MW01528086

PERIMETER AND AREA FORMULAS

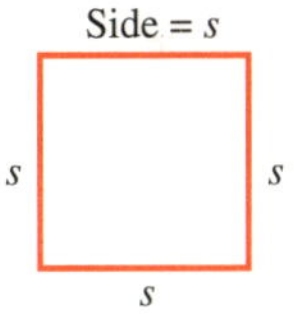

Square
$P = 4s$
$A = s^2$

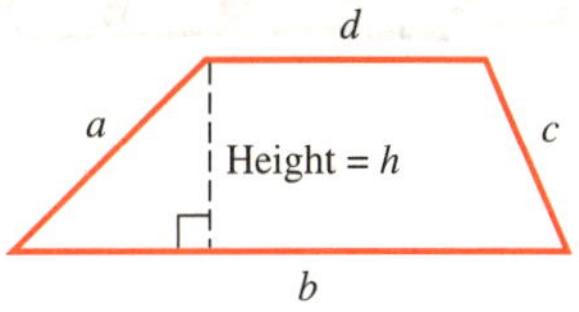

Trapezoid
$P = a + b + c + d$
$A = \frac{1}{2}h(b + d)$

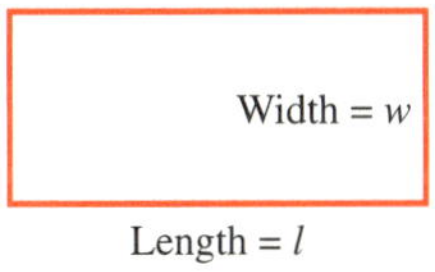

Rectangle
$P = 2l + 2w$
$A = lw$

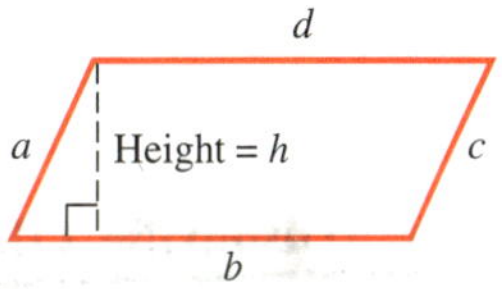

Parallelogram
$P = a + b + c + d$
$A = bh$

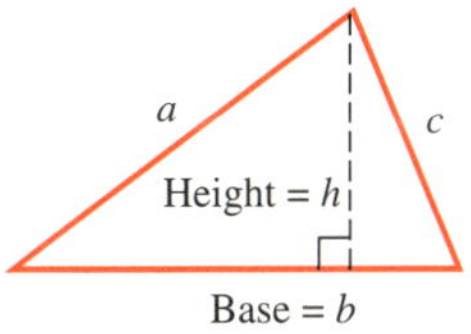

Triangle
$P = a + b + c$
$A = \frac{1}{2}bh$

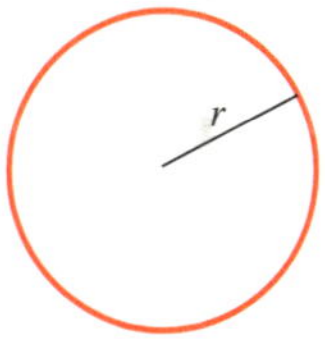

Circle
$C = 2\pi r$ or $C = \pi D$
where $\pi = 3.14$
$A = \pi r^2$

VOLUME FORMULAS

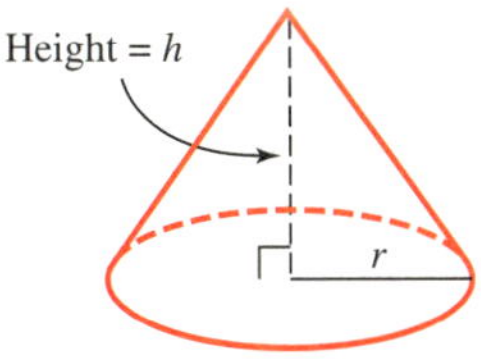

Cone
$V = \frac{1}{3}\pi r^2 h$

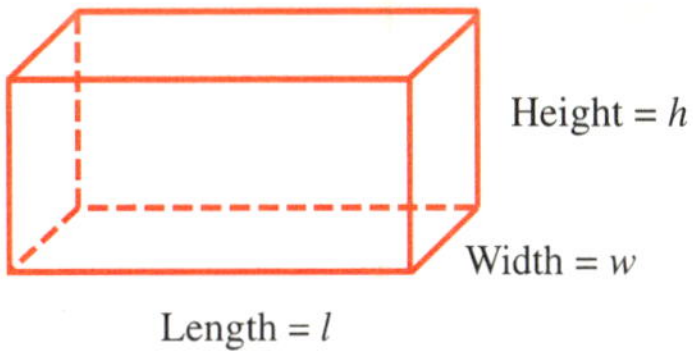

Rectangular solid
$V = lwh$

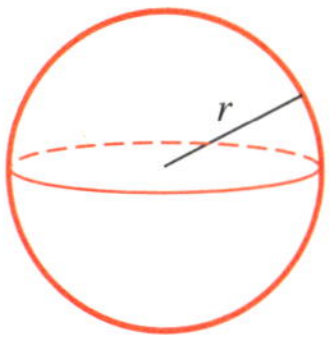

Sphere
$V = \frac{4}{3}\pi r^3$

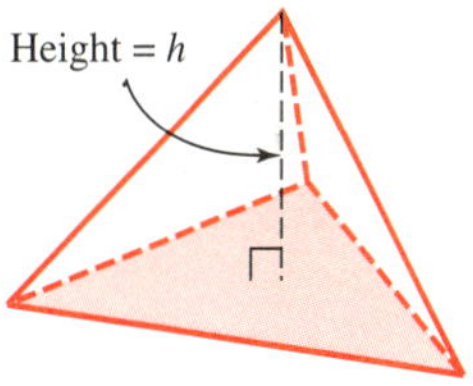

Pyramid
$V = \frac{1}{3}Bh$*
*B represents the area of the base.

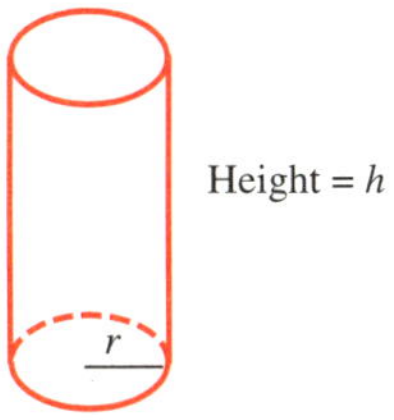

Cylinder
$V = \pi r^2 h$

Property of
Pioneer Pacific
COLLEGE

Pioneer Pacific College 14

Intermediate College Algebra

00013786

REMOVED FROM
INVENTORY
PPC / WIL

27501 SW Parkway Ave. Wilsonville, OR 97070

www.brookscole.com

www.brookscole.com is the World Wide Web site for Thomson Brooks/Cole and is your direct source to dozens of online resources.

At *www.brookscole.com* you can find out about supplements, demonstration software, and student resources. You can also send email to many of our authors and preview new publications and exciting new technologies.

www.brookscole.com
Changing the way the world learns®

Books in the Tussy and Gustafson Series

In hardcover:

Elementary Algebra, Third Edition
Intermediate Algebra, Third Edition
Elementary and Intermediate Algebra, Third Edition

In paperback:

Basic Mathematics for College Students, Second Edition
Basic Geometry for College Students
Prealgebra, Second Edition
Introductory Algebra, Second Edition
Intermediate Algebra, Second Edition
Developmental Mathematics

For more information, please visit www.brookscole.com

Edition

3

Intermediate Algebra

Alan S. Tussy
Citrus College

R. David Gustafson
Rock Valley College

THOMSON
BROOKS/COLE

Australia • Canada • Mexico • Singapore • Spain
United Kingdom • United States

THOMSON

BROOKS/COLE

Executive Editor: Jennifer Huber
Executive Publisher: Curt Hinrichs
Development Editor: Kirsten Markson
Assistant Editor: Rebecca Subity
Editorial Assistant: Sarah Woicicki
Technology Project Manager: Rachael Sturgeon
Marketing Manager: Greta Kleinert
Marketing Assistant: Jessica Bothwell
Advertising Project Manager: Bryan Vann
Project Manager, Editorial Production: Hal Humphrey
Senior Art Director: Vernon T. Boes
Print/Media Buyer: Barbara Britton
Permissions Editor: Kiely Sexton
Production Service: Helen Walden
Text Designer: Kim Rokusek
Photo Researcher: Helen Walden
Illustrator: Lori Heckelman Illustration
Cover Designer: Patrick Devine
Cover Image: Kevin Tolman
Compositor: Graphic World, Inc.
Text and Cover Printer: Quebecor World/Taunton

© 2005 Thomson Brooks/Cole, a part of the Thomson Corporation. Thomson, the Star logo, and Brooks/Cole are trademarks used herein under license.

ALL RIGHTS RESERVED. No part of this work covered by the copyright hereon may be reproduced or used in any form or by any means—graphic, electronic, or mechanical, including but not limited to photocopying, recording, taping, Web distribution, information networks, or information storage and retrieval systems—without the written permission of the publisher.

Printed in the United States of America
1 2 3 4 5 6 7 08 07 06 05 04

For more information about our products, contact us at:
Thomson Learning Academic Resource Center
1-800-423-0563
For permission to use material from this text or product, submit a request online at http://www.thomson.com.
Any additional questions about permissions can be submitted by e-mail to thomsonrights@thomson.com.

ExamView® and *ExamView Pro*® are registered trademarks of FSCreations, Inc. Windows is a registered trademark of the Microsoft Corporation used herein under license. Macintosh and Power Macintosh are registered trademarks of Apple Computer, Inc. Used herein under license.

© 2005 Thomson Learning, Inc. All Rights Reserved. Thomson Learning WebTutor™ is a trademark of Thomson Learning, Inc.

Library of Congress Control Number: 2004110436

Student Edition: ISBN 0-534-41923-2

Annotated Instructor's Edition: ISBN 0-534-41924-0

Thomson Higher Education
10 Davis Drive
Belmont, CA 94002
USA

Asia
Thomson Learning
5 Shenton Way #01-01
UIC Building
Singapore 068808

Australia/New Zealand
Thomson Learning
102 Dodds Street
Southbank, Victoria 3006
Australia

Canada
Nelson
1120 Birchmount Road
Toronto, Ontario M1K 5G4
Canada

Europe/Middle East/Africa
Thomson Learning
High Holborn House
50/51 Bedford Row
London WC1R 4LR
United Kingdom

Latin America
Thomson Learning
Seneca, 53
Colonia Polanco
11560 Mexico D.F.
Mexico

Spain/Portugal
Paraninfo
Calle Magallanes, 25
28015 Madrid, Spain

QA154.3
T88
2005

To Helen Walden,

Thank you for your tireless devotion
to this series.

AST
RDG

Contents

Chapter 6 Rational Expressions and Equations 412

Chapter 7 Radical Expressions and Equations 513

Chapter 8 Quadratic Equations, Functions, and Inequalities 598

Preface

Algebra is a language in its own right. The purpose of this textbook is to teach students how to read, write, and think mathematically using the language of algebra. It presents all the topics associated with a second course in algebra. *Intermediate Algebra,* Third Edition, employs a variety of instructional methods that reflect the recommendations of NCTM and AMATYC. In this book, you will find the vocabulary, practice, and well-defined pedagogy of a traditional approach. You will also find that we emphasize the reasoning, modeling, and communicating skills that are part of today's reform movement.

The third edition retains the basic philosophy of the second edition. However, we have made several improvements as a direct result of the comments and suggestions we received from instructors and students. Our goal has been to make the book more enjoyable to read, easier to understand, and more relevant.

NEW TO THIS EDITION

- New chapter openers reference the *TLE* computer lessons that accompany each chapter.
- The new Language of Algebra features, along with Success Tips, Notation, Calculator Boxes, and Cautions, are presented in the margins to promote understanding and increased clarity.
- Many additional applications involving real-life data have been added.
- Answers to the popular *Self Check* feature have been relocated to the end of each section, right before the *Study Set.*
- Several higher-level Challenge Problems have been added to each *Study Set.*
- The Accent on Teamwork feature has been redesigned to offer the instructor two or three collaborative activities per chapter that can be assigned as group work.
- More illustrations, diagrams, and color have been added for the visual learner.

REVISED TABLE OF CONTENTS

Chapter 1: *A Review of Basic Algebra* Section 1.5, *Solving Linear Equations and Formulas,* now includes a more detailed discussion of identities and contradictions.

Chapter 2: *Graphs, Equations of Lines, and Functions* Section 2.5, *An Introduction to Functions,* and Section 2.6, *Graphs of Functions,* were rewritten. The concept of function is introduced using a real-world example and mapping diagrams. More emphasis is placed on reading and interpreting the graphs of functions.

Chapter 3: *Systems of Equations* In Section 3.4, *Solving Systems Using Matrices,* a more in-depth explanation of matrix solutions of systems of three equations is presented.

Chapter 4: *Inequalities* Section 4.3, *Solving Absolute Value Equations and Inequalities,* was rewritten to better relate the geometric, graphic, and algebraic interpretations of absolute value.

Chapter 5: *Exponents, Polynomials, and Polynomial Functions* The section on *Factoring Trinomials* was relocated so that it now appears before factoring differences of two squares and sums and differences of two cubes.

Chapter 6: *Rational Expressions and Equations* The topic of *Synthetic Division,* formerly in the Appendix, is now Section 6.6.

Chapter 7: *Radical Expressions and Equations* The sections in this chapter have been reordered so that *Solving Radical Equations* follows the sections that discuss simplifying radical expressions. The topic of *Complex Numbers,* formerly in Chapter 8, is now the final section of Chapter 7.

Chapter 8: *Quadratic Equations, Functions, and Inequalities* The section, *The Discriminant and Equations That Can Be Written in Quadratic Form,* was relocated so that it now follows Section 8.2, *The Quadratic Formula.* Section 8.5, *Quadratic and Other Nonlinear Inequalities,* was rewritten. It now includes a more detailed discussion of the interval testing method.

Chapter 9: *Exponential and Logarithmic Functions* Several sections have been edited to improve clarity. Section 9.8, *Exponential and Logarithmic Equations,* has been reorganized: Exponential equations whose sides can be written as a power of the same base appear first, followed by equations that can be solved by taking the logarithm of both sides.

Chapter 10: *Conic Sections; More Graphing* More detailed drawings are included in Section 10.1, *The Circle and the Parabola,* where conic sections and their applications are introduced.

Chapter 11: *Miscellaneous Topics* In Section 11.3, *Geometric Sequences and Series,* more in-depth explanations of solution methods are presented.

ACKNOWLEDGMENTS

We are grateful to the following people who reviewed the manuscript at various stages of its development. They all had valuable suggestions that have been incorporated into the text.

The following people reviewed the first and second editions:

Sally Copeland
Johnson County Community College

Ben Cornelius
Oregon Institute of Technology

Mary Lou Hammond
Spokane Community College

Judith Jones
Valencia Community College

Therese Jones
Amarillo College

Janice McFatter
Gulf Coast Community College

June Strohm
Pennsylvania State Community College–DuBois

Jo Anne Temple
Texas Technical University

Sharon Testone
Onondaga Community College

Marilyn Treder
Rochester Community College

Betty Weissbecker
J. Sargeant Reynolds Community College

Cathleen Zucco
SUNY-New Paltz

The following people reviewed the third edition:

Mike Adams
Modesto Junior College

Ray Brinker
Western Illinois University

Cynthia J. Broughton
Arizona Western College

Don K. Brown
Macon State College

Light Bryant
Arizona Western College

Warren S. Butler
Daytona Beach Community College

John Scott Collins
Pima Community College

Lucy H. Edwards
Ohlone College

Hajrudin Fejzic
California State University, San Bernardino

Lee Gibbs
Arizona Western College

Barry T. Gibson
Daytona Beach Community College

Haile K. Haile
Minneapolis Community and Technical College

Suzanne Harris-Smith
Albuquerque Technical Vocational Institute

Kamal Hennayake
Chesapeake College

Doreen Kelly
Mesa Community College

Lynn Marecek
Santa Ana College

Michael Marzinske
Inver Hills Community College

Jamie McGill
East Tennessee State University

Margaret Michener
University of Nebraska, Kearney

Micheal Montano
Riverside Community College

Brian W. Moudry
Davis & Elkins College

William Peters
San Diego Mesa College

Bernard J. Pina
Dona Ana Branch Community College

Carol Purcell
Century Community College

Daniel Russow
Arizona Western College

Donald W. Solomon
University of Wisconsin, Milwaukee

John Thoo
Yuba College

Susan M. Twigg
Wor-Wic Community College

Gloria Upson
Winston-Salem State University

Gizelle Worley
California State University, Stanislaus

We want to express our gratitude to Bob Billups, George Carlson, Robin Carter, Jim Cope, Terry Damron, Marion Hammond, Karl Hunsicker, Doug Keebaugh, Arnold Kondo, John McKeown, Kent Miller, Donna Neff, Steve Odrich, Eric Rabitoy, Maryann Rachford, Dave Ryba, Chris Scott, Rob Everest, Bill Tussy, Liz Tussy, and the Citrus College Library Staff (including Barbara Rugeley) for their help with some of the application problems in the textbook.

Without the talents and dedication of the editorial, marketing, and production staff of Brooks/Cole, this revision of *Intermediate Algebra* could not have been so well accomplished. We express our sincere appreciation for the hard work of Bob Pirtle, Jennifer Huber, Helen Walden, Lori Heckleman, Vernon Boes, Kim Rokusek, Sarah Woicicki, Greta Kleinert, Jessica Bothwell, Bryan Vann, Kirsten Markson, Rebecca Subity, Hal Humphrey, Tammy Fisher-Vasta, Christine Davis, Ellen Brownstein, Diane Koenig, Ian Crewe, and Graphic World for their help in creating the book.

Alan S. Tussy
R. David Gustafson

For the Student

SUCCESS IN ALGEBRA

To be successful in mathematics, you need to know how to study it. The following checklist will help you develop your own personal strategy to study and learn the material. The suggestions below require some time and self-discipline on your part, but it will be worth the effort. This will help you get the most out of the course.

As you read each of the following statements, place a check mark in the box if you can truthfully answer Yes. If you can't answer Yes, think of what you might do to make the suggestion part of your personal study plan. You should go over this checklist several times during the semester to be sure you are following it.

Preparing for the Class

- ❑ I have made a commitment to myself to give this course my best effort.
- ❑ I have the proper materials: a pencil with an eraser, paper, a notebook, a ruler, a calculator, and a calendar or day planner.
- ❑ I am willing to spend a minimum of two hours doing homework for every hour of class.
- ❑ I will try to work on this subject every day.
- ❑ I have a copy of the class syllabus. I understand the requirements of the course and how I will be graded.
- ❑ I have scheduled a free hour after the class to give me time to review my notes and begin the homework assignment.

Class Participation

- ❑ I know my instructor's name.
- ❑ I will regularly attend the class sessions and be on time.
- ❑ When I am absent, I will find out what the class studied, get a copy of any notes or handouts, and make up the work that was assigned when I was gone.
- ❑ I will sit where I can hear the instructor and see the board.
- ❑ I will pay attention in class and take careful notes.
- ❑ I will ask the instructor questions when I don't understand the material.
- ❑ When tests, quizzes, or homework papers are passed back and discussed in class, I will write down the correct solutions for the problems I missed so that I can learn from my mistakes.

Study Sessions

- ❑ I will find a comfortable and quiet place to study.
- ❑ I realize that reading a math book is different from reading a newspaper or a novel. Quite often, it will take more than one reading to understand the material.
- ❑ After studying an example in the textbook, I will work the accompanying Self Check.
- ❑ I will begin the homework assignment only after reading the assigned section.
- ❑ I will try to use the mathematical vocabulary mentioned in the book and used by my instructor when I am writing or talking about the topics studied in this course.
- ❑ I will look for opportunities to explain the material to others.
- ❑ I will check all my answers to the problems with those provided in the back of the book (or with the *Student Solutions Manual*) and resolve any differences.
- ❑ My homework will be organized and neat. My solutions will show all the necessary steps.
- ❑ I will work some review problems every day.

- ❑ After completing the homework assignment, I will read the next section to prepare for the coming class session.
- ❑ I will keep a notebook containing my class notes, homework papers, quizzes, tests, and any handouts—all in order by date.

Special Help

- ❑ I know my instructor's office hours and am willing to go in to ask for help.
- ❑ I have formed a study group with classmates that meets regularly to discuss the material and work on problems.
- ❑ When I need additional explanation of a topic, I use the tutorial videos and the interactive CD, as well as the Web site.
- ❑ I make use of extra tutorial assistance that my school offers for mathematics courses.
- ❑ I have purchased the *Students Solutions Manual* that accompanies this text, and I use it.

To follow each of these suggestions will take time. It takes a lot of practice to learn mathematics, just as with any other skill.

No doubt, you will sometimes become frustrated along the way. This is natural. When it occurs, take a break and come back to the material after you have had time to clear your thoughts. Keep in mind that the skills and discipline you learn in this course will help make for a brighter future. Gook luck!

iLrn Tutorial Quick Start Guide

iLrn CAN HELP YOU SUCCEED IN MATH

iLrn is an online program that facilitates math learning by providing resources and practice to help you succeed in your math course. Your instructor chose to use iLrn because it provides online opportunities for learning (Explanations found by clicking **Read Book**), practice (Exercises), and evaluating (Quizzes). It also gives you a way to keep track of your own progress and manage your assignments.

The mathematical notation in iLrn is the same as that you see in your textbooks, in class, and when using other math tools like a graphing calculator. iLrn can also help you run calculations, plot graphs, enter expressions, and grasp difficult concepts. You will encounter various problem types as you work through iLrn, all of which are designed to strengthen your skills and engage you in learning in different ways.

LOGGING IN TO iLrn

Registering with the PIN Code on the iLrn Card *Situation:* Your instructor has not given you a PIN code for an online course, but you have a textbook with an iLrn product PIN code.

Initial Log-in

1. Go to **http://iLrn.com.**
2. In the menu at the left, click on **Student Tutorial.**
3. Make sure that the name of your school appears in the "School" field. If your school name does not appear, follow steps a–d below. If your school is listed, go to step 4.
   a. Click on **Find Your School.**
   b. In the "State" field, select your state from the drop-down menu.

c. In the “Name of school” field, type the first few letters of your school’s name; then click on **Search.** The school list will appear at the right.

d. Click on your school. The “First Time Users” screen will open.

4. In the “PIN Code” field, type the iLrn PIN code supplied on your iLrn card.
5. In the “ISBN” field, type the ISBN of your book (from the textbook’s back cover), for example, 0-534-41914-3.
6. Click on **Register.**
7. Enter the appropriate information. Fields marked with a red asterisk must be filled in.
8. Click on **Register and Enter iLrn.**

You will be asked to select a user name and password. Save your user name and password in a safe place. You will need them to log in the next time you use iLrn. Only your user name and password will allow you to reenter iLrn.

Subsequent Log-in

1. Go to **http://iLrn.com.**
2. Click on **Login.**
3. Make sure the name of your school appears in the “School” field. If not, then follow steps 3a–d under “Initial Login” to identify your school.
4. Type your user name and password (see boxed information above); then click on **Login.** The “My Assignments” page will open.

NAVIGATING THROUGH iLrn

To navigate between chapters and sections, use the drop-down menu below the top navigation bar. This will give you access to the study activities available for each section.

The view of a tutorial in iLrn looks like this.

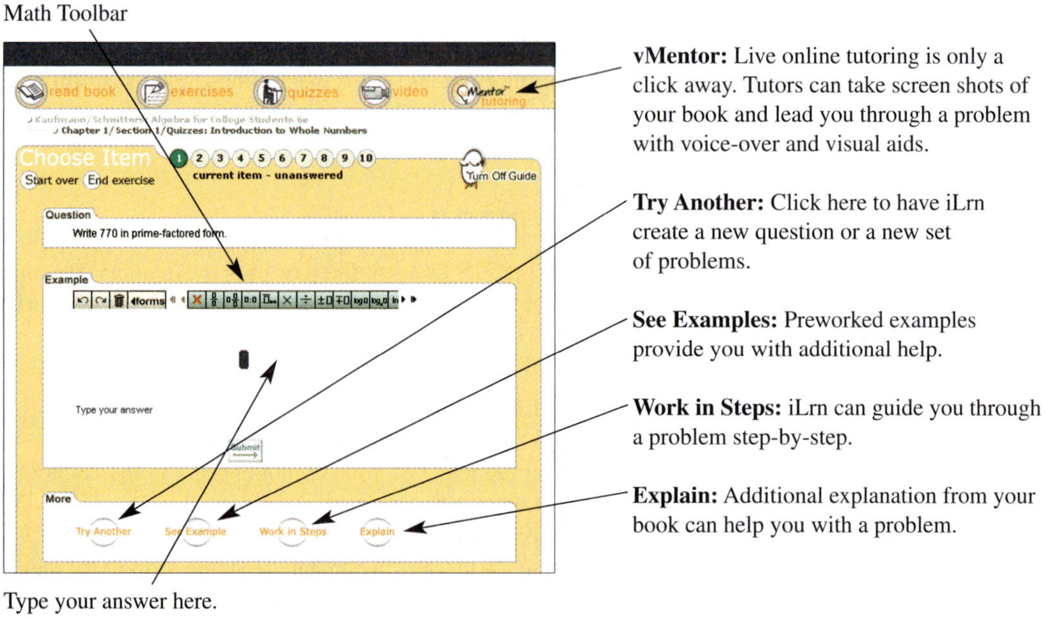
Math Toolbar
read book
exercises
quizzes
video
Choose Item
Start over
End exercise
current item - unanswered
Turn Off Guide
Question
Write 770 in prime-factored form.
Example
Type your answer
More
Try Another
See Example
Work in Steps
Explain
vMentor: Live online tutoring is only a click away. Tutors can take screen shots of your book and lead you through a problem with voice-over and visual aids.
Try Another: Click here to have iLrn create a new question or a new set of problems.
See Examples: Preworked examples provide you with additional help.
Work in Steps: iLrn can guide you through a problem step-by-step.
Explain: Additional explanation from your book can help you with a problem.
Type your answer here.

ONLINE TUTORING WITH vMENTOR

Access to iLrn also means access to online tutors and support through vMentor, which provides live homework help and tutorials. To access vMentor while you are working in the Exercises or "Tutorial" areas in iLrn, click on the **vMentor Tutoring** button at the top right of the navigation bar above the problem or exercise.

Next, click on the **vMentor** button; you will be taken to a Web page that lists the steps for entering a vMentor classroom. If you are a first-time user of vMentor, you might need to download Java software before entering the class for the first class. You can either take an Orientation Session or log in to a vClass from the links at the bottom of the opening screen.

All vMentor Tutoring is done through a vClass, an Internet-based virtual classroom that features two-way audio, a shared whiteboard, chat, messaging, and experienced tutors.

You can access vMentor Sunday through Thursday, as follows:

5 p.m. to 9 p.m. Pacific Time
6 p.m. to 10 p.m. Mountain Time
7 p.m. to 11 p.m. Central Time
8 p.m. to midnight Eastern Time

If you need additional help using vMentor, you can access the Participant Guide at this Web site: **http://www.elluminate.com/support/guide/pdf.**

INTERACT WITH TLE ONLINE LABS

If your text came with TLE Online Labs, use the labs to explore and reinforce key concepts introduced in this text. These electronic labs give you access to additional instruction and practice problems, so you can explore each concept interactively, at your own pace. Not only will you be better prepared, but you will also perform better in the class overall.

To access TLE Online Labs:

1. Go to http://tle.brookscole.com.
2. In the "Pin Code" field, type the TLE PIN code supplied on the TLE card that came shrink-wrapped with your book.
3. Click on **Register.**
4. Enter the appropriate information. Fields marked with a red asterisk must be filled in.
5. Click on **Register** and **Begin TLE.**

> You will be asked to select a user name and password. Save your user name and password in a safe place. You will need them to log in the next time you use TLE. Only your user name and password will allow you to reenter TLE.

Subsequent Log-in

1. Go to **http://tle.brookscole.com.**
2. Click on **Login.**
3. Make sure the name of your school appears in the "School" field. If not, then follow steps 3a–d under "Initial Login" found on page xv to identify your school.
4. Type your user name and password (see boxed information above); then click on **Login.** The "My Assignments" page will open.

Applications Index

Examples that are applications are shown in **boldface page numbers.**
Exercises that are applications are shown in lightface page numbers.

Chapter

1 A Review of Basic Algebra

Getty/Brand X Pictures

Today, banks and lending institutions offer a variety of financial services. To manage our money wisely, we need to be familiar with the terms and conditions of our checking accounts, credit cards, and loans. Financial transactions are described using positive and negative whole numbers, fractions, and decimals. These numbers belong to a set that we call the *real numbers.*

To learn more about real numbers and how they are used in the financial world, visit *The Learning Equation* on the Internet at http://tle.brookscole.com. (The log-in instructions are in the Preface.) For Chapter 1, the following online lesson is available:

- *TLE* Lesson 1: The Real Numbers

This chapter reviews many of the fundamental concepts that are studied in an elementary algebra course.

1.1 The Language of Algebra

- Variables, algebraic expressions, and equations
- Verbal models
- Constructing tables
- Graphical models
- Formulas

Algebra is the result of contributions from many cultures over thousands of years. The word *algebra* comes from the title of the book *Al-jabr wa'l muquabalah,* written by the Arabian mathematician al-Khwarizmi around A.D. 800. Using the vocabulary and notation of algebra we can mathematically describe many situations in the real world. In this section, we will review some of the basic components of the language of algebra.

VARIABLES, ALGEBRAIC EXPRESSIONS, AND EQUATIONS

The following rental agreement shows that two operations need to be performed to calculate the cost of renting a banquet hall.

- First, we must *multiply* the hourly rental cost of \$100 by the number of hours the hall is to be rented.
- To that result, we must then *add* the cleanup fee of \$200.

Rental Agreement
ROYAL VISTA BANQUET HALL
Wedding Receptions•Dances•Reunions•Fashion Shows

Rented To____________________Date____________
Lessee's Address____________________________

Rental Charges
- \$100 per hour
- Nonrefundable \$200 cleanup fee

Terms and conditions
Lessor leases the undersigned lessee the above described property upon the terms and conditions set forth on this page and on the back of this page. Lessee promises to pay rental cost stated herein.

In words, we can describe the process as follows:

The cost of renting the hall	is	100	times	the number of hours it is rented	plus	200.

We can also describe the procedure for calculating the cost using *variables* and mathematical symbols. A **variable** is a letter that is used to stand for a number. If we let C stand for the cost of renting the hall and h stand for the number of hours it is rented, the words can be translated to form a **mathematical model.**

The cost of renting the hall	is	100	times	the number of hours it is rented	plus	200.
C	$=$	100	$\cdot$	h	$+$	200

Language of Algebra

Words such as *is, was, gives,* and *yields* translate to an = symbol.

The statement $C = 100h + 200$ is called an **equation.** The = symbol indicates that two quantities, C and $100h + 200$, are equal.

Equations An **equation** is a mathematical sentence that contains an = symbol.

On the right-hand side of the equation $C = 100h + 200$, the notation $100h + 200$ is called an **algebraic expression**, or more simply, an **expression.**

Algebraic Expressions Variables and/or numbers can be combined with the operations of addition, subtraction, multiplication, division, raising to a power, and finding a root to create **expressions.**

Here are some examples of expressions.

$5a - 12$ This expression involves the operations of multiplication and subtraction.

$\frac{50 - y}{3y^3}$ This expression involves the operations of subtraction, division, multiplication, and raising to a power.

$\sqrt{a^2 + b^2}$ This expression involves the operations of addition, raising to a power, and finding a root.

VERBAL MODELS

The following table lists some words and phrases that are often used in mathematics to denote the operations of addition, subtraction, multiplication, and division.

Addition +	Subtraction −	Multiplication ·	Division ÷
added to	subtracted from	multiplied by	divided by
sum	difference	product	quotient
plus	less than	times	ratio
more than	decreased by	percent (or fraction) of	half
increased by	reduced by	twice	into
greater than	minus	triple	per

In the banquet hall example, the equation $C = 100h + 200$ was used to describe a procedure to calculate the cost of renting the hall. Using vocabulary from the table, we can write a **verbal model** that also describes this procedure. One such model is:

The cost (in dollars) of renting the hall is the *product* of 100 and the number of hours it is rented, *increased by* 200.

Here is another example of creating a verbal and a mathematical model of a real-life situation.

EXAMPLE 1

Catering. It costs \$6 per person to have a dinner catered. For groups of more than 200, a \$100 discount is given. Write a verbal and a mathematical model that describe the relationship between the catering cost and the number of people being served, for groups larger than 200.

Solution To find the catering cost C (in dollars) for groups larger than 200, we need to *multiply* the number n of people served by \$6 and then *subtract* the \$100 discount.

A verbal model is:

The catering cost (in dollars) is the *product* of 6 and the number of people served, *decreased by 100.*

In symbols, the mathematical model is:

$$C = 6n - 100$$

Self Check 1 After winning a lottery, three friends split the prize equally. Each person then had to pay \$2,000 in taxes on his or her share. Write a verbal model and a mathematical model that relate the amount of each person's share, after taxes, to the amount of the lottery prize.

CONSTRUCTING TABLES

In the banquet hall example, the equation $C = 100h + 200$ can be used to determine the cost of renting the banquet hall for *any* number of hours.

EXAMPLE 2

Find the cost of renting the banquet hall for 3 hours and for 4 hours. Write the results in a table.

Solution We begin by constructing the table below with the appropriate column headings: h for the number of hours the hall is rented and C for the cost (in dollars) to rent the hall. Then we enter the number of hours of each rental time in the left column.

Next, we use the equation $C = 100h + 200$ to find the total rental cost for 3 hours and for 4 hours.

$$\begin{aligned} C &= 100h + 200 \\ C &= 100(3) + 200 && \text{Replace } h \text{ with 3.} \\ &= 300 + 200 && \text{Multiply.} \\ &= 500 \end{aligned}$$

$$\begin{aligned} C &= 100h + 200 \\ C &= 100(4) + 200 && \text{Replace } h \text{ with 4.} \\ &= 400 + 200 && \text{Multiply.} \\ &= 600 \end{aligned}$$

Finally, we enter these results in the right-hand column of the table: \$500 for a 3-hour rental and \$600 for a 4-hour rental.

h	C
3	500
4	600

Self Check 2 Find the cost of renting the hall for 6 hours and for 7 hours. Write the results in the table.

GRAPHICAL MODELS

The cost of renting the banquet hall for various lengths of time can also be presented graphically. The following **bar graph** has a **horizontal axis** labeled "Number of hours the hall is rented." The **vertical axis,** labeled "Cost to rent the hall ($)," is scaled in units of 50 dollars. The bars above each of the times (1, 2, 3, 4, 5, 6, and 7 hours) extend to a height that gives the corresponding cost to rent the hall. For example, if the hall is rented for 5 hours, the bar indicates that the cost is $700.

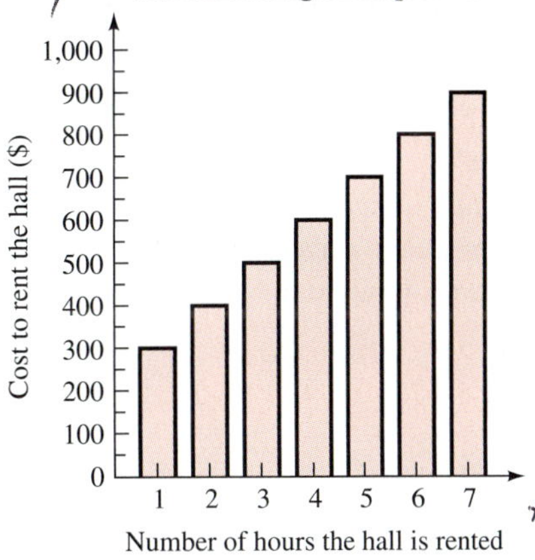

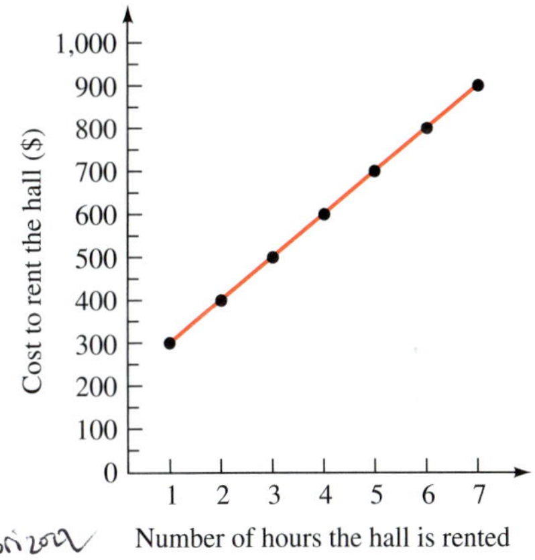

The **line graph** above also shows the rental costs. This type of graph consists of a series of dots drawn at the correct height, connected with line segments. We can use the line graph to find the cost of renting the banquet hall for lengths of time not shown in the bar graph.

EXAMPLE 3

Use the line graph shown above to determine the cost of renting the hall for $4\frac{1}{2}$ hours.

Solution In the figure to the right, we locate $4\frac{1}{2}$ on the horizontal axis and draw a vertical line upward to intersect the graph. From the point of intersection with the graph, we draw a horizontal line to the left that intersects the vertical axis. On the vertical axis, we can read that the rental cost is $650 for $4\frac{1}{2}$ hours.

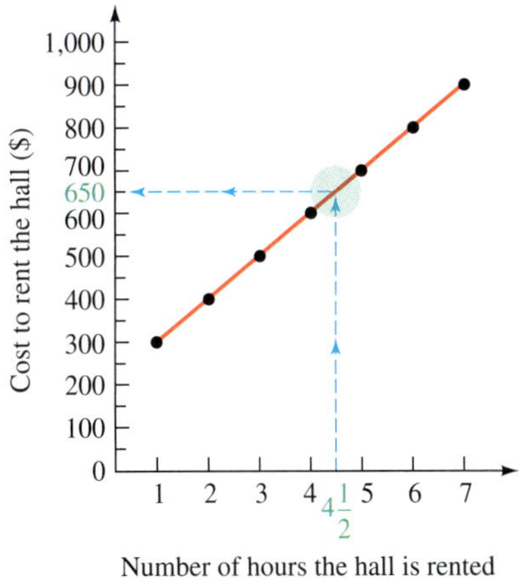

Success Tip

The video icons (see above) show which examples are taught on tutorial video tapes or disks.

Self Check 3 Use the figure to find the cost of renting the banquet hall for $6\frac{1}{2}$ hours.

FORMULAS

Equations that express a relationship between two or more quantities, represented by variables, are called **formulas.** Formulas are used in many fields, such as automotive technology, economics, medicine, retail sales, and banking.

EXAMPLE 4

Use variables to express each relationship as a formula.

a. The distance in miles traveled by a vehicle is the product of its average rate of speed in mph and the time in hours it travels at that rate.

b. The sale price of an item is the difference between the regular price and the discount.

Solution

a. The word *product* indicates multiplication. If we let d stand for the distance traveled in miles, r for the vehicle's average rate of speed in mph, and t for the length of time traveled in hours, we can write the formula as

$$d = rt$$

b. The word *difference* indicates subtraction. If we let s stand for the sale price of the item, p for the regular price, and d for the discount, we have

$$s = p - d$$

Self Check 4 Express the following relationship as a formula: the simple interest earned by a deposit is the product of the principal, the annual rate of interest, and the time.

Some commonly used geometric formulas are presented inside the front and back covers of this book. For example, to find the **perimeter** of a rectangle (the distance around it), the appropriate formula to use is $P = 2l + 2w$, where l is the length and w is the width of the rectangle.

EXAMPLE 5

Landscaping. Find the number of feet of redwood edging needed to outline a square flower bed having sides that are 6.5 feet long.

Solution

To find the amount of redwood edging needed, we need to find the perimeter of the square flower bed.

$$P = 4s \quad \text{This is the formula for the perimeter of a square.}$$

$$P = 4(6.5) \quad \text{Substitute 6.5 for } s \text{, the length of one side of the square.}$$

$$= 26$$

26 feet of redwood edging is needed to outline the flower bed.

Self Check 5 Find the amount of fencing needed to enclose a triangular lot with sides that are 150 ft, 205.5 ft, and 165 ft long.

Answers to Self Checks

1. Each person's share, after taxes, is the quotient of the lottery prize and 3, decreased by 2,000; $S = \frac{p}{3} - 2{,}000$.

2.

h	C
6	800
7	900

3. \$850 4. $I = Prt$ 5. 520.5 ft

1.1 STUDY SET

VOCABULARY Fill in the blanks.

1. An ________ is a mathematical sentence that contains an = symbol.
2. A ________ is a letter that is used to stand for a number.
3. Variables and/or numbers can be combined with mathematical operations to create algebraic ________.
4. Phrases such as *increased by* and *more than* are used to indicate the operation of ________.
5. Phrases such as *decreased by* and *less than* are used to indicate the operation of ________.
6. Words such as *is, was, gives,* and *yields* translate to an ____ symbol.
7. A ________ is an equation that expresses a relationship between two or more quantities represented by variables.
8. The distance around a geometric figure is its ________.

CONCEPTS

9. a. What type of graph is shown?
   b. What units are used to scale the horizontal axis? The vertical axis?
   c. Estimate the height of the candle after it has burned for $3\frac{1}{2}$ hours. For 8 hours.

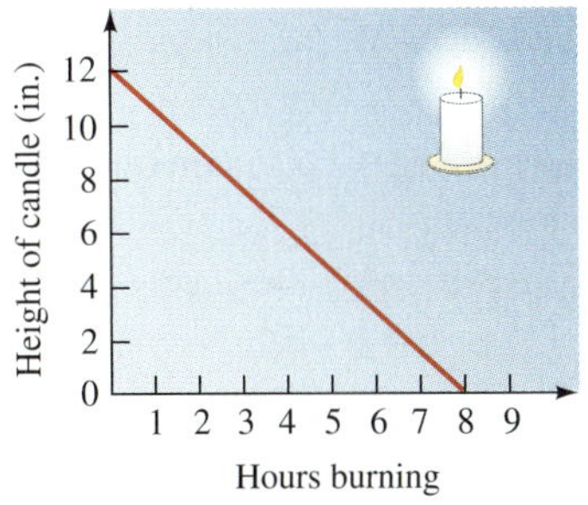

10. a. What type of graph is shown?
    b. What units are used to scale the horizontal axis? The vertical axis?
    c. In what year was the average expenditure on auto insurance the least? Estimate the amount. In what year was it the greatest? Estimate the amount.

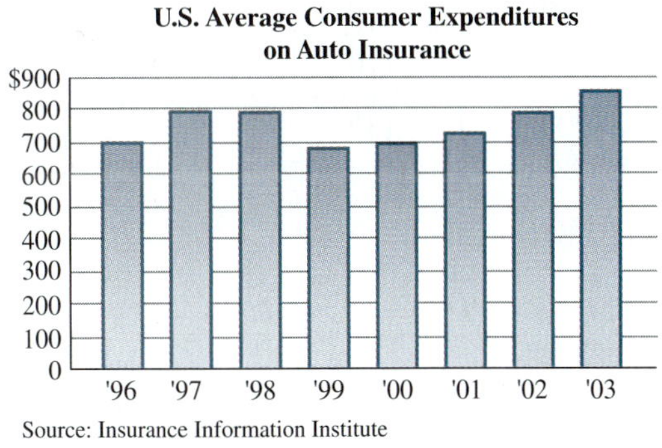

Translate each verbal model into a mathematical model.

11. The cost each semester is \$13 times the number of units taken plus a student services fee of \$24.
12. The yearly salary is \$25,000 plus \$75 times the number of years of experience.
13. The quotient of the number of clients and seventy-five gives the number of social workers needed.
14. The difference between 500 and the number of people in a theater gives the number of unsold tickets.
15. Each test score was increased by 15 points to give a new adjusted test score.
16. The weight of a super-size order of French fries is twice that of a regular-size order.
17. The product of the number of boxes of crayons in a case and 12 gives the number of crayons in a case.

18. The perimeter of an equilateral triangle can be found by tripling the length of one of its sides.

Use the data in each table to find a formula that mathematically describes the relationship between the two quantities. Then state the relationship in words. (Answers may vary.)

19.

Tower height (ft)	Height of base (ft)
15.5	5.5
22	12
25.25	15.25
45.125	35.125

20.

Seasonal employees	Employees
25	75
50	100
60	110
80	130

NOTATION

21. Classify each of the following as an expression or an equation.

a. $6x - 5$

b. $4s - 5 = 5$

c. $P = a + b + c$

d. $\frac{x + y}{8}$

e. $\sqrt{2x^2}$

f. Prt

22. Translate the verbal model into a mathematical model.

7	times	the age of a dog in years	gives	the dog's equivalent human age.

PRACTICE Use the given formula to complete each table.

23. $c = \frac{p}{12}$

Number of packages p	Cartons c
24	
72	
180	

24. $y = 100c$

Number of centuries c	Years y
1	
6	
21	

25. $n = 22.44 - K$

K	n
0	
1.01	
22.44	

26. $y = x + 15$

x	y
0	
15	
30	

27. The lengths of the two parallel sides of a trapezoid are 10 inches and 15 inches. The other two sides are each 6 inches long. Find the perimeter of the trapezoid.

28. Find the perimeter of a parallelogram if two of its adjacent sides are 50 meters and 100 meters long.

29. Find the perimeter of a square quilt that has sides 2 yards long.

30. Find the perimeter of a triangular postage stamp with sides 1.8, 1.8, and 1.5 centimeters long.

APPLICATIONS

31. CARPET CLEANING See the following ad.

a. Write a verbal model that states the relationship between the cost C of renting the carpet-cleaning system and the number of hours h it is rented.

b. Translate the verbal model written in part a to a mathematical model.

c. Use your result from part b to complete the table, and then draw a line graph.

h	C
1	
2	
3	
4	
8	

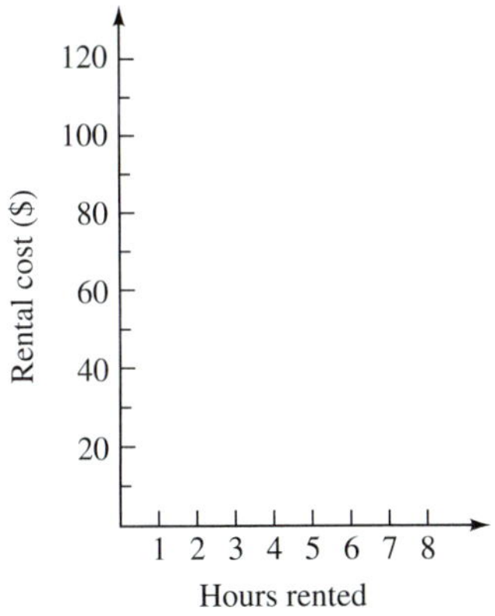

32. FLOOR MATS What geometric concept applies when finding the length of the plastic trim around the cargo area floor mat? Estimate the amount of trim used.

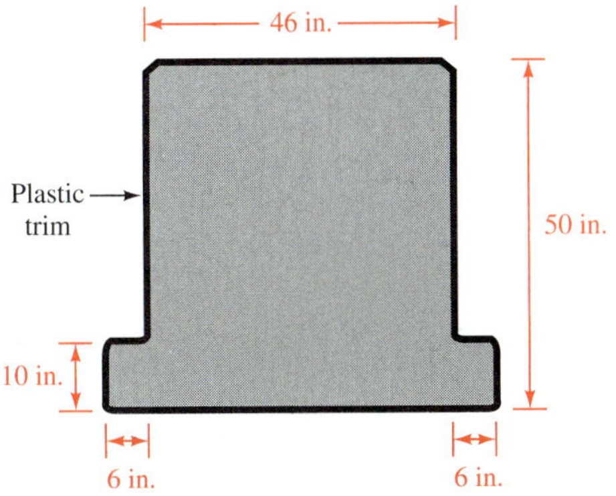

33. CARPENTRY A miter saw can pivot 180° to make angled cuts on molding. The formula that relates the angle measure s on the scrap piece of molding and the angle measure f on the finish piece of molding is $s = 180 - f$. Complete the following table and then draw a line graph.

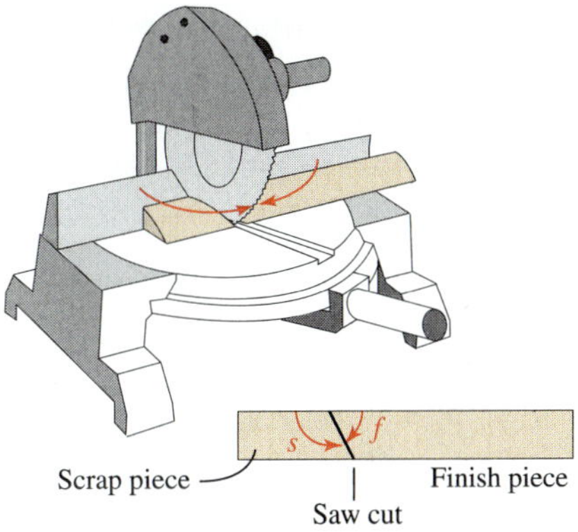

f	s
30	
45	
90	
135	
150	

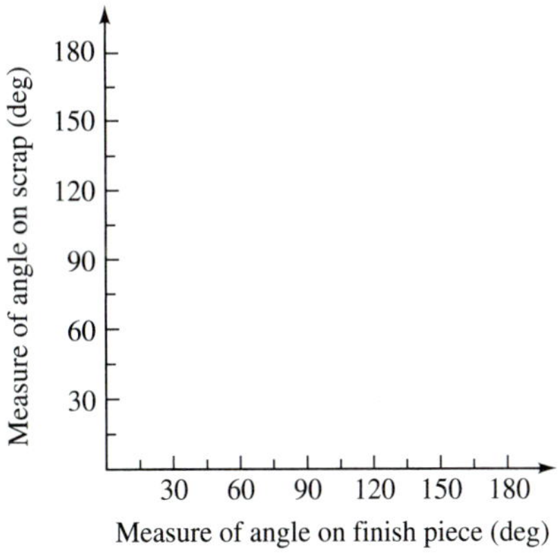

34. PRODUCTION PLANNING Suppose r towel racks are to be manufactured. Complete the four formulas that planners could use to order the necessary number of oak mounting plates p, bar holders b, chrome bars c, and wood screws s.

$p = \underline{\quad\quad}$ $b = \underline{\quad\quad}\,r$ $c = \underline{\quad\quad}$ $s = \underline{\quad\quad}\,r$

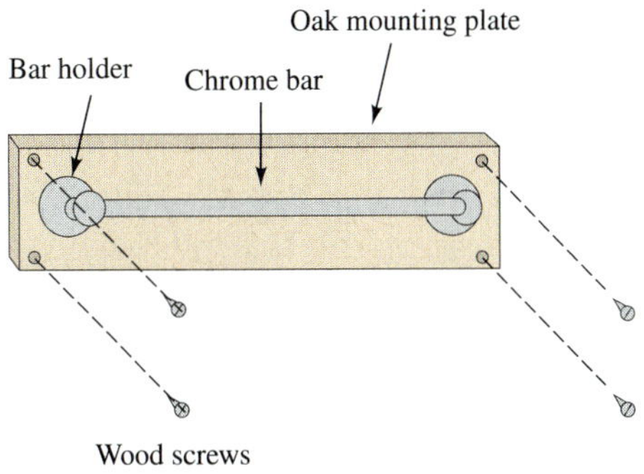

WRITING

35. Explain the difference between an expression and an equation. Give examples.

36. Use each word below in a sentence that indicates a mathematical operation. If you are unsure of the meaning of a word, look it up in a dictionary.

quadrupled	deleted	bisected
confiscated	annexed	docked

CHALLENGE PROBLEMS

37. Use the formula $F = \frac{9}{5}C + 32$ to complete the table.

C	F
5	
	50
15	

38. Fill in the blank: If $T = 16s$ and $s = \frac{r}{2}$ then $T = \underline{\quad\quad}\, r$.

1.2 The Real Number System

- Natural numbers, whole numbers, and integers
- Rational numbers
- Irrational numbers
- Real numbers
- The real number line
- Inequality symbols
- Opposites
- Absolute value

In this course, we will work with *real numbers.* The set of real numbers is a collection of several other important sets of numbers.

NATURAL NUMBERS, WHOLE NUMBERS, AND INTEGERS

The following graph shows the daily low temperatures in Anchorage, Alaska, for the first seven days of January. On the horizontal axis, 1, 2, 3, 4, 5, 6, and 7 denote the days of the month. This collection of numbers is called a **set,** and the **members** or **elements** of the set can be listed within **braces** { }.

$$\{1, 2, 3, 4, 5, 6, 7\}$$

This is read as "the set containing the elements 1, 2, 3, 4, 5, 6, and 7."

Each of these numbers also belongs to a more extensive set of numbers that we use to count with, called the *natural numbers.*

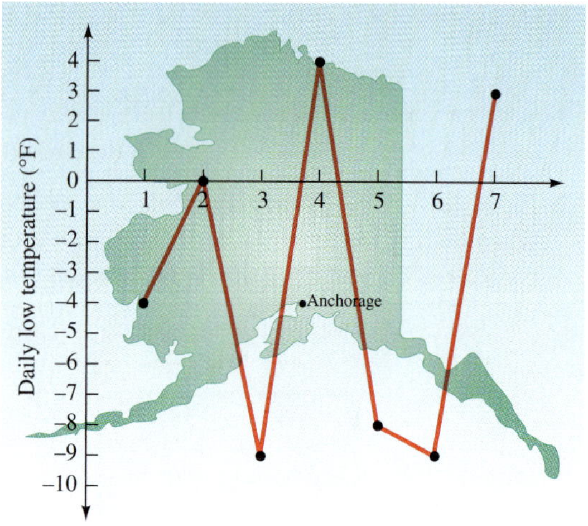

Natural Numbers The set of **natural numbers** is $\{1, 2, 3, 4, 5, 6, 7, 8, 9, 10, \ldots\}$.

The three dots . . . in the previous definition mean that the established pattern continues forever.

The natural numbers, together with 0, form the set of *whole numbers.*

Whole Numbers The set of **whole numbers** is $\{0, 1, 2, 3, 4, 5, \ldots\}$.

When all the members of one set are members of a second set, we say the first set is a **subset** of the second set. Since every natural number is also a whole number, the set of natural numbers is a subset of the set of whole numbers.

Two other important subsets of the whole numbers are the *prime numbers* and the *composite numbers.*

Prime Numbers and Composite Numbers A **prime number** is a whole number greater than 1 that has only itself and 1 as factors. The first ten prime numbers are 2, 3, 5, 7, 11, 13, 17, 19, 23, and 29.

A **composite number** is a whole number, greater than 1, that is not prime. The first ten composite numbers are 4, 6, 8, 9, 10, 12, 14, 15, 16, and 18.

Recall from arithmetic that every composite number can be written as the product of prime numbers. For example,

$$6 = 2 \cdot 3, \quad 25 = 5 \cdot 5, \quad \text{and} \quad 168 = 2 \cdot 2 \cdot 2 \cdot 3 \cdot 7$$

The graph of the daily low temperatures contains both **positive numbers,** numbers greater than 0, and **negative numbers,** numbers less than 0. For example, on January 7, the low was 3°F (3 degrees above zero) and on January 3 it was -9°F (9 degrees below zero). On January 2, the low temperature was 0°F. Zero is neither positive nor negative. These numbers, 3, -9, and 0, are examples of *integers.*

Integers The set of **integers** is $\{\ldots, -4, -3, -2, -1, 0, 1, 2, 3, 4, \ldots\}$.

The Language of Algebra

The *positive integers* are: 1, 2, 3, 4, 5, ...
The *negative integers* are: $-1, -2, -3, -4, -5, \ldots$.

Integers that are divisible by 2 are called *even integers,* and integers that are not divisible by 2 are called *odd integers.*

Even integers: $\ldots, -6, -4, -2, 0, 2, 4, 6, \ldots$

Odd integers: $\ldots, -5, -3, -1, 1, 3, 5, \ldots$

Since every whole number is also an integer, the set of whole numbers is a subset of the set of integers.

RATIONAL NUMBERS

In this course, we will work with positive and negative fractions. For example, the slope of a line might be $\frac{7}{12}$ or a tank might drain at a rate of $-\frac{40}{3}$ gallons per minute. We will also work with mixed numbers. For instance, we might speak of $5\frac{7}{8}$ cups of flour or of a river that is $3\frac{1}{2}$ feet below flood stage ($-3\frac{1}{2}$ ft). These fractions and mixed numbers are examples of *rational numbers.*

Rational Numbers A **rational number** is any number that can be written as $\frac{a}{b}$, where a and b represent integers and $b \neq 0$.

Some other examples of rational numbers are

$$\frac{3}{4}, \quad \frac{25}{25}, \quad \text{and} \quad \frac{19}{6}$$

To show that negative fractions are rational numbers, we use the following fact.

Negative Fractions Let a and b represent numbers, where b is not 0,

$$-\frac{a}{b} = \frac{-a}{b} = \frac{a}{-b}$$

The Language of Algebra

Rational numbers are so named because they can be expressed as the ratio (quotient) of two integers: $\frac{\text{integer}}{\text{integer}}$.

To illustrate this rule, consider $-\frac{40}{3}$. It is a rational number because it can be written as $\frac{-40}{3}$, or as $\frac{40}{-3}$.

Positive and negative mixed numbers such as $5\frac{7}{8}$ and $-3\frac{1}{2}$ are rational numbers because they can be expressed as fractions.

$$5\frac{7}{8} = \frac{47}{8} \quad \text{and} \quad -3\frac{1}{2} = -\frac{7}{2} = \frac{-7}{2}$$

Any natural number, whole number, or integer can be expressed as a fraction with a denominator of 1. For example, $5 = \frac{5}{1}$, $0 = \frac{0}{1}$, and $-3 = \frac{-3}{1}$. Therefore, every natural number, whole number, and integer is also a rational number.

Throughout the book we will work with decimals. Some examples of uses of decimals are:

- The interest rate of a loan was $11\% = 0.11$.
- In baseball, the distance from home plate to second base is 127.279 feet.
- The third-quarter loss for a business was -2.7 million dollars.

Terminating decimals such as 0.11, 127.279, and -2.7 are rational numbers, because they can be written as fractions with integer numerators and nonzero integer denominators.

$$0.11 = \frac{11}{100} \qquad 127.279 = 127\frac{279}{1{,}000} = \frac{127{,}279}{1{,}000} \qquad -2.7 = -2\frac{7}{10} = \frac{-27}{10}$$

Examples of **repeating decimals** are 0.333 . . . and 4.252525. . . . Any repeating decimal can be expressed as a fraction with an integer numerator and a nonzero integer denominator. For example, $0.333\ldots = \frac{1}{3}$ and $4.252525\ldots = 4\frac{25}{99} = \frac{421}{99}$. Since every repeating decimal can be written as a fraction, repeating decimals are also rational numbers.

Rational Numbers The set of **rational numbers** is the set of all terminating and all repeating decimals.

EXAMPLE 1

Change each fraction to a decimal to determine whether the decimal terminates or repeats:

a. $\frac{4}{5}$ and **b.** $\frac{17}{6}$.

Solution **a.** To change $\frac{4}{5}$ to a decimal, we divide the numerator by the denominator.

$$\begin{array}{r} .8 \\ 5\overline{)4.0} \\ \underline{4\,0} \\ 0 \end{array}$$

Write a decimal point and a 0 to the right of 4.

In decimal form, $\frac{4}{5}$ is 0.8. This is a terminating decimal.

b. To change $\frac{17}{6}$ to a decimal, we perform the division and obtain 2.8333. . . . This is a repeating decimal, because the digit 3 repeats forever. It can be written as $2.8\overline{3}$, where the overbar indicates that the 3 repeats.

Self Check 1 Change each fraction to a decimal to determine whether it terminates or repeats:

a. $\frac{25}{990}$ and **b.** $\frac{47}{50}$.

The set of rational numbers is too extensive to list in the same way that we listed the other sets in this section. Instead, we use the following **set-builder** notation to describe it. The set of rational numbers is

$$\left\{\frac{a}{b} \,\middle|\, a \text{ and } b \text{ are integers, with } b \neq 0.\right\}$$

Read as "the set of all numbers of the form $\frac{a}{b}$, such that a and b represent integers, with $b \neq 0$."

IRRATIONAL NUMBERS

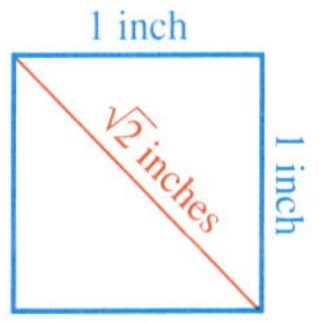

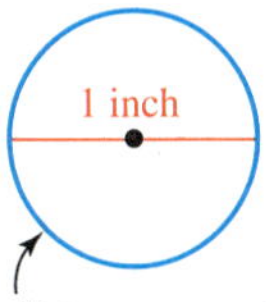

The distance around the circle is π inches

Some numbers cannot be expressed as fractions with an integer numerator and a nonzero integer denominator. Such numbers are called **irrational numbers.** One example of an irrational number is $\sqrt{2}$. It can be shown that a square, with sides of length 1 inch, has a diagonal that is $\sqrt{2}$ inches long.

The number represented by the Greek letter π (pi) is another example of an irrational number. It can be shown that a circle, with a 1-inch diameter, has a circumference of π inches.

Expressed in decimal form,

$$\sqrt{2} = 1.414213562\ldots \quad \text{and} \quad \pi = 3.141592654\ldots$$

These decimals neither terminate nor repeat.

Irrational Numbers An **irrational number** is a nonterminating, nonrepeating decimal. An irrational number cannot be expressed as a fraction with an integer numerator and a nonzero integer denominator.

ACCENT ON TECHNOLOGY: APPROXIMATING IRRATIONAL NUMBERS

We can approximate the value of irrational numbers with a scientific calculator. To find the value of π, we press the [π] key.

[π] (You may have to use a [2nd] or [Shift] key first.) — 3.141592654

We see that $\pi \approx 3.141592654$. (Read $\approx$ as "is approximately equal to.") To the nearest thousandth, $\pi \approx 3.142$.

To approximate $\sqrt{2}$, we enter 2 and press the square root key [√].

2 [√] — 1.414213562

We see that $\sqrt{2} \approx 1.414213562$. To the nearest hundredth, $\sqrt{2} \approx 1.41$.

To find π and $\sqrt{2}$ with a graphing calculator, we proceed as follows.

[2nd] π [ENTER] — π
3.141592654

[2nd] [√] 2 [)] [ENTER] — √(2)
1.414213562

Some other examples of irrational numbers are

$$\sqrt{97} = 9.848857802\ldots$$
$$-\sqrt{7} = -2.64575131\ldots \quad \text{This is a negative irrational number.}$$
$$2\pi = 6.283185307\ldots \quad 2\pi \text{ means } 2 \cdot \pi.$$

Caution

Don't classify a number such as 4.12122122212222 . . . as a repeating decimal. Although it exhibits a pattern, no block of digits repeats forever. It is a nonterminating, nonrepeating decimal—an irrational number.

Not all square roots are irrational numbers. When we simplify square roots such as $\sqrt{9}$, $\sqrt{36}$, and $\sqrt{400}$, it is apparent that they are rational numbers: $\sqrt{9} = 3$, $\sqrt{36} = 6$, and $\sqrt{400} = 20$.

REAL NUMBERS

The set of rational numbers together with the set of irrational numbers form the set of **real numbers.** This means that every real number can be written as either a terminating decimal, a repeating decimal, or a nonterminating, nonrepeating decimal. Thus, the set of real numbers is the set of all decimals.

The Real Numbers A **real number** is any number that is either a rational or an irrational number. All the points on a number line represent the set of real numbers.

The figure shows how the sets of numbers introduced in this section are related; it also gives some specific examples of each type of number. Note that a number can belong to more than one set. For example, -6 is an integer, a rational number, and a real number.

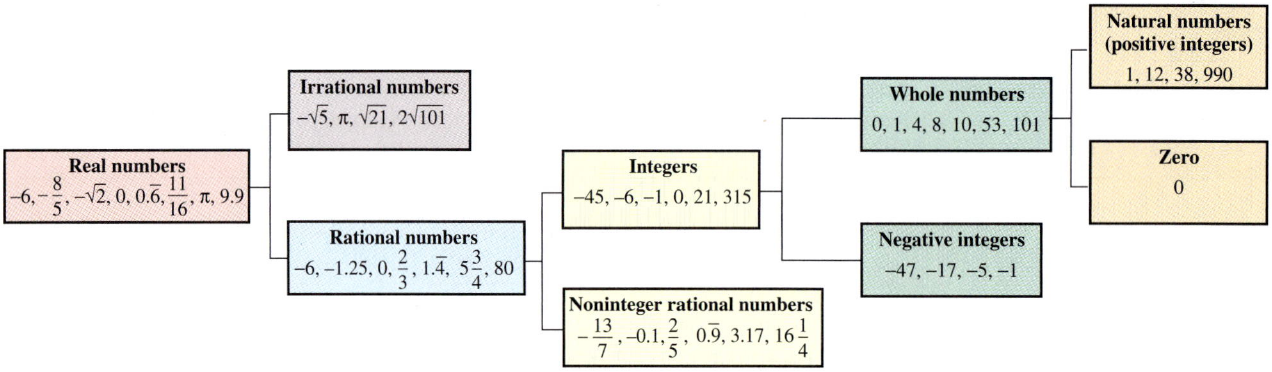

EXAMPLE 2

Classifying real numbers. Which numbers in the following set are natural numbers, whole numbers, integers, rational numbers, irrational numbers, and real numbers?

$$\left\{\frac{5}{8}, -0.03, 45, -9, \sqrt{7}, 5\frac{2}{3}, 0, -1.727227222\ldots, 0.\overline{25}\right\}$$

Solution

Natural numbers: 45

Whole numbers: 45, 0

Integers: 45, -9, 0

Rational numbers: $\frac{5}{8}$, 45, -9, $5\frac{2}{3}$, and 0 are rational numbers because each of them can be expressed as a fraction: $45 = \frac{45}{1}$, $-9 = \frac{-9}{1}$, $5\frac{2}{3} = \frac{17}{3}$, and $0 = \frac{0}{1}$.

The terminating decimal $-0.03 = \frac{-3}{100}$ and the repeating decimal $0.\overline{25} = \frac{25}{99}$ are also rational numbers.

Irrational numbers: The nonterminating, nonrepeating decimals $\sqrt{7} = 2.645751311\ldots$ and $-1.727227222\ldots$ are irrational numbers.

Real numbers: $\frac{5}{8}, -0.03, 45, -9, \sqrt{7}, 5\frac{2}{3}, 0, -1.727227222\ldots, 0.\overline{25}$

Self Check 2 Use the instructions for Example 2 with the following set:

$$\left\{-\pi, -5, 3.4, \sqrt{19}, 1, \frac{16}{5}, 9.\overline{7}\right\}$$

THE REAL NUMBER LINE

We can illustrate real numbers using a **number line.** To each real number, there corresponds a point on the line. Furthermore, to each point on the line, there corresponds a number, called its **coordinate.**

EXAMPLE 3

Graph the set $\left\{-\frac{8}{3}, -1.1, 0.\overline{56}, \frac{\pi}{2}, -\sqrt{15}, \text{and } 2\sqrt{2}\right\}$ on a number line.

Solution To help locate the graph of each number, we make some observations.

- Expressed as a mixed number, $-\frac{8}{3} = -2\frac{2}{3}$.
- Since -1.1 is less than -1, its graph is to the *left* of -1.
- $0.\overline{56} \approx 0.6$
- From a calculator, $\frac{\pi}{2} \approx 1.6$.
- From a calculator, $-\sqrt{15} \approx -3.9$.
- $2\sqrt{2}$ means $2 \cdot \sqrt{2}$. From a calculator, $2\sqrt{2} \approx 2.8$.

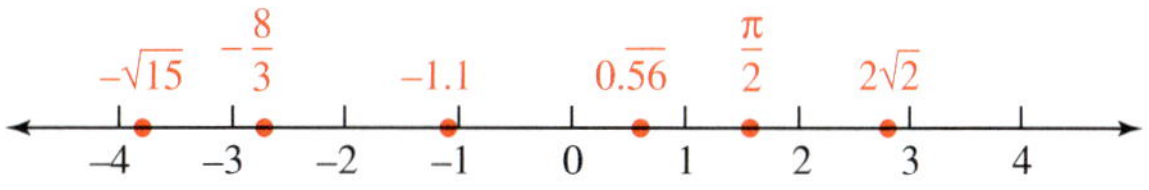

The Language of Algebra

An example of a number that is not on the real number line is $\sqrt{-4}$. It is called an *imaginary number.* We will discuss such numbers in Chapter 8.

Self Check 3 Graph the set $\left\{\pi, -2.\overline{1}, \sqrt{3}, \frac{11}{4}, \text{and } -0.9\right\}$ on a number line.

INEQUALITY SYMBOLS

To show that two quantities are not equal, we can use one of the **inequality symbols** shown in the following table.

The Language of Algebra

If a real number x is positive, then $x > 0$. If a real number x is *nonnegative,* then $x \geq 0$. If a real number x is a *negative* number, then $x < 0$.

Symbol	Read as	Examples
$\neq$	"is not equal to"	$6 \neq 9$ and $0.33 \neq \frac{3}{5}$
$<$	"is less than"	$\frac{22}{3} < \frac{23}{3}$ and $-7 < -6$
$>$	"is greater than"	$19 > 5$ and $\frac{1}{2} > 0.3$
$\leq$	"is less than or equal to"	$3.5 \leq 3.\overline{5}$ and $1\frac{4}{5} \leq 1.8$
$\geq$	"is greater than or equal to"	$29 \geq 29$ and $-15.2 \geq -16.7$

It is always possible to write an equivalent inequality with the inequality symbol pointing in the opposite direction. For example,

$-3 < 4$	is equivalent to	$4 > -3$
$5.3 \geq 2.9$	is equivalent to	$2.9 \leq 5.3$

EXAMPLE 4

Use one of the symbols $>$ or $<$ to make each statement true: **a.** $-24 \quad\square\quad -25$ and **b.** $\frac{3}{4} \quad\square\quad 0.76$.

Solution **a.** Since -24 is to the right of -25 on the number line, $-24 > -25$.

b. If we express the fraction $\frac{3}{4}$ as a decimal, we can easily compare it to 0.76.

Since $\frac{3}{4} = 0.75$, $\frac{3}{4} < 0.76$.

Self Check 4 Use one of the symbols $\geq$ or $\leq$ to make each statement true: **a.** $\frac{2}{3} \quad\square\quad \frac{4}{3}$ and **b.** $8\frac{1}{2} \quad\square\quad 8.4$.

OPPOSITES

In the figure, we can see that -3 and 3 are both a distance of 3 units away from zero on the number line. Because of this, we say that -3 and 3 are **opposites** or **additive inverses.**

Parentheses are used to express the opposite of a negative number. For example, the opposite of -3 is written as $-(-3)$. Since -3 and 3 are the same distance from zero, the opposite of -3 is 3. Symbolically, this can be written $-(-3) = 3$. In general, we have the following.

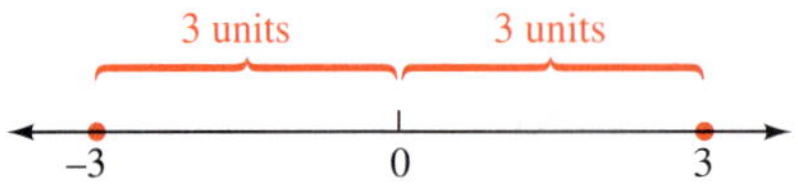

Opposites The **opposite** of a number a is the number $-a$. If a is a real number, then $-(-a) = a$.

ABSOLUTE VALUE

The **absolute value** of any real number is the distance between the number and zero on a number line. To indicate absolute value, the number is inserted between two vertical bars. For example, the points shown in the previous figure with coordinates of 3 and -3 both lie 3 units from zero. Thus, $|3| = 3$ and $|-3| = 3$.

The absolute value of a number can be defined more formally as follows.

Absolute Value For any real number a, $\begin{cases} \text{If } a \geq 0, \text{ then } |a| = a. \\ \text{If } a < 0, \text{ then } |a| = -a. \end{cases}$

EXAMPLE 5

Find the value of each expression: **a.** $|34|$, **b.** $\left|-\frac{4}{5}\right|$, **c.** $|0|$, and **d.** $-|-1.8|$.

Solution

a. Since 34 is a distance of 34 from 0 on a number line, $|34| = 34$.

b. $-\frac{4}{5}$ is a distance of $\frac{4}{5}$ from 0 on a number line. Therefore, $\left|-\frac{4}{5}\right| = \frac{4}{5}$.

c. $|0| = 0$

d. The negative sign outside the absolute value bars means to find the opposite of $|-1.8|$.

$$-|\mathbf{-1.8}| = -(\mathbf{1.8}) \quad \text{Find } |-1.8| \text{ first: } |-1.8| = 1.8.$$
$$= -1.8$$

Self Check 5 Find the value of each expression: **a.** $|-9.6|$, **b.** $-|-12|$, **c.** $\left|\frac{3}{2}\right|$.

Answers to Self Checks

1. a. $\frac{25}{990} = 0.0\overline{25}$, repeating decimal, **b.** $\frac{47}{50} = 0.94$, terminating decimal

2. natural numbers: 1; whole numbers: 1; integers: -5, 1; rational numbers: -5, 3.4, 1, $\frac{16}{5}$, $9.\overline{7}$; irrational numbers: $-\pi$, $\sqrt{19}$; real numbers; all

3.

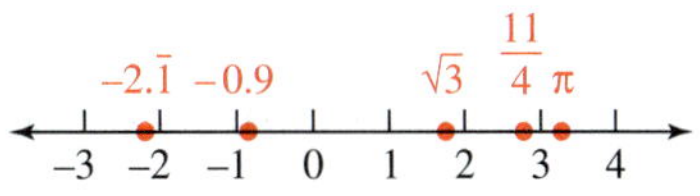

4. a. $\le$, **b.** $\ge$

5. a. 9.6, **b.** -12, **c.** $\frac{3}{2}$

1.2 STUDY SET

VOCABULARY **Fill in the blanks.**

1. A ________ number is any number that can be written as a fraction with an integer numerator and a nonzero integer denominator.

2. A _______ number is a whole number greater than 1 that has only itself and 1 as factors. A _________ number is a whole number greater than 1 that is not prime.

3. The ____________ of any real number is the distance between the number and zero on a number line.

4. The set of rational numbers together with the set of irrational numbers form the set of _____ numbers.

5. _________ numbers are nonterminating, nonrepeating decimals.

6. ________ numbers are greater than 0 and ________ numbers are less than 0.

7. When all the members of one set are members of a second set, we say the first set is a _______ of the second set.

8. ______ is neither positive nor negative.

CONCEPTS **List the elements of $\{-3, -\frac{8}{5}, 0, \frac{2}{3}, 1, 2, \sqrt{3}, \pi, 4.75, 9, 16.\overline{6}\}$ that belong to the following sets.**

9. Natural numbers

10. Whole numbers

11. Integers

12. Rational numbers

13. Irrational numbers

14. Real numbers

15. Even natural numbers

16. Odd integers

17. Prime numbers

18. Composite numbers

19. Odd composite numbers

20. Odd prime numbers

Decide whether each number is a repeating or a nonrepeating decimal, and whether it is a rational or an irrational number.

21. 0.090090009. . .

22. $0.\overline{09}$

23. 5.41414141. . .

24. 1.414213562. . .

25. Show that each of the following numbers is a rational number by expressing it as a fraction with an integer numerator and a nonzero integer denominator.

$$7, -7\tfrac{3}{5}, 0.007, 700.1$$

26. Decide whether each statement is true or false.

a. All prime numbers are odd numbers.

b. $6 \geq 6$

c. 0 is neither even nor odd.

d. Every real number is a rational number.

27. Fill in the blanks:

For any real number a, $\begin{cases} \text{If } a \geq 0, \text{ then } |a| = \text{\_\_\_\_} \\ \text{If } a < 0, \text{ then } |a| = \text{\_\_\_\_} \end{cases}$

28. Name two numbers that are 6 units away from −2 on the number line.

29. The following diagram can be used to show how the natural numbers, whole numbers, integers, rational numbers, and irrational numbers make up the set of real numbers. If the natural numbers can be represented as shown, label each of the other sets.

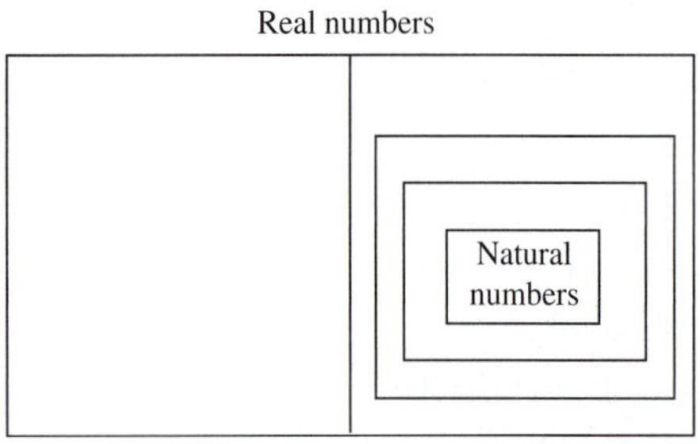

30. The formula $C = \pi D$ gives the circumference C of a circle, where D is the length of its diameter. Find the circumference of the wedding ring. Give an *exact* answer and then an *approximate* answer, rounded to the nearest hundredth of an inch.

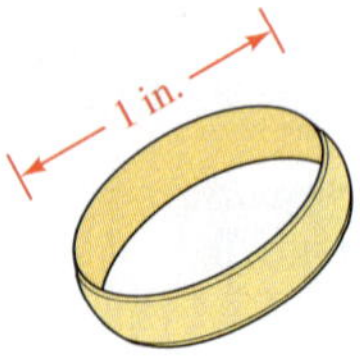

NOTATION **Fill in the blanks.**

31. The symbol < means "__________."

32. $|-2|$ is read as "the ________ value _____ −2."

33. The symbols { } are called ________.

34. The symbol ≥ means "____________________."

35. Describe the set of rational numbers using set-builder notation.

36. List two other ways that the fraction $-\frac{2}{3}$ can be written.

PRACTICE **Change each fraction into a decimal and classify the result as a terminating or a repeating decimal.**

37. $\frac{7}{8}$

38. $\frac{8}{3}$

39. $-\frac{11}{15}$

40. $-\frac{19}{16}$

Graph each set on a number line.

41. $\left\{-\frac{5}{2}, -0.1, 2.142765. . . , \frac{\pi}{3}, -\sqrt{11}, 2\sqrt{3}\right\}$

42. $\left\{2\frac{1}{9}, -3.821134. . . , -\frac{\pi}{2}, \sqrt{15}, -0.9, \frac{\sqrt{2}}{2}\right\}$

43. $\{3.\overline{15}, \frac{22}{7}, 3\frac{1}{8}, \pi, \sqrt{10}, 3.1\}$

44. $\{-0.\overline{331}, -0.331, -\frac{1}{3}, -\sqrt{0.11}\}$

45. The set of prime numbers less than 8

46. The set of integers between −7 and 0

47. The set of odd integers between 10 and 18

48. The set of composite numbers less than 10

49. The set of positive odd integers less than 12

50. The set of negative even integers greater than −7

Insert either a < or a > symbol to make a true statement.

51. 8 ☐ 9

52. 9 ☐ 0

53. $-(-5)$ ☐ -10

54. $|-3|$ ☐ $-(-6)$

55. -7.999 ☐ -7.1

56. $4\frac{1}{2}$ ☐ $\frac{7}{2}$

57. $6.\overline{1}$ ☐ $-(-6)$

58. -6.07 ☐ $-\frac{17}{6}$

Write each statement with the inequality symbol pointing in the opposite direction.

59. $19 > 12$

60. $-3 \geq -5$

61. $-6 \leq -5$

62. $-10 < 0$

Find the value of each expression.

63. $|20|$

64. $|-20|$

65. $-|-6|$

66. $-|-8|$

67. $|-5.9|$

68. $-|1.\overline{27}|$

69. $\left|\frac{5}{4}\right|$

70. $\left|-\frac{5}{16}\right|$

APPLICATIONS

71. DRAFTING Express each dimension in the drawing of a bracket as a four-place decimal.

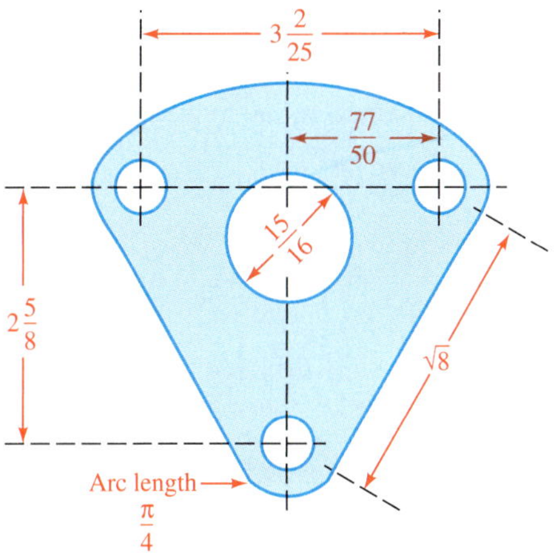

72. pH SCALE The pH scale is used to measure the strength of acids and bases (alkalines) in chemistry. It can be thought of as a number line. On the scale, graph and label each pH measurement given in the table.

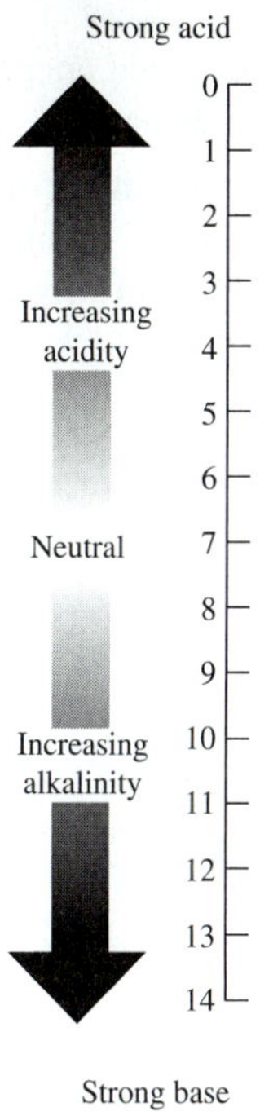

Solution	pH
Seawater	8.5
Cola	2.9
Battery acid	1.0
Milk	6.6
Blood	7.4
Ammonia	11.9
Saliva	6.1
Oven cleaner	13.2
Black coffee	5.0
Toothpaste	9.9
Tomato juice	4.1

WRITING

73. Explain why the whole numbers are a subset of the integers.

74. What is a real number? Give examples.

75. Explain why there are no even prime numbers greater than 2.

76. Explain why every integer is a rational number, but not every rational number is an integer.

REVIEW

77. Is $\frac{3x-4}{2}$ an equation or an expression?

78. Translate into mathematical symbols: The weight of an object in ounces is 16 times its weight in pounds.

Complete each table.

79. $T = x - 1.5$

x	T
3.7	
10	
30.6	

80. $j = 3m$

m	j
0	
15	
300	

CHALLENGE PROBLEMS

81. How many integers have an absolute value that is less than 50?

82. How many odd integers have an absolute value between 20 and 40?

83. The **trichotomy property** of real numbers states that: If a and b are real numbers, then $a < b$, $a = b$, or $a > b$. Explain why this is true.

84. Which of the following statements are always true?

a. $|a + b| = |a| + |b|$

b. $|a \cdot b| = |a| \cdot |b|$

c. $|a + b| \leq |a| + |b|$

1.3 Operations with Real Numbers

- Adding real numbers
- Subtracting real numbers
- Multiplying real numbers
- Dividing real numbers
- Raising a real number to a power
- Finding a square root
- Order of operations
- Evaluating algebraic expressions
- Area and volume

Six operations can be performed with real numbers: addition, subtraction, multiplication, division, raising to a power, and finding a root. In this section, we will review the rules for performing these operations. We will also discuss how to evaluate numerical expressions involving several operations.

ADDING REAL NUMBERS

When two numbers are added, the result is their **sum.** The rules for adding real numbers are as follows:

Adding Two Real Numbers

To add two positive numbers, add them in the usual way. The answer is positive.

To add two negative numbers, add their absolute values and make the answer negative.

To add a positive number and a negative number, subtract the smaller absolute value from the larger.

1. If the positive number has the larger absolute value, the answer is positive.
2. If the negative number has the larger absolute value, make the answer negative.

EXAMPLE 1

Add: **a.** $-5 + (-3)$, **b.** $8.9 + (-5.1)$, **c.** $-\frac{13}{15} + \frac{3}{5}$, and **d.** $6 + (-10) + (-1)$.

Solution

a. $-5 + (-3) = -8$

Both numbers are negative. Add their absolute values, 5 and 3, to get 8, and make the answer negative.

b. $8.9 + (-5.1) = 3.8$

One number is positive and the other is negative. Subtract their absolute values, 5.1 from 8.9, to get 3.8. Because 8.9 has the larger absolute value, the answer is positive.

c. $-\frac{13}{15} + \frac{3}{5} = -\frac{13}{15} + \frac{9}{15}$

Express $\frac{3}{5}$ in terms of the lowest common denominator, 15: $\frac{3}{5} = \frac{3 \cdot 3}{5 \cdot 3} = \frac{9}{15}$.

$= -\frac{4}{15}$

Subtract the absolute values, $\frac{9}{15}$ from $\frac{13}{15}$, to get $\frac{4}{15}$, and make the answer negative because $-\frac{13}{15}$ has the larger absolute value.

d. To add three or more real numbers, add from left to right.

$$6 + (-10) + (-1) = -4 + (-1)$$
$$= -5$$

Self Check 1. Add: **a.** $-34 + 25$, **b.** $-70.4 + (-21.2)$, **c.** $\frac{7}{4} + \left(-\frac{3}{2}\right)$, and **d.** $-16 + 17 + (-5)$.

SUBTRACTING REAL NUMBERS

When two numbers are subtracted, the result is their **difference.** To find a difference, we can change the subtraction into an equivalent addition. For example, the subtraction $7 - 4$ is equivalent to the addition $7 + (-4)$, because they have the same answer:

$$7 - 4 = 3 \quad \text{and} \quad 7 + (-4) = 3$$

This suggests that to subtract two numbers, we can change the sign of the number being subtracted and add.

Subtracting Two Real Numbers

To subtract two real numbers, add the first number to the opposite (additive inverse) of the number to be subtracted.

Let a and b represent real numbers,

$$a - b = a + (-b)$$

EXAMPLE 2

Subtract: **a.** $2 - 8$ **b.** $-1.3 - 5.5$, **c.** $-\frac{14}{3} - \left(-\frac{7}{3}\right)$, **d.** Subtract 9 from -6, and **e.** $-11 - (-1) - 5$.

Solution

a. $2 - 8 = 2 + (-8)$ (Add ... the opposite)

$= -6$

Here, 8 is being subtracted, so we change the sign of 8 and add. Do not change the sign of 2.

b. $-1.3 - 5.5 = -1.3 + (-5.5)$

$= -6.8$

Change the sign of 5.5 and add. Do not change the sign of -1.3.

The Language of Algebra

The rule for subtracting real numbers is often summarized as: *Subtraction is the same as adding the opposite.*

c. $-\frac{14}{3} - \left(-\frac{7}{3}\right) = -\frac{14}{3} + \frac{7}{3}$

$= -\frac{7}{3}$

Change the sign of $-\frac{7}{3}$ and add.

d. The number to be subtracted is 9. When we translate, we must reverse the order in which 9 and -6 appear in the sentence.

Subtract 9 from -6.

$$\begin{aligned} -6 - 9 &= -6 + (-9) \quad \text{Add the opposite of 9.} \\ &= -15 \end{aligned}$$

e. To subtract three or more real numbers, subtract from left to right.

$$\begin{aligned} \mathbf{-11 - (-1)} - 5 &= \mathbf{-10} - 5 \\ &= -15 \end{aligned}$$

Self Check 2 Subtract: **a.** $-15 - 4$, **b.** $-12.1 - (-7.6)$, **c.** $\frac{5}{9} - \frac{7}{9}$, **d.** Subtract 1 from -5, and **e.** $5 - 4 - (-15)$.

MULTIPLYING REAL NUMBERS

When two numbers are multiplied, we call the numbers **factors** and the result is their **product.** The rules for multiplying real numbers are as follows:

Multiplying Two Numbers with Unlike Signs To multiply a positive number and a negative number, multiply their absolute values and make the answer negative.

Multiplying Two Numbers with Like Signs To multiply two real numbers with the same sign, multiply their absolute values. The product is positive.

EXAMPLE 3

Multiply: **a.** $4(-7)$, **b.** $-5.2(-3)$, **c.** $-\frac{7}{9}\left(\frac{3}{16}\right)$, and **d.** $8(-2)(-3)$.

Solution

a. $4(-7) = -28$ — Multiply the absolute values, 4 and 7, to get 28. Since the signs are unlike, make the answer negative.

b. $-5.2(-3) = 15.6$ — Multiply the absolute values 5.2 and 3, to get 15.6. Since the signs are like, the answer is positive.

c. $-\frac{7}{9}\left(\frac{3}{16}\right) = -\frac{7 \cdot 3}{9 \cdot 16}$ — Multiply the numerators and multiply the denominators. Since the signs of the factors are unlike, the product is negative.

$= -\frac{7 \cdot \overset{1}{\cancel{3}}}{\underset{1}{\cancel{3}} \cdot 3 \cdot 16}$ — Factor 9 as $3 \cdot 3$ and simplify the fraction: $\frac{3}{3} = 1$.

$= -\frac{7}{48}$ — Multiply in the numerator and denominator.

d. To multiply three or more real numbers, multiply from left to right.

$$8(-2)(-3) = -16(-3)$$
$$= 48$$

Self Check 3 Multiply: **a.** $(-6)(5)$, **b.** $(-4.1)(-8)$, **c.** $\left(\frac{4}{3}\right)\left(-\frac{1}{8}\right)$, and **d.** $-4(-9)(-3)$.

DIVIDING REAL NUMBERS

When two numbers are divided, the result is their **quotient.** In the division $\frac{x}{y} = q$, the quotient q is a number such that $y \cdot q = x$. We can use this relationship to find rules for dividing real numbers.

$$\frac{10}{2} = 5, \text{ because } 2(5) = 10 \qquad \frac{-10}{-2} = 5, \text{ because } -2(5) = -10$$

$$\frac{-10}{2} = -5, \text{ because } 2(-5) = -10 \qquad \frac{10}{-2} = -5, \text{ because } -2(-5) = 10$$

These results suggest the following rules for dividing real numbers. Note that they are similar to those for multiplying real numbers.

Dividing Two Real Numbers

To divide two real numbers, divide their absolute values.

1. The quotient of two numbers with *like* signs is positive.
2. The quotient of two numbers with *unlike* signs is negative.

EXAMPLE 4

Divide: **a.** $\frac{-44}{11}$ and **b.** $\frac{-2.7}{-9}$.

Solution **a.** $\frac{-44}{11} = -4$ Divide the absolute values, 44 by 11, to get 4. Since the signs are unlike, make the answer negative.

b. $\frac{-2.7}{-9} = 0.3$ Divide the absolute values, 2.7 by 9, to get 0.3. Since the signs are like, the quotient is positive.

Self Check 4 Divide: **a.** $\frac{55}{-5}$ and **b.** $\frac{-7.2}{-6}$.

To divide two fractions, we multiply the first fraction by the **reciprocal** of the second fraction. In symbols, if a, b, c, and d are real numbers, and no denominators are 0, then

$$\frac{a}{b} \div \frac{c}{d} = \frac{a}{b} \cdot \frac{d}{c}$$ $\frac{d}{c}$ is the reciprocal of $\frac{c}{d}$.

EXAMPLE 5

Divide: **a.** $\frac{2}{3} \div \left(-\frac{3}{5}\right)$ and **b.** $-\frac{1}{2} \div (-6)$.

Solution **a.** $\frac{2}{3} \div \left(-\frac{3}{5}\right) = \frac{2}{3} \cdot \left(-\frac{5}{3}\right)$ Multiply by the reciprocal of $-\frac{3}{5}$, which is $-\frac{5}{3}$.

$= -\frac{10}{9}$ Since the factors have unlike signs, the answer is negative.

b. $-\frac{1}{2} \div (-6) = -\frac{1}{2} \cdot \left(-\frac{1}{6}\right)$ Multiply by the reciprocal of -6, which is $-\frac{1}{6}$.

$= \frac{1}{12}$ Since the factors have like signs, the answer is positive.

Self Check 5 Divide: **a.** $-\frac{7}{8} \div \frac{2}{3}$ and **b.** $-\frac{1}{10} \div (-5)$.

The Language of Algebra

When we say a division by 0, such as $\frac{4}{0}$, is *undefined,* we mean it is not allowed or it is not defined. That is, $\frac{4}{0}$ does not represent a number.

Students often confuse division problems such as $\frac{0}{4}$ and $\frac{4}{0}$. We know that $\frac{0}{4} = 0$, because $4 \cdot 0 = 0$. However, $\frac{4}{0}$ is undefined, because there is no real number q such that $0 \cdot q = 4$. In general, if $x \neq 0$, $\frac{0}{x} = 0$ and $\frac{x}{0}$ is undefined.

RAISING A REAL NUMBER TO A POWER

Exponents indicate repeated multiplication. For example,

$3^2 = 3 \cdot 3$ Read 3^2 as "3 to the second power" or "3 squared."

$(-9.1)^3 = (-9.1)(-9.1)(-9.1)$ Read $(-9.1)^3$ as "-9.1 to the third power" or "-9.1 cubed."

$\left(\frac{2}{3}\right)^4 = \left(\frac{2}{3}\right)\left(\frac{2}{3}\right)\left(\frac{2}{3}\right)\left(\frac{2}{3}\right)$ Read $\left(\frac{2}{3}\right)^4$ as "$\frac{2}{3}$ to the fourth power."

These examples suggest the following definition.

Natural-number Exponents A natural-number exponet tells how many times its base is to be used as a factor. For any real number x and any natural number n,

$$x^n = \underbrace{x \cdot x \cdot x \cdot \cdots \cdot x}_{n \text{ factors of } x}$$

The exponential expression x^n is called a **power of x,** and we read it as "x to the nth power." In this expression, x is called the **base,** and n is called the **exponent.** A natural-number exponent tells how many times the base of an exponential expression is to be used as a factor in a product.

Base $\longrightarrow x^n \longleftarrow$ Exponent

EXAMPLE 6

Find each power: **a.** $(-2)^4$, **b.** $\left(\frac{3}{4}\right)^2$, and **c.** -0.1 cubed.

Solution In each case, we use the fact that an exponent tells how many times the base is to be used as a factor in a product.

a. $(-2)^4 = (-2)(-2)(-2)(-2) = 16$ The base is -2. The exponent is 4.

b. $\left(\frac{3}{4}\right)^2 = \frac{3}{4}\left(\frac{3}{4}\right) = \frac{9}{16}$ The base is $\frac{3}{4}$. The exponent is 2.

c. -0.1 cubed means $(-0.1)^3$. The base is -0.1. The exponent is 3.

$$(-0.1)^3 = (-0.1)(-0.1)(-0.1) = -0.001$$

Success Tip

When multiplying signed numbers, an odd number of negative factors gives a negative product. An even number of negative factors gives a positive product.

Self Check 6 Find each power: **a.** $(-3)^3$, **b.** $(0.8)^2$, **c.** 2^4, and **d.** $\frac{7}{5}$ squared.

ACCENT ON TECHNOLOGY: THE SQUARING AND EXPONENTIAL KEYS

A homeowner plans to install a cooking island in her kitchen. (See the figure.) To find the number of square feet of floor space that will be lost, we substitute 3.25 for s in the formula for the area of a square, $A = s^2$. Using the squaring key $\boxed{x^2}$ on a scientific calculator, we can evaluate $(3.25)^2$ as follows:

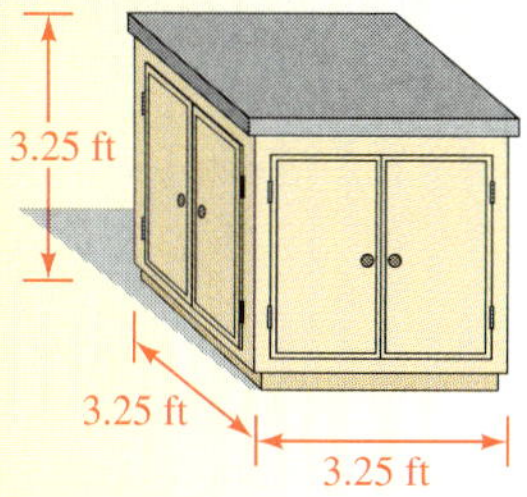

3.25 $\boxed{x^2}$

```
10.5625
```

On a graphing calculator, we have:

3.25 $\boxed{x^2}$ $\boxed{\text{ENTER}}$

```
3.25²
               10.5625
```

About 10.6 square feet of floor space will be lost.

The number of cubic feet of storage space that the cooking island will add can be found by substituting 3.25 for s in the formula for the volume of a cube, $V = s^3$. Using the exponential key $\boxed{y^x}$ ($\boxed{x^y}$ on some calculators), we can evaluate $(3.25)^3$ on a scientific calculator as follows.

3.25 $\boxed{y^x}$ 3 $\boxed{=}$

```
34.328125
```

On a graphing calculator, we have:

3.25 $\boxed{\wedge}$ 3 $\boxed{\text{ENTER}}$

```
3.25^3
               34.328125
```

The cooking island will add about 34.3 cubic feet of storage space.

Although the expressions $(-3)^2$ and -3^2 look alike, they are not. In $(-3)^2$, the base is -3. In -3^2, the base is 3. The $-$ sign in front of 3^2 means the opposite of 3^2. When we evaluate them, we see that the results are different:

$$(-3)^2 = (-3)(-3) = 9 \qquad -3^2 = -(3 \cdot 3) = -9$$

Different results

ACCENT ON TECHNOLOGY: THE PARENTHESES AND NEGATIVE KEYS

To compute $(-3)^2$ with a scientific calculator, use the *parentheses* keys [( ] [ )] and the *negative* key [+/−]. Notice that the negative key is different from the subtraction key [−]. To enter -3, press [+/−] *after* entering 3.

[( ] 3 [+/−] [ )] [x^2] [=] Display: 9

If a graphing calculator is used to find $(-3)^2$, press the negative key [(−)] *before* entering 3.

[( ] [(−)] 3 [ )] [x^2] [ENTER] Display: (−3)² 9

To compute -3^2 with a scientific calculator, think of the expression as $-1 \cdot 3^2$. First, find 3^2. Then press [+/−], which is equivalent to multiplying 3^2 by -1.

3 [x^2] [+/−] Display: −9

A graphing calculator recognizes -3^2 as $-1 \cdot 3^2$, so we can find -3^2 by entering the following:

[(−)] 3 [x^2] [ENTER] Display: −3² −9

FINDING A SQUARE ROOT

Since the product $3 \cdot 3$ can be denoted by the exponential expression 3^2, we say that 3 is squared. The opposite of squaring a number is called finding its **square root.**

All positive numbers have two square roots, one positive and one negative. For example, the two square roots of 9 are 3 and -3. The number 3 is a square root of 9, because $3^2 = 9$, and -3 is a square root of 9, because $(-3)^2 = 9$.

The symbol $\sqrt{\ }$, called a **radical symbol,** is used to represent the positive (or *principal*) square root of a number.

Principal Square Root

A number b is a square root of a if $b^2 = a$.

If $a > 0$, the expression $\sqrt{a}$ represents the **principal** (or positive) **square root** of a. The principal square root of 0 is 0: $\sqrt{0} = 0$.

The principal square root of a positive number is always positive. Although 3 and -3 are both square roots of 9, only 3 is the principal square root. The symbol $\sqrt{9}$ represents 3. To represent -3, we place a $-$ sign in front of the radical:

$$\sqrt{9} = 3 \quad \text{and} \quad -\sqrt{9} = -3$$

EXAMPLE 7 Find each square root: **a.** $\sqrt{121}$, **b.** $-\sqrt{49}$, **c.** $\sqrt{\frac{1}{4}}$, and **d.** $\sqrt{0.09}$.

Solution **a.** $\sqrt{121} = 11$, because $11^2 = 121$. **b.** Since $\sqrt{49} = 7$, $-\sqrt{49} = -7$.

c. $\sqrt{\frac{1}{4}} = \frac{1}{2}$, because $\left(\frac{1}{2}\right)^2 = \frac{1}{4}$. **d.** $\sqrt{0.09} = 0.3$, because $(0.3)^2 = 0.09$.

Self Check 7 Find each square root: **a.** $\sqrt{64}$, **b.** $-\sqrt{100}$, **c.** $\sqrt{\frac{4}{25}}$, **d.** $\sqrt{1}$, **e.** $\sqrt{0.81}$, and **f.** $-\sqrt{400}$.

ORDER OF OPERATIONS

We will often have to evaluate expressions involving several operations. For example, consider the expression $3 + 2 \cdot 5$. To evaluate it, we can perform the addition first and then the multiplication. Or we can perform the multiplication first and then the addition. However, we get different results.

Method 1: Add first

$3 + 2 \cdot 5 = 5 \cdot 5$ Add 3 and 2 first.
$= 25$ Multiply.

Method 2: Multiply first

$3 + 2 \cdot 5 = 3 + 10$ Multiply 2 and 5 first.
$= 13$ Add.

Different results

This example shows that we need to establish an order of operations. Otherwise, the same expression can have two different values. To guarantee that calculations will have one correct result, we will use the following set of priority rules.

Rules for the Order of Operations

1. Perform all calculations within parentheses and other grouping symbols, following the order listed in steps 2–4 and working from the innermost pair to the outermost pair.
2. Evaluate all exponential expressions (powers) and roots.
3. Perform all multiplications and divisions as they occur from left to right.
4. Perform all additions and subtractions as they occur from left to right.

When all grouping symbols have been removed, repeat steps 2–4 to complete the calculation.

If a fraction bar is present, evaluate the expression above the bar (the *numerator*) and the expression below the bar (the *denominator*) separately. Then perform the division indicated by the fraction bar, if possible.

To evaluate $3 + 2 \cdot 5$ correctly, we follow steps 2, 3, and 4 of the rules for the order of operations. Since the expression does not contain any powers or roots, we perform the multiplication first, followed by the addition.

$3 + 2 \cdot 5 = 3 + 10$ Ignore the addition for now and multiply 2 and 5.
$= 13$ Next, perform the addition.

We see that the correct answer is 13.

EXAMPLE 8

Evaluate: **a.** $-5 + 4(-3)^2$ and **b.** $-10 \div 5 - 5(3) + 6$.

Solution **a.** Although the expression contains parentheses, there are no operations to perform *within* the parentheses. So we proceed with steps 2, 3, and 4 of the rules for the order of operations.

$$\begin{aligned} -5 + 4(-3)^2 &= -5 + 4(9) && \text{First, evaluate the power: } (-3)^2 = 9. \\ &= -5 + 36 && \text{Multiply.} \\ &= 31 && \text{Add.} \end{aligned}$$

The Language of Algebra

Sometimes, the word *simplify* is used in the place of the word *evaluate*. For instance, Example 8a could read:

Simplify: $-5 + 4(-3)^2$

b. Since the expression does not contain any powers, we perform the multiplications and divisions, working from left to right.

$$\begin{aligned} -10 \div 5 - 5(3) + 6 &= -2 - 5(3) + 6 && \text{Divide: } -10 \div 5 = -2. \\ &= -2 - 15 + 6 && \text{Multiply.} \\ &= -17 + 6 && \text{Working from left to right, subtract: } -2 - 15 = -17. \\ &= -11 && \text{Add.} \end{aligned}$$

Self Check 8 Evaluate: **a.** $-9 + 2(-4)^2$ and **b.** $20 \div (-5) - (-6)(-5) + (-12)$.

Grouping symbols serve as mathematical punctuation marks. They help determine the order in which an expression is evaluated. Examples of grouping symbols are parentheses ( ), brackets [ ], and the fraction bar ——.

EXAMPLE 9

Evaluate: **a.** $3 - (4 - 8)^2$ and **b.** $2 + 3[-2 - 8(4 - 3^2)]$.

Solution **a.** We begin by performing the operation within the parentheses.

$$\begin{aligned} 3 - (4 - 8)^2 &= 3 - (-4)^2 && \text{Perform the subtraction: } 4 - 8 = -4. \\ &= 3 - 16 && \text{Evaluate the power: } (-4)^2 = 16. \\ &= -13 && \text{Subtract.} \end{aligned}$$

b. First, we work within the innermost grouping symbols, the parentheses.

$$\begin{aligned} 2 + 3[-2 - 8(4 - 3^2)] &= 2 + 3[-2 - 8(4 - 9)] && \text{Find the power: } 3^2 = 9. \\ &= 2 + 3[-2 - 8(-5)] && \text{Subtract: } 4 - 9 = -5. \end{aligned}$$

Next, we work within the brackets.

$$\begin{aligned} &= 2 + 3[-2 - (-40)] && \text{Multiply: } 8(-5) = -40. \\ &= 2 + 3(-2 + 40) \end{aligned}$$

Since only one set of grouping symbols was needed, we wrote $-2 + 40$ within parentheses.

$$= 2 + 3(38) \quad \text{Add: } -2 + 40 = 38.$$
$$= 2 + 114 \quad \text{Multiply.}$$
$$= 116 \quad \text{Add.}$$

Self Check 9 Evaluate: **a.** $(5 - 3)^3 - 40$ and **b.** $-3[-2(5^3 - 3) + 4] - 1$.

EXAMPLE 10

Evaluate: $|-45 + 30|(2 - 7)$.

Solution Since the absolute value bars are grouping symbols, we perform the operations within the absolute value bars and the parentheses first.

$$|-45 + 30|(2 - 7) = |-15|(-5) \quad \text{Perform the addition within the absolute value bars and the subtraction within the parentheses.}$$
$$= 15(-5) \quad \text{Find the absolute value: } |-15| = 15.$$
$$= -75 \quad \text{Multiply.}$$

Self Check 10 Evaluate: $2|-25 - (-6)(3)|$.

ACCENT ON TECHNOLOGY: ORDER OF OPERATIONS

Scientific and graphing calculators are programmed to follow the rules for the order of operations. For example, when finding $3 + 2 \cdot 5$, both types of calculators give the correct answer, 13.

3 [+] 2 [×] 5 [=] — display: 13

3 [+] 2 [×] 5 [ENTER] — display: 3+2*5 / 13

Both types of calculators use parentheses keys [(] [)] when grouping symbols are needed. To evaluate $3 - (4 - 8)^2$, we proceed as follows.

3 [−] [(] 4 [−] 8 [)] [x^2] [=] — display: −13

3 [−] [(] 4 [−] 8 [)] [x^2] [ENTER] — display: 3−(4−8)² / −13

Both types of calculators require that we group the terms in the numerator together and the terms in the denominator together when calculating the value of an expression such as $\frac{200 + 120}{20 - 16}$.

[(] 200 + 120 [)] [÷] [(] 20 [−] 16 [)] [=] — display: 80

[(] 200 + 120 [)] [÷] [(] 20 [−] 16 [)] [ENTER] — display: (200+120)/(20−16) / 80

If parentheses aren't used when finding $\frac{200 + 120}{20 - 16}$, you will obtain an incorrect result of 190. That is because the calculator will interpret the entry as $200 + \frac{120}{20} - 16$.

EVALUATING ALGEBRAIC EXPRESSIONS

Recall that an algebraic expression is a combination of variables and numbers with the operations of arithmetic. To evaluate these expressions, we substitute specific numbers for the variables and then apply the rules for the order of operations.

EXAMPLE 11

If $a = -2$, $b = 9$, and $c = -1$, evaluate **a.** $-\frac{1}{2}a^2$ and **b.** $\frac{-a\sqrt{b} + 3c^3}{c(c - b)}$.

Solution **a.** We substitute -2 for a and use the rules for the order of operations.

$$-\frac{1}{2}a^2 = -\frac{1}{2}(-2)^2$$

Substitute -2 for a. Write parentheses around -2 so that it is squared.

$$= -\frac{1}{2}(4)$$

Evaluate the power: $(-2)^2 = 4$.

$$= -2$$

Multiply.

b. $$\frac{-a\sqrt{b} + 3c^3}{c(c - b)} = \frac{-(-2)\sqrt{9} + 3(-1)^3}{-1(-1 - 9)}$$

Substitute -2 for a, 9 for b, and -1 for c.

$$= \frac{-(-2)(3) + 3(-1)}{-1(-10)}$$

In the numerator, evaluate the square root and the power: $\sqrt{9} = 3$ and $(-1)^3 = -1$. In the denominator, subtract.

$$= \frac{2(3) + 3(-1)}{-1(-10)}$$

In the numerator, simplify: $-(-2) = 2$.

$$= \frac{6 + (-3)}{10}$$

In the numerator, multiply. In the denominator, multiply.

$$= \frac{3}{10}$$

In the numerator, add.

Self Check 11 If $r = 2$, $s = -5$, and $t = 3$, evaluate: **a.** $-\frac{1}{3}s^3t$ and **b.** $\frac{\sqrt{-5s}}{(s + t)r^2}$.

ACCENT ON TECHNOLOGY: EVALUATING ALGEBRAIC EXPRESSIONS

Graphing calculators can evaluate algebraic expressions. For example, to evaluate

$$\frac{-a\sqrt{b} + 3c^3}{c(c - b)}$$

(Example 11, part b) using a TI-83 Plus calculator, we first enter the values of $a = -2$, $b = 9$, and $c = -1$, using the store key [STO] and the [ALPHA] key. See figure (a).

[(−)] 2 [STO] [ALPHA] [A] [ALPHA] [:] This enters $a = -2$.

9 [STO] [ALPHA] [B] [ALPHA] [:] This enters $b = 9$.

[(−)] 1 [STO] [ALPHA] [C] [ALPHA] [:] This enters $c = -1$.

Next, enter the expression as shown in figure (b) and press [ENTER] to find that the value of the expression is 0.3. To express the result as a fraction, press [MATH], highlight Frac, and then press [ENTER] [ENTER]. See figure (c).

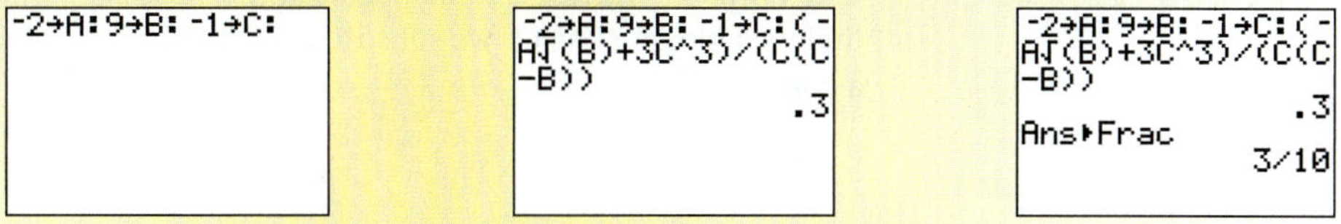

(a) **(b)** **(c)**

AREA AND VOLUME

The **area** of a two-dimensional geometric figure is a measure of the surface it encloses. Several commonly used area formulas are shown inside the front cover of this book.

EXAMPLE 12

Band-aids®. Find the amount of skin covered by a rectangular bandage $\frac{5}{8}$ inches wide and $3\frac{1}{2}$ inches long.

Solution To find the amount of skin covered by the bandage, we need to find its area.

$$A = lw \quad \text{This is the formula for the area of a rectangle.}$$

$$A = 3\frac{1}{2}\left(\frac{5}{8}\right) \quad \text{Substitute } 3\frac{1}{2} \text{ for } l \text{ and } \frac{5}{8} \text{ for } w.$$

$$A = \frac{7}{2}\left(\frac{5}{8}\right) \quad \text{Write } 3\frac{1}{2} \text{ as a fraction: } 3\frac{1}{2} = \frac{7}{2}.$$

$$A = \frac{35}{16}$$

The bandage covers $\frac{35}{16}$ or $2\frac{3}{16}$ in.^{2} (square inches) of skin.

Self Check 12 A solar panel is in the shape of a trapezoid. Its upper and lower bases measure $53\frac{1}{2}$ centimeters and $79\frac{1}{2}$ centimeters, respectively, and its height is 47 centimeters. In square centimeters, how large a surface do the sun's rays strike?

The **volume** of a three-dimensional geometric figure is a measure of its capacity. Several commonly used volume formulas are shown inside the back cover of this book.

EXAMPLE 13

Finding volume. Find the amount of sand in the hourglass.

Solution The sand is in the shape of a cone. The radius of the cone is one-half the diameter of the base of the hourglass, and the height of the cone is one-half the height of the hourglass. To find the amount of sand, we substitute 1 for r and 2.5 for h in the formula for the volume of a cone.

$$V = \frac{1}{3}\pi r^2 h$$

$$V = \frac{1}{3}\pi(1)^2(2.5)$$

$$V = \frac{2.5\pi}{3}$$

$$V \approx 2.617993878 \quad \text{Use a calculator.}$$

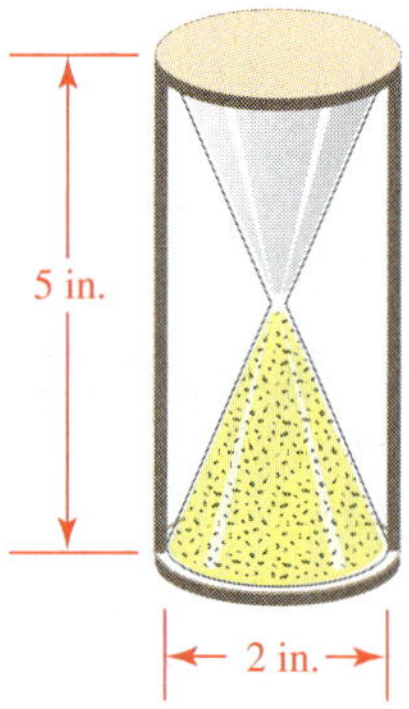

There are about 2.6 in.^{3} (cubic inches) of sand in the hourglass.

Caution

When finding area, remember to write the appropriate *square units* in your answer. For volume problems, write the appropriate *cubic units* in your answer.

Self Check 13 Find the volume of a drinking straw that is 250 millimeters long with an inside diameter of 6 millimeters.

Answers to Self Checks **1. a.** -9, **b.** -91.6, **c.** $\frac{1}{4}$, **d.** -4 **2. a.** -19, **b.** -4.5, **c.** $-\frac{2}{9}$, **d.** -6, **e.** 16 **3. a.** -30 **b.** 32.8, **c.** $-\frac{1}{6}$, **d.** -108 **4. a.** -11, **b.** 1.2 **5. a.** $-\frac{21}{16}$, **b.** $\frac{1}{50}$ **6. a.** -27, **b.** 0.64, **c.** 16, **d.** $\frac{49}{25}$ **7. a.** 8, **b.** -10, **c.** $\frac{2}{5}$, **d.** 1, **e.** 0.9, **f.** -20 **8. a.** 23, **b.** -46 **9. a.** -32, **b.** 719 **10.** 14 **11. a.** 125, **b.** $-\frac{5}{8}$ **12.** $3{,}125\frac{1}{2}$ cm^2 **13.** about 7,069 mm^3

1.3 STUDY SET

VOCABULARY **Fill in the blanks.**

1. When we add two numbers, the result is called the sum. When we subtract two numbers, the result is called the different.
2. When we multiply two numbers, the result is called the product. When we divide two numbers, the result is called the quotient.
3. To evaluate an algebraic expression, we substitute values for the variables and then apply the rules for the order of operations.
4. In the expression $9 + 6[22 - (6 - 1)]$, the __________ are the innermost grouping symbols, and the brackets are the __________ grouping symbols.
5. 6^2 can be read as "six square," and 6^3 can be read as "six cubed."

6. 4^5 is the fifth ________ of four.

7. In the exponential expression x^2, x is the ________, and 2 is the ________.

8. An ________ is used to represent repeated multiplication.

9. Subtraction is the same as adding the ________ of the number being subtracted.

10. The principal ________ ________ of 16 is 4.

CONCEPTS

11. Consider the expression $6 + 3 \cdot 2$.
    a. In what two different ways *might* we evaluate the given expression?
    b. Which result from part (a) is correct and why?

12. a. What operations does the expression $60 - (-9)^2 + 5(-1)$ contain?
    b. In what order should they be performed?

13. What are we finding when we calculate
    a. the amount of surface a circle encloses?
    b. the capacity of a cylinder?

14. a. What is the related multiplication statement for the division statement $\frac{0}{6} = 0$?
    b. Why isn't there a related multiplication statement for $\frac{6}{0}$?

NOTATION

15. a. In the expression $(-6)^2$, what is the base?
    b. In the expression -6^2, what is the base?

16. Translate each expression into symbols, and then evaluate it.
    a. Negative four squared.
    b. The opposite of four squared.

17. What is the name of the symbol $\sqrt{\ }$?

18. What is the one number that a fraction cannot have as its denominator?

PRACTICE Perform the operations.

19. $-3 + (-5)$
20. $-2 + (-8)$
21. $-7.1 + 2.8$
22. $3.1 + (-5.2)$
23. $-9 + (-8) + 4$
24. $2 + (-6) + (-3)$
25. $-3 - 4$
26. $-11 - (-17)$
27. $-3.3 - (-3.3)$
28. $0.14 - (-0.13)$
29. $-1 - 5 - (-4)$
30. $5 - (-3) - 2$
31. $-2(6)$
32. $-3(-7)$
33. $-0.3(5)$
34. $-0.4(-0.6)$
35. $-5(6)(-2)$
36. $-9(-1)(-3)$
37. $\dfrac{-8}{4}$
38. $\dfrac{-16}{-4}$
39. $\dfrac{1}{2} + \left(-\dfrac{1}{3}\right)$
40. $-\dfrac{3}{4} + \left(-\dfrac{1}{5}\right)$
41. Subtract $-\dfrac{3}{5}$ from $\dfrac{1}{2}$
42. Subtract $\dfrac{11}{13}$ from $\dfrac{1}{26}$
43. $\left(-\dfrac{3}{5}\right)\left(\dfrac{10}{7}\right)$
44. $\left(-\dfrac{6}{7}\right)\left(-\dfrac{5}{12}\right)$
45. $-\dfrac{16}{5} \div \left(-\dfrac{10}{3}\right)$
46. $-\dfrac{5}{24} \div \dfrac{10}{3}$

Evaluate each expression.

47. 12^2
48. 9^2
49. -5^2
50. $(-5)^2$
51. $(-8)^2$
52. -8^2
53. $4 \cdot 2^3$
54. $(4 \cdot 2)^3$
55. $(1.3)^2$
56. $\left(\dfrac{3}{5}\right)^2$
57. $\sqrt{64}$
58. $\sqrt{121}$
59. $-\sqrt{\dfrac{9}{16}}$
60. $-\sqrt{0.16}$
61. $3 - 5 \cdot 4$
62. $12 - 2 \cdot 3$
63. $\left(-3 - \sqrt{25}\right)^2$
64. $4^2 - (-2)^2$
65. $2 + 3\left(\dfrac{25}{5}\right) + (-4)$
66. $(-2)^3\left(\dfrac{-6}{-2}\right)(-1)$
67. $\dfrac{-\sqrt{49} - 3^2}{2 \cdot 4}$
68. $\dfrac{1}{2}\left(\dfrac{1}{8}\right) + \left(-\dfrac{1}{4}\right)^2$
69. $-2|4 - 8|$
70. $|\sqrt{49} - 8(4 - 7)|$
71. $(4 + 2 \cdot 3)^4$
72. $|9 - 5(1 - 8)|$

73. $3 + 2[-1 - 4(5)]$

74. $-3[5^2 - (7 - 3)^2]$

75. $30 + 6[-4 - 5(6 - 4)^2]$

76. $7 - 12[7^2 - 4(2 - 5)^2]$

77. $3 - [3^3 + (3 - 1)^3]$

78. $8 - 4\,|-(3 \cdot 5 - 2 \cdot 6)^2|$

79. $\dfrac{|-25| - 2(-5)}{2^4 - 9}$

80. $\dfrac{2[-4 - 2(3 - 1)]}{3(3)(2)}$

81. $\dfrac{3[-9 + 2(7 - 3)]}{(8 - 5)(9 - 7)}$

82. $\dfrac{5 \cdot 4 \cdot 3 \cdot 2 \cdot 1}{1 \cdot 2 \cdot 3 \cdot 4}$

83. $\dfrac{(6 - 5)^4 + 21}{27 - (\sqrt{16})^2}$

84. $\dfrac{3(3{,}246 - 1{,}111)}{561 - 546}$

85. $54^3 - 16^4 + 19(3)$

86. $\dfrac{36^2 - 2(48)}{(25)^2 - \sqrt{105{,}625}}$

Evaluate each expression for the given values.

87. $-\dfrac{2}{3}a^2$ for $a = -6$

88. $\left(-\dfrac{2}{3}a\right)^2$ for $a = -6$

89. $\dfrac{y_2 - y_1}{x_2 - x_1}$ for $x_1 = -3, x_2 = 5, y_1 = 12, y_2 = -4$

90. $P_0\left(1 + \dfrac{r}{k}\right)^{kt}$ for $P_0 = 500, r = 4, k = 2, t = 3$

91. $(x + y)(x^2 - xy + y^2)$ for $x = -4, y = 5$

92. $\dfrac{-b + \sqrt{b^2 - 4ac}}{2a}$ for $a = 1, b = 2, c = -3$

93. $\dfrac{x^2}{a^2} + \dfrac{y^2}{b^2}$ for $x = -3, y = -4, a = 5, b = -5$

94. $\dfrac{n}{2}[2a_1 + (n - 1)d]$ for $n = 50, a_1 = -4, d = 5$

95. $\sqrt{(x_2 - x_1)^2 + (y_2 - y_1)^2}$ for $x_1 = -2, x_2 = 4,$ $y_1 = 4, y_2 = -4$

96. $\dfrac{|Ax_0 + By_0 + C|}{\sqrt{A^2 + B^2}}$ for $A = 3, B = 4, C = -5,$ $x_0 = 2$, and $y_0 = -1$

Find each area to the nearest tenth.

97. The area of a triangle with a base of 2.75 centimeters (cm) and a height of 8.25 cm

98. The area of a circle with a radius of 5.7 meters

Find each volume to the nearest hundredth.

99. The volume of a rectangular solid with dimensions of 2.5 cm, 3.7 cm, and 10.2 cm

100. The volume of a pyramid whose base is a square with each side measuring 2.57 cm and with a height of 12.32 cm

101. The volume of a sphere with a radius of 5.7 meters

102. The volume of a cone whose base has a radius of 5.5 in. and whose height is 8.52 in.

APPLICATIONS

103. ALUMINUM FOIL Find the number of *square feet* of aluminum foil on a roll if the dimensions printed on the box are $8\frac{1}{3}$ yards × 12 inches.

104. HOCKEY A goal is scored in hockey when the puck, a vulcanized rubber disk 2.5 cm (1 in.) thick and 7.6 cm (3 in.) in diameter, is driven into the opponent's goal. Find the volume of a puck in cubic centimeters and cubic inches. Round to the nearest tenth.

105. PAPER PRODUCTS When folded, the paper sheet shown in the illustration forms a rectangular-shaped envelope. The formula

$$A = \frac{1}{2}h_1(b_1 + b_2) + b_3h_3 + \frac{1}{2}b_1h_2 + b_1b_3$$

gives the amount of paper (in square units) used in the design. Explain what each of the four terms in the formula finds. Then evaluate the formula for $b_1 = 6$, $b_2 = 2$, $b_3 = 3$, $h_1 = 2$, $h_2 = 2.5$, and $h_3 = 3$. All dimensions are in inches.

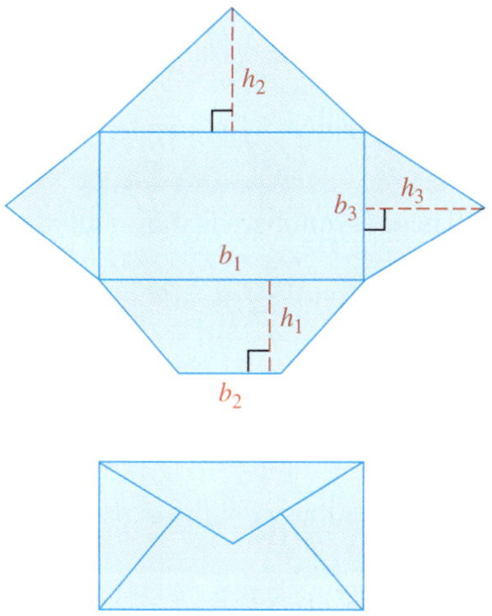

106. INVESTMENT IN BONDS In the following graph, positive numbers represent new cash *inflow* into U.S. bond funds. Negative numbers represent cash *outflow* from bond funds. Was there a net inflow or outflow over the 10-year period? What was it?

New Net Cash Flow to U.S. Bond Funds (in billions of dollars)

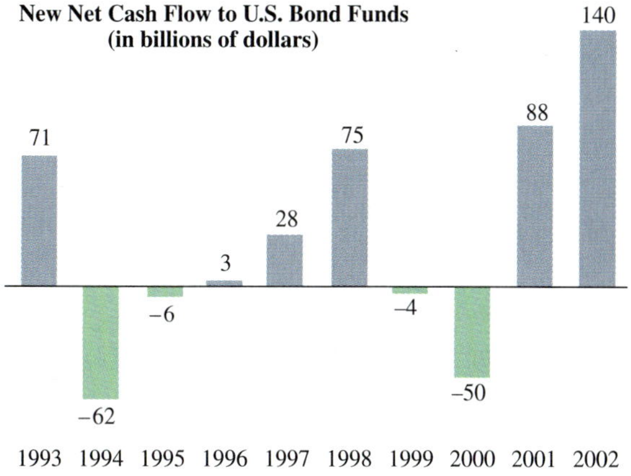

Source: Investment Company Institute

107. ACCOUNTING On a financial balance sheet, debts (negative numbers) are denoted within parentheses. Assets (positive numbers) are written without parentheses. What is the 2003 fund balance for the preschool whose financial records are shown in the table?

Community Care Preschool Balance Sheet, June 2003	
Fund balances	
Classroom supplies	\$ 5,889
Emergency needs	927
Holiday program	(2,928)
Insurance	1,645
Janitorial	(894)
Licensing	715
Maintenance	(6,321)
BALANCE	?

108. TEMPERATURE EXTREMES The highest and lowest temperatures ever recorded in several cities are shown in the table. List the cities in order, from the smallest to the largest range in temperature extremes.

	Extreme temperatures	
City	**Highest**	**Lowest**
Atlanta, Georgia	105	−8
Boise, Idaho	111	−25
Helena, Montana	105	−42
New York, New York	107	−3
Omaha, Nebraska	114	−23

109. ICE CREAM If the two equal-sized scoops of ice cream melt completely into the cone, will they overflow the cone?

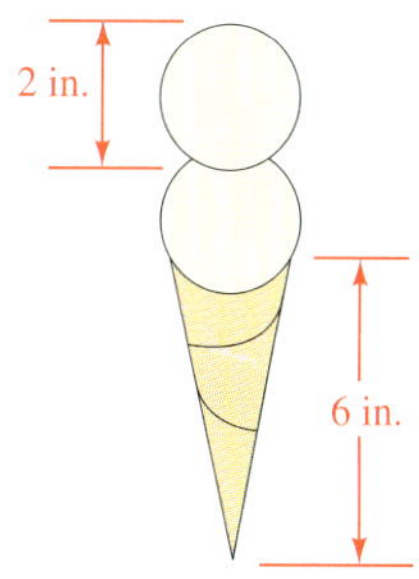

110. PHYSICS Waves are motions that carry energy from one place to another. The illustration shows an example of a wave called a *standing wave.* What is the difference in the height of the crest of the wave and the depth of the trough of the wave?

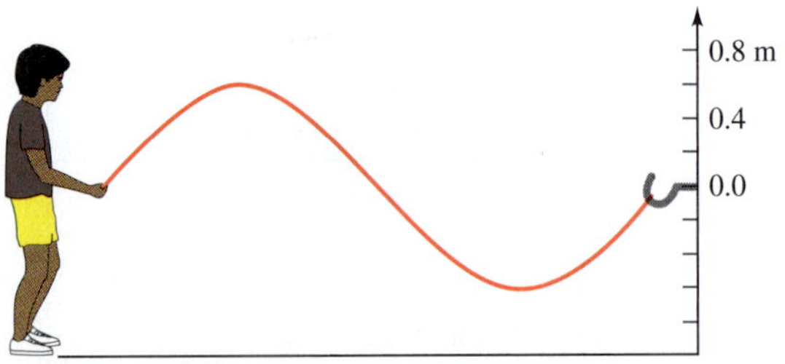

111. PEDIATRICS Young's rule, shown below, is used by some doctors to calculate dosage for infants and children.

$$\frac{\text{Age of child}}{\text{Age of child} + 12}\left(\begin{array}{c}\text{average}\\ \text{adult dose}\end{array}\right) = \begin{array}{c}\text{child's}\\ \text{dose}\end{array}$$

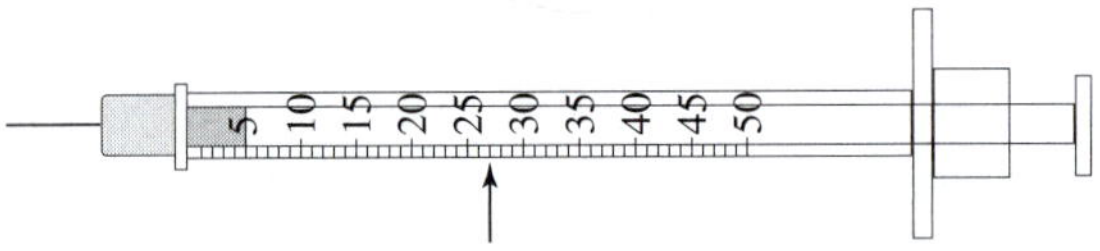

The syringe shows the adult dose of a certain medication. Use Young's rule to determine the dosage for a 6-year-old child. Then use an arrow to locate the dosage on the calibration.

112. DOSAGES The adult dosage of procaine penicillin is 300,000 units daily. Calculate the dosage for a 12-year-old child using Young's rule. (See Exercise 111.)

WRITING

113. Explain what the statement $x - y = x + (-y)$ means.

114. Explain why rules for the order of operations are necessary.

REVIEW

115. What two numbers are a distance of 5 away from -2 on the number line?

116. Place the proper symbol (> or <) in the blank: -4.6 ☐ -4.5.

117. List the set of integers.

118. Translate into mathematical symbols: ten less than twice x.

119. True or false: The real numbers is the set of all decimals.

120. True or false: Irrational numbers are nonterminating, nonrepeating decimals.

CHALLENGE PROBLEMS

121. Insert one pair of parentheses in the expression so that its value is 0.

$$71 - 1 - 2 \cdot 5^2 + 10$$

122. Point C is the center of the largest circle in the figure. Find the area of the shaded region. Round to the nearest tenth.

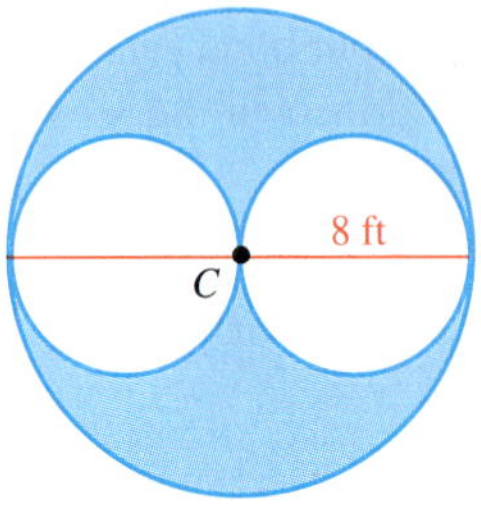

1.4 Simplifying Algebraic Expressions

- Properties of real numbers
- Properties of 0 and 1
- More properties of real numbers
- Simplifying algebraic expressions
- The distributive property
- Combining like terms

Suppose we want to find the total dollar amount of the checks that are recorded in the following register. The *commutative* and *associative* properties of addition guarantee that we will obtain the same result whether we add them in their original order or in the more convenient way suggested by the notes. In this section, we will use these properties and others to simplify expressions containing variables.

Number	Date	Description of Transaction	Payment/Debit	
101	3/6	DR. OKAMOTO, DDS	$64	00
102	3/6	UNION OIL CO.	$25	00
103	3/8	STATER BROS.	$16	00
104	3/9	LITTLE LEAGUE	$75	00

Add $64.00 and $16.00 to get $80.00.

Add $25.00 and $75.00 to get $100.00.

Now add the two subtotals to get the total dollar amount of the checks: $80.00 + $100.00 = $180.00.

PROPERTIES OF REAL NUMBERS

When working with real numbers, we will use the following properties.

Properties of Real Numbers

If a, b, and c represent real numbers, then we have

The associative properties of addition and multiplication

$$(a + b) + c = a + (b + c) \qquad (ab)c = a(bc)$$

The commutative properties of addition and multiplication

$$a + b = b + a \qquad ab = ba$$

The *associative properties* enable us to group the numbers in a sum or a product any way that we wish and get the same result.

EXAMPLE 1

Evaluate $(14 + 94) + 6$ in two ways.

Solution

$$\begin{aligned}(14 + 94) + 6 &= 108 + 6 && \text{Work within the parentheses first.}\\ &= 114\end{aligned}$$

To evaluate the expression another way, we use the associative property of addition.

$$\begin{aligned}(14 + 94) + 6 &= 14 + (94 + 6) && \text{Use parentheses to group 94 with 6.}\\ &= 14 + 100 && \text{Add within the parentheses.}\\ &= 114\end{aligned}$$

Notice that the results are the same.

The Language of Algebra

Associative is a form of the word *associate,* meaning to join a group. The National Basketball *Association* (NBA) is a group of professional basketball players.

Self Check 1 Evaluate $2 \cdot (50 \cdot 37)$ in two ways.

Subtraction and division are not associative, because different groupings give different results. For example,

$$(8 - 4) - 2 = 4 - 2 = 2 \quad \text{but} \quad 8 - (4 - 2) = 8 - 2 = 6$$

$$(8 \div 4) \div 2 = 2 \div 2 = 1 \quad \text{but} \quad 8 \div (4 \div 2) = 8 \div 2 = 4$$

The *commutative properties* enable us to add or multiply two numbers in either order and obtain the same result. Here are two examples.

The Language of Algebra
Commutative is a form of the word *commute*, meaning to go back and forth. *Commuter* trains take people to and from work.

$$3 + (-5) = -2 \quad \text{and} \quad -5 + 3 = -2$$
$$-2.6(-8) = 20.8 \quad \text{and} \quad -8(-2.6) = 20.8$$

Subtraction and division are not commutative, because performing these operations in different orders will give different results. For example,

$$8 - 4 = 4 \quad \text{but} \quad 4 - 8 = -4$$
$$8 \div 4 = 2 \quad \text{but} \quad 4 \div 8 = \frac{1}{2}$$

PROPERTIES OF 0 AND 1

The real numbers 0 and 1 have important special properties.

Properties of 0 and 1

Additive identity: The sum of 0 and any number is the number itself.

$$0 + a = a + 0 = a$$

Multiplicative identity: The product of 1 and any number is the number itself.

$$1 \cdot a = a \cdot 1 = a$$

Multiplication property of 0: The product of any number and 0 is 0.

$$a \cdot 0 = 0 \cdot a = 0$$

For example,

$$7 + 0 = 7, \quad 1(5.4) = 5.4, \quad \left(-\frac{7}{3}\right)1 = -\frac{7}{3}, \quad \text{and} \quad -19(0) = 0$$

MORE PROPERTIES OF REAL NUMBERS

If the sum of two numbers is 0, they are called **additive inverses,** or **opposites** of each other. For example, 6 and -6 are additive inverses, because $6 + (-6) = 0$.

The Additive Inverse Property

For every real number a, there exists a real number $-a$ such that

$$a + (-a) = -a + a = 0$$

If the product of two numbers is 1, the numbers are called **multiplicative inverses** or **reciprocals** of each other.

The Multiplicative Inverse Property

For every nonzero real number a, there exists a real number $\frac{1}{a}$ such that

$$a \cdot \frac{1}{a} = \frac{1}{a} \cdot a = 1$$

Some examples of reciprocals (multiplicative inverses) are

- 5 and $\frac{1}{5}$ are reciprocals, because $5\left(\frac{1}{5}\right) = 1$.
- $\frac{3}{2}$ and $\frac{2}{3}$ are reciprocals, because $\frac{3}{2}\left(\frac{2}{3}\right) = 1$.
- -0.25 and -4 are reciprocals, because $-0.25(-4) = 1$.

Caution

The reciprocal of 0 does not exist, because $\frac{1}{0}$ is undefined.

Recall that when a number is divided by 1, the result is the number itself, and when a nonzero number is divided by itself, the result is 1.

Division Properties

Division by 1: If a represents any real number, then $\frac{a}{1} = a$.

Division of a number by itself: For any nonzero real number a, $\frac{a}{a} = 1$.

There are three possible cases to consider when discussing division involving 0.

Division with 0

Division of 0: For any nonzero real number a, $\frac{0}{a} = 0$.

Division by 0: For any nonzero real number a, $\frac{a}{0}$ is undefined.

Division of 0 *by* 0: $\frac{0}{0}$ is indeterminate.

To show that division of zero by zero doesn't have a single result, we consider $\frac{0}{0} = ?$ and its equivalent multiplication fact $0(?) = 0$.

Multiplication fact	***Division fact***
$0(?) = 0$	$\frac{0}{0}$ = indeterminate
↑ Any number multiplied by 0 gives 0.	↑ We cannot determine this—it could be any number.

We say that zero divided by zero is *indeterminate.*

SIMPLIFYING ALGEBRAIC EXPRESSIONS

To **simplify algebraic expressions,** we write the expressions in a simpler form. As an example, let's consider $6(5x)$ and simplify it.

$$
\begin{aligned}
6(5x) &= 6 \cdot (5 \cdot x) \\
&= (6 \cdot 5) \cdot x && \text{Use the associative property of multiplication to group 5 with 6.} \\
&= 30x && \text{Multiply within the parentheses.}
\end{aligned}
$$

Since $6(5x) = 30x$, we say that $6(5x)$ simplifies to $30x$.

EXAMPLE 2

Simplify: **a.** $9(10t)$, **b.** $-5.3r(-2s)$, and **c.** $-\frac{21}{2}a\left(\frac{1}{3}\right)$.

Solution **a.** $9(10t) = (9 \cdot 10)t$ — Use the associative property of multiplication to regroup the factors.

$= 90t$ — Multiply inside the parentheses: $9 \cdot 10 = 90$.

b. $-5.3r(-2s) = [-5.3(-2)](r \cdot s)$ — Use the commutative and associative properties to group the numbers and group the variables.

$= 10.6rs$ — Multiply.

c. $-\frac{21}{2}a\left(\frac{1}{3}\right) = -\frac{21}{2}\left(\frac{1}{3}\right)a$ — Use the commutative property of multiplication to change the order of the factors a and $\frac{1}{3}$.

$= -\frac{7}{2}a$ — Multiply: $-\frac{21}{2} \cdot \frac{1}{3} = -\frac{21 \cdot 1}{2 \cdot 3} = -\frac{7 \cdot \overset{1}{\cancel{3}} \cdot 1}{2 \cdot \underset{1}{\cancel{3}}} = -\frac{7}{2}$.

Self Check 2 Simplify: **a.** $14 \cdot 3s$, **b.** $-1.6b(3t)$, and **c.** $-\frac{2}{3}x(-9)$.

THE DISTRIBUTIVE PROPERTY

The *distributive property* enables us to evaluate many expressions involving a multiplication and an addition. For example, let's consider $4(5 + 3)$, which can be evaluated in two ways.

Method 1: Rules for the Order of Operations

In this method, we compute the sum within the parentheses first.

$4(5 + 3) = 4(8)$ — Add inside the parentheses first.

$= 32$ — Multiply.

Method 2: The Distributive Property

In this method, we distribute the multiplication by 4 to 5 and to 3, find each product separately, and add the results.

First product — Second product

$4(5 + 3) = 4 \cdot 5 + 4 \cdot 3$ — Multiply each term inside the parentheses by the factor outside the parentheses.

$= 20 + 12$

$= 32$

Notice that each method gives a result of 32. We now state the distributive property in symbols.

The Distributive Property

The distributive property of multiplication over addition
If a, b, and c represent real numbers,

$$a(b + c) = ab + ac$$

The Language of Algebra

When we use the distributive property to write a product, such as $5(x + 2)$, as the sum, $5x + 10$, we say that we have *removed* or *cleared* parentheses.

To illustrate one use of the distributive property, let's consider the expression $5(x + 2)$. Since we are not given the value of x, we cannot add x and 2 within the parentheses. However, we can distribute the multiplication by the factor of 5 that is outside the parentheses to x and to 2 and simplify.

$$\begin{aligned} 5(x + 2) &= 5 \cdot x + 5 \cdot 2 && \text{Distribute the multiplication by 5.} \\ &= 5x + 10 \end{aligned}$$

Since subtraction is the same as adding the opposite, the distributive property also holds for subtraction.

$$\begin{aligned} 5(x - 2) &= 5 \cdot x - 5 \cdot 2 && \text{Distribute the multiplication by 5.} \\ &= 5x - 10 \end{aligned}$$

EXAMPLE 3

Use the distributive property to remove parentheses: **a.** $6(a + 9)$ and **b.** $-15(4b - 1)$.

Solution

a. $$\begin{aligned} 6(a + 9) &= 6 \cdot a + 6 \cdot 9 && \text{Distribute the multiplication by 6.} \\ &= 6a + 54 \end{aligned}$$

b. $$\begin{aligned} -15(4b - 1) &= -15(4b) - (-15)(1) && \text{Distribute the multiplication by } -15. \\ &= -60b + 15 \end{aligned}$$

Self Check 3 Remove parentheses: **a.** $9(r + 4)$ and **b.** $-11(-3x - 5)$.

A more general form of the distributive property is the **extended distributive property.**

$$a(b + c + d + e + \cdots) = ab + ac + ad + ae + \cdots$$

EXAMPLE 4

Remove parentheses: $-0.5(7 - 5y + 6z)$.

Solution

$$\begin{aligned} &-0.5(7 - 5y + 6z) \\ &= -0.5(7) - (-0.5)(5y) + (-0.5)(6z) && \text{Distribute the multiplication by } -0.5. \\ &= -3.5 + 2.5y - 3z \end{aligned}$$

Self Check 4 Remove parentheses: $\frac{1}{3}(-6t + 3s - 9)$.

Since multiplication is commutative, we can write the distributive property in the following forms.

$$(b + c)a = ba + ca, \quad (b - c)a = ba - ca, \quad (b + c + d)a = ba + ca + da$$

EXAMPLE 5

Remove parentheses: $(-8 - 3y)(-30)$.

Solution

$$(-8 - 3y)(-30) = -8(-30) - 3y(-30) \quad \text{Distribute the multiplication by } -30.$$
$$= 240 + 90y \quad \text{Multiply.}$$

Self Check 5 Remove parentheses: $(-5s + 4t)(-10)$.

To use the distributive property to simplify $-(x + 3)$, we interpret the $-$ symbol as a factor of -1, and proceed as follows.

$$-(x + 3) = -1(x + 3)$$
$$= -1(x) + (-1)(3) \quad \text{Distribute the multiplication by } -1.$$
$$= -x - 3$$

EXAMPLE 6

Simplify: $-(-21 - 20m)$.

Solution

$$-(-21 - 20m) = -1(-21 - 20m) \quad \text{Write the } - \text{ sign in front of the parentheses as } -1.$$
$$= -1(-21) - (-1)(20m) \quad \text{Distribute the multiplication by } -1.$$
$$= 21 + 20m$$

Self Check 6 Simplify: $-(-27k + 15)$.

COMBINING LIKE TERMS

Addition signs separate algebraic expressions into parts called **terms.** For example, the expression $3x^2 + 2x + 4$ has three terms: $3x^2$, $2x$, and 4. A term may be

- a number (called a **constant**); examples are -6, 45.7, 35, and $\frac{2}{3}$.
- a variable or a product of variables (which may be raised to powers); examples are x, bh, s^2, Prt, and a^3bc^4.
- a product of a number and one or more variables (which may be raised to powers); examples are $3x$, $-7y$, $2.5y^2$, and πr^2h.

Since subtraction can be written as addition of the opposite, the expression $6a - 5b$ can be written in the equivalent form $6a + (-5b)$. We can then see that the expression $6a - 5b$ contains two terms, $6a$ and $-5b$.

The **numerical coefficients,** or simply the **coefficients,** of the terms of the expression $x^3 - 5x^2 - x + 28$ are 1, -5, -1, and 28, respectively.

Like Terms **Like terms** are terms with exactly the same variables raised to exactly the same powers. Any constant terms in an expression are considered to be like terms. Terms that are not like terms are called **unlike terms.**

Here are some examples of like and unlike terms.

$5x$ and $6x$ are like terms.

$27x^2y^3$ and $-326x^2y^3$ are like terms.

$4x$ and $-17y$ are unlike terms, because they have different variables.

$15x^2y$ and $6xy^2$ are unlike terms, because the variables have different exponents.

If we are to add (or subtract) objects, they must have the same units. For example, we can add dollars to dollars and inches to inches, but we cannot add dollars to inches. The same is true when working with terms of an expression. They can be added or subtracted only when they are like terms.

This expression can be simplified, because it contains like terms.

$$5x + 6x$$

These are like terms; the variable parts are the same.

This expression cannot be simplified, because its terms are not like terms.

$$5x + 6y$$

These are unlike terms; the variable parts are not the same.

Simplifying the sum or difference of like terms is called **combining like terms.** To simplify expressions containing like terms, we use the distributive property. For example,

$$5x + 6x = (5 + 6)x = 11x \quad \text{and} \quad 32y - 16y = (32 - 16)y = 16y$$

These examples suggest the following rule.

Combining Like Terms To add or subtract like terms, combine their coefficients and keep the same variables with the same exponents.

EXAMPLE 7

Simplify each expression: **a.** $-8f + (-12f)$, **b.** $0.56s^3 - 0.2s^3$, and **c.** $-\frac{1}{2}ab + \frac{1}{3}ab$.

Solution

a. $-8f + (-12f) = -20f$ — Add the coefficients of the like terms: $-8 + (-12) = -20$. Keep the variable f.

b. $0.56s^3 - 0.2s^3 = 0.36s^3$ — Subtract: $0.56 - 0.2 = 0.36$. Keep s^3.

c. $-\frac{1}{2}ab + \frac{1}{3}ab = -\frac{1}{2} \cdot \frac{3}{3}ab + \frac{1}{3} \cdot \frac{2}{2}ab$ — Express each fraction in terms of the LCD, 6.

$= -\frac{3}{6}ab + \frac{2}{6}ab$ — Multiply.

$= -\frac{1}{6}ab$ — Add the coefficients: $-\frac{3}{6} + \frac{2}{6} = -\frac{1}{6}$. Keep ab.

Self Check 7 Simplify by combining like terms: **a.** $5k + 8k$, **b.** $-600a^2 - (-800a^2)$, and **c.** $\frac{2}{3}xy - \frac{3}{4}xy$.

EXAMPLE 8

Simplify: $9b - B - 14b + 34B$.

Solution Since the uppercase B and lowercase b are different variables, the first and third terms are like terms, and the second and fourth terms are like terms.

$$9b - B - 14b + 34B = -5b + 33B$$ Combine like terms: $9b - 14b = -5b$ and $-B + 34B = 33B$.

Self Check 8 Simplify: $8R + 7r - 14R - 21r$.

EXAMPLE 9

Simplify: $9(x + 1) - 3(7x - 1)$.

Solution We use the distributive property and combine like terms.

$$\begin{aligned} 9(x + 1) - 3(7x - 1) &= 9x + 9 - 21x + 3 && \text{Use the distributive property twice.} \\ &= -12x + 12 && \text{Combine like terms: } 9x - 21x = -12x \text{ and } 9 + 3 = 12. \end{aligned}$$

Self Check 9 Simplify: $-5(y - 4) + 2(4y + 8)$.

Answers to Self Checks **1.** 3,700 **2. a.** $42s$, **b.** $-4.8bt$, **c.** $6x$ **3. a.** $9r + 36$, **b.** $33x + 55$ **4.** $-2t + s - 3$ **5.** $50s - 40t$ **6.** $27k - 15$ **7. a.** $13k$, **b.** $200a^2$, **c.** $-\frac{1}{12}xy$ **8.** $-6R - 14r$ **9.** $3y + 36$

1.4 STUDY SET

VOCABULARY Fill in the blanks.

1. ______ terms are terms with exactly the same variables raised to exactly the same powers.

2. To add or subtract like terms, combine their __________ and keep the same variables and exponents.

3. A number or the product of numbers and variables is called a ______.

4. $\frac{1}{3}$ and 3 are __________, because $\frac{1}{3} \cdot 3 = 1$.

5. To ________ expressions, we use properties of real numbers to write the expressions in a less complicated form.

6. We can use the __________ property to remove or clear parentheses in the expression $2(x + 8)$.

7. Division by 0 is __________.

8. The __________ of the term $-8c$ is -8.

CONCEPTS

9. Using the variables x, y, and z, write the associative property of addition.

10. Using the variables x and y, write the commutative property of multiplication.

11. Using the variables r, s, and t, write the distributive property of multiplication over addition.

12. a. What is the additive identity?
    b. What is the multiplicative identity?
    c. Simplify: $-(-10)$.
13. What number should be
    a. subtracted from 5 to obtain 0?
    b. added to 5 to obtain 0?
14. By what number should
    a. 5 be divided to obtain 1?
    b. 5 be multiplied to obtain 1?
15. Give the reciprocal.
    a. $\frac{15}{16}$ b. -20
    c. 0.5 d. x
16. Does the distributive property apply?
    a. 2(3)(5) b. $2(3 \cdot 5)$
    c. $2(3x)$ d. $2(x - 3)$
17. Consider the expression $2x^2 - x + 6$.
    a. What are the terms of the expression?
    b. Give the coefficient of each term.
18. Which properties of real numbers involve changing *order* and which involve changing *grouping?*

Decide whether the terms are like terms. If they are, combine them.

19. $2x, 6x$
20. $-3x, 5y$
21. $-5xy, -7yz$
22. $-3t^2, 12t^2$
23. $3x^2, -5x^2$
24. $5y^2, 7xy$
25. $xy, 3xt$
26. $-4x, -5x$

NOTATION

27. In $-(x - 7)$, what does the negative sign in front of the parentheses represent?
28. Perform each division, if possible.
    a. $\frac{0}{8}$ b. $\frac{8}{0}$

PRACTICE

Fill in the blanks by applying the given property of the real numbers.

29. $3 + 7 =$ ______ Commutative property of addition
30. $2(5 \cdot 97) =$ ________ Associative property of multiplication
31. $3(2 + d) =$ ________ Distributive property
32. $1 \cdot y =$ ______ Commutative property of multiplication
33. $c + 0 =$ ____ Additive identity property
34. $-4(x - 2) =$ ________ Distributive property and simplifying
35. $25 \cdot \frac{1}{25} =$ ____ Multiplicative inverse property
36. $z + (9 - 27) =$ __________ Commutative property of addition
37. $8 + (7 + a) =$ __________ Associative property of addition
38. ____ $\cdot 3 = 3$ Multiplicative identity property
39. $(x + y)2 =$ ________ Commutative property of multiplication
40. $h + (-h) =$ ____ Additive inverse property

Evaluate each side of the equation separately to show that the same result is obtained. Identify the property of real numbers that is being illustrated.

41. $(37.9 + 25.2) + 14.3 = 37.9 + (25.2 + 14.3)$
42. $7.1(3.9 + 8.8) = 7.1 \cdot 3.9 + 7.1 \cdot 8.8$
43. $2.73(4.534 + 57.12) = 2.73 \cdot 4.534 + 2.73 \cdot 57.12$
44. $(6.789 + 345.1) + 27.347 = (345.1 + 6.789) + 27.347$

Remove parentheses.

45. $-4(t - 3)$
46. $-4(-t + 3)$
47. $-(t - 3)$
48. $-(-t + 3)$
49. $(y - 2)(-3)$
50. $(2t + 5)(-2)$
51. $\frac{2}{3}(3s - 9)$
52. $\frac{1}{5}(5s - 15)$
53. $0.7(s + 2)$
54. $2.5(6s - 8)$
55. $3\left(\frac{4}{3}x - \frac{5}{3}y + \frac{1}{3}\right)$
56. $6\left(-\frac{4}{3} + \frac{7}{6}s + \frac{16}{3}t\right)$

Simplify each expression.

57. $9(8m)$
58. $12n(4)$
59. $5(-9q)$
60. $-7(2t)$

61. $(-5p)(-6b)$
62. $(-7d)(-7e)$
63. $-5(8r)(-2y)$
64. $-7s(-4t)(-1)$
65. $3x + 15x$
66. $12y - 17y$
67. $18x^2 - 5x^2$
68. $37x^2 + 3x^2$
69. $-9x + 9x$
70. $-26y + 26y$
71. $-b^2 + b^2$
72. $-3c^3 + 3c^3$
73. $8x + 5x - 7x$
74. $-y + 3y + 6y$
75. $3x^2 + 2x^2 - 5x^2$
76. $8x^3 - x^3 + 2x^3$
77. $3.8h - 0.7h$
78. $-5.7m + 5.3m$
79. $\frac{2}{5}ab - \left(-\frac{1}{2}ab\right)$
80. $-\frac{3}{4}st - \frac{1}{3}st$
81. $\frac{3}{5}t + \frac{1}{3}t$
82. $\frac{3}{16}x - \frac{5}{4}x$
83. $4(y + 9) - 8y$
84. $-(4 + z) + 2z$
85. $2z + 5(z - 4)$
86. $12(2m + 11) - 11$
87. $8(2c + 7) - 2(c - 3)$
88. $9(z + 3) - 5(3 - z)$
89. $2x^2 + 4(3x - x^2) + 3x$
90. $3p^2 - 6(5p^2 + p) + p^2$
91. $-(a + 2A) - (a - A)$
92. $3T - 2(t - T) + t$
93. $-3(p - 2) + 2(p + 3) - 5(p - 1)$
94. $5(q + 7) - 3(q - 1) - (q + 2)$
95. $36\left(\frac{2}{9}x - \frac{3}{4}\right) + 36\left(\frac{1}{2}\right)$
96. $40\left(\frac{3}{8}y - \frac{1}{4}\right) + 40\left(\frac{4}{5}\right)$
97. $3[2(x + 2)] - 5[3(x - 5)]$
98. $-5[3(x - 4) - 2(x + 2)] - 7(x - 3)$

APPLICATIONS

99. PARKING AREAS Refer to the illustration in the next column.

a. Express the area of the entire parking lot as the product of its length and width.

b. Express the area of the entire lot as the sum of the areas of the self-parking space and the valet parking space.

c. Write an equation that shows that your answers to parts (a) and (b) are equal. What property of real numbers is illustrated by this example?

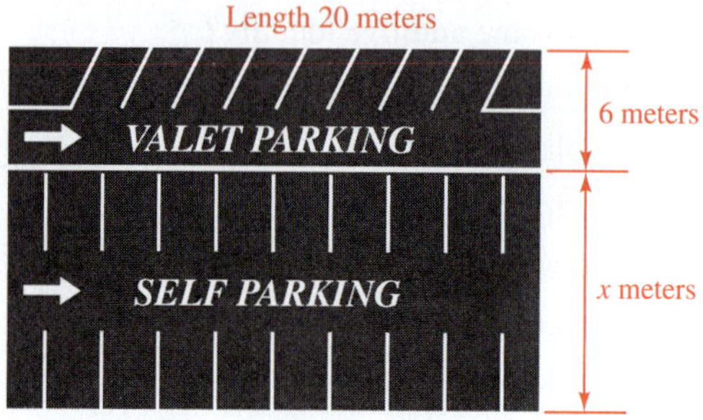

100. CROSS SECTION OF A CASTING When the steel casting shown in the illustration is cut down the middle, it has a uniform cross section consisting of two identical trapezoids. Find the area of the cross section. (The measurements are in inches.)

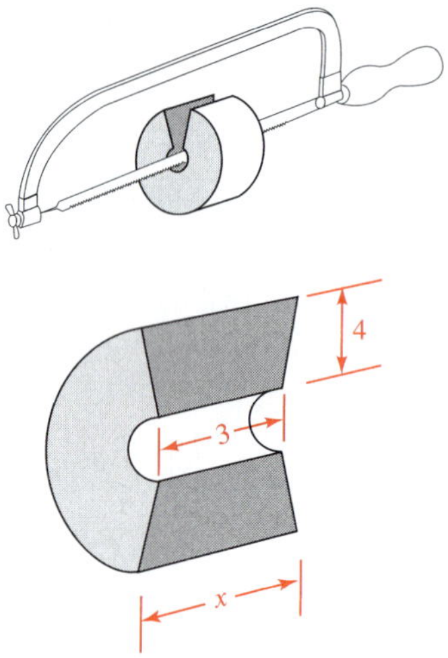

WRITING

101. Explain why the distributive property does not apply when simplifying $6(2 \cdot x)$.

102. In each case, explain what you can conclude about one or both of the numbers.

a. When the two numbers are added, the result is 0.

b. When the two numbers are subtracted, the result is 0.

c. When the two numbers are multiplied, the result is 0.

d. When the two numbers are divided, the result is 0.

103. What are like terms?

104. Use each of the words *commute, associate,* and *distribute* in a sentence in which the context is nonmathematical.

REVIEW **Evaluate each expression.**

105. $-5.6 - (-5.6)$

106. $\left(-\frac{3}{2}\right)\left(\frac{7}{12}\right)$

107. $(4 + 2 \cdot 3)^3$

108. $-3\,|4 - 8|$

109. $\dfrac{-\sqrt{64} - 5^2}{2 \cdot 4 + 3}$

110. $\frac{1}{2} - \left(-\frac{4}{5}\right)$

CHALLENGE PROBLEMS

111. Simplify: $\frac{x}{2} + \frac{x}{3} + \frac{x}{4} + \frac{x}{5} + \frac{x}{6}$.

112. Fill in the blank:

$\underline{\qquad}(0.05x + 0.2y - 0.003z) = 50x + 200y - 3z$

1.5 Solving Linear Equations and Formulas

- Solutions of equations
- Linear equations
- Properties of equality
- Solving linear equations
- Simplifying expressions to solve equations
- Identities and contradictions
- Solving formulas

To solve problems, we often begin by letting a variable stand for an unknown quantity. Then we write an equation involving the variable to describe the situation mathematically. Finally, we perform a series of steps on the equation to find the value represented by the variable. The process of determining the values represented by a variable is called *solving the equation.* In this section, we will discuss an equation-solving strategy for *linear equations* in one variable.

SOLUTIONS OF EQUATIONS

An **equation** is a statement that two expressions are equal. The equation $2 + 4 = 6$ is true, and the equation $2 + 5 = 6$ is false. If an equation contains a variable (say, x), it can be either true or false, depending on the value of x. For example, if x is 1, then the equation $7x - 3 = 4$ is true.

$7x - 3 = 4$

$7(1) - 3 \stackrel{?}{=} 4$ Substitute 1 for x. At this stage, we don't know whether the left- and right-hand sides of the equation are equal, so we use an "is possibly equal to" symbol $\stackrel{?}{=}$.

$7 - 3 \stackrel{?}{=} 4$

$4 = 4$ We obtain a true statement.

Since 1 makes the equation true, we say that 1 *satisfies* the equation. However, the equation is false for all other values of x.

The set of numbers that satisfy an equation is called its **solution set.** The elements of the solution set are called **solutions** of the equation. Finding all of the solutions of an equation is called **solving the equation.**

EXAMPLE 1

Determine whether 2 is a solution of $3x + 2 = 2x + 5$.

Solution We substitute 2 for x where it appears in the equation and see whether it satisfies the equation.

$$3x + 2 = 2x + 5 \quad \text{This is the original equation.}$$
$$3(2) + 2 \stackrel{?}{=} 2(2) + 5 \quad \text{Substitute 2 for } x.$$
$$6 + 2 \stackrel{?}{=} 4 + 5$$
$$8 = 9 \quad \text{False.}$$

Since $8 = 9$ is a false statement, the number 2 does not satisfy the equation. It is not a solution of $3x + 2 = 2x + 5$.

Self Check 1 Is -5 a solution of $2x - 5 = 3x$?

LINEAR EQUATIONS

Usually, we do not know the solutions of an equation—we need to find them. In this text, we will discuss how to solve many different types of equations. The easiest equations to solve are **linear equations.**

Linear Equations A **linear equation in one variable** can be written in the form

$$ax + b = c \quad \text{where } a, b, \text{ and } c \text{ are real numbers, and } a \neq 0.$$

Some examples of linear equations are

$$-2x - 8 = 0 \qquad \frac{3}{4}y = -7 \qquad 4b - 7 + 2b = 1 + 2b + 8$$

Linear equations are also called **first-degree equations,** since the highest power on the variable is 1.

PROPERTIES OF EQUALITY

When solving linear equations, the objective is to *isolate* the variable on one side of the equation. This is achieved by undoing the operations performed on the variable. As we undo the operations, we produce a series of simpler equations, all having the same solutions. Such equations are called *equivalent equations.*

Equivalent Equations Equations with the same solutions are called **equivalent equations.**

The solution of the equation $x = 2$ is obviously 2, because replacing x with 2 yields a true statement, $2 = 2$. The equation $x + 4 = 6$ also has a solution of 2. Since $x = 2$ and $x + 4 = 6$ have the same solution, they are equivalent equations.

The following properties are used to isolate a variable on one side of an equation.

Properties of Equality

Adding the same number to, or subtracting the same number from, both sides of an equation does not change the solution.

If a, b, and c are real numbers and $a = b$,

$a + c = b + c$ **Addition property of equality**

$a - c = b - c$ **Subtraction property of equality**

Multiplying or dividing both sides of an equation by the same nonzero number does not change the solution.

If a, b, and c are real numbers with $c \neq 0$, and $a = b$,

$ca = cb$ **Multiplication property of equality**

$\frac{a}{c} = \frac{b}{c}$ **Division property of equality**

SOLVING LINEAR EQUATIONS

We use the properties of equality to solve equations.

EXAMPLE 2

Solve: $2x - 8 = 0$.

Solution

We note that x is multiplied by 2 and then 8 is subtracted from that product. To isolate x on the left-hand side of the equation, we use the rules for the order of operations in reverse.

- To undo the subtraction of 8, we add 8 to both sides.
- To undo the multiplication by 2, we divide both sides by 2.

$2x - 8 + 8 = 0 + 8$ Use the addition property of equality: Add 8 to both sides.

$2x = 8$ Simplify both sides of the equation.

$\frac{2x}{2} = \frac{8}{2}$ Divide both sides by 2.

$x = 4$ Simplify both sides of the equation.

Check: We substitute 4 for x to verify that it satisfies the original equation.

$2x - 8 = 0$

$2(4) - 8 \stackrel{?}{=} 0$ Substitute 4 for x.

$8 - 8 \stackrel{?}{=} 0$ Multiply.

$0 = 0$ True.

Since we obtain a true statement, 4 is the solution of $2x - 8 = 0$ and the solution set is $\{4\}$.

Success Tip

Since division by 2 is the same as multiplication by $\frac{1}{2}$, we can also solve $2x = 8$ using the multiplication property of equality:

$$2x = 8$$
$$\frac{1}{2} \cdot 2x = \frac{1}{2} \cdot 8$$
$$x = 4$$

Self Check 2 Solve: $3a + 15 = 0$.

EXAMPLE 3

Solve: $\frac{3}{4}y = -7$.

Solution On the left-hand side, y is multiplied by $\frac{3}{4}$. We can undo the multiplication by dividing both sides by $\frac{3}{4}$. Since division by $\frac{3}{4}$ is equivalent to multiplication by its **reciprocal,** we can isolate y by multiplying both sides by $\frac{4}{3}$.

$$\frac{3}{4}y = -7$$

$$\frac{4}{3}\left(\frac{3}{4}y\right) = \frac{4}{3}(-7)$$ Use the multiplication property of equality: Multiply both sides by the reciprocal of $\frac{3}{4}$, which is $\frac{4}{3}$.

$$\left(\frac{4}{3} \cdot \frac{3}{4}\right)y = \frac{4}{3}(-7)$$ Use the associative property of multiplication to regroup.

$$1y = \frac{4}{3}(-7)$$ The product of a number and its reciprocal is 1: $\frac{4}{3} \cdot \frac{3}{4} = 1$.

$$y = -\frac{28}{3}$$ On the right-hand side, multiply.

Check:

$$\frac{3}{4}y = -7$$ This is the original equation.

$$\frac{3}{4}\left(-\frac{28}{3}\right) \stackrel{?}{=} -7$$ Substitute $-\frac{28}{3}$ for y.

$$-\frac{\overset{1}{\cancel{3}} \cdot \overset{1}{\cancel{4}} \cdot 7}{\underset{1}{\cancel{4}} \cdot \underset{1}{\cancel{3}}} \stackrel{?}{=} -7$$ Multiply the numerators and the denominators. Factor 28 and simplify.

$$-7 = -7$$ True.

The solution is $-\frac{28}{3}$ and the solution set is $\left\{-\frac{28}{3}\right\}$.

Self Check 3 Solve: $\frac{2}{3}b - 3 = -15$.

The equation in Example 3 can be solved using an alternate two-step approach.

$$\frac{3}{4}y = -7$$

$$4\left(\frac{3}{4}y\right) = 4(-7)$$ Multiply both sides by 4 to undo the division by 4.

$$3y = -28$$ Simplify: $4\left(\frac{3}{4}y\right) = \frac{4}{1}\left(\frac{3}{4}y\right) = \frac{\overset{1}{\cancel{4}} \cdot 3}{1 \cdot \underset{1}{\cancel{4}}}y = 3y$.

$$\frac{3y}{3} = -\frac{28}{3}$$ To undo the multiplication by 3, divide both sides by 3.

$$y = -\frac{28}{3}$$

SIMPLIFYING EXPRESSIONS TO SOLVE EQUATIONS

To solve more complicated equations, we often need to use the distributive property and combine like terms.

EXAMPLE 4

Solve: $-7(a - 2) = 8$.

Solution We begin by using the distributive property to remove parentheses.

$$-7(a - 2) = 8$$

$$-7a + 14 = 8 \quad \text{Distribute the multiplication by } -7.$$

$$-7a + 14 - 14 = 8 - 14 \quad \text{To undo the addition of 14, subtract 14 from both sides.}$$

$$-7a = -6$$

$$\frac{-7a}{-7} = \frac{-6}{-7} \quad \text{To undo the multiplication by } -7\text{, divide both sides by } -7.$$

$$a = \frac{6}{7}$$

Check:

$$-7(a - 2) = 8 \quad \text{This is the original equation.}$$

$$-7\left(\frac{6}{7} - 2\right) \stackrel{?}{=} 8 \quad \text{Substitute } \frac{6}{7} \text{ for } a.$$

$$-7\left(\frac{6}{7} - \frac{14}{7}\right) \stackrel{?}{=} 8 \quad \text{Get a common denominator: } 2 = \frac{14}{7}.$$

$$-7\left(-\frac{8}{7}\right) \stackrel{?}{=} 8 \quad \text{Subtract the fractions.}$$

$$8 = 8 \quad \text{True.}$$

Caution

When checking solutions, always use the original equation.

The solution is $\frac{6}{7}$ and the solution set is $\left\{\frac{6}{7}\right\}$.

Self Check 4 Solve: $-2(x + 3) = 18$.

EXAMPLE 5

Solve: $4b - 7 + 2b = 1 + 2b + 8$.

Solution First, we combine like terms on each side of the equation.

$$4b - 7 + 2b = 1 + 2b + 8$$

$$6b - 7 = 2b + 9 \quad \text{Combine like terms: } 4b + 2b = 6b \text{ and } 1 + 8 = 9.$$

We note that terms involving b appear on both sides of the equation. To isolate b on the left-hand side, we need to eliminate $2b$ on the right-hand side.

$$6b - 7 = 2b + 9$$

$$6b - 7 - 2b = 2b + 9 - 2b \quad \text{Subtract } 2b \text{ from both sides.}$$

$$4b - 7 = 9 \quad \text{Combine like terms on each side: } 6b - 2b = 4b \text{ and } 2b - 2b = 0.$$

$$4b - 7 + 7 = 9 + 7 \quad \text{To undo the subtraction of 7, add 7 to both sides.}$$

$$4b = 16 \quad \text{Simplify each side of the equation.}$$

$$b = 4 \quad \text{Divide both sides by 4.}$$

Check:

$4b - 7 + 2b = 1 + 2b + 8$	This is the original equation.
$4(4) - 7 + 2(4) \stackrel{?}{=} 1 + 2(4) + 8$	Substitute 4 for b.
$16 - 7 + 8 \stackrel{?}{=} 1 + 8 + 8$	
$17 = 17$	True.

The solution is 4.

Self Check 5 Solve: $-6t - 12 - 6t = 1 + 2t - 5$.

In general, we will follow these steps to solve linear equations in one variable.

Solving Linear Equations

1. If the equation contains fractions, multiply both sides of the equation by a nonzero number that will eliminate the denominators.
2. Use the distributive property to remove all sets of parentheses and then combine like terms.
3. Use the addition and subtraction properties to get all variable terms on one side of the equation and all constants on the other side. Combine like terms, if necessary.
4. Use the multiplication and division properties to make the coefficient of the variable equal to 1.
5. Check the result by replacing the variable with the possible solution and verifying that the number satisfies the equation.

EXAMPLE 6

Solve: $\frac{1}{3}(6x + 15) = \frac{3}{2}(x + 2) - 2$.

Solution

Step 1: We can clear the equation of fractions by multiplying both sides by the least common denominator (LCD) of $\frac{1}{3}$ and $\frac{3}{2}$. The LCD of these fractions is the smallest number that can be divided by both 2 and 3 exactly. That number is 6.

Success Tip

Before multiplying both sides of an equation by the LCD, frame the left-hand side and frame the right-hand side with parentheses or brackets.

$\frac{1}{3}(6x + 15) = \frac{3}{2}(x + 2) - 2$	
$6\left[\frac{1}{3}(6x + 15)\right] = 6\left[\frac{3}{2}(x + 2) - 2\right]$	To eliminate the fractions, multiply both sides by the LCD, 6.
$2(6x + 15) = 6 \cdot \frac{3}{2}(x + 2) - 6 \cdot 2$	On the left-hand side, multiply: $6 \cdot \frac{1}{3} = 2$. On the right-hand side, distribute the multiplication by 6.
$2(6x + 15) = 9(x + 2) - 12$	Perform the multiplications on the right-hand side.

Step 2: We remove parentheses and then combine like terms.

$12x + 30 = 9x + 18 - 12$	Distribute the multiplication by 2 and the multiplication by 9.
$12x + 30 = 9x + 6$	Combine like terms.

Step 3: To get the variable term on the left-hand side and the constant on the right-hand side, subtract $9x$ and 30 from both sides.

$$12x + 30 - 9x - 30 = 9x + 6 - 9x - 30$$

$$3x = -24 \quad \text{On each side, combine like terms.}$$

Step 4: The coefficient of the variable x is 3. To undo the multiplication by 3, we divide both sides by 3.

$$\frac{3x}{3} = \frac{-24}{3} \quad \text{Divide both sides by 3.}$$

$$x = -8$$

Step 5: We check by substituting -8 for x in the original equation and simplifying:

$$\frac{1}{3}(6x + 15) = \frac{3}{2}(x + 2) - 2$$

$$\frac{1}{3}[6(-8) + 15] \stackrel{?}{=} \frac{3}{2}(-8 + 2) - 2$$

$$\frac{1}{3}(-48 + 15) \stackrel{?}{=} \frac{3}{2}(-6) - 2$$

$$\frac{1}{3}(-33) \stackrel{?}{=} -9 - 2$$

$$-11 = -11 \quad \text{True.}$$

The solution is -8.

Self Check 6 Solve: $\frac{1}{3}(2x - 2) = \frac{1}{4}(5x + 1) + 2$.

EXAMPLE 7

Solve: $\frac{x+2}{5} - 4x = \frac{8}{5} - \frac{x+9}{2}$.

Solution Some of the steps used to solve an equation can be done in your head, as you will see in this example.

$$\frac{x+2}{5} - 4x = \frac{8}{5} - \frac{x+9}{2}$$

$$10\left(\frac{x+2}{5} - 4x\right) = 10\left(\frac{8}{5} - \frac{x+9}{2}\right) \quad \text{To eliminate the fractions, multiply both sides by the LCD, 10.}$$

$$10 \cdot \frac{x+2}{5} - 10 \cdot 4x = 10 \cdot \frac{8}{5} - 10 \cdot \frac{x+9}{2} \quad \text{On each side, distribute the 10.}$$

$$2(x + 2) - 40x = 2(8) - 5(x + 9) \quad \text{Perform each multiplication by 10.}$$

$$2x + 4 - 40x = 16 - 5x - 45 \quad \text{On each side, remove parentheses.}$$

$$-38x + 4 = -5x - 29 \quad \text{On each side, combine like terms.}$$

$$33 = 33x \quad \text{Add 38x and 29 to both sides. These steps are done in your head—we don't show them.}$$

$$1 = x \quad \text{Divide both sides by 33. This step is also done in your head.}$$

Check by substituting 1 for x in the original equation.

Success Tip

In step 6, we could have eliminated $-5x$ from the right-hand side by adding $5x$ to both sides:

$$-38x + 4 + 5x = -5x - 29 + 5x$$

$$-33x + 4 = -29$$

However, it is usually easier to isolate the variable term on the side that will result in a *positive* coefficient, as we did.

Self Check 7 Solve: $\frac{a+3}{2} + 2a = \frac{3}{2} - \frac{a+27}{5}$.

EXAMPLE 8

Solve: $-35.6 = 77.89 - x$.

Solution

$$-35.6 = 77.89 - x$$

$$-35.6 - 77.89 = 77.89 - x - 77.89 \quad \text{Subtract 77.89 from both sides.}$$

$$-113.49 = -x \quad \text{Simplify each side of the equation.}$$

$$-113.49 = -1x \quad -x = -1x.$$

$$\frac{-113.49}{-1} = \frac{-1x}{-1} \quad \text{To isolate } x \text{, multiply both sides by } -1 \text{ or divide both sides by } -1.$$

$$113.49 = x \quad \text{Simplify each side of the equation.}$$

$$x = 113.49$$

Check that 113.49 satisfies the equation.

Self Check 8 Solve: $-1.3 = -2.6 - x$.

For more complicated equations involving decimals, we can multiply both sides of the equation by a power of 10 to clear the equation of decimals.

EXAMPLE 9

Solve: $0.04(12) + 0.01x = 0.02(12 + x)$.

Solution The equation contains the decimals 0.04, 0.01, and 0.02. Multiplying both sides by $10^2 = 100$ changes the decimals in the equation to integers, which are easier to work with.

$$0.04(12) + 0.01x = 0.02(12 + x)$$

$$100[0.04(12) + 0.01x] = 100[0.02(12 + x)] \quad \text{To make 0.04, 0.01, and 0.02 integers, multiply both sides by 100.}$$

$$100 \cdot 0.04(12) + 100 \cdot 0.01x = 100 \cdot 0.02(12 + x) \quad \text{On the left-hand side, distribute the multiplication by 100.}$$

$$4(12) + 1x = 2(12 + x) \quad \text{Perform the multiplication by 100.}$$

$$48 + x = 24 + 2x \quad \text{Remove parentheses.}$$

$$48 + x - 24 - x = 24 + 2x - 24 - x \quad \text{Subtract 24 and } x \text{ from both sides.}$$

$$24 = x \quad \text{Simplify each side.}$$

$$x = 24$$

Check by substituting 24 for x in the original equation.

Self Check 9 Solve: $0.08x + 0.07(15{,}000 - x) = 1{,}110$.

IDENTITIES AND CONTRADICTIONS

The equations discussed so far are called **conditional equations.** For these equations, some numbers satisfy the equation and others do not. An **identity** is an equation that is satisfied by every number for which both sides of the equation are defined.

EXAMPLE 10

Solve: $-2(x - 1) - 4 = -4(1 + x) + 2x + 2$.

Solution

$$-2(x - 1) - 4 = -4(1 + x) + 2x + 2$$

$$-2x + 2 - 4 = -4 - 4x + 2x + 2 \quad \text{Use the distributive property.}$$

$$-2x - 2 = -2x - 2 \quad \text{On each side, combine like terms.}$$

$$-2 = -2 \quad \text{True.}$$

The terms involving x drop out. The resulting true statement indicates that the original equation is true for every value of x. The solution set is the set of real numbers denoted $\mathbb{R}$. The equation is an identity.

Self Check 10 Solve: $3(a + 4) + 5 = 2(a - 1) + a + 19$ and give the solution set.

A **contradiction** is an equation that is never true.

EXAMPLE 11

Solve: $-6.2(-x - 1) - 4 = 4.2x - (-2x)$.

Solution

$$-6.2(-x - 1) - 4 = 4.2x - (-2x)$$

$$6.2x + 6.2 - 4 = 4.2x + 2x \quad \text{On the left-hand side, remove parentheses. On the right-hand side, write the subtraction as addition of the opposite.}$$

$$6.2x + 2.2 = 6.2x \quad \text{On each side, combine like terms.}$$

$$6.2x + 2.2 - 6.2x = 6.2x - 6.2x \quad \text{Subtract } 6.2x \text{ from both sides.}$$

$$2.2 = 0 \quad \text{False.}$$

The Language of Algebra

Contradiction is a form of the word *contradict,* meaning conflicting ideas. During a trial, evidence might be introduced that *contradicts* the testimony of a witness.

The terms involving x drop out. The resulting false statement indicates that no value for x makes the original equation true. The solution set contains no elements and can be denoted as the **empty set** $\{\ \}$ or the **null set** $\varnothing$. The equation is a contradiction.

Self Check 11 Solve: $3(a + 4) + 2 = 2(a - 1) + a + 19$.

SOLVING FORMULAS

To solve a formula for a variable means to isolate that variable on one side of the equation and have all other quantities on the other side. We can use the skills discussed in this section to solve many types of formulas for a specified variable.

EXAMPLE 12

Solve: $A = \frac{1}{2}bh$ for h.

Solution

$$A = \frac{1}{2}bh \quad \text{This is the formula for the area of a triangle.}$$

$$2A = bh \quad \text{To eliminate the fraction, multiply both sides by 2.}$$

$$\frac{2A}{b} = h \quad \text{To isolate } h\text{, multiply both sides by } \frac{1}{b} \text{ or divide both sides by } b.$$

$$h = \frac{2A}{b} \quad \text{Write the equation with } h \text{ on the left-hand side.}$$

Self Check 12 Solve: $A = \frac{1}{2}bh$ for b.

EXAMPLE 13

For simple interest, the formula $A = P + Prt$ gives the amount of money in an account at the end of a specific time. A represents the amount, P the principal, r the rate of interest, and t the time. We can solve the formula for t as follows:

Solution

$$A = P + Prt$$

$$A - P = Prt \quad \text{To isolate the term involving } t\text{, subtract } P \text{ from both sides.}$$

$$\frac{A - P}{Pr} = t \quad \text{To isolate } t\text{, multiply both sides by } \frac{1}{Pr} \text{ or divide both sides by } Pr.$$

$$t = \frac{A - P}{Pr} \quad \text{Write the equation with } t \text{ on the left-hand side.}$$

Self Check 13 Solve $A = P + Prt$ for r.

EXAMPLE 14

The formula $F = \frac{9}{5}C + 32$ converts degrees Celsius to degrees Fahrenheit. Solve it for C.

Solution

$$F = \frac{9}{5}C + 32$$

$$F - 32 = \frac{9}{5}C \quad \text{To isolate the term involving } C\text{, subtract 32 from both sides.}$$

$$\frac{5}{9}(F - 32) = \frac{5}{9}\left(\frac{9}{5}C\right) \quad \text{To isolate } C\text{, multiply both sides by } \frac{5}{9}.$$

$$\frac{5}{9}(F - 32) = C \quad \frac{5}{9} \cdot \frac{9}{5} = 1.$$

$$C = \frac{5}{9}(F - 32)$$

To convert degrees Fahrenheit to degrees Celsius, we can use the formula $C = \frac{5}{9}(F - 32)$.

Self Check 14 Solve: $S = \frac{180(t - 2)}{7}$ for t.

Answers to Self Checks **1.** yes **2.** -5 **3.** -18 **4.** -12 **5.** $-\frac{4}{7}$ **6.** -5 **7.** -2 **8.** -1.3 **9.** 6,000 **10.** all real numbers, $\mathbb{R}$ **11.** no solution, $\varnothing$ **12.** $b = \frac{2A}{h}$ **13.** $r = \frac{A - P}{Pt}$ **14.** $t = \frac{7S + 360}{180}$ or $t = \frac{7S}{180} + 2$

1.5 STUDY SET

VOCABULARY **Fill in the blanks.**

1. An ________ is a statement that two expressions are equal.
2. $2x + 1 = 4$ and $5(y - 3) = 8$ are examples of ______ equations in one variable.
3. If a number is substituted for a variable in an equation and the equation is true, we say that the number ________ the equation.
4. If two equations have the same solution set, they are called __________ equations.
5. An equation that is true for all values of its variable is called an ________.
6. An equation that is not true for any values of its variable is called a ____________.

CONCEPTS **Fill in the blanks.**

7. If $a = b$, then $a + c = b +$ ☐ and $a - c = b -$ ☐. ________ (or subtracting) the same number to (or from) ______ sides of an equation does not change the solution.
8. If $a = b$, then $ca =$ ☐ and $\frac{a}{c} = \frac{b}{☐}$. ___________ (or dividing) both sides of an equation by the ______ nonzero number does not change the solution.
9. **a.** Simplify: $5y + 2 - 3y$.
   b. Solve: $5y + 2 - 3y = 8$.
   c. Evaluate $5y + 2 - 3y$ for $y = 8$.
10. When solving $\frac{x+1}{3} - \frac{2}{15} = \frac{x-1}{5}$, why would we multiply both sides by 15?
11. When solving $1.45x - 0.5(1 - x) = 0.7x$, why would we multiply both sides by 100?
12. **a.** Suppose you solve a linear equation in one variable, the variable drops out, and you obtain $8 = 8$. What is the solution set?
    b. Suppose you solve a linear equation in one variable, the variable drops out, and you obtain $8 = 7$. What is the solution set?

NOTATION **Complete the solution to solve the equation. Then check the result.**

13.
$$-2(x + 7) = 20$$
$$☐ - 14 = 20$$
$$-2x - 14 + ☐ = 20 + ☐$$
$$-2x = 34$$
$$\frac{-2x}{☐} = \frac{34}{☐}$$
$$x = -17$$

Check:
$$-2(x + 7) = 20$$
$$-2(☐ + 7) \stackrel{?}{=} 20$$
$$-2(☐)\ ☐\ 20$$
$$☐ = 20$$

The solution is ☐.

14. Fill in the blanks to make the statements true.
    a. $-x = ☐\,x$ **b.** $\frac{2t}{3} = ☐\,t$
15. When checking a solution of an equation, the symbol $\stackrel{?}{=}$ is used. What does it mean?
16. **a.** What does the symbol $\mathbb{R}$ denote?
    b. What symbol denotes a set with no members?

PRACTICE **Determine whether 5 is a solution of each equation.**

17. $3x + 2 = 17$

18. $7x - 2 = 53 - 5x$

19. $3(2m - 3) = 15$

20. $\frac{3}{5}p - 5 = -2$

Solve each equation. Check each result.

21. $2x - 12 = 0$

22. $3x - 24 = 0$

23. $5y + 6 = 0$

24. $7y + 3 = 0$

25. $2x + 2(1) = 6$

26. $3x - 4(1) = 8$

27. $4(3) + 2y = -6$

28. $5(2) + 10y = -10$

29. $3x + 1 = 3$

30. $8k - 2 = 13$

31. $\frac{x}{4} = 7$

32. $-\frac{x}{6} = 8$

33. $-\frac{4}{5}s = 16$

34. $-3 = -\frac{9}{8}s$

35. $1.6a = 4.032$

36. $0.52 = 0.05y$

37. $\frac{x}{6} - 7 = -12$

38. $\frac{a}{8} + 1 = -10$

39. $3(k - 4) = -36$

40. $4(x + 6) = 84$

41. $8x = x$

42. $-z = 5z$

43. $4j + 12.54 = 18.12$

44. $9.8 - 15r = -15.7$

45. $4a - 22 - a = -2a - 7$

46. $a + 18 = 5a - 3 + a$

47. $2(2x + 1) = x + 15 + 2x$

48. $-2(x + 5) = x + 30 - 2x$

49. $2(a - 5) - (3a + 1) = 0$

50. $8(3a - 5) - 4(2a + 3) = 12$

51. $9(x + 2) = -6(4 - x) + 18$

52. $3(x + 2) - 2 = -(5 + x) + x$

53. $12 + 3(x - 4) - 21 = 5[5 - 4(4 - x)]$

54. $1 + 3[-2 + 6(4 - 2x)] = -(x + 3)$

55. $\frac{1}{2}x - 4 = -1 + 2x$

56. $2x + 3 = \frac{2}{3}x - 1$

57. $\frac{b}{2} - \frac{b}{3} = 4$

58. $\frac{w}{2} + \frac{w}{3} = 10$

59. $\frac{a + 1}{3} + \frac{a - 1}{5} = \frac{2}{15}$

60. $\frac{2z + 3}{3} + \frac{3z - 4}{6} = \frac{z - 2}{2}$

61. $\frac{5a}{2} - 12 = \frac{a}{3} + 1$

62. $5 - \frac{x + 2}{3} = 7 - x$

63. $\frac{3 + p}{3} - 4p = 1 - \frac{p + 7}{2}$

64. $\frac{4 - t}{2} - \frac{3t}{5} = 2 + \frac{t + 1}{3}$

65. $\frac{4}{5}(x + 5) = \frac{7}{8}(3x + 23) - 7$

66. $\frac{2}{3}(2x + 2) + 4 = \frac{1}{6}(5x + 29)$

67. $0.45 = 16.95 - 0.25(75 - 3x)$

68. $0.02x + 0.0175(15{,}000 - x) = 277.5$

69. $0.04(12) + 0.01t - 0.02(12 + t) = 0$

70. $0.25(t + 32) = 3.2 + t$

Solve each equation. If the equation is an identity or a contradiction, so indicate.

71. $4(2 - 3t) + 6t = -6t + 8$

72. $2x - 6 = -2x + 4(x - 2)$

73. $3(x - 4) + 6 = -2(x + 4) + 5x$

74. $2(x - 3) = \frac{3}{2}(x - 4) + \frac{x}{2}$

75. $2y + 1 = 5(0.2y + 1) - (4 - y)$

76. $-3x = -2x + 1 - (5 + x)$

Solve each formula for the indicated variable.

77. $V = \frac{1}{3}Bh$ for B

78. $A = \frac{1}{2}bh$ for b

79. $I = Prt$ for t

80. $E = mc^2$ for m

81. $P = 2l + 2w$ for w

82. $T - W = ma$ for W

83. $A = \frac{1}{2}h(B + b)$ for B

84. $\ell = a + (n - 1)d$ for n

85. $y = mx + b$ for x

86. $\lambda = Ax + AB$ for B

87. $\bar{v} = \frac{1}{2}(v + v_0)$ for v_0

88. $\ell = a + (n - 1)d$ for d

89. $S = \frac{a - \ell r}{1 - r}$ for ℓ

90. $s = \frac{1}{2}gt^2 + vt$ for g

91. $S = \frac{n(a + \ell)}{2}$ for ℓ

92. $K = \frac{Mv_0^2}{2} + \frac{Iw^2}{2}$ for I

APPLICATIONS

93. CONVERTING TEMPERATURES In preparing an American almanac for release in Europe, editors need to convert temperature ranges for the planets from degrees Fahrenheit to degrees Celsius. Solve the formula $F = \frac{9}{5}C + 32$ for C. Then use your result to make the conversions for the data shown in the table. Round to the nearest degree.

Planet	High °F	Low °F	High °C	Low °C
Mercury	810	−290		
Earth	136	−129		
Mars	63	−87		

94. THERMODYNAMICS In thermodynamics, the Gibbs free-energy function is given by the formula $G = U - TS + pV$. Solve for S.

95. WIPER DESIGN The area cleaned by the windshield wiper assembly shown in the illustration in the next column is given by the formula

$$A = \frac{d\pi(r_1^2 - r_2^2)}{360}$$

Engineers have determined the amount of windshield area that needs to be cleaned by the wiper for two different vehicles. Solve the equation for d and use your result to find the number of degrees d the wiper arm must swing in each case. Round to the nearest degree.

Vehicle	Area cleaned	d (deg)
Luxury car	513 in.^{2}	
Sport utility vehicle	586 in.^{2}	

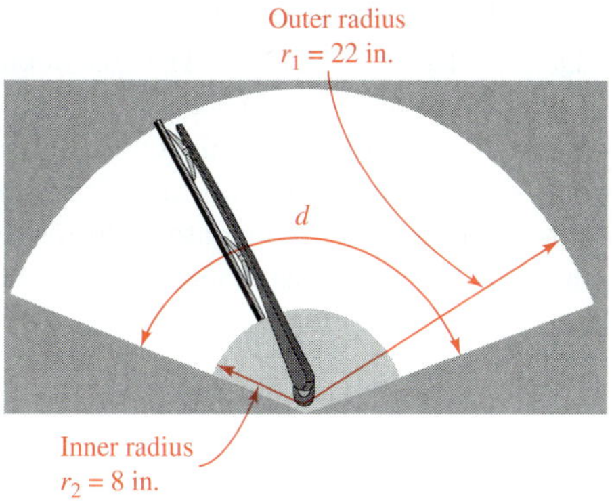
Outer radius
$r_1 = 22$ in.
d
Inner radius
$r_2 = 8$ in.

96. ELECTRONICS The illustration is a schematic diagram of a resistor connected to a voltage source of 60 volts. As a result, the resistor dissipates power in the form of heat. The power P lost when a voltage E is placed across a resistance R (in ohms) is given by the formula

$$P = \frac{E^2}{R}$$

Solve for R. If P is 4.8 watts and E is 60 volts, find R.

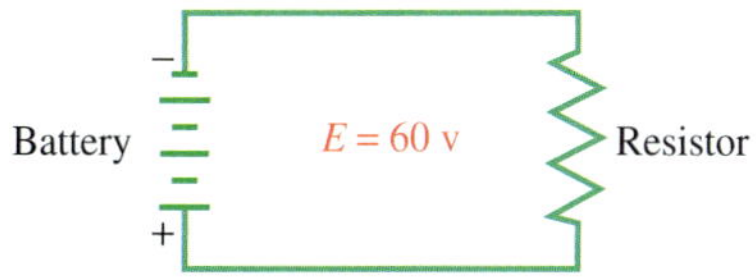
Battery
E = 60 v
Resistor
−
+

97. CHEMISTRY LAB In chemistry, the ideal gas law equation is $PV = nR(T + 273)$, where P is the pressure, V the volume, T the temperature, and n the number of moles of a gas. R is a constant, 0.082. Solve the equation for n. Then use your result and the data from the student lab notebook in the illustration to find the value of n to the nearest thousandth for trial 1 and trial 2.

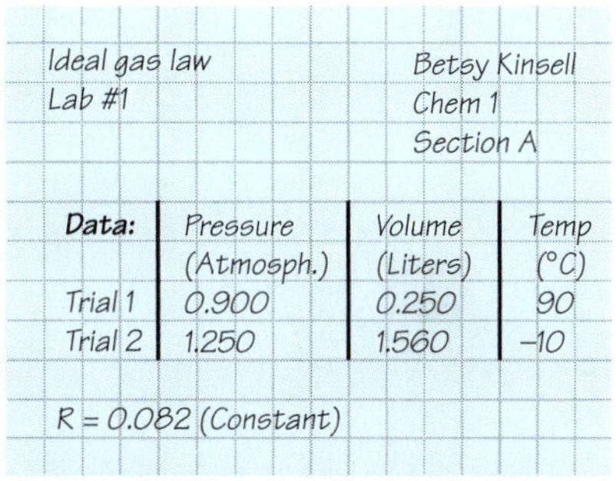
Ideal gas law
Lab #1
Betsy Kinsell
Chem 1
Section A
Data: Pressure (Atmosph.) Volume (Liters) Temp (°C)
Trial 1 0.900 0.250 90
Trial 2 1.250 1.560 −10
R = 0.082 (Constant)

98. INVESTMENTS An amount P, invested at a simple interest rate r, will grow to an amount A in t years according to the formula $A = P(1 + rt)$. Solve for P. Suppose a man invested some money at 5.5%. If after 5 years, he had \$6,693.75 on deposit, what amount did he originally invest?

99. COST OF ELECTRICITY The cost of electricity in a city is given by the formula $C = 0.07n + 6.50$, where C is the cost and n is the number of kilowatt hours used. Solve for n. Then find the number of kilowatt hours used each month by the homeowner whose checks to pay the monthly electric bills are shown in the illustration.

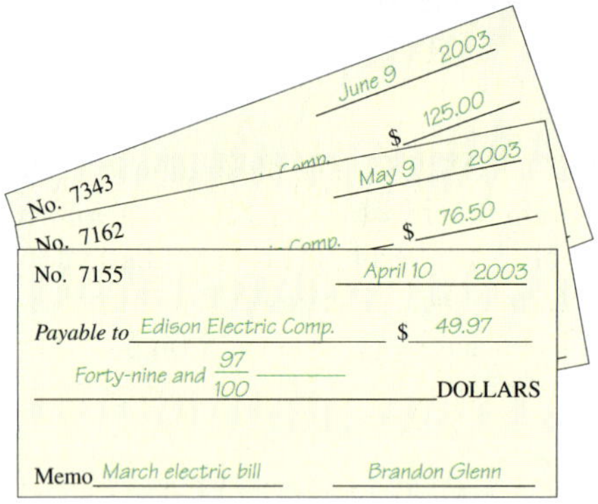

100. COST OF WATER A monthly water bill in a certain city is calculated by using the formula $n = \frac{5{,}000C - 17{,}500}{6}$, where n is the number of gallons used and C is the monthly cost. Solve for C and compute the bill for quantities of 500, 1,200, and 2,500 gallons.

101. SURFACE AREA To find the amount of tin needed to make the coffee can shown in the illustration, we use the formula for the surface area of a right circular cylinder,

$$A = 2\pi r^2 + 2\pi rh$$

Solve the formula for h.

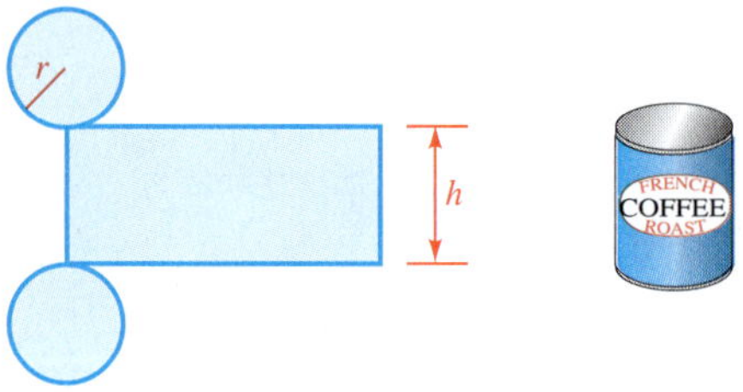

102. CARPENTRY A regular polygon has n equal sides and n equal angles. The measure a of an interior angle in degrees is given by $a = 180\left(1 - \frac{2}{n}\right)$. Solve for n. How many sides does the outdoor bandstand shown below have if the performance platform is a regular polygon with interior angles measuring 135°?

WRITING

103. What does it mean to *solve an equation?*

104. Why doesn't the equation $x = x + 1$ have a real-number solution?

105. What is an identity? Give an example.

106. When solving a linear equation in one variable, the objective is to isolate the variable on one side of the equation. What does that mean?

REVIEW **Simplify each expression.**

107. $-(4 + t) + 2t$

108. $12(2r + 11) - 11 - 3$

109. $4(b + 8) - 8b$

110. $-2(m - 3) + 8(2m + 7)$

111. $3.8b - 0.9b$

112. $-5.7p + 5.1p$

113. $\frac{3}{5}t + \frac{2}{5}t$

114. $-\frac{3}{16}x - \frac{5}{16}x$

CHALLENGE PROBLEMS

115. Find the value of k that makes 4 a solution of the following linear equation in x.

$$k + 3x - 6 = 3kx - k + 16$$

116. Solve for x: $\frac{2n + 3x}{6} - \frac{4n - x}{2} = \frac{5n + x}{4} + 2.$

1.6 Using Equations to Solve Problems

- A problem-solving strategy • Translating words to form an equation
- Analyzing a problem • Number–value problems • Geometry problems
- Using formulas to solve problems

A major objective of this course is to improve your problem-solving abilities. In the next two sections, you will have the opportunity to do that as we discuss how to use equations to solve many different types of problems.

A PROBLEM-SOLVING STRATEGY

The key to problem solving is understanding the problem and devising a plan for solving it. The following list provides a strategy for solving problems.

Problem Solving

1. ***Analyze the problem*** by reading it carefully to understand the given facts. What information is given? What are you asked to find? What vocabulary is given? Often a diagram or table will help you visualize the facts of the problem.
2. ***Form an equation*** by picking a variable to represent the quantity to be found. Then express all other quantities mentioned as expressions involving the variable. Finally, translate the words of the problem into an equation.
3. ***Solve the equation.***
4. ***State the conclusion.***
5. ***Check the result*** in the words of the problem.

TRANSLATING WORDS TO FORM AN EQUATION

In order to solve problems, which are almost always given in words, we must translate those words into mathematical symbols. In the next example, we use translation to write an equation that mathematically models the situation.

EXAMPLE 1

Leading U. S. employers. In 2003, McDonald's and Wal-Mart were the nation's top two employers. Their combined work forces totaled 2,900,000 people. If Wal-Mart employed 100,000 fewer people than McDonald's, how many employees did each company have?

Analyze the Problem

- The phrase *combined work forces totaled 2,900,000* suggests that if we add the number of employees of each company, the result will be 2,900,000.
- The phrase *Wal-Mart employed 100,000 fewer people than McDonald's* suggests that the number of employees of Wal-Mart can be found by subtracting 100,000 from the number of employees of McDonald's.
- We are to find the number of employees of each company.

Form an Equation

If we let $x =$ the number of employees of McDonald's, then $x - 100{,}000 =$ the number of employees of Wal-Mart. We can now translate the words of the problem into an equation.

The number of employees of McDonald's	plus	the number of employees of Wal-Mart	is	2,900,000
x	$+$	$x - 100{,}000$	$=$	$2{,}900{,}000$

Solve the Equation

$$x + x - 100{,}000 = 2{,}900{,}000$$
$$2x - 100{,}000 = 2{,}900{,}000 \quad \text{Combine like terms.}$$
$$2x = 3{,}000{,}000 \quad \text{Add 100,000 to both sides.}$$
$$x = 1{,}500{,}000 \quad \text{Divide both sides by 2.}$$

Recall that x represents the number of employees of McDonald's. To find the number of employees of Wal-Mart, we evaluate $x - 100{,}000$ for $x = 1{,}500{,}000$.

$$x - 100{,}000 = \mathbf{1{,}500{,}000} - 100{,}000$$
$$= 1{,}400{,}000$$

State the Conclusion In 2003, McDonald's had 1,500,000 employees and Wal-Mart had 1,400,000 employees.

Check the Result Since $1{,}500{,}000 + 1{,}400{,}000 = 2{,}900{,}000$, and since 1,400,000 is 100,000 less than 1,500,000, the answers check.

When solving problems, diagrams are often helpful, because they allow us to visualize the facts of the problem.

EXAMPLE 2

Triathlons. A triathlon in Hawaii includes swimming, long-distance running, and cycling. The long-distance run is 11 times longer than the distance the competitors swim. The distance they cycle is 85.8 miles longer than the run. Overall, the competition covers 140.6 miles. Find the length of each part of the triathlon and round each length to the nearest tenth of a mile.

Analyze the Problem The entire triathlon course covers a distance of 140.6 miles. We note that the distance the competitors run is related to the distance they swim, and the distance they cycle is related to the distance they run.

Form an Equation If x = the distance the competitors swim, then $11x$ = the length of the long-distance run, and $11x + 85.8$ = the distance they cycle. From the diagram, we can see that the sum of the individual parts of the triathlon must equal the total distance covered.

We can now form the equation.

The distance they swim	plus	the distance they run	plus	the distance they cycle	is	the total length of the course.
x	$+$	$11x$	$+$	$11x + 85.8$	$=$	140.6

Solve the Equation

$$x + 11x + 11x + 85.8 = 140.6$$
$$23x + 85.8 = 140.6 \quad \text{Combine like terms.}$$
$$23x = 54.8 \quad \text{Subtract 85.8 from both sides.}$$
$$x \approx 2.382608696 \quad \text{Divide both sides by 23.}$$

State the Conclusion To the nearest tenth, the distance the competitors swim is 2.4 miles. The distance they run is $11x$, or approximately $11(2.382608696) = 26.20869565$ miles. To the nearest tenth, that is 26.2 miles. The distance they cycle is $11x + 85.8$, or approximately $26.20869565 + 85.8 = 112.0086957$ miles. To the nearest tenth, that is 112.0 miles.

Check the Result If we add the lengths of the three parts of the triathlon and round to the nearest tenth, we get 140.6 miles. The answers check.

ANALYZING A PROBLEM

The wording of a problem doesn't always contain key phrases that translate directly to an equation. In such cases, an analysis of the problem often gives clues that help us write an equation.

EXAMPLE 3

Travel promotions. The price of a 7-day Alaskan cruise, normally \$2,752 per person, is reduced by \$1.75 per person for large groups traveling together. How large a group is needed for the price to be \$2,500 per person?

Analyze the Problem For each member of the group, the cost is reduced by \$1.75. For a group of 20 people, the \$2,752 price is reduced by 20(\$1.75) = \$35.

$$\text{The per-person price of the cruise} = \$2{,}752 - 20(\$1.75)$$

For a group of 30 people, the \$2,752 cost is reduced by 30(\$1.75) = \$52.50.

$$\text{The per-person price of the cruise} = \$2{,}752 - 30(\$1.75)$$

Form an Equation If we let x = the group size necessary for the price of the cruise to be \$2,500 per person, we can form the following equation:

The price of the cruise	is	\$2,752	minus	the number of people in the group	times	\$1.75.
2,500	=	2,752	−	x	·	1.75

Solve the Equation

$$2{,}500 = 2{,}752 - 1.75x$$
$$2{,}500 - 2{,}752 = 2{,}752 - 1.75x - 2{,}752 \quad \text{Subtract 2,752 from both sides.}$$
$$-252 = -1.75x \quad \text{Simplify each side.}$$
$$144 = x \quad \text{Divide both sides by } -1.75.$$

State the Conclusion If 144 people travel together, the price will be \$2,500 per person.

Check the Result For 144 people, the cruise cost of \$2,752 will be reduced by 144(\$1.75) = \$252. If we subtract, \$2,752 − \$252 = \$2,500. The answer checks.

NUMBER–VALUE PROBLEMS

Some problems deal with quantities that have a value. In these problems, we must distinguish between the *number of* and the *value of* the unknown quantity. For problems such as these, we will use the relationship

$$\text{Number} \cdot \text{value} = \text{total value}$$

EXAMPLE 4

Portfolio analysis. A college foundation owns stock in Kodak (selling at \$25 per share), Coca-Cola (selling at \$50 per share), and IBM (selling at \$100 per share). The foundation owns an equal number of shares of Kodak and Coca-Cola stock, but five times as many shares of IBM stock. If this portfolio is worth \$402,500, how many shares of each stock does the foundation own?

Analyze the Problem The value of the Kodak stock plus the value of the Coca-Cola stock plus the value of the IBM stock must equal \$402,500. We need to find the number of shares of each of these stocks held by the foundation.

Form an Equation If we let x = the number of shares of Kodak stock, then x = the number of shares of Coca-Cola stock. Since the foundation owns five times as many shares of IBM stock as Kodak or Coca-Cola stock, $5x$ = the number of shares of IBM. The value of the shares of each stock is the *product* of the number of shares of that stock and its per-share value. See the table.

Success Tip

We can let x represent the number of shares of Kodak stock *and* the number of shares of Coca-Cola stock because the foundation owns an *equal number* of shares of these stocks.

Stock	Number of shares ·	Value per share =	Total value of the stock
Kodak	x	25	$25x$
Coca-Cola	x	50	$50x$
IBM	$5x$	100	$100(5x)$

We can now form the equation.

The value of Kodak stock	plus	the value of Coca-Cola stock	plus	the value of IBM stock	is	the total value of all of the stock.
$25x$	$+$	$50x$	$+$	$100(5x)$	$=$	402,500

Solve the Equation

$$25x + 50x + 500x = 402{,}500$$

$$575x = 402{,}500 \quad \text{Combine like terms on the left-hand side.}$$

$$x = 700 \quad \text{Divide both sides by 575.}$$

State the Conclusion The foundation owns 700 shares of Kodak, 700 shares of Coca-Cola, and $5(700) = 3{,}500$ shares of IBM.

Check the Result The value of 700 shares of Kodak stock is 700(\$25) = \$17,500. The value of 700 shares of Coca-Cola is 700(\$50) = \$35,000. The value of 3,500 shares of IBM is 3,500(\$100) = \$350,000. The sum is \$17,500 + \$35,000 + \$350,000 = \$402,500. The answers check.

GEOMETRY PROBLEMS

Sometimes a geometric fact or formula is helpful in solving a problem. The following illustration shows several geometric figures. A **right angle** is an angle whose measure is 90°. A **straight angle** is an angle whose measure is 180°. An **acute angle** is an angle whose measure is greater than 0° and less than 90°. An angle whose measure is greater than 90° and less than 180° is called an **obtuse angle.**

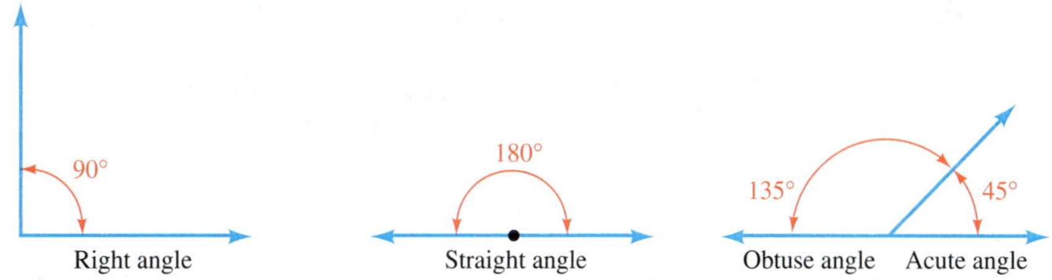

If the sum of two angles equals 90°, the angles are called **complementary,** and each angle is called the **complement** of the other. If the sum of two angles equals 180°, the angles are called **supplementary,** and each angle is the **supplement** of the other.

A **right triangle** is a triangle with one right angle. An **isosceles triangle** is a triangle with two sides of equal measure that meet to form the **vertex angle.** The angles opposite the equal sides, called the **base angles,** are also equal. An **equilateral triangle** is a triangle with three equal sides and three equal angles.

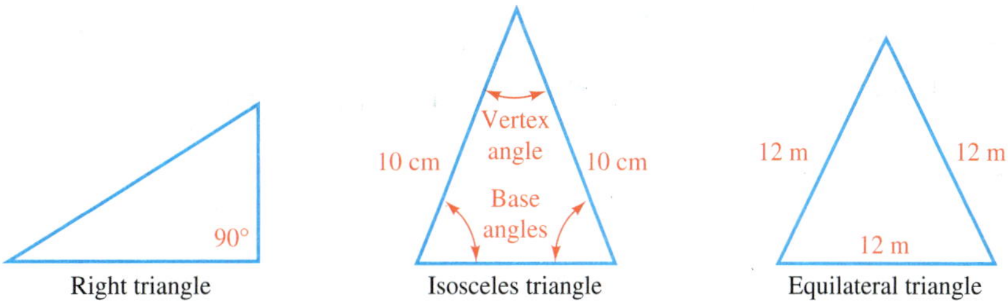

EXAMPLE 5

Flag design. The flag of Guyana, a republic on the northern coast of South America, is one isosceles triangle superimposed over another on a field of green, as shown. The measure of a base angle of the larger triangle is 14° more than the measure of a base angle of the smaller triangle. The measure of the vertex angle of the larger triangle is 34°. Find the measure of each base angle of the smaller triangle.

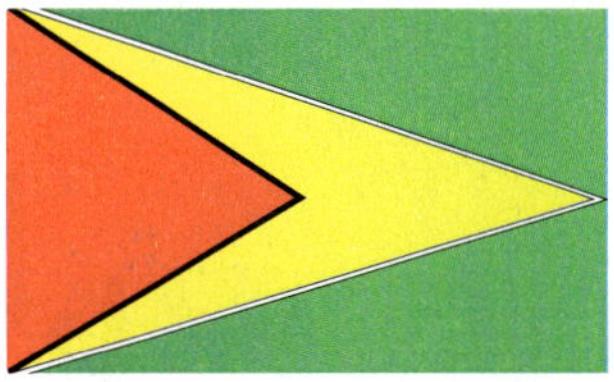

Analyze the Problem We are working with isosceles triangles. Therefore, the base angles of the smaller triangle have the same measure, and the base angles of the larger triangle have the same measure.

Form an Equation If we let $x =$ the measure in degrees of one base angle of the smaller isosceles triangle, then the measure of its other base angle is also x. (See the figure.)

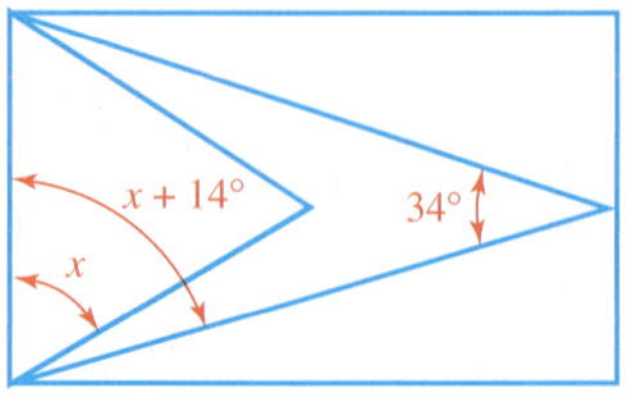

The measure of a base angle of the larger isosceles triangle is $x + 14°$, since its measure is 14° more than the measure of a base angle of the smaller triangle. We are given that the vertex angle of the larger triangle measures 34°.

The sum of the measures of the angles of any triangle (in this case, the larger triangle) is 180°.

We can now form the equation.

The measure of one base angle	plus	the measure of the other base angle	plus	the measure of the vertex angle	is	180°.
$x + 14$	$+$	$x + 14$	$+$	34	$=$	180

Solve the Equation

$$x + 14 + x + 14 + 34 = 180$$
$$2x + 62 = 180 \quad \text{Combine like terms.}$$
$$2x = 118 \quad \text{Subtract 62 from both sides.}$$
$$x = 59 \quad \text{Divide both sides by 2.}$$

State the Conclusion The measure of each base angle of the smaller triangle is 59°.

Check the Result If $x = 59$, then $x + 14 = 73$. The sum of the measures of each base angle and the vertex angle of the *larger* triangle is $73° + 73° + 34° = 180°$. The answer checks.

USING FORMULAS TO SOLVE PROBLEMS

When preparing to write an equation to solve a problem, the given facts of the problem often suggest a formula that can be used to model the situation mathematically.

EXAMPLE 6

Kennels. A man has a 50-foot roll of fencing to make a rectangular kennel. If he wants the kennel to be 6 feet longer than it is wide, find its dimensions.

Analyze the Problem The perimeter P of the rectangular kennel is 50 feet. Recall that the formula for the perimeter of a rectangle is $P = 2l + 2w$. We need to find its length and width.

Form an Equation We let $w =$ the width of the kennel shown below. Then the length, which is 6 feet more than the width, is represented by the expression $w + 6$.

We can now form the equation by substituting 50 for P and $w + 6$ for the length in the formula for the perimeter of a rectangle.

$$P = 2l + 2w$$
$$50 = 2(w + 6) + 2w$$

Solve the Equation

$50 = 2(w + 6) + 2w$

$50 = 2w + 12 + 2w$ Use the distributive property to remove parentheses.

$50 = 4w + 12$ Combine like terms.

$38 = 4w$ Subtract 12 from both sides.

$9.5 = w$ Divide both sides by 4.

State the Conclusion The width of the kennel is 9.5 feet. The length is 6 feet more than this, or 15.5 feet.

Check the Result If a rectangle has a width of 9.5 feet and a length of 15.5 feet, its length is 6 feet more than its width, and the perimeter is 2(9.5) feet + 2(15.5) feet = 50 feet.

1.6 STUDY SET

VOCABULARY Fill in the blanks.

1. An ______ angle has a measure of more than 0° and less than 90°.
2. A ______ angle is an angle whose measure is 90°.
3. If the sum of the measures of two angles equals 90°, the angles are called ______________ angles.
4. If the sum of the measures of two angles equals 180°, the angles are called ______________ angles.
5. If a triangle has a right angle, it is called a ______ triangle.
6. If a triangle has two sides with equal measures, it is called an ________ triangle.
7. The sum of the measures of the _______ of a triangle is 180°.
8. An __________ triangle has three sides of equal length and three angles of equal measure.

CONCEPTS

9. The unit used to measure the intensity of sound is called the *decibel.* In the table, translate the comments in the right-hand column into mathematical symbols to complete the decibel column.

	Decibels	Compared to conversation
Conversation	d	—
Vacuum cleaner		15 decibels more
Circular saw		10 decibels less than twice
Jet takeoff		20 decibels more than twice
Whispering		10 decibels less than half
Rock band		Twice the decibel level

10. The following table shows the four types of problems an instructor put on a history test.

a. Complete the table.

b. Which type of question appears the most on the test?

c. Write an algebraic expression that represents the total number of points on the test.

Type of question	Number ·	Value =	Total value
Multiple choice	x	5	
True/false	$3x$	2	
Essay	$x - 2$	10	
Fill-in	x	5	

APPLICATIONS

11. INSTRUMENTS The flute consists of three pieces. Write an algebraic expression that represents

a. the length of the shortest piece.

b. the length of the longest piece.

c. the length of the flute.

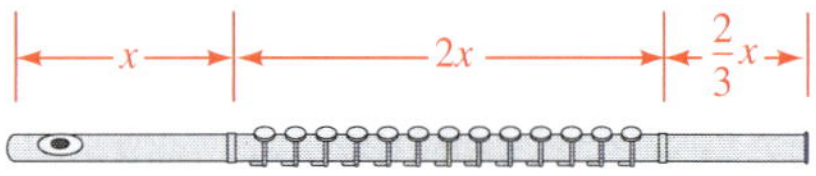

12. For each of the two pictures shown, what geometric concept studied in this section is illustrated?

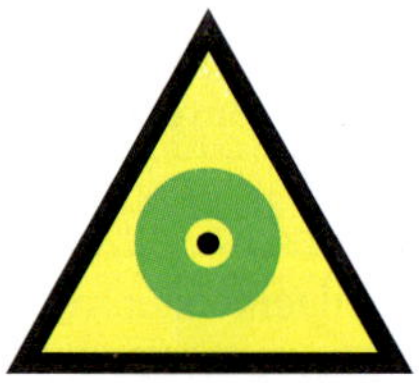

Radiation warning

13. TENNIS Write an algebraic expression that represents the length in inches of the head of the tennis racquet.

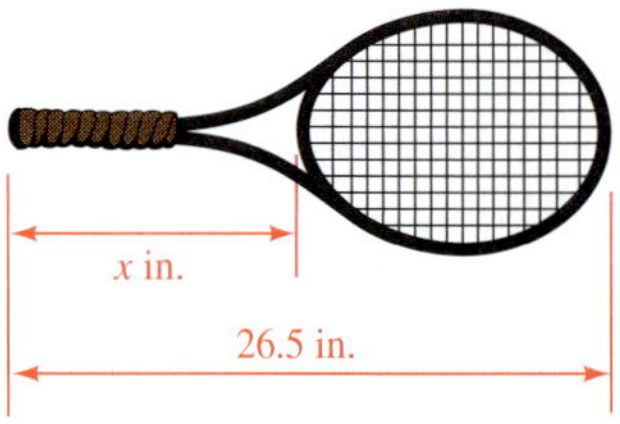

14. GEOGRAPHY The surface area of the Earth is 510,066,000 square kilometers (km^2). If we let x represent the number of km^2 covered by water, what would the algebraic expression $510{,}066{,}000 - x$ represent?

15. CEREAL SALES In 2001, the two top-selling cereals were General Mills' Cheerios and Kellogg's Frosted Flakes, with combined sales of $1,003 million. Frosted Flakes sales were $324 million less than sales of Cheerios. What were the 2001 sales for each brand?

16. FILMS As of March 2004, Denzel Washington's three top grossing films, *Remember the Titans, The Pelican Brief,* and *Crimson Tide,* had earned $307.8 million. If *Remember the Titans* earned $14.8 million more than *The Pelican Brief,* and if *The Pelican Brief* earned $9.4 million more than *Crimson Tide,* how much did each film earn as of that date?

17. SPRING TOURS A group of junior high students will be touring Washington, D.C. Their chaperons will have the $1,810 cost of the tour reduced by $15.50 for each student they personally supervise. How many students will a chaperon have to supervise so that his or her cost to take the tour will be $1,500?

18. MACHINING Each pass through a lumber plane shaves off 0.015 inch of thickness from a board. How many times must a board, originally 0.875 inch thick, be run through the planer if a board of thickness 0.74 inch is desired?

19. MOVING EXPENSES To help move his furniture, a man rents a truck for $41.50 per day plus 35¢ per mile. If he has budgeted $150 for transportation expenses, how many miles will he be able to drive the truck if the move takes 1 day?

20. COMPUTING SALARIES A student working for a delivery company earns $57.50 per day plus $4.75 for each package she delivers. How many deliveries must she make each day to earn $200 a day?

21. VALUE OF AN IRA In an Individual Retirement Account (IRA) valued at $53,900, a couple has 500 shares of stock, some in Big Bank Corporation and some in Safe Savings and Loan. If Big Bank sells for $115 per share and Safe Savings sells for $97 per share, how many shares of each does the couple own?

22. ASSETS OF A PENSION FUND A pension fund owns 2,000 fewer shares in mutual stock funds than mutual bond funds. Currently, the stock funds sell for \$12 per share, and the bond funds sell for \$15 per share. How many shares of each does the pension fund own if the value of the securities is \$165,000?

23. SELLING CALCULATORS Last month, a bookstore ran the following ad. Sales of \$4,980 were generated, with 15 more graphing calculators sold than scientific calculators. How many of each type of calculator did the bookstore sell?

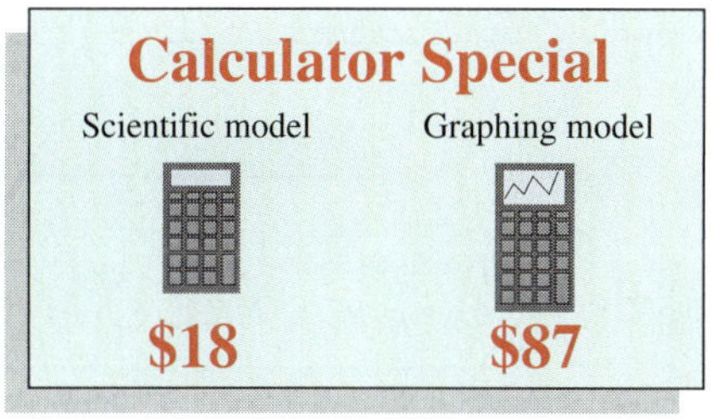

24. SELLING SEED A seed company sells two grades of grass seed. A 100-pound bag of a mixture of rye and Kentucky bluegrass sells for \$245, and a 100-pound bag of bluegrass sells for \$347. How many bags of each are sold in a week when the receipts for 19 bags are \$5,369?

25. WOODWORKING The carpenter saws a board that is 22 feet long into two pieces. One piece is to be 1 foot longer than twice the length of the shorter piece. Find the length of each piece.

26. STATUE OF LIBERTY From the foundation of the large pedestal on which it sits to the top of the torch, the Statue of Liberty National Monument measures 305 feet. The pedestal is 3 feet taller than the statue. Find the height of the pedestal and the height of the statue.

27. NURSING The illustration in the next column shows the angle a needle should make with the skin when administering a certain type of intradermal injection. Find the measure of both angles labeled.

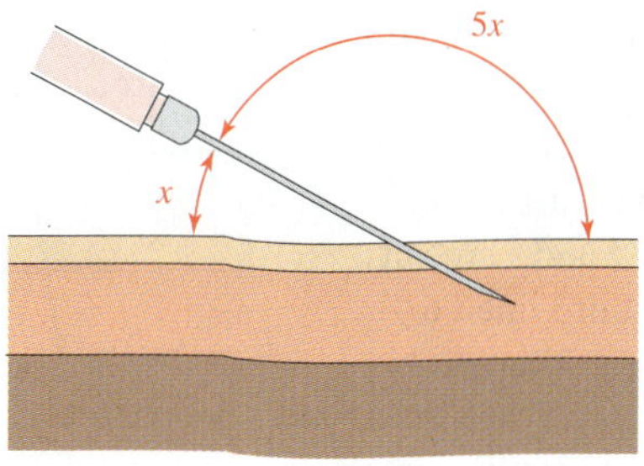

28. SUPPLEMENTARY ANGLES Refer to the illustration and find x.

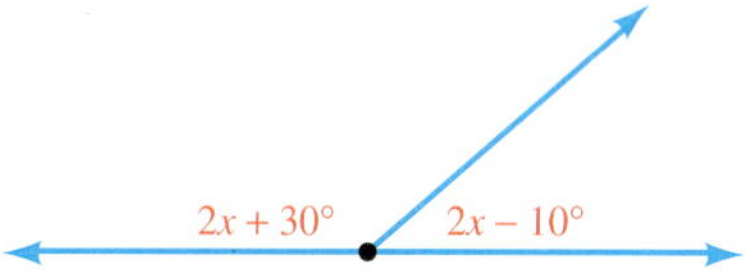

29. ARCHITECTURE Because of soft soil and a shallow foundation, the Leaning Tower of Pisa in Italy is not vertical. How many degrees from vertical is the tower?

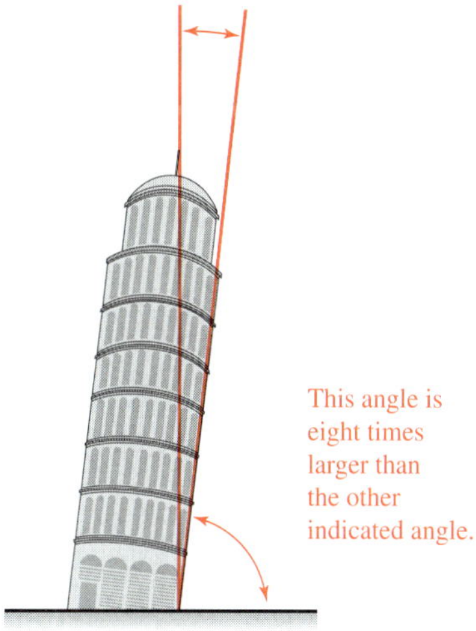

30. STEPSTOOLS The sum of the measures of the three angles of any triangle is 180°. In the illustration, the measure of ∠ 2 (angle 2) is 10° larger than the measure of ∠ 1. The measure of ∠ 3 is 10° larger than the measure of ∠ 2. Find each angle measure.

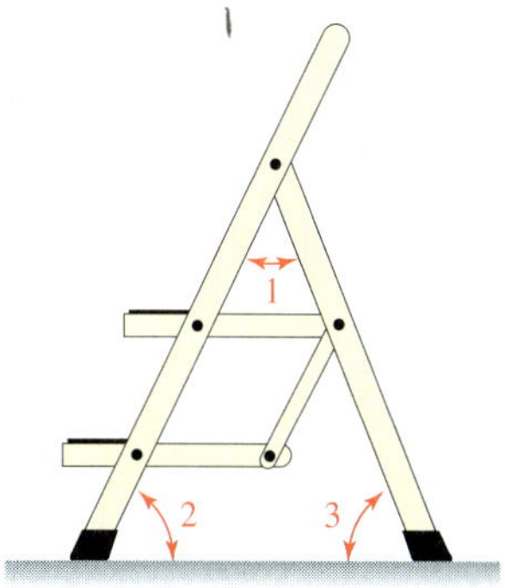

31. SUPPLEMENTARY ANGLES AND PARALLEL LINES In the illustration, lines r and s are cut by a third line l to form $\angle 1$ (angle 1) and $\angle 2$. When lines r and s are parallel, $\angle 1$ and $\angle 2$ are supplementary. If $\angle 1 = x + 50°$, $\angle 2 = 2x - 20°$, and lines r and s are parallel, find x.

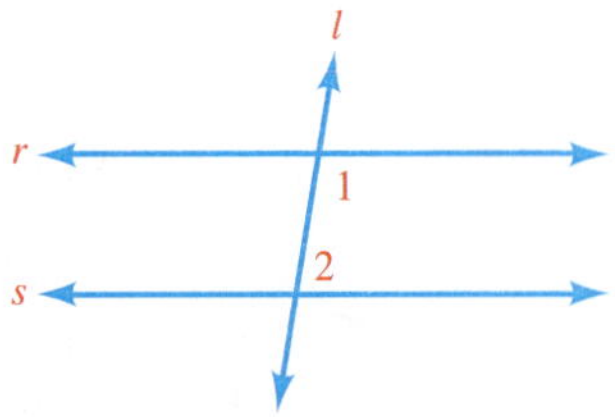

32. ANGLES AND PARALLEL LINES In the illustration, $r \parallel s$ (read as "line r is parallel to line s"), and $a = 103$. Find b, c, and d. (*Hint:* See Problem 31.)

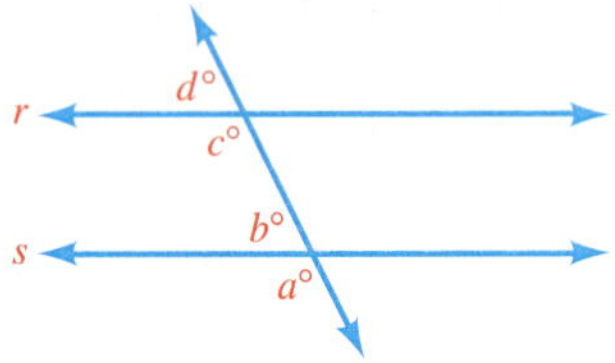

33. VERTICAL ANGLES When two lines intersect, four angles are formed. Angles that are side-by-side, such as $\angle 1$ (angle 1) and $\angle 2$, are called **adjacent angles.** Angles that are nonadjacent, such as $\angle 1$ and $\angle 3$ or $\angle 2$ and $\angle 4$, are called **vertical angles.** From geometry, we know that if two lines intersect, vertical angles have the same measure. If $\angle 1 = 3x + 10°$ and $\angle 3 = 5x - 10°$, find x.

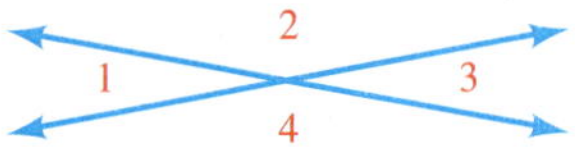

34. ANGLES AND PARALLEL LINES In the illustration, $r \parallel s$ (read as "line r is parallel to line s"), and $b = 137$. Find a and c.

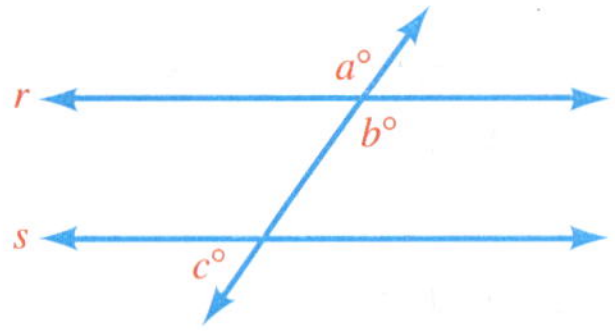

35. ANGLES OF A QUADRILATERAL The sum of the angles of any four-sided figure (called a *quadrilateral*) is 360°. The quadrilateral shown has two equal base angles. Find x.

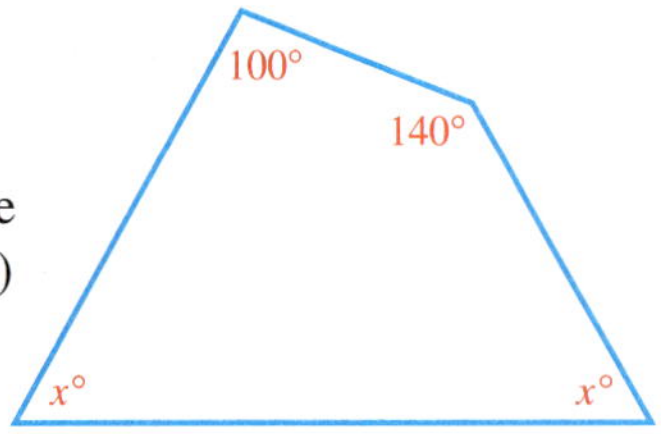

36. HEIGHT OF A TRIANGLE If the height of a triangle with a base of 8 inches is tripled, its area is increased by 96 square inches. Find the height of the triangle.

37. GOLDEN RECTANGLES Throughout history, most artists and designers have felt that rectangles with a length 1.618 times as long as their width have the most visually attractive shape. Such rectangles are known as *golden rectangles.* Measure the length and width of the rectangles in the illustration. Which of the rectangles is closest to being a golden rectangle?

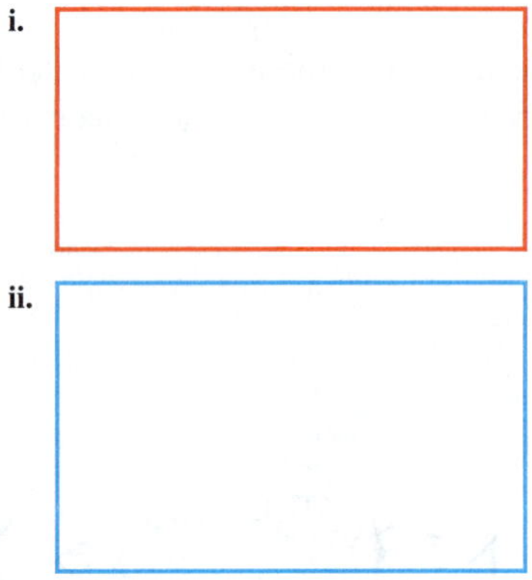

38. QUILTING A woman is planning to make a quilt in the shape of a golden rectangle. (See Problem 37.) She has exactly 22 feet of a special lace that she plans to sew around the edge of the quilt. What should the length and width of the quilt be? Round both answers up to the nearest hundredth.

39. FENCING PASTURES A farmer has 624 feet of fencing to enclose a pasture. Because a river runs along one side, fencing will be needed on only three sides. Find the dimensions of the pasture if its length is double its width.

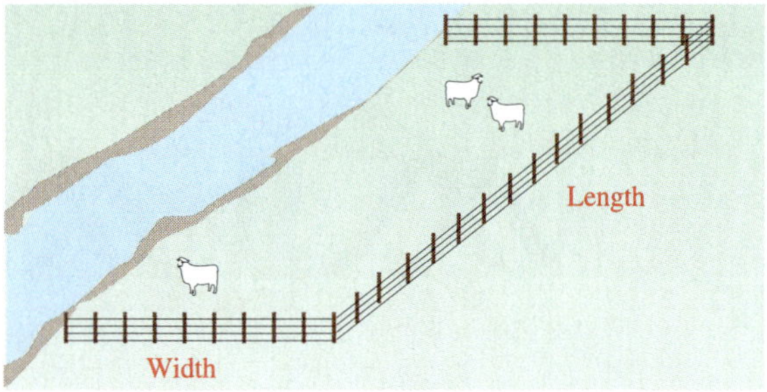

40. FENCING PENS A man has 150 feet of fencing to build the pen shown in the illustration. If one end is a square, find the outside dimensions.

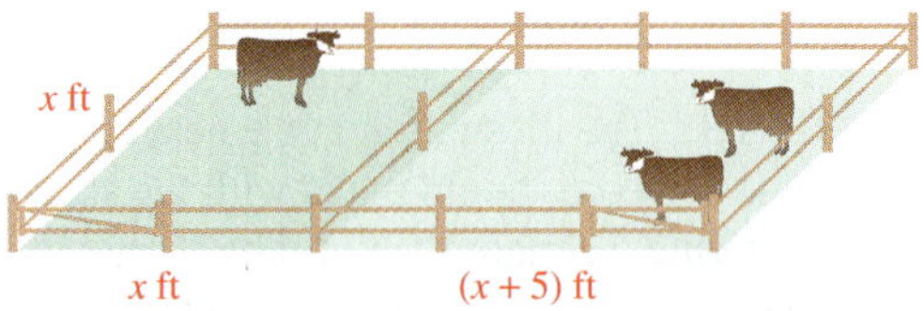

41. SWIMMING POOLS A woman wants to enclose the pool shown and have a walkway of uniform width all the way around. How wide will the walkway be if the woman uses 180 feet of fencing?

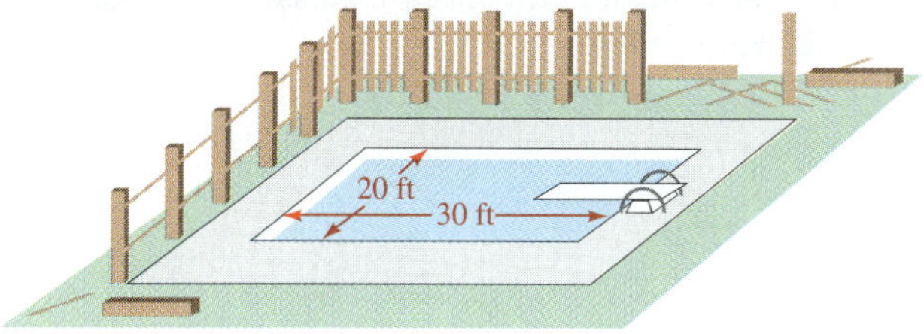

42. INSTALLING SOLAR HEATING One solar panel in the illustration is to be 3 feet wider than the other. To be equally efficient, they must have the same area. Find the width of each.

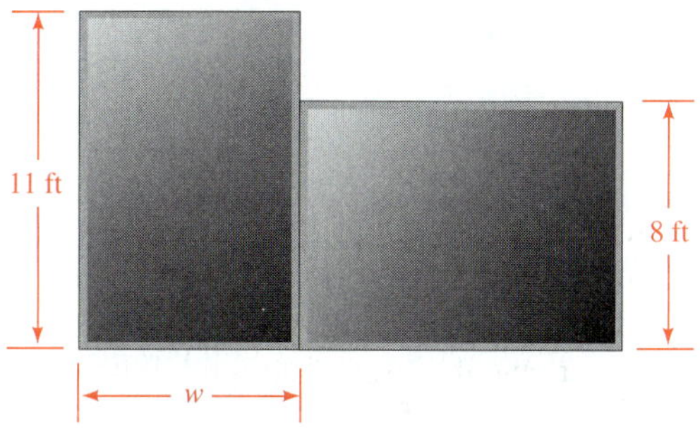

43. MAKING FURNITURE A woodworker wants to put two partitions crosswise in a drawer that is 28 inches deep, as shown in the illustration. He wants to place the partitions so that the spaces created increase by 3 inches from front to back. If the thickness of each partition is $\frac{1}{2}$ inch, how far from the front end should he place the first partition?

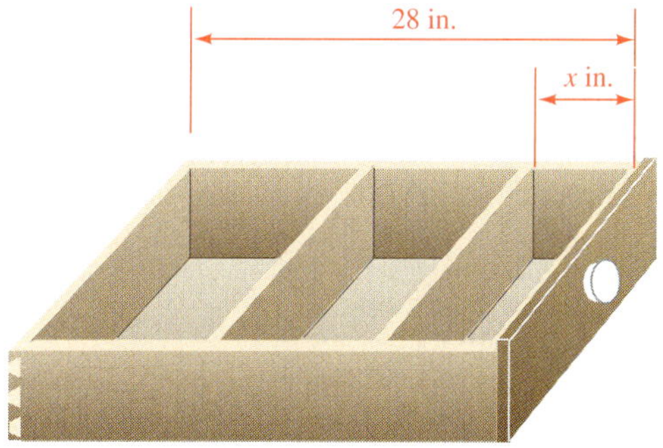

44. BUILDING SHELVES See the illustration in the next column. A carpenter wants to put four shelves on an 8-foot wall so that the five spaces created decrease by 6 inches as we move up the wall. If the thickness of each shelf is $\frac{3}{4}$ inch, how far will the bottom shelf be from the floor?

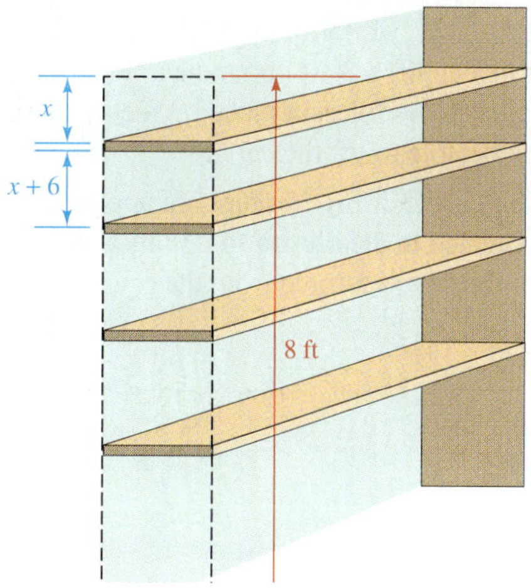

WRITING

45. Briefly explain what should be accomplished in each of the steps (*analyze, form, solve, state,* and *check*) of the problem-solving strategy used in this section.

46. Write a problem that can be represented by the following verbal model.

Measure of 1st angle	plus	measure of 2nd angle	plus	measure of 3rd angle	is 180°.
x	$+$	$2x$	$+$	$x + 10$	$= 180$

REVIEW

47. When expressed as a decimal, is $\frac{7}{9}$ a terminating or repeating decimal?

48. Solve: $x + 20 = 4x - 1 + 2x$.

49. List the integers.

50. Solve: $2x + 2 = \frac{2}{3}x - 2$.

51. Evaluate $2x^2 + 5x - 3$ for $x = -3$.

52. Solve: $T - R = ma$ for R.

CHALLENGE PROBLEMS

A lever will be in balance when the sum of the products of the forces on one side of a fulcrum and their respective distances from the fulcrum is equal to the sum of the products of the forces on the other side of the fulcrum and their respective distances from the fulcrum.

53. MOVING A STONE A woman uses a 10-foot bar to lift a 210-pound stone. If she places another rock 3 feet from the stone to act as the fulcrum, how much force must she exert to move the stone?

54. LIFTING A CAR A 350-pound football player brags that he can lift a 2,500-pound car. If he uses a 12-foot bar with the fulcrum placed 3 feet from the car, will he be able to lift the car?

55. BALANCING A LEVER Forces are applied to a lever as indicated in the illustration. Find x, the distance of the smallest force from the fulcrum.

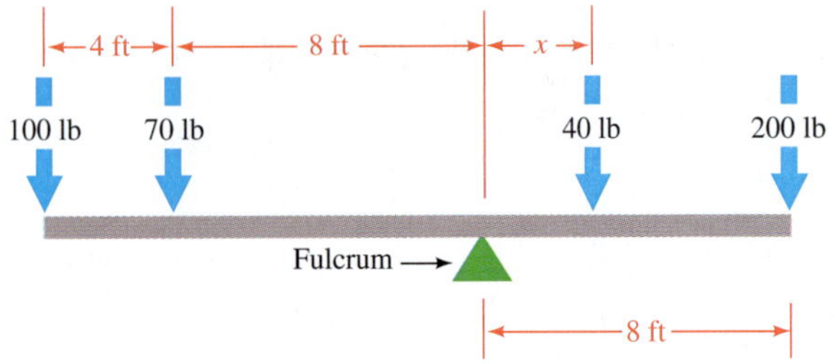

56. BALANCING A SEESAW Jim and Bob sit at opposite ends of an 18-foot seesaw, with the fulcrum at its center. Jim weighs 160 pounds, and Bob weighs 200 pounds. Kim sits 4 feet in front of Jim, and the seesaw balances. How much does Kim weigh?

1.7 More Applications of Equations

- Percent problems
- Statistics problems
- Investment problems
- Uniform motion problems
- Mixture problems

In this section, we will again use equations as we solve a variety of problems.

The Language of Algebra

The names of the parts of a percent sentence are:

5 is 50% of 10.
amount percent base

They are related by the formula:

Amount = percent · base

PERCENT PROBLEMS

Percents are often used to present numeric information. **Percent** means parts per one hundred. One method to solve applied percent problems is to use the given facts to write a **percent sentence** of the form:

_____ is _____ % of _____ ?

We enter the appropriate numbers in two of the blanks and the word "what" in the remaining blank. Then we translate the sentence to mathematical symbols and solve the resulting equation.

EXAMPLE 1

Gold mining. Use the following data about South Africa, the world's largest producer of gold, to determine the total world gold production in 2002.

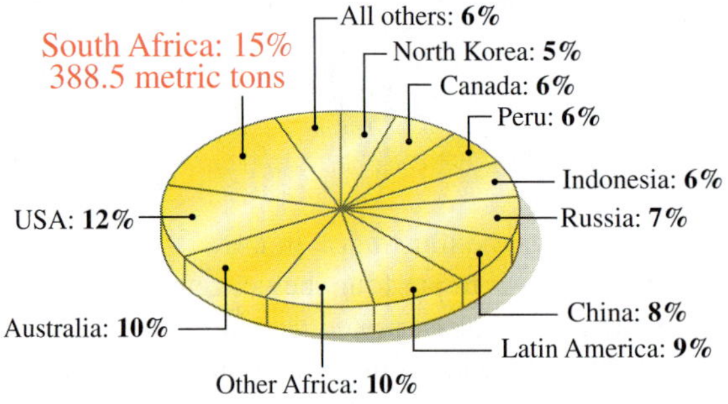

Source: World Gold Council

Analyze the Problem In the **circle graph,** we see that the 388.5 metric tons of gold produced by South Africa was 15% of the world's production in 2002.

Form an Equation Let x = world gold production (in metric tons) for 2002. First, we write a percent sentence using the given data. Then we translate to form an equation.

388.5	is	15%	of	what?	
388.5	=	15%	·	x	The amount is 388.5, the percent is 15%, and the base is x.

Solve the Equation

$$388.5 = 15\% \cdot x$$

$$388.5 = 0.15x \quad \text{Write 15\% as a decimal: 15\% = 0.15.}$$

$$\frac{388.5}{\mathbf{0.15}} = \frac{0.15x}{\mathbf{0.15}} \quad \text{Divide both sides by 0.15.}$$

$$2{,}590 = x \quad \text{Divide.}$$

State the Conclusion The world produced 2,590 metric tons of gold in 2002.

Check the Result If 2,590 metric tons of gold were produced, then the 388.5 metric tons produced by South Africa were $\frac{388.5}{2{,}590} = 0.15 = 15\%$ of the world's production. The answer checks.

When the regular price of merchandise is reduced, the amount of reduction is called **markdown** (or discount).

Sale price = regular price − markdown

Usually, the markdown is expressed as a percent of the regular price.

Markdown = percent of markdown · regular price

EXAMPLE 2

Wedding gowns. At a bridal shop, a wedding gown that normally sells for $397.98 is on sale for $265.32. Find the percent of markdown.

Analyze the Problem In this case, $265.32 is the sale price, $397.98 is the regular price, and the markdown is the *product* of $397.98 and the percent of markdown.

Form an Equation We let r = the percent of markdown, expressed as a decimal. We then substitute $265.32 for the sale price and $397.98 for the regular price in the formula.

Sale price	is	regular price	minus	markdown.	
265.32	=	397.98	−	$r \cdot 397.98$	Markdown = percent of markdown · regular price

Solve the Equation

$$265.32 = 397.98 - r \cdot 397.98$$

$$265.32 = 397.98 - 397.98r \quad \text{Rewrite } r \cdot 397.98 \text{ as } 397.98r.$$

$$-132.66 = -397.98r \quad \text{Subtract 397.98 from both sides.}$$

$$\frac{-132.66}{-397.98} = r \quad \text{Divide both sides by } -397.98.$$

$$0.333333\ldots = r \quad \text{Do the division using a calculator.}$$

$$33.3333\ldots\% = r \quad \text{To write the decimal as a percent, multiply 0.333333... by 100 and insert a \% sign.}$$

State the Conclusion The percent of markdown on the wedding gown is $33.3333\ldots\%$ or $33\frac{1}{3}\%$.

Check the Result The markdown is $33\frac{1}{3}\%$ of \$397.98, or \$132.66. The sale price is \$397.98 − \$132.66, or \$265.32. The answer checks.

Percents are often used to describe how a quantity has changed. To describe such changes, we use **percent of increase** or **percent of decrease.**

EXAMPLE 3

Entertainment. Use the following data to determine the percent of increase in the number of movie theater screens in the United States from 1990 to 2002. Round to the nearest one percent.

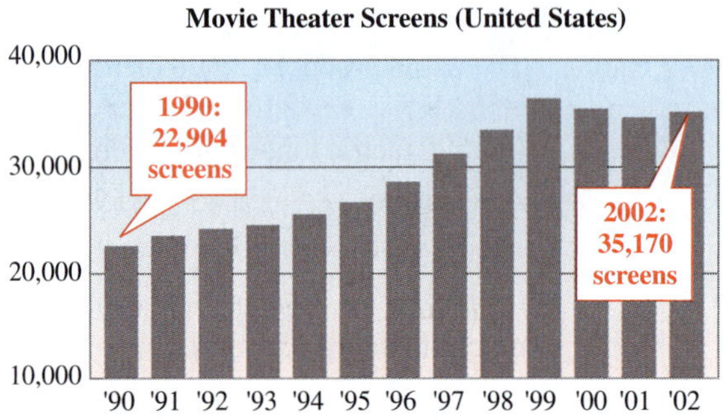

Source: National Association of Theater Owners

Analyze the Problem To find the percent of increase, we first find the *amount of increase* by subtracting the number of screens in 1990 from the number of screens in 2002.

$$35{,}170 - 22{,}904 = 12{,}266$$

Form an Equation Next, we find what percent of the original 22,904 screens the 12,266 increase represents. We let x = the unknown percent and translate the words into an equation.

Caution

Always find the percent of increase (or decrease) with respect to the *original* amount.

12,266	is	what percent	of	22,904?
12,266	$=$	x	$\cdot$	22,904

The amount is 12,266, the percent is x, and the base is 22,904.

Solve the Equation

$$12{,}266 = x \cdot 22{,}904$$

$$12{,}266 = 22{,}904x$$

$$\frac{12{,}266}{22{,}904} = \frac{22{,}904x}{22{,}904} \quad \text{Divide both sides by 22,904.}$$

$$0.535539644\ldots = x \quad \text{Divide.}$$

$$53.5539644\ldots\% \approx x \quad \text{Write the decimal as a percent.}$$

$$54\% \approx x \quad \text{Round to the nearest one percent.}$$

State the Conclusion There was a 54% increase in the number of movie screens in the United States from 1990 to 2002.

Check the Result A 50% increase from 22,904 screens would be approximately 11,000 additional screens. It seems reasonable that 12,266 more screens would be a 54% increase.

STATISTICS PROBLEMS

Statistics is a branch of mathematics that deals with the analysis of numerical data. Three types of averages are commonly used in statistics as measures of central tendency of a collection of data: **mean,** the **median,** and the **mode.**

Mean, Median, and Mode

The **mean** $\bar{x}$ of a collection of values is the sum S of those values divided by the number of values n.

$$\bar{x} = \frac{S}{n}$$

Read $\bar{x}$ as "x bar."

The **median** of a collection of values is the middle value. To find the median,

1. Arrange the values in increasing order.
2. If there are an odd number of values, choose the middle value.
3. If there are an even number of values, add the middle two values and divide by 2.

The **mode** of a collection of values is the value that occurs most often.

EXAMPLE 4

Physiology. As a project for a physiology class, a student measured ten people's reaction times to a visual stimulus. Their reaction times, in seconds, are listed below. Find **a.** the mean, **b.** the median, and **c.** the mode of the collection of data.

0.29, 0.22, 0.19, 0.36, 0.28, 0.23, 0.16, 0.28, 0.33, 0.26

Solution **a.** To find the mean, we add the values and divide by the number of values, which is 10.

$$\bar{x} = \frac{0.29 + 0.22 + 0.19 + 0.36 + 0.28 + 0.23 + 0.16 + 0.28 + 0.33 + 0.26}{10}$$

$$= 0.26 \text{ second}$$

b. To find the median, we first arrange the values in increasing order:

0.16, 0.19, 0.22, 0.23, 0.26, 0.28, 0.28, 0.29, 0.33, 0.36

Because there are an even number of measurements, the median will be the sum of the middle two values, 0.26 and 0.28, divided by 2. Thus, the median is

$$\text{Median} = \frac{0.26 + 0.28}{2} = 0.27 \text{ second}$$

c. Since the time 0.28 second occurs most often, it is the mode.

The Language of Algebra

In statistics, the mean, median, and mode are classified as types of *averages.* In daily life, when the word *average* is used, it most often is referring to the mean.

EXAMPLE 5

Bank service charges. When the average (mean) daily balance of a customer's checking account falls below \$500 in any week, the bank assesses a \$15 service charge. What minimum balance will the account shown need to have on Friday to avoid the service charge?

Security Savings

☐ Weekly Statement ☐

Acct: 201-234-002 Type: checking

Day	Date	Daily balance	Comments
Mon	3/11	\$730.70	
Tue	3/12	\$350.19	
Wed	3/13	–\$50.19	overdrawn
Thu	3/14	\$275.55	
Fri	3/15		

Analyze the Problem We can find the average (mean) daily balance for the week by adding the daily balances and dividing by 5. We want the mean to be \$500 so that there is no service charge.

Form an Equation We will let x = the minimum balance needed on Friday. Then we translate the words into mathematical symbols.

The sum of the five daily balances	divided by	5	is	\$500.

$$\frac{730.70 + 350.19 + (-50.19) + 275.55 + x}{5} = 500$$

Solve the Equation

$$\frac{730.70 + 350.19 + (-50.19) + 275.55 + x}{5} = 500$$

$$\frac{1{,}306.25 + x}{5} = 500 \qquad \text{Combine like terms in the numerator.}$$

$$5\left(\frac{1{,}306.25 + x}{5}\right) = 5(500) \qquad \text{Multiply both sides by 5.}$$

$$1{,}306.25 + x = 2{,}500$$

$$x = 1{,}193.75 \qquad \text{Subtract 1,306.25 from both sides.}$$

State the Conclusion On Friday, the account balance needs to be \$1,193.75 to avoid a service charge.

Check the Result Check the result by adding the five daily balances and dividing by 5.

INVESTMENT PROBLEMS

The money an investment earns is called *interest.* **Simple interest** is computed by the formula $I = Prt$, where I is the interest earned, P is the principal (amount invested), r is the annual interest rate, and t is the length of time the principal is invested.

EXAMPLE 6

Interest income. To protect against a major loss, a financial analyst suggests a diversified plan for a client who has \$50,000 to invest for 1 year.

1. Alco Development, Inc. Builds mini-malls. High yield: 12% per year. Risky!
2. Certificate of deposit (CD). Insured, safe. Low yield: 4.5% annual interest.

If the client puts some money in each investment and wants to earn \$3,600 in interest, how much should be invested at each rate?

Analyze the Problem In this case, we are working with two investments made at two different rates for 1 year. If we add the interest from the two investments, the sum should equal \$3,600.

Form an Equation If we let $x =$ the number of dollars invested at 12%, the interest earned is $I = Prt = \$x(12\%)(1) = \$0.12x$. If \$$x$ is invested at 12%, there is \$$(50{,}000 - x)$ to invest at 4.5%, which will earn \$$0.045(50{,}000 - x)$ in interest. These facts are listed in the table.

	P ·	r ·	t =	I
Alco Development, Inc.	x	0.12	1	$0.12x$
Certificate of deposit	$50{,}000 - x$	0.045	1	$0.045(50{,}000 - x)$

The sum of the two amounts of interest should equal \$3,600. We now translate the words into an equation.

The interest earned at 12%	plus	the interest earned at 4.5%	is	the total interest earned.
$0.12x$	$+$	$0.045(50{,}000 - x)$	$=$	$3{,}600$

Solve the Equation

$$0.12x + 0.045(50{,}000 - x) = 3{,}600$$

$$1{,}000[0.12x + 0.045(50{,}000 - x)] = 1{,}000(3{,}600)$$ To eliminate the decimals, multiply both sides by 1,000.

$$120x + 45(50{,}000 - x) = 3{,}600{,}000$$ Distribute the 1,000 and simplify both sides.

$$120x + 2{,}250{,}000 - 45x = 3{,}600{,}000$$ Remove parentheses.

$$75x + 2{,}250{,}000 = 3{,}600{,}000$$ Combine like terms.

$$75x = 1{,}350{,}000$$ Subtract 2,250,000 from both sides.

$$x = 18{,}000$$ Divide both sides by 75.

State the Conclusion \$18,000 should be invested at 12% and \$(50,000 − 18,000) = \$32,000 should be invested at 4.5%.

Check the Result The annual interest on \$18,000 is 0.12(\$18,000) = \$2,160. The interest earned on \$32,000 is 0.045(\$32,000) = \$1,440. The total interest is \$2,160 + \$1,440 = \$3,600. The answers check.

UNIFORM MOTION PROBLEMS

Problems that involve an object traveling at a constant rate for a specified period of time over a certain distance are called **uniform motion** problems. To solve these problems, we use the formula $d = rt$, where d is distance, r is rate, and t is time.

EXAMPLE 7

Travel time. After a stay on her grandparents' farm, a girl is to return home, 385 miles away. To split up the drive, the parents and grandparents start at the same time and drive toward each other, planning to meet somewhere along the way. If the parents travel at an average rate of 60 mph and the grandparents at 50 mph, how long will it take them to meet?

Analyze the Problem The vehicles are traveling toward each other as shown in the following figure. We know the rates the cars are traveling (60 mph and 50 mph). We also know that they will travel for the same amount of time.

Form an Equation We can let t = the time that each vehicle travels. Then the distance traveled by the parents is $60t$ miles, and the distance traveled by the grandparents is $50t$ miles. This information is organized in the table in the figure. The sum of the distances traveled by the parents and grandparents is 385 miles.

The distance the parents travel	plus	the distance the grandparents travel	is	the distance between the child's home and the grandparent's farm.
$60t$	$+$	$50t$	$=$	385

Caution

When using $d = rt$, make sure the units are consistent. For example, if the rate is given in miles per hour, the time must be expressed in hours.

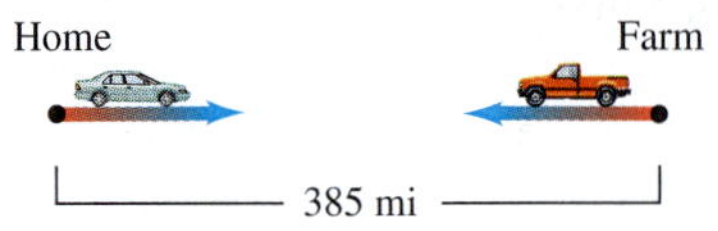

	r	$\cdot\ t$	$= d$
Parents	60	t	$60t$
Grandparents	50	t	$50t$

Solve the Equation

$$60t + 50t = 385$$

$$110t = 385 \quad \text{Combine like terms.}$$

$$t = 3.5 \quad \text{Divide both sides by 110.}$$

State the Conclusion The parents and grandparents will meet in $3\frac{1}{2}$ hours.

Check the Result The parents travel $3.5(60) = 210$ miles. The grandparents travel $3.5(50) = 175$ miles. The total distance traveled is $210 + 175 = 385$ miles. The answer checks.

MIXTURE PROBLEMS

We now discuss two types of mixture problems. In the first example, a *dry mixture* of a specified value is created from two differently priced components.

EXAMPLE 8

Mixing nuts. The owner of a candy store notices that 20 pounds of gourmet cashews did not sell because of their high price of \$12 per pound. The owner decides to mix peanuts with the cashews to lower the price per pound. If peanuts sell for \$3 per pound, how many pounds of peanuts must be mixed with the cashews to make a mixture that could be sold for \$6 per pound?

Analyze the Problem To solve this problem, we will use the formula $v = np$, where v represents value, n represents the number of pounds, and p represents the price per pound.

Form an Equation We can let x = the number of pounds of peanuts to be used. Then $20 + x$ = the number of pounds in the mixture. We enter the known information in the following table. The value of the cashews plus the value of the peanuts will be equal to the value of the mixture.

	n	$\cdot\ p$	$= v$
Cashews	20	12	240
Peanuts	x	3	$3x$
Mixture	$20 + x$	6	$6(20 + x)$

We can now form the equation.

The value of the cashews	plus	the value of the peanuts	is	the value of the mixture.
240	+	$3x$	=	$6(20 + x)$

Solve the Equation

$$240 + 3x = 6(20 + x)$$

$$240 + 3x = 120 + 6x \quad \text{Use the distributive property to remove parentheses.}$$

$$120 = 3x \quad \text{Subtract } 3x \text{ and 120 from both sides.}$$

$$40 = x \quad \text{Divide both sides by 3.}$$

State the Conclusion The owner should mix 40 pounds of peanuts with the 20 pounds of cashews.

Check the Result The cashews are valued at \$12(20) = \$240, and the peanuts are valued at \$3(40) = \$120. The mixture is valued at \$6(60) = \$360. Since the value of the cashews plus the value of the peanuts equals the value of the mixture, the answer checks.

In the next example, a *liquid mixture* of a desired strength is to be made from two solutions with different concentrations.

EXAMPLE 9

Milk production. Owners of a dairy find that milk with a 2% butterfat content is their best seller. Suppose the dairy has large quantities of whole milk having a 4% butterfat content and milk having a 1% butterfat content. How much of each type of milk should be mixed to obtain 120 gallons of milk that is 2% butterfat?

Analyze the Problem We are to find the amount of 4% milk to mix with 1% milk to get 120 gallons of a milk that has a 2% butterfat content. In the figure, if we let g = the number of gallons of the 4% milk used in the mixture, then $120 - g$ = the number of gallons of the 1% milk needed to obtain the desired concentration.

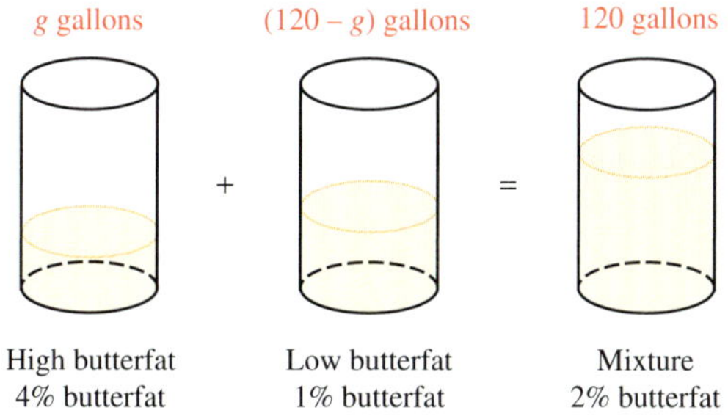

	Gallons of milk	· Percent butterfat	= Gallons of butterfat
High butterfat	g	0.04	$0.04g$
Low butterfat	$120 - g$	0.01	$0.01(120 - g)$
Mixture	120	0.02	$0.02(120)$

Form an Equation The amount of butterfat in a tank is the *product* of the percent butterfat and the number of gallons of milk in the tank. In the first tank shown in the figure, 4% of the g gallons, or $0.04g$ gallons, is butterfat. In the second tank, 1% of the $(120 - g)$ gallons, or $0.01(120 - g)$ gallons, is butterfat. Upon mixing, the third tank will have 0.02(120) gallons of butterfat in it. These results are recorded in the last column of the table.

The sum of the amounts of butterfat in the first two tanks should equal the amount of butterfat in the third tank.

The amount of butterfat in g gallons of 4% milk	plus	the amount of butterfat in $(120 - g)$ gallons of 1% milk	is	the amount of butterfat in 120 gallons of the mixture.
$0.04(g)$	$+$	$0.01(120 - g)$	$=$	$0.02(120)$

Solve the Equation

$$0.04(g) + 0.01(120 - g) = 0.02(120)$$

$$4(g) + 1(120 - g) = 2(120)$$ Multiply both sides by 100 to clear the equation of decimals.

$$4g + 120 - g = 240$$

$$3g + 120 = 240$$ Combine like terms.

$$3g = 120$$ Subtract 120 from both sides.

$$g = 40$$ Divide both sides by 3.

State the Conclusion 40 gallons of 4% milk and $120 - 40 = 80$ gallons of 1% milk should be mixed to get 120 gallons of milk with a 2% butterfat content.

Check the Result The 40 gallons of 4% milk contains $0.04(40) = 1.6$ gallons of butterfat, and 80 gallons of 1% milk contains $0.01(80) = 0.8$ gallons of butterfat—a total of $1.6 + 0.8 = 2.4$ gallons of butterfat. The 120 gallons of the 2% mixture contains 2.4 gallons of butterfat. The answers check.

1.7 STUDY SET

VOCABULARY Fill in the blanks.

1. When an investment is made, the amount of money invested is called the ________.
2. The value that occurs the most in a collection of data is called the ______.
3. The middle value of a collection of data is called the ________.
4. The ______ of several values is the sum of those values divided by the number of values.
5. In the statement, "10 is 20% of 50," 10 is the ________, and 50 is the ______.
6. When the regular price of an item is reduced, the amount of reduction is called the __________.

CONCEPTS

7. One method to solve applied percent problems is to use the given facts to write a percent sentence. What is the basic form of a percent sentence?

8.

Total Paid Circulation *Seventeen* Magazine	
1995: 2,172,923	**2002:** 2,445,539

a. Find the *amount* of increase in circulation of *Seventeen* magazine.

b. Fill in the percent sentence that can be used to find the percent of increase in circulation.

 is % of ?

9. a. Write 4.5% as a decimal.

 b. Write 0.06 as a percent.

10. For each collection of values, give the median.

 a. 8, 9, 11, 15, 17

 b. 1, 3, 8, 16, 21, 44

11. Complete the following table for each 1-year investment.

	Principal ·	Rate ·	Time =	Interest
CD	1,500	0.06	1	
Bonds	x	0.0565	1	
Stocks	$850 - x$	0.07	1	

12. The following table shows how a retired teacher invested a total of \$8,000 in two accounts for 1 year. Complete the table.

	P ·	r ·	t =	I
S & L	x	0.03		
Credit Union		0.04		

13. A banker invested the *same* amount in two money-making opportunities for 1 year. Complete the table that describes the investments.

	P ·	r ·	t =	I
Cattle futures	x	0.15		
Soybeans		0.18		

14. Complete the following table given the speed of light and sound through air and water.

	Rate ·	Time =	Distance
Light (air)	186,224 mi/sec	60 sec	
Light (water)	140,060 mi/sec	60 sec	
Sound (air)	1,088 ft/sec	x sec	
Sound (water)	4,870 ft/sec	$(x - 3)$ sec	

15. Complete the following table.

	Pounds ·	Price =	Value
M & M plain	30	7.45	
M & M peanut	p	8.25	
Mixture	$p + 30$	7.75	

16. Complete the following table that could be used to solve this problem: How many pints of punch from the orange cooler must be mixed with the entire contents of the blue cooler to get a 12% punch mixture?

	Amount ·	Strength =	Pure concentrate
Too strong			
Too weak			
Mixture			

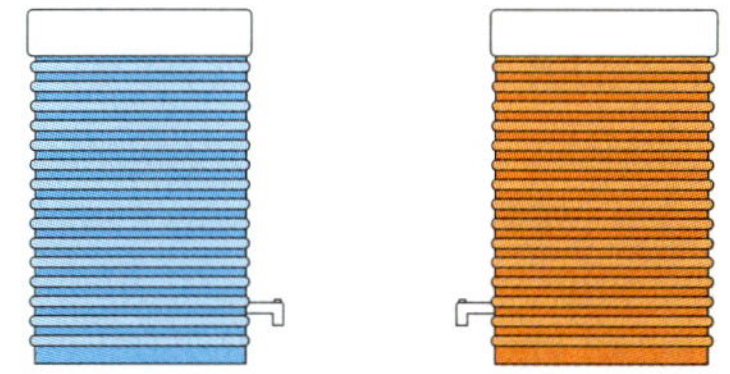

20 pints of punch, 20% concentrate

x pints of punch, 10% concentrate

NOTATION Translate each statement into mathematical symbols.

17. What number is 5% of 10.56?

18. 16 is what percent of 55?

19. 32.5 is 74% of what number?

20. What is 83.5% of 245?

21. What formula is used to solve simple interest problems?

22. What formula is used to solve uniform motion problems?

23. What formula is used to solve dry mixture problems?

24. What formula is used to find the mean of a collection of values?

APPLICATIONS

25. ENERGY In 2002, the United States alone accounted for 23.5% of the world's total energy consumption, using 97.6 quadrillion British thermal units (Btu). What was the world's energy consumption in 2002?

26. COMPUTERS In 2001, 61 million, or 56.5%, of U.S. households had personal computers. How many U.S. households were there in 2001? Round to the nearest one million.

27. BUYING A WASHER AND A DRYER Use the following ad to find the percent of markdown of the sale.

28. BUYING FURNITURE A bedroom set regularly sells for \$983. If it is on sale for \$737.25, what is the percent of markdown?

29. FLEA MARKETS A vendor sells tool chests at a flea market for \$65. If she makes a profit of 30% on each unit sold, what does she pay the manufacturer for each tool chest? (*Hint:* The retail price = the wholesale price + the markup.)

30. BOOKSTORES A bookstore sells a textbook for \$39.20. If the bookstore makes a profit of 40% on each sale, what does the bookstore pay the publisher for each book? (*Hint:* The retail price = the wholesale price + the markup.)

31. IMPROVING PERFORMANCE The following graph shows how the installation of a special computer chip increases the horsepower of a truck. What is the percent of increase in horsepower for the engine running at 4,000 revolutions per minute (rpm)? Round to the nearest tenth of one percent.

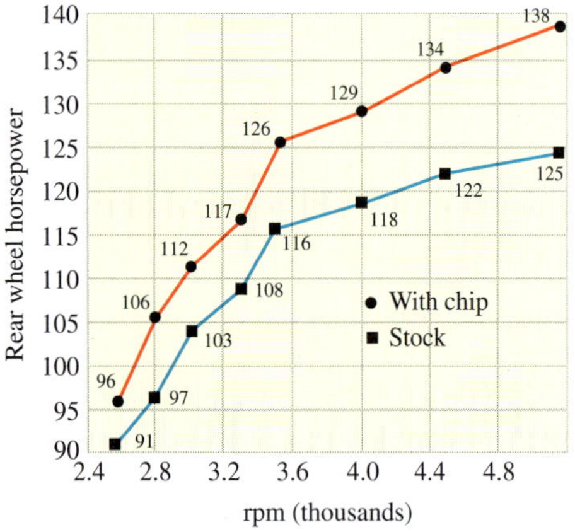

32. GREENHOUSE GASES The total U.S. greenhouse gas emissions in 2000 were 1,906 million metric tons carbon equivalent. In 2001, they were 1,883 million metric tons carbon equivalent. What percent of decrease is this? Round to the nearest tenth of one percent.

33. BROADWAY SHOWS Complete the table to find the percent of increase or decrease in attendance at Broadway shows for each season compared to the previous season. Round to the nearest tenth.

Season	Broadway attendance	% of increase or decrease
2000–01	11.89 million	—
2001–02	10.95 million	
2002–03	11.42 million	

Source: LiveBroadway.com

34. DRIVE-INS The number of drive-in movie theaters in the United States peaked in 1958. Since then, the numbers have steadily declined. Determine the percent of decrease in the number of drive-ins. Round to the nearest one percent.

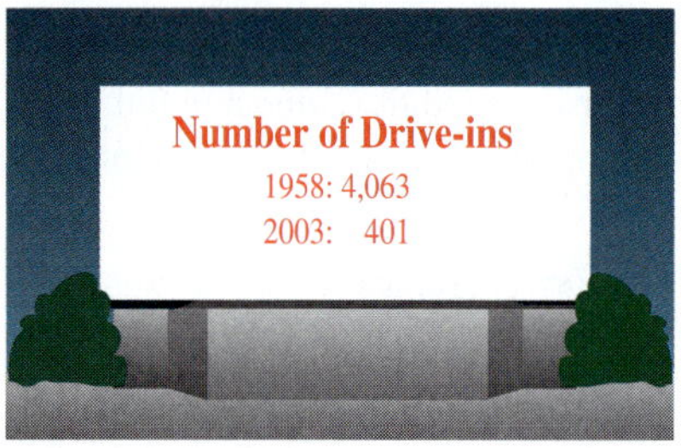

Source: Drive-in Theater Owners Association

35. FUEL EFFICIENCY The ten most fuel-efficient cars in 2002, based on manufacturer's estimated city and highway average miles per gallon (mpg), are shown in the table. Find the mean, median, and mode of both sets of data.

Model	mpg city/hwy
Honda Insight	61/68
Toyota Prius	52/45
Honda Civic Hybrid	47/51
VW Jetta Wagon	42/50
VW Golf	42/49
VW Jetta Sedan	42/49
VW Beetle	42/49
Honda Civic Coupe	36/44
Toyota Echo	34/41
Chevy Prizm	32/41

Source: edmonds.com

36. SPORT FISHING The report shown below lists the fishing conditions at Pyramid Lake for a Saturday in January. Find the median and the mode of the weights of the striped bass caught at the lake.

> **Pyramid Lake**—Some striped bass are biting but are on the small side. Striking jigs and plastic worms. Water is cold: 38°. Weights of fish caught (lb): 6, 9, 4, 7, 4, 3, 3, 5, 6, 9, 4, 5, 8, 13, 4, 5, 4, 6, 9

37. JOB TESTING To be accepted into a police training program, a recruit must have an average score of 85 on a battery of four tests. If a candidate scored 78 on the oral test, 91 on the physical fitness test, and 87 on the psychological test, what is the lowest score she can obtain on the written test and still be accepted into the training program?

38. WNBA CHAMPIONS The results of each 2003 playoff game for the Detroit Shock are shown below. If they averaged 71.5 points per game in the playoffs, how many points did they score in game 3 of the finals against the Sparks?

First Round	*Second Round*
Shock 76, Rockers 74	Shock 73, Sun 63
Rockers 66, Shock 59	Shock 79, Sun 73
Shock 77, Rockers 63	

WNBA Finals

Sparks 75, Shock 63

Shock 62, Sparks 61

Shock **?**, Sparks 78

39. HIGHEST RATES Based on the information in the table, a woman invested $12,000, some in an account paying the highest rate and the rest in an account paying the second highest rate. How much was invested in each account if the interest from both investments is $1,060 per year?

First Republic Savings and Loan	
Account	**Rate**
NOW	5.5%
Savings	7.5%
Money market	8.0%
Checking	4.0%
5-year CD	9.0%

40. ENTREPRENEURS Last year, a women's professional organization made two small-business loans totaling $28,000 to young women beginning their own businesses. The money was lent at 7% and 10% simple interest rates. If the annual income the organization received from these loans was $2,560, what was each loan amount?

41. INHERITANCES Paula split an inheritance between two investments, one a certificate of deposit paying 7% annual interest, and the other a promising biotech company offering an annual return of 10%. She invested twice as much in the 10% investment as she did in the 7% investment. If her combined annual income from the two investments was $4,050, how much did she inherit?

42. TAX RETURNS On a federal income tax form, Schedule B, a taxpayer forgot to write in the amount of interest income he earned for the year. From what is written on the form, determine the amount of interest earned from each investment and the amount he invested in stocks.

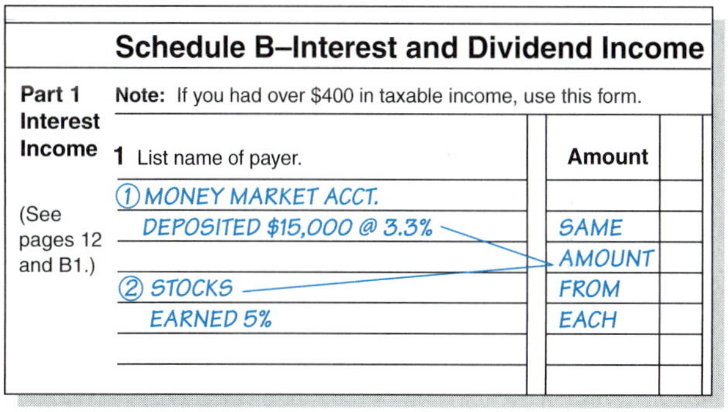
Schedule B–Interest and Dividend Income
Part 1 Interest Income
Note: If you had over $400 in taxable income, use this form.
(See pages 12 and B1.)
1 List name of payer.
Amount
① MONEY MARKET ACCT.
DEPOSITED $15,000 @ 3.3%
② STOCKS
EARNED 5%
SAME
AMOUNT
FROM
EACH

43. MONEY-LAUNDERING Use the evidence compiled by investigators to determine how much money a suspect deposited in the Cayman Islands bank.

- On 6/1/03, the suspect electronically transferred $300,000 to a Swiss bank account paying an 8% annual yield.
- That same day, the suspect opened another account in a Cayman Islands bank that offered a 5% annual yield.
- A document dated 6/3/04 was seized during a raid of the suspect's home. It stated, "The total interest earned in one year from the two overseas accounts was 7.25% of the total amount deposited."

44. FINANCIAL PRESENTATIONS A financial planner showed her client the following investment plan. Find the total amount the client will have to invest to earn $2,700 in interest.

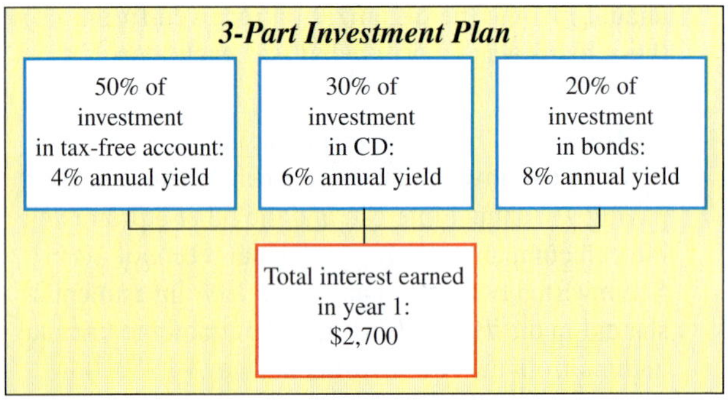

3-Part Investment Plan
50% of investment in tax-free account: 4% annual yield
30% of investment in CD: 6% annual yield
20% of investment in bonds: 8% annual yield
Total interest earned in year 1: $2,700

45. TRAVEL TIMES A man called his wife to tell her that they needed to switch vehicles so he could use the family van to pick up some building materials after work. The wife left their home, traveling toward his office in their van at 35 mph. At the same time, the husband left his office in his car, traveling toward their home at 45 mph. If his office is 20 miles from their home, how long will it take them to meet so they can switch vehicles?

46. AIR TRAFFIC CONTROL An airplane leaves Los Angeles bound for Caracas, Venezuela, flying at an average rate of 500 mph. At the same time, another airplane leaves Caracas bound for Los Angeles, averaging 550 mph. If the airports are 3,675 miles apart, when will the air traffic controllers have to make the pilots aware that the planes are passing each other?

47. CYCLING A cyclist leaves his training base for a morning workout, riding at the rate of 18 mph. One hour later, her support staff leaves the base in a car going 45 mph in the same direction. How long will it take the support staff to catch up with the cyclist?

48. RUNNING MARATHONS Two marathon runners leave the starting gate, one running 12 mph and the other 10 mph. If they maintain the pace, how long will it take for them to be one-quarter of a mile apart?

49. RADIO COMMUNICATIONS At 2 P.M., two military convoys leave Eagle River, WI, one headed north and one headed south. The convoy headed north averages 50 mph, and the convoy headed south averages 40 mph. They will lose radio contact when the distance between them is more than 135 miles. When will this occur?

50. SEARCH AND RESCUE Two search-and-rescue teams leave base camp at the same time, looking for a lost child. The first team, on horseback, heads north at 3 mph, and the other team, on foot, heads south at 1.5 mph. How long will it take them to search a distance of 18 miles between them?

51. JET SKIING A jet ski can go 12 mph in still water. If a rider goes upstream for 3 hours against a current of 4 mph, how long will it take the rider to return? (*Hint:* Upstream speed is $(12 - 4)$ mph; how far can the rider go in 3 hours?)

52. PHYSICAL FITNESS For her workout, Sarah walks north at the rate of 3 mph and returns at the rate of 4 mph. How many miles does she walk if the round trip takes 3.5 hours?

53. MIXING CANDY The owner of a candy store wants to make a 30-pound mixture of two candies to sell for $1 per pound. If red licorice bits sell for 95¢ per pound and lemon gumdrops sell for $1.10 per pound, how many pounds of each should be used?

54. HEALTH FOODS A pound of dried pineapple bits sells for $6.19, a pound of dried banana chips sells for $4.19, and a pound of raisins sells for $2.39 a pound. Two pounds of raisins are to be mixed with equal amounts of pineapple and banana to create a trail mix that will sell for $4.19 a pound. How many pounds of pineapple and banana chips should be used?

55. GARDENING A wholesaler of premium organic planting mix notices that the retail garden centers are not buying her product because of its high price of $1.57 per cubic foot. She decides to mix sawdust with the planting mix to lower the price per cubic foot. If the wholesaler can buy the sawdust for $0.10 per cubic foot, how many cubic feet of each must be mixed to have 6,000 cubic feet of planting mix that could be sold to retailers for $1.08 per cubic foot?

56. METALLURGY A 1-ounce commemorative coin is to be made of a combination of pure gold, costing $380 an ounce, and a gold alloy that costs $140 an ounce. If the cost of the coin is to be $200, and 500 are to be minted, how many ounces of gold and gold alloy are needed to make the coins?

57. DILUTING SOLUTIONS How much water should be added to 20 ounces of a 15% solution of alcohol to dilute it to a 10% solution?

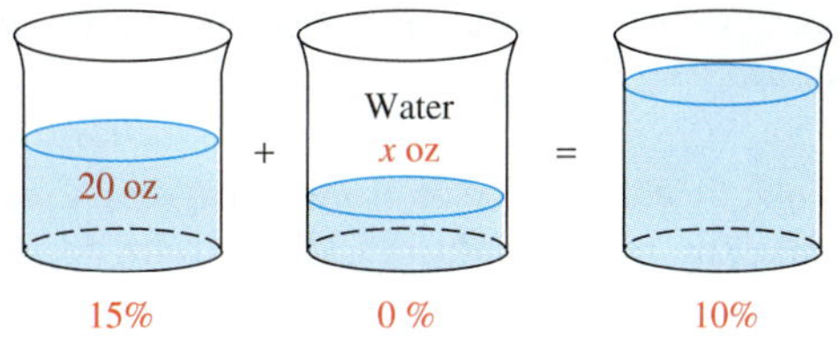

58. INCREASING CONCENTRATIONS The beaker shown below contains a 2% saltwater solution.

a. How much water must be boiled away to increase the concentration of the salt solution from 2% to 3%?

b. Where on the beaker would the new water level be?

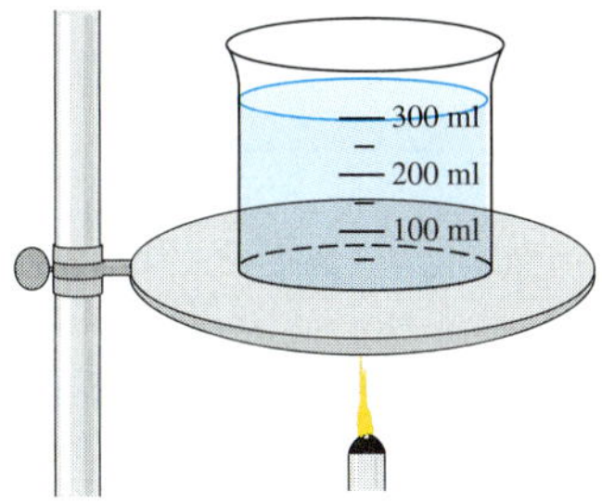

59. DAIRY FOODS Cream is approximately 22% butterfat. How many gallons of cream must be mixed with milk testing at 2% butterfat to get 20 gallons of milk containing 4% butterfat?

60. LOWERING FAT How many pounds of extra-lean hamburger that is 7% fat must be mixed with 30 pounds of lean hamburger that is 15% fat to obtain a mixture that is 10% fat?

WRITING

61. If a car travels at 60 mph for 30 minutes, explain why the distance traveled is *not* $60 \cdot 30 = 1{,}800$ miles.

62. If a mixture is to be made from solutions with concentrations of 12% and 30%, can the mixture have a concentration less than 12% or greater than 30%? Explain.

63. Write a mixture problem that can be represented by the following verbal model and equation.

The value of the regular coffee	plus	the value of the gourmet coffee	equals	the value of the blend.
$4x$	$+$	$7(40 - x)$	$=$	$5(40)$

64. Write a uniform motion problem that can be represented by the following verbal model and equation.

The distance traveled by the 1st train	plus	the distance traveled by the 2nd train	equals	330 miles.
$45t$	$+$	$55t$	$=$	330

REVIEW **Solve each equation.**

65. $9x = 6x$

66. $7a + 2 = 12 - 4(a - 3)$

67. $\dfrac{8(y - 5)}{3} = 2(y - 4)$

68. $\dfrac{t - 1}{3} = \dfrac{t + 2}{6} + 2$

CHALLENGE PROBLEMS

69. Determine a set of 5 values such that the mean is 10, the median is 8, and the mode is 2.

70. Solve the following problem. Then explain why the solution does not make sense.

Adult tickets cost \$4, and student tickets cost \$2. Sales of 71 tickets bring in \$245. How many of each were sold?

ACCENT ON TEAMWORK

WRITING FRACTIONS AS DECIMALS AND AS PERCENTS

Overview: This is a good activity to try at the beginning of the course. You can become acquainted with other students in your class while you review some important arithmetic skills.

Instructions: Form groups of 5 students. Select one person from your group to record the group's responses on the questionnaire. Express the results in fraction form, decimal form, and as percents.

What fraction, decimal, and percent of the students in your group . . .	Fraction	Decimal	Percent
have the letter *a* in their first names?			
have a birthday in January or February?			
say that vanilla is their favorite flavor of ice cream?			
have ever been on television?			
live more than 20 miles from campus?			
say they enjoy rainy days?			
work full-time or part-time?			

MAKING LEMONADE

Overview: This activity will give you a better understanding of the liquid-mixture problems discussed in this chapter.

Instructions: Form groups of 2 or 3 students. Each group needs a dozen 3-ounce paper cups, 2 pitchers, two 8-ounce cans of thawed lemonade concentrate, a large spoon, and a half-gallon of bottled water.

In one pitcher, make a 10% lemonade/water solution by mixing 1 paper cup of lemonade concentrate with 9 paper cups of water. In another pitcher, make a 40% lemonade/water solution by mixing 4 paper cups of lemonade concentrate with 6 paper cups of water.

Use algebra to determine how many paper cups of the 40% solution must be mixed with all of the 10% lemonade to get a 20% lemonade solution. Measure out the appropriate amount of 40% lemonade and pour it into the 10% lemonade to make the 20% mixture. Taste the 20% and 40% lemonades. Can you tell the difference in the concentrations?

KEY CONCEPT: LET x =

In Chapter 1, we discussed one of the most important problem-solving techniques used in algebra. In this method, the first step is to let a variable represent the unknown quantity. Then we use the variable in writing an equation that mathematically describes the situation. Finally, we solve the equation to find the value represented by the variable. Let's review some key parts of this problem-solving method.

ANALYZING THE PROBLEM

In Exercises 1 and 2, what do we know and what are we asked to find?

1. GEOGRAPHY Of the 48 contiguous states, 4 more lie east of the Mississippi River than lie west of the Mississippi. How many states are west of the Mississippi River?

2. GEOMETRY In a right triangle, the measure of one acute angle is 5° more than twice that of the other angle. Find the measure of the smallest angle.

LETTING A VARIABLE REPRESENT AN UNKNOWN QUANTITY

In Exercises 3 and 4, what quantity should the variable represent? State your response in the form "Let $x = \ldots$".

3. CAMPING To make anchor lines for a tent, a 60-foot rope is cut into four pieces, each successive piece twice as long as the previous one. Find the length of each anchor line.

4. INSURANCE COVERAGE While waiting for his van to be repaired, a man rents a car for \$25 per day and 30 cents per mile. His insurance company will pay up to \$100 of the rental fee. If he needs the car for two days, how many miles of driving will his policy cover?

FORMING AN EQUATION

As Exercises 5–8 show, several methods can be used to help form an equation.

5. TRANSLATION For each phrase, what operation is indicated?

a. less than

b. of

c. increased by

d. ratio

6. FORMULAS What formula is suggested by each type of problem?

a. Uniform motion

b. Simple interest

c. Dry mixture

d. Perimeter of a rectangle

7. TABLES Complete the table. What equation is suggested?

	Amount	· Strength	= Amount alcohol
Too weak	15 oz	0.15	
Too strong	x oz	0.50	
Mixture	$(15 + x)$ oz	0.40	

8. DIAGRAMS What equation is suggested by the diagram below?

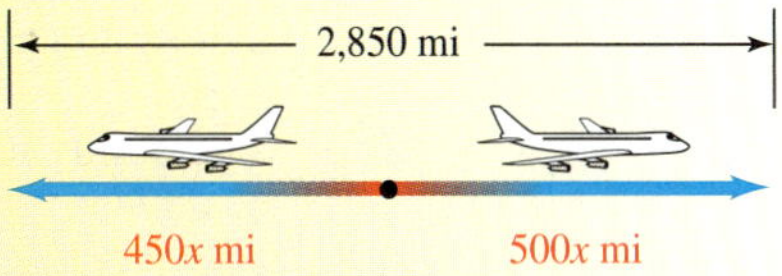

CHAPTER REVIEW

SECTION 1.1 The Language of Algebra

CONCEPTS

A *variable* is a letter that stands for a number.

An *equation* is a mathematical sentence that contains an = symbol.

Formulas are equations that express a relationship between two or more quantities represented by variables.

Bar graphs and *line graphs* display numerical relationships.

The *perimeter* of a geometric figure is the distance around it.

REVIEW EXERCISES

Translate each verbal model into a mathematical model.

1. The cost C (in dollars) to rent t tables is \$15 more than the product of \$2 and t.

2. A rectangle has an area of 25 in.^{2}. The length of the rectangle is the quotient of its area and its width.

3. The waiting period for a business license is now 3 weeks less than it used to be.

4. To determine the cooking time for prime rib, a cookbook suggests using the formula $T = 30p$, where T is the cooking time in minutes and p is the weight of the prime rib in pounds. Use this formula to complete the table.

p	T
6.0	
6.5	
7.0	
7.5	
8.0	

5. Use the data from the table in Exercise 4 to draw each type of graph.

a. Bar graph

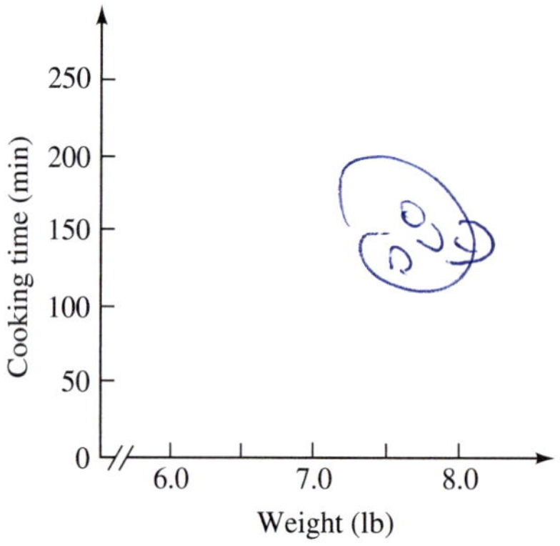

b. Line graph

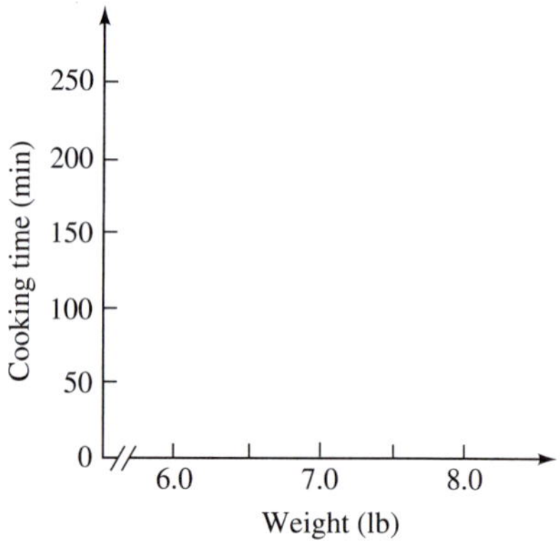

6. The owner of a new business wants to frame the first dollar bill her business ever received. How long a piece of molding will she need if a dollar is 2.625 inches wide and 6.125 inches long?

SECTION 1.2 The Real Number System

Natural numbers: $\{1, 2, 3, \dots\}$

Whole numbers: $\{0, 1, 2, 3, \dots\}$

Integers: $\{\dots, -2, -1, 0, 1, 2, \dots\}$

List the numbers in $\{-5, 0, -\sqrt{3}, 2.4, 7, -\frac{2}{3}, -3.\overline{6}, \pi, \frac{15}{4}, 0.13242368\dots\}$ that belong to the following sets.

7. Natural numbers

8. Whole numbers

9. Integers

10. Rational numbers

Prime numbers: $\{2, 3, 5, 7, 11, 13, \ldots\}$
Composite numbers: $\{4, 6, 8, 9, 10, 12, \ldots\}$
Integers divisible by 2 are *even integers.* Integers not divisible by 2 are *odd integers.*

Rational numbers are numbers that can be written as $\frac{a}{b}$, where a and b are integers and $b \neq 0$. Terminating and repeating decimals are rational numbers.

Irrational numbers are nonterminating, nonrepeating decimals.

A *real number* is any number that is either a rational or an irrational number. All points on the number line represent the set of real numbers.

For any real number x:

$$\begin{cases} \text{If } x \geq 0, \text{ then } |x| = x. \\ \text{If } x < 0, \text{ then } |x| = -x. \end{cases}$$

11. Irrational numbers

12. Real numbers

13. Negative numbers

14. Positive numbers

15. Prime numbers

16. Composite numbers

17. Even integers

18. Odd integers

19. Use one of the symbols > or < to make each statement true.

a. $-16 \quad -17$

b. $-(-1.8) \quad 2\frac{1}{2}$

20. Tell whether each statement is true or false.

a. $23.000001 \geq 23.1$

b. $-11 \leq -11$

21. Graph the prime numbers between 20 and 30 on the number line.

22. Graph the set $\left\{2.75, 2.\overline{3}, \sqrt{7}, \frac{8}{3}, \frac{3\pi}{4}\right\}$ on the number line.

Find the value of each expression.

23. $|-18|$

24. $-|-6.26|$

SECTION 1.3 Operations with Real Numbers

Adding real numbers: With like signs, add the absolute values and keep the common sign. With unlike signs, subtract the absolute values and keep the sign of the number that has the greatest absolute value.

Subtracting real numbers:

$$x - y = x + (-y)$$

Multiplying and dividing real numbers: With like signs, multiply (or divide) their absolute values. The product is positive.

Perform the operations.

25. $-3 + (-4)$

26. $-70.5 + 80.6$

27. $-\frac{1}{2} - \frac{1}{4}$

28. $-6 - (-8)$

29. $(-4.2)(-3.0)$

30. $-\frac{1}{10} \cdot \frac{5}{16}$

31. $\frac{-2.2}{-11}$

32. $-\frac{9}{8} \div 21$

33. $15 - 25 - 23$

34. $-3.5 + (-7.1) + 4.9$

35. $-3(-5)(-8)$

36. $-1(-1)(-1)(-1)$

With unlike signs, multiply (or divide) their absolute values and make the answer negative.

x^n is a *power of x*. x is the base, and n is the *exponent*.

$$x^n = \overbrace{x \cdot x \cdot x \cdot \cdots \cdot x}^{n \text{ factors of } x}$$

A number b is a *square root* of a if $b^2 = a$. $\sqrt{a}$ represents the *principal* (positive) *square root* of a.

Order of operations:

1. Work from the innermost pair to the outermost pair of grouping symbols in the following order.
2. Evaluate all powers and roots.
3. Perform all multiplications and divisions, working from left to right.
4. Perform all additions and subtractions, working from left to right.

When the grouping symbols have been removed, repeat steps 2–4 to finish the calculation. In a fraction, simplify the numerator and denominator separately, and then simplify the fraction.

To *evaluate an algebraic expression,* substitute the values for the variables and then apply the rules for the order of operations.

The *area* of a figure is the amount of surface it encloses. The *volume* of an object is its capacity.

Evaluate each expression.

37. $(-3)^5$

38. $\left(-\frac{2}{3}\right)^2$

39. 0.3 cubed

40. -5^2

Evaluate each expression.

41. $\sqrt{4}$

42. $-\sqrt{100}$

43. $\sqrt{\frac{9}{25}}$

44. $\sqrt{0.64}$

Evaluate each expression.

45. $-6 + 2(-5)^2$

46. $\frac{-20}{4} - (-3)(-2)\left(-\sqrt{1}\right)$

47. $4 - (5 - 9)^2$

48. $4 + 6[-1 - 5(25 - 3^3)]$

49. $2\left|-1.3 + (-2.7)\right|$

50. $\frac{(7-6)^4 + 32}{36 - \left(\sqrt{16} + 1\right)^2}$

51. $(-10)^3\left(\frac{-6}{-2}\right)(-1)$

52. $-(-2 \cdot 4)^2$

Evaluate the algebraic expression for the given values of the variables.

53. $(x + y)(x^2 - xy + y^2)$ for $x = -2$ and $y = 4$

54. $\frac{-b - \sqrt{b^2 - 4ac}}{2a}$ for $a = 2$, $b = -3$, and $c = -2$

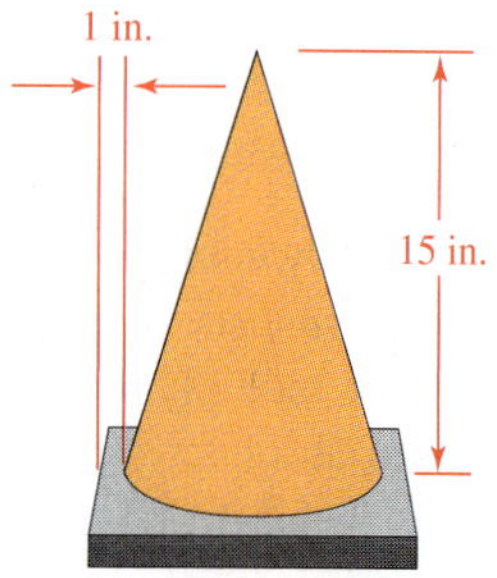

55. SAFETY CONES Find the area covered by the square rubber base if its sides are 10 inches long.

56. SAFETY CONES Find the volume of the cone that is centered atop the base. Round to the nearest tenth.

SECTION 1.4 Simplifying Algebraic Expressions

Properties of real numbers:

1. *Associative properties:* $(a + b) + c = a + (b + c)$ $(ab)c = a(bc)$
2. *Commutative properties:* $a + b = b + a$ $ab = ba$
3. *Distributive property:* $a(b + c) = ab + ac$
4. 0 is the *additive identity:* $a + 0 = 0 + a = a$
5. 1 is the *multiplicative identity:* $1 \cdot a = a \cdot 1 = a$
6. *Multiplication property of 0:* $a \cdot 0 = 0 \cdot a = 0$
7. $-a$ is the *opposite* (or *additive inverse*) of a: $a + (-a) = 0$
8. If $a \neq 0$, then $\frac{1}{a}$ is the *reciprocal* (or *multiplicative inverse* of a): $a \cdot \frac{1}{a} = \frac{1}{a} \cdot a = 1$

To *simplify algebraic expressions,* we use properties of real numbers to write them in a less complicated form.

Terms with exactly the same variables raised to exactly the same powers are called *like* (similar) *terms.*

To add or subtract *like terms,* combine their coefficients and keep the same variables with the same exponents.

Fill in the blanks by applying the indicated property of the real numbers.

57. $3(x + 7) =$ ________ Distributive property (and simplify)

58. $t \cdot 5 =$ ____ Commutative property of multiplication

59. $-x + x =$ ___ Additive inverse property

60. $(27 + 1) + 99 =$ ____________ Associative property of addition

61. $\frac{1}{8} \cdot 8 =$ ___ Multiplicative inverse property

62. $0 + m =$ ____ Additive identity property

63. ___ $\cdot\, 9.87 = 9.87$ Multiplicative identity property

64. $5(-9)(0)(2{,}345) =$ ___ Multiplication property of 0

65. $(-3 \cdot 5)2 =$ _________ Associative property of multiplication

66. $(t + z) \cdot t =$ _________ Commutative property of addition

Perform each division, if possible.

67. $\dfrac{102}{102}$

68. $\dfrac{0}{6}$

69. $\dfrac{-25}{1}$

70. $\dfrac{5.88}{0}$

Remove the parentheses and simplify.

71. $8(x + 6)$

72. $-6(x - 2)$

73. $-(-4 + 3y)$

74. $(3x - 2y)1.2$

75. $\frac{3}{4}(8c^2 - 4c + 1)$

76. $\frac{2}{3}(3t + 9)$

Simplify each expression.

77. $8(6k)$

78. $(-7x)(-10y)$

79. $-9(-3p)(-7)$

80. $15a + 30a + 7$

81. $3g^2 - 3g^2$

82. $-m + 4(m - 12)$

83. $\frac{7}{8}x + \frac{1}{8}x$

84. $21.45l - 45.99l$

85. $4[-2(a^3 - 1) - 2(3 - 6a^3)]$

86. $\frac{3}{4}(2h + 9) - \frac{5}{4}(h - 1)$

SECTION 1.5 Solving Linear Equations and Formulas

The set of numbers that satisfy an equation is called its *solution set.*

Properties of equality: If $a = b$, then

$$a + c = b + c$$
$$a - c = b - c$$
$$ca = cb \quad (c \neq 0)$$
$$\frac{a}{c} = \frac{b}{c} \quad (c \neq 0)$$

To solve a linear equation:

1. Clear the equation of any fractions.
2. Remove all parentheses and combine like terms.
3. Get all variables on one side and all constants on the other. Combine like terms.
4. Make the coefficient of the variable 1.
5. Check the result.

An *identity* is an equation that is satisfied by every number for which both sides are defined. A *contradiction* is an equation that is never true.

To *solve a formula* for a variable, isolate that variable on one side and get all other quantities on the other side.

Determine whether −6 is a solution of each equation.

87. $6 - x = 2x + 24$

88. $\frac{5}{3}(x - 3) = -12$

Solve each equation and give the solution set.

89. $\frac{x}{5} = -45$

90. $t - 3.67 = 4.23$

91. $0.0035 = 0.25g$

92. $0 = x + 4$

Solve each equation.

93. $5x + 12 = 0$

94. $-3x - 7 + x = 6x + 20 - 5x$

95. $4(y - 1) = 28$

96. $2 - 13(x - 1) = 4 - 6x$

97. $\frac{8}{3}(x - 5) = \frac{2}{5}(x - 4)$

98. $\frac{3y}{4} - 14 = -\frac{y}{3} - 1$

99. $-k = -0.06$

100. $\frac{5}{4}p = -10$

101. $\frac{4t + 1}{3} - \frac{t + 5}{6} = \frac{t - 3}{6}$

102. $33.9 - 0.5(75 - 3x) = 0.9$

Solve each equation. If the equation is an identity or a contradiction, so state.

103. $2(x - 6) = 10 + 2x$

104. $-5x + 2x - 1 = -(3x + 1)$

Solve each formula for the indicated variable.

105. $V = \pi r^2 h$ for h

106. $Y + 2g = m$ for g

107. $\frac{T}{6} = \frac{1}{6}ab(x + y)$ for x

108. $V = \frac{4}{3}\pi r^3$ for r^3

SECTION 1.6 Using Equations to Solve Problems

Problem-solving strategy:

1. Analyze the problem.
2. Form an equation.
3. Solve the equation.
4. State the conclusion.
5. Check the result.

Number · value = total value

109. AIRPORTS The world's two busiest airports are Hartsfield Atlanta International and Chicago O'Hare International. Together they served 143 million passengers in 2002, with Atlanta handling 10 million more than O'Hare. How many passengers did each airport serve?

110. TUITION A private school reduces the monthly tuition cost of $245 by $5 per child if a family has more than one child attending the school. Write an algebraic expression that gives the monthly tuition cost per child for a family having c children.

111. TREASURY BILLS What is the value of five $1,000 T-bills? What is the value of x $1,000 T-bills?

112. CABLE TV A 186-foot television cable is to be cut into four pieces. Find the length of each piece if each successive piece is 3 feet longer than the previous one.

113. TOOLING The illustration shows the angle at which a drill is to be held when drilling a hole into a piece of aluminum. Find the measures of both labeled angles.

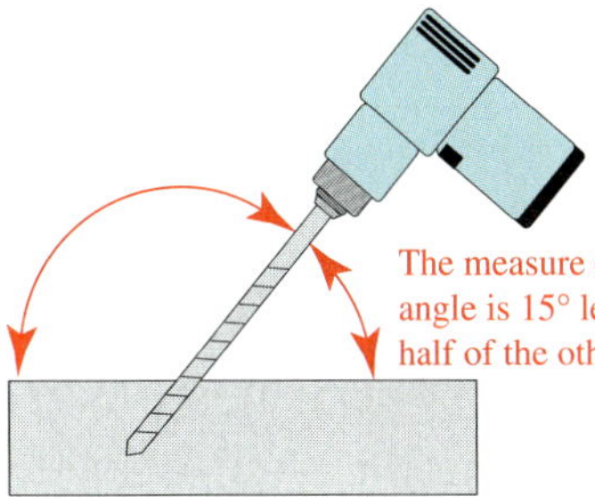
The measure of this angle is 15° less than half of the other angle.

SECTION 1.7 More Applications of Equations

Interest = Principal · rate · time

Percent means parts per one hundred.

Amount = percent · base

114. INVESTMENTS Sally has $25,000 to invest. She invests some money at 10% interest and the rest at 9%. If her total annual income from these two investments is $2,430, how much does she invest at each rate?

115. **a.** Determine the percent of increase in the number of Toyota Camrys sold in 2002 compared to 2001. Round to the nearest tenth of one percent.

b. Determine the percent of decrease in the number of Honda Accords sold in 2002 compared to 2001. Round to the nearest tenth of one percent.

The Two Top-Selling Passenger Cars in the U.S.

2002	
1. Toyota Camry	434,145
2. Honda Accord	398,980
2001	
1. Honda Accord	414,718
2. Toyota Camry	390,449

Source: *The World Almanac 2004*

Mean = $\dfrac{\text{sum of the values}}{\text{number of values}}$

The *mode* is the value that occurs most often.

The *median* is the middle value after the values have been arranged in increasing order.

116. SCHOLASTIC APTITUDE TEST The mean SAT verbal test scores of college-bound seniors for the years 1993–2003 are listed below. Find the mean, median, and mode. Round to the nearest one point.

1993	1994	1995	1996	1997	1998	1999	2000	2001	2002	2003
500	499	504	505	505	505	505	505	506	504	507

Source: *The World Almanac,* 2004

Distance = rate · time

117. PAPARAZZI A celebrity leaves a nightclub in his car and travels at 1 mile per minute (60 mph) trying to avoid a tabloid photographer. One minute later, the photographer leaves the nightclub on his motorcycle, traveling at 1.5 miles per minute (90 mph) in pursuit of the celebrity. How long will it take the photographer to catch up with the celebrity?

118. PEST CONTROL How much water must be added to 20 gallons of a 12% pesticide/water solution to dilute it to an 8% solution?

$v = np$, where v is the value, n is the number of pounds, and p is the price per pound.

119. CANDY SALES Write an algebraic expression that gives the value of a mixture of 3 pounds of Tootsie Rolls with x pounds of Bit O'Honey, if the mix is to sell for \$3.95 per pound.

CHAPTER 1 TEST

Translate each verbal model into a mathematical model.

1. Each test score T was increased by 10 points to give a new adjusted test score s.
2. The area A of a triangle is the product of one-half the length of the base b and the height h.

Consider the set $\left\{-2, \pi, 0, -3\frac{3}{4}, 9.2, \frac{14}{5}, 5, -\sqrt{7}\right\}$.

3. Which numbers are integers?
4. Which numbers are rational numbers?
5. Which numbers are irrational numbers?
6. Which numbers are real numbers?

Graph each set on the number line.

7. $\left\{\dfrac{7}{6}, \dfrac{\pi}{2}, 1.8234503\ldots, \sqrt{3}, 1.\overline{91}\right\}$
8. The set of prime numbers less than 12

Evaluate each expression.

9. $-|8|$
10. $|-5.5|$
11. $7 - (-5.3)$
12. $-\dfrac{5}{3}\left(-\dfrac{4}{25}\right)$
13. $\dfrac{1}{2} - \left(-\dfrac{3}{5}\right)$
14. $(-4)^3$
15. $\dfrac{2[-4 - 2(3 - 1)]}{3(\sqrt{9})(2)}$
16. $7 + 2[-1 - 4(5)]$
17. Evaluate the expression for $a = 2$, $b = -3$, and $c = 4$.
$$\frac{(-3b + c)^2 - 17a}{-b + a^2bc}$$
18. PEDIATRICS Some doctors use Young's rule in calculating dosage for infants and children.
$$\frac{\text{Age of child}}{\text{Age of child} + 12}\left(\begin{array}{c}\text{average}\\ \text{adult dose}\end{array}\right) = \text{child's dose}$$
The adult dose of Achromycin is 250 mg. What is the dose for an 8-year-old child?

Determine which property of real numbers justifies each statement.

19. $3 + 5 = 5 + 3$

20. $x(yz) = (xy)z$

Simplify each expression.

21. $-y + 3y + 9y$

22. $-(4 + t) + t$

23. $8(2h + H) - 5(h + 5H) - 11$

24. $\frac{2}{5}\left(x^2 + 2\right) + \frac{3}{5}\left(x^2 - 3\right)$

Solve each equation.

25. $9(x + 4) + 4 = 4(x - 5)$

26. $\frac{y - 1}{5} + 2 = \frac{2y - 3}{3}$

27. $6 - (x - 3) - 5x = 3[1 - 2(x + 2)]$

For Exercises 28–30, refer to the data in the table.

U.S. Unemployment Rate (1990–2002) in percent						
1990	**1991**	**1992**	**1993**	**1994**	**1995**	**1996**
5.6	6.8	7.5	6.9	6.1	5.6	5.4
1997	**1998**	**1999**	**2000**	**2001**	**2002**	
4.9	4.5	4.2	4.0	4.7	5.8	

Source: U.S. Department of Labor

28. Find the mean. Round to the nearest tenth.

29. Find the median.

30. Find the mode.

31. Solve $P = L + \frac{s}{f}i$ for i.

32. CALCULATORS The viewing window of a calculator has a perimeter of 26 centimeters and is 5 centimeters longer than it is wide. Find the dimensions of the window.

33. WOMEN'S TENNIS Determine the percent of increase in prize money earned by Serena Williams in 2002 compared to 2001. Round to the nearest one percent.

Serena Williams–WTA Tennis Tour
Prize Money 2002: \$3,935,668
Prize Money 2001: \$2,136,263

Source: AcemanTennis.com

34. INVESTING An investment club invested part of \$10,000 at 9% annual interest and the rest at 8%. If the annual income from these investments was \$860, how much was invested at 8%?

35. MIXING ALLOYS How many ounces of a 40% gold alloy must be mixed with 10 ounces of a 10% gold alloy to obtain an alloy that is 25% gold?

36. What does it mean when we say that 3 is a solution of the equation $2x - 3 = x$?

Chapter

2 Graphs, Equations of Lines, and Functions

Getty Images/Thinkstock

An increasing number of people today are choosing to lease rather than buy a car. When deciding whether to lease or buy, one needs to weigh the advantages and disadvantages of each option. Graphs can be helpful in making these comparisons. In this chapter, we will draw graphs that display paired data using a rectangular coordinate system. This type of graph can be used to show the annual costs of leasing versus buying a vehicle over an extended period of time.

To learn more about the rectangular coordinate system, visit *The Learning Equation* on the Internet at http://tle.brookscole.com. (The log-in instructions are in the Preface.) For Chapter 2, the online lessons are:

- *TLE* Lesson 2: The Rectangular Coordinate System
- *TLE* Lesson 3: Rate of Change and the Slope of a Line
- *TLE* Lesson 4: Function Notation

Many relationships between two quantities can be described by using a table, a graph, or an equation.

2.1 The Rectangular Coordinate System

- The rectangular coordinate system
- Graphing mathematical relationships
- Reading graphs
- Step graphs
- The midpoint formula

It is often said that a picture is worth a thousand words. In this section, we will show how numerical relationships can be described by mathematical pictures called *graphs.* We will draw the graphs on a *rectangular coordinate system.*

THE RECTANGULAR COORDINATE SYSTEM

Many cities are laid out on a rectangular grid as shown below. For example, on the east side of Rockford, Illinois, all streets run north and south, and all avenues run east and west. If we agree to list the street numbers first, every address can be identified by using an ordered pair of numbers. If Jose Quevedo lives on the corner of Third Street and Sixth Avenue, his address is given by the ordered pair (3, 6).

This is the street. → (3, 6) ← This is the avenue.

If Lisa Kumar has an address of (6, 3), we know that she lives on the corner of Sixth Street and Third Avenue. From the figure, we can see that

- Bob Anderson's address is (4, 1).
- Rosa Vang's address is (7, 5).
- The address of the store is (8, 2).

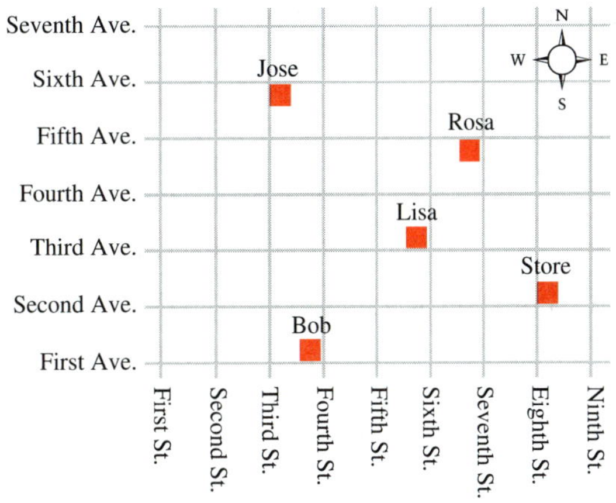

The idea of associating an ordered pair of numbers with points on a grid is attributed to the 17th-century French mathematician René Descartes. The grid is often called a **rectangular coordinate system,** or **Cartesian coordinate system** after its inventor.

In general, a rectangular coordinate system is formed by two intersecting perpendicular number lines, as shown in the figure. The horizontal number line is usually called the ***x*-axis.** The vertical number line is usually called the ***y*-axis.**

The positive direction on the x-axis is to the right, and the positive direction on the y-axis is upward. If no scale is indicated on the axes, we assume that the axes are scaled in units of 1.

The Language of Algebra

The prefix *quad* means four, as in quadrilateral (4 sides), quadraphonic sound (4 speakers), and quadruple (4 times).

The point where the axes cross is called the **origin.** This is the 0 point on each axis. The two axes form a **coordinate plane** and divide it into four regions called **quadrants,** which are numbered using Roman numerals.

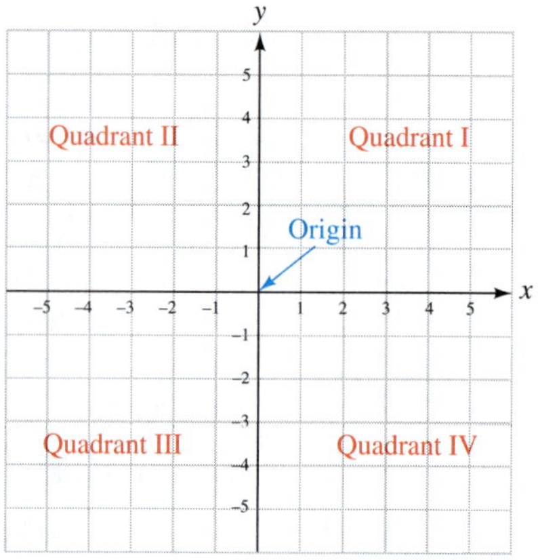

Every point on a coordinate plane can be identified by an **ordered pair** of real numbers x and y, written as (x, y). The first number in the pair is the ***x*-coordinate,** and the second number is the ***y*-coordinate.** The numbers are called the **coordinates** of the point. Some examples are $(-4, 6)$, $(2, 3)$, and $(6, -4)$.

$(-4, 6)$

In an ordered pair, the x-coordinate is listed first. The y-coordinate is listed second.

Caution

If no scale is indicated on the axes, we assume that they are scaled in units of 1.

The process of locating a point in the coordinate plane is called **graphing** or **plotting** the point. Below, we use red arrows to graph the point with coordinates $(6, -4)$. Since the x-coordinate, 6, is positive, we start at the origin and move 6 units to the *right* along the x-axis. Since the y-coordinate, -4, is negative, we then move *down* 4 units, and draw a dot. This locates the point $(6, -4)$, which lies in quadrant IV.

In the figure, blue arrows are used to show how to plot $(-4, 6)$. We start at the origin, move 4 units to the *left* along the x-axis, and then 6 units *up* and draw a dot. This locates the point $(-4, 6)$, which lies in quadrant II.

Caution

Note that the point $(-4, 6)$ is not the same as the point $(6, -4)$. This illustrates that the order of the coordinates of a point is important.

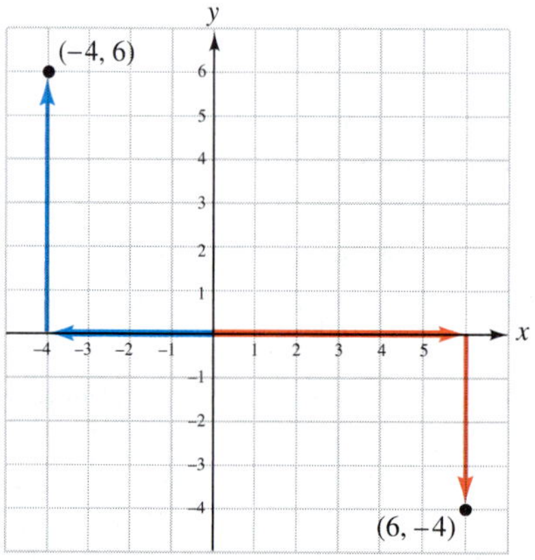

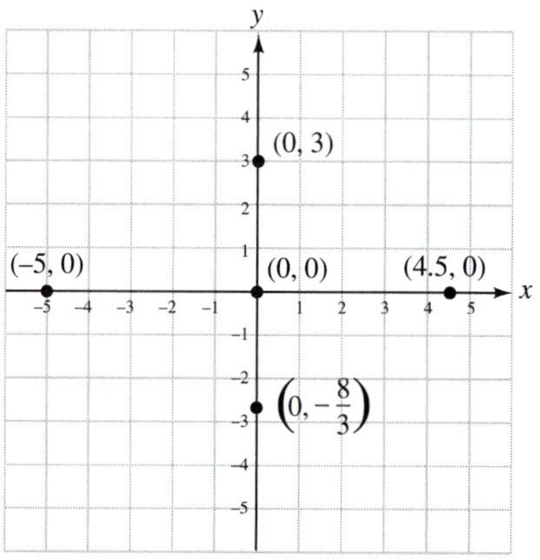

In the figure on the right, we see that the points $(-5, 0)$, $(0, 0)$, and $(4.5, 0)$ all lie on the x-axis. In fact, every point with a y-coordinate of 0 will lie on the x-axis. We also see that the points $\left(0, -\frac{8}{3}\right)$, $(0, 0)$, and $(0, 3)$ all lie on the y-axis. In fact, every point with an x-coordinate of 0 will lie on the y-axis. Note that the coordinates of the origin are $(0, 0)$.

A point may lie in one of the four quadrants or it may lie on one of the axes, in which case the point is not considered to be in any quadrant. For points in quadrant I, the x- and y-coordinates are positive. Points in quadrant II have a negative x-coordinate and a positive y-coordinate. In quadrant III, both coordinates are negative. In quadrant IV, the x-coordinate is positive and the y-coordinate is negative.

EXAMPLE 1

Astronomy. Halley's comet passes Earth every 76 years as it travels in an orbit about the sun. Use the graph to determine the comet's position for the years 1912, 1930, 1948, 1966, 1978, and the most recent time it passed by the Earth, 1986.

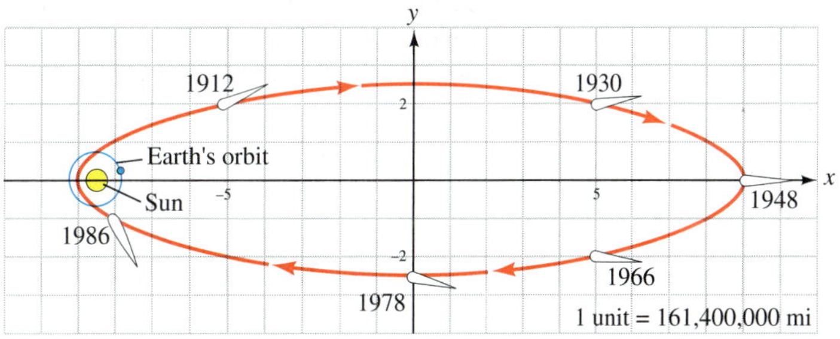

Solution To find the coordinates of each position, we start at the origin and move left or right along the x-axis to find the x-coordinate and then up or down to find the y-coordinate.

Year	Position of comet on graph	Coordinates
1912	5 units to the *left,* then 2 units *up*	$(-5, 2)$
1930	5 units to the *right,* then 2 units *up*	$(5, 2)$
1948	9 units to the *right,* no units *up* or *down*	$(9, 0)$
1966	5 units to the *right,* then 2 units *down*	$(5, -2)$
1978	No units *left* or *right,* then 2.5 units *down*	$(0, -2.5)$
1986	8 units to the *left,* then 1 unit *down*	$(-8, -1)$

GRAPHING MATHEMATICAL RELATIONSHIPS

Every day, we deal with quantities that are related.

- The distance we travel depends on how fast we are going.
- Your test score depends on the amount of time you study.
- The height of a toy rocket depends on the time since it was launched.

Graphs are often used to show relationships between two quantities. For example, suppose we know the height of a toy rocket at 1-second intervals from 0 to 6 seconds after it is launched. We can list this information in a table and write each data pair as an ordered pair of the form (time, height).

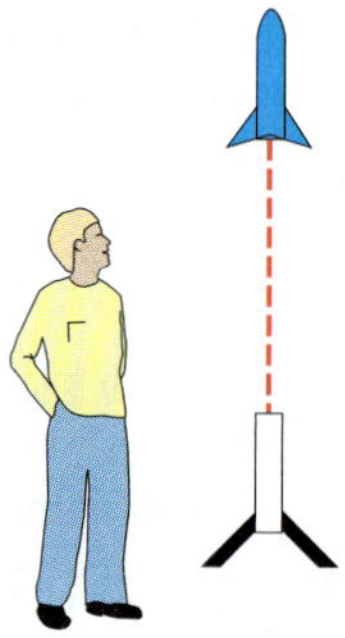

Time (seconds)	Height of rocket (feet)	
0	0	→ (0, 0)
1	80	→ (1, 80)
2	128	→ (2, 128)
3	144	→ (3, 144)
4	128	→ (4, 128)
5	80	→ (5, 80)
6	0	→ (6, 0)

↑ x-coordinate ↑ y-coordinate ↑ The data in the table can be expressed as ordered pairs.

The ordered pairs in the table can then be plotted on a rectangular coordinate system and a smooth curve drawn through the points.

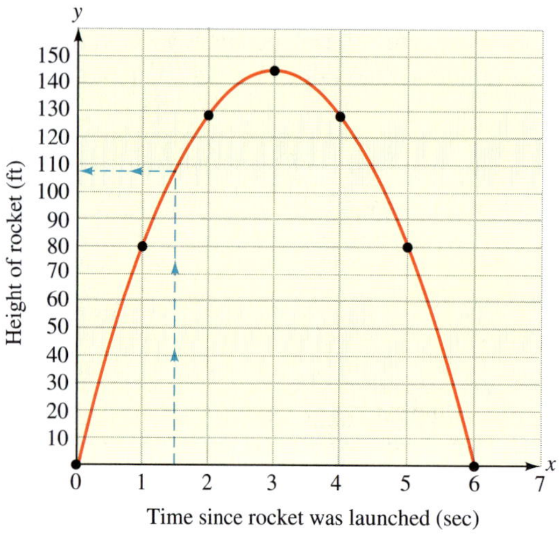

This graph shows the height of the rocket in relation to the time since it was launched. It does not show the path of the rocket.

From the graph, we can see that the height of the rocket increases as the time increases from 0 second to 3 seconds. Then the height decreases until the rocket hits the ground in 6 seconds. We can also use the graph to make observations about the height of the rocket at other times. For example, the dashed blue lines on the graph show that in 1.5 seconds, the height of the rocket will be approximately 108 feet.

READING GRAPHS

Since graphs are becoming an increasingly popular way to present information, the ability to read and interpret them is becoming ever more important.

EXAMPLE 2

Water management. The graph below shows the water level of a reservoir before, during, and after a storm. On the x-axis, zero represents the day the storm began. On the y-axis, zero represents the normal water level that operators try to maintain.

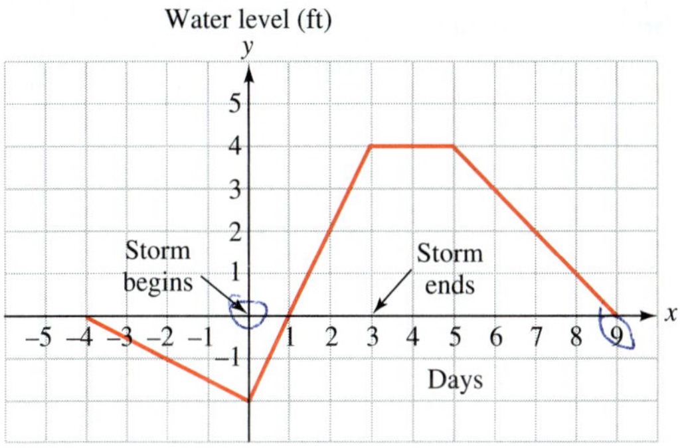

a. In anticipation of the storm, operators released water to lower the level of the reservoir. By how many feet was the water lowered prior to the storm?

b. After the storm ended, on what day did the water level begin to fall?

c. When was the water level 2 feet above normal?

Solution **a.** The graph starts at the point $(-4, 0)$. This means that four days before the storm began, the water level was at the normal level. If we look below zero on the y-axis, we see that the point $(0, -2)$ is on the graph. So the day the storm began, the water level had been lowered 2 feet.

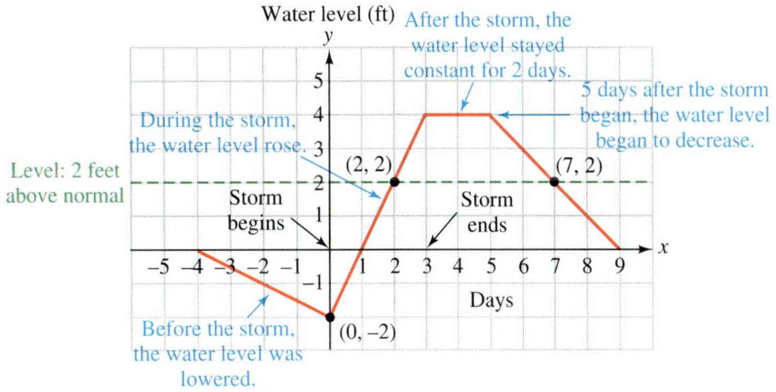

b. If we look at the x-axis, we see that the storm lasted 3 days. From the third to the fifth day, the water level remained constant, 4 feet above normal. The graph does not begin to decrease until day 5.

c. We can draw a horizontal line passing through 2 on the y-axis. This line intersects the graph in two places—at the points $(2, 2)$ and $(7, 2)$. This means that 2 days and 7 days after the storm began, the water level was 2 feet above normal.

Self Check 2 Refer to the graph. **a.** When was the water at the normal level? **b.** By how many feet did the water level rise during the storm? **c.** After the storm ended and the water level began to fall, how long did it take for the water level to return to normal?

STEP GRAPHS

The graph on page 102 shows the cost of renting a rototiller for different periods of time. The horizontal axis is labeled with the variable d. This reinforces the fact that it is associated with the number of *days* the rototiller is rented. The vertical axis is labeled with the variable c, for *cost.* In this case, ordered pairs on the graph will be of the form (d, c). For

Notation

Variables other than x and y can be used to label the horizontal and vertical axes of a rectangular coordinate graph.

example, the point (3, 50) on the graph tells us that the cost of renting a rototiller for 3 days is \$50. The cost of renting the rototiller for more than 4 days up to 5 days is \$70. We call this type of graph a **step graph.**

The point at the end of each step indicates the rental cost for 1, 2, 3, 4, 5, and 6 days. Each open circle indicates that that point is not on the graph.

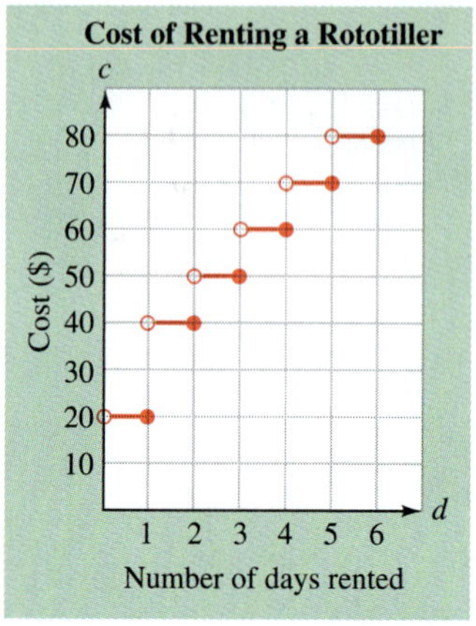

EXAMPLE 3

Rental costs. Use the graph to answer the following questions. **a.** Find the cost of renting the rototiller for 2 days. **b.** Find the cost of renting the rototiller for $5\frac{1}{2}$ days. **c.** How long can you rent the rototiller if you have budgeted \$60 for the rental? **d.** Is the cost of renting the rototiller the same each day?

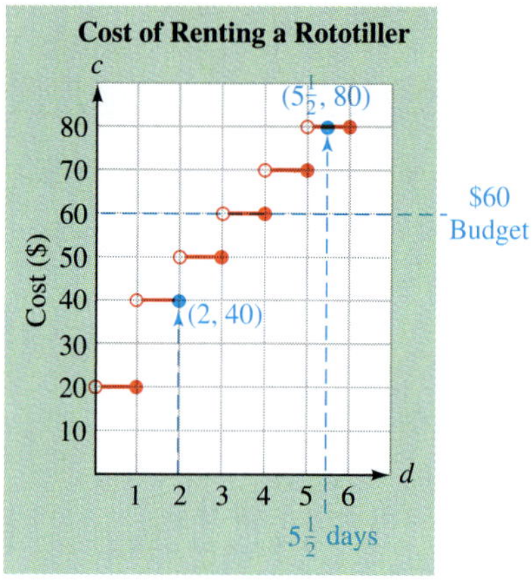

Solution

a. We locate 2 on the d-axis and move up to locate the point on the graph directly above the 2. Since that point has coordinates (2, 40), a two-day rental costs \$40.

b. We locate $5\frac{1}{2}$ on the d-axis and move straight up to locate the point on the graph with coordinates $(5\frac{1}{2}, 80)$, which indicates that a $5\frac{1}{2}$-day rental would cost \$80.

c. We draw a horizontal line through the point labeled 60 on the c-axis. Since this line intersects one of the steps of the graph, we can look down to the d-axis to find the d-values that correspond to a c-value of 60. We see that the rototiller can be rented for more than 3 and up to 4 days for \$60.

d. The cost each day is not the same. If we look at how the c-coordinates change, we see that the first-day rental fee is \$20. The second day, the cost jumps another \$20. The third day, and all subsequent days, the cost jumps \$10.

Self Check 3 Use the graph to find the cost of renting the rototiller for **a.** 4 days and **b.** $2\frac{1}{2}$ days.

THE MIDPOINT FORMULA

Notation

Points are labeled with capital letters. The notation $P(-2, 5)$ indicates that point P has coordinates $(-2, 5)$.

If point M in the figure on the next page lies midway between point P with coordinates $(-2, 5)$ and point Q with coordinates $(4, -2)$, it is called the **midpoint** of line segment PQ. To find the coordinates of point M, we find the mean of the x-coordinates and the mean of the y-coordinates of points P and Q.

$$x = \frac{-2 + 4}{2} \quad \text{and} \quad y = \frac{5 + (-2)}{2}$$

$$= \frac{2}{2} \qquad\qquad = \frac{3}{2}$$

$$= 1$$

Thus, point M has coordinates $\left(1, \frac{3}{2}\right)$.

To distinguish between the coordinates of two general points on a line segment, we often use **subscript notation.** In the right-hand figure, point $P(x_1, y_1)$ is read as "point P with coordinates x sub 1 and y sub 1," and point $Q(x_2, y_2)$ is read as "point Q with coordinates x sub 2 and y sub 2." Using this notation, we can write the midpoint formula in the following way.

The Language of Algebra

The prefix *sub* means below or beneath, as in submarine or subway. In x_2, the *subscript* 2 is written lower than the variable.

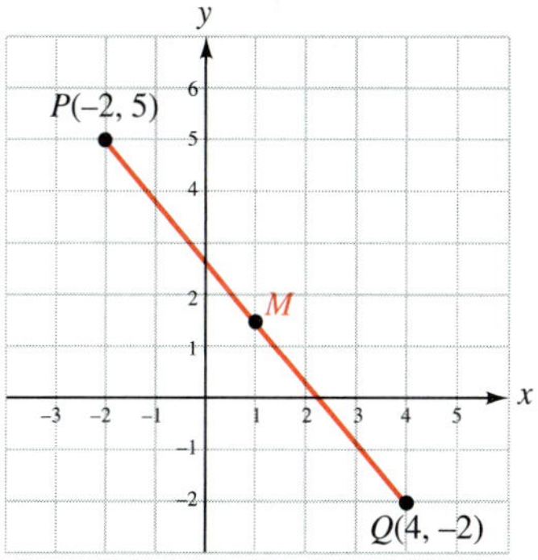

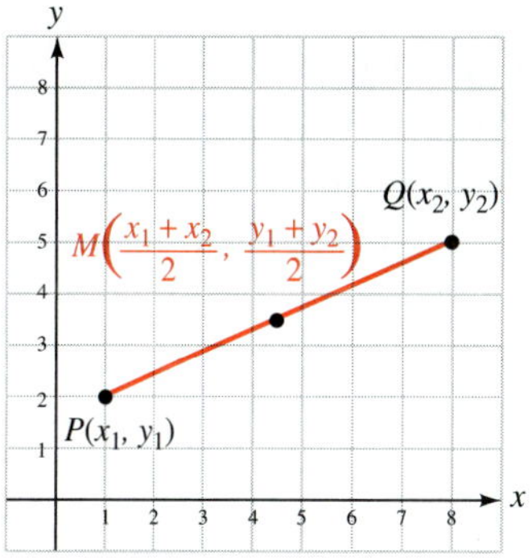

The Midpoint Formula

The **midpoint** of a line segment with endpoints at (x_1, y_1) and (x_2, y_2) is the point with coordinates

$$\left(\frac{x_1 + x_2}{2}, \frac{y_1 + y_2}{2}\right)$$

EXAMPLE 4

The midpoint of the line segment joining $P(-5, -3)$ and $Q(x_2, y_2)$ is the point $(-1, 2)$. Find the coordinates of point Q.

Solution

We can let $P(x_1, y_1) = P(-5, -3)$ and $(x_M, y_M) = (-1, 2)$, where x_M represents the x-coordinate and y_M represents the y-coordinate of the midpoint. We can then find the coordinates of point Q using the midpoint formula.

Caution

Don't confuse x^2 with x_2. Recall that $x^2 = x \cdot x$. When working with two ordered pairs, x_2 represents the x-coordinate of the second ordered pair.

$$x_M = \frac{x_1 + x_2}{2} \quad \text{and} \quad y_M = \frac{y_1 + y_2}{2}$$

Read x_M as "x sub M" and y_M as "y sub M."

$$-1 = \frac{-5 + x_2}{2} \qquad 2 = \frac{-3 + y_2}{2}$$

$$-2 = -5 + x_2 \qquad 4 = -3 + y_2$$

Multiply both sides by 2.

$$3 = x_2 \qquad 7 = y_2$$

Since $x_2 = 3$ and $y_2 = 7$, the coordinates of point Q are $(3, 7)$.

Self Check 4 If the midpoint of a segment PQ is $(-2, 5)$ and one endpoint is $Q(6, -2)$, find the coordinates of point P.

Answers to Self Checks **2. a.** 4 days before the storm began, 1 day and 9 days after the storm began, **b.** 6 ft, **c.** 4 days **3. a.** \$60, **b.** \$50 **4.** $(-10, 12)$

2.1 STUDY SET

VOCABULARY Fill in the blanks.

1. The pair of numbers (6, −2) is called an ________ pair.
2. In the ordered pair (−2, −9), −9 is called the ____coordinate.
3. The point (0, 0) is the ________.
4. The x- and y-axes divide the coordinate plane into four regions called __________.
5. Ordered pairs of numbers can be graphed on a __________ coordinate system.
6. The process of locating a point on a coordinate plane is called ________________ the point.
7. If a point is midway between two points P and Q, it is called the _________ of segment PQ.
8. If a line segment joins points P and Q, points P and Q are called _______________ of the segment.

CONCEPTS Fill in the blanks.

9. To plot (6, −3.5), we start at the ______ and move 6 units to the ______ and then 3.5 units ______.
10. To plot $\left(-6, \frac{3}{2}\right)$, we start at the ______ and move 6 units to the ______ and then $\frac{3}{2}$ units ______.
11. In which quadrant do points with a negative x-coordinate and a positive y-coordinate lie?
12. In which quadrant do points with a positive x-coordinate and a negative y-coordinate lie?

NOTATION

13. Do these ordered pairs name the same point? $\left(5.25, -\frac{3}{2}\right), \left(5\frac{1}{4}, -1.5\right), \left(\frac{21}{4}, -1\frac{1}{2}\right)$
14. For the ordered pair (t, d), which variable is associated with the horizontal axis?
15. What type of letters are used to label points?
16. Fill in the blank: The expression x_1 is read as ________.
17. Explain the difference between x^2 and x_2.
18. Fill in the blanks: The x-coordinate of the midpoint of the line segment joining (x_1, y_1) and (x_2, y_2) is ________ and the y-coordinate is ________.

PRACTICE Plot each point on the rectangular coordinate system.

19. (4, 3)
20. (−2, 1)
21. (3.5, −2)
22. (−2.5, −3)
23. (5, 0)
24. (−4, 0)
25. $\left(\frac{8}{3}, 0\right)$
26. $\left(0, \frac{10}{3}\right)$

Give the coordinates of each point.

27. A
28. B
29. C
30. D
31. E
32. F
33. G
34. H

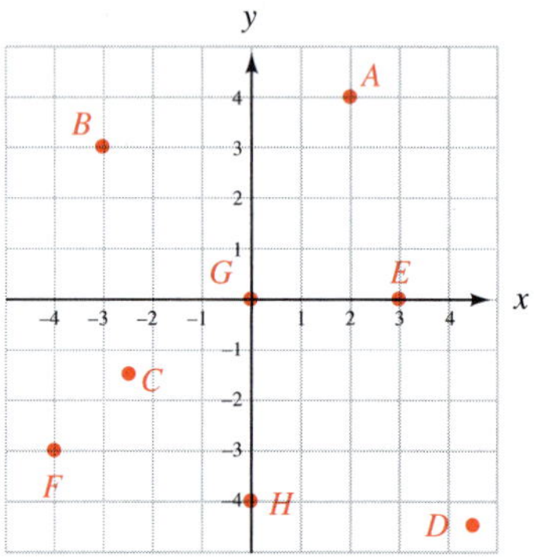

35. The graph in the following illustration shows the depths of a submarine at certain times.
    a. Where is the sub when $t = 2$?
    b. What is the sub doing as t increases from $t = 2$ to $t = 3$?
    c. How deep is the sub when $t = 4$?
    d. How large an ascent does the sub begin to make when $t = 6$?

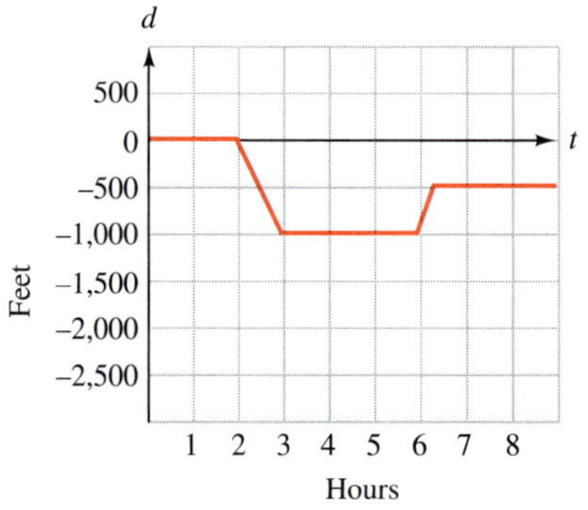

56. TRAMPOLINES The table shows the distance a girl is from the ground (in relation to time) as she bounds into the air and back down to the trampoline. Plot the ordered pairs in the table on a rectangular coordinate system and then draw a smooth curve through the points.

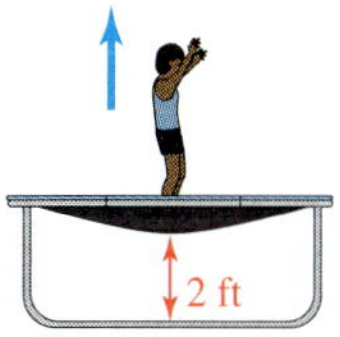

Time (sec)	Height (ft)
0	2
0.25	9
0.5	14
1.0	18
1.5	14
1.75	9
2.0	2

57. GOLF Tiger Woods came back in the final 18 holes of the 2000 AT&T Pebble Beach National Pro-Am golf tournament to overtake the leader, Matt Gogel. See the graph. (In golf, the player with the score that is the farthest *under* par is the winner.)

a. At the beginning of the final round, by how many strokes did Gogel lead Woods?

b. What was the largest lead that Gogel had over Woods in the final round?

c. On what hole did Woods tie up the match?

d. On what hole did Woods take the lead?

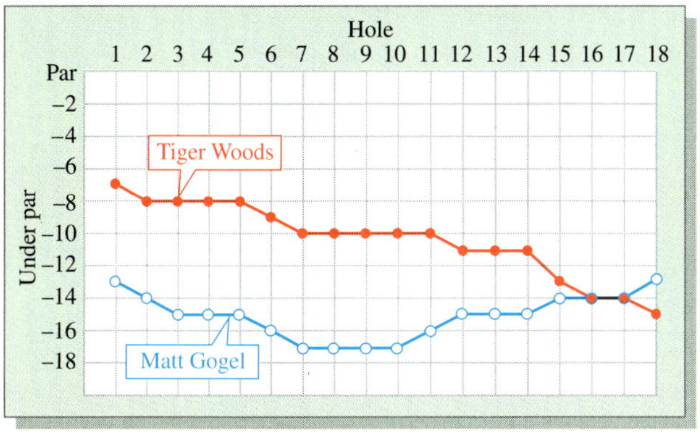

58. VIDEO RENTALS The charges for renting a video are shown in the following graph.

a. Find the 1-day rental charge.

b. Find the 2-day rental charge.

c. Find the 5-day rental charge.

d. Find the charge if the video is kept for a week.

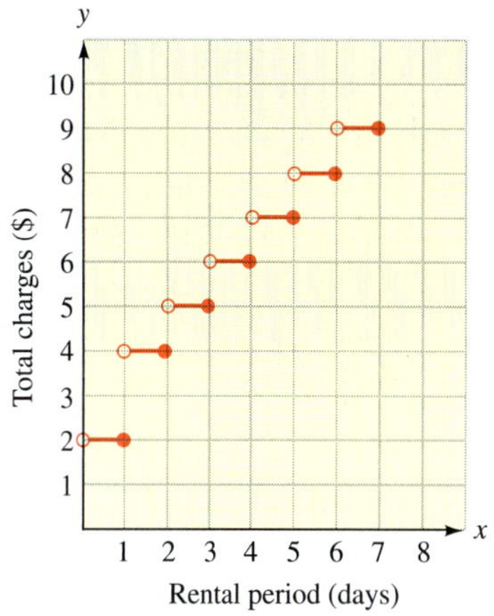

59. POSTAGE The graph shown below gives the first-class postage rates in 2004 for mailing parcels weighing up to 5 ounces.

a. Find the cost of postage to mail a 3-oz letter.

b. Find the difference in cost for a 2.75-oz letter and a 3.75-oz letter.

c. What is the heaviest letter that can be mailed first class for $1?

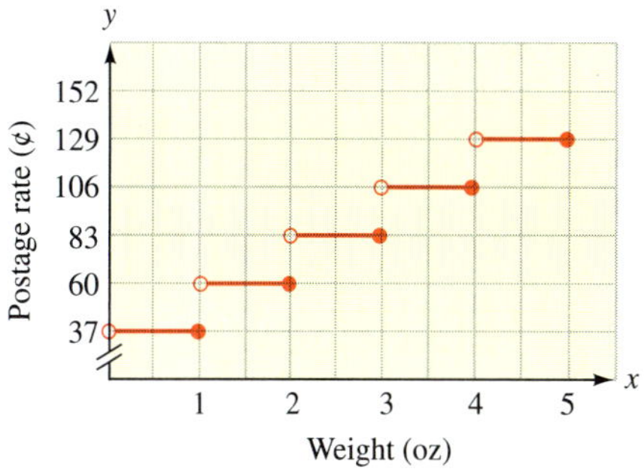

60. ROAST TURKEY Guidelines that appear on the label of a frozen turkey are listed in the table. Draw a step graph that illustrates these instructions.

Size	Time thawing in refrigerator
10 lb to just under 18 lb	3 days
18 lb to just under 22 lb	4 days
22 lb to just under 24 lb	5 days
24 lb to just under 30 lb	6 days

61. MULTICULTURAL STUDIES Social scientists use the following diagram to classify cultures. The amount of group/family loyalty in a culture is measured on the horizontal *group* axis. The amount of social mobility is measured on the vertical *social grid* axis. In the diagram, four cultures are classified. In which culture, R, S, T, or U, would you expect that. . .

a. anyone can grow up to be president, and parents expect their children to get out on their own as soon as possible?

b. only the upper class attends college, and people must marry within their own social class?

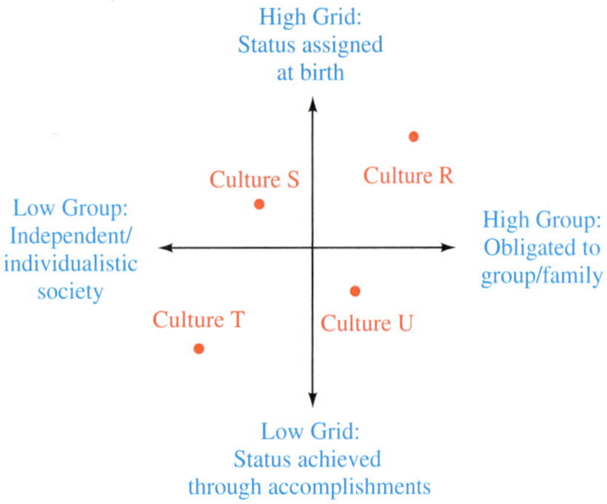

62. PSYCHOLOGY The results of a personal profile test taken by an employee are plotted as an ordered pair on the grid in the illustration. The test shows whether the employee is more task oriented or people oriented. From the results, would you expect the employee to agree or disagree with each of the following statements?

a. Completing a project is almost an obsession with me, and I cannot be content until I am finished.

b. Even if I'm in a hurry while running errands, I will stop to talk with a friend.

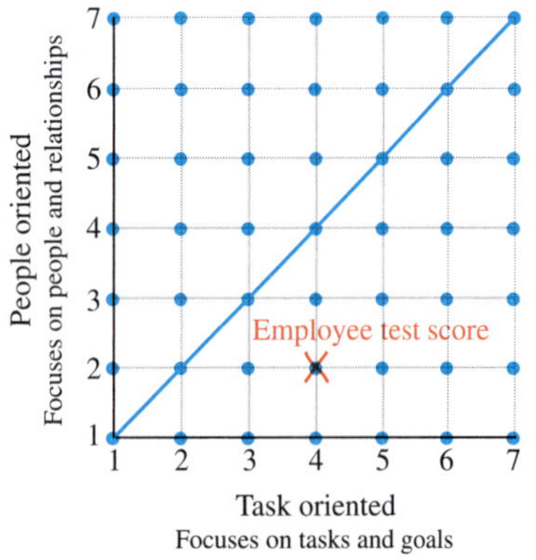

WRITING

63. Explain how to plot the point with coordinates of $(-2, 5)$.

64. Explain why the coordinates of the origin are $(0, 0)$.

REVIEW Evaluate each expression.

65. $-5 - 5(-5)$

66. $(-5)^2 + (-5)$

67. $\dfrac{-3 + 5(2)}{9 + 5}$

68. $|-1 - 9|$

69. Solve: $-4x + 0.7 = -2.1$.

70. Solve $P = 2l + 2w$ for w.

CHALLENGE PROBLEMS

71. What are the coordinates of the three points that divide the line segment joining $P(a, b)$ and $Q(c, d)$ into four equal parts?

72. AIRPLANES When designing an airplane, engineers use a coordinate system with 3 axes, as shown. Any point on the airplane can be described by an *ordered triple* of the form (x, y, z). The coordinates of three points on the plane are $(0, 181, 56)$, $(-46, 48, 19)$, and $(84, 94, 24)$. Which highlighted part of the plane corresponds with which ordered triple?

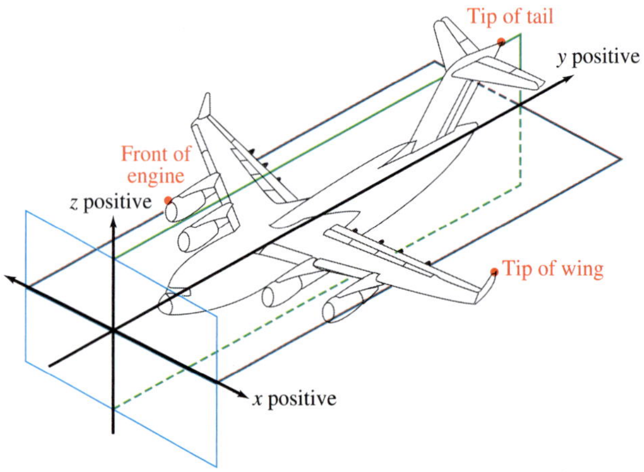

2.2 Graphing Linear Equations

- Solutions of equations in two variables
- Graphing linear equations
- The intercept method
- Graphing horizontal and vertical lines
- Linear models

In this section, we will discuss equations that contain two variables. Such equations are used to describe algebraic relationships between two quantities.

SOLUTIONS OF EQUATIONS IN TWO VARIABLES

We will now extend our equation-solving skills to find solutions of **equations in two variables.** To begin, let's consider $y = -\frac{1}{2}x + 3$, an equation in x and y. In general, a solution of an equation in two variables is an ordered pair of numbers that make a true statement when substituted into the equation.

EXAMPLE 1

Is $(-4, 5)$ a solution of $y = -\dfrac{1}{2}x + 3$?

Solution The ordered pair $(-4, 5)$ has an x-coordinate of -4 and a y-coordinate of 5. We substitute these values for the variables and see whether the resulting equation is true.

The Language of Algebra

We say that a solution of an equation in two variables *satisfies* the equation.

$$y = -\frac{1}{2}x + 3$$

$$5 \stackrel{?}{=} -\frac{1}{2}(-4) + 3 \quad \text{Substitute 5 for } y \text{ and } -4 \text{ for } x.$$

$$5 \stackrel{?}{=} 2 + 3$$

$$5 = 5$$

Since the result is a true statement, $(-4, 5)$ is a solution of $y = -\dfrac{1}{2}x + 3$.

Self Check 1 Is $(4, -1)$ a solution of $y = -\dfrac{1}{2}x + 3$?

To find a solution of an equation in two variables, we can select a number for one of the variables and find the corresponding value of the other variable. For example, to find a solution of $y = -\frac{1}{2}x + 3$, we can select a value of x, say 6, and find the corresponding value of y.

$$y = -\frac{1}{2}x + 3$$

$$y = -\frac{1}{2}(6) + 3 \quad \text{Substitute 6 for } x.$$

$$y = -3 + 3$$

$$y = 0$$

Thus, $(6, 0)$ is a solution of $y = -\dfrac{1}{2}x + 3$.

Since we can choose any real number for x, and since any choice of x will give a corresponding value for y, the equation $y = -\frac{1}{2}x + 3$ has *infinitely many solutions.* It would be impossible to list all of the solutions. Instead, we can draw a mathematical picture of the solutions, called a **graph** of the equation.

GRAPHING LINEAR EQUATIONS

Equations in two variables can be graphed in several ways. If an equation in x and y is solved for y, we can graph it by selecting values of x and calculating the corresponding values of y.

EXAMPLE 2

Graph $y = -\dfrac{1}{2}x + 3$.

Solution To graph $y = -\frac{1}{2}x + 3$, we construct a **table of solutions** by choosing several values of x and finding the corresponding values of y. For example, if x is -2, we have

$$y = -\frac{1}{2}x + 3$$

$$y = -\frac{1}{2}(-2) + 3 \qquad \text{Substitute } -2 \text{ for } x.$$

$$y = 1 + 3$$

$$y = 4$$

Success Tip

Choose x-values that are multiples of 2 to make the computations easier when multiplying x by $-\frac{1}{2}$.

Thus, $(-2, 4)$ is a solution. In a similar manner, we find corresponding y-values for x-values of 0, 2, and 4, and enter the solutions in a table.

When we plot the ordered-pair solutions on a rectangular coordinate system, we see that they lie in a straight line. Using a straight edge or ruler, we then draw a line through the points because the graph of any solution of $y = -\frac{1}{2}x + 3$ will lie on this line. Furthermore, every point of this line represents a solution. We call the line the **graph of the equation.** It represents all of the solutions of $y = -\frac{1}{2}x + 3$.

$y = -\frac{1}{2}x + 3$

x	y	(x, y)
-2	4	$(-2, 4)$
0	3	$(0, 3)$
2	2	$(2, 2)$
4	1	$(4, 1)$

Choose values for x. Compute each value of y.

Construct a table of solutions.

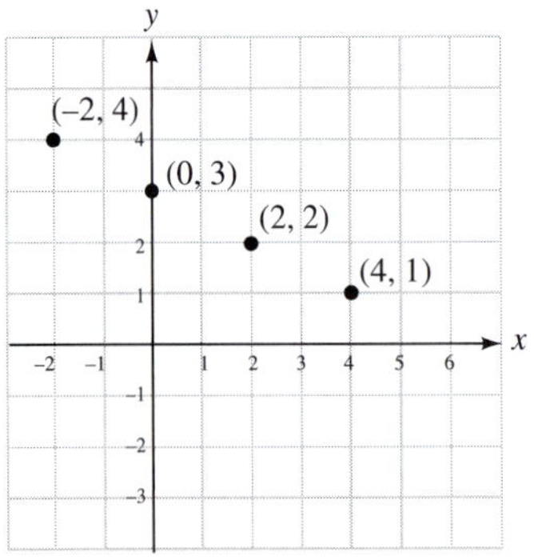

Plot the ordered pairs.

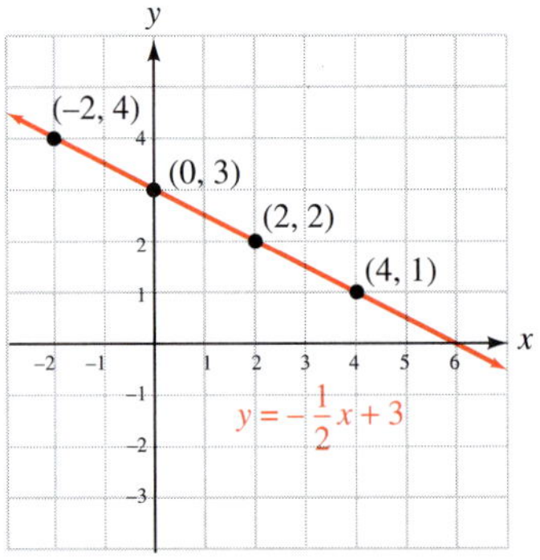

Draw a straight line through the points. This is the *graph of the equation.*

Self Check 2 Graph: $y = \dfrac{1}{3}x + 1$.

When the graph of an equation is a line, we call the equation a *linear equation.*

General (Standard) Form of a Linear Equation

A **linear equation in two variables** is an equation that can be written in the form

$$Ax + By = C$$

where A, B, and C are real numbers and A and B are not both 0.

Some examples of linear equations are

$$y = -\frac{1}{2}x + 3, \quad 2x - 5y = 10, \quad y = 3, \quad \text{and} \quad x = 2$$

THE INTERCEPT METHOD

In Example 2, the graph intersected the y-axis at the point (0, 3) (called the ***y*-intercept**) and intersected the x-axis at the point (6, 0) (called the ***x*-intercept**). In general, we have the following definitions.

Intercepts of a Line

The ***y*-intercept** of a line is the point $(0, b)$, where the line intersects the y-axis. To find b, substitute 0 for x in the equation of the line and solve for y.

The ***x*-intercept** of a line is the point $(a, 0)$, where the line intersects the x-axis. To find a, substitute 0 for y in the equation of the line and solve for x.

EXAMPLE 3

Use the x- and y-intercepts to graph $2x - 5y = 10$.

Solution To find the y-intercept, we substitute 0 for x and solve for y:

$$\begin{aligned} 2x - 5y &= 10 \\ 2(0) - 5y &= 10 && \text{Substitute 0 for } x. \\ -5y &= 10 \\ y &= -2 && \text{Divide both sides by } -5. \end{aligned}$$

The y-intercept is the point $(0, -2)$. To find the x-intercept, we substitute 0 for y and solve for x:

$$\begin{aligned} 2x - 5y &= 10 \\ 2x - 5(0) &= 10 && \text{Substitute 0 for } y. \\ 2x &= 10 \\ x &= 5 && \text{Divide both sides by 2.} \end{aligned}$$

The x-intercept is the point (5, 0).

Success Tip

The exponent on each variable of a linear equation is an understood 1. For example, $2x - 5y = 10$ can be thought of as $2x^1 - 5y^1 = 10$.

Although two points are enough to draw the line, it is a good idea to find and plot a third point as a check. To find the coordinates of a third point, we can substitute any convenient number (such as -5) for x and solve for y:

The Language of Algebra

For any two points, exactly one line passes through them. We say two points *determine* a line.

$$\begin{aligned} 2x - 5y &= 10 \\ 2(-5) - 5y &= 10 && \text{Substitute } -5 \text{ for } x. \\ -10 - 5y &= 10 \\ -5y &= 20 && \text{Add 10 to both sides.} \\ y &= -4 && \text{Divide both sides by } -5. \end{aligned}$$

The line will also pass through the point $(-5, -4)$.

A table of solutions and the graph of $2x - 5y = 10$ are shown below.

$2x - 5y = 10$

x	y	(x, y)
0	-2	$(0, -2)$
5	0	$(5, 0)$
-5	-4	$(-5, -4)$

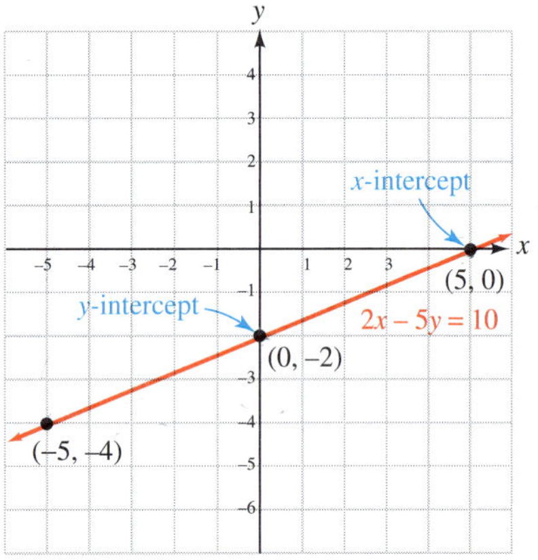

Self Check 3 Find the x- and y-intercepts and graph $5x + 15y = -15$.

ACCENT ON TECHNOLOGY: GENERATING TABLES OF SOLUTIONS

Courtesy of Texas Instruments

If an equation in x and y is solved for y, we can use a graphing calculator to generate a table of solutions. The instructions in this discussion are for a TI-83 or a TI-83 Plus graphing calculator. For specific details about other brands, please consult the owner's manual.

To construct a table of solutions for $2x - 5y = 10$, we first solve for y.

$$\begin{aligned} 2x - 5y &= 10 \\ -5y &= -2x + 10 && \text{Subtract } 2x \text{ from both sides.} \\ y &= \frac{2}{5}x - 2 && \text{Divide both sides by } -5 \text{ and simplify.} \end{aligned}$$

To enter $y = \frac{2}{5}x - 2$, we press [Y =] and enter $(2/5)x - 2$, as shown in figure (a). (Ignore the subscript 1 on y; it is not relevant at this time.)

To enter the x-values that are to appear in the table, we press [2nd] [TBLSET] and enter the first value for x on the line labeled TblStart =. In figure (b), -5 has been entered on this line. Other values for x that are to appear in the table are determined by setting an **increment value** on the line labeled $\triangle$Tbl =. Figure (b) shows that an increment of 1 was entered. This means that each x-value in the table will be 1 unit larger than the previous x-value.

The final step is to press the keys [2nd] [TABLE]. This displays a table of solutions, as shown in figure (c).

(a)

(b)

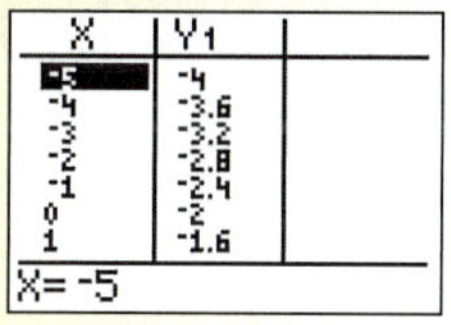

(c)

GRAPHING HORIZONTAL AND VERTICAL LINES

Equations such as $y = 3$ and $x = -2$ are linear equations, because they can be written in the form $Ax + By = C$.

$$y = 3 \quad \text{is equivalent to} \quad 0x + 1y = 3$$
$$x = -2 \quad \text{is equivalent to} \quad 1x + 0y = -2$$

EXAMPLE 4

Graph: **a.** $y = 3$ and **b.** $x = -2$

Solution

a. Since the equation $y = 3$ does not contain x, the numbers chosen for x have no effect on y. The value of y is always 3.

After plotting the ordered pairs shown in the table, we see that the graph (shown on the next page) is a horizontal line, parallel to the x-axis, with a y-intercept of (0, 3). The line has no x-intercept.

b. Since the equation $x = -2$ does not contain y, the value of y can be any number.

After plotting the ordered pairs shown in the table, we see that the graph (on the next page) is a vertical line, parallel to the y-axis, with an x-intercept of $(-2, 0)$. The line has no y-intercept.

$y = 3$

x	y	(x, y)
-3	3	$(-3, 3)$
0	3	(0, 3)
2	3	(2, 3)
4	3	(4, 3)

↑ The value of x can be any number.

$x = -2$

x	y	(x, y)
-2	-2	$(-2, -2)$
-2	0	$(-2, 0)$
-2	2	$(-2, 2)$
-2	6	$(-2, 6)$

↑ The value of y can be any number.

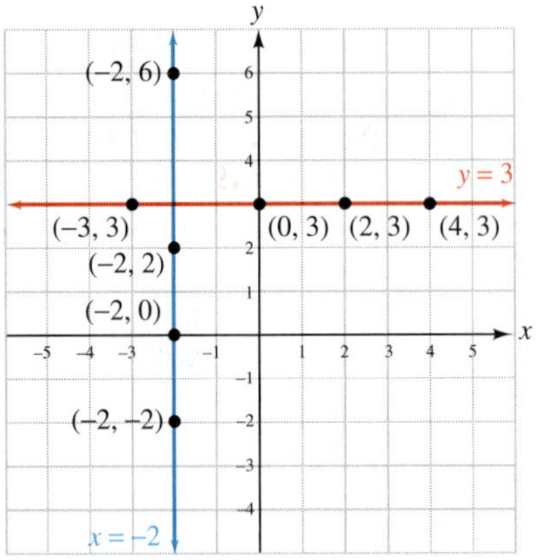

Self Check 4 Graph: **a.** $x = 4$ and **b.** $y = -3$.

The results of Example 4 suggest the following facts.

Equations of Horizontal and Vertical Lines	The equation $y = b$ represents the horizontal line that intersects the y-axis at $(0, b)$. The equation $x = a$ represents the vertical line that intersects the x-axis at $(a, 0)$.

The graph of the equation $y = 0$ has special significance; it is the x-axis. Similarly, the graph of the equation $x = 0$ is the y-axis.

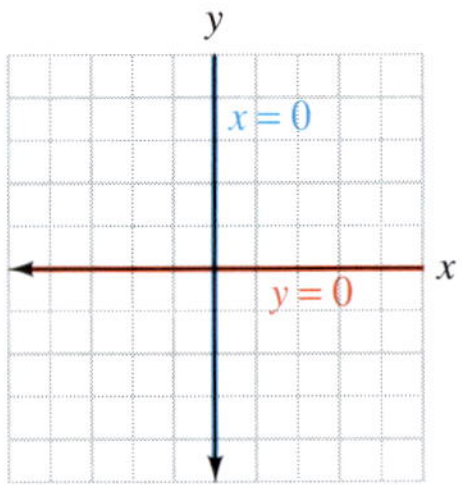

LINEAR MODELS

In the next examples, we will see how linear equations can model real-life situations. In each case, the equations describe a *linear relationship* between two quantities; when they are graphed, the result is a line. We can make observations about what has taken place in the past and what might take place in the future by carefully inspecting the graph.

EXAMPLE 5

U.S. labor statistics. The linear equation $p = 0.6t + 38$ models the percent of women 16 years or older who were part of the civilian labor force for each of the years 1960–2000. In the equation, t represents the number of years after 1960, and p represents the percent. Graph this equation.

Solution The variables t and p are used in the equation. We will associate t with the horizontal axis and p with the vertical axis. Ordered pairs will be of the form (t, p).

To graph the equation, we pick three values for t, substitute them into the equation, and find each corresponding value of p. The results are listed in the following table.

For $t = 0$	***For $t = 10$***	***For $t = 20$***
(The year 1960)	(The year 1970)	(The year 1980)
$p = 0.6t + 38$	$p = 0.6t + 38$	$p = 0.6t + 38$
$p = 0.6(0) + 38$	$p = 0.6(10) + 38$	$p = 0.6(20) + 38$
$p = 38$	$p = 6 + 38$	$p = 12 + 38$
	$p = 44$	$p = 50$

The pairs (0, 38), (10, 44), and (20, 50) satisfy the equation. Next, we plot these points and draw a line through them. From the graph, we see that there has been a steady increase in the percent of the female population 16 years or older that is part of the labor force.

$p = 0.6t + 38$

t	p	(t, p)
0	38	(0, 38)
10	44	(10, 44)
20	50	(20, 50)

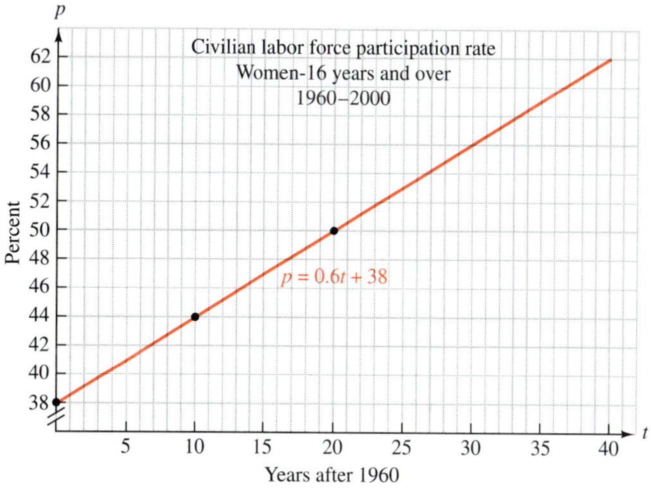

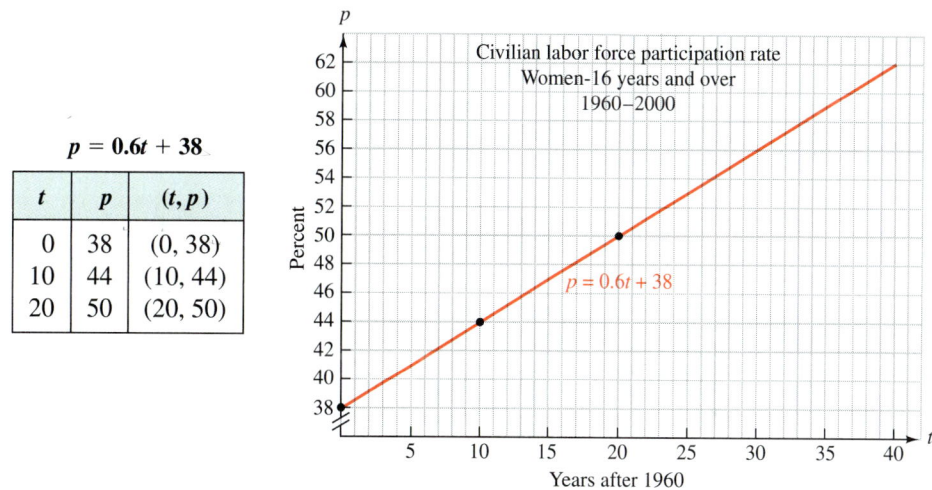

Source: Bureau of Labor Statistics

Self Check 5 **a.** Use the equation $p = 0.6t + 38$ to determine the percent of women 16 years or older who were part of the labor force in 1975. **b.** Use the graph to determine the percent of women who were part of the labor force in 2000.

EXAMPLE 6

Linear depreciation. A copy machine that was purchased for \$6,750 is expected to depreciate according to the formula $y = -950x + 6{,}750$, where y is the value of the copier after x years. When will the copier have no value?

Solution The copier will have no value when y is 0. To find x when $y = 0$, we substitute 0 for y and solve for x.

$$y = -950x + 6{,}750$$

$$0 = -950x + 6{,}750$$

$$-6{,}750 = -950x \quad \text{Subtract 6,750 from both sides.}$$

$$7.105263158 \approx x \quad \text{Divide both sides by } -950.$$

The copier will have no value in about 7.1 years.

The Language of Algebra

Depreciation is a form of the word *depreciate,* meaning to lose value. You've probably heard that the minute you drive a new car off the lot, it has depreciated.

The equation $y = -950x + 6{,}750$ is graphed below. Important information can be obtained from the intercepts of the graph.

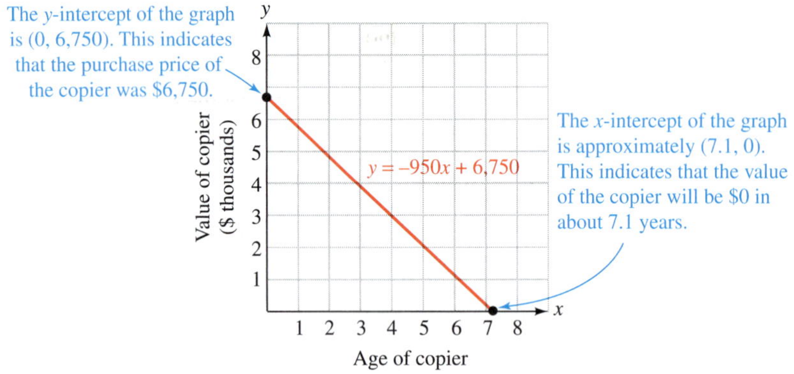
The y-intercept of the graph is (0, 6,750). This indicates that the purchase price of the copier was $6,750.
The x-intercept of the graph is approximately (7.1, 0). This indicates that the value of the copier will be $0 in about 7.1 years.
y = –950x + 6,750
Value of copier ($ thousands)
Age of copier
y
x
1 2 3 4 5 6 7 8
1 2 3 4 5 6 8

Self Check 6 **a.** Use the equation $y = -950x + 6{,}750$ to determine when the copier will be worth $3,900.
b. Use the graph in Example 6 to determine when the copier will be worth $2,000.

ACCENT ON TECHNOLOGY: GRAPHING LINES

We have graphed linear equations by finding solutions, plotting points, and drawing lines through those points. Graphing is often easier using a graphing calculator.

Window settings Graphing calculators have a window to display graphs. To see the proper picture of a graph, we must decide on the minimum and maximum values for the x- and y-coordinates. A window with standard settings of

$$\text{Xmin} = -10 \quad \text{Xmax} = 10 \quad \text{Ymin} = -10 \quad \text{Ymax} = 10$$

will produce a graph where the values of x and the values of y are between -10 and 10, inclusive. We can use the notation $[-10, 10]$ to describe such intervals.

Graphing lines To graph $5x - 2y = 4$, we must first solve the equation for y.

$$y = \frac{5}{2}x - 2 \quad \text{Subtract } 5x \text{ from both sides and then divide both sides by } -2.$$

Next, we press [Y =] and enter the right-hand side of the equation after the symbol $Y_1 =$. See figure (a). We then press the [GRAPH] key to get the graph shown in figure (b). To show more detail, we can draw the graph using window settings of $[-2, 5]$ for x and $[-4, 5]$ for y. See figure (c).

Plot1 Plot2 Plot3
\Y1=(5/2)X-2
\Y2=
\Y3=
\Y4=
\Y5=
\Y6=
\Y7=
(a)

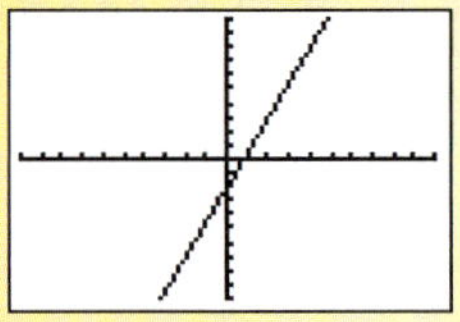
(b)

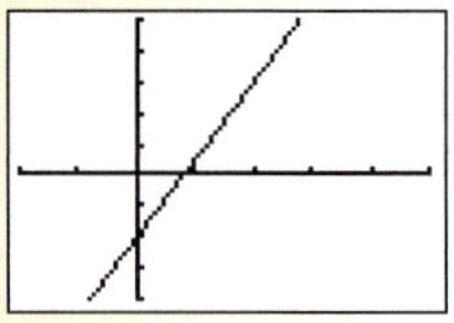
(c)

Finding the coordinates of a point on the graph

If we reenter the standard window settings of $[-10, 10]$ for x and for y, press GRAPH, and press the TRACE key, we get the display shown in figure (d). The y-intercept of the graph is highlighted by the flashing cursor, and the x- and y-coordinates of that point are given at the bottom of the screen. We can use the ▶ and ◀ keys to move the cursor along the line to find the coordinates of any point on the line. After pressing the ▶ key 12 times, we will get the display in figure (e).

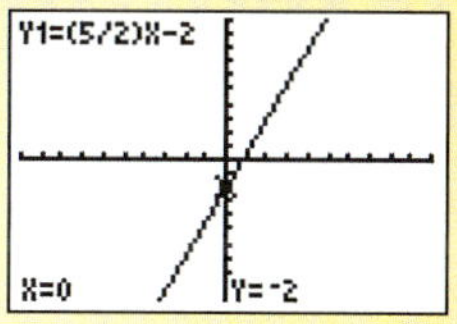

(d)

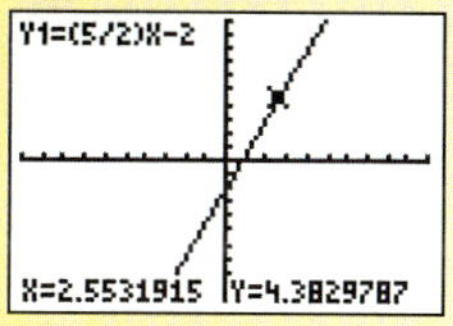

(e)

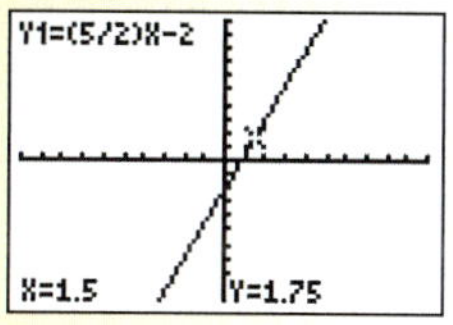

(f)

To find the y-coordinate of any point on the line, given its x-coordinate, we press 2nd CALC and select the *value* option. We enter the x-coordinate of the point and press ENTER. The y-coordinate is then displayed. In figure (f), 1.5 was entered for the x-coordinate, and its corresponding y-coordinate, 1.75, was found.

The *table* feature, discussed on pages 112–113, gives us a third way of finding the coordinates of a point on the line.

Determining the x-intercepts of a graph

To determine the x-intercept of the graph of $y = \frac{5}{2}x - 2$, we can use the *zero* option, found under the CALC menu. (Be sure to reenter the standard window settings for x and y before using CALC.) After we guess left and right bounds, as shown in figure (g), the cursor automatically moves to the x-intercept of the graph when we press ENTER. Figure (h) shows how the coordinates of the x-intercept are then displayed at the bottom of the screen.

We can also use the *trace* and *zoom* features to determine the x-intercept of the graph of $y = \frac{5}{2}x - 2$. After graphing the equation using the standard window settings, we press TRACE. Then we move the cursor along the line toward the x-intercept until we arrive at a point with the coordinates shown in figure (i). To get better results, we press ZOOM, select the *zoom in* option, and press ENTER to get a magnified picture. We press TRACE again and move the cursor to the point with coordinates shown in figure (j). Since the y-coordinate is nearly 0, this point is nearly the x-intercept. We can achieve better results with more zooms and traces.

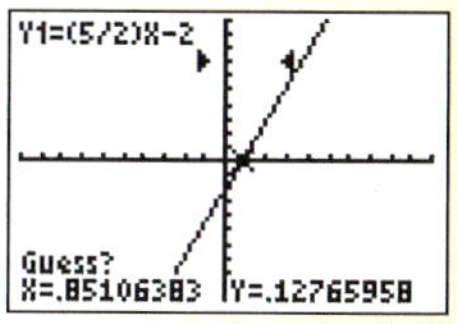

(g)

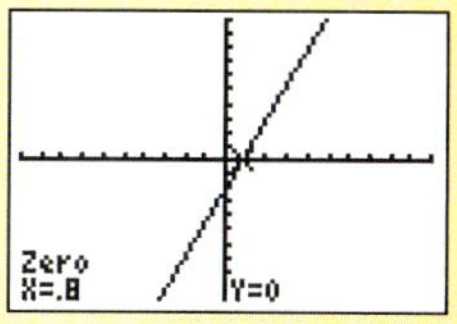

(h)

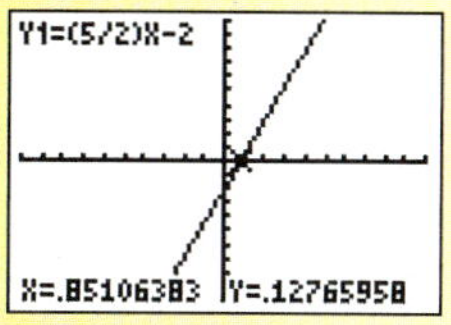

(i)

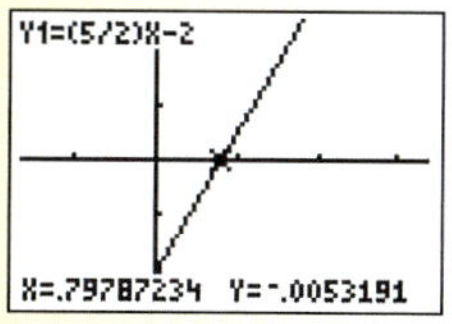

(j)

Answers to Self Checks **1.** no **2.**

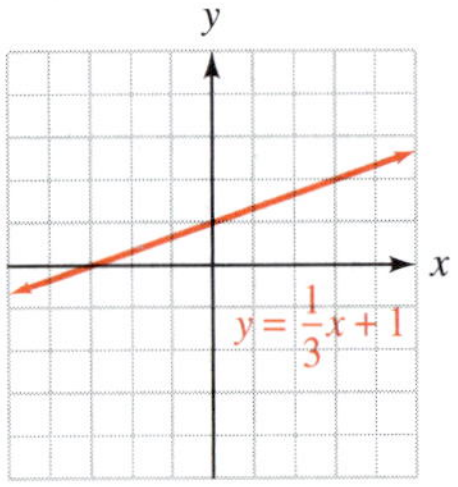

3.

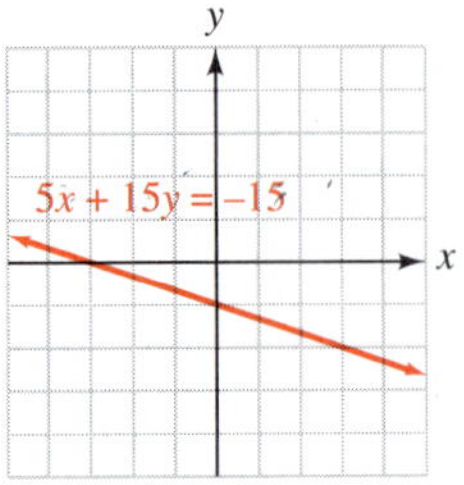

4.

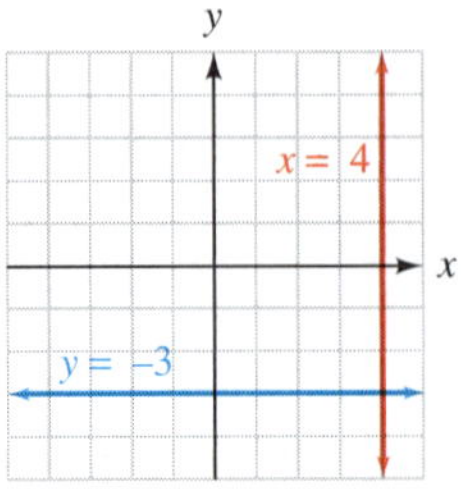

5. a. 47%, **b.** 62% **6. a.** 3 years, **b.** 5 years

2.2 STUDY SET

VOCABULARY Fill in the blanks.

1. A solution of an equation in two variables is an ________ ______ of numbers that make a true statement when substituted into the equation.
2. The graph of an equation is the graph of all points (x, y) on the rectangular coordinate system whose coordinates ________ the equation.
3. Any equation whose graph is a line is called a ________ equation.
4. The point where the graph of an equation intersects the y-axis is called the ____________, and the point where it intersects the x-axis is called the ____________.
5. The graph of any equation of the form $x = a$ is a ________ line.
6. The graph of any equation of the form $y = b$ is a __________ line.

CONCEPTS

7. Determine whether the given ordered pair is a solution of $y = -5x - 2$.
   a. $(-1, 3)$ **b.** $(3, -13)$
8. Determine whether the given ordered pair is a solution of $2x - 5y = 9$.
   a. $(-4, 2)$ **b.** $(2, -1)$
9. **a.** Consider the equation $2x + 4 = 8$, studied in Chapter 1. How many variables does it contain? How many solutions does it have?
   b. Consider $2x + 4y = 8$. How many variables does it contain? How many solutions does it have?
10. Consider the linear equation $6x - 4y = -12$.
    a. Find the x-intercept of its graph.
    b. Find the y-intercept of its graph.
    c. Does its graph pass through $(2, 6)$?
11. Fill in the blanks: The exponent on each variable of a linear equation is an understood ____. For example, $4x + 7y = 3$ can be thought of as $4x^{\_\_} + 7y^{\_\_} = 3$.
12. A table of solutions for a linear equation is given below. From the table, determine the x-intercept and the y-intercept of the graph of the equation.

x	y	(x, y)
−6	0	(−6, 0)
−4	1	(−4, 1)
−2	2	(−2, 2)
0	3	(0, 3)

13. Refer to the graph.

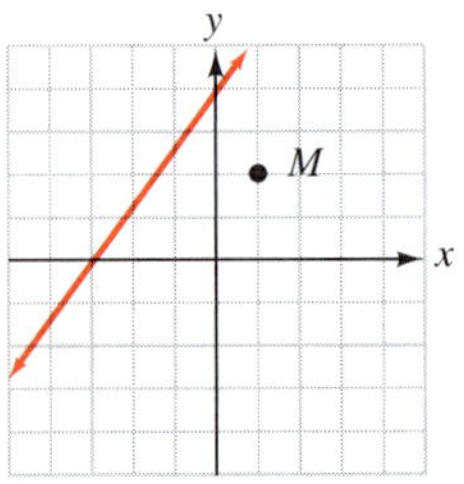

    a. What is the x-intercept and what is the y-intercept of the line?
    b. If the coordinates of point M are substituted into the equation of the line that is graphed here, will a true or a false statement result?

14. Use the graph to determine three solutions of $2x + 3y = 9$.

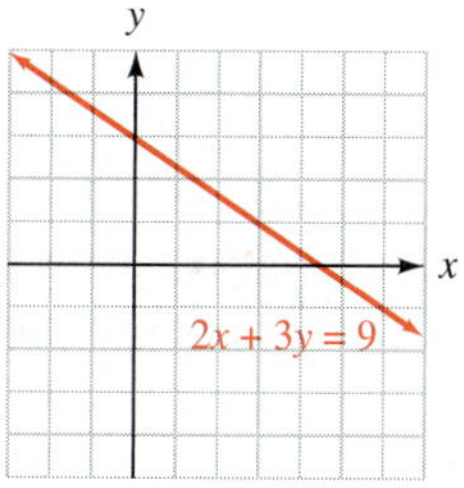

15. A graphing calculator display is shown below. It is a table of solutions for which one of the following linear equations?

$$y = -2x - 1, \quad y = -3x - 1, \quad \text{or} \quad y = -4x - 1$$

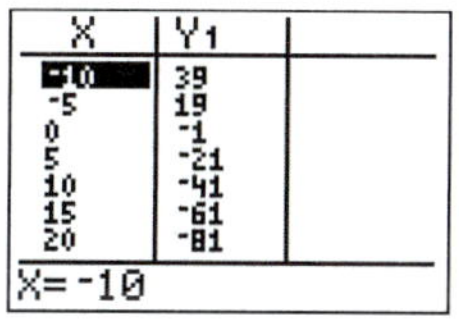

X	Y1
-10	39
-5	19
0	-1
5	-21
10	-41
15	-61
20	-81

X=-10

16. The graphing calculator displays below show the graph of $y = -2x - \frac{5}{4}$.

a. In figure (a), what important feature of the line is highlighted by the cursor?

b. In figure (b), what important feature of the line is highlighted by the cursor?

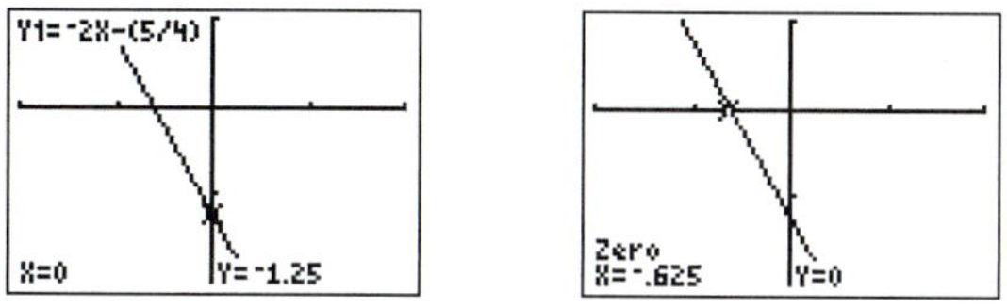

(a) (b)

NOTATION

17. a. The graph of the equation $x = 0$ is which axis?

b. The graph of the equation $y = 0$ is which axis?

18. A linear equation in two variables is an equation that can be written in the form $Ax + By = C$. For $x - 5y = 4$, determine A, B, and C.

PRACTICE **Complete each table of solutions.**

19. $y = -x + 4$

x	y
-1	
0	
2	

20. $y = x - 2$

x	y
-2	
0	
4	

21. $y = -\frac{1}{3}x - 1$

x	y
-3	
0	
3	

22. $y = -\frac{1}{2}x + \frac{5}{2}$

x	y
-1	
3	
5	

Use the results from Exercises 19–22 to graph each equation.

23. $y = -x + 4$

24. $y = x - 2$

25. $y = -\frac{1}{3}x - 1$

26. $y = -\frac{1}{2}x + \frac{5}{2}$

Graph each equation.

27. $y = x$

28. $y = -2x$

29. $y = -3x + 2$

30. $y = 2x - 3$

31. $y = 3 - x$

32. $y = 5 - x$

33. $y = \frac{x}{4} - 1$

34. $y = -\frac{x}{4} + 2$

35. $x = 3$

36. $y = -4$

Write each equation in $y = b$ or $x = a$ form. Then graph it.

37. $y - 2 = 0$

38. $x + 1 = 0$

39. $-3y + 2 = 5$

40. $-2x + 3 = 11$

Graph each equation using the intercept method. Label the intercepts on each graph.

41. $3x + 4y = 12$

42. $4x - 3y = 12$

43. $3y = 6x - 9$

44. $2x = 4y - 10$

45. $2y + x = -2$

46. $4y + 2x = -8$

47. $3x + 4y - 8 = 0$

48. $-2y - 3x + 9 = 0$

49. $3x = 4y - 11$

50. $-5x + 3y = 11$

Use a graphing calculator to graph each equation, and then find the x-coordinate of the x-intercept to the nearest hundredth.

51. $y = 3.7x - 4.5$

52. $y = \frac{3}{5}x + \frac{5}{4}$

53. $1.5x - 3y = 7$

54. $0.3x + y = 7.5$

APPLICATIONS

55. BUYING TICKETS Tickets to a circus cost \$10 each from Ticketron plus a \$2 service fee for each block of tickets purchased.

a. Write a linear equation that gives the cost c when t tickets are purchased.

b. Complete the table and graph the equation.

c. Use the graph to estimate the cost of buying 6 tickets.

t	c
1	
2	
3	
4	

56. TELEPHONE COSTS In a community, the monthly cost of local telephone service is \$5 per month, plus 25¢ per call.

a. Write a linear equation that gives the cost c for a person making n calls. Then graph the equation.

b. Complete the table.

c. Use the graph to estimate the cost of service in a month when 20 calls were made.

n	c
4	
8	
12	
16	

57. U.S. SPORTS PARTICIPATION The equation $s = -0.9t + 65.5$ gives the approximate number of people 7 years of age and older who went swimming during a given year, where s is the annual number of swimmers (in millions) and t is the number of years since 1990. Graph the equation. (Source: National Sporting Goods Association)

a. What information can be obtained from the s-intercept of the graph?

b. From the graph, estimate the number of swimmers in 2002.

58. FARMING The equation

$$a = -3{,}700{,}000t + 983{,}000{,}000$$

gives the approximate number of acres a of farmland in the United States, t years after 1990. Graph the equation. (Source: U.S. Department of Agriculture)

a. What information can be obtained from the a-intercept of the graph?

b. From the graph, estimate the number of acres of farmland in 1998.

59. LIVING LONGER According to the National Center for Health Statistics, life expectancy in the United States is increasing. The equation $y = 0.13t + 74$ is a linear model that approximates life expectancy; y is the number of years of life expected for a child born t years after 1980. Graph the equation.

a. What information can be obtained from the y-intercept of the graph?

b. From the graph, estimate the life expectancy for someone born in 1998.

60. LABOR The equation $p = 0.45t + 11$ gives the approximate average hourly pay p (in dollars) of a U.S. production worker, t years after 1994. Graph the equation. (Source: U.S. Department of Labor)

a. What information can be obtained from the p-intercept of the graph?

b. From the graph, estimate the average hourly pay for production workers in 2002.

61. DEPRECIATION The graph shows how the value of a computer decreased over the age of the computer. What information can be obtained from the x-intercept? The y-intercept?

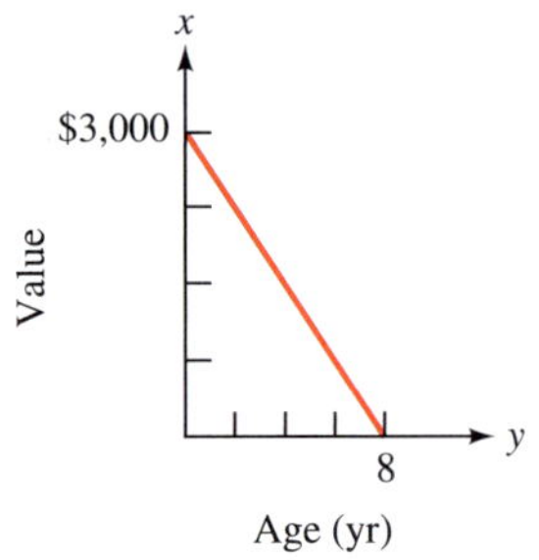

62. CAR DEPRECIATION A car purchased for \$17,000 is expected to depreciate (lose value) according to the formula $y = -1{,}360x + 17{,}000$. When will the car have no value?

63. DEMAND EQUATION The number of television sets that consumers buy depends on price. The higher the price, the fewer TVs people will buy. The equation that relates price to the number of TVs sold at that price is called a **demand equation.** If the demand equation for a 25-inch TV is $p = -\frac{1}{10}q + 170$, where p is the price and q is the number of TVs sold at that price, how many TVs will be sold at a price of \$150?

64. SUPPLY EQUATION The number of television sets that manufacturers produce depends on price. The higher the price, the more TVs manufacturers will produce. The equation that relates price to the number of TVs produced at that price is called a **supply equation.** If the supply equation for a 25-inch TV is $p = \frac{1}{10}q + 130$, where p is the price and q is the number of TVs produced for sale at that price, how many TVs will be produced if the price is \$150?

WRITING

65. Explain how to graph a line using the intercept method.

66. When graphing a line by plotting points, why is it a good practice to find three solutions instead of two?

REVIEW

67. List the prime numbers between 10 and 30.

68. Write the first ten composite numbers.

69. In what quadrant does the point $(-2, -3)$ lie?

70. What is the formula that gives the area of a circle?

71. Simplify: $-4(-20s)$.

72. Approximate π to the nearest thousandth.

73. Remove parentheses: $-(-3x - 8)$.

74. Simplify: $\frac{1}{3}b + \frac{1}{3}b + \frac{1}{3}b$.

CHALLENGE PROBLEMS Graph each equation.

75. $0.2x + 0.3y = 6$

76. $\frac{x}{2} - \frac{y}{3} - 4 = 0$

2.3 Rate of Change and the Slope of a Line

- Average rate of change
- Slope of a line
- Applications of slope
- Horizontal and vertical lines
- Slopes of parallel lines
- Slopes of perpendicular lines

Our world is one of constant change. In this section, we will show how to describe the amount of change in one quantity in relation to the amount of change in another by finding an *average rate of change.*

AVERAGE RATE OF CHANGE

The following line graphs model the approximate number of morning and evening newspapers published in the United States for the years 1990–1999. We see that the number of morning newspapers increased and the number of evening newspapers decreased over this time span.

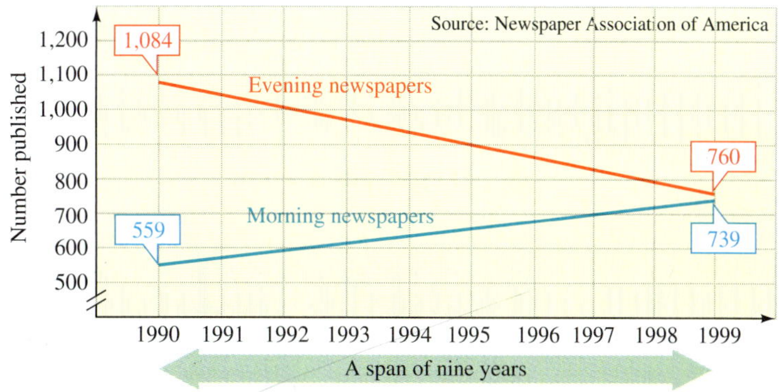

If we want to know the rate at which the number of morning newspapers increased or the rate at which the number of evening newspapers decreased, we can do so by finding an **average rate of change.** To find an average rate of change, we find the *ratio* of the change in the number of newspapers to the length of time in which that change took place.

Ratios and Rates

A **ratio** is a comparison of two numbers by their indicated quotient. In symbols, if a and b are two numbers, the ratio of a to b is $\frac{a}{b}$. Ratios that are used to compare quantities with different units are called **rates.**

In the previous figure, we see that in 1990, the number of morning newspapers published was 559. In 1999, the number grew to 739. This is a change of 739 − 559 or 180 over a 9-year time span. So we have

$$\text{Average rate of change} = \frac{\text{change in number of morning newspapers}}{\text{change in time}}$$

The rate of change is a ratio that includes units.

$$= \frac{180 \text{ newspapers}}{9 \text{ years}}$$

$$= \frac{\overset{1}{\cancel{9}} \cdot 20 \text{ newspapers}}{\underset{1}{\cancel{9}} \text{ years}}$$

Factor 180 as 9 · 20 and simplify: $\frac{9}{9} = 1$.

$$= \frac{20 \text{ newspapers}}{1 \text{ year}}$$

The number of morning newspapers published in the United States increased, on average, at a rate of 20 newspapers per year (written 20 newspapers/year) from 1990 through 1999.

Success Tip

In general, to find the change in a quantity, we subtract the earlier value from the later value.

In the previous graph, we see that in 1990 the number of evening newspapers published was 1,084. In 1999, the number fell to 760. To find the change, we subtract: 760 − 1,084 = −324. The negative result indicates a decline in the number of evening newspapers over the 9-year time span. So we have

$$\text{Average rate of change} = \frac{-324 \text{ newspapers}}{9 \text{ years}}$$

$$= \frac{-36 \cdot \overset{1}{\cancel{9}} \text{ newspapers}}{\underset{1}{\cancel{9}} \text{ years}}$$

Factor −324 as −36 · 9 and simplify: $\frac{9}{9} = 1$.

$$= \frac{-36 \text{ newspapers}}{1 \text{ year}}$$

The number of evening newspapers changed at a rate of −36 newspapers/year. That is, on average, there were 36 fewer evening newspapers per year, every year, from 1990 through 1999.

The Language of Algebra

m is used to denote the slope of a line. Historians credit this to the fact that it is the first letter of the French word *monter,* meaning to ascend or to climb.

SLOPE OF A LINE

In the newspaper example, we measured the steepness of the two lines in a graph to determine the average rates of change. In doing so, we found the **slope** of each line. The slope of a nonvertical line is a number that measures the line's steepness.

To calculate the slope of a line (usually denoted by the letter m), we first pick two points on the line. To distinguish between the coordinates of the points, we use **subscript**

notation. The first point can be denoted as (x_1, y_1), and the second point as (x_2, y_2). After picking two points on the line, we write the ratio of the vertical change to the corresponding horizontal change as we move from one point to the other.

Slope of a Line

The **slope** of a line passing through points (x_1, y_1) and (x_2, y_2) is

$$m = \frac{\text{change in } y}{\text{change in } x} = \frac{y_2 - y_1}{x_2 - x_1} \quad \text{where } x_2 \neq x_1$$

EXAMPLE 1

Find the slope of the line passing through $(-2, 4)$ and $(3, -4)$.

Solution We can let $(x_1, y_1) = (-2, 4)$ and $(x_2, y_2) = (3, -4)$. Then we have

$$m = \frac{\text{change in } y}{\text{change in } x}$$

$$= \frac{y_2 - y_1}{x_2 - x_1} \quad \text{This is the slope formula.}$$

$$= \frac{-4 - 4}{3 - (-2)} \quad \text{Substitute } -4 \text{ for } y_2, 4 \text{ for } y_1, 3 \text{ for } x_2, \text{ and } -2 \text{ for } x_1.$$

$$= \frac{-8}{5}$$

$$= -\frac{8}{5}$$

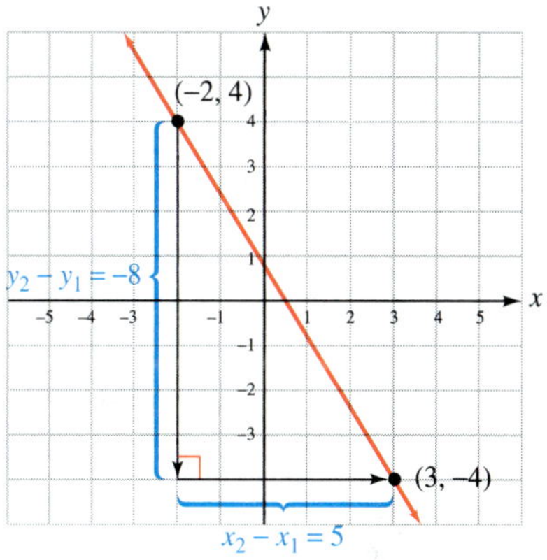

Notation

In the figure, the symbol ⌝ denotes a right angle.

The slope of the line is $-\frac{8}{5}$.

Self Check 1 Find the slope of the line passing through $(-3, 6)$ and $(4, -8)$.

When calculating slope, it doesn't matter which point we call (x_1, y_1) and which point we call (x_2, y_2). We will obtain the same result in Example 1 if we let $(x_1, y_1) = (3, -4)$ and $(x_2, y_2) = (-2, 4)$.

$$m = \frac{y_2 - y_1}{x_2 - x_1} = \frac{4 - (-4)}{-2 - 3} = \frac{8}{-5} = -\frac{8}{5}$$

Caution When using the slope formula, we must be careful to subtract the y-coordinates and the x-coordinates in the same order. For instance, in Example 1 with $(x_1, y_1) = (-2, 4)$ and $(x_2, y_2) = (3, -4)$, it would be incorrect to write

This is $y_2 - y_1$. The subtraction is not in the same order.

$$m = \frac{-4 - 4}{-2 - 3} \quad \text{(crossed out)}$$

This is $x_1 - x_2$.

This is $y_1 - y_2$. The subtraction is not in the same order.

$$m = \frac{4 - (-4)}{3 - (-2)} \quad \text{(crossed out)}$$

This is $x_2 - x_1$.

The Language of Algebra
The symbol Δ is the letter *delta* from the Greek alphabet.

The change in y (denoted as Δy and read as "delta y") is the **rise** of the line between two points on the line. The change in x (denoted as Δx and read as "delta x") is the **run.** Using this terminology, we can define slope as the ratio of the rise to the run:

$$m = \frac{\Delta y}{\Delta x} = \frac{\text{rise}}{\text{run}} \quad \text{where } \Delta x \neq 0$$

EXAMPLE 2

Find the slope of the line on the following graph.

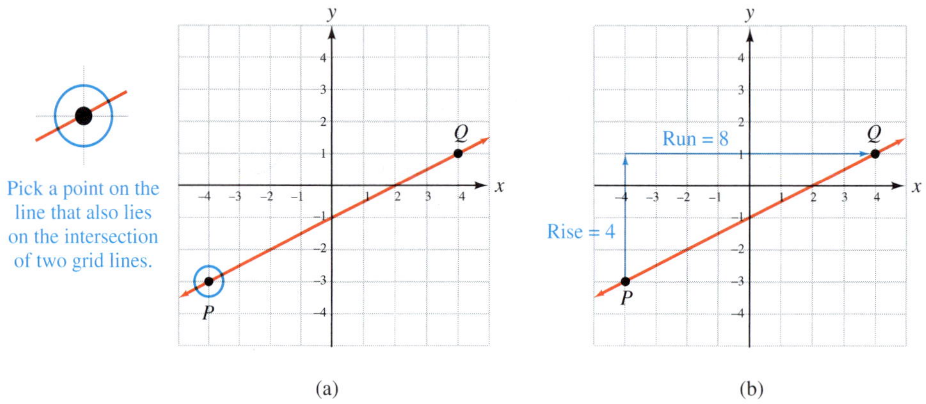

Solution We begin by choosing two points on the line, P and Q, as shown in illustration (a). One way to move from P to Q is shown in illustration (b). Starting at P, we move upward, a rise of 4, and then to the right, a run of 8, to reach Q. These steps create a right triangle called a **slope triangle.**

Success Tip
When drawing a slope triangle, remember that upward movements are positive, downward movements are negative, movements to the right are positive, and movements to the left are negative.

$$m = \frac{\text{rise}}{\text{run}} = \frac{4}{8} = \frac{1}{2} \quad \text{Simplify the fraction.}$$

The slope of the line is $\frac{1}{2}$.

The two-step process to move from P to Q can be reversed. Starting at P, we can move to the right, a run of 8; and then upward, a rise of 4, to reach Q. With this approach, the slope triangle is below the line. When we form the ratio to find the slope, we get the same result as before:

$$m = \frac{\text{rise}}{\text{run}} = \frac{4}{8} = \frac{1}{2}$$

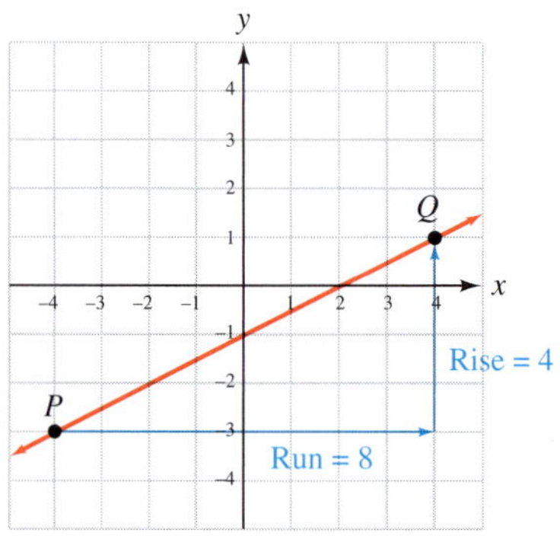

Self Check 2 Find the slope of the line shown above using two points different from those used in the solution of Example 2.

The identical answers from Example 2 and its Self Check illustrate an important fact: the same value will be obtained no matter which two points on a line are used to find the slope.

APPLICATIONS OF SLOPE

The concept of slope has many applications. For example, architects use slope when designing ramps and determining the *pitch* of roofs. Truckers must be aware of the slope, or *grade,* of the roads. Mountain resorts rate the difficulty level of ski runs by the degree of steepness.

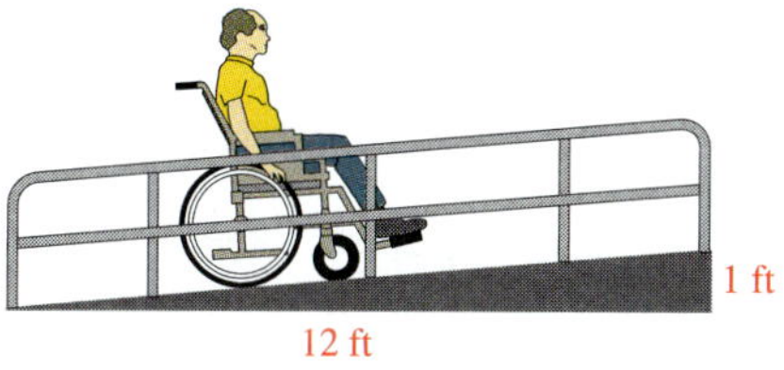

The maximum slope for a wheelchair ramp is 1 foot of rise for every 12 feet of run: $m = \frac{1}{12}$.

A 6% **grade** means a vertical change of 6 feet for every horizontal change of 100 feet: $m = \frac{6}{100}$.

EXAMPLE 3

Building stairs. The slope of a staircase is defined to be the ratio of the total rise to the total run, as shown in the illustration. What is the slope of the staircase?

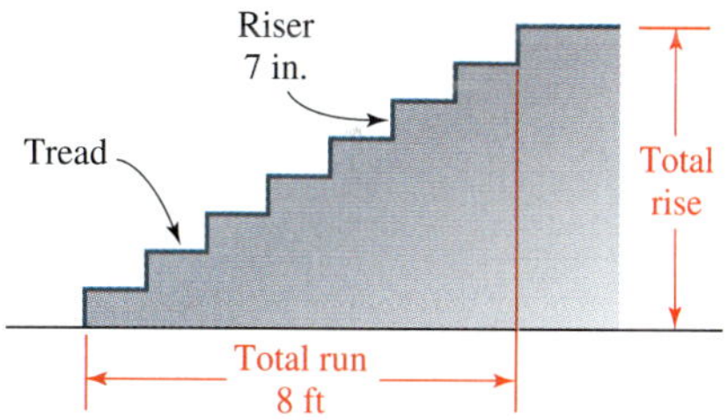

Solution Since the design has eight 7-inch risers, the total rise is $8 \cdot 7 = 56$ inches. The total run is 8 feet, or 96 inches. With these quantities expressed in the same units, we can now form their ratio.

$$m = \frac{\text{total rise}}{\text{total run}}$$

$$= \frac{56}{96}$$

$$= \frac{7}{12} \qquad \text{Simplify the fraction: } \frac{56}{96} = \frac{\overset{1}{\cancel{8}} \cdot 7}{\underset{1}{\cancel{8}} \cdot 12} = \frac{7}{12}.$$

The slope of the staircase is $\frac{7}{12}$.

Self Check 3 Find the slope of the staircase if the riser height is changed to 6.5 inches.

EXAMPLE 4

Rate of descent. It takes a skier 25 minutes to complete the course shown in the illustration. Find her average rate of descent in feet per minute.

Solution We can write the information about the skier's position as ordered pairs of the form (time, elevation). To find the average rate of descent, we must find the ratio of the change in elevation ΔE to the change in time Δt. To find this ratio, we calculate the slope of the line passing through the points (0, 12,000) and (25, 8,500).

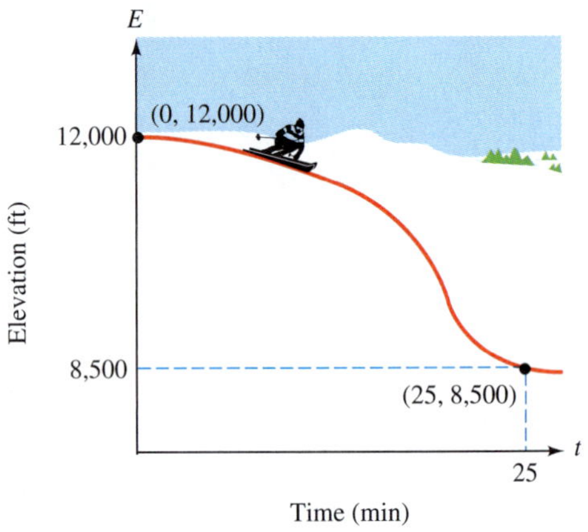

$$\text{Average rate of descent} = \frac{\Delta E}{\Delta t}$$

$$= \frac{8,500 - 12,000}{25 - 0}$$ In the numerator, write the change in altitude; in the denominator, the change in time.

$$= \frac{-3,500}{25}$$ Perform the subtractions.

$$= -140$$ Simplify.

The average rate of descent is 140 feet per minute.

Self Check 4 Find the average rate of descent if the skier completes the course in 20 minutes.

HORIZONTAL AND VERTICAL LINES

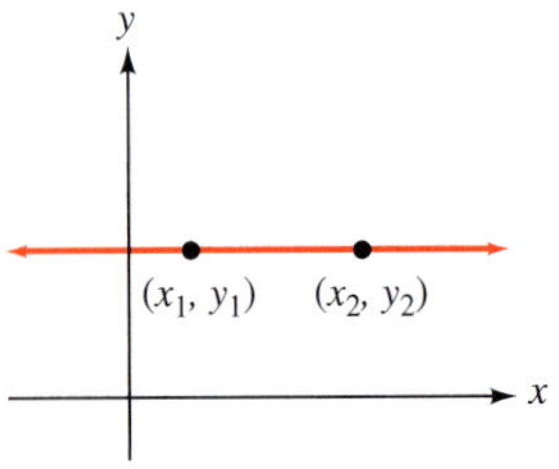

If (x_1, y_1) and (x_2, y_2) are distinct points on the horizontal line to the left, then $y_1 = y_2$, and the numerator of the fraction

$$\frac{y_2 - y_1}{x_2 - x_1}$$ On a horizontal line, $x_2 \neq x_1$.

is 0. Thus, the value of the fraction is 0, and the slope of the horizontal line is 0.

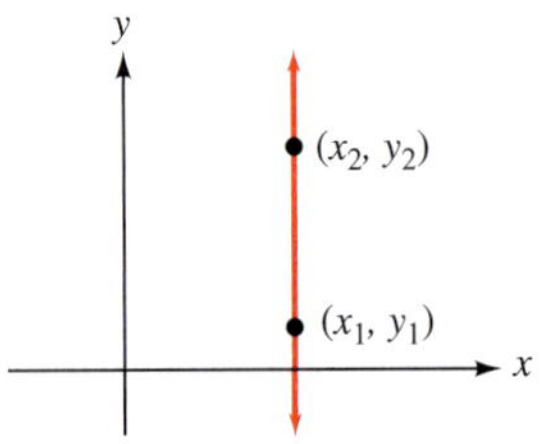

If (x_1, y_1) and (x_2, y_2) are distinct points on the vertical line to the left, then $x_1 = x_2$, and the denominator of the fraction

$$\frac{y_2 - y_1}{x_2 - x_1}$$ On a vertical line, $y_2 \neq y_1$.

is 0. Since the denominator of a fraction cannot be 0, a vertical line has no defined slope.

Slopes of Horizontal and Vertical Lines

Horizontal lines (lines with equations of the form $y = b$) have a slope of 0.

Vertical lines (lines with equations of the form $x = a$) have no defined slope.

If a line rises as we follow it from left to right, its slope is positive. If a line drops as we follow it from left to right, its slope is negative. If a line is horizontal, its slope is 0. If a line is vertical, it has undefined slope.

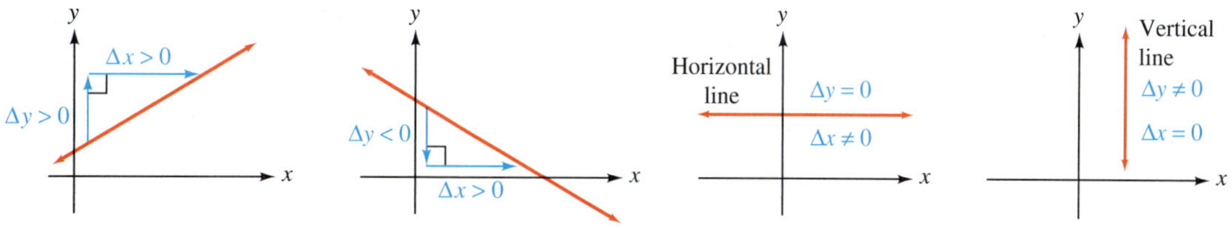

Positive slope   Negative slope   Zero slope   Undefined slope

SLOPES OF PARALLEL LINES

Caution

Note that zero slope and undefined slope do not mean the same thing.

To see a relationship between parallel lines and their slopes, we refer to the parallel lines l_1 and l_2 shown below, with slopes of m_1 and m_2, respectively. Because right triangles *ABC* and *DEF* are similar, it follows that

$$m_1 = \frac{\Delta y \text{ of } l_1}{\Delta x \text{ of } l_1} \qquad \text{Read } l_1 \text{ as "line } l \text{ sub 1."}$$

$$= \frac{\Delta y \text{ of } l_2}{\Delta x \text{ of } l_2} \qquad \text{Since the triangles are similar, corresponding sides of } \Delta ABC \text{ and } \Delta DEF \text{ are proportional: } \frac{CB}{BA} = \frac{FE}{ED}.$$

$$= m_2$$

Thus, if two nonvertical lines are parallel, they have the same slope. It is also true that when two lines have the same slope, they are parallel.

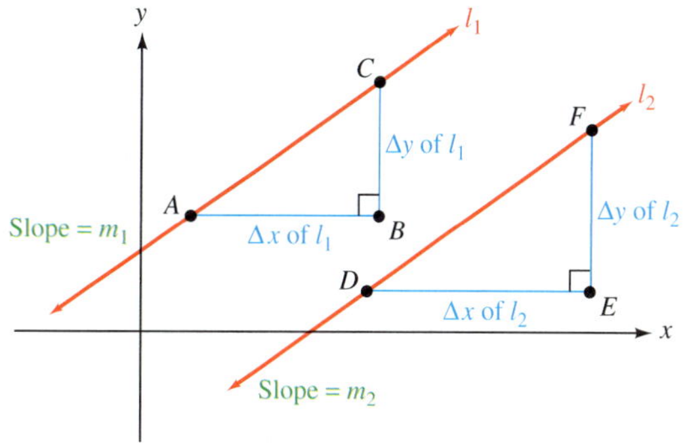

Slopes of Parallel Lines Nonvertical parallel lines have the same slope, and different lines having the same slope are parallel.

EXAMPLE 5

Slopes of parallel lines. Determine whether the line that passes through the points $(-6, 2)$ and $(3, -1)$ is parallel to a line with a slope $-\frac{1}{3}$.

Solution We can use the slope formula to find the slope of the line that passes through $(-6, 2)$ and $(3, -1)$.

$$m = \frac{y_2 - y_1}{x_2 - x_1}$$

$$m = \frac{-1 - 2}{3 - (-6)}$$ Substitute -1 for y_2, 2 for y_1, 3 for x_2, and -6 for x_1.

$$= \frac{-3}{9}$$

$$= -\frac{1}{3}$$

Both lines have slope $-\frac{1}{3}$, and therefore they are parallel.

Self Check 5 Determine whether the line that passes through the points $(4, -8)$ and $(1, -2)$ is parallel to a line with slope 2.

SLOPES OF PERPENDICULAR LINES

The two lines shown in the figure meet at right angles and are called **perpendicular lines.** Each of the four angles that are formed has a measure of 90°.

The product of the slopes of two (nonvertical) perpendicular lines is -1. For example, the perpendicular lines shown in the figure have slopes of $\frac{3}{2}$ and $-\frac{2}{3}$. If we find the product of their slopes, we have

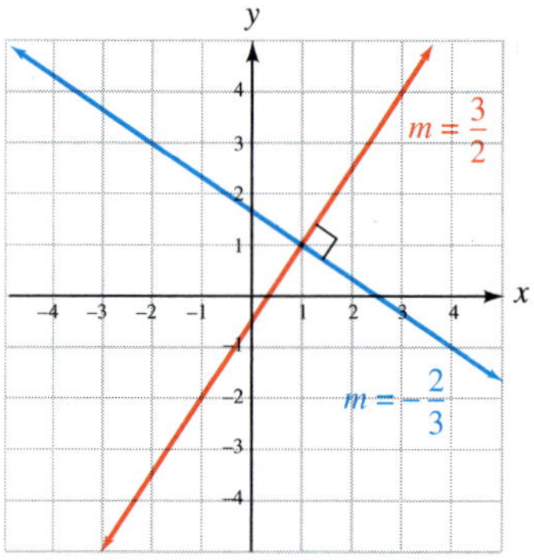

$$\frac{3}{2}\left(-\frac{2}{3}\right) = -\frac{6}{6} = -1$$

Two numbers whose product is -1, such as $\frac{3}{2}$ and $-\frac{2}{3}$, are called **negative reciprocals.** The term *negative reciprocal* can be used to relate perpendicular lines and their slopes.

Slopes of Perpendicular Lines

If two nonvertical lines are perpendicular, their slopes are negative reciprocals.

If the slopes of two lines are negative reciprocals, the lines are perpendicular.

We can also state the fact given above symbolically: If the slopes of two nonvertical lines are m_1 and m_2, then the lines are perpendicular if

$$m_1 \cdot m_2 = -1 \quad \text{or} \quad m_2 = -\frac{1}{m_1}$$

Because a horizontal line is perpendicular to a vertical line, a line with a slope of 0 is perpendicular to a line with no defined slope.

EXAMPLE 6

Slopes of perpendicular lines. Are the lines l_1 and l_2 shown in the figure perpendicular?

Solution We find the slopes of the lines and see whether they are negative reciprocals.

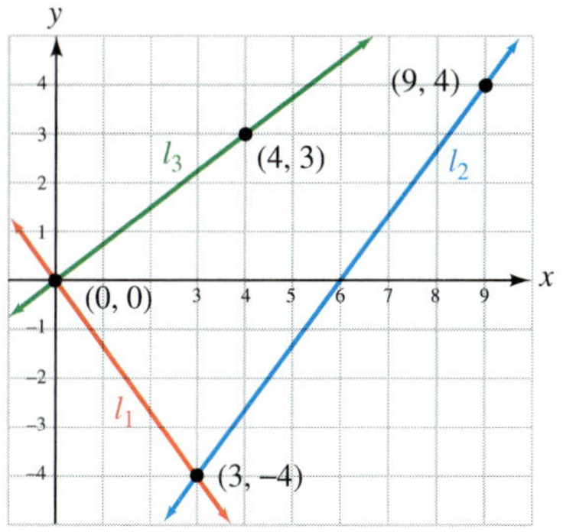

$$m_1 = \frac{y_2 - y_1}{x_2 - x_1} \quad \text{This is the slope of } l_1.$$
$$= \frac{-4 - 0}{3 - 0}$$
$$= -\frac{4}{3}$$

$$m_2 = \frac{y_2 - y_1}{x_2 - x_1} \quad \text{This is the slope of } l_2.$$
$$= \frac{4 - (-4)}{9 - 3}$$
$$= \frac{8}{6}$$
$$= \frac{4}{3}$$

Since their slopes are not negative reciprocals $\left(-\frac{4}{3} \cdot \frac{4}{3} \neq -1\right)$, the lines are not perpendicular.

Self Check 6 Is l_1 perpendicular to l_3?

Answers to Self Checks **1.** -2 **2.** $\frac{1}{2}$ **3.** $\frac{13}{24}$ **4.** 175 feet per minute **5.** They are not parallel. **6.** yes

2.3 STUDY SET

VOCABULARY Fill in the blanks.

1. ______ is defined as the change in y divided by the change in x.

2. A slope is an average ______ of change.

3. The ________ in x (denoted as Δx) is the horizontal run of the line between two points on the line.

4. The change in y (denoted as Δy) is the vertical ______ of the line between two points on the line.

5. $\frac{7}{8}$ and $-\frac{8}{7}$ are negative __________.

6. ________ lines have the same slope. The slopes of ___________ lines are negative reciprocals.

CONCEPTS

7. Refer to the graph.

a. Which line is horizontal? What is its slope?

b. Which line is vertical? What is its slope?

c. Which line has a positive slope? What is it?

d. Which line has a negative slope? What is it?

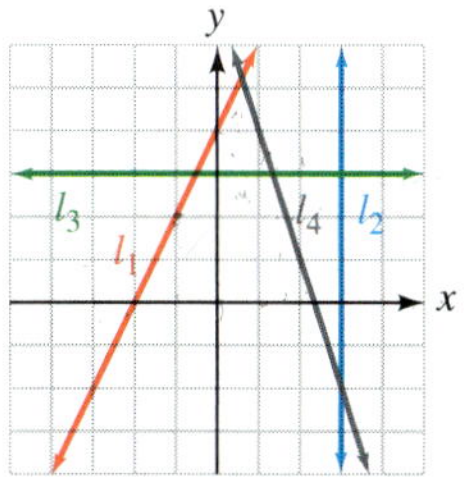

8. Refer to the graph.

a. Find the slopes of lines l_1 and l_2. Are they parallel?

b. Find the slopes of lines l_2 and l_3. Are they perpendicular?

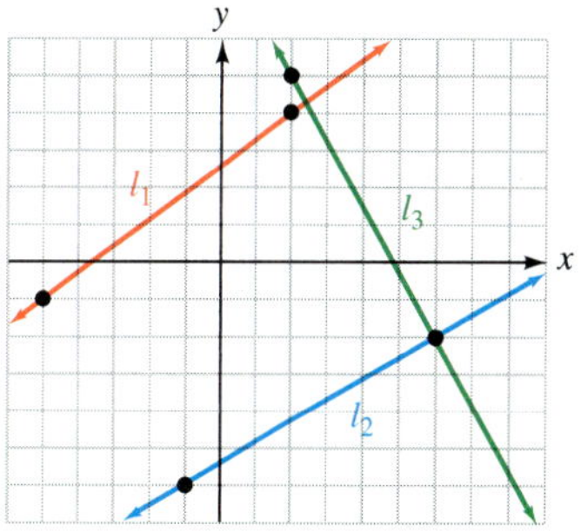

9. THE RECORDING INDUSTRY The graphs on the next page model the approximate number of CDs and cassettes that were shipped for sale from 1990 through 1999.

a. What was the rate of increase in the number of CDs shipped?

b. What was the rate of decrease in the number of cassettes shipped?

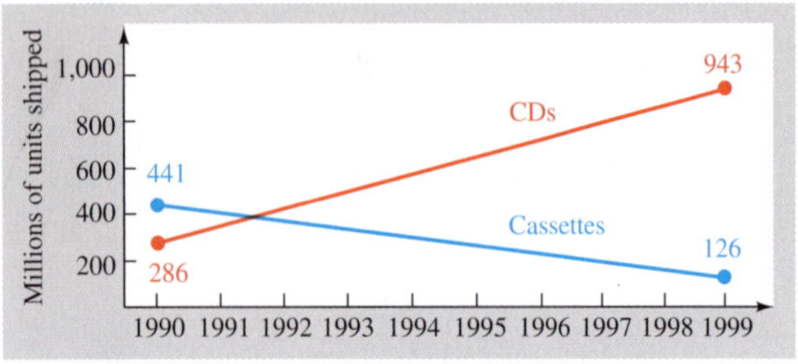

Source: *Statistical Abstract of the United States* (2003)

10. HALLOWEEN A couple kept records of the number of trick-or-treaters who came to their door on Halloween night. (See the graph.) Find the rate of change in the number of trick-or-treaters.

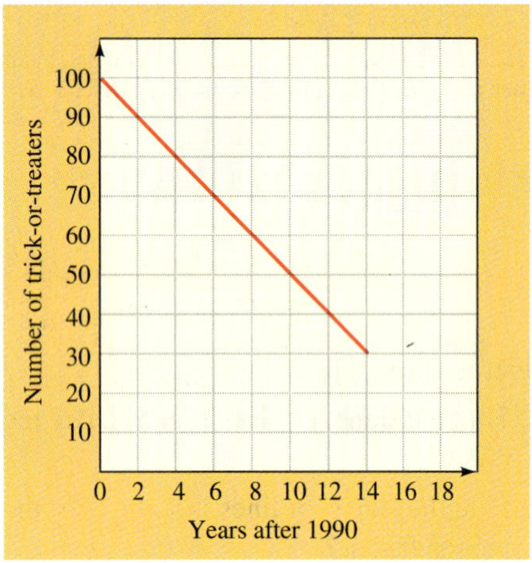

11. Determine the slope of each line.

a.

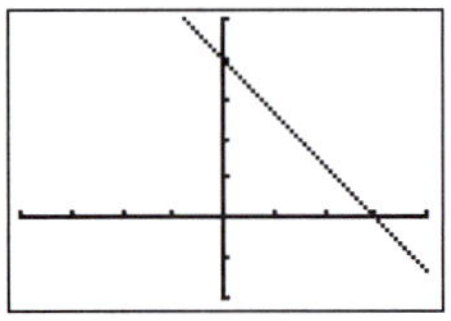

b.

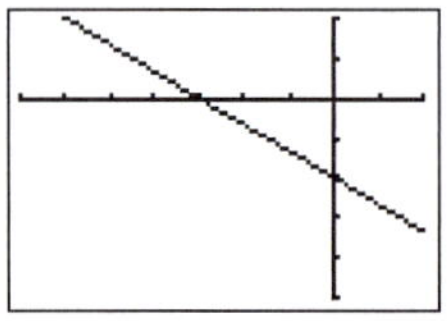

12. A table of solutions for a linear equation is shown here. Find the slope of the graph of the equation.

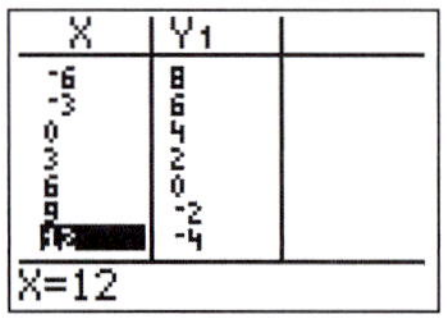

NOTATION

13. What formula is used to find the slope of a line?

14. Explain the difference between x^2 and x_2.

15. Refer to the graph.

a. What is Δy?

b. What is Δx?

c. What is $\frac{\Delta y}{\Delta x}$?

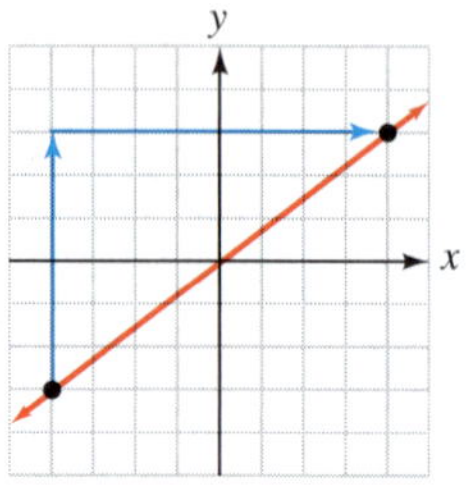

16. Refer to the graph.

a. What is the rise?

b. What is the run?

c. What is $\frac{\text{rise}}{\text{run}}$?

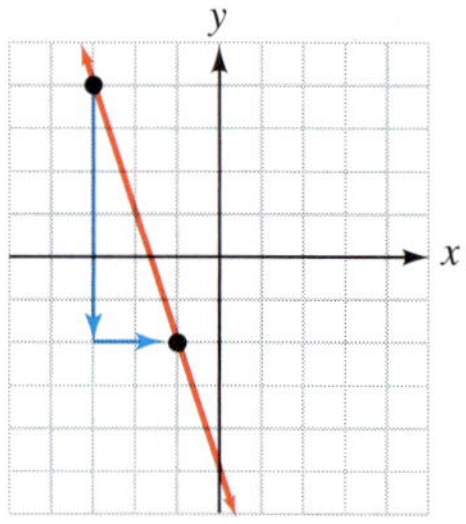

PRACTICE Find the slope of each line.

17.

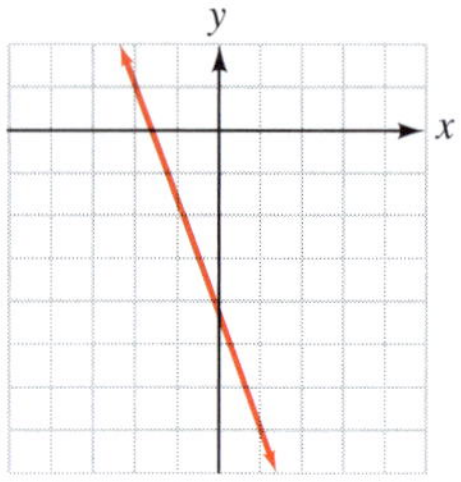

18.

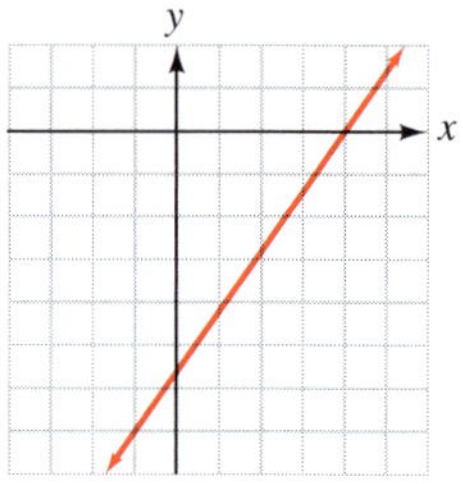

19.

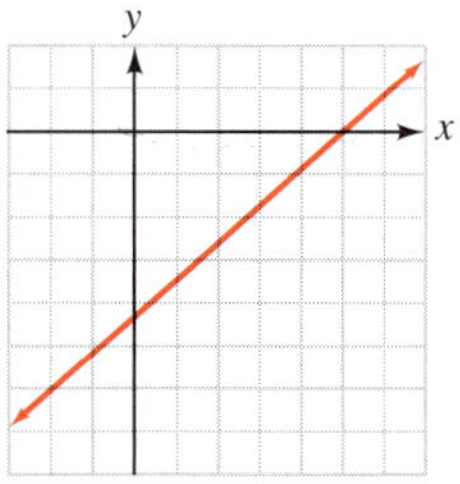

20.

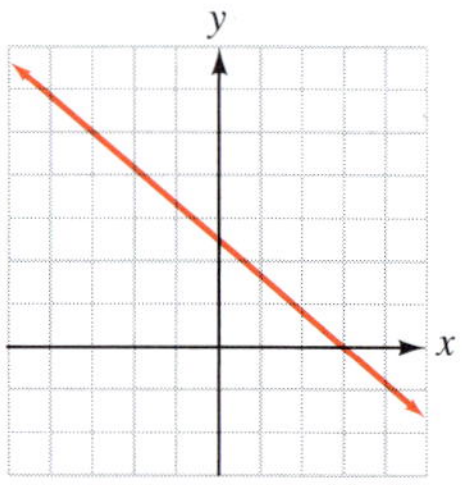

Find the slope of the line that passes through the given points, if possible.

21. (0, 0), (3, 9)

22. (9, 6), (0, 0)

23. (−1, 8), (6, 1)

24. (−5, −8), (3, 8)

25. (3, −1), (−6, 2)

26. (0, −8), (−5, 0)

27. (7, 5), (−9, 5)

28. (2, −8), (3, −8)

29. $(-7, -5), (-7, -2)$

30. $(3, -5), (3, 14)$

31. $\left(\frac{1}{4}, \frac{1}{2}\right), \left(-\frac{3}{4}, 0\right)$

32. $\left(\frac{1}{8}, \frac{3}{4}\right), \left(\frac{3}{8}, -\frac{1}{4}\right)$

33. $(0.7, -0.6), (-0.9, 0.2)$

34. $(-1.2, 8.6), (-1.1, 7.6)$

35. $(a, b), (b, a)$

36. $(a, b), (-b, -a)$

Determine whether the lines with the given slopes are parallel, perpendicular, or neither.

37. $m_1 = 3, m_2 = -\frac{1}{3}$

38. $m_1 = \frac{1}{4}, m_2 = 4$

39. $m_1 = 4, m_2 = 0.25$

40. $m_1 = -5, m_2 = -\frac{1}{0.2}$

41. $m_1 = \frac{1}{a}, m_2 = a$

42. $m_1 = a, m_2 = -\frac{1}{a}$

Determine whether the line that passes through the two given points is parallel or perpendicular (or neither) to a line with a slope of -2.

43. $(3, 4), (4, 2)$

44. $(6, 4), (8, 5)$

45. $(-2, 1), (6, 5)$

46. $(3, 4), (-3, -5)$

47. $(5, 4), (6, 6)$

48. $(-2, 3), (4, -9)$

APPLICATIONS

49. LANDING PLANES A jet descends in a stairstep pattern, as shown in the illustration in the next column. The required elevations of the plane's path are given. Find the slope of the descent in each of the three parts of its landing that are labeled. Which part is the steepest?

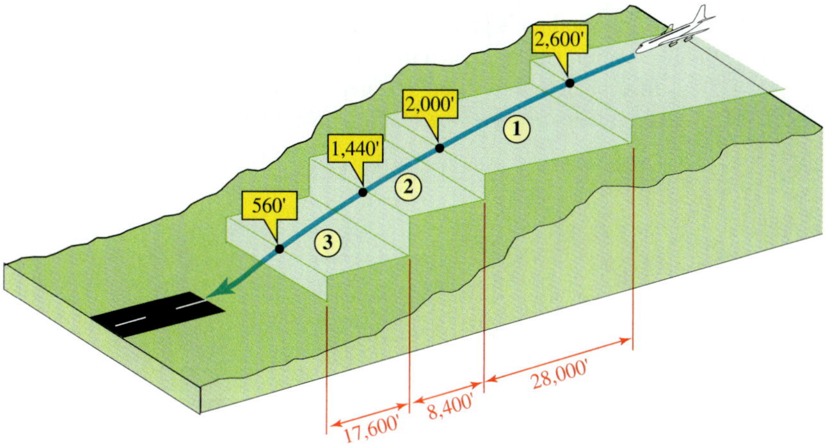

Based on data from *Los Angeles Times* (August 7, 1997), p. A8

50. COMPUTERS The price of computers has been dropping for the past ten years. If a desktop PC cost \$5,700 ten years ago, and the same computing power cost \$400 two years ago, find the rate of decrease per year. (Assume a straight-line model.)

51. MAPS Topographic maps have contour lines that connect points of equal elevation on a mountain. The vertical distance between contour lines in the illustration is 50 feet. Find the slope of the west face and the slope of the east face of the mountain peak.

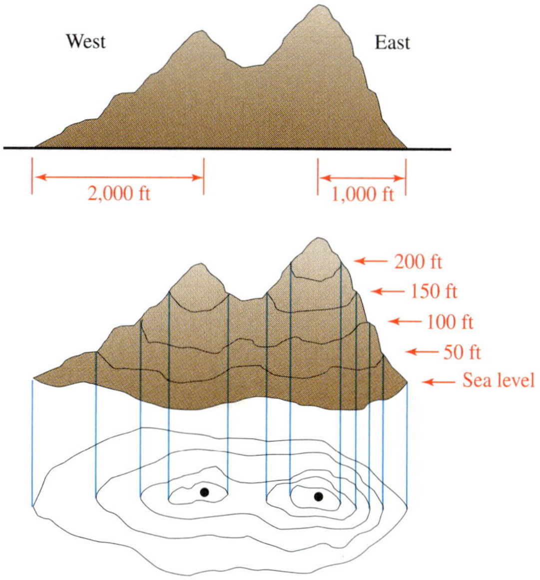

52. SKIING The men's giant slalom course shown in the illustration is longer than the women's course. Does this mean that the men's course is steeper? Use the concept of the slope of a line to explain.

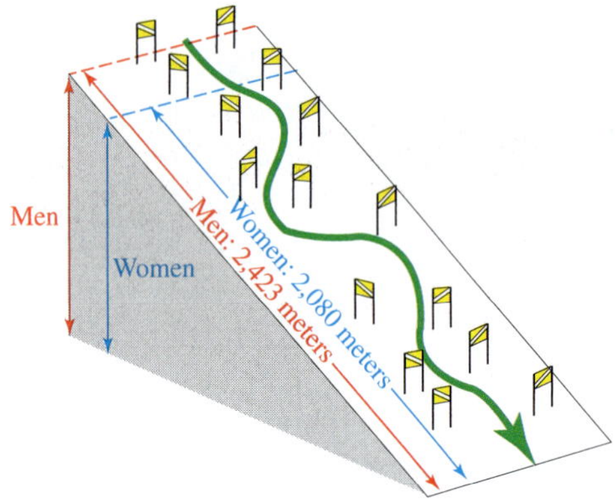

53. STEEP GRADES Find the grade of the road shown in the illustration. (*Hint:* 1 mi = 5,280 ft.)

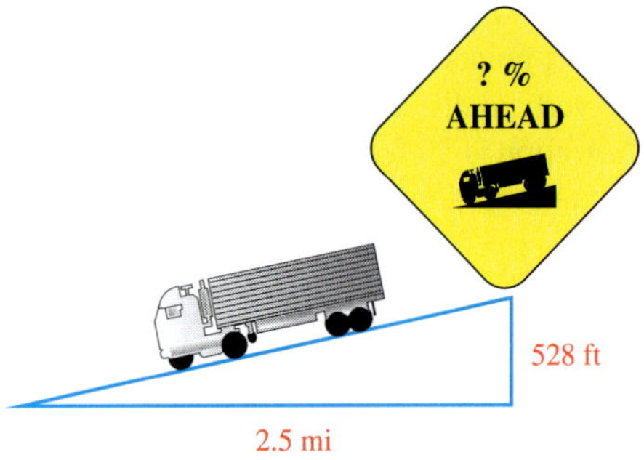

54. GREENHOUSE EFFECT The graphs below are estimates of future average global temperature rise due to the greenhouse effect. Assume that the models are straight lines. Estimate the average rate of change of each model. Express your answers as fractions.

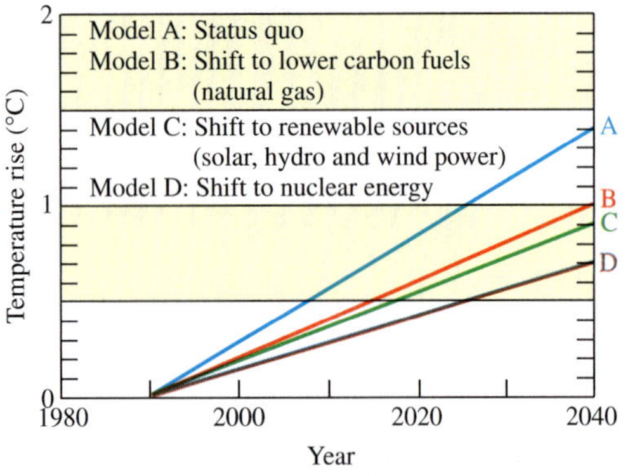

Based on data from *The Blue Planet* (Wiley, 1995)

55. DECK DESIGNS See the illustration. Find the slopes of the cross-brace and the supports. Is the cross-brace perpendicular to either support?

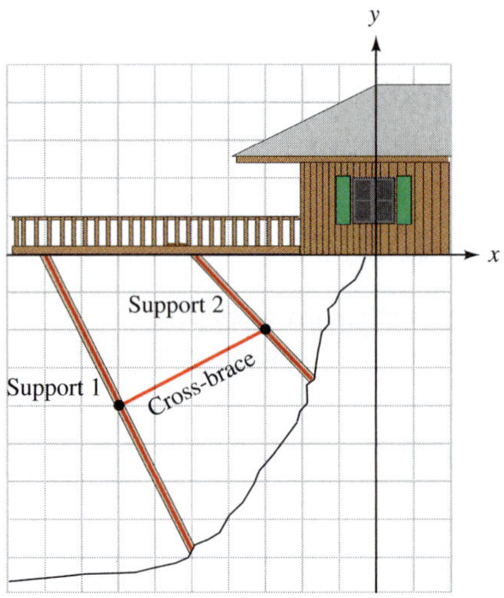

56. AIR PRESSURE Air pressure, measured in units called pascals (Pa), decreases with altitude. Find the rate of change in Pascals for the fastest and the slowest decreasing steps of the following graph.

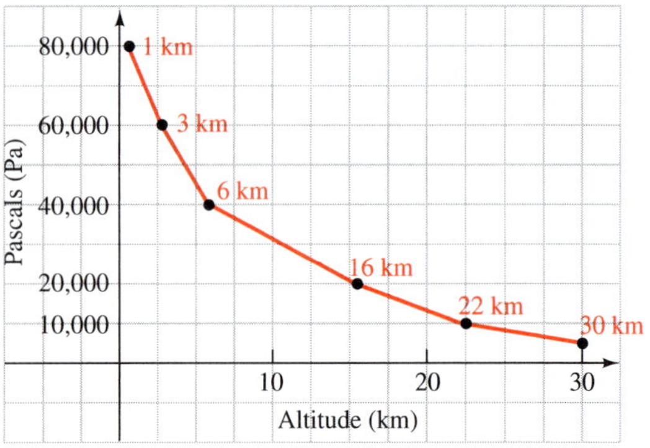

Based on data from *The Blue Planet* (Wiley, 1995)

WRITING

57. POLITICS The following illustration shows how federal Medicare spending would have continued if the Republican-sponsored Balanced Budget Act hadn't become law in 1997. Explain why Democrats could argue that the budget act "cut spending." Then explain why Republicans could respond by saying, "There was no cut in spending—only a reduction in the rate of growth of spending."

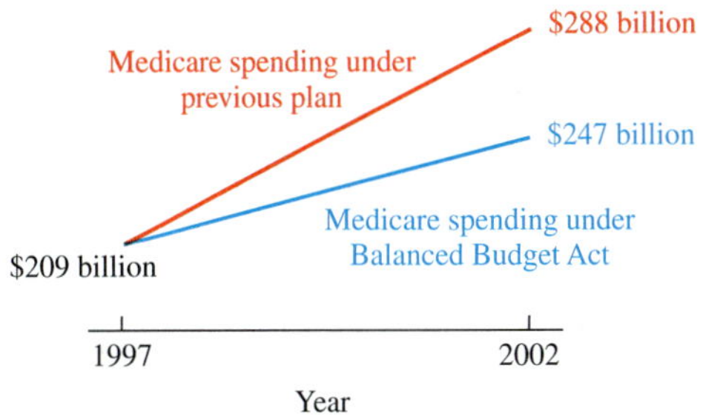

Based on information supplied by Congressman David Drier's office

58. NUCLEAR ENERGY Since 1998, the number of nuclear reactors licensed for operation in the United States has remained the same. Knowing this, what can be said about the rate of change in the number of reactors since 1998? Explain your answer.

59. Explain why a vertical line has no defined slope.

60. Explain how to determine from their slopes whether two lines are parallel, perpendicular, or neither.

REVIEW

61. HALLOWEEN CANDY A candy maker wants to make a 60-pound mixture of two candies to sell for \$2 per pound. If black licorice bits sell for \$1.90 per pound and orange gumdrops sell for \$2.20 per pound, how many pounds of each should be used?

62. CIRCLE GRAPHS In the illustration, each part of the circle represents the amount of money spent in each of five categories. Approximately what percent was spent on rent?

Monthly Expenses of Joe Sigueri

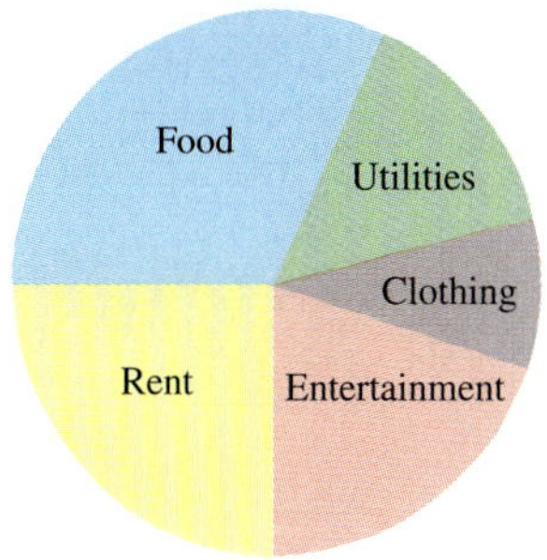

CHALLENGE PROBLEMS

63. The two lines graphed in the illustration are parallel. Find x and y.

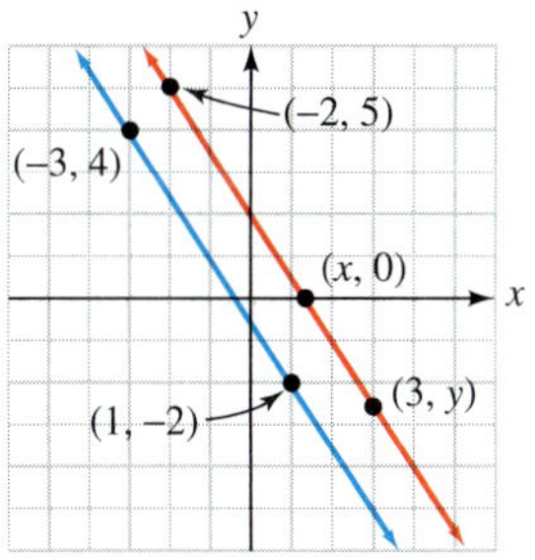

64. The line passing through (1, 3) and (−2, 7) is perpendicular to the line passing through points (4, b) and (8, −1). Without graphing, find b.

2.4 Writing Equations of Lines

- Point–slope form of the equation of a line
- Slope–intercept form of the equation of a line
- Using slope as an aid when graphing
- Parallel and perpendicular lines
- Curve fitting

We have seen that linear relationships are often presented in graphs. In this section, we begin a discussion of how to write an equation to model a linear relationship.

POINT–SLOPE FORM OF THE EQUATION OF A LINE

Suppose that line l in the figure has a slope of m and passes through (x_1, y_1). If (x, y) is a second point on line l, we have

$$m = \frac{y - y_1}{x - x_1}$$

or if we multiply both sides by $x - x_1$, we have

(1) $y - y_1 = m(x - x_1)$

Because Equation 1 displays the coordinates of the point (x_1, y_1) on the line and the slope m of the line, it is called the **point–slope form** of the equation of a line.

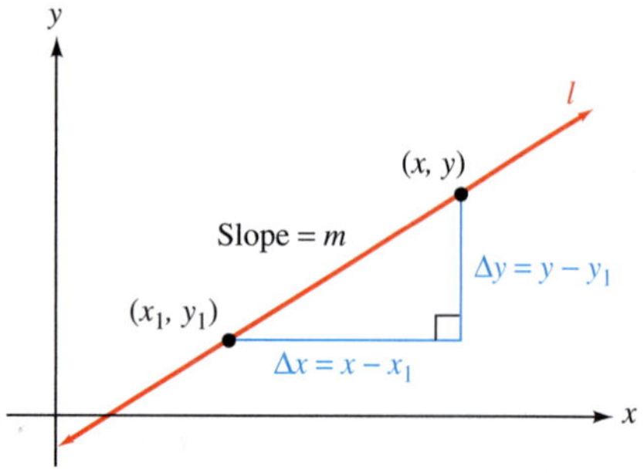

Point–Slope Form The equation of the line passing through (x_1, y_1) and with slope m is

$$y - y_1 = m(x - x_1)$$

EXAMPLE 1

Write an equation of the line that has slope $-\frac{2}{3}$ and passes through $(-4, 5)$.

Solution We substitute $-\frac{2}{3}$ for m, -4 for x_1, and 5 for y_1 into the point–slope form and simplify.

$y - y_1 = m(x - x_1)$	This is the point–slope form.
$y - 5 = -\frac{2}{3}[x - (-4)]$	Substitute $-\frac{2}{3}$ for m, -4 for x_1, and 5 for y_1.
$y - 5 = -\frac{2}{3}(x + 4)$	Simplify the expression within the brackets.
$y - 5 = -\frac{2}{3}x - \frac{8}{3}$	Use the distributive property to remove the parentheses.
$y = -\frac{2}{3}x + \frac{7}{3}$	To solve for y, add 5 in the form of $\frac{15}{3}$ to both sides and simplify.

The equation of the line is $y = -\frac{2}{3}x + \frac{7}{3}$.

Self Check 1 Write an equation of the line that has slope $\frac{5}{4}$ and passes through $(0, 5)$.

EXAMPLE 2

Write an equation of the line passing through $(-5, 4)$ and $(8, -6)$.

Solution First we find the slope of the line.

$m = \frac{y_2 - y_1}{x_2 - x_1}$	This is the slope formula.
$= \frac{-6 - 4}{8 - (-5)}$	Substitute -6 for y_2, 4 for y_1, 8 for x_2, and -5 for x_1.
$= -\frac{10}{13}$	

Since the line passes through $(-5, 4)$ and $(8, -6)$, we can choose either point and substitute its coordinates into the point–slope form. If we select $(-5, 4)$, we substitute -5 for x_1, 4 for y_1, and $-\frac{10}{13}$ for m and proceed as follows.

Success Tip

In Example 2, either of the given points can be used as (x_1, y_1) when writing the point–slope equation.

Looking ahead, we usually choose the point whose coordinates will make the computations the easiest.

$y - y_1 = m(x - x_1)$	This is the point–slope form.
$y - 4 = -\frac{10}{13}[x - (-5)]$	Substitute $-\frac{10}{13}$ for m, -5 for x_1, and 4 for y_1.
$y - 4 = -\frac{10}{13}(x + 5)$	Simplify the expression inside the brackets.
$y - 4 = -\frac{10}{13}x - \frac{50}{13}$	Remove the parentheses: distribute $-\frac{10}{13}$.
$y = -\frac{10}{13}x + \frac{2}{13}$	To solve for y, add 4 in the form of $\frac{52}{13}$ to both sides and simplify.

The equation of the line is $y = -\frac{10}{13}x + \frac{2}{13}$.

Self Check 2 Write an equation of the line passing through $(-2, 5)$ and $(4, -3)$.

Linear models can be used to describe certain types of financial gain or loss. For example, **straight-line depreciation** is used when aging equipment declines in value and **straight-line appreciation** is used when property or collectibles increase in value.

EXAMPLE 3

Accounting. After purchasing a new drill press, a machine shop owner had his accountant prepare a depreciation worksheet for tax purposes. See the illustration.

Depreciation Worksheet

Drill press $1,970
(new)

Salvage value $270
(in 10 years)

a. Assuming straight-line depreciation, write an equation that gives the value v of the drill press after x years of use.

b. Find the value of the drill press after $2\frac{1}{2}$ years of use.

c. What is the economic meaning of the v-intercept of the line?

d. What is the economic meaning of the slope of the line?

Solution

a. The facts presented in the worksheet can be expressed as ordered pairs of the form

$$(x, v)$$

number of years of use → x; value of the drill press → v

- When purchased, the new $1,970 drill press had been used 0 years: (0, 1,970).
- After 10 years of use, the value of the drill press will be $270: (10, 270).

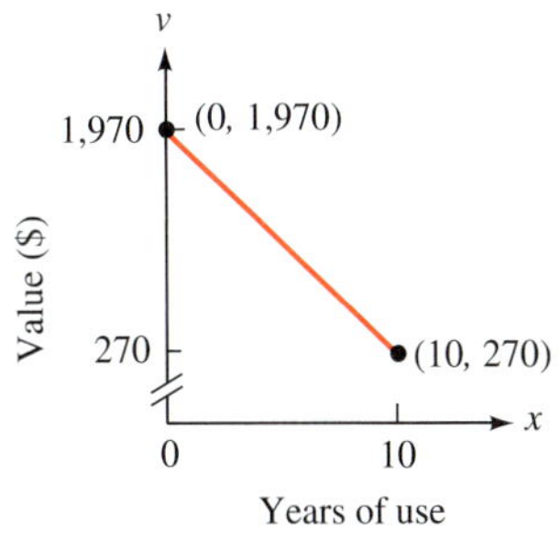

A simple sketch showing these ordered pairs and the line of depreciation is helpful in visualizing the situation.

Since we know two points that lie on the line, we can write its equation using the point–slope form. First, we find the slope of the line.

$$m = \frac{v_2 - v_1}{x_2 - x_1}$$ This is the slope formula written in terms of x and v.

$$= \frac{270 - 1{,}970}{10 - 0}$$ $(x_1, v_1) = (0, 1{,}970)$ and $(x_2, v_2) = (10, 270)$.

$$= \frac{-1{,}700}{10}$$

$$= -170$$

To find the equation of the line, we substitute -170 for m, 0 for x_1, and 1,970 for v_1 in the point–slope form and simplify.

$$v - v_1 = m(x - x_1)$$ This is the point–slope form written in terms x and v.

$$v - 1{,}970 = -170(x - 0)$$

$$v = -170x + 1{,}970$$ This is the straight-line depreciation equation.

The value v of the drill press after x years of use is given by the linear model $v = -170x + 1{,}970$.

b. To find the value of the drill press after $2\frac{1}{2}$ years of use, we substitute 2.5 for x in the depreciation equation and find v.

$$\begin{aligned} v &= -170x + 1{,}970 \\ &= -170(2.5) + 1{,}970 \\ &= -425 + 1{,}970 \\ &= 1{,}545 \end{aligned}$$

In $2\frac{1}{2}$ years, the drill press will be worth \$1,545.

c. From the sketch, we see that the v-intercept of the graph of the depreciation line is (0, 1,970). This gives the original cost of the drill press, \$1,970.

d. Each year, the value of the drill press decreases by \$170, because the slope of the line is -170. The slope of the line is the *annual depreciation rate.*

SLOPE–INTERCEPT FORM OF THE EQUATION OF A LINE

Since the y-intercept of the line l shown in the figure is the point $(0, b)$, we can write its equation by substituting 0 for x_1 and b for y_1 in the point–slope form and simplifying.

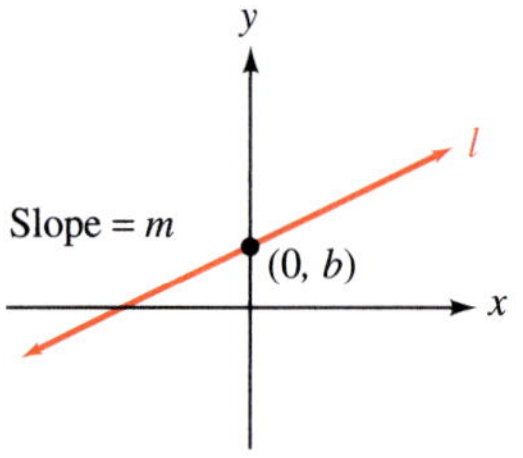

$$\begin{aligned} y - y_1 &= m(x - x_1) \\ y - b &= m(x - 0) \\ y - b &= mx \end{aligned}$$

(2) $$y = mx + b$$ To solve for y, add b to both sides.

Because Equation 2 displays the slope m and the y-coordinate b of the y-intercept, it is called the **slope–intercept form** of the equation of a line.

Slope–Intercept Form The equation of the line with slope m and y-intercept $(0, b)$ is

$$y = mx + b$$

EXAMPLE 4

Use the slope–intercept form to write an equation of the line that has slope 4 and passes through (5, 9).

Solution Since we are given that $m = 4$ and that (5, 9) satisfies the equation, we can substitute 5 for x, 9 for y, and 4 for m in the equation $y = mx + b$ and solve for b.

Success Tip

If a point lies on a line, the coordinates of the point satisfy the equation.

$y = mx + b$	This is the slope–intercept form.
$9 = 4(5) + b$	Substitute 9 for y, 4 for m, and 5 for x.
$9 = 20 + b$	Perform the multiplication.
$-11 = b$	To solve for b, subtract 20 from both sides.

Because $m = 4$ and $b = -11$, the equation is $y = 4x - 11$.

Self Check 4 Use the slope–intercept form to write an equation of the line that has slope -2 and passes through $(-2, 8)$.

EXAMPLE 5

School supplies. Each turn of the handle of a pencil sharpener shaves off 0.05 inch from a 7.25-inch-long pencil.

a. Write a linear equation in slope–intercept form that gives the new length L of the pencil after the sharpener handle has been turned t times.

b. How long is the pencil after the sharpener handle has been turned 20 times?

Solution

a. Since the length L of the pencil depends on the number of turns t of the handle, the equation will have the form $L = mt + b$. We need to determine m and b.

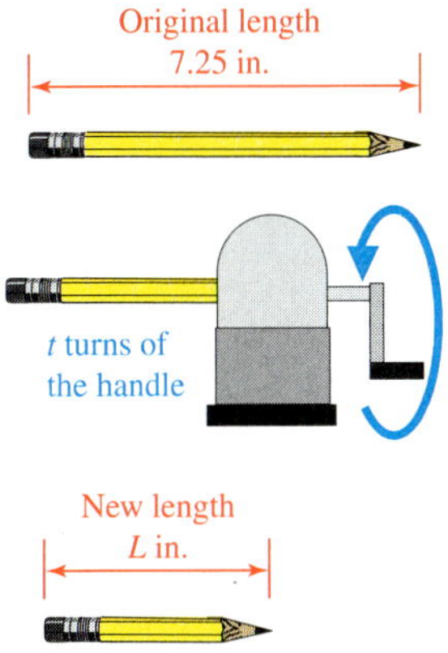

- The length of the pencil *decreases* as the handle is turned. This rate of change, -0.05 inch per turn, is the slope of the graph of the equation. Thus, $m = -0.05$.
- Before any turns of the handle are made (when $t = 0$), the length of the pencil is 7.25 inches. Written as an ordered pair of the form (t, L), we have (0, 7.25). When graphed, this would be the L-intercept of the graph. Thus, $b = 7.25$.

Substituting for m and b, we have the linear equation that models this situation.

$$L = -0.05t + 7.25$$

The slope is the rate of change of the length of the pencil. The intercept is the original length of the pencil.

b. To find the pencil's length after the handle is turned 20 times, we proceed as follows:

$$L = -0.05t + 7.25$$
$$L = -0.05(20) + 7.25$$
$$L = -1 + 7.25$$
$$= 6.25$$

If the sharpener handle is turned 20 times, the pencil will be 6.25 inches long.

USING SLOPE AS AN AID WHEN GRAPHING

If we know the slope and the y-intercept of a line, we can graph the line without having to construct a table of solutions.

EXAMPLE 6

Find the slope and the y-intercept of the line with the equation $2x + 3y = -9$. Then graph the line.

Solution To find the slope and y-intercept of the line, we write the equation in slope–intercept form: $y = mx + b$.

$$2x + 3y = -9 \quad \text{The given equation is in general form.}$$
$$3y = -2x - 9 \quad \text{Subtract } 2x \text{ from both sides.}$$
$$\frac{3y}{3} = \frac{-2x}{3} - \frac{9}{3} \quad \text{To solve for } y\text{, divide both sides by 3.}$$
$$y = -\frac{2}{3}x - 3 \quad \text{Simplify both sides. We see that } m = -\frac{2}{3} \text{ and } b = -3.$$

Caution

When using the y-intercept and the slope to graph a line, remember to draw the slope triangle from the y-intercept, *not* from the origin.

The slope of the line is $-\frac{2}{3}$, which can be expressed as $\frac{-2}{3}$. After plotting the y-intercept, $(0, -3)$, we move 2 units downward (rise) and then 3 units to the right (run). This locates a second point on the line, $(3, -5)$. From this point, we move another 2 units downward and 3 units to the right to locate a *third point* on the line, $(6, -7)$. Then we draw a line through the points to obtain the graph shown in the figure.

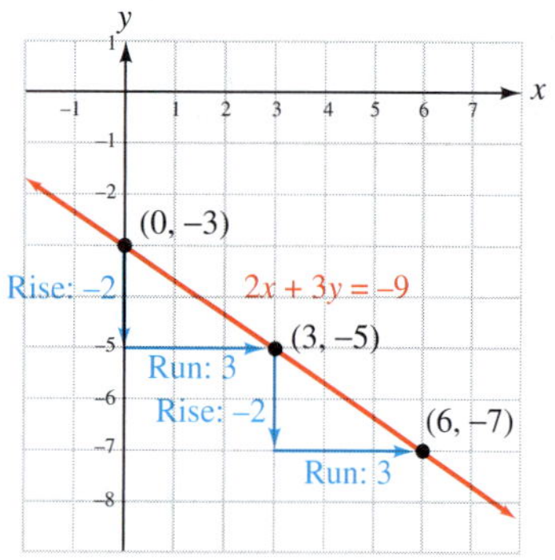

Self Check 6 Find the slope and the y-intercept of the line with the equation $3x - 2y = -4$. Then graph the line.

PARALLEL AND PERPENDICULAR LINES

EXAMPLE 7

a. Show that the lines represented by $4x + 8y = 10$ and $2x = 12 - 4y$ are parallel.

b. Show that the lines represented by $4x + 8y = 10$ and $4x - 2y = 21$ are perpendicular.

Solution **a.** We solve each equation for y to see that the lines are distinct and that their slopes are equal.

$$4x + 8y = 10 \qquad\qquad 2x = 12 - 4y$$
$$8y = -4x + 10 \qquad\qquad 4y = -2x + 12$$
$$y = -\frac{1}{2}x + \frac{5}{4} \qquad\qquad y = -\frac{1}{2}x + 3$$

Since the values of b in these equations are different $\left(\frac{5}{4} \text{ and } 3\right)$, the lines are distinct. Since the slope of each line is $-\frac{1}{2}$, they are parallel.

b. We solve each equation for y to see that the slopes of their straight-line graphs are negative reciprocals.

$$4x + 8y = 10 \qquad\qquad 4x - 2y = 21$$
$$8y = -4x + 10 \qquad\qquad -2y = -4x + 21$$
$$y = -\frac{1}{2}x + \frac{5}{4} \qquad\qquad y = 2x - \frac{21}{2}$$

Since the slopes are negative reciprocals ($-\frac{1}{2}$ and 2), the lines are perpendicular.

Self Check 7 **a.** Are the lines represented by $3x - 2y = 4$ and $2x = 5(y + 1)$ parallel?

b. Are the lines represented by $3x + 2y = 6$ and $2x - 3y = 6$ perpendicular?

EXAMPLE 8

Write an equation of the line that passes through $(-2, 5)$ and is parallel to the line $y = 8x - 3$.

Solution Since the slope of the line given by $y = 8x - 3$ is the coefficient of x, the slope is 8. Since the desired equation is to have a graph that is parallel to the graph of $y = 8x - 3$, its slope must also be 8.

We substitute -2 for x_1, 5 for y_1, and 8 for m in the point–slope form and simplify.

$$y - y_1 = m(x - x_1)$$
$$y - 5 = 8[x - (-2)] \quad \text{Substitute 5 for } y_1\text{, 8 for } m\text{, and } -2 \text{ for } x_1.$$
$$y - 5 = 8(x + 2) \quad \text{Simplify the expression inside the brackets.}$$
$$y - 5 = 8x + 16 \quad \text{Use the distributive property to remove the parentheses.}$$
$$y = 8x + 21 \quad \text{To solve for } y\text{, add 5 to both sides.}$$

The equation is $y = 8x + 21$.

Self Check 8 Write an equation of the line that is parallel to the line $y = 8x - 3$ and passes through the origin.

When asked to *write the equation of a line,* determine what you know about the graph of the line: its slope, its y-intercept, points it passes through, and so on. Then substitute the appropriate numbers into one of the following forms of a linear equation.

Forms of a Linear Equation

General form	$Ax + By = C$	A and B cannot both be 0.
Slope–intercept form	$y = mx + b$	The slope is m, and the y-intercept is $(0, b)$.
Point–slope form	$y - y_1 = m(x - x_1)$	The slope is m, and the line passes through (x_1, y_1).
A **horizontal line**	$y = b$	The slope is 0, and the y-intercept is $(0, b)$.
A **vertical line**	$x = a$	There is no defined slope, and the x-intercept is $(a, 0)$.

CURVE FITTING

In statistics, the process of using one variable to predict another is called **regression.** For example, if we know a man's height, we can usually make a good prediction about his weight because taller men tend to weigh more than shorter men.

The table in figure (a) shows the results of sampling twelve men at random and recording the height h and weight w of each. In figure (b), the ordered pairs (h, w) from the table are plotted to form a **scatter diagram.** Notice that the data points fall more or less along an imaginary straight line, indicating a linear relationship between h and w.

Height *h* in.	Weight *w* lb
66	145
67	150
68	150
68	165
70	180
70	165
71	175
72	200
73	190
74	190
75	205
75	215

(a)

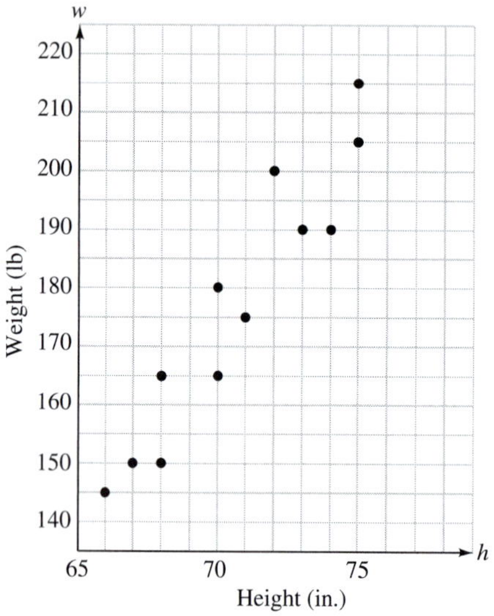

(b)

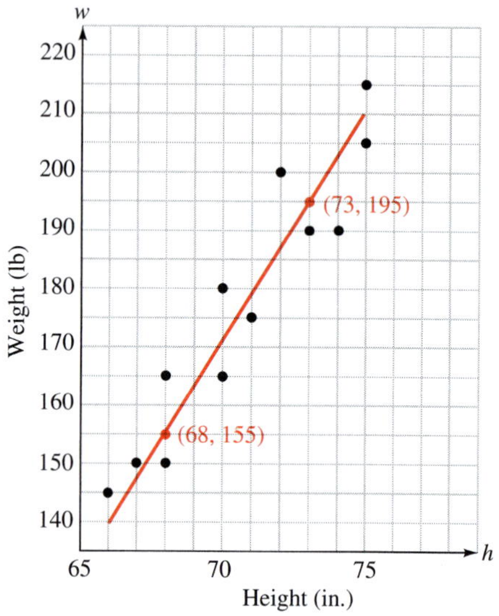

(c)

To write a *prediction equation* (sometimes called a *regression equation*) that relates height and weight, we must find the equation of the line that comes closer to all of the data points in the scatter diagram than any other possible line. In statistics, there are exact methods to find this equation; however, they are beyond the scope of this book. In this course, we will draw "by eye" a line that we feel best fits the data points.

In figure (c), a straight edge was placed on the scatter diagram and a line was drawn that seemed to best fit all of the data points. Note that it passes through (68, 155) and (73, 195). To write the equation of that line, we first need to find its slope.

Caution

When drawing a line through the data points of a scatter diagram by eye, the results could vary from person to person. Graphing calculators have a program that finds the line of best fit for a collection of data. Look in the owner's manual under **linear regression.**

$$m = \frac{w_2 - w_1}{h_2 - h_1}$$ This is the slope formula written in terms of h and w.

$$= \frac{195 - 155}{73 - 68}$$ Choose $(h_1, w_1) = (68, 155)$ and $(h_2, w_2) = (73, 195)$.

$$= \frac{40}{5}$$

$$= 8$$

We then use the point–slope form to find the equation of the line. Since the line passes through (68, 155) and (73, 195), we can use either one to write its equation.

$w - w_1 = m(h - h_1)$	This is the point–slope form written in terms of h and w.
$w - 155 = 8(h - 68)$	Choose (68, 155) for (h_1, w_1).
$w - 155 = 8h - 544$	Distribute the multiplication by 8.
$w = 8h - 389$	To solve for w, add 155 to both sides.

The equation of the line that was drawn through the data points in the scatter diagram is $w = 8h - 389$. We can use this equation to predict the weight of a man who is 72 inches tall.

$w = 8h - 389$

$w = 8(72) - 389$ Substitute 72 for h.

$w = 576 - 389$

$w = 187$

We predict that a 72-inch-tall man chosen at random will weigh about 187 pounds.

Answers to Self Checks **1.** $y = \frac{5}{4}x + 5$ **2.** $y = -\frac{4}{3}x + \frac{7}{3}$ **4.** $y = -2x + 4$ **6.** $m = \frac{3}{2}$, (0, 2)

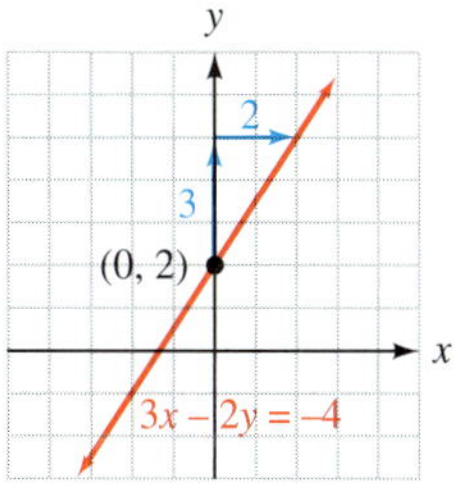

7. a. no, **b.** yes **8.** $y = 8x$

2.4 STUDY SET

VOCABULARY Fill in the blanks.

1. The point–slope form of the equation of a line is ________________.

2. The ______________ form of the equation of a line is $y = mx + b$.

3. Two lines are ______________ when their slopes are negative reciprocals.

4. Two lines are _________ when they have the same slope.

CONCEPTS

5. If you know the slope of a line, is that enough information about the line to write its equation?

6. If you know a point that a line passes through, is that enough information about the line to write its equation?

7. The line in the illustration passes through the point $(-2, -3)$. Find its slope. Then write its equation in point–slope form.

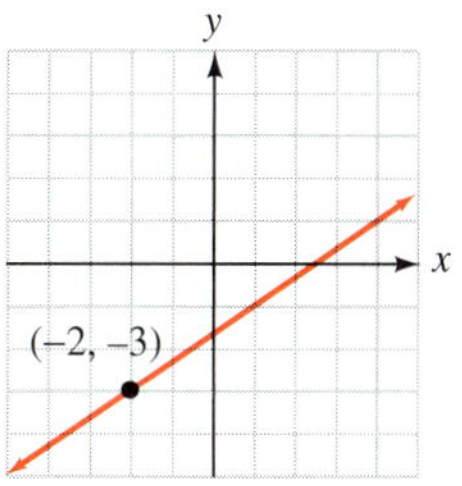

8. For the line in the illustration, find the slope and the y-intercept. Then write the equation of the line in slope–intercept form.

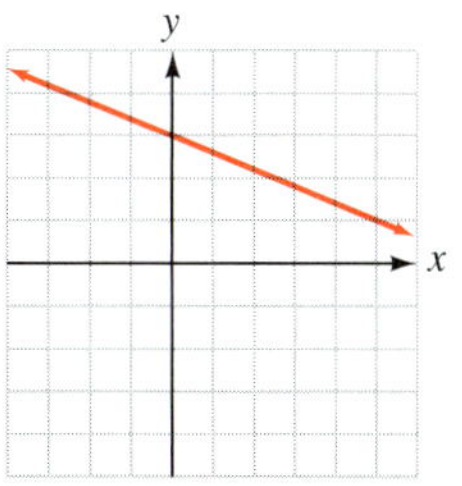

9. When the graph of the line $y = -\frac{2}{3}x + 1$ is drawn, what slope and y-intercept will the line have?

10. When the graph of the line $y - 3 = -\frac{2}{3}(x + 1)$ is drawn, what slope will it have? What point does the equation indicate it will pass through?

11. Do the equations $y - 2 = 3(x - 2)$, $y = 3x - 4$, and $3x - y = 4$ all describe the same line?

12. See the linear model graphed below.

a. What information does the y-intercept give?

b. What information does the slope give?

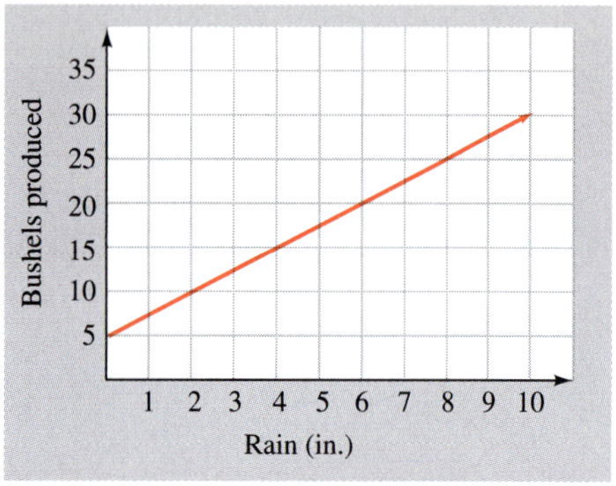

13. When each equation is graphed, what will the y-intercept be?

a. $y = 2x$ **b.** $x = -3$

14. When each equation is graphed, what will the slope of the line be?

a. $y = -x$ **b.** $x = -3$

15. The two lines graphed as follows appear to be perpendicular. Their equations are also displayed. Are the lines actually perpendicular? Explain.

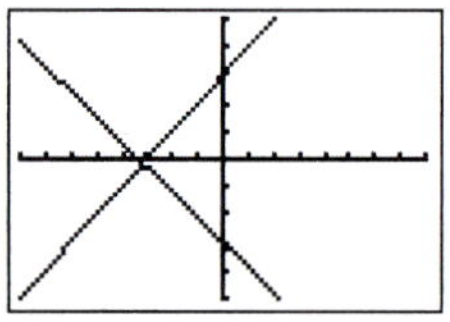

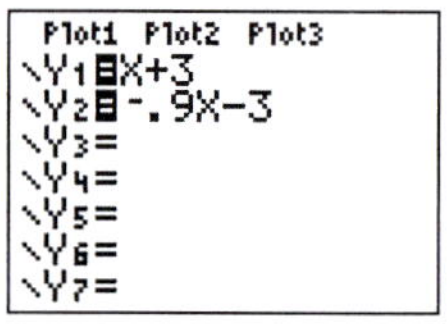

16. The two lines graphed as follows appear to be parallel. Their equations are also displayed below. Are the lines actually parallel? Explain.

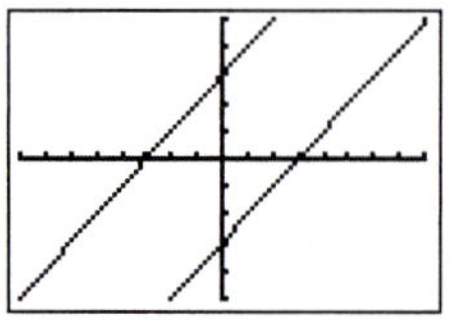

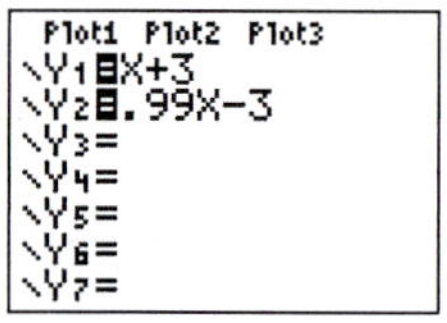

NOTATION **Complete each solution.**

17. Write $y + 2 = \frac{1}{3}(x + 3)$ in slope–intercept form.

$$y + 2 = \frac{1}{3}(x + 3)$$

$$y + 2 = \underline{\quad} + 1$$

$$y + 2 - \underline{\quad} = \frac{1}{3}x + 1 - \underline{\quad}$$

$$y = \frac{1}{3}x - \underline{\quad}$$

$$m = \underline{\quad}, b = \underline{\quad}$$

18. Write an equation of the line that has slope -2 and passes through the point (3, 1).

$$y - y_1 = m(x - x_1)$$

$$y - \underline{\quad} = -2(x - \underline{\quad})$$

$$y - 1 = \underline{\quad} + 6$$

$$y = -2x + \underline{\quad}$$

PRACTICE **Use the point–slope form to write an equation of the line with the given properties. Then write each equation in slope–intercept form.**

19. $m = 5$, passing through (0, 7)

20. $m = -8$, passing through (0, −2)

21. $m = -3$, passing through (2, 0)

22. $m = 4$, passing through (−5, 0)

Use the point–slope form to write an equation of the line passing through the two given points. Then write each equation in slope–intercept form.

23. (0, 0), (4, 4)

24. (−5, 5), (0, 0)

25.

x	y
3	4
0	−3

26.

x	y
4	0
6	−8

27.

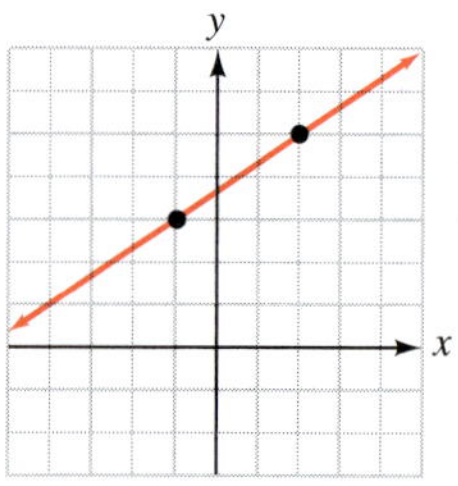

28.

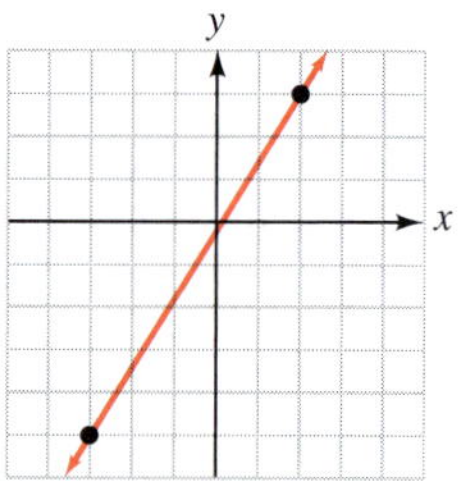

Use the slope–intercept form to write an equation of the line with the given properties.

29. $m = 3, b = 17$

30. $m = -2, b = 11$

31. $m = -7$, passing through $(7, 5)$

32. $m = 3$, passing through $(-2, -5)$

33. $m = 0$, passing through $(2, -4)$

34. $m = -7$, passing through the origin

35. passing through $(6, 8)$ and $(2, 10)$

36. passing through $(-4, 5)$ and $(2, -6)$

Write each equation in slope–intercept form. Then find the slope and the *y*-intercept of the line determined by the equation.

37. $3x - 2y = 8$

38. $-2x + 4y = 12$

39. $-2(x + 3y) = 5$

40. $5(2x - 3y) = 4$

Find the slope and *y*-intercept and use them to draw the graph of the line.

41. $y = x - 1$

42. $y = -x + 2$

43. $y = \frac{2}{3}x + 2$

44. $y = -\frac{5}{4}x + \frac{5}{2}$

45. $4y - 3 = -3x - 11$

46. $-2x + 4y = 12$

Determine whether the graphs of each pair of equations are parallel, perpendicular, or neither.

47. $y = 3x + 4, y = 3x - 7$

48. $y = 4x - 13, y = \frac{1}{4}x + 13$

49. $x + y = 2, y = x + 5$

50. $x = y + 2, y = x + 3$

51. $3x + 6y = 1, y = \frac{1}{2}x$

52. $2x + 3y = 9, 3x - 2y = 5$

53. $y = 3, x = 4$

54. $y = -3, y = -7$

Write an equation of the line that passes through the given point and is parallel to the given line. Write the answer in slope–intercept form.

55. $(0, 0), y = 4x - 7$

56. $(0, 0), x = -3y - 12$

57. $(2, 5), 4x - y = 7$

58. $(-6, 3), y + 3x = -12$

59. $(4, -2), x = \frac{5}{4}y - 2$

60. $(1, -5), x = -\frac{3}{4}y + 5$

Write an equation of the line that passes through the given point and is perpendicular to the given line. Write the answer in slope–intercept form.

61. $(0, 0), y = 4x - 7$

62. $(0, 0), x = -3y - 12$

63. $(2, 5), 4x - y = 7$

64. $(-6, 3), y + 3x = -12$

65. $(4, -2), x = \frac{5}{4}y - 2$

66. $(1, -5), x = -\frac{3}{4}y + 5$

APPLICATIONS

67. BIG-SCREEN TV Find the straight-line depreciation equation for the TV in the following want ad.

> *For Sale*: 3-year-old 45-inch TV, with matrix surround sound & picture within picture, remote. $1,750 new. Asking $800. Call 875-5555. Ask for Mike.

68. SALVAGE VALUES A truck was purchased for \$19,984. Its salvage value at the end of 8 years is expected to be \$1,600. Find the straight-line depreciation equation.

69. ART In 1987, the painting *Rising Sunflowers* by Vincent van Gogh sold for \$36,225,000. Suppose that an appraiser expected the painting to double in value in 20 years. Let x represent the time in years after 1987. Find the straight-line appreciation equation.

70. REAL ESTATE LISTINGS Use the information given in the following description of the property to write a straight-line appreciation equation for the house.

Vacation Home
\$122,000
Only 2 years old

- Great investment property!
- Expected to appreciate \$4,000/yr

Sq ft: 1,635	**Fam rm:** yes	**Den:** no
Bdrm: 3	**Ba:** 1.5	**Gar:** enclosed
A/C: yes	**Firepl:** yes	**Kit:** built-ins

71. CRIMINOLOGY City growth and the number of burglaries for a certain city are related by a linear equation. Records show that 575 burglaries were reported in a year when the local population was 77,000 and that the rate of increase in the number of burglaries was 1 for every 100 new residents.

a. Using the variables p for population and B for burglaries, write an equation (in slope–intercept form) that police can use to predict future burglary statistics.

b. How many burglaries can be expected when the population reaches 110,000?

72. CABLE TV Since 1990, when the average monthly basic cable TV rate in the United States was \$15.81, the cost has risen by about \$1.52 a year.

a. Write an equation in slope–intercept form to predict cable TV costs in the future. Use t to represent time in years after 1990 and C to represent the average basic monthly cost. (Source: Kagan World Media)

b. If the equation in part a were graphed, what would be the meaning of the C-intercept and the slope of the line?

73. PSYCHOLOGY EXPERIMENTS The scattergram in the following illustration shows the performance of a rat in a maze.

a. Draw a line through (1, 10) and (19, 1). Write its equation using the variables t and E. In psychology, this equation is called the *learning curve* for the rat.

b. What does the slope of the line tell us?

c. What information does the t-intercept of the graph give?

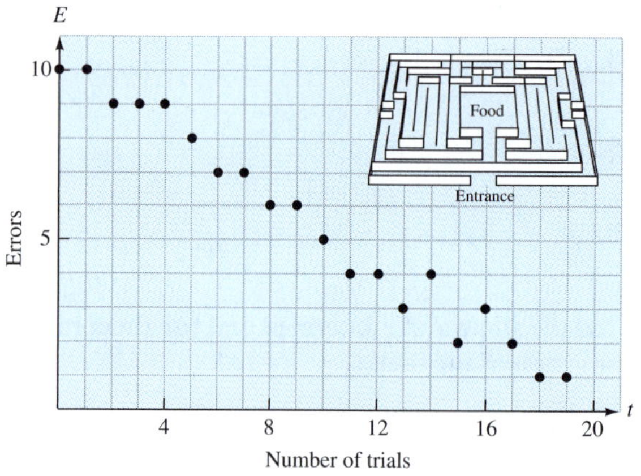

74. UNDERSEA DIVING The illustration on the next page shows that the pressure p that divers experience is related to the depth d of the dive. A linear model can be used to describe this relationship.

a. Write the linear model in slope–intercept form.

b. Pearl and sponge divers often reach depths of 100 feet. What pressure do they experience? Round to the nearest tenth.

c. Scuba divers can safely dive to depths of 250 feet. What pressure do they experience? Round to the nearest tenth.

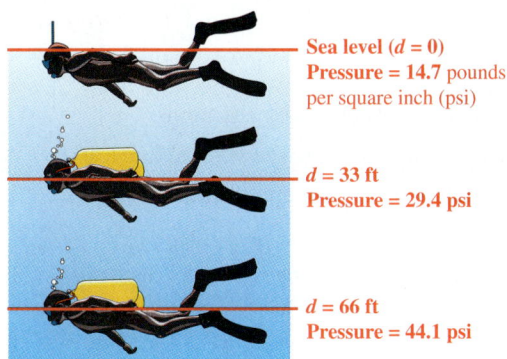

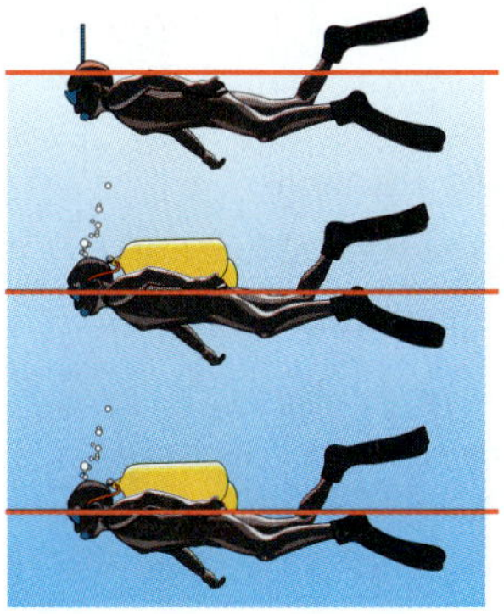

75. WIND-CHILL A combination of cold and wind makes a person feel colder than the actual temperature. The table shows what temperatures of 35°F and 15°F feel like when a 15-mph wind is blowing. The relationship between the actual temperature and the wind-chill temperature can be modeled with a linear equation.

a. Write the equation that models this relationship. Answer in slope–intercept form.

b. What information is given by the y-intercept of the graph of the equation found in part a?

Actual temperature	Wind-chill temperature
35°F	16°F
15°F	−11°F

76. COMPUTER-AIDED DRAFTING The illustration shows a computer-generated drawing of an airplane part. When the designer clicks the mouse on a line on the drawing, the computer finds the equation of the line. Use a calculator to determine whether the angle where the weld is to be made is a right angle.

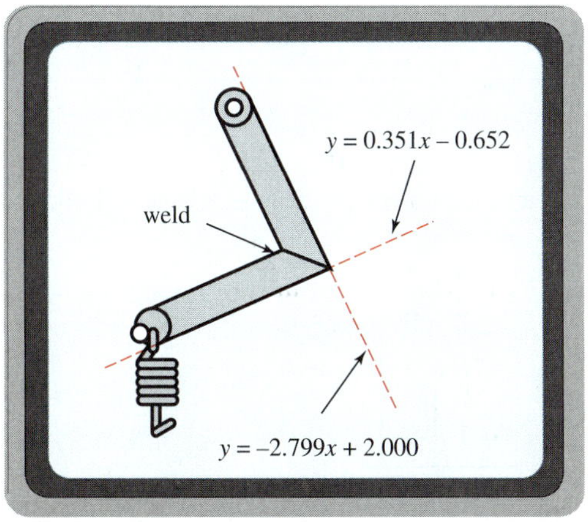

WRITING

77. Explain how to find the equation of a line passing through two given points.

78. Explain what m, x_1, and y_1 represent in the point–slope form of the equation of a line.

79. A student was asked to determine the slope of the graph of the line $y = 6x - 4$. His answer was $m = 6x$. Explain his error.

80. Linear relationships between two quantities can be described by an equation or a graph. Which do you think is the more informative? Why?

REVIEW

81. INVESTMENTS Equal amounts are invested at 6%, 7%, and 8% annual interest. The three investments yield a total of $2,037 annual interest. Find the total amount of money invested.

82. MEDICATIONS A doctor prescribes an ointment that is 2% hydrocortisone. A pharmacist has 1% and 5% concentrations in stock. How many ounces of each should the pharmacist use to make a 1-ounce tube?

CHALLENGE PROBLEMS **Investigate the properties of the slope and the y-intercept by experimenting with the following problems.**

83. a. Graph $y = mx + 2$ for several positive values of m. What do you notice?

b. Graph $y = mx + 2$ for several negative values of m. What do you notice?

84. a. Graph $y = 2x + b$ for several increasing positive values of b. What do you notice?

b. Graph $y = 2x + b$ for several decreasing negative values of b. What do you notice?

85. If the graph of $y = ax + b$ passes through quadrants I, II, and IV, what can be known about the constants a and b?

86. The graph of $Ax + By = C$ passes only through quadrants I and IV. What is known about the constants A, B, and C?

2.5 An Introduction to Functions

- Functions; domain and range
- Functions defined by equations
- Function notation
- The graph of a function
- The vertical line test
- Finding the domain and range of a function
- An application

The concept of a *function* is one of the most important ideas in all of mathematics. To introduce this topic, let's look at a table that one might see on television or printed in a newspaper.

FUNCTIONS; DOMAIN AND RANGE

The following table shows the number of women serving in the House of Representatives during the most recent sessions of Congress.

Women in the U.S. House of Representatives						
Session of Congress	103rd	104th	105th	106th	107th	108th
Women members	47	48	54	56	59	59

For each session of Congress, there corresponds exactly one number of women representatives. Such a correspondence is an example of a *function.*

Functions A **function** is a rule (or correspondence) that assigns to each value of one variable (called the **independent variable**) exactly one value of another variable (called the **dependent variable**).

Domain and Range The set of all possible values that can be used for the independent variable is called the **domain.** The set of all values of the dependent variable is called the **range.**

An **arrow** or **mapping diagram** can be used to show how a function assigns to each member of the domain exactly one member of the range. For the House of Representatives example, we have the diagram shown on the right.

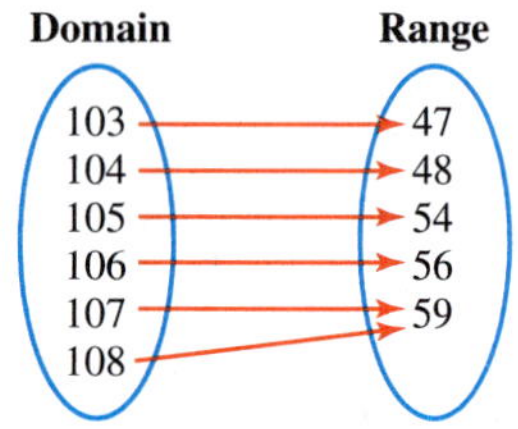

We can restate the definition of a function using the variables x and y.

y is a Function of x If to each value of x in the domain there is assigned exactly one value of y in the range, then y is said to be a function of x.

EXAMPLE 1

Determine whether the arrow diagram and the tables define y as a function of x:

a.

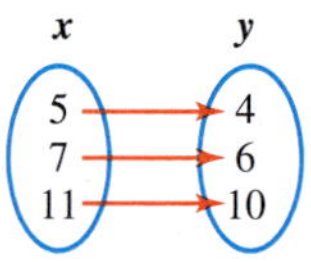

b.

x	y
8	2
1	4
8	3
9	9

c.

x	y
−2	3
−1	3
0	3
1	3

Solution **a.** The arrow diagram defines a function because each x-value is assigned exactly one y-value: $5 \rightarrow 4$, $7 \rightarrow 6$, and $11 \rightarrow 10$.

b. This table does not define a function, because to the x-value 8 there is assigned more than one y-value. In the first row, 2 is assigned to 8, and in the third row, 3 is also assigned to 8.

c. Since the table assigns to each x-value exactly one y-value, it defines a function. It also illustrates an important fact about functions: *Different values of x may be assigned the same value of y.* In this case, each x-value is assigned the y-value 3.

Self Check 1 Determine whether the arrow diagram and the table define y as a function of x.

a.

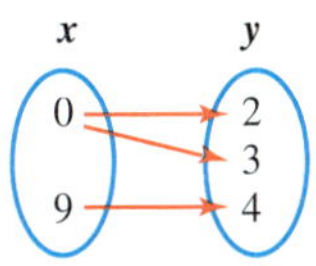

b.

x	y
−1	−60
0	55
3	0

FUNCTIONS DEFINED BY EQUATIONS

A function can also be defined by an equation. For example, $y = \frac{1}{2}x + 3$ is a rule that assigns to each value of x exactly one value of y. To find the y-value (called an **output**) that is assigned to the x-value 4 (called an **input**), we substitute 4 for x and evaluate the right-hand side of the equation.

$$\begin{aligned} y &= \frac{1}{2}x + 3 \\ &= \frac{1}{2}(4) + 3 && \text{Substitute 4 for } x. \\ &= 2 + 3 \\ &= 5 && \text{The output is 5.} \end{aligned}$$

The function $y = \frac{1}{2}x + 3$ assigns the y-value 5 to the x-value 4.

Not all equations define functions, as we see in the following example.

EXAMPLE 2

Determine whether each equation defines y to be a function of x: **a.** $y = 2x - 5$ and **b.** $y^2 = x$.

Solution **a.** To find the output value y that is assigned to an input value x, we *multiply x by 2 and then subtract 5.* Since this arithmetic gives one result, to each value of x there is assigned exactly one y-value. Thus, $y = 2x - 5$ defines y to be a function of x.

b. The equation $y^2 = x$ does not define y to be a function of x, because we can find an input value x that is assigned more than one output value y. For example, consider $x = 16$. It is assigned two values of y, 4 and -4, because $4^2 = 16$ and $(-4)^2 = 16$.

x	y
16	4
16	−4

Self Check 2 Determine whether each equation defines y to be a function of x: **a.** $y = -2x + 5$ and **b.** $|y| = x$.

FUNCTION NOTATION

A special notation is used to name functions that are defined by equations.

Function Notation The notation $y = f(x)$ denotes that the variable y is a function of x.

In Example 2a, we saw that $y = 2x - 5$ defines y to be a function of x. To write this equation using function notation, we replace y with $f(x)$, to get $f(x) = 2x - 5$. This is read as "f of x is equal to $2x$ minus 5."

Caution

The symbol $f(x)$ denotes a function. It does not mean $f \cdot x$ (f times x).

This variable represents the input.

$$f(x) = 2x - 5$$

This is the name of the function.

This expression shows how to obtain an output from a given input.

Function notation provides a compact way of denoting the output value that is assigned to some input value x. For example, if $f(x) = 2x - 5$, the value that is assigned to an x-value 6 is represented by $f(6)$.

$f(x) = 2x - 5$

$f(6) = 2(6) - 5$ Substitute 6 for each x. (The input is 6.)

$= 12 - 5$ Evaluate the right-hand side.

$= 7$

The Language of Algebra

Another way to read $f(6) = 7$ is to say "the value of the function at 6 is 7."

Thus, $f(6) = 7$. The output 7 is called a **function value.**

To see why function notation is helpful, consider these equivalent sentences:

1. If $y = 2x - 5$, find the value of y when x is 6.
2. If $f(x) = 2x - 5$, find $f(6)$.

Statement 2, which uses $f(x)$ notation, is much more concise.

EXAMPLE 3

Let $f(x) = 4x + 3$. Find **a.** $f(3)$, **b.** $f(-1)$, **c.** $f(0)$, and **d.** $f(r + 1)$.

Solution **a.** To find $f(3)$, we replace x with 3:

$f(x) = 4x + 3$

$f(3) = 4(3) + 3$

$= 12 + 3$

$= 15$

b. To find $f(-1)$, we replace x with -1:

$f(x) = 4x + 3$

$f(-1) = 4(-1) + 3$

$= -4 + 3$

$= -1$

c. To find $f(0)$, we replace x with 0:

$$f(x) = 4x + 3$$
$$f(0) = 4(0) + 3$$
$$= 3$$

d. To find $f(r + 1)$, we replace x with $r + 1$:

$$f(x) = 4x + 3$$
$$f(r + 1) = 4(r + 1) + 3$$
$$= 4r + 4 + 3$$
$$= 4r + 7$$

Self Check 3 If $f(x) = -2x - 1$, find **a.** $f(2)$, **b.** $f(-3)$, and **c.** $f(-t)$.

The letter f used in the notation $y = f(x)$ represents the word *function.* However, other letters can be used to represent functions. For example, the notations $y = g(x)$ and $y = h(x)$ are often used to denote functions involving the independent variable x.

EXAMPLE 4

Let $g(x) = x^2 - 2x$. Find **a.** $g\left(\frac{2}{5}\right)$ and **b.** $g(-2.4)$.

Solution **a.** To find $g\left(\frac{2}{5}\right)$, we replace x with $\frac{2}{5}$:

$$g(x) = x^2 - 2x$$
$$g\left(\frac{2}{5}\right) = \left(\frac{2}{5}\right)^2 - 2\left(\frac{2}{5}\right)$$
$$= \frac{4}{25} - \frac{4}{5}$$
$$= -\frac{16}{25}$$

b. To find $g(-2.4)$, we replace x with -2.4:

$$g(x) = x^2 - 2x$$
$$g(-2.4) = (-2.4)^2 - 2(-2.4)$$
$$= 5.76 + 4.8$$
$$= 10.56$$

Self Check 4 Let $h(x) = -\frac{x^2 + 2}{2}$. Find **a.** $h(4)$ and **b.** $h(-0.6)$.

EXAMPLE 5

Archery. The area of a circle with a diameter of length d is given by the function $A(d) = \pi\left(\frac{d}{2}\right)^2$. Find the area of the archery target.

Solution Since the diameter of the circular target is 48 inches, $A(48)$ gives the area of the target. To find $A(48)$, we replace d with 48.

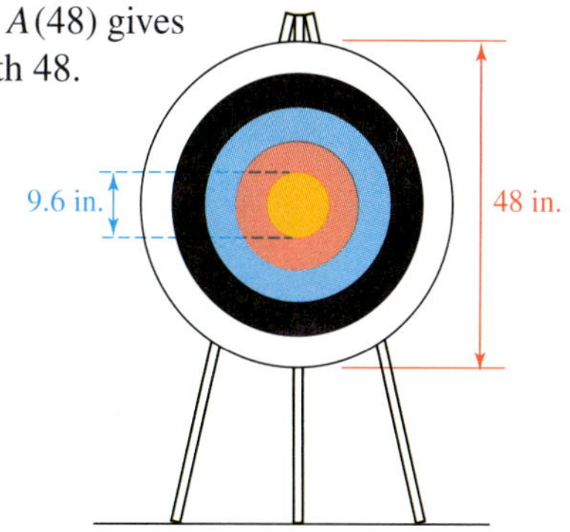

$$A(d) = \pi\left(\frac{d}{2}\right)^2$$
$$A(48) = \pi\left(\frac{48}{2}\right)^2 \quad \text{Substitute 48 for } d.$$
$$= \pi(24)^2$$
$$= 576\pi$$
$$\approx 1{,}809.557368 \quad \text{Use a calculator.}$$

To the nearest tenth, the area of the target is 1,809.6 in.^{2}.

The Language of Algebra

The function was named A because it finds the *area* of a circle. Since the area of a circle is a function of its *diameter,* d was used for the independent variable.

Self Check 5 Find the area of the "bull's eye" to the nearest tenth of a square inch.

If we are given an output of a function, we can work in reverse to find the corresponding input(s).

EXAMPLE 6

Let: $f(x) = \frac{1}{3}x + 4$. For what value(s) of x is $f(x) = 2$?

Solution To find the value(s) where $f(x) = 2$, we substitute 2 for $f(x)$ and solve for x.

$$f(x) = \frac{1}{3}x + 4$$

$$2 = \frac{1}{3}x + 4 \quad \text{Substitute 2 for } f(x).$$

$$-2 = \frac{1}{3}x \quad \text{Subtract 4 from both sides.}$$

$$-6 = x \quad \text{Multiply both sides by 3.}$$

To check, we can substitute -6 for x and verify that $f(-6) = 2$.

$$f(x) = \frac{1}{3}x + 4$$

$$f(-6) = \frac{1}{3}(-6) + 4$$

$$= -2 + 4$$

$$= 2$$

Self Check 6 For what value(s) of x is $f(x) = -5$?

THE GRAPH OF A FUNCTION

We have seen that a function assigns to each value of x a single value $f(x)$. The "input-output" pairs that a function generates can be plotted on a rectangular coordinate system to get the graph of the function.

EXAMPLE 7

Graph: $f(x) = \frac{1}{2}x + 3$.

Solution We begin by constructing a table of function values. To make a table, we choose several values for x and find the corresponding values of $f(x)$. If x is -2, we have

$$f(x) = \frac{1}{2}x + 3 \quad \text{This is the function to graph.}$$

$$f(-2) = \frac{1}{2}(-2) + 3 \quad \text{Substitute } -2 \text{ for each } x.$$

$$= -1 + 3$$

$$= 2$$

Thus, $f(-2) = 2$. This means that, when x is -2, $f(x)$ is 2, and it indicates that the ordered pair $(-2, 2)$ lies on the graph of f.

In a similar manner, we find the corresponding values of $f(x)$ for x-values of 0, 2, and 4 and record them in the table. Then we plot the ordered pairs and draw a straight line through the points to get the graph of $f(x) = \frac{1}{2}x + 3$.

$f(x) = \frac{1}{2}x + 3$

x	$f(x)$	
−2	2	→ (−2, 2)
0	3	→ (0, 3)
2	4	→ (2, 4)
4	5	→ (4, 5)

↑ Since $y = f(x)$, this column may be labeled $f(x)$ or y.

This axis can be labeled y or $f(x)$.

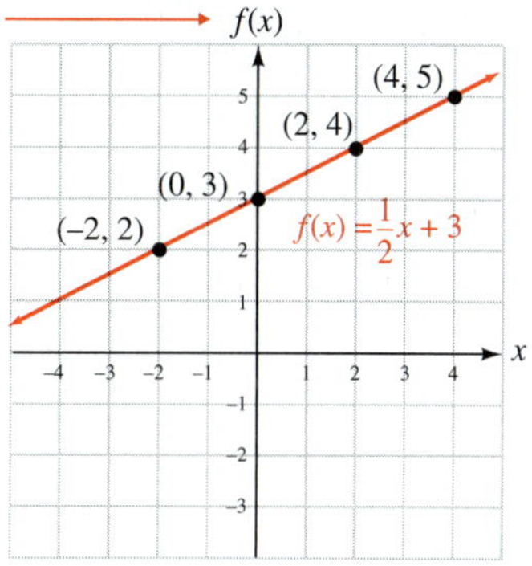

Self Check 7 Graph: $f(x) = -3x - 2$.

We call $f(x) = \frac{1}{2}x + 3$ from Example 7 a **linear function** because its graph is a nonvertical straight line. Any linear equation, except those of the form $x = a$, can be written in function notation by writing it in slope–intercept form ($y = mx + b$) and then replacing y with $f(x)$.

THE VERTICAL LINE TEST

Some graphs define functions and some do not. If any vertical line intersects a graph more than once, the graph cannot represent a function, because to one value of x there would be assigned more than one value of y.

The Vertical Line Test If a vertical line intersects a graph in more than one point, the graph is not the graph of a function.

The Language of Algebra

Graphs that do not represent functions are called *relations.* A *relation* is simply a set of ordered pairs.

The graph shown in red in figure (a) is not the graph of a function because the vertical line intersects the graph at more than one point. The points of intersection indicate that the x-value 3 is assigned two y-values, 2.5 and −2.5.

The graph shown in red in figure (b) represents a function, because no vertical line intersects the graph at more than one point. Several vertical lines are drawn to illustrate this.

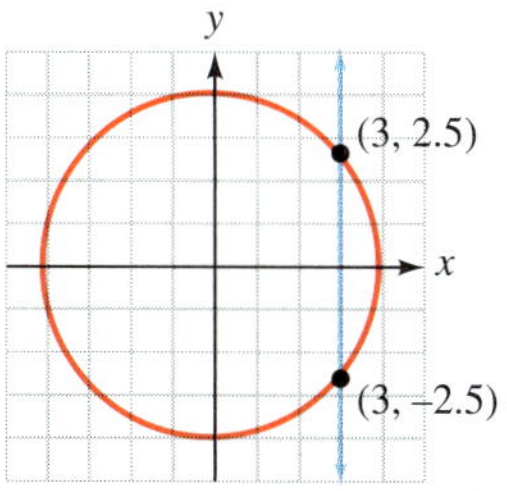

x	y
3	2.5
3	−2.5

(a)

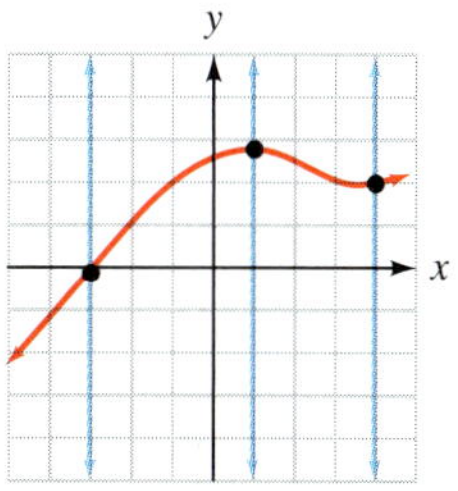

(b)

FINDING THE DOMAIN AND RANGE OF A FUNCTION

We can think of a function as a machine that takes some input x and turns it into some output $f(x)$, as shown in figure (a). The machine shown in figure (b) turns the input -6 into the output -11. The set of numbers that we put into the machine is the domain of the function, and the set of numbers that comes out is the range.

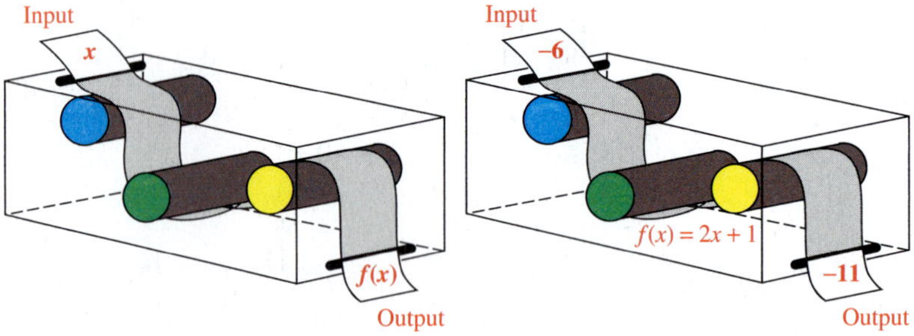

(a) (b)

EXAMPLE 8

Find the domain and range of each function: **a.** $\{(-2, 4), (0, 6), (2, 8)\}$,

b. $f(x) = 3x + 1$, and **c.** $f(x) = \dfrac{1}{x - 2}$.

Solution

a. This function consists of only three ordered pairs. The ordered pairs set up a correspondence between x (the input) and y (the output), where a single value of y is assigned to each x.

- The domain is the set of first coordinates in the set of ordered pairs: $\{-2, 0, 2\}$.
- The range is the set of second coordinates in the set of ordered pairs: $\{4, 6, 8\}$.

b. We will be able to evaluate $3x + 1$ for any real-number input x. So the domain of the function is the set of real numbers. Since the output y can be any real number, the range is the set of real numbers.

c. To find the domain of $f(x) = \frac{1}{x - 2}$, we exclude any real-number x inputs for which we would be unable to compute $\frac{1}{x - 2}$. The number 2 cannot be substituted for x, because that would make the denominator equal to zero. Since any real number except 2 can be substituted for x in the equation $f(x) = \frac{1}{x - 2}$, the domain is the set of all real numbers except 2.

Since a fraction with a numerator of 1 cannot be 0, the range is the set of all real numbers except 0.

Self Check 8 Find the domain and range of each function: **a.** $\{(-3, 5), (-2, 7), (1, 11)\}$ and **b.** $f(x) = \dfrac{2}{x + 3}$.

AN APPLICATION

Functions are used to mathematically describe certain relationships where one quantity depends upon another. Letters other than f and x are often chosen to more clearly describe these situations.

EXAMPLE 9

Cosmetology. A cosmetologist rents a station from the owner of a beauty salon for \$18 a day. She expects to make \$12 profit from each customer she serves. Write a linear function describing her daily income if she serves c customers per day. Then graph the function.

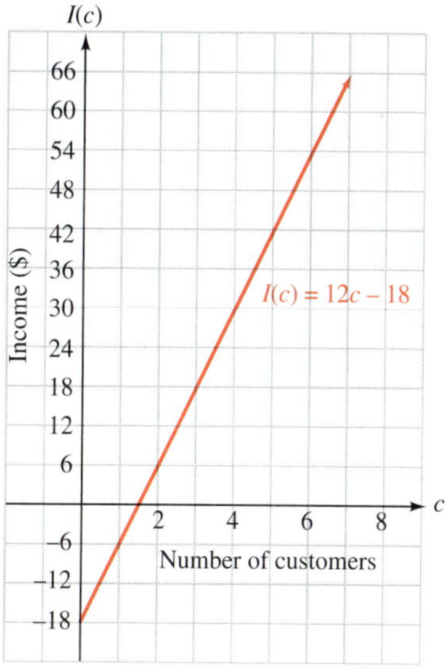

Solution The cosmetologist makes a profit of \$12 per customer, so if she serves c customers a day, she will make \$12c. To find her income, we must *subtract* the \$18 rental fee from the profit. Therefore, the income function is $I(c) = 12c - 18$.

The graph of this linear function is a line with slope 12 and intercept $(0, -18)$. Since the cosmetologist cannot have a negative number of customers, we do not extend the line into quadrant III.

ACCENT ON TECHNOLOGY: EVALUATING FUNCTIONS

We can use a graphing calculator to find function values.

For example, to find the income earned by the cosmetologist in Example 9 for different numbers of customers, we first graph the income function $I(c) = 12c - 18$ as $y = 12x - 18$, using window settings of [0, 10] for x and [0, 100] for y to obtain figure (a). To find her income when she serves seven customers, we trace and move the cursor until the x-coordinate on the screen is nearly 7, as in figure (b). From the screen, we see that her income is about \$66.25.

To find her income when she serves nine customers, we trace and move the cursor until the x-coordinate is nearly 9, as in figure (c). From the screen, we see that her income is about \$90.51.

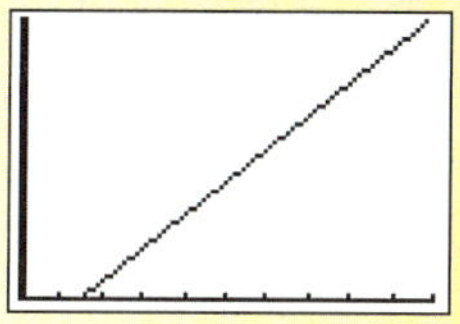

(a)

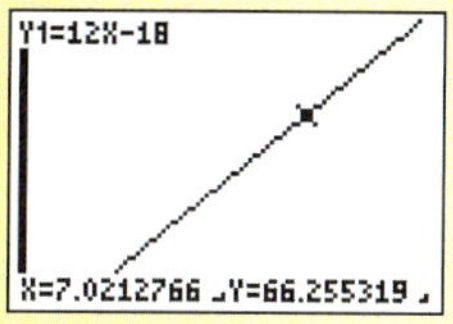

(b)

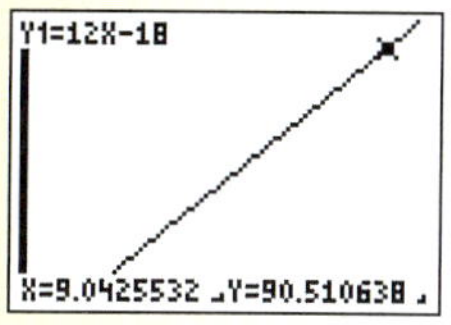

(c)

With some graphing calculator models, we can evaluate a function by entering function notation. To find $I(15)$, the income earned by the cosmetologist of Example 9 if she serves 15 customers, we use the following steps on a TI-83 Plus calculator.

With $I(c) = 12c - 18$ entered as $Y_1 = 12x - 18$, we call up the home screen by pressing [2nd] [QUIT]. Then we enter [VARS] [▶] [1] [ENTER]. The symbolism Y_1 will be displayed. See figure (a). Next, we enter the input value 15, as shown in figure (b), and press [ENTER]. In figure (c) we see that $Y_1(15) = 162$. That is, $I(15) = 162$. The cosmetologist will earn \$162 if she serves 15 customers in one day.

(a)

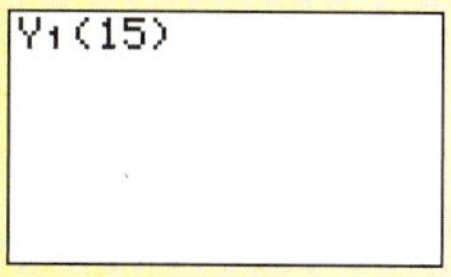

(b)

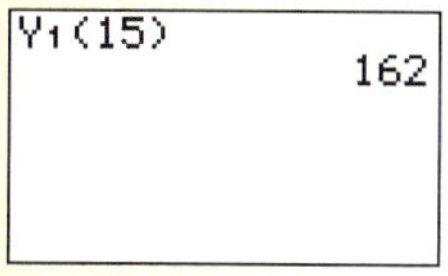

(c)

Answers to Self Checks **1. a.** no, **b.** yes **2. a.** yes, **b.** no; $x = 3$ is assigned two y-values, 3 and -3. **3. a.** -5, **b.** 5, **c.** $2t - 1$ **4. a.** -9, **b.** -1.18 **5.** 72.4 in.^{2} **6.** -27

7.

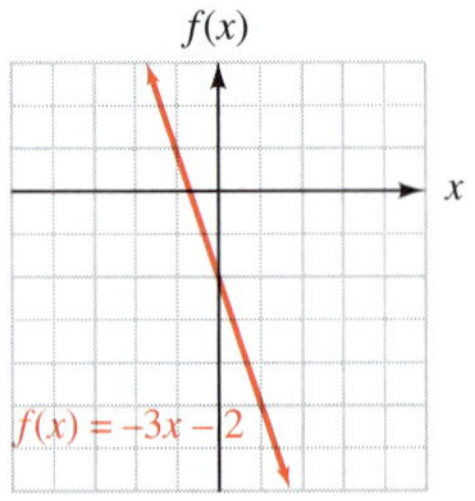

8. a. $\{-3, -2, 1\}$, $\{5, 7, 11\}$, **b.** D: the set of all real numbers except -3; R: the set of all real numbers except 0

2.5 STUDY SET

VOCABULARY Fill in the blanks.

1. A ________ is a rule (or correspondence) that assigns to each value of one variable (called the independent variable) exactly ______ value of another variable (called the dependent variable).

2. The set of all possible values that can be used for the independent variable is called the ________. The set of all values of the dependent variable is called the ________.

3. We can think of a function as a machine that takes some ________ x and turns it into some ________ $f(x)$.

4. The notation $y = f(x)$ denotes that the variable y is a ________ of x.

5. If $f(2) = -1$, we call -1 a function ________.

6. We call $f(x) = 2x + 1$ a ________ function because its graph is a straight line.

CONCEPTS

7. RECYCLING The following table gives the annual average price (in cents) paid for one pound of aluminum cans. Use an arrow diagram to show how members of the domain are assigned members of the range.

Year	1996	1997	1998	1999	2000	2001
Cents per lb	33	35	30	30	32	35

Source: *Midwest Assistance Program*

8. The arrow diagram describes a function.

a. What is the domain of the function?

b. What is the range of the function?

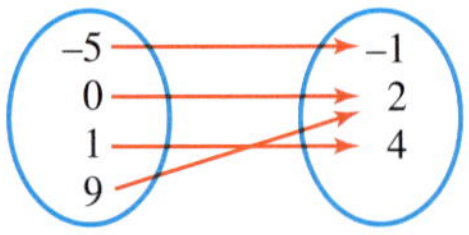

9. Fill in the blank so that the statements are equivalent:

- If $y = 5x + 1$, find the value of y when $x = 8$.
- If $f(x) = 5x + 1$, find ______.

10. For the given input, what value will the function machine output?

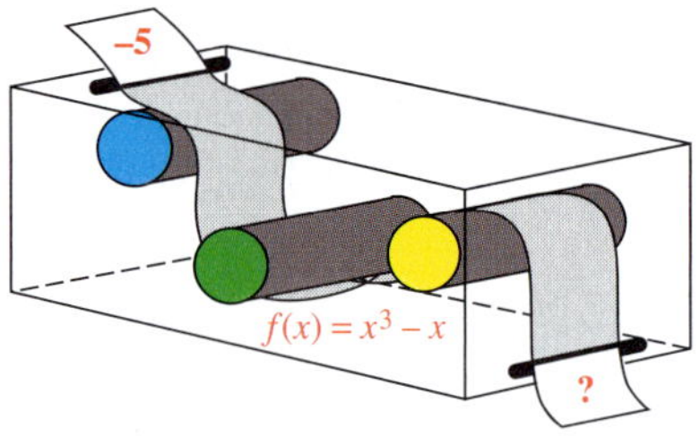

11. Complete the table of function values. Then give the corresponding ordered pairs.

$f(x) = 2x^2 - 1$

x	y	
-3		→
0		→
2		→

12. Fill in the blank: If a ________ line intersects a graph in more than one point, the graph is not the graph of a function.

13. a. Give the coordinates of the points where the given vertical line intersects the graph.

b. Is this the graph of a function? Explain your answer.

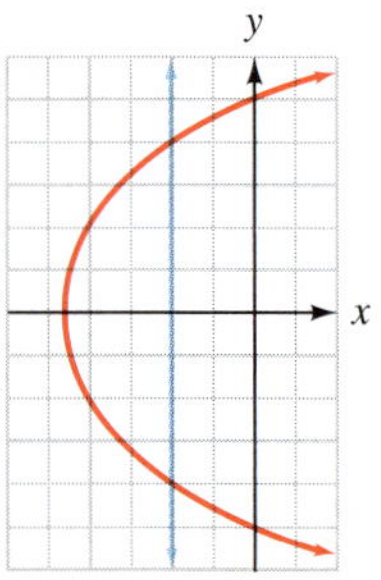

14. Explain why -4 isn't in the domain of $f(x) = \dfrac{1}{x+4}$.

NOTATION **Fill in the blanks.**

15. We read $f(x) = 5x - 6$ as "f ___ x is $5x$ minus 6."

16. This variable represents the ___.

$$f(x) = 2x - 5$$

This is the ___ of the function.

Use this expression to find the ___.

17. Since $y =$ ___, the equations $y = 3x + 2$ and $f(x) = 3x + 2$ are equivalent.

18. The notation $f(2) = 7$ indicates that when the x-value ___ is input into a function rule, the output is ___. This fact can be shown graphically by plotting the ordered pair (___ , ___).

19. When graphing the function $f(x) = -x + 5$, the vertical axis of the rectangular coordinate system can be labeled ___ or ___.

20. The graphing calculator display shows a table of values for a function f.

$f(-1) =$ ___ $\qquad f(3) =$ ___

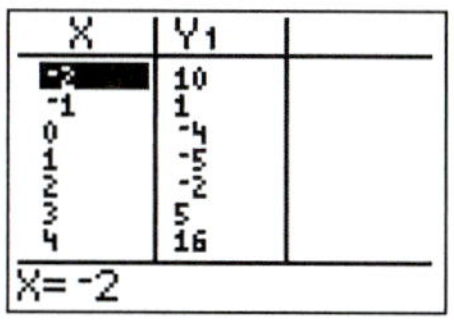

PRACTICE **Determine whether each arrow diagram and table defines y as a function of x. If it does not, indicate a value of x that is assigned more than one value of y.**

21.

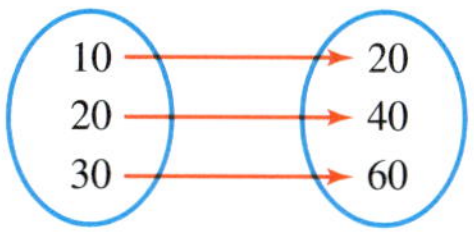

22.

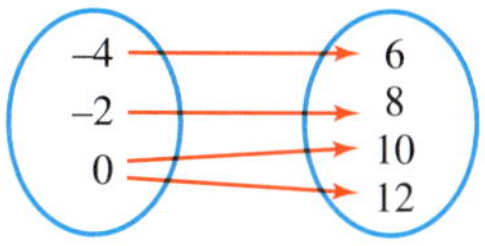

23.

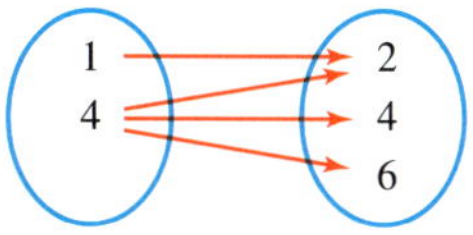

24.

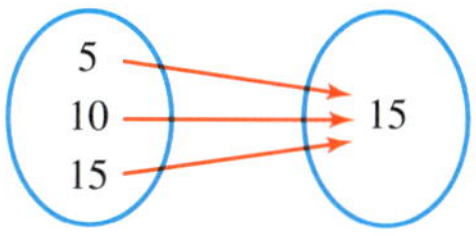

25.

x	y
1	7
2	15
3	23
4	16
5	8

26.

x	y
30	2
30	4
30	6
30	8
30	10

27.

x	y
-4	6
-1	0
0	-3
2	4
-1	2

28.

x	y
1	1
2	2
3	3
4	4

29.

x	y
3	4
3	-4
4	3
4	-3

30.

x	y
-1	1
-3	1
-5	1
-7	1
-9	1

Decide whether the equation defines y as a function of x.

31. $y = 2x + 3$

32. $y = 4x - 1$

33. $y = 2x^2$

34. $y^2 = x + 1$

35. $y^2 = 3 - 2x$

36. $y = 3 + 7x^2$

37. $x = |y|$

38. $y = |x|$

Find $f(3)$ and $f(-1)$.

39. $f(x) = 3x$

40. $f(x) = -4x$

41. $f(x) = 2x - 3$

42. $f(x) = 3x - 5$

43. $f(x) = 7 + 5x$

44. $f(x) = 3 + 3x$

45. $f(x) = 9 - 2x$

46. $f(x) = 12 + 3x$

Find $g(2)$ and $g(3)$.

47. $g(x) = x^2$

48. $g(x) = x^2 - 2$

49. $g(x) = x^3 - 1$

50. $g(x) = x^3$

51. $g(x) = (x + 1)^2$

52. $g(x) = (x - 3)^2$

53. $g(x) = 2x^2 - x$

54. $g(x) = 5x^2 + 2x$

Find $h(2)$ and $h(-2)$.

55. $h(x) = |x| + 2$

56. $h(x) = |x| - 5$

57. $h(x) = x^2 - 2$

58. $h(x) = x^2 + 3$

59. $h(x) = \dfrac{1}{x + 3}$

60. $h(x) = \dfrac{3}{x - 4}$

61. $h(x) = \dfrac{x}{x - 3}$

62. $h(x) = \dfrac{x}{x^2 + 2}$

Complete each table.

63. $f(t) = |t - 2|$

t	$f(t)$
−1.7	
0.9	
5.4	

64. $f(r) = -2r^2 + 1$

Input	Output
−1.7	
0.9	
5.4	

65. $g(x) = x^3$

Input	Output
$-\frac{3}{4}$	
$\frac{1}{6}$	
$\frac{5}{2}$	

66. $g(x) = 2\left(-x - \frac{1}{4}\right)$

x	$g(x)$
$-\frac{3}{4}$	
$\frac{1}{8}$	
$\frac{5}{2}$	

Find $g(w)$ and $g(w + 1)$.

67. $g(x) = 2x$

68. $g(x) = -3x$

69. $g(x) = 3x - 5$

70. $g(x) = 2x - 7$

Let $f(x) = -2x + 5$. For what value(s) of x is

71. $f(x) = 5$?

72. $f(x) = -7$?

Let $f(x) = \frac{3}{2}x - 2$. For what value(s) of x is

73. $f(x) = -\frac{1}{2}$?

74. $f(x) = \frac{2}{3}$?

Find the domain and range of each function.

75. $\{(-2, 3), (4, 5), (6, 7)\}$

76. $\{(0, 2), (1, 2), (3, 4)\}$

77. $s(x) = 3x + 6$

78. $h(x) = \frac{4}{5}x - 8$

79. $f(x) = x^2$

80. $g(x) = x^3$

81. $s(x) = |x - 7|$

82. $t(x) = \left|\frac{2x}{3} + 1\right|$

83. $f(x) = \dfrac{1}{x - 4}$

84. $f(x) = \dfrac{5}{x + 1}$

Use the vertical line test to decide whether the given graph represents a function.

85.

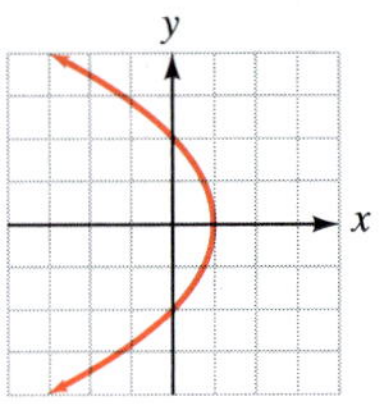

86.

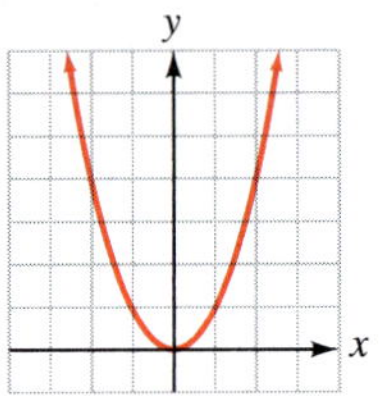

87.

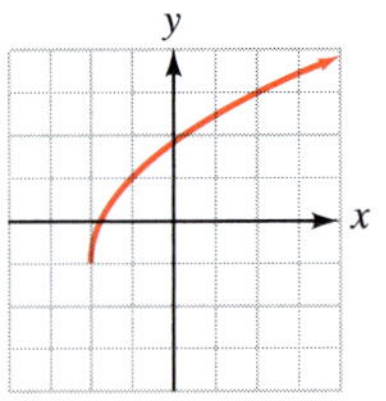

88.

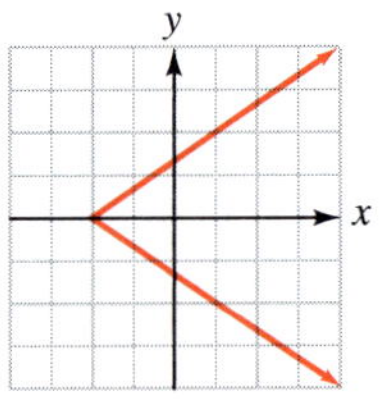

89.

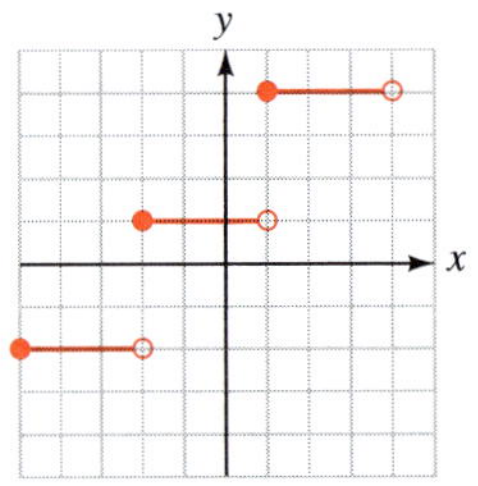

90.

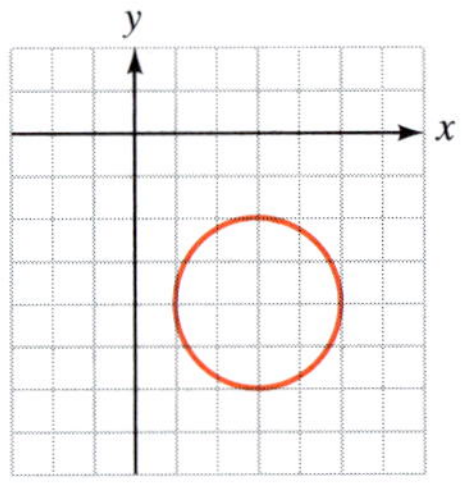

91.

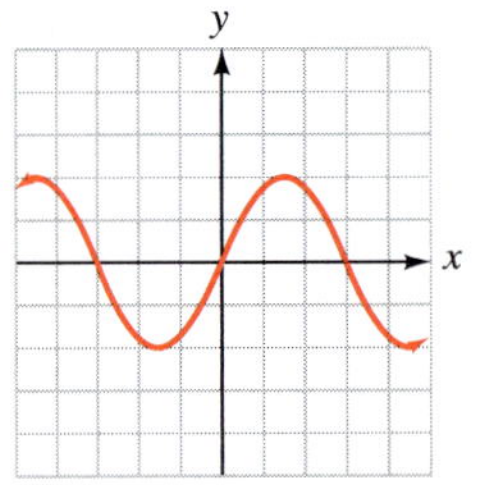

92.

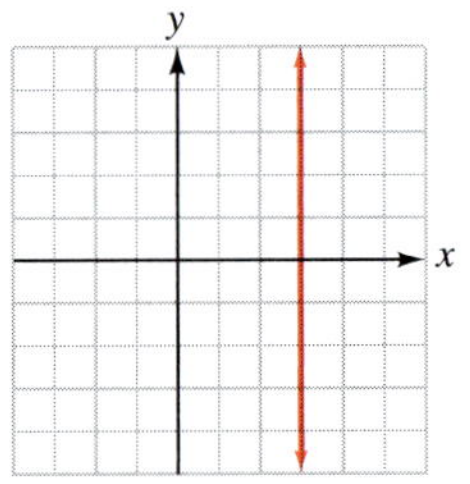

Graph each function.

93. $f(x) = 2x - 1$

94. $f(x) = -x + 2$

95. $f(x) = \frac{2}{3}x - 2$

96. $f(x) = -\frac{3}{2}x - 3$

APPLICATIONS

97. DECONGESTANTS The temperature in degrees Celsius that is equivalent to a temperature in degrees Fahrenheit is given by the linear function $C(F) = \frac{5}{9}(F - 32)$. Use this function to find the temperature range, in degrees Celsius, at which a bottle of Dimetapp should be stored. The label directions follow.

> **DIRECTIONS:** Adults and children 12 years of age and over: Two teaspoons every 4 hours. DO NOT EXCEED 6 DOSES IN A 24-HOUR PERIOD. Store at a controlled room temperature between 68°F and 77°F.

98. BODY TEMPERATURES The temperature in degrees Fahrenheit that is equivalent to a temperature in degrees Celsius is given by the linear function $F(C) = \frac{9}{5}C + 32$. Convert each of the temperatures in the following excerpt from *The Good Housekeeping Family Health and Medical Guide* to degrees Fahrenheit. (Round to the nearest degree.)

> *In disease, the temperature of the human body may vary from about 32.2°C to 43.3°C for a time, but there is grave danger to life should it drop and remain below 35°C or rise and remain at or above 41°C.*

99. CONCESSIONAIRES A baseball club pays a peanut vendor $50 per game for selling bags of peanuts for $1.75 each.

a. Write a linear function that describes the income the vendor makes for the baseball club during a game if she sells b bags of peanuts.

b. Find the income the baseball club will make if the vendor sells 110 bags of peanuts during a game.

100. HOME CONSTRUCTION In a proposal to some prospective clients, a housing contractor listed the following costs.

Fees, permits, miscellaneous	$12,000
Construction, per square foot	$75

a. Write a linear function that the clients could use to determine the cost of building a home having f square feet.

b. Find the cost to build a home having 1,950 square feet.

101. EARTH'S ATMOSPHERE The illustration shows a graph of the temperatures of the atmosphere at various altitudes above the Earth's surface. The temperature is expressed in degrees Kelvin, a scale widely used in scientific work.

a. Estimate the coordinates of three points on the graph that have an x-coordinate of 200.

b. Explain why this is not the graph of a function.

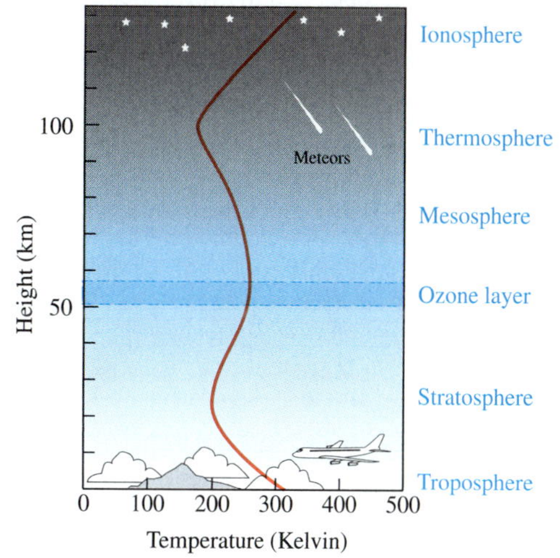

102. CHEMICAL REACTIONS When students in a chemistry laboratory mixed solutions of acetone and chloroform, they found that heat was immediately generated. As time went by, the mixture cooled down. The illustration on the next page shows a graph of data points of the form (time, temperature) taken by the students.

a. The linear function $T(t) = -\frac{t}{240} + 30$ models the relationship between the elapsed time t since the solutions were combined and the temperature $T(t)$ of the mixture. Graph the function.

b. Predict the temperature of the mixture immediately after the two solutions are combined.

c. Is $T(180)$ more or less than the temperature recorded by the students for $t = 300$?

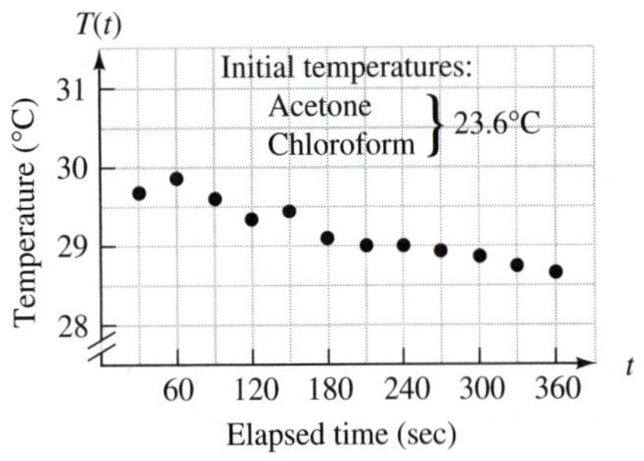

103. TAXES The function

$$T(a) = 700 + 0.15(a - 7{,}000)$$

(where a is adjusted gross income) is a model of the instructions given on the first line of the following tax rate Schedule X.

a. Find $T(25{,}000)$ and interpret the result.

b. Write a function that models the second line on Schedule X.

Schedule X–Use if your filing status is **Single**				**2003**
If your adjusted gross income is: *Over—*	*But not over—*	Your tax is		*of the amount over—*
\$ 7,000	\$28,400	**\$ 700 +**	**15%**	**\$ 7,000**
\$28,400	\$68,800	**\$3,910 +**	**25%**	**\$28,400**

104. COST FUNCTIONS An electronics firm manufactures DVD recorders, receiving \$120 for each recorder it makes. If x represents the number of recorders produced, the income received is determined by the *revenue function* $R(x) = 120x$. The manufacturer has fixed costs of \$12,000 per month and variable costs of \$57.50 for each recorder manufactured. Thus, the *cost function* is $C(x) = 57.50x + 12{,}000$. How many recorders must the company sell for revenue to equal cost? (*Hint:* Set $R(x) = C(x)$.)

WRITING

105. What is a function?

106. Explain why we can think of a function as a machine.

REVIEW **Show that each number is a rational number by expressing it as a ratio of two integers.**

107. $-3\frac{3}{4}$

108. 4.7

109. 0.333. . .

110. $-0.\overline{6}$

CHALLENGE PROBLEMS **Let $f(x) = 2x + 1$ and $g(x) = x^2$.**

111. Is $f(x) + g(x)$ equal to $g(x) + f(x)$?

112. Is $f(x) - g(x)$ equal to $g(x) - f(x)$?

2.6 Graphs of Functions

- Finding function values graphically
- Finding domain and range graphically
- Graphs of nonlinear functions
- Translations of graphs
- Reflections of graphs

Since a graph is often the best way to describe a function, we need to know how to construct and interpret graphs of functions.

FINDING FUNCTION VALUES GRAPHICALLY

Recall that the graph of a function is a picture of the ordered pairs $(x, f(x))$ that define the function. From the graph of a function, we can determine function values.

EXAMPLE 1

Refer to the graph of function f in figure (a). **a.** Find $f(-3)$, and **b.** find the value of x for which $f(x) = -2$.

Solution

a. To find $f(-3)$, we need to find the y-value that f assigns to the x-value -3. If we draw a vertical line through -3 on the x-axis, as shown in figure (b), the line intersects the graph of f at $(-3, 5)$. Therefore, 5 is assigned to -3, and it follows that $f(-3) = 5$.

b. We need to find the input value x that is assigned the output value -2. If we draw a horizontal line through -2 on the y-axis, as shown in figure (c), it intersects the graph of f at $(4, -2)$. Therefore, the function assigns -2 to 4, and it follows that $f(4) = -2$.

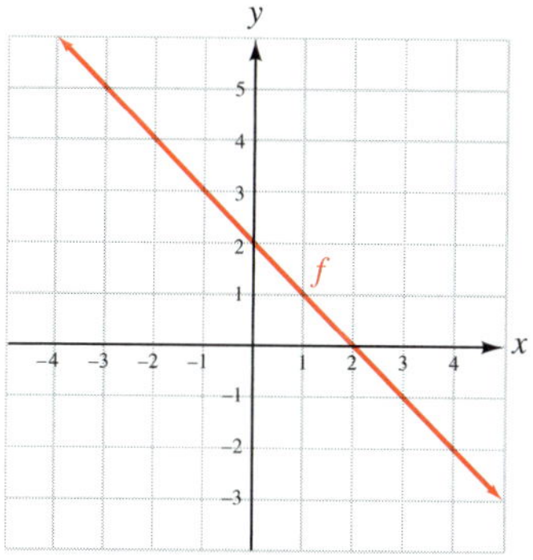

(a)

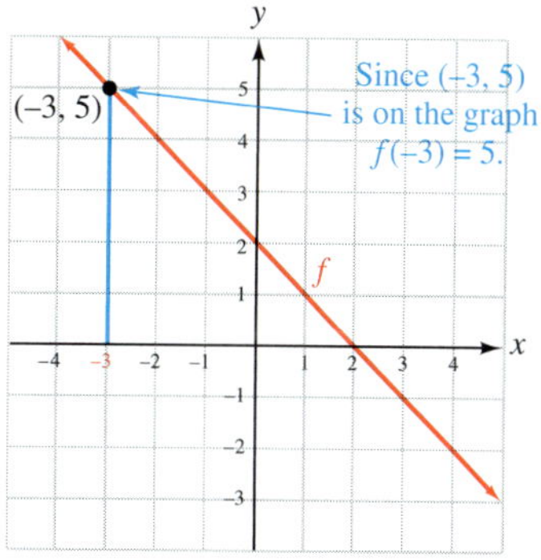

(b)

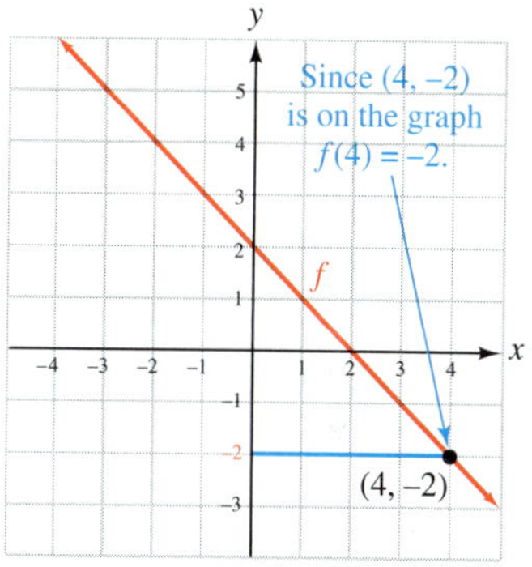

(c)

Self Check 1 From the graph of function g: **a.** find $g(-3)$, and **b.** find the x-value for which $g(x) = 4$.

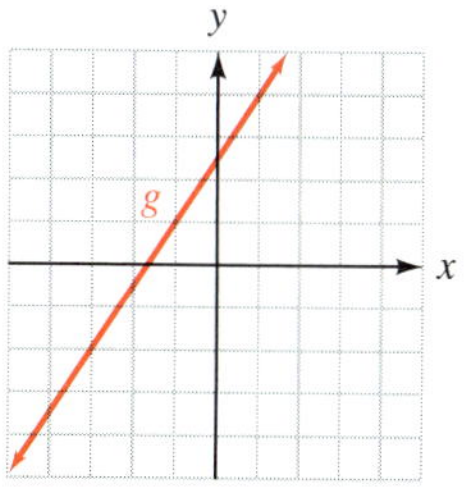

FINDING DOMAIN AND RANGE GRAPHICALLY

We can determine the domain and range of a function from its graph. For example, to find the domain of the linear function graphed in figure (a), we *project* the graph onto the x-axis. Because the graph of the function extends indefinitely to the left and to the right, the projection includes all the real numbers. Therefore, the domain of the function is the set of real numbers.

To determine the range of the same linear function, we project the graph onto the y-axis, as shown in figure (b). Because the graph of the function extends indefinitely upward and downward, the projection includes all the real numbers. Therefore, the range of the function is the set of real numbers.

The Language of Algebra

Think of the *projection* of a graph on an axis as the "shadow" that the graph makes on the axis.

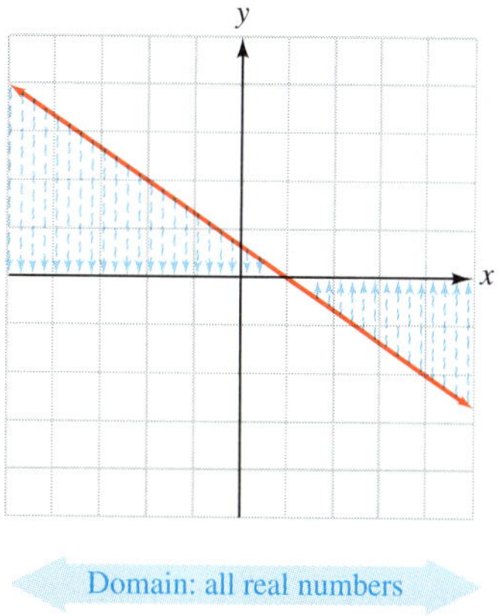

(a)

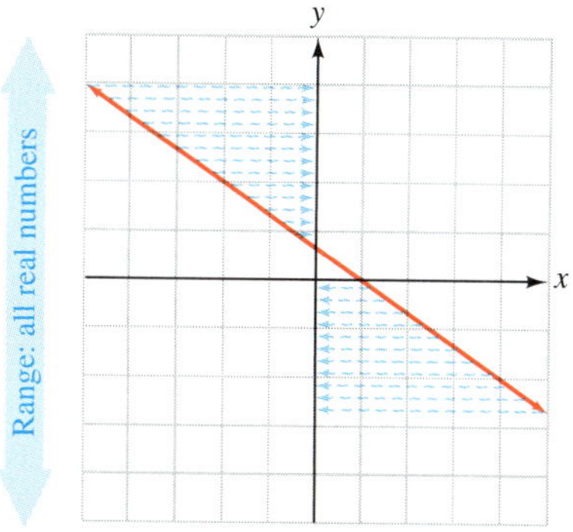

(b)

GRAPHS OF NONLINEAR FUNCTIONS

We have seen that the graph of a linear function is a straight line. We will now consider several examples of **nonlinear functions.** Their graphs are not straight lines. We will begin with $f(x) = x^2$, called the **squaring function.**

EXAMPLE 2

Graph $f(x) = x^2$ and find its domain and range.

Solution To graph the function, we select numbers for x and find the corresponding values for $f(x)$. For example, if we choose -3 for x, we have

$$f(x) = x^2$$
$$f(-3) = (-3)^2 \quad \text{Substitute } -3 \text{ for } x.$$
$$= 9$$

Since $f(-3) = 9$, the ordered pair $(-3, 9)$ lies on the graph of f. In a similar manner, we find the corresponding values of $f(x)$ for other x-values and list the ordered pairs in the table of values. Then we plot the points and draw a smooth curve through them to get the graph, called a **parabola.**

The Language of Algebra

The cup-like shape of a *parabola* has many real-life applications. For example, a satellite TV dish is often called a *parabolic* dish.

$f(x) = x^2$

x	$f(x)$	
-3	9	→ $(-3, 9)$
-2	4	→ $(-2, 4)$
-1	1	→ $(-1, 1)$
0	0	→ $(0, 0)$
1	1	→ $(1, 1)$
2	4	→ $(2, 4)$
3	9	→ $(3, 9)$

↑ Choose values for x. ↑ Compute each $f(x)$. ↑ Plot these points.

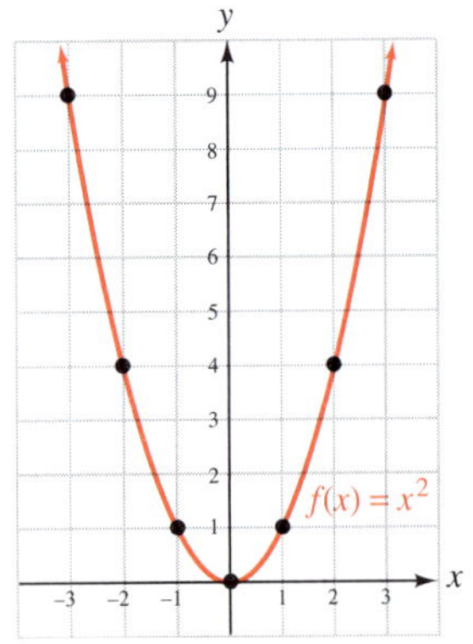

The Language of Algebra

The set of **nonnegative real numbers** is the set of real numbers greater than or equal to 0.

Because the graph extends indefinitely to the left and to the right, the projection of the graph onto the x-axis includes all the real numbers. This means that the domain of the squaring function is the set of real numbers.

Because the graph extends upward indefinitely from the point (0, 0), the projection of the graph on the y-axis includes only positive real numbers and zero. This means that the range of the squaring function is the set of nonnegative real numbers.

Self Check 2 Graph $g(x) = x^2 - 2$ and find its domain and range. Compare the graph to the graph of $f(x) = x^2$.

Another important nonlinear function is $f(x) = x^3$, called the **cubing function.**

EXAMPLE 3

Graph $f(x) = x^3$ and find its domain and range.

Solution To graph the function, we select numbers for x and find the corresponding values for $f(x)$. For example, if we choose -2 for x, we have

$$\begin{aligned} f(x) &= x^3 \\ f(-2) &= (-2)^3 \quad \text{Substitute } -2 \text{ for } x. \\ &= -8 \end{aligned}$$

Since $f(-2) = -8$, the ordered pair $(-2, -8)$ lies on the graph of f. In a similar manner, we find the corresponding values of $f(x)$ for other x-values and list the ordered pairs in the table. Then we plot the points and draw a smooth curve through them to get the graph.

$f(x) = x^3$

x	$f(x)$	
-2	-8	→ $(-2, -8)$
-1	-1	→ $(-1, -1)$
0	0	→ $(0, 0)$
1	1	→ $(1, 1)$
2	8	→ $(2, 8)$

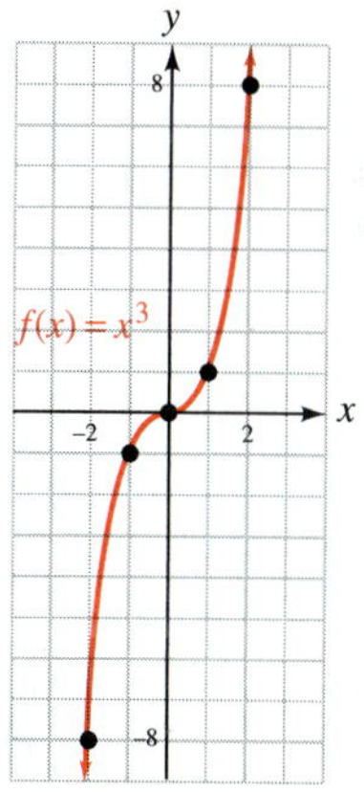

Because the graph of the function extends indefinitely to the left and to the right, the projection includes all the real numbers. Therefore, the domain of the cubing function is the set of real numbers.

Because the graph of the function extends indefinitely upward and downward, the projection includes all the real numbers. Therefore, the range of the cubing function is the set of real numbers.

Self Check 3 Graph $g(x) = x^3 + 1$ and find its domain and range. Compare the graph to the graph of $f(x) = x^3$.

A third nonlinear function is $f(x) = |x|$, called the **absolute value function.**

EXAMPLE 4

Graph $f(x) = |x|$ and find its domain and range.

Solution To graph the function, we select numbers for x and find the corresponding values for $f(x)$. For example, if we choose -3 for x, we have

$$\begin{aligned} f(x) &= |x| \\ f(-3) &= |-3| \quad \text{Substitute } -3 \text{ for } x. \\ &= 3 \end{aligned}$$

Since $f(-3) = 3$, the ordered pair $(-3, 3)$ lies on the graph of f. In a similar manner, we find the corresponding values of $f(x)$ for other x-values and list the ordered pairs in the table. Then we plot the points and connect them to get the following V-shaped graph.

$f(x) = |x|$

x	$f(x)$	
−3	3	→ (−3, 3)
−2	2	→ (−2, 2)
−1	1	→ (−1, 1)
0	0	→ (0, 0)
1	1	→ (1, 1)
2	2	→ (2, 2)
3	3	→ (3, 3)

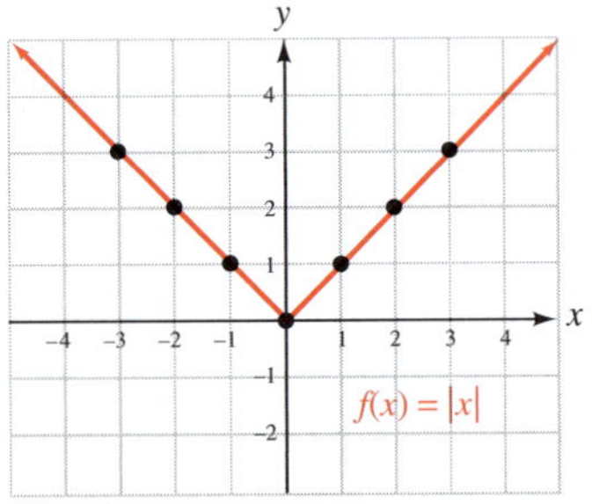

Because the graph extends indefinitely to the left and to the right, the projection of the graph onto the x-axis includes all the real numbers. The domain of the absolute value function is the set of real numbers.

Because the graph extends upward indefinitely from the point (0, 0), the projection of the graph on the y-axis includes only positive real numbers and zero. The range of the absolute value function is the set of nonnegative real numbers.

Self Check 4 Graph $g(x) = |x - 2|$ and find its domain and range. Compare the graph to the graph of $f(x) = |x|$.

ACCENT ON TECHNOLOGY: GRAPHING FUNCTIONS

We can graph nonlinear functions with a graphing calculator. For example, to graph $f(x) = x^2$ in a standard window of $[-10, 10]$ for x and $[-10, 10]$ for y, we press [Y =], enter the function by typing [x^2], and press the [GRAPH] key. We will obtain the graph shown in figure (a).

To graph $f(x) = x^3$, we enter the function by typing x ^ 3 and then press the [GRAPH] key to obtain the graph in figure (b). To graph $f(x) = |x|$, we enter the function by selecting abs from the NUM option within the MATH menu, typing x, and pressing the [GRAPH] key to obtain the graph in figure (c).

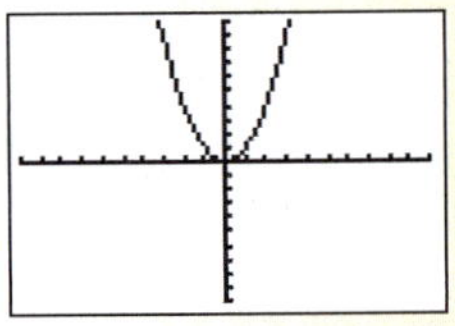

(a)

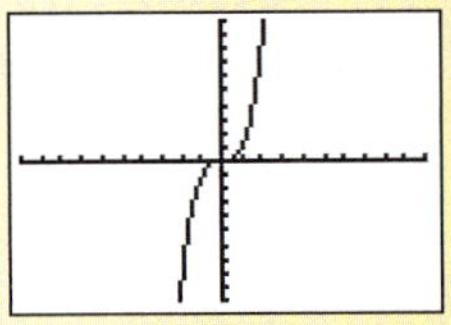

(b)

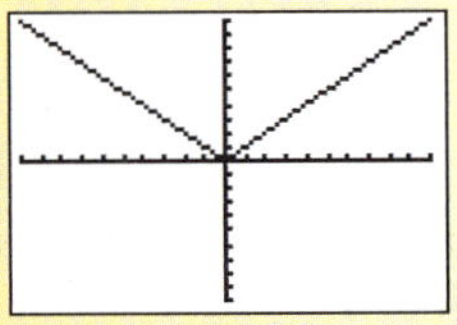

(c)

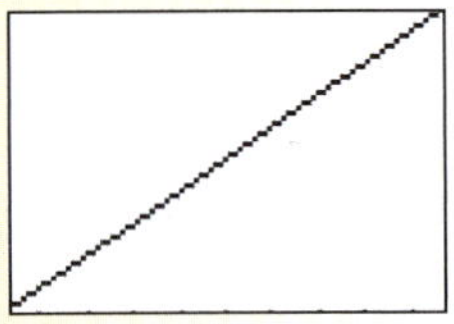

(d)

When using a graphing calculator, we must be sure that the viewing window does not show a misleading graph. For example, if we graph $f(x) = |x|$ in the window [0, 10] for x and [0, 10] for y, we will obtain a misleading graph that looks like a line. See figure (d). This is not correct. The proper graph is the V-shaped graph shown in figure (c). One of the challenges of using graphing calculators is finding an appropriate viewing window.

TRANSLATIONS OF GRAPHS

Examples 2, 3, and 4 and their Self Checks suggest that the graphs of different functions may be identical except for their positions in the *xy*-plane. For example, the figure shows the graph of $f(x) = x^2 + k$ for three different values of k. If $k = 0$, we get the graph of $f(x) = x^2$. If $k = 3$, we get the graph of $f(x) = x^2 + 3$, which is identical to the graph of $f(x) = x^2$ except that it is shifted 3 units upward. If $k = -4$, we get the graph of $f(x) = x^2 - 4$, which is identical to the graph of $f(x) = x^2$ except that it is shifted 4 units downward. These shifts are called **vertical translations.**

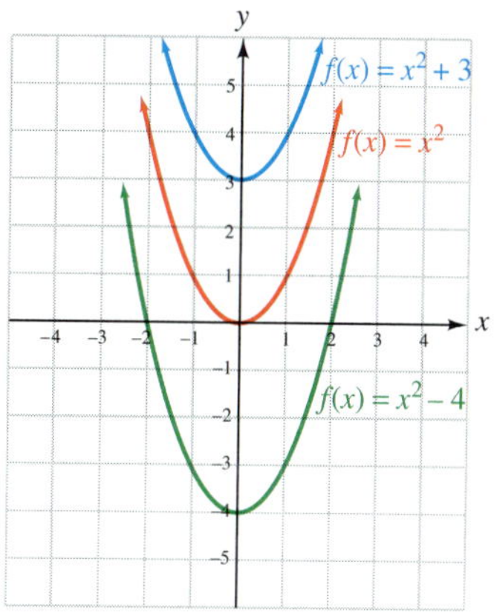

In general, we can make these observations.

Vertical Translations

If f is a function and k represents a positive number, then

- The graph of $y = f(x) + k$ is identical to the graph of $y = f(x)$ except that it is translated k units upward.
- The graph of $y = f(x) - k$ is identical to the graph of $y = f(x)$ except that it is translated k units downward.

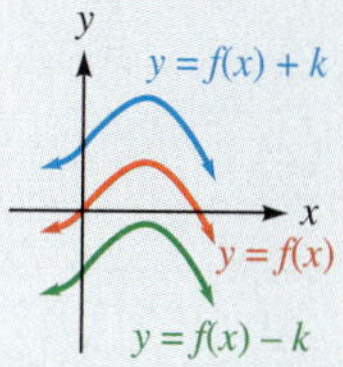

EXAMPLE 5

Graph: $g(x) = |x| + 2$.

Solution The graph of $g(x) = |x| + 2$ will be the same V-shaped graph as $f(x) = |x|$, except that it is shifted 2 units upward.

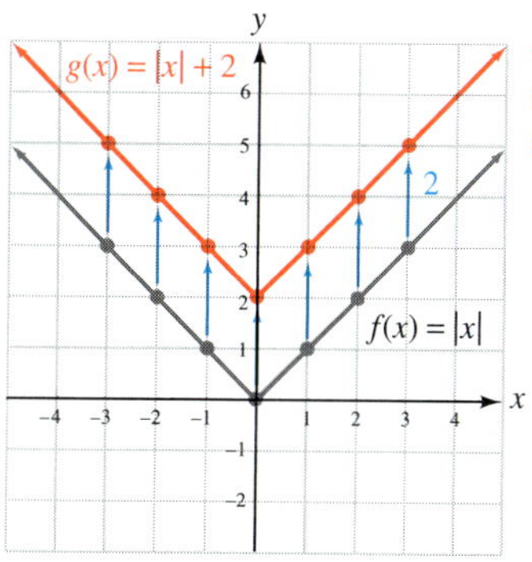

To graph $g(x) = |x| + 2$, translate each point on the graph of $f(x) = |x|$ up 2 units.

Self Check 5 Graph: $g(x) = |x| - 3$.

The figure on the next page shows the graph of $f(x) = (x + h)^2$ for three different values of h. If $h = 0$, we get the graph of $f(x) = x^2$. The graph of $f(x) = (x - 3)^2$ is identical to the graph of $f(x) = x^2$ except that it is shifted 3 units to the right. The graph of $f(x) = (x + 2)^2$

is identical to the graph of $f(x) = x^2$ except that it is shifted 2 units to the left. These shifts are called **horizontal translations.**

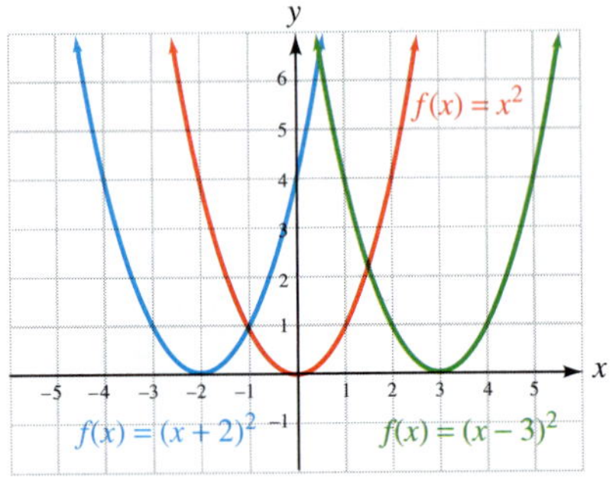

In general, we can make these observations.

Horizontal Translations

If f is a function and h is a positive number, then

- The graph of $y = f(x - h)$ is identical to the graph of $y = f(x)$ except that it is translated h units to the right.
- The graph of $y = f(x + h)$ is identical to the graph of $y = f(x)$ except that it is translated h units to the left.

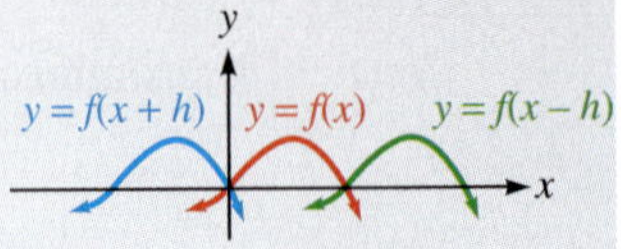

EXAMPLE 6

Graph: $g(x) = (x + 3)^3$.

Solution The graph of $g(x) = (x + 3)^3$ will be the same shape as the graph of $f(x) = x^3$ except that it is shifted 3 units to the left.

Success Tip

To determine the direction of the horizontal translation, find the value of x that makes the expression within the parentheses, $x + 3$, equal to 0. Since -3 makes $x + 3 = 0$, the translation is 3 units to the *left*.

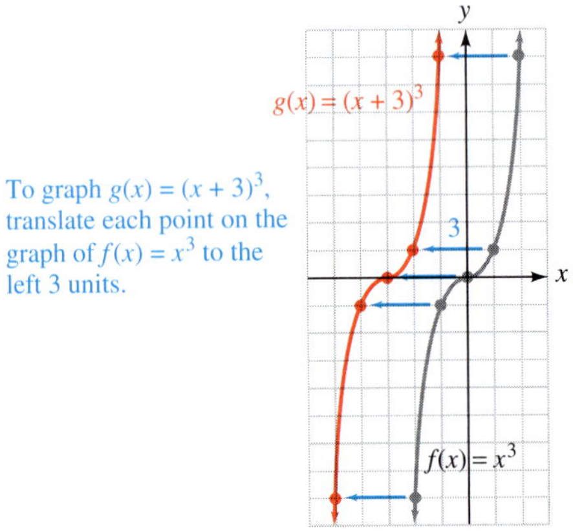

Self Check 6 Graph $g(x) = (x - 2)^2$.

EXAMPLE 7

Graph: $g(x) = (x - 3)^2 + 2$.

Solution Two translations are made to a basic graph. We can graph this function by translating the graph of $f(x) = x^2$ to the right 3 units and then 2 units up, as follows.

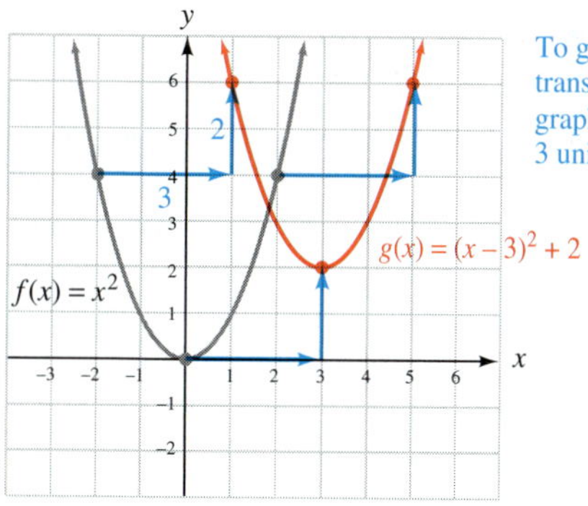

To graph $g(x) = (x - 3)^2 + 2$, translate each point on the graph of $f(x) = x^2$ to the right 3 units and then 2 units up..

Self Check 7 Graph: $g(x) = |x + 2| - 3$.

REFLECTIONS OF GRAPHS

The following figure shows a table of values for $f(x) = x^2$ and for $g(x) = -x^2$. We note that for a given value of x, the corresponding y-values in the tables are opposites. When graphed, we see that the $-$ in $g(x) = -x^2$ has the effect of flipping the graph of $f(x) = x^2$ over the x-axis so that the parabola opens downward. We say that the graph of $g(x) = -x^2$ is a **reflection** of the graph of $f(x) = x^2$ about the x-axis.

$f(x) = x^2$

x	$f(x)$	
−2	4	→ (−2, 4)
−1	1	→ (−1, 1)
0	0	→ (0, 0)
1	1	→ (1, 1)
2	4	→ (2, 4)

$g(x) = -x^2$

x	$f(x)$	
−2	−4	→ (−2, −4)
−1	−1	→ (−1, −1)
0	0	→ (0, 0)
1	−1	→ (1, −1)
2	−4	→ (2, −4)

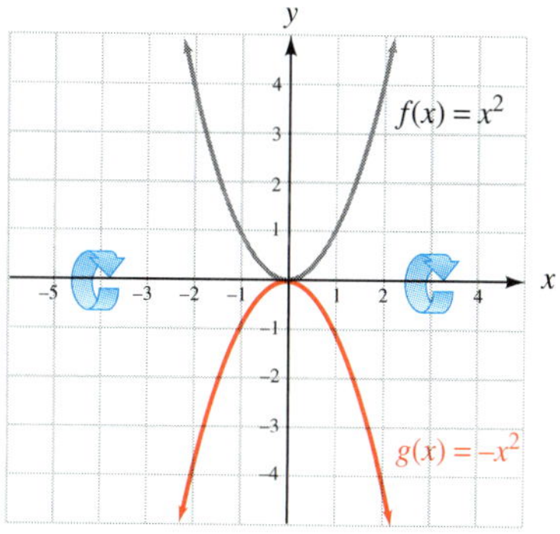

EXAMPLE 8

Graph: $g(x) = -x^3$.

Solution To graph $g(x) = -x^3$, we use the graph of $f(x) = x^3$ from Example 3. First, we reflect the portion of the graph of $f(x) = x^3$ in quadrant I to quadrant IV, as shown. Then we reflect the portion of the graph of $f(x) = x^3$ in quadrant III to quadrant II.

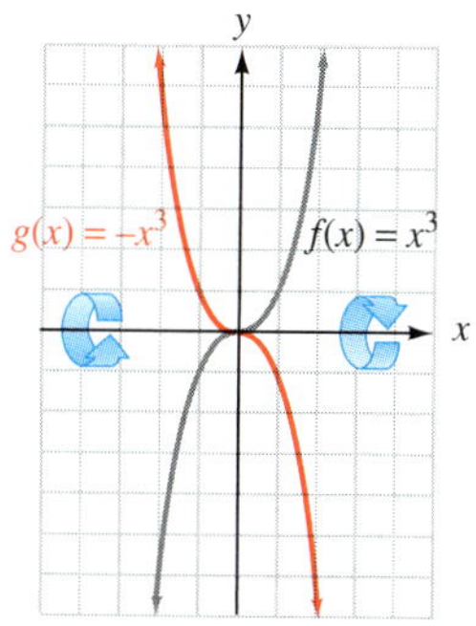

Self Check 8 Graph: $g(x) = -|x|$.

Reflection of a Graph	The graph of $y = -f(x)$ is the graph of $y = f(x)$ reflected about the x-axis.

Answers to Self Checks

1. **a.** −2, **b.** 1

2. D: the set of real numbers, R: the set of all real numbers greater than or equal to −2; the graph has the same shape but is 2 units lower.

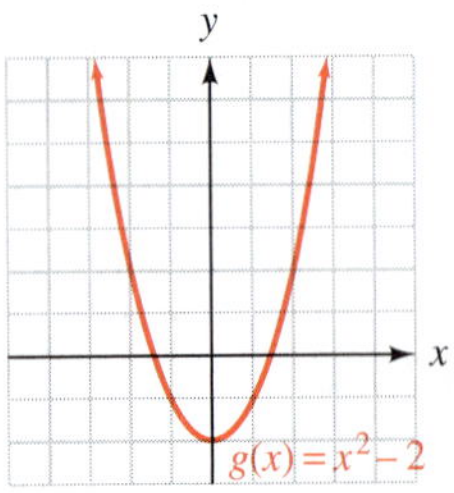

3. D: the set of real numbers, R: the set of real numbers; the graph has the same shape but is 1 unit higher.

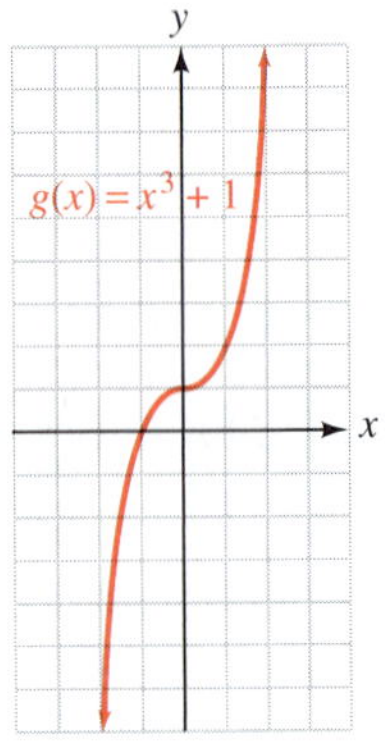

4. D: the set of real numbers, R: the set of nonnegative real numbers; the graph has the same shape but is 2 units to the right.

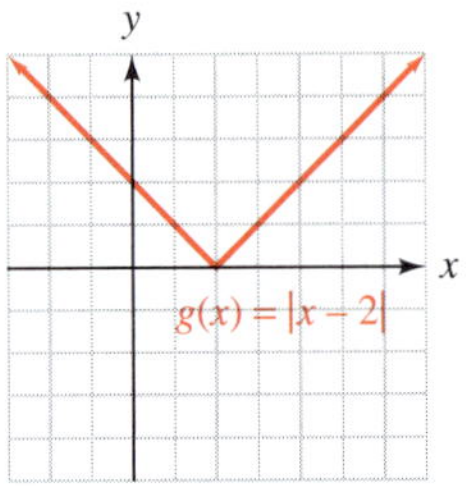

5.

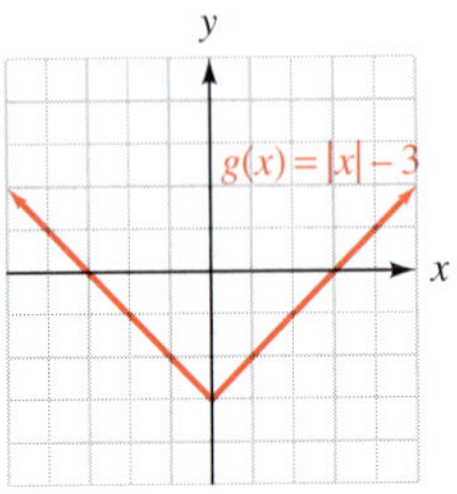

6.

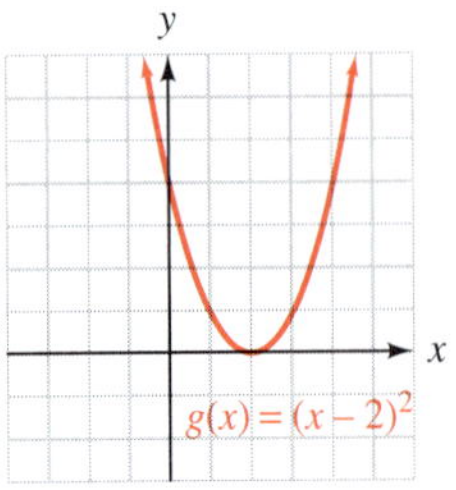

7.

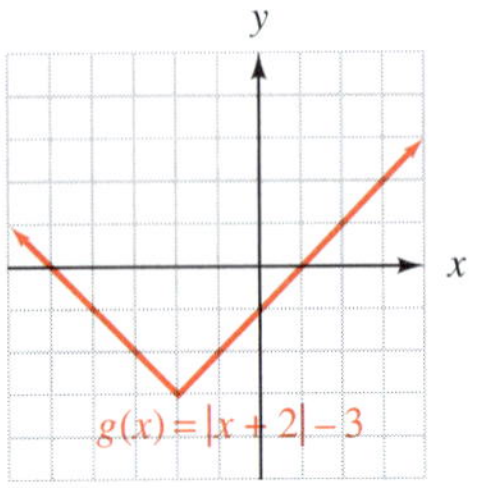

8.

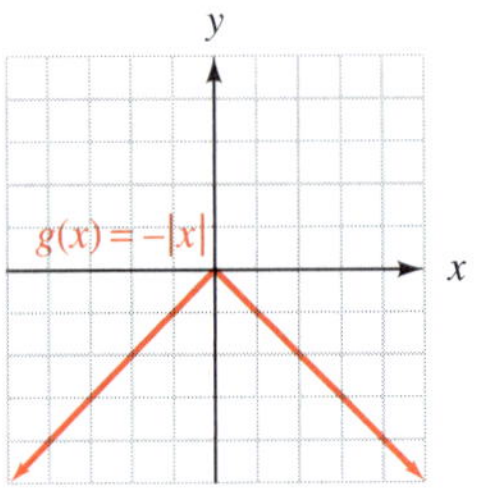

2.6 STUDY SET

VOCABULARY Fill in the blanks.

1. Functions whose graphs are not straight lines are called __________ functions.
2. The function $f(x) = x^2$ is called the __________ function.
3. The graph of $f(x) = x^2$ is a cup-like shape called a __________.
4. The set of __________ real numbers is the set of real numbers greater than or equal to 0.
5. The function $f(x) = x^3$ is called the __________ function.
6. The function $f(x) = |x|$ is called the __________ function.

CONCEPTS

7. Use the graph of function f to find each of the following.

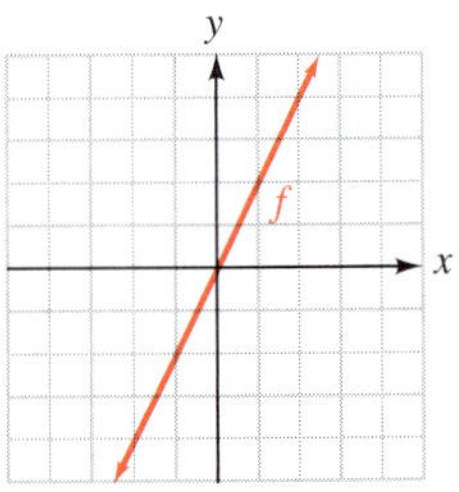

   a. $f(-2)$ **b.** $f(0)$

   c. The value of x for which $f(x) = 4$.

   d. The value of x for which $f(x) = -2$.

8. Use the graph of function g to find each of the following.

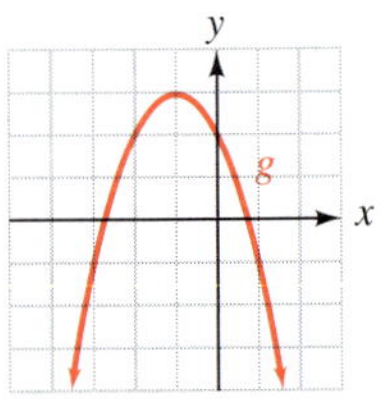

   a. $g(-2)$ **b.** $g(0)$

   c. The value of x for which $g(x) = 3$.

   d. The values of x for which $g(x) = -1$.

9. Use the graph of function h to find each of the following.

a. $h(-3)$ **b.** $h(4)$

c. The value(s) of x for which $h(x) = 1$.

d. The value(s) of x for which $h(x) = 0$.

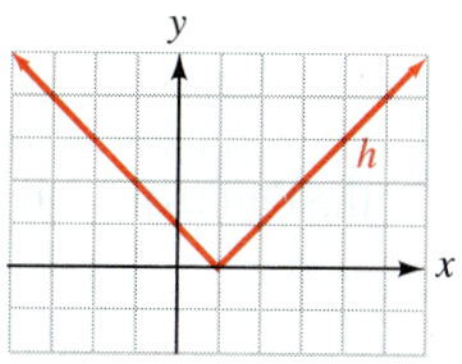

10. Fill in the blanks. The illustration shows the projection of the graph of function f on the ______. We see that the ______ of f is the set of real numbers less than or equal to 0.

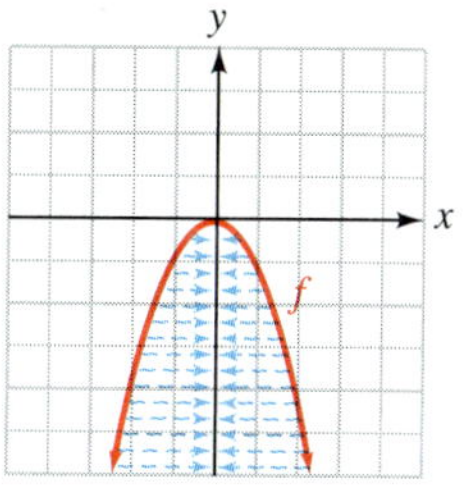

11. Consider the graph of the function f.

a. Label each arrow in the illustration with the appropriate term: *domain* or *range*.

b. Give the domain and range of f.

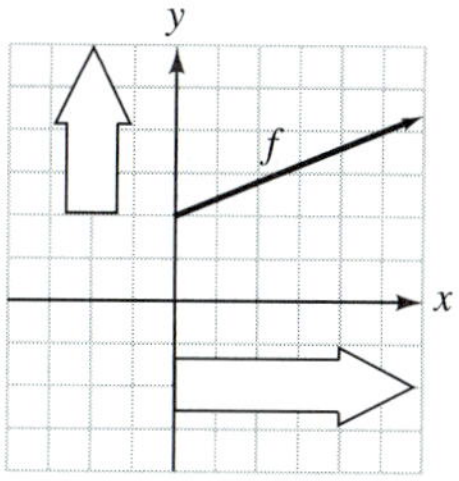

12. The illustration shows the graph of $f(x) = x^2 + k$ for three values of k. What are the three values?

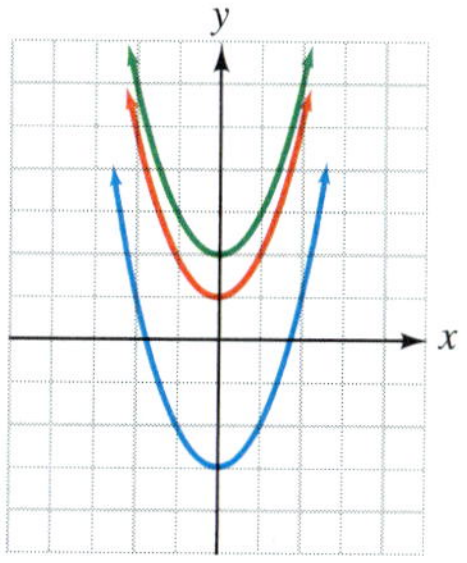

13. The illustration shows the graph of $f(x) = |x + h|$ for three values of h. What are the three values?

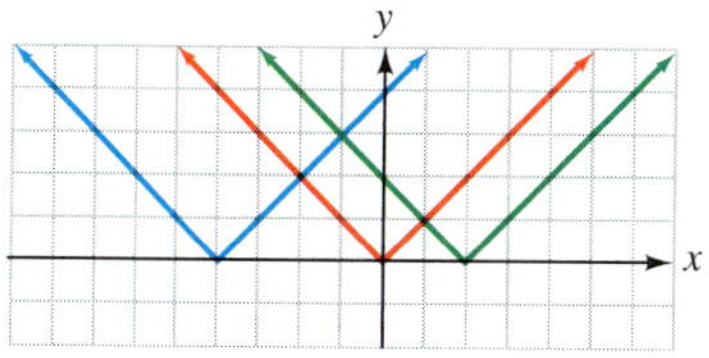

14. Translate each point plotted on the graph to the left 5 units and then up 1 unit.

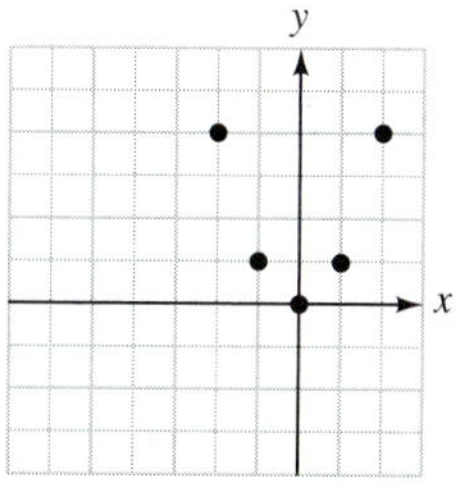

15. Translate each point plotted on the graph to the right 4 units and then down 3 units.

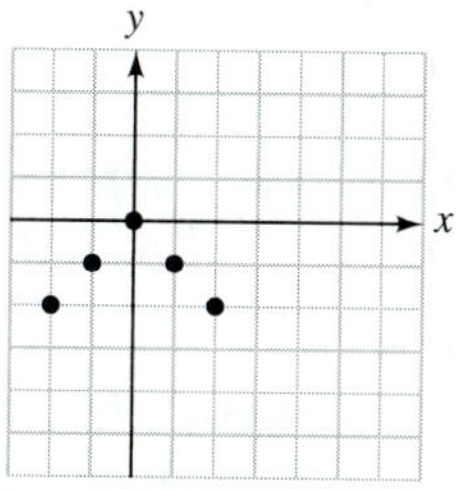

16. Use a graphing calculator to sketch the reflection of the following graph.

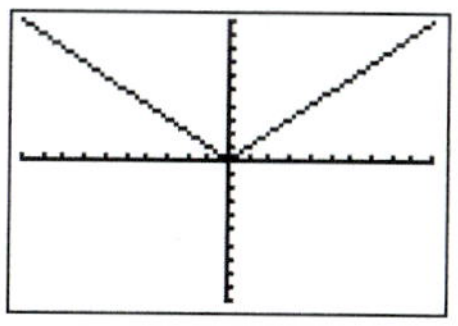

NOTATION **Fill in the blanks.**

17. The graph of $f(x) = (x + 4)^3$ is the same as the graph of $f(x) = x^3$ except that it is shifted ___ units to the ______.

18. The graph of $f(x) = x^3 - 2$ is the same as the graph of $f(x) = x^3$ except that it is shifted ___ units ______.

19. The graph of $f(x) = x^2 + 5$ is the same as the graph of $f(x) = x^2$ except that it is shifted ___ units _____.

20. The graph of $f(x) = |x - 5|$ is the same as the graph of $f(x) = |x|$ except that it is shifted ___ units to the ______.

PRACTICE **Graph each function by plotting points. Give the domain and range.**

21. $f(x) = x^2 - 3$

22. $f(x) = x^2 + 2$

23. $f(x) = (x - 1)^3$

24. $f(x) = (x + 1)^3$

25. $f(x) = |x| - 2$

26. $f(x) = |x| + 1$

27. $f(x) = |x - 1|$

28. $f(x) = |x + 2|$

29. $f(x) = -3x$

30. $f(x) = \frac{1}{4}x + 4$

Graph each function using window settings of [−4, 4] for x and [−4, 4] for y. The graph is not what it appears to be. Pick a better viewing window and find a better representation of the true graph.

31. $f(x) = x^2 + 8$

32. $f(x) = x^3 - 8$

33. $f(x) = |x + 5|$

34. $f(x) = |x - 5|$

35. $f(x) = (x - 6)^2$

36. $f(x) = (x + 9)^2$

37. $f(x) = x^3 + 8$

38. $f(x) = x^3 - 12$

For each function, first sketch the graph of its associated function, $f(x) = x^2$, $f(x) = x^3$, or $f(x) = |x|$. Then draw each graph using a translation or a reflection.

39. $f(x) = x^2 - 5$

40. $f(x) = x^3 + 4$

41. $f(x) = (x - 1)^3$

42. $f(x) = (x + 4)^2$

43. $f(x) = |x - 2| - 1$

44. $f(x) = (x + 2)^2 - 1$

45. $f(x) = (x + 1)^3 - 2$

46. $f(x) = |x + 4| + 3$

47. $f(x) = -x^3$

48. $f(x) = -|x|$

49. $f(x) = -x^2$

50. $f(x) = -(x + 1)^2$

APPLICATIONS

51. OPTICS See the illustration. The law of reflection states that the angle of reflection is equal to the angle of incidence. What function studied in this section models the path of the reflected light beam with an angle of incidence measuring 45°?

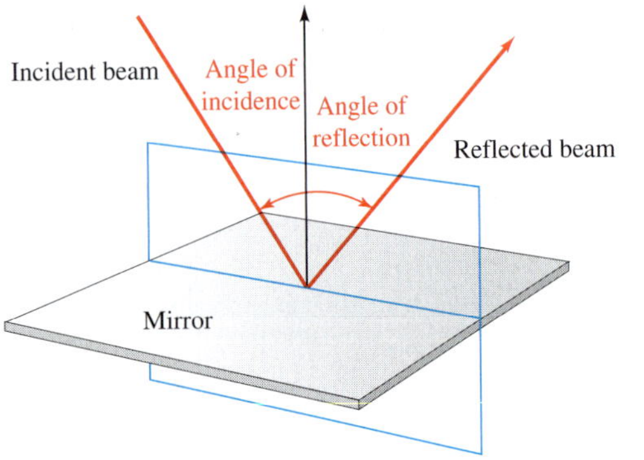

52. BILLIARDS In the illustration, a rectangular coordinate system has been superimposed over a billiard table. Write a function that models the path of the ball that is shown banking off of the far cushion.

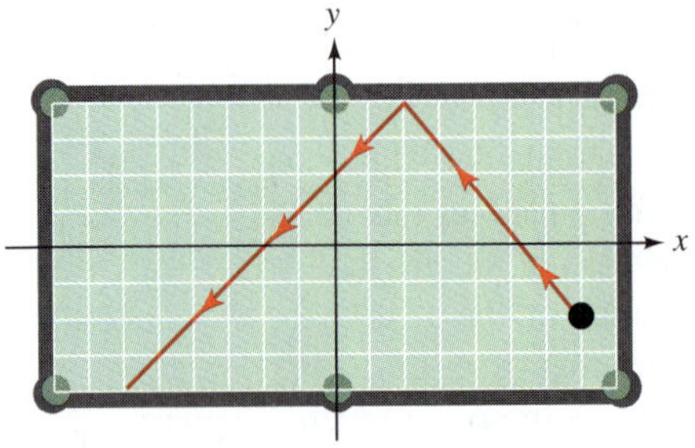

53. CENTER OF GRAVITY See the illustration. As a diver performs a $1\frac{1}{2}$-somersault in the tuck position, her center of gravity follows a path that can be described by a graph shape studied in this section. What graph shape is that?

54. FLASHLIGHTS Light beams coming from a bulb are reflected outward by a parabolic mirror as parallel rays.

a. The cross-sectional view of a parabolic mirror is given by the function $f(x) = x^2$ for the following values of x: −0.7, −0.6, −0.5, −0.4, −0.3, −0.2, −0.1, 0, 0.1, 0.2, 0.3, 0.4, 0.5, 0.6, 0.7. Sketch the parabolic mirror using the following graph.

b. From the lightbulb filament at (0, 0.25), draw a line segment representing a beam of light that strikes the mirror at (−0.4, 0.16) and then reflects outward, parallel to the y-axis.

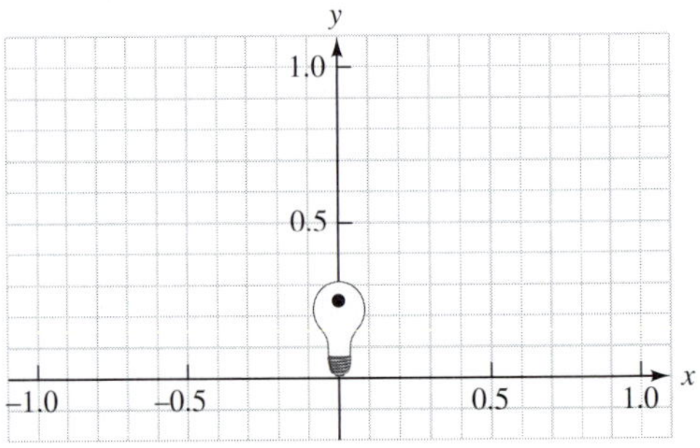

WRITING

55. Explain how to graph a function by plotting points.

56. What does it mean when we say that the domain of a function is the set of all real numbers?

57. What does it mean to vertically translate a graph?

58. Explain why the correct choice of window settings is important when using a graphing calculator.

REVIEW **Solve each formula for the indicated variable.**

59. $T - W = ma$ for W

60. $a + (n - 1)d = l$ for n

61. $s = \frac{1}{2}gt^2 + vt$ for g

62. $e = mc^2$ for m

63. BUDGETING Last year, Rock Valley College had an operating budget of \$4.5 million. Due to salary increases and a new robotics program, the budget was increased by 20%. Find the operating budget for this year.

64. In the illustration, the line passing through points R, C, and S is parallel to line segment AB. Find the measure of $\angle ACB$. (Read $\angle ACB$ as "angle ACB." *Hint:* Recall from geometry that alternate interior angles have the same measure.)

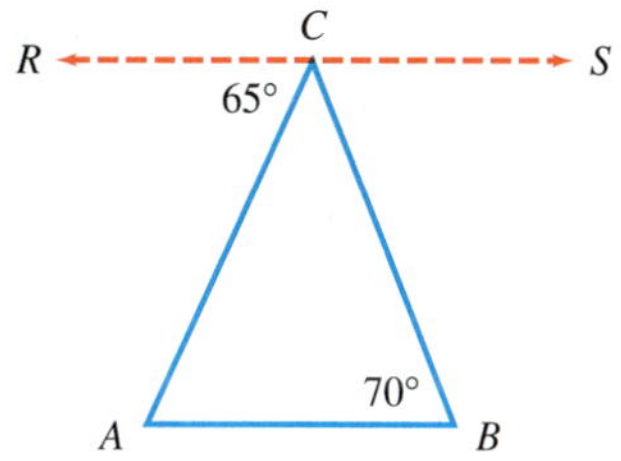

CHALLENGE PROBLEMS **Graph each function.**

65. $f(x) = \begin{cases} |x| & \text{for } x \geq 0 \\ x^3 & \text{for } x < 0 \end{cases}$

66. $f(x) = \begin{cases} x^2 & \text{for } x \geq 0 \\ |x| & \text{for } x < 0 \end{cases}$

ACCENT ON TEAMWORK

MEASURING SLOPE

Overview: This hands-on activity will give you a better understanding of slope.

Instructions: Form groups of 2 or 3 students. Use a ruler and a level to find the slopes of five ramps or inclines by measuring $\frac{\text{rise}}{\text{run}}$, as shown in the illustration. Record your results in a table, listing the slopes in order from smallest to largest.

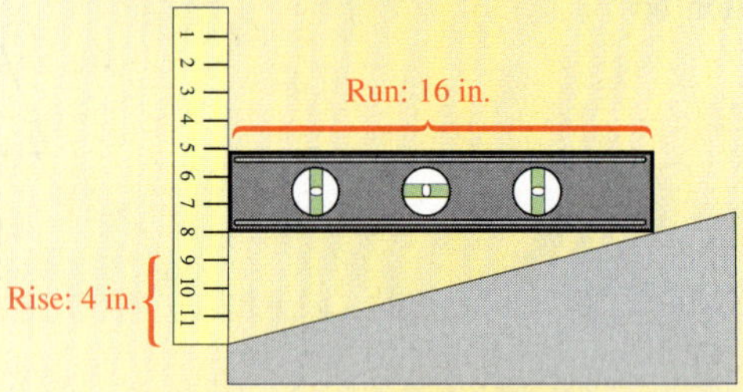

Object/location	Slope
Ramp outside the cafeteria	$\frac{\text{Rise}}{\text{Run}} = \frac{4 \text{ in.}}{16 \text{ in.}} = \frac{1}{4}$

WRITING A LINEAR MODEL

Overview: In this hands-on activity, you will write an equation that mathematically models a real-life situation.

Instructions: Form groups of 2 or 3 students. You will need a 5-gallon pail (it needs to have vertical sides), a yardstick, a watch that shows seconds, and access to a garden hose.

Tape the yardstick to the pail as shown in the illustration. Turn on the water, leaving it running at a constant rate, and begin to fill the pail. Keep track of the time (in seconds) that it takes to fill the pail to reach heights of 2, 4, 6, 8, 10, and 12 inches.

Create a rectangular coordinate graph with the horizontal axis labeled *time in seconds* and the vertical axis labeled *height of the water in inches.* Plot your data as ordered pairs of the form (time, height). Draw a straight line that best fits the data and determine the equation of the line. Then use the equation to predict the height of the water column if it were allowed to run in an infinitely tall pail for 24 hours (86,400 seconds).

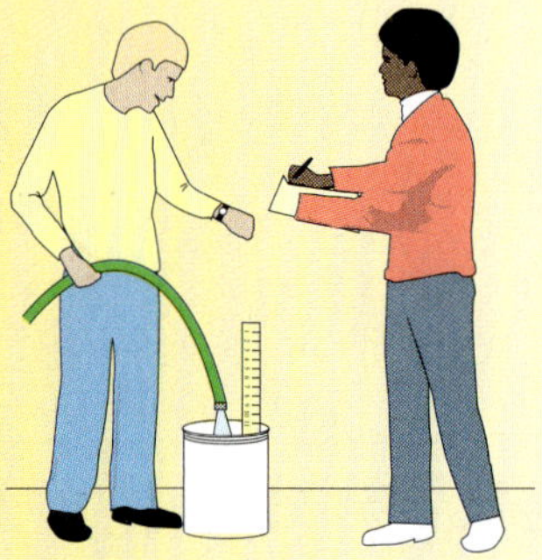

KEY CONCEPT: FUNCTIONS

In Chapter 2, we introduced one of the most important concepts in mathematics, that of a **function.**

1. Fill in the blanks.

a. A _________ is a rule (or correspondence) that assigns to each value of one variable (called the ___________ variable) exactly one value of another variable (called the ___________ variable).

b. The set of all possible values that can be used for the independent variable is called the _________. The set of all values of the dependent variable is called the _______.

2. We can think of a function as a machine. Using the words *input, output, domain,* and *range,* explain how the following function machine works.

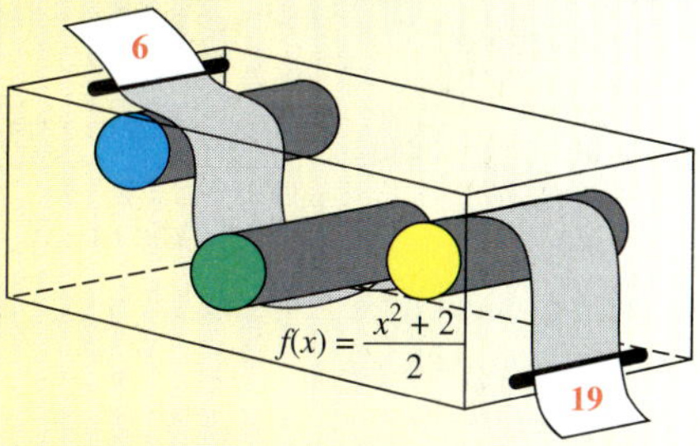

FOUR WAYS TO REPRESENT A FUNCTION

Functions can be described in words, with an equation, with a table, or with a graph.

3. The equation $y = 2x + 3$ determines a correspondence between the values of x and y. Find the value of y that corresponds to the x-value -10.

4. The area of a circle is the product of π and the radius squared. Use the variables A and r to describe this relationship with an equation.

5. Use the following table to determine the height of a projectile 1.5 seconds after it was shot vertically into the air.

Time t (seconds)	Height h (feet)
0	0
0.5	28
1.0	48
1.5	60
2.0	64

6. Use the following graph to determine what y-value the function f assigns to the x-value 1.

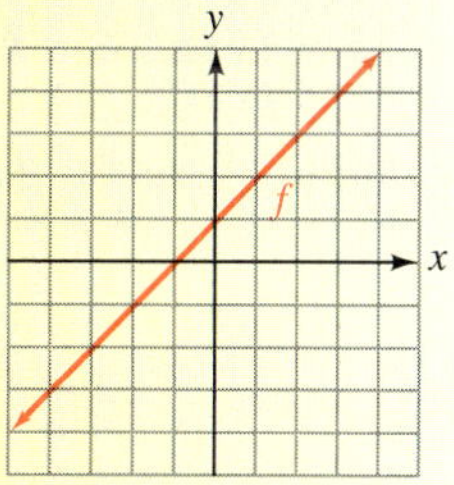

FUNCTION NOTATION

The notation $y = f(x)$ denotes that the variable y is a function of x.

7. Write the equation in Problem 3 using function notation. Find $f(0)$.

8. Use function notation to represent the relationship described in Problem 4.

9. If the function $h(t) = -16t^2 + 64t$ gives the height of the projectile described in Problem 5, find $h(4)$ and interpret the result.

10. Refer to the graph in Problem 6.

a. Find $f(-3)$.

b. Find the x-value for which $f(x) = 3$.

CHAPTER REVIEW

SECTION 2.1 The Rectangular Coordinate System

CONCEPTS

A *rectangular coordinate system* is formed by two intersecting perpendicular number lines called the *x-axis* and the *y-axis,* which divide the plane into four *quadrants.*

The process of locating a point in the coordinate plane is called *plotting* or *graphing* that point.

REVIEW EXERCISES

Plot each point on the given rectangular coordinate system.

1. (0, 3)

2. (−2, −4)

3. $\left(\frac{5}{2}, -1.75\right)$

4. the origin

5. (2.5, 0)

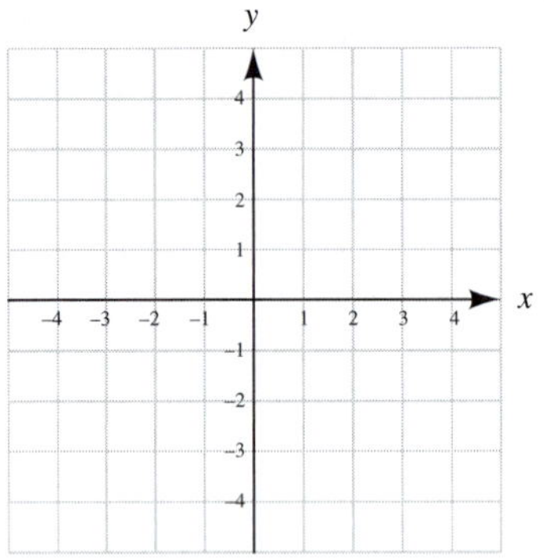

Graphs can be used to visualize relationships between two quantities.

6. The given graph shows how the height of the water in a flood control channel changed over a 7-day period.

a. Describe the height of the water at the beginning of day 2.

b. By how much did the water level increase or decrease from day 4 to day 5?

c. During what time period did the water level stay the same?

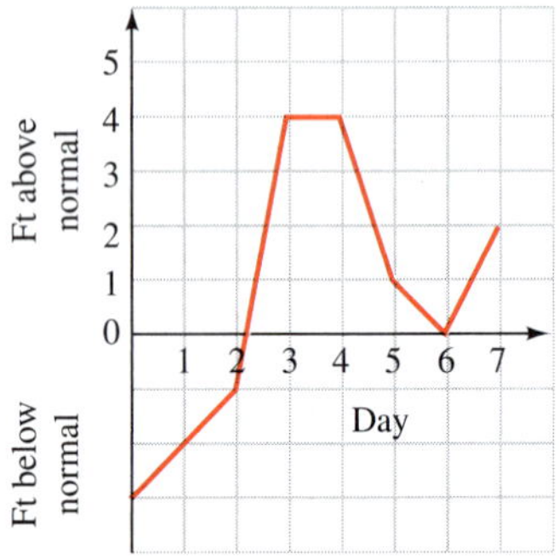

7. AUCTIONS The dollar increments used by an auctioneer during the bidding process depend on what initial price the auctioneer began with for the item. See the step graph in the illustration.

a. What increments are used by the auctioneer if the bidding on an item began at $150?

b. If the first bid on an item being auctioned is $750, what will be the next price asked for by the auctioneer?

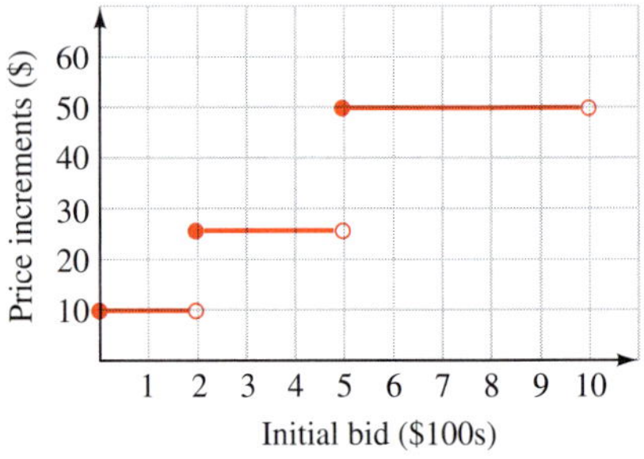

Midpoint formula: The midpoint of a line segment with endpoints at (x_1, y_1) and (x_2, y_2) is the point M with coordinates

$$\left(\frac{x_1 + x_2}{2}, \frac{y_1 + y_2}{2}\right)$$

8. Find the midpoint of the line segment joining (8, −2) and (6, −4).

SECTION 2.2 Graphing Linear Equations

A *solution* of an equation in two variables is an ordered pair of numbers that makes the equation a true statement.

9. Is $(3, -6)$ a solution of $y = -5x + 9$?

10. The graph of a linear equation is shown in the illustration.

a. If the coordinates of point A are substituted in the equation, will a true or false statement result?

b. If the coordinates of point B are substituted in the equation, will a true or false statement result?

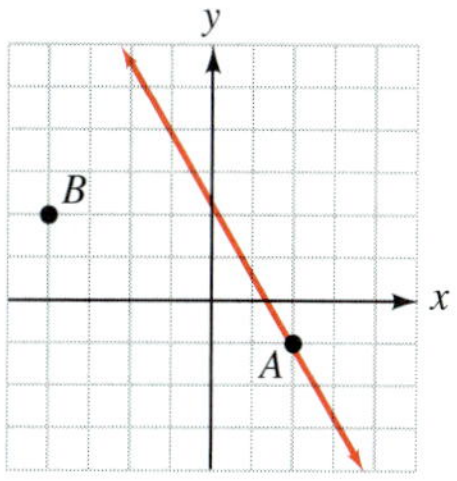

The *graph of an equation* is the graph of all points on the rectangular coordinate system whose coordinates satisfy the equation.

Complete the table of solutions for each equation.

11. $y = -3x$

x	y
-3	
0	
3	

12. $y = \frac{1}{2}x - \frac{5}{2}$

x	y
-3	
0	
3	

Graph each equation.

13. $y = 3x + 4$

14. $y = -\frac{1}{3}x - 1$

To find the *y-intercept* of a line, substitute 0 for x in the equation and solve for y. To find the *x-intercept* of a line, substitute 0 for y in the equation and solve for x.

Graph each equation using the intercept method.

15. $2x + y = 4$

16. $3x - 4y - 8 = 0$

The graph of the equation $x = a$ is a *vertical line* with x-intercept at $(a, 0)$.

The graph of the equation $y = b$ is a *horizontal line* with y-intercept at $(0, b)$.

Graph each equation.

17. $y = 4$

18. $x = -2$

A *linear equation in two variables* is an equation that can be written in the form $Ax + By = C$.

19. Fill in the blanks. The exponent on each variable of a linear equation is an understood 1. For example, $3x + 2y = 5$ can be thought of as $3x^{\ \ } + 2y^{\ \ } = 5$.

20. RECYCLING It takes more aluminum cans to weigh one pound than it used to because manufacturers continue to use thinner materials. The equation $n = 0.45t + 24.25$ gives the approximate number n of empty aluminum cans needed to weigh one pound, where t is the number of years since 1980. Graph the equation. (Source: The Aluminum Association)

a. What information can be obtained from the n-intercept of the graph?

b. From the graph, estimate the number of cans it took to weigh one pound in 2004.

SECTION 2.3 Rate of Change and the Slope of a Line

The *slope* of a nonvertical line is defined to be

$$m = \frac{\text{rise}}{\text{run}} = \frac{\Delta y}{\Delta x}$$

The *slope formula:*

$$m = \frac{y_2 - y_1}{x_2 - x_1}$$

The slope of a line with the appropriate units attached gives the *average rate of change.*

21. Find the slope of lines l_1 and l_2 in the illustration.

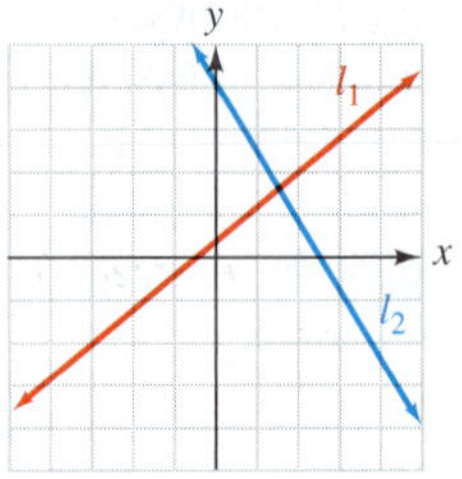

22. U.S. VEHICLE SALES On the graph in the illustration, draw a line through the points (0, 21.2) and (20, 48.6). Use this linear model to estimate the rate of increase in the market share of minivans, sport utility vehicles, and light trucks over the years 1980–2000.

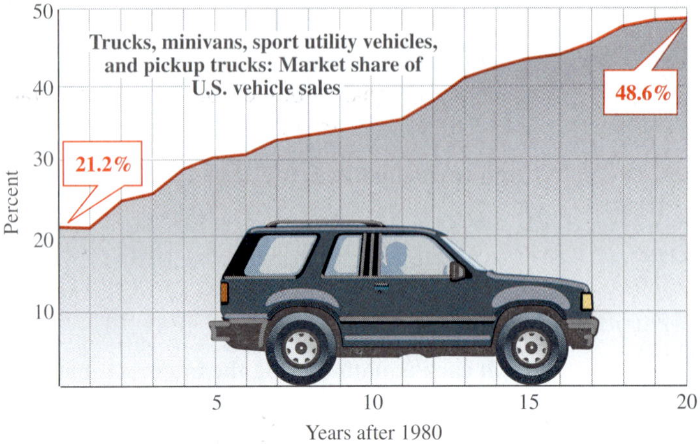

Source: American Automotive Association and U.S. Bureau of Economic Analysis

Horizontal lines have a slope of 0. Vertical lines have no defined slope.

Find the slope of the line passing through the given points.

23. (2, 5) and (5, 8)

24. (3, −2) and (−6, 12)

25. (−2, 4) and (8, 4)

26. (−5, −4) and (−5, 8)

Parallel lines have the same slope. The slopes of two nonvertical *perpendicular lines* are negative reciprocals.

Determine whether the lines with the given slopes are parallel, perpendicular, or neither.

27. $m_1 = 4, m_2 = -\frac{1}{4}$

28. $m_1 = 0.5, m_2 = \frac{1}{2}$

29. Determine whether a line that passes through (−2, 1) and (6, 5) is parallel or perpendicular to a line with slope −2.

30. SAN FRANCISCO According to the city Bureau of Engineering, the steepest street in San Francisco is 22nd Street between Church and Vicksburg. For a run of 200 feet, the street rises 63 feet. Find the grade of the street.

SECTION 2.4

Writing Equations of Lines

Equations of a line:
Point–slope form:
$y - y_1 = m(x - x_1)$

Slope–intercept form:
$y = mx + b$

Write an equation of the line with the given properties. Express the result in slope–intercept form.

31. Slope of 3; passing through $(-8, 5)$

32. Passing through $(-2, 4)$ and $(6, -9)$

33. Passing through $(-3, -5)$; parallel to the graph of $3x - 2y = 7$

34. Passing through $(-3, -5)$; perpendicular to the graph of $3x - 2y = 7$

35. Write $3x + 4y = -12$ in slope–intercept form. Give the slope and y-intercept of the graph of the equation. Then use this information to graph the line.

Write the equation of each line.

36. The x-axis

37. The y-axis

38. Write the equation of the line shown in the graph. Answer in slope–intercept form.

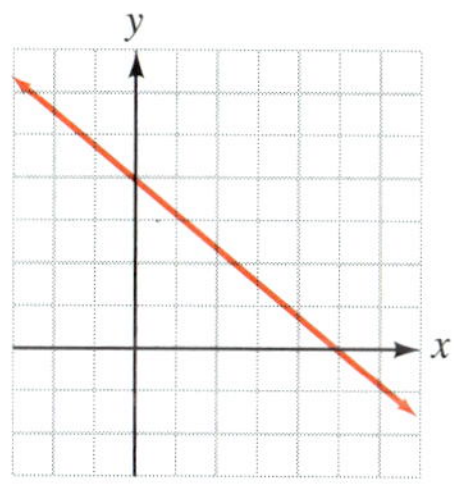

39. BUSINESS GROWTH City growth and the number of business licenses issued by a certain city are related by a linear equation. Records show that 250 licenses had been issued when the local population was 21,000, and that the rate of increase in the number of licenses issued was 1 for every 150 new residents. Use the variables p for population and L for the number of business licenses to write an equation (in slope–intercept form) that city officials can use to predict future business growth.

Many real-life situations can be modeled by linear equations.

40. DEPRECIATION A manufacturing company purchased a new diamond-tipped saw blade for \$8,700 and will depreciate it on a straight-line basis over the next 5 years. At the end of its useful life, it will be sold for scrap for \$100.

a. Write a depreciation equation for the saw blade using the variables x and y.

b. If the depreciation equation is graphed, explain the significance of the y-intercept.

SECTION 2.5 An Introduction to Functions

A *function* is a rule (or correspondence) that assigns to each value of one variable (called the independent variable) exactly one value of another variable (called the dependent variable).

The notation $y = f(x)$ denotes that the variable y (the *dependent* variable) is a function of x (the *independent* variable).

The *domain* of a function is the set of input values. The *range* is the set of output values.

Determine whether the arrow diagram or the table defines y as a function of x.

41.

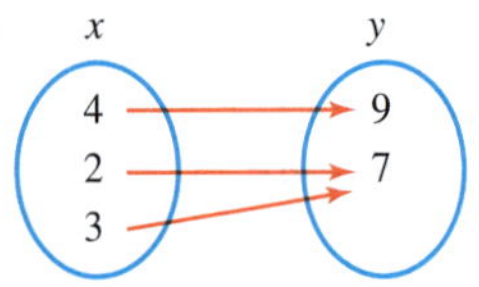

42.

x	y
-1	8
0	5
4	1
-1	9

Determine whether each equation determines y to be a function of x.

43. $y = 6x - 4$

44. $y = 4 - x^2$

45. $y^2 = x$

46. $|y| = x$

Let $f(x) = 3x + 2$ and $g(x) = \dfrac{x^2 - 4x + 4}{2}$. Find each function value.

47. $f(-3)$

48. $g(8)$

49. $g(-2)$

50. $f(t)$

51. Let $f(x) = -5x + 7$. For what value of x is $f(x) = -8$?

52. Let $g(x) = \dfrac{3}{4}x - 1$. For what value of x is $g(x) = 0$?

Find the domain and range of each function.

53. $f(x) = 4x - 1$

54. $f(x) = x^2 + 1$

55. $f(x) = \dfrac{4}{2 - x}$

56. $y = -|4x|$

The *vertical line test* can be used to determine whether a graph represents a function.

Determine whether each graph represents a function.

57.

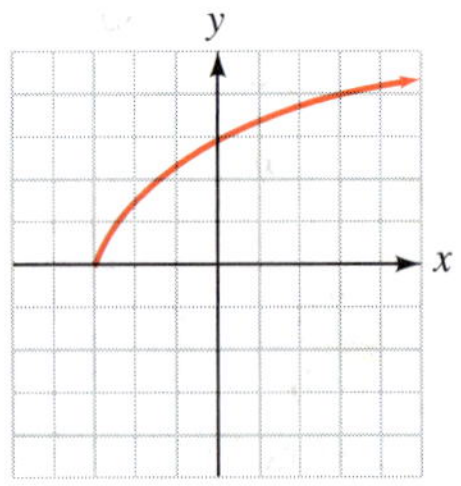

58.

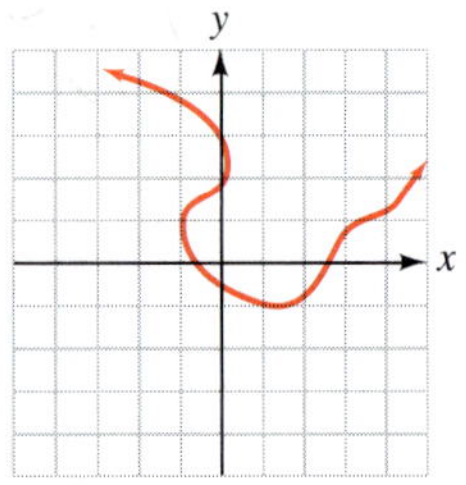

59. MARKET SHARE In Exercise 22, a line was drawn on the graph to estimate the rate of increase in the market share of minivans, sport utility vehicles, and light trucks. Use this information to write an equation of the line. Express your result using function notation. Then use the function to predict the market share for the year 2004 if the trend continues.

60. Graph: $f(x) = \dfrac{2}{3}x - 2$.

SECTION 2.6 Graphs of Functions

Graphs of *nonlinear functions* are not lines.

The *squaring function:*

$f(x) = x^2$

The *cubing function:*

$f(x) = x^3$

The *absolute value function:*

$f(x) = |x|$

A *horizontal translation* shifts a graph left or right. A *vertical translation* shifts a graph upward or downward. A *reflection* "flips" a graph about the x-axis.

The domain of a function is the *projection* of its graph onto the x-axis. The range of a function is the *projection* of its graph onto the y-axis.

61. Use the graph in the illustration to find each value.

a. $f(-2)$ **b.** $f(3)$

c. The value of x for which $f(x) = 0$.

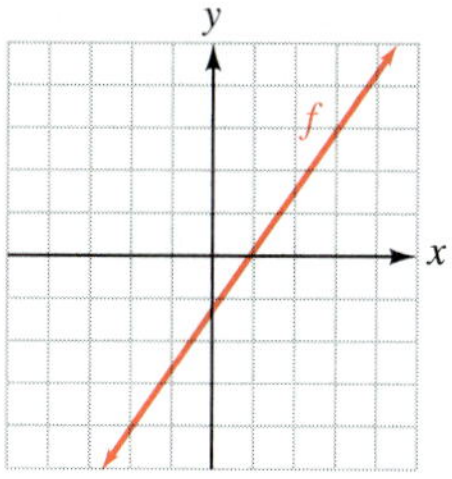

62. Graph $f(x) = |x + 2|$, $g(x) = |x| - 3$, and $h(x) = -|x|$ on the same coordinate system.

Graph each function.

63. $f(x) = x^2 - 3$

64. $f(x) = (x - 2)^3 + 1$

Give the domain and range of each function graphed below.

65.

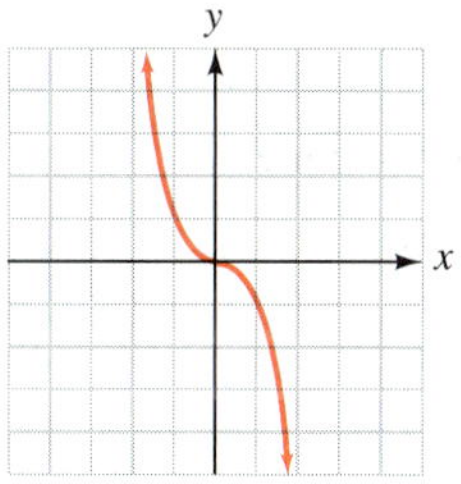

66.

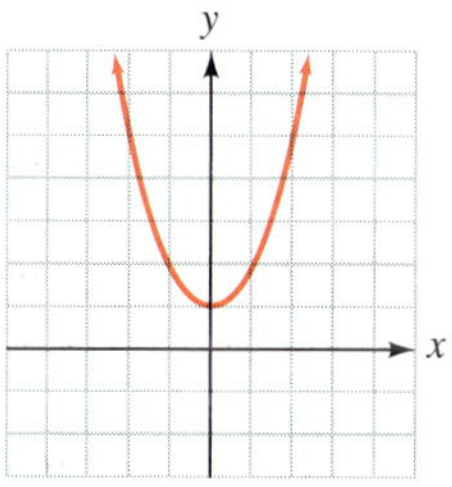

CHAPTER 2 TEST

Refer to the graph, which shows the height of an object at different times after it was shot straight up into the air.

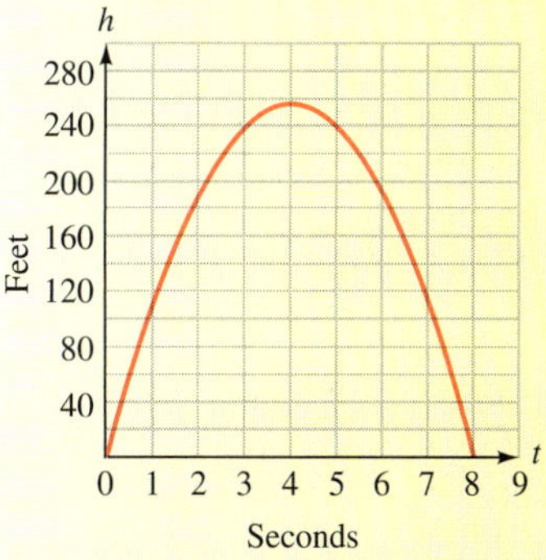

1. How high was the object 3 seconds into the flight?

2. At what times was the object about 110 feet above the ground?

3. What was the maximum height reached by the object?

4. How long did the flight take?

5. Find the coordinates of the midpoint of the line segment joining $(-2, -5)$ and $(7, 8)$.

6. Graph: $y = -\frac{2}{3}x - 2$.

7. Find the x- and y-intercepts of the graph of $2x - 5y = 10$. Then graph the equation.

8. Graph: $y = -2$.

9. Find the slope of the line shown in the illustration.

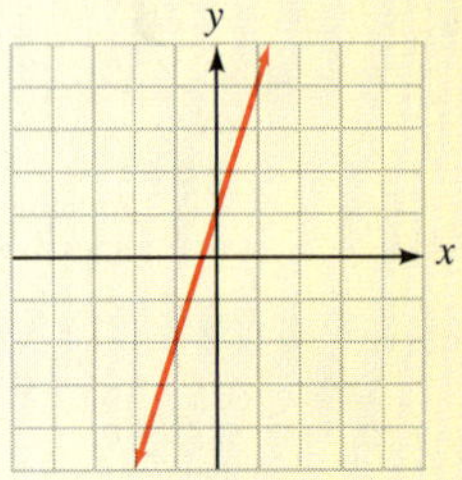

10. Find the rate of change of the temperature for the period of time shown in the graph.

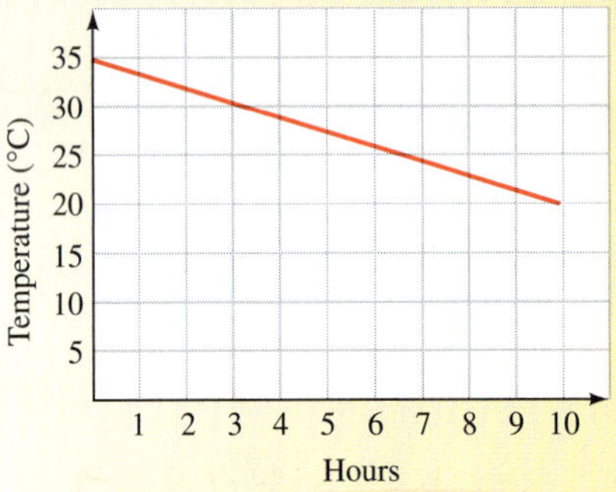

Find the slope of each line, if possible.

11. The line through $(-2, 4)$ and $(6, 8)$

12. The graph of $2x - 3y = 8$

13. The graph of $x = 12$

14. The graph of $y = 12$

15. Write an equation of the line shown in Exercise 9. Give the answer in slope–intercept form.

16. Write an equation of the line that passes through $(-2, 6)$ and $(-4, -10)$. Give the answer in slope–intercept form.

17. Find the slope and the y-intercept of the graph of $-2x - 9 = 6y$.

18. Determine whether the graphs of $4x - y = 12$ and $y = \frac{1}{4}x + 3$ are parallel, perpendicular, or neither.

19. Write an equation of the line that passes through the origin and is parallel to the graph of $y = -\frac{3}{2}x - 7$.

20. ACCOUNTING After purchasing a new color copier, a business owner had his accountant prepare a depreciation worksheet for tax purposes. (See the illustration.)

a. Assuming straight-line depreciation, write an equation that gives the value v of the copier after x years of use.

b. If the depreciation equation is graphed, explain the significance of its v-intercept.

Depreciation Worksheet

Color copier . . . . . . (new)	$4,000
Salvage value . . . . . . (in 6 years)	$400

21. Does $|y| = x$ define y to be a function of x?

22. Does the table define y as a function of x?

x	y
-3	4
4	-3
1	4
2	5

23. Find the domain and range of the function $f(x) = |x|$.

24. Let $f(x) = -\frac{4}{5}x - 12$. For what value of x is $f(x) = 4$?

Let $f(x) = 3x + 1$ and $g(x) = x^2 - 2x - 1$. Find each value.

25. $f(3)$

26. $g(0)$

27. $f\left(\frac{2}{3}\right)$

28. $g(r)$

For Exercises 29 and 30, refer to the graph of function f on the next page.

29. Find $f(-2)$.

30. Find the value of x for which $f(x) = 3$.

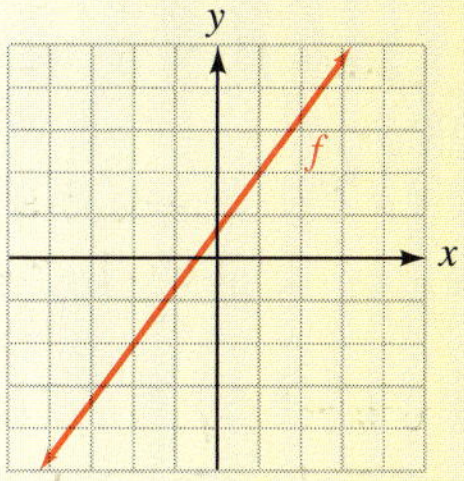

Determine whether each graph represents a function.

31.

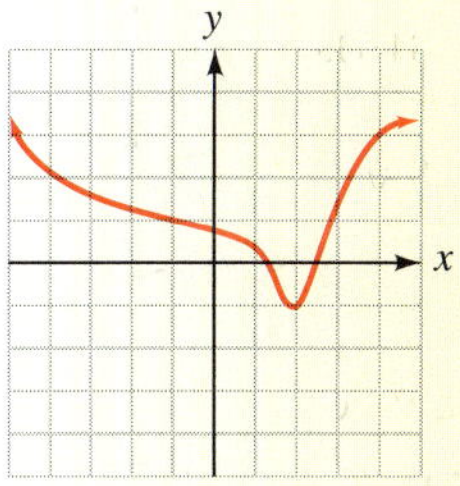

32.

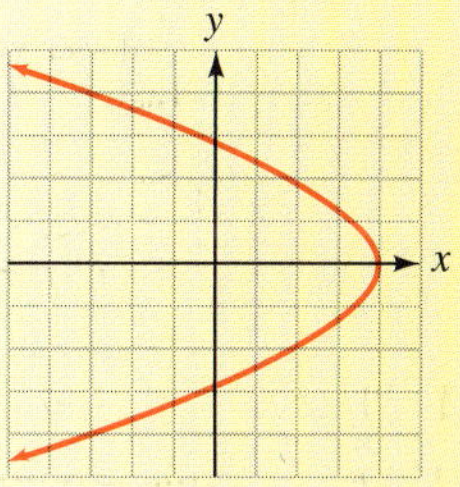

33. Graph: $f(x) = x^2 + 3$.

34. Graph: $g(x) = -|x + 2|$.

35. Give the domain and range of function f graphed as follows.

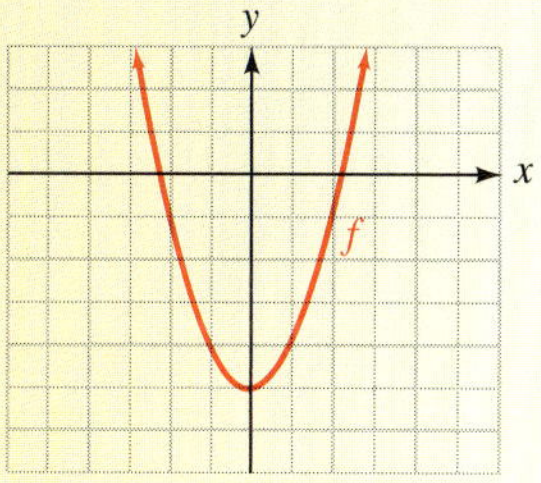

36. Explain why the graph of a circle does not represent a function.

CHAPTERS 1–2 CUMULATIVE REVIEW EXERCISES

Determine which numbers in the set

$$\{-2, 0, 1, 2, \tfrac{13}{12}, 6, 7, \sqrt{5}, \pi\}$$

are in each category.

1. Natural numbers

2. Whole numbers

3. Rational numbers

4. Irrational numbers

5. Negative numbers

6. Real numbers

7. Prime numbers

8. Composite numbers

9. Even numbers

10. Odd numbers

Evaluate each expression.

11. $-|5| + |-3|$

12. $\dfrac{|-5| + |-3|}{-|4|}$

13. $2 + 4 \cdot 5$

14. $\dfrac{8 - 4}{2 - 4}$

15. $-\dfrac{16}{5} \div \left(-\dfrac{10}{3}\right)$

16. $\dfrac{(9 - 8)^4 + 21}{3^3 - (\sqrt{16})^2}$

Evaluate each expression for $x = 2$ and $y = -3$.

17. $-x - 2y$

18. $\dfrac{x^2 - y^2}{2x + y}$

Determine which property of real numbers justifies each statement.

19. $(a + b) + c = a + (b + c)$

20. $3(x + y) = 3x + 3y$

21. $(a + b) + c = c + (a + b)$

22. $(ab)c = a(bc)$

Simplify each expression.

23. $12y - 17y$

24. $-7s(-4t)(-1)$

25. $3x^2 + 2x^2 - 5x^2$

26. $-(4 + z) + 2z$

Solve each equation.

27. $2x - 5 = 11$

28. $\dfrac{2x - 6}{3} = x + 7$

29. $4(y - 3) + 4 = -3(y + 5)$

30. $2x - \dfrac{3(x - 2)}{2} = 7 - \dfrac{x - 3}{3}$

31. $-3 = -\dfrac{9}{8}s$

32. $0.04(24) + 0.02x = 0.04(12 + x)$

Solve each formula for the indicated variable.

33. $-Tx + 3By = c$ for B

34. $A = \dfrac{1}{2}h(b_1 + b_2)$ for h

35. INVESTMENTS A woman invested part of \$20,000 at 6% and the rest at 7%. If her annual interest is \$1,260, how much did she invest at 6%?

36. DRIVING RATES John drove to a distant city in 5 hours. When he returned, there was less traffic, and the trip took only 3 hours. If he drove 26 mph faster on the return trip, how fast did he drive each way?

37. Graph: $2x - 3y = 6$.

38. Find the slope of the following line.

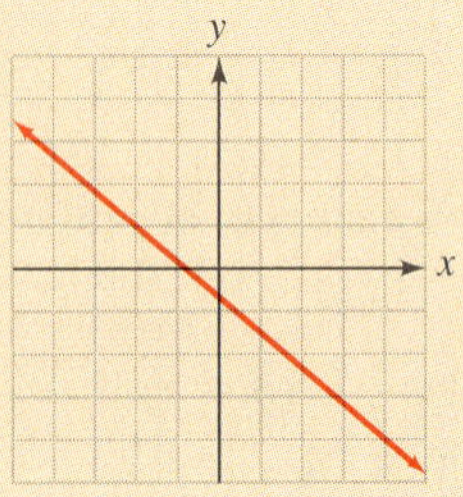

39. Write an equation of the line passing through $(-2, 5)$ and $(8, -9)$.

40. Write an equation of the line passing through $(-2, 3)$ and parallel to the graph of $3x + y = 8$.

Refer to the following graph of function *f*.

41. Find $f(1)$.

42. Find the value of x for which $f(x) = 1$.

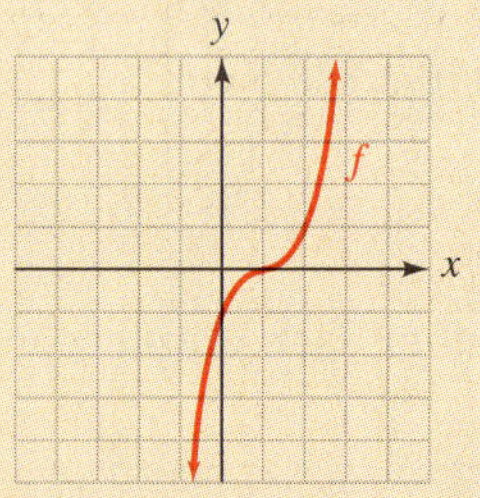

43. Determine whether the graph represents a function.

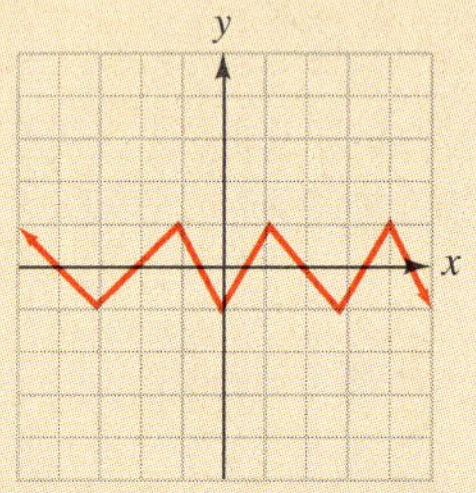

44. Does the arrow diagram define y as a function of x?

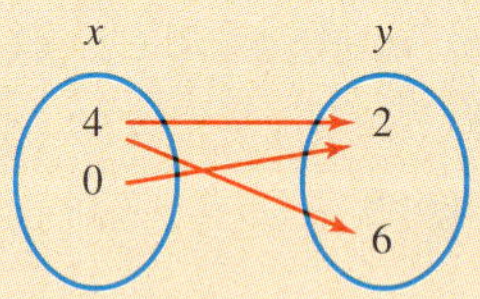

45. Does $y^2 = x$ define y as a function of x?

46. Let $h(x) = -\dfrac{1}{5}x - 12$. For what value of x is $h(x) = 0$?

Let $f(x) = 3x^2 + 2$ and $g(x) = -2x - 1$. Find each function value.

47. $f(-1)$

48. $g(0)$

49. $g(-2)$

50. $f(-r)$

Graph each equation and determine whether it is a function. If it is a function, give the domain and range.

51. $y = -x^2 + 1$

52. $y = |x - 3|$

53. See the illustration. Explain why there is not a linear relationship between the height of the antenna and its maximum range of reception.

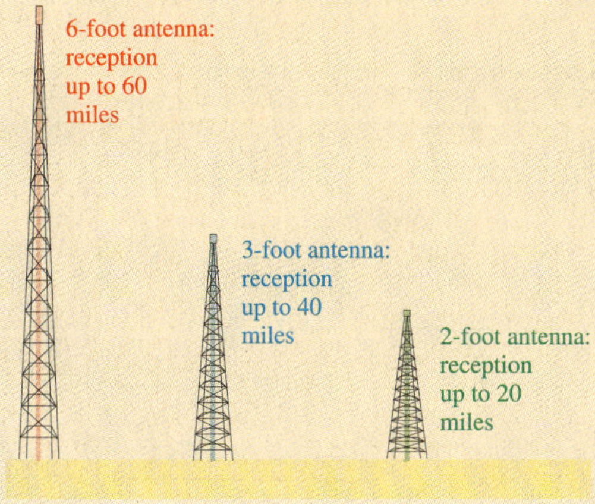

54. ELECTRICITY The electrical resistance R of a coil of wire and its temperature t are related by a linear equation.

a. Use the following data to write an equation that models this relationship.

b. Use answer to part a to predict the resistance if the temperature of the coil of wire is 100° Celsius.

t (in degrees Celsius)	10	30
R (in milliohms)	5.25	5.65

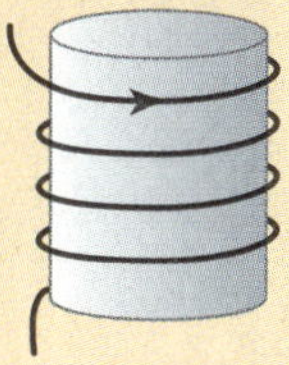

Chapter 3 Systems of Equations

Getty Images/Gabriel M. Covian

Manufacturing companies regularly consider replacing old machinery with newer, more efficient models to increase productivity and expand profits. Financial decisions such as this can be made by writing and then graphing a system of equations to compare the costs associated with the purchase of new equipment to the costs of continuing with the equipment currently in use.

To learn more about solving systems of linear equations graphically, visit *The Learning Equation* on the Internet at http://tle.brookscole.com. (The log-in instructions are in the Preface.) For Chapter 3, the online lesson is:

- *TLE* Lesson 5: Solving Systems of Equations by Graphing

To solve many problems, we must use two and sometimes three variables. This requires that we solve a system of equations.

3.1 Solving Systems by Graphing

- The graphing method
- Consistent systems
- Inconsistent systems
- Dependent equations
- Solving equations graphically

The red line in the following graph shows the cost for a company to produce a given number of skateboards. The blue line shows the revenue the company will receive for selling a given number of those skateboards. The graph offers the company important financial information.

- The production costs exceed the revenue earned if fewer than 400 skateboards are sold. In this case, the company loses money.
- The revenue earned exceeds the production costs if more than 400 skateboards are sold. In this case, the company makes a profit.
- Production costs equal revenue earned if exactly 400 skateboards are sold. This fact is indicated by the point of intersection of the two lines, (400, 20,000), which is called the **break-even point.**

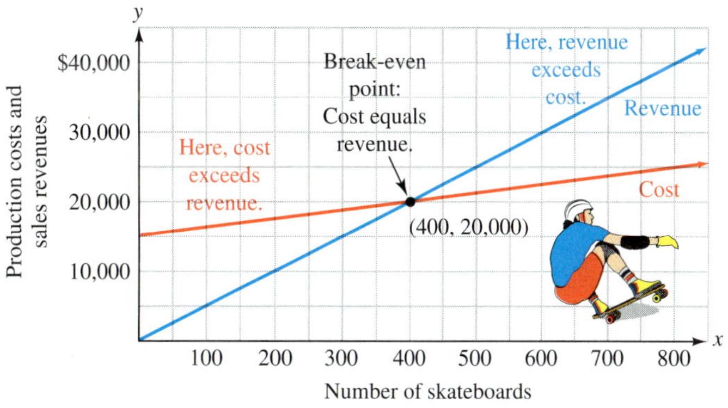

This example shows that important information can be learned by finding the point of intersection of two lines. In this section, we will discuss how to use the *graphing method* to do this.

THE GRAPHING METHOD

In Chapter 2, we discussed linear equations in two variables, x and y. We found that such equations had infinitely many solutions (x, y), and that we could graph them on a rectangular coordinate system. In this chapter, we will discuss **systems of linear equations** involving two or three equations. To write a system of equations, we use a left brace {.

In the pair of equations

$$\begin{cases} x + 2y = 4 \\ 2x - y = 3 \end{cases}$$ This is called a system of two linear equations.

there are infinitely many pairs (x, y) that satisfy the first equation and infinitely many pairs (x, y) that satisfy the second equation. However, there is only one pair (x, y) that satisfies both equations. The process of finding this pair is called *solving the system.*

To solve a system of two equations in two variables by graphing, we use the following steps.

The Graphing Method

1. On a single set of coordinate axes, graph each equation.
2. Find the coordinates of the point (or points) where the graphs intersect. These coordinates give the solution of the system.
3. If the graphs have no point in common, the system has no solution.
4. Check the proposed solution in both of the original equations.

CONSISTENT SYSTEMS

When a system of equations (as in Example 1) has at least one solution, the system is called a **consistent system.**

EXAMPLE 1

Solve the system: $\begin{cases} x + 2y = 4 \\ 2x - y = 3 \end{cases}$.

Solution We graph both equations on one set of coordinate axes, as shown.

$x + 2y = 4$

x	y	(x, y)
4	0	(4, 0)
0	2	(0, 2)
−2	3	(−2, 3)

$2x - y = 3$

x	y	(x, y)
$\frac{3}{2}$	0	$\left(\frac{3}{2}, 0\right)$
0	−3	(0, −3)
−1	−5	(−1, −5)

Use the intercept method to graph each line.

Success Tip

Since accuracy is crucial when using the graphing method to solve a system:

- Use graph paper.
- Use a sharp pencil.
- Use a straightedge.

Although infinitely many ordered pairs (x, y) satisfy $x + 2y = 4$, and infinitely many ordered pairs (x, y) satisfy $2x - y = 3$, only the coordinates of the point where the graphs intersect satisfy both equations. From the graph, it appears that the intersection point has coordinates (2, 1). To verify that it is the solution, we substitute 2 for x and 1 for y in both equations and verify that (2, 1) satisfies each one.

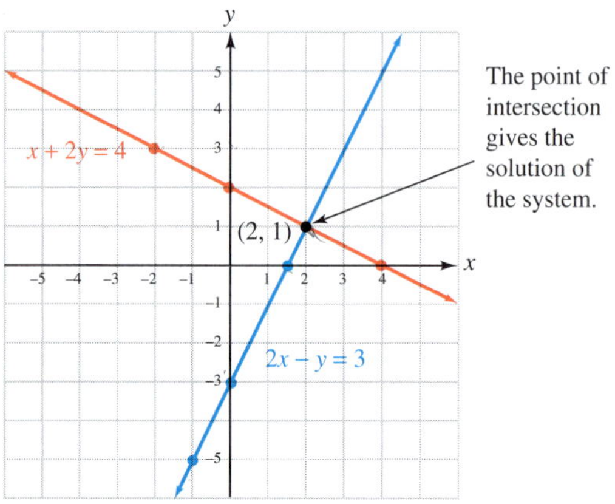

Check: *The first equation*

$$x + 2y = 4$$
$$2 + 2(1) \stackrel{?}{=} 4$$
$$2 + 2 \stackrel{?}{=} 4$$
$$4 = 4 \quad \text{True.}$$

The second equation

$$2x - y = 3$$
$$2(2) - 1 \stackrel{?}{=} 3$$
$$4 - 1 \stackrel{?}{=} 3$$
$$3 = 3 \quad \text{True.}$$

Since (2, 1) makes both equations true, it is the solution of the system.

Self Check 1 Solve: $\begin{cases} x - 3y = -5 \\ 2x + y = 4 \end{cases}$.

INCONSISTENT SYSTEMS

When a system has no solution (as in Example 2), it is called an **inconsistent system.**

EXAMPLE 2 Solve: $\begin{cases} 2x + 3y = 6 \\ 4x + 6y = 24 \end{cases}$, if possible.

Solution Using the intercept method, we graph both equations on one set of coordinate axes, as shown in the figure.

$2x + 3y = 6$

x	y	(x, y)
3	0	(3, 0)
0	2	(0, 2)
−3	4	(−3, 4)

$4x + 6y = 24$

x	y	(x, y)
6	0	(6, 0)
0	4	(0, 4)
−3	6	(−3, 6)

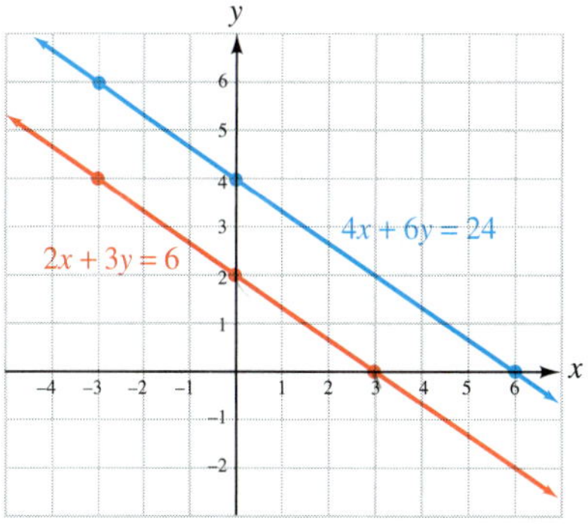

In this example, the graphs are parallel, because the slopes of the two lines are equal and they have different y-intercepts. We can see that the slope of each line is $-\frac{2}{3}$ by writing each equation in slope–intercept form.

$$2x + 3y = 6 \qquad\qquad 4x + 6y = 24$$
$$3y = -2x + 6 \qquad\qquad 6y = -4x + 24$$
$$y = -\frac{2}{3}x + 2 \qquad\qquad y = -\frac{2}{3}x + 4$$

Since the graphs are parallel lines, the lines do not intersect, and the system does not have a solution. It is an inconsistent system.

Self Check 2 Solve: $\begin{cases} 3y - 2x = 6 \\ 2x - 3y = 6 \end{cases}$.

DEPENDENT EQUATIONS

When the equations of a system have different graphs (as in Examples 1 and 2), the equations are called **independent equations.** Two equations with the same graph are called **dependent equations.**

EXAMPLE 3 Solve the system: $\begin{cases} y = \frac{1}{2}x + 2 \\ 2x + 8 = 4y \end{cases}$.

Solution We graph each equation on one set of coordinate axes, as shown in the figure. Since the graphs coincide, the system has infinitely many solutions. Any ordered pair (x, y) that satisfies one equation also satisfies the other. From the graph, we see that $(-4, 0)$, $(0, 2)$, and

(2, 3) are three of infinitely many solutions. Because the two equations have the same graph, they are dependent equations.

Graph by using the slope and y-intercept.

$$y = \frac{1}{2}x + 2$$

$m = \frac{1}{2}$ $\quad b = 2$

Slope $= \frac{1}{2}$ $\quad$ y-intercept: (0, 2)

Graph by using the intercept method.

$$2x + 8 = 4y$$

x	y	(x, y)
−4	0	(−4, 0)
0	2	(0, 2)
2	3	(2, 3)

The Language of Algebra

Here the graphs of the lines *coincide.* That is, they occupy the same location. To illustrate this concept, think of a clock. At noon and midnight, the hands of the clock *coincide.*

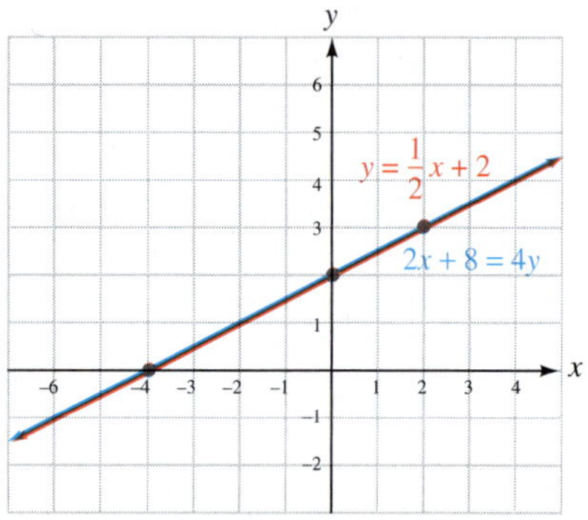

Self Check 3 Solve: $\begin{cases} 2x - y = 4 \\ y = 2x - 4 \end{cases}$.

We now summarize the possibilities that can occur when two linear equations, each with two variables, are graphed.

Solving a System of Equations by the Graphing Method

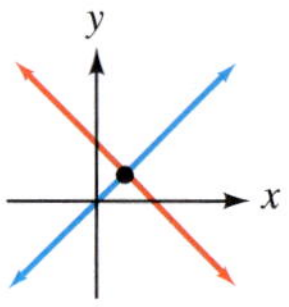

If the lines are different and intersect, the equations are independent, and the system is consistent. **One solution exists.** It is the point of intersection.

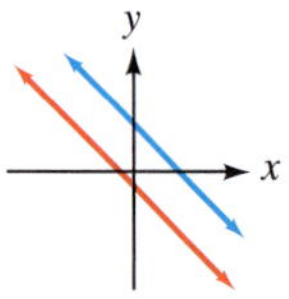

If the lines are different and parallel, the equations are independent, and the system is inconsistent. **No solution exists.**

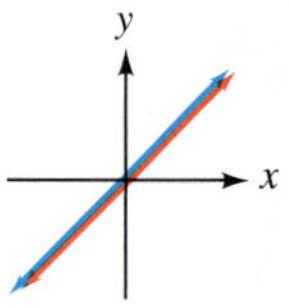

If the lines coincide, the equations are dependent, and the system is consistent. **Infinitely many solutions exist.** Any point on the line is a solution.

If each equation in one system is equivalent to a corresponding equation in another system, the systems are called **equivalent.**

EXAMPLE 4

Solve the system: $\begin{cases} \frac{3}{2}x - y = \frac{5}{2} \\ x + \frac{1}{2}y = 4 \end{cases}$.

Solution We multiply both sides of $\frac{3}{2}x - y = \frac{5}{2}$ by 2 to eliminate the fractions and obtain the equation $3x - 2y = 5$. We multiply both sides of $x + \frac{1}{2}y = 4$ by 2 to eliminate the fractions and obtain the equation $2x + y = 8$.

The new system

$$\begin{cases} 3x - 2y = 5 \\ 2x + y = 8 \end{cases}$$

is equivalent to the original system and is easier to solve, since it has no fractions. If we graph each equation in the new system, it appears that the coordinates of the point where the lines intersect are (3, 2).

$3x - 2y = 5$

x	y	(x, y)
$\frac{5}{3}$	0	$\left(\frac{5}{3}, 0\right)$
0	$-\frac{5}{2}$	$\left(0, -\frac{5}{2}\right)$
1	-1	$(1, -1)$

$2x + y = 8$

x	y	(x, y)
4	0	(4, 0)
0	8	(0, 8)
1	6	(1, 6)

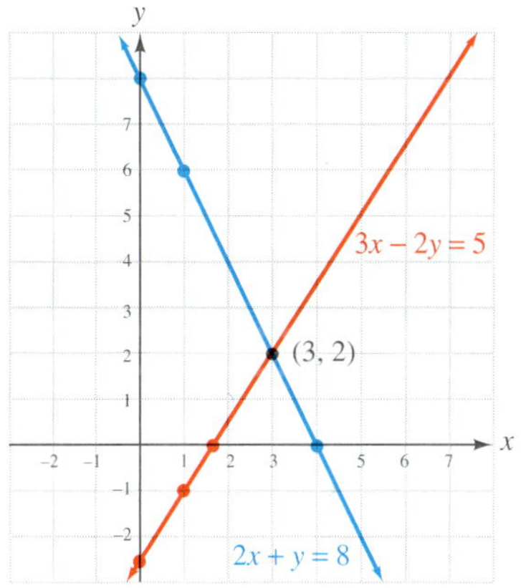

To verify that (3, 2) is the solution, we substitute 3 for x and 2 for y in each equation of the original system.

Caution

When checking the solution of a system of equations, always substitute the values of the variables into the original equations.

Check:

$$\frac{3}{2}x - y = \frac{5}{2} \qquad\qquad x + \frac{1}{2}y = 4$$

$$\frac{3}{2}(3) - 2 \stackrel{?}{=} \frac{5}{2} \qquad\qquad 3 + \frac{1}{2}(2) \stackrel{?}{=} 4$$

$$\frac{9}{2} - 2 \stackrel{?}{=} \frac{5}{2} \qquad\qquad 3 + 1 \stackrel{?}{=} 4$$

$$\frac{5}{2} = \frac{5}{2} \quad \text{True.} \qquad\qquad 4 = 4 \quad \text{True.}$$

Self Check 4 Solve: $\begin{cases} \frac{5}{2}x - y = 2 \\ x + \frac{1}{3}y = 3 \end{cases}$ by the graphing method.

ACCENT ON TECHNOLOGY: SOLVING SYSTEMS BY GRAPHING

The graphing method is limited to equations with two variables. Systems with three or more variables cannot be solved graphically. Also, it is often difficult to find exact solutions graphically. However, the TRACE and ZOOM capabilities of graphing calculators enable us to get very good approximations of such solutions.

To solve the system $\begin{cases} 3x + 2y = 12 \\ 2x - 3y = 12 \end{cases}$

with a graphing calculator, we must first solve each equation for y so that we can enter the equations into the calculator. After solving for y, we obtain the following equivalent system:

$$\begin{cases} y = -\frac{3}{2}x + 6 \\ y = \frac{2}{3}x - 4 \end{cases}$$

If we use window settings of $[-10, 10]$ for x and for y, the graphs of the equations will look like those in figure (a). If we zoom in on the intersection point of the two lines and trace, we will get an approximate solution like the one shown in figure (b). To get better results, we can do more zooms. We would then find that, to the nearest hundredth, the solution is $(4.63, -0.94)$. Verify that this is reasonable.

We can also find the intersection of two lines by using the INTERSECT feature found on most graphing calculators. After graphing the lines and using INTERSECT, we obtain a graph similar to figure (c). The display shows the approximate coordinates of the point of intersection.

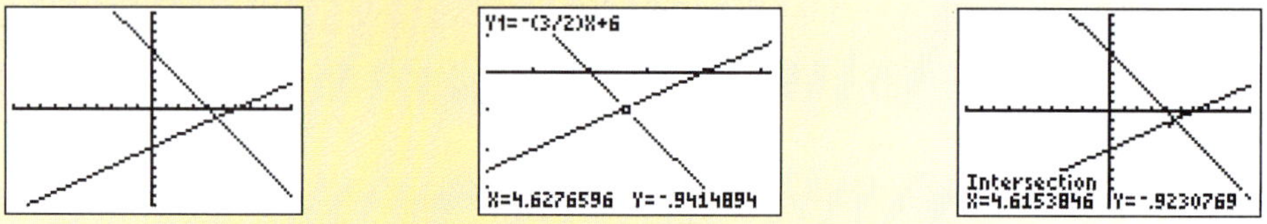

(a) (b) (c)

SOLVING EQUATIONS GRAPHICALLY

The graphing method discussed in this section can be used to solve equations in one variable.

EXAMPLE 5 Solve $2x + 4 = -2$ graphically.

Solution The graphs of $y = 2x + 4$ and $y = -2$ are shown in the figure. To solve $2x + 4 = -2$, we need to find the value of x that makes $2x + 4$ equal -2. The point of intersection of the graphs is $(-3, -2)$. This tells us that if x is -3, the expression $2x + 4$ equals -2. So the solution of $2x + 4 = -2$ is -3. Check this result.

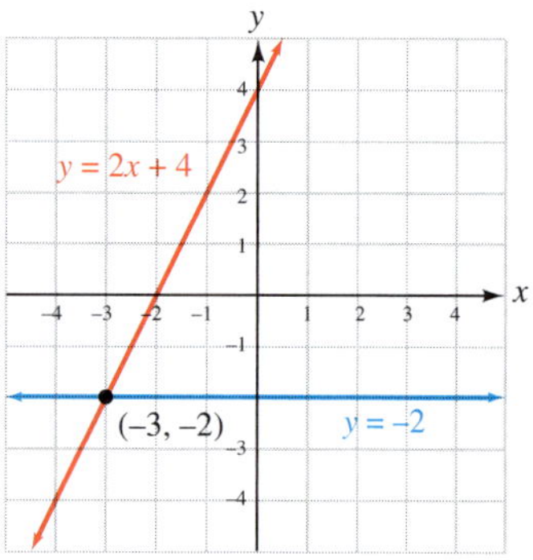

Self Check 5 Solve $2x + 4 = 2$ graphically.

ACCENT ON TECHNOLOGY: SOLVING EQUATIONS GRAPHICALLY

To solve $2(x - 3) + 3 = 7$ with a graphing calculator, we graph the left-hand side and the right-hand side of the equation in the same window by entering

$$Y_1 = 2(x - 3) + 3$$
$$Y_2 = 7$$

Figure (a) shows the graphs, generated using settings of $[-10, 10]$ for x and for y.

The coordinates of the point of intersection of the graphs can be determined using the INTERSECT feature found on most graphing calculators. With this feature, the cursor automatically highlights the intersection point, and the x- and y-coordinates are displayed.

In figure (b), we see that the point of intersection is (5, 7), which indicates that 5 is a solution of $2(x - 3) + 3 = 7$.

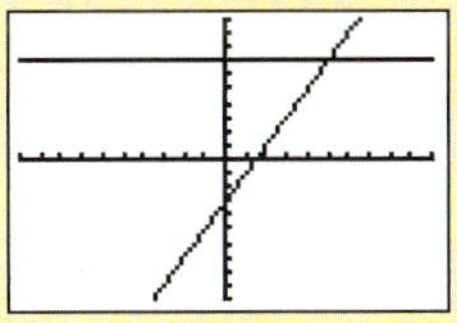

(a)

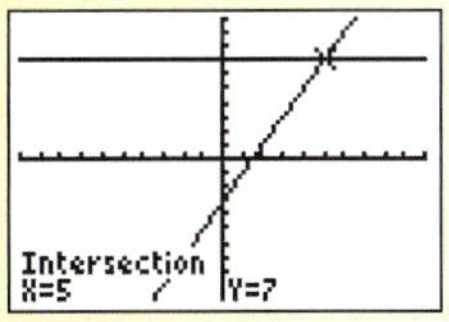

(b)

Answers to Self Checks

1. (1, 2)

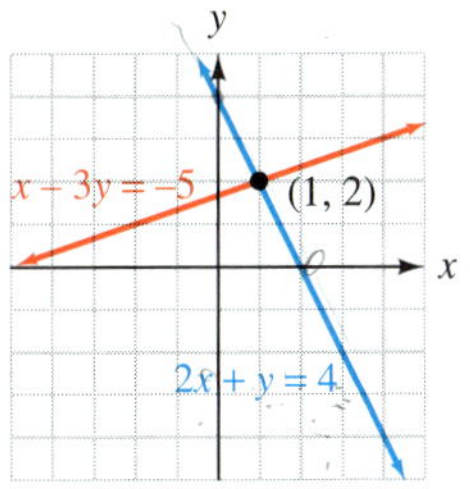

2. no solution

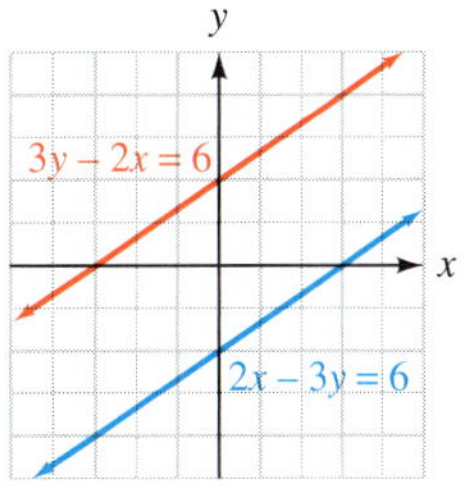

3. There are infinitely many solutions; three of them are (0, −4), (2, 0), and (4, 4).

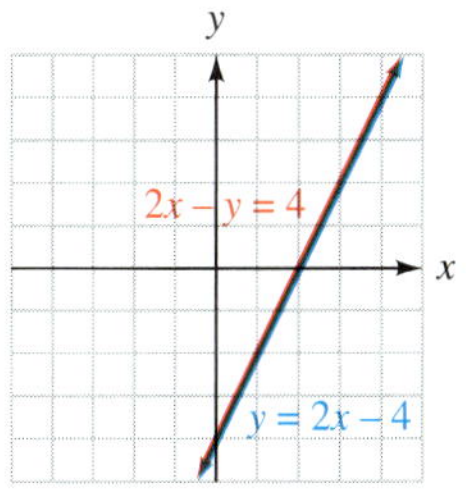

4. (2, 3)

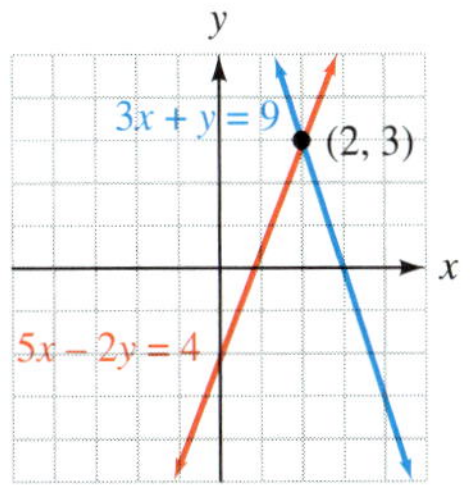

5. −1

3.1 STUDY SET

VOCABULARY Fill in the blanks.

1. $\begin{cases} x - 2y = 4 \\ 2x - y = 3 \end{cases}$ is called a ________ of linear equations.
2. When a system of equations has at least one solution, it is called a __________ system.
3. If a system has no solutions, it is called an ___________ system.
4. If two equations have different graphs, they are called ____________ equations.
5. Two equations with the same graph are called __________ equations.
6. When solving a system of two linear equations by the graphing method, we look for the point of ___________ of the two lines.

CONCEPTS

7. Refer to the illustration. Decide whether a true or a false statement would be obtained when the coordinates of

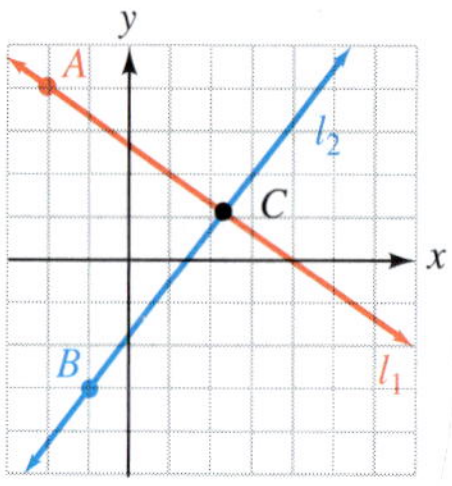

   a. Point A are substituted into the equation for line l_1.
   b. Point B are substituted into the equation for line l_1.
   c. Point C are substituted into the equation for line l_1.
   d. Point C are substituted into the equation for line l_2.

8. Refer to the illustration.

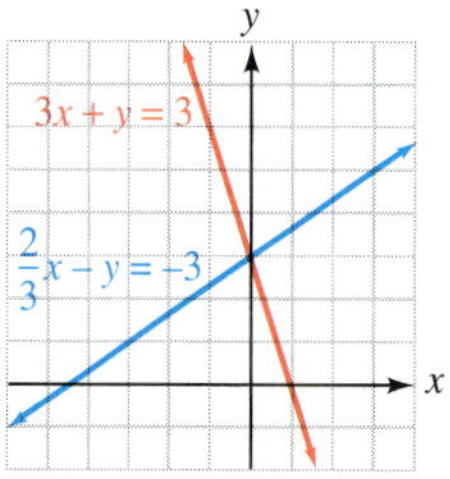

   a. How many ordered pairs satisfy the equation $3x + y = 3$? Name three.
   b. How many ordered pairs satisfy the equation $\frac{2}{3}x - y = -3$? Name three.
   c. How many ordered pairs satisfy both equations? Name it or them.

9. a. The intercept method can be used to graph $2x - 4y = -8$. Complete the following table.

x	y	(x, y)
	0	
0		
2		

   b. What is the x-intercept of the graph of $2x - 4y = -8$? What is the y-intercept?

10. a. To graph $y = 3x + 1$, we can pick three numbers for x and find the corresponding values of y. Complete the table on the right.

x	y	(x, y)
−1		
0		
2		

   b. We can also graph $y = 3x + 1$ if we know the slope and the y-intercept of the line. What are they?

11. How many solutions does the system of equations graphed on the right have? Are the equations dependent or independent?

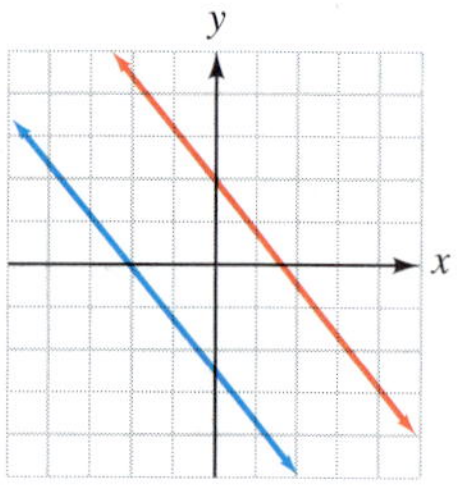

12. How many solutions does the system of equations graphed on the right have? Give three of the solutions. Is the system consistent or inconsistent?

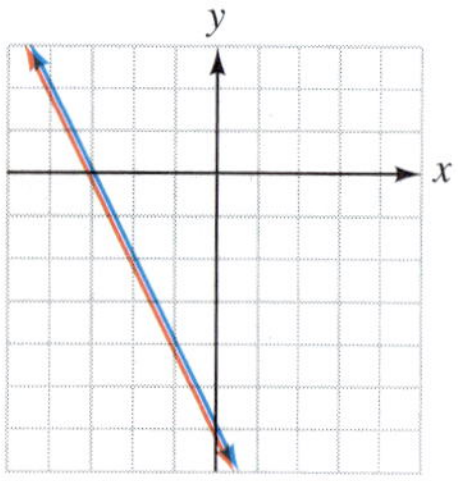

13. Estimate the solution of the system of linear equations shown in the following display.

```
Plot1 Plot2 Plot3
\Y1=(2/3)X+(1/3)
\Y2=(-3/2)X-4
\Y3=
\Y4=
\Y5=
\Y6=
```

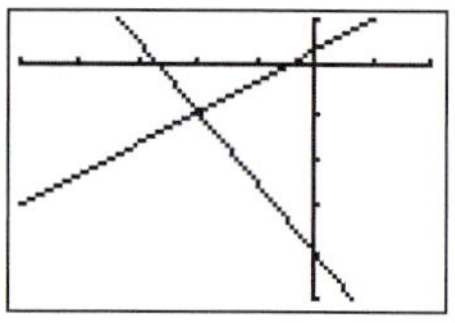

14. Use the graphs in the illustration to solve each equation.

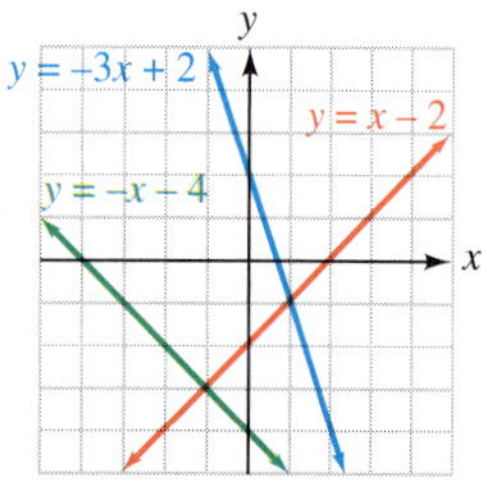

   a. $-3x + 2 = x - 2$
   b. $-x - 4 = x - 2$

NOTATION Fill in the blanks.

15. The symbol { is called a left ______. It is used when writing a system of equations.

16. The solution of a system of two linear equations is written as an ordered ______.

PRACTICE Tell whether the ordered pair is a solution of the system of equations.

17. (1, 2); $\begin{cases} 2x - y = 0 \\ y = \frac{1}{2}x + \frac{3}{2} \end{cases}$

18. (−1, 2); $\begin{cases} y = 3x + 5 \\ y = x + 4 \end{cases}$

19. (2, −3); $\begin{cases} y + 2 = \frac{1}{2}x \\ 3x + 2y = 0 \end{cases}$

20. (−4, 3); $\begin{cases} 4x - y = -19 \\ 3x + 2y = -6 \end{cases}$

Solve each system by graphing, if possible. If a system is inconsistent or if the equations are dependent, so indicate.

21. $\begin{cases} x + y = 6 \\ x - y = 2 \end{cases}$

22. $\begin{cases} x - y = 4 \\ 2x + y = 5 \end{cases}$

23. $\begin{cases} 2x + y = 1 \\ x - 2y = -7 \end{cases}$

24. $\begin{cases} 3x - y = -3 \\ 2x + y = -7 \end{cases}$

25. $\begin{cases} x + y = 0 \\ y = 2x - 6 \end{cases}$

26. $\begin{cases} 4x - 3y = 5 \\ y = -2x \end{cases}$

27. $\begin{cases} 3x + y = 3 \\ 3x + 2y = 0 \end{cases}$

28. $\begin{cases} 2x + 2y = -1 \\ 3x + 4y = 0 \end{cases}$

29. $\begin{cases} x = 13 - 4y \\ 3x = 4 + 2y \end{cases}$

30. $\begin{cases} 3x = 7 - 2y \\ 2x = 2 + 4y \end{cases}$

31. $\begin{cases} x = 3 - 2y \\ 2x + 4y = 6 \end{cases}$

32. $\begin{cases} 3x = 5 - 2y \\ 3x + 2y = 7 \end{cases}$

33. $\begin{cases} x = 2 \\ y = -\frac{1}{2}x \end{cases}$

34. $\begin{cases} y = -2 \\ y = \frac{2}{3}x - \frac{4}{3} \end{cases}$

35. $\begin{cases} y = 3 \\ x = 2 \end{cases}$

36. $\begin{cases} 2x + 3y = -15 \\ 2x + y = -9 \end{cases}$

37. $\begin{cases} x = \dfrac{11 - 2y}{3} \\ y = \dfrac{11 - 6x}{4} \end{cases}$

38. $\begin{cases} x = \dfrac{1 - 3y}{4} \\ y = \dfrac{12 + 3x}{2} \end{cases}$

39. $\begin{cases} y = -\frac{5}{2}x + \frac{1}{2} \\ 2x - \frac{3}{2}y = 5 \end{cases}$

40. $\begin{cases} \frac{5}{2}x + 3y = 6 \\ y = -\frac{5}{6}x + 2 \end{cases}$

41. $\begin{cases} x = \dfrac{5y - 4}{2} \\ x - \frac{5}{3}y + \frac{1}{3} = 0 \end{cases}$

42. $\begin{cases} 2x = 5y - 11 \\ 3x = 2y \end{cases}$

43. $\begin{cases} x = -\frac{3}{2}y \\ x = \frac{3}{2}y - 2 \end{cases}$

44. $\begin{cases} 4x = 3y - 1 \\ 3y = 4 - 8x \end{cases}$

Use a graphing calculator to solve each system. Give all answers to the nearest hundredth.

45. $\begin{cases} y = 3.2x - 1.5 \\ y = -2.7x - 3.7 \end{cases}$

46. $\begin{cases} y = -0.45x + 5 \\ y = 5.55x - 13.7 \end{cases}$

47. $\begin{cases} 1.7x + 2.3y = 3.2 \\ y = 0.25x + 8.95 \end{cases}$

48. $\begin{cases} 2.75x = 12.9y - 3.79 \\ 7.1x - y = 35.76 \end{cases}$

Use a graphing calculator to solve each equation.

49. $4(x - 1) = 3x$

50. $4(x - 3) - x = x - 6$

51. $11x + 6(3 - x) = 3$

52. $2(x + 2) = 2(1 - x) + 10$

APPLICATIONS

53. MAPS In the following map, what New Mexico city lies on the intersection of Interstate 25 and Interstate 40?

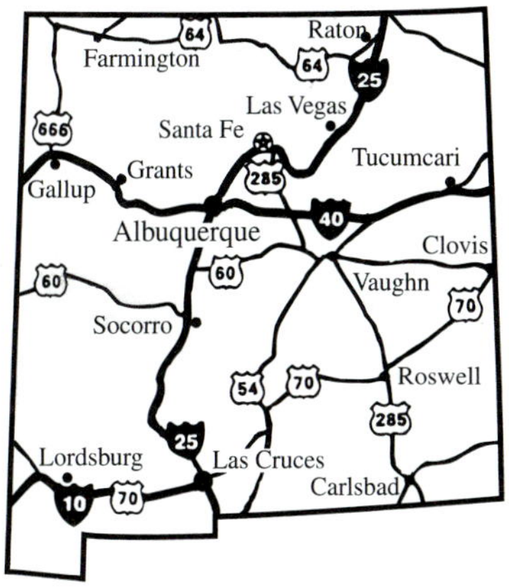

54. BUSINESS Estimate the break-even point (where cost = revenue) on the graph in the illustration. Then determine why is it called the break-even point.

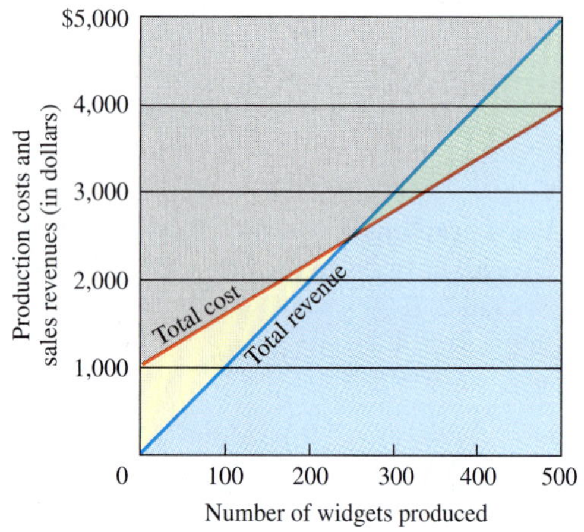

55. HEARING TESTS See the illustration. At what frequency and decibel level were the hearing test results the same for the left and right ear? Write your answer as an ordered pair.

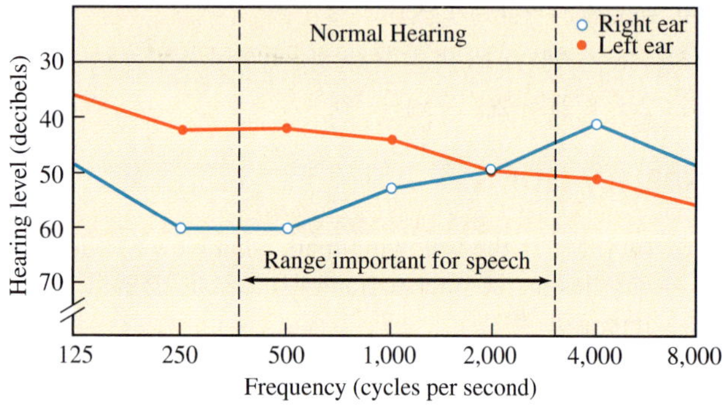

56. LAW OF SUPPLY AND DEMAND The demand function, graphed in the next column, describes the relationship between the price x of a certain camera and the demand for the camera.

a. The supply function, $S(x) = \frac{25}{4}x - 525$, describes the relationship between the price x of the camera and the number of cameras the manufacturer is willing to supply. Graph this function in the illustration.

b. For what price will the supply of cameras equal the demand?

c. As the price of the camera is increased, what happens to supply and what happens to demand?

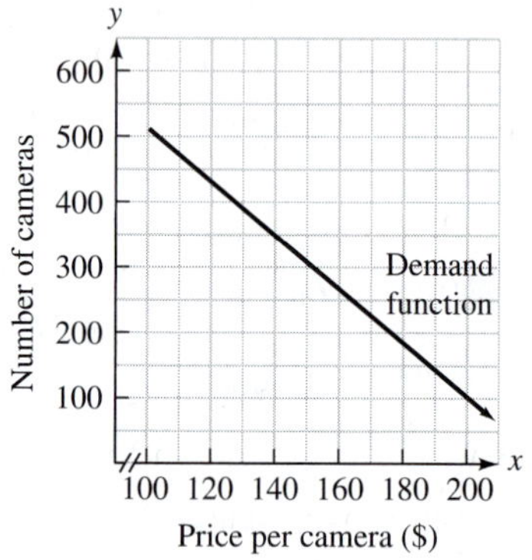

57. LEISURE TIME The graph shows how the leisure activities of Americans are changing. When was the time spent on the following activities (or when will it be) the same? Approximately how many hours for each?

a. Internet and reading magazines

b. Internet and reading newspapers

c. Video games and reading books

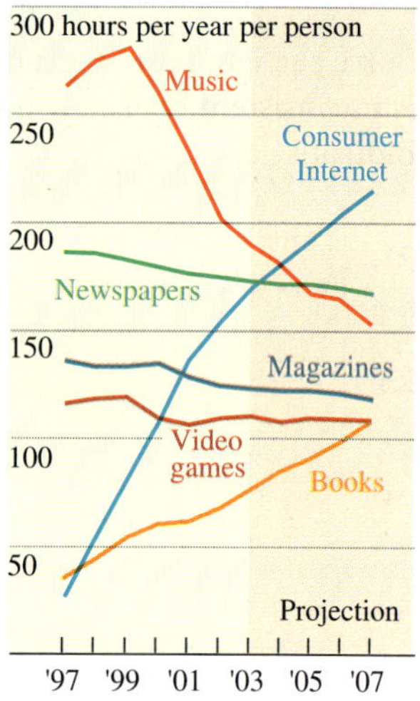

Source: *Fortune* magazine, September 15, 2003

58. COST AND REVENUE The function $C(x) = 200x + 400$ gives the cost for a college to offer x sections of an introductory class in CPR (cardio-pulmonary resuscitation). The function $R(x) = 280x$ gives the amount of revenue the college brings in when offering x sections of CPR.

a. Find the break-even point (where cost = revenue) by graphing each function on the same coordinate system.

b. How many sections does the college need to offer to make a profit on the CPR training course?

59. NAVIGATION The paths of two ships are tracked on the same coordinate system. One ship is following a path described by the equation $2x + 3y = 6$, and the other is following a path described by the equation $y = \frac{2}{3}x - 3$.

a. Is there a possibility of a collision?

b. What are the coordinates of the danger point?

c. Is a collision a certainty?

60. AIR TRAFFIC CONTROL Two airplanes are tracked using the same coordinate system on a radar screen. One plane is following a path described by the equation $y = \frac{2}{5}x - 2$, and the other is following a path described by the equation $2x = 5y + 7$. Is there a possibility of a collision?

WRITING

61. Suppose the solution of a system of two linear equations is $\left(\frac{14}{5}, -\frac{8}{3}\right)$. Knowing this, explain any drawbacks with solving the system by the graphing method.

62. Can a system of two linear equations have exactly two solutions? Why or why not?

REVIEW **Let $f(x) = -x^3 + 2x - 2$ and $g(x) = \frac{2 - x}{9 + x}$ and find each value.**

63. $f(-1)$

64. $f(10)$

65. $g(2)$

66. $g(-20)$

67. Determine the domain and range of $f(x) = x^2 - 2$.

68. Find the slope of the line passing through the points $(-4, 8)$ and $(3, 8)$.

69. In the illustration, the area of the square is 81 square centimeters. Find the area of the shaded triangle.

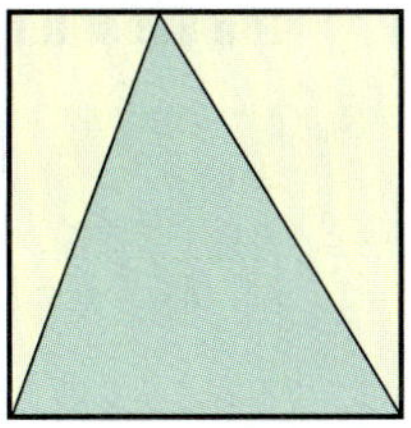

70. If the area of the circle in the illustration is 49π square centimeters, find the area of the square.

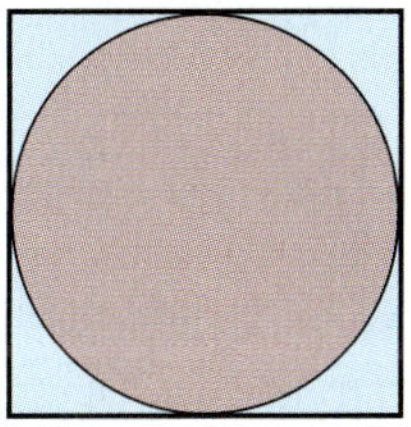

CHALLENGE PROBLEMS

71. Write an independent system of equations with a solution of $(-5, 2)$.

72. Write a dependent system of equations with a solution of $(-5, 2)$.

3.2 Solving Systems Algebraically

- The substitution method
- The elimination method
- An inconsistent system
- A system with infinitely many solutions
- Problem solving

The graphing method enables us to visualize the process of solving systems of equations. However, it can often be difficult to determine the exact coordinates of the point of intersection. In this section, we will discuss two other methods, called the *substitution* and the *elimination* methods. They can be used to find the exact solutions of systems of equations algebraically.

THE SUBSTITUTION METHOD

To solve a system of two equations in two variables by the **substitution method,** we follow these steps.

The Substitution Method

1. If necessary, solve one equation for one of its variables—preferably a variable with a coefficient of 1 or -1. We call the equation found in step 1 the **substitution equation.**
2. Substitute the resulting expression for that variable into the other equation and solve it.
3. Find the value of the other variable by substituting the value of the variable found in step 2 into the equation found in step 1.
4. State the solution.
5. Check the proposed solution in both of the original equations. Write the solution as an ordered pair.

EXAMPLE 1

Solve the system: $\begin{cases} 4x + y = 13 \\ -2x + 3y = -17 \end{cases}$.

Solution

Step 1: We solve the first equation for y, because y has a coefficient of 1.

$$4x + y = 13$$

$$y = -4x + 13$$ To isolate y, subtract $4x$ from both sides. This is the substitution equation.

Success Tip

With this method, the objective is to use an appropriate substitution to obtain *one* equation in *one* variable.

Step 2: We then substitute $-4x + 13$ for y in the second equation to eliminate the variable y from that equation. The result will be an equation containing only one variable, x.

$$-2x + 3y = -17$$ This is the second equation.

$$-2x + 3(\mathbf{-4x + 13}) = -17$$ Substitute $-4x + 13$ for y. The variable y is eliminated from the equation.

$$-2x - 12x + 39 = -17$$ Distribute the multiplication by 3.

$$-14x = -56$$ To solve for x, first combine like terms and then subtract 39 from both sides.

$$x = 4$$ Divide both sides by -14.

Language of Algebra

The phrase *back-substitute* can also be used to describe step 3 of the substitution method. To find y, we *back-substitute* 4 for x in the equation $y = -4x + 13$.

Step 3: To find y, we substitute 4 for x in the substitution equation and simplify:

$$y = -4x + 13$$

$$= -4(\mathbf{4}) + 13$$ Substitute 4 for x.

$$= -3$$

Step 4: The solution is $(4, -3)$. If graphed, the equations of the given system would intersect at the point $(4, -3)$.

Step 5: To verify that this solution satisfies both equations, we substitute 4 for x and -3 for y into each equation in the system and simplify.

Check: ***The first equation*** ***The second equation***

$$4x + y = 13 \qquad -2x + 3y = -17$$
$$4(4) + (-3) \stackrel{?}{=} 13 \qquad -2(4) + 3(-3) \stackrel{?}{=} -17$$
$$16 - 3 \stackrel{?}{=} 13 \qquad -8 - 9 \stackrel{?}{=} -17$$
$$13 = 13 \quad \text{True.} \qquad -17 = -17 \quad \text{True.}$$

Since (4, −3) satisfies both equations of the system, it checks.

Self Check 1 Solve: $\begin{cases} x + 3y = 9 \\ 2x - y = -10 \end{cases}$.

EXAMPLE 2

Solve the system: $\begin{cases} \frac{2}{9}x - \frac{2}{9}y = \frac{2}{3} \\ 0.1x = 0.2 - 0.1y \end{cases}$.

Solution First we find an equivalent system without fractions or decimals. To do this, we multiply both sides of the first equation by 9, which is the lowest common denominator of the fractions in the equation. Then we multiply both sides of the second equation by 10.

Notation

To clarify the solution process, we number the equations (1) and (2).

(1) $\begin{cases} 2x - 2y = 6 \\ x = 2 - y \end{cases}$
(2) This is the substitution equation.

Since the variable x is isolated in Equation 2, we will substitute $2 - y$ for x in Equation 1. This step will eliminate x from Equation 1, leaving an equation containing only one variable, y. We then solve for y.

$$2x - 2y = 6 \quad \text{This is Equation 1.}$$
$$2(2 - y) - 2y = 6 \quad \text{Substitute } 2 - y \text{ for } x.$$
$$4 - 2y - 2y = 6 \quad \text{Distribute the multiplication by 2.}$$
$$-4y = 2 \quad \text{To solve for } y \text{, combine like terms and then subtract 4 from both sides.}$$
$$y = -\frac{1}{2} \quad \text{Divide both sides by } -4 \text{ and then simplify the fraction.}$$

Caution

Always use the *original* equations when checking a solution. Do not use a substitution equation or an equivalent equation that you found algebraically. If an error was made, a proposed solution that would not satisfy the original system might appear to be correct.

We can find x by substituting $-\frac{1}{2}$ for y in Equation 2 and simplifying:

$$x = 2 - y \quad \text{This is Equation 2.}$$
$$x = 2 - \left(-\frac{1}{2}\right) \quad \text{Substitute } -\frac{1}{2} \text{ for } y.$$
$$= 2 + \frac{1}{2}$$
$$= \frac{5}{2} \quad 2 + \frac{1}{2} = \frac{4}{2} + \frac{1}{2} = \frac{5}{2}.$$

The solution is $\left(\frac{5}{2}, -\frac{1}{2}\right)$. Verify that it satisfies both equations in the original system.

Self Check 2 Solve: $\begin{cases} \frac{x}{8} + \frac{y}{4} = \frac{1}{2} \\ 0.01y = -0.02x + 0.04 \end{cases}$.

THE ELIMINATION METHOD

Another method for solving a system of linear equations is the **elimination** or **addition method.** In this method, we combine the equations in a way that will eliminate the terms involving one of the variables.

The Elimination Method

1. Write both equations of the system in general form: $Ax + By = C$.
2. Multiply the terms of one or both of the equations by constants chosen to make the coefficients of x (or y) differ only in sign.
3. Add the equations and solve the resulting equation, if possible.
4. Substitute the value obtained in step 3 into either of the original equations and solve for the remaining variable.
5. State the solution obtained in steps 3 and 4.
6. Check the proposed solution in both of the original equations. Write the solution as an ordered pair.

EXAMPLE 3

Solve the system: $\begin{cases} 4x + y = 13 \\ -2x + 3y = -17 \end{cases}$.

Solution

Step 1: This is the system discussed in Example 1. In this example, we will solve it by the elimination method. Since both equations are already written in general form, step 1 is unnecessary.

Step 2: We note that the coefficient of x in the first equation is 4. If we multiply both sides of the second equation by 2, the coefficient of x in that equation will be -4. Then the coefficients of x will differ only in sign.

$$\begin{cases} 4x + y = 13 \\ -4x + 6y = -34 \end{cases}$$

Step 3: When these equations are added, the terms involving x drop out (or are eliminated), and we get an equation that contains only the variable y. We then proceed by solving for y.

$$\begin{array}{r} 4x + y = 13 \\ + \quad -4x + 6y = -34 \\ \hline 7y = -21 \end{array}$$

Add the like terms, column by column: $4x + (-4x) = 0$, $y + 6y = 7y$, and $13 + (-34) = -21$.

$$y = -3$$

To solve for y, divide both sides by 7.

Step 4: To find x, we substitute -3 for y in either of the original equations and solve for x. If we use $4x + y = 13$, we have

$$4x + y = 13$$

$$4x + (-3) = 13$$ Substitute -3 for y.

$$4x = 16$$ To solve for x, add 3 to both sides.

$$x = 4$$ Divide both sides by 4.

Step 5: The solution is $(4, -3)$.

Step 6: The check was completed in Example 1.

Self Check 3 Solve: $\begin{cases} 3x + 2y = 0 \\ 2x - y = -7 \end{cases}$.

EXAMPLE 4

Solve the system: $\begin{cases} 4x = 3(2 + y) \\ 3(x - 10) = -2y \end{cases}$.

Solution To use the elimination method, we must write each equation in general form. In each case, the first step is to remove the parentheses.

The first equation	***The second equation***
$4x = 3(2 + y)$	$3(x - 10) = -2y$
$4x = 6 + 3y$	$3x - 30 = -2y$
$4x - 3y = 6$	$3x + 2y = 30$

We now solve the equivalent system

$$\begin{array}{ll} \textbf{(1)} & \\ \textbf{(2)} & \end{array} \begin{cases} 4x - 3y = 6 \\ 3x + 2y = 30 \end{cases}$$

Since the coefficients of y already have opposite signs, we choose to eliminate y. To make the y-terms drop out when we add the equations, we multiply both sides of Equation 1 by 2 and both sides of Equation 2 by 3 to get

$$\begin{cases} 8x - 6y = 12 \\ 9x + 6y = 90 \end{cases}$$

When these equations are added, the y-terms drop out, and we get

$17x = 102$ $\quad$ $8x + 9x = 17x$, $-6y + 6y = 0$, and $12 + 90 = 102$.

$x = 6$ $\quad$ To solve for x, divide both sides by 17.

To find y, we can substitute 6 for x in either of the two original equations, or in Equation 1 or Equation 2. If we substitute 6 for x in Equation 2, we get

$3x + 2y = 30$

$3(6) + 2y = 30$ $\quad$ Substitute 6 for x.

$18 + 2y = 30$ $\quad$ Perform the multiplication.

$2y = 12$ $\quad$ Subtract 18 from both sides.

$y = 6$ $\quad$ Divide both sides by 2.

The solution is (6, 6).

Self Check 4 Solve: $\begin{cases} 4(2x - y) = 18 \\ 3(x - 3) = 2y - 1 \end{cases}$.

AN INCONSISTENT SYSTEM

EXAMPLE 5

Solve the system $\begin{cases} y = 2x + 4 \\ 8x - 4y = 7 \end{cases}$, if possible.

Solution Because the first equation is already solved for y, we use the substitution method.

$$8x - 4y = 7 \quad \text{This is the second equation.}$$
$$8x - 4(2x + 4) = 7 \quad \text{Substitute } 2x + 4 \text{ for } y.$$

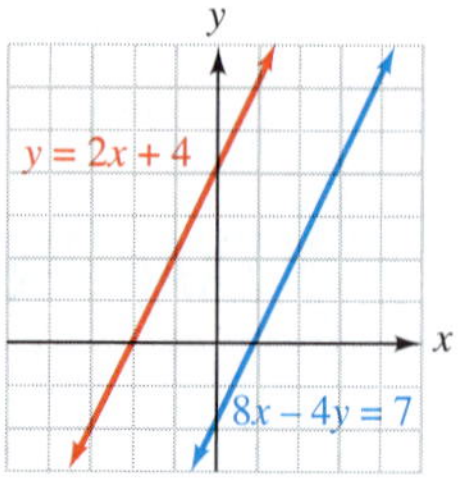

We then solve this equation for x:

$$8x - 8x - 16 = 7 \quad \text{Distribute the multiplication by } -4.$$
$$-16 = 7$$

Here, the terms involving x drop out, and we get $-16 = 7$. This false statement indicates that the system has *no solution* and is, therefore, inconsistent. The graphs of the equations of the system verify this—they are parallel lines.

Self Check 5 Solve: $\begin{cases} x = -2.5y + 8 \\ y = -0.4x + 2 \end{cases}$.

A SYSTEM WITH INFINITELY MANY SOLUTIONS

EXAMPLE 6 Solve the system: $\begin{cases} 4x + 6y = 12 \\ -2x - 3y = -6 \end{cases}$.

Solution Since the equations are written in general form, we use the elimination method. We copy the first equation and multiply both sides of the second equation by 2 to get

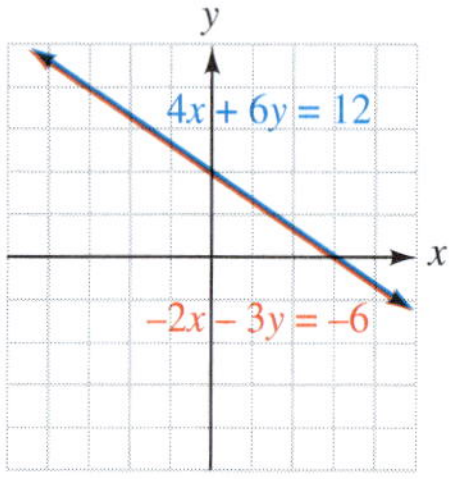

$$\begin{aligned} 4x + 6y &= 12 \\ -4x - 6y &= -12 \end{aligned}$$

After adding the left-hand sides and the right-hand sides, we get

$$0x + 0y = 0$$
$$0 = 0$$

Here, both the x- and y-terms drop out. The resulting true statement $0 = 0$ indicates that the equations are dependent and that the system has an *infinitely many solutions.*

Note that the equations of the system are equivalent, because when the second equation is multiplied by -2, it becomes the first equation. The graphs of these equations would coincide. Any ordered pair that satisfies one of the equations also satisfies the other. To find some solutions, we can substitute 0, 3, and -3 for x in either equation to obtain (0, 2), (3, 0), and $(-3, 4)$.

Self Check 6 Solve: $\begin{cases} x - \frac{5}{2}y = \frac{19}{2} \\ -\frac{2}{5}x + y = -\frac{19}{5} \end{cases}$.

PROBLEM SOLVING

To solve problems using two variables, we follow the same problem-solving strategy discussed in Chapters 1 and 2, except that we form two equations using two variables instead of one using one variable.

EXAMPLE 7

Wedding pictures. A professional photographer offers two different packages for wedding pictures. Use the information in the figure to determine the cost of an 8×10-in. and a 5×7-in. photograph.

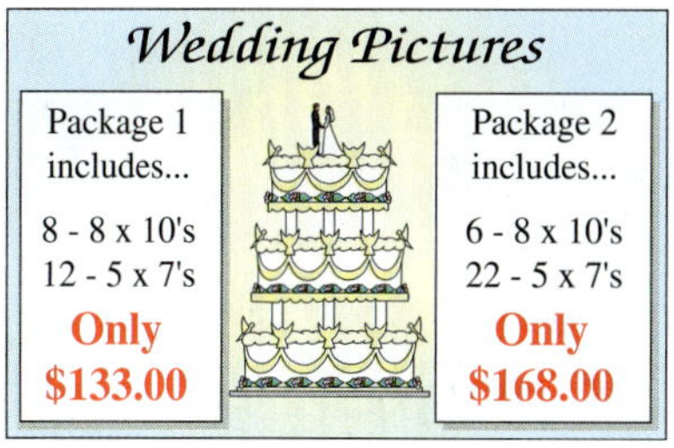

Analyze the Problem

- Eight 8×10 and twelve 5×7 pictures cost \$133.
- Six 8×10 and twenty-two 5×7 pictures cost \$168.
- Find the cost of an 8×10 and a 5×7 photograph.

Form Two Equations

We can let x = the cost of an 8×10 photograph and let y = the cost of a 5×7 photograph. For the first package, the cost of eight 8×10 pictures is $8 \cdot \$x = \$8x$, and the cost of twelve 5×7 pictures is $12 \cdot \$y = \$12y$. For the second package, the cost of six 8×10 pictures is $\$6x$, and the cost of twenty-two 5×7 pictures is $\$22y$. To find x and y, we must write and solve two equations.

Caution

If two variables are used to represent two unknown quantities, we must form a system of two equations to find the unknowns.

The cost of eight 8×10 photographs	plus	the cost of twelve 5×7 photographs	is	the cost of the first package.
$8x$	$+$	$12y$	$=$	133
The cost of six 8×10 photographs	plus	the cost of twenty-two 5×7 photographs	is	the cost of the second package.
$6x$	$+$	$22y$	$=$	168

Solve the System

To find the cost of the 8×10 and the 5×7 photographs, we must solve the following system:

$$\begin{array}{ll} \textbf{(1)} & \left\{\begin{array}{l} 8x + 12y = 133 \\ 6x + 22y = 168 \end{array}\right. \\ \textbf{(2)} & \end{array}$$

We will use the elimination method to solve this system. To make the x-terms drop out, we multiply both sides of Equation 1 by 3. Then we multiply both sides of Equation 2 by -4. We then add the resulting equations and solve for y:

$$\begin{aligned} 24x + 36y &= 399 \\ -24x - 88y &= -672 \\ \hline -52y &= -273 && \text{Add like terms, column by column. The } x\text{-terms drop out.} \\ y &= 5.25 && \text{Divide both sides by } -52. \end{aligned}$$

To find x, we substitute 5.25 for y in Equation 1 and solve for x:

$$\begin{aligned} 8x + 12y &= 133 \\ 8x + 12(5.25) &= 133 && \text{Substitute 5.25 for } y. \\ 8x + 63 &= 133 && \text{Perform the multiplication.} \\ 8x &= 70 && \text{Subtract 63 from both sides.} \\ x &= 8.75 && \text{Divide both sides by 8.} \end{aligned}$$

State the Conclusion The cost of an 8×10 photo is \$8.75, and the cost of a 5×7 photo is \$5.25.

Check the Result If the first package contains eight 8×10 and twelve 5×7 photographs, the value of the package is 8(\$8.75) + 12(\$5.25) = \$70 + \$63 = \$133. If the second package contains six 8×10 and twenty-two 5×7 photographs, the value of the package is 6(\$8.75) + 22(\$5.25) = \$52.50 + \$115.50 = \$168. The answers check.

EXAMPLE 8

Water treatment. A technician determines that 50 fluid ounces of a 15% muriatic acid solution needs to be added to the water in a swimming pool to kill a growth of algae. If the technician has 5% and 20% muriatic solutions on hand, how many ounces of each must be combined to create the 15% solution?

Analyze the Problem We need to find the number of ounces of a 5% solution and the number of ounces of a 20% solution that must be combined to obtain 50 ounces of a 15% solution.

Form Two Equations We can let x = the number of ounces of the 5% solution and let y = the number of ounces of the 20% solution that are to be mixed. Then the amount of muriatic acid in the 5% solution is $0.05x$ ounces, and the amount of muriatic acid in the 20% solution is $0.20y$ ounces. The sum of these amounts is also the amount of muriatic acid in the final mixture, which is 15% of 50 ounces. This information is shown in the figure.

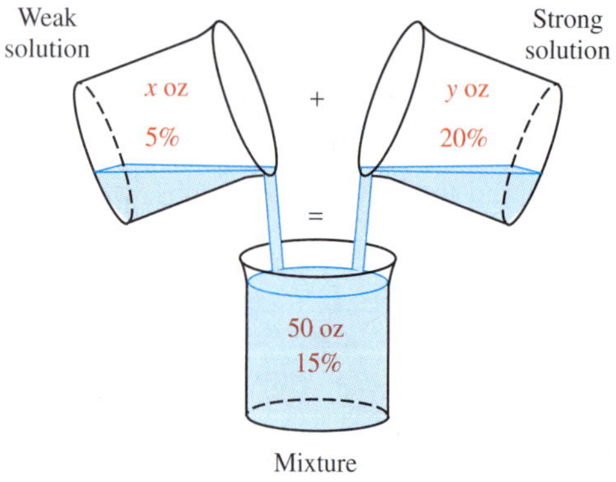

	Ounces ·	Strength =	Amount of acid
Weak	x	0.05	$0.05x$
Strong	y	0.20	$0.20y$
Mixture	50	0.15	0.15(50)

One equation comes from the information in this column. (Ounces)

Another equation comes from the information in this column. (Amount of acid)

The facts of the problem give the following two equations:

The number of ounces of 5% solution	plus	the number of ounces of 20% solution	is	the total number of ounces in the 15% mixture.
x	$+$	y	$=$	50
The acid in the 5% solution	plus	the acid in the 20% solution	is	the acid in the 15% mixture.
$0.05x$	$+$	$0.20y$	$=$	$0.15(50)$

Solve the System To find out how many ounces of each are needed, we solve the following system:

$$\begin{cases} x + y = 50 & (1) \\ 0.05x + 0.20y = 7.5 & (2) \end{cases} \quad 0.15(50) = 7.5.$$

To solve this system by substitution, we can solve the first equation for y:

$$x + y = 50$$
$$y = 50 - x \quad \text{Subtract } x \text{ from both sides. This is the substitution equation.}$$

Then we substitute $50 - x$ for y in Equation 1 and solve for x.

$$0.05x + 0.20y = 7.5 \quad \text{This is Equation 1.}$$
$$0.05x + 0.20(50 - x) = 7.5 \quad \text{Substitute } 50 - x \text{ for } y.$$
$$5x + 20(50 - x) = 750 \quad \text{Multiply both sides by 100.}$$
$$5x + 1{,}000 - 20x = 750 \quad \text{Use the distributive property to remove parentheses.}$$
$$-15x = -250 \quad \text{Combine like terms and subtract 1,000 from both sides.}$$
$$x = \frac{-250}{-15} \quad \text{Divide both sides by } -15.$$
$$x = \frac{50}{3} \quad \text{Simplify: } \frac{250}{15} = \frac{\overset{1}{\cancel{5}} \cdot 50}{\underset{1}{\cancel{5}} \cdot 3}.$$

To find y, we can substitute $\frac{50}{3}$ for x in the substitution equation:

$$y = 50 - x$$
$$= 50 - \frac{50}{3} \quad \text{Substitute } \frac{50}{3} \text{ for } x.$$
$$= \frac{100}{3} \quad 50 = \frac{150}{3}.$$

State the Conclusion To obtain 50 ounces of a 15% solution, the technician must mix $\frac{50}{3}$ or $16\frac{2}{3}$ ounces of the 5% solution with $\frac{100}{3}$ or $33\frac{1}{3}$ ounces of the 20% solution.

Check the Result We note that $16\frac{2}{3}$ ounces of solution plus $33\frac{1}{3}$ ounces of solution equals the required 50 ounces of solution. We also note that 5% of $16\frac{2}{3} \approx 0.83$ and 20% of $33\frac{1}{3} \approx 6.67$, giving a total of 7.5, which is 15% of 50. The answers check.

EXAMPLE 9

Parallelograms. Refer to the parallelogram and find the values of x and y.

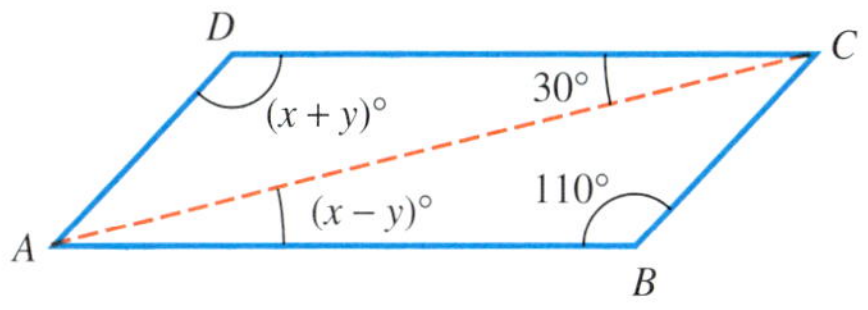

Solution To solve this problem, we will use two important facts about parallelograms.

- When a diagonal intersects two parallel sides of a parallelogram, pairs of *alternate interior angles* have the same measure. In the figure, $\angle BAC$ and $\angle DCA$ are alternate interior angles and therefore have the same measure. Thus, $(x - y)° = 30°$.
- *Opposite angles* of a parallelogram have the same measure. Since $\angle B$ and $\angle D$ in the figure are opposite angles of the parallelogram, $(x + y)° = 110°$.

We can form the following system of equations and solve it by elimination.

$$\begin{aligned} x - y &= 30 \\ x + y &= 110 \\ \hline 2x \quad &= 140 \quad && \text{Add the equations. The } y\text{-terms drop out.} \\ x &= 70 && \text{Divide both sides by 2.} \end{aligned}$$

We can substitute 70 for x in the second equation and solve for y.

$$\begin{aligned} x + y &= 110 \\ 70 + y &= 110 \quad && \text{Substitute 70 for } x. \\ y &= 40 && \text{Subtract 70 from both sides.} \end{aligned}$$

Thus, $x = 70$ and $y = 40$.

Running a machine involves both *setup costs* and *unit costs.* Setup costs include the cost of preparing a machine to do a certain job. The costs to make one item are unit costs. They depend on the number of items to be manufactured, including costs of raw materials and labor.

EXAMPLE 10

Break point. The setup cost of a machine that makes wooden coathangers is \$400. After setup, it costs \$1.50 to make each hanger (the unit cost). Management is considering the purchase of a new machine that can manufacture the same type of coathanger at a cost of \$1.25 per hanger. If the setup cost of the new machine is \$500, find the number of coathangers that the company would need to manufacture to make the cost the same using either machine. This is called the **break point.**

Analyze the Problem

We are to find the number of coathangers that will cost equal amounts to produce on either machine. The machines have different setup costs and different unit costs.

Form Two Equations

The cost C_1 of manufacturing x coathangers on the machine currently in use is $\$1.50x + \$400$ (the number of coathangers manufactured times \$1.50, plus the setup cost of \$400). The cost C_2 of manufacturing the same number of coathangers on the new machine is $\$1.25x + \$500$. The break point occurs when the costs to make the same number of hangers using either machine are equal ($C_1 = C_2$).

If $x =$ the number of coathangers to be manufactured, the cost C_1 using the machine currently in use is

The cost of using the current machine	is	the cost of manufacturing x coathangers	plus	the setup cost.
C_1	$=$	$1.5x$	$+$	400

The cost C_2 using the new machine to make x coathangers is

The cost of using the new machine	is	the cost of manufacturing x coathangers	plus	the setup cost.
C_2	$=$	$1.25x$	$+$	500

Solve the System To find the break point, we must solve the system $\begin{cases} C_1 = 1.5x + 400 \\ C_2 = 1.25x + 500 \end{cases}$.

Since the break point occurs when $C_1 = C_2$, we can substitute $1.5x + 400$ for C_2 in the second equation to get

$$
\begin{aligned}
1.5x + 400 &= 1.25x + 500 \\
1.5x &= 1.25x + 100 && \text{Subtract 400 from both sides.} \\
0.25x &= 100 && \text{Subtract } 1.25x \text{ from both sides.} \\
x &= 400 && \text{Divide both sides by 0.25.}
\end{aligned}
$$

State the Conclusion The break point is 400 coathangers.

Check the Result To make 400 coathangers, the cost on the current machine would be \$400 + \$1.50(400) = \$400 + \$600 = \$1,000. The cost using the new machine would be \$500 + \$1.25(400) = \$500 + \$500 = \$1,000. Since the costs are equal, the break point is 400.

Answers to Self Checks **1.** $(-3, 4)$ **2.** $\left(\frac{4}{3}, \frac{4}{3}\right)$ **3.** $(-2, 3)$ **4.** $\left(1, -\frac{5}{2}\right)$ **5.** no solution
6. There are infinitely many solutions; three of them are $(2, -3)$, $(12, 1)$, and $\left(\frac{19}{2}, 0\right)$.

3.2 STUDY SET

VOCABULARY Fill in the blanks.

1. $Ax + By = C$ is the ________ form of a linear equation.

2. In the equation $x + 3y = -1$, the x-term has an understood __________ of 1.

3. When we add the two equations of the system $\begin{cases} x + y = 5 \\ x - y = -3 \end{cases}$, the y-terms are __________.

4. To solve $\begin{cases} y = 3x \\ x + y = 4 \end{cases}$, we can _________ $3x$ for y in the second equation.

CONCEPTS

5. If the system $\begin{cases} 4x - 3y = 7 \\ 3x - 2y = 6 \end{cases}$ is to be solved using the elimination method, by what constant should each equation be multiplied if

a. the x-terms are to drop out?

b. the y-terms are to drop out?

6. If the system $\begin{cases} 4x - 3y = 7 \\ 3x + y = 6 \end{cases}$ is to be solved using the substitution method, what variable in what equation would it be easier to solve for?

7. Can the system $\begin{cases} 2x + 5y = 7 \\ 4x - 3y = 16 \end{cases}$ be solved more easily by the substitution or the elimination method?

8. Given the equation $3x + y = -4$.

a. Solve for x.

b. Solve for y.

c. Which variable was easier to solve for? Explain why.

9. The substitution method was used to solve three systems of linear equations. The results after y was eliminated and the remaining equation was solved for x are listed below. Match each result with one of the possible graphs shown.

a. $-2 = 3$ **b.** $x = 3$ **c.** $3 = 3$

Possible graphs

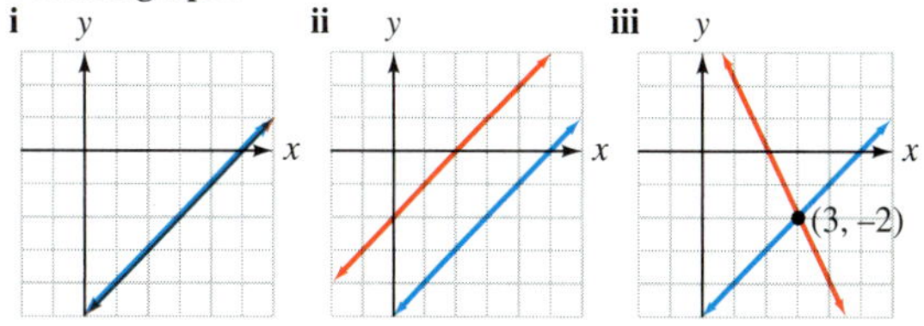

10. Consider the system: $\begin{cases} \frac{2}{3}x - \frac{y}{6} = \frac{16}{9} \\ 0.03x + 0.02y = 0.03 \end{cases}$.

a. What algebraic step should be performed to clear the first equation of fractions?

b. What algebraic step should be performed to clear the second equation of decimals?

NOTATION **Write each equation of the system in general form: $Ax + By = C$.**

11. $\begin{cases} 4y = 8 - 7x \\ 2x = y - 3 \end{cases}$

12. $\begin{cases} 2(x - 4y) = \frac{9}{2} \\ 3x - y = 2(x + 4) \end{cases}$

PRACTICE **Solve each system by substitution, if possible. If a system is inconsistent or if the equations are dependent, so indicate.**

13. $\begin{cases} y = x \\ x + y = 4 \end{cases}$

14. $\begin{cases} y = x + 2 \\ x + 2y = 16 \end{cases}$

15. $\begin{cases} x = 2 + y \\ 2x + y = 13 \end{cases}$

16. $\begin{cases} x = -4 + y \\ 3x - 2y = -5 \end{cases}$

17. $\begin{cases} x + 2y = 6 \\ 3x - y = -10 \end{cases}$

18. $\begin{cases} 2x - y = -21 \\ 4x + 5y = 7 \end{cases}$

19. $\begin{cases} \frac{3}{2}x + 2 = y \\ 0.6x - 0.4y = -0.4 \end{cases}$

20. $\begin{cases} 2x - \frac{5}{2} = y \\ 0.04x - 0.02y = 0.05 \end{cases}$

Solve each system by elimination, if possible. If a system is inconsistent or if the equations are dependent, so indicate.

21. $\begin{cases} x - y = 3 \\ x + y = 7 \end{cases}$

22. $\begin{cases} x + y = 1 \\ x - y = 7 \end{cases}$

23. $\begin{cases} 2x + y = -10 \\ 2x - y = -6 \end{cases}$

24. $\begin{cases} x + 2y = -9 \\ x - 2y = -1 \end{cases}$

25. $\begin{cases} 2x + 3y = 8 \\ 3x - 2y = -1 \end{cases}$

26. $\begin{cases} 5x - 2y = 19 \\ 3x + 4y = 1 \end{cases}$

27. $\begin{cases} 4(x - 2) = -9y \\ 2(x - 3y) = -3 \end{cases}$

28. $\begin{cases} 2(2x + 3y) = 5 \\ 8x = 3(1 + 3y) \end{cases}$

Solve each system by any method, if possible. If a system is inconsistent or if the equations are dependent, so indicate.

29. $\begin{cases} 3x - 4y = 9 \\ x + 2y = 8 \end{cases}$

30. $\begin{cases} 3x - 2y = -10 \\ 6x + 5y = 25 \end{cases}$

31. $\begin{cases} 2(x + y) + 1 = 0 \\ 3x + 4y = 0 \end{cases}$

32. $\begin{cases} 5x + 3y = -7 \\ 3(x - y) - 7 = 0 \end{cases}$

33. $\begin{cases} 0.16x - 0.08y = 0.32 \\ 2x - 4 = y \end{cases}$

34. $\begin{cases} 0.6y - 0.9x = -3.9 \\ 3x - 17 = 4y \end{cases}$

35. $\begin{cases} x = \frac{3}{2}y + 5 \\ 2x - 3y = 8 \end{cases}$

36. $\begin{cases} x = \frac{2}{3}y \\ y = 4x + 5 \end{cases}$

37. $\begin{cases} 0.5x + 0.5y = 6 \\ \frac{x}{2} - \frac{y}{2} = -2 \end{cases}$

38. $\begin{cases} \frac{x}{2} - \frac{y}{3} = -4 \\ \frac{x}{2} + \frac{y}{9} = 0 \end{cases}$

39. $\begin{cases} \frac{3}{4}x + \frac{2}{3}y = 7 \\ \frac{3}{5}x - \frac{1}{2}y = 18 \end{cases}$

40. $\begin{cases} \frac{2}{3}x - \frac{1}{4}y = -8 \\ 0.5x - 0.375y = -9 \end{cases}$

41. $\begin{cases} \frac{3x}{2} - \frac{2y}{3} = 0 \\ \frac{3x}{4} + \frac{4y}{3} = \frac{5}{2} \end{cases}$

42. $\begin{cases} \frac{3x}{5} + \frac{5y}{3} = 2 \\ \frac{6x}{5} - \frac{5y}{3} = 1 \end{cases}$

43. $\begin{cases} 12x - 5y - 21 = 0 \\ \frac{3}{4}x - \frac{2}{3}y = \frac{19}{8} \end{cases}$

44. $\begin{cases} 4y + 5x - 7 = 0 \\ \frac{10}{7}x - \frac{4}{9}y = \frac{17}{21} \end{cases}$

Solve each system. When writing the solution as an ordered pair, write the values for the variables in alphabetical order.

45. $\begin{cases} \frac{3}{2}p + \frac{1}{3}q = 2 \\ \frac{2}{3}p + \frac{1}{9}q = 1 \end{cases}$

46. $\begin{cases} a + \frac{b}{3} = \frac{5}{3} \\ \frac{a+b}{3} = 3 - a \end{cases}$

47. $\begin{cases} \frac{m-n}{5} + \frac{m+n}{2} = 6 \\ \frac{m-n}{2} - \frac{m+n}{4} = 3 \end{cases}$

48. $\begin{cases} \frac{r-2}{5} + \frac{s+3}{2} = 5 \\ \frac{r+3}{2} + \frac{s-2}{3} = 6 \end{cases}$

Solve each system. To do so, substitute a for $\frac{1}{x}$ and b for $\frac{1}{y}$ and solve for a and b. Then find x and y using the fact that $a = \frac{1}{x}$ and $b = \frac{1}{y}$.

49. $\begin{cases} \frac{1}{x} + \frac{1}{y} = \frac{5}{6} \\ \frac{1}{x} - \frac{1}{y} = \frac{1}{6} \end{cases}$

50. $\begin{cases} \frac{1}{x} + \frac{1}{y} = \frac{9}{20} \\ \frac{1}{x} - \frac{1}{y} = \frac{1}{20} \end{cases}$

51. $\begin{cases} \frac{1}{x} + \frac{2}{y} = -1 \\ \frac{2}{x} - \frac{1}{y} = -7 \end{cases}$

52. $\begin{cases} \frac{3}{x} - \frac{2}{y} = -30 \\ \frac{2}{x} - \frac{3}{y} = -30 \end{cases}$

APPLICATIONS **Use two variables to solve each problem.**

53. ADVERTISING Use the information in the ad to find the cost of a 15-second and a 30-second radio commercial on radio station KLIZ.

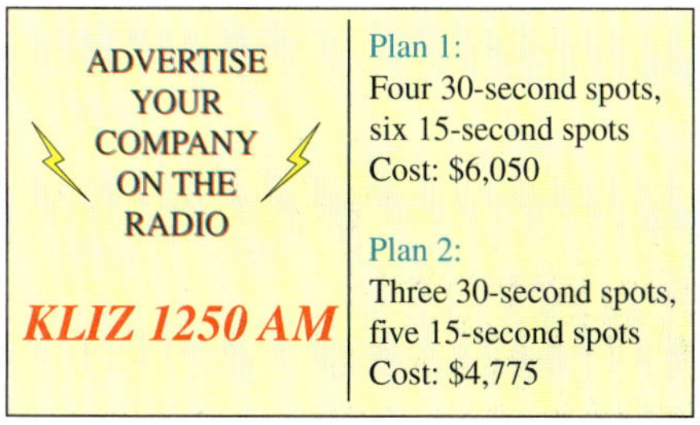
ADVERTISE YOUR COMPANY ON THE RADIO
KLIZ 1250 AM
Plan 1:
Four 30-second spots, six 15-second spots
Cost: $6,050
Plan 2:
Three 30-second spots, five 15-second spots
Cost: $4,775

54. TEMPORARY HELP A law firm hired several workers to help finish a large project. From the following billing records, determine the daily fee charged by the employment agency for a clerk-typist and for a computer programmer.

TEMPORARY EMPLOYMENT, INC.
We meet your employment needs!
Billed to: Archer Law Offices Attn: B. Kinsell

Day	Position/Employee Name	Total cost
Mon. 3/22	*Clerk-typists:* K. Amad, B. Tran, S. Smith *Programmers:* T. Lee, C. Knox	$685
Tues. 3/23	*Clerk-typists:* K. Amad, B. Tran, S. Smith, W. Morada *Programmers:* T. Lee, C. Knox, B. Morales	$975

55. PETS According to the Pet Food Institute, in 2003 there were an estimated 135 million dogs and cats in the United States. If there were 15 million more cats than dogs, how many of each type of pet were there in 2003?

56. ELECTRONICS In the illustration, two resistors in the voltage divider circuit have a total resistance of 1,375 ohms. To provide the required voltage, R_1 must be 125 ohms greater than R_2. Find both resistances.

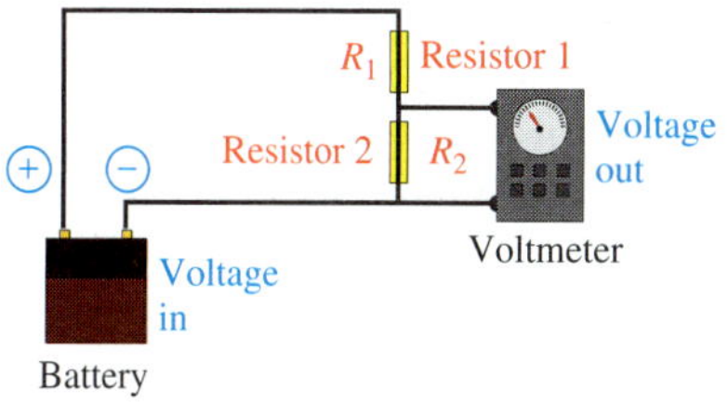

57. FENCING A FIELD The rectangular field is surrounded by 72 meters of fencing. If the field is partitioned as shown, a total of 88 meters of fencing is required. Find the dimensions of the field.

58. GEOMETRY In a right triangle, one acute angle is 15° greater than two times the other acute angle. Find the difference between the measures of the angles.

59. BRACING The bracing of a basketball backboard forms a parallelogram. Find the values of x and y.

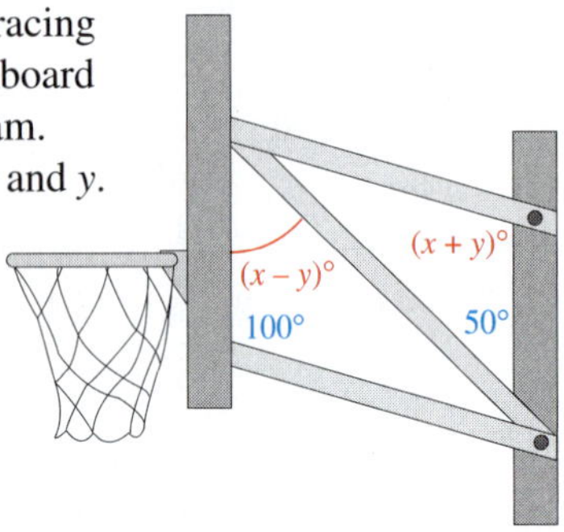

60. TRAFFIC SIGNALS In the illustration, braces A and B are perpendicular. Find the values of x and y.

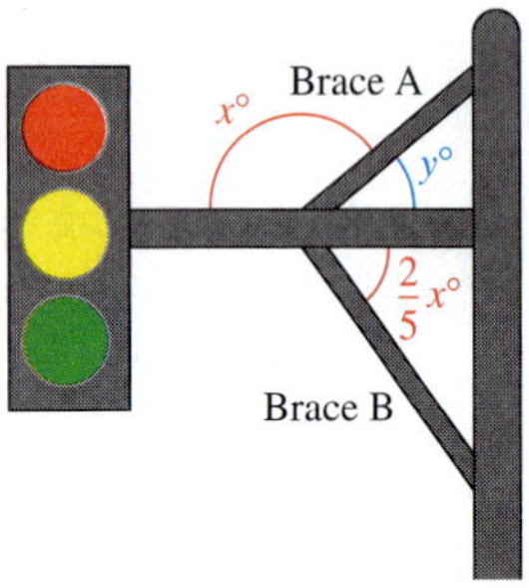

61. INVESTMENT CLUBS Part of \$8,000 was invested by an investment club at 10% interest and the rest at 12%. If the annual income from these investments is \$900, how much was invested at each rate?

62. RETIREMENT INCOME A retired couple invested part of \$12,000 at 6% interest and the rest at 7.5%. If their annual income from these investments is \$810, how much was invested at each rate?

63. TV NEWS A news van and a helicopter left a TV station parking lot at the same time, headed in opposite directions to cover breaking news stories that were 145 miles apart. If the helicopter had to travel 55 miles farther than the van, how far did the van have to travel to reach the location of the news story?

64. DELIVERY SERVICE A delivery truck travels 50 miles in the same time that a cargo plane travels 180 miles. The speed of the plane is 143 mph faster than the speed of the truck. Find the speed of the delivery truck.

65. PRODUCTION PLANNING A manufacturer builds racing bikes and mountain bikes, with the per unit manufacturing costs shown in the table. The company has budgeted \$15,900 for labor and \$13,075 for materials. How many bicycles of each type can be built?

Model	Cost of materials	Cost of labor
Racing	\$55	\$60
Mountain	\$70	\$90

66. FARMING A farmer keeps some animals on a strict diet. Each animal is to receive 15 grams of protein and 7.5 grams of carbohydrates. The farmer uses two food mixes, with nutrients as shown in the table. How many grams of each mix should be used to provide the correct nutrients for each animal?

Mix	Protein	Carbohydrates
Mix A	12%	9%
Mix B	15%	5%

67. DERMATOLOGY Tests of an antibacterial face-wash cream showed that a mixture containing 0.3% Triclosan (active ingredient) gave the best results. How many grams of cream from each tube shown in the illustration should be used to make an equal-size tube of the 0.3% cream?

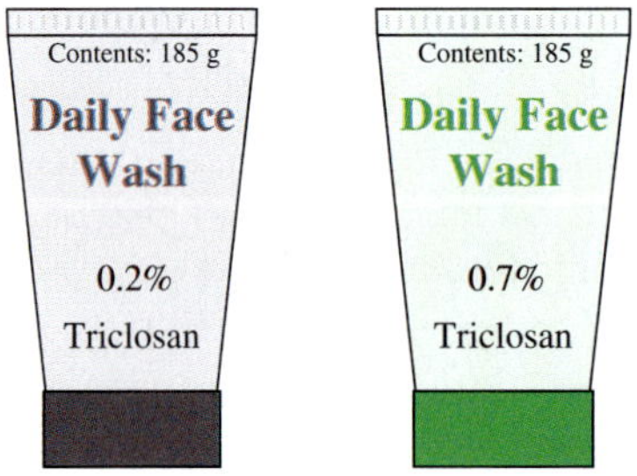

68. MIXING SOLUTIONS How many ounces of the two alcohol solutions must be mixed to obtain 100 ounces of a 12.2% solution?

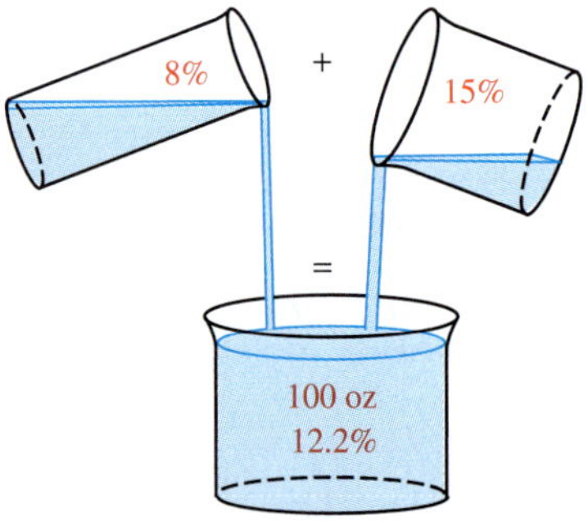

69. MIXING CANDY How many pounds of each candy shown in the illustration must be mixed to obtain 60 pounds of candy that would be worth $4 per pound?

70. RECORDING COMPANIES Three people invest a total of $105,000 to start a recording company that will produce reissues of classic jazz. Each release will be a set of 3 CDs that will retail for $45 per set. If each set can be produced for $18.95, how many sets must be sold for the investors to make a profit?

71. MACHINE SHOPS Two machines can mill a brass plate. One machine has a setup cost of $300 and a cost per plate of $2. The other machine has a setup cost of $500 and a cost per plate of $1. Find the break point.

72. PUBLISHING A printer has two presses. One has a setup cost of $210 and can print the pages of a certain book for $5.98. The other press has a setup cost of $350 and can print the pages of the same book for $5.95. Find the break point.

73. COSMETOLOGY A beauty shop specializing in permanents has fixed costs of $2,101.20 per month. The owner estimates that the cost for each permanent is $23.60, which covers labor, chemicals, and electricity. If her shop can give as many permanents as she wants at a price of $44 each, how many must be given each month to break even?

74. PRODUCTION PLANNING A paint manufacturer can choose between two processes for manufacturing house paint, with monthly costs as shown in the table. Assume that the paint sells for $18 per gallon.

Process	Fixed costs	Unit cost (per gallon)
A	$32,500	$13
B	$80,600	$5

a. Find the break point for process A.

b. Find the break point for process B.

c. If expected sales are 7,000 gallons per month, which process should the company use?

75. MANUFACTURING A manufacturer of automobile water pumps is considering retooling for one of two manufacturing processes, with monthly fixed costs and unit costs as indicated in the table. Each water pump can be sold for $50.

Process	Fixed costs	Unit cost
A	$12,390	$29
B	$20,460	$17

a. Find the break point for process A.

b. Find the break point for process B.

c. If expected sales are 550 per month, which process should be used?

76. SALARY OPTIONS A sales clerk can choose from two salary plans: a straight 7% commission, or $150 + 2% commission. How much would the clerk have to sell for each plan to produce the same monthly paycheck?

WRITING

77. Which method would you use to solve the system $\begin{cases} 4x + 6y = 5 \\ 8x - 3y = 3 \end{cases}$? Explain why.

78. Which method would you use to solve the system $\begin{cases} x - 2y = 2 \\ 2x + 3y = 11 \end{cases}$? Explain why.

79. When solving a problem using two variables, why must we write two equations?

80. Write a problem that can be solved by solving the system $\begin{cases} x + y = 36 \\ \$1.29x + \$2.29y = \$72.44 \end{cases}$.

81. Write a problem to fit the information given in the table.

	Ounces ·	Strength =	Amount of insecticide
Weak	x	0.02	$0.02x$
Strong	y	0.10	$0.10y$
Mixture	80	0.07	0.07(80)

82. Complete the following table, listing one advantage and one disadvantage for each of the methods that can be used to solve a system of two linear equations.

Method	Advantage	Disadvantage
Graphing		
Substitution		
Elimination		

REVIEW **Find the slope of each line.**

83.

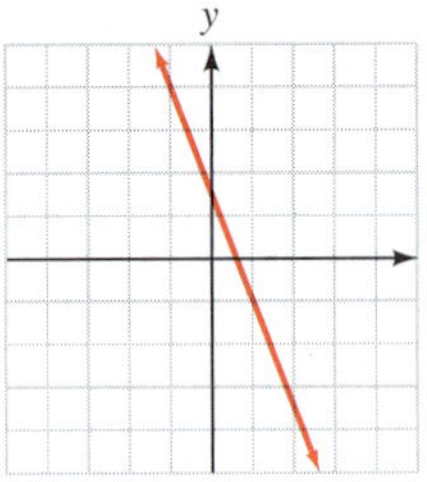

84.

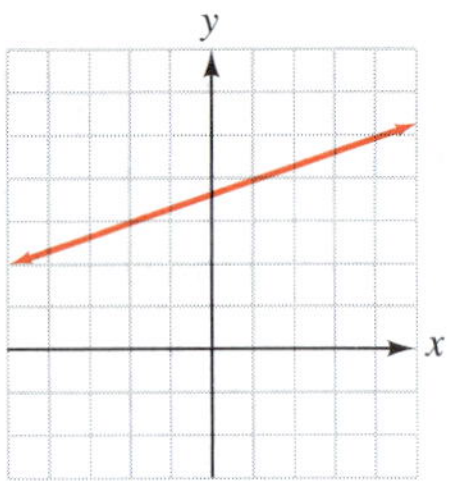

85. The line passing through $(0, -8)$ and $(-5, 0)$

86. The line with equation $y = -3x + 4$

87. The line with equation $4x - 3y = -3$

88. The line with equation $y = 3$

CHALLENGE PROBLEMS

89. If the solution of the system $\begin{cases} Ax + By = -2 \\ Bx - Ay = -26 \end{cases}$ is $(-3, 5)$, what are the values of the constants A and B?

90. Solve: $\begin{cases} 2ab - 3cd = 1 \\ 3ab - 2cd = 1 \end{cases}$ for a and c. Assume that b and d are constants.

3.3 Systems with Three Variables

- Solving three equations with three variables
- Consistent systems
- An inconsistent system
- Systems with dependent equations
- Problem solving
- Curve fitting

In the preceding sections, we solved systems of two linear equations with two variables. In this section, we will solve systems of linear equations with three variables by using a combination of the elimination method and the substitution method. We will then use that procedure to solve problems involving three variables.

SOLVING THREE EQUATIONS WITH THREE VARIABLES

We now extend the definition of a linear equation to include equations of the form $Ax + By + Cz = D$. The solution of a system of three linear equations with three variables is an **ordered triple** of numbers. For example, the solution of the system

$$\begin{cases} 2x + 3y + 4z = 20 \\ 3x + 4y + 2z = 17 \\ 3x + 2y + 3z = 16 \end{cases}$$

is the triple $(1, 2, 3)$, since each equation is satisfied if $x = 1$, $y = 2$, and $z = 3$.

$$\begin{aligned} 2x + 3y + 4z &= 20 \\ 2(1) + 3(2) + 4(3) &\stackrel{?}{=} 20 \\ 2 + 6 + 12 &\stackrel{?}{=} 20 \\ 20 &= 20 \end{aligned} \qquad \begin{aligned} 3x + 4y + 2z &= 17 \\ 3(1) + 4(2) + 2(3) &\stackrel{?}{=} 17 \\ 3 + 8 + 6 &\stackrel{?}{=} 17 \\ 17 &= 17 \end{aligned} \qquad \begin{aligned} 3x + 2y + 3z &= 16 \\ 3(1) + 2(2) + 3(3) &\stackrel{?}{=} 16 \\ 3 + 4 + 9 &\stackrel{?}{=} 16 \\ 16 &= 16 \end{aligned}$$

The Language of Algebra

A system of equations is two (or more) equations that we consider *simultaneously*—at the same time. Some professional sports teams *simulcast* their games. That is, the announcer's play-by-play description is broadcast on radio and television at the same time.

The graph of an equation of the form $Ax + By + Cz = D$ is a flat surface called a *plane.* A system of three linear equations with three variables is consistent or inconsistent, depending on how the three planes corresponding to the three equations intersect. The following illustration shows some of the possibilities.

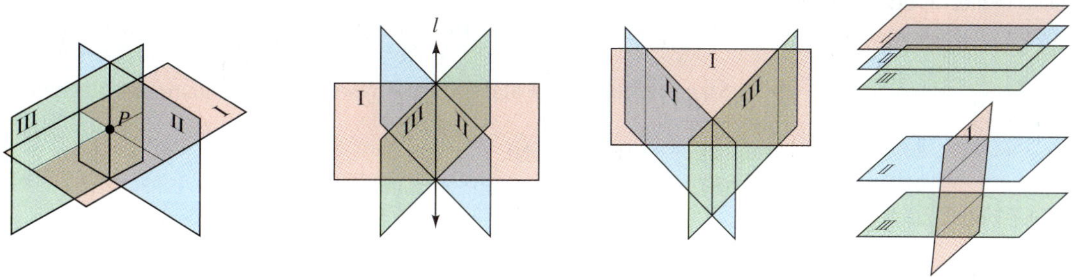

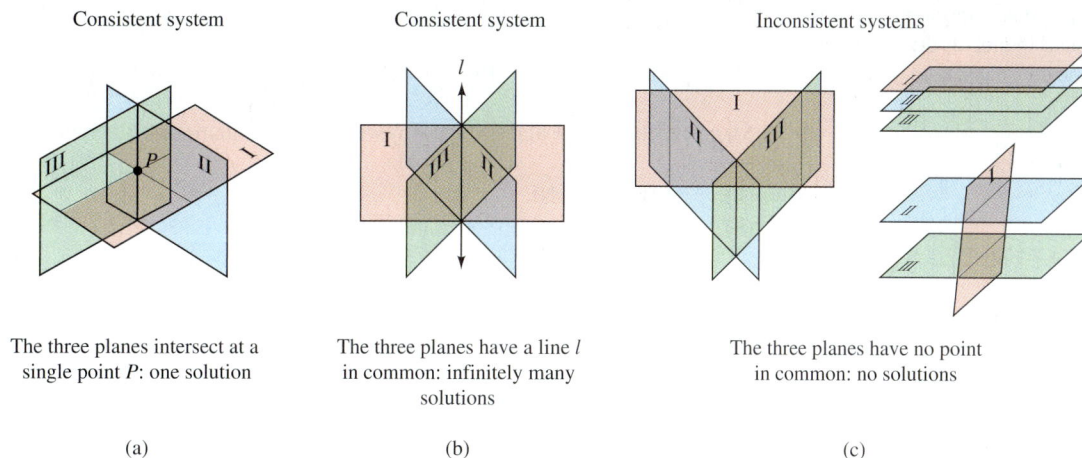

Consistent system — The three planes intersect at a single point P: one solution (a)

Consistent system — The three planes have a line l in common: infinitely many solutions (b)

Inconsistent systems — The three planes have no point in common: no solutions (c)

To solve a system of three linear equations with three variables, we follow these steps.

Solving Three Equations with Three Variables

1. Pick any two equations and eliminate a variable.
2. Pick a different pair of equations and eliminate the same variable as in step 1.
3. Solve the resulting pair of two equations in two variables.
4. To find the value of the third variable, substitute the values of the two variables found in step 3 into any equation containing all three variables and solve the equation.
5. Check the proposed solution in all three of the original equations. Write the solution as an ordered triple.

CONSISTENT SYSTEMS

Recall that when a system has a solution, it is called a *consistent* system.

EXAMPLE 1

Solve: $\begin{cases} 2x + y + 4z = 12 \\ x + 2y + 2z = 9 \\ 3x - 3y - 2z = 1 \end{cases}$.

Solution *Step 1:* We are given the system

$$\begin{aligned} &(1) \quad \\ &(2) \quad \\ &(3) \quad \end{aligned}\begin{cases} 2x + y + 4z = 12 \\ x + 2y + 2z = 9 \\ 3x - 3y - 2z = 1 \end{cases}$$

Notation

To clarify the solution process, we number the equations.

If we pick Equations 2 and 3 and add them, the variable z is eliminated.

$$\begin{array}{lrcl} \textbf{(2)} & x + 2y + 2z &=& 9 \\ \textbf{(3)} & 3x - 3y - 2z &=& 1 \\ \hline \textbf{(4)} & 4x - y &=& 10 \end{array}$$

This equation does not contain z.

Caution

In step 2, choose a different pair of equations than those used in step 1, but eliminate the same variable.

Step 2: We now pick a different pair of equations (Equations 1 and 3) and eliminate z again. If each side of Equation 3 is multiplied by 2 and the resulting equation is added to Equation 1, z is eliminated.

(1)	$2x + y + 4z = 12$	
	$6x - 6y - 4z = 2$	Multiply both sides of Equation 3 by 2.
(5)	$8x - 5y = 14$	This equation does not contain z.

Step 3: Equations 4 and 5 form a system of two equations with two variables, x and y.

$$\begin{cases} 4x - y = 10 & (4) \\ 8x - 5y = 14 & (5) \end{cases}$$

Success Tip

With this method, we use elimination to reduce a system of three equations in three variables to a system of two equations in two variables.

To solve this system, we multiply Equation 4 by -5 and add the resulting equation to Equation 5 to eliminate y:

	$-20x + 5y = -50$	Multiply both sides of Equation 4 by -5.
(5)	$8x - 5y = 14$	
	$-12x = -36$	
	$x = 3$	To find x, divide both sides by -12.

To find y, we substitute 3 for x in any equation containing x and y (such as Equation 5) and solve for y:

(5)	$8x - 5y = 14$	
	$8(3) - 5y = 14$	Substitute 3 for x.
	$24 - 5y = 14$	Simplify.
	$-5y = -10$	Subtract 24 from both sides.
	$y = 2$	Divide both sides by -5.

Step 4: To find z, we substitute 3 for x and 2 for y in any equation containing x, y, and z (such as Equation 1) and solve for z:

(1)	$2x + y + 4z = 12$	
	$2(3) + 2 + 4z = 12$	Substitute 3 for x and 2 for y.
	$8 + 4z = 12$	Simplify.
	$4z = 4$	Subtract 8 from both sides
	$z = 1$	Divide both sides by 4.

The solution of the system is (3, 2, 1). Because this system has a solution, it is a consistent system.

Step 5: Verify that these values satisfy each equation in the original system.

Self Check 1 Solve: $\begin{cases} 2x + y + 4z = 16 \\ x + 2y + 2z = 11 \\ 3x - 3y - 2z = -9 \end{cases}$.

When one or more of the equations of a system is missing a term, the elimination of a variable that is normally performed in step 1 of the solution process can be skipped.

EXAMPLE 2

Solve: $\begin{cases} 3x = 6 - 2y + z \\ -y - 2z = -8 - x \\ x = 1 - 2z \end{cases}$.

Solution ***Step 1:*** First, we write each equation in the form $Ax + By + Cz = D$.

$$\begin{array}{l} \textbf{(1)} \\ \textbf{(2)} \\ \textbf{(3)} \end{array} \quad \begin{cases} 3x + 2y - z = 6 \\ x - y - 2z = -8 \\ x + 2z = 1 \end{cases}$$

Since Equation 3 does not have a y-term, we can proceed to step 2, where we will find another equation that does not contain a y-term.

Step 2: If each side of Equation 2 is multiplied by 2 and the resulting equation is added to Equation 1, y is eliminated.

$$\begin{array}{lrl} \textbf{(1)} & 3x + 2y - z = 6 & \\ & 2x - 2y - 4z = -16 & \text{Multiply both sides of Equation 2 by 2.} \\ \hline \textbf{(4)} & 5x \qquad - 5z = -10 & \end{array}$$

Step 3: Equations 3 and 4 form a system of two equations with two variables, x and z:

$$\begin{array}{l} \textbf{(3)} \\ \textbf{(4)} \end{array} \quad \begin{cases} x + 2z = 1 \\ 5x - 5z = -10 \end{cases}$$

To solve this system, we multiply Equation 3 by -5 and add the resulting equation to Equation 4 to eliminate x:

$$\begin{array}{lrl} & -5x - 10z = -5 & \text{Multiply both sides of Equation 3 by } -5. \\ \textbf{(4)} & 5x - 5z = -10 & \\ \hline & -15z = -15 & \\ & z = 1 & \text{To find } z\text{, divide both sides by } -15. \end{array}$$

To find x, we substitute 1 for z in Equation 3.

$$\begin{array}{lrl} \textbf{(3)} & x + 2z = 1 & \\ & x + 2(1) = 1 & \text{Substitute 1 for } z. \\ & x + 2 = 1 & \text{Multiply.} \\ & x = -1 & \text{Subtract 2 from both sides.} \end{array}$$

Step 4: To find y, we substitute -1 for x and 1 for z in Equation 1:

$$\begin{array}{lrl} \textbf{(1)} & 3x + 2y - z = 6 & \\ & 3(-1) + 2y - 1 = 6 & \text{Substitute } -1 \text{ for } x \text{ and 1 for } z. \\ & -3 + 2y - 1 = 6 & \text{Multiply.} \\ & 2y = 10 & \text{Add 4 to both sides.} \\ & y = 5 & \text{Divide both sides by 2.} \end{array}$$

The solution of the system is $(-1, 5, 1)$.

Step 5: Check the proposed solution in all three of the original equations.

Self Check 2 Solve: $\begin{cases} x + 2y - z = 1 \\ 2x - y + z = 3 \\ x + z = 3 \end{cases}$.

AN INCONSISTENT SYSTEM

Recall that when a system has no solution, it is called an *inconsistent system.*

EXAMPLE 3

Solve: $\begin{cases} 2a + b - 3c = -3 \\ 3a - 2b + 4c = 2 \\ 4a + 2b - 6c = -7 \end{cases}$.

Solution We can multiply the first equation of the system by 2 and add the resulting equation to the second equation to eliminate b:

$$\begin{aligned} 4a + 2b - 6c &= -6 \quad \text{Multiply both sides of the first equation by 2.} \\ 3a - 2b + 4c &= 2 \\ \hline \textbf{(1)} \quad 7a \qquad - 2c &= -4 \end{aligned}$$

Now add the second and third equations of the system to eliminate b again:

$$\begin{aligned} 3a - 2b + 4c &= 2 \\ 4a + 2b - 6c &= -7 \\ \hline \textbf{(2)} \quad 7a \qquad - 2c &= -5 \end{aligned}$$

Equations 1 and 2 form the system

$$\begin{aligned} &\textbf{(1)} \\ &\textbf{(2)} \end{aligned} \quad \begin{cases} 7a - 2c = -4 \\ 7a - 2c = -5 \end{cases}$$

Since $7a - 2c$ cannot equal both -4 and -5, the system is inconsistent and has no solution.

Self Check 3 Solve: $\begin{cases} 2a + b - 3c = 8 \\ 3a - 2b + 4c = 10 \\ 4a + 2b - 6c = -5 \end{cases}$.

SYSTEMS WITH DEPENDENT EQUATIONS

When the equations in a system of two equations with two variables are dependent, the system has infinitely many solutions. This is not always true for systems of three equations with three variables. In fact, a system can have dependent equations and still be inconsistent. The following illustration shows the different possibilities.

Consistent system | Consistent system | Inconsistent system

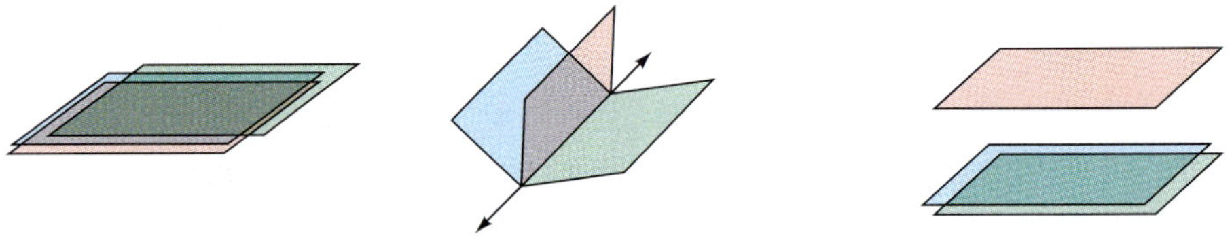

(a) When three planes coincide, the equations are dependent, and there are infinitely many solutions.

(b) When three planes intersect in a common line, the equations are dependent, and there are infinitely many solutions.

(c) When two planes coincide and are parallel to a third plane, the system is inconsistent, and there are no solutions.

EXAMPLE 4

Solve: $\begin{cases} 3x - 2y + z = -1 \\ 2x + y - z = 5 \\ 5x - y = 4 \end{cases}$.

Solution We can add the first two equations to get

$$\begin{array}{lrcr} & 3x - 2y + z &=& -1 \\ & 2x + \ \ y - z &=& 5 \\ \hline \textbf{(1)} & 5x - \ \ y \qquad &=& 4 \end{array}$$

Since Equation 1 is the same as the third equation of the system, the equations of the system are dependent, and there are infinitely many solutions. From a graphical perspective, the equations represent three planes that intersect in a common line.

To write the general solution of this system, we can solve Equation 1 for y to get

$$\begin{aligned} 5x - y &= 4 \\ -y &= -5x + 4 && \text{Subtract } 5x \text{ from both sides.} \\ y &= 5x - 4 && \text{Multiply both sides by } -1. \end{aligned}$$

We can then substitute $5x - 4$ for y in the first equation of the system and solve for z to get

$$\begin{aligned} 3x - 2y + z &= -1 \\ 3x - 2(5x - 4) + z &= -1 && \text{Substitute } 5x - 4 \text{ for } y. \\ 3x - 10x + 8 + z &= -1 && \text{Use the distributive property to remove parentheses.} \\ -7x + 8 + z &= -1 && \text{Combine like terms.} \\ z &= 7x - 9 && \text{Add } 7x \text{ and } -8 \text{ to both sides.} \end{aligned}$$

Since we have found the values of y and z in terms of x, every solution of the system has the form $(x, 5x - 4, 7x - 9)$, where x can be any real number. For example,

If $x = 1$, a solution is $(1, 1, -2)$. $5(1) - 4 = 1$, and $7(1) - 9 = -2$.
If $x = 2$, a solution is $(2, 6, 5)$. $5(2) - 4 = 6$, and $7(2) - 9 = 5$.
If $x = 3$, a solution is $(3, 11, 12)$. $5(3) - 4 = 11$, and $7(3) - 9 = 12$.

Self Check 4 Solve: $\begin{cases} 3x + 2y + z = -1 \\ 2x - y - z = 5 \\ 5x + y = 4 \end{cases}$.

PROBLEM SOLVING

EXAMPLE 5

Tool manufacturing. A company makes three types of hammers, which are marketed as "good," "better," and "best." The cost of manufacturing each type of hammer is \$4, \$6, and \$7, respectively, and the hammers sell for \$6, \$9, and \$12. Each day, the cost of manufacturing 100 hammers is \$520, and the daily revenue from their sale is \$810. How many hammers of each type are manufactured?

Analyze the Problem We need to find how many of each type of hammer are manufactured daily. We must write three equations to find three unknowns.

Form Three Equations If we let x represent the number of good hammers, y represent the number of better hammers, and z represent the number of best hammers, we know that

The total number of hammers is $x + y + z$.

The cost of manufacturing the good hammers is \$4x (\$4 times x hammers).

The cost of manufacturing the better hammers is \$6y (\$6 times y hammers).

The cost of manufacturing the best hammers is \$7z (\$7 times z hammers).

The revenue received by selling the good hammers is \$6x (\$6 times x hammers).

The revenue received by selling the better hammers is \$9y (\$9 times y hammers).

The revenue received by selling the best hammers is \$12z (\$12 times z hammers).

We can assemble the facts of the problem to write three equations.

The number of good hammers	plus	the number of better hammers	plus	the number of best hammers	is	the total number of hammers.
x	$+$	y	$+$	z	$=$	100
The cost of good hammers	plus	the cost of better hammers	plus	the cost of best hammers	is	the total cost.
$4x$	$+$	$6y$	$+$	$7z$	$=$	520
The revenue from good hammers	plus	the revenue from better hammers	plus	the revenue from best hammers	is	the total revenue.
$6x$	$+$	$9y$	$+$	$12z$	$=$	810

Solve the System We must now solve the system

$$\begin{array}{ll} \textbf{(1)} & \\ \textbf{(2)} & \\ \textbf{(3)} & \end{array}\begin{cases} x + y + z = 100 \\ 4x + 6y + 7z = 520 \\ 6x + 9y + 12z = 810 \end{cases}$$

If we multiply Equation 1 by -7 and add the result to Equation 2, we get

$$\begin{array}{lrl} & -7x - 7y - 7z = & -700 \\ & 4x + 6y + 7z = & 520 \\ \hline \textbf{(4)} & -3x - y \qquad = & -180 \end{array}$$

If we multiply Equation 1 by -12 and add the result to Equation 3, we get

$$\begin{array}{lrl} & -12x - 12y - 12z = & -1{,}200 \\ & 6x + 9y + 12z = & 810 \\ \hline \textbf{(5)} & -6x - 3y \qquad = & -390 \end{array}$$

If we multiply Equation 4 by -3 and add it to Equation 5, we get

$$\begin{array}{rl} 9x + 3y = & 540 \\ -6x - 3y = & -390 \\ \hline 3x \qquad = & 150 \end{array}$$

$x = 50$ To find x, divide both sides by 3.

To find y, we substitute 50 for x in Equation 4:

$$-3x - y = -180$$
$$-3(50) - y = -180 \quad \text{Substitute 50 for } x.$$
$$-150 - y = -180 \quad -3(50) = -150.$$
$$-y = -30 \quad \text{Add 150 to both sides.}$$
$$y = 30 \quad \text{Divide both sides by } -1.$$

To find z, we substitute 50 for x and 30 for y in Equation 1:

$$x + y + z = 100$$
$$50 + 30 + z = 100$$
$$z = 20 \quad \text{Subtract 80 from both sides.}$$

State the Conclusion The company manufactures 50 good hammers, 30 better hammers, and 20 best hammers each day.

Check the Result Check the proposed solution in each equation in the original system.

CURVE FITTING

EXAMPLE 6

The equation of a parabola opening upward or downward is of the form $y = ax^2 + bx + c$. Find the equation of the parabola shown to the right by determining the values of a, b, and c.

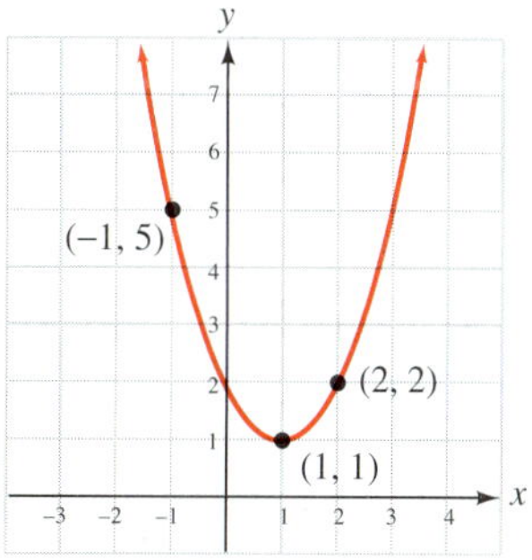

Solution Since the parabola passes through the points $(-1, 5)$, $(1, 1)$, and $(2, 2)$, each pair of coordinates must satisfy the equation $y = ax^2 + bx + c$. If we substitute the x- and y-coordinates of each point into the equation and simplify, we obtain the following system of three equations with three variables.

$$\begin{cases} a - b + c = 5 & (1) \quad \text{Substitute the coordinates of } (-1, 5) \text{ into } y = ax^2 + bx + c \text{ and simplify.} \\ a + b + c = 1 & (2) \quad \text{Substitute the coordinates of } (1, 1) \text{ into } y = ax^2 + bx + c \text{ and simplify.} \\ 4a + 2b + c = 2 & (3) \quad \text{Substitute the coordinates of } (2, 2) \text{ into } y = ax^2 + bx + c \text{ and simplify.} \end{cases}$$

If we add Equations 1 and 2, we obtain

$$\begin{aligned} a - b + c &= 5 \\ a + b + c &= 1 \\ \hline (4) \quad 2a \quad + 2c &= 6 \end{aligned}$$

If we multiply Equation 1 by 2 and add the result to Equation 3, we get

$$\begin{aligned} 2a - 2b + 2c &= 10 \\ 4a + 2b + c &= 2 \\ \hline (5) \quad 6a \quad + 3c &= 12 \end{aligned}$$

We can then divide both sides of Equation 4 by 2 to get Equation 6 and divide both sides of Equation 5 by 3 to get Equation 7. We now have the system

$$\begin{array}{ll} (6) & \\ (7) & \end{array} \begin{cases} a + c = 3 \\ 2a + c = 4 \end{cases}$$

To eliminate c, we multiply Equation 6 by -1 and add the result to Equation 7. We get

$$\begin{array}{r} -a - c = -3 \\ 2a + c = 4 \\ \hline a = 1 \end{array}$$

To find c, we can substitute 1 for a in Equation 6 and find that $c = 2$. To find b, we can substitute 1 for a and 2 for c in Equation 2 and find that $b = -2$.

After we substitute these values of a, b, and c into the equation $y = ax^2 + bx + c$, we have the equation of the parabola.

$$y = ax^2 + bx + c$$
$$y = 1x^2 - 2x + 2$$
$$y = x^2 - 2x + 2$$

Answers to Self Checks **1.** (1, 2, 3) **2.** (1, 1, 2) **3.** no solution **4.** There are infinitely many solutions. A general solution is $(x, 4 - 5x, -9 + 7x)$. Three solutions are $(1, -1, -2)$, $(2, -6, 5)$, and $(3, -11, 12)$.

3.3 STUDY SET

VOCABULARY Fill in the blanks.

1. $\begin{cases} 2x + y - 3z = 0 \\ 3x - y + 4z = 5 \\ 4x + 2y - 6z = 0 \end{cases}$ is called a ________ of three linear equations.
2. If the first two equations of the system in Exercise 1 are added, the variable y is ________.
3. The equation $2x + 3y + 4z = 5$ is a linear equation with ______ variables.
4. The graph of the equation $2x + 3y + 4z = 5$ is a flat surface called a ________.
5. When three planes coincide, the equations of the system are ________, and there are infinitely many solutions.
6. When three planes intersect in a line, the system will have ________ many solutions.

CONCEPTS

7. For each graph of a system of three equations, determine whether the solution set contains one solution, infinitely many solutions, or no solution.

a. 

b. 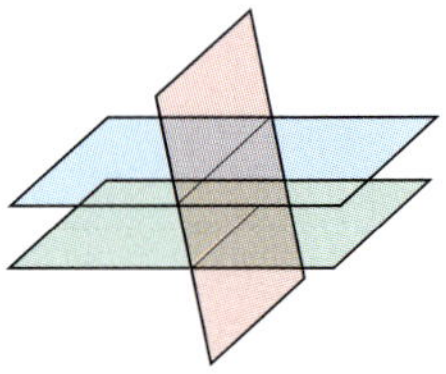

8. Consider the system $\begin{cases} -2x + y + 4z = 3 \\ x - y + 2z = 1 \\ x + y - 3z = 2 \end{cases}$.
   a. What is the result if Equation 1 and Equation 2 are added?
   b. What is the result if Equation 2 and Equation 3 are added?
   c. What variable was eliminated in the steps performed in parts a and b?

NOTATION

9. Write the equation $3z - 2y = x + 6$ in $Ax + By + Cz = D$ form.

10. Fill in the blank: Solutions of a system of three equations in three variables, x, y, and z, are written in the form (x, y, z) and are called ordered ________.

PRACTICE

Determine whether the given ordered triple is a solution of the given system.

11. $(2, 1, 1)$, $\begin{cases} x - y + z = 2 \\ 2x + y - z = 4 \\ 2x - 3y + z = 2 \end{cases}$

12. $(-3, 2, -1)$, $\begin{cases} 2x + 2y + 3z = -1 \\ 3x + y - z = -6 \\ x + y + 2z = 1 \end{cases}$

Solve each system. If a system is inconsistent or if the equations are dependent, so indicate.

13. $\begin{cases} x + y + z = 4 \\ 2x + y - z = 1 \\ 2x - 3y + z = 1 \end{cases}$

14. $\begin{cases} x + y + z = 4 \\ x - y + z = 2 \\ x - y - z = 0 \end{cases}$

15. $\begin{cases} 2x + 2y + 3z = 10 \\ 3x + y - z = 0 \\ x + y + 2z = 6 \end{cases}$

16. $\begin{cases} x - y + z = 4 \\ x + 2y - z = -1 \\ x + y - 3z = -2 \end{cases}$

17. $\begin{cases} b + 2c = 7 - a \\ a + c = 8 - 2b \\ 2a + b + c = 9 \end{cases}$

18. $\begin{cases} 2a = 2 - 3b - c \\ 4a + 6b + 2c - 5 = 0 \\ a + c = 3 + 2b \end{cases}$

19. $\begin{cases} 2x + y - z = 1 \\ x + 2y + 2z = 2 \\ 4x + 5y + 3z = 3 \end{cases}$

20. $\begin{cases} 4x + 3z = 4 \\ 2y - 6z = -1 \\ 8x + 4y + 3z = 9 \end{cases}$

21. $\begin{cases} a + b + c = 180 \\ \frac{a}{4} + \frac{b}{2} + \frac{c}{3} = 60 \\ 2b + 3c - 330 = 0 \end{cases}$

22. $\begin{cases} 2a + 3b - 2c = 18 \\ 5a - 6b + c = 21 \\ 4b - 2c - 6 = 0 \end{cases}$

23. $\begin{cases} 0.5a + 0.3b = 2.2 \\ 1.2c - 8.5b = -24.4 \\ 3.3c + 1.3a = 29 \end{cases}$

24. $\begin{cases} 4a - 3b = 1 \\ 6a - 8c = 1 \\ 2b - 4c = 0 \end{cases}$

25. $\begin{cases} 2x + 3y + 4z = 6 \\ 2x - 3y - 4z = -4 \\ 4x + 6y + 8z = 12 \end{cases}$

26. $\begin{cases} x - 3y + 4z = 2 \\ 2x + y + 2z = 3 \\ 4x - 5y + 10z = 7 \end{cases}$

27. $\begin{cases} x + \frac{1}{3}y + z = 13 \\ \frac{1}{2}x - y + \frac{1}{3}z = -2 \\ x + \frac{1}{2}y - \frac{1}{3}z = 2 \end{cases}$

28. $\begin{cases} x - \frac{1}{5}y - z = 9 \\ \frac{1}{4}x + \frac{1}{5}y - \frac{1}{2}z = 5 \\ 2x + y + \frac{1}{6}z = 12 \end{cases}$

APPLICATIONS

29. MAKING STATUES An artist makes three types of ceramic statues at a monthly cost of \$650 for 180 statues. The manufacturing costs for the three types are \$5, \$4, and \$3. If the statues sell for \$20, \$12, and \$9, respectively, how many of each type should be made to produce \$2,100 in monthly revenue?

30. POTPOURRI The owner of a home decorating shop wants to mix dried rose petals selling for \$6 per pound, dried lavender selling for \$5 per pound, and buckwheat hulls selling for \$4 per pound to get 10 pounds of a mixture that would sell for \$5.50 per pound. She wants to use twice as many pounds of rose petals as lavender. How many pounds of each should she use?

31. NUTRITION A dietitian is to design a meal that will provide a patient with exactly 14 grams (g) of fat, 9 g of carbohydrates, and 9 g of protein. She is to use a combination of the three foods listed in the table on the next page. If one ounce of each of the foods has the nutrient content shown in the table, how many ounces of each food should be used?

Food	Fat	Carbohydrates	Protein
A	2 g	1 g	2 g
B	3 g	2 g	1 g
C	1 g	1 g	2 g

32. NUTRITIONAL PLANNING One ounce of each of three foods has the vitamin and mineral content shown in the table. How many ounces of each must be used to provide exactly 22 milligrams (mg) of niacin, 12 mg of zinc, and 20 mg of vitamin C?

Food	Niacin	Zinc	Vitamin C
A	1 mg	1 mg	2 mg
B	2 mg	1 mg	1 mg
C	2 mg	1 mg	2 mg

33. CHAINSAW SCULPTING A wood sculptor carves three types of statues with a chainsaw. The number of hours required for carving, sanding, and painting a totem pole, a bear, and a deer are shown in the table. How many of each should be produced to use all available labor hours?

	Totem pole	Bear	Deer	Time available
Carving	2 hr	2 hr	1 hr	14 hr
Sanding	1 hr	2 hr	2 hr	15 hr
Painting	3 hr	2 hr	2 hr	21 hr

34. MAKING CLOTHES A clothing manufacturer makes coats, shirts, and slacks. The time required for cutting, sewing, and packaging each item is shown in the table. How many of each should be made to use all available labor hours?

	Coats	Shirts	Slacks	Time available
Cutting	20 min	15 min	10 min	115 hr
Sewing	60 min	30 min	24 min	280 hr
Packaging	5 min	12 min	6 min	65 hr

35. EARTH'S ATMOSPHERE Use the information in the circle graph to determine what percent of Earth's atmosphere is nitrogen, is oxygen, and is other gases.

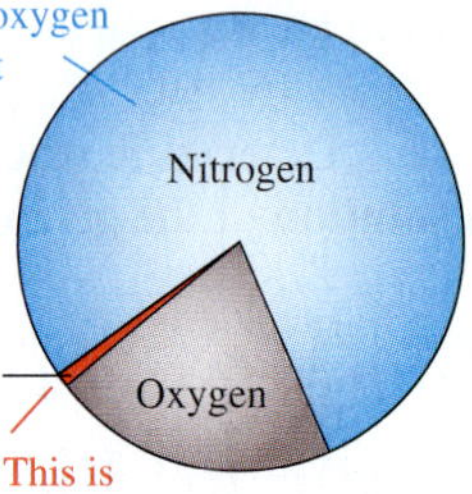
Nitrogen: This is 12% more than three times the sum of the percent oxygen and the percent other gases.
Nitrogen
Other gases
Oxygen
Other gases: This is 20% less than the percent oxygen.

36. NFL RECORDS Jerry Rice, who played with the San Francisco 49ers and the Oakland Raiders, holds the all-time record for touchdown passes caught. Here are some interesting facts about this feat.

- He caught 30 more TD passes from Steve Young than he did from Joe Montana.
- He caught 39 more TD passes from Joe Montana than he did from Rich Gannon.
- He caught a total of 156 TD passes from Young, Montana, and Gannon.

Determine the number of touchdown passes Rice has caught from Young, from Montana, and from Gannon as of 2003.

37. GRAPHS OF SYSTEMS Explain how each of the following pictures could be thought of as an example of the graph of a system of three equations. Then describe the solution, if there is any.

a.

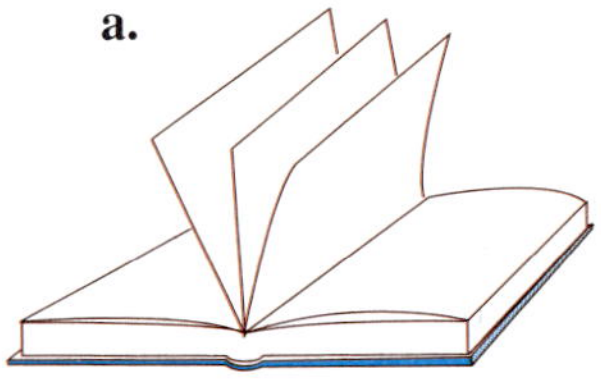

b.

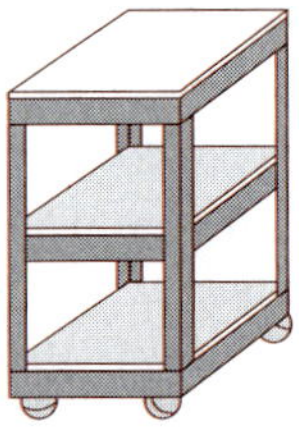

c.

d.

38. ZOOLOGY An X-ray of a mouse revealed a cancerous tumor located at the intersection of the coronal, sagittal, and transverse planes. From this description, would you expect the tumor to be at the base of the tail, on the back, in the stomach, on the tip of the right ear, or in the mouth of the mouse?

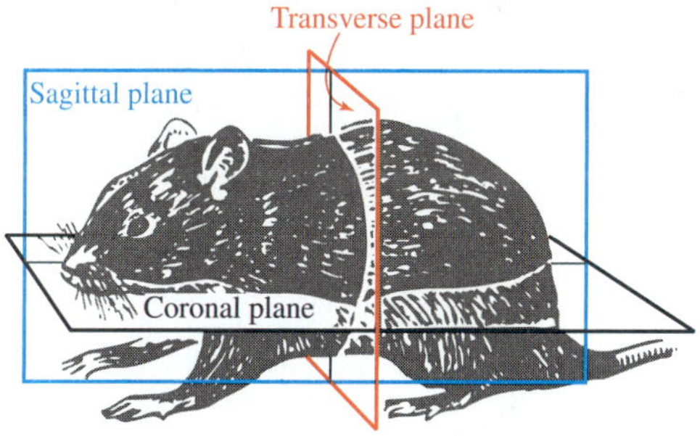

39. ASTRONOMY Comets have elliptical orbits, but the orbits of some comets are so vast that they are indistinguishable from parabolas. Find the equation of the parabola that describes the orbit of the comet shown in the illustration.

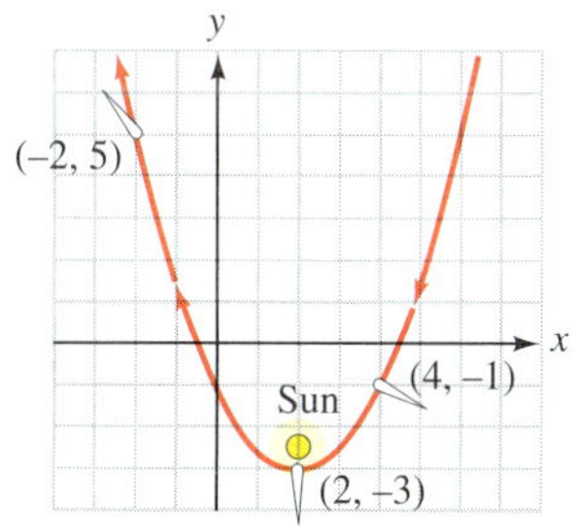

40. CURVE FITTING Find the equation of the parabola shown in the illustration.

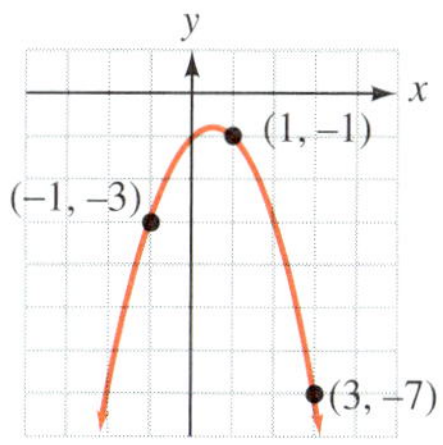

41. WALKWAYS A circular sidewalk is to be constructed in a city park. The walk is to pass by three particular areas of the park, as shown in the illustration in the next column. If an equation of a circle is of the form $x^2 + y^2 + Cx + Dy + E = 0$, find the equation that describes the path of the sidewalk by determining C, D, and E.

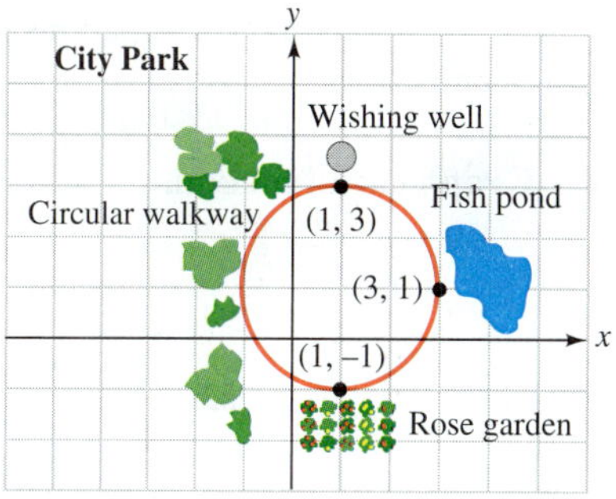

42. CURVE FITTING The equation of a circle is of the form $x^2 + y^2 + Cx + Dy + E = 0$. Find the equation of the circle shown in the illustration by determining C, D, and E.

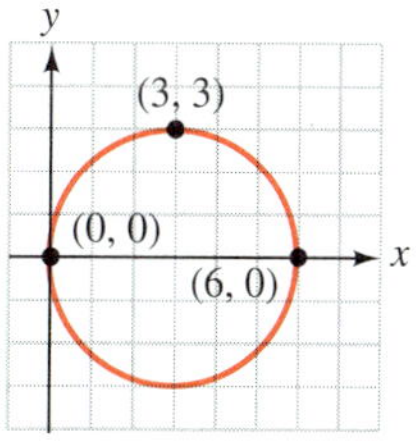

43. TRIANGLES The sum of the measures of the angles of any triangle is 180°. In $\triangle ABC$, $\angle A$ measures 100° less than the sum of the measures of $\angle B$ and $\angle C$, and the measure of $\angle C$ is 40° less than twice the measure of $\angle B$. Find the measure of each angle of the triangle.

44. QUADRILATERALS A quadrilateral is a four-sided polygon. The sum of the measures of the angles of any quadrilateral is 360°. In the illustration below, the measures of $\angle A$ and $\angle B$ are the same. The measure of $\angle C$ is 20° greater than the measure of $\angle A$, and $\angle D$ measures 40°. Find the measure of $\angle A$, $\angle B$, and $\angle C$.

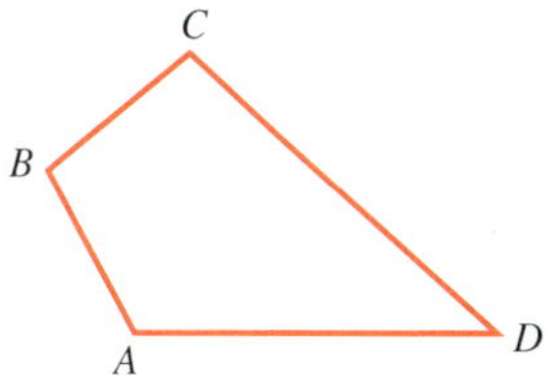

45. INTEGER PROBLEM The sum of three integers is 48. If the first integer is doubled, the sum is 60. If the second integer is doubled, the sum is 63. Find the integers.

46. INTEGER PROBLEM The sum of three integers is 18. The third integer is four times the second, and the second integer is 6 more than the first. Find the integers.

WRITING

47. Explain how a system of three equations with three variables can be reduced to a system of two equations with two variables.

48. What makes a system of three equations with three variables inconsistent?

REVIEW **Graph each function.**

49. $f(x) = |x|$

50. $g(x) = x^2$

51. $h(x) = x^3$

52. $S(x) = x$

CHALLENGE PROBLEMS **Solve each system.**

53. $\begin{cases} w + x + y + z = 3 \\ w - x + y + z = 1 \\ w + x - y + z = 1 \\ w + x + y - z = 3 \end{cases}$

54. $\begin{cases} w + 2x + y + z = 3 \\ w + x - 2y - z = -3 \\ w - x + y + 2z = 3 \\ 2w + x + y - z = 4 \end{cases}$

3.4 Solving Systems Using Matrices

- Matrices
- Augmented matrices
- Gaussian elimination
- Solving a system of three equations
- Inconsistent systems and dependent equations

In this section, we will discuss another method for solving systems of linear equations. This technique uses a mathematical tool called a *matrix* in a series of steps that are based on the addition method.

MATRICES

Another method of solving systems of equations involves rectangular arrays of numbers called *matrices* (plural for *matrix*).

Matrices A **matrix** is any rectangular array of numbers arranged in rows and columns, written within brackets.

The Language of Algebra

An *array* is an orderly arrangement. For example, a jewelry store might display an impressive *array* of gemstones.

Some examples of matrices are

$$A = \begin{bmatrix} 1 & -3 & 8 \\ 2 & 5 & -1 \end{bmatrix} \begin{matrix} \leftarrow \text{Row 1} \\ \leftarrow \text{Row 2} \end{matrix}$$

(↑ Column 1, ↑ Column 2, ↑ Column 3)

$$B = \begin{bmatrix} 1 & 4 & -2 & -4 \\ 6 & -2 & 6 & 1 \\ 3 & 8 & -3 & 12 \end{bmatrix} \begin{matrix} \leftarrow \text{Row 1} \\ \leftarrow \text{Row 2} \\ \leftarrow \text{Row 3} \end{matrix}$$

(↑ Column 1, ↑ Column 2, ↑ Column 3, ↑ Column 4)

The numbers in each matrix are called **elements.** Because matrix A has two rows and three columns, it is called a 2×3 matrix (read "2 by 3" matrix). Matrix B is a 3×4 matrix (three rows and four columns).

AUGMENTED MATRICES

To show how to use matrices to solve systems of linear equations, we consider the system

$$\begin{cases} x - y = 4 \\ 2x + y = 5 \end{cases}$$

which can be represented by the following matrix, called an **augmented matrix:**

$$\left[\begin{array}{cc|c} 1 & -1 & 4 \\ 2 & 1 & 5 \end{array}\right]$$

Each row of the augmented matrix represents one equation of the system. The first two columns of the augmented matrix are determined by the coefficients of x and y in the equations of the system. The last column is determined by the constants in the equations.

$$\left[\begin{array}{cc|c} 1 & -1 & 4 \\ 2 & 1 & 5 \end{array}\right]$$

← This row represents the equation $x - y = 4$.
← This row represents the equation $2x + y = 5$.

Coefficients of x — Coefficients of y — Constants

EXAMPLE 1

Represent each system using an augmented matrix:

a. $\begin{cases} 3x + y = 11 \\ x - 8y = 0 \end{cases}$ and **b.** $\begin{cases} 2a + b - 3c = -3 \\ 9a + 4c = 2 \\ a - b - 6c = -7 \end{cases}$.

Solution

a. $\begin{cases} 3x + y = 11 \\ x - 8y = 0 \end{cases} \begin{matrix} \leftrightarrow \\ \leftrightarrow \end{matrix} \left[\begin{array}{cc|c} 3 & 1 & 11 \\ 1 & -8 & 0 \end{array}\right]$

b. $\begin{cases} 2a + b - 3c = -3 \\ 9a + 4c = 2 \\ a - b - 6c = -7 \end{cases} \begin{matrix} \leftrightarrow \\ \leftrightarrow \\ \leftrightarrow \end{matrix} \left[\begin{array}{ccc|c} 2 & 1 & -3 & -3 \\ 9 & 0 & 4 & 2 \\ 1 & -1 & -6 & -7 \end{array}\right]$

Self Check 1 Represent each system using an augmented matrix:

a. $\begin{cases} 2x - 4y = 9 \\ 5x - y = -2 \end{cases}$ and **b.** $\begin{cases} a + b - c = -4 \\ -2b + 7c = 0 \\ 10a + 8b - 4c = 5 \end{cases}$.

GAUSSIAN ELIMINATION

To solve a 2×2 system of equations by **Gaussian elimination,** we transform the augmented matrix into a matrix that has 1's down its main diagonal and a 0 below the 1 in the first column.

$$\left[\begin{array}{cc|c} 1 & a & b \\ 0 & 1 & c \end{array}\right]$$

Main diagonal

a, b, and c represent real numbers.

To write the augmented matrix in this form, we use three operations called **elementary row operations.**

Elementary Row Operations

Type 1: Any two rows of a matrix can be interchanged.
Type 2: Any row of a matrix can be multiplied by a nonzero constant.
Type 3: Any row of a matrix can be changed by adding a nonzero constant multiple of another row to it.

- A type 1 row operation corresponds to interchanging two equations of the system.
- A type 2 row operation corresponds to multiplying both sides of an equation by a nonzero constant.
- A type 3 row operation corresponds to adding a nonzero multiple of one equation to another.

None of these row operations will change the solution of the given system of equations.

EXAMPLE 2

Consider the augmented matrices

$$A = \left[\begin{array}{cc|c} 2 & 4 & -3 \\ 1 & -8 & 0 \end{array}\right] \qquad B = \left[\begin{array}{cc|c} 1 & -1 & 2 \\ 4 & -8 & 0 \end{array}\right] \qquad C = \left[\begin{array}{ccc|c} 2 & 1 & -8 & 4 \\ 0 & 1 & 4 & -2 \\ 0 & 0 & -6 & 24 \end{array}\right]$$

a. Interchange rows 1 and 2 of matrix A.

b. Multiply row 3 of matrix C by $-\frac{1}{6}$.

c. To the numbers in row 2 of matrix B, add the results of multiplying each number in row 1 by -4.

Solution **a.** Interchanging the rows of matrix A, we obtain $\left[\begin{array}{cc|c} 1 & -8 & 0 \\ 2 & 4 & -3 \end{array}\right]$.

b. We multiply each number in row 3 by $-\frac{1}{6}$. Rows 1 and 2 remain unchanged.

$$\left[\begin{array}{ccc|c} 2 & 1 & -8 & 4 \\ 0 & 1 & 4 & -2 \\ 0 & 0 & 1 & -4 \end{array}\right]$$

We can represent the instruction to multiply the third row by $-\frac{1}{6}$ with the symbolism $-\frac{1}{6}R_3$.

c. If we multiply each number in row 1 of matrix B by -4, we get

$$-4 \quad 4 \quad -8$$

We then add these numbers to row 2. (Note that row 1 remains unchanged.)

$$\left[\begin{array}{cc|c} 1 & -1 & 2 \\ 4 + (-4) & -8 + 4 & 0 + (-8) \end{array}\right]$$

We can abbreviate this procedure using the notation $-4R_1 + R_2$, which means "Multiply row 1 by -4 and add the result to row 2."

After simplifying, we have the matrix

$$\left[\begin{array}{cc|c} 1 & -1 & 2 \\ 0 & -4 & -8 \end{array}\right]$$

Self Check 2 Refer to Example 2.

a. Interchange the rows of matrix B.

b. To the numbers in row 1 of matrix A, add the results of multiplying each number in row 2 by -2.

c. Interchange rows 2 and 3 of matrix C.

We now solve a system of two linear equations using the **Gaussian elimination** process, which involves a series of elementary row operations.

EXAMPLE 3

Solve the system: $\begin{cases} 2x + y = 5 \\ x - y = 4 \end{cases}$.

Solution We can represent the system with the following augmented matrix:

$$\left[\begin{array}{cc|c} 2 & 1 & 5 \\ 1 & -1 & 4 \end{array}\right]$$

First, we want to get a 1 in the top row of the first column where the red 2 is. This can be achieved by applying a type 1 row operation: Interchange rows 1 and 2.

$$\left[\begin{array}{cc|c} 1 & -1 & 4 \\ 2 & 1 & 5 \end{array}\right]$$ Interchanging row 1 and row 2 can be abbreviated as $R_1 \leftrightarrow R_2$.

To get a 0 under the 1 in the first column where the red 2 is, we use a type 3 row operation. To row 2, we add the results of multiplying each number in row 1 by -2.

$$\left[\begin{array}{cc|c} 1 & -1 & 4 \\ 0 & 3 & -3 \end{array}\right] \quad -2R_1 + R_2$$

To get a 1 in the bottom row of the second column where the red 3 is, we use a type 2 row operation: Multiply row 2 by $\frac{1}{3}$.

$$\left[\begin{array}{cc|c} 1 & -1 & 4 \\ 0 & 1 & -1 \end{array}\right] \quad \tfrac{1}{3}R_2$$

This augmented matrix represents the equations

$$1x - 1y = 4$$
$$0x + 1y = -1$$

Writing the equations without the coefficients of 1 and -1, we have

(1) $x - y = 4$

(2) $y = -1$

From Equation 2, we see that $y = -1$. We can *back-substitute* -1 for y in Equation 1 to find x.

$$x - y = 4$$
$$x - (-1) = 4 \qquad \text{Substitute } -1 \text{ for } y.$$
$$x + 1 = 4 \qquad -(-1) = 1.$$
$$x = 3 \qquad \text{Subtract 1 from both sides.}$$

The solution of the system is $(3, -1)$. Verify that this ordered pair satisfies the original system.

Self Check 3 Solve: $\begin{cases} 3x - 2y = -5 \\ x - y = -4 \end{cases}$.

In general, if a system of linear equations has a single solution, we can use the following steps to solve the system using matrices.

Solving Systems of Linear Equations Using Matrices

1. Write an augmented matrix for the system.
2. Use elementary row operations to transform the augmented matrix into a matrix with 1's down its main diagonal and 0's under the 1's.
3. When step 2 is complete, write the resulting system. Then use back substitution to find the solution.
4. Check the proposed solution in the equations of the original system.

SOLVING A SYSTEM OF THREE EQUATIONS

To show how to use matrices to solve systems of three linear equations containing three variables, we consider the system

$$\begin{cases} x - 2y - z = 6 \\ 2x + 2y - z = 1 \\ -x - y + 2z = 1 \end{cases}$$

which can be represented by the augmented matrix

$$\left[\begin{array}{ccc:c} 1 & -2 & -1 & 6 \\ 2 & 2 & -1 & 1 \\ -1 & -1 & 2 & 1 \end{array}\right]$$

To solve a 3×3 system of equations by Gaussian elimination, we transform the augmented matrix into a matrix with 1's down its main diagonal and 0's below its main diagonal.

$$\left[\begin{array}{ccc:c} 1 & a & b & c \\ 0 & 1 & d & e \\ 0 & 0 & 1 & f \end{array}\right]$$

$a, b, c, \ldots, f$ represent real numbers.

Main diagonal

EXAMPLE 4 Solve the system:

$$\begin{cases} 3x + y + 5z = 8 \\ 2x + 3y - z = 6 \\ x + 2y + 2z = 10 \end{cases}$$

Solution This system can be represented by the augmented matrix

$$\left[\begin{array}{ccc:c} 3 & 1 & 5 & 8 \\ 2 & 3 & -1 & 6 \\ 1 & 2 & 2 & 10 \end{array}\right]$$

Success Tip

Follow this order in getting 1's and 0's in the proper positions of the augmented matrix.

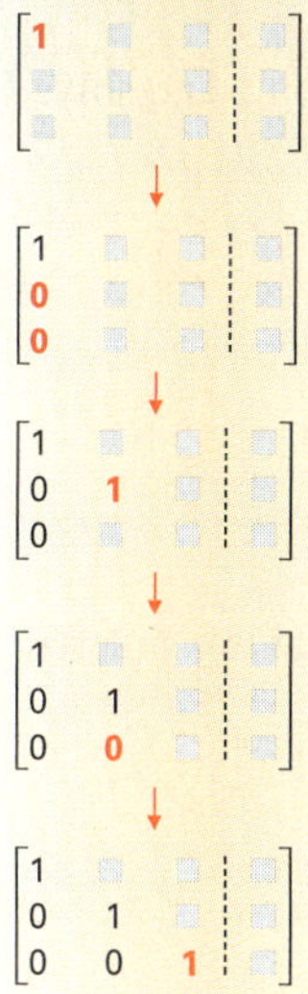

To get a 1 in the first column where the red 3 is, we perform a type 1 row operation: Interchange rows 1 and 3.

$$\left[\begin{array}{ccc|c} 1 & 2 & 2 & 10 \\ 2 & 3 & -1 & 6 \\ 3 & 1 & 5 & 8 \end{array}\right] \quad R_1 \leftrightarrow R_3$$

To get a 0 under the 1 in the first column where the red 2 is, we perform a type 3 row operation: Multiply row 1 by -2 and add the results to row 2. Note that row 1 remains the same.

$$\left[\begin{array}{ccc|c} 1 & 2 & 2 & 10 \\ 0 & -1 & -5 & -14 \\ 3 & 1 & 5 & 8 \end{array}\right] \quad -2R_1 + R_2$$

To get a 0 under the 0 in the first column where the red 3 is, we perform another type 3 row operation: Multiply row 1 by -3 and add the results to row 3. Again, row 1 remains the same.

$$\left[\begin{array}{ccc|c} 1 & 2 & 2 & 10 \\ 0 & -1 & -5 & -14 \\ 0 & -5 & -1 & -22 \end{array}\right] \quad -3R_1 + R_3$$

To get a 1 under the 2 in the second column where the red -1 is, we perform a type 2 row operation: Multiply row 2 by -1.

$$\left[\begin{array}{ccc|c} 1 & 2 & 2 & 10 \\ 0 & 1 & 5 & 14 \\ 0 & -5 & -1 & -22 \end{array}\right] \quad -1R_2$$

To get a 0 under the 1 in the second column where the red -5 is, we perform a type 3 row operation: Multiply row 2 by 5 and add the results to row 3. Row 2 remains the same.

$$\left[\begin{array}{ccc|c} 1 & 2 & 2 & 10 \\ 0 & 1 & 5 & 14 \\ 0 & 0 & 24 & 48 \end{array}\right] \quad 5R_2 + R_3$$

To get a 1 under the 5 in the third column where the red 24 is, we perform a type 2 row operation: Multiply row 3 by $\frac{1}{24}$.

$$\left[\begin{array}{ccc|c} 1 & 2 & 2 & 10 \\ 0 & 1 & 5 & 14 \\ 0 & 0 & 1 & 2 \end{array}\right] \quad \tfrac{1}{24}R_3$$

The final matrix represents the system

$$\begin{cases} 1x + 2y + 2z = 10 \\ 0x + 1y + 5z = 14 \\ 0x + 0y + 1z = 2 \end{cases}$$

which can be written without the coefficients of 0 and 1 as

$$\begin{cases} x + 2y + 2z = 10 & \textbf{(1)} \\ y + 5z = 14 & \textbf{(2)} \\ z = 2 & \textbf{(3)} \end{cases}$$

From Equation 3, we can read that z is 2. To find y, we back substitute 2 for z in Equation 2 and solve for y:

$$\begin{aligned} y + 5z &= 14 && \text{This is Equation 2.} \\ y + 5(2) &= 14 && \text{Substitute 2 for } z. \\ y + 10 &= 14 \\ y &= 4 && \text{Subtract 10 from both sides.} \end{aligned}$$

Thus, y is 4. To find x, we back substitute 2 for z and 4 for y in Equation 1 and solve for x:

$$\begin{aligned} x + 2y + 2z &= 10 && \text{This is Equation 1.} \\ x + 2(4) + 2(2) &= 10 && \text{Substitute 2 for } z \text{ and 4 for } y. \\ x + 8 + 4 &= 10 \\ x + 12 &= 10 \\ x &= -2 && \text{Subtract 12 from both sides.} \end{aligned}$$

Thus, x is -2. The solution of the given system is $(-2, 4, 2)$. Verify that this ordered triple satisfies each equation of the original system.

Self Check 4 Solve: $\begin{cases} 2x - y + z = 5 \\ x + y - z = -2 \\ -x + 2y + 2z = 1 \end{cases}$.

INCONSISTENT SYSTEMS AND DEPENDENT EQUATIONS

In the next example, we consider a system with no solution.

EXAMPLE 5 Using matrices, solve the system: $\begin{cases} x + y = -1 \\ -3x - 3y = -5 \end{cases}$.

Solution This system can be represented by the augmented matrix

$$\left[\begin{array}{cc|c} 1 & 1 & -1 \\ -3 & -3 & -5 \end{array}\right]$$

Since the matrix has a 1 in the top row of the first column, we proceed to get a 0 under it by multiplying row 1 by 3 and adding the results to row 2.

$$\left[\begin{array}{cc|c} 1 & 1 & -1 \\ 0 & 0 & -8 \end{array}\right] \quad 3R_1 + R_2$$

This matrix represents the system

$$\begin{cases} x + y = -1 \\ 0 + 0 = -8 \end{cases}$$

This system has no solution, because the second equation is never true. Therefore, the system is inconsistent. It has no solutions.

Self Check 5 Solve: $\begin{cases} 4x - 8y = 9 \\ x - 2y = -5 \end{cases}$

In the next example, we consider a system with infinitely many solutions.

EXAMPLE 6

Using matrices, solve the system:

$$\begin{cases} 2x + 3y - 4z = 6 \\ 4x + 6y - 8z = 12 \\ -6x - 9y + 12z = -18 \end{cases}$$

Solution This system can be represented by the augmented matrix

$$\left[\begin{array}{ccc|c} 2 & 3 & -4 & 6 \\ 4 & 6 & -8 & 12 \\ -6 & -9 & 12 & -18 \end{array}\right]$$

To get a 1 in the top row of the first column, we multiply row 1 by $\frac{1}{2}$.

$$\left[\begin{array}{ccc|c} 1 & \frac{3}{2} & -2 & 3 \\ 4 & 6 & -8 & 12 \\ -6 & -9 & 12 & -18 \end{array}\right] \quad \tfrac{1}{2}R_1$$

Next, we want to get 0's under the 1 in the first column. This can be achieved by multiplying row 1 by -4 and adding the results to row 2, and multiplying row 1 by 6 and adding the results to row 3.

$$\left[\begin{array}{ccc|c} 1 & \frac{3}{2} & -2 & 3 \\ 0 & 0 & 0 & 0 \\ 0 & 0 & 0 & 0 \end{array}\right] \quad \begin{array}{l} \\ -4R_1 + R_2 \\ 6R_1 + R_3 \end{array}$$

The last matrix represents the system

$$\begin{cases} x + \frac{3}{2}y - 2z = 3 \\ 0x + 0y + 0z = 0 \\ 0x + 0y + 0z = 0 \end{cases}$$

If we clear the first equation of fractions, we have the system

$$\begin{cases} 2x + 3y - 4z = 6 \\ 0 = 0 \\ 0 = 0 \end{cases}$$

This system has dependent equations and infinitely many solutions. Solutions of this system would be any triple (x, y, z) that satisfies the equation $2x + 3y - 4z = 6$. Two such solutions would be $(0, 2, 0)$ and $(1, 0, -1)$.

Self Check 6 Solve: $\begin{cases} 5x - 10y + 15z = 35 \\ -3x + 6y - 9z = -21. \\ 2x - 4y + 6z = 14 \end{cases}$

Answers to Self Checks

1. a. $\left[\begin{array}{cc|c} 2 & -4 & 9 \\ 5 & -1 & -2 \end{array}\right]$, **b.** $\left[\begin{array}{ccc|c} 1 & 1 & -1 & -4 \\ 0 & -2 & 7 & 0 \\ 10 & 8 & -4 & 5 \end{array}\right]$

2. a. $\left[\begin{array}{cc|c} 4 & -8 & 0 \\ 1 & -1 & 2 \end{array}\right]$, **b.** $\left[\begin{array}{cc|c} 0 & 20 & -3 \\ 1 & -8 & 0 \end{array}\right]$, **c.** $\left[\begin{array}{ccc|c} 2 & 1 & -8 & 4 \\ 0 & 0 & -6 & 24 \\ 0 & 1 & 4 & -2 \end{array}\right]$

3. (3, 7) **4.** (1, −1, 2) **5.** no solution

6. There are infinitely many solutions—any triple satisfying the equation $x - 2y + 3z = 7$.

3.4 STUDY SET

VOCABULARY Fill in the blanks.

1. A ________ is a rectangular array of numbers.

2. The numbers in a matrix are called its ________.

3. A 3 × 4 matrix has 3 ______ and 4 ________.

4. Elementary ______ operations are used to produce new matrices that lead to the solution of a system.

5. A matrix that represents the equations of a system is called an __________ matrix.

6. The augmented matrix $\left[\begin{array}{cc|c} 1 & 3 & -2 \\ 0 & 1 & 4 \end{array}\right]$ has 1's down its main ________.

CONCEPTS

7. For each matrix, determine the number of rows and the number of columns.

a. $\left[\begin{array}{cc|c} 4 & 6 & -1 \\ \frac{1}{2} & 9 & -3 \end{array}\right]$

b. $\left[\begin{array}{ccc|c} 1 & -2 & 3 & 1 \\ 0 & 1 & 6 & 4 \\ 0 & 0 & 1 & \frac{1}{3} \end{array}\right]$

8. For each augmented matrix, give the system of equations it represents.

a. $\left[\begin{array}{cc|c} 1 & 6 & 7 \\ 0 & 1 & 4 \end{array}\right]$

b. $\left[\begin{array}{ccc|c} 2 & -2 & 9 & 1 \\ 3 & 1 & 1 & 0 \\ 2 & -6 & 8 & -7 \end{array}\right]$

9. Write the system of equations represented by the augmented matrix and use back substitution to find the solution.

$$\left[\begin{array}{cc|c} 1 & -1 & -10 \\ 0 & 1 & 6 \end{array}\right]$$

10. Write the system of equations represented by the augmented matrix and use back substitution to find the solution.

$$\left[\begin{array}{ccc|c} 1 & -2 & 1 & -16 \\ 0 & 1 & 2 & 8 \\ 0 & 0 & 1 & 4 \end{array}\right]$$

11. Matrices were used to solve a system of two linear equations. The final matrix is shown here. Explain what the result tells about the system.

$$\left[\begin{array}{cc|c} 1 & 2 & -4 \\ 0 & 0 & 2 \end{array}\right]$$

12. Matrices were used to solve a system of two linear equations. The final matrix is shown here. Explain what the result tells about the equations.

$$\left[\begin{array}{cc|c} 1 & 2 & -4 \\ 0 & 0 & 0 \end{array}\right]$$

NOTATION

13. Consider the matrix

$$A = \left[\begin{array}{ccc|c} 3 & 6 & -9 & 0 \\ 1 & 5 & -2 & 1 \\ -2 & 2 & -2 & 5 \end{array}\right].$$

a. Explain what is meant by $\frac{1}{3}R_1$. Then perform the operation on matrix A.

b. Explain what is meant by $-R_1 + R_2$. Then perform the operation on the answer to part a.

14. Consider the matrix $B = \left[\begin{array}{cc|c} -3 & 1 & -6 \\ 1 & -4 & 4 \end{array}\right]$.

a. Explain what is meant by $R_1 \leftrightarrow R_2$. Then perform the operation on matrix B.

b. Explain what is meant by $3R_1 + R_2$. Then perform the operation on the answer to part a.

Complete each solution.

15. Solve: $\begin{cases} 4x - y = 14 \\ x + y = 6 \end{cases}$.

$$\left[\begin{array}{cc|c} 4 & & 14 \\ 1 & 1 & 6 \end{array}\right]$$

$$\left[\begin{array}{cc|c} & 1 & 6 \\ 4 & -1 & 14 \end{array}\right] \quad R_1 \leftrightarrow R_2$$

$$\left[\begin{array}{cc|c} 1 & 1 & 6 \\ 0 & & -10 \end{array}\right] \quad -4R_1 + R_2$$

$$\left[\begin{array}{cc|c} 1 & 1 & 6 \\ 0 & 1 & \end{array}\right] \quad -\tfrac{1}{5}R_2$$

This matrix represents the system

$$\begin{cases} x + y = 6 \\ = 2 \end{cases}$$

The solution is (, 2).

16. Solve: $\begin{cases} 2x + 2y = 18 \\ x - y = 5 \end{cases}$.

$$\left[\begin{array}{cc|c} 2 & 2 & 18 \\ & -1 & 5 \end{array}\right]$$

$$\left[\begin{array}{cc|c} 1 & 1 & 9 \\ & -1 & 5 \end{array}\right] \quad \tfrac{1}{2}R_1$$

$$\left[\begin{array}{cc|c} 1 & 1 & 9 \\ 0 & & -4 \end{array}\right] \quad -R_1 + R_2$$

$$\left[\begin{array}{cc|c} 1 & 1 & 9 \\ 0 & 1 & \end{array}\right] \quad -\tfrac{1}{2}R_2$$

This matrix represents the system

$$\begin{cases} x + y = \\ y = 2 \end{cases}$$

The solution is (, 2).

PRACTICE

Use matrices to solve each system of equations. If the equations of a system are dependent or if a system is inconsistent, so indicate.

17. $\begin{cases} x + y = 2 \\ x - y = 0 \end{cases}$

18. $\begin{cases} x + y = 3 \\ x - y = -1 \end{cases}$

19. $\begin{cases} 2x + y = 1 \\ x + 2y = -4 \end{cases}$

20. $\begin{cases} 5x - 4y = 10 \\ x - 7y = 2 \end{cases}$

21. $\begin{cases} 2x - y = -1 \\ x - 2y = 1 \end{cases}$

22. $\begin{cases} 2x - y = 0 \\ x + y = 3 \end{cases}$

23. $\begin{cases} 3x + 4y = -12 \\ 9x - 2y = 6 \end{cases}$

24. $\begin{cases} 2x - 3y = 16 \\ -4x + y = -22 \end{cases}$

25. $\begin{cases} x + y + z = 6 \\ x + 2y + z = 8 \\ x + y + 2z = 9 \end{cases}$

26. $\begin{cases} x - y + z = 2 \\ x + 2y - z = 6 \\ 2x - y - z = 3 \end{cases}$

27. $\begin{cases} 3x + y - 3z = 5 \\ x - 2y + 4z = 10 \\ x + y + z = 13 \end{cases}$

28. $\begin{cases} 2x + y - 3z = -1 \\ 3x - 2y - z = -5 \\ x - 3y - 2z = -12 \end{cases}$

29. $\begin{cases} 3x - 2y + 4z = 4 \\ x + y + z = 3 \\ 6x - 2y - 3z = 10 \end{cases}$

30. $\begin{cases} 2x + 3y - z = -8 \\ x - y - z = -2 \\ -4x + 3y + z = 6 \end{cases}$

31. $\begin{cases} 2a + b + 3c = 3 \\ -2a - b + c = 5 \\ 4a - 2b + 2c = 2 \end{cases}$

32. $\begin{cases} 3a + 2b + c = 8 \\ 6a - b + 2c = 16 \\ -9a + b - c = -20 \end{cases}$

33. $\begin{cases} 2x + y - 3z = -7 \\ 3x - y + 2z = -9 \\ -2x - y - z = 3 \end{cases}$

34. $\begin{cases} -2x + 3y + z = -12 \\ 3x + y - z = 12 \\ 3x - y - z = 14 \end{cases}$

35. $\begin{cases} 2x + y - 2z = 6 \\ 4x - y + z = -1 \\ 6x - 2y + 3z = -5 \end{cases}$

36. $\begin{cases} 2x - 3y + 3z = 14 \\ 3x + 3y - z = 2 \\ -2x + 6y + 5z = 9 \end{cases}$

37. $\begin{cases} x - 3y = 9 \\ -2x + 6y = 18 \end{cases}$

38. $\begin{cases} -6x + 12y = 10 \\ 2x - 4y = 8 \end{cases}$

39. $\begin{cases} 4x + 4y = 12 \\ -x - y = -3 \end{cases}$

40. $\begin{cases} 5x - 15y = 10 \\ 2x - 6y = 4 \end{cases}$

41. $\begin{cases} 6x + y - z = -2 \\ x + 2y + z = 5 \\ 5y - z = 2 \end{cases}$

42. $\begin{cases} 2x + 3y - 2z = 18 \\ 5x - 6y + z = 21 \\ 4y - 2z = 6 \end{cases}$

43. $\begin{cases} 2x + y - z = 1 \\ x + 2y + 2z = 2 \\ 4x + 5y + 3z = 3 \end{cases}$

44. $\begin{cases} x - 3y + 4z = 2 \\ 2x + y + 2z = 3 \\ 4x - 5y + 10z = 7 \end{cases}$

45. $\begin{cases} 5x + 3y = 4 \\ 3y - 4z = 4 \\ x + z = 1 \end{cases}$

46. $\begin{cases} y + 2z = -2 \\ x + y = 1 \\ 2x - z = 0 \end{cases}$

47. $\begin{cases} x - y = 1 \\ 2x - z = 0 \\ 2y - z = -2 \end{cases}$

48. $\begin{cases} x + y - 3z = 4 \\ 2x + 2y - 6z = 5 \\ -3x + y - z = 2 \end{cases}$

Remember these facts from geometry: The sum of the measures of complementary angles is 90°, and the sum of the measures of supplementary angles is 180°.

49. One angle measures 46° more than the measure of its complement. Find the measure of each angle.

50. One angle measures 14° more than the measure of its supplement. Find the measure of each angle.

51. In the illustration, $\angle B$ measures 25° more than the measure of $\angle A$, and the measure of $\angle C$ is 5° less than twice the measure of $\angle A$. Find the measure of each angle of the triangle.

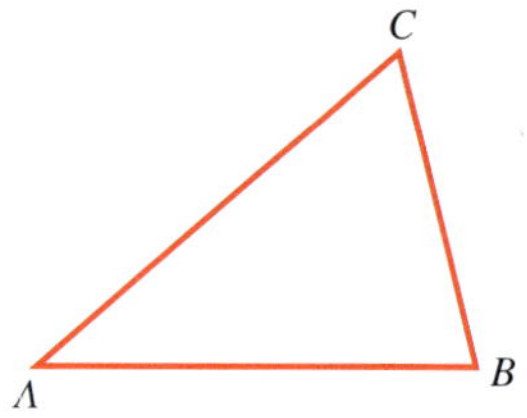

51. In the illustration, $\angle A$ measures 10° less than the measure of $\angle B$, and the measure of $\angle B$ is 10° less than the measure of $\angle C$. Find the measure of each angle of the triangle.

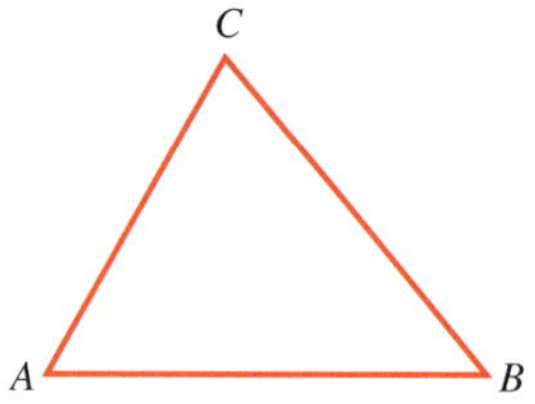

APPLICATIONS

53. DIGITAL PHOTOGRAPHY A digital camera stores the black and white photograph shown below as a 512 × 512 matrix. Each element of the matrix corresponds to a small dot of grey scale shading, called a *pixel,* in the picture. How many elements does a 512 × 512 matrix have?

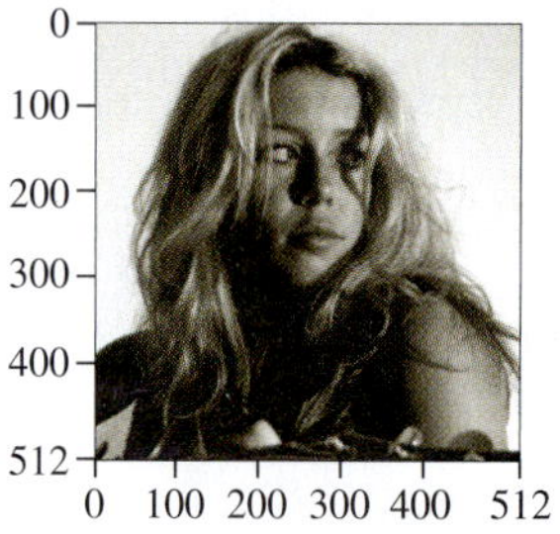

54. DIGITAL IMAGING A scanner stores a black and white photograph as a matrix that has a total of 307,200 elements. If the matrix has 480 rows, how many columns does it have?

Write a system of equations to solve each problem. Use matrices to solve the system.

55. PHYSICAL THERAPY After an elbow injury, a volleyball player has restricted movement of her arm. Her range of motion (the measure of $\angle 1$) is 28° less than the measure of $\angle 2$. Find the measure of each angle.

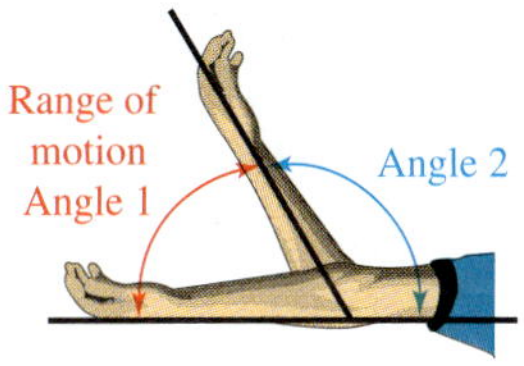

56. PIGGY BANKS When a child breaks open her piggy bank, she finds a total of 64 coins, consisting of nickels, dimes, and quarters. The total value of the coins is \$6. If the nickels were dimes, and the dimes were nickels, the value of the coins would be \$5. How many nickels, dimes, and quarters were in the piggy bank?

57. THEATER SEATING The illustration shows the cash receipts and the ticket prices from two sold-out performances of a play. Find the number of seats in each of the three sections of the 800-seat theater.

Sunday Ticket Receipts

Matinee	\$13,000
Evening	\$23,000

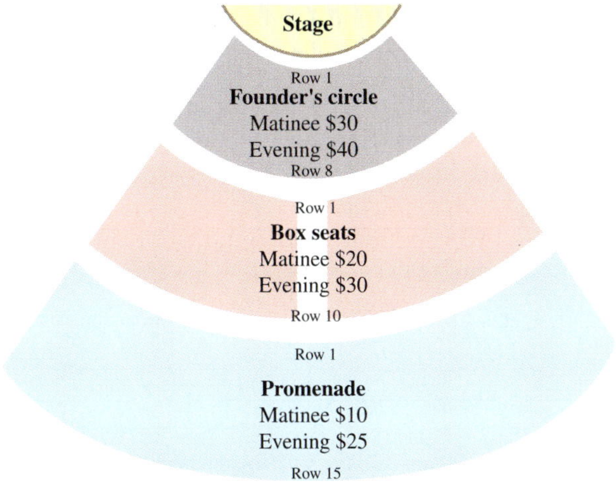

58. ICE SKATING Three circles are traced out by a figure skater during her performance. If the centers of the circles are the given distances apart, find the radius of each circle.

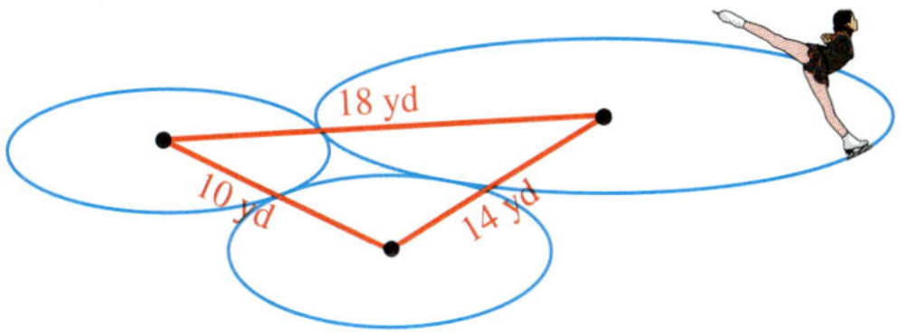

WRITING

59. Explain what is meant by the phrase *back substitution.*

60. Explain how a type 3 row operation is similar to the elimination method of solving a system of equations.

REVIEW

61. What is the formula used to find the slope of a line, given two points on the line?

62. What is the form of the equation of a horizontal line? Of a vertical line?

63. What is the point-slope form of the equation of a line?

64. What is the slope-intercept form of the equation of a line?

CHALLENGE PROBLEMS

65. If the system represented by

$$\left[\begin{array}{ccc|c} 1 & 1 & 0 & 1 \\ 0 & 0 & 1 & 2 \\ 0 & 0 & 0 & k \end{array}\right]$$

has no solution, what do you know about k?

66. Use matrices to solve the system.

$$\begin{cases} w + x + y + z = 0 \\ w - 2x + y - 3z = -3 \\ 2w + 3x + y - 2z = -1 \\ 2w - 2x - 2y + z = -12 \end{cases}$$

3.5 Solving Systems Using Determinants

- Determinants • Evaluating a determinant
- Using Cramer's rule to solve a system of two equations
- Using Cramer's rule to solve a system of three equations

In this section, we will discuss another method for solving systems of linear equations. With this method, called *Cramer's rule,* we work with combinations of the coefficients and the constants of the equations written as *determinants.*

DETERMINANTS

An idea related to the concept of matrix is the **determinant.** A determinant is a number that is associated with a **square matrix,** a matrix that has the same number of rows and columns. For any square matrix A, the symbol $|A|$ represents the determinant of A. To write a determinant, we put the elements of a square matrix between two vertical lines.

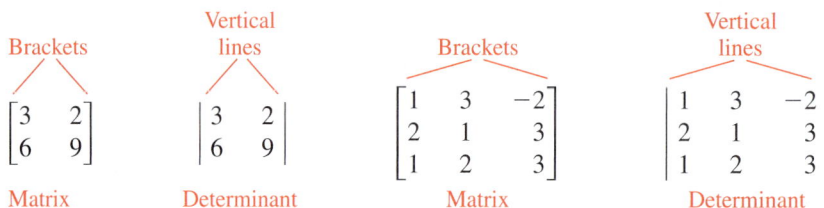

Brackets: $\begin{bmatrix} 3 & 2 \\ 6 & 9 \end{bmatrix}$ Matrix

Vertical lines: $\begin{vmatrix} 3 & 2 \\ 6 & 9 \end{vmatrix}$ Determinant

Brackets: $\begin{bmatrix} 1 & 3 & -2 \\ 2 & 1 & 3 \\ 1 & 2 & 3 \end{bmatrix}$ Matrix

Vertical lines: $\begin{vmatrix} 1 & 3 & -2 \\ 2 & 1 & 3 \\ 1 & 2 & 3 \end{vmatrix}$ Determinant

Like matrices, determinants are classified according to the number of rows and columns they contain. The determinant on the left is a 2×2 determinant. The other is a 3×3 determinant.

EVALUATING A DETERMINANT

The determinant of a 2×2 matrix is the number that is equal to the product of the numbers on the main diagonal minus the product of the numbers on the other diagonal.

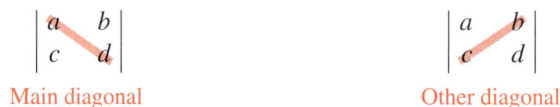

$\begin{vmatrix} a & b \\ c & d \end{vmatrix}$ Main diagonal

$\begin{vmatrix} a & b \\ c & d \end{vmatrix}$ Other diagonal

Value of a 2×2 Determinant

If a, b, c, and d are numbers, the **determinant** of the matrix $\begin{bmatrix} a & b \\ c & d \end{bmatrix}$ is

$$\begin{vmatrix} a & b \\ c & d \end{vmatrix} = ad - bc$$

EXAMPLE 1

Find each value: **a.** $\begin{vmatrix} 3 & 2 \\ 6 & 9 \end{vmatrix}$ and **b.** $\begin{vmatrix} -5 & \frac{1}{2} \\ -1 & 0 \end{vmatrix}$.

Solution From the product of the numbers along the main diagonal, we subtract the product of the numbers along the other diagonal.

a. $\begin{vmatrix} 3 & 2 \\ 6 & 9 \end{vmatrix} = 3(9) - 2(6)$

$= 27 - 12$

$= 15$

b. $\begin{vmatrix} -5 & \frac{1}{2} \\ -1 & 0 \end{vmatrix} = -5(0) - \frac{1}{2}(-1)$

$= 0 + \frac{1}{2}$

$= \frac{1}{2}$

Self Check 1 Evaluate: $\begin{vmatrix} 4 & -3 \\ 2 & 1 \end{vmatrix}$.

A 3×3 determinant is evaluated by **expanding by minors.**

Value of a 3 × 3 Determinant

Minor of a_1 ↓ Minor of b_1 ↓ Minor of c_1 ↓

$$\begin{vmatrix} a_1 & b_1 & c_1 \\ a_2 & b_2 & c_2 \\ a_3 & b_3 & c_3 \end{vmatrix} = a_1\begin{vmatrix} b_2 & c_2 \\ b_3 & c_3 \end{vmatrix} - b_1\begin{vmatrix} a_2 & c_2 \\ a_3 & c_3 \end{vmatrix} + c_1\begin{vmatrix} a_2 & b_2 \\ a_3 & b_3 \end{vmatrix}$$

To find the minor of a_1, we cross out the elements of the determinant that are in the same row and column as a_1:

$$\begin{vmatrix} a_1 & b_1 & c_1 \\ a_2 & b_2 & c_2 \\ a_3 & b_3 & c_3 \end{vmatrix}$$ The minor of a_1 is $\begin{vmatrix} b_2 & c_2 \\ b_3 & c_3 \end{vmatrix}$.

To find the minor of b_1, we cross out the elements of the determinant that are in the same row and column as b_1:

$$\begin{vmatrix} a_1 & b_1 & c_1 \\ a_2 & b_2 & c_2 \\ a_3 & b_3 & c_3 \end{vmatrix}$$ The minor of b_1 is $\begin{vmatrix} a_2 & c_2 \\ a_3 & c_3 \end{vmatrix}$.

To find the minor of c_1, we cross out the elements of the determinant that are in the same row and column as c_1:

$$\begin{vmatrix} a_1 & b_1 & c_1 \\ a_2 & b_2 & c_2 \\ a_3 & b_3 & c_3 \end{vmatrix}$$ The minor of c_1 is $\begin{vmatrix} a_2 & b_2 \\ a_3 & b_3 \end{vmatrix}$.

EXAMPLE 2

Find the value of $\begin{vmatrix} 1 & 3 & -2 \\ 2 & 0 & 3 \\ 1 & 2 & 3 \end{vmatrix}$.

Solution We evaluate this determinant by expanding by minors along the first row of the determinant.

Minor of 1 ↓ — Minor of 3 ↓ — Minor of −2 ↓

$$\begin{vmatrix} 1 & 3 & -2 \\ 2 & 0 & 3 \\ 1 & 2 & 3 \end{vmatrix} = 1\begin{vmatrix} 0 & 3 \\ 2 & 3 \end{vmatrix} - 3\begin{vmatrix} 2 & 3 \\ 1 & 3 \end{vmatrix} + (-2)\begin{vmatrix} 2 & 0 \\ 1 & 2 \end{vmatrix}$$

$= 1(0 - 6) - 3(6 - 3) - 2(4 - 0)$ Evaluate each 2×2 determinant.

$= 1(-6) - 3(3) - 2(4)$

$= -6 - 9 - 8$

$= -23$

Self Check 2 Evaluate: $\begin{vmatrix} 2 & 3 & -1 \\ 0 & 2 & 4 \\ -2 & 5 & 6 \end{vmatrix}$.

We can evaluate a 3×3 determinant by expanding it along any row or column. To determine the signs between the terms of the expansion of a 3×3 determinant, we use the following array of signs.

Array of Signs for a 3×3 Determinant

$$\begin{matrix} + & - & + \\ - & + & - \\ + & - & + \end{matrix}$$

This array of signs is commonly referred to as the **checkerboard pattern.**

EXAMPLE 3

Evaluate the determinant $\begin{vmatrix} 1 & 3 & -2 \\ 2 & 0 & 3 \\ 1 & 2 & 3 \end{vmatrix}$ by expanding on the middle column.

Solution This is the determinant of Example 2. To expand it along the middle column, we use the signs of the middle column of the array of signs:

Minor of 3 ↓ — Minor of 1 ↓ — Minor of 2 ↓

$$\begin{vmatrix} 1 & 3 & -2 \\ 2 & 0 & 3 \\ 1 & 2 & 3 \end{vmatrix} = -3\begin{vmatrix} 2 & 3 \\ 1 & 3 \end{vmatrix} + 0\begin{vmatrix} 1 & -2 \\ 1 & 3 \end{vmatrix} - 2\begin{vmatrix} 1 & -2 \\ 2 & 3 \end{vmatrix}$$

Use the middle column of the checkerboard pattern:

$$\begin{matrix} + & - & + \\ - & + & - \\ + & - & + \end{matrix}$$

$= -3(6 - 3) + 0 - 2[3 - (-4)]$ Evaluate each 2×2 determinant.

$= -3(3) + 0 - 2(7)$

$= -9 + 0 - 14$

$= -23$

As expected, we get the same value as in Example 2.

Success Tip

When evaluating a determinant, expanding along a row or column that contains 0's can simplify the computations.

Self Check 3 Evaluate $\begin{vmatrix} 2 & 3 & -1 \\ 0 & 2 & 4 \\ -2 & 5 & 6 \end{vmatrix}$ by expanding along the first column.

ACCENT ON TECHNOLOGY: EVALUATING DETERMINANTS

It is possible to use a graphing calculator to evaluate determinants. For example, to evaluate the determinant in Example 3, we first enter the matrix by pressing the [MATRIX] key, selecting EDIT, and pressing the [ENTER] key. Next, we enter the dimensions and the elements of the matrix to get figure (a). We then press [2nd] [QUIT] to clear the screen, press [MATRIX], select MATH, and press 1 to get figure (b). We then press [MATRIX], select NAMES, press 1, and press [)] and [ENTER] to get the value of the determinant. Figure (c) shows that the value of the determinant is -23.

(a) (b) (c)

USING CRAMER'S RULE TO SOLVE A SYSTEM OF TWO EQUATIONS

The method of using determinants to solve systems of linear equations is called **Cramer's rule,** named after the 18th-century mathematician Gabriel Cramer. To develop Cramer's rule, we consider the system

$$\begin{cases} ax + by = e \\ cx + dy = f \end{cases}$$

where x and y are variables and a, b, c, d, e, and f are constants.

If we multiply both sides of the first equation by d and multiply both sides of the second equation by $-b$, we can add the equations and eliminate y:

$$\begin{array}{rcl} adx + bdy &=& ed \\ -bcx - bdy &=& -bf \\ \hline adx - bcx &=& ed - bf \end{array}$$

To solve for x, we use the distributive property to write $adx - bcx$ as $(ad - bc)x$ on the left-hand side and divide each side by $ad - bc$:

$$(ad - bc)x = ed - bf$$

$$x = \frac{ed - bf}{ad - bc} \quad \text{where } ad - bc \neq 0$$

We can find y in a similar manner. After eliminating the variable x, we get

$$y = \frac{af - ec}{ad - bc} \quad \text{where } ad - bc \neq 0$$

Determinants provide an easy way of remembering these formulas. Note that the denominator for both x and y is

$$\begin{vmatrix} a & b \\ c & d \end{vmatrix} = ad - bc$$

The numerators can be expressed as determinants also:

$$x = \frac{ed - bf}{ad - bc} = \frac{\begin{vmatrix} e & b \\ f & d \end{vmatrix}}{\begin{vmatrix} a & b \\ c & d \end{vmatrix}} \quad \text{and} \quad y = \frac{af - ec}{ad - bc} = \frac{\begin{vmatrix} a & e \\ c & f \end{vmatrix}}{\begin{vmatrix} a & b \\ c & d \end{vmatrix}}$$

If we compare these formulas with the original system

$$\begin{cases} ax + by = e \\ cx + dy = f \end{cases}$$

we note that in the expressions for x and y above, the denominator determinant is formed by using the coefficients a, b, c, and d of the variables in the equations. The numerator determinants are the same as the denominator determinant, except that the column of coefficients of the variable for which we are solving is replaced with the column of constants e and f.

Cramer's Rule for Two Equations in Two Variables

The solution of the system $\begin{cases} ax + by = e \\ cx + dy = f \end{cases}$ is given by

$$x = \frac{D_x}{D} = \frac{\begin{vmatrix} e & b \\ f & d \end{vmatrix}}{\begin{vmatrix} a & b \\ c & d \end{vmatrix}} \quad \text{and} \quad y = \frac{D_y}{D} = \frac{\begin{vmatrix} a & e \\ c & f \end{vmatrix}}{\begin{vmatrix} a & b \\ c & d \end{vmatrix}}$$

If every determinant is 0, the system is consistent, but the equations are dependent.

If $D = 0$ and D_x or D_y is nonzero, the system is inconsistent. If $D \neq 0$, the system is consistent, and the equations are independent.

EXAMPLE 4

Use Cramer's rule to solve $\begin{cases} 4x - 3y = 6 \\ -2x + 5y = 4 \end{cases}$.

Solution The value of x is the quotient of two determinants, D_x and D. The denominator determinant D is made up of the coefficients of x and y:

$$D = \begin{vmatrix} 4 & -3 \\ -2 & 5 \end{vmatrix}$$

To solve for x, we form the numerator determinant D_x from D by replacing its first column (the coefficients of x) with the column of constants (6 and 4).

To solve for y, we form the numerator determinant D_y from D by replacing the second column (the coefficients of y) with the column of constants (6 and 4).

To find the values of x and y, we evaluate each determinant:

$$x = \frac{D_x}{D} = \frac{\begin{vmatrix} 6 & -3 \\ 4 & 5 \end{vmatrix}}{\begin{vmatrix} 4 & -3 \\ -2 & 5 \end{vmatrix}} = \frac{6(5) - (-3)(4)}{4(5) - (-3)(-2)} = \frac{30 + 12}{20 - 6} = \frac{42}{14} = 3$$

$$y = \frac{D_y}{D} = \frac{\begin{vmatrix} 4 & 6 \\ -2 & 4 \end{vmatrix}}{\begin{vmatrix} 4 & -3 \\ -2 & 5 \end{vmatrix}} = \frac{4(4) - 6(-2)}{14} = \frac{16 + 12}{14} = \frac{28}{14} = 2$$

The solution of this system is (3, 2). Verify that it satisfies both equations.

Self Check 4 Use Cramer's rule to solve $\begin{cases} 2x - 3y = -16 \\ 3x + 5y = 14 \end{cases}$.

EXAMPLE 5

Use Cramer's rule to solve $\begin{cases} 7x = 8 - 4y \\ 2y = 3 - \frac{7}{2}x \end{cases}$.

Solution We multiply both sides of the second equation by 2 to eliminate the fraction and write the system in the form

$$\begin{cases} 7x + 4y = 8 \\ 7x + 4y = 6 \end{cases}$$

When we attempt to use Cramer's rule to solve this system for x, we obtain

$$x = \frac{D_x}{D} = \frac{\begin{vmatrix} 8 & 4 \\ 6 & 4 \end{vmatrix}}{\begin{vmatrix} 7 & 4 \\ 7 & 4 \end{vmatrix}} = \frac{8}{0} \quad \text{which is undefined}$$

Success Tip

If any two rows or any two columns of a determinant are identical, the value of the determinant is 0.

Since the denominator determinant D is 0 and the numerator determinant D_x is not 0, the system is inconsistent. It has no solution.

We can see directly from the system that it is inconsistent. For any values of x and y, it is impossible that 7 times x plus 4 times y could be both 8 and 6.

Self Check 5 Use Cramer's rule to solve $\begin{cases} 3x = 8 - 4y \\ y = \frac{5}{2} - \frac{3}{4}x \end{cases}$.

USING CRAMER'S RULE TO SOLVE A SYSTEM OF THREE EQUATIONS

Cramer's rule can be extended to solve systems of three linear equations with three variables.

Cramer's Rule for Three Equations with Three Variables

The solution of the system $\begin{cases} ax + by + cz = j \\ dx + ey + fz = k \\ gx + hy + iz = l \end{cases}$ is given by

$$x = \frac{D_x}{D}, \quad y = \frac{D_y}{D}, \quad \text{and} \quad z = \frac{D_z}{D}$$

where

$$D = \begin{vmatrix} a & b & c \\ d & e & f \\ g & h & i \end{vmatrix} \qquad D_x = \begin{vmatrix} j & b & c \\ k & e & f \\ l & h & i \end{vmatrix}$$

$$D_y = \begin{vmatrix} a & j & c \\ d & k & f \\ g & l & i \end{vmatrix} \qquad D_z = \begin{vmatrix} a & b & j \\ d & e & k \\ g & h & l \end{vmatrix}$$

If every determinant is 0, the system is consistent, but the equations are dependent.

If $D = 0$ and D_x or D_y or D_z is nonzero, the system is inconsistent. If $D \neq 0$, the system is consistent, and the equations are independent.

EXAMPLE 6

Use Cramer's rule to solve $\begin{cases} 2x + y + 4z = 12 \\ x + 2y + 2z = 9 \\ 3x - 3y - 2z = 1 \end{cases}$.

Solution The denominator determinant D is the determinant formed by the coefficients of the variables. The numerator determinants, D_x, D_y, and D_z, are formed by replacing the coefficients of the variable being solved for by the column of constants. We form the quotients for x, y, and z and evaluate each determinant by expanding by minors about the first row:

$$x = \frac{D_x}{D} = \frac{\begin{vmatrix} 12 & 1 & 4 \\ 9 & 2 & 2 \\ 1 & -3 & -2 \end{vmatrix}}{\begin{vmatrix} 2 & 1 & 4 \\ 1 & 2 & 2 \\ 3 & -3 & -2 \end{vmatrix}} = \frac{12\begin{vmatrix} 2 & 2 \\ -3 & -2 \end{vmatrix} - 1\begin{vmatrix} 9 & 2 \\ 1 & -2 \end{vmatrix} + 4\begin{vmatrix} 9 & 2 \\ 1 & -3 \end{vmatrix}}{2\begin{vmatrix} 2 & 2 \\ -3 & -2 \end{vmatrix} - 1\begin{vmatrix} 1 & 2 \\ 3 & -2 \end{vmatrix} + 4\begin{vmatrix} 1 & 2 \\ 3 & -3 \end{vmatrix}}$$

$$= \frac{12(2) - 1(-20) + 4(-29)}{2(2) - 1(-8) + 4(-9)} = \frac{-72}{-24} = 3$$

$$y = \frac{D_y}{D} = \frac{\begin{vmatrix} 2 & 12 & 4 \\ 1 & 9 & 2 \\ 3 & 1 & -2 \end{vmatrix}}{\begin{vmatrix} 2 & 1 & 4 \\ 1 & 2 & 2 \\ 3 & -3 & -2 \end{vmatrix}} = \frac{2\begin{vmatrix} 9 & 2 \\ 1 & -2 \end{vmatrix} - 12\begin{vmatrix} 1 & 2 \\ 3 & -2 \end{vmatrix} + 4\begin{vmatrix} 1 & 9 \\ 3 & 1 \end{vmatrix}}{-24}$$

$$= \frac{2(-20) - 12(-8) + 4(-26)}{-24} = \frac{-48}{-24} = 2$$

$$z = \frac{D_z}{D} = \frac{\begin{vmatrix} 2 & 1 & 12 \\ 1 & 2 & 9 \\ 3 & -3 & 1 \end{vmatrix}}{\begin{vmatrix} 2 & 1 & 4 \\ 1 & 2 & 2 \\ 3 & -3 & -2 \end{vmatrix}} = \frac{2\begin{vmatrix} 2 & 9 \\ -3 & 1 \end{vmatrix} - 1\begin{vmatrix} 1 & 9 \\ 3 & 1 \end{vmatrix} + 12\begin{vmatrix} 1 & 2 \\ 3 & -3 \end{vmatrix}}{-24}$$

$$= \frac{2(29) - 1(-26) + 12(-9)}{-24} = \frac{-24}{-24} = 1$$

The solution of this system is (3, 2, 1). Verify that it satisfies the three original equations.

Self Check 6 Use Cramer's rule to solve $\begin{cases} x + y + 2z = 6 \\ 2x - y + z = 9 \\ x + y - 2z = -6 \end{cases}$.

Answers to Self Checks **1.** 10 **2.** −44 **3.** −44 **4.** (−2, 4) **5.** no solution **6.** (2, −2, 3)

3.5 STUDY SET

VOCABULARY Fill in the blanks.

1. A determinant is a ________ that is associated with a square matrix.

2. $\begin{vmatrix} 2 & 1 \\ -6 & 1 \end{vmatrix}$ is a 2×2 ________.

3. The ________ of b_1 in $\begin{vmatrix} a_1 & b_1 & c_1 \\ a_2 & b_2 & c_2 \\ a_3 & b_3 & c_3 \end{vmatrix}$ is $\begin{vmatrix} a_2 & c_2 \\ a_3 & c_3 \end{vmatrix}$.

4. In $\begin{vmatrix} 7 & -3 \\ 1 & 2 \end{vmatrix}$, 7 and 2 lie along the main ________.

5. A 3×3 determinant has 3 ________ and 3 ________.

6. ________ rule uses determinants to solve systems of linear equations.

CONCEPTS Fill in the blanks.

7. If the denominator determinant D for a system of equations is zero, the equations of the system are ________ or the system is ________.

8. To find the minor of 5, we ________ the elements of the determinant that are in the same row and column as 5.

$$\begin{vmatrix} 3 & 5 & 1 \\ 6 & -2 & 2 \\ 8 & -1 & 4 \end{vmatrix}$$

9. $\begin{vmatrix} a & b \\ c & d \end{vmatrix} =$ ____ − ____

10. $\begin{vmatrix}5&1&-1\\8&7&4\\9&7&6\end{vmatrix}=-1\begin{vmatrix}8&7\\9&7\end{vmatrix}-4\begin{vmatrix}5&1\\9&7\end{vmatrix}+6\begin{vmatrix}5&1\\8&7\end{vmatrix}$

In evaluating this determinant, about what row or column was it expanded?

11. What is the denominator determinant D for the system $\begin{cases}3x+4y=7\\2x-3y=5\end{cases}$?

12. What is the denominator determinant D for the system $\begin{cases}x+2y=-8\\3x+y-z=-2\\8x+4y-z=6\end{cases}$?

13. For the system $\begin{cases}3x+2y=1\\4x-y=3\end{cases}$, $D_x=-7$, $D_y=5$, and $D=-11$. What is the solution of the system?

14. For the system $\begin{cases}2x+3y-z=-8\\x-y-z=-2\\-4x+3y+z=6\end{cases}$, $D_x=-28$, $D_y=-14$, $D_z=14$, and $D=14$. What is the solution?

NOTATION **Complete the evaluation of each determinant.**

15. $\begin{vmatrix}5&-2\\-2&6\end{vmatrix}=5(\square)-(-2)(-2)$

$=\square-4$

$=26$

16. $\begin{vmatrix}2&1&3\\3&4&2\\1&5&3\end{vmatrix}$

$=2\begin{vmatrix}4&\square\\5&3\end{vmatrix}\square 1\begin{vmatrix}3&2\\\square&3\end{vmatrix}+3\begin{vmatrix}3&4\\1&\square\end{vmatrix}$

$=2(\square-10)-1(9-\square)+3(15-\square)$

$=2(2)-1(\square)+\square(11)$

$=4-7+\square$

$=30$

PRACTICE **Evaluate each determinant.**

17. $\begin{vmatrix}2&3\\-2&1\end{vmatrix}$

18. $\begin{vmatrix}3&-2\\-2&4\end{vmatrix}$

19. $\begin{vmatrix}-1&2\\3&-4\end{vmatrix}$

20. $\begin{vmatrix}-1&-2\\-3&-4\end{vmatrix}$

21. $\begin{vmatrix}10&0\\1&20\end{vmatrix}$

22. $\begin{vmatrix}1&15\\15&0\end{vmatrix}$

23. $\begin{vmatrix}-6&-2\\15&4\end{vmatrix}$

24. $\begin{vmatrix}3&-2\\12&-8\end{vmatrix}$

25. $\begin{vmatrix}1&2&0\\0&1&2\\0&0&1\end{vmatrix}$

26. $\begin{vmatrix}-1&2&1\\2&1&-3\\1&1&1\end{vmatrix}$

27. $\begin{vmatrix}1&-2&3\\-2&1&1\\-3&-2&1\end{vmatrix}$

28. $\begin{vmatrix}1&1&2\\2&1&-2\\3&1&3\end{vmatrix}$

29. $\begin{vmatrix}1&0&1\\0&1&0\\1&1&1\end{vmatrix}$

30. $\begin{vmatrix}3&5&1\\6&-2&2\\8&-1&4\end{vmatrix}$

31. $\begin{vmatrix}1&2&1\\-3&7&3\\-4&3&-5\end{vmatrix}$

32. $\begin{vmatrix}1&4&7\\2&5&8\\3&6&9\end{vmatrix}$

Use Cramer's rule to solve each system of equations, if possible. If a system is inconsistent or if the equations are dependent, so indicate.

33. $\begin{cases}x+y=6\\x-y=2\end{cases}$

34. $\begin{cases}x-y=4\\2x+y=5\end{cases}$

35. $\begin{cases}2x+3y=0\\4x-6y=-4\end{cases}$

36. $\begin{cases}4x-3y=-1\\8x+3y=4\end{cases}$

37. $\begin{cases}3x+2y=11\\6x+4y=11\end{cases}$

38. $\begin{cases}5x+6y=12\\10x+12y=24\end{cases}$

39. $\begin{cases}y=\dfrac{-2x+1}{3}\\3x-2y=8\end{cases}$

40. $\begin{cases}2x+3y=-1\\x=\dfrac{y-9}{4}\end{cases}$

41. $\begin{cases}x+y+z=4\\x+y-z=0\\x-y+z=2\end{cases}$

42. $\begin{cases}x+y+z=4\\x-y+z=2\\x-y-z=0\end{cases}$

43. $\begin{cases}x+y+2z=7\\x+2y+z=8\\2x+y+z=9\end{cases}$

44. $\begin{cases}x+2y+2z=10\\2x+y+2z=9\\2x+2y+z=1\end{cases}$

45. $\begin{cases}2x+y+z=5\\x-2y+3z=10\\x+y-4z=-3\end{cases}$

46. $\begin{cases}3x+2y-z=-8\\2x-y+7z=10\\2x+2y-3z=-10\end{cases}$

47. $\begin{cases}4x-3y=1\\6x-8z=1\\2y-4z=0\end{cases}$

48. $\begin{cases}4x+3z=4\\2y-6z=-1\\8x+4y+3z=9\end{cases}$

49. $\begin{cases} 2x + 3y + 4z = 6 \\ 2x - 3y - 4z = -4 \\ 4x + 6y + 8z = 12 \end{cases}$

50. $\begin{cases} x - 3y + 4z - 2 = 0 \\ 2x + y + 2z - 3 = 0 \\ 4x - 5y + 10z - 7 = 0 \end{cases}$

51. $\begin{cases} 2x + y - z - 1 = 0 \\ x + 2y + 2z - 2 = 0 \\ 4x + 5y + 3z - 3 = 0 \end{cases}$

52. $\begin{cases} 2x - y + 4z + 2 = 0 \\ 5x + 8y + 7z = -8 \\ x + 3y + z + 3 = 0 \end{cases}$

53. $\begin{cases} x + y = 1 \\ \frac{1}{2}y + z = \frac{5}{2} \\ x - z = -3 \end{cases}$

54. $\begin{cases} \frac{1}{2}x + y + z + \frac{3}{2} = 0 \\ x + \frac{1}{2}y + z - \frac{1}{2} = 0 \\ x + y + \frac{1}{2}z + \frac{1}{2} = 0 \end{cases}$

APPLICATIONS

Write a system of equations to solve each problem. Then use Cramer's rule to solve the system.

55. INVENTORIES The table shows an end-of-the-year inventory report for a warehouse that supplies electronics stores. If the warehouse stocks two models of cordless telephones, one valued at \$67 and the other at \$100, how many of each model of phone did the warehouse have at the time of the inventory?

Item	Number	Merchandise value
Television	800	\$1,005,450
Radios	200	\$15,785
Cordless phones	360	\$29,400

56. SIGNALING A system of sending signals uses two flags held in various positions to represent letters of the alphabet. The illustration shows how the letter U is signaled. Find x and y, if y is to be 30° more than x.

57. INVESTING A student wants to average a 6.6% return by investing \$20,000 in the three stocks listed in the table. Because HiTech is a high-risk investment, he wants to invest three times as much in SaveTel and OilCo combined as he invests in HiTech. How much should he invest in each stock?

Stock	Rate of return
HiTech	10%
SaveTel	5%
OilCo	6%

58. INVESTING A woman wants to average a $7\frac{1}{3}\%$ return by investing \$30,000 in three certificates of deposit. (See the table.) She wants to invest five times as much in the 8% CD as in the 6% CD. How much should she invest in each CD?

Type of CD	Rate of return
12-month	6%
24-month	7%
36-month	8%

Use a calculator with matrix capabilities to evaluate each determinant.

59. $\begin{vmatrix} 2 & -3 & 4 \\ -1 & 2 & 4 \\ 3 & -3 & 1 \end{vmatrix}$

60. $\begin{vmatrix} -3 & 2 & -5 \\ 3 & -2 & 6 \\ 1 & -3 & 4 \end{vmatrix}$

61. $\begin{vmatrix} 2 & 1 & -3 \\ -2 & 2 & 4 \\ 1 & -2 & 2 \end{vmatrix}$

62. $\begin{vmatrix} 4 & 2 & -3 \\ 2 & -5 & 6 \\ 2 & 5 & -2 \end{vmatrix}$

WRITING

63. Explain how to find the minor of an element of a determinant.

64. Explain how to find x when solving a system of three linear equations by Cramer's rule. Use the words *coefficients* and *constants* in your explanation.

65. Explain how the following checkerboard pattern is used when evaluating a 3 × 3 determinant.

$$\begin{matrix} + & - & + \\ - & + & - \\ + & - & + \end{matrix}$$

66. Explain the difference between a matrix and a determinant. Give an example of each.

REVIEW

67. Are the lines $y = 2x - 7$ and $x - 2y = 7$ perpendicular?

68. Are the lines $y = 2x - 7$ and $2x - y = 10$ parallel?

69. How are the graphs of $f(x) = x^2$ and $g(x) = x^2 - 2$ related?

70. Is the graph of a circle the graph of a function?

71. The graph of a line passes through $(0, -3)$. Is this the x-intercept or the y-intercept of the line?

72. What is the name of the function $f(x) = |x|$?

73. For the function $y = 2x^2 + 6x + 1$, what is the independent variable and what is the dependent variable?

74. If $f(x) = x^3 - x$, what is $f(-1)$?

CHALLENGE PROBLEMS

75. Show that

$$\begin{vmatrix} x & y & 1 \\ -2 & 3 & 1 \\ 3 & 5 & 1 \end{vmatrix} = 0$$

is the equation of the line passing through $(-2, 3)$ and $(3, 5)$.

76. Show that

$$\frac{1}{2}\begin{vmatrix} 0 & 0 & 1 \\ 3 & 0 & 1 \\ 0 & 4 & 1 \end{vmatrix}$$

is the area of the triangle with vertices at $(0, 0)$, $(3, 0)$, and $(0, 4)$.

ACCENT ON TEAMWORK

INTERSECTION POINTS ON GRAPHS

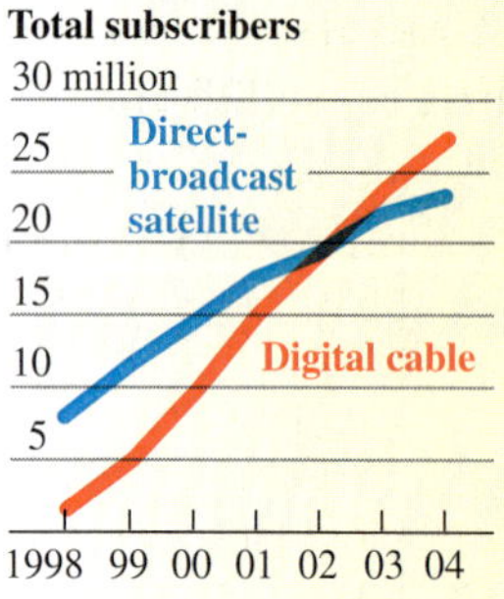

Source: Forrester Research

Overview: This activity will improve your ability to read and interpret graphs.

Instructions: Each student in the class should find a graph that involves intersecting lines. (See the example shown here.) Your school library is a good resource to find such graphs. Ask to look through the collection of recent magazines and newspapers, or scan encyclopedias and books from other disciplines such as nursing and science. You might also use the Internet to find a graph.

Form groups of 5 or 6 students. Have each student show his or her graph to the group and explain the information that is given by the point (or points) of intersection of the lines in the graph. After everyone has taken their turn, vote to determine which graph is the most interesting. The winner from each group should then present his or her graph to the entire class.

BREAK-POINT ANALYSIS

Overview: In this activity, you are to interpret a graph that contains a break point and submit your observations in writing in the form of a financial report.

Instructions: Form groups of 2 or 3 students. Suppose you are a financial analyst for the coathanger company mentioned in Example 10 of Section 3.2. It is your job to decide whether the company should purchase the new machine. First, graph the equations

$$\begin{cases} C = 1.5x + 400 \\ C = 1.25x + 500 \end{cases}$$

on the same coordinate system. Then write a brief report that could be given to company managers, explaining their options concerning the purchase of the new machine. Under what conditions should they keep the machine currently in use? Under what conditions should they buy the new machine?

METHODS OF SOLUTION

Overview: In this activity, you will explore the advantages and disadvantages of several methods for solving a system of linear equations.

Instructions: Form groups of 5 students. Have each member of your group solve the system

$$\begin{cases} x - y = 4 \\ 2x + y = 5 \end{cases}$$

in a different way. The methods to use are graphing, substitution, elimination, matrices, and Cramer's rule. Have each person briefly explain his or her method of solution to the group. After everyone has presented a solution, discuss the advantages and drawbacks of each method. Then rank the five methods, from most desirable to least desirable.

KEY CONCEPT: SYSTEMS OF EQUATIONS

In Chapter 3, we have solved problems involving two and three variables by writing and solving a **system of equations.**

SOLUTIONS OF A SYSTEM OF EQUATIONS

A solution of a system of equations involving two or three variables is an ordered pair or an ordered triple whose coordinates satisfy each equation of the system. In Exercises 1 and 2, decide whether the given ordered pair or ordered triple is a solution of the system.

1. $\begin{cases} 2x - y = 1 \\ 4x + 2y = 0 \end{cases}, \quad \left(\frac{1}{4}, -\frac{1}{2}\right)$

2. $\begin{cases} 2x - y + z = 9 \\ 3x + y - 4z = 8 \\ 2x - 7z = -1 \end{cases}, \quad (4, 0, 1)$

METHODS OF SOLVING SYSTEMS OF LINEAR EQUATIONS

There are several methods for solving systems of two and three linear equations.

3. Solve $\begin{cases} 2x + 5y = 8 \\ y = 3x + 5 \end{cases}$ using the *graphing method.*

4. Solve $\begin{cases} 9x - 8y = 1 \\ 6x + 12y = 5 \end{cases}$ using the *elimination method.*

5. Solve $\begin{cases} 4x - y - 10 = 0 \\ 3x + 5y = 19 \end{cases}$ using the *substitution method.*

6. Solve $\begin{cases} -x + 3y + 2z = 5 \\ 3x + 2y + z = -1 \\ 2x - y + 3z = 4 \end{cases}$ using the *elimination method.*

7. Solve $\begin{cases} x - 6y = 3 \\ x + 3y = 21 \end{cases}$ using *matrices.*

8. Solve $\begin{cases} x + 2z = 7 \\ 2x - y + 3z = 9 \\ y - z = 1 \end{cases}$ using *Cramer's rule.*

DEPENDENT EQUATIONS AND INCONSISTENT SYSTEMS

If the equations in a system of two linear equations are dependent, the system has infinitely many solutions. An inconsistent system has no solutions.

9. Suppose you are solving a system of two equations by the elimination method, and you obtain the following.

$$\begin{array}{r} 2x - 3y = 4 \\ \underline{-2x + 3y = -4} \\ 0 = 0 \end{array}$$

What can you conclude?

10. Suppose you are solving a system of two equations by the substitution method, and you obtain

$$\begin{aligned} -2(x - 3) + 2x &= 7 \\ -2x + 6 + 2x &= 7 \\ 6 &= 7 \end{aligned}$$

What can you conclude?

CHAPTER REVIEW

SECTION 3.1 Solving Systems by Graphing

CONCEPTS

The graph of a linear equation is the graph of all points (x, y) on the rectangular coordinate system whose coordinates satisfy the equation.

A *solution* of a system of equations is an ordered pair that satisfies both equations of the system.

To solve a system *graphically:*

1. Graph each equation on the same rectangular coordinate system.
2. Determine the coordinates of the point of intersection of the graphs. That ordered pair is the solution.
3. Check the proposed solution in each equation of the original system.

A system of equations that has at least one solution is called a *consistent system.* If the graphs are parallel lines, the system has no solution, and it is called an *inconsistent system.*

Equations with different graphs are called *independent equations.* If the graphs are the same line, the system has infinitely many solutions. The equations are called *dependent equations.*

REVIEW EXERCISES

1. See the illustration.

a. Give three points that satisfy the equation $2x + y = 5$.

b. Give three points that satisfy the equation $x - y = 4$.

c. What is the solution of $\begin{cases} 2x + y = 5 \\ x - y = 4 \end{cases}$?

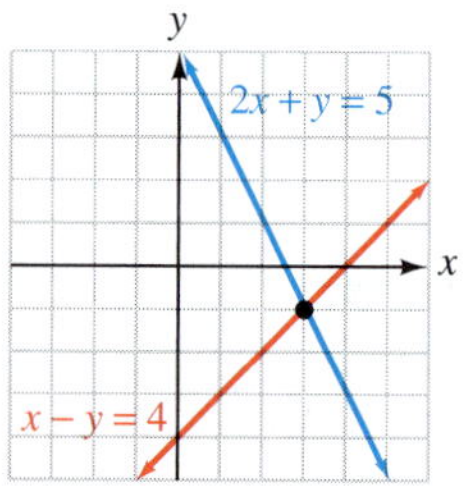

2. POLITICS Explain the importance of the points of intersection of the graphs shown below.

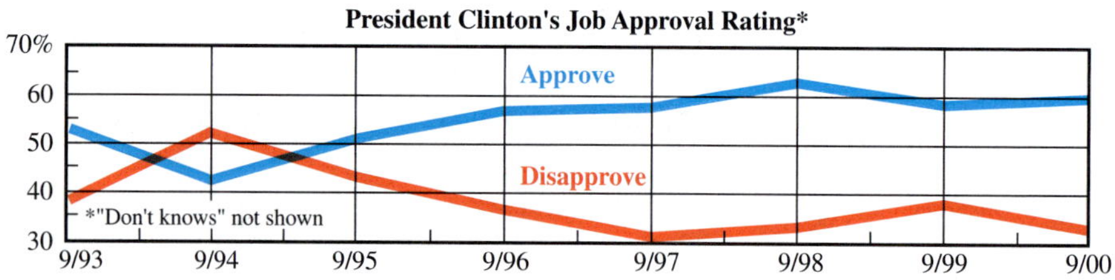

Solve each system by the graphing method, if possible. If a system is inconsistent or if the equations are dependent, so indicate.

3. $\begin{cases} 2x + y = 11 \\ -x + 2y = 7 \end{cases}$

4. $\begin{cases} y = -\frac{3}{2}x \\ 2x - 3y + 13 = 0 \end{cases}$

5. $\begin{cases} \frac{1}{2}x + \frac{1}{3}y = 2 \\ y = 6 - \frac{3}{2}x \end{cases}$

6. $\begin{cases} \frac{x}{3} - \frac{y}{2} = 1 \\ 6x - 9y = 3 \end{cases}$

Use the graphs in the illustration to solve each equation. Check each answer.

7. $2(2 - x) + x = x$

8. $2(2 - x) + x = 5$

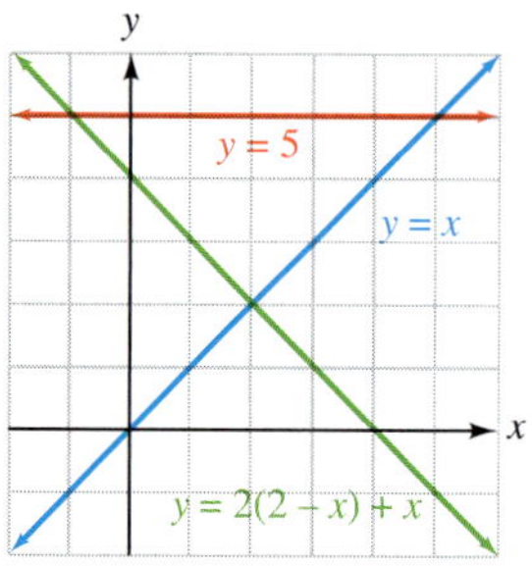

SECTION 3.2 Solving Systems Algebraically

To solve a system by the *substitution method:*

1. Solve one equation for one of its variables.
2. Substitute the resulting expression for that variable into the other equation and solve that equation.
3. Find the value of the other variable by substituting the value of the variable found in step 2 into the equation from step 1.

To solve a system by the *elimination method:*

1. Write both equations in general form: $Ax + By = C$.
2. Multiply the terms of one or both equations by constants so that the coefficients of one variable differ only in sign.
3. Add the equations from step 2 and solve the resulting equation.
4. Substitute the value obtained in step 3 into either original equation and solve for the remaining variable.

Solve each system using the substitution method, if possible. If a system is inconsistent or if the equations are dependent, so indicate.

9. $\begin{cases} x = y - 4 \\ 2x + 3y = 7 \end{cases}$

10. $\begin{cases} y = 2x + 5 \\ 3x - 5y = -4 \end{cases}$

11. $\begin{cases} 0.1x + 0.2y = 1.1 \\ 2x - y = 2 \end{cases}$

12. $\begin{cases} x = -2 - 3y \\ -2x - 6y = 4 \end{cases}$

Solve each system using the elimination method, if possible.

13. $\begin{cases} x + y = -2 \\ 2x + 3y = -3 \end{cases}$

14. $\begin{cases} 2x - 3y = 5 \\ 2x - 3y = 8 \end{cases}$

15. $\begin{cases} x + \frac{1}{2}y = 7 \\ -2x = 3y - 6 \end{cases}$

16. $\begin{cases} y = \dfrac{x - 3}{2} \\ x = \dfrac{2y + 7}{2} \end{cases}$

17. To solve $\begin{cases} 5x - 2y = 19 \\ 3x + 4y = 1 \end{cases}$, which method, elimination or substitution, would you use? Explain why.

18. Estimate the solution of the system $\begin{cases} y = -\frac{2}{3}x \\ 2x - 3y = -4 \end{cases}$ from the graphs in the illustration. Then solve the system algebraically.

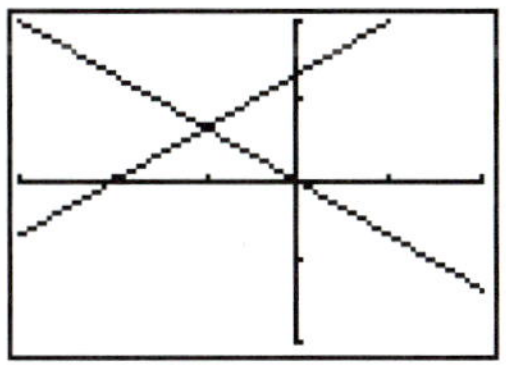

Use two equations to solve each problem.

19. MAPS See the illustration. The distance between Austin and Houston is 4 miles less than twice the distance between Austin and San Antonio. The round trip from Houston to Austin to San Antonio and back to Houston is 442 miles. Determine the mileages between Austin and Houston and between Austin and San Antonio.

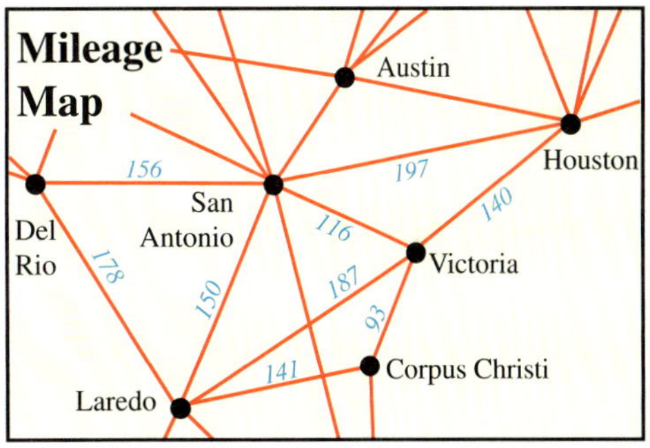

20. RIVERBOATS A Mississippi riverboat travels 30 miles downstream in three hours and then makes the return trip upstream in five hours. Find the speed of the riverboat in still water and the speed of the current.

21. BREAK POINTS A bottling company is considering purchasing a new piece of equipment for their production line. The machine they currently use has a setup cost of \$250 and a cost of \$0.04 per bottle. The new machine has a setup cost of \$600 and a cost of \$0.02 per bottle. Find the break point.

SECTION 3.3 Systems with Three Variables

The solution of a system of three linear equations is an *ordered triple.*

To solve a system of linear equations with three variables:

1. Pick any two equations and eliminate a variable.
2. Pick a different pair of equations and eliminate the same variable.
3. Solve the resulting pair of equations.
4. Use substitution to find the value of the third variable.

22. Determine whether (2, −1, 1) is a solution of the system $\begin{cases} x - y + z = 4 \\ x + 2y - z = -1 \\ x + y - 3z = -1 \end{cases}$.

Solve each system, if possible.

23. $\begin{cases} x + y + z = 6 \\ x - y - z = -4 \\ -x + y - z = -2 \end{cases}$

24. $\begin{cases} 2x + 3y + z = -5 \\ -x + 2y - z = -6 \\ 3x + y + 2z = 4 \end{cases}$

25. $\begin{cases} x + y - z = -3 \\ x + z = 2 \\ 2x - y + 2z = 3 \end{cases}$

26. $\begin{cases} 3x + 3y + 6z = -6 \\ -x - y - 2z = 2 \\ 2x + 2y + 4z = -4 \end{cases}$

27. A system of three linear equations in three variables is graphed on the right. Does the system have a solution? If so, how many solutions does it have?

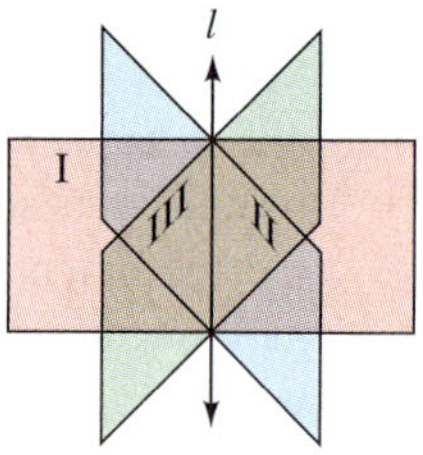

28. MIXING NUTS The owner of a produce store wanted to mix peanuts selling for \$3 per pound, cashews selling for \$9 per pound, and Brazil nuts selling for \$9 per pound to get 50 pounds of a mixture that would sell for \$6 per pound. She used 15 fewer pounds of cashews than peanuts. How many pounds of each did she use?

SECTION 3.4 Solving Systems Using Matrices

A *matrix* is a rectangular array of numbers.

A system of linear equations can be represented by an *augmented matrix.*

Represent each system of equations using an augmented matrix.

29. $\begin{cases} 5x + 4y = 3 \\ x - y = -3 \end{cases}$

30. $\begin{cases} x + 2y + 3z = 6 \\ x - 3y - z = 4 \\ 6x + y - 2z = -1 \end{cases}$

Systems of linear equations can be solved using *Gaussian elimination* and *elementary row operations:*

1. Any two rows can be interchanged.
2. Any row can be multiplied by a nonzero constant.
3. Any row can be changed by adding a nonzero constant multiple of another row to it.

Solve each system using matrices, if possible.

31. $\begin{cases} x - y = 4 \\ 3x + 7y = -18 \end{cases}$

32. $\begin{cases} x + 2y - 3z = 5 \\ x + y + z = 0 \\ 3x + 4y + 2z = -1 \end{cases}$

33. $\begin{cases} 16x - 8y = 32 \\ -2x + y = -4 \end{cases}$

34. $\begin{cases} x + 2y - z = 4 \\ x + 3y + 4z = 1 \\ 2x + 4y - 2z = 3 \end{cases}$

35. INVESTING One year, a couple invested a total of \$10,000 in two projects. The first investment, a mini-mall, made a 6% profit. The other investment, a skateboard park, made a 12% profit. If their investments made \$960, how much was invested at each rate? To answer this question, write a system of two equations and solve it using matrices.

SECTION 3.5 Solving Systems Using Determinants

A *determinant* of a *square matrix* is a number.

To evaluate a 2 × 2 determinant:

$$\begin{vmatrix} a & b \\ c & d \end{vmatrix} = ad - bc$$

To evaluate a 3 × 3 determinant, we expand it by *minors* along any row or column using the *array of signs.*

Cramer's rule can be used to solve systems of linear equations.

Evaluate each determinant.

36. $\begin{vmatrix} 2 & 3 \\ -4 & 3 \end{vmatrix}$

37. $\begin{vmatrix} -3 & -4 \\ 5 & -6 \end{vmatrix}$

38. $\begin{vmatrix} -1 & 2 & -1 \\ 2 & -1 & 3 \\ 1 & -2 & 2 \end{vmatrix}$

39. $\begin{vmatrix} 3 & -2 & 2 \\ 1 & -2 & -2 \\ 2 & 1 & -1 \end{vmatrix}$

Use Cramer's rule to solve each system, if possible.

40. $\begin{cases} 3x + 4y = 10 \\ 2x - 3y = 1 \end{cases}$

41. $\begin{cases} -6x - 4y = -6 \\ 3x + 2y = 5 \end{cases}$

42. $\begin{cases} x + 2y + z = 0 \\ 2x + y + z = 3 \\ x + y + 2z = 5 \end{cases}$

43. $\begin{cases} 2x + 3y + z = 2 \\ x + 3y + 2z = 7 \\ x - y - z = -7 \end{cases}$

44. VETERINARY MEDICINE The daily requirements of a balanced diet for an animal are shown in the nutritional pyramid. The number of grams per cup of nutrients in three food mixes are shown in the table. How many cups of each mix should be used to meet the daily requirements for protein, carbohydrates, and essential fatty acids in the animal's diet? To answer this problem, write a system of three equations and solve it using Cramer's rule.

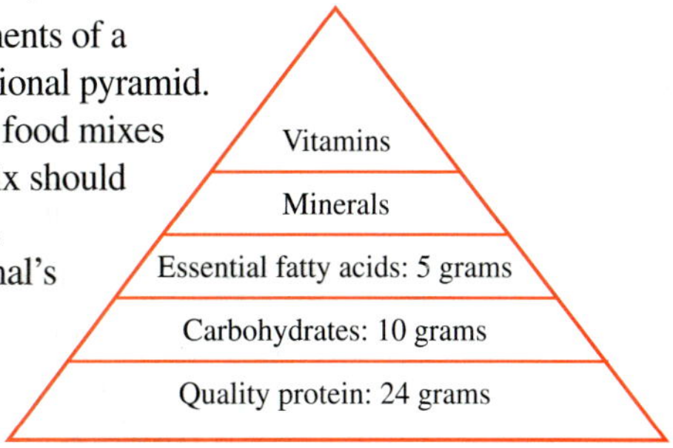

	Grams per cup		
	Protein	**Carbohydrates**	**Fatty Acids**
Mix A	5	2	1
Mix B	6	3	2
Mix C	8	3	1

CHAPTER 3 TEST

1. Solve $\begin{cases} 2x + y = 5 \\ y = 2x - 3 \end{cases}$ by graphing.

2. Use the graphs in the illustration to solve $3(x - 2) - 2(-2 + x) = 1$.

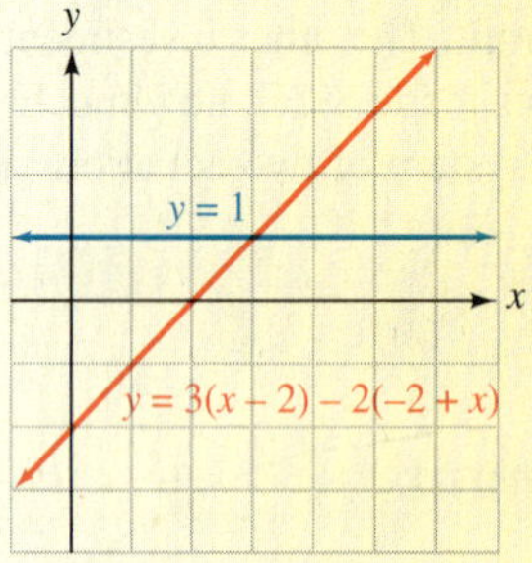

3. Use substitution to solve $\begin{cases} 2x - 4y = 14 \\ x + 2y = 7 \end{cases}$.

4. Use elimination to solve $\begin{cases} 2x + 3y = -5 \\ 3x - 2y = 12 \end{cases}$.

5. Are the equations of the system

$$\begin{cases} 3(x + y) = x - 3 \\ -y = \dfrac{2x + 3}{3} \end{cases}$$

dependent or independent?

6. Is $\left(-1, -\frac{1}{2}, 5\right)$ a solution of $\begin{cases} x - 2y + z = 5 \\ 2x + 4y = -4 \\ -6y + 4z = 22 \end{cases}$?

7. Solve the system $\begin{cases} x + y + z = 4 \\ x + y - z = 6 \\ 2x - 3y + z = -1 \end{cases}$ using elimination.

Write a system of equations to solve each problem.

8. In the sign, find x and y, if y is 15 more than x.

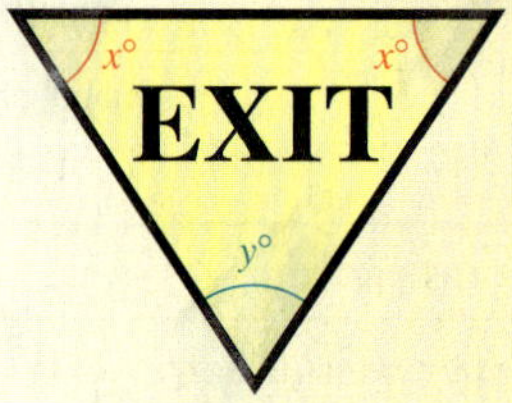

9. ANTIFREEZE How much of a 40% antifreeze solution must a mechanic mix with an 80% antifreeze solution if 20 gallons of a 50% antifreeze solution are needed?

10. BREAK POINTS A metal stamping plant is considering purchasing a new piece of equipment. The machine they currently use has a setup cost of \$1,775 and a cost of \$5.75 per impression. The new machine has a setup cost of \$3,975 and a cost of \$4.15 per impression. Find the break point.

Use matrices to solve each system, if possible.

11. $\begin{cases} x + y = 4 \\ 2x - y = 2 \end{cases}$

12. $\begin{cases} x - 3y + 2z = 1 \\ x - 2y + 3z = 5 \\ 2x - 6y + 4z = 3 \end{cases}$

Evaluate each determinant.

13. $\begin{vmatrix} 2 & -3 \\ 4 & 5 \end{vmatrix}$

14. $\begin{vmatrix} 1 & 2 & 0 \\ 2 & 0 & 3 \\ 1 & -2 & 2 \end{vmatrix}$

Consider the system $\begin{cases} x - y = -6 \\ 3x + y = -6 \end{cases}$, which is to be solved using Cramer's rule.

15. a. When solving for x, what is the numerator determinant D_x? **(Don't evaluate it.)**

b. When solving for y, what is the denominator determinant D? **(Don't evaluate it.)**

16. Solve the system for x:

17. Solve the system for y:

18. Solve the following system for z only, using Cramer's rule.

$$\begin{cases} x + y + z = 4 \\ x + y - z = 6 \\ 2x - 3y + z = -1 \end{cases}$$

19. MOVIE TICKETS The receipts for one showing of a movie were \$410 for an audience of 100 people. The ticket prices are given in the table. If twice as many children's tickets as general admission tickets were purchased, how many of each type of ticket were sold?

Ticket prices	
Children	\$3
General Admission	\$6
Seniors	\$5

20. Which method, substitution or elimination, would you use to solve the following system? Explain your reasoning.

$$\begin{cases} \dfrac{x}{2} - \dfrac{y}{3} = -4 \\ y = -2 - x \end{cases}$$

21. What does it mean to say that a system of two linear equations in two variables is an *inconsistent* system?

22. POPULATION PROJECTIONS See the illustration on the right. If the population trends for the years 2010–2020 continue as projected, estimate the point of intersection of the graphs. Interpret your answer.

Children under age 18 and adults 65 and older as a percent of the U.S. population

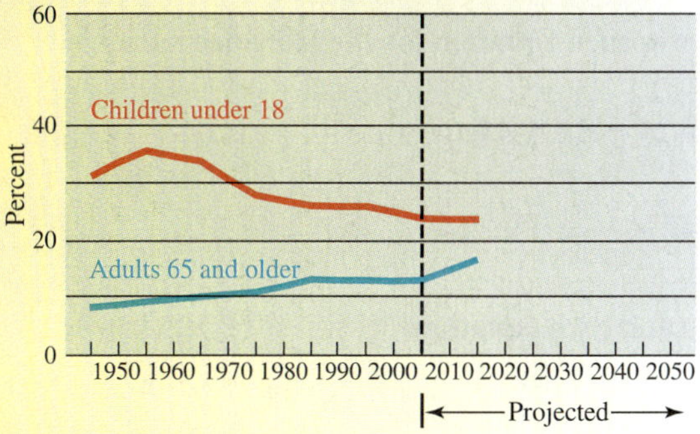

Source: U.S. Bureau of the Census

CHAPTERS 1–3 CUMULATIVE REVIEW EXERCISES

1. Complete the illustration by labeling the rational numbers, irrational numbers, integers, and whole numbers.

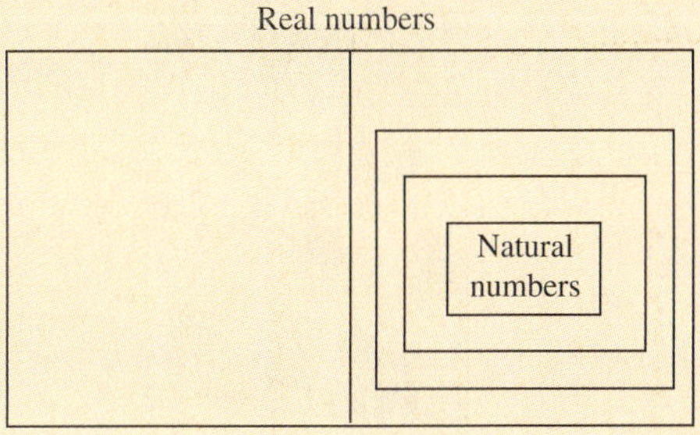

2. FEDERAL BUDGET President Bush's proposed budget for the fiscal year 2005 was \$2.4 trillion. The illustration shows how a typical dollar of the budget was to be spent. Determine the amount he proposed to spend on Social Security.

Source: Budget of the United States Government FY 2005

Evaluate each expression for $a = -3$ and $b = -5$.

3. $-|b| - ab^2$

4. $\dfrac{14 + 2[2a - (b - a)]}{-b - 2}$

Simplify each expression.

5. $0.5x^2 - 6(2.1x^2 - x) + 6.7x$

6. $-(c + 2) - (2 - c)$

7. COMMUTING Use the following facts to determine a commuter's average speed when she drives to work.

- If she drives her car, it takes a quarter of an hour to get to work.
- If she rides the bus, it takes half an hour to get to work.
- When she drives, her average speed is 10 miles per hour faster than that of the bus.

8. DRIED FRUITS Dried apple slices cost \$4.60 per pound, and dried banana chips sell for \$3.40 per pound. How many pounds of each should be used to create a 10-pound mixture that sells for \$4 per pound?

Solve each equation, if possible. If an equation is an identity, so indicate.

9. $\dfrac{3}{4}x + 1.5 = -19.5$

10. $7 - x - x - x = 8$

11. $\dfrac{x+7}{3} = \dfrac{x-2}{5} - \dfrac{x}{15} + \dfrac{7}{3}$

12. $3p - 6 = 4(p - 2) + 2 - p$

Solve each equation for the indicated variable.

13. $\lambda = Ax + AB$ for B

14. $v = \dfrac{d_1 - d_2}{t}$ for d_2

Graph each equation.

15. $3x = 4y - 11$

16. $y = -4$

17. Write an equation of the line that passes through (4, 5) and is parallel to the graph of $y = -3x$. Answer in slope–intercept form.

18. Find the slope of the line.

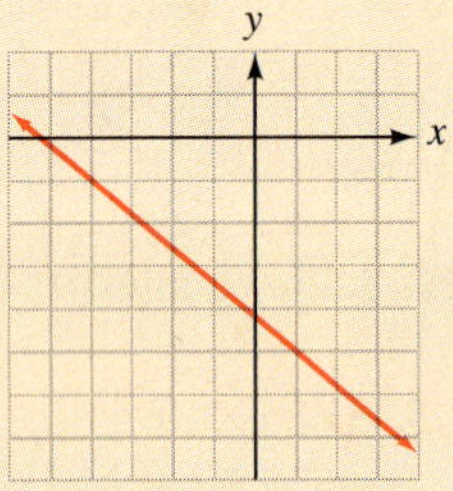

If $f(x) = -x^2 - \dfrac{x}{2}$, find each value.

19. $f(10)$

20. $f(-10)$

21. We can think of a function as a machine. (See the illustration.) Write a function that turns the given input into the given output.

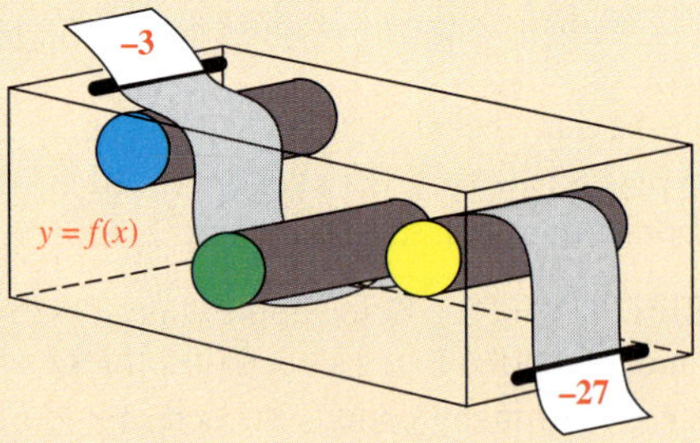

22. Does the table define y as a function of x?

x	y
−2	5
−1	2
0	2
5	5

23. Graph $f(x) = (x + 4)^2$. Give the domain and range.

24. Use the graph of function f to find each of the following.

a. $f(-2)$

b. The value for x for which $f(x) = 3$

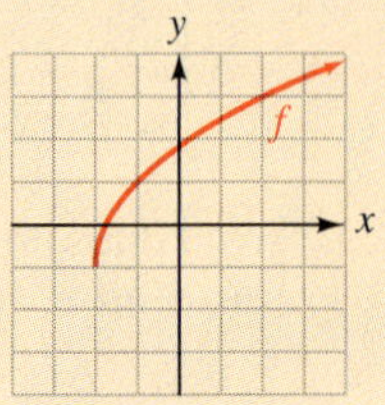

25. Determine whether the graph on the right is the graph of a function. Explain why or why not.

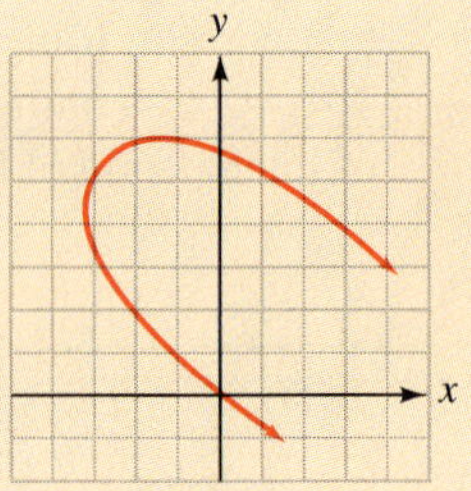

26. COLLECTIBLES A collector buys the Hummel figurine shown in the illustration anticipating that it will be worth \$650 in 20 years. Assuming straight-line appreciation, write an equation that gives the value v of the figurine x years after it is purchased.

27. Solve: $\begin{cases} y = \dfrac{-2x+1}{3} \\ 3x - 2y = 8 \end{cases}$.

28. Solve: $\begin{cases} -x + 3y + 2z = 5 \\ 3x + 2y + z = -1 \\ 2x - y + 3z = 4 \end{cases}$.

Evaluate each determinant.

29. $\begin{vmatrix} 5 & -2 \\ -2 & 6 \end{vmatrix}$

30. $\begin{vmatrix} 2 & 1 & -3 \\ -2 & 2 & 4 \\ 1 & -2 & 2 \end{vmatrix}$

Chapter

4 Inequalities

C. McIntyre/Photolink/Getty Images

There are many ways to measure distance. The above signpost in Maine measures the distances in miles to these towns and lake areas. On long trips, motorists use the car's odometer to measure the distance traveled. When building a staircase, carpenters use a tape measure to make sure the distances between the vertical posts are the same. Scientists use a beam of light to measure the distances from Earth to the planets. In mathematics, we use absolute value to measure distance. Recall that the absolute value of a real number is its distance from zero on the number line. In this chapter, we will define absolute value more formally and we will solve equations and inequalities that contain the absolute value of a variable expression.

To learn more about absolute value, visit *The Learning Equation* on the Internet at http://tle.brookscole.com. (The log-in instructions are in the Preface.) For Chapter 4, the online lesson is:

- *TLE* Lesson 6: Absolute Value Equations

When working with unequal quantities, we use inequalities instead of equations to describe the situation mathematically.

4.1 Solving Linear Inequalities

- Inequalities
- Graphs, intervals, and set-builder notation
- Solving linear inequalities
- Problem solving

Traffic signs like the one shown here often appear in front of schools. From the sign, a motorist knows that

- A speed *greater than* 25 miles per hour breaks the law and could possibly result in a ticket for speeding.
- A speed *less than or equal to* 25 miles per hour is within the posted speed limit.

Statements such as these can be expressed mathematically using *inequality symbols.*

INEQUALITIES

Inequalities are statements indicating that two quantities are unequal. Inequalities contain one or more of the following symbols.

Inequality Symbols

$a \neq b$	means	"a is not equal to b."
$a < b$	means	"a is less than b."
$a > b$	means	"a is greater than b."
$a \leq b$	means	"a is less than or equal to b."
$a \geq b$	means	"a is greater than or equal to b."

The Language of Algebra

Because $<$ requires one number to be strictly less than another number and $>$ requires one number to be strictly greater than another number, $<$ and $>$ are called *strict inequalities.*

By definition, $a < b$ means that "a is less than b," but it also means that $b > a$. Furthermore, if a is to the left of b on a number line, then $a < b$. If a is to the right of b on a number line, then $a > b$.

By definition, $a \leq b$ is true if a is less than b or if a is equal to b. For example, the inequality $-2 \leq 4$ is true, and so is $4 \leq 4$.

We can use a variable and inequality symbols to describe the warning that the traffic sign shown above gives to drivers. If x represents the motorist's speed in miles per hour, the driver is in danger of receiving a speeding ticket if $x > 25$. The driver is observing the posted speed limit if $x \leq 25$.

GRAPHS, INTERVALS, AND SET-BUILDER NOTATION

The graph of a set of real numbers that is a portion of a number line is called an **interval.** The graph shown on the next page represents all real numbers that are greater than -5. This interval contains numbers that satisfy the inequality $x > -5$, such as -4.99, -3, -1.8, 0, $2\frac{3}{4}$, π, and 1,050. The left **parenthesis** at -5 indicates that -5 is not included in the interval.

We can also express this interval in **interval notation** as $(-5, \infty)$, where ∞ (read as **positive infinity**) indicates that the interval extends indefinitely to the right. The left parenthesis is used to show that the endpoint -5 is not included.

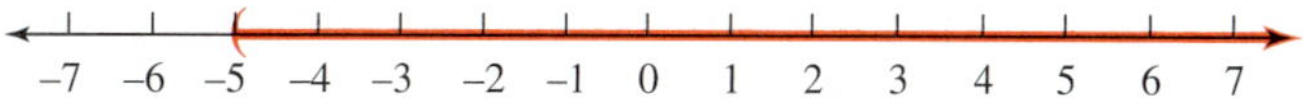

Set-builder notation is another way of describing the set of real numbers graphed in the figure above. With this notation, the condition for membership in the set is specified using a variable. For example, the set of real numbers greater than -5 is written in set-builder notation as

$$\{x \mid x > -5\}$$

the set of all real numbers x — such that — x is greater than -5

Notation

Note that a parenthesis rather than a bracket is written next to an infinity symbol.

$(-5, \infty)$ $(-\infty, 7]$

The interval shown in the following figure is the graph of the real numbers less than or equal to 7. It contains the numbers that satisfy the inequality $x \leq 7$. The right **bracket** at 7 indicates that 7 is included in the interval. To express this interval in interval notation, we write $(-\infty, 7]$, where $-\infty$ (read as **negative infinity**) indicates that the interval extends indefinitely to the left. The bracket is used to show that 7 is included in the interval. To describe the interval using set-builder notation, we write $\{x \mid x \leq 7\}$.

EXAMPLE 1

Represent the set of real numbers greater than or equal to 8 using interval notation, with a graph, and using set-builder notation.

Solution All real numbers that are greater than or equal to 8 are included in the interval $[8, \infty)$. The graph is shown below. Using set-builder notation, we write $\{x \mid x \geq 8\}$.

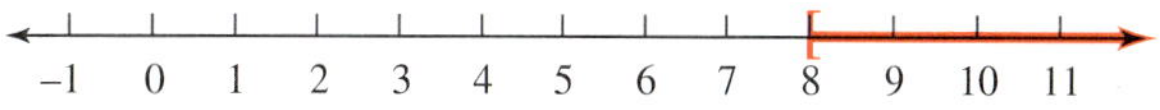

Self Check 1 Represent the set of negative real numbers using interval notation, with a graph, and using set-builder notation.

If an interval extends forever in one direction, as in the previous examples, it is called an **unbounded interval.** The following chart illustrates the various types of unbounded intervals and shows how they are described using an inequality and a graph.

Unbounded Intervals

The interval (a, ∞) includes all real numbers x such that $x > a$.

The interval $[a, \infty)$ includes all real numbers x such that $x \geq a$.

The interval $(-\infty, a)$ includes all real numbers x such that $x < a$.

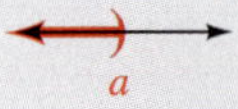

The interval $(-\infty, a]$ includes all real numbers x such that $x \leq a$.

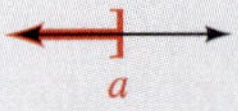

The interval $(-\infty, \infty)$ includes all real numbers x. The graph of this interval is the entire number line.

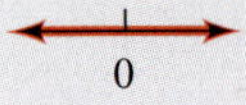

When graphing intervals, an open circle can be used to show that a point is not included in a graph, and a solid circle can be used to show that a point *is* included. For example,

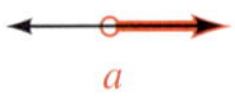

is equivalent to

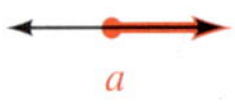

is equivalent to

We will use parentheses and brackets when graphing intervals, because they are consistent with interval notation.

Notation

The symbols ∞ and $-\infty$ do not represent numbers. Instead, ∞ indicates that an interval extends indefinitely to the right and $-\infty$ indicates that an interval extends indefinitely to the left.

SOLVING LINEAR INEQUALITIES

In this section, we will work with **linear inequalities** in one variable.

Linear Inequalities

A **linear inequality** in one variable (say, x) is any inequality that can be expressed in one of the following forms, where a, b, and c represent real numbers and $a \neq 0$.

$$ax + b < c \qquad ax + b \leq c \qquad ax + b > c \qquad \text{or} \qquad ax + b \geq c$$

Some examples of linear inequalities are

$$3x < 0, \qquad 3(2x - 9) < 9, \qquad \text{and} \qquad -12x - 8 \geq 16$$

To **solve a linear inequality** means to find all the values that, when substituted for the variable, make the inequality true. The set of all solutions of an inequality is called its **solution set.** Most of the inequalities we will solve have infinitely many solutions. We will use the following properties to solve inequalities.

Addition and Subtraction Properties of Inequality

Adding the same number to, or subtracting the same number from, both sides of an inequality does not change the solutions.

For any real numbers a, b, and c,

If $a < b$, then $a + c < b + c$.

If $a < b$, then $a - c < b - c$.

Similar statements can be made for the symbols $\leq$, $>$, or $\geq$.

As with equations, there are properties for multiplying and dividing both sides of an inequality by the same number. To develop what is called *the multiplication property of inequality,* consider the true statement $2 < 5$. If both sides are multiplied by a positive number, such as 3, another true inequality results.

$$2 < 5$$
$$3 \cdot 2 < 3 \cdot 5 \quad \text{Multiply both sides by 3.}$$
$$6 < 15 \quad \text{This is a true inequality.}$$

However, if we multiply both sides of $2 < 5$ by a negative number, such as -3, the direction of the inequality symbol is reversed to produce another true inequality.

$$2 < 5$$
$$-3 \cdot 2 > -3 \cdot 5 \quad \text{Multiply both sides by the negative number } -3 \text{ and reverse the direction of the inequality.}$$
$$-6 > -15 \quad \text{This is a true inequality.}$$

The inequality $-6 > -15$ is true because -6 is to the right of -15 on the number line.

Dividing both sides of an inequality by the same negative number also requires that the direction of the inequality symbol be reversed.

$$-4 < 6 \quad \text{This is a true inequality.}$$
$$\frac{-4}{-2} > \frac{6}{-2} \quad \text{Divide both sides by } -2 \text{ and change } < \text{ to } >.$$
$$2 > -3 \quad \text{This is a true inequality.}$$

These examples illustrate the multiplication and division properties of inequality.

Multiplication and Division Properties of Inequality

Multiplying or dividing both sides of an inequality by the same positive number does not change the solutions.

For any real numbers a, b, and c, where c is positive,

If $a < b$, then $ac < bc$.

If $a < b$, then $\frac{a}{c} < \frac{b}{c}$.

If we multiply or divide both sides of an inequality by a negative number, the direction of the inequality symbol must be reversed for the inequalities to have the same solutions.

For any real numbers a, b, and c, where c is negative,

If $a < b$, then $ac > bc$.

If $a < b$, then $\frac{a}{c} > \frac{b}{c}$.

Similar statements can be made for the symbols $\leq$, $>$, or $\geq$.

After applying one of the properties of inequality, the resulting inequality is equivalent to the original one. Like equivalent equations, **equivalent inequalities** have the same solution set.

EXAMPLE 2

Solve: $3(2x - 9) < 9$. Write the solution set in interval notation and graph it.

Solution We want to isolate x on one side of the inequality symbol. To do that, we use the same strategy as we used to solve equations.

$$3(2x - 9) < 9$$

$$6x - 27 < 9 \quad \text{Distribute the multiplication by 3.}$$

$$6x < 36 \quad \text{To undo the subtraction of 27, add 27 to both sides.}$$

$$x < 6 \quad \text{To undo the multiplication by 6, divide both sides by 6.}$$

The solution set is the interval $(-\infty, 6)$, whose graph is shown. We can also write the solution set using set-builder notation: $\{x \mid x < 6\}$.

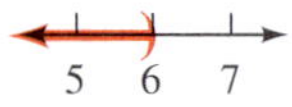

The solution set contains infinitely many real numbers. We cannot check to see whether all of them satisfy the original inequality. As an informal check, we pick one number in the graph, such as 4, and see whether it satisfies the inequality.

Check:

$$3(2x - 9) < 9 \quad \text{This is the original inequality.}$$

$$3[2(4) - 9] \stackrel{?}{<} 9 \quad \text{Substitute 4 for } x\text{. Read } \stackrel{?}{<} \text{ as "is possibly less than."}$$

$$3(8 - 9) \stackrel{?}{<} 9 \quad \text{Multiply: } 2(4) = 8.$$

$$3(-1) \stackrel{?}{<} 9 \quad \text{Subtract: } 8 - 9 = -1.$$

$$-3 < 9 \quad \text{This is a true statement.}$$

Since $-3 < 9$, 4 satisfies the inequality. The solution appears to be correct.

Self Check 2 Solve: $2(3x + 2) > -44$.

EXAMPLE 3

Solve: $-12x - 8 \le 16$. Write the solution set in interval notation and graph it.

Solution To solve this inequality, we isolate x.

$$-12x - 8 \le 16$$

$$-12x \le 24 \quad \text{To undo the subtraction of 8, add 8 to both sides.}$$

$$x \ge -2 \quad \text{To undo the multiplication by } -12\text{, divide both sides by } -12\text{. Because we are dividing by a negative number, we reverse the } \le \text{ symbol.}$$

The solution set is $\{x \mid x \ge -2\}$ or the interval $[-2, \infty)$, whose graph is shown.

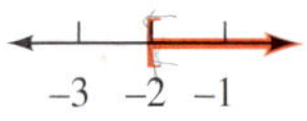

Self Check 3 Solve: $-6x + 6 \le 0$.

EXAMPLE 4

Solve: $\frac{2}{3}(x + 2) > \frac{4}{5}(x - 3)$.

Solution To clear the inequality of fractions, we multiply both sides by the LCD of $\frac{2}{3}$ and $\frac{4}{5}$.

$$\frac{2}{3}(x+2) > \frac{4}{5}(x-3)$$

$$\mathbf{15} \cdot \frac{2}{3}(x+2) > \mathbf{15} \cdot \frac{4}{5}(x-3)$$ Multiply both sides by the LCD of $\frac{2}{3}$ and $\frac{4}{5}$, which is 15.

$$10(x+2) > 12(x-3)$$ Simplify: $15 \cdot \frac{2}{3} = 10$ and $15 \cdot \frac{4}{5} = 12$.

$$10x + 20 > 12x - 36$$ Distribute the multiplication by 10 and 12.

$$-2x + 20 > -36$$ To eliminate $12x$ on the right-hand side, subtract $12x$ from both sides.

$$-2x > -56$$ Subtract 20 from both sides.

$$x < 28$$ Divide both sides by -2 and reverse the $>$ symbol.

The solution set is the interval $(-\infty, 28)$, as shown in the graph.

Self Check 4 Solve: $\frac{3}{2}(x+2) < \frac{3}{5}(x-3)$.

Caution When solving inequalities, the variable can end up on the right-hand side. For example, if we solve an inequality and obtain $-3 < x$, we can write the inequality in the equivalent form $x > -3$.

EXAMPLE 5

Solve: $3a - 4 < 3(a + 5)$. Write the solution set in interval notation and graph it.

Solution

$$3a - 4 < 3(a+5)$$

$$3a - 4 < 3a + 15$$ Distribute the multiplication by 3.

$$3a - 4 - 3a < 3a + 15 - 3a$$ Subtract $3a$ from both sides.

$$-4 < 15$$ This is a true statement.

The terms involving a drop out. The resulting true statement indicates that the original inequality is true for all values of a. Therefore, the solution set is the set of real numbers, denoted $(-\infty, \infty)$ or $\mathbb{R}$, and its graph is as shown.

Success Tip

When solving an inequality, if the variables drop out and the result is false, the solution set has no elements and is denoted $\varnothing$. The graph is an unshaded number line.

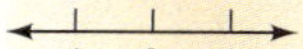

Self Check 5 Solve: $-8n + 10 \geq 1 - 2(4n - 2)$.

ACCENT ON TECHNOLOGY: SOLVING LINEAR INEQUALITIES

There are several ways to solve linear inequalities graphically. For example, to solve $3(2x - 9) < 9$ we can subtract 9 from both sides and solve instead the equivalent inequality $3(2x - 9) - 9 < 0$. Using standard window settings of $[-10, 10]$ for x and $[-10, 10]$ for y, we graph $y = 3(2x - 9) - 9$ and then use TRACE. Moving the cursor closer and closer to the x-axis, as shown in figure (a), we see that the graph is below the x-axis for x-values in the interval $(-\infty, 6)$. This interval is the solution, because in this interval, $3(2x - 9) - 9 < 0$.

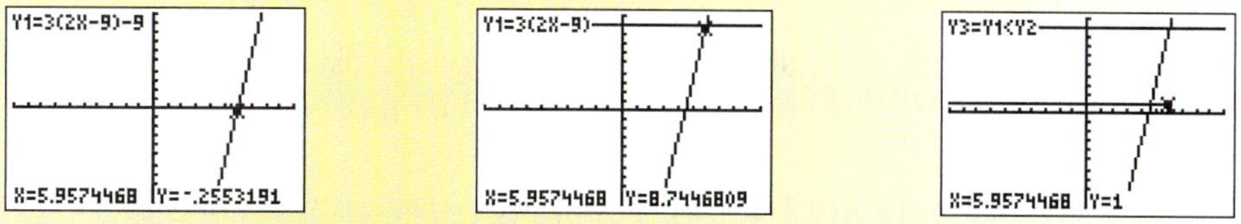

(a) (b) (c)

Another way to solve $3(2x - 9) < 9$ is to graph $y = 3(2x - 9)$ and $y = 9$. We can then trace to see that the graph of $y = 3(2x - 9)$ is below the graph of $y = 9$ for x-values in the interval $(-\infty, 6)$. See figure (b). This interval is the solution, because in this interval, $3(2x - 9) < 9$.

A third approach is to enter and then graph

$Y_1 = 3(2x - 9)$

$Y_2 = 9$

$Y_3 = Y_1 < Y_2$ To do this, use the VARS key. Consult your owner's manual for the specific directions.

The graphs of $y = 3(2x - 9)$, $y = 9$, and a horizontal line 1 unit above the x-axis will be displayed, as shown in figure (c). In the TRACE mode, we then move the cursor to the rightmost endpoint of the horizontal line to determine that the interval $(-\infty, 6)$ is the solution of $3(2x - 9) < 9$.

PROBLEM SOLVING

We have used a five-step problem-solving strategy to solve problems. This process involved writing and solving equations. We will now show how inequalities can be used to solve problems. To decide whether to use an equation or an inequality to solve a problem, you must look for key words and phrases. Here are some statements that translate to inequalities.

The statement	Translates to
a does not exceed *b*.	$a \leq b$
a is at most *b*.	$a \leq b$
a is no more than *b*.	$a \leq b$

The statement	Translates to
a is at least *b*.	$a \geq b$
a is not less than *b*.	$a \geq b$
a will exceed *b*.	$a > b$

EXAMPLE 6

Translate the sentence to mathematical symbols: *The instructor said that the test would take no more than 50 minutes.*

Solution Since the test will take no more than 50 minutes, it will take 50 minutes or less to complete. If we let t represent the time it takes to complete the test, then $t \leq 50$.

Self Check 6 Translate the sentence to mathematical symbols: *A PG-13 movie rating means that you must be at least 13 years old to see the movie.*

EXAMPLE 7

Political contributions. Some volunteers are making long-distance telephone calls to solicit contributions for their candidate. The calls are billed at the rate of 25¢ for the first three minutes and 7¢ for each additional minute or part thereof. If the campaign chairperson has ordered that the cost of each call is not to exceed \$1.00, for how many minutes can a volunteer talk to a prospective donor on the phone?

Analyze the problem We are given the rate at which a call is billed. Since the cost of a call is not to exceed \$1.00, the cost must be *less than or equal to* \$1.00. This phrase indicates that we should write an inequality to find how long a volunteer can talk to a prospective donor.

Form an inequality We will let $x =$ the total number of minutes that a call can last. Then the cost of a call will be 25¢ for the first three minutes plus 7¢ times the number of additional minutes, where the number of *additional* minutes is $x - 3$ (the total number of minutes minus the first 3 minutes). With this information, we can form an inequality.

The cost of the first three minutes	plus	the cost of the additional minutes	is not to exceed	\$1.00.
0.25	$+$	$0.07(x - 3)$	$\leq$	1

Solve the inequality To simplify the computations, we first clear the inequality of decimals.

$$0.25 + 0.07(x - 3) \leq 1$$

$$25 + 7(x - 3) \leq 100 \quad \text{To eliminate the decimals, multiply both sides by 100.}$$

$$25 + 7x - 21 \leq 100 \quad \text{Distribute the multiplication by 7.}$$

$$7x + 4 \leq 100 \quad \text{Combine like terms.}$$

$$7x \leq 96 \quad \text{Subtract 4 from both sides.}$$

$$x \leq 13.\overline{714285} \quad \text{Divide both sides by 7.}$$

State the conclusion Since the phone company doesn't bill for part of a minute, the longest time a call can last is 13 minutes. If a call lasts for $13.\overline{714285}$ minutes, it will be charged as a 14-minute call, and the cost will be \$0.25 + \$0.07(11) = \$1.02.

Check the result If the call lasts 13 minutes, the cost will be \$0.25 + \$0.07(10) = \$0.95. This is less than \$1.00. The result checks.

Answers to Self Checks

1. $(-\infty, 0)$ 

$\{x \mid x < 0\}$

2. $(-8, \infty)$ 

3. $[1, \infty)$ 

4. $\left(-\infty, -\frac{16}{3}\right)$ 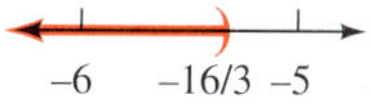

5. $(-\infty, \infty)$ 

6. $a \geq 13$

4.1 STUDY SET

VOCABULARY **Fill in the blanks.**

1. $<, >, \leq$, and $\geq$ are __________ symbols.

2. $(-\infty, 5)$ is an example of an unbounded __________.

3. The __________ on the right of the interval notation $(-\infty, 5)$ indicates that 5 is not included in the interval.

4. To ________ an inequality means to find all values of the variable that make the inequality true.

5. $3x + 2 \geq 7$ is an example of a ______ inequality.

6. ∞ is a symbol representing positive ________.

7. The symbol for "__________" is $<$. The symbol for "____________________" is $\geq$.

8. We read the __________ notation $\{x \mid x < 1\}$ as "the set of all real numbers x __________ x is less than 1."

CONCEPTS

9. Classify each of the following as an equation, an expression, or an inequality.
   a. $-6 - 5x = 8$
   b. $5 - 2x$
   c. $7x - 5x > -4x$
   d. $-(7x - 9)$

10. In the illustration, which of the following are true?
   i. $b > 0$
   ii. $a - b < 0$
   iii. $ab > 0$

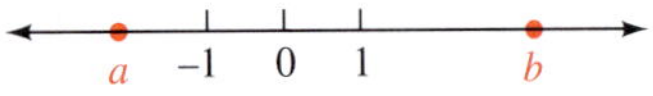

11. In the illustration above, which of the following are true?
   i. $b - a > 0$
   ii. $ab < 0$
   iii. $|a| > |b|$

12. What inequality is suggested by each sentence?
   a. As many as 16 people were seriously injured.
   b. There are no fewer than 10 references to carpools in the speech.

13. Perform each step listed below on the inequality $4 > -2$ and give the resulting true inequality.
   a. Add 2 to both sides.
   b. Subtract 4 from both sides.
   c. Multiply both sides by 4.
   d. Divide both sides by -2.

14. Write an equivalent inequality with the variable on the left-hand side.
   a. $-10 > x$
   b. $\frac{7}{8} < x$
   c. $0 \leq x$

15. Consider the linear inequality $3x + 6 \leq 6$. Decide whether each value is a solution of the inequality.
   a. 0 **b.** $\frac{2}{3}$
   c. -10 **d.** 1.5

16. The solution set of a linear inequality in x is graphed below. Tell whether a true or false statement results when

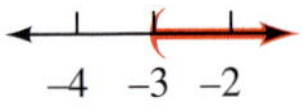

   a. -4 is substituted for x.
   b. -3 is substituted for x.
   c. 0 is substituted for x.

17. Suppose that when solving a linear inequality, the variables drop out, and the result is $6 \leq 10$. Write the solution set in interval notation and graph it.

18. Suppose that when solving a linear inequality, the variables drop out, and the result is $7 < -1$. What symbol is used to represent the solution set? Graph the solution set.

NOTATION **Complete the solution to solve the inequality.**

19. $-5x - 1 \geq -11$
$$-5x \geq \square$$
$$\frac{-5x}{-5} \;\square\; \frac{-10}{\square}$$
$$x \leq 2$$
The solution set is $(\square, 2]$. Using set-builder notation, it is $\{x \mid \square\}$.

20. Describe each set of real numbers using interval notation and set-builder notation, and then graph it.
   a. All real numbers greater than 4
   b. All real numbers less than -4
   c. All real numbers less than or equal to 4

21. Match each interval with its graph.
   a. $(-\infty, -1]$ **i.**
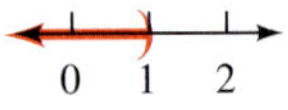

   b. $(-\infty, 1)$ **ii.**

   c. $[-1, \infty)$ **iii.**
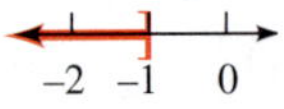

22. In each case, tell what is wrong with the interval notation.

a. $(\infty, -3)$

b. $[-\infty, -3)$

PRACTICE **Solve each inequality. Write the solution set in interval notation and then graph it.**

23. $3x > -9$

24. $4x < -36$

25. $-30y \le -600$

26. $-6y \ge -600$

27. $0.6x \ge 36$

28. $0.2x < 8$

29. $3 > -\frac{9}{10}x$

30. $-\frac{2}{5} < -\frac{4}{5}x$

31. $x + 4 < 5$

32. $x - 5 > 2$

33. $-5t + 3 \le 5$

34. $-9t + 6 \ge 16$

35. $-3x - 1 \le 5$

36. $-2y + 6 < 16$

37. $7 < \frac{5}{3}a - 3$

38. $5 > \frac{7}{2}a - 9$

39. $-7y + 5 > -5y - 1$

40. $-2s - 105 \le -7s - 205$

41. $10x - 12 > 4x - 15 + 6x$

42. $5x + 2 < 6x + 1 - x$

43. $\frac{6 - d}{-2} \le -6$

44. $\frac{9 - 3b}{-8} < 3$

45. $0.4x + 0.4 \le 0.1x + 0.85$

46. $0.05 - 0.5x \le -0.7 - 0.8x$

47. $3(z - 2) \le 2(z + 7)$

48. $5(3 + z) > -3(z + 3)$

49. $-11(2 - b) < 4(2b + 2)$

50. $-9(h - 3) + 2h \le 8(4 - h)$

51. $\frac{1}{2}x + 6 \ge 4 + 2x$

52. $\frac{1}{3}x + 1 < 4 + 5x$

53. $5(2n + 2) - n > 3n - 3(1 - 2n)$

54. $-1 + 4(y - 1) + 2y \le \frac{1}{2}(12y - 30) + 15$

55. $\frac{3b + 7}{3} \le \frac{2b - 9}{2}$

56. $-\frac{5x}{4} > \frac{3 - 5x}{4}$

57. $\frac{x - 7}{2} - \frac{x - 1}{5} \ge -\frac{x}{4}$

58. $\frac{3a + 1}{3} - \frac{4 - 3a}{5} \ge -\frac{1}{15}$

59. $\frac{1}{2}y + 2 \ge \frac{1}{3}y - 4$

60. $\frac{1}{4}x - \frac{1}{3} \le x + 2$

61. $\frac{2}{3}x + \frac{3}{2}(x - 5) \le x$

62. $\frac{5}{9}(x + 3) - \frac{4}{3}(x - 3) \ge x - 1$

63. $5[3t - (t - 4)] - 11 \le -12(t - 6) - (-t)$

64. $2 - 2[3h - (7 - h)] > 6[-(19 + h) - (1 - h)]$

Use a graphing calculator to solve each inequality.

65. $2x + 3 < 5$

66. $3x - 2 > 4$

67. $5x + 2 \geq 4x - 2$

68. $3x - 4 \leq 2x + 4$

APPLICATIONS

69. REAL ESTATE Refer to the illustration. For which regions of the country was the following inequality true in the year 2003?

Median sales price < U.S. median price

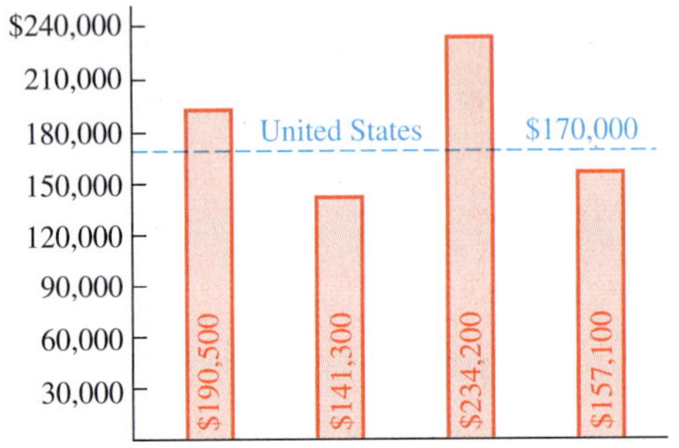

Source: National Association of Realtors

70. PUBLIC EDUCATION Refer to the illustration. For which years is the following inequality true?

$$\text{Enrollment in grade 4} \geq \text{Enrollment in grade 1}$$

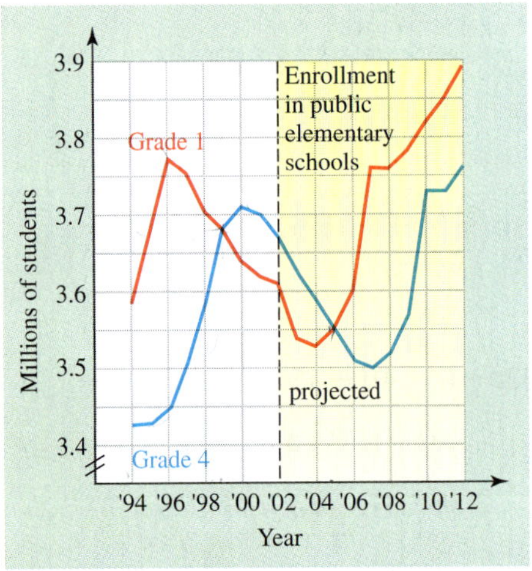

Source: National Center for Education Statistics

71. GEOMETRY The **triangle inequality** states an important relationship between the sides of any triangle:

$$\text{The sum of the lengths of two sides of a triangle} > \text{the length of the third side.}$$

Use the triangle inequality to show that the dimensions of the shuffleboard court shown in the illustration must be mislabeled.

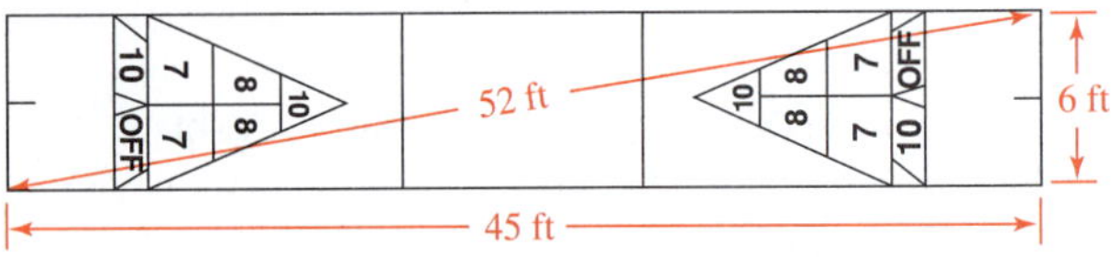

72. COMPUTER PROGRAMMING Flowcharts like the one shown are used by programmers to show the step-by-step instructions of a computer program. For row 1 in the table, work through the steps of the flow chart using the values of a, b, and c, and tell what the computer printout would be. Now do the same for row 2, and then for row 3.

	a	b	c
Row 1	1	1	1
Row 2	9	−12	4
Row 3	11	−25	−24

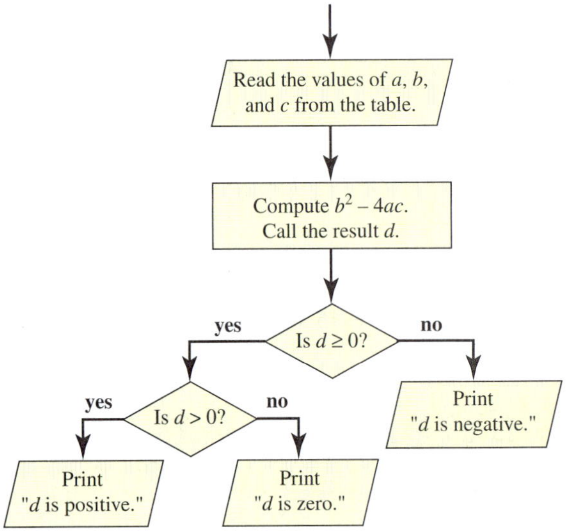

73. FUNDRAISING A school PTA wants to rent a dunking tank for its annual school fundraising carnival. The cost is \$85.00 for the first three hours and then \$19.50 for each additional hour or part thereof. How long can the tank be rented if up to \$185 is budgeted for this expense?

74. INVESTMENTS If a woman has invested \$10,000 at 8% annual interest, how much more must she invest at 9% so that her annual income will exceed \$1,250?

75. BUYING A COMPUTER A student who can afford to spend up to \$2,000 sees the ad shown in the illustration. If she decides to buy the computer, find the greatest number of CD-ROMs that she can also purchase. (Disregard sales tax.)

76. AVERAGING GRADES A student has scores of 70, 77, and 85 on three government exams. What score does she need on a fourth exam to give her an average of 80 or better?

77. WORK SCHEDULES A student works two part-time jobs. He earns \$7 an hour for working at the college library and \$12 an hour for construction work. To save time for study, he limits his work to 20 hours a week. If he enjoys the work at the library more, how many hours can he work at the library and still earn at least \$175 a week?

78. SCHEDULING EQUIPMENT An excavating company charges \$300 an hour for the use of a backhoe and \$500 an hour for the use of a bulldozer. (Part of an hour counts as a full hour.) The company employs one operator for 40 hours per week to operate the machinery. If the company wants to bring in at least \$18,500 each week from equipment rental, how many hours per week can it schedule the operator to use a backhoe?

79. MEDICAL PLANS A college provides its employees with a choice of the two medical plans shown in the following table. For what size hospital bills is Plan 2 better for the employee than Plan 1? (*Hint:* The cost to the employee includes both the deductible payment and the employee's coinsurance payment.)

Plan 1	Plan 2
Employee pays \$100	Employee pays \$200
Plan pays 70% of the rest	Plan pays 80% of the rest

80. MEDICAL PLANS To save costs, the college in Exercise 75 raised the employee deductible, as shown in the following table. For what size hospital bills is Plan 2 better for the employee than Plan 1? (*Hint:* The cost to the employee includes both the deductible payment and the employee's coinsurance payment.)

Plan 1	Plan 2
Employee pays \$200	Employee pays \$400
Plan pays 70% of the rest	Plan pays 80% of the rest

WRITING

81. The techniques for solving linear equations and linear inequalities are similar, yet different. Explain.

82. Explain how the symbol ∞ is used in this section. Is ∞ a real number?

83. Explain how to use the following graph to solve $2x + 1 < 3$.

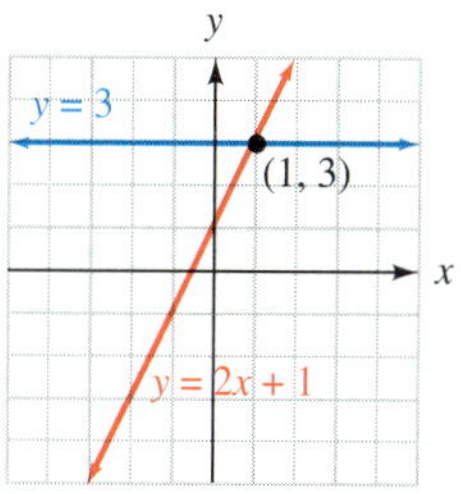

84. Explain what is wrong with the following statement:

When solving inequalities involving negative numbers, the direction of the inequality symbol must be reversed.

REVIEW **Use the graph of the function to find $f(-1)$, $f(0)$, and $f(2)$.**

85.

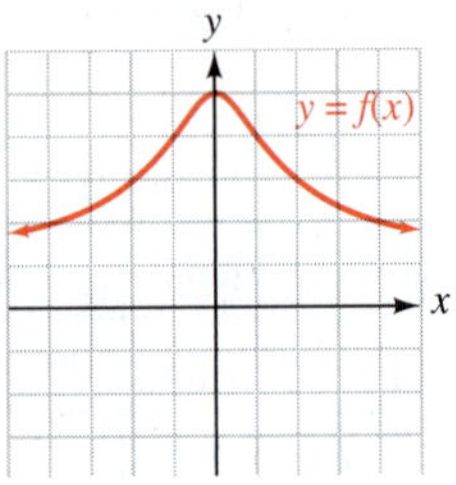

86.

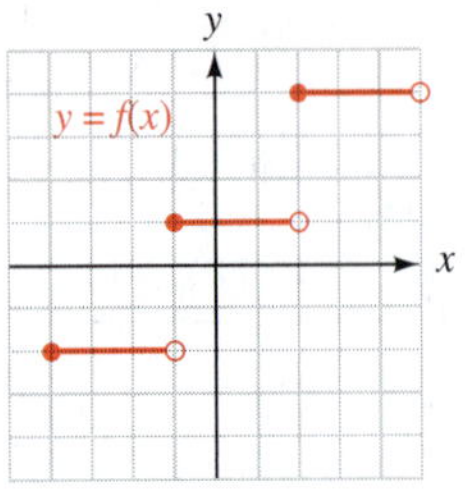

Complete each input/output table.

87. $f(x) = x - x^3$

Input	Output
−2	
2	

88. $g(t) = \dfrac{t^2 - 1}{5}$

Input	Output
−6	
4	

CHALLENGE PROBLEMS

89. The **trichotomy property** states that for any real numbers a and b, exactly one of the following statements is true:

$$a < b, \qquad a = b, \qquad \text{or} \qquad a > b$$

Explain this property and give examples to illustrate it.

90. The **transitive property** of $<$ states that if a, b, and c are real numbers with $a < b$ and $b < c$, then $a < c$. Explain this property and give examples to illustrate it.

91. Which of the relations is transitive?

a. $=$ **b.** $\leq$ **c.** $\ngtr$ **d.** $\neq$

92. Find the error in the following solution.

Solve: $\frac{1}{3} > \frac{1}{x}$.

$$\frac{1}{3} > \frac{1}{x}$$

$$3x\left(\frac{1}{3}\right) > 3x\left(\frac{1}{x}\right)$$

$$x > 3$$

4.2 Solving Compound Inequalities

- Solving compound inequalities containing the word *and*
- Double linear inequalities
- Compound inequalities containing the word *or*
- Solving compound inequalities containing the word *or*

A label on a tube of antibiotic ointment advises the user about the temperature at which the medication should be stored. A careful reading reveals that the storage instructions consist of two parts:

The storage temperature should be at least 59°F

and

The storage temperature should be at most 77°F

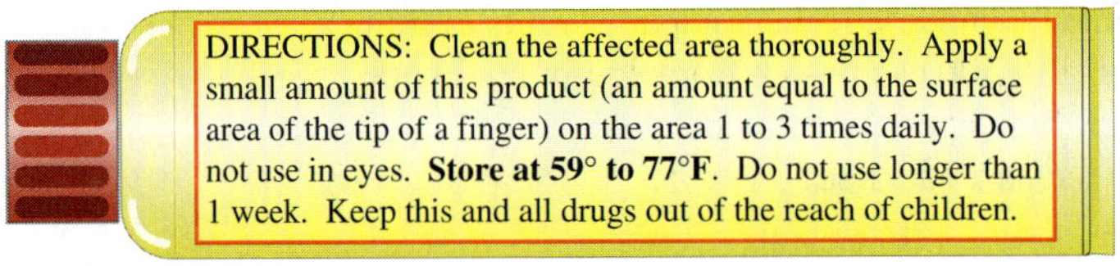
DIRECTIONS: Clean the affected area thoroughly. Apply a small amount of this product (an amount equal to the surface area of the tip of a finger) on the area 1 to 3 times daily. Do not use in eyes. Store at 59° to 77°F. Do not use longer than 1 week. Keep this and all drugs out of the reach of children.

When *and* or *or* are used to connect pairs of inequalities, we call the statements *compound inequalities.* In this section, we will discuss the procedures used to solve three types of compound inequalities, as well as the notation used to express their solution sets.

SOLVING COMPOUND INEQUALITIES CONTAINING THE WORD *AND*

The Language of Algebra

Compound means composed of two or more parts, as in *compound* inequalities, chemical *compounds,* and *compound* sentences.

When two inequalities are joined with the word *and,* we call the statement a **compound inequality.** Some examples are

$$x \geq -3 \quad \text{and} \quad x \leq 6$$

$$\frac{x}{2} + 1 > 0 \quad \text{and} \quad 2x - 3 < 5$$

$$x + 3 \leq 2x - 1 \quad \text{and} \quad 3x - 2 < 5x - 4$$

The solution set of a compound inequality containing the word *and* includes all numbers that make *both* of the inequalities true. For example, we can find the solution set of the compound inequality $x \geq -3$ and $x \leq 6$ by graphing the solution sets of each inequality on the same number line and looking for the numbers common to both graphs.

In the following figure, the graph of the solution set of $x \geq -3$ is shown in red, and the graph of the solution set of $x \leq 6$ is shown in blue.

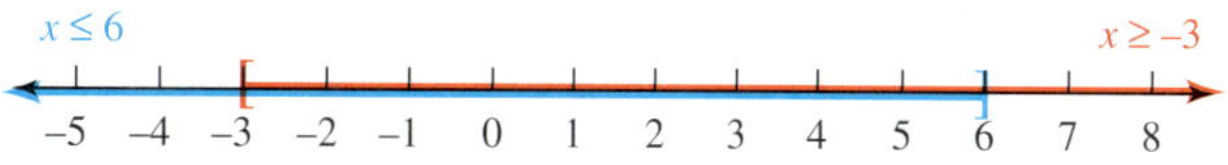

The figure below shows the graph of the solution of $x \geq -3$ and $x \leq 6$. The purple shaded interval is where the red and blue graphs overlap. It represents the numbers that are common to the graphs of $x \geq -3$ and $x \leq 6$.

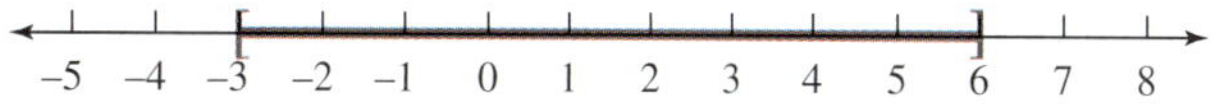

The Language of Algebra

The *intersection* of two sets is the collection of elements that they have in common. When two streets cross, we call the area of pavement that they have in common an *intersection.*

The solution set of $x \geq -3$ and $x \leq 6$ can be denoted by the **bounded interval** $[-3, 6]$, where the brackets indicate that the endpoints, -3 and 6, are included. It represents all real numbers between -3 and 6, including -3 and 6. Intervals such as this, which contain both endpoints, are called **closed intervals.**

When solving a compound inequality containing *and,* the solution set is the *intersection* of the solution sets of the two inequalities. The **intersection** of two sets is the set of elements that are common to both sets. We can denote the intersection of two sets using the symbol $\cap$, which is read as "intersection." For the compound inequality $x \geq -3$ and $x \leq 6$, we can write

$$[-3, \infty) \cap (-\infty, 6] = [-3, 6]$$

The solution set of the compound inequality $x \geq -3$ and $x \leq 6$ can be expressed in several ways:

1. *As a graph:* 

2. *In interval notation:* $[-3, 6]$
3. *In words:* all real numbers between -3 and 6, including -3 and 6
4. *Using set-builder notation:* $\{x \mid x \geq -3 \text{ and } x \leq 6\}$

EXAMPLE 1

Solve: $\frac{x}{2} + 1 > 0$ and $2x - 3 < 5$. Graph the solution set.

Solution We solve each linear inequality separately.

$$\begin{array}{rcl} \frac{x}{2} + 1 > 0 & \text{and} & 2x - 3 < 5 \\ \frac{x}{2} > -1 & \Big| & 2x < 8 \\ x > -2 & \Big| & x < 4 \end{array}$$

Next, we graph the solutions of each inequality on the same number line and determine their intersection.

> **Notation**
> When graphing on a number line, (−2, 4) represents an *interval.* When graphing on a rectangular coordinate system, (−2, 4) is an *ordered pair* that gives the coordinates of a point.

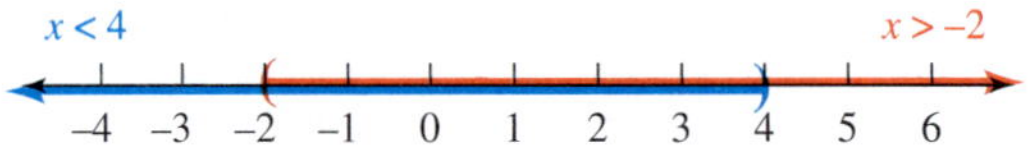

The intersection of the graphs is the set of all real numbers between −2 and 4. The solution set of the compound inequality is the interval (−2, 4), whose graph is shown below. This bounded interval, which does not include either endpoint, is called an **open interval.**

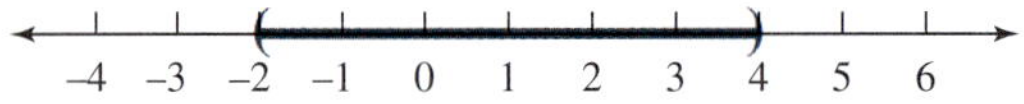

Self Check 1 Solve: $3x > -18$ and $\frac{x}{5} - 1 \le 1$. Graph the solution set.

The solution of the compound inequality in the Self Check of Example 1 is the interval (−6, 10]. A bounded interval such as this, which includes only one endpoint, is called a **half-open interval.** The following chart shows the various types of bounded intervals, along with the inequalities and interval notation that describe them.

Intervals			
	Open intervals	The interval (a, b) includes all real numbers x such that $a < x < b$.	
	Half-open intervals	The interval $[a, b)$ includes all real numbers x such that $a \le x < b$.	
		The interval $(a, b]$ includes all real numbers x such that $a < x \le b$.	
	Closed intervals	The interval $[a, b]$ includes all real numbers x such that $a \le x \le b$.	

EXAMPLE 2

Solve: $x + 3 \leq 2x - 1$ and $3x - 2 < 5x - 4$. Graph the solution set.

Solution We solve each inequality separately.

$$
\begin{array}{rcl}
x + 3 \leq 2x - 1 & \text{and} & 3x - 2 < 5x - 4 \\
4 \leq x & & 2 < 2x \\
x \geq 4 & & 1 < x \\
 & & x > 1
\end{array}
$$

The graph of $x \geq 4$ is shown below in red and the graph of $x > 1$ is shown below in blue.

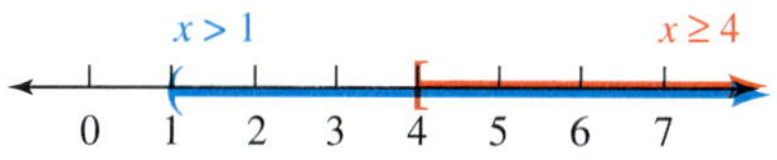

Only those x where $x \geq 4$ and $x > 1$ are in the solution set of the compound inequality. Since all numbers greater than or equal to 4 are also greater than 1, the solutions are the numbers x where $x \geq 4$. The solution set is the interval $[4, \infty)$, whose graph is shown below.

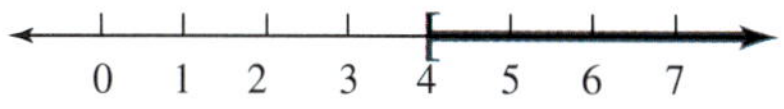

Self Check 2 Solve $2x + 3 < 4x + 2$ and $3x + 1 < 5x + 3$. Graph the solution set.

EXAMPLE 3

Solve: $x - 1 > -3$ and $2x < -8$.

Solution We solve each inequality separately.

$$
\begin{array}{rcl}
x - 1 > -3 & \text{and} & 2x < -8 \\
x > -2 & & x < -4
\end{array}
$$

We note that the graphs of the solution sets shown below do not intersect.

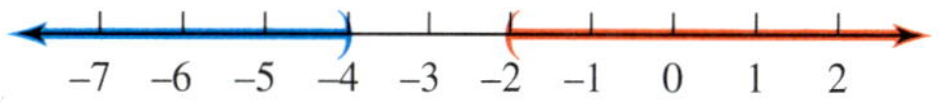

This means there are no numbers that make both parts of the original compound inequality true. The solution set of the compound inequality is the empty set, which can be denoted $\varnothing$.

Notation

The graphs of two linear inequalities can intersect at a single point, as shown below. The interval notation used to describe this point of intersection is [3, 3].

Self Check 3 Solve: $2x - 3 < x - 2$ and $0 < x - 3.5$.

DOUBLE LINEAR INEQUALITIES

Inequalities containing two inequality symbols are called **double inequalities.** An example is

$-3 \le 2x + 5 < 7$ Read as "-3 is less than or equal to $2x + 5$ and $2x + 5$ is less than 7."

Any double linear inequality can be written as a compound inequality containing the word *and.* In general, the following is true.

Double Linear Inequalities The compound inequality $c < x < d$ is equivalent to $c < x$ and $x < d$.

EXAMPLE 4

Solve: $-3 \le 2x + 5 < 7$. Graph the solution set.

Solution This double inequality $-3 \le 2x + 5 < 7$ means that

$$-3 \le 2x + 5 \quad \text{and} \quad 2x + 5 < 7$$

We could solve each linear inequality separately, but we note that each solution would involve the same steps: subtracting 5 from both sides and dividing both sides by 2. We can solve the double inequality more efficiently by leaving it in its original form and applying these steps to each of its three parts to isolate x in the middle.

$$-3 \le 2x + 5 < 7$$

$$-3 - 5 \le 2x + 5 - 5 < 7 - 5$$ To undo the addition of 5, subtract 5 from all three parts.

$$-8 \le 2x < 2$$ Perform the subtractions.

$$\frac{-8}{2} \le \frac{2x}{2} < \frac{2}{2}$$ To undo the multiplication by 2, divide all three parts by 2.

$$-4 \le x < 1$$ Perform the divisions.

The solution set of the double linear inequality is the half-open interval $[-4, 1)$, whose graph is shown below.

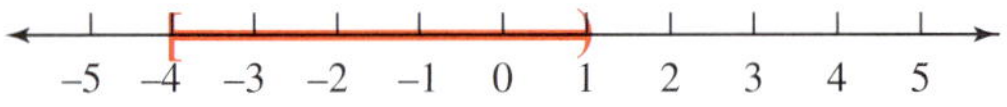

Self Check 4 Solve: $-5 \le 3x - 8 \le 7$. Graph the solution set.

Caution When multiplying or dividing all three parts of a double inequality by a negative number, don't forget to reverse the direction of *both* inequalities. As an example, we solve $-15 < -5x \le 25$.

$$-15 < -5x \le 25$$

$$\frac{-15}{-5} > \frac{-5x}{-5} \ge \frac{25}{-5}$$ Divide all three parts by -5 to isolate x in the middle. Reverse both inequality signs.

$$3 > x \ge -5$$ Perform the divisions.

$$-5 \le x < 3$$ Write an equivalent compound inequality with the smaller number, -5, on the left.

COMPOUND INEQUALITIES CONTAINING THE WORD *OR*

A warning on the water temperature gauge of a commercial dishwasher cautions the operator to shut down the unit if

The water temperature goes below 140°

or

The water temperature goes above 160°

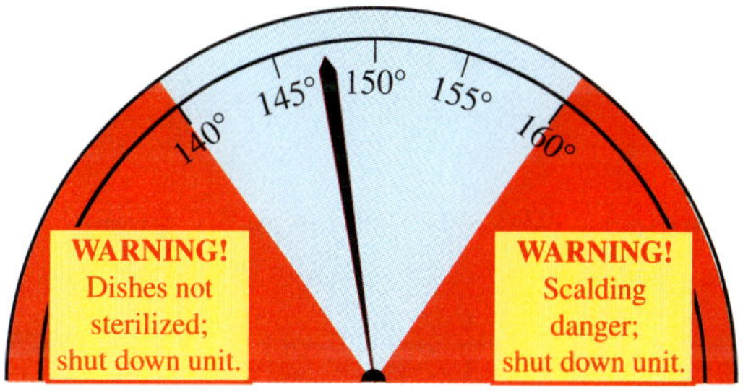

When two inequalities are joined with the word *or,* we also call the statement a compound inequality. Some examples are

$$x < 140 \quad \text{or} \quad x > 160$$
$$x \leq -3 \quad \text{or} \quad x \geq 2$$
$$\frac{x}{3} > \frac{2}{3} \quad \text{or} \quad -(x - 2) > 3$$

SOLVING COMPOUND INEQUALITIES CONTAINING THE WORD *OR*

Caution

It is incorrect to write the statement $x \leq -3$ or $x \geq 2$ as the double inequality $2 \leq x \leq -3$, because that would imply that $2 \leq -3$, which is false.

The solution set of a compound inequality containing the word *or* includes all numbers that make *one or the other or both* inequalities true. For example, we can find the solution set of $x \leq -3$ or $x \geq 2$ by putting the graphs of each inequality on the same number line.

In the following figure, the graph of the solution set of $x \leq -3$ is shown in red, and the graph of the solution set of $x \geq 2$ is shown in blue.

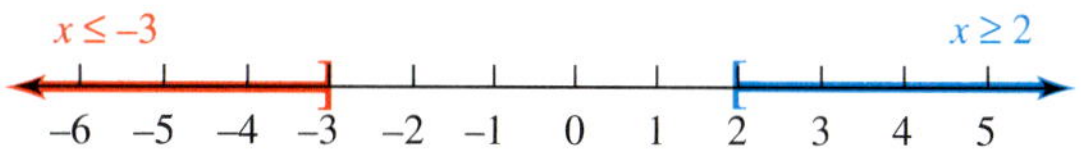

The figure below shows the graph of the solution set of $x \leq -3$ or $x \geq 2$. This graph is a combination of the graph of $x \leq -3$ with the graph of $x \geq 2$.

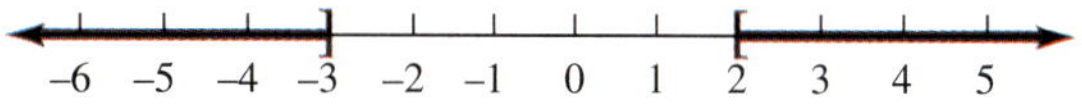

The Language of Algebra

The *union* of two sets is the collection of elements that belong to either set. The concept is similar to that of a family *reunion,* which brings together the members of several families.

When solving a compound inequality containing *or,* the solution set is the *union* of the solution sets of the two inequalities. The **union** of two sets is the set of elements that are in either of the sets or both. We can denote the union of two sets using the symbol $\cup$, which is read as "union." For the compound inequality $x \leq -3$ or $x \geq 2$, we can write the solution set using interval notation:

$$(-\infty, -3] \cup [2, \infty)$$

We can express the solution set of the compound inequality $x \le -3$ or $x \ge 2$ in several ways:

1. *As a graph:* 

2. *In interval notation:* $(-\infty, -3] \cup [2, \infty)$
3. *In words:* all real numbers less than or equal to -3 or greater than or equal to 2
4. *Using set-builder notation:* $\{x \mid x \le -3 \text{ or } x \ge 2\}$

EXAMPLE 5

Solve: $\frac{x}{3} > \frac{2}{3}$ or $-(x - 2) > 3$. Graph the solution set.

Solution We solve each inequality separately.

$$\frac{x}{3} > \frac{2}{3} \quad \text{or} \quad -(x-2) > 3$$

$$x > 2 \qquad\qquad -x + 2 > 3$$

$$-x > 1$$

$$x < -1$$

The Language of Algebra

The meaning of the word *or* in a compound inequality differs from our everyday use of the word. For example, when we say, "I will go shopping today *or* tomorrow," we mean that we will go one day or the other, but *not* both. With compound inequalities, *or* includes one possibility, or the other, or both.

Next, we graph the solutions of each inequality on the same number line and determine their union.

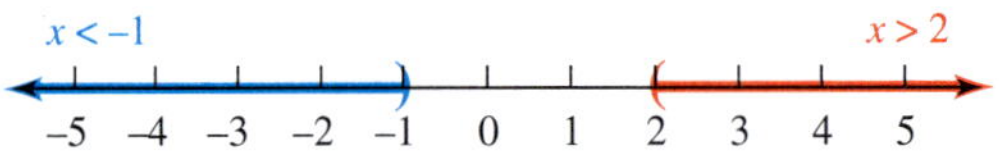

The union of the two solution sets consists of all real numbers less than -1 or greater than 2. The solution set of the compound inequality is the interval $(-\infty, -1) \cup (2, \infty)$. Its graph appears below.

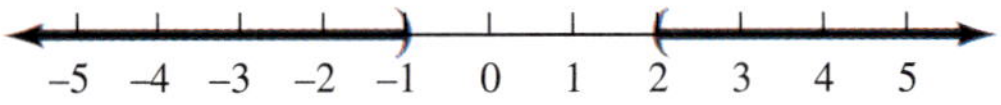

Self Check 5 Solve: $\frac{x}{2} > 2$ or $-3(x - 2) > 0$. Graph the solution set.

EXAMPLE 6

Solve: $x + 3 \ge -3$ or $-x > 0$. Graph the solution set.

Solution We solve each inequality separately.

$$x + 3 \ge -3 \quad \text{or} \quad -x > 0$$

$$x \ge -6 \qquad\qquad x < 0$$

We graph the solution set of each inequality on the same number line and determine their union.

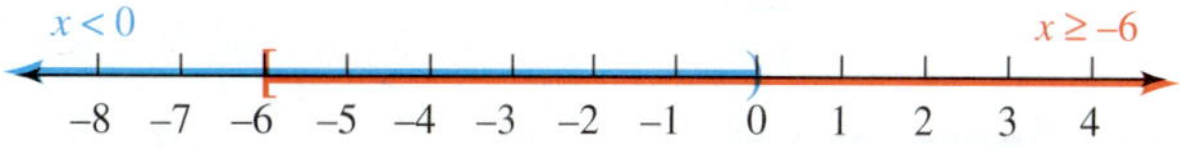

Since the entire number line is shaded, all real numbers satisfy the original compound inequality and the solution set is denoted as $(-\infty, \infty)$ or $\mathbb{R}$. Its graph is shown below.

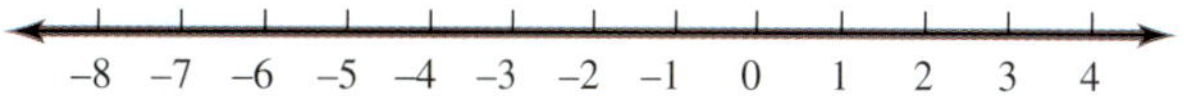

Self Check 6 Solve: $x - 1 < 5$ or $-2x \leq 10$. Graph the solution set.

Answers to Self Checks

1. $(-6, 10]$ 

2. $\left(\frac{1}{2}, \infty\right)$ 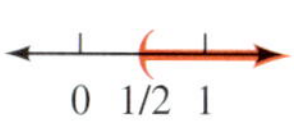

3. no solution

4. $[1, 5]$ 

5. $(-\infty, 2) \cup (4, \infty)$ 

6. $(-\infty, \infty)$ 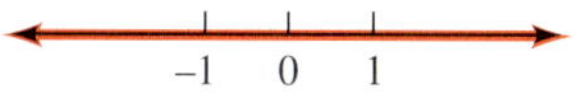

4.2 STUDY SET

VOCABULARY **Fill in the blanks.**

1. $x \geq 3$ and $x < 4$ is a __________ inequality.
2. $3x < x - 1$ or $x \leq 6$ is a compound __________.
3. $-6 < x + 1 \leq 1$ is a ________ linear inequality.
4. $(2, 8)$ is an example of an open ________.
   $[-4, 0]$ is an example of a ________ interval.
   $(0, 9]$ is an example of a __________ interval.
5. The __________ of two sets is the set of elements that are common to both sets.
6. The ________ of two sets is the set of elements that are in one set, or the other, or both.

CONCEPTS **Fill in the blanks.**

7. The solution set of a compound inequality containing the word *and* includes all numbers that make ______ inequalities true.
8. The solution set of a compound inequality containing the word *or* includes all numbers that make ______, or the other, or ______ inequalities true.
9. The double inequality $4 < 3x + 5 \leq 15$ is equivalent to $4 < 3x + 5$ ______ $3x + 5 \leq 15$.
10. When solving a compound inequality containing the word *and,* the solution set is the __________ of the solution sets of the inequalities.
11. When solving a compound inequality containing the word *or*, the solution set is the _______ of the solution sets of the inequalities.
12. When multiplying or dividing all three parts of a double inequality by a negative number, the direction of both inequality symbols must be _________.
13. In each case, decide whether -3 is a solution of the compound inequality.

    a. $\frac{x}{3} + 1 \geq 0$ and $2x - 3 < -10$

    b. $2x \leq 0$ or $-3x < -5$
14. In each case, decide whether -3 is a solution of the double linear inequality.

    a. $-1 < -3x + 4 < 12$

    b. $-1 < -3x + 4 < 14$
15. Use interval notation, if possible, to describe the intersection of each pair of graphs.

    a. 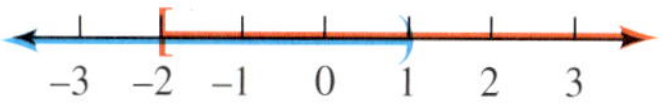

    b. 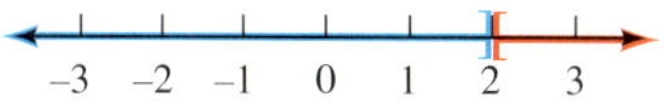

    c. 

16. Use interval notation to describe the union of each pair of graphs.

a. 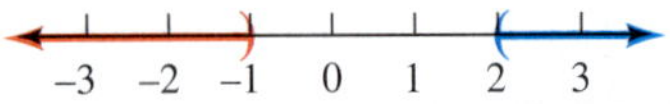

b. 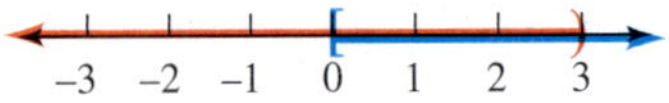

c. 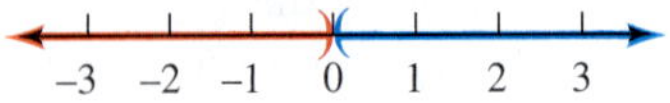

NOTATION

17. Fill in the blanks: We read $\cup$ as ______ and $\cap$ as ______.

18. Match each interval with its corresponding graph.

a. $[2, 3)$ **i.** 

b. $(2, 3)$ **ii.** 

c. $[2, 3]$ **iii.** 

19. Give the interval notation that describes each set and graph it.

a. The real numbers between -3 and 3

b. The real numbers less than -3 or greater than 3

c. The real numbers between -3 and 3, including 3

20. What set is denoted by the interval notation $(-\infty, \infty)$? Graph it.

21. a. Graph: $(-\infty, 2) \cup [3, \infty)$.

b. Graph: $(-\infty, 3) \cap [-2, \infty)$.

22. What is incorrect about the double inequality $3 < -3x + 4 < -3$?

PRACTICE **Solve each compound inequality. Write the solution set (if one exists) in interval notation and graph it.**

23. $x > -2$ and $x \le 5$

24. $x \le -4$ and $x \ge -7$

25. $2.2x < -19.8$ and $-4x < 40$

26. $\frac{1}{2}x \le 2$ and $0.75x \ge -6$

27. $x + 3 < 3x - 1$ and $4x - 3 \le 3x$

28. $4x \ge -x + 5$ and $6 \ge 4x - 3$

29. $x + 2 < -\frac{1}{3}x$ and $-6x < 9x$

30. $\frac{3}{2}x + \frac{1}{5} < 5$ and $2x + 1 > 9$

31. $5(x - 2) \ge 0$ and $-3x < 9$

32. $x - 1 \le 2(x + 2)$ and $x \le 2x - 5$

33. $5(x + 1) \le 4(x + 3)$ and $x + 12 < -3$

34. $4 \le x + 3 \le 7$

35. $-5.3 < x - 2.3 < -1.3$

36. $1.5 > 2x - 0.7 > 0.9$

37. $25 > 3x - 2 > 7$

38. $-2 < -b + 3 < 5$

39. $2 < -t - 2 < 9$

40. $-6 < -3(x - 4) \le 24$

41. $-4 \le -2(x + 8) < 8$

42. $2x + 1 \ge 5$ and $-3(x + 1) \ge -9$

43. $2(-2) \le 3x - 1$ and $3x - 1 \le -1 - 3$

44. $\frac{x}{0.7} + 5 > 4$ and $-4.8 \le \frac{3x}{-0.125}$

45. $-2 \le 2 - t \le -4$

46. $-x < -2x$ and $3x > 2x$

47. $-4 > \frac{2}{3}x - 2 > -6$

48. $-6 \le \frac{1}{3}a + 1 < 0$

49. $0 \le \frac{4 - x}{3} \le 2$

50. $-2 \le \frac{5 - 3x}{2} \le 2$

51. $x \le 6 - \frac{1}{2}x$ and $\frac{1}{2}x + 1 \ge 3$

52. $3\left(x + \frac{2}{3}\right) \le -7$ and $2(x + 2) \ge -2$

53. $x \le -2$ or $x > 6$

54. $x \ge -1$ or $x \le -3$

55. $x - 3 < -4$ or $x - 2 > 0$

56. $4x < -12$ or $\frac{x}{2} > 4$

57. $3x + 2 < 8$ or $2x - 3 > 11$

58. $3x + 4 < -2$ or $3x + 4 > 10$

59. $x > 3$ or $x < 5$

60. $x < -15$ or $x > -100$

61. $-4(x + 2) \geq 12$ or $3x + 8 < 11$

62. $4.5x - 1 < -10$ or $6 - 2x \geq 12$

63. $4.5x - 2 > 2.5$ or $\frac{1}{2}x \leq 1$

64. $0 < x$ or $3x - 5 > 4x - 7$

APPLICATIONS

65. BABY FURNITURE See the illustration. A company manufactures various sizes of playpens having perimeters between 128 and 192 inches, inclusive.

a. Complete the double inequality that describes the range of the perimeters of the playpens.

$? \leq 4s \leq ?$

b. Solve the double inequality to find the range of the side lengths of the playpens.

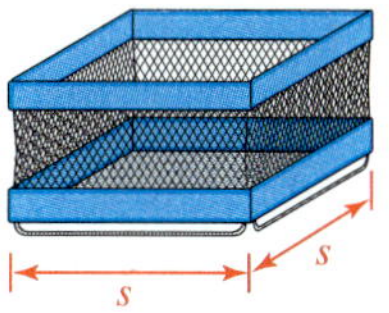
s
s

66. TRUCKING The distance that a truck can travel in 8 hours, at a constant rate of r mph, is given by $8r$. A trucker wants to travel at least 350 miles, and company regulations don't allow him to exceed 450 miles in one 8-hour shift.

a. Complete the double inequality that describes the mileage range of the truck.

$? \leq 8r \leq ?$

b. Solve the double inequality to find the range of the average rate (speed) of the truck for the 8-hour trip.

67. TREATING A FEVER Use the flow chart in the next column to determine what action should be taken for a 13-month-old child who has had a 99.8° temperature for 3 days and is not suffering any other symptoms. T represents the child's temperature, A the child's age in months, and S the number of hours the child has experienced the symptoms.

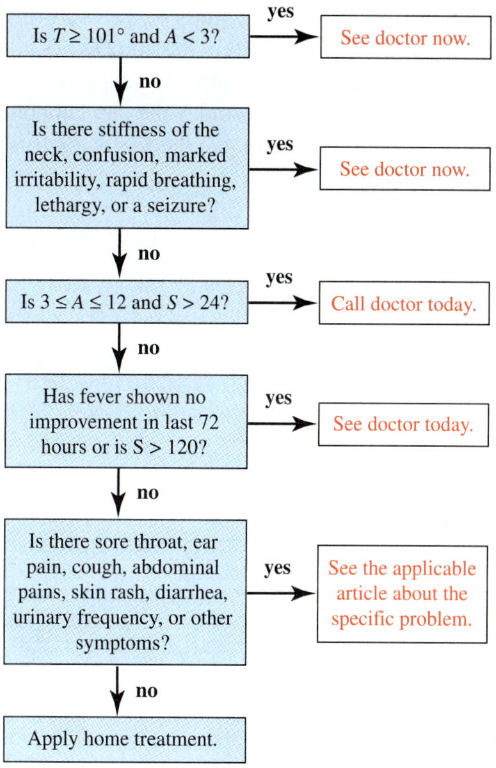
Is T ≥ 101° and A < 3?
yes
See doctor now.
no
Is there stiffness of the neck, confusion, marked irritability, rapid breathing, lethargy, or a seizure?
yes
See doctor now.
no
Is 3 ≤ A ≤ 12 and S > 24?
yes
Call doctor today.
no
Has fever shown no improvement in last 72 hours or is S > 120?
yes
See doctor today.
no
Is there sore throat, ear pain, cough, abdominal pains, skin rash, diarrhea, urinary frequency, or other symptoms?
yes
See the applicable article about the specific problem.
no
Apply home treatment.

Based on information from *Take Care of Yourself* (Addison-Wesley, 1993)

68. THERMOSTATS The *Temp range* control on the thermostat shown below directs the heater to come on when the room temperature gets 5 degrees below the *Temp setting;* it directs the air conditioner to come on when the room temperature gets 5 degrees above the *Temp setting.* Use interval notation to describe

a. the temperature range for the room when neither the heater nor the air conditioner will be on.

b. the temperature range for the room when either the heater or air conditioner will be on. (*Note:* The lowest temperature theoretically possible is −460° F, called *absolute zero.*)

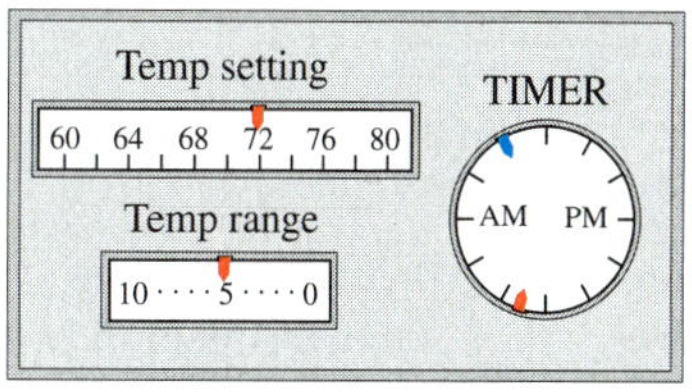
Temp setting
60 64 68 72 76 80
Temp range
10 · · · · 5 · · · · 0
TIMER
AM PM

69. U.S. HEALTH CARE Refer to the following illustration. Let P represent the percent of children covered by private insurance, M the percent covered by Medicare/Medicaid, and N the percent not covered. For what years are the following true?

a. $P \geq 68$ and $M \geq 18$

b. $P \geq 68$ or $M \geq 18$

c. $P \geq 67$ and $N \leq 12.5$

d. $P \geq 67$ or $N \leq 12.5$

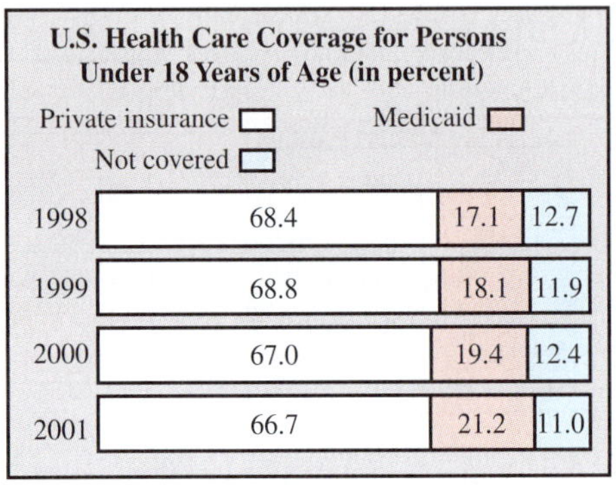

Source: U.S. Department of Health and Human Services

70. POLLS For each response to the poll question shown below, the *margin of error* is +/− (read as "plus or minus") 3.2%. This means that for the statistical methods used to do the polling, the actual response could be as much as 3.2 points more or 3.2 points less than shown. Use interval notation to describe the possible interval (in percent) for each response.

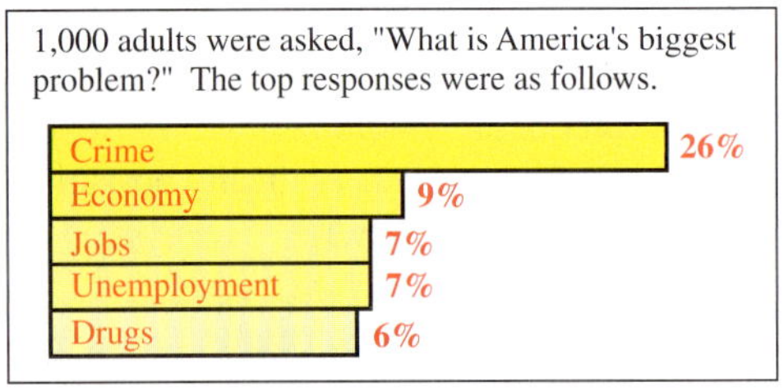

71. STREET INTERSECTIONS

a. Shade the area that represents the intersection of the two streets shown in the next column.

b. Shade the area that represents the union of the two streets.

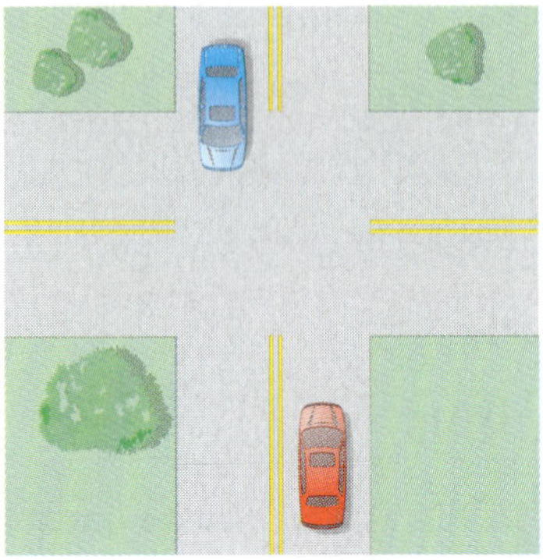

72. TRAFFIC SIGNS The pair of signs shown below are a real-life example of which concept discussed in this section?

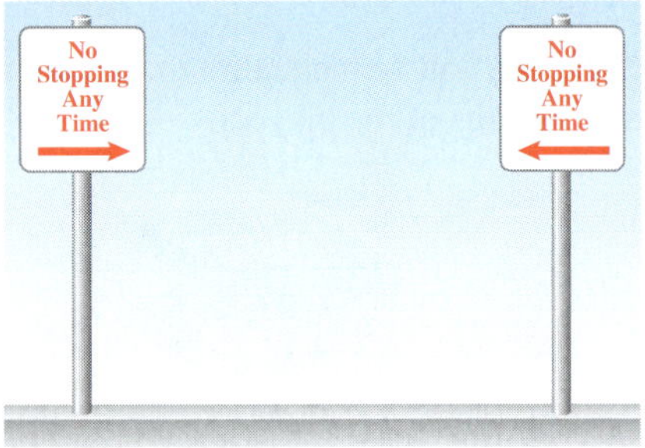

WRITING

73. Explain how to find the union and how to find the intersection of $(-\infty, 5)$ and $(-2, \infty)$ graphically.

74. Explain why the double inequality

$$2 < x < 8$$

can be written in the equivalent form

$$2 < x \quad \text{and} \quad x < 8$$

75. Explain the meaning of (2, 3) for each type of graph.

a.

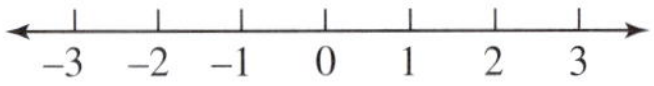

b.

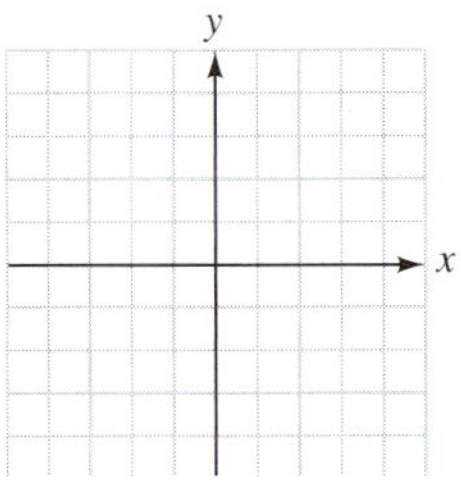

76. The meaning of the word *or* in a compound inequality differs from our everyday use of the word. Explain the difference.

REVIEW **Refer to the illustration, which shows the results of each of the games of the eventual champion, the University of Kentucky, in the 1998 NCAA Men's Basketball Tournament. Round to the nearest tenth when necessary.**

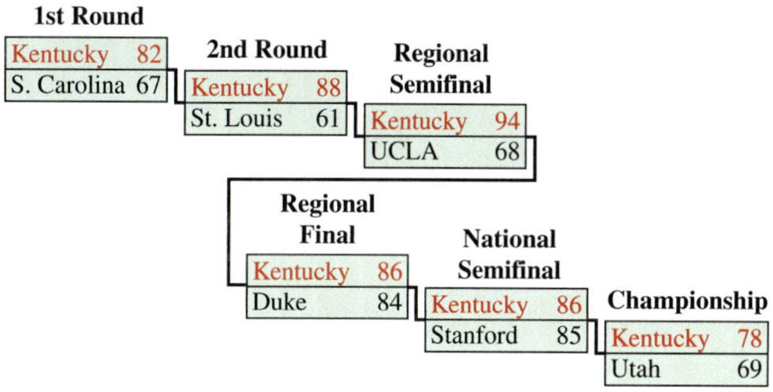

77. What are the mean, median, and mode of the set of Kentucky scores?

78. What are the mean and the median of the set of scores of Kentucky's opponents?

79. Find the margin of victory for Kentucky in each of its games. Then find the average (mean) margin of victory for Kentucky in the tournament.

80. What was the average (mean) combined score for Kentucky and its opponents in the tournament?

CHALLENGE PROBLEMS **Solve each compound inequality. Write the solution set in interval notation and graph it.**

81. $-5 < \frac{x+2}{-2} < 0$ or $2x + 10 \geq 30$

82. $-2 \leq \frac{x-4}{3} \leq 0$ and $\frac{x-5}{2} \geq -3$

4.3 Solving Absolute Value Equations and Inequalities

- Equations of the form $|X| = k$
- Equations with two absolute values
- Inequalities of the form $|X| < k$
- Inequalities of the form $|X| > k$

Many quantities studied in mathematics, science, and engineering are expressed as positive numbers. To guarantee that a quantity is positive, we often use absolute value. In this section, we will consider equations and inequalities involving the absolute value of an algebraic expression. Some examples are

$$|3x - 2| = 5, \qquad |2x - 3| < 9, \qquad \text{and} \qquad \left|\frac{3 - x}{5}\right| \geq 6$$

To solve these *absolute value equations* and *inequalities,* we write and then solve equivalent compound equations and inequalities.

EQUATIONS OF THE FORM $|X| = k$

Recall that the absolute value of a real number is its distance from 0 on a number line. To solve the **absolute value equation** $|x| = 5$, we must find all real numbers x whose distance from 0 on the number line is 5. There are two such numbers: 5 and -5. We say that the solutions of $|x| = 5$ are 5 and -5 and the solution set is $\{5, -5\}$.

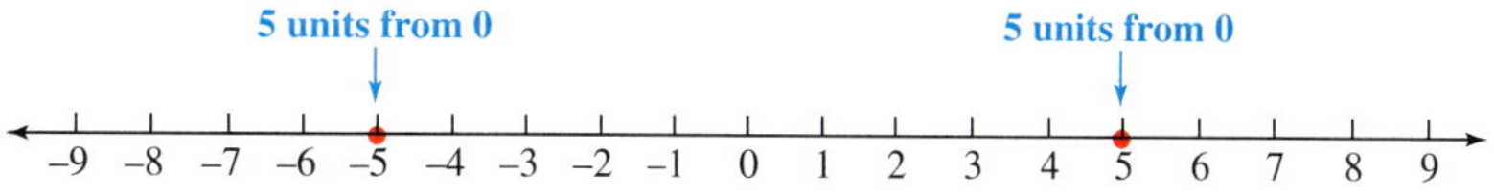

EXAMPLE 1

Solve: **a.** $|x| = 8$, **b.** $|s| = 0.003$, and **c.** $|c| = -15$.

Solutions **a.** To solve $|x| = 8$, we must find all real numbers x whose distance from 0 on the number line is 8. Therefore, the solutions are 8 and -8 and the solution set is $\{8, -8\}$.

b. To solve $|s| = 0.003$, we must find all real numbers s whose distance from 0 on the number line is 0.003. Therefore, the solutions are 0.003 and -0.003.

c. Recall that the absolute value of a number is either positive or zero, but never negative. Therefore, there is no value for c for which $|c| = -15$. The equation has no solution and the solution set is $\varnothing$.

Self Check 1 Solve: **a.** $|y| = 24$, **b.** $|x| = \frac{1}{2}$, and **c.** $|a| = -1.1$.

The results from Example 1 suggest the following approach for solving absolute value equations.

Solving Absolute Value Equations

For any positive number k and any algebraic expression X:

To solve $|X| = k$, solve the equivalent compound equation $X = k$ or $X = -k$.

EXAMPLE 2

Solve: **a.** $|3x - 2| = 5$ and **b.** $|10 - x| = -40$.

Solution **a.** To solve $|3x - 2| = 5$, we write and then solve an equivalent compound equation.

$$|3x - 2| = 5$$

means

$$3x - 2 = 5 \quad \text{or} \quad 3x - 2 = -5$$

Now we solve each equation for x:

$$\begin{array}{rcl|rcl} 3x - 2 &=& 5 & 3x - 2 &=& -5 \\ 3x &=& 7 & 3x &=& -3 \\ x &=& \frac{7}{3} & x &=& -1 \end{array}$$

The Language of Algebra

When two equations are joined with the word *or,* we call the statement a *compound equation.*

The results must be checked separately to see whether each of them produces a true statement. We substitute $\frac{7}{3}$ for x and then -1 for x in the original equation.

The Language of Algebra

When we say that the absolute value equation and a compound equation are *equivalent,* we mean that they have the same solution(s).

Check: ***For*** $x = \frac{7}{3}$ ***For*** $x = -1$

$$\begin{array}{rcl} |3x - 2| &=& 5 \\ \left|3\left(\frac{7}{3}\right) - 2\right| &\stackrel{?}{=}& 5 \\ |7 - 2| &\stackrel{?}{=}& 5 \\ |5| &\stackrel{?}{=}& 5 \\ 5 &=& 5 \end{array} \qquad \begin{array}{rcl} |3x - 2| &=& 5 \\ |3(-1) - 2| &\stackrel{?}{=}& 5 \\ |-3 - 2| &\stackrel{?}{=}& 5 \\ |-5| &\stackrel{?}{=}& 5 \\ 5 &=& 5 \end{array}$$

The resulting true statements indicate that the equation has two solutions: $\frac{7}{3}$ and -1.

b. Since an absolute value can never be negative, there are no real numbers x that make $|10 - x| = -40$ true. The equation has no solution. The solution set is $\varnothing$.

Self Check 2 Solve: **a.** $|2x - 3| = 7$ and **b.** $\left|\frac{x}{4} - 1\right| = -3$.

Caution When solving absolute value equations (or inequalities), isolate the absolute value expression on one side *before* writing the equivalent compound statement.

EXAMPLE 3

Solve: $\left|\frac{2}{3}x + 3\right| + 4 = 10$.

Solution We can isolate $\left|\frac{2}{3}x + 3\right|$ on the left-hand side by subtracting 4 from both sides.

$$\left|\frac{2}{3}x + 3\right| + 4 = 10$$

$$\left|\frac{2}{3}x + 3\right| = 6 \quad \text{Subtract 4 from both sides. The equation is in the form } |X| = k.$$

With the absolute value now isolated, we can solve $\left|\frac{2}{3}x + 3\right| = 6$ by writing and solving an equivalent compound equation.

$$\left|\frac{2}{3}x + 3\right| = 6$$

means

$$\frac{2}{3}x + 3 = 6 \quad \text{or} \quad \frac{2}{3}x + 3 = -6$$

Now we solve each equation for x:

$$\frac{2}{3}x + 3 = 6 \quad \text{or} \quad \frac{2}{3}x + 3 = -6$$

$$\frac{2}{3}x = 3 \quad \Big| \quad \frac{2}{3}x = -9$$

$$2x = 9 \quad \Big| \quad 2x = -27$$

$$x = \frac{9}{2} \quad \Big| \quad x = -\frac{27}{2}$$

Verify that both solutions check.

Caution

A common error when solving absolute value equations is to forget to isolate the absolute value expression first. Note:

$$\left|\frac{2}{3}x + 3\right| + 4 = 10$$

does not mean

$$\frac{2}{3}x + 3 + 4 = 10$$

or

$$\frac{2}{3}x + 3 + 4 = -10$$

Self Check 3 Solve: $|0.4x - 2| - 0.6 = 0.4$.

EXAMPLE 4

Solve: $3\left|\frac{1}{2}x - 5\right| - 4 = -4$.

Solution We first isolate $\left|\frac{1}{2}x - 5\right|$ on the left-hand side.

$$3\left|\frac{1}{2}x - 5\right| - 4 = -4$$

$$3\left|\frac{1}{2}x - 5\right| = 0 \quad \text{Add 4 to both sides.}$$

$$\left|\frac{1}{2}x - 5\right| = 0 \quad \text{Divide both sides by 3. The equation is in the form } |X| = k.$$

Since 0 is the only number whose absolute value is 0, the expression $\frac{1}{2}x - 5$ must be 0, and we have

$$\frac{1}{2}x - 5 = 0$$

$$\frac{1}{2}x = 5 \quad \text{Add 5 to both sides.}$$

$$x = 10 \quad \text{Multiply both sides by 2.}$$

Verify that 10 satisfies the original equation.

Success Tip

To solve most absolute value equations, we must consider two cases. However, if an absolute value is equal to 0, we need only consider one: the case when the expression within the absolute value bars is equal to 0.

Self Check 4 Solve: $-5\left|\frac{2x}{3} + 4\right| + 1 = 1$.

In Section 2.6 we discussed absolute value functions and their graphs. If we are given an output of an absolute value function, we can work in reverse to find the corresponding input(s).

EXAMPLE 5

Let: $f(x) = |x + 4|$. For what value(s) of x is $f(x) = 20$?

Solution To find the value(s) where $f(x) = 20$, we substitute 20 for $f(x)$ and solve for x.

$$f(x) = |x + 4|$$

$$20 = |x + 4| \quad \text{Substitute 20 for } f(x).$$

To solve $20 = |x + 4|$, we write and then solve an equivalent compound equation.

$$20 = |x + 4|$$

means

$$20 = x + 4 \quad \text{or} \quad -20 = x + 4$$

Now we solve each equation for x:

$$20 = x + 4 \quad \text{or} \quad -20 = x + 4$$

$$16 = x \quad \Big| \quad -24 = x$$

To check, substitute 16 and then -24 for x, and verify that $f(x) = 20$ in each case.

Self Check 5 For what value(s) of x is $f(x) = 11$?

EQUATIONS WITH TWO ABSOLUTE VALUES

Equations can contain two absolute value expressions. To develop a strategy to solve them, consider the following example.

$|3| = |3|$ (The same number.) or $|-3| = |-3|$ (The same number.) or $|3| = |-3|$ (These numbers are opposites.) or $|-3| = |3|$ (These numbers are opposites.)

Look closely to see that these four possible cases are really just two cases: *For two expressions to have the same absolute value, they must either be equal or be opposites of each other.* This observation suggests the following approach for solving equations having two absolute value expressions.

Solving Equations with Two Absolute Values

For any algebraic expressions X and Y:

To solve $|X| = |Y|$, solve the equivalent compound equation $X = Y$ or $X = -Y$.

EXAMPLE 6

Solve: $|5x + 3| = |3x + 25|$.

Solution To solve $|5x + 3| = |3x + 25|$, we write and then solve an equivalent compound equation.

$$|5x + 3| = |3x + 25|$$

means

The expressions within the absolute value symbols are equal:

$$5x + 3 = 3x + 25$$
$$2x = 22$$
$$x = 11$$

or

The expressions within the absolute value symbols are opposites:

$$5x + 3 = -(3x + 25)$$
$$5x + 3 = -3x - 25$$
$$8x = -28$$
$$x = -\frac{28}{8}$$
$$x = -\frac{7}{2}$$

Verify that both solutions check.

Self Check 6 Solve: $|2x - 3| = |4x + 9|$.

INEQUALITIES OF THE FORM $|X| < k$

To solve the **absolute value inequality** $|x| < 5$, we must find all real numbers x whose distance from 0 on the number line is less than 5. From the graph, we see that there are many such numbers. For example, -4.999, -3, -2.4, $-1\frac{7}{8}$, $-\frac{3}{4}$, 0, 1, 2.8, 3.001, and 4.999 all meet this requirement. We conclude that the solution set is all numbers between -5 and 5, which can be written $(-5, 5)$.

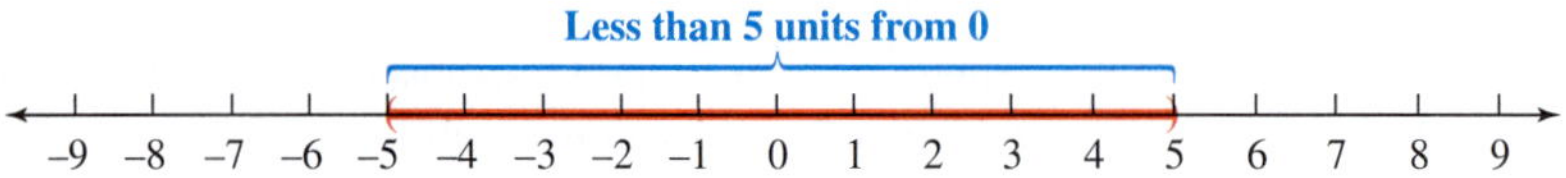

Since x is between -5 and 5, it follows that $|x| < 5$ is equivalent to $-5 < x < 5$. The observation suggests the following approach for solving absolute value inequalities of the form $|X| < k$ and $|X| \le k$.

Solving $|X| < k$ and $|X| \le k$

For any positive number k and any algebraic expression X:

To solve $|X| < k$, solve the equivalent double inequality $-k < X < k$.

To solve $|X| \le k$, solve the equivalent double inequality $-k \le X \le k$.

EXAMPLE 7

Solve $|2x - 3| < 9$ and graph the solution set.

Solution To solve $|2x - 3| < 9$, we write and then solve an equivalent double inequality.

$$|2x - 3| < 9 \quad \text{means} \quad -9 < 2x - 3 < 9$$

Now we solve for x:

$$-9 < 2x - 3 < 9$$
$$-6 < 2x < 12 \quad \text{Add 3 to all three parts.}$$
$$-3 < x < 6 \quad \text{Divide all parts by 2.}$$

Any number between -3 and 6 is in the solution set. This is the interval $(-3, 6)$; its graph is shown on the right.

Self Check 7 Solve $|3x + 2| < 4$ and graph the solution set.

EXAMPLE 8

Tolerances. When manufactured parts are inspected by a quality control engineer, they are classified as acceptable if each dimension falls within a given *tolerance range* of the dimensions listed on the blueprint. For the bracket shown in the figure, the distance between the two drilled holes is given as 2.900 inches. Because the tolerance is ± 0.015 inch, this distance can be as much as 0.015 inch longer or 0.015 inch shorter, and the part will be considered acceptable. The acceptable distance d between holes can be represented by the absolute value inequality $|d - 2.900| \le 0.015$. Solve the inequality and explain the result.

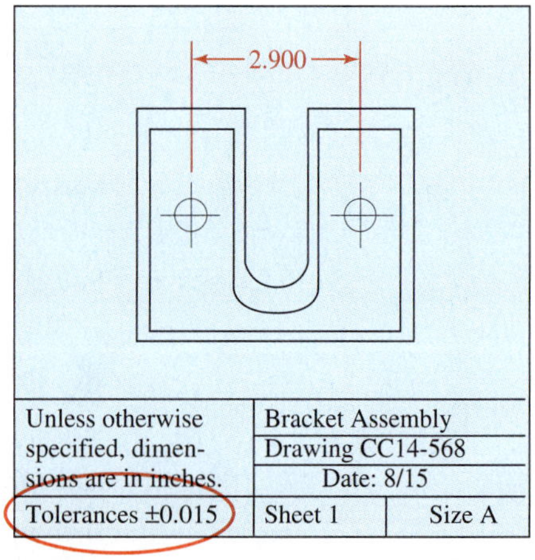

Solution To solve the absolute value inequality, we write and then solve an equivalent double inequality.

$$|d - 2.900| \le 0.015 \quad \text{means} \quad -0.015 \le d - 2.900 \le 0.015$$

Now we solve for d:

$$-0.015 \le d - 2.900 \le 0.015$$
$$2.885 \le d \le 2.915 \qquad \text{Add 2.900 to all three parts.}$$

The solution set is the interval [2.885, 2.915]. This means that the distance between the two holes should be between 2.885 and 2.915 inches, inclusive. If the distance is less than 2.885 inches or more than 2.915 inches, the part should be rejected.

EXAMPLE 9

Solve: $|4x - 5| < -2$.

Solution Since $|4x - 5|$ is always greater than or equal to 0 for any real number x, this absolute value inequality has no solution. The solution set is $\varnothing$.

Self Check 9 Solve: $|6x + 24| < -51$.

INEQUALITIES OF THE FORM $|X| > k$

To solve the absolute value inequality $|x| > 5$, we must find all real numbers x whose distance from 0 on the number line is greater than 5. From the following graph, we see that there are many such numbers. For example, -5.001, -6, -7.5, $-8\frac{3}{8}$, 5.001, 6.2, 7, 8, and $9\frac{1}{2}$ all meet this requirement. We conclude that the solution set is all numbers less than -5 or greater than 5, which can be written $(-\infty, -5) \cup (5, \infty)$.

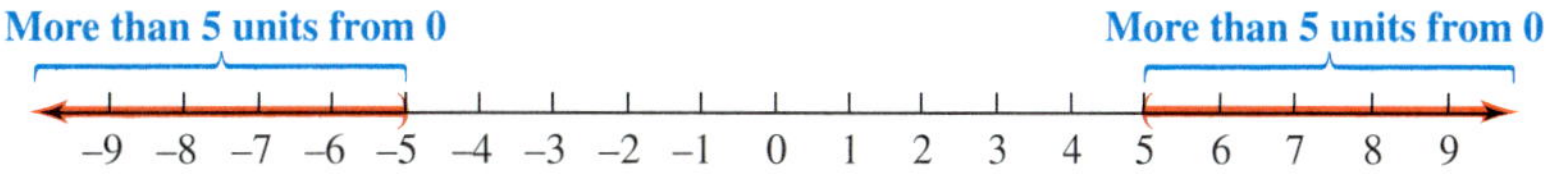

Since x is less than -5 or greater than 5, it follows that $|x| > 5$ is equivalent to $x < -5$ or $x > 5$. The observation suggests the following approach for solving absolute value inequalities of the form $|X| > k$ and $|X| \ge k$.

Solving $|X| > k$ and $|X| \ge k$

For any positive number k and any algebraic expression X:

To solve $|X| > k$, solve the equivalent compound inequality $X < -k$ or $X > k$.

To solve $|X| \ge k$, solve the equivalent compound inequality $X \le -k$ or $X \ge k$.

EXAMPLE 10

Solve $\left|\frac{3 - x}{5}\right| \ge 6$ and graph the solution set.

Solution To solve $\left|\frac{3 - x}{5}\right| \ge 6$, we write and then solve an equivalent compound inequality.

$$\left|\frac{3-x}{5}\right| \geq 6$$

means

$$\frac{3-x}{5} \leq -6 \quad \text{or} \quad \frac{3-x}{5} \geq 6$$

Then we solve each inequality for x:

$$\frac{3-x}{5} \leq -6 \quad \text{or} \quad \frac{3-x}{5} \geq 6$$

$3 - x \leq -30$	$3 - x \geq 30$	Multiply both sides by 5.
$-x \leq -33$	$-x \geq 27$	Subtract 3 from both sides.
$x \geq 33$	$x \leq -27$	Divide both sides by -1 and reverse the direction of the inequality symbol.

The solution set is the interval $(-\infty, -27] \cup [33, \infty)$. Its graph appears on the left.

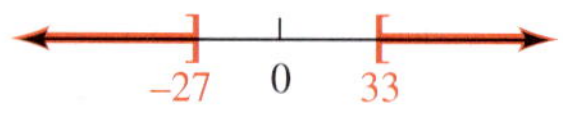

Self Check 10 Solve $\left|\frac{2-x}{4}\right| \geq 1$ and graph the solution set.

EXAMPLE 11

Solve $\left|\frac{2}{3}x - 2\right| - 3 > 6$ and graph the solution set.

Solution We add 3 to both sides to isolate the absolute value on the left-hand side.

$$\left|\frac{2}{3}x - 2\right| - 3 > 6$$

$$\left|\frac{2}{3}x - 2\right| > 9 \qquad \text{Add 3 to both sides to isolate the absolute value.}$$

To solve this absolute value inequality, we write and then solve an equivalent compound inequality.

$$\frac{2}{3}x - 2 < -9 \quad \text{or} \quad \frac{2}{3}x - 2 > 9$$

$\frac{2}{3}x < -7$	$\frac{2}{3}x > 11$	Add 2 to both sides.
$2x < -21$	$2x > 33$	Multiply both sides by 3.
$x < -\frac{21}{2}$	$x > \frac{33}{2}$	Divide both sides by 2.

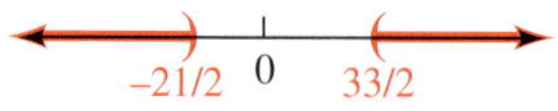

The solution set is $\left(-\infty, -\frac{21}{2}\right) \cup \left(\frac{33}{2}, \infty\right)$. Its graph appears on the left.

Self Check 11 Solve $\left|\frac{3}{4}x + 2\right| - 1 > 3$ and graph the solution set.

EXAMPLE 12

Solve: $\left|\frac{x}{8} - 1\right| \geq -4$.

Solution Since $\left|\frac{x}{8} - 1\right|$ is always greater than or equal to 0 for any real number x, this absolute value inequality is true for all real numbers. The solution set is $(-\infty, \infty)$ or $\mathbb{R}$.

Self Check 12 Solve: $|-x - 9| > -0.5$.

The following summary shows how we can interpret absolute value in three ways. Assume $k > 0$.

Geometric description	Graphic description	Algebraic description
1. $\lvert x \rvert = k$ means that x is k units from 0 on the number line.	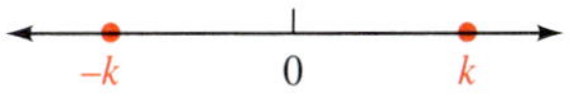	$\lvert x \rvert = k$ is equivalent to $x = k$ or $x = -k$.
2. $\lvert x \rvert < k$ means that x is less than k units from 0 on the number line.	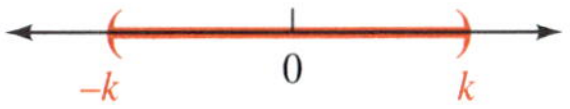	$\lvert x \rvert < k$ is equivalent to $-k < x < k$.
3. $\lvert x \rvert > k$ means that x is more than k units from 0 on the number line.	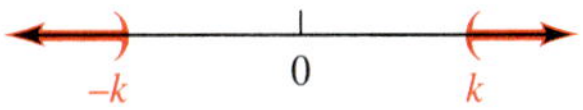	$\lvert x \rvert > k$ is equivalent to $x > k$ or $x < -k$.

ACCENT ON TECHNOLOGY: SOLVING ABSOLUTE VALUE EQUATIONS AND INEQUALITIES

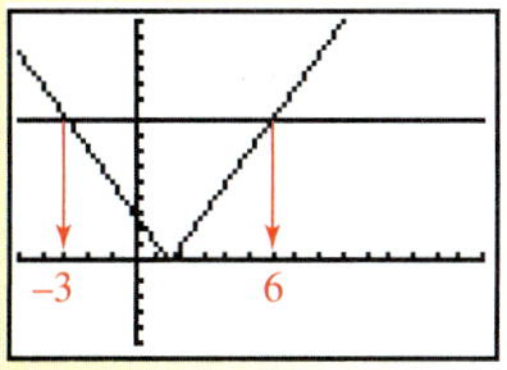

We can solve absolute value equations and inequalities with a graphing calculator. For example, to solve $|2x - 3| = 9$, we graph the equations $y = |2x - 3|$ and $y = 9$ on the same coordinate system, as shown in the figure. The equation $|2x - 3| = 9$ will be true for all x-coordinates of points that lie on *both* graphs. Using the TRACE or INTERSECT feature, we can see that the graphs intersect at the points $(-3, 9)$ and $(6, 9)$. Thus, the solutions of the absolute value equation are -3 and 6.

The inequality $|2x - 3| < 9$ will be true for all x-coordinates of points that lie on the graph of $y = |2x - 3|$ and *below* the graph of $y = 9$. We see that these values of x are between -3 and 6. Thus, the solution set is the interval $(-3, 6)$.

The inequality $|2x - 3| > 9$ will be true for all x-coordinates of points that lie on the graph of $y = |2x - 3|$ and *above* the graph of $y = 9$. We see that these values of x are less than -3 or greater than 6. Thus, the solution set is the interval $(-\infty, -3) \cup (6, \infty)$.

Answers to Self Checks **1. a.** 24, -24, **b.** $\frac{1}{2}, -\frac{1}{2}$, **c.** no solution **2. a.** 5, -2, **b.** no solution **3.** 7.5, 2.5

4. -6 **5.** 7, -15 **6.** $-1, -6$ **7.** $\left(-2, \frac{2}{3}\right)$ 

9. no solution

10. $(-\infty, -2] \cup [6, \infty)$ 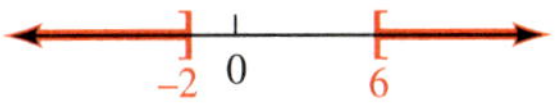

11. $(-\infty, -8) \cup \left(\frac{8}{3}, \infty\right)$ 

12. $(-\infty, \infty)$

4.3 STUDY SET

VOCABULARY Fill in the blanks.

1. The __________ of a number is its distance from 0 on a number line.
2. $|2x - 1| = 10$ is an absolute value ________.
3. $|2x - 1| > 10$ is an absolute value ________.
4. To ________ the absolute value in $|3 - x| - 4 = 5$, we add 4 to both sides.
5. $-(2x + 9)$ is the ________ of $2x + 9$.
6. When we say that the absolute value equation and a compound equation are equivalent, we mean that they have the same ________.
7. When two equations are joined by the word *or*, such as $x + 1 = 5$ or $x + 1 = -5$, we call the statement a ________ equation.
8. $f(x) = |6x - 2|$ is called an absolute value ________.

CONCEPTS Fill in the blanks.

9. The absolute value of a real number is greater than or equal to 0, but never ________.
10. For two expressions to have the same absolute value, they must either be equal or ________ of each other.
11. To solve $|x| > 5$, we must find the coordinates of all points on a number line that are ________ 5 units from the origin.
12. To solve $|x| < 5$, we must find the coordinates of all points on a number line that are ________ 5 units from the origin.
13. To solve $|x| = 5$, we must find the coordinates of all points on a number line that are ___ units from the origin.
14. To solve these absolute value equations and inequalities, we write and then solve equivalent ________ equations and inequalities.
15. Consider the following real numbers: $-4, -3, -2.01, -2, -1.99, -1, 0, 1, 1.99, 2, 2.01, 3, 4$
    a. Which of them make $|x| = 2$ true?
    b. Which of them make $|x| < 2$ true?
    c. Which of them make $|x| > 2$ true?
16. Decide whether -3 is a solution of the given equation or inequality.
    a. $|x - 1| = 4$ b. $|x - 1| > 4$
    c. $|x - 1| \leq 4$ d. $|5 - x| = |x + 12|$

For each absolute value equation, write an equivalent compound equation.

17. a. $|x - 7| = 8$ means
    ____ = ____ or ____ = ____
    b. $|x + 10| = |x - 3|$ means
    ____ = ____ or ____ = ____

For each absolute value inequality, write an equivalent compound inequality.

18. a. $|x + 5| < 1$ means
    ____ < ____ < ____
    b. $|x - 6| \geq 3$ means
    ____ ≥ ____ or ____ ≤ ____
19. For each absolute value equation or inequality, write an equivalent compound equation or inequality.
    a. $|x| = 8$ b. $|x| \geq 8$
    c. $|x| \leq 8$ d. $|5x - 1| = |x + 3|$
20. Perform the necessary steps to isolate the absolute value expression on one side of the equation. **Do not solve.**
    a. $|3x + 2| - 7 = -5$
    b. $6 + |5x - 19| \leq 40$

NOTATION

21. Match each equation or inequality with its graph.
    a. $|x| = 1$ i. 

    b. $|x| > 1$ ii. 

    c. $|x| < 1$ iii. 

22. Match each graph with its corresponding equation or inequality.
    a. 

    i. $|x| \geq 2$
    b. 

    ii. $|x| \leq 2$
    c. 

    iii. $|x| = 2$

23. Describe the set graphed below using interval notation.

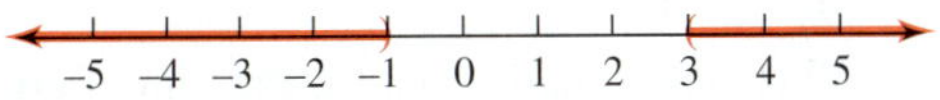

24. a. If an absolute value inequality has no solution, what symbol is used to represent the solution set?

b. If the solution set of an absolute value inequality is all real numbers, what notation is used to represent the solution set?

PRACTICE **Solve each equation, if possible.**

25. $|x| = 23$

26. $|x| = 90$

27. $|5x| = 20$

28. $|6x| = 12$

29. $|x - 3.1| = 6$

30. $|x + 4.3| = 8.9$

31. $|3x + 2| = 16$

32. $|5x - 3| = 22$

33. $|x| - 3 = 9$

34. $|x| + 6 = 11$

35. $\left|\frac{7}{2}x + 3\right| = -5$

36. $5|x - 21| = -8$

37. $|3 - 4x| + 1 = 6$

38. $|8 - 5x| - 8 = 10$

39. $2|3x + 24| = 0$

40. $\left|\frac{2x}{3} + 10\right| = 0$

41. $\left|\frac{3x + 48}{3}\right| = 12$

42. $\left|\frac{4x - 64}{4}\right| = 32$

43. $-7 = 2 - |0.3x - 3|$

44. $-1 = 1 - |0.1x + 8|$

45. $\frac{6}{5} = \left|\frac{3x}{5} + \frac{x}{2}\right|$

46. $\frac{11}{12} = \left|\frac{x}{3} - \frac{3x}{4}\right|$

47. $|2x + 1| = |3(x + 1)|$

48. $|5x - 7| = |4(x + 1)|$

49. $|2 - x| = |3x + 2|$

50. $|4x + 3| = |9 - 2x|$

51. $\left|\frac{x}{2} + 2\right| = \left|\frac{x}{2} - 2\right|$

52. $|7x + 12| = |x - 6|$

53. $\left|x + \frac{1}{3}\right| = |x - 3|$

54. $\left|x - \frac{1}{4}\right| = |x + 4|$

55. Let $f(x) = |x + 3|$. For what value(s) of x is $f(x) = 3$?

56. Let $g(x) = |2 - x|$. For what value(s) of x is $g(x) = 2$?

57. Let $f(x) = \frac{1}{2}|3x| - 1$. For what value(s) of x is $f(x) = \frac{1}{4}$?

58. Let $h(x) = 5\left|-\frac{x}{3}\right| - 2$. For what value(s) of x is $h(x) = -\frac{1}{3}$?

Solve each inequality. Write the solution set in interval notation (if possible) and graph it.

59. $|x| < 4$

60. $|x| < 9$

61. $|x + 9| \le 12$

62. $|x - 8| \le 12$

63. $|3x - 2| < 10$

64. $|4 - 3x| \le 13$

65. $|3x + 2| \le -3$

66. $|5x - 12| < -5$

67. $|x| > 3$

68. $|x| > 7$

69. $|x - 12| > 24$

70. $|x + 5| \ge 7$

71. $|3x + 2| > 14$

72. $|2x - 5| > 25$

73. $|4x + 3| > -5$

74. $|7x + 2| > -8$

75. $|2 - 3x| \ge 8$

76. $|-1 - 2x| > 5$

77. $-|2x - 3| < -7$

78. $-|3x + 1| < -8$

79. $\left|\frac{x - 2}{3}\right| \le 4$

80. $\left|\frac{x - 2}{3}\right| > 4$

81. $|3x + 1| + 2 < 6$

82. $1 + \left|\frac{1}{7}x + 1\right| \le 1$

83. $\left|\frac{1}{3}x + 7\right| + 5 > 6$

84. $-2|3x - 4| < 16$

85. $|0.5x + 1| + 2 \le 0$

86. $15 \ge 7 - |1.4x + 9|$

87. Let $f(x) = |2(x - 1) + 4|$. For what value(s) of x is $f(x) < 4$?

88. Let $g(x) = |4(3x + 2) - 2|$. For what value(s) of x is $g(x) \le 6$?

89. Let $f(x) = \left|\frac{x}{4} + \frac{1}{3}\right|$. For what value(s) of x is $f(x) \ge \frac{1}{12}$?

90. Let $h(x) = \left|\frac{x}{5} - \frac{1}{2}\right|$. For what value(s) of x is $h(x) > \frac{9}{10}$?

APPLICATIONS

91. TEMPERATURE RANGES The temperatures on a sunny summer day satisfied the inequality $|t - 78°| \le 8°$, where t is a temperature in degrees Fahrenheit. Solve this inequality and express the range of temperatures as a double inequality.

92. OPERATING TEMPERATURES A car CD player has an operating temperature of $|t - 40°| < 80°$, where t is a temperature in degrees Fahrenheit. Solve the inequality and express this range of temperatures as an interval.

93. AUTO MECHANICS On most cars, the bottoms of the front wheels are closer together than the tops, creating a *camber angle.* This lessens road shock to the steering system. (See the illustration.) The specifications for a certain car state that the camber angle c of its wheels should be $0.6° \pm 0.5°$.

a. Express the range with an inequality containing absolute value symbols.

b. Solve the inequality and express this range of camber angles as an interval.

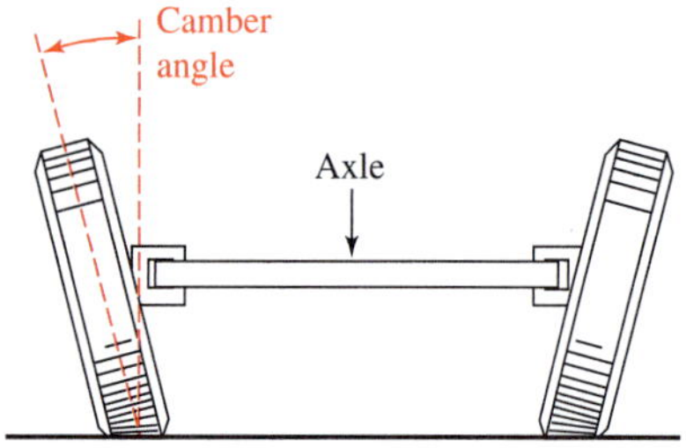

94. STEEL PRODUCTION A sheet of steel is to be 0.250 inch thick with a tolerance of 0.025 inch.

a. Express this specification with an inequality containing absolute value symbols, using x to represent the thickness of a sheet of steel.

b. Solve the inequality and express the range of thickness as an interval.

95. ERROR ANALYSIS In a lab, students measured the percent of copper p in a sample of copper sulfate. The students know that copper sulfate is actually 25.46% copper by mass. They are to compare their results to the actual value and find the amount of *experimental error.*

Lab 4 Section A
Title:
"Percent copper (Cu) in copper sulfate ($CuSO_4 \cdot 5H_2O$)"

Results

	% Copper
Trial #1:	22.91%
Trial #2:	26.45%
Trial #3:	26.49%
Trial #4:	24.76%

a. Which measurements shown in the illustration satisfy the absolute value inequality $|p - 25.46| \le 1.00$?

b. What can be said about the amount of error for each of the trials listed in part a?

96. ERROR ANALYSIS See Exercise 95.

a. Which measurements satisfy the absolute value inequality $|p - 25.46| > 1.00$?

b. What can be said about the amount of error for each of the trials listed in part a?

WRITING

97. Explain the error.

Solve: $|x| + 2 = 6$

$x + 2 = 6$ or $x + 2 = -6$

$x = 4$ | $x = -8$

98. Explain why the equation $|x - 4| = -5$ has no solutions.

99. Explain the differences between the solution sets of $|x| < 8$ and $|x| > 8$.

100. Explain how to use the graph in the illustration to solve the following.

a. $|x - 2| = 3$

b. $|x - 2| \le 3$

c. $|x - 2| \ge 3$

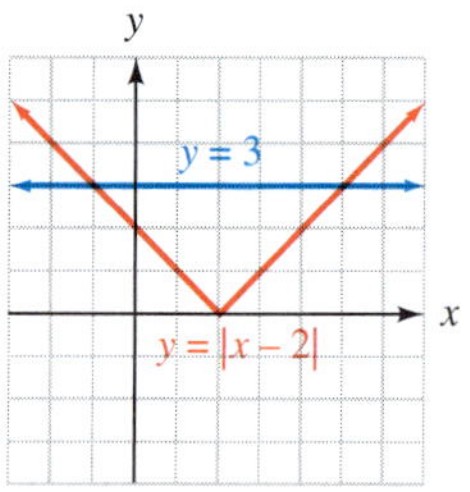

REVIEW

101. RAILROAD CROSSINGS The warning sign in the illustration is to be painted on the street in front of a railroad crossing. If y is 30° more than twice x, find x and y.

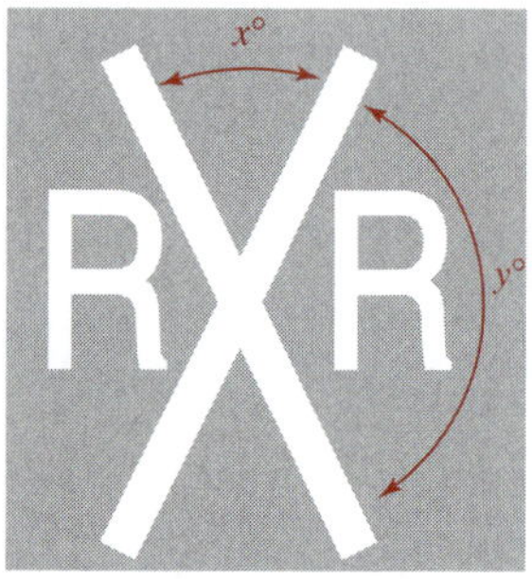

102. GEOMETRY Refer to the illustration. What is $2x + 2y$?

CHALLENGE PROBLEMS

103. For what values of k does $|x| + k = 0$ have exactly two solutions?

104. For what values of k does $|x| + k = 0$ have exactly one solution?

105. Under what conditions is $|x| + |y| > |x + y|$?

106. Under what conditions is $|x| + |y| = |x + y|$? (Assume that x and y are nonzero.)

4.4 Linear Inequalities in Two Variables

- Graphing linear inequalities
- Horizontal and vertical boundary lines
- Problem solving

In the first three sections of this chapter, we worked with linear inequalities in one variable. Some examples are

$$x \geq -7, \quad 5 < \frac{7}{2}a - 9, \quad \text{and} \quad 5(3 + z) > -3(z + 3)$$

These inequalities have infinitely many solutions. When their solutions are graphed on a real number line, we obtain an interval.

In this section, we will discuss **linear inequalities in two variables.** Some examples are

$$y > 3x + 2, \quad 2x - 3y \leq 6, \quad \text{and} \quad y < 2x$$

Linear Inequalities in Two Variables

A **linear inequality** in x and y is any inequality that can be written in the form

$$Ax + By < C \quad \text{or} \quad Ax + By > C \quad \text{or} \quad Ax + By \leq C \quad \text{or} \quad Ax + By \geq C$$

where A, B, and C are real numbers and A and B are not both 0.

The solutions of these inequalities are ordered pairs. We can graph their solutions on a rectangular coordinate system.

GRAPHING LINEAR INEQUALITIES

The **graph of a linear inequality** in x and y is the graph of all ordered pairs (x, y) that *satisfy* the inequality.

EXAMPLE 1

Graph: $y > 3x + 2$.

Solution

To graph the linear inequality $y > 3x + 2$, we begin by graphing the linear *equation* $y = 3x + 2$. The graph of $y = 3x + 2$, shown in figure (a), is a **boundary line** that separates the rectangular coordinate plane into two regions called **half-planes.** It is drawn with a dashed line to show that it is not part of the graph of $y > 3x + 2$.

To find which half-plane is the graph of $y > 3x + 2$, we can substitute the coordinates of any point in either half-plane. We will choose the origin as the test point because its coordinates, (0, 0), make the computations easy. We substitute 0 for x and 0 for y into the inequality and simplify.

Caution

When using a test point to determine which half-plane to shade, remember to substitute the coordinates into the given inequality, not the equation for the boundary.

Check the test point (0, 0):

$y > 3x + 2$	This is the original inequality.
$0 \overset{?}{>} 3(0) + 2$	Substitute 0 for y and 0 for x.
$0 > 2$	This statement is false.

Since the coordinates of the origin don't satisfy $y > 3x + 2$, the origin is not part of the graph of the inequality. The half-plane on the other side of the dashed line is the graph. We then shade that region, as shown in figure (b).

The Language of Algebra

The boundary line is also called an *edge* of the half-plane.

Success Tip

As an *informal* check, pick several points that lie in the shaded region. Substitute their coordinates into the inequality. In each case, you should obtain a true statement.

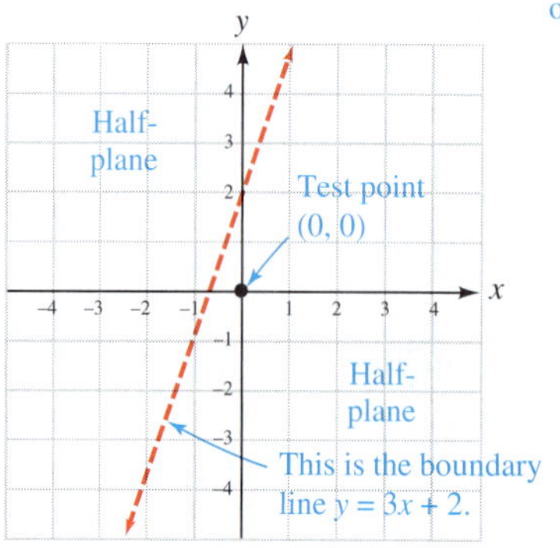

Half-plane
Test point (0, 0)
Half-plane
This is the boundary line y = 3x + 2.

(a)

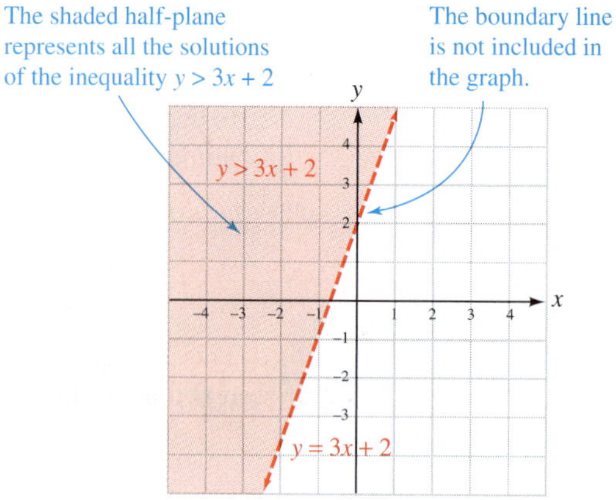

The shaded half-plane represents all the solutions of the inequality y > 3x + 2
The boundary line is not included in the graph.
y > 3x + 2
y = 3x + 2

(b)

Self Check 1 Graph: $y > 2x - 4$.

EXAMPLE 2

Graph: $2x - 3y \le 6$.

Solution This inequality is the combination of the inequality $2x - 3y < 6$ and the equation $2x - 3y = 6$.

We begin by graphing $2x - 3y = 6$ to find the boundary line that separates the two half-planes. This time, we draw the solid line shown in figure (a), because equality is permitted by the symbol $\le$. To decide which half-plane to shade, we check to see whether the coordinates of the origin satisfy the inequality.

Success Tip

Draw a solid boundary line if the inequality has $\le$ or $\ge$. Draw a dashed line if the inequality has $<$ or $>$.

Check the test point (0, 0):

$$2x - 3y \le 6$$

$$2(0) - 3(0) \overset{?}{\le} 6 \quad \text{Substitute 0 for } x \text{ and 0 for } y.$$

$$0 \le 6 \quad \text{This statement is true.}$$

The coordinates of the origin satisfy the inequality. In fact, the coordinates of every point on the same side of the boundary line as the origin satisfy the inequality. We then shade that half-plane to complete the graph of $2x - 3y \le 6$, shown in figure (b).

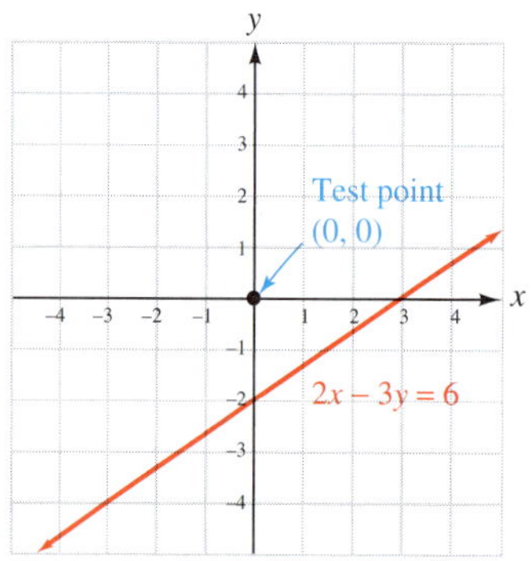

Test point (0, 0)
2x – 3y = 6

(a)

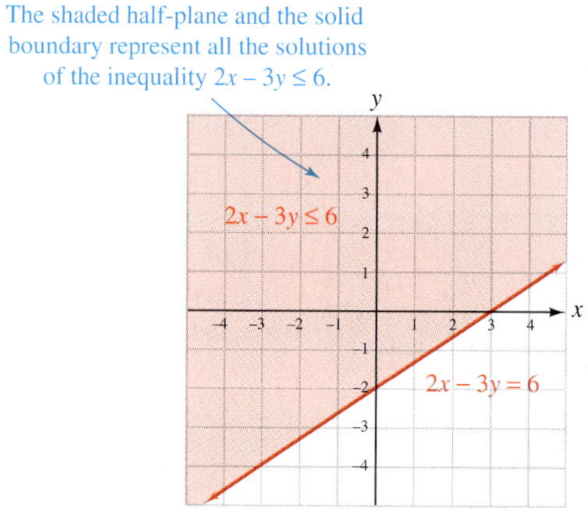

The shaded half-plane and the solid boundary represent all the solutions of the inequality 2x – 3y ≤ 6.
2x – 3y ≤ 6
2x – 3y = 6

(b)

Self Check 2 Graph: $3x - 2y \geq 6$.

EXAMPLE 3

Graph: $y < 2x$.

Solution To graph $y = 2x$, we use the fact that the equation is in slope–intercept form and that $m = 2 = \frac{2}{1}$ and $b = 0$. Since the symbol $<$ does not include an equal symbol, the points on the graph of $y = 2x$ are not on the graph of $y < 2x$. We draw the boundary line as a dashed line to show this, as in figure (a).

To decide which half-plane is the graph of $y < 2x$, we check to see whether the coordinates of some fixed point satisfy the inequality. We cannot use the origin as a test point, because the boundary line passes through the origin. However, we can choose a different point—say, (2, 0).

Success Tip

The origin (0, 0) is a smart choice for a test point because computations involving 0 are usually easy. If the origin is on the boundary, choose a test point not on the boundary that has one coordinate that is 0, such as (0, 1) or (2, 0).

Check the test point (2, 0):

$y < 2x$

$0 \stackrel{?}{<} 2(2)$ Substitute 2 for x and 0 for y.

$0 < 4$ This is a true statement.

Since $0 < 4$ is a true inequality, the point (2, 0) satisfies the inequality and is in the graph of $y < 2x$. We then shade the half-plane containing (2, 0), as shown in figure (b).

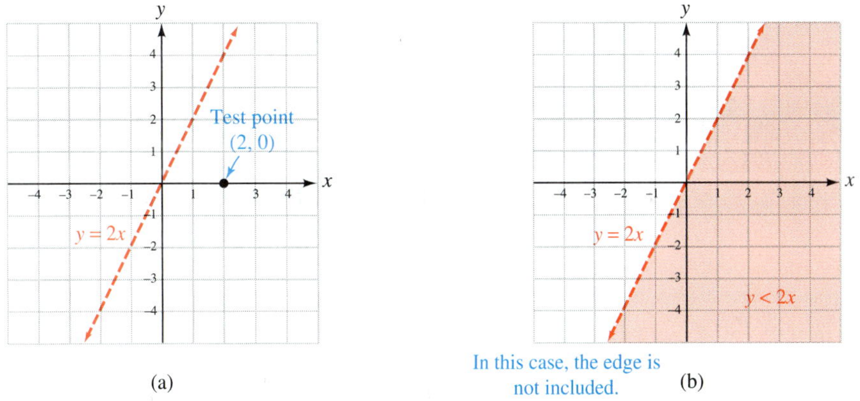

(a) (b)

Self Check 3 Graph: $y > -x$.

The following is a summary of the procedure for graphing linear inequalities.

Graphing Linear Inequalities in Two Variables

1. Graph the boundary line of the region. If the inequality allows equality (the symbol is either $\leq$ or $\geq$), draw the boundary line as a solid line. If equality is not allowed ($<$ or $>$), draw the boundary line as a dashed line.
2. Pick a test point that is on one side of the boundary line. (Use the origin if possible.) Replace x and y in the original inequality with the coordinates of that point. If the inequality is satisfied, shade the side that contains that point. If the inequality is not satisfied, shade the other side of the boundary.

HORIZONTAL AND VERTICAL BOUNDARY LINES

Recall that the graph of $x = a$ is a vertical line with x-intercept at $(a, 0)$, and the graph of $y = b$ is a horizontal line with y-intercept at $(0, b)$.

EXAMPLE 4

Graph: $x \geq -1$.

Solution The graph of the boundary $x = -1$ is a vertical line passing through $(-1, 0)$. We draw the boundary as a solid line to show that it is part of the solution. See figure (a).

In this case, we need not pick a test point. The inequality $x \geq -1$ is satisfied by points with an x-coordinate greater than or equal to -1. Points satisfying this condition lie to the right of the boundary. We shade that half-plane, as shown in figure (b), to complete the graph of $x \geq -1$.

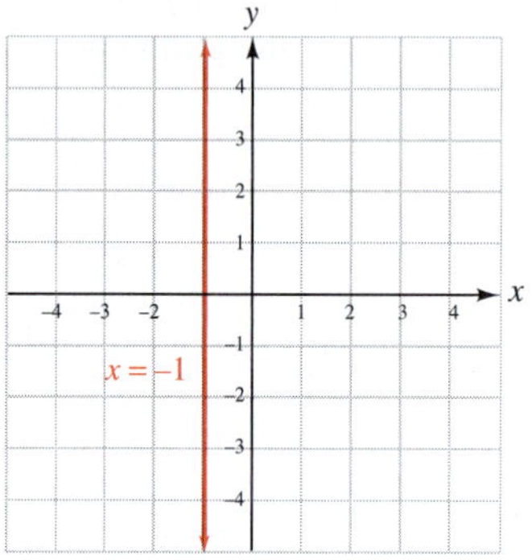

(a)

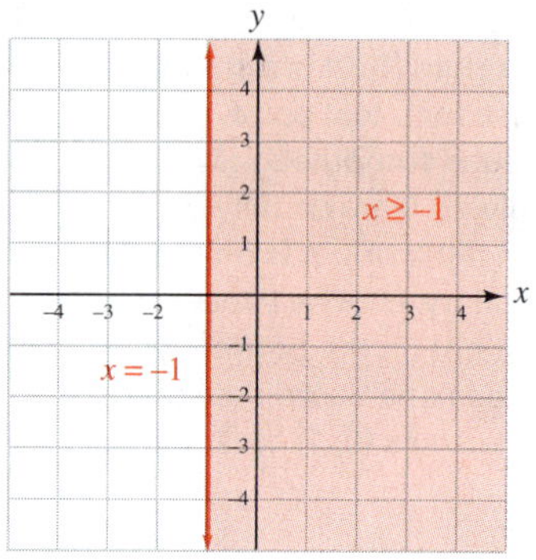

(b)

Self Check 4 Graph: $y < 4$.

PROBLEM SOLVING

In the next example, we solve a problem by writing a linear inequality in two variables to model a situation mathematically.

EXAMPLE 5

Social Security. Retirees, ages 62–65, can earn as much as \$11,640 and still receive their full Social Security benefits. If their annual earnings exceed \$11,640, their benefits are reduced. A 64-year-old retired woman receiving Social Security works two part-time jobs: one at the library, paying \$485 per week and another at a pet store, paying \$388 per week. Write an inequality representing the number of weeks the woman can work at each job during the year without losing any of her benefits.

Analyze the Problem We need to find the various combinations of weeks she can work at the library and at the pet store so that her annual income is less than or equal to \$11,640.

Form an Inequality If we let $x =$ the number of weeks she works at the library, she will earn \$485x annually. If we let $y =$ the number of weeks she works at the pet store, she will earn \$388y annually. Combining the income from these jobs, the total is not to exceed \$11,640.

The weekly rate on the library job	·	the weeks worked on the library job	plus	the weekly rate on the pet store job	·	the weeks worked on the pet store job	should not exceed	\$11,640
\$485	·	x	+	\$388	·	y	$\leq$	\$11,640

Solve the Inequality The graph of $485x + 388y \leq 11{,}640$ is shown in the figure. Since she cannot work a negative number of weeks, the graph has no meaning when x or y is negative, so only the first quadrant is used. Any point in the shaded region indicates a way that she can schedule her work weeks and earn \$11,640 or less annually. For example, if she works 8 weeks at the library and 16 weeks at the pet store, represented by the ordered pair (8, 16), she will earn

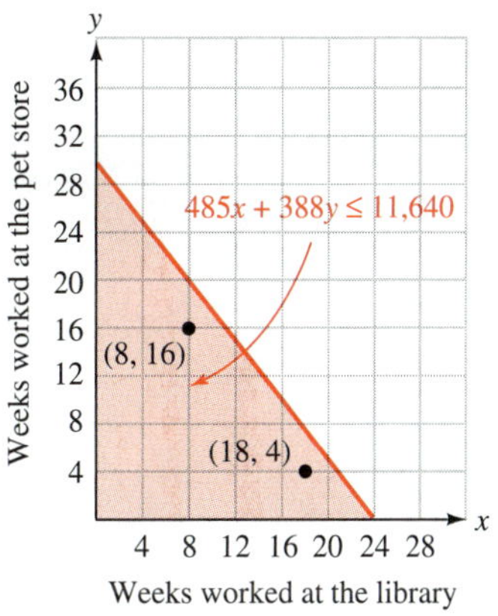

$$\begin{aligned} \$485(8) + \$388(16) &= \$3{,}880 + \$6{,}208 \\ &= \$10{,}088 \end{aligned}$$

If she works 18 weeks at the library and 4 weeks at the pet store, represented by (18, 4), she will earn

$$\begin{aligned} \$485(18) + \$388(4) &= \$8{,}730 + \$1{,}552 \\ &= \$10{,}282 \end{aligned}$$

ACCENT ON TECHNOLOGY: GRAPHING INEQUALITIES

Some graphing calculators (such as the TI-83 Plus) have a graph style icon in the $y =$ editor. Some of the different graph styles are

\	line	A straight line or curved graph is shown.	$\backslash Y_1 =$
◥	above	Shading covers the area above a graph.	◥$Y_1 =$
◣	below	Shading covers the area below a graph.	◣$Y_1 =$

We can change the icon in front of Y_1 by placing the cursor on it and pressing the ENTER key.

To graph $2x - 3y \leq 6$ of Example 2, we first write it in an equivalent form, with y isolated on the left-hand side.

$$\begin{aligned} 2x - 3y &\leq 6 \\ -3y &\leq -2x + 6 && \text{Subtract } 2x \text{ from both sides.} \\ y &\geq \frac{2}{3}x - 2 && \text{Divide both sides by } -3. \text{ Change the direction of the inequality symbol.} \end{aligned}$$

We then change the graph style icon to above (◥), because the inequality $y \geq \frac{2}{3}x - 2$ contains a $\geq$ symbol. Using window settings of $[-10, 10]$ for x and $[-10, 10]$ for y, we enter the boundary equation $y = \frac{2}{3}x - 2$. See figure (a). Finally, we press the GRAPH key to get figure (b).

To graph $y < 2x$ from Example 3, we change the graph style icon to below (◣), because the inequality contains a $<$ symbol. Using window settings of $[-10, 10]$ for x and $[-10, 10]$ for y, we enter the boundary equation $y = 2x$ and press the GRAPH key to get figure (c).

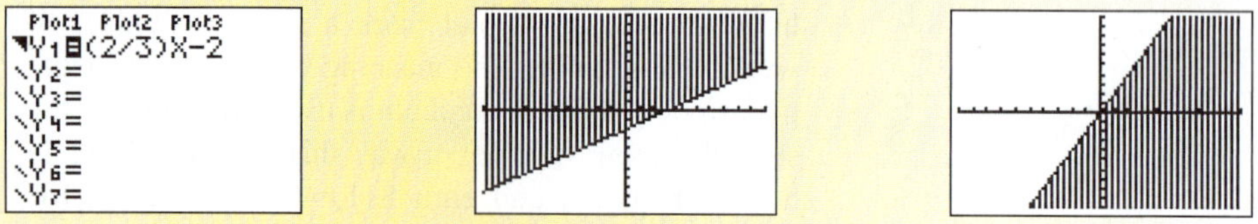

(a) (b) (c)

If your calculator does not have a graph style icon, you can graph linear inequalities with a SHADE feature. Graphing calculators do not distinguish between solid and dashed lines to show whether the edge of a region is included in the graph.

Answers to Self Checks

1.

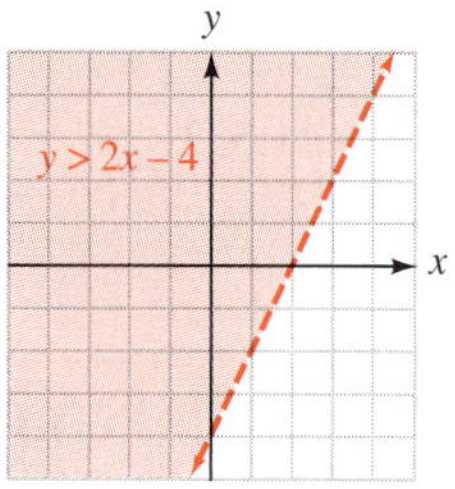

2.

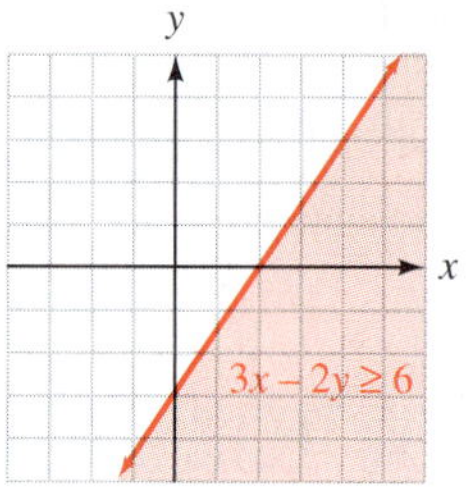

3.

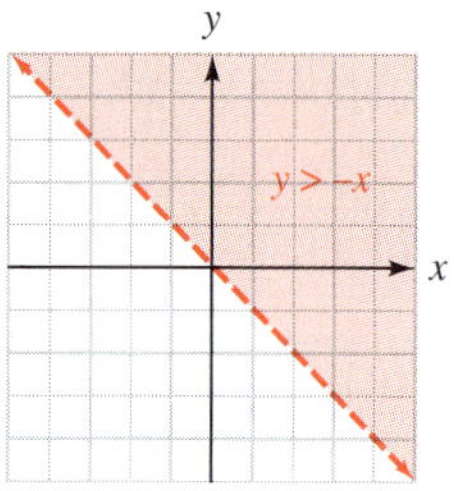

4.

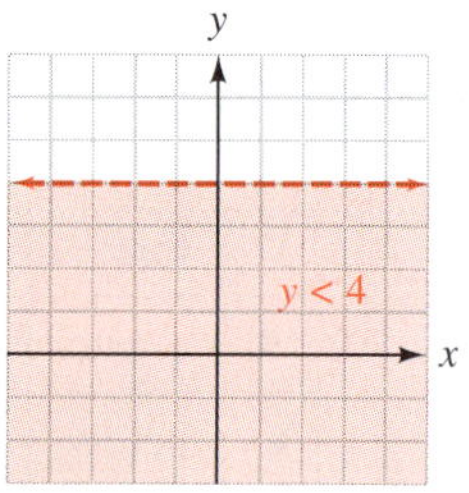

4.4 STUDY SET

VOCABULARY Fill in the blanks.

1. $4x - 2y \geq -8$ is an example of a ______ inequality in ______ variables.
2. Graphs of linear inequalities are ___________.
3. The boundary line of a half-plane is called an ______.
4. The graph of a linear inequality in x and y contains the points (x, y) whose coordinates _______ the inequality.

CONCEPTS

5. Decide whether each ordered pair is a solution of $3x - 2y \geq 5$.

   a. $(3, 1)$
   b. $(0, 3)$
   c. $(-1, -4)$
   d. $\left(1, \frac{1}{2}\right)$

6. The graph of a linear inequality is shown. Tell whether each point satisfies the inequality.

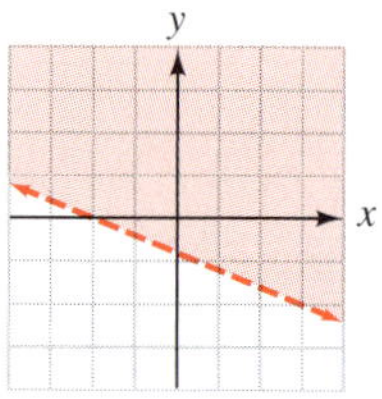

   a. $(-1, 4)$
   b. $(3, -2)$
   c. $(0, 0)$
   d. $(-3, -3)$

7. To graph the inequality $y > 3x - 1$, we begin by graphing the boundary line $y = 3x - 1$. What is the slope m of the line? What is its y-intercept?
8. To graph the inequality $2x + 3y \leq -6$, we begin by graphing the boundary line $2x + 3y = -6$. What are its x- and y-intercepts?

9. ZOOS To determine the allowable number of juvenile chimpanzees x and adult chimpanzees y that can live in an enclosure, a zookeeper refers to the illustration. Can 7 juvenile and 4 adult chimps be kept in the enclosure?

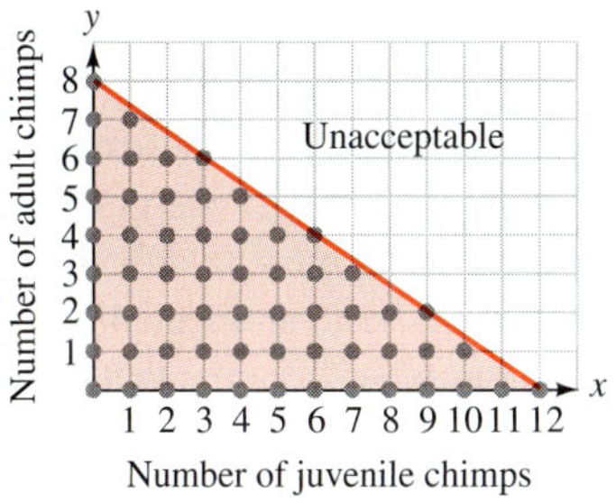

10. The boundary for the graph of a linear inequality is shown in the illustration. Why can't the origin be used as a test point to decide which side to shade?

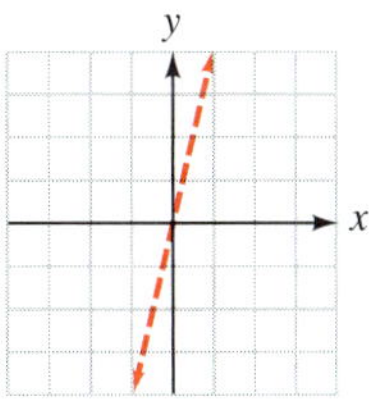

NOTATION

11. a. Solve the inequality in one variable and graph its solution set: $2x + 4 \geq 8$.

b. Graph the inequality in two variables: $2x + 4y \geq 8$.

12. Tell whether the graph of each inequality includes the boundary line. In each case, would the boundary be a solid or a dashed line?

a. $y < 3x - 1$ **b.** $2x + 3y \geq -6$

c. $y \leq -10$ **d.** $x > 1$

PRACTICE Graph each inequality.

13. $y > x + 1$ **14.** $y < 2x - 1$

15. $y \geq x$ **16.** $y \leq 2x$

17. $2x + y \leq 6$ **18.** $x - 2y \geq 4$

19. $3x \geq -y + 3$ **20.** $2x \leq -3y - 12$

21. $y \geq 1 - \frac{3}{2}x$ **22.** $y < \frac{x}{3} - 1$

23. $3x + y > 2 + x$ **24.** $3x - y > 6 + y$

25. $y < -\frac{x}{2}$ **26.** $y > \frac{x}{3}$

27. $\frac{x}{2} + \frac{y}{2} \leq 2$ **28.** $\frac{x}{3} - \frac{y}{2} \geq 1$

29. $x < 4$ **30.** $y \geq -2$

31. $y < 0$ **32.** $x \geq 0$

Find the equation of the boundary line. Then give the inequality whose graph is shown.

33.

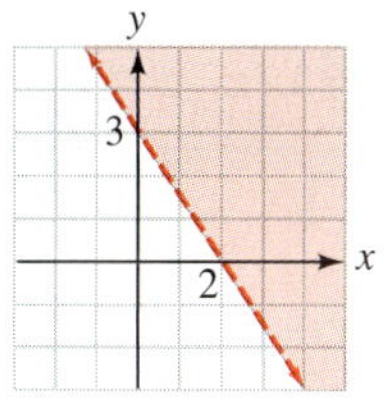

34.

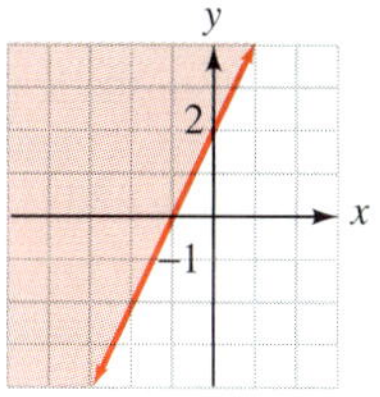

35.

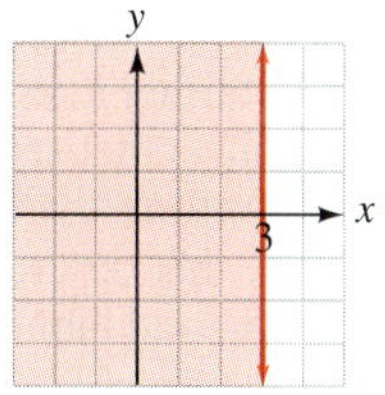

36.

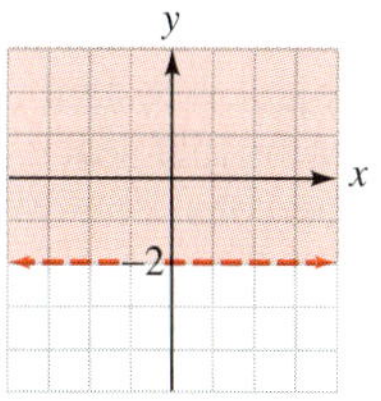

Use a graphing calculator to graph each inequality.

37. $y < 0.27x - 1$ **38.** $y > -3.5x + 2.7$

39. $y \geq -2.37x + 1.5$ **40.** $y \leq 3.37x - 1.7$

APPLICATIONS

41. GEOGRAPHY A region of the continental United States is shaded in the following map.

a. What is the boundary that separates the shaded and unshaded regions?

b. In words, describe the shaded area with respect to the boundary.

42. THE KOREAN WAR After World War II, the 38th parallel of north latitude was established as the boundary between North Korea and South Korea. The Korean War began on June 25, 1950, when the North Korean army crossed this line and invaded South Korea. In the illustration, shade the region of the Korean Peninsula south of the 38th parallel.

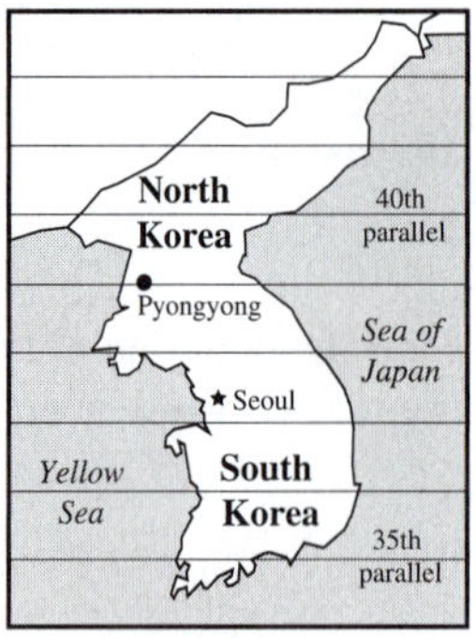

Write a linear inequality that models the situation. Then graph each inequality for nonnegative values of *x* and *y* and give three ordered pairs that satisfy the inequality.

43. RESTAURANT SEATING As part of a remodeling project, a restaurant owner will be installing new booths that seat 4 persons, and new tables that seat 6 persons. The overall seating must conform to the sign shown below. Write an inequality that describes the possible combinations of booths (x) and tables (y) that the owner can install.

MAXIMUM OCCUPANCY
NOT TO EXCEED
120
By order of Clake County Fire Marshal

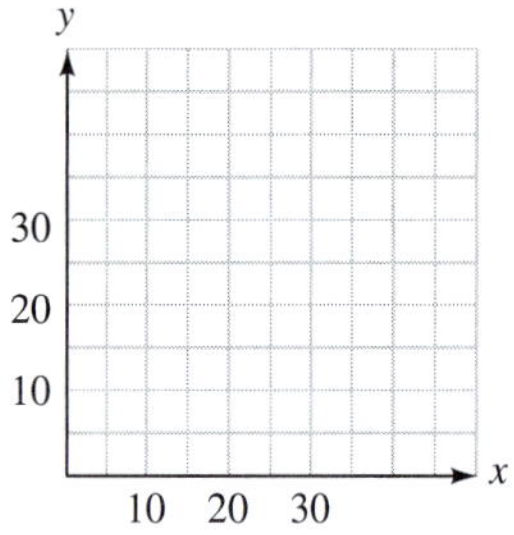

44. GARDENING During an Arbor Day sale, a garden store sold more than $2,000 worth of trees. If a 6-foot maple costs $100 and a 5-foot pine costs $125, write an inequality that shows the possible ways that maple trees (x) and pine trees (y) were sold.

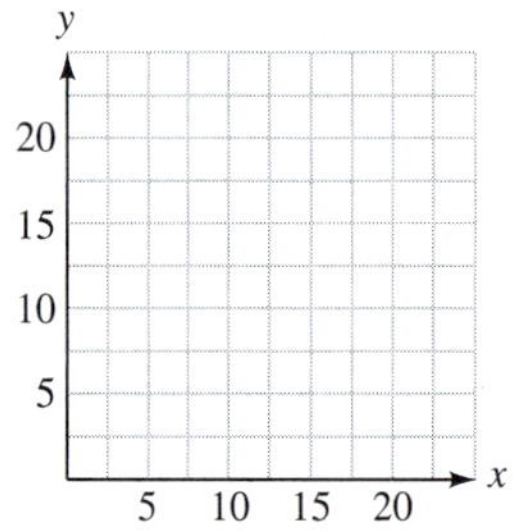

45. SPORTING GOODS A sporting goods manufacturer allocates at least 1,200 units of time per day to make fishing rods and reels. If it takes 10 units of time to make a rod and 15 units of time to make a reel, write an inequality that describes the possible ways to schedule the time to make rods (x) and reels (y).

46. HOUSEKEEPING One housekeeper charges $12 per hour, and another charges $9 per hour. If Sarah can afford no more than $54 per month to clean her house, write an inequality that describes the possible number of hours that she can hire the first housekeeper (x) and the second housekeeper (y).

WRITING

47. Explain how to decide whether the boundary of the graph of a linear inequality should be drawn as a solid or a dashed line.

48. Explain how to decide which side of the boundary of the graph of a linear inequality should be shaded.

REVIEW

Decide whether the ordered pair (−4, 3) is a solution of the system of linear equations.

49. $\begin{cases} 4x - y = -19 \\ 3x + 2y = -6 \end{cases}$

50. $\begin{cases} y = 2x + 11 \\ \frac{x}{2} + y = 0 \end{cases}$

Solve each system of equations.

51. $\begin{cases} x = \frac{2}{3}y \\ y = 4x + 5 \end{cases}$

52. $\begin{cases} x - \frac{y}{2} = -2 \\ 0.01x + 0.02y = 0.03 \end{cases}$

CHALLENGE PROBLEMS

53. Can an inequality be an identity, one that is satisfied by all pairs (x, y)? Illustrate.

54. Can an inequality have no solutions? Illustrate.

4.5 Systems of Linear Inequalities

- Solving systems of linear inequalities
- Compound inequalities
- Problem solving

We have discussed how to solve systems of linear equations by the graphing method. For example, to solve

$$\begin{cases} y = -x + 1 \\ 2x - y = 2 \end{cases}$$

we graph both equations on the same set of coordinate axes and then find the coordinates of the point of intersection of the straight lines. We will now discuss how to solve **systems of linear inequalities** graphically, such as

$$\begin{cases} y \leq -x + 1 \\ 2x - y > 2 \end{cases}$$

SOLVING SYSTEMS OF LINEAR INEQUALITIES

When the solution of a linear inequality in x and y is graphed, the result is a half-plane. To solve a system of linear inequalities, we graph each of the inequalities on one set of coordinate axes and look for the intersection of the shaded half-planes.

EXAMPLE 1

Graph the solution set of $\begin{cases} y \leq -x + 1 \\ 2x - y > 2 \end{cases}$.

Solution We graph each inequality on one set of coordinate axes, as shown in the figure.

To graph $y \leq -x + 1$, we graph the boundary $y = -x + 1$, as shown in figure (a). Since the edge is to be included, we draw it as a solid line. To determine which half-plane to shade, we use the origin as a test point. Because the coordinates of the origin satisfy $y \leq -x + 1$, we shade (in red) the half-plane containing the origin.

In figure (b), we superimpose the graph of $2x - y > 2$ on the graph of $y \leq -x + 1$ so that we can determine the points that the graphs have in common. To graph $2x - y > 2$, we graph the boundary $2x - y = 2$ as a dashed line. Since the test point (0, 0) does not satisfy $2x - y > 2$, we then shade (in blue) the half-plane that does not contain (0, 0).

The area that is shaded twice represents the solutions of the given system. Any point in the doubly shaded region in purple (including the purple portion of one of the boundaries) has coordinates that satisfy both inequalities.

The Language of Algebra

To solve a system of linear inequalities, we *superimpose* the graphs of the inequalities. That is, we place one graph over the other. Most video camcorders can *superimpose* the date and time over the picture being recorded.

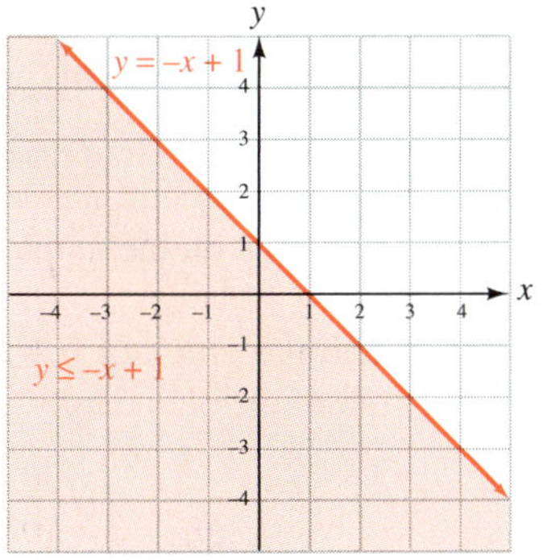

The graph of $y \leq -x + 1$ is shaded in red.

(a)

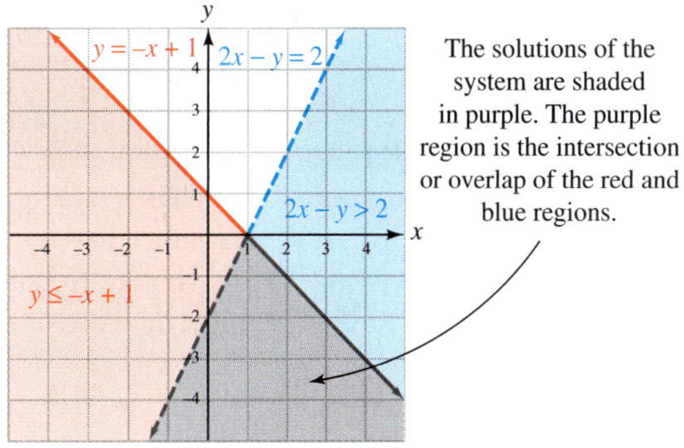

The graph of $2x - y > 2$ is shaded in blue. It is drawn over the graph of $y \leq -x + 1$.

(b)

Since there are an infinite number of solutions, we cannot check each of them. However, as an informal check, we can select several points that lie in the doubly shaded region and show that their coordinates satisfy both inequalities of the system.

Self Check 1 Graph: $\begin{cases} x + y \geq 1 \\ 2x - y < 2 \end{cases}$.

ACCENT ON TECHNOLOGY: SOLVING SYSTEMS OF INEQUALITIES

To solve the system of Example 1 with a graphing calculator, we can use window settings of $x = [-10, 10]$ and $y = [-10, 10]$. To graph $y \leq -x + 1$, we enter the boundary equation $y = -x + 1$ and change the graph style icon to below (◣). See figure (a). To graph $2x - y > 2$, we first write it in equivalent form as $y < 2x - 2$. Then we enter the boundary equation $y = 2x - 2$ and change the graph style icon to below (◣). See figure (a). Finally, we press the GRAPH key to obtain figure (b).

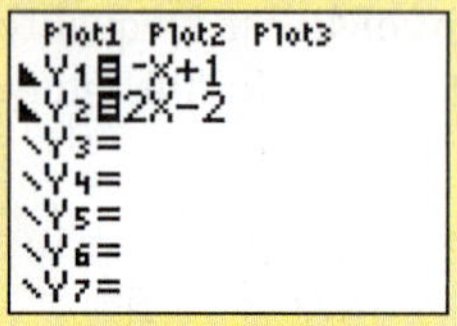

(a)

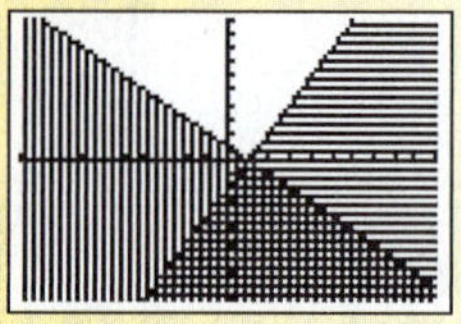

(b)

In general, to solve systems of linear inequalities, we will follow these steps.

Solving Systems of Linear Inequalities

1. Graph each inequality on the same rectangular coordinate system.
2. Use shading to highlight the intersection of the graphs (the region where the graphs overlap). The points in this region are the solutions of the system.
3. As an informal check, pick a point from the region and verify that its coordinates satisfy each inequality of the original system.

EXAMPLE 2

Graph the solution set: $\begin{cases} x \geq 1 \\ y \geq x \\ 4x + 5y < 20 \end{cases}$.

Solution We will find the graph of the solution set of the system in stages, using several graphs.

The graph of $x \geq 1$ includes the points that lie on the graph of $x = 1$ and to the right, as shown in red in figure (a).

Figure (b) shows the graph of $x \geq 1$ and the graph of $y \geq x$. The graph of $y \geq x$, in blue, includes the points that lie on the graph of the boundary $y = x$ and above it.

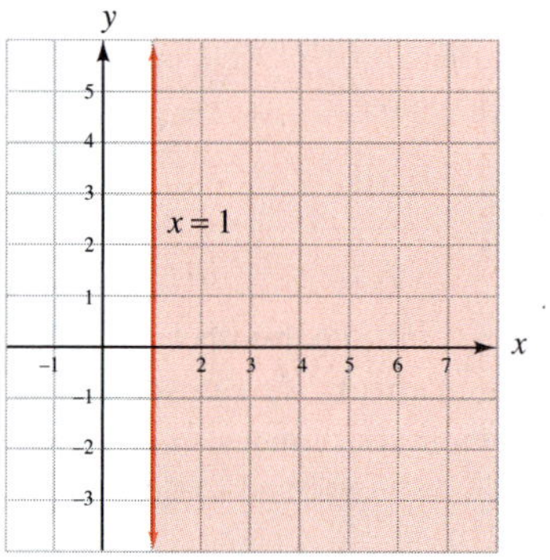

The graph of $x \geq 1$ is shaded in red.

(a)

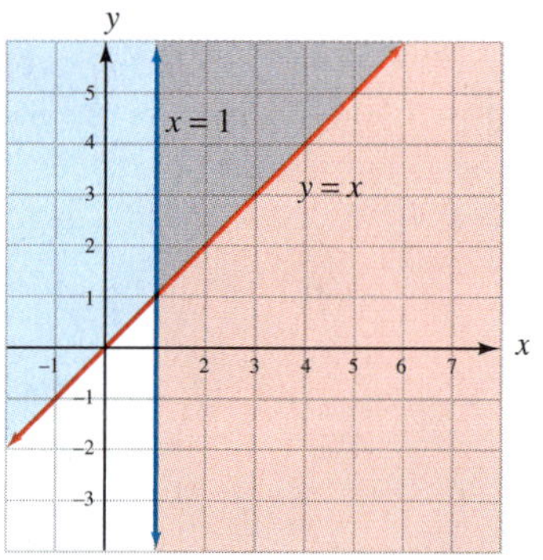

The graph of $y \geq x$ is shaded in blue.

(b)

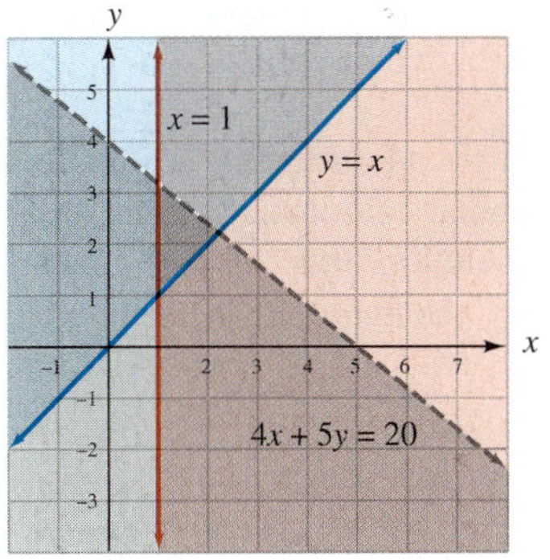

The graph of $4x + 5y < 20$ is shaded in grey.

(c)

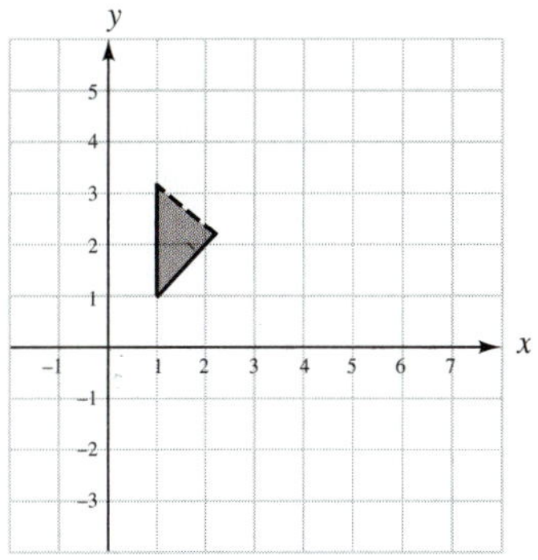

This is the graph of the solution of the system.

(d)

Figure (c) shows the graphs of $x \geq 1$, $y \geq x$, and $4x + 5y < 20$. The graph of $4x + 5y < 20$ includes the points that lie below the graph of the boundary $4x + 5y = 20$.

The graph of the solution of the system includes the points that lie within the shaded triangle together with the points on the two sides of the triangle that are drawn with solid line segments, as shown in figure (d).

Check: Pick a point in the shaded region, such as (1.5, 2), and show that it satisfies each inequality of the system.

Self Check 2 Graph: $\begin{cases} x \geq 0 \\ y \leq 0 \\ y \geq -2 \end{cases}$.

COMPOUND INEQUALITIES

We have graphed the solution set of double linear inequalities, such as $2 < x \leq 5$, on a number line. In the next example, we will graph the solution set of $2 < x \leq 5$ in the context of two variables.

EXAMPLE 3

Graph $2 < x \leq 5$ on the rectangular coordinate plane.

Solution The compound inequality $2 < x \leq 5$ is equivalent to the following system of two linear inequalities:

$$\begin{cases} 2 < x \\ x \leq 5 \end{cases}$$

The graph of $2 < x$, shown in the figure in red, is the half-plane to the right of the vertical line $x = 2$. The graph of $x \leq 5$, shown in the figure in blue, includes the line $x = 5$ and the half-plane to its left. The graph of $2 < x \leq 5$ will contain all points in the plane that satisfy the inequalities $2 < x$ and $x \leq 5$ simultaneously. These points are in the purple-shaded region of the figure.

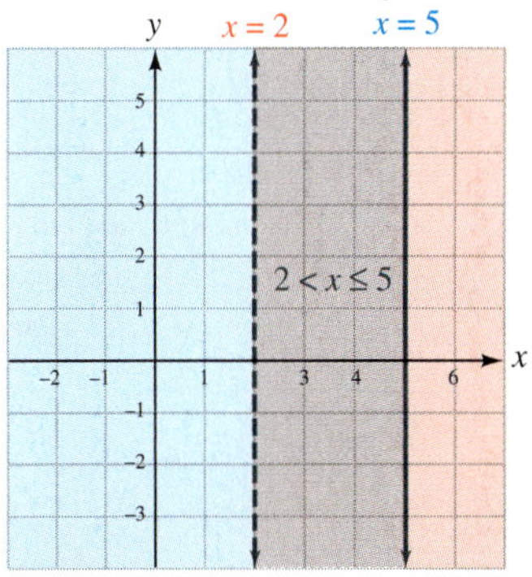

Success Tip

Colored pencils are often used to graph systems of inequalities. A standard pencil can also be used. Just draw different patterns of lines instead of shading.

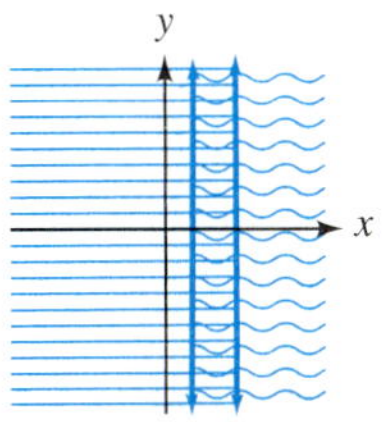

Self Check 3 Graph $-2 \le y < 3$ on the rectangular coordinate plane.

To graph a compound inequality containing the word *or* in the rectangular coordinate system, we sketch the *union* of the solution sets of the inequalities involved. For example, the figure shows the graph of the compound inequality

$$x \le -2 \quad \text{or} \quad x > 3$$

in the rectangular coordinate system.

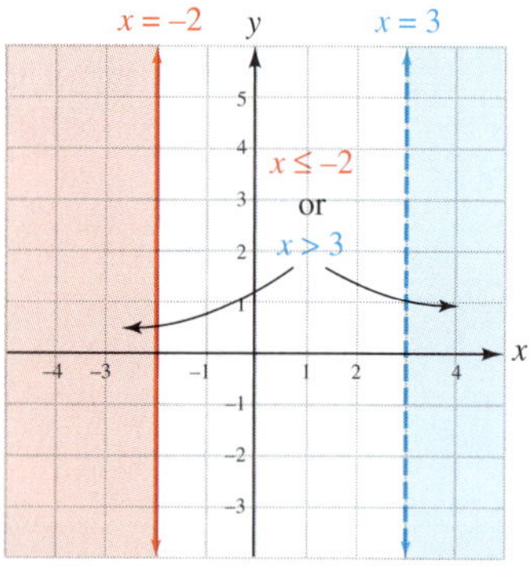

PROBLEM SOLVING

EXAMPLE 4

Landscaping. A homeowner has a budget of \$300 to \$600 for trees and bushes to landscape his yard. After shopping, he finds that good trees cost \$150 and mature bushes cost \$75. What combinations of trees and bushes can he afford to buy?

Analyze the Problem We must find the number of trees and bushes that the homeowner can afford. This suggests we should use two variables. We know that he is willing to spend *at least* \$300 and *at most* \$600 for trees and bushes. These phrases suggest that we should write two inequalities that model the situation.

Form Two Inequalities If x = the number of trees purchased, then \$150x will be the cost of the trees. If y = the number of bushes purchased, then \$75y will be the cost of the bushes. The homeowner wants the sum of these costs to be from \$300 to \$600. Using this information, we can form the following system of linear inequalities.

The cost of a tree	times	the number of trees purchased	plus	the cost of a bush	times	the number of bushes purchased	should be at least	\$300.
\$150	$\cdot$	x	$+$	\$75	$\cdot$	y	$\ge$	\$300
The cost of a tree	times	the number of trees purchased	plus	the cost of a bush	times	the number of bushes purchased	should be at most	\$600.
\$150	$\cdot$	x	$+$	\$75	$\cdot$	y	$\le$	\$600

Solve the System We graph the system

$$\begin{cases} 150x + 75y \ge 300 \\ 150x + 75y \le 600 \end{cases}$$

as in the following figure. The coordinates of each point shown in the graph give a possible combination of trees (x) and bushes (y) that can be purchased.

State the Conclusion The possible combinations of trees and bushes that can be purchased are given by

(0, 4), (0, 5), (0, 6), (0, 7), (0, 8)
(1, 2), (1, 3), (1, 4), (1, 5), (1, 6)
(2, 0), (2, 1), (2, 2), (2, 3), (2, 4)
(3, 0), (3, 1), (3, 2), (4, 0)

The ordered pair (1, 6), for example, indicates that the homeowner can afford 1 tree and 6 bushes.

Only these points can be used, because the homeowner cannot buy a portion of a tree or a bush.

Check the Result Check some of the ordered pairs to verify that they satisfy both inequalities.

Because the homeowner cannot buy a negative number of trees or bushes, we graph the system for $x \geq 0$ and $y \geq 0$.

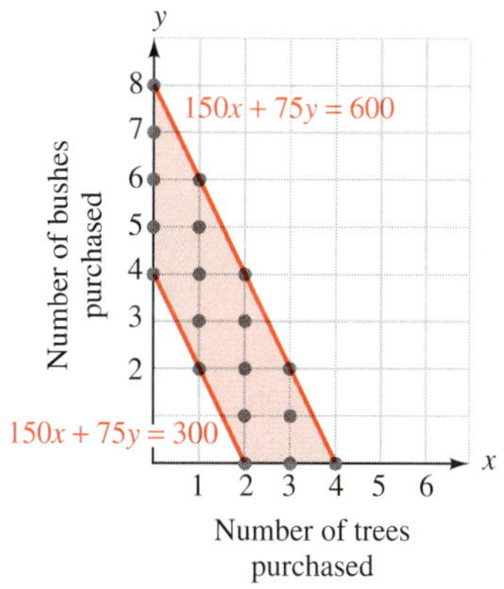

Answers to Self Checks

1.

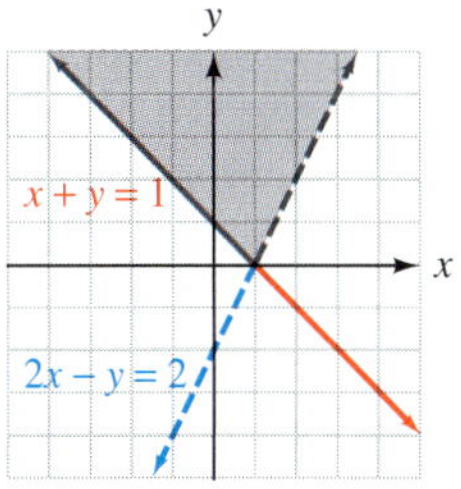

2.

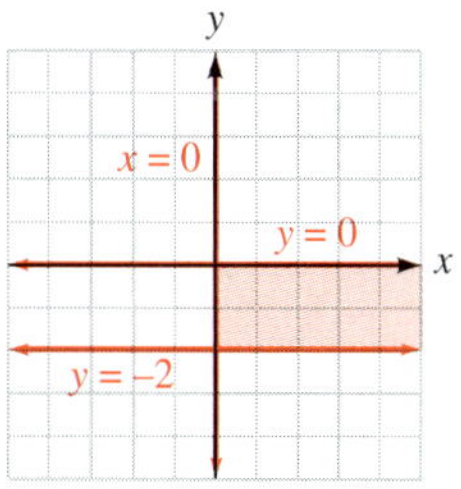

3.

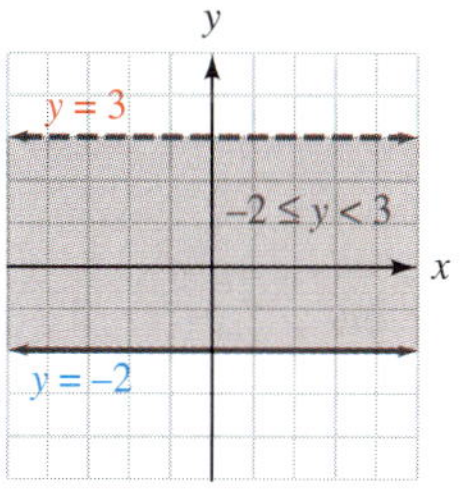

4.5 STUDY SET

VOCABULARY Fill in the blanks.

1. $\begin{cases} x + y \leq 2 \\ x - 3y > 10 \end{cases}$ is a system of linear ________ in two variables.
2. If an edge is included in the graph of an inequality, we draw it as a ______ line.
3. To solve a system of inequalities by graphing, we graph each inequality. The solution is the region where the graphs overlap or ________.
4. To determine which half-plane to shade when graphing a linear inequality, we see whether the coordinates of a test ______ satisfy the inequality.

CONCEPTS

5. Tell whether each point satisfies the system of linear inequalities $\begin{cases} x + y \leq 2 \\ x - 3y > 10 \end{cases}$.
   a. $(2, -3)$ b. $(12, -1)$
   c. $(0, -3)$ d. $(-0.5, -5)$
6. a. Decide whether $(-3, 10)$ satisfies the compound inequality $-5 < x \leq 8$ in the rectangular coordinate system.
   b. Decide whether $(-3, 3)$ satisfies the compound inequality $y \leq 0$ or $y > 4$ in the rectangular coordinate system.

7. In the illustration, the solution of one linear inequality is shaded in red, and the solution of a second is shaded in blue. Decide whether a true or false statement results if the coordinates of the given point are substituted into the given inequality.

a. A, inequality 1 **b.** A, inequality 2

c. B, inequality 1 **d.** B, inequality 2

e. C, inequality 1 **f.** C, inequality 2

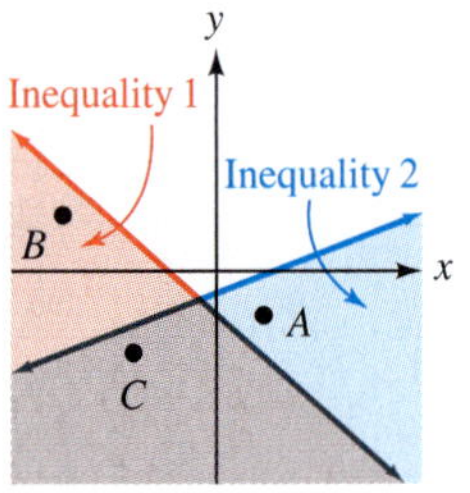

8. Match each equation, inequality, or system with the graph of its solution in the illustration.

a. $2x + y = 2$ **b.** $2x + y \geq 2$

c. $\begin{cases} 2x + y = 2 \\ 2x - y = 2 \end{cases}$ **d.** $\begin{cases} 2x + y \geq 2 \\ 2x - y \leq 2 \end{cases}$

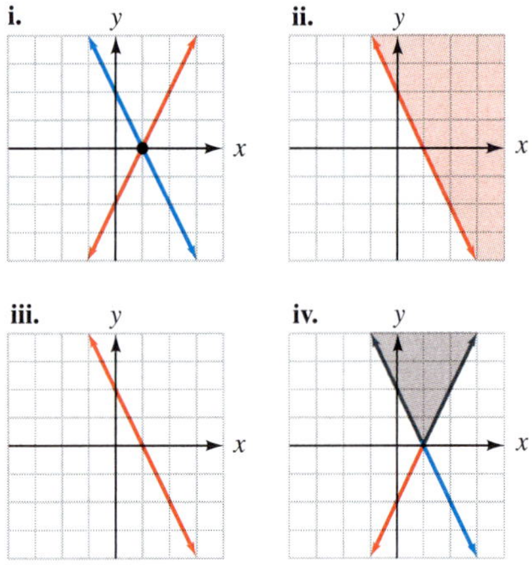

PRACTICE **Graph the solution set of each system of inequalities.**

9. $\begin{cases} y < 3x + 2 \\ y < -2x + 3 \end{cases}$ **10.** $\begin{cases} y \leq x - 2 \\ y \geq 2x + 1 \end{cases}$

11. $\begin{cases} 3x + 2y > 6 \\ x + 3y \leq 2 \end{cases}$ **12.** $\begin{cases} x + y < 2 \\ x + y \leq 1 \end{cases}$

13. $\begin{cases} 3x + y \leq 1 \\ -x + 2y \geq 6 \end{cases}$ **14.** $\begin{cases} x + 2y < 3 \\ 2x + 4y < 8 \end{cases}$

15. $\begin{cases} x > 0 \\ y > 0 \end{cases}$ **16.** $\begin{cases} x \leq 0 \\ y < 0 \end{cases}$

17. $\begin{cases} 2x + 3y \leq 6 \\ 3x + y \leq 1 \\ x \leq 0 \end{cases}$ **18.** $\begin{cases} 2x + y \leq 2 \\ y \geq x \\ x \geq 0 \end{cases}$

19. $\begin{cases} x - y < 4 \\ y \leq 0 \\ x \geq 0 \end{cases}$ **20.** $\begin{cases} x \geq 0 \\ y \geq 0 \\ 9x + 3y \leq 18 \\ 3x + 6y \leq 18 \end{cases}$

Graph each inequality in the rectangular coordinate system.

21. $-2 \leq x < 0$ **22.** $-3 < y \leq -1$

23. $y < -2$ or $y > 3$ **24.** $-x \leq 1$ or $x \geq 2$

Use a graphing calculator to solve each system.

25. $\begin{cases} y < 3x + 2 \\ y < -2x + 3 \end{cases}$ **26.** $\begin{cases} y > -x + 2 \\ y < -x + 4 \end{cases}$

27. $\begin{cases} 2x + y \geq 6 \\ y \leq 2(2x - 3) \end{cases}$ **28.** $\begin{cases} 3x + y < -2 \\ y > 3(1 - x) \end{cases}$

APPLICATIONS

29. FOOTBALL In 2003, the Green Bay Packers scored either a touchdown or a field goal 65.4% of the time when their offense was in the *red zone.* This was the best record in the NFL! If x represents the yard line the football is on, a team's red zone is an area on their opponent's half of the field that can be described by the system

$$\begin{cases} x > 0 \\ x \leq 20 \end{cases}$$

Shade the red zone on the field shown below.

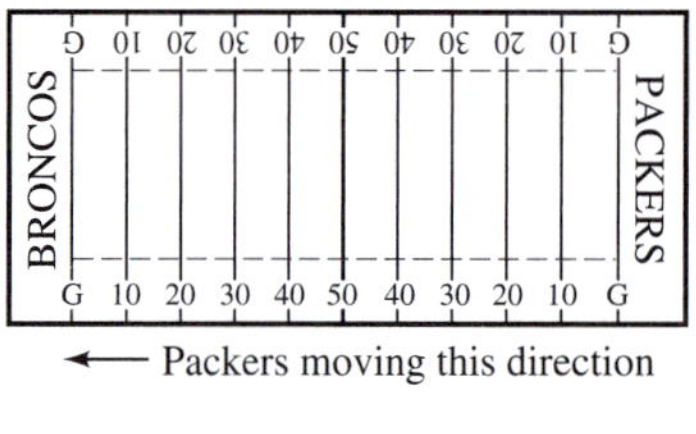

30. TRACK AND FIELD In the shot put, the solid metal ball must land in a marked sector for it to be a fair throw. In the illustration, graph the system of inequalities that describes the region in which a shot must land.

$$\begin{cases} y \leq \frac{3}{8}x \\ y \geq -\frac{3}{8}x \\ x \geq 1 \end{cases}$$

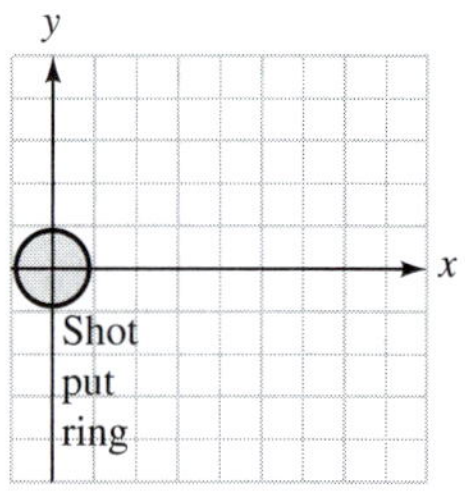

31. NO-FLY ZONES After the Gulf War, U.S. and Allied forces enforced northern and southern "no-fly" zones over Iraq. Iraqi aircraft was prohibited from flying in this air space. If y represents the north latitude parallel measurement, the no-fly zones can be described by

$$y \geq 36 \quad \text{or} \quad y \leq 33$$

On the map, shade the regions of Iraq over which there was a no-fly zone.

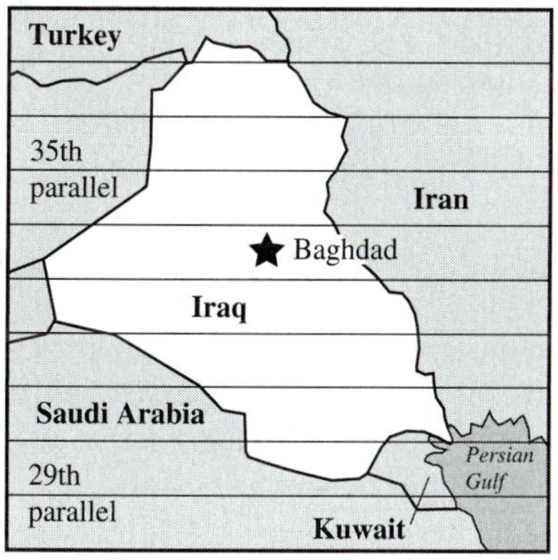

32. CARDIOVASCULAR FITNESS The graph in the illustration shows the range of pulse rates that persons ages 20–90 should maintain during aerobic exercise to get the most benefit from the training. The shaded region "Effective Training Heart Rate Zone" can be described by a system of linear inequalities. Determine what inequality symbol should be inserted in each blank.

$$\begin{cases} x \; \underline{\quad} \; 20 \\ x \; \underline{\quad} \; 90 \\ y \; \underline{\quad} \; -0.87x + 191 \\ y \; \underline{\quad} \; -0.72x + 158 \end{cases}$$

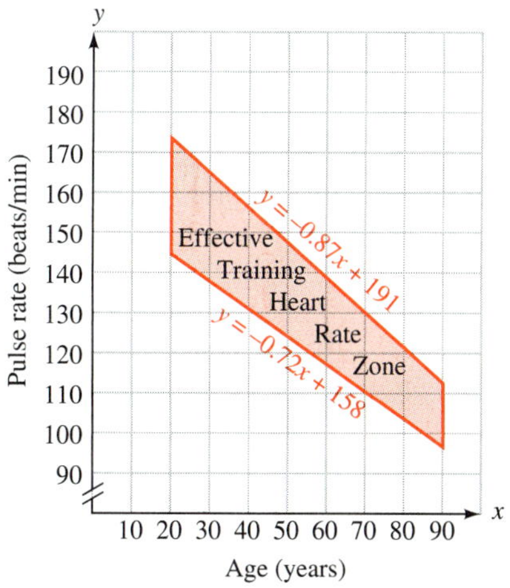

Graph each system of inequalities and give two possible solutions.

33. COMPACT DISCS Melodic Music has compact discs on sale for either \$10 or \$15. If a customer wants to spend at least \$30 but no more than \$60 on CDs, use the illustration to graph a system of inequalities that will show the possible ways a customer can buy \$10 CDs (x) and \$15 CDs (y).

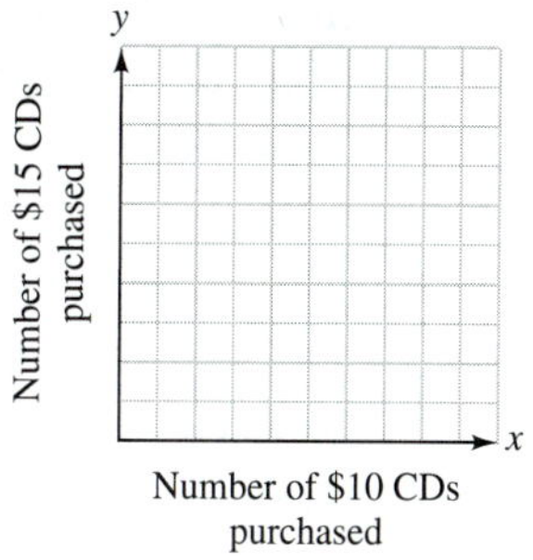

34. BOAT SALES Dry Boat Works wholesales aluminum boats for \$800 and fiberglass boats for \$600. Northland Marina wants to order at least \$2,400 worth but no more than \$4,800 worth of boats. Use the illustration to graph a system of inequalities that will show the possible combinations of aluminum boats (x) and fiberglass boats (y) that can be ordered.

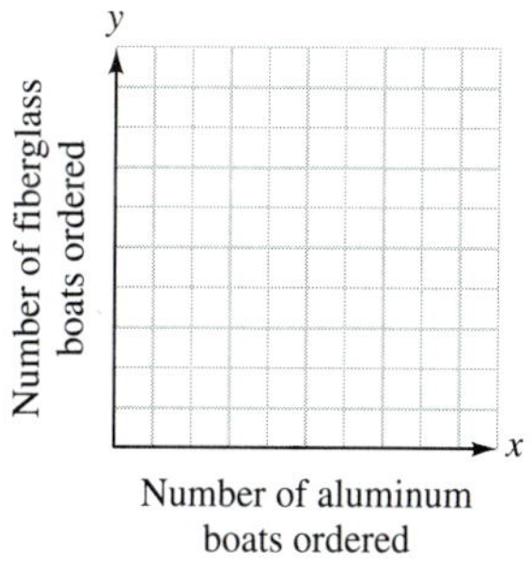

35. FURNITURE SALES A distributor wholesales desk chairs for \$150 and side chairs for \$100. Best Furniture wants to order no more than \$900 worth of chairs, including more side chairs than desk chairs. Use the illustration to graph a system of inequalities that will show the possible combinations of desk chairs (x) and side chairs (y) that can be ordered.

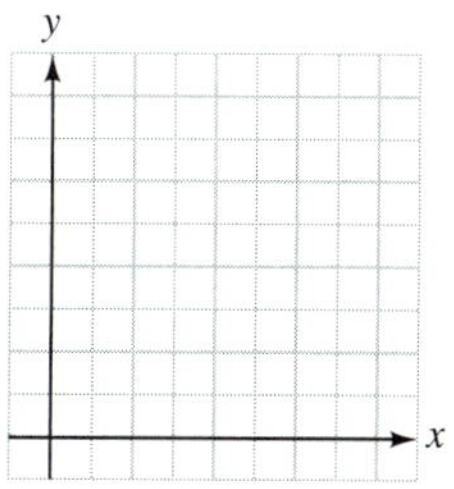

36. FURNACE EQUIPMENT J. Bolden Heating Company wants to order no more than \$2,000 worth of electronic air cleaners and humidifiers from a wholesaler that charges \$500 for air cleaners and \$200 for humidifiers. If Bolden wants more humidifiers than air cleaners, use the illustration to graph a system of inequalities that will show the possible combinations of air cleaners (x) and humidifiers (y) that can be ordered.

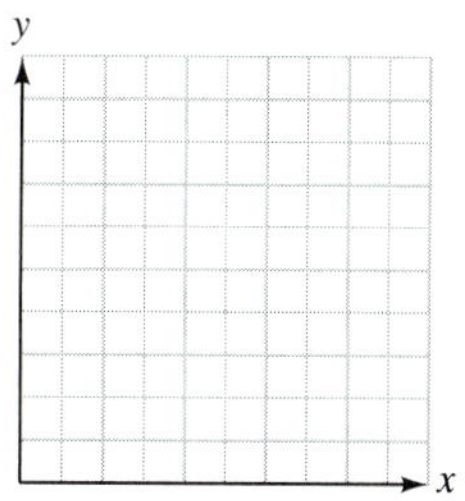

WRITING

37. Explain how to solve a system of two linear inequalities graphically.

38. Explain how a system of two linear inequalities might have no solution.

39. A student graphed the system

$$\begin{cases} -x + 3y > 0 \\ x + 3y < 3 \end{cases}$$

as shown. Explain how to informally check the result.

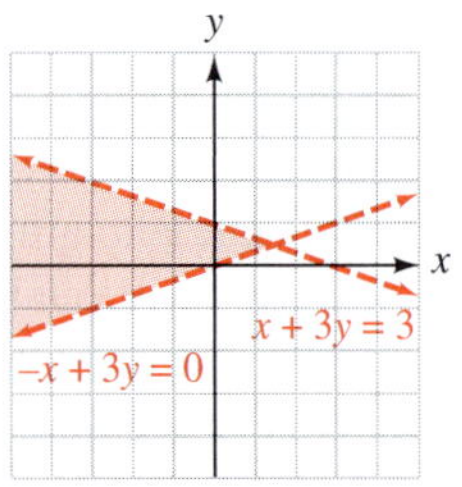

40. Describe the result when $-3 \le x < 4$ is graphed on a number line. Describe the result when $-3 \le x < 4$ is graphed on the rectangular coordinate plane.

REVIEW

Use the given conditions to determine in which quadrant of a rectangular coordinate system each point (x, y) is located.

41. $x > 0$ and $y < 0$

42. $x < 0$ and $y < 0$

43. $x < 0$ and $y > 0$

44. $x > 0$ and $y > 0$

CHALLENGE PROBLEMS

Write a system of linear inequalities in two variables whose graph is shown.

45.

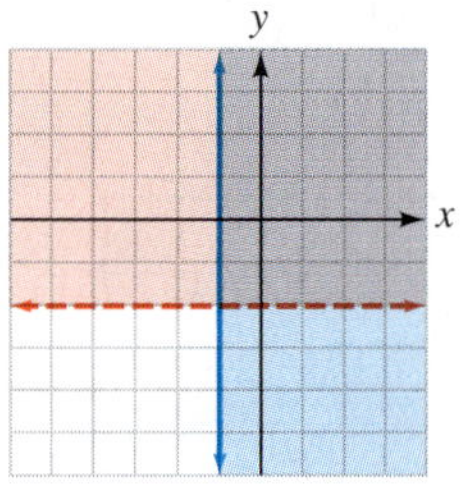

46.

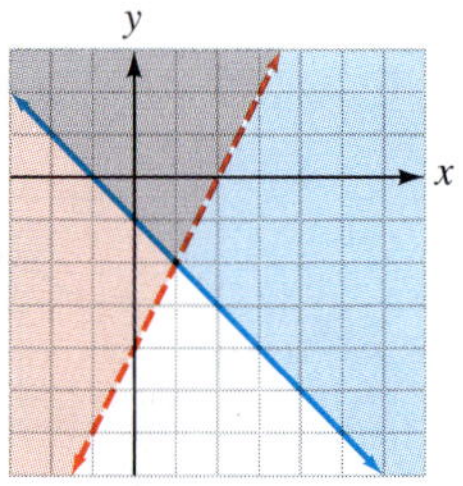

47. The solution of a system of inequalities in two variables is *bounded* if it is possible to draw a circle around the solution. Can the solution of two linear inequalities be bounded?

48. The solution of $\begin{cases} y \ge |x| \\ y \le k \end{cases}$ has an area of 25. Find k.

ACCENT ON TEAMWORK

VENN DIAGRAMS

Overview: In this activity, we will discuss several of the fundamental concepts of what is known as *set theory.*

Instructions: Venn diagrams are a convenient way to visualize relationships between sets and operations on sets. They were invented by the English mathematician John Venn (1834–1923). To draw a Venn diagram, we begin with a large rectangle, called the *universal set.* Ovals or circles are then drawn in the interior of the rectangle to represent subsets of the universal set.

Form groups of 2 or 3 students. Study the following figures, which illustrate three set operations: union, intersection, and complement.

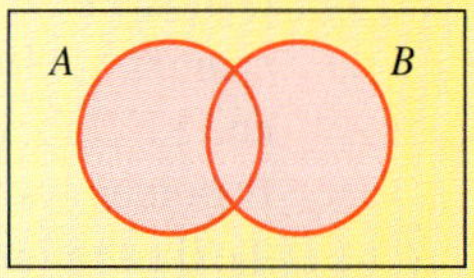

$A \cup B$
The shaded region is the *union* of set A and set B.

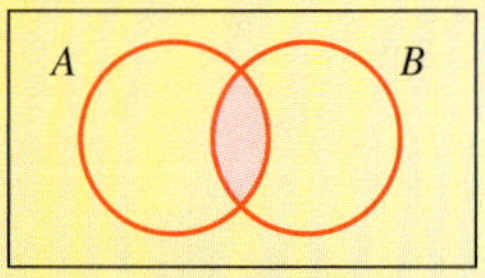

$A \cap B$
The shaded region is the *intersection* of set A and set B.

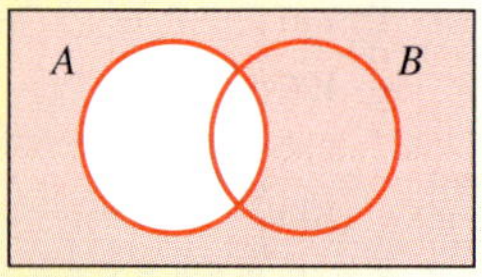

$\overline{A}$
The shaded region is the complement of set A.

For each of the following exercises, sketch the following blank Venn diagram and then shade the indicated region.

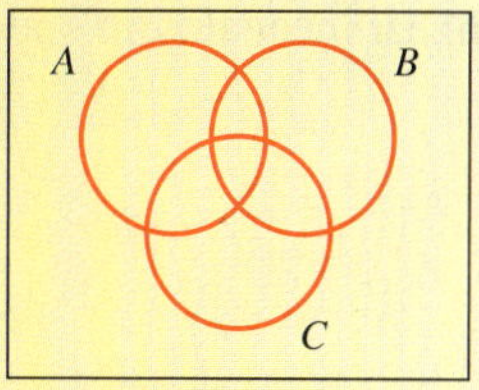

1. $A \cup B$
2. $A \cap B$
3. $A \cap C$
4. $A \cup C$
5. $A \cup B \cup C$
6. $A \cap B \cap C$
7. $(B \cup C) \cap A$
8. $C \cup (A \cap B)$
9. $\overline{A}$
10. $\overline{B} \cup \overline{C}$
11. $\overline{B} \cap \overline{C}$
12. $\overline{A \cup B}$

KEY CONCEPT: INEQUALITIES

TYPES OF INEQUALITIES

An **inequality** is a statement indicating that quantities are unequal. In Chapter 4, we have worked with several different types of inequalities and combinations of inequalities.

Classify each statement as one of the following: linear inequality in one variable, compound inequality, double linear inequality, absolute value inequality, linear inequality in two variables, system of linear inequalities.

1. $x - 3 < -4$ or $x - 2 > 0$
2. $\begin{cases} y < 3x + 2 \\ y < -2x + 3 \end{cases}$
3. $|x - 8| \leq 12$
4. $y < \frac{x}{3} - 1$
5. $\frac{1}{2}x + 2 \geq \frac{1}{3}x - 4$
6. $-6 < -3(x - 4) \leq 24$
7. $5(x - 2) \geq 0$ and $-3x < 9$
8. $|-1 - 2x| > 5$
9. $y > -x$

SOLUTIONS OF INEQUALITIES

A solution of a linear inequality in one variable is a value that, when substituted for the variable, makes the inequality true. A solution of a linear inequality in two variables (or a system of linear inequalities) is an ordered pair whose coordinates satisfy the inequality (or inequalities).

Decide whether -2 is a solution of the inequalities in one variable. Determine whether $(-1, 3)$ is a solution of the inequalities (or system of inequalities) in two variables.

10. $x - 3 < -4$ or $x - 2 > 0$
11. $\begin{cases} y < 3x + 2 \\ y < -2x + 3 \end{cases}$
12. $|x - 8| \leq 12$
13. $y < \frac{x}{3} - 1$
14. $\frac{1}{2}x + 2 \geq \frac{1}{3}x - 4$
15. $-6 < -3(x - 4) \leq 24$
16. $5(x - 2) \geq 0$ and $-3x < 9$
17. $|-1 - 2x| > 5$
18. $y > -x$

GRAPHS OF INEQUALITIES

To graph the solution set of an inequality in one variable, we use a number line. To graph the solution set of an inequality in two variables, we use a rectangular coordinate system.

19. Graph the solution set of the linear inequality in one variable: $2x + 1 > 4$.

20. Graph the solution set of the linear inequality in two variables: $2x + y \geq 4$.

CHAPTER REVIEW

SECTION 4.1 Solving Linear Inequalities

CONCEPTS

To solve an inequality, apply the *properties of inequalities.* If both sides of an inequality are multiplied (or divided) by a negative number, another inequality results, but with the opposite direction from the original inequality.

The graph of a set of real numbers that is a portion of a number line is called an *interval.*

REVIEW EXERCISES

Solve each inequality. Give each solution set in interval notation and graph it.

1. $5(x - 2) \leq 5$

2. $0.3x - 0.4 \geq 1.2 - 0.1x$

3. $-16 < -\frac{4}{5}x$

4. $\frac{7}{4}(x + 3) < \frac{3}{8}(x - 3)$

5. $7 - [6t - 5(t - 3)] > 2(t - 3) - 3(t + 1)$

6. $\frac{2b + 7}{2} \leq \frac{3b - 1}{3}$

7. Explain how to use the graph of $y = 1$ and $y = x - 3$ to solve $x - 3 \leq 1$.

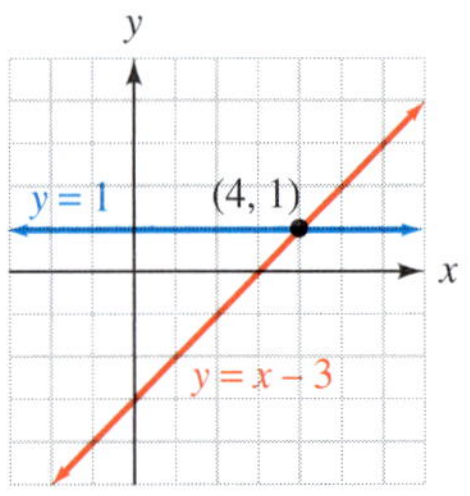

8. INVESTMENTS A woman has invested $10,000 at 6% annual interest. How much more must she invest at 7% so that her annual income is at least $2,000?

SECTION 4.2 Solving Compound Inequalities

A solution of a compound inequality containing *and* makes both of the inequalities true.

Determine whether −4 is a solution of the compound inequality.

9. $x < 0$ and $x > -5$

10. $x + 3 < -3x - 1$ and $4x - 3 > 3x$

Graph each set.

11. $(-3, 3) \cup [1, 6]$

12. $(-\infty, 2] \cap [1, 4)$

The solution set of a compound inequality containing *and* is the *intersection* of the two solution sets.

$\cap$ means intersection.

Double linear inequalities:
$c < x < d$
is equivalent to
$c < x$ and $x < d$

A solution of a compound inequality containing the word *or* makes one, or the other, or both inequalities true.

The solution set of a compound inequality containing *or* is the *union* of the two solution sets.

$\cup$ means union.

Solve each compound inequality. Give the result in interval notation and graph the solution set.

13. $-2x > 8$ and $x + 4 \geq -6$

14. $5(x + 2) \leq 4(x + 1)$ and $11 + x < 0$

15. $\frac{2}{5}x - 2 < -\frac{4}{5}$ and $\frac{x}{-3} < -1$

16. $4\left(x - \frac{1}{4}\right) \leq 3x - 1$ and $x \geq 0$

Solve each double inequality. Give the result in interval notation and graph the solution set.

17. $3 < 3x + 4 < 10$

18. $-2 \leq \frac{5 - x}{2} \leq 2$

Determine whether −4 is a solution of the compound inequality.

19. $x < 1.6$ or $x > -3.9$

20. $x + 1 < 2x - 1$ or $4x - 3 > 3x$

Solve each compound inequality. Give the result in interval notation and graph the solution set.

21. $x + 1 < -4$ or $x - 4 > 0$

22. $\frac{x}{2} + 3 > -2$ or $4 - x > 4$

23. INTERIOR DECORATING A manufacturer makes a line of decorator rugs that are 4 feet wide and of varying lengths l (in feet). The floor area covered by the rugs ranges from 17 ft^2 to 25 ft^2. Write and then solve a double linear inequality to find the range of the lengths of the rugs.

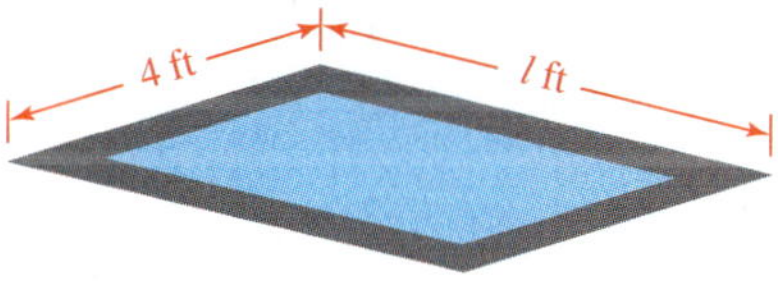

24. Match each word in Column 1 with *two* items in Column II.

Column 1	*Column II*
a. or	**i.** $\cap$
	ii. $\cup$
b. and	**iii.** intersection
	iv. union

SECTION 4.3 Solving Absolute Value Equations and Inequalities

Absolute value equations: For $k > 0$ and any algebraic expressions X and Y:

$|X| = k$ is equivalent to $X = k$ or $X = -k$.

$|X| = |Y|$ is equivalent to $X = Y$ or $X = -Y$

Solve each absolute value equation.

25. $|4x| = 8$

26. $2|3x + 1| - 1 = 19$

27. $\left|\frac{3}{2}x - 4\right| - 10 = -1$

28. $\left|\frac{2 - x}{3}\right| = -4$

29. $|3x + 2| = |2x - 3|$

30. $\left|\frac{2(1 - x) + 1}{2}\right| = \left|\frac{3x - 2}{3}\right|$

Absolute value inequalities: For $k > 0$ and any algebraic expression X:

$|X| < k$ is equivalent to $-k < X < k$

$|X| > k$ is equivalent to $X < -k$ or $X > k$

Solve each absolute value inequality. Give the solution in interval notation and graph it.

31. $|x| \le 3$

32. $|2x + 7| < 3$

33. $2|5 - 3x| \le 28$

34. $\left|\frac{2}{3}x + 14\right| + 6 < 6$

35. $|x| > 1$

36. $\left|\frac{1 - 5x}{3}\right| \ge 7$

37. $|3x - 8| - 4 > 0$

38. $\left|\frac{3}{2}x - 14\right| \ge 0$

39. Explain why $|0.04x - 8.8| < -2$ has no solution.

40. Explain why the solution set of $\left|\frac{3x}{50} + \frac{1}{45}\right| \ge -\frac{4}{5}$ is the set of all real numbers.

41. PRODUCE Before packing, freshly picked tomatoes are weighed on the scale shown. Tomatoes having a weight w (in ounces) that falls within the highlighted range are sold to grocery stores.

a. Express this acceptable weight range using an absolute value inequality.

b. Solve the inequality and express this range as an interval.

42. Let $f(x) = \frac{1}{3}|6x| - 1$. For what value(s) of x is $f(x) = 5$?

SECTION 4.4 Linear Inequalities in Two Variables

To graph a *linear inequality* in x and y, graph the *boundary line,* and then use a *test point* to decide which side of the boundary should be shaded.

Graph each inequality in the rectangular coordinate system.

43. $2x + 3y > 6$

44. $y \le 4 - x$

45. $y < \frac{1}{2}x$

46. $x \ge -\frac{3}{2}$

47. CONCERT TICKETS Tickets to a concert cost \$6 for reserved seats and \$4 for general admission. If receipts must be at least \$10,200 to meet expenses, find an inequality that shows the possible ways that the box office can sell reserved seats (x) and general admission tickets (y). Then graph the inequality for nonnegative values of x and y and give three ordered pairs that satisfy the inequality.

48. Find the equation of the boundary line. Then give the inequality whose graph is shown.

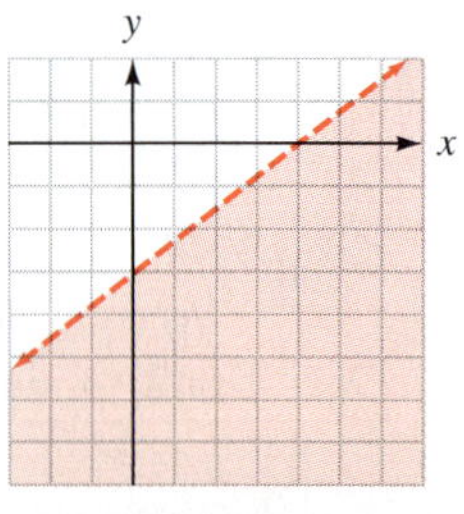

SECTION 4.5 Systems of Linear Inequalities

To solve a *system of linear inequalities,* graph each of the inequalities on the same set of coordinate axes and look for the intersection of the shaded *half-planes.*

Compound inequalities can be graphed in the rectangular coordinate system.

Graph the solution set of each system of inequalities.

49. $\begin{cases} y \geq x + 1 \\ 3x + 2y < 6 \end{cases}$

50. $\begin{cases} x - y < 3 \\ y \leq 0 \\ x \geq 0 \end{cases}$

Graph each compound inequality in the rectangular coordinate system.

51. $-2 < x < 4$

52. $y \leq -2$ or $y > 1$

53. PETROLEUM EXPLORATION Organic matter converts to oil and gas within a specific range of temperature and depth called the *petroleum window.* The petroleum window shown can be described by a system of linear inequalities, where x is the temperature in °C of the soil at a depth of y meters. Determine what inequality symbol should be inserted in each blank.

$\begin{cases} x \;\square\; 35 \\ x \;\square\; 130 \\ y \;\square\; -56x + 280 \\ y \;\square\; -18x + 90 \end{cases}$

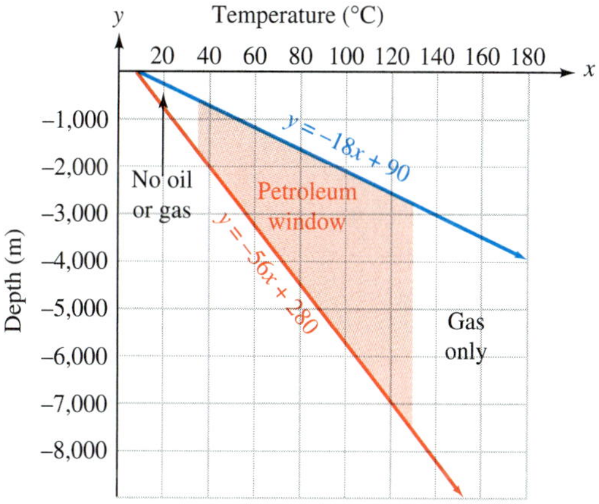

Based on data from *The Blue Planet* (Wiley, 1995)

54. In the illustration, the solution of one linear inequality is shaded in red, and the solution of a second is shaded in blue. Decide whether a true or false statement results if the coordinates of the given point are substituted into the given inequality.

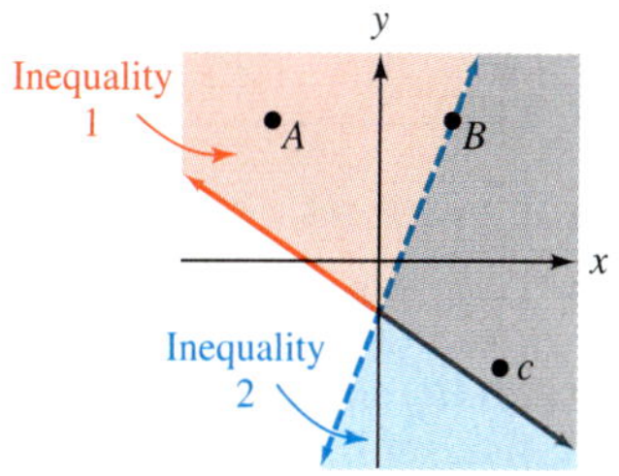

a. A, inequality 1

b. A, inequality 2

c. B, inequality 1

d. B, inequality 2

e. C, inequality 1

f. C, inequality 2

CHAPTER 4 TEST

1. Decide whether the statement is true or false.
$-5.67 \geq -5$

2. Decide whether -2 is a solution of the inequality.
$3(x - 2) \leq 2(x + 7)$

Graph the solution set of each inequality and give the solution in interval notation.

3. $7 < \frac{2}{3}t - 1$

4. $-2(2x + 3) \geq 14$

5. $\frac{x}{4} - \frac{1}{3} > \frac{5}{6} + \frac{x}{3}$

6. $4 - 4[3t - 2(3 - t)] \leq -15t - (5t - 28)$

7. AVERAGING GRADES Use the information from the gradebook to determine what score Karen Nelson-Sims needs on the fifth exam so that her exam average exceeds 80.

Sociology 101 8:00-10:00 pm MW	Exam 1	Exam 2	Exam 3	Exam 4	Exam 5
Nelson-Sims, Karen	70	79	85	88	

Solve each compound inequality. Give the result in interval notation, if possible, and graph the solution set.

8. $3x \geq -2x + 5$ and $7 \geq 4x - 2$

9. $3x < -9$ or $-\frac{x}{4} < -2$

10. $-2 < \frac{x - 4}{3} < 4$

11. $\frac{4}{5}(x + 1) > 1$ and $-(0.3x + 1.5) > 2.9 - 0.2x$

Solve each equation.

12. $|4 - 3x| = 19$

13. $|3x + 4| = |x + 12|$

14. $10 = 4\left|\frac{3x}{8} - \frac{3x}{2}\right| + 6$

15. $|16x| = -16$

Graph each set.

16. $(-3, 6) \cup [5, \infty)$

17. $[-2, 7] \cap (-\infty, 1)$

Graph the solution set of each inequality and give the solution set in interval notation.

18. $|x + 3| \leq 4$

19. $\left|\frac{x - 2}{2}\right| > 5.5$

20. $|4 - 2x| + 48 > 50$

21. $2|3(x - 2)| \leq 4$

22. $|4.5x - 0.9| \geq -0.7$

23. Let $f(x) = |2x + 9|$. For what value(s) of x is $f(x) < 3$?

Graph each solution set.

24. $3x + 2y \geq 6$

25. $y < x$

26. $\begin{cases} x - \frac{3}{2}y \geq 3 \\ y \leq -x + 1 \end{cases}$

27. $-2 \leq y < 5$

28. ACCOUNTING On average, it takes an accountant 1 hour to complete a simple tax return and 3 hours to complete a complicated return. If the accountant wants to work less than 9 hours per day, find an inequality that shows the number of possible ways that simple returns (x) and complicated returns (y) can be completed each day. Then graph the inequality and give three ordered pairs that satisfy it.

29. Two linear inequalities are graphed on the same coordinate axes in the illustration. The solution set of the first inequality is shaded in red, and the solution set of the second in blue.

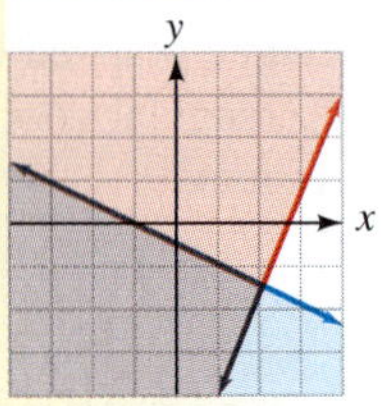

a. Determine from the graph whether $(3, -4)$ is a solution of either inequality.

b. Is $(3, -4)$ a solution of the system of two linear inequalities? Explain your answer.

30. INDOOR CLIMATES The general zone of comfort acceptable to most people when working in an office can be described by a system of linear inequalities where x is the dry bulb temperature and y is the percent relative humidity. See the illustration. Determine what inequality symbol should be inserted in each blank.

$$\begin{cases} y \ \square \ 60 \\ y \ \square \ 27 \\ y \ \square \ -11x + 852 \\ y \ \square \ -5x + 445 \end{cases}$$

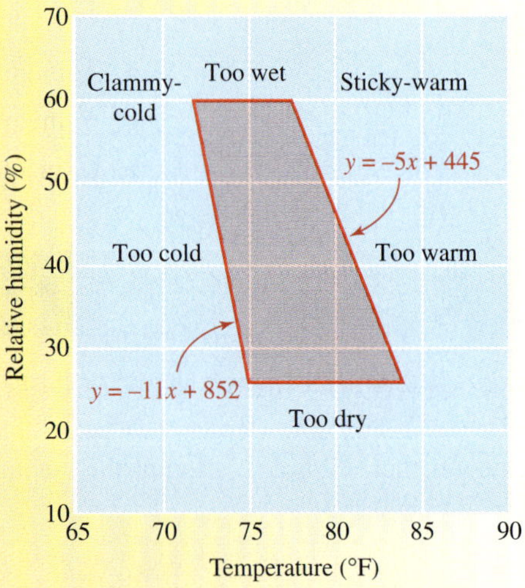

CHAPTERS 1–4 CUMULATIVE REVIEW EXERCISES

1. The diagram shows the sets that compose the set of real numbers. Which of the indicated sets make up the *rational numbers* and the *irrational numbers?*

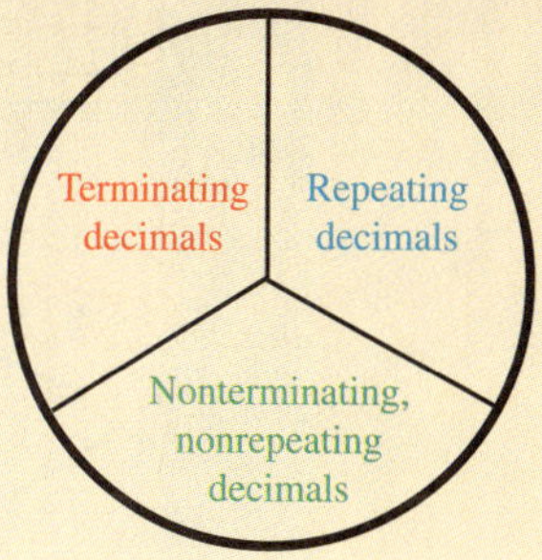

2. HARDWARE The thread profile of a screw is determined by the distance between threads. This distance, indicated by the letter p, is known as the *pitch.* If $p = 0.125$, find each of the dimensions labeled in the illustration.

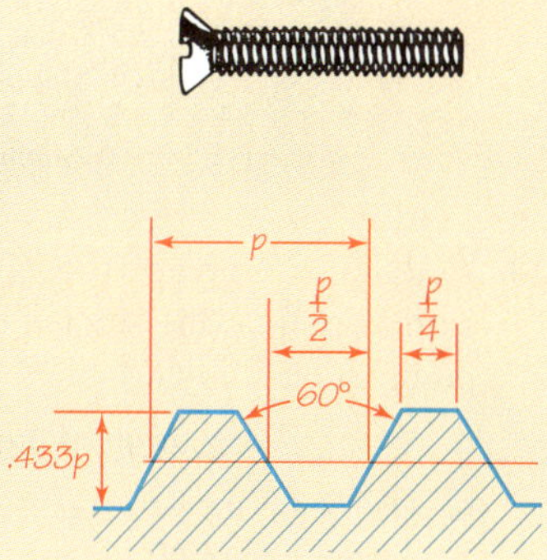

Evaluate each expression for $x = 2$ and $y = -4$.

3. $|x| - xy$

4. $\dfrac{x^2 - y^2}{3x + y}$

Simplify each expression.

5. $-(a + 2) - (a - b)$

6. $36\left(\dfrac{2}{9}t - \dfrac{3}{4}\right) + 36\left(\dfrac{1}{2}\right)$

7. PLASTIC WRAP Estimate the number of *square feet* of plastic wrap on a roll if the dimensions printed on the box describe the roll as 205 feet long by $11\frac{3}{4}$ inches wide.

8. INVESTMENTS Find the amount of money that was invested at $8\frac{7}{8}\%$ if it earned \$1,775 in simple interest in one year.

Solve each equation, if possible.

9. $6(x - 1) = 2(x + 3)$

10. $\dfrac{5b}{2} - 10 = \dfrac{b}{3} + 3$

11. $2a - 5 = -2a + 4(a - 2) + 1$

12. $\dfrac{2z + 3}{3} + \dfrac{3z - 4}{6} = \dfrac{z - 2}{2}$

13. Solve $\ell = a + (n - 1)d$ for d.

14. Determine whether the lines represented by the equations are parallel, perpendicular, or neither.

$$3x = y + 4$$
$$y = 3(x - 4) - 1$$

15. Write the equation of the line that passes through $(-2, 3)$ and is perpendicular to the graph of $3x + y = 8$. Answer in slope–intercept form.

16. Find the slope of the line that passes through $(0, -8)$ and $(-5, 0)$.

17. PRISONS The following graph shows the growth of the U.S. prison population from 1970 to 2000. Find the rate of change in the prison population from 1970 to 1975.

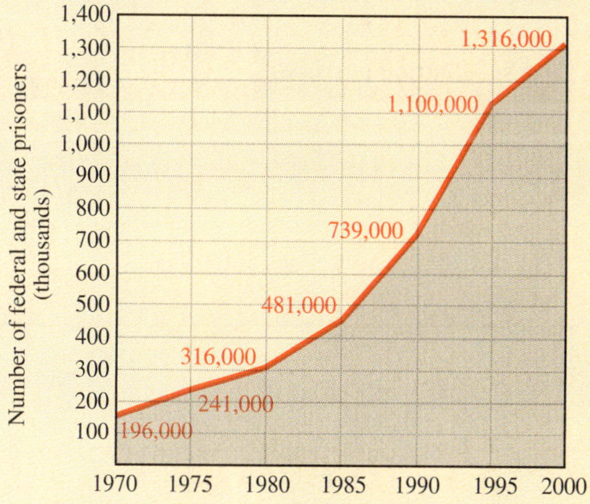

Source: *U.S. Statistical Abstract and Time Almanac* 2004

18. PRISONS Refer to the graph above. During what five-year period was the rate of change in the U.S. prison population the greatest? Find the rate of change.

Let $f(x) = 3x^2 - x$ and find each value.

19. $f(2)$

20. $f(-2)$

21. Graph $f(x) = |x| - 2$ and give its domain and range.

22. BOATING The graph in the illustration shows the vertical distance from a point on the tip of a propeller to the centerline as the propeller spins. Is this the graph of a function?

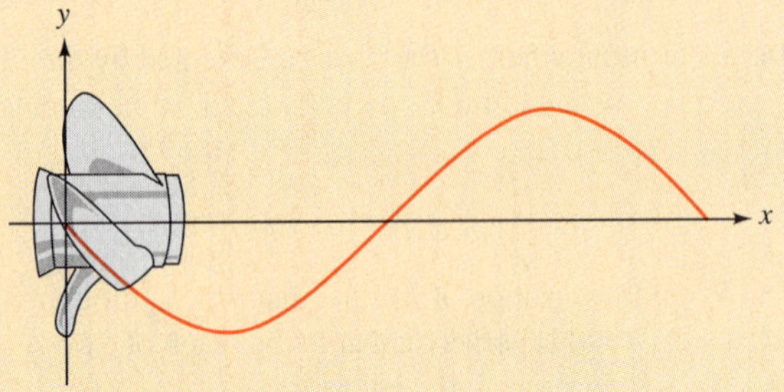

23. Use graphing to solve $\begin{cases} 2x + y = 5 \\ x - 2y = 0 \end{cases}$.

24. Use elimination to solve $\begin{cases} \dfrac{x}{10} + \dfrac{y}{5} = \dfrac{1}{2} \\ \dfrac{x}{2} - \dfrac{y}{5} = \dfrac{13}{10} \end{cases}$.

25. Use substitution to solve $\begin{cases} 3x = 4 - y \\ 4x - 3y = -1 + 2x \end{cases}$.

26. Solve: $\begin{cases} x + y + z = 1 \\ 2x - y - z = -4. \\ x - 2y + z = 4 \end{cases}$

27. Use matrices to solve the system.

$$\begin{cases} 4x - 3y = -1 \\ 3x + 4y = -7 \end{cases}$$

28. Use Cramer's rule to solve the system.

$$\begin{cases} x - 2y - z = -2 \\ 3x + y - z = 6 \\ 2x - y + z = -1 \end{cases}$$

29. U.S. WORKERS The illustration in the next column shows how the makeup of the U.S. workforce changed over the years 1900–1990. Estimate the coordinates of the points of intersection in the graph. Explain their significance.

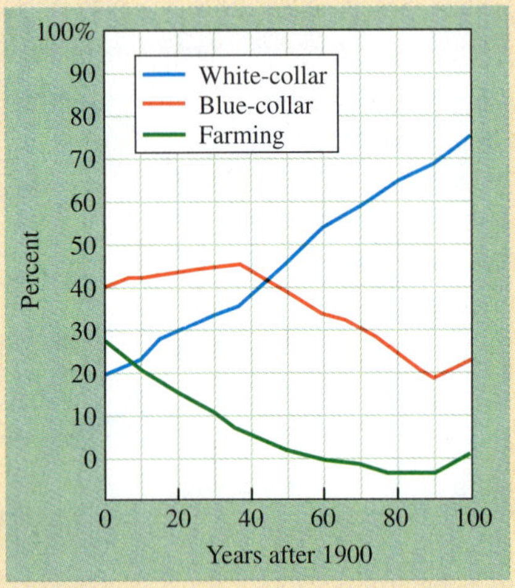

Source: *U.S. Statistical Abstract*

30. AGING The graph shows the effects of aging on cardiac output (the amount of blood that the heart can pump in one minute).

a. Write the equation of the line.

b. Use your answer to part a to determine the cardiac output at age 90.

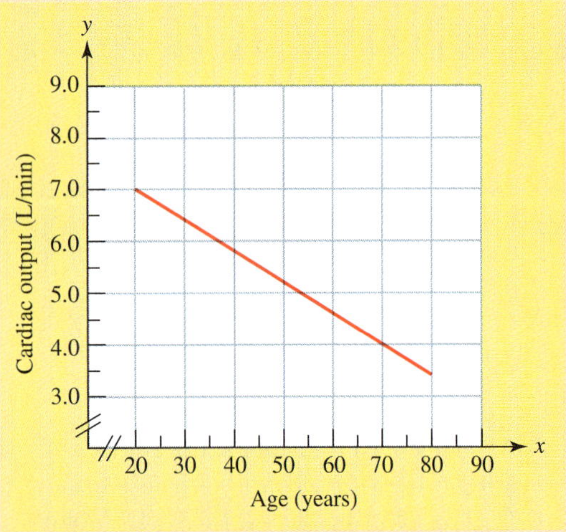

Based on data from *Cardiopulmonary Anatomy and Physiology, Essentials for Respiratory Care*, 2nd ed. (Delmar Publishers, 1994)

31. ENTREPRENEURS A person invests \$18,375 to set up a small business producing a piece of computer software that will sell for \$29.95. If each piece can be produced for \$5.45, how many pieces must be sold to break even?

32. CONCERT TICKETS Tickets for a concert cost \$5, \$3, and \$2. Twice as many \$5 tickets were sold as \$2 tickets. The receipts for 750 tickets were \$2,625. How many tickets were sold at each price?

Solve each equation.

33. $2|4x - 3| + 1 = 19$

34. $|2x - 1| = |3x + 4|$

Solve each inequality. Give the solution in interval notation and graph it.

35. $-3(x - 4) \geq x - 32$

36. $-8 < -3x + 1 < 10$

37. $|3x - 2| \leq 4$

38. $|2x + 3| - 1 > 4$

Use graphing to solve each inequality or system of inequalities.

39. $2x - 3y \leq 12$

40. $\begin{cases} y < x + 2 \\ 3x + y \leq 6 \end{cases}$

Chapter

5 Exponents, Polynomials, and Polynomial Functions

Getty Images/David Noton

Over the past twenty years, the federal government, and many state and local governments, have increased their efforts to clean up hazardous waste sites that threaten public health and the environment. Environmental engineers are often hired to oversee these projects that deal with leaking underground storage tanks, chemical spills, asbestos, and lead paint. They use principles of biology, chemistry, and mathematics to develop plans to restore the sites to their original condition.

To learn more about the role of mathematics in environmental cleanup, visit *The Learning Equation* on the Internet at http://tle.brookscole.com. (The log-in instructions are in the Preface.) For Chapter 5, the online lessons are:

- *TLE* Lesson 7: The Greatest Common Factor and Factoring by Grouping
- *TLE* Lesson 8: Factoring Trinomials and the Difference of Squares

Polynomials are algebraic expressions that are used to model many real-world situations. They often contain terms in which the variables have exponents.

5.1 Exponents

- Exponents
- Rules for exponents
- Zero exponents
- Negative exponents
- More rules for exponents

We have evaluated exponential expressions having natural-number exponents. In this section, we will extend the definition of exponent to include negative-integer exponents, as in 3^{-2}, and zero exponents, as in 3^0. We will also develop several rules that simply work with exponents.

EXPONENTS

The **exponential expression** x^n is called a **power of *x*,** and we read it as "x to the nth power." In this expression, x is called the **base,** and n is called the **exponent.**

Base → x^n ← Exponent

Exponents provide a way to write products of *repeated factors* in compact form.

Natural-Number Exponents

A natural-number exponent tells how many times its base is to be used as a factor. For any number x and any natural number n,

$$x^n = \overbrace{x \cdot x \cdot x \cdot \cdots \cdot x}^{n \text{ factors of } x}$$

EXAMPLE 1

Identify the base and the exponent in each expression: **a.** $(5x)^3$, **b.** $5x^3$, **c.** $-a^4$, **d.** $\left(\frac{2b^8}{9c}\right)^4$, and **e.** $(x - 7)^2$.

Solution

a. When an exponent is written outside parentheses, the expression within the parentheses is the base. For $(5x)^3$, $5x$ is the base and 3 is the exponent: $(5x)^3 = (5x)(5x)(5x)$.

$(5x)^3$ ← Exponent; $(5x)$ ← Base

Notation

An exponent of 1 means the base is to be used as a factor 1 time. For example, $x^1 = x$.

b. $5x^3$ means $5 \cdot x^3$. Thus, x is the base and 3 is the exponent: $5x^3 = 5 \cdot x \cdot x \cdot x$.

c. $-a^4$ means $-1 \cdot a^4$. Thus, a is the base and 4 is the exponent: $-a^4 = -1(a \cdot a \cdot a \cdot a)$.

d. Because of the parentheses, $\left(\frac{2b^8}{9c}\right)$ is the base and 4 is the exponent:

$$\left(\frac{2b^8}{9c}\right)^4 = \left(\frac{2b^8}{9c}\right)\left(\frac{2b^8}{9c}\right)\left(\frac{2b^8}{9c}\right)\left(\frac{2b^8}{9c}\right)$$

e. Because of the parentheses, $x - 7$ is the base and 2 is the exponent: $(x - 7)^2 = (x - 7)(x - 7)$.

Self Check 1 Identify the base and the exponent in each expression: **a.** $(kt)^4$, **b.** πr^2, **c.** $-h^8$, **d.** $\left(\frac{3n}{2m^5}\right)^5$, and **e.** $(y+1)^3$.

RULES FOR EXPONENTS

Several rules for exponents come directly from the definition of exponent. To develop the first rule, we consider $x^5 \cdot x^3$, the product of two exponential expressions having the same base. Since x^5 means that x is to be used as a factor five times, and since x^3 means that x is to be used as a factor three times, $x^5 \cdot x^3$ means that x will be used as a factor eight times.

$$x^5x^3 = \underbrace{x \cdot x \cdot x \cdot x \cdot x}_{\text{5 factors of } x} \cdot \underbrace{x \cdot x \cdot x}_{\text{3 factors of } x} = \underbrace{x \cdot x \cdot x \cdot x \cdot x \cdot x \cdot x \cdot x}_{\text{8 factors of } x} = x^8$$

In general,

$$x^mx^n = \underbrace{x \cdot x \cdot x \cdot \cdots \cdot x}_{m \text{ factors of } x} \cdot \underbrace{x \cdot x \cdot \cdots \cdot x}_{n \text{ factors of } x} = \underbrace{x \cdot x \cdot x \cdot x \cdot \cdots \cdot x}_{m+n \text{ factors of } x} = x^{m+n}$$

This result is called the **product rule for exponents.**

Product Rule for Exponents

To multiply exponential expressions with the same base, keep the common base and add the exponents.

For any real number x and any natural numbers m and n,

$$x^m \cdot x^n = x^{m+n}$$

EXAMPLE 2

Simplify each expression: **a.** $x^{11}x^5$, **b.** y^5y^4y, **c.** $a^2b^3a^3b^2$, **d.** $-8x^4(x^3)$.

Solution

a. $x^{11}x^5 = x^{11+5}$ Keep the common base x. Add the exponents.

$= x^{16}$

b. $y^5y^4y = (y^5y^4)y$

$= y^9y^1$

$= y^{10}$

c. $a^2b^3a^3b^2 = a^2a^3b^3b^2$

$= a^5b^5$

d. $-8x^4(x^3) = -8(x^4x^3)$

$= -8x^7$

Self Check 2 Simplify each expression: **a.** 2^32^5, **b.** $k \cdot k^4$, **c.** $a^2b^3a^3b^4$, and **d.** $-8a^4(a^2b)$.

Caution Here are examples of two common errors associated with the product rule:

$3^2 \cdot 3^4 \neq 9^6$

Do not multiply the common bases. Keep the common base and add exponents to get 3^6.

$2^3 \cdot 5^2 \neq 10^5$

The power rule does not apply. The bases are not the same.

The Language of Algebra

An exponential expression raised to a power, such as $(x^4)^3$, is called a *power of a power.*

To develop another rule, we consider $(x^4)^3$, which means x^4 cubed.

$$(x^4)^3 = x^4 \cdot x^4 \cdot x^4 = \underbrace{x \cdot x \cdot x \cdot x}_{x^4} \cdot \underbrace{x \cdot x \cdot x \cdot x}_{x^4} \cdot \underbrace{x \cdot x \cdot x \cdot x}_{x^4} = x^{12}$$

In general, we have

$$(x^m)^n = \underbrace{x^m \cdot x^m \cdot x^m \cdot \cdots \cdot x^m}_{n \text{ factors of } x^m} = \underbrace{x \cdot x \cdot x \cdot x \cdot x \cdot \cdots \cdot x}_{mn \text{ factors of } x} = x^{mn}$$

This result is called the **power rule for exponents.**

Power Rule for Exponents To raise an exponential expression to a power, keep the base and multiply the exponents.

For any real number x and any natural numbers m and n,

$$(x^m)^n = x^{mn}$$

EXAMPLE 3

Simplify each expression: **a.** $(3^2)^3$, **b.** $(x^{11})^5$, **c.** $(x^2x^3)^6$, and **d.** $(x^2)^4(x^3)^2$.

a. $(3^2)^3 = 3^{2\cdot3}$ Keep the base. Multiply the exponents.

$= 3^6$

$= 729$

b. $(x^{11})^5 = x^{11\cdot5}$

$= x^{55}$

c. $(x^2x^3)^6 = (x^5)^6$ Within the parentheses, keep the common base and add the exponents.

$= x^{30}$ Keep the base. Multiply the exponents.

d. $(x^2)^4(x^3)^2 = x^8x^6$

$= x^{14}$

Self Check 3 Simplify each expression: **a.** $(a^5)^8$, **b.** $(6^3)^5$, **c.** $(a^4a^3)^3$, and **d.** $(a^3)^3(a^2)^3$.

To develop a third rule, we consider $(3x)^2$, which means $3x$ squared.

$$(3x)^2 = (3x)(3x) = 3 \cdot 3 \cdot x \cdot x = 3^2x^2 = 9x^2$$

In general, we have

$$(xy)^n = \underbrace{(xy)(xy)(xy) \cdot \cdots \cdot (xy)}_{n \text{ factors of } xy} = \underbrace{xxx \cdot \cdots \cdot x}_{n \text{ factors of } x} \cdot \underbrace{yyy \cdot \cdots \cdot y}_{n \text{ factors of } y} = x^ny^n$$

To develop a fourth rule, we consider $\left(\frac{x}{3}\right)^3$, which means $\frac{x}{3}$ cubed.

$$\left(\frac{x}{3}\right)^3 = \frac{x}{3} \cdot \frac{x}{3} \cdot \frac{x}{3} = \frac{x \cdot x \cdot x}{3 \cdot 3 \cdot 3} = \frac{x^3}{3^3} = \frac{x^3}{27}$$

In general, we have

$$\left(\frac{x}{y}\right)^n = \underbrace{\left(\frac{x}{y}\right)\left(\frac{x}{y}\right)\left(\frac{x}{y}\right)\cdot\cdots\cdot\left(\frac{x}{y}\right)}_{n \text{ factors of } \frac{x}{y}} \quad \text{where } y \neq 0$$

$$= \frac{\overbrace{xxx\cdot\cdots\cdot x}^{n \text{ factors of } x}}{\underbrace{yyy\cdot\cdots\cdot y}_{n \text{ factors of } y}} \qquad \text{Multiply the numerators and multiply the denominators.}$$

$$= \frac{x^n}{y^n}$$

Caution

There is no rule for the power of a sum or power of a difference. To show why, consider this example:

$$(3+2)^2 \stackrel{?}{=} 3^2 + 2^2$$
$$5^2 \stackrel{?}{=} 9 + 4$$
$$25 \neq 13$$

The previous results are called the **power of a product** and the **power of a quotient rules.**

Powers of a Product and a Quotient

To raise a product to a power, raise each factor of the product to that power. To raise a quotient to a power, raise the numerator and the denominator to that power.

For any real numbers x and y, and any natural number n,

$$(xy)^n = x^n y^n \quad \text{and} \quad \left(\frac{x}{y}\right)^n = \frac{x^n}{y^n}, \quad \text{where } y \neq 0$$

EXAMPLE 4

Simplify each expression. Assume that no denominators are zero: **a.** $(x^2y)^3$, **b.** $(2y^4)^5$, **c.** $\left(\frac{x}{y^2}\right)^4$, and **d.** $\left(\frac{6x^3}{5y^4}\right)^2$.

a. $(x^2y)^3 = (x^2)^3y^3 = x^6y^3$ — Raise each factor of the product x^2y to the 3rd power.

b. $(2y^4)^5 = (2)^5(y^4)^5 = 32y^{20}$

c. $\left(\frac{x}{y^2}\right)^4 = \frac{x^4}{(y^2)^4} = \frac{x^4}{y^8}$ — Raise the numerator and denominator to the 4th power.

d. $\left(\frac{6x^3}{5y^4}\right)^2 = \frac{6^2(x^3)^2}{5^2(y^4)^2} = \frac{36x^6}{25y^8}$

Self Check 4 Simplify each expression: **a.** $(a^4b^5)^2$, **b.** $\left(\frac{-6a^5}{b^7}\right)^3$, and **c.** $(-2d^5)^4$.

ZERO EXPONENTS

To develop the definition of a zero exponent, we consider the expression $x^0 \cdot x^n$, where x is not 0. By the product rule,

$$x^0 \cdot x^n = x^{0+n} = x^n = 1x^n$$

For the product rule to hold true for 0 exponents, $x^0 \cdot x^n$ must equal $1x^n$. Comparing factors, it follows that $x^0 = 1$. This result suggests the following definition.

Zero Exponents A nonzero base raised to the 0 power is 1.

For any nonzero base x,

$$x^0 = 1$$

The Language of Algebra
Note that 0^0 is undefined. This expression is said to be an *indeterminate form.*

For example, if no variables are zero, then

$$3^0 = 1 \qquad (-7)^0 = 1 \qquad (3ax^3)^0 = 1 \qquad \left(\frac{1}{2}x^5y^7z^9\right)^0 = 1$$

EXAMPLE 5

Simplify each expression: **a.** $(5x)^0$, **b.** $5x^0$, **c.** $-(5xy)^0$, and **d.** $-5x^0y$.

Solution

a. $(5x)^0 = 1$ The base is $5x$ and the exponent is 0.

b. $5x^0 = 5 \cdot x^0 = 5 \cdot 1 = 5$ The base is x and the exponent is 0.

c. $-(5xy)^0 = -1$ The base is $5xy$ and the exponent is 0.

d. $-5x^0y = -5 \cdot x^0 \cdot y = -5 \cdot 1 \cdot y = -5y$

Self Check 5 Simplify each expression: **a.** $2xy^0$ and **b.** $-(xy)^0$.

NEGATIVE EXPONENTS

To develop the definition of a negative integer exponent, we consider the expression $x^{-n} \cdot x^n$, where x is not 0. By the product rule,

$$x^{-n} \cdot x^n = x^{-n+n} = x^0 = 1$$

Since their product is 1, x^{-n} and x^n must be reciprocals. It is also true that $\frac{1}{x^n}$ and x^n are reciprocals and their product is 1.

$$x^{-n} \cdot x^n = 1 \qquad \frac{1}{x^n} \cdot x^n = 1$$

Comparing factors, it follows that x^{-n} must equal $\frac{1}{x^n}$. This result suggests the following definition.

Negative Exponents For any nonzero real number x and any integer n,

$$x^{-n} = \frac{1}{x^n}$$

In words, x^{-n} is the reciprocal of x^n.

From the definition, we see that another way to write x^{-n} is to write its reciprocal and change the sign of the exponent. For example,

Caution

A negative exponent does not indicate a negative number. It indicates a reciprocal.

$$3^{-2} = \frac{1}{3^2}$$ First, write the reciprocal of 3^{-2}, which is $\frac{1}{3^{-2}}$. Then change the sign of the exponent.

$$= \frac{1}{9}$$

EXAMPLE 6

Write each expression using positive exponents only. Simplify, if possible: **a.** 4^{-3}, **b.** $(-2)^{-5}$, **c.** $7m^{-8}$, and **d.** $-n^{-4}$.

Solution

a. $4^{-3} = \frac{1}{4^3}$ Write the reciprocal of 4^{-3} and change the sign of the exponent.

$= \frac{1}{64}$ Evaluate: $4^3 = 64$.

b. $(-2)^{-5} = \frac{1}{(-2)^5}$

$= -\frac{1}{32}$

Caution

Don't confuse negative numbers with negative exponents. For example, the expressions -2 and 2^{-1} are not the same.

$$2^{-1} = \frac{1}{2^1} = \frac{1}{2}$$

c. $7m^{-8} = 7 \cdot m^{-8}$ Since there are no parentheses, the base is m.

$= 7 \cdot \frac{1}{m^8}$ Write the reciprocal of m^{-8} and change the sign of the exponent.

$= \frac{7}{m^8}$

d. $-n^{-4} = -1 \cdot n^{-4}$

$= -1 \cdot \frac{1}{n^4}$

$= -\frac{1}{n^4}$

Self Check 6 Write each expression using positive exponents only. Simplify, if possible: **a.** 8^{-2}, **b.** $(-3)^{-3}$, **c.** $12h^{-9}$, and $-c^{-1}$.

The rules for exponents involving products and powers are also true for negative exponents.

EXAMPLE 7

Simplify each expression. Write answers using positive exponents only. **a.** $x^{-5}x^3$ and **b.** $(x^{-3})^{-2}$.

a. $x^{-5}x^3 = x^{-5+3}$ Keep the common base x and add the exponents.

$= x^{-2}$

$= \frac{1}{x^2}$

b. $(x^{-3})^{-2} = x^{(-3)(-2)}$ Keep the base and multiply the exponents.

$= x^6$

Self Check 7 Simplify each expression using positive exponents only: **a.** $a^{-7}a^3$ and **b.** $(a^{-5})^{-3}$.

Negative exponents can appear in the numerator and/or the denominator of a fraction. To develop rules to apply to such situations, consider the following example.

$$\frac{x^{-4}}{y^{-3}} = \frac{\frac{1}{x^4}}{\frac{1}{y^3}} = \frac{1}{x^4} \cdot \frac{y^3}{1} = \frac{y^3}{x^4}$$

We can obtain this result in a simpler way. Beginning with $\frac{x^{-4}}{y^{-3}}$, move x^{-4} to the denominator and change the sign of its exponent. Then move y^{-3} to the numerator and change the sign of its exponent.

$$\frac{x^{-4}}{y^{-3}} = \frac{y^3}{x^4}$$

This example suggests the following rules.

Changing from Negative to Positive Exponents

A factor can be moved from the denominator to the numerator or from the numerator to the denominator of a fraction if the sign of its exponent is changed.

For any nonzero real numbers x and y, and any integers m and n,

$$\frac{1}{x^{-n}} = x^n \quad \text{and} \quad \frac{x^{-m}}{y^{-n}} = \frac{y^n}{x^m}$$

EXAMPLE 8

Write each expression using positive exponents only. Simplify, if possible: **a.** $\frac{1}{c^{-10}}$, **b.** $\frac{2^{-3}}{3^{-4}}$, and **c.** $-\frac{s^{-2}}{5t^{-9}}$.

Solution

a. $\frac{1}{c^{-10}} = c^{10}$ Move c^{-10} to the numerator and change the sign of the exponent.

b. $\frac{2^{-3}}{3^{-4}} = \frac{3^4}{2^3}$ Move 2^{-3} to the denominator and change the sign of the exponent. Move 3^{-4} to the numerator and change the sign of the exponent.

$= \frac{81}{8}$ Evaluate 3^4 and 2^3.

c. $-\frac{s^{-2}}{5t^{-9}} = -\frac{t^9}{5s^2}$ Move s^{-2} to the denominator and change the sign of the exponent. Since $5t^{-9}$ has no parentheses, t is the base. Move t^{-9} to the numerator and change the sign of the exponent.

Caution

This rule does not allow us to move *terms* that have negative exponents. For example,

$$\frac{3^{-2} + 8}{5} \neq \frac{8}{3^2 \cdot 5}$$

Self Check 8 Write each expression using positive exponents only. Simplify, if possible: **a.** $\frac{1}{t^{-9}}$, **b.** $\frac{5^{-2}}{4^{-3}}$, and **c.** $-\frac{h^{-6}}{8r^{-7}}$.

MORE RULES FOR EXPONENTS

To develop a rule for dividing exponential expressions, we proceed as follows:

$$\frac{x^m}{x^n} = x^m\left(\frac{1}{x^n}\right) = x^m x^{-n} = x^{m+(-n)} = x^{m-n}$$

This result is called the **quotient rule for exponents.**

Quotient Rule for Exponents

To divide exponential expressions with the same base, keep the common base and subtract the exponents.

For any nonzero number x and any integers m and n,

$$\frac{x^m}{x^n} = x^{m-n}$$

EXAMPLE 9

Simplify each expression. Write answers using positive exponents only. **a.** $\frac{a^5}{a^3}$ and **b.** $\frac{2x^{-5}}{x^{11}}$.

a. $\frac{a^5}{a^3} = a^{5-3}$ Keep the common base a. Subtract the exponents.

$= a^2$

b. $\frac{2x^{-5}}{x^{11}} = 2x^{-5-11}$

$= 2x^{-16}$

$= \frac{2}{x^{16}}$

Self Check 9 Simplify each expression: **a.** $\frac{b^7}{b^5}$ and **b.** $\frac{3b^{-3}}{b^3}$.

EXAMPLE 10

Simplify each expression. Write answers using positive exponents only.

a. $\frac{x^4x^3}{x^{-5}} = \frac{x^7}{x^{-5}}$

$= x^{7-(-5)}$

$= x^{12}$

b. $\frac{(x^2)^3}{(x^3)^2} = \frac{x^6}{x^6}$

$= x^{6-6}$

$= x^0$

$= 1$

c. $\frac{x^2y^3}{7xy^4} = \frac{x^{2-1}y^{3-4}}{7}$

$= \frac{xy^{-1}}{7}$

$= \frac{x}{7y}$

d. $\left(\frac{2a^{-2}b^3}{3a^5b^4}\right)^3 = \left(\frac{2a^{-2-5}b^{3-4}}{3}\right)^3$

$= \left(\frac{2a^{-7}b^{-1}}{3}\right)^3$

$= \left(\frac{2}{3a^7b}\right)^3$

$= \frac{8}{27a^{21}b^3}$

Success Tip

When more than one rule for exponents is involved in a simplification, more than one approach can often be used. In Example 10a, we obtain the same result with this alternate approach:

$\frac{x^4x^3}{x^{-5}} = x^4x^3x^5$

$= x^{12}$

Self Check 10 Simplify each expression: **a.** $\frac{(a^{-2})^3}{(a^2)^{-3}}$ and **b.** $\left(\frac{a^{-2}b^5}{5b^8}\right)^{-3}$.

To develop another rule, we consider the following simplification:

$$\left(\frac{2}{3}\right)^{-4} = \frac{1}{\left(\frac{2}{3}\right)^4} = \frac{1}{\frac{2^4}{3^4}} = 1 \div \frac{2^4}{3^4} = 1 \cdot \frac{3^4}{2^4} = \frac{3^4}{2^4} = \left(\frac{3}{2}\right)^4$$

This result suggests the following rule for exponents.

Fractions to Negative Powers

To raise a fraction to the negative nth power, invert the fraction and then raise it to the nth power.

For any nonzero real numbers x and y, and any integer n,

$$\left(\frac{x}{y}\right)^{-n} = \left(\frac{y}{x}\right)^n$$

EXAMPLE 11

Write each expression without using parentheses or negative exponents.

a. $\left(\frac{2}{3}\right)^{-4} = \left(\frac{3}{2}\right)^4$ Invert $\frac{2}{3}$. Change the exponent to positive 4.

$= \frac{3^4}{2^4}$

$= \frac{81}{16}$

b. $\left(\frac{y^2}{x^3}\right)^{-3} = \left(\frac{x^3}{y^2}\right)^3$

$= \frac{x^9}{y^6}$

c. $\left(\frac{a^{-2}b^3}{a^2a^3b^4}\right)^{-3} = \left(\frac{a^2a^3b^4}{a^{-2}b^3}\right)^3$ Invert the fraction within the parentheses. Change the exponent to positive 3.

$= \left(\frac{a^5b^4}{a^{-2}b^3}\right)^3$

$= (a^{5-(-2)}b^{4-3})^3$

$= (a^7b)^3$

$= a^{21}b^3$

d. $\left(\frac{2x^2}{3y^{-3}}\right)^{-4} = \left(\frac{3y^{-3}}{2x^2}\right)^4$

$= \frac{3^4y^{-12}}{2^4x^8}$

$= \frac{81}{16x^8y^{12}}$

Self Check 11 Write $\left(\frac{3a^3}{2b^{-2}}\right)^{-5}$ without using parentheses or negative exponents.

We summarize the rules for exponents as follows.

Rules for exponents

If there are no divisions by zero, then for any integers m and n,

Product rule	*Quotient rule*	*Power rule*
$x^m \cdot x^n = x^{m+n}$	$\frac{x^m}{x^n} = x^{m-n}$	$(x^m)^n = x^{mn}$
Power of a product	*Power of a quotient*	*Zero exponent*
$(xy)^n = x^ny^n$	$\left(\frac{x}{y}\right)^n = \frac{x^n}{y^n}$	$x^0 = 1$
Negative exponent	*Negative exponent*	*Negative exponent*
$x^{-m} = \frac{1}{x^m}$	$\frac{x^{-n}}{y^{-n}} = \frac{y^n}{x^n}$	$\left(\frac{x}{y}\right)^{-n} = \left(\frac{y}{x}\right)^n$

Answers to Self Checks **1. a.** kt; 4, **b.** r; 2, **c.** h; 8, **d.** $\frac{3n}{2m^5}$; 5, **e.** $y + 1$; 3 **2. a.** $2^8 = 256$, **b.** k^5, **c.** a^5b^7, **d.** $-8a^6b$ **3. a.** a^{40}, **b.** 6^{15}, **c.** a^{21}, **d.** a^{15} **4. a.** a^8b^{10}, **b.** $-\frac{216a^{15}}{b^{21}}$, **c.** $16d^{20}$ **5. a.** $2x$, **b.** -1 **6. a.** $\frac{1}{64}$, **b.** $-\frac{1}{27}$, **c.** $\frac{12}{h^9}$, **d.** $-\frac{1}{c}$ **7. a.** $\frac{1}{a^4}$, **b.** a^{15} **8. a.** t^9, **b.** $\frac{64}{25}$, **c.** $-\frac{r^7}{8h^6}$ **9. a.** b^2, **b.** $\frac{3}{b^6}$ **10. a.** 1, **b.** $125a^6b^9$ **11.** $\frac{32}{243a^{15}b^{10}}$

5.1 STUDY SET

VOCABULARY Fill in the blanks.

1. Expressions such as x^4, 10^3, and $(5t)^2$ are called ________ expressions.
2. In the exponential expression x^n, x is called the ______, and n is called the ________.
3. The expression x^4 represents a repeated multiplication where x is to be written as a ______ four times.
4. $3^4 \cdot 3^8$ is a ________ of exponential expressions with the same base and $\frac{x^4}{x^2}$ is a ________ of exponential expressions with the same base.
5. $(h^3)^7$ is a ______ of an exponential expression.
6. In the expression 5^{-1}, the exponent is a ________ integer.

CONCEPTS Complete the rules for exponents. Assume that $x \neq 0$ and $y \neq 0$.

7. **a.** $x^mx^n = \square$ **b.** $(x^m)^n = \square$
   c. $(xy)^n = \square$ **d.** $\left(\frac{x}{y}\right)^n = \square$
   e. $x^0 = \square$ **f.** $x^{-n} = \square$
   g. $\frac{x^m}{x^n} = \square$ **h.** $\left(\frac{x}{y}\right)^{-n} = \left(\frac{y}{x}\right)^{\square}$
   i. $\frac{x^{-m}}{y^{-n}} = \square$
8. **a.** To multiply exponential expressions with the same base, keep the common base and _____ the exponents.
   b. To divide exponential expressions with the same base, keep the common base and ________ the exponents.
9. To raise an exponential expression to a power, keep the base and ________ the exponents.
10. **a.** To raise a product to a power, raise each ______ of the product to that power.
    b. To raise a quotient to a power, raise the ________ and the ________ to that power.
11. **a.** Any nonzero base raised to the 0 power is ___.
    b. x^{-n} is the ________ of x^n.
12. **a.** A factor can be moved from the denominator to the numerator or from the numerator to the denominator of a fraction if the _____ of its exponent is changed.
    b. To raise a fraction to the negative nth power, ______ the fraction and then raise it to the nth power.

NOTATION Complete each simplification.

13. $\frac{x^5x^4}{x^{-2}} = \frac{x^{\square}}{x^{-2}}$
    $= x^{9-\square}$
    $= x^{\square}$

14. $\left(\frac{a^{-4}}{a^3}\right)^2 = (a^{-4-3})^2$
    $= (a^{\square})^2$
    $= a^{\square}$
    $= \frac{\square}{a^{14}}$

PRACTICE Identify the base and the exponent.

15. $(6x)^3$
16. -7^2
17. $-x^5$
18. $(-t)^4$
19. $2b^6$
20. $12a^2$
21. $\left(\frac{n}{4}\right)^3$
22. $\left(\frac{3x}{y}\right)^0$
23. $(m - 8)^6$
24. $(t + 7)^8$

Evaluate each expression.

25. -3^2
26. -3^4
27. $(-3)^2$
28. $(-3)^3$
29. 5^{-2}
30. 5^{-4}
31. -9^{-2}
32. -2^{-4}
33. $(-6)^{-2}$
34. $(-4)^{-4}$
35. -8^0
36. -9^0
37. $(-8)^0$
38. $(-9)^0$
39. $\left(\frac{3}{4}\right)^3$
40. $\left(\frac{2}{5}\right)^2$
41. $\frac{1}{7^{-2}}$
42. $\frac{1}{4^{-3}}$
43. $\frac{2^{-4}}{1^{-10}}$
44. $\frac{3^{-4}}{1^{-9}}$
45. $\frac{-3}{4^{-2}}$
46. $\frac{-2}{6^{-2}}$
47. $\left(\frac{2}{3}\right)^{-2}$
48. $\left(\frac{4}{5}\right)^{-3}$

Simplify each expression. Assume that variables represent nonzero real numbers. Write answers using positive exponents only.

49. x^2x^3
50. y^3y^4
51. $y^3y^7y^2$
52. $x^2x^3x^5$
53. $x^8x^{11}x$
54. k^0k^7k
55. $h^{-3} \cdot h^8$
56. $s^{-10} \cdot s^{12}$
57. $m^{-4} \cdot m^{-6}$
58. $n^{-9} \cdot n^{-2}$
59. $2aba^3b^4$
60. $2x^2y^3x^3y^2$
61. $3p^9pp^0$
62. $4z^7z^0z$
63. $(-x)^2y^4x^3$
64. $-x^2y^7y^3x^{-2}$
65. $(b^{-8})^9$
66. $(z^{12})^2$
67. $(x^4)^7$
68. $(y^7)^5$
69. $(-2x)^5$
70. $(-3a)^3$
71. $r^{-10} \cdot r^{12} \cdot r$
72. $t^{-3} \cdot t^8 \cdot t$
73. $m^{-4} \cdot m^2 \cdot m^{-8}$
74. $n^{-9} \cdot n^5 \cdot n^{-7}$
75. $-5r^{-5}(r^6)^3$
76. $-8d^{-8}(d^9)^2$
77. $(r^{-3}s)^3$
78. $(m^5n^2)^{-3}$
79. $(2a^2a^3)^4$
80. $(3bb^2b^3)^4$
81. $(-3d^2)^3(d^{-3})^3$
82. $(c^3)^2(2c^4)^{-2}$
83. $(3x^3y^4)^3$
84. $\left(\frac{1}{2}a^2b^5\right)^4$
85. $(-s^2)^{-3}$
86. $(-t^2)^{-5}$
87. $\left(-\frac{1}{3}mn^2\right)^6$
88. $(-3p^2q^3)^5$
89. $\left(\frac{a^3}{b^2}\right)^5$
90. $\left(\frac{a^2}{b^3}\right)^4$
91. $\frac{1}{a^{-4}}$
92. $\frac{3}{b^{-5}}$
93. $\frac{a^{-3}}{a^{-21}}$
94. $\frac{n^{-5}}{n^{-8}}$
95. $\left(\frac{a^{-3}}{b^{-2}}\right)^{-2}$
96. $\left(\frac{k^{-3}}{k^{-4}}\right)^{-1}$
97. $\frac{a^8}{a^3}$
98. $\frac{c^7}{c^2}$
99. $\frac{c^{12}c^5}{c^{10}}$
100. $\frac{a^{33}}{a^2a^3}$
101. $\frac{8t^{-3} \cdot t^{-11}}{t^{-14}}$
102. $\frac{4x^{-9} \cdot x^{-3}}{x^{-12}}$
103. $\frac{(3x^2)^{-2}}{x^3x^{-4}x^0}$
104. $\frac{y^{-3}y^{-4}y^0}{(2y^{-2})^3}$
105. $\frac{3(-d^{-1})^{-5}}{8(d^{-4})^{-2}}$
106. $\frac{(-c^{-2})^{-4}}{15(c^{-3})^{-7}}$
107. $\frac{(3x^2)^{-2}}{x^3x^{-4}x^0}$
108. $\frac{y^{-3}y^{-4}y^0}{(-2y^{-2})^3}$
109. $\left(\frac{4a^{-2}b}{3ab^{-3}}\right)^3$
110. $\left(\frac{2ab^{-3}}{3a^{-2}b^2}\right)^2$
111. $\frac{b^0 - (4d)^0}{25(d+3)^0}$
112. $\frac{a^0 + b^0}{2(a+b)^0}$
113. $\left(\frac{-2a^4b}{a^{-3}b^2}\right)^3$
114. $\left(\frac{-3x^4y^2}{-9x^5y^{-2}}\right)^2$
115. $\left(-\frac{2a^3b^2}{3a^{-3}b^2}\right)^{-3}$
116. $\left(\frac{3x^5y^2}{6x^5y^{-2}}\right)^{-4}$
117. $\left(\frac{4a^2b^3z^{-4}}{3a^{-2}b^{-1}z^3}\right)^{-3}$
118. $\left(\frac{-3pqr^{-4}}{2p^2q^{-3}r^2}\right)^{-2}$

Use a calculator to verify that each statement is true by showing that the values on either side of the equation are equal.

119. $(3.68)^0 = 1$

120. $(2.1)^4(2.1)^3 = (2.1)^7$

121. $\left(\frac{5.4}{2.7}\right)^{-4} = \left(\frac{2.7}{5.4}\right)^4$

122. $(7.23)^{-3} = \frac{1}{(7.23)^3}$

APPLICATIONS

123. MICROSCOPES The illustration shows the relative sizes of some chemical and biological structures, expressed as fractions of a meter (m). Express each fraction shown in the illustration as a power of 10, from the largest to the smallest.

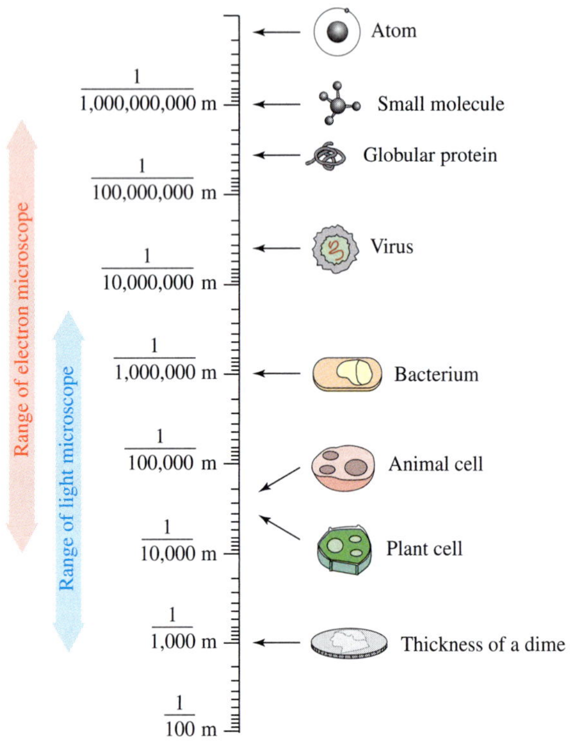

124. ASTRONOMY See the illustration in the next column. The distance d, in miles, of the nth planet from the sun is given by the formula

$$d = 9{,}275{,}200[3(2^{n-2}) + 4]$$

Find the distance of Earth and the distance of Mars from the sun.

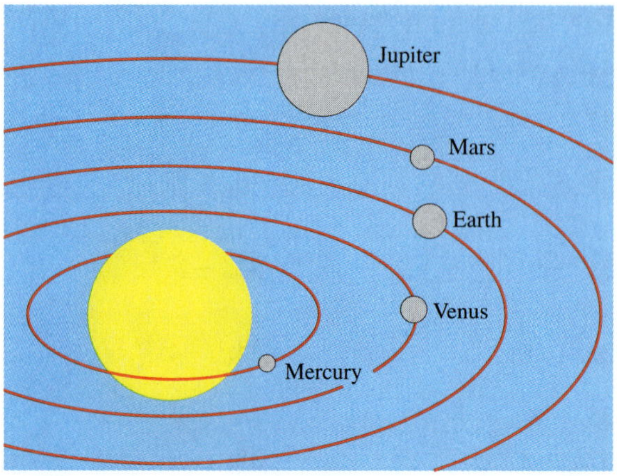

125. LICENSE PLATES The number of different license plates of the form three digits followed by three letters, as in the illustration, is $10 \cdot 10 \cdot 10 \cdot 26 \cdot 26 \cdot 26$. Write this expression using exponents. Then evaluate it.

126. PHYSICS Albert Einstein's work in the area of special relativity resulted in the observation that the total energy E of a body is equal to its total mass m times the square of the speed of light c. This relationship is given by the famous equation $E = mc^2$. Identify the base and exponent on the right-hand side.

127. GEOMETRY A cube is shown on the right.

a. Find the area of its base.

b. Find its volume.

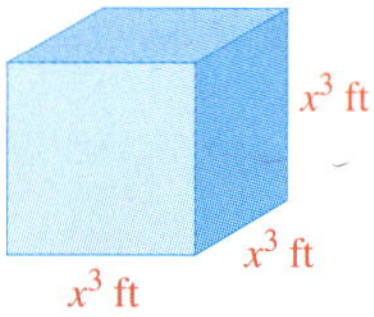

128. GEOMETRY A rectangular solid is shown on the right.

a. Find the area of its base.

b. Find its volume.

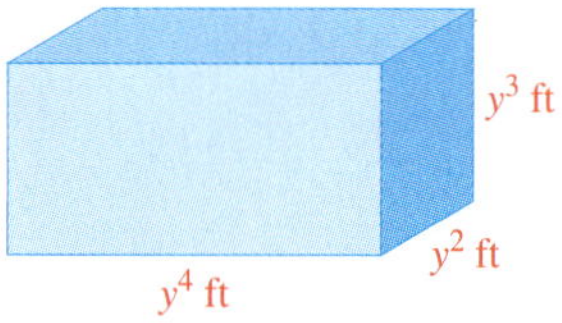

WRITING

129. Explain how an exponential expression with a negative exponent can be expressed as an equivalent expression with a positive exponent. Give an example.

130. Explain the error in the following solution.

Write $-8ab^{-3}$ using positive exponents only.

$$-8ab^{-3} = \frac{a}{8b^3}$$

REVIEW

Solve each inequality. Find the solution set in interval notation and then graph it.

131. $-9x + 5 \geq 15$

132. $\frac{1}{4}p - \frac{1}{3} \leq p + 2$

CHALLENGE PROBLEMS

Evaluate each expression.

133. $(2^{-1} + 3^{-1} - 4^{-1})^{-1}$

134. $(3^{-1} + 4^{-1})^{-2}$

Simplify each expression. Assume there are no divisions by 0.

135. $\dfrac{8^{5a}(8^{6a})^5}{8^{-2a} \cdot 8^a \cdot 8^{4a}}$

136. $\left(\dfrac{(y^{5x})^2(y^{4x})^4}{(y^{2x} \cdot y^x)^{-3}}\right)^{-2}$

5.2 Scientific Notation

- Writing numbers in scientific notation
- Converting from scientific notation
- Using scientific notation to simplify computations

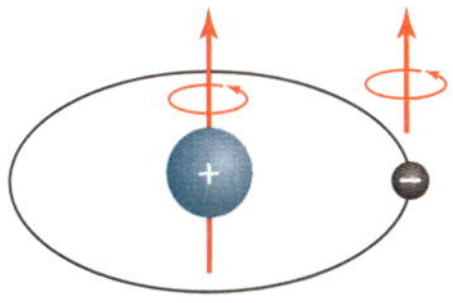

Hydrogen atom

Very large and very small numbers occur in science and other disciplines. For example, the star nearest to the Earth (excluding the sun) is Proxima Centauri, about 24,793,000,000,000 miles away, and the mass of a hydrogen atom is approximately 0.000000000000000000000001673 gram.

These numbers, written in **standard notation,** are difficult to read and cumbersome to work with in computations because they contain many zeros. In this section, we will discuss a notation that enables us to express such numbers in a more manageable form.

WRITING NUMBERS IN SCIENTIFIC NOTATION

Scientific notation provides a compact way of writing very large or very small numbers.

Scientific Notation A positive number is written in **scientific notation** when it is written in the form $N \times 10^n$, where $1 \leq N < 10$ and n is an integer.

Notation

A raised dot · is sometimes used when writing scientific notation.

$3.67 \times 10^6 = 3.67 \cdot 10^6$

Some examples of numbers written in scientific notation are

$$3.67 \times 10^6 \qquad 2.24 \times 10^{-4} \qquad 9.875 \times 10^{22}$$

Every positive number written in scientific notation is the product of a decimal number that is at least 1, but less than 10, and a power of 10.

A decimal that is at least 1, but less than 10 — An integer exponent

$$\square.\square \times 10^{\square}$$

EXAMPLE 1

Write each number in scientific notation: **a.** 24,793,000,000,000 and **b.** 0.00000000000000000000000001673.

Solution **a.** The number 2.4793 is between 1 and 10. To get 24,793,000,000,000, the decimal point in 2.4793 must be moved 13 places to the *right.*

2.4,793,000,000,000.
13 places

We can move the decimal point 13 places to the right by multiplying 2.4793 by 10^{13}.

$$24{,}793{,}000{,}000{,}000 = 2.4793 \times 10^{13}$$

b. The number 1.673 is between 1 and 10. To get 0.00000000000000000000000001673, the decimal point in 1.673 must be moved 24 places to the *left.*

0.000000000000000000000000001.673
24 places

We can move the decimal point 24 places to the left by multiplying 1.673 by 10^{-24}.

$$0.00000000000000000000000001673 = 1.673 \times 10^{-24}$$

Self Check 1 Write each italicized number in scientific notation. **a.** In 2002, the country earning the most money from tourism was the United States, *$66,500,000,000.* **b.** DNA molecules contain and transmit the information that allows cells to reproduce. They are only *0.000000002* meter wide.

Numbers such as 47.2×10^3 and 0.063×10^{-2} appear to be written in scientific notation, because they are the product of a number and a power of 10. However, they are not. Their first factors (47.2 and 0.063) are not between 1 and 10.

EXAMPLE 2

Write **a.** 47.2×10^3 and **b.** 0.063×10^{-2} in scientific notation.

Solution Since the first factors are not between 1 and 10, neither number is in scientific notation. However, we can change them to scientific notation as follows:

Notation

When writing numbers in scientific notation, keep the negative exponents. Don't apply the negative exponent rule.

6.3×10^{-4} ~~$6.3 \times \frac{1}{10^4}$~~

a. $47.2 \times 10^3 = (4.72 \times 10^1) \times 10^3$ Write 47.2 in scientific notation.

$= 4.72 \times (10^1 \times 10^3)$ Group the powers of 10 together.

$= 4.72 \times 10^4$ Apply the product rule for exponents: $10^1 \times 10^3 = 10^{1+3} = 10^4$.

b. $0.063 \times 10^{-2} = (6.3 \times 10^{-2}) \times 10^{-2}$ Write 0.063 in scientific notation.

$= 6.3 \times (10^{-2} \times 10^{-2})$

$= 6.3 \times 10^{-4}$

Self Check 2 Write **a.** 27.3×10^2 and **b.** 0.0025×10^{-3} in scientific notation.

CONVERTING FROM SCIENTIFIC NOTATION

Each of the following numbers is written in scientific and standard notation. In each case, the exponent gives the number of places that the decimal point moves, and the sign of the exponent indicates the direction that it moves:

$5.32 \times 10^4 = 5.3\,2\,0\,0.$
4 places to the right

$6.45 \times 10^7 = 6.4\,5\,0\,0\,0\,0\,0.$
7 places to the right

$2.37 \times 10^{-4} = 0.0\,0\,0\,2.3\,7$
4 places to the left

$9.234 \times 10^{-2} = 0.0\,9.2\,3\,4$
2 places to the left

$4.8 \times 10^0 = 4.8$
No movement of the decimal point

Success Tip

Since $10^0 = 1$, scientific notation involving 10^0 is easily simplified. For example,

$$4.8 \times 10^0 = 4.8 \times 1 = 4.8$$

These results suggest the following steps for changing a number written in scientific notation to standard notation.

Converting from Scientific to Standard Notation

1. If the exponent is positive, move the decimal point the same number of places to the right as the exponent.
2. If the exponent is negative, move the decimal point the same number of places to the left as the absolute value of the exponent.

EXAMPLE 3

Change **a.** 8.706×10^5 and **b.** 1.1×10^{-3} to standard notation.

Solution

a. Multiplication by 10^5, which is 100,000, moves the decimal point 5 places to the right:

$$8.706 \times 10^5 = 8.7\,0\,6\,0\,0. = 870{,}600$$

b. Multiplication by 10^{-3}, which is 0.001, moves the decimal point 3 places to the left:

$$1.1 \times 10^{-3} = 0.0\,0\,1\,.\,1 = 0.0011$$

Self Check 3 Change each number in scientific notation to standard notation. **a.** The country with the largest area of forest is Russia, with 1.9×10^9 acres. **b.** The average distance between molecules of air in a room is 3.937×10^{-7} inch.

USING SCIENTIFIC NOTATION TO SIMPLIFY COMPUTATIONS

Scientific notation is useful when multiplying and dividing very large or very small numbers.

EXAMPLE 4

Astronomy. The wheel-shaped galaxy in which we live is called the Milky Way. This system of some 10^{11} stars, one of which is the sun, has a diameter of approximately 100,000 light years. (A light year is the distance light travels in a vacuum in one year: 9.46×10^{15} meters.) Find the diameter of the Milky Way in meters.

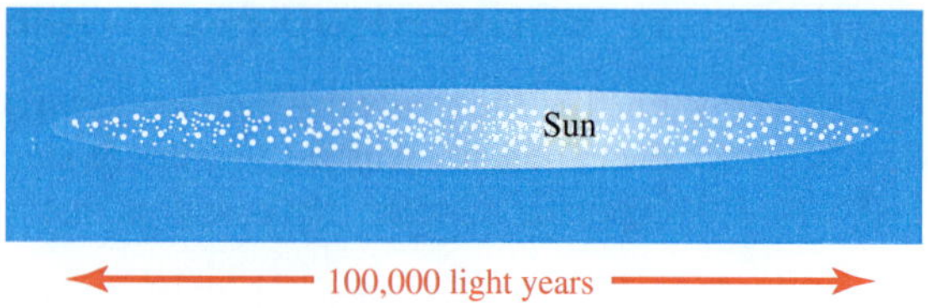

A cross-sectional representation of the Milky Way Galaxy

Solution To find the diameter of the Milky Way, in meters, we multiply its diameter, expressed in light years, by the number of meters in a light year. To perform the calculation, we write 100,000 in scientific notation as 1.0×10^5.

$$1.0 \times 10^5 \cdot 9.46 \times 10^{15}$$

$= (1.0 \cdot 9.46) \times (10^5 \cdot 10^{15})$ Apply the commutative and associative properties of multiplication to group the first factors together and the powers of 10 together.

$= 9.46 \times 10^{5+15}$ Perform the multiplication: $1.0 \cdot 9.46 = 9.46$. For the powers of 10, keep the base and add the exponents.

$= 9.46 \times 10^{20}$ Perform the addition.

The Milky Way Galaxy is about 9.46×10^{20} meters in diameter.

Self Check 4 A light year is 5.88×10^{12} miles. Find the diameter of the Milky Way in miles.

EXAMPLE 5

World oil reserves/production. According to estimates in the *Oil and Gas Journal,* there were 1.03×10^{12} barrels of oil reserves in the ground at the start of 2003. At that time, world production was 2.89×10^{10} barrels per year. If annual production remains the same and if no new oil discoveries are made, when will the world's oil supply run out?

Solution If we divide the estimated number of barrels of oil in reserve, 1.03×10^{12}, by the number of barrels produced each year, 2.89×10^{10}, we can find the number of years of oil supply left.

$\dfrac{1.03 \times 10^{12}}{2.89 \times 10^{10}} = \dfrac{1.03}{2.89} \times \dfrac{10^{12}}{10^{10}}$ Divide the first factors and the second factors in the numerator and denominator separately.

$\approx 0.36 \times 10^{12-10}$ Perform the division: $\frac{1.03}{2.89} \approx 0.36$. For the powers of 10, keep the base and subtract the exponents.

$\approx 0.36 \times 10^2$ Perform the subtraction.

≈ 36 Write 0.36×10^2 in standard notation.

According to industry estimates, as of 2003, there were 36 years of oil reserves left. Under these conditions, the world's oil supply will run out in the year 2039.

EXAMPLE 6

Use scientific notation to evaluate $\dfrac{(0.00000064)(24{,}000{,}000{,}000)}{(400{,}000{,}000)(0.0000000012)}$.

Solution After writing each number in scientific notation, we can perform the arithmetic on the numbers and the exponential expressions separately.

$$\begin{aligned}\frac{(0.00000064)(24{,}000{,}000{,}000)}{(400{,}000{,}000)(0.0000000012)} &= \frac{(6.4 \times 10^{-7})(2.4 \times 10^{10})}{(4.0 \times 10^{8})(1.2 \times 10^{-9})} \\ &= \frac{(6.4)(2.4)}{(4)(1.2)} \times \frac{10^{-7}10^{10}}{10^{8}10^{-9}} \\ &= \frac{15.36}{4.8} \times 10^{-7+10-8-(-9)} \\ &= 3.2 \times 10^{4}\end{aligned}$$

The result is 3.2×10^4. In standard notation, this is 32,000.

Self Check 6 Use scientific notation to evaluate $\frac{(320)(25{,}000)}{0.00004}$.

ACCENT ON TECHNOLOGY: USING SCIENTIFIC NOTATION

Scientific and graphing calculators often give answers in scientific notation. For example, if we use a calculator to find 301.2^8, the display will read

6.77391496 19	On a scientific calculator
301.2 ^ 8 6.773914961E19	On a graphing calculator

In either case, the answer is given in scientific notation and means

$$6.77391496 \times 10^{19}$$

Numbers can be entered into a calculator in scientific notation. For example, to enter 24,000,000,000 (which is 2.4×10^{10} in scientific notation), we enter these numbers and press these keys:

2.4 [EXP] 10 On most scientific calculators

2.4 [EE] 10 On a graphing calculator and on some scientific calculators

To use a scientific calculator to evaluate

$$\frac{(24{,}000{,}000{,}000)(0.00000006495)}{0.00000004824}$$

we enter each number in scientific notation, because each number has too many digits to be entered directly. In scientific notation, the three numbers are

$$2.4 \times 10^{10} \qquad 6.495 \times 10^{-8} \qquad 4.824 \times 10^{-8}$$

Using a scientific calculator, we enter these numbers and press these keys:

2.4 [EXP] 10 [×] 6.495 [EXP] 8 [+/−] [÷] 4.824 [EXP] 8 [+/−] [=]

The display will read [3.231343284 10]. In standard notation, the answer is 32,313,432,840. The steps are similar on a graphing calculator.

Answers to Self Checks **1. a.** 6.65×10^{10}, **b.** 2.0×10^{-9} **2. a.** 2.73×10^{3}, **b.** 2.5×10^{-6} **3. a.** 1,900,000,000, **b.** 0.0000003937 **4.** 5.88×10^{17} mi **6.** $2.0 \times 10^{11} = 200{,}000{,}000{,}000$

5.2 STUDY SET

VOCABULARY **Fill in the blanks.**

1. 7.4×10^{10} is written in ________ notation and 7,400,000 is written in ________ notation.
2. 10^{-3}, 10^{0}, 10^{1}, and 10^{4} are ________ of 10.

CONCEPTS **Fill in the blanks.**

3. A positive number is written in scientific notation when it is written in the form $N \times$ _____, where $1 \leq N < 10$ and n is an ________.
4. Insert > or <: 5.3×10^{2} _____ 5.3×10^{-2}.
5. To change 6.31×10^{-4} to standard notation, we move the decimal point four places to the _____.
6. To change 9.7×10^{3} to standard notation, we move the decimal point three places to the ______.

NOTATION

7. **a.** Explain why the number 60.22×10^{22} is not written in scientific notation.

   b. Explain why the number 0.6022×10^{24} is not written in scientific notation.

8. Determine what type of *exponent* must be used when writing each of the three categories of real numbers in scientific notation.

   a. For real numbers between 0 and 1:

   ■ × 10

   b. For numbers at least 1, but less than 10:

   ■ × 10

   c. For real numbers greater than or equal to 10:

   ■ × 10

PRACTICE **Write each number in scientific notation.**

9. 3,900
10. 1,700
11. 0.0078
12. 0.068
13. 173,000,000,000,000
14. 89,800,000,000
15. 0.0000096
16. 0.000000046
17. 323×10^{5}
18. 689×10^{9}
19. $6{,}000 \times 10^{-7}$
20. 765×10^{-5}
21. 0.0527×10^{5}
22. 0.0298×10^{3}
23. 0.0317×10^{-2}
24. 0.0012×10^{-3}

Write each number in standard notation.

25. 2.7×10^{2}
26. 7.2×10^{3}
27. 3.23×10^{-3}
28. 6.48×10^{-2}
29. 7.96×10^{5}
30. 9.67×10^{6}
31. 3.7×10^{-4}
32. 4.12×10^{-5}
33. 5.23×10^{0}
34. 8.67×10^{0}
35. 23.65×10^{6}
36. 75.6×10^{-5}

Perform the operations. Give all answers in scientific notation.

37. $(7.9 \times 10^{5})(2.3 \times 10^{6})$
38. $(6.1 \times 10^{8})(3.9 \times 10^{5})$
39. $(9.1 \times 10^{-5})(5.5 \times 10^{12})$
40. $(8.4 \times 10^{-13})(4.8 \times 10^{9})$
41. $(9.0 \times 10^{-1})(8.0 \times 10^{-6})$
42. $(8.1 \times 10^{-4})(2.4 \times 10^{-15})$
43. $\dfrac{4.2 \times 10^{-12}}{8.4 \times 10^{-5}}$
44. $\dfrac{1.21 \times 10^{-15}}{1.1 \times 10^{2}}$
45. $\dfrac{(3.9 \times 10^{-9})(9.5 \times 10^{-4})}{1.95 \times 10^{-2}}$
46. $\dfrac{(4.9 \times 10^{60})(2.7 \times 10^{30})}{6.3 \times 10^{40}}$

Write each number in scientific notation and perform the operations. Give all answers in scientific notation and in standard notation.

47. (89,000,000,000)(4,500,000,000)

48. (0.000000061)(3,500,000,000)

49. $\dfrac{0.00000129}{0.0003}$

50. $\dfrac{4{,}400{,}000{,}000{,}000}{0.0002}$

51. $\dfrac{(220{,}000)(0.000009)}{0.00033}$

52. $\dfrac{(640{,}000)(2{,}700{,}000)}{120{,}000}$

53. $\dfrac{(0.00024)(96{,}000{,}000)}{640{,}000{,}000}$

54. $\dfrac{(0.0000013)(0.00009)}{0.00039}$

APPLICATIONS

55. FIVE-CARD POKER The odds against being dealt the hand shown in the illustration are about 2.6×10^6 to 1. Express the odds using standard notation.

56. ENERGY See the illustration. Express each of the following using scientific notation. (1 quadrillion is 10^{15}.)

a. U.S. energy consumption

b. U.S. energy production

c. The difference in 2001 consumption and production

2001 U.S. Energy Consumption and Production

(petroleum, natural gas, coal, hydroelectric, nuclear, geothermal, solar, wind)

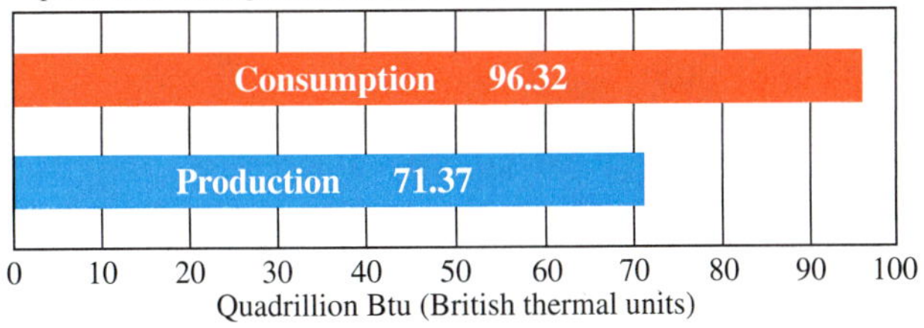

Source: Energy Information Administration, United States Department of Energy

57. THE YEAR 2000 Prior to January 1, 2000, the U.S. government spent a large sum to reprogram its computers to protect them from the Y2K bug. Express in scientific notation each of the dollar amounts mentioned in the following article from the *Federal Computer Week* Web page (February 1998).

> President Clinton's fiscal 1999 budget proposal of \$1.7 trillion includes expenditures of about \$3.9 billion to ensure that federal computers can accept dates after Dec. 31, 1999. Clinton has proposed spending \$275 million at the Defense Department and \$312 million at the Treasury Department to fix the year 2000 problem.

58. TV TRIVIA In the series *Star Trek,* the *U.S.S. Enterprise* traveled at warp speeds. To convert a warp speed, W, to an equivalent velocity in miles per second, v, we can use the equation

$$v = W^3c$$

where c is the speed of light, 1.86×10^5 miles per second. Find the velocity of a spacecraft traveling at warp 2.

59. ATOMS A simple model of a helium atom is shown. If a proton has a mass of 1.7×10^{-24} grams, and if the mass of an electron is only about $\frac{1}{2{,}000}$ that of a proton, find the mass of an electron.

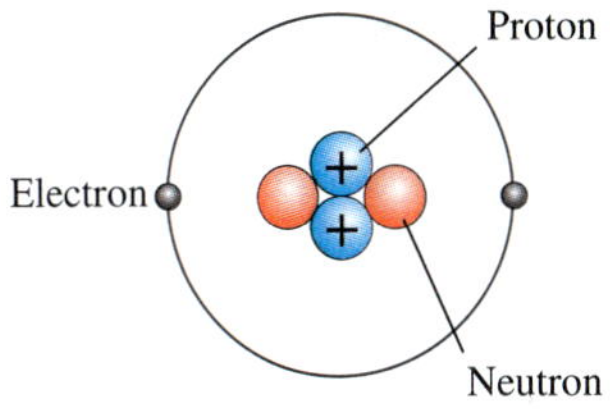

60. OCEANS The mass of the Earth's oceans is only about $\frac{1}{4{,}400}$ that of the Earth. If the mass of the Earth is 6.578×10^{21} tons, find the mass of the oceans.

61. LIGHT YEAR Light travels about 300,000,000 meters per second. A **light year** is the distance that light can travel in one year. Estimate the number of meters in one light year.

62. AQUARIUMS Express the volume of the fish tank in scientific notation.

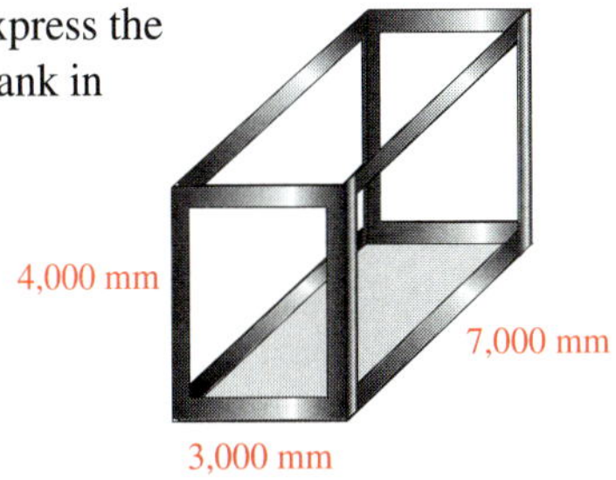

63. THE BIG DIPPER One star in the Big Dipper is named Merak. It is approximately 4.65×10^{14} miles from the Earth.

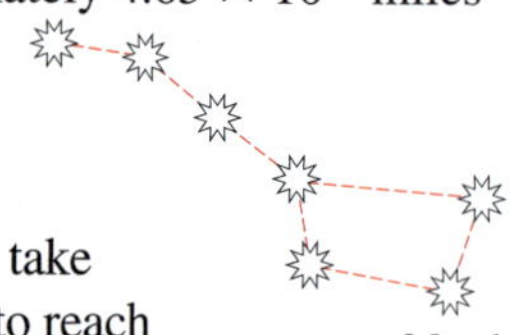

a. If light travels about 1.86×10^5 miles/sec, how many seconds does it take light emitted from Merak to reach the Earth? (*Hint:* Use the formula $t = \frac{d}{r}$.)

b. Convert your result from part a to years.

64. BIOLOGY A paramecium is a single-celled organism that propels itself with hair-like projections called *cilia.* Use the scale in the illustration to estimate the length of the paramecium. Express the result in scientific and in standard notation.

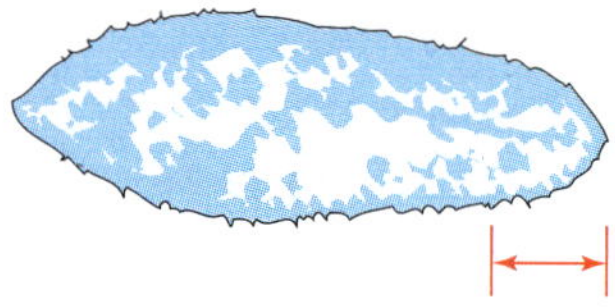

65. COMETS On March 23, 1997, Comet Hale-Bopp made its closest approach to Earth, coming within 1.3 **astronomical units.** One astronomical unit (AU) is the distance from the Earth to the sun—about 9.3×10^7 miles. Express this distance in miles, using scientific notation.

66. DIAMONDS The approximate number of atoms of carbon in a $\frac{1}{2}$-carat diamond can be found by computing

$$\frac{6.0 \times 10^{23}}{1.2 \times 10^2}$$

Express the number of carbon atoms in scientific and in standard notation.

67. ATOMS A hydrogen atom is so small that a single drop of water contains more than a million million billion hydrogen atoms. Express this number in scientific notation.

68. ASTRONOMY The American Physical Society recently honored first-year graduate student Gwen Bell for coming up with what it considers the most accurate estimate of the mass of the Milky Way. In pounds, her estimate is a 3 with 42 zeros after it. Express this number in scientific notation.

WRITING

69. Explain how to change a number from standard notation to scientific notation.

70. Explain how to change a number from scientific notation to standard notation.

71. Explain why 9.99×10^n represents a number less than 1 but greater than 0 if n is a negative integer.

72. Explain the advantages of writing very large and very small numbers in scientific notation.

REVIEW

Solve each compound inequality. Give the result in interval notation and graph the solution set.

73. $4x \geq -x + 5$ and $6 \geq 4x - 3$

74. $15 > 2x - 7 > 9$

75. $3x + 2 < 8$ or $2x - 3 > 11$

76. $-4(x + 2) \geq 12$ or $3x + 8 < 11$

CHALLENGE PROBLEMS

77. What is the reciprocal of the opposite of 2.5×10^{-24}? Write the result in scientific notation.

78. Solve: $(1.1 \times 10^{-16})x - (1.2 \times 10^{10}) = (6.5 \times 10^{10})$. Write the solution in scientific notation.

5.3 Polynomials and Polynomial Functions

- Polynomials
- Degree of a polynomial
- Polynomial functions
- Evaluating polynomial functions
- Graphing polynomial functions
- Simplifying polynomials by combining like terms
- Adding and subtracting polynomials

In arithmetic, we add, subtract, multiply, divide, and find powers of real numbers. In algebra, we perform these operations on algebraic expressions called *polynomials.*

POLYNOMIALS

The Language of Algebra

The prefix *poly* means many. A *poly*gon is a many-sided figure and *poly*unsaturated fats are molecules having many strong chemical bonds.

A **term** is a number or a product of a number and a variable (or variables) raised to a power. Some examples are

$$17, \quad 9x, \quad \frac{15}{16}y^2, \quad \text{and} \quad -2.4x^4y^5$$

If a term contains only a number, such as 17, it is called a **constant term,** or simply a **constant.**

The **numerical coefficient,** or simply the **coefficient,** is the numerical factor of a term. For example, the coefficient of $9x$ is 9 and the coefficient of $-2.4x^4y^5$ is -2.4. The coefficient of a constant term is that constant.

Polynomials

A **polynomial** is a single term or the sum of terms in which all variables have whole-number exponents. No variable appears in a denominator.

Notation

Since $3x^2 - 2x$ can be written as $3x^2 + (-2x)$, it can be thought of as a sum of terms and is, therefore, a polynomial.

The following expressions are polynomials in x:

$$6x, \quad 3x^2 - 2x, \quad \frac{3}{2}x^5 + \frac{7}{3}x^4 + \frac{8}{3}x^3, \quad \text{and} \quad 19x^{20} + \sqrt{3}x^{14} + 4.5x^{11} - x^2$$

Caution The following expressions are not polynomials:

$$\frac{2x}{x^2+1}, \quad x^{1/2} - 8, \quad \text{and} \quad x^{-3} + 2x + 24$$

The first expression is a quotient and has a variable in the denominator. The last two have exponents that are not whole numbers.

If any terms of a polynomial contain more than one variable, we say that the polynomial is in more than one variable. Some examples are

$$3xy, \quad 5x^2y^2 + 2xy - 3y, \quad \text{and} \quad u^2v^2w^2 + uv + 1$$

The Language of Algebra

We say that $5x^2y^2 + 2xy - 3y$ is a polynomial *in x and y.*

Polynomials can be classified according to the number of terms they have. A polynomial with one term is called a **monomial,** a polynomial with two terms is called a **binomial,** and a polynomial with three terms is called a **trinomial.**

Monomials	***Binomials***	***Trinomials***
$2x^3$	$2x + 5$	$2x^2 + 4x + 3$
a^2b	$-17x^4 - \frac{3}{5}x$	$3mn^3 - m^2n^3 + 7n$
$3x^3y^5z^2$	$32x^{13}y^5z^3 + 47x^3yz$	$-12x^5y^2 + 13x^4y^3 - 7x^3y^3$

DEGREE OF A POLYNOMIAL

Because x occurs three times as a factor in the monomial $2x^3$, it is called a *third-degree monomial* or a *monomial of degree 3.* The monomial $3x^3y^5z^2$ is called a *monomial of degree 10,* because the variables x, y, and z occur as factors a total of ten times $(3 + 5 + 2)$. These examples illustrate the following definition.

Degree of a Monomial

The **degree** of a monomial with one variable is the exponent on the variable. The degree of a monomial in several variables is the sum of the exponents on those variables. If the monomial is a nonzero constant, its degree is 0. The constant 0 has no defined degree.

EXAMPLE 1

Find the degree of **a.** $3x^4$, **b.** $-4x^2y^3$, **c.** t, **d.** $\left(\frac{1}{2}\right)^2 m^6$, and **e.** 3.

Solution

a. $3x^4$ is a monomial of degree 4, because the exponent on the variable is 4.

b. $-4x^2y^3$ is a monomial of degree 5, because the sum of the exponents on the variables is 5.

c. t is a monomial of degree 1, because the exponent on the variable is an understood 1: $t = t^1$.

d. $\left(\frac{1}{2}\right)^2 m^6$ is a monomial of degree 6 because the exponent on the variable is 6.

e. 3 is a monomial of degree 0, because $3 = 3x^0$.

The Language of Algebra

The word *degree* is also used in other disciplines for classification. Doctors speak of *third-degree* burns.

Self Check 1 Find the degree of **a.** $-12a^2$, **b.** $8a^3b^2$, **c.** s, and **d.** $\frac{1}{2}x^3y^2z^{12}$.

We determine the degree of a polynomial by considering the degrees of each of its terms.

Degree of a Polynomial

The **degree of a polynomial** is the same as the degree of the term in the polynomial with largest degree.

EXAMPLE 2

Find the degree of each polynomial: **a.** $3x^5 + 4x^2 + 7$, **b.** $7x^2y^8 - 3x^2y^2$, and **c.** $3x + 2y - xy + 15$

Solution

a. The terms of $3x^5 + 4x^2 + 7$ have degree 5, 2, and 0, respectively. This trinomial is of degree 5, because the largest degree of the three terms is 5.

b. $7x^2y^8 - 3x^2y^2$ is a binomial of degree 10.

c. $3x + 2y - xy + 15$ is a polynomial of degree 2. (Recall that $xy = x^1y^1$.)

Self Check 2 Find the degree of **a.** $x^2 - x + 1$ and **b.** $x^2y^3 - 12x^7y^2 + 3x^9y^3 - 3$.

If there is exactly one term of a polynomial with the highest degree, that term is called the **lead term** and its coefficient is called the **lead coefficient.** For the polynomial $2x^2 - 4x - 6$, the lead term is $2x^2$ and the lead coefficient is 2.

If the terms of a polynomial in one variable are written so that the exponents decrease as we move from left to right, we say that the terms are written with their exponents in descending order. If the terms are written so that the exponents increase as we move from left to right, we say that the terms are written with their exponents in ascending order.

$-5x^4 + 2x^3 + 7x^2 + 3x - 1$ This polynomial is written in descending powers of x.

$-1 + 3x + 7x^2 + 2x^3 - 5x^4$ The same polynomial is now written in ascending powers of x.

POLYNOMIAL FUNCTIONS

We have seen that linear functions are defined by equations of the form $f(x) = mx + b$. Some examples of linear functions are

$$f(x) = 3x + 1 \qquad g(x) = -\frac{1}{2}x - 1 \qquad h(x) = 5x$$

In each case, the right-hand side of the equation is a polynomial. For this reason, linear functions are members of a larger class of functions known as *polynomial functions.*

Polynomial Functions A **polynomial function** is a function whose equation is defined by a polynomial in one variable.

Another example of a polynomial function is $f(x) = -x^2 + 6x - 8$. This is a second-degree polynomial function, called a **quadratic function.** Quadratic functions are of the form $f(x) = ax^2 + bx + c$, where $a \neq 0$.

An example of a third-degree polynomial function is $f(x) = x^3 - 3x^2 - 9x + 2$. Third-degree polynomial functions, also called **cubic functions,** are of the form $f(x) = ax^3 + bx^2 + cx + d$, where $a \neq 0$.

EVALUATING POLYNOMIAL FUNCTIONS

Polynomial functions can be used to model many real-life situations. If we are given a polynomial function model, we can learn more about the situation by evaluating the function at specific values.

EXAMPLE 3

Rocketry. If a toy rocket is shot straight up with an initial velocity of 128 feet per second, its height, in feet, t seconds after being launched is given by the function

$$h(t) = -16t^2 + 128t$$

Find the height of the rocket **a.** 2 seconds after being launched and **b.** 7.9 seconds after being launched.

Solution

a. To find the height of the rocket 2 seconds after being launched, we need to evaluate the function at $t = 2$. That is, we need to find $h(2)$.

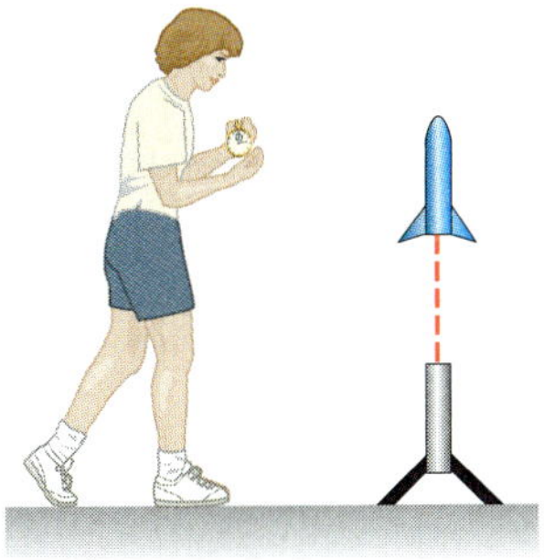

$h(t) = -16t^2 + 128t$	This is the given function.
$h(2) = -16(2)^2 + 128(2)$	Substitute 2 for each t. (The input is 2.)
$= -16(4) + 256$	Evaluate the right-hand side.
$= -64 + 256$	
$= 192$	The output is 192.

We have found that $h(2) = 192$. Thus, 2 seconds after it is launched, the height of the rocket is 192 feet.

b. To find the height of the rocket 7.9 seconds after it is launched, we need to find $h(7.9)$.

$h(t) = -16t^2 + 128t$	This is the given function.
$h(7.9) = -16(7.9)^2 + 128(7.9)$	Substitute 7.9 for each t. (The input is 7.9.)
$= -16(62.41) + 1{,}011.2$	Evaluate the right-hand side.
$= -998.56 + 1{,}011.2$	
$= 12.64$	The output is 12.64.

At 7.9 seconds, the height of the rocket is 12.64 feet. It has almost fallen back to Earth.

Self Check 3 Find the height of the rocket 4 seconds after it is launched.

EXAMPLE 4

Packaging. To make boxes, a maufacturer cuts equal-sized squares from each corner of the 10 in. × 12 in. piece of cardboard shown below and then folds up the sides. The polynomial function $f(x) = 4x^3 - 44x^2 + 120x$ gives the volume (in cubic inches) of the resulting box when a square with sides x inches long is cut from each corner. Find the volume of a box if 3-inch squares are cut out.

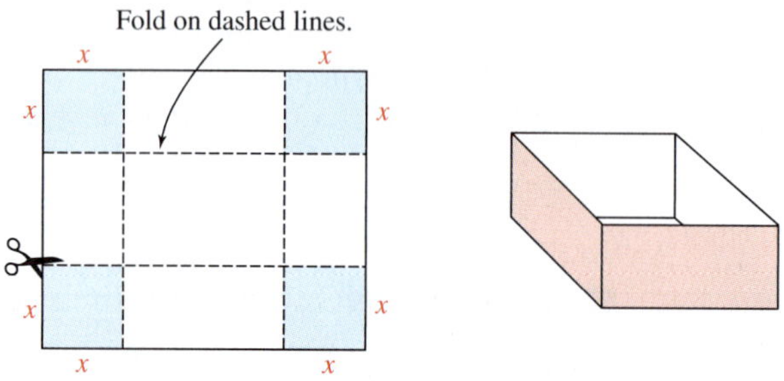

Solution To find the volume of the box, we evaluate the function for $x = 3$.

$$f(x) = 4x^3 - 44x^2 + 120x \quad \text{This is the given function.}$$
$$f(3) = 4(3)^3 - 44(3)^2 + 120(3) \quad \text{Substitute 3 for each } x.$$
$$= 4(27) - 44(9) + 120(3) \quad \text{Evaluate the right-hand side.}$$
$$= 108 - 396 + 360$$
$$= 72$$

If 3-inch squares are cut out, the box will have a volume of 72 in.^{3}.

Self Check 4 Find the volume of the resulting box if 2-inch squares are cut from each corner of the cardboard.

GRAPHING POLYNOMIAL FUNCTIONS

The Language of Algebra

$f(x) = x$ is called the **identity function** because it assigns each real number to itself. Note that the graph passes through $(-2, -2)$, $(0, 0)$, $(1, 1)$, and so on.

The graphs of three basic polynomial functions are shown below. The domain and range of the functions are expressed in interval notation.

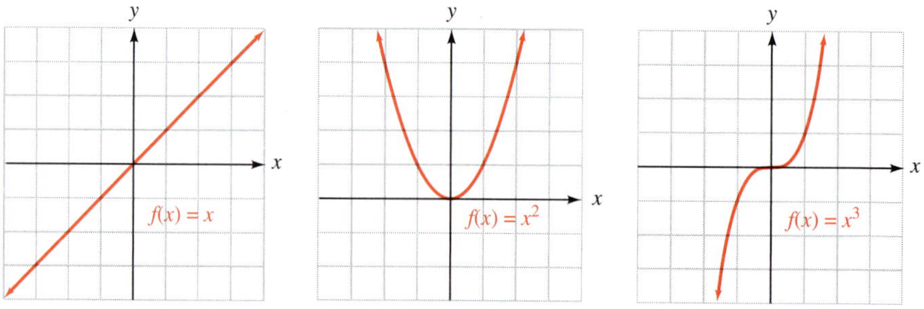

The identity function The domain is $(-\infty, \infty)$. The range is $(-\infty, \infty)$.

The squaring function The domain is $(-\infty, \infty)$. The range is $[0, \infty)$.

The cubing function The domain is $(-\infty, \infty)$. The range is $(-\infty, \infty)$.

When graphing a linear function, we need to plot only two points, because the graph is a straight line. The graphs of polynomial functions of degree greater than 1 are smooth, continuous curves. To graph them, we must plot more points.

EXAMPLE 5

Graph: $f(x) = x^3 - 3x^2 - 9x + 2$.

Solution To graph this cubic function, we begin by evaluating it for $x = -3$.

$$\begin{aligned} f(x) &= x^3 - 3x^2 - 9x + 2 \\ f(-3) &= (-3)^3 - 3(-3)^2 - 9(-3) + 2 \quad \text{Substitute } -3 \text{ for each } x. \\ &= -27 - 3(9) - 9(-3) + 2 \\ &= -27 - 27 + 27 + 2 \\ &= -25 \end{aligned}$$

In the following table, we enter the ordered pair $(-3, -25)$. We continue the evaluation process for $x = -2, -1, 0, 1, 2, 3, 4$, and 5, and list the results in the table. After plotting the ordered pairs, we draw a smooth curve through the points to get the graph of function f.

Success Tip

The graphs of many polynomial functions of degree 3 and higher have these characteristic *peaks* and *valleys*.

$f(x) = x^3 - 3x^2 - 9x + 2$

x	$f(x)$	
−3	−25	→ (−3, −25)
−2	0	→ (−2, 0)
−1	7	→ (−1, 7)
0	2	→ (0, 2)
1	−9	→ (1, −9)
2	−20	→ (2, −20)
3	−25	→ (3, −25)
4	−18	→ (4, −18)
5	7	→ (5, 7)

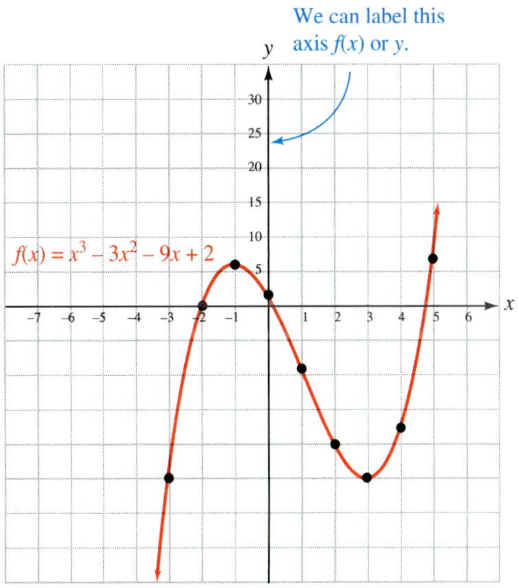

Self Check 5 What are the domain and the range of the function graphed above? Express each in interval notation.

EXAMPLE 6

Labor statistics. The number of manufacturing jobs in the United States, in millions, is approximated by the polynomial function

$$J(x) = 0.000003x^4 - 0.0121x^3 + 0.204x^2 - 0.893x + 17.902$$

where x is the number of years after 1990. Use the graph of the function in figure (a) on the next page to answer the following questions.

a. Find $J(6)$. Explain what the result means.

b. Find the value(s) of x for which $J(x) = 16.8$. Explain what the results mean.

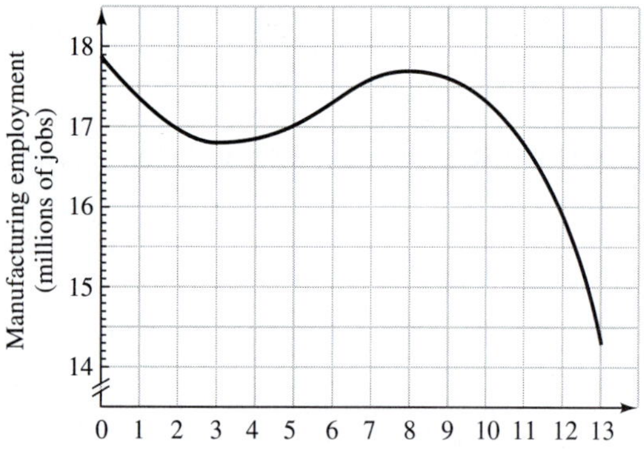

Source: Congressional Budget Office

(a)

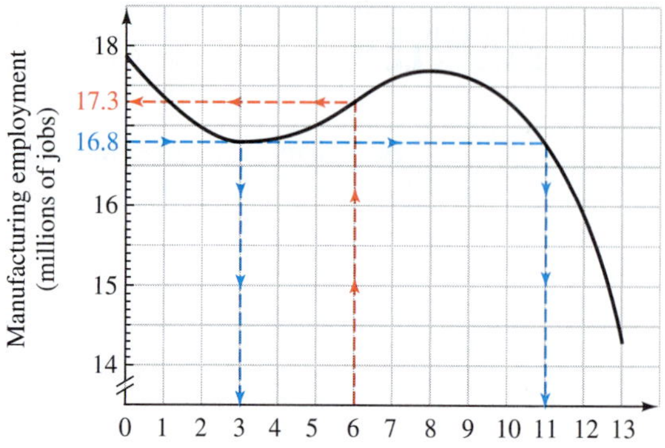

(b)

Solution **a.** Refer to figure (b). To find $J(6)$, we use the dashed red lines to determine that $J(6) \approx 17.3$. This means 6 years after 1990, or in 1996, there were approximately 17.3 million manufacturing jobs in the United States.

b. Refer again to figure (b). To find the input values x that are assigned the output value 16.8, we used the dashed blue lines to determine that $J(3) \approx 16.8$ and $J(11) \approx 16.8$. This means 3 years after 1990, or in 1993, and 11 years after 1990, or in 2001, there were approximately 16.8 million manufacturing jobs in the United States.

ACCENT ON TECHNOLOGY: GRAPHING POLYNOMIAL FUNCTIONS

We can graph polynomial functions with a graphing calculator. For example, to graph $f(x) = x^3 - 3x^2 - 9x + 2$ from Example 5, we enter it as shown in figure (a). Using window settings of $[-8, 8]$ for x and $[-50, 50]$ for y, we get the graph shown in figure (b).

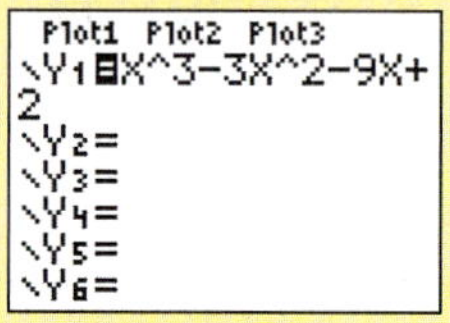

(a)

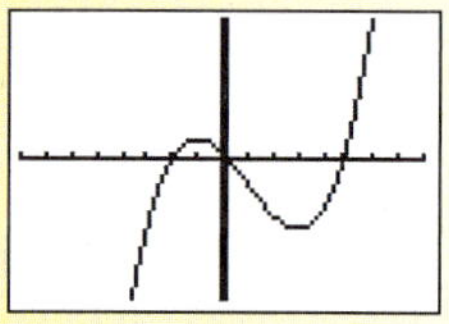
(b)

SIMPLIFYING POLYNOMIALS BY COMBINING LIKE TERMS

Recall that **like terms** have the same variables with the same exponents:

Like terms	*Unlike terms*
$-7x$ and $15x$	$-7x$ and $15a$
$4y^3$ and $16y^3$	$4y^3$ and $16y^2$
$\frac{1}{2}xy^2$ and $-\frac{1}{3}xy^2$	$\frac{1}{2}xy^2$ and $-\frac{1}{3}x^2y$

Also recall that to **combine like terms,** we combine their coefficients and keep the same variables with the same exponents. For example,

$$4y + 5y = (4 + 5)y \qquad 8x^2 - x^2 = (8 - 1)x^2$$
$$= 9y \qquad\qquad = 7x^2$$

Polynomials with like terms can be simplified by combining like terms.

EXAMPLE 7

Simplify each polynomial: **a.** $4x^4 + 81x^4$, **b.** $17x^2y^2 + 2x^2y - 6x^2y^2$, **c.** $-3r - 4r + 6r$, and **d.** $ab + 8 - 15 + 4ab$.

Solution

a. $4x^4 + 81x^4 = 85x^4$ $\quad (4 + 81)x^4 = 85x^4$.

b. The first and third terms are like terms.

$$17x^2y^2 + 2x^2y - 6x^2y^2 = 11x^2y^2 + 2x^2y \qquad (17 - 6)x^2y^2 = 11x^2y^2.$$

c. $-3r - 4r + 6r = -r$ $\quad (-3 - 4 + 6)r = -1r = -r$.

d. The first and fourth terms are like terms, and the second and third terms are like terms.

$$ab + 8 - 15 + 4ab = 5ab - 7 \qquad (1 + 4)ab = 5ab \text{ and } 8 - 15 = -7.$$

Self Check 7 Simplify each polynomial: **a.** $6m^4 + 3m^4$, **b.** $17s^3t + 3s^2t - 6s^3t$, **c.** $-19x + 21x - x$, and **d.** $rs + 3r - 5rs + 4r$.

ADDING AND SUBTRACTING POLYNOMIALS

Adding Polynomials To add polynomials, combine their like terms.

EXAMPLE 8

Add: **a.** $(3x^2 - 2x + 4) + (2x^2 + 4x - 3)$ and **b.** $(-5x^3y^2 - 4x^2y^3) + (2x^3y^2 + x^3y + 5x^2y^3)$.

Solution

a.
$$\begin{aligned} &(3x^2 - 2x + 4) + (2x^2 + 4x - 3) && \text{We are to add two trinomials.}\\ &= 3x^2 - 2x + 4 + 2x^2 + 4x - 3 && \text{Remove the parentheses.}\\ &= 5x^2 + 2x + 1 && \text{Combine like terms.}\end{aligned}$$

b.
$$\begin{aligned} &(-5x^3y^2 - 4x^2y^3) + (2x^3y^2 + x^3y + 5x^2y^3) && \text{We are to add a binomial and a trinomial.}\\ &= -5x^3y^2 - 4x^2y^3 + 2x^3y^2 + x^3y + 5x^2y^3 && \text{Remove the parentheses.}\\ &= -3x^3y^2 + x^3y + x^2y^3 && \text{Combine like terms.}\end{aligned}$$

Notation

When performing operations on polynomials, it is standard practice to write the terms of a result in descending powers of one variable.

Self Check 8 Add: **a.** $(2a^2 - 3a + 5) + (5a^2 + 4a - 2)$ and **b.** $(-6a^2b^3 - 5a^3b^2) + (3a^2b^3 + 2a^3b^2 + ab^2)$.

The additions in Example 8 can be done by aligning the terms vertically and combining like terms column by column.

$$\begin{array}{r} 3x^2 - 2x + 4 \\ +\ \ 2x^2 + 4x - 3 \\ \hline 5x^2 + 2x + 1 \end{array} \qquad \begin{array}{r} -5x^3y^2 \qquad\quad - 4x^2y^3 \\ +\ \ 2x^3y^2 + x^3y + 5x^2y^3 \\ \hline -3x^3y^2 + x^3y + x^2y^3 \end{array}$$

Because of the distributive property, we can remove parentheses enclosing several terms when the sign preceding the parentheses is a $-$ sign. We simply drop the $-$ sign and the parentheses, and *change the sign of every term within the parentheses.*

$$\begin{aligned}-(3x^2 + 3x - 2) &= -(3x^2 + 3x - 2)\\ &= -1(3x^2) + (-1)(3x) + (-1)(-2)\\ &= -3x^2 + (-3x) + 2\\ &= -3x^2 - 3x + 2\end{aligned}$$

This suggests a way to subtract polynomials.

Subtracting Polynomials To subtract two polynomials, change the signs of the terms of the polynomial being subtracted, drop the parentheses, and combine like terms.

EXAMPLE 9

Subtract: **a.** $(8x^3y + 2x^2y) - (2x^3y - 3x^2y)$ and **b.** $(3rt^2 + 4r^2t^2) - (8rt^2 - 4r^2t^2 + r^3t^2)$.

Solution **a.** $(8x^3y + 2x^2y) - (2x^3y - 3x^2y)$

$= 8x^3y + 2x^2y - 2x^3y + 3x^2y$ Change the sign of each term of $2x^3y - 3x^2y$ and drop the parentheses.

$= 6x^3y + 5x^2y$ Combine like terms.

b. $(3rt^2 + 4r^2t^2) - (8rt^2 - 4r^2t^2 + r^3t^2)$

$= 3rt^2 + 4r^2t^2 - 8rt^2 + 4r^2t^2 - r^3t^2$ Change the signs of the terms of the polynomial being subtracted.

$= -5rt^2 + 8r^2t^2 - r^3t^2$ Combine like terms.

Self Check 9 Subtract: $(6a^2b^3 - 2a^2b^2) - (-2a^2b^3 + a^2b^2)$.

Just as real numbers have opposites, polynomials have opposites as well. To find the opposite of a polynomial, multiply each of its terms by -1. This changes the sign of each term of the polynomial.

A polynomial		***Its opposite***
$2x^2 - 4x + 5$	Multiply by -1 $\longrightarrow$	$-(2x^2 - 4x + 5)$ or $-2x^2 + 4x - 5$

To subtract polynomials in vertical form, we add the opposite of the polynomial that is being subtracted.

$$\begin{array}{r} 8x^3y + 2x^2y \\ -\ \ \underline{2x^3y - 3x^2y} \end{array} \quad \rightarrow \quad \begin{array}{r} 8x^3y + 2x^2y \\ +\ \ \underline{-2x^3y + 3x^2y} \\ 6x^3y + 5x^2y \end{array}$$

This is the opposite of $2x^3y - 3x^2y$.

Answers to Self Checks **1. a.** 2, **b.** 5, **c.** 1, **d.** 17 **2. a.** 2, **b.** 12 **3.** 256 ft **4.** 96 in.^{3} **5.** domain: $(-\infty, \infty)$, range: $(-\infty, \infty)$ **7. a.** $9m^4$, **b.** $11s^3t + 3s^2t$, **c.** x, **d.** $-4rs + 7r$ **8. a.** $7a^2 + a + 3$, **b.** $-3a^2b^3 - 3a^3b^2 + ab^2$ **9.** $8a^2b^3 - 3a^2b^2$

5.3 STUDY SET

VOCABULARY **Fill in the blanks.**

1. A __________ is the sum of one or more algebraic terms whose variables have whole-number exponents.
2. A __________ is a polynomial with one term. A __________ is a polynomial with two terms. A __________ is a polynomial with three terms.
3. The ________ of a monomial with one variable is the exponent on the variable.
4. A second-degree polynomial function is also called a __________ function. A third-degree polynomial function is also called a ________ function.
5. The __________ of the term $-15x^2y^3$ is -15. The ________ of the term is 5.
6. For $9y^3 + y^2 - 6y - 17$, the lead term is $9y^3$ and the lead __________ is 9.
7. Terms having the same variables with the same exponents are called ______ terms.
8. The ____________________ of $x^2 + x - 3$ is $-x^2 - x + 3$.

Classify each polynomial as a monomial, binomial, trinomial, or none of these. Then determine the degree of the polynomial.

9. $3x^2$
10. $2y^3 + 4y^2$
11. $3x^2y - 2x + 3y$
12. $a^2 + ab + b^2$
13. $x^2 - y^2$
14. $\frac{17}{2}x^3 + 3x^2 - x - 4$
15. 5
16. $\left(\frac{1}{4}\right)^2 x^3y^5$
17. $9x^2y^4 - x - y^{10} + 1$
18. x^{17}
19. $4x^9 + 3x^2y^4$
20. -12

Decide whether the terms are like or unlike terms. If they are like terms, combine them.

21. $3x, 7x$
22. $-8x, 3y$
23. $7x, 7y$
24. $3mn, 5mn$
25. $3r^2t^3, -8r^2t^3$
26. $9u^2v, 10u^2v$
27. $9x^2y^3, 3x^2y^2$
28. $27x^6y^4z, 8x^6y^4z^2$

29. Write a polynomial that represents the perimeter of the following triangle.

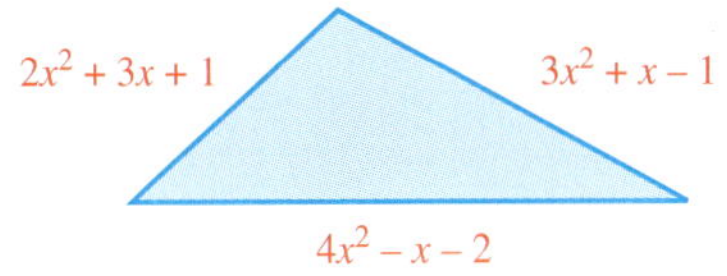

30. Use the graph of function *f* to find each of the following
   a. $f(-1)$
   b. $f(1)$
   c. The values of x for which $f(x) = 0$.
   d. The domain and range of f.

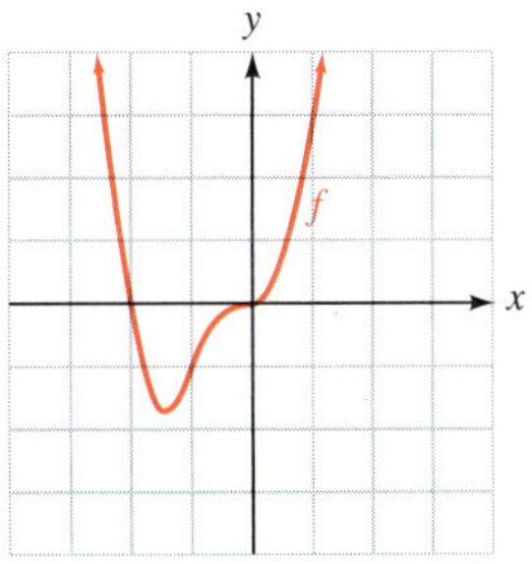

NOTATION **Complete the evaluation.**

31. If $h(t) = -t^3 - t^2 + 2t + 1$, find $h(3)$.

$$h(t) = -t^3 - t^2 + 2t + 1$$
$$h(\quad) = -(\quad)^3 - (\quad)^2 + 2(3) + 1$$
$$= \quad - 9 + 6 + 1$$
$$= \quad$$

32. Determine whether each expression is a polynomial.
   a. $\frac{3}{x^2} + \frac{4}{x} + 2$
   b. $\frac{4}{3}\pi r^3$
   c. $y^{-2} - 5y^{-1}$
33. Write each polynomial with the exponents on x in descending order.
   a. $3x - 2x^4 + 7 - 5x^2$
   b. $a^2x - ax^3 + 7a^3x^5 - 5a^3x^2$
34. Write each polynomial with the exponents on y in ascending order.
   a. $4y^2 - 2y^5 + 7y - 5y^3$
   b. $x^3y^2 + x^2y^3 - 2x^3y + x^7y^6 - 3x^6$

PRACTICE **Complete each table of values. Then graph each polynomial function.**

35. $f(x) = 2x^2 - 4x + 2$

x	$f(x)$
-1	
0	
1	
2	
3	

36. $f(x) = -x^2 + 2x + 6$

x	$f(x)$
-2	
-1	
0	
1	
2	
3	
4	

37. $f(x) = 2x^3 - 3x^2 - 11x + 6$

x	$f(x)$
-3	
-2	
-1	
0	
1	
2	
3	
4	

38. $f(x) = -x^3 - x^2 + 6x$

x	$f(x)$
-4	
-3	
-2	
-1	
0	
1	
2	
3	

Use a graphing calculator to graph each polynomial function. Use window settings of [−4, 6] for *x* and [−5, 5] for *y*.

39. $f(x) = 2.75x^2 - 4.7x + 1.5$

40. $f(x) = 0.37x^3 - 1.4x + 1.5$

Simplify each polynomial.

41. $15x^2 + 4x - 5 + 5x^2$

42. $8m^3 + 5m - 5 + 3m$

43. $-11y^3 - 7y - 4 + y^3$

44. $-3y^2 + 2y - 5 + 2y^2$

45. $ab^2 - 4ab - a + 5ab$

46. $8c^4d + 5cd - 7cd + 1$

47. $9rst^2 - 5 - rst^2 + 4$

48. $m^4n - 5mn - m^4n$

Perform each operation.

49. $(3x^2 + 2x + 1) + (-2x^2 - 7x + 5)$

50. $(-2a^2 - 5a - 7) + (-3a^2 + 7a + 1)$

51. $(-a^2 + 2a + 3) - (4a^2 - 2a - 1)$

52. $(x^2 - 3x + 8) - (3x^2 + x + 3)$

53. $(2a^2 + 4ab - 7) + (3a^2 - ab - 2)$

54. $(6x^3 + 3xy - 2) - (2x^3 + 3x^2 + 5)$

55. $(7y^3 + 4y^2 + y + 3) + (-8y^3 - y + 3)$

56. $(-8p^3 - 2p - 4) - (2p^3 + p^2 - p)$

57. $(3p^2q^2 + p - q) + (-p^2q^2 - p - q)$

58. $(-2m^2n^2 + 2m - n) - (-2m^2n^2 - 2m + n)$

59. $(-2x^2y^3 + 6xy + 5y^2) - (-4x^2y^3 - 7xy + 2y^2)$

60. $(3ax^3 - 2ax^2 + 3a^3) + (4ax^3 + 3ax^2 - 2a^3)$

61. $(3x^2 + 4x - 3) + (2x^2 - 3x - 1) - (x^2 + x + 7)$

62. $(-2x^2 + 6x + 5) - (-4x^2 - 7x + 2) - (4x^2 + 10x + 5)$

63. $\left(\frac{1}{3}y^6 - \frac{1}{6}y^4 - \frac{4}{3}y^2\right) + \left(-\frac{1}{6}y^6 - \frac{1}{2}y^4 + \frac{5}{6}y^2\right)$

64. $(0.2xy^7 + 0.8xy^5) - (0.5xy^7 - 0.6xy^5 + 0.2xy)$

65.
$$\begin{array}{rrrrr} & 3x^3 & -\ 2x^2 & +\ 4x & -\ 3 \\ & -2x^3 & +\ 3x^2 & +\ 3x & -\ 2 \\ + & 5x^3 & -\ 7x^2 & +\ 7x & -\ 12 \\ \hline \end{array}$$

66.
$$\begin{array}{rrrrr} & 7a^3 & & +\ 3a & +\ 7 \\ & -2a^3 & +\ 4a^2 & & -\ 13 \\ + & 3a^3 & -\ 3a^2 & +\ 4a & +\ 5 \\ \hline \end{array}$$

67.
$$\begin{array}{rrrr} & 3x^2 & -\ 4x & +\ 17 \\ - & 2x^2 & +\ 4x & -\ 5 \\ \hline \end{array}$$

68.
$$\begin{array}{rrrr} & -2y^2 & -\ 4y & +\ 3 \\ - & 3y^2 & +\ 10y & -\ 5 \\ \hline \end{array}$$

69.
$$\begin{array}{rrrrr} & -5y^3 & +\ 4y^2 & -\ 11y & +\ 3 \\ - & -2y^3 & -\ 14y^2 & +\ 17y & -\ 32 \\ \hline \end{array}$$

70.
$$\begin{array}{rrrrr} & 17x^4 & -\ 3x^2 & -\ 65x & -\ 12 \\ - & 23x^4 & +\ 14x^2 & +\ 3x & -\ 23 \\ \hline \end{array}$$

71.
$$\begin{array}{rrrr} & 4x^3 & & +\ 6a \\ - & 4x^3 & -\ 2x^2 & -\ a \\ \hline \end{array}$$

72.
$$\begin{array}{rrrr} & -2a^3 & & -\ 7a \\ - & -2a^3 & +\ 3a^2 & -\ 6a \\ \hline \end{array}$$

73. Find the difference when $3x^2y^3 + 4xy^2 - 3x^2$ is subtracted from the sum of $-2x^2y^3 - xy^2 + 7x^2$ and $5x^2y^3 + 3xy^2 - x^2$.

74. Find the difference when $8m^3n^3 + 2m^2n - n^2$ is subtracted from the sum of $m^2n + mn^2 + 2n^2$ and $2m^3n^3 - mn^2 + 9n^2$.

75. Find the sum when the difference of $2x^2 - 4x + 3$ and $8x^2 + 5x - 3$ is added to $-2x^2 + 7x - 4$.

76. Find the sum when the difference of $7x^3 - 4x$ and $x^2 + 2$ is added to $x^2 + 3x + 5$.

APPLICATIONS

77. JUGGLING During a performance, a juggler tosses one ball straight upward while continuing to juggle three others. The height $f(t)$, in feet, of the ball is given by the polynomial function $f(t) = -16t^2 + 32t + 4$, where t is the time in seconds since the ball was thrown. Find the height of the ball 1 second after it is tossed upward.

78. STOPPING DISTANCES The number of feet that a car travels before stopping depends on the driver's reaction time and the braking distance. For one driver, the stopping distance $d(v)$, in feet, is given by the polynomial function $d(v) = 0.04v^2 + 0.9v$, where v is the velocity of the car. Find the stopping distance at 60 mph.

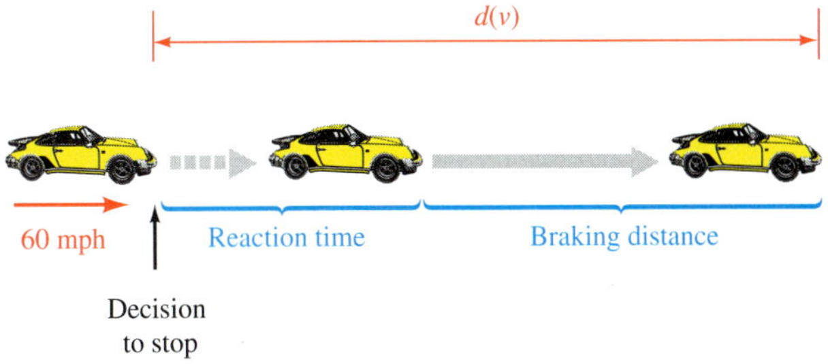

79. STORAGE TANKS The volume $V(r)$ of the gasoline storage tank, in cubic feet, is given by the polynomial function $V(r) = 4.2r^3 + 37.7r^2$, where r is the radius in feet of the cylindrical part of the tank. What is the capacity of the tank if its radius is 4 feet?

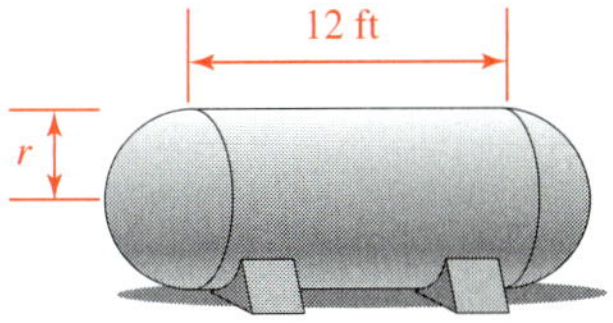

80. ROLLER COASTERS The polynomial function $f(x) = 0.001x^3 - 0.12x^2 + 3.6x + 10$ models the path of a portion of the track of a roller coaster. Find the height of the track for $x = 0, 20, 40,$ and 60.

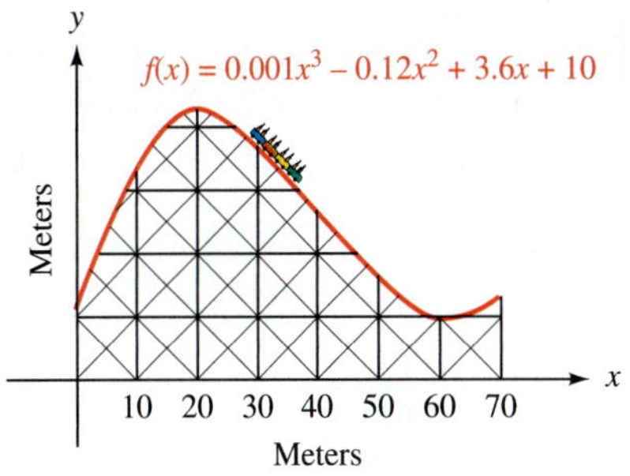

81. RAIN GUTTERS A rectangular sheet of metal will be used to make a rain gutter by bending up its sides, as shown. If the ends are covered, the capacity $f(x)$ of the gutter is a polynomial function of x: $f(x) = -240x^2 + 1{,}440x$. Find the capacity of the gutter if x is 3 inches.

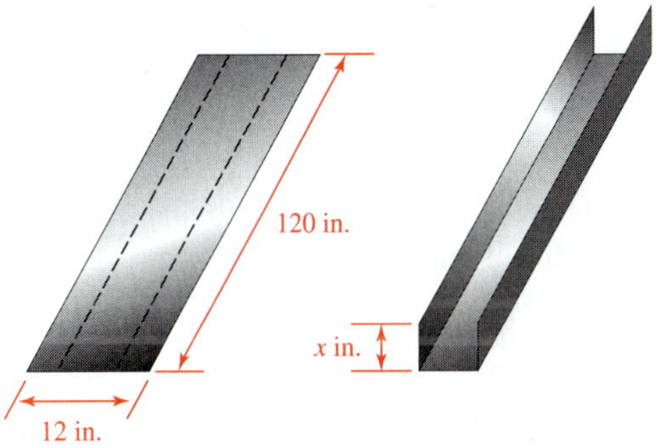

82. CUSTOMER SERVICE A software service hotline has found that on Mondays, the polynomial function $C(t) = -0.0625t^4 + t^3 - 6t^2 + 16t$ approximates the number of callers to the hotline at any one time. Here, t represents the time, in hours, since the hotline opened at 8:00 A.M. How many service technicians should be on duty on Mondays at noon if the company doesn't want any callers to the hotline waiting to be helped by a technician?

83. TRANSPORTATION ENGINEERING The polynomial function

$$A(x) = -0.000000000002x^3 + 0.00000008x^2 - 0.0006x + 2.45$$

approximates the number of accidents per mile in one year on a 4-lane interstate, where x is the average daily traffic in number of vehicles. Use the graph of the function on the next page to answer the following questions.

a. Find $A(20{,}000)$. Explain what the result means.

b. Find the value of x for which $A(x) = 2$. Explain what the result means.

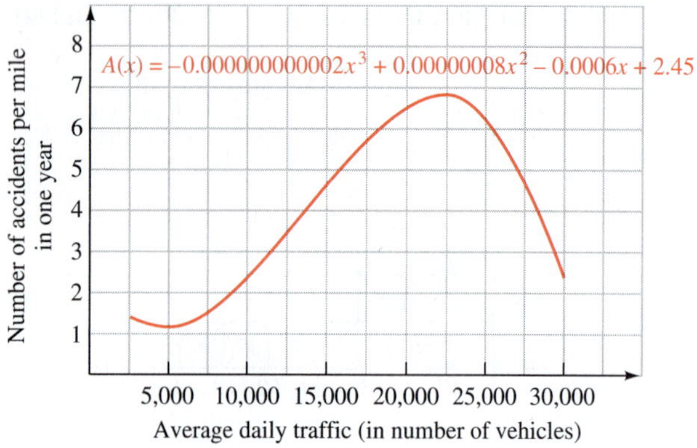

Source: Highway Safety Manual, Colorado Department of Transportation

84. BUSINESS EXPENSES A company purchased two cars for its sales force to use. The following functions give the respective values of the vehicles after x years.

Toyota Camry LE: $T(x) = -2{,}100x + 16{,}600$

Ford Explorer Sport: $F(x) = -2{,}700x + 19{,}200$

a. Find one polynomial function V that will give the value of both cars after x years.

b. Use your answer in part a to find the combined value of the two cars after 3 years.

WRITING

85. Explain why the terms x^2y and xy^2 are not like terms.

86. Explain why the identity function, the squaring function, and the cubic function belong to the family of polynomial functions.

87. Explain the error in the following solution.

Subtract $2x - 3$ from $3x + 4$.

$$\begin{aligned}(2x - 3) - (3x + 4) &= 2x - 3 - 3x - 4\\ &= -x - 7\end{aligned}$$

88. Explain why $f(x) = \dfrac{1}{x + 1}$ is not a polynomial function.

89. Use the word *descending* in a sentence in which the context is not mathematical. Do the same for the word *ascending*.

90. Look up the meaning of the prefix *poly* in a dictionary. Why do you think the name *polynomial* was given to expressions such as $x^3 - x^2 + 2x + 15$?

REVIEW

Solve each inequality. Write the solution set in interval notation.

91. $|x| \le 5$

92. $|x| > 7$

93. $|x - 4| < 5$

94. $|2x + 1| \ge 7$

CHALLENGE PROBLEMS

95. What polynomial should be subtracted from $5x^3y - 5xy + 2x$ to obtain the polynomial $8x^3y - 7xy + 11x$?

96. Find two trinomials such that their sum is a binomial and their difference is a monomial.

5.4 Multiplying Polynomials

- Multiplying monomials
- Multiplying a polynomial by a monomial
- Multiplying a polynomial by a polynomial
- The FOIL method
- Multiplying three polynomials
- Special products
- Simplifying expressions
- Applications of multiplying polynomials

In this section, we discuss the procedures used to multiply polynomials. These procedures involve the application of several algebraic concepts introduced in earlier chapters, such as the commutative and associative properties of multiplication, the rules for exponents, and the distributive property.

MULTIPLYING MONOMIALS

We begin by considering the simplest case of polynomial multiplication, multiplying two monomials.

Multiplying Monomials To multiply two monomials, multiply the numerical factors (the coefficients) and then multiply the variable factors.

EXAMPLE 1

Find each product: **a.** $(3x^2)(6x^3)$, **b.** $(-8x)(2y)(xy)$, and **c.** $(2a^3b)(-7b^2c)(-12ac^4)$.

Solution We can use the commutative and associative properties of multiplication to rearrange the terms and regroup the factors.

a. $(3x^2)(6x^3) = 3 \cdot x^2 \cdot 6 \cdot x^3$

$= (3 \cdot 6)(x^2 \cdot x^3)$

$= 18x^5$ To simplify $x^2 \cdot x^3$, keep the base and add the exponents.

Success Tip

In this section, you will see that every polynomial multiplication is a series of monomial multiplications.

b. $(-8x)(2y)(xy) = -8 \cdot x \cdot 2 \cdot y \cdot x \cdot y$

$= (-8 \cdot 2) \cdot x \cdot x \cdot y \cdot y$

$= -16x^2y^2$

c. $(2a^3b)(-7b^2c)(-12ac^4) = 2 \cdot a^3 \cdot b \cdot (-7) \cdot b^2 \cdot c \cdot (-12) \cdot a \cdot c^4$

$= 2(-7)(-12) \cdot a^3 \cdot a \cdot b \cdot b^2 \cdot c \cdot c^4$

$= 168a^4b^3c^5$

Self Check 1 Multiply: **a.** $(-2a^3)(4a^2)$ and **b.** $(-5b^3)(-3a)(a^2b)$.

MULTIPLYING A POLYNOMIAL BY A MONOMIAL

To multiply a polynomial by a monomial, we use the distributive property.

Multiplying Polynomials by Monomials To multiply a monomial and a polynomial, multiply each term of the polynomial by the monomial.

EXAMPLE 2

Find each product: **a.** $3x^2(6xy + 3y^2)$, **b.** $5x^3y^2(xy^3 - 2x^2y)$, and **c.** $-2ab^2(3bz - 2az + 4z^3)$.

Solution We can use the distributive property to remove parentheses.

a. $3x^2(6xy + 3y^2) = 3x^2(6xy) + 3x^2(3y^2)$ Distribute the multiplication by $3x^2$.

$= 18x^3y + 9x^2y^2$ Do the multiplications.

Since $18x^3y$ and $9x^2y^2$ are not like terms, we cannot add them.

b. $5x^3y^2(xy^3 - 2x^2y) = 5x^3y^2(xy^3) - 5x^3y^2(2x^2y)$ Distribute $5x^3y^2$.

$= 5x^4y^5 - 10x^5y^3$

c. $-2ab^2(3bz - 2az + 4z^3)$

$= -2ab^2 \cdot 3bz - (-2ab^2) \cdot 2az + (-2ab^2) \cdot 4z^3$

$= -6ab^3z + 4a^2b^2z - 8ab^2z^3$

Self Check 2 Multiply: $-2a^2(a^2 - a + 3)$.

MULTIPLYING A POLYNOMIAL BY A POLYNOMIAL

To multiply a polynomial by a polynomial, we use the distributive property repeatedly.

EXAMPLE 3

Find each product: **a.** $(3x + 2)(4x + 9)$ and **b.** $(2a - b)(3a^2 - 4ab + b^2)$.

Solution We can use the distributive property to remove parentheses.

a.

$$\begin{aligned}(3x + 2)(4x + 9) &= (3x + 2) \cdot 4x + (3x + 2) \cdot 9 && \text{Distribute } 3x + 2.\\ &= 12x^2 + 8x + 27x + 18 && \text{Distribute } 4x \text{ and distribute } 9.\\ &= 12x^2 + 35x + 18 && \text{Combine like terms.}\end{aligned}$$

b.

$$\begin{aligned}&(2a - b)(3a^2 - 4ab + b^2)\\ &= (2a - b)3a^2 - (2a - b)4ab + (2a - b)b^2 && \text{Distribute } 2a - b.\\ &= 6a^3 - 3a^2b - 8a^2b + 4ab^2 + 2ab^2 - b^3 && \text{Distribute } 3a^2, -4ab, \text{ and } b^2.\\ &= 6a^3 - 11a^2b + 6ab^2 - b^3 && \text{Combine like terms.}\end{aligned}$$

Self Check 3 Multiply: $(2a + b)(3a - 2b)$.

The results of Example 3 suggest the following rule.

Multiplying Polynomials To multiply two polynomials, multiply each term of one polynomial by each term of the other polynomial, and then combine like terms.

In the next example, we organize the work done in Example 3 vertically.

EXAMPLE 4

Find each product: **a.** $(3x + 2)(4x + 9)$ and **b.** $(3a^2 - 4ab + b^2)(2a - b)$.

Solution

a.

$$\begin{array}{rrrl} & 3x & +\ 2 & \\ & 4x & +\ 9 & \\ \hline 12x^2 & +\ 8x & & \leftarrow \text{This is the result of } 4x(3x + 2). \\ & +\ 27x & +\ 18 & \leftarrow \text{This is the result of } 9(3x + 2). \\ \hline 12x^2 & +\ 35x & +\ 18 & \text{Combine like terms, column by column.} \end{array}$$

b.

$$\begin{array}{rrrrl} 3a^2 & -\ 4ab & +\ b^2 & & \\ & 2a & -\ b & & \\ \hline 6a^3 & -\ 8a^2b & +\ 2ab^2 & & \leftarrow \text{This row is } 2a(3a^2 - 4ab + b^2) \\ & -\ 3a^2b & +\ 4ab^2 & -\ b^3 & \leftarrow \text{This row is } -b(3a^2 - 4ab + b^2) \\ \hline 6a^3 & -\ 11a^2b & +\ 6ab^2 & -\ b^3 & \end{array}$$

Success Tip

If we multiply each term of a three-term polynomial by each term of a two-term polynomial, there will be $3 \cdot 2 = 6$ multiplications to perform.

Self Check 4 Multiply:

$$\begin{array}{r} 3x^2 + 2x - 5 \\ 2x + 1 \\ \hline \end{array}$$

EXAMPLE 5

Multiply: $(-2y^3 - 6y^2 + 1)(5y^2 - 10y - 2)$.

Solution To multiply these expressions, we must multiply each term of one polynomial by each term of the other polynomial.

$$(-2y^3 - 6y^2 + 1)(5y^2 - 10y - 2)$$

For lengthy multiplications like this, we can use the vertical form. We begin by multiplying $-2y^3 - 6y^2 + 1$ by -2; then we multiply $-2y^3 - 6y^2 + 1$ by $-10y$; and finally we multiply $-2y^3 - 6y^2 + 1$ by $5y^2$. Then we combine like terms, column by column.

$$\begin{array}{rrrrrrl}
 & & -2y^3 & -\ 6y^2 & & +\ 1 & \\
 & & & 5y^2 & -\ 10y & -\ 2 & \\
\hline
 & & 4y^3 & +\ 12y^2 & & -\ 2 & \text{There is no } y\text{-term; leave a space.} \\
 & 20y^4 & +\ 60y^3 & & -\ 10y & & \text{There is no } y^2\text{-term; leave a space.} \\
-10y^5 & -\ 30y^4 & & +\ 5y^2 & & & \text{There is no } y^3\text{-term; leave a space.} \\
\hline
-10y^5 & -\ 10y^4 & +\ 64y^3 & +\ 17y^2 & -\ 10y & -\ 2 &
\end{array}$$

Self Check 5 Multiply: $(2a^2 + 6a - 1)(3a^2 + 9a - 5)$.

THE FOIL METHOD

When we multiply two binomials, each term of one binomial must be multiplied by each term of the other binomial. This fact can be emphasized by drawing arrows to show the indicated products. For example, to multiply $3x + 2$ and $x + 4$, we can write

First terms   Last terms

$$(3x + 2)(x + 4) = 3x(x) + 3x(4) + 2(x) + 2(4)$$

$$= 3x^2 + 12x + 2x + 8$$

$$= 3x^2 + 14x + 8 \qquad \text{Combine like terms: } 12x + 2x = 14x.$$

Inner terms

Outer terms

We note that

- the product of the **F**irst terms is $3x \cdot x = 3x^2$,
- the product of the **O**uter terms is $3x \cdot 4 = 12x$,
- the product of the **I**nner terms is $2 \cdot x = 2x$, and
- the product of the **L**ast terms is $2 \cdot 4 = 8$.

The procedure is called the **FOIL** method of multiplying two binomials. FOIL is an acronym for **F**irst terms, **O**uter terms, **I**nner terms, and **L**ast terms. The resulting terms of the product must be combined, if possible.

It is easy to multiply binomials by sight using the FOIL method. We find the product of the first terms, then find the products of the outer terms and the inner terms and add them (when possible), and then find the product of the last terms.

EXAMPLE 6

Find each product: **a.** $(2x - 3)(3x + 2)$, **b.** $(3x + 1)(3x + 4)$, and **c.** $(4xy - 5)(2x^2 - 3y)$.

Solution We can use the FOIL method to perform each multiplication.

a. $(2x - 3)(3x + 2) = 6x^2 - 5x - 6$

The product of the first terms is $2x \cdot 3x = 6x^2$. The middle term in the result comes from combining the outer and inner products of $4x$ and $-9x$:

$$4x + (-9x) = -5x$$

The product of the last terms is $-3 \cdot 2 = -6$.

b. $(3x + 1)(3x + 4) = 9x^2 + 15x + 4$

The product of the first terms is $3x \cdot 3x = 9x^2$. The middle term in the result comes from combining the products $12x$ and $3x$:

$$12x + 3x = 15x$$

The product of the last terms is $1 \cdot 4 = 4$.

c. $(4xy - 5)(2x^2 - 3y) = 8x^3y - 12xy^2 - 10x^2 + 15y$

The product of the first terms is $4xy \cdot 2x^2 = 8x^3y$. The product of the outer terms is $4xy(-3y) = -12xy^2$. The product of the inner terms is $-5(2x^2) = -10x^2$. The product of the last terms is $-5(-3y) = 15y$. Since the terms of $8x^3y - 12xy^2 - 10x^2 + 15y$ are unlike, we cannot simplify this result.

The Language of Algebra

The acronym FOIL helps us remember the order to follow when multiplying two binomials. Another popular acronym is PEMDAS. It represents the order of operations rules: **P**arentheses, **E**xponents, **M**ultiply, **D**ivide, **A**dd, **S**ubtract.

Self Check 6 Multiply: **a.** $(3a + 4b)(2a - b)$, **b.** $(a^2b - 3)(a^2b - 1)$, and **c.** $(6c^4 - d)(3c + d)$.

MULTIPLYING THREE POLYNOMIALS

When finding the product of three polynomials, we begin by multiplying *any* two of them, and then we multiply that result by the third polynomial.

EXAMPLE 7

Multiply: $3cd(c + 2d)(3c - d)$.

Solution First, we find the product of the two binomials. Then we multiply that result by $3cd$.

$3cd(c + 2d)(3c - d) = 3cd(3c^2 - cd + 6cd - 2d^2)$	Use the FOIL method to find $(c + 2d)(3c - d)$.
$= 3cd(3c^2 + 5cd - 2d^2)$	Combine like terms: $-cd + 6cd = 5cd$.
$= 9c^3d + 15c^2d^2 - 6cd^3$	Distribute the multiplication by $3cd$.

Self Check 7 Multiply: $-2r(r - 2s)(5r - 4s)$.

SPECIAL PRODUCTS

We often must find the square of a binomial. To do so, we can use the FOIL method. For example, to find $(x + y)^2$ and $(x - y)^2$, we proceed as follows.

$$\begin{aligned}(x + y)^2 &= (x + y)(x + y) \\ &= x^2 + xy + xy + y^2 \\ &= x^2 + 2xy + y^2\end{aligned} \qquad \begin{aligned}(x - y)^2 &= (x - y)(x - y) \\ &= x^2 - xy - xy + y^2 \\ &= x^2 - 2xy + y^2\end{aligned}$$

In each case, we see that the square of the binomial is the square of its first term, twice the product of its two terms, and the square of its last term.

The figure shows how $(x + y)^2$ can be found graphically.

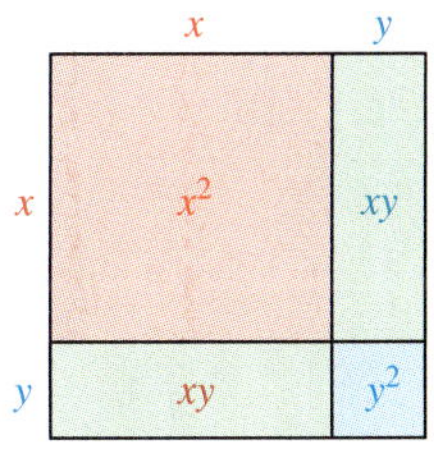

The area of the largest square is the product of its length and width: $(x + y)(x + y) = (x + y)^2$.

The area of the largest square is also the sum of its four parts: $x^2 + xy + xy + y^2 = x^2 + 2xy + y^2$.

Thus, $(x + y)^2 = x^2 + 2xy + y^2$.

Another common binomial product is the product of the sum and difference of the same two terms. An example of such a product is $(x + y)(x - y)$. To find this product, we use the FOIL method.

$$\begin{aligned}(x + y)(x - y) &= x^2 - xy + xy - y^2 \\ &= x^2 - y^2 \qquad \text{Combine like terms: } -xy + xy = 0.\end{aligned}$$

We see that the product of the sum and the difference of the same two terms is the square of the first term minus the square of the second term.

The square of a binomial and the product of the sum and the difference of the same two terms are called **special products.** Because special products occur so often, it is useful to learn their forms.

Special Product Formulas

$(x + y)^2 = (x + y)(x + y) = x^2 + 2xy + y^2$	The square of a sum.
$(x - y)^2 = (x - y)(x - y) = x^2 - 2xy + y^2$	The square of a difference.
$(x + y)(x - y) = x^2 - y^2$	The product of the sum and difference of two terms.

Caution Remember that the square of a binomial is a *trinomial.* A common error when squaring a binomial is to forget the middle term of the product. For example,

$(x + y)^2 \neq x^2 + y^2$ (Missing $2xy$) and $(x - y)^2 \neq x^2 - y^2$ (Missing $-2xy$; Should be + symbol)

Also remember that the product $(x + y)(x - y)$ is the binomial $x^2 - y^2$. And since $(x + y)(x - y) = (x - y)(x + y)$ by the commutative property of multiplication,

$$(x - y)(x + y) = x^2 - y^2$$

EXAMPLE 8

Multiply: **a.** $(5c + 3d)^2$, **b.** $\left(\frac{1}{2}a^4 - b^2\right)^2$, and **c.** $(0.2m^3 + 2.5n)(0.2m^3 - 2.5n)$.

Solution **a.** To find $(5c + 3d)^2$ using a special product formula, we begin by noting that the first term of the binomial is $5c$ and the last term is $3d$.

The square of the first term $5c$ ↓ | Twice the product of the two terms ↓ | The square of the last term $3d$ ↓

$$(5c + 3d)^2 = (5c)^2 + 2(5c)(3d) + (3d)^2$$
$$= 25c^2 + 30cd + 9d^2$$

The Language of Algebra

When squaring a binomial, the result is called a *perfect square trinomial.* For example,

$$(t + 9)^2 = t^2 + 18t + 81$$

Perfect square trinomial

b. To find $\left(\frac{1}{2}a^4 - b^2\right)^2$ using a special product formula, we begin by noting that the first term of the binomial is $\frac{1}{2}a^4$ and the last term is $-b^2$.

The square of the first term ↓ | Twice the product of the two terms ↓ | The square of the last term ↓

$$\left(\frac{1}{2}a^4 - b^2\right)^2 = \left(\frac{1}{2}a^4\right)^2 + 2\left(\frac{1}{2}a^4\right)(-b^2) + (-b^2)^2$$
$$= \frac{1}{4}a^8 - a^4b^2 + b^4$$

Success Tip

We can use the FOIL method to find each of the special products discussed in this section. However, these forms occur so often, it is worthwhile to learn the special product rules.

c. $(0.2m^3 + 2.5n)(0.2m^3 - 2.5n)$ is the product of the sum and the difference of the same two terms: $0.2m^3$ and $2.5n$. Using a special product formula, we proceed as follows.

The square of the first term ↓ | The square of the last term ↓

$$(0.2m^3 + 2.5n)(0.2m^3 - 2.5n) = (0.2m^3)^2 - (2.5n)^2$$
$$= 0.04m^6 - 6.25n^2$$

Self Check 8 Multiply: **a.** $(8r + 2s)^2$, **b.** $\left(\frac{1}{3}a^3 - b^6\right)^2$, and **c.** $(0.4x + 1.2y^4)(0.4x - 1.2y^4)$.

SIMPLIFYING EXPRESSIONS

The procedures discussed in this section are often useful when we simplify algebraic expressions that involve the multiplication of polynomials.

EXAMPLE 9

Simplify: $(5x - 4)^2 - (x - 7)(x + 1)$.

Solution Before doing the subtraction, we use a special product formula to find $(5x - 4)^2$ and the FOIL method to find $(x - 7)(x + 1)$.

$$(5x - 4)^2 - (x - 7)(x + 1)$$

$= 25x^2 - 40x + 16 - (x^2 - 6x - 7)$ $(5x - 4)^2 = (5x)^2 + 2(5x)(-4) + (-4)^2$.

$= 25x^2 - 40x + 16 - x^2 + 6x + 7$ To subtract $(x^2 - 6x - 7)$, remove the parentheses and change the sign of each term within the parentheses.

$= 24x^2 - 34x + 23$ Combine like terms.

Self Check 9 Simplify: $(y - 7)(y + 7) - (4y + 3)^2$.

APPLICATIONS OF MULTIPLYING POLYNOMIALS

Profit, revenue, and cost are terms used in the business world. The profit earned on the sale of one or more items is given by the formula

$$\text{Profit} = \text{revenue} - \text{cost}$$

If a salesperson has 12 vacuum cleaners and sells them for \$225 each, the revenue will be $r = \$(12 \cdot 225) = \$2{,}700$. This illustrates the following formula for finding the revenue r:

$$r = \boxed{\begin{matrix}\text{number of}\\ \text{items sold } x\end{matrix}} \cdot \boxed{\begin{matrix}\text{selling price}\\ \text{of each item } p\end{matrix}} = xp = px$$

EXAMPLE 10

Selling vacuum cleaners. Over the years, a saleswoman has found that the number of vacuum cleaners she can sell depends on price. The lower the price, the more she can sell. She has determined that the number of vacuums x that she can sell at a price p is related by the equation $x = -\frac{2}{25}p + 28$.

a. Find a formula for the revenue r.

b. How much revenue will she take in if the vacuums are priced at \$250?

Solution **a.** To find a formula for revenue, we substitute $-\frac{2}{25}p + 28$ for x in the formula $r = px$ and multiply.

$r = px$ This is the formula for revenue.

$r = p\left(-\frac{2}{25}p + 28\right)$ Substitute $-\frac{2}{25}p + 28$ for x.

$r = -\frac{2}{25}p^2 + 28p$ Multiply the polynomials.

b. To find how much revenue she will take in if the vacuums are priced at \$250, we substitute 250 for p in the formula for revenue.

$r = -\frac{2}{25}p^2 + 28p$ This is the formula for revenue.

$r = -\frac{2}{25}(250)^2 + 28(250)$ Substitute 250 for p.

$= -5{,}000 + 7{,}000$

$= 2{,}000$

The revenue will be \$2,000.

Answers to Self Checks **1. a.** $-8a^5$, **b.** $15a^3b^4$ **2.** $-2a^4 + 2a^3 - 6a^2$ **3.** $6a^2 - ab - 2b^2$ **4.** $6x^3 + 7x^2 - 8x - 5$ **5.** $6a^4 + 36a^3 + 41a^2 - 39a + 5$ **6. a.** $6a^2 + 5ab - 4b^2$, **b.** $a^4b^2 - 4a^2b + 3$, **c.** $18c^5 + 6c^4d - 3cd - d^2$ **7.** $-10r^3 + 28r^2s - 16rs^2$ **8. a.** $64r^2 + 32rs + 4s^2$, **b.** $\frac{1}{9}a^6 - \frac{2}{3}a^3b^6 + b^{12}$, **c.** $0.16x^2 - 1.44y^8$ **9.** $-15y^2 - 24y - 58$

5.4 STUDY SET

VOCABULARY **Fill in the blanks.**

1. The expression $(2x^3)(3x^4)$ is the product of two __________ and the expression $(x + 4)(x - 5)$ is the product of two __________.
2. $(x + 4)^2$ is the ________ of a sum and $(m - 9)^2$ is the square of a __________.
3. $(b + 1)(b - 1)$ is the product of the ______ and difference of two terms.
4. Since $x^2 + 16x + 64$ is the square of $x + 8$, it is called a ________ square trinomial.

CONCEPTS **Fill in the blanks.**

5. To multiply a monomial by a monomial, we multiply the numerical ________ and then multiply the variable factors.
6. To multiply a polynomial by a monomial, we multiply each ______ of the polynomial by the monomial.
7. To multiply a polynomial by a polynomial, we multiply each ______ of one polynomial by each term of the other polynomial.
8. FOIL is an acronym for ______ terms, _______ terms, _______ terms, and ______ terms.
9. $(x + y)^2 = (x + y)(x + y) =$ ______________
10. $(x - y)^2 = (x - y)(x - y) =$ ______________
11. $(x + y)(x - y) =$ _________
12. **a.** The square of a binomial is the ________ of its first term, _______ the product of its two terms, plus the ________ of its last term.
    b. The product of the sum and difference of the same two terms is the ________ of the first term minus the ________ of the second term.
13. Write a polynomial that represents the area of the rectangle shown in the illustration.

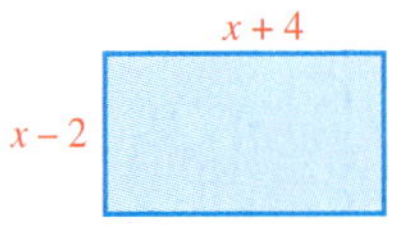

14. Write a polynomial that represents the area of the triangle shown in the illustration.

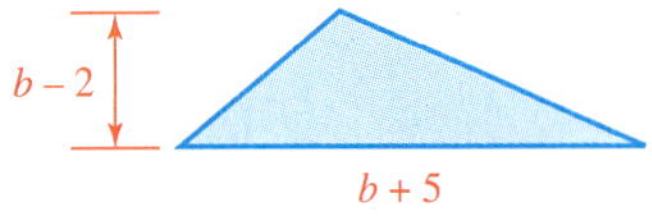

15. Write a polynomial that represents the area of the square shown in the illustration.

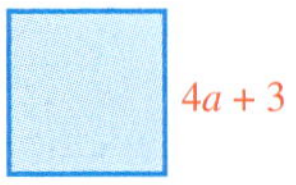

16. Write a polynomial that represents the area of the rectangle shown in the illustration.

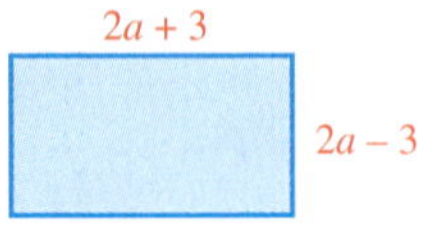

17. Consider $(2x + 4)(4x - 3)$. Give the
    a. First terms **b.** Outer terms
    c. Inner terms **d.** Last terms
18. Find
    a. $(4b - 1) + (2b - 1)$
    b. $(4b - 1)(2b - 1)$
    c. $(4b - 1) - (2b - 1)$

PRACTICE **Find each product.**

19. $(2a^2)(-3ab)$
20. $(-3x^2y)(3xy)$
21. $(-3ab^2c)(5ac^2)$
22. $(-2m^2n)(-4mn^3)$
23. $(4a^2b)(-5a^3b^2)(6a^4)$
24. $(2x^2y^3)(4xy^5)(-5y^6)$
25. $(-5xx^2)(-3xy)^4$
26. $(-2a^2ab^2)^3(-3ab^2b^2)$
27. $3(x + 2)$
28. $-5(a + b)$
29. $3x(x^2 + 3x)$
30. $-2x(3x^2 - 2)$
31. $-2x(3x^2 - 3x + 2)$
32. $3a(4a^2 + 3a - 4)$
33. $7rst(r^2 + s^2 - t^2)$
34. $3x^2yz(x^2 - 2y + 3z^2)$
35. $4m^2n(-3mn)(m + n)$
36. $-3a^2b^3(2b)(3a + b)$
37. $(x + 2)(x + 3)$
38. $(y - 3)(y + 4)$
39. $(3t - 2)(2t + 3)$
40. $(p + 3)(3p - 4)$
41. $(3y - z)(2y - z)$
42. $(2m - n)(3m - n)$
43. $\left(\frac{1}{2}b + 8\right)(4b + 6)$
44. $\left(\frac{2}{3}x + 1\right)(15x - 9)$
45. $(0.4t - 3)(0.5t - 3)$
46. $(0.7d - 2)(0.1d + 3)$
47. $(b^3 - 1)(b + 1)$
48. $(c^3 + 1)(1 - c)$

49. $(3tu - 1)(-2tu + 3)$

50. $(-5st + 1)(10st - 7)$

51. $(9b^3 - c)(3b^2 - c)$

52. $(h^5 - k)(4h^3 - k)$

53. $(11m^2 + 3n^3)(5m + 2n^2)$

54. $(50m^4 - 3n^4)(2m + 2n^3)$

55. $6p^2(3p - 4)(p + 3)$

56. $4a(2a + 3)(3a - 2)$

57. $(3m - y)(4my)(2m - y)$

58. $(2h - z)(-3hz)(3h - z)$

59. $(x + 2)^2$

60. $(x - 3)^2$

61. $(3a - 4)^2$

62. $(2y + 5)^2$

63. $(2a + b)^2$

64. $(a - 2b)^2$

65. $(5r^2 + 6)^2$

66. $(6p^2 - 3)^2$

67. $(9ab^2 - 4)^2$

68. $(2yz^2 + 5)^2$

69. $\left(\frac{1}{4}b + 2\right)^2$

70. $\left(\frac{2}{3}y - 7\right)^2$

71. $(4k - 1.3)^2$

72. $(0.5k + 6)^2$

73. $(x + 2)(x - 2)$

74. $(z + 3)(z - 3)$

75. $(y^3 + 2)(y^3 - 2)$

76. $(y^4 + 3)(y^4 - 3)$

77. $(xy - 6)(xy + 6)$

78. $(a^4b - c)(a^4b + c)$

79. $\left(\frac{1}{2}x - 16\right)\left(\frac{1}{2}x + 16\right)$

80. $\left(\frac{3}{4}h^2 - \frac{2}{3}\right)\left(\frac{3}{4}h^2 + \frac{2}{3}\right)$

81. $(2.4 + y)(2.4 - y)$

82. $(3.5t + 4.1u)(3.5t - 4.1u)$

83. $(x - y)(x^2 + xy + y^2)$

84. $(x + y)(x^2 - xy + y^2)$

85. $(3y + 1)(2y^2 + 3y + 2)$

86. $(a + 2)(3a^2 + 4a - 2)$

87. $(2a - b)(4a^2 + 2ab + b^2)$

88. $(x - 3y)(x^2 + 3xy + 9y^2)$

89. $(a + b)(a - b)(a - 3b)$

90. $(x - y)(x + 2y)(x - 2y)$

91. $(a + b + c)(2a - b - 2c)$

92. $(x + 2y + 3z)^2$

93. $(r + s)^2(r - s)^2$

94. $r(r + s)(r - s)^2$

Simplify each expression.

95. $3x(2x + 4) - 3x^2$

96. $2y - 3y(y^2 + 4)$

97. $3pq - p(p - q)$

98. $-4rs(r - 2) + 4rs$

99. $(x + 3)(x - 3) + (2x - 1)(x + 2)$

100. $(2b + 3)(b - 1) - (b + 2)(3b - 1)$

101. $(3x - 4)^2 - (2x + 3)^2$

102. $(3y + 1)^2 + (2y - 4)^2$

Use a calculator to help find each product.

103. $(3.21x - 7.85)(2.87x + 4.59)$

104. $(7.44y + 56.7)(-2.1y - 67.3)$

105. $(-17.3y + 4.35)^2$

106. $(-0.31x + 29.3)(-0.31x - 29.3)$

APPLICATIONS

107. THE YELLOW PAGES Refer to the illustration.

a. Describe the area occupied by the ads for movers by using a product of two binomials.

b. Describe the area occupied by the ad for Budget Moving Co. by using a product. Then perform the multiplication.

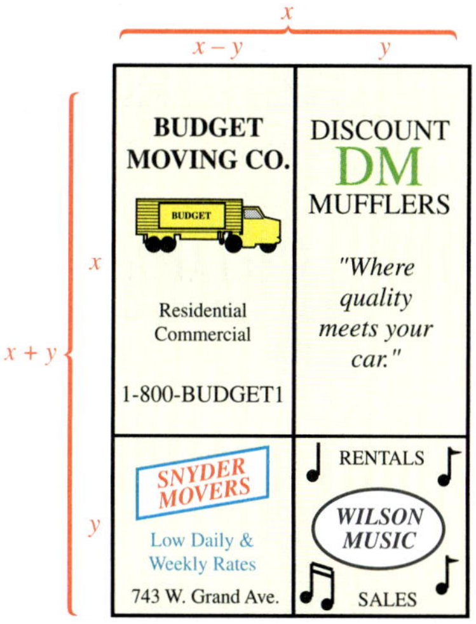

c. Describe the area occupied by the ad for Snyder Movers by using a product. Then perform the multiplication.

d. Explain why your answer to part a is equal to the sum of your answers to parts b and c. What special product does this exercise illustrate?

108. HELICOPTER PADS To determine the amount of fluorescent paint needed to paint the circular ring on the landing pad design shown in the illustration, painters must find its area. The area of the ring is given by the expression $\pi(R + r)(R - r)$.

a. Find the product $\pi(R + r)(R - r)$.

b. If $R = 25$ feet and $r = 20$ feet, find the area to be painted. Round to the nearest tenth.

c. If a quart of fluorescent paint covers 65 ft^2, how many quarts will be needed to paint the ring?

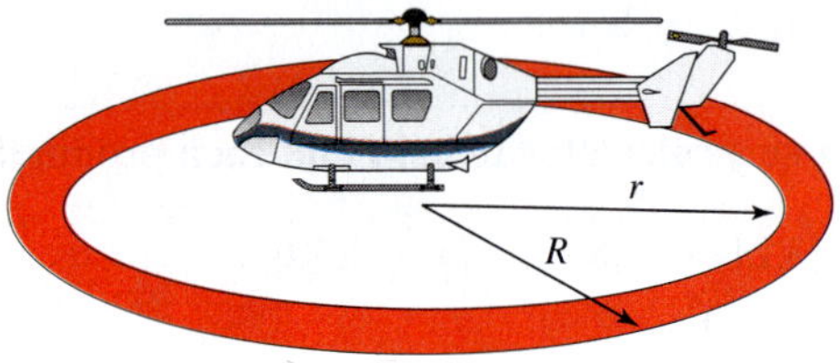

109. GIFT BOXES The corners of a 12-in.-by-12-in. piece of cardboard are creased, folded inward, and glued to make a gift box. (See the illustration.) Write a polynomial that gives the volume of the resulting box.

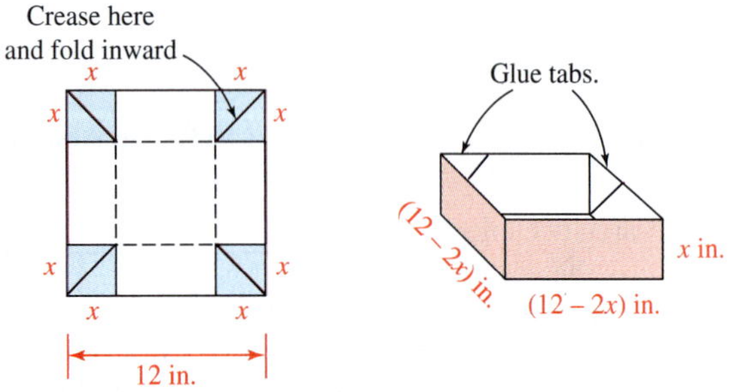

110. CALCULATING REVENUE A salesperson has found that the number x of televisions she can sell at a certain price p is related by the equation $x = -\frac{1}{5}p + 90$.

a. Find the number of TVs she will sell if the price is \$375.

b. Write a formula for the revenue when x TVs are sold.

c. Find the revenue generated by TV sales if they are priced at \$400 each.

WRITING

111. Explain how to use the FOIL method.

112. Explain how you would multiply two trinomials.

113. On a test, when asked to find $(x - y)^2$, a student answered $x^2 - y^2$. What error did the student make?

114. Describe each expression in words:
$(x + y)^2 \quad (x - y)^2 \quad (x + y)(x - y)$

REVIEW **Graph each inequality or system of inequalities.**

115. $2x + y \leq 2$

116. $x \geq 2$

117. $\begin{cases} y - 2 < 3x \\ y + 2x < 3 \end{cases}$

118. $\begin{cases} y < 0 \\ x < 0 \end{cases}$

CHALLENGE PROBLEMS **Find each product. Write all answers without negative exponents.**

119. $ab^{-2}c^{-3}(a^{-4}bc^3 + a^{-3}b^4c^3)$

120. $(5x^{-4} - 4y^2)(5x^2 - 4y^{-4})$

Find each product. Assume n is a natural number.

121. $a^{2n}(a^n + a^{2n})$

122. $(a^{3n} - b^{3n})(a^{3n} + b^{3n})$

5.5 The Greatest Common Factor and Factoring by Grouping

- The greatest common factor (GCF)
- Factoring out the greatest common factor
- Factoring by grouping
- Formulas

In Section 5.4, we discussed ways of multiplying polynomials. In this section, we will discuss the reverse process—*factoring* polynomials. When factoring a polynomial, the first step is to determine whether its terms have any common factors.

THE GREATEST COMMON FACTOR (GCF)

If one number a divides a second number b exactly, then a is called a **factor** of b. For example, because 3 divides 24 exactly, it is a factor of 24. Each number in the following list is a factor of 24.

1, 2, 3, 4, 6, 8, 12, and 24

To factor a natural number, we write it as a product of other natural numbers. If each factor is a prime number, the natural number is said to be written in **prime-factored form.** Example 1 shows how to find the prime-factored forms of 60, 84, and 180, respectively.

EXAMPLE 1

Find the prime factorization of each number: **a.** 60, **b.** 84, and **c.** 180.

Solution

a.
$$\begin{aligned} 60 &= 6 \cdot 10 \\ &= 2 \cdot 3 \cdot 2 \cdot 5 \\ &= 2^2 \cdot 3 \cdot 5 \end{aligned}$$

b.
$$\begin{aligned} 84 &= 4 \cdot 21 \\ &= 2 \cdot 2 \cdot 3 \cdot 7 \\ &= 2^2 \cdot 3 \cdot 7 \end{aligned}$$

c.
$$\begin{aligned} 180 &= 10 \cdot 18 \\ &= 2 \cdot 5 \cdot 3 \cdot 6 \\ &= 2 \cdot 5 \cdot 3 \cdot 3 \cdot 2 \\ &= 2^2 \cdot 3^2 \cdot 5 \end{aligned}$$

Self Check 1 Find the prime factorization of 120.

The Language of Algebra
Recall that the *prime numbers* are: 2, 3, 5, 7, 11, 13, 17, 19, ...

The largest natural number that divides 60, 84, and 180 is called their **greatest common factor (GCF).** Because 60, 84, and 180 all have two factors of 2 and one factor of 3, their GCF is $2^2 \cdot 3 = 12$. We note that

$$\frac{60}{12} = 5, \quad \frac{84}{12} = 7, \quad \text{and} \quad \frac{180}{12} = 15$$

There is no natural number greater than 12 that divides 60, 84, and 180.

The Greatest Common Factor (GCF) The **greatest common factor (GCF)** of a list of integers is the largest common factor of those integers.

To find the greatest common factor of a list of terms, we can use the following approach.

Strategy for Finding the GCF

1. Write each coefficient as a product of prime factors.
2. Identify the numerical and variable factors common to each term.
3. Multiply the common factors identified in Step 2 to obtain the GCF. If there are no common factors, the GCF is 1.

EXAMPLE 2

Find the GCF of $6a^2b^3c$, $9a^3b^2c$, and $18a^4c^3$.

Solution We begin by factoring each term.

$$6a^2b^3c = 3 \cdot 2 \cdot a \cdot a \cdot b \cdot b \cdot b \cdot c$$
$$9a^3b^2c = 3 \cdot 3 \cdot a \cdot a \cdot a \cdot b \cdot b \cdot c$$
$$18a^4c^3 = 2 \cdot 3 \cdot 3 \cdot a \cdot a \cdot a \cdot a \cdot c \cdot c \cdot c$$

Success Tip

The exponent on any variable in a GCF is the *smallest* exponent that appears on that variable in all of the terms under consideration.

Since each term has one factor of 3, two factors of a, and one factor of c in common, the GCF is

$$3^1 \cdot a^2 \cdot c^1 = 3a^2c$$

Self Check 2 Find the GCF of $24x^2y^3$, $3x^3y$, and $18x^2y^2$.

FACTORING OUT THE GREATEST COMMON FACTOR

We have seen that the distributive property provides a method for multiplying a polynomial by a monomial. For example,

Multiplication →

$$2x^3(3x^2 + 4y^3) = 2x^3 \cdot 3x^2 + 2x^3 \cdot 4y^3$$
$$= 6x^5 + 8x^3y^3$$

If the product of a multiplication is $6x^5 + 8x^3y^3$, we can use the distributive property in reverse to find the individual factors.

Factoring →

$$6x^5 + 8x^3y^3 = 2x^3 \cdot 3x^2 + 2x^3 \cdot 4y^3$$
$$= 2x^3(3x^2 + 4y^3)$$

Since $2x^3$ is the GCF of the terms of $6x^5 + 8x^3y^3$, this process is called **factoring out the greatest common factor.** *When we factor a polynomial, we write a sum of terms as a product of factors.*

$$\underbrace{6x^5 + 8x^3y^3}_{\text{Sum of terms}} = \underbrace{2x^3(3x^2 + 4y^3)}_{\text{Product of factors}}$$

EXAMPLE 3

Factor: $25a^3b + 15ab^3$.

Solution We begin by factoring each monomial:

$$25a^3b = 5 \cdot 5 \cdot a \cdot a \cdot a \cdot b$$
$$15ab^3 = 5 \cdot 3 \cdot a \cdot b \cdot b \cdot b$$

Since each term has one factor of 5, one factor of a, and one factor of b in common, and there are no other common factors, $5ab$ is the GCF of the two terms. We can use the distributive property to factor it out.

Success Tip

Always verify a factorization by performing the indicated multiplication. The result should be the original polynomial.

$$25a^3b + 15ab^3 = \mathbf{5ab} \cdot 5a^2 + \mathbf{5ab} \cdot 3b^2$$
$$= \mathbf{5ab}(5a^2 + 3b^2)$$

To check, we multiply: $5ab(5a^2 + 3b^2) = 25a^3b + 15ab^3$.

Since we obtain the original polynomial, $25a^3b + 15ab^3$, the factorization is correct.

Self Check 3 Factor: $9x^4y^2 - 12x^3y^3$.

EXAMPLE 4

Factor: $3xy^2z^3 + 6xyz^3 - 3xz^2$.

Solution We begin by factoring each term:

$$3xy^2z^3 = 3 \cdot x \cdot y \cdot y \cdot z \cdot z \cdot z$$
$$6xyz^3 = 3 \cdot 2 \cdot x \cdot y \cdot z \cdot z \cdot z$$
$$-3xz^2 = -3 \cdot x \cdot z \cdot z$$

Success Tip

On the game show Jeopardy!, answers are revealed and contestants respond with the appropriate questions. Factoring is similar. Answers to multiplications are given. You are to respond by telling what factors were multiplied.

Since each term has one factor of 3, one factor of x, and two factors of z in common, and because there are no other common factors, $3xz^2$ is the GCF of the three terms. We can use the distributive property to factor it out.

$$3xy^2z^3 + 6xyz^3 - 3xz^2 = \mathbf{3xz^2} \cdot y^2z + \mathbf{3xz^2} \cdot 2yz - \mathbf{3xz^2} \cdot 1$$
$$= \mathbf{3xz^2}(y^2z + 2yz - 1)$$

When the $3xz^2$ is factored out, remember to write the -1.

Self Check 4 Factor: $2a^4b^2 + 6a^3b^2 - 4a^2b$.

A polynomial that cannot be factored is called a **prime polynomial** or an **irreducible polynomial.**

EXAMPLE 5

Factor $3x^2 + 4y + 7$, if possible.

Solution We factor each term:

$$3x^2 = 3 \cdot x \cdot x \qquad 4y = 2 \cdot 2 \cdot y \qquad 7 = 7$$

Since there are no common factors other than 1, this polynomial cannot be factored. It is a prime polynomial.

Self Check 5 Factor: $6a^3 + 7b^2 + 5$.

EXAMPLE 6

Factor -1 out of $-a^3 + 2a^2 - 4$.

Solution First, we write each term of the polynomial as the product of -1 and another factor. Then we factor out the common factor, -1.

$$-a^3 + 2a^2 - 4 = (-1)a^3 + (-1)(-2a^2) + (-1)4$$

$$= -1(a^3 - 2a^2 + 4) \quad \text{Factor out } -1.$$

$$= -(a^3 - 2a^2 + 4) \quad \text{The coefficient of 1 need not be written.}$$

Success Tip

To factor out -1, simply change the sign of each term of $-a^3 + 2a^2 - 4$ and write a $-$ symbol in front of the parentheses.

Self Check 6 Factor -1 out of $-b^4 - 3b^2 + 2$.

EXAMPLE 7

Factor the opposite of the GCF from $-6u^2v^3 + 8u^3v^2$.

Solution Because the GCF of the two terms is $2u^2v^2$, the opposite of the GCF is $-2u^2v^2$. To factor out $-2u^2v^2$, we proceed as follows:

$$-6u^2v^3 + 8u^3v^2 = -2u^2v^2 \cdot 3v + 2u^2v^2 \cdot 4u$$

$$= -2u^2v^2 \cdot 3v - (-2u^2v^2)4u$$

$$= -2u^2v^2(3v - 4u)$$

Self Check 7 Factor out the opposite of the GCF from $-8a^2b^2 - 12ab^3$.

A common factor can have more than one term.

EXAMPLE 8

Factor: **a.** $x(x + 1) + y(x + 1)$ and **b.** $a(x - y + z) - b(x - y + z) + 3(x - y + z)$.

Solution **a.** The binomial $x + 1$ is a factor of both terms. We can factor it out to get

$$x(x + 1) + y(x + 1) = (x + 1)x + (x + 1)y \quad \text{Use the commutative property of multiplication.}$$

$$= (x + 1)(x + y)$$

b. We can factor out the GCF of the three terms, which is $(x - y + z)$.

$$a(x - y + z) - b(x - y + z) + 3(x - y + z)$$

$$= (x - y + z)a - (x - y + z)b + (x - y + z)3$$

$$= (x - y + z)(a - b + 3)$$

Self Check 8 Factor: **a.** $c(y^2 + 1) + d(y^2 + 1) + e(y^2 + 1)$, and **b.** $x(a + b - c) - y(a + b - c)$.

FACTORING BY GROUPING

Suppose that we wish to factor

$$ac + ad + bc + bd$$

Although there is no factor common to all four terms, there is a common factor of a in the first two terms and a common factor of b in the last two terms. We can factor out these common factors to get

$$ac + ad + bc + bd = a(c + d) + b(c + d)$$

We can now factor out the common factor of $c + d$ on the right-hand side:

$$ac + ad + bc + bd = (c + d)(a + b)$$

Caution

Don't think that

$$a(c + d) + b(c + d)$$

is in factored form and stop. It is still a *sum* of two terms. A factorization of a polynomial must be a *product.*

The grouping in this type of problem is not always unique. For example, if we write the polynomial $ac + ad + bc + bd$ in the form

$$ac + bc + ad + bd$$

and factor c from the first two terms and d from the last two terms, we obtain

$$\begin{aligned} ac + bc + ad + bd &= c(a + b) + d(a + b) \\ &= (a + b)(c + d) \end{aligned}$$

This is equivalent to $(c + d)(a + b)$.

The method used in the previous examples is called **factoring by grouping.**

Factoring by Grouping

1. Group the terms of the polynomial so that each group has a common factor.
2. Factor out the common factor from each group.
3. Factor out the resulting common factor. If there is no common factor, regroup the terms of the polynomial and repeat steps 2 and 3.

EXAMPLE 9

Factor: **a.** $2c - 2d + cd - d^2$ and **b.** $3ax^2 + 3bx^2 + a + 5ax + b + 5bx$.

Solution

a. The first two terms have a common factor of 2 and the last two terms have a common factor of d.

$$2c - 2d + cd - d^2 = 2(c - d) + d(c - d)$$

Factor out 2 from $2c - 2d$ and d from $cd - d^2$.

$$= (c - d)(2 + d)$$

Factor out the common binomial factor, $c - d$.

We check by multiplying:

$$\begin{aligned} (c - d)(2 + d) &= 2c + cd - 2d - d^2 \\ &= 2c - 2d + cd - d^2 \end{aligned}$$

Rearrange the terms to get the original polynomial.

Caution

Factoring by grouping can be attempted on any polynomial with four or more terms. However, not every such polynomial can be factored in this way.

b. Although there is no factor common to all six terms, $3x^2$ is common to the first two terms, and $5x$ is common to the fourth and sixth terms.

$$3ax^2 + 3bx^2 + a + 5ax + b + 5bx$$

GCF = $3x^2$ (first and second terms) GCF = $5x$ (fourth and sixth terms)

These observations suggest that it would be beneficial to rearrange the terms of the polynomial. By the commutative property of addition, we have

$$3ax^2 + 3bx^2 + a + 5ax + b + 5bx = 3ax^2 + 3bx^2 + 5ax + 5bx + a + b$$

Caution

If the GCF of the terms of a polynomial is the same as one of its terms, remember to include a term of 1 within the parentheses of the factored form.

Now we factor by grouping.

$$3ax^2 + 3bx^2 + a + 5ax + b + 5bx = 3x^2(a + b) + 5x(a + b) + 1(a + b)$$

Since $a + b$ is common to all three terms, it can be factored out to get

$$3ax^2 + 3bx^2 + a + 5ax + b + 5bx = (a + b)(3x^2 + 5x + 1)$$

Self Check 9 Factor: **a.** $7m - 7n + mn - n^2$ and **b.** $2x^3 + 8x^2 + x + 2x^2y + y + 8xy$.

To **factor a polynomial completely,** it is often necessary to factor more than once. When factoring a polynomial, *always look for a common factor first.*

EXAMPLE 10

Factor: $3x^3y - 4x^2y^2 - 6x^2y + 8xy^2$.

Solution We begin by factoring out the common factor of xy.

$$3x^3y - 4x^2y^2 - 6x^2y + 8xy^2 = xy(3x^2 - 4xy - 6x + 8y)$$

Caution

The instruction "Factor" means for you to factor the given expression *completely.* Each factor of a completely factored expression will be prime.

We can now factor $3x^2 - 4xy - 6x + 8y$ by grouping:

$$\begin{aligned} &3x^3y - 4x^2y^2 - 6x^2y + 8xy^2 \\ &\quad = xy(3x^2 - 4xy - 6x + 8y) \\ &\quad = xy[x(3x-4y) - 2(3x-4y)] && \text{Factor } x \text{ from } 3x^2 - 4xy \text{ and } -2 \text{ from } -6x + 8y. \\ &\quad = xy(3x-4y)(x - 2) && \text{Factor out } 3x - 4y. \end{aligned}$$

Because no more factoring can be done, the factorization is complete.

Self Check 10 Factor: $3a^3b + 3a^2b - 2a^2b^2 - 2ab^2$.

FORMULAS

Factoring is often required to solve a formula for one of its variables.

EXAMPLE 11

Electronics. The formula $r_1r_2 = rr_2 + rr_1$ is used in electronics to relate the combined resistance, r, of two resistors wired in parallel. The variable r_1 represents the resistance of the first resistor, and the variable r_2 represents the resistance of the second. Solve for r_2.

Solution To isolate r_2 on one side of the equation, we get all terms involving r_2 on the left-hand side and all terms not involving r_2 on the right-hand side. We proceed as follows:

$$\begin{aligned} r_1r_2 &= rr_2 + rr_1 \\ r_1r_2 - rr_2 &= rr_1 && \text{Subtract } rr_2 \text{ from both sides.} \\ r_2(r_1 - r) &= rr_1 && \text{Factor out } r_2 \text{ on the left-hand side.} \\ r_2 &= \frac{rr_1}{r_1 - r} && \text{Divide both sides by } r_1 - r. \end{aligned}$$

Self Check 11 Solve $A = p + prt$ for p.

Answers to Self Checks **1.** $2^3 \cdot 3 \cdot 5$ **2.** $3x^2y$ **3.** $3x^3y^2(3x - 4y)$ **4.** $2a^2b(a^2b + 3ab - 2)$ **5.** a prime polynomial **6.** $-(b^4 + 3b^2 - 2)$ **7.** $-4ab^2(2a + 3b)$ **8. a.** $(y^2 + 1)(c + d + e)$, **b.** $(a + b - c)(x - y)$ **9. a.** $(m - n)(7 + n)$, **b.** $(x + y)(2x^2 + 8x + 1)$ **10.** $ab(3a - 2b)(a + 1)$ **11.** $p = \dfrac{A}{1 + rt}$

5.5 STUDY SET

VOCABULARY **Fill in the blanks.**

1. When we write $2x + 4$ as $2(x + 2)$, we say that we have ________ $2x + 4$.

2. When we ________ a polynomial, we write a sum of terms as a product of factors.

3. The abbreviation GCF stands for ____________________.

4. If a polynomial cannot be factored, it is called a ________ polynomial or an irreducible polynomial.

5. To factor means to factor ____________. Each factor of a completely factored expression will be ________.

6. To factor $ab + 6a + 2b + 12$ by __________, we begin by factoring out a from the first two terms and 2 from the last two terms.

CONCEPTS

7. The prime factorizations of three terms are shown here. Find their GCF.

$2 \cdot 2 \cdot 3 \cdot x \cdot x \cdot y \cdot y \cdot y$

$2 \cdot 3 \cdot 3 \cdot x \cdot y \cdot y \cdot y \cdot y$

$2 \cdot 3 \cdot 3 \cdot 7 \cdot x \cdot x \cdot x \cdot y \cdot y$

8. a. What property is illustrated here?

$4a^2b(2ab^3 - 3a^2b^4) = 4a^2b \cdot 2ab^3 - 4a^2b \cdot 3a^2b^4$

b. Explain how we use the distributive property in reverse to factor $8a^3b^4 - 12a^4b^5$.

9. Explain why each factorization of $30t^2 - 20t^3$ is not complete.

a. $5t^2(6 - 4t)$

b. $10t(3t - 2t^2)$

10. a. Factor $-5y^3 - 10y^2 + 15y$ by factoring out the positive GCF.

b. Factor $-5y^3 - 10y^2 + 15y$ by factoring out the opposite of the GCF.

NOTATION **Complete each factorization.**

11. $3a - 12 = 3(a - \quad)$

12. $8z^3 + 4z^2 + 2z = 2z(4z^2 + 2z + \quad)$

13. $x^3 - x^2 + 2x - 2 = \quad(x - 1) + \quad(x - 1)$
$= (\quad)(x^2 + 2)$

14. $-24a^3b^2 + 12ab^2 = -12ab^2(2a^2 \quad 1)$

PRACTICE **Find the prime-factored form of each number.**

15. 6 **16.** 10

17. 135 **18.** 98

19. 128 **20.** 357

21. 325 **22.** 288

Find the GCF of each list.

23. 36, 48 **24.** 45, 75

25. 42, 36, 98 **26.** 16, 40, 60

27. $4a^2b$, $8a^3c$ **28.** $6x^3y^2z$, $9xyz^2$

29. $18x^4y^3z^2$, $-12xy^2z^3$ **30.** $6x^2y^3$, $24xy^3$, $40x^2y^2z^3$

Factor, if possible.

31. $2x + 8$ **32.** $3y - 9$

33. $2x^2 - 6x$ **34.** $3y^3 + 3y^2$

35. $5xy + 12ab^2$ **36.** $7x^2 + 14x$

37. $15x^2y - 10x^2y^2$ **38.** $11m^3n^2 - 12x^2y$

39. $14r^2s^3 + 15t^6$

40. $13ab^2c^3 - 26a^3b^2c$

41. $27z^3 + 12z^2 + 3z$

42. $25t^6 - 10t^3 + 5t^2$

43. $45x^{10}y^3 - 63x^7y^7 + 81x^{10}y^{10}$

44. $48u^6v^6 - 16u^4v^4 - 3u^6v^3$

45. $\frac{3}{5}ax^4 + \frac{1}{5}bx^2 - \frac{4}{5}ax^3$

46. $\frac{3}{2}t^2y^4 - \frac{1}{2}ty^4 - \frac{5}{2}ry^3$

Factor out −1 from each polynomial.

47. $-a - b$

48. $-2x - y$

49. $-5xy + y - 4$

50. $-7m - 12n + 16$

51. $-60P^2 - 17$

52. $-2x^3 - 1$

Factor each polynomial by factoring out the opposite of the GCF.

53. $-3a - 6$

54. $-6b + 12$

55. $-3x^2 - x$

56. $-4a^3 + a^2$

57. $-6x^2 - 3xy$

58. $-15y^3 + 25y^2$

59. $-18a^2b - 12ab^2$

60. $-21t^5 + 28t^3$

61. $-63u^3v^6z^9 + 28u^2v^7z^2 - 21u^3v^3z^4$

62. $-56x^4y^3z^2 - 72x^3y^4z^5 + 80xy^2z^3$

Factor.

63. $4(x + y) + t(x + y)$

64. $5(a - b) - t(a - b)$

65. $(a - b)r - (a - b)s$

66. $(x + y)u + (x + y)v$

67. $3(m + n + p) + x(m + n + p)$

68. $x(x - y - z) + y(x - y - z)$

69. $(u + v)^2 - (u + v)$

70. $a(x - y) - (x - y)^2$

71. $-a(x + y) - b(x + y)$

72. $-bx(a - b) - cx(a - b)$

73. $4(x^2 + 1)^2 + 2(x^2 + 1)^3$

74. $6(x^3 - 7x + 1)^2 - 3(x^3 - 7x + 1)^3$

Factor by grouping.

75. $ax + bx + ay + by$

76. $ar - br + as - bs$

77. $x^2 + yx + x + y$

78. $c + d + cd + d^2$

79. $3c - cd + 3d - c^2$

80. $x^2 + 4y - xy - 4x$

81. $1 - m + mn - n$

82. $a^2x^2 - 10 - 2x^2 + 5a^2$

83. $2ax^2 - 4 + a - 8x^2$

84. $a^3b^2 - 3 + a^3 - 3b^2$

85. $a^2 - 4b + ab - 4a$

86. $7u + v^2 - 7v - uv$

87. $a^2x + bx - a^2 - b$

88. $x^2y - ax - xy + a$

89. $x^2 + xy + xz + xy + y^2 + zy$

90. $ab - b^2 - bc + ac - bc - c^2$

Factor by grouping. Factor out the GCF first.

91. $mpx + mqx + npx + nqx$

92. $abd - abe + acd - ace$

93. $x^2y + xy^2 + 2xyz + xy^2 + y^3 + 2y^2z$

94. $a^3 - 2a^2b + a^2c - a^2b + 2ab^2 - abc$

95. $2n^4p - 2n^2 - n^3p^2 + np + 2mn^3p - 2mn$

96. $a^2c^3 + ac^2 + a^3c^2 - 2a^2bc^2 - 2bc^2 + c^3$

Solve for the indicated variable.

97. $r_1r_2 = rr_2 + rr_1$ for r_1

98. $r_1r_2 = rr_2 + rr_1$ for r

99. $d_1d_2 = fd_2 + fd_1$ for f

100. $d_1d_2 = fd_2 + fd_1$ for d_1

101. $b^2x^2 + a^2y^2 = a^2b^2$ for a^2

102. $b^2x^2 + a^2y^2 = a^2b^2$ for b^2

103. $S(1 - r) = a - \ell r$ for r

104. $Sn = (n - 2)180°$ for n

APPLICATIONS

105. GEOMETRIC FORMULAS

a. Write an expression that gives the area of the part of the figure that is shaded red.

b. Do the same for the part of the figure that is shaded blue.

c. Add the results from parts a and b and then factor that expression. What important formula from geometry do you obtain?

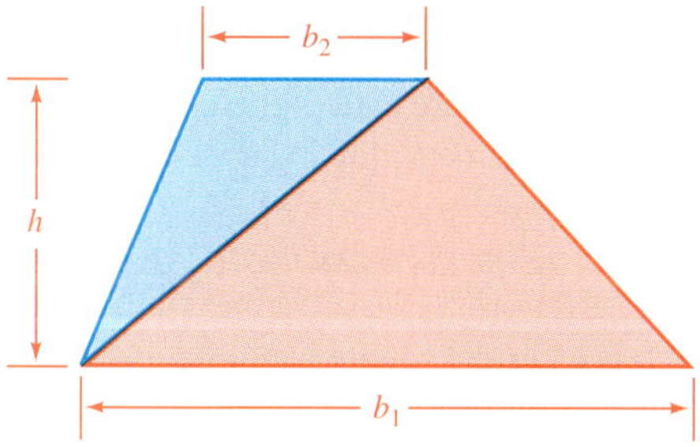

106. PACKAGING The amount of cardboard needed to make the cereal box shown below can be found by computing the area A, which is given by the formula

$$A = 2wh + 4wl + 2lh$$

where w is the width, h the height, and l the length. Solve the equation for the width.

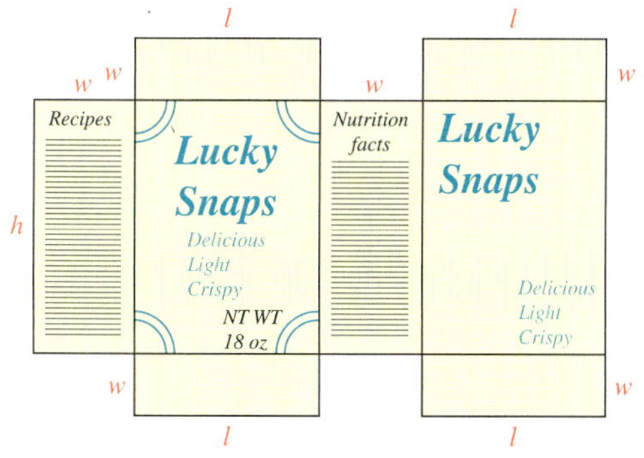

107. LANDSCAPING The combined area of the portions of a square lot that the sprinkler doesn't reach is given by $4r^2 - \pi r^2$, where r is the radius of the circular spray. Factor this expression.

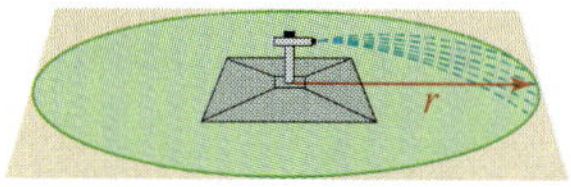

108. CRAYONS The amount of colored wax used to make the crayon shown below can be found by computing its volume using the formula

$$V = \pi r^2 h_1 + \frac{1}{3}\pi r^2 h_2$$

Factor the expression on the right-hand side of this equation.

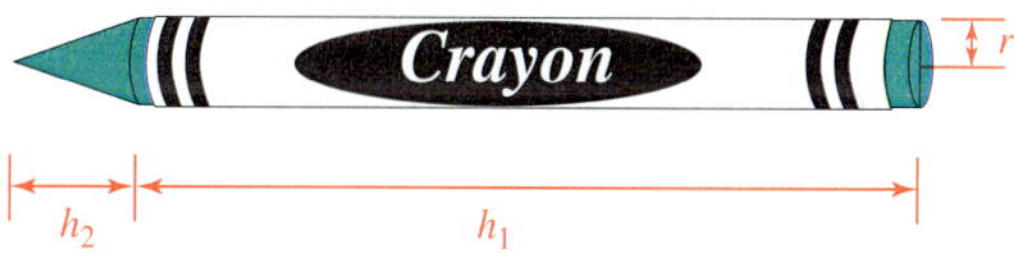

WRITING

109. How are factorizations of polynomials checked? Give an example.

110. Explain why the following factorization is not complete. Then finish the solution.

$$ax + ay + x + y = a(x + y) + x + y$$

111. What is a prime polynomial?

112. Explain the error in the following solution.

Solve for r_1: $\quad r_1r_2 = rr_2 + rr_1.$

$$\frac{r_1r_2}{r_2} = \frac{rr_2 + rr_1}{r_2}$$

$$r_1 = \frac{rr_2 + rr_1}{r_2}$$

REVIEW

113. INVESTMENTS Equal amounts are invested in each of three accounts paying 7%, 8%, and 10.5% annually. If one year's combined interest income is \$1,249.50, how much is invested in each account?

114. SEARCH AND RESCUE Two search-and-rescue teams leave base at the same time looking for a lost boy. The first team, on foot, heads north at 2 mph and the other, on horseback, south at 4 mph. How long will it take them to search a distance of 21 miles between them?

CHALLENGE PROBLEMS **Factor out the designated factor.**

115. x^2 from $x^{n+2} + x^{n+3}$

116. y^n from $2y^{n+2} - 3y^{n+3}$

117. x^{-2} from $x^4 - 5x^6$

118. t^{-3} from $t^5 + 4t^{-6}$

119. $4y^{-2n}$ from $8y^{2n} + 12 + 16y^{-2n}$

120. $7x^{-3n}$ from $21x^{6n} + 7x^{3n} + 14$

5.6 Factoring Trinomials

- Perfect square trinomials
- Factoring trinomials with a lead coefficient of 1
- Factoring trinomials with lead coefficients other than 1
- Test for factorability
- Using substitution to factor trinomials
- The grouping method

We will now discuss techniques for factoring trinomials. These techniques are based on the fact that the product of two binomials is often a trinomial. With that observation in mind, we begin the study of trinomial factoring by considering two special products.

PERFECT SQUARE TRINOMIALS

Recall that trinomials that are squares of a binomial are called **perfect square trinomials.** Perfect square trinomials can be factored by using the following special product formulas.

$$(x + y)^2 = x^2 + 2xy + y^2 \qquad (x - y)^2 = x^2 - 2xy + y^2$$

To factor $n^2 + 20n + 100$, we note that it is a perfect square trinomial because

- The first term n^2 is the square of n.
- The last term 100 is the square of 10: $10^2 = 100$.
- The middle term $20n$ is twice the product of n and 10: $2(n)(10) = 20n$.

To find the factorization, we match the given trinomial to the proper special product formula.

$$\begin{array}{c} x^2 + 2\ x\ y + y^2 = (x + y)^2 \\ \downarrow \quad \downarrow \quad \downarrow \quad \downarrow \quad \downarrow \quad \downarrow \quad \downarrow \\ n^2 + 20n + 100 = n^2 + 2 \cdot n \cdot 10 + 10^2 = (n + 10)^2 \end{array}$$

Thus, $n^2 + 20n + 100 = (n + 10)^2$. We check the factorization as follows.

Check: $\quad (n + 10)^2 = (n + 10)(n + 10)$
$\quad\quad\quad\quad\quad\quad\quad = n^2 + 20n + 100$

EXAMPLE 1 Factor: $9a^2 - 30ab^2 + 25b^4$.

Solution $9a^2 - 30ab + 25b^4$ is a perfect square trinomial because

- The first term $9a^2$ is the square of $3a$: $(3a)^2 = 9a^2$.
- The last term $25b^4$ is the square of $-5b^2$: $(-5b^2)^2 = 25b^4$.
- The middle term $-30ab^2$ is twice the product of $3a$ and $-5b^2$: $2(3a)(-5b^2) = -30ab^2$.

We can match the trinomial to the special product formula $x^2 - 2xy + y^2 = (x - y)^2$ to find the factorization.

$$9a^2 - 30ab^2 + 25b^4 = (3a)^2 + 2 \cdot 3a \cdot (-5b^2) + (-5b^2)^2$$
$$= (3a - 5b^2)^2$$

Thus, $9a^2 - 30ab^2 + 25b^4 = (3a - 5b^2)^2$. Check by multiplying.

Self Check 1 Factor: $49b^4 - 28b^2c + 4c^2$.

We begin our discussion of general trinomials by considering trinomials with lead coefficients of 1.

FACTORING TRINOMIALS WITH A LEAD COEFFICIENT OF 1

To develop a method for factoring trinomials, we will find the product of $x + 6$ and $x + 4$ and make some observations about the result.

$$(x + 6)(x + 4) = x \cdot x + 4x + 6x + 6 \cdot 4 \quad \text{Use the FOIL method to multiply.}$$
$$= \underbrace{x^2}_{\text{First term}} + \underbrace{10x}_{\text{Middle term}} + \underbrace{24}_{\text{Last term}}$$

The result is a trinomial, where

- The first term, x^2, is the product of x and x.
- The last term, 24, is the product of 6 and 4.
- The coefficient of the middle term, 10, is the sum of 6 and 4.

These observations suggest a strategy to use to factor trinomials with a lead coefficient of 1.

Factoring Trinomials with a Lead Coefficient of 1

To factor a trinomial of the form $x^2 + bx + c$, find two numbers whose product is c and whose sum is b.

1. If c is positive, the numbers have the same sign.
2. If c is negative, the numbers have different signs.

Then write the trinomial as a product of two binomials. You can check by multiplying.

The product of these numbers must be c.

$$x^2 + bx + c = (x \; \square)(x \; \square)$$

The sum of these numbers must be b.

EXAMPLE 2

Factor: $x^2 - 6x + 8$.

Solution We assume that $x^2 - 6x + 8$ factors as the product of two binomials. Since the first term of the trinomial is x^2, we enter x and x as the first terms of the binomial factors.

$$x^2 - 6x + 8 = (x \; \square)(x \; \square) \quad \text{Because } x \cdot x \text{ will give } x^2.$$

The second terms of the binomials must be two integers whose product is 8 and whose sum is −6. All possible integer-pair factors of 8 are listed in the table.

Factors of 8	Sum of factors
1(8)	$1 + 8 = 9$
2(4)	$2 + 4 = 6$
−1(−8)	$-1 + (-8) = -9$
−2(−4)	**$-2 + (-4) = -6$** ← This is the one to choose.

The Language of Algebra

Make sure you understand the following vocabulary: *Many trinomials factor as the product of two binomials.*

Trinomial: $x^2 - 6x + 8$ = Product of two binomials: $(x - 2)(x - 4)$

The fourth row of the table contains the correct pair of integers −2 and −4, whose product is 8 and whose sum is −6. To complete the factorization, we enter −2 and −4 as the second terms of the binomial factors.

$$x^2 - 6x + 8 = (x - 2)(x - 4)$$

Check: We can verify this result by multiplication:

$$(x - 2)(x - 4) = x^2 - 4x - 2x + 8 \quad \text{Use the FOIL method.}$$
$$= x^2 - 6x + 8$$

Self Check 2 Factor: $a^2 - 7a + 12$.

EXAMPLE 3

Factor: $-4x + x^2 - 12$.

Solution We begin by writing the trinomial in descending powers of x:

$$-4x + x^2 - 12 = x^2 - 4x - 12$$

Caution

Always write a trinomial in descending powers of one variable before beginning the factoring process.

The possible factorizations of the third term are

This is the one to choose. ↓ (at 2(−6))

1(−12) 2(−6) 3(−4) 4(−3) 6(−2) 12(−1)

In the trinomial, the coefficient of the middle term is −4. The only factorization of −12 whose sum of factors is −4 is 2(−6). Thus, 2 and −6 are the second terms of the binomial factors.

$$x^2 - 4x - 12 = (x + 2)(x - 6)$$

Self Check 3 Factor: $-3a + a^2 - 10$.

EXAMPLE 4

Factor: $2xy^2 + 4xy - 30x$.

Solution Each term in this trinomial has a common factor of $2x$, which we will factor out.

$$2xy^2 + 4xy - 30x = 2x(y^2 + 2y - 15) \quad \text{The GCF is } 2x.$$

Caution

When factoring a polynomial, always factor out the GCF first. For multistep factorizations, remember to write the GCF in the final factored form:

$$\underset{\underset{\text{GCF}}{\uparrow}}{2x}(y-3)(y+5)$$

To factor $y^2 + 2y - 15$, we list the factors of -15 and find the pair whose sum is 2.

$$1(-15) \qquad 3(-5) \qquad -1(15) \qquad \overset{\overset{\text{This is the one to choose.}}{\downarrow}}{-3(5)}$$

The only factorization where the sum of the factors is 2 is $-3(5)$. Thus, $y^2 + 2y - 15$ factors as $(y - 3)(y + 5)$.

$$\begin{aligned} 2xy^2 + 4xy - 30x &= 2x(\mathbf{y^2 + 2y - 15}) \\ &= 2x(\mathbf{y - 3})(\mathbf{y + 5}) \end{aligned}$$

Self Check 4 Factor: $3ab^2 + 6ab - 105a$.

FACTORING TRINOMIALS WITH LEAD COEFFICIENTS OTHER THAN 1

There are more combinations of factors to consider when factoring trinomials with lead coefficients other than 1. To factor $5x^2 + 7x + 2$, for example, we assume that it factors as the product of two binomials.

$$5x^2 + 7x + 2 = (\square \quad \square)(\square \quad \square)$$

Since the first term of the trinomial $5x^2 + 7x + 2$ is $5x^2$, the first terms of the binomial factors must be $5x$ and x.

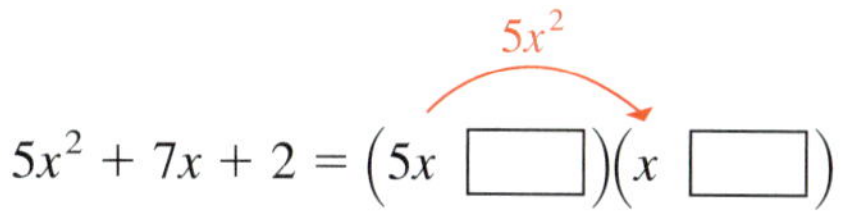

Since the product of the last terms must be 2, and the sum of the products of the outer and inner terms must be $7x$, we must find two numbers whose product is 2 that will give a middle term of $7x$.

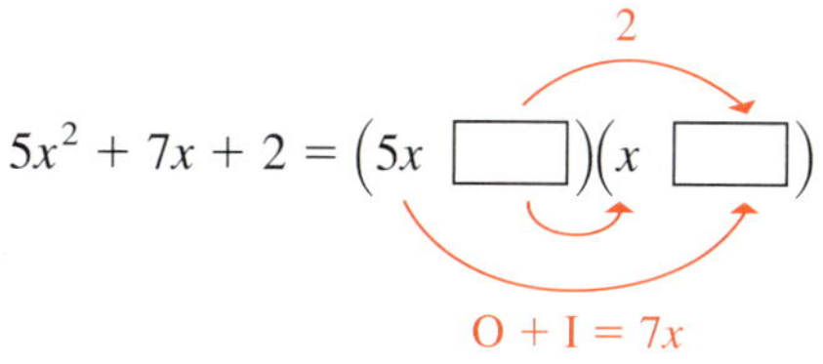

Since $2(1)$ and $(-2)(-1)$ give a product of 2, there are four possible combinations to consider:

$$\begin{array}{ll} (5x + 2)(x + 1) & (5x - 2)(x - 1) \\ (5x + 1)(x + 2) & (5x - 1)(x - 2) \end{array}$$

Of these possibilities, only the one in blue gives the correct middle term of $7x$.

$$5x^2 + 7x + 2 = (5x + 2)(x + 1)$$

We can verify this result by multiplication:

Check: $\begin{aligned} (5x + 2)(x + 1) &= 5x^2 + 5x + 2x + 2 \\ &= 5x^2 + 7x + 2 \end{aligned}$

TEST FOR FACTORABILITY

If a trinomial has the form $ax^2 + bx + c$, with integer coefficients and $a \neq 0$, we can test to see whether it is factorable.

The Language of Algebra

A number that is the square of an integer is called a **perfect integer square.** Some examples are: 1, 4, 9, 16, 25, 36, 49, 64, 81, and 100.

- If the value of $b^2 - 4ac$ is a perfect integer square, the trinomial can be factored using only integers.
- If the value of $b^2 - 4ac$ is not a perfect integer square, the trinomial cannot be factored using only integers.

For example, $5x^2 + 7x + 2$ is a trinomial in the form $ax^2 + bx + c$ with $a = 5$, $b = 7$, and $c = 2$. For this trinomial, the value of $b^2 - 4ac$ is

$$\begin{aligned} b^2 - 4ac &= 7^2 - 4(5)(2) && \text{Substitute 5 for } a\text{, 7 for } b\text{, and 2 for } c. \\ &= 49 - 40 \\ &= 9 \end{aligned}$$

Since 9 is a perfect integer square, the trinomial is factorable. Earlier, we found that $5x^2 + 7x + 2$ factors as $(5x + 2)(x + 1)$.

Test for Factorability

A trinomial of the form $ax^2 + bx + c$, with integer coefficients and $a \neq 0$, will factor into two binomials with integer coefficients if the value of $b^2 - 4ac$ is a perfect square. If $b^2 - 4ac = 0$, the factors will be the same.

EXAMPLE 5

Factor: $3p^2 - 4p - 4$.

Solution In the trinomial, $a = 3$, $b = -4$, and $c = -4$. To see whether it factors, we evaluate $b^2 - 4ac$.

$$\begin{aligned} b^2 - 4ac &= (-4)^2 - 4(3)(-4) \\ &= 16 + 48 \\ &= 64 \end{aligned}$$

Since 64 is a perfect square, the trinomial is factorable.

To factor the trinomial, we note that the first terms of the binomial factors must be $3p$ and p to give the first term of $3p^2$.

$$3p^2 \quad\text{(product of first terms)}$$

$$3p^2 - 4p - 4 = (3p \ \square)(p \ \square)$$

The product of the last terms must be -4, and the sum of the products of the outer terms and the inner terms must be $-4p$.

$$-4 \quad\text{(product of last terms)}$$

$$3p^2 - 4p - 4 = (3p \ \square)(p \ \square)$$

$$\text{O} + \text{I} = -4p$$

Because $1(-4)$, $-1(4)$, and $-2(2)$ all give a product of -4, there are six possible combinations to consider:

$$(3p + 1)(p - 4) \qquad (3p - 4)(p + 1)$$
$$(3p - 1)(p + 4) \qquad (3p + 4)(p - 1)$$
$$(3p - 2)(p + 2) \qquad (3p + 2)(p - 2)$$

Of these possibilities, only the one in blue gives the required middle term of $-4p$.

$$3p^2 - 4p - 4 = (3p + 2)(p - 2)$$

Notation

By the commutative property of multiplication, the factors of a trinomial can be written in either order. Thus, we could also write:

$3p^2 - 4p - 4 = (p - 2)(3p + 2)$

Self Check 5 Factor: $4q^2 - 9q - 9$.

EXAMPLE 6

Factor: $4t^2 - 3t - 5$, if possible.

Solution In the trinomial, $a = 4$, $b = -3$, and $c = -5$. To see whether the trinomial is factorable, we evaluate $b^2 - 4ac$ by substituting the values of a, b, and c.

$$\begin{aligned} b^2 - 4ac &= (-3)^2 - 4(4)(-5) \\ &= 9 + 80 \\ &= 89 \end{aligned}$$

Since 89 is not a perfect square, the trinomial is not factorable using only integer coefficients.

The Language of Algebra

When a trinomial is not factorable using only integers, we say it is **prime** and that it does not factor *over the integers.*

Self Check 6 Factor $5a^2 - 8a + 2$, if possible.

It is not easy to give specific rules for factoring general trinomials. However, the following hints are helpful. This approach is called the **trial-and-check method.**

Factoring Trinomials with Lead Coefficients Other Than 1

To factor trinomials with lead coefficients other than 1:

1. Factor out any GCF (including -1 if that is necessary to make $a > 0$ in a trinomial of the form $ax^2 + bx + c$).
2. Write the trinomial as a product of two binomials. The coefficients of the first terms of each binomial factor must be factors of a, and the last terms must be factors of c).

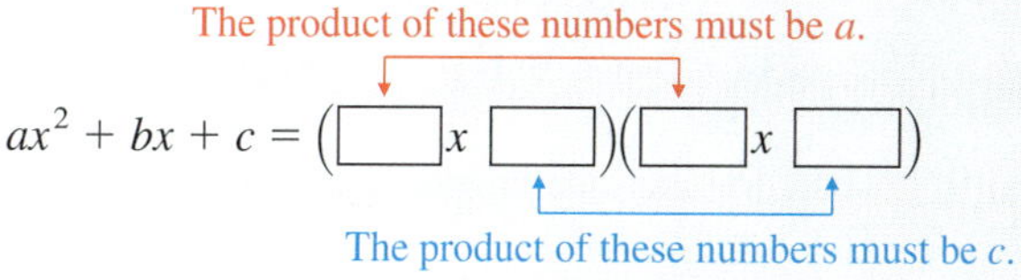
The product of these numbers must be a.
ax² + bx + c = (x)(x)
The product of these numbers must be c.

3. If c is positive, the signs within the binomial factors match the sign of b. If c is negative, the signs within the binomial factors are opposites.
4. Try combinations of first terms and second terms of the binomial factors until you find the one that gives the proper middle term. If no combination works, the trinomial is prime.
5. Check by multiplying.

EXAMPLE 7

Factor: $-15x^2 + 25xy + 60y^2$.

Solution Factor out -5 from each term.

$$-15x^2 + 25xy + 60y^2 = -5(3x^2 - 5xy - 12y^2) \quad \text{The opposite of the GCF is } -5.$$

Caution

When factoring a trinomial, be sure to factor it completely. Always check to see whether any of the factors of your result can be factored further.

A test for factorability using $a = 3$, $b = -5$, and $c = -12$ will show that $3x^2 - 5xy - 12y^2$ will factor.

To factor $3x^2 - 5xy - 12y^2$, we examine its terms.

- Since the first term is $3x^2$, the first terms of the binomial factors must be $3x$ and x.
- Since the sign of the first term of the trinomial is positive and the sign of the last term is negative, the signs within the binomial factors will be opposites.
- Since the last term of the trinomial contains y^2, the second terms of the binomial factors must contain y.

$$-5(3x^2 - 5xy - 12y^2) = -5\left(3x \,\square\, y\right)\left(x \,\square\, y\right) \quad (3x^2)$$

The product of the last terms must be $-12y^2$, and the sum of the product of the outer terms and the product of the inner terms must be $-5xy$.

$$-15x^2 + 25xy + 60y^2 = -5\left(3x \,\square\, y\right)\left(x \,\square\, y\right) \quad (-12;\ \text{O} + \text{I} = -5xy)$$

Since $1(-12)$, $2(-6)$, $3(-4)$, $-1(12)$, $-2(6)$ and $-3(4)$ all give a product of -12, there are 12 possible combinations to consider.

$(3x + 1y)(x - 12y)$	$(3x - 12y)(x + 1y)$	← $3x - 12y$ has a common factor 3.
$(3x + 2y)(x - 6y)$	$(3x - 6y)(x + 2y)$	
$(3x + 3y)(x - 4y)$	$(3x - 4y)(x + 3y)$	
$(3x - 1y)(x + 12y)$	$(3x + 12y)(x - 1y)$	
$(3x - 2y)(x + 6y)$	$(3x + 6y)(x - 2y)$	
$(3x - 3y)(x + 4y)$	$(3x + 4y)(x - 3y)$	← This is the one to choose.

Success Tip

If the terms of a trinomial do not have a common factor, the terms of each of its binomial factors will not have a common factor.

The combinations in blue cannot work, because one of the factors has a common factor. This implies that $3x^2 - 5xy - 12y^2$ would have a common factor, which it doesn't.

After mentally trying the remaining combinations, we find that only $(3x + 4y)(x - 3y)$ gives the proper middle term of $-5xy$.

$$\begin{aligned} -15x^2 + 25xy + 60y^2 &= -5(3x^2 - 5xy - 12y^2) \\ &= -5(3x + 4y)(x - 3y) \end{aligned}$$

Self Check 7 Factor: $-6x^2 - 15xy - 6y^2$.

EXAMPLE 8

Factor: $6y^3 + 13x^2y^3 + 6x^4y^3$.

Solution We write the expression in descending powers of x and then factor out the common factor y^3.

$$\begin{aligned}6y^3 + 13x^2y^3 + 6x^4y^3 &= 6x^4y^3 + 13x^2y^3 + 6y^3\\ &= y^3(6x^4 + 13x^2 + 6)\end{aligned}$$

A test for factorability will show that $6x^4 + 13x^2 + 6$ will factor.

To factor $6x^4 + 13x^2 + 6$, we examine its terms.

- Since the first term is $6x^4$, the first terms of the binomial factors must be either $2x^2$ and $3x^2$ or x^2 and $6x^2$.

$$6x^4 + 13x^2 + 6 = (2x^2 \;\square)(3x^2 \;\square) \quad \text{or} \quad (x^2 \;\square)(6x^2 \;\square)$$

- Since the signs of the middle term and the last term of the trinomial are positive, the signs within each binomial factor will be positive.
- Since the product of the last terms of the binomial factors must be 6, we must find two numbers whose product is 6 that will lead to a middle term of $13x^2$.

After trying some combinations, we find the one that works.

$$\begin{aligned}6x^4y^3 + 13x^2y^3 + 6y^3 &= y^3(6x^4 + 13x^2 + 6)\\ &= y^3(2x^2 + 3)(3x^2 + 2)\end{aligned}$$

Self Check 8 Factor: $4b + 11a^2b + 6a^4b$.

USING SUBSTITUTION TO FACTOR TRINOMIALS

For more complicated expressions, a substitution sometimes helps to simplify the factoring process.

EXAMPLE 9

Factor: $(x + y)^2 + 7(x + y) + 12$.

Solution We rewrite the trinomial $(x + y)^2 + 7(x + y) + 12$ as $z^2 + 7z + 12$, where $z = x + y$. The trinomial $z^2 + 7z + 12$ factors as $(z + 4)(z + 3)$.

To find the factorization of $(x + y)^2 + 7(x + y) + 12$, we substitute $x + y$ for z in the expression $(z + 4)(z + 3)$ to obtain

$$z^2 + 7z + 12 = (z + 4)(z + 3)$$

$$(x + y)^2 + 7(x + y) + 12 = (x + y + 4)(x + y + 3) \quad \text{Replace each } z \text{ with } x + y.$$

Self Check 9 Factor: $(a + b)^2 - 3(a + b) - 10$.

THE GROUPING METHOD

Another way to factor trinomials is to write them as equivalent four-termed polynomials and factor by grouping. For example, to factor $2x^2 + 5x + 3$ in this way, we proceed as follows.

Identify the values of a, b and c. Then, find the product ac, called the **key number:** $ac = 2(3) = 6$.

$$\left.\begin{array}{c} ax^2 + bx + c \\ \downarrow \quad \downarrow \quad \downarrow \\ 2x^2 + 5x + 3 \end{array}\right\} a = 2, b = 5, \text{ and } c = 3$$

Next, find two numbers whose product is $ac = 6$ and whose sum is $b = 5$. Since the numbers must have a positive product and a positive sum, we consider only positive factors of 6.

Key number = 6

Positive factors of 6	Sum of the factors of 6
$1 \cdot 6 = 6$	$1 + 6 = 7$
$2 \cdot 3 = 6$	**$2 + 3 = 5$**

The second row of the table contains the correct pair of integers 2 and 3, whose product is 6 and whose sum is 5. Use the factors 2 and 3 as coefficients of two terms, $2x$ and $3x$, to be placed between $2x^2$ and 3.

$$2x^2 + 5x + 3 = 2x^2 + 2x + 3x + 3$$ Express $5x$ as $2x + 3x$.

Factor the four-termed polynomial by grouping:

$$2x^2 + 2x + 3x + 3 = 2x(x + 1) + 3(x + 1)$$ Factor $2x$ out of $2x^2 + 2x$ and 3 out of $3x + 3$.

$$= (x + 1)(2x + 3)$$ Factor out $x + 1$.

The factorization is $(x + 1)(2x + 3)$. Check by multiplying.

Factoring by grouping is especially useful when the lead coefficient, a, and the constant term, c, have many factors.

Factoring Trinomials by Grouping

To factor a trinomial by grouping:

1. Factor out any GCF (including -1 if that is necessary to make $a > 0$ in a trinomial of the form $ax^2 + bx + c$).
2. Identify a, b, and c, and find the key number ac.
3. Find two numbers whose product is the key number and whose sum is b.
4. Enter the two numbers as coefficients of x between the first and last terms and factor the polynomial by grouping.

The product of these numbers must be ac.

$$ax^2 + \square x + \square x + c$$

The sum of these numbers must be b.

5. Check by multiplying.

EXAMPLE 10

Factor by grouping: **a.** $x^2 + 8x + 15$ and **b.** $10x^2 + 13x - 3$.

Solution **a.** Since $x^2 + 8x + 15 = 1x^2 + 8x + 15$, we identify a as 1, b as 8, and c as 15. The key number is $ac = 1(15) = 15$. We must find two integers whose product is 15 and whose sum is $b = 8$. Since the integers must have a positive product and a positive sum, we consider only positive factors of 15.

Key number = 15

Positive factors of 15	Sum of the factors
$1 \cdot 15 = 15$	$1 + 15 = 16$
$\mathbf{3 \cdot 5 = 15}$	$\mathbf{3 + 5 = 8}$

The second row of the table contains the correct pair of integers 3 and 5, whose product is 15 and whose sum is 8. They serve as the coefficients of $3x$ and $5x$ that we place between x^2 and 15.

$$
\begin{aligned}
x^2 + 8x + 15 &= x^2 + 3x + 5x + 15 && \text{Express } 8x \text{ as } 3x + 5x.\\
&= x(x + 3) + 5(x + 3) && \text{Factor } x \text{ out of } x^2 + 3x \text{ and 5 out of } 5x + 15.\\
&= (x + 3)(x + 5) && \text{Factor out } x + 3.
\end{aligned}
$$

The factorization is $(x + 3)(x + 5)$. Check by multiplying.

b. In $10x^2 + 13x - 3$, $a = 10$, $b = 13$, and $c = -3$. The key number is $ac = 10(-3) = -30$. We must find a factorization of -30 in which the sum of the factors is $b = 13$. Since the factors must have a negative product, their signs must be different. The possible factor pairs are listed in the table.

The seventh row contains the correct pair of numbers 15 and -2, whose product is -30 and whose sum is 13. They serve as the coefficients of the two terms, $15x$ and $-2x$, that we place between $10x^2$ and -3.

Key number = −30

Factors of −30	Sum of the factors
$1(-30) = -30$	$1 + (-30) = -29$
$2(-15) = -30$	$2 + (-15) = -13$
$3(-10) = -30$	$3 + (-10) = -7$
$5(-6) = -30$	$5 + (-6) = -1$
$6(-5) = -30$	$6 + (-5) = 1$
$10(-3) = -30$	$10 + (-3) = 7$
$\mathbf{15(-2) = -30}$	$\mathbf{15 + (-2) = 13}$
$30(-1) = -30$	$30 + (-1) = 29$

$$10x^2 + 13x - 3 = 10x^2 + 15x - 2x - 3 \qquad \text{Express } 13x \text{ as } 15x - 2x.$$

Finally, we factor by grouping.

$$
\begin{aligned}
10x^2 + 15x - 2x - 3 &= 5x(2x + 3) - 1(2x + 3) && \text{Factor out } 5x \text{ from } 10x^2 + 15x. \text{ Factor out } -1 \text{ from } -2x - 3.\\
&= (2x + 3)(5x - 1) && \text{Factor out } 2x + 3.
\end{aligned}
$$

So $10x^2 + 13x - 3 = (2x + 3)(5x - 1)$. Check by multiplying.

Notation

In Example 10b, the middle term, $13x$, may be expressed as $15x - 2x$ or as $-2x + 15x$ when using factoring by grouping. The resulting factorizations will be equivalent.

Self Check 10 Factor by grouping: **a.** $m^2 + 13m + 42$ and **b.** $15a^2 + 17a - 4$.

EXAMPLE 11 Factor by grouping: $12x^5 - 17x^4 + 6x^3$.

Solution First, we factor out the GCF, which is x^3.

$$12x^5 - 17x^4 + 6x^3 = x^3(12x^2 - 17x + 6)$$

To factor $12x^2 - 17x + 6$, we must find two integers whose product is $12(6) = 72$ and whose sum is -17. Two such numbers are -8 and -9.

$$\begin{aligned} 12x^2 - 17x + 6 &= 12x^2 - 8x - 9x + 6 && \text{Express } -17x \text{ as } -8x - 9x. \\ &= 4x(3x - 2) - 3(3x - 2) && \text{Factor out } 4x \text{ and factor out } -3. \\ &= (3x - 2)(4x - 3) && \text{Factor out } 3x - 2. \end{aligned}$$

The complete factorization of the original trinomial is

$$12x^5 - 17x^4 + 6x^3 = x^3(3x - 2)(4x - 3)$$

Check by multiplying.

Self Check 11 Factor by grouping: $21a^4 - 13a^3 + 2a^2$.

Answers to Self Checks **1.** $(7b^2 - 2c)^2$ **2.** $(a - 4)(a - 3)$ **3.** $(a + 2)(a - 5)$ **4.** $3a(b - 5)(b + 7)$ **5.** $(4q + 3)(q - 3)$ **6.** a prime polynomial **7.** $-3(x + 2y)(2x + y)$ **8.** $b(2a^2 + 1)(3a^2 + 4)$ **9.** $(a + b + 2)(a + b - 5)$ **10. a.** $(m + 7)(m + 6)$, **b.** $(3a + 4)(5a - 1)$ **11.** $a^2(7a - 2)(3a - 1)$

5.6 STUDY SET

VOCABULARY Fill in the blanks.

1. A polynomial with three terms, such as $3x^2 - 2x + 4$, is called a ________.
2. Since $y^2 + 2y + 1$ is the square of $y + 1$, we call $y^2 + 2y + 1$ a _______ square trinomial.
3. The _____ coefficient of the trinomial $x^2 - 3x + 2$ is 1, the __________ of the middle term is -3, and the last term is ___.
4. The trinomial $4a^2 - 5a - 6$ is written in __________ powers of a.
5. The numbers 6 and -2 are two integers whose _______ is -12 and whose _____ is 4.
6. A trinomial is factored __________ when no factor can be factored further.
7. A ______ polynomial cannot be factored by using only integers.
8. The statement $x^2 - x - 12 = (x - 4)(x + 3)$ shows that $x^2 - x - 12$ factors into the _______ of two binomials.

CONCEPTS

9. Consider $3x^2 - x + 16$. What is the sign of the
   a. First term?
   b. Middle term?
   c. Last term?

10. If $b^2 - 4ac$ is a perfect integer square, the trinomial $ax^2 + bx + c$ can be factored using integers. Give three examples of perfect integer squares.

11. Fill in the blanks. When factoring a trinomial, we write it in ___________ powers of the variable. Then we factor out any _______ (including -1 if that is necessary to make the lead coefficient _________).

12. Check to see whether $(3t - 1)(5t - 6)$ is the correct factorization of $15t^2 - 19t + 6$.

13. Complete the table.

Factors of 8	Sum of the factors of 8
$1(8) = 8$	
$2(4) = 8$	
$-1(-8) = 8$	
$-2(-4) = 8$	

14. Find two integers whose

a. product is 10 and whose sum is 7.

b. product is 8 and whose sum is -6.

c. product is -6 and whose sum is 1.

d. product is -9 and whose sum is -8.

15. Complete the key number table.

Key number = 12

Negative factors of 12	Sum of factors of 12
$-1(-12) = 12$	
	$-2 + (-6) = -8$
$-3(-4) = 12$	

16. Use the substitution $x = a + b$ to rewrite the trinomial $6(a + b)^2 - 17(a + b) - 3$.

NOTATION

17. The trinomial $4m^2 - 4m + 1$ is written in the form $ax^2 + bx + c$. Identify a, b, and c.

18. Consider the trinomial $15s^2 + 4s - 4$. Is $b^2 - 4ac$ a perfect square?

PRACTICE **Complete each factorization.**

19. $x^2 + 5x + 6 = (x + 3)$

20. $x^2 - 6x + 8 = (x - 4)$

21. $x^2 + 2x - 15 = (x + 5)$

22. $x^2 - 3x - 18 = (x - 6)$

23. $2a^2 + 9a + 4 = \qquad (a + 4)$

24. $6p^2 - 5p - 4 = \qquad (2p + 1)$

Use a special product formula to factor each perfect square trinomial.

25. $x^2 + 2x + 1$

26. $y^2 - 2y + 1$

27. $a^2 - 18a + 81$

28. $b^2 + 12b + 36$

29. $4y^2 + 4y + 1$

30. $9x^2 + 6x + 1$

31. $9b^2 - 12bc^2 + 4c^4$

32. $4a^2 - 12ab + 9b^2$

33. $y^4 + 10y^2 + 25$

34. $a^4 - 14a^2 + 49$

35. $25m^8 - 60m^4n + 36n^2$

36. $49s^6 + 84s^3n^2 + 36n^4$

Test each trinomial for factorability and factor it, if possible.

37. $x^2 - 5x + 6$

38. $y^2 + 7y + 6$

39. $x^2 - 7x + 10$

40. $c^2 - 7c + 12$

41. $b^2 + 8b + 18$

42. $x^2 + 4x - 28$

43. $-x + x^2 - 30$

44. $a^2 - 45 + 4a$

45. $-50 + a^2 + 5a$

46. $-36 + b^2 + 9b$

47. $x^2 - 4xy - 21y^2$

48. $a^2 + 4ab - 5b^2$

49. $s^2 - 10st + 16t^2$

50. $h^2 - 8hk + 15k^2$

51. $y^4 - 13y^2 + 30$

52. $y^4 - 13y^2 + 42$

53. $g^6 - 2g^3 - 63$

54. $d^8 - d^4 - 90$

Factor each trinomial. If the lead coefficient is negative, begin by factoring out -1.

55. $3x^2 + 12x - 63$

56. $2y^2 + 4y - 48$

57. $b^4x^2 - 12b^2x^2 + 35x^2$

58. $c^3x^4 + 11c^3x^2 - 42c^3$

59. $32 - a^2 + 4a$

60. $15 - x^2 - 2x$

61. $-3a^2x^2 + 15a^2x - 18a^2$

62. $-2bcy^2 - 16bcy + 40bc$

63. $-2p^2 - 2pq + 4q^2$

64. $-6m^2 + 3mn + 3n^2$

Factor each expression. Factor out all common monomials first (including −1 if the lead coefficient is negative). If a trinomial is prime, so indicate.

65. $6y^2 + 7y + 2$

66. $6x^2 - 11x + 3$

67. $8a^2 + 6a - 9$

68. $15b^2 + 4b - 4$

69. $6x^2 - 5xy - 4y^2$

70. $18y^2 - 3yz - 10z^2$

71. $5x^2 + 4x + 1$

72. $6z^2 + 17z + 12$

73. $3 - 10x + 8x^2$

74. $3 + 4a^2 + 20a$

75. $64h^6 + 24h^5 - 4h^4$

76. $27x^2yz + 90xyz - 72yz$

77. $3x^3 - 11x^2 + 8x$

78. $3t^3 - 3t^2 + t$

79. $-3a^2 + ab + 2b^2$

80. $-2x^2 + 3xy + 5y^2$

81. $20a^2 - 60b^2 + 45ab$

82 $-4x^2 - 9 + 12x$

83. $21x^4 - 10x^3 - 16x^2$

84. $16x^3 - 50x^2 + 36x$

85. $12y^6 + 23y^3 + 10$

86. $5m^8 + 29m^4 - 42$

87. $6a^2(m + n) + 13a(m + n) - 15(m + n)$

88. $15n^2(q - r) - 17n(q - r) - 18(q - r)$

Use a substitution to help factor each expression.

89. $(x + a)^2 + 2(x + a) + 1$

90. $(a + b)^2 - 2(a + b) + 1$

91. $(a + b)^2 - 2(a + b) - 24$

92. $(x - y)^2 + 3(x - y) - 10$

93. $14(q - r)^2 - 17(q - r) - 6$

94. $8(h + s)^2 + 34(h + s) + 35$

APPLICATIONS

95. ICE The surface area of a cubical block of ice is $6x^2 + 36x + 54$. Find the length of an edge of the block.

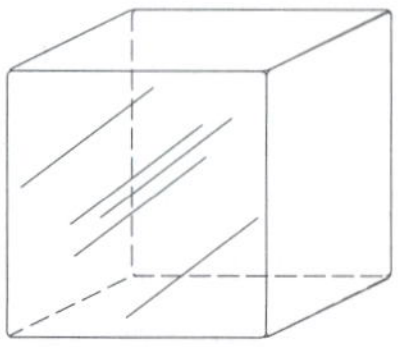

96. CHECKERS The area of a square checkerboard is $25x^2 - 40x + 16$. Find the length of a side.

WRITING

97. Explain the error.

Factor: $2x^2 - 4x - 6$.

$$2x^2 - 4x - 6 = (2x + 2)(x - 3)$$

98. How do you know when a polynomial has been factored completely?

99. How was substitution used in this section?

100. How does one determine whether a trinomial is a perfect square trinomial?

REVIEW

101. If $f(x) = |2x - 1|$, find $f(-2)$.

102. If $g(x) = 2x^2 - 1$, find $g(-2)$.

103. Solve: $-3 = -\frac{9}{8}s$.

104. Solve: $2x + 3 = \frac{2}{3}x - 1$.

105. Simplify: $3p^2 - 6(5p^2 + p) + p^2$.

106. Solve the system: $\begin{cases} 2(2x + 3y) = 5 \\ 8x = 3(1 + 3y) \end{cases}$.

CHALLENGE PROBLEMS

107. What are the only integer values of b for which $9m^2 + bx - 1$ can be factored?

108. Find the missing factors:

$$17y^2 + 1{,}496y - 11{,}305 = ?(y - 7)?$$

Factor. Assume that n is a natural number.

109. $x^{2n} + 2x^n + 1$

110. $2a^{6n} - 3a^{3n} - 2$

111. $x^{4n} + 2x^{2n}y^{2n} + y^{4n}$

112. $6x^{2n} + 7x^n - 3$

5.7 The Difference of Two Squares; the Sum and Difference of Two Cubes

- Perfect squares
- The difference of two squares
- Perfect cubes
- The sum and difference of two cubes

We will now discuss some special rules of factoring. These rules are applied to polynomials that can be written as the difference of two squares or as the sum or difference of two cubes. To use these factoring methods, we must first be able to recognize such polynomials. We begin with a discussion that will help you recognize polynomials with terms that are *perfect squares.*

PERFECT SQUARES

To factor the difference of two squares, it is helpful to know the first twenty **perfect integer squares.**

1, 4, 9, 16, 25, 36, 49, 64, 81, 100, 121, 144, 169, 196, 225, 256, 289, 324, 361, 400

Expressions such as x^6y^4 are also perfect squares, because they can be written as the square of another quantity:

$$x^6y^4 = (x^3y^2)^2$$

THE DIFFERENCE OF TWO SQUARES

The Language of Algebra

The expression $x^2 - y^2$ is a *difference of two squares,* whereas $(x - y)^2$ is the *square of a difference.* They are not equivalent because $(x - y)^2 \neq x^2 - y^2$.

In Section 5.4, we developed the special product formula

(1) $$(x + y)(x - y) = x^2 - y^2$$

The binomial $x^2 - y^2$ is called the **difference of two squares,** because x^2 represents the square of x, y^2 represents the square of y, and $x^2 - y^2$ represents the difference of these squares.

Equation 1 can be written in reverse order to give a formula for factoring the difference of two squares.

Factoring the Difference of Two Squares

$$x^2 - y^2 = (x + y)(x - y)$$

If we think of the difference of two squares as the square of a **F**irst quantity minus the square of a **L**ast quantity, we have the formula

$$F^2 - L^2 = (F + L)(F - L)$$

and we say: *To factor the square of a **F**irst quantity minus the square of a **L**ast quantity, we multiply the **F**irst plus the **L**ast by the **F**irst minus the **L**ast.*

EXAMPLE 1

Factor: $49x^2 - 16$.

Solution We begin by rewriting the binomial $49x^2 - 16$ as a difference of two squares: $(7x)^2 - (4)^2$. Then we use the formula for factoring the difference of two squares:

$$\begin{array}{ccccc} F^2 - L^2 &=& (F + L)(F - L) \\ \downarrow \quad \downarrow & & \downarrow \quad \downarrow \quad \downarrow \quad \downarrow \\ (7x)^2 - 4^2 &=& (7x + 4)(7x - 4) \end{array}$$

We can verify this result using the FOIL method to do the multiplication.

$$\begin{aligned}(7x + 4)(7x - 4) &= 49x^2 - 28x + 28x - 16 \\ &= 49x^2 - 16\end{aligned}$$

Success Tip

Always verify a factorization by doing the indicated multiplication. The result should be the original polynomial.

Self Check 1 Factor: $81p^2 - 25$.

EXAMPLE 2

Factor: $64a^4 - 25b^2$.

Solution We can write $64a^4 - 25b^2$ in the form $(8a^2)^2 - (5b)^2$ and use the formula for factoring the difference of two squares.

$$\begin{array}{ccccc} F^2 - L^2 &=& (F + L)(F - L) \\ \downarrow \quad \downarrow & & \downarrow \quad \downarrow \quad \downarrow \quad \downarrow \\ (8a^2)^2 - (5b)^2 &=& (8a^2 + 5b)(8a^2 - 5b) \end{array}$$

Notation

By the commutative property of multiplication, this factorization could be written

$(8a^2 - 5b)(8a^2 + 5b)$

Self Check 2 Factor: $36r^4 - s^2$.

EXAMPLE 3

Factor: $x^4 - 1$.

Solution Because the binomial is the difference of the squares of x^2 and 1, it factors into the sum of x^2 and 1 and the difference of x^2 and 1.

$$\begin{aligned}x^4 - 1 &= (x^2)^2 - (1)^2 \\ &= (x^2 + 1)(x^2 - 1)\end{aligned}$$

The factor $x^2 + 1$ is the sum of two quantities and is prime. However, the factor $x^2 - 1$ is the difference of two squares and can be factored as $(x + 1)(x - 1)$. Thus,

$$\begin{aligned}x^4 - 1 &= (x^2 + 1)(x^2 - 1) \\ &= (x^2 + 1)(x + 1)(x - 1)\end{aligned}$$

Caution

The binomial $x^2 + 1$ is the **sum of two squares.** In general, after removing any common factor, a sum of two squares cannot be factored using real numbers.

Self Check 3 Factor: $a^4 - 81$.

EXAMPLE 4

Factor: $(x + y)^4 - z^4$.

Solution This expression is the difference of two squares and can be factored:

$$\begin{aligned}(x + y)^4 - z^4 &= [(x + y)^2]^2 - (z^2)^2 \\ &= [(x + y)^2 + z^2][(x + y)^2 - z^2]\end{aligned}$$

Caution

When factoring a polynomial, be sure to factor it completely. Always check to see whether any of the factors of your result can be factored further.

The factor $(x + y)^2 + z^2$ is the sum of two squares and is prime. However, the factor $(x + y)^2 - z^2$ is the difference of two squares and can be factored as $(x + y + z)(x + y - z)$. Thus,

$$\begin{aligned}(x + y)^4 - z^4 &= [(x + y)^2 + z^2][(x + y)^2 - z^2] \\ &= [(x + y)^2 + z^2](x + y + z)(x + y - z)\end{aligned}$$

Self Check 4 Factor: $(a - b)^4 - c^4$.

When possible, we always factor out a common factor before factoring the difference of two squares. The factoring process is easier when all common factors are factored out first.

EXAMPLE 5

Factor: $2x^4y - 32y$.

Solution

$$\begin{aligned}2x^4y - 32y &= 2y(x^4 - 16) && \text{Factor out the GCF, which is } 2y. \\ &= 2y(x^2 + 4)(x^2 - 4) && \text{Factor } x^4 - 16. \\ &= 2y(x^2 + 4)(x + 2)(x - 2) && \text{Factor } x^2 - 4.\end{aligned}$$

Self Check 5 Factor: $3a^4 - 3$.

EXAMPLE 6

Factor: $x^2 - y^2 + x - y$.

Solution If we group the first two terms and factor the difference of two squares, we have

$$\begin{aligned}x^2 - y^2 + x - y &= (x + y)(x - y) + (x - y) && \text{Factor } x^2 - y^2. \\ &= (x - y)(x + y + 1) && \text{Factor out } x - y.\end{aligned}$$

Self Check 6 Factor: $a^2 - b^2 + a + b$.

EXAMPLE 7

Factor: $x^2 + 6x + 9 - z^2$.

Solution We group the first three terms together and factor the trinomial to get

$$\begin{aligned}x^2 + 6x + 9 - z^2 &= (x + 3)(x + 3) - z^2 \\ &= (x + 3)^2 - z^2\end{aligned}$$

We can now factor the difference of two squares to get

$$x^2 + 6x + 9 - z^2 = (x + 3 + z)(x + 3 - z)$$

Self Check 7 Factor: $a^2 + 4a + 4 - b^2$.

PERFECT CUBES

The number 64 is called a perfect cube, because $4^3 = 64$. To factor the sum or difference of two cubes, it is helpful to know the first ten **perfect integer cubes:**

1, 8, 27, 64, 125, 216, 343, 512, 729, 1,000

Expressions such as x^9y^6 are also perfect cubes, because they can be written as the cube of another quantity:

$$x^9y^6 = (x^3y^2)^3$$

THE SUM AND DIFFERENCE OF TWO CUBES

To find formulas for factoring the sum or difference of two cubes, we use the following product formulas:

(2) $(x + y)(x^2 - xy + y^2) = x^3 + y^3$

(3) $(x - y)(x^2 + xy + y^2) = x^3 - y^3$

The Language of Algebra

The expression $x^3 + y^3$ is a *sum of two cubes,* whereas $(x + y)^3$ is the *cube of a sum.* If you expand $(x + y)^3$, you will see that they are not equivalent.

To verify Equation 2, we multiply $x^2 - xy + y^2$ by $x + y$.

$$\begin{aligned}(x + y)(x^2 - xy + y^2) &= (x + y)x^2 - (x + y)xy + (x + y)y^2 \\ &= x \cdot x^2 + y \cdot x^2 - x \cdot xy - y \cdot xy + x \cdot y^2 + y \cdot y^2 \\ &= x^3 + x^2y - x^2y - xy^2 + xy^2 + y^3 \\ &= x^3 + y^3\end{aligned}$$

Equation 3 can also be verified by multiplication.

The binomial $x^3 + y^3$ is called the **sum of two cubes,** because x^3 represents the cube of x, y^3 represents the cube of y, and $x^3 + y^3$ represents the sum of these cubes. Similarly, $x^3 - y^3$ is called the **difference of two cubes.**

If we write Equations 2 and 3 in reverse order, we have the formulas for factoring the sum and difference of two cubes.

Sum and Difference of Two Cubes

$$x^3 + y^3 = (x + y)(x^2 - xy + y^2)$$
$$x^3 - y^3 = (x - y)(x^2 + xy + y^2)$$

If we think of the sum of two cubes as the sum of the cube of a **F**irst quantity plus the cube of a **L**ast quantity, we have the formula

$$F^3 + L^3 = (F + L)(F^2 - FL + L^2)$$

*To factor the cube of a **First** quantity plus the cube of a **Last** quantity, we multiply the sum of the **First** and **Last** by*

- *the **First** squared*
- *minus the **First** times the **Last***
- *plus the **Last** squared.*

The formula for the difference of two cubes is

$$F^3 - L^3 = (F - L)(F^2 + FL + L^2)$$

*To factor the cube of a **First** quantity minus the cube of a **Last** quantity, we multiply the difference of the **First** and **Last** by*

- *the **First** squared*
- *plus the **First** times the **Last***
- *plus the **Last** squared.*

EXAMPLE 8

Factor: $a^3 + 8$.

Solution Since $a^3 + 8$ can be written as $a^3 + 2^3$, we have the sum of two cubes, which factors as follows:

$$F^3 + L^3 = (F + L)(F^2 - FL + L^2)$$
$$\downarrow \quad \downarrow \quad \downarrow \quad \downarrow \quad \downarrow \quad \downarrow\downarrow \quad \downarrow$$
$$a^3 + 2^3 = (a + 2)(a^2 - a2 + 2^2)$$
$$= (a + 2)(a^2 - 2a + 4)$$

$a^2 - 2a + 4$ does not factor.

Caution

In Example 8, a common error is to try to factor $a^2 - 2a + 4$. It is not a perfect square trinomial, because the middle term needs to be $-4a$. Furthermore, it cannot be factored by the methods of Section 5.7. It is prime.

Therefore, $a^3 + 8 = (a + 2)(a^2 - 2a + 4)$. We can check by multiplying.

$$(a + 2)(a^2 - 2a + 4) = a^3 + 2a^2 - 2a^2 - 4a + 4a + 8$$
$$= a^3 + 8$$

Self Check 8 Factor: $p^3 + 27$.

You should memorize the formulas for factoring the sum and the difference of two cubes. Note that each has the form

(a binomial)(a trinomial)

and that there is a relationship between the signs that appear in these forms.

$$F^3 + L^3 = (F + L)(F^2 - FL + L^2) \qquad F^3 - L^3 = (F - L)(F^2 + FL + L^2)$$

In each formula: same (sign between F^3, L^3 and sign in the binomial); opposite (sign in the binomial and sign of the FL term); positive (sign before L^2).

EXAMPLE 9

Factor: $27a^3 - 64b^3$.

Solution Since $27a^3 - 64b^3$ can be written as $(3a)^3 - (4b)^3$, we have the difference of two cubes, which factors as follows:

$$\mathbf{F}^3 - \mathbf{L}^3 = (\mathbf{F} - \mathbf{L})(\mathbf{F}^2 + \mathbf{F}\ \mathbf{L} + \mathbf{L}^2)$$

$$\begin{aligned}(3a)^3 - (4b)^3 &= (3a - 4b)[(3a)^2 + (3a)(4b) + (4b)^2]\\ &= (3a - 4b)(9a^2 + 12ab + 16b^2)\end{aligned}$$

Thus, $27a^3 - 64b^3 = (3a - 4b)(9a^2 + 12ab + 16b^2)$.

Self Check 9 Factor: $8c^3 - 125d^3$.

EXAMPLE 10

Factor: $a^3 - (c + d)^3$.

Solution $a^3 - (c + d)^3 = [a - (c + d)][a^2 + a(c + d) + (c + d)^2]$

Now we simplify the expressions inside both sets of brackets.

$$a^3 - (c + d)^3 = (a - c - d)(a^2 + ac + ad + c^2 + 2cd + d^2)$$

Self Check 10 Factor: $(p + q)^3 - r^3$.

EXAMPLE 11

Factor: $x^6 - 64$.

Solution This expression is both the difference of two squares and the difference of two cubes. It is easier to factor it as the difference of two squares first.

$$\begin{aligned}x^6 - 64 &= (x^3)^2 - 8^2\\ &= (x^3 + 8)(x^3 - 8)\end{aligned}$$

Each of these factors can be factored further. One is the sum of two cubes and the other is the difference of two cubes.

$$x^6 - 64 = (x + 2)(x^2 - 2x + 4)(x - 2)(x^2 + 2x + 4)$$

Self Check 11 Factor: $x^6 - 1$.

EXAMPLE 12

Factor: $2a^5 + 250a^2$.

Solution We first factor out the common monomial factor of $2a^2$ to obtain

$$2a^5 + 250a^2 = 2a^2(a^3 + 125)$$

Then we factor $a^3 + 125$ as the sum of two cubes to obtain

$$2a^5 + 250a^2 = 2a^2(a + 5)(a^2 - 5a + 25)$$

Self Check 12 Factor: $3x^5 + 24x^2$.

Answers to Self Checks
1. $(9p + 5)(9p - 5)$ **2.** $(6r^2 + s)(6r^2 - s)$ **3.** $(a^2 + 9)(a + 3)(a - 3)$
4. $[(a - b)^2 + c^2](a - b + c)(a - b - c)$ **5.** $3(a^2 + 1)(a + 1)(a - 1)$
6. $(a + b)(a - b + 1)$ **7.** $(a + 2 + b)(a + 2 - b)$ **8.** $(p + 3)(p^2 - 3p + 9)$
9. $(2c - 5d)(4c^2 + 10cd + 25d^2)$ **10.** $(p + q - r)(p^2 + 2pq + q^2 + pr + qr + r^2)$
11. $(x + 1)(x^2 - x + 1)(x - 1)(x^2 + x + 1)$ **12.** $3x^2(x + 2)(x^2 - 2x + 4)$

5.7 STUDY SET

VOCABULARY Fill in the blanks.

1. When the polynomial $4x^2 - 25$ is written as $(2x)^2 - (5)^2$, we see that it is the difference of two ________.
2. When the polynomial $8x^3 + 125$ is written as $(2x)^3 + (5)^3$, we see that it is the sum of two ________.

CONCEPTS

3. Write the first ten perfect integer squares.
4. Write the first ten perfect integer cubes.
5. **a.** Use multiplication to verify that the sum of two squares $x^2 + 25$ does not factor as $(x + 5)(x + 5)$.
   b. Use multiplication to verify that the difference of two squares $x^2 - 25$ factors as $(x + 5)(x - 5)$.
6. Explain the error.
   a. Factor: $4g^2 - 16 = (2g + 4)(2g - 4)$
   b. Factor: $1 - t^8 = (1 + t^4)(1 - t^4)$
7. When asked to factor $81t^2 - 16$, one student answered $(9t - 4)(9t + 4)$, and another answered $(9t + 4)(9t - 4)$. Explain why both students are correct.
8. Factor each polynomial.
   a. $5p^2 + 20$
   b. $5p^2 - 20$
   c. $5p^3 + 20$
   d. $5p^3 + 40$

Complete each factorization.

9. $p^2 - q^2 = (p + q)$ ______
10. $36y^2 - 49m^2 = (\quad)^2 - (7m)^2$
    $= (6y \quad 7m)(6y - \quad)$
11. $p^2q + pq^2 = \quad (p + q)$
12. $p^3 + q^3 = (p + q)$ ______
13. $p^3 - q^3 = (p - q)$ ______
14. $h^3 - 27k^3 = (h)^3 - (\quad)^3$
    $= (h \quad 3k)(h^2 + \quad + 9k^2)$

NOTATION

15. Give an example of each.
    a. a difference of two squares
    b. a square of a difference
    c. a sum of two squares
    d. a sum of two cubes
    e. a cube of a sum
16. Fill in the blanks.
    a. $x^2 - y^2 = (x \quad y)(x \quad y)$
    b. $x^3 + y^3 = (x \quad y)(x^2 \quad xy \quad y^2)$
    c. $x^3 - y^3 = (x \quad y)(x^2 \quad xy \quad y^2)$

PRACTICE **Factor, if possible.**

17. $x^2 - 4$

18. $y^2 - 9$

19. $9y^2 - 64$

20. $16x^4 - 81y^2$

21. $x^2 + 25$

22. $144a^2 - b^4$

23. $400 - c^2$

24. $900 - t^2$

25. $625a^2 - 169b^4$

26. $4y^2 + 9z^4$

27. $81a^4 - 49b^2$

28. $64r^6 - 121s^2$

29. $36x^4y^2 - 49z^4$

30. $4a^2b^4c^6 - 9d^8$

31. $(x + y)^2 - z^2$

32. $a^2 - (b - c)^2$

33. $(a - b)^2 - c^2$

34. $(m + n)^2 - p^4$

35. $x^4 - y^4$

36. $16a^4 - 81b^4$

37. $256x^4y^4 - z^8$

38. $225a^4 - 16b^8c^{12}$

39. $\frac{1}{36} - y^4$

40. $\frac{4}{81} - m^4$

41. $2x^2 - 288$

42. $8x^2 - 72$

43. $2x^3 - 32x$

44. $3x^3 - 243x$

45. $5x^3 - 125x$

46. $6x^4 - 216x^2$

47. $r^2s^2t^2 - t^2x^4y^2$

48. $16a^4b^3c^4 - 64a^2bc^6$

49. $a^2 - b^2 + a + b$

50. $x^2 - y^2 - x - y$

51. $a^2 - b^2 + 2a - 2b$

52. $m^2 - n^2 + 3m + 3n$

53. $2x + y + 4x^2 - y^2$

54. $m - 2n + m^2 - 4n^2$

55. $x^3 - xy^2 - 4x^2 + 4y^2$

56. $m^2n - 9n + 9m^2 - 81$

57. $x^2 + 4x + 4 - y^2$

58. $x^2 - 6x + 9 - 4y^2$

59. $x^2 + 2x + 1 - 9z^2$

60. $x^2 + 10x + 25 - 16z^2$

61. $c^2 - 4a^2 + 4ab - b^2$

62. $4c^2 - a^2 - 6ab - 9b^2$

63. $r^3 + s^3$

64. $t^3 - v^3$

65. $x^3 - 8y^3$

66. $27a^3 + b^3$

67. $64a^3 - 125b^6$

68. $8x^6 + 125y^3$

69. $125x^3y^6 + 216z^9$

70. $1{,}000a^6 - 343b^3c^6$

71. $x^6 + y^6$

72. $x^9 + y^9$

73. $5x^3 + 625$

74. $2x^3 - 128$

75. $4x^5 - 256x^2$

76. $2x^6 + 54x^3$

77. $128u^2v^3 - 2t^3u^2$

78. $56rs^2t^3 + 7rs^2v^6$

79. $(a + b)x^3 + 27(a + b)$

80. $(c - d)r^3 - (c - d)s^3$

81. $x^9 - y^{12}z^{15}$

82. $r^{12} + s^{18}t^{24}$

83. $(a + b)^3 + 27$

84. $(b - c)^3 - 1{,}000$

85. $y^3(y^2 - 1) - 27(y^2 - 1)$

86. $z^3(y^2 - 4) + 8(y^2 - 4)$

Factor each expression completely. Factor a difference of two squares first.

87. $x^6 - 1$

88. $x^6 - y^6$

89. $x^{12} - y^6$

90. $a^{12} - 64$

Factor each trinomial.

91. $a^4 - 13a^2 + 36$

92. $b^4 - 17b^2 + 16$

APPLICATIONS

93. CANDY To find the amount of chocolate used in the outer coating of the malted-milk ball shown, we can find the volume V of the chocolate shell using the formula $V = \frac{4}{3}\pi r_1^3 - \frac{4}{3}\pi r_2^3$. Factor the expression on the right-hand side of the formula.

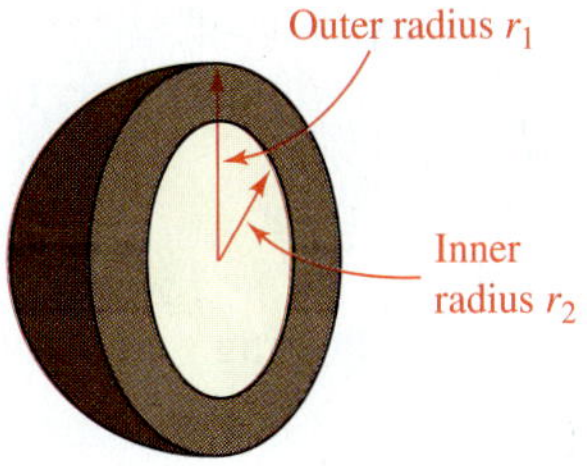

94. MOVIE STUNTS The function that gives the distance a stuntwoman is above the ground t seconds after she falls over the side of a 144-foot tall building is $h(t) = 144 - 16t^2$. Factor the right-hand side.

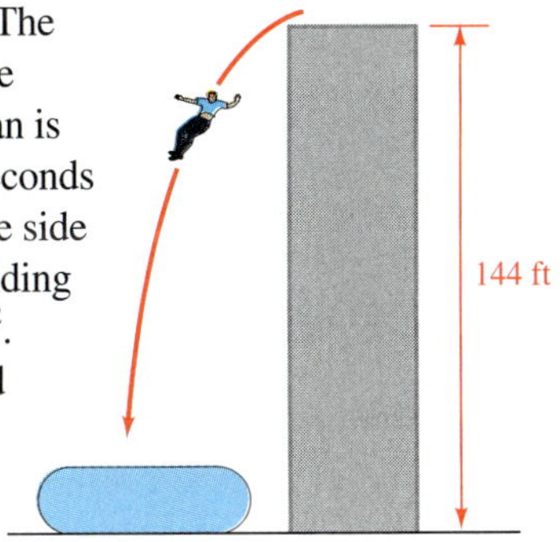

WRITING

95. Describe the pattern used to factor the difference of two squares.

96. Describe the patterns used to factor the sum and the difference of two cubes.

REVIEW **Graph the line with the given characteristics.**

97. Passing through $(-2, -1)$; slope $= -\frac{2}{3}$

98. y-intercept $(0, -4)$; slope $= 3$

99. Write the equation of line l shown to the right.

100. Write the equation of line r shown to the right.

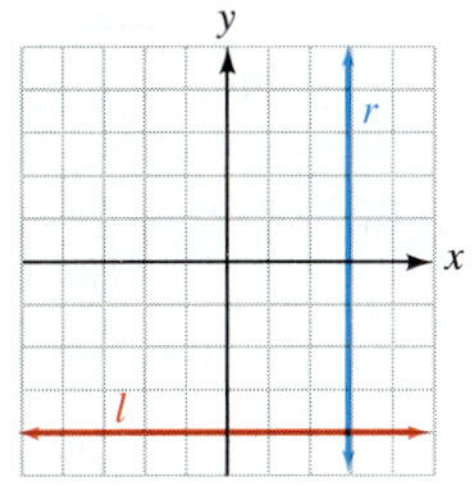

CHALLENGE PROBLEMS **Factor. Assume all variables represent natural numbers.**

101. $4x^{2n} - 9y^{2n}$

102. $25 - x^{6n}$

103. $a^{3b} - c^{3b}$

104. $8 - x^{3n}$

105. $27x^{3n} + y^{3n}$

106. $a^{3b} + b^{3c}$

107. Factor $x^{32} - y^{32}$.

108. Find the error in this proof that $2 = 1$.

$$x = y$$
$$x^2 = xy$$
$$x^2 - y^2 = xy - y^2$$
$$(x + y)(x - y) = y(x - y)$$
$$\frac{(x + y)\cancel{(x - y)}}{\cancel{(x - y)}} = \frac{y\cancel{(x - y)}}{\cancel{x - y}}$$
$$x + y = y$$
$$y + y = y$$
$$2y = y$$
$$\frac{2y}{y} = \frac{y}{y}$$
$$2 = 1$$

5.8 Summary of Factoring Techniques

- A general factoring strategy

When we factor a polynomial, we write a sum of terms as an equivalent product of factors. For many polynomials, this process involves several steps in which two or more factoring techniques must be used. In such cases, it is helpful to follow a general factoring strategy.

A GENERAL FACTORING STRATEGY

The following strategy is helpful when factoring polynomials.

Steps for Factoring a Polynomial

1. Is there a common factor? If so, factor out the GCF.
2. How many terms does the polynomial have?
   If it has *two terms,* look for the following problem types:
   a. The difference of two squares
   b. The sum of two cubes
   c. The difference of two cubes
   If it has *three terms,* look for the following problem types:
   a. A perfect square trinomial
   b. If the trinomial is not a perfect square, use the trial-and-check method or the grouping method.
   If it has *four* or more terms, try to factor by grouping.
3. Can any factors be factored further? If so, factor them completely.
4. Does the factorization check? Check by multiplying.

EXAMPLE 1

Factor: $12x^2y^2z^3 - 2xy^2z^3 - 4y^2z^3$.

Solution ***Is there a common factor?*** Yes. Factor out the greatest common factor $2y^2z^3$.

$$12x^2y^2z^3 - 2xy^2z^3 - 4y^2z^3 = 2y^2z^3(6x^2 - x - 2)$$

How many terms does it have? The polynomial within the parentheses has three terms. We can factor $6x^2 - x - 2$, using the trial-and-check method or the key number method, to get

$$12x^2y^2z^3 - 2xy^2z^3 - 4y^2z^3 = 2y^2z^3(6x^2 - x - 2)$$
$$= 2y^2z^3(3x - 2)(2x + 1)$$

Don't forget to write the GCF from the first step.

The Language of Algebra

Remember that the instruction to *factor* means to *factor completely.* A polynomial is *factored completely* when no factor can be factored further.

Is it factored completely? Yes. Since each of the individual factors is prime, the factorization is complete.

Does it check? To check, we multiply.

$$2y^2z^3(3x - 2)(2x + 1) = 2y^2z^3(6x^2 - x - 2) \quad \text{Multiply the binomials first.}$$
$$= 12x^2y^2z^3 - 2xy^2z^3 - 4y^2z^3 \quad \text{Distribute the multiplication by } 2y^2z^3.$$

Since we obtain the original polynomial, the factorization is correct.

Self Check 1 Factor: $30a^2b^3c - 27ab^3c + 6b^3c$.

EXAMPLE 2

Factor: $48a^4c^3 - 3b^4c^3$.

Solution ***Is there a common factor?*** Yes. Factor out the greatest common factor $3c^3$.

$$48a^4c^3 - 3b^4c^3 = 3c^3(16a^4 - b^4)$$

How many terms does it have? The polynomial within the parentheses, $16a^4 - b^4$, has two terms. It is the difference of two squares and factors as $(4a^2 + b^2)(4a^2 - b^2)$.

$$48a^4c^3 - 3b^4c^3 = 3c^3(16a^4 - b^4)$$
$$= 3c^3(4a^2 + b^2)(4a^2 - b^2)$$

Is it factored completely? No. The binomial $4a^2 + b^2$ is the sum of two squares and is prime. However, $4a^2 - b^2$ is the difference of two squares and factors as $(2a + b)(2a - b)$.

$$\begin{aligned} 48a^4c^3 - 3b^4c^3 &= 3c^3(16a^4 - b^4) \\ &= 3c^3(4a^2 + b^2)(4a^2 - b^2) \\ &= 3c^3(4a^2 + b^2)(2a + b)(2a - b) \end{aligned}$$

Since each of the individual factors is prime, the factorization is now complete.

Does it check? Multiply to verify that this factorization is correct.

Self Check 2 Factor: $3p^4r^3 - 3q^4r^3$.

EXAMPLE 3

Factor: $x^5y + x^2y^4 - x^3y^3 - y^6$.

Solution ***Is there a common factor?*** Yes. Factor out the greatest common factor of y.

$$x^5y + x^2y^4 - x^3y^3 - y^6 = y(x^5 + x^2y^3 - x^3y^2 - y^5)$$

How many terms does it have? The polynomial $x^5 + x^2y^3 - x^3y^2 - y^5$ has four terms. We try factoring by grouping to obtain

$$\begin{aligned} &x^5y + x^2y^4 - x^3y^3 - y^6 \\ &\quad = y(x^5 + x^2y^3 - x^3y^2 - y^5) && \text{Factor out } y. \\ &\quad = y[x^2(x^3 + y^3) - y^2(x^3 + y^3)] && \text{Factor by grouping.} \\ &\quad = y(x^3 + y^3)(x^2 - y^2) && \text{Factor out } x^3 + y^3. \end{aligned}$$

Is it factored completely? No. We can factor $x^3 + y^3$ (the sum of two cubes) and $x^2 - y^2$ (the difference of two squares) to obtain

$$x^5y + x^2y^4 - x^3y^3 - y^6 = y(x + y)(x^2 - xy + y^2)(x + y)(x - y)$$

Because each of the individual factors is prime, the factorization is complete.

Does it check? Multiply to verify that this factorization is correct.

Self Check 3 Factor: $a^5p - a^3b^2p + a^2b^3p - b^5p$.

EXAMPLE 4

Factor: $x^3 + 5x^2 + 6x + x^2y + 5xy + 6y$.

Solution ***Is there a common factor?*** No. There is no common factor (other than 1).

How many terms does it have? Since there are more than three terms, we try factoring by grouping. We can factor x from the first three terms and y from the last three terms.

$$\begin{aligned} &x^3 + 5x^2 + 6x + x^2y + 5xy + 6y \\ &\quad = x(x^2 + 5x + 6) + y(x^2 + 5x + 6) \\ &\quad = (x^2 + 5x + 6)(x + y) && \text{Factor out } x^2 + 5x + 6. \end{aligned}$$

Is it factored completely? No. We can factor the trinomial $x^2 + 5x + 6$ to obtain

$$x^3 + 5x^2 + 6x + x^2y + 5xy + 6y = (x + 3)(x + 2)(x + y)$$

Since each of the individual factors is prime, the factorization is now complete.

Does it check? Multiply to verify that the factorization is correct.

Self Check 4 Factor: $a^3 - 5a^2 + 6a + a^2b - 5ab + 6b$.

EXAMPLE 5

Factor: $4x^4 + 4x^3 + x^2 + 2x + 1$.

Solution ***Is there a common factor?*** There is no common factor (other than 1).

How many terms does it have? Since there are more than three terms, we try factoring by grouping. We can factor x^2 from the first three terms, and group the last two terms together.

$$4x^4 + 4x^3 + x^2 + 2x + 1 = x^2(4x^2 + 4x + 1) + (2x + 1)$$

Is it factored completely? No. We recognize $4x^2 + 4x + 1$ as a perfect square trinomial, because $4x^2 = (2x)^2$, $1 = (1)^2$, and $4x = 2 \cdot 2x \cdot 1$. Therefore, it factors as $(2x + 1)(2x + 1)$.

$$\begin{aligned} 4x^4 + 4x^3 + x^2 + 2x + 1 &= x^2(\mathbf{4x^2 + 4x + 1}) + (2x + 1) \\ &= x^2(\mathbf{2x + 1})(\mathbf{2x + 1}) + (2x + 1) \end{aligned}$$

Finally, we factor out the common factor $2x + 1$.

$$\begin{aligned} 4x^4 + 4x^3 + x^2 + 2x + 1 &= x^2(4x^2 + 4x + 1) + (2x + 1) \\ &= x^2(\mathbf{2x + 1})(2x + 1) + (\mathbf{2x + 1}) \\ &= (\mathbf{2x + 1})[x^2(2x + 1) + 1] \\ &= (2x + 1)(2x^3 + x^2 + 1) \end{aligned}$$

Within the brackets, distribute the multiplication by x^2.

Since each of the individual factors is prime, the factorization is complete.

Does it check? Multiply to verify that this factorization is correct.

Self Check 5 Factor: $a^4 - a^3 - 2a^2 + a - 2$.

Answers to Self Checks

1. $3b^3c(5a - 2)(2a - 1)$ **2.** $3r^3(p^2 + q^2)(p + q)(p - q)$
3. $p(a + b)(a^2 - ab + b^2)(a + b)(a - b)$ **4.** $(a - 2)(a - 3)(a + b)$
5. $(a - 2)(a^3 + a^2 + 1)$

5.8 STUDY SET

VOCABULARY Fill in the blanks.

1. To factor means to factor ________. Each factor of a completely factored expression will be ______.
2. When we factor a polynomial, we write a sum of ______ as an equivalent product of _______.
3. $x^3 + y^3$ is called a sum of two ______ and $x^3 - y^3$ is called a difference of two ______.
4. $x^2 - y^2$ is called a ________ of two squares.

CONCEPTS Fill in the blanks.

5. In any factoring problem, always factor out any ________ factors first.
6. If a polynomial has two terms, check to see whether the problem type is the ________ of two squares, the sum of two ______, or the ________ of two cubes.
7. If a polynomial has three terms, try to factor it as a ________.
8. If a polynomial has four or more terms, try factoring it by ________.
9. Explain how to verify that $y^2z^3(x + 6)(x + 1)$ is the factored form of $x^2y^2z^3 + 7xy^2z^3 + 6y^2z^3$.
10. Why is the polynomial $x + 6$ classified as prime?

NOTATION Complete each factorization.

11. $18a^3b + 3a^2b^2 - 6ab^3 = \square(6a^2 + ab - 2b^2)$
$= 3ab(3a + \square)(\square - b)$
12. $2x^4 - 1{,}250 = 2(\square)$
$= 2(\square)(x^2 - 25)$
$= 2(x^2 + 25)(x + 5)(\square)$

PRACTICE Factor, if possible.

13. $4a^2bc + 4abcd - 120bcd^2$
14. $8x^3y^4 - 27y$
15. $3x^2y + 6xy^2 - 12xy$
16. $xy - ty + xs^2 - ts^2$
17. $9x^4 + 6x^3 + x^2 + 3x + 1$
18. $b^2c + b^2 + bcd + bd$
19. $25x^2 - 16y^2$
20. $27x^9 - y^3$
21. $16c^2g^2 + h^4$
22. $12x^2 + 14x - 6$
23. $6x^2 - 14x + 8$
24. $m^4 - 13m^2 + 36$
25. $4x^2y^2 + 4xy^2 + y^2$
26. $x^3 + (a^2y)^3$
27. $4x^2y^2z^2 - 26x^2y^2z^3$
28. $2x^3 - 54$
29. $4(xy)^3 + 256$
30. $a^2 + b^2 + 25$
31. $a^2x^2 + b^2y^2 + b^2x^2 + a^2y^2$
32. $2(x + y)^2 + (x + y) - 3$
33. $(x - y)^3 + 125$
34. $625x^4 - 256y^4$
35. $2(a - b)^2 + 5(a - b) + 3$
36. $36x^4 - 36$
37. $6x^2 - 63 - 13x$
38. $a^4b^2 - 20ab^2 + 64b^2$
39. $x^4 - 17x^2 + 16$
40. $x^2 + 6x + 9 - y^2$
41. $x^2 + 10x + 25 - y^8$
42. $4x^2 + 4x + 1 - 4y^2$
43. $9x^2 - 6x + 1 - 25y^2$
44. $x^2 - y^2 - 2y - 1$
45. $a^2 - b^2 + 4b - 4$
46. $60q^2r^2s^4 + 78qr^2s^4 - 18r^2s^4$
47. $32x^{10} + 48x^9 + 18x^8$
48. $3ax^2 - 3axy + 3ay^2 - 3x^2 + 3xy - 3y^2$
49. $\frac{81}{16}x^4 - y^{40}$
50. $\frac{d^2x^2}{2} - \frac{f^2x^2}{2} - \frac{c^2d^2}{2} + \frac{c^2f^2}{2}$
51. $16m^{16} - 16$
52. $8(4 - a^2) - x^3(4 - a^2)$
53. $9y^5 + 6y^4 + y^3 + 3y + 1$
54. $25m^4 - 10m^3 + m^2 + 5m - 1$

WRITING

55. What is your strategy for factoring a polynomial?

56. For the factorization below, explain why the polynomial is not factored completely.

$$48a^4c^3 - 3b^4c^3 = 3c^3(16a^4 - b^4)$$

REVIEW Evaluate each determinant.

57. $\begin{vmatrix} -6 & -2 \\ 15 & 4 \end{vmatrix}$

58. $\begin{vmatrix} 3 & -2 \\ 12 & -8 \end{vmatrix}$

59. $\begin{vmatrix} -1 & 2 & 1 \\ 2 & 1 & -3 \\ 1 & 1 & 1 \end{vmatrix}$

60. $\begin{vmatrix} 1 & 0 & 1 \\ 0 & 1 & 0 \\ 1 & 1 & 1 \end{vmatrix}$

CHALLENGE PROBLEMS

61. Factor: $x^4 + x^2 + 1$. (*Hint:* Add and subtract x^2.)

62. Factor: $x^4 + 7x^2 + 16$. (*Hint:* Add and subtract x^2.)

Factor. Assume that *n* is a natural number.

63. $2a^{2n} + 2a^n - 24$

64. $ma^{2n} - mb^{2n}$

65. $54a^{3n} + 16b^{3n}$

66. $-a^{2n} - 2a^nb^n - b^{2n}$

67. $12m^{4n} + 10m^{2n} + 2$

68. $nx^{4n} + 2nx^{2n}y^{2n} + ny^{4n}$

5.9 Solving Equations by Factoring

- Solving quadratic equations
- Solving higher-degree polynomial equations
- Problem solving

We have previously solved *linear equations in one variable* such as $3x - 1 = 5$ and $10y + 9 = y + 4$. These equations are also called **polynomial equations** because they involve two polynomials that are set equal to each other. Some other examples of polynomial equations are

$$3x^2 = -6x, \quad 6x^3 - x^2 = 2x, \quad \text{and} \quad x^4 + 4 - 5x^2 = 0$$

In this section, we will discuss a method for solving polynomial equations like these.

SOLVING QUADRATIC EQUATIONS

A second-degree polynomial equation in one variable is called a *quadratic equation.*

Quadratic Equations A **quadratic equation** is any equation that can be written in **standard form**

$$ax^2 + bx + c = 0$$

where a, b, and c represent real numbers and $a \neq 0$.

Of the following examples of quadratic equations, only the last one is written in standard form.

$$2x^2 + 6x = 15 \qquad y^2 = -16y \qquad 4m^2 - 7m + 2 = 0$$

Many quadratic equations can be solved by factoring and then by using the *zero-factor property.*

The Zero-Factor Property

When the product of two real numbers is 0, at least one of them is 0.

If a and b represent real numbers, and

$$\text{if } ab = 0, \text{ then } a = 0 \text{ or } b = 0$$

This property also applies to three or more factors.

EXAMPLE 1

Solve: $x^2 + 5x + 6 = 0$.

Solution To solve this quadratic equation, we factor its left-hand side to obtain

$$(x + 3)(x + 2) = 0$$

Since the product of $x + 3$ and $x + 2$ is 0, at least one of the factors must be 0. Thus, we can set each factor equal to 0 and solve each resulting linear equation for x:

$$x + 3 = 0 \quad \text{or} \quad x + 2 = 0$$
$$x = -3 \quad \Big| \quad x = -2$$

To check these solutions, we substitute -3 and -2 for x in the equation and verify that each number satisfies the equation.

Check:

$$x^2 + 5x + 6 = 0 \qquad \text{or} \qquad x^2 + 5x + 6 = 0$$
$$(-3)^2 + 5(-3) + 6 \stackrel{?}{=} 0 \qquad (-2)^2 + 5(-2) + 6 \stackrel{?}{=} 0$$
$$9 - 15 + 6 \stackrel{?}{=} 0 \qquad 4 - 10 + 6 \stackrel{?}{=} 0$$
$$0 = 0 \qquad 0 = 0$$

Both -3 and -2 are solutions, because they satisfy the equation. The solution set is $\{-3, -2\}$.

The Language of Algebra

A *solution* of a quadratic equation is a value of the variable that makes the equation true. *To solve a quadratic equation* means to find all of its solutions.

Self Check 1 Solve: $x^2 - 4x - 45 = 0$

EXAMPLE 2

Solve: $3x^2 = 6x$.

Solution To use the zero-factor property, we need 0 on one side of the equation. To get 0 on the right-hand side, we subtract $6x$ from both sides.

$$3x^2 = 6x$$
$$3x^2 - 6x = 0$$

To solve the equation, we factor the left-hand side, set each factor equal to 0, and solve each resulting equation for x.

$$3x^2 - 6x = 0$$

$$3x(x - 2) = 0 \qquad \text{Factor out the common factor } 3x.$$

$$3x = 0 \quad \text{or} \quad x - 2 = 0 \qquad \text{By the zero-factor property, at least one of the factors must be equal to zero.}$$

$$x = 0 \quad \Big| \quad x = 2 \qquad \text{Solve each linear equation.}$$

The solutions are 0 and 2. Check each in the original equation.

Caution

A creative, but incorrect, approach is to divide both sides of $3x^2 = 6x$ by $3x$

$$\frac{3x^2}{3x} = \frac{6x}{3x}$$

You will obtain $x = 2$. However, you will lose the second solution, 0.

Self Check 2 Solve: $4p^2 = 12p$.

EXAMPLE 3

Let $f(x) = x^2 - 71$. For what value(s) of x is $f(x) = 10$?

Solution To find the value(s) where $f(x) = 10$, we substitute 10 for $f(x)$ and solve for x.

$$f(x) = x^2 - 71$$
$$10 = x^2 - 71$$

To solve this quadratic equation, we first write it in standard form. Then we factor the difference of two squares on the right-hand side, set each factor equal to 0, and solve each resulting equation.

$$0 = x^2 - 81 \quad \text{Subtract 10 from both sides.}$$
$$0 = (x + 9)(x - 9) \quad \text{Factor } x^2 - 81.$$
$$x + 9 = 0 \quad \text{or} \quad x - 9 = 0$$
$$x = -9 \quad \Big| \quad x = 9$$

If x is -9 or 9, then $f(x) = 10$. Check these results by substituting -9 and 9 for x in $f(x) = x^2 - 71$.

Self Check 3 Let $g(x) = x^2 - 21$. For what value(s) of x is $g(x) = 100$?

The following steps can be used to solve a quadratic equation by factoring.

Solving a Quadratic Equation by the Factoring Method

1. Write the equation in standard form: $ax^2 + bx + c = 0$.
2. Factor the polynomial.
3. Use the zero-factor property to set each factor equal to zero.
4. Solve each resulting equation.
5. Check the results in the original equation.

EXAMPLE 4

Solve: $x = \frac{6}{5} - \frac{6}{5}x^2$.

Solution We must write the equation in standard form. To clear the equation of fractions, we multiply both sides by 5.

$$x = \frac{6}{5} - \frac{6}{5}x^2$$
$$5x = 6 - 6x^2 \quad \text{Multiply both sides by 5.}$$

To use factoring to solve this quadratic equation, one side of the equation must be 0. Since it is easier to factor a second-degree polynomial if the coefficient of the squared term is positive, we add $6x^2$ to both sides and subtract 6 from both sides to obtain

The Language of Algebra

Quadratic equations involve the square of a variable, not the fourth power as *quad* might suggest. Why the inconsistency? A closer look at the origin of the word *quadratic* reveals that it comes from the Latin word *quadratus,* meaning square.

$$6x^2 + 5x - 6 = 0$$

$$(3x - 2)(2x + 3) = 0 \quad \text{Factor the trinomial.}$$

$$3x - 2 = 0 \quad \text{or} \quad 2x + 3 = 0 \quad \text{Set each factor equal to 0 and solve for } x.$$

$$3x = 2 \quad \Big| \quad 2x = -3$$

$$x = \frac{2}{3} \quad \Big| \quad x = -\frac{3}{2}$$

The solutions are $\frac{2}{3}$ and $-\frac{3}{2}$. Check them in the original equation.

Self Check 4 Solve: $x = \frac{6}{7}x^2 - \frac{3}{7}$.

Caution To solve a quadratic equation by factoring, set the quadratic polynomial equal to 0 before factoring and applying the zero-factor property. Do not make the following error:

$$6x^2 + 5x = 6$$

$$x(6x + 5) = 6$$

$$x = 6 \quad \text{or} \quad 6x + 5 = 6$$

$$\Big| \quad x = \frac{1}{6}$$

If the product of two numbers is 6, neither number need be 6. For example, $2 \cdot 3 = 6$.

Neither solution checks.

Many equations that don't appear to be quadratic can be put into standard form and then solved by factoring.

EXAMPLE 5

Solve: $(x + 5)^2 = 9(2x + 1)$.

Solution To put the equation in standard form, we square the binomial on the left-hand side and distribute the multiplication by 9 on the right-hand side.

$$(x + 5)^2 = 9(2x + 1)$$

$$x^2 + 10x + 25 = 18x + 9$$

$$x^2 - 8x + 16 = 0 \quad \text{Subtract } 18x \text{ and 9 from both sides to get 0 on the right-hand side.}$$

$$(x - 4)(x - 4) = 0 \quad \text{Factor the trinomial.}$$

$$x - 4 = 0 \quad \text{or} \quad x - 4 = 0 \quad \text{Set each factor equal to 0.}$$

$$x = 4 \quad \Big| \quad x = 4$$

We see that the two solutions are the same. We call 4 a *repeated solution.* Check by substituting into the original equation.

Self Check 5 Solve: $(x + 6)^2 = -4(x + 7)$.

ACCENT ON TECHNOLOGY: SOLVING QUADRATIC EQUATIONS

To solve a quadratic equation such as $x^2 + 4x - 5 = 0$ with a graphing calculator, we can use window settings of $[-10, 10]$ for x and $[-10, 10]$ for y and graph the quadratic function $y = x^2 + 4x - 5$, as shown in figure (a). We can then trace to find the x-coordinates of the x-intercepts of the parabola. See figures (b) and (c). For better results, we can zoom in. Since these are the numbers x that make $y = 0$, they are the solutions of the equation.

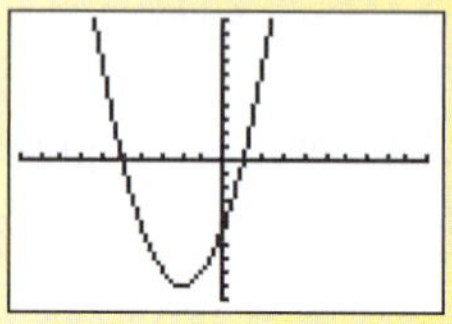

(a)

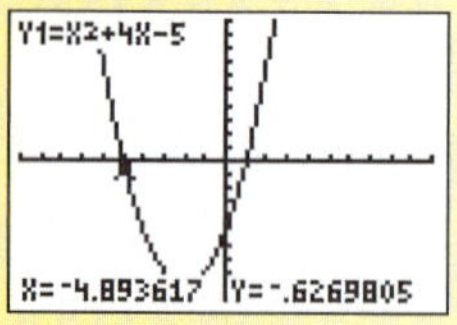

(b)

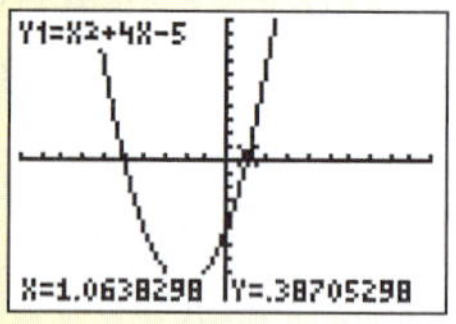

(c)

We can also find the x-intercepts of the graph of $y = x^2 + 4x - 5$ by using the ZERO feature found on most graphing calculators. Figures (d) and (e) show how this feature locates the x-intercept and displays its coordinates.

From the displays, we can conclude that $x = -5$ and $x = 1$ are solutions of $x^2 + 4x - 5 = 0$.

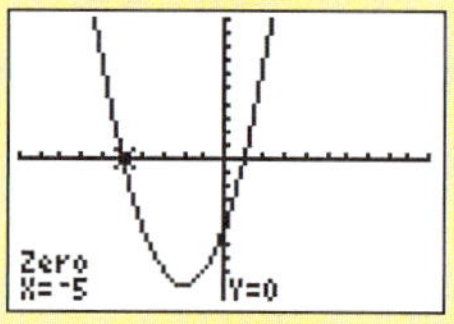

(d)

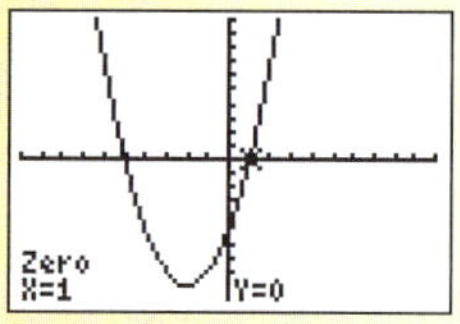

(e)

SOLVING HIGHER-DEGREE POLYNOMIAL EQUATIONS

We can solve many polynomial equations with degree greater than 2 by factoring and applying an extension of the zero-factor property.

EXAMPLE 6

Solve: $6x^3 - x^2 = 2x$.

Solution First, we subtract $2x$ from both sides so that the right-hand side of the equation is 0.

$$6x^3 - x^2 - 2x = 0 \quad \text{This is a third-degree polynomial equation.}$$

Then we factor x from the third-degree polynomial on the left-hand side and proceed as follows:

$$6x^3 - x^2 - 2x = 0$$
$$x(6x^2 - x - 2) = 0 \quad \text{Factor out } x.$$
$$x(3x - 2)(2x + 1) = 0 \quad \text{Factor } 6x^2 - x - 2.$$
$$x = 0 \quad \text{or} \quad 3x - 2 = 0 \quad \text{or} \quad 2x + 1 = 0 \quad \text{Set each of the three factors equal to 0.}$$
$$x = \frac{2}{3} \qquad x = -\frac{1}{2} \quad \text{Solve each equation.}$$

The solutions are 0, $\frac{2}{3}$, and $-\frac{1}{2}$. Check each one.

Self Check 6 Solve: $5x^3 + 13x^2 = 6x$.

EXAMPLE 7

Solve: $-x^4 + 5x^2 - 4 = 0$.

Solution To make the lead coefficient positive, we multiply both sides of the equation by -1. Then we factor the resulting trinomial and set the factors equal to 0.

$$-x^4 + 5x^2 - 4 = 0$$
$$-1(-x^4 + 5x^2 - 4) = -1(0)$$
$$x^4 - 5x^2 + 4 = 0 \quad \text{This is a fourth-degree polynomial equation.}$$
$$(x^2 - 1)(x^2 - 4) = 0$$
$$(x + 1)(x - 1)(x + 2)(x - 2) = 0 \quad \text{Factor } x^2 - 1 \text{ and } x^2 - 4.$$

$$x + 1 = 0 \quad \text{or} \quad x - 1 = 0 \quad \text{or} \quad x + 2 = 0 \quad \text{or} \quad x - 2 = 0$$
$$x = -1 \qquad x = 1 \qquad x = -2 \qquad x = 2$$

The solutions are -1, 1, -2, and 2. Check each one in the original equation.

Self Check 7 Solve: $-a^4 - 36 + 13a^2 = 0$.

ACCENT ON TECHNOLOGY: SOLVING EQUATIONS

To solve $x^4 - 5x^2 + 4 = 0$ with a graphing calculator, we can use window settings of $[-6, 6]$ for x and $[-5, 10]$ for y and graph the polynomial function $y = x^4 - 5x^2 + 4$ as shown in the figure. We can then read the values of x that make $y = 0$. They are $x = -2, -1, 1,$ and 2. If the x-coordinates of the x-intercepts were not obvious, we could approximate their values by using TRACE and ZOOM or by using the ZERO feature.

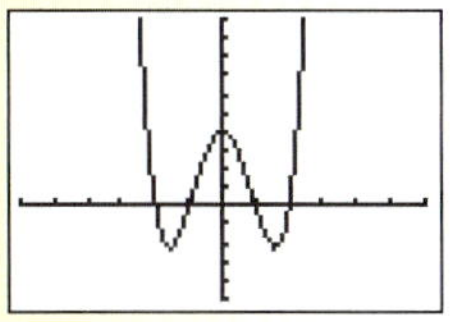

PROBLEM SOLVING

EXAMPLE 8

Stained glass. The window in the illustration is to be installed in a chapel. The length of the base of the window is 3 times its height, and its area is 96 square feet. Find its base and height.

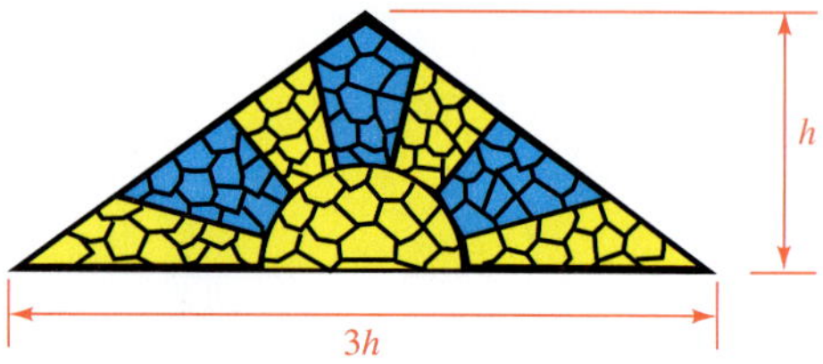

Analyze the Problem We are to find the length of the base and height of the window. The formula that gives the area of a triangle is $A = \frac{1}{2}bh$, where b is the length of the base and h the height.

Form an Equation We can let h = the height of the window. Then $3h$ = the length of the base. To form an equation in terms of h, we can substitute $3h$ for b and 96 for A in the formula for the area of a triangle.

$$A = \frac{1}{2}bh$$

$$96 = \frac{1}{2}(3h)h$$

Solve the Equation To solve this equation, we must write it in standard form.

$$96 = \frac{1}{2}(3h)h$$

$$192 = 3h^2 \quad \text{To clear the equation of the fraction, multiply both sides by 2.}$$

$$64 = h^2 \quad \text{Divide both sides by 3.}$$

$$0 = h^2 - 64 \quad \text{To obtain 0 on the left-hand side, subtract 64 from both sides.}$$

$$0 = (h + 8)(h - 8) \quad \text{Factor the difference of two squares.}$$

$$h + 8 = 0 \quad \text{or} \quad h - 8 = 0$$

$$h = -8 \quad \Big| \quad h = 8$$

State the Conclusion Since the height cannot be negative, we discard the negative solution. Thus, the height of the window is 8 feet, and the length of its base is 3(8), or 24 feet.

Check the Result The area of a triangle with a base of 24 feet and a height of 8 feet is 96 square feet:

$$A = \frac{1}{2}bh = \frac{1}{2}(24)(8) = 12(8) = 96$$

The result checks.

EXAMPLE 9

Ballistics. If the initial velocity of an object thrown from the ground straight up into the air is 176 feet per second, when will the object strike the ground?

Analyze the Problem The height of an object thrown straight up into the air from the ground with an initial velocity of v feet per second is given by the formula

$$h = -16t^2 + vt$$

The height h is in feet, and t represents the number of seconds since the object was released. When the object hits the ground, its height will be 0.

Form an Equation In the formula, we set h equal to 0 and set v equal to 176.

$$h = -16t^2 + vt$$
$$0 = -16t^2 + 176t$$

Solve the Equation To solve this equation, we will use the factoring method.

$$0 = -16t^2 + 176t$$
$$0 = -16t(t - 11) \quad \text{Factor out } -16t.$$
$$-16t = 0 \quad \text{or} \quad t - 11 = 0 \quad \text{Set each factor equal to 0.}$$
$$t = 0 \quad | \quad t = 11$$

State the Conclusion When t is 0, the object's height above the ground is 0 feet, because it has not been released. When t is 11, the height is again 0 feet, and the object has returned to the ground. The solution is 11 seconds.

Check the Result Verify that $h = 0$ when $t = 11$.

EXAMPLE 10

Recreation. A rectangular-shaped spa, 5 feet wide and 6 feet long, is surrounded by decking of uniform width, as shown in the illustration. If the total area of the deck is 60 ft^2, how wide is the decking?

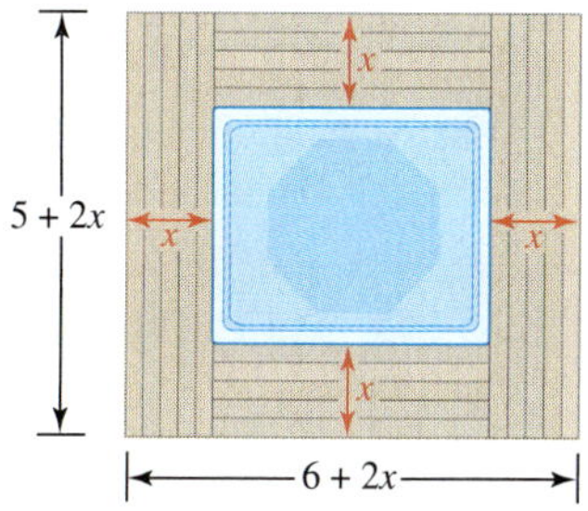

Analyze the Problem Since the dimensions of the rectangular spa are 5 feet by 6 feet, its surface area is $6 \cdot 5 = 30$ ft^2. The decking has an area of 60 ft^2.

Form an Equation Let x = the width of the decking in feet. Then the width of the outer rectangle (the pool and the decking) is $x + 5 + x$ or $(5 + 2x)$ feet, and the length of the outer rectangle is $x + 6 + x$ or $(6 + 2x)$ feet. The area of the outer rectangle is the product of its length and width: $(6 + 2x)(5 + 2x)$ ft^2. We can now form an equation.

The area of the outer rectangle	minus	the area of the spa	equals	the area of the decking.
$(6 + 2x)(5 + 2x)$	$-$	30	$=$	60

Solve the Equation

$$(6 + 2x)(5 + 2x) - 30 = 60$$
$$30 + 12x + 10x + 4x^2 - 30 = 60 \quad \text{Multiply the binomials on the left-hand side.}$$
$$4x^2 + 22x = 60 \quad \text{Simplify the left-hand side.}$$
$$4x^2 + 22x - 60 = 0 \quad \text{Subtract 60 from both sides to get 0 on the right-hand side.}$$
$$2x^2 + 11x - 30 = 0 \quad \text{Divide both sides by 2.}$$
$$(2x + 15)(x - 2) = 0 \quad \text{Factor the trinomial.}$$

$2x + 15 = 0$	or $x - 2 = 0$	Set each factor equal to 0.
$2x = -15$	$x = 2$	
$x = -\dfrac{15}{2}$		

State the Conclusion The decking is 2 feet wide. (Since x represents width, we discard the negative solution.)

Check the Result The illustration shows that if the decking is 2 feet wide, then the total area of the decking is $18 + 12 + 18 + 12 = 60$ ft^2. The result checks.

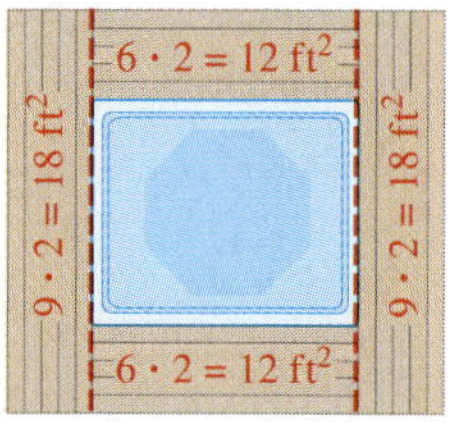

Answers to Self Checks **1.** 9, −5 **2.** 0, 3 **3.** −11, 11 **4.** $\frac{3}{2}, -\frac{1}{3}$ **5.** −8, −8 **6.** 0, $\frac{2}{5}$, −3 **7.** 2, −2, 3, −3

5.9 STUDY SET

VOCABULARY Fill in the blanks.

1. A ________ equation is any equation that can be written in the form $ax^2 + bx + c = 0$, where $a \neq 0$.

2. To ______ an equation means to find all the values of the variable that make the equation true.

3. When a quadratic equation is written in ________ form, 0 is on one side of the equation.

4. $2x^2 - 4x = 0$, $3x^3 - x^2 - 6x = 0$, and $x^4 - 5x^2 + 4 = 0$ are examples of __________ equations.

CONCEPTS

5. If the product of two numbers is 0, what must be true about at least one of the numbers?

6. Use a check to determine whether −5 and 4 are solutions of $a^2 - 9a + 20 = 0$.

7. Determine whether each equation is a quadratic equation.

a. $w^2 + 7w + 12 = 0$

b. $6t + 11 = 0$

c. $x(x + 3) = -2$

d. $k^3 - 4k^2 + k - 15 = 0$

8. Write each equation in standard form.

a. $20 - 10x = -x^2$

b. $5x^3 - 10x^2 = 20x$

9. What first step should be performed to solve each quadratic equation?

a. $x^2 + 24 = -11x$

b. $x^2 + x + \frac{1}{4} = 0$

c. $-2x^2 + 7x + 4 = 0$

d. $m(m + 3) = 2$

10. Divide both sides of the equation by 4.

$4y^2 - 40y + 96 = 0$

11. Use the zero-factor property to solve each equation.

a. $(x - 3)(2x + 5) = 0$

b. $6x(x - 1)(2x + 15) = 0$

12. a. Write an expression that represents the width of the outer rectangle.

b. Write an expression that represents the length of the outer rectangle.

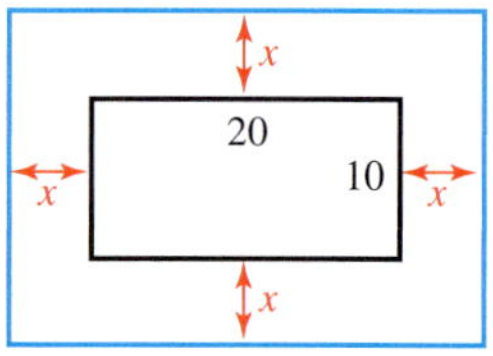

13. Use the graph to solve $x^2 - 2x - 3 = 0$.

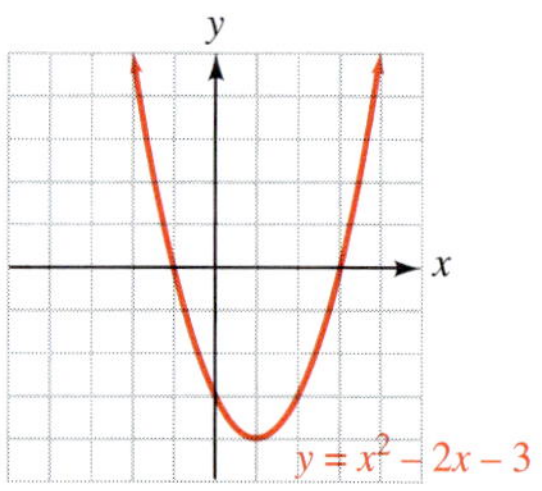

14. Use the graph to solve $x^3 - 4x^2 + 4x = 0$

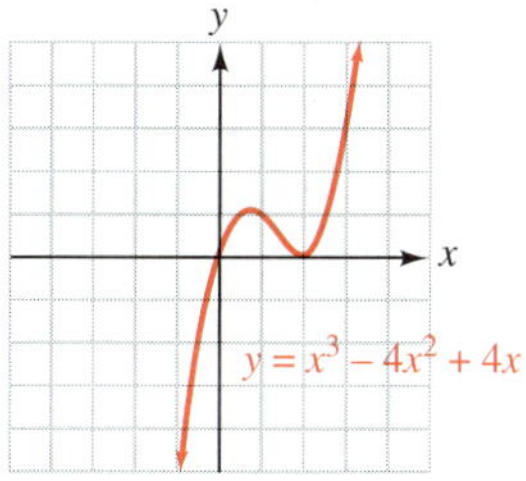

NOTATION **Complete each solution.**

15. Solve: $y^2 - 3y - 54 = 0$.

$(y - 9)(\quad) = 0$

$\quad = 0$ or $y + 6 = 0$

$y = 9$ | $y = \quad$

16. Solve: $2x^2 - 3x - 1 = 1$.

$2x^2 - 3x - 2 = \quad$

$(\quad)(x - 2) = 0$

$2x + 1 = \quad$ or $\quad = 0$

$2x = -1$ | $x = 2$

$x = \quad$

PRACTICE **Solve each equation.**

17. $4x^2 + 8x = 0$

18. $x^2 - 9 = 0$

19. $y^2 - 16 = 0$

20. $y^2 - 25 = 0$

21. $x^2 + x = 0$

22. $x^2 - 3x = 0$

23. $5y^2 - 25y = 0$

24. $y^2 - 36 = 0$

25. $z^2 + 8z + 15 = 0$

26. $w^2 + 7w + 12 = 0$

27. $x^2 + 6x + 8 = 0$

28. $x^2 + 9x + 20 = 0$

29. $3m^2 + 10m + 3 = 0$

30. $2r^2 + 5r + 3 = 0$

31. $2y^2 - 5y = -2$

32. $2x^2 - 3x = -1$

33. $2x^2 = x + 1$

34. $2x^2 = 3x + 5$

35. $x(x - 6) + 9 = 0$

36. $x^2 + 8(x + 2) = 0$

37. $8a^2 = 3 - 10a$

38. $5z^2 = 6 - 13z$

39. $b(6b - 7) = 10$

40. $2y(4y + 3) = 9$

41. $\frac{3a^2}{2} = \frac{1}{2} - a$

42. $x^2 = \frac{1}{2}(x + 1)$

43. $x^2 + 1 = \frac{5}{2}x$

44. $\frac{3}{5}(x^2 - 4) = -\frac{9}{5}x$

45. $x\left(3x + \frac{22}{5}\right) = 1$

46. $x\left(\frac{x}{11} - \frac{1}{7}\right) = \frac{6}{77}$

47. $x^3 + x^2 = 0$

48. $2x^4 + 8x^3 = 0$

49. $y^3 - 49y = 0$

50. $2z^3 - 200z = 0$

51. $x^3 - 4x^2 - 21x = 0$

52. $x^3 + 8x^2 - 9x = 0$

53. $z^4 - 13z^2 + 36 = 0$

54. $y^4 - 10y^2 + 9 = 0$

55. $3a(a^2 + 5a) = -18a$

56. $7t^3 = 2t\left(t + \frac{5}{2}\right)$

57. $\frac{x^2(6x + 37)}{35} = x$

58. $x^2 = -\frac{4x^3(3x + 5)}{3}$

59. $\frac{a^2}{2} + \frac{5a}{12} = \frac{7}{4}$

60. $\frac{r^2}{15} + \frac{r}{20} = \frac{1}{6}$

61. $(m + 4)(2m + 3) - 22 = 10m$

62. $(d - 2)(d + 1) - d = 1$

63. $(x + 7)^2 = -2(x + 7) - 1$

64. $2(7x + 18) - 1 = (x + 6)^2$

65. $n(3n - 4) = n^2 - n + 35$

66. $s(2s + 7) = s^2 + s + 72$

67. $3x^3 + 3x^2 = 12(x + 1)$

68. $9(y + 4) = y^3 + 4y^2$

69. Let $f(x) = x^2 - 3x + 3$. For what value(s) of x is $f(x) = 1$?

70. Let $f(x) = 6x^2 + 5x + 2$. For what value(s) of x is $f(x) = 6$?

71. Let $f(x) = x^3 - 6x^2 + 8x + 2$. For what value(s) of x is $f(x) = 2$?

72. Let $f(x) = x^3 - 2x^2 - 8x + 10$. For what value(s) of x is $f(x) = 10$?

Use a graphing calculator to find the solutions of each equation, if one exists. If an answer is not exact, give the answer to the nearest hundredth.

73. $2x^2 - 7x + 4 = 0$

74. $x^2 - 4x + 7 = 0$

75. $-3x^3 - 2x^2 + 5 = 0$

76. $-2x^3 - 3x - 5 = 0$

APPLICATIONS

77. INTEGER PROBLEM The product of two consecutive even integers is 288. Find the integers. (*Hint:* Let $x =$ the first even integer. Then represent the second even integer in terms of x.)

78. INTEGER PROBLEM The product of two consecutive odd integers is 143. Find the integers. (*Hint:* Let $x =$ the first odd integer. Then represent the second odd integer in terms of x.)

79. COOKING A griddle has a cooking surface of 160 square inches. Find its length and width.

80. ENGINEERING The formula for the area of a trapezoid is $A = \frac{h(B + b)}{2}$. The area of a trapezoidal truss is 44 square feet. Find the length of the truss if the length of the shorter base is the same as the height.

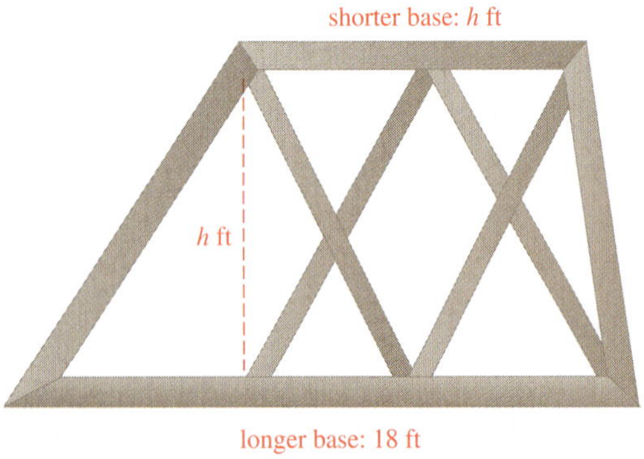

81. SWIMMING POOL DESIGN Building codes require that a rectangular swimming pool be surrounded by a uniform-width walkway of at least 516 square feet. The length of the pool illustrated in the next column is 10 feet less than twice the width. How wide should the border be?

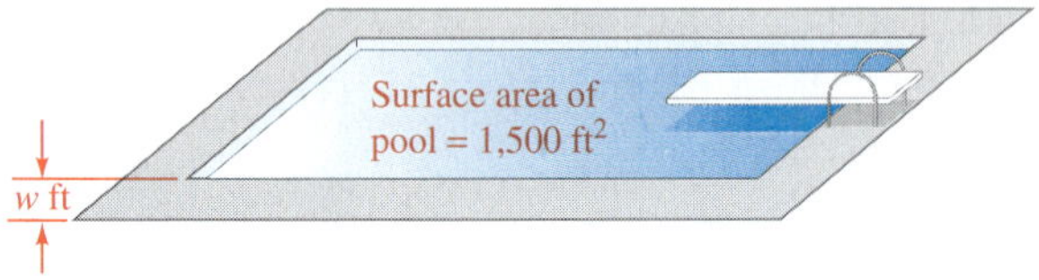

82. FINE ARTS An artist intends to paint a 60-square-foot mural on a large wall. Find the dimensions of the mural if the artist leaves a border of uniform width around it.

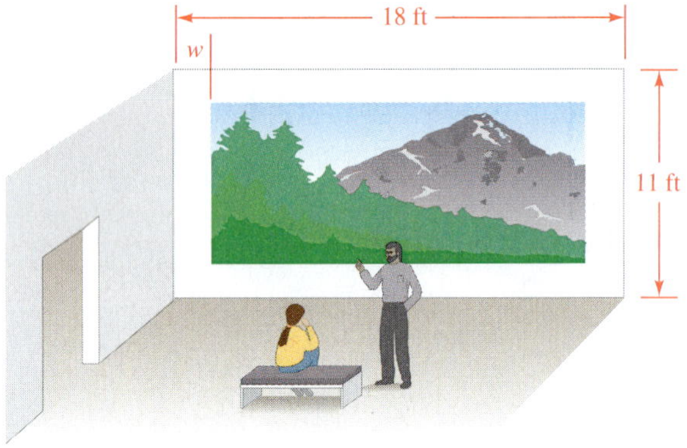

83. ARCHITECTURE The following rectangular room is twice as long as it is wide. It is divided into two rectangular parts by a partition. If the larger part of the room contains 560 square feet, find the dimensions of the entire room.

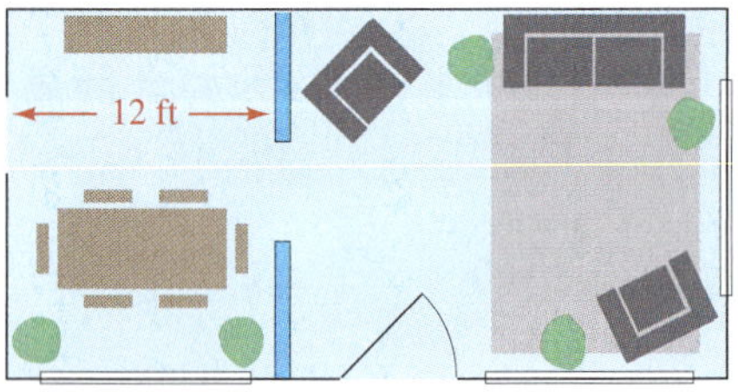

84. WINTER RECREATION The length of a rectangular ice-skating rink is 20 meters greater than twice its width. Find the width.

85. BALLISTICS The muzzle velocity of a cannon is 480 feet per second. If a cannonball is fired vertically, at what times will it be at a height of 3,344 feet? (See Example 9.)

86. SLINGSHOTS A slingshot can provide an initial velocity of 128 feet per second. At what times will a stone, shot vertically upward, be 192 feet above the ground? (See Example 9.)

87. BASEBALL In 1998, Roger Clemens, then of the Toronto Blue Jays, threw a fastball clocked at 97 mph. This is approximately 144 feet per second. If he could throw the baseball vertically into the air with this velocity, how long would it take for the ball to fall to the ground? (See Example 9.)

88. BUNGEE JUMPING The formula $h = -16t^2 + 212$ gives the distance a bungee jumper is from the ground for the free-fall portion of the jump, t seconds after leaping off a bridge. We can find the number of seconds it takes the jumper to reach the point in the fall where the 64-foot bungee cord starts to stretch by substituting 148 for h and solving for t. Find t.

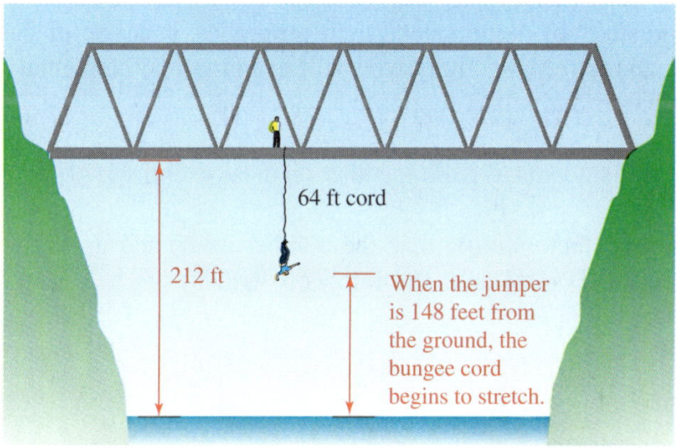

89. FORENSIC MEDICINE The kinetic energy E of a moving object is given by $E = \frac{1}{2}mv^2$, where m is the mass of the object (in kilograms) and v is the object's velocity (in meters per second). Kinetic energy is measured in joules. By measuring the damage done to a victim who has been struck by a 3-kilogram club, a pathologist finds that the energy at impact was 54 joules. Find the velocity of the club at impact.

90. TRAFFIC ACCIDENTS Investigators at a traffic accident used the function $d(v) = 0.04v^2 + 0.8v$, where v is the velocity of the car (in mph) and $d(v)$ is the stopping distance of the car (in feet), to reconstruct the events leading up to a collision. From physical evidence, it was concluded that it took one car 32 feet to stop. At what velocity was the car traveling prior to the accident?

91. BREAK-EVEN POINT The cost for a guitar maker to hand-craft x guitars is given by the function $C(x) = \frac{1}{8}x^2 - x + 6$. The revenue taken in with the sale of x guitars is given by the function $R(x) = \frac{1}{4}x^2$. Find the number of guitars that must be sold so that the cost equals the revenue.

92. REVENUE Over the years, the manager of a store has found that the number of scented candles x she can sell in a month depends on the price p according to the formula $x = 200 - 10p$. At what price should she sell the candles if she needs to bring in $750 in revenue a month? (*Hint:* Revenue = price · number sold = px.)

WRITING

93. Explain the zero-factor property.

94. In the work shown below, explain why the student has not solved for x.

Solve: $x^2 + x - 6 = 0$.

$$x^2 + x = 6$$

$$\boxed{x = 6 - x^2}$$

95. Explain what is wrong with the following solution.

Solve: $x^2 = x$.

$$\frac{x^2}{x} = \frac{x}{x}$$

$$x = 1$$

96. Explain what is wrong with the following solution.

Solve: $x^2 - x = 6$.

$$x(x - 1) = 6$$

$$\boxed{x = 6} \quad \text{or} \quad x - 1 = 6$$

$$\boxed{x = 7}$$

97. The following graphs of two polynomial functions $f(x) = 2x^3 - 8x$ and $f(x) = 2x(x + 2)(x - 2)$ appear to be the same. After examining their equations, explain why we know that they are identical graphs.

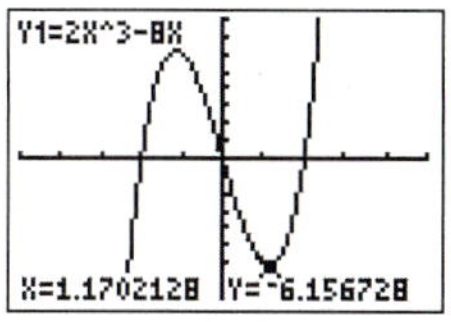

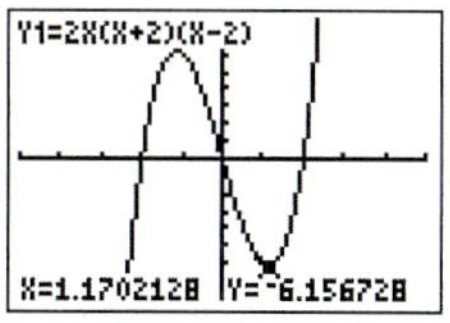

98. Explain why the x-coordinate of the x-intercept in the following graph of $y = 8x^2 + 10x - 3$ is a solution of $8x^2 + 10x - 3 = 0$.

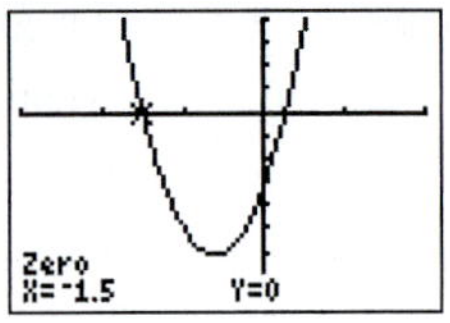

REVIEW

99. ALUMINUM FOIL Find the number of square feet of aluminum foil on a roll if it has dimensions of $8\frac{1}{3}$ yards × 12 inches.

100. HOCKEY A hockey puck is a vulcanized rubber disk 2.5 cm (1 in.) thick and 7.6 cm (3 in.) in diameter. Find the volume of a puck in cubic centimeters and cubic inches. Round to the nearest tenth.

CHALLENGE PROBLEMS **Find a quadratic equation with the following solutions.**

101. $-2, 6$

102. $\frac{1}{4}, -\frac{4}{3}$

Find a polynomial equation of degree 3 with the given solutions.

103. $0, 2, 4$

104. $-3, -2, 3$

ACCENT ON TEAMWORK

NUMBER THEORY

Overview: Number theory is the study of numbers and their properties. It is one of the oldest branches of pure mathematics. In this activity, you will examine two elementary number theory topics: perfect numbers and relatively prime numbers.

Instructions: Form groups of 2 or 3 students to work together on these exercises.

1. A number whose **proper factors** (factors other than the number itself) add up to the number is called a **perfect number.** For example, the number 6, with factors 1, 2, 3, and 6, is a perfect number because

$$1 + 2 + 3 = 6$$

Show that 28 and 496 are perfect numbers. (If you feel ambitious, the next perfect number is 8,128.)

Notice that each of these perfect numbers is even. No odd perfect numbers are known at this time. An exhaustive computer search has shown that there are no odd perfect numbers smaller than 10^{300}.

2. If the greatest common factor of two (or more) whole numbers is 1, the numbers are called **relatively prime.** For example, 12 and 13 are relatively prime because their greatest common factor is 1. The numbers 12 and 14 are not relatively prime, because their greatest common factor is 2. Determine whether the numbers in each list are relatively prime.

a. 14, 45 **b.** 33, 57 **c.** 116, 145
d. 60, 28, 36 **e.** 55, 49, 78 **f.** 30, 42, 70, 105

A FACTORING FLOWCHART

Overview: This activity will improve your ability to factor polynomials.

Instructions: Form groups of 2 or 3 students.

A **flowchart** is a diagram that illustrates the steps of a particular process. When completed, the flowchart shown below can be used to identify the type(s) of factoring necessary for any given polynomial having two or more terms.

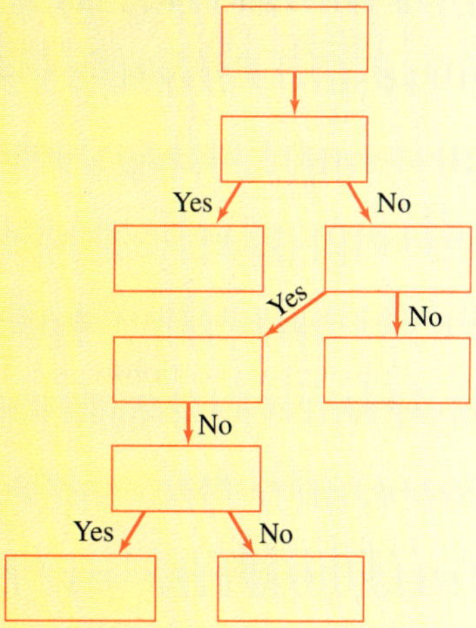

Redraw a larger version of the flowchart and write the correct statement from the list below in each box.

Can the polynomial be factored by grouping?

Does it factor as a perfect square trinomial?

Can the trinomial be factored
1. using the trial-and-check method?
2. by grouping?

Does the polynomial have exactly 2 terms?

Does it factor as a
1. difference of two squares?
2. sum of two cubes?
3. difference of two cubes?

Can you find two integers whose product is the last term and whose sum is the coefficient of the middle term?

Does the polynomial have exactly 3 terms?

Is the lead coefficient 1?

1. Write the polynomial in descending powers of a variable.
2. Factor out the GCF.
3. If necessary, factor out -1 so that the lead coefficient is positive.

Use the completed flowchart to help you factor each of the following polynomials.

1. $-3x^2 + 21x - 36$
2. $rt + 2r + st + 2s$
3. $46w - 6 + 16w^2$
4. $v^3 - 8$
5. $x^2 - 121y^2$
6. $25y^2 - 20y + 4$

KEY CONCEPT: POLYNOMIALS

A **polynomial** is an algebraic term or the sum of two or more algebraic terms whose variables have whole-number exponents. No variable appears in a denominator.

OPERATIONS WITH POLYNOMIALS

In arithmetic, we learned how to add, subtract, multiply, divide, and find powers of numbers. In algebra, we need to be able to perform these operations on polynomials.

Perform each operation

1. $(2x^3 - 6x^2 + 8x - 4) + (-6x^3 + 2x^2 - 3x + 2)$

2. $(6s^3t + 3s - 2t) - (2s^3t + 3s^2 + 5t)$

3. $(3m - 4)(m + 3)$

4. $3r^2st(r^2 - 2s + 3t^2)$

5. $(a - 2d)^2$

6. $-3x(x - 1)(x - 3)$

7. $(3b + 1)(2b^2 + 3b + 2)$

8. $(2y + 3)(y - 1) - (y + 2)(3y - 1)$

POLYNOMIAL FUNCTIONS AND THEIR GRAPHS

Polynomial functions can be used to model many real-world situations.

9. WINDOW WASHERS A man on a scaffold, washing the outside windows of a skyscraper, drops a squeegee. As it falls, its distance in feet from the ground $d(t)$, t seconds after being dropped, is given by the polynomial function $d(t) = -16t^2 + 576$. Find $d(6)$ and explain the result.

10. Write a polynomial function $V(x)$ that gives the volume of the ice chest shown. Then find $V(3)$.

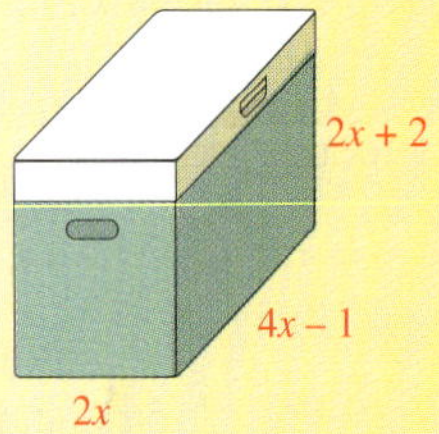

11. If $f(x) = x^3 - x^2 - 4x + 1$, find $f(10)$.

12. The graph of $f(x) = x^3 - x^2 - 4x + 4$ is shown below. Use the graph to find each of the following.

a. $f(-1)$ **b.** $f(3)$

c. The values of x for which $f(x) = 0$

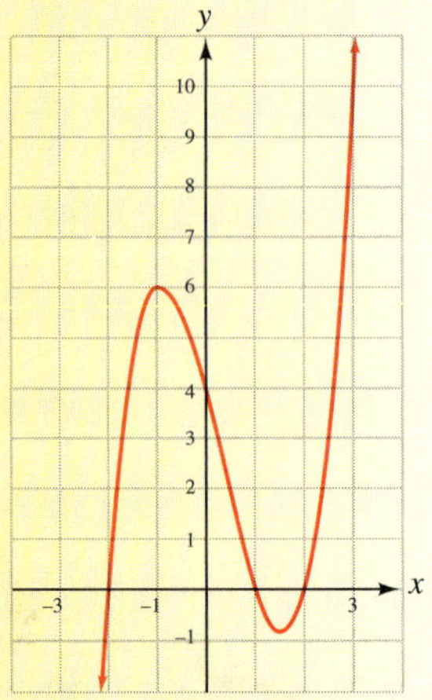

SOLVING POLYNOMIAL EQUATIONS BY FACTORING

Quadratic equations are polynomial equations of degree 2. They can be written in the form $ax^2 + bx + c = 0$. Many quadratic equations can be solved by factoring the polynomial $ax^2 + bx + c$ and applying the zero-factor property. Some higher-degree polynomial equations can also be solved by using an extension of this procedure.

Solve each equation by factoring.

13. $x^2 - 81 = 0$

14. $5x^2 - 25x = 0$

15. $z^2 + 8z + 15 = 0$

16. $3(x + 1) = 7 + (5x - 2)(x + 2)$

17. $\dfrac{3t^2}{2} + t = \dfrac{1}{2}$

18. $m^3 = 9m - 8m^2$

CHAPTER REVIEW

SECTION 5.1 Exponents

CONCEPTS

If n is a natural number,

$$x^n = \underbrace{x \cdot x \cdot x \cdot \cdots \cdot x}_{n \text{ factors of } x}$$

where x is the *base* and n the *exponent.*

Rules for exponents:
If there are no divisions by 0, then for all integers m and n,

$x^m x^n = x^{m+n}$ $(x^m)^n = x^{mn}$

$(xy)^n = x^n y^n$ $\left(\dfrac{x}{y}\right)^n = \dfrac{x^n}{y^n}$

$x^0 = 1$ $x^{-n} = \dfrac{1}{x^n}$

$\dfrac{x^m}{x^n} = x^{m-n}$

$\left(\dfrac{x}{y}\right)^{-n} = \left(\dfrac{y}{x}\right)^n$

REVIEW EXERCISES

Evaluate each expression.

1. 3^6
2. -2^5
3. $(-4)^3$
4. $\left(\dfrac{2}{3}\right)^{-2}$

Simplify each expression and write all answers without negative exponents.

5. $x^4 \cdot x^2$
6. $m^{-3}n^{-4}m^6n^{-1}$
7. $(m^6)^3$
8. $(-t^2)^2(t^3)^3$
9. $(3x^2y^3)^2$
10. $\left(\dfrac{x^4}{b}\right)^4$
11. $-3x^0$
12. $(x^2)^{-5}$
13. $\left(\dfrac{2a}{b}\right)^{-1}$
14. $\dfrac{70}{x^{-4}}$
15. $(3x^{-3})^{-2}$
16. $\dfrac{2x^{-4}x^3}{9}$
17. $-\left(\dfrac{c^{-3}}{c^{-5}}\right)^5$
18. $\left(\dfrac{4}{5}\right)^{-2}$
19. $\dfrac{y^{-3}}{y^4 y}$
20. $\left(\dfrac{-2a^4b}{a^{-3}b^2}\right)^{-3}$

SECTION 5.2 Scientific Notation

Scientific notation is a compact way of writing large and small numbers. Positive numbers are written in the form

$N \times 10^n$

where $1 \le N < 10$ and n is an integer.

Write each number in scientific notation.

21. 19,300,000,000
22. 0.00000002735

Write each number in standard notation.

23. 7.277×10^7
24. 8.3×10^{-9}

Write each number in scientific notation and do the operations. Give answers in scientific notation.

25. SPEED OF LIGHT Light travels at about 300,000 kilometers per second. If the average distance from the sun to the planet Mars is approximately 228,000,000 kilometers, how long does it take light from the sun to reach Mars?

26. PROTONS If the mass of 1 proton is 0.00000000000000000000000167248 gram, find the mass of 1 million protons.

27. Evaluate: $\dfrac{(616{,}000{,}000)(0.000009)}{0.00066}$.

SECTION 5.3 Polynomials and Polynomial Functions

A *polynomial* is a single term or the sum of terms in which all variables have whole-number exponents. No variable appears in a denominator.

Tell whether each expression is a polynomial.

28. $\dfrac{2x^2}{x+1}$

29. $-5x^3 + x^2 - 5x - 4$

30. $2.8y^{15} - y^{10} + y^8 - \dfrac{3}{2}y^6$

31. $x^{-3} + x^{-2} - x^{-1} - 1$

The *degree of a polynomial* is the degree of the term with the highest degree contained within the polynomial.

Classify each polynomial as a monomial, binomial, trinomial, or none of these. Then determine the degree of the polynomial.

32. $x^2 - 8$

33. $-15a^3b$

34. $x^4 + x^3 - x^2 + x - 4$

35. $9x^2y + 13x^3y^2 + 8x^4y^4$

To *evaluate a polynomial* function, we replace the variable in the defining equation with its value, called the *input.* Then we simplify to find the *output.*

36. SQUIRT GUNS The volume of the reservoir on top of the squirt gun is given by the polynomial function $V(r) = 4.19r^3 + 25.13r^2$, where r is the radius in inches. Find $V(2)$ to the nearest cubic inch.

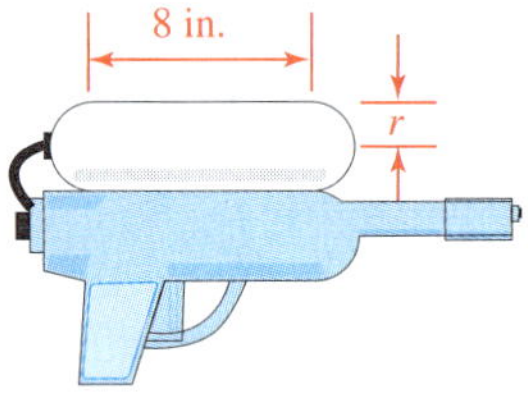

Graph each polynomial function.

37. $f(x) = x^2 - 2x$

38. $f(x) = x^3 - 3x^2 + 4$

To add polynomials, remove parentheses and *combine like terms* (terms having the same variables with the same exponents).

Perform each operation.

39. $(2x^2y^3 - 5x^2y + 9y) + (x^2y^3 - 3x^2y - y)$

40.
$$\begin{array}{r} -10k^4 - 4k^3 + 5k^2 - k + 1 \\ -\ \underline{-16k^4 + 2k^3 - 4k^2 - k + 3} \end{array}$$

To subtract polynomials, add the first polynomial and the opposite of the second polynomial.

41. Subtract $6c^2d^2 + 4c^2d - 5cd^2$ from the sum of $-c^2d^2 + 5c^2d - 10cd^2$ and $11c^2d^2 - c^2d + 9cd^2$.

42. Use the graph of function f to find each of the following

a. $f(0)$

b. The values of x for which $f(x) = 0$

c. The domain and range of f

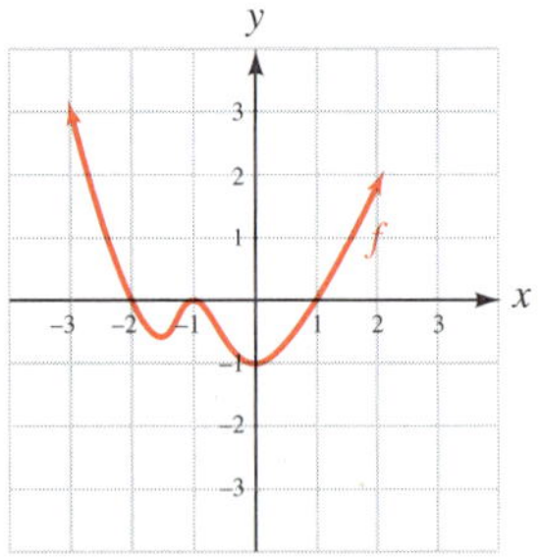

SECTION 5.4 Multiplying Polynomials

To *multiply monomials,* multiply their numerical factors and multiply their variable factors.

To *multiply a polynomial by a monomial,* multiply each term of the polynomial by the monomial.

The *FOIL method* is used to multiply two binomials.

To *multiply polynomials,* multiply each term of one polynomial by each term of the other polynomial.

Find each product.

43. $(8a^2)\left(-\frac{1}{2}a\right)$

44. $(-3xy^2z)(-2xz^3)(xz)$

45. $2xy^2(x^3y - 4xy^5)$

46. $-a^2b(-a^2 - 2ab + b^2)$

47. $(3x^2 + 2)(2x - 4)$

48. $(5at - 6)^2$

49. $(7c^2d^2 - d)(7c^2d^2 + d)$

50. $(5x^2 - 4x)(3x^2 - 2x + 10)$

51. $(r + s)(r - s)(r - 3s)$

52. $\left(3c - \frac{3}{4}\right)^2$

53. SHAVING A razor blade is made from a thin piece of platinum steel. Before its center is punched out, the blade has the shape shown in the illustration. Write a polynomial that gives the area of the front of the blade shown here.

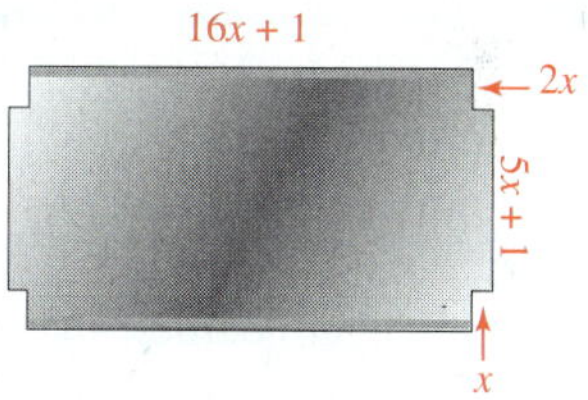

54. GEOMETRY The length, width, and height of the rectangular solid shown in the illustration are consecutive integers.

a. Write a polynomial function that gives the volume of the solid.

b. What is the volume of the solid if the shortest dimension is 5 inches?

SECTION 5.5 The Greatest Common Factor and Factoring by Grouping

To *prime factor* a natural number means to write it as a product of prime numbers.

The largest natural number that divides each number in a set of numbers is called their *greatest common factor (GCF).*

When we factor a polynomial, we write a sum of terms as a product of factors.

Always factor out *common factors* as the first step in a factoring problem.

55. Find the prime factorization of 350.

Find the GCF of each list.

56. a. 42, 36, 54

b. $6x^2y^5$, $15xy^3$

Factor, if possible.

57. $4x^4 + 8$

58. $\frac{3x^3}{5} - \frac{6x^2}{5} + \frac{x}{5}$

59. $6x^2y^3 - 11mn^2$

60. $7a^4b^2 + 49a^3b$

61. $5x^2(x + y + 1) - 15x^3(x + y + 1)$

62. $27x^3y^3z^3 + 81x^4y^5z^2 - 90x^2y^3z^7$

A polynomial that cannot be factored is a *prime polynomial.*

If an expression has four or more terms, try to factor the expression by *grouping.*

Factor out the opposite of the greatest common factor.

63. $-7b^3 + 14c$

64. $-49a^3b^2(a - b)^4 + 63a^2b^4(a - b)^3$

Factor by grouping.

65. $xy + 2y + 4x + 8$

66. $r^2y - ar - ry + a + r - 1$

67. Solve $m_1m_2 = mm_2 + mm_1$ for m_1.

68. GEOMETRY The formula for the surface area of a cylinder is $A = 2\pi r^2 + 2\pi rh$. Rewrite the formula with the right-hand side in factored form.

SECTION 5.6 Factoring Trinomials

Test for factorability: A trinomial of the form $ax^2 + bx + c$ will factor with integer coefficients if $b^2 - 4ac$ is a perfect square.

Perfect square trinomials are the squares of binomials:

$$x^2 + 2xy + y^2 = (x + y)^2$$
$$x^2 - 2xy + y^2 = (x - y)^2$$

To factor trinomials with a *lead coefficient of 1,* list the factorizations of the third term.

To factor trinomials with *lead coefficients other than 1,* factor by trial-and-check or by the key number/grouping method.

Use the test for factorability to determine if each trinomial is factorable.

69. $h^2 + 8h + 18$

70. $9c^2 - 12cd + 4d^2$

Factor, if possible.

71. $x^2 + 10x + 25$

72. $49a^6 + 84a^3b^2 + 36b^4$

73. $y^2 + 21y + 20$

74. $z^2 + 30 - 11z$

75. $-x^2 - 3x + 28$

76. $a^2 - 24b^2 - 5ab$

77. $4a^2 - 5a + 1$

78. $3b^2 + 2b + 1$

79. $y^3 + y^2 - 2y$

80. $27r^2st + 90rst - 72st$

81. $6t^2(r + s) + 13t(r + s) - 15(r + s)$

82. $v^4 - 13v^2 + 42$

83. $w^8 - w^4 - 90$

84. Use a substitution to factor $(s + t)^2 - 2(s + t) + 1$.

SECTION 5.7 The Difference of Two Squares; the Sum and Difference of Two Cubes

Factoring the *difference of two squares:*

$$x^2 - y^2 = (x + y)(x - y)$$

Factor, if possible.

85. $z^2 - 16$

86. $x^2y^4 - 64z^6$

87. $a^2b^2 + c^2$

88. $c^2 - (a + b)^2$

Factoring the *sum of two cubes:*

$x^3 + y^3$
$= (x + y)(x^2 - xy + y^2)$

Factoring the *difference of two cubes:*

$x^3 - y^3$
$= (x - y)(x^2 + xy + y^2)$

89. $m^4 - 16$

90. $m^2 - n^2 - m - n$

91. $32a^4c - 162b^4c$

92. $k^2 + 2k + 1 - 9m^2$

93. $t^3 + 64$

94. $8a^3 - 125b^9$

SECTION 5.8

Summary of Factoring Techniques

Use these steps to factor a random expression:

1. Factor out all common factors.
2. If an expression has two terms, check to see whether it is
   - a. The difference of two squares: $(x^2 - y^2) = (x + y)(x - y)$
   - b. The sum of two cubes: $(x^3 + y^3) = (x + y)(x^2 - xy + y^2)$
   - c. The difference of two cubes: $(x^3 - y^3) = (x - y)(x^2 + xy + y^2)$
3. If an expression has three terms, attempt to factor it as a *general trinomial.*
4. If an expression has four or more terms, try factoring by *grouping.*
5. Continue until each individual factor is prime.
6. Check the results by multiplying.

Factor, if possible.

95. $4q^2rs + 4qrst - 120rst^2$

96. $2(m + n)^2 + (m + n) - 3$

97. $z^2 - 4 + zx - 2x$

98. $m^4 + 16n^2$

99. $x^2 + 4x + 4 - 4p^4$

100. $y^2 + 3y + 2 + 2x + xy$

101. $4a^3b^3c^2 + 256c^2$

102. $-13a^2 + 36 + a^4$

103. $4x^4 + 12x^3 + 9x^2 + 2x + 3$

104. SPANISH ROOF TILES The amount of clay used to make a roof tile is given by

$$V = \frac{\pi}{2}r_1^2h - \frac{\pi}{2}r_2^2h$$

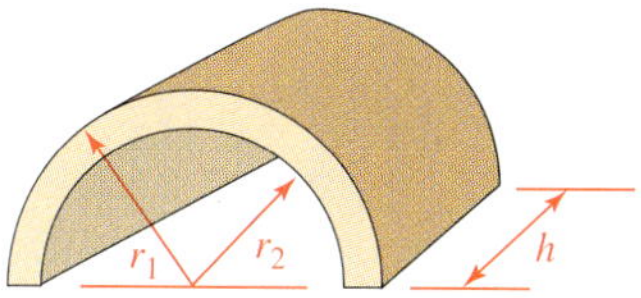

Factor the right-hand side of the formula completely.

SECTION 5.9 Solving Equations by Factoring

To solve a *quadratic equation* by factoring:

1. Write the equation in the form $ax^2 + bx + c = 0$.
2. Factor the polynomial.
3. Use the zero-factor property to set each factor equal to zero.
4. Solve each resulting equation.
5. Check each solution.

The *zero-factor property:* If a and b are real numbers, then

If $ab = 0$, then $a = 0$ or $b = 0$.

Solve each equation by factoring.

105. $4x^2 - 3x = 0$

106. $x^2 - 36 = 0$

107. $12x^2 = 5 - 4x$

108. $d^4 - 10d^2 = -9$

109. $t^2(15t - 2) = 8t$

110. $u^3 = \frac{u}{3}(19u + 14)$

111. $(y + 7)^2 + 8 = -2(y + 7) + 7$

112. PYRAMIDS The volume of a pyramid is given by the formula $V = \frac{Bh}{3}$, where B is the area of its base and h is its height. The volume of the pyramid is 1,020 cubic meters. Find the dimensions of its rectangular base if one edge of the base is 3 meters longer than the other, and the height of the pyramid is 9 meters.

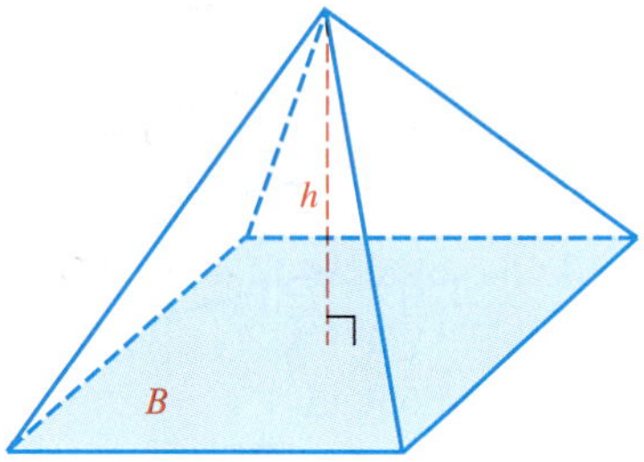

113. Use the graph of $y = 2x^2 - x - 1$, shown in the illustration, to estimate the solutions of $2x^2 - x - 1 = 0$.

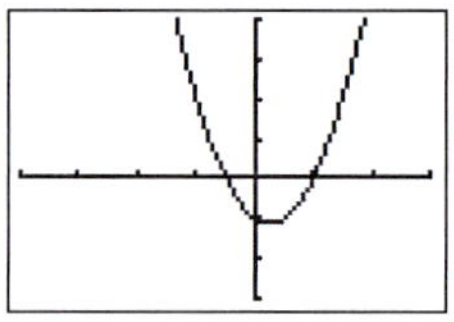

114. Let $f(x) = x^2 - 4x + 2$. For what value(s) of x is $f(x) = -1$?

CHAPTER 5 TEST

Simplify each expression. Write all answers without using negative exponents. Assume that no denominators are zero.

1. $x^3 \cdot x^5 \cdot x$

2. $\left(\frac{-2x^2y^3}{5}\right)^3$

3. $12\,m^3(m^{-4})^2$

4. $\left(\frac{3m^2n^3}{m^4n^{-2}}\right)^{-2}$

5. $\frac{2^{-3}}{3^{-2}}$

6. $\left(\frac{44t^0}{st}\right)^{-1}$

7. Write 4,706,000,000,000 in scientific notation.

8. Write 2.45×10^{-4} in standard notation.

9. Evaluate: $\frac{3.19 \times 10^{15}}{2.2 \times 10^{-4}}$. Express the answer in scientific notation.

10. SPEED OF LIGHT Light travels 1.86×10^5 miles per second. How far does it travel in a minute? Express the answer in scientific notation.

Find the degree of each polynomial.

11. $3x^3 - 4x^5 - 3x^2 - \frac{5}{3}$

12. $3x^5y^3 - x^8y^2 + 2x^9y^4 - 3x^2y^5 + 4$

13. BOATING The height (in feet) of a warning flare from the surface of the ocean t seconds after being shot into the air is given by the polynomial function $h(t) = -16t^2 + 80t + 10$. What is the height of the flare 2.5 seconds after being fired?

14. STRUCTURAL ENGINEERING Write a polynomial function that gives the cross-sectional area of the wooden beam shown in the illustration.

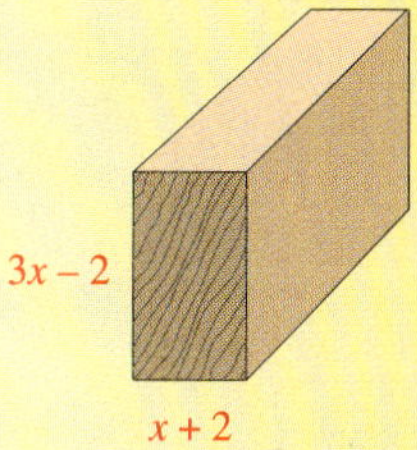

15. Graph the function: $f(x) = x^2 + 2x$.

16. **a.** Graph the function: $f(x) = x^3 + 4x^2 + 4x$.

b. From the graph, determine the solutions of $x^3 + 4x^2 + 4x = 0$.

17. Use the graph of function f to find each of the following.

a. $f(4)$

b. The values of x for which $f(x) = 2$.

c. The domain and range of f.

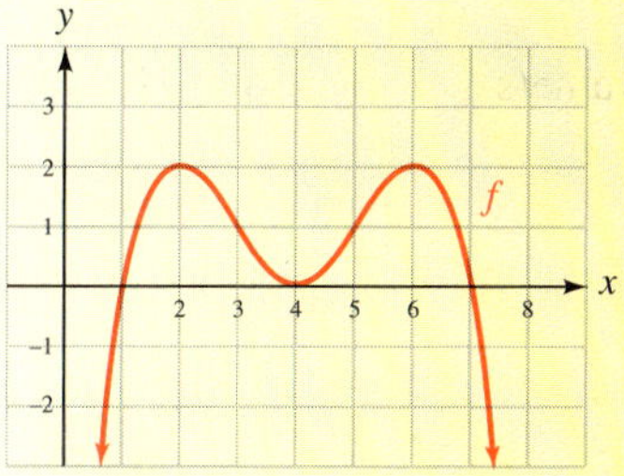

Perform the operations.

18. $(7y^3 + 4y^2 + y + 3) + (-8y^3 - y + 3)$

19. $(-2x^2y^3 + 6xy + 5y^2) - (-4x^2y^3 - 7xy + 2y^2)$

20. $-5a^2b(3ab^3 - 2ab^4)$

21. $(3y + 1)(2y^2 + 3y + 2)$

22. $(0.6d - 2)(0.1d + 3)$

23. $(4t^4 - 9)^2$

24. $2s(s - t)(s + t)$

Factor, if possible.

25. $12a^3b^2c - 3a^2b^2c^2 + 6abc^3$

26. $x^2 + xy + xz + xy + y^2 + zy$

27. $25m^8 - 60m^4n + 36n^2$

28. $21x^4 - 10x^3 - 16x^2$

29. $s^4 - 13s^2 + 36$

30. $144b^2 + 25$

31. $5x^3 + 625$

32. $64a^3 - 125b^6$

33. $(x - y)^2 + 3(x - y) - 10$

34. $6b^2 + bc - 2c^2$

35. $x^2 + 6x + 9 - y^2$

Solve each equation.

36. $5m^2 - 25m = 0$

37. $2x(4x + 3) = 9$

38. $x^3 + 8x^2 - 9x = 0$

39. $\frac{m^2}{18} + \frac{m}{3} + \frac{1}{2} = 0$

40. Solve for v: $v_1v_3 - v_3v = v_1v$

41. PREFORMED CONCRETE A slab of concrete is twice as long as it is wide. The area in which it is placed includes a 1-foot-wide border that has an area of 70 square feet. Find the dimensions of the slab.

42. Explain how a factorization of a polynomial can be verified using a check. Give an example.

Chapter

6 Rational Expressions and Equations

©Bettmann/CORBIS

TLE

Air travel has come a long way since the Wright brothers made the world's first powered airplane flight on December 17, 1903. With Orville at the controls and Wilbur watching on the ground, the *Wright Flyer* traveled 120 feet in 12 seconds. That's an average rate of speed of $\frac{120 \text{ feet}}{12 \text{ seconds}}$, or 10 feet per second. Amazingly, today's passenger jets have cruising speeds of 500 mph, which is about 730 feet per second!

When calculating average rates of speed, we use the formula $r = \frac{d}{t}$, where r is the rate, d is the distance traveled, and t is the time traveled at that rate. This formula is an example of a *rational equation.* To learn more about rational equations, visit *The Learning Equation* on the Internet at http://tle.brookscole.com. (The log-in instructions are in the Preface.) For Chapter 6, the online lesson is:

- *TLE* Lesson 9: Solving Rational Equations

In this chapter, we extend the concept of fractions to include quotients of two polynomials. These algebraic fractions are called rational expressions.

6.1 Rational Functions and Simplifying Rational Expressions

- Rational expressions
- Rational functions
- Graphing rational functions
- Finding the domain of a rational function
- Simplifying rational expressions
- Simplifying rational expressions by factoring out -1

Linear and polynomial functions can be used to model many real-world situations. In this section, we introduce another family of functions known as *rational functions.* Rational functions get their name from the fact that their defining equation contains a *ratio* (fraction) of two polynomials.

RATIONAL EXPRESSIONS

Fractions that are the quotient of two integers are *rational numbers.* Fractions that are the quotient of two polynomials are called *rational expressions.*

Rational Expressions A **rational expression** is an expression of the form $\frac{P}{Q}$, where P and Q are polynomials and Q does not equal 0.

Some examples of rational expressions are

$$\frac{-8y^3z^5}{6y^4z^3}, \quad \frac{3x}{x-7}, \quad \frac{5m+n}{8m+16}, \quad \text{and} \quad \frac{6a^2-13a+6}{3a^2+a-2}$$

Caution Since division by 0 is undefined, the value of a polynomial in the denominator of a rational expression cannot be 0. For example, x cannot be 7 in the rational expression $\frac{3x}{x-7}$, because the value of the denominator would be 0. In the rational expression $\frac{5m+n}{8m+16}$, m cannot be -2, because the value of the denominator would be 0.

RATIONAL FUNCTIONS

Rational expressions often define functions. For example, if the cost of subscribing to an online information network is \$6 per month plus \$1.50 per hour of access time, the average (mean) hourly cost of the service is the total monthly cost, divided by the number of hours of access time used that month:

$$\frac{C}{n} = \frac{1.50n + 6}{n}$$ C is the total monthly cost, and n is the number of hours the service is used that month.

The right-hand side of this equation is a rational expression: the quotient of the binomial $1.50n + 6$ and the monomial n.

The rational function that gives the average hourly cost of using the network for n hours per month can be written

$$f(n) = \frac{1.50n + 6}{n}$$

We are assuming that at least one access call will be made each month, so the function is defined for $n > 0$.

Rational Functions A **rational function** is a function whose equation is defined by a rational expression in one variable, where the value of the polynomial in the denominator is never zero.

EXAMPLE 1

Use the function $f(n) = \frac{1.50n + 6}{n}$ to find the average hourly cost when the network described earlier is used for **a.** 1 hour and **b.** 9 hours.

Solution **a.** To find the average hourly cost for 1 hour of access time, we find $f(1)$:

$$f(1) = \frac{1.50(1) + 6}{1} = 7.5 \quad \text{Input 1 for } n \text{ and simplify.}$$

The average hourly cost for 1 hour of access time is \$7.50.

b. To find the average hourly cost for 9 hours of access time, we find $f(9)$:

$$f(9) = \frac{1.50(9) + 6}{9} = 2.166666666\ldots \quad \text{Input 9 for } n \text{ and simplify.}$$

The average hourly cost for 9 hours of access time is approximately \$2.17.

Self Check 1 Find the average hourly cost when the network is used for **a.** 3 hours and **b.** 100 hours.

GRAPHING RATIONAL FUNCTIONS

To graph the rational function $f(n) = \frac{1.50n + 6}{n}$, we substitute values for n (the inputs) in the equation, compute the corresponding values of $f(n)$ (the outputs), and express the results as ordered pairs. From the evaluations in Example 1 and its Self Check, we know four ordered pairs that satisfy the equation: (1, 7.50), (3, 3.50), (9, 2.17), and (100, 1.56). Those pairs and others are listed in the table below. We then plot the points and draw a smooth curve through them to get the graph.

$$f(n) = \frac{1.50n + 6}{n}$$

n	$f(n)$	
1	7.50	→ (1, 7.50)
2	4.50	→ (2, 4.50)
3	3.50	→ (3, 3.50)
4	3.00	→ (4, 3.00)
5	2.70	→ (5, 2.70)
6	2.50	→ (6, 2.50)
7	2.36	→ (7, 2.36)
8	2.25	→ (8, 2.25)
9	2.17	→ (9, 2.17)
10	2.10	→ (10, 2.10)
100	1.56	→ (100, 1.56)

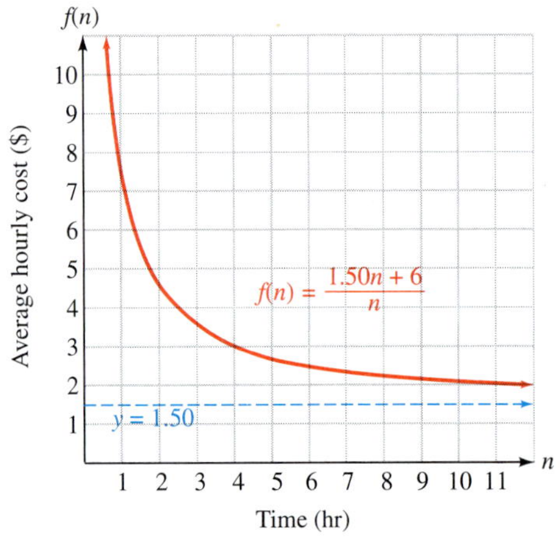

As the access time increases, the graph approaches the line $y = 1.50$, which indicates that the average hourly cost approaches \$1.50 as the hours of use increase.

From the graph, we can see that the average hourly cost decreases as the number of hours of access time increases. Since the cost of each extra hour of access time is \$1.50, the average hourly cost can approach \$1.50 but never drop below it. Thus, the graph of the function approaches the line $y = 1.5$ as n increases. When a graph approaches a line, we call the line an **asymptote.** The line $y = 1.5$ is a **horizontal asymptote** of the graph.

As n gets smaller and approaches 0, the graph approaches the y-axis. The y-axis is a **vertical asymptote** of the graph.

FINDING THE DOMAIN OF A RATIONAL FUNCTION

Since division by 0 is undefined, any values that make the denominator 0 in a rational function must be excluded from the domain of the function.

EXAMPLE 2

Find the domain of $f(x) = \dfrac{3x + 2}{x^2 + x - 6}$.

Solution From the set of real numbers, we must exclude all values of x that make the denominator 0. To find these values, we set $x^2 + x - 6$ equal to 0 and solve for x.

$$x^2 + x - 6 = 0$$

$$(x + 3)(x - 2) = 0 \quad \text{Factor the trinomial.}$$

$$x + 3 = 0 \quad \text{or} \quad x - 2 = 0 \quad \text{Set each factor equal to 0.}$$

$$x = -3 \quad \Big| \quad x = 2 \quad \text{Solve each linear equation.}$$

Thus, the domain of the function is the set of all real numbers except -3 and 2. In interval notation, the domain is $(-\infty, -3) \cup (-3, 2) \cup (2, \infty)$.

The Language of Algebra

Another way that Example 2 could be phrased is: State the *restrictions* on the variable. For $\frac{3x + 2}{x^2 + x - 6}$, we can write $x \neq -3$ and $x \neq 2$.

Self Check 2 Find the domain of $f(x) = \dfrac{x^2 + 1}{x - 2}$.

ACCENT ON TECHNOLOGY: FINDING THE DOMAIN AND RANGE OF A RATIONAL FUNCTION

We can find the domain and range of the function in Example 2 by looking at its graph. If we use window settings of $[-10, 10]$ for x and $[-10, 10]$ for y and graph the function

$$f(x) = \frac{3x + 2}{x^2 + x - 6}$$

we will obtain the graph in figure (a).

From the figure, we can see that

- As x approaches -3 from the left, the values of y decrease, and the graph approaches the vertical line $x = -3$. As x approaches -3 from the right, the values of y increase, and the graph approaches the vertical line $x = -3$.

From the figure, we can also see that

- As x approaches 2 from the left, the values of y decrease, and the graph approaches the vertical line $x = 2$. As x approaches 2 from the right, the values of y increase, and the graph approaches the vertical line $x = 2$.

The lines $x = -3$ and $x = 2$ are vertical asymptotes. Although the vertical lines in the graph appear to be the graphs of $x = -3$ and $x = 2$, they are not. Graphing calculators draw graphs by connecting dots whose x-coordinates are close together. Often when two such points straddle a vertical asymptote and their y-coordinates are far apart, the calculator draws a line between them anyway, producing what appears to be a vertical asymptote. If you set your calculator to dot mode instead of connected mode, the vertical lines will not appear.

From figure (a), we can also see that

- As x increases to the right of 2, the values of y decrease and approach the line $y = 0$.
- As x decreases to the left of -3, the values of y increase and approach the line $y = 0$.

The line $y = 0$ (the x-axis) is a horizontal asymptote. Graphing calculators do not draw lines that appear to be horizontal asymptotes.

From the graph, we can see that every real number x, except -3 and 2, gives a value of y. This observation confirms that the domain of the function is $(-\infty, -3) \cup (-3, 2) \cup (2, \infty)$. We can also see that y can be any value. Thus, the range is $(-\infty, \infty)$.

To find the domain and range of the function $f(x) = \frac{2x+1}{x-1}$, we use a calculator to draw the graph shown in figure (b). From this graph, we can see that the line $x = 1$ is a vertical asymptote and that the line $y = 2$ is a horizontal asymptote. Since x can be any real number except 1, the domain is the interval $(-\infty, 1) \cup (1, \infty)$. Since y can be any value except 2, the range is $(-\infty, 2) \cup (2, \infty)$.

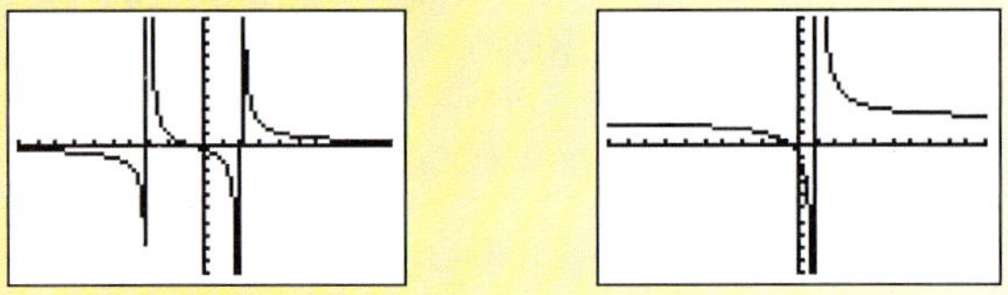

(a) (b)

SIMPLIFYING RATIONAL EXPRESSIONS

When working with rational expressions, we will use some familiar rules from arithmetic.

Properties of Fractions If a, b, c, d, and k represent real numbers, and if there are no divisions by 0, then

1. $\frac{a}{b} = \frac{c}{d}$ if and only if $ad = bc$ **2.** $\frac{a}{1} = a$ and $\frac{a}{a} = 1$

3. $\frac{ak}{bk} = \frac{a}{b} \cdot \frac{k}{k} = \frac{a}{b}$ **4.** $-\frac{a}{b} = \frac{-a}{b} = \frac{a}{-b}$

Property 3 of fractions is used to simplify rational expressions. It is true because any number times 1 is the number.

$$\frac{ak}{bk} = \frac{a}{b} \cdot \frac{k}{k} = \frac{a}{b} \cdot 1 = \frac{a}{b} \quad \text{where } b \neq 0 \text{ and } k \neq 0$$

To streamline this process, we can replace $\frac{k}{k}$ in $\frac{ak}{bk}$ with the equivalent fraction $\frac{1}{1}$.

The Language of Algebra

Property 3 is known as the **fundamental property of fractions.** Stated in another way, it enables us to *divide out* factors that are common to the numerator and denominator of a fraction.

$$\frac{ak}{bk} = \frac{a\overset{1}{\cancel{k}}}{b\underset{1}{\cancel{k}}} = \frac{a}{b} \qquad \frac{k}{k} = \frac{1}{1} = 1.$$

We say that we have simplified $\frac{ak}{bk}$ by *removing a factor equal to 1.*

To **simplify a rational expression** means to write it so that the numerator and denominator have no common factors other than 1.

Simplifying Rational Expressions

1. Factor the numerator and denominator completely to determine their common factors.
2. Remove factors equal to 1 by replacing each pair of factors common to the numerator and denominator with the equivalent fraction $\frac{1}{1}$.

EXAMPLE 3

Simplify: **a.** $\frac{10k}{25k^2}$ and **b.** $\frac{-8y^3z^5}{6y^4z^3}$.

Solution To simplify these expressions, we factor each numerator and denominator and remove common factors.

The Language of Algebra

When a rational expression is simplified, the result is an *equivalent expression.* In Example 3a, this means that $\frac{10k}{25k^2}$ has the same value as $\frac{2}{5k}$ for all values of k, except those that make the denominator 0.

a.

$$\frac{10k}{25k^2} = \frac{5 \cdot 2 \cdot k}{5 \cdot 5 \cdot k \cdot k}$$

$$= \frac{\overset{1}{\cancel{5}} \cdot 2 \cdot \overset{1}{\cancel{k}}}{\underset{1}{\cancel{5}} \cdot 5 \cdot \underset{1}{\cancel{k}} \cdot k}$$ Replace $\frac{5}{5}$ and $\frac{k}{k}$ with the equivalent fraction $\frac{1}{1}$. This removes the factor $\frac{5 \cdot k}{5 \cdot k} = 1$.

$$= \frac{2}{5k}$$ Do the multiplications in the numerator and the denominator.

b.

$$\frac{-8y^3z^5}{6y^4z^3} = \frac{-2 \cdot 4 \cdot y \cdot y \cdot y \cdot z \cdot z \cdot z \cdot z \cdot z}{2 \cdot 3 \cdot y \cdot y \cdot y \cdot y \cdot z \cdot z \cdot z}$$

$$= \frac{-\overset{1}{\cancel{2}} \cdot 4 \cdot \overset{1}{\cancel{y}} \cdot \overset{1}{\cancel{y}} \cdot \overset{1}{\cancel{y}} \cdot \overset{1}{\cancel{z}} \cdot \overset{1}{\cancel{z}} \cdot \overset{1}{\cancel{z}} \cdot z \cdot z}{\underset{1}{\cancel{2}} \cdot 3 \cdot \underset{1}{\cancel{y}} \cdot \underset{1}{\cancel{y}} \cdot \underset{1}{\cancel{y}} \cdot y \cdot \underset{1}{\cancel{z}} \cdot \underset{1}{\cancel{z}} \cdot \underset{1}{\cancel{z}}}$$ $\frac{2}{2} = 1$, $\frac{y}{y} = 1$, and $\frac{z}{z} = 1$.

$$= -\frac{4z^2}{3y}$$

Self Check 3 Simplify: $\frac{-12a^4b^2}{-3ab^4}$.

The fractions in Example 3 can also be simplified using the rules of exponents:

$$\begin{aligned}\frac{10k}{25k^2} &= \frac{5 \cdot 2}{5 \cdot 5}k^{1-2} \\ &= \frac{2}{5} \cdot k^{-1} \\ &= \frac{2}{5} \cdot \frac{1}{k} \\ &= \frac{2}{5k}\end{aligned} \qquad \begin{aligned}\frac{-8y^3z^5}{6y^4z^3} &= \frac{-2 \cdot 4}{2 \cdot 3}y^{3-4}z^{5-3} \\ &= \frac{-4}{3} \cdot y^{-1}z^2 \\ &= -\frac{4}{3} \cdot \frac{1}{y} \cdot \frac{z^2}{1} \\ &= -\frac{4z^2}{3y}\end{aligned}$$

EXAMPLE 4

Simplify: $\frac{x^2 - 16}{x + 4}$.

Solution

$$\frac{x^2 - 16}{x + 4} = \frac{\overset{1}{\cancel{(x + 4)}}(x - 4)}{\underset{1}{\cancel{(x + 4)}}} \quad \text{Factor the difference of two squares. } \frac{x+4}{x+4} = 1.$$

$$= \frac{x - 4}{1}$$

$$= x - 4$$

Self Check 4 Simplify: $\frac{x^2 - 9}{x - 3}$.

ACCENT ON TECHNOLOGY: CHECKING AN ALGEBRAIC SIMPLIFICATION

After simplifying an expression, we can use a scientific calculator to check the answer. One way to check whether $\frac{x^2 - 16}{x + 4} = x - 4$ is correct in Example 4 is to evaluate $\frac{x^2 - 16}{x + 4}$ and $x - 4$ for a value of x (say, 3). The expressions should give identical results.

(3 x^2 − 16) ÷ (3 + 4) =   −1

Since $x - 4$ is -1 when $x = 3$, the results of the evaluations are the same. Evaluate the expressions for several other values of x. If the results differ for any given value, the original expression was not simplified correctly.

A graphing calculator can also be used to show that the simplification in Example 4 is correct. To do this, we enter the functions $f(x) = \frac{x^2 - 16}{x + 4}$ and $g(x) = x - 4$ as Y_1 and Y_2, respectively. See figure (a). Then select the TABLE feature. Reading across the table, the values of Y_1 and Y_2 should be the same for each value of x as shown in figure (b).

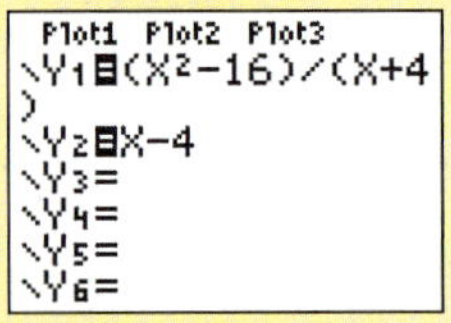

(a)

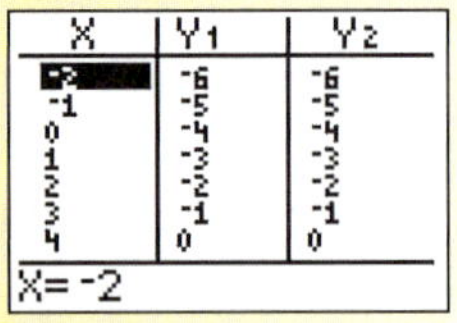

(b)

To use a third method, we can compare the graphs of $f(x) = \frac{x^2 - 16}{x + 4}$, shown in figure (c), and $g(x) = x - 4$, shown in figure (d). Except for the point where $x = -4$, the graphs are the same. The point where $x = -4$ is excluded from the graph of $f(x) = \frac{x^2 - 16}{x + 4}$, because -4 is not in the domain of f. However, using a standard window setting, graphing calculators do not show that this point is excluded. The point where $x = -4$ is included in the graph of $g(x) = x - 4$, because -4 is in the domain of g.

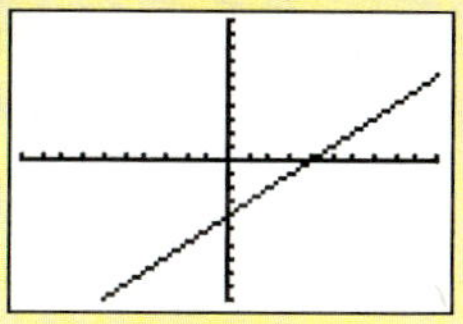

(c)

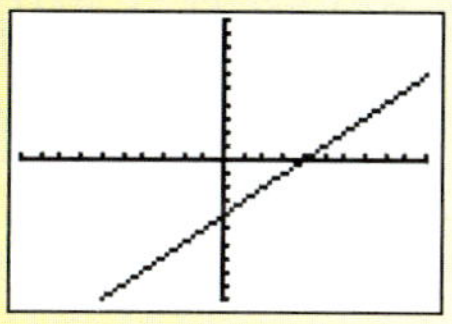

(d)

EXAMPLE 5

Simplify: $\dfrac{6a^2 - 13a + 6}{3a^2 + a - 2}$.

Solution We factor the trinomials in the numerator and the denominator and then remove the common factor, $3a - 2$.

$$\frac{6a^2 - 13a + 6}{3a^2 + a - 2} = \frac{\overset{1}{\cancel{(3a - 2)}}(2a - 3)}{\underset{1}{\cancel{(3a - 2)}}(a + 1)} \qquad \frac{3a - 2}{3a - 2} = 1.$$

$$= \frac{2a - 3}{a + 1} \qquad \text{This expression does not simplify further.}$$

Caution Do not remove the a's in $\frac{2a - 3}{a + 1}$. The a in the numerator is a factor of the first term only, not a factor of the entire numerator. Likewise, the a in the denominator is a factor of the first term only, not a factor of the entire denominator.

When simplifying rational expressions, we can only remove factors common to the entire numerator and denominator. It is incorrect to remove terms common to the numerator and denominator.

$$\frac{\overset{1}{\cancel{x}} + 1}{\underset{1}{\cancel{x}}} \qquad \frac{a^2 - 3a + \overset{1}{\cancel{2}}}{a + \underset{1}{\cancel{2}}} \qquad \frac{\overset{1}{\cancel{y^2}} - 36}{\underset{1}{\cancel{y^2}} - y - 7}$$

x is a term of $x + 1$. 2 is a term of $a^2 - 3a + 2$ and a term of $a + 2$. y^2 is a term of $y^2 - 36$ and a term of $y^2 - y - 7$.

Self Check 5 Simplify: $\dfrac{2b^2 + 7b - 15}{2b^2 + 13b + 15}$.

We will encounter fractions that are already in simplified form. For example, to attempt to simplify

$$\frac{x^2 + xa + 2x + 2a}{x^2 + x - 6}$$

we factor the numerator and denominator:

$$\frac{x^2 + xa + 2x + 2a}{x^2 + x - 6} = \frac{x(x + a) + 2(x + a)}{(x - 2)(x + 3)} = \frac{(x + a)(x + 2)}{(x - 2)(x + 3)}$$

Since there are no common factors in the numerator and denominator, the fraction is in *lowest terms.* It cannot be simplified.

SIMPLIFYING RATIONAL EXPRESSIONS BY FACTORING OUT −1

If the terms of two polynomials are the same, except that they are opposite in sign, the polynomials are **opposites.** For example, $b - a$ and $a - b$ are opposites.

To simplify $\frac{b-a}{a-b}$, the quotient of opposites, we factor -1 from the numerator and remove any factors common to both the numerator and the denominator:

Success Tip

When a difference is reversed, the original binomial and the resulting binomial are opposites. Here are some pairs of opposites:

$b - a$	and	$a - b$
$y - 6$	and	$6 - y$
$x^2 - 4$	and	$4 - x^2$

$$\frac{b - a}{a - b} = \frac{-a + b}{a - b}$$ Rewrite the numerator.

$$= \frac{-1\overset{1}{\cancel{(a - b)}}}{\underset{1}{\cancel{(a - b)}}}$$ Factor out -1 from each term in the numerator. $\frac{a-b}{a-b} = 1$.

$$= \frac{-1}{1}$$

$$= -1$$

In general, we have the following principle.

The Quotient of Opposites The quotient of any nonzero polynomial and its opposite is -1.

EXAMPLE 6

Simplify: $\dfrac{3x^2 - 10xy - 8y^2}{4y^2 - xy}$.

Solution We factor the numerator and denominator. Because $x - 4y$ and $4y - x$ are opposites, their quotient is -1.

Caution

A − symbol preceding a fraction may be applied to the numerator or to the denominator, but not to both. For example,

$$-\frac{3x + 2y}{y} \neq \frac{-3x - 2y}{-y}$$

$$\frac{3x^2 - 10xy - 8y^2}{4y^2 - xy} = \frac{(3x + 2y)\overset{-1}{\cancel{(x - 4y)}}}{y\underset{1}{\cancel{(4y - x)}}}$$ Since $x - 4y$ and $4y - x$ are opposites, simplify by replacing $\frac{x - 4y}{4y - x}$ with the equivalent fraction $\frac{-1}{1} = -1$.

$$= \frac{-(3x + 2y)}{y}$$

$$= \frac{-3x - 2y}{y}$$

This result can also be written as $\dfrac{-(3x + 2y)}{y}$ or $-\dfrac{3x + 2y}{y}$.

Self Check 6 Simplify: $\dfrac{2a^2 - 3ab - 9b^2}{3b^2 - ab}$.

Answers to Self Checks **1. a.** \$3.50, **b.** \$1.56 **2.** $(-\infty, 2) \cup (2, \infty)$ **3.** $\dfrac{4a^3}{b^2}$ **4.** $x + 3$ **5.** $\dfrac{2b - 3}{2b + 3}$ **6.** $-\dfrac{2a + 3b}{b}$ or $\dfrac{-2a - 3b}{b}$

6.1 STUDY SET

VOCABULARY Fill in the blanks.

1. A quotient of two polynomials, such as $\frac{x^2 + x}{x^2 - 3x}$, is called a ________ expression.
2. In the rational expression $\frac{(x + 2)(3x - 1)}{(x + 2)(4x + 2)}$, $x + 2$ is a common _______ of the numerator and the denominator.
3. To ________ a rational expression, we remove factors common to the numerator and denominator.
4. Because of the division by 0, the expression $\frac{8}{0}$ is __________.
5. The binomials $x - 15$ and $15 - x$ are called __________, because their terms are the same, except that they are opposite in sign.
6. In Exercise 7, the graph of the function approaches the positive x-axis. When a graph approaches a line, we call the line an __________.

CONCEPTS

7. The graph of rational function f for $x > 0$ is shown in the illustration. Find each of the following.
   a. $f(1)$ **b.** $f(2)$
   c. The value(s) of x for which $f(x) = 4$
   d. The domain and range of f

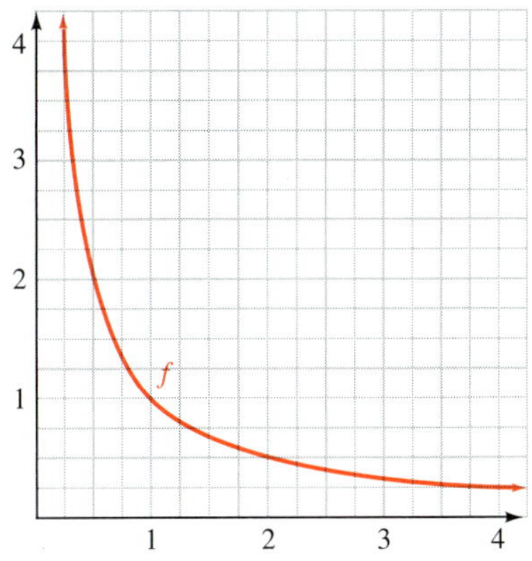

8. Show that $\frac{x - y}{y - x} = -1$ by factoring out -1 from each term in the numerator.

9. Simplify each rational expression.
   a. $\frac{x + 8}{x + 8}$ **b.** $\frac{x + 8}{8 + x}$
   c. $\frac{x - 8}{x - 8}$ **d.** $\frac{x - 8}{8 - x}$

10. Simplify each rational expression, if possible.
   a. $\frac{x + 8}{x}$ **b.** $\frac{x + 8}{8}$
   c. $\frac{a^3 + 8}{2}$ **d.** $\frac{x^2 + 5x + 6}{x^2 + x - 12}$

In Exercises 11–12, refer to the following graphs. Each graph shows the average cost to manufacture a certain item for a given number of units produced.

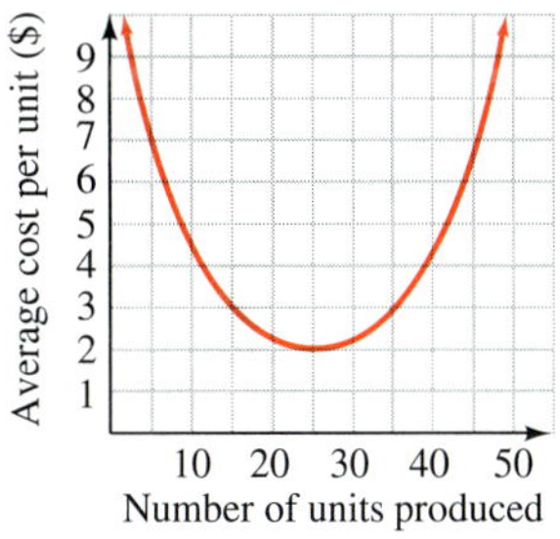

Item 1

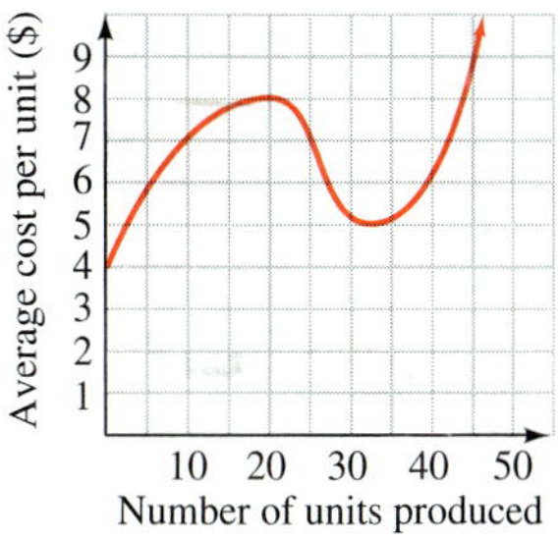

Item 2

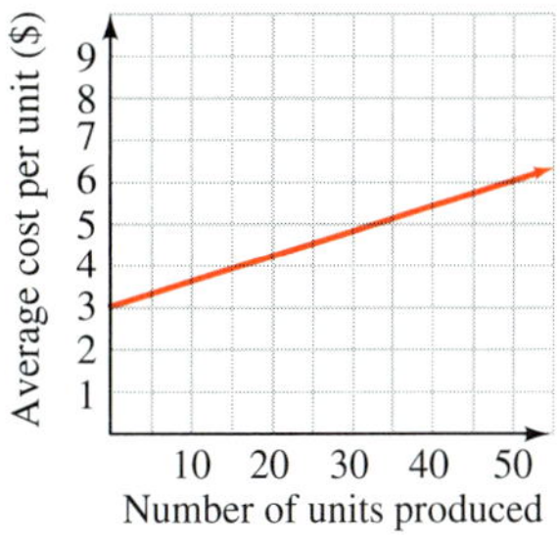

Item 3

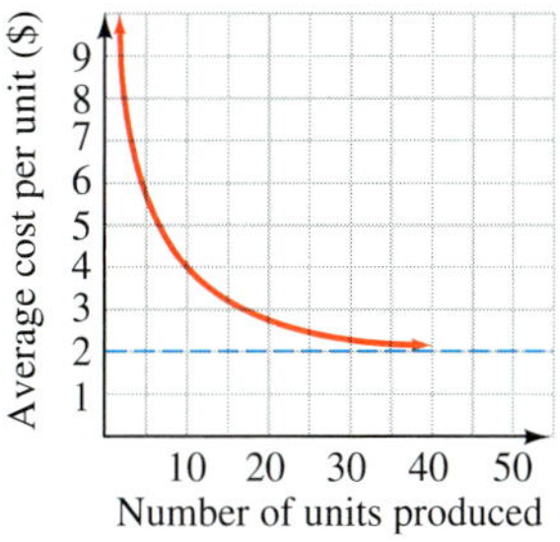

Item 4

11. MANUFACTURING For each graph on the previous page, briefly describe how the average cost per unit changes as the number of units produced increases.

12. Which graph on the previous page is best described as the graph of a

a. linear function?

b. quadratic function?

c. rational function?

d. polynomial function?

NOTATION

13. In the following table, the answers to five homework problems are compared to the answers in the back of the book. Are the answers equivalent?

Answer	Book's answer	Equivalent?
$\frac{-3}{x+3}$	$-\frac{3}{x+3}$	
$\frac{-x+4}{6x+1}$	$\frac{-(x-4)}{6x+1}$	
$\frac{x+7}{(x-4)(x+2)}$	$\frac{x+7}{(x+2)(x-4)}$	
$-\frac{x-4}{x+4}$	$\frac{4-x}{x+4}$	
$\frac{a-3b}{2b-a}$	$\frac{3b-a}{a-2b}$	

14. a. In $\frac{(x+5)\overset{1}{\cancel{(x-5)}}}{x\underset{1}{\cancel{(x-5)}}}$, what do the slashes show?

b. In $\frac{(x-3)\overset{-1}{\cancel{(x-7)}}}{(x+3)\underset{1}{\cancel{(7-x)}}}$, what do the slashes show?

PRACTICE

Complete the table of values for each rational function (round to the nearest hundredth when applicable). Then graph it. Each function is defined for $x > 0$. Label the horizontal asymptote.

15. $f(x) = \frac{6}{x}$

x	$f(x)$
1	6
2	3
4	3/2
6	1
8	3/4
10	3/5
12	1/2

16. $f(x) = \frac{12}{x}$

x	$f(x)$
1	
4	
8	
12	
16	
20	
24	

17. $f(x) = \frac{x+2}{x}$

x	$f(x)$
1	
2	
4	
6	
8	
10	
12	

18. $f(x) = \frac{2x+4}{x}$

x	$f(x)$
1	
4	
8	
12	
16	
20	
24	

Find the domain of each rational function. Use interval notation.

19. $f(x) = \frac{2}{x}$

20. $f(x) = \frac{8}{x-1}$

21. $f(x) = \frac{2x}{x+2}$

22. $f(x) = \frac{2x+1}{x^2-2x}$

23. $f(x) = \frac{3x-1}{x-x^2}$

24. $f(x) = \frac{x^2+36}{x^2-36}$

25. $f(x) = \dfrac{x^2 + 3x + 2}{x^2 - x - 56}$

26. $f(x) = \dfrac{2x^2 - 3x - 2}{x^2 + 2x - 24}$

Simplify each rational expression when possible.

27. $\dfrac{12}{18}$

28. $\dfrac{25}{55}$

29. $-\dfrac{112}{36}$

30. $-\dfrac{49}{21}$

31. $\dfrac{12x^3}{3x}$

32. $-\dfrac{15a^2}{25a^3}$

33. $\dfrac{-24x^3y^4}{18x^4y^3}$

34. $\dfrac{15a^5b^4}{21b^3c^2}$

35. $-\dfrac{11x(x-y)^3}{22(x-y)^4}$

36. $\dfrac{x(x-2)^2}{(x-2)^3}$

37. $\dfrac{(a-b)(d-c)}{(c-d)(a-b)}$

38. $\dfrac{(p+q)(p-r)}{(r-p)(p+q)}$

39. $\dfrac{y+x}{x^2-y^2}$

40. $\dfrac{x-y}{x^2-y^2}$

41. $\dfrac{5x-10}{x^2-4x+4}$

42. $\dfrac{y-xy}{xy-x}$

43. $\dfrac{12-3x^2}{x^2-x-2}$

44. $\dfrac{x^2+2x-15}{25-x^2}$

45. $\dfrac{x^2+y^2}{x+y}$

46. $\dfrac{3x+6y}{2y+x}$

47. $\dfrac{x^3+8}{x^2-2x+4}$

48. $\dfrac{x^2+3x+9}{x^3-27}$

49. $\dfrac{x^2+2x+1}{x^2+4x+3}$

50. $\dfrac{6x^2+x-2}{8x^2+2x-3}$

51. $\dfrac{sx+4s-3x-12}{sx+4s+6x+24}$

52. $\dfrac{ax+by+ay+bx}{a^2-b^2}$

53. $\dfrac{4x^2+24x+32}{16x^2+8x-48}$

54. $\dfrac{a^2-4}{a^3-8}$

55. $\dfrac{3x^2-3y^2}{x^2+2y+2x+yx}$

56. $\dfrac{x^2+x-30}{x^2-x-20}$

57. $\dfrac{4x^2+8x+3}{6+x-2x^2}$

58. $\dfrac{6x^2+13x+6}{6-5x-6x^2}$

59. $\dfrac{a^3+27}{4a^2-36}$

60. $\dfrac{a-b}{b^2-a^2}$

61. $\dfrac{2x^2-3x-9}{2x^2+3x-9}$

62. $\dfrac{6x^2-7x-5}{2x^2+5x+2}$

63. $\dfrac{(m+n)^3}{m^2+2mn+n^2}$

64. $\dfrac{x^3-27}{3x^2-8x-3}$

65. $\dfrac{9g-gx+18-2x}{gx-9g+2x-18}$

66. $\dfrac{4ac-4ad+c-d}{4ad-4ac+d-c}$

67. $\dfrac{m^3-mn^2}{mn^2+m^2n-2m^3}$

68. $\dfrac{p^3+p^2q-2pq^2}{pq^2+p^2q-2p^3}$

69. $\dfrac{x^4-y^4}{(x^2+2xy+y^2)(x^2+y^2)}$

70. $\dfrac{(x^2-1)(x+1)}{(x^2-2x+1)^2}$

71. $\dfrac{10+10(t-3)}{3(t-3)+3}$

72. $\dfrac{6(m+3)-6}{7-7(m+3)}$

73. $\dfrac{6xy - 4x - 9y + 6}{6y^2 - 13y + 6}$

74. $\dfrac{x^2 + 2xy}{x + 2y + x^2 - 4y^2}$

75. $\dfrac{(2x^2 + 3xy + y^2)(3a + b)}{(x + y)(2xy + 2bx + y^2 + by)}$

76. $\dfrac{(x - 1)(6ax + 9x + 4a + 6)}{(3x + 2)(2ax - 2a + 3x - 3)}$

77. $\dfrac{(x^2 + 2x + 1)(x^2 - 2x + 1)}{(x^2 - 1)^2}$

78. $\dfrac{2x^2 + 2x - 12}{x^3 + 3x^2 - 4x - 12}$

Use a graphing calculator to graph each rational function. From the graph, determine its domain and range.

79. $f(x) = \dfrac{x}{x - 2}$

80. $f(x) = \dfrac{x + 2}{x}$

81. $f(x) = \dfrac{x + 1}{x^2 - 4}$

82. $f(x) = \dfrac{x - 2}{x^2 - 3x - 4}$

APPLICATIONS

83. ENVIRONMENTAL CLEANUP Suppose the cost (in dollars) of removing $p\%$ of the pollution in a river is given by the rational function

$$f(p) = \frac{50{,}000p}{100 - p} \quad \text{where } 0 \le p < 100$$

Find the cost of removing each percent of pollution.

a. 50%
b. 80%

84. DIRECTORY COSTS The average (mean) cost for a service club to publish a directory of its members is given by the rational function

$$f(x) = \frac{1.25x + 700}{x}$$

where x is the number of directories printed. Find the average cost per directory if

a. 500 directories are printed.
b. 2,000 directories are printed.

85. UTILITY COSTS An electric company charges \$7.50 per month plus 9¢ for each kilowatt hour (kwh) of electricity used.

a. Find a linear function that gives the total cost of n kwh of electricity.
b. Find a rational function that gives the average cost per kwh when using n kwh.
c. Find the average cost per kwh when 775 kwh are used.

86. SCHEDULING WORK CREWS The rational function

$$f(t) = \frac{t^2 + 2t}{2t + 2}$$

gives the number of days it would take two construction crews, working together, to frame a house that crew 1 (working alone) could complete in t days and crew 2 (working alone) could complete in $t + 2$ days.

a. If crew 1 could frame a certain house in 15 days, how long would it take both crews working together?
b. If crew 2 could frame a certain house in 20 days, how long would it take both crews working together?

87. FILLING A POOL The rational function

$$f(t) = \frac{t^2 + 3t}{2t + 3}$$

gives the number of hours it would take two pipes, working together, to fill a pool that the larger pipe (working alone) could fill in t hours and the smaller pipe (working alone) could fill in $t + 3$ hours.

a. If the smaller pipe could fill a pool in 7 hours, how long would it take both pipes to fill the pool?
b. If the larger pipe could fill a pool in 8 hours, how long would it take both pipes to fill the pool?

88. RETENTION STUDY After learning a list of words, two subjects were tested over a 28-day period to see what percent of the list they remembered. In

both cases, their percent recall could be modeled by rational functions, as shown in the illustration.

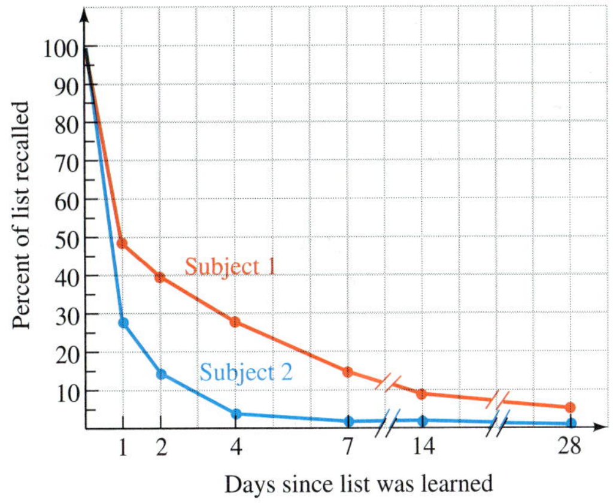

a. Use the graphs to complete the table.

Days since learning	0	1	2	4	7	14	28
% recall—subject 1							
% recall—subject 2							

b. After 28 days, which subject had the better recall?

WRITING

89. A student simplified $\frac{6x^2 - 7x - 5}{2x^2 + 5x + 2}$ and obtained $\frac{3x - 5}{x - 2}$. As a check, she graphed $Y_1 = \frac{6x^2 - 7x - 5}{2x^2 + 5x + 2}$ and $Y_2 = \frac{3x - 5}{x - 2}$. What conclusion can be drawn from the graphs? Explain your answer.

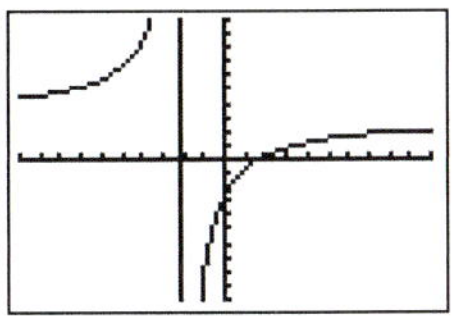

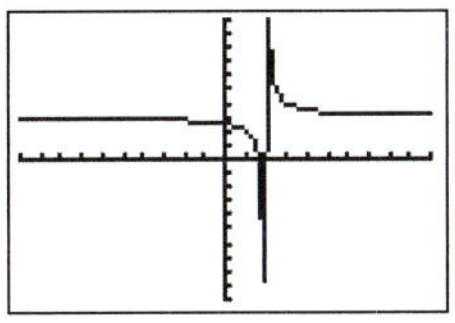

90. Simplify: $\frac{6x^2 + x - 2}{8x^2 + 2x - 3}$. Then explain how the table of values for $Y_1 = \frac{6x^2 + x - 2}{8x^2 + 2x - 3}$ and $Y_2 = \frac{3x + 2}{4x + 3}$ shown in the illustration can be used to check your result.

X	Y_1	Y_2
-2	.8	.8
-1	1	1
0	.66667	.66667
1	.71429	.71429
2	.72727	.72727
3	.73333	.73333
4	.73684	.73684

X= -2

REVIEW Perform each operation.

91. $(a^2 - 4a - 3)(a - 2)$

92. $(3c^2 + 5c) + (7 - c^2 - 5c)$

93. $-3mn^2(m^3 - 7mn - 2m^2)$

94. $(4u^2 + z^2 - 3u^2z^2) - (u^3 + 3z^2 - 3u^2z^2)$

CHALLENGE PROBLEMS Simplify each expression.

95. $\dfrac{x^{32} - 1}{x^{16} - 1}$

96. $\dfrac{20m^2(m^2 - 1) - 47m(1 - m^2) + 24(m^2 - 1)}{4m^2 - m - 3}$

97. $\dfrac{a^6 - 64}{(a^2 + 2a + 4)(a^2 - 2a + 4)}$

98. $\dfrac{(p + q)^3 + 64}{(p + q)^2 - 16}$

6.2 Multiplying and Dividing Rational Expressions

- Multiplying rational expressions
- Finding powers of rational expressions
- Dividing rational expressions
- Mixed operations

In this section, we review the rules for multiplying and dividing arithmetic fractions—fractions whose numerators and denominators are integers. Then we use these rules, in combination with the simplification skills learned in Section 6.1, to multiply and divide rational expressions.

MULTIPLYING RATIONAL EXPRESSIONS

Recall that to multiply fractions, we multiply the numerators and multiply the denominators. For example,

$$\frac{3}{5} \cdot \frac{2}{7} = \frac{3 \cdot 2}{5 \cdot 7} = \frac{6}{35}$$

$$\frac{4}{7} \cdot \frac{5}{8} = \frac{4 \cdot 5}{7 \cdot 8} = \frac{\overset{1}{\cancel{2}} \cdot \overset{1}{\cancel{2}} \cdot 5}{7 \cdot \underset{1}{\cancel{2}} \cdot \underset{1}{\cancel{2}} \cdot 2} = \frac{5}{14}$$ Simplify.

We use the same procedure to multiply rational expressions.

Multiplying Rational Expressions

Let A, B, C, and D represent polynomials, where B and D are not 0,

$$\frac{A}{B} \cdot \frac{C}{D} = \frac{AC}{BD}$$

Then simplify, if possible.

EXAMPLE 1

Multiply: $\frac{7x^2y}{t} \cdot \frac{xy^3}{t^3}$.

Solution

$$\frac{7x^2y}{t} \cdot \frac{xy^3}{t^3} = \frac{7x^2y \cdot xy^3}{tt^3}$$ Multiply the numerators and multiply the denominators.

$$= \frac{7x^2x \cdot yy^3}{t^4}$$

$$= \frac{7x^3y^4}{t^4}$$ Use the rules for exponents to simplify the numerator.

Self Check 1 Multiply: $\frac{a^3b^8}{m} \cdot \frac{3a^4b}{m^5}$.

EXAMPLE 2

Multiply: $\frac{x^2 - 6x + 9}{20x} \cdot \frac{5x^2}{x - 3}$.

Solution We multiply the numerators and multiply the denominators and then factor to simplify the resulting fraction.

$$\frac{x^2 - 6x + 9}{20x} \cdot \frac{5x^2}{x - 3} = \frac{(x^2 - 6x + 9)5x^2}{20x(x - 3)}$$ Multiply the numerators and multiply the denominators.

$$= \frac{(x - 3)(x - 3)5xx}{4 \cdot 5 \cdot x(x - 3)}$$ Factor the numerator. In the denominator factor 20 as $4 \cdot 5$.

$$= \frac{\overset{1}{\cancel{(x - 3)}}(x - 3)\overset{1}{\cancel{5}}\overset{1}{\cancel{x}}x}{4 \cdot \underset{1}{\cancel{5}} \cdot \underset{1}{\cancel{x}}\underset{1}{\cancel{(x - 3)}}}$$ Simplify.

$$= \frac{x(x - 3)}{4}$$

Caution

When multiplying rational expressions, always write the result in simplest form by removing any factors common to the numerator and denominator.

Self Check 2 Multiply: $\frac{a^2 + 6a + 9}{18a} \cdot \frac{3a^3}{a + 3}$.

ACCENT ON TECHNOLOGY: CHECKING AN ALGEBRAIC SIMPLIFICATION

We can check the simplification in Example 2 by graphing the functions $f(x) = \left(\frac{x^2 - 6x + 9}{20x}\right)\left(\frac{5x^2}{x - 3}\right)$, shown in figure (a), and $g(x) = \frac{x(x - 3)}{4}$, shown in figure (b), and observing that the graphs are the same, except that 0 and 3 are not included in the domain of the first function.

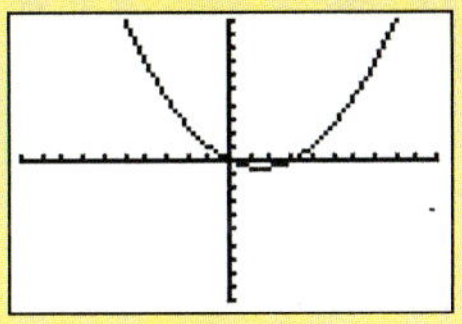

(a)

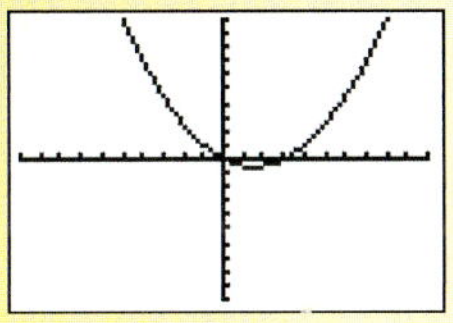

(b)

The split-screen G-T (graph, table) mode can also be used to check a simplification. If we enter $Y_3 = Y_1 - Y_2$, use the cursor to highlight the = sign as shown in figure (c), and then press GRAPH, we get the display shown in figure (d). The zeros under the Y_3 column indicate that the value of $\left(\frac{x^2 - 6x + 9}{20x}\right)\left(\frac{5x^2}{x - 3}\right)$ and the value of $\frac{x(x - 3)}{4}$ are the same for different values of x. (The error message is given because when $x = 0$, $\left(\frac{x^2 - 6x + 9}{20x}\right)\left(\frac{5x^2}{x - 3}\right)$ is undefined.)

The graph of $Y_3 = Y_1 - Y_2$ is difficult to see because it lies on the x-axis. The graph indicates that for all x-values (except those that make the fractions undefined), $Y_3 = 0$, or more specifically, $\left(\frac{x^2 - 6x + 9}{20x}\right)\left(\frac{5x^2}{x - 3}\right) = \frac{x(x - 3)}{4}$.

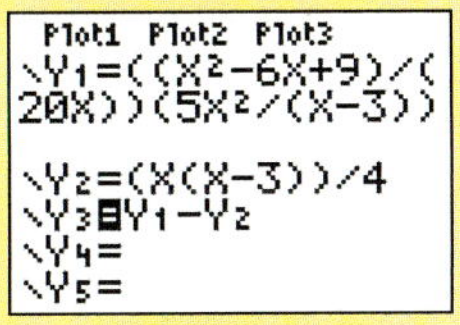

(c)

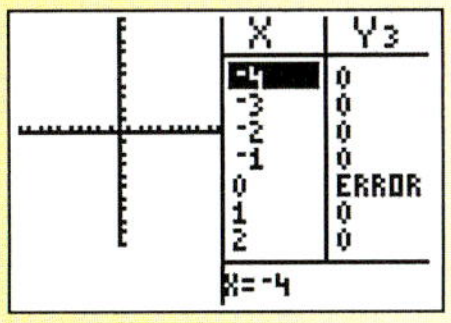

(d)

EXAMPLE 3

Multiply: $\frac{x^2 - x - 6}{x^2 - 4} \cdot \frac{x^2 + x - 6}{x^2 - 9}$.

Solution

$$\frac{x^2 - x - 6}{x^2 - 4} \cdot \frac{x^2 + x - 6}{x^2 - 9}$$

$$= \frac{(x^2 - x - 6)(x^2 + x - 6)}{(x^2 - 4)(x^2 - 9)}$$ Multiply the numerators and multiply the denominators.

$$= \frac{(x - 3)(x + 2)(x + 3)(x - 2)}{(x + 2)(x - 2)(x + 3)(x - 3)}$$ Factor the polynomials.

$$= \frac{\overset{1}{\cancel{(x - 3)}}\overset{1}{\cancel{(x + 2)}}\overset{1}{\cancel{(x + 3)}}\overset{1}{\cancel{(x - 2)}}}{\underset{1}{\cancel{(x + 2)}}\underset{1}{\cancel{(x - 2)}}\underset{1}{\cancel{(x + 3)}}\underset{1}{\cancel{(x - 3)}}}$$ Simplify.

$$= 1$$

Self Check 3 Multiply: $\dfrac{a^2 + a - 56}{a^2 - 49} \cdot \dfrac{a^2 - a - 56}{a^2 - 64}$.

Caution Note that when all factors divide out, the result is 1 and not 0.

EXAMPLE 4

Multiply: $\dfrac{6x^2 + 5xy - 4y^2}{2x^2 + 5xy + 3y^2} \cdot \dfrac{8x^2 + 6xy - 9y^2}{12x^2 + 7xy - 12y^2}$

Solution

$$\frac{6x^2 + 5xy - 4y^2}{2x^2 + 5xy + 3y^2} \cdot \frac{8x^2 + 6xy - 9y^2}{12x^2 + 7xy - 12y^2}$$

$$= \frac{(6x^2 + 5xy - 4y^2)(8x^2 + 6xy - 9y^2)}{(2x^2 + 5xy + 3y^2)(12x^2 + 7xy - 12y^2)}$$ Multiply the numerators and multiply the denominators.

$$= \frac{(3x + 4y)(2x - y)(4x - 3y)(2x + 3y)}{(2x + 3y)(x + y)(3x + 4y)(4x - 3y)}$$ Factor the polynomials.

$$= \frac{\overset{1}{\cancel{(3x + 4y)}}(2x - y)\overset{1}{\cancel{(4x - 3y)}}\overset{1}{\cancel{(2x + 3y)}}}{\underset{1}{\cancel{(2x + 3y)}}(x + y)\underset{1}{\cancel{(3x + 4y)}}\underset{1}{\cancel{(4x - 3y)}}}$$ Simplify.

$$= \frac{2x - y}{x + y}$$

Self Check 4 Multiply: $\dfrac{2a^2 + 5ab - 12b^2}{2a^2 + 11ab + 12b^2} \cdot \dfrac{2a^2 - 3ab - 9b^2}{2a^2 - ab - 3b^2}$

EXAMPLE 5

Multiply: $(2x - x^2) \cdot \dfrac{x}{x^2 - xb - 2x + 2b}$.

Solution

$$(2x - x^2) \cdot \frac{x}{x^2 - xb - 2x + 2b}$$

$$= \frac{2x - x^2}{1} \cdot \frac{x}{x^2 - xb - 2x + 2b}$$ Write $2x - x^2$ as the fraction $\dfrac{2x - x^2}{1}$.

$$= \frac{(2x - x^2)x}{1(x^2 - xb - 2x + 2b)}$$ Multiply the fractions.

$$= \frac{x\overset{-1}{\cancel{(2 - x)}}x}{1(x - b)\underset{1}{\cancel{(x - 2)}}}$$ Factor out x in the numerator. In the denominator, factor by grouping. Recall that the quotient of any nonzero quantity and its opposite is -1: $\frac{2 - x}{x - 2} = -1$.

$$= \frac{-x^2}{x - b}$$

Success Tip

In Examples 2–5, we would obtain the same answer if we had factored the numerators and denominators first and simplified before we multiplied.

Since the − sign can be written in front of the fraction, this result can be written as

$$-\frac{x^2}{x - b}$$

Self Check 5 Multiply: $\dfrac{x^2 + 5x + 6}{4x + 8 - x^2 - 2x}(x^2 - 4x)$.

FINDING POWERS OF RATIONAL EXPRESSIONS

EXAMPLE 6 Find: $\left(\dfrac{x^2 + x - 1}{2x + 3}\right)^2$.

Solution To square the expression, we write it as a factor twice and perform the multiplication.

$$\begin{aligned}\left(\frac{x^2 + x - 1}{2x + 3}\right)^2 &= \left(\frac{x^2 + x - 1}{2x + 3}\right)\left(\frac{x^2 + x - 1}{2x + 3}\right)\\ &= \frac{(x^2 + x - 1)(x^2 + x - 1)}{(2x + 3)(2x + 3)}\\ &= \frac{x^4 + 2x^3 - x^2 - 2x + 1}{4x^2 + 12x + 9}\end{aligned}$$

Self Check 6 Find: $\left(\dfrac{x + 5}{x^2 - 6x}\right)^2$.

DIVIDING RATIONAL EXPRESSIONS

Recall that one number is called the **reciprocal** of another if their product is 1. To find the reciprocal of a fraction, we invert its numerator and denominator. We have seen that to divide fractions, we multiply the first fraction by the reciprocal of the second fraction.

$$\begin{aligned}\frac{3}{5} \div \frac{2}{7} &= \frac{3}{5} \cdot \frac{7}{2}\\ &= \frac{3 \cdot 7}{5 \cdot 2}\\ &= \frac{21}{10}\end{aligned}$$

$$\begin{aligned}\frac{4}{7} \div \frac{2}{21} &= \frac{4}{7} \cdot \frac{21}{2}\\ &= \frac{4 \cdot 21}{7 \cdot 2}\\ &= \frac{\overset{1}{\cancel{2}} \cdot 2 \cdot 3 \cdot \overset{1}{\cancel{7}}}{\underset{1}{\cancel{7}} \cdot \underset{1}{\cancel{2}}} \quad \text{Simplify.}\\ &= 6\end{aligned}$$

We use the same procedure to divide rational expressions.

Dividing Rational Expressions

Let A, B, C, and D represent polynomials, where B, C, and D are not 0,

$$\frac{A}{B} \div \frac{C}{D} = \frac{A}{B} \cdot \frac{D}{C} = \frac{AD}{BC}$$

Then simplify, if possible.

EXAMPLE 7

Divide: $\dfrac{6x^2}{y^3z^2} \div \dfrac{x^2}{yz^3}$.

Solution

$$\frac{6x^2}{y^3z^2} \div \frac{x^2}{yz^3} = \frac{6x^2}{y^3z^2} \cdot \frac{yz^3}{x^2}$$ Multiply the first rational expression by the reciprocal of the second.

$$= \frac{6x^2yz^3}{x^2y^3z^2}$$ Multiply the numerators and the denominators.

$$= 6x^{2-2}y^{1-3}z^{3-2}$$ To divide exponential expressions with the same base, keep the base and subtract the exponents.

$$= 6x^0y^{-2}z^1$$ Simplify the exponents.

$$= 6 \cdot 1 \cdot y^{-2} \cdot z$$ $x^0 = 1$.

$$= \frac{6z}{y^2}$$ Write the result without the negative exponent.

Success Tip

To find the reciprocal of a rational expression, invert its numerator and denominator. For example, the reciprocal of $\frac{yz^3}{x^2}$ is $\frac{x^2}{yz^3}$.

Self Check 7 Divide: $\dfrac{8a^4}{t^5s^7} \div \dfrac{a^3s^5}{t^2}$.

EXAMPLE 8

Divide: $\dfrac{x^3+8}{4x+4} \div \dfrac{x^2-2x+4}{2x^2-2}$.

Solution

$$\frac{x^3+8}{4x+4} \div \frac{x^2-2x+4}{2x^2-2}$$

$$= \frac{x^3+8}{4x+4} \cdot \frac{2x^2-2}{x^2-2x+4}$$ Multiply the first rational expression by the reciprocal of the second.

$$= \frac{(x^3+8)(2x^2-2)}{(4x+4)(x^2-2x+4)}$$ Multiply the numerators and the denominators.

$$= \frac{(x+2)\overset{1}{\cancel{(x^2-2x+4)}}\overset{1}{\cancel{2}}\overset{1}{\cancel{(x+1)}}(x-1)}{2 \cdot \underset{1}{\cancel{2}}\underset{1}{\cancel{(x+1)}}\underset{1}{\cancel{(x^2-2x+4)}}}$$ Factor x^3+8, $2x^2-2$, and $4x+4$. The polynomial x^2-2x+4 does not factor. Then simplify.

$$= \frac{(x+2)(x-1)}{2}$$

Caution

When dividing rational expressions, always write the result in simplest form by removing any factors common to the numerator and denominator.

Self Check 8 Divide: $\dfrac{x^3-8}{9x-9} \div \dfrac{x^2+2x+4}{3x^2-3x}$.

EXAMPLE 9

Divide: $\dfrac{b^3-4b}{x-1} \div (b-2)$.

Solution

$$\frac{b^3-4b}{x-1} \div (b-2) = \frac{b^3-4b}{x-1} \div \frac{b-2}{1}$$ Write $b-2$ as a fraction with a denominator of 1.

$$= \frac{b^3-4b}{x-1} \cdot \frac{1}{b-2}$$ Multiply the first rational expression by the reciprocal of the second.

$$= \frac{b^3-4b}{(x-1)(b-2)}$$ Multiply the numerators and the denominators.

$$= \frac{b(b+2)\overset{1}{\cancel{(b-2)}}}{(x-1)\underset{1}{\cancel{(b-2)}}} \quad \text{Factor } b^3 - 4b \text{ and then simplify.}$$

$$= \frac{b(b+2)}{x-1}$$

Self Check 9 Divide: $\frac{m^4 - 9m^2}{a^2 - 3a} \div (m^2 + 3m)$.

MIXED OPERATIONS

EXAMPLE 10 Simplify: $\frac{x^2 + 2x - 3}{6x^2 + 5x + 1} \div \frac{2x^2 - 2}{2x^2 - 5x - 3} \cdot \frac{6x^2 + 4x - 2}{x^2 - 2x - 3}$.

Solution Since multiplications and divisions are done in order from left to right, we begin by focusing on the division. We introduce grouping symbols to emphasize this. To divide the expressions in the parentheses, we invert $\frac{2x^2 - 2}{2x^2 - 5x - 3}$ and multiply.

$$\left(\frac{x^2 + 2x - 3}{6x^2 + 5x + 1} \div \frac{2x^2 - 2}{2x^2 - 5x - 3}\right)\frac{6x^2 + 4x - 2}{x^2 - 2x - 3} = \left(\frac{x^2 + 2x - 3}{6x^2 + 5x + 1} \cdot \frac{2x^2 - 5x - 3}{2x^2 - 2}\right)\frac{6x^2 + 4x - 2}{x^2 - 2x - 3}$$

Next, we multiply the three fractions and simplify the result.

$$= \frac{(x^2 + 2x - 3)(2x^2 - 5x - 3)(6x^2 + 4x - 2)}{(6x^2 + 5x + 1)(2x^2 - 2)(x^2 - 2x - 3)}$$

$$= \frac{(x+3)\overset{1}{\cancel{(x-1)}}\overset{1}{\cancel{(2x+1)}}\overset{1}{\cancel{(x-3)}}\overset{1}{\cancel{2}}(3x-1)\overset{1}{\cancel{(x+1)}}}{(3x+1)\underset{1}{\cancel{(2x+1)}}\underset{1}{\cancel{2}}(x+1)\underset{1}{\cancel{(x-1)}}\underset{1}{\cancel{(x-3)}}\underset{1}{\cancel{(x+1)}}}$$

$$= \frac{(x+3)(3x-1)}{(3x+1)(x+1)}$$

Self Check 10 Simplify: $\frac{x^2 - 25}{4x^2 + 12x + 9} \div \frac{x^2 - 5x}{3x - 1} \cdot \frac{2x + 3}{3x^2 + 14x - 5}$.

Answers to Self Checks **1.** $\frac{3a^7b^9}{m^6}$ **2.** $\frac{a^2(a+3)}{6}$ **3.** 1 **4.** $\frac{a-3b}{a+b}$ **5.** $-x(x+3)$ **6.** $\frac{x^2+10x+25}{x^4-12x^3+36x^2}$ **7.** $\frac{8a}{t^3s^{12}}$ **8.** $\frac{x(x-2)}{3}$ **9.** $\frac{m(m-3)}{a(a-3)}$ **10.** $\frac{1}{x(2x+3)}$

6.2 STUDY SET

VOCABULARY **Fill in the blanks.**

1. $\frac{a^2 - 9}{a^2 - 49} \cdot \frac{a - 7}{a + 3}$ is the product of two ________ expressions.

2. The ________ of $\frac{a+3}{a+7}$ is $\frac{a+7}{a+3}$.

3. To find the reciprocal of a rational expression, we ________ its numerator and denominator.

4. To simplify a rational expression, remove any factors ________ to the numerator and denominator.

CONCEPTS **Fill in the blanks.**

5. To multiply rational expressions, multiply their ________ and multiply their ________.
In symbols.
$$\frac{A}{B} \cdot \frac{C}{D} = \square$$

6. To divide two rational expressions, multiply the first by the ________ of the second. In symbols,
$$\frac{A}{B} \div \frac{C}{D} = \square$$

7. The product of a rational expression and its reciprocal is $\square$.

8. The quotient of ________ is -1. For example,
$$\frac{x-8}{8-x} = -1$$

NOTATION **Complete each solution.**

9. $$\frac{x^2+3x}{5x-25} \cdot \frac{x-5}{x+3} = \frac{(x^2+3x)\square}{\square(x+3)} = \frac{\square(x-5)}{\square(x+3)} = \frac{\square}{5}$$

10. $$\frac{x^2-x-6}{4x^2+16x} \div \frac{x-3}{x+4} = \frac{x^2-x-6}{4x^2+16x} \cdot \square = \frac{\square(x+4)}{(4x^2+16x)\square} = \frac{\square(x+2)(x+4)}{\square(x-3)} = \frac{x+2}{\square}$$

11. A student checks her answers with those in the back of her textbook. Tell whether they are equivalent.

Student's answer	Book's answer	Equivalent?
$\frac{-x^{10}}{y^2}$	$-\frac{x^{10}}{y^2}$	
$\frac{x-3}{x+3}$	$\frac{3-x}{3+x}$	
$\frac{b+a}{(2-x)(d+c)}$	$-\frac{a+b}{(x-2)(c+d)}$	

12. a. Write $5x^2 + 35x$ as a fraction.

b. What is the reciprocal of $5x^2 + 35x$?

PRACTICE **Perform the operations and simplify, if possible.**

13. $\frac{3}{4} \cdot \frac{5}{3}$

14. $-\frac{5}{6} \cdot \frac{3}{7}$

15. $-\frac{6}{11} \div \frac{36}{55}$

16. $\frac{17}{12} \div \frac{34}{3}$

17. $\frac{2x^2y^2}{cd} \cdot \frac{cd^2}{4c^2x}$

18. $\frac{b^2x}{6a^2y} \cdot \frac{3a^4b^4}{x^2y^3}$

19. $-\frac{x^3y^3}{y^2} \div \frac{y^3}{x^7}$

20. $(a^3)^2b \div \frac{b}{(a^2)^3}$

21. $15x\left(\frac{x+1}{15x}\right)$

22. $30t\left(\frac{t-7}{30t}\right)$

23. $12y\left(\frac{y+8}{6y}\right)$

24. $16x\left(\frac{3x+8}{4x}\right)$

25. $10(h+9)\frac{h-3}{h+9}$

26. $r(r-25)\frac{r+4}{r-25}$

27. $\frac{x^2+2x+1}{9x} \cdot \frac{2x^2-2x}{2x^2-2}$

28. $\frac{a+6}{16-a^2} \cdot \frac{3a-12}{3a+18}$

29. $\frac{2x^2-x-3}{x^2-1} \cdot \frac{x^2+x-2}{2x^2+x-6}$

30. $\frac{9x^2+3x-20}{3x^2-7x+4} \cdot \frac{3x^2-5x+2}{9x^2+18x+5}$

31. $\frac{x^2-16}{x^2-25} \div \frac{5x+20}{10x-50}$

32. $\frac{a^2-9}{a^2-49} \div \frac{9a+27}{3a+21}$

33. $-\frac{a^2+2a-35}{12x} \div \frac{ax-3x}{a^2+4a-21}$

34. $\frac{x^2-4}{2b-bx} \div \frac{x^2+4x+4}{2b+bx}$

35. $\frac{3t^2-t-2}{6t^2-5t-6} \cdot \frac{4t^2-9}{2t^2+5t+3}$

36. $\frac{2p^2-5p-3}{p^2-9} \cdot \frac{2p^2+5p-3}{2p^2+5p+2}$

37. $\frac{3n^2+5n-2}{12n^2-13n+3} \div \frac{n^2+3n+2}{4n^2+5n-6}$

38. $\dfrac{8y^2 - 14y - 15}{6y^2 - 11y - 10} \div \dfrac{4y^2 - 9y - 9}{3y^2 - 7y - 6}$

39. $\dfrac{2x^2 + 5xy + 3y^2}{3x^2 - 5xy + 2y^2} \div \dfrac{2x^2 + xy - 3y^2}{-3x^2 + 5xy - 2y^2}$

40. $\dfrac{2p^2 - 5pq - 3q^2}{p^2 - 9q^2} \div \dfrac{2p^2 + 5pq + 2q^2}{2p^2 + 5pq - 3q^2}$

41. $(x + 1) \cdot \dfrac{1}{x^2 + 2x + 1}$

42. $-\dfrac{x - 2}{x} \div (x^2 - 4)$

43. $(x^2 + x - 2cx - 2c) \cdot \dfrac{x^2 + 3x + 2}{x^2 - 4c^2}$

44. $(2ax - 10x + a - 5) \cdot \dfrac{x}{2x^2 + x}$

45. $\dfrac{a^2 + 2a - 35}{12x} \div \dfrac{ax - 3x}{a^2 + 4a - 21}$

46. $\dfrac{x^3 + 1}{4} \div \dfrac{x + 1}{2}$

47. $\dfrac{x^3 + y^3}{x^3 - y^3} \div \dfrac{x^2 - xy + y^2}{x^2 + xy + y^2}$

48. $\dfrac{x^2 - 6x + 9}{4 - x^2} \div \dfrac{x^2 - 9}{x^2 - 8x + 12}$

49. $\dfrac{ax + ay + bx + by}{x^3 - 27} \div \dfrac{xc + xd + yc + yd}{x^2 + 3x + 9}$

50. $\dfrac{x^2 + 3x + yx + 3y}{x^2 - 9} \div \dfrac{x + 3}{x - 3}$

51. $\dfrac{x^2 - x - 6}{x^2 - 4} \cdot \dfrac{x^2 - x - 2}{9 - x^2}$

52. $\dfrac{p^3 - q^3}{q^2 - p^2} \cdot \dfrac{q^2 + pq}{p^3 + p^2q + pq^2}$

53. $(4x + 12) \cdot \dfrac{x^2}{2x - 6} \div \dfrac{2}{x - 3}$

54. $(4x^2 - 9) \div \dfrac{2x^2 + 5x + 3}{x + 2} \div (2x - 3)$

55. $(x^2 - x - 6) \div (x - 3) \div (x - 2)$

56. $(x^2 - x - 6) \div [(x - 3) \div (x - 2)]$

57. $\dfrac{2x^2 - 2x - 4}{x^2 + 2x - 8} \cdot \dfrac{3x^2 + 15x}{x + 1} \div \dfrac{4x^2 - 100}{x^2 - x - 20}$

58. $\dfrac{6a^2 - 7a - 3}{a^2 - 1} \div \dfrac{4a^2 - 12a + 9}{a^2 - 1} \cdot \dfrac{2a^2 - a - 3}{3a^2 - 2a - 1}$

59. $\dfrac{x^2 - x - 12}{x^2 + x - 2} \div \dfrac{x^2 - 6x + 8}{x^2 - 3x - 10} \cdot \dfrac{x^2 - 3x + 2}{x^2 - 2x - 15}$

60. $\dfrac{4a^2 - 10a + 6}{a^4 - 3a^3} \div \dfrac{3 - 2a}{2a^3} \cdot \dfrac{a - 3}{2a - 2}$

Find each power.

61. $\left(\dfrac{x - 3}{x^3 + 4}\right)^2$

62. $\left(\dfrac{2t^2 + t}{t - 1}\right)^2$

63. $\left(\dfrac{2m^2 - m - 3}{x^2 - 1}\right)^2$

64. $\left(\dfrac{-k - 3}{x^2 - x + 1}\right)^2$

APPLICATIONS

65. PHYSICS EXPERIMENTS The following table contains data from a physics experiment. Complete the table.

Trial	Rate (m/sec)	Time (sec)	Distance (m)
1	$\dfrac{k_1^2 + 3k_1 + 2}{k_1 - 3}$	$\dfrac{k_1^2 - 3k_1}{k_1 + 1}$	
2	$\dfrac{k_2^2 + 6k_2 + 5}{k_2 + 1}$		$k_2^2 + 11k_2 + 30$

66. TRUNK CAPACITY The shape of the storage space in the trunk of a car is approximately a rectangular solid. Write a simplified rational expression that gives the number of cubic units of storage space in the trunk.

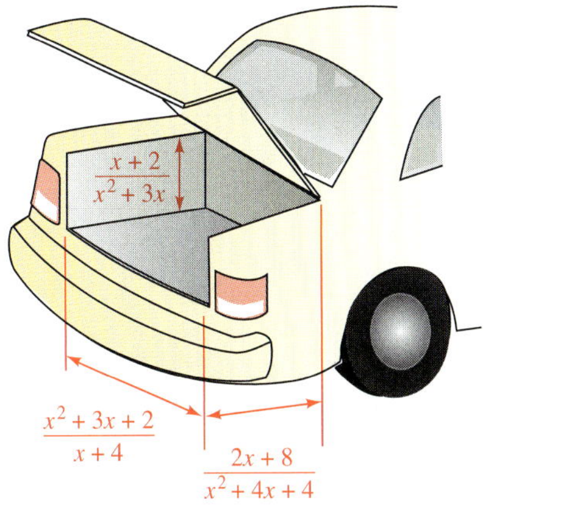

WRITING

67. Explain how to multiply two rational expressions.

68. Write some comments to the student who wrote the following solution, explaining the error.

$$\frac{x^2+x-2}{x^2-4}\cdot\frac{x-2}{x-1}=\frac{\cancel{(x+2)}\cancel{(x-1)}\cancel{(x-2)}}{\cancel{(x+2)}\cancel{(x-2)}\cancel{(x-1)}}=0$$

69. The graph of $Y_3 = Y_1 - Y_2$, where

$$Y_1=\frac{2x^2-5x-3}{x^2-9}\cdot\frac{2x^2+5x-3}{2x^2+5x+2}$$

$$Y_2=\frac{2x-1}{x+2}$$

is shown below. Explain how the graph and table can be used to verify that

$$\frac{2x^2-5x-3}{x^2-9}\cdot\frac{2x^2+5x-3}{2x^2+5x+2}=\frac{2x-1}{x+2}$$

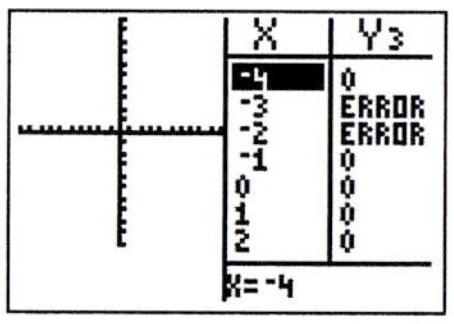

70. A student obtained an answer of $\frac{x+3}{x+7}$ after performing $\frac{x^2-9}{x^2-49}\div\frac{x+3}{x+7}$. As a check, he graphed $Y_3 = Y_1 - Y_2$, where

$$Y_1=\left(\frac{x^2-9}{x^2-49}\right)\div\left(\frac{x+3}{x+7}\right)$$

$$Y_2=\frac{x+3}{x+7}$$

The graph is shown in the next column.

Explain what conclusion can be drawn from the graph and the table.

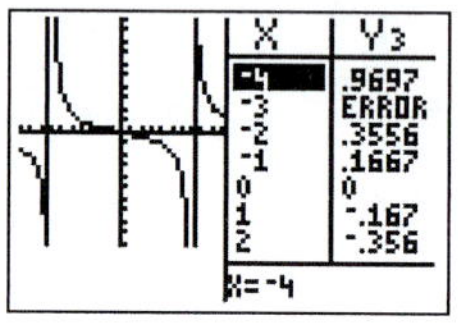

REVIEW **Complete the rules for exponents. Assume that $x \neq 0$ and $y \neq 0$.**

71. $x^m x^n =$ ▢

72. $(x^m)^n =$ ▢

73. $(xy)^n =$ ▢

74. $\left(\frac{x}{y}\right)^n =$ ▢

75. $x^0 =$ ▢

76. $x^{-n} =$ ▢

77. $\frac{x^m}{x^n} =$ ▢

78. $\left(\frac{x}{y}\right)^{-n} =$ ▢

79. $\frac{x^{-m}}{y^{-n}} =$ ▢

80. $x^1 =$ ▢

CHALLENGE PROBLEMS **Insert either a multiplication or a division symbol in each box to make a true statement.**

81. $\frac{x^2}{y}\ \square\ \frac{x}{y^2}\ \square\ \frac{x^2}{y^2}=\frac{x^3}{y}$

82. $\frac{x^2}{y}\ \square\ \frac{x}{y^2}\ \square\ \frac{x^2}{y^2}=\frac{y^3}{x}$

6.3 Adding and Subtracting Rational Expressions

- Adding and subtracting rational expressions with like denominators
- Adding and subtracting rational expressions with unlike denominators
- Finding the least common denominator
- Mixed operations

The methods used to add and subtract rational expressions are based on the rules for adding and subtracting arithmetic fractions. In this section, we will add and subtract rational expressions with *like* and *unlike* denominators.

ADDING AND SUBTRACTING RATIONAL EXPRESSIONS WITH LIKE DENOMINATORS

To add or subtract fractions that have a common denominator, we add or subtract their numerators and write the sum or difference over the common denominator. For example,

$$\frac{3}{7} + \frac{2}{7} = \frac{3+2}{7} = \frac{5}{7} \qquad \frac{3}{7} - \frac{2}{7} = \frac{3-2}{7} = \frac{1}{7}$$

We use the same procedure to add and subtract rational expressions with like denominators.

Adding and Subtracting Rational Expressions

If $\frac{A}{D}$ and $\frac{B}{D}$ are rational expressions,

$$\frac{A}{D} + \frac{B}{D} = \frac{A+B}{D} \qquad \frac{A}{D} - \frac{B}{D} = \frac{A-B}{D}$$

Then simplify, if possible.

EXAMPLE 1

Perform the operations. Simplify the result when possible: **a.** $\frac{4}{3x} + \frac{7}{3x}$ and **b.** $\frac{a^2}{a^2-1} - \frac{a}{a^2-1}$.

Solution

a. $\frac{4}{3x} + \frac{7}{3x} = \frac{4+7}{3x}$ Add the numerators. Write the sum over the common denominator $3x$.

$= \frac{11}{3x}$ Perform the addition in the numerator.

b. $\frac{a^2}{a^2-1} - \frac{a}{a^2-1} = \frac{a^2-a}{a^2-1}$ Subtract the numerators. Write the difference over the common denominator $a^2 - 1$.

We note that the polynomials factor in the numerator and the denominator of the result.

$$\frac{a^2}{a^2-1} - \frac{a}{a^2-1} = \frac{a(a-1)}{(a+1)(a-1)}$$

$$= \frac{a\overset{1}{\cancel{(a-1)}}}{(a+1)\underset{1}{\cancel{(a-1)}}} \quad \text{Simplify.}$$

$$= \frac{a}{a+1}$$

Caution

When adding or subtracting rational expressions, always write the result in simplest form, by removing any factors common to the numerator and denominator.

Self Check 1 Perform the operations: **a.** $\frac{17}{22} + \frac{13}{22}$, **b.** $\frac{1}{6a} - \frac{7}{6a}$, and **c.** $\frac{3a}{a-2} + \frac{2a}{a-2}$.

ACCENT ON TECHNOLOGY: CHECKING ALGEBRA

We can check the subtraction in part b of Example 1 by graphing the rational functions $f(a) = \frac{a^2}{a^2 - 1} - \frac{a}{a^2 - 1}$, shown in figure (a), and $g(a) = \frac{a}{a + 1}$, shown in figure (b), and observing that the graphs are the same. Note that -1 and 1 are not in the domain of the first function and that -1 is not in the domain of the second function.

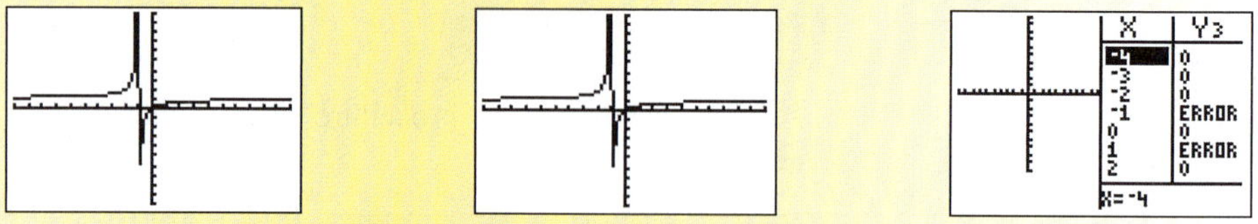

(a) (b) (c)

Figure (c) shows the display when the G-T mode is used to check the simplification. Here, $Y_3 = Y_1 - Y_2$, where $Y_1 = \frac{x^2}{x^2 - 1} - \frac{x}{x^2 - 1}$ and $Y_2 = \frac{x}{x + 1}$.

ADDING AND SUBTRACTING RATIONAL EXPRESSIONS WITH UNLIKE DENOMINATORS

Recall that writing a fraction as an equivalent fraction with a larger denominator is called building the fraction. For example, to write $\frac{3}{5}$ as an equivalent fraction with a denominator of 35, we multiply it by 1 in the form of $\frac{7}{7}$. When a number is multiplied by 1, its value does not change.

$$\frac{3}{5} = \frac{3}{5} \cdot \frac{7}{7} = \frac{21}{35}$$

To add and subtract rational expressions with different denominators, we must write them as equivalent expressions having a common denominator. To do so, we build rational expressions.

Building Rational Expressions

To build a rational expression, multiply it by 1 in the form of $\frac{c}{c}$, where c is any nonzero number or expression.

The following steps summarize how to add or subtract rational expressions with different denominators.

Adding and Subtracting Rational Expressions with Unlike Denominators

1. Find the LCD.
2. Write each rational expression as an equivalent expression whose denominator is the LCD.
3. Add or subtract the numerators and write the sum or difference over the LCD.
4. Simplify the resulting rational expression, if possible.

EXAMPLE 2

Add: $\dfrac{3}{x} + \dfrac{4}{y}$.

Solution A common denominator for the fractions is xy. We multiply each numerator and denominator by the appropriate factor so that each denominator builds to xy.

$$\frac{3}{x} + \frac{4}{y} = \frac{3}{x} \cdot \frac{y}{y} + \frac{4}{y} \cdot \frac{x}{x}$$

Build the rational expressions so that each has a denominator of xy.

$$= \frac{3y}{xy} + \frac{4x}{xy}$$

Multiply the numerators. Multiply the denominators.

$$= \frac{3y + 4x}{xy}$$

Add the numerators. Write the sum over the common denominator xy.

Self Check 2 Add: $\dfrac{5}{a} + \dfrac{7}{b}$.

EXAMPLE 3

Subtract: $\dfrac{4x}{x+2} - \dfrac{7x}{x-2}$.

Solution By inspection, we see that a common denominator is $(x + 2)(x - 2)$. We multiply the numerator and denominator of each expression by the appropriate factor, so that each one has a denominator of $(x + 2)(x - 2)$.

Success Tip

We use the distributive property to multiply the numerators of $\frac{4x}{x+2}$ and $\frac{x-2}{x-2}$. We don't multiply out the denominators.

$$\frac{4x}{x+2} \cdot \frac{x-2}{x-2}$$

$$\frac{4x}{x+2} - \frac{7x}{x-2}$$

$$= \frac{4x}{x+2} \cdot \frac{x-2}{x-2} - \frac{7x}{x-2} \cdot \frac{x+2}{x+2}$$

Build each rational expression.

$$= \frac{4x^2 - 8x}{(x+2)(x-2)} - \frac{7x^2 + 14x}{(x+2)(x-2)}$$

Multiply the numerators. Multiply the denominators.

This numerator is written within parentheses to make sure that we subtract both of its terms.

$$= \frac{(4x^2 - 8x) - (7x^2 + 14x)}{(x+2)(x-2)}$$

Subtract the numerators. Write the difference over the common denominator.

$$= \frac{4x^2 - 8x - 7x^2 - 14x}{(x+2)(x-2)}$$

In the numerator, to subtract the polynomials, add the first and the opposite of the second.

$$= \frac{-3x^2 - 22x}{(x+2)(x-2)}$$

Combine like terms in the numerator.

Notation

The numerator of the result may be written in two forms:

Not factored	Factored
$\dfrac{-3x^2 - 22x}{(x+2)(x-2)}$	$\dfrac{-x(3x+22)}{(x+2)(x+2)}$

If the common factor of $-x$ is factored out of the terms in the numerator, this result can be written in two other equivalent forms.

$$\frac{-3x^2 - 22x}{(x+2)(x-2)} = \frac{-x(3x+22)}{(x+2)(x-2)} = -\frac{x(3x+22)}{(x+2)(x-2)}$$

Self Check 3 Subtract: $\dfrac{3a}{a+3} - \dfrac{5a}{a-3}$.

We can use the following fact to add or subtract rational expressions whose denominators are opposites.

Multiplying by -1	When a polynomial is multiplied by -1, the result is its opposite.

EXAMPLE 4

Add: $\dfrac{x}{x-y}+\dfrac{y}{y-x}$.

Solution We note that the denominators are opposites. Either can serve as the LCD; we will choose $x-y$.

We must multiply the denominator of $\frac{y}{y-x}$ by -1 to obtain the LCD. It follows that $\frac{-1}{-1}$ should be the form of 1 that is used to build an equivalent rational expression.

$$\frac{x}{x-y}+\frac{y}{y-x}=\frac{x}{x-y}+\frac{y}{y-x}\cdot\frac{-1}{-1}$$
Build $\frac{y}{y-x}$ so that it has a denominator of $x-y$.

$$=\frac{x}{x-y}+\frac{-y}{-y+x}$$
Multiply the numerators. Multiply the denominators.

$$=\frac{x}{x-y}+\frac{-y}{x-y}$$
Rewrite the second denominator, $-y+x$, as $x-y$. The fractions now have a common denominator.

$$=\frac{x-y}{x-y}$$
Add the numerators. Write the difference over the common denominator $x-y$.

$$=1$$
Simplify.

Self Check 4 Add: $\dfrac{2a}{a-b}+\dfrac{b}{b-a}$.

EXAMPLE 5

Subtract: $3-\dfrac{7}{x-2}$.

Solution If we write 3 as $\frac{3}{1}$ and multiply the numerator and denominator by $x-2$, the fractions will have a common denominator of $x-2$.

$$3-\frac{7}{x-2}=\frac{3}{1}-\frac{7}{x-2}$$
$3=\frac{3}{1}$.

$$=\frac{3}{1}\cdot\frac{x-2}{x-2}-\frac{7}{x-2}$$
Build $\frac{3}{1}$ to a fraction with denominator $x-2$.

$$=\frac{3x-6}{x-2}-\frac{7}{x-2}$$
Distribute the 3.

$$=\frac{3x-6-7}{x-2}$$
Subtract the numerators. Write the difference over the common denominator $x-2$.

$$=\frac{3x-13}{x-2}$$
Combine like terms in the numerator. The result does not simplify.

Self Check 5 Subtract: $6-\dfrac{5y}{6-y}$.

FINDING THE LEAST COMMON DENOMINATOR

When adding or subtracting rational expressions with unlike denominators, it is easiest if we write the rational expressions in terms of the smallest common denominator possible, called the **least** (or lowest) **common denominator (LCD).** To find the least common denominator of several rational expressions, we follow these steps.

Finding the LCD

1. Factor each denominator completely.
2. The LCD is a product that uses each different factor obtained in step 1 the greatest number of times it appears in any one factorization.

EXAMPLE 6

Find the LCD of: **a.** $\frac{5a}{24b}$ and $\frac{11a}{18b^2}$ and **b.** $\frac{1}{x^2 - 12x + 36}$ and $\frac{3 - x}{x^2 - 6x}$.

Solution

a. We write each denominator as the product of prime numbers and variables.

$$24b = 2 \cdot 2 \cdot 2 \cdot 3 \cdot b = 2^3 \cdot 3 \cdot b$$
$$18b^2 = 2 \cdot 3 \cdot 3 \cdot b \cdot b = 2 \cdot 3^2 \cdot b^2$$

To find the LCD, we form a product using each of these factors the greatest number of times it appears in any one factorization.

The greatest number of times the factor 2 appears is three times.
The greatest number of times the factor 3 appears is twice.
The greatest number of times the factor b appears is twice.

$$\text{LCD} = 2 \cdot 2 \cdot 2 \cdot 3 \cdot 3 \cdot b \cdot b$$
$$= 72b^2$$

Success Tip

Note that the highest power of each factor is used to form the LCD:

$$24b = 2^{③} \cdot 3 \cdot b$$
$$18b^2 = 2 \cdot 3^{②} \cdot b^{②}$$
$$\text{LCD} = 2^3 \cdot 3^2 \cdot b^2 = 72b^2$$

b. We factor each denominator completely:

$$x^2 - 12x + 36 = (x - 6)^2 = (x - 6)(x - 6)$$
$$x^2 - 6x = x(x - 6)$$

To find the LCD, we form a product using the highest power of each of the factors:

The greatest number of times the factor x appears is once.
The greatest number of times the factor $x - 6$ appears is twice.

$$\text{LCD} = x(x - 6)^2$$

Self Check 6 Find the LCD of: **a.** $\frac{3y}{28z^3}$ and $\frac{5x}{21z}$ and **b.** $\frac{a - 1}{a^2 - 25}$ and $\frac{3 - a^2}{a^2 + 7a + 10}$.

EXAMPLE 7

Add: $\frac{5a}{24b} + \frac{11a}{18b^2}$.

Solution In Example 6, we saw that the LCD of these rational expressions is $72b^2$. We multiply each numerator and denominator by whatever it takes to build the denominator to $72b^2$.

$$\frac{5a}{24b} + \frac{11a}{18b^2} = \frac{5a}{24b} \cdot \frac{3b}{3b} + \frac{11a}{18b^2} \cdot \frac{4}{4}$$

Build each rational expression.

$$= \frac{15ab}{72b^2} + \frac{44a}{72b^2}$$

Multiply the numerators. Multiply the denominators.

$$= \frac{15ab + 44a}{72b^2}$$

Add the numerators. Write the sum over the common denominator. The result does not simplify.

Self Check 7 Add: $\frac{3y}{28z^3} + \frac{5x}{21z}$.

EXAMPLE 8

Subtract: $\frac{x + 1}{x^2 - 2x + 1} - \frac{x - 4}{x^2 - 1}$.

Solution We factor each denominator to find the LCD:

$$x^2 - 2x + 1 = (x - 1)(x - 1) = (x - 1)^2$$
$$x^2 - 1 = (x + 1)(x - 1)$$

The greatest number of times $x - 1$ appears is twice. The greatest number of times $x + 1$ appears is once.

The LCD is $(x - 1)^2(x + 1)$ or $(x - 1)(x - 1)(x + 1)$.

We now write each rational expression with its denominator in factored form. Then we multiply each numerator and denominator by the missing factor, so that each rational expression has a denominator of $(x - 1)(x - 1)(x + 1)$.

$$\frac{x + 1}{x^2 - 2x + 1} - \frac{x - 4}{x^2 - 1}$$

$$= \frac{x + 1}{(x - 1)(x - 1)} - \frac{x - 4}{(x + 1)(x - 1)}$$

Write each denominator in factored form.

$$= \frac{x + 1}{(x - 1)(x - 1)} \cdot \frac{x + 1}{x + 1} - \frac{x - 4}{(x + 1)(x - 1)} \cdot \frac{x - 1}{x - 1}$$

Build each rational expression.

$$= \frac{x^2 + 2x + 1}{(x - 1)(x - 1)(x + 1)} - \frac{x^2 - 5x + 4}{(x + 1)(x - 1)(x - 1)}$$

Multiply the numerators using the FOIL method. Multiply the denominators.

$$= \frac{(x^2 + 2x + 1) - (x^2 - 5x + 4)}{(x - 1)(x - 1)(x + 1)}$$

Subtract the numerators. Write the difference over the common denominator.

$$= \frac{x^2 + 2x + 1 - x^2 + 5x - 4}{(x - 1)(x - 1)(x + 1)}$$

In the numerator, subtract the polynomials.

$$= \frac{7x - 3}{(x - 1)(x - 1)(x + 1)}$$

Combine like terms. The result does not simplify.

Self Check 8 Subtract: $\frac{a + 2}{a^2 - 4a + 4} - \frac{a - 3}{a^2 - 4}$.

MIXED OPERATIONS

EXAMPLE 9 Combine: $\dfrac{2x}{x^2 - 4} - \dfrac{1}{x^2 - 3x + 2} + \dfrac{x + 1}{x^2 + x - 2}$.

Solution We factor each denominator to find the LCD and note that the greatest number of times each factor appears is once.

$$\left.\begin{aligned} x^2 - 4 &= (x - 2)(x + 2) \\ x^2 - 3x + 2 &= (x - 2)(x - 1) \\ x^2 + x - 2 &= (x - 1)(x + 2) \end{aligned}\right\} \text{LCD} = (x - 2)(x + 2)(x - 1)$$

We then write each rational expression as an equivalent rational expression with the LCD as its denominator and do the subtraction and addition.

$$\frac{2x}{x^2 - 4} - \frac{1}{x^2 - 3x + 2} + \frac{x + 1}{x^2 + x - 2}$$

$$= \frac{2x}{(x - 2)(x + 2)} - \frac{1}{(x - 2)(x - 1)} + \frac{x + 1}{(x - 1)(x + 2)}$$ Factor the denominators.

$$= \frac{2x}{(x - 2)(x + 2)} \cdot \frac{x - 1}{x - 1} - \frac{1}{(x - 2)(x - 1)} \cdot \frac{x + 2}{x + 2} + \frac{(x + 1)}{(x - 1)(x + 2)} \cdot \frac{x - 2}{x - 2}$$

$$= \frac{2x(x - 1) - 1(x + 2) + (x + 1)(x - 2)}{(x + 2)(x - 2)(x - 1)}$$ Write the sum and difference over the common denominator.

$$= \frac{2x^2 - 2x - x - 2 + x^2 - x - 2}{(x + 2)(x - 2)(x - 1)}$$

$$= \frac{3x^2 - 4x - 4}{(x + 2)(x - 2)(x - 1)}$$ Combine like terms.

$$= \frac{(3x + 2)\overset{1}{\cancel{(x - 2)}}}{(x + 2)\underset{1}{\cancel{(x - 2)}}(x - 1)}$$ Factor the trinomial and simplify.

$$= \frac{3x + 2}{(x + 2)(x - 1)}$$

Self Check 9 Combine: $\dfrac{5a}{a^2 - 25} - \dfrac{7}{a - 5} + \dfrac{2}{a + 5}$.

Answers to Self Checks **1. a.** $\dfrac{15}{11}$, **b.** $-\dfrac{1}{a}$, **c.** $\dfrac{5a}{a - 2}$ **2.** $\dfrac{5b + 7a}{ab}$ **3.** $\dfrac{-2a(a + 12)}{(a + 3)(a - 3)}$ **4.** $\dfrac{2a - b}{a - b}$

5. $\dfrac{-11y + 36}{6 - y}$ **6. a.** $84z^3$, **b.** $(a - 5)(a + 5)(a + 2)$ **7.** $\dfrac{9y + 20xz^2}{84z^3}$

8. $\dfrac{9a - 2}{(a - 2)(a - 2)(a + 2)}$ **9.** $-\dfrac{45}{(a + 5)(a - 5)}$

6.3 STUDY SET

VOCABULARY Fill in the blanks.

1. The rational expressions $\frac{7}{6n}$ and $\frac{n+1}{6n}$ have a ________ denominator of $6n$.
2. The ______ __________ _____________ of $\frac{x-8}{x+6}$ and $\frac{6-5x}{x}$ is $x(x+6)$.
3. To ______ a rational expression, we multiply it by a form of 1. For example, $\frac{2}{n^2} \cdot \frac{8}{8} = \frac{16}{8n^2}$.
4. Two polynomials are _________ if their terms are the same but are opposite in sign.

CONCEPTS Fill in the blanks.

5. To add or subtract rational expressions that have the same denominator, add or subtract the __________, and write the sum or difference over the common __________.

In symbols, if $\frac{A}{D}$ and $\frac{B}{D}$ are rational expressions,

$$\frac{A}{D} + \frac{B}{D} = \frac{}{D} \quad \text{or} \quad \frac{A}{D} - \frac{B}{D} = \frac{}{D}$$

6. When a number is multiplied by ___, its value does not change.
7. To find the least common denominator of several rational expressions, ______ each denominator completely. The LCD is a product that uses each different factor the _______ number of times it appears in any one factorization.
8. $\frac{x^2+3x}{x-1} - \frac{2x-1}{x-1} = \frac{x^2+3x-(\qquad)}{x-1}$
9. Consider the following two procedures.

i. $\frac{x^2-2x}{x^2+4x-12} = \frac{\overset{1}{x(\cancel{x-2})}}{(x+6)\underset{1}{\cancel{(x-2)}}} = \frac{x}{x+6}$

ii. $\frac{x}{x+6} = \frac{x}{x+6} \cdot \frac{x-2}{x-2} = \frac{x^2-2x}{x^2+4x-12}$

   a. In which of these procedures are we *building* a rational expression?
   b. For what type of problem is this procedure often necessary?
   c. What name is used to describe the other procedure?

10. The LCD for $\frac{2x+1}{x^2+5x+6}$ and $\frac{3x}{x^2-4}$ is

$$\text{LCD} = (x+2)(x+3)(x-2)$$

If we want to subtract these rational expressions, what form of 1 should be used

   a. to build $\frac{2x+1}{x^2+5x+6}$?
   b. to build $\frac{3x}{x^2-4}$?

11. Consider the following factorizations.

$2 \cdot 3 \cdot 3 \cdot (x-2)$

$3(x-2)(x+1)$

   a. What is the greatest number of times the factor 3 appears in any one factorization?
   b. What is the greatest number of times the factor $x-2$ appears in any one factorization?

12. The factorizations of the denominators of two rational expressions follow. Find the LCD.

$2 \cdot 3 \cdot a \cdot a \cdot a$

$2 \cdot 3 \cdot 3 \cdot a \cdot a$

13. Factor each denominator completely.
   a. $\frac{17}{40x^2}$
   b. $\frac{x+25}{2x^2-6x}$
   c. $\frac{n^2+3n-4}{n^2-64}$

14. By what must $y-4$ be multiplied to obtain $4-y$?

NOTATION Complete each solution.

15. $\frac{6x-1}{3x-1} + \frac{3x-2}{3x-1} = \frac{6x-1+\qquad}{3x-1}$

$$= \frac{9x-\quad}{3x-1}$$

$$= \frac{3(\qquad)}{3x-1}$$

$$= \quad$$

16. $\frac{8}{3v} - \frac{1}{4v^2} = \frac{8}{3v} \cdot \square - \frac{1}{4v^2} \cdot \square$

$= \frac{\square}{12v^2} - \frac{3}{\square}$

$= \frac{32v - 3}{\square}$

PRACTICE **Perform the operations and simplify the result when possible.**

17. $\frac{3}{4y} + \frac{8}{4y}$

18. $\frac{5}{3z^2} - \frac{6}{3z^2}$

19. $\frac{3x}{2x+2} + \frac{x+4}{2x+2}$

20. $\frac{4y}{y-4} - \frac{16}{y-4}$

21. $\frac{3x}{x-3} - \frac{9}{x-3}$

22. $\frac{9x}{x-y} - \frac{9y}{x-y}$

23. $\frac{5x}{x+1} + \frac{3}{x+1} - \frac{2x}{x+1}$

24. $\frac{4}{a+4} - \frac{2a}{a+4} + \frac{3a}{a+4}$

25. $\frac{3(x^2+x)}{x^2-5x+6} + \frac{-3(x^2-x)}{x^2-5x+6}$

26. $\frac{2x+4}{x^2+13x+12} - \frac{x+3}{x^2+13x+12}$

27. $\frac{2x+1}{x^4-81} + \frac{2-x}{x^4-81}$

28. $\frac{2m^2-7}{m^4-9} + \frac{4-m^2}{m^4-9}$

29. $\frac{3bx-x^2}{x-1} - \frac{3b-x}{x-1}$

30. $\frac{4ct+c^2}{c-7} - \frac{28t+7c}{c-7}$

The denominators of several fractions are given. Find the LCD.

31. $12x, 18x^2$

32. $15ab^2, 27a^2b$

33. x^2+3x, x^2-9

34. $3y^2-6y, 3y(y-4)$

35. x^3+27, x^2+6x+9

36. x^3-8, x^2-4x+4

37. $2x^2+5x+3, 4x^2+12x+9, x^2+2x+1$

38. $2x^2+5x+3, 4x^2+12x+9, 4x+6$

Perform the operations and simplify the result when possible.

39. $\frac{3}{4x} + \frac{2}{3x}$

40. $\frac{2}{5a} + \frac{3}{2b}$

41. $\frac{3a}{2b} - \frac{2b}{3a}$

42. $\frac{5m}{2n} - \frac{3n}{4m}$

43. $\frac{3}{ab^2} - \frac{5}{a^2b}$

44. $\frac{1}{xy^3} - \frac{2}{x^2y}$

45. $\frac{r}{4b^2} + \frac{s}{6b}$

46. $\frac{t}{12c^3} + \frac{t}{15c^2}$

47. $\frac{a+b}{3} + \frac{a-b}{7}$

48. $\frac{x-y}{2} + \frac{x+y}{3}$

49. $\frac{3}{x+2} + \frac{5}{x-4}$

50. $\frac{2}{a+4} - \frac{6}{a+3}$

51. $\frac{x+2}{x+5} - \frac{x-3}{x+7}$

52. $\frac{7}{x+3} + \frac{4x}{x+6}$

53. $\frac{2x+1}{3x+9} - \frac{x-2}{4x+12}$

54. $\frac{x-1}{4x-24} - \frac{3x-2}{5x-30}$

55. $4 + \frac{1}{x}$

56. $2 - \frac{1}{x+1}$

57. $\frac{x+8}{x-3} - \frac{x-14}{3-x}$

58. $\frac{3-x}{2-x} + \frac{x-1}{x-2}$

59. $\frac{2a+1}{3a-2} - \frac{a-4}{2-3a}$

60. $\frac{4}{x-3} + \frac{5}{3-x}$

61. $\frac{x}{x^2+5x+6} + \frac{x}{x^2-4}$

62. $\dfrac{x}{3x^2 - 2x - 1} + \dfrac{4}{3x^2 + 10x + 3}$

63. $\dfrac{a^2 + ab}{a^3 - b^3} - \dfrac{b^2}{b^3 - a^3}$ **64.** $\dfrac{y^2 - 3xy}{x^3 - y^3} - \dfrac{x^2 + 4xy}{y^3 - x^3}$

65. $\dfrac{4}{x^2 - 2x - 3} - \dfrac{x}{3x^2 - 7x - 6}$

66. $\dfrac{2a}{a^2 - 2a - 8} + \dfrac{3}{a^2 - 5a + 4}$

67. $2x + 3 + \dfrac{1}{x + 1}$ **68.** $x + 1 + \dfrac{1}{x - 1}$

69. $1 + x - \dfrac{x}{x - 5}$ **70.** $2 - x + \dfrac{3}{x - 9}$

71. $\dfrac{8}{x^2 - 9} + \dfrac{2}{x - 3} - \dfrac{6}{x}$

72. $\dfrac{x}{x^2 - 4} - \dfrac{x}{x + 2} + \dfrac{2}{x}$

73. $\dfrac{s + 7}{s + 3} - \dfrac{s - 3}{s + 7}$

74. $\dfrac{t + 5}{t - 5} - \dfrac{t - 5}{t + 5}$

75. $\dfrac{3x}{2x - 1} + \dfrac{x + 1}{3x + 2} - \dfrac{2x}{6x^3 + x^2 - 2x}$

76. $\dfrac{x + 3}{2x^2 - 5x + 2} - \dfrac{3x - 1}{x^2 - x - 2}$

77. $\dfrac{3}{x + 1} - \dfrac{2}{x - 1} + \dfrac{x + 3}{x^2 - 1}$

78. $\dfrac{2}{x - 2} + \dfrac{3}{x + 2} - \dfrac{x - 1}{x^2 - 4}$

79. $\dfrac{a}{a - b} + \dfrac{b}{a + b} + \dfrac{a^2 + b^2}{b^2 - a^2}$

80. $\dfrac{1}{x + y} - \dfrac{1}{x - y} - \dfrac{2y}{y^2 - x^2}$

81. $\dfrac{7n^2}{m - n} + \dfrac{3m}{n - m} - \dfrac{3m^2 - n}{m^2 - 2mn + n^2}$

82. $\dfrac{3b}{2a - b} + \dfrac{2a - 1}{b - 2a} - \dfrac{3a^2 + b}{b^2 - 4ab + 4a^2}$

83. $\dfrac{m + 1}{m^2 + 2m + 1} + \dfrac{m - 1}{m^2 - 2m + 1} + \dfrac{2}{m^2 - 1}$

(*Hint:* Simplify first.)

84. $\dfrac{a + 2}{a^2 + 3a + 2} + \dfrac{a - 1}{a^2 - 1} + \dfrac{3}{a + 1}$

(*Hint:* Simplify first.)

APPLICATIONS

85. DRAFTING Among the tools used in drafting are the 45°-45°-90° and the 30°-60°-90° triangles shown. Find the perimeter of each triangle. Express each result as a single rational expression.

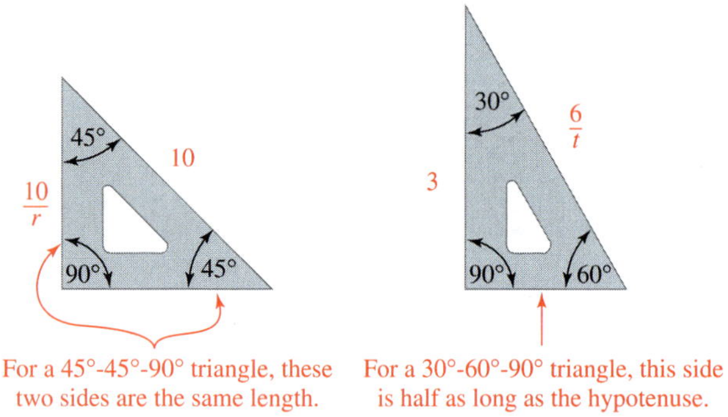

86. THE AMAZON The Amazon River flows in an easterly direction to the Atlantic Ocean. In Brazil, when the river is at low stage, the rate of flow is about 5 mph. Suppose that a river guide can canoe in still water at a rate of r mph.

a. Complete the table to find rational expressions that represent the time it would take the guide to canoe 3 miles downriver and to canoe 3 miles upriver on the Amazon.

	Rate (mph)	Time (hr)	Distance (mi)
Downriver	$r + 5$		3
Upriver	$r - 5$		3

b. Find the difference in the times for the trips upriver and downriver. Express the result as a single rational expression.

WRITING

87. Explain how to find the least common denominator of a set of fractions.

88. Add the fractions by expressing them in terms of a common denominator $24b^3$. (*Note:* This is not the LCD.)

$$\frac{r}{4b^2} + \frac{s}{6b}$$

An extra step had to be performed because the lowest common denominator was not used. What was the step?

89. Write some comments to the student who wrote the following solution, explaining his misunderstanding.

$$\text{Multiply: } \frac{1}{x} \cdot \frac{3}{2} = \frac{1 \cdot 2}{x \cdot 2} \cdot \frac{3 \cdot x}{2 \cdot x}$$
$$= \frac{2}{2x} \cdot \frac{3x}{2x}$$
$$= \frac{6x}{2x}$$

90. Write some comments to the student who wrote the following solution, pointing out where she made an error.

$$\text{Subtract: } \frac{1}{x} - \frac{x+1}{x} = \frac{1 - x + 1}{x}$$
$$= \frac{2 - x}{x}$$

REVIEW **Solve each equation.**

91. $a(a - 6) = -9$

92. $x^2 - \frac{1}{2}(x + 1) = 0$

93. $y^3 + y^2 = 0$

94. $5x^2 = 6 - 13x$

CHALLENGE PROBLEMS

95. Find two rational expressions, each with denominator $x^2 + 5x + 6$, such that their sum is $\frac{1}{x + 2}$.

96. Add: $x^{-1} + x^{-2} + x^{-3} + x^{-4} + x^{-5}$.

6.4 Simplifying Complex Fractions

- Two methods for simplifying complex fractions

A rational expression whose numerator and/or denominator contain rational expressions is called a **complex rational expression** or, more simply, a **complex fraction.** The expression above the main fraction bar of a complex fraction is the numerator, and the expression below the main fraction bar is the denominator. Two examples are:

$$\frac{\frac{3a}{b}}{\frac{6ac}{b^2}}, \qquad \frac{\frac{1}{x} + \frac{1}{y}}{\frac{1}{x} - \frac{1}{y}}$$

← Numerator →
← Main fraction bar →
← Denominator →

In this section, we will simplify complex fractions.

TWO METHODS FOR SIMPLIFYING COMPLEX FRACTIONS

To *simplify a complex fraction* means to express it in the form $\frac{P}{Q}$, where P and Q are polynomials that have no common factors. We can use two methods to simplify the complex fraction

$$\frac{\frac{3a}{b}}{\frac{6ac}{b^2}}$$

In the first method, we eliminate the fractions in the numerator and denominator by writing the complex fraction as a division and using the division rule for fractions:

Success Tip

Simplifying using division works well when a complex fraction is written, or can be easily written, as a quotient of two single rational expressions.

$$\frac{\frac{3a}{b}}{\frac{6ac}{b^2}} = \frac{3a}{b} \div \frac{6ac}{b^2}$$ The main fraction bar of the complex fraction indicates division.

$$= \frac{3a}{b} \cdot \frac{b^2}{6ac}$$ Multiply by the reciprocal of $\frac{6ac}{b^2}$.

$$= \frac{b}{2c}$$ Multiply the rational expressions and then simplify.

In the second method, we eliminate the fractions in the numerator and denominator by multiplying the complex fraction by 1, written in the form $\frac{b^2}{b^2}$. We use $\frac{b^2}{b^2}$ because b^2 is the LCD of $\frac{3a}{b}$ and $\frac{6ac}{b^2}$.

$$\frac{\frac{3a}{b}}{\frac{6ac}{b^2}} = \frac{\frac{3a}{b}}{\frac{6ac}{b^2}} \cdot \frac{b^2}{b^2}$$ Multiply the complex fraction by a form of 1: $\frac{b^2}{b^2} = 1$.

$$= \frac{\frac{3ab^2}{b}}{\frac{6acb^2}{b^2}}$$ Multiply the numerators: $\frac{3a}{b} \cdot b^2$. Multiply the denominators: $\frac{6ac}{b^2} \cdot b^2$.

$$= \frac{3ab}{6ac}$$ Simplify the numerator and the denominator of the complex fraction.

$$= \frac{b}{2c}$$ Simplify.

With either method, the result is the same.

Methods for Simplifying Complex Fractions

Method 1

1. Write the numerator and the denominator of the complex fraction as single rational expressions.
2. Perform the division by multiplying the numerator of the complex fraction by the reciprocal of the denominator.
3. Simplify the result, if possible.

Method 2

1. Find the LCD of all rational expressions in the complex fraction.
2. Multiply the complex fraction by 1 in the form $\frac{\text{LCD}}{\text{LCD}}$.
3. Perform the operations in the numerator and denominator. No fractional expressions should remain within the complex fraction.
4. Simplify the result, if possible.

EXAMPLE 1

Simplify: $\dfrac{\frac{2}{x} + 1}{x + 3}$.

Solution ***Method 1***

We add in the numerator to make it a single rational expression.

$$\frac{\frac{2}{x} + 1}{x + 3} = \frac{\frac{2}{x} + \frac{x}{x}}{\frac{x + 3}{1}}$$ Write 1 as $\frac{x}{x}$ and $x + 3$ as $\frac{x + 3}{1}$. The denominator is now a single rational expression.

$$= \frac{\frac{2 + x}{x}}{\frac{x + 3}{1}}$$ Add $\frac{2}{x}$ and $\frac{x}{x}$ to get $\frac{2 + x}{x}$.

$$= \frac{2 + x}{x} \div \frac{x + 3}{1}$$ Write the division indicated by the main fraction bar using a ÷ symbol.

$$= \frac{2 + x}{x} \cdot \frac{1}{x + 3}$$ Multiply by the reciprocal of $\frac{x + 3}{1}$.

$$= \frac{2 + x}{x^2 + 3x}$$ Multiply the numerators and multiply the denominators.

The Language of Algebra

The second step of this method can be phrased: Perform the division by *inverting the denominator of the complex fraction and multiplying.*

Method 2

The LCD of all rational expressions in the complex fraction is x.

$$\frac{\frac{2}{x} + 1}{x + 3} = \frac{\frac{2}{x} + 1}{x + 3} \cdot \frac{x}{x}$$ Multiply the complex fraction by 1 in the form $\frac{x}{x}$.

$$= \frac{\left(\frac{2}{x} + 1\right)x}{(x + 3)x}$$ Multiply the numerators. Multiply the denominators.

$$= \frac{\frac{2}{x} \cdot x + 1 \cdot x}{x \cdot x + 3 \cdot x}$$ Distribute the multiplication by x.

$$= \frac{2 + x}{x^2 + 3x}$$

Notation

The result after simplifying a complex fraction can often have several forms. This result could be written:

$$\frac{x + 2}{x(x + 3)}$$

Self Check 1 Simplify: $\dfrac{\frac{3}{a} + 2}{a + 2}$.

ACCENT ON TECHNOLOGY: CHECKING ALGEBRA

A check of the simplification done in Example 1 can be performed using a scientific calculator. If

$$\frac{\frac{2}{x}+1}{3+x}=\frac{2+x}{x^2+3x}$$

then the expressions on each side will have identical values when evaluated for a given value of x (say, $x = 5$). To evaluate the expression on the left-hand side, we enter these numbers and press these keys.

[( ] 2 [÷] 5 [+] 1 [ )] [÷] [( ] 3 [+] 5 [ )] [=]   0.175

To evaluate the expression on the right-hand side, we enter these numbers and press these keys.

[( ] 2 [+] 5 [ )] [÷] [( ] 5 [x^2] [+] 3 [×] 5 [ )] [=]   0.175

The results are the same, so it appears that the simplification is correct. We say "appears" because checking for only a single value of x is not definitive. The expressions should yield identical values when evaluated for any value of x for which the fractions are defined.

We can also check the simplification in Example 1 by graphing the functions

$f(x) = \dfrac{\frac{2}{x}+1}{3+x}$ shown in figure (a), and

$g(x) = \dfrac{2+x}{x^2+3x}$ shown in figure (b),

and observing that the graphs are the same. Each graph has window settings of $[-10, 10]$ for x and $[-10, 10]$ for y.

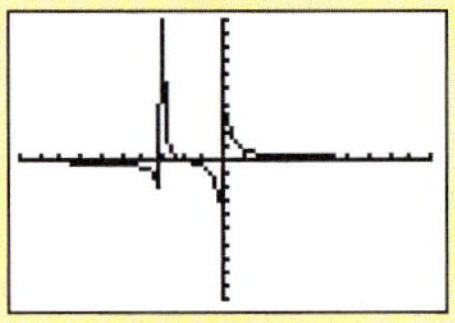

(a)

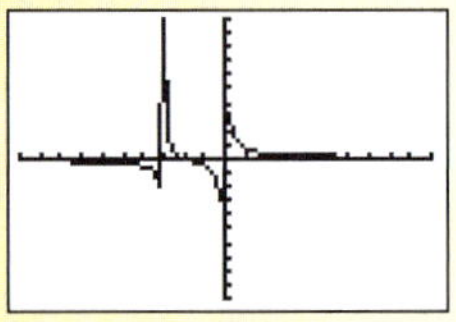

(b)

EXAMPLE 2

Simplify: $\dfrac{\frac{1}{x}+\frac{1}{y}}{\frac{1}{x}-\frac{1}{y}}$.

Solution ***Method 1***

To write the numerator and denominator of the complex fraction as single fractions, we add the rational expressions in the numerator and subtract the rational expressions in the denominator.

$$\frac{\frac{1}{x}+\frac{1}{y}}{\frac{1}{x}-\frac{1}{y}}=\frac{\frac{1}{x}\cdot\frac{y}{y}+\frac{1}{y}\cdot\frac{x}{x}}{\frac{1}{x}\cdot\frac{y}{y}-\frac{1}{y}\cdot\frac{x}{x}}$$

Build the rational expressions in the numerator to have the LCD xy. Do the same in the denominator.

$$=\frac{\frac{y+x}{xy}}{\frac{y-x}{xy}}$$

Add the rational expressions in the numerator, and subtract the rational expressions in the denominator.

$$=\frac{y+x}{xy}\div\frac{y-x}{xy}$$

Write the complex fraction as a division.

$$=\frac{y+x}{xy}\cdot\frac{xy}{y-x}$$

Multiply by the reciprocal of $\frac{y-x}{xy}$.

$$=\frac{(y+x)\overset{1}{\cancel{x}}\overset{1}{\cancel{y}}}{\underset{1}{\cancel{x}}\underset{1}{\cancel{y}}(y-x)}$$

Multiply the rational expressions and then simplify.

$$=\frac{y+x}{y-x}$$

Method 2

The LCD of the fractions appearing in the complex fraction is xy. We multiply the complex fraction by 1 in the form of $\frac{\text{LCD}}{\text{LCD}}$.

$$\frac{\frac{1}{x}+\frac{1}{y}}{\frac{1}{x}-\frac{1}{y}}=\frac{\frac{1}{x}+\frac{1}{y}}{\frac{1}{x}-\frac{1}{y}}\cdot\frac{xy}{xy}$$

Multiply the complex fraction by a form of 1: $\frac{xy}{xy}$.

$$=\frac{\left(\frac{1}{x}+\frac{1}{y}\right)xy}{\left(\frac{1}{x}-\frac{1}{y}\right)xy}$$

Multiply the numerators.
Multiply the denominators.

$$=\frac{\frac{xy}{x}+\frac{xy}{y}}{\frac{xy}{x}-\frac{xy}{y}}$$

Distribute the multiplication by xy.

$$=\frac{y+x}{y-x}$$

Simplify each of the four rational expressions.

Success Tip

Simplifying using the LCD works well when the complex fraction has sums and/or differences in the numerator or denominator.

Self Check 2 Simplify: $\frac{\frac{1}{a}-\frac{1}{b}}{\frac{1}{a}+\frac{1}{b}}$.

EXAMPLE 3

Simplify: $\dfrac{x^{-1}+y^{-1}}{x^{-2}-y^{-2}}$.

Solution We write the complex fraction without using negative exponents. Then we use Method 2 to simplify.

$$\frac{x^{-1}+y^{-1}}{x^{-2}-y^{-2}} = \frac{\frac{1}{x}+\frac{1}{y}}{\frac{1}{x^2}-\frac{1}{y^2}} \qquad \text{Write the fraction without negative exponents.}$$

$$= \frac{\frac{1}{x}+\frac{1}{y}}{\frac{1}{x^2}-\frac{1}{y^2}} \cdot \frac{x^2y^2}{x^2y^2} \qquad \text{The LCD of all rational expressions in the complex fraction is } x^2y^2.$$

$$= \frac{\left(\frac{1}{x}+\frac{1}{y}\right)x^2y^2}{\left(\frac{1}{x^2}-\frac{1}{y^2}\right)x^2y^2} \qquad \text{Multiply the numerators. Multiply the denominators.}$$

$$= \frac{xy^2+yx^2}{y^2-x^2} \qquad \text{Distribute the multiplication by } x^2y^2.$$

$$= \frac{xy(y+x)}{(y+x)(y-x)} \qquad \text{Factor the numerator and denominator.}$$

$$= \frac{xy}{y-x} \qquad \text{Simplify.}$$

Caution

Recall that a *factor* can be moved from the numerator to the denominator if the sign of its exponent is changed. However, in this case, x^{-1}, y^{-1}, x^{-2}, and $-y^{-2}$ are *terms*. Thus,

$$\frac{x^{-1}+y^{-1}}{x^{-2}-y^{-2}} \neq \frac{x^2+y^2}{x-y}$$

Self Check 3 Simplify: $\dfrac{a^{-2}+b^{-2}}{a^{-1}-b^{-1}}$.

EXAMPLE 4

Simplify: $\dfrac{\frac{1}{a^2-3a+2}}{\frac{3}{a-2}-\frac{2}{a-1}}$.

Solution We will use Method 2 to simplify. To determine the LCD for all the fractions appearing in the complex fraction, we must factor a^2-3a+2.

$$\frac{\frac{1}{a^2-3a+2}}{\frac{3}{a-2}-\frac{2}{a-1}} = \frac{\frac{1}{(a-2)(a-1)}}{\frac{3}{a-2}-\frac{2}{a-1}}$$

The LCD of the fractions in the numerator and denominator of the complex fraction is $(a-2)(a-1)$. We multiply the numerator and the denominator by the LCD.

$$= \frac{\frac{1}{(a-2)(a-1)}}{\frac{3}{a-2}-\frac{2}{a-1}} \cdot \frac{(a-2)(a-1)}{(a-2)(a-1)}$$

$$= \frac{\left(\frac{1}{(a-2)(a-1)}\right)(a-2)(a-1)}{\left(\frac{3}{a-2}-\frac{2}{a-1}\right)(a-2)(a-1)}$$

$$= \frac{\frac{(a-2)(a-1)}{(a-2)(a-1)}}{\frac{3(a-2)(a-1)}{a-2}-\frac{2(a-2)(a-1)}{a-1}}$$ Perform the multiplication in the numerator. In the denominator, distribute the LCD.

$$= \frac{1}{3(a-1)-2(a-2)}$$ Simplify each of the three rational expressions.

$$= \frac{1}{3a-3-2a+4}$$ In the denominator, remove parentheses.

$$= \frac{1}{a+1}$$ Combine like terms.

The Language of Algebra

After multiplying a complex fraction by $\frac{\text{LCD}}{\text{LCD}}$ and performing the multiplications, the numerator and denominator of the complex fraction will be cleared of fractions.

Self Check 4 Simplify: $\dfrac{\frac{b}{b+4}+\frac{3}{b+3}}{\frac{b}{b^2+7b+12}}$.

If a fraction has a complex fraction in its numerator or denominator, it is often called a **continued fraction.**

EXAMPLE 5

Simplify: $\dfrac{\frac{2x}{1-\frac{1}{x}}+3}{3-\frac{2}{x}}$.

Solution We begin by multiplying the numerator and denominator of

$$\frac{2x}{1-\frac{1}{x}}$$

by x to eliminate the complex fraction in the numerator of the continued fraction.

$$\frac{\dfrac{2x}{1-\dfrac{1}{x}}+3}{3-\dfrac{2}{x}} = \frac{\dfrac{2x}{1-\dfrac{1}{x}}\cdot\dfrac{x}{x}+3}{3-\dfrac{2}{x}}$$

$$= \frac{\dfrac{2x^2}{x-1}+3}{3-\dfrac{2}{x}}$$

We then multiply the numerator and denominator of the previous fraction by $x(x-1)$, the LCD of $\frac{2x^2}{x-1}$, 3, and $\frac{2}{x}$, and simplify.

$$\frac{\dfrac{2x}{1-\dfrac{1}{x}}+3}{3-\dfrac{2}{x}} = \frac{\dfrac{2x^2}{x-1}+3}{3-\dfrac{2}{x}}\cdot\frac{x(x-1)}{x(x-1)}$$

$$= \frac{2x^3+3x(x-1)}{3x(x-1)-2(x-1)}$$ Distribute the multiplication by $x(x-1)$.

$$= \frac{2x^3+3x^2-3x}{3x^2-5x+2}$$

This result does not simplify.

Self Check 5 Simplify: $\dfrac{3+\dfrac{2}{a}}{\dfrac{2a}{1+\dfrac{1}{a}}+3}$.

Answers to Self Checks **1.** $\dfrac{2a+3}{a^2+2a}$ **2.** $\dfrac{b-a}{b+a}$ **3.** $\dfrac{b^2+a^2}{ab(b-a)}$ **4.** $\dfrac{b^2+6b+12}{b}$ **5.** $\dfrac{(3a+2)(a+1)}{a(2a^2+3a+3)}$

6.4 STUDY SET

VOCABULARY Fill in the blanks.

1. $\dfrac{\frac{x}{y}+\frac{1}{x}}{\frac{1}{y}+\frac{2}{x}}$ and $\dfrac{\frac{5a^2}{b}}{\frac{b}{2a^3}}$ are examples of ________ rational expressions, or more simply, complex ________.

2. To ________ a complex fraction means to express it in the form $\frac{P}{Q}$, where P and Q are polynomials with no common factors.

CONCEPTS

3. To simplify the complex fraction shown below, it is multiplied by a form of 1. What form of 1 is used?

$$\frac{\frac{4}{t^2}+\frac{b}{t}}{\frac{3b}{t}} = \frac{\frac{4}{t^2}+\frac{b}{t}}{\frac{3b}{t}} \cdot \frac{t^2}{t^2}$$

4. Determine the LCD of the rational expressions appearing in each complex fraction.

a. $\dfrac{1+\frac{4}{c}}{\frac{2}{c}+c}$

b. $\dfrac{\frac{6}{m^2}+\frac{1}{2m}}{\frac{m^2-1}{4}}$

c. $\dfrac{\frac{p}{p+2}+\frac{12}{p+3}}{\frac{p-1}{p^2+5p+6}}$

d. $\dfrac{2+\frac{3}{x+1}}{\frac{1}{x}+x+x^2}$

NOTATION Complete each solution.

5.
$$\frac{\frac{5m^2}{6}}{\frac{25m}{3}} = \frac{5m^2}{6}\ \square\ \frac{25m}{3}$$
$$= \frac{5m^2}{6} \cdot \square$$
$$= \frac{5 \cdot \square \cdot m \cdot \square}{2 \cdot \square \cdot 5 \cdot 5 \cdot \square}$$
$$= \frac{m}{\square}$$

6.
$$\frac{\frac{2}{a^2}-\frac{1}{b}}{\frac{2}{a}+\frac{1}{b^2}} = \frac{\frac{2}{a^2}-\frac{1}{b}}{\frac{2}{a}+\frac{1}{b^2}} \cdot \frac{\square}{\square}$$
$$= \frac{\frac{2a^2b^2}{a^2} - \square}{\frac{2a^2b^2}{a} + \square}$$
$$= \frac{\square - a^2b}{2ab^2 + \square}$$

7. The fraction $\dfrac{\frac{a}{b}}{\frac{c}{d}}$ is equivalent to $\frac{a}{b}\ \square\ \frac{c}{d}$.

8. What is the numerator and what is the denominator of the following complex fraction?

$$\frac{6-k-\frac{5}{k}}{\frac{6}{k^2}+\frac{4}{k}-4}$$

PRACTICE Simplify each complex fraction.

9. $\dfrac{\frac{1}{2}}{\frac{3}{4}}$

10. $\dfrac{\frac{3}{4}}{\frac{1}{2}}$

11. $\dfrac{\frac{1}{2}-\frac{2}{3}}{\frac{2}{3}+\frac{1}{2}}$

12. $\dfrac{\frac{1}{4}-\frac{1}{5}}{\frac{1}{3}}$

13. $\dfrac{\frac{4x}{y}}{\frac{6xz}{y^2}}$

14. $\dfrac{\frac{5t^4}{9x}}{\frac{2t}{18x}}$

15. $\dfrac{5ab^2}{\frac{ab}{25}}$

16. $\dfrac{\frac{6a^2b}{4t}}{3a^2b^2}$

17. $\dfrac{\frac{x-y}{xy}}{\frac{y-x}{x}}$

18. $\dfrac{\frac{x^2+5x+6}{3xy}}{\frac{x^2-9}{6xy}}$

19. $\dfrac{\frac{1}{x}-\frac{1}{y}}{xy}$

20. $\dfrac{xy}{\frac{1}{x}-\frac{1}{y}}$

21. $\dfrac{\frac{1}{a}+\frac{1}{b}}{\frac{1}{a}}$

22. $\dfrac{\frac{1}{b}}{\frac{1}{a}-\frac{1}{b}}$

23. $\dfrac{1+\frac{x}{y}}{1-\frac{x}{y}}$

24. $\dfrac{\frac{x}{y}+1}{1-\frac{x}{y}}$

25. $\dfrac{\frac{ac-ad-c+d}{a^3-1}}{\frac{c^2-2cd+d^2}{a^2+a+1}}$

26. $\dfrac{\frac{2x-tx+2y-ty}{x^2+2xy+y^2}}{\frac{t^3-8}{15x+15y}}$

27. $\dfrac{\frac{y}{x}-\frac{x}{y}}{\frac{1}{x}+\frac{1}{y}}$

28. $\dfrac{\frac{y}{x}-\frac{x}{y}}{\frac{1}{y}-\frac{1}{x}}$

29. $\dfrac{\frac{1}{a}-\frac{1}{b}}{\frac{a}{b}-\frac{b}{a}}$

30. $\dfrac{\frac{1}{a}+\frac{1}{b}}{\frac{a}{b}-\frac{b}{a}}$

31. $\dfrac{x+1-\frac{6}{x}}{\frac{1}{x}}$

32. $\dfrac{x-1-\frac{2}{x}}{\frac{x}{3}}$

33. $\dfrac{5xy}{1+\frac{1}{xy}}$

34. $\dfrac{3a}{a+\frac{1}{a}}$

35. $\dfrac{a-4+\frac{1}{a}}{-\frac{1}{a}-a+4}$

36. $\dfrac{a+1+\frac{1}{a^2}}{\frac{1}{a^2}+a-1}$

37. $\dfrac{1+\frac{6}{x}+\frac{8}{x^2}}{1+\frac{1}{x}-\frac{12}{x^2}}$

38. $\dfrac{1-x-\frac{2}{x}}{\frac{6}{x^2}+\frac{1}{x}-1}$

39. $\dfrac{\frac{1}{a+1}+1}{\frac{3}{a-1}+1}$

40. $\dfrac{2+\frac{4}{y-7}}{\frac{4}{y-7}}$

41. $\dfrac{2+\frac{3}{x+1}}{\frac{1}{x}+x}$

42. $\dfrac{\frac{1}{x}-\frac{4}{x-1}}{\frac{3}{x-1}+\frac{2}{x}}$

43. $\dfrac{y}{x^{-1}-y^{-1}}$

44. $\dfrac{x^{-1}+y^{-1}}{(x+y)^{-1}}$

45. $\dfrac{x-y^{-2}}{y-x^{-2}}$

46. $\dfrac{x^{-2}-y^{-2}}{x^{-1}-y^{-1}}$

47. $\dfrac{\frac{t}{x^2-y^2}}{\frac{t}{x+y}}$

48. $\dfrac{\frac{7}{a-b}}{\frac{b}{a^3-b^3}}$

49. $\dfrac{\frac{2}{x+3}-\frac{1}{x-3}}{\frac{3}{x^2-9}}$

50. $\dfrac{2+\frac{1}{x^2-1}}{1+\frac{1}{x-1}}$

51. $\dfrac{\frac{h}{h^2+3h+2}}{\frac{4}{h+2}-\frac{4}{h+1}}$

52. $\dfrac{\frac{1}{r^2+4r+4}}{\frac{r}{r+2}+\frac{r}{r+2}}$

53. $a+\dfrac{a}{1+\frac{a}{a+1}}$

54. $b+\dfrac{b}{1-\frac{b+1}{b}}$

55. $\dfrac{x - \dfrac{1}{1 - \dfrac{x}{2}}}{\dfrac{3}{x + \dfrac{2}{3}} - x}$

56. $\dfrac{3x - \dfrac{1}{3 - \dfrac{x}{2}}}{\dfrac{3}{\dfrac{x}{2} - 3} + x}$

APPLICATIONS

57. ENGINEERING The stiffness k of the shaft shown is given by the formula

$$k = \frac{1}{\dfrac{1}{k_1} + \dfrac{1}{k_2}}$$

where k_1 and k_2 are the individual stiffnesses of each section. Simplify the complex fraction.

58. TRANSPORTATION If a bus travels a distance d_1 at a speed s_1, and then travels a distance d_2 at a speed s_2, the average (mean) speed $\bar{s}$ is given by the formula

$$\bar{s} = \frac{d_1 + d_2}{\dfrac{d_1}{s_1} + \dfrac{d_2}{s_2}}$$

Simplify the complex fraction.

59. KITCHEN UTENSILS What is the ratio of the width of the opening of the ice tongs to the width of the opening of the handles? Express the result in simplest form.

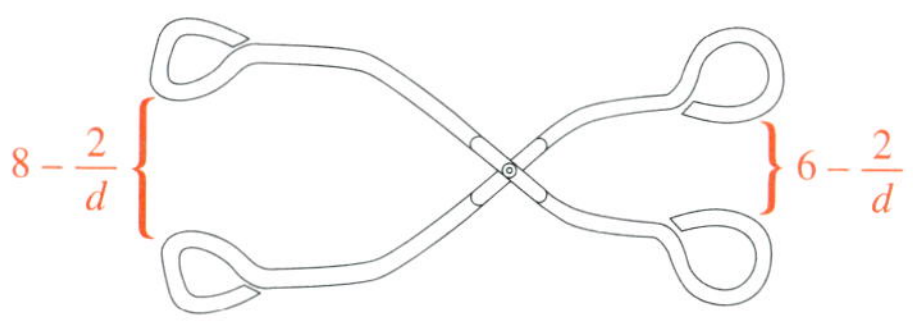

60. DATA ANALYSIS Use the data in the table to find the average measurement for the three-trial experiment. Express the answer as a rational expression.

	Trial 1	Trial 2	Trial 3
Measurement	$\frac{k}{3}$	$\frac{k}{5}$	$\frac{k}{6}$

WRITING

61. What is a complex fraction?

62. Two methods can be used to simplify a complex fraction. Which method do you think is simpler? Why?

63. Evaluate each expression for $x = 2$.

$$\frac{\dfrac{4}{3} + x}{\dfrac{x}{2} + \dfrac{2}{x}} \quad \text{and} \quad \frac{x^2 + 4}{x + 2}$$

Determine whether the result verifies that

$$\frac{\dfrac{4}{3} + x}{\dfrac{x}{2} + \dfrac{2}{x}} = \frac{x^2 + 4}{x + 2}$$

64. To check a simplification to verify that

$$\frac{\dfrac{4}{3} + x}{\dfrac{x}{2} + \dfrac{2}{x}} = \frac{x^2 + 4}{x + 2}$$

a student graphed each expression's associated function. What conclusion can be made about the simplification from the graphs?

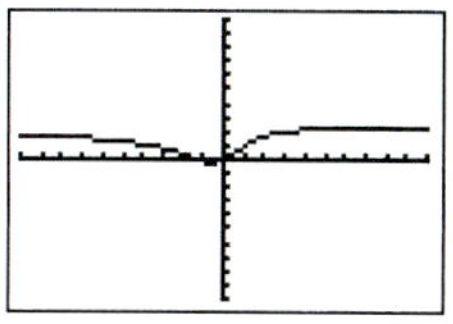

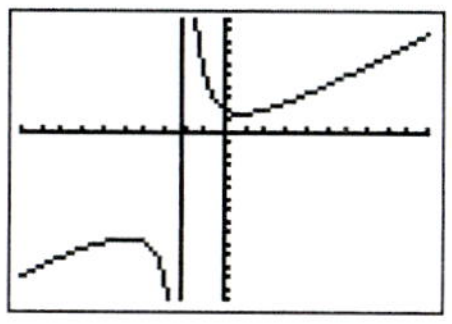

REVIEW Solve each equation.

65. $\dfrac{8(a - 5)}{3} = 2(a - 4)$

66. $\dfrac{3t^2}{5} + \dfrac{7t}{10} = \dfrac{3t + 6}{5}$

67. $a^4 - 13a^2 + 36 = 0$

68. $|2x - 1| = 9$

CHALLENGE PROBLEMS

69. Simplify: $(x^{-1}y^{-1})(x^{-1} + y^{-1})^{-1}$.

70. Simplify: $[(x^{-1} + 1)^{-1} + 1]^{-1}$.

6.5 Dividing Polynomials

- Dividing a monomial by a monomial
- Dividing a polynomial by a monomial
- Dividing a polynomial by a polynomial
- Missing terms

We have discussed addition, subtraction, and multiplication of polynomials. We will now introduce the procedures used to divide polynomials. This topic appears in a chapter about rational expressions because rational expressions indicate division of polynomials. For example,

$$\frac{x^2 - 3x + 7}{x + 1} = (x^2 - 3x + 7) \div (x + 1)$$

We begin the discussion of division of polynomials with the simplest case, a monomial divided by a monomial.

DIVIDING A MONOMIAL BY A MONOMIAL

To divide monomials, we can use the method for simplifying fractions or the quotient rule for exponents.

EXAMPLE 1

Simplify: **a.** $\frac{21x^5}{7x^2}$ and **b.** $\frac{10r^6s}{6rs^3}$.

Solution

By simplifying fractions

a. $\frac{21x^5}{7x^2} = \frac{3 \cdot \overset{1}{\cancel{7}} \cdot \overset{1}{\cancel{x}} \cdot \overset{1}{\cancel{x}} \cdot x \cdot x \cdot x}{\underset{1}{\cancel{7}} \cdot \underset{1}{\cancel{x}} \cdot \underset{1}{\cancel{x}}}$

$= 3x^3$

b. $\frac{10r^6s}{6rs^3} = \frac{\overset{1}{\cancel{2}} \cdot 5 \cdot \overset{1}{\cancel{r}} \cdot r \cdot r \cdot r \cdot r \cdot r \cdot \overset{1}{\cancel{s}}}{\underset{1}{\cancel{2}} \cdot 3 \cdot \underset{1}{\cancel{r}} \cdot \underset{1}{\cancel{s}} \cdot s \cdot s}$

$= \frac{5r^5}{3s^2}$

Using the rules for exponents

$\frac{21x^5}{7x^2} = 3x^{5-2}$ Divide the coefficients. Keep the common base x and subtract exponents.

$= 3x^3$

$\frac{10r^6s}{6rs^3} = \frac{5}{3}r^{6-1}s^{1-3}$ Simplify $\frac{10}{6}$.

$= \frac{5}{3}r^5s^{-2}$

$= \frac{5r^5}{3s^2}$ Move s^{-2} to the denominator and change the sign of the exponent.

Success Tip

In this section, you will see that regardless of the number of terms involved, every polynomial division is a series of monomial divisions.

Self Check 1 Simplify: **a.** $\frac{30y^4}{5y^2}$ and **b.** $\frac{8c^2d^6}{32c^5d^2}$.

DIVIDING A POLYNOMIAL BY A MONOMIAL

Recall that to add two fractions with the same denominator, we add their numerators and keep the common denominator.

$$\frac{a}{d} + \frac{b}{d} = \frac{a+b}{d}$$

We can use this rule in reverse to divide polynomials by monomials.

Dividing a Polynomial by a Monomial

To divide a polynomial by a monomial, divide each term of the polynomial by the monomial.

Let a, b, and d represent monomials, where d is not 0,

$$\frac{a+b}{d} = \frac{a}{d} + \frac{b}{d}$$

EXAMPLE 2

Divide: **a.** $\dfrac{9x^2 + 6x}{3x}$ and **b.** $\dfrac{12a^4b^3 - 18a^3b^2 + 2a^2}{6a^2b^2}$.

Solution

a. Here, we have a binomial divided by a monomial.

$$\frac{9x^2 + 6x}{3x} = \frac{9x^2}{3x} + \frac{6x}{3x}$$ Divide each term of the numerator by the denominator, $3x$.

$$= 3x^{2-1} + 2x^{1-1}$$ Perform each monomial division. Divide the coefficients. Subtract the exponents.

$$= 3x + 2$$ Recall that $x^0 = 1$.

Check: We multiply the divisor, $3x$, and the quotient, $3x + 2$. The result should be the dividend, $9x^2 + 6x$.

$$3x(3x + 2) = 9x^2 + 6x$$ The answer checks.

b. Here, we have a trinomial divided by a monomial.

$$\frac{12a^4b^3 - 18a^3b^2 + 2a^2}{6a^2b^2} = \frac{12a^4b^3}{6a^2b^2} - \frac{18a^3b^2}{6a^2b^2} + \frac{2a^2}{6a^2b^2}$$ Divide each term of the numerator by the denominator, $6a^2b^2$.

$$= 2a^{4-2}b^{3-2} - 3a^{3-2}b^{2-2} + \frac{a^{2-2}}{3b^2}$$ Perform each monomial division. Simplify: $\frac{2}{6} = \frac{1}{3}$.

$$= 2a^2b - 3a + \frac{1}{3b^2}$$

Recall that the variables in a polynomial must have whole-number exponents. Therefore, the result, $2a^2b - 3a + \frac{1}{3b^2}$, is not a polynomial because the last term can be written $\frac{1}{3}b^{-2}$.

Check: $$6a^2b^2\left(2a^2b - 3a + \frac{1}{3b^2}\right) = 12a^4b^3 - 18a^3b^2 + 2a^2$$ The answer checks.

The Language of Algebra

The names of the parts of a division statement are

Dividend

$$\frac{9x^2 + 6x}{3x} = 3x + 2$$

Divisor

Quotient

Success Tip

The sum, difference, and product of two polynomials are always polynomials. However, as seen in Example 2b, the quotient of two polynomials is not always a polynomial.

Self Check 2 Divide: **a.** $\dfrac{50h^3 + 15h^2}{5h^2}$ and **b.** $\dfrac{22s^5t^2 - s^4t^3 + 44s^2t}{11s^2t^2}$.

DIVIDING A POLYNOMIAL BY A POLYNOMIAL

There is an **algorithm** (a repeating series of steps) to use when the divisor is not a monomial. To use the division algorithm to divide $x^2 + 7x + 12$ (the **dividend**) by $x + 4$ (the **divisor**), we write the division in long division form and proceed as follows:

$$x + 4\overline{)x^2 + 7x + 12} \quad \text{quotient: } x$$

How many times does x divide x^2? $\frac{x^2}{x} = x$. Place the x in the quotient.

$$\begin{array}{r} x \\ x + 4\overline{)x^2 + 7x + 12} \\ \underline{x^2 + 4x} \\ 3x + 12 \end{array}$$

Multiply each term in the divisor by x to get $x^2 + 4x$, subtract $x^2 + 4x$ from $x^2 + 7x$, and bring down the 12.

$$\begin{array}{r} x + 3 \\ x + 4\overline{)x^2 + 7x + 12} \\ \underline{x^2 + 4x} \\ 3x + 12 \end{array}$$

How many times does x divide $3x$? $\frac{3x}{x} = 3$. Place the 3 in the quotient.

$$\begin{array}{r} x + 3 \\ x + 4\overline{)x^2 + 7x + 12} \\ \underline{x^2 + 4x} \\ 3x + 12 \\ \underline{3x + 12} \\ 0 \end{array}$$

Multiply each term in the divisor by 3 to get $3x + 12$, and subtract $3x + 12$ from $3x + 12$ to get 0.

Success Tip

The long division process is a series of four steps that are repeated: Divide, multiply, subtract, and bring down.

The division process stops when the result of the subtraction is a constant or a polynomial with degree less than the degree of the divisor. Here, the quotient is $x + 3$ and the remainder is 0.

We can check the answer using the fact that for any division:

divisor · quotient + remainder = dividend

$$\underbrace{(x + 4)}_{\text{Divisor}} \cdot \underbrace{(x + 3)}_{\text{quotient}} + \underbrace{0}_{\text{remainder}} = \underbrace{x^2 + 7x + 12}_{\text{dividend}}$$

Check: $(x + 4)(x + 3) + 0 = x^2 + 7x + 12$ The quotient checks.

EXAMPLE 3 Divide: $\dfrac{2a^3 + 9a^2 + 5a - 6}{2a + 3}$.

Solution

$$\begin{array}{r} a^2 \\ 2a + 3\overline{)2a^3 + 9a^2 + 5a - 6} \end{array}$$

How many times does $2a$ divide $2a^3$? $\frac{2a^3}{2a} = a^2$. Place a^2 in the quotient.

$$\begin{array}{r} a^2 \\ 2a + 3\overline{)2a^3 + 9a^2 + 5a - 6} \\ \underline{2a^3 + 3a^2} \\ 6a^2 + 5a \end{array}$$

Multiply each term in the divisor by a^2 to get $2a^3 + 3a^2$, subtract $2a^3 + 3a^2$ from $2a^3 + 9a^2$, and bring down the $5a$.

$$\begin{array}{r}
a^2 + 3a \phantom{{}+5a-6} \\
2a + 3\overline{)2a^3 + 9a^2 + 5a - 6} \\
\underline{2a^3 + 3a^2} \phantom{{}+5a-6} \\
6a^2 + 5a \phantom{{}-6}
\end{array}$$

How many times does $2a$ divide $6a^2$? $\frac{6a^2}{2a} = 3a$. Place the $+3a$ in the quotient.

$$\begin{array}{r}
a^2 + 3a \phantom{{}+5a-6} \\
2a + 3\overline{)2a^3 + 9a^2 + 5a - 6} \\
\underline{2a^3 + 3a^2} \phantom{{}+5a-6} \\
6a^2 + 5a \phantom{{}-6} \\
\underline{6a^2 + 9a} \phantom{{}-6} \\
-4a - 6
\end{array}$$

Multiply each term in the divisor by $3a$ to get $6a^2 + 9a$, subtract $6a^2 + 9a$ from $6a^2 + 5a$, and bring down the -6.

$$\begin{array}{r}
a^2 + 3a - 2 \phantom{{}-6} \\
2a + 3\overline{)2a^3 + 9a^2 + 5a - 6} \\
\underline{2a^3 + 3a^2} \phantom{{}+5a-6} \\
6a^2 + 5a \phantom{{}-6} \\
\underline{6a^2 + 9a} \phantom{{}-6} \\
-4a - 6
\end{array}$$

How many times does $2a$ divide $-4a$? $\frac{-4a}{2a} = -2$. Place the -2 in the quotient.

$$\begin{array}{r}
a^2 + 3a - 2 \phantom{{}-6} \\
2a + 3\overline{)2a^3 + 9a^2 + 5a - 6} \\
\underline{2a^3 + 3a^2} \phantom{{}+5a-6} \\
6a^2 + 5a \phantom{{}-6} \\
\underline{6a^2 + 9a} \phantom{{}-6} \\
-4a - 6 \\
\underline{-4a - 6} \\
0
\end{array}$$

Multiply each term in the divisor by -2 to get $-4a - 6$; subtract $-4a - 6$ from $-4a - 6$ to get 0.

Since the remainder is 0, the quotient is $a^2 + 3a - 2$. We can check the quotient by verifying that

$$\underbrace{\text{Divisor}}\cdot\underbrace{\text{quotient}} = \underbrace{\text{dividend}}$$

Check: $\quad (2a + 3)(a^2 + 3a - 2) = 2a^3 + 9a^2 + 5a - 6$

Self Check 3 Divide: $\dfrac{4x^3 - 5x^2 - 2x + 3}{4x + 3}$.

EXAMPLE 4

Divide: $\dfrac{3x^3 + 2x^2 - 3x - 28}{x - 2}$.

Solution

$$\begin{array}{r}
3x^2 + 8x\ + 13 \phantom{{}-28} \\
x - 2\overline{)3x^3 + 2x^2 - \ 3x - 28} \\
\underline{3x^3 - 6x^2} \phantom{{}-\ 3x - 28} \\
8x^2 - \ 3x \phantom{{}-28} \\
\underline{8x^2 - 16x} \phantom{{}-28} \\
13x - 28 \\
\underline{13x - 26} \\
-2
\end{array}$$

Subtract: $-28 - (-26) = -2$.

The remainder is negative.

This division gives a quotient of $3x^2 + 8x + 13$ and a remainder of -2. It is common to form a fraction with the remainder as the numerator and the divisor as the denominator and to write the result as

Notation

Because the remainder is negative, we can also write the result as:

$$3x^2 + 8x + 13 - \frac{2}{x - 2}$$

$$3x^2 + 8x + 13 + \frac{-2}{x - 2}$$

To check, we verify that

$$(x - 2)\left(3x^2 + 8x + 13 + \frac{-2}{x - 2}\right) = 3x^3 + 2x^2 - 3x - 28.$$

Self Check 4 Divide: $\dfrac{2a^3 + 3a^2 - a - 85}{a - 3}$.

EXAMPLE 5

Divide: $(-9x + 8x^3 + 10x^2 - 9) \div (3 + 2x)$.

Solution The division algorithm works best when the polynomials in the dividend and the divisor are written in descending powers of x. We can use the commutative property of addition to rearrange the terms. Then the division is routine:

$$\begin{array}{r@{}l}
 & 4x^2 - x - 3 \\
2x + 3 & \overline{)8x^3 + 10x^2 - 9x - 9} \\
 & \underline{8x^3 + 12x^2} \\
 & \quad - 2x^2 - 9x \\
 & \quad \underline{- 2x^2 - 3x} \\
 & \qquad - 6x - 9 \\
 & \qquad \underline{- 6x - 9} \\
 & \qquad\qquad 0
\end{array}$$

Thus,

$$\frac{-9x + 8x^3 + 10x^2 - 9}{3 + 2x} = 4x^2 - x - 3$$

Self Check 5 Divide: $2 + 3a\overline{)-4a + 15a^2 + 18a^3 - 4}$.

MISSING TERMS

If a power of the variable is missing in the dividend, it is helpful to insert placeholder terms, because they aid in the subtraction step of the long division procedure.

EXAMPLE 6

Divide $8x^3 + 1$ by $2x + 1$.

Solution When we write the terms in the dividend in descending powers of x, we see that the terms involving x^2 and x are missing. We can introduce the terms $0x^2$ and $0x$ in the dividend or leave spaces for them. Then the division is routine.

$$\begin{array}{r} 4x^2 - 2x + 1 \\ 2x + 1\overline{)8x^3 + 0x^2 + 0x + 1} \\ \underline{8x^3 + 4x^2} \\ -4x^2 + 0x \\ \underline{-4x^2 - 2x} \\ 2x + 1 \\ \underline{2x + 1} \\ 0 \end{array}$$

Thus,

$$\frac{8x^3 + 1}{2x + 1} = 4x^2 - 2x + 1$$

Self Check 6 Divide $27a^3 - 1$ by $3a - 1$.

EXAMPLE 7

Divide $-17x^2 + 5x + x^4 + 2$ by $x^2 - 1 + 4x$.

Solution We write the problem with the divisor and the dividend in descending powers of x. After introducing $0x^3$ for the missing term in the dividend, we proceed as follows:

$$\begin{array}{r} x^2 - 4x \\ x^2 + 4x - 1\overline{)x^4 + 0x^3 - 17x^2 + 5x + 2} \\ \underline{x^4 + 4x^3 - x^2} \\ -4x^3 - 16x^2 + 5x \\ \underline{-4x^3 - 16x^2 + 4x} \\ x + 2 \end{array}$$

This division gives a quotient of $x^2 - 4x$ and a remainder of $x + 2$.

$$\frac{-17x^2 + 5x + x^4 + 2}{x^2 - 1 + 4x} = x^2 - 4x + \frac{x + 2}{x^2 + 4x - 1}$$

Self Check 7 Divide: $\dfrac{2a^2 + 3a^3 + a^4 - 6 + a}{a^2 - 2a + 1}$.

Answers to Self Checks **1. a.** $6y^2$, **b.** $\dfrac{d^4}{4c^3}$ **2. a.** $10h + 3$, **b.** $2s^3 - \dfrac{s^2t}{11} + \dfrac{4}{t}$ **3.** $x^2 - 2x + 1$

4. $2a^2 + 9a + 26 - \dfrac{7}{a - 3}$ **5.** $6a^2 + a - 2$ **6.** $9a^2 + 3a + 1$

7. $a^2 + 5a + 11 + \dfrac{18a - 17}{a^2 - 2a + 1}$

6.5 STUDY SET

VOCABULARY Fill in the blanks.

1. The expression $\frac{18x^7}{9x^4}$ is a monomial divided by a ________. The expression $\frac{6x^3y - 4x^2y^2 + 8xy^3 - 2y^4}{2x^4}$ is a ________ divided by a monomial. The expression $\frac{x^2 - 8x + 12}{x - 6}$ is a trinomial divided by a ________.

2. The powers of x in $2x^4 + 3x^3 + 4x^2 - 7x - 8$ are written in ________ order.

3.
$$\begin{array}{rl}
\square & \\
\downarrow \qquad x - 2 & \leftarrow \square \\
x - 6\overline{)x^2 - 8x - 4} & \leftarrow \square \\
\underline{x^2 - 6x} & \\
-2x - 4 & \\
\underline{-2x + 12} & \\
-16 & \leftarrow \square
\end{array}$$

4. The expression $5x^2 + 6$ is missing an x-term. We can insert a ________ $0x$ term and write it as $5x^2 + 0x + 6$.

CONCEPTS Fill in the blanks.

5. **a.** To divide a polynomial by a monomial, divide each ______ of the polynomial by the monomial.

   b. $\frac{18x + 9}{9} = \frac{18x}{\square} + \frac{9}{\square}$

   c. $\frac{30x^2 + 12x - 24}{6} = \frac{30x^2}{\square} + \frac{12x}{\square} - \frac{24}{\square}$

6. Divisor · ________ + remainder = dividend

7. Suppose that after dividing $2x^3 + 5x^2 - 11x + 4$ by $2x - 1$, you obtain $x^2 + 3x - 4$. Show how multiplication can be used to check the result.

8. Consider the first step of the division process for
$$2x^2 - 1\overline{)4x^4 + 0x^3 + 0x^2 + 0x - 1}$$
How many times does $2x^2$ divide $4x^4$?

NOTATION Complete each solution.

9.
$$\begin{array}{r}
2x + 1 \\
x + 4\overline{)2x^2 + 9x + 4} \\
\underline{\square + 8x} \\
\square + 4 \\
\underline{x + \square} \\
0
\end{array}$$

10.
$$\begin{array}{r}
2x - 1 \\
3x + 4\overline{)6x^2 + 5x - 4} \\
\underline{6x^2 + \square} \\
\square - 4 \\
\underline{-3x - \square} \\
\square
\end{array}$$

11. If a polynomial is divided by $3a - 2$ and the quotient is $3a^2 + 5$ with a remainder of 6, how do we write the result?

12. A polynomial is divided by $3a - 2$. The quotient is $3a^2 + 5$ with a remainder of -6. Write the answer to the division in two ways.

13. List three ways we can use symbols to write $x^2 - x - 12$ divided by $x - 4$.

14. Is the following statement true or false? Justify your answer.
$$2x^3 - 9 = 2x^3 + 0x^2 + 0x - 9$$

PRACTICE Perform each division. Write each answer using positive exponents only.

15. $\frac{4x^2y^3}{8x^5y^2}$

16. $\frac{25x^4y^7}{5xy^9}$

17. $-\frac{33a^2b^2}{44a^4b^2}$

18. $\frac{-63a^4}{81a^6b^3}$

19. $\frac{4x + 6}{2}$

20. $\frac{11a^3 - 11a^2}{11}$

21. $\frac{4x^2 - x^3}{-6x}$

22. $\frac{5y^4 + 45y^3}{-15y^2}$

23. $\frac{12x^2y^3 + x^3y^2}{6xy}$

24. $\frac{54a^3y^2 - 18a^4y^3}{27a^2y^2}$

25. $\frac{24x^6y^7 - 12x^5y^{12} + 36xy}{48x^2y^3}$

26. $\frac{9x^4y^3 + 18x^2y - 27xy^4}{-9x^3y^3}$

Perform each division.

27. $\dfrac{x^2 + 5x + 6}{x + 3}$

28. $\dfrac{x^2 - 5x + 6}{x - 3}$

29. $\dfrac{x^2 + 10x + 21}{x + 3}$

30. $\dfrac{x^2 + 10x + 21}{x + 7}$

31. $\dfrac{6x^2 - x - 12}{2x + 3}$

32. $\dfrac{6x^2 - x - 12}{2x - 3}$

33. $\dfrac{3x^3 - 2x^2 + x - 6}{x - 1}$

34. $\dfrac{4a^3 + a^2 - 3a + 7}{a + 1}$

35. $\dfrac{6x^3 + 11x^2 - x - 2}{3x - 2}$

36. $\dfrac{6x^3 + 11x^2 - 9x - 5}{2x + 3}$

37. $\dfrac{6x^3 - x^2 - 6x - 9}{2x - 3}$

38. $\dfrac{16x^3 + 16x^2 - 9x - 5}{4x + 5}$

39. $(2a + 1 + a^2) \div (a + 1)$

40. $(a - 15 + 6a^2) \div (2a - 3)$

41. $(6y - 4 + 10y^2) \div (5y - 2)$

42. $(-10x + x^2 + 16) \div (x - 2)$

43. $\dfrac{-18x + 12 + 6x^2}{x - 1}$

44. $\dfrac{27x + 23x^2 + 6x^3}{2x + 3}$

45. $\dfrac{13x + 16x^4 + 3x^2 + 3}{4x + 3}$

46. $\dfrac{3x^2 + 9x^3 + 4x + 4}{3x + 2}$

47. $a^3 + 1$ divided by $a - 1$

48. $27a^3 - 8$ divided by $3a - 2$

49. $\dfrac{15a^3 - 29a^2 + 16}{3a - 4}$

50. $\dfrac{4x^3 - 12x^2 + 17x - 12}{2x - 3}$

51. $y - 2\overline{)-24y + 24 + 6y^2}$

52. $3 - a\overline{)21a - a^2 - 54}$

53. $x^2 - 2\overline{)x^6 - x^4 + 2x^2 - 8}$

54. $x^2 + 3\overline{)x^6 + 2x^4 - 6x^2 - 9}$

55. $\dfrac{x^4 + 2x^3 + 4x^2 + 3x + 2}{x^2 + x + 2}$

56. $\dfrac{2x^4 + 3x^3 + 3x^2 - 5x - 3}{2x^2 - x - 1}$

57. $\dfrac{x^3 + 3x + 5x^2 + 6 + x^4}{x^2 + 3}$

58. $\dfrac{x^5 + 3x + 2}{x^3 + 1 + 2x}$

59. $\dfrac{-9a^4 - 6a^3 - 55a^2 - 18a - 81}{3a^2 + a + 9}$

60. $\dfrac{-8x^4 + 6x^3 - 11x^2 + 4x - 3}{4x^2 - x + 3}$

61. $\dfrac{a^8 + a^6 - 4a^4 + 5a^2 - 3}{a^4 + 2a^2 - 3}$

62. $\dfrac{c^{24} + c^{18} - 3c^{12} + 7c^6 - 6}{c^{12} + 2c^6 - 3}$

63. $3x^2 - 7x + 4\overline{)6x^3 - 5x^2 + 7x - 1}$

64. $3m^2 - m + 4\overline{)9m^3 - 6m^2 + 5m - 11}$

Use a calculator to help find each quotient.

65. $x - 2\overline{)9.8x^2 - 3.2x - 69.3}$

66. $2.5x - 3.7\overline{)-22.25x^2 - 38.9x - 16.65}$

APPLICATIONS

67. ADVERTISING Find the length of one of the longer sides of the billboard if its area is given by $x^3 - 4x^2 + x + 6$.

68. MASONRY The trowel shown below is in the shape of an isosceles triangle. Find the height if its area is given by $6 + 18t + t^2 + 3t^3$.

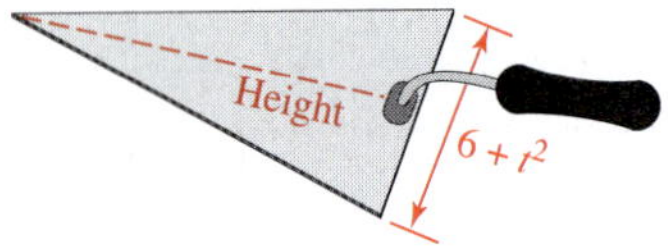

69. WINTER TRAVEL Complete the following table, which lists the rate (mph), time traveled (hr), and distance traveled (mi) by an Alaskan trail guide using two different means of travel.

	r ·	t =	d
Dog sled		$4x + 7$	$12x^2 + 13x - 14$
Snowshoes	$3x + 4$		$3x^2 + 19x + 20$

70. PRICING Complete the table for two items sold at a produce store.

	Price per lb ·	**Number of lb =**	**Value**
Cashews	$x^2 + 2x + 4$		$x^4 + 4x^2 + 16$
Sunflower seeds		$x^2 + 6$	$x^4 - x^2 - 42$

WRITING

71. Explain how to divide a monomial by a monomial.

72. Explain how to check the result of a division problem if there is a nonzero remainder.

REVIEW **Simplify each expression.**

73. $2(x^2 + 4x - 1) + 3(2x^2 - 2x + 2)$

74. $3(2a^2 - 3a + 2) - 4(2a^2 + 4a - 7)$

75. $-2(3y^3 - 2y + 7) - (y^2 + 2y - 4) + 4(y^3 + 2y - 1)$

76. $3(4y^3 + 3y - 2) + 2(3y^2 - y + 3) - 5(2y^3 - y^2 - 2)$

CHALLENGE PROBLEMS **Perform each division.**

77. $\left(3c^2 - \frac{7}{4}c - 3\right) \div (4c + 3)$

78. $x - 1\overline{)x^5 - 1}$

79. $\dfrac{c^4 - c^2d^2 + 10c^2 - 6d^2 + 23}{c^2 + 6}$

80. $\dfrac{0.03a^2 + 0.17a + 0.1}{0.03a + 0.02}$

(*Hint:* Think of a way to simplify the division.)

6.6 Synthetic Division

• Synthetic division • The remainder theorem • The factor theorem

We have discussed how to divide polynomials by polynomials using a long division process. We will now discuss a shortcut method, called **synthetic division,** that we can use to divide a polynomial by a binomial of the form $x - k$.

SYNTHETIC DIVISION

To see how synthetic division works, we consider the division of $4x^3 - 5x^2 - 11x + 20$ by $x - 2$.

$$
\begin{array}{rl}
 & 4x^2 + 3x - 5 \\
x - 2 & \overline{)4x^3 - 5x^2 - 11x + 20} \\
 & \underline{4x^3 - 8x^2} \\
 & 3x^2 - 11x \\
 & \underline{3x^2 - 6x} \\
 & -5x + 20 \\
 & \underline{-5x + 10} \\
 & 10 \quad \text{(remainder)}
\end{array}
\qquad
\begin{array}{rl}
 & 4 \quad 3 - 5 \\
1 - 2 & \overline{)4 - 5 - 11 \quad 20} \\
 & \underline{4 - 8} \\
 & 3 - 11 \\
 & \underline{3 - 6} \\
 & -5 \quad 20 \\
 & \underline{-5 \quad 10} \\
 & 10 \quad \text{(remainder)}
\end{array}
$$

On the left is the long division, and on the right is the same division with the variables and their exponents removed. The various powers of x can be remembered without actually writing them, because the exponents of the terms in the divisor, dividend, and quotient were written in descending order.

We can further shorten the version on the right. The numbers printed in color need not be written, because they are duplicates of the numbers above them. Thus, we can write the division in the following form:

The Language of Algebra

Synthetic means devised to imitate something natural. You've probably heard of *synthetic* fuels or *synthetic* fibers. *Synthetic* division imitates the long division process.

$$\begin{array}{r|rrrr}
 & 4 & 3 & -5 & \\
\hline
1-2 & 4 & -5 & -11 & 20 \\
 & -8 & & & \\
\hline
 & & 3 & & \\
 & & -6 & & \\
\hline
 & & & -5 & \\
 & & & & 10 \\
\hline
 & & & & 10
\end{array}$$

We can shorten the process further by compressing the work vertically and eliminating the 1 (the coefficient of x in the divisor):

$$\begin{array}{r|rrrr}
 & 4 & 3 & -5 & \\
\hline
-2 & 4 & -5 & -11 & 20 \\
 & & -8 & -6 & 10 \\
\hline
 & & 3 & -5 & 10
\end{array}$$

If we write the 4 in the quotient on the bottom line, the bottom line gives the coefficients of the quotient and the remainder. If we eliminate the top line, the division appears as follows:

$$\begin{array}{r|rrrr}
-2 & 4 & -5 & -11 & 20 \\
 & & -8 & -6 & 10 \\
\hline
 & 4 & 3 & -5 & 10
\end{array}$$

The bottom line was obtained by subtracting the middle line from the top line. If we replace the -2 in the divisor by 2, the division process will reverse the signs of every entry in the middle line, and then the bottom line can be obtained by addition. This gives the final form of the synthetic division.

The Language of Algebra

Synthetic division is used to divide a polynomial by a binomial of the form $x - k$. We call k the **synthetic divisor.** In this example, we are dividing by $x - 2$, so k is 2.

$$\begin{array}{r|rrrr}
2 & 4 & -5 & -11 & 20 \\
 & & 8 & 6 & -10 \\
\hline
 & 4 & 3 & -5 & 10
\end{array}$$

These are the coefficients of the dividend.

These are the coefficients of the quotient and the remainder.

$$4x^2 + 3x - 5 + \frac{10}{x-2}$$

Read the result from the bottom row.

Thus,

$$\frac{4x^3 - 5x^2 - 11x + 20}{x - 2} = 4x^2 + 3x - 5 + \frac{10}{x - 2}$$

EXAMPLE 1

Use synthetic division to find $(6x^2 + 5x - 2) \div (x - 5)$.

Solution We write the coefficients in the dividend and the 5 in the divisor in the following form:

Since we are dividing the polynomial by $x - 5$, the synthetic divisor is 5. →

$$\begin{array}{r|rrr} 5 & 6 & 5 & -2 \\ \hline & & & \end{array}$$

← This represents the dividend $6x^2 + 5x - 2$.

Then we follow these steps:

Success Tip

In his process, numbers below the line are multiplied by the synthetic divisor and that product is carried above the line to the next column. Numbers above the horizontal line are added.

$$\begin{array}{r|rrr} 5 & 6 & 5 & -2 \\ & \downarrow & & \\ \hline & 6 & & \end{array}$$

Begin by bringing down the 6.

$$\begin{array}{r|rrr} 5 & 6 & 5 & -2 \\ & & 30 & \\ \hline & 6 & & \end{array}$$

Multiply 5 by 6 to get 30.

$$\begin{array}{r|rrr} 5 & 6 & 5 & -2 \\ & & 30 & \\ \hline & 6 & 35 & \end{array}$$

Add 5 and 30 to get 35.

$$\begin{array}{r|rrr} 5 & 6 & 5 & -2 \\ & & 30 & 175 \\ \hline & 6 & 35 & \end{array}$$

Multiply 35 by 5 to get 175.

$$\begin{array}{r|rrr} 5 & 6 & 5 & -2 \\ & & 30 & 175 \\ \hline & 6 & 35 & 173 \end{array}$$

Add -2 and 175 to get 173.

The numbers 6 and 35 represent the quotient $6x + 35$, and 173 is the remainder. Thus,

$$\frac{6x^2 + 5x - 2}{x - 5} = 6x + 35 + \frac{173}{x - 5}$$

Self Check 1 Divide $5x^2 - 4x + 2$ by $x - 3$.

EXAMPLE 2

Use synthetic division to find $\dfrac{x^3 + x^2 - 1}{x - 3}$.

Solution We begin by writing

$$\begin{array}{r|rrrr} 3 & 1 & 1 & 0 & -1 \\ \hline & & & & \end{array}$$

Write 0 for the coefficient of x, the missing term.

and complete the division as follows.

3	1	1	0	−1
		3		
	1	4		

Multiply, then add.

3	1	1	0	−1
		3	12	
	1	4	12	

Multiply, then add.

3	1	1	0	−1
		3	12	36
	1	4	12	35

Multiply, then add.

Thus,

$$\frac{x^3 + x^2 - 1}{x - 3} = x^2 + 4x + 12 + \frac{35}{x - 3}$$

Self Check 2 Use synthetic division to find $\dfrac{x^3 + 3x - 90}{x - 4}$.

EXAMPLE 3

Use synthetic division to divide $5x^2 + 6x^3 + 2 - 4x$ by $x + 2$.

Solution First, we write the dividend with the exponents in descending order.

$$6x^3 + 5x^2 - 4x + 2$$

Then we write the divisor in $x - k$ form: $x - (\mathbf{-2})$. Thus, $k = -2$. Using synthetic division, we begin by writing

This represents division by $x + 2$. →

−2	6	5	−4	2

and complete the division.

−2	6	5	−4	2
		−12	14	−20
	6	−7	10	−18

The remainder is negative.

Notation

Because the remainder is negative, we can also write the result as

$$6x^2 - 7x + 10 + \frac{-18}{x + 2}$$

Thus,

$$\frac{5x^2 + 6x^3 + 2 - 4x}{x + 2} = 6x^2 - 7x + 10 - \frac{18}{x + 2}$$

Self Check 3 Divide $2x - 4x^2 + 3x^3 - 3$ by $x + 1$.

THE REMAINDER THEOREM

Synthetic division is important because of the **remainder theorem.**

Remainder Theorem If a polynomial $P(x)$ is divided by $x - k$, the remainder is $P(k)$.

It follows from the remainder theorem that we can evaluate polynomials using synthetic division. We illustrate this in the following example.

EXAMPLE 4

Let $P(x) = 2x^3 - 3x^2 - 2x + 1$. Find **a.** $P(3)$ and **b.** the remainder when $P(x)$ is divided by $x - 3$.

Solution To find $P(3)$ we evaluate the function for $x = 3$.

Notation

Naming the function with the letter *P*, instead of *f*, stresses that we are working with a polynomial function.

a.
$$\begin{aligned} P(3) &= 2(3)^3 - 3(3)^2 - 2(3) + 1 \quad \text{Substitute 3 for } x.\\ &= 2(27) - 3(9) - 6 + 1\\ &= 54 - 27 - 6 + 1\\ &= 22 \end{aligned}$$

Thus, $P(3) = 22$.

b. We use synthetic division to find the remainder when $2x^3 - 3x^2 - 2x + 1$ is divided by $x - 3$.

Success Tip

It is often easier to find $P(k)$ by using synthetic division than by substituting k for x in $P(x)$. This is especially true if k is a decimal.

$$\begin{array}{r|rrrr} 3 & 2 & -3 & -2 & 1\\ & & 6 & 9 & 21\\ \hline & 2 & 3 & 7 & 22 \end{array}$$

Thus, the remainder is 22.

The same results in parts a and b show that rather than substituting 3 for x in $P(x) = 2x^3 - 3x^2 - 2x + 1$, we can divide the polynomial $2x^3 - 3x^2 - 2x + 1$ by $x - 3$ to find $P(3)$.

Self Check 4 Let $P(x) = 5x^3 - 3x^2 + x + 6$. Find **a.** $P(1)$ and **b.** use synthetic division to find the remainder when $P(x)$ is divided by $x - 1$.

THE FACTOR THEOREM

If two quantities are multiplied, each is called a **factor** of the product. Thus, $x - 2$ is a factor of $6x - 12$, because $6(x - 2) = 6x - 12$. A theorem, called the **factor theorem,** tells us how to find one factor of a polynomial if the remainder of a certain division is 0.

Factor Theorem If $P(x)$ is a polynomial in x, then

$$P(k) = 0 \quad \text{if and only if} \quad x - k \text{ is a factor of } P(x)$$

If $P(x)$ is a polynomial in x and if $P(k) = 0$, k is called a **zero of the polynomial function.**

EXAMPLE 5

Let $P(x) = 3x^3 - 5x^2 + 3x - 10$. Show that **a.** $P(2) = 0$ and **b.** $x - 2$ is a factor of $P(x)$.

Solution **a.** Use the remainder theorem to evaluate $P(2)$ by dividing $P(x) = 3x^3 - 5x^2 + 3x - 10$ by $x - 2$.

$$\begin{array}{r|rrrr} 2 & 3 & -5 & 3 & -10 \\ & & 6 & 2 & 10 \\ \hline & 3 & 1 & 5 & \mathbf{0} \end{array}$$

The remainder in this division is 0. By the remainder theorem, the remainder is $P(2)$. Thus, $P(2) = 0$, and 2 is a zero of the polynomial.

b. Because the remainder is 0, the numbers 3, 1, and 5 in the synthetic division in part a represent the quotient $3x^2 + x + 5$. Thus,

$$\underbrace{(x - 2)}_{\text{Divisor}} \cdot \underbrace{(3x^2 + x + 5)}_{\text{quotient}} + \underbrace{0}_{\text{remainder}} = \underbrace{3x^3 - 5x^2 + 3x - 10}_{\text{the dividend, } P(x)}$$

or

$$(x - 2)(3x^2 + x + 5) = 3x^3 - 5x^2 + 3x - 10$$

Thus, $x - 2$ is a factor of $3x^3 - 5x^2 + 3x - 10$.

The Language of Algebra

The phrase *if and only if* in the factor theorem means:

If $P(2) = 0$, then $x - 2$ is a factor of $P(x)$

and

If $x - 2$ is a factor of $P(x)$, then $P(2) = 0$.

Self Check 5 Let $P(x) = x^3 - 4x^2 + x + 6$. Show that $x + 1$ is a factor of $P(x)$ using synthetic division.

The result in Example 5 is true, because the remainder, $P(2)$, is 0. If the remainder had not been 0, then $x - 2$ would not have been a factor of $P(x)$.

ACCENT ON TECHNOLOGY: APPROXIMATING ZEROS OF POLYNOMIALS

We can use a graphing calculator to approximate the real zeros of a polynomial function $f(x)$. For example, to find the real zeros of $f(x) = 2x^3 - 6x^2 + 7x - 21$, we graph the function as in the figure.

It is clear from the figure that the function f has a zero at $x = 3$.

$$\begin{aligned} f(3) &= 2(3)^3 - 6(3)^2 + 7(3) - 21 && \text{Substitute 3 for } x. \\ &= 2(27) - 6(9) + 21 - 21 \\ &= 0 \end{aligned}$$

From the factor theorem, we know that $x - 3$ is a factor of the polynomial. To find the other factor, we can synthetically divide by 3.

$$\begin{array}{r|rrrr} 3 & 2 & -6 & 7 & -21 \\ & & 6 & 0 & 21 \\ \hline & 2 & 0 & 7 & 0 \end{array}$$

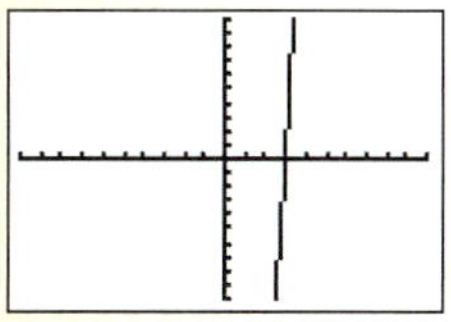

Thus, $f(x) = (x - 3)(2x^2 + 7)$. Since $2x^2 + 7$ cannot be factored over the real numbers, we can conclude that 3 is the only real zero of the polynomial function.

Answers to Self Checks **1.** $5x + 11 + \frac{35}{x - 3}$ **2.** $x^2 + 4x + 19 - \frac{14}{x - 4}$ **3.** $3x^2 - 7x + 9 - \frac{12}{x + 1}$ **4.** 9 **5.** Since $P(-1) = 0$, $x + 1$ is a factor of $P(x)$.

6.6 STUDY SET

VOCABULARY Fill in the blanks.

1. The method of dividing $x^2 + 2x - 9$ by $x - 4$ shown below is called ________ division.

$$\begin{array}{r|rrr} 4 & 1 & 2 & -9 \\ & & 4 & 24 \\ \hline & 1 & 6 & 15 \end{array}$$

2. Synthetic division is used to divide a polynomial by a ________ of the form $x - k$.
3. In Exercise 1, the synthetic ________ is 4.
4. By the ________ theorem, if a polynomial $P(x)$ is divided by $x - k$, the remainder is $P(k)$.
5. The factor ________ tells us how to find one factor of a polynomial if the remainder of a certain division is 0.
6. If $P(x)$ is a polynomial and if $P(k) = 0$, then k is called a ______ of the polynomial.

CONCEPTS

7. **a.** What division is represented below?

 b. What is the answer?

$$\begin{array}{r|rrrr} -2 & 5 & 0 & 1 & -3 \\ & & -10 & 20 & -42 \\ \hline & 5 & -10 & 21 & -45 \end{array}$$

Fill in the blanks.

8. In the synthetic division process, numbers below the line are ________ by the synthetic divisor and that product is carried above the line to the next column. Numbers above the horizontal line are ______.
9. Rather than substituting 8 for x in $P(x) = 6x^3 - x^2 - 17x + 9$, we can divide the polynomial ______________ by ______ to find $P(8)$.
10. For $P(x) = x^3 - 4x^2 + x + 6$, suppose we know that $P(3) = 0$. Then ______ is a factor of $x^3 - 4x^2 + x + 6$.

NOTATION Complete each synthetic division.

11. Divide $6x^3 + x^2 - 23x + 2$ by $x - 2$.

$$\begin{array}{r|rrrr} \square & 6 & \square & -23 & 2 \\ & & \square & \square & 6 \\ \hline & \square & 13 & 3 & \square \end{array}$$

12. Divide $2x^3 - 4x^2 - 25x + 15$ by $x + 3$.

$$\begin{array}{r|rrrr} \square & 2 & -4 & \square & 15 \\ & & \square & 30 & \square \\ \hline & \square & \square & \square & 0 \end{array}$$

PRACTICE Use synthetic division to perform each division.

13. $\frac{x^2 + x - 2}{x - 1}$
14. $\frac{x^2 + x - 6}{x - 2}$
15. $\frac{x^2 - 7x + 12}{x - 4}$
16. $\frac{x^2 - 6x + 5}{x - 5}$
17. $\frac{x^2 + 8 + 6x}{x + 4}$
18. $\frac{x^2 - 15 - 2x}{x + 3}$
19. $\frac{x^2 - 5x + 14}{x + 2}$
20. $\frac{x^2 + 13x + 42}{x + 6}$
21. $\frac{3x^3 - 10x^2 + 5x - 6}{x - 3}$
22. $\frac{2x^3 - 9x^2 + 10x - 3}{x - 3}$
23. $\frac{2x^3 - 6 - 5x}{x - 2}$
24. $\frac{4x^3 - 1 + 5x^2}{x + 2}$
25. $\frac{5x^2 + 4 + 6x^3}{x + 1}$
26. $\frac{4 - 3x^2 + x}{x - 4}$
27. $\frac{t^3 + t^2 + t + 2}{t + 1}$
28. $\frac{m^3 - m^2 - m - 1}{m - 1}$
29. $\frac{a^5 - 1}{a - 1}$
30. $\frac{b^4 - 81}{b - 3}$

31. $\dfrac{-5x^5 + 4x^4 + 30x^3 + 2x^2 + 20x + 3}{x - 3}$

32. $\dfrac{-6c^5 + 14c^4 + 38c^3 + 4c^2 + 25c - 36}{c - 4}$

33. $\dfrac{8t^3 - 4t^2 + 2t - 1}{t - \frac{1}{2}}$

34. $\dfrac{9a^3 + 3a^2 - 21a - 7}{a + \frac{1}{3}}$

35. $\dfrac{x^4 - x^3 - 56x^2 - 2x + 16}{x - 8}$

36. $\dfrac{x^4 - 9x^3 + x^2 - 7x - 20}{x - 9}$

Use a calculator and synthetic division to perform each division.

37. $\dfrac{7.2x^2 - 2.1x + 0.5}{x - 0.2}$

38. $\dfrac{2.7x^2 + x - 5.2}{x + 1.7}$

39. $\dfrac{9x^3 - 25}{x + 57}$

40. $\dfrac{0.5x^3 + x}{x - 2.3}$

Let $P(x) = 2x^3 - 4x^2 + 2x - 1$. Evaluate $P(x)$ by substituting the given value of x into the polynomial and simplifying. Then evaluate the polynomial by using the remainder theorem and synthetic division.

41. $P(1)$
42. $P(2)$
43. $P(-2)$
44. $P(-1)$
45. $P(3)$
46. $P(-4)$
47. $P(0)$
48. $P(4)$

Let $Q(x) = x^4 - 3x^3 + 2x^2 + x - 3$. Evaluate $Q(x)$ by substituting the given value of x into the polynomial and simplifying. Then evaluate the polynomial by using the remainder theorem and synthetic division.

49. $Q(-1)$
50. $Q(1)$
51. $Q(2)$
52. $Q(-2)$
53. $Q(3)$
54. $Q(0)$
55. $Q(-3)$
56. $Q(-4)$

Use the remainder theorem and synthetic division to find $P(k)$.

57. $P(x) = x^3 - 4x^2 + x - 2;\ k = 2$
58. $P(x) = x^3 - 3x^2 + x + 1;\ k = 1$
59. $P(x) = 2x^3 + x + 2;\ k = 3$
60. $P(x) = x^3 + x^2 + 1;\ k = -2$
61. $P(x) = x^4 - 2x^3 + x^2 - 3x + 2;\ k = -2$
62. $P(x) = x^5 + 3x^4 - x^2 + 1;\ k = -1$
63. $P(x) = 3x^5 + 1;\ k = -\dfrac{1}{2}$
64. $P(x) = 5x^7 - 7x^4 + x^2 + 1;\ k = 2$

Use the factor theorem and tell whether the first expression is a factor of $P(x)$.

65. $x - 3;\ P(x) = x^3 - 3x^2 + 5x - 15$
66. $x + 1;\ P(x) = x^3 + 2x^2 - 2x - 3$
(*Hint:* Write $x + 1$ as $x - (-1)$.)
67. $x + 2;\ P(x) = 3x^2 - 7x + 4$
(*Hint:* Write $x + 2$ as $x - (-2)$.)
68. $x;\ P(x) = 7x^3 - 5x^2 - 8x$
(*Hint:* $x = x - 0$.)

Use a calculator to work each problem.

69. Find 2^6 by using synthetic division to evaluate the polynomial $P(x) = x^6$ at $x = 2$. Then check the answer by evaluating 2^6 with a calculator.
70. Find $(-3)^5$ by using synthetic division to evaluate the polynomial $P(x) = x^5$ at $x = -3$. Then check the answer by evaluating $(-3)^5$ with a calculator.

WRITING

71. When dividing a polynomial by a binomial of the form $x - k$, synthetic division is considered to be faster than long division. Explain why.
72. Let $P(x) = x^3 - 6x^2 - 9x + 4$. You now know two ways to find $P(6)$. What are they? Which method do you prefer?
73. Explain the factor theorem.
74. What is a *zero* of a polynomial function?

REVIEW

Evaluate each expression for $x = -3$, $y = -5$, and $z = 0$.

75. $x^2z(y^3 - z)$
76. $|y^3 - z|$
77. $\dfrac{x - y^2}{2y - 1 + x}$
78. $\dfrac{2y + 1}{x} - x$

CHALLENGE PROBLEMS

Suppose that $P(x) = x^{100} - x^{99} + x^{98} - x^{97} + \cdots + x^2 - x + 1$.

79. Find the remainder when $P(x)$ is divided by $x - 1$.
80. Find the remainder when $P(x)$ is divided by $x + 1$.

6.7 Solving Rational Equations

- Solving rational equations
- Extraneous solutions
- Solving formulas for a specified variable
- Problem solving

In this section, we will solve problems from disciplines such as business, photography, aviation, and electronics. When we write mathematical models of these situations, we will encounter a new type of equation, called a *rational equation.*

SOLVING RATIONAL EQUATIONS

If an equation contains one or more rational expressions, it is called a **rational equation.** Some examples are

$$\frac{3}{5} + \frac{7}{x+2} = 2, \quad \frac{x+3}{x-3} = \frac{2}{x^2-4}, \quad \text{and} \quad \frac{-x^2+10}{x^2-1} + \frac{3x}{x-1} = \frac{2x}{x+1}$$

To solve a rational equation, we must find all values of the variable that make the equation true. We do this by multiplying both sides of the equation by the LCD of the rational expressions in the equation to clear it of fractions.

EXAMPLE 1 Solve: $\frac{3}{5} + \frac{7}{x+2} = 2.$

Solution We note that x cannot be -2, because this would give a 0 in the denominator of $\frac{7}{x+2}$. If $x \neq -2$, we can multiply both sides of the equation by $5(x+2)$, the LCD of $\frac{3}{5}$ and $\frac{7}{x+2}$.

$$5(x+2)\left(\frac{3}{5} + \frac{7}{x+2}\right) = 5(x+2)(2) \quad \text{Multiply both sides by the LCD.}$$

$$5(x+2)\left(\frac{3}{5}\right) + 5(x+2)\left(\frac{7}{x+2}\right) = 5(x+2)2 \quad \text{On the left-hand side, distribute } 5(x+2).$$

$$\overset{1}{\cancel{5}}(x+2)\left(\frac{3}{\underset{1}{\cancel{5}}}\right) + 5\overset{1}{\cancel{(x+2)}}\left(\frac{7}{\underset{1}{\cancel{(x+2)}}}\right) = 5(x+2)2 \quad \text{On the left-hand side, simplify: } \frac{5}{5} = 1 \text{ and } \frac{x+2}{x+2} = 1.$$

$$3(x+2) + 5(7) = 10(x+2) \quad \text{Simplify each side.}$$

The resulting equation does not contain any fractions. We now solve this *linear equation* for x.

$$\begin{aligned} 3x + 6 + 35 &= 10x + 20 && \text{Use the distributive property and simplify.} \\ 3x + 41 &= 10x + 20 && \text{Combine like terms.} \\ -7x &= -21 && \text{Subtract } 10x \text{ and 41 from both sides.} \\ x &= 3 && \text{Divide both sides by } -7. \end{aligned}$$

Success Tip

Don't confuse procedures. To *simplify the expression* $\frac{3}{5} + \frac{7}{x+2}$, we build each fraction to have the LCD $5(x+2)$, add the numerators, and write the sum over the LCD. To *solve the equation* $\frac{3}{5} + \frac{7}{x+2} = 2$, we multiply both sides by the LCD $5(x+2)$ to eliminate the denominators.

The solution is 3. To check, we substitute 3 for x in the original equation and simplify:

Check:

$$\frac{3}{5} + \frac{7}{x+2} = 2$$

$$\frac{3}{5} + \frac{7}{3+2} \stackrel{?}{=} 2$$

$$\frac{3}{5} + \frac{7}{5} \stackrel{?}{=} 2$$

$$2 = 2$$

Self Check 1 Solve: $\frac{2}{5} + \frac{5}{x-2} = \frac{29}{10}$.

ACCENT ON TECHNOLOGY: SOLVING RATIONAL EQUATIONS GRAPHICALLY

To use a graphing calculator to solve $\frac{3}{5} + \frac{7}{x+2} = 2$, we graph the functions $f(x) = \frac{3}{5} + \frac{7}{x+2}$ and $g(x) = 2$. If we trace and move the cursor closer to the intersection point of the two graphs, we will get the approximate value of x shown in figure (a). If we zoom twice and trace again, we get the results shown in figure (b). Algebra will show that the exact solution is 3.

An alternate way of finding the point of intersection of the two graphs is to use the INTERSECT feature. In figure (c), the display shows that the graphs intersect at the point (3, 2). This implies that the solution of the rational equation is 3.

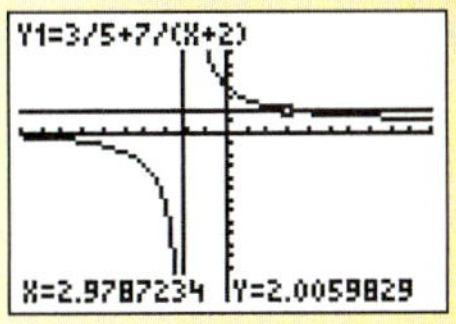

(a)

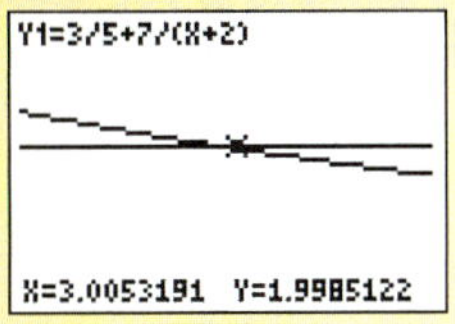

(b)

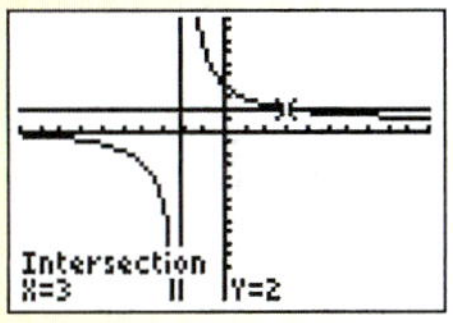

(c)

EXAMPLE 2

Solve: $\frac{-x^2 + 10}{x^2 - 1} + \frac{3x}{x-1} = \frac{2x}{x+1}$.

Solution We note that x cannot be 1 or -1, because this would give a 0 in the denominator of a fraction. If $x \neq 1$ and $x \neq -1$, we can clear the equation of fractions by multiplying both sides by the LCD of the three rational expressions and proceeding as follows:

$$\frac{-x^2 + 10}{x^2 - 1} + \frac{3x}{x-1} = \frac{2x}{x+1}$$

$$\frac{-x^2 + 10}{(x+1)(x-1)} + \frac{3x}{x-1} = \frac{2x}{x+1}$$ Factor the denominator $x^2 - 1$. We determine the LCD to be $(x+1)(x-1)$.

$$(x+1)(x-1)\left(\frac{-x^2 + 10}{(x+1)(x-1)} + \frac{3x}{x-1}\right) = (x+1)(x-1)\frac{2x}{x+1}$$ Multiply both sides by the LCD.

$$\frac{(x+1)(x-1)(-x^2+10)}{(x+1)(x-1)} + \frac{3x(x+1)(x-1)}{x-1} = \frac{2x(x+1)(x-1)}{x+1}$$ Distribute the multiplication by $(x + 1)(x - 1)$.

$$\frac{\overset{1}{\cancel{(x+1)}}\overset{1}{\cancel{(x-1)}}(-x^2+10)}{\underset{1}{\cancel{(x+1)}}\underset{1}{\cancel{(x-1)}}} + \frac{3x(x+1)\overset{1}{\cancel{(x-1)}}}{\underset{1}{\cancel{(x-1)}}} = \frac{2x\overset{1}{\cancel{(x+1)}}(x-1)}{\underset{1}{\cancel{(x+1)}}}$$ Simplify each rational expression.

$$-x^2 + 10 + 3x(x + 1) = 2x(x - 1)$$ Simplify each side. The resulting equation does not contain any fractions.

$$-x^2 + 10 + 3x^2 + 3x = 2x^2 - 2x$$ Remove parentheses.

$$2x^2 + 10 + 3x = 2x^2 - 2x$$ Combine like terms.

$$10 + 3x = -2x$$ Subtract $2x^2$ from both sides.

$$10 + 5x = 0$$ Add $2x$ to both sides.

$$5x = -10$$ Subtract 10 from both sides.

$$x = -2$$ Divide both sides by 5.

Verify that -2 is a solution of the original equation.

Self Check 2 Solve: $\dfrac{5x}{x-2} - \dfrac{4x}{x+2} = \dfrac{x^2-54}{x^2-4}$.

ACCENT ON TECHNOLOGY: CHECKING APPARENT SOLUTIONS

We can use a scientific calculator to check the solution -2 found in Example 2 by evaluating

$$\frac{-x^2+10}{x^2-1} + \frac{3x}{x-1} \quad \text{and} \quad \frac{2x}{x+1}$$

In each case, the result is 4. Since the results are the same, -2 is a solution of the equation.

A check can also be done using a graphing calculator. One way of doing this is to enter

$$Y_1 = \frac{-x^2+10}{x^2-1} + \frac{3x}{x-1} \quad \text{and} \quad Y_2 = \frac{2x}{x+1}$$

and compare the values of the expressions when $x = -2$ in the table mode. See the figure. We know that -2 is a solution of

$$\frac{-x^2+10}{x^2-1} + \frac{3x}{x-1} = \frac{2x}{x+1}$$

because the value of Y_1 and Y_2 are the same (namely, 4) for $x = -2$.

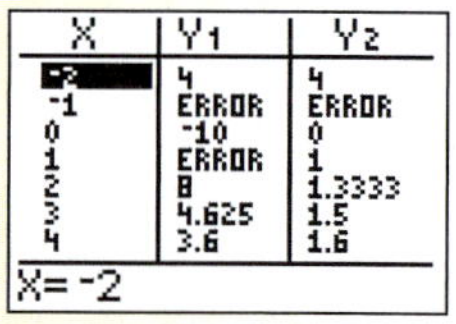

EXAMPLE 3

Solve: $\frac{x+1}{5} - 2 = -\frac{4}{x}$.

Solution We note that x cannot be 0, because this would give a 0 in a denominator. If $x \neq 0$, we can clear the equation of fractions by multiplying both sides by $5x$, the LCD of $\frac{x+1}{5}$ and $\frac{4}{x}$.

$$\frac{x+1}{5} - 2 = -\frac{4}{x}$$

$$5x\left(\frac{x+1}{5} - 2\right) = 5x\left(-\frac{4}{x}\right) \quad \text{Multiply both sides by the LCD, } 5x.$$

$$5x\left(\frac{x+1}{5}\right) - 5x(2) = 5x\left(-\frac{4}{x}\right) \quad \text{Distribute } 5x.$$

$$\overset{1}{\cancel{5}}x\left(\frac{x+1}{\underset{1}{\cancel{5}}}\right) - 5x(2) = \overset{1}{5\cancel{x}}\left(-\frac{4}{\underset{1}{\cancel{x}}}\right) \quad \text{Simplify: } \frac{5}{5} = 1 \text{ and } \frac{x}{x} = 1.$$

$$x(x+1) - 10x = -20$$

$$x^2 + x - 10x = -20 \quad \text{Remove parentheses.}$$

To use factoring to solve this quadratic equation, we must write it in standard quadratic form $ax^2 + bx + c = 0$.

$$x^2 - 9x + 20 = 0 \quad \text{Combine like terms and add 20 to both sides.}$$

$$(x-5)(x-4) = 0 \quad \text{Factor } x^2 - 9x + 20.$$

$$x - 5 = 0 \quad \text{or} \quad x - 4 = 0 \quad \text{Set each factor equal to 0.}$$

$$x = 5 \quad \Big| \quad x = 4$$

Verify that 4 and 5 both satisfy the original equation.

Self Check 3 Solve: $a + \frac{2}{3} = \frac{2a - 12}{3(a-3)}$.

We can now summarize the procedure used to solve rational equations.

> **Caution**
> When solving rational equations, each term on both sides must be multiplied by the LCD.

Solving Rational Equations

1. Factor all denominators.
2. Determine which numbers cannot be solutions of the equation.
3. Multiply both sides of the equation by the LCD of all rational expressions in the equation.
4. Use the distributive property to remove parentheses, remove any factors equal to 1, and write the result in simplified form.
5. Solve the resulting equation.
6. Check all possible solutions in the original equation.

EXTRANEOUS SOLUTIONS

When we multiply both sides of an equation by a quantity that contains a variable, we can get false solutions, called **extraneous solutions.** This happens when we multiply both sides of an equation by 0 and get a solution that gives a 0 in the denominator of a rational expression. Extraneous solutions must be discarded.

EXAMPLE 4

Solve: $\dfrac{2(x+1)}{x-3} = \dfrac{x+5}{x-3}$.

Solution We note that x cannot be 3, because this would give a 0 in a denominator. If $x \neq 3$, we can clear the equation of fractions by multiplying both sides by $x - 3$.

$$\frac{2(x+1)}{x-3} = \frac{x+5}{x-3}$$

$$(x-3)\frac{2(x+1)}{x-3} = (x-3)\frac{x+5}{x-3} \quad \text{Multiply both sides by the LCD.}$$

$$2(x+1) = x+5 \quad \text{Simplify: } \tfrac{x-3}{x-3} = 1.$$

$$2x + 2 = x + 5 \quad \text{Remove parentheses.}$$

$$x + 2 = 5 \quad \text{Subtract } x \text{ from both sides.}$$

$$x = 3 \quad \text{Subtract 2 from both sides.}$$

Since x cannot be 3, it is an extraneous solution and must be discarded. This equation has no solutions.

The Language of Algebra

Extraneous means not a vital part. Mathematicians speak of *extraneous* solutions. Rock groups don't want any *extraneous* sounds (like humming or feedback) coming from their amplifiers. Artists erase any *extraneous* marks on their sketches.

Self Check 4 Solve: $\dfrac{5}{a} - \dfrac{3}{2} = 2 + \dfrac{5}{a}$.

SOLVING FORMULAS FOR A SPECIFIED VARIABLE

Many formulas involve rational expressions.

EXAMPLE 5

Physics. The *law of gravitation,* formulated by Sir Isaac Newton in 1684, states that if two masses, m_1 and m_2, are separated by a distance of r, the force F exerted by one mass on the other is

$$F = \frac{Gm_1m_2}{r^2}$$

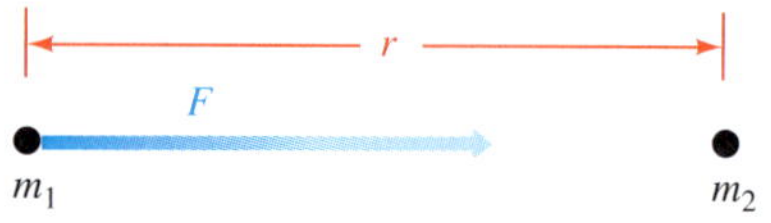

where G is the gravitational constant. Solve for m_2.

Solution

$$F = \frac{Gm_1m_2}{r^2}$$

$$r^2(F) = \overset{1}{\cancel{r^2}}\left(\frac{Gm_1m_2}{\underset{1}{\cancel{r^2}}}\right) \quad \text{Multiply both sides by the LCD, } r^2\text{. Simplify: } \frac{r^2}{r^2} = 1.$$

$$\frac{r^2F}{Gm_1} = \frac{Gm_1m_2}{Gm_1} \quad \text{To isolate } m_2\text{, divide both sides by } Gm_1.$$

$$\frac{r^2F}{Gm_1} = m_2 \quad \text{Simplify the right-hand side.}$$

$$m_2 = \frac{r^2F}{Gm_1}$$

Self Check 5 Solve the law of gravitation formula for r^2.

EXAMPLE 6

Electronics. In electronic circuits, resistors oppose the flow of an electric current. The total resistance R of a parallel combination of two resistors as shown is given by

$$\frac{1}{R} = \frac{1}{R_1} + \frac{1}{R_2}$$

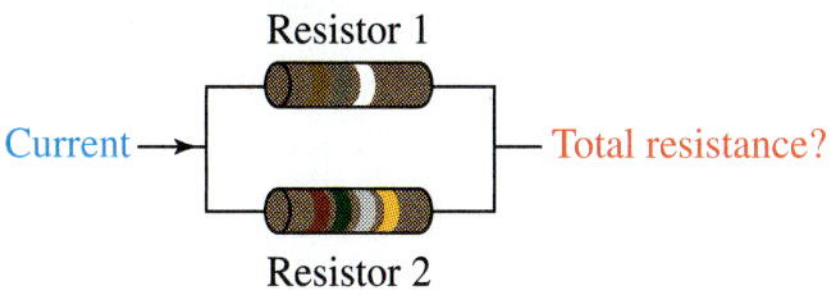

where R_1 is the resistance of the first resistor and R_2 is the resistance of the second resistor. Solve for R.

Solution We begin by clearing the equation of fractions by multiplying both sides by the LCD, which is RR_1R_2.

$$\frac{1}{R} = \frac{1}{R_1} + \frac{1}{R_2}$$

$$RR_1R_2\left(\frac{1}{R}\right) = RR_1R_2\left(\frac{1}{R_1} + \frac{1}{R_2}\right) \qquad \text{Multiply both sides by the LCD.}$$

$$\frac{\overset{1}{\cancel{R}}R_1R_2}{\underset{1}{\cancel{R}}} = \frac{R\overset{1}{\cancel{R_1}}R_2}{\underset{1}{\cancel{R_1}}} + \frac{RR_1\overset{1}{\cancel{R_2}}}{\underset{1}{\cancel{R_2}}} \qquad \text{Distribute } RR_1R_2\text{. Simplify each rational expression.}$$

$$R_1R_2 = RR_2 + RR_1 \qquad \text{Simplify.}$$

$$R_1R_2 = R(R_2 + R_1) \qquad \text{Factor out } R \text{ on the right-hand side.}$$

$$\frac{R_1R_2}{R_2 + R_1} = R \qquad \text{To isolate } R\text{, divide both sides by } R_2 + R_1.$$

$$R = \frac{R_1R_2}{R_2 + R_1}$$

Self Check 6 The *two-intercept form* for the equation of a line is $\frac{x}{a} + \frac{y}{b} = 1$. Solve for b.

PROBLEM SOLVING

Problems in which two or more people (or machines) work together to complete a job are called *shared-work problems.* To solve such problems, we must determine the **rate of work** for each person or machine involved. For example, suppose it takes you 4 hours to clean your house. Your rate of work can be expressed as $\frac{1}{4}$ of the job is completed per hour. If someone else takes 5 hours to clean the same house, they complete $\frac{1}{5}$ of the job per hour. In general, a rate of work can be determined in the following way.

Rate of Work If a job can be completed in x hours, the rate of work can be expressed as:

$\frac{1}{x}$ of the job is completed per hour.

If a job is completed in some other unit of time, such as x minutes or x days, then the rate of work is expressed in that unit.

To solve shared-work problems, we must also determine the *amount of work* completed. To do this, we use a formula similar to the distance formula $d = rt$ used for motion problems.

$$\text{Work completed} = \text{rate of work} \cdot \text{time worked} \qquad \text{or} \qquad W = rt$$

EXAMPLE 7

Home construction. One crew can drywall a house in 4 days and another crew can drywall the same house in 5 days. If both crews work together, how long will it take to drywall the house?

Analyze the Problem It is helpful to organize the facts of a shared-work problem in a table.

Form an Equation Let $x =$ the number of days it will take to drywall the house if both crews work together. Since the crews will be working for the same amount of time, enter x as the time worked for each crew.

If the first crew can drywall the house in 4 days, its rate working alone is $\frac{1}{4}$ of the job per day. If the second crew can drywall the house in 5 days, its rate working alone is $\frac{1}{5}$ of the job per day. To determine the work completed by each crew, multiply the rate by the time.

	Rate ·	Time =	Work completed
1st crew	$\frac{1}{4}$	x	$\frac{x}{4}$
2nd crew	$\frac{1}{5}$	x	$\frac{x}{5}$

Enter this information first. (Rate and Time columns)

Multiply to get each of these entries; $W = rt$. (Work completed column)

In shared-work problems, the number 1 represents one whole job completed. So we have

The part of job done by 1st crew	plus	part of job done by 2nd crew	equals	1 job completed.
$\frac{x}{4}$	$+$	$\frac{x}{5}$	$=$	1

Solve the Equation

$$\frac{x}{4} + \frac{x}{5} = 1$$

$$20\left(\frac{x}{4} + \frac{x}{5}\right) = 20(1)$$ Clear the equation of fractions by multiplying both sides by the LCD, 20.

$$20\left(\frac{x}{4}\right) + 20\left(\frac{x}{5}\right) = 20$$ Distribute the multiplication by 20.

$$\frac{\overset{1}{\cancel{4}} \cdot 5 \cdot x}{\underset{1}{\cancel{4}}} + \frac{4 \cdot \overset{1}{\cancel{5}} \cdot x}{\underset{1}{\cancel{5}}} = 20$$ Factor 20 as $4 \cdot 5$, and simplify each rational expression.

$$5x + 4x = 20$$

$$9x = 20$$ Combine like terms.

$$x = \frac{20}{9}$$ Divide both sides by 9.

State the Conclusion If both crews work together, it will take $\frac{20}{9}$ or $2\frac{2}{9}$ days to drywall the house.

Check the Result In $\frac{20}{9}$ days, the first crew drywalls $\frac{1}{4} \cdot \frac{20}{9} = \frac{5}{9}$ of the house and the second crew drywalls $\frac{1}{5} \cdot \frac{20}{9} = \frac{4}{9}$ of the house. The sum of these efforts, $\frac{4}{9} + \frac{5}{9}$, is $\frac{9}{9}$ or 1 house drywalled. The result checks.

Example 7 can be solved in a different way by considering *the amount of work done by each crew in 1 day.* As before, if we let $x =$ the number of days it will take to drywall the house if both crews work together, then together, in 1 day, they will complete $\frac{1}{x}$ of the job. If we add what the first crew can do in 1 day to what the second crew can do in 1 day, the sum is what they can do together in 1 day.

What crew 1 can do in one day	plus	what crew 2 can do in one day	is	what they can do together in one day.
$\frac{1}{4}$	$+$	$\frac{1}{5}$	$=$	$\frac{1}{x}$

To solve the equation, begin by clearing it of fractions.

$$\mathbf{20x}\left(\frac{1}{4} + \frac{1}{5}\right) = \mathbf{20x}\left(\frac{1}{x}\right) \quad \text{Multiply both sides by the LCD, } 20x.$$

$$5x + 4x = 20 \quad \text{Distribute the multiplication by } 20x \text{ and simplify.}$$

$$9x = 20 \quad \text{Combine like terms.}$$

$$x = \frac{20}{9} \quad \text{Divide both sides by 9.}$$

This is the same as the solution obtained in Example 7.

In the next two examples, rational equations are used to model situations involving *uniform motion.*

EXAMPLE 8

Driving to a convention. A doctor drove 200 miles to attend a national convention. Because of poor weather, her average speed on the return trip was 10 mph less than her average speed going to the convention. If the return trip took 1 hour longer, how fast did she drive in each direction?

Analyze the Problem We need to find her rates of speed going to and returning from the convention. They can be represented using a variable. The distance traveled was 200 miles each way. To describe the travel times, we note that

$$rt = d \quad r \text{ is the rate of speed, } t \text{ is the time, and } d \text{ is the distance.}$$

or

$$t = \frac{d}{r} \quad \text{Divide both sides by } r.$$

Form an Equation Let $r =$ the average rate of speed going to the meeting. Then $r - 10 =$ the average rate of speed on the return trip. We can organize the facts of the problem in the table.

	Rate	· Time	= Distance
Going	r	$\frac{200}{r}$	200
Returning	$r - 10$	$\frac{200}{r-10}$	200

Enter this information first.

We obtained these entries by dividing the distance by the rate: $t = \frac{d}{r}$.

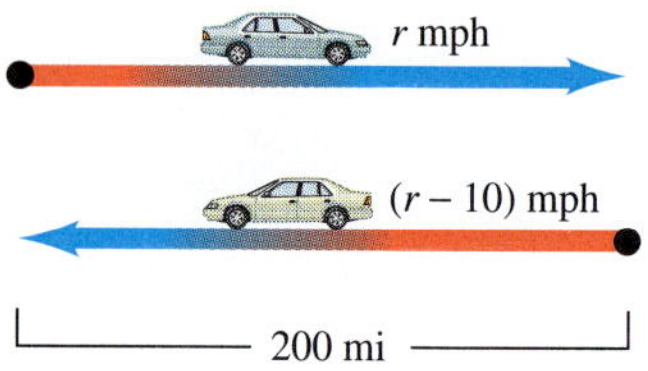

Because the return trip took 1 hour longer, we can form the following equation:

The time it took to travel to the convention	plus	1	is	the time it took to return
$\frac{200}{r}$	$+$	1	$=$	$\frac{200}{r-10}$

Solve the Equation We can solve the equation as follows:

$$r(r-10)\left(\frac{200}{r}+1\right) = r(r-10)\left(\frac{200}{r-10}\right)$$ Multiply both sides by $r(r - 10)$.

$$\overset{1}{\cancel{r}}(r-10)\frac{200}{\underset{1}{\cancel{r}}} + r(r-10)1 = r\overset{1}{\cancel{(r-10)}}\left(\frac{200}{\underset{1}{\cancel{r-10}}}\right)$$ Distribute $r(r - 10)$ and then simplify.

$$200(r-10) + r(r-10) = 200r$$

$$200r - 2{,}000 + r^2 - 10r = 200r$$ Distribute 200 and r. This is a quadratic equation.

$$r^2 - 10r - 2{,}000 = 0$$ Subtract $200r$ from both sides.

$$(r-50)(r+40) = 0$$ Factor $r^2 - 10r - 2{,}000$.

$$r - 50 = 0 \quad \text{or} \quad r + 40 = 0$$ Set each factor equal to 0.

$$r = 50 \quad \Big| \quad r = -40$$

State the Conclusion We must exclude the solution of -40, because a speed cannot be negative. Thus, the doctor averaged 50 mph going to the convention, and she averaged $50 - 10$ or 40 mph returning.

Check the Result At 50 mph, the 200-mile trip took 4 hours. At 40 mph, the return trip took 5 hours, which is 1 hour longer.

EXAMPLE 9

Riverboat cruises. The Forest City Queen can make a 9-mile trip down the Rock River and return in a total of 1.6 hours. If the riverboat travels 12 mph in still water, find the speed of the current in the Rock River.

Analyze the Problem We can represent the upstream and downstream rates of speed using a variable. In each case, the distance traveled is 9 miles. To write an expression for the time traveled, divide the distance by the rate of speed.

Form an Equation We can let c = the speed of the current. Since the boat travels 12 mph and a current of c mph pushes the boat while it is going downstream, the speed of the boat going downstream is $(12 + c)$ mph. On the return trip, the current pushes against the boat, and its speed is $(12 - c)$ mph. Since $t = \frac{d}{r}$ $\left(\text{time} = \frac{\text{distance}}{\text{rate}}\right)$, the time required for the downstream leg of the trip is $\frac{9}{12 + c}$ hours, and the time required for the upstream leg of the trip is $\frac{9}{12 - c}$ hours. We can organize this information in the table.

	Rate	· Time	= Distance
Going downstream	$12 + c$	$\frac{9}{12 + c}$	9
Going upstream	$12 - c$	$\frac{9}{12 - c}$	9

Enter this information first. (Rate and Distance columns)

Divide the distance by the rate. (Time column)

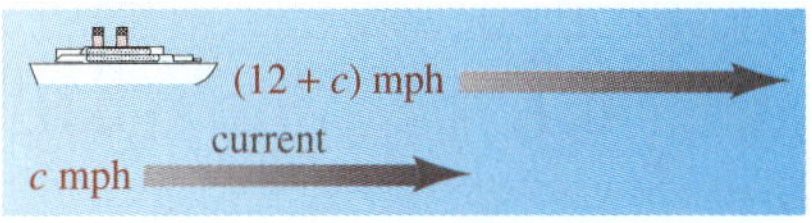

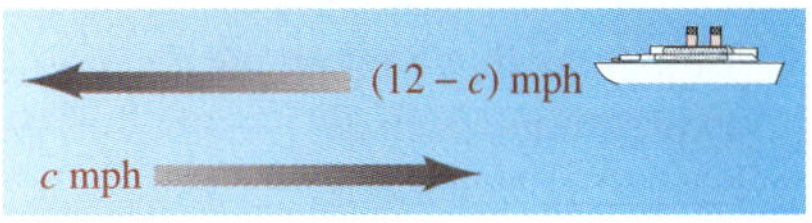

We also know that the total time required for the round trip is 1.6 or $\frac{8}{5}$ hours.

The time it takes to travel downstream	plus	the time it takes to travel upstream	is	the total time for the round trip.
$\frac{9}{12 + c}$	$+$	$\frac{9}{12 - c}$	$=$	$\frac{8}{5}$

Solve the Equation Multiply both sides of this equation by $5(12 + c)(12 - c)$ to clear it of fractions.

$$5(12 + c)(12 - c)\left(\frac{9}{12 + c} + \frac{9}{12 - c}\right) = 5(12 + c)(12 - c)\left(\frac{8}{5}\right)$$

$$\frac{5\overset{1}{\cancel{(12 + c)}}(12 - c)9}{\underset{1}{\cancel{12 + c}}} + \frac{5(12 + c)\overset{1}{\cancel{(12 - c)}}9}{\underset{1}{\cancel{12 - c}}} = \frac{\overset{1}{\cancel{5}}(12 + c)(12 - c)8}{\underset{1}{\cancel{5}}}$$ Distribute, and then simplify.

$$45(12 - c) + 45(12 + c) = 8(12 + c)(12 - c)$$

$$540 - 45c + 540 + 45c = 8(144 - c^2)$$ On the left-hand side, distribute. On the right-hand side, use the FOIL method.

$$1{,}080 = 1{,}152 - 8c^2$$ Combine like terms and multiply. This is a quadratic equation.

$$8c^2 - 72 = 0$$ Add $8c^2$ and subtract 1,152 from both sides.

$$c^2 - 9 = 0$$ Divide both sides by 8.

$$(c + 3)(c - 3) = 0$$ Factor $c^2 - 9$.

$$c + 3 = 0 \quad \text{or} \quad c - 3 = 0$$ Set each factor equal to 0.

$$c = -3 \quad | \quad c = 3$$

State the Conclusion Since the current cannot be negative, the solution -3 must be discarded. The current in the Rock River is 3 mph.

Check the Result The downstream trip is at $12 + 3 = 15$ mph for $\frac{9}{12 + 3} = \frac{3}{5}$ hr. Thus, the distance traveled is $15 \cdot \frac{3}{5} = 9$ miles. The upstream trip is at $12 - 3 = 9$ mph for $\frac{9}{12 - 3} = 1$ hr. Thus, the distance traveled is $9 \cdot 1 = 9$ miles. Since both distances are 9 miles, the result checks.

Answers to Self Checks 1. 4 2. −3 3. 1, 2 4. no solutions; 0 is extraneous 5. $r^2 = \frac{Gm_1m_2}{F}$ 6. $b = -\frac{ay}{x-a}$ or $b = \frac{ay}{a-x}$

6.7 STUDY SET

VOCABULARY Fill in the blanks.

1. Equations that contain one or more rational expressions, such as $\frac{x}{x+2} = 4 + \frac{10}{x+2}$, are called ________ equations.
2. To ______ a rational equation we find all the values of the variable that make the equation true.
3. To ______ a rational equation of fractions, multiply both sides by the LCD of all rational expressions in the equation.
4. When solving a rational equation, if we obtain a number that does not satisfy the original equation, the number is called an __________ solution.

CONCEPTS

5. Is 2 a solution of the following equations?
   a. $\frac{x+2}{x+3} + \frac{1}{x^2+2x-3} = 1$
   b. $\frac{x+2}{x-2} + \frac{1}{x^2-4} = 1$
6. To clear the equation
   $$\frac{4}{10} + y = \frac{4y-50}{5y-25}$$
   of fractions, by what should both sides be multiplied?
7. Complete the table.

	r ·	t =	d
Running	x		12
Bicycling	$x + 15$		12

8. Complete the table.

	Rate ·	Time =	Work completed
1st crew	$\frac{1}{15}$	x	
2nd crew	$\frac{1}{8}$	x	

9. Consider the rational equation $\frac{x}{x-3} = \frac{1}{x} + \frac{2}{x-3}$.
   a. What values of x make a denominator 0?
   b. What values of x make a rational expression undefined?
   c. What numbers can't be solutions of the equation?
10. Perform each multiplication.
    a. $4x\left(\frac{3}{4x}\right)$
    b. $(x+6)(x-2)\left(\frac{3}{x-2}\right)$

NOTATION Complete each solution.

11.
$$\frac{10}{3y} - \frac{7}{30} = \frac{9}{2y}$$
$$\boxed{}\left(\frac{10}{3y} - \frac{7}{30}\right) = 30y\left(\boxed{}\right)$$
$$30y\left(\boxed{}\right) - 30y\left(\frac{7}{30}\right) = 30y\left(\frac{9}{2y}\right)$$
$$100 - \boxed{} = 135$$
$$-7y = \boxed{}$$
$$y = -5$$

12.
$$\frac{2}{u-1} + \frac{1}{u} = \frac{1}{u^2-u}$$
$$\boxed{} + \frac{1}{u} = \frac{1}{u(u-1)}$$
$$\boxed{}\left(\frac{2}{u-1} + \frac{1}{u}\right) = u(u-1)\left[\frac{1}{u(u-1)}\right]$$
$$u(u-1)\left(\frac{2}{u-1}\right) + u(u-1)\left(\boxed{}\right) = u(u-1)\left[\frac{1}{u(u-1)}\right]$$
$$2u + (\boxed{}) = \boxed{}$$
$$\boxed{} = 2$$
$$u = \boxed{}$$

PRACTICE **Solve each equation. If a solution is extraneous, so indicate.**

13. $\frac{1}{4} + \frac{9}{x} = 1$

14. $\frac{1}{3} - \frac{10}{x} = -3$

15. $\frac{34}{x} - \frac{3}{2} = -\frac{13}{20}$

16. $\frac{1}{2} + \frac{7}{x} = 2 + \frac{1}{x}$

17. $\frac{3}{y} + \frac{7}{2y} = 13$

18. $\frac{2}{x} + \frac{1}{2} = \frac{7}{2x}$

19. $\frac{x+1}{x} - \frac{x-1}{x} = 0$

20. $\frac{2}{x} + \frac{1}{2} = \frac{9}{4x} - \frac{1}{2x}$

21. $\frac{7}{5x} - \frac{1}{2} = \frac{5}{6x} + \frac{1}{3}$

22. $\frac{x-3}{x-1} - \frac{2x-4}{x-1} = 0$

23. $\frac{3-5y}{2+y} = \frac{3+5y}{2-y}$

24. $\frac{x}{x-2} = 1 + \frac{1}{x-3}$

25. $\frac{a+2}{a+1} - \frac{a-4}{a-3} = 0$

26. $\frac{z+2}{z+8} - \frac{z-3}{z-2} = 0$

27. $\frac{x+2}{x+3} - 1 = \frac{-1}{x^2+2x-3}$

28. $\frac{x-3}{x-2} - \frac{1}{x} = \frac{x-3}{x}$

29. $\frac{m+6}{3m-12} + \frac{5}{4-m} = \frac{2}{3}$

30. $\frac{21}{x^2-4} - \frac{14}{x+2} = \frac{3}{2-x}$

31. $\frac{4t^2+36}{t^2-9} - \frac{4t}{t+3} = \frac{12}{3-t}$

32. $\frac{2c}{4-c^2} - \frac{5}{c+2} = -\frac{3}{c-2}$

33. $\frac{x}{x+2} = 1 - \frac{3x+2}{x^2+4x+4}$

34. $\frac{3+2a}{a^2+6+5a} + \frac{2-5a}{a^2-4} = \frac{2-3a}{a^2-6+a}$

35. $\frac{2}{x-2} + \frac{1}{x+1} = \frac{1}{x^2-x-2}$

36. $\frac{5}{y-1} + \frac{3}{y-3} = \frac{8}{y-2}$

37. $\frac{a-1}{a+3} - \frac{1-2a}{3-a} = \frac{2-a}{a-3}$

38. $\frac{5}{2z^2+z-3} - \frac{2}{2z+3} = \frac{z+1}{z-1} - 1$

39. $\frac{5}{x+4} + \frac{1}{x+4} = x - 1$

40. $\frac{2}{5x-5} + \frac{x-2}{15} = \frac{4}{5x-5}$

41. $\frac{3}{2x+4} - \frac{x-2}{2} = \frac{x-5}{2x+4}$

42. $\frac{x-4}{x-3} + \frac{x-2}{x-3} = x - 3$

43. $\frac{2}{x-3} + \frac{3}{4} = \frac{17}{2x}$

44. $\frac{30}{y-2} + \frac{24}{y-5} = 13$

45. $\frac{x+4}{2x+14} - \frac{x}{2x+6} = \frac{3}{16}$

46. $\frac{5}{3x+12} - \frac{1}{9} = \frac{x-1}{3x}$

47. $\frac{3}{r} = \frac{12}{4r-r^2} - \frac{7}{r-4}$

48. $\frac{1}{y+5} = \frac{1}{3y+6} - \frac{y+2}{y^2+7y+10}$

49. $\frac{3}{s-2} + \frac{s-14}{2s^2-3s-2} - \frac{4}{2s+1} = 0$

50. $\frac{1}{y^2-2y-3} + \frac{1}{y^2-4y+3} - \frac{1}{y^2-1} = 0$

Solve each formula for the indicated variable.

51. $I = \frac{E}{R_L + r}$ for r (from physics)

52. $P = \frac{R-C}{n}$ for C (from business)

53. $S = \frac{a - \ell r}{1-r}$ for r (from mathematics)

54. $P = \frac{Q_1}{Q_2 - Q_1}$ for Q_1 (from refrigeration/heating)

55. $\mu_R = \frac{n_1(n_1 + n_2 + 1)}{2}$ for n_2 (from statistics)

56. $\frac{P_1V_1}{T_1} = \frac{P_2V_2}{T_2}$ for T_2 (from chemistry)

57. $\dfrac{1}{R} = \dfrac{1}{R_1} + \dfrac{1}{R_2} + \dfrac{1}{R_3}$ for R (from electronics)

58. $P + \dfrac{a}{V^2} = \dfrac{RT}{V - b}$ for b (from physics)

59. Let $f(x) = \dfrac{x^3 - 3x^2 + 12}{x}$. For what values of x is $f(x) = 4$?

60. Let $f(x) = \dfrac{x^3 + 2x^2 - 32}{x}$. For what values of x is $f(x) = 16$?

61. Let $f(x) = \dfrac{2x^3 + x^2}{98x + 49}$. For what values of x is $f(x) = 1$?

62. Let $f(x) = \dfrac{x^3 + 4x^2}{25x + 100}$. For what values of x is $f(x) = 1$?

APPLICATIONS

63. PHOTOGRAPHY The illustration shows the relationship between distances when taking a photograph. The design of a camera lens uses the equation

$$\frac{1}{f} = \frac{1}{s_1} + \frac{1}{s_2}$$

which relates the focal length f of a lens to the image distance s_1 and the object distance s_2. Find the focal length of the lens in the illustration. (*Hint:* Convert feet to inches.)

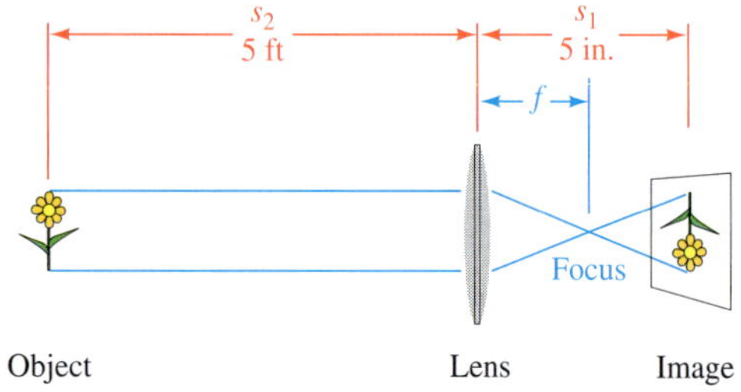

64. OPTICS The focal length, f, of a lens is given by the lensmaker's formula,

$$\frac{1}{f} = 0.6\left(\frac{1}{r_1} + \frac{1}{r_2}\right)$$

where f is the focal length of the lens and r_1 and r_2 are the radii of the two circular surfaces. See the illustration in the next column. Find the focal length of the lens in the illustration.

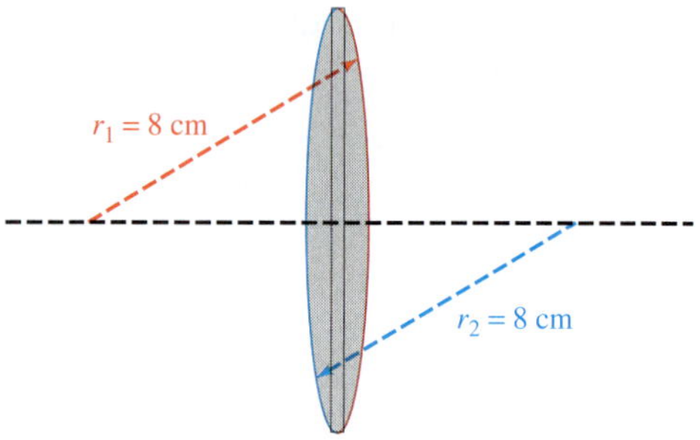

65. ACCOUNTING As a piece of equipment gets older, its value usually lessens. One way to calculate *depreciation* is to use the formula

$$V = C - \left(\frac{C - S}{L}\right)N$$

where V denotes the value of the equipment at the end of year N, L is its useful lifetime (in years), C is its cost new, and S is its salvage value at the end of its useful life. Solve for L. Then determine what an accountant considered the useful lifetime of a forklift that cost $25,000 new, was worth $13,000 after 4 years, and has a salvage value of $1,000.

66. ENGINEERING The equation

$$a = \frac{9.8m_2 - f}{m_2 + m_1}$$

models the system shown below, where a is the acceleration of the suspended block, m_1 and m_2 are the masses of the blocks, and f is the friction force. Solve for m_2.

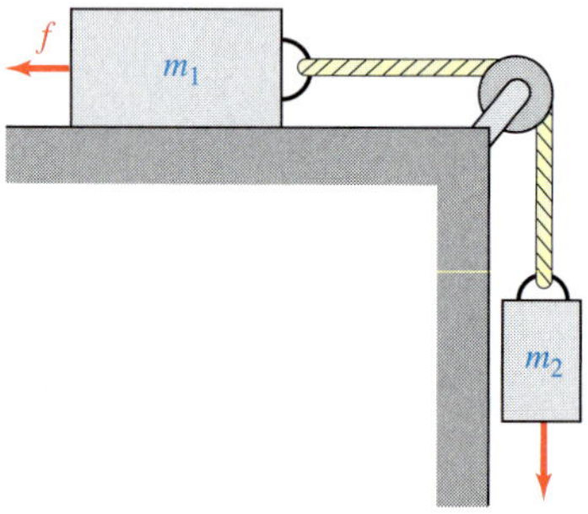

67. HOUSEPAINTING The illustration on the next page shows two bids to paint a house.

a. To get the job done quicker, the homeowner hired both the painters who submitted bids. How long will it take them to paint the house working together?

b. What will the homeowner have to pay each painter?

68. ROOFING HOUSES A homeowner estimates that it will take him 7 days to roof his house. A professional roofer estimates that he could roof the house in 4 days. How long will it take if the homeowner helps the roofer?

69. OYSTERS According to the *Guinness Book of World Records,* the record for opening oysters is 100 in 140 seconds by Mike Racz in Invercargill, New Zealand on July 16, 1990. If it would take a novice $8\frac{1}{2}$ minutes to perform the same task, how long would it take them working together to open 100 oysters?

70. FARMING In 10 minutes, a conveyor belt can move 1,000 bushels of corn into the storage bin shown below. A smaller belt can move 1,000 bushels to the storage bin in 14 minutes. If both belts are used, how long will it take to move 1,000 bushels to the storage bin?

71. FILLING PONDS One pipe can fill a pond in 3 weeks, and a second pipe can fill it in 5 weeks. However, evaporation and seepage can empty the pond in 10 weeks. If both pipes are used, how long will it take to fill the pond?

72. HOUSECLEANING Sally can clean the house in 6 hours, her father can clean the house in 4 hours, and her younger brother, Dennis, can completely mess up the house in 8 hours. If Sally and her father clean and Dennis plays, how long will it take to clean the house?

73. DETAILING A CAR It takes a man 3 hours to wash and wax the family car. If his teenage son helps him, it only takes 1 hour. How long would it take the son, working alone, to wash and wax the car?

74. CLEANUP CREWS It takes one crew 4 hours to clean an auditorium after an event. If a second crew helps, it only takes 1.5 hours. How long would it take the second crew, working alone, to clean the auditorium?

75. BOXING For his morning workout, a boxer bicycles for 8 miles and then jogs back to camp along the same route. If he bicycles 6 mph faster than he jogs, and the entire workout lasts 2 hours, how fast does he jog?

76. DELIVERIES A FedEx delivery van traveled from Rockford to Chicago in 3 hours less time than it took a second van to travel from Rockford to St. Louis. If the vans traveled at the same average speed, use the information in the map to help determine how long the first driver was on the road.

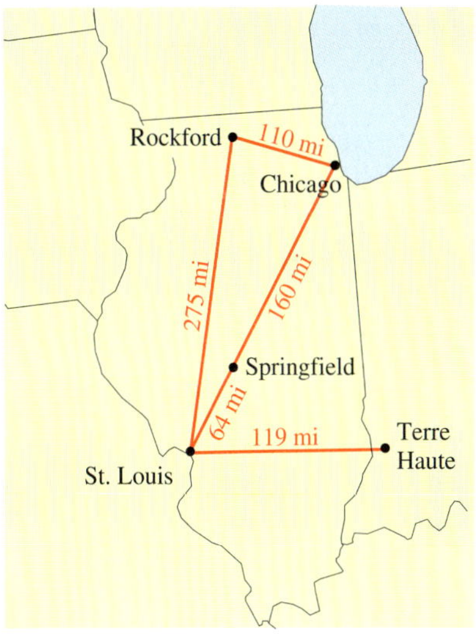

77. RATES OF SPEED Two trains made the same 315-mile run. Since one train traveled 10 mph faster than the other, it arrived 2 hours earlier. Find the speed of each train.

78. TRAIN TRAVEL A train traveled 120 miles from Freeport to Chicago and returned the same distance in a total time of 5 hours. If the train traveled 20 mph slower on the return trip, how fast did the train travel in each direction?

79. CROP DUSTING A helicopter spraying fertilizer over a field can fly 0.5 mile downwind in the same time as it can fly 0.4 mile upwind. Find the speed of the wind if the helicopter travels 45 mph in still air when dusting crops.

80. BOATING A man can drive a motorboat 45 miles down the Rock River in the same amount of time that he can drive 27 miles upstream. Find the speed of the current if the speed of the boat is 12 mph in still water.

81. UNIT COST One month, a store manager bought several microwave ovens for a total of $1,800. The next month, because the unit cost of the same model of microwave increased by $25, she bought one fewer oven for the same total price. How many ovens did she buy the first month? (*Hint:* Write an expression for the unit cost of a microwave for the second month, then use the formula: Unit cost · number = total cost.)

82. EXTENDED VACATION Use the facts in the e-mail message to determine how long the student had originally planned to stay in Europe. (*Hint:* Unit cost · number = total cost.)

> **E-Mail**
>
> Hi Mom and Dad,
> After working so hard to earn $1,200 to take this trip to Europe, it was all worth it! I've been very frugal and been able to cut $20 from my daily expenses. Because of this, I'll be able to stay three extra days. Please pick me up on Friday instead. Love, Liz

WRITING

83. Why is it necessary to check the solutions of a rational equation?

84. Explain what it means to *clear* a rational equation of fractions. Give an example.

85. Explain how to solve the rational equation graphically:

$$\frac{3x}{x-2} + \frac{1}{5} = 2$$

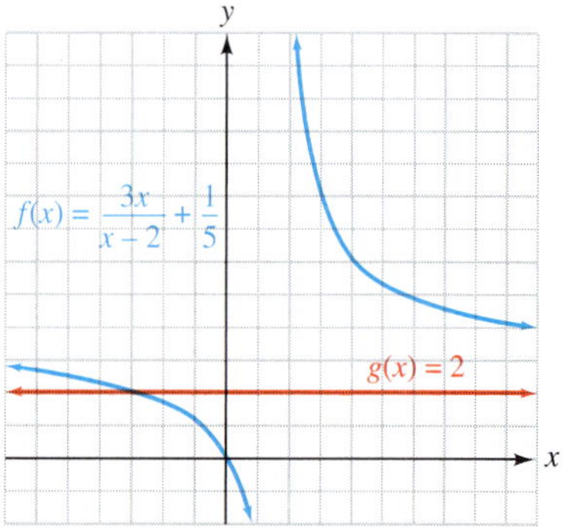

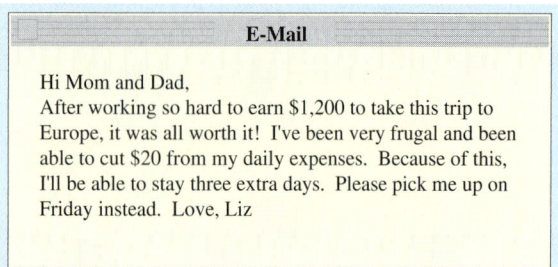

86. Would you use the same approach to answer the following problems? Explain why or why not.

$$\text{Simplify: } \frac{x^2-10}{x^2-1} - \frac{3x}{x-1} - \frac{2x}{x+1}$$

$$\text{Solve: } \frac{x^2-10}{x^2-1} - \frac{3x}{x-1} = -\frac{2x}{x+1}$$

87. In Example 7, one crew could drywall a house in 4 days, and another crew could drywall the same house in 5 days. We were asked to find how long it would take them to drywall the house working together. Explain why each of the following approaches is incorrect.

The time it would take to drywall the house

- is the *sum* of the lengths of time it takes each crew to drywall the house: 4 days + 5 days = 9 days.
- is the *difference* in lengths of time it takes each crew to drywall the house: 5 days − 4 days = 1 day.
- is the *average* of the lengths of time it takes each crew to drywall the house: $\frac{4 \text{ days} + 5 \text{ days}}{2} = \frac{9}{2}$ days $= 4\frac{1}{2}$ days.

88. Write a shared-work problem that can be modeled by the equation

$$\frac{x}{3} + \frac{x}{4} = 1$$

REVIEW **Write each italicized number in scientific notation.**

89. OIL The total cost of the Alaskan pipeline, running 800 miles from Prudhoe Bay to Valdez, was *$9,000,000,000*.

90. NATURAL GAS The TransCanada Pipeline transported a record *2,352,000,000,000* cubic feet of gas in 1995.

91. RADIOACTIVITY The least stable radioactive isotope is lithium 5, which decays in *0.00000000000000000000044* second.

92. BALANCES The finest balances in the world are made in Germany. They can weigh objects to an accuracy of 35×10^{-11} ounce.

CHALLENGE PROBLEMS

93. Solve: $\left(\frac{1}{2}\right)^{-1} = \frac{5b^{-1}}{2} + 2b(b+1)^{-1}$.

94. Invent a rational equation that has an extraneous solution of 3.

6.8 Proportion and Variation

- Ratios and rates
- Proportions
- Solving proportions
- Similar triangles
- Direct variation
- Inverse variation
- Joint variation
- Combined variation

We will now discuss five mathematical models that have a variety of applications. First, we show how a *ratio-proportion model* can be used to solve shopping problems and to determine the height of a tree given the length of its shadow. Then we introduce four types of *variation models,* each of which expresses a special relationship between two or more quantities. We will use these models to solve problems involving travel, lighting, geometry, and highway construction.

RATIOS AND RATES

The quotient of two numbers or two quantities with the same units is often referred to as a **ratio.** For example, $\frac{2}{3}$ can be read as "the ratio of 2 to 3." The notation 2:3 (read as "2 is to 3") is another common way to denote a ratio. Some more examples of ratios are

$$\frac{4x}{7y} \quad \text{The ratio of } 4x \text{ to } 7y. \qquad \text{and} \qquad \frac{x-2}{3x} \quad \text{The ratio of } x-2 \text{ to } 3x.$$

When we compare two quantities having different units, we call the comparison a **rate,** and we can write it as a fraction. One example is an average rate of speed.

A distance traveled → in a period of time →

$$\frac{372 \text{ miles}}{6 \text{ hours}} = 62 \text{ mph}$$

← The average rate of speed.

Rates are often used to express **unit costs,** such as the cost per pound of ground beef.

The cost of a package of ground beef → The weight of the package →

$$\frac{\$7.47}{5 \text{ lb}} \approx \$1.49 \text{ per lb}$$

← The cost per pound

PROPORTIONS

An equation indicating that two ratios or rates are equal is called a **proportion.** Two examples are

$$\frac{1}{4} = \frac{2}{8} \quad \text{and} \quad \frac{4}{7} = \frac{12}{21}$$

The Language of Algebra

The word *proportion* implies a comparative relationship in size. For a picture to appear realistic, the artist must draw the shapes in the proper *proportion.* Sometimes the news media is accused of blowing things way *out of proportion.*

In the proportion $\frac{a}{b} = \frac{c}{d}$, the **terms** a and d are called the **extremes** of the proportion, and the terms b and c are called the **means.**

To develop a fundamental property of proportions, we suppose that

$$\frac{a}{b} = \frac{c}{d}$$

is a proportion and multiply both sides by bd to obtain

$$bd\left(\frac{a}{b}\right) = bd\left(\frac{c}{d}\right)$$

$$\frac{\cancel{b}da}{\cancel{b}} = \frac{b\cancel{d}c}{\cancel{d}} \qquad \text{Simplify: } \frac{b}{b} = 1 \text{ and } \frac{d}{d} = 1.$$

$$ad = bc$$

Since $ad = bc$, the product of the extremes equals the product of the means.

The same products ad and bc can be found by multiplying diagonally in the proportion $\frac{a}{b} = \frac{c}{d}$. We call ad and bc **cross products.**

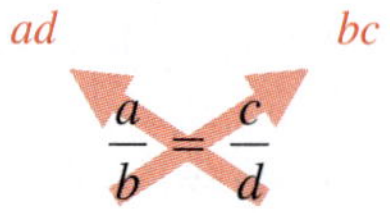

The Fundamental Property of Proportions

In a proportion, the product of the extremes is equal to the product of the means.

$$\text{If } \frac{a}{b} = \frac{c}{d}, \text{ then } ad = bc \text{ and if } ad = bc, \text{ then } \frac{a}{b} = \frac{c}{d}.$$

SOLVING PROPORTIONS

We can solve many problems by writing and then solving a proportion. To solve a proportion, we apply the fundamental property of proportions.

EXAMPLE 1

Solve for x: $\dfrac{x+3}{x} = \dfrac{x}{x+6}$.

Solution

$$\frac{x+3}{x} = \frac{x}{x+6}$$

$(x+3)(x+6) = x \cdot x$ — In a proportion, the product of the extremes equals the product of the means.

$x^2 + 9x + 18 = x^2$ — Perform the multiplications.

$9x + 18 = 0$ — Subtract x^2 from both sides.

$x = -2$ — Subtract 18 from both sides and then divide by 9.

Caution

The expression $\frac{x+3}{x}$ is undefined if x is 0, because division by 0 would be indicated. Similarly, $\frac{x}{x+6}$ is undefined if x is -6. Thus, we can rule out 0 and -6 as possible solutions of $\frac{x+3}{x} = \frac{x}{x+6}$.

The solution is -2. To check, we substitute -2 for x in the proportion and simplify each side.

Check:

$$\frac{x+3}{x} = \frac{x}{x+6}$$

$$\frac{-2+3}{-2} \stackrel{?}{=} \frac{-2}{-2+6}$$

$$\frac{-1}{2} \stackrel{?}{=} \frac{-2}{4}$$

$$-\frac{1}{2} = -\frac{1}{2}$$

Self Check 1 Solve for x: $\dfrac{x-1}{x} = \dfrac{x}{x+3}$.

EXAMPLE 2

Solve: $\dfrac{5a+2}{2a} = \dfrac{18}{a+4}$.

Solution Because $\frac{5a+2}{2a}$ is undefined if $a = 0$, we state the restriction that $a \neq 0$. Because $\frac{18}{a+4}$ is undefined if $a = -4$, we also note that $a \neq -4$.

$$\frac{5a+2}{2a} = \frac{18}{a+4}$$

$(5a+2)(a+4) = 2a(18)$ — In a proportion, the product of the extremes equals the product of the means.

$5a^2 + 22a + 8 = 36a$ — Multiply.

$5a^2 - 14a + 8 = 0$ — Subtract $36a$ from both sides.

$(5a-4)(a-2) = 0$ — Factor to solve the quadratic equation.

$5a - 4 = 0$ or $a - 2 = 0$ — Set each factor equal to 0.

$5a = 4$ | $a = 2$ — Solve each linear equation.

$a = \frac{4}{5}$

The solutions are $\frac{4}{5}$ and 2. Check each of them.

Caution

Remember that a cross product is the product of the means and the extremes of a *proportion.* For example, it would be incorrect to try to compute cross products to solve

$$\frac{5a+2}{2a} = \frac{18}{a+4} + \frac{1}{a}$$

It is not a proportion. The right-hand side is not a ratio.

Self Check 2 Solve: $\frac{3x+1}{12} = \frac{x}{x+2}$.

EXAMPLE 3

Gourmet cooking. To make a dessert of Pears Hélène, a chef needs to purchase 14 pears. If they are on sale at 6 for \$2.34, what will 14 cost?

Solution First, we let c = the cost of 14 pears. The price per pear when purchasing 6 pears is $\frac{\$2.34}{6}$, and the price per pear when purchasing 14 pears is $\frac{\$c}{14}$. Since these ratios are equal, we have the following proportion: *\$2.34 is to 6 pears as \$c is to 14 pears.*

$$\frac{2.34}{6} = \frac{c}{14}$$

(Cost of 6 pears → 2.34; 6 pears → 6; c ← Cost of 14 pears; 14 ← 14 pears)

$14(2.34) = 6c$ — In a proportion, the product of the extremes is equal to the product of the means.

$32.76 = 6c$ — Multiply.

$\frac{32.76}{6} = c$ — Divide both sides by 6.

$c = 5.46$ — Simplify.

Fourteen pears will cost \$5.46.

Success Tip

Since proportions are rational equations, they can also be solved by multiplying both sides by the LCD. For Example 3, an alternate approach is to multiply both sides by the LCD of 6 and 14, which is 42.

$$42\left(\frac{2.34}{6}\right) = 42\left(\frac{c}{14}\right)$$

Self Check 3 A model railroad engine is 9 inches long. The scale is 87 feet to 1 foot. How long is a real engine? (*Hint:* Note the difference in units.)

SIMILAR TRIANGLES

If two angles of one triangle have the same measure as two angles of a second triangle, the triangles will have the same shape. In this case, we call the triangles **similar triangles.** Here are some facts about similar triangles.

Similar Triangles If two triangles are similar, then

1. the three angles of the first triangle have the same measure, respectively, as the three angles of the second triangle.
2. the lengths of all corresponding sides are in proportion.

The following triangles are similar triangles.

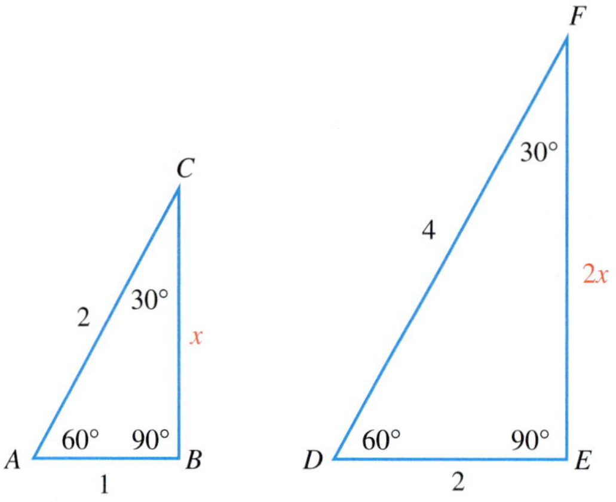

Corresponding angles have the same measure.

The corresponding sides are in proportion:

$$\frac{2}{4} = \frac{x}{2x} \qquad \frac{x}{2x} = \frac{1}{2} \qquad \frac{1}{2} = \frac{2}{4}$$

The properties of similar triangles often enable us to determine the lengths of the sides of triangles indirectly. For example, we can find the height of a tree and stay safely on the ground.

EXAMPLE 4

Height of a tree. A tree casts a shadow of 29 feet at the same time as a vertical yardstick casts a shadow of 2.5 feet. Find the height of the tree.

Solution Refer to the figure, which shows the triangles determined by the tree and its shadow and the yardstick and its shadow. Because the triangles have the same shape, they are similar, and the measures of their corresponding sides are in proportion. If we let h = the height of the tree, we can find h by setting up and solving the following proportion: *h is to 3 as 29 is to 2.5.*

$$\frac{h}{3} = \frac{29}{2.5}$$

height of tree → h; height of yardstick → 3; 29 ← length of tree's shadow; 2.5 ← length of yardstick's shadow

$2.5h = 3(29)$ In a proportion, the product of the extremes is equal to the product of the means.

$2.5h = 87$ Multiply.

$h = 34.8$ Divide both sides by 2.5.

The tree is about 35 feet tall.

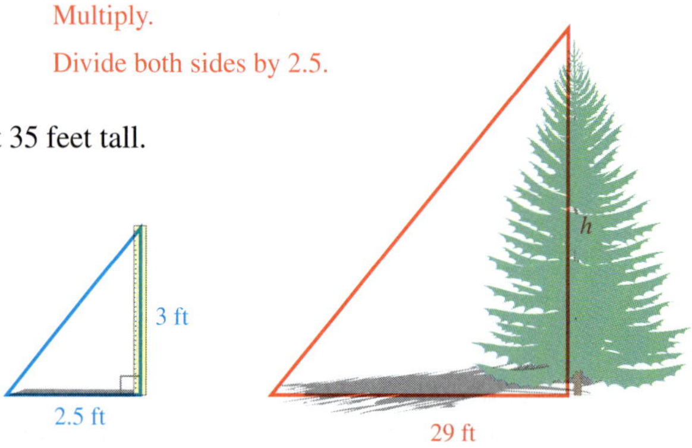

Self Check 4 Suppose the tree casts a shadow of 32 feet at the same time the yardstick casts a shadow of 4 feet. Find the height of the tree.

DIRECT VARIATION

To introduce direct variation, we consider the formula for the circumference of a circle

$$C = \pi D$$

where C is the circumference, D is the diameter, and $\pi \approx 3.14159$. If we double the diameter of a circle, we determine another circle with a larger circumference C_1 such that

$$C_1 = \pi(2D) = 2\pi D = 2C$$

Thus, doubling the diameter results in doubling the circumference. Likewise, if we triple the diameter, we will triple the circumference.

In this formula, we say that the variables C and D *vary directly,* or that they are *directly proportional.* This is because C is always found by multiplying D by a constant. In this example, the constant π is called the *constant of variation* or the *constant of proportionality.*

Direct Variation The words "y varies directly with x" or "y is directly proportional to x" means that $y = kx$ for some nonzero constant k. The constant k is called the **constant of variation** or the **constant of proportionality.**

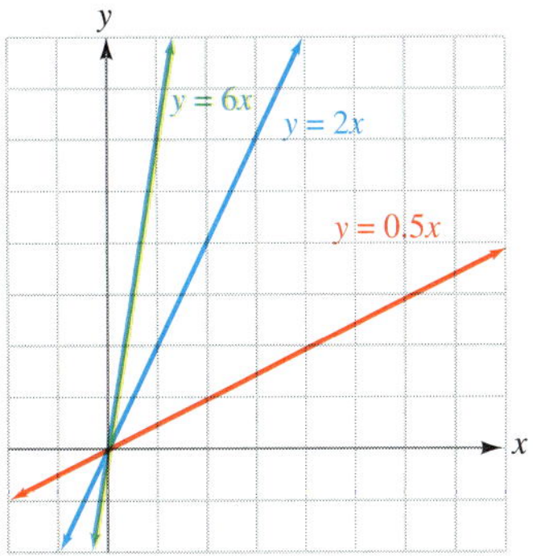

Since the formula for direct variation ($y = kx$) defines a linear function, its graph is always a line with a y-intercept at the origin. The graph of $y = kx$ appears in the figure for three positive values of k.

One example of direct variation is Hooke's law from physics. Hooke's law states that the distance a spring will stretch varies directly with the force that is applied to it.

If d represents a distance and f represents a force, this verbal model of Hooke's law can be expressed as

$d = kf$ This direct variation model can also be read as "d is directly proportional to f."

where k is the constant of variation. Suppose we know that a certain spring stretches 10 inches when a weight of 6 pounds is attached (see the figure). We can find k as follows:

$d = kf$

$10 = k(6)$ Substitute 10 for d and 6 for f.

$\frac{5}{3} = k$

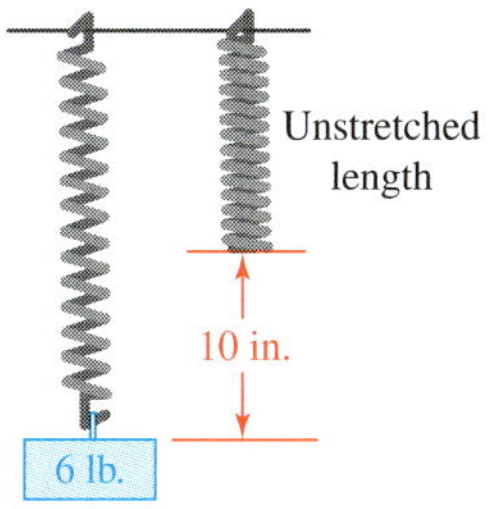

To find the force required to stretch the spring a distance of 35 inches, we can solve the equation $d = kf$ for f, with $d = 35$ and $k = \frac{5}{3}$.

$d = kf$

$35 = \frac{5}{3}f$ Substitute 35 for d and $\frac{5}{3}$ for k.

$105 = 5f$ Multiply both sides by 3.

$21 = f$ Divide both sides by 5.

Thus, the force required to stretch the spring a distance of 35 inches is 21 pounds.

EXAMPLE 5

Currency exchange. The currency calculator shown on the right converts from U.S. dollars to Japanese yen. When exchanging these currencies, the number of yen received is directly proportional to the number of dollars to be exchanged. How many yen will an exchange of \$1,200 bring?

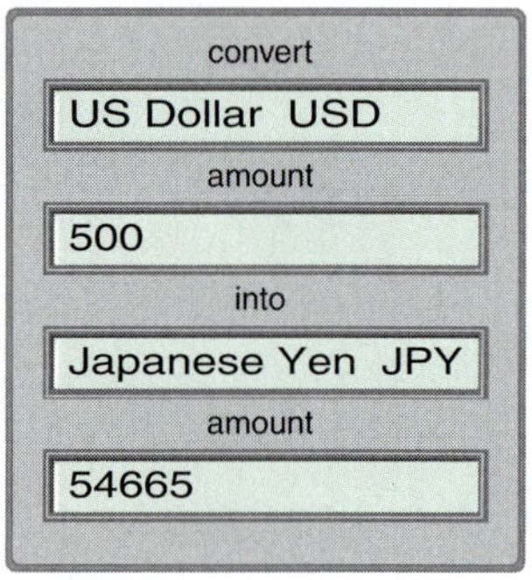

Solution The verbal model *the number of yen is directly proportional to the number of dollars* can be expressed by the equation

$y = kd$ This is a direct variation model.

where y is the number of yen, k is the constant of variation, and d is the number of dollars. From the illustration, we see that an exchange of \$500 brings 54,665 yen. To find k, we substitute 500 for d and 54,665 for y, and then we solve for k.

$$y = kd$$
$$54{,}665 = k(500)$$
$$109.33 = k \quad \text{Divide both sides by 500.}$$

To find how many yen an exchange of \$1,200 will bring, we substitute 109.33 for k and 1,200 for d in the direct variation model, and then we evaluate the right-hand side.

$$y = kd$$
$$y = 109.33(1{,}200)$$
$$y = 131{,}196$$

An exchange of \$1,200 will bring 131,196 yen.

Self Check 5 When exchanging currencies, the number of British pounds received is directly proportional to the number of U.S. dollars to be exchanged. If \$800 converts to 440 pounds, how many pounds will be received if \$1,500 is exchanged?

Solving Variation Problems To solve a variation problem:

1. Translate the verbal model into an equation.
2. Substitute the first set of values into the equation from step 1 to determine the value of k.
3. Substitute the value of k into the equation from step 1.
4. Substitute the remaining set of values into the equation from step 3 and solve for the unknown.

INVERSE VARIATION

In the formula $w = \frac{12}{l}$, w gets smaller as l gets larger, and w gets larger as l gets smaller. Since these variables vary in opposite directions in a predictable way, we say that the variables *vary inversely,* or that they are *inversely proportional.* The constant 12 is the constant of variation.

Inverse Variation The words "y varies inversely with x" or "y is inversely proportional to x" mean that $y = \frac{k}{x}$ for some nonzero constant k. The constant k is called the **constant of variation.**

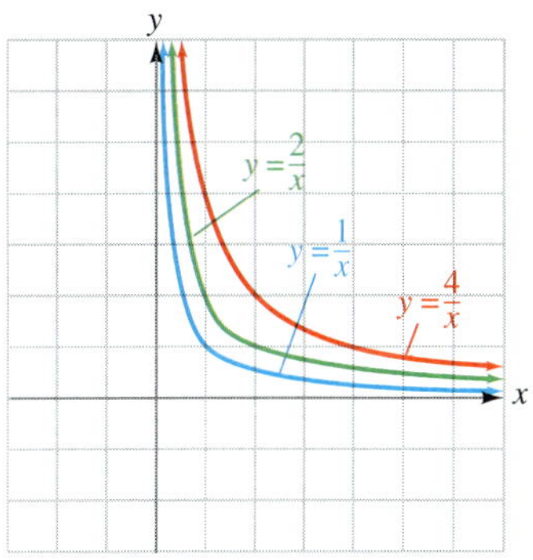

The formula for inverse variation, $y = \frac{k}{x}$, defines a rational function whose graph will have the x- and y-axes as asymptotes. The graph of $y = \frac{k}{x}$ appears in the figure for three positive values of k.

In an elevator, the amount of floor space per person varies inversely with the number of people in the elevator. If f represents the amount of floor space per person and n the number of people in the elevator, the relationship between f and n can be expressed by the equation.

$$f = \frac{k}{n}$$ This inverse variation model can also be read as "f is inversely proportional to n."

The figure shows 6 people in an elevator; each has 8.25 square feet of floor space. To determine how much floor space each person would have if 15 people were in the elevator, we begin by determining k.

$$f = \frac{k}{n}$$

$$8.25 = \frac{k}{6}$$ Substitute 8.25 for f and 6 for n.

$$k = 49.5$$ Multiply both sides by 6 to solve for k.

To find the amount of floor space per person if 15 people are in the elevator, we proceed as follows:

$$f = \frac{k}{n}$$

$$f = \frac{49.5}{15}$$ Substitute 49.5 for k and 15 for n.

$$f = 3.3$$ Do the division.

If 15 people were in the elevator, each would have 3.3 square feet of floor space.

EXAMPLE 6

Photography. The intensity I of light received from a light source varies inversely with the square of the distance from the light source. If a photographer, 16 feet away from his subject, has a light meter reading of 4 foot-candles of illuminance, what will the meter read if the photographer moves in for a close-up 4 feet away from the subject?

Solution The words *intensity varies inversely with the square of the distance d* can be expressed by the equation

$$I = \frac{k}{d^2}$$ This inverse variation model can also be read as "I is inversely proportional to d^2."

To find k, we substitute 4 for I and 16 for d and solve for k.

Success Tip

The constant of variation is usually positive, because most real-life applications involve only positive quantities. However, the definitions of direct, inverse, joint, and combined variation allow for a negative constant of variation.

$$I = \frac{k}{d^2}$$

$$4 = \frac{k}{16^2}$$

$$4 = \frac{k}{256}$$

$$1{,}024 = k$$

To find the intensity when the photographer is 4 feet away from the subject, we substitute 4 for d and 1,024 for k and simplify.

$$I = \frac{k}{d^2}$$

$$I = \frac{1{,}024}{4^2}$$

$$= 64$$

The intensity at 4 feet is 64 foot-candles.

Self Check 6 Find the intensity when the photographer is 8 feet away from the subject.

JOINT VARIATION

There are times when one variable varies with the product of several variables. For example, the area of a triangle varies directly with the product of its base and height:

$$A = \frac{1}{2}bh$$

Such variation is called *joint variation.*

Joint Variation If one variable varies directly with the product of two or more variables, the relationship is called **joint variation.** If y varies jointly with x and z, then $y = kxz$. The nonzero constant k is called the **constant of variation.**

EXAMPLE 7

Force of the wind. The force of the wind on a billboard varies jointly with the area of the billboard and the square of the wind velocity. When the wind is blowing at 20 mph, the force on a billboard 30 feet wide and 18 feet high is 972 pounds. Find the force on a billboard having an area of 300 square feet caused by a 40-mph wind.

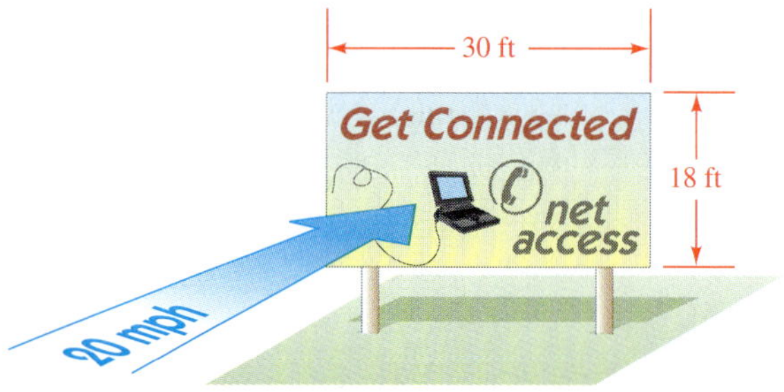

Solution We will let f represent the force of the wind, A the area of the billboard, and v the velocity of the wind. The words *the force of the wind on a billboard varies jointly with the area of the billboard and the square of the wind velocity* mean that *f varies directly as the product of A and v^2*. Thus,

$$f = kAv^2$$ The joint variation model can also be read as "f is directly proportional to the product of A and v^2."

Since the billboard is 30 feet wide and 18 feet high, it has an area of $30 \cdot 18 = 540$ square feet. We can find k by substituting 972 for f, 540 for A, and 20 for v.

$$f = kAv^2$$
$$972 = k(540)(20)^2$$
$$972 = k(216{,}000)$$ First, find the power: $(20)^2 = 400$. Then do the multiplication.
$$0.0045 = k$$ Divide both sides by 216,000 to solve for k.

To find the force exerted on a 300-square-foot billboard by a 40-mph wind, we use the formula $f = 0.0045Av^2$ and substitute 300 for A and 40 for v.

$$f = 0.0045Av^2$$
$$= 0.0045(300)(40)^2$$
$$= 2{,}160$$

The 40-mph wind exerts a force of 2,160 pounds on the billboard.

COMBINED VARIATION

Many applied problems involve a combination of direct and inverse variation. Such variation is called **combined variation.**

EXAMPLE 8

Highway construction. The time it takes to build a highway varies directly with the length of the road, but inversely with the number of workers. If it takes 100 workers 4 weeks to build 2 miles of highway, how long will it take 80 workers to build 10 miles of highway?

Solution We can let t represent the time in weeks to build a highway, ℓ represent the length of the highway in miles, and w represent the number of workers. The relationship between these variables can be expressed by the equation

$$t = \frac{k\ell}{w}$$ This is a combined variation model.

We substitute 4 for t, 100 for w, and 2 for l to find k:

$$4 = \frac{k(2)}{100}$$
$$400 = 2k$$ Multiply both sides by 100.
$$200 = k$$ Divide both sides by 2.

We now substitute 80 for w, 10 for ℓ, and 200 for k in the equation $t = \frac{k\ell}{w}$ and simplify:

$$t = \frac{k\ell}{w}$$

$$t = \frac{200(10)}{80}$$

$$= 25$$

It will take 25 weeks for 80 workers to build 10 miles of highway.

Self Check 8 How long will it take 60 workers to build 6 miles of highway?

Answers to Self Checks **1.** $\frac{3}{2}$ **2.** $\frac{2}{3}$, 1 **3.** $65\frac{1}{4}$ ft **4.** 24 ft **5.** 825 British pounds **6.** 16 foot-candles **8.** 20 weeks

6.8 STUDY SET

VOCABULARY **Fill in the blanks.**

1. A ______ is the quotient of two numbers or two quantities with the same units.
2. An equation that states that two ratios are equal, such as $\frac{1}{2} = \frac{4}{8}$, is called a __________.
3. In a proportion, the product of the ________ is equal to the product of the _______.
4. If two angles of one triangle have the same measure as two angles of a second triangle, the triangles are _______.
5. The equation $y = kx$ defines _______ variation, and $y = \frac{k}{x}$ defines _______ variation.
6. The equation $y = kxz$ defines ______ variation, and $y = \frac{kx}{z}$ defines ___________ variation.
7. ________ variation is represented by a rational function.
8. _______ variation is represented by a linear function.

CONCEPTS **Decide whether direct or inverse variation applies and sketch a possible graph for the situation.**

9.

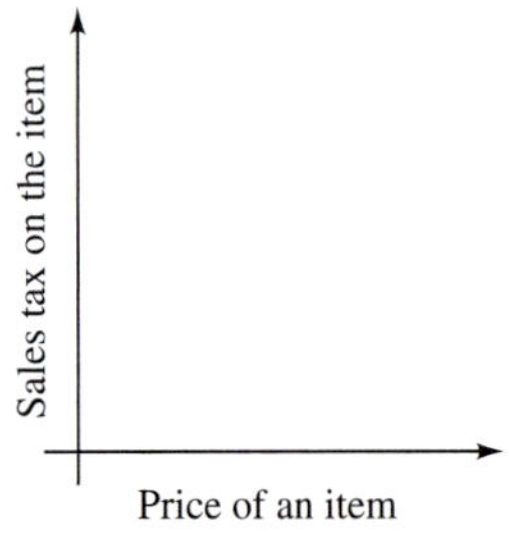

10.

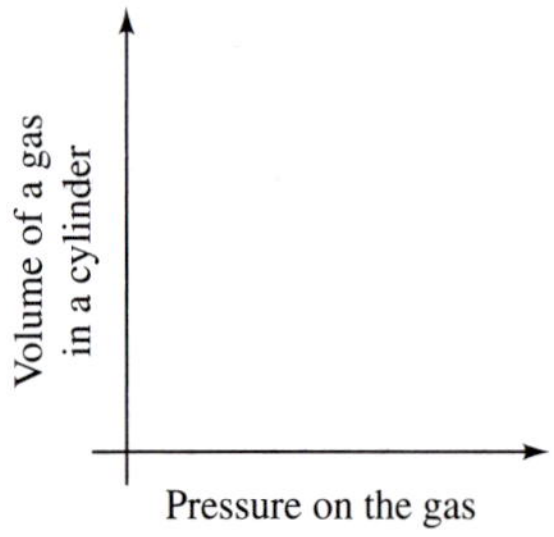

11.

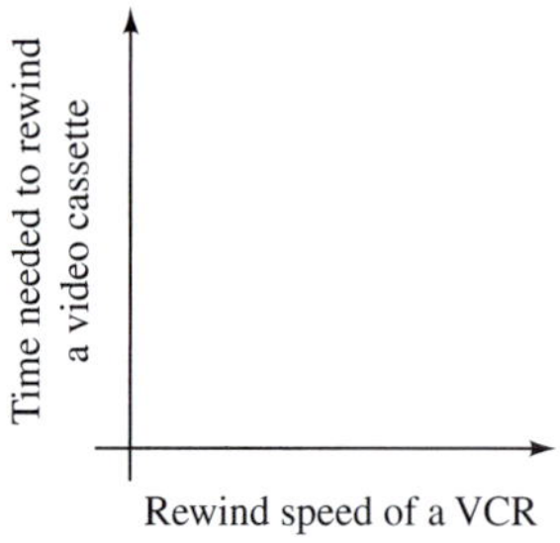

12.

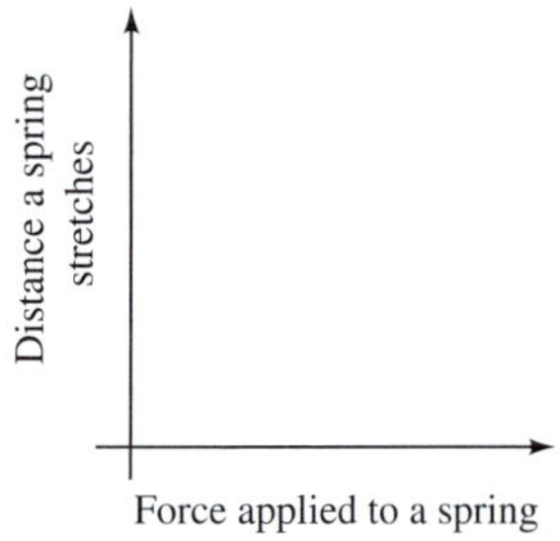

NOTATION **Complete each solution.**

13. Solve: $\frac{-7}{6} = \frac{x+3}{12}$

$$\square(12) = \square(x+3)$$

$$-84 = 6x + \square$$

$$\square = 6x$$

$$\square = x$$

14. Solve: $\frac{18}{2x+1} = \frac{3}{14}$

$$\square(14) = (\square)3$$

$$252 = \square + 3$$

$$249 = \square$$

$$\square = x$$

PRACTICE **Solve each proportion, if possible.**

15. $\frac{x}{5} = \frac{15}{25}$

16. $\frac{4}{y} = \frac{6}{27}$

17. $\dfrac{r-2}{3} = \dfrac{r}{5}$

18. $\dfrac{x+1}{x-1} = \dfrac{6}{4}$

19. $\dfrac{5}{5z+3} = \dfrac{2z}{2z^2+6}$

20. $\dfrac{9t+6}{t} = \dfrac{7}{3}$

21. $\dfrac{2(y+3)}{3} = \dfrac{4(y-4)}{5}$

22. $\dfrac{b+4}{5} = \dfrac{3(b-2)}{3}$

23. $\dfrac{2}{3x} = \dfrac{6x}{36}$

24. $\dfrac{y}{4} = \dfrac{4}{y}$

25. $\dfrac{2}{c} = \dfrac{c-3}{2}$

26. $\dfrac{2}{x+6} = \dfrac{-2x}{5}$

27. $\dfrac{1}{x+3} = \dfrac{-2x}{x+5}$

28. $\dfrac{x-1}{x+1} = \dfrac{2}{3x}$

29. $\dfrac{9z+6}{z(z+3)} = \dfrac{7}{z+3}$

30. $\dfrac{3}{n(n+3)} = \dfrac{2}{(n+1)(n+3)}$

31. $\dfrac{h^2}{5} = \dfrac{h}{2h-9}$

32. $\dfrac{b^2}{5} = \dfrac{b}{6b-13}$

33. $\dfrac{t^2-1}{5} = \dfrac{1-t^2}{2t}$

34. $\dfrac{n^2}{6} = \dfrac{n}{n-1}$

Express each verbal model in symbols.

35. A varies directly with the square of p.

36. z varies inversely with the cube of t.

37. v varies inversely with the square of r.

38. C varies jointly with x, y, and z.

39. P varies directly with the square of a and inversely with the cube of j.

40. M varies inversely with the cube of n and jointly with x and the square of z.

Express each variation model in words. In each equation, k is the constant of variation.

41. $L = kmn$

42. $P = \dfrac{km}{n}$

43. $R = \dfrac{kL}{d^2}$

44. $U = krs^2t$

APPLICATIONS **Set up and solve the required proportion.**

45. CAFFEINE Many convenience stores sell super-size 44-ounce soft drinks in refillable cups. For each of the products listed in the table, find the amount of caffeine contained in one of the large cups. Round to the nearest milligram.

Soft drink, 12 oz	Caffeine (mg)
Mountain Dew	55
Coca-Cola Classic	47
Pepsi	37

Based on data from *Los Angeles Times* (November 11, 1997) p. S4

46. TELEPHONES As of 2003, Iceland had 221 mobile cellular telephones per 250 inhabitants—the highest rate of any country in the world. If Iceland's population is about 280,000, how many mobile cellular telephones does the country have?

47. WALLPAPERING Read the instructions on the label of wallpaper adhesive. Estimate the amount of adhesive needed to paper 500 square feet of kitchen walls if a heavy wallpaper will be used.

COVERAGE: One-half gallon will hang approximately 4 single rolls (140 sq ft), depending on the weight of the wall covering and the condition of the wall.

48. RECOMMENDED DOSAGES The recommended child's dose of the sedative hydroxine is 0.006 gram per kilogram of body mass. Find the dosage for a 30-kg child in milligrams.

49. ERGONOMICS The science of ergonomics coordinates the design of working conditions with the requirements of the worker. The illustration gives guidelines for the dimensions (in inches) of a computer workstation to be used by a person whose height is 69 inches. Find a set of workstation dimensions for a person 5 feet 11 inches tall. Round to the nearest tenth.

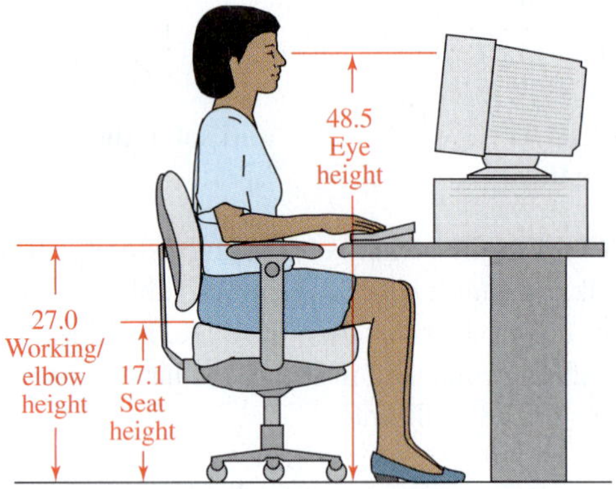

50. SHOPPING A recipe for guacamole dip calls for 5 avocados. If they are advertised at 3 for \$1.98, what will 5 avocados cost?

51. DRAWING See the illustration. To make an enlargement of the sailboat, an artist drew a grid over the smaller picture and transferred the contents of each small box to its corresponding larger box on another sheet of paper. If the smaller picture is 3 in. × 5 in. and if the width of the enlargement is 7.5 in., what is the length of the enlargement?

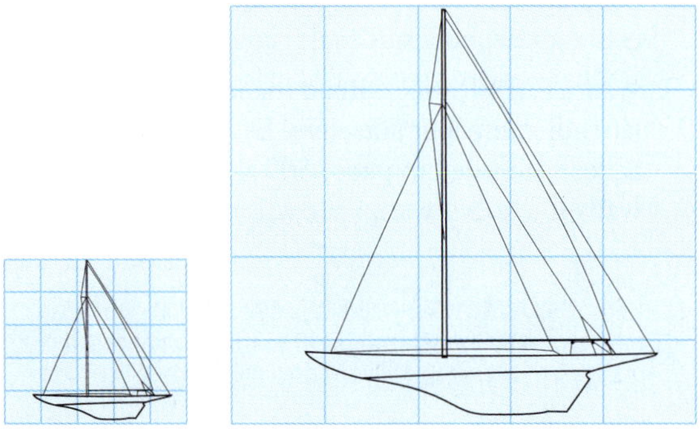

52. DRAFTING In a scale drawing, a 280-foot antenna tower is drawn $7\frac{1}{2}$ inches high. The building next to it is drawn $2\frac{1}{4}$ inches high. How tall is the actual building?

Use similar triangles to help solve each problem.

53. WASHINGTON, D.C. The Washington Monument casts a shadow of $166\frac{1}{2}$ feet at the same time as a 5-foot-tall tourist casts a shadow of $1\frac{1}{2}$ feet. Find the height of the monument.

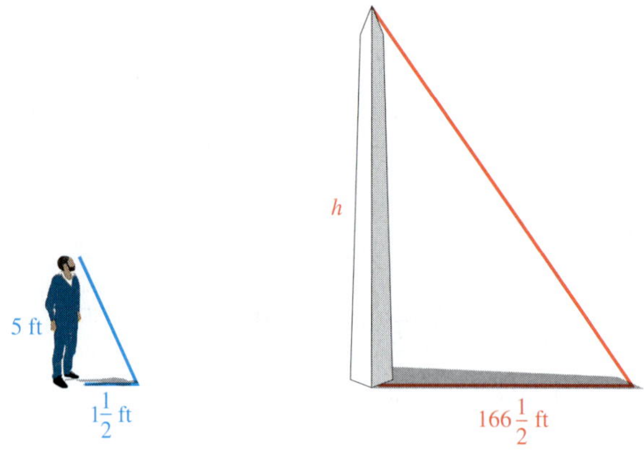

54. FLAGPOLES A man places a mirror on the ground and sees the reflection of the top of a flagpole, as in the illustration. The two triangles in the illustration are similar. Find the height h of the flagpole.

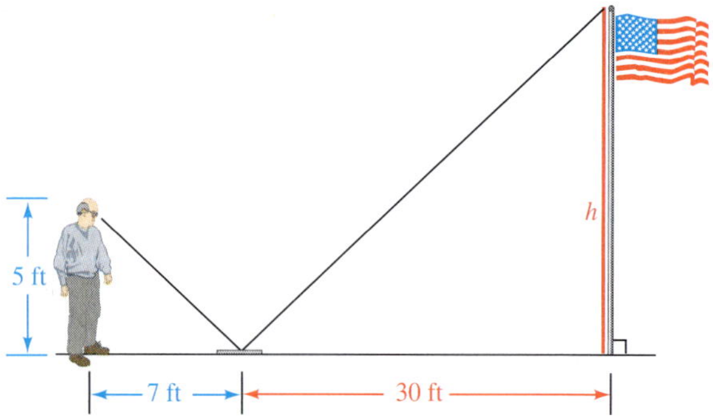

55. WIDTH OF A RIVER Use the dimensions in the illustration to find w, the width of the river. The two triangles in the illustration are similar.

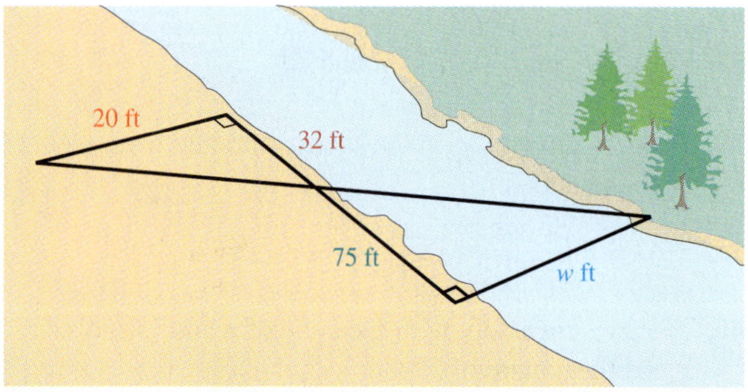

56. FLIGHT PATHS An airplane ascends 150 feet as it flies a horizontal distance of 1,000 feet. How much altitude will it gain as it flies a horizontal distance of 1 mile? (*Hint:* 5,280 feet = 1 mile.)

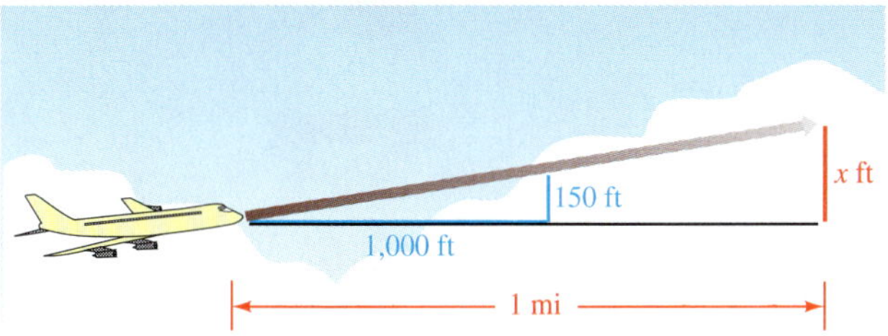

57. SKI RUNS A ski course with $\frac{1}{2}$ mile of horizontal run falls 100 feet in every 300 feet of run. Find the height of the hill.

58. GRAPHIC ARTS The compass in the illustration is used to draw circles with different radii (plural for radius). For the setting shown, what radius will the resulting circle have?

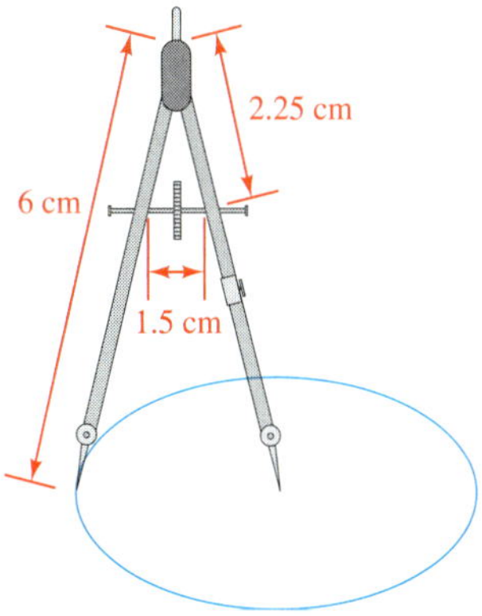

Solve each problem by writing a variation model.

59. FREE FALL An object in free fall travels a distance s that is directly proportional to the square of the time t. If an object falls 1,024 feet in 8 seconds, how far will it fall in 10 seconds?

60. FINDING DISTANCE The distance that a car can go varies directly with the number of gallons of gasoline it consumes. If a car can go 288 miles on 12 gallons of gasoline, how far can it go on a full tank of 18 gallons?

61. FARMING The number of days that a given number of bushels of corn will last when feeding cattle varies inversely with the number of animals. If x bushels will feed 25 cows for 10 days, how long will the feed last for 10 cows?

62. ORGAN PIPES The frequency of vibration of air in an organ pipe is inversely proportional to the length of the pipe. If a pipe 2 feet long vibrates 256 times per second, how many times per second will a 6-foot pipe vibrate?

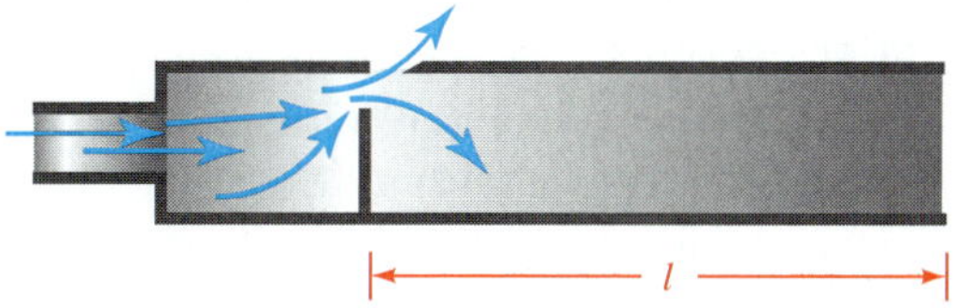

63. GAS PRESSURE Under constant temperature, the volume occupied by a gas varies inversely to the pressure applied. If the gas occupies a volume of 20 cubic inches under a pressure of 6 pounds per square inch, find the volume when the gas is subjected to a pressure of 10 pounds per square inch.

64. REAL ESTATE The following table shows the listing price for three homes in the same general locality. Write the variation model (direct or inverse) that describes the relationship between the listing price and the number of square feet of a house in this area.

Number of square feet	Listing price
1,720	$180,600
1,205	126,525
1,080	113,400

65. TRUCKING COSTS The costs of a trucking company vary jointly with the number of trucks in service and the number of hours they are used. When 4 trucks are used for 6 hours each, the costs are $1,800. Find the costs of using 10 trucks, each for 12 hours.

66. OIL STORAGE The number of gallons of oil that can be stored in a cylindrical tank varies jointly with the height of the tank and the square of the radius of its base. The constant of proportionality is 23.5. Find the number of gallons that can be stored in the cylindrical tank shown.

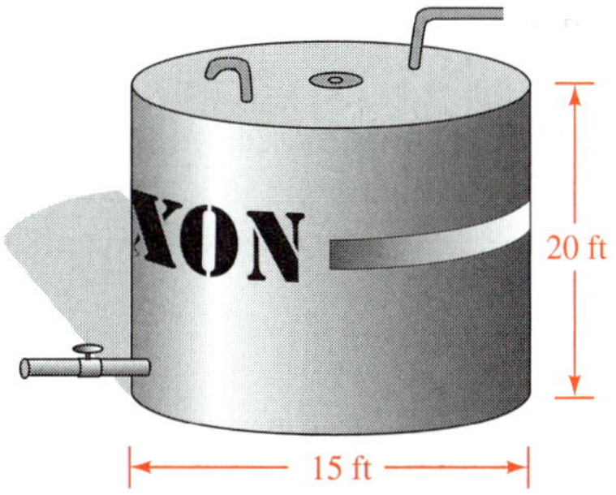

67. ELECTRONICS The voltage (in volts) measured across a resistor is directly proportional to the current (in amperes) flowing through the resistor. The constant of variation is the **resistance** (in ohms). If 6 volts is measured across a resistor carrying a current of 2 amperes, find the resistance.

68. ELECTRONICS The power (in watts) lost in a resistor (in the form of heat) varies directly with the square of the current (in amperes) passing through it. The constant of proportionality is the resistance (in ohms). What power is lost in a 5-ohm resistor carrying a 3-ampere current?

69. STRUCTURAL ENGINEERING The deflection of a beam is inversely proportional to its width and the cube of its depth. If the deflection of a 4-inch-by-4-inch beam is 1.1 inches, find the deflection of a 2-inch-by-8-inch beam positioned as in the illustration.

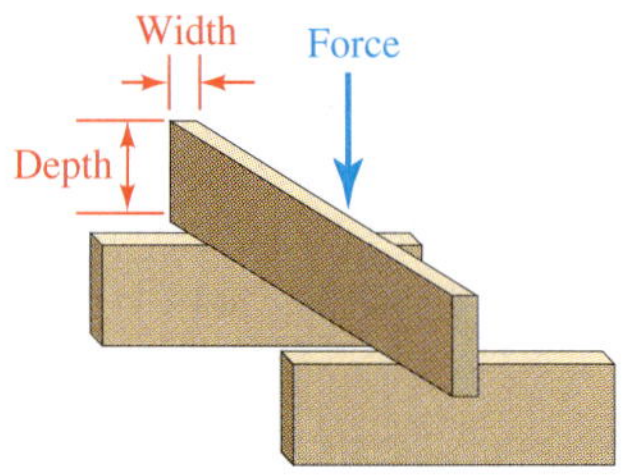

70. STRUCTURAL ENGINEERING Find the deflection of the beam in Exercise 69 when the beam is positioned as in the illustration.

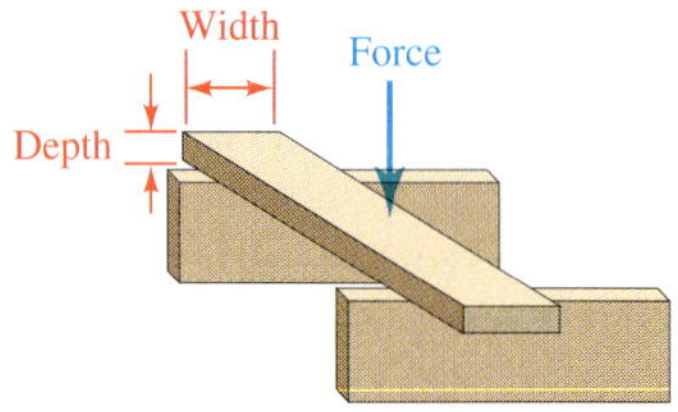

71. TENSION IN A STRING When playing with a Skip It toy, a child swings a weighted ball on the end of a string in a circular motion around one leg while jumping over the revolving string with the other leg. See the illustration in the next column. The tension T in the string is directly proportional to the square of the speed s of the ball and inversely proportional to the radius r of the circle. If the tension in the string is 6 pounds when the speed of the ball is 6 feet per second and the radius is 3 feet, find the tension when the speed is 8 feet per second and the radius is 2.5 feet.

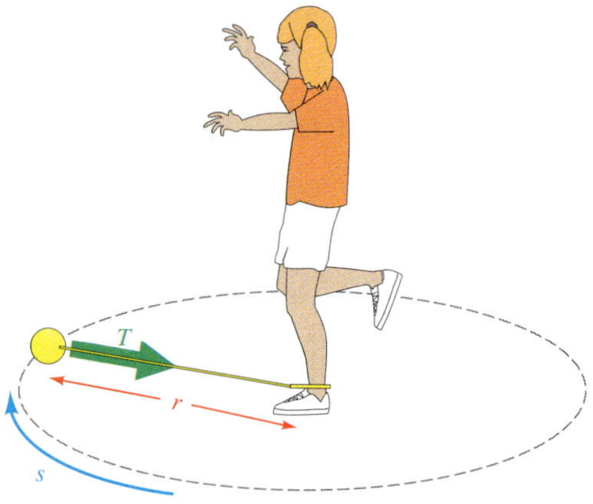

72. GAS PRESSURE The pressure of a certain amount of gas is directly proportional to the temperature (measured on the Kelvin scale) and inversely proportional to the volume. A sample of gas at a pressure of 1 atmosphere occupies a volume of 1 cubic meter at a temperature of 273 Kelvin. When heated, the gas expands to twice its volume, but the pressure remains constant. To what temperature is it heated?

WRITING

73. Distinguish between a *ratio* and a *proportion.*

74. From everyday life, give examples of two quantities that vary directly and two quantities that vary inversely.

REVIEW **Simplify each expression. Write each answer using positive exponents only.**

75. $(c^3)^2(c^4)^{-2}$

76. $\left(\dfrac{a^3a^5}{a^{-2}}\right)^3$

77. $\dfrac{b^0 - 2b^0}{b^0}$

78. $\left(\dfrac{2r^{-2}r^{-3}}{4r^{-5}}\right)^{-3}$

CHALLENGE PROBLEMS

79. As the cost of a purchase that is less than \$5 increases, the amount of change received from a five-dollar bill decreases. Is this inverse variation? Explain.

80. You've probably heard of Murphy's first law: *If anything can go wrong, it will.* Another of Murphy's laws is: *The chances of a piece of bread falling with the grape jelly side down varies directly with the cost of the carpet.* Write one of your own witty sayings using the phrase *varies directly.*

ACCENT ON TEAMWORK

USING PROPORTIONS WHEN COOKING

Overview: In this activity, you will gain experience with proportions.

Instructions: Each student should bring to class the recipe for their favorite food written on the front of a 3×5 card. Form groups of 2 or 3 students. Working together, the group should write proportions to determine the amount of each ingredient needed to make the recipe for the exact number of people in the class. For instance, suppose there are 32 students in your class. If a recipe that serves 8 calls for 3 cups of flour, the amount of flour needed for the recipe to serve 32 is given by the proportion

$$\begin{matrix}\text{Cups of flour} \rightarrow \\ \text{People served} \rightarrow\end{matrix} \frac{3}{8} = \frac{x}{32} \begin{matrix}\leftarrow \text{Cups of flour} \\ \leftarrow \text{People served}\end{matrix}$$

Write the new ingredient list on the back of each recipe card. Since the divisions involved might not be exact, be prepared to round when making the calculations. After all of the student's recipes have been adjusted, exchange recipe cards with someone in your class.

CONTINUED FRACTIONS

Overview: In this activity, as you gain experience simplifying complex fractions, you will make an interesting discovery about continued fractions.

Instructions: Form groups of 2 or 3 students. Working as a group, simplify each expression in the following list. Note that the third, fourth, fifth, and all the subsequent fractions in the list have a complex fraction in their denominator. These expressions are called **continued fractions.**

$$1 + \frac{1}{2}, \qquad 1 + \cfrac{1}{1 + \cfrac{1}{2}}, \qquad 1 + \cfrac{1}{1 + \cfrac{1}{1 + \cfrac{1}{2}}},$$

$$1 + \cfrac{1}{1 + \cfrac{1}{1 + \cfrac{1}{1 + \cfrac{1}{2}}}}, \qquad 1 + \cfrac{1}{1 + \cfrac{1}{1 + \cfrac{1}{1 + \cfrac{1}{1 + \cfrac{1}{2}}}}}, \ldots$$

Each of these expressions can be simplified by using the value of the expression preceding it. For example, to simplify the second expression in the list, replace $1 + \frac{1}{2}$ with $\frac{3}{2}$. Show that the expressions in the list simplify to $\frac{3}{2}, \frac{5}{3}, \frac{8}{5}, \frac{13}{8}, \frac{21}{13}, \ldots$. Then write the next 3 continued fractions in the list. From what you have learned, *predict* the answers if each of them were simplified.

KEY CONCEPT: EXPRESSIONS AND EQUATIONS

In this chapter, we have discussed procedures for working with **rational expressions** and procedures for solving **rational equations.**

RATIONAL EXPRESSIONS

To simplify rational expressions and when multiplying and dividing rational expressions, we remove factors equal to 1 by replacing each pair of factors common to the numerator and denominator with the equivalent fraction $\frac{1}{1}$.

1. a. Simplify: $\dfrac{6x^2 + x - 2}{8x^2 + 2x - 3}$.

b. What common factor was removed?

2. a. Multiply: $\dfrac{3d^2 - d - 2}{6d^2 - 5d - 6} \cdot \dfrac{4d^2 - 9}{2d^2 + 5d + 3}$.

b. What common factors were removed?

When adding and subtracting rational expressions and when simplifying complex fractions, we must often multiply by 1 in the form of $\frac{c}{c}$, where c is a nonzero expression.

3. a. Add: $\dfrac{3}{x + 2} + \dfrac{5}{x - 4}$.

b. By what did you multiply the first fraction to rewrite it in terms of the LCD? By what did you multiply the second fraction?

4. a. Simplify: $\dfrac{n - 1 - \frac{2}{n}}{\frac{n}{3}}$.

b. By what did you multiply the numerator and denominator to simplify the complex fraction?

RATIONAL EQUATIONS

The multiplication property of equality states that *if equal quantities are multiplied by the same nonzero number, the results will be equal quantities.* We use this property when solving rational equations. If we multiply both sides of the equation by the LCD of the rational expressions in the equation, we can clear it of fractions.

5. a. Solve: $\dfrac{t - 3}{t - 2} - \dfrac{t - 3}{t} = \dfrac{1}{t}$.

b. By what did you multiply both sides to clear the equation of fractions?

6. a. Solve: $\dfrac{5}{2x^2 + x - 3} - \dfrac{x + 1}{x - 1} = \dfrac{2}{2x + 3} - 1$.

b. By what did you multiply both sides to clear the equation of fractions?

7. a. Solve: $\dfrac{x + 1}{x + 2} = \dfrac{x - 3}{x - 4}$.

b. By what did you multiply both sides to clear the equation of fractions?

8. a. Solve: $\dfrac{x^2}{a^2} - \dfrac{y^2}{b^2} = 1$ for a^2.

b. By what did you multiply both sides to clear the equation of fractions?

CHAPTER REVIEW

SECTION 6.1 Rational Functions and Simplifying Rational Expressions

CONCEPTS

A *rational expression* is an expression of the form $\frac{P}{Q}$, where P and Q are polynomials and Q does not equal 0.

Since division by 0 is undefined, we must make sure that the denominator of a rational expression is not 0.

To *simplify* a rational expression:

1. Factor the numerator and denominator completely.
2. Remove factors equal to 1 by replacing each pair of factors common to the numerator and denominator with the equivalent fraction $\frac{1}{1}$.

The quotient of any nonzero expression and its opposite is -1.

REVIEW EXERCISES

1. Complete the table of values for the rational function $f(x) = \frac{4}{x}$ where $x > 0$. Then graph it. Label the horizontal asymptote.

x	$f(x)$
$\frac{1}{2}$	
1	
2	
3	
4	
5	
6	
7	
8	

2. Use the graph of function f to find each of the following:

a. $f(12)$

b. The value(s) of x for which $f(x) = 6$

c. The domain and range of f

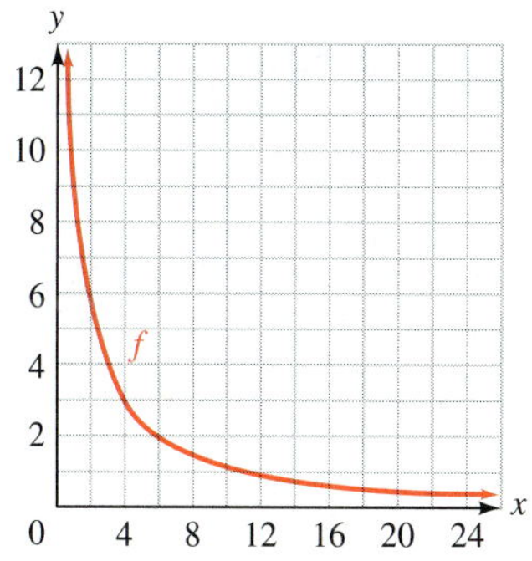

3. Find the domain of the rational function $f(x) = \frac{2x^2 + 8x}{x^2 + 2x - 24}$. Use interval notation.

4. Use a graphing calculator to graph the rational function $f(x) = \frac{3x + 2}{x}$. From the graph, determine the equations of the horizontal and vertical asymptotes and the domain and range.

Simplify each rational expression.

5. $\frac{48x^2y}{76xy^8}$

6. $\frac{x^2 - 49}{x^2 + 14x + 49}$

7. $\frac{x^2 - 2x + 4}{2x^3 + 16}$

8. $\frac{x^2 + 6x + 36}{x^3 - 216}$

9. $\frac{ac - ad + bc - bd}{d^2 - c^2}$

10. $\frac{m^3 + m^2n - 2mn^2}{2m^3 - mn^2 - m^2n}$

11. $\frac{6x^2 - 5x - 4}{(3x - 4)^3}$

12. $\frac{2m - 2n}{n - m}$

SECTION 6.2 Multiplying and Dividing Rational Expressions

To *multiply* rational expressions, multiply the numerators and multiply the denominators.

$$\frac{A}{B} \cdot \frac{C}{D} = \frac{AC}{BD}$$

Then simplify, if possible.

To divide rational expressions, multiply the first by the reciprocal of the second.

$$\frac{A}{B} \div \frac{C}{D} = \frac{A}{B} \cdot \frac{D}{C} = \frac{AD}{BC}$$

Then simplify, if possible.

Perform the operations and simplify:

13. $\dfrac{3x^3y^4}{c^2d} \cdot \dfrac{c^3d^2}{21x^5y^4}$

14. $\dfrac{x^2+4x+4}{x^2-x-6} \cdot \dfrac{9-x^2}{x^2+5x+6}$

15. $\dfrac{2a^2-5a-3}{a^2-9} \div \dfrac{2a^2+5a+2}{2a^2+5a-3}$

16. $\dfrac{t^4-4t^2}{t} \div (t^3+2t^2)$

17. $\left(\dfrac{h-2}{h^3+4}\right)^2$

18. $\dfrac{m^2+3m+9}{m^2+mp+mr+pr} \div \dfrac{m^3-27}{am+ar+bm+br}$

19. $\dfrac{8m^2+6mn-9n^2}{2m^2+5mn+3n^2} \cdot \dfrac{6m^2+5mn-4n^2}{12m^2+7mn-12n^2}$

20. $\dfrac{x^3+3x^2+2x}{2x^2-2x-12} \cdot \dfrac{3x^2-3x}{x^3-3x^2-4x} \div \dfrac{x^2+3x+2}{2x^2-4x-16}$

SECTION 6.3 Adding and Subtracting Rational Expressions

To add (or subtract) two rational expressions with *like denominators,* add (or subtract) the numerators and keep the common denominator.

To find the LCD of several rational expressions, factor each denominator and use each factor the greatest number of times that it appears in any one denominator. The product of these factors is the LCD.

To add or subtract rational expressions with *unlike denominators,* find the LCD and express each rational expression with a denominator that is the LCD. Add (or subtract) the resulting fractions and simplify the result if possible.

Perform the operations and simplify.

21. $\dfrac{5y}{x-y} - \dfrac{3}{x-y}$

22. $\dfrac{d^2}{c^3-d^3} + \dfrac{c^2+cd}{c^3-d^3}$

23. $\dfrac{4}{t-3} + \dfrac{6}{3-t}$

24. $\dfrac{p+3}{p^2+13p+12} - \dfrac{2p+4}{p^2+13p+12}$

The denominators of some rational expressions are given. Find the LCD.

25. $15a^2h, 20ah^3$

26. ab^2-ab, ab^2, b^2-b

27. x^2-4x-5, x^2-25

28. m^2-4m+4, m^3-8

Perform the operations and simplify.

29. $9 - \dfrac{1}{a+1}$

30. $\dfrac{5x}{14z^2} + \dfrac{y^2}{16z}$

31. $\dfrac{4x}{x-4} - \dfrac{3}{x+3}$

32. $\dfrac{4}{3xy-6y} - \dfrac{4}{10-5x}$

33. $\dfrac{y+7}{y+3} - \dfrac{y-3}{y+7}$

34. $\dfrac{-2(3+x)}{x^2+6x+9} + \dfrac{3(x+2)}{x^2-6x+9} - \dfrac{1}{x^2-9}$

SECTION 6.4 Simplifying Complex Fractions

To *simplify a complex fraction:*

Method 1: Write the numerator and denominator as single fractions. Then divide the fractions and simplify.

Method 2: Multiply the numerator and denominator by the LCD of the fractions in the numerator and denominator. Then simplify the results.

Simplify each complex fraction.

35. $\dfrac{\frac{4a^3b^2}{9c}}{\frac{14a^3b}{9c^4}}$

36. $\dfrac{\frac{p^2-9}{6pt}}{\frac{p^2+5p+6}{3pt}}$

37. $\dfrac{\frac{1}{a}+\frac{2}{b}}{\frac{2}{a}-\frac{1}{b}}$

38. $\dfrac{1-\frac{1}{x}-\frac{2}{x^2}}{1+\frac{4}{x}+\frac{3}{x^2}}$

39. $\dfrac{(x-y)^{-2}}{x^{-2}-y^{-2}}$

40. $\dfrac{3x-\frac{1}{3-\frac{x}{2}}}{\frac{3}{x-6}+x}$

SECTION 6.5 Dividing Polynomials

To divide monomials, use the method for simplifying fractions or use the rules for exponents.

To divide a polynomial by a monomial, divide each term of the numerator by the denominator.

Long division is used to divide one polynomial by another. It works best when the exponents of the terms of the divisor and the dividend are written in descending order.

When the dividend is missing a term, write it with a coefficient of zero or leave a blank space.

Perform each division. Write each answer using positive exponents only.

41. $\dfrac{25h^4k^7}{55hk^9}$

42. $(-5x^3y^3z^{10}) \div (10x^3y^6z^{20})$

Find each quotient.

43. $\dfrac{36a+32}{6}$

44. $\dfrac{30x^3y^2-15x^2y-10xy^2}{-10xy}$

Find each quotient using long division.

45. $b+5\overline{)b^2+9b+20}$

46. $\dfrac{-33v-8v^2+3v^3-10}{1+3v}$

47. $x+2\overline{)x^3+8}$

48. $\dfrac{m^8+m^6-4m^4+5m^2-1}{m^4+2m^2-3}$

SECTION 6.6 Synthetic Division

Synthetic division is used to divide a polynomial by a binomial of the form $x - k$.

Remainder theorem: If a polynomial $P(x)$ is divided by $x - k$, then the remainder is $P(k)$.

It follows from the remainder theorem that a polynomial can be evaluated using synthetic division.

Factor theorem: If $P(x)$ is divided by $x - k$, then $P(k) = 0$, if and only if $x - k$ is a factor of $P(x)$.

Use synthetic division to perform each division.

49. $(x^2 + 13x + 42) \div (x + 6)$

50. $\dfrac{4x^3 + 5x^2 - 1}{x + 2}$

51. $\dfrac{-5n^5 + 4n^4 + 30n^3 + 2n^2 + 20n + 3}{n - 3}$

52. Let $P(x) = x^4 - 2x^3 + x^2 - 3x + 12$. Use the remainder theorem and synthetic division to find $P(-2)$.

Use the factor theorem to decide whether the first expression is a factor of $P(x)$.

53. $x - 5$; $P(x) = x^3 - 3x^2 - 8x - 10$

54. $x + 5$; $P(x) = x^3 + 4x^2 - 5x + 5$ (*Hint:* Write $x + 5$ as $x - (-5)$.)

SECTION 6.7 Solving Rational Equations

To solve a *rational equation,* multiply both sides by the LCD of the rational expressions in the equation to clear it of fractions.

Multiplying both sides of an equation by a quantity that contains a variable can lead to *extraneous* solutions.

All possible solutions of a rational equation must be checked.

Solve each equation, if possible.

55. $\dfrac{4}{x} - \dfrac{1}{10} = \dfrac{7}{2x}$

56. $\dfrac{5}{7 + t} - 1 = \dfrac{-4}{t + 7}$

57. $\dfrac{2}{7y^2} + \dfrac{1}{14y} = \dfrac{9}{28y^2} - \dfrac{1}{14y^2}$

58. $\dfrac{2}{3x + 15} - \dfrac{1}{18} = \dfrac{1}{3x + 12}$

59. $\dfrac{2(x - 5)}{x - 2} = \dfrac{6x + 12}{4 - x^2}$

60. $\dfrac{x + 3}{x - 5} + \dfrac{6 + 2x^2}{x^2 - 7x + 10} = \dfrac{3x}{x - 2}$

Solve each formula for the indicated variable.

61. $\dfrac{x^2}{a^2} - \dfrac{y^2}{b^2} = 1$ for y^2

62. $H = \dfrac{2ab}{a + b}$ for b

63. ADVERTISING A small plane pulling a banner can fly at a rate of 75 mph in calm air. Flying down the coast, with a tailwind, the plane flew 40 miles in the same time that it took to fly 35 miles up the coast, into a headwind. Find the rate of the wind.

64. TRIP LENGTH Traffic reduced a driver's usual speed by 10 mph, which lengthened her 200-mile trip by 1 hour. Find the driver's usual speed.

65. **a.** If a painter can complete a job in 10 hours, what is the painter's rate of work?

b. If the painter works for x hours, how much of the job is completed?

To solve *shared-work* problems, we use the formula

$$W = rt$$

where W is the amount of work completed, r is the rate of work, and t is the time worked.

66. DRAINING A TANK If one outlet pipe can drain a tank in 24 hours and another pipe can drain the tank in 36 hours, how long will it take for both pipes to drain the tank?

67. INSTALLING SIDING Two men have estimated that they can side a house in 8 days. If one of them, who could have sided the house alone in 14 days, gets sick, how long will it take the other man to side the house alone?

68. METALLURGY The stiffness of the flagpole is given by the formula

$$k = \frac{1}{\frac{1}{k_1} + \frac{1}{k_2}}$$

where k_1 and k_2 are the individual stiffnesses of each section. If the design specifications require that the stiffness k of the entire pole be 1,900,000 in. lb/rad, what must the stiffness of Section 1 be?

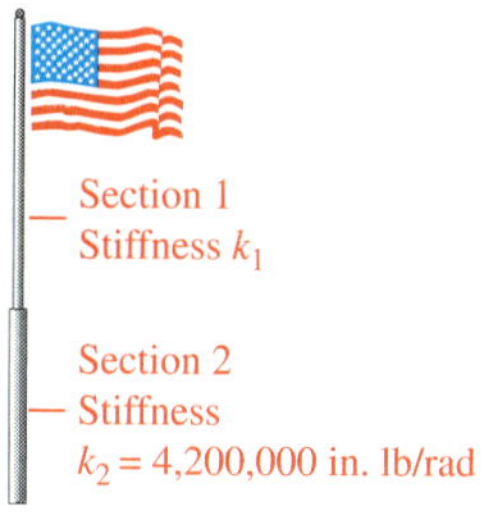

SECTION 6.8 Proportion and Variation

In a *proportion,* the product of the *extremes* is equal to the product of the *means.*

If two angles of one triangle have the same measure as two angles of a second triangle, the triangles are *similar.* The lengths of corresponding sides of similar triangles are proportional.

Direct variation: As one variable gets larger, the other gets larger as described by the equation $y = kx$, where k is the *constant of proportionality.*

Inverse variation: As one variable gets larger, the other gets smaller as described by the equation

$$y = \frac{k}{x} \quad (k \text{ is a constant})$$

Solve each proportion.

69. $\dfrac{x+1}{8} = \dfrac{4x-2}{24}$

70. $\dfrac{1}{x+6} = \dfrac{x+10}{12}$

71. SIMILAR TRIANGLES Find the height of a tree if it casts a 44-foot shadow when a 4-foot shrub casts a $2\frac{1}{2}$-foot shadow.

72. COOKING A recipe for spaghetti sauce requires four 16-ounce bottles of ketchup to make 2 gallons of sauce. How many bottles of ketchup are needed to make 10 gallons of sauce?

73. SCALE MODELS A model of a playhouse was made using a 1/12th scale. If the scale model is 5.5 inches tall, how tall is the playhouse?

74. PROPERTY TAX The property tax in a certain county varies directly as assessed valuation. If a tax of $1,575 is levied on a single-family home assessed at $90,000, determine the property tax on an apartment complex assessed at $312,000.

75. ELECTRICITY For a fixed voltage, the current in an electrical circuit varies inversely as the resistance in the circuit. If a certain circuit has a current of $2\frac{1}{2}$ amps when the resistance is 150 ohms, find the current in the circuit when the resistance is doubled.

Joint variation: One variable varies with the product of several variables. For example, $y = kxz$ (k is a constant).

Combined variation: a combination of direct and inverse variation. For example,

$$y = \frac{kx}{z} \quad (k \text{ is a constant})$$

76. HURRICANE WINDS The wind force on a vertical surface varies jointly as the area of the surface and the square of the wind's velocity. If a 10-mph wind exerts a force of 1.98 pounds on the sign shown in the illustration, find the force on the sign if the wind is blowing at 80 mph.

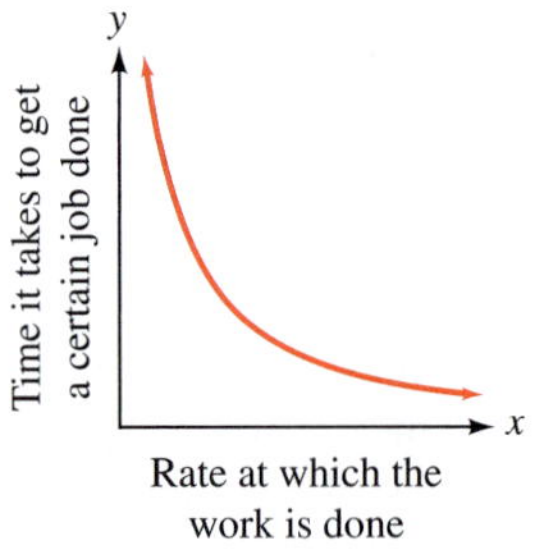

77. Does the graph above show direct or inverse variation?

78. Assume that x_1 varies directly with the third power of t and inversely with x_2. Find the constant of variation if $x_1 = 1.6$ when $t = 8$ and $x_2 = 64$.

CHAPTER 6 TEST

Simplify each rational expression.

1. $\dfrac{12x^2y^3z^2}{18x^3y^4z^2}$

2. $\dfrac{2x + 4}{x^2 - 4}$

3. $\dfrac{3y - 6z}{2z - y}$

4. $\dfrac{2x^2 + 7xy + 3y^2}{4xy + 12y^2}$

5. Graph the rational function $f(x) = \frac{2}{x}$ for $x > 0$. Label the horizontal asymptote.

6. Find the domain of the rational function $f(x) = \frac{x^2 + 6x + 5}{x - x^2}$. Use interval notation.

Perform the operations and simplify, if necessary. Write all answers using positive exponents only.

7. $\dfrac{x^2}{x^3z^2y^2} \cdot \dfrac{x^2z^4}{y^2z}$

8. $\dfrac{a^2 + 5a + 6}{a^2 - 4} \cdot \dfrac{a^2 - 5a + 6}{a^2 - 9}$

9. $\dfrac{xu + 2u + 3x + 6}{u^2 - 9} \cdot \dfrac{13u - 39}{x^2 + 3x + 2}$

10. $\dfrac{x^3 + y^3}{16x^2} \div \dfrac{x^2 - xy + y^2}{8x^2 + 8xy}$

11. $\dfrac{a^2 + 7a + 12}{a + 3} \div \dfrac{16 - a^2}{a - 4}$

12. $\dfrac{(2x - 3)^3}{x^2 - 2x + 1} \div \dfrac{3x^2 + 7x + 2}{3x^2 - 2x - 1} \cdot \dfrac{x^2 + x - 2}{2x^7 - 3x^6}$

13. $\dfrac{-3t + 4}{t^2 + t - 20} + \dfrac{6 + 5t}{t^2 + t - 20}$

14. $\dfrac{3wx}{wx - 5} + \dfrac{wx + 10}{5 - wx}$

15. $8b - 5 + \dfrac{5b + 4}{3b + 1}$

16. $\dfrac{a + 3}{a^2 - a - 2} - \dfrac{a - 4}{a^2 - 2a - 3}$

Simplify each complex fraction.

17. $\dfrac{\dfrac{2u^2w^3}{v^2}}{\dfrac{4uw^4}{uv}}$

18. $\dfrac{\dfrac{4}{3k} + \dfrac{k}{k+1}}{\dfrac{k}{k+1} - \dfrac{3}{k}}$

19. Divide: $\dfrac{18x^2y^3 - 12x^3y^2 + 9xy}{-3xy^4}$

20. Divide: $(y^3 - 48) \div (y + 2)$.

21. Let $P(x) = 4x^3 + 3x^2 + 2x - 7$. Use synthetic division to find $P(2)$.

22. Use the factor theorem to decide whether $x + 3$ is a factor of $P(x) = x^4 + 3x^3 - 16x^2 - 27x + 63$.

Solve each equation.

23. $\dfrac{34}{x^2} + \dfrac{13}{20x} = \dfrac{3}{2x}$

24. $\dfrac{u-2}{u-3} + 3 = u + \dfrac{u-4}{3-u}$

25. $\dfrac{3}{x-2} = \dfrac{x+3}{2x}$

26. $\dfrac{4}{m^2-9} + \dfrac{5}{m^2-m-12} = \dfrac{7}{m^2-7m+12}$

Solve each formula for the indicated variable.

27. $\dfrac{x^2}{a^2} + \dfrac{y^2}{b^2} = 1$ for a^2

28. $\dfrac{1}{r} = \dfrac{1}{r_1} + \dfrac{1}{r_2}$ for r_2

29. ROOFING One crew can finish a 2,800-square-foot roof in 12 hours, and another crew can do the job in 10 hours. If they work together, can they finish before a predicted rain in 5 hours? If not, how long will they have to work in the rain?

30. TOURING THE COUNTRYSIDE A man bicycles 5 mph faster than he can walk. He bicycles a distance of 24 miles and then hikes back along the same route. If the entire trip takes 11 hours, how fast does he walk?

31. SHADOWS Refer to the illustration. Find the height of the tree.

32. ANNIVERSARY GIFTS A florist sells a dozen long-stemmed red roses for $57.99. In honor of their 16th wedding anniversary, a man wants to buy 16 roses for his wife. What will the roses cost?

33. SOUND Sound intensity (loudness) varies inversely as the square of the distance from the source. If a rock band has a sound intensity of 100 decibels 30 feet away from the amplifier, find the sound intensity 60 feet away from the amplifier.

34. Draw a possible graph showing that the weekly salary of a person *varies directly* with the number of hours worked during the week. Label the axes.

35. Explain how to find the LCD for several rational expressions.

36. Explain the error that was made in the solution shown below.

$$\text{Simplify: } \frac{2(x+2) + 3(x-3)}{x+2} = \frac{2(\overset{1}{\cancel{x+2}}) + 3(x-3)}{\underset{1}{\cancel{x+2}}}$$

$$= \frac{2 + 3x - 9}{1}$$

$$= 3x - 7$$

CHAPTERS 1–6 CUMULATIVE REVIEW EXERCISES

1. Solve: $-3 = -\frac{9}{8}t$.

2. Solve: $\frac{3x - 4}{6} - \frac{x - 2}{2} = \frac{-2x - 3}{3}$.

3. AUTO SALES See the following graph. An automobile dealership is going to order 80 new Ford Escorts. According to the survey, exactly how many green Escorts should be purchased to meet the expected customer demand?

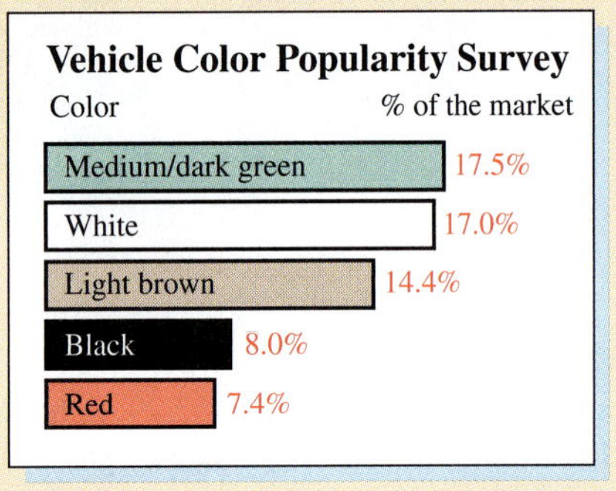

4. LIFE EXPECTANCY Determine the predicted rate of change in the life expectancy of females during the years 2000–2050, as shown in the graph.

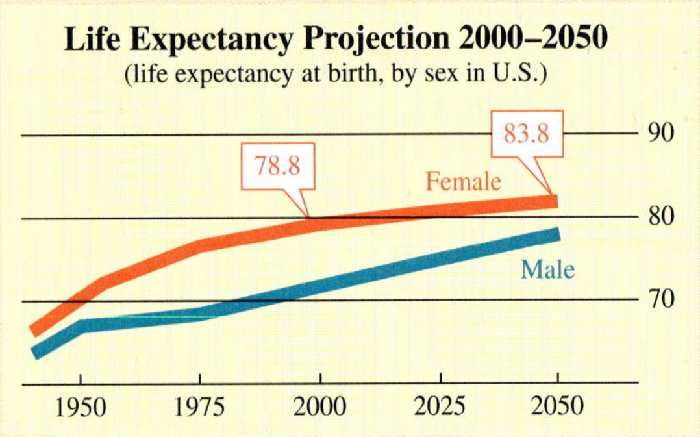

Based on data from the Social Security Administration, Office of Chief Actuary

Write the equation of the line with the given properties. Express your answer in slope–intercept form.

5. Slope of -7, passing through $(7, 5)$
6. Passing through $(-4, 5)$ and $(2, -6)$

7. Graph the linear function $f(x) = \frac{2}{3}x - 2$. Then use interval notation to specify the domain and range.

8. ENGINEERING The tensions T_1 and T_2 (in pounds) in each of the ropes shown in the illustration can be found by solving the system
$$\begin{cases} 0.6T_1 - 0.8T_2 = 0 \\ 0.8T_1 + 0.6T_2 = 100 \end{cases}$$
Find T_1 and T_2.

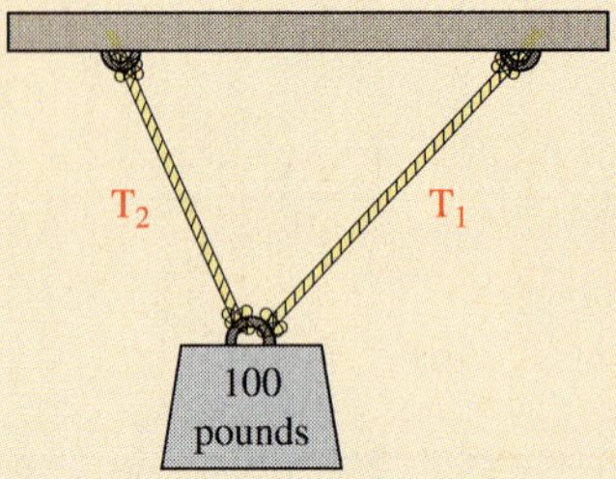

Solve. Write the solution set in interval notation and then graph it.

9. $\frac{1}{2}x + 6 \geq 4 + 2x$
10. $\left| \frac{3a}{5} - 2 \right| + 1 \geq \frac{6}{5}$
11. $5(x + 2) \leq 4(x + 1)$ and $11 + x < 0$
12. $-4(x + 2) \geq 12$ or $3x + 8 < 11$

Simplify each expression. Write answers using positive exponents only.

13. $a^3b^2a^5b^2$
14. $\frac{a^3b^6}{a^7b^2}$
15. $\frac{1}{3^{-4}}$
16. $\left(\frac{2x^{-2}y^3}{x^2x^3y^4} \right)^{-3}$

Write each number in standard notation.

17. 4.25×10^4
18. 7.12×10^{-4}

19. Express as a formula: *y varies directly with the product of x and z, and inversely with r.*

20. Evaluate the determinant: $\begin{vmatrix} 3 & -2 \\ -2 & 4 \end{vmatrix}$.

21. Graph: $y < 4 - x$.
22. Graph: $f(x) = 2x^2 - 3$. Then use interval notation to specify the domain and range.

23. If $g(x) = -3x^3 + x - 4$, find $g(-2)$.

24. Find the degree of $3 + x^2y + 17x^3y^4$.

25. The graph of a line is shown on the graphing calculator screen below.

a. Give the x- and y-intercepts of the line.

b. What is the slope of the line?

c. The line doesn't lie in one quadrant. Which quadrant is that?

d. What is the equation of the line?

e. Does the line pass through $(15, -42)$?

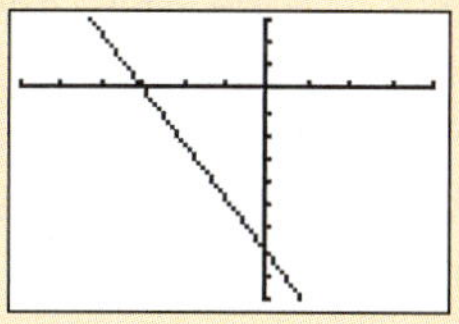

26. Express the dollar value of each type of United States coin and currency shown in the illustration, using a power of 10. For example, one hundred dollars can be expressed as $\$10^2$.

Perform the operations and simplify.

27. $(x^3 + 3x^2 - 2x + 7) + (x^3 - 2x^2 + 2x + 5)$

28. $(-5x^2 + 3x + 4) - (-2x^2 + 3x + 7)$

29. $(a + b + c)(2a - b - 2c)$

30. $(2x^3 - 1)^2$

For Exercises 31–32, refer to the following illustration. The graph shows the correction that must be made to a sundial reading to obtain accurate clock time. The difference is caused by the Earth's orbit and tilted axis.

31. a. Is this the graph of a function?

b. During the year, what is the maximum number of minutes the sundial reading gets ahead of a clock?

c. During the year, what is the maximum number of minutes the sundial reading falls behind a clock?

32. How many times during a year is the sundial reading exactly the same as a clock?

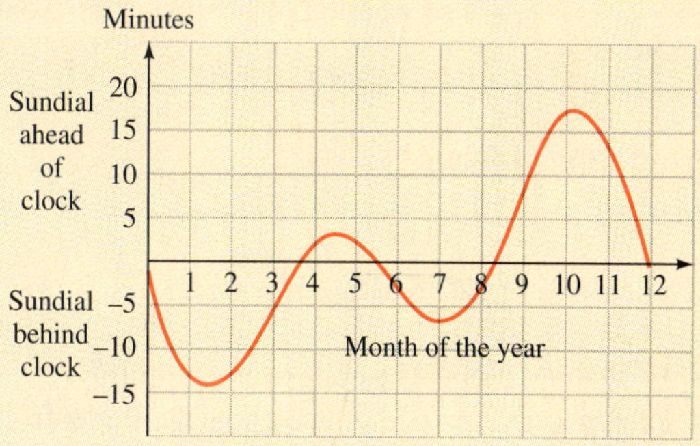

Factor each expression.

33. $3r^2s^3 - 6rs^4$

34. $5(x - y) - a(x - y)$

35. $xu + yv + xv + yu$

36. $81x^4 - 16y^4$

37. $8x^3 - 27y^6$

38. $3 - 10x + 8x^2$

39. $x^2 + 10x + 25 - 16z^2$

40. $(x - y)^2 + 3(x - y) - 10$

41. Solve: $6x^2 + 7 = -23x$.

42. Solve: $x^3 - 4x = 0$.

43. Solve: $b^2x^2 + a^2y^2 = a^2b^2$ for b^2.

44. CAMPING The rectangular-shaped cooking surface of a small camping stove is 108 in.^{2}. If its length is 3 inches longer than its width, what are its dimensions?

Simplify each expression.

45. $\dfrac{2x^2y + xy - 6y}{3x^2y + 5xy - 2y}$

46. $\dfrac{p^3 - q^3}{q^2 - p^2} \cdot \dfrac{q^2 + pq}{p^3 + p^2q + pq^2}$

47. $\dfrac{2}{x + y} + \dfrac{3}{x - y} - \dfrac{x - 3y}{x^2 - y^2}$

48. $\dfrac{\frac{a}{b} + b}{a - \frac{b}{a}}$

49. Solve: $\dfrac{5x - 3}{x + 2} = \dfrac{5x + 3}{x - 2}$.

50. Solve: $\dfrac{3}{x - 2} + \dfrac{x^2}{(x + 3)(x - 2)} = \dfrac{x + 4}{x + 3}$.

Perform the division.

51. $(x^2 + 9x + 20) \div (x + 5)$

52. $(3x^2 + 9x - 2x^3 + 3) \div (2x - 1)$

53. ECONOMICS The controversial Phillips curve shown in the next column depicts the tradeoff between unemployment and inflation as seen by one school of economists. If unemployment drops to very low levels, what does the theoretical model predict will happen to the inflation rate?

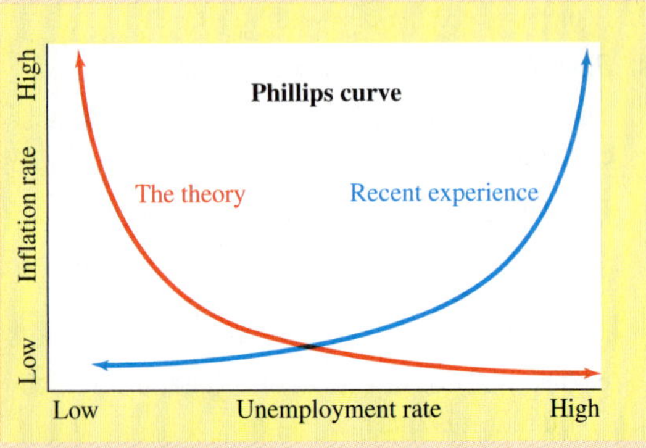

54. ECONOMICS See Exercise 53. The graph shows that economic factors have not followed the Phillips model in recent years. As unemployment has dropped to very low levels, what has happened to the inflation rate?

55. MAKING BROWNIES A recipe for brownies calls for 4 eggs and $1\frac{1}{2}$ cups of flour. If the recipe makes 15 brownies, how many cups of flour will be needed to make 130 brownies?

56. FARMING The number of days a given number of bushels of corn will last when feeding chickens varies inversely with the number of animals. If a certain number of bushels will feed 300 chickens for 4 days, how long will the feed last for 1,200 chickens?

Chapter

7 Radical Expressions and Equations

CORBIS

When investigating an automobile accident, police and insurance professionals use clues from the scene to reconstruct the events that led to the collision. They estimate the speed of a vehicle prior to braking using a formula that involves the length of the skid marks and the condition of the road surface. This formula, $s = \sqrt{30Df}$, contains a square root. Square roots are more formally known as *radical expressions.*

To learn more about radical expressions and radical equations, visit *The Learning Equation* on the Internet at http://tle.brookscole.com. (The log-in instructions are in the Preface.) For Chapter 7, the online lessons are:

- *TLE* Lesson 10: Simplifying Radical Expressions
- *TLE* Lesson 11: Radical Equations

Radical expressions have the form $\sqrt[n]{a}$. In this chapter, we will see how they are used to model many real-world situations.

7.1 Radical Expressions and Radical Functions

- Square roots
- Square roots of expressions containing variables
- The square root function
- Cube roots
- The cube root function
- *n*th roots

In this section, we will reverse the squaring process and learn how to find *square roots* of numbers. Then we will generalize the concept of root and consider cube roots, fourth roots, and so on. We will also discuss a new family of functions, called *radical functions.*

SQUARE ROOTS

When we raise a number to the second power, we are squaring it, or finding its **square.**

- The square of 5 is 25 because $5^2 = 25$.
- The square of -5 is 25, because $(-5)^2 = 25$.

We can reverse the squaring process to find **square roots** of numbers. For example, to find the square roots of 25, we ask ourselves "What number, when squared, is equal to 25?" There are two possible answers.

- 5 is a square root of 25, because $5^2 = 25$.
- -5 is a square root of 25, because $(-5)^2 = 25$.

In general, we have the following definition.

Square Root of *a* The number b is a **square root** of the number a if $b^2 = a$.

Every positive number has two square roots, one positive and one negative. For example, the two square roots of 9 are 3 and -3, and the two square roots of 144 are 12 and -12. The number 0 is the only real number with exactly one square root. In fact, it is its own square root, because $0^2 = 0$.

A **radical symbol** $\sqrt{}$ represents the **positive** or **principal square root** of a number. Since 3 is the positive square root of 9, we can write

$$\sqrt{9} = 3$$

The Language of Algebra
We can read $\sqrt{9}$ as "the square root of 9" or as "radical 9."

The symbol $-\sqrt{}$ represents the **negative square root** of a number. It is the opposite of the principal square root. Since -12 is the negative square root of 144, we can write

$$-\sqrt{144} = -12$$ Read as "the negative square root of 144 is -12" or "the opposite of the square root of 144 is -12."

Square Root Notation If a is a positive real number,

1. $\sqrt{a}$ represents the **positive** or **principal square root** of a. It is the positive number we square to get a.
2. $-\sqrt{a}$ represents the **negative square root** of a. It is the opposite of the principal square root of a: $-\sqrt{a} = -1 \cdot \sqrt{a}$.
3. The principal square root of 0 is 0: $\sqrt{0} = 0$.

The number or variable expression under a radical symbol is called the **radicand,** and the radical symbol and radicand are called a **radical.** An algebraic expression containing a radical is called a **radical expression.**

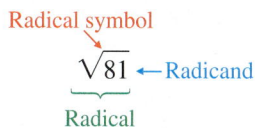

Radical symbol

$\sqrt{81}$ ← Radicand

Radical

EXAMPLE 1

Find each square root: **a.** $\sqrt{81}$, **b.** $-\sqrt{225}$, **c.** $\sqrt{\frac{49}{4}}$, and **d.** $\sqrt{0.36}$.

Solution

a. $\sqrt{81} = 9$ Because $9^2 = 81$.

b. $-\sqrt{225} = -15$ Because $(15)^2 = 225$.

c. $\sqrt{\frac{49}{4}} = \frac{7}{2}$ Because $\left(\frac{7}{2}\right)^2 = \frac{49}{4}$.

d. $\sqrt{0.36} = 0.6$ Because $(0.6)^2 = 0.36$.

Self Check 1 Find each square root: **a.** $\sqrt{64}$, **b.** $-\sqrt{1}$, **c.** $\sqrt{\frac{1}{16}}$, and **d.** $\sqrt{0.09}$.

A table of square roots

n	$\sqrt{n}$
1	1.000
2	1.414
3	1.732
4	2.000
5	2.236
6	2.449
7	2.646
8	2.828
9	3.000
10	3.162

A number such as 81, 225, $\frac{1}{4}$, and 0.36, that is the square of some rational number, is called a **perfect square.** In Example 1, we saw that the square root of a perfect square is a rational number.

If a positive number is not a perfect square, its square root is irrational. For example, $\sqrt{5}$ is an irrational number because 5 is not a perfect square. Since $\sqrt{5}$ is irrational, its decimal representation is nonterminating and nonrepeating. We can find an approximate value of $\sqrt{5}$ using the square root key $\sqrt{}$ on a calculator or from the table of square roots found in Appendix II.

$$\sqrt{5} \approx 2.236067978$$

Caution Square roots of negative numbers are not real numbers. For example, $\sqrt{-9}$ is not a real number, because no real number squared equals -9. Square roots of negative numbers come from a set called the **imaginary numbers,** which we will discuss later in this chapter. If we attempt to evaluate $\sqrt{-9}$ using a calculator, we will get an error message.

Scientific calculator

Graphing calculator

Caution

Although they look similar, these radical expressions have very different meanings.

$-\sqrt{9} = -3$

$\sqrt{-9}$ is not a real number.

SQUARE ROOTS OF EXPRESSIONS CONTAINING VARIABLES

If $x \neq 0$, the positive number x^2 has x and $-x$ for its two square roots. To denote the positive square root of $\sqrt{x^2}$, we must know whether x is positive or negative.

If x is positive, we can write

$\sqrt{x^2} = x$ $\sqrt{x^2}$ represents the positive square root of x^2, which is x.

If x is negative, then $-x > 0$, and we can write

$\sqrt{x^2} = -x$ $\sqrt{x^2}$ represents the positive square root of x^2, which is $-x$.

If we don't know whether x is positive or negative, we can use absolute value symbols to guarantee that $\sqrt{x^2}$ is positive.

Definition of $\sqrt{x^2}$ For any real number x,

$$\sqrt{x^2} = |x|$$

We use this definition to *simplify* square root radical expressions.

EXAMPLE 2

Simplify: **a.** $\sqrt{16x^2}$, **b.** $\sqrt{x^2 + 2x + 1}$, and **c.** $\sqrt{m^4}$.

Solution If x can be any real number, we have

a. $\sqrt{16x^2} = \sqrt{(4x)^2}$ Write the radicand $16x^2$ as $(4x)^2$.

$= |4x|$ Because $(|4x|)^2 = 16x^2$. Since x could be negative, absolute value symbols are needed.

$= 4|x|$ Since 4 is a positive constant in the product $4x$, we can write it outside the absolute value symbols.

b. $\sqrt{x^2 + 2x + 1} = \sqrt{(x + 1)^2}$ Factor the radicand: $x^2 + 2x + 1 = (x + 1)^2$.

$= |x + 1|$ Since $x + 1$ can be negative (for example, when $x = -5$, $x + 1$ is -4), absolute value symbols are needed.

c. $\sqrt{m^4} = m^2$ Because $(m^2)^2 = m^4$. Since $m^2 \geq 0$, no absolute value symbols are needed.

Self Check 2 Simplify: **a.** $\sqrt{25a^2}$ and **b.** $\sqrt{16a^4}$.

If we know that x is positive in parts a and b of Example 2, we don't need to use absolute value symbols. For example, if $x > 0$, then

$\sqrt{16x^2} = 4x$ If x is positive, $4x$ is positive.

$\sqrt{x^2 + 2x + 1} = x + 1$ If x is positive, $x + 1$ is positive.

THE SQUARE ROOT FUNCTION

Since there is one principal square root for every nonnegative real number x, the equation $f(x) = \sqrt{x}$ determines a function, called a **square root function.** Square root functions belong to a larger family of functions known as **radical functions.**

EXAMPLE 3

Graph $f(x) = \sqrt{x}$ and find its domain and range.

Solution To graph this square root function, we will evaluate it for several values of x. We begin with $x = 0$, since 0 is the smallest input for which $\sqrt{x}$ is defined.

$f(\mathbf{x}) = \sqrt{\mathbf{x}}$

$f(\mathbf{0}) = \sqrt{\mathbf{0}}$ Substitute 0 for x.

$f(0) = 0$

We enter 0 for x and 0 for $f(x)$ in the table. Then we let $x = 1$, 4, 9, and 16, and list each corresponding function value in the table. After plotting the ordered pairs, we draw a smooth curve through the points to get the graph shown in figure (a). Since the equation defines a function, its graph passes the vertical line test.

We can use a graphing calculator to get the graph shown in figure (b). From either graph, we can see that the domain and the range are the set of nonnegative real numbers. Expressed in interval notation, the domain is $[0, \infty)$, and the range is $[0, \infty)$.

$f(x) = \sqrt{x}$

x	$f(x)$	
0	**0**	→ (0, 0)
1	1	→ (1, 1)
4	2	→ (4, 2)
9	3	→ (9, 3)
16	4	→ (16, 4)

↑ Select values of x that are perfect squares.

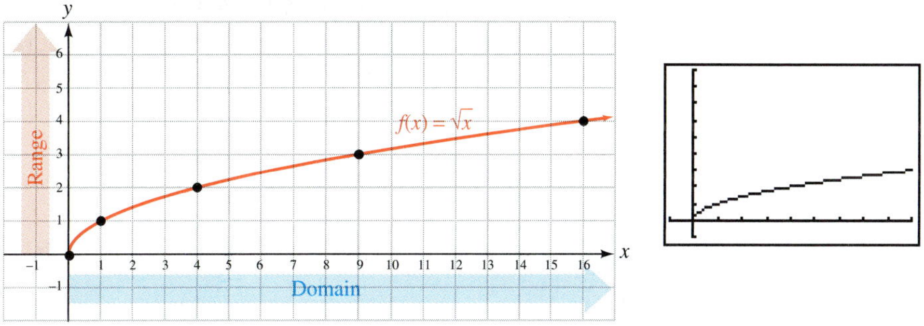

(a) (b)

Self Check 3 Graph: $g(x) = \sqrt{x} + 2$. Then give its domain and range and compare it to the graph of $f(x) = \sqrt{x}$.

EXAMPLE 4

Consider the function $g(x) = \sqrt{x + 3}$. **a.** Find its domain, **b.** graph the function, and **c.** find its range.

Solution

a. We can determine the domain algebraically. Since the expression $\sqrt{x + 3}$ is not a real number when $x + 3$ is negative, we must require that

$$x + 3 \geq 0$$

To solve for x, we subtract 3 from both sides,

$$x \geq -3 \quad \text{The } x\text{-inputs must be real numbers greater than or equal to } -3.$$

Thus, the domain of $g(x) = \sqrt{x + 3}$ is the interval $[-3, \infty)$.

b. To graph the function, we construct a table of function values. We begin by selecting $x = -3$, since -3 is the smallest input for which $\sqrt{x + 3}$ is defined.

$$g(x) = \sqrt{x + 3}$$
$$g(-3) = \sqrt{-3 + 3}$$
$$g(-3) = \sqrt{0} = 0$$

We enter -3 for x and 0 for $g(x)$ in the table. Then we let $x = -2$, 1, and 6 and list each corresponding function value in the table. After plotting the ordered pairs, we draw a smooth curve through the points to get the graph shown in figure (a) on the next page. In figure (b), we see that the graph of $g(x) = \sqrt{x + 3}$ is the graph of $f(x) = \sqrt{x}$, translated 3 units to the left.

$g(x) = \sqrt{x + 3}$

x	$g(x)$	
−3	**0**	→ (−3, 0)
−2	1	→ (−2, 1)
1	2	→ (1, 2)
6	3	→ (6, 3)

↑ Select values of x that make $x + 3$ a perfect square.

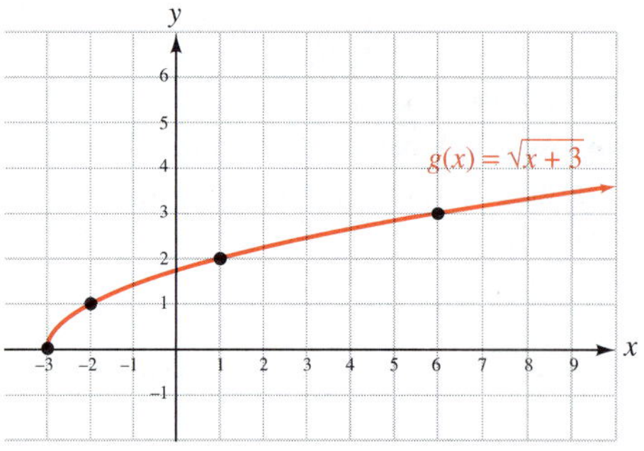

(a)

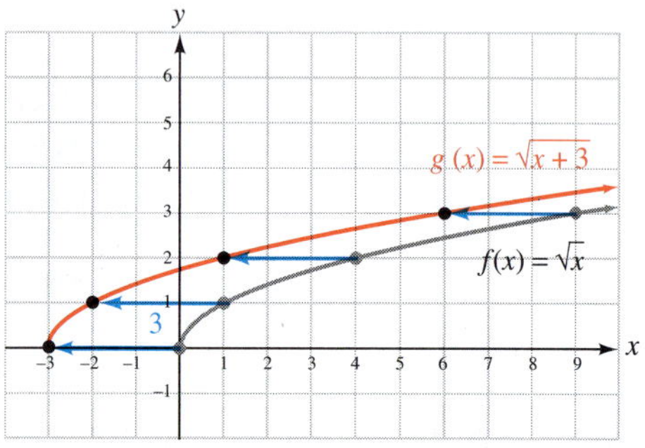

(b)

c. From the graph, we see that the range of $g(x) = \sqrt{x + 3}$ is $[0, \infty)$.

Self Check 4 Consider $h(x) = \sqrt{x - 2}$. **a.** Find its domain, **b.** graph the function, and **c.** find its range.

EXAMPLE 5

Pendulums. The **period of a pendulum** is the time required for the pendulum to swing back and forth to complete one cycle. The period (in seconds) is a function of the pendulum's length L (in feet) and is given by

$$f(L) = 2\pi\sqrt{\frac{L}{32}}$$

Find the period of the 5-foot-long pendulum of a clock.

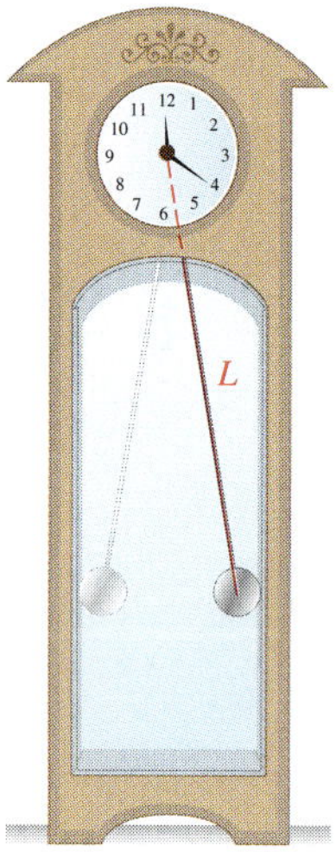

Solution To determine the period, we substitute 5 for L.

$$f(L) = 2\pi\sqrt{\frac{L}{32}}$$

$$f(5) = 2\pi\sqrt{\frac{5}{32}}$$

$$\approx 2.483647066 \quad \text{Use a calculator to find an approximation.}$$

The period is approximately 2.5 seconds.

Notation

$2\pi\sqrt{\frac{5}{32}}$ means $2 \cdot \pi \cdot \sqrt{\frac{5}{32}}$.

Self Check 5 To the nearest hundredth, find the period of a pendulum that is 3 feet long.

ACCENT ON TECHNOLOGY: EVALUATING A SQUARE ROOT FUNCTION

To solve Example 5 with a graphing calculator, we graph the function $f(x) = 2\pi\sqrt{\frac{x}{32}}$, as in figure (a). We then trace and move the cursor toward an x-value of 5 until we see the coordinates shown in figure (b). The pendulum's period is given by the y-value shown on the screen. By zooming in, we can get better results.

After entering $Y_1 = 2\pi\sqrt{\frac{x}{32}}$, we can also use the TABLE mode to find $f(5)$. See figure (c).

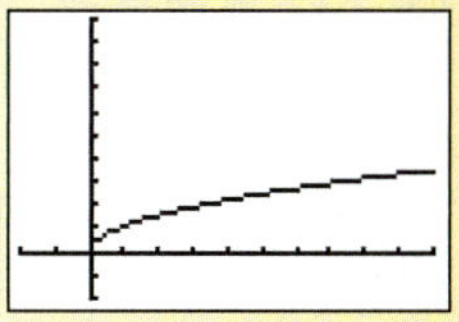

(a)

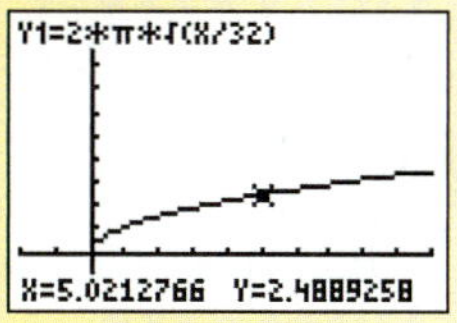

(b)

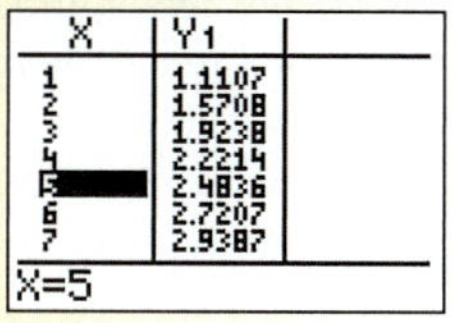
X Y1
1 1.1107
2 1.5708
3 1.9238
4 2.2214
5 2.4836
6 2.7207
7 2.9387
X=5

(c)

CUBE ROOTS

When we raise a number to the third power, we are cubing it, or finding its **cube.** We can reverse the cubing process to find **cube roots** of numbers. To find the cube root of 8, we ask "What number, when cubed, is equal to 8?" It follows that 2 is a cube root of 8, because $2^3 = 8$.

In general, we have this definition.

Cube Root of *a* The number b is a **cube root** of the number a if $b^3 = a$.

All real numbers have one real cube root. A positive number has a positive cube root, a negative number has a negative cube root, and the cube root of 0 is 0.

Cube Root Notation The **cube root of *a*** is denoted by $\sqrt[3]{a}$. By definition,

$$\sqrt[3]{a} = b \quad \text{if} \quad b^3 = a$$

Earlier, we determined that the cube root of 8 is 2. In symbols, we can write: $\sqrt[3]{8} = 2$. The number 3 is called the **index.**

Index → $\sqrt[3]{8}$

Notation

For the square root symbol $\sqrt{}$, the unwritten index is understood to be 2.

$$\sqrt{a} = \sqrt[2]{a}$$

A number such as 125, $\frac{1}{64}$, -27, and -8, that is the cube of some rational number, is called a **perfect cube.** To simplify cube root radical expressions, we look for perfect cubes and apply the following definition.

Definition of $\sqrt[3]{x^3}$ For any real number x,

$$\sqrt[3]{x^3} = x$$

EXAMPLE 6

Simplify: **a.** $\sqrt[3]{125}$, **b.** $\sqrt[3]{\frac{1}{64}}$, **c.** $\sqrt[3]{-27x^3}$, **d.** $\sqrt[3]{-\frac{8a^3}{b^3}}$, and **e.** $\sqrt[3]{0.216x^3y^6}$.

Solution **a.** $\sqrt[3]{125} = 5$ Because $5^3 = 5 \cdot 5 \cdot 5 = 125$.

b. $\sqrt[3]{\dfrac{1}{64}} = \dfrac{1}{4}$ Because $\left(\dfrac{1}{4}\right)^3 = \dfrac{1}{4} \cdot \dfrac{1}{4} \cdot \dfrac{1}{4} = \dfrac{1}{64}$.

c. $\sqrt[3]{-27x^3} = -3x$ Because $(-3x)^3 = (-3x)(-3x)(-3x) = -27x^3$.

d. $\sqrt[3]{-\dfrac{8a^3}{b^3}} = -\dfrac{2a}{b}$ Because $\left(-\dfrac{2a}{b}\right)^3 = \left(-\dfrac{2a}{b}\right)\left(-\dfrac{2a}{b}\right)\left(-\dfrac{2a}{b}\right) = -\dfrac{8a^3}{b^3}$.

e. $\sqrt[3]{0.216x^3y^6} = 0.6xy^2$ Because $(0.6xy^2)^3 = (0.6xy^2)(0.6xy^2)(0.6xy^2) = 0.216x^3y^6$.

Success Tip

Since every real number has exactly one real cube root, it is unnecessary to use absolute value symbols when simplifying cube roots.

Self Check 6 Simplify: **a.** $\sqrt[3]{1{,}000}$, **b.** $\sqrt[3]{\dfrac{1}{27}}$, and **c.** $\sqrt[3]{125a^3}$.

THE CUBE ROOT FUNCTION

Since there is one cube root for every real number x, the equation $f(x) = \sqrt[3]{x}$ defines a function, called the **cube root function.** Like square root functions, cube root functions belong to the family of radical functions.

EXAMPLE 7

Consider $f(x) = \sqrt[3]{x}$. **a.** Graph the function, **b.** find its domain and range, and **c.** graph $g(x) = \sqrt[3]{x} - 2$.

Solution **a.** To graph this cube root function, we will evaluate it for several values of x. We begin with $x = -8$.

$$f(x) = \sqrt[3]{x}$$
$$f(-8) = \sqrt[3]{-8} \quad \text{Substitute } -8 \text{ for } x.$$
$$f(-8) = -2$$

We enter -8 for x and -2 for $f(x)$ in the table. Then we let $x = -1$, 0, 1, and 8, and list each corresponding function value in the table. After plotting the ordered pairs, we draw a smooth curve through the points to get the graph shown in figure (a).

$f(x) = \sqrt[3]{x}$

x	$f(x)$	
-8	-2	→ $(-8, -2)$
-1	-1	→ $(-1, -1)$
0	0	→ $(0, 0)$
1	1	→ $(1, 1)$
8	2	→ $(8, 2)$

↑ Select values of x that are perfect cubes.

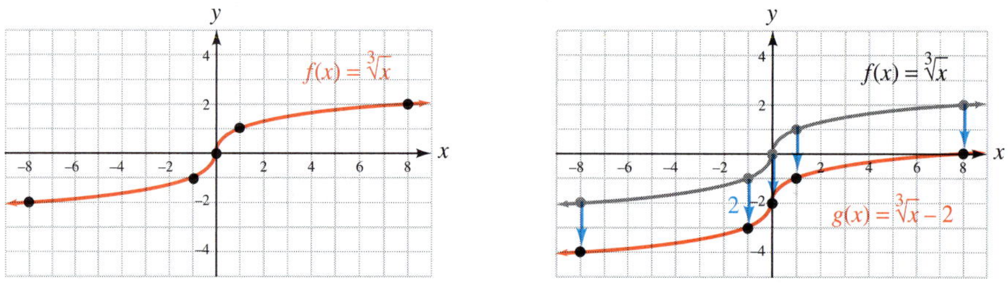

(a) (b)

b. From the graph in figure (a), we see that the domain and the range of function f are the set of real numbers. Thus, the domain is $(-\infty, \infty)$ and the range is $(-\infty, \infty)$.

c. Refer to figure (b). The graph of $g(x) = \sqrt[3]{x} - 2$ is the graph of $f(x) = \sqrt[3]{x}$, translated 2 units downward.

Self Check 7 Consider $f(x) = \sqrt[3]{x} + 1$. **a.** Graph the function and **b.** find its domain and range.

*n*TH ROOTS

Just as there are square roots and cube roots, there are fourth roots, fifth roots, sixth roots, and so on. In general, we have the following definition.

nth Roots of a

The ***n*th root of *a*** is denoted by $\sqrt[n]{a}$, and

$$\sqrt[n]{a} = b \quad \text{if} \quad b^n = a$$

The number n is called the **index** (or **order**) of the radical. If n is an even natural number, a must be positive or zero, and b must be positive.

When n is an odd natural number, the expression $\sqrt[n]{x}$ $(n > 1)$ represents an **odd root.** Since every real number has just one real nth root when n is odd, we don't need absolute value symbols when finding odd roots. For example,

$$\sqrt[5]{243} = \sqrt[5]{3^5} = 3 \qquad \text{Because } 3^5 = 243.$$

$$\sqrt[7]{-128x^7} = \sqrt[7]{(-2x)^7} = -2x \qquad \text{Because } (-2x)^7 = -128x^7.$$

When n is an even natural number, the expression $\sqrt[n]{x}$, where $n > 1, x > 0$, represents an **even root.** In this case, there will be one positive and one negative real nth root. For example, the real sixth roots of 729 are 3 and -3, because $3^6 = 729$ and $(-3)^6 = 729$. When finding even roots, we can use absolute value symbols to guarantee that the nth root is positive.

$$\sqrt[4]{(-3)^4} = |-3| = 3$$

We could also simplify this as follows: $\sqrt[4]{(-3)^4} = \sqrt[4]{81} = 3.$

$$\sqrt[6]{729x^6} = \sqrt[6]{(3x)^6} = |3x| = 3|x|$$

The absolute value symbols guarantee that the sixth root is positive.

In general, we have the following rules.

Rules for $\sqrt[n]{x^n}$

If x is a real number and $n > 1$, then

If n is an odd natural number, $\sqrt[n]{x^n} = x$.

If n is an even natural number, $\sqrt[n]{x^n} = |x|$.

EXAMPLE 8

Simplify each radical expression, if possible: **a.** $\sqrt[4]{625}$, **b.** $\sqrt[4]{-1}$, **c.** $\sqrt[5]{-32}$, **d.** $\sqrt[6]{\frac{1}{64}}$, and **e.** $\sqrt[7]{10^7}$.

Solution

a. $\sqrt[4]{625} = 5$, because $5^4 = 625$. Read $\sqrt[4]{625}$ as "the fourth root of 625."

b. $\sqrt[4]{-1}$ is not a real number. This is an even root of a negative number.

c. $\sqrt[5]{-32} = -2$, because $(-2)^5 = -32$. Read $\sqrt[5]{-32}$ as "the fifth root of -32."

d. $\sqrt[6]{\frac{1}{64}} = \frac{1}{2}$, because $\left(\frac{1}{2}\right)^6 = \frac{1}{64}$. Read $\sqrt[6]{\frac{1}{64}}$ as "the sixth root of $\frac{1}{64}$."

e. $\sqrt[7]{10^7} = 10$, because $10^7 = 10^7$. Read $\sqrt[7]{10^7}$ as "the seventh root of 10^7."

Caution

When n is even $(n > 1)$ and $x < 0$, $\sqrt[n]{x}$ is not a real number. For example, $\sqrt[4]{-81}$ is not a real number, because no real number raised to the fourth power is -81.

Self Check 8 Simplify: **a.** $\sqrt[4]{\frac{1}{81}}$, **b.** $\sqrt[5]{10^5}$, and **c.** $\sqrt[6]{-64}$.

ACCENT ON TECHNOLOGY: FINDING ROOTS

The square root key $\boxed{\sqrt{\ }}$ on a scientific calculator can be used to evaluate square roots. To evaluate roots with an index greater than 2, we can use the root key $\boxed{\sqrt[x]{y}}$. For example, the function

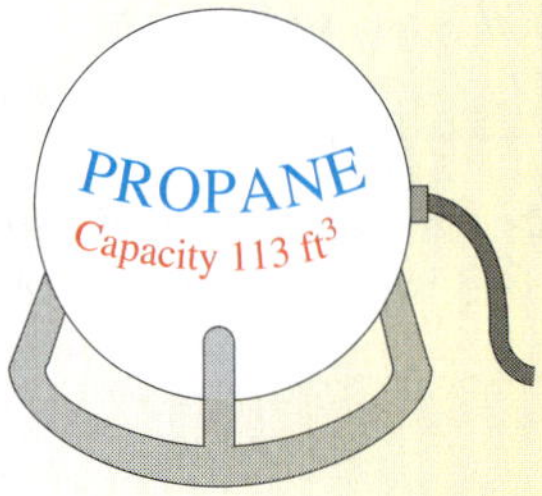

$$r(V) = \sqrt[3]{\frac{3V}{4\pi}}$$

gives the radius of a sphere with volume V. To find the radius of the spherical propane tank shown on the left, we substitute 113 for V to get

$$r(V) = \sqrt[3]{\frac{3V}{4\pi}}$$

$$r(113) = \sqrt[3]{\frac{3(113)}{4\pi}}$$

To evaluate a root, we enter the radicand and press the root key $\boxed{\sqrt[x]{y}}$ followed by the index of the radical, which in this case is 3.

3 $\boxed{\times}$ 113 $\boxed{\div}$ $\boxed{(}$ 4 $\boxed{\times}$ $\boxed{\pi}$ $\boxed{)}$ $\boxed{=}$ $\boxed{\text{2nd}}$ $\boxed{\sqrt[x]{y}}$ $\boxed{3}$ $\boxed{=}$

2.999139118

To evaluate the cube root of $\frac{3(113)}{4\pi}$ with a graphing calculator, we enter

$\boxed{\text{MATH}}$ 4 $\boxed{(}$ 3 $\boxed{\times}$ 113 $\boxed{)}$ $\boxed{\div}$ $\boxed{(}$ 4 $\boxed{\times}$ $\boxed{\text{2nd}}$ $\boxed{\pi}$ $\boxed{)}$ $\boxed{)}$ $\boxed{\text{ENTER}}$

```
³√((3*113)/(4*π)
)
                2.999139118
```

The radius of the propane tank is about 3 feet.

EXAMPLE 9

Simplify each radical expression. Assume that x can be any real number. **a.** $\sqrt[5]{x^5}$, **b.** $\sqrt[4]{16x^4}$, **c.** $\sqrt[6]{(x+4)^6}$, **d.** $\sqrt[3]{(x+1)^3}$, and **e.** $\sqrt{9x^4}$.

Solution

a. $\sqrt[5]{x^5} = x$ — Since n is odd, absolute value symbols aren't needed.

b. $\sqrt[4]{16x^4} = |2x| = 2|x|$ — Since n is even and x can be negative, absolute value symbols are needed to guarantee that the result is positive.

c. $\sqrt[6]{(x+4)^6} = |x+4|$ — Absolute value symbols are needed to guarantee that the result is positive.

d. $\sqrt[3]{(x+1)^3} = x + 1$ — Since n is odd, absolute value symbols aren't needed.

e. $\sqrt{9x^4} = 3x^2$ — Since x^2 is always nonnegative, we don't need absolute value symbols.

The Language of Algebra

Another way to say that x can be any real number is to say that the variable is *unrestricted.*

Self Check 9 Simplify: **a.** $\sqrt[6]{x^6}$, **b.** $\sqrt[5]{(a+5)^5}$, **c.** $\sqrt{(x^2+4x+4)^2}$, and **d.** $\sqrt[4]{16a^8}$.

If we know that x is positive in parts b and c of Example 9, we don't need to use absolute value symbols. For example, if $x > 0$, then

$$\sqrt[4]{16x^4} = 2x \qquad \text{If } x \text{ is positive, } 2x \text{ is positive.}$$

$$\sqrt[6]{(x+4)^6} = x + 4 \qquad \text{If } x \text{ is positive, } x + 4 \text{ is positive.}$$

We summarize the definitions concerning $\sqrt[n]{x}$ as follows.

Summary of the Definitions of $\sqrt[n]{x}$

If n is a natural number greater than 1 and x is a real number, then

If $x > 0$, then $\sqrt[n]{x}$ is the positive number such that $\left(\sqrt[n]{x}\right)^n = x$.

If $x = 0$, then $\sqrt[n]{x} = 0$.

If $x < 0$ $\begin{cases} \text{and } n \text{ is odd, then } \sqrt[n]{x} \text{ is the negative number such that } \left(\sqrt[n]{x}\right)^n = x. \\ \text{and } n \text{ is even, then } \sqrt[n]{x} \text{ is not a real number.} \end{cases}$

Answers to Self Checks

1. a. 8, **b.** -1, **c.** $\frac{1}{4}$, **d.** 0.3 **2. a.** $5|a|$, **b.** $4a^2$

3.

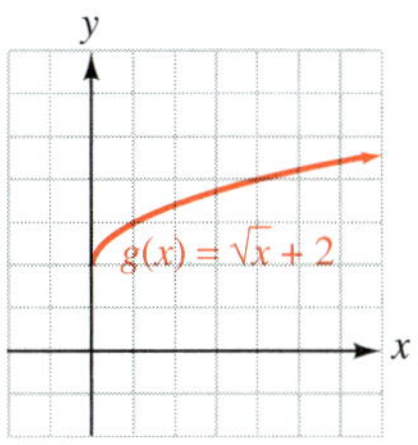

D: $[0, \infty)$, R: $[2, \infty)$; the graph is 2 units higher

4. a. $[2, \infty)$ **b.**

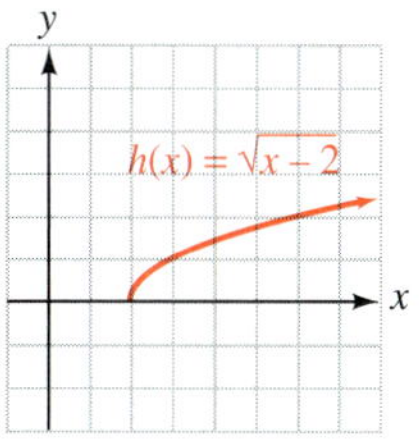

c. $[0, \infty)$

5. 1.92 sec **6. a.** 10, **b.** $\frac{1}{3}$, **c.** $5a$

7. a.

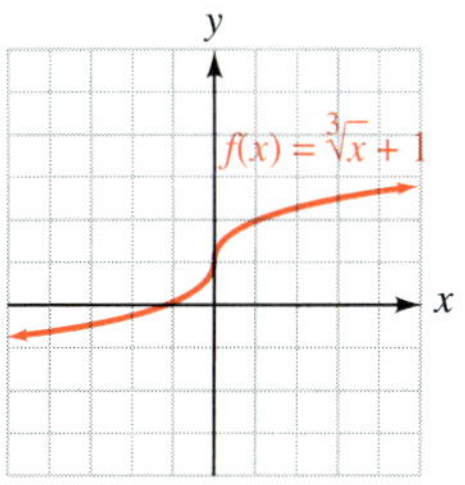

b. D: $(-\infty, \infty)$; R: $(-\infty, \infty)$

8. a. $\frac{1}{3}$, **b.** 10, **c.** not a real number

9. a. $|x|$, **b.** $a + 5$, **c.** $(x + 2)^2$, **d.** $2a^2$

7.1 STUDY SET

VOCABULARY **Fill in the blanks.**

1. $5x^2$ is the ________ root of $25x^4$, because $(5x^2)^2 = 25x^4$. The ______ root of 216 is 6 because $6^3 = 216$.
2. The symbol $\sqrt{}$ is called a ________ symbol.
3. In the expression $\sqrt[3]{27x^6}$, the ______ is 3 and $27x^6$ is the ________.
4. When we write $\sqrt{b^4} = b^2$, we say that we have ________ the radical expression.
5. When n is an odd number, $\sqrt[n]{x}$ represents an ______ root. When n is an ______ number, $\sqrt[n]{x}$ represents an even root.
6. $f(x) = \sqrt{x}$ and $g(x) = \sqrt[3]{x}$ are ________ functions.

CONCEPTS **Fill in the blanks.**

7. b is a square root of a if $b^2 =$ ___.
8. $\sqrt{0} =$ ___ and $\sqrt[3]{0} =$ ___.

9. The number 25 has ______ square roots. The principal square root of 25 is ______.

10. $\sqrt{-4}$ is not a real number, because no real number ______ equals -4.

11. $\sqrt[3]{x} = y$ if $y^3 =$ ______.

12. $\sqrt{x^2} =$ ______ and $\sqrt[3]{x^3} =$ ______.

13. The graph of $g(x) = \sqrt{x} + 3$ is the graph of $f(x) = \sqrt{x}$ translated ______ units ______.

14. The graph of $g(x) = \sqrt{x + 5}$ is the graph of $f(x) = \sqrt{x}$ translated ______ units to the ______.

15. The graph of a square root function f is shown below. Find each of the following.

a. $f(11)$ **b.** $f(2)$

c. The value(s) of x for which $f(x) = 2$

d. The domain and range of f

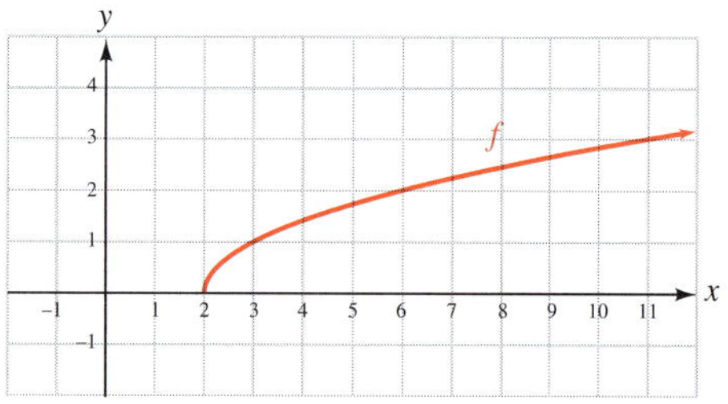

16. The graph of a cube root function f is shown below. Find each of the following.

a. $f(-8)$ **b.** $f(0)$

c. The value(s) of x for which $f(x) = -2$

d. The domain and range of f

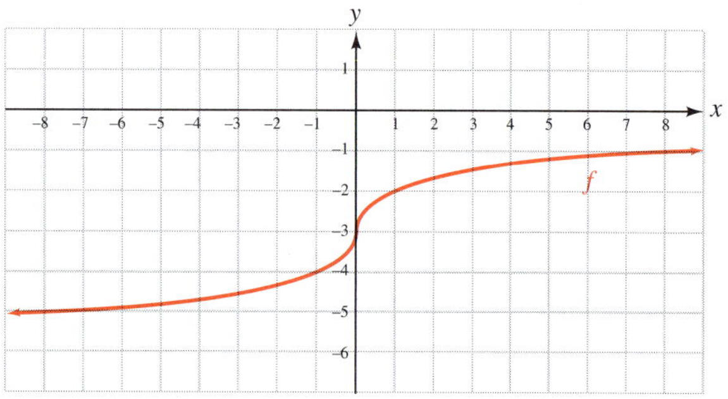

Complete each table of values and graph the function. Then find the domain and range.

17. $f(x) = \sqrt{x}$

x	y
0	
1	
4	
9	
16	

18. $f(x) = \sqrt[3]{x}$

x	y
-8	
-1	
0	
1	
8	

NOTATION **Translate each sentence into mathematical symbols.**

19. The square root of x squared is the absolute value of x.

20. The cube root of x cubed is x.

21. f of x equals the square root of the quantity x minus five.

22. The fifth root of negative thirty-two is negative two.

PRACTICE **Find each square root, if possible.**

23. $\sqrt{121}$

24. $\sqrt{144}$

25. $-\sqrt{64}$

26. $-\sqrt{1}$

27. $\sqrt{\frac{1}{9}}$

28. $-\sqrt{\frac{4}{25}}$

29. $\sqrt{0.25}$

30. $\sqrt{0.16}$

31. $\sqrt{-25}$

32. $-\sqrt{-49}$

33. $\sqrt{(-4)^2}$

34. $\sqrt{(-9)^2}$

Use a calculator to find each square root. Give each answer to four decimal places.

35. $\sqrt{12}$

36. $\sqrt{340}$

37. $\sqrt{679.25}$

38. $\sqrt{0.0063}$

Simplify each expression. Assume that all variables are unrestricted and use absolute value symbols when necessary.

39. $\sqrt{4x^2}$

40. $\sqrt{16y^4}$

41. $\sqrt{(t + 5)^2}$

42. $\sqrt{(a + 6)^2}$

43. $\sqrt{(-5b)^2}$

44. $\sqrt{(-8c)^2}$

45. $\sqrt{a^2 + 6a + 9}$

46. $\sqrt{x^2 + 10x + 25}$

Find each value given that $f(x) = \sqrt{x - 4}$ and $g(x) = \sqrt[3]{x - 4}$.

47. $f(4)$
48. $f(8)$
49. $f(20)$
50. $f(29)$
51. $g(12)$
52. $g(-4)$
53. $g(-996)$
54. $g(1{,}004)$

Find each value given that $f(x) = \sqrt{x^2 + 1}$ and $g(x) = \sqrt[3]{x^2 + 1}$. Give each answer to four decimal places.

55. $f(4)$
56. $f(2.35)$
57. $g(6)$
58. $g(21.57)$

Complete each table and graph the function. Find the domain and range.

59. $f(x) = -\sqrt{x}$

x	y
0	
1	
4	
9	
16	

60. $f(x) = -\sqrt[3]{x}$

x	y
-8	
-1	
0	
1	
8	

Graph each function and find its domain and range.

61. $f(x) = \sqrt{x + 4}$
62. $f(x) = \sqrt{x} - 1$
63. $f(x) = \sqrt[3]{x} + 3$
64. $f(x) = \sqrt[3]{x - 3}$

Simplify each cube root.

65. $\sqrt[3]{1}$
66. $\sqrt[3]{-8}$
67. $\sqrt[3]{-125}$
68. $\sqrt[3]{512}$
69. $\sqrt[3]{-\frac{8}{27}}$
70. $\sqrt[3]{\frac{125}{216}}$
71. $\sqrt[3]{0.064}$
72. $\sqrt[3]{0.001}$
73. $\sqrt[3]{8a^3}$
74. $\sqrt[3]{-27x^6}$
75. $\sqrt[3]{-1{,}000p^3q^3}$
76. $\sqrt[3]{343a^6b^3}$
77. $\sqrt[3]{-\frac{1}{8}m^6n^3}$
78. $\sqrt[3]{0.008z^9}$
79. $\sqrt[3]{-0.064s^9t^6}$
80. $\sqrt[3]{\frac{27}{1{,}000}a^6b^6}$

Simplify each radical, if possible. Assume that all variables represent positive real numbers.

81. $\sqrt[4]{81}$
82. $\sqrt[6]{64}$
83. $-\sqrt[5]{243}$
84. $-\sqrt[4]{625}$
85. $\sqrt[4]{-256}$
86. $\sqrt[6]{-729}$
87. $\sqrt[4]{\frac{16}{625}}$
88. $\sqrt[5]{-\frac{243}{32}}$
89. $-\sqrt[5]{-\frac{1}{32}}$
90. $-\sqrt[4]{\frac{81}{256}}$
91. $\sqrt[5]{32a^5}$
92. $\sqrt[5]{-32x^5}$
93. $\sqrt[4]{16a^4}$
94. $\sqrt[8]{x^{24}}$
95. $\sqrt[4]{k^{12}}$
96. $\sqrt[6]{64b^6}$
97. $\sqrt[4]{\frac{1}{16}m^4}$
98. $\sqrt[4]{\frac{1}{81}x^8}$
99. $\sqrt[25]{(x + 2)^{25}}$
100. $\sqrt[44]{(x + 4)^{44}}$

APPLICATIONS **Use a calculator to solve each problem. Round answers to the nearest tenth.**

101. EMBROIDERY The radius r of a circle is given by the formula

$$r = \sqrt{\frac{A}{\pi}}$$

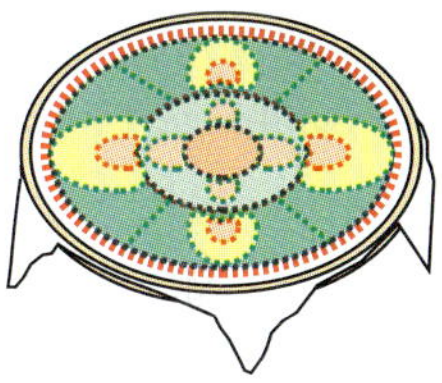

where A is its area. Find the diameter of the embroidery hoop if there are 38.5 in.^{2} of stretched fabric on which to embroider.

102. BASEBALL The length of a diagonal of a square is given by the function $d(s) = \sqrt{2s^2}$, where s is the length of a side of the square. Find the distance from home plate to second base on a softball diamond and on a baseball diamond. The illustration gives the dimensions of each type of infield.

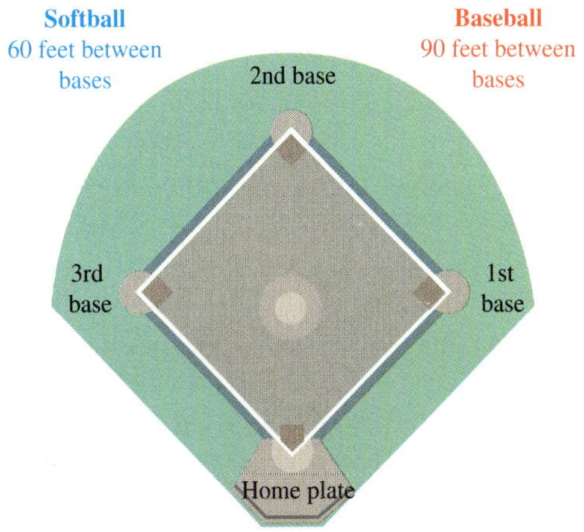

103. PULSE RATES The approximate pulse rate (in beats per minute) of an adult who is t inches tall is given by the function

$$p(t) = \frac{590}{\sqrt{t}}$$

The *Guinness Book of World Records 1998* lists Ri Myong-hun of North Korea as the tallest living man, at 7 ft $8\frac{1}{2}$ in. Find his approximate pulse rate as predicted by the function.

104. THE GRAND CANYON The time t (in seconds) that it takes for an object to fall a distance of s feet is given by the formula

$$t = \frac{\sqrt{s}}{4}$$

In some places, the Grand Canyon is one mile (5,280 feet) deep. How long would it take a stone dropped over the edge of the canyon to hit bottom?

105. BIOLOGY Scientists will place five rats inside a controlled environment to study the rats' behavior. The function

$$d(V) = \sqrt[3]{12\left(\frac{V}{\pi}\right)}$$

gives the diameter of a hemisphere with volume V. Use the function to determine the diameter of the base of the hemisphere, if each rat requires 125 cubic feet of living space.

106. AQUARIUMS The function

$$s(g) = \sqrt[3]{\frac{g}{7.5}}$$

determines how long (in feet) an edge of a cube-shaped tank must be if it is to hold g gallons of water. What dimensions should a cube-shaped aquarium have if it is to hold 1,250 gallons of water?

107. COLLECTIBLES The *effective rate of interest r* earned by an investment is given by the formula

$$r = \sqrt[n]{\frac{A}{P}} - 1$$

where P is the initial investment that grows to value A after n years. Determine the effective rate of interest earned by a collector on a Lladró porcelain figurine purchased for \$800 and sold for \$950 five years later.

108. LAW ENFORCEMENT The graphs of the two radical functions shown in the illustration in the next column can be used to estimate the speed (in mph) of a car involved in an accident. Suppose a police accident report listed skid marks to be 220 feet long but failed to give the road conditions. Estimate the possible speeds the car was traveling prior to the brakes being applied.

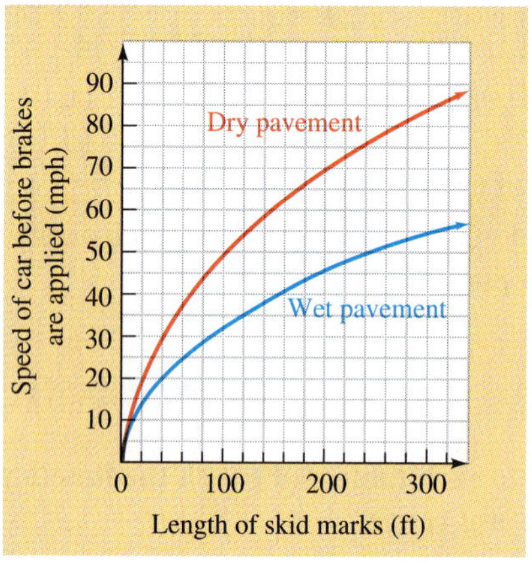

WRITING

109. Explain why 36 has two square roots, but $\sqrt{36}$ is just 6, and not −6.

110. If x is any real number, then $\sqrt{x^2} = x$ is not correct. Explain.

111. Explain what is wrong with the graph in the illustration if it is supposed to be the graph of $f(x) = \sqrt{x}$.

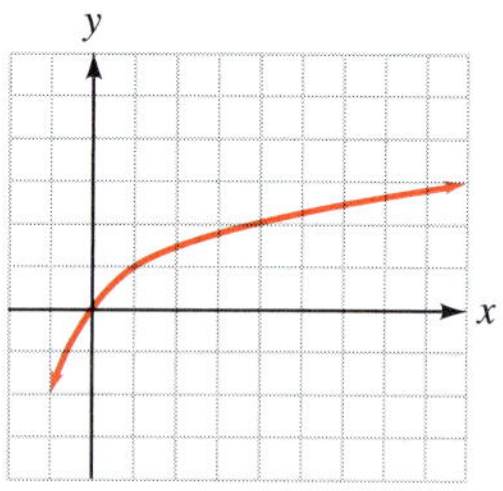

112. Explain how to estimate the domain and range of the radical function shown below.

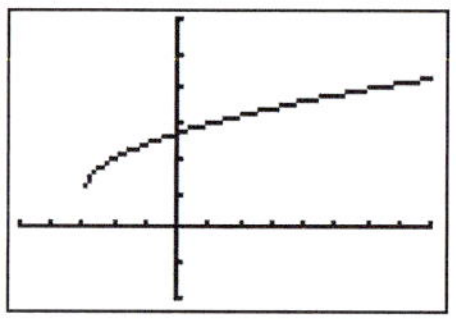

REVIEW **Perform the operations.**

113. $\dfrac{x^2 - x - 6}{x^2 - 2x - 3} \cdot \dfrac{x^2 - 1}{x^2 + x - 2}$

114. $\dfrac{x^2 - 3xy - 4y^2}{x^2 + cx - 2yx - 2cy} \div \dfrac{x^2 - 2xy - 3y^2}{x^2 + cx - 4yx - 4cy}$

115. $\dfrac{3}{m+1} + \dfrac{3m}{m-1}$

116. $\dfrac{2x+3}{3x-1} - \dfrac{x-4}{2x+1}$

CHALLENGE PROBLEMS

117. Graph: $f(x) = -\sqrt{x-2} + 3$. Find the domain and range.

118. Simplify: $\sqrt{9a^{16} + 12a^8b^{25} + 4b^{50}}$. Assume that $a > 0$ and $b > 0$.

7.2 Rational Exponents

- Rational exponents
- Exponential expressions with variables in their bases
- Rational exponents with numerators other than 1
- Negative rational exponents
- Applying the rules for exponents
- Simplifying radical expressions

In this section, we will extend the definition of exponent to include rational (fractional) exponents. We will see how expressions such as $9^{1/2}$, $\left(\frac{1}{16}\right)^{3/4}$, and $(-32x^5)^{-2/5}$ can be simplified by writing them in an equivalent radical form using two new rules for exponents.

RATIONAL EXPONENTS

The Language of Algebra

Rational exponents are also called *fractional exponents.*

It is possible to raise numbers to fractional powers. To give meaning to rational exponents, we first consider $\sqrt{7}$. Because $\sqrt{7}$ is the positive number whose square is 7, we have

$$(\sqrt{7})^2 = 7$$

We now consider the notation $7^{1/2}$. If rational exponents are to follow the same rules as integer exponents, the square of $7^{1/2}$ must be 7, because

$$\begin{aligned} (7^{1/2})^2 &= 7^{1/2 \cdot 2} && \text{Keep the base and multiply the exponents.} \\ &= 7^1 && \text{Do the multiplication: } \tfrac{1}{2} \cdot 2 = 1 \\ &= 7 \end{aligned}$$

Since the square of $7^{1/2}$ and the square of $\sqrt{7}$ are both equal to 7, we define $7^{1/2}$ to be $\sqrt{7}$. Similarly,

$$7^{1/3} = \sqrt[3]{7}, \quad 7^{1/4} = \sqrt[4]{7}, \quad \text{and} \quad 7^{1/5} = \sqrt[5]{7}$$

In general, we have the following definition.

The Definition of $x^{1/n}$

A **rational exponent** of $\frac{1}{n}$ indicates the nth root of its base.

If n represents a positive integer greater than 1 and $\sqrt[n]{x}$ represents a real number,

$$x^{1/n} = \sqrt[n]{x}$$

We can use this definition to simplify exponential expressions that have rational exponents with a numerator of 1. For example, to simplify $8^{1/3}$, we write it as an equivalent expression in radical form and proceed as follows:

Index

$$8^{1/3} = \sqrt[3]{8} = 2$$

Radicand

The base of the exponential expression, 8, is the radicand of the radical expression. The denominator of the fractional exponent, 3, is the index of the radical.

Thus, $8^{1/3} = 2$.

EXAMPLE 1

Write each expression in radical form and simplify, if possible: **a.** $9^{1/2}$, **b.** $(-64)^{1/3}$, **c.** $-\left(\frac{1}{32}\right)^{1/5}$, **d.** $16^{1/4}$, and **e.** $(2x^5)^{1/6}$.

Solution

a. $9^{1/2} = \sqrt{9} = 3$ — Because the denominator of the exponent is 2, find the square root of the base, 9.

b. $(-64)^{1/3} = \sqrt[3]{-64} = -4$ — Because the denominator of the exponent is 3, find the cube root of the base, -64.

c. $-\left(\frac{1}{32}\right)^{1/5} = -\sqrt[5]{\frac{1}{32}} = -\frac{1}{2}$ — Because the denominator of the exponent is 5, find the fifth root of the base, $\frac{1}{32}$.

d. $16^{1/4} = \sqrt[4]{16} = 2$ — Because the denominator of the exponent is 4, find the fourth root of the base, 16.

e. $(2x^5)^{1/6} = \sqrt[6]{2x^5}$ — Because the denominator of the exponent is 6, find the sixth root of the base, $2x^5$. This expression does not simplify further.

Self Check 1 Write each expression in radical form and simplify, if possible: **a.** $16^{1/2}$, **b.** $\left(-\frac{27}{8}\right)^{1/3}$, and **c.** $-(6x^3)^{1/4}$.

EXAMPLE 2

Write $\sqrt{5xyz}$ as an exponential expression with a rational exponent.

Solution The radicand is $5xyz$, so the base of the exponential expression is $5xyz$. The index of the radical is an understood 2, so the denominator of the fractional exponent is 2.

$$\sqrt{5xyz} = (5xyz)^{1/2}$$ Recall: $\sqrt{5xyz} = \sqrt[2]{5xyz}$.

Self Check 2 Write the radical with a fractional exponent: $\sqrt[6]{7ab}$.

Rational exponents appear in formulas used in many disciplines, such as science and engineering.

EXAMPLE 3

Satellites. The formula

$$r = \left(\frac{GMP^2}{4\pi^2}\right)^{1/3}$$

gives the orbital radius (in meters) of a satellite circling the Earth, where G and M are constants and P is the time in seconds for the satellite to make one complete revolution. Write the formula using a radical.

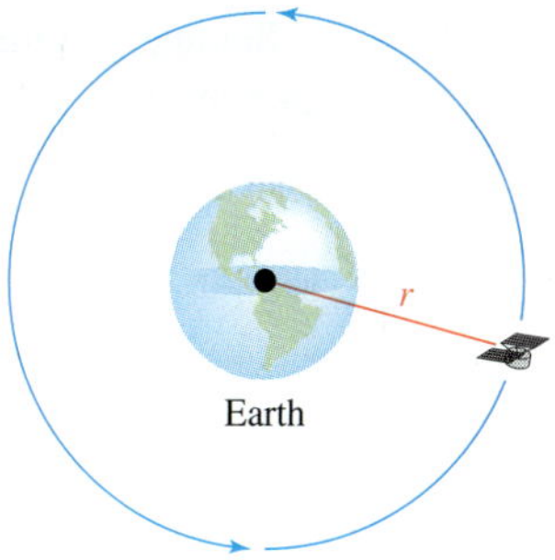

Solution The fractional exponent $\frac{1}{3}$ has a denominator of 3, which indicates that we are to find the cube root of the base of the exponential expression. So we have

$$r = \sqrt[3]{\frac{GMP^2}{4\pi^2}}$$

EXPONENTIAL EXPRESSIONS WITH VARIABLES IN THEIR BASES

As with radicals, when n is an *odd natural number* in the expression $x^{1/n}$ $(n > 1)$, there is exactly one real nth root, and we don't need to use absolute value symbols.

When n is an *even natural number,* there are two nth roots. Since we want the expression $x^{1/n}$ to represent the positive nth root, we must often use absolute value symbols to guarantee that the simplified result is positive. Thus, if n is even,

$$(x^n)^{1/n} = |x|$$

When n is even and x is negative, the expression $x^{1/n}$ is not a real number.

EXAMPLE 4

Simplify each expression. Assume that the variables can be any real number.

a. $(-27x^3)^{1/3}$, **b.** $(256a^8)^{1/8}$, **c.** $[(y + 4)^2]^{1/2}$, **d.** $(25b^4)^{1/2}$, and **e.** $(-256x^4)^{1/4}$.

Solution

a. $(-27x^3)^{1/3} = -3x$ Because $(-3x)^3 = -27x^3$. Since n is odd, no absolute value symbols are needed.

b. $(256a^8)^{1/8} = 2|a|$ Because $\left(2|a|\right)^8 = 256a^8$. Since n is even and a can be any real number, $2a$ can be negative. Thus, absolute value symbols are needed.

c. $[(y + 4)^2]^{1/2} = |y + 4|$ Because $|y + 4|^2 = (y + 4)^2$. Since n is even and y can be any real number, $y + 4$ can be negative. Thus, absolute value symbols are needed.

d. $(25b^4)^{1/2} = 5b^2$ Because $(5b^2)^2 = 25b^4$. Since $b^2 \geq 0$, no absolute value symbols are needed.

e. $(-256x^4)^{1/4}$ is not a real number Because no real number raised to the 4th power is $-256x^4$.

Self Check 4 Simplify each expression: **a.** $(625a^4)^{1/4}$ and **b.** $(b^4)^{1/2}$.

If we are told that the variables represent positive real numbers in parts b and c of Example 4, the absolute value symbols in the answers are not needed.

$(256a^8)^{1/8} = 2a$ If a represents a positive number, then $2a$ is positive.

$[(y + 4)^2]^{1/2} = y + 4$ If y represents a positive number, then $y + 4$ is positive.

We summarize the cases as follows.

Summary of the Definitions of $x^{1/n}$

If n is a natural number greater than 1 and x is a real number,

If $x > 0$, then $x^{1/n}$ is the positive number such that $(x^{1/n})^n = x$.
If $x = 0$, then $x^{1/n} = 0$.
If $x < 0$ $\begin{cases} \text{and } n \text{ is odd, then } x^{1/n} \text{ is the negative number such that } (x^{1/n})^n = x. \\ \text{and } n \text{ is even, then } x^{1/n} \text{ is not a real number.} \end{cases}$

RATIONAL EXPONENTS WITH NUMERATORS OTHER THAN 1

We can extend the definition of $x^{1/n}$ to include fractional exponents with numerators other than 1. For example, since $8^{2/3}$ can be written as $(8^{1/3})^2$, we have

$8^{2/3} = (8^{1/3})^2$
$= (\sqrt[3]{8})^2$ Write $8^{1/3}$ in radical form.
$= 2^2$ Find the cube root first: $\sqrt[3]{8} = 2$.
$= 4$ Then find the power.

Thus, we can simplify $8^{2/3}$ by finding the second power of the cube root of 8.

The numerator of the rational exponent is the power.

$8^{2/3} = (\sqrt[3]{8})^2$ The base of the exponential expression is the radicand.

The denominator of the exponent is the index of the radical.

We can also simplify $8^{2/3}$ by taking the cube root of 8 squared.

$8^{2/3} = (8^2)^{1/3}$
$= 64^{1/3}$ Find the power first: $8^2 = 64$.
$= \sqrt[3]{64}$ Write $64^{1/3}$ in radical form.
$= 4$ Now find the cube root.

In general, we have the following definition.

The Definition of $x^{m/n}$

If m and n represent positive integers ($n \neq 1$) and $\sqrt[n]{x}$ represents a real number,

$$x^{m/n} = (\sqrt[n]{x})^m \quad \text{and} \quad x^{m/n} = \sqrt[n]{x^m}$$

Because of the previous definition, we can interpret $x^{m/n}$ in two ways:

1. $x^{m/n}$ means the nth root of the mth power of x.
2. $x^{m/n}$ means the mth power of the nth root of x.

We can use this definition to evaluate exponential expressions that have rational exponents with a numerator that is not 1. To avoid large numbers, we usually find the root of the base first and then calculate the power using the relationship $x^{m/n} = \left(\sqrt[n]{x}\right)^m$.

EXAMPLE 5

Evaluate: **a.** $32^{2/5}$, **b.** $81^{3/4}$, and **c.** $-25^{3/2}$.

Solution

a. To evaluate $32^{2/5}$, we write it in an equivalent radical form. The denominator of the rational exponent is the same as the index of the corresponding radical. The numerator of the rational exponent indicates the power to which the radical base is raised.

Power

Root

$$32^{2/5} = \left(\sqrt[5]{32}\right)^2 = (2)^2 = 4$$

Because the exponent is 2/5, find the fifth root of the base, 32, to get 2. Then find the second power of 2.

b. $81^{3/4} = \left(\sqrt[4]{81}\right)^3 = (3)^3 = 27$

Because the exponent is 3/4, find the fourth root of the base, 81, to get 3. Then find the third power of 3.

c. For $-25^{3/2}$, the base is 25, not -25.

$$-25^{3/2} = -(25^{3/2}) = -\left(\sqrt{25}\right)^3 = -(5)^3 = -125$$

Because the exponent is 3/2, find the square root of the base, 25, to get 5. Then find the third power of 5.

Caution

We can also evaluate $x^{m/n}$ using $\sqrt[n]{x^m}$, however the resulting radicand is often extremely large. For example,

$$81^{3/4} = \sqrt[4]{81^3} = \sqrt[4]{531{,}441} = 27$$

Self Check 5 Simplify: **a.** $16^{3/2}$, **b.** $125^{4/3}$, and **c.** $-32^{4/5}$.

EXAMPLE 6

Simplify each expression. Assume that the variables can represent any real number.

a. $(36m^4)^{3/2}$, **b.** $(-8x^3)^{4/3}$, and **c.** $(x^5y^5)^{2/5}$.

Solution

Power

Root

a. $(36m^4)^{3/2} = \left(\sqrt{36m^4}\right)^3 = (6m^2)^3 = 216m^6$

Because the exponent is $\frac{3}{2}$, find the square root of the base, $36m^4$, to get $6m^2$. Then find the third power of $6m^2$.

b. $(-8x^3)^{4/3} = \left(\sqrt[3]{-8x^3}\right)^4 = (-2x)^4 = 16x^4$

Because the exponent is $\frac{4}{3}$, find the cube root of the base, $-8x^3$, to get $-2x$. Then find the fourth power of $-2x$.

c. $(x^5y^5)^{2/5} = \left(\sqrt[5]{x^5y^5}\right)^2 = (xy)^2 = x^2y^2$

Because the exponent is $\frac{2}{5}$, find the fifth root of the base, x^5y^5, to get xy. Then find the second power of xy.

Self Check 6 Simplify: **a.** $(4c^4)^{3/2}$ and **b.** $(-27m^3n^3)^{2/3}$.

ACCENT ON TECHNOLOGY: RATIONAL EXPONENTS

We can evaluate expressions containing rational exponents using the exponential key $\boxed{y^x}$ or $\boxed{x^y}$ on a scientific calculator. For example, to evaluate $10^{2/3}$, we enter

10 $\boxed{y^x}$ $\boxed{(}$ 2 $\boxed{\div}$ 3 $\boxed{)}$ $\boxed{=}$

4.641588834

Note that parentheses were used when entering the power. Without them, the calculator would interpret the entry as $10^2 \div 3$.

To evaluate the exponential expression using a graphing calculator, we use the [^] key, which raises a base to a power. Again, we use parentheses when entering the power.

10 [^] [(] 2 [÷] 3 [)] [ENTER]

```
10^(2/3)
        4.641588834
```

To the nearest hundredth, $10^{2/3} \approx 4.64$.

NEGATIVE RATIONAL EXPONENTS

To be consistent with the definition of negative-integer exponents, we define $x^{-m/n}$ as follows.

Definition of $x^{-m/n}$

If m and n are positive integers, $\frac{m}{n}$ is in simplified form, and $x^{1/n}$ is a real number, then

$$x^{-m/n} = \frac{1}{x^{m/n}} \quad \text{and} \quad \frac{1}{x^{-m/n}} = x^{m/n} \quad (x \neq 0)$$

From the definition, we see that another way to write $x^{-m/n}$ is to write its reciprocal and change the sign of the exponent.

EXAMPLE 7

Simplify each expression. Assume that x can represent any nonzero real number.

a. $64^{-1/2}$, **b.** $(-16)^{-5/4}$, **c.** $-625^{-3/4}$, **d.** $(-32x^5)^{-2/5}$, and **e.** $\dfrac{1}{25^{-3/2}}$.

Solution

a.

Reciprocal

$$64^{-1/2} = \frac{1}{64^{1/2}} = \frac{1}{\sqrt{64}} = \frac{1}{8}$$

Change sign

Because the exponent is negative, write the reciprocal of $64^{-1/2}$, and change the sign of the exponent.

b. $(-16)^{-5/4}$ is not a real number because $(-16)^{5/4}$ is not a real number.

c. In $-625^{-3/4}$, the base is 625.

$$-625^{-3/4} = -\frac{1}{625^{3/4}} = -\frac{1}{(\sqrt[4]{625})^3} = -\frac{1}{(5)^3} = -\frac{1}{125}$$

d. $$(-32x^5)^{-2/5} = \frac{1}{(-32x^5)^{2/5}} = \frac{1}{(\sqrt[5]{-32x^5})^2} = \frac{1}{(-2x)^2} = \frac{1}{4x^2}$$

e.

Reciprocal

$$\frac{1}{25^{-3/2}} = 25^{3/2} = (\sqrt{25})^3 = (5)^3 = 125$$

Change sign

Because the exponent is negative, write the reciprocal of $\frac{1}{25^{-3/2}}$, and change the sign of the exponent.

Caution

A negative exponent does not indicate a negative number. For example,

$$64^{-1/2} = \frac{1}{8}$$

Caution

A base of 0 raised to a negative power is undefined. For example, $0^{-2} = \frac{1}{0^2}$ is undefined because we cannot divide by 0.

Self Check 7 Simplify: **a.** $(36)^{-3/2}$ and **b.** $(-27a^3)^{-2/3}$.

APPLYING THE RULES FOR EXPONENTS

We can use the rules for exponents to simplify many expressions with fractional exponents. If all variables represent positive numbers, absolute value symbols are not needed.

EXAMPLE 8

Simplify each expression. Assume that all variables represent positive numbers. Write all answers using positive exponents only.

a. $5^{2/7}5^{3/7}$, **b.** $(5^{2/7})^3$, **c.** $(a^{2/3}b^{1/2})^6$, and **d.** $\dfrac{a^{8/3}a^{1/3}}{a^2}$.

Solution

a. $5^{2/7}5^{3/7} = 5^{2/7+3/7}$ — Use the rule $x^m x^n = x^{m+n}$.

$= 5^{5/7}$ — Add: $\frac{2}{7} + \frac{3}{7} = \frac{5}{7}$.

b. $(5^{2/7})^3 = 5^{(2/7)(3)}$ — Use the rule $(x^m)^n = x^{mn}$.

$= 5^{6/7}$ — Multiply: $\frac{2}{7}(3) = \frac{6}{7}$.

c. $(a^{2/3}b^{1/2})^6 = (a^{2/3})^6(b^{1/2})^6$ — Use the rule $(xy)^n = x^n y^n$.

$= a^{12/3}b^{6/2}$ — Use the rule $(x^m)^n = x^{mn}$ twice.

$= a^4b^3$ — Simplify the exponents.

d. $\dfrac{a^{8/3}a^{1/3}}{a^2} = a^{8/3+1/3-2}$ — Use the rules $x^m x^n = x^{m+n}$ and $\dfrac{x^m}{x^n} = x^{m-n}$.

$= a^{8/3+1/3-6/3}$ — $2 = \frac{6}{3}$.

$= a^{3/3}$ — $\frac{8}{3} + \frac{1}{3} - \frac{6}{3} = \frac{3}{3}$.

$= a$ — $\frac{3}{3} = 1$.

Self Check 8 Simplify: **a.** $(x^{1/3}y^{3/2})^6$ and **b.** $\dfrac{x^{5/3}x^{2/3}}{x^{1/3}}$.

EXAMPLE 9

Perform each multiplication and then simplify if possible. Assume all variables represent positive numbers. Write all answers using positive exponents only. **a.** $a^{4/5}(a^{1/5} + a^{3/5})$ and **b.** $x^{1/2}(x^{-1/2} + x^{1/2})$.

Solution

a. $a^{4/5}(a^{1/5} + a^{3/5}) = a^{4/5}a^{1/5} + a^{4/5}a^{3/5}$ — Use the distributive property.

$= a^{4/5+1/5} + a^{4/5+3/5}$ — Use the rule $x^m x^n = x^{m+n}$.

$= a^{5/5} + a^{7/5}$ — Simplify the exponents.

$= a + a^{7/5}$ — We cannot add these terms because they are not like terms.

b. $x^{1/2}(x^{-1/2} + x^{1/2}) = x^{1/2}x^{-1/2} + x^{1/2}x^{1/2}$ — Use the distributive property.

$= x^{1/2+(-1/2)} + x^{1/2+1/2}$ — Use the rule $x^m x^n = x^{m+n}$.

$= x^0 + x^1$ — Simplify each exponent.

$= 1 + x$ — $x^0 = 1$.

Self Check 9 Simplify: $t^{5/8}(t^{3/8} + t^{-5/8})$.

SIMPLIFYING RADICAL EXPRESSIONS

We can simplify many radical expressions by using the following steps.

Using Rational Exponents to Simplify Radicals

1. Change the radical expression into an exponential expression.
2. Simplify the rational exponents.
3. Change the exponential expression back into a radical.

EXAMPLE 10

Simplify: **a.** $\sqrt[4]{3^2}$, **b.** $\sqrt[8]{x^6}$, **c.** $\sqrt[9]{27x^6y^3}$, and **d.** $\sqrt[5]{\sqrt[3]{t}}$.

Solution

a.

$\sqrt[4]{3^2} = (3^2)^{1/4}$ Change the radical to an exponential expression.

$= 3^{2/4}$ Use the rule $(x^m)^n = x^{mn}$.

$= 3^{1/2}$ $\frac{2}{4} = \frac{1}{2}$.

$= \sqrt{3}$ Change back to radical form.

b.

$\sqrt[8]{x^6} = (x^6)^{1/8}$ Change the radical to an exponential expression.

$= x^{6/8}$ Use the rule $(x^m)^n = x^{mn}$.

$= x^{3/4}$ $\frac{6}{8} = \frac{3}{4}$.

$= (x^3)^{1/4}$ $\frac{3}{4} = 3\left(\frac{1}{4}\right)$.

$= \sqrt[4]{x^3}$ Change back to radical form.

c.

$\sqrt[9]{27x^6y^3} = (3^3x^6y^3)^{1/9}$ Write 27 as 3^3 and change the radical to an exponential expression.

$= 3^{3/9}x^{6/9}y^{3/9}$ Raise each factor to the $\frac{1}{9}$ power by multiplying the fractional exponents.

$= 3^{1/3}x^{2/3}y^{1/3}$ Simplify each fractional exponent.

$= (3x^2y)^{1/3}$ Use the rule $(xy)^n = x^ny^n$.

$= \sqrt[3]{3x^2y}$ Change back to radical form.

d.

$\sqrt[5]{\sqrt[3]{t}} = \sqrt[5]{t^{1/3}}$ Change the radical $\sqrt[3]{t}$ to exponential notation.

$= (t^{1/3})^{1/5}$ Change the radical $\sqrt[5]{t^{1/3}}$ to exponential notation.

$= t^{1/15}$ Use the rule $(x^m)^n = x^{mn}$. Multiply: $\frac{1}{3} \cdot \frac{1}{5} = \frac{1}{15}$.

$= \sqrt[15]{t}$ Change back to radical form.

Self Check 10 Simplify: **a.** $\sqrt[6]{3^3}$, **b.** $\sqrt[4]{49x^2y^2}$, and **c.** $\sqrt[3]{\sqrt[4]{m}}$.

Answers to Self Checks

1. a. 4, **b.** $-\frac{3}{2}$, **c.** $-\sqrt[4]{6x^3}$ **2.** $(7ab)^{1/6}$ **4. a.** $5|a|$, **b.** b^2
5. a. 64, **b.** 625, **c.** -16 **6. a.** $8c^6$, **b.** $9m^2n^2$ **7. a.** $\frac{1}{216}$, **b.** $\frac{1}{9a^2}$
8. a. x^2y^9, **b.** x^2 **9.** $t + 1$ **10. a.** $\sqrt{3}$, **b.** $\sqrt{7xy}$ **c.** $\sqrt[12]{m}$

7.2 STUDY SET

VOCABULARY **Fill in the blanks.**

1. The expressions $4^{1/2}$ and $(-8)^{-2/3}$ have ________________ exponents.
2. In the exponential expression $27^{4/3}$, 27 is the ______, and 4/3 is the __________.
3. In the radical expression $\sqrt[3]{4{,}096x^{12}}$, 3 is the ______, and $4{,}096x^{12}$ is the _________.
4. $32^{4/5}$ means the fourth _______ of the fifth ______ of 32.

CONCEPTS

5. Complete the table by writing the given expression in the alternate form.

Radical form	Exponential form
$\sqrt[5]{25}$	
	$(-27)^{2/3}$
$(\sqrt[4]{16})^{-3}$	
	$81^{3/2}$
$-\sqrt{\dfrac{9}{64}}$	

6. In your own words, explain the three rules for rational exponents illustrated in the diagrams below.

a. $(-32)^{1/5} = \sqrt[5]{-32}$

b. $125^{4/3} = (\sqrt[3]{125})^4$

c. $8^{-1/3} = \dfrac{1}{8^{1/3}}$

7. Graph each number on the number line.

$$\left\{8^{2/3}, (-125)^{1/3}, -16^{-1/4}, 4^{3/2}, -\left(\frac{9}{100}\right)^{-1/2}\right\}$$

8. Evaluate $25^{3/2}$ in two ways. Which way is easier?

Complete each rule for exponents.

9. $x^{1/n} =$ ____
10. $x^{m/n} =$ ____ $= \sqrt[n]{x^m}$
11. $x^{-m/n} =$ ____
12. $\dfrac{1}{x^{-m/n}} =$ ____

NOTATION **Complete each solution.**

13. Simplify: $(100a^4)^{3/2}$.

$$(100a^4)^{3/2} = (\sqrt{})^3$$
$$= ()^3$$
$$= 1{,}000a^6$$

14. Simplify: $(m^{1/3}n^{1/2})^6$.

$$(m^{1/3}n^{1/2})^6 = ()^6(n^{1/2})^6$$
$$= m^{}n^{6/2}$$
$$= m^2n^3$$

PRACTICE **Write each expression in radical form.**

15. $x^{1/3}$
16. $b^{1/2}$
17. $(3x)^{1/4}$
18. $(4ab)^{1/6}$
19. $(6x^3y)^{1/4}$
20. $(7a^2b^2)^{1/5}$
21. $(x^2 + y^2)^{1/2}$
22. $(x^3 + y^3)^{1/3}$

Change each radical to an exponential expression.

23. $\sqrt{m}$
24. $\sqrt[3]{r}$
25. $\sqrt[4]{3a}$
26. $3\sqrt[5]{a}$
27. $\sqrt[6]{8abc}$
28. $\sqrt[7]{7p^2q}$
29. $\sqrt[3]{a^2 - b^2}$
30. $\sqrt{x^2 + y^2}$

Simplify each expression, if possible.

31. $4^{1/2}$
32. $25^{1/2}$
33. $125^{1/3}$
34. $8^{1/3}$
35. $81^{1/4}$
36. $625^{1/4}$
37. $32^{1/5}$
38. $0^{1/5}$
39. $\left(\dfrac{1}{4}\right)^{1/2}$
40. $\left(\dfrac{1}{16}\right)^{1/2}$
41. $-16^{1/4}$
42. $-125^{1/3}$

43. $(-64)^{1/2}$

44. $(-216)^{1/2}$

45. $(-216)^{1/3}$

46. $(-1{,}000)^{1/3}$

Simplify each expression, if possible. Assume that all variables are unrestricted and use absolute value symbols when necessary.

47. $(x^2)^{1/2}$

48. $(x^3)^{1/3}$

49. $(m^4)^{1/2}$

50. $(a^4)^{1/4}$

51. $(n^9)^{1/3}$

52. $(t^{10})^{1/5}$

53. $(25y^2)^{1/2}$

54. $(-27x^3)^{1/3}$

55. $(16x^4)^{1/4}$

56. $(-16x^4)^{1/2}$

57. $(-64x^8)^{1/4}$

58. $(243x^{10})^{1/5}$

59. $[(x+1)^4]^{1/4}$

60. $[(x+5)^3]^{1/3}$

Simplify each expression. Assume that all variables represent positive numbers.

61. $36^{3/2}$

62. $27^{2/3}$

63. $81^{3/4}$

64. $100^{3/2}$

65. $144^{3/2}$

66. $1{,}000^{2/3}$

67. $\left(\frac{1}{8}\right)^{2/3}$

68. $\left(\frac{4}{9}\right)^{3/2}$

69. $(25x^4)^{3/2}$

70. $(27a^3b^3)^{2/3}$

71. $\left(\frac{8x^3}{27}\right)^{2/3}$

72. $\left(\frac{27}{64y^6}\right)^{2/3}$

Write each expression without using negative exponents. Assume that all variables represent positive numbers.

73. $4^{-1/2}$

74. $8^{-1/3}$

75. $25^{-5/2}$

76. $4^{-3/2}$

77. $(16x^2)^{-3/2}$

78. $(81c^4)^{-3/2}$

79. $(-27y^3)^{-2/3}$

80. $(-8z^9)^{-2/3}$

81. $\left(\frac{27}{8}\right)^{-4/3}$

82. $\left(\frac{25}{49}\right)^{-3/2}$

83. $\left(-\frac{8x^3}{27}\right)^{-1/3}$

84. $\left(\frac{16}{81y^4}\right)^{-3/4}$

Perform the operations. Write the answers without negative exponents. Assume that all variables represent positive numbers.

85. $9^{3/7}9^{2/7}$

86. $4^{2/5}4^{2/5}$

87. $6^{-2/3}6^{-4/3}$

88. $5^{1/3}5^{-5/3}$

89. $\frac{3^{4/3}3^{1/3}}{3^{2/3}}$

90. $\frac{2^{5/6}2^{1/3}}{2^{1/2}}$

91. $a^{2/3}a^{1/3}$

92. $b^{3/5}b^{1/5}$

93. $(a^{2/3})^{1/3}$

94. $(t^{4/5})^{10}$

95. $(a^{1/2}b^{1/3})^{3/2}$

96. $(mn^{-2/3})^{-3/5}$

97. $(27x^{-3})^{-1/3}$

98. $(16a^{-2})^{-1/2}$

Perform the multiplications. Assume that all variables are positive.

99. $y^{1/3}(y^{2/3} + y^{5/3})$

100. $y^{2/5}(y^{-2/5} + y^{3/5})$

101. $x^{3/5}(x^{7/5} - x^{2/5} + 1)$

102. $x^{4/3}(x^{2/3} + 3x^{5/3} - 4)$

Use rational exponents to simplify each radical. Assume that all variables represent positive numbers.

103. $\sqrt[6]{p^3}$

104. $\sqrt[8]{q^2}$

105. $\sqrt[4]{25b^2}$

106. $\sqrt[9]{-8x^6}$

107. $\sqrt[10]{x^2y^2}$

108. $\sqrt[6]{x^2y^2}$

109. $\sqrt[9]{\sqrt{c}}$

110. $\sqrt[4]{\sqrt{x}}$

111. $\sqrt[5]{\sqrt[3]{7m}}$

112. $\sqrt[3]{\sqrt[4]{21x}}$

Use a calculator to evaluate each expression. Round to the nearest hundredth.

113. $\sqrt[3]{15}$

114. $\sqrt[4]{50.5}$

115. $\sqrt[5]{1.045}$

116. $\sqrt[5]{-1{,}000}$

APPLICATIONS

117. BALLISTIC PENDULUMS The formula

$$v = \frac{m + M}{m}(2gh)^{1/2}$$

gives the velocity (in ft/sec) of a bullet with weight *m* fired into a block with weight *M*, that raises the height of the block h feet after the collision. The letter g represents the constant, 32. Find the velocity of the bullet to the nearest ft/sec.

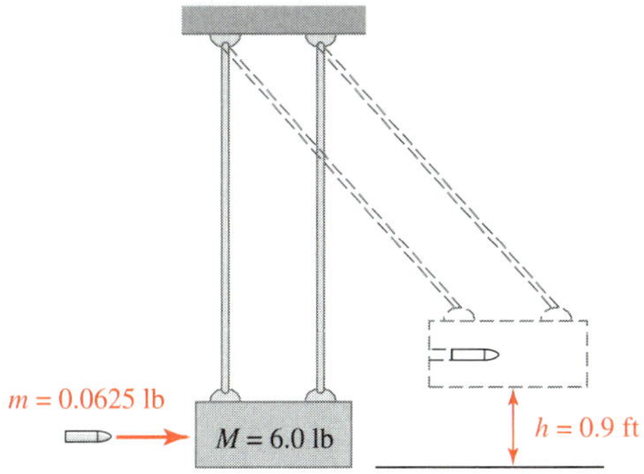

118. GEOGRAPHY The formula

$$A = [s(s - a)(s - b)(s - c)]^{1/2}$$

gives the area of a triangle with sides of length a, b, and c, where s is one-half of the perimeter. Estimate the area of Virginia (to the nearest square mile) using the data given in the illustration.

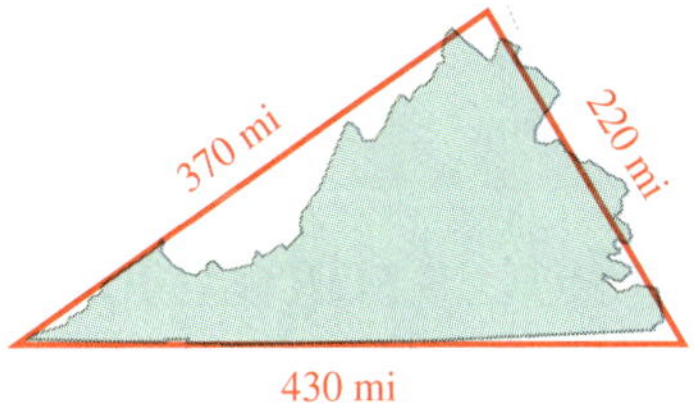

119. RELATIVITY One concept of relativity theory is that an object moving past an observer at a speed near the speed of light appears to have a larger mass because of its motion. If the mass of the object is m_0 when the object is at rest relative to the observer, its mass m will be given by the formula

$$m = m_0\left(1 - \frac{v^2}{c^2}\right)^{-1/2}$$

when it is moving with speed v (in miles per second) past the observer. The variable c is the speed of light, 186,000 mi/sec. If a proton with a rest mass of 1 unit is accelerated by a nuclear accelerator to a speed of 160,000 mi/sec, what mass will the technicians observe it to have? Round to the nearest hundredth.

120. LOGGING The width w and height h of the strongest rectangular beam that can be cut from a cylindrical log of radius a are given by

$$w = \frac{2a}{3}(3^{1/2}) \qquad h = a\left(\frac{8}{3}\right)^{1/2}$$

Find the width, height, and cross-sectional area of the strongest beam that can be cut from a log with *diameter* 4 feet. Round to the nearest hundredth.

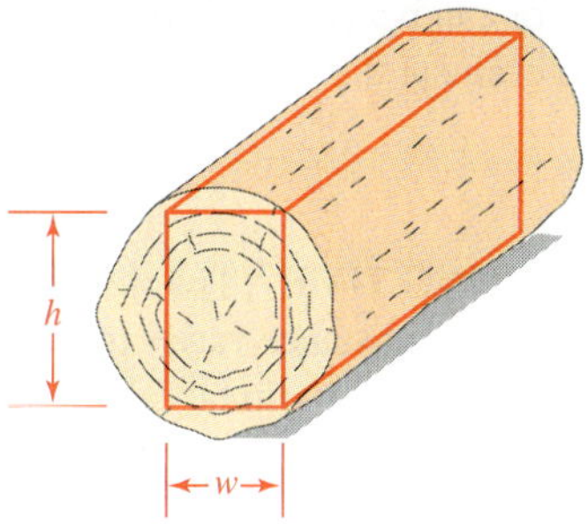

121. CUBICLES The area of the base of a cube is given by the function $A(V) = V^{2/3}$, where V is the volume of the cube. In a preschool room, 18 children's cubicles like the one shown are placed on the floor around the room. Estimate how much floor space is lost to the cubicles. Give your answer in square inches and in square feet.

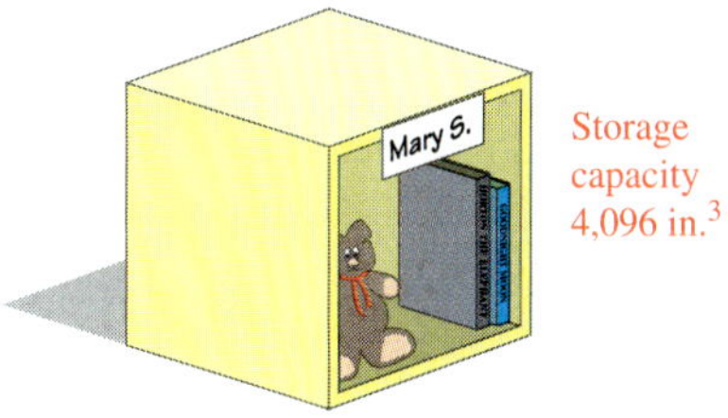

122. CARPENTRY The length L of the longest board that can be carried horizontally around the right-angle corner of two intersecting hallways is given by the formula

$$L = (a^{2/3} + b^{2/3})^{3/2}$$

where a and b represent the widths of the hallways. Find the longest shelf that a carpenter can carry around the corner if $a = 40$ in. and $b = 64$ in. Give your result in inches and in feet. In each case, round to the nearest tenth.

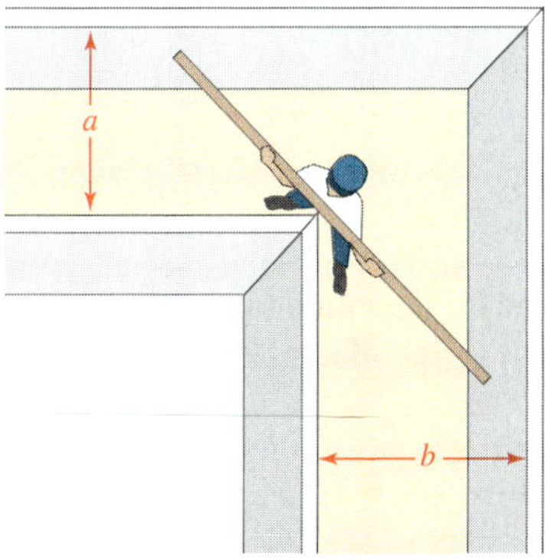

WRITING

123. What is a rational exponent? Give some examples.

124. Explain how the root key $\sqrt[x]{y}$ on a scientific calculator can be used in combination with other keys to evaluate the expression $16^{3/4}$.

REVIEW

125. COMMUTING TIME The time it takes a car to travel a certain distance varies inversely with its rate of speed. If a certain trip takes 3 hours at 50 miles per hour, how long will the trip take at 60 miles per hour?

126. BANKRUPTCY After filing for bankruptcy, a company was able to pay its creditors only 15 cents on the dollar. If the company owed a lumberyard \$9,712, how much could the lumberyard expect to be paid?

CHALLENGE PROBLEMS

127. The fraction $\frac{2}{4}$ is equal to $\frac{1}{2}$. Is $16^{2/4}$ equal to $16^{1/2}$? Explain.

128. How would you evaluate an expression with a mixed-number exponent? For example, what is $8^{1\frac{1}{3}}$? What is $25^{2\frac{1}{2}}$? Discuss.

7.3 Simplifying and Combining Radical Expressions

- The product and quotient rules for radicals
- Simplifying radical expressions
- Adding and subtracting radical expressions

In algebra, it is often helpful to replace an expression with a simpler equivalent expression. This is certainly true when working with radicals. In most cases, radical expressions should be written in simplified form. We use two rules for radicals to do this.

THE PRODUCT AND QUOTIENT RULES FOR RADICALS

To introduce the product rule for radicals, we will find $\sqrt{4 \cdot 25}$ and $\sqrt{4}\sqrt{25}$, and compare the results.

Square root of a product

$$\sqrt{4 \cdot 25} = \sqrt{100} = 10$$

Product of square roots

$$\sqrt{4}\sqrt{25} = 2 \cdot 5 = 10$$

In each case, the answer is 10. Thus, $\sqrt{4 \cdot 25} = \sqrt{4}\sqrt{25}$.

Notation

The products $\sqrt{4}\sqrt{25}$ and $\sqrt[3]{8}\sqrt[3]{27}$ can also be written using a raised dot:

$\sqrt{4} \cdot \sqrt{25}$ $\quad$ $\sqrt[3]{8} \cdot \sqrt[3]{27}$

Similarly, we will find $\sqrt[3]{8 \cdot 27}$ and $\sqrt[3]{8}\sqrt[3]{27}$, and compare the results.

Cube root of a product

$$\sqrt[3]{8 \cdot 27} = \sqrt[3]{216} = 6$$

Product of cube roots

$$\sqrt[3]{8}\sqrt[3]{27} = 2 \cdot 3 = 6$$

In each case, the answer is 6. Thus, $\sqrt[3]{8 \cdot 27} = \sqrt[3]{8}\sqrt[3]{27}$. These results illustrate the *product rule for radicals.*

The Product Rule for Radicals

The nth root of the product of two numbers is equal to the product of their nth roots. If $\sqrt[n]{a}$ and $\sqrt[n]{b}$ are real numbers,

$$\sqrt[n]{ab} = \sqrt[n]{a}\sqrt[n]{b}$$

Caution The product rule for radicals applies to the nth root of a product. There is no such property for sums or differences. For example,

$$\sqrt{9+4} \neq \sqrt{9} + \sqrt{4} \qquad \sqrt{9-4} \neq \sqrt{9} - \sqrt{4}$$
$$\sqrt{13} \neq 3 + 2 \qquad \sqrt{5} \neq 3 - 2$$
$$\sqrt{13} \neq 5 \qquad \sqrt{5} \neq 1$$

Thus, $\sqrt{a+b} \neq \sqrt{a} + \sqrt{b}$ and $\sqrt{a-b} \neq \sqrt{a} - \sqrt{b}$.

To introduce the quotient rule for radicals, we will find $\sqrt{\dfrac{100}{4}}$ and $\dfrac{\sqrt{100}}{\sqrt{4}}$, and compare the results.

Square root of a quotient

$$\sqrt{\frac{100}{4}} = \sqrt{25}$$
$$= 5$$

Quotient of square roots

$$\frac{\sqrt{100}}{\sqrt{4}} = \frac{10}{2}$$
$$= 5$$

Since the answer is 5 in each case, $\sqrt{\dfrac{100}{4}} = \dfrac{\sqrt{100}}{\sqrt{4}}$.

Similarly, we will find $\sqrt[3]{\dfrac{64}{8}}$ and $\dfrac{\sqrt[3]{64}}{\sqrt[3]{8}}$, and compare the results.

Cube root of a quotient

$$\sqrt[3]{\frac{64}{8}} = \sqrt[3]{8}$$
$$= 2$$

Quotient of cube roots

$$\frac{\sqrt[3]{64}}{\sqrt[3]{8}} = \frac{4}{2}$$
$$= 2$$

Since the answer is 2 in each case, $\sqrt[3]{\dfrac{64}{8}} = \dfrac{\sqrt[3]{64}}{\sqrt[3]{8}}$. These results illustrate the *quotient rule for radicals.*

The Quotient Rule for Radicals

The nth root of the quotient of two numbers is equal to the quotient of their nth roots.

If $\sqrt[n]{a}$ and $\sqrt[n]{b}$ are real numbers, and b is not 0,

$$\sqrt[n]{\frac{a}{b}} = \frac{\sqrt[n]{a}}{\sqrt[n]{b}}$$

SIMPLIFYING RADICAL EXPRESSIONS

When a radical expression is written in **simplified form,** each of the following is true.

Simplified Form of a Radical Expression

1. Each factor in the radicand is to a power that is less than the index of the radical.
2. The radicand contains no fractions or negative numbers.
3. No radicals appear in the denominator of a fraction.

To simplify radical expressions, we must often factor the radicand using two natural-number factors. To simplify square root and cube root radicals, it is helpful to have the following lists memorized.

Perfect squares: **1, 4, 9, 16, 25, 36, 49, 64, 81, 100, 121, 144, 169, 196, 225, . . .**

Perfect cubes: **1, 8, 27, 64, 125, 216, 343, 512, 729, 1,000, . . .**

EXAMPLE 1

Simplify: **a.** $\sqrt{12}$, **b.** $\sqrt{98}$, **c.** $\sqrt[3]{54}$, and **d.** $-\sqrt[4]{48}$.

Solution **a.** To simplify $\sqrt{12}$, we first factor 12 so that one factor is the largest perfect square that divides 12. Since 4 is the largest perfect square factor of 12, we write 12 as $4 \cdot 3$, use the multiplication property of radicals, and simplify.

$\sqrt{12} = \sqrt{4 \cdot 3}$ Write 12 as $12 = 4 \cdot 3$.

Write the perfect square factor first.

$= \sqrt{4}\sqrt{3}$ The square root of a product is equal to the product of the square roots.

$= 2\sqrt{3}$ Find the square root of the perfect square factor: $\sqrt{4} = 2$. Read as "2 times the square root of 3" or as "2 radical 3."

We say that $2\sqrt{3}$ is the simplified form of $\sqrt{12}$.

b. The largest perfect square factor of 98 is 49. Thus,

$\sqrt{98} = \sqrt{49 \cdot 2}$ Write 98 in factored form: $98 = 49 \cdot 2$.

$= \sqrt{49}\sqrt{2}$ The square root of a product is equal to the product of the square roots: $\sqrt{49 \cdot 2} = \sqrt{49}\sqrt{2}$.

$= 7\sqrt{2}$ Simplify: $\sqrt{49} = 7$.

c. Since the largest perfect cube factor of 54 is 27, we have

$\sqrt[3]{54} = \sqrt[3]{27 \cdot 2}$ Write 54 as $27 \cdot 2$.

$= \sqrt[3]{27}\sqrt[3]{2}$ The cube root of a product is equal to the product of the cube roots: $\sqrt[3]{27 \cdot 2} = \sqrt[3]{27}\sqrt[3]{2}$.

$= 3\sqrt[3]{2}$ Simplify: $\sqrt[3]{27} = 3$.

d. The largest perfect fourth-power factor of 48 is 16. Thus,

$-\sqrt[4]{48} = -\sqrt[4]{16 \cdot 3}$ Write 48 as $16 \cdot 3$.

$= -\sqrt[4]{16}\sqrt[4]{3}$ The fourth root of a product is equal to the product of the fourth roots: $\sqrt[4]{16 \cdot 3} = \sqrt[4]{16} \cdot \sqrt[4]{3}$.

$= -2\sqrt[4]{3}$ Simplify: $\sqrt[4]{16} = 2$.

The Language of Algebra

Perfect fourth-powers are

1, 16, 81, 256, 625, . . .

Self Check 1 Simplify: **a.** $\sqrt{20}$, **b.** $\sqrt[3]{24}$, and **c.** $\sqrt[5]{-128}$.

Variable expressions can also be perfect squares, perfect cubes, perfect fourth-powers, and so on. For example,

Perfect squares: $x^2, x^4, x^6, x^8, x^{10}, \ldots$ Perfect cubes: $x^3, x^6, x^9, x^{12}, x^{15}, \ldots$

EXAMPLE 2

Simplify: **a.** $\sqrt{m^9}$, **b.** $\sqrt{128a^5}$, **c.** $\sqrt[3]{24x^5}$, and **d.** $\sqrt[5]{a^9b^5}$. Assume that all variables represent positive real numbers.

Solution

a. The largest perfect square factor of m^9 is m^8.

$$\begin{aligned} \sqrt{m^9} &= \sqrt{m^8 \cdot m} && \text{Write } m^9 \text{ in factored form as } m^8 \cdot m. \\ &= \sqrt{m^8}\sqrt{m} && \text{Use the product rule for radicals.} \\ &= m^4\sqrt{m} && \text{Simplify: } \sqrt{m^8} = m^4. \end{aligned}$$

b. Since the largest perfect square factor of 128 is 64 and the largest perfect square factor of a^5 is a^4, the largest perfect square factor of $128a^5$ is $64a^4$. We write $128a^5$ as $64a^4 \cdot 2a$ and proceed as follows:

$$\begin{aligned} \sqrt{128a^5} &= \sqrt{64a^4 \cdot 2a} && \text{Write } 128a^5 \text{ in factored form as } 64a^4 \cdot 2a. \\ &= \sqrt{64a^4}\sqrt{2a} && \text{Use the product rule for radicals.} \\ &= 8a^2\sqrt{2a} && \text{Simplify: } \sqrt{64a^4} = 8a^2. \end{aligned}$$

c. We write $24x^5$ as $8x^3 \cdot 3x^2$ and proceed as follows:

$$\begin{aligned} \sqrt[3]{24x^5} &= \sqrt[3]{8x^3 \cdot 3x^2} && 8x^3 \text{ is the largest perfect cube factor of } 24x^5. \\ &= \sqrt[3]{8x^3}\sqrt[3]{3x^2} && \text{Use the product rule for radicals.} \\ &= 2x\sqrt[3]{3x^2} && \text{Simplify: } \sqrt[3]{8x^3} = 2x. \end{aligned}$$

d. The largest perfect fifth-power factor of a^9 is a^5, and b^5 is a perfect fifth power.

$$\begin{aligned} \sqrt[5]{a^9b^5} &= \sqrt[5]{a^5b^5 \cdot a^4} && a^5b^5 \text{ is the largest perfect fifth-power factor of } a^9b^5. \\ &= \sqrt[5]{a^5b^5}\sqrt[5]{a^4} && \text{Use the product rule for radicals.} \\ &= ab\sqrt[5]{a^4} && \text{Simplify: } \sqrt[5]{a^5b^5} = ab. \end{aligned}$$

The Language of Algebra

Perfect fifth-powers of *a* are

$a^5, a^{10}, a^{15}, a^{20}, a^{25}, \ldots$

Self Check 2 Simplify: **a.** $\sqrt{98b^3}$, **b.** $\sqrt[3]{54y^5}$, and **c.** $\sqrt[4]{t^8u^{15}}$.

EXAMPLE 3

Simplify each expression:

a. $\sqrt{\dfrac{7}{64}}$, **b.** $\sqrt{\dfrac{15}{49x^2}}$, and **c.** $\sqrt[3]{\dfrac{10x^2}{27y^6}}$. Assume that the variables represent positive real numbers.

Solution

a. We can use the quotient rule for radicals to simplify each expression.

$$\begin{aligned} \sqrt{\frac{7}{64}} &= \frac{\sqrt{7}}{\sqrt{64}} && \text{The square root of a quotient is equal to the quotient of the square roots.} \\ &= \frac{\sqrt{7}}{8} && \text{Simplify the denominator: } \sqrt{64} = 8. \end{aligned}$$

Success Tip

In Example 3, a radical of a quotient is written as a quotient of radicals.

$$\sqrt[n]{\frac{a}{b}} = \frac{\sqrt[n]{a}}{\sqrt[n]{b}}$$

b.

$$\begin{aligned} \sqrt{\frac{15}{49x^2}} &= \frac{\sqrt{15}}{\sqrt{49x^2}} && \text{The square root of a quotient is equal to the quotient of the square roots.} \\ &= \frac{\sqrt{15}}{7x} && \text{Simplify the denominator: } \sqrt{49x^2} = 7x. \end{aligned}$$

c. $\sqrt[3]{\dfrac{10x^2}{27y^6}} = \dfrac{\sqrt[3]{10x^2}}{\sqrt[3]{27y^6}}$ The cube root of a quotient is equal to the quotient of the cube roots.

$= \dfrac{\sqrt[3]{10x^2}}{3y^2}$ Simplify the denominator.

Self Check 3 Simplify: **a.** $\sqrt{\dfrac{11}{36a^2}}$ $(a > 0)$ and **b.** $\sqrt[4]{\dfrac{a^3}{625y^{12}}}$ $(a > 0, y > 0)$.

EXAMPLE 4

Simplify each expression. Assume that all variables represent positive numbers.

a. $\dfrac{\sqrt{45xy^2}}{\sqrt{5x}}$ and **b.** $\dfrac{\sqrt[3]{-432x^5}}{\sqrt[3]{8x}}$.

Success Tip

In Example 4, a quotient of radicals is written as a radical of a quotient.

$$\frac{\sqrt[n]{a}}{\sqrt[n]{b}} = \sqrt[n]{\frac{a}{b}}$$

Solution **a.** We can write the quotient of the square roots as the square root of a quotient.

$\dfrac{\sqrt{45xy^2}}{\sqrt{5x}} = \sqrt{\dfrac{45xy^2}{5x}}$ Use the quotient rule for radicals.

$= \sqrt{9y^2}$ Simplify the radicand: $\dfrac{45xy^2}{5x} = \dfrac{\overset{1}{\cancel{5}} \cdot 9 \cdot \overset{1}{\cancel{x}} \cdot y^2}{\underset{1}{\cancel{5}} \cdot \underset{1}{\cancel{x}}} = 9y^2$.

$= 3y$ Simplify the radical.

b. We can write the quotient of the cube roots as the cube root of a quotient.

$\dfrac{\sqrt[3]{-432x^5}}{\sqrt[3]{8x}} = \sqrt[3]{\dfrac{-432x^5}{8x}}$ Use the quotient rule for radicals.

$= \sqrt[3]{-54x^4}$ Simplify the radicand: $\dfrac{-432x^5}{8x} = -54x^4$.

$= \sqrt[3]{-27x^3 \cdot 2x}$ $-27x^3$ is the largest perfect cube that divides $-54x^4$.

$= \sqrt[3]{-27x^3}\sqrt[3]{2x}$ Use the product rule for radicals.

$= -3x\sqrt[3]{2x}$ Simplify: $\sqrt[3]{-27x^3} = -3x$.

Self Check 4 Simplify each expression (assume that all variables represent positive numbers):

a. $\dfrac{\sqrt{50ab^2}}{\sqrt{2a}}$ and **b.** $\dfrac{\sqrt[3]{-2{,}000x^5y^3}}{\sqrt[3]{2x}}$.

ADDING AND SUBTRACTING RADICAL EXPRESSIONS

Radical expressions with the same index and the same radicand are called **like** or **similar radicals.** For example, $3\sqrt{2}$ and $2\sqrt{2}$ are like radicals. However,

- $3\sqrt{5}$ and $4\sqrt{2}$ are not like radicals, because the radicands are different.
- $3\sqrt[4]{5}$ and $2\sqrt[3]{5}$ are not like radicals, because the indices are different.

Success Tip

Combining like radicals is similar to combining like terms.

$3\sqrt{2} + 2\sqrt{2} = 5\sqrt{2}$

$3x + 2x = 5x$

For an expression with two or more radical terms, we should attempt to combine like radicals, if possible. For example, to simplify the expression $3\sqrt{2} + 2\sqrt{2}$, we use the distributive property to factor out $\sqrt{2}$ and simplify.

$$3\sqrt{2} + 2\sqrt{2} = (3 + 2)\sqrt{2}$$
$$= 5\sqrt{2}$$

Radicals with the same index but different radicands can often be written as like radicals. For example, to simplify the expression $\sqrt{27} - \sqrt{12}$, we simplify both radicals first and then combine the like radicals.

$$\begin{aligned} \sqrt{27} - \sqrt{12} &= \sqrt{9 \cdot 3} - \sqrt{4 \cdot 3} && \text{Write 27 and 12 in factored form.} \\ &= \sqrt{9}\sqrt{3} - \sqrt{4}\sqrt{3} && \text{Use the product rule for radicals.} \\ &= 3\sqrt{3} - 2\sqrt{3} && \text{Simplify } \sqrt{9} \text{ and } \sqrt{4}. \\ &= (3 - 2)\sqrt{3} && \text{Factor out } \sqrt{3}. \\ &= \sqrt{3} && 1\sqrt{3} = \sqrt{3}. \end{aligned}$$

As the previous examples suggest, we can add or subtract radicals as follows.

Adding and Subtracting Radicals

1. To add or subtract radicals, simplify each radical expression and combine all like radicals.
2. To add or subtract like radicals, combine the coefficients and keep the common radical.

EXAMPLE 5

Simplify: $2\sqrt{12} - 3\sqrt{48} + 3\sqrt{3}$.

Solution We simplify $2\sqrt{12}$ and $3\sqrt{48}$ and then combine like radicals.

$$\begin{aligned} 2\sqrt{12} - 3\sqrt{48} + 3\sqrt{3} &= 2\sqrt{4 \cdot 3} - 3\sqrt{16 \cdot 3} + 3\sqrt{3} \\ &= 2\sqrt{4}\sqrt{3} - 3\sqrt{16}\sqrt{3} + 3\sqrt{3} \\ &= 2(2)\sqrt{3} - 3(4)\sqrt{3} + 3\sqrt{3} \\ &= 4\sqrt{3} - 12\sqrt{3} + 3\sqrt{3} && \text{All three expressions have the same index and radicand.} \\ &= (4 - 12 + 3)\sqrt{3} && \text{Combine the coefficients of these like radicals and keep } \sqrt{3}. \\ &= -5\sqrt{3} \end{aligned}$$

Self Check 5 Simplify: $3\sqrt{75} - 2\sqrt{12} + 2\sqrt{48}$.

EXAMPLE 6

Simplify: $\sqrt[3]{16} - \sqrt[3]{54} + \sqrt[3]{24}$.

Solution We begin by simplifying each radical expression:

$$\begin{aligned} \sqrt[3]{16} - \sqrt[3]{54} + \sqrt[3]{24} &= \sqrt[3]{8 \cdot 2} - \sqrt[3]{27 \cdot 2} + \sqrt[3]{8 \cdot 3} \\ &= \sqrt[3]{8}\sqrt[3]{2} - \sqrt[3]{27}\sqrt[3]{2} + \sqrt[3]{8}\sqrt[3]{3} \\ &= 2\sqrt[3]{2} - 3\sqrt[3]{2} + 2\sqrt[3]{3} \end{aligned}$$

Notation

Just as $-1x = -x$,

$-1\sqrt[n]{x} = -\sqrt[n]{x}$

And just as $1x = x$,

$1\sqrt[n]{x} = \sqrt[n]{x}$

Now we combine the two radical expressions that have the same index and radicand.

$$\sqrt[3]{16} - \sqrt[3]{54} + \sqrt[3]{24} = -\sqrt[3]{2} + 2\sqrt[3]{3} \qquad 2\sqrt[3]{2} - 3\sqrt[3]{2} = -1\sqrt[3]{2} = -\sqrt[3]{2}.$$

Self Check 6 Simplify: $\sqrt[3]{24} - \sqrt[3]{16} + \sqrt[3]{54}$.

Caution Even though the expressions $-\sqrt[3]{2}$ and $2\sqrt[3]{3}$ in the last line of Example 6 have the same index, we cannot combine them, because their radicands are different. Neither can we combine radical expressions having the same radicand but a different index. For example, the expression $\sqrt[3]{2} + \sqrt[4]{2}$ cannot be simplified.

EXAMPLE 7

Simplify: $\sqrt[3]{16x^4} + \sqrt[3]{54x^4} - \sqrt[3]{-128x^4}$.

Solution We simplify each expression and then combine like radicals.

$$\begin{aligned}
&\sqrt[3]{16x^4} + \sqrt[3]{54x^4} - \sqrt[3]{-128x^4} \\
&\quad = \sqrt[3]{8x^3 \cdot 2x} + \sqrt[3]{27x^3 \cdot 2x} - \sqrt[3]{-64x^3 \cdot 2x} \\
&\quad = \sqrt[3]{8x^3}\sqrt[3]{2x} + \sqrt[3]{27x^3}\sqrt[3]{2x} - \sqrt[3]{-64x^3}\sqrt[3]{2x} \\
&\quad = 2x\sqrt[3]{2x} + 3x\sqrt[3]{2x} + 4x\sqrt[3]{2x} && \text{All three radicals have the same index and radicand.} \\
&\quad = (2x + 3x + 4x)\sqrt[3]{2x} \\
&\quad = 9x\sqrt[3]{2x} && \text{Within the parentheses, combine like terms.}
\end{aligned}$$

Self Check 7 Simplify: $\sqrt{32x^3} + \sqrt{50x^3} - \sqrt{18x^3}$.

Answers to Self Checks **1. a.** $2\sqrt{5}$, **b.** $2\sqrt[3]{3}$, **c.** $-2\sqrt[5]{4}$ **2. a.** $7b\sqrt{2b}$, **b.** $3y\sqrt[3]{2y^2}$, **c.** $t^2u^3\sqrt[4]{u^3}$ **3. a.** $\frac{\sqrt{11}}{6a}$, **b.** $\frac{\sqrt[4]{a^3}}{5y^3}$ **4. a.** $5b$, **b.** $-10xy\sqrt[3]{x}$ **5.** $19\sqrt{3}$ **6.** $2\sqrt[3]{3} + \sqrt[3]{2}$ **7.** $6x\sqrt{2x}$

7.3 STUDY SET

VOCABULARY **Fill in the blanks.**

1. Radical expressions such as $\sqrt[3]{4}$ and $6\sqrt[3]{4}$ with the same index and the same radicand are called ______ radicals.
2. Numbers such as 1, 4, 9, 16, 25, and 36 are called perfect ________. Numbers such as 1, 8, 27, 64, and 125 are called perfect _______.
3. The largest perfect square _______ of 27 is 9.
4. "To ________ $\sqrt{24}$" means to write it as $2\sqrt{6}$.

CONCEPTS **Fill in the blanks.**

5. $\sqrt[n]{ab} =$

   In words, the nth root of the ________ of two numbers is equal to the product of their nth ______.
6. $\sqrt[n]{\frac{a}{b}} =$

   In words, the nth root of the ________ of two numbers is equal to the quotient of their nth ______.
7. Consider the expressions

   $\sqrt{4 \cdot 5}$ and $\sqrt{4}\sqrt{5}$

   Which expression is
   a. the square root of a product?
   b. the product of square roots?
   c. How are these two expressions related?
8. Consider the expressions

   $\frac{\sqrt[3]{a}}{\sqrt[3]{x^2}}$ and $\sqrt[3]{\frac{a}{x^2}}$

   Which expression is
   a. the cube root of a quotient?
   b. the quotient of cube roots?
   c. How are these two expressions related?

9. a. Write two radical expressions that have the same radicand but a different index. Can the expressions be added?

   b. Write two radical expressions that have the same index but a different radicand. Can the expressions be added?

10. Explain the mistake in the student's solution shown below.

Simplify: $\sqrt[3]{54}$.

$$\begin{aligned}\sqrt[3]{54} &= \sqrt[3]{27+27}\\ &= \sqrt[3]{27} + \sqrt[3]{27}\\ &= 3 + 3\\ &= 6\end{aligned}$$

NOTATION **Complete each solution.**

11. Simplify:

$$\begin{aligned}\sqrt[3]{32k^4} &= \sqrt[3]{\square \cdot 4k}\\ &= \sqrt[3]{\square}\sqrt[3]{4k}\\ &= 2k\sqrt[3]{\square}\end{aligned}$$

12. Simplify:

$$\begin{aligned}\frac{\sqrt{80s^2t^4}}{\sqrt{5s^2}} &= \sqrt{\frac{80s^2t^4}{\square}}\\ &= \sqrt{\square}\\ &= 4t^2\end{aligned}$$

PRACTICE **Simplify each expression.**

13. $\sqrt{20}$
14. $\sqrt{8}$
15. $\sqrt{200}$
16. $\sqrt{250}$
17. $\sqrt[3]{80}$
18. $\sqrt[3]{270}$
19. $\sqrt[3]{-81}$
20. $\sqrt[3]{-72}$
21. $\sqrt[4]{32}$
22. $\sqrt[4]{48}$
23. $-\sqrt[5]{96}$
24. $-\sqrt[7]{256}$
25. $\sqrt[6]{320}$
26. $\sqrt[6]{192}$
27. $\sqrt{\frac{7}{9}}$
28. $\sqrt{\frac{3}{4}}$
29. $\sqrt[3]{\frac{7}{64}}$
30. $\sqrt[3]{\frac{4}{125}}$
31. $\sqrt[4]{\frac{3}{10{,}000}}$
32. $\sqrt[5]{\frac{4}{243}}$
33. $\sqrt[5]{\frac{3}{32}}$
34. $\sqrt[6]{\frac{5}{64}}$
35. $\frac{\sqrt{500}}{\sqrt{5}}$
36. $\frac{\sqrt{128}}{\sqrt{2}}$
37. $\frac{\sqrt[3]{48}}{\sqrt[3]{6}}$
38. $\frac{\sqrt[3]{64}}{\sqrt[3]{8}}$

Simplify each radical expression. All variables represent positive numbers.

39. $\sqrt{50x^2}$
40. $\sqrt{75a^2}$
41. $\sqrt{32b}$
42. $\sqrt{80c}$
43. $-\sqrt{112a^3}$
44. $\sqrt{147a^5}$
45. $\sqrt{175a^2b^3}$
46. $\sqrt{128a^3b^5}$
47. $-\sqrt{300xy}$
48. $\sqrt{200x^2y}$
49. $\sqrt[3]{-54x^6}$
50. $-\sqrt[3]{-81a^3}$
51. $\sqrt[3]{16x^{12}y^3}$
52. $\sqrt[3]{40a^3b^6}$
53. $\sqrt[4]{32x^{12}y^4}$
54. $\sqrt[5]{64x^{10}y^5}$
55. $\sqrt[5]{a^7}$
56. $\sqrt[5]{b^8}$
57. $\sqrt[6]{m^{11}}$
58. $\sqrt[6]{n^{13}}$
59. $\sqrt[5]{32t^{11}}$
60. $\sqrt[5]{243r^{22}}$
61. $\frac{\sqrt[3]{189a^4}}{\sqrt[3]{7a}}$
62. $\frac{\sqrt[3]{243x^7}}{\sqrt[3]{9x}}$
63. $\frac{\sqrt{98x^3}}{\sqrt{2x}}$
64. $\frac{\sqrt{75y^5}}{\sqrt{3y}}$
65. $\sqrt{\frac{z^2}{16x^2}}$
66. $\sqrt{\frac{b^4}{64a^8}}$
67. $\sqrt[4]{\frac{5x}{16z^4}}$
68. $\sqrt[3]{\frac{11a^2}{125b^6}}$

Simplify and combine like radicals. All variables represent positive numbers.

69. $4\sqrt{2x} + 6\sqrt{2x}$
70. $6\sqrt[3]{5y} + 3\sqrt[3]{5y}$
71. $8\sqrt[5]{7a^2} - 7\sqrt[5]{7a^2}$
72. $10\sqrt[6]{12xyz} - \sqrt[6]{12xyz}$
73. $\sqrt{2} - \sqrt{8}$
74. $\sqrt{20} - \sqrt{125}$
75. $\sqrt{98} - \sqrt{50}$
76. $\sqrt{72} - \sqrt{200}$
77. $3\sqrt{24} + \sqrt{54}$
78. $\sqrt{18} + 2\sqrt{50}$
79. $\sqrt[3]{24x} + \sqrt[3]{3x}$
80. $\sqrt[3]{16y} + \sqrt[3]{128y}$
81. $\sqrt[3]{32} - \sqrt[3]{108}$
82. $\sqrt[3]{80} - \sqrt[3]{10{,}000}$

83. $2\sqrt[3]{125} - 5\sqrt[3]{64}$

84. $3\sqrt[3]{27} + 12\sqrt[3]{216}$

85. $14\sqrt[4]{32} - 15\sqrt[4]{162}$

86. $23\sqrt[4]{768} + \sqrt[4]{48}$

87. $3\sqrt[4]{512} + 2\sqrt[4]{32}$

88. $4\sqrt[4]{243} - \sqrt[4]{48}$

89. $\sqrt{98} - \sqrt{50} - \sqrt{72}$

90. $\sqrt{20} + \sqrt{125} - \sqrt{80}$

91. $\sqrt{18t} + \sqrt{300t} - \sqrt{243t}$

92. $\sqrt{80m} - \sqrt{128m} + \sqrt{288m}$

93. $2\sqrt[3]{16} - \sqrt[3]{54} - 3\sqrt[3]{128}$

94. $\sqrt[4]{48} - \sqrt[4]{243} - \sqrt[4]{768}$

95. $\sqrt{25y^2z} - \sqrt{16y^2z}$

96. $\sqrt{25yz^2} + \sqrt{9yz^2}$

97. $\sqrt{36xy^2} + \sqrt{49xy^2}$

98. $3\sqrt{2x} - \sqrt{8x}$

99. $2\sqrt[3]{64a} + 2\sqrt[3]{8a}$

100. $3\sqrt[4]{x^4y} - 2\sqrt[4]{x^4y}$

101. $\sqrt{y^5} - \sqrt{9y^5} - \sqrt{25y^5}$

102. $\sqrt{8y^7} + \sqrt{32y^7} - \sqrt{2y^7}$

103. $\sqrt[5]{x^6y^2} + \sqrt[5]{32x^6y^2} + \sqrt[5]{x^6y^2}$

104. $\sqrt[3]{xy^4} + \sqrt[3]{8xy^4} - \sqrt[3]{27xy^4}$

APPLICATIONS

First give the exact answer, expressed as a simplified radical expression. Then give an approximation, rounded to the nearest tenth.

105. UMBRELLAS The surface area of a cone is given by the formula $S = \pi r\sqrt{r^2 + h^2}$, where r is the radius of the base and h is its height. Use this formula to find the number of square feet of waterproof cloth used to make the umbrella shown below.

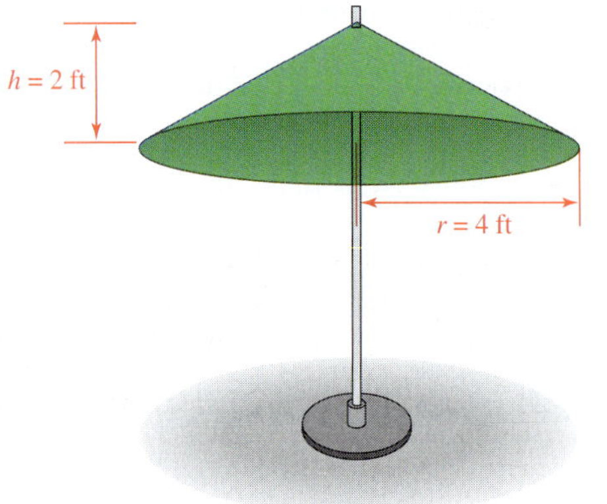

106. STRUCTURAL ENGINEERING Engineers have determined that two additional supports need to be added to strengthen the truss shown. Find the length L of each new support using the formula

$$L = \sqrt{\frac{b^2}{2} + \frac{c^2}{2} - \frac{a^2}{4}}$$

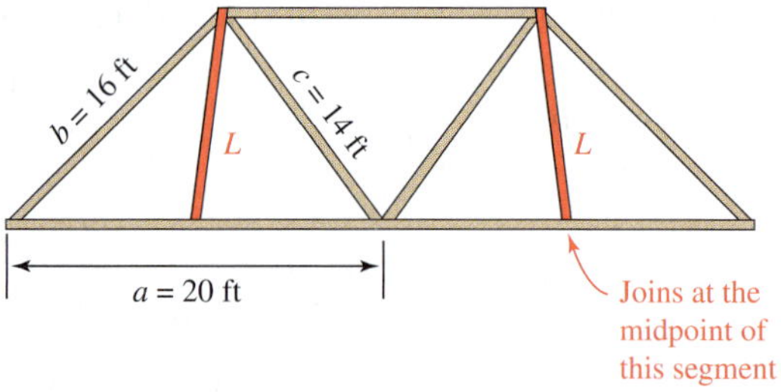

107. BLOW DRYERS The current I (in amps), the power P (in watts), and the resistance R (in ohms) are related by the formula $I = \sqrt{\frac{P}{R}}$. What current is needed for a 1,200-watt hair dryer if the resistance is 16 ohms?

108. COMMUNICATIONS SATELLITES Engineers have determined that a spherical communications satellite needs to have a capacity of 565.2 cubic feet to house all of its operating systems. The volume V of a sphere is related to its radius r by the formula $r = \sqrt[3]{\frac{3V}{4\pi}}$. What radius must the satellite have to meet the engineer's specification? Use 3.14 for π.

109. DUCTWORK The following pattern is laid out on a sheet of galvanized tin. Then it is cut out with snips and bent to make an air conditioning duct connection. Find the total length of the cut that must be made with the tin snips. (All measurements are in inches.)

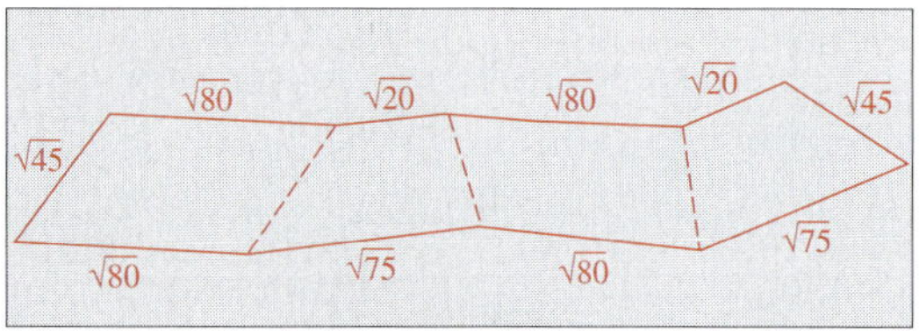

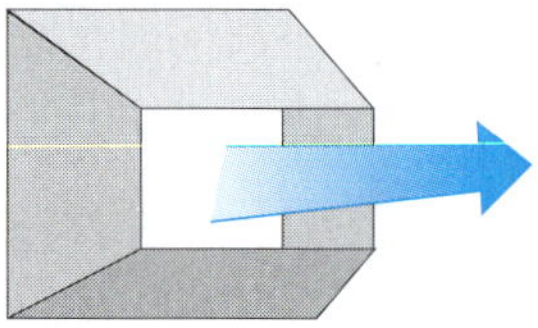

110. OUTDOOR COOKING The diameter of a circle is given by the function $d(A) = 2\sqrt{\frac{A}{\pi}}$, where A is the area of the circle. Find the difference between the diameters of the barbecue grills on the next page.

WRITING

111. Explain why $\sqrt[3]{9x^4}$ is not in simplified form.

112. How are the procedures used to simplify $3x + 4x$ and $3\sqrt{x} + 4\sqrt{x}$ similar?

113. Explain how the graphs of $Y_1 = 3\sqrt{24x} + \sqrt{54x}$ (on the left) and $Y_2 = 9\sqrt{6x}$ (on the right) can be used to verify the simplification $3\sqrt{24x} + \sqrt{54x} = 9\sqrt{6x}$. In each graph, settings of $[-5, 20]$ for x and $[-5, 100]$ for y were used.

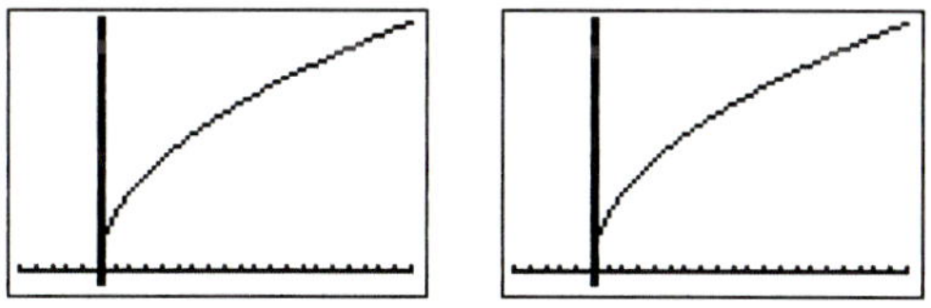

114. Explain how to verify algebraically that

$$\sqrt{200x^3y^5} = 10xy^2\sqrt{2xy}$$

REVIEW Perform each operation.

115. $3x^2y^3(-5x^3y^{-4})$

116. $(2x^2 - 9x - 5) \cdot \dfrac{x}{2x^2 + x}$

117. $2p - 5\overline{)6p^2 - 7p - 25}$

118. $\dfrac{xy}{\frac{1}{x} - \frac{1}{y}}$

CHALLENGE PROBLEMS

119. Can you find any numbers a and b such that $\sqrt{a + b} = \sqrt{a} + \sqrt{b}$?

120. Find the sum:

$$\sqrt{3} + \sqrt{3^2} + \sqrt{3^3} + \sqrt{3^4} + \sqrt{3^5}$$

7.4 Multiplying and Dividing Radical Expressions

- Multiplying radical expressions
- Powers of radical expressions
- Rationalizing denominators
- Rationalizing two-term denominators
- Rationalizing numerators

In this section, we will discuss the methods used to multiply and divide radical expressions.

MULTIPLYING RADICAL EXPRESSIONS

We have used the *product rule for radicals* to write radical expressions in simplified form. We can also use this rule to multiply radical expressions that have the same index.

The Product Rule for Radicals

The product of the nth roots of two nonnegative numbers is equal to the nth root of the product of those numbers.

If $\sqrt[n]{a}$ and $\sqrt[n]{b}$ are real numbers,

$$\sqrt[n]{a} \cdot \sqrt[n]{b} = \sqrt[n]{a \cdot b}$$

EXAMPLE 1

Multiply and then simplify, if possible: **a.** $\sqrt{5}\sqrt{10}$, **b.** $3\sqrt{6}(2\sqrt{3})$, and **c.** $-2\sqrt[3]{7x}\cdot 6\sqrt[3]{49x^2}$.

Solution

a.

$$\begin{aligned}\sqrt{5}\sqrt{10} &= \sqrt{5\cdot 10} && \text{Use the product rule for radicals.}\\ &= \sqrt{50} && \text{Multiply under the radical. Note that } \sqrt{50} \text{ can be simplified.}\\ &= \sqrt{25\cdot 2} && \text{Begin the process of simplifying } \sqrt{50} \text{ by factoring 50.}\\ &= 5\sqrt{2} && \sqrt{25\cdot 2} = \sqrt{25}\sqrt{2} = 5\sqrt{2}.\end{aligned}$$

Caution

Note that to multiply radical expressions, they must have the same index.

$$\sqrt[n]{a}\cdot\sqrt[n]{b} = \sqrt[n]{a\cdot b}$$

b. We use the commutative and associative properties of multiplication to multiply the coefficients and the radicals separately. Then we simplify any radicals in the product, if possible.

$$\begin{aligned}3\sqrt{6}(2\sqrt{3}) &= 3(2)\sqrt{6}\sqrt{3} && \text{Multiply the coefficients and multiply the radicals.}\\ &= 6\sqrt{18} && \text{Use the product rule for radicals.}\\ &= 6\sqrt{9}\sqrt{2} && \text{Simplify: } \sqrt{18} = \sqrt{9\cdot 2} = \sqrt{9}\sqrt{2}.\\ &= 6(3)\sqrt{2} && \sqrt{9} = 3.\\ &= 18\sqrt{2} && \text{Multiply.}\end{aligned}$$

c.

$$\begin{aligned}-2\sqrt[3]{7x}\cdot 6\sqrt[3]{49x^2} &= -2(6)\sqrt[3]{7x}\sqrt[3]{49x^2} && \text{Write the coefficients together and the radicals together.}\\ &= -12\sqrt[3]{7x\cdot 49x^2} && \text{Multiply the coefficients and multiply the radicals.}\\ &= -12\sqrt[3]{7x\cdot 7^2x^2} && \text{Write 49 as } 7^2.\\ &= -12\sqrt[3]{7^3x^3} && \text{Write } 7x\cdot 7^2x^2 \text{ as } 7^3x^3.\\ &= -12(7x) && \text{Simplify: } \sqrt[3]{7^3x^3} = 7x.\\ &= -84x && \text{Multiply.}\end{aligned}$$

Self Check 1 Multiply: **a.** $\sqrt{7}\sqrt{14}$, **b.** $-2\sqrt{7}(5\sqrt{2})$, and **c.** $\sqrt[4]{4x^3}\cdot 9\sqrt[4]{8x^2}$.

Recall that to multiply a polynomial by a monomial, we use the distributive property. We use the same technique to multiply a radical expression that has two or more terms by a radical expression that has only one term.

EXAMPLE 2

Multiply and then simplify, if possible: $3\sqrt{3}(4\sqrt{8} - 5\sqrt{10})$.

Solution

$$\begin{aligned}&3\sqrt{3}(4\sqrt{8} - 5\sqrt{10})\\ &= 3\sqrt{3}\cdot 4\sqrt{8} - 3\sqrt{3}\cdot 5\sqrt{10} && \text{Distribute the multiplication by } 3\sqrt{3}.\\ &= 12\sqrt{24} - 15\sqrt{30} && \text{Multiply the coefficients and multiply the radicals.}\\ &= 12\sqrt{4}\sqrt{6} - 15\sqrt{30} && \text{Simplify: } \sqrt{24} = \sqrt{4\cdot 6} = \sqrt{4}\sqrt{6}.\\ &= 12(2)\sqrt{6} - 15\sqrt{30} && \sqrt{4} = 2.\\ &= 24\sqrt{6} - 15\sqrt{30}\end{aligned}$$

Self Check 2 Simplify: $4\sqrt{2}(3\sqrt{5} - 2\sqrt{8})$.

Recall that to multiply two binomials, we multiply each term of one binomial by each term of the other binomial and simplify. We multiply two radical expressions, each having two terms, in the same way.

EXAMPLE 3

Multiply and then simplify, if possible: **a.** $(\sqrt{7} + \sqrt{2})(\sqrt{7} - 3\sqrt{2})$ and **b.** $(\sqrt[3]{x^2} - 4\sqrt[3]{5})(\sqrt[3]{x} + \sqrt[3]{2})$.

Solution **a.** $(\sqrt{7} + \sqrt{2})(\sqrt{7} - 3\sqrt{2})$

$= \sqrt{7}\sqrt{7} - 3\sqrt{7}\sqrt{2} + \sqrt{2}\sqrt{7} - 3\sqrt{2}\sqrt{2}$ Use the FOIL method.

$= 7 - 3\sqrt{14} + \sqrt{14} - 3(2)$ Perform each multiplication.

$= 7 - 2\sqrt{14} - 6$ Combine like radicals.

$= 1 - 2\sqrt{14}$ Combine like terms.

b. $(\sqrt[3]{x^2} - 4\sqrt[3]{5})(\sqrt[3]{x} + \sqrt[3]{2})$

$= \sqrt[3]{x^2}\sqrt[3]{x} + \sqrt[3]{x^2}\sqrt[3]{2} - 4\sqrt[3]{5}\sqrt[3]{x} - 4\sqrt[3]{5}\sqrt[3]{2}$ Use the FOIL method.

$= \sqrt[3]{x^3} + \sqrt[3]{2x^2} - 4\sqrt[3]{5x} - 4\sqrt[3]{10}$ Perform each multiplication.

$= x + \sqrt[3]{2x^2} - 4\sqrt[3]{5x} - 4\sqrt[3]{10}$ Simplify the first term.

Self Check 3 Multiply: **a.** $(\sqrt{5} + 2\sqrt{3})(\sqrt{5} - \sqrt{3})$ and **b.** $(\sqrt[3]{a} + 9\sqrt[3]{2})(\sqrt[3]{a^2} - \sqrt[3]{3})$.

POWERS OF RADICAL EXPRESSIONS

To find the power of a radical expression, such as $(\sqrt{5})^2$ or $(\sqrt[3]{2})^3$, we can use the definition of exponent and the product rule for radicals.

$$(\sqrt{5})^2 = \sqrt{5} \cdot \sqrt{5} = \sqrt{25} = 5 \qquad (\sqrt[3]{2})^3 = \sqrt[3]{2} \cdot \sqrt[3]{2} \cdot \sqrt[3]{2} = \sqrt[3]{8} = 2$$

These results illustrate the following property of radicals.

The *n*th Power of the *n*th Root

If $\sqrt[n]{a}$ is a real number,

$$(\sqrt[n]{a})^n = a$$

EXAMPLE 4

Find: **a.** $(\sqrt{5})^2$, **b.** $(2\sqrt[3]{7x^2})^3$, and **c.** $(\sqrt{m+1} + 2)^2$, where $m > 0$.

Solution **a.** $(\sqrt{5})^2 = 5$ Because the square of the square root of a positive number is that number.

b. We can use the power of a product rule for exponents to find $(2\sqrt[3]{7x^2})^3$.

$(2\sqrt[3]{7x^2})^3 = 2^3(\sqrt[3]{7x^2})^3$ Raise each factor of $2\sqrt[3]{7x^2}$ to the 3rd power.

$= 8(7x^2)$ Evaluate: $2^3 = 8$. Use $(\sqrt[n]{a})^n = a$.

$= 56x^2$

c. We can use the FOIL method to find the product.

$$\begin{aligned}(\sqrt{m+1}+2)^2 &= (\sqrt{m+1}+2)(\sqrt{m+1}+2)\\ &= (\sqrt{m+1})^2 + 2\sqrt{m+1} + 2\sqrt{m+1} + 2\cdot 2\\ &= m+1 + 2\sqrt{m+1} + 2\sqrt{m+1} + 4 \quad \text{Use } (\sqrt[n]{a})^n = a.\\ &= m + 4\sqrt{m+1} + 5 \quad \text{Combine like terms.}\end{aligned}$$

Self Check 4 Find: **a.** $(\sqrt{11})^2$, **b.** $(3\sqrt[3]{4y})^3$, and **c.** $(\sqrt{x-8}-5)^2$.

RATIONALIZING DENOMINATORS

We have seen that when a radical expression is written in simplified form, each of the following statements is true.

1. Each factor in the radicand is to a power that is less than the index of the radical.
2. The radicand contains no fractions or negative numbers.
3. No radicals appear in the denominator of a fraction.

We now consider radical expressions that do not satisfy requirement 2 and those that do not satisfy requirement 3. We will introduce an algebraic technique, called *rationalizing the denominator,* that is used to write such expressions in an equivalent simplified form.

To divide radical expressions, we **rationalize the denominator** of a fraction to replace the denominator with a rational number. For example, to divide $\sqrt{5}$ by $\sqrt{3}$, we write the division as the fraction

$$\frac{\sqrt{5}}{\sqrt{3}} \quad \text{This radical expression is not in simplified form, because a radical appears in the denominator.}$$

The Language of Algebra
Since $\sqrt{3}$ is an irrational number, the fraction $\frac{\sqrt{5}}{\sqrt{3}}$ has an *irrational* denominator.

We want to find a fraction equivalent to $\frac{\sqrt{5}}{\sqrt{3}}$ that does not have a radical in its denominator. If we multiply $\frac{\sqrt{5}}{\sqrt{3}}$ by $\frac{\sqrt{3}}{\sqrt{3}}$, the denominator becomes $\sqrt{3}\cdot\sqrt{3}=3$, a rational number.

$$\begin{aligned}\frac{\sqrt{5}}{\sqrt{3}} &= \frac{\sqrt{5}}{\sqrt{3}}\cdot\frac{\sqrt{3}}{\sqrt{3}} \quad \text{To build an equivalent fraction, multiply by } \frac{\sqrt{3}}{\sqrt{3}} = 1.\\ &= \frac{\sqrt{15}}{3} \quad \text{Multiply the numerators: } \sqrt{5}\cdot\sqrt{3}=\sqrt{15}. \text{ Multiply the denominators: } \sqrt{3}\cdot\sqrt{3}=(\sqrt{3})^2=3.\end{aligned}$$

Success Tip
As an informal check, we can use a calculator to evaluate each expression.

$$\frac{\sqrt{5}}{\sqrt{3}} \approx 1.290994449$$
$$\frac{\sqrt{15}}{3} \approx 1.290994449$$

Thus, $\frac{\sqrt{5}}{\sqrt{3}} = \frac{\sqrt{15}}{3}$. These equivalent fractions represent the same number, but have different forms. Since there is no radical in the denominator, and $\sqrt{15}$ is in simplest form, the division is complete.

EXAMPLE 5 Simplify by rationalizing the denominator:

a. $\sqrt{\frac{20}{7}}$ and **b.** $\frac{4}{\sqrt[3]{2}}$.

Solution **a.** This radical expression is not in simplified form, because the radicand contains a fraction. We begin by writing the square root of the quotient as the quotient of two square roots:

$$\sqrt{\frac{20}{7}} = \frac{\sqrt{20}}{\sqrt{7}} \qquad \text{Use the division property of radicals: } \sqrt[n]{\frac{a}{b}} = \frac{\sqrt[n]{a}}{\sqrt[n]{b}}.$$

To rationalize the denominator, we proceed as follows:

$$\frac{\sqrt{20}}{\sqrt{7}} = \frac{\sqrt{20}}{\sqrt{7}} \cdot \frac{\sqrt{7}}{\sqrt{7}} \qquad \text{To build an equivalent fraction, multiply by } \frac{\sqrt{7}}{\sqrt{7}} = 1.$$

$$= \frac{\sqrt{140}}{7} \qquad \text{Multiply the numerators. Multiply the denominators: } (\sqrt{7})^2 = 7.$$

$$= \frac{2\sqrt{35}}{7} \qquad \text{Simplify: } \sqrt{140} = \sqrt{4 \cdot 35} = \sqrt{4}\sqrt{35} = 2\sqrt{35}.$$

Caution

Do not attempt to remove a common factor of 7 from the numerator and denominator of $\frac{2\sqrt{35}}{7}$. The numerator, $2\sqrt{35}$, does not have a factor of 7.

$$\frac{2\sqrt{35}}{7} = \frac{2 \cdot \sqrt{5 \cdot 7}}{7}$$

b. This expression is not in simplified form because a radical appears in the denominator of a fraction. Here, we must rationalize a denominator that is a cube root. We multiply the numerator and the denominator by a number that will give a perfect cube under the radical. Since $2 \cdot 4 = 8$ is a perfect cube, $\sqrt[3]{4}$ is such a number.

$$\frac{4}{\sqrt[3]{2}} = \frac{4}{\sqrt[3]{2}} \cdot \frac{\sqrt[3]{4}}{\sqrt[3]{4}} \qquad \text{To build an equivalent fraction, multiply by } \frac{\sqrt[3]{4}}{\sqrt[3]{4}} = 1.$$

$$= \frac{4\sqrt[3]{4}}{\sqrt[3]{8}} \qquad \text{Multiply the numerators. Multiply the denominators. This radicand is now a perfect cube.}$$

$$= \frac{4\sqrt[3]{4}}{2} \qquad \text{Simplify: } \sqrt[3]{8} = 2.$$

$$= 2\sqrt[3]{4} \qquad \text{Simplify: } \frac{4\sqrt[3]{4}}{2} = \frac{\overset{1}{\cancel{2}} \cdot 2\sqrt[3]{4}}{\underset{1}{\cancel{2}}} = 2\sqrt[3]{4}.$$

Caution

Note that multiplying $\frac{4}{\sqrt[3]{2}}$ by $\frac{\sqrt[3]{2}}{\sqrt[3]{2}}$ does not rationalize the denominator.

$$\frac{4}{\sqrt[3]{2}} \cdot \frac{\sqrt[3]{2}}{\sqrt[3]{2}} = \frac{4\sqrt[3]{2}}{\sqrt[3]{4}}$$

Since 4 is not a perfect cube, this radical does not simplify.

Self Check 5 Rationalize the denominator: **a.** $\sqrt{\frac{24}{5}}$ and **b.** $\frac{5}{\sqrt[4]{3}}$.

EXAMPLE 6

Rationalize the denominator:

$$\frac{\sqrt{5xy^2}}{\sqrt{xy^3}}$$

Solution Two possible methods for rationalizing the denominator are shown on the next page. In each case, we simplify the expression first.

Caution

We will assume that all of the variables appearing in the following examples represent positive numbers.

Method 1

$$\frac{\sqrt{5xy^2}}{\sqrt{xy^3}} = \sqrt{\frac{5xy^2}{xy^3}}$$

$$= \sqrt{\frac{5}{y}}$$

$$= \frac{\sqrt{5}}{\sqrt{y}}$$

$$= \frac{\sqrt{5}}{\sqrt{y}} \cdot \frac{\sqrt{y}}{\sqrt{y}} \quad \text{Multiply outside the radical.}$$

$$= \frac{\sqrt{5y}}{y}$$

Method 2

$$\frac{\sqrt{5xy^2}}{\sqrt{xy^3}} = \sqrt{\frac{5xy^2}{xy^3}}$$

$$= \sqrt{\frac{5}{y}}$$

$$= \sqrt{\frac{5}{y} \cdot \frac{y}{y}} \quad \text{Multiply within the radical.}$$

$$= \frac{\sqrt{5y}}{\sqrt{y^2}}$$

$$= \frac{\sqrt{5y}}{y}$$

Self Check 6 Rationalize the denominator: $\frac{\sqrt{4ab^3}}{\sqrt{2a^2b^2}}$.

EXAMPLE 7

Rationalize the denominator: $\frac{11}{\sqrt{20q^5}}$.

Solution We could begin by multiplying $\frac{11}{\sqrt{20q^5}}$ by $\frac{\sqrt{20q^5}}{\sqrt{20q^5}}$. However, to work with smaller numbers, it is easier if we simplify $\sqrt{20q^5}$ first, and then rationalize the denominator.

Success Tip

We usually simplify a radical expression before rationalizing the denominator.

$$\frac{11}{\sqrt{20q^5}} = \frac{11}{\sqrt{4q^4 \cdot 5q}} \quad \text{To simplify } \sqrt{20q^5}\text{, factor } 20q^5 \text{ as } 4q^4 \cdot 5q.$$

$$= \frac{11}{2q^2\sqrt{5q}} \quad \sqrt{4q^4 \cdot 5q} = \sqrt{4q^4}\sqrt{5q} = 2q^2\sqrt{5q}.$$

$$= \frac{11}{2q^2\sqrt{5q}} \cdot \frac{\sqrt{5q}}{\sqrt{5q}} \quad \text{To rationalize the denominator, multiply by } \frac{\sqrt{5q}}{\sqrt{5q}} = 1.$$

$$= \frac{11\sqrt{5q}}{2q^2(5q)} \quad \text{Multiply the numerators. Multiply the denominators: } (\sqrt{5q})^2 = 5q.$$

$$= \frac{11\sqrt{5q}}{10q^3} \quad \text{Multiply in the denominator.}$$

Self Check 7 Rationalize the denominator: $\sqrt[3]{\frac{1}{16h^4}}$.

EXAMPLE 8

Rationalize each denominator:

a. $\frac{5}{\sqrt[3]{6mn^2}}$ and **b.** $\frac{\sqrt[4]{b}}{\sqrt[4]{9a}}$.

Solution **a.** To rationalize the denominator $\sqrt[3]{6mn^2}$, we need the radicand to be a perfect cube. Since $6mn^2 = 6 \cdot m \cdot n \cdot n$, the radicand needs two more factors of 6, two more factors of m, and one more factor of n. It follows that we should multiply the given expression by $\frac{\sqrt[3]{36m^2n}}{\sqrt[3]{36m^2n}}$.

$$\frac{5}{\sqrt[3]{6mn^2}} = \frac{5}{\sqrt[3]{6mn^2}} \cdot \frac{\sqrt[3]{36m^2n}}{\sqrt[3]{36m^2n}}$$ Multiply by a form of 1 to rationalize the denominator.

$$= \frac{5\sqrt[3]{36m^2n}}{\sqrt[3]{216m^3n^3}}$$ Multiply the numerators. Multiply the denominators. This radicand is now a perfect cube.

$$= \frac{5\sqrt[3]{36m^2n}}{6mn}$$ Simplify: $\sqrt[3]{216m^3n^3} = 6mn$.

b. To rationalize the denominator $\sqrt[4]{9a}$, we need the radicand to be a perfect fourth power. Since $9a = 3 \cdot 3 \cdot a$, the radicand needs two more factors of 3 and three more factors of a. It follows that we should multiply the given expression by $\frac{\sqrt[4]{9a^3}}{\sqrt[4]{9a^3}}$.

$$\frac{\sqrt[4]{b}}{\sqrt[4]{9a}} = \frac{\sqrt[4]{b}}{\sqrt[4]{9a}} \cdot \frac{\sqrt[4]{9a^3}}{\sqrt[4]{9a^3}}$$ Multiply by a form of 1 to rationalize the denominator.

$$= \frac{\sqrt[4]{9a^3b}}{\sqrt[4]{81a^4}}$$ Multiply the numerators. Multiply the denominators. This radicand is now a perfect fourth power.

$$= \frac{\sqrt[4]{9a^3b}}{3a}$$ Simplify: $\sqrt[4]{81a^4} = 3a$.

Self Check 8 Rationalize each denominator: **a.** $\frac{11}{\sqrt[3]{100ab^2}}$ and **b.** $\frac{\sqrt[4]{s}}{\sqrt[4]{x^3y^2}}$.

RATIONALIZING TWO-TERM DENOMINATORS

So far, we have rationalized denominators that had only one term. We will now discuss a method to rationalize denominators that have two terms.

One-term denominators

$$\frac{\sqrt{5}}{\sqrt{3}}, \quad \frac{11}{\sqrt{20q^5}}, \quad \frac{4}{\sqrt[3]{2}}$$

Two-term denominators

$$\frac{1}{\sqrt{2}+1}, \quad \frac{\sqrt{x}+\sqrt{2}}{\sqrt{x}-\sqrt{2}}$$

Success Tip

Recall the special product formula for finding the product of the sum and difference of two terms:

$$(x + y)(x - y) = x^2 - y^2$$

To rationalize the denominator of $\frac{1}{\sqrt{2}+1}$, for example, we multiply the numerator and denominator by $\sqrt{2} - 1$, because the product $(\sqrt{2} + 1)(\sqrt{2} - 1)$ contains no radicals.

$$(\sqrt{2} + 1)(\sqrt{2} - 1) = (\sqrt{2})^2 - (1)^2$$ Use a special product formula.

$$= 2 - 1$$

$$= 1$$

Radical expressions that involve the sum and difference of the same two terms, such as $\sqrt{2} + 1$ and $\sqrt{2} - 1$, are called **conjugates**.

EXAMPLE 9

Rationalize the denominator:

a. $\dfrac{1}{\sqrt{2}+1}$ and **b.** $\dfrac{\sqrt{x}+\sqrt{2}}{\sqrt{x}-\sqrt{2}}$.

Solution **a.** To find a fraction equivalent to $\frac{1}{\sqrt{2}+1}$ that does not have a radical in its denominator, we multiply $\frac{1}{\sqrt{2}+1}$ by a form of 1 that uses the conjugate of $\sqrt{2}+1$.

$$\frac{1}{\sqrt{2}+1} = \frac{1}{\sqrt{2}+1}\cdot\frac{\sqrt{2}-1}{\sqrt{2}-1}$$

$$= \frac{\sqrt{2}-1}{(\sqrt{2})^2-(1)^2}$$ Multiply the numerators. Multiply the denominators.

$$= \frac{\sqrt{2}-1}{2-1}$$

$$= \frac{\sqrt{2}-1}{1}$$

$$= \sqrt{2}-1$$

b. We multiply the numerator and denominator by $\sqrt{x}+\sqrt{2}$, which is the conjugate of $\sqrt{x}-\sqrt{2}$, and simplify.

$$\frac{\sqrt{x}+\sqrt{2}}{\sqrt{x}-\sqrt{2}} = \frac{\sqrt{x}+\sqrt{2}}{\sqrt{x}-\sqrt{2}}\cdot\frac{\sqrt{x}+\sqrt{2}}{\sqrt{x}+\sqrt{2}}$$

$$= \frac{x+\sqrt{2x}+\sqrt{2x}+2}{(\sqrt{x})^2-(\sqrt{2})^2}$$ Multiply the numerators. Multiply the denominators.

$$= \frac{x+\sqrt{2x}+\sqrt{2x}+2}{x-2}$$

$$= \frac{x+2\sqrt{2x}+2}{x-2}$$ In the numerator, combine like radicals.

Self Check 9 Rationalize the denominator: $\dfrac{\sqrt{x}-\sqrt{2}}{\sqrt{x}+\sqrt{2}}$.

RATIONALIZING NUMERATORS

In calculus, we sometimes have to rationalize a numerator by multiplying the numerator and denominator of the fraction by the conjugate of the numerator.

EXAMPLE 10

Rationalize the numerator:

$$\frac{\sqrt{x}-3}{\sqrt{x}}$$

Solution We multiply the numerator and denominator by $\sqrt{x} + 3$, which is the conjugate of the numerator.

$$\frac{\sqrt{x} - 3}{\sqrt{x}} = \frac{\sqrt{x} - 3}{\sqrt{x}} \cdot \frac{\sqrt{x} + 3}{\sqrt{x} + 3}$$ Multiply by a form of 1 to rationalize the numerator.

$$= \frac{(\sqrt{x})^2 - (3)^2}{x + 3\sqrt{x}}$$ Multiply the numerators. Multiply the denominators.

$$= \frac{x - 9}{x + 3\sqrt{x}}$$

The final expression is not in simplified form. However, this nonsimplified form is often desirable in calculus.

Self Check 10 Rationalize the numerator: $\frac{\sqrt{x} + 3}{\sqrt{x}}$.

Answers to Self Checks **1. a.** $7\sqrt{2}$, **b.** $-10\sqrt{14}$, **c.** $18x\sqrt[4]{2x}$ **2.** $12\sqrt{10} - 32$ **3. a.** $-1 + \sqrt{15}$, **b.** $a - \sqrt[3]{3a} + 9\sqrt[3]{2a^2} - 9\sqrt[3]{6}$ **4. a.** 11, **b.** $108y$, **c.** $x - 10\sqrt{x - 8} + 17$ **5. a.** $\frac{2\sqrt{30}}{5}$, **b.** $\frac{5\sqrt[4]{27}}{3}$ **6.** $\frac{\sqrt{2ab}}{a}$ **7.** $\frac{\sqrt[3]{4h^2}}{4h^2}$ **8. a.** $\frac{11\sqrt[3]{10a^2b}}{10ab}$, **b.** $\frac{\sqrt[4]{sxy^2}}{xy}$ **9.** $\frac{x - 2\sqrt{2x} + 2}{x - 2}$ **10.** $\frac{x - 9}{x - 3\sqrt{x}}$

7.4 STUDY SET

VOCABULARY Fill in the blanks.

1. To multiply $(\sqrt{3} + \sqrt{2})(\sqrt{3} - 2\sqrt{2})$, we can use the _______ method.

2. To multiply $2\sqrt{5}(3\sqrt{8} + \sqrt{3})$, use the __________ property to remove parentheses.

3. The denominator of the fraction $\frac{4}{\sqrt{5}}$ is an _________ number.

4. The __________ of $\sqrt{x} + 1$ is $\sqrt{x} - 1$.

5. To obtain a _______ cube radicand in the denominator of $\frac{\sqrt[3]{7}}{\sqrt[3]{5n}}$, we multiply the fraction by $\frac{\sqrt[3]{25n^2}}{\sqrt[3]{25n^2}}$.

6. To __________ the denominator of $\frac{4}{\sqrt{5}}$, we multiply the fraction by $\frac{\sqrt{5}}{\sqrt{5}}$.

CONCEPTS

7. Perform each operation, if possible.

a. $4\sqrt{6} + 2\sqrt{6}$ **b.** $4\sqrt{6}(2\sqrt{6})$

c. $3\sqrt{2} - 2\sqrt{3}$ **d.** $3\sqrt{2}(-2\sqrt{3})$

8. Perform each operation, if possible.

a. $5 + 6\sqrt[3]{6}$ **b.** $5(6\sqrt[3]{6})$

c. $\frac{30\sqrt[3]{15}}{5}$ **d.** $\frac{\sqrt[3]{15}}{5}$

9. Consider $\frac{\sqrt{3}}{\sqrt{7}} = \frac{\sqrt{3}}{\sqrt{7}} \cdot \frac{\sqrt{7}}{\sqrt{7}}$. Explain why the expressions on the left-hand side and the right-hand side of the equation are equal.

10. To rationalize the denominator of $\frac{\sqrt[4]{2}}{\sqrt[4]{3}}$, why wouldn't we multiply the numerator and denominator by $\frac{\sqrt[4]{3}}{\sqrt[4]{3}}$?

11. Explain why $\frac{\sqrt[3]{12}}{\sqrt[3]{5}}$ is not in simplified form.

12. Explain why $\sqrt{\frac{3a}{11k}}$ is not in simplified form.

NOTATION **Fill in the blanks.**

13. Multiply: $5\sqrt{8} \cdot 7\sqrt{6}$.

$$5\sqrt{8} \cdot 7\sqrt{6} = 5(7)\sqrt{8\,\square}$$
$$= 35\sqrt{\square}$$
$$= 35\sqrt{\square \cdot 3}$$
$$= 35(\square)\sqrt{3}$$
$$= 140\sqrt{3}$$

14. Rationalize the denominator: $\dfrac{9}{\sqrt[3]{4a^2}}$.

$$\frac{9}{\sqrt[3]{4a^2}} = \frac{9}{\sqrt[3]{4a^2}} \cdot \frac{\sqrt[3]{2a}}{\square}$$
$$= \frac{9\sqrt[3]{2a}}{\sqrt[3]{\square}}$$
$$= \frac{9\sqrt[3]{2a}}{\square}$$

PRACTICE **Perform each multiplication and simplify, if possible. All variables represent positive real numbers.**

15. $\sqrt{11}\sqrt{11}$
16. $\sqrt{35}\sqrt{35}$
17. $(\sqrt{7})^2$
18. $(\sqrt{23})^2$
19. $\sqrt{2}\sqrt{8}$
20. $\sqrt{3}\sqrt{27}$
21. $\sqrt{5} \cdot \sqrt{10}$
22. $\sqrt{7} \cdot \sqrt{35}$
23. $2\sqrt{3}\sqrt{6}$
24. $-3\sqrt{11}\sqrt{33}$
25. $\sqrt[3]{5}\sqrt[3]{25}$
26. $-\sqrt[3]{7}\sqrt[3]{49}$
27. $(3\sqrt{2})^2$
28. $(2\sqrt{5})^2$
29. $(-2\sqrt{2})^2$
30. $(-3\sqrt{10})^2$
31. $(3\sqrt[3]{9})(2\sqrt[3]{3})$
32. $(2\sqrt[3]{16})(-\sqrt[3]{4})$
33. $\sqrt[3]{2} \cdot \sqrt[3]{12}$
34. $\sqrt[3]{3} \cdot \sqrt[3]{18}$
35. $\sqrt{ab^3}(\sqrt{ab})$
36. $\sqrt{8x}(\sqrt{2x^3y})$
37. $\sqrt{5ab}\sqrt{5a}$
38. $\sqrt{15rs^2}\sqrt{10r}$
39. $-4\sqrt[3]{5r^2s}(5\sqrt[3]{2r})$
40. $-\sqrt[3]{3xy^2}(-\sqrt[3]{9x^3})$
41. $\sqrt{x(x+3)}\sqrt{x^3(x+3)}$
42. $\sqrt{y^2(x+y)}\sqrt{(x+y)^3}$
43. $(\sqrt[3]{9b})^3$
44. $(\sqrt[3]{5c^2})^3$
45. $\sqrt[4]{2a^3} \cdot \sqrt[4]{3a^2b}$
46. $\sqrt[4]{9m^3n} \cdot \sqrt[4]{2mn^5}$
47. $\sqrt[5]{2t}(\sqrt[5]{16t})$
48. $\sqrt[5]{27y^3}(\sqrt[5]{9y^4})$

Perform each multiplication and simplify. All variables represent positive real numbers.

49. $3\sqrt{5}(4 - \sqrt{5})$
50. $2\sqrt{7}(3\sqrt{7} - 1)$
51. $3\sqrt{2}(4\sqrt{6} + 2\sqrt{7})$
52. $-\sqrt{3}(\sqrt{7} - \sqrt{15})$
53. $-2\sqrt{5x}(4\sqrt{2x} - 3\sqrt{3})$
54. $3\sqrt{7t}(2\sqrt{7t} + 3\sqrt{3t^2})$
55. $(\sqrt{2} + 1)(\sqrt{2} - 3)$
56. $(2\sqrt{3} + 1)(\sqrt{3} - 1)$
57. $(\sqrt[3]{5z} + \sqrt[3]{3})(\sqrt[3]{5z} + 2\sqrt[3]{3})$
58. $(\sqrt[3]{3p} - 2\sqrt[3]{2})(\sqrt[3]{3p} + \sqrt[3]{2})$
59. $(\sqrt{3x} - \sqrt{2y})(\sqrt{3x} + \sqrt{2y})$
60. $(\sqrt{3m} + \sqrt{2n})(\sqrt{3m} - \sqrt{2n})$
61. $(2\sqrt{3a} - \sqrt{b})(\sqrt{3a} + 3\sqrt{b})$
62. $(5\sqrt{p} - \sqrt{3q})(\sqrt{p} + 2\sqrt{3q})$
63. $(3\sqrt{2r} - 2)^2$
64. $(2\sqrt{3t} + 5)^2$
65. $-2(\sqrt{3x} + \sqrt{3})^2$
66. $3(\sqrt{5x} - \sqrt{3})^2$
67. $(2\sqrt[3]{4a^2} + 1)(\sqrt[3]{4a^2} - 3)$
68. $(\sqrt[3]{9b^2} - 5)(\sqrt[3]{9b^2} - 2)$

Simplify each radical expression by rationalizing the denominator. All variables represent positive real numbers.

69. $\sqrt{\dfrac{1}{7}}$
70. $\sqrt{\dfrac{5}{3}}$
71. $\dfrac{6}{\sqrt{30}}$
72. $\dfrac{8}{\sqrt{10}}$
73. $\dfrac{\sqrt{5}}{\sqrt{8}}$
74. $\dfrac{\sqrt{3}}{\sqrt{50}}$
75. $\dfrac{1}{\sqrt[3]{2}}$
76. $\dfrac{2}{\sqrt[3]{6}}$
77. $\dfrac{3}{\sqrt[3]{9}}$
78. $\dfrac{2}{\sqrt[3]{a}}$
79. $\dfrac{\sqrt[3]{2}}{\sqrt[3]{9}}$
80. $\dfrac{\sqrt[3]{9}}{\sqrt[3]{54}}$
81. $\dfrac{\sqrt{8}}{\sqrt{xy}}$
82. $\dfrac{\sqrt{9xy}}{\sqrt{3x^2y}}$

83. $\dfrac{\sqrt{10xy^2}}{\sqrt{2xy^3}}$

84. $\dfrac{\sqrt{5ab^2c}}{\sqrt{10abc}}$

85. $\dfrac{\sqrt[3]{4a^2}}{\sqrt[3]{2ab}}$

86. $\dfrac{\sqrt[3]{9x}}{\sqrt[3]{3xy}}$

87. $\dfrac{1}{\sqrt[4]{4}}$

88. $\dfrac{1}{\sqrt[5]{2}}$

89. $\dfrac{1}{\sqrt[5]{16}}$

90. $\dfrac{4}{\sqrt[4]{32}}$

91. $\dfrac{\sqrt[4]{s}}{\sqrt[4]{3t^2}}$

92. $\dfrac{\sqrt[4]{c^2}}{\sqrt[4]{5b^3}}$

93. $\dfrac{t}{\sqrt[5]{27a}}$

94. $\dfrac{n}{\sqrt[5]{8m^2}}$

Rationalize each denominator. All variables represent positive real numbers.

95. $\dfrac{\sqrt{2}}{\sqrt{5}+3}$

96. $\dfrac{\sqrt{3}}{\sqrt{3}-2}$

97. $\dfrac{\sqrt{7}-\sqrt{2}}{\sqrt{2}+\sqrt{7}}$

98. $\dfrac{\sqrt{3}+\sqrt{2}}{\sqrt{3}-\sqrt{2}}$

99. $\dfrac{3\sqrt{2}-5\sqrt{3}}{2\sqrt{3}-3\sqrt{2}}$

100. $\dfrac{3\sqrt{6}+5\sqrt{5}}{2\sqrt{5}-3\sqrt{6}}$

101. $\dfrac{2}{\sqrt{x}+1}$

102. $\dfrac{3}{\sqrt{x}-2}$

103. $\dfrac{2z-1}{\sqrt{2z}-1}$

104. $\dfrac{3t-1}{\sqrt{3t}+1}$

105. $\dfrac{\sqrt{x}-\sqrt{y}}{\sqrt{x}+\sqrt{y}}$

106. $\dfrac{\sqrt{x}+\sqrt{y}}{\sqrt{x}-\sqrt{y}}$

107. $\dfrac{\sqrt{a}+3\sqrt{b}}{\sqrt{a}-3\sqrt{b}}$

108. $\dfrac{4\sqrt{m}-2\sqrt{n}}{\sqrt{n}+4\sqrt{m}}$

Rationalize each numerator. All variables represent positive numbers.

109. $\dfrac{\sqrt{x}+3}{x}$

110. $\dfrac{2+\sqrt{x}}{5x}$

111. $\dfrac{\sqrt{x}+\sqrt{y}}{\sqrt{x}}$

112. $\dfrac{\sqrt{x}-\sqrt{y}}{\sqrt{x}+\sqrt{y}}$

APPLICATIONS

113. STATISTICS An example of a normal distribution curve, or *bell-shaped* curve, is shown below. A fraction that is part of the equation that models this curve is

$$\frac{1}{\sigma\sqrt{2\pi}}$$

where σ is a letter from the Greek alphabet. Rationalize the denominator of the fraction.

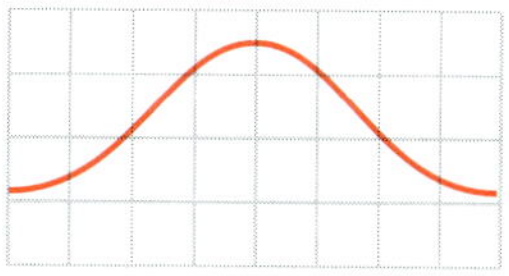

114. ANALYTIC GEOMETRY The length of the perpendicular segment drawn from $(-2, 2)$ to the line with equation $2x - 4y = 4$ is given by

$$L = \frac{|2(-2) + (-4)(2) + (-4)|}{\sqrt{(2)^2 + (-4)^2}}$$

Find L. Express the result in simplified radical form. Then give an approximation to the nearest tenth.

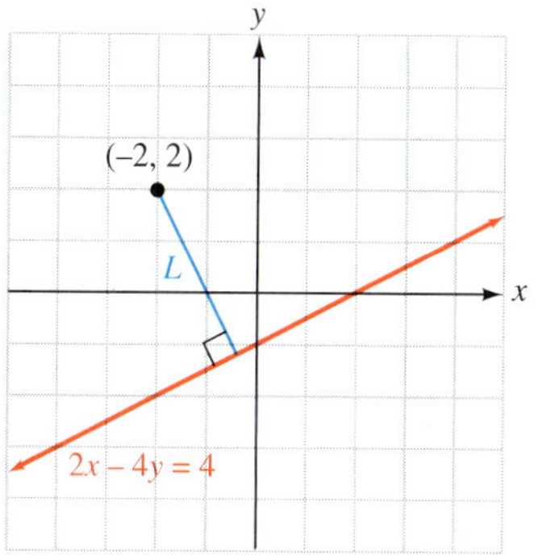

115. TRIGONOMETRY In trigonometry, we must often find the ratio of the lengths of two sides of right triangles. Use the information in the illustration to find the ratio

$$\frac{\text{length of side } AC}{\text{length of side } AB}$$

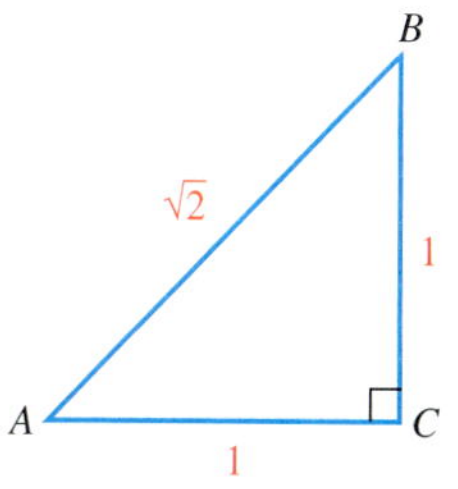

Write the result in simplified radical form.

116. ENGINEERING A measure of how fast the block shown below will oscillate when the system is set in motion is given by the formula

$$\omega = \sqrt{\frac{k_1 + k_2}{m}}$$

where k_1 and k_2 indicate the stiffness of the springs and m is the mass of the block. Rationalize the right-hand side and restate the formula.

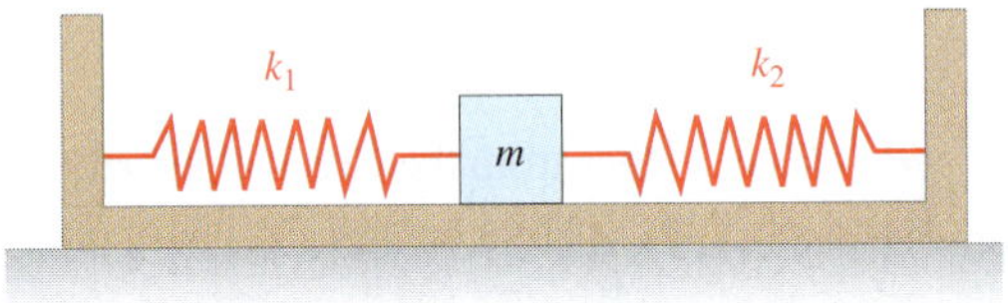

WRITING

117. Explain why $\sqrt{m} \cdot \sqrt{m} = m$ but $\sqrt[3]{m} \cdot \sqrt[3]{m} \neq m$. (Assume that $m > 0$.)

118. Explain why the product of $\sqrt{m} + 3$ and $\sqrt{m} - 3$ does not contain a radical.

REVIEW **Solve each equation.**

119. $\dfrac{8}{b-2} + \dfrac{3}{2-b} = -\dfrac{1}{b}$

120. $\dfrac{2}{x-2} + \dfrac{1}{x+1} = \dfrac{1}{(x+1)(x-2)}$

CHALLENGE PROBLEMS

121. Multiply: $\sqrt{2} \cdot \sqrt[3]{2}$. (*Hint:* Keep in mind two things. The indices (plural for index) must be the same to use the product rule for radicals, and radical expressions can be written using rational exponents.)

122. Show that $\dfrac{\sqrt[3]{a^2} + \sqrt[3]{a}\sqrt[3]{b} + \sqrt[3]{b^2}}{\sqrt[3]{a^2} + \sqrt[3]{a}\sqrt[3]{b} + \sqrt[3]{b^2}}$ can be used to rationalize the denominator of $\dfrac{1}{\sqrt[3]{a} - \sqrt[3]{b}}$.

7.5 Solving Radical Equations

- The power rule
- Equations containing one radical
- Equations containing two radicals
- Solving formulas containing radicals

When we solve equations containing fractions, we clear them of the fractions by multiplying both sides by the LCD. To solve equations containing radical expressions, we take a similar approach. The first step is to clear them of the radicals. To do this, we raise both sides to a power.

THE POWER RULE

Radical equations contain a radical expression with a variable radicand. Some examples are

$$\sqrt{x+3} = 4 \qquad \sqrt[3]{x^3+7} = x + 1 \qquad \sqrt{x} + \sqrt{x+2} = 2$$

To solve equations containing radicals, we will use the *power rule.*

The Power Rule If we raise two equal quantities to the same power, the results are equal quantities.

If x, y, and n are real numbers and $x = y$, then

$$x^n = y^n$$

If both sides of an equation are raised to the same power, all solutions of the original equation are also solutions of the new equation. However, the resulting equation might not be equivalent to the original equation. For example, if we square both sides of the equation

(1) $x = 3$

with a solution set of $\{3\}$

we obtain the equation

(2) $x^2 = 9$

with a solution set of $\{3, -3\}$.

Equations 1 and 2 are not equivalent, because they have different solution sets, and the solution -3 of Equation 2 does not satisfy Equation 1. Since raising both sides of an equation to the same power can produce an equation with apparent solutions that don't satisfy the original equation, we must always check each apparent solution in the original equation and discard any *extraneous solutions.*

EQUATIONS CONTAINING ONE RADICAL

To develop a method for solving any radical equation, let's see how the power rule can be used to solve an equation that contains a square root.

EXAMPLE 1

Solve: $\sqrt{x+3} = 4$.

Solution

To eliminate the radical, we use the power rule by squaring both sides of the equation and proceeding as follows:

The Language of Algebra

When we square both sides of an equation, we are *raising both sides* to the second power.

$$\sqrt{x+3} = 4$$
$$\left(\sqrt{x+3}\right)^2 = (4)^2 \quad \text{Square both sides.}$$
$$x + 3 = 16$$
$$x = 13 \quad \text{Subtract 3 from both sides.}$$

We must check the apparent solution 13 to see whether it satisfies the original equation.

Check:

$$\sqrt{x+3} = 4$$
$$\sqrt{13+3} \stackrel{?}{=} 4 \quad \text{Substitute 13 for } x.$$
$$\sqrt{16} \stackrel{?}{=} 4$$
$$4 = 4$$

Since 13 satisfies the original equation, it is the solution. The solution set is $\{13\}$.

Self Check 1 Solve: $\sqrt{a-2} = 3$.

The method used in Example 1 to solve a radical equation containing a square root can be generalized, as follows.

Solving an Equation Containing Radicals

1. Isolate one radical expression on one side of the equation.
2. Raise both sides of the equation to the power that is the same as the index of the radical.
3. Solve the resulting equation. If it still contains a radical, go back to step 1.
4. Check the results to eliminate extraneous solutions.

EXAMPLE 2

Amusement park rides. The distance d in feet that an object will fall in t seconds is given by the formula

$$t = \sqrt{\frac{d}{16}}$$

If the designers of the amusement park attraction want the riders to experience 3 seconds of vertical free fall, what length of vertical drop is needed?

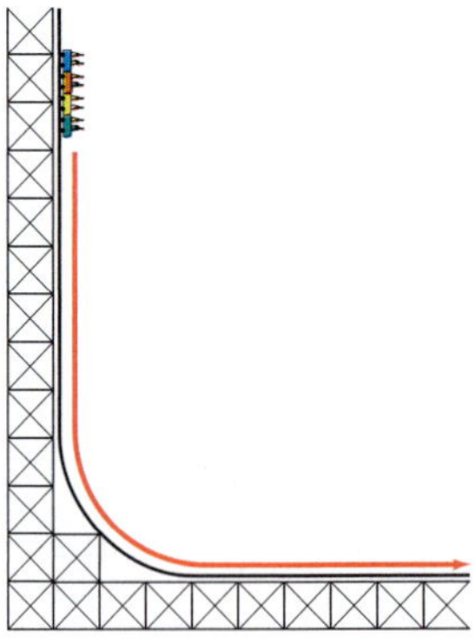

Solution We substitute 3 for t in the formula and solve for d.

$$t = \sqrt{\frac{d}{16}}$$

$$3 = \sqrt{\frac{d}{16}}$$ Here the radical is isolated on the right-hand side.

$$(3)^2 = \left(\sqrt{\frac{d}{16}}\right)^2$$ Use the power rule: Raise both sides to the second power.

$$9 = \frac{d}{16}$$ Simplify.

$$144 = d$$ Solve the resulting equation by multiplying both sides by 16.

The amount of vertical drop needs to be 144 feet.

Caution

When using the power rule, don't forget to raise both sides to the same power. For this example, a common error would be to write

$$3 = \left(\sqrt{\frac{d}{16}}\right)^2$$

Self Check 2 How long a vertical drop is needed if the riders are to free fall for 3.5 seconds?

EXAMPLE 3

Solve: $\sqrt{3x + 1} + 1 = x$.

Solution We first subtract 1 from both sides to isolate the radical. Then, to eliminate the radical, we square both sides of the equation and proceed as follows:

$$\sqrt{3x + 1} + 1 = x$$

$$\sqrt{3x + 1} = x - 1$$ Subtract 1 from both sides.

$$\left(\sqrt{3x + 1}\right)^2 = (x - 1)^2$$ Square both sides to eliminate the square root.

$$3x + 1 = x^2 - 2x + 1$$ On the right-hand side, use the FOIL method: $(x - 1)^2 = (x - 1)(x - 1) = x^2 - x - x + 1 = x^2 - 2x + 1$.

$$0 = x^2 - 5x$$ Subtract $3x$ and 1 from both sides. This is a quadratic equation. Use factoring to solve it.

$$0 = x(x - 5)$$ Factor $x^2 - 5x$.

$$x = 0 \quad \text{or} \quad x - 5 = 0$$ Set each factor equal to 0.

$$x = 0 \quad \Big| \quad x = 5$$

We must check each apparent solution to see whether it satisfies the original equation.

Notation

We have seen that expressions with rational exponents can be written as radical expressions. The equation in this example,

$$\sqrt{3x + 1} + 1 = x$$

could also be written as

$$(3x + 1)^{1/2} + 1 = x$$

$$\textit{Check:}\quad \begin{aligned} \sqrt{3x+1}+1 &= x \\ \sqrt{3(0)+1}+1 &\stackrel{?}{=} 0 \\ \sqrt{1}+1 &\stackrel{?}{=} 0 \\ 2 &\neq 0 \end{aligned} \qquad \begin{aligned} \sqrt{3x+1}+1 &= x \\ \sqrt{3(5)+1}+1 &\stackrel{?}{=} 5 \\ \sqrt{16}+1 &\stackrel{?}{=} 5 \\ 5 &= 5 \end{aligned}$$

Since 0 does not check, it must be discarded. The only solution is 5.

Self Check 3 Solve: $\sqrt{4x+1}+1=x$.

ACCENT ON TECHNOLOGY: SOLVING RADICAL EQUATIONS

To find solutions for $\sqrt{3x+1}+1=x$ with a graphing calculator, we graph the functions $f(x)=\sqrt{3x+1}+1$ and $g(x)=x$, as in figure (a). We then trace to find the approximate x-coordinate of their intersection point, as in figure (b). After repeated zooms, we will see that x is 5.

We can also use the INTERSECT feature to approximate the point of intersection of the graphs. See figure (c). The intersection point of (5, 5) implies that $x=5$ is a solution of the radical equation.

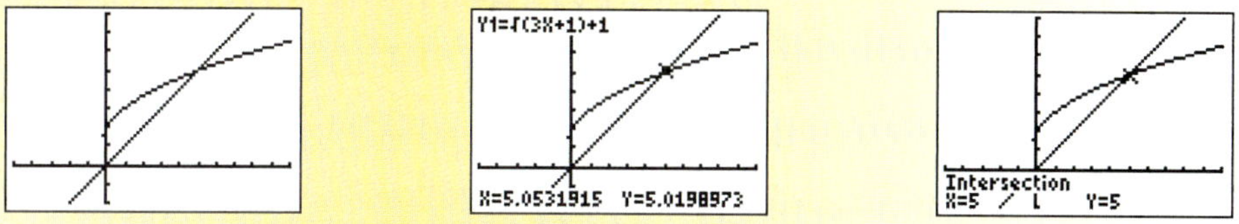

(a) (b) (c)

EXAMPLE 4

Solve: $\sqrt{3x}+6=0$.

Solution To isolate the radical, we subtract 6 from both sides and proceed as follows:

$$\begin{aligned} \sqrt{3x}+6 &= 0 \\ \sqrt{3x} &= -6 \\ \left(\sqrt{3x}\right)^2 &= (-6)^2 \quad \text{Square both sides to eliminate the square root.} \\ 3x &= 36 \\ x &= 12 \end{aligned}$$

We check the proposed solution 12 in the original equation.

$$\begin{aligned} \sqrt{3x}+6 &= 0 \\ \sqrt{3(12)}+6 &\stackrel{?}{=} 0 \quad \text{Substitute 12 for } x. \\ \sqrt{36}+6 &\stackrel{?}{=} 0 \\ 12 &\neq 0 \end{aligned}$$

Since 12 does not satisfy the original equation, it is extraneous. The equation has no solution. The solution set is $\varnothing$.

Success Tip

After isolating the radical, we obtained the equation $\sqrt{3x}=-6$. Since $\sqrt{3x}$ cannot be negative, we immediately know that $\sqrt{3x}+6=0$ has no solution.

Self Check 4 Solve: $\sqrt{4x+1}+5=0$.

EXAMPLE 5

Solve: $\sqrt[3]{x^3 + 7} = x + 1$.

Solution To eliminate the radical, we cube both sides of the equation and proceed as follows:

$$\begin{aligned}
\sqrt[3]{x^3 + 7} &= x + 1 \\
\left(\sqrt[3]{x^3 + 7}\right)^3 &= (x + 1)^3 && \text{Cube both sides to eliminate the cube root.} \\
x^3 + 7 &= x^3 + 3x^2 + 3x + 1 && (x + 1)^3 = (x + 1)(x + 1)(x + 1). \\
0 &= 3x^2 + 3x - 6 && \text{Subtract } x^3 \text{ and 7 from both sides.} \\
0 &= x^2 + x - 2 && \text{Divide both sides by 3. To solve this quadratic equation, use factoring.} \\
0 &= (x + 2)(x - 1) && \text{Factor the trinomial.}
\end{aligned}$$

$$x + 2 = 0 \quad \text{or} \quad x - 1 = 0$$
$$x = -2 \quad \Big| \quad x = 1$$

Success Tip

After raising both sides of a radical equation to a power, we use $(\sqrt[n]{a})^n = a$ to simplify one side. For example:

$$\left(\sqrt[3]{x^3 + 7}\right)^3 = x^3 + 7$$

We check each apparent solution to see whether it satisfies the original equation.

Check:

$$\begin{aligned}
\sqrt[3]{x^3 + 7} &= x + 1 \\
\sqrt[3]{(-2)^3 + 7} &\stackrel{?}{=} -2 + 1 \\
\sqrt[3]{-8 + 7} &\stackrel{?}{=} -1 \\
\sqrt[3]{-1} &\stackrel{?}{=} -1 \\
-1 &= -1
\end{aligned}
\qquad\qquad
\begin{aligned}
\sqrt[3]{x^3 + 7} &= x + 1 \\
\sqrt[3]{1^3 + 7} &\stackrel{?}{=} 1 + 1 \\
\sqrt[3]{1 + 7} &\stackrel{?}{=} 2 \\
\sqrt[3]{8} &\stackrel{?}{=} 2 \\
2 &= 2
\end{aligned}$$

Both -2 and 1 satisfy the original equation.

Self Check 5 Solve: $\sqrt[3]{x^3 + 8} = x + 2$.

EXAMPLE 6

Let $f(x) = \sqrt[4]{2x + 1}$. For what value(s) of x is $f(x) = 5$?

Solution To find the value(s) where $f(x) = 5$, we substitute 5 for $f(x)$ and solve for x.

$$\begin{aligned}
f(x) &= \sqrt[4]{2x + 1} \\
5 &= \sqrt[4]{2x + 1}
\end{aligned}$$

Since the equation contains a fourth root, we raise both sides to the fourth power to solve for x.

$$\begin{aligned}
(5)^4 &= \left(\sqrt[4]{2x + 1}\right)^4 && \text{Use the power rule to eliminate the radical.} \\
625 &= 2x + 1 \\
624 &= 2x \\
312 &= x
\end{aligned}$$

If x is 312, then $f(x) = 5$. Verify this by evaluating $f(312)$.

Self Check 6 Let $g(x) = \sqrt[5]{10x + 1}$. For what value(s) of x is $g(x) = 1$?

EQUATIONS CONTAINING TWO RADICALS

EXAMPLE 7

Solve: $\sqrt{5x+9} = 2\sqrt{3x+4}$.

Solution Each radical is isolated on one side of the equation, so we square both sides to eliminate them.

Caution

When finding $(2\sqrt{3x+4})^2$, remember to square both 2 and $\sqrt{3x+4}$ to get:

$$2^2(\sqrt{3x+4})^2$$

$$\sqrt{5x+9} = 2\sqrt{3x+4}$$

$$(\sqrt{5x+9})^2 = (2\sqrt{3x+4})^2 \quad \text{Square both sides.}$$

$$5x + 9 = 4(3x+4) \quad \text{On the right-hand side: } (2\sqrt{3x+4})^2 = 2^2(\sqrt{3x+4})^2 = 4(3x+4).$$

$$5x + 9 = 12x + 16 \quad \text{Remove parentheses.}$$

$$-7 = 7x \quad \text{Subtract } 5x \text{ and 16 from both sides.}$$

$$-1 = x \quad \text{Divide both sides by 7.}$$

We check the solution by substituting -1 for x in the original equation.

$$\sqrt{5x+9} = 2\sqrt{3x+4}$$

$$\sqrt{5(-1)+9} \stackrel{?}{=} 2\sqrt{3(-1)+4} \quad \text{Substitute } -1 \text{ for } x.$$

$$\sqrt{4} \stackrel{?}{=} 2\sqrt{1}$$

$$2 = 2$$

The solution checks.

Self Check 7 Solve: $\sqrt{x-4} = 2\sqrt{x-16}$.

When more than one radical appears in an equation, we can use the power rule more than once.

EXAMPLE 8

Solve: $\sqrt{x} + \sqrt{x+2} = 2$.

Solution To remove the radicals, we square both sides of the equation. Since this is easier to do if one radical is on each side of the equation, we subtract $\sqrt{x}$ from both sides to isolate $\sqrt{x+2}$ on the left-hand side.

Caution

When finding $(2-\sqrt{x})^2$, remember to use FOIL or a special product formula:

$$(2-\sqrt{x})^2 \neq 2^2 - (\sqrt{x})^2$$

$$\sqrt{x} + \sqrt{x+2} = 2$$

$$\sqrt{x+2} = 2 - \sqrt{x} \quad \text{Subtract } \sqrt{x} \text{ from both sides.}$$

$$(\sqrt{x+2})^2 = (2-\sqrt{x})^2 \quad \text{Square both sides to eliminate the square root.}$$

$$x + 2 = 4 - 4\sqrt{x} + x \quad \text{Use FOIL: } (2-\sqrt{x})^2 = (2-\sqrt{x})(2-\sqrt{x}) = 4 - 2\sqrt{x} - 2\sqrt{x} + x = 4 - 4\sqrt{x} + x.$$

$$2 = 4 - 4\sqrt{x} \quad \text{Subtract } x \text{ from both sides.}$$

$$-2 = -4\sqrt{x} \quad \text{Subtract 4 from both sides.}$$

$$\frac{1}{2} = \sqrt{x} \quad \text{Divide both sides by } -4 \text{ and simplify.}$$

$$\frac{1}{4} = x \quad \text{Square both sides.}$$

Check:

$$
\begin{aligned}
\sqrt{x} + \sqrt{x+2} &= 2 \\
\sqrt{\frac{1}{4}} + \sqrt{\frac{1}{4} + 2} &\stackrel{?}{=} 2 \\
\frac{1}{2} + \sqrt{\frac{9}{4}} &\stackrel{?}{=} 2 \\
\frac{1}{2} + \frac{3}{2} &\stackrel{?}{=} 2 \\
2 &= 2
\end{aligned}
$$

The result $\frac{1}{4}$ checks.

Self Check 8 Solve: $\sqrt{a} + \sqrt{a+3} = 3$.

ACCENT ON TECHNOLOGY: SOLVING RADICAL EQUATIONS

To find solutions for $\sqrt{x} + \sqrt{x+2} = 4$ (an equation similar to Example 8) with a graphing calculator, we graph the functions $f(x) = \sqrt{x} + \sqrt{x+2}$ and $g(x) = 4$. We then trace to find an approximation of the x-coordinate of their intersection point, as in figure (a). From the figure, we can see that $x \approx 2.98$. We can zoom to get better results.

Figure (b) shows that the INTERSECT feature gives the approximate coordinates of the point of intersection of the two graphs as (3.06, 4). Therefore, an approximate solution of the radical equation is 3.06. Check its reasonableness.

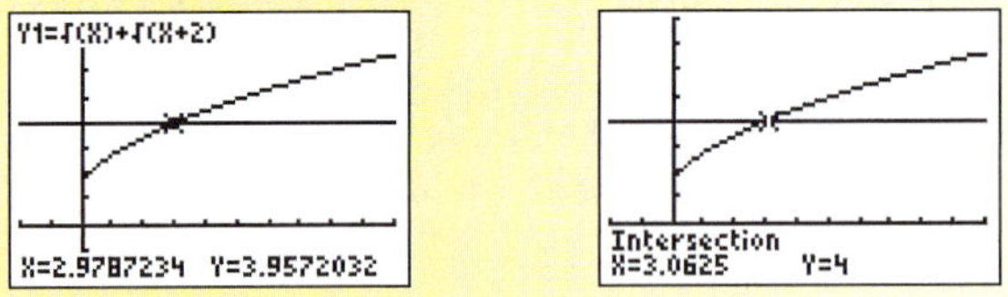

(a) **(b)**

SOLVING FORMULAS CONTAINING RADICALS

To *solve a formula for a variable* means to isolate that variable on one side of the equation, with all other quantities on the other side.

EXAMPLE 9

Depreciation rates. Some office equipment that is now worth V dollars originally cost C dollars 3 years ago. The rate r at which it has depreciated is given by

$$r = 1 - \sqrt[3]{\frac{V}{C}}$$

Solve the formula for C.

Solution We begin by isolating the cube root on the right-hand side of the equation.

$$r = 1 - \sqrt[3]{\frac{V}{C}}$$

$$r - 1 = -\sqrt[3]{\frac{V}{C}} \qquad \text{Subtract 1 from both sides.}$$

$$(r-1)^3 = \left[-\sqrt[3]{\frac{V}{C}}\right]^3 \qquad \text{To eliminate the radical, cube both sides.}$$

$$(r-1)^3 = -\frac{V}{C} \qquad \text{Simplify the right-hand side.}$$

$$C(r-1)^3 = -V \qquad \text{Multiply both sides by } C.$$

$$C = -\frac{V}{(r-1)^3} \qquad \text{Divide both sides by } (r-1)^3.$$

Self Check 9 A formula used in statistics to determine the size of a sample to obtain a desired degree of accuracy is

$$E = z_0\sqrt{\frac{pq}{n}}$$

Solve the formula for n.

Answers to Self Checks **1.** 11 **2.** 196 ft **3.** 6, 0 is extraneous **4.** 6 is extraneous, no solution **5.** 0, −2 **6.** 0 **7.** 20 **8.** 1 **9.** $n = \frac{z_0^2 pq}{E^2}$

7.5 STUDY SET

VOCABULARY Fill in the blanks.

1. Equations such as $\sqrt{x+4} - 4 = 5$ and $\sqrt[3]{x+1} = 12$ are called _______ equations.
2. When solving equations containing radicals, try to _______ one radical expression on one side of the equation.
3. Squaring both sides of an equation can introduce __________ solutions.
4. To _______ an apparent solution means to substitute it into the original equation and see whether a true statement results.

CONCEPTS

5. Fill in the blanks: The power rule states that if x, y, and n are real numbers and $x = y$, then

   $x^{\quad} = y^{\quad}$

6. Determine whether 6 is a solution of each radical equation.

   a. $\sqrt{x+3} = x - 3$

   b. $\sqrt{4x+1} = \sqrt{6x} - 1$

   c. $\sqrt[3]{5x-3} = x - 9$

7. What is the first step in solving each equation?

a. $\sqrt{x+11}=5$

b. $2=\sqrt[3]{x-2}$

c. $\sqrt[3]{5x+4}+3=30$

8. What is the first step in solving each equation?

a. $\sqrt{5x+4}+\sqrt{x+4}=0$

b. $\sqrt{x+8}-\sqrt{2x+9}=1$

9. Simplify each expression.

a. $(\sqrt{x})^2$ **b.** $(\sqrt{x-5})^2$

c. $(4\sqrt{2x})^2$ **d.** $(-\sqrt{x+3})^2$

10. Simplify each expression.

a. $(\sqrt[3]{x})^3$ **b.** $(\sqrt[4]{x})^4$

c. $(-\sqrt[3]{2x})^3$ **d.** $(2\sqrt[3]{x+3})^3$

11. Multiply.

a. $(\sqrt{x}-3)^2$

b. $(\sqrt{2y+1}+5)^2$

12. Perform the necessary steps to isolate the radical on the right-hand side of the equation.

$$3x-6=2x+\sqrt{2y+1}-9$$

13. Explain why it is immediately apparent that $\sqrt{8x-7}=-2$ has no solution.

14. Solve $\sqrt{x-2}+2=4$ graphically, using the graphs in the illustration.

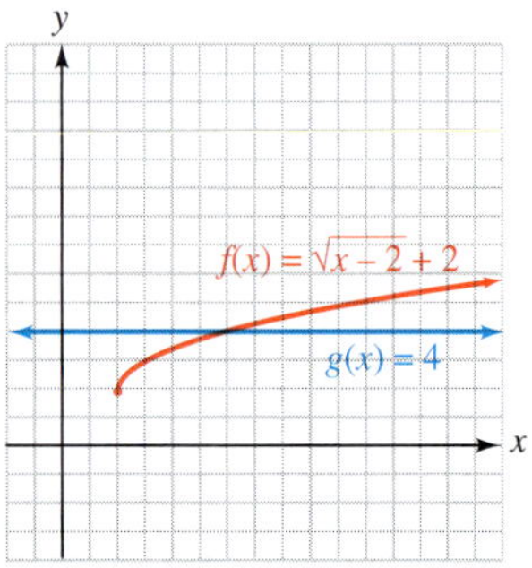

NOTATION **Complete each solution.**

15. Solve: $\sqrt{3x}-1=5$

$$\sqrt{3x}=\square$$
$$(\sqrt{3x})^{\square}=(6)^{\square}$$
$$\square=36$$
$$x=12$$

Does 12 check?

16. Solve: $\sqrt{1-2x}=\sqrt{x+10}$.

$$(\square)^2=(\sqrt{x+10})^{\square}$$
$$\square=x+10$$
$$\square=9$$
$$x=-3$$

Does -3 check?

PRACTICE **Solve each equation. Write all apparent solutions. Cross out those that are extraneous.**

17. $\sqrt{5x-6}=2$

18. $\sqrt{7x-10}=12$

19. $\sqrt{6x+1}+2=7$

20. $\sqrt{6x+13}-2=5$

21. $2\sqrt{4x+1}=\sqrt{x+4}$

22. $\sqrt{3(x+4)}=\sqrt{5x-12}$

23. $\sqrt[3]{7n-1}=3$

24. $\sqrt[3]{12m+4}=4$

25. $\sqrt[4]{10p+1}=\sqrt[4]{11p-7}$

26. $\sqrt[4]{10y+6}=2\sqrt[4]{y}$

27. $(6x+2)^{1/2}=(5x+3)^{1/2}$

28. $(5x+3)^{1/2}=(x+11)^{1/2}$

29. $(x+8)^{1/3}=-2$

30. $(x+4)^{1/3}=-1$

31. $\sqrt{5-x}+10=9$

32. $1=2+\sqrt{4x+75}$

33. $x=\dfrac{\sqrt{12x-5}}{2}$

34. $x=\dfrac{\sqrt{16x-12}}{2}$

35. $\sqrt{x+2}-\sqrt{4-x}=0$

36. $\sqrt{6-x}-\sqrt{2x+3}=0$

37. $2\sqrt{x}=\sqrt{5x-16}$

38. $3\sqrt{x}=\sqrt{3x+54}$

39. $r-9=\sqrt{2r-3}$

40. $-s-3=2\sqrt{5-s}$

41. $(m^4+m^2-25)^{1/4}=m$

42. $n=(n^3+n^2-1)^{1/3}$

43. $\sqrt{-5x+24}=6-x$

44. $\sqrt{-x+2}=x-2$

45. $\sqrt{y+2}=4-y$

46. $\sqrt{22y+86}=y+9$

47. $\sqrt[3]{x^3-7}=x-1$

48. $\sqrt[3]{x^3+56}-2=x$

49. $\sqrt[4]{x^4+4x^2-4}=-x$

50. $u=\sqrt[4]{u^4-6u^2+24}$

51. $\sqrt[4]{12t+4}+2=0$

52. $\sqrt[4]{8x-8}+2=0$

53. $\sqrt{2y+1}=1-2\sqrt{y}$

54. $\sqrt{u}+3=\sqrt{u-3}$

55. $\sqrt{n^2+6n+3}=\sqrt{n^2-6n-3}$

56. $\sqrt{m^2-12m-3}=\sqrt{m^2+12m+3}$

57. $\sqrt{7t^2+4}=\sqrt{17t-8t^2}$

58. $\sqrt{b^2+b}=\sqrt{3-b^2}$

59. $\sqrt{y+7}+3=\sqrt{y+4}$

60. $1+\sqrt{z}=\sqrt{z+3}$

61. $2+\sqrt{u}=\sqrt{2u+7}$

62. $5r+4=\sqrt{5r+20}+4r$

63. $\sqrt{6t+1}-3\sqrt{t}=-1$

64. $\sqrt{4s+1}-\sqrt{6s}=-1$

65. $\sqrt{2x+5}+\sqrt{x+2}=5$

66. $\sqrt{2x+5}+\sqrt{2x+1}+4=0$

67. $\sqrt{x-5}-\sqrt{x+3}=4$

68. $\sqrt{x+8}-\sqrt{x-4}=-2$

69. Let $f(x)=\sqrt{3x-6}$. For what value(s) of x is $f(x)=3$?

70. Let $f(x)=\sqrt{2x^2-7x}$. For what value(s) of x is $f(x)=2$?

71. Let $f(x)=\sqrt{x+8}-\sqrt{x}$. For what value(s) of x is $f(x)=2$?

72. Let $f(x)=\sqrt{x}-\sqrt{x+5}$. For what value(s) of x is $f(x)=-1$?

Solve each equation for the indicated variable.

73. $v=\sqrt{2gh}$ for h

74. $d=1.4\sqrt{h}$ for h

75. $T=2\pi\sqrt{\frac{\ell}{32}}$ for ℓ

76. $d=\sqrt[3]{\frac{12V}{\pi}}$ for V

77. $r=\sqrt[3]{\frac{A}{P}}-1$ for A

78. $r=\sqrt[3]{\frac{A}{P}}-1$ for P

79. $L_A=L_B\sqrt{1-\frac{v^2}{c^2}}$ for v^2

80. $R_1=\sqrt{\frac{A}{\pi}-R_2{}^2}$ for A

APPLICATIONS

81. HIGHWAY DESIGN A curved road will accommodate traffic traveling s mph if the radius of the curve is r feet, according to the formula $s=3\sqrt{r}$. If engineers expect 40-mph traffic, what radius should they specify? Give the result to the nearest foot.

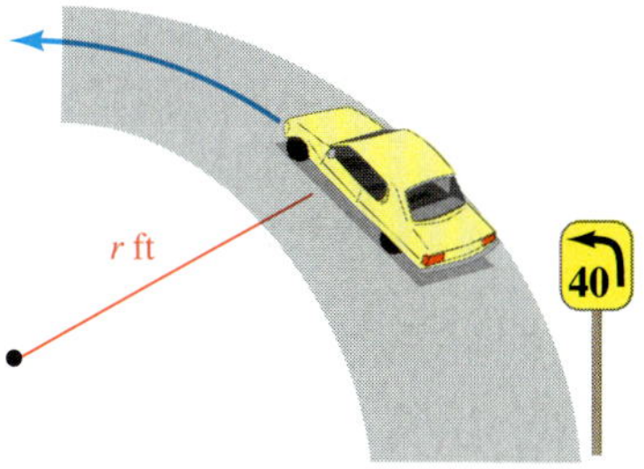

82. FORESTRY The taller a lookout tower, the farther an observer can see. That distance d (called the *horizon distance,* measured in miles) is related to the height h of the observer (measured in feet) by the formula $d=1.22\sqrt{h}$. How tall must a lookout tower be to see the edge of the forest, 25 miles away? (Round to the nearest foot.)

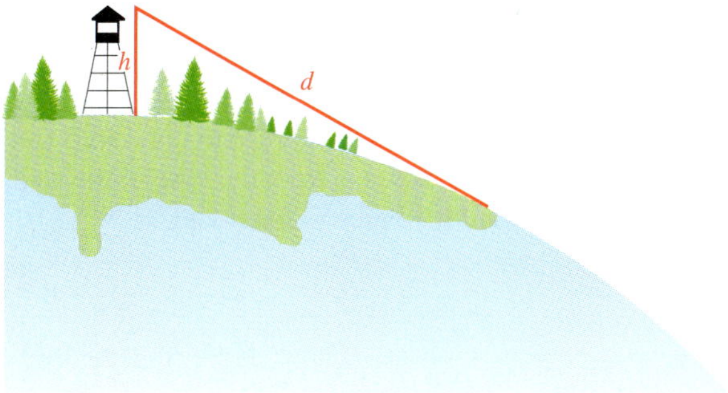

83. WIND POWER The power generated by a windmill is related to the velocity of the wind by the formula

$$v=\sqrt[3]{\frac{P}{0.02}}$$

where P is the power (in watts) and v is the velocity of the wind (in mph). Find how much power the windmill is generating when the wind is 29 mph.

84. DIAMONDS The *effective rate of interest r* earned by an investment is given by the formula

$$r = \sqrt[n]{\frac{A}{P}} - 1$$

where P is the initial investment that grows to value A after n years. If a diamond buyer got \$4,000 for a 1.73-carat diamond that he had purchased 4 years earlier, and earned an annual rate of return of 6.5% on the investment, what did he originally pay for the diamond?

85. THEATER PRODUCTIONS The ropes, pulleys, and sandbags shown in the illustration are part of a mechanical system used to raise and lower scenery for a stage play. For the scenery to be in the proper position, the following formula must apply:

$$w_2 = \sqrt{w_1^{\,2} + w_3^{\,2}}$$

If $w_2 = 12.5$ lb and $w_3 = 7.5$ lb, find w_1.

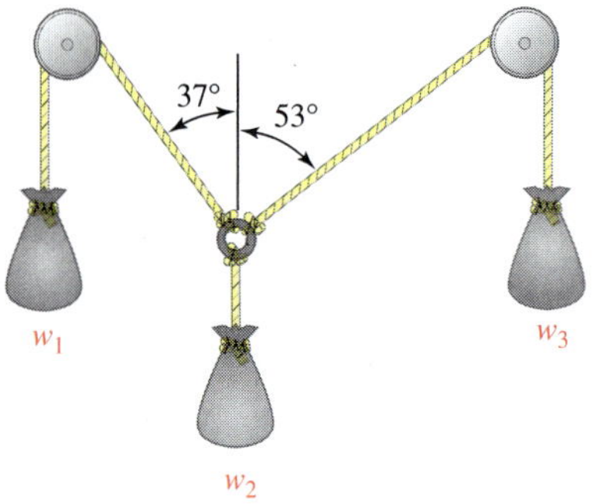

86. CARPENTRY During construction, carpenters often brace walls as shown in the illustration, where the length ℓ of the brace is given by the formula

$$\ell = \sqrt{f^2 + h^2}$$

If a carpenter nails a 10-ft brace to the wall 6 feet above the floor, how far from the base of the wall should he nail the brace to the floor?

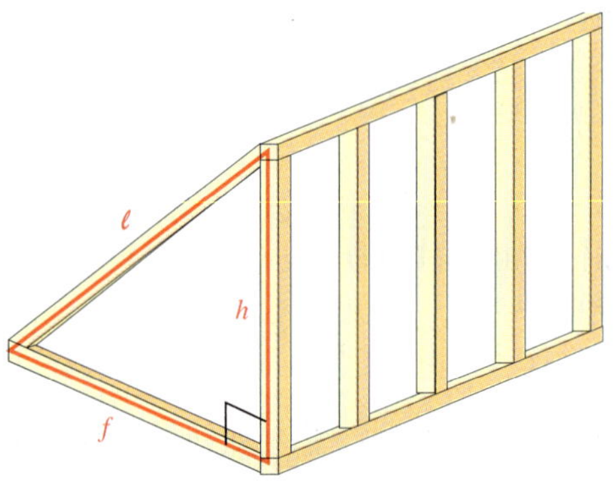

87. SUPPLY AND DEMAND The number of wrenches that will be produced at a given price can be predicted by the formula $s = \sqrt{5x}$, where s is the supply (in thousands) and x is the price (in dollars). The demand d for wrenches can be predicted by the formula $d = \sqrt{100 - 3x^2}$. Find the equilibrium price—that is, find the price at which supply will equal demand.

88. SUPPLY AND DEMAND The number of mirrors that will be produced at a given price can be predicted by the formula $s = \sqrt{23x}$, where s is the supply (in thousands) and x is the price (in dollars). The demand d for mirrors can be predicted by the formula $d = \sqrt{312 - 2x^2}$. Find the equilibrium price—that is, find the price at which supply will equal demand.

WRITING

89. If both sides of an equation are raised to the same power, the resulting equation might not be equivalent to the original equation. Explain.

90. What is wrong with the student's work shown below?

Solve: $\sqrt{x + 1} - 3 = 8$.

$$\sqrt{x + 1} = 11$$
$$\left(\sqrt{x + 1}\right)^2 = 11$$
$$x + 1 = 11$$
$$x = 10$$

91. The first step of a student's solution is shown below. What is a better way to begin the solution?

Solve: $\sqrt{x} + \sqrt{x + 22} = 12$.

$$\left(\sqrt{x} + \sqrt{x + 22}\right)^2 = 102^2$$

92. Explain how $\sqrt{2x - 1} = x$ can be solved graphically.

93. Explain how the table can be used to solve $\sqrt{4x - 3} - 2 = \sqrt{2x - 5}$ if $Y_1 = \sqrt{4x - 3} - 2$ and $Y_2 = \sqrt{2x - 5}$.

X	Y1	Y2
2	.23607	ERROR
3	1	1
4	1.6056	1.7321
5	2.1231	2.2361
6	2.5826	2.6458
7	3	3
8	3.3852	3.3166

X=2

94. Explain how to use the graph of $f(x) = \sqrt[3]{x - 0.5} - 1$, shown in the illustration, to approximate the solution of $\sqrt[3]{x - 0.5} = 1$.

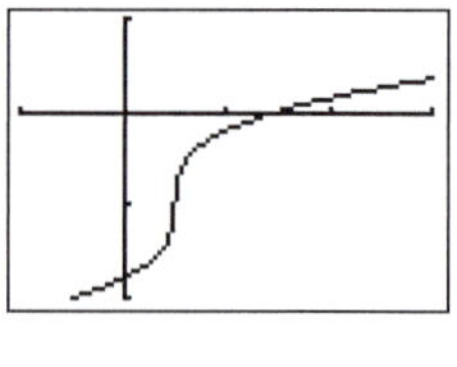

REVIEW

95. LIGHTING The intensity of light from a light bulb varies inversely as the square of the distance from the bulb. If you are 5 feet away from a bulb and the intensity is 40 foot-candles, what will the intensity be if you move 20 feet away from the bulb?

96. COMMITTEES What type of variation is shown in the illustration? As the number of people on this committee increased, what happened to its effectiveness?

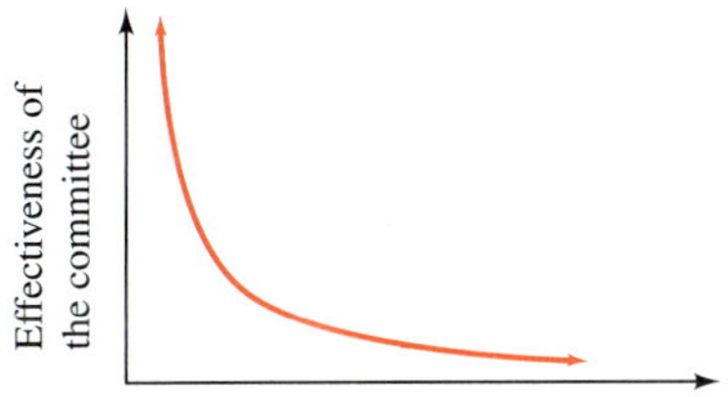

97. TYPESETTING If 12-point type is 0.166044 inch tall, how tall is 30-point type?

98. GUITAR STRINGS The frequency of vibration of a string varies directly as the square root of the tension and inversely as the length of the string. Suppose a string 2.5 feet long, under a tension of 16 pounds, vibrates 25 times per second. Find k, the constant of proportionality.

CHALLENGE PROBLEMS **Solve each equation.**

99. $\sqrt[3]{2x} = \sqrt{x}$ (*Hint:* Square and then cube both sides.)

100. $\sqrt[4]{x} = \sqrt{\dfrac{x}{4}}$

101. $\sqrt{x+2} + \sqrt{2x} = \sqrt{18-x}$

102. $\sqrt{8-x} - \sqrt{3x-8} = \sqrt{x-4}$

7.6 Geometric Applications of Radicals

- The Pythagorean theorem
- 45°–45°–90° triangles
- 30°–60°–90° triangles
- The distance formula

The Language of Algebra

A *theorem* is a mathematical statement that can be proved. The *Pythagorean theorem* is named after *Pythagoras,* a Greek mathematician who lived about 2,500 years ago. He is thought to have been the first to prove the theorem.

We will now consider applications of square roots in geometry. Then we will find the distance between two points on a rectangular coordinate system, using a formula that contains a square root. We begin by considering an important theorem about right triangles.

THE PYTHAGOREAN THEOREM

If we know the lengths of two legs of a right triangle, we can find the length of the **hypotenuse** (the side opposite the 90° angle) by using the **Pythagorean theorem.**

The Pythagorean Theorem

If a and b are the lengths of the legs of a right triangle and c is the length of the hypotenuse.

$$a^2 + b^2 = c^2$$

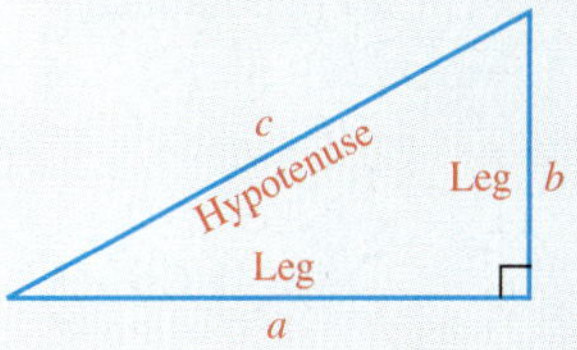

In words, the Pythagorean theorem is expressed as follows:

In any right triangle, the square of the hypotenuse is equal to the sum of the squares of the two legs.

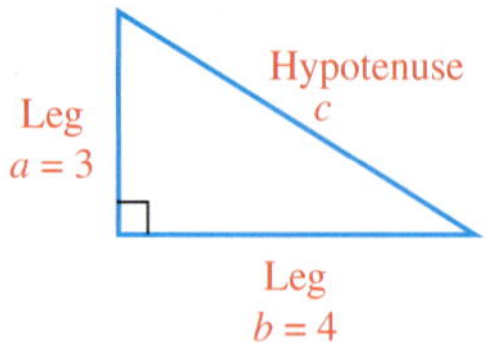

Suppose the right triangle shown in the figure has legs of length 3 and 4 units. To find the length of the hypotenuse, we use the Pythagorean theorem.

$$a^2 + b^2 = c^2$$
$$3^2 + 4^2 = c^2 \quad \text{Substitute 3 for } a \text{ and 4 for } b.$$
$$9 + 16 = c^2$$
$$25 = c^2$$

To find c, we ask "What number, when squared, is equal to 25?" There are two such numbers: the positive square root of 25 and the negative square root of 25. Since c represents the length of the hypotenuse, and it cannot be negative, it follows that c is the positive square root of 25.

$$\sqrt{25} = c \quad \text{Recall that a radical symbol } \sqrt{\ } \text{ is used to represent the positive, or principal square root of a number.}$$
$$5 = c$$

The length of the hypotenuse is 5 units.

EXAMPLE 1

Firefighting. To fight a fire, the forestry department plans to clear a rectangular fire break around the fire, as shown in the illustration. Crews are equipped with mobile communications that have a 3,000-yard range. Can crews at points A and B remain in radio contact?

Solution Points A, B, and C form a right triangle. To find the distance c from point A to point B, we can use the Pythagorean theorem, substituting 2,400 for a and 1,000 for b and solving for c.

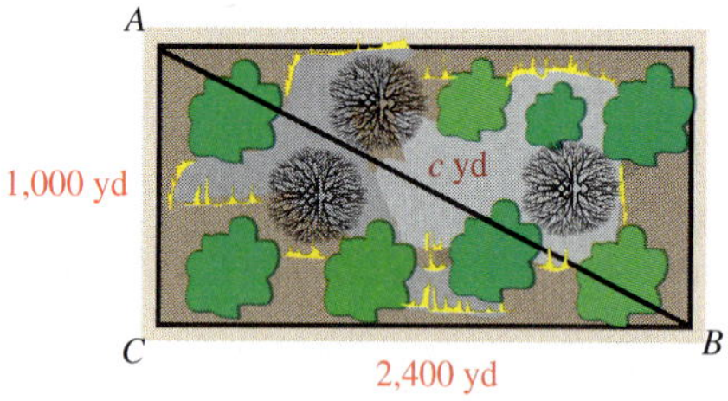

Caution

When using the Pythagorean theorem $a^2 + b^2 = c^2$, we can let a represent the length of either leg of the right triangle. We then let b represent the length of the other leg. The variable c must always represent the length of the hypotenuse.

$$a^2 + b^2 = c^2$$
$$2{,}400^2 + 1{,}000^2 = c^2$$
$$5{,}760{,}000 + 1{,}000{,}000 = c^2$$
$$6{,}760{,}000 = c^2$$
$$\sqrt{6{,}760{,}000} = c \quad \text{If } c^2 = 6{,}760{,}000\text{, then } c \text{ must be a square root of 6,760,000. Because } c \text{ represents a length, it must be the positive square root of 6,760,000.}$$
$$2{,}600 = c \quad \text{Use a calculator to find the square root.}$$

The two crews are 2,600 yards apart. Because this distance is less than the range of the radios, they can communicate by radio.

Self Check 1 Can the crews communicate if $b = 1{,}500$ yards?

45°–45°–90° TRIANGLES

An **isosceles right triangle** is a right triangle with two legs of equal length. Isosceles right triangles have angle measures of 45°, 45°, and 90°. If we know the length of one leg of an isosceles right triangle, we can use the Pythagorean theorem to find the length of the hypotenuse. Since the triangle shown in the figure is a right triangle, we have

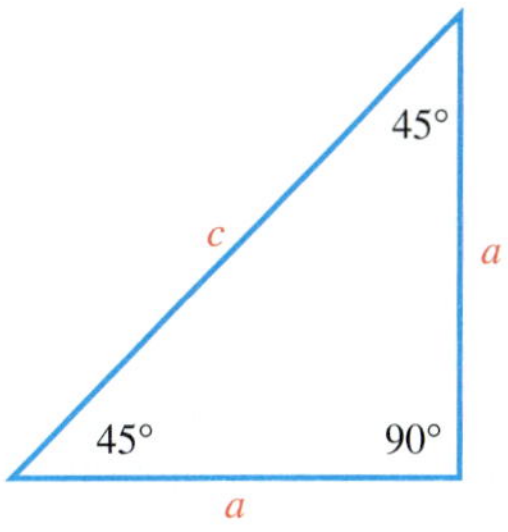

$c^2 = a^2 + b^2$

$c^2 = a^2 + a^2$ Both legs are a units long, so replace b with a.

$c^2 = 2a^2$ Combine like terms.

$c = \sqrt{2a^2}$ If $c^2 = 2a^2$, then c must be a square root of $2a^2$. Because c represents a length, it must be the positive square root of $2a^2$.

$c = a\sqrt{2}$ Simplify the radical: $\sqrt{2a^2} = \sqrt{2}\sqrt{a^2} = \sqrt{2}a = a\sqrt{2}$.

Thus, *in an isosceles right triangle, the length of the hypotenuse is the length of one leg times* $\sqrt{2}$.

EXAMPLE 2

If one leg of the isosceles right triangle shown above is 10 feet long, find the length of the hypotenuse.

Solution Since the length of the hypotenuse is the length of a leg times $\sqrt{2}$, we have

$$c = 10\sqrt{2}$$

The length of the hypotenuse is $10\sqrt{2}$ feet. To two decimal places, the length is 14.14 feet.

Self Check 2 Find the length of the hypotenuse of an isosceles right triangle if one leg is 12 meters long.

If the length of the hypotenuse of an isosceles right triangle is known, we can use the Pythagorean theorem to find the length of each leg.

EXAMPLE 3

Find the exact length of each leg of the following isosceles right triangle.

Solution We use the Pythagorean theorem.

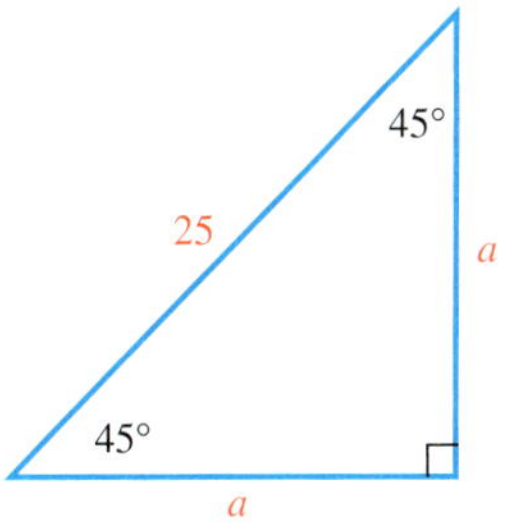

$c^2 = a^2 + b^2$

$25^2 = a^2 + a^2$ Since both legs are a units long, substitute a for b. The hypotenuse is 25 units long. Substitute 25 for c.

$25^2 = 2a^2$ Combine like terms.

$\frac{625}{2} = a^2$ Square 25 and divide both sides by 2.

$\sqrt{\frac{625}{2}} = a$ If $a^2 = \frac{625}{2}$, then a must be the positive square root of $\frac{625}{2}$.

$\frac{\sqrt{625}}{\sqrt{2}} \cdot \frac{\sqrt{2}}{\sqrt{2}} = a$ Write $\sqrt{\frac{625}{2}}$ as $\frac{\sqrt{625}}{\sqrt{2}}$. Then rationalize the denominator.

$\frac{25\sqrt{2}}{2} = a$ In the numerator, simplify the radical: $\sqrt{625} = 25$. In the denominator, do the multiplication: $\sqrt{2} \cdot \sqrt{2} = 2$.

The exact length of each leg is $\frac{25\sqrt{2}}{2}$ units. To two decimal places, the length is 17.68 units.

Self Check 3 Find the exact length of each leg of an isosceles right triangle if the length of the hypotenuse is 9 inches.

30°–60°–90° TRIANGLES

From geometry, we know that an **equilateral triangle** is a triangle with three sides of equal length and three 60° angles. Each side of the following equilateral triangle is $2a$ units long. If an **altitude** is drawn to its base the altitude bisects the base and divides the equilateral triangle, into two 30°–60°–90° triangles. We can see that the shorter leg of each 30°–60°–90° triangle (the side *opposite* the 30° angle) is a units long. Thus,

The length of the shorter leg of a 30°–60°–90° right triangle is half as long as the hypotenuse.

We can discover another important relationship between the legs of a 30°–60°–90° triangle if we find the length of the altitude h in the figure. We begin by applying the Pythagorean theorem to one of the 30°–60°–90° triangles.

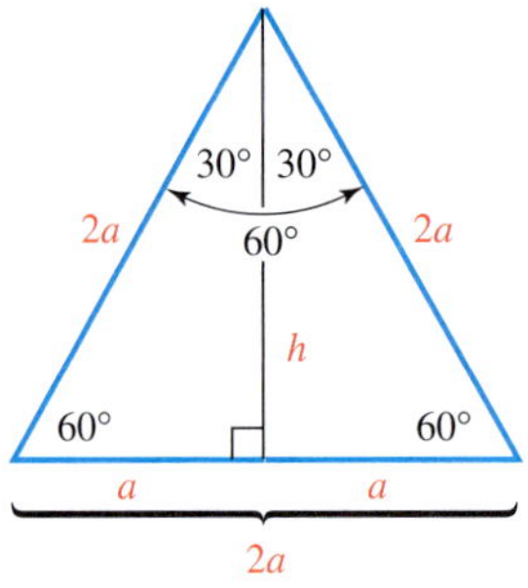

$a^2 + b^2 = c^2$

$a^2 + h^2 = (2a)^2$ One leg is h units long, so replace b with h. The hypotenuse is $2a$ units long, so replace c with $2a$.

$a^2 + h^2 = 4a^2$ $(2a)^2 = (2a)(2a) = 4a^2$.

$h^2 = 3a^2$ Subtract a^2 from both sides.

$h = \sqrt{3a^2}$ If $h^2 = 3a^2$, then h must be the positive square root of $3a^2$.

$h = a\sqrt{3}$ Simplify the radical: $\sqrt{3a^2} = \sqrt{3}\sqrt{a^2} = a\sqrt{3}$.

We see that the altitude—the longer leg of the 30°–60°–90° triangle—is $\sqrt{3}$ times as long as the shorter leg. Thus,

The length of the longer leg of a 30°–60°–90° triangle is the length of the shorter leg times $\sqrt{3}$.

EXAMPLE 4 Find the length of the hypotenuse and the longer leg of the right triangle.

Solution Since the shorter leg of a 30°–60°–90° triangle is half as long as the hypotenuse, the hypotenuse is 12 centimeters long.

Since the length of the longer leg is the length of the shorter leg times $\sqrt{3}$, the longer leg is $6\sqrt{3}$ (about 10.39) centimeters long.

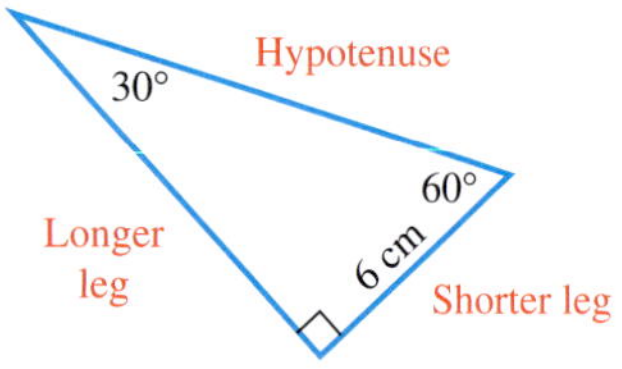

Self Check 4 Find the length of the hypotenuse and the longer leg of a 30°–60°–90° triangle if the shorter leg is 8 centimeters long.

EXAMPLE 5

Stretching exercises. A doctor prescribed the exercise shown in figure (a). The patient was instructed to raise his leg to an angle of 60° and hold the position for 10 seconds. If the patient's leg is 36 inches long, how high off the floor will his foot be when his leg is held at the proper angle?

Solution In figure (b), we see that a 30°–60°–90° triangle, which we will call triangle *ABC*, models the situation. Since the side opposite the 30° angle of a 30°–60°–90° triangle is half as long as the hypotenuse, side *AC* is 18 inches long.

Since the length of the side opposite the 60° angle is the length of the side opposite the 30° angle times $\sqrt{3}$, side *BC* is $18\sqrt{3}$, or about 31 inches long. So the patient's foot will be about 31 inches from the floor when his leg is in the proper position.

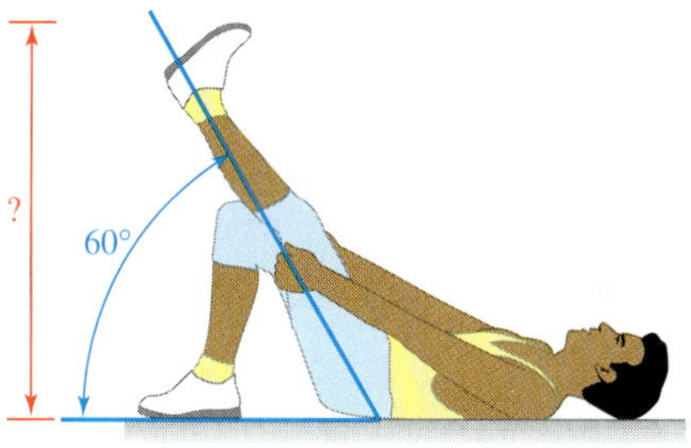

(a)

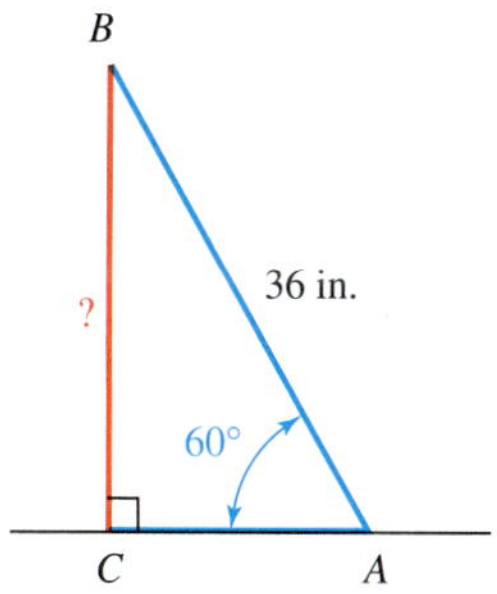

(b)

THE DISTANCE FORMULA

With the *distance formula,* we can find the distance between any two points graphed on a rectangular coordinate system.

To find the distance *d* between points $P(x_1, y_1)$ and $Q(x_2, y_2)$ shown in the figure, we construct the right triangle *PRQ*. The distance between *P* and *R* is $|x_2 - x_1|$, and the distance between *R* and *Q* is $|y_2 - y_1|$. We apply the Pythagorean theorem to the right triangle *PRQ* to get

$$\begin{aligned} d^2 &= |x_2 - x_1|^2 + |y_2 - y_1|^2 \\ &= (x_2 - x_1)^2 + (y_2 - y_1)^2 \end{aligned}$$

Because $|x_2 - x_1|^2 = (x_2 - x_1)^2$ and $|y_2 - y_1|^2 = (y_2 - y_1)^2$.

Because *d* represents the distance between two points, it must be equal to the positive square root of $(x_2 - x_1)^2 + (y_2 - y_1)^2$.

$$d = \sqrt{(x_2 - x_1)^2 + (y_2 - y_1)^2}$$

We call this result the *distance formula.*

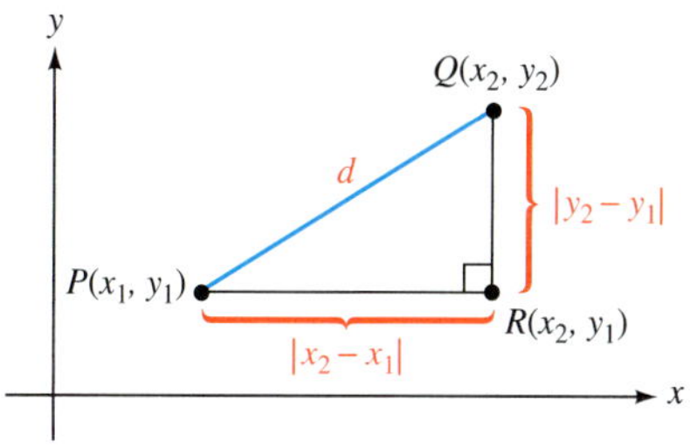

Distance Formula The distance *d* between two points with coordinates (x_1, y_1) and (x_2, y_2) is given by

$$d = \sqrt{(x_2 - x_1)^2 + (y_2 - y_1)^2}$$

EXAMPLE 6

Find the distance between the points $(-2, 3)$ and $(4, -5)$.

Solution To find the distance, we can use the distance formula by substituting 4 for x_2, -2 for x_1, -5 for y_2, and 3 for y_1.

$$\begin{aligned} d &= \sqrt{(x_2 - x_1)^2 + (y_2 - y_1)^2} \\ &= \sqrt{[4 - (-2)]^2 + (-5 - 3)^2} \\ &= \sqrt{(4 + 2)^2 + (-5 - 3)^2} \\ &= \sqrt{6^2 + (-8)^2} \\ &= \sqrt{36 + 64} \\ &= \sqrt{100} \\ &= 10 \end{aligned}$$

The distance between the points is 10 units.

Self Check 6 Find the distance between $(-2, -2)$ and $(3, 10)$.

EXAMPLE 7

Robotics. Robots are used to weld parts of an automobile chassis on an automated production line. To do this, an imaginary coordinate system is superimposed on the side of the vehicle, and the robot is programmed to move to specific positions to make each weld. See the figure, which is scaled in inches. If the welder unit moves from point to point at an average rate of speed of 48 in./sec, how long will it take it to move from position 1 to position 2?

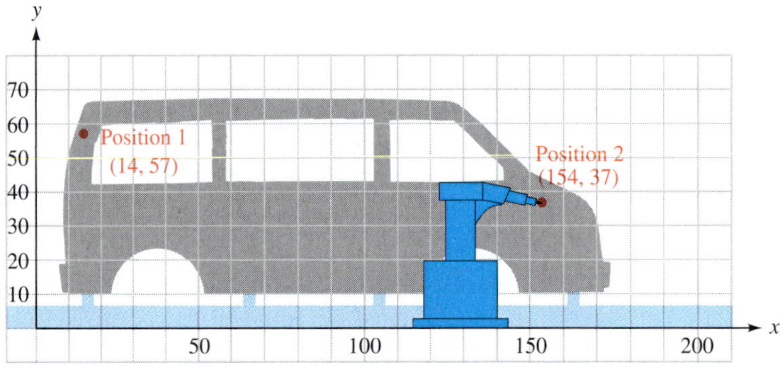

Solution This is a uniform motion problem. We can use the formula $t = \frac{d}{r}$ to find the time it takes for the welder to move from position 1 at $(14, 57)$ to position 2 at $(154, 37)$.

We can use the distance formula to find the distance d that the welder unit moves.

$$d = \sqrt{(x_2 - x_1)^2 + (y_2 - y_1)^2}$$

$$d = \sqrt{(154 - 14)^2 + (37 - 57)^2} \quad \text{Substitute 154 for } x_2\text{, 14 for } x_1\text{, 37 for } y_2\text{, and 57 for } y_1.$$

$$= \sqrt{140^2 + (-20)^2}$$

$$= \sqrt{20{,}000} \quad 140^2 + (-20)^2 = 19{,}600 + 400 = 20{,}000.$$

$$= 100\sqrt{2} \quad \text{Simplify: } \sqrt{20{,}000} = \sqrt{100 \cdot 100 \cdot 2} = 100\sqrt{2}.$$

The welder travels $100\sqrt{2}$ inches as it moves from position 1 to position 2. To find the time this will take, we divide the distance by the average rate of speed, 48 in./sec.

$$t = \frac{d}{r}$$

$$t = \frac{100\sqrt{2}}{48} \quad \text{Substitute } 100\sqrt{2} \text{ for } d \text{ and 48 for } r.$$

$$t \approx 2.9 \quad \text{Use a calculator to find an approximation to the nearest tenth.}$$

It will take the welder about 2.9 seconds to travel from position 1 to position 2.

Answers to Self Checks **1.** yes **2.** $12\sqrt{2}$ m **3.** $\frac{9\sqrt{2}}{2}$ in. **4.** 16 cm, $8\sqrt{3}$ cm **6.** 13

7.6 STUDY SET

VOCABULARY Fill in the blanks.

1. In a right triangle, the side opposite the 90° angle is called the ___________.
2. An _________ right triangle is a right triangle with two legs of equal length.
3. The ___________ theorem states that in any right triangle, the square of the hypotenuse is equal to the sum of the squares of the lengths of the two legs.
4. An _________ triangle has three sides of equal length and three 60° angles.

CONCEPTS Fill in the blanks.

5. If a and b are the lengths of the legs of a right triangle and c is the length of the hypotenuse, then ___________.
6. In any right triangle, the square of the hypotenuse is equal to the _____ of the squares of the two _____.
7. In an isosceles right triangle, the length of the hypotenuse is the length of one leg times ____.
8. The shorter leg of a 30°–60°–90° triangle is _____ as long as the hypotenuse.
9. The length of the longer leg of a 30°–60°–90° triangle is the length of the shorter leg times ____.
10. The formula to find the distance between two points (x_1, y_1) and (x_2, y_2) is $d =$ ___________.
11. In a right triangle, the shorter leg is opposite the _____ angle, and the longer leg is opposite the _____ angle.
12. An isosceles triangle has _____ sides of equal length.
13. Solve for c, where c represents the length of the hypotenuse of a right triangle.
    a. $c^2 = 64$
    b. $c^2 = 15$
    c. $c^2 = 24$
14. When the lengths of the sides of a certain triangle are substituted into the equation $a^2 + b^2 = c^2$, the result is a false statement. Explain why.

$$a^2 + b^2 = c^2$$
$$2^2 + 4^2 = 5^2$$
$$4 + 16 = 25$$
$$20 = 25$$

NOTATION Complete each solution.

15. Evaluate:

$$\begin{aligned}\sqrt{(-1-3)^2 + [2-(-4)]^2} &= \sqrt{(-4)^2 + [\quad]^2}\\ &= \sqrt{\quad}\\ &= \sqrt{\quad \cdot 13}\\ &= \quad\sqrt{13}\\ &\approx 7.21\end{aligned}$$

16. Solve: $8^2 + 4^2 = c^2$. Assume $c > 0$.

$$\square + 16 = c^2$$
$$\square = c^2$$
$$\sqrt{\square} = \square$$
$$\sqrt{\square \cdot 5} = c$$
$$\square\sqrt{5} = c$$
$$c \approx 8.94$$

PRACTICE **The lengths of two sides of the right triangle *ABC* are given. Find the length of the missing side.**

17. $a = 6$ ft and $b = 8$ ft

18. $a = 10$ cm and $c = 26$ cm

19. $b = 18$ m and $c = 82$ m

20. $a = 14$ in. and $c = 50$ in.

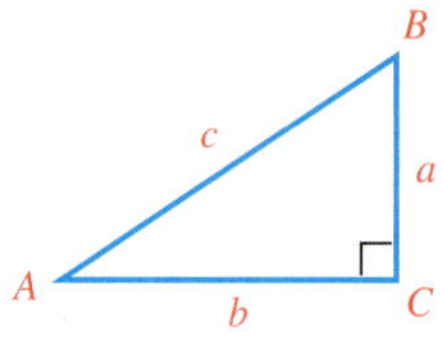

Find the missing lengths in each triangle. Give the exact answer and then an approximation to two decimal places, when applicable.

21.

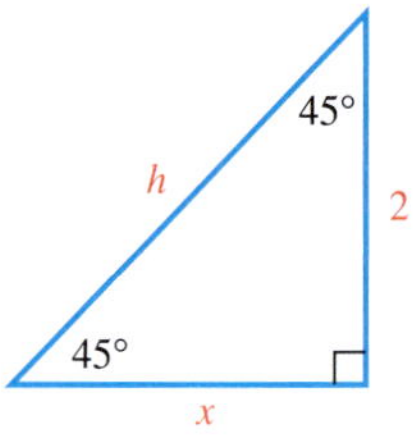

22.

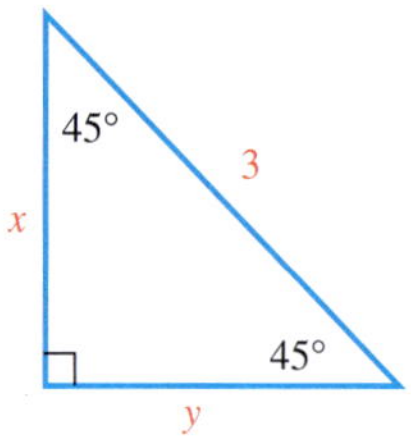

23.

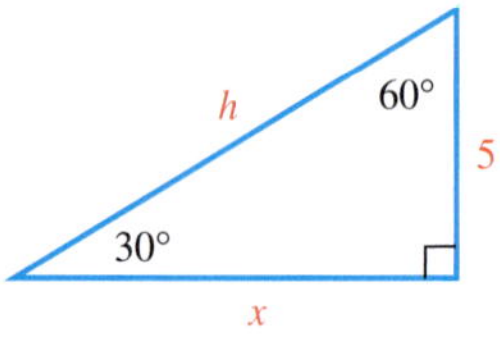

24.

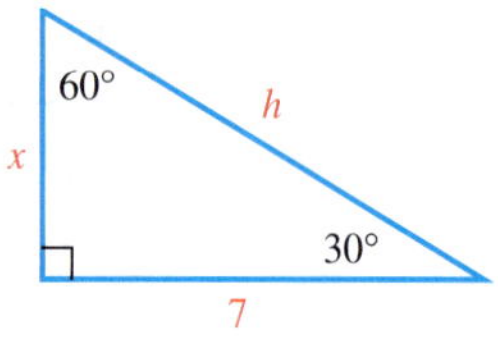

Find the missing lengths in each triangle. Give the answer to two decimal places.

25.

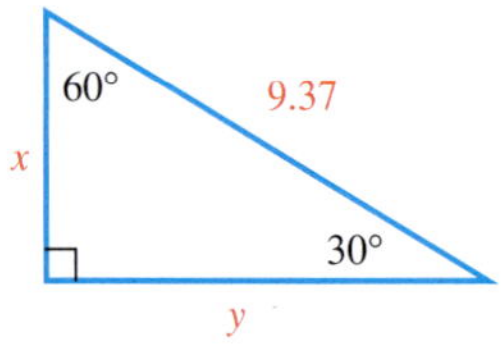

26.

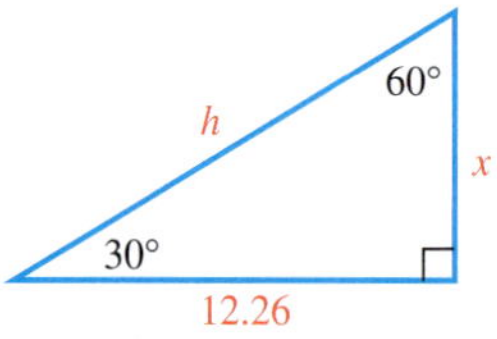

27.

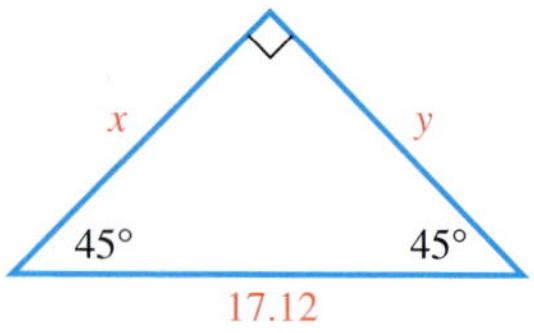

28.

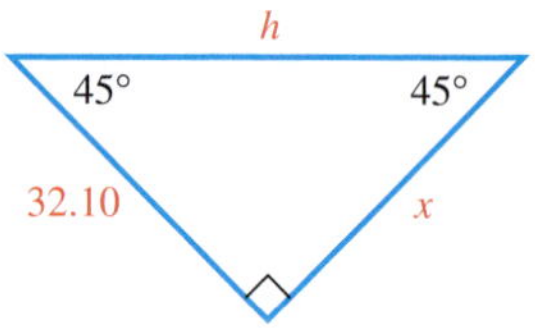

29. GEOMETRY Find the exact length of the diagonal (in blue) of one of the *faces* of the cube shown below.

30. GEOMETRY Find the exact length of the diagonal (in green) of the cube shown below.

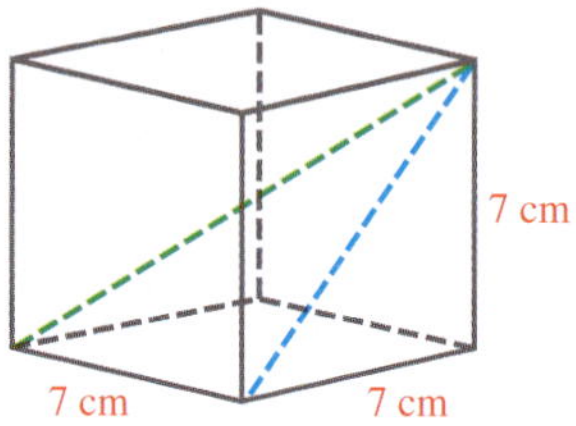

Find the distance between each pair of points.

31. (0, 0), (3, −4)

32. (0, 0), (−12, 16)

33. (−2, −8), (3, 4)

34. (−5, −2), (7, 3)

35. (6, 8), (12, 16)

36. (10, 4), (2, −2)

37. (−3, 5), (−5, −5)

38. (2, −3), (4, −8)

39. ISOSCELES TRIANGLES Use the distance formula to show that a triangle with vertices (−2, 4), (2, 8), and (6, 4) is isosceles.

40. RIGHT TRIANGLES Use the distance formula and the Pythagorean theorem to show that a triangle with vertices (2, 3), (−3, 4), and (1, −2) is a right triangle.

APPLICATIONS **Find the exact answer. Then give an approximation to two decimal places.**

41. WASHINGTON, D.C. The square in the map shows the 100-square-mile site selected by George Washington in 1790 to serve as a permanent capital for the United States. In 1847, the part of the district lying on the west bank of the Potomac was returned to Virginia. Find the coordinates of each corner of the original square that outlined the District of Columbia.

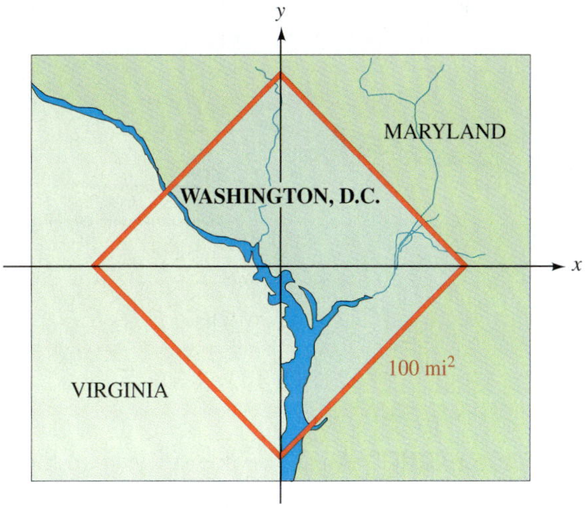

42. PAPER AIRPLANES The illustration gives the directions for making a paper airplane from a square piece of paper with sides 8 inches long. Find the length ℓ of the plane when it is completed.

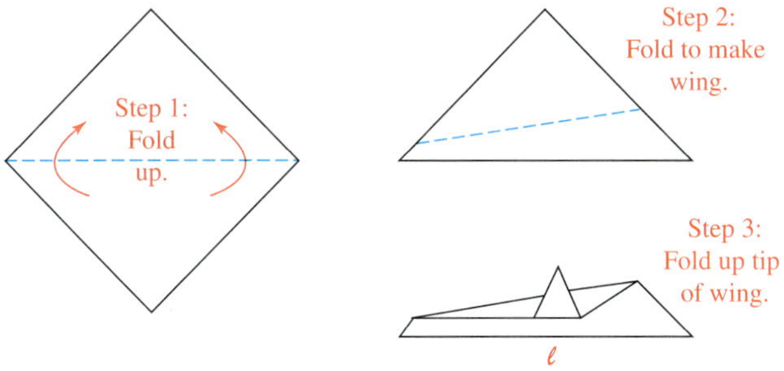

43. HARDWARE The sides of a regular hexagonal nut are 10 millimeters long. Find the height h of the nut.

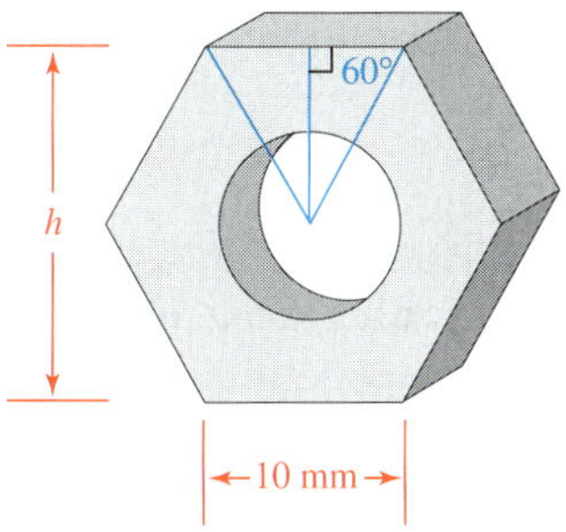

44. IRONING BOARDS Find the height h of the ironing board shown in the illustration in the next column.

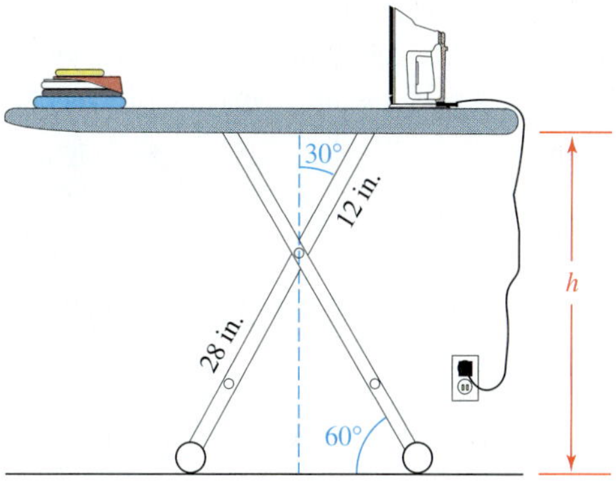

45. BASEBALL A baseball diamond is a square, 90 feet on a side. If the third baseman fields a ground ball 10 feet directly behind third base, how far must he throw the ball to throw a runner out at first base?

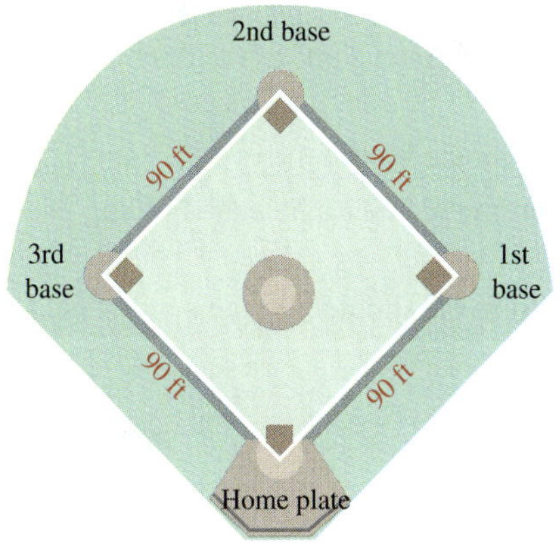

46. BASEBALL A shortstop fields a grounder at a point one-third of the way from second base to third base. How far will he have to throw the ball to make an out at first base?

47. CLOTHESLINES A pair of damp jeans are hung on a clothesline to dry. They pull the center down 1 foot. By how much is the line stretched?

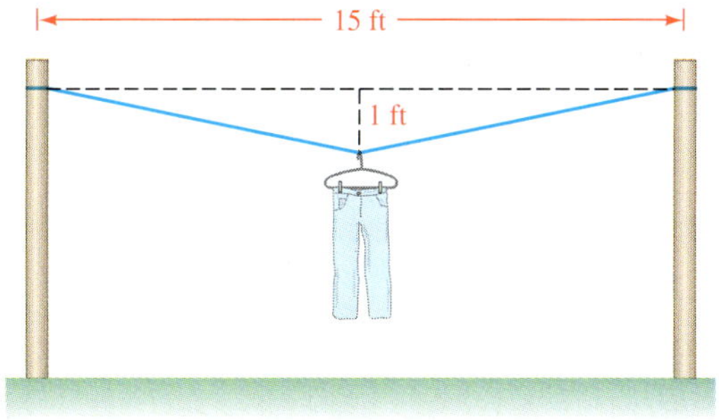

48. FIREFIGHTING The base of the 37-foot ladder is 9 feet from the wall. Will the top reach a window ledge that is 35 feet above the ground? Verify your result.

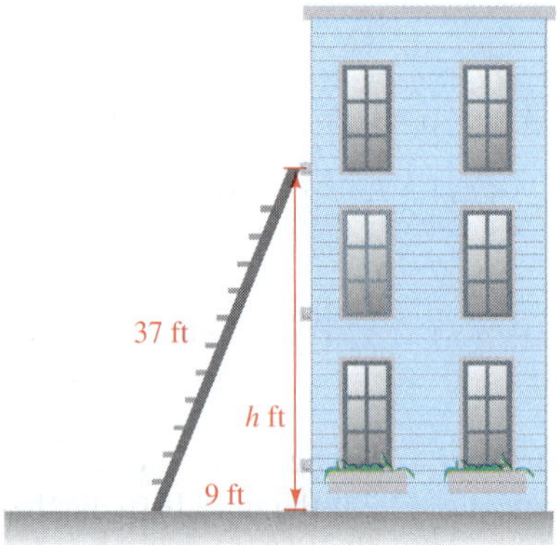

49. ART HISTORY A figure displaying some of the characteristics of Egyptian art is shown in the illustration. Use the distance formula to find the following dimensions of the drawing. Round your answers to two decimal places.

a. From the foot to the eye

b. From the belt to the hand holding the staff

c. From the shoulder to the symbol held in the hand

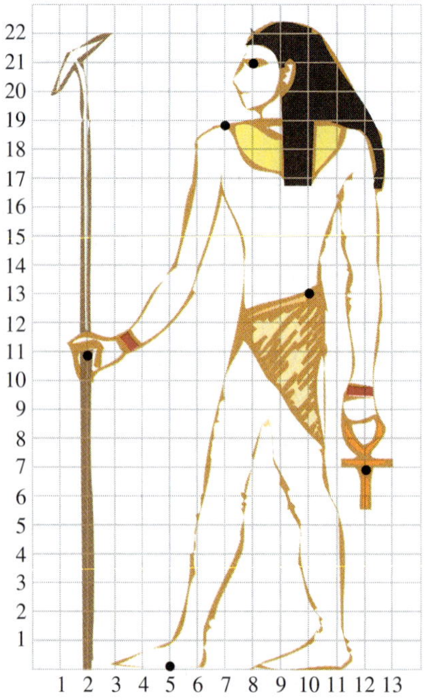

50. PACKAGING The diagonal d of a rectangular box with dimensions $a \times b \times c$ is given by

$$d = \sqrt{a^2 + b^2 + c^2}$$

Will the umbrella fit in the shipping carton in the illustration? Verify your result.

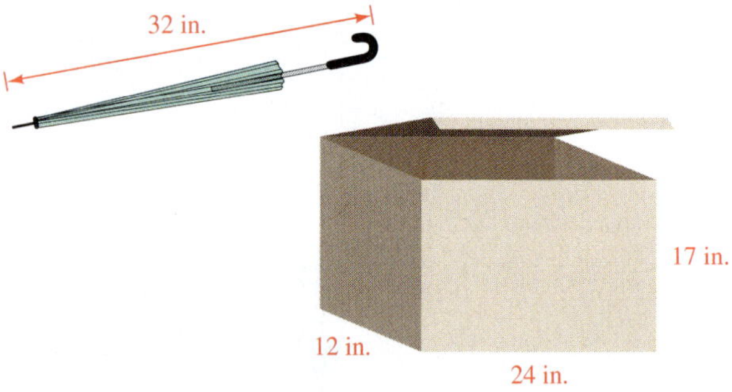

51. PACKAGING An archaeologist wants to ship a 34-inch femur bone. Will it fit in a 4-inch-tall box that has a 24-inch-square base? (See Exercise 50.) Verify your result.

52. TELEPHONE SERVICE The telephone cable in the illustration runs from A to B to C to D. How much cable is required to run from A to D directly?

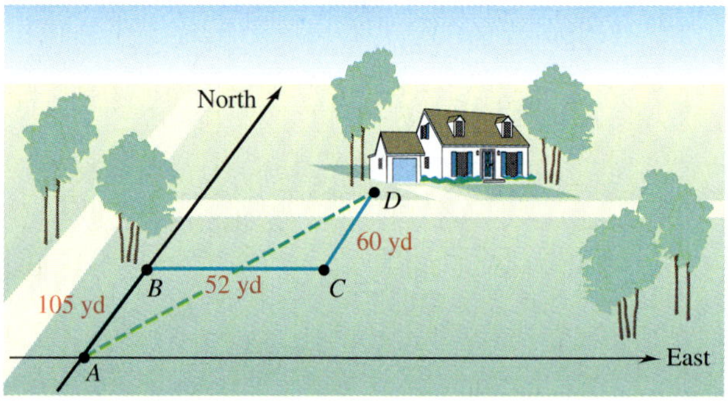

WRITING

53. State the Pythagorean theorem in words.

54. List the facts that you learned about special right triangles in this section.

REVIEW

55. DISCOUNT BUYING A repairman purchased some washing-machine motors for a total of \$224. When the unit cost decreased by \$4, he was able to buy one extra motor for the same total price. How many motors did he buy originally?

56. AVIATION An airplane can fly 650 miles with the wind in the same amount of time as it can fly 475 miles against the wind. If the wind speed is 40 mph, find the speed of the plane in still air.

57. Find the mean of 16, 6, 10, 4, 5, 13

58. Find the median of 16, 6, 10, 4, 5

CHALLENGE PROBLEMS

59. Find the length of the diagonal of the cube.

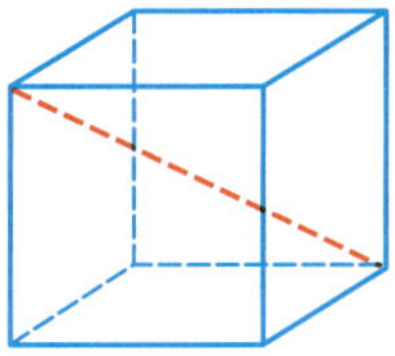

60. Show that the length of the diagonal of the rectangular solid shown is $\sqrt{a^2 + b^2 + c^2}$ cm.

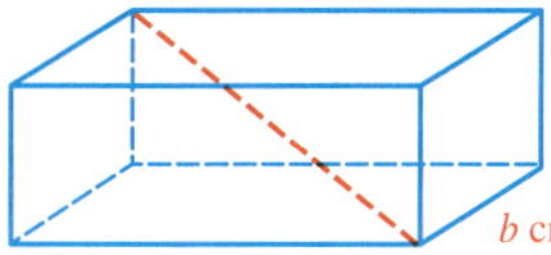

7.7 Complex Numbers

- The imaginary number i
- Simplifying square roots of negative numbers
- Complex numbers
- Arithmetic of complex numbers
- Complex conjugates
- Division of complex numbers
- Powers of i

The Language of Algebra

For years, mathematicians thought numbers like $\sqrt{-9}$ and $\sqrt{-25}$ were useless. In the 17th century, René Descartes (1596–1650) called them *imaginary numbers.* Today they have important uses such as describing alternating electric current.

Recall that the square root of a negative number is not a real number. However, an expanded number system, called the *complex number system,* has been devised to give meaning to $\sqrt{-9}$, $\sqrt{-25}$, and the like. To define complex numbers, we use a new type of number that is denoted by the letter i.

THE IMAGINARY NUMBER i

Some equations do not have real-number solutions. For example, $x^2 = -1$ has no real-number solutions because the square of a real number is never negative. To provide a solution to this equation, mathematicians have defined the number i in such a way that $i^2 = -1$.

The Number i

The **imaginary number i** is defined as

$$i = \sqrt{-1}$$

From the definition, it follows that $i^2 = -1$.

This definition enables us to write the square root of any negative number in terms of i.

SIMPLIFYING SQUARE ROOTS OF NEGATIVE NUMBERS

We can use extensions of the product and quotient rules for radicals to write the square root of a negative number as the product of a real number and i.

EXAMPLE 1

Write each expression in terms of i: **a.** $\sqrt{-9}$, **b.** $\sqrt{-7}$, **c.** $-\sqrt{-18}$, and **d.** $\sqrt{-\frac{24}{49}}$.

Solution We write each negative radicand as the product of -1 and a positive number and use the product rule for radicals. (The product rule, $\sqrt{ab} = \sqrt{a}\sqrt{b}$, holds when a is a negative real number and b is a positive real number.) Then we replace $\sqrt{-1}$ with i.

a. $\sqrt{-9} = \sqrt{-1 \cdot 9} = \sqrt{-1}\sqrt{9} = i \cdot 3 = 3i$

b. $\sqrt{-7} = \sqrt{-1 \cdot 7} = \sqrt{-1}\sqrt{7} = i\sqrt{7}$ or $\sqrt{7}i$

c. $-\sqrt{-18} = -\sqrt{-1 \cdot 9 \cdot 2} = -\sqrt{-1}\sqrt{9}\sqrt{2} = -i \cdot 3 \cdot \sqrt{2} = -3i\sqrt{2}$ or $-3\sqrt{2}i$

d. $\sqrt{-\frac{24}{49}} = \sqrt{-1 \cdot \frac{24}{49}} = \frac{\sqrt{-1 \cdot 24}}{\sqrt{49}} = \frac{\sqrt{-1}\sqrt{4}\sqrt{6}}{\sqrt{49}} = \frac{2i\sqrt{6}}{7}$ or $\frac{2\sqrt{6}}{7}i$

Self Check 1 Write each expression in terms of i: **a.** $\sqrt{-25}$, **b.** $-\sqrt{-19}$, **c.** $\sqrt{-45}$, and **d.** $\sqrt{-\frac{50}{81}}$.

The results from Example 1 illustrate a rule for simplifying square roots of negative numbers.

Square Root of a Negative Number

For any positive real number b,

$$\sqrt{-b} = i\sqrt{b}$$

Notation

Since it is easy to confuse $\sqrt{b}i$ with $\sqrt{bi}$, we write i first so that it is clear that the i is not under the radical symbol and part of the radicand. However, both $i\sqrt{b}$ and $\sqrt{b}i$ are correct.

To justify this rule, we use the fact that $\sqrt{-1} = i$.

$$\begin{aligned}\sqrt{-b} &= \sqrt{-1 \cdot b}\\ &= \sqrt{-1}\sqrt{b}\\ &= i\sqrt{b}\end{aligned}$$

COMPLEX NUMBERS

The imaginary number i is used to define *complex numbers.*

Complex Numbers

A **complex number** is any number that can be written in the form $a + bi$, where a and b are real numbers and $i = \sqrt{-1}$.

Complex numbers of the form $a + bi$, where $b \neq 0$, are also called **imaginary numbers.***

Notation

It is acceptable to use $a - bi$ as a substitute for the form $a + (-b)i$. For example: $6 - 9i = 6 + (-9)i$.

For a complex number written in the **standard form** $a + bi$, we call a the **real part** and b the **imaginary part.** Some examples of complex numbers written in standard form are

$$2 + 11i \qquad 6 - 9i \qquad -\frac{1}{2} + 0i \qquad 0 + i\sqrt{3}$$

Two complex numbers are equal if and only if their real parts are equal and their imaginary parts are equal. Thus, $0.5 + 0.9i = \frac{1}{2} + \frac{9}{10}i$ because $0.5 = \frac{1}{2}$ and $0.9 = \frac{9}{10}$.

EXAMPLE 2

Write each number in the form $a + bi$: **a.** 6, **b.** $\sqrt{-64}$, and **c.** $-2 + \sqrt{-63}$.

Solution

a. $6 = 6 + 0i$ — The imaginary part is 0.

b. $\sqrt{-64} = 0 + 8i$ — The real part is 0. $\sqrt{-64} = \sqrt{-1}\sqrt{64} = 8i$.

c. $-2 + \sqrt{-63} = -2 + 3i\sqrt{7}$ — $\sqrt{-63} = \sqrt{-1}\sqrt{63} = \sqrt{-1}\sqrt{9}\sqrt{7} = 3i\sqrt{7}$.

*Some textbooks define imaginary numbers as complex numbers with $a = 0$ and $b \neq 0$.

Self Check 2 Write each number in the form $a + bi$: **a.** -18, **b.** $\sqrt{-36}$, and **c.** $1 + \sqrt{-24}$.

The following illustration shows the relationship between the real numbers, the imaginary numbers, and the complex numbers.

Success Tip

Just as real numbers are either rational or irrational, but not both, complex numbers are either real or imaginary, but not both.

Complex numbers

Real numbers				Imaginary numbers		
-6	$\frac{5}{16}$	-1.75	π	$9 + 7i$	$-2i$	$\frac{1}{4} - \frac{3}{4}i$
$48 + 0i$	0	$-\sqrt{10}$	$-\frac{7}{2}$	$0.56i$	$\sqrt{-10}$	$6 + i\sqrt{3}$

ARITHMETIC OF COMPLEX NUMBERS

We now consider how to add, subtract, multiply, and divide complex numbers.

Addition and Subtraction of Complex Numbers To add (or subtract) two complex numbers, add (or subtract) their real parts and add (or subtract) their imaginary parts.

EXAMPLE 3

Perform each operation: **a.** $(8 + 4i) + (12 + 8i)$, **b.** $(-6 + i) - (3 + 2i)$, and **c.** $\left(7 - \sqrt{-16}\right) + \left(9 + \sqrt{-4}\right)$.

Solution

a. $(8 + 4i) + (12 + 8i) = (8 + 12) + (4 + 8)i$ Add the real parts. Add the imaginary parts.

$= 20 + 12i$

b. $(-6 + i) - (3 + 2i) = (-6 - 3) + (1 - 2)i$ Subtract the real parts. Subtract the imaginary parts.

$= -9 - i$

c. $\left(7 - \sqrt{-16}\right) + \left(9 + \sqrt{-4}\right)$

$= (7 - 4i) + (9 + 2i)$ Write $\sqrt{-16}$ and $\sqrt{-4}$ in terms of i.

$= (7 + 9) + (-4 + 2)i$ Add the real parts. Add the imaginary parts.

$= 16 - 2i$ Write $16 + (-2i)$ in the form $16 - 2i$.

Success Tip

Always change complex numbers to $a + bi$ form before performing any arithmetic.

Self Check 3 Perform the operations: **a.** $(3 - 5i) + (-2 + 7i)$ and **b.** $\left(3 - \sqrt{-25}\right) - \left(-2 + \sqrt{-49}\right)$.

Imaginary numbers are not real numbers; some properties of real numbers do not apply to imaginary numbers. For example, we cannot use the product rule for radicals to multiply two imaginary numbers.

Caution If a and b are both negative, then $\sqrt{a}\sqrt{b} \neq \sqrt{ab}$. For example, if $a = -4$ and $b = -9$,

$$\cancel{\sqrt{-4}\sqrt{-9} = \sqrt{-4(-9)} = \sqrt{36} = 6} \qquad \sqrt{-4}\sqrt{-9} = 2i(3i) = 6i^2 = 6(-1) = -6$$

EXAMPLE 4

Multiply: $\sqrt{-2}\sqrt{-20}$.

Solution We first express $\sqrt{-2}$ and $\sqrt{-20}$ in terms of i. Then, we can multiply the radical expressions as usual because the radicands are positive numbers.

$$\begin{aligned} \sqrt{-2}\sqrt{-20} &= \left(i\sqrt{2}\right)\left(2i\sqrt{5}\right) && \text{Simplify: } \sqrt{-20} = i\sqrt{20} = 2i\sqrt{5}. \\ &= 2i^2\sqrt{10} && i \cdot 2i = 2i^2 \text{ and } \sqrt{2}\sqrt{5} = \sqrt{10}. \\ &= -2\sqrt{10} && \text{Simplify: } i^2 = -1. \end{aligned}$$

Self Check 4 Multiply: $\sqrt{-3}\sqrt{-32}$.

EXAMPLE 5

Multiply: **a.** $6(2 + 9i)$ and **b.** $-5i(4 - 8i)$.

Solution **a.** To multiply a complex number by a real number, we use the distributive property to remove parentheses and then simplify. For example,

$$\begin{aligned} 6(2 + 9i) &= 6(2) + 6(9i) && \text{Use the distributive property.} \\ &= 12 + 54i && \text{Simplify.} \end{aligned}$$

b. To multiply a complex number by an imaginary number, we use the distributive property to remove parentheses and then simplify. For example,

$$\begin{aligned} -5i(4 - 8i) &= -5i(4) - (-5i)8i && \text{Use the distributive property.} \\ &= -20i + 40i^2 && \text{Simplify.} \\ &= -40 - 20i && \text{Since } i^2 = -1,\ 40i^2 = 40(-1) = -40. \end{aligned}$$

Caution

A common mistake is to replace i with -1. Remember, $i \neq -1$. By definition, $i = \sqrt{-1}$ and $i^2 = -1$.

Self Check 5 Multiply: **a.** $-2(-9 - i)$ and **b.** $10i(7 + 4i)$.

To multiply two complex numbers, we can use the FOIL method.

EXAMPLE 6

Multiply: **a.** $(2 + 3i)(3 - 2i)$ and **b.** $(-4 + 2i)(2 + i)$.

Solution **a.**

$$\begin{aligned} (2 + 3i)(3 - 2i) &= 6 - 4i + 9i - 6i^2 && \text{Use the FOIL method.} \\ &= 6 + 5i - (-6) && \text{Combine the imaginary terms: } -4i + 9i = 5i. \text{ Simplify: } i^2 = -1, \text{ so } 6i^2 = -6. \\ &= 6 + 5i + 6 \\ &= 12 + 5i && \text{Combine like terms.} \end{aligned}$$

Success Tip

i is not a variable, but you can think of it as one when adding, subtracting, and multiplying. For example:

$-4i + 9i = 5i$
$6i - 2i = 4i$
$i \cdot i = i^2$

b. $(-4 + 2i)(2 + i) = -8 - 4i + 4i + 2i^2$ Use the FOIL method.
$= -8 + 0i - 2$ $-4i + 4i = 0i$. Since $i^2 = -1$, $2i^2 = -2$.
$= -10 + 0i$

Self Check 6 Multiply: $(-2 + 3i)(3 - 2i)$.

COMPLEX CONJUGATES

Before we can discuss division of complex numbers, we must introduce an important fact about *complex conjugates.*

Complex Conjugates The complex numbers $a + bi$ and $a - bi$ are called **complex conjugates.**

For example,

- $7 + 4i$ and $7 - 4i$ are complex conjugates.
- $5 - i$ and $5 + i$ are complex conjugates.
- $-6i$ and $6i$ are complex conjugates, because $-6i = 0 - 6i$ and $6i = 0 + 6i$.

The Language of Algebra

Recall that the word *conjugate* was used earlier when we rationalized the denominators of radical expressions such as $\frac{5}{\sqrt{6} - 1}$.

In general, the product of the complex number $a + bi$ and its complex conjugate $a - bi$ is the real number $a^2 + b^2$, as the following work shows:

$(a + bi)(a - bi) = a^2 - abi + abi - b^2i^2$ Use the FOIL method.
$= a^2 - b^2(-1)$ $-abi + abi = 0$. Replace i^2 with -1.
$= a^2 + b^2$

EXAMPLE 7

Find the product of $3 + i$ and its complex conjugate.

Solution The complex conjugate of $3 + i$ is $3 - i$. We can find the product as follows:

$(3 + i)(3 - i) = 9 - 3i + 3i - i^2$ Use the FOIL method.
$= 9 - i^2$ Combine like terms: $-3i + 3i = 0$.
$= 9 - (-1)$ $i^2 = -1$.
$= 10$

The product of the complex numbers $3 + i$ and $3 - i$ is the real number 10.

Self Check 7 Multiply: $(2 + 3i)(2 - 3i)$.

DIVISION OF COMPLEX NUMBERS

Recall that to divide *radical expressions,* we rationalize the denominator. We use a similar approach to divide complex numbers.

Division of Complex Numbers To divide complex numbers, multiply the numerator and denominator by the complex conjugate of the denominator.

EXAMPLE 8

Find the quotient: **a.** $\dfrac{1}{3+i}$ and **b.** $\dfrac{3-i}{2+i}$.

Solution **a.** We multiply the numerator and the denominator of the fraction by the complex conjugate of the denominator, which is $3 - i$.

$$\frac{1}{3+i} = \frac{1}{3+i} \cdot \frac{3-i}{3-i}$$
Multiply $\frac{1}{3+i}$ by a form of 1: $\frac{3-i}{3-i} = 1$.

$$= \frac{3-i}{9-3i+3i-i^2}$$
Multiply the numerators and multiply the denominators.

$$= \frac{3-i}{9-(-1)}$$
$i^2 = -1$. Note that the denominator no longer contains i.

$$= \frac{3-i}{10}$$
Simplify in the denominator.

$$= \frac{3}{10} - \frac{1}{10}i$$
Write the complex number in $a + bi$ form.

b.
$$\frac{3-i}{2+i} = \frac{3-i}{2+i} \cdot \frac{2-i}{2-i}$$
The complex conjugate of the denominator of the fraction is $2 - i$. Multiply by $\frac{2-i}{2-i} = 1$.

$$= \frac{6-3i-2i+i^2}{4-2i+2i-i^2}$$
Multiply the numerators and multiply the denominators.

$$= \frac{5-5i}{4-(-1)}$$
$i^2 = -1$. The denominator is now a real number.

$$= \frac{\overset{1}{\cancel{5}}(1-i)}{\underset{1}{\cancel{5}}}$$
Factor out 5 in the numerator and remove the common factor of 5.

$$= 1 - i$$

Self Check 8 Find the quotient: **a.** $\dfrac{1}{5-i}$ and **b.** $\dfrac{5+4i}{3+2i}$.

EXAMPLE 9

Find the quotient: $\dfrac{4+\sqrt{-16}}{2+\sqrt{-4}}$. Write the result in $a + bi$ form.

Solution

$$\frac{4+\sqrt{-16}}{2+\sqrt{-4}} = \frac{4+4i}{2+2i}$$
Write the numerator and denominator in $a + bi$ form.

$$= \frac{2\overset{1}{\cancel{(2+2i)}}}{\underset{1}{\cancel{2+2i}}}$$
Factor out 2 in the numerator and remove the common factor of $2 + 2i$.

$$= 2 + 0i$$
Write 2 in the form $a + bi$.

Self Check 9 Find the quotient: $\dfrac{3+\sqrt{-9}}{4+\sqrt{-16}}$.

EXAMPLE 10

Find the quotient: $\dfrac{7}{2i}$. Write the result in $a + bi$ form.

Solution The denominator can be expressed as $0 + 2i$. Its conjugate is $0 - 2i$, or just $-2i$.

$$\frac{7}{2i} = \frac{7}{2i} \cdot \frac{-2i}{-2i} \qquad \text{Multiply } \frac{7}{2i} \text{ by a form of 1 using the complex conjugate of the denominator: } \frac{-2i}{-2i} = 1.$$

$$= \frac{-14i}{-4i^2}$$

$$= \frac{-14i}{4} \qquad i^2 = -1. \text{ The denominator is now a real number.}$$

$$= \frac{-7i}{2} \qquad \text{Simplify.}$$

$$= 0 - \frac{7}{2}i \qquad \text{Write in } a + bi \text{ form.}$$

Success Tip

In this example, the denominator of $\frac{7}{2i}$ is of the form bi. In such cases, we can eliminate i in the denominator by simply multiplying by $\frac{i}{i}$.

$$\frac{7}{2i} = \frac{7}{2i} \cdot \frac{i}{i} = \frac{7i}{2i^2} = -\frac{7}{2}i$$

Self Check 10 Divide: $\dfrac{5}{-i}$.

POWERS OF *i*

The powers of i produce an interesting pattern:

$$i = \sqrt{-1} = i \qquad i^5 = i^4 i = 1i = i$$

$$i^2 = \left(\sqrt{-1}\right)^2 = -1 \qquad i^6 = i^4 i^2 = 1(-1) = -1$$

$$i^3 = i^2 i = -1i = -i \qquad i^7 = i^4 i^3 = 1(-i) = -i$$

$$i^4 = i^2 i^2 = (-1)(-1) = 1 \qquad i^8 = i^4 i^4 = (1)(1) = 1$$

The pattern continues: $i, -1, -i, 1, \ldots$.

Success Tip

Note that the powers of i cycle through four possible outcomes:

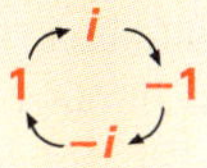

Larger powers of i can be simplified by using the fact that $i^4 = 1$. For example, to simplify i^{29}, we note that 29 divided by 4 gives a quotient of 7 and a remainder of 1. Thus, $29 = 4 \cdot 7 + 1$ and

$$i^{29} = i^{4\cdot 7+1} \qquad 4 \cdot 7 = 28.$$

$$= (i^4)^7 \cdot i^1$$

$$= 1^7 \cdot i \qquad i^4 = 1.$$

$$= i \qquad 1 \cdot i = i.$$

The result of this example illustrates the following fact.

Powers of *i* If n is a natural number that has a remainder of r when divided by 4, then

$$i^n = i^r$$

EXAMPLE 11

Simplify: i^{55}.

Solution We divide 55 by 4 and get a remainder of 3. Therefore,

$$i^{55} = i^3 = -i$$

Self Check 11 Simplify: i^{62}.

Answers to Self Checks **1. a.** $5i$, **b.** $-i\sqrt{19}$, **c.** $3i\sqrt{5}$, **d.** $\frac{5\sqrt{2}}{9}i$ **2. a.** $-18 + 0i$, **b.** $0 + 6i$, **c.** $1 + 2i\sqrt{6}$ **3. a.** $1 + 2i$, **b.** $5 - 12i$ **4.** $-4\sqrt{6}$ **5. a.** $18 + 2i$, **b.** $-40 + 70i$ **6.** $0 + 13i$ **7.** 13 **8. a.** $\frac{5}{26} + \frac{1}{26}i$, **b.** $\frac{23}{13} + \frac{2}{13}i$ **9.** $\frac{3}{4} + 0i$ **10.** $0 + 5i$ **11.** -1

7.7 STUDY SET

VOCABULARY Fill in the blanks.

1. The ________ number i is defined as $i = \sqrt{-1}$.
2. A ________ number is any number that can be written in the form $a + bi$, where a and b are real numbers and $i = \sqrt{-1}$.
3. For the complex number $2 + 5i$, we call 2 the ____ part and 5 the ________ part.
4. Complex numbers such as $5i$ and $-10i$ are called ____ imaginary numbers.
5. $6 + 3i$ and $6 - 3i$ are called complex ________.
6. i^{25} is called a ______ of i.

CONCEPTS Fill in the blanks.

7. **a.** $i =$ ____ **b.** $i^2 =$ ____ **c.** $i^3 =$ ____ **d.** $i^4 =$ ____
8. Simplify:
$$\sqrt{-36} = \sqrt{\quad \cdot 36} = \sqrt{\quad}\sqrt{36} = 6\ \ $$
9. To add (or subtract) complex numbers, add (or subtract) their ____ parts and add (or subtract) their ________ parts.
10. We ________ two complex numbers by using the FOIL method.
11. To divide complex numbers, multiply the numerator and denominator by the complex conjugate of the ________
12. The powers of i cycle through ____ possible outcomes.
13. Explain the error. Then find the correct result.
    a. Add: $\sqrt{-16} + \sqrt{-9} = \sqrt{-25}$.
    b. Multiply: $\sqrt{-2}\sqrt{-3} = \sqrt{-2(-3)} = \sqrt{6}$.
14. Give the complex conjugate of each number.
    a. $2 - 3i$ **b.** 2 **c.** $-3i$
15. Complete the illustration. Label the real numbers, the imaginary numbers, the complex numbers, the rational numbers, and the irrational numbers.

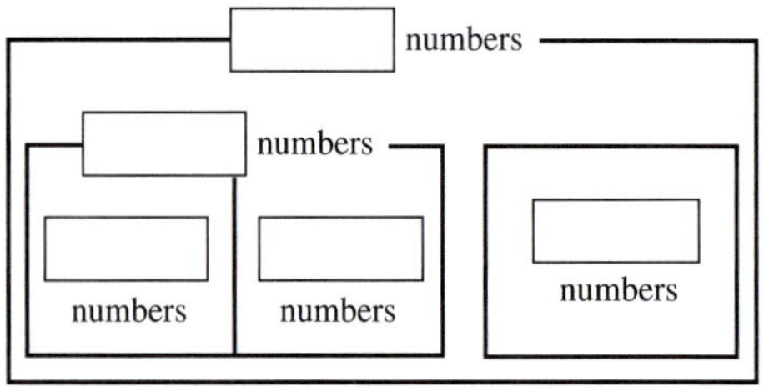

16. Decide whether each statement is true or false.
    a. Every complex number is a real number.
    b. Every real number is a complex number.
    c. i is a real number.
    d. The square root of a negative number is an imaginary number.

17. Decide whether the complex numbers are equal.

a. $4 - \frac{2}{5}i, \frac{8}{2} - 0.4i$

b. $0.25 + 0.7i, \frac{1}{4} + \frac{7}{10}i$

18. To divide $6 + 7i$ by $1 - 8i$, we multiply $\frac{6 + 7i}{1 - 8i}$ by a form of 1. What form of 1 do we use?

NOTATION **Complete each operation.**

19. $(3 + 2i)(3 - i) = \square - 3i + \square - 2i^2$
$= 9 + 3i + \square$
$= \square + 3i$

20. $\frac{3}{2 - i} = \frac{3}{2 - i} \cdot \frac{3}{\square}$
$= \frac{6 + \square}{4 + \square - 2i - i^2}$
$= \frac{6 + 3i}{\square}$
$= \square + \frac{3}{5}i$

21. Decide whether each statement is true or false.

a. $\sqrt{6}i = i\sqrt{6}$ **b.** $\sqrt{8}i = \sqrt{8i}$

c. $\sqrt{-25} = -\sqrt{25}$ **d.** $-i = i$

22. Write each number in the form $a + bi$.

a. $\frac{9 + 11i}{4}$ **b.** $\frac{1 - i}{18}$

PRACTICE **Express each number in terms of *i*.**

23. $\sqrt{-9}$ **24.** $\sqrt{-4}$

25. $\sqrt{-7}$ **26.** $\sqrt{-11}$

27. $\sqrt{-24}$ **28.** $\sqrt{-28}$

29. $-\sqrt{-24}$ **30.** $-\sqrt{-72}$

31. $5\sqrt{-81}$ **32.** $6\sqrt{-49}$

33. $\sqrt{-\frac{25}{9}}$ **34.** $-\sqrt{-\frac{121}{144}}$

Simplify each expression.

35. $\sqrt{-1}\sqrt{-36}$ **36.** $\sqrt{-9}\sqrt{-100}$

37. $\sqrt{-2}\sqrt{-6}$ **38.** $\sqrt{-3}\sqrt{-6}$

39. $\frac{\sqrt{-25}}{\sqrt{-64}}$ **40.** $\frac{\sqrt{-4}}{\sqrt{-1}}$

41. $-\frac{\sqrt{-400}}{\sqrt{-1}}$ **42.** $-\frac{\sqrt{-225}}{\sqrt{-16}}$

Perform the operations. Write all answers in $a + bi$ form.

43. $(3 + 4i) + (5 - 6i)$ **44.** $(5 + 3i) - (6 - 9i)$

45. $(7 - 3i) - (4 + 2i)$ **46.** $(8 + 3i) + (-7 - 2i)$

47. $(6 - i) + (9 + 3i)$ **48.** $(5 - 4i) - (3 + 2i)$

49. $\left(8 + \sqrt{-25}\right) + \left(7 + \sqrt{-4}\right)$

50. $\left(-7 + \sqrt{-81}\right) - \left(-2 - \sqrt{-64}\right)$

51. $3(2 - i)$ **52.** $-4(3 + 4i)$

53. $-5i(5 - 5i)$ **54.** $2i(7 + 2i)$

55. $(2 + i)(3 - i)$ **56.** $(4 - i)(2 + i)$

57. $(3 - 2i)(2 + 3i)$ **58.** $(3 - i)(2 + 3i)$

59. $(4 + i)(3 - i)$ **60.** $(1 - 5i)(1 - 4i)$

61. $\left(2 - \sqrt{-16}\right)\left(3 + \sqrt{-4}\right)$

62. $\left(3 - \sqrt{-4}\right)\left(4 - \sqrt{-9}\right)$

63. $\left(2 + i\sqrt{2}\right)\left(3 - i\sqrt{2}\right)$ **64.** $\left(5 + i\sqrt{3}\right)\left(2 - i\sqrt{3}\right)$

65. $(2 + i)^2$ **66.** $(3 - 2i)^2$

67. $(3i)^2$ **68.** $(5i)^2$

69. $\left(i\sqrt{6}\right)^2$ **70.** $\left(i\sqrt{2}\right)^2$

71. $\frac{1}{i}$ **72.** $\frac{1}{i^3}$

73. $\frac{4}{5i^3}$ **74.** $\frac{3}{2i}$

75. $\frac{3i}{8\sqrt{-9}}$ **76.** $\frac{5i^3}{2\sqrt{-4}}$

77. $\frac{-3}{5i^5}$ **78.** $\frac{-4}{6i^7}$

79. $\frac{5}{2 - i}$ **80.** $\frac{3}{5 + i}$

81. $\frac{-12}{7 - \sqrt{-1}}$ **82.** $\frac{-4}{3 + \sqrt{-1}}$

83. $\frac{5i}{6 + 2i}$ **84.** $\frac{3i}{6 - i}$

85. $\dfrac{-2i}{3+2i}$

86. $\dfrac{-4i}{2-6i}$

87. $\dfrac{3-2i}{3+2i}$

88. $\dfrac{2+3i}{2-3i}$

89. $\dfrac{3+2i}{3+i}$

90. $\dfrac{2-5i}{2+5i}$

91. $\dfrac{\sqrt{5}-i\sqrt{3}}{\sqrt{5}+i\sqrt{3}}$

92. $\dfrac{\sqrt{3}+i\sqrt{2}}{\sqrt{3}-i\sqrt{2}}$

Simplify each expression.

93. i^{21}

94. i^{19}

95. i^{27}

96. i^{22}

97. i^{100}

98. i^{42}

99. i^{97}

100. i^{200}

APPLICATIONS

101. FRACTALS Complex numbers are fundamental in the creation of the intricate geometric shape shown below, called a *fractal.* The process of creating this image is based on the following sequence of steps, which begins by picking any complex number, which we will call z.

1. Square z, and then add that result to z.
2. Square the result from step 1, and then add it to z.
3. Square the result from step 2, and then add it to z.

If we begin with the complex number i, what is the result after performing steps 1, 2, and 3?

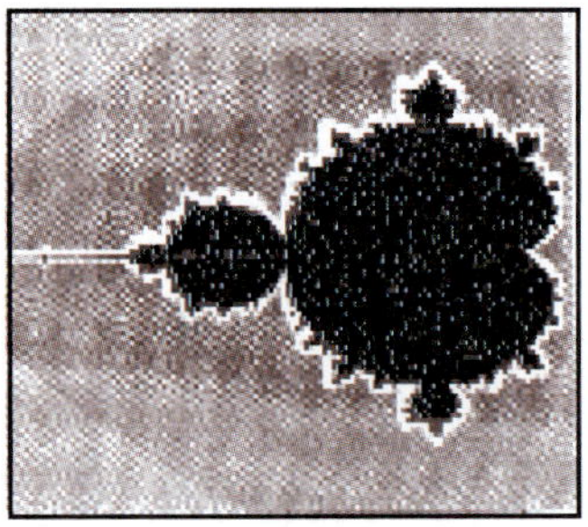

102. ELECTRONICS The impedance Z in an ac (alternating current) circuit is a measure of how much the circuit impedes (hinders) the flow of current through it. The impedance is related to the voltage V and the current I by the formula

$$V = IZ$$

If a circuit has a current of $(0.5 + 2.0i)$ amps and an impedance of $(0.4 - 3.0i)$ ohms, find the voltage.

WRITING

103. What is an imaginary number? What is a complex number?

104. The method used to divide complex numbers is similar to the method used to divide radical expressions. Explain why. Give an example.

REVIEW

105. WIND SPEEDS A plane that can fly 200 mph in still air makes a 330-mile flight with a tail wind and returns, flying into the same wind. Find the speed of the wind if the total flying time is $3\frac{1}{3}$ hours.

106. FINDING RATES A student drove a distance of 135 miles at an average speed of 50 mph. How much faster would she have to drive on the return trip to save 30 minutes of driving time?

CHALLENGE PROBLEMS

107. Rationalize the numerator of $\dfrac{2+3i}{2-3i}$.

108. Simplify: $(2+3i)^{-2}$. Write the result in the form $a + bi$.

ACCENT ON TEAMWORK

A SPIRAL OF ROOTS

Overview: In this activity, you will create a visual representation of a collection of square roots.

Instructions: Form groups of 2 or 3 students. You will need a piece of unlined paper, a protractor, a ruler, and a pencil. Begin by drawing an isosceles right triangle with legs of length 1 inch in the middle of the paper. (See the illustration.) Use the Pythagorean theorem to determine the length of the hypotenuse. Draw a second right triangle using the hypotenuse of the first right triangle as one leg. Draw its second leg with length 1 inch. Find the length of the hypotenuse of the second triangle.

Continue creating right triangles, using the previous hypotenuse as one leg and drawing a new second leg of length 1 inch each time. Calculate the length of each resulting hypotenuse. When the figure begins to spiral onto itself, you may stop the process. Make a list of the lengths of each hypotenuse. What pattern do you see?

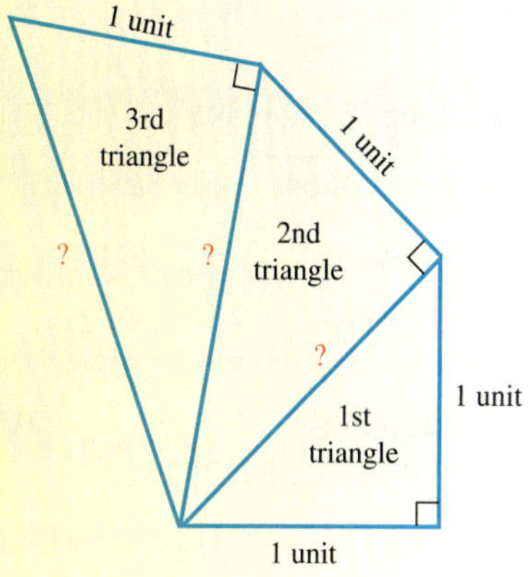

GRAPHING IN THREE DIMENSIONS

Overview: In this activity, you will find the distance between two points that lie in three-dimensional space.

Instructions: Form groups of 2 or 3 students. In a three-dimensional Cartesian coordinate system, the positive x-axis is horizontal and pointing toward the viewer (out of the page), the positive y-axis is also horizontal and pointing to the right, and the positive z-axis is vertical, pointing up. A point is located by plotting an ordered triple of numbers (x, y, z). In the illustration, the point $(3, 2, 4)$ is plotted.

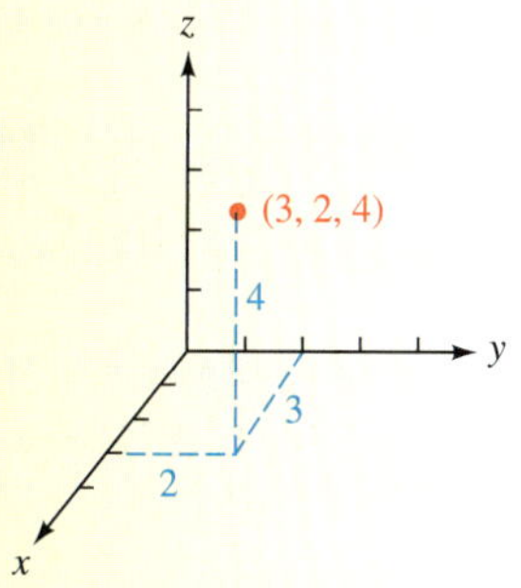

In three dimensions, the distance formula is

$$d = \sqrt{(x_2 - x_1)^2 + (y_2 - y_1)^2 + (z_2 - z_1)^2}$$

1. Copy the illustration shown above. Then plot the point $(1, 4, 3)$. Use the distance formula to find the distance between these two points.
2. Draw another three-dimensional coordinate system and plot the points $(-3, 3, -4)$ and $(2, -3, 2)$. Use the distance formula to find the distance between these two points.

KEY CONCEPT: RADICALS

The expression $\sqrt[n]{a}$ is called a **radical expression.** In this chapter, we have discussed the properties and procedures used when simplifying radical expressions, solving radical equations, and writing radical expressions using rational exponents.

EXPRESSIONS CONTAINING RADICALS

When working with expressions containing radicals, we must often apply the product rule and/or the quotient rule for radicals to simplify the expression. Recall that if a and b are both nonnegative,

$$\sqrt[n]{ab} = \sqrt[n]{a}\sqrt[n]{b} \qquad \sqrt[n]{\frac{a}{b}} = \frac{\sqrt[n]{a}}{\sqrt[n]{b}}, \quad \text{where } b \neq 0$$

Perform each operation and simplify the expression.

1. Simplify: $\sqrt[3]{-54h^6}$.
2. Add: $2\sqrt[3]{64e} + 3\sqrt[3]{8e}$.
3. Subtract: $\sqrt{72} - \sqrt{200}$.
4. Multiply: $-4\sqrt[3]{5r^2s}\left(5\sqrt[3]{2r}\right)$.
5. Multiply: $\left(\sqrt{3s} - \sqrt{2t}\right)\left(\sqrt{3s} + \sqrt{2t}\right)$.
6. Multiply: $-\sqrt{3}\left(\sqrt{7} - \sqrt{5}\right)$.
7. Find the power: $\left(3\sqrt{2n} - 2\right)^2$.
8. Rationalize the denominator: $\dfrac{\sqrt[3]{9j}}{\sqrt[3]{3jk}}$.

EQUATIONS CONTAINING RADICALS

When solving radical equations, our objective is to rid the equation of the radical. This is achieved by using the *power rule:*

If x, y, and n are real numbers and $x = y$, then $x^n = y^n$.

If we raise both sides of an equation to the same power, the resulting equation might not be equivalent to the original equation. We must always check for extraneous solutions.

Solve each equation, if possible.

9. $\sqrt{1 - 2g} = \sqrt{g + 10}$
10. $4 - \sqrt[3]{4 + 12x} = 0$
11. $\sqrt{y + 2} - 4 = -y$
12. $\sqrt[4]{12t + 4} + 2 = 0$

RADICALS AND RATIONAL EXPONENTS

Radicals can be written using rational (fractional) exponents, and exponential expressions having fractional exponents can be written in radical form. To do this, we use two rules for exponents introduced in this chapter.

$$x^{1/n} = \sqrt[n]{x} \qquad x^{m/n} = \sqrt[n]{x^m} = \left(\sqrt[n]{x}\right)^m$$

13. Express using a rational exponent: $\sqrt[3]{3}$.
14. Express in radical form: $5a^{2/5}$.

CHAPTER REVIEW

SECTION 7.1 Radical Expressions and Radical Functions

CONCEPTS

The number b is a *square root of a* if $b^2 = a$.

If $x > 0$, the *principal square root of x* is the positive square root of x, denoted $\sqrt{x}$. If x can be any real number, then $\sqrt{x^2} = |x|$.

The *cube root of x* is denoted as $\sqrt[3]{x}$ and is defined by

$\sqrt[3]{x} = y$ if $y^3 = x$

REVIEW EXERCISES

Simplify each expression, if possible. Assume that x and y can be any real number.

1. $\sqrt{49}$
2. $-\sqrt{121}$
3. $\sqrt{\dfrac{225}{49}}$
4. $\sqrt{-4}$
5. $\sqrt{0.01}$
6. $\sqrt{25x^2}$
7. $\sqrt{x^8}$
8. $\sqrt{x^2 + 4x + 4}$
9. $\sqrt[3]{-27}$
10. $-\sqrt[3]{216}$
11. $\sqrt[3]{64x^6y^3}$
12. $\sqrt[3]{\dfrac{x^9}{125}}$
13. $\sqrt[4]{625}$
14. $\sqrt[5]{-32}$

If n is an even natural number,

$\sqrt[n]{a^n} = |a|$

If n is an odd natural number,

$\sqrt[n]{a^n} = a$

If n is a natural number greater than 1 and x is a real number, then

- If $x > 0$, then $\sqrt[n]{x}$ is the positive number such that $(\sqrt[n]{x})^n = x$.
- If $x = 0$, then $\sqrt[n]{x} = 0$.
- If $x < 0$, and n is odd, $\sqrt[n]{x}$ is the negative number such that $(\sqrt[n]{x})^n = x$.
- If $x < 0$, and n is even, $\sqrt[n]{x}$ is not a real number.

15. $\sqrt[4]{256x^8y^4}$

16. $\sqrt[15]{(x+1)^{15}}$

17. $-\sqrt[4]{\dfrac{1}{16}}$

18. $\sqrt[6]{-1}$

19. $\sqrt{0}$

20. $\sqrt[3]{0}$

21. GEOMETRY The side of a square with area A square feet is given by the function $s(A) = \sqrt{A}$. Find the *perimeter* of a square with an area of 144 square feet.

22. VOLUME OF A CUBE The total surface area of a cube is related to its volume V by the function $A(V) = 6\sqrt[3]{V^2}$. Find the surface area of a cube with a volume of 8 cm^3.

Graph each function. Find the domain and range.

23. $f(x) = \sqrt{x+2}$

24. $f(x) = -\sqrt[3]{x} + 3$

SECTION 7.2 Rational Exponents

If n is a natural number greater than 1 and $\sqrt[n]{x}$ is a real number, then

$x^{1/n} = \sqrt[n]{x}$

If n is a natural number greater than 1 and x is a real number,

- If $x > 0$, then $x^{1/n}$ is the positive number such that $(x^{1/n})^n = x$.
- If $x = 0$, then $x^{1/n} = 0$.
- If $x < 0$, and n is odd, then $x^{1/n}$ is the negative number such that $(x^{1/n})^n = x$.
- If $x < 0$ and n is even, then $x^{1/n}$ is not a real number.

If m and n are positive integers, $x > 0$, and $\frac{m}{n}$ is in simplest form,

$x^{m/n} = \sqrt[n]{x^m} = (\sqrt[n]{x})^m$

$x^{-m/n} = \dfrac{1}{x^{m/n}}$

$\dfrac{1}{x^{-m/n}} = x^{m/n}$

Write each expression in radical form.

25. $t^{1/2}$

26. $(5xy^3)^{1/4}$

Simplify each expression, if possible. Assume that all variables represent positive real numbers.

27. $25^{1/2}$

28. $-36^{1/2}$

29. $(-36)^{1/2}$

30. $1^{1/5}$

31. $\left(\dfrac{9}{x^2}\right)^{1/2}$

32. $(-8)^{1/3}$

33. $625^{1/4}$

34. $(81c^4d^4)^{1/4}$

35. $9^{3/2}$

36. $8^{-2/3}$

37. $-49^{5/2}$

38. $\dfrac{1}{100^{-1/2}}$

39. $\left(\dfrac{4}{9}\right)^{-3/2}$

40. $\dfrac{1}{25^{5/2}}$

41. $(25x^2y^4)^{3/2}$

42. $(8u^6v^3)^{-2/3}$

The *rules for exponents* can be used to simplify expressions with fractional exponents.

Perform the operations. Write answers without negative exponents. Assume that all variables represent positive real numbers.

43. $5^{1/4}5^{1/2}$

44. $a^{3/7}a^{-2/7}$

45. $(k^{4/5})^{10}$

46. $\dfrac{3^{5/6}3^{1/3}}{3^{1/2}}$

Perform the multiplications. Assume all variables represent positive real numbers.

47. $u^{1/2}(u^{1/2} - u^{-1/2})$

48. $v^{2/3}(v^{1/3} + v^{4/3})$

Use rational exponents to simplify each radical. All variables represent positive real numbers.

49. $\sqrt[4]{a^2}$

50. $\sqrt[3]{\sqrt{c}}$

51. VISIBILITY The distance d in miles a person in an airplane can see to the horizon on a clear day is given by the formula $d = 1.22a^{1/2}$, where a is the altitude of the plane in feet. Find d.

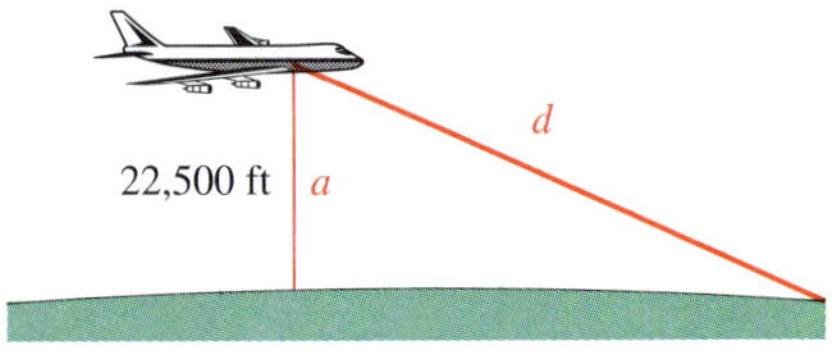

52. Substitute the x- and y-coordinates of each point labeled in the graph into the equation

$$x^{2/3} + y^{2/3} = 32$$

Show that each one satisfies the equation.

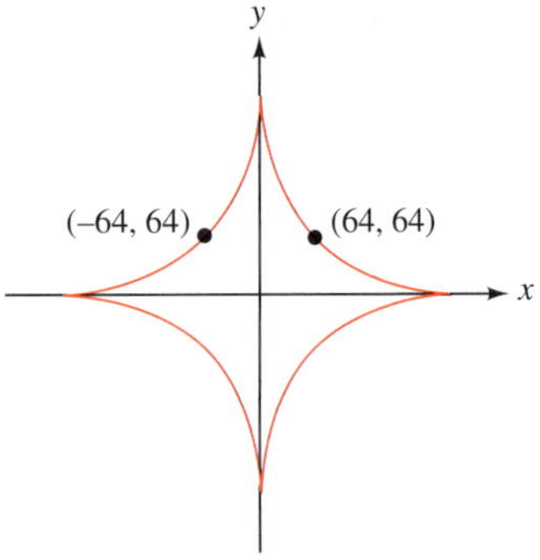

SECTION 7.3 Simplifying and Combining Radical Expressions

A radical is in *simplest form* when:

1. Each factor in the radicand appears to a power less than the index.
2. The radicand contains no fractions or negative numbers.
3. No radicals appear in a denominator.

Simplify each expression. Assume that all variables represent positive real numbers.

53. $\sqrt{240}$

54. $\sqrt[3]{54}$

55. $\sqrt[4]{32}$

56. $-2\sqrt[5]{-96}$

57. $\sqrt{8x^5}$

58. $\sqrt[3]{r^{17}}$

59. $\sqrt[3]{16x^5y^4}$

60. $3\sqrt[3]{27j^7k}$

61. $\dfrac{\sqrt{32x^3}}{\sqrt{2x}}$

62. $\sqrt{\dfrac{17xy}{64a^4}}$

Rules for radicals:

Multiplication:

$$\sqrt[n]{ab} = \sqrt[n]{a}\sqrt[n]{b}$$

Division:

$$\sqrt[n]{\frac{a}{b}} = \frac{\sqrt[n]{a}}{\sqrt[n]{b}} \quad (b \neq 0)$$

Like radicals can be combined by addition and subtraction.

Radicals that are not like can often be converted to radicals that are and then combined.

Simplify and combine like radicals. Assume that all variables represent positive real numbers.

63. $\sqrt{2} + 2\sqrt{2}$

64. $6\sqrt{20} - \sqrt{5}$

65. $2\sqrt[3]{3} - \sqrt[3]{24}$

66. $-\sqrt[4]{32a^5} - 2\sqrt[4]{162a^5}$

67. $2x\sqrt{8} + 2\sqrt{200x^2} + \sqrt{50x^2}$

68. $\sqrt[3]{54x^3} - 3\sqrt[3]{16x^3} + 4\sqrt[3]{128x^3}$

69. Explain the error that was made in each simplification.

a. $2\sqrt{5x} + 3\sqrt{5x} = 5\sqrt{10x}$

b. $30 + 30\sqrt[4]{2} = 60\sqrt[4]{2}$

c. $7\sqrt[3]{y^2} - 5\sqrt[3]{y^2} = 2$

d. $6\sqrt{11ab} - 3\sqrt{5ab} = 3\sqrt{6ab}$

70. SEWING A corner of fabric is folded over to form a collar and stitched down as shown. From the dimensions given in the figure, determine the exact number of inches of stitching that must be made. Then give an approximation to one decimal place. (All measurements are in inches.)

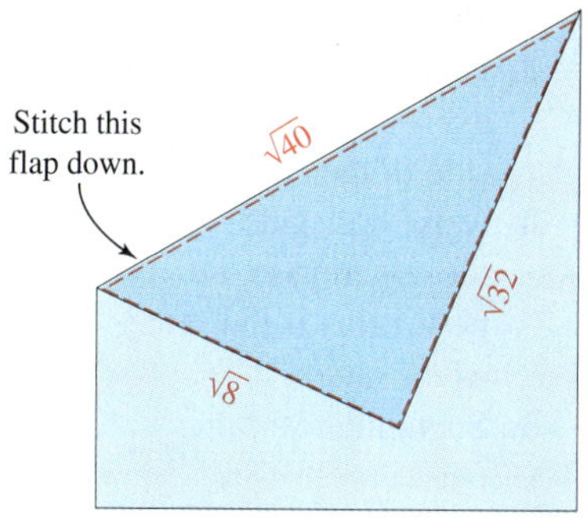

SECTION 7.4 Multiplying and Dividing Radical Expressions

If two radicals have the same index, they can be multiplied:

$$\sqrt[n]{a}\sqrt[n]{b} = \sqrt[n]{ab}$$

If $\sqrt[n]{a}$ is a real number,

$$\left(\sqrt[n]{a}\right)^n = a$$

If a radical appears in a denominator of a fraction, or if a radicand contains a fraction, we can write the radical in simplest form by *rationalizing the denominator.*

To *rationalize a two-term denominator* of a fraction, multiply the numerator and the denominator by the conjugate of the denominator.

Simplify each expression. Assume that all variables represent positive real numbers.

71. $\sqrt{7}\sqrt{7}$

72. $(2\sqrt{5})(3\sqrt{2})$

73. $(-2\sqrt{8})^2$

74. $2\sqrt{6}\sqrt{216}$

75. $\sqrt{9x}\sqrt{x}$

76. $\left(\sqrt[3]{x+1}\right)^3$

77. $-\sqrt[3]{2x^2}\sqrt[3]{4x^8}$

78. $\sqrt[5]{9} \cdot \sqrt[5]{27}$

79. $3\sqrt{7t}(2\sqrt{7t} + 3\sqrt{3t^2})$

80. $-\sqrt[4]{256x^5y^{11}}\sqrt[4]{625x^9y^3}$

81. $(\sqrt{3b} + \sqrt{3})^2$

82. $(\sqrt[3]{3p} - 2\sqrt[3]{2})(\sqrt[3]{3p} + \sqrt[3]{2})$

Rationalize each denominator.

83. $\dfrac{10}{\sqrt{3}}$

84. $\sqrt{\dfrac{3}{5xy}}$

85. $\dfrac{\sqrt[3]{uv}}{\sqrt[3]{u^5v^7}}$

86. $\dfrac{\sqrt[4]{a}}{\sqrt[4]{3b^2}}$

87. $\dfrac{2}{\sqrt{2} - 1}$

88. $\dfrac{4\sqrt{x} - 2\sqrt{z}}{\sqrt{z} + 4\sqrt{x}}$

89. Rationalize the numerator: $\dfrac{\sqrt{a} - \sqrt{b}}{\sqrt{a}}$.

90. VOLUME The formula relating the radius r of a sphere and its volume V is $r = \sqrt[3]{\dfrac{3V}{4\pi}}$. Write the radical in simplest form.

SECTION 7.5 Solving Radical Equations

The power rule:

If $x = y$, then $x^n = y^n$.

Solving equations containing radicals:

1. Isolate one radical expression on one side of the equation.
2. Raise both sides of the equation to the power that is the same as the index.
3. Solve the resulting equation. If it still contains a radical, go back to step 1.
4. Check the solutions to eliminate *extraneous* solutions.

Solve each equation. Write all solutions. Cross out those that are extraneous.

91. $\sqrt{7x - 10} - 1 = 11$

92. $u = \sqrt{25u - 144}$

93. $2\sqrt{y - 3} = \sqrt{2y + 1}$

94. $\sqrt{z + 1} + \sqrt{z} = 2$

95. $\sqrt[3]{x^3 + 56} - 2 = x$

96. $\sqrt[4]{8x - 8} + 2 = 0$

97. $(x + 2)^{1/2} - (4 - x)^{1/2} = 0$

98. $\sqrt{b^2 + b} = \sqrt{3 - b^2}$

99. Let $f(x) = \sqrt{2x^2 - 7x}$. For what value(s) of x is $f(x) = 2$?

100. Using the graphs of $f(x) = \sqrt{2x - 3}$ and $g(x) = -2x + 5$, estimate the solution of

$$\sqrt{2x - 3} = -2x + 5$$

Check the result.

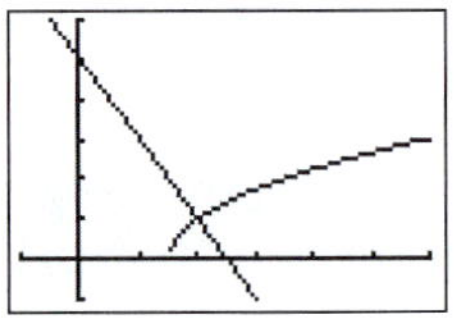

Solve each equation for the indicated variable.

101. $r = \sqrt{\dfrac{A}{P}} - 1$ for P

102. $h = \sqrt[3]{\dfrac{12I}{b}}$ for I

SECTION 7.6 Geometric Applications of Radicals

The Pythagorean theorem:

If a and b are the lengths of the *legs* of a right triangle and c is the length of the *hypotenuse*, then $a^2 + b^2 = c^2$.

103. CARPENTRY The gable end of the roof shown below is divided in half by a vertical brace, 8 feet in height. Find the length of the roof line.

104. SAILING A technique called *tacking* allows a sailboat to make progress into the wind. A sailboat follows the course shown below. Find d, the distance the boat advances into the wind after tacking.

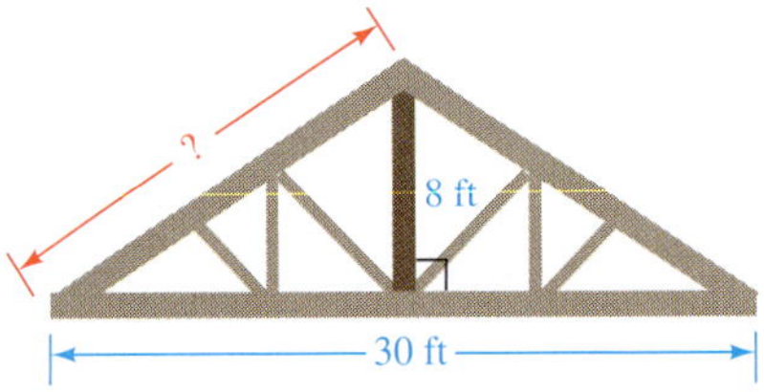

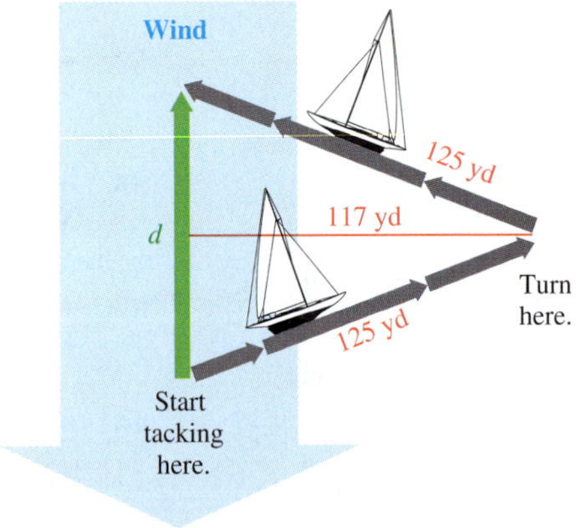

In an *isosceles right triangle,* the length of the hypotenuse is the length of one leg times $\sqrt{2}$.

The shorter leg of a *30°–60°–90° triangle* (the side opposite the 30° angle) is half as long as the hypotenuse. The longer leg (the side opposite the 60° angle) is the length of the shorter leg times $\sqrt{3}$.

105. Find the length of the hypotenuse of an isosceles right triangle whose legs measure 7 meters.

106. The hypotenuse of a 30°–60°–90° triangle measures $12\sqrt{3}$ centimeters. Find the length of each leg.

Find *x* to two decimal places.

107.

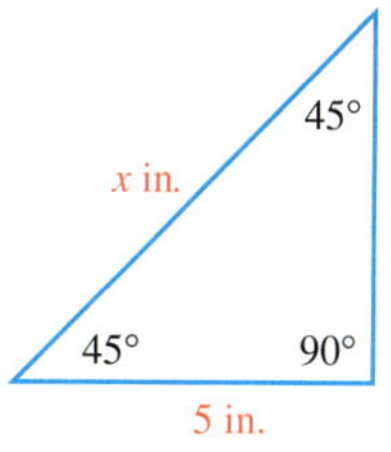

108.

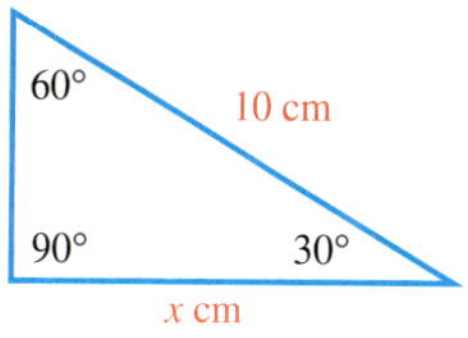

The distance formula:

$$d = \sqrt{(x_2 - x_1)^2 + (y_2 - y_1)^2}$$

Find the distance between the points.

109. (0, 0) and (5, −12)

110. (−4, 6) and (−2, 8)

SECTION 7.7 Complex Numbers

The *imaginary number i* is defined as

$$i = \sqrt{-1}$$

From the definition, it follows that $i^2 = -1$.

A *complex number* is any number that can be written in the form $a + bi$, where a and b are real numbers and $i = \sqrt{-1}$. We call a the *real part* and b the *imaginary part.*

Complex numbers of the form $a + bi$, where $b \neq 0$, are also called *imaginary numbers.*

The complex numbers $a + bi$ and $a - bi$ are called *complex conjugates.*

To add complex numbers, add their real parts and add their imaginary parts.

Write each expression in terms of *i*.

111. $\sqrt{-25}$

112. $4\sqrt{-18}$

113. $-\sqrt{-6}$

114. $\sqrt{-\dfrac{9}{64}}$

115. Complete the diagram.

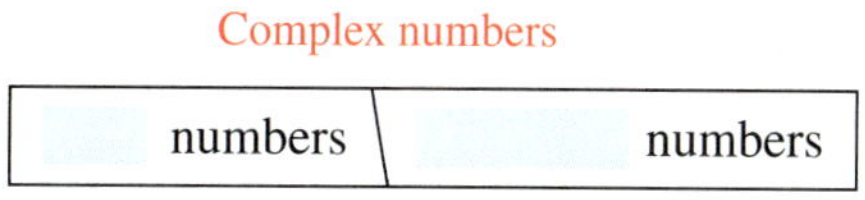

116. Determine whether each statement is true or false.

a. Every real number is a complex number.

b. $3 - 4i$ is a complex number.

c. $\sqrt{-4}$ is a real number.

d. i is a real number.

Give the complex conjugate of each number.

117. $3 + 6i$

118. $-1 - 7i$

119. $19i$

120. $-i$

Perform the operations. Write all answers in the form $a + bi$.

121. $(3 + 4i) + (5 - 6i)$

122. $\left(7 - \sqrt{-9}\right) - \left(4 + \sqrt{-4}\right)$

123. $3i(2 - i)$

124. $(2 - 7i)(-3 + 4i)$

To subtract complex numbers, add the opposite of the complex number being subtracted.

Multiplying complex numbers is similar to multiplying polynomials.

To divide complex numbers, multiply the numerator and denominator by the complex conjugate of the denominator.

125. $\sqrt{-3} \cdot \sqrt{-9}$

126. $(9i)^2$

127. $\dfrac{3}{4i}$

128. $\dfrac{2 + 3i}{2 - 3i}$

The powers of i cycle through four possible outcomes: i, -1, $-i$, and 1.

Simplify each expression.

129. i^{42}

130. i^{97}

CHAPTER 7 TEST

1. Graph: $f(x) = \sqrt{x - 1}$. Find the domain and range.

2. DIVING The velocity v of an object in feet per second after it has fallen a distance of d feet is approximated by the function $v(d) = \sqrt{64.4d}$. Olympic diving platforms are 10 meters (32.8 feet) tall. Estimate the velocity at which a diver hits the water from this height. Round to the nearest foot per second.

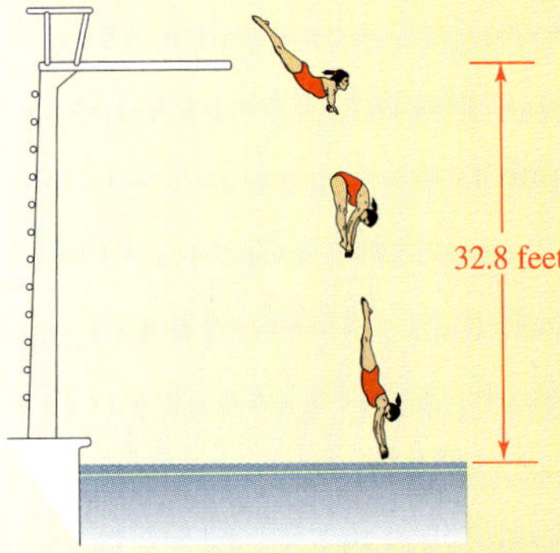

3. Use the graph in the next column to find each of the following.

a $f(-1)$

b. $f(8)$

c. The value(s) of x for which $f(x) = 1$

d. The domain and range of f

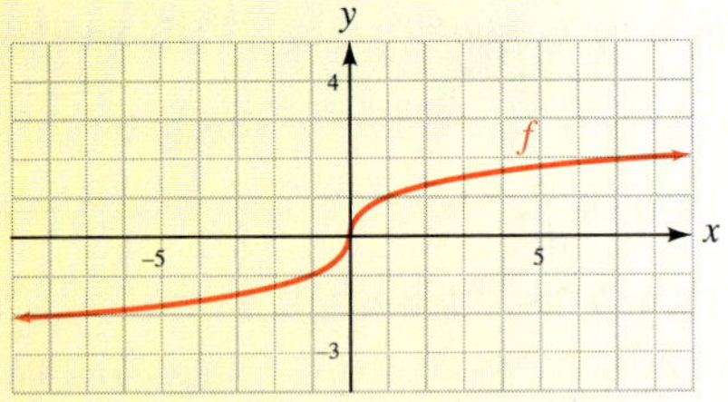

4. Explain why $\sqrt[4]{-16}$ is not a real number.

Simplify each expression. Assume that all variables represent positive real numbers and write answers without using negative exponents.

5. $(49x^4)^{1/2}$

6. $-27^{2/3}$

7. $36^{-3/2}$

8. $\left(-\dfrac{8}{125n^6}\right)^{-2/3}$

9. $\dfrac{2^{5/3}2^{1/6}}{2^{1/2}}$

10. $(a^{2/3})^{1/6}$

Simplify each expression. Assume that the variables are unrestricted.

11. $\sqrt{x^2}$

12. $\sqrt{y^2 + 10y + 25}$

Simplify each expression. Assume that all variables represent positive real numbers.

13. $\sqrt[3]{-64x^3y^6}$

14. $\sqrt{\dfrac{4a^2}{9}}$

15. $\sqrt[5]{(t+8)^5}$

16. $\sqrt{250x^3y^5}$

17. $\dfrac{\sqrt[3]{24x^{15}y^4}}{\sqrt[3]{y}}$

18. $\sqrt[7]{256}$

Perform the operations and simplify. Assume that all variables represent positive real numbers.

19. $2\sqrt{48y^5} - 3y\sqrt{12y^3}$

20. $2\sqrt[3]{40} - \sqrt[3]{5{,}000} + 4\sqrt[3]{625}$

21. $\sqrt[4]{243z^{13}} + z\sqrt[4]{48z^9}$

22. $-2\sqrt{xy}\left(3\sqrt{x} + \sqrt{xy^3}\right)$

23. $\left(3\sqrt{2} + \sqrt{3}\right)\left(2\sqrt{2} - 3\sqrt{3}\right)$

24. $\left(\sqrt[3]{2a} + 9\right)^2$

25. $\dfrac{8}{\sqrt{10}}$

26. $\dfrac{3t-1}{\sqrt{3t}-1}$

27. $\sqrt[3]{\dfrac{9}{4a}}$

28. Rationalize the numerator: $\dfrac{\sqrt{5}+3}{-4\sqrt{2}}$.

Solve each equation and check each result.

29. $2\sqrt{x} = \sqrt{x+1}$

30. $\sqrt[3]{6n+4} - 4 = 0$

stop here!

31. $1 - \sqrt{u} = \sqrt{u-3}$

32. $(2m^2 - 9)^{1/2} = m$

33. Explain why, without having to perform any algebraic steps, it is obvious that the equation

$$\sqrt{x-8} = -10$$

has no solutions.

34. Solve $r = \sqrt[3]{\dfrac{GMt^2}{4\pi^2}}$ for G.

Find x to two decimal places.

35.

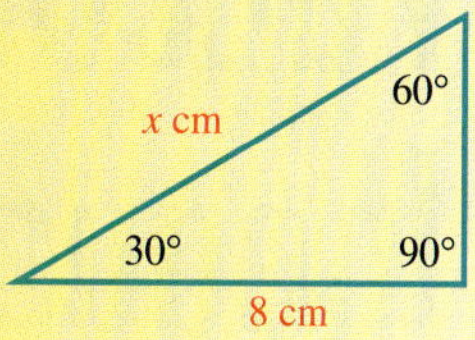

36.

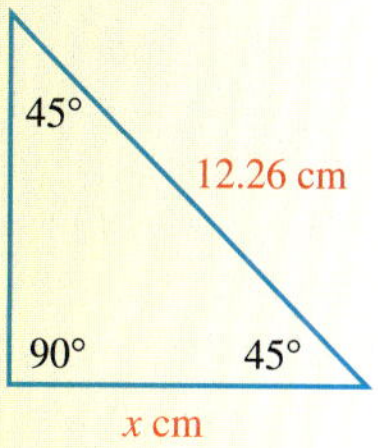

37. Find the distance between $(-2, 5)$ and $(22, 12)$.

38. SHIPPING CRATES The diagonal brace on the shipping crate in the illustration is 53 inches. Find the height h of the crate.

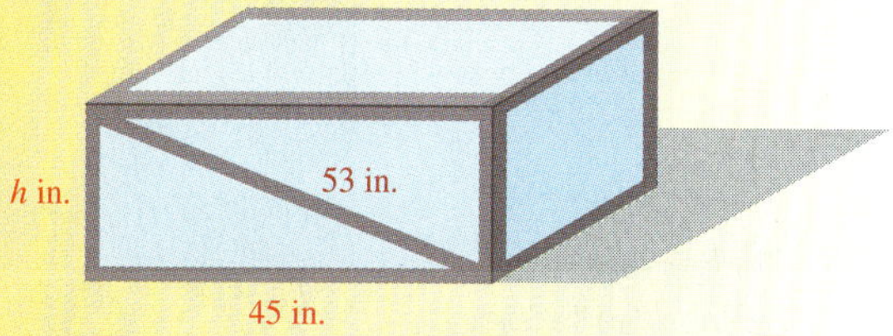

39. Express $\sqrt{-5}$ in terms of i.

40. Simplify: i^{22}.

Perform the operations. Give answers in $a + bi$ form.

41. $(2 + 4i) + (-3 + 7i)$

42. $\left(3 - \sqrt{-9}\right) - \left(-1 + \sqrt{-16}\right)$

43. $2i(3 - 4i)$

44. $(3 + 2i)(-4 - i)$

45. $\dfrac{1}{i\sqrt{2}}$

46. $\dfrac{2+i}{3-i}$

Chapter

8 Quadratic Equations, Functions, and Inequalities

©Bob Krist, CORBIS

TLE In a watercolor class, students learn that light, shadow, color, and perspective are fundamental components of an attractive painting. They also learn that the appropriate matting and frame can enhance their work. In this section, we will use mathematics to determine the dimensions of a uniform matting that is to have the same area as the picture it frames. To do this, we will write a quadratic equation and then solve it by *completing the square.* The technique of completing the square can be used to derive *the quadratic formula.* This formula is a valuable algebraic tool for solving any quadratic equation.

To learn more about quadratic equations, visit *The Learning Equation* on the Internet at http://tle.brookscole.com. (The log-in instructions are in the Preface.) For Chapter 8, the online lesson is:

- *TLE* Lesson 12: The Quadratic Formula

We have previously solved quadratic equations by factoring. In this chapter, we will discuss more general methods for solving quadratic equations, and we will consider the graphs of quadratic functions.

8.1 The Square Root Property and Completing the Square

- The square root property
- Completing the square
- Solving equations by completing the square
- Problem solving

Recall that a *quadratic equation* is an equation of the form $ax^2 + bx + c = 0$ where a, b, and c are real numbers and $a \neq 0$. We have solved quadratic equations using factoring and the zero-factor property. For example, to solve $6x^2 - 7x - 3 = 0$, we proceed as follows:

$$6x^2 - 7x - 3 = 0$$

$$(2x - 3)(3x + 1) = 0 \qquad \text{Factor.}$$

$$2x - 3 = 0 \quad \text{or} \quad 3x + 1 = 0 \qquad \text{Set each factor equal to 0.}$$

$$x = \frac{3}{2} \quad \Big| \quad x = -\frac{1}{3} \qquad \text{Solve each linear equation.}$$

Many expressions do not factor as easily as $6x^2 - 7x - 3$. For example, it would be difficult to solve $2x^2 + 4x + 1 = 0$ by factoring, because $2x^2 + 4x + 1$ cannot be factored by using only integers. With this in mind, we will now develop a more general method that enables us to solve *any* quadratic equation. It is based on the *square root property.*

THE SQUARE ROOT PROPERTY

To develop general methods for solving all quadratic equations, we first consider the equation $x^2 = c$. If $c \geq 0$, we can find the real solutions of $x^2 = c$ as follows:

$$x^2 = c$$

$$x^2 - c = 0 \qquad \text{Subtract } c \text{ from both sides.}$$

$$x^2 - (\sqrt{c})^2 = 0 \qquad \text{Replace } c \text{ with } (\sqrt{c})^2\text{, since } c = (\sqrt{c})^2.$$

$$(x + \sqrt{c})(x - \sqrt{c}) = 0 \qquad \text{Factor the difference of two squares.}$$

$$x + \sqrt{c} = 0 \quad \text{or} \quad x - \sqrt{c} = 0 \qquad \text{Set each factor equal to 0.}$$

$$x = -\sqrt{c} \quad \Big| \quad x = \sqrt{c} \qquad \text{Solve each linear equation.}$$

The solutions of $x^2 = c$ are $\sqrt{c}$ and $-\sqrt{c}$.

The Square Root Property For any nonnegative real number c, if $x^2 = c$, then

$$x = \sqrt{c} \quad \text{or} \quad x = -\sqrt{c}$$

EXAMPLE 1

Solve: $x^2 - 12 = 0$.

Solution We isolate x^2 on the left-hand side and use the square root property.

$$x^2 - 12 = 0$$

$$x^2 = 12 \quad \text{Add 12 to both sides.}$$

$$x = \sqrt{12} \quad \text{or} \quad x = -\sqrt{12} \quad \text{Use the square root property.}$$

$$x = 2\sqrt{3} \quad | \quad x = -2\sqrt{3} \quad \text{Simplify: } \sqrt{12} = \sqrt{4}\sqrt{3} = 2\sqrt{3}.$$

Notation

We can use **double-sign notation** $\pm$ to write the solutions in more compact form as $\pm 2\sqrt{3}$. Read $\pm$ as "positive or negative."

Check:

$$x^2 - 12 = 0 \qquad\qquad x^2 - 12 = 0$$

$$(2\sqrt{3})^2 - 12 \stackrel{?}{=} 0 \qquad\qquad (-2\sqrt{3})^2 - 12 \stackrel{?}{=} 0$$

$$12 - 12 \stackrel{?}{=} 0 \qquad\qquad 12 - 12 \stackrel{?}{=} 0$$

$$0 = 0 \qquad\qquad 0 = 0$$

The solutions are $2\sqrt{3}$ and $-2\sqrt{3}$ and the solution set is $\{2\sqrt{3}, -2\sqrt{3}\}$.

Self Check 1 Solve: $x^2 - 18 = 0$.

EXAMPLE 2

Phonograph records. Before compact discs, one way of recording music was by engraving grooves on thin vinyl discs called records. The vinyl discs used for long-playing records had a surface area of about 111 square inches per side and were played at $33\frac{1}{3}$ revolutions per minute on a turntable. What is the radius of a long-playing record?

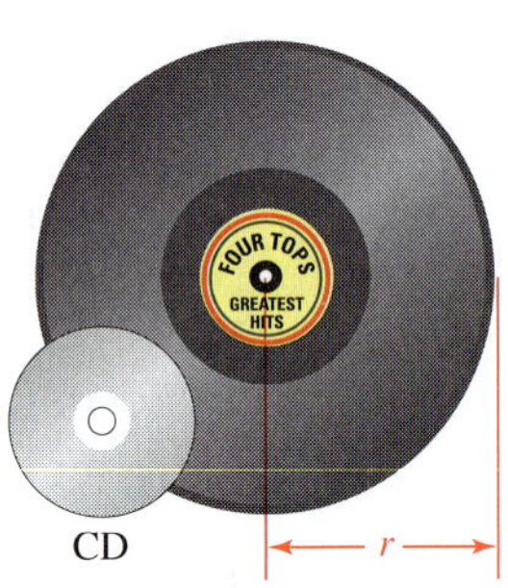

Solution The relationship between the area of a circle and its radius is given by the formula $A = \pi r^2$. We can find the radius of a record by substituting 111 for A and solving for r.

$$A = \pi r^2 \quad \text{This is the formula for the area of a circle.}$$

$$111 = \pi r^2 \quad \text{Substitute 111 for } A.$$

$$\frac{111}{\pi} = r^2 \quad \text{To undo the multiplication by } \pi \text{, divide both sides by } \pi.$$

$$r = \sqrt{\frac{111}{\pi}} \quad \text{or} \quad r = -\sqrt{\frac{111}{\pi}}$$

Use the square root property. Since the radius of the record cannot be negative, discard the second solution.

The radius of a record is $\sqrt{\frac{111}{\pi}}$ inches—to the nearest tenth, 5.9 inches.

EXAMPLE 3

Solve: $(x - 1)^2 = 16$.

Solution

$$
\begin{aligned}
(x-1)^2 &= 16 \\
x - 1 &= \pm\sqrt{16} && \text{Use the square root property.} \\
x - 1 &= \pm 4 && \text{Simplify: } \sqrt{16} = 4. \\
x &= 1 \pm 4 && \text{Add 1 to both sides.}
\end{aligned}
$$

$$
\begin{array}{lcl}
x = 1 + 4 & \text{or} & x = 1 - 4 \qquad \text{To find one solution use } +. \text{ To find the other use } -. \\
x = 5 & \vert & x = -3 \qquad \text{Add (subtract).}
\end{array}
$$

Verify that 5 and -3 satisfy the original equation.

Notation

We read 1 ± 4 as "one plus or minus four."

Self Check 3 Solve: $(x + 2)^2 = 9$.

Some quadratic equations have solutions that are not real numbers.

EXAMPLE 4

Solve: $4x^2 + 25 = 0$.

Solution

$$
\begin{aligned}
4x^2 + 25 &= 0 \\
x^2 &= -\frac{25}{4} && \text{To isolate } x^2\text{, subtract 25 from both sides and divide both sides by 4.} \\
x &= \pm\sqrt{-\frac{25}{4}} && \text{Use the square root property.}
\end{aligned}
$$

Since

$$\sqrt{-\frac{25}{4}} = \sqrt{-1 \cdot \frac{25}{4}} = \sqrt{-1}\,\frac{\sqrt{25}}{\sqrt{4}} = \frac{5}{2}i$$

we have

$$x = \pm\frac{5}{2}i$$

The solutions are $\frac{5}{2}i$ and $-\frac{5}{2}i$.

Check:

$$
\begin{array}{rcl}
4x^2 + 25 = 0 & \qquad & 4x^2 + 25 = 0 \\
4\left(\frac{5}{2}i\right)^2 + 25 \stackrel{?}{=} 0 & & 4\left(-\frac{5}{2}i\right)^2 + 25 \stackrel{?}{=} 0 \\
4\left(\frac{25}{4}\right)i^2 + 25 \stackrel{?}{=} 0 & & 4\left(\frac{25}{4}\right)i^2 + 25 \stackrel{?}{=} 0 \\
25(-1) + 25 \stackrel{?}{=} 0 & & 25(-1) + 25 \stackrel{?}{=} 0 \\
0 = 0 & & 0 = 0
\end{array}
$$

The Language of Algebra

The $\pm$ symbol is often seen in surveys and polls. Suppose a poll predicts a candidate will receive 48% of the vote, $\pm$4%. That means the candidate's support could be anywhere between $48 - 4 = 44\%$ and $48 + 4 = 52\%$.

Self Check 4 Solve: $16x^2 + 49 = 0$.

COMPLETING THE SQUARE

When the polynomial in a quadratic equation doesn't factor easily, we can solve the equation by *completing the square.* This method is based on the following special products:

$$x^2 + 2bx + b^2 = (x + b)^2 \quad \text{and} \quad x^2 - 2bx + b^2 = (x - b)^2$$

In each of these perfect square trinomials, the third term is the square of one-half of the coefficient of x.

The Language of Algebra
Recall that trinomials that are the square of a binomial are called **perfect square trinomials.**

- In $x^2 + 2bx + b^2$, the coefficient of x is $2b$. If we find $\frac{1}{2} \cdot 2b$, which is b, and square it, we get the third term, b^2.
- In $x^2 - 2bx + b^2$, the coefficient of x is $-2b$. If we find $\frac{1}{2}(-2b)$, which is $-b$, and square it, we get the third term: $(-b)^2 = b^2$.

We can use these observations to change certain binomials into perfect square trinomials. For example, to change $x^2 + 12x$ into a perfect square trinomial, we find one-half of the coefficient of x, square the result, and add the square to $x^2 + 12x$.

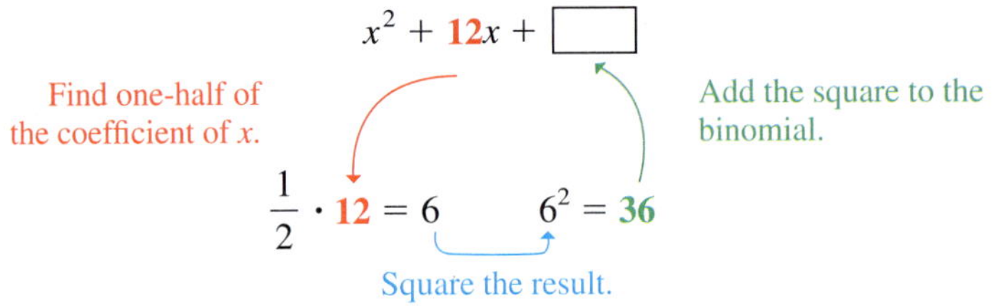

x² + 12x + □
Find one-half of the coefficient of x.
Add the square to the binomial.
1/2 · 12 = 6
6² = 36
Square the result.

We obtain the perfect square trinomial $x^2 + 12x + 36$ that factors as $(x + 6)^2$. By adding 36 to $x^2 + 12x$, we **completed the square** on $x^2 + 12x$.

Completing the Square

To complete the square on $x^2 + bx$, add the square of one-half of the coefficient of x:

$$x^2 + bx + \left(\frac{1}{2}b\right)^2$$

EXAMPLE 5

Complete the square and factor the resulting perfect square trinomial: **a.** $x^2 + 10x$ and **b.** $x^2 - 11x$.

Solution

a. To make $x^2 + 10x$ a perfect square trinomial, we find one-half of 10, square it, and add that result to $x^2 + 10x$.

$$x^2 + 10x + 25 \qquad \frac{1}{2} \cdot 10 = 5 \text{ and } 5^2 = 25. \text{ Add 25 to the binomial.}$$

This trinomial factors as $(x + 5)^2$.

b. To make $x^2 - 11x$ a perfect square trinomial, we find one-half of -11, square it, and add that result to $x^2 - 11x$.

$$x^2 - 11x + \frac{121}{4} \qquad \frac{1}{2}(-11) = \frac{-11}{2} \text{ and } \left(-\frac{11}{2}\right)^2 = \frac{121}{4}. \text{ Add } \frac{121}{4} \text{ to the binomial.}$$

This trinomial factors as $\left(x - \frac{11}{2}\right)^2$.

Self Check 5 Complete the square on $a^2 - 5a$ and factor the resulting trinomial.

SOLVING EQUATIONS BY COMPLETING THE SQUARE

To solve an equation of the form $ax^2 + bx + c = 0$ by completing the square, we use the following steps.

Completing the Square to Solve a Quadratic Equation

1. If the coefficient of x^2 is 1, go to step 2. If it is not 1, make it 1 by dividing both sides of the equation by the coefficient of x^2.
2. Get all variable terms on one side of the equation and constants on the other side.
3. Complete the square by finding one-half of the coefficient of x, squaring the result, and adding the square to both sides of the equation.
4. Factor the perfect square trinomial as the square of a binomial.
5. Solve the resulting equation using the square root property.
6. Check the answers in the original equation.

EXAMPLE 6

Use completing the square to solve $x^2 + 8x + 7 = 0$.

Solution

Step 1: In this example, the coefficient of x^2 is an understood 1.

Step 2: We subtract 7 from both sides so that the variable terms are on one side and the constant is on the other side of the equation.

$$x^2 + 8x + 7 = 0$$
$$x^2 + 8x = -7$$

The Language of Algebra

In $x^2 + 8x + 7 = 0$, x^2 and $8x$ are called *variable terms* and 7 is called the *constant term.*

Step 3: The coefficient of x is 8, one-half of 8 is 4, and $4^2 = 16$. To complete the square, we add 16 to both sides.

$$x^2 + 8x + 16 = 16 - 7$$
$$\text{(1)} \quad x^2 + 8x + 16 = 9 \qquad \text{Simplify: } 16 - 7 = 9.$$

Step 4: Since the left-hand side of Equation 1 is a perfect square trinomial, we can factor it to get $(x + 4)^2$.

$$x^2 + 8x + 16 = 9$$
$$\text{(2)} \quad (x + 4)^2 = 9$$

Step 5: We solve Equation 2 by using the square root property.

$$x + 4 = \pm\sqrt{9}$$
$$x + 4 = \pm 3 \qquad \text{Simplify: } \sqrt{9} = 3.$$
$$x = -4 \pm 3 \qquad \text{Subtract 4 from both sides.}$$

Caution

When using the square root property to solve an equation, always write the ± symbol, or you will lose one of the solutions.

This result represents two solutions. To find the first, we add 3, and to find the second, we subtract 3.

$$x = -4 + 3 \quad \text{or} \quad x = -4 - 3 \qquad \pm \text{ represents } + \text{ or } -.$$

$$x = -1 \quad\quad\quad x = -7 \qquad \text{Add and subtract.}$$

Step 6: The solutions are -1 and -7. Verify that they satisfy the original equation.

Self Check 6 Solve: $x^2 + 12x + 11 = 0$.

EXAMPLE 7

Solve: $6x^2 + 5x - 6 = 0$.

Solution ***Step 1:*** To make the coefficient of x^2 equal to 1, we divide both sides of the equation by 6.

$$6x^2 + 5x - 6 = 0$$

$$\frac{6x^2}{6} + \frac{5}{6}x - \frac{6}{6} = \frac{0}{6} \qquad \text{Divide both sides by 6.}$$

$$x^2 + \frac{5}{6}x - 1 = 0 \qquad \text{Simplify.}$$

Success Tip

When solving a quadratic equation using the factoring method, one side of the equation must be 0. When we complete the square, we are not concerned with that requirement.

Step 2: To have the constant term on one side of the equation and the variable terms on the other, add 1 to both sides.

$$x^2 + \frac{5}{6}x = 1$$

Step 3: The coefficient of x is $\frac{5}{6}$, one-half of $\frac{5}{6}$ is $\frac{5}{12}$, and $\left(\frac{5}{12}\right)^2 = \frac{25}{144}$. To complete the square, we add $\frac{25}{144}$ to both sides.

$$x^2 + \frac{5}{6}x + \frac{25}{144} = 1 + \frac{25}{144}$$

(3) $$x^2 + \frac{5}{6}x + \frac{25}{144} = \frac{169}{144} \qquad \text{Simplify: } 1 + \frac{25}{144} = \frac{144}{144} + \frac{25}{144} = \frac{169}{144}.$$

Step 4: Factor the left-hand side of Equation 3.

(4) $$\left(x + \frac{5}{12}\right)^2 = \frac{169}{144} \qquad x^2 + \frac{5}{6}x + \frac{25}{144} \text{ is a perfect square trinomial.}$$

Step 5: We can solve Equation 4 by using the square root property.

$$x + \frac{5}{12} = \pm\sqrt{\frac{169}{144}}$$

$$x + \frac{5}{12} = \pm\frac{13}{12} \qquad \text{Simplify: } \sqrt{\frac{169}{144}} = \frac{13}{12}.$$

$$x = -\frac{5}{12} \pm \frac{13}{12} \qquad \text{To isolate } x\text{, subtract } \frac{5}{12} \text{ from both sides.}$$

$$x = -\frac{5}{12} + \frac{13}{12} \quad \text{or} \quad x = -\frac{5}{12} - \frac{13}{12}$$

$$x = \frac{8}{12} \quad \Big| \quad x = -\frac{18}{12} \qquad \text{Add (subtract) the fractions.}$$

$$x = \frac{2}{3} \quad \Big| \quad x = -\frac{3}{2} \qquad \text{Simplify each fraction.}$$

Step 6: Verify that $\frac{2}{3}$ and $-\frac{3}{2}$ satisfy the original equation.

Self Check 7 Solve: $3x^2 + 2x - 8 = 0$.

EXAMPLE 8

Solve: $2x^2 + 4x + 1 = 0$.

Solution

$$2x^2 + 4x + 1 = 0$$

$$x^2 + 2x + \frac{1}{2} = 0 \qquad \text{Divide both sides by 2 to make the coefficient of } x^2 \text{ equal to 1.}$$

$$x^2 + 2x = -\frac{1}{2} \qquad \text{Subtract } \frac{1}{2} \text{ from both sides.}$$

$$x^2 + 2x + 1 = 1 - \frac{1}{2} \qquad \text{Square one-half of the coefficient of } x \text{ and add it to both sides.}$$

$$(x + 1)^2 = \frac{1}{2} \qquad \text{Factor and combine like terms.}$$

$$x + 1 = \pm\sqrt{\frac{1}{2}} \qquad \text{Use the square root property.}$$

$$x = -1 \pm \sqrt{\frac{1}{2}} \qquad \text{To isolate } x \text{, subtract 1 from both sides.}$$

Caution

A common error is to add a constant to one side of an equation to complete the square and forget to add it to the other side.

To write $\sqrt{\frac{1}{2}}$ in simplified radical form, we write it as a quotient of square roots and then rationalize the denominator.

$$x = -1 + \frac{\sqrt{2}}{2} \quad \text{or} \quad x = -1 - \frac{\sqrt{2}}{2} \qquad \sqrt{\frac{1}{2}} = \frac{\sqrt{1}}{\sqrt{2}} = \frac{1 \cdot \sqrt{2}}{\sqrt{2}\sqrt{2}} = \frac{\sqrt{2}}{2}.$$

We can express each solution in an alternate form if we write -1 as a fraction with a denominator of 2.

$$x = -\frac{2}{2} + \frac{\sqrt{2}}{2} \quad \text{or} \quad x = -\frac{2}{2} - \frac{\sqrt{2}}{2} \qquad \text{Write } -1 \text{ as } -\frac{2}{2}.$$

$$x = \frac{-2 + \sqrt{2}}{2} \quad \Big| \quad x = \frac{-2 - \sqrt{2}}{2} \qquad \text{Add (subtract) the numerators and keep the common denominator of 2.}$$

Caution

Recall that to simplify a fraction, we divide out common *factors* of the numerator and denominator. Since -2 is a *term* of the numerator of $\frac{-2 + \sqrt{2}}{2}$, no further simplification of this expression can be made.

The exact solutions are $\frac{-2 + \sqrt{2}}{2}$ and $\frac{-2 - \sqrt{2}}{2}$, or simply, $\frac{-2 \pm \sqrt{2}}{2}$. We can use a calculator to approximate them. To the nearest hundredth, they are -0.29 and -1.71.

Self Check 8 Solve: $3x^2 + 6x + 1 = 0$.

ACCENT ON TECHNOLOGY: CHECKING SOLUTIONS OF QUADRATIC EQUATIONS

We can use a graphing calculator to check the solutions of the quadratic equation $2x^2 + 4x + 1 = 0$ found in Example 8. After entering $Y_1 = 2x^2 + 4x + 1$, we call up the home screen by pressing [2nd] QUIT. Then we press the [VARS] key, arrow [▶] to Y-VARS, and enter 1 to get the display shown in figure (a). We evaluate $2x^2 + 4x + 1$ for $x = \frac{-2 + \sqrt{2}}{2}$ by entering the solution using function notation, as shown in figure (b). When [ENTER] is pressed, the result of 0 is confirmation that $x = \frac{-2 + \sqrt{2}}{2}$ is a solution of the equation.

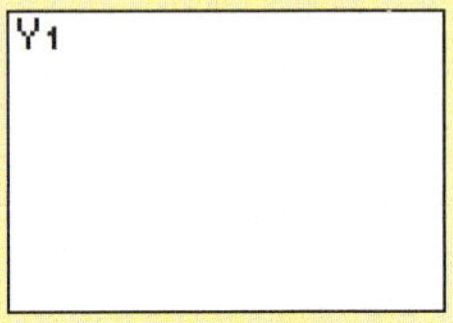

(a)

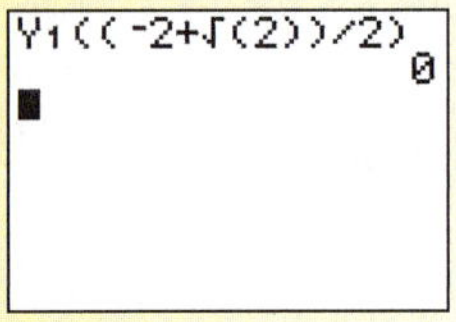

(b)

In the next example, the solutions of the equation are two complex numbers that contain i.

EXAMPLE 9

Solve: $3x^2 + 2x + 2 = 0$.

Solution

$$3x^2 + 2x + 2 = 0$$

$$x^2 + \frac{2}{3}x + \frac{2}{3} = \frac{0}{3}$$ Divide both sides by 3 to make the coefficient of x^2 equal to 1.

$$x^2 + \frac{2}{3}x = -\frac{2}{3}$$ Subtract $\frac{2}{3}$ from both sides.

$$x^2 + \frac{2}{3}x + \frac{1}{9} = \frac{1}{9} - \frac{2}{3}$$ $\frac{1}{2} \cdot \frac{2}{3} = \frac{1}{3}$ and $\left(\frac{1}{3}\right)^2 = \frac{1}{9}$. Add $\frac{1}{9}$ to both sides.

$$\left(x + \frac{1}{3}\right)^2 = -\frac{5}{9}$$ Factor and combine terms: $\frac{1}{9} - \frac{2}{3} = \frac{1}{9} - \frac{6}{9} = -\frac{5}{9}$.

$$x + \frac{1}{3} = \pm\sqrt{-\frac{5}{9}}$$ Use the square root property.

$$x = -\frac{1}{3} \pm \sqrt{-\frac{5}{9}}$$ Subtract $\frac{1}{3}$ from both sides.

Since

$$\sqrt{-\frac{5}{9}} = \sqrt{-1 \cdot \frac{5}{9}} = \sqrt{-1}\frac{\sqrt{5}}{\sqrt{9}} = \frac{\sqrt{5}}{3}i$$

we have

$$x = -\frac{1}{3} \pm \frac{\sqrt{5}}{3}i$$

The solutions are $-\frac{1}{3} + \frac{\sqrt{5}}{3}i$ and $-\frac{1}{3} - \frac{\sqrt{5}}{3}i$.

Self Check 9 Solve: $x^2 + 4x + 6 = 0$.

Success Tip

Perhaps you noticed that Examples 6 and 7 can be solved by factoring. However, that is not the case for Examples 8 and 9. These observations illustrate that completing the square can be used to solve *any* quadratic equation.

Notation

The solutions are written in complex number form $a + bi$. They could also be written as

$$\frac{-1 \pm i\sqrt{5}}{3}$$

PROBLEM SOLVING

EXAMPLE 10

Graduation announcements. To create the announcement shown to the right, a graphic artist must follow two design requirements:

- A border of uniform width should surround the text.
- Equal areas should be devoted to the text and to the border.

To meet these requirements, how wide should the border be?

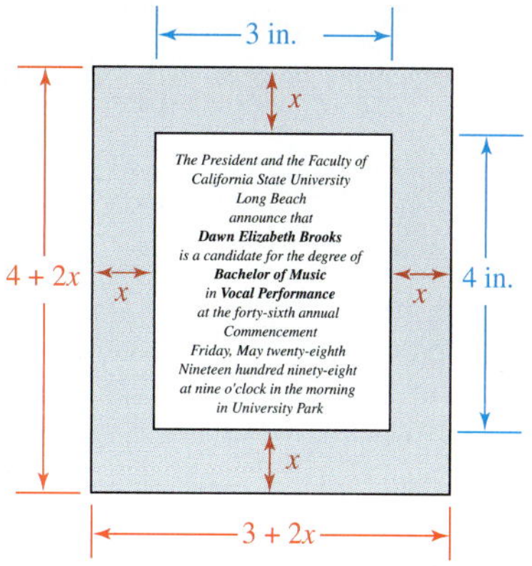

Analyze the Problem The text occupies $4 \cdot 3 = 12$ in.^{2} of space. The border must also have an area of 12 in.^{2}.

Form an Equation If we let x = the width of the border, the length of the announcement is $(4 + 2x)$ inches and the width is $(3 + 2x)$ inches. We can now form the equation.

The area of the announcement	minus	the area of the text	equals	the area of the border.
$(4 + 2x)(3 + 2x)$	$-$	12	$=$	12

Solve the Equation

$$(4 + 2x)(3 + 2x) - 12 = 12$$

$$12 + 8x + 6x + 4x^2 - 12 = 12$$ On the left-hand side, use the FOIL method.

$$4x^2 + 14x = 12$$ Combine like terms.

$$2x^2 + 7x - 6 = 0$$ Subtract 12 from both sides. Then divide both sides of $4x^2 + 14x - 12 = 0$ by 2.

We note that the trinomial on the left-hand side does not factor. We will solve the equation by completing the square.

$$x^2 + \frac{7}{2}x - 3 = 0$$ Divide both sides by 2 so that the coefficient of x^2 is 1.

$$x^2 + \frac{7}{2}x = 3$$ Add 3 to both sides.

$$x^2 + \frac{7}{2}x + \frac{49}{16} = 3 + \frac{49}{16}$$ One-half of $\frac{7}{2}$ is $\frac{7}{4}$. Square $\frac{7}{4}$, which is $\frac{49}{16}$, and add it to both sides.

$$\left(x + \frac{7}{4}\right)^2 = \frac{97}{16}$$ On the left-hand side, factor the trinomial. On the right-hand side, $3 = \frac{3 \cdot 16}{1 \cdot 16} = \frac{48}{16}$ and $\frac{48}{16} + \frac{49}{16} = \frac{97}{16}$.

$$x + \frac{7}{4} = \pm\sqrt{\frac{97}{16}}$$ Apply the square root property.

$$x = -\frac{7}{4} \pm \frac{\sqrt{97}}{4}$$ Subtract $\frac{7}{4}$ from both sides and simplify: $\sqrt{\frac{97}{16}} = \frac{\sqrt{97}}{\sqrt{16}} = \frac{\sqrt{97}}{4}$.

$$x = \frac{-7 + \sqrt{97}}{4} \quad \text{or} \quad x = \frac{-7 - \sqrt{97}}{4}$$ Write each expression as a single fraction.

State the Conclusion The width of the border should be $\frac{-7 + \sqrt{97}}{4} \approx 0.71$ inch. (We discard the solution $\frac{-7 - \sqrt{97}}{4}$, since it is negative.)

Check the Result If the border is 0.71 inch wide, the announcement has an area of about $5.42 \cdot 4.42 \approx 23.96$ in.^{2}. If we subtract the area of the text from the area of the announcement, we get $23.96 - 12 = 11.96$ in.^{2}. This represents the area of the border, which was to be 12 in.^{2}. The answer seems reasonable.

Answers to Self Checks **1.** $\pm 3\sqrt{2}$ **3.** $1, -5$ **4.** $\frac{7}{4}i, -\frac{7}{4}i$ **5.** $\left(a - \frac{5}{2}\right)^2$ **6.** $-1, -11$ **7.** $\frac{4}{3}, -2$ **8.** $\frac{-3 \pm \sqrt{6}}{3}$ **9.** $-2 \pm i\sqrt{2}$

8.1 STUDY SET

VOCABULARY Fill in the blanks.

1. An equation of the form $ax^2 + bx + c = 0$, where $a \neq 0$, is called a ______ equation.
2. We read $8 \pm \sqrt{3}$ as "eight ______ the square root of 3."
3. $x^2 + 6x + 9$ is called a ______ square trinomial because it factors as $(x + 3)^2$.
4. The ______ of x^2 in $x^2 - 12x + 36 = 0$ is 1, and the ______ term is 36.

CONCEPTS Fill in the blanks.

5. For any nonnegative number c, if $x^2 = c$, then $x =$ ______ or $x =$ ______.
6. To complete the square on x in $x^2 + 6x$, find one-half of ______, square it to get ______, and add ______ to get ______.
7. $\frac{1}{5} \pm \frac{\sqrt{2}}{5} = \frac{1 \pm \sqrt{2}}{}$ ______
8. Suppose $x = 5 \pm 9$. Then $x =$ ______ or $x =$ ______.
9. Check to see whether $-3\sqrt{2}$ is a solution of $x^2 - 18 = 0$.
10. Check to see whether $-2 + \sqrt{2}$ is a solution of $x^2 + 4x + 2 = 0$.
11. Find one-half of the coefficient of x and then square it.

    a. $x^2 + 12x$ **b.** $x^2 - 5x$

    c. $x^2 - \frac{x}{2}$ **d.** $x^2 + \frac{3}{4}x$

12. Add a number to make each binomial a perfect square trinomial. Then factor the result.

    a. $x^2 + 8x$

    b. $x^2 - 4x$

    c. $x^2 - x$

13. What is the first step in solving the equation $x^2 + 12x = 35$

    a. by the factoring method?

    b. by completing the square?

14. Solve $x^2 = 16$

    a. by the factoring method.

    b. by the square root method.

15. Solve for x: $x + 7 = \pm\sqrt{6}$.
16. **a.** In the expression $8 \pm \frac{\sqrt{15}}{2}$, replace 8 with an equivalent fraction that has a denominator of 2.

    b. Write your answer to part a as a single fraction with denominator 2.

17. Explain the error in the work shown below.

    a. $$\frac{4 \pm \sqrt{3}}{8} = \frac{\overset{1}{\cancel{4}} \pm \sqrt{3}}{\underset{1}{\cancel{4}} \cdot 2} = \frac{1 \pm \sqrt{3}}{2}$$

b. $\dfrac{1 \pm \sqrt{5}}{5} = \dfrac{1 \pm \sqrt{\cancel{5}}^{\,1}}{\cancel{5}_{\,1}}$

$= \dfrac{1 \pm 1}{1}$

18. a. Write an expression that represents the width of the larger rectangle shown below.

b. Write an expression that represents the length of the larger rectangle shown below.

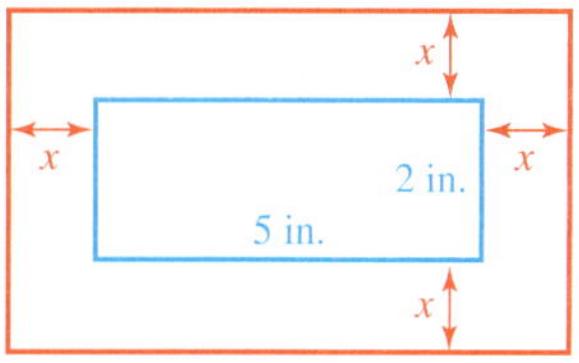

NOTATION

19. When solving a quadratic equation, a student obtains $x = \pm 2\sqrt{5}$.

a. How many solutions are represented by this notation? List them.

b. Approximate the solutions to the nearest hundredth.

20. When solving a quadratic equation, a student obtains $x = \dfrac{-5 \pm \sqrt{7}}{3}$.

a. How many solutions are represented by this notation? List them.

b. Approximate the solutions to the nearest hundredth.

PRACTICE **Use factoring to solve each equation.**

21. $6x^2 + 12x = 0$ **22.** $5x^2 + 11x = 0$

23. $y^2 - 25 = 0$ **24.** $y^2 - 16 = 0$

25. $r^2 + 6r + 8 = 0$ **26.** $x^2 + 9x + 20 = 0$

27. $2z^2 = -2 + 5z$ **28.** $3x^2 = 8 - 10x$

Use the square root property to solve each equation.

29. $x^2 = 36$ **30.** $x^2 = 144$

31. $z^2 = 5$ **32.** $u^2 = 24$

33. $3x^2 - 16 = 0$ **34.** $5x^2 - 49 = 0$

35. $(x + 1)^2 = 1$ **36.** $(x - 1)^2 = 4$

37. $(s - 7)^2 = 9$ **38.** $(t + 4)^2 = 16$

39. $(x + 5)^2 - 3 = 0$ **40.** $(x + 3)^2 - 7 = 0$

41. $(a + 2)^2 = 8$ **42.** $(c + 2)^2 = 12$

43. $(3x - 1)^2 = 25$ **44.** $(5x - 2)^2 = 64$

45. $p^2 = -16$ **46.** $q^2 = -25$

47. $4m^2 + 81 = 0$ **48.** $9n^2 + 121 = 0$

49. $(x - 3)^2 = -5$ **50.** $(x + 2)^2 = -3$

Use the square root property to solve for the indicated variable. Assume that all variables represent positive numbers. Express all radicals in simplified form.

51. $2d^2 = 3h$ for d **52.** $2x^2 = d^2$ for d

53. $E = mc^2$ for c **54.** $A = \pi r^2$ for r

Use completing the square to solve each equation.

55. $x^2 + 2x - 8 = 0$ **56.** $x^2 + 6x + 5 = 0$

57. $k^2 - 8k + 12 = 0$ **58.** $p^2 - 4p + 3 = 0$

59. $g^2 + 5g - 6 = 0$ **60.** $s^2 + 5s - 14 = 0$

61. $x^2 - 3x - 4 = 0$ **62.** $x^2 - 7x + 12 = 0$

63. $x^2 + 8x + 6 = 0$ **64.** $x^2 + 6x + 4 = 0$

65. $x^2 - 2x = 17$ **66.** $x^2 + 10x = 7$

67. $m^2 - 7m + 3 = 0$ **68.** $m^2 - 5m + 3 = 0$

69. $a^2 - a = 3$ **70.** $b^2 - 3b = 5$

71. $2x^2 - x - 1 = 0$ **72.** $2x^2 - 5x + 2 = 0$

73. $3x^2 - 6x = 1$ **74.** $2x^2 - 6x = -3$

75. $4x^2 - 4x = 7$

76. $4x^2 - 4x = 1$

77. $2x^2 + 5x - 2 = 0$

78. $2x^2 - 8x + 5 = 0$

79. $\frac{7x + 1}{5} = -x^2$

80. $\frac{3x^2}{8} = \frac{1}{8} - x$

81. $p^2 + 2p + 2 = 0$

82. $x^2 - 6x + 10 = 0$

83. $y^2 + 8y + 18 = 0$

84. $t^2 + t + 3 = 0$

85. $3m^2 - 2m + 3 = 0$

86. $4p^2 + 2p + 3 = 0$

APPLICATIONS

87. FLAGS In 1912, an order by President Taft fixed the width and length of the U.S. flag in the ratio 1 to 1.9. If 100 square feet of cloth are to be used to make a U.S. flag, estimate its dimensions to the nearest $\frac{1}{4}$ foot.

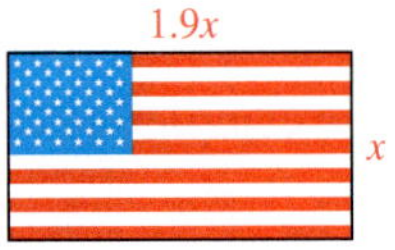

88. MOVIE STUNTS According to the *Guinness Book of World Records, 1998,* stuntman Dan Koko fell a distance of 312 feet into an airbag after jumping from the Vegas World Hotel and Casino. The distance d in feet traveled by a free-falling object in t seconds is given by the formula $d = 16t^2$. To the nearest tenth of a second, how long did the fall last?

89. ACCIDENTS The height h (in feet) of an object that is dropped from a height of s feet is given by the formula $h = s - 16t^2$, where t is the time the object has been falling. A 5-foot-tall woman on a sidewalk looks directly overhead and sees a window washer drop a bottle from 4 stories up. How long does she have to get out of the way? Round to the nearest tenth. (A story is 12 feet.)

90. GEOGRAPHY The surface area S of a sphere is given by the formula $S = 4\pi r^2$, where r is the radius of the sphere. An almanac lists the surface area of the Earth as 196,938,800 square miles. Assuming the Earth to be spherical, what is its radius to the nearest mile?

91. AUTOMOBILE ENGINES As the piston shown moves upward, it pushes a cylinder of a gasoline/air mixture that is ignited by the spark plug. The formula that gives the volume of a cylinder is $V = \pi r^2 h$, where r is the radius and h the height. Find the radius of the piston (to the nearest hundredth of an inch) if it displaces 47.75 cubic inches of gasoline/air mixture as it moves from its lowest to its highest point.

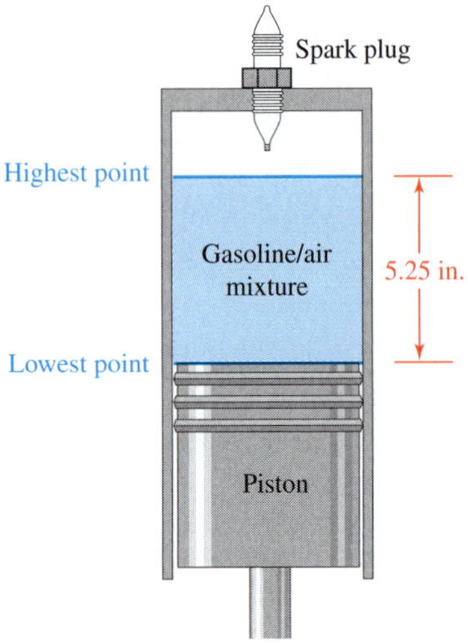

92. INVESTMENTS If P dollars are deposited in an account that pays an annual rate of interest r, then in n years, the amount of money A in the account is given by the formula $A = P(1 + r)^n$. A savings account was opened on January 3, 1996, with a deposit of \$10,000 and closed on January 2, 1998, with an ending balance of \$11,772.25. Find the rate of interest.

93. PICTURE FRAMING The matting around the picture has a uniform width. How wide is the matting if its area equals the area of the picture? Round to the nearest hundredth of an inch.

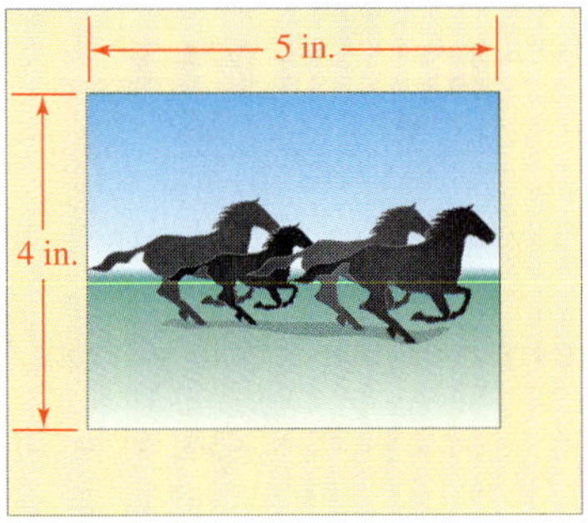

94. SWIMMING POOLS In the advertisement shown, how wide will the free concrete decking be if a uniform width is constructed around the perimeter of the pool? Round to the nearest hundredth of a yard. (*Hint:* Note the difference in units.)

SAHARA POOL & SPA
SUMMER SPECIAL
This 18 ft x 30 ft pool: only $18,500
Buy now and receive 28 square yards of concrete decking FREE!

95. DIMENSIONS OF A RECTANGLE A rectangle is 4 feet longer than it is wide, and its area is 20 square feet. Find its dimensions to the nearest tenth of a foot.

96. DIMENSIONS OF A TRIANGLE The height of a triangle is 4 meters longer than twice its base. Find the base and height if the area of the triangle is 10 square meters. Round to the nearest hundredth of a meter.

WRITING

97. Give an example of a perfect square trinomial. Why do you think the word "perfect" is used to describe it?

98. Explain why completing the square on $x^2 + 5x$ is more difficult than completing the square on $x^2 + 4x$.

REVIEW

Simplify each expression. All variables represent positive real numbers.

99. $\sqrt[3]{40a^3b^6}$

100. $\sqrt[3]{-27x^6}$

101. $\sqrt[8]{x^{24}}$

102. $\sqrt[4]{\dfrac{16}{625}}$

103. $\sqrt{175a^2b^3}$

104. $\sqrt{\dfrac{z^2}{16x^2}}$

CHALLENGE PROBLEMS

105. What number must be added to $x^2 + \sqrt{3}x$ to make a perfect square trinomial?

106. Solve $x^2 + \sqrt{3}x - \frac{1}{4} = 0$ by completing the square.

8.2 The Quadratic Formula

- The quadratic formula
- Solving quadratic equations using the quadratic formula
- Solving an equivalent equation
- Problem solving

We can solve any quadratic equation by the method of completing the square, but the work is often tedious. In this section, we will develop a formula, called the *quadratic formula,* that lets us solve quadratic equations with less effort.

THE QUADRATIC FORMULA

To develop a formula that will produce the solutions of any given quadratic equation, we start with the **general quadratic equation** $ax^2 + bx + c = 0$, with $a > 0$, and solve it for x by completing the square.

$$ax^2 + bx + c = 0$$

$$\frac{ax^2}{a} + \frac{bx}{a} + \frac{c}{a} = \frac{0}{a}$$

Divide both sides by a so that the coefficient of x^2 is 1.

$$x^2 + \frac{b}{a}x + \frac{c}{a} = 0$$

Simplify: $\frac{ax^2}{a} = x^2$. Write $\frac{bx}{a}$ as $\frac{b}{a}x$.

$$x^2 + \frac{b}{a}x = -\frac{c}{a}$$

Subtract $\frac{c}{a}$ from both sides so that only the terms involving x are on the left-hand side of the equation.

We can complete the square on $x^2 + \frac{b}{a}x$ by adding the square of one-half of the coefficient of x. Since the coefficient of x is $\frac{b}{a}$, we have $\frac{1}{2} \cdot \frac{b}{a} = \frac{b}{2a}$ and $\left(\frac{b}{2a}\right)^2 = \frac{b^2}{4a^2}$.

$$x^2 + \frac{b}{a}x + \frac{b^2}{4a^2} = -\frac{c}{a} + \frac{b^2}{4a^2}$$ To complete the square, add $\frac{b^2}{4a^2}$ to both sides.

$$x^2 + \frac{b}{a}x + \frac{b^2}{4a^2} = -\frac{4ac}{4aa} + \frac{b^2}{4a^2}$$ Multiply $-\frac{c}{a}$ by $\frac{4a}{4a}$. Now the fractions on the right side have the common denominator $4a^2$.

$$\left(x + \frac{b}{2a}\right)^2 = \frac{b^2 - 4ac}{4a^2}$$ On the left-hand side, factor. On the right-hand side, add the fractions.

$$x + \frac{b}{2a} = \pm\sqrt{\frac{b^2 - 4ac}{4a^2}}$$ Use the square root property.

$$x + \frac{b}{2a} = \pm\frac{\sqrt{b^2 - 4ac}}{\sqrt{4a^2}}$$ The square root of a quotient is the quotient of square roots.

$$x + \frac{b}{2a} = \pm\frac{\sqrt{b^2 - 4ac}}{2a}$$ Since $a > 0$, $\sqrt{4a^2} = 2a$.

$$x = -\frac{b}{2a} \pm \frac{\sqrt{b^2 - 4ac}}{2a}$$ To isolate x, subtract $\frac{b}{2a}$ from both sides.

$$x = \frac{-b \pm \sqrt{b^2 - 4ac}}{2a}$$ Combine the fractions.

The Language of Algebra

Your instructor may ask you to *derive* the quadratic formula. That means to solve $ax^2 + bx + c = 0$ for x, using the series of steps shown here, to obtain

$$x = \frac{-b \pm \sqrt{b^2 - 4ac}}{2a}$$

This result is called the *quadratic formula*. To develop this formula, we assumed that a was positive. If a is negative, similar steps are used, and we obtain the same result.

Quadratic Formula The solutions of $ax^2 + bx + c = 0$, with $a \neq 0$, are

$$x = \frac{-b \pm \sqrt{b^2 - 4ac}}{2a}$$

SOLVING QUADRATIC EQUATIONS USING THE QUADRATIC FORMULA

EXAMPLE 1 Solve: $2x^2 - 5x - 3 = 0$.

Solution To use the quadratic formula, we need to identify a, b, and c. We do this by comparing the given equation to the general quadratic equation $ax^2 + bx + c = 0$.

$$2x^2 - 5x - 3 = 0$$
$$\uparrow \quad \uparrow \quad \uparrow$$
$$ax^2 + bx + c = 0$$

We see that $a = 2$, $b = -5$, and $c = -3$. To find the solutions of the equation, we substitute these values into the formula and evaluate the right-hand side.

$$x = \frac{-b \pm \sqrt{b^2 - 4ac}}{2a}$$ This is the quadratic formula.

$$x = \frac{-(-5) \pm \sqrt{(-5)^2 - 4(2)(-3)}}{2(2)}$$ Substitute 2 for a, -5 for b, and -3 for c.

$$x = \frac{5 \pm \sqrt{25 - (-24)}}{4}$$ Simplify: $-(-5) = 5$. Evaluate the power and multiply within the radical. Multiply in the denominator.

$$x = \frac{5 \pm \sqrt{49}}{4}$$ Simplify within the radical.

$$x = \frac{5 \pm 7}{4}$$ Simplify: $\sqrt{49} = 7$.

Caution

When writing the quadratic formula, be careful to draw the fraction bar so that it includes the entire numerator. Do not write

$$x = -b \pm \frac{\sqrt{b^2 - 4ac}}{2a}$$ (crossed out)

or

$$x = -b \pm \sqrt{\frac{b^2 - 4ac}{2a}}$$ (crossed out)

To find the first solution, evaluate the expression using the + symbol. To find the second solution, evaluate the expression using the − symbol.

$$x = \frac{5 + 7}{4} \quad \text{or} \quad x = \frac{5 - 7}{4}$$

$$x = \frac{12}{4} \qquad\qquad x = \frac{-2}{4}$$

$$x = 3 \qquad\qquad x = -\frac{1}{2}$$

The solutions are 3 and $-\frac{1}{2}$. Verify that both satisfy the original equation.

Self Check 1 Solve: $4x^2 - 7x - 2 = 0$.

When using the quadratic formula, we should write the equation in $ax^2 + bx + c = 0$ form (called **quadratic form**) so that a, b, and c can be determined.

EXAMPLE 2

Solve: $2x^2 = -4x - 1$.

Solution We begin by writing the equation in quadratic form.

$$2x^2 = -4x - 1$$

$$2x^2 + 4x + 1 = 0$$ Add $4x$ and 1 to both sides.

In this equation, $a = 2$, $b = 4$, and $c = 1$.

$$x = \frac{-b \pm \sqrt{b^2 - 4ac}}{2a}$$ This is the quadratic formula.

$$x = \frac{-4 \pm \sqrt{4^2 - 4(2)(1)}}{2(2)}$$ Substitute 2 for a, 4 for b, and 1 for c.

$$x = \frac{-4 \pm \sqrt{16 - 8}}{4}$$ Evaluate the expression within the radical. Multiply in the denominator.

$$x = \frac{-4 \pm \sqrt{8}}{4}$$

$$x = \frac{-4 \pm 2\sqrt{2}}{4}$$ Simplify: $\sqrt{8} = \sqrt{4 \cdot 2} = 2\sqrt{2}$.

Success Tip

Perhaps you noticed that Example 1 could be solved by factoring. However, that is not the case for Example 2. These observations illustrate that the quadratic formula can be used to solve any quadratic equation.

We can write the solutions in simpler form by factoring out 2 from the terms in the numerator and then removing the common factor of 2 in the numerator and denominator.

$$x = \frac{-4 \pm 2\sqrt{2}}{4} = \frac{2(-2 \pm \sqrt{2})}{4} = \frac{\overset{1}{\cancel{2}}(-2 \pm \sqrt{2})}{\underset{1}{\cancel{2}} \cdot 2} = \frac{-2 \pm \sqrt{2}}{2}$$

Notation

The solutions can also be written as

$$-\frac{2}{2} \pm \frac{\sqrt{2}}{2} = -1 \pm \frac{\sqrt{2}}{2}$$

The solutions are $\frac{-2 + \sqrt{2}}{2}$ and $\frac{-2 - \sqrt{2}}{2}$. We can approximate the solutions using a calculator. To two decimal places, they are -0.29 and -1.71.

Self Check 2 Solve $3x^2 = 2x + 3$. Approximate the solutions to two decimal places.

The solutions to the next example are imaginary numbers.

EXAMPLE 3

Solve: $x^2 + x = -1$.

Solution We begin by writing the equation in quadratic form before identifying a, b, and c.

$$x^2 + x + 1 = 0$$

In this equation, $a = 1$, $b = 1$, and $c = 1$:

$$x = \frac{-b \pm \sqrt{b^2 - 4ac}}{2a}$$

$$x = \frac{-1 \pm \sqrt{1^2 - 4(1)(1)}}{2(1)}$$ Substitute 1 for a, 1 for b, and 1 for c.

$$x = \frac{-1 \pm \sqrt{1 - 4}}{2}$$ Evaluate the expression within the radical.

$$x = \frac{-1 \pm \sqrt{-3}}{2}$$

$$x = \frac{-1 \pm i\sqrt{3}}{2}$$ $\sqrt{-3} = \sqrt{-1 \cdot 3} = \sqrt{-1}\sqrt{3} = i\sqrt{3}$.

Notation

The solutions are written in complex number form $a + bi$. They could also be written as

$$\frac{-1 \pm i\sqrt{3}}{2}$$

The solutions are $-\frac{1}{2} + \frac{\sqrt{3}}{2}i$ and $-\frac{1}{2} - \frac{\sqrt{3}}{2}i$.

Self Check 3 Solve: $a^2 + 3a + 5 = 0$.

SOLVING AN EQUIVALENT EQUATION

When solving a quadratic equation by the quadratic formula, we can often simplify the computations by solving an equivalent equation.

EXAMPLE 4

For each equation, write an equivalent equation so that the quadratic formula computations will be easier:

a. $-2x^2 + 4x - 1 = 0$, **b.** $x^2 + \frac{4}{5}x - \frac{1}{3} = 0$, and **c.** $20x^2 - 60x - 40 = 0$.

Solution **a.** It is often easier to solve a quadratic equation using the quadratic formula if a is positive. If we multiply (or divide) both sides of $-2x^2 + 4x - 1 = 0$ by -1, we obtain an equivalent equation with $a > 0$.

$$-2x^2 + 4x - 1 = 0 \qquad \text{Here, } a = -2.$$
$$(-1)(-2x^2 + 4x - 1) = (-1)(0)$$
$$2x^2 - 4x + 1 = 0 \qquad \text{Now } a = 2.$$

b. For $x^2 + \frac{4}{5}x - \frac{1}{3} = 0$, two coefficients are fractions: $b = \frac{4}{5}$ and $c = -\frac{1}{3}$. We can multiply both sides of the equation by their least common denominator, 15, to obtain an equivalent equation having coefficients that are integers.

$$x^2 + \frac{4}{5}x - \frac{1}{3} = 0 \qquad \text{Here, } a = 1, b = \frac{4}{5}, \text{ and } c = -\frac{1}{3}.$$
$$15\left(x^2 + \frac{4}{5}x - \frac{1}{3}\right) = 15(0)$$
$$15x^2 + 12x - 5 = 0 \qquad \text{Now } a = 15, b = 12, \text{ and } c = -5.$$

c. For $20x^2 - 60x - 40 = 0$, the coefficients 20, -60, and -40 have a common factor of 20. If we divide both sides of the equation by their GCF, we obtain an equivalent equation having smaller coefficients.

$$20x^2 - 60x - 40 = 0 \qquad \text{Here, } a = 20, b = -60, \text{ and } c = -40.$$
$$\frac{20x^2}{20} - \frac{60x}{20} - \frac{40}{20} = \frac{0}{20}$$
$$x^2 - 3x - 2 = 0 \qquad \text{Now } a = 1, b = -3, \text{ and } c = -2.$$

Self Check 4 For each equation, write an equivalent equation so that the quadratic formula computations will be simpler: **a.** $-6x^2 + 7x - 9 = 0$, **b.** $\frac{1}{3}x^2 - \frac{2}{3}x - \frac{5}{6} = 0$, and **c.** $44x^2 + 66x - 55 = 0$.

PROBLEM SOLVING

EXAMPLE 5

Shortcuts. Instead of using the hallways, students are wearing a path through a planted quad area to walk 195 feet directly from the classrooms to the cafeteria. If the length of the hallway from the office to the cafeteria is 105 feet longer than the hallway from the office to the classrooms, how much walking are the students saving by taking the shortcut?

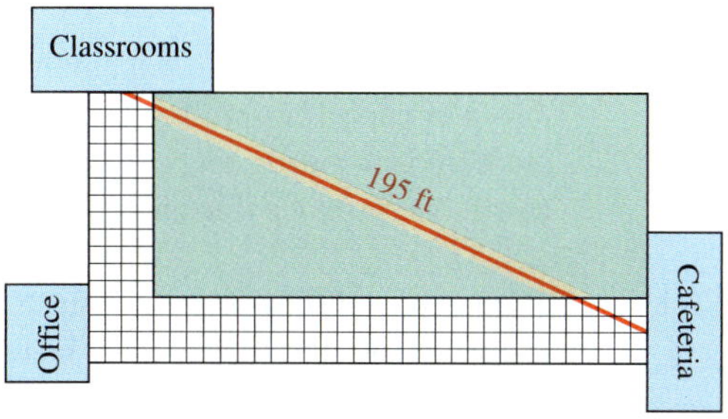

Analyze the Problem The two hallways and the shortcut form a right triangle with a hypotenuse 195 feet long. We will use the Pythagorean theorem to solve this problem.

Form an Equation If we let $x =$ the length (in feet) of the hallway from the classrooms to the office, then the length of the hallway from the office to the cafeteria is $(x + 105)$ feet. Substituting these lengths into the Pythagorean theorem, we have

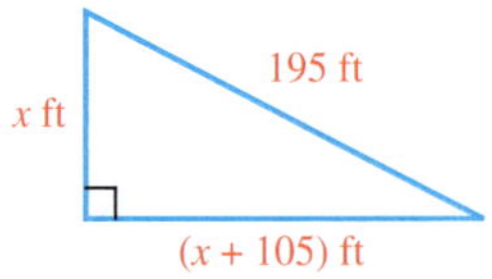

$$a^2 + b^2 = c^2 \quad \text{This is the Pythagorean theorem.}$$

$$x^2 + (x + 105)^2 = 195^2 \quad \text{Substitute } x \text{ for } a, (x + 105) \text{ for } b, \text{ and 195 for } c.$$

$$x^2 + x^2 + 105x + 105x + 11{,}025 = 38{,}025 \quad \text{Find } (x + 105)^2.$$

$$2x^2 + 210x + 11{,}025 = 38{,}025 \quad \text{Combine like terms.}$$

$$2x^2 + 210x - 27{,}000 = 0 \quad \text{Subtract 38,025 from both sides.}$$

$$x^2 + 105x - 13{,}500 = 0 \quad \text{Divide both sides by 2.}$$

Solve the Equation To solve $x^2 + 105x - 13{,}500 = 0$, we will use the quadratic formula with $a = 1$, $b = 105$, and $c = -13{,}500$.

$$x = \frac{-b \pm \sqrt{b^2 - 4ac}}{2a}$$

$$x = \frac{-105 \pm \sqrt{(105)^2 - 4(1)(-13{,}500)}}{2(1)}$$

$$x = \frac{-105 \pm \sqrt{65{,}025}}{2} \quad \text{Simplify: } (105)^2 - 4(1)(-13{,}500) = 11{,}025 + 54{,}000 = 65{,}025.$$

$$x = \frac{-105 \pm 255}{2} \quad \text{Use a calculator: } \sqrt{65{,}025} = 255.$$

$$x = \frac{150}{2} \quad \text{or} \quad x = \frac{-360}{2}$$

$$x = 75 \quad \Big| \quad x = -180 \quad \text{Since the length of the hallway can't be negative, discard the solution } x = -180.$$

State the Conclusion The length of the hallway from the classrooms to the office is 75 feet. The length of the hallway from the office to the cafeteria is $75 + 105 = 180$ feet. Instead of using the hallways, a distance of $75 + 180 = 255$ feet, the students are taking the 195-foot shortcut to the cafeteria, a savings of $(255 - 195)$, or 60 feet.

Check the Result The length of the 180-foot hallway is 105 feet longer than the length of the 75-foot hallway. The sum of the squares of the lengths of the hallways is $75^2 + 180^2 = 38{,}025$. This equals the square of the length of the 195-foot shortcut. The result checks.

EXAMPLE 6

Mass transit. A bus company has 4,000 passengers daily, each currently paying a 75¢ fare. For each 15¢ fare increase, the company estimates that it will lose 50 passengers. If the company needs to bring in \$6,570 per day to stay in business, what fare must be charged to produce this amount of revenue?

Analyze the Problem To understand how a fare increase affects the number of passengers, let's consider what happens if there are two fare increases. We organize the data in a table. The fares are expressed in terms of dollars.

Number of increases	New fare	Number of passengers
One \$0.15 increase	\$0.75 + \$0.15(1) = \$0.90	4,000 − 50(1) = 3,950
Two \$0.15 increases	\$0.75 + \$0.15(2) = \$1.05	4,000 − 50(2) = 3,900

In general, the new fare will be the old fare (\$0.75) plus the number of fare increases times \$0.15. The number of passengers who will pay the new fare is 4,000 minus 50 times the number of \$0.15 fare increases.

Form an Equation If we let x = the number of \$0.15 fare increases necessary to bring in \$6,570 daily, then $\$(0.75 + 0.15x)$ is the fare that must be charged. The number of passengers who will pay this fare is $4{,}000 - 50x$. We can now form the equation.

The bus fare	times	the number of passengers who will pay that fare	equals	\$6,570.
$(0.75 + 0.15x)$	$\cdot$	$(4{,}000 - 50x)$	$=$	$6{,}570$

Solve the Equation

$$(0.75 + 0.15x)(4{,}000 - 50x) = 6{,}570$$

$$3{,}000 - 37.5x + 600x - 7.5x^2 = 6{,}570$$ Multiply the binomials.

$$-7.5x^2 + 562.5x + 3{,}000 = 6{,}570$$ Combine like terms: $-37.5x + 600x = 562.5x$.

$$-7.5x^2 + 562.5x - 3{,}570 = 0$$ Subtract 6,570 from both sides.

$$7.5x^2 - 562.5x + 3{,}570 = 0$$ Multiply both sides by -1 so that a, 7.5, is positive.

To solve this equation, we will use the quadratic formula.

$$x = \frac{-b \pm \sqrt{b^2 - 4ac}}{2a}$$

$$x = \frac{-(-562.5) \pm \sqrt{(-562.5)^2 - 4(7.5)(3{,}570)}}{2(7.5)}$$ Substitute 7.5 for a, -562.5 for b, and 3,570 for c.

$$x = \frac{562.5 \pm \sqrt{209{,}306.25}}{15}$$ Simplify: $(-562.5)^2 - 4(7.5)(3{,}570) = 316{,}406.25 - 107{,}100 = 209{,}306.25$.

$$x = \frac{562.5 \pm 457.5}{15}$$ Use a calculator: $\sqrt{209{,}306.25} = 457.5$.

$$x = \frac{1{,}020}{15} \quad \text{or} \quad x = \frac{105}{15}$$

$$x = 68 \quad \Big| \quad x = 7$$

State the Conclusion If there are 7 fifteen-cent increases in the fare, the new fare will be \$0.75 + \$0.15(7) = \$1.80. If there are 68 fifteen-cent increases in the fare, the new fare will be \$0.75 + \$0.15(68) = \$10.95. Although this fare would bring in the necessary revenue, a \$10.95 bus fare is unreasonable, so we discard it.

Check the Result A fare of \$1.80 will be paid by [4,000 − 50(7)] = 3,650 bus riders. The amount of revenue brought in would be \$1.80(3,650) = \$6,570. The result checks.

EXAMPLE 7

Lawyers. The number of lawyers N in the United States each year from 1980 to 2002 is approximated by $N = 222x^2 + 17{,}630x + 571{,}178$, where $x = 0$ corresponds to the year 1980, $x = 1$ corresponds to 1981, $x = 2$ corresponds to 1982, and so on. (Thus, $0 \le x \le 22$). In what year does this model indicate that the United States had one million lawyers?

Solution We will substitute 1,000,000 for N in the equation. Then we can solve for x, which will give the number of years after 1980 that the United States had approximately 1,000,000 lawyers.

$$N = 222x^2 + 17{,}630x + 571{,}178$$

$$1{,}000{,}000 = 222x^2 + 17{,}630x + 571{,}178 \quad \text{Replace } N \text{ with 1,000,000.}$$

$$0 = 222x^2 + 17{,}630x - 428{,}822 \quad \text{Subtract 1,000,000 from both sides so that the equation is in quadratic form.}$$

We can simplify the computations by dividing both sides of the equation by 2, which is the greatest common factor of 222, 17,630, and 428,822.

$$111x^2 + 8{,}815x - 214{,}411 = 0 \quad \text{Divide both sides by 2.}$$

We solve this equation using the quadratic formula.

$$x = \frac{-b \pm \sqrt{b^2 - 4ac}}{2a}$$

$$x = \frac{-8{,}815 \pm \sqrt{(8{,}815)^2 - 4(111)(-214{,}411)}}{2(111)} \quad \text{Substitute 111 for } a\text{, 8,815 for } b\text{, and } -214{,}411 \text{ for } c.$$

$$x = \frac{-8{,}815 \pm \sqrt{172{,}902{,}709}}{222} \quad \text{Evaluate the expression within the radical.}$$

$$x \approx \frac{4{,}334}{222} \quad \text{or} \quad x \approx \frac{-21{,}964}{222} \quad \text{Use a calculator.}$$

$$x \approx 19.5 \quad \Big| \quad x \approx -98.9 \quad \text{Since the model is defined only for } 0 \le x \le 22\text{, we discard the second solution.}$$

In 19.5 years after 1980, or midway through 1999, the United States had approximately 1,000,000 lawyers.

Answers to Self Checks **1.** $2, -\frac{1}{4}$ **2.** $\frac{1 \pm \sqrt{10}}{3}$; $-0.72, 1.39$ **3.** $-\frac{3}{2} \pm \frac{\sqrt{11}}{2}i$

4. a. $6x^2 - 7x + 9 = 0$, **b.** $2x^2 - 4x - 5 = 0$, **c.** $4x^2 + 6x - 5 = 0$

8.2 STUDY SET

VOCABULARY **Fill in the blanks.**

1. An equation of the form $ax^2 + bx + c = 0$, with $a \ne 0$, is a _________ equation.

2. The formula

$$x =$$

is called the quadratic formula.

CONCEPTS

3. Write each equation in quadratic form.

a. $x^2 + 2x = -5$ **b.** $3x^2 = -2x + 1$

4. For each quadratic equation, find a, b, and c.

a. $x^2 + 5x + 6 = 0$ **b.** $8x^2 - x = 10$

5. Decide whether each statement is true or false.

a. Any quadratic equation can be solved by using the quadratic formula.

b. Any quadratic equation can be solved by completing the square.

6. What is wrong with the beginning of the solution shown below?

Solve: $x^2 - 3x = 2$.

$a = 1 \quad b = -3 \quad c = 2$

Evaluate each expression.

7. a. $\dfrac{-2 \pm \sqrt{2^2 - 4(1)(-8)}}{2(1)}$

b. $\dfrac{-(-1) \pm \sqrt{(-1)^2 - 4(2)(-4)}}{2(2)}$

8. A student used the quadratic formula to solve a quadratic equation and obtained $x = \dfrac{-2 \pm \sqrt{3}}{2}$.

a. How many solutions does the equation have? What are they exactly?

b. Graph the solutions on a number line.

9. Simplify each of the following.

a. $\dfrac{3 \pm 6\sqrt{2}}{3}$

b. $\dfrac{-12 \pm 4\sqrt{7}}{8}$

10. For each of the following, write an equivalent equation so that the quadratic formula computations will be easier to perform.

a. $-5x^2 + 9x - 2 = 0$

b. $\dfrac{1}{8}x^2 + \dfrac{1}{2}x - \dfrac{3}{4} = 0$

c. $45x^2 + 30x - 15 = 0$

NOTATION

11. On a quiz, students were asked to write the quadratic formula. What is wrong with each answer shown below?

a. $x = -b \pm \dfrac{\sqrt{b^2 - 4ac}}{2a}$

b. $x = \dfrac{-b\sqrt{b^2 - 4ac}}{2a}$

12. In reading $\dfrac{-b \pm \sqrt{b^2 - 4ac}}{2a}$, we say, "The ________ of b, plus or ______ the square ______ of b ________ minus times a times c, all ______________ $2a$."

PRACTICE **Use the quadratic formula to solve each equation.**

13. $x^2 + 3x + 2 = 0$

14. $x^2 - 3x + 2 = 0$

15. $x^2 + 12x = -36$

16. $y^2 - 18y = -81$

17. $2x^2 + 5x - 3 = 0$

18. $6x^2 - x - 1 = 0$

19. $5x^2 + 5x + 1 = 0$

20. $4w^2 + 6w + 1 = 0$

21. $8u = -4u^2 - 3$

22. $4t + 3 = 4t^2$

23. $-16y^2 - 8y + 3 = 0$

24. $-16x^2 - 16x - 3 = 0$

25. $x^2 - \dfrac{14}{15}x = \dfrac{8}{15}$

26. $x^2 = -\dfrac{5}{4}x + \dfrac{3}{2}$

27. $\dfrac{x^2}{2} + \dfrac{5}{2}x = -1$

28. $\dfrac{x^2}{8} - \dfrac{x}{4} = \dfrac{1}{2}$

29. $2x^2 - 1 = 3x$

30. $-9x = 2 - 3x^2$

31. $-x^2 + 10x = 18$

32. $-3x = \dfrac{x^2}{2} + 2$

33. $x^2 - 6x = 391$

34. $-x^2 + 27x = -280$

35. $x^2 - \frac{5}{3} = -\frac{11}{6}x$

36. $x^2 - \frac{1}{2} = \frac{2}{3}x$

37. $x^2 + 2x + 2 = 0$

38. $x^2 + 3x + 3 = 0$

39. $2x^2 + x + 1 = 0$

40. $3x^2 + 2x + 1 = 0$

41. $3x^2 - 4x = -2$

42. $2x^2 + 3x = -3$

43. $3x^2 - 2x = -3$

44. $5x^2 = 2x - 1$

45. $\frac{x^2}{8} - \frac{x}{2} + 1 = 0$

46. $\frac{x^2}{2} + 3x + \frac{13}{2} = 0$

47. $\frac{a^2}{10} - \frac{3a}{5} + \frac{7}{5} = 0$

48. $\frac{c^2}{4} + c + \frac{11}{4} = 0$

49. $50x^2 + 30x - 10 = 0$

50. $120b^2 + 120b - 40 = 0$

51. $900x^2 - 8{,}100x = 1{,}800$

52. $-14x^2 + 21x = -49$

53. $-0.6x^2 - 0.03 = -0.4x$

54. $2x^2 + 0.1x = 0.04$

Use the quadratic formula and a scientific calculator to solve each equation. Give all answers to the nearest hundredth.

55. $x^2 + 8x + 5 = 0$

56. $2x^2 - x - 9 = 0$

57. $3x^2 - 2x - 2 = 0$

58. $81x^2 + 12x - 80 = 0$

59. $0.7x^2 - 3.5x - 25 = 0$

60. $-4.5x^2 + 0.2x + 3.75 = 0$

APPLICATIONS

61. IMAX SCREENS The largest permanent movie screen is in the Panasonic Imax theater at Darling Harbor, Sydney, Australia. The rectangular screen has an area of 11,349 square feet. Find the dimensions of the screen if it is 20 feet longer than it is wide.

62. ROCK CONCERTS During a 1997 tour, the rock group U2 used an LED (light emitting diode) electronic screen as part of the stage backdrop. The rectangular screen, with an area of 9,520 square feet, had a length that was 2 feet more than three times its width. Find the dimensions of the LED screen.

63. PARKS Central Park is one of New York's best-known landmarks. Rectangular in shape, its length is 5 times its width. When measured in miles, its perimeter numerically exceeds its area by 4.75. Find the dimensions of Central Park if we know that its width is less than 1 mile.

64. HISTORY One of the important cities of the ancient world was Babylon. Greek historians wrote that the city was square-shaped. Measured in miles, its area numerically exceeded its perimeter by about 124. Find its dimensions. (Round to the nearest tenth.)

65. BADMINTON The person who wrote the instructions for setting up the badminton net shown below forgot to give the specific dimensions for securing the pole. How long is the support string?

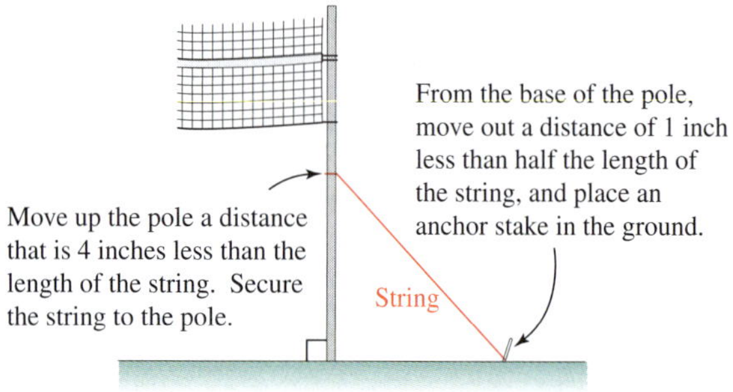
From the base of the pole, move out a distance of 1 inch less than half the length of the string, and place an anchor stake in the ground.
Move up the pole a distance that is 4 inches less than the length of the string. Secure the string to the pole.
String

66. RIGHT TRIANGLES The hypotenuse of a right triangle is 2.5 units long. The longer leg is 1.7 units longer than the shorter leg. Find the lengths of the sides of the triangle.

67. DANCES Tickets to a school dance cost \$4, and the projected attendance is 300 persons. It is further projected that for every 10¢ increase in ticket price, the average attendance will decrease by 5. At what ticket price will the receipts from the dance be \$1,248?

68. TICKET SALES A carnival usually sells three thousand 75¢ ride tickets on a Saturday. For each 15¢ increase in price, management estimates that 80 fewer tickets will be sold. What increase in ticket price will produce \$2,982 of revenue on Saturday?

69. MAGAZINE SALES The *Gazette's* profit is \$20 per year for each of its 3,000 subscribers. Management estimates that the profit per subscriber will increase by 1¢ for each additional subscriber over the current 3,000. How many subscribers will bring a total profit of \$120,000?

70. POLYGONS A five-sided polygon, called a *pentagon,* has 5 diagonals. The number of diagonals d of a polygon of n sides is given by the formula

$$d = \frac{n(n-3)}{2}$$

Find the number of sides of a polygon if it has 275 diagonals.

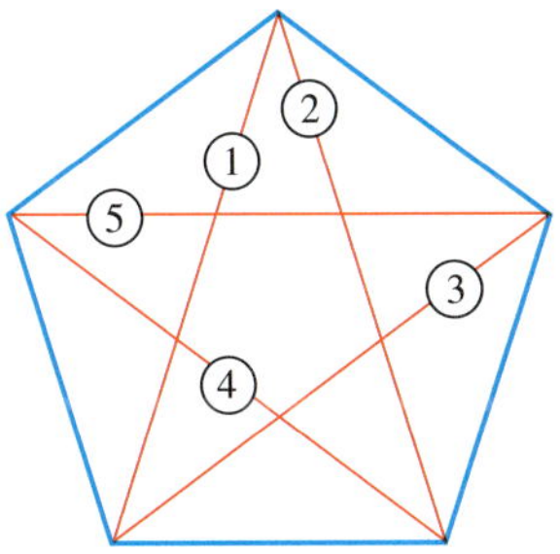

71. INVESTMENT RATES A woman invests \$1,000 in a fund for which interest is compounded annually at a rate r. After one year, she deposits an additional \$2,000. After two years, the balance in the account is

$$\$1{,}000(1 + r)^2 + \$2{,}000(1 + r)$$

If this amount is \$3,368.10, find r.

72. METAL FABRICATION A box with no top is to be made by cutting a 2-inch square from each corner of the square sheet of metal. After bending up the sides, the volume of the box is to be 220 cubic inches. How large should the piece of metal be? Round to the nearest hundredth.

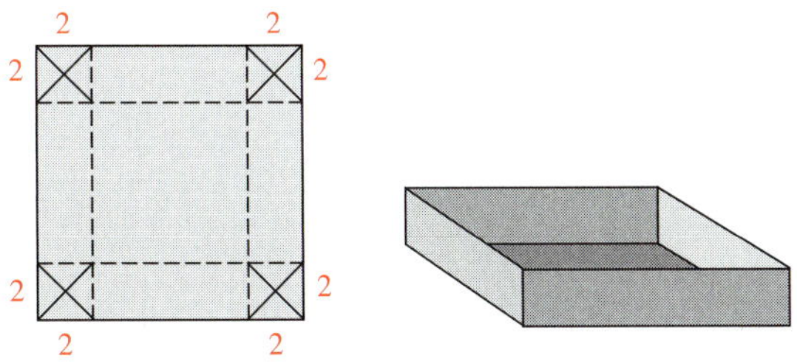

73. RETIREMENT The labor force participation rate P (in percent) for men ages 55–64 from 1970 to 2000 is approximated by the quadratic equation

$$P = 0.03x^2 - 1.37x + 82.51$$

where $x = 0$ corresponds to the year 1970, $x = 1$ corresponds to 1971, $x = 2$ corresponds to 1972, and so on. (Thus, $0 \le x \le 30$.) When does the model indicate that 75% of the men ages 55–64 were part of the workforce?

74. SPACE PROGRAM The yearly budget B (in billions of dollars) for the National Aeronautics and Space Administration (NASA) is approximated by the quadratic equation

$$B = 0.0596x^2 - 0.3811x + 14.2709$$

where x is the number of years since 1995 and $0 \le x \le 9$. In what year does the model indicate that NASA's budget was about \$15 billion?

WRITING

75. Explain why the quadratic formula, in most cases, is less tedious to use in solving a quadratic equation than is the method of completing the square.

76. On an exam, a student was asked to solve the equation $-4w^2 - 6w - 1 = 0$. Her first step was to multiply both sides of the equation by -1. She then used the quadratic formula to solve $4w^2 + 6w + 1 = 0$ instead. Is this a valid approach? Explain.

REVIEW **Change each radical to an exponential expression.**

77. $\sqrt{n}$

78. $\sqrt[7]{\frac{3}{8}r^2s}$

79. $\sqrt[4]{3b}$

80. $3\sqrt[3]{c^2 - d^2}$

Write each expression in radical form.

81. $t^{1/3}$

82. $\left(\frac{3}{4}m^2n^2\right)^{1/5}$

83. $(3t)^{1/4}$

84. $(c^2 + d^2)^{1/2}$

CHALLENGE PROBLEMS **All of the equations we have solved so far have had rational-number coefficients. However, the quadratic formula can be used to solve quadratic equations with irrational or even imaginary coefficients. Solve each equation.**

85. $x^2 + 2\sqrt{2}x - 6 = 0$

86. $\sqrt{2}x^2 + x - \sqrt{2} = 0$

87. $x^2 - 3ix - 2 = 0$

88. $100ix^2 + 300x - 200i = 0$

8.3 The Discriminant and Equations That Can Be Written in Quadratic Form

- The discriminant
- Equations that are quadratic in form
- Problem solving

In this section, we will discuss how to predict what type of solutions a quadratic equation will have without solving the equation. We will then solve some special equations that can be written in quadratic form. Finally, we will use the equation-solving methods of this chapter to solve a shared-work problem.

THE DISCRIMINANT

We can predict what type of solutions a particular quadratic equation will have without solving it. To see how, we suppose that the coefficients a, b, and c in the equation $ax^2 + bx + c = 0$ represent real numbers and $a \neq 0$. Then the solutions of the equation are given by the quadratic formula

$$x = \frac{-b \pm \sqrt{b^2 - 4ac}}{2a}$$

If $b^2 - 4ac \geq 0$, the solutions are real numbers. If $b^2 - 4ac < 0$, the solutions are not real numbers. Thus, the value of $b^2 - 4ac$, called the **discriminant,** determines the type of solutions for a particular quadratic equation.

The Discriminant If a, b, and c represent real numbers and

if $b^2 - 4ac$ is . . .	**the solutions are . . .**
positive,	two different real numbers.
0,	two real numbers that are equal.
negative,	two different complex numbers containing i that are complex conjugates.

If a, b, and c represent rational numbers and

if $b^2 - 4ac$ is . . .	**the solutions are . . .**
a perfect square,	two different rational numbers.
positive and not a perfect square,	two different irrational numbers.

EXAMPLE 1

Determine the type of solutions for each equation: **a.** $x^2 + x + 1 = 0$ and **b.** $3x^2 + 5x + 2 = 0$.

Solution **a.** We calculate the discriminant for $x^2 + x + 1 = 0$:

$$b^2 - 4ac = 1^2 - 4(1)(1) \quad a = 1, b = 1, \text{ and } c = 1.$$
$$= -3 \quad \text{The result is a negative number.}$$

Since $b^2 - 4ac < 0$, the solutions of $x^2 + x + 1 = 0$ are two complex numbers containing i that are complex conjugates.

b. For $3x^2 + 5x + 2 = 0$,

$$
\begin{aligned}
b^2 - 4ac &= 5^2 - 4(3)(2) && a = 3, b = 5, \text{ and } c = 2. \\
&= 25 - 24 \\
&= 1 && \text{The result is a positive number.}
\end{aligned}
$$

Since $b^2 - 4ac > 0$ and $b^2 - 4ac$ is a perfect square, the two solutions of $3x^2 + 5x + 2 = 0$ are rational and unequal.

Self Check 1 Determine the type of solutions for **a.** $x^2 + x - 1 = 0$ and **b.** $4x^2 - 10x + 25 = 0$.

EQUATIONS THAT ARE QUADRATIC IN FORM

Many equations that are not quadratic can be written in quadratic form ($ax^2 + bx + c = 0$) and then solved using the techniques discussed in previous sections. For example, a careful inspection of the equation $x^4 - 5x^2 + 4 = 0$ leads to the following observations:

$$x^4 - 5x^2 + 4 = 0$$

The lead term, x^4, is the square of the expression x^2, here in the middle term: $x^4 = (x^2)^2$.

The last term is a constant.

Equations that contain an expression, the same expression squared, and a constant term are said to be *quadratic in form.* One method used to solve such equations is to make a substitution.

EXAMPLE 2

Solve: $x^4 - 3x^2 - 4 = 0$.

Solution If we write x^4 as $(x^2)^2$, then the equation takes the form

$$(x^2)^2 - 3x^2 - 4 = 0$$

and it is said to be *quadratic in* x^2. We can solve this equation by letting $y = x^2$.

$$y^2 - 3y - 4 = 0 \qquad \text{Replace each } x^2 \text{ with } y.$$

We can solve this quadratic equation by factoring.

$$
\begin{aligned}
&(y - 4)(y + 1) = 0 && \text{Factor } y^2 - 3y - 4. \\
&y - 4 = 0 \quad \text{or} \quad y + 1 = 0 && \text{Set each factor equal to 0.} \\
&y = 4 \quad\quad\quad\;\; | \quad\; y = -1
\end{aligned}
$$

These are *not* the solutions for x. To find x, we now undo the earlier substitutions by replacing each y with x^2. Then we solve for x.

$$
\begin{aligned}
&x^2 = 4 \quad \text{or} \quad x^2 = -1 && \text{Substitute } x^2 \text{ for } y. \\
&x = \pm\sqrt{4} \quad | \quad x = \pm\sqrt{-1} && \text{Use the square root property.} \\
&x = \pm 2 \quad\;\; | \quad x = \pm i
\end{aligned}
$$

Notation

The choice of the letter y for the substitution is arbitrary. We could just as well let $b = x^2$.

Caution

If you are solving an equation in x, you can't answer with values of y. Remember to undo any substitutions, and solve for the variable in the original equation.

This equation has four solutions: 2, -2, i, and $-i$. Verify that each of them satisfies the original equation.

Self Check 2 Solve: $x^4 - 5x^2 - 36 = 0$.

EXAMPLE 3

Solve: $x - 7\sqrt{x} + 12 = 0$.

Solution We examine the lead term and middle term.

The lead term, x, is the square of the expression $\sqrt{x}$, here in the middle term: $x = (\sqrt{x})^2$.

$$x - 7\sqrt{x} + 12 = 0$$

> **The Language of Algebra**
>
> Equations such as $x - 7\sqrt{x} + 12 = 0$ that are quadratic in form are also said to be *reducible to a quadratic.*

If we write x as $(\sqrt{x})^2$, then the equation takes the form

$$(\sqrt{x})^2 - 7\sqrt{x} + 12 = 0$$

and it is said to be *quadratic in* $\sqrt{x}$. We can solve this equation by letting $y = \sqrt{x}$ and factoring.

$$y^2 - 7y + 12 = 0 \qquad \text{Replace each } \sqrt{x} \text{ with } y.$$
$$(y - 3)(y - 4) = 0 \qquad \text{Factor } y^2 - 7y + 12.$$
$$y - 3 = 0 \quad \text{or} \quad y - 4 = 0 \qquad \text{Set each factor equal to 0.}$$
$$y = 3 \quad | \quad y = 4$$

To find x, we undo the substitutions by replacing each y with $\sqrt{x}$. Then we solve the radical equations by squaring both sides.

$$\sqrt{x} = 3 \quad \text{or} \quad \sqrt{x} = 4$$
$$x = 9 \quad | \quad x = 16$$

The solutions are 9 and 16. Verify that both satisfy the original equation.

Self Check 3 Solve: $x + \sqrt{x} - 6 = 0$.

EXAMPLE 4

Solve: $2m^{2/3} - 2 = 3m^{1/3}$.

Solution After writing the equation in descending powers of m, we see that

$$2m^{2/3} - 3m^{1/3} - 2 = 0$$

is *quadratic in* $m^{1/3}$, because $m^{2/3} = (m^{1/3})^2$. We will use the substitution $y = m^{1/3}$ to write this equation in quadratic form.

$$2m^{2/3} - 3m^{1/3} - 2 = 0$$

$$2(m^{1/3})^2 - 3m^{1/3} - 2 = 0 \quad \text{Write } m^{2/3} \text{ as } (m^{1/3})^2.$$

$$2y^2 - 3y - 2 = 0 \quad \text{Replace each } m^{1/3} \text{ with } y.$$

$$(2y + 1)(y - 2) = 0 \quad \text{Factor } 2y^2 - 3y - 2.$$

$$2y + 1 = 0 \quad \text{or} \quad y - 2 = 0 \quad \text{Set each factor equal to 0.}$$

$$y = -\frac{1}{2} \quad \Big| \quad y = 2$$

To find m, we undo the substitutions by replacing each y with $m^{1/3}$. Then we solve the equations by cubing both sides.

$$m^{1/3} = -\frac{1}{2} \quad \text{or} \quad m^{1/3} = 2$$

$$(m^{1/3})^3 = \left(-\frac{1}{2}\right)^3 \quad \Big| \quad (m^{1/3})^3 = (2)^3 \quad \text{Recall that } m^{1/3} = \sqrt[3]{m}. \text{ To solve for } m, \text{ cube both sides.}$$

$$m = -\frac{1}{8} \quad \Big| \quad m = 8$$

The solutions are $-\frac{1}{8}$ and 8. Verify that both satisfy the original equation.

Self Check 4 Solve: $a^{2/3} = -3a^{1/3} + 10$.

EXAMPLE 5

Solve: $(4t + 2)^2 - 30(4t + 2) + 224 = 0$.

Solution This equation is *quadratic in* $4t + 2$. If we make the substitution $y = 4t + 2$, we have

$$y^2 - 30y + 224 = 0$$

which can be solved by using the quadratic formula.

$$y = \frac{-b \pm \sqrt{b^2 - 4ac}}{2a}$$

$$y = \frac{-(-30) \pm \sqrt{(-30)^2 - 4(1)(224)}}{2(1)} \quad \text{Substitute 1 for } a, -30 \text{ for } b, \text{ and 224 for } c.$$

$$y = \frac{30 \pm \sqrt{900 - 896}}{2} \quad \text{Simplify within the radical.}$$

$$y = \frac{30 \pm 2}{2} \quad \sqrt{900 - 896} = \sqrt{4} = 2.$$

$$y = 16 \quad \text{or} \quad y = 14$$

To find t, we replace y with $4t + 2$ and solve for t.

$$4t + 2 = 16 \quad \text{or} \quad 4t + 2 = 14$$

$$4t = 14 \quad \Big| \quad 4t = 12$$

$$t = 3.5 \quad \Big| \quad t = 3$$

Verify that 3.5 and 3 satisfy the original equation.

Self Check 5 Solve: $(n + 3)^2 - 6(n + 3) = -8$.

EXAMPLE 6

Solve: $15a^{-2} - 8a^{-1} + 1 = 0$.

Solution When we write the terms $15a^{-2}$ and $-8a^{-1}$ using positive exponents, we see that this equation is *quadratic in* $\frac{1}{a}$.

$$\frac{15}{a^2} - \frac{8}{a} + 1 = 0 \qquad \text{Think of this equation as } 15 \cdot \left(\frac{1}{a}\right)^2 - 8 \cdot \frac{1}{a} + 1 = 0.$$

If we let $y = \dfrac{1}{a}$, the resulting quadratic equation can be solved by factoring.

$$15y^2 - 8y + 1 = 0 \qquad \text{Substitute } \tfrac{1}{a} \text{ for } y \text{ and } \tfrac{1}{a^2} \text{ for } y^2.$$
$$(5y - 1)(3y - 1) = 0 \qquad \text{Factor } 15y^2 - 8y + 1 = 0.$$
$$5y - 1 = 0 \quad \text{or} \quad 3y - 1 = 0$$
$$y = \frac{1}{5} \quad \Big| \quad y = \frac{1}{3}$$

To find y, we undo the substitution by replacing each y with $\dfrac{1}{a}$.

$$\frac{1}{a} = \frac{1}{5} \quad \text{or} \quad \frac{1}{a} = \frac{1}{3}$$
$$5 = a \quad \Big| \quad 3 = a \qquad \text{Solve the proportions.}$$

The solutions are 5 and 3. Verify that they satisfy the original equation.

Self Check 6 Solve: $28c^{-2} - 3c^{-1} - 1 = 0$.

PROBLEM SOLVING

EXAMPLE 7

Household appliances. The illustration shows a water temperature control on a washing machine. When the "warm" setting is selected, both the hot and cold water inlets open to fill the tub in 2 minutes 15 seconds. When the "cold" temperature setting is chosen, the cold water inlet fills the tub 45 seconds faster than when the "hot" setting is used. How long does it take to fill the washing machine with hot water?

Electronic Temperature Control

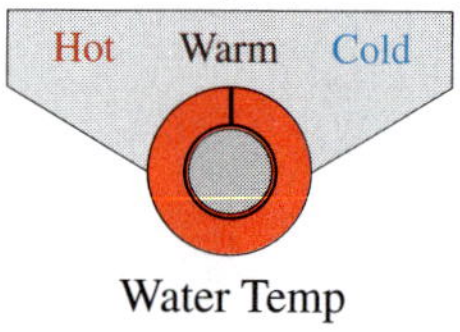

Analyze the Problem It is helpful to organize the facts of this shared-work problem in a table.

Form an Equation Let x = the number of seconds it takes to fill the tub with hot water. Since the cold water inlet fills the tub in 45 seconds less time, $x - 45$ = the number of seconds it takes to fill the tub with cold water. The hot and cold water inlets will be open for the same time: 2 minutes 15 seconds, or 135 seconds.

To determine the work completed by each inlet, multiply the rate by the time.

	Rate	· Time	= Work completed
Hot water	$\frac{1}{x}$	135	$\frac{135}{x}$
Cold water	$\frac{1}{x-45}$	135	$\frac{135}{x-45}$

Enter this information first. Multiply to get each of these entries: $W = rt$.

Success Tip

An alternate way to form an equation is to note that what the hot water inlet can do in 1 second plus what the cold water inlet can do in 1 second equals what they can do together in 1 second:

$$\frac{1}{x} + \frac{1}{x-45} = \frac{1}{135}$$

In shared-work problems, the number 1 represents one whole job completed. So we have,

The fraction of tub filled with hot water	plus	the fraction of the tub filled with cold water	equals	1 tub filled.
$\frac{135}{x}$	$+$	$\frac{135}{x-45}$	$=$	1

Solve the Equation

$$\frac{135}{x} + \frac{135}{x-45} = 1$$

$$x(x-45)\left(\frac{135}{x} + \frac{135}{x-45}\right) = x(x-45)(1)$$ Multiply both sides by the LCD $x(x-45)$ to clear the equation of fractions.

$$135(x-45) + 135x = x(x-45)$$

$$135x - 6{,}075 + 135x = x^2 - 45x$$ Distribute the multiplication by 135 and by x.

$$270x - 6{,}075 = x^2 - 45x$$ Combine like terms.

$$0 = x^2 - 315x + 6{,}075$$ Subtract $270x$ from both sides. Add 6,075 to both sides.

To solve this equation, we will use the quadratic formula, with $a = 1$, $b = -315$, and $c = 6{,}075$.

$$x = \frac{-b \pm \sqrt{b^2 - 4ac}}{2a}$$

$$x = \frac{-(-315) \pm \sqrt{(-315)^2 - 4(1)(6{,}075)}}{2(1)}$$ Substitute 1 for a, -315 for b, and 6,075 for c.

$$x = \frac{315 \pm \sqrt{99{,}225 - 24{,}300}}{2}$$ Simplify within the radical.

$$x = \frac{315 \pm \sqrt{74{,}925}}{2}$$

$$x \approx \frac{589}{2} \quad \text{or} \quad x \approx \frac{41}{2}$$

$$x \approx 294 \quad \Big| \quad x \approx 21$$

State the Conclusion We can discard the solution of 21 seconds, because this would imply that the cold water inlet fills the tub in a negative number of seconds $(21 - 45 = -24)$. Therefore, the hot water inlet fills the washing machine tub in about 294 seconds, which is 4 minutes 54 seconds.

Check the Result Use estimation to check the result.

Answers to Self Checks **1. a.** real numbers that are irrational and unequal, **b.** two complex numbers containing i that are complex conjugates **2.** $3, -3, 2i, -2i$ **3.** 4 **4.** $-125, 8$ **5.** $-1, 1$ **6.** $-7, 4$

8.3 STUDY SET

VOCABULARY **Fill in the blanks.**

1. For the quadratic equation $ax^2 + bx + c = 0$, the __________ is $b^2 - 4ac$.

2. We can solve $x - 2\sqrt{x} - 8 = 0$ by making a __________: Let $y = \sqrt{x}$.

CONCEPTS **Consider the quadratic equation $ax^2 + bx + c = 0$, where a, b, and c represent rational numbers, and fill in the blanks.**

3. If $b^2 - 4ac < 0$, the solutions of the equation are two complex numbers containing i that are complex __________.

4. If $b^2 - 4ac = \square$, the solutions of the equation are equal real numbers.

5. If $b^2 - 4ac$ is a perfect square, the solutions are __________ numbers and __________.

6. If $b^2 - 4ac$ is positive and not a perfect square, the solutions are __________ numbers and __________.

7. For each equation, determine the substitution that should be made to write the equation in quadratic form.

a. $x^4 - 12x^2 + 27 = 0$ Let $y = \square$

b. $x - 13\sqrt{x} + 40 = 0$ Let $y = \square$

c. $x^{2/3} + 2x^{1/3} - 3 = 0$ Let $y = \square$

d. $x^{-2} - x^{-1} - 30 = 0$ Let $y = \square$

e. $(x + 1)^2 - (x + 1) - 6 = 0$ Let $y = \square$

8. Fill in the blanks.

a. $x^4 = (\square)^2$ **b.** $x = (\square)^2$

c. $x^{2/3} = (\square)^2$ **d.** $\frac{1}{x^2} = \left(\square\right)^2$

NOTATION **Complete each solution.**

9. To find the type of solutions for the equation $x^2 + 5x + 6 = 0$, we compute the discriminant.

$$b^2 - \square = \square^2 - 4(1)(\square)$$
$$= 25 - \square$$
$$= 1$$

Since a, b, and c are rational numbers and the value of the discriminant is a perfect square, the solutions are __________ numbers and unequal.

10. Change $\frac{3}{4} + x = \frac{3x - 50}{4(x - 6)}$ to quadratic form.

$$\square\left(\frac{3}{4} + x\right) = \square\frac{3x - 50}{4(x - 6)}$$
$$3(x - 6) + 4x(\square) = 3x - 50$$
$$3x - \square + 4x^2 - \square = 3x - 50$$
$$4x^2 - 24x + \square = 0$$
$$\square - 6x + 8 = 0$$

PRACTICE **Use the discriminant to determine what type of solutions exist for each equation. Do not solve the equation.**

11. $4x^2 - 4x + 1 = 0$

12. $6x^2 - 5x - 6 = 0$

13. $5x^2 + x + 2 = 0$

14. $3x^2 + 10x - 2 = 0$

15. $2x^2 = 4x - 1$

16. $9x^2 = 12x - 4$

17. $x(2x - 3) = 20$

18. $x(x - 3) = -10$

19. Use the discriminant to determine whether the solutions of $1{,}492x^2 + 1{,}776x - 2{,}000 = 0$ are real numbers.

20. Use the discriminant to determine whether the solutions of $1{,}776x^2 - 1{,}492x + 2{,}000 = 0$ are real numbers.

Solve each equation.

21. $x^4 - 17x^2 + 16 = 0$

22. $x^4 - 10x^2 + 9 = 0$

23. $x^4 = 6x^2 - 5$

24. $2x^4 + 24 = 26x^2$

25. $t^4 + 3t^2 = 28$

26. $3h^4 + h^2 - 2 = 0$

27. $x^4 + 19x^2 + 18 = 0$

28. $t^4 + 4t^2 - 5 = 0$

29. $2x + \sqrt{x} - 3 = 0$

30. $2x - \sqrt{x} - 1 = 0$

31. $3x + 5\sqrt{x} + 2 = 0$

32. $3x - 4\sqrt{x} + 1 = 0$

33. $x - 6x^{1/2} = -8$

34. $x - 5x^{1/2} + 4 = 0$

35. $2x - \sqrt{x} = 3$

36. $3x + 4\sqrt{x} = 4$

37. $x^{2/3} + 5x^{1/3} + 6 = 0$

38. $x^{2/3} - 7x^{1/3} + 12 = 0$

39. $a^{2/3} - 2a^{1/3} - 3 = 0$

40. $r^{2/3} + 4r^{1/3} - 5 = 0$

41. $2x^{2/5} - 5x^{1/5} = -3$

42. $2x^{2/5} + 3x^{1/5} = -1$

43. $2(2x + 1)^2 - 7(2x + 1) + 6 = 0$

44. $3(2 - x)^2 + 10(2 - x) - 8 = 0$

45. $(c + 1)^2 - 4(c + 1) + 8 = 0$

46. $(k - 7)^2 + 6(k - 7) + 10 = 0$

47. $(a^2 - 4)^2 - 4(a^2 - 4) - 32 = 0$

48. $(y^2 - 9)^2 + 2(y^2 - 9) - 99 = 0$

49. $9\left(\dfrac{3m + 2}{m}\right)^2 - 30\left(\dfrac{3m + 2}{m}\right) + 25 = 0$

50. $4\left(\dfrac{c - 7}{c}\right)^2 - 12\left(\dfrac{c - 7}{c}\right) + 9 = 0$

51. $\left(8 - \sqrt{a}\right)^2 + 6\left(8 - \sqrt{a}\right) - 7 = 0$

52. $\left(10 - \sqrt{t}\right)^2 - 4\left(10 - \sqrt{t}\right) - 45 = 0$

53. $8x^{-2} - 10x^{-1} - 3 = 0$

54. $2x^{-2} - 5x^{-1} - 3 = 0$

55. $8(t + 1)^{-2} - 30(t + 1)^{-1} + 7 = 0$

56. $2(s - 2)^{-2} + 3(s - 2)^{-1} - 5 = 0$

57. $x^{-4} - 2x^{-2} + 1 = 0$

58. $4x^{-4} + 1 = 5x^{-2}$

59. $x + \dfrac{2}{x - 2} = 0$

60. $x + \dfrac{x + 5}{x - 3} = 0$

61. $x + 5 + \dfrac{4}{x} = 0$

62. $x - 4 + \dfrac{3}{x} = 0$

63. $\dfrac{1}{x + 2} + \dfrac{24}{x + 3} = 13$

64. $\dfrac{3}{x} + \dfrac{4}{x + 1} = 2$

65. $\dfrac{2}{x - 1} + \dfrac{1}{x + 1} = 3$

66. $\dfrac{3}{x - 2} - \dfrac{1}{x + 2} = 5$

APPLICATIONS

67. FLOWER ARRANGING A florist needs to determine the height h of the flowers shown in the illustration. The radius r, the width w, and the height h of the circular-shaped arrangement are related by the formula

$$r = \frac{4h^2 + w^2}{8h}$$

If w is to be 34 inches and r is to be 18 inches, find h to the nearest tenth of an inch.

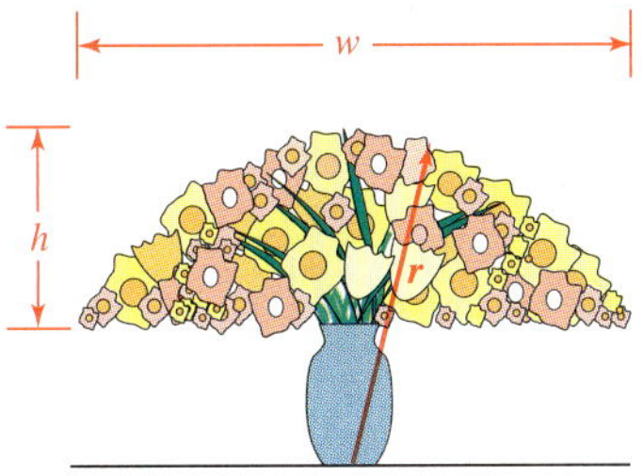

68. ARCHITECTURE A **golden rectangle** is one of the most visually appealing of all geometric forms. The Parthenon, built by the Greeks in the 5th century B.C., fits into a golden rectangle if its ruined triangular pediment is included. See the illustration on the next page.

In a golden rectangle, the length ℓ and width w must satisfy the equation

$$\frac{\ell}{w} = \frac{w}{\ell - w}$$

If a rectangular billboard is to have a width of 20 feet, what should its length be so that it is a golden rectangle? Round to the nearest tenth.

69. SNOWMOBILES A woman drives her snowmobile 150 miles at a rate of r mph. She could have gone the same distance in 2 hours less time if she had increased her speed by 20 mph. Find r.

70. BICYCLING Tina bicycles 160 miles at the rate of r mph. The same trip would have taken 2 hours longer if she had decreased her speed by 4 mph. Find r.

71. CROWD CONTROL After a performance at a county fair, security guards have found that the grandstand area can be emptied in 6 minutes if both the east and west exits are opened. If just the east exit is used, it takes 4 minutes longer to clear the grandstand than it does if just the west exit is opened. How long does it take to clear the grandstand if everyone must file through the east exit?

72. PAPER ROUTES When a father, in a car, and his son, on a bicycle, work together to distribute the morning newspaper, it takes them 35 minutes to complete the route. Working alone, it takes the son 25 minutes longer than the father. To the nearest minute, how long does it take the son to cover the route on his bicycle?

WRITING

73. Describe how to predict what type of solutions the equation $3x^2 - 4x + 5 = 0$ will have.

74. What error is made in the following solution?

Solve: $x^4 - 12x^2 + 27 = 0$

Let $y = x^2$

$y^2 - 12y + 27 = 0$

$(y - 9)(y - 3) = 0$

$y - 9 = 0$ or $y - 3 = 0$

$y = 9$ | $y = 3$

The solutions are 9 and 3.

REVIEW

75. Write an equation of the vertical line that passes through (3, 4).

76. Write an equation of the line that passes through $(-1, -6)$ and $(-2, -1)$. Express the result in slope–intercept form.

77. Write an equation of the line with slope $\frac{2}{3}$ that passes through the origin.

78. Write an equation of the line that passes through $(2, -3)$ and is perpendicular to the line whose equation is $y = \frac{x}{5} + 6$. Express the result in slope–intercept form.

CHALLENGE PROBLEMS

79. Solve: $x^6 + 17x^3 + 16 = 0$.

80. Find the real-number solutions of $x^4 - 3x^2 - 2 = 0$. Rationalize the denominators of the solutions.

8.4 Quadratic Functions and Their Graphs

- Graphing $f(x) = ax^2$
- Graphing $f(x) = ax^2 + k$
- Graphing $f(x) = a(x - h)^2$
- Graphing $f(x) = a(x - h)^2 + k$
- Graphing $f(x) = ax^2 + bx + c$ by completing the square
- A formula to find the vertex
- Determining minimum and maximum values
- Solving quadratic equations graphically

In this section, we will discuss methods for graphing *quadratic functions.*

Quadratic Functions

A **quadratic function** is a second-degree polynomial function that can be written in the form

$$f(x) = ax^2 + bx + c$$

where a, b, and c are real numbers and $a \neq 0$.

Quadratic functions are often written in an alternate form, called **standard form,**

$$f(x) = a(x - h)^2 + k$$

where a, h, and k are real numbers and $a \neq 0$. This form is useful because a, h, and k give us important information about the graph of the function. To develop a strategy for graphing quadratic functions written in standard form, we will begin by considering the simplest case, $f(x) = ax^2$.

Notation

Since $y = f(x)$, quadratic functions can also be written as $y = a(x - h)^2 + k$ and $y = ax^2 + bx + c$.

GRAPHING $f(x) = ax^2$

One way to graph quadratic functions is to plot points.

EXAMPLE 1

Graph: **a.** $f(x) = x^2$, **b.** $g(x) = 3x^2$, and **c.** $h(x) = \frac{1}{3}x^2$.

Solution We can make a table of values for each function, plot each point, and join them with a smooth curve. We note that the graph of $g(x) = 3x^2$ is narrower than the graph of $f(x) = x^2$, and the graph of $h(x) = \frac{1}{3}x^2$ is wider than the graph of $f(x) = x^2$. For $f(x) = ax^2$, the smaller the value of $|a|$, the wider the graph.

$f(x) = x^2$

x	$f(x)$
−2	4
−1	1
0	0
1	1
2	4

$g(x) = 3x^2$

x	$g(x)$
−2	12
−1	3
0	0
1	3
2	12

$h(x) = \frac{1}{3}x^2$

x	$h(x)$
−2	$\frac{4}{3}$
−1	$\frac{1}{3}$
0	0
1	$\frac{1}{3}$
2	$\frac{4}{3}$

The values of $g(x)$ increase faster than the values of $f(x)$, making its graph steeper.

The values of $h(x)$ increase more slowly than the values of $f(x)$, making its graph flatter.

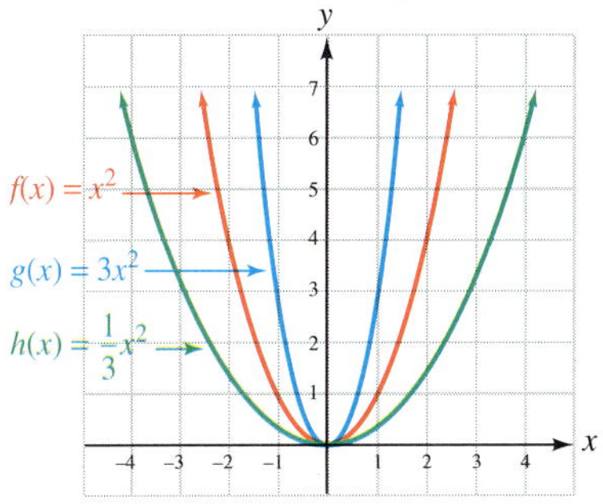

EXAMPLE 2

Graph: $f(x) = -3x^2$.

Solution We make a table of values for the function, plot each point, and join them with a smooth curve. We see that the parabola opens downward and has the same shape as the graph of $g(x) = 3x^2$ that was graphed in Example 1.

$f(x) = -3x^2$

x	$f(x)$	
−2	−12	→ (−2, −12)
−1	−3	→ (−1, −3)
0	0	→ (0, 0)
1	−3	→ (1, −3)
2	−12	→ (2, −12)

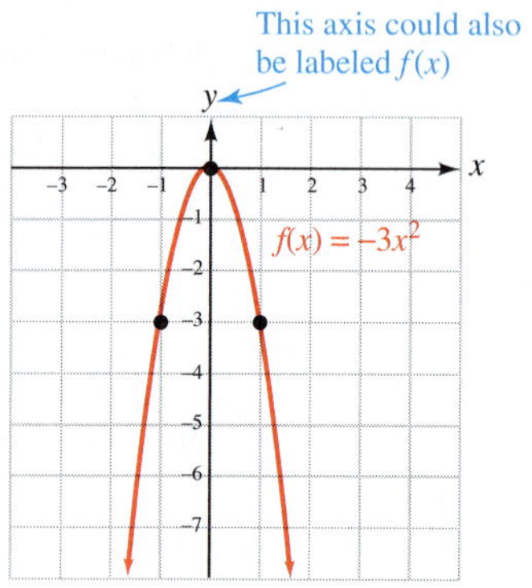

Self Check 2 Graph: $f(x) = -\frac{1}{3}x^2$.

The graphs of functions of the form $f(x) = ax^2$ are **parabolas.** The lowest point on a parabola that opens upward, or the highest point on a parabola that opens downward, is called the **vertex** of the parabola. The vertical line, called an **axis of symmetry,** that passes through the vertex divides the parabola into two congruent halves. If we fold the paper along the axis of symmetry, the two sides of the parabola will match.

The Language of Algebra

An axis of symmetry divides a parabola into two matching sides. The sides are said to be *mirror images* of each other.

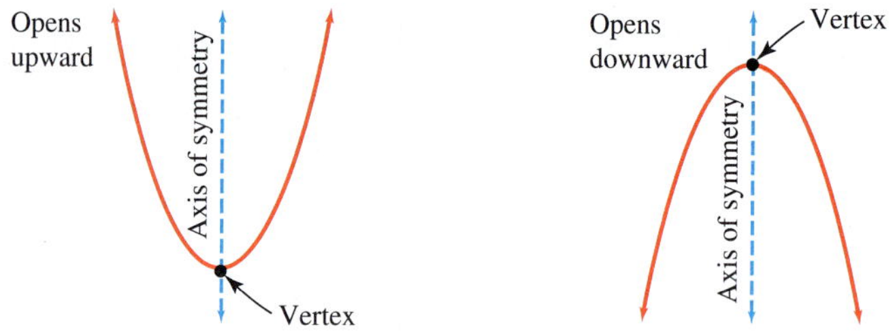

The results from Examples 1 and 2 confirm the following facts.

The Graph of $f(x) = ax^2$ The graph of $f(x) = ax^2$ is a parabola opening upward when $a > 0$ and downward when $a < 0$, with vertex at the point (0, 0) and axis of symmetry the line $x = 0$.

GRAPHING $f(x) = ax^2 + k$

EXAMPLE 3 Graph: **a.** $f(x) = 2x^2$, **b.** $g(x) = 2x^2 + 3$, and **c.** $h(x) = 2x^2 - 3$.

Solution We make a table of values for each function, plot each point, and join them with a smooth curve. We note that the graph of $g(x) = 2x^2 + 3$ is identical to the graph of $f(x) = 2x^2$, except that it has been translated 3 units upward. The graph of $h(x) = 2x^2 - 3$ is identical to the graph of $f(x) = 2x^2$, except that it has been translated 3 units downward. In each case, the axis of symmetry is the line $x = 0$.

$f(x) = 2x^2$

x	$f(x)$
−2	8
−1	2
0	0
1	2
2	8

$g(x) = 2x^2 + 3$

x	$g(x)$
−2	11
−1	5
0	3
1	5
2	11

$h(x) = 2x^2 - 3$

x	$h(x)$
−2	5
−1	−1
0	−3
1	−1
2	5

For each x-value, $g(x)$ is 3 more than $f(x)$.

For each x-value, $h(x)$ is 3 less than $f(x)$.

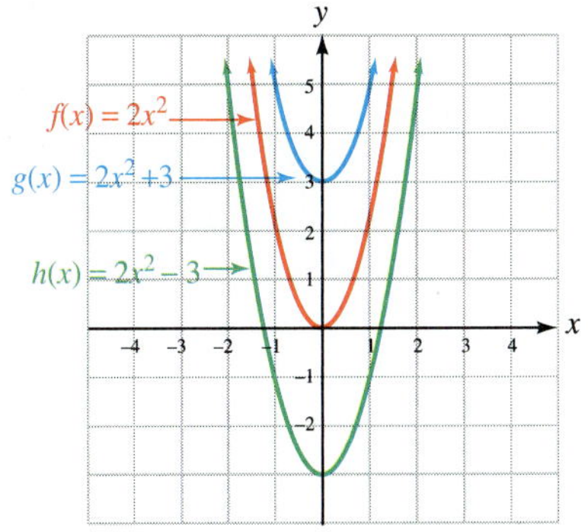

The results of Example 3 confirm the following facts.

The Graph of $f(x) = ax^2 + k$ The graph of $f(x) = ax^2 + k$ is a parabola having the same shape as $f(x) = ax^2$ but translated upward k units if k is positive and downward $|k|$ units if k is negative. The vertex is at the point $(0, k)$, and the axis of symmetry is the line $x = 0$.

GRAPHING $f(x) = a(x - h)^2$

EXAMPLE 4 Graph: **a.** $f(x) = 2x^2$, **b.** $g(x) = 2(x - 3)^2$, and **c.** $h(x) = 2(x + 3)^2$.

Solution We make a table of values for each function, plot each point, and join them with a smooth curve. We note that the graph of $g(x) = 2(x - 3)^2$ is identical to the graph of $f(x) = 2x^2$, except that it has been translated 3 units to the right. The graph of $h(x) = 2(x + 3)^2$ is identical to the graph of $f(x) = 2x^2$, except that it has been translated 3 units to the left.

$f(x) = 2x^2$

x	$f(x)$
−2	8
−1	2
0	0
1	2
2	8

$g(x) = 2(x - 3)^2$

x	$g(x)$
1	8
2	2
3	0
4	2
5	8

$h(x) = 2(x + 3)^2$

x	$h(x)$
−5	8
−4	2
−3	0
−2	2
−1	8

When an x-value is increased by 3, the function's outputs are the same.

When an x-value is decreased by 3, the function's outputs are the same.

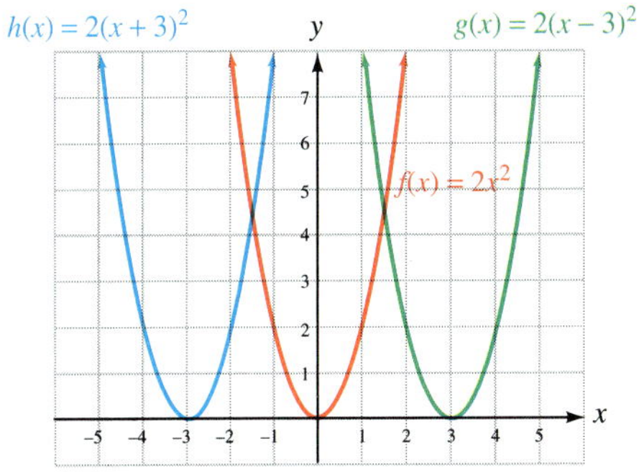

The results of Example 4 confirm the following facts.

The Graph of $f(x) = a(x - h)^2$ The graph of $f(x) = a(x - h)^2$ is a parabola having the same shape as $f(x) = ax^2$ but translated h units to the right if h is positive and $|h|$ units to the left if h is negative. The vertex is at the point $(h, 0)$, and the axis of symmetry is the line $x = h$.

GRAPHING $f(x) = a(x - h)^2 + k$

The results of Examples 1–4 suggest a general strategy for graphing quadratic functions that are written in the form $f(x) = a(x - h)^2 + k$.

Graphing a Quadratic Function in Standard Form

The graph of the quadratic function

$$f(x) = a(x - h)^2 + k \quad \text{where } a \neq 0$$

is a parabola with vertex at (h, k). The axis of symmetry is the line $x = h$. The parabola opens upward when $a > 0$ and downward when $a < 0$.

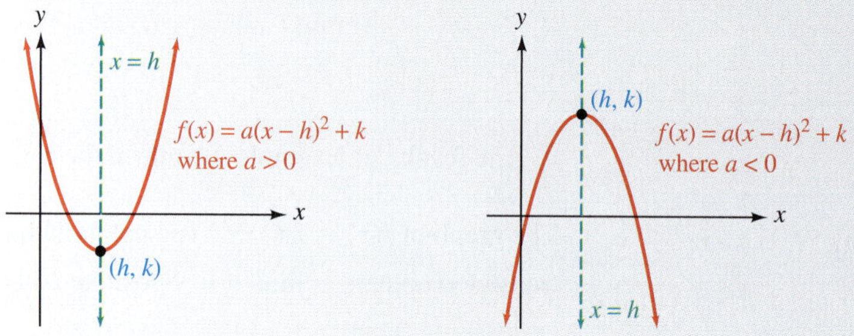

EXAMPLE 5

Graph: $f(x) = 2(x - 3)^2 - 4$. Label the vertex and draw the axis of symmetry.

Solution The graph of $f(x) = 2(x - 3)^2 - 4$ is identical to the graph of $g(x) = 2(x - 3)^2$, except that it has been translated 4 units downward. The graph of $g(x) = 2(x - 3)^2$ is identical to the graph of $h(x) = 2x^2$, except that it has been translated 3 units to the right.

We can learn more about the graph of $f(x) = 2(x - 3)^2 - 4$ by determining a, h, and k.

$$\left.\begin{array}{l} f(x) = 2(x - 3)^2 - 4 \\ f(x) = a(x - h)^2 + k \end{array}\right\} \quad a = 2,\ h = 3, \text{ and } k = -4$$

Upward/downward: Since $a = 2$ and $2 > 0$, the parabola opens upward.

Vertex: The vertex of the parabola is $(h, k) = (3, -4)$, as shown below.

Axis of symmetry: Since $h = 3$, the axis of symmetry is the line $x = 3$, as shown below.

Plotting points: We can construct a table of values to determine several points on the parabola. Since the x-coordinate of the vertex is 3, we choose the x-values of 4 and 5, find $f(4)$ and $f(5)$, and record the results in a table. Then we plot $(4, -2)$ and $(5, 4)$, and use symmetry to locate two other points on the parabola: $(2, -2)$ and $(1, 4)$. Finally, we draw a smooth curve through the points to get the graph.

$f(x) = 2(x - 3)^2 - 4$

x	$f(x)$	
4	−2	→ (4, −2)
5	4	→ (5, 4)

The x-coordinate of the vertex is 3. Choose values for x close to 3 and on the same side of the axis of symmetry.

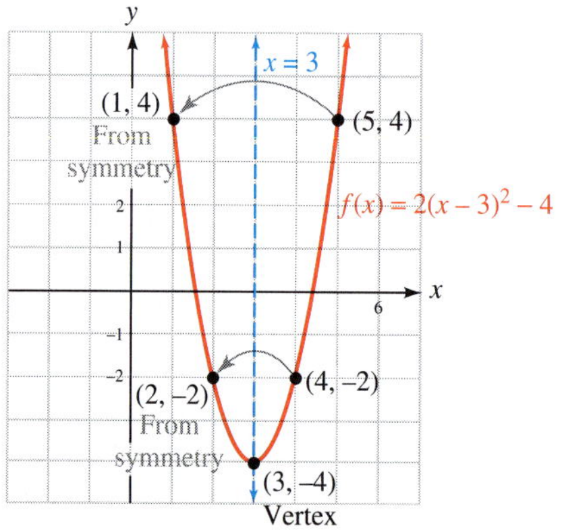

Self Check 5 Graph: $f(x) = 2(x - 1)^2 - 2$. Label the vertex and draw the axis of symmetry.

GRAPHING $f(x) = ax^2 + bx + c$ BY COMPLETING THE SQUARE

To graph functions of the form $f(x) = ax^2 + bx + c$, we can complete the square to write the function in standard form $f(x) = a(x - h)^2 + k$.

EXAMPLE 6

Determine the vertex and the axis of symmetry of the graph of $f(x) = x^2 + 8x + 21$. Will the graph open upward or downward?

Solution To determine the vertex and the axis of symmetry of the graph, we complete the square on the right-hand side so that we can write the function in $f(x) = a(x - h)^2 + k$ form.

$$f(x) = x^2 + 8x + 21$$

$$f(x) = (x^2 + 8x \quad) + 21$$ Prepare to complete the square on x by writing parentheses around $x^2 + 8x$.

To complete the square on $x^2 + 8x$, we note that one-half of the coefficient of x is $\frac{1}{2} \cdot 8 = 4$, and $4^2 = 16$. If we add 16 to $x^2 + 8x$, we obtain a perfect square trinomial within the parentheses. Since this step adds 16 to the right-hand side, we must also subtract 16 from the right-hand side so that it remains in an equivalent form.

Success Tip

When a number is added to and that same number is subtracted from one side of an equation, the value of that side of the equation remains the same.

Add 16 to the right-hand side. Subtract 16 from the right-hand side.

$$f(x) = (x^2 + 8x + 16) + 21 - 16$$

$$f(x) = (x + 4)^2 + 5$$ Factor $x^2 + 8x + 16$ and combine like terms.

The function is now written in standard form and we can determine a, h, and k.

The standard form requires a minus symbol here.

$$f(x) = \underset{a}{\uparrow}\left[(x - \underset{h}{(-4)})\right]^2 + \underset{k}{5}$$ Write $x + 4$ as $x - (-4)$ to determine h. $a = 1$, $h = -4$, and $k = 5$.

The vertex is $(h, k) = (-4, 5)$ and the axis of symmetry is the line $x = -4$. Since $a = 1$ and $1 > 0$, the parabola opens upward.

Self Check 6 Determine the vertex and the axis of symmetry of the graph of $f(x) = x^2 + 4x + 10$. Will the graph open upward or downward?

EXAMPLE 7

Graph: $f(x) = 2x^2 - 4x - 1$.

Solution Recall that to complete the square on $2x^2 - 4x$, the coefficient of x^2 must be equal to 1. Therefore, we factor 2 from $2x^2 - 4x$.

$$f(x) = 2x^2 - 4x - 1$$

$$f(x) = 2(x^2 - 2x \quad) - 1$$

To complete the square on $x^2 - 2x$, we note that one-half of the coefficient of x is $\frac{1}{2}(-2) = -1$, and $(-1)^2 = 1$. If we add 1 to $x^2 - 2x$, we obtain a perfect square trinomial within the parentheses. Since this step adds 2 to the right-hand side, we must also subtract 2 from the right-hand side so that it remains in an equivalent form.

By the distributive property, when 1 is added to the expression within the parentheses, $2 \cdot 1 = 2$ is added to the right-hand side.

Subtract 2 to counteract the addition of 2.

$$f(x) = 2(x^2 - 2x + 1) - 1 - 2$$

$$f(x) = 2(x - 1)^2 - 3 \qquad \text{Factor } x^2 - 2x + 1 \text{ and combine like terms.}$$

We see that $a = 2$, $h = 1$, and $k = -3$. Thus, the vertex is at the point $(1, -3)$, and the axis of symmetry is $x = 1$. Since $a = 2$ and $2 > 0$, the parabola opens upward. We plot the vertex and axis of symmetry as shown below.

Finally, we construct a table of values, plot the points, use symmetry to plot the corresponding points, and then draw the graph.

$$f(x) = 2x^2 - 4x - 1$$
or
$$f(x) = 2(x - 1)^2 - 3$$

x	$f(x)$	
2	−1	→ (2, −1)
3	5	→ (3, 5)

↑ The x-coordinate of the vertex is 1. Choose values for x close to 1 and on the same side of the axis of symmetry.

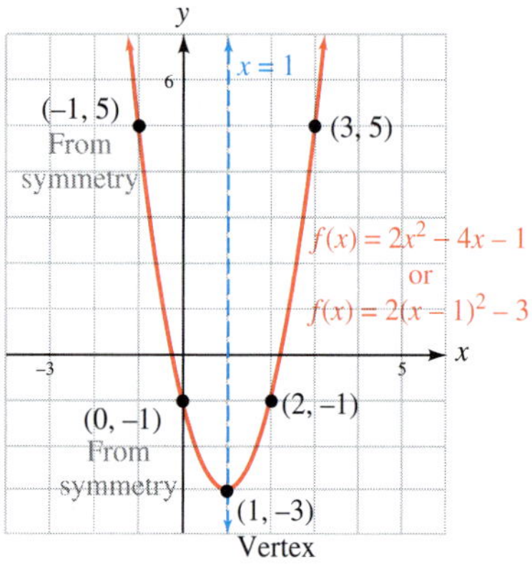

Self Check 7 Graph: $f(x) = 3x^2 - 12x + 8$.

A FORMULA TO FIND THE VERTEX

We can derive a formula for the vertex of the graph of $f(x) = ax^2 + bx + c$ by completing the square in the same manner as we did in Example 7. After using similar steps, the result is

$$f(x) = a\left[x - \left(-\frac{b}{2a}\right)\right]^2 + \frac{4ac - b^2}{4a}$$

(↑ h under $-\frac{b}{2a}$; ↑ k under $\frac{4ac - b^2}{4a}$)

The x-coordinate of the vertex is $-\frac{b}{2a}$. The y-coordinate of the vertex is $\frac{4ac - b^2}{4a}$. However, we can also find the y-coordinate of the vertex by substituting the x-coordinate, $-\frac{b}{2a}$, for x in the quadratic function.

Formula for the Vertex of a Parabola

The vertex of the graph of the quadratic function $f(x) = ax^2 + bx + c$ is

$$\left(-\frac{b}{2a}, f\left(-\frac{b}{2a}\right)\right)$$

and the axis of symmetry of the parabola is the line $x = -\frac{b}{2a}$.

EXAMPLE 8

Find the vertex of the graph of $f(x) = 2x^2 - 4x - 1$.

Solution The function is written in $f(x) = ax^2 + bx + c$ form, where $a = 2$, $b = -4$, and $c = -1$. To find the vertex of its graph, we compute

$$-\frac{b}{2a} = -\frac{-4}{2(2)} = -\frac{-4}{4} = 1 \qquad f\left(-\frac{b}{2a}\right) = f(1) = 2(1)^2 - 4(1) - 1 = -3$$

> **Success Tip**
>
> We can find the vertex of the graph of a quadratic function by completing the square or by using the formula.

The vertex is the point $(1, -3)$. This agrees with the result we obtained in Example 7 by completing the square.

Self Check 8 Find the vertex of the graph of $f(x) = 3x^2 - 12x + 8$.

ACCENT ON TECHNOLOGY: FINDING THE VERTEX

We can use a graphing calculator to graph the function $f(x) = 2x^2 + 6x - 3$ and find the coordinates of the vertex and the axis of symmetry of the parabola. If we enter the function, we will obtain the graph shown in figure (a).

We then trace to move the cursor to the lowest point on the graph, as shown in figure (b). By zooming in, we can see that the vertex is the point $(-1.5, -7.5)$, or $\left(-\frac{3}{2}, -\frac{15}{2}\right)$, and that the line $x = -\frac{3}{2}$ is the axis of symmetry.

Some calculators have an fmin or fmax feature that can also be used to find the vertex.

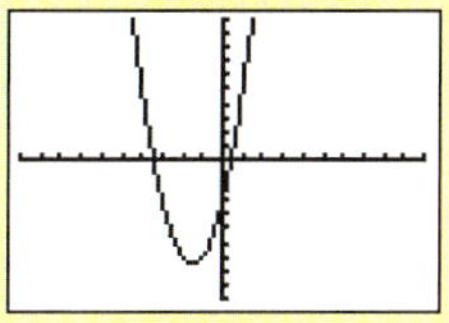

(a)

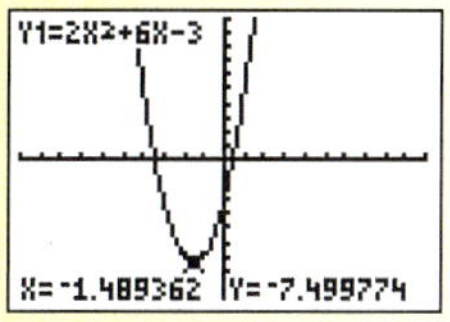

(b)

Much can be determined about the graph of $f(x) = ax^2 + bx + c$ from the coefficients a, b, and c. This information is summarized as follows:

Graphing a Quadratic Function $f(x) = ax^2 + bx + c$

Determine whether the parabola opens upward or downward by examining a.

The x-coordinate of the vertex of the parabola is $x = -\frac{b}{2a}$.

To find the y-coordinate of the vertex, substitute $-\frac{b}{2a}$ for x and find $f\left(-\frac{b}{2a}\right)$.

The axis of symmetry is the vertical line passing through the vertex.

The y-intercept is determined by the value of $f(x)$ when $x = 0$: the y-intercept is $(0, c)$.

The x-intercepts (if any) are determined by the values of x that make $f(x) = 0$. To find them, solve the quadratic equation $ax^2 + bx + c = 0$.

EXAMPLE 9

Graph: $f(x) = -2x^2 - 8x - 8$.

Solution ***Step 1:*** *Determine whether the parabola opens upward or downward.* The function is in the form $f(x) = ax^2 + bx + c$, with $a = -2$, $b = -8$, and $c = -8$. Since $a < 0$, the parabola opens downward.

Step 2: *Find the vertex and draw the axis of symmetry.* To find the coordinates of the vertex, we compute

$$x = -\frac{b}{2a} \qquad f\left(-\frac{b}{2a}\right) = f(-2)$$

$$x = -\frac{-8}{2(-2)} \quad \text{Substitute } -2 \text{ for } a \text{ and } -8 \text{ for } b. \qquad = -2(-2)^2 - 8(-2) - 8$$

$$= -2 \qquad\qquad = -8 + 16 - 8$$

$$= 0$$

The vertex of the parabola is the point $(-2, 0)$. This point is in blue on the graph. The axis of symmetry is the line $x = -2$.

Step 3: *Find the x- and y-intercepts.* Since $c = -8$, the y-intercept of the parabola is $(0, -8)$. The point $(-4, -8)$, two units to the left of the axis of symmetry, must also be on the graph. We plot both points in black on the graph.

To find the x-intercepts, we set $f(x)$ equal to 0 and solve the resulting quadratic equation.

$$f(x) = -2x^2 - 8x - 8$$

$$0 = -2x^2 - 8x - 8 \quad \text{Set } f(x) = 0.$$

$$0 = x^2 + 4x + 4 \quad \text{Divide both sides by } -2.$$

$$0 = (x + 2)(x + 2) \quad \text{Factor the trinomial.}$$

$$x + 2 = 0 \quad \text{or} \quad x + 2 = 0 \quad \text{Set each factor equal to 0.}$$

$$x = -2 \quad \Big| \quad x = -2$$

Since the solutions are the same, the graph has only one x-intercept: $(-2, 0)$. This point is the vertex of the parabola and has already been plotted.

Step 4: *Plot another point.* Finally, we find another point on the parabola. If $x = -3$, then $f(-3) = -2$. We plot $(-3, -2)$ and use symmetry to determine that $(-1, -2)$ is also on the graph. Both points are in green.

Step 5: Draw a smooth curve through the points, as shown.

$f(x) = -2x^2 - 8x - 8$

x	$f(x)$	
-3	-2	$\rightarrow (-3, -2)$

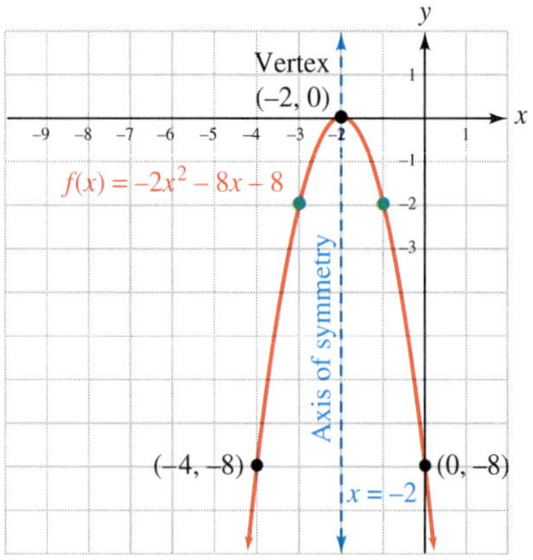

DETERMINING MINIMUM AND MAXIMUM VALUES

It is often useful to know the smallest or largest possible value a quantity can assume. For example, companies try to minimize their costs and maximize their profits. If the quantity is expressed by a quadratic function, the y-coordinate of the vertex of the graph of the function gives its minimum or maximum value.

EXAMPLE 10

Minimizing costs. A glassworks that makes lead crystal vases has daily production costs given by the function $C(x) = 0.2x^2 - 10x + 650$, where x is the number of vases made each day. How many vases should be produced to minimize the per-day costs? What will the costs be?

Solution The graph of $C(x) = 0.2x^2 - 10x + 650$ is a parabola opening upward. The vertex is the lowest point on the graph. To find the vertex, we compute

$$-\frac{b}{2a} = -\frac{-10}{2(0.2)} \quad b = -10 \text{ and } a = 0.2.$$
$$= -\frac{-10}{0.4}$$
$$= 25$$

$$f\left(-\frac{b}{2a}\right) = f(25)$$
$$= 0.2(25)^2 - 10(25) + 650$$
$$= 525$$

The Language of Algebra

We say that 25 is the value for which the function $C(x) = 0.2x^2 - 10x + 650$ is a minimum.

The vertex is (25, 525), and it indicates that the costs are a minimum of \$525 when 25 vases are made daily.

To solve this problem with a graphing calculator, we graph the function $C(x) = 0.2x^2 - 10x + 650$. By using TRACE and ZOOM, we can locate the vertex of the graph. The coordinates of the vertex indicate that the minimum cost is \$525 when the number of vases produced is 25.

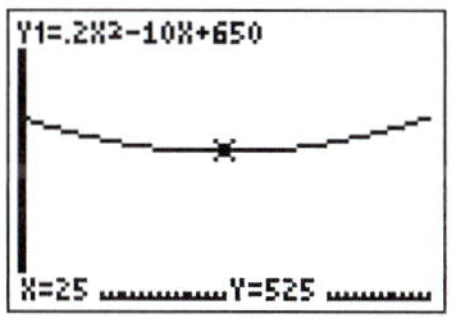

SOLVING QUADRATIC EQUATIONS GRAPHICALLY

When solving quadratic equations graphically, we must consider three possibilities. If the graph of the associated quadratic function has two x-intercepts, the quadratic equation has two real-number solutions. Figure (a) shows an example of this. If the graph has one x-intercept, as shown in figure (b), the equation has one real-number solution. Finally, if the graph does not have an x-intercept, as shown in figure (c), the equation does not have any real-number solutions.

Success Tip

Note that the solutions of each quadratic equation are given by the x-coordinate of the x-intercept of each respective graph.

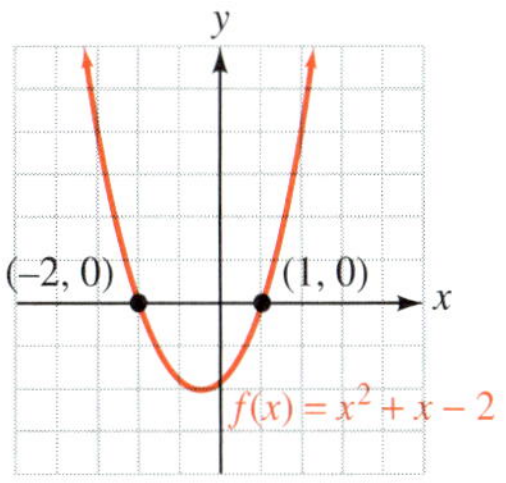

$x^2 + x - 2 = 0$ has two solutions, -2 and 1.

(a)

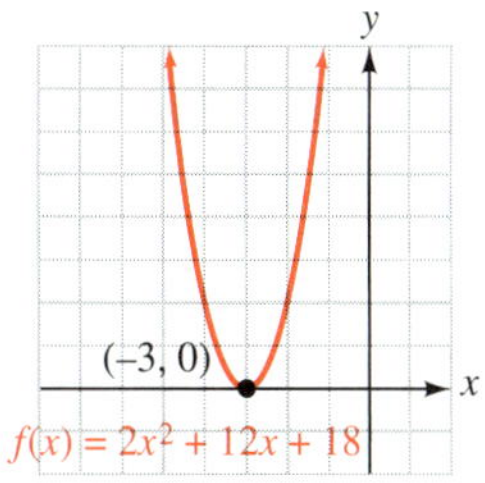

$2x^2 + 12x + 18 = 0$ has one solution, -3.

(b)

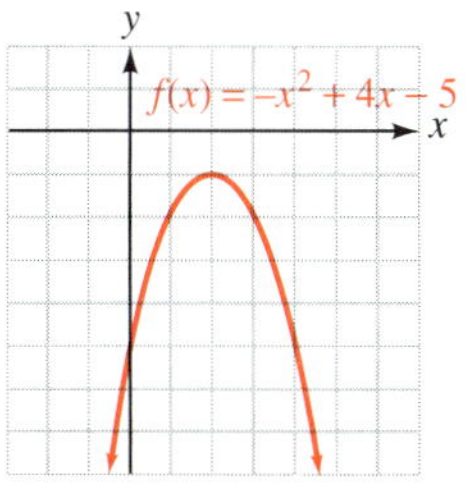

$-x^2 + 4x - 5 = 0$ has no real-number solutions.

(c)

ACCENT ON TECHNOLOGY: SOLVING QUADRATIC EQUATIONS GRAPHICALLY

We can use a graphing calculator to find approximate solutions of quadratic equations. For example, the solutions of $0.7x^2 + 2x - 3.5 = 0$ are the numbers x that will make $y = 0$ in the quadratic function $f(x) = 0.7x^2 + 2x - 3.5$. To approximate these numbers, we graph the quadratic function and read the x-intercepts from the graph using the ZERO feature. In the figure, we see that the x-coordinate of the left-most x-intercept of the graph is given as -4.082025. This means that an approximate solution of the equation is -4.08. To find the positive x-intercept, we use similar steps.

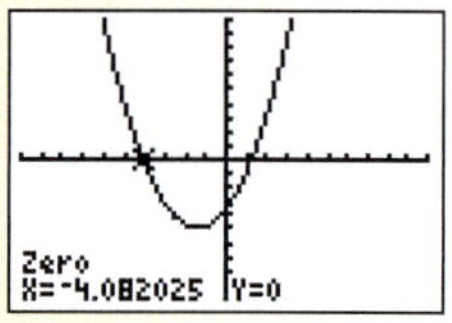

Answers to Self Checks

2.

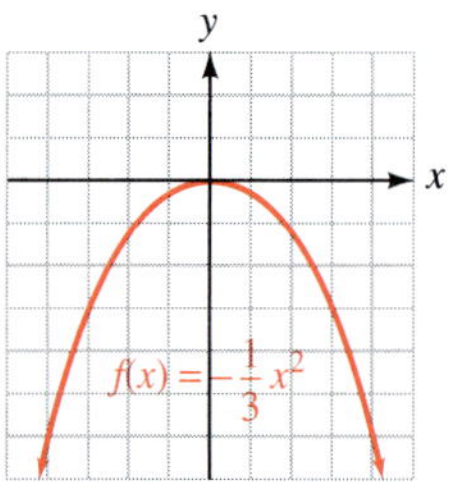

5.

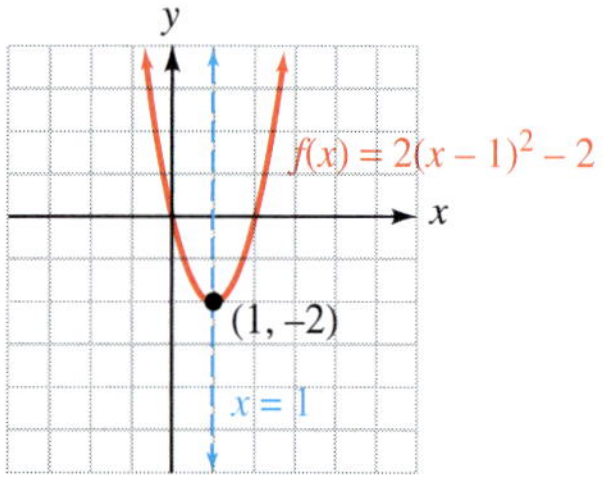

6. $(-2, 6)$; $x = -2$; opens upward

7.

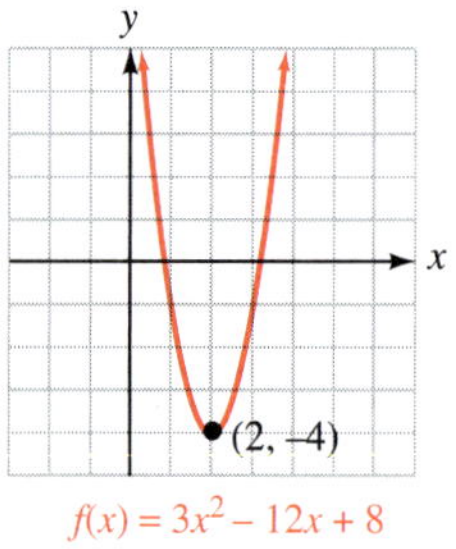

8. $(2, -4)$

8.4 STUDY SET

VOCABULARY **Refer to the graph. Fill in the blanks.**

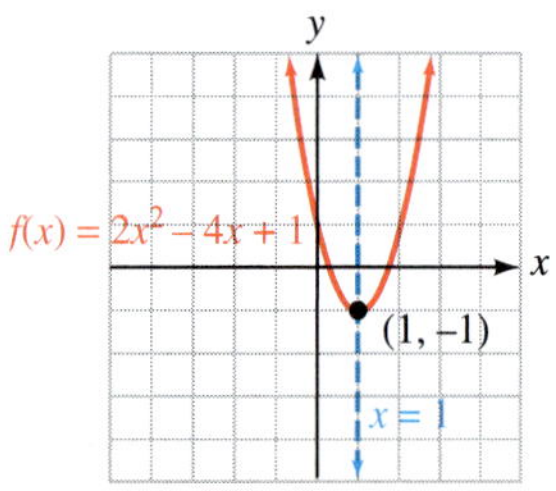

1. $f(x) = 2x^2 - 4x + 1$ is called a __________ function. Its graph is a cup-shaped figure called a __________.
2. The lowest point on the graph is $(1, -1)$. This is called the ________ of the parabola.
3. The vertical line $x = 1$ divides the parabola into two halves. This line is called the _________________.
4. $f(x) = a(x - h)^2 + k$ is called the __________ form of the equation of a quadratic function.

CONCEPTS

5. Refer to the graph below.

a. What do we call the curve shown there?

b. What are the *x*-intercepts of the graph?

c. What is the *y*-intercept of the graph?

d. What is the vertex?

e. What is the axis of symmetry?

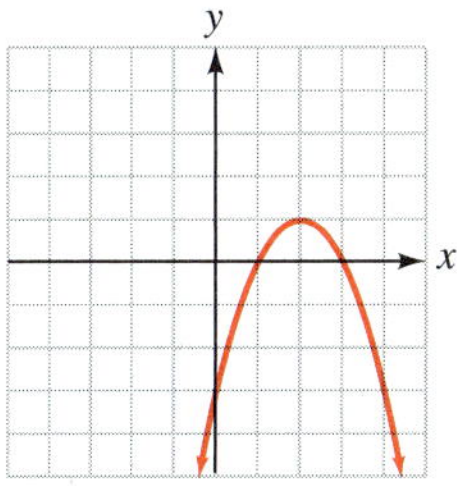

6. The vertex of a parabola is at $(1, -3)$, its *y*-intercept is $(0, -2)$, and it passes through the point $(3, 1)$, as shown in the illustration. Draw the axis of symmetry and use it to help determine two other points on the parabola.

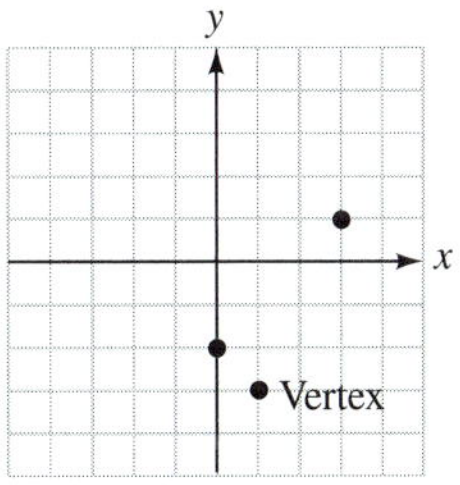

7. Draw the graph of a quadratic function using the given facts about its graph.

- Opens upward
- *y*-intercept: $(0, -3)$
- Vertex: $(-1, -4)$
- *x*-intercepts: $(-3, 0), (1, 0)$
-

x	$f(x)$
2	5

8. For $f(x) = -x^2 + 6x - 7$, the value of $-\frac{b}{2a}$ is 3. Find the *y*-coordinate of the vertex of the graph of this function.

9. a. To complete the square on the right-hand side of $f(x) = 2x^2 + 12x + 11$, what should be factored from the first two terms?

$$f(x) = \quad (x^2 + 6x) + 11$$

b. To complete the square on $x^2 + 6x$ shown below, what should be added within the parentheses and what should be subtracted outside the parentheses?

$$f(x) = 2(x^2 + 6x + \quad) + 11 - \quad$$

10. To complete the square on $x^2 + 4x$ shown below, what should be added within the parentheses and what should be added outside the parentheses?

$$f(x) = -5(x^2 + 4x + \quad) + 7 + \quad$$

11. Use the graph of $f(x) = \frac{x^2}{10} - \frac{x}{5} - \frac{3}{2}$, shown below, to estimate the solutions of the equation $\frac{x^2}{10} - \frac{x}{5} - \frac{3}{2} = 0$.

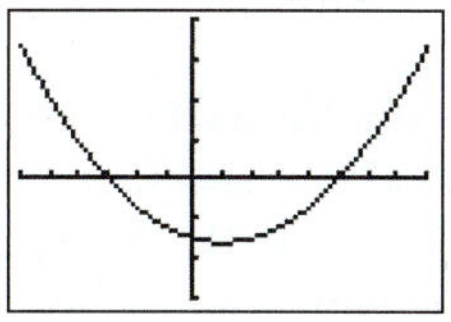

12. Three quadratic equations are to be solved graphically. The graphs of their associated quadratic functions are shown below. Decide which graph indicates that the equation has

a. two real solutions.

b. one real solution.

c. no real solutions.

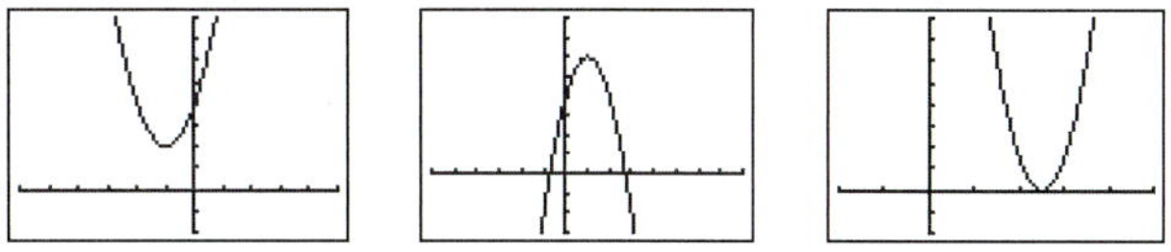

(i) **(ii)** **(iii)**

NOTATION

13. The function $f(x) = 2(x + 1)^2 + 6$ is written in the form $f(x) = a(x - h)^2 + k$. Is $h = -1$ or is $h = 1$? Explain.

14. Consider the function $f(x) = 2x^2 + 4x - 8$.

a. What are a, b, and c?

b. Find $-\frac{b}{2a}$.

Make a table of values for each function. Then graph them on the same coordinate system.

15. $f(x) = x^2, g(x) = 2x^2, h(x) = \frac{1}{2}x^2$

16. $f(x) = -x^2, g(x) = -\frac{1}{4}x^2, h(x) = -4x^2$

Make a table of values to graph function *f*. Then use a translation to graph the other two functions on the same coordinate system.

17. $f(x) = 4x^2, g(x) = 4x^2 + 3, h(x) = 4x^2 - 2$

18. $f(x) = 3x^2, g(x) = 3(x + 2)^2, h(x) = 3(x - 3)^2$

Make a table of values to graph function f. Then use a series of translations to graph function g on the same coordinate system.

19. $f(x) = -3x^2$, $g(x) = -3(x-2)^2 - 1$

20. $f(x) = -\frac{1}{2}x^2$, $g(x) = -\frac{1}{2}(x+1)^2 + 2$

Find the vertex and the axis of symmetry of the graph of each function. If necessary, complete the square on x to write the equation in the form $f(x) = a(x-h)^2 + k$. Do not graph the equation, but tell whether the graph will open upward or downward.

21. $f(x) = (x-1)^2 + 2$

22. $f(x) = 2(x-2)^2 - 1$

23. $f(x) = 2(x+3)^2 - 4$

24. $f(x) = -3(x+1)^2 + 3$

25. $f(x) = 0.5(x-7.5)^2 + 8.5$

26. $f(x) = -\frac{3}{2}\left(x+\frac{1}{4}\right)^2 + \frac{7}{8}$

27. $f(x) = 2x^2 - 4x$

28. $f(x) = 3x^2 - 3$

29. $f(x) = -4x^2 + 16x + 5$

30. $f(x) = 5x^2 + 20x + 25$

31. $f(x) = 3x^2 + 4x + 2$

32. $f(x) = -6x^2 + 5x - 7$

First determine the vertex and the axis of symmetry of the graph of the function. Then plot several points and complete the graph. (See Example 5.)

33. $f(x) = (x-3)^2 + 2$ **34.** $f(x) = (x+1)^2 - 2$

35. $f(x) = -(x-2)^2$ **36.** $f(x) = -(x+2)^2$

37. $f(x) = -2(x+3)^2 + 4$ **38.** $f(x) = 2(x-2)^2 - 4$

39. $f(x) = \frac{1}{2}(x+1)^2 - 3$ **40.** $f(x) = \frac{1}{3}(x-1)^2 + 2$

Complete the square to write each function in $f(x) = a(x-h)^2 + k$ form. Determine the vertex and the axis of symmetry of the graph of the function. Then plot several points and complete the graph. (See Examples 6 and 7.)

41. $f(x) = x^2 + 2x - 3$ **42.** $f(x) = x^2 + 6x + 5$

43. $f(x) = 3x^2 - 12x + 10$ **44.** $f(x) = 4x^2 + 24x + 37$

45. $f(x) = 2x^2 + 8x + 6$ **46.** $f(x) = 3x^2 - 12x + 9$

47. $f(x) = x^2 + x - 6$ **48.** $f(x) = x^2 - x - 6$

49. $f(x) = -x^2 - 8x - 17$ **50.** $f(x) = -x^2 + 6x - 8$

51. $f(x) = -4x^2 + 16x - 10$ **52.** $f(x) = -2x^2 + 4x + 3$

Find the x- and y-intercepts of the graph of the quadratic function.

53. $f(x) = x^2 - 2x + 1$

54. $f(x) = 2x^2 - 4x$

55. $f(x) = -x^2 - 10x - 21$

56. $f(x) = 3x^2 + 6x - 9$

First determine the coordinates of the vertex and the axis of symmetry of the graph of the function using the vertex formula. Then determine the x- and y-intercepts of the graph. Finally, plot several points and complete the graph. (See Example 9.)

57. $f(x) = x^2 + 4x + 4$ **58.** $f(x) = x^2 - 6x + 9$

59. $f(x) = -x^2 + 2x - 1$ **60.** $f(x) = -x^2 - 2x - 1$

61. $f(x) = x^2 - 2x$ **62.** $f(x) = x^2 + x$

63. $f(x) = 4x^2 - 12x + 9$ **64.** $f(x) = 3x^2 - 12x + 12$

65. $f(x) = 2x^2 - 8x + 6$ **66.** $f(x) = 4x^2 + 4x - 3$

67. $f(x) = -6x^2 - 12x - 8$ **68.** $f(x) = -2x^2 + 8x - 10$

Use a graphing calculator to find the coordinates of the vertex of the graph of each quadratic function. Round to the nearest hundredth.

69. $f(x) = 2x^2 - x + 1$ **70.** $f(x) = x^2 + 5x - 6$

71. $f(x) = 7 + x - x^2$ **72.** $f(x) = 2x^2 - 3x + 2$

Use a graphing calculator to solve each equation. If an answer is not exact, round to the nearest hundredth.

73. $x^2 + x - 6 = 0$

74. $2x^2 - 5x - 3 = 0$

75. $0.5x^2 - 0.7x - 3 = 0$

76. $2x^2 - 0.5x - 2 = 0$

APPLICATIONS

77. CROSSWORD PUZZLES Darken the appropriate squares to the right of the dashed red line so that the puzzle has symmetry with respect to that line.

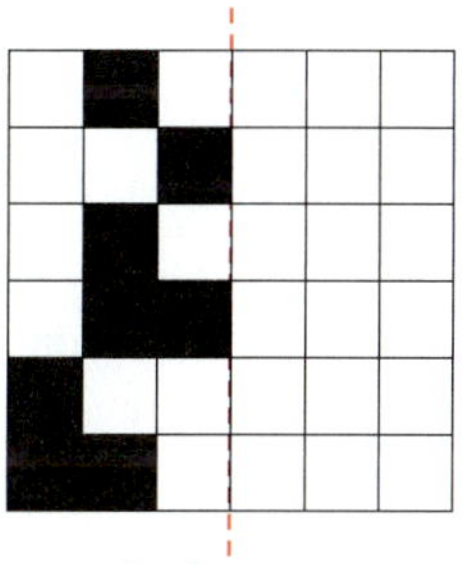

78. GRAPHIC ARTS Draw an axis of symmetry over the letter shown.

M

79. FIREWORKS A fireworks shell is shot straight up with an initial velocity of 120 feet per second. Its height s after t seconds is given by the equation $s = 120t - 16t^2$. If the shell is designed to explode when it reaches its maximum height, how long after being fired, and at what height, will the fireworks appear in the sky?

80. BALLISTICS From the top of the building in the illustration, a ball is thrown straight up with an initial velocity of 32 feet per second. The equation

$$s = -16t^2 + 32t + 48$$

gives the height s of the ball t seconds after it is thrown. Find the maximum height reached by the ball and the time it takes for the ball to hit the ground.

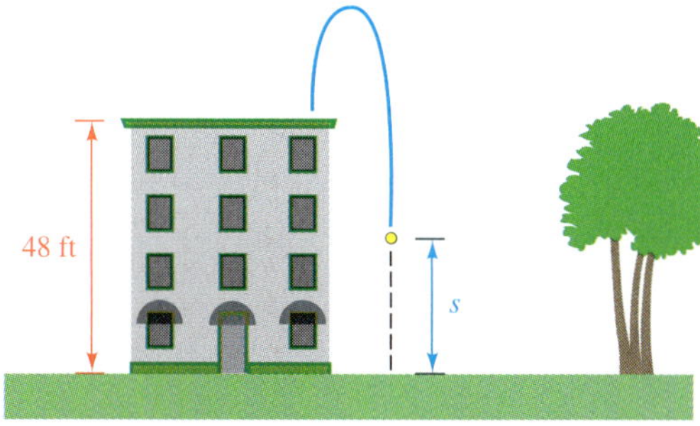

81. FENCING A FIELD See the illustration in the next column. A farmer wants to fence in three sides of a rectangular field with 1,000 feet of fencing. The other side of the rectangle will be a river. If the enclosed area is to be maximum, find the dimensions of the field.

82. POLICE INVESTIGATIONS A police officer seals off the scene of a car collision using a roll of yellow police tape that is 300 feet long, as shown in the illustration. What dimensions should be used to seal off the maximum rectangular area around the collision? What is the maximum area?

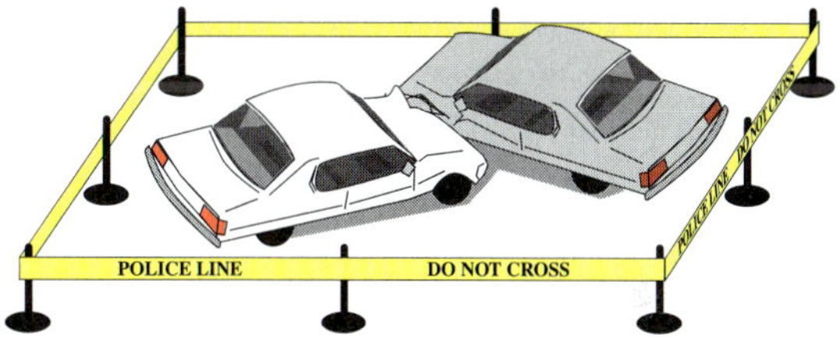

83. OPERATING COSTS The cost C in dollars of operating a certain concrete-cutting machine is related to the number of minutes n the machine is run by the function

$$C(n) = 2.2n^2 - 66n + 655$$

For what number of minutes is the cost of running the machine a minimum? What is the minimum cost?

84. WATER USAGE The height (in feet) of the water level in a reservoir over a 1-year period is modeled by the function

$$H(t) = 3.3(t - 9)^2 + 14$$

where $t = 1$ represents January, $t = 2$ represents February, and so on. How low did the water level get that year, and when did it reach the low mark?

85. U.S. ARMY The function

$$N(x) = -0.0534x^2 + 0.337x + 0.97$$

gives the number of active-duty military personnel in the United States Army (in millions) for the years 1965–1972, where $x = 0$ corresponds to 1965, $x = 1$ corresponds to 1966, and so on. For this period, when was the army's personnel strength level at its highest, and what was it? Historically, can you explain why?

86. SCHOOL ENROLLMENT The total annual enrollment (in millions) in U.S. elementary and secondary schools for the years 1975–1996 is given by the model

$$E(x) = 0.058x^2 - 1.162x + 50.604$$

where $x = 0$ corresponds to 1975, $x = 1$ corresponds to 1976, and so on. For this period, when was enrollment the lowest? What was the enrollment?

87. MAXIMIZING REVENUE The revenue R received for selling x stereos is given by the formula

$$R = -\frac{x^2}{5} + 80x - 1{,}000$$

How many stereos must be sold to obtain the maximum revenue? Find the maximum revenue.

88. MAXIMIZING REVENUE When priced at \$30 each, a toy has annual sales of 4,000 units. The manufacturer estimates that each \$1 increase in cost will decrease sales by 100 units. Find the unit price that will maximize total revenue. (*Hint:* Total revenue = price · the number of units sold.)

WRITING

89. Use the example of a stream of water from a drinking fountain to explain the concepts of the vertex and the axis of symmetry of a parabola.

90. What are some quantities that are good to maximize? What are some quantities that are good to minimize?

91. A mirror is held against the y-axis of the graph of a quadratic function. What fact about parabolas does this illustrate?

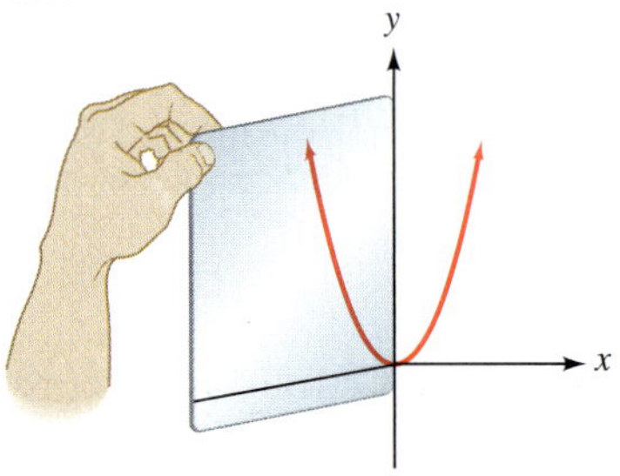

92. The vertex of a quadratic function $f(x) = ax^2 + bx + c$ is given by the formula $\left(-\frac{b}{2a}, f\left(-\frac{b}{2a}\right)\right)$. Explain what is meant by the notation $f\left(-\frac{b}{2a}\right)$.

93. A table of values for $f(x) = 2x^2 - 4x + 3$ is shown. Explain why it appears that the vertex of the graph of f is the point (1, 1).

X	Y1
.25	2.125
.5	1.5
.75	1.125
1	1
1.25	1.125
1.5	1.5
1.75	2.125

X=.25

94. The illustration shows the graph of the quadratic function $f(x) = -4x^2 + 12x$ with domain [0, 3]. Explain how the value of $f(x)$ changes as the value of x increases from 0 to 3.

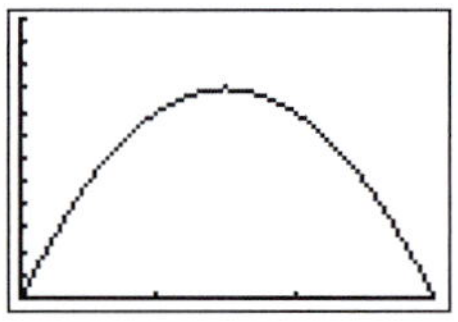

REVIEW **Simplify each expression. Assume all variables represent positive numbers.**

95. $\sqrt{8a}\sqrt{2a^3b}$

96. $\left(\sqrt{23}\right)^2$

97. $\frac{\sqrt{3}}{\sqrt{50}}$

98. $\frac{3}{\sqrt[3]{9}}$

99. $3\left(\sqrt{5b} - \sqrt{3}\right)^2$

100. $-2\sqrt{5b}\left(4\sqrt{2b} - 3\sqrt{3}\right)$

CHALLENGE PROBLEMS

101. Use completing the square to show that the vertex of the graph of the quadratic function $f(x) = ax^2 + bx + c$ is $\left(-\frac{b}{2a}, \frac{4ac - b^2}{4a}\right)$.

102. Determine a quadratic function whose graph has x-intercepts of (2, 0) and (−4, 0).

8.5 Quadratic and Other Nonlinear Inequalities

- Solving quadratic inequalities
- Solving rational inequalities
- Graphs of nonlinear inequalities in two variables

We have previously solved *linear* inequalities in one variable such as $2x + 3 > 8$ and $6x - 7 < 4x - 9$. To find their solution sets, we used properties of inequalities to isolate the variable on one side of the inequality.

In this section, we will solve *quadratic* inequalities in one variable such as $x^2 + x - 6 < 0$ and $x^2 + 4x \geq 5$. We will use an interval testing method on the number line to determine their solution sets.

SOLVING QUADRATIC INEQUALITIES

Recall that a quadratic equation can be written in the form $ax^2 + bx + c = 0$. If we replace the = symbol with an inequality symbol, we have a quadratic inequality.

Quadratic Inequalities

A **quadratic inequality** can be written in one of the standard forms

$$ax^2 + bx + c < 0 \quad ax^2 + bx + c > 0 \quad ax^2 + bx + c \leq 0 \quad ax^2 + bx + c \geq 0$$

where a, b, and c are real numbers and $a \neq 0$.

To solve a quadratic inequality in one variable, we find all values of the variable that make the inequality true using the following steps.

Solving Quadratic Inequalities

1. Write the inequality in standard form and then solve its related quadratic equation.
2. Locate the solutions (called **critical numbers**) of the related quadratic equation on the number line.
3. Test each interval created in step 2 by choosing a test value from the interval and determining whether it satisfies the inequality. The solution set includes the interval(s) whose test value makes the inequality true.
4. Determine whether the endpoints of the intervals are included in the solution set.

EXAMPLE 1

Solve: $x^2 + x - 6 < 0$.

Solution The expression $x^2 + x - 6$ can be positive, or negative, or 0, depending on what value is substituted for x. Solutions of the inequality are x-values that make $x^2 + x - 6$ less than 0. To find them, we first find the values of x that make $x^2 + x - 6$ equal to 0.

Step 1: *Solve the related quadratic equation.* For the quadratic inequality $x^2 + x - 6 < 0$, the related quadratic equation is $x^2 + x - 6 = 0$.

$$x^2 + x - 6 = 0$$

$$(x + 3)(x - 2) = 0 \qquad \text{Factor the trinomial.}$$

$$x + 3 = 0 \quad \text{or} \quad x - 2 = 0 \qquad \text{Set each factor equal to 0.}$$

$$x = -3 \quad \Big| \quad x = 2$$

The solutions of $x^2 + x - 6 = 0$ are -3 and 2. These solutions are the critical numbers.

The Language of Algebra

We say that the critical numbers *partition* the real-number line into test intervals. Interior decorators use freestanding screens to *partition* off parts of a room.

Step 2: *Locate the critical numbers.* When we highlight -3 and 2 on a number line, we see that they separate it into three intervals:

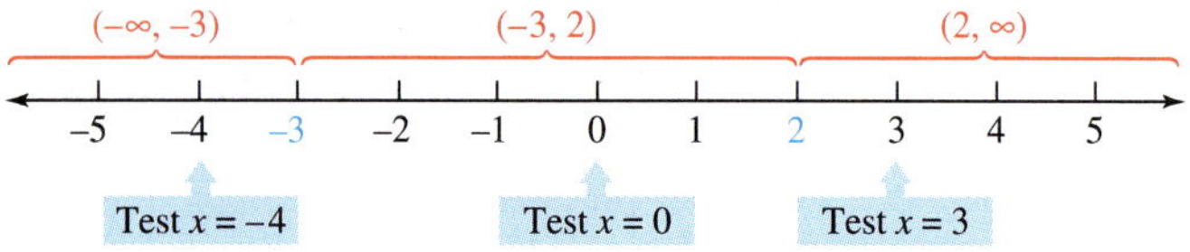

Step 3: *Test each interval.* To determine whether the numbers in $(-\infty, -3)$ are solutions of the inequality, we choose any number from that interval, substitute it for x, and see whether it satisfies $x^2 + x - 6 < 0$. *If one number in that interval satisfies the inequality, all numbers in that interval will satisfy the inequality.*

If we choose -4 from $(-\infty, -3)$, we have:

$$x^2 + x - 6 < 0 \quad \text{This is the original inequality.}$$
$$(-4)^2 + (-4) - 6 \stackrel{?}{<} 0 \quad \text{Substitute } -4 \text{ for } x.$$
$$16 + (-4) - 6 \stackrel{?}{<} 0$$
$$6 < 0 \quad \text{False.}$$

Since -4 does not satisfy the inequality, none of the numbers in $(-\infty, -3)$ are solutions.

To test the second interval, $(-3, 2)$, we choose $x = 0$.

$$x^2 + x - 6 < 0 \quad \text{This is the original inequality.}$$
$$0^2 + 0 - 6 \stackrel{?}{<} 0 \quad \text{Substitute 0 for } x.$$
$$-6 < 0 \quad \text{True.}$$

Since 0 satisfies the inequality, all of the numbers in $(-3, 2)$ are solutions.

To test the third interval, $(2, \infty)$, we choose $x = 3$.

$$x^2 + x - 6 < 0 \quad \text{This is the original inequality.}$$
$$3^2 + 3 - 6 \stackrel{?}{<} 0 \quad \text{Substitute 3 for } x.$$
$$9 + 3 - 6 \stackrel{?}{<} 0$$
$$6 < 0 \quad \text{False.}$$

Since 3 does not satisfy the inequality, none of the numbers in $(2, \infty)$ are solutions.

Success Tip

If a quadratic inequality contains $\le$ or $\ge$, the endpoints of the intervals are included in the solution set. If the inequality contains $<$ or $>$, they are not.

Step 4: *Are the endpoints included?* From the interval testing, we see that only numbers from $(-3, 2)$ satisfy $x^2 + x - 6 < 0$. The endpoints -3 and 2 are not included in the solution set because they do not satisfy the inequality. (Recall that -3 and 2 make $x^2 + x - 6$ equal to 0.) The solution set is $(-3, 2)$ as graphed below.

Self Check 1 Solve: $x^2 + x - 12 < 0$.

EXAMPLE 2

Solve: $x^2 + 4x \ge 5$.

Solution The inequality is not in standard form. To get 0 on the right-hand side, we subtract 5 from both sides.

$$x^2 + 4x \geq 5$$
$$x^2 + 4x - 5 \geq 0 \quad \text{Write the inequality in the form } ax^2 + bx + c \geq 0.$$

Now we solve the related quadratic equation $x^2 + 4x - 5 = 0$.

$$x^2 + 4x - 5 = 0$$
$$(x + 5)(x - 1) = 0 \quad \text{Factor the trinomial.}$$
$$x + 5 = 0 \quad \text{or} \quad x - 1 = 0 \quad \text{Set each factor equal to 0.}$$
$$x = -5 \quad \Big| \quad x = 1$$

The critical numbers -5 and 1 separate the number line into three intervals. We pick a test value from each interval to see whether it satisfies $x^2 + 4x - 5 \geq 0$.

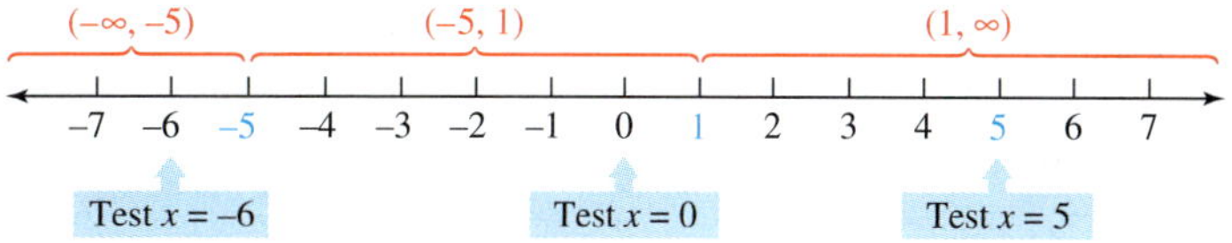

$$x^2 + 4x - 5 \geq 0 \qquad x^2 + 4x - 5 \geq 0 \qquad x^2 + 4x - 5 \geq 0$$
$$(-6)^2 + 4(-6) - 5 \overset{?}{\geq} 0 \qquad 0^2 + 4(0) - 5 \overset{?}{\geq} 0 \qquad 5^2 + 4(5) - 5 \overset{?}{\geq} 0$$
$$7 \geq 0 \ \text{True.} \qquad -5 \geq 0 \ \text{False.} \qquad 40 \geq 0 \ \text{True.}$$

Success Tip

When choosing a test value from an interval, pick a convenient number that makes the computations easy. When applicable, 0 is an obvious choice.

The numbers in the intervals $(-\infty, -5)$ and $(1, \infty)$ satisfy the inequality. Since the endpoints -5 and 1 also satisfy $x^2 + 4x - 5 \geq 0$, they are included in the solution set. (Recall that -5 and 1 make $x^2 + 4x - 5$ equal to 0.) Thus, the solution set is the union of two intervals: $(-\infty, -5] \cup [1, \infty)$. The graph of the solution set is shown below.

Self Check 2 Solve: $x^2 + 3x \geq 40$.

SOLVING RATIONAL INEQUALITIES

Rational inequalities in one variable such as $\frac{9}{x} < 8$ and $\frac{x^2 + x - 2}{x - 4} \geq 0$ can also be solved using the interval testing method.

Solving Rational Inequalities

1. Write the inequality in standard form and then solve its related rational equation.
2. Set the denominator equal to zero and solve that equation.
3. Locate the solutions (called critical numbers) found in steps 1 and 2 on the number line.
4. Test each interval on the number line created in step 3 by choosing a test value from the interval and determining whether it satisfies the inequality. The solution set includes the interval(s) whose test point makes the inequality true.
5. Determine whether the endpoints of the intervals are included in the solution set. Exclude any values that make the denominator 0.

EXAMPLE 3

Solve: $\dfrac{9}{x} < 8$.

Solution The inequality is not in standard form. To get 0 on the right-hand side, we subtract 8 from both sides. We then find a common denominator to simplify the left-hand side.

Caution

When solving rational inequalities such as $\frac{9}{x} < 8$, a common error is to multiply both sides by x to clear it of the fraction. However, we don't know whether x is positive or negative, so we don't know whether or not to reverse the inequality symbol.

$$\frac{9}{x} < 8$$

$$\frac{9}{x} - 8 < 0 \quad \text{Subtract 8 from both sides.}$$

$$\frac{9}{x} - 8 \cdot \frac{x}{x} < 0 \quad \text{Build 8 to a fraction with denominator } x.$$

$$\frac{9}{x} - \frac{8x}{x} < 0$$

$$\frac{9 - 8x}{x} < 0 \quad \text{Subtract the numerators and keep the common denominator, } x.$$

Now we solve the related rational equation.

$$\frac{9 - 8x}{x} = 0$$

$$9 - 8x = 0 \quad \text{To clear the equation of the fraction, multiply both sides by } x.$$

$$-8x = -9 \quad \text{Subtract 9 from both sides.}$$

$$x = \frac{9}{8} \quad \text{This is a critical number.}$$

If we set the denominator of $\frac{9 - 8x}{x}$ equal to 0, we obtain a second critical number, $x = 0$. When graphed, the critical numbers 0 and $\frac{9}{8}$ separate the number line into three intervals. We pick a test value from each interval to see whether it satisfies $\frac{9 - 8x}{x} < 0$.

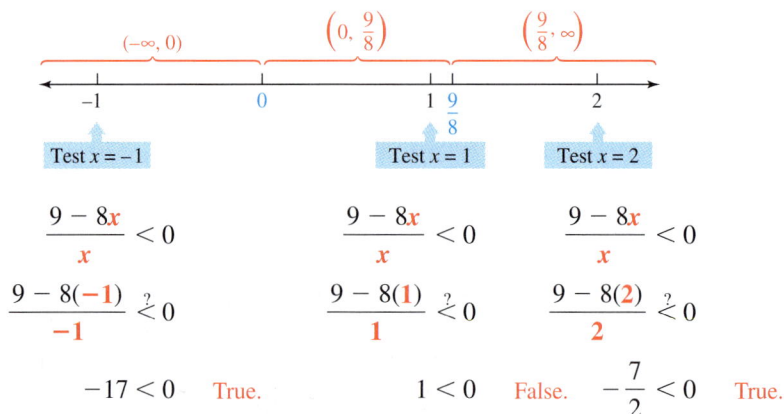

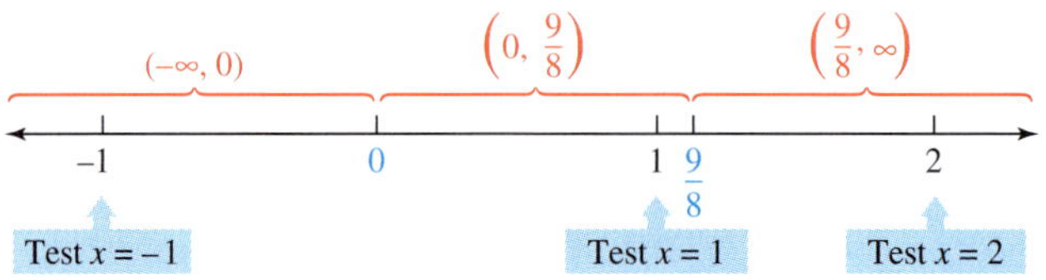

$$\frac{9 - 8x}{x} < 0 \qquad \frac{9 - 8x}{x} < 0 \qquad \frac{9 - 8x}{x} < 0$$

$$\frac{9 - 8(-1)}{-1} \stackrel{?}{<} 0 \qquad \frac{9 - 8(1)}{1} \stackrel{?}{<} 0 \qquad \frac{9 - 8(2)}{2} \stackrel{?}{<} 0$$

$$-17 < 0 \quad \text{True.} \qquad 1 < 0 \quad \text{False.} \qquad -\frac{7}{2} < 0 \quad \text{True.}$$

The numbers in the intervals $(-\infty, 0)$ and $\left(\frac{9}{8}, \infty\right)$ satisfy the inequality. We do not include the endpoint 0 in the solution set, because it makes the denominator of the original inequality 0. Neither do we include $\frac{9}{8}$, because it does not satisfy $\frac{9 - 8x}{x} < 0$. (Recall that $\frac{9}{8}$ makes

$\frac{9-8x}{x}$ equal to 0.) Thus, the solution set is the union of two intervals: $(-\infty, 0) \cup (\frac{9}{8}, \infty)$. Its graph is shown below.

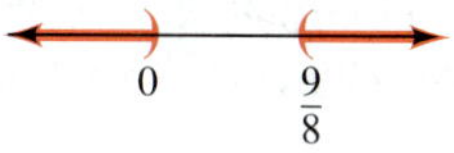

Self Check 3 Solve: $\frac{3}{x} < 5$.

EXAMPLE 4

Solve: $\frac{x^2 + x - 2}{x - 4} \geq 0$.

Solution We first solve the related rational equation to find the critical numbers.

$$\frac{x^2 + x - 2}{x - 4} = 0$$

$$x^2 + x - 2 = 0 \quad \text{To clear the equation of the fraction, multiply both sides by } x - 4.$$

$$(x + 2)(x - 1) = 0 \quad \text{Factor the trinomial.}$$

$$x + 2 = 0 \quad \text{or} \quad x - 1 = 0 \quad \text{Set each factor equal to 0.}$$

$$x = -2 \quad | \quad x = 1 \quad \text{These are critical numbers.}$$

If we set the denominator of $\frac{x^2 + x - 2}{x - 4}$ equal to 0, we see that $x = 4$ is also a critical number. When graphed, the critical numbers, -2, 1, and 4, separate the number line into four intervals. We pick a test value from each interval to see whether it satisfies $\frac{x^2 + x - 2}{x - 4} \geq 0$.

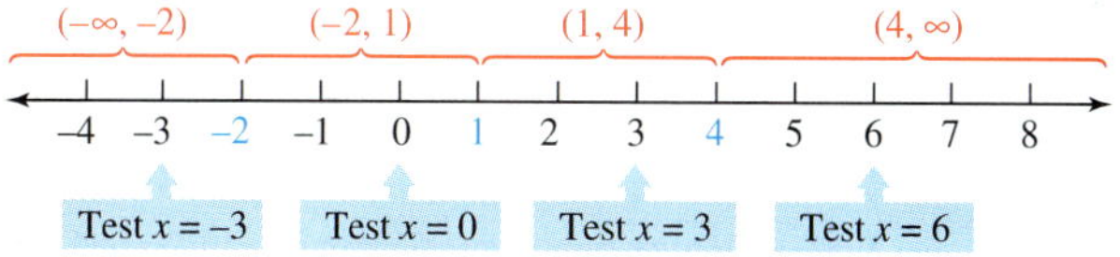

$$\frac{(-3)^2 + (-3) - 2}{-3 - 4} \overset{?}{\geq} 0 \qquad \frac{0^2 + 0 - 2}{0 - 4} \overset{?}{\geq} 0 \qquad \frac{3^2 + 3 - 2}{3 - 4} \overset{?}{\geq} 0 \qquad \frac{6^2 + 6 - 2}{6 - 4} \overset{?}{\geq} 0$$

$$-\frac{4}{7} \geq 0 \qquad \frac{1}{2} \geq 0 \qquad -10 \geq 0 \qquad 20 \geq 0$$

False. True. False. True.

The numbers in the intervals $(-2, 1)$ and $(4, \infty)$ satisfy the inequality. We include the endpoints -2 and 1 in the solution set, because they satisfy the inequality. We do not include 4, because it makes the denominator of the inequality 0. Thus, the solution set is $[-2, 1] \cup (4, \infty)$ as graphed below.

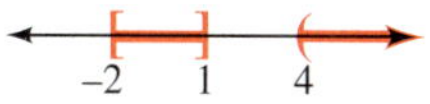

Self Check 4 Solve: $\frac{x + 2}{x^2 - 2x - 3} \geq 0$.

EXAMPLE 5

Solve: $\dfrac{3}{x-1} < \dfrac{2}{x}$.

Solution We subtract $\dfrac{2}{x}$ from both sides to get 0 on the right-hand side and proceed as follows:

$$\frac{3}{x-1} < \frac{2}{x}$$

$$\frac{3}{x-1} - \frac{2}{x} < 0 \qquad \text{Subtract } \frac{2}{x} \text{ from both sides.}$$

$$\frac{3}{x-1}\cdot\frac{x}{x} - \frac{2}{x}\cdot\frac{x-1}{x-1} < 0 \qquad \text{Build each rational expression to have the common denominator } x(x-1).$$

$$\frac{3x - 2x + 2}{x(x-1)} < 0 \qquad \text{Subtract the numerators and keep the common denominator.}$$

$$\frac{x+2}{x(x-1)} < 0 \qquad \text{Combine like terms.}$$

The only solution of the related rational equation $\frac{x+2}{x(x-1)}$ is -2. Thus, -2 is a critical number. When we set the denominator equal to 0 and solve $x(x-1) = 0$, we find two more critical numbers, 0 and 1. These three critical numbers create four intervals to test.

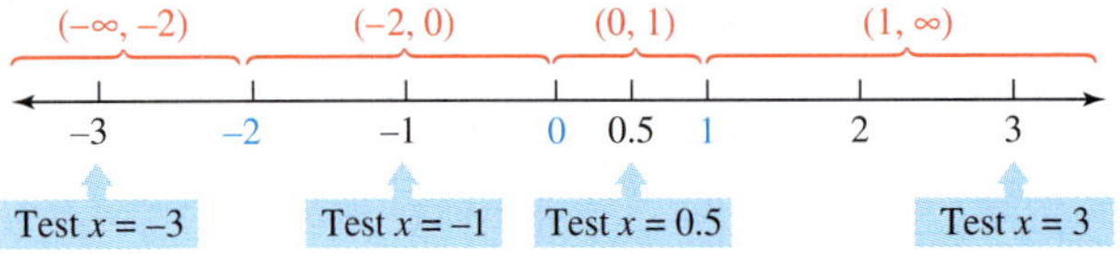

$$\frac{-3+2}{-3(-3-1)} \overset{?}{<} 0 \qquad \frac{-1+2}{-1(-1-1)} \overset{?}{<} 0 \qquad \frac{0.5+2}{0.5(0.5-1)} \overset{?}{<} 0 \qquad \frac{3+2}{3(3-1)} \overset{?}{<} 0$$

True. False. True. False.

Success Tip

When the endpoints of an interval are consecutive integers, such as with the third interval (0, 1), we cannot choose an integer as a test value. For these cases, choose a fraction or decimal that lies within the interval.

The numbers 0 and 1 are not included in the solution set because they make the denominator 0, and the number -2 is not included because it does not satisfy the inequality. The solution set is $(-\infty, -2) \cup (0, 1)$ as graphed below.

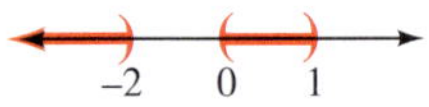

Self Check 5 Solve: $\dfrac{2}{x+1} > \dfrac{1}{x}$.

ACCENT ON TECHNOLOGY: SOLVING INEQUALITIES GRAPHICALLY

We can solve $x^2 + 4x \geq 5$ (Example 2) graphically by first writing the inequality as $x^2 + 4x - 5 \geq 0$, and then graphing the quadratic function $f(x) = x^2 + 4x - 5$, as shown in figure (a). The solution set of the inequality will be those values of x for which the graph lies on or above the x-axis. We can trace to determine that this interval is $(-\infty, -5] \cup [1, \infty)$.

To solve $\frac{3}{x-1} < \frac{2}{x}$ (Example 5) graphically, we first write the inequality in the form $\frac{x+2}{x(x-1)} < 0$, and then we graph the rational function $f(x) = \frac{x+2}{x(x-1)}$, as shown in figure (b). The solution of the inequality will be those values of x for which the graph lies below the axis.

We can trace to see that the graph is below the x-axis when x is less than -2. Since we cannot see the graph in the interval $0 < x < 1$, we redraw the graph using window settings of $[-1, 2]$ for x and $[-25, 10]$ for y, as shown in figure (c).

Now we see that the graph is below the x-axis in the interval $(0, 1)$. Thus, the solution set of the inequality is the union of the two intervals: $(-\infty, -2) \cup (0, 1)$.

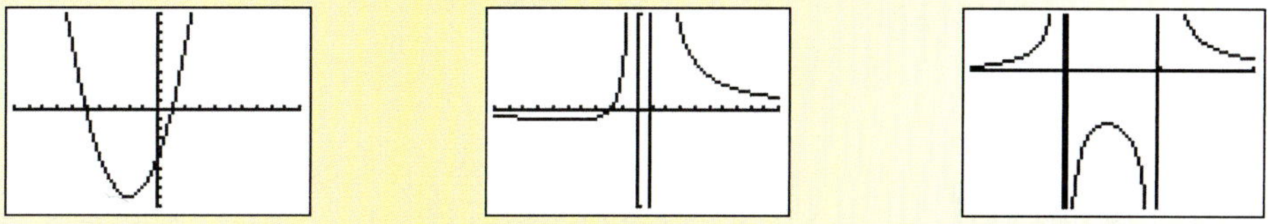

(a) **(b)** **(c)**

GRAPHS OF NONLINEAR INEQUALITIES IN TWO VARIABLES

We have previously graphed linear inequalities in two variables such as $y > 3x + 2$ and $2x - 3y \leq 6$ using the following steps.

Graphing Inequalities in Two Variables

1. Graph the boundary line of the region. If the inequality allows equality (the symbol is either $\leq$ or $\geq$), draw the boundary line as a solid line. If equality is not allowed ($<$ or $>$), draw the boundary line as a dashed line.
2. Pick a test point that is on one side of the boundary line. (Use the origin if possible.) Replace x and y in the original inequality with the coordinates of that point. If the inequality is satisfied, shade the side that contains that point. If the inequality is not satisfied, shade the other side of the boundary.

We use the same procedure to graph *nonlinear* inequalities in two variables.

EXAMPLE 6

Graph: $y < -x^2 + 4$.

Solution The graph of the boundary $y = -x^2 + 4$ is a parabola opening downward, with vertex at $(0, 4)$ and axis of symmetry $x = 0$ (the y-axis). Since the inequality contains a $<$ symbol, and equality is not allowed, we draw the parabola using a dashed line.

To determine which region to shade, we pick the test point $(0, 0)$ and substitute its coordinates into the inequality. We shade the region containing $(0, 0)$ because its coordinates satisfy $y < -x^2 + 4$.

Graph the boundary

$y = -x^2 + 4$

Compare to $y = a(x - h)^2 + k$

$a = -1$: Opens downward

$h = 0$ and $k = 4$: Vertex (0, 4)

Axis of symmetry $x = 0$

x	y
1	3
2	0

***Shading: Use the test point* (0, 0)**

$y < -x^2 + 4$

$0 \overset{?}{<} -0^2 + 4$

$0 < 4$ True.

Since $0 < 4$ is true, (0, 0) is a solution of $y < -x^2 + 4$.

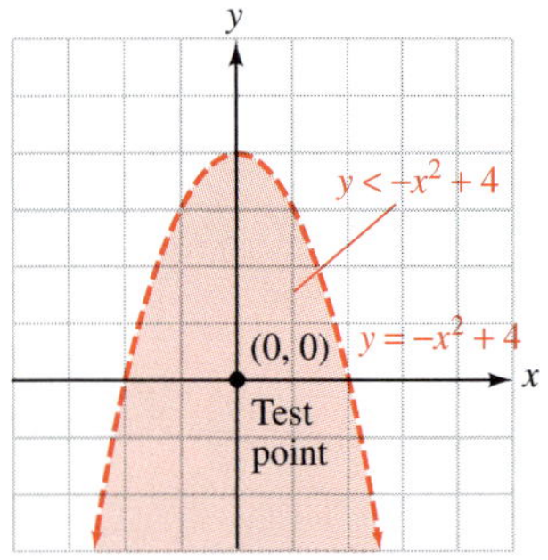

Self Check 6 Graph: $y \geq -x^2 + 4$.

EXAMPLE 7

Graph: $x \leq |y|$.

Solution To graph the boundary, $x = |y|$, we construct a table of solutions, as shown in figure (a). In figure (b), the boundary is graphed using a solid line because the inequality contains a $\leq$ symbol and equality is permitted. Since the origin is on the graph, we cannot use it as a test point. However, any other point, such as (1, 0), will do. We substitute 1 for x and 0 for y into the inequality to get

$x \leq |y|$

$1 \overset{?}{\leq} |0|$

$1 \leq 0$ False.

Since $1 \leq 0$ is a false statement, the point (1, 0) does not satisfy the inequality and is not part of the graph. Thus, the graph of $x \leq |y|$ is to the left of the boundary.

The complete graph is shown in figure (c).

$x = |y|$

x	y
0	0
1	1
1	−1
2	2
2	−2

(a)

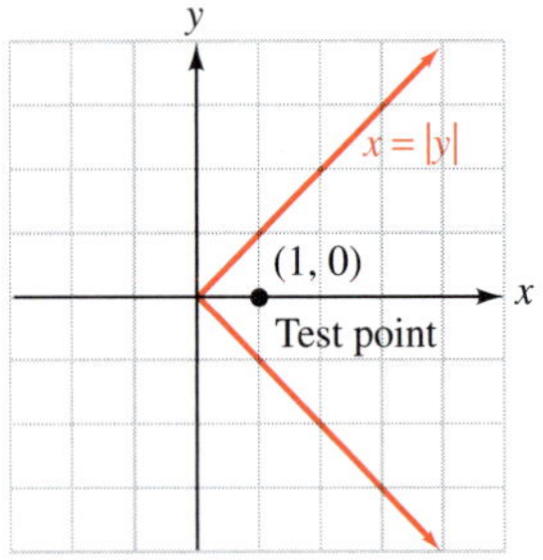

(b)

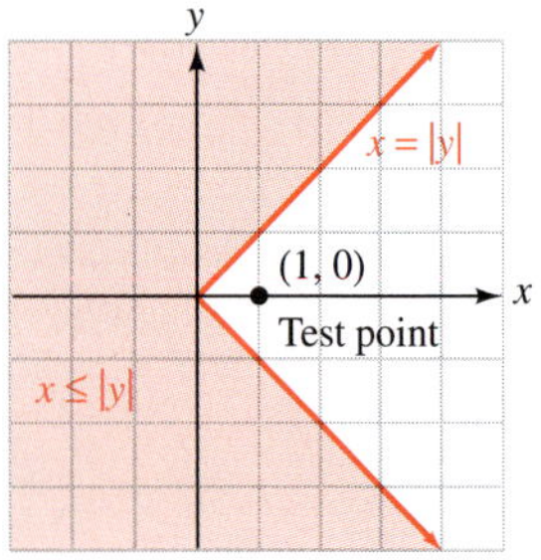

(c)

Self Check 7 Graph: $x \geq -|y|$.

Answers to Self Checks

1. $(-4, 3)$

2. $(-\infty, -8] \cup [5, \infty)$

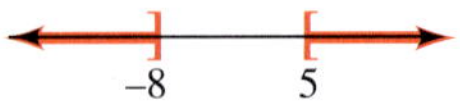

3. $(-\infty, 0) \cup \left(\frac{3}{5}, \infty\right)$

4. $[-2, -1) \cup (3, \infty)$

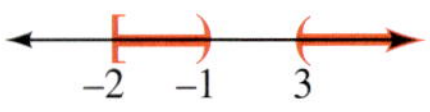

5. $(-1, 0) \cup (1, \infty)$

6.

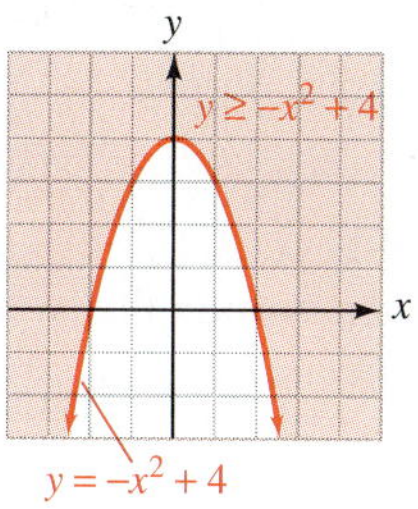

7.

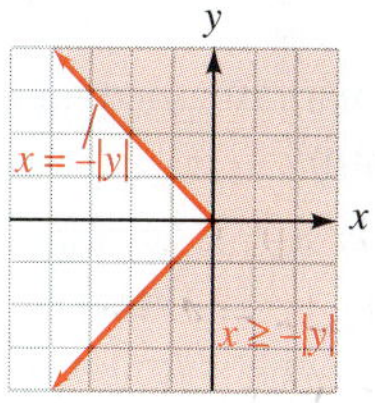

8.5 STUDY SET

VOCABULARY Fill in the blanks.

1. $x^2 + 3x - 18 < 0$ is an example of a ________ inequality in one variable.
2. $\frac{x-1}{x^2-x-20} \leq 0$ is an example of a ________ inequality in one variable.
3. $y \leq x^2 - 4x + 3$ is an example of a nonlinear inequality in _____ variables.
4. The set of real numbers greater than 3 can be represented using the ________ notation $(3, \infty)$.

CONCEPTS

5. The critical numbers of a quadratic inequality are highlighted in red on the number line shown below. Use interval notation to represent each interval that must be tested to solve the inequality.

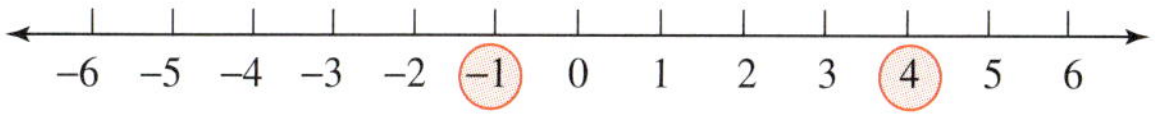

6. The graph of the solution set of a rational inequality in one variable is shown below. Determine whether each of the following numbers is a solution of the inequality.

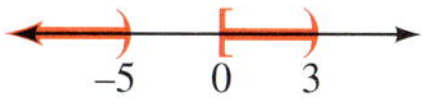

a. -10 **b.** -5

c. 0 **d.** 4

7. Graph each of the following solution sets.

a. $(-2, 4)$

b. $(-\infty, -2) \cup (3, 5]$

8. What are the critical numbers for each inequality?

a. $x^2 - 2x - 48 \geq 0$

b. $\frac{x-3}{x(x+4)} > 0$

9. **a.** The results after interval testing for a quadratic inequality containing a $>$ symbol are shown below. (The critical numbers are highlighted in red.) What is the solution set?

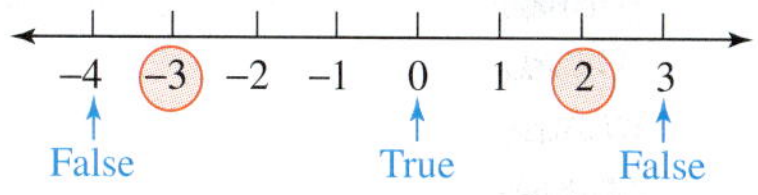

b. The results after interval testing for a quadratic inequality containing a $\leq$ symbol are shown below. (The critical numbers are highlighted in red.) What is the solution set?

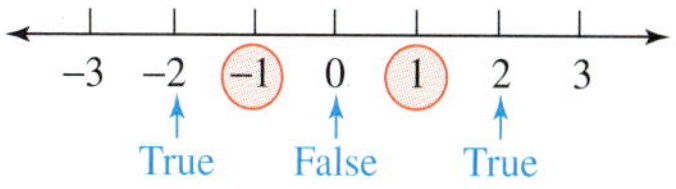

10. Fill in the blank to complete this important fact about the interval testing method discussed in this section: *If one number in an interval satisfies the inequality, _____ numbers in that interval will satisfy the inequality.*

11. **a.** When graphing the solution of $y \leq x^2 + 2x + 1$, should the boundary be solid or dashed?

b. Does the test point (0, 0) satisfy the inequality?

12. **a.** Estimate the solution of $x^2 - x - 6 > 0$ using the graph of $y = x^2 - x - 6$ shown in figure (a) below.

b. Estimate the solution of $\frac{x-3}{x} \leq 0$ using the graph of $y = \frac{x-3}{x}$ shown in figure (b) below.

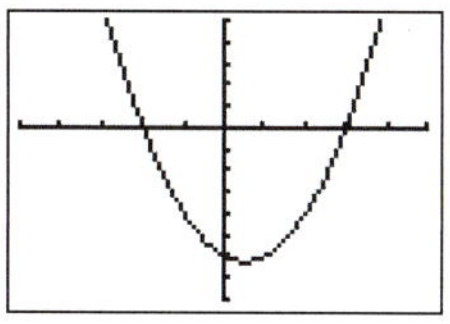

(a)

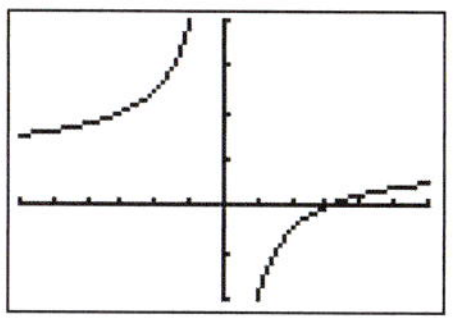

(b)

NOTATION

13. Write the quadratic inequality $x^2 - 6x \geq 7$ in standard form.

14. The solution set of a rational inequality consists of the intervals, $(-1, 4]$ and $(7, \infty)$. When writing the solution set, what symbol is used between the two intervals?

PRACTICE

Solve each inequality. Write the solution set in interval notation and graph it.

15. $x^2 - 5x + 4 < 0$

16. $x^2 - 3x - 4 > 0$

17. $x^2 - 8x + 15 > 0$

18. $x^2 + 2x - 8 < 0$

19. $x^2 + x - 12 \leq 0$

20. $x^2 - 8x \leq -15$

21. $x^2 + 8x < -16$

22. $x^2 + 6x \geq -9$

23. $x^2 \geq 9$

24. $x^2 \geq 16$

25. $2x^2 - 50 < 0$

26. $3x^2 - 243 < 0$

27. $\frac{1}{x} < 2$

28. $\frac{1}{x} > 3$

29. $-\frac{5}{x} < 3$

30. $\frac{4}{x} \geq 8$

31. $\frac{x^2 - x - 12}{x - 1} < 0$

32. $\frac{x^2 + x - 6}{x - 4} \geq 0$

33. $\frac{6x^2 - 5x + 1}{2x + 1} > 0$

34. $\frac{6x^2 + 11x + 3}{3x - 1} < 0$

35. $\frac{3}{x - 2} < \frac{4}{x}$

36. $\frac{-6}{x + 1} \geq \frac{1}{x}$

37. $\frac{7}{x - 3} \geq \frac{2}{x + 4}$

38. $\frac{-5}{x - 4} < \frac{3}{x + 1}$

39. $\frac{x}{x + 4} \leq \frac{1}{x + 1}$

40. $\frac{x}{x + 9} \geq \frac{1}{x + 1}$

41. $(x + 2)^2 > 0$

42. $(x - 3)^2 < 0$

Use a graphing calculator to solve each inequality. Write the solution set in interval notation.

43. $x^2 - 2x - 3 < 0$

44. $x^2 + x - 6 > 0$

45. $\frac{x + 3}{x - 2} > 0$

46. $\frac{3}{x} < 2$

Graph each inequality.

47. $y < x^2 + 1$

48. $y > x^2 - 3$

49. $y \leq x^2 + 5x + 6$

50. $y \geq x^2 + 5x + 4$

51. $y < |x + 4|$

52. $y \geq |x - 3|$

53. $y \leq -|x| + 2$

54. $y > |x| - 2$

APPLICATIONS

55. BRIDGES If an x-axis is superimposed over the roadway of the Golden Gate Bridge, with the origin at the center of the bridge, the length L in feet of a vertical support cable can be approximated by the formula

$$L = \frac{1}{9{,}000}x^2 + 5$$

For the Golden Gate Bridge, $-2{,}100 < x < 2{,}100$. For what intervals along the x-axis are the vertical cables more than 95 feet long?

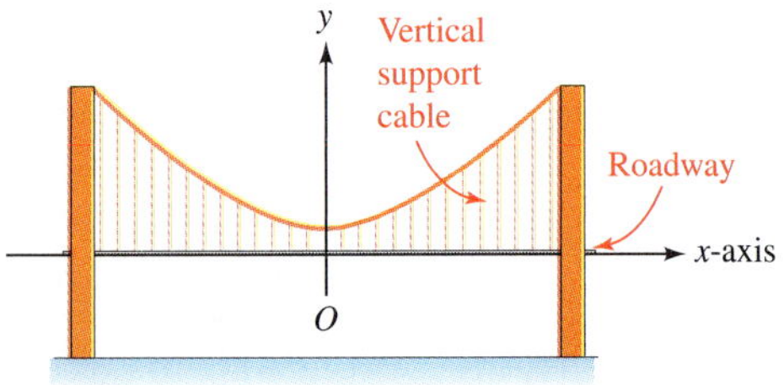

56. MALLS The number of people n in a mall is modeled by the formula

$$n = -100x^2 + 1{,}200x$$

where x is the number of hours since the mall opened. If the mall opened at 9 A.M., when were there 2,000 or more people in it?

WRITING

57. How are critical numbers used when solving a quadratic inequality in one variable?

58. Explain how to find the graph of $y \geq x^2$.

59. The graph of $f(x) = x^2 - 3x + 4$ is shown below. Explain why the quadratic inequality $x^2 - 3x + 4 < 0$ has no solution.

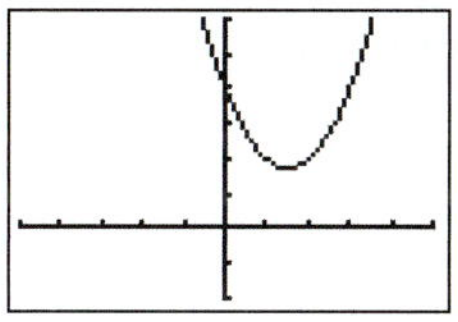

60. Describe the following solution set of a rational inequality in words: $(-\infty, 4] \cup (6, 7)$.

REVIEW **Translate each statement into an equation.**

61. x varies directly with y.

62. y varies inversely with t.

63. t varies jointly with x and y.

64. d varies directly with t and inversely with u^2.

CHALLENGE PROBLEMS

65. **a.** Solve: $x^2 - x - 12 > 0$.

b. Find a rational inequality in one variable that has the same solution set as the quadratic inequality in part a.

66. **a.** Solve: $\frac{1}{x} < 1$.

b. Now incorrectly "solve" $\frac{1}{x} < 1$ by multiplying both sides by x to clear it of the fraction. What part of the solution set is not obtained with this faulty approach?

ACCENT ON TEAMWORK

PICTURE FRAMING

Overview: When framing pictures, mats are often used to enhance the images and give them a sense of depth. In this activity, you will use the quadratic formula to design the matting for several pictures.

Instructions: Form groups of 3 students. Each person in your group is to bring a picture to class. You can use a picture from a magazine or newspaper, a picture postcard, or a photograph that is no larger than 5 in. × 7 in. You will also need a pair of scissors, a ruler, glue, and three pieces of construction paper (12 in. × 18 in.).

Select one of the pictures and find its area. A mat of *uniform* width is to be placed around the picture. The area of the mat should equal the area of the picture. To determine the proper width of the matting, follow the steps of Example 10 in Section 8.1. However, use the quadratic formula, instead of completing the square, to solve the equation. Once you have determined the proper width, cut out the mat from the construction paper and glue it to the picture.

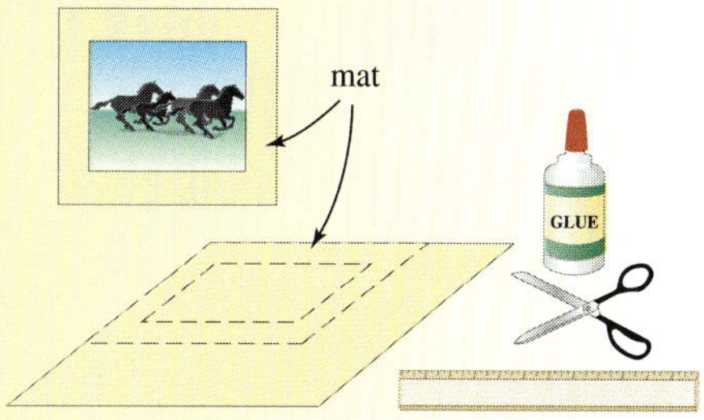

Then, choose another picture and find its area. Determine the uniform width that a matting should have so that its area is double that of the picture. Cut out the proper size matting from the construction paper and glue it to the second picture.

Finally, find the area of the third picture and determine the uniform width that a matting should have so that its area is one-half that of the picture. Cut out the proper-size matting from the construction paper and glue it to the third picture.

Is one size matting more visually appealing than another? Discuss this among the members of your group.

SOLUTIONS OF QUADRATIC EQUATIONS

Overview: In this activity, you will learn of an interesting fact about the solutions of a quadratic equation.

Instructions: Form groups of 2 or 3 students. Consider the following property about the sum and product of the solutions of a quadratic equation:

If $r_1 + r_2 = -\frac{b}{a}$ and $r_1 \cdot r_2 = \frac{c}{a}$, then r_1 and r_2 are the solutions of the quadratic equation $ax^2 + bx + c = 0$, where $a \neq 0$.

Use this property to show that

1. $\frac{3}{2}$ and $-\frac{1}{3}$ are solutions of $6x^2 - 7x - 3 = 0$.
2. $1 + 3\sqrt{2}$ and $1 - 3\sqrt{2}$ are solutions of $x^2 - 2x - 17 = 0$.
3. $i\sqrt{51}$ and $-i\sqrt{51}$ are solutions of $x^2 + 51 = 0$.

KEY CONCEPT: SOLVING QUADRATIC EQUATIONS

We have discussed five methods for solving **quadratic equations.** Let's review each of them and list an advantage and a drawback of each method.

FACTORING

- Factoring can be very quick and simple if the factoring pattern is evident.
- Much of the time, $ax^2 + bx + c$ cannot be factored or is not easily factored.

Solve each equation by factoring.

1. $4k^2 + 8k = 0$
2. $z^2 + 8z + 15 = 0$
3. $2r^2 + 5r = -3$

THE SQUARE ROOT METHOD

- If the equation can be written in the form $x^2 = a$ or $(x + d)^2 = a$, where a is a constant, the square root method is a fast method, requiring few computations.
- Most quadratic equations that we must solve are not written in either of these forms.

Solve each equation by the square root method.

4. $u^2 = 24$
5. $(s - 7)^2 - 9 = 0$
6. $3x^2 + 16 = 0$

COMPLETING THE SQUARE

- Completing the square can be used to solve any quadratic equation.
- Most often, it involves more steps than the other methods.

Solve each equation by completing the square.

7. $x^2 + 10x - 7 = 0$
8. $4x^2 - 4x - 1 = 0$
9. $x^2 + 2x + 2 = 0$

THE QUADRATIC FORMULA

- The quadratic formula simply involves an evaluation of the expression $\frac{-b \pm \sqrt{b^2 - 4ac}}{2a}$.
- If applicable, the factoring method and the square root method are usually faster.

Solve each equation by using the quadratic formula.

10. $2x^2 - 1 = 3x$ **11.** $x^2 - 6x - 391 = 0$ **12.** $3x^2 + 2x + 1 = 0$

THE GRAPHING METHOD

- We can solve the equation using a graphing calculator. It doesn't require any computations.
- It usually gives only approximations of the solutions.

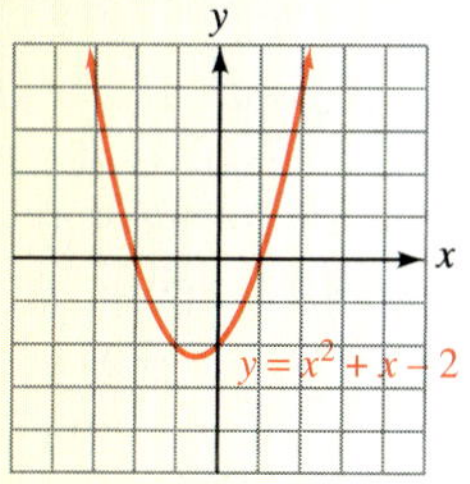

13. Use the graph of $y = x^2 + x - 2$ to solve $x^2 + x - 2 = 0$.

CHAPTER REVIEW

SECTION 8.1 The Square Root Property and Completing the Square

CONCEPTS

The square root property: If $c > 0$, the equation $x^2 = c$ has two real solutions:

$x = \sqrt{c}$ and $x = -\sqrt{c}$

To *complete the square:*

1. Make sure the coefficient of x^2 is 1.
2. Make sure the constant term is on the right-hand side of the equation.
3. Add the square of one-half of the coefficient of x to both sides.
4. Factor the trinomial.
5. Use the square root property.
6. Check the answers.

REVIEW EXERCISES

Solve each equation by factoring or using the square root property.

1. $x^2 + 9x + 20 = 0$ **2.** $6x^2 + 17x + 5 = 0$

3. $x^2 = 28$ **4.** $(t + 2)^2 = 36$

5. $5a^2 + 11a = 0$ **6.** $5x^2 - 49 = 0$

7. $a^2 + 25 = 0$

8. What number must be added to $x^2 - x$ to make a perfect square trinomial?

Solve each equation by completing the square.

9. $x^2 + 6x + 8 = 0$ **10.** $2x^2 - 6x + 3 = 0$

11. $x^2 - 2x + 13 = 0$

12. Solve $A = \pi r^2$ for r. Assume that all variables represent positive numbers. Express any radical in simplified form.

13. Explain the error:

$$\frac{2 \pm \sqrt{7}}{2} = \frac{\overset{1}{\cancel{2}} \pm \sqrt{7}}{\underset{1}{\cancel{2}}} = 1 \pm \sqrt{7}$$

14. HAPPY NEW YEAR As part of a New Year's Eve celebration, a huge ball is to be dropped from the top of a 605-foot-tall building at the proper moment so that it strikes the ground at exactly 12:00 midnight. The distance d in feet traveled by a free-falling object in t seconds is given by the formula $d = 16t^2$. To the nearest second, when should the ball be dropped from the building?

SECTION 8.2 The Quadratic Formula

The quadratic formula: The solutions of $ax^2 + bx + c = 0$ where $a \neq 0$ are given by

$$x = \frac{-b \pm \sqrt{b^2 - 4ac}}{2a}$$

When solving a quadratic equation by the quadratic formula, we can often simplify the computations by solving an equivalent equation that does not involve fractions or decimals, and whose lead coefficient is positive.

Solve each equation using the quadratic formula.

15. $x^2 - 10x = 0$

16. $-x^2 + 10x - 18 = 0$

17. $2x^2 + 13x = 7$

18. $26y - 3y^2 = 2$

19. $\frac{p^2}{3} + \frac{p}{2} + \frac{1}{2} = 0$

20. $3{,}000t^2 - 4{,}000t = -2{,}000$

21. $0.5x^2 + 0.3x - 0.1 = 0$

22. TUTORING A private tutoring company charges \$20 for a 1-hour session. Currently, 300 students are tutored each week. Since the company is losing money, the owner has decided to increase the price. For each 50¢ increase, she estimates that 5 fewer students will participate. If the center needs to bring in \$6,240 per week to stay in business, what price must be charged for a 1-hour tutoring session to produce this amount of revenue?

23. POSTERS The specifications for a poster of Cesar Chavez call for a 615-square-inch photograph to be surrounded by a blue border. (See below.) The borders on the sides of the poster are to be half as wide as those at the top and bottom. Find the width of each border.

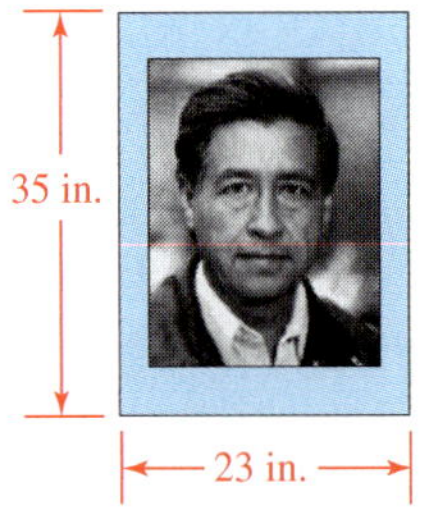

24. ACROBATS To begin his routine on a trapeze, an acrobat is catapulted upward as shown in the illustration above. His distance d (in feet) from the arena floor during this maneuver is given by the formula $d = -16t^2 + 40t + 5$, where t is the time in seconds since being launched. If the trapeze bar is 25 feet in the air, at what two times will he be able to grab it? Round to the nearest tenth.

SECTION 8.3 The Discriminant and Equations That Can Be Written in Quadratic Form

The *discriminant* predicts the type of solutions of $ax^2 + bx + c = 0$:

1. If $b^2 - 4ac > 0$, the solutions are unequal real numbers.
2. If $b^2 - 4ac = 0$, the solutions are equal real numbers.
3. If $b^2 - 4ac < 0$, the solutions are complex conjugates.

Equations that contain an expression to a power, the same expression to that power squared, and a constant term are said to be *quadratic in form.* One method used to solve such equations is to make a substitution.

Use the discriminant to determine the type of solutions for each equation.

25. $3x^2 + 4x - 3 = 0$

26. $4x^2 - 5x + 7 = 0$

27. $9x^2 - 12x + 4 = 0$

28. $m(2m - 3) = 20$

Solve each equation.

29. $x - 13\sqrt{x} + 12 = 0$

30. $a^{2/3} + a^{1/3} - 6 = 0$

31. $3x^4 + x^2 - 2 = 0$

32. $\dfrac{6}{x+2} + \dfrac{6}{x+1} = 5$

33. $(x - 7)^2 + 6(x - 7) + 10 = 0$

34. $m^{-4} - 2m^{-2} + 1 = 0$

35. $4\left(\dfrac{x+1}{x}\right)^2 + 12\left(\dfrac{x+1}{x}\right) + 9 = 0$

36. WEEKLY CHORES Working together, two sisters can do the yard work at their house in 45 minutes. When the older girl does it all herself, she can complete the job in 20 minutes less time than it takes the younger girl working alone. How long does it take the older girl to do the yard work?

SECTION 8.4 Quadratic Functions and Their Graphs

A *quadratic function* is a second-degree polynomial function of the form $f(x) = ax^2 + bx + c$.

The graph of $f(x) = ax^2$ is a *parabola* opening upward when $a > 0$ and downward when $a < 0$, with *vertex* at the point $(0, 0)$ and *axis of symmetry* the line $x = 0$.

Each of the following functions has a graph that is the same shape as $f(x) = ax^2$ but involves a vertical or horizontal translation.

1. $f(x) = ax^2 + k$: translated upward if $k > 0$, downward if $k < 0$.

37. AEROSPACE INDUSTRY The annual sales of the Boeing Company in billions of dollars for the years 1993–1998 can be modeled by the quadratic function

$$S(x) = 2.2x^2 - 7.7x + 39.9$$

where x is the number of years since 1993. Use the function to determine the annual sales for 1997.

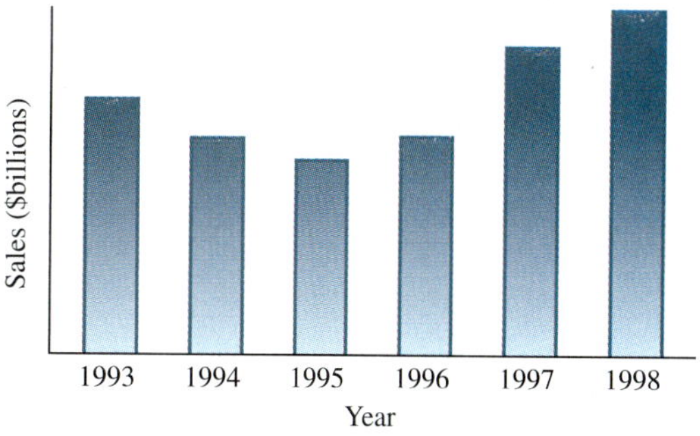

Source: The Boeing Company Annual Report, 1999 and 2002

Make a table of values to graph function f. Then use a series of translations to graph function g on the same coordinate system.

38. $f(x) = 2x^2$ $g(x) = 2x^2 - 3$

39. $f(x) = -4x^2$ $g(x) = -4(x - 2)^2 + 1$

40. Find the vertex and the axis of symmetry of the graph of $f(x) = -2(x - 1)^2 + 4$. Then plot several points and complete the graph.

2. $f(x) = a(x - h)^2$: translated right if $h > 0$, and left if $h < 0$.

The graph of $f(x) = a(x - h)^2 + k$ is a parabola with vertex at (h, k). It opens upward when $a > 0$ and downward when $a < 0$. The axis of symmetry is the line $x = h$.

The vertex of the graph of $f(x) = ax^2 + bx + c$ is

$$\left(-\frac{b}{2a}, f\left(-\frac{b}{2a}\right)\right)$$

and the axis of symmetry is the line

$$x = -\frac{b}{2a}$$

The *y-intercept* is determined by the value of $f(x)$ when $x = 0$: the y-intercept is $(0, c)$. To find the *x-intercepts,* let $f(x) = 0$ and solve $ax^2 + bx + c = 0$.

The y-coordinate of the vertex of the graph of a quadratic function gives the *minimum* or *maximum* value of the function.

41. Complete the square to write $f(x) = 4x^2 + 16x + 9$ in the form $f(x) = a(x - h)^2 + k$. Determine the vertex and the axis of symmetry of the graph. Then plot several points and complete the graph.

42. First determine the coordinates of the vertex and the axis of symmetry of the graph of $f(x) = x^2 + x - 2$ using the vertex formula. Then determine the x- and y-intercepts of the graph. Finally, plot several points and complete the graph.

43. FARMING The number of farms in the United States for the years 1870–1970 is approximated by

$$N(x) = -1{,}526x^2 + 155{,}652x + 2{,}500{,}200$$

where $x = 0$ represents 1870, $x = 1$ represents 1871, and so on. For this period, when was the number of U.S. farms a maximum? How many farms were there?

44. Estimate the solutions of $-3x^2 - 5x + 2 = 0$ from the graph of $f(x) = -3x^2 - 5x + 2$, shown on the right.

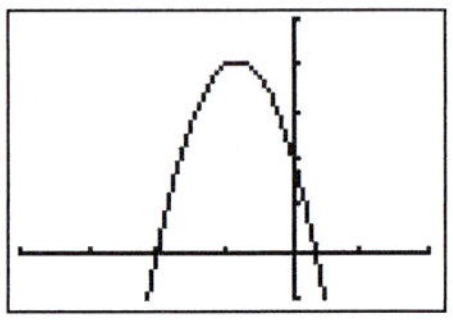

SECTION 8.5 Quadratic and Other Nonlinear Inequalities

To solve a *quadratic inequality,* get 0 on one side and solve the related quadratic equation. Locate the *critical numbers* on a number line and test each interval. Finally, check the endpoints.

Solve each inequality. Write the solution set in interval notation and graph it.

45. $x^2 + 2x - 35 > 0$

46. $x^2 \leq 81$

47. $\frac{3}{x} \leq 5$

48. $\frac{2x^2 - x - 28}{x - 1} > 0$

To solve a *rational inequality,* get 0 on one side and solve the related rational equation. Locate the *critical numbers* (including any values that make the denominator 0) on a number line and test each interval. Finally, check the endpoints.

To graph a *nonlinear inequality in two variables,* first graph the boundary. Then use a test point to determine which half-plane to shade.

49. Estimate the solution set of $3x^2 + 10x - 8 \le 0$ from the graph of $f(x) = 3x^2 + 10x - 8$ shown in figure (a) below.

50. Estimate the solution set of $\frac{x-1}{x} > 0$ from the graph of $f(x) = \frac{x-1}{x}$ shown in figure (b).

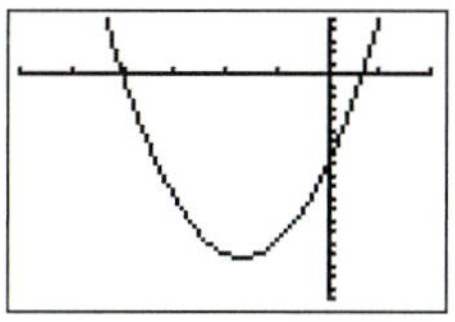

(a)

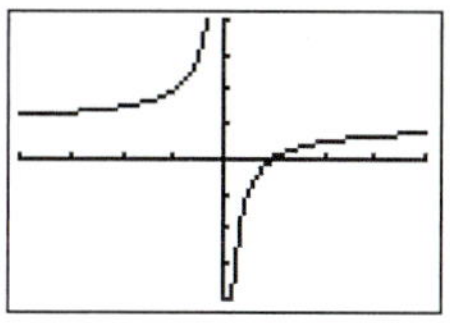

(b)

Graph each inequality.

51. $y < \frac{1}{2}x^2 - 1$

52. $y \ge -|x|$

CHAPTER 8 TEST

Solve each equation by factoring or using the square root property.

1. $3x^2 + 18x = 0$

2. $m^2 + 4 = 0$

3. $(a + 7)^2 = 50$

4. $x(6x + 19) = -15$

5. Determine what number must be added to $x^2 + 24x$ to make it a perfect square trinomial.

6. Solve $3x^2 + x - 24 = 0$ by completing the square.

Use the quadratic formula to solve each equation.

7. $2x^2 - 8x + 5 = 0$

8. $\frac{t^2}{8} - \frac{t}{4} = \frac{1}{2}$

9. $-t^2 + 4t - 13 = 0$

10. $0.01x^2 = -0.08x - 0.15$

Solve by any method.

11. $2y - 3\sqrt{y} + 1 = 0$

12. $m^{-2} + m^{-1} = -1$

13. $x^4 - x^2 - 12 = 0$

14. $4\left(\frac{x+2}{3x}\right)^2 - 4\left(\frac{x+2}{3x}\right) - 3 = 0$

15. Solve $E = mc^2$ for c. Assume that all variables represent positive numbers. Express any radical in simplified form.

16. Use the discriminant to determine the type of solutions for each equation.

a. $3x^2 + 5x + 17 = 0$

b. $9m^2 - 12m = -4$

17. TABLECLOTHS In 1990, Sportex of Highland, Illinois, made what was at the time the world's longest tablecloth. Find the dimensions of the rectangular tablecloth if it covered an area of 6,759 square feet and its length was 8 feet more than 332 times its width.

18. COOKING Working together, a chef and his assistant can make a pastry dessert in 25 minutes. When the chef makes it himself, it takes him 8 minutes less time than it takes his assistant working alone. How long does it take the chef to make the dessert?

19. DRAWING An artist uses four equal-sized right triangles to block out a perspective drawing of an old hotel. For each triangle, one leg is 14 inches longer than the other, and the hypotenuse is 26 inches. On the centerline of the drawing, what is the length of the segment extending from the ground to the top of the building?

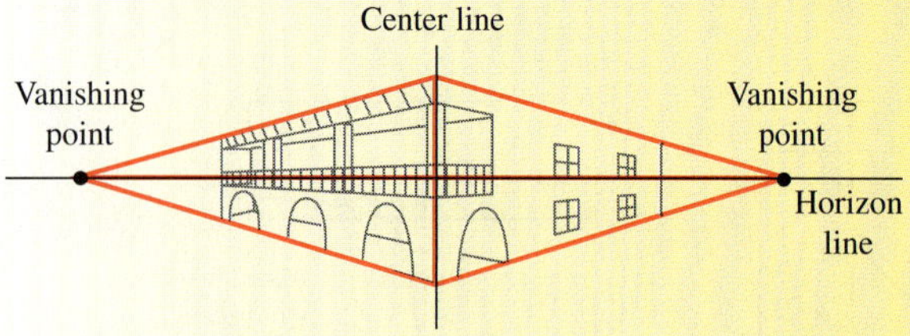

20. ANTHROPOLOGY Anthropologists refer to the shape of the human jaw as a *parabolic dental arcade.* Which function is the best mathematical model of the parabola shown in the illustration?

i. $f(x) = -\frac{3}{8}(x - 4)^2 + 6$

ii. $f(x) = -\frac{3}{8}(x - 6)^2 + 4$

iii. $f(x) = -\frac{3}{8}x^2 + 6$

iv. $f(x) = \frac{3}{8}x^2 + 6$

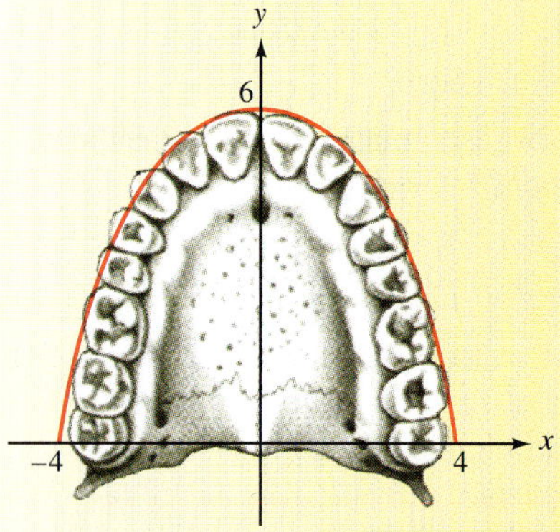

21. Find the vertex and the axis of symmetry of the graph of $f(x) = -3(x - 1)^2 - 2$. Then plot several points and complete the graph.

22. Complete the square to write the function $f(x) = 5x^2 + 10x - 1$ in the form $f(x) = a(x - h)^2 + k$. Determine the vertex and the axis of symmetry of the graph. Then plot several points and complete the graph.

23. First determine the coordinates of the vertex and the axis of symmetry of the graph of $f(x) = 2x^2 + x - 1$ using the vertex formula. Then determine the x- and y-intercepts of the graph. Finally, plot several points and complete the graph.

24. DISTRESS SIGNALS A flare is fired directly upward into the air from a boat that is experiencing engine problems. The height of the flare (in feet) above the water, t seconds after being fired, is given by the formula $h = -16t^2 + 112t + 15$. If the flare is designed to explode when it reaches its highest point, at what height will this occur?

Solve each inequality. Write the solution set in interval notation and then graph it.

25. $x^2 - 2x > 8$

26. $\frac{x - 2}{x + 3} \le 0$

27. Explain the error.

$$\text{Solve: } \frac{12}{x} > 6$$
$$12 > 6x$$
$$2 > x$$
$$x < 2$$
$$(-\infty, 2)$$

28. Graph: $y \le -x^2 + 3$.

29. The graph of a quadratic function of the form $f(x) = ax^2 + bx + c$ is shown. Estimate the solutions of the corresponding quadratic equation $ax^2 + bx + c = 0$.

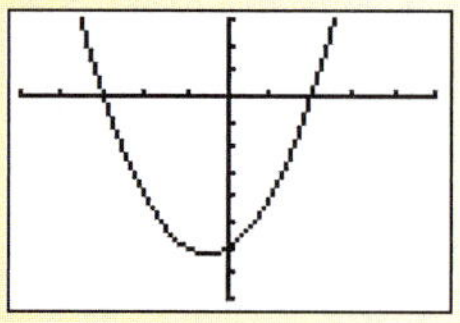

30. See Problem 29. Estimate the solution of the quadratic inequality $ax^2 + bx + c \le 0$.

CHAPTERS 1–8 CUMULATIVE REVIEW EXERCISES

Write an equation of the line with the given properties.

1. $m = 3$, passing through $(-2, -4)$

2. Parallel to the graph of $2x + 3y = 6$ and passing through $(0, -2)$

3. AIRPORT TRAFFIC From the graph in the illustration, determine the projected average rate of change in the number of takeoffs and landings at Los Angeles International Airport for the years 2000–2015.

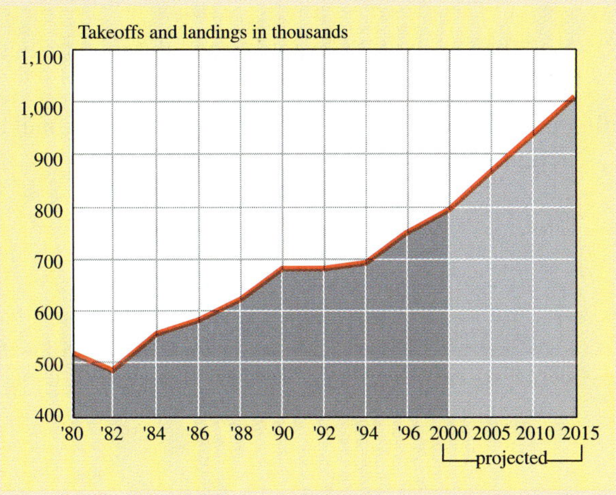

Based on data from *Los Angeles Times* (July 6, 1998) p. 83

4. Solve the system by graphing.

$$\begin{cases} y = -\frac{5}{2}x + \frac{1}{2} \\ 2x - \frac{3}{2}y = 5 \end{cases}$$

5. Solve the system using Cramer's rule.

$$\begin{cases} x - y + z = 4 \\ x + 2y - z = -1 \\ x + y - 3z = -2 \end{cases}$$

6. Graph the solution set of the system

$$\begin{cases} 3x + 2y > 6 \\ x + 3y \le 2 \end{cases}$$

Solve each inequality. Write the solution set in interval notation and then graph it.

7. $5(-2x + 2) > 20 - x$

8. $|2x - 5| \ge 25$

9. Simplify: $\left(\dfrac{-3x^4y^2}{-9x^5y^{-2}}\right)^2$.

10. TIDES The illustration shows the graph of a function f, which gives the height of the tide for a 24-hour period in Seattle, Washington. (Note that military time is used on the x-axis: 3 A.M. = 3, noon = 12, 3 P.M. = 15, 9 P.M. = 21, and so on.)

a. Find the domain of the function.

b. Find $f(3)$.

c. Find $f(6)$.

d. Find $f(15)$.

e. What information does $f(12)$ give?

f. Estimate the values of x for which $f(x) = 0$.

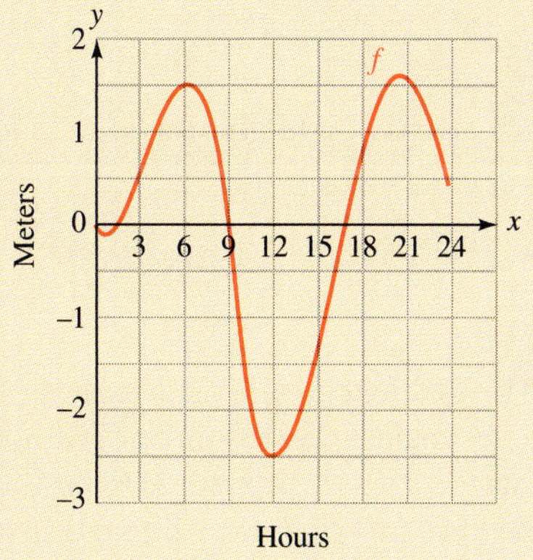

Perform the operations.

11. $(a + 2)(3a^2 + 4a - 2)$

12. $\dfrac{x^3 + y^3}{x^3 - y^3} \div \dfrac{x^2 - xy + y^2}{x^2 + xy + y^2}$

13. $\dfrac{1}{x + y} - \dfrac{1}{x - y} - \dfrac{2y}{y^2 - x^2}$

14. Simplify: $\dfrac{\frac{1}{r^2 + 4r + 4}}{\frac{r}{r + 2} + \frac{r}{r + 2}}$.

Factor each expression.

15. $x^4 - 16y^4$

16. $30a^4 - 4a^3 - 16a^2$

17. $x^2 + 4y - xy - 4x$

18. $8x^6 + 125y^3$

19. $x^2 + 10x + 25 - y^8$

20. $49s^6 - 84s^3n^2 + 36n^4$

Solve each equation.

21. $(m + 4)(2m + 3) - 22 = 10m$

22. $6a^3 - 2a = a^2$

23. $\dfrac{x - 4}{x - 3} + \dfrac{x - 2}{x - 3} = x - 3$

24. $P + \dfrac{a}{V^2} = \dfrac{RT}{V - b}$ solve for b

Graph each function and give its domain and range.

25. $f(x) = x^3 + x^2 - 6x$

26. $f(x) = \dfrac{4}{x}$ for $x > 0$

27. LIGHT As light energy radiates away from its source, its intensity varies inversely as the square of the distance from the source. The illustration shows that the light energy passing through an area 1 foot from the source spreads out over 4 units of area 2 feet from the source. That energy is therefore less intense 2 feet from the source than it was 1 foot from the source. Over how many units of area will the light energy spread out 3 feet from the source?

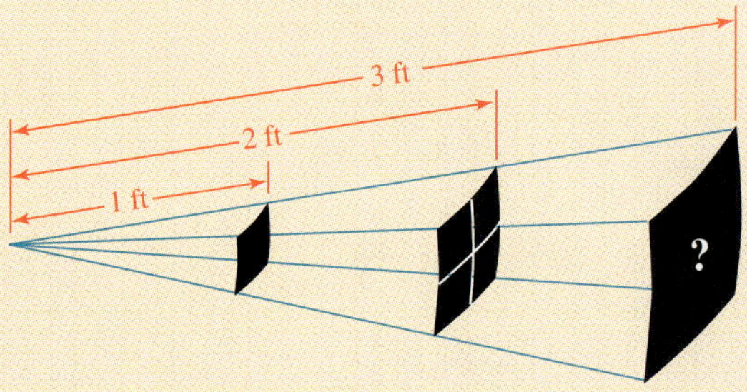

28. Graph the function $f(x) = \sqrt{x - 2}$ and give its domain and range.

Simplify each expression.

29. $\sqrt[3]{-27x^3}$

30. $\sqrt{48t^3}$

31. $64^{-2/3}$

32. $\dfrac{x^{5/3}x^{1/2}}{x^{3/4}}$

33. $-3\sqrt[4]{32} - 2\sqrt[4]{162} + 5\sqrt[4]{48}$

34. $3\sqrt{2}\left(2\sqrt{3} - 4\sqrt{12}\right)$

35. $\dfrac{\sqrt{x} + 2}{\sqrt{x} - 1}$

36. $\dfrac{5}{\sqrt[3]{x}}$

Solve each equation.

37. $5\sqrt{x + 2} = x + 8$

38. $\sqrt{x} + \sqrt{x + 2} = 2$

39. Find the length of the hypotenuse of the right triangle in figure (a).

40. Find the length of the hypotenuse of the right triangle in figure (b).

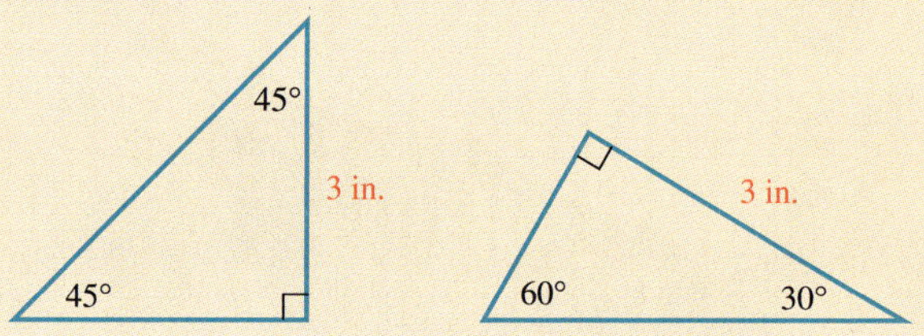

(a) **(b)**

41. Find the distance between $(-2, 6)$ and $(4, 14)$.

42. Simplify: i^{43}.

Perform the operations. Write each result in $a + bi$ form.

43. $\left(-7 + \sqrt{-81}\right) - \left(-2 - \sqrt{-64}\right)$

44. $\dfrac{5}{3 - i}$

45. $(2 + i)^2$

46. $\dfrac{-4}{6i^7}$

47. What number must be added to $x^2 + 6x$ to make a perfect square trinomial?

48. Use the method of completing the square to solve $2x^2 + x - 3 = 0$.

49. Use the quadratic formula to solve $\dfrac{a^2}{8} - \dfrac{a}{2} = -1$.

50. COMMUNITY GARDENS Residents of a community can work their own 16 ft × 24 ft plot of city-owned land if they agree to the following stipulations:

- The area of the garden cannot exceed 180 square feet.
- A path of uniform width must be maintained around the garden.

Find the dimensions of the largest possible garden.

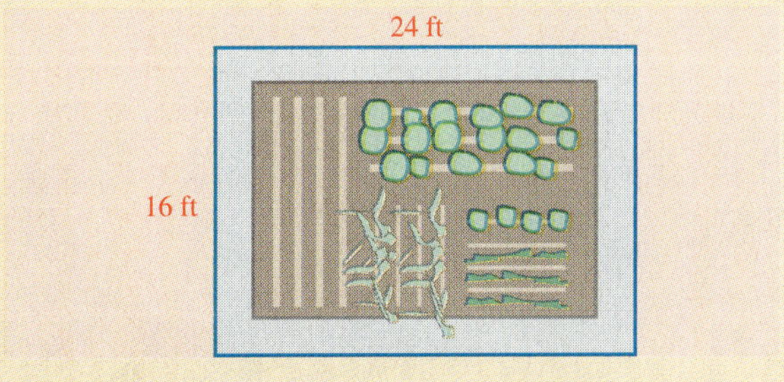

51. INSTALLING A SIDEWALK A 170-meter-long sidewalk from the mathematics building M to the student center C is shown in red in the illustration. However, students prefer to walk directly from M to C. How long are the two segments of the existing sidewalk?

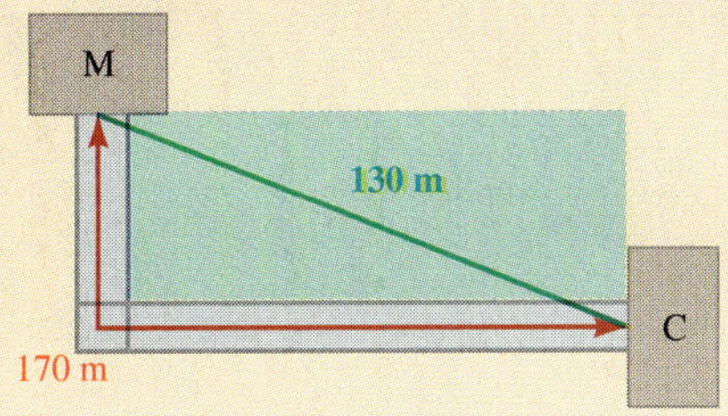

52. First determine the vertex and the axis of symmetry of the graph of $f(x) = -x^2 - 4x$ using the vertex formula. Then determine the x- and y-intercepts of the graph. Finally, plot several points and complete the graph.

Solve each equation.

53. $a - 7a^{1/2} + 12 = 0$

54. $x^{-4} - 2x^{-2} + 1 = 0$

55. The graph of $f(x) = 16x^2 + 24x + 9$ is shown below. Estimate the solution(s) of $16x^2 + 24x + 9 = 0$.

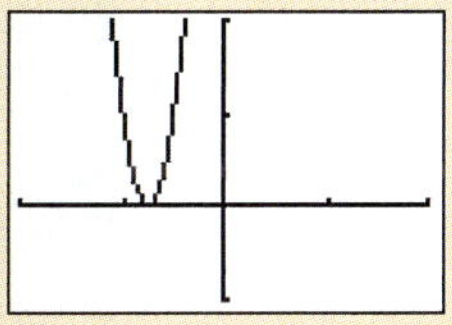

56. Use the graph above to determine the solution of $16x^2 + 24x + 9 < 0$.

Chapter

9 Exponential and Logarithmic Functions

CORBIS

Financial planners advise us that we should begin saving for our retirement at a young age. However, it's difficult to make saving a budgeting priority with today's high cost of living. Fortunately, we don't have to choose between paying our current financial obligations and saving for retirement. Thanks to the power of compounding, a modest amount of money invested wisely can turn into a large sum over time. That is because compounding calculates interest on the principal and on the prior period's interest.

To learn more about compound interest, visit *The Learning Equation* on the Internet at http://tle.brookscole.com. (The log-in instructions are in the Preface.) For Chapter 9, the online lessons are:

- *TLE* Lesson 13: Exponential Functions
- *TLE* Lesson 14: Properties of Logarithms

In this chapter, we discuss the concept of function in more depth. We also introduce two new families of functions—exponential and logarithmic functions—which have applications in many areas.

9.1 Algebra and Composition of Functions

- Algebra of functions
- Composition of functions
- The identity function
- Writing composite functions

Just as it is possible to perform arithmetic operations on real numbers, it is also possible to perform those operations on functions. We call the process of adding, subtracting, multiplying, and dividing functions the *algebra of functions.*

ALGEBRA OF FUNCTIONS

The sum, difference, product, and quotient of two functions are themselves functions.

Operations on Functions

If the domains and ranges of functions f and g are subsets of the real numbers, then

The **sum** of f and g, denoted as $\boldsymbol{f + g}$, *is defined by*

$$(f + g)(x) = f(x) + g(x)$$

The **difference** of f and g, denoted as $\boldsymbol{f - g}$, is defined by

$$(f - g)(x) = f(x) - g(x)$$

The **product** of f and g, denoted as $\boldsymbol{f \cdot g}$, is defined by

$$(f \cdot g)(x) = f(x)g(x)$$

The **quotient** of f and g, denoted as $\boldsymbol{f/g}$, is defined by

$$(f/g)(x) = \frac{f(x)}{g(x)} \text{ where } g(x) \neq 0$$

The domain of each of these functions is the set of real numbers x that are in the domain of both f and g. In the case of the quotient, there is the further restriction that $g(x) \neq 0$.

EXAMPLE 1

Let $f(x) = 2x^2 + 1$ and $g(x) = 5x - 3$. Find each function and its domain: **a.** $f + g$ and **b.** $f - g$.

Solution **a.**

$$\begin{aligned}(f + g)(x) &= f(x) + g(x) \\ &= (2x^2 + 1) + (5x - 3) && \text{Replace } f(x) \text{ with } 2x^2 + 1 \text{ and } g(x) \text{ with } 5x - 3. \\ &= 2x^2 + 5x - 2 && \text{Combine like terms.}\end{aligned}$$

The domain of $f + g$ is the set of real numbers that are in the domain of both f and g. Since the domain of both f and g is the interval $(-\infty, \infty)$, the domain of $f + g$ is also the interval $(-\infty, \infty)$.

b. $(f - g)(x) = f(x) - g(x)$
$= (2x^2 + 1) - (5x - 3)$
$= 2x^2 + 1 - 5x + 3$ Remove parentheses.
$= 2x^2 - 5x + 4$ Combine like terms.

Since the domain of both f and g is $(-\infty, \infty)$, the domain of $f - g$ is also the interval $(-\infty, \infty)$.

Self Check 1 Let $f(x) = 3x - 2$ and $g(x) = 2x^2 + 3x$. Find **a.** $f + g$ and **b.** $f - g$.

EXAMPLE 2

Let $f(x) = 2x^2 + 1$ and $g(x) = 5x - 3$. Find each function and its domain: **a.** $f \cdot g$ and **b.** f/g.

Solution **a.** $(f \cdot g)(x) = f(x)g(x)$ Replace $f(x)$ with $2x^2 + 1$ and $g(x)$ with $5x - 3$.
$= (2x^2 + 1)(5x - 3)$
$= 10x^3 - 6x^2 + 5x - 3$ Use the FOIL method.

The domain of $f \cdot g$ is the set of real numbers that are in the domain of both f and g. Since the domain of both f and g is the interval $(-\infty, \infty)$, the domain of $f \cdot g$ is also the interval $(-\infty, \infty)$.

b. $(f/g)(x) = \dfrac{f(x)}{g(x)}$
$= \dfrac{2x^2 + 1}{5x - 3}$

Since the denominator of the fraction cannot be 0, $x \neq \dfrac{3}{5}$. Thus, the domain of f/g is the interval $\left(-\infty, \frac{3}{5}\right) \cup \left(\frac{3}{5}, \infty\right)$.

Self Check 2 Let $f(x) = 2x^2 - 3$ and $g(x) = x^2 + 1$. Find **a.** $f \cdot g$ and **b.** f/g.

COMPOSITION OF FUNCTIONS

We have seen that a function can be represented by a machine: We put in a number from the domain, and a number from the range comes out. For example, if we put the number 2 into the machine shown in figure (a) on the next page, the number $f(2) = 8$ comes out. In general, if we put x into the machine shown in figure (b), the value $f(x)$ comes out.

Often one quantity is a function of a second quantity that depends, in turn, on a third quantity. For example, the cost of a car trip is a function of the gasoline consumed. The amount of gasoline consumed, in turn, is a function of the number of miles driven. Such chains of dependence can be analyzed mathematically as **compositions of functions.**

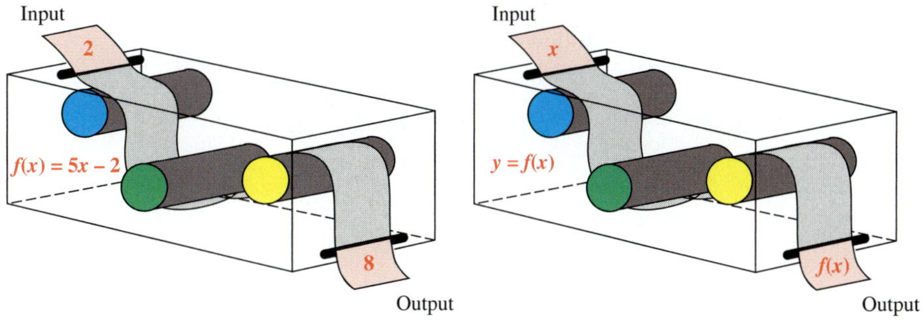

(a) (b)

> **The Language of Algebra**
>
> A *composite* is something that is formed by bringing together distinct parts. For example, a police artist creates a *composite* sketch from the eyewitnesses' descriptions of the suspect.

Suppose that $y = f(x)$ and $y = g(x)$ define two functions. Any number x in the domain of g will produce the corresponding value $g(x)$ in the range of g. If $g(x)$ is in the domain of function f, then $g(x)$ can be substituted into f, and a corresponding value $f(g(x))$ will be determined. This two-step process defines a new function, called a **composite function,** denoted by $f \circ g$. (This is read as "f composed with g.")

The function machines shown below illustrate the composition $f \circ g$. When we put a number x into the function g, $g(x)$ comes out. The value $g(x)$ goes into function f, which transforms $g(x)$ into $f(g(x))$. (This is read as "f of g of x.") If the function machines for g and f were connected to make a single machine, that machine would be named $f \circ g$.

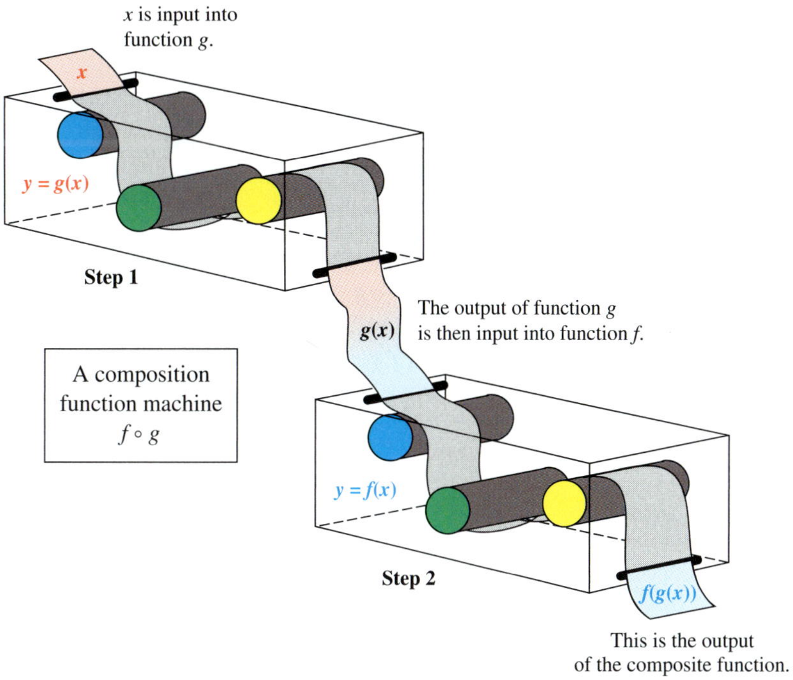

To be in the domain of the composite function $f \circ g$, a number x has to be in the domain of g and the output of g must be in the domain of f. Thus, the domain of $f \circ g$ consists of those numbers x that are in the domain of g, and for which $g(x)$ is in the domain of f.

Composite Functions The **composite function $f \circ g$** is defined by

$$(f \circ g)(x) = f(g(x))$$

For example, if $f(x) = 4x$ and $g(x) = 3x + 2$, to find $f \circ g$ and $g \circ f$, we proceed as follows.

$$\begin{aligned}(f \circ g)(x) &= f(g(x)) \\ &= f(3x + 2) \\ &= 4(3x + 2) \\ &= 12x + 8\end{aligned} \qquad \begin{aligned}(g \circ f)(x) &= g(f(x)) \\ &= g(4x) \\ &= 3(4x) + 2 \\ &= 12x + 2\end{aligned}$$

Different results

The different results illustrate that the composition of functions is not commutative: $(f \circ g)(x) \neq (g \circ f)(x)$.

EXAMPLE 3

Let $f(x) = 2x + 1$ and $g(x) = x - 4$. Find **a.** $(f \circ g)(9)$, **b.** $(f \circ g)(x)$, and **c.** $(g \circ f)(-2)$.

Solution

a. $(f \circ g)(9)$ means $f(g(9))$. In figure (a) on the next page, function g receives the number 9, subtracts 4, and releases the number $g(9) = 5$. Then 5 goes into the f function, which doubles 5 and adds 1. The final result, 11, is the output of the composite function $f \circ g$:

Read as "f of g of 9."

$$(f \circ g)(9) = f(g(9)) = f(5) = 2(5) + 1 = 11$$

Thus, $(f \circ g)(9) = 11$.

Notation

The notation $f \circ g$ can also be read as "f circle g." Remember, it means that the function g is applied first and function f is applied second.

b. $(f \circ g)(x)$ means $f(g(x))$. In figure (a) on the next page, function g receives the number x, subtracts 4, and releases the number $x - 4$. Then $x - 4$ goes into the f function, which doubles $x - 4$ and adds 1. The final result, $2x - 7$, is the output of the composite function $f \circ g$.

Read as "f of g of x."

$$(f \circ g)(x) = f(g(x)) = f(x - 4) = 2(x - 4) + 1 = 2x - 7$$

Thus, $(f \circ g)(x) = 2x - 7$.

c. $(g \circ f)(-2)$ means $g(f(-2))$. In figure (b) on the next page, function f receives the number -2, doubles it and adds 1, and releases -3 into the g function. Function g subtracts 4 from -3 and outputs a final result of -7. Thus,

Read as "g of f of -2."

$$(g \circ f)(-2) = g(f(-2)) = g(-3) = -3 - 4 = -7$$

Thus, $(g \circ f)(-2) = -7$.

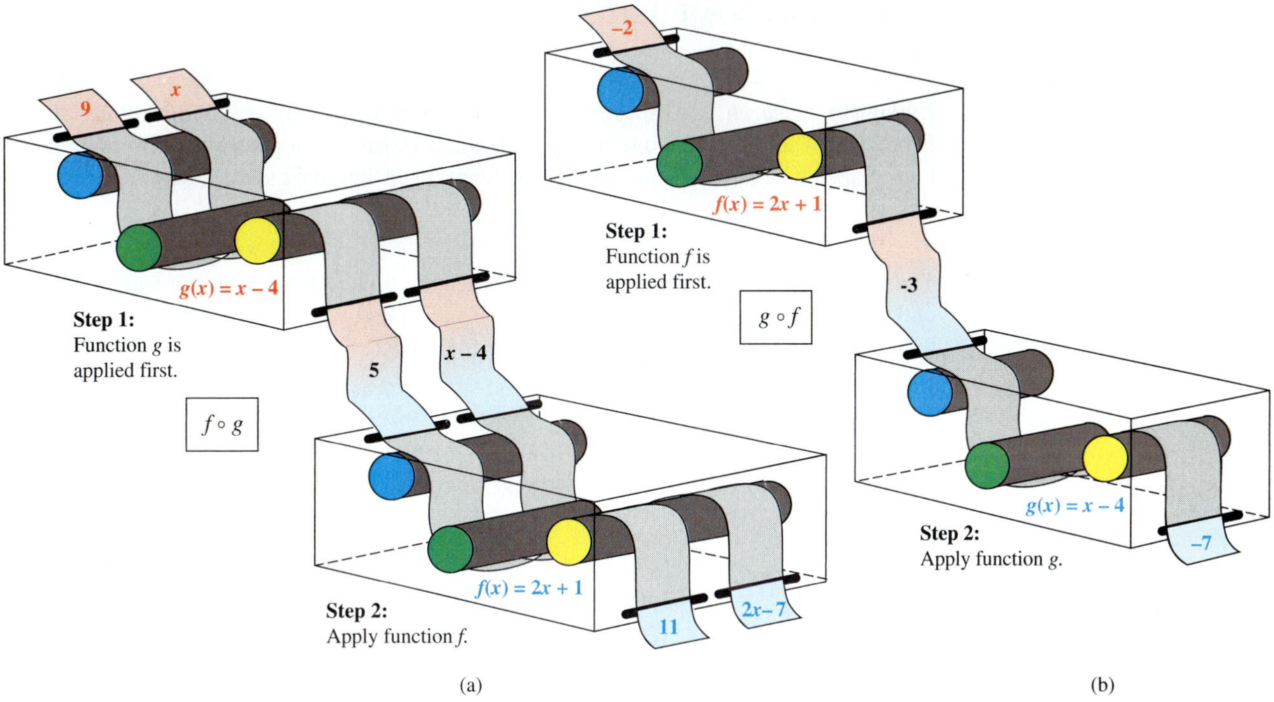

Self Check 3 Let $f(x) = x^3$ and $g(x) = 6 - x$. Find **a.** $(f \circ g)(8)$, **b.** $(g \circ f)(1)$, and **c.** $(g \circ f)(x)$.

THE IDENTITY FUNCTION

The **identity function** is defined by the equation $I(x) = x$. Under this function, the value that is assigned to any real number x is x itself. For example $I(2) = 2$, $I(-3) = -3$, and $I(7.5) = 7.5$. If f is any function, the composition of f with the identity function is just the function f:

$$(f \circ I)(x) = (I \circ f)(x) = f(x)$$

EXAMPLE 4

Let f be any function and let I be the identity function, $I(x) = x$. Show that **a.** $(f \circ I)(x) = f(x)$ and **b.** $(I \circ f)(x) = f(x)$.

Solution **a.** $(f \circ I)(x)$ means $f(I(x))$. Because $I(x) = x$, we have

$$(f \circ I)(x) = f(I(x)) = f(x)$$

b. $(I \circ f)(x)$ means $I(f(x))$. Because I passes any number through unchanged, we have

$$(I \circ f)(x) = I(f(x)) = f(x)$$

The Language of Algebra

The identity function pairs each real number with itself such that each output is *identical* to its corresponding input.

WRITING COMPOSITE FUNCTIONS

EXAMPLE 5

Biological research. A specimen is stored in refrigeration at a temperature of 15° Fahrenheit. Biologists remove the specimen and warm it at a controlled rate of 3° F per hour. Express its Celsius temperature as a function of the time t since it was removed from refrigeration.

Solution The temperature of the specimen is 15° F when the time $t = 0$. Because it warms at a rate of 3° F per hour, its initial temperature of 15° increases by $3t$° F in t hours. The Fahrenheit temperature of the specimen is given by the function

$$F(t) = 3t + 15$$

The Celsius temperature C is a function of this Fahrenheit temperature F, given by the function

$$C(F) = \frac{5}{9}(F - 32)$$

To express the specimen's Celsius temperature as a function of *time,* we find the composite function

$$\begin{aligned}(C \circ F)(t) &= C(F(t)) \\ &= C(3t + 15) && \text{Substitute } 3t + 15 \text{ for } F(t). \\ &= \frac{5}{9}[(3t + 15) - 32] && \text{Substitute } 3t + 15 \text{ for } F \text{ in } \tfrac{5}{9}(F - 32). \\ &= \frac{5}{9}(3t - 17) && \text{Simplify.} \\ &= \frac{15}{9}t + \frac{85}{9} \\ &= \frac{5}{3}t + \frac{85}{9}\end{aligned}$$

The composite function, $C(t) = \frac{5}{3}t + \frac{85}{9}$, gives the temperature of the specimen in degrees Celsius t hours after it is removed from refrigeration.

Answers to Self Checks **1. a.** $(f + g)(x) = 2x^2 + 6x - 2$, **b.** $(f - g)(x) = -2x^2 - 2$ **2. a.** $(f \cdot g)(x) = 2x^4 - x^2 - 3$, **b.** $(f/g)(x) = \dfrac{2x^2 - 3}{x^2 + 1}$ **3. a.** -8, **b.** 5, **c.** $(g \circ f)(x) = 6 - x^3$

9.1 STUDY SET

VOCABULARY **Fill in the blanks.**

1. The ______ of f and g, denoted as $f + g$, is defined by $(f + g)(x) =$ ______ and the ______ of f and g, denoted as $f - g$, is defined by $(f - g)(x) =$ ______.

2. The ______ of f and g, denoted as $f \cdot g$, is defined by $(f \cdot g)(x) =$ ______ and the ______ of f and g, denoted as f/g, is defined by $(f/g)(x) =$ ______.

3. The ________ of the function $f + g$ is the set of real numbers x that are in the domain of both f and g.
4. The __________ function $f \circ g$ is defined by $(f \circ g)(x) = $ ________.
5. Under the ________ function, the value that is assigned to any real number x is x itself.
6. When reading the notation $f(g(x))$, we say "f ___ g ___ x."

CONCEPTS

7. Fill in the blanks to make the statements true.
   a. $(f \circ g)(3) = f(\quad)$
   b. To find $f(g(3))$, we first find ____ and then substitute that value for x in $f(x)$.
8. a. If $f(x) = 3x + 1$ and $g(x) = 1 - 2x$, find $f(g(3))$ and $g(f(3))$.
   b. Is the composition of functions commutative?
9. Fill in the blanks in the drawing of the function machines that show how to compute $g(f(-2))$.

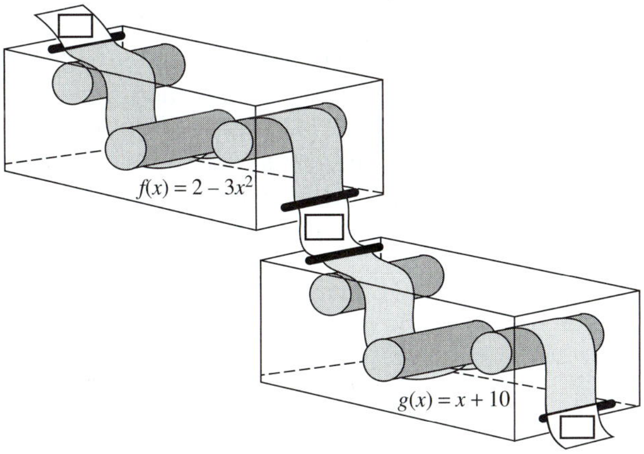

10. Complete the table of values for the identity function, $I(x) = x$. Then graph it.

x	$I(x)$
−3	
−2	
−1	
0	
1	
2	
3	

11. Use the table of values for functions f and g to find each of the following.
   a. $(f + g)(1)$ b. $(f - g)(5)$
   c. $(f \cdot g)(1)$ d. $(g/f)(5)$

x	$f(x)$
1	3
5	8

x	$g(x)$
1	4
5	0

12. Use the table of values for functions f and g to find each of the following.
   a. $(f \circ g)(1)$ b. $(g \circ f)(2)$

x	$f(x)$
2	5
4	7

x	$g(x)$
1	2
5	−3

NOTATION **Complete each solution.**

13. Let $f(x) = 3x - 1$ and $g(x) = 2x + 3$. Find $f \cdot g$.

$$(f \cdot g)(x) = f(x) \cdot \square$$
$$= \square(2x + 3)$$
$$= 6x^2 + \square - \square - 3$$
$$(f \cdot g)(x) = 6x^2 + 7x - 3$$

14. Let $f(x) = 3x - 1$ and $g(x) = 2x + 3$. Find $f \circ g$.

$$(f \circ g)(x) = f(\square)$$
$$= f(\square)$$
$$= 3(\square) - 1$$
$$= \square + \square - 1$$
$$(f \circ g)(x) = 6x + 8$$

PRACTICE **Let $f(x) = 3x$ and $g(x) = 4x$. Find each function and its domain.**

15. $f + g$
16. $f - g$
17. $f \cdot g$
18. f/g
19. $g - f$
20. $g + f$
21. g/f
22. $g \cdot f$

Let $f(x) = 2x + 1$ and $g(x) = x - 3$. Find each function and its domain.

23. $f + g$

24. $f - g$

25. $f \cdot g$

26. f/g

27. $g - f$

28. $g + f$

29. g/f

30. $g \cdot f$

Let $f(x) = 3x - 2$ and $g(x) = 2x^2 + 1$. Find each function and its domain.

31. $f - g$

32. $f + g$

33. f/g

34. $f \cdot g$

Let $f(x) = x^2 - 1$ and $g(x) = x^2 - 4$. Find each function and its domain.

35. $f - g$

36. $f + g$

37. g/f

38. $g \cdot f$

Let $f(x) = 2x + 1$ and $g(x) = x^2 - 1$. Find each composition.

39. $(f \circ g)(2)$

40. $(g \circ f)(2)$

41. $(g \circ f)(-3)$

42. $(f \circ g)(-3)$

43. $(f \circ g)(0)$

44. $(g \circ f)(0)$

45. $(f \circ g)\left(\frac{1}{2}\right)$

46. $(g \circ f)\left(\frac{1}{3}\right)$

47. $(f \circ g)(x)$

48. $(g \circ f)(x)$

49. $(g \circ f)(2x)$

50. $(f \circ g)(2x)$

Let $f(x) = 3x - 2$ and $g(x) = x^2 + x$. Find each composition.

51. $(f \circ g)(4)$

52. $(g \circ f)(4)$

53. $(g \circ f)(-3)$

54. $(f \circ g)(-3)$

55. $(g \circ f)(0)$

56. $(f \circ g)(0)$

57. $(g \circ f)(x)$

58. $(f \circ g)(x)$

Let $f(x) = \dfrac{1}{x}$ and $g(x) = \dfrac{1}{x^2}$. Find each composition.

59. $(f \circ g)(4)$

60. $(f \circ g)(6)$

61. $(g \circ f)\left(\frac{1}{3}\right)$

62. $(g \circ f)\left(\frac{1}{10}\right)$

63. $(g \circ f)(8x)$

64. $(f \circ g)(5x)$

65. If $f(x) = x + 1$ and $g(x) = 2x - 5$, show that $(f \circ g)(x) \neq (g \circ f)(x)$.

66. If $f(x) = x^2 + 1$ and $g(x) = 3x^2 - 2$, show that $(f \circ g)(x) \neq (g \circ f)(x)$.

APPLICATIONS **In Exercises 67–70, refer to the following illustration. The graph of function f gives the average score on the verbal portion of the SAT college entrance exam, the graph of function g gives the average score on the mathematics portion, and x represents the number of years since 1990.**

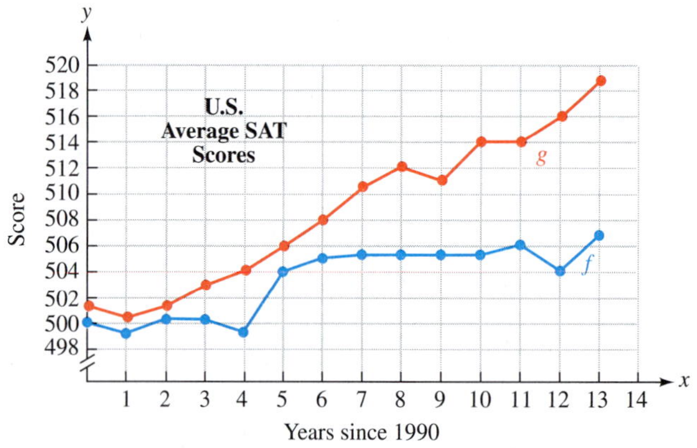

Source: *The World Almanac*, 1999 and 2004

67. From the graph, determine $f(3)$ and $g(3)$. Then use those results to find $(f + g)(3)$.

68. From the graph, determine $f(6)$ and $g(6)$. Then use those results to find $(g - f)(6)$.

69. Find $(f + g)(10)$ and explain what information about SAT scores it gives.

70. Find $(g - f)(12)$ and explain what information about SAT scores it gives.

71. METALLURGY A molten alloy must be cooled slowly to control crystallization. When removed from the furnace, its temperature is 2,700° F, and it will be cooled at 200° per hour. Express the Celsius temperature as a function of the number of hours t since cooling began.

72. WEATHER FORECASTING A high-pressure area promises increasingly warmer weather for the next 48 hours. The temperature is now 34° Celsius and is expected to rise 1° every 6 hours. Express the Fahrenheit temperature as a function of the number of hours from now. $\left(\textit{Hint: } F = \frac{9}{5}C + 32.\right)$

73. VACATION MILEAGE COSTS

a. Use the following graphs to determine the cost of the gasoline consumed if a family drove 500 miles on a summer vacation.

b. Write a composition function that expresses the cost of the gasoline consumed on the vacation as a function of the miles driven.

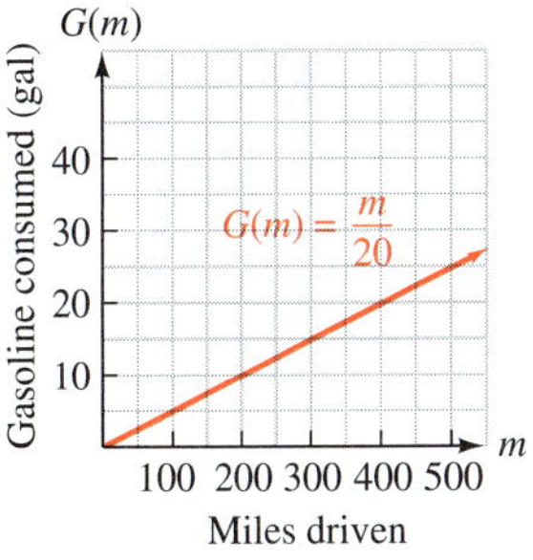

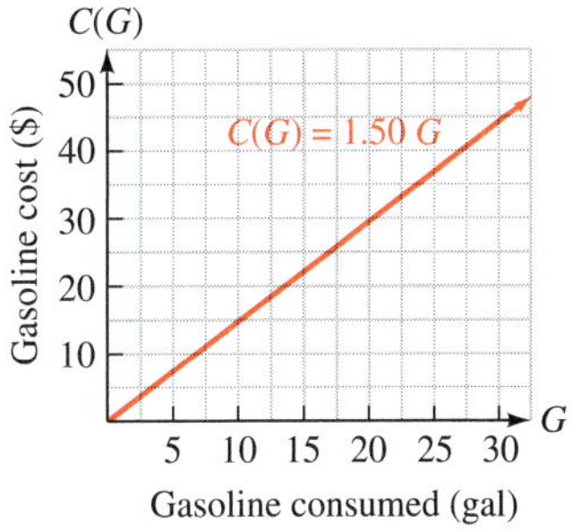

74. HALLOWEEN COSTUMES The tables on the back of a pattern package (see the next column) can be used to determine the number of yards of material needed to make a rabbit costume for a child.

a. How many yards of material are needed if the child's chest measures 29 inches?

b. In this exercise, one quantity is a function of a second quantity that depends, in turn, on a third quantity. Explain this dependence.

PATTERN 9810 **Simplicity**

Costumes have front zipper, long raglan sleeves, elastic sleeve and leg casings, hood. Fabrics: Fleece or suede

BODY MEASUREMENTS

Chest (in.)	21	22	23	25	26	27	$28\frac{1}{2}$	29	30
Pattern Size	2	3	4	6	7	8	10	11	12

YARDAGE NEEDED

Pattern Size	2-4	6-8	10-12
Yards	$2\frac{5}{8}$	$3\frac{3}{8}$	$3\frac{3}{4}$

WRITING

75. Exercise 73 illustrates a chain of dependence between the cost of the gasoline, the gasoline consumed, and the miles driven. Describe another chain of dependence that could be represented by a composition function.

76. In this section, what operations are performed on functions? Give an example of each.

77. Write out in words how to say each of the following:

$(f \circ g)(2)$ $\quad g(f(-8))$

78. If $Y_1 = f(x)$ and $Y_2 = g(x)$, explain how to use the following tables to find $g(f(2))$.

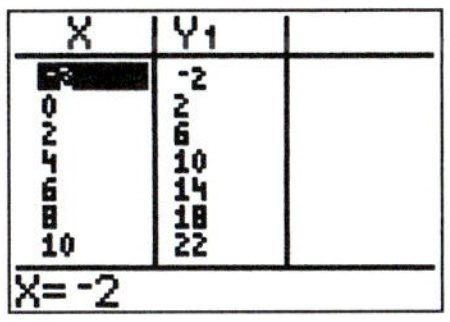

(a)

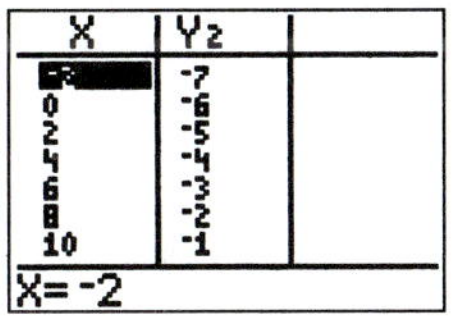

(b)

REVIEW **Simplify each complex fraction.**

79. $$\frac{\dfrac{ac - ad - c + d}{a^3 - 1}}{\dfrac{c^2 - 2cd + d^2}{a^2 + a + 1}}$$

80. $$\frac{2 + \dfrac{1}{x^2 - 1}}{1 + \dfrac{1}{x - 1}}$$

CHALLENGE PROBLEMS **Fill in the blanks.**

81. If $f(x) = x^2$ and $g(x) =$ ______, then $(f \circ g)(x) = 4x^2 + 20x + 25$.

82. If $f(x) = \sqrt{3x}$ and $g(x) =$ ______, then $(g \circ f)(x) = 9x^2 + 7$.

9.2 Inverse Functions

- The inverse of a function
- One-to-one functions
- The horizontal line test
- Finding the inverse of a function
- The composition of a function and its inverse
- Graphing a function and its inverse

In the previous section, we used the operations of arithmetic and composition to create new functions from given functions. Another way that a new function can be created from a given function is to find its *inverse*.

THE INVERSE OF A FUNCTION

In figure (a) below, the arrow diagram defines a function f. If we reverse the arrows, as shown in figure (b), the domain of f becomes the range, and the range of f becomes the domain, of a new correspondence. The new correspondence is a function, because it assigns to each member of the domain exactly one member of the range. We call this new correspondence the **inverse** of f, or f inverse.

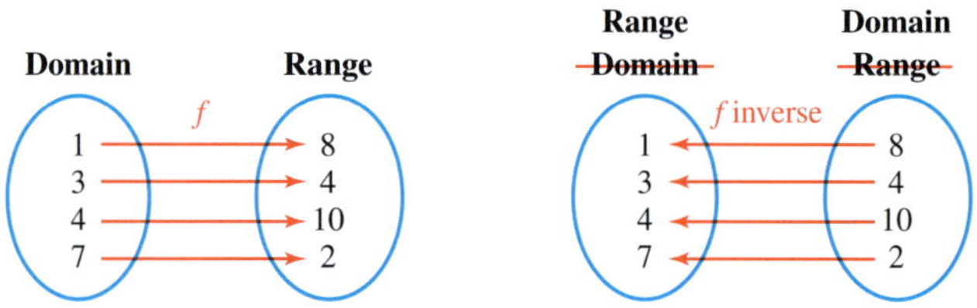

(a) (b)

The reversing process does not always produce a function. Consider the function g defined by the diagram in figure (a) below. When we reverse the arrows, the resulting correspondence is not a function, because it assigns two members of the range, 8 and 4, to the number 2.

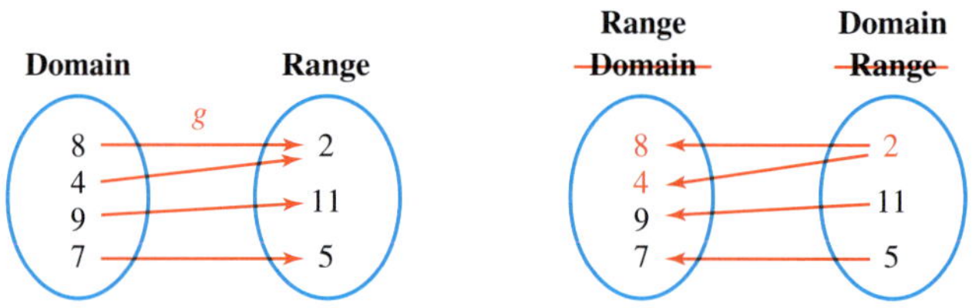

(a) (b)

The question that arises is, "What must be true of the original function to guarantee that the reversing process produces a function?" The answer to that question is: *the original function must be one-to-one.*

ONE-TO-ONE FUNCTIONS

We have seen that a function assigns to each input exactly one output. For some functions, different inputs are assigned different outputs, as shown in figure (a) on the next page. For other functions, different inputs are assigned the *same* output, as in figure (b). When each output corresponds to exactly one input, as in figure (a), we say the function is *one-to-one.*

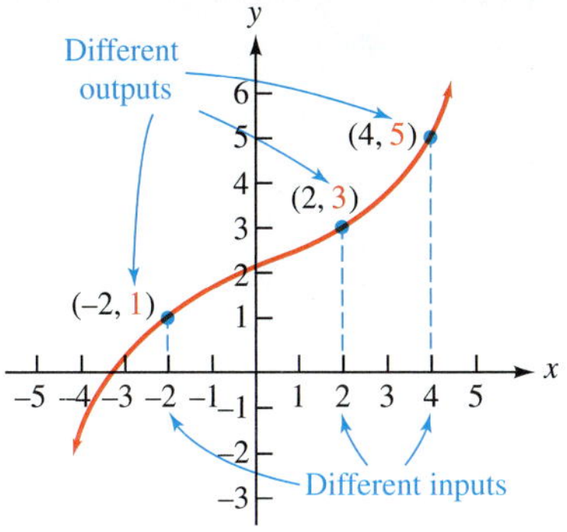

A one-to-one function

(a)

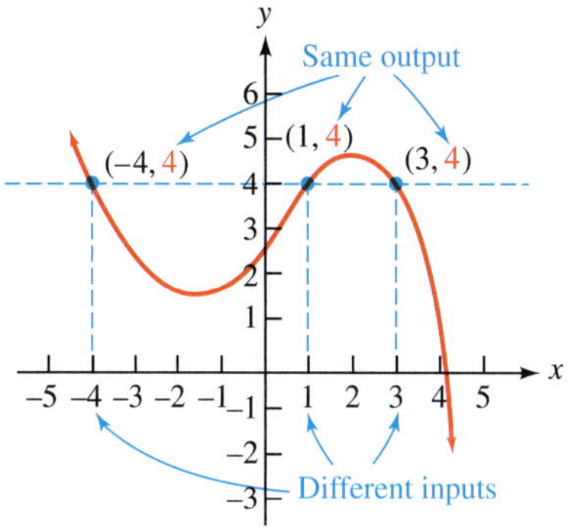

Not a one-to-one function

(b)

One-to-One Functions For a **one-to-one function,** each input is assigned exactly one output, and each output corresponds to exactly one input.

EXAMPLE 1

Determine whether each function is one-to-one: **a.** $f(x) = x^2$ and **b.** $f(x) = x^3$.

Solution

a. Since the output 9 corresponds to two different inputs, −3 and 3, $f(x) = x^2$ is not one-to-one.

Success Tip

Example 1 illustrates that not every function is one-to-one.

$f(-3) = (-3)^2 = 9$ and $f(3) = 3^2 = 9$

x	$f(x)$
−3	9
3	9

The output 9 does not correspond to exactly one input.

b. Since different numbers have different cubes, each output of $f(x) = x^3$ corresponds to exactly one input. The function is one-to-one.

Self Check 1 Determine whether each function is one-to-one. If it is not, find an output that corresponds to more than one input: **a.** $f(x) = 2x + 3$ and **b.** $f(x) = x^4$.

THE HORIZONTAL LINE TEST

It is often easier to examine the graph of a function, rather than its defining equation, to determine whether the function is one-to-one. If two (or more) points on the graph of a function have the same y-coordinate, the function is not one-to-one. This observation suggests the following *horizontal line test.*

The Horizontal Line Test A function is one-to-one if every horizontal line intersects the graph of the function at most once.

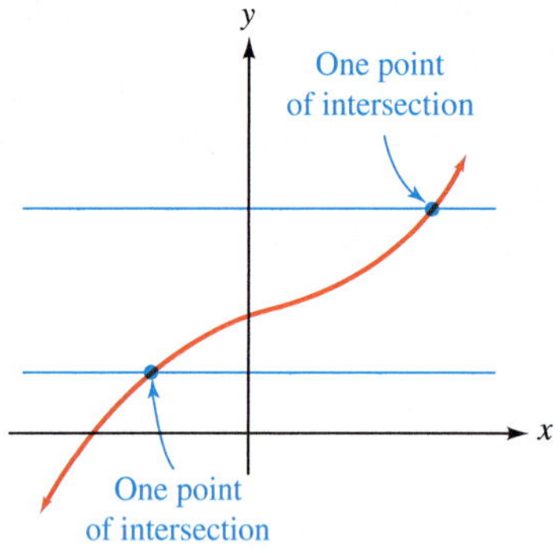

A one-to-one function

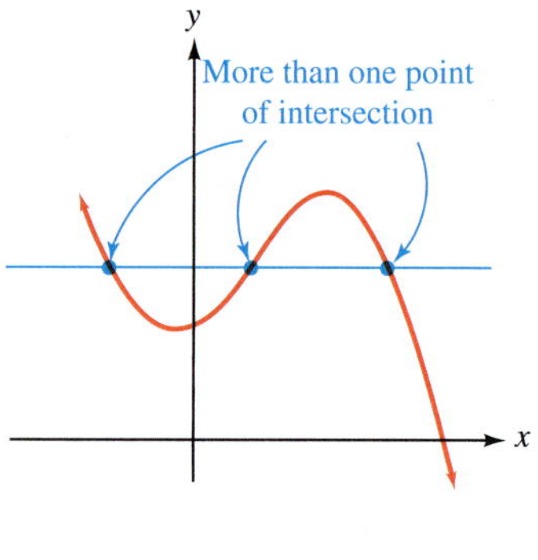

Not a one-to-one function

EXAMPLE 2

Use the horizontal line test to decide whether the following graphs represent one-to-one functions.

Solution

a. Because we can draw a horizontal line that intersects the graph shown in figure (a) twice, the graph does not represent a one-to-one function.

b. Because every horizontal line that intersects the graph in figure (b) does so exactly once, the graph represents a one-to-one function.

Success Tip

Recall that we use the *vertical line test* to determine whether a graph is the graph of a function. We use the *horizontal line test* to determine whether the function that is graphed is one-to-one.

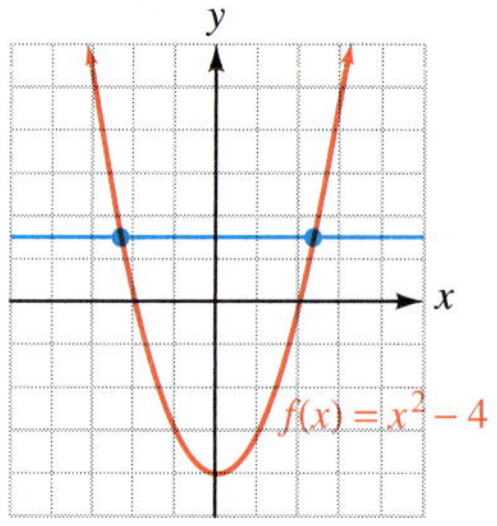

(a)

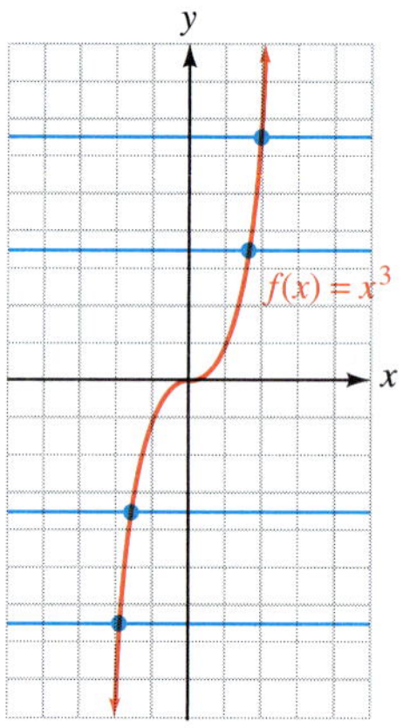

(b)

Self Check 2 Determine whether the following graphs represent one-to-one functions.

a.

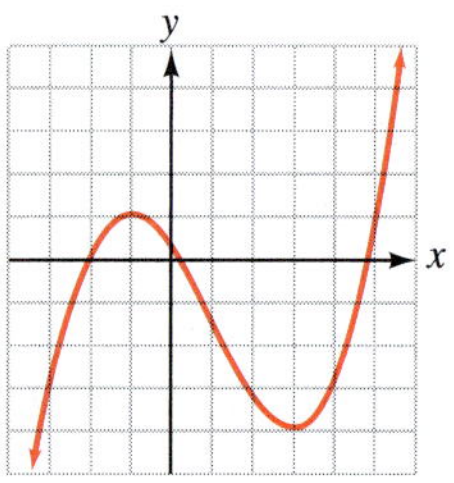

b.

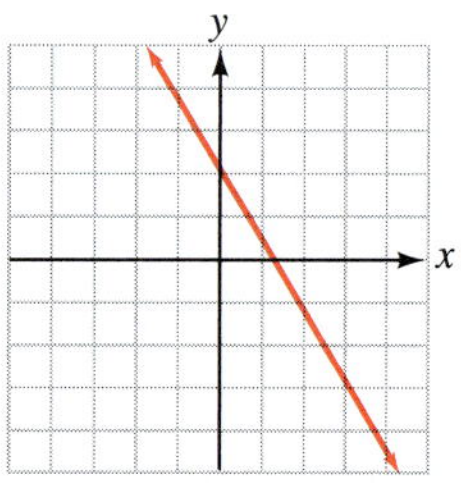

FINDING THE INVERSE OF A FUNCTION

If f is the one-to-one function defined by the arrow diagram in figure (a), it turns the number 1 into 10, 2 into 20, and 3 into 30. The ordered pairs that determine f can be listed in a table. Since the inverse of f must turn 10 back into 1, 20 back into 2, and 30 back into 3, it consists of the ordered pairs shown in the table in figure (b).

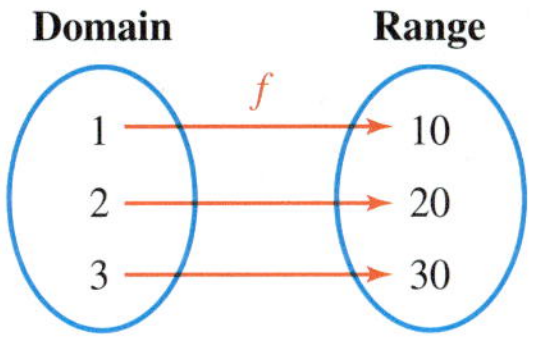

Function f

x	y	(x, y)
1	10	(1, 10)
2	20	(2, 20)
3	30	(3, 30)

(a)

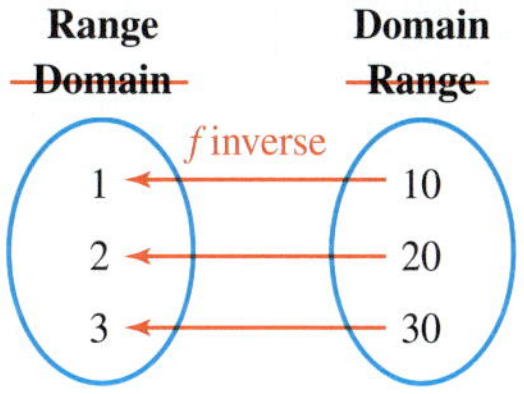

f inverse

x	y	(x, y)
10	1	(10, 1)
20	2	(20, 2)
30	3	(30, 3)

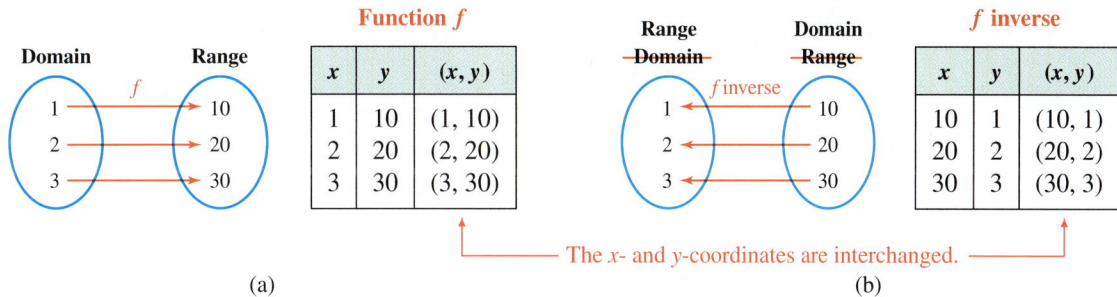

The x- and y-coordinates are interchanged.

(b)

Caution

The -1 in the notation $f^{-1}(x)$ is not an exponent:

$$f^{-1}(x) \neq \frac{1}{f(x)}$$

We note that the domain of f and the range of its inverse is $\{1, 2, 3\}$. The range of f and the domain of its inverse is $\{10, 20, 30\}$.

This example suggests that to form the inverse of a function f, we simply interchange the coordinates of each ordered pair that determines f. When the inverse of a function is also a function, we call it ***f* inverse** and denote it with the symbol f^{-1}. The symbol $f^{-1}(x)$ is read as "the inverse of $f(x)$" or "f inverse of x."

The Inverse of a Function

If f is a one-to-one function consisting of ordered pairs of the form (x, y), the **inverse** of f, denoted f^{-1}, is the one-to-one function consisting of all ordered pairs of the form (y, x).

When a one-to-one function is defined by an equation, we use the following method to find its inverse.

Finding the Inverse of a Function

If a function is one-to-one, we find its inverse as follows:

1. If the function is written using function notation, replace $f(x)$ with y.
2. Interchange the variables x and y.
3. Solve the resulting equation for y.
4. We can substitute $f^{-1}(x)$ for y.

EXAMPLE 3

Determine whether each function is one-to-one. If it is, find the equation of its inverse: **a.** $f(x) = 4x + 2$ and **b.** $f(x) = x^3$.

Solution

a. We recognize $f(x) = 4x + 2$ as a linear function whose graph is a straight line with slope 4 and y-intercept (0, 2). Since such a graph would pass the horizontal line test, we conclude that f is one-to-one.

To find the inverse, we proceed as follows:

Success Tip

Every linear function, except those of the form $f(x) = \text{constant}$, is one-to-one.

$$f(x) = 4x + 2$$

$$y = 4x + 2 \quad \text{Replace } f(x) \text{ with } y.$$

$$\mathbf{x} = 4\mathbf{y} + 2 \quad \text{Interchange the variables } x \text{ and } y.$$

$$x - 2 = 4y \quad \text{Subtract 2 from both sides.}$$

$$\frac{x - 2}{4} = y \quad \text{Divide both sides by 4.}$$

$$y = \frac{x - 2}{4} \quad \text{Write the equation with } y \text{ on the left-hand side.}$$

To denote that this equation is the inverse of function f, we replace y with $f^{-1}(x)$.

$$f^{-1}(x) = \frac{x - 2}{4}$$

b. In Example 2, we used the horizontal line test to determine that $f(x) = x^3$ is a one-to-one function.

To find the inverse, we proceed as follows:

Caution

Only one-to-one functions have inverse functions.

$$y = x^3 \quad \text{Replace } f(x) \text{ with } y.$$

$$\mathbf{x} = \mathbf{y}^3 \quad \text{Interchange the variables } x \text{ and } y.$$

$$\sqrt[3]{x} = y \quad \text{Take the cube root of both sides.}$$

$$y = \sqrt[3]{x}$$

Replacing y with $f^{-1}(x)$, we have

$$f^{-1}(x) = \sqrt[3]{x}$$

Self Check 3 Determine whether each function is one-to-one. If it is, find the equation of its inverse: **a.** $f(x) = -5x - 3$ and **b.** $f(x) = x^5$.

THE COMPOSITION OF A FUNCTION AND ITS INVERSE

To emphasize an important relationship between a function and its inverse, we substitute some number x, such as $x = 3$, into the function $f(x) = 4x + 2$ of Example 3. The corresponding value of y that is produced is

$$f(3) = 4(3) + 2 = 14 \quad f \text{ determines the ordered pair } (3, 14).$$

If we substitute 14 into the inverse function, $f^{-1}(x) = \frac{x-2}{4}$, the corresponding value of y that is produced is

$$f^{-1}(14) = \frac{14 - 2}{4} = 3 \quad f^{-1} \text{ determines the ordered pair } (14, 3).$$

Thus, the function f turns 3 into 14, and the inverse function f^{-1} turns 14 back into 3.

In general, the composition of a function and its inverse function is the identity function such that any input x is assigned the output x. This fact can be stated symbolically as follows.

The Composition of Inverse Functions For any one-to-one function f and its inverse, f^{-1},

$$(f \circ f^{-1})(x) = x \quad \text{and} \quad (f^{-1} \circ f)(x) = x$$

We can use this property to determine whether two functions are inverses.

EXAMPLE 4

Prove that $f(x) = 4x + 2$ and $f^{-1}(x) = \frac{x-2}{4}$ are inverses.

Solution To prove that $f(x) = 4x + 2$ and $f^{-1}(x) = \frac{x-2}{4}$ are inverses, we must show that for each composition, an input x is assigned an output of x.

$$\begin{aligned}(f \circ f^{-1})(x) &= f(f^{-1}(x)) \\ &= f\left(\frac{x-2}{4}\right) \\ &= 4\left(\frac{x-2}{4}\right) + 2 \\ &= x - 2 + 2 \\ &= x\end{aligned} \qquad \begin{aligned}(f^{-1} \circ f)(x) &= f^{-1}(f(x)) \\ &= f^{-1}(4x + 2) \\ &= \frac{4x + 2 - 2}{4} \\ &= \frac{4x}{4} \\ &= x\end{aligned}$$

Because $(f \circ f^{-1})(x) = x$ and $(f^{-1} \circ f)(x) = x$, the functions are inverses.

Self Check 4 Use composition to determine whether $f(x) = x - 4$ and $g(x) = x + 4$ are inverses.

GRAPHING A FUNCTION AND ITS INVERSE

Success Tip

Recall that the line $y = x$ passes through points whose x- and y-coordinates are equal: $(-1, -1)$, $(0, 0)$, $(1, 1)$, $(2, 2)$, and so on.

If a point (a, b) is on the graph of function f, it follows that the point (b, a) is on the graph of f^{-1}, and vice versa. There is a geometric relationship between a pair of points whose coordinates are interchanged. For example, in the graph, we see that the line segment between $(1, 3)$ and $(3, 1)$ is perpendicular to and cut in half by the line $y = x$. We say that $(1, 3)$ and $(3, 1)$ are mirror images of each other with respect to $y = x$.

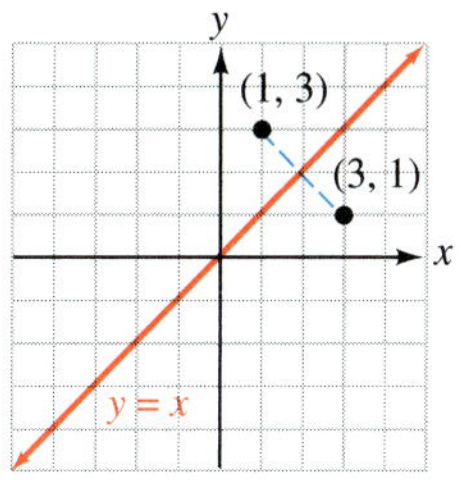

Since each point on the graph of f^{-1} is a mirror image of a point on the graph of f, and vice versa, the graphs of f and f^{-1} must be mirror images of each other with respect to $y = x$.

EXAMPLE 5

Find the inverse of $f(x) = -\frac{3}{2}x + 3$. Then graph f and its inverse on one coordinate system.

Solution Since $f(x) = -\frac{3}{2}x + 3$ is a linear function, it is one-to-one and has an inverse. To find the inverse function, we replace $f(x)$ with y, and interchange x and y to obtain

$$x = -\frac{3}{2}y + 3$$

Then we solve for y to get

$$x - 3 = -\frac{3}{2}y \qquad \text{Subtract 3 from both sides.}$$

$$-\frac{2}{3}x + 2 = y \qquad \text{Multiply both sides by } -\frac{2}{3}.$$

Success Tip

To graph f^{-1}, we don't need to construct a table of values. We can simply interchange the coordinates of the ordered pairs in the table for f and use them to graph f^{-1}.

When we replace y with $f^{-1}(x)$, we have $f^{-1}(x) = -\frac{2}{3}x + 2$.

To graph f and f^{-1}, we construct tables of values and plot points. Because the functions are inverses of each other, their graphs are mirror images about the line $y = x$.

$f(x) = -\frac{3}{2}x + 3$

x	$f(x)$	
0	3	→ (0, 3)
2	0	→ (2, 0)
4	−3	→ (4, −3)

$f^{-1}(x) = -\frac{2}{3}x + 2$

x	$f^{-1}(x)$	
3	0	→ (3, 0)
0	2	→ (0, 2)
−3	4	→ (−3, 4)

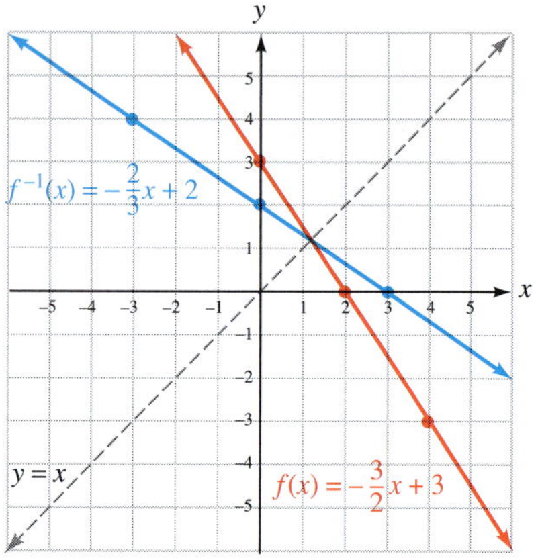

Self Check 5 Find the inverse of $f(x) = \frac{2}{3}x - 2$. Then graph the function and its inverse on one coordinate system.

ACCENT ON TECHNOLOGY: GRAPHING THE INVERSE OF A FUNCTION

We can use a graphing calculator to check the result found in Example 5. First, we enter $f(x) = -\frac{3}{2}x + 3$. Then we enter what we believe to be the inverse function, $f^{-1}(x) = -\frac{2}{3}x + 2$, as well as the equation $y = x$. See figure (a). Before graphing, we adjust the display so that the graphing grid will be composed of squares. The axis of symmetry is then at a 45° angle to the positive x-axis.

In figure (b), it appears that the two graphs are symmetric about the line $y = x$. Although it is not definitive, this visual check does help to validate the result of Example 5.

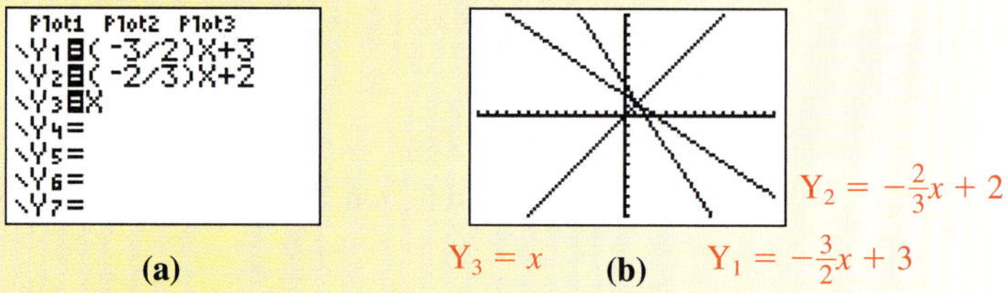

(a) (b)

EXAMPLE 6

Graph the inverse of function f shown in figure (a).

Solution To graph the inverse, we determine the coordinates of several points on the graph of f, interchange their coordinates, and plot them, as shown in figure (b). Then we draw a smooth curve with those points to get the graph of f^{-1}. We also graph the line $y = x$ to emphasize the symmetry.

The Language of Algebra

We can also say that the graphs of f and f^{-1} are *reflections* of each other about the line $y = x$, or they are *symmetric* about $y = x$.

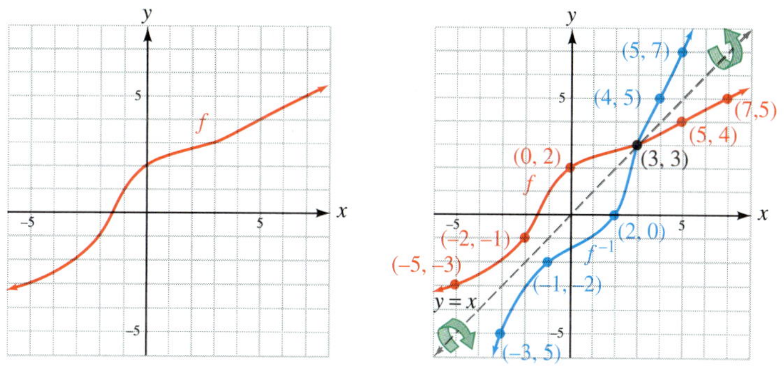

(a) (b)

Answers to Self Checks **1. a.** yes, **b.** no, (−1, 1), (1, 1) **2. a.** no, **b.** yes **3. a.** $f^{-1}(x) = \dfrac{-x-3}{5}$, **b.** $f^{-1}(x) = \sqrt[5]{x}$ **4.** They are inverses.

5.

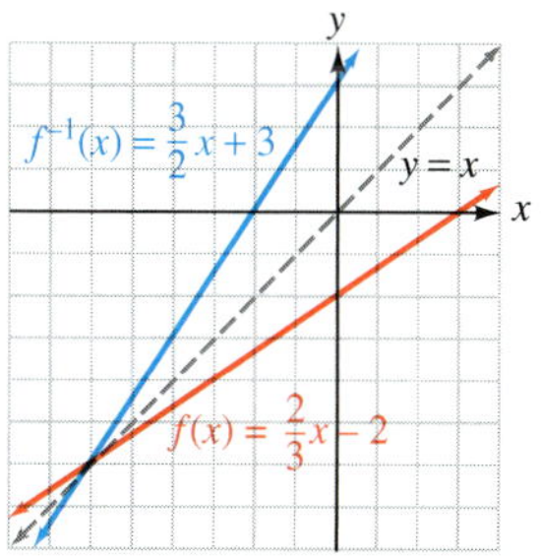

9.2 STUDY SET

VOCABULARY **Fill in the blanks.**

1. For a __________ function, each input is assigned exactly one output, and each output corresponds to exactly one input.

2. The __________ line test can be used to decide whether the graph of a function represents a one-to-one function.

3. The functions f and f^{-1} are ________.

4. When we __________ the coordinates of the point (2, 5) we get the point (5, 2).

5. The graphs of a function and its inverse are mirror ________ of each other with respect to $y = x$. We also say that their graphs are __________ with respect to the line $y = x$.

6. $(f \circ f^{-1})(x)$ is the __________ of a function f and its inverse f^{-1}.

CONCEPTS **Fill in the blanks.**

7. a. If every horizontal line that intersects the graph of a function does so only ______, the function is one-to-one.

b. If any horizontal line that intersects the graph of a function does so more than once, the function is not __________.

8. If a function turns an input of 2 into an output of 5, the inverse function will turn an input of 5 into an output of ____.

9. The graphs of a function and its inverse are symmetrical about the line $y =$ ____.

10. $(f \circ f^{-1})(x) =$ ____ and $(f^{-1} \circ f)(x) =$ ____.

11. To find the inverse of the function $f(x) = 2x - 3$, we begin by replacing $f(x)$ with ____, and then we ____________ x and y.

12. If f is a one-to-one function, the domain of f is the ________ f^{-1}, and the range of f is the ________ f^{-1}.

13. Is the correspondence defined by the following arrow diagram a function? If it is, is the function one-to-one?

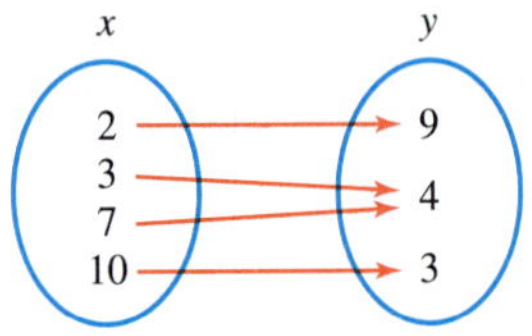

14. How can we tell that function f is not one-to-one from the table of values?

x	$f(x)$
-2	4
-1	1
0	0
2	4
3	9

15. Is the inverse of a one-to-one function always a function?

16. Name four points that the line $y = x$ passes through.

17. If f is a one-to-one function, and if $f(2) = 6$, then what is $f^{-1}(6)$?

18. If the point $(2, -4)$ is on the graph of the one-to-one function f, then what point is on the graph of f^{-1}?

19. Two functions are graphed on the following square grid along with the line $y = x$. Explain why we know that the functions are not inverses of each other.

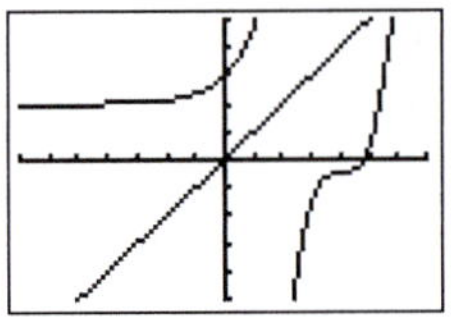

20. Use the table of values of the one-to-one function f to complete a table of values for f^{-1}.

x	$f(x)$
-6	-3
-4	-2
0	0
2	1
8	4

x	$f^{-1}(x)$
-3	
-2	
0	
1	
4	

21. Redraw the graph of function f. Then graph f^{-1} and the axis of symmetry on the same coordinate system.

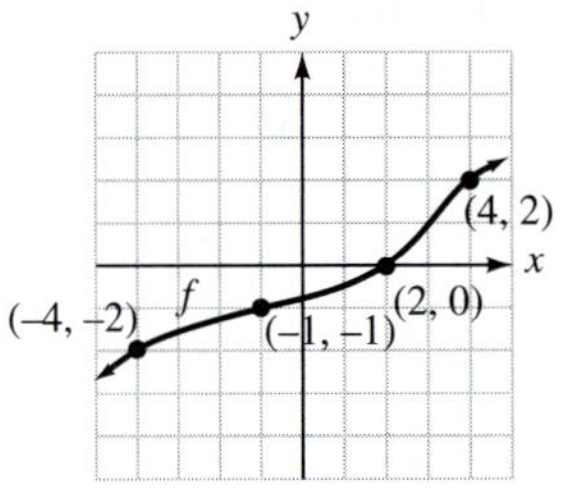

22. A table of values for a function f is shown in figure (a). A table of values for f^{-1} is shown in figure (b). Use the tables to find $f^{-1}(f(4))$ and $f(f^{-1}(2))$.

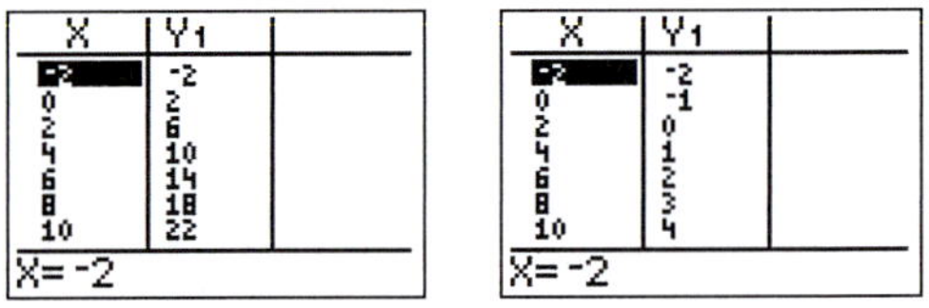

X	Y1
-2	-2
0	2
2	6
4	10
6	14
8	18
10	22

X=-2

(a)

X	Y1
-2	-2
0	-1
2	0
4	1
6	2
8	3
10	4

X=-2

(b)

NOTATION Complete each solution.

23. Find the inverse of $f(x) = 2x - 3$.

$$\square = 2x - 3$$
$$x = \square - 3$$
$$x + \square = 2y$$
$$\frac{x+3}{2} = \square$$

The inverse of $f(x) = 2x - 3$ is $\square = \dfrac{x+3}{2}$.

24. Find the inverse of $f(x) = \sqrt[3]{x} + 2$.

$$\square = \sqrt[3]{x} + 2$$
$$x = \sqrt[3]{\square} + 2$$
$$x - \square = \sqrt[3]{y}$$
$$(x-2)^3 = \square$$

The inverse of $f(x) = \sqrt[3]{x} + 2$ is $\square = (x - 2)^3$.

25. The symbol f^{-1} is read as "the ________ f" or "f ________."

26. Explain the difference in the meaning of the -1 in the notation $f^{-1}(x)$ as compared to x^{-1}.

PRACTICE **Determine whether each function is one-to-one.**

27. $f(x) = 2x$

28. $f(x) = |x|$

29. $f(x) = x^4$

30. $f(x) = x^3 + 1$

31. $f(x) = -x^2 + 3x$

32. $f(x) = \frac{2}{3}x + 8$

33. $\{(1, 1), (2, 1), (3, 1), (4, 1)\}$

34. $\{(3, 2), (2, 1), (1,0)\}$

Each graph represents a function. Use the horizontal line test to decide whether the function is one-to-one.

35.

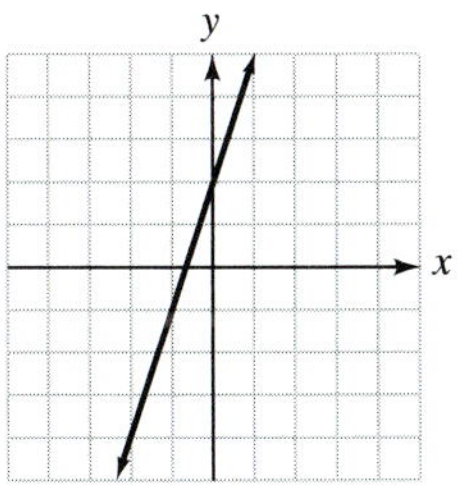

36.

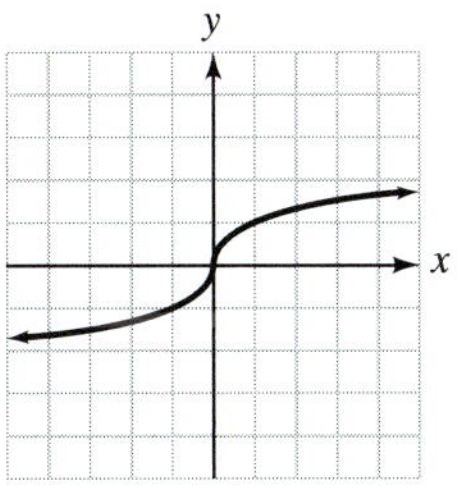

37.

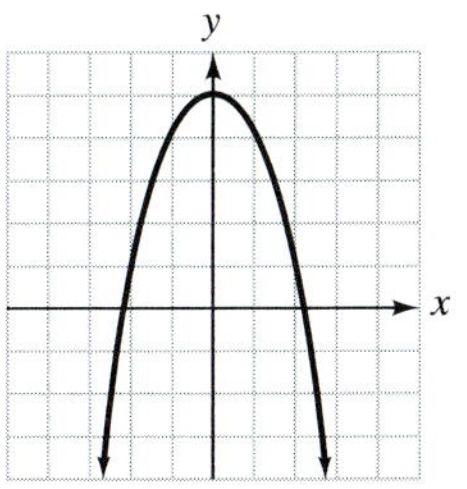

38.

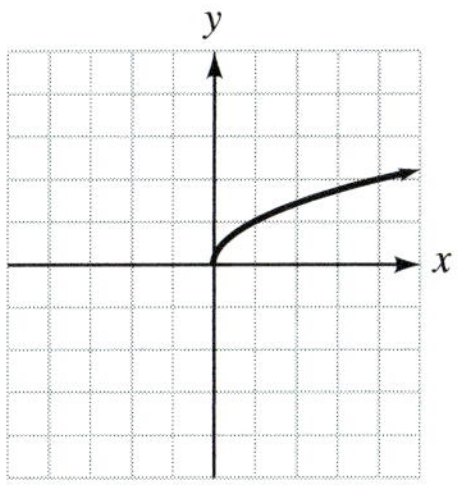

39.

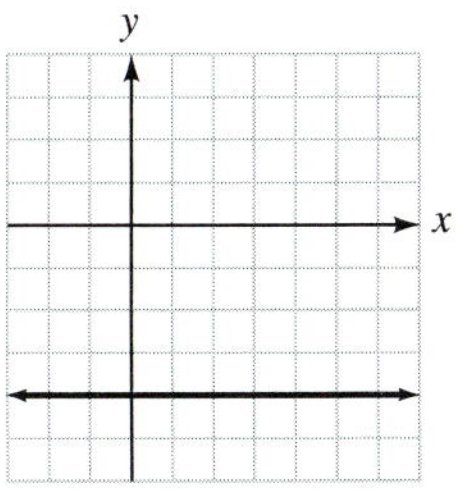

40.

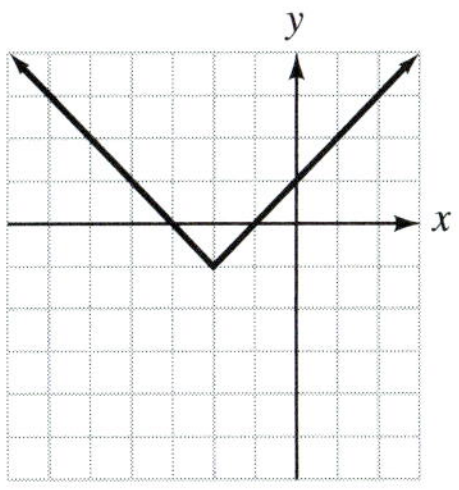

Find the inverse of the function and express it using $f^{-1}(x)$ notation.

41. $f(x) = 2x + 4$

42. $f(x) = 5x - 1$

43. $f(x) = \frac{x}{5} + \frac{4}{5}$

44. $f(x) = \frac{x}{3} - \frac{1}{3}$

45. $f(x) = \frac{x - 4}{5}$

46. $f(x) = \frac{2x + 6}{3}$

47. $f(x) = \frac{2}{x - 3}$

48. $f(x) = \frac{3}{x + 1}$

49. $f(x) = \frac{4}{x}$

50. $f(x) = \frac{1}{x}$

51. $f(x) = x^3 + 8$

52. $f(x) = x^3 - 4$

53. $f(x) = \sqrt[3]{x}$

54. $f(x) = \sqrt[3]{x - 5}$

55. $f(x) = (x + 10)^3$

56. $f(x) = (x - 9)^3$

57. $f(x) = 2x^3 - 3$

58. $f(x) = \frac{3}{x^3} - 1$

Use composition to show that each pair of functions are inverses. (See Example 4.)

59. $f(x) = 2x + 9$, $f^{-1}(x) = \frac{x - 9}{2}$

60. $f(x) = 5x - 1$, $f^{-1}(x) = \frac{x + 1}{5}$

61. $f(x) = \frac{2}{x - 3}$, $f^{-1}(x) = \frac{2}{x} + 3$

62. $f(x) = \sqrt[3]{x - 6}$, $f^{-1}(x) = x^3 + 6$

Find the inverse of each function. Then graph the function and its inverse on one coordinate system. Show the line of symmetry on the graph.

63. $f(x) = 2x$

64. $f(x) = -3x$

65. $f(x) = 4x + 3$

66. $f(x) = \frac{x}{3} + \frac{1}{3}$

67. $f(x) = -\frac{2}{3}x + 3$

68. $f(x) = -\frac{1}{3}x + \frac{4}{3}$

69. $f(x) = x^3$

70. $f(x) = x^3 + 1$

71. $f(x) = x^2 - 1 \quad (x \geq 0)$

72. $f(x) = x^2 + 1 \quad (x \geq 0)$

APPLICATIONS

73. INTERPERSONAL RELATIONSHIPS Feelings of anxiety in a relationship can increase or decrease, depending on what is going on in the relationship. The graph shows how a person's anxiety might vary as a relationship develops over time.

a. Is this the graph of a function? Is its inverse a function?

b. Does each anxiety level correspond to exactly one point in time? Use the dashed lined labeled *Maximum threshold* to explain.

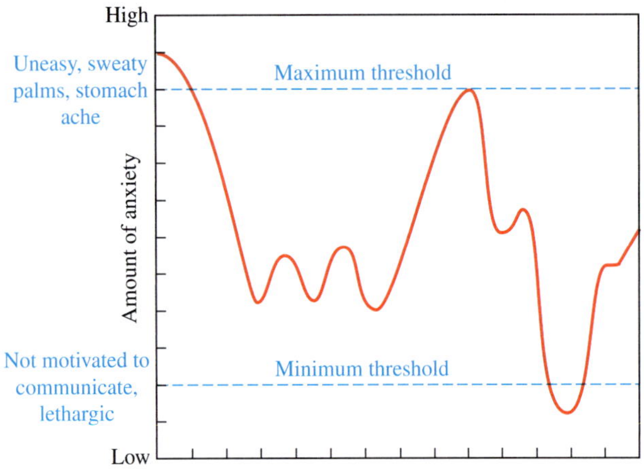

Source: Gudykunst, Building Bridges: Interpersonal Skills for a Changing World (Houghton Mifflin, 1994)

74. LIGHTING LEVELS The ability of the eye to see detail increases as the level of illumination increases. This relationship can be modeled by a function E, whose graph is shown here.

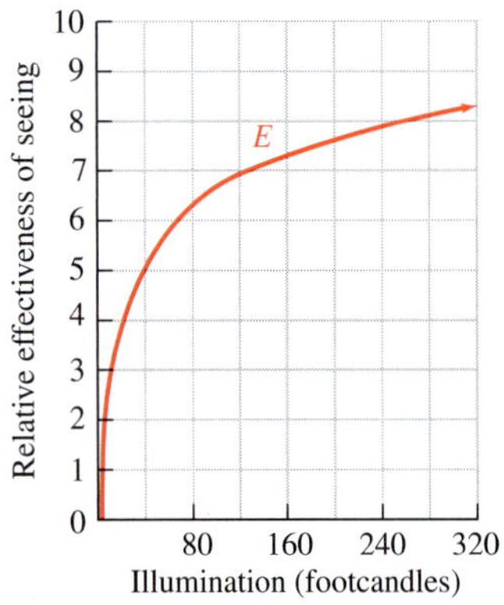

a. From the graph, determine $E(240)$.

b. Is function E one-to-one? Does E have an inverse?

c. If the effectiveness of seeing in an office is 7, what is the illumination in the office? How can this question be asked using inverse function notation?

WRITING

75. In your own words, what is a one-to-one function?

76. Explain the purpose of the horizontal line test.

77. In the illustration, a function f and its inverse f^{-1} have been graphed on the same coordinate system. Explain what concept can be demonstrated by folding the graph paper on the dashed line.

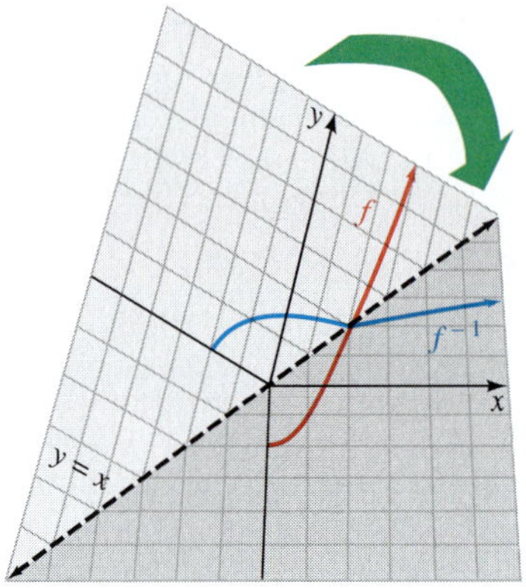

78. Write in words how to read the notation.

a. $f^{-1}(x) = \frac{1}{2}x - 3$

b. $(f \circ f^{-1})(x) = x$

REVIEW **Simplify. Write the result in $a + bi$ form.**

79. $3 - \sqrt{-64}$

80. $(2 - 3i) + (4 + 5i)$

81. $(3 + 4i)(2 - 3i)$

82. $\dfrac{6 + 7i}{3 - 4i}$

83. $(6 - 8i)^2$

84. i^{100}

CHALLENGE PROBLEMS

85. Find the inverse of $f(x) = \dfrac{x + 1}{x - 1}$.

86. Using the functions of Exercise 85, show that $(f \circ f^{-1})(x) = x$ and $(f^{-1} \circ f)(x) = x$.

9.3 Exponential Functions

- Irrational exponents
- Exponential functions
- Graphing exponential functions
- Vertical and horizontal translations
- Compound interest
- Exponential functions as models

The graph below shows the balance in a bank account in which \$10,000 was invested in 1998 at 9%, compounded monthly. The graph shows that in the year 2008, the value of the account will be approximately \$25,000, and in the year 2028, the value will be approximately \$147,000. The red curve is the graph of a function called an *exponential function.*

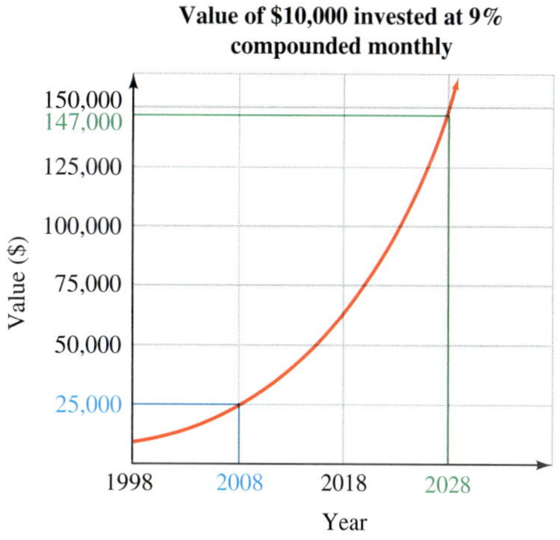

Exponential functions are also used to model many other situations, such as population growth, the spread of an epidemic, the temperature of a heated object as it cools, and radioactive decay. Before we can discuss exponential functions in more detail, we must define irrational exponents.

IRRATIONAL EXPONENTS

We have discussed expressions of the form b^x, where x is a rational number.

$8^{1/2}$ means "the square root of 8."

$5^{1/3}$ means "the cube root of 5."

$3^{-2/5} = \dfrac{1}{3^{2/5}}$ means "the reciprocal of the fifth root of 3^2."

To give meaning to b^x when x is an irrational number, we consider the expression

$5^{\sqrt{2}}$ where $\sqrt{2}$ is the irrational number 1.414213562 . . .

Each number in the following list is defined, because each exponent is a rational number.

$$5^{1.4}, \quad 5^{1.41}, \quad 5^{1.414}, \quad 5^{1.4142}, \quad 5^{1.41421}, \quad \ldots$$

Since the exponents are getting closer to $\sqrt{2}$, the numbers in this list are successively better approximations of $5^{\sqrt{2}}$. We can use a calculator to obtain a very good approximation.

ACCENT ON TECHNOLOGY: EVALUATING EXPONENTIAL EXPRESSIONS

To find the value of $5^{\sqrt{2}}$ with a scientific calculator, we enter these numbers and press these keys:

5 [y^x] 2 [$\sqrt{\ }$] [=] 9.738517742

With a graphing calculator, we enter these numbers and press these keys:

5 [^] [$\sqrt{\ }$] 2 [)] [Enter]

5^√(2)
9.738517742

If $b > 0$ and x is a real number, b^x represents a positive number. It can be shown that all of the familiar rules of exponents are also true for irrational exponents.

EXAMPLE 1

Use the rules of exponents to simplify **a.** $\left(5^{\sqrt{2}}\right)^{\sqrt{2}}$ and **b.** $b^{\sqrt{3}} \cdot b^{\sqrt{12}}$.

Solution

a. $\left(5^{\sqrt{2}}\right)^{\sqrt{2}} = 5^{\sqrt{2}\sqrt{2}}$ Keep the base and multiply the exponents.

$= 5^2$ Multiply: $\sqrt{2}\sqrt{2} = \sqrt{4} = 2$.

$= 25$

b. $b^{\sqrt{3}} \cdot b^{\sqrt{12}} = b^{\sqrt{3}+\sqrt{12}}$ Keep the base and add the exponents.

$= b^{\sqrt{3}+2\sqrt{3}}$ Simplify: $\sqrt{12} = \sqrt{4}\sqrt{3} = 2\sqrt{3}$.

$= b^{3\sqrt{3}}$ Combine like radicals. $\sqrt{3} + 2\sqrt{3} = 3\sqrt{3}$.

Self Check 1 Simplify: **a.** $\left(3^{\sqrt{2}}\right)^{\sqrt{8}}$ and **b.** $b^{\sqrt{2}} \cdot b^{\sqrt{18}}$.

EXPONENTIAL FUNCTIONS

If $b > 0$ and $b \neq 1$, the function $f(x) = b^x$ is called an **exponential function.** Since x can be any real number, its domain is the set of real numbers. This is the interval $(-\infty, \infty)$.

Because b is positive, the value of $f(x)$ is positive, and the range is the set of positive numbers. This is the interval $(0, \infty)$.

Since $b \neq 1$, an exponential function *cannot* be the constant function $f(x) = 1^x$, in which $f(x) = 1$ for every real number x.

Exponential Functions An **exponential function with base *b*** is defined by the equation

$$f(x) = b^x \quad \text{or} \quad y = b^x \quad \text{where } b > 0,\ b \neq 1, \text{ and } x \text{ is a real number}$$

The domain of $f(x) = b^x$ is the interval $(-\infty, \infty)$, and the range is the interval $(0, \infty)$.

GRAPHING EXPONENTIAL FUNCTIONS

Since the domain and range of $f(x) = b^x$ are sets of real numbers, we can graph exponential functions on a rectangular coordinate system.

EXAMPLE 2

Graph: $f(x) = 2^x$.

Solution To graph $f(x) = 2^x$, we construct a table of function values by choosing several values for x and finding the corresponding values of $f(x)$. If x is -1, we have

$$f(x) = 2^x$$
$$f(-1) = 2^{-1} \quad \text{Substitute } -1 \text{ for } x.$$
$$= \frac{1}{2}$$

Notation

We have previously graphed the linear function $f(x) = 2x$ and the squaring function $f(x) = x^2$. For the exponential function $f(x) = 2^x$, note that the variable is in the exponent.

The point $\left(-1, \frac{1}{2}\right)$ is on the graph of $f(x) = 2^x$. In a similar manner, we find the corresponding values of $f(x)$ for x values of 0, 1, 2, 3, and 4 and record them in a table. Then we plot the ordered pairs and draw a smooth curve through them.

$f(x) = 2^x$

x	$f(x)$	
-1	$\frac{1}{2}$	$\to (-1, \frac{1}{2})$
0	1	$\to (0, 1)$
1	2	$\to (1, 2)$
2	4	$\to (2, 4)$
3	8	$\to (3, 8)$
4	16	$\to (4, 16)$

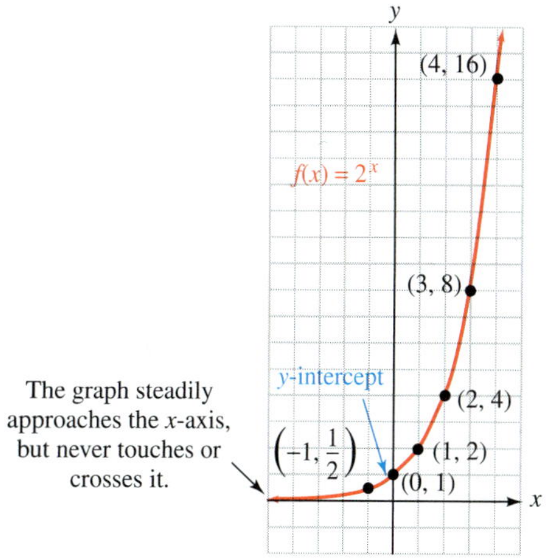

The Language of Algebra

We encountered the word *asymptote* earlier, when we graphed rational functions. Recall that an asymptote is not part of the graph. It is a line that the graph approaches and, in this case, never touches.

From the graph, we can verify that the domain of $f(x) = 2^x$ is the interval $(-\infty, \infty)$ and the range is the interval $(0, \infty)$. Since the graph passes the horizontal line test, the function is one-to-one.

Note that as x decreases, the values of $f(x)$ decrease and approach 0. Thus, the x-axis is a horizontal asymptote of the graph. The graph does not have an x-intercept, the y-intercept is $(0, 1)$, and the graph passes through the point $(1, 2)$.

Self Check 2 Graph: $f(x) = 4^x$.

EXAMPLE 3

Graph: $f(x) = \left(\frac{1}{2}\right)^x$.

Solution We make a table of values for the function. For example, if $x = -4$, we have

$$\begin{aligned} f(x) &= \left(\frac{1}{2}\right)^x \\ f(-4) &= \left(\frac{1}{2}\right)^{-4} \\ &= \left(\frac{2}{1}\right)^4 \quad \text{Recall: } \left(\frac{x}{y}\right)^{-n} = \left(\frac{y}{x}\right)^n. \\ &= 16 \end{aligned}$$

The point $(-4, 16)$ is on the graph of $f(x) = \left(\frac{1}{2}\right)^x$. In a similar manner, we find the corresponding values of $f(x)$ for other x-values and record them in a table.

$f(x) = \left(\frac{1}{2}\right)^x$

x	$f(x)$	
−4	16	→ (−4, 16)
−3	8	→ (−3, 8)
−2	4	→ (−2, 4)
−1	2	→ (−1, 2)
0	1	→ (0, 1)
1	$\frac{1}{2}$	→ $\left(1, \frac{1}{2}\right)$

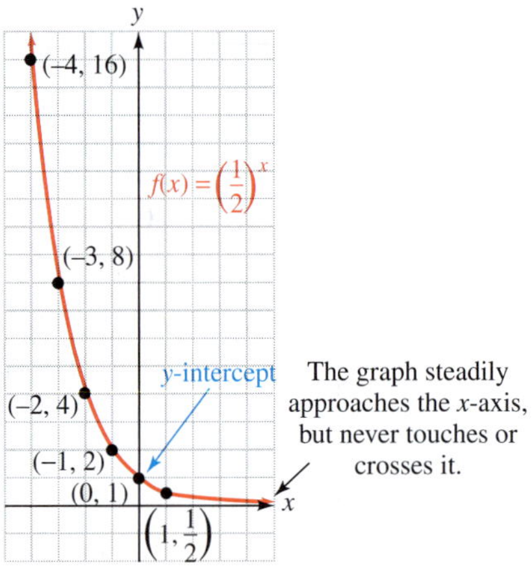

From the graph, we can verify that the domain of $f(x) = \left(\frac{1}{2}\right)^x$ is the interval $(-\infty, \infty)$ and the range is the interval $(0, \infty)$. Since the graph passes the horizontal line test, the function is one-to-one.

Note that as x increases, the values of $f(x)$ decrease and approach 0. Thus, the x-axis is a horizontal asymptote of the graph. The graph does not have an x-intercept, the y-intercept is $(0, 1)$, and the graph passes through the point $\left(1, \frac{1}{2}\right)$.

Self Check 3 Graph: $g(x) = \left(\frac{1}{3}\right)^x$.

Examples 2 and 3 illustrate the following properties of exponential functions.

Properties of Exponential Functions

The domain of the exponential function $f(x) = b^x$ is the interval $(-\infty, \infty)$.

The range is the interval $(0, \infty)$.

The graph has a y-intercept of $(0, 1)$.

The x-axis is an asymptote of the graph.

The graph of $f(x) = b^x$ passes through the point $(1, b)$.

In Example 2 (where $b = 2$), the values of y increase as the values of x increase. Since the graph rises as we move to the right, we call the function an *increasing function.* When $b > 1$, the larger the value of b, the steeper the curve.

In Example 3 $\left(\text{where } b = \frac{1}{2}\right)$, the values of y decrease as the values of x increase. Since the graph drops as we move to the right, we call the function a *decreasing function.* When $0 < b < 1$, the smaller the value of b, the steeper the curve.

In general, the following is true.

Increasing and Decreasing Functions

If $b > 1$, then $f(x) = b^x$ is an **increasing function.**

If $0 < b < 1$, then $f(x) = b^x$ is a **decreasing function.**

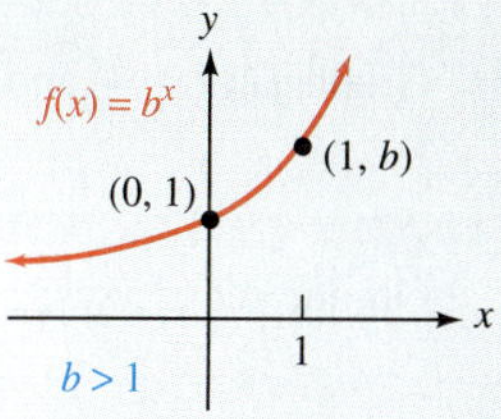

Increasing function

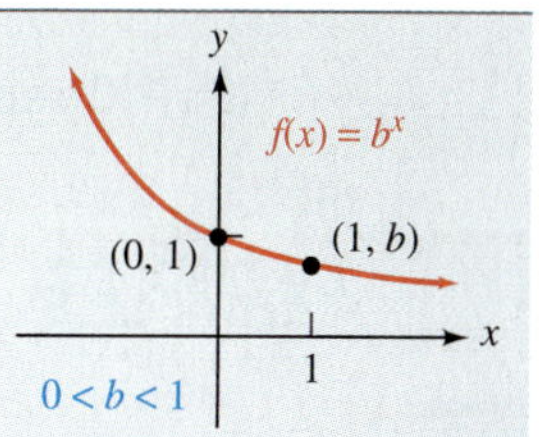

Decreasing function

An exponential function with base b is either increasing (for $b > 1$) or decreasing ($0 < b < 1$). Since different real numbers x determine different values of b^x, exponential functions are one-to-one.

ACCENT ON TECHNOLOGY: GRAPHING EXPONENTIAL FUNCTIONS

To use a graphing calculator to graph $f(x) = \left(\frac{2}{3}\right)^x$ and $g(x) = \left(\frac{3}{2}\right)^x$, we enter the right-hand sides of the equations after the symbols $Y_1 =$ and $Y_2 =$. The screen will show the following equations.

$Y_1 = (2/3) \wedge X$

$Y_2 = (3/2) \wedge X$

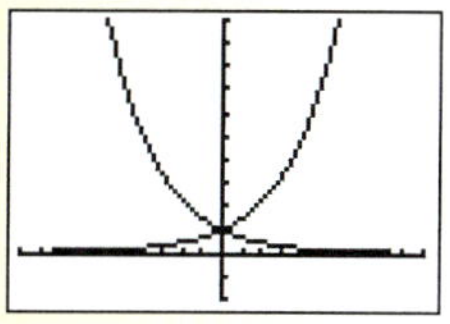

If we press the [GRAPH] key, we will obtain the display shown.

We note that the graph of $f(x) = \left(\frac{2}{3}\right)^x$ passes through the points (0, 1) and $\left(1, \frac{2}{3}\right)$. Since $\frac{2}{3} < 1$, the function is decreasing. The graph of $g(x) = \left(\frac{3}{2}\right)^x$ passes through the points (0, 1) and $\left(1, \frac{3}{2}\right)$. Since $\frac{3}{2} > 1$, the function is increasing.

Since both graphs pass the horizontal line test, each function is one-to-one.

VERTICAL AND HORIZONTAL TRANSLATIONS

EXAMPLE 4 On one set of axes, graph $f(x) = 2^x$ and $g(x) = 2^x + 3$, and describe the translation.

Solution The graph of $g(x) = 2^x + 3$ is identical to the graph of $f(x) = 2^x$, except that it is translated 3 units upward.

$f(x) = 2^x$

x	$f(x)$	
-4	$\frac{1}{16}$	$\to \left(-4, \frac{1}{16}\right)$
0	1	$\to (0, 1)$
2	4	$\to (2, 4)$

$g(x) = 2^x + 3$

x	$g(x)$	
-4	$3\frac{1}{16}$	$\to \left(-4, 3\frac{1}{16}\right)$
0	4	$\to (0, 4)$
2	7	$\to (2, 7)$

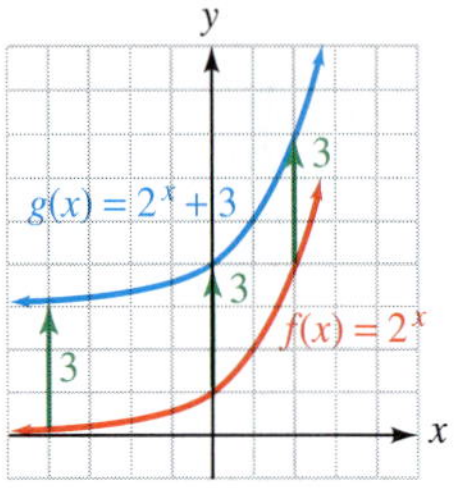

Self Check 4 Graph $f(x) = 4^x$ and $g(x) = 4^x - 3$, and describe the translation.

EXAMPLE 5

On one set of axes, graph $f(x) = 2^x$ and $g(x) = 2^{x+3}$, and describe the translation.

Solution The graph of $g(x) = 2^{x+3}$ is identical to the graph of $f(x) = 2^x$, except that it is translated 3 units to the left.

$f(x) = 2^x$

x	$f(x)$	
-1	$\frac{1}{2}$	$\to \left(-1, \frac{1}{2}\right)$
0	1	$\to (0, 1)$
1	2	$\to (1, 2)$

$g(x) = 2^{x+3}$

x	$g(x)$	
-1	4	$\to (-1, 4)$
0	8	$\to (0, 8)$
1	16	$\to (1, 16)$

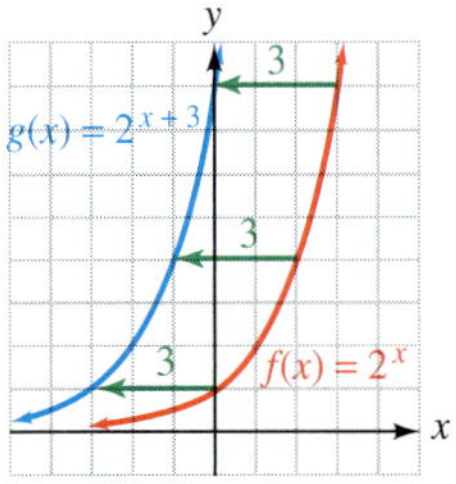

Self Check 5 On one set of axes, graph $f(x) = 4^x$ and $g(x) = 4^{x-3}$, and describe the translation.

The graphs of $f(x) = kb^x$ and $f(x) = b^{kx}$ are vertical and horizontal stretchings of the graph of $f(x) = b^x$. To graph these functions, we can plot several points and join them with a smooth curve, or we can use a graphing calculator.

ACCENT ON TECHNOLOGY: GRAPHING EXPONENTIAL FUNCTIONS

To use a graphing calculator to graph the exponential function $f(x) = 3(2^{x/3})$, we enter the right-hand side of the equation after the symbol $Y_1 =$. The display will show the equation

$$Y_1 = 3(2 \wedge (X/3))$$

If we press the graph key, we will obtain the display shown.

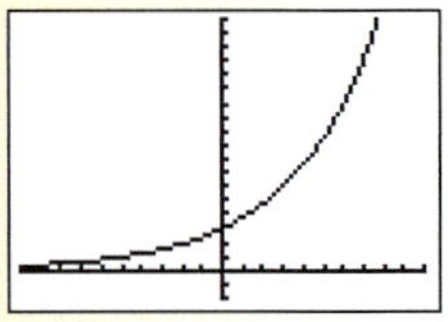

COMPOUND INTEREST

If we deposit $\$P$ in an account paying an annual interest rate r, we can find the amount A in the account at the end of t years by using the formula $A = P + Prt$, or $A = P(1 + rt)$.

Suppose that we deposit \$500 in such an account that pays interest every 6 months. Then $P = 500$, and after 6 months $\left(\frac{1}{2}\text{ year}\right)$, the amount in the account will be

$$A = 500(1 + rt)$$

$$= 500\left(1 + r \cdot \frac{1}{2}\right) \quad \text{Substitute } \frac{1}{2} \text{ for } t.$$

$$= 500\left(1 + \frac{r}{2}\right)$$

The account will begin the second 6-month period with a value of $\$500\left(1 + \frac{r}{2}\right)$. After the second 6-month period, the amount in the account will be

$$A = P(1 + rt)$$

$$A = \left[500\left(1 + \frac{r}{2}\right)\right]\left(1 + r \cdot \frac{1}{2}\right) \quad \text{Substitute } 500\left(1 + \frac{r}{2}\right) \text{ for } P \text{ and } \frac{1}{2} \text{ for } t.$$

$$= 500\left(1 + \frac{r}{2}\right)\left(1 + \frac{r}{2}\right)$$

$$= 500\left(1 + \frac{r}{2}\right)^2$$

At the end of a third 6-month period, the amount in the account will be

$$A = 500\left(1 + \frac{r}{2}\right)^3$$

In this discussion, the earned interest is deposited back in the account and also earns interest, and we say that the account is earning **compound interest.** The preceding discussion suggests the following formula for compound interest.

Formula for Compound Interest

If $\$P$ is deposited in an account and interest is paid k times a year at an annual rate r, the amount A in the account after t years is given by

$$A = P\left(1 + \frac{r}{k}\right)^{kt}$$

EXAMPLE 6

Saving for college. To save for college, parents invest \$12,000 for their newborn child in a mutual fund that should average a 10% annual return. If the quarterly dividends are reinvested, how much will be available in 18 years?

Solution We substitute 12,000 for P, 0.10 for r, and 18 for t in the formula for compound interest and find A. Since interest is paid quarterly, $k = 4$.

$$A = P\left(1 + \frac{r}{k}\right)^{kt}$$

$$A = 12{,}000\left(1 + \frac{0.10}{4}\right)^{4(18)} \quad \text{Express } r = 10\% \text{ as a decimal.}$$

$$= 12{,}000(1 + 0.025)^{72}$$

$$= 12{,}000(1.025)^{72}$$

$$= 71{,}006.74$$

Use a scientific calculator and press these keys: 1.025 y^x 72 = × 12,000 = .

In 18 years, the account will be worth \$71,006.74.

Self Check 6 How much would be available if the parents invested \$20,000?

In business applications, the initial amount of money deposited is often called the **present value** (PV). The amount to which the money will grow is called the **future value** (FV). The interest rate used for each compounding period is the **periodic interest rate** (i), and the number of times interest is compounded is the number of **compounding periods** (n). Using these definitions, we have an alternate formula for compound interest.

Formula for Compound Interest

$$FV = PV(1 + i)^n$$

This alternate formula appears on business calculators. To use this formula to solve Example 6, we proceed as follows:

$$FV = PV(1 + i)^n$$

$$FV = 12{,}000(1 + 0.025)^{72} \qquad i = \frac{0.10}{4} = 0.025 \text{ and } n = 4(18) = 72.$$

$$\approx 71{,}006.74 \qquad \text{Use a calculator to evaluate the expression.}$$

ACCENT ON TECHNOLOGY: SOLVING INVESTMENT PROBLEMS

Suppose \$1 is deposited in an account earning 6% annual interest, compounded monthly. To use a graphing calculator to estimate how much will be in the account in 100 years, we can substitute 1 for P, 0.06 for r, and 12 for k in the formula

$$A = P\left(1 + \frac{r}{k}\right)^{kt}$$

$$A = 1\left(1 + \frac{0.06}{12}\right)^{12t}$$

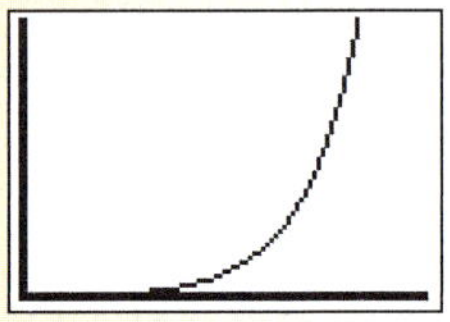

and simplify to get

$$A = (1.005)^{12t}$$

We now graph $A = (1.005)^{12t}$ using window settings of [0, 120] for t and [0, 400] for A to obtain the graph shown. We can then trace and zoom to estimate that \$1 grows to be approximately \$397 in 100 years. From the graph, we can see that the money grows slowly in the early years and rapidly in the later years.

EXPONENTIAL FUNCTIONS AS MODELS

EXAMPLE 7

Cellular phones. For the years 1990–1997, the U.S. cellular telephone industry experienced exponential growth. The exponential function $S(n) = 5.74(1.39)^n$ approximates the number of cellular telephone subscribers in millions, where n is the number of years since 1990 and $0 \le n \le 7$.

Use the function to answer the following.

a. How many subscribers were there in 1990?

b. How many subscribers were there in 1997?

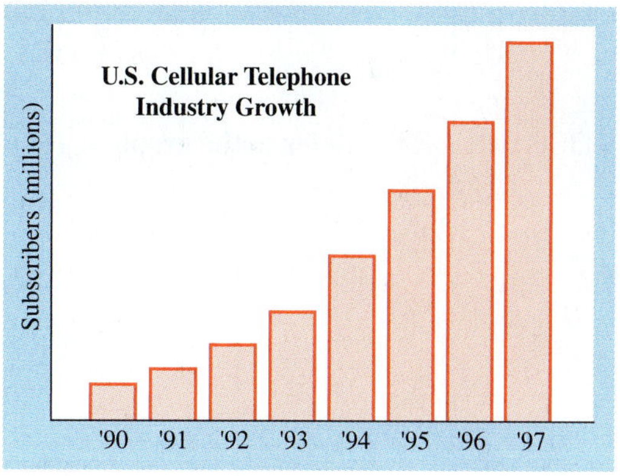

Source: *New York Times Almanac*, 1998

Solution

The Language of Algebra

The word *exponential* is used in many settings to describe rapid growth. For example, we hear that the processing power of computers is growing *exponentially.*

a. The year 1990 is 0 years after 1990. To find the number of subscribers in 1990, we substitute 0 for n in the function and find $S(0)$.

$$S(n) = 5.74(1.39)^n$$
$$S(0) = 5.74(1.39)^0$$
$$= 5.74 \cdot 1 \quad (1.39)^0 = 1.$$
$$= 5.74$$

In 1990, there were approximately 5.74 million cellular telephone subscribers.

b. The year 1997 is 7 years after 1990. We need to find $S(7)$.

$$S(n) = 5.74(1.39)^n$$
$$S(7) = 5.74(1.39)^7 \quad \text{Substitute 7 for } n.$$
$$\approx 57.54604675 \quad \text{Use a calculator to find an approximation.}$$

In 1997, there were approximately 57.55 million cellular telephone subscribers.

Answers to Self Checks

1. a. 81, **b.** $b^{4\sqrt{2}}$

2.

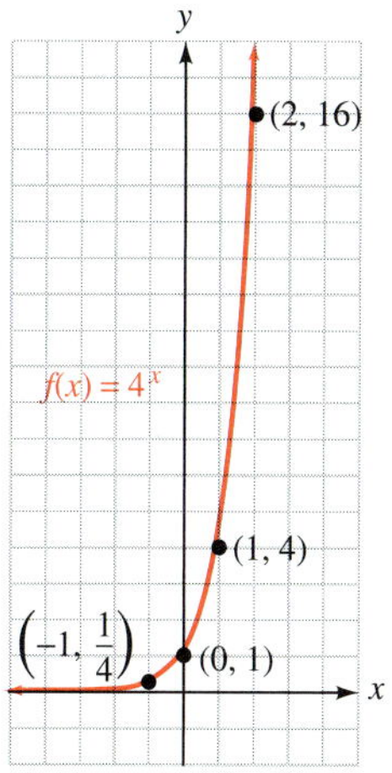

3.

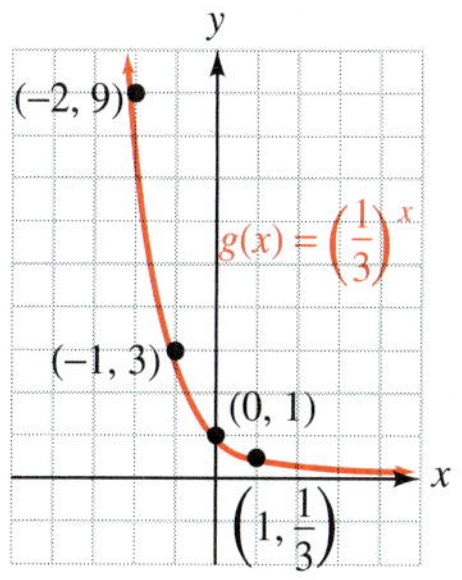

4. The graph of $f(x) = 4^x$ is translated 3 units downward.

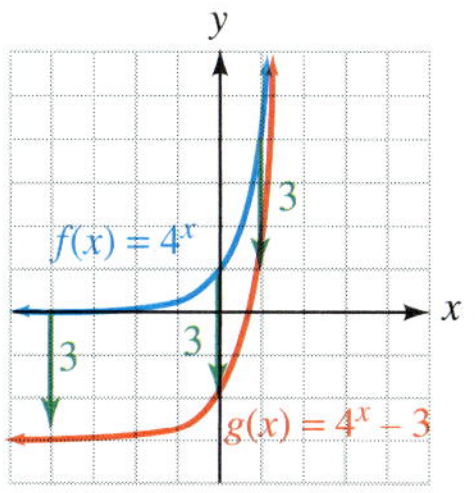

5. The graph of $f(x) = 4^x$ is translated 3 units to the right.

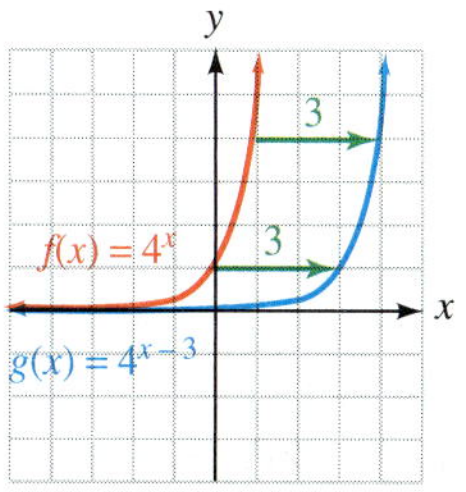

6. $118,344.56

9.3 STUDY SET

VOCABULARY

Refer to the graph of $f(x) = 3^x$.

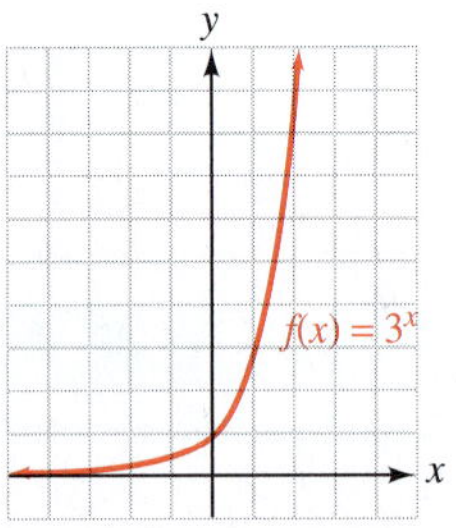

1. What type of function is $f(x) = 3^x$?
2. What is the domain of the function?
3. What is the range of the function?
4. **a.** What is the y-intercept of the graph?
   b. What is the x-intercept of the graph?
5. Is the function one-to-one?
6. What is an asymptote of the graph?
7. Is f an increasing or a decreasing function?
8. The graph passes through the point $(1, y)$. What is y?

CONCEPTS

9. Graph $f(x) = x^2$ and $g(x) = 2^x$ on the same set of coordinate axes.
10. Graph $y = x^{1/2}$ and $y = \left(\frac{1}{2}\right)^x$ on the same set of coordinate axes.
11. What are the two formulas that are used to determine the amount of money in a savings account that is earning compound interest?
12. Explain the order in which the expression $20{,}000(1.036)^{72}$ should be evaluated.
13. The illustration shows the graph of $f(x) = 2^x$, as well as two vertical translations of that graph. Using the notation $g(x)$ for one translation and $h(x)$ for the other, write the defining equation for each function.

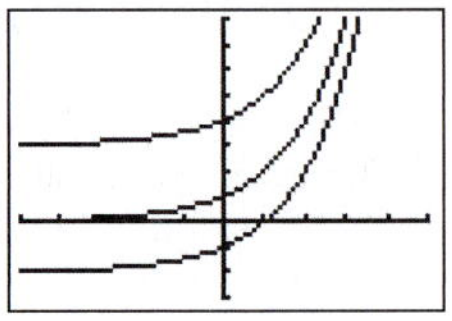

14. The illustration shows the graph of $f(x) = 2^x$, as well as a horizontal translation of that graph. Using the notation $g(x)$ for the translation, write its defining equation.

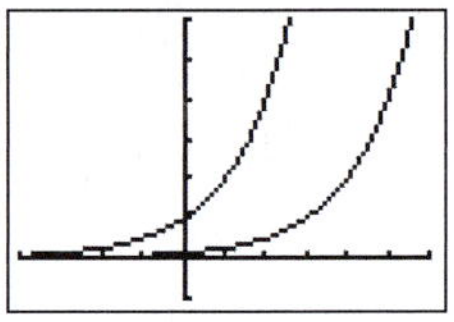

NOTATION

15. In $A = P\left(1 + \frac{r}{k}\right)^{kt}$, what is the base, and what is the exponent?
16. For an exponential function of the form $f(x) = b^x$, what are the restrictions on b?

PRACTICE

Find each value to four decimal places.

17. $2^{\sqrt{2}}$
18. $7^{\sqrt{2}}$
19. $5^{\sqrt{5}}$
20. $6^{\sqrt{3}}$

Simplify each expression.

21. $\left(2^{\sqrt{3}}\right)^{\sqrt{3}}$
22. $3^{\sqrt{2}}3^{\sqrt{18}}$
23. $7^{\sqrt{3}}7^{\sqrt{12}}$
24. $\left(3^{\sqrt{5}}\right)^{\sqrt{5}}$

Graph each function.

25. $f(x) = 5^x$
26. $f(x) = 6^x$
27. $y = \left(\dfrac{1}{4}\right)^x$
28. $y = \left(\dfrac{1}{5}\right)^x$
29. $f(x) = 3^x - 2$
30. $y = 2^x + 1$
31. $f(x) = 3^{x-1}$
32. $f(x) = 2^{x+1}$

Use a graphing calculator to graph each function. Determine whether the function is an increasing or a decreasing function.

33. $f(x) = \dfrac{1}{2}(3^{x/2})$
34. $f(x) = -3(2^{x/3})$
35. $y = 2(3^{-x/2})$
36. $y = -\dfrac{1}{4}(2^{-x/2})$

APPLICATIONS **In Exercises 37–42, assume that there are no deposits or withdrawals.**

37. COMPOUND INTEREST An initial deposit of \$10,000 earns 8% interest, compounded quarterly. How much will be in the account after 10 years?

38. COMPOUND INTEREST An initial deposit of \$10,000 earns 8% interest, compounded monthly. How much will be in the account after 10 years?

39. COMPARING INTEREST RATES How much more interest could \$1,000 earn in 5 years, compounded quarterly, if the annual interest rate were $5\frac{1}{2}\%$ instead of 5%?

40. COMPARING SAVINGS PLANS Which institution in the ads provides the better investment?

Fidelity Savings & Loan
Earn 5.25%
compounded monthly

Union Trust
Money Market Account
paying 5.35%
compounded annually

41. COMPOUND INTEREST If \$1 had been invested on July 4, 1776, at 5% interest, compounded annually, what would it be worth on July 4, 2076?

42. FREQUENCY OF COMPOUNDING \$10,000 is invested in each of two accounts, both paying 6% annual interest. In the first account, interest compounds quarterly, and in the second account, interest compounds daily. Find the difference between the accounts after 20 years.

43. WORLD POPULATION See the following graph.

a. Estimate when the world's population reached $\frac{1}{2}$ billion and when it reached 1 billion.

b. Estimate the world's population in the year 2000.

c. What type of function does it appear could be used to model the population growth?

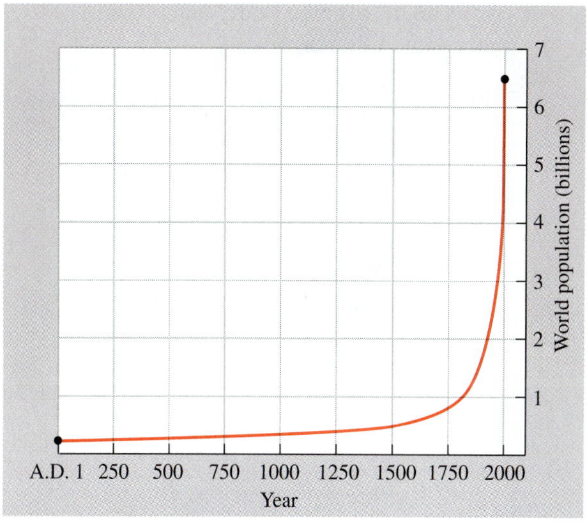

Source: *The Blue Planet* (Wiley, 1995)

44. THE STOCK MARKET The Dow Jones Industrial Average is a measure of how well the stock market is doing. Graph the following Dow milestones as ordered pairs of the form (year, average). What type of function could be used to model the growth of the stock market between 1906 and 1999?

Dow Jones Milestones			
Year	**Average**	**Year**	**Average**
Jan. 1906	100	Nov. 1995	5,000
Mar. 1956	500	Oct. 1996	6,000
Nov. 1972	1,000	Feb. 1997	7,000
Jan. 1987	2,000	July 1997	8,000
Apr. 1991	3,000	April 1998	9,000
Feb. 1995	4,000	March 1999	10,000
		May 1999	11,000

Source: finfacts.com

45. VALUE OF A CAR The graph on the next page shows how the value of the average car depreciates as a percent of its original value over a 10-year period. It also shows the yearly maintenance costs as a percent of the car's value.

a. When is the car worth half of its purchase price?

b. When is the car worth a quarter of its purchase price?

c. When do the average yearly maintenance costs surpass the value of the car?

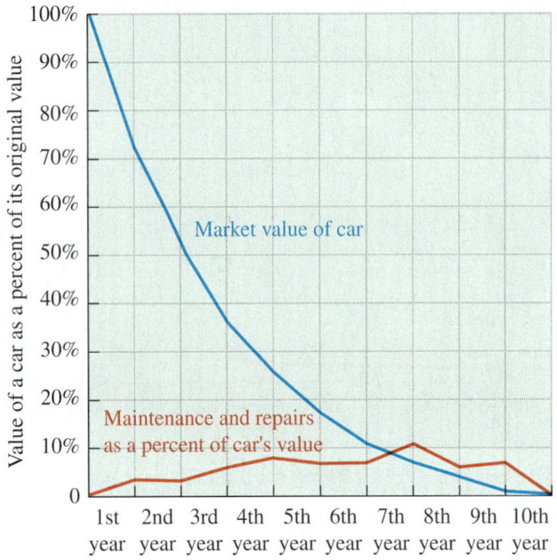

Source: U.S. Department of Transportation

46. DIVING *Bottom time* is the time a scuba diver spends descending plus the actual time spent at a certain depth. Graph the bottom time limits given in the table.

Bottom time limits			
Depth (ft)	**Bottom time (min)**	**Depth (ft)**	**Bottom time (min)**
30	no limit	80	40
35	310	90	30
40	200	100	25
50	100	110	20
60	60	120	15
70	50	130	10

47. BACTERIAL CULTURES A colony of 6 million bacteria is growing in a culture medium. The population P after t hours is given by the formula $P = (6 \times 10^6)(2.3)^t$. Find the population after 4 hours.

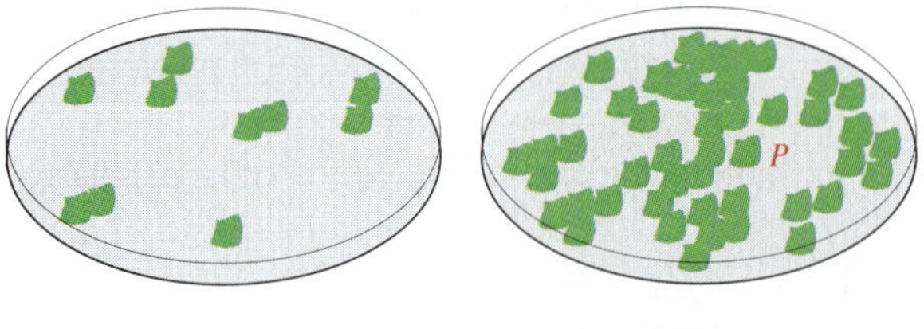

48. RADIOACTIVE DECAY A radioactive material decays according to the formula $A = A_0\left(\frac{2}{3}\right)^t$, where A_0 is the initial amount present and t is measured in years. Find the amount present in 5 years.

49. DISCHARGING A BATTERY The charge remaining in a battery decreases as the battery discharges. The charge C (in coulombs) after t days is given by the formula $C = (3 \times 10^{-4})(0.7)^t$. Find the charge after 5 days.

50. POPULATION GROWTH The population of North Rivers is decreasing exponentially according to the formula $P = 3{,}745(0.93)^t$, where t is measured in years from the present date. Find the population in 6 years, 9 months.

51. SALVAGE VALUES A small business purchased a computer for \$4,700. It is expected that its value each year will be 75% of its value in the preceding year. If the business disposes of the computer after 5 years, find its salvage value (the value after 5 years).

52. THE LOUISIANA PURCHASE In 1803, the United States negotiated the Louisiana Purchase with France. The country doubled its territory by adding 827,000 square miles of land for \$15 million. If the land appreciated at the rate of 6% each year, what would one square mile of land be worth in 2005?

WRITING

53. If world population is increasing exponentially, why is there cause for concern?

54. How do the graphs of $f(x) = 3^x$ and $g(x) = \left(\frac{1}{3}\right)^x$ differ? How are they similar?

55. A snowball rolling downhill grows *exponentially* with time. Explain what this means. Sketch a simple graph that models the situation.

56. Explain why the graph of $f(x) = 3^x$ gets closer and closer to the x-axis as the values of x decrease. Does the graph ever cross the x-axis? Explain why or why not.

REVIEW **In Exercises 57–60, refer to the illustration below in which lines *r* and *s* are parallel.**

57. Find x.

58. Find the measure of $\angle 1$.

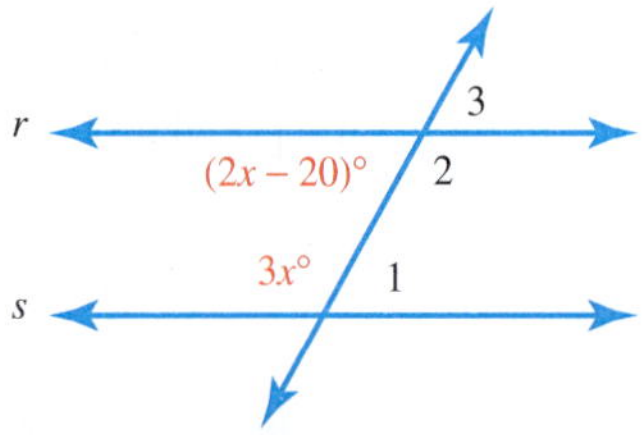

59. Find the measure of $\angle 2$.

60. Find the measure of $\angle 3$.

CHALLENGE PROBLEMS

61. In the definition of the exponential function, b could not be negative. Why?

62. Graph $f(x) = 3^x$. Then use the graph to estimate the value of $3^{1.5}$.

9.4 Base-*e* Exponential Functions

- Continuous compound interest
- The natural exponential function
- Graphing the natural exponential function
- Vertical and horizontal translations
- Malthusian population growth
- Base-*e* exponential function models

An exponential function that has many real-life applications is the base-*e* exponential function. In this section, we will show how to evaluate *e*, graph the base-*e* exponential function, and discuss one of its applications in analyzing population growth.

CONTINUOUS COMPOUND INTEREST

If a bank pays interest twice a year, we say that interest is compounded semiannually. If it pays interest four times a year, we say that interest is compounded quarterly. If it pays interest continuously (infinitely many times in a year), we say that interest is compounded continuously.

To develop the formula for continuous compound interest, we start with the formula

$$A = P\left(1 + \frac{r}{k}\right)^{kt}$$

This is the formula for compound interest: r is the annual rate and k is the number of times per year interest is paid.

and let $rn = k$. Since r and k are positive numbers, so is n.

$$A = P\left(1 + \frac{r}{rn}\right)^{rnt}$$

We can then simplify the fraction $\frac{r}{rn}$ and use the commutative property of multiplication to change the order of the exponents.

$$A = P\left(1 + \frac{1}{n}\right)^{nrt}$$

Finally, we can use a property of exponents to write the formula as

$$\text{(1)} \quad A = P\left[\left(1 + \frac{1}{n}\right)^{n}\right]^{rt}$$

Use the property $a^{mn} = (a^m)^n$.

To find the value of $\left(1 + \frac{1}{n}\right)^n$, we evaluate it for several values of n, as shown in the table.

n	$\left(1 + \frac{1}{n}\right)^n$
1	2
2	2.25
4	2.44140625. . .
12	2.61303529. . .
365	2.71456748. . .
1,000	2.71692393. . .
100,000	2.71826830. . .
1,000,000	2.71828137. . .

Notation

Swiss born Leonhard Euler (1707–1783) is said to have published more than any mathematician in history. He had a great influence on the notation that we use today. Through his work, the symbol *e* came into common use.

The results suggest that as n gets larger, the value of $\left(1 + \frac{1}{n}\right)^n$ approaches the number 2.71828. . . . This number is called e, which has the following value.

$$e = 2.718281828459\ldots$$

Like π, the number e is irrational. Its decimal representation is nonterminating and nonrepeating. Rounded to four decimal places, $e \approx 2.7183$.

In continuous compound interest, k (the number of compoundings) is infinitely large. Since k, r, and n are all positive, r is a fixed rate, and $k = rn$, as k gets very large (approaches infinity), so does n. Therefore, we can replace $\left(1 + \frac{1}{n}\right)^n$ in Equation 1 with e to get

$$A = Pe^{rt}$$

Formula for Exponential Growth

If a quantity P increases or decreases at an annual rate r, compounded continuously, the amount A after t years is given by

$$A = Pe^{rt}$$

If time is measured in years, then r is called the **annual growth rate.** If r is negative, the growth represents a decrease.

THE NATURAL EXPONENTIAL FUNCTION

Of all possible bases for an exponential function, e is the most convenient for applied problems involving growth or decay. Since these situations occur often in natural settings, we call $f(x) = e^x$ the *natural exponential function.*

The Natural Exponential Function

The function defined by $f(x) = e^x$ is the **natural exponential function** (or the **base-*e* exponential function**) where $e = 2.71828\ldots$ The domain of $f(x) = e^x$ is the interval $(-\infty, \infty)$. The range is the interval $(0, \infty)$.

The e^x key on a calculator is used to evaluate the natural exponential function.

ACCENT ON TECHNOLOGY: THE NATURAL EXPONENTIAL FUNCTION KEY

To compute the amount to which \$12,000 will grow if invested for 18 years at 10% annual interest, compounded continuously, we substitute 12,000 for P, 0.10 for r, and 18 for t in the formula for continuous compound interest.

$$A = Pe^{rt}$$

$$A = 12{,}000e^{0.10(18)} \quad \text{Write 10\% as 0.10.}$$

$$= 12{,}000e^{1.8}$$

To evaluate this expression using a scientific calculator, we enter

1.8 [e^x] [×] 12000 [=] — display: 72595.76957

Using a graphing calculator, we enter

12000 [×] [2nd] [e^x] 1.8 [)] [ENTER] — display: 12000*e^(1.8) 72595.76957

After 18 years, the account will contain \$72,595.77. This is \$1,589.03 more than the result in Example 6 in the previous section, where interest was compounded quarterly.

EXAMPLE 1

Investing. If \$25,000 accumulates interest at an annual rate of 8%, compounded continuously, find the balance in the account in 50 years.

Solution

$$A = Pe^{rt} \quad \text{This is the formula for continuous compound interest.}$$

$$A = 25{,}000e^{0.08(50)} \quad \text{Write 8\% as 0.08.}$$

$$= 25{,}000e^{4}$$

$$\approx 1{,}364{,}953.75 \quad \text{Use a calculator.}$$

In 50 years, the balance will be \$1,364,953.75—more than a million dollars.

Self Check 1 Find the balance in 60 years.

GRAPHING THE NATURAL EXPONENTIAL FUNCTION

To graph $f(x) = e^x$, we construct a table of function values by choosing several values for x and finding the corresponding values of $f(x)$. For example, if $x = -2$, we have

$$f(x) = e^x$$

$$f(-2) = e^{-2}$$

$$= 0.135335283\ldots \quad \text{Use a calculator.}$$

$$\approx 0.1 \quad \text{Round to the nearest tenth.}$$

We enter $(-2, 0.1)$ in the table. Similarly, we find $f(-1)$, $f(0)$, $f(1)$ and $f(2)$, enter each result in the table, and plot the ordered pairs. We draw a smooth curve through the points to get the graph on the next page.

From the graph, we can verify that the domain of $f(x) = e^x$ is the interval $(-\infty, \infty)$ and the range is the interval $(0, \infty)$. Since the graph passes the horizontal line test, the function is one-to-one.

Note that as x decreases, the values of $f(x)$ decrease and approach 0. Thus, the x-axis is a horizontal asymptote of the graph. The graph does not have an x-intercept, the y-intercept is $(0, 1)$, and the graph passes through the point $(1, e)$.

$f(x) = e^x$

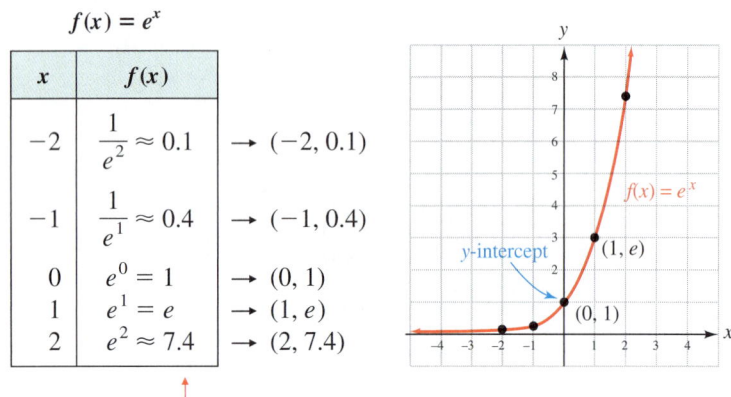

x	$f(x)$	
-2	$\frac{1}{e^2} \approx 0.1$	→ $(-2, 0.1)$
-1	$\frac{1}{e^1} \approx 0.4$	→ $(-1, 0.4)$
0	$e^0 = 1$	→ $(0, 1)$
1	$e^1 = e$	→ $(1, e)$
2	$e^2 \approx 7.4$	→ $(2, 7.4)$

The outputs can be found using the e^x key on a calculator. Some are rounded to the nearest tenth to make point-plotting easier.

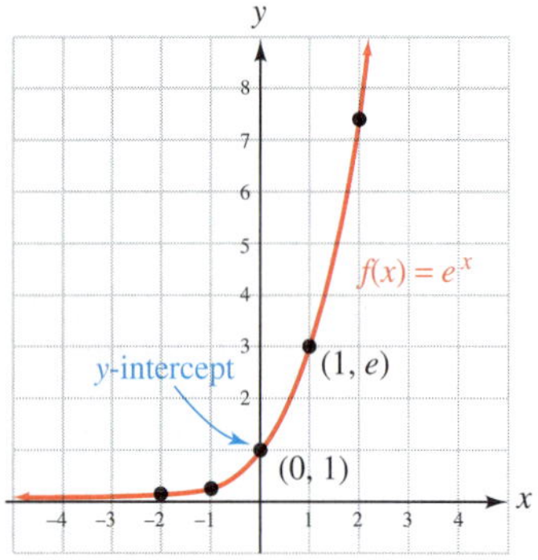

To graph more complicated natural exponential functions, point-plotting can be tedious. In such cases, a graphing calculator is a useful tool.

ACCENT ON TECHNOLOGY: GRAPHING EXPONENTIAL FUNCTIONS

The figure shows the calculator graph of $f(x) = 3e^{-x/2}$. To graph this function, we enter the right-hand side of the equation after the symbol $Y_1 =$. The display will show the equation

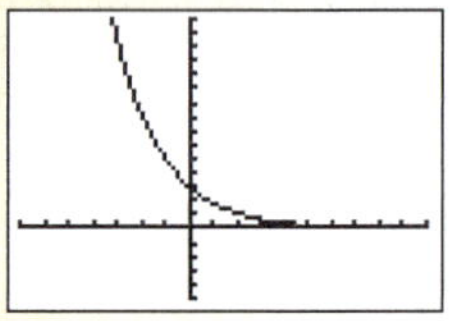

$$Y_1 = 3(e \wedge (-X/2))$$

VERTICAL AND HORIZONTAL TRANSLATIONS

We can illustrate the effects of vertical and horizontal translations of the natural exponential function by using a graphing calculator.

ACCENT ON TECHNOLOGY: TRANSLATIONS OF THE NATURAL EXPONENTIAL FUNCTION

Figure (a) on the next page shows the calculator graphs of $f(x) = e^x$, $g(x) = e^x + 5$, and $h(x) = e^x - 3$. To graph these functions, we enter the right-hand sides of the equations after the symbols $Y_1 =$, $Y_2 =$, and $Y_3 =$. The display will show

$$Y_1 = e \wedge (X) \qquad Y_2 = e \wedge (X) + 5 \qquad Y_3 = e \wedge (X) - 3$$

After graphing these functions, we can see that the graph of $g(x) = e^x + 5$ is 5 units above the graph of $f(x) = e^x$, and that the graph of $h(x) = e^{x-3}$ is 3 units to the right of the graph of $f(x) = e^x$.

Figure (b) shows the calculator graphs of $f(x) = e^x$, $g(x) = e^{x+5}$ and $h(x) = e^{x-3}$. To graph these functions, we enter the right-hand sides of the equations after the symbols $Y_1 =$, $Y_2 =$, and $Y_3 =$. The display will show

$$Y_1 = e\wedge(X) \qquad Y_2 = e\wedge(X + 5) \qquad Y_3 = e\wedge(X - 3)$$

After graphing these functions, we can see that the graph of $g(x) = e^{x+5}$ is 5 units to the left of the graph of $f(x) = e^x$, and that the graph of $h(x) = e^{x-3}$ is 3 units to the right of the graph of $f(x) = e^x$.

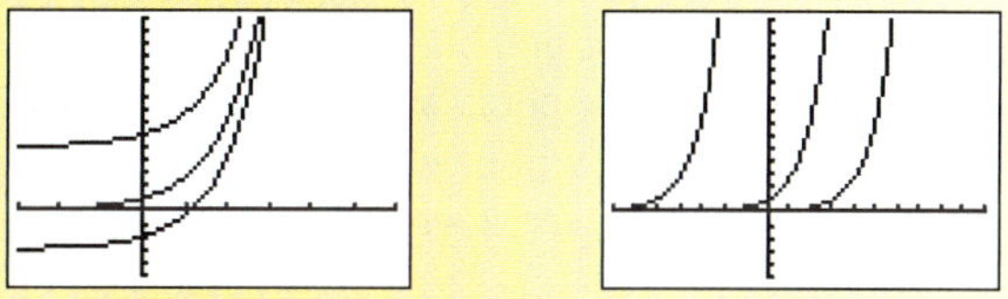

(a) (b)

MALTHUSIAN POPULATION GROWTH

An equation based on the natural exponential function provides a model for **population growth.** In the **Malthusian model for population growth,** the future population of a colony is related to the present population by the formula $A = Pe^{rt}$.

EXAMPLE 2

City planning. The population of a city is currently 15,000, but economic conditions are causing the population to decrease 3% each year. If this trend continues, find the population in 30 years.

Solution Since the population is decreasing 3% each year, the annual growth rate is -3%, or -0.03. We can substitute -0.03 for r, 30 for t, and 15,000 for P in the formula for exponential growth and find A.

$$A = Pe^{rt}$$
$$A = 15{,}000e^{-0.03(30)}$$
$$= 15{,}000e^{-0.9}$$
$$\approx 6{,}099$$

In 30 years, the expected population will be 6,099.

Success Tip

For quantities that are decreasing, remember to enter a negative value for r, the annual rate, in the formula $A = Pe^{rt}$.

Self Check 2 Find the population in 50 years.

The English economist Thomas Robert Malthus (1766–1834) was a pioneer in studying population. He believed that poverty and starvation were unavoidable, because the human population tends to grow exponentially but the food supply tends to grow linearly.

EXAMPLE 3

The Malthusian model. Suppose that a country with a population of 1,000 people is growing exponentially according to the formula

$$P = 1{,}000e^{0.02t} \quad \text{The annual growth rate is } 2\% = 0.02.$$

where t is in years. Furthermore, assume that the food supply F, measured in adequate food per day per person, is growing linearly according to the formula

$$F = 30.625t + 2{,}000 \quad (t \text{ is time in years})$$

In how many years will the population outstrip the food supply?

Solution We can use a graphing calculator with window settings of [0, 100] for x and [0, 10,000] for y. After graphing the functions, we obtain figure (a). If we trace, as in figure (b), we can find the point where the two graphs intersect. From the graph, we can see that the food supply will be adequate for about 71 years. At that time, the population of approximately 4,200 people will begin to have problems.

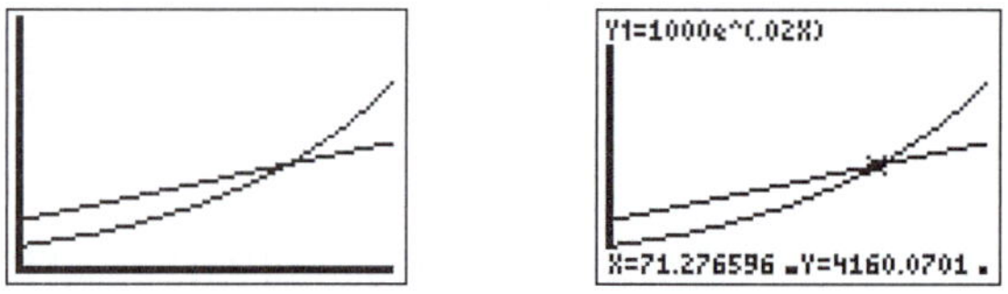

(a) (b)

Self Check 3 Suppose that the population grows at a 2.5% rate. Use a graphing calculator to determine for how many years the food supply will be adequate.

BASE-*e* EXPONENTIAL FUNCTION MODELS

EXAMPLE 4

Baking. A mother takes a cake out of the oven and sets it on a rack to cool. The function $T(t) = 68 + 220e^{-0.18t}$ gives the cake's temperature in degrees Fahrenheit after it has cooled for t minutes. If her children will be home from school in 20 minutes, will the cake have cooled enough for the children to eat it?

Solution When the children arrive home, the cake will have cooled for 20 minutes. To find the temperature of the cake at that time, we need to find $T(20)$.

$$T(t) = 68 + 220e^{-0.18t}$$
$$T(20) = 68 + 220e^{-0.18(20)} \quad \text{Substitute 20 for } t.$$
$$= 68 + 220e^{-3.6}$$
$$\approx 74.0 \quad \text{Use a calculator.}$$

When the children return home, the temperature of the cake will be about 74°, and it can be eaten.

Answers to Self Checks **1.** \$3,037,760.44 **2.** 3,347 **3.** about 51 years

9.4 STUDY SET

VOCABULARY **Refer to the graph of $f(x) = e^x$.**

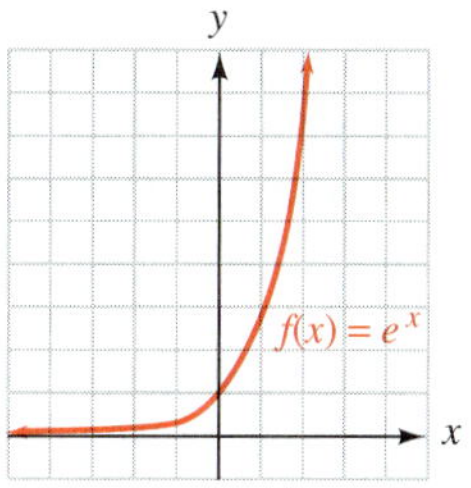

1. What is the name of the function $f(x) = e^x$?
2. What is the domain of the function?
3. What is the range of the function?
4. **a.** What is the y-intercept of the graph?
   b. What is the x-intercept of the graph?
5. Is the function one-to-one?
6. What is an asymptote of the graph?
7. Is f an increasing or a decreasing function?
8. The graph passes through the point $(1, y)$. What is y?

CONCEPTS **Fill in the blanks.**

9. In __________ compound interest, the number of compoundings is infinitely large.
10. The formula for continuous compound interest is $A =$ ____.
11. To two decimal places, the value of e is ____.
12. If n gets larger and larger, the value of $\left(1 + \frac{1}{n}\right)^n$ approaches the value of ____.
13. Graph each irrational number on the number line. $\{\pi, e, \sqrt{2}\}$

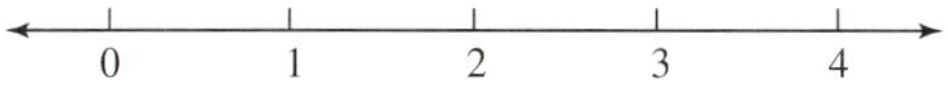

14. Complete the table of values. Round to the nearest hundredth.

x	-2	-1	0	1	2
e^x					

15. POPULATION OF THE UNITED STATES Graph the U.S. census population figures shown in the table (in millions). What type of function does it appear could be used to model the population?

Year	Population	Year	Population
1790	3.9	1900	76.0
1800	5.3	1910	92.2
1810	7.2	1920	106.0
1820	9.6	1930	123.2
1830	12.9	1940	132.1
1840	17.0	1950	151.3
1850	23.1	1960	179.3
1860	31.4	1970	203.3
1870	38.5	1980	226.5
1880	50.1	1990	248.7
1890	62.9	2000	281.4

16. What is the Malthusian population growth formula?
17. The function $f(x) = e^x$ is graphed to the right and the TRACE feature is used. What is the y-coordinate of the point on the graph having an x-coordinate of 1? What is the name given this number?

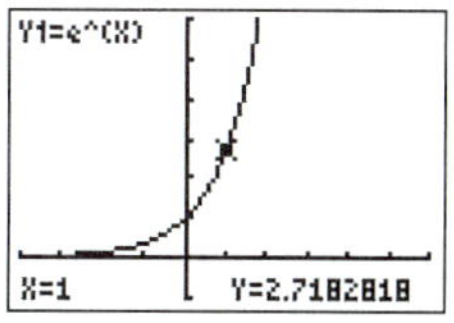

18. The illustration shows a table of values for $f(x) = e^x$. As x decreases, what happens to the values of $f(x)$ listed in the Y_1 column? Will the value of $f(x)$ ever be 0 or negative?

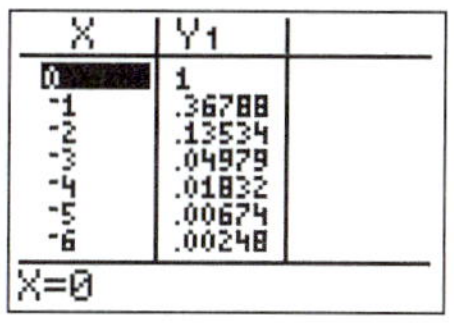

NOTATION **Evaluate A in the formula $A = Pe^{rt}$ for the following values of P, r, and t.**

19. $P = 1{,}000$, $r = 0.09$ and $t = 10$

$$A = \_\_\_\_\, e^{(0.09)(\_\_)}$$
$$= 1{,}000e^{\_\_}$$
$$\approx \_\_\_\_(2.459603111)$$
$$\approx 2{,}459.603111$$

20. $P = 1{,}000$, $r = 0.12$ and $t = 50$

$$A = 1{,}000e^{(\qquad)(50)}$$
$$= \qquad e^{6}$$
$$\approx 1{,}000(\qquad\qquad)$$
$$\approx 403{,}428.7935$$

PRACTICE **Graph each function.**

21. $f(x) = e^{x} + 1$

22. $f(x) = e^{x} - 2$

23. $y = e^{x+3}$

24. $y = e^{x-5}$

25. $f(x) = -e^{x}$

26. $f(x) = -e^{x} + 1$

27. $f(x) = 2e^{x}$

28. $f(x) = \frac{1}{2}e^{x}$

APPLICATIONS **In Exercises 29–34, assume that there are no deposits or withdrawals.**

29. CONTINUOUS COMPOUND INTEREST An initial investment of \$5,000 earns 8.2% interest, compounded continuously. What will the investment be worth in 12 years?

30. CONTINUOUS COMPOUND INTEREST An initial investment of \$2,000 earns 8% interest, compounded continuously. What will the investment be worth in 15 years?

31. COMPARISON OF COMPOUNDING METHODS An initial deposit of \$5,000 grows at an annual rate of 8.5% for 5 years. Compare the final balances resulting from annual compounding and continuous compounding.

32. COMPARISON OF COMPOUNDING METHODS An initial deposit of \$30,000 grows at an annual rate of 8% for 20 years. Compare the final balances resulting from annual compounding and continuous compounding.

33. DETERMINING THE INITIAL DEPOSIT An account now contains \$11,180 and has been accumulating interest at 7% annual interest, compounded continuously, for 7 years. Find the initial deposit.

34. DETERMINING THE PREVIOUS BALANCE An account now contains \$3,610 and has been accumulating interest at 8% annual interest, compounded continuously. How much was in the account 4 years ago?

35. WORLD POPULATION GROWTH The population of the Earth is approximately 6.1 billion people and is growing at an annual rate of 1.4%. Assuming a Malthusian growth model, find the world population in 30 years.

36. HIGHS AND LOWS Somalia, in eastern Africa, has one of the greatest population growth rates in the world. Bulgaria, in southeastern Europe, has one of the smallest. Assuming a Malthusian growth model, complete the table.

Country	Population 2003	Annual growth rate	Estimated population 2015
Somalia	8,025,190	3.43%	
Bulgaria	7,537,929	−1.09%	

Source: nationmaster.com

37. POPULATION GROWTH The growth of a population is modeled by

$$P = 173e^{0.03t}$$

How large will the population be when $t = 20$?

38. POPULATION DECLINE The decline of a population is modeled by

$$P = (1.2 \times 10^{6})e^{-0.008t}$$

How large will the population be when $t = 30$?

39. EPIDEMICS The spread of hoof and mouth disease through a herd of cattle can be modeled by the formula

$$P = P_0e^{0.27t} \qquad (t \text{ is in days})$$

If a rancher does not act quickly to treat two cases, how many cattle will have the disease in 12 days?

40. OCEANOGRAPHY The width w (in millimeters) of successive growth spirals of the sea shell *Catapulus voluto,* shown below, is given by the exponential function

$$w = 1.54e^{0.503n}$$

where n is the spiral number. Find the width, to the nearest tenth of a millimeter, of the sixth spiral.

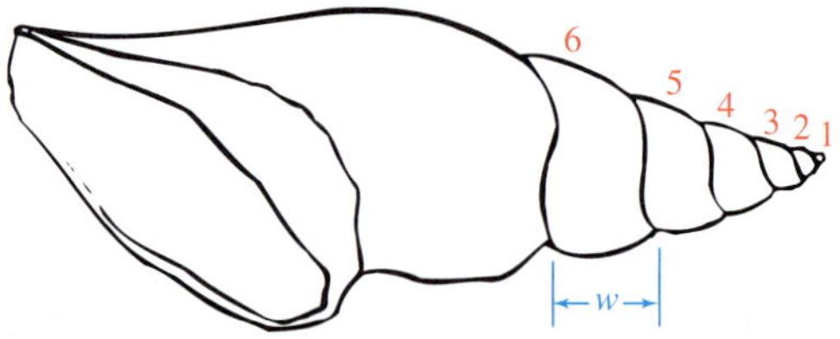

41. HALF-LIFE OF A DRUG The quantity of a prescription drug in the bloodstream of a patient t hours after it is administered can be modeled by an exponential function. (See the graph.) Determine the time it takes to eliminate half of the initial dose from the body.

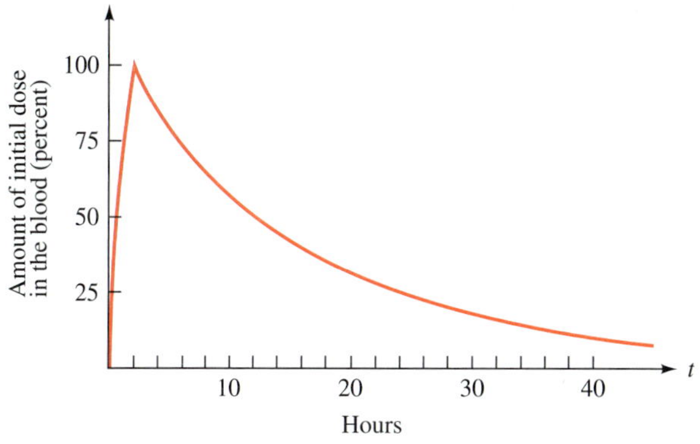

42. MEDICINE The concentration x of a certain prescription drug in an organ after t minutes is given by

$$x = 0.08(1 - e^{-0.1t})$$

Find the concentration of the drug at 30 minutes.

43. SKYDIVING Before the parachute opens, a skydiver's velocity v in meters per second is given by

$$v = 50(1 - e^{-0.2t})$$

Find the velocity after 20 seconds of free fall.

44. FREE FALL After t seconds a certain falling object has a velocity v given by

$$v = 50(1 - e^{-0.3t})$$

Which is falling faster after 2 seconds—the object or the skydiver in Exercise 43?

Use a graphing calculator to solve each problem.

45. THE MALTHUSIAN MODEL In Example 3, suppose that better farming methods changed the formula for food growth to $F = 31t + 2{,}000$. How long would the food supply be adequate?

46. THE MALTHUSIAN MODEL In Example 3, suppose that a birth-control program changed the formula for population growth to $P = 1{,}000e^{0.01t}$. How long would the food supply be adequate?

WRITING

47. Explain why the graph of $y = e^x - 5$ is five units below the graph of $y = e^x$.

48. A feature article in a newspaper stated that the sport of snowboarding was growing *exponentially.* Explain what the author of the article meant by that.

49. As of 2003, the population growth rate for Russia was -0.3% annually. What are some of the consequences for a country that has a negative population growth?

50. What is e?

REVIEW

Simplify each expression. Assume that all variables represent positive numbers.

51. $\sqrt{240x^5}$

52. $\sqrt[3]{-125x^5y^4}$

53. $4\sqrt{48y^3} - 3y\sqrt{12y}$

54. $\sqrt[4]{48z^5} + \sqrt[4]{768z^5}$

CHALLENGE PROBLEMS

55. Is the statement $e^e > e^3$ true or false? Explain your reasoning.

56. Graph the function defined by the equation

$$f(x) = \frac{e^x + e^{-x}}{2}$$

from $x = -2$ to $x = 2$. The graph will look like a parabola, but it is not. The graph, called a **catenary,** is important in the design of power distribution networks, because it represents the shape of a uniform flexible cable whose ends are suspended from the same height. The function is called the **hyperbolic cosine function.**

57. If $e^{t+5} = ke^t$, find k.

58. If $e^{5t} = k^t$, find k.

9.5 Logarithmic Functions

- The definition of logarithm
- Exponential and logarithmic form
- Base-10 logarithms
- Logarithmic functions and their graphs
- Vertical and horizontal translations
- Applications of logarithms

In this section, we will study inverses of exponential functions. These functions are called *logarithmic* functions and they can be used to solve applied problems from fields such as electronics, seismology (the study of earthquakes), and business.

THE DEFINITION OF LOGARITHM

The graph of the exponential function $f(x) = 2^x$ is shown in red below. Since it passes the horizontal line test, it is a one-to-one function, and therefore, has an inverse. To graph f^{-1}, we interchange the coordinates of the ordered pairs in the table, plot those points, and draw a smooth curve through them, as shown below in blue. As expected, the graphs of f and f^{-1} are symmetric with respect to the line $y = x$.

$f(x) = 2^x$

To graph f^{-1}, interchange each pair of coordinates.

x	$f(x)$			
-3	$\frac{1}{8}$	→	$\left(-3, \frac{1}{8}\right)$	$\left(\frac{1}{8}, -3\right)$
-2	$\frac{1}{4}$	→	$\left(-2, \frac{1}{4}\right)$	$\left(\frac{1}{4}, -2\right)$
-1	$\frac{1}{2}$	→	$\left(-1, \frac{1}{2}\right)$	$\left(\frac{1}{2}, -1\right)$
0	1	→	(0, 1)	(1, 0)
1	2	→	(1, 2)	(2, 1)
2	4	→	(2, 4)	(4, 2)
3	8	→	(3, 8)	(8, 3)

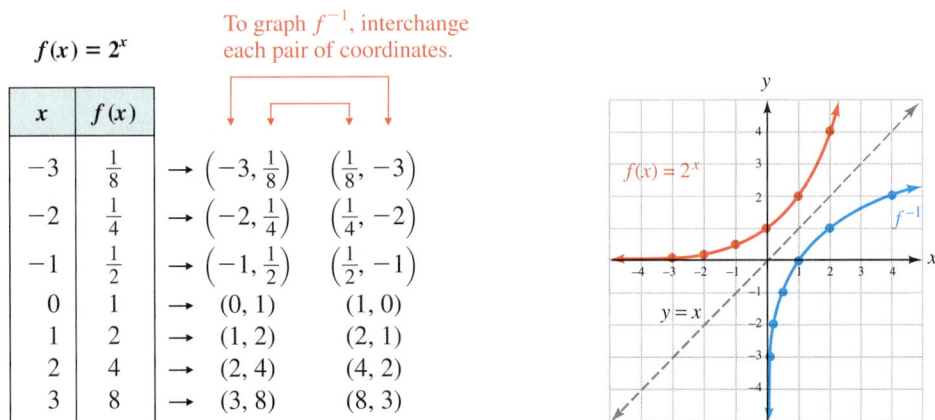

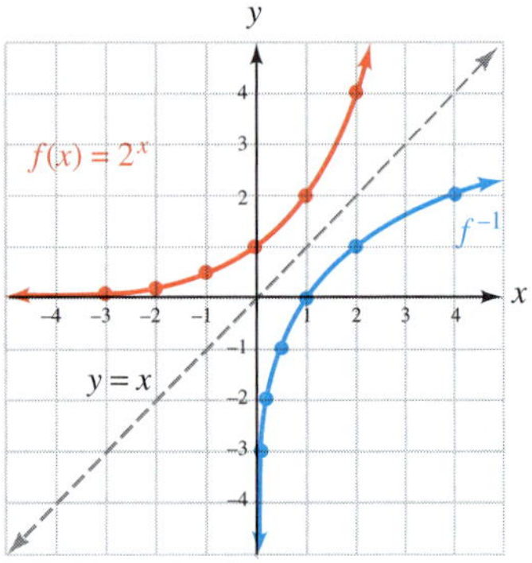

To write an equation for the inverse of $f(x) = 2^x$, we proceed as follows:

$$f(x) = 2^x$$
$$y = 2^x \quad \text{Replace } f(x) \text{ with } y.$$
$$x = 2^y \quad \text{Interchange the variables } x \text{ and } y.$$

Now we must solve for y. However, we have not studied methods for solving for a variable located in an exponent. Instead, we translate the relationship $x = 2^y$ into words:

$y =$ the power to which we raise 2 to get x

Finally, we substitute $f^{-1}(x)$ for y.

$f^{-1}(x) =$ the power to which we raise 2 to get x

A new notation, called **logarithmic notation,** enables us to write the inverse in simpler form. If we define the symbol $\log_2 x$ to mean *the power to which we raise 2 to get x,* we can write the equation for the inverse as

$$f^{-1}(x) = \log_2 x \quad \text{Read } \log_2 x \text{ as "the logarithm, base 2, of } x\text{" or "log, base 2, of } x\text{."}$$

We have found that the inverse of the exponential function $f(x) = 2^x$ is $f^{-1}(x) = \log_2 x$. To find the inverse of exponential functions with other bases, such as $f(x) = 3^x$ and $f(x) = 10^x$, we define logarithm in the following way.

Definition of Logarithm

For all positive numbers b, where $b \neq 1$, and all positive numbers x,

$$y = \log_b x \quad \text{is equivalent to} \quad x = b^y$$

This definition guarantees that any pair (x, y) that satisfies the logarithmic equation $y = \log_b x$ also satisfies the exponential equation $x = b^y$. Because of this relationship, a statement written in logarithmic form can be written in an equivalent exponential form, and vice versa. The following diagram will help you remember the respective positions of the exponent and base in each form.

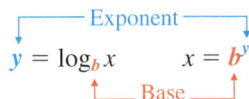

Exponent

$$y = \log_b x \qquad x = b^y$$

Base

EXPONENTIAL AND LOGARITHMIC FORM

Success Tip

Study the relationship between exponential and logarithmic statements carefully. Your success with the material in the rest of this chapter depends greatly on your understanding of this definition.

The following table shows several pairs of equivalent statements.

Logarithmic form	***Exponential form***
$\log_2 8 = 3$	$2^3 = 8$
$\log_3 81 = 4$	$3^4 = 81$
$\log_4 4 = 1$	$4^1 = 4$
$\log_5 \frac{1}{125} = -3$	$5^{-3} = \frac{1}{125}$

EXAMPLE 1

Write as an exponential equation: **a.** $\log_4 64 = 3$, **b.** $\log_7 \sqrt{7} = \frac{1}{2}$, and **c.** $\log_6 \frac{1}{36} = -2$.

Solution **a.** $\log_4 64 = 3$ is equivalent to $4^3 = 64$.

b. $\log_7 \sqrt{7} = \frac{1}{2}$ is equivalent to $7^{1/2} = \sqrt{7}$.

c. $\log_6 \frac{1}{36} = -2$ is equivalent to $6^{-2} = \frac{1}{36}$.

Self Check 1 Write $\log_2 128 = 7$ as an exponential equation.

EXAMPLE 2

Write as a logarithmic equation: **a.** $8^0 = 1$, **b.** $6^{1/3} = \sqrt[3]{6}$, and **c.** $\left(\frac{1}{4}\right)^2 = \frac{1}{16}$.

Solution **a.** $8^0 = 1$ is equivalent to $\log_8 1 = 0$

b. $6^{1/3} = \sqrt[3]{6}$ is equivalent to $\log_6 \sqrt[3]{6} = \frac{1}{3}$

c. $\left(\frac{1}{4}\right)^2 = \frac{1}{16}$ is equivalent to $\log_{1/4} \frac{1}{16} = 2$

Self Check 2 Write $9^{-1} = \frac{1}{9}$ as a logarithmic equation.

Certain logarithmic equations can be solved by writing them as exponential equations.

EXAMPLE 3

Solve each equation for x: **a.** $\log_x 25 = 2$, **b.** $\log_3 x = -3$, and **c.** $\log_{1/2} \frac{1}{16} = x$.

Solution

a. Since $\log_x 25 = 2$ is equivalent to $x^2 = 25$, we can solve $x^2 = 25$ to find x.

$$x^2 = 25$$
$$x = \pm\sqrt{25} \quad \text{Use the square root property.}$$
$$x = \pm 5$$

In the expression $\log_x 25$, the base of the logarithm is x. Because the base must be positive, we discard -5 and we have

$$x = 5$$

The solution is 5. To check, verify that $\log_5 25 = 2$.

b. Since $\log_3 x = -3$ is equivalent to $3^{-3} = x$, we can solve $3^{-3} = x$ to find x.

$$3^{-3} = x$$
$$\frac{1}{3^3} = x$$
$$x = \frac{1}{27}$$

The solution is $\frac{1}{27}$. To check, verify that $\log_3 \frac{1}{27} = -3$.

Success Tip

To solve this equation, we note that if the bases are equal, the exponents must be equal.

$$\left(\frac{1}{2}\right)^x = \left(\frac{1}{2}\right)^4$$

c. Since $\log_{1/2} \frac{1}{16} = x$ is equivalent to $\left(\frac{1}{2}\right)^x = \frac{1}{16}$, we can solve $\left(\frac{1}{2}\right)^x = \frac{1}{16}$ to find x.

$$\left(\frac{1}{2}\right)^x = \frac{1}{16}$$
$$\left(\frac{1}{2}\right)^x = \left(\frac{1}{2}\right)^4 \quad \text{Write } \frac{1}{16} \text{ as a power of } \frac{1}{2} \text{ to match the bases: } \frac{1}{2}\cdot\frac{1}{2}\cdot\frac{1}{2}\cdot\frac{1}{2} = \frac{1}{16}.$$
$$x = 4$$

Since the bases are the same, and since exponential functions are one-to-one, the exponents must be equal.

The solution is 4. To check, verify that $\log_{1/2} \frac{1}{16} = 4$.

Self Check 3 Solve each equation for x: **a.** $\log_x 49 = 2$, **b.** $\log_{1/3} x = 2$, and **c.** $\log_6 216 = x$.

In the previous examples, we have seen that the logarithm of a number is an exponent. In fact,

$\log_b x$ is the exponent to which b is raised to get x.

Translating this statement into symbols, we have

$$b^{\log_b x} = x$$

EXAMPLE 4

Evaluate each logarithmic expression: **a.** $\log_8 64$, **b.** $\log_3 \frac{1}{3}$, and **c.** $\log_4 2$.

Solution

a. $\log_8 64 = 2$ Ask: "To what power must we raise 8 to get 64?" Since $8^2 = 64$, the answer is: the 2nd power.

b. $\log_3 \frac{1}{3} = -1$ Ask: "To what power must we raise 3 to get $\frac{1}{3}$?" Since $3^{-1} = \frac{1}{3}$, the answer is: the -1 power.

c. $\log_4 2 = \frac{1}{2}$ Ask: "To what power must we raise 4 to get 2?" Since $\sqrt{4} = 4^{1/2} = 2$, the answer is: the $\frac{1}{2}$ power.

Self Check 4 Evaluate each expression: **a.** $\log_9 81$, **b.** $\log_4 \frac{1}{16}$, and **c.** $\log_9 3$.

BASE-10 LOGARITHMS

For computational purposes and in many applications, we will use base-10 logarithms (also called **common logarithms**). When the base b is not indicated in the notation $\log x$, we assume that $b = 10$:

$\log x$ means $\log_{10} x$

The table below shows several pairs of equivalent statements involving base-10 logarithms.

Logarithmic form	*Exponential form*	
$\log 100 = 2$	$10^2 = 100$	Read log 100 as "log of 100."
$\log \frac{1}{10} = -1$	$10^{-1} = \frac{1}{10}$	
$\log 1 = 0$	$10^0 = 1$	

In general, we have

$$\log_{10} 10^x = x$$

The Language of Algebra

London professor Henry Briggs (1561–1630) and Scottish lord John Napier (1550–1617) are credited with developing the concept of *common logarithms.* Their tables of logarithms were useful tools at that time for those performing large calculations.

EXAMPLE 5

Evaluate each logarithmic expression, if possible: **a.** $\log 1{,}000$, **b.** $\log \frac{1}{100}$, **c.** $\log 10$, and **d.** $\log(-10)$.

Solution

a. $\log 1{,}000 = 3$ Ask: "To what power must we raise 10 to get 1,000?" Since $10^3 = 1{,}000$, the answer is: the 3rd power.

b. $\log \frac{1}{100} = -2$ Ask: "To what power must we raise 10 to get $\frac{1}{100}$?" Since $10^{-2} = \frac{1}{100}$, the answer is: the -2 power.

c. $\log 10 = 1$ Ask: "To what power must we raise 10 to get 10?" Since $10^1 = 10$, the answer is: the 1st power.

d. To find $\log(-10)$, we must find a power of 10 such that $10^? = -10$. There is no such number. Thus,

$$\log(-10) \text{ is undefined}$$

Self Check 5 Evaluate each expression: **a.** $\log 10{,}000$, **b.** $\log \frac{1}{1{,}000}$, and **c.** $\log 0$.

Many logarithmic expressions are not as easy to evaluate as those in the previous example. For instance, to find $\log 2.34$, we ask, "To what power must we raise 10 to get 2.34?" The answer certainly isn't obvious. In such cases, we use a calculator.

ACCENT ON TECHNOLOGY: EVALUATING LOGARITHMS

To find $\log 2.34$ with a scientific calculator we enter

2.34 [LOG] .369215857

On some calculators, the [10^x] key also serves as the [log] key when [2nd] or [SHIFT] is pressed. This is because $f(x) = 10^x$ and $f(x) = \log x$ are inverses.

To use a graphing calculator, we enter

[LOG] 2.34 [)] [ENTER]

```
log(2.34)
        .3692158574
```

To four decimal places, $\log 2.34 = 0.3692$. This means, $10^{0.3692} \approx 2.34$.

If we attempt to evaluate logarithmic expressions such as log 0, or the logarithm of a negative number, such as $\log(-5)$, the following error statements are displayed.

[Error]

```
ERR:DOMAIN
1:QUIT
2:Goto
```

```
ERR:NONREAL ANS
1:QUIT
2:Goto
```

EXAMPLE 6

Solve: $\log x = 0.3568$. Round to four decimal places.

Solution The equation $\log x = 0.3568$ is equivalent to $10^{0.3568} = x$. To find x with a scientific calculator, we enter

10 y^x .3568 =

The display will read 2.274049951. To four decimal places,

$$x = 2.2740$$

If your calculator has a 10^x key, enter .3568 and press it to get the same result. The solution is 2.2740. To check, use your calculator to verify that $\log 2.2740 \approx 0.3568$.

Self Check 6 Solve: $\log x = 1.87737$. Round to four decimal places.

LOGARITHMIC FUNCTIONS AND THEIR GRAPHS

Because an exponential function defined by $f(x) = b^x$ is one-to-one, it has an inverse function that is defined by $x = b^y$. When we write $x = b^y$ in the equivalent form $y = \log_b x$, the result is called a *logarithmic function*.

Logarithmic Functions If $b > 0$ and $b \neq 1$, the **logarithmic function with base b** is defined by

$$f(x) = \log_b x \qquad \text{or} \qquad y = \log_b x$$

The domain of $f(x) = \log_b x$ is the interval $(0, \infty)$ and the range is the interval $(-\infty, \infty)$.

Since every logarithmic function is the inverse of a one-to-one exponential function, logarithmic functions are one-to-one.

We can use the point-plotting method to graph logarithmic functions. For example, to graph $f(x) = \log_2 x$, we construct a table of function values, plot the resulting ordered pairs, and then draw a smooth curve through the points to get the graph, as shown in figure (a). To graph $f(x) = \log_{1/2} x$, we use the same procedure, as shown in figure (b).

$f(x) = \log_2 x$

x	$f(x)$	
$\frac{1}{4}$	-2	→ $\left(\frac{1}{4}, -2\right)$
$\frac{1}{2}$	-1	→ $\left(\frac{1}{2}, -1\right)$
1	0	→ $(1, 0)$
2	1	→ $(2, 1)$
4	2	→ $(4, 2)$
8	3	→ $(8, 3)$

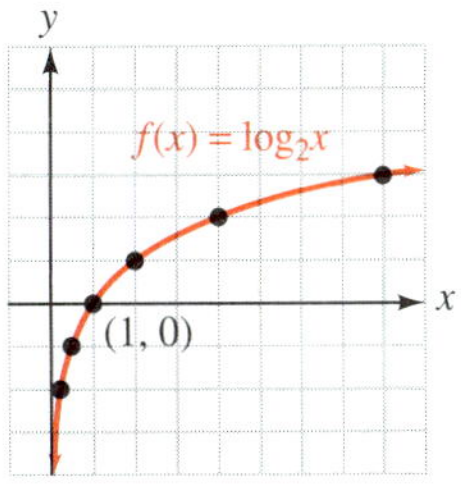

Because the base of the function is 2, choose values for x that are integer powers of 2.

(a)

$f(x) = \log_{1/2} x$

x	$f(x)$	
$\frac{1}{4}$	2	→ $\left(\frac{1}{4}, 2\right)$
$\frac{1}{2}$	1	→ $\left(\frac{1}{2}, 1\right)$
1	0	→ $(1, 0)$
2	-1	→ $(2, -1)$
4	-2	→ $(4, -2)$
8	-3	→ $(8, -3)$

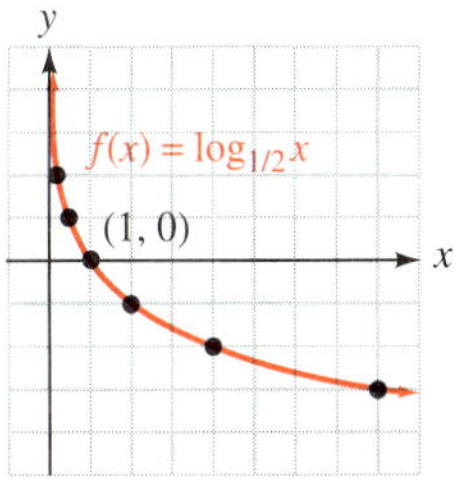

Because the base of the function is $\frac{1}{2}$, choose values for x that are integer powers of $\frac{1}{2}$.

(b)

The graphs of all logarithmic functions are similar to those shown below. If $b > 1$, the logarithmic function is increasing, as in figure (a). If $0 < b < 1$, the logarithmic function is decreasing, as in figure (b).

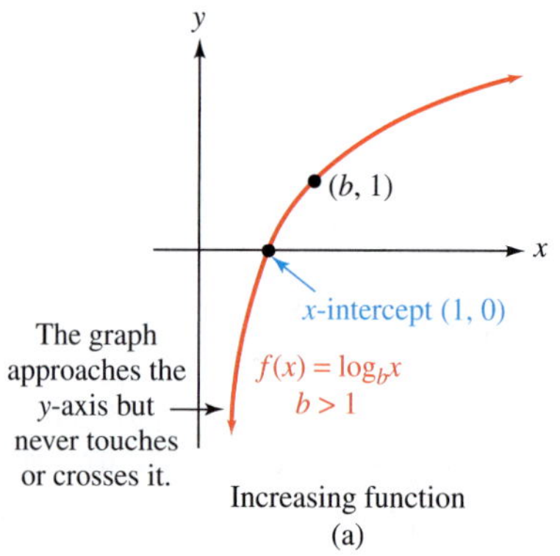

Increasing function
(a)

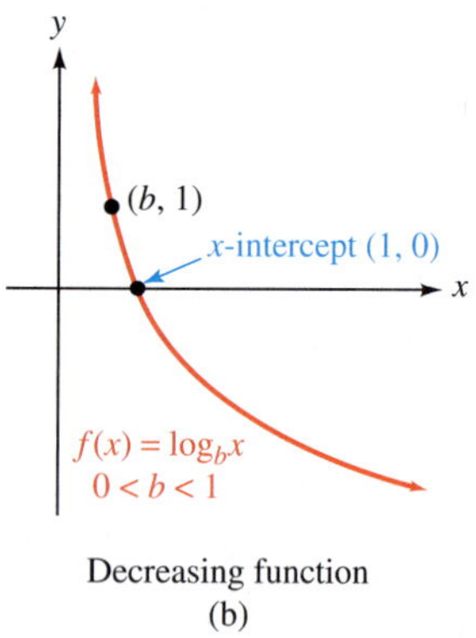

Decreasing function
(b)

Properties of Logarithmic Functions

The graph of $f(x) = \log_b x$ (or $y = \log_b x$) has the following properties.

1. It passes through the point (1, 0).
2. It passes through the point $(b, 1)$.
3. The y-axis (the line $x = 0$) is an asymptote.
4. The domain is $(0, \infty)$ and the range is $(-\infty, \infty)$.

Caution

Since the domain of the logarithmic function is the set of positive real numbers, it is impossible to find the logarithm of 0 or the logarithm of a negative number. For example,

$\log_2(-4)$ and $\log_2 0$

are undefined.

The exponential and logarithmic functions are inverses of each other, so their graphs have symmetry about the line $y = x$. The graphs of $f(x) = \log_b x$ and $g(x) = b^x$ are shown in figure (a) when $b > 1$ and in figure (b) when $0 < b < 1$.

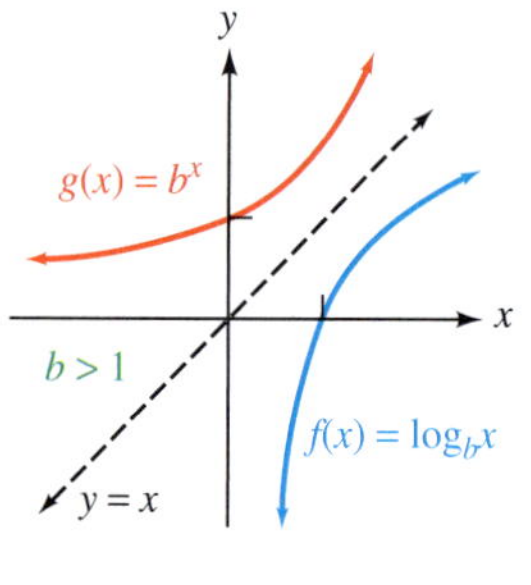

(a)

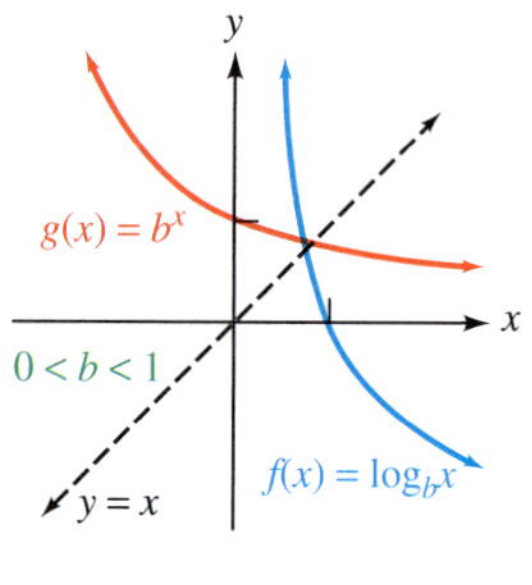

(b)

VERTICAL AND HORIZONTAL TRANSLATIONS

The graphs of many functions involving logarithms are translations of the basic logarithmic graphs.

EXAMPLE 7

Graph the function $f(x) = 3 + \log_2 x$ and describe the translation.

The graph of $f(x) = 3 + \log_2 x$ is identical to the graph of $g(x) = \log_2 x$, except that it is translated 3 units upward.

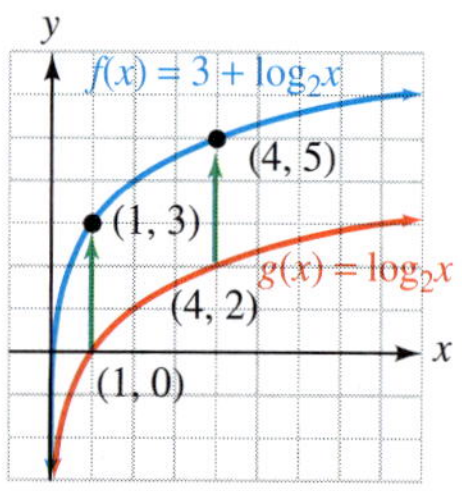

Notation

Since $y = f(x)$, we can write $f(x) = 3 + \log_2 x$ as $y = 3 + \log_2 x$.

Self Check 7 Graph $y = (\log_3 x) - 2$ and describe the translation.

EXAMPLE 8

Graph $f(x) = \log_{1/2}(x - 1)$ and describe the translation.

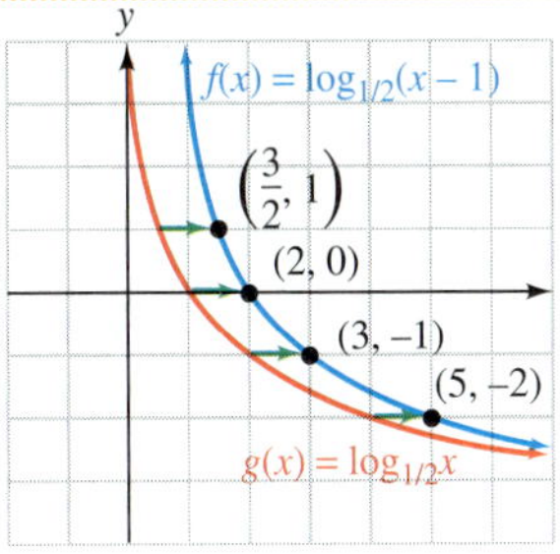

Solution The graph of $f(x) = \log_{1/2}(x - 1)$ is identical to the graph of $g(x) = \log_{1/2} x$, except that it is translated 1 unit to the right.

Notation

Parentheses are used to write the function in Example 8, because $f(x) = \log_{1/2} x - 1$ could be interpreted as $f(x) = \log_{1/2}(x - 1)$ or as $f(x) = (\log_{1/2} x) - 1$.

Self Check 8 Graph $f(x) = \log_{1/3}(x + 2)$ and describe the translation.

ACCENT ON TECHNOLOGY: GRAPHING LOGARITHMIC FUNCTIONS

To use a calculator to graph the logarithmic function $f(x) = -2 + \log_{10} \frac{x}{2}$, we enter the right-hand side of the equation after the symbol $Y_1 =$. The display will show the equation

$$Y_1 = -2 + \log(X/2)$$

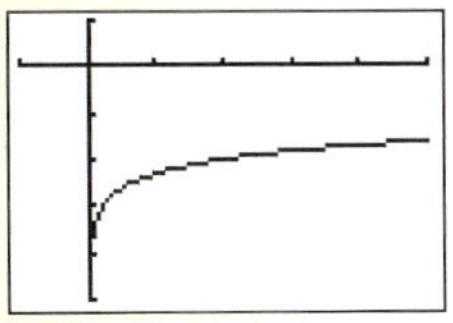

If we use window settings of $[-1, 5]$ for x and $[-4, 1]$ for y and press the GRAPH key, we will obtain the graph shown.

APPLICATIONS OF LOGARITHMS

Common logarithms are used in electrical engineering to express the voltage gain (or loss) of an electronic device such as an amplifier. The unit of gain (or loss), called the **decibel,** is defined by a logarithmic relation.

Decibel Voltage Gain If E_O is the output voltage of a device and E_I is the input voltage, the decibel voltage gain of the device (db gain) is given by

$$\text{db gain} = 20 \log \frac{E_O}{E_I}$$

EXAMPLE 9

db gain. If the input to an amplifier is 0.5 volt and the output is 40 volts, find the decibel voltage gain of the amplifier.

Solution We can find the decibel voltage gain by substituting 0.5 for E_I and 40 for E_O into the formula for db gain:

$$\text{db gain} = 20 \log \frac{E_O}{E_I}$$

$$\text{db gain} = 20 \log \frac{40}{0.5}$$

$$= 20 \log 80 \qquad \text{Divide: } \frac{40}{0.5} = 80.$$

$$\approx 38 \qquad \text{Use a calculator: } 20 \log 80 \text{ means } 20 \cdot \log 80.$$

The amplifier provides a 38-decibel voltage gain.

In seismology, common logarithms are used to measure the intensity of earthquakes on the **Richter scale.** The intensity of an earthquake is given by the following logarithmic function.

Richter Scale If R is the intensity of an earthquake, A is the amplitude (measured in micrometers), and P is the period (the time of one oscillation of the Earth's surface measured in seconds), then

$$R = \log \frac{A}{P}$$

EXAMPLE 10

Earthquakes. Find the measure on the Richter scale of an earthquake with an amplitude of 5,000 micrometers (0.5 centimeter) and a period of 0.1 second.

Solution We substitute 5,000 for A and 0.1 for P in the Richter scale formula and simplify:

$$R = \log \frac{A}{P}$$

$$R = \log \frac{5{,}000}{0.1}$$

$$= \log 50{,}000 \qquad \text{Divide: } \frac{5{,}000}{0.1} = 50{,}000.$$

$$\approx 4.698970004 \qquad \text{Use a calculator.}$$

The earthquake measures about 4.7 on the Richter scale.

The Language of Algebra

The *Richter* scale was developed in 1935 by Charles F. Richter of the California Institute of Technology.

Answers to Self Checks **1.** $2^7 = 128$ **2.** $\log_9 \frac{1}{9} = -1$ **3. a.** 7, **b.** $\frac{1}{9}$, **c.** 3 **4. a.** 2, **b.** -2, **c.** $\frac{1}{2}$

5. a. 4, **b.** -3, **c.** undefined **6.** 75.3998

7. The graph of $y = \log_3 x$ is translated 2 units downward.

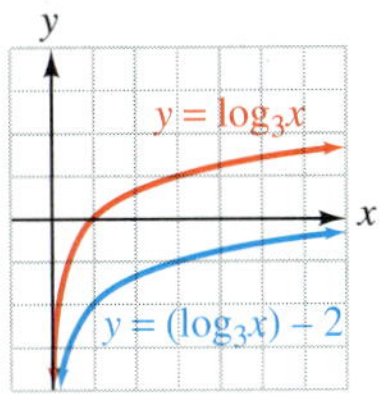

8. The graph of $g(x) = \log_{1/3} x$ is translated 2 units to the left.

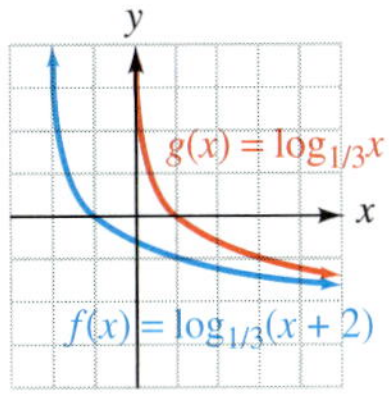

9.5 STUDY SET

VOCABULARY **Refer to the graph of $f(x) = \log_4 x$.**

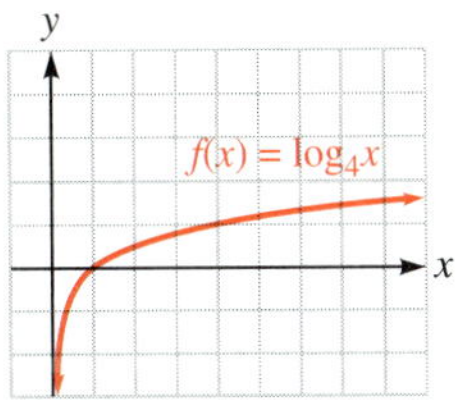

1. What type of function is $f(x) = \log_4 x$?
2. What is the domain of the function?
3. What is the range of the function?
4. **a.** What is the y-intercept of the graph?
   b. What is the x-intercept of the graph?
5. Is f a one-to-one function?
6. What is an asymptote of the graph?
7. Is f an increasing or a decreasing function?
8. The graph passes through the point $(4, y)$. What is y?

CONCEPTS **Fill in the blanks.**

9. The equation $y = \log_b x$ is equivalent to the exponential equation ______.
10. $\log_b x$ is the ________ to which b is raised to get x.
11. The functions $f(x) = \log_{10} x$ and $f(x) = 10^x$ are ________ functions.
12. The inverse of an exponential function is called a __________ function.

Complete the table of values, where possible.

13. $y = \log x$

x	y
$\frac{1}{100}$	
$\frac{1}{10}$	
1	
10	
100	

14. $f(x) = \log_5 x$

x	$f(x)$
$\frac{1}{25}$	
$\frac{1}{5}$	
1	
5	
25	

15. $f(x) = \log_6 x$

Input	Output
-6	
0	
$\frac{1}{216}$	
$\sqrt{6}$	
6^8	

16. $f(x) = \log_8 x$

x	$f(x)$
-8	
0	
$\frac{1}{8}$	
$\sqrt{8}$	
64	

17. Use a calculator to complete the table of values for $f(x) = \log x$. Round to the nearest hundredth.

x	$f(x)$
0.5	
1	
2	
4	
6	
8	
10	

18. Graph $f(x) = \log x$ in the illustration. (See Exercise 17.) Note that the units on the x- and y-axes are different.

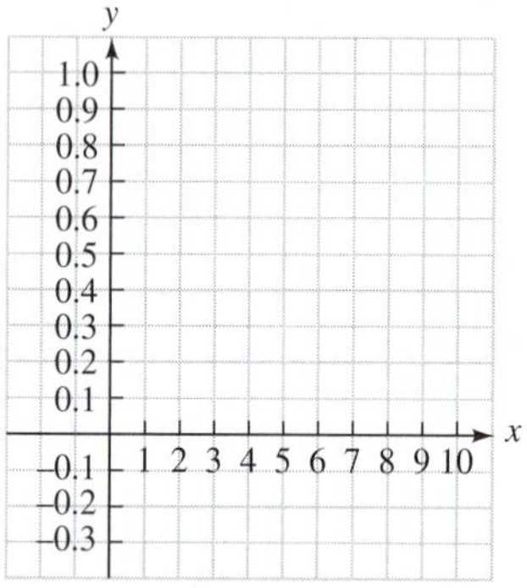

19. A table of solutions for $f(x) = \log x$ is shown below. As x decreases and gets close to 0, what happens to the values of $f(x)$?

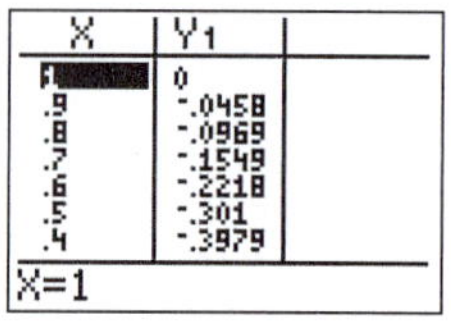

20. For each function, determine $f^{-1}(x)$.

a. $f(x) = 10^x$ **b.** $f(x) = 3^x$

c. $f(x) = \log x$ **d.** $f(x) = \log_2 x$

NOTATION Fill in the blanks.

21. a. $\log x = \log_{\_\_} x$ **b.** $\log_{10} 10^x =$ ____

22. a. Read $\log_5 25$ as "log, ____ 5, ____ 25."

b. Read $\log x$ as "____ of x."

PRACTICE Write each equation in exponential form.

23. $\log_3 81 = 4$ **24.** $\log_7 7 = 1$

25. $\log 10 = 1$ **26.** $\log 100 = 2$

27. $\log_4 \frac{1}{64} = -3$ **28.** $\log_6 \frac{1}{36} = -2$

29. $\log_5 \sqrt{5} = \frac{1}{2}$ **30.** $\log_8 \sqrt[3]{8} = \frac{1}{3}$

Write each equation in logarithmic form.

31. $8^2 = 64$ **32.** $10^3 = 1{,}000$

33. $4^{-2} = \frac{1}{16}$ **34.** $3^{-4} = \frac{1}{81}$

35. $\left(\frac{1}{2}\right)^{-5} = 32$ **36.** $\left(\frac{1}{3}\right)^{-3} = 27$

37. $x^y = z$ **38.** $m^n = p$

Evaluate each expression.

39. $\log_2 8$ **40.** $\log_3 9$

41. $\log_4 16$ **42.** $\log_6 216$

43. $\log_{1/2} \frac{1}{32}$ **44.** $\log_{1/3} \frac{1}{81}$

45. $\log_9 3$ **46.** $\log_{125} 5$

47. $\log \frac{1}{10}$ **48.** $\log \frac{1}{100}$

49. $\log 1{,}000{,}000$ **50.** $\log 100{,}000$

Solve for x.

51. $\log_8 x = 2$ **52.** $\log_7 x = 0$

53. $\log_{25} x = \frac{1}{2}$ **54.** $\log_4 x = \frac{1}{2}$

55. $\log_5 x = -2$ **56.** $\log_3 x = -4$

57. $\log_{36} x = -\frac{1}{2}$ **58.** $\log_{27} x = -\frac{1}{3}$

59. $\log_x 0.01 = -2$ **60.** $\log_x 0.001 = -3$

61. $\log_{27} 9 = x$ **62.** $\log_{12} x = 0$

63. $\log_x 5^3 = 3$ **64.** $\log_x 5 = 1$

65. $\log_x \frac{\sqrt{3}}{3} = \frac{1}{2}$ **66.** $\log_x \frac{9}{4} = 2$

67. $\log_{100} x = \frac{3}{2}$ **68.** $\log_x \frac{1}{1{,}000} = -\frac{3}{2}$

69. $\log_x \frac{1}{64} = -3$ **70.** $\log_x \frac{1}{100} = -2$

71. $\log_8 x = 0$ **72.** $\log_4 8 = x$

Use a calculator to find each value. Give answers to four decimal places.

73. $\log 3.25$
74. $\log 0.57$
75. $\log 0.00467$
76. $\log 375.876$

Use a calculator to find each value of x. Give answers to two decimal places.

77. $\log x = 1.4023$
78. $\log x = 0.926$
79. $\log x = -1.71$
80. $\log x = -0.5$

Graph each function. Decide whether each function is an increasing or a decreasing function.

81. $f(x) = \log_3 x$
82. $f(x) = \log_{1/3} x$
83. $y = \log_{1/2} x$
84. $y = \log_4 x$

Graph each function.

85. $f(x) = 3 + \log_3 x$
86. $f(x) = \log_{1/3} x - 1$
87. $y = \log_{1/2}(x - 2)$
88. $y = \log_4(x + 2)$

Graph each pair of inverse functions on a single coordinate system. Draw the axis of symmetry.

89. $f(x) = 6^x$
$f^{-1}(x) = \log_6 x$
90. $f(x) = 3^x$
$f^{-1}(x) = \log_3 x$
91. $f(x) = 5^x$
$f^{-1}(x) = \log_5 x$
92. $f(x) = 8^x$
$f^{-1}(x) = \log_8 x$

APPLICATIONS

93. INPUT VOLTAGE Find the db gain of an amplifier if the input voltage is 0.71 volt when the output voltage is 20 volts.

94. OUTPUT VOLTAGE Find the db gain of an amplifier if the output voltage is 2.8 volts when the input voltage is 0.05 volt.

95. db GAIN Find the db gain of the amplifier shown below.

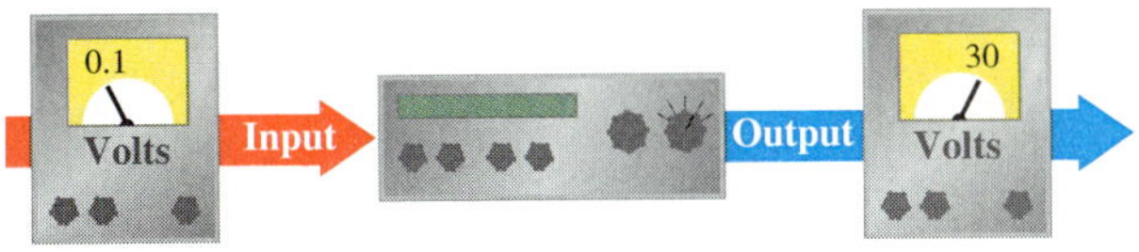

96. db GAIN An amplifier produces an output of 80 volts when driven by an input of 0.12 volt. Find the amplifier's db gain.

97. THE RICHTER SCALE An earthquake has amplitude of 5,000 micrometers and a period of 0.2 second. Find its measure on the Richter scale.

98. EARTHQUAKES Find the period of an earthquake with amplitude of 80,000 micrometers that measures 6 on the Richter scale.

99. EARTHQUAKES An earthquake with a period of $\frac{1}{4}$ second measures 4 on the Richter scale. Find its amplitude.

100. EARTHQUAKES In 1985, Mexico City experienced an earthquake of magnitude 8.1 on the Richter scale. In 1989, the San Francisco Bay area was rocked by an earthquake measuring 7.1. By what factor must the amplitude of an earthquake change to increase its severity by 1 point on the Richter scale? (Assume that the period remains constant.)

101. DEPRECIATION In business, equipment is often depreciated using the double declining-balance method. In this method, a piece of equipment with a life expectancy of N years, costing $\$C$, will depreciate to a value of $\$V$ in n years, where n is given by the formula

$$n = \frac{\log V - \log C}{\log\left(1 - \frac{2}{N}\right)}$$

A computer that cost \$37,000 has a life expectancy of 5 years. If it has depreciated to a value of \$8,000, how old is it?

102. DEPRECIATION A printer worth \$470 when new had a life expectancy of 12 years. If it is now worth \$189, how old is it? (See Exercise 101.)

103. INVESTING If $\$P$ is invested at the end of each year in an annuity earning annual interest at a rate r, the amount in the account will be $\$A$ after n years, where

$$n = \frac{\log\left[\frac{Ar}{P} + 1\right]}{\log(1 + r)}$$

If \$1,000 is invested each year in an annuity earning 12% annual interest, how long will it take for the account to be worth \$20,000?

104. GROWTH OF MONEY If \$5,000 is invested each year in an annuity earning 8% annual interest, how long will it take for the account to be worth \$50,000? (See Exercise 103.)

WRITING

105. Explain the mathematical relationship between $f(x) = \log x$ and $g(x) = 10^x$.

106. Explain why it is impossible to find the logarithm of a negative number.

REVIEW Solve each equation.

107. $\sqrt[3]{6x + 4} = 4$

108. $\sqrt{3x + 4} = \sqrt{7x + 2}$

109. $\sqrt{a + 1} - 1 = 3a$

110. $3 - \sqrt{t - 3} = \sqrt{t}$

CHALLENGE PROBLEMS

111. Without graphing, determine the domain of the function $f(x) = \log_5 (x^2 - 1)$. Express the result in interval notation.

112. Evaluate: $\log_6 (\log_5 (\log_4 1{,}024))$.

9.6 Base-*e* Logarithms

- Base-*e* logarithms
- The natural logarithmic function and its graph
- An application of base-*e* logarithms

We have seen the importance of e in mathematically modeling the growth and decay of natural events. Just as $f(x) = e^x$ is called the natural exponential function, its inverse, the base-e logarithmic function, is called the *natural logarithmic function.* Natural logarithmic functions have many applications. They play a very important role in advanced mathematics courses, such as calculus.

BASE-*e* LOGARITHMS

Base-e logarithms are called **natural logarithms** or **Napierian logarithms** after John Napier (1550–1617). They are usually written as $\ln x$ rather than $\log_e x$:

$\ln x$ means $\log_e x$ Read $\ln x$ letter-by-letter as "ℓ-n of x."

Caution

Because of the font used to print the natural log of *x*, some students initially misread the notation as In *x*. In handwriting, ln *x* should look like

ℓn x

In general, the logarithm of a number is an exponent. For natural logarithms,

ln x is the exponent to which e is raised to get x.

Translating this statement into symbols, we have

$$e^{\ln x} = x$$

EXAMPLE 1

Evaluate each natural logarithmic expression: **a.** $\ln e$, **b.** $\ln \frac{1}{e^2}$, **c.** $\ln 1$, and **d.** $\ln \sqrt{e}$.

Solution

a. $\ln e = 1$ Ask: "To what power must we raise e to get e?" Since $e^1 = e$, the answer is: the 1st power.

b. $\ln \frac{1}{e^2} = -2$ Ask: "To what power must we raise e to get $\frac{1}{e^2}$?" Since $e^{-2} = \frac{1}{e^2}$, the answer is: the -2 power.

c. $\ln 1 = 0$ Ask: "To what power must we raise e to get 1?" Since $e^0 = 1$, the answer is: the 0 power.

d. $\ln \sqrt{e} = \dfrac{1}{2}$ Ask: "To what power must we raise e to get $\sqrt{e}$?" Since $e^{1/2} = \sqrt{e}$, the answer is: the $\frac{1}{2}$ power.

Self Check 1 Evaluate each expression: **a.** $\ln e^3$, **b.** $\ln \dfrac{1}{e}$, and **c.** $\ln \sqrt[3]{e}$.

Many natural logarithmic expressions are not as easy to evaluate as those in the previous example. For instance, to find ln 2.34, we ask, "To what power must we raise e to get 2.34?" The answer certainly isn't obvious. In such cases, we use a calculator.

ACCENT ON TECHNOLOGY: EVALUATING BASE-*e* (NATURAL) LOGARITHMS

To find ln 2.34 with a scientific calculator, we enter

2.34 [LN] .8501509294

On some calculators, the [e^x] key also serves as the [LN] key when [2nd] or [SHIFT] is pressed. This is because $f(x) = e^x$ and $g(x) = \ln x$ are inverses.

To use a graphing calculator, we enter

[LN] 2.34 [)] [ENTER]

```
ln(2.34)
     .8501509294
```

To four decimal places, $\ln 2.34 = 0.8502$. This means $e^{0.8502} \approx 2.34$.

If we attempt to evaluate logarithmic expressions such as $\ln 0$, or the logarithm of a negative number, such as $\ln(-5)$, the following error statements are displayed.

[Error]

```
ERR:DOMAIN
1:QUIT
2:Goto
```

```
ERR:NONREAL ANS
1:QUIT
2:Goto
```

Certain natural logarithmic equations can be solved by writing them as natural exponential equations.

EXAMPLE 2

Solve each equation: **a.** $\ln x = 1.335$ and **b.** $\ln x = -5.5$. Give each result to four decimal places.

Solution **a.** Since the base of the natural logarithmic function is e, the equation $\ln x = 1.335$ is equivalent to $e^{1.335} = x$. To use a scientific calculator to find x, enter:

1.335 e^x

The display will read 3.799995946. To four decimal places,

$$x = 3.8000$$

The solution is 3.8000. To check, use your calculator to verify that $\ln 3.8000 \approx 1.335$.

b. The equation $\ln x = -5.5$ is equivalent to $e^{-5.5} = x$. To use a scientific calculator to find x, enter:

5.5 +/− e^x

The display will read 0.004086771. To four decimal places,

$$x = 0.0041$$

The solution is 0.0041. To check, use your calculator to verify that $\ln 0.0041 \approx -5.5$.

Self Check 2 Solve: **a.** $\ln x = 1.9344$ and **b.** $-3 = \ln x$. Give each result to four decimal places.

THE NATURAL LOGARITHMIC FUNCTION AND ITS GRAPH

Because the natural exponential function defined by $f(x) = e^x$ is one-to-one, it has an inverse function that is defined by $x = e^y$. When we write $x = e^y$ in the equivalent form $y = \ln x$, the result is called the *natural logarithmic function.*

The Natural Logarithmic Function

The **natural logarithmic function with base *e*** is defined by

$$f(x) = \ln x \quad \text{or} \quad y = \ln x, \quad \text{where } \ln x = \log_e x.$$

The domain of $f(x) = \ln x$ is the interval $(0, \infty)$ and the range is the interval $(-\infty, \infty)$.

Since the natural logarithmic function is the inverse of the one-to-one natural exponential function, the natural logarithmic function is one-to-one.

We can use the point-plotting method to graph $f(x) = \ln x$. We construct a table of function values, plot the resulting ordered pairs, and then draw a smooth curve through the points to get the graph, as shown in figure (a). Figure (b) shows the calculator graph of $f(x) = \ln x$.

$f(x) = \ln x$

To plot these ordered pairs, use a calculator to approximate some x-coordinates.

x	$f(x)$		
$\frac{1}{e}$	-1	→ $\left(\frac{1}{e}, -1\right)$	→ $(0.4, -1)$
1	0	→ $(1, 0)$	→ $(1,0)$
e	1	→ $(e, 1)$	→ $(2.7, 1)$
e^2	2	→ $(e^2, 2)$	→ $(7.4, 2)$

Since the base of the natural logarithmic function is e, choose x-values that are integer powers of e.

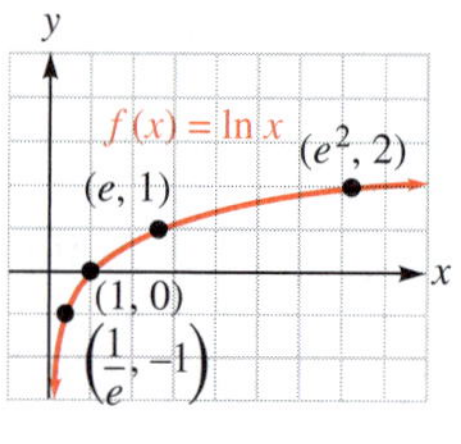

(a)

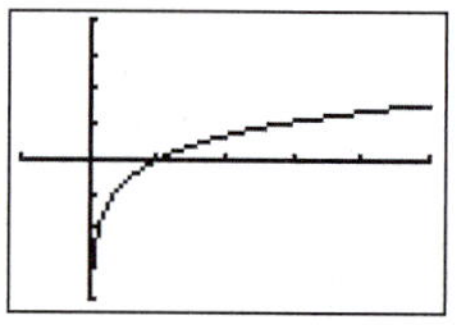

(b)

The exponential function and the natural logarithm function are inverse functions. The figure shows that their graphs are symmetric to the line $y = x$.

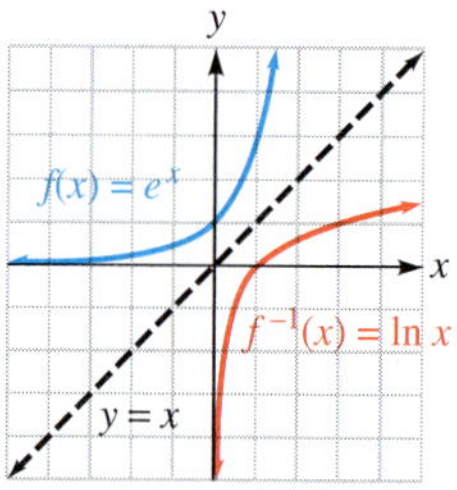

ACCENT ON TECHNOLOGY: GRAPHING BASE-e LOGARITHMIC FUNCTIONS

Many graphs of logarithmic functions involve translations of the graph of $f(x) = \ln x$. For example, the figure below shows calculator graphs of the functions $f(x) = \ln x$, $g(x) = (\ln x) + 2$, and $h(x) = (\ln x) - 3$.

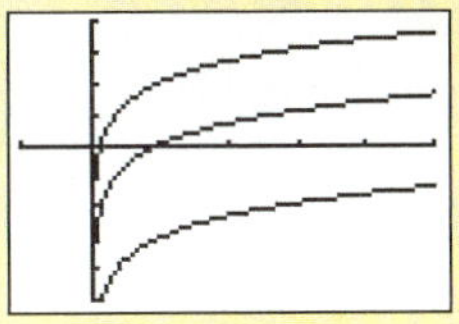

The graph of $g(x) = (\ln x) + 2$ is 2 units above the graph of $f(x) = \ln x$.

The graph of $h(x) = (\ln x) - 3$ is 3 units below the graph of $f(x) = \ln x$.

The next figure shows the calculator graph of the functions $f(x) = \ln x$, $g(x) = \ln(x - 2)$, and $h(x) = \ln(x + 3)$.

The graph of $h(x) = \ln(x + 3)$ is 3 units to the left of the graph of $f(x) = \ln x$.

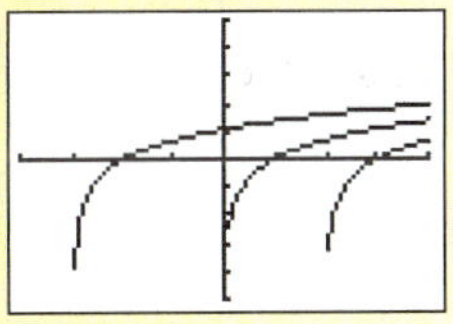

The graph of $g(x) = \ln(x - 2)$ is 2 units to the right of the graph of $f(x) = \ln x$.

AN APPLICATION OF BASE-e LOGARITHMS

Base-e logarithms have many applications. If a population grows exponentially at a certain annual rate, the time required for the population to double is called the **doubling time.** It is given by the following formula.

Formula for Doubling Time If r is the annual rate, compounded continuously, and t is time required for a population to double, then

$$t = \frac{\ln 2}{r}$$

EXAMPLE 3

Doubling time. The population of the Earth is growing at the approximate rate of 2% per year. If this rate continues, how long will it take for the population to double?

Solution Because the population is growing at the rate of 2% per year, we substitute 0.02 for r in the formula for doubling time and simplify.

$$t = \frac{\ln 2}{r}$$

$$t = \frac{\ln 2}{0.02}$$

$$\approx 34.65735903$$ Use a calculator. Find ln 2 first, then divide the result by 0.02.

The population will double in about 35 years.

Self Check 3 If the population's annual growth rate could be reduced to 1.5% per year, what would be the doubling time?

EXAMPLE 4

Doubling time. How long will it take $1,000 to double at an annual rate of 8%, compounded continuously?

Solution We substitute 0.08 for r and simplify:

$$t = \frac{\ln 2}{r}$$

$$t = \frac{\ln 2}{0.08}$$

$$\approx 8.664339757$$ Use a calculator. Find ln 2 first, then divide the result by 0.08.

It will take about $8\frac{2}{3}$ years for the money to double.

Self Check 4 How long will it take at 9%, compounded continuously?

Answers to Self Checks **1. a.** 3, **b.** −1, **c.** $\frac{1}{3}$ **2. a.** 6.9199, **b.** 0.0498 **3.** about 46 years **4.** about 7.7 years

9.6 STUDY SET

VOCABULARY Fill in the blanks.

1. $f(x) = \ln x$ is called the ________ logarithmic function. The base is ___.

2. $f(x) = \ln x$ and $g(x) = e^x$ are ________ functions.

CONCEPTS

3. Use a calculator to complete the table of values for $f(x) = \ln x$. Round to the nearest hundredth.

x	$f(x)$
0.5	
1	
2	
4	
6	
8	
10	

4. Graph $f(x) = \ln x$ in the illustration. (See Exercise 3.) Note that the units on the x- and y-axes are different.

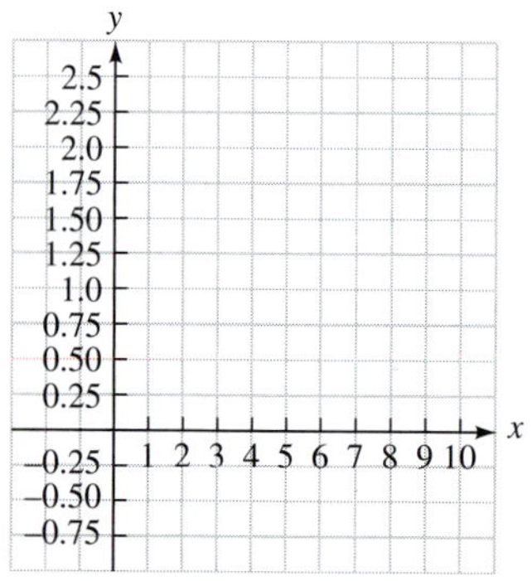

Fill in the blanks.

5. The graph of $f(x) = \ln x$ has the ________ as an asymptote.

6. The domain of the function $f(x) = \ln x$ is the interval ___.

7. The range of the function $f(x) = \ln x$ is the interval ____.

8. To find $\ln e^2$, we ask, "To what power must we raise ____ to get e^2?" Since the answer is the 2nd power, $\ln e^2 =$ ____.

9. The graph of $f(x) = \ln x$ has the x-intercept (____, 0). The y-axis is an ____ of the graph.

10. The logarithmic statement $\ln x = 1.5318$ is equivalent to the exponential statement ____.

11. A table of values for $f(x) = \ln x$ is shown below. Explain why ERROR appears in the Y_1 column for the first three entries.

X	Y1
-2	ERROR
-1	ERROR
0	ERROR
1	0
2	.69315
3	1.0986
4	1.3863

X= -2

12. The illustration shows the graph of $f(x) = \ln x$, as well as a vertical translation of that graph. Using the notation $g(x)$ for the translation, write the defining equation for the function.

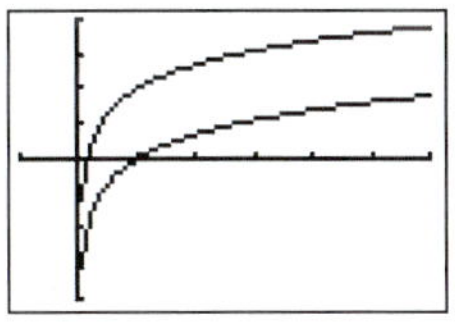

13. In the illustration, $f(x) = \ln x$ was graphed, and the TRACE feature was used. What is the x-coordinate of the point on the graph having a y-coordinate of 1? What is the name given this number?

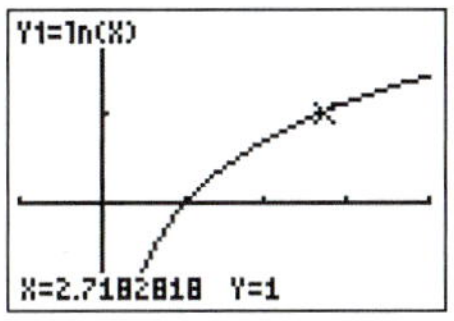

14. The graphs of $f(x) = \ln x$, $g(x) = e^x$, and $y = x$ are shown below. What phrase is used to describe the relationship between the graphs?

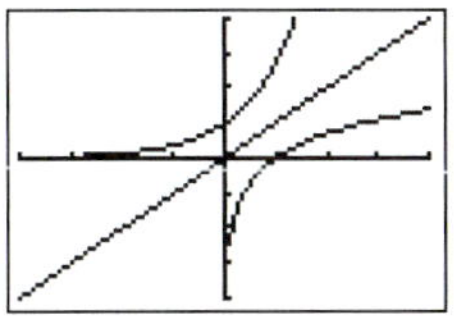

NOTATION **Fill in the blanks.**

15. We read $\ln x$ letter-by-letter as "____-____ of x."

16. **a.** $\ln 2$ means $\log_{\_\_}$ 2.
b. $\log 2$ means $\log_{\_\_}$ 2.

17. If a population grows exponentially at a rate r, the time it will take the population to double is given by the formula $t =$ ____.

18. To evaluate a base-10 logarithm with a calculator, use the ____ key. To evaluate the base-e logarithm, use the ____ key.

Evaluate each expression without using a calculator.

19. $\ln e$
20. $\ln e^2$
21. $\ln e^6$
22. $\ln e^4$
23. $\ln \frac{1}{e}$
24. $\ln \frac{1}{e^3}$
25. $\ln \sqrt[4]{e}$
26. $\ln \sqrt[5]{e}$

PRACTICE **Use a calculator to find each value, if possible. Express all answers to four decimal places.**

27. $\ln 35.15$
28. $\ln 0.675$
29. $\ln 0.00465$
30. $\ln 378.96$
31. $\ln 1.72$
32. $\ln 2.7$
33. $\ln(-0.1)$
34. $\ln(-10)$

Solve each equation. Express all answers to four decimal places.

35. $\ln x = 1.4023$
36. $\ln x = 2.6490$
37. $\ln x = 4.24$
38. $\ln x = 0.926$
39. $\ln x = -3.71$
40. $\ln x = -0.28$
41. $1.001 = \ln x$
42. $\ln x = -0.001$

Use a graphing calculator to graph each function.

43. $y = \ln\left(\frac{1}{2}x\right)$
44. $y = \ln x^2$
45. $f(x) = \ln(-x)$
46. $f(x) = \ln(3x)$

APPLICATIONS **Use a calculator to solve each problem.**

47. POPULATION GROWTH How long will it take the population of River City to double? 5.78

River City
A growing community
• 6 parks • 12% annual growth
• Good schools • Low crime rate

48. DOUBLING MONEY How long will it take \$1,000 to double if it is invested at an annual rate of 5% compounded continuously? 13.86

49. POPULATION GROWTH A population growing continuously at an annual rate r will triple in a time t given by the formula

$$t = \frac{\ln 3}{r}$$

How long will it take the population of a town to triple if it is growing at the rate of 12% per year?

50. TRIPLING MONEY Find the length of time for \$25,000 to triple when it is invested at 6% annual interest, compounded continuously. See Exercise 49.

51. FORENSIC MEDICINE To estimate the number of hours t that a murder victim had been dead, a coroner used the formula

$$t = \frac{1}{0.25} \ln \frac{98.6 - T_s}{82 - T_s}$$

where T_s is the temperature of the surroundings where the body was found. If the crime took place in an apartment where the thermostat was set at 70° F, approximately how long ago did the murder occur?

52. MAKING JELL-O® After the contents of a package of JELL-O® are combined with boiling water, the mixture is placed in a refrigerator whose temperature remains a constant 42° F. Estimate the number of hours t that it will take for the JELL-O® to cool to 50° F using the formula

$$t = -\frac{1}{0.9} \ln \frac{50 - T_r}{200 - T_r}$$

where T_r is the temperature of the refrigerator.

WRITING

53. Explain the difference between the functions $f(x) = \log x$ and $g(x) = \ln x$.

54. How are the functions $f(x) = \ln x$ and $g(x) = e^x$ related?

55. Explain why $\ln e = 1$.

56. Why is $f(x) = \ln x$ called the natural logarithmic function?

REVIEW **Write the equation of the required line.**

57. Parallel to $y = 5x - 8$ and passing through the origin

58. Having a slope of 7 and a y-intercept of 3

59. Passing through the point (3, 2) and perpendicular to the line $y = \frac{2}{3}x - 12$

60. Parallel to the line $3x + 2y = 9$ and passing through the point $(-3, 5)$

61. Vertical line through the point (2, 3)

62. Horizontal line through the point (2, 3)

CHALLENGE PROBLEMS

63. Use the formula $P = P_0e^{rt}$ to verify that P will be twice P_0 when $t = \frac{\ln 2}{r}$.

64. Use the formula $P = P_0e^{rt}$ to verify that P will be three times as large as P_0 when $t = \frac{\ln 3}{r}$.

65. Find a formula to find how long it will take money to quadruple.

Use a graphing calculator to graph the following function and discuss its graph.

66. $$f(x) = \frac{1}{1 + e^{-2x}}$$

9.7 Properties of Logarithms

- Basic properties for logarithms
- The product rule for logarithms
- The quotient rule for logarithms
- The power rule for logarithms
- Writing expressions as a single logarithm
- The change-of-base formula
- An application from chemistry

Since a logarithm is an exponent, we would expect there to be properties of logarithms just as there are properties of exponents. In this section, we will introduce seven properties of logarithms and use them to simplify logarithmic expressions.

BASIC PROPERTIES FOR LOGARITHMS

The first four properties of logarithms follow directly from the definition of logarithm.

Properties of Logarithms For all positive numbers b, where $b \neq 1$,

1. $\log_b 1 = 0$ **2.** $\log_b b = 1$ **3.** $\log_b b^x = x$ **4.** $b^{\log_b x} = x \quad (x > 0)$

We can use properties of exponents to prove that these properties of logarithms are true.

1. $\log_b 1 = 0$, because $b^0 = 1$.
2. $\log_b b = 1$, because $b^1 = b$.
3. $\log_b b^x = x$, because $b^x = b^x$.
4. $b^{\log_b x} = x$, because $\log_b x$ is the exponent to which b is raised to get x.

Properties 3 and 4 also indicate that the composition of the exponential and logarithmic functions (in both directions) is the identity function. This is expected, because the exponential and logarithmic functions are inverse functions.

EXAMPLE 1

Simplify each expression: **a.** $\log_5 1$, **b.** $\log_3 3$, **c.** $\ln e^3$, and **d.** $b^{\log_b 7}$.

Solution

a. By property 1, $\log_5 1 = 0$, because $5^0 = 1$.
b. By property 2, $\log_3 3 = 1$, because $3^1 = 3$.
c. By property 3, $\ln e^3 = 3$, because $e^3 = e^3$.
d. By property 4, $b^{\log_b 7} = 7$, because $\log_b 7$ is the power to which b is raised to get 7.

Self Check 1 Simplify: **a.** $\log_4 1$, **b.** $\log_4 4$, **c.** $\log_2 2^4$, and **d.** $5^{\log_5 2}$.

THE PRODUCT RULE FOR LOGARITHMS

The next property of logarithms is related to the product rule for exponents: $x^m \cdot x^n = x^{m+n}$.

The Product Rule for Logarithms

The logarithm of a product is equal to the sum of the logarithms.
For all positive numbers b, where $b \neq 1$,

$$\log_b MN = \log_b M + \log_b N$$

EXAMPLE 2

Use the product rule for logarithms to write each expression as a sum of logarithms. Then simplify, if possible. **a.** $\log_2(2 \cdot 7)$, **b.** $\log 100x$, and **c.** $\log_5 125yz$.

Solution

a. $\log_2(2 \cdot 7) = \log_2 2 + \log_2 7$ — The log of a product is the sum of the logs.
$= 1 + \log_2 7$ — By property 2, $\log_2 2 = 1$.

b. $\log 100x = \log 100 + \log x$ — The log of a product is the sum of the logs.
$= 2 + \log x$ — By property 3, $\log 100 = \log 10^2 = 2$.

c. $\log_5 125yz = \log_5(125y)z$ — Group the first two factors together.
$= \log_5(125y) + \log_5 z$ — The log of a product is the sum of the logs.
$= \log_5 125 + \log_5 y + \log_5 z$ — The log of a product is the sum of the logs.
$= 3 + \log_5 y + \log_5 z$ — By property 3, $\log_5 125 = \log_5 5^3 = 3$.

Caution

As we apply properties of logarithms to rewrite expressions, assume that all variables represent positive numbers.

Self Check 2 Write each expression as the sum of logarithms. Then simplify, if possible. **a.** $\log_3(4 \cdot 3)$, **b.** $\log 1{,}000y$, and **c.** $\log_5 25cd$.

PROOF

To prove the product rule for logarithms, we let $x = \log_b M$ and $y = \log_b N$. We use the definition of logarithm to write each equation in exponential form.

$$M = b^x \quad \text{and} \quad N = b^y$$

Then $MN = b^x b^y$, and a property of exponents gives

$MN = b^{x+y}$ — Keep the base and add the exponents: $b^x b^y = b^{x+y}$.

We write this exponential equation in logarithmic form as

$$\log_b MN = x + y$$

Substituting the values of x and y completes the proof.

$$\log_b MN = \log_b M + \log_b N$$

□

By the product rule, the logarithm of a *product* is equal to the *sum* of the logarithms. The logarithm of a sum or a difference usually does not simplify. In general,

$$\log_b(M + N) \neq \log_b M + \log_b N \quad \text{and} \quad \log_b(M - N) \neq \log_b M - \log_b N$$

For example,

$$\log_2(2 + 7) \neq \log_2 2 + \log_2 7 \quad \text{and} \quad \log(100 - y) \neq \log 100 - \log y$$

Caution

The log of a sum *does not* equal the sum of the logs. The log of a difference *does not* equal the difference of the logs.

ACCENT ON TECHNOLOGY: VERIFYING PROPERTIES OF LOGARITHMS

We can use a calculator to illustrate the product rule for logarithms by showing that

$$\ln[(3.7)(15.9)] = \ln 3.7 + \ln 15.9$$

We calculate the left- and right-hand sides of the equation separately and compare the results. To use a scientific calculator to find $\ln[(3.7)(15.9)]$, we enter

3.7 [×] 15.9 [=] [LN] — 4.074651929

To find $\ln 3.7 + \ln 15.9$, we enter

3.7 [LN] [+] 15.9 [LN] [=] — 4.074651929

Since the left- and right-hand sides are equal, the equation $\ln[(3.7)(15.9)] = \ln 3.7 + \ln 15.9$ is true.

THE QUOTIENT RULE FOR LOGARITHMS

The next property of logarithms is related to the quotient rule for exponents: $\frac{x^m}{x^n} = x^{m-n}$.

The Quotient Rule for Logarithms

The logarithm of a quotient is equal to the difference of the logarithms.

For all positive numbers b, where $b \neq 1$,

$$\log_b \frac{M}{N} = \log_b M - \log_b N$$

The proof of the quotient rule for logarithms is similar to the proof for the product rule for logarithms.

EXAMPLE 3

Use the quotient rule for logarithms to write each expression as a difference of logarithms. Then simplify, if possible. **a.** $\ln \frac{10}{7}$ and **b.** $\log_4 \frac{x}{64}$.

Solution

a. $\ln \frac{10}{7} = \ln 10 - \ln 7$ — The log of a quotient is the difference of the logs.

b. $\log_4 \frac{x}{64} = \log_4 x - \log_4 64$ — The log of a quotient is the difference of the logs.

$= \log_4 x - 3$ — $\log_4 64 = \log_4 4^3 = 3$.

Self Check 3 Write each expression as a difference of logarithms. Then simplify, if possible. **a.** $\log_6 \frac{6}{5}$ and **b.** $\ln \frac{y}{100}$.

By the quotient rule, the logarithm of a *quotient* is equal to the *difference* of the logarithms. The logarithm of a quotient is not the quotient of the logarithms:

$$\log_b \frac{M}{N} \neq \frac{\log_b M}{\log_b N}$$

For example,

$$\ln \frac{10}{7} \neq \frac{\ln 10}{\ln 7} \quad \text{and} \quad \log_4 \frac{x}{64} \neq \frac{\log_4 x}{\log_4 64}$$

In the next example, the product and quotient rules for logarithms are used in combination to rewrite an expression.

EXAMPLE 4

Use properties of logarithms to rewrite $\log \frac{xy}{10z}$.

Solution We begin by recognizing that the expression is the logarithm of the quotient $\frac{xy}{10z}$.

$$\log \frac{xy}{10z} = \log xy - \log 10z$$ The log of a quotient is the difference of the logs.

$$= \log x + \log y - (\log 10 + \log z)$$ The log of a product is the sum of the logs.

Write parentheses here so that the sum, $\log 10 + \log z$, is subtracted.

$$= \log x + \log y - \log 10 - \log z$$ Remove parentheses.

$$= \log x + \log y - 1 - \log z$$ Simplify: $\log 10 = 1$.

Self Check 4 Use properties of logarithms to rewrite $\log_b \frac{x}{yz}$.

THE POWER RULE FOR LOGARITHMS

The next property of logarithms is related to the power rule for exponents: $(x^m)^n = x^{mn}$.

The Power Rule for Logarithms

The logarithm of a power is equal to the power times the logarithm.

For all positive numbers b, where $b \neq 1$, and any real number p,

$$\log_b M^p = p \log_b M$$

EXAMPLE 5

Use the power rule for logarithms to rewrite each of the following: **a.** $\log_5 6^2$ and **b.** $\log \sqrt{10}$.

Solution

a. $\log_5 6^2 = 2 \log_5 6$ The log of a power is equal to the power times the log.

b. $\log \sqrt{10} = \log (10)^{1/2}$ Write $\sqrt{10}$ using a fractional exponent: $\sqrt{10} = (10)^{1/2}$.

$= \frac{1}{2} \log 10$ The log of a power is equal to the power times the log.

$= \frac{1}{2}$ Simplify: $\log 10 = 1$.

Self Check 5 Use the power rule for logarithms to rewrite **a.** $\ln x^4$ and **b.** $\log_2 \sqrt[3]{3}$.

PROOF

To prove the power rule, we let $x = \log_b M$, write the expression in exponential form, and raise both sides to the pth power:

$$\begin{aligned} M &= b^x \\ (M)^p &= (b^x)^p && \text{Raise both sides to the } p\text{th power.} \\ M^p &= b^{px} && \text{Keep the base and multiply the exponents.} \end{aligned}$$

Using the definition of logarithms gives

$$\log_b M^p = px$$

Substituting the value for x completes the proof.

$$\log_b M^p = p \log_b M$$

□

EXAMPLE 6

Use logarithm properties to rewrite each expression: **a.** $\log_b x^2y^3z$ and **b.** $\ln \frac{y^3\sqrt{x}}{z}$.

Solution

a. We begin by recognizing that $\log_b x^2y^3z$ is the logarithm of a product.

$$\begin{aligned} \log_b x^2y^3z &= \log_b x^2 + \log_b y^3 + \log_b z && \text{The log of a product is the sum of the logs.} \\ &= 2\log_b x + 3\log_b y + \log_b z && \text{The log of a power is the power times the log.} \end{aligned}$$

b. The expression $\ln \frac{y^3\sqrt{x}}{z}$ is the logarithm of a quotient.

$$\begin{aligned} \ln \frac{y^3\sqrt{x}}{z} &= \ln (y^3\sqrt{x}) - \ln z && \text{The log of a quotient is the difference of the logs.} \\ &= \ln y^3 + \ln \sqrt{x} - \ln z && \text{The log of a product is the sum of the logs.} \\ &= \ln y^3 + \ln x^{1/2} - \ln z && \text{Write } \sqrt{x} \text{ as } x^{1/2}. \\ &= 3\ln y + \frac{1}{2}\ln x - \ln z && \text{The log of a power is the power times the log.} \end{aligned}$$

The Language of Algebra

In Examples 2, 3, 4, and 6, we use properties of logarithms to *expand* logarithmic expressions.

Self Check 6 Expand: $\log \sqrt[4]{\frac{x^3y}{z}}$.

WRITING EXPRESSIONS AS A SINGLE LOGARITHM

EXAMPLE 7

Write each of the given expressions as one logarithm: **a.** $3\log_5 x + \frac{1}{2}\log_5 y$ and **b.** $\frac{1}{2}\log_b (x - 2) - \log_b y + 3\log_b z$.

Solution

a. We begin by applying the power rule to each term of the expression.

$$\begin{aligned} 3\log_5 x + \frac{1}{2}\log_5 y &= \log_5 x^3 + \log_5 y^{1/2} && \text{A power times a log is the log of the power.} \\ &= \log_5 x^3y^{1/2} && \text{The sum of two logs is the log of the product.} \end{aligned}$$

The Language of Algebra

In these examples, we use properties of logarithms to *condense* the given expression into a single logarithmic expression. To *condense* means to make more compact. Summer school is a *condensed* version of the regular semester.

b. The first and third terms of this expression can be rewritten using the power rule of logarithms.

$$\frac{1}{2}\log_b(x-2) - \log_b y + 3\log_b z$$

$= \log_b(x-2)^{1/2} - \log_b y + \log_b z^3$ — A power times a log is the log of the power.

$= \log_b \dfrac{(x-2)^{1/2}}{y} + \log_b z^3$ — The difference of two logs is the log of the quotient.

$= \log_b \dfrac{z^3\sqrt{x-2}}{y}$ — The sum of two logs is the log of the product. Write $(x-2)^{1/2}$ as $\sqrt{x-2}$.

Self Check 7 Write the expression as one logarithm: $2\log_a x + \frac{1}{2}\log_a y - 2\log_a(x-y)$.

The properties of logarithms can be used when working with numerical values.

EXAMPLE 8

Given that $\log 2 \approx 0.3010$ and $\log 3 \approx 0.4771$, find approximations for **a.** $\log 6$ and **b.** $\log 18$.

Solution

a. $\log 6 = \log(2 \cdot 3)$ — Write 6 using the factors 2 and 3.

$= \log 2 + \log 3$ — The log of a product is the sum of the logs.

$\approx 0.3010 + 0.4771$ — Substitute the value of each logarithm.

≈ 0.7781

b. $\log 18 = \log(2 \cdot 3^2)$ — Write 18 using the factors 2 and 3.

$= \log 2 + \log 3^2$ — The log of a product is the sum of the logs.

$= \log 2 + 2\log 3$ — The log of a power is the power times the log.

$\approx 0.3010 + 2(0.4771)$ — Substitute the value of each logarithm.

≈ 1.2552

Self Check 8 Find: **a.** $\log 1.5$ and **b.** $\log 0.75$.

We summarize the properties of logarithms as follows.

Properties of Logarithms

If b, M, and N are positive numbers, $b \neq 1$, and p is any real number,

1. $\log_b 1 = 0$ **2.** $\log_b b = 1$

3. $\log_b b^x = x$ **4.** $b^{\log_b x} = x$

5. $\log_b MN = \log_b M + \log_b N$ **6.** $\log_b \dfrac{M}{N} = \log_b M - \log_b N$

7. $\log_b M^p = p\log_b M$

THE CHANGE-OF-BASE FORMULA

Most calculators can find common logarithms and natural logarithms. If we need to find a logarithm with some other base, we use a conversion formula.

If we know the base-a logarithm of a number, we can find its logarithm to some other base b by using a formula called the **change-of-base formula.**

Change-of-Base Formula If a, b, and x are positive, $a \neq 1$, and $b \neq 1$,

$$\log_b x = \frac{\log_a x}{\log_a b}$$

If we know logarithms to base a (for example, $a = 10$), we can find the logarithm of x to a new base b. We simply divide the base-a logarithm of x by the base-a logarithm of b.

EXAMPLE 9

Find $\log_3 5$.

Solution We can use base-10 logarithms to find a base-3 logarithm. To do this, we substitute 3 for b, 10 for a, and 5 for x in the change-of-base formula:

$$\log_b x = \frac{\log_a x}{\log_a b}$$

$$\log_3 5 = \frac{\log_{10} 5}{\log_{10} 3} \quad b = 3,\ x = 5,\text{ and } a = 10.$$

$$\approx 1.464973521 \quad \text{Use a scientific calculator and enter 5 log } \div \text{ 3 log } = .$$

To four decimal places, $\log_3 5 = 1.4650$.

Caution

Don't misapply the quotient rule:

$\frac{\log_{10} 5}{\log_{10} 3}$ means $\log_{10} 5 \div \log_{10} 3$.

It is the expression $\log_{10} \frac{5}{3}$ that means $\log_{10} 5 - \log_{10} 3$.

We can also use the natural logarithm function (base e) in the change-of-base formula to find a base-3 logarithm.

$$\log_b x = \frac{\log_a x}{\log_a b}$$

$$\log_3 5 = \frac{\ln 5}{\ln 3} \quad b = 3,\ x = 5,\text{ and } a = e.\ \log_e 5 = \ln 5 \text{ and } \log_e 3 = \ln 3.$$

$$\approx 1.464973521 \quad \text{Use a calculator.}$$

We obtain the same result.

Caution

Wait until the final calculation has been made to round. Don't round any values when performing intermediate calculations. That could make the final result incorrect because of a build-up of rounding error.

Self Check 9 Find $\log_5 3$ to four decimal places.

PROOF

To prove the change-of-base formula, we begin with the equation $\log_b x = y$.

$$y = \log_b x$$

$$x = b^y \quad \text{Change the equation from logarithmic to exponential form.}$$

$$\log_a x = \log_a b^y \quad \text{Take the base-}a\text{ logarithm of both sides.}$$

$$\log_a x = y \log_a b \quad \text{The log of a power is the power times the log.}$$

$$y = \frac{\log_a x}{\log_a b} \quad \text{Divide both sides by } \log_a b.$$

$$\log_b x = \frac{\log_a x}{\log_a b} \quad \text{Refer to the first equation and substitute } \log_b x \text{ for } y.$$

□

AN APPLICATION FROM CHEMISTRY

In chemistry, common logarithms are used to express the acidity of solutions. The more acidic a solution, the greater the concentration of hydrogen ions. This concentration is indicated indirectly by the *pH scale,* or *hydrogen ion index.* The pH of a solution is defined as follows.

pH of a Solution If $[H^+]$ is the hydrogen ion concentration in gram-ions per liter, then

$$pH = -\log [H^+]$$

EXAMPLE 10

pH meters. One of the most accurate ways to measure pH is with a probe and meter. What reading should the meter give for pure water if water has a hydrogen ion concentration $[H^+]$ of approximately 10^{-7} gram-ions per liter?

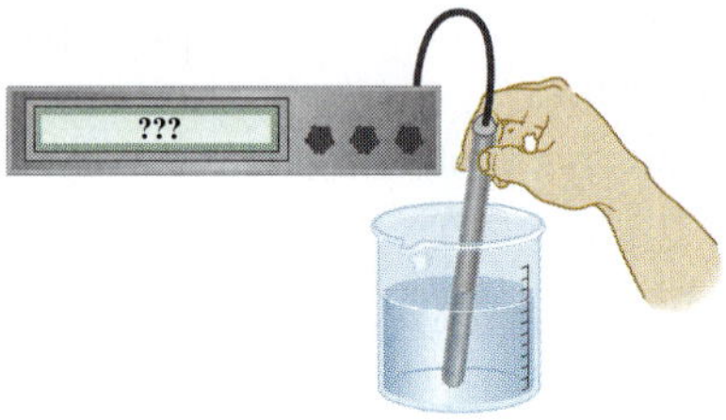

Solution Since pure water has approximately 10^{-7} gram-ions per liter, its pH is

$$\begin{aligned} pH &= -\log [H^+] && \text{This is the formula for pH.} \\ pH &= -\log 10^{-7} \\ &= -(-7)\log 10 && \text{The log of a power is the power times the log.} \\ &= -(-7)\cdot 1 && \text{Simplify: } \log 10 = 1. \\ &= 7 \end{aligned}$$

The meter should give a reading of 7.

EXAMPLE 11

Hydrogen ion concentration. Find the hydrogen ion concentration of seawater if its pH is 8.5.

Solution To find its hydrogen ion concentration, we solve the following equation for $[H^+]$.

$$\begin{aligned} pH &= -\log [H^+] && \text{This is the formula for pH.} \\ 8.5 &= -\log [H^+] && \text{Substitute 8.5 for pH.} \\ -8.5 &= \log [H^+] && \text{Multiply both sides by } -1. \\ [H^+] &= 10^{-8.5} && \text{Write the equation in the equivalent exponential form.} \end{aligned}$$

We can use a calculator to find that

$$[H^+] \approx 3.2 \times 10^{-9} \text{ gram-ions per liter}$$

Answers to Self Checks **1. a.** 0, **b.** 1, **c.** 4, **d.** 2 **2. a.** $\log_3 4 + 1$, **b.** $3 + \log y$, **c.** $2 + \log_5 c + \log_5 d$ **3. a.** $1 - \log_6 5$, **b.** $\ln y - \ln 100$ **4.** $\log_b x - \log_b y - \log_b z$ **5. a.** $4 \ln x$, **b.** $\frac{1}{3}\log_2 3$ **6.** $\frac{1}{4}(3\log x + \log y - \log z)$ **7.** $\log_a \frac{x^2\sqrt{y}}{(x-y)^2}$ **8. a.** 0.1761, **b.** −0.1249 **9.** 0.6826

9.7 STUDY SET

VOCABULARY **Fill in the blanks.**

1. The expression $\log_3(4x)$ is the logarithm of a ________.
2. The expression $\log_2 \frac{5}{x}$ is the logarithm of a ________.
3. The expression $\log 4^x$ is the logarithm of a ________.
4. In the expression $\log_5 4$, the number 5 is the ______ of the logarithm.

CONCEPTS **Fill in the blanks.**

5. $\log_b 1 =$ ____
6. $\log_b b =$ ____
7. $\log_b MN = \log_b$ ____ $+ \log_b$ ____
8. $b^{\log_b x} =$ ____
9. $\log_b \frac{M}{N} = \log_b M$ ____ $\log_b N$
10. $\log_b M^p = p \log_b$ ____
11. $\log_b b^x =$ ____
12. $\log_b (A + B)$ ____ $\log_b A + \log_b B$
13. $\log_b \frac{M}{N}$ ____ $\frac{\log_b M}{\log_b N}$
14. $\log_b AB$ ____ $\log_b A + \log_b B$
15. $\log_b x = \frac{\log_a x}{\text{____}}$
16. pH = ____

17. Three logarithmic expressions have been evaluated, and the results are shown on the calculator display. Show that each result is correct by writing the equivalent base-10 exponential statement.

```
log(1)
               0
log(10)
               1
log(10²)
               2
```

18. Three logarithmic expressions have been evaluated, and the results are shown on the calculator display. Show that each result is correct by writing the equivalent base-e exponential statement. (The notation ln(e ∧ (2)) means $\ln e^2$.)

```
ln(e^(2))
               2
ln(e^(3))
               3
ln(e^(4))
               4
```

Evaluate each expression.

19. $\log_4 1$
20. $\log_4 4$
21. $\log_4 4^7$
22. $\ln e^8$
23. $5^{\log_5 10}$
24. $8^{\log_8 10}$
25. $\log_5 5^2$
26. $\log_4 4^2$
27. $\ln e$
28. $\log_7 1$
29. $\log_3 3^7$
30. $5^{\log_5 8}$

NOTATION **Complete each solution.**

31. $\log_b rst = \log_b(\text{__})t$
$= \log_b(rs) + \log_b$ __
$= \log_b$ __ $+ \log_b$ __ $+ \log_b t$

32. $\log \frac{r}{st} = \log r - \log(\text{__})$
$= \log r - (\log \text{__} + \log t)$
$= \log r - \log s$ __ $\log t$

PRACTICE **Use a calculator to verify that each equation is true.**

33. $\log[(2.5)(3.7)] = \log 2.5 + \log 3.7$
34. $\log 45.37 = \frac{\ln 45.37}{\ln 10}$
35. $\ln(2.25)^4 = 4 \ln 2.25$
36. $\ln \frac{11.3}{6.1} = \ln 11.3 - \ln 6.1$
37. $\log \sqrt{24.3} = \frac{1}{2} \log 24.3$
38. $\ln 8.75 = \frac{\log 8.75}{\log e}$

Use the properties of logarithms to rewrite each expression. Assume that x, y, and z are positive numbers.

39. $\log_2(4 \cdot 5)$
40. $\log_3(27 \cdot 5)$
41. $\log_6 \frac{x}{36}$
42. $\log_8 \frac{y}{8}$
43. $\ln y^4$
44. $\ln z^9$
45. $\log \sqrt{5}$
46. $\log \sqrt[3]{7}$

Assume that x, y, z, and b are positive numbers and $b \neq 1$. Use the properties of logarithms to write each expression in terms of the logarithms of x, y, and z.

47. $\log xyz$

48. $\log 4xz$

49. $\log_2 \dfrac{2x}{y}$

50. $\log_3 \dfrac{x}{yz}$

51. $\log x^3y^2$

52. $\log xy^2z^3$

53. $\log_b (xy)^{1/2}$

54. $\log_b x^3y^{1/2}$

55. $\log_a \dfrac{\sqrt[3]{x}}{\sqrt[4]{yz}}$

56. $\log_b \sqrt[4]{\dfrac{x^3y^2}{z^4}}$

57. $\ln x\sqrt{z}$

58. $\ln \sqrt{xy}$

Assume that x, y, z, and b are positive numbers and $b \neq 1$. Use the properties of logarithms to write each expression as the logarithm of a single quantity.

59. $\log_2 (x + 1) - \log_2 x$

60. $\log_3 x + \log_3 (x + 2) - \log_3 8$

61. $2 \log x + \dfrac{1}{2} \log y$

62. $-2 \log x - 3 \log y + \log z$

63. $-3 \log_b x - 2 \log_b y + \dfrac{1}{2} \log_b z$

64. $3 \log_b (x + 1) - 2 \log_b (x + 2) + \log_b x$

65. $\ln\left(\dfrac{x}{z} + x\right) - \ln\left(\dfrac{y}{z} + y\right)$

66. $\ln (xy + y^2) - \ln (xz + yz) + \ln z$

Determine whether the given statement is true. If a statement is false, explain why.

67. $\log xy = (\log x)(\log y)$

68. $\log ab = \log a + 1$

69. $\log_b (A - B) = \dfrac{\log_b A}{\log_b B}$

70. $\dfrac{\log_b A}{\log_b B} = \log_b A - \log_b B$

71. $\log_b \dfrac{A}{B} = \log_b A - \log_b B$

72. $\log_b 2 = \log_2 b$

Assume that $\log_b 4 = 0.6021$, $\log_b 7 = 0.8451$, and $\log_b 9 = 0.9542$. Use these values and the properties of logarithms to find each value.

73. $\log_b 28$

74. $\log_b \dfrac{7}{4}$

75. $\log_b \dfrac{4}{63}$

76. $\log_b 36$

77. $\log_b \dfrac{63}{4}$

78. $\log_b 2.25$

79. $\log_b 64$

80. $\log_b 49$

Use the change-of-base formula to find each logarithm to four decimal places.

81. $\log_3 7$

82. $\log_7 3$

83. $\log_{1/3} 3$

84. $\log_{1/2} 6$

85. $\log_3 8$

86. $\log_5 10$

87. $\log_{\sqrt{2}} \sqrt{5}$

88. $\log_{\pi} e$

APPLICATIONS

89. pH OF A SOLUTION Find the pH of a solution with a hydrogen ion concentration of 1.7×10^{-5} gram-ions per liter.

90. HYDROGEN ION CONCENTRATION Find the hydrogen ion concentration of a saturated solution of calcium hydroxide whose pH is 13.2.

91. AQUARIUMS To test for safe pH levels in a freshwater aquarium, a test strip is compared with the scale shown below. Find the corresponding range in the hydrogen ion concentration.

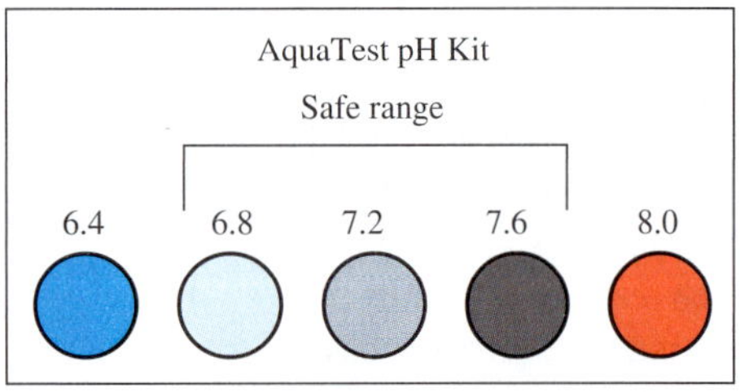

92. pH OF PICKLES The hydrogen ion concentration of sour pickles is 6.31×10^{-4}. Find the pH.

WRITING

93. Explain the difference between a logarithm of a product and the product of logarithms.

94. How can the LOG key on a calculator be used to find $\log_2 7$?

REVIEW

Consider the line that passes through $P(-2, 3)$ and $Q(4, -4)$.

95. Find the slope of line PQ.

96. Find the distance PQ.

97. Find the midpoint of line segment PQ.

98. Write the equation of line PQ.

CHALLENGE PROBLEMS

99. Explain why $e^{\ln x} = x$.

100. If $\log_b 3x = 1 + \log_b x$, find b.

101. Show that $\log_{b^2} x = \frac{1}{2} \log_b x$.

102. Show that $e^{x \ln a} = a^x$.

9.8 Exponential and Logarithmic Equations

- Solving exponential equations
- Solving logarithmic equations
- Radioactive decay
- Population growth

An **exponential equation** is an equation that contains a variable in one of its exponents. Some examples of exponential equations are

$$3^{x+1} = 81, \quad 6^{x-3} = 2^x, \quad \text{and} \quad e^{0.9t} = 8$$

A **logarithmic equation** is an equation with a logarithmic expression that contains a variable. Some examples of logarithmic equations are

$$\log 5x = 3, \quad \log(3x + 2) = \log(2x - 3), \quad \text{and} \quad \frac{\log_2(5x - 6)}{\log_2 x} = 2$$

In this section, we will learn how to solve exponential and logarithmic equations.

SOLVING EXPONENTIAL EQUATIONS

If both sides of an exponential equation can be expressed as a power of the same base, we can use the following property to solve the equation.

Exponent Property of Equality

If two exponential expressions with the same base are equal, their exponents are equal. For any real number b, where $b \neq -1, 0$, or 1,

$$b^x = b^y \text{ is equivalent to } x = y$$

EXAMPLE 1

Solve: $3^{x+1} = 81$.

Solution Since $81 = 3^4$, each side of the equation can be expressed as a power of 3.

$$3^{x+1} = 81$$

$$3^{x+1} = 3^4 \qquad \text{Write 81 as } 3^4.$$

$$x + 1 = 4 \qquad \text{If two exponential expressions with the same base are equal, their exponents are equal.}$$

$$x = 3$$

The solution is 3. To check this result, we substitute 3 for x in the original equation.

Check:
$$3^{x+1} = 81$$
$$3^{3+1} \stackrel{?}{=} 81$$
$$3^{4} \stackrel{?}{=} 81$$
$$81 = 81$$

Self Check 1 Solve: $5^{3x-4} = 25$.

EXAMPLE 2

Solve: $2^{x^2+2x} = \frac{1}{2}$.

Solution Since $\frac{1}{2} = 2^{-1}$, each side of the equation can be expressed as a power of 2.

$$2^{x^2+2x} = \frac{1}{2}$$

$$2^{x^2+2x} = 2^{-1} \quad \text{Write } \tfrac{1}{2} \text{ as } 2^{-1}.$$

$$x^2 + 2x = -1 \quad \text{If two exponential expressions with the same base are equal, their exponents are equal.}$$

$$x^2 + 2x + 1 = 0 \quad \text{Add 1 to both sides.}$$

$$(x + 1)(x + 1) = 0 \quad \text{Factor the trinomial.}$$

$$x + 1 = 0 \quad \text{or} \quad x + 1 = 0 \quad \text{Set each factor equal to 0.}$$

$$x = -1 \quad\quad x = -1$$

The solution is -1. Verify that it satisfies the original equation.

Self Check 2 Solve: $3^{x^2-2x} = \frac{1}{3}$.

ACCENT ON TECHNOLOGY: SOLVING EXPONENTIAL EQUATIONS GRAPHICALLY

To use a graphing calculator to approximate the solutions of $2^{x^2+2x} = \frac{1}{2}$ (see Example 2), we can subtract $\frac{1}{2}$ from both sides of the equation to get

$$2^{x^2+2x} - \frac{1}{2} = 0$$

and graph the corresponding function

$$y = 2^{x^2+2x} - \frac{1}{2}$$

as shown in figure (a) on the next page.

The solutions of $2^{x^2+2x} - \frac{1}{2} = 0$ are the x-coordinates of the x-intercepts of the graph of $y = 2^{x^2+2x} - \frac{1}{2}$. Using the ZERO feature, we see in figure (a) that the graph has only one x-intercept, $(-1, 0)$. Therefore, $x = -1$ is the only solution of $2^{x^2+2x} - \frac{1}{2} = 0$.

We can also solve $2^{x^2+2x} = \frac{1}{2}$ using the INTERSECT feature found on most graphing calculators. After graphing $Y_1 = 2^{x^2+2x}$ and $Y_2 = \frac{1}{2}$, we select INTERSECT, which approximates the coordinates of the point of intersection of the two graphs. From the display shown in figure (b), we can conclude that the solution is $x = -1$. Verify this by checking.

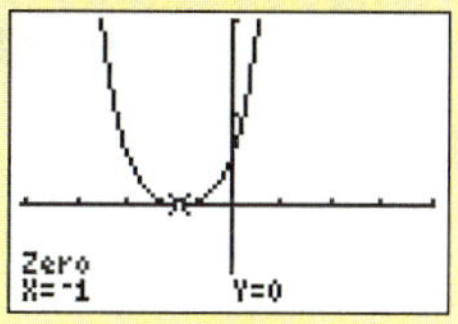

(a)

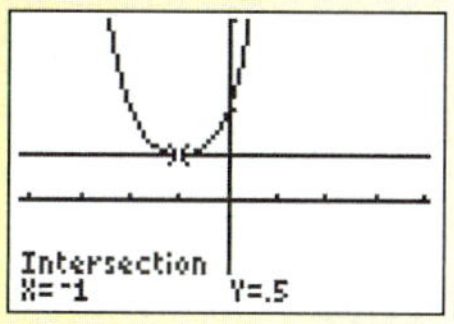

(b)

When it is difficult or impossible to write each side of an exponential equation as a power of the same base, we can often use the following property of logarithms to solve the equation.

The Logarithm Property of Equality

If two positive numbers are equal, then the logarithms base-b of the numbers are equal. For any positive number b, where $b \neq 1$, and positive numbers x and y,

$$\log_b x = \log_b y \quad \text{is equivalent to} \quad x = y$$

EXAMPLE 3

Solve: $3^x = 5$.

Solution

Unlike Example 1, $3^{x+1} = 81$, it is not possible to write each side of $3^x = 5$ as a power of the same base 3. Instead, we use the logarithm property of equality and *take the logarithm of both sides* to solve the equation. Any base logarithm can be chosen, however, the computations with a calculator are usually simplest if the common or the natural logarithm is used.

$$3^x = 5$$

$$\log 3^x = \log 5 \quad \text{Take the common logarithm of each side.}$$

$$x \log 3 = \log 5 \quad \text{The log of a power is the power times the log: } \log 3^x = x \log 3.$$

$$x = \frac{\log 5}{\log 3} \quad \text{Divide both sides by log 3. This is the exact solution.}$$

$$x \approx 1.464973521 \quad \text{Use a calculator.}$$

The exact solution is $\frac{\log 5}{\log 3}$. To four decimal places, the solution is 1.4650.

Success Tip

The power rule of logarithms provides a way of moving the variable x from its position in the exponent to a position as a factor of $x \log 3$.

$$\log 3^x = \log 5$$
$$x \log 3 = \log 5$$

We can also take the natural logarithm of each side of the equation to solve for x.

$$3^x = 5$$

$$\ln 3^x = \ln 5 \quad \text{Take the natural logarithm of each side.}$$

$$x \ln 3 = \ln 5 \quad \text{Use the power rule of logarithms: } \ln 3^x = x \ln 3.$$

$$x = \frac{\ln 5}{\ln 3} \quad \text{Divide both sides by ln 3.}$$

$$x \approx 1.464973521 \quad \text{Use a calculator.}$$

Caution

Don't misapply the quotient rule:

$\frac{\ln 5}{\ln 3}$ means $\ln 5 \div \ln 3$.

It is the expression $\ln \frac{5}{3}$ that means $\ln 5 - \ln 3$.

The result is the same using the natural logarithm. To check, we substitute 1.4650 for x in 3^x and see if $3^{1.4650}$ is close to 5.

Check:

$$\begin{aligned} 3^x &= 5 \\ 3^{1.4650} &\stackrel{?}{=} 5 \\ 5.000145454 &\approx 5 && \text{Use a calculator: Enter 3 } \boxed{y^x} \text{ 1.4650 } \boxed{=}. \end{aligned}$$

Self Check 3 Solve $5^x = 4$. Give the answer to four decimal places.

EXAMPLE 4

Solve: $6^{x-3} = 2^x$.

Solution

$$\begin{aligned} 6^{x-3} &= 2^x \\ \log 6^{x-3} &= \log 2^x && \text{Take the common logarithm of each side.} \\ (x-3)\log 6 &= x \log 2 && \text{The log of a power is the power times the log.} \\ x\log 6 - 3\log 6 &= x\log 2 && \text{Use the distributive property.} \\ x\log 6 - x\log 2 &= 3\log 6 && \text{On both sides, add } 3\log 6 \text{ and subtract } x\log 2. \\ x(\log 6 - \log 2) &= 3\log 6 && \text{Factor out } x \text{ on the left-hand side.} \\ x &= \frac{3\log 6}{\log 6 - \log 2} && \text{Divide both sides by } \log 6 - \log 2. \\ x &\approx 4.892789261 && \text{Use a calculator.} \end{aligned}$$

To four decimal places, the solution is 4.8928.

The Language of Algebra

$\dfrac{3\log 6}{\log 6 - \log 2}$ is the *exact* solution of $6^{x-3} = 2^x$. An *approximate* solution is 4.8928.

Self Check 4 Solve: $5^{x-2} = 3^x$.

EXAMPLE 5

Solve: $e^{0.9t} = 10$.

Solution The exponential expression on the left-hand side has base e. In such cases, the computations are somewhat simpler if we take the natural logarithm of each side.

$$\begin{aligned} e^{0.9t} &= 10 \\ \ln e^{0.9t} &= \ln 10 && \text{Take the natural logarithm of each side.} \\ 0.9t \ln e &= \ln 10 && \text{Use the power rule of logarithms: } \ln e^{0.9t} = 0.9t \ln e. \\ 0.9t \cdot 1 &= \ln 10 && \text{Simplify: } \ln e = 1. \\ 0.9t &= \ln 10 \\ t &= \frac{\ln 10}{0.9} && \text{Divide both sides by 0.9.} \\ t &\approx 2.558427881 && \text{Use a calculator.} \end{aligned}$$

To four decimal places, the solution is 2.5584

Self Check 5 Solve: $e^{2.1t} = 35$.

SOLVING LOGARITHMIC EQUATIONS

We can solve many logarithmic equations using properties of logarithms.

EXAMPLE 6

Solve: $\log 5x = 3$.

Solution Recall that $\log 5x = \log_{10} 5x$. We can write the equation $\log 5x = 3$ as an equivalent base-10 exponential equation $10^3 = 5x$ and solve for x.

$$\log 5x = 3$$
$$10^3 = 5x \quad \text{Write using exponential form.}$$
$$1{,}000 = 5x \quad \text{Simplify: } 10^3 = 1{,}000.$$
$$200 = x \quad \text{Divide both sides by 5.}$$

The solution is 200.

Caution

Always check your solutions to a logarithmic equation.

Check:

$$\log 5x = 3$$
$$\log 5(200) \stackrel{?}{=} 3 \quad \text{Substitute 200 for } x.$$
$$\log 1{,}000 \stackrel{?}{=} 3$$
$$3 = 3 \quad \log 1{,}000 = \log 10^3 = 3.$$

Self Check 6 Solve: $\log_2 (x - 3) = -1$.

EXAMPLE 7

Solve: $\log (3x + 2) = \log (2x - 3)$.

Solution We can use the logarithm property of equality to solve the equation.

Caution

Don't make this error of trying to "distribute" log:

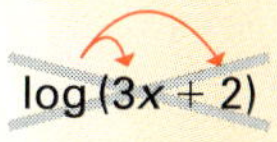

The notation *log* is not a number, it is the name of a function and cannot be distributed.

$$\log (3x + 2) = \log (2x - 3)$$
$$(3x + 2) = (2x - 3) \quad \text{If the logarithms of two numbers are equal, the numbers are equal.}$$
$$x + 2 = -3 \quad \text{Subtract } 2x \text{ from both sides.}$$
$$x = -5$$

Check:

$$\log (3x + 2) = \log (2x - 3)$$
$$\log [3(-5) + 2] \stackrel{?}{=} \log [2(-5) - 3] \quad \text{Substitute } -5 \text{ for } x.$$
$$\log (-13) \stackrel{?}{=} \log (-13)$$

Since the logarithm of a negative number does not exist, the proposed solution of -5 must be discarded. This equation has no solutions.

Self Check 7 Solve: $\log (5x + 2) = \log (7x - 2)$.

EXAMPLE 8

Solve: $\log x + \log(x - 3) = 1$.

Solution

$$\log x + \log(x - 3) = 1$$

$\log x(x - 3) = 1$ — Use the product rule of logarithms.

$\log_{10} x(x-3) = 1$ — The base is 10.

$x(x-3) = 10^1$ — Write the equation in exponential form.

$x^2 - 3x - 10 = 0$ — Remove parentheses. Subtract 10 from both sides.

$(x + 2)(x - 5) = 0$ — Factor the trinomial.

$x + 2 = 0$ or $x - 5 = 0$ — Set each factor equal to 0.

$x = -2$ | $x = 5$

Check: The number -2 is not a solution, because it does not satisfy the equation (a negative number does not have a logarithm). We will check the other result, 5.

$$\log x + \log(x - 3) = 1$$

$\log 5 + \log(5 - 3) \stackrel{?}{=} 1$ — Substitute 5 for x.

$\log 5 + \log 2 \stackrel{?}{=} 1$

$\log 10 \stackrel{?}{=} 1$ — Use the product rule of logarithms: $\log 5 + \log 2 = \log(5 \cdot 2) = \log 10$.

$1 = 1$ — Simplify: $\log 10 = 1$.

Since 5 satisfies the equation, it is the solution.

Caution

Proposed solutions of a logarithmic equation must be checked to see whether they produce undefined logarithms in the original equation.

Self Check 8 Solve: $\log x + \log(x + 3) = 1$.

ACCENT ON TECHNOLOGY: SOLVING LOGARITHMIC EQUATIONS GRAPHICALLY

To use a graphing calculator to approximate the solutions of the logarithmic equation $\log x + \log(x - 3) = 1$ (see Example 8), we can subtract 1 from both sides of the equation to get

$$\log x + \log(x - 3) - 1 = 0$$

and graph the corresponding function

$$y = \log x + \log(x - 3) - 1$$

as shown in figure (a). Since the solution of the equation is the x-coordinate of the x-intercept, we can find the solutions using the ZERO feature.

We can also solve $\log x + \log(x - 3) = 1$ using the INTERSECT feature. After graphing $Y_1 = \log x + \log(x - 3)$ and $Y_2 = 1$, we select INTERSECT, which approximates the coordinates of the point of intersection of the two graphs. From the display shown in figure (b), we can conclude that the solution is $x = 5$.

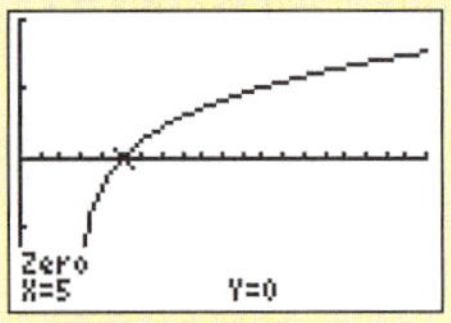

(a)

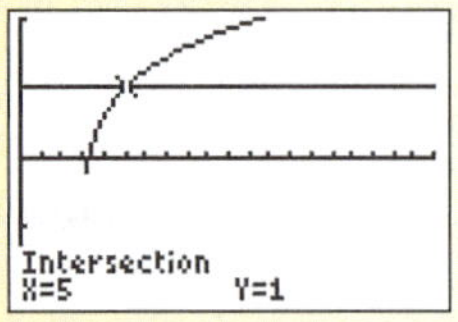

(b)

EXAMPLE 9

Solve: $\dfrac{\log_2 (5x - 6)}{\log_2 x} = 2.$

Solution We can multiply both sides of the equation by $\log_2 x$ to get

$$\log_2 (5x - 6) = 2 \log_2 x$$

and apply the power rule of logarithms to get

$$\log_2 (5x - 6) = \log_2 x^2$$

By the logarithm property of equality, $5x - 6 = x^2$, because they have equal logarithms. Thus,

$$\begin{aligned} 5x - 6 &= x^2 \\ 0 &= x^2 - 5x + 6 \\ 0 &= (x - 3)(x - 2) \end{aligned}$$

$$x - 3 = 0 \quad \text{or} \quad x - 2 = 0$$
$$x = 3 \quad \Big| \quad x = 2$$

Verify that both 2 and 3 satisfy the equation.

Self Check 9 Solve: $\dfrac{\log_3 (5x + 6)}{\log_3 x} = 2.$

RADIOACTIVE DECAY

Experiments have determined the time it takes for half of a sample of a given radioactive material to decompose. This time is a constant, called the material's **half-life.**

When living organisms die, the oxygen/carbon dioxide cycle common to all living things ceases, and carbon-14, a radioactive isotope with a half-life of 5,700 years, is no longer absorbed. By measuring the amount of carbon-14 present in an ancient object, archaeologists can estimate the object's age by using the radioactive decay formula.

Radioactive Decay Formula If A is the amount of radioactive material present at time t, A_0 was the amount present at $t = 0$, and h is the material's half-life, then

$$A = A_0 2^{-t/h}$$

EXAMPLE 10

Carbon-14 dating. How old is a wooden statue that retains only one-third of its original carbon-14 content?

Solution To find the time t when $A = \frac{1}{3}A_0$, we substitute $\frac{A_0}{3}$ for A and 5,700 for h in the radioactive decay formula and solve for t:

$$A = A_0 2^{-t/h}$$

$$\frac{A_0}{3} = A_0 2^{-t/5{,}700}$$ The half-life of carbon-14 is 5,700 years.

$$1 = 3\left(2^{-t/5{,}700}\right)$$ Divide both sides by A_0 and multiply both sides by 3.

$$\log 1 = \log 3\left(2^{-t/5{,}700}\right)$$ Take the common logarithm of each side.

$$0 = \log 3 + \log 2^{-t/5{,}700}$$ $\log 1 = 0$, and use the product rule of logarithms.

$$-\log 3 = -\frac{t}{5{,}700}\log 2$$ Subtract log 3 from both sides and use the power rule of logarithms.

$$5{,}700\left(\frac{\log 3}{\log 2}\right) = t$$ Multiply both sides by $-\frac{5{,}700}{\log 2}$.

$$t \approx 9{,}034.286254$$ Use a calculator.

Notation

The initial amount of radioactive material is represented by A_0, and it is read as "A sub 0."

The statue is approximately 9,000 years old.

Self Check 10 How old is a statue that retains 25% of its original carbon-14 content?

POPULATION GROWTH

Recall that when there is sufficient food and space, populations of living organisms tend to increase exponentially according to the Malthusian growth model.

Malthusian Growth Model If P is the population at some time t, P_0 is the initial population at $t = 0$, and k depends on the rate of growth, then

$$P = P_0 e^{kt}$$

EXAMPLE 11

Population growth. The bacteria in a laboratory culture increased from an initial population of 500 to 1,500 in 3 hours. How long will it take for the population to reach 10,000?

Solution We substitute 500 for P_0, 1,500 for P, and 3 for t and simplify to find k:

Notation

The initial population of bacteria is represented by P_0, and it is read as "P sub 0."

$$P = P_0 e^{kt}$$ This is the population growth formula.

$$1{,}500 = 500(e^{k3})$$ Substitute 1,500 for P, 500 for P_0, and 3 for t.

$$3 = e^{3k}$$ Divide both sides by 500.

$$3k = \ln 3$$ Change the equation from exponential to logarithmic form.

$$k = \frac{\ln 3}{3}$$ Divide both sides by 3.

To find when the population will reach 10,000, we substitute 10,000 for P, 500 for P_0, and $\frac{\ln 3}{3}$ for k in the equation $P = P_0 e^{kt}$ and solve for t:

$$P = P_0 e^{kt}$$
$$10{,}000 = 500e^{[(\ln 3)/3]t}$$
$$20 = e^{[(\ln 3)/3]t} \quad \text{Divide both sides by 500.}$$
$$\left(\frac{\ln 3}{3}\right)t = \ln 20 \quad \text{Change the equation to logarithmic form.}$$
$$t = \frac{3 \ln 20}{\ln 3} \quad \text{Multiply both sides by } \frac{3}{\ln 3}.$$
$$\approx 8.180499084 \quad \text{Use a calculator.}$$

The culture will reach 10,000 bacteria in about 8 hours.

Self Check 11 How long will it take the population to reach 20,000?

EXAMPLE 12

Generation time. If a medium is inoculated with a bacterial culture that contains 1,000 cells per milliliter, how many generations will pass by the time the culture has grown to a population of 1 million cells per milliliter?

Solution During bacterial reproduction, the time required for a population to double is called the *generation time.* If b bacteria are introduced into a medium, then after the generation time of the organism has elapsed, there are $2b$ cells. After another generation, there are $2(2b)$ or $4b$ cells, and so on. After n generations, the number of cells present will be

(1) $$B = b \cdot 2^n$$

To find the number of generations that have passed while the population grows from b bacteria to B bacteria, we solve Equation 1 for n. To do this, we can take the common logarithm or the natural logarithm of each side.

$$\ln B = \ln (b \cdot 2^n) \quad \text{Take the natural logarithm of each side.}$$
$$\ln B = \ln b + n \ln 2 \quad \text{Apply the product and power rules of logarithms.}$$
$$\ln B - \ln b = n \ln 2 \quad \text{Subtract } \ln b \text{ from both sides.}$$
$$n = \frac{1}{\ln 2}(\ln B - \ln b) \quad \text{Multiply both sides by } \frac{1}{\ln 2}.$$

(2) $$n = \frac{1}{\ln 2}\left(\ln \frac{B}{b}\right) \quad \text{Use the quotient rule of logarithms.}$$

Equation 2 is a formula that gives the number of generations that will pass as the population grows from b bacteria to B bacteria.

To find the number of generations that have passed while a population of 1,000 cells per milliliter has grown to a population of 1 million cells per milliliter, we substitute 1,000 for b and 1,000,000 for B in Equation 2 and solve for n.

$$n = \frac{1}{\ln 2} \ln \frac{1{,}000{,}000}{1{,}000}$$
$$= \frac{1}{\ln 2} \ln 1{,}000 \quad \text{Simplify.}$$
$$n \approx 9.965784285 \quad \text{Use a calculator:}$$

Approximately 10 generations will have passed.

Answers to Self Checks 1. 2 2. 1, 1 3. 0.8614 4. $\frac{2 \log 5}{\log 5 - \log 3} \approx 6.3013$ 5. $\frac{\ln 35}{2.1} \approx 1.6930$ 6. $\frac{7}{2}$ 7. 2 8. 2 9. 6 10. about 11,400 years 11. about 10 hours

9.8 STUDY SET

VOCABULARY Fill in the blanks.

1. An equation with a variable in its exponent, such as $3^{2x} = 8$, is called a(n) __________ equation.
2. An equation with a logarithmic expression that contains a variable, such as $\log_5(2x - 3) = \log_5(x + 4)$, is a(n) __________ equation.

CONCEPTS

3. Perform a check to determine whether -2 is a solution of $5^{2x+3} = \frac{1}{5}$.
4. Perform a check to determine whether -4 is a solution of $\log_5(x + 3) = \frac{1}{5}$.
5. Use a calculator to determine whether 2.5646 is an approximate solution of $2^{2x+1} = 70$.
6. a. The exponential property of equality: If two exponential expressions with the same base are equal, their exponents are ______.

   $b^x = b^y$ is equivalent to ____

   b. The logarithm property of equality: If the logarithms base-b of two numbers are equal, the numbers are ______

   $\log_b x = \log_b y$ is equivalent to ____
7. Both sides of the exponential equation $5^{x-3} = 125$ can be written as a power of ____.
8. If $6^{4x} = 6^{-2}$, then $4x =$ ____.
9. a. Write the equivalent base-10 exponential equation for $\log(x + 1) = 2$.

   b. Write the equivalent base-e exponential equation for $\ln(x + 1) = 2$.
10. Fill in the blanks: To solve $5^x = 21$, we can take the __________ of both sides of the equation to get

    $\log 5^x = \log 21$

    The power rule for logarithms then provides a way of moving the variable x from its position as an __________ to a position as a factor of $x \log 5$.
11. If $2^x = 9$, then $\log 2^x =$ ____.
12. If $e^{x+2} = 4$, then $\ln e^{x+2} =$ ____.
13. Apply the power rule for logarithms to the left-hand side of the equation

    $\log 7^x = 12$
14. How do we solve $x \ln 3 = \ln 5$ for x?
15. a. Find $\frac{\log 8}{\log 5}$. Round to four decimal places.

    b. Find $\frac{2 \ln 12}{\ln 9}$. Round to four decimal places.
16. Does $\frac{\log 7}{\log 3} = \log 7 - \log 3$?
17. Complete each formula.

    a. Radioactive decay: $A =$ ____.

    b. Population growth: $P =$ ____.
18. Use the graphs to estimate the solution of each equation.

    a. $2^x = 3^{-x+3}$ b. $3 \log(x - 1) = 2 \log x$

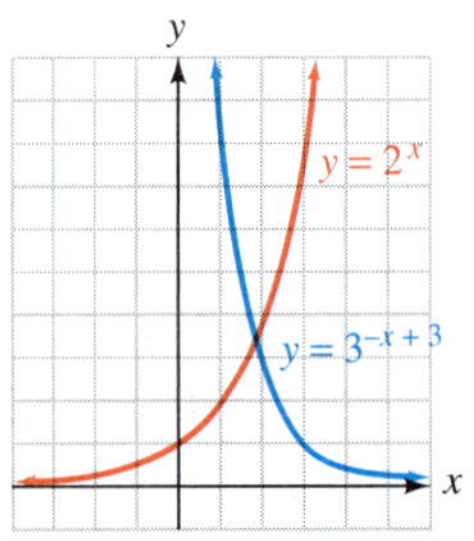

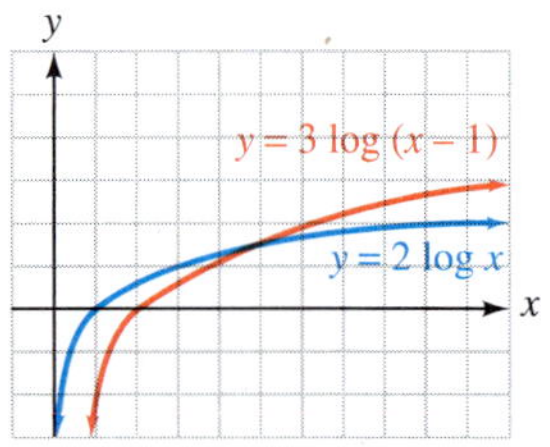

NOTATION Complete each solution.

19. Solve: $2^x = 7$.

    $2^x = 7$

    ____ $2^x = \log 7$

    x ____ $= \log 7$

    $x = \frac{\log 7}{\log 2}$

20. Solve: $\log_2(2x - 3) = \log_2(x + 4)$.

$$\log_2(2x - 3) = \log_2(x + 4)$$
$$\underline{\qquad\qquad} = x + 4$$
$$x = 7$$

PRACTICE **Solve each equation. Give answers to four decimal places when necessary.**

21. $2^{x-2} = 64$

22. $3^{-3x+1} = 243$

23. $5^{4x} = \dfrac{1}{125}$

24. $8^{-x+1} = \dfrac{1}{64}$

25. $2^{x^2-2x} = 8$

26. $3^{x^2-3x} = 81$

27. $3^{x^2+4x} = \dfrac{1}{81}$

28. $7^{x^2+3x} = \dfrac{1}{49}$

29. $4^x = 5$

30. $7^x = 12$

31. $13^{x-1} = 2$

32. $5^{x+1} = 3$

33. $2^{x+1} = 3^x$

34. $5^{x-3} = 3^{2x}$

35. $2^x = 3^x$

36. $3^{2x} = 4^x$

37. $7^{x^2} = 10$

38. $8^{x^2} = 11$

39. $8^{x^2} = 9^x$

40. $5^{x^2} = 2^{5x}$

41. $e^{3x} = 9$

42. $e^{4x} = 60$

43. $e^{-0.2t} = 14.2$

44. $e^{0.3t} = 9.1$

Use a graphing calculator to solve each equation. Give all answers to the nearest tenth.

45. $2^{x+1} = 7$

46. $3^{x-1} = 2^x$

47. $3^x - 10 = 3^{-x}$

48. $4(2^{x^2}) = 8^{3x}$

Solve each equation.

49. $\log(x + 2) = 4$

50. $\log 5x = 4$

51. $\log(7 - x) = 2$

52. $\log(2 - x) = 3$

53. $\ln x = 1$

54. $\ln x = 5$

55. $\ln(x + 1) = 3$

56. $\ln 2x = 5$

57. $\log 2x = \log 4$

58. $\log 3x = \log 9$

59. $\ln(3x + 1) = \ln(x + 7)$

60. $\ln(x^2 + 4x) = \ln(x^2 + 16)$

61. $\log(3 - 2x) - \log(x + 24) = 0$

62. $\log(3x + 5) - \log(2x + 6) = 0$

63. $\log \dfrac{4x + 1}{2x + 9} = 0$

64. $\log \dfrac{2 - 5x}{2(x + 8)} = 0$

65. $\log x^2 = 2$

66. $\log x^3 = 3$

67. $\log x + \log(x - 48) = 2$

68. $\log x + \log(x + 9) = 1$

69. $\log x + \log(x - 15) = 2$

70. $\log x + \log(x + 21) = 2$

71. $\log(x + 90) = 3 - \log x$

72. $\log(x - 90) = 3 - \log x$

73. $\log(x - 6) - \log(x - 2) = \log \dfrac{5}{x}$

74. $\log(3 - 2x) - \log(x + 9) = 0$

75. $\dfrac{\log(3x - 4)}{\log x} = 2$

76. $\dfrac{\log(8x - 7)}{\log x} = 2$

77. $\dfrac{\log(5x + 6)}{2} = \log x$

78. $\dfrac{1}{2}\log(4x + 5) = \log x$

79. $\log_3 x = \log_3\left(\dfrac{1}{x}\right) + 4$

80. $\log_5(7 + x) + \log_5(8 - x) - \log_5 2 = 2$

81. $2\log_2 x = 3 + \log_2(x - 2)$

82. $2\log_3 x - \log_3(x - 4) = 2 + \log_3 2$

83. $\log(7y + 1) = 2\log(y + 3) - \log 2$

84. $2\log(y + 2) = \log(y + 2) - \log 12$

Use a graphing calculator to solve each equation. If an answer is not exact, round to the nearest tenth.

85. $\log x + \log(x - 15) = 2$

86. $\log x + \log(x + 3) = 1$

87. $\ln(2x + 5) - \ln 3 = \ln(x - 1)$

88. $2\log(x^2 + 4x) = 1$

APPLICATIONS

89. TRITIUM DECAY The half-life of tritium is 12.4 years. How long will it take for 25% of a sample of tritium to decompose?

90. RADIOACTIVE DECAY In two years, 20% of a radioactive element decays. Find its half-life.

91. THORIUM DECAY An isotope of thorium, ^{227}Th, has a half-life of 18.4 days. How long will it take for 80% of the sample to decompose?

92. LEAD DECAY An isotope of lead, ^{201}Pb, has a half-life of 8.4 hours. How many hours ago was there 30% more of the substance?

93. CARBON-14 DATING A bone fragment analyzed by archaeologists contains 60% of the carbon-14 that it is assumed to have had initially. How old is it?

94. CARBON-14 DATING Only 10% of the carbon-14 in a small wooden bowl remains. How old is the bowl?

95. COMPOUND INTEREST If \$500 is deposited in an account paying 8.5% annual interest, compounded semiannually, how long will it take for the account to increase to \$800?

96. CONTINUOUS COMPOUND INTEREST In Exercise 95, how long will it take if the interest is compounded continuously?

97. COMPOUND INTEREST If \$1,300 is deposited in a savings account paying 9% interest, compounded quarterly, how long will it take the account to increase to \$2,100?

98. COMPOUND INTEREST A sum of \$5,000 deposited in an account grows to \$7,000 in 5 years. Assuming annual compounding, what interest rate is being paid?

99. RULE OF SEVENTY A rule of thumb for finding how long it takes an investment to double is called the **rule of seventy.** To apply the rule, divide 70 by the interest rate written as a percent. At 5%, an investment takes $\frac{70}{5} = 14$ years to double. At 7%, it takes $\frac{70}{7} = 10$ years. Explain why this formula works.

100. BACTERIAL GROWTH A bacterial culture grows according to the formula

$$P = P_0a^t$$

If it takes 5 days for the culture to triple in size, how long will it take to double in size?

101. RODENT CONTROL The rodent population in a city is currently estimated at 30,000. If it is expected to double every 5 years, when will the population reach 1 million?

102. POPULATION GROWTH The population of a city is expected to triple every 15 years. When can the city planners expect the present population of 140 persons to double?

103. BACTERIA CULTURE A bacteria culture doubles in size every 24 hours. By how much will it have increased in 36 hours?

104. OCEANOGRAPHY The intensity I of a light a distance x meters beneath the surface of a lake decreases exponentially. Use the illustration in the next column to find the depth at which the intensity will be 20%.

105. MEDICINE If a medium is inoculated with a bacteria culture containing 500 cells per milliliter, how many generations will have passed by the time the culture contains 5×10^6 cells per milliliter?

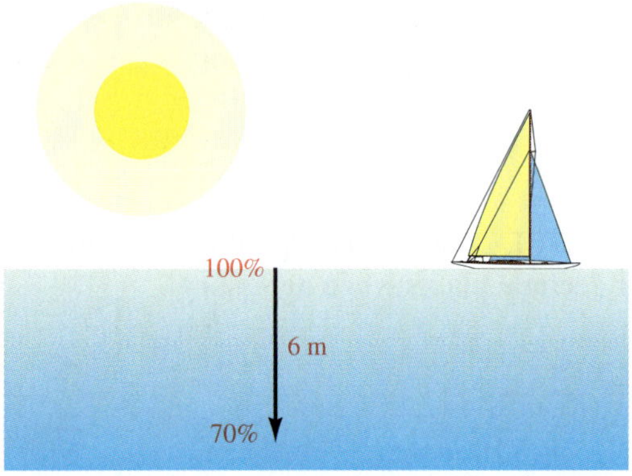

106. MEDICINE If a medium is inoculated with a bacteria culture containing 800 cells per milliliter, how many generations will have passed by the time the culture contains 6×10^7 cells per milliliter?

107. NEWTON'S LAW OF COOLING Water initially at 100°C is left to cool in a room at temperature 60°C. After 3 minutes, the water temperature is 90°. The water temperature T is a function of time t given by

$$T = 60 + 40e^{kt}$$

Find k.

108. NEWTON'S LAW OF COOLING Refer to Exercise 107 and find the time for the water temperature to reach 70°C.

WRITING

109. Explain how to solve the equation $2^{x+1} = 31$.

110. Explain how to solve the equation $2^{x+1} = 32$.

111. Write a justification for each step of the solution.

$15^x = 9$	This is the equation to solve.
$\log 15^x = \log 9$	____________.
$x \log 15 = \log 9$	____________.
$x = \dfrac{\log 9}{\log 15}$	____________.

112. What is meant by the term *half-life*?

REVIEW

113. Find the length of leg AC.

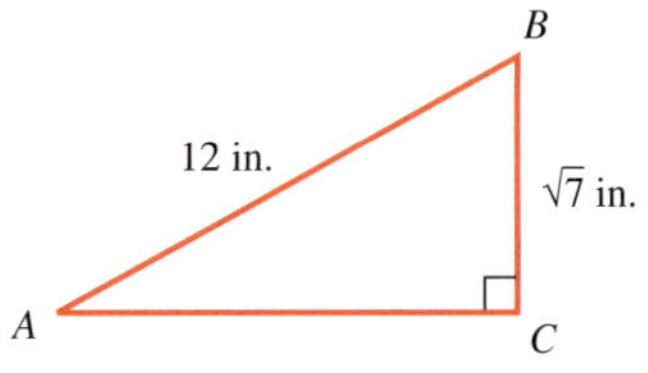

114. The amount of medicine a patient should take is often proportional to his or her weight. If a patient weighing 83 kilograms needs 150 milligrams of medicine, how much will be needed by a person weighing 99.6 kilograms?

CHALLENGE PROBLEMS

115. Without solving the following equation, find the values of x that cannot be a solution:

$$\log(x-3) - \log(x^2+2) = 0$$

116. Solve the equation: $x^{\log x} = 10{,}000$.

ACCENT ON TEAMWORK

THE NUMBER *e*

Overview: In this activity, you will use a calculator to find progressively more accurate approximations of e.

Instructions: Form groups of 2 students. Each student will need a scientific calculator. Begin by finding an approximation of e using the $\boxed{e^x}$ key on your calculator. Copy the table shown below, and write the number displayed on the calculator screen at the top of the table.

The value of e can be calculated to any degree of accuracy by adding the terms of the following pattern:

$$e = 1 + 1 + \frac{1}{2} + \frac{1}{2 \cdot 3} + \frac{1}{2 \cdot 3 \cdot 4} + \frac{1}{2 \cdot 3 \cdot 4 \cdot 5} + \cdots$$

The more terms that are added, the closer the sum will be to e.

You are to add as many terms as necessary until you obtain a sum that matches the value of e given by the $\boxed{e^x}$ key. Work together as a team. One member of the group should compute the fractional form of the term to be added. (See the middle column of the table.) The other member should take that information and calculate the cumulative sum. (See the right-hand column of the table.)

How many terms must be added so that the cumulative sum approximation and the $\boxed{e^x}$ key approximation match in each decimal place?

Approximation of e found using the $\boxed{e^x}$ key: $e \approx$ ________________

Number of terms in the sum	Term (Expressed as a fraction)	Cumulative sum (An approximation of e)
1	1	1
2	1	2
3	$\frac{1}{2}$	2.5
4	$\frac{1}{2 \cdot 3} = \frac{1}{6}$	2.666666667
⋮	⋮	⋮

KEY CONCEPT: INVERSE FUNCTIONS

ONE-TO-ONE FUNCTIONS

For a one-to-one function, each input is assigned exactly one output, and each output corresponds to exactly one input. Determine whether each function is one-to-one.

1. $f(x) = x^2$

2. $f(x) = |x|$

3.

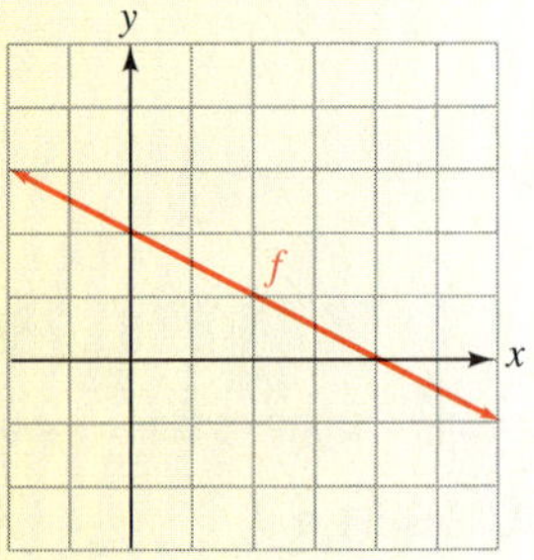

THE INVERSE OF A FUNCTION

If a function is one-to-one, its inverse is a function. To find the inverse of a function, replace $f^{-1}(x)$ with y, interchange x and y, and solve for y.

4. Find $f^{-1}(x)$ if $f(x) = -2x - 1$.

5. Given the table of values for a one-to-one function f, complete the table of values for f^{-1}.

x	-2	1	3
$f(x)$	4	-2	-6

x	4	-2	-6
$f^{-1}(x)$			

EXPONENTIAL AND LOGARITHMIC FUNCTIONS

The exponential function $f(x) = b^x$ and logarithmic function $g(x) = \log_b x$ (where $b > 0$ and $b \neq 1$) are inverse functions.

6. Write the exponential statement $10^3 = 1{,}000$ in logarithmic form.

7. Write the logarithmic statement $\log_2 \frac{1}{8} = -3$ in exponential form.

8. If $\log_4 x = \frac{1}{2}$, what is x?

9. If $\log_x \frac{9}{4} = 2$, what is x?

THE NATURAL EXPONENTIAL AND NATURAL LOGARITHMIC FUNCTIONS

A special exponential function that is used in many real-life applications involving growth and decay is the base-e exponential function, $f(x) = e^x$. Its inverse is the natural logarithm function $g(x) = \ln x$.

10. What is an approximate value of e?

11. What is the base of the logarithmic function $f(x) = \ln x$?

12. Use a calculator to find x: $\ln x = -0.28$.

13. Graph $f(x) = e^x$ and $g(x) = \ln x$ on the coordinate system shown. Label the axis of symmetry.

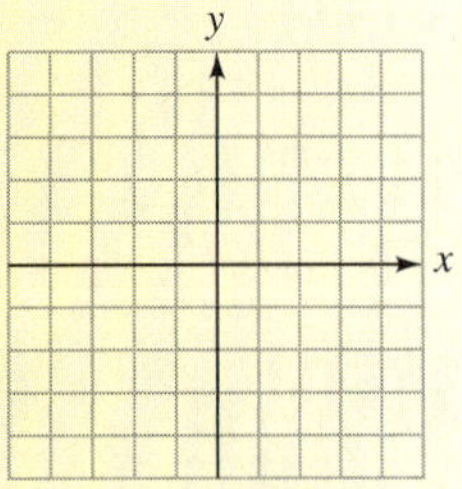

CHAPTER REVIEW

SECTION 9.1 Algebra and Composition of Functions

CONCEPTS

Sum, difference, product, and quotient functions are defined as:

$(f + g)(x) = f(x) + g(x)$

$(f - g)(x) = f(x) - g(x)$

$(f \cdot g)(x) = f(x)g(x)$

$(f/g)(x) = \dfrac{f(x)}{g(x)}$, with $g(x) \neq 0$

Composition of functions:

$(f \circ g)(x) = f(g(x))$

REVIEW EXERCISES

Let $f(x) = 2x$ and $g(x) = x + 1$. Find each function and its domain.

1. $f + g$

2. $f - g$

3. $f \cdot g$

4. f/g

Let $f(x) = x^2 + 2$ and $g(x) = 2x + 1$. Find each composition.

5. $(f \circ g)(-1)$

6. $(g \circ f)(0)$

7. $(f \circ g)(x)$

8. $(g \circ f)(x)$

9. Use the table of values for functions f and g to find each of the following.

a. $(f + g)(2)$ **b.** $(f \cdot g)(2)$ **c.** $(f \circ g)(2)$ **d.** $(g \circ f)(2)$

x	$f(x)$
2	3
9	7

x	$g(x)$
2	9
3	0

10. MILEAGE COSTS The function $f(m) = \frac{m}{8}$ gives the number of gallons of fuel consumed if a bus travels m miles. The function $C(f) = 1.85f$ gives the cost (in dollars) of f gallons of fuel. Write a composition function that expresses the cost of the fuel consumed as a function of the number of miles driven.

SECTION 9.2 Inverse Functions

For a *one-to-one function,* each input is assigned exactly one output, and each output corresponds to exactly one input.

Horizontal line test: A function is one-to-one if every horizontal line intersects the graph of the function at most once.

In Exercises 11–16, determine whether the function is one-to-one.

11. $f(x) = x^2 + 3$

12. $f(x) = \dfrac{1}{3}x - 8$

13. $\{(3, 4), (5, 10), (10, -1), (6, 6)\}$

14.

x	$f(x)$
0	−5
2	10
4	−5
6	15

15.

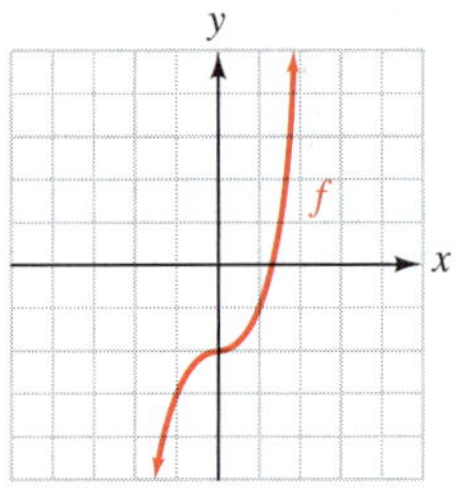

16.

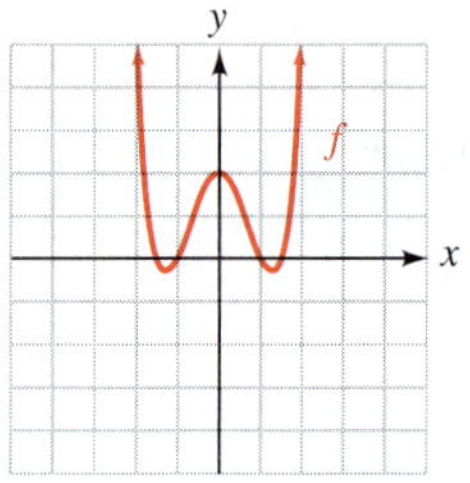

The graph of a function and its inverse are symmetric about the line $y = x$.

17. Use the table of values of the one-to-one function f to complete a table of values for f^{-1}.

x	$f(x)$
−6	−6
−1	−3
7	12
20	3

x	$f^{-1}(x)$
−6	
−3	
12	
3	

18. Given the graph of function f, graph f^{-1} on the same coordinate axes. Label the axis of symmetry.

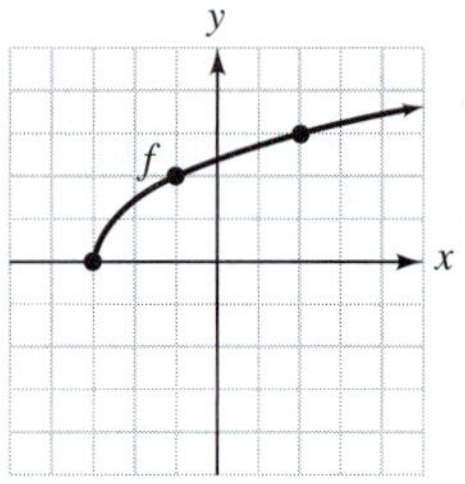

To *find the inverse of a function,* replace $f(x)$ with y, interchange the variables x and y, solve for y, and replace y with $f^{-1}(x)$.

Find the inverse of each function.

19. $f(x) = 6x - 3$

20. $f(x) = \dfrac{4}{x - 1}$

21. $f(x) = (x + 2)^3$

22. $f(x) = \dfrac{x}{6} - \dfrac{1}{6}$

For any one-to-one function f and its inverse, f^{-1},

$$(f \circ f^{-1})(x) = x$$
$$(f^{-1} \circ f)(x) = x$$

23. Find the inverse of $f(x) = \sqrt[3]{x - 1}$. Then graph the function and its inverse on one coordinate system. Show the line of symmetry on the graph.

24. Use composition to show that $f(x) = 5 - 4x$ and $f^{-1}(x) = -\dfrac{x - 5}{4}$ are inverse functions.

SECTION 9.3 Exponential Functions

An *exponential function* with base b is defined by the equation

$f(x) = b^x$, with $b > 0$, $b \neq 1$

If $b > 1$, then $f(x) = b^x$ is an *increasing function.*

If $0 < b < 1$, then $f(x) = b^x$ is a *decreasing function.*

Use properties of exponents to simplify each expression.

25. $5^{\sqrt{6}} \cdot 5^{3\sqrt{6}}$

26. $\left(2^{\sqrt{14}}\right)^{\sqrt{2}}$

Graph each function and give the domain and the range. Label the *y*-intercept.

27. $f(x) = 3^x$

28. $f(x) = \left(\dfrac{1}{3}\right)^x$

29. $f(x) = \left(\dfrac{1}{2}\right)^x - 2$

30. $f(x) = 3^{x-1}$

31. In Exercise 30, what is the asymptote of the graph of $f(x) = 3^{x-1}$?

Exponential functions are suitable models for describing many situations involving the *growth* or *decay* of a quantity.

32. COAL PRODUCTION The table gives the number of tons of coal produced in the United States for the years 1800–1920. Graph the data. What type of function does it appear could be used to model coal production over this period?

Year	Tons
1800	108,000
1810	178,000
1820	881,000
1830	1,334,000
1840	2,474,000
1850	8,356,000
1860	20,041,000

Year	Tons
1870	40,429,000
1880	79,407,000
1890	157,771,000
1900	269,684,000
1910	501,596,000
1920	658,265,000

Source: *World Book Encyclopedia*

Compound interest: If $P is the deposit, and interest is paid k times a year at an annual rate r, the amount A in the account after t years is given by

$$A = P\left(1 + \frac{r}{k}\right)^{kt}$$

33. COMPOUND INTEREST How much will $10,500 become if it earns 9% annual interest, compounded quarterly, for 60 years?

34. DEPRECIATION The value (in dollars) of a certain model car is given by the function $V(t) = 12{,}000\left(10^{-0.155t}\right)$, where t is the number of years from the present. What will the value of the car be in 5 years?

SECTION 9.4 Base-e Exponential Functions

The function defined by $f(x) = e^x$ is the *natural exponential function* where $e = 2.718281828459\ldots$

If a quantity increases or decreases at an annual rate r, *compounded continuously,* then the amount A after t years is given by

$$A = Pe^{rt}$$

If r is negative, the quantity is decreasing.

35. MORTGAGE RATES The average annual interest rate on a 30-year fixed-rate home mortgage for the years 1980–1996 can be approximated by the function $r(t) = 13.9e^{-0.035t}$, where t is the number of years since 1980. To the nearest hundredth of a percent, what does this model predict was the 30-year fixed rate in 1995?

36. INTEREST COMPOUNDED CONTINUOUSLY If $10,500 accumulates interest at an annual rate of 9%, compounded continuously, how much will be in the account in 60 years?

Graph each function, and give the domain and the range.

37. $f(x) = e^x + 1$

38. $f(x) = e^{x-3}$

Malthusian population growth is modeled by the formula

$A = Pe^{rt}$

39. POPULATION GROWTH As of July 2003, the population of the United States was estimated to be 290,340,000, with an annual growth rate of 0.92%. If the growth rate remains the same, how large will the population be in 50 years?

40. MEDICAL TESTS A radioactive dye is injected into a patient as part of a test to detect heart disease. The amount of dye remaining in his bloodstream t hours after the injection is given by the function $f(t) = 10e^{-0.27t}$. How can you tell from the function that the amount of dye in the bloodstream is decreasing?

SECTION 9.5 Logarithmic Functions

If $b > 0$ and $b \neq 1$, then the *logarithmic function with base b* is defined by $f(x) = \log_b x$.

$y = \log_b x$ means $x = b^y$.

41. Give the domain and range of $f(x) = \log x$.

42. When a student used a calculator to evaluate $\log 0$, she obtained the following message:

Error

Explain why.

43. Write the statement $\log_4 64 = 3$ in exponential form.

44. Write the statement $7^{-1} = \frac{1}{7}$ in logarithmic form.

$\log_b x$ is the exponent to which b is raised to get x.

$b^{\log_b x} = x$

Find each value, if possible.

45. $\log_3 9$ **46.** $\log_9 \frac{1}{81}$ **47.** $\log_{1/2} 1$

48. $\log_5 (-25)$ **49.** $\log_6 \sqrt{6}$ **50.** $\log 1{,}000$

For computational purposes and in many applications, we use base-10 logarithms called *common logarithms.*

$\log x$ means $\log_{10} x$

Solve each equation.

51. $\log_2 x = 5$ **52.** $\log_3 x = -4$ **53.** $\log_x 16 = 2$

54. $\log_x \frac{1}{100} = -2$ **55.** $\log_9 3 = x$ **56.** $\log_{27} x = \frac{2}{3}$

Use a calculator to find the value of x to four decimal places.

57. $\log 4.51 = x$ **58.** $\log x = 1.43$

If $b > 1$, then $f(x) = \log_b x$ is an *increasing function.* If $0 < b < 1$, then $f(x) = \log_b x$ is a *decreasing function.*

Graph each pair of equations on one set of coordinate axes. Draw the axis of symmetry.

59. $f(x) = \log_4 x$ and $g(x) = 4^x$ **60.** $f(x) = \log_{1/3} x$ and $g(x) = \left(\frac{1}{3}\right)^x$

The exponential function $f(x) = b^x$ and the logarithmic function $f(x) = \log_b x$ are inverses of each other.

Graph each function. Label the x-intercept.

61. $f(x) = \log(x - 2)$

62. $f(x) = 3 + \log x$

Decibel voltage gain:

$$\text{db gain} = 20 \log \frac{E_O}{E_I}$$

63. ELECTRICAL ENGINEERING An amplifier has an output of 18 volts when the input is 0.04 volt. Find the db gain.

The Richter scale:

$$R = \log \frac{A}{P}$$

64. EARTHQUAKES An earthquake had a period of 0.3 second and an amplitude of 7,500 micrometers. Find its measure on the Richter scale.

SECTION 9.6 Base-*e* Logarithms

Natural logarithms:

$\ln x$ means $\log_e x$

$\ln x$ is the exponent to which e is raised to get x.

$e^{\ln x} = x$

The *natural logarithmic function* with base e is defined by

$f(x) = \ln x$

The domain of $f(x) = \ln x$ is the interval $(0, \infty)$ and the range is the interval $(-\infty, \infty)$.

Evaluate each expression, if possible. Do not use a calculator.

65. $\ln e$

66. $\ln e^2$

67. $\ln \frac{1}{e^5}$

68. $\ln \sqrt{e}$

69. $\ln(-e)$

70. $\ln 0$

71. $\ln 1$

72. $\ln e^{-7}$

Use a calculator to find each value to four decimal places.

73. $\ln 452$

74. $\ln 0.85$

Use a calculator to find the value of x to four decimal places.

75. $\ln x = 2.336$

76. $\ln x = -8.8$

77. Explain the difference between the functions $f(x) = \log x$ and $g(x) = \ln x$.

78. What function is the inverse of $f(x) = \ln x$?

Graph each function.

79. $f(x) = 1 + \ln x$

80. $f(x) = \ln(x + 1)$

Population doubling time:

$$t = \frac{\ln 2}{r}$$

81. POPULATION GROWTH How long will it take the population of Mexico to double if the growth rate is currently about 1.43%?

82. BOTANY The height (in inches) of a certain plant is approximated by the function $H(a) = 13 + 20.03 \ln a$, where a is its age in years. How tall will it be when it is 19 years old?

SECTION 9.7 Properties of Logarithms

Properties of logarithms:
If b is a positive number and $b \neq 1$,

1. $\log_b 1 = 0$
2. $\log_b b = 1$
3. $\log_b b^x = x$
4. $b^{\log_b x} = x$
5. $\log_b MN = \log_b M + \log_b N$
6. $\log_b \frac{M}{N} = \log_b M - \log_b N$
7. $\log_b M^P = p \log_b M$

Properties of logarithms can be used to condense expressions.

Change-of-base formula:

$$\log_b x = \frac{\log_a x}{\log_a b}$$

pH scale:

$$\text{pH} = -\log\,[\text{H}^+]$$

Simplify each expression.

83. $\log_2 1$

84. $\log_9 9$

85. $\log 10^3$

86. $7^{\log_7 4}$

Use the properties of logarithms to rewrite each expression.

87. $\log_3 27x$

88. $\log \frac{100}{x}$

89. $\log_5 \sqrt{27}$

90. $\log_b 10ab$

Write each expression in terms of the logarithms of x, y and z.

91. $\log_b \frac{x^2y^3}{z}$

92. $\ln \sqrt{\frac{x}{yz^2}}$

Write each expression as the logarithm of one quantity.

93. $3 \log_2 x - 5 \log_2 y + 7 \log_2 z$

94. $-3 \log_b y - 7 \log_b z + \frac{1}{2} \log_b (x + 2)$

Assume that $\log_b 5 = 1.1609$ and $\log_b 8 = 1.5000$. Use these values and the properties of logarithms to find each value to four decimal places.

95. $\log_b 40$

96. $\log_b 64$

97. Find $\log_5 17$ to four decimal places.

98. pH OF GRAPEFRUIT The pH of grapefruit juice is about 3.1. Find its hydrogen ion concentration.

SECTION 9.8 Exponential and Logarithmic Equations

If two exponential expressions with the same base are equal, their exponents are equal.

$$b^x = b^y$$

is equivalent to $x = y$

Solve each equation for x. Give answers to four decimal places when necessary.

99. $5^{x+2} = 625$

100. $2^{x^2+4x} = \frac{1}{8}$

101. $3^x = 7$

102. $2^x = 3^{x-1}$

103. $e^x = 7$

104. $e^{-0.4t} = 25$

When it is difficult to write each side of an exponential equation as a power of the same base, take the logarithm of each side.

If two positive numbers are equal, then the logarithms base-b of the numbers are equal.

$$\log_b x = \log_b y$$

is equivalent to $x = y$

Carbon dating:

$$A = A_0 2^{-t/h}$$

Population growth

$$P = P_0 e^{kt}$$

Solve each equation for x.

105. $\log(x - 4) = 2$

106. $\ln(2x - 3) = \ln 15$

107. $\log x + \log(29 - x) = 2$

108. $\log_2 x + \log_2(x - 2) = 3$

109. $\dfrac{\log(7x - 12)}{\log x} = 2$

110. $\log_2(x + 2) + \log_2(x - 1) = 2$

111. $\log x + \log(x - 5) = \log 6$

112. $\log 3 - \log(x - 1) = -1$

113. Evaluate both sides of the statement $\frac{\log 8}{\log 15} \neq \log 8 - \log 15$ to show that the sides are indeed not equal.

114. CARBON-14 DATING A wooden statue found in Egypt has a carbon-14 content that is two-thirds of that found in living wood. If the half-life of carbon-14 is 5,700 years, how old is the statue?

115. ANTS The number of ants in a colony is estimated to be 800. If the ant population is expected to triple every 14 days, how long will it take for the population to reach one million?

116. The approximate coordinates of the points of intersection of the graphs of $f(x) = \log x$ and $g(x) = 1 - \log(7 - x)$ are shown in parts (a) and (b) of the illustration. Use the graphs to estimate the solutions of the logarithmic equation $\log x = 1 - \log(7 - x)$. Then check your answers.

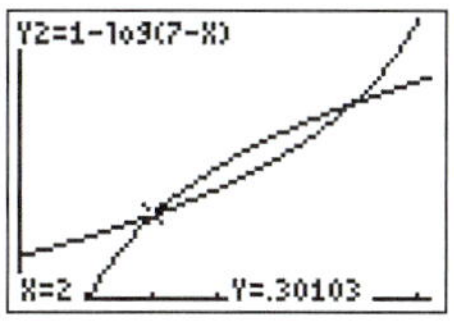

(a)

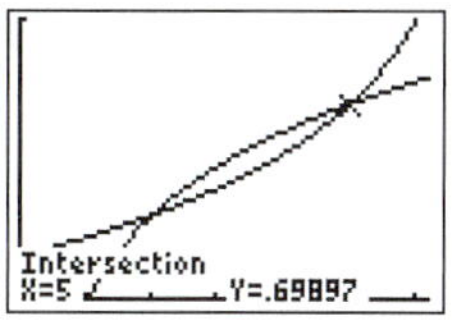

(b)

CHAPTER 9 TEST

Let $f(x) = x + 9$ and $g(x) = 4x^2 - 3x + 2$. Find each function and give its domain.

1. $f + g$

2. g/f

Let $f(x) = 2x^2 + 3$ and $g(x) = 4x - 8$. Find each composition.

3. $(g \circ f)(-3)$

4. $(f \circ g)(x)$

Use the tables of values for functions f and g to find each of the following.

5. $(f \cdot g)(9)$

6. $(f \circ g)(-3)$

x	$f(x)$
9	-1
10	17

x	$g(x)$
-3	10
9	16

Determine whether each function is one-to-one.

7. $f(x) = x$

8.

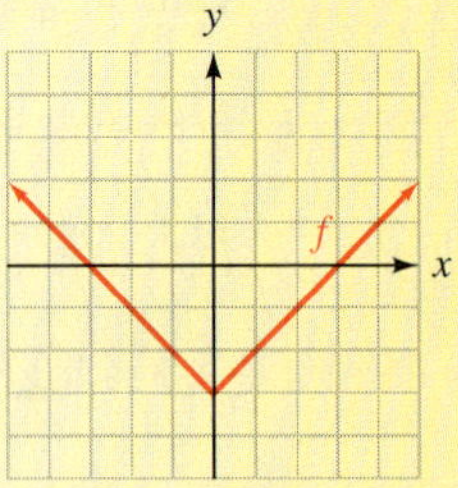

9. Find the inverse of $f(x) = -\frac{1}{3}x$ and then graph f and its inverse on the same coordinate axes.

10. Find the inverse of $f(x) = (x - 15)^3$.

11. Use composition to show that $f(x) = 4x + 4$ and $f^{-1}(x) = \dfrac{x - 4}{4}$ are inverse functions.

12. Consider the following graph of the function f.
   a. Is f a one-to-one function?
   b. Is its inverse a function?
   c. What is $f^{-1}(260)$? What information does it give?

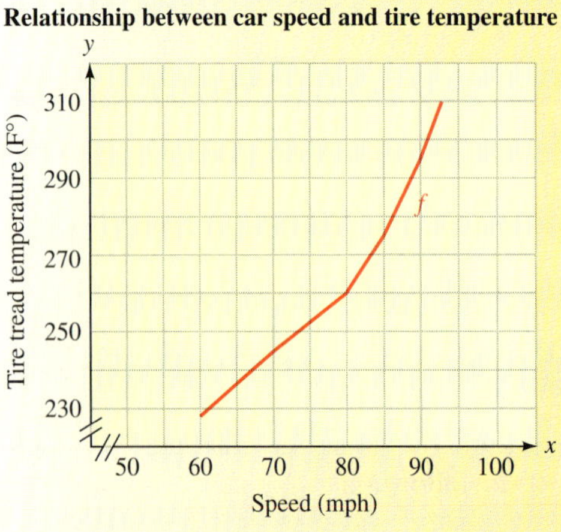

Graph each function and give the domain and the range.

13. $f(x) = 2^x + 1$

14. $f(x) = 3^{-x}$

15. RADIOACTIVE DECAY A radioactive material decays according to the formula $A = A_0(2)^{-t}$. How much of a 3-gram sample will be left in 6 years?

16. COMPOUND INTEREST An initial deposit of \$1,000 earns 6% interest, compounded twice a year. How much will be in the account in one year?

17. Graph: $f(x) = e^x$. Label the y-intercept and the asymptote of the graph.

18. POPULATION GROWTH As of July 2003, the population of India was estimated to be 1,050,000,000, with an annual growth rate of 1.47%. If the growth rate remains the same, how large will the population be in 30 years?

19. Write the statement $\log_6 \frac{1}{36} = -2$ in exponential form.

20. a. What are the domain and range of the function $f(x) = \log x$?
   b. What is the inverse of $f(x) = \log x$?

Evaluate each logarithmic expression, if possible.

21. $\log_5 25$

22. $\log_9 \dfrac{1}{81}$

23. $\log(-100)$

24. $\ln \dfrac{1}{e^6}$

25. $\log_4 2$

26. $\log_{1/3} 1$

Find x.

27. $\log_x 32 = 5$

28. $\log_8 x = \dfrac{4}{3}$

29. $\log_3 x = -3$

30. $\ln x = 1$

Graph each function.

31. $f(x) = -\log_3 x$

32. $f(x) = \ln x$

33. pH Find the pH of a solution with a hydrogen ion concentration of 3.7×10^{-7}. (*Hint:* $\text{pH} = -\log[\text{H}^+]$.)

34. ELECTRONICS Find the db gain of an amplifier when $E_O = 60$ volts and $E_I = 0.3$ volt.
   Hint: db gain $= 20 \log\left(\dfrac{E_O}{E_I}\right)$.

35. Use a calculator to find x to four decimal places.
   $\log x = -1.06$

36. Use the change-of-base formula to find $\log_7 3$ to four decimal places.

37. Write the expression $\log_b a^2bc^3$ in terms of the logarithms of a, b, and c.

38. Write the expression $\frac{1}{2}\ln(a + 2) + \ln b - 3\ln c$ as a logarithm of a single quantity.

Solve each equation. Round to four decimal places when necessary.

39. $5^x = 3$

40. $3^{x-1} = 27$

41. $\ln(5x + 2) = \ln(2x + 5)$

42. $\log x + \log(x - 9) = 1$

43. The illustration shows the graphs of $y = \frac{1}{2}\ln(x - 1)$ and $y = \ln 2$ and the approximate coordinates of their point of intersection. Estimate the solution of the logarithmic equation $\frac{1}{2}\ln(x - 1) = \ln 2$. Then check the result.

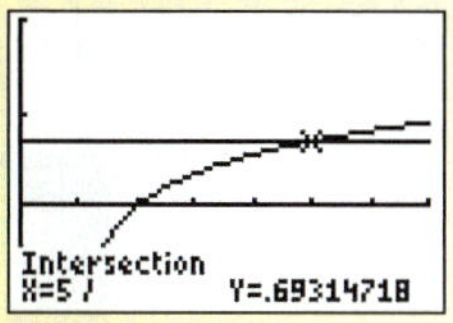

44. INSECTS The number of insects attracted to a bright light is currently 5. If the number is expected to quadruple every 6 minutes, how long will it take for the number to reach 500?

Chapter

10 Conic Sections; More Graphing

Getty Images/Image Bank

The solar system consists of the sun, the nine planets, more than 130 satellites of the planets, and a large number of comets and asteroids. The orbits of the planets are ellipses, though all except Mercury and Pluto are very nearly circular. The orbit of a comet can be an ellipse or a hyperbola, depending on the total energy of the comet. Circles, ellipses, and hyperbolas belong to a family of curves called *conic sections.* They are so named because they can be formed by the intersection of a plane and right-circular cone.

To learn more about conic sections, visit *The Learning Equation* on the Internet at http://tle.brookscole.com. (The log-in instructions are in the Preface.) For Chapter 10, the online lesson is:

- *TLE* Lesson 15: Introduction to Conic Sections

In this chapter, we will discuss graphs that do not represent functions. Since these curves are cross sections of cones, they are called conic sections.

10.1 The Circle and the Parabola

- Conic sections
- The circle
- Problem solving using circles
- The parabola

We have previously graphed first-degree equations in two variables such as $y = 3x + 8$ and $4x - 3y = 12$. Their graphs are lines. In this section, we will graph second-degree equations in two variables such as $x^2 + y^2 = 25$ and $x = -3y^2 - 12y - 13$. The graphs of these equations are *conic sections.*

CONIC SECTIONS

The curves formed by the intersection of a plane with an infinite right-circular cone are called **conic sections.** Those curves have four basic shapes, called **circles, parabolas, ellipses,** and **hyperbolas,** as shown below.

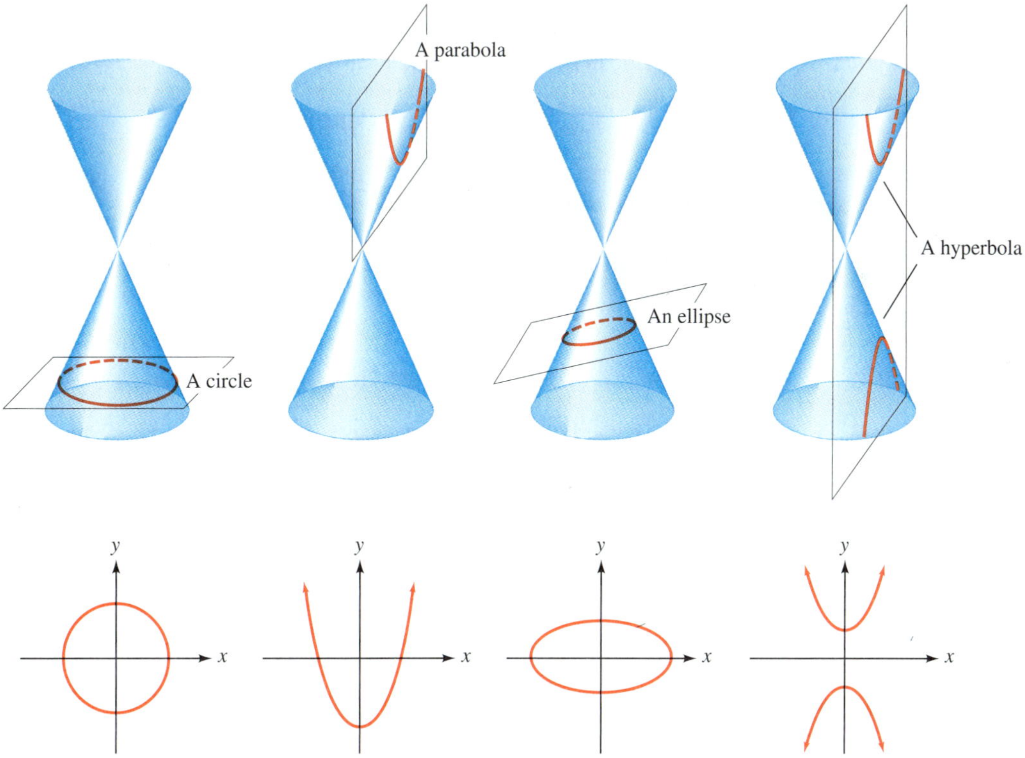

Conic sections have many applications. For example, everyone is familiar with circular wheels and gears, pizza cutters, and hula hoops.

Parabolas can be rotated to generate dish-shaped surfaces called **paraboloids.** Any light or sound placed at the **focus** of a paraboloid is reflected outward in parallel paths. This property makes parabolic surfaces ideal for flashlight and headlight reflectors. It also makes parabolic surfaces good antennas, because signals captured by such antennas are concentrated at the focus. Parabolic mirrors are capable of concentrating the rays of the sun at a single point, thereby generating tremendous heat. This property is used in the design of solar furnaces.

The Language of Algebra

Conic sections are often simply called *conics.*

Any object thrown upward and outward travels in a parabolic path. An example of this is a stream of water flowing from a drinking fountain. In architecture, many arches are parabolic in shape, because this gives them strength. Cables that support suspension bridges hang in the shape of a parabola.

Parabolas

Radar dish

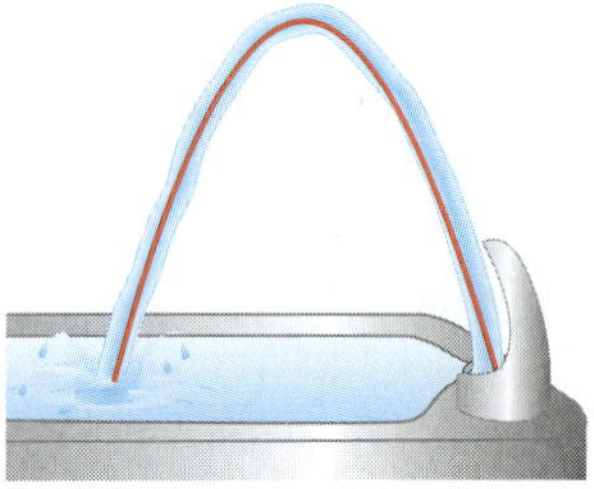

Stream of water

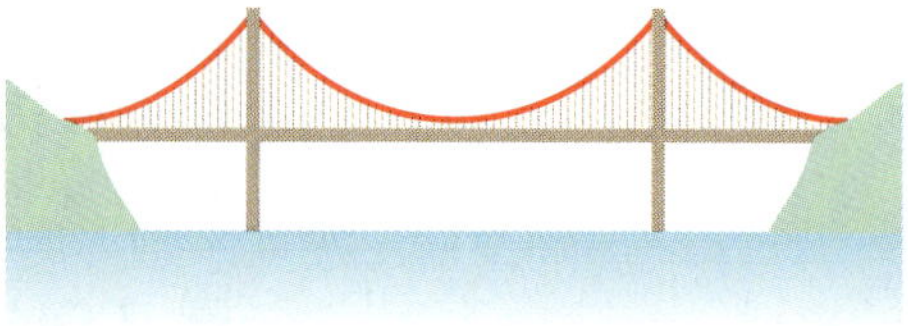

Support cables

Ellipses have optical and acoustical properties that are useful in architecture and engineering. Many arches are portions of an ellipse, because the shape is pleasing to the eye. The planets and many comets have elliptical orbits. Certain gears have elliptical shapes to provide nonuniform motion.

Ellipses

Arches

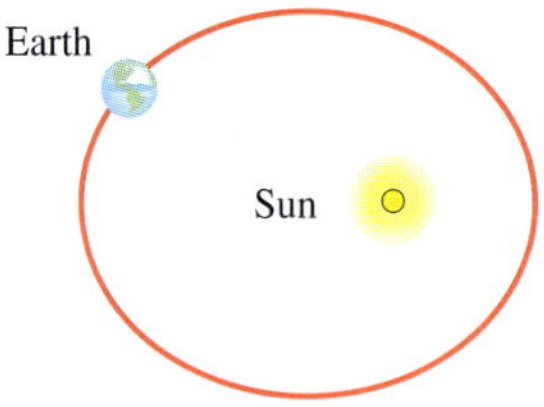

Earth's orbit

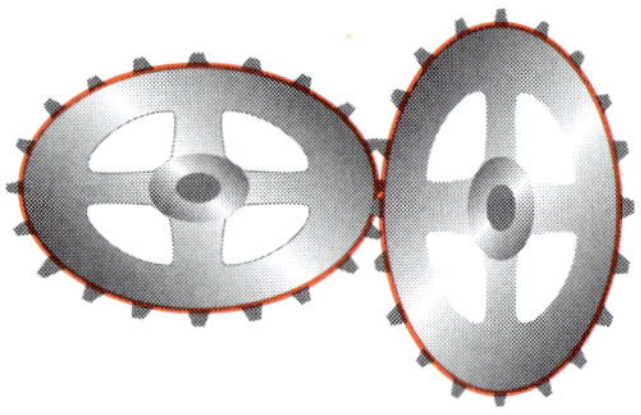

Gears

Hyperbolas serve as the basis of a navigational system known as LORAN (LOng RAnge Navigation). They are also used to find the source of a distress signal, are the basis for the design of hypoid gears, and describe the orbits of some comets.

A sonic shock wave created by a jet has the shape of a cone. The sound wave intersects the ground as one branch of a hyperbola, as shown below. The sonic boom is heard by anyone on the branch at the same time.

Hyperbolas

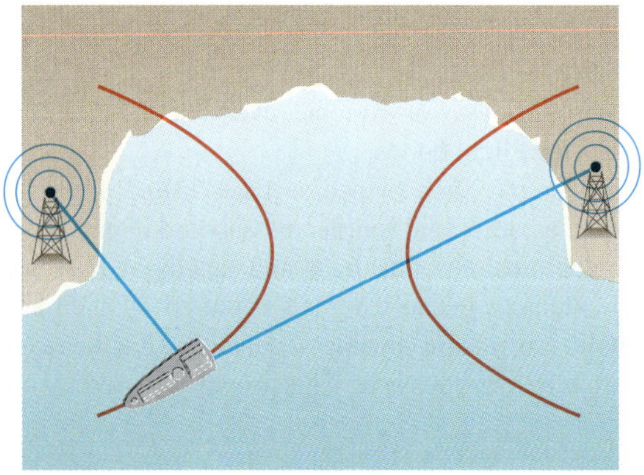

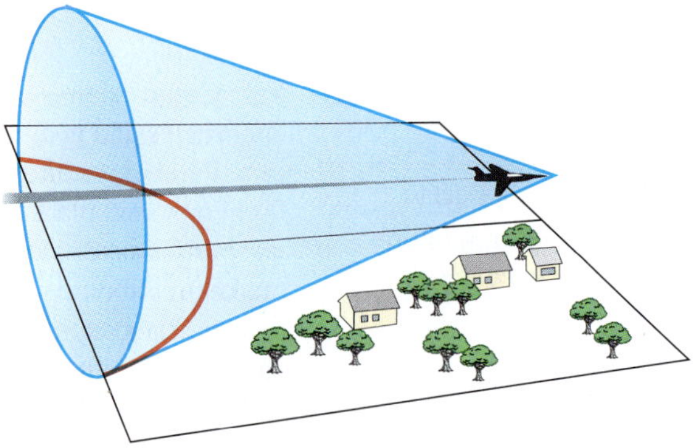

THE CIRCLE

Every conic section can be represented by a second-degree equation in x and y. To find the form of an equation of a circle, we use the following definition.

Definition of a Circle

A **circle** is the set of all points in a plane that are a fixed distance from a fixed point called its **center.** The fixed distance is called the **radius** of the circle.

On a rectangular coordinate system, if we let (h, k) be the center of a circle, and (x, y) be some point on the circle, then the distance from (h, k) to (x, y) is the radius of the circle, r units. We can use the distance formula to find r.

$$r = \sqrt{(x - h)^2 + (y - k)^2}$$

We then square both sides to eliminate the radical and obtain

$$r^2 = (x - h)^2 + (y - k)^2$$

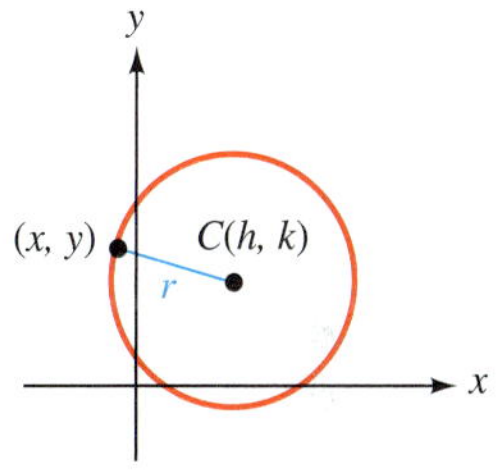

This result is called the *standard form of the equation of a circle* with radius r and center at (h, k).

Equation of a Circle

The **standard form of the equation of a circle** with radius r and center at (h, k) is

$$(x - h)^2 + (y - k)^2 = r^2$$

EXAMPLE 1

Find the center and the radius of each circle and then graph it: **a.** $(x - 4)^2 + (y - 1)^2 = 9$, **b.** $x^2 + y^2 = 25$, and **c.** $(x + 3)^2 + y^2 = 12$.

Solution

a. It is easy to determine the center and the radius of a circle when its equation is written in standard form.

$$(x - 4)^2 + (y - 1)^2 = 9$$
$$\uparrow \quad \uparrow \quad \uparrow$$
$$(x - h)^2 + (y - k)^2 = r^2$$

$h = 4$, $k = 1$, and $r^2 = 9$. Since the radius of a circle must be positive, $r = 3$.

The center of the circle is $(h, k) = (4, 1)$ and the radius is 3.

To plot four points on the circle, we move up, down, left, and right 3 units from the center, as shown in figure (a). Then we draw a circle through the points to get the graph of $(x - 4)^2 + (y - 1)^2 = 9$, as shown in figure (b).

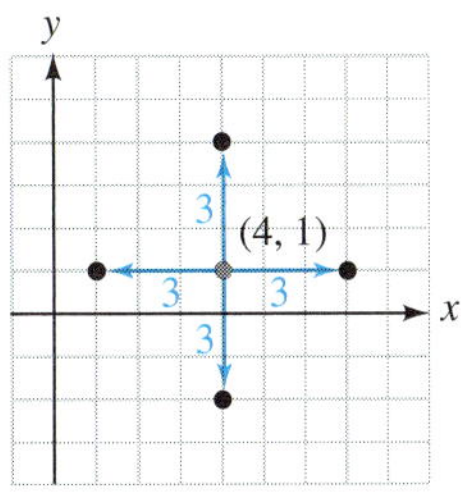

(a)

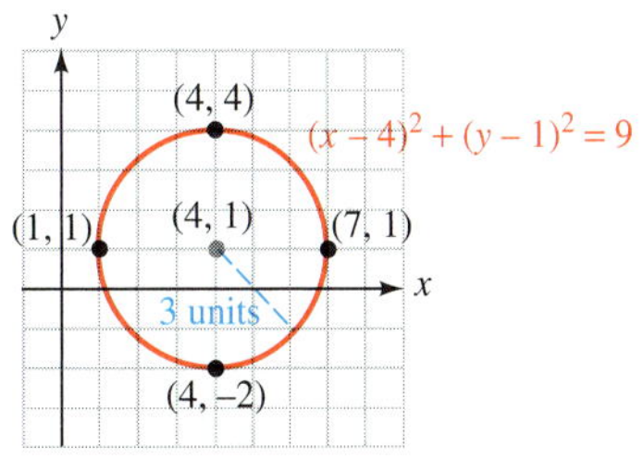

(b)

Success Tip

A compass can be used to draw a circle. The distance between the compass point and pencil should be set to equal the radius.

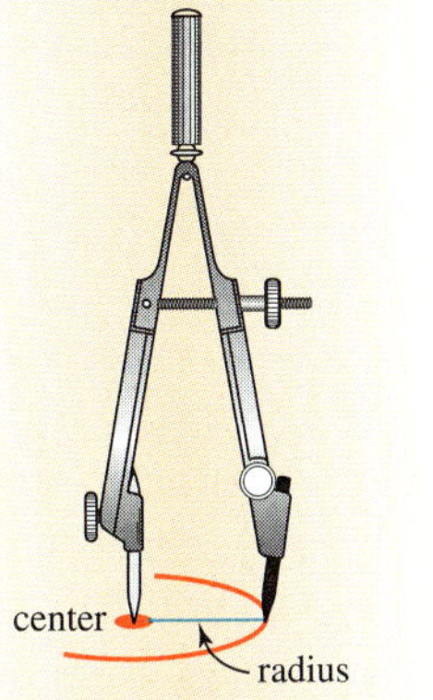

b. To determine h and k, it is helpful to write $x^2 + y^2 = 25$ in the following way:

$$(x - 0)^2 + (y - 0)^2 = 25$$

$h = 0$, $k = 0$, and $r^2 = 25$. Since the radius must be positive, $r = 5$.

Success Tip

An equation of the form

$$x^2 + y^2 = r^2$$

has a graph that is a circle with radius r and center at (0, 0).

The center of the circle is at (0, 0) and the radius is 5.

To plot four points on the circle, we move up, down, left, and right 5 units from the center. Then we draw a circle through the points to get the graph of $x^2 + y^2 = 25$, as shown.

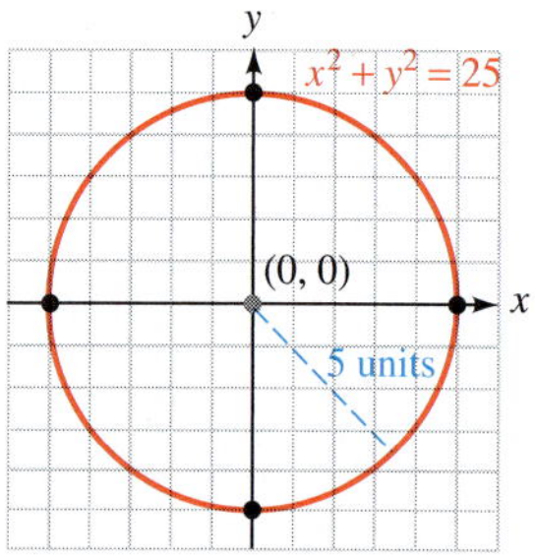

c. To determine h, it is helpful to write $x + 3$ as $x - (-3)$.

Standard form requires a minus symbol here.

$$[x - (-3)]^2 + (y - 0)^2 = 12$$

$h = -3$, $k = 0$, and $r^2 = 12$.

If $r^2 = 12$, by the square root property

$$r = \pm\sqrt{12} = \pm 2\sqrt{3}$$

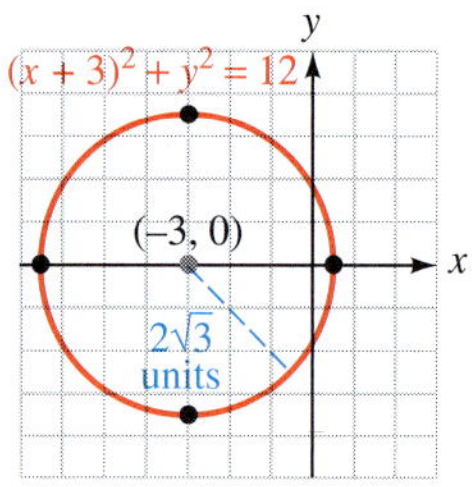

Since the radius can't be negative, we get $r = 2\sqrt{3}$. The center of the circle is at $(-3, 0)$ and the radius is $2\sqrt{3}$.

To plot four points on the circle, we move up, down, left, and right $2\sqrt{3} \approx 3.5$ units from the center. Then we draw a circle through the points to get the graph of $(x + 3)^2 + y^2 = 12$, as shown.

Self Check 1 Find the center and the radius of each circle and then graph it: **a.** $(x - 3)^2 + (y + 4)^2 = 4$ and **b.** $x^2 + y^2 = 8$.

EXAMPLE 2

Find the equation of the circle with radius 9 and center at $(6, -5)$.

Solution We substitute 9 for r, 6 for h, and -5 for k in the standard form and simplify.

$$(x - h)^2 + (y - k)^2 = r^2$$
$$(x - 6)^2 + [y - (-5)]^2 = 9^2$$
$$(x - 6)^2 + (y + 5)^2 = 9^2 \quad \text{Write } y - (-5) \text{ as } y + 5.$$

Notation

Standard form can be written

$$(x - 6)^2 + (y + 5)^2 = 9^2$$
or
$$(x - 6)^2 + (y + 5)^2 = 81$$

If we express 9^2 as 81, we have

$$(x - 6)^2 + (y + 5)^2 = 81$$

Self Check 2 Find the equation of the circle with radius 10 and center at $(-7, 1)$.

In Example 2, the result was written in standard form: $(x-6)^2+(y+5)^2=81$. If we square $x-6$ and $y+5$, we obtain a different form for the circle's equation.

$$(x-6)^2+(y+5)^2=9^2$$
$$x^2-12x+36+y^2+10y+25=81$$
$$x^2-12x+y^2+10y-20=0 \quad \text{Subtract 81 from both sides. Combine like terms.}$$
$$x^2+y^2-12x+10y-20=0 \quad \text{Rearrange the terms, writing the squared terms first.}$$

This result is written in the *general form of the equation of a circle.*

Equation of a Circle The **general form of the equation of a circle** is

$$x^2+y^2+Dx+Ey+F=0$$

We can convert from the general form to the standard form of the equation of a circle by completing the square.

EXAMPLE 3

Write in standard form: $x^2+y^2-4x+2y-11=0$.

Solution To write the equation in standard form, we complete the square twice.

$$x^2+y^2-4x+2y-11=0$$
$$x^2-4x+y^2+2y-11=0 \quad \text{Write the } x\text{-terms together and the } y\text{-terms together.}$$
$$x^2-4x \quad +y^2+2y \quad =11 \quad \text{Add 11 to both sides.}$$

To complete the square on x^2-4x, we note that $\frac{1}{2}(-4)=-2$ and $(-2)^2=4$. To complete the square on y^2+2y, we note that $\frac{1}{2}(2)=1$ and $1^2=1$. We add 4 and 1 to both sides of the equation.

$$x^2-4x+4+y^2+2y+1=11+4+1$$
$$(x-2)^2+(y+1)^2=16 \quad \text{Factor } x^2-4x+4 \text{ and } y^2+2y+1.$$

The equation could also be written as $(x-2)^2+(y+1)^2=4^2$.

Self Check 3 Write in standard form: $x^2+y^2+12x-6y-4=0$.

ACCENT ON TECHNOLOGY: GRAPHING CIRCLES

Since the graphs of circles fail the vertical line test, their equations do not represent functions. It is somewhat more difficult to use a graphing calculator to graph equations that are not functions. For example, to graph the circle described by $(x-1)^2+(y-2)^2=4$, we must split the equation into two functions and graph each one separately. We begin by solving the equation for y.

$$(x-1)^2+(y-2)^2=4$$
$$(y-2)^2=4-(x-1)^2 \quad \text{Subtract } (x-1)^2 \text{ from both sides.}$$
$$y-2=\pm\sqrt{4-(x-1)^2} \quad \text{Use the square root property.}$$
$$y=2\pm\sqrt{4-(x-1)^2} \quad \text{Add 2 to both sides.}$$

This equation defines two functions. If we graph

$$y = 2 + \sqrt{4 - (x - 1)^2} \quad \text{and} \quad y = 2 - \sqrt{4 - (x - 1)^2}$$

we get the distorted circle shown in figure (a). To get a better circle, we can use the graphing calculator's square window feature, which gives an equal unit distance on both the x- and y-axes. Using this feature, we get the circle shown in figure (b). Sometimes the two arcs will not join because of approximations made by the calculator at each endpoint.

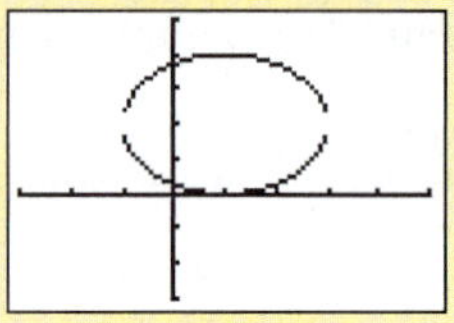

(a)

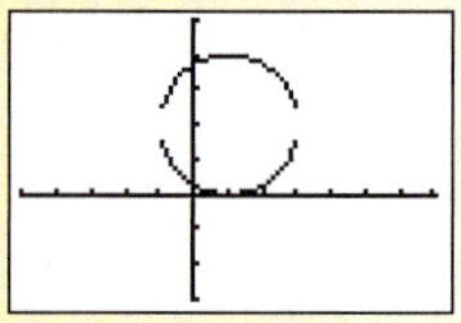

(b)

PROBLEM SOLVING USING CIRCLES

EXAMPLE 4

Radio translators. The broadcast area of a television station is bounded by the circle $x^2 + y^2 = 3{,}600$, where x and y are measured in miles. A translator station picks up the signal and retransmits it from the center of a circular area bounded by

$$(x + 30)^2 + (y - 40)^2 = 1{,}600$$

Find the location of the translator and the greatest distance from the main transmitter that the signal can be received.

Solution The coverage of the TV station is bounded by $x^2 + y^2 = 60^2$, a circle centered at the origin with a radius of 60 miles, as shown in yellow in the figure. Because the translator is at the center of the circle $(x + 30)^2 + (y - 40)^2 = 1{,}600$, it is located at $(-30, 40)$, a point 30 miles west and 40 miles north of the TV station. The radius of the translator's coverage is $\sqrt{1{,}600}$, or 40 miles.

As shown in the figure, the greatest distance of reception is the sum of d, the distance from the translator to the television station, and 40 miles, the radius of the translator's coverage.

To find d, we use the distance formula to find the distance between $(x_1, y_1) = (-30, 40)$ and the origin, $(x_2, y_2) = (0, 0)$.

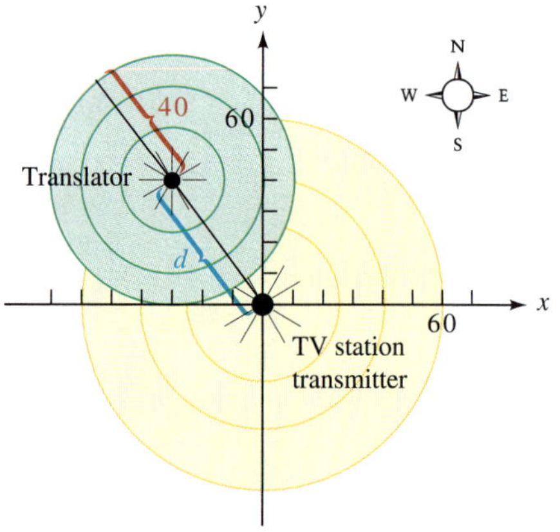

$$d = \sqrt{(x_1 - x_2)^2 + (y_1 - y_2)^2}$$

$$d = \sqrt{(-30 - 0)^2 + (40 - 0)^2}$$

$$d = \sqrt{(-30)^2 + 40^2}$$

$$= \sqrt{900 + 1{,}600}$$

$$= \sqrt{2{,}500}$$

$$= 50$$

The translator is located 50 miles from the television station, and it broadcasts the signal 40 miles. The greatest reception distance from the main transmitter signal is, therefore, 50 + 40, or 90 miles.

THE PARABOLA

Another type of conic section is the parabola. We have previously discussed the equations of parabolas whose graphs open upward or downward. Parabolas can also open to the right and to the left, but the equations for these parabolas are not functions because their graphs fail the vertical line test.

The two *general forms of the equation of a parabola* are similar.

Equation of a Parabola

The **general forms of the equation of a parabola** are:

1. $y = ax^2 + bx + c$ The graph opens upward if $a > 0$ and downward if $a < 0$.
2. $x = ay^2 + by + c$ The graph opens to the right if $a > 0$ and to the left if $a < 0$.

Recall from Chapter 8 that equations written in the **standard form** $y = a(x - h)^2 + k$ represent parabolas with vertex at (h, k) and axis of symmetry $x = h$. They open upward when $a > 0$ and downward when $a < 0$.

EXAMPLE 5

Graph: $y = -2x^2 + 12x - 15$.

Solution Because the equation is not in standard form, the coordinates of the vertex are not obvious. To write the equation in standard form, we complete the square.

$$y = -2x^2 + 12x - 15$$

$$y = -2(x^2 - 6x \quad\ ) - 15 \qquad \text{Factor out } -2 \text{ from } -2x^2 + 12x.$$

This step adds $-2 \cdot 9$ or -18 to this side. Add 18 to counteract the addition of -18.

$$y = -2(x^2 - 6x + 9) - 15 + 18 \qquad \text{Complete the square on } x^2 - 6x.$$

$$y = -2(x - 3)^2 + 3 \qquad \text{Factor } x^2 - 6x + 9 \text{ and combine like terms.}$$

Success Tip

Recall that we can find the x-coordinate of the vertex using

$$x = -\frac{b}{2a} = -\frac{12}{2(-2)} = 3$$

To find the y-coordinate, substitute:

$$y = -2(3)^2 + 12(3) - 15 = 3$$

The vertex is at (3, 3).

This equation is written in the form $y = a(x - h)^2 + k$, where $a = -2$, $h = 3$, and $k = 3$. Thus, the graph of the equation is a parabola that opens downward with vertex at (3, 3) and an axis of symmetry $x = 3$. We can construct a table of solutions and use symmetry to plot several points on the parabola. Then we draw a smooth curve through the points to get the graph of $y = -2x^2 + 12x - 15$, as shown below.

$y = -2x^2 + 12x - 15$

x	y	
1	−5	→ (1, −5)
2	1	→ (2, 1)

The x-coordinate of the vertex is 3. Choose values for x close to 3 on the same side of the axis of symmetry.

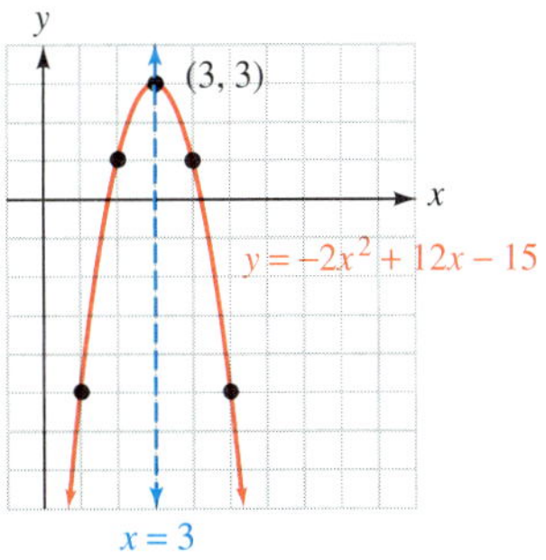

Self Check 5 Graph: $y = 2x^2 + 4x + 5$.

The *standard form* for the equation of a parabola that opens to the right or left is similar to $y = a(x - h)^2 + k$, except that the variables, x and y, exchange positions as do the constants, h and k.

Standard Form of the Equation of a Parabola

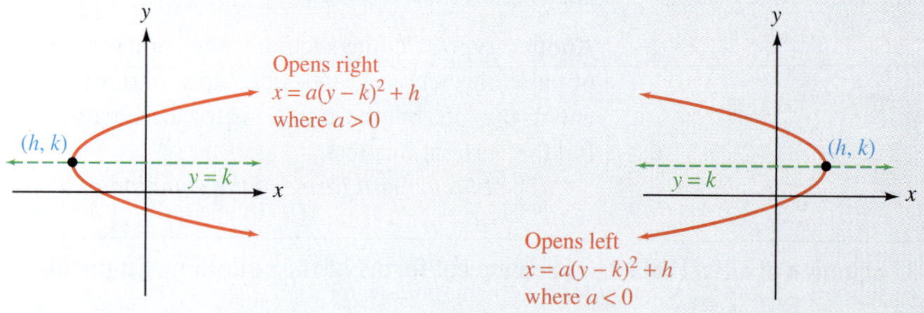

EXAMPLE 6

Graph: $x = \frac{1}{2}y^2$.

Solution This equation is written in the form $x = a(y - k)^2 + h$, where $a = \frac{1}{2}$, $k = 0$, and $h = 0$. The graph of the equation is a parabola that opens to the right with vertex at (0, 0) and an axis of symmetry $y = 0$.

To construct a table of solutions, we choose *values of y* and find their corresponding values of x. For example, if $y = 1$, we have

$$x = \frac{1}{2}y^2$$

$$x = \frac{1}{2}(1)^2 \quad \text{Substitute 1 for } y.$$

$$x = \frac{1}{2}$$

The point $\left(\frac{1}{2}, 1\right)$ is on the parabola.

We plot the ordered pairs from the table and use symmetry to plot three more points on the parabola. Then we draw a smooth curve through the points to get the graph of $x = \frac{1}{2}y^2$, as shown below.

$x = \frac{1}{2}y^2$

x	y	
$\frac{1}{2}$	1	→ $\left(\frac{1}{2}, 1\right)$
2	2	→ (2, 2)
8	4	→ (8, 4)

The y-coordinate of the vertex is 0. Choose values for y close to 0 on the same side of the axis of symmetry.

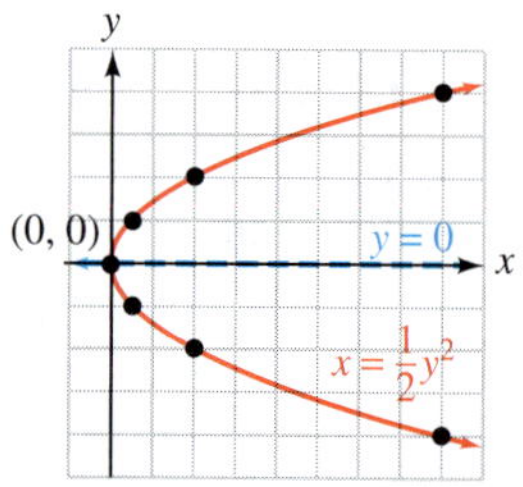

Self Check 6 Graph: $x = -\frac{2}{3}y^2$.

EXAMPLE 7

Graph: $x = -3y^2 - 12y - 13$.

Solution To write the equation in standard form, we complete the square.

$x = -3y^2 - 12y - 13$	
$x = -3(y^2 + 4y \quad\ ) - 13$	Factor out -3 from $-3y^2 - 12y$.
$x = \mathbf{-3}(y^2 + 4y \mathbf{+ 4}) - 13 \mathbf{+ 12}$	Complete the square on $y^2 + 4y$. Then add 12 to the right-hand side to counteract $-3 \cdot 4 = -12$.
$x = -3(y + 2)^2 - 1$	Factor $y^2 + 4y + 4$ and combine like terms.

This equation is in the standard form $x = a(y - k)^2 + h$, where $a = -3$, $k = -2$, and $h = -1$. The graph of the equation is a parabola that opens to the left with vertex at $(-1, -2)$ and an axis of symmetry $y = -2$.

We can construct a table of solutions and use symmetry to plot several points on the parabola. Then we draw a smooth curve through the points to get the graph of $x = -3y^2 - 12y - 13$, as shown below.

Success Tip

When an equation is of the form $x = ay^2\ by + c$, we can find the y-coordinate of the vertex using

$$y = -\frac{b}{2a} = -\frac{-12}{2(-3)} = \mathbf{-2}$$

To find the x-coordinate, substitute:

$$x = -3(\mathbf{-2})^2 - 12(\mathbf{-2}) - 13 = -1$$

The vertex is at $(-1, -2)$.

$$\mathbf{x = -3y^2 - 12y - 13}$$

or

$$\mathbf{x = -3(y + 2)^2 - 1}$$

x	y
-4	-1
-13	0

Choose values for y, and find the corresponding x-values.

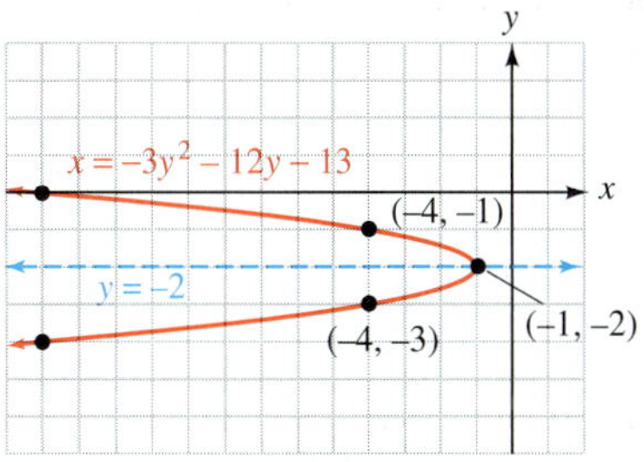

Success Tip

The equation of a circle contains an x^2 and a y^2 term. The equation of a parabola has either an x^2 term or a y^2 term, but not both.

Self Check 7 Graph: $x = 3y^2 - 6y - 1$.

Answers to Self Checks

1. a.

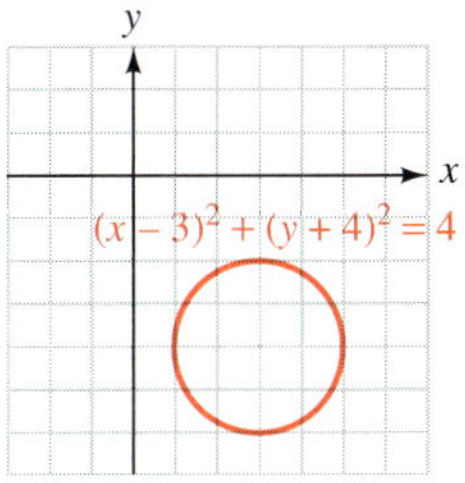

b.

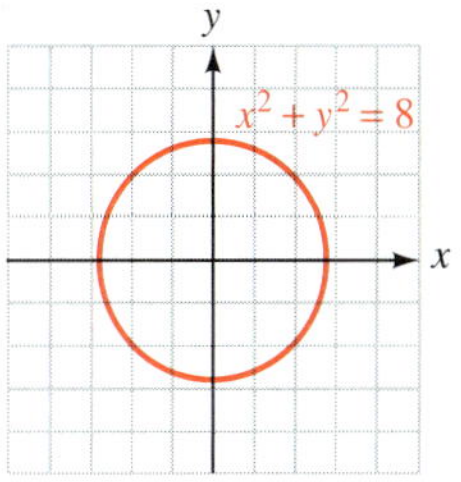

2. $(x + 7)^2 + (y - 1)^2 = 100$

3. $(x + 6)^2 + (y - 3)^2 = 49$

5.

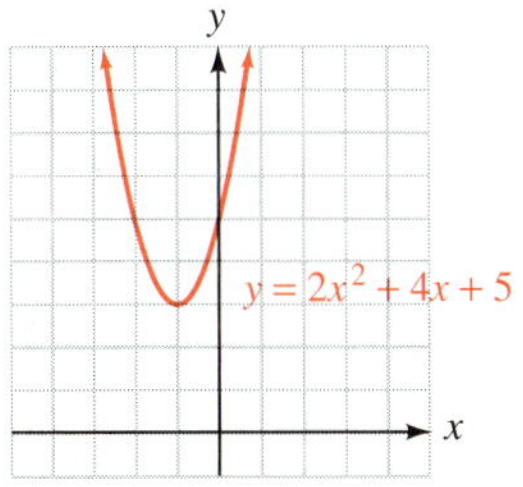

6.

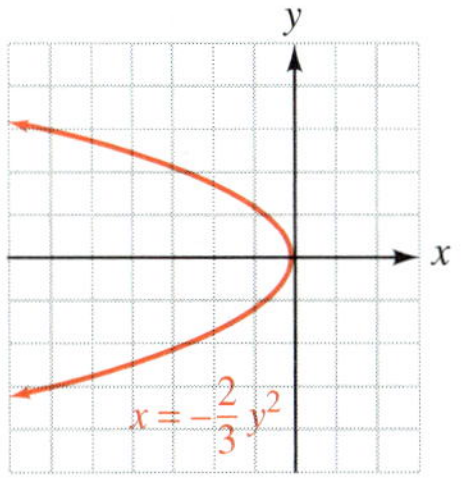

7.

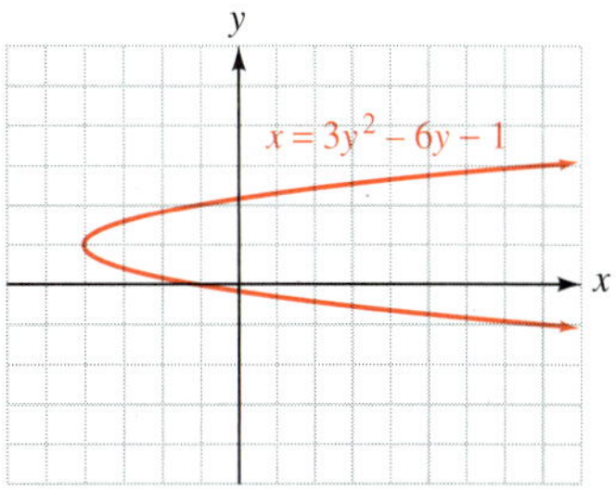

10.1 STUDY SET

VOCABULARY Fill in the blanks.

1. The curves formed by the intersection of a plane with an infinite right-circular cone are called _____ _______.

2. Give the name of each curve shown below.

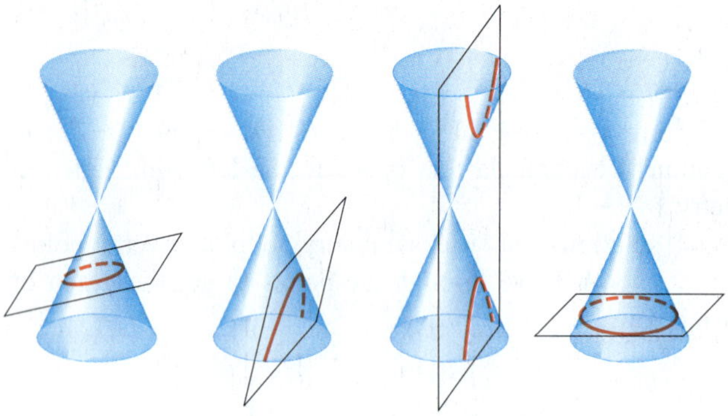

3. A _______ is the set of all points in a plane that are a fixed distance from a fixed point called its center. The fixed distance is called the _______.

4. The line that divides a parabola into two identical halves is called the axis of _________.

CONCEPTS

5. **a.** What is the standard form for the equation of a circle?

   b. What is the standard form for the equation of a circle with the center at the origin?

6. **a.** What is the center and the radius of the circle graphed below?

   b. What is the equation of the circle?

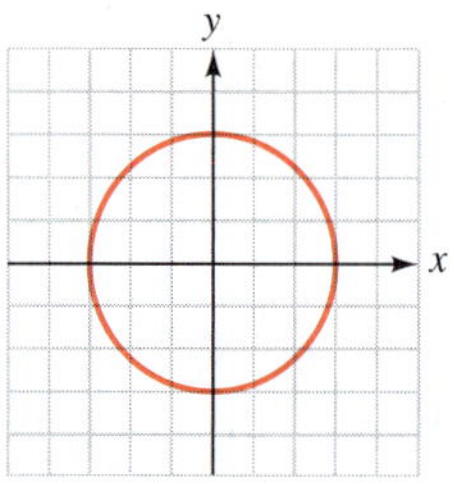

7. **a.** What are the center and the radius of the circle graphed below?

   b. What is the equation of the circle?

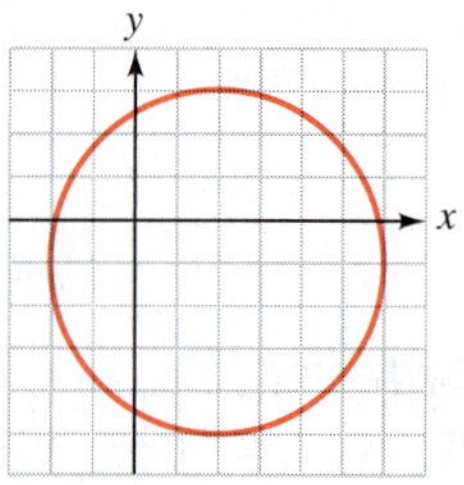

8. Fill in the blanks. To complete the square on $x^2 + 2x$ and on $y^2 - 6y$, what numbers must be added to each side of the equation?

   $$x^2 + 2x + \square + y^2 - 6y + \square = 2 + \square + \square$$

9. **a.** What is the standard form for the equation of a parabola opening upward or downward?

   b. What is the standard form for the equation of a parabola opening to the right or left?

10. Fill in the blanks.

    a. To complete the square on the right-hand side, what should be factored from the first two terms?

    $$x = 4y^2 + 16y + 9$$
    $$x = \square(y^2 + 4y \qquad) + 9$$

    b. To complete the square on $y^2 + 4y$, what should be added within the parentheses, and what should be subtracted outside the parentheses?

    $$x = 4\left(y^2 + 4y + \square\right) + 9 - \square$$

11. Determine whether the graph of each equation is a circle or a parabola.

    a. $x^2 + y^2 - 6x + 8y - 10 = 0$

    b. $y^2 - 2x + 3y - 9 = 0$

    c. $x^2 + 5x - y = 0$

    d. $x^2 + 12x + y^2 = 0$

12. Draw a parabola using the given facts.

    - Opens right
    - Vertex $(-3, 2)$
    - Passes through $(-2, 1)$
    - x-intercept $(1, 0)$

NOTATION

13. Find h, k, and r: $(x - 6)^2 + (y + 2)^2 = 9$.

14. **a.** Find a, h, and k: $y = 6(x - 5)^2 - 9$.

b. Find a, h, and k: $x = -3(y + 2)^2 + 1$.

PRACTICE Graph each equation.

15. $x^2 + y^2 = 9$

16. $x^2 + y^2 = 16$

17. $(x - 2)^2 + y^2 = 9$

18. $x^2 + (y - 3)^2 = 4$

19. $(x - 2)^2 + (y - 4)^2 = 4$

20. $(x - 3)^2 + (y - 2)^2 = 4$

21. $(x + 3)^2 + (y - 1)^2 = 16$

22. $(x - 1)^2 + (y + 4)^2 = 9$

23. $x^2 + (y + 3)^2 = 1$

24. $(x + 4)^2 + y^2 = 1$

25. $x^2 + y^2 = 6$

26. $x^2 + y^2 = 10$

27. $(x - 1)^2 + (y - 3)^2 = 15$

28. $(x + 1)^2 + (y + 1)^2 = 8$

Use a graphing calculator to graph each equation.

29. $x^2 + y^2 = 7$

30. $x^2 + y^2 = 5$

31. $(x + 1)^2 + y^2 = 16$

32. $x^2 + (y - 2)^2 = 4$

Write the equation of the circle with the following properties.

33. Center at the origin; radius 1

34. Center at the origin; radius 4

35. Center at (6, 8); radius 5

36. Center at (5, 3); radius 2

37. Center at (−2, 6); radius 12

38. Center at (5, −4); radius 6

39. Center (0, 0); radius $\frac{1}{4}$

40. Center (0, 0); radius $\frac{1}{3}$

41. Center $\left(\frac{2}{3}, -\frac{7}{8}\right)$; radius $\sqrt{2}$

42. Center (−0.7, −0.2); radius $\sqrt{11}$

43. Center at the origin; diameter $4\sqrt{2}$

44. Center at the origin; diameter $8\sqrt{3}$

Graph each circle and give the coordinates of the center and the radius.

45. $x^2 + y^2 - 2x + 4y = -1$

46. $x^2 + y^2 + 6x - 4y = -12$

47. $x^2 + y^2 + 4x + 2y = 4$

48. $x^2 + y^2 + 8x + 2y = -13$

49. $x^2 + y^2 + 2x - 8 = 0$

50. $x^2 + y^2 - 4y = 12$

51. $x^2 + y^2 - 6x + 8y + 18 = 0$

52. $x^2 + y^2 - 4x + 4y - 3 = 0$

Graph each parabola and give the coordinates of the vertex.

53. $x = y^2$

54. $x = -y^2 + 1$

55. $x = -\frac{1}{4}y^2$

56. $x = 4y^2$

57. $x = 2(y + 1)^2 + 3$

58. $x = 3(y - 2)^2 - 1$

59. $x = -3y^2 + 18y - 25$

60. $x = -2y^2 + 4y + 1$

61. $x = \frac{1}{2}y^2 + 2y$

62. $x = -\frac{1}{3}y^2 - 2y$

63. $y = 2x^2 - 4x + 5$

64. $y = -2x^2 - 4x$

65. $y = -x^2 - 2x + 3$

66. $y = x^2 + 4x + 5$

67. $y^2 + 4x - 6y = -1$

68. $x^2 - 2y - 2x = -7$

Use a graphing calculator to graph each equation. (*Hint:* Solve for *y* and graph two functions when necessary.)

69. $x = 2y^2$

70. $x = y^2 - 4$

71. $x^2 - 2x + y = 6$

72. $x = -2(y - 1)^2 + 2$

APPLICATIONS

73. MESHING GEARS For design purposes, the large gear is described by the circle $x^2 + y^2 = 16$. The smaller gear is a circle centered at (7, 0) and tangent to the larger circle. Find the equation of the smaller gear.

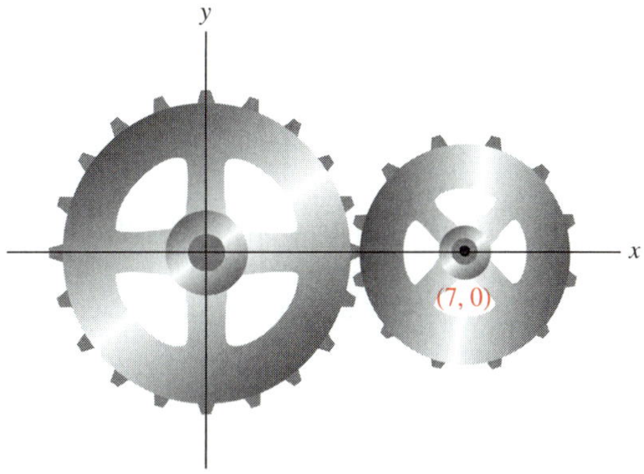

74. WALKWAYS The walkway shown is bounded by the two circles $x^2 + y^2 = 2{,}500$ and $(x - 10)^2 + y^2 = 900$, measured in feet. Find the largest and the smallest width of the walkway.

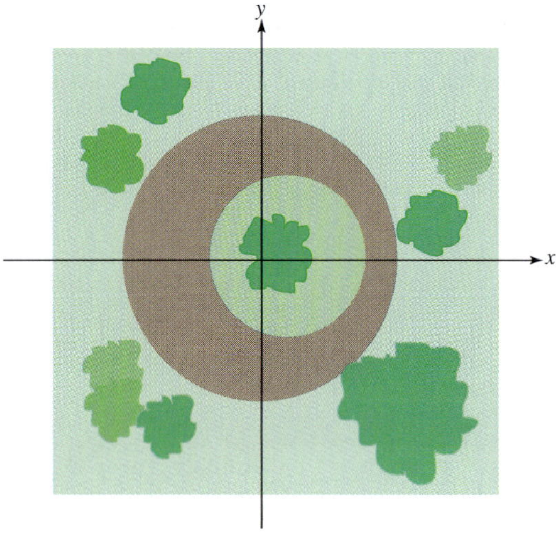

75. BROADCAST RANGES Radio stations applying for licensing may not use the same frequency if their broadcast areas overlap. One station's coverage is bounded by $x^2 + y^2 - 8x - 20y + 16 = 0$, and the other's by $x^2 + y^2 + 2x + 4y - 11 = 0$. May they be licensed for the same frequency?

76. HIGHWAY DESIGN Engineers want to join two sections of highway with a curve that is one-quarter of a circle, as shown. The equation of the circle is $x^2 + y^2 - 16x - 20y + 155 = 0$, where distances are measured in kilometers. Find the locations (relative to the center of town) of the intersections of the highway with State and with Main.

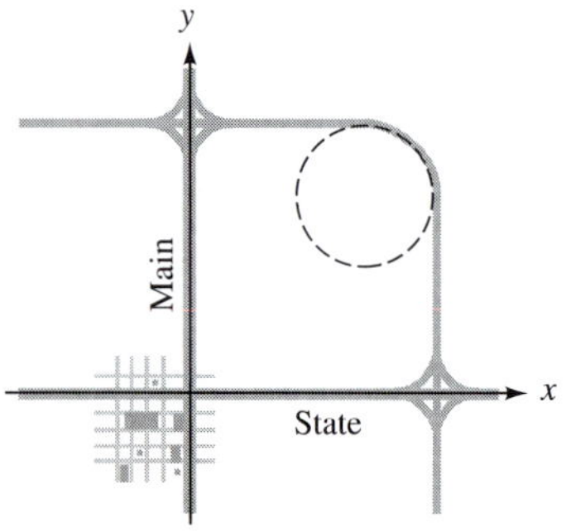

77. PROJECTILES The cannonball in the illustration in the next column follows the parabolic trajectory $y = 30x - x^2$. How far short of the castle does it land?

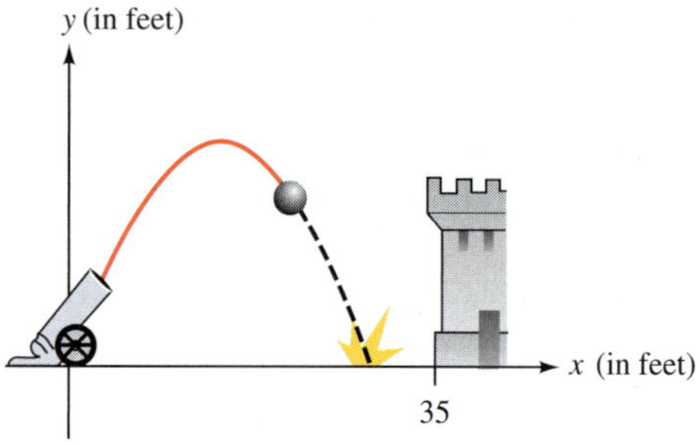

78. PROJECTILES In Exercise 77, how high does the cannonball get?

79. COMETS If the orbit of the comet is approximated by the equation $2y^2 - 9x = 18$, how far is it from the sun at the vertex of the orbit? Distances are measured in astronomical units (AU).

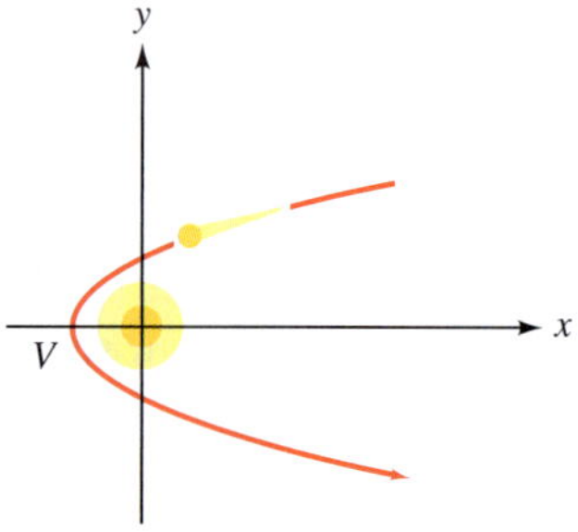

80. SATELLITE ANTENNAS The cross section of the satellite antenna in the illustration is a parabola given by the equation $y = \frac{1}{16}x^2$, with distances measured in feet. If the dish is 8 feet wide, how deep is it?

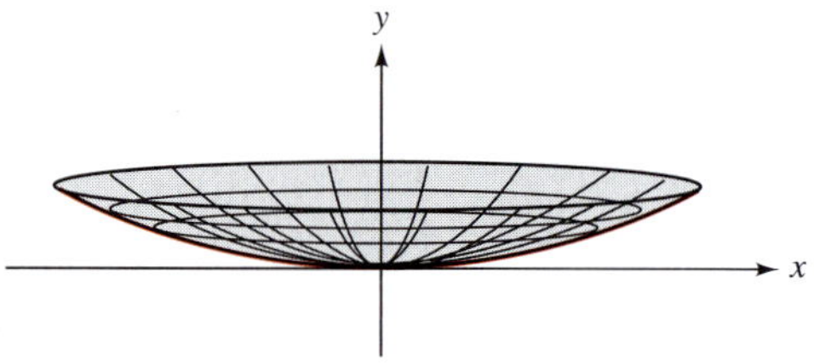

WRITING

81. Explain how to decide from its equation whether the graph of a parabola opens up, down, right, or left.

82. From the equation of a circle, explain how to determine the radius and the coordinates of the center.

83. On the day of an election, the following warning was posted in front of a school. Explain what it means.

No electioneering within a 1,000-foot radius of this polling place.

84. What is meant by the *turning radius* of a truck?

REVIEW **Solve each equation.**

85. $|3x - 4| = 11$

86. $\left|\frac{4 - 3x}{5}\right| = 12$

87. $|3x + 4| = |5x - 2|$

88. $|6 - 4x| = |x + 2|$

CHALLENGE PROBLEMS

89. Could the intersection of a plane with a pair of right-circular cones be a single point? If so, draw a picture that illustrates this.

90. Under what conditions will the graph of $x = a(y - k)^2 + h$ have no y-intercepts?

91. Write the equation of a circle with a diameter whose endpoints are at $(-2, -6)$ and $(8, 10)$.

92. Write the equation of a circle with a diameter whose endpoints are at $(-5, 4)$ and $(7, -3)$.

10.2 The Ellipse

- The definition of an ellipse
- Graphing ellipses centered at the origin
- Graphing ellipses centered at (h, k)
- Problem solving using ellipses

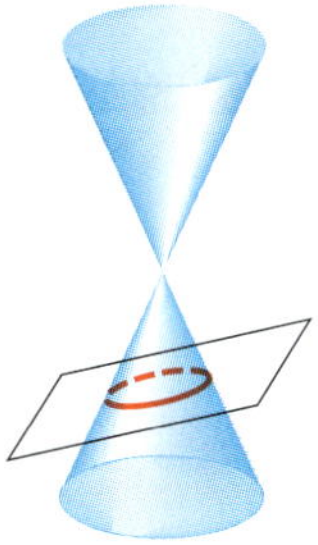

A third conic section is an oval-shaped curve called an *ellipse.* Ellipses can be nearly round and look almost like a circle, or they can be long and narrow. In this section, we will learn how to construct ellipses and how to graph equations that represent ellipses.

THE DEFINITION OF AN ELLIPSE

To define a circle, we considered a fixed distance from a fixed point. The definition of an ellipse involves *two* distances from *two* fixed points.

Definition of an Ellipse An **ellipse** is the set of all points in a plane for which the sum of the distances from two fixed points is a constant.

The figure below illustrates that any point on an ellipse is a constant distance $d_1 + d_2$ from two fixed points, each of which is called a **focus.** Midway between the **foci** is the **center** of the ellipse.

The Language of Algebra

The word *foci* (pronounced *foe-sigh*) is the plural form of the word *focus.* In the illustration on the right, the foci are labeled using subscript notation. One focus is F_1 and the other is F_2.

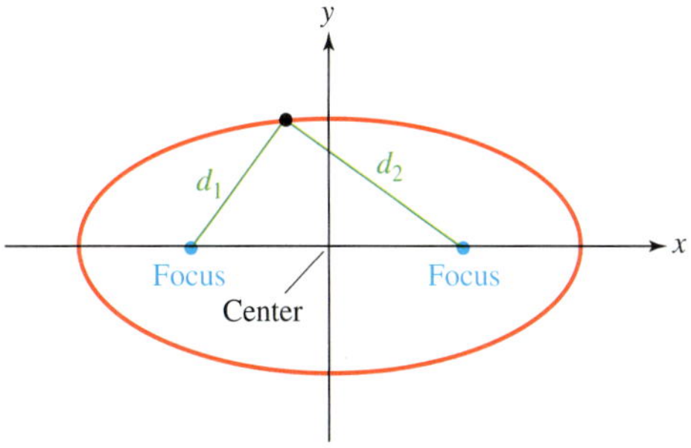

We can construct an ellipse by placing two thumbtacks fairly close together to serve as foci. We then tie each end of a piece of string to a thumbtack, catch the loop with the point of a pencil, and (keeping the string taut) draw the ellipse.

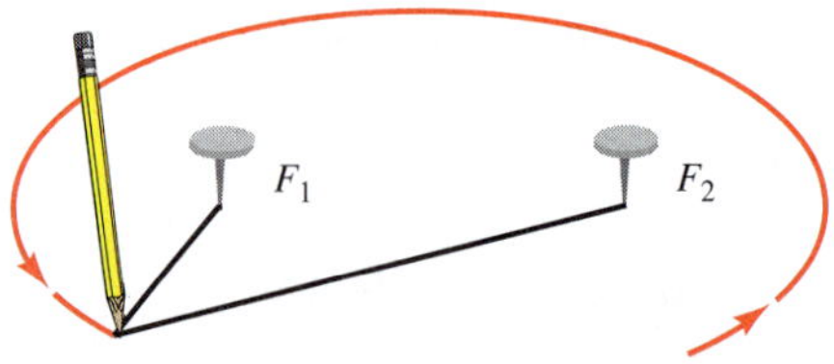

GRAPHING ELLIPSES CENTERED AT THE ORIGIN

The definition of an ellipse can be used to develop the standard form for the equation of an ellipse. To learn more about the derivation, see Problem 60 in the Challenge Problem section of the Study Set.

Equation of an Ellipse Centered at the Origin

The **standard form of the equation of an ellipse** that is symmetric with respect to both axes and centered at (0, 0) is

$$\frac{x^2}{a^2} + \frac{y^2}{b^2} = 1 \quad \text{where } a > 0 \text{ and } b > 0$$

To graph an ellipse centered at the origin, it is helpful to know the intercepts of the graph. To find the x-intercepts of the graph of

$$\frac{x^2}{a^2} + \frac{y^2}{b^2} = 1$$

we let $y = 0$ and solve for x.

$$\frac{x^2}{a^2} + \frac{\mathbf{0}^2}{b^2} = 1 \quad \text{Substitute 0 for } y.$$

$$\frac{x^2}{a^2} + 0 = 1 \quad \frac{0^2}{b^2} = 0.$$

$$x^2 = a^2 \quad \text{Simplify and multiply both sides by } a^2.$$

$$x = \pm a \quad \text{Use the square root property.}$$

The x-intercepts are $(a, 0)$ and $(-a, 0)$.

To find the y-intercepts of the graph, we can let $x = 0$ and solve for y.

$$\frac{\mathbf{0}^2}{a^2} + \frac{y^2}{b^2} = 1$$

$$0 + \frac{y^2}{b^2} = 1$$

$$y^2 = b^2 \quad \text{Simplify and multiply both sides by } b^2.$$

$$y = \pm b$$

The y-intercepts are $(0, b)$ and $(0, -b)$.

In general, we have the following results.

The Intercepts of an Ellipse

The graph of $\frac{x^2}{a^2} + \frac{y^2}{b^2} = 1$ is an ellipse, centered at the origin, with x-intercepts $(a, 0)$ and $(-a, 0)$ and y-intercepts $(0, b)$ and $(0, -b)$.

For $\frac{x^2}{a^2} + \frac{y^2}{b^2} = 1$, if $a > b$, the ellipse is horizontal, as shown in figure (a). If $b > a$, the ellipse is vertical, as shown in figure (b). The points V_1 and V_2 are the **vertices** of the ellipse. The line segment joining the vertices is called the **major axis,** and its midpoint is the **center** of the ellipse. The line segment perpendicular to the major axis at the center is the **minor axis** of the ellipse.

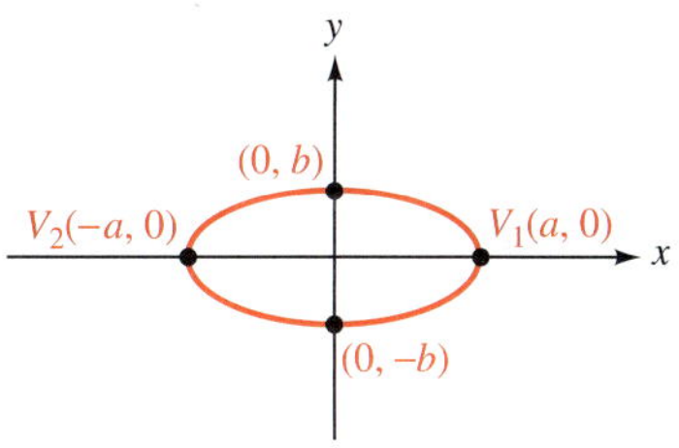

Horizontal ellipse
(a)

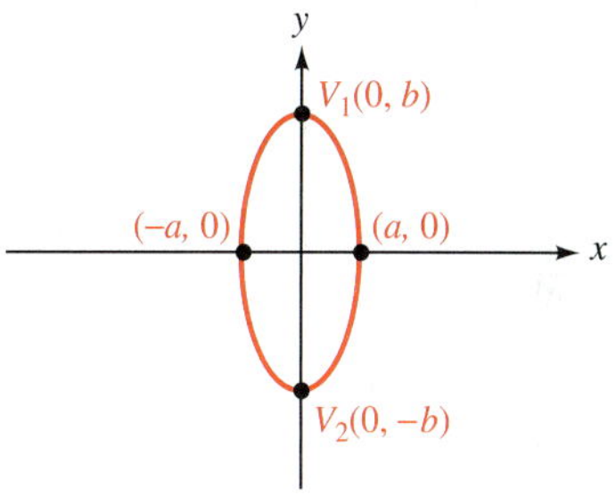

Vertical ellipse
(b)

EXAMPLE 1

Graph: $\frac{x^2}{36} + \frac{y^2}{9} = 1$.

Solution This is the equation of an ellipse centered at the origin. To determine the intercepts of the graph, we find a and b.

$$\frac{x^2}{36} + \frac{y^2}{9} = 1 \qquad \frac{x^2}{a^2} + \frac{y^2}{b^2} = 1$$

Since $a^2 = 36$, it follows that $a = 6$.

Since $b^2 = 9$, it follows that $b = 3$.

The Language of Algebra

The word *vertices* is the plural form of the word *vertex.* From the graph, we see that one vertex of this horizontal ellipse is the point (6, 0) and the other vertex is the point (−6, 0).

The x-intercepts are $(a, 0)$ and $(-a, 0)$, or $(6, 0)$ and $(-6, 0)$. The y-intercepts are $(0, b)$ and $(0, -b)$, or $(0, 3)$ and $(0, -3)$. Using these four points as a guide, we draw an oval-shaped curve through them, as shown in figure (a). The result is a horizontal ellipse.

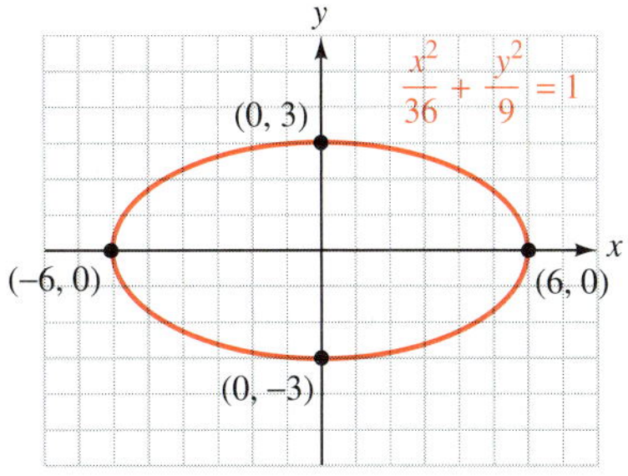

(a)

$$\frac{x^2}{36} + \frac{y^2}{9} = 1$$

x	y	
2	$\pm 2\sqrt{2}$	→ $(2, \pm 2\sqrt{2})$
4	$\pm\sqrt{5}$	→ $(4, \pm\sqrt{5})$

Approximate the radicals to graph.

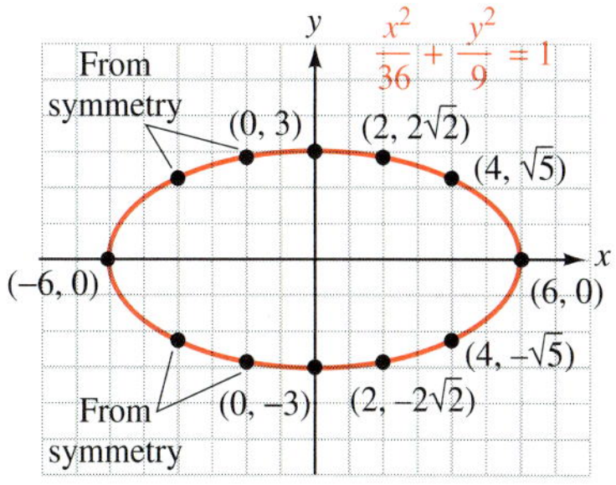

(b)

To increase the accuracy of the graph, we can find additional ordered pairs that satisfy the equation and plot them. For example, if $x = 2$, we have

$$\frac{2^2}{36} + \frac{y^2}{9} = 1$$ Substitute 2 for x in the equation of the ellipse.

$$36\left(\frac{4}{36} + \frac{y^2}{9}\right) = 36(1)$$ To clear the equation of fractions, multiply both sides by the LCD, 36.

$$4 + 4y^2 = 36$$ Distribute the multiplication by 36.

$$y^2 = 8$$ Subtract 4 from both sides, then divide both sides by 4.

$$y = \pm\sqrt{8}$$ Use the square root property.

$$y = \pm 2\sqrt{2}$$ Simplify the radical.

Since two values of y, $2\sqrt{2}$ and $-2\sqrt{2}$, correspond to the x-value 2, we have found two points on the ellipse: $(2, 2\sqrt{2})$ and $(2, -2\sqrt{2})$.

In a similar manner, we can find the corresponding values of y for the x-value 4. In figure (b) we record these ordered pairs in a table, plot them, use symmetry with respect to the y-axis to plot four other points, and then draw the graph of the ellipse.

Self Check 1 Graph: $\frac{x^2}{49} + \frac{y^2}{25} = 1$.

EXAMPLE 2

Graph: $16x^2 + y^2 = 16$.

Solution This equation is not in standard form. To write it in standard form with 1 on the right-hand side, we divide both sides by 16.

Success Tip

Although the term $\frac{16x^2}{16}$ simplifies to x^2, we write it as the fraction $\frac{x^2}{1}$ so that it has the form $\frac{x^2}{a^2}$.

$$16x^2 + y^2 = 16$$

$$\frac{16x^2}{16} + \frac{y^2}{16} = \frac{16}{16}$$ Divide both sides by 16.

$$\frac{x^2}{1} + \frac{y^2}{16} = 1$$ Simplify: $\frac{16x^2}{16} = x^2 = \frac{x^2}{1}$, and $\frac{16}{16} = 1$.

To determine a and b, we can write the equation in the form

$$\frac{x^2}{1^2} + \frac{y^2}{4^2} = 1$$ To find a, write 1 as 1^2. To find b, write 16 as 4^2.

Since a^2 (the denominator of x^2) is 1^2, it follows that $a = 1$, and since b^2 (the denominator of y^2) is 4^2, it follows that $b = 4$. Thus, the x-intercepts of the graph are $(1, 0)$ and $(-1, 0)$ and the y-intercepts are $(0, 4)$ and $(0, -4)$. We use these four points as guides to sketch the graph of the ellipse, as shown on the right. The result is a vertical ellipse.

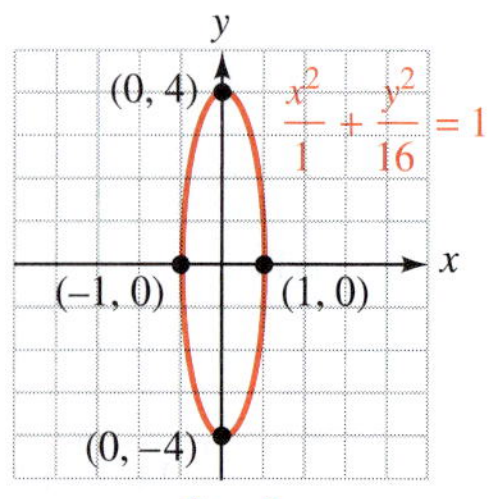

$16x^2 + y^2 = 16$

Self Check 2 Graph: $9x^2 + y^2 = 9$.

GRAPHING ELLIPSES CENTERED AT (*h, k*)

Not all ellipses are centered at the origin. As with the graphs of circles and parabolas, the graph of an ellipse can be translated horizontally and vertically.

The Equation of an Ellipse Centered at (h, k)

The **standard form of the equation of a horizontal or vertical ellipse** centered at (h, k) is

$$\frac{(x-h)^2}{a^2} + \frac{(y-k)^2}{b^2} = 1 \qquad \text{where } a > 0 \text{ and } b > 0$$

For a horizontal ellipse, a is the distance from the center to a vertex. For a vertical ellipse, b is the distance from the center to a vertex.

EXAMPLE 3

Graph: $\frac{(x-2)^2}{16} + \frac{(y+3)^2}{25} = 1$.

Solution To determine h, k, a, and b, we write the equation in the form

$$\frac{(x-2)^2}{4^2} + \frac{[y-(-3)]^2}{5^2} = 1$$

To find k, write $y + 3$ as $y - (-3)$. To find a, write 16 as 4^2. To find b, write 25 as 5^2.

We find the center of the ellipse in the same way we would find the center of a circle, by examining $(x-2)^2$ and $(y+3)^2$. Since $h = 2$ and $k = -3$, this is the equation of an ellipse centered at $(h, k) = (2, -3)$. From the denominators, 4^2 and 5^2, we find that $a = 4$ and $b = 5$. Because $b > a$, it is a vertical ellipse.

We first plot the center, as shown below. Since b is the distance from the center to a vertex for a vertical ellipse, we can locate the vertices by counting 5 units above and 5 units below the center. The vertices are the points $(2, 2)$ and $(2, -8)$.

To locate two more points on the ellipse, we use the fact that a is 4 and count 4 units to the left and to the right of the center. We see that the points $(-2, -3)$ and $(6, -3)$ are also on the graph.

Using these four points as guides, we draw the graph shown below.

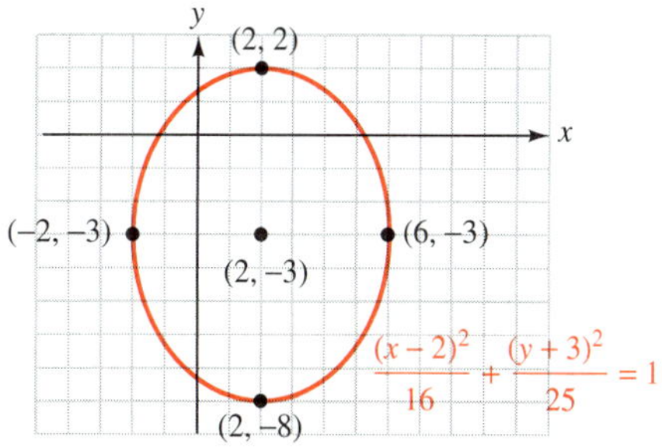

Self Check 3 Graph: $\frac{(x-1)^2}{9} + \frac{(y+2)^2}{16} = 1$.

ACCENT ON TECHNOLOGY: GRAPHING ELLIPSES

To use a graphing calculator to graph the equation from Example 3,

$$\frac{(x-2)^2}{16} + \frac{(y+3)^2}{25} = 1$$

we first clear the equation of fractions and then we solve for y.

$$25(x-2)^2 + 16(y+3)^2 = 400$$ Multiply both sides by 400.

$$16(y+3)^2 = 400 - 25(x-2)^2$$ Subtract $25(x-2)^2$ from both sides.

$$(y+3)^2 = \frac{400 - 25(x-2)^2}{16}$$ Divide both sides by 16.

$$y + 3 = \pm\frac{\sqrt{400 - 25(x-2)^2}}{4}$$ Use the square root property.

$$y = -3 \pm \frac{\sqrt{400 - 25(x-2)^2}}{4}$$ Subtract 3 from both sides.

If we graph the two functions that $y = -3 \pm \frac{\sqrt{400 - 25(x-2)^2}}{4}$ represents in a square window, we get the graph of the ellipse shown below.

$$y = -3 + \frac{\sqrt{400 - 25(x-2)^2}}{4} \quad \text{and} \quad y = -3 - \frac{\sqrt{400 - 25(x-2)^2}}{4}$$

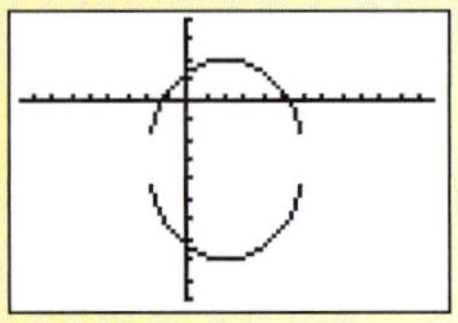

As we saw with circles, the two portions of the ellipse do not quite connect. This is because the graphs are nearly vertical there.

EXAMPLE 4

Graph: $4(x-2)^2 + 9(y-1)^2 = 36$.

Solution This equation is not in standard form. To write it in standard form with 1 on the right-hand side, we divide both sides by 36.

$$4(x-2)^2 + 9(y-1)^2 = 36$$

$$\frac{4(x-2)^2}{\mathbf{36}} + \frac{9(y-1)^2}{\mathbf{36}} = \frac{36}{\mathbf{36}}$$ Divide both sides by 36.

$$\frac{(x-2)^2}{9} + \frac{(y-1)^2}{4} = 1$$ Simplify: $\frac{4}{36} = \frac{1}{9}$, $\frac{9}{36} = \frac{1}{4}$, and $\frac{36}{36} = 1$.

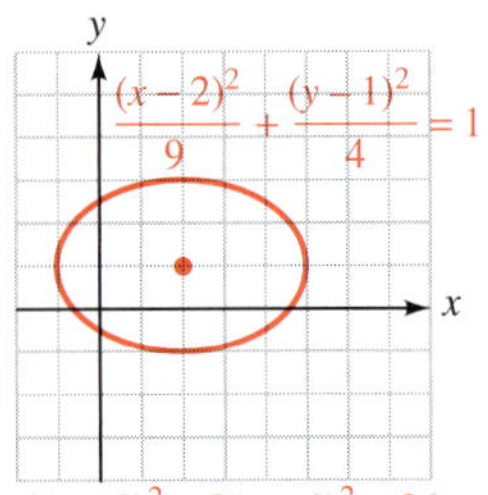

$4(x-2)^2 + 9(y-1)^2 = 36$

This is the standard form of the equation of a horizontal ellipse, centered at (2, 1), with $a = 3$ and $b = 2$. The graph of the ellipse is shown in the margin.

Self Check 4 Graph: $12(x-1)^2 + 3(y+1)^2 = 48$.

PROBLEM SOLVING USING ELLIPSES

EXAMPLE 5

Landscape design. A landscape architect is designing an elliptical pool that will fit in the center of a 20-by-30-foot rectangular garden, leaving at least 5 feet of space on all sides. Find the equation of the ellipse.

Solution We place the rectangular garden in the coordinate system shown below.

To maintain 5 feet of clearance at the ends of the ellipse, the x-intercepts must be the points $(10, 0)$ and $(-10, 0)$. Similarly, the y-intercepts are the points $(0, 5)$ and $(0, -5)$.

Since the ellipse is centered at the origin, its equation has the form

$$\frac{x^2}{a^2} + \frac{y^2}{b^2} = 1$$

with $a = 10$ and $b = 5$. Thus, the equation of the boundary of the pool is

$$\frac{x^2}{100} + \frac{y^2}{25} = 1$$

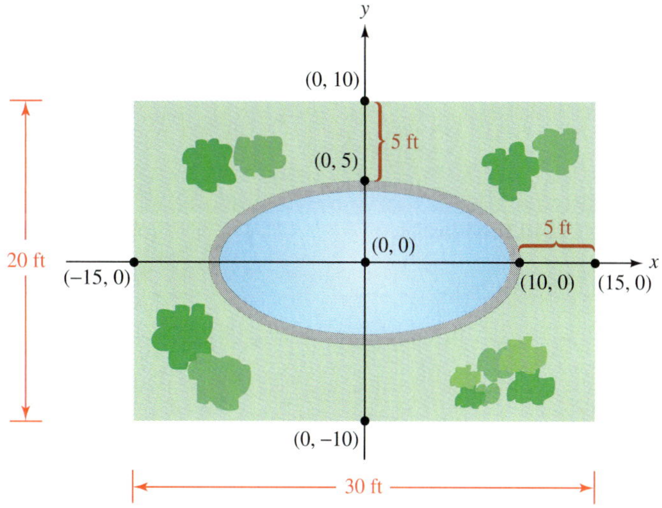

Ellipses, like parabolas, have reflective properties that are used in many practical applications. For example, any light or sound originating at one focus of an ellipse is reflected by the interior of the figure to the other focus.

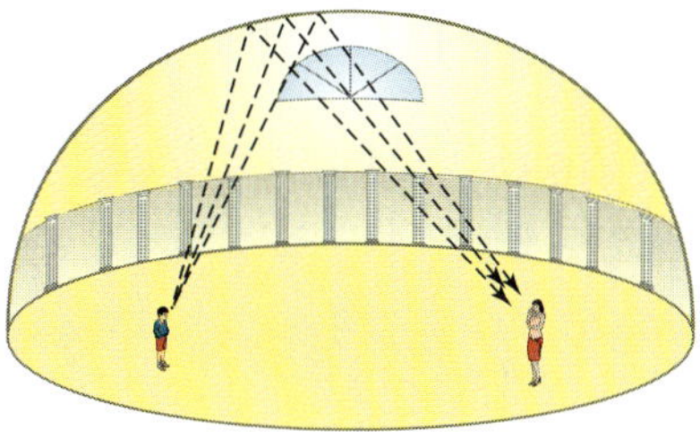

Whispering galleries

In an elliptical dome, even the slightest whisper made by a person standing at one focus can be heard by a person standing at the other focus.

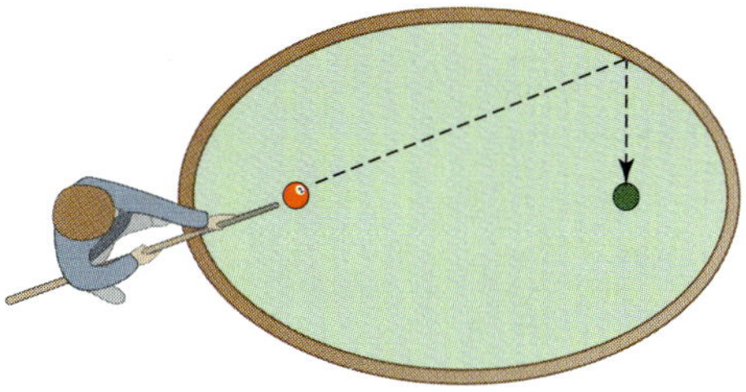

Elliptical billiards tables

When a ball is shot from one focus, it will rebound off the side of the table into a pocket located at the other focus.

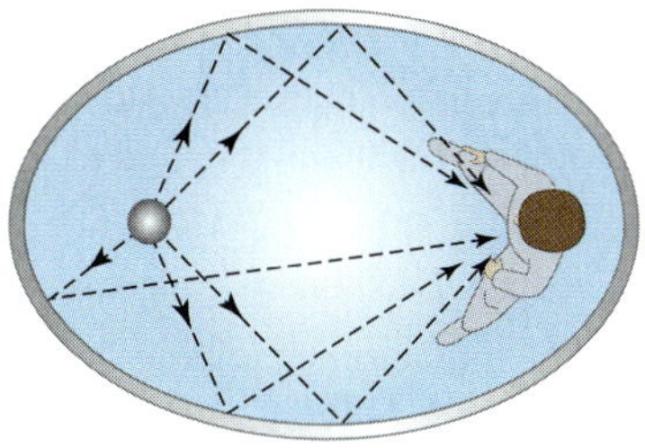

Treatment for kidney stones

The patient is positioned in an elliptical tank of water so that the kidney stone is at one focus. High-intensity sound waves generated at another focus are reflected to the stone to shatter it.

Answers to Self Checks

1.

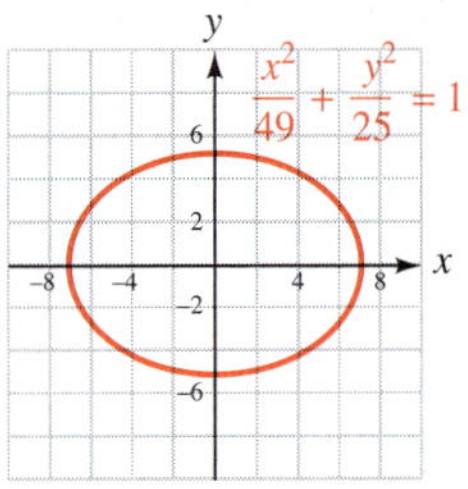

2.

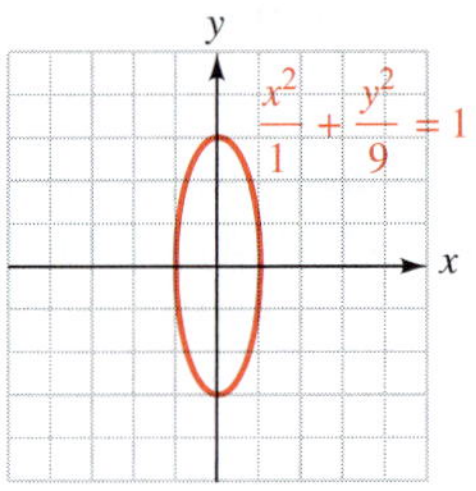

$9x^2 + y^2 = 9$

3.

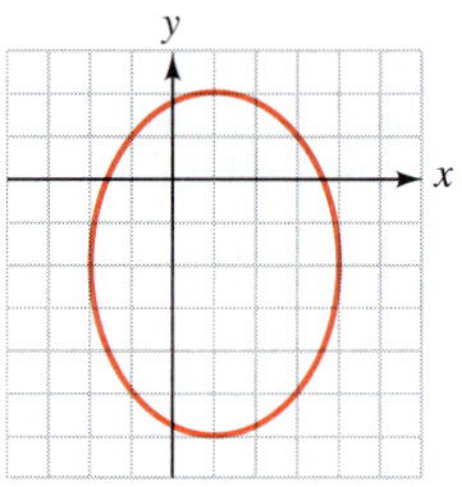

$\frac{(x-1)^2}{9} + \frac{(y+2)^2}{16} = 1$

4.

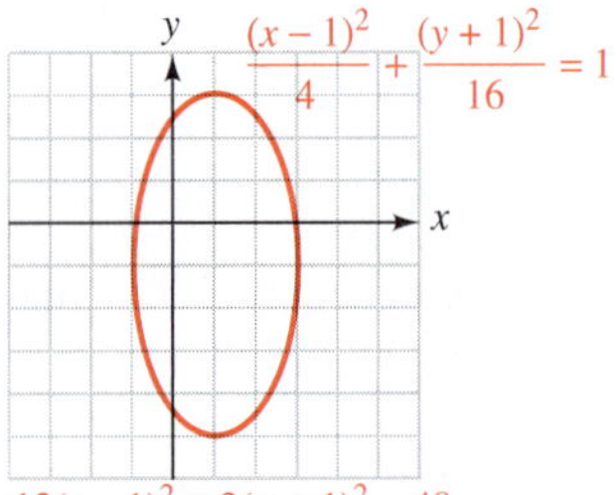

$12(x-1)^2 + 3(y+1)^2 = 48$

10.2 STUDY SET

VOCABULARY Fill in the blanks.

1. The curve graphed below is an ________.

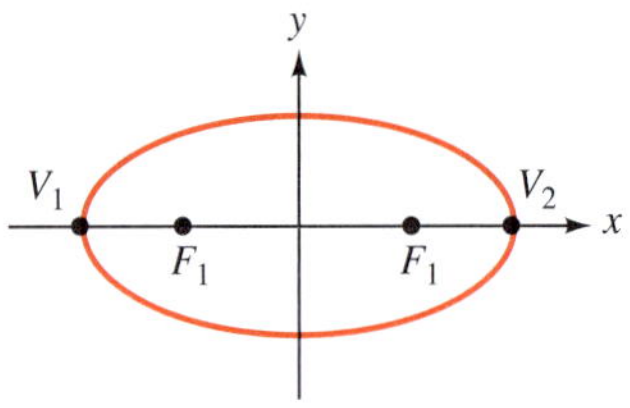

2. An ________ is the set of all points in a plane for which the sum of the distances from two fixed points is a constant.
3. In the graph above, F_1 and F_2 are the ______ of the ellipse. Each one is called a _______ of the ellipse.
4. In the graph above, V_1 and V_2 are the _________ of the ellipse. Each one is called a _______ of the ellipse.
5. The line segment joining the vertices of an ellipse is called the _______ _______ of the ellipse.
6. The midpoint of the major axis of an ellipse is the _______ of the ellipse.

CONCEPTS

7. What is the standard form for the equation of an ellipse centered at the origin and symmetric to both axes?
8. What is the standard form for the equation of a horizontal or vertical ellipse centered at (h, k)?
9. What are the x- and the y-intercepts of the graph of $\frac{x^2}{a^2} + \frac{y^2}{b^2} = 1$?
10. **a.** What is the center of the ellipse graphed below? What are a and b?
    b. Is the ellipse horizontal or vertical?
    c. What is the equation of the ellipse?

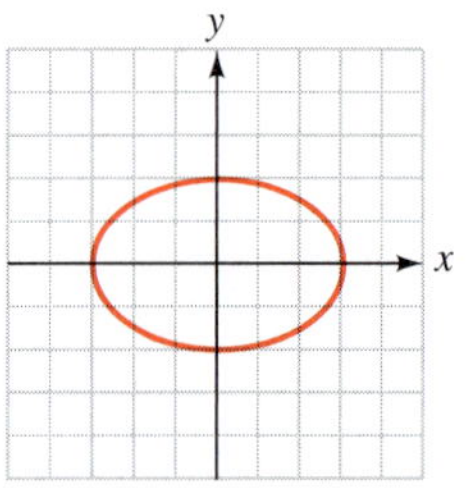

11. **a.** What is the center of the ellipse graphed below? What are a and b?

b. Is the ellipse horizontal or vertical?

c. What is the equation of the ellipse?

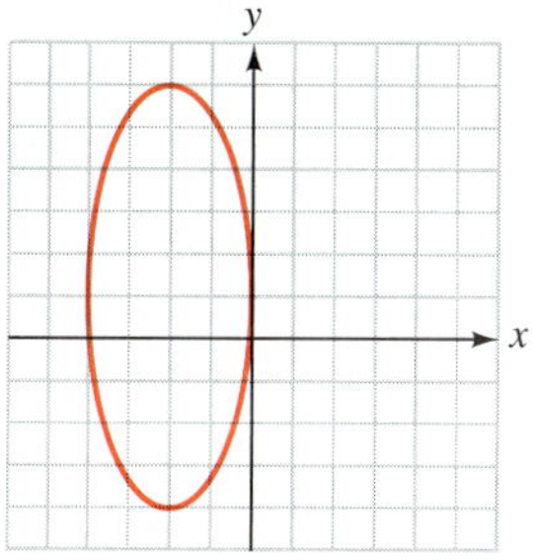

12. Find two points on the graph of $\frac{x^2}{16} + \frac{y^2}{4} = 1$ by letting $x = 2$ and finding the corresponding values of y.

13. Divide both sides of the equation by 64 and write the equation in standard form:

$$4(x - 1)^2 + 64(y + 5)^2 = 64$$

14. Determine whether the equation, when graphed, will be a circle, a parabola, or an ellipse.

a. $x = y^2 - 2y + 10$ **b.** $\frac{x^2}{49} + \frac{y^2}{64} = 1$

c. $(x - 3)^2 + (y + 4)^2 = 25$

d. $2(x - 1)^2 + 8(y + 5)^2 = 32$

NOTATION

15. Find h, k, a, and b: $\frac{(x + 8)^2}{100} + \frac{(y - 6)^2}{144} = 1$.

16. Write each denominator in the equation $\frac{x^2}{81} + \frac{y^2}{49} = 1$ as the square of a number.

PRACTICE Graph each equation.

17. $\frac{x^2}{25} + \frac{y^2}{4} = 1$

18. $\frac{x^2}{16} + \frac{y^2}{9} = 1$

19. $\frac{x^2}{4} + \frac{y^2}{9} = 1$

20. $\frac{x^2}{16} + \frac{y^2}{25} = 1$

21. $\frac{x^2}{16} + \frac{y^2}{1} = 1$

22. $\frac{x^2}{1} + \frac{y^2}{9} = 1$

23. $x^2 + 9y^2 = 9$

24. $25x^2 + 9y^2 = 225$

25. $16x^2 + 4y^2 = 64$

26. $4x^2 + 9y^2 = 36$

27. $x^2 = 100 - 4y^2$

28. $x^2 = 36 - 4y^2$

29. $\frac{(x - 2)^2}{9} + \frac{(y - 1)^2}{4} = 1$

30. $\frac{(x - 1)^2}{9} + \frac{(y - 3)^2}{4} = 1$

31. $\frac{(x + 2)^2}{64} + \frac{(y - 2)^2}{100} = 1$

32. $\frac{(x - 6)^2}{36} + \frac{(y + 6)^2}{144} = 1$

33. $(x + 1)^2 + 4(y + 2)^2 = 4$

34. $25(x + 1)^2 + 9y^2 = 225$

35. $(x - 2)^2 + 4(y + 1)^2 = 4$

36. $(x - 1)^2 + 4(y - 2)^2 = 4$

37. $9(x - 1)^2 = 36 - 4(y + 2)^2$

38. $16(x - 5)^2 = 400 - 25(y - 4)^2$

Use a graphing calculator to graph each equation.

39. $\frac{x^2}{9} + \frac{y^2}{4} = 1$

40. $x^2 + 16y^2 = 16$

41. $\frac{x^2}{4} + \frac{(y - 1)^2}{9} = 1$

42. $\frac{(x + 1)^2}{9} + \frac{(y - 2)^2}{4} = 1$

APPLICATIONS

43. FITNESS EQUIPMENT With elliptical cross-training equipment, the feet move through the natural elliptical pattern that one experiences when walking, jogging, or running. Write the equation of the elliptical pattern shown below.

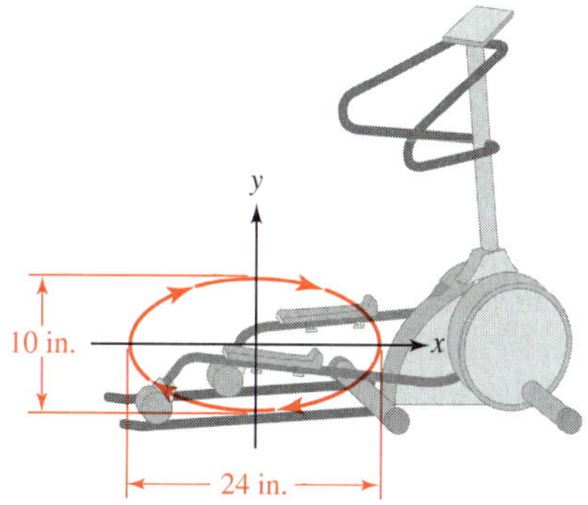

44. DESIGNING AN UNDERPASS The arch of an underpass is a part of an ellipse. Find the equation of the ellipse.

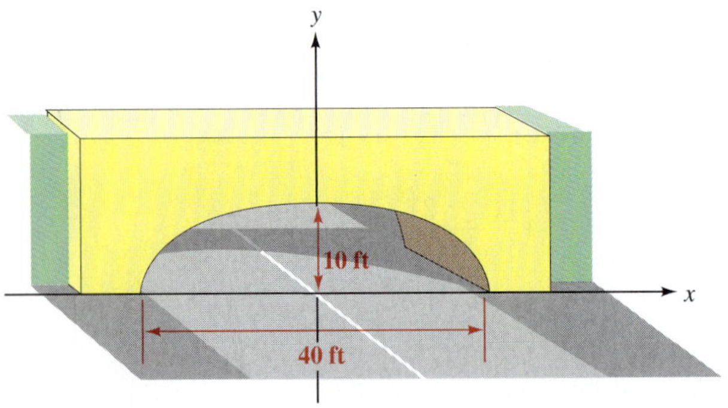

45. CALCULATING CLEARANCE Find the height of the elliptical arch in Exercise 44 at a point 10 feet from the center of the roadway.

46. POOL TABLES Find the equation of the outer edge of the elliptical pool table shown below.

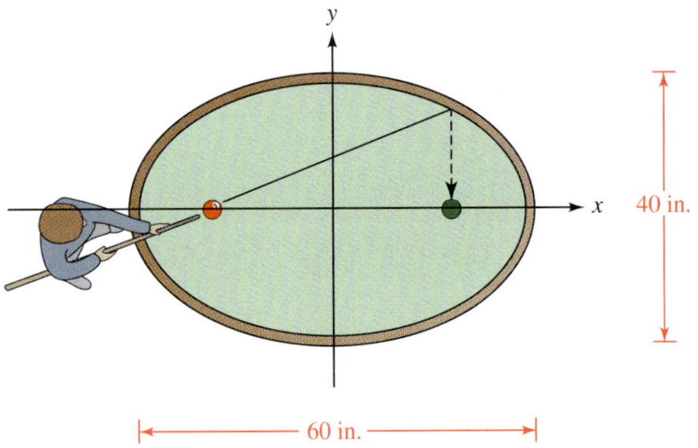

47. AREA OF AN ELLIPSE The area A of the ellipse

$$\frac{x^2}{a^2} + \frac{y^2}{b^2} = 1$$

is given by $A = \pi ab$. Find the area of the ellipse $9x^2 + 16y^2 = 144$.

48. AREA OF A TRACK The elliptical track shown in the next column is bounded by the ellipses $4x^2 + 9y^2 = 576$ and $9x^2 + 25y^2 = 900$. Find the area of the track. (See Exercise 47.)

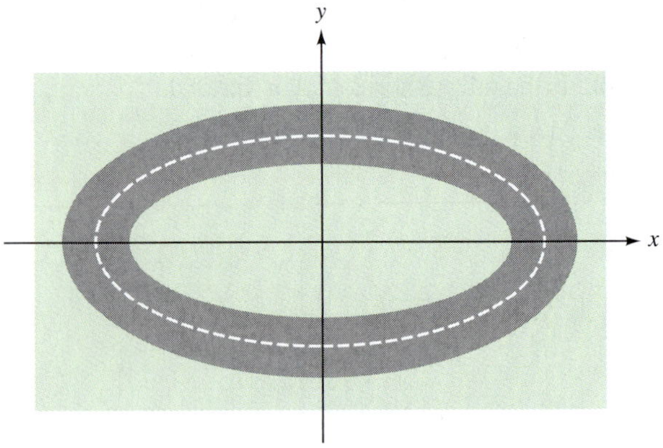

WRITING

49. What is an ellipse?

50. Explain the difference between the focus of an ellipse and the vertex of an ellipse.

51. Compare the graphs of $\frac{x^2}{81} + \frac{y^2}{64} = 1$ and $\frac{x^2}{64} + \frac{y^2}{81} = 1$. Do they have any similarities?

52. What are the reflective properties of an ellipse?

REVIEW **Find each product.**

53. $3x^{-2}y^2(4x^2 + 3y^{-2})$

54. $(2a^{-2} - b^{-2})(2a^{-2} + b^{-2})$

Simplify each complex fraction.

55. $\dfrac{x^{-2} + y^{-2}}{x^{-2} - y^{-2}}$

56. $\dfrac{2x^{-3} - 2y^{-3}}{4x^{-3} + 4y^{-3}}$

CHALLENGE PROBLEMS

57. What happens to the graph of

$$\frac{x^2}{a^2} + \frac{y^2}{b^2} = 1$$

when $a = b$?

58. Graph: $9x^2 + 4y^2 = 1$.

59. Write the equation $9x^2 + 4y^2 - 18x + 16y = 11$ in the standard form of the equation of an ellipse.

60. Let the foci of an ellipse be $(c, 0)$ and $(-c, 0)$. Suppose that the sum of the distances from any point (x, y) on the ellipse to the two foci is the constant $2a$. Show that the equation for the ellipse is

$$\frac{x^2}{a^2} + \frac{y^2}{a^2 - c^2} = 1$$

Then let $b^2 = a^2 - c^2$ to obtain the standard form of the equation of an ellipse.

10.3 The Hyperbola

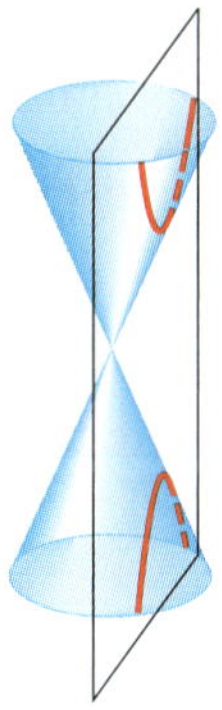

- The definition of a hyperbola
- Graphing hyperbolas centered at the origin
- Graphing hyperbolas centered at (h, k)
- Problem solving using hyperbolas

The final conic section that we will discuss, the *hyperbola,* is a curve that has two branches. In this section, we will learn how to graph equations that represent hyperbolas.

THE DEFINITION OF A HYPERBOLA

Ellipses and hyperbolas have completely different shapes, but their definitions are similar. Instead of the *sum* of distances, the definition of a hyperbola involves a *difference* of distances.

Definition of a Hyperbola A **hyperbola** is the set of all points in a plane for which the difference of the distances from two fixed points is a constant.

The figure below illustrates that any point on the hyperbola is a constant distance $d_1 - d_2$ from two fixed points, each of which is called a **focus.** Midway between the **foci** is the **center** of the hyperbola.

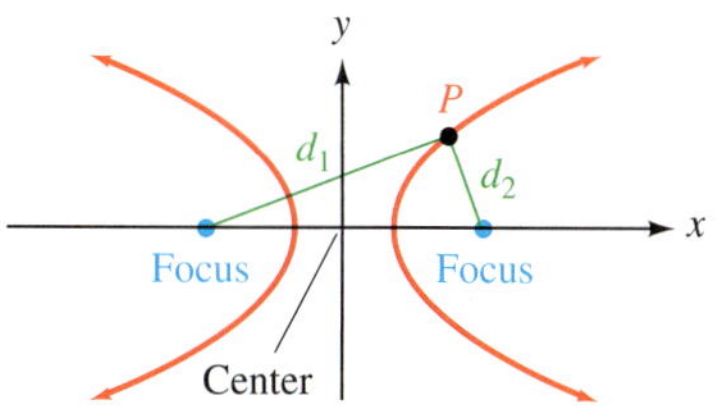

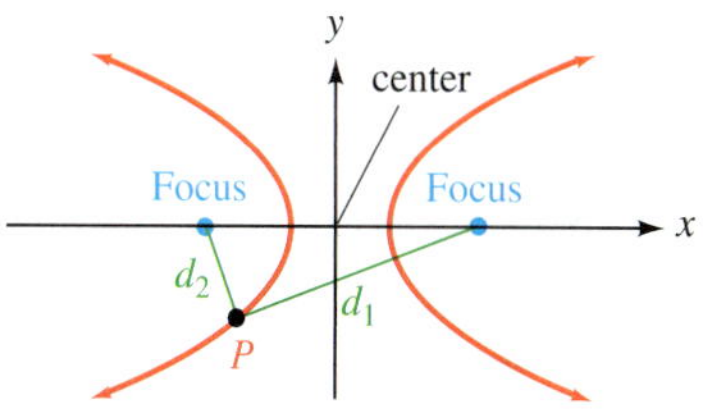

GRAPHING HYPERBOLAS CENTERED AT THE ORIGIN

The graph of the equation

$$\frac{x^2}{25} - \frac{y^2}{9} = 1$$

is a hyperbola. To graph the equation, we make a table of solutions that satisfy the equation, plot each point, and join them with a smooth curve.

$$\frac{x^2}{25} - \frac{y^2}{9} = 1$$

x	y	
−7	±2.9	→ (−7, ±2.9)
−6	±2.0	→ (−6, ±2.0)
−5	0	→ (−5, 0)
5	0	→ (5, 0)
6	±2.0	→ (6, ±2.0)
7	±2.9	→ (7, ±2.9)

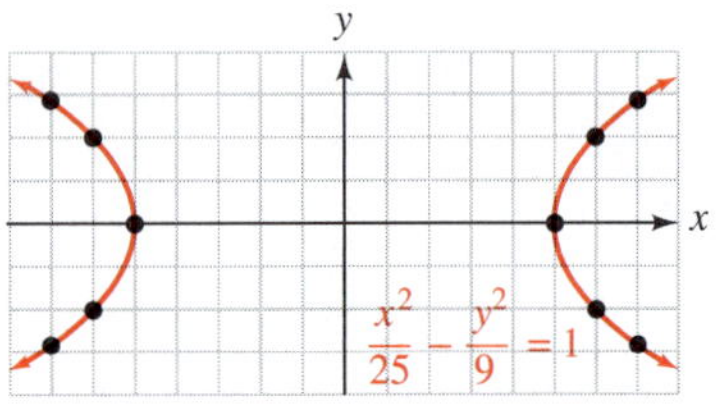

Caution

Although the two branches of a hyperbola look like parabolas, they are not parabolas.

This graph is centered at the origin and intersects the x-axis at (5, 0) and (−5, 0). We also note that the graph does not intersect the y-axis.

It is possible to draw a hyperbola without plotting points. For example, if we want to graph the hyperbola with an equation of

$$\frac{x^2}{a^2} - \frac{y^2}{b^2} = 1$$

we first find the x- and y-intercepts. To find the x-intercepts, we let $y = 0$ and solve for x:

$$\frac{x^2}{a^2} - \frac{0^2}{b^2} = 1$$

$$x^2 = a^2$$

$$x = \pm a \quad \text{Use the square root property.}$$

The hyperbola crosses the x-axis at the points $V_1(a, 0)$ and $V_2(-a, 0)$, called the **vertices** of the hyperbola.

To attempt to find the y-intercepts, we let $x = 0$ and solve for y:

$$\frac{0^2}{a^2} - \frac{y^2}{b^2} = 1$$

$$y^2 = -b^2$$

$$y = \pm\sqrt{-b^2}$$

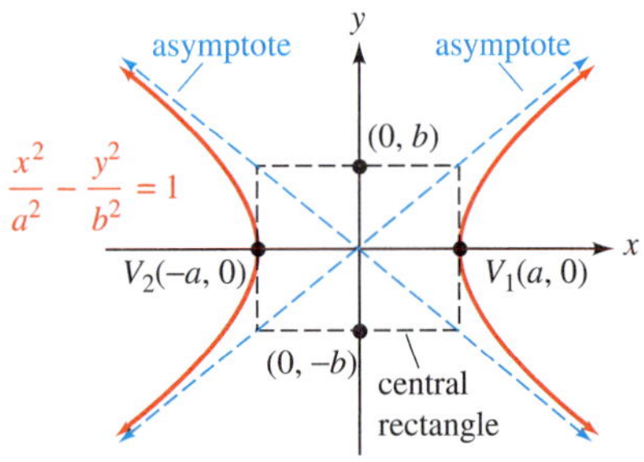

Since b^2 is always positive, $\sqrt{-b^2}$ is an imaginary number. This means that the hyperbola does not intersect the y-axis.

If we construct a rectangle, called the **central rectangle,** whose sides pass horizontally through $\pm b$ on the y-axis and vertically through $\pm a$ on the x-axis, the extended diagonals of the rectangle will be **asymptotes** of the hyperbola.

The Language of Algebra

The central rectangle is also called the **fundamental rectangle.**

Standard Form for the Equation of a Hyperbola Centered at the Origin

Any equation that can be written in the form

$$\frac{x^2}{a^2} - \frac{y^2}{b^2} = 1$$

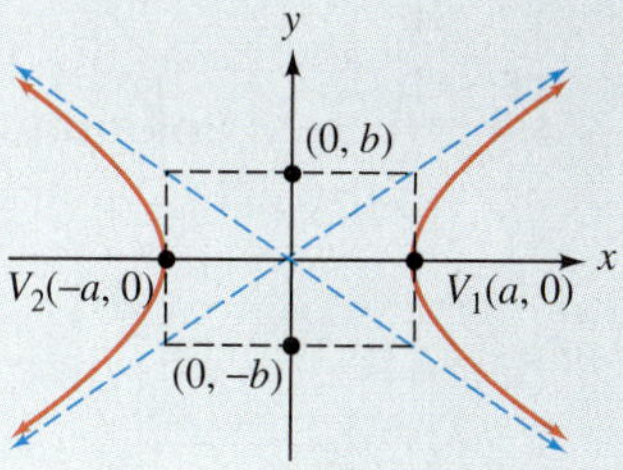

has a graph that is a hyperbola centered at the origin. The x-intercepts are the vertices $V_1(a, 0)$ and $V_2(-a, 0)$. There are no y-intercepts.

The asymptotes of the hyperbola are the extended diagonals of the central rectangle.

The branches of the hyperbola in previous discussions open to the left and to the right. It is possible for hyperbolas to have different orientations with respect to the x- and y-axes. For example, the branches of a hyperbola can open upward and downward. In that case, the following equation applies.

Standard Form for the Equation of a Hyperbola Centered at the Origin

Any equation that can be written in the form

$$\frac{y^2}{a^2} - \frac{x^2}{b^2} = 1$$

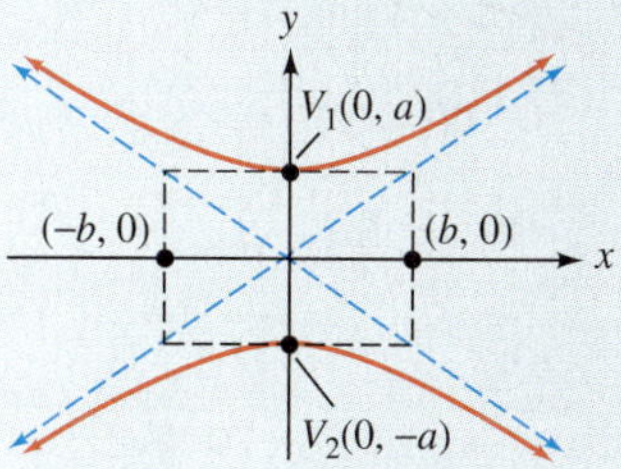

has a graph that is a hyperbola centered at the origin. The y-intercepts are the vertices $V_1(0, a)$ and $V_2(0, -a)$. There are no x-intercepts.

The asymptotes of the hyperbola are the extended diagonals of the central rectangle.

EXAMPLE 1

Graph: $\frac{x^2}{9} - \frac{y^2}{16} = 1$.

Solution This is the standard form of the equation of a hyperbola, centered at the origin, that opens left and right. To determine the vertices and the central rectangle of the graph, we find a and b.

$$\frac{x^2}{9} - \frac{y^2}{16} = 1 \qquad \frac{x^2}{a^2} - \frac{y^2}{b^2} = 1$$

Since $a^2 = 9$, it follows that $a = 3$

Since $b^2 = 16$, it follows that $b = 4$.

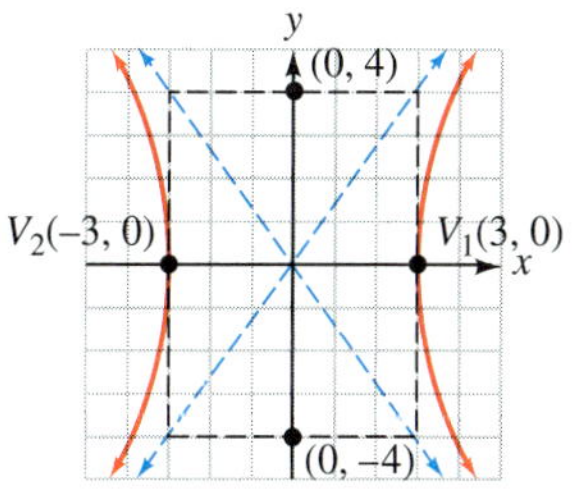

$\frac{x^2}{9} - \frac{y^2}{16} = 1$

The x-intercepts are $(a, 0)$ and $(-a, 0)$, or $(3, 0)$ and $(-3, 0)$. They are also the vertices of the hyperbola.

To construct the central rectangle, we use the values of $a = 3$ and $b = 4$. The rectangle passes through $(3, 0)$ and $(-3, 0)$ on the x-axis, and $(0, 4)$ and $(0, -4)$ on the y-axis. We draw extended diagonal dashed lines through the rectangle to obtain the asymptotes. Then we draw a smooth curve through each vertex that gets close to the asymptotes.

Self Check 1 Graph: $\frac{x^2}{25} - \frac{y^2}{4} = 1$.

EXAMPLE 2

Graph: $9y^2 - 4x^2 = 36$.

Solution To write the equation in standard from, we divide both sides by 36 to obtain

$$\frac{9y^2}{36} - \frac{4x^2}{36} = 1$$

$$\frac{y^2}{4} - \frac{x^2}{9} = 1 \quad \text{Simplify each fraction.}$$

This is the standard form of the equation of a hyperbola, centered at the origin, that opens up and down. To determine the vertices and the central rectangle of the graph, we find a and b.

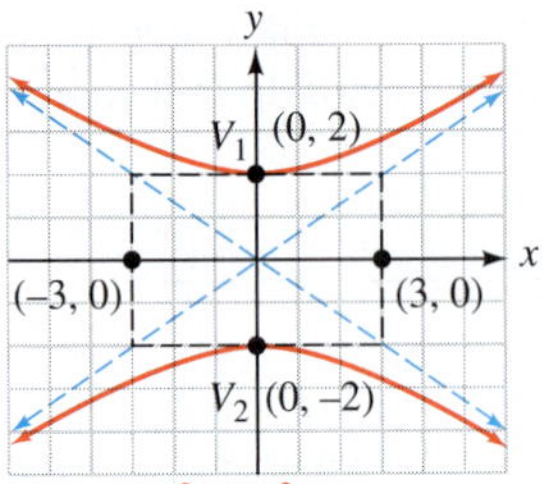

$9y^2 - 4x^2 = 36$
or
$\frac{y^2}{4} - \frac{x^2}{9} = 1$

$$\frac{y^2}{\mathbf{4}} - \frac{x^2}{\mathbf{9}} = 1 \qquad \frac{y^2}{a^2} - \frac{x^2}{b^2} = 1$$

Since $a^2 = 4$, it follows that $a = 2$.

Since $b^2 = 9$, it follows that $b = 3$.

The y-intercepts are $(0, a)$ and $(0, -a)$, or $(0, 2)$ and $(0, -2)$. They are also the vertices of the hyperbola.

Since $a = 2$ and $b = 3$, the central rectangle passes through $(0, 2)$ and $(0, -2)$, and $(3, 0)$ and $(-3, 0)$. We draw its extended diagonals and sketch the hyperbola.

Self Check 2 Graph: $16y^2 - x^2 = 16$.

ACCENT ON TECHNOLOGY: GRAPHING HYPERBOLAS

To graph $\frac{x^2}{9} - \frac{y^2}{16} = 1$ from Example 1 using a graphing calculator, we follow the same procedure that we used for circles and ellipses. To write the equation as two functions, we solve for y to get $y = \pm\frac{\sqrt{16x^2 - 144}}{3}$. Then we graph the following two functions in a square window setting to get the graph of the hyperbola shown below.

$$y = \frac{\sqrt{16x^2 - 144}}{3} \quad \text{and} \quad y = -\frac{\sqrt{16x^2 - 144}}{3}$$

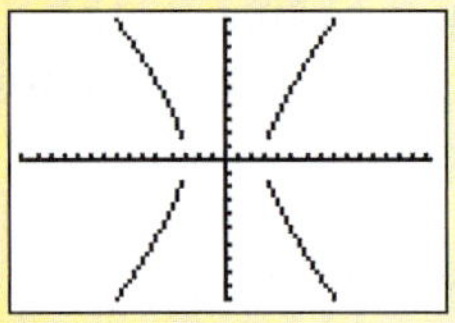

GRAPHING HYPERBOLAS CENTERED AT (h, k)

If a hyperbola is centered at a point with coordinates (h, k), the following equations apply.

Standard Forms for the Equations of Hyperbolas Centered at (h, k)

Any equation that can be written in the form

$$\frac{(x-h)^2}{a^2} - \frac{(y-k)^2}{b^2} = 1$$

is a hyperbola that has its center at (h, k) and opens left and right.

Any equation of the form

$$\frac{(y-k)^2}{a^2} - \frac{(x-h)^2}{b^2} = 1$$

is a hyperbola that has its center at (h, k) and opens up and down.

EXAMPLE 3

Graph: $\frac{(x-3)^2}{16} - \frac{(y+1)^2}{4} = 1$.

Solution We write the equation in the form

$$\frac{(x-3)^2}{4^2} - \frac{[y-(-1)]^2}{2^2} = 1$$

to see that its graph will be a hyperbola centered at the point $(h, k) = (3, -1)$. Its vertices are located at $a = 4$ units to the right and left of the center, at $(7, -1)$ and $(-1, -1)$. Since $b = 2$, we can count 2 units above and below the center to locate points $(3, 1)$ and $(3, -3)$. With these four points, we can draw the central rectangle along with its extended diagonals. We can then sketch the hyperbola, as shown.

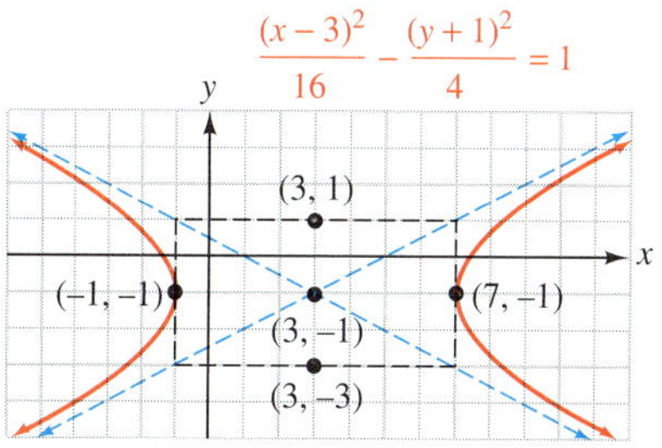

Self Check 3 Graph: $\frac{(x+2)^2}{9} - \frac{(y-1)^2}{4} = 1$.

EXAMPLE 4

Write the equation $x^2 - y^2 - 2x + 4y = 12$ in standard form to show that the equation represents a hyperbola. Then graph it.

Solution We proceed as follows.

$$x^2 - y^2 - 2x + 4y = 12$$

$$x^2 - 2x - y^2 + 4y = 12 \quad \text{Use the commutative property to rearrange terms.}$$

$$x^2 - 2x - (y^2 - 4y) = 12 \quad \text{Factor } -1 \text{ from } -y^2 + 4y.$$

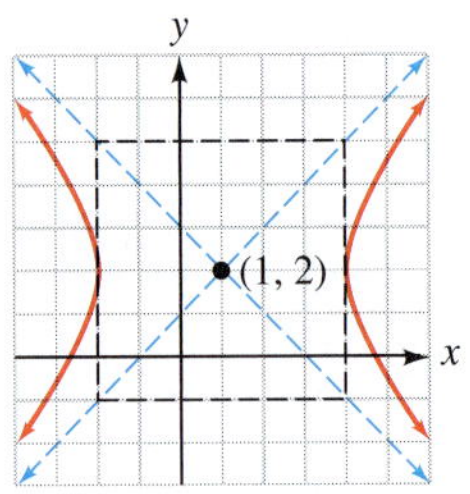

$x^2 - y^2 - 2x + 4y = 12$
or
$\frac{(x-1)^2}{9} - \frac{(y-2)^2}{9} = 1$

We then complete the square on x and y to make $x^2 - 2x$ and $y^2 - 4y$ perfect square trinomials.

$$x^2 - 2x + 1 - (y^2 - 4y + 4) = 12 + 1 - 4$$ Add 1 to both sides and subtract 4 from both sides.

We then factor $x^2 - 2x + 1$ and $y^2 - 4y + 4$ to get

$$(x - 1)^2 - (y - 2)^2 = 9$$

$$\frac{(x - 1)^2}{9} - \frac{(y - 2)^2}{9} = 1$$ Divide both sides by 9.

This is the equation of a hyperbola with center at (1, 2). Its graph is shown in the figure.

Self Check 4 Graph: $x^2 - 4y^2 + 2x - 8y = 7$.

There is a special type of hyperbola (also centered at the origin) that does not intersect either the x- or the y-axis. These hyperbolas have equations of the form $xy = k$, where $k \neq 0$.

EXAMPLE 5

Graph: $xy = -8$.

Solution To make a table of solutions, we can solve the equation $xy = -8$ for y:

$$y = \frac{-8}{x}$$

Then we choose several values for x, find the corresponding values of y, and record the results in the table below. We plot the ordered pairs and join them with a smooth curve to obtain the graph of the hyperbola.

The Language of Algebra

The asymptotes of this hyperbola are the x- and y-axes. A hyperbola for which the asymptotes are perpendicular is called a **rectangular hyperbola.**

$xy = -8$ or $y = \frac{-8}{x}$

x	y	
1	−8	→ (1, −8)
2	−4	→ (2, −4)
4	−2	→ (4, −2)
8	−1	→ (8, −1)
−1	8	→ (−1, 8)
−2	4	→ (−2, 4)
−4	2	→ (−4, 2)
−8	1	→ (−8, 1)

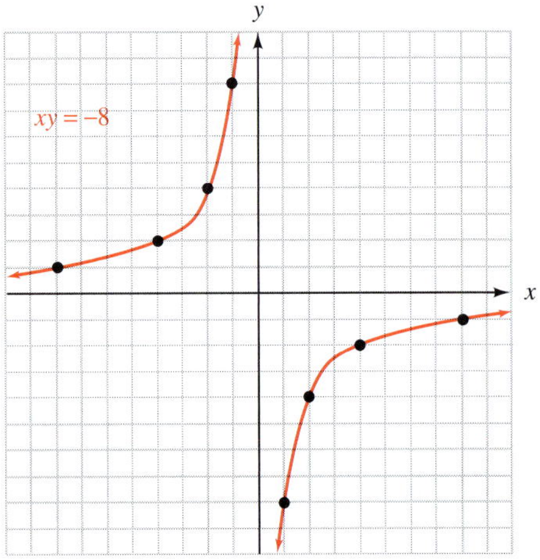

Self Check 5 Graph: $xy = 6$.

The result in Example 5 illustrates the following general equation.

Equations of Hyperbolas of the Form $xy = k$

Any equation of the form $xy = k$, where $k \neq 0$, has a graph that is a **hyperbola,** which does not intersect either the x- or the y-axis.

PROBLEM SOLVING USING HYPERBOLAS

EXAMPLE 6

Atomic structure. In an experiment that led to the discovery of the atomic structure of matter, Lord Rutherford (1871–1937) shot high-energy alpha particles toward a thin sheet of gold. Many of them were reflected, and Rutherford showed the existence of the nucleus of a gold atom. An alpha particle is repelled by the nucleus at the origin; it travels along the hyperbolic path given by $4x^2 - y^2 = 16$. How close does the particle come to the nucleus?

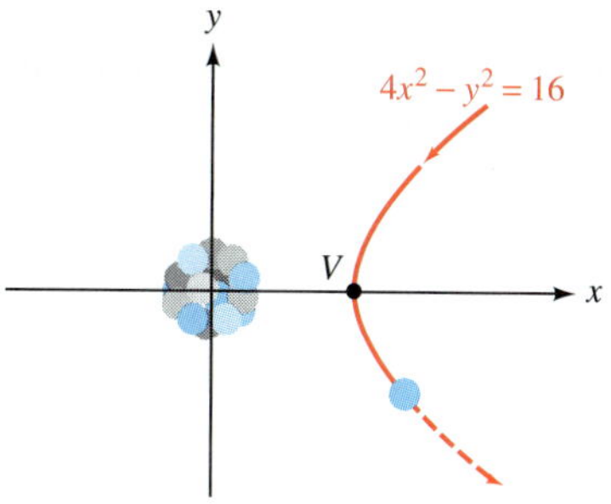

Solution To find the distance from the nucleus at the origin, we must find the coordinates of the vertex V. To do so, we write the equation of the particle's path in standard form:

$$4x^2 - y^2 = 16$$

$$\frac{4x^2}{\mathbf{16}} - \frac{y^2}{\mathbf{16}} = \frac{16}{\mathbf{16}} \quad \text{Divide both sides by 16.}$$

$$\frac{x^2}{4} - \frac{y^2}{16} = 1 \quad \text{Simplify.}$$

$$\frac{x^2}{2^2} - \frac{y^2}{4^2} = 1 \quad \text{To determine } a \text{ and } b\text{, write 4 as } 2^2 \text{ and 16 as } 4^2.$$

This equation is in the form

$$\frac{x^2}{a^2} - \frac{y^2}{b^2} = 1$$

with $a = 2$. Thus, the vertex of the path is (2, 0). The particle is never closer than 2 units from the nucleus

Answers to Self Checks

1.

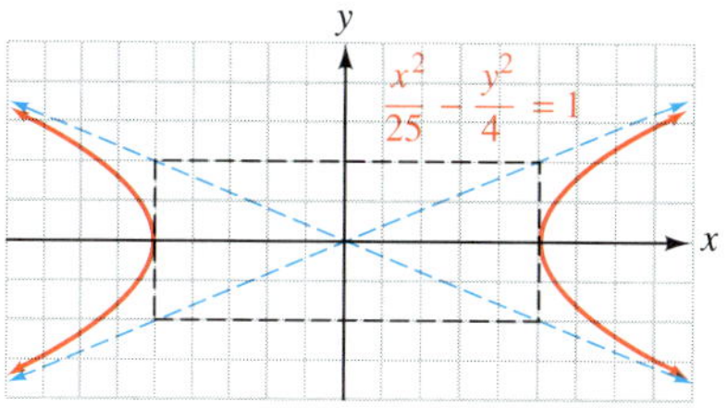

2.

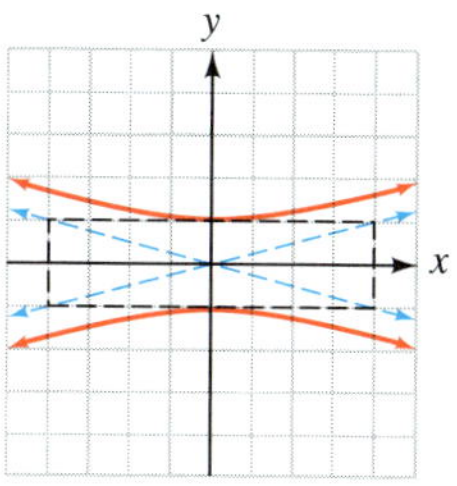

$16y^2 - x^2 = 16$ or $\frac{y^2}{1} - \frac{x^2}{16} = 1$

3.

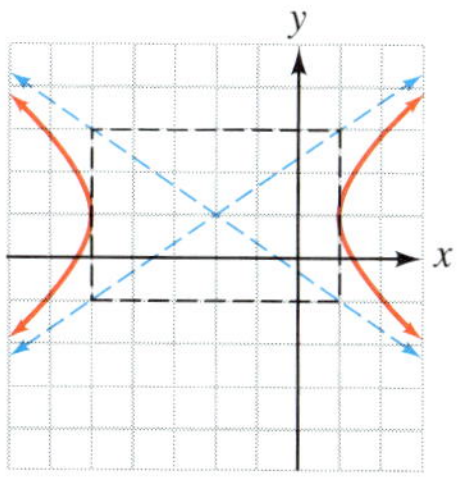

$\frac{(x+2)^2}{9} - \frac{(y-1)^2}{4} = 1$

4.

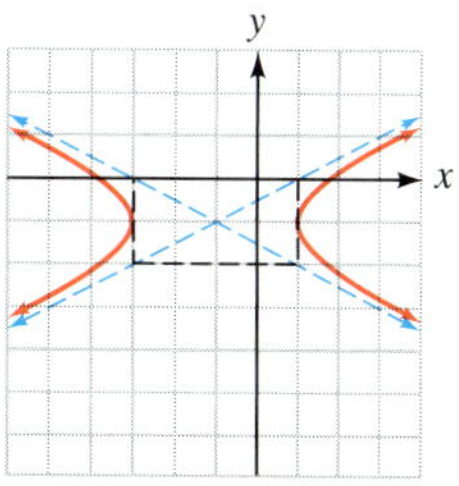

$x^2 - 4y^2 + 2x - 8y = 7$ or $\frac{(x+1)^2}{4} - \frac{(y+1)^2}{1} = 1$

5.

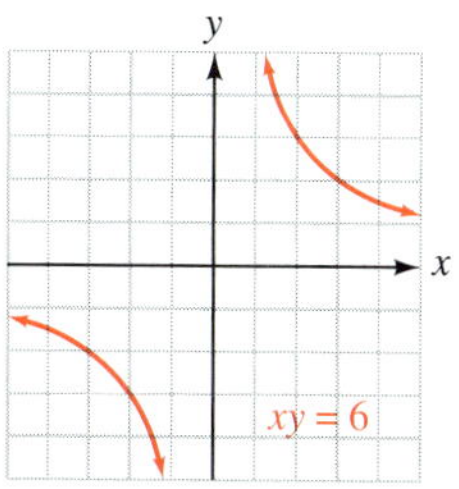

10.3 STUDY SET

VOCABULARY Fill in the blanks.

1. The two-branch curve graphed below is a ________.

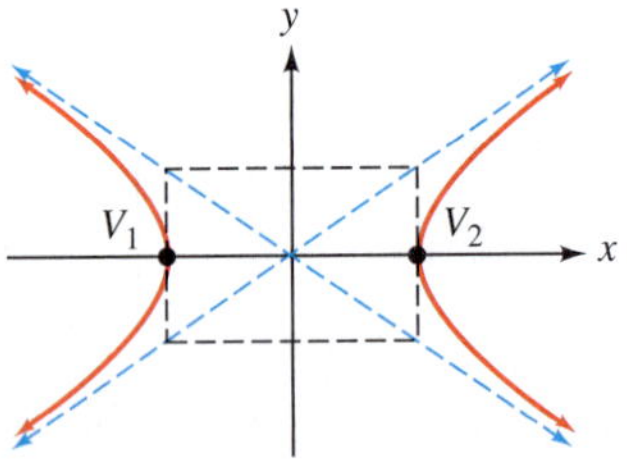

2. A ________ is the set of all points in a plane for which the difference of the distances from two fixed points is a constant.
3. In the graph above, V_1 and V_2 are the ________ of the hyperbola.
4. In the graph above, the figure drawn using dashed black lines is called the ________ ________.
5. The extended ________ of the central rectangle are asymptotes of the hyperbola.
6. To write $9x^2 - 4y^2 = 36$ in ________ form, we divide both sides by 36.

CONCEPTS

7. What is the standard form for the equation of a hyperbola centered at the origin that opens left and right?
8. What is the standard form for the equation of a hyperbola centered at (h, k) that opens up and down?
9. What is the standard form for the equation of a hyperbola centered at (h, k) that opens left and right?
10. **a.** What is the center of the hyperbola graphed below? What are a and b?

    b. What are the x-intercepts of the graph? What are the y-intercepts of the graph?

    c. What is the equation of the hyperbola?

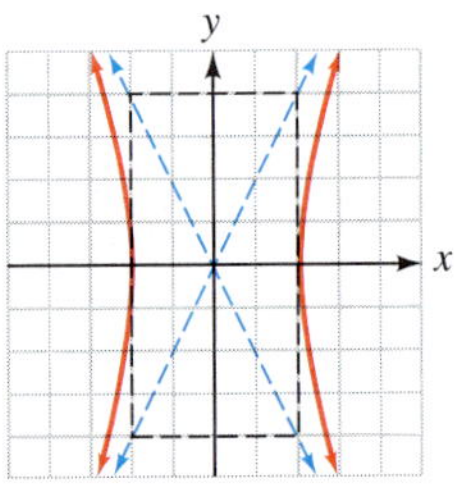

11. **a.** What is the center of the hyperbola graphed below? What are a and b?

    b. What is the equation of the hyperbola?

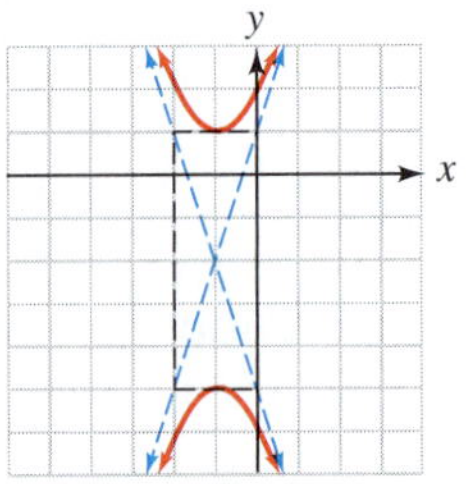

12. **a.** Fill in the blank: An equation of the form $xy = k$, where $k \neq 0$, has a graph that is a ________ that does not intersect either the x-axis or the y-axis.

    b. Complete the table of solutions for $xy = 10$.

x	y
−2	
5	

13. Divide both sides of the equation by 100 and write the equation in standard form:

$$100(x + 1)^2 - 25(y - 5)^2 = 100$$

14. Determine whether the equation, when graphed, will be a circle, a parabola, an ellipse, or a hyperbola.

    a. $(x - 8)^2 + y^2 = 10$ **b.** $\dfrac{y^2}{16} - \dfrac{x^2}{9} = 1$

    c. $x = y^2 - 3y + 6$

    d. $\dfrac{(x - 4)^2}{1} + \dfrac{(y + 12)^2}{49} = 1$

NOTATION

15. Find h, k, a, and b: $\dfrac{(x - 5)^2}{25} - \dfrac{(y + 11)^2}{36} = 1$.
16. Write each denominator in the equation $\dfrac{x^2}{36} - \dfrac{y^2}{81} = 1$ as the square of a number.

PRACTICE **Graph each hyperbola.**

17. $\frac{x^2}{9} - \frac{y^2}{4} = 1$
18. $\frac{x^2}{4} - \frac{y^2}{4} = 1$
19. $\frac{y^2}{4} - \frac{x^2}{9} = 1$
20. $\frac{y^2}{4} - \frac{x^2}{64} = 1$
21. $25x^2 - y^2 = 25$
22. $9x^2 - 4y^2 = 36$
23. $\frac{(x-2)^2}{9} - \frac{y^2}{16} = 1$
24. $\frac{(x+2)^2}{16} - \frac{(y-3)^2}{25} = 1$
25. $\frac{(y+1)^2}{1} - \frac{(x-2)^2}{4} = 1$
26. $\frac{(y-2)^2}{4} - \frac{(x+1)^2}{1} = 1$
27. $4(x+3)^2 - (y-1)^2 = 4$
28. $(x+5)^2 - 16y^2 = 16$
29. $\frac{y^2}{25} - \frac{(x-2)^2}{4} = 1$
30. $\frac{y^2}{36} - \frac{(x+2)^2}{4} = 1$

Write each equation in standard form and graph it.

31. $x^2 + 2x - y^2 - 2y = 9$
32. $x^2 - 4x - y^2 + 2y = 13$
33. $x^2 - y^2 - 6y = 34$
34. $y^2 - 2y - x^2 - 8 = 0$
35. $xy = 8$
36. $xy = -10$

Use a graphing calculator to graph each equation.

37. $\frac{x^2}{9} - \frac{y^2}{4} = 1$
38. $y^2 - 16x^2 = 16$
39. $\frac{x^2}{4} - \frac{(y-1)^2}{9} = 1$
40. $\frac{(y+1)^2}{9} - \frac{(x-2)^2}{4} = 1$

APPLICATIONS

41. ALPHA PARTICLES The particle in the illustration approaches the nucleus at the origin along the path $9y^2 - x^2 = 81$ in the coordinate system shown. How close does the particle come to the nucleus?

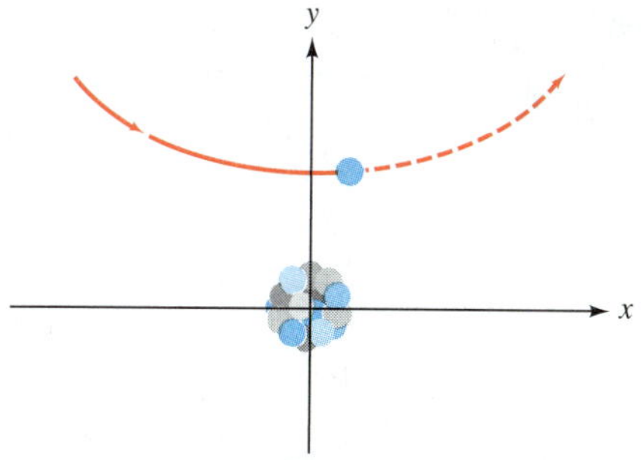

42. LORAN By determining the difference of the distances between the ship in the illustration and two radio transmitters, the LORAN navigation system places the ship on the hyperbola $x^2 - 4y^2 = 576$ in the coordinate system shown. If the ship is 5 miles out to sea, find its coordinates.

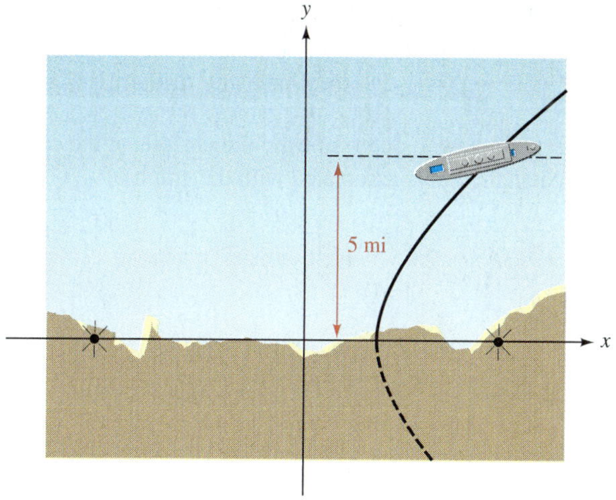

43. SONIC BOOM The position of a sonic boom caused by the faster-than-sound aircraft is one branch of the hyperbola $y^2 - x^2 = 25$ in the coordinate system shown. How wide is the hyperbola 5 miles from its vertex?

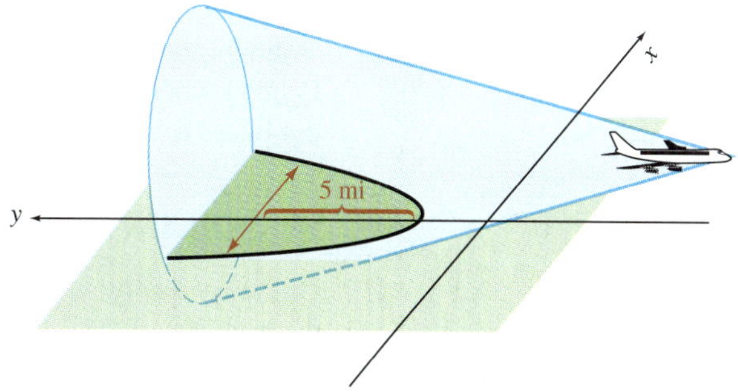

44. FLUIDS See the illustration below. Two glass plates in contact at the left, and separated by about 5 millimeters on the right, are dipped in beet juice, which rises by capillary action to form a hyperbola. The hyperbola is modeled by an equation of the form $xy = k$. If the curve passes through the point (12, 2), what is k?

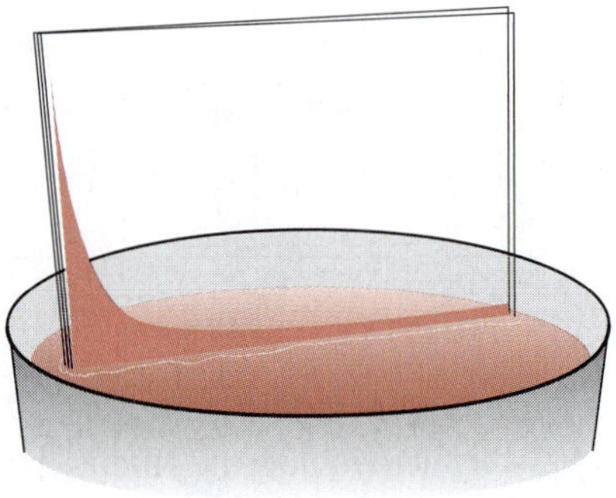

WRITING

45. What is a hyperbola?

46. Compare the graphs of $\frac{x^2}{81} - \frac{y^2}{64} = 1$ and $\frac{y^2}{81} - \frac{x^2}{64} = 1$. Do they have any similarities?

47. Explain how to determine the dimensions of the central rectangle that is associated with the graph of

$$\frac{x^2}{36} - \frac{y^2}{25} = 1$$

48. Explain why the graph of the hyperbola

$$\frac{x^2}{a^2} - \frac{y^2}{b^2} = 1$$

has no y-intercept.

REVIEW **Find each value of x.**

49. $\log_8 x = 2$

50. $\log_{25} x = \frac{1}{2}$

51. $\log_{1/2} \frac{1}{8} = x$

52. $\log_{12} x = 0$

53. $\log_x \frac{9}{4} = 2$

54. $\log_6 216 = x$

CHALLENGE PROBLEMS

55. Write $36x^2 - 25y^2 - 72x - 100y = 964$ in the standard form of the equation of a hyperbola.

56. Explain how a plane could intersect two right-circular cones to form two intersecting lines. Draw a picture to illustrate this.

57. Write an equation of a hyperbola whose graph has the following characteristics:

- vertices $(\pm 1, 0)$
- equations of asymptotes: $y = \pm 5x$

58. Graph: $16x^2 - 25y^2 = 1$.

10.4 Solving Nonlinear Systems of Equations

- Solving systems by graphing
- Solving systems by substitution
- Solving systems by elimination

In Chapter 3, we discussed how to solve systems of linear equations by the graphing, substitution, and elimination methods. In this section, we will use these methods to solve systems where at least one of the equations is nonlinear. Recall that equations are classified nonlinear because their graphs are not straight lines.

SOLVING SYSTEMS BY GRAPHING

One way to solve a system of two equations in two variables is to graph the equations on the same set of axes.

EXAMPLE 1

Solve $\begin{cases} x^2 + y^2 = 25 \\ 2x + y = 10 \end{cases}$ by graphing.

Solution The graph of $x^2 + y^2 = 25$ is a circle with center at the origin and radius of 5. The graph of $2x + y = 10$ is a line. Depending on whether the line is a **secant** (intersecting the circle at two points) or a **tangent** (intersecting the circle at one point) or does not intersect the circle at all, there are two, one, or no solutions to the system, respectively.

After graphing the circle and the line, it appears that the points of intersection are (5, 0) and (3, 4). To verify that they are solutions of the system, we need to check each one.

Success Tip

It is helpful to sketch the possibilities before solving the system:

2 points of intersection: (2 real solutions)

1 point of intersection: (1 real solution)

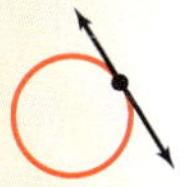

No points of intersection: (0 real solutions)

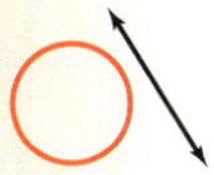

Check:

***For* (5, 0)**		***For* (3, 4)**	
$2x + y = 10$	$x^2 + y^2 = 25$	$2x + y = 10$	$x^2 + y^2 = 25$
$2(5) + 0 \stackrel{?}{=} 10$	$5^2 + 0^2 \stackrel{?}{=} 25$	$2(3) + 4 \stackrel{?}{=} 10$	$3^2 + 4^2 \stackrel{?}{=} 25$
$10 = 10$	$25 = 25$	$10 = 10$	$25 = 25$

The ordered pair (5, 0) satisfies both equations of the system, and so does (3, 4). Thus, there are two solutions, (5, 0) and (3, 4), and the solution set is $\{(5, 0), (3, 4)\}$.

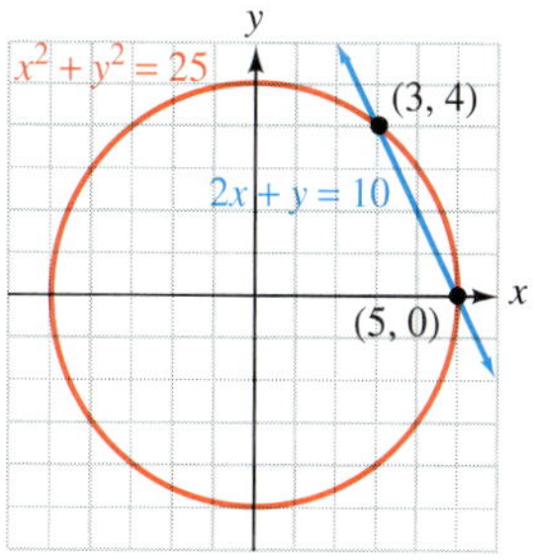

Self Check 1 Solve: $\begin{cases} x^2 + y^2 = 13 \\ y = -\dfrac{1}{5}x + \dfrac{13}{5}. \end{cases}$

ACCENT ON TECHNOLOGY: SOLVING SYSTEMS OF EQUATIONS

To solve Example 1 with a graphing calculator, we graph the circle and the line on one set of coordinate axes. (See figure (a).) We then trace to find the coordinates of the intersection points of the graphs. (See figures (b) and (c).)

We can zoom for better results.

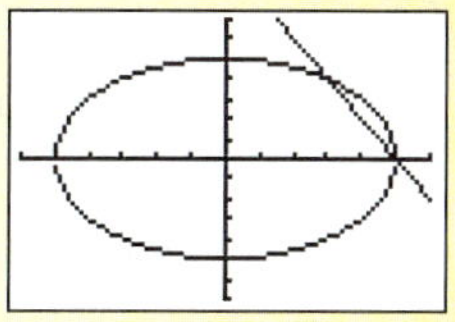

(a)

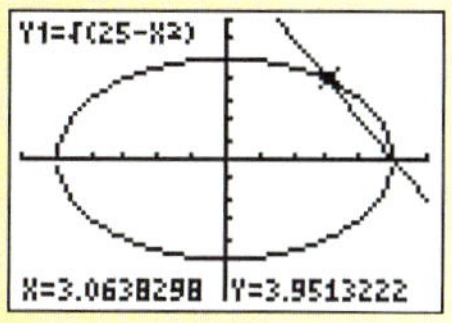

(b)

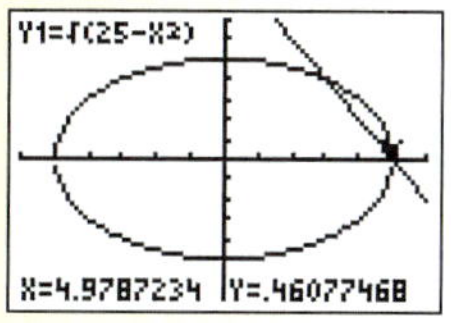

(c)

SOLVING SYSTEMS BY SUBSTITUTION

When solving a system by graphing, it is often difficult to determine the coordinates of the intersection points. A more precise algebraic method called the **substitution method** can be used to solve certain systems involving nonlinear equations.

EXAMPLE 2

Solve $\begin{cases} x^2 + y^2 = 2 \\ 2x - y = 1 \end{cases}$ by substitution.

Solution This system has one second-degree equation and one first-degree equation. We can solve this type of system by substitution. Solving the linear equation for y gives

$$2x - y = 1$$

$$-y = -2x + 1 \quad \text{Subtract } 2x \text{ from both sides.}$$

$$y = 2x - 1 \quad \text{Multiply both sides by } -1. \text{ We call this the substitution equation.}$$

We can substitute $2x - 1$ for y in the second-degree equation and solve the resulting quadratic equation for x:

$$x^2 + y^2 = 2$$

$$x^2 + (2x - 1)^2 = 2$$

$$x^2 + 4x^2 - 4x + 1 = 2 \quad \text{Find } (2x - 1)^2.$$

$$5x^2 - 4x - 1 = 0 \quad \text{Combine like terms and subtract 2 from both sides.}$$

$$(5x + 1)(x - 1) = 0 \quad \text{Factor.}$$

$$5x + 1 = 0 \quad \text{or} \quad x - 1 = 0 \quad \text{Set each factor equal to 0.}$$

$$x = -\frac{1}{5} \quad \bigg| \quad x = 1$$

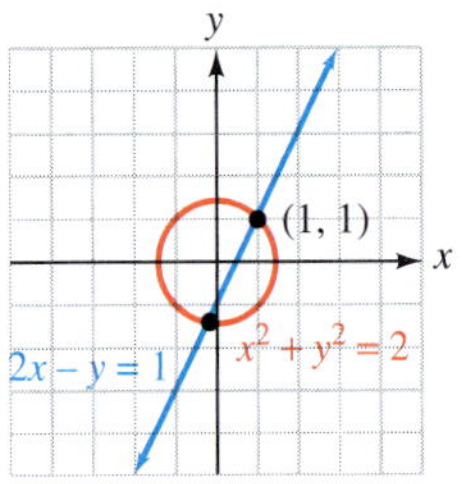

If we substitute $-\frac{1}{5}$ for x in the equation $y = 2x - 1$, we get $y = -\frac{7}{5}$. If we substitute 1 for x in $y = 2x - 1$, we get $y = 1$. Thus, the system has two solutions, $\left(-\frac{1}{5}, -\frac{7}{5}\right)$ and $(1, 1)$. Verify that each ordered pair satisfies both equations of the original system.

The graph in the margin confirms that the system has two solutions, and that one of them is $(1, 1)$. However, it would be virtually impossible to determine that the coordinates of the second point of intersection are $\left(-\frac{1}{5}, -\frac{7}{5}\right)$ from the graph.

Self Check 2 Solve $\begin{cases} x^2 + y^2 = 10 \\ y = x + 2 \end{cases}$ by substitution.

EXAMPLE 3

Solve: $\begin{cases} 4x^2 + 9y^2 = 5 \\ y = x^2 \end{cases}$.

Solution We can solve this system by substitution.

Success Tip

$4x^2 + 9y^2 = 5$ is the equation of an ellipse centered at (0, 0), and $y = x^2$ is the equation of a parabola with vertex at (0, 0), opening upward. We would expect two solutions.

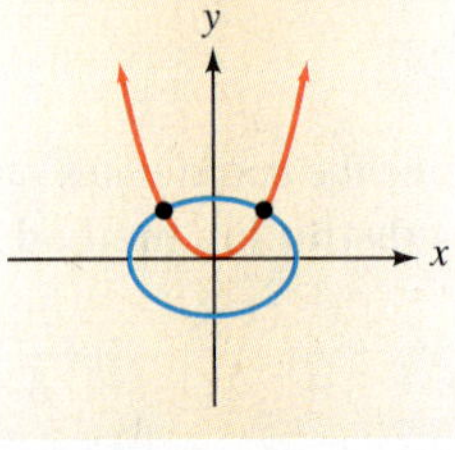

$$4x^2 + 9y^2 = 5$$

$$4y + 9y^2 = 5 \quad \text{Substitute } y \text{ for } x^2.$$

$$9y^2 + 4y - 5 = 0 \quad \text{Subtract 5 from both sides.}$$

$$(9y - 5)(y + 1) = 0 \quad \text{Factor } 9y^2 + 4y - 5.$$

$$9y - 5 = 0 \quad \text{or} \quad y + 1 = 0 \quad \text{Set each factor equal to 0.}$$

$$y = \frac{5}{9} \quad \bigg| \quad y = -1$$

Since $y = x^2$, the values of x are found by solving the equations

$$x^2 = \frac{5}{9} \quad \text{and} \quad \cancel{x^2 = -1}$$

Because $x^2 = -1$ has no real solutions, this possibility is discarded. The solutions of $x^2 = \frac{5}{9}$ are

$$x = \frac{\sqrt{5}}{3} \quad \text{or} \quad x = -\frac{\sqrt{5}}{3}$$

Caution

In this section, we are solving for only the *real* values of *x* and *y*.

Thus, the solutions of the system are

$$\left(\frac{\sqrt{5}}{3}, \frac{5}{9}\right) \quad \text{and} \quad \left(-\frac{\sqrt{5}}{3}, \frac{5}{9}\right)$$

Self Check 3 Solve: $\begin{cases} x^2 + y^2 = 20 \\ y = x^2 \end{cases}$.

SOLVING SYSTEMS BY ELIMINATION

EXAMPLE 4 Solve: $\begin{cases} 3x^2 + 2y^2 = 36 \\ 4x^2 - y^2 = 4 \end{cases}$.

Solution To solve this system of two second-degree equations, we can use either the substitution or the elimination method. We will use the elimination method because the y^2-terms can be eliminated by multiplying the second equation by 2 and adding it to the first equation.

Success Tip

The elimination method is generally better than the substitution method when both equations of the system are of the form $Ax^2 + By^2 = C$.

$$\begin{cases} 3x^2 + 2y^2 = 36 \\ 4x^2 - y^2 = 4 \end{cases} \quad \begin{array}{c} \xrightarrow{\text{Unchanged}} \\ \xrightarrow[\text{Multiply by 2}]{} \end{array} \quad \begin{array}{l} 3x^2 + 2y^2 = 36 \\ 8x^2 - 2y^2 = 8 \end{array}$$

We add the two equations on the right to eliminate y^2 and solve the resulting equation for x:

$$11x^2 = 44$$
$$x^2 = 4$$
$$x = 2 \quad \text{or} \quad x = -2$$

Success Tip

$3x^2 + 2y^2 = 36$ is the equation of an ellipse, centered at (0, 0), and $4x^2 - y^2 = 4$ is the equation of a hyperbola, centered at (0, 0), opening left and right. It is possible to have four solutions.

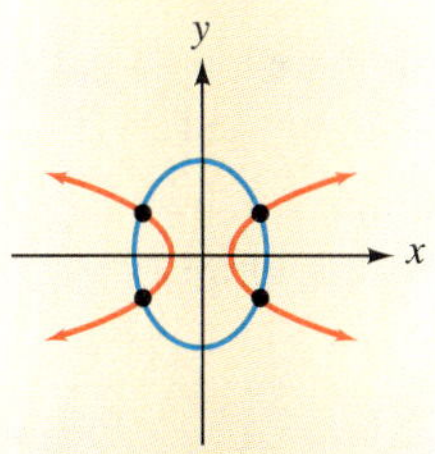

To find y, we can substitute 2 for x and then -2 for x into any equation containing both variables. It appears that the computations will be simplest if we use $3x^2 + 2y^2 = 36$.

For x = 2

$$3x^2 + 2y^2 = 36$$
$$3(2)^2 + 2y^2 = 36$$
$$12 + 2y^2 = 36$$
$$2y^2 = 24$$
$$y^2 = 12$$
$$y = \sqrt{12} \quad \text{or} \quad y = -\sqrt{12}$$
$$y = 2\sqrt{3} \quad | \quad y = -2\sqrt{3}$$

For x = −2

$$3x^2 + 2y^2 = 36$$
$$3(-2)^2 + 2y^2 = 36$$
$$12 + 2y^2 = 36$$
$$2y^2 = 24$$
$$y^2 = 12$$
$$y = \sqrt{12} \quad \text{or} \quad y = -\sqrt{12}$$
$$y = 2\sqrt{3} \quad | \quad y = -2\sqrt{3}$$

The four solutions of this system are

$$\left(2, 2\sqrt{3}\right), \quad \left(2, -2\sqrt{3}\right), \quad \left(-2, 2\sqrt{3}\right), \quad \text{and} \quad \left(-2, -2\sqrt{3}\right)$$

Self Check 4 Solve: $\begin{cases} x^2 + 4y^2 = 16 \\ x^2 - y^2 = 1 \end{cases}$.

Answers to Self Checks **1.** (3, 2), (−2, 3)

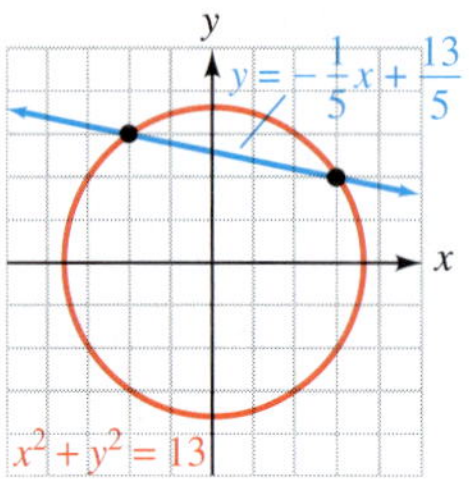

2. (1, 3), (−3, −1) **3.** (2, 4), (−2, 4) **4.** $\left(2, \sqrt{3}\right), \left(2, -\sqrt{3}\right), \left(-2, \sqrt{3}\right), \left(-2, -\sqrt{3}\right)$

10.4 STUDY SET

VOCABULARY Fill in the blanks.

1. $\begin{cases} 4x^2 + 6y^2 = 24 \\ 9x^2 - y^2 = 9 \end{cases}$ is a ________ of two nonlinear equations.
2. __________ equations have graphs that are not straight lines.
3. When solving a system by __________, it is often difficult to determine the coordinates of the intersection points.
4. Two algebraic methods for solving systems of nonlinear equations are the ___________ method and the ___________ method.
5. A _______ is a line that intersects a circle at two points.
6. A ________ is a line that intersects a circle at one point.

CONCEPTS

7. **a.** At most, a line can intersect an ellipse at _____ points.
   b. At most, an ellipse can intersect a parabola at _____ points.
   c. At most, an ellipse can intersect a circle at _____ points.
   d. At most, a hyperbola can intersect a circle at _____ points.
8. Check to determine whether (1, −1) is a solution of the system $\begin{cases} 2x + y - 1 = 0 \\ x^2 + y^2 = 3 \end{cases}$.
9. Determine the solutions of the system $\begin{cases} x^2 + 4y^2 = 25 \\ x^2 - 2y^2 = 1 \end{cases}$ that is graphed below.

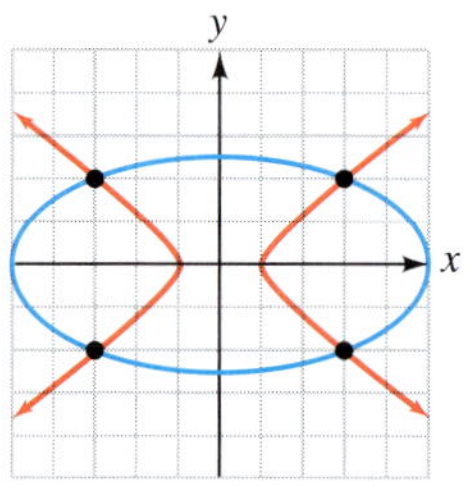

10. What should be used as the substitution equation to solve the system $\begin{cases} 2x^2 - y^2 = 6 \\ x^2 - y = 3 \end{cases}$?
11. Consider the system $\begin{cases} 6x^2 + y^2 = 9 \\ 3x^2 + 4y^2 = 36 \end{cases}$.
    a. If the y^2-terms are to be eliminated, by what should the first equation be multiplied?
    b. If the x^2-terms are to be eliminated, by what should the second equation be multiplied?

12. Suppose you begin to solve the system $\begin{cases} x^2 + y^2 = 10 \\ 4x^2 + y^2 = 13 \end{cases}$ and find that x is ± 1. Use the first equation to find the corresponding y-values for $x = 1$ and $x = -1$. State the solutions as ordered pairs.

NOTATION Complete each solution.

13. Solve: $\begin{cases} x^2 + y^2 = 5 \\ y = 2x \end{cases}$.

$$x^2 + y^2 = 5$$
$$x^2 + (\square)^2 = 5$$
$$x^2 + 4x^2 = \square$$
$$\square x^2 = 5$$
$$x^2 = \square$$
$$x = 1 \quad \text{or} \quad x = -1$$

If $x = 1$, then $y = 2(\square) = 2$.

If $x = -1$, then $y = 2(\square) = -2$.

The solutions are $(1, 2)$ and $(-1, \square)$.

14. Solve: $\begin{cases} x^2 + y^2 = 13 \\ x^2 - y^2 = 5 \end{cases}$

$$2x^2 = \square \qquad \text{Add the equations.}$$
$$x^2 = \square$$
$$x = \square \quad \text{or} \quad x = \square$$

$$2y^2 = \square \qquad \text{Subtract the equations.}$$
$$y^2 = \square$$
$$y = \square \quad \text{or} \quad y = \square$$

The solutions are $(3, \square), (3, \square), (-3, 2), (-3, -2)$

PRACTICE Solve each system of equations by graphing.

15. $\begin{cases} 8x^2 + 32y^2 = 256 \\ x = 2y \end{cases}$

16. $\begin{cases} x^2 + y^2 = 2 \\ x + y = 2 \end{cases}$

17. $\begin{cases} x^2 + y^2 = 10 \\ y = 3x^2 \end{cases}$

18. $\begin{cases} x^2 + y^2 = 5 \\ x + y = 3 \end{cases}$

19. $\begin{cases} x^2 + y^2 = 25 \\ 12x^2 + 64y^2 = 768 \end{cases}$

20. $\begin{cases} x^2 + y^2 = 13 \\ y = x^2 - 1 \end{cases}$

21. $\begin{cases} x^2 - 13 = -y^2 \\ y = \frac{2}{3}x \end{cases}$

22. $\begin{cases} x^2 + y^2 = 20 \\ y = x^2 \end{cases}$

Use a graphing calculator to solve each system.

23. $\begin{cases} x^2 - 6x - y = -5 \\ x^2 - 6x + y = -5 \end{cases}$

24. $\begin{cases} x^2 - y^2 = -5 \\ 3x^2 + 2y^2 = 30 \end{cases}$

Solve each system of equations algebraically for real values of *x* and *y*.

25. $\begin{cases} 25x^2 + 9y^2 = 225 \\ 5x + 3y = 15 \end{cases}$

26. $\begin{cases} x^2 + y^2 = 20 \\ y = x^2 \end{cases}$

27. $\begin{cases} x^2 + y^2 = 2 \\ x + y = 2 \end{cases}$

28. $\begin{cases} x^2 + y^2 = 36 \\ 49x^2 + 36y^2 = 1{,}764 \end{cases}$

29. $\begin{cases} x^2 + y^2 = 5 \\ x + y = 3 \end{cases}$

30. $\begin{cases} x^2 - x - y = 2 \\ 4x - 3y = 0 \end{cases}$

31. $\begin{cases} x^2 + y^2 = 13 \\ y = x^2 - 1 \end{cases}$

32. $\begin{cases} x^2 + y^2 = 25 \\ 2x^2 - 3y^2 = 5 \end{cases}$

33. $\begin{cases} x^2 + y^2 = 30 \\ y = x^2 \end{cases}$

34. $\begin{cases} 9x^2 - 7y^2 = 81 \\ x^2 + y^2 = 9 \end{cases}$

35. $\begin{cases} x^2 + y^2 = 13 \\ x^2 - y^2 = 5 \end{cases}$

36. $\begin{cases} 2x^2 + y^2 = 6 \\ x^2 - y^2 = 3 \end{cases}$

37. $\begin{cases} x^2 + y^2 = 20 \\ x^2 - y^2 = -12 \end{cases}$

38. $\begin{cases} xy = -\dfrac{9}{2} \\ 3x + 2y = 6 \end{cases}$

39. $\begin{cases} y^2 = 40 - x^2 \\ y = x^2 - 10 \end{cases}$

40. $\begin{cases} x^2 - 6x - y = -5 \\ x^2 - 6x + y = -5 \end{cases}$

41. $\begin{cases} y = x^2 - 4 \\ x^2 - y^2 = -16 \end{cases}$

42. $\begin{cases} 6x^2 + 8y^2 = 182 \\ 8x^2 - 3y^2 = 24 \end{cases}$

43. $\begin{cases} x^2 - y^2 = -5 \\ 3x^2 + 2y^2 = 30 \end{cases}$

44. $\begin{cases} \dfrac{1}{x} + \dfrac{1}{y} = 5 \\ \dfrac{1}{x} - \dfrac{1}{y} = -3 \end{cases}$

45. $\begin{cases} \dfrac{1}{x} + \dfrac{2}{y} = 1 \\ \dfrac{2}{x} - \dfrac{1}{y} = \dfrac{1}{3} \end{cases}$

46. $\begin{cases} \dfrac{1}{x} + \dfrac{3}{y} = 4 \\ \dfrac{2}{x} - \dfrac{1}{y} = 7 \end{cases}$

47. $\begin{cases} 3y^2 = xy \\ 2x^2 + xy - 84 = 0 \end{cases}$

48. $\begin{cases} x^2 + y^2 = 10 \\ 2x^2 - 3y^2 = 5 \end{cases}$

49. $\begin{cases} xy = \dfrac{1}{6} \\ y + x = 5xy \end{cases}$

50. $\begin{cases} xy = \dfrac{1}{12} \\ y + x = 7xy \end{cases}$

51. INTEGER PROBLEM The product of two integers is 32, and their sum is 12. Find the integers.

52. NUMBER PROBLEM The sum of the squares of two numbers is 221, and the sum of the numbers is 9. Find the numbers.

APPLICATIONS

53. GEOMETRY The area of a rectangle is 63 square centimeters, and its perimeter is 32 centimeters. Find the dimensions of the rectangle.

54. INVESTING Grant receives \$225 annual income from one investment. Jeff invested \$500 more than Grant, but at an annual rate of 1% less. Jeff's annual income is \$240. What are the amount and rate of Grant's investment?

55. INVESTING Carol receives \$67.50 annual income from one investment. John invested \$150 more than Carol at an annual rate of $1\frac{1}{2}$% more. John's annual income is \$94.50. What are the amount and rate of Carol's investment? (*Hint:* There are two answers.)

56. ARTILLERY See the illustration in the next column. A shell fired from the base of a hill follows the parabolic path $y = -\frac{1}{6}x^2 + 2x$, with distances measured in miles. The hill has a slope of $\frac{1}{3}$. How far from the cannon is the point of impact? (*Hint:* Find the coordinates of the point and then the distance.)

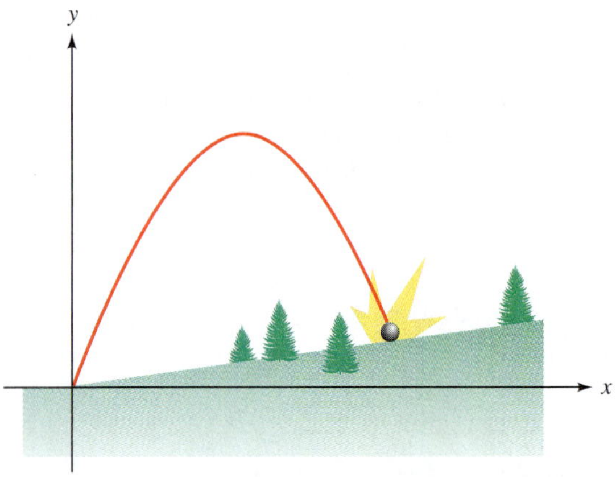

57. DRIVING RATES Jim drove 306 miles. Jim's brother made the same trip at a speed 17 mph slower than Jim did and required an extra $1\frac{1}{2}$ hours. What was Jim's rate and time?

58. FENCING PASTURES The rectangular pasture shown below is to be fenced in along a riverbank. If 260 feet of fencing is to enclose an area of 8,000 square feet, find the dimensions of the pasture.

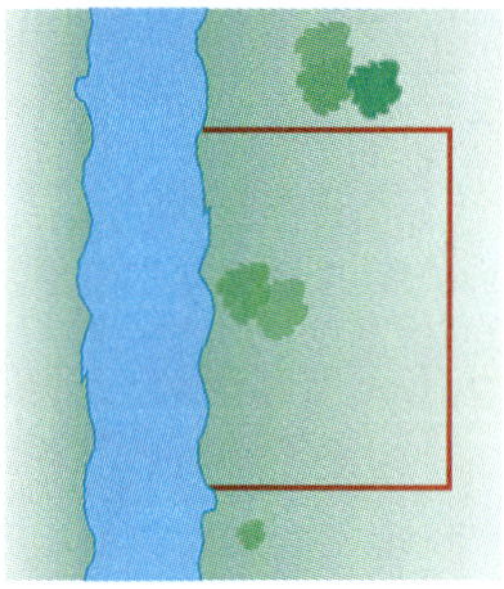

WRITING

59. **a.** Describe the benefits of the graphical method for solving a system of equations.

b. Describe the drawbacks of the graphical method.

60. Explain why the elimination method, not the substitution method, is the better method to solve the system

$$\begin{cases} 4x^2 + 9y^2 = 52 \\ 9x^2 + 4y^2 = 52 \end{cases}$$

REVIEW **Solve each equation.**

61. $\log 5x = 4$

62. $\log 3x = \log 9$

63. $\dfrac{\log(8x - 7)}{\log x} = 2$

64. $\log x + \log(x + 9) = 1$

CHALLENGE PROBLEMS

65. a. The graphs of the two independent equations of a system are parabolas. How many solutions might the system have?

b. The graphs of the two independent equations of a system are hyperbolas. How many solutions might the system have?

66. Solve the system $\begin{cases} x^2 - y^2 = 16 \\ x^2 + y^2 = 9 \end{cases}$ over the complex numbers.

ACCENT ON TEAMWORK

CONIC SECTIONS

Overview: In this activity, you will construct the four basic conic sections from clay.

Instructions: Form groups of 3 students. Mold some clay into the shape of a right-circular cone, as shown. With both hands, one student should pull a thin wire through the clay to slice it in such a way that a circular shape results. Then he or she should slice it to get an elliptical shape, a parabolic shape, and one branch of a hyperbolic shape.

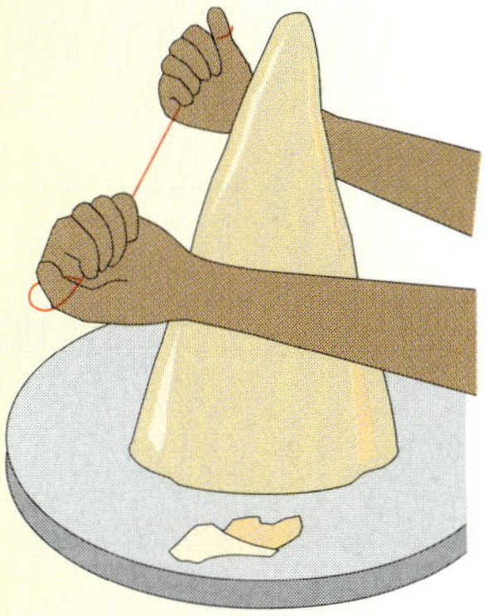

The second student should then mold the pieces back together into a right-circular cone whose base is wider and whose height is shorter than the first model, and slice it to create the four conics.

Finally, the third student should mold the pieces back together into a right-circular cone whose base is narrower and whose height is taller than the first model and create the four conics.

PARABOLAS

Overview: In this activity, you will construct several models of parabolas.

Instructions: Form groups of 2 or 3 students. You will need a T-square, string, paper, pencil, and thumbtack. To construct a parabola, secure one end of a piece of string that is as long as the T-square to a large piece of paper using a brad or thumbtack, as shown below. Attach the other end of the string to the upper end of the T-square. Hold the string taut against the T-square with a pencil and slide the T-square along the edge of the table. As the T-square moves, the pencil will trace a parabola.

Each point on the parabola is the same distance away from a point as it is from a given line. With this model, what is the given point, and what is the given line?

Make other models by moving the fixed point closer and further away from the edge of the table. How is the shape of the parabola affected?

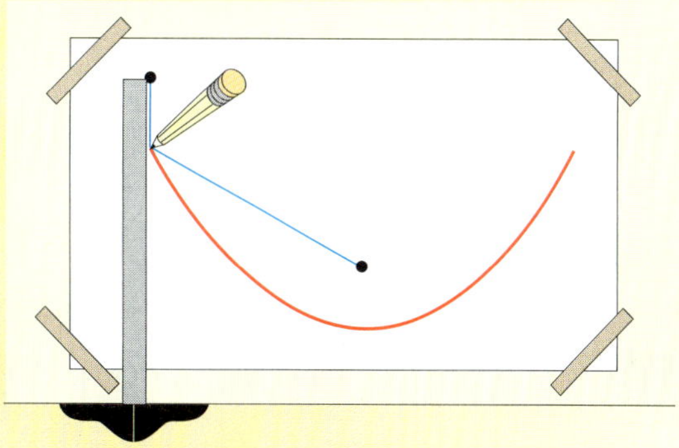

ELLIPSES

Overview: In this activity, you will construct several ellipses.

Instructions: Form groups of 2 or 3 students. You will need two thumbtacks, a pencil, and a length of string with a loop tied at one end. To construct an ellipse, place two thumbtacks (or brads) fairly close together, as shown in the illustration. Catch the loop of the string with the point of the pencil and, keeping the string taut, draw the ellipse.

Make several models by moving one of the thumbtacks farther away and then closer to the other thumbtack. How does the shape of the ellipse change?

For each point on the ellipse, the sum of the distances of the point from two given points is a constant. With this method of construction, what are the two points? What is the constant distance?

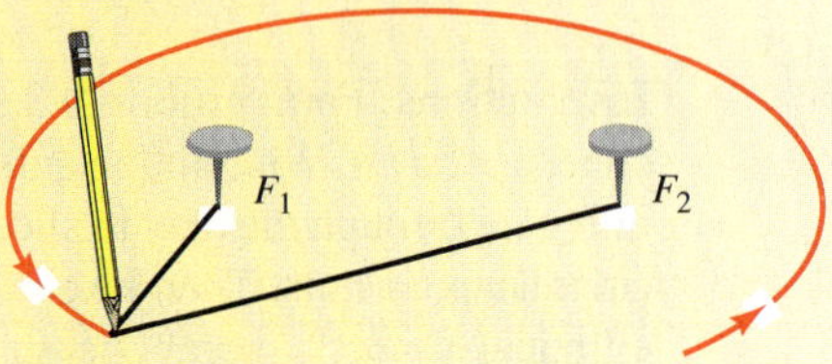

KEY CONCEPT: CONIC SECTIONS

In this chapter, we have studied four conic sections: the circle, the parabola, the ellipse, and the hyperbola. They are called conic sections because they are formed by the intersection of a plane and a pair of right-circular cones.

CLASSIFYING CONICS

In Exercises 1–9, classify the graph of each equation as a circle, a parabola, an ellipse, or a hyperbola.

1. $\dfrac{(x-2)^2}{9} + \dfrac{(y-1)^2}{4} = 1$
2. $(x-3)^2 + (y-2)^2 = 4$
3. $y = 4x^2 - 2x + 3$
4. $\dfrac{x^2}{4} - \dfrac{(y-1)^2}{9} = 1$
5. $4(x-7)^2 - (y+10)^2 = 4$
6. $\dfrac{x^2}{4} + \dfrac{y^2}{9} = 1$
7. $3(x-1)^2 + 12(y+2)^2 = 48$
8. $x^2 - 2y - 2x = -7$
9. $x^2 + y^2 + 8x + 2y = -13$

EQUATIONS OF CONIC SECTIONS

When the equation of a conic section is written in standard form, important features of its graph are apparent.

10. Consider $(x+1)^2 + (y-2)^2 = 16$.
    - **a.** What are the coordinates of the center of the circle?
    - **b.** What is the radius of the circle?
11. Consider $x = \dfrac{1}{2}(y-1)^2 - 2$.
    - **a.** What are the coordinates of the vertex of the parabola?
    - **b.** In which direction does the parabola open?

12. Consider $\frac{x^2}{4} + \frac{y^2}{16} = 1$.

a. What are the coordinates of the center of the ellipse?

b. Is the ellipse horizontal or vertical?

c. What are the vertices of the ellipse?

13. Consider $\frac{(x + 2)^2}{9} - \frac{(y - 1)^2}{4} = 1$.

a. What are the coordinates of the center of the hyperbola?

b. In which direction do the branches of the hyperbola open?

c. What are the dimensions of the central rectangle?

GRAPHING CONIC SECTIONS

Use the results from Exercises 10–13 to graph each conic section.

14. $(x + 1)^2 + (y - 2)^2 = 16$

15. $x = \frac{1}{2}(y - 1)^2 - 2$

16. $\frac{x^2}{4} + \frac{y^2}{16} = 1$

17. $\frac{(x + 2)^2}{9} - \frac{(y - 1)^2}{4} = 1$

CHAPTER REVIEW

SECTION 10.1 The Circle and the Parabola

CONCEPTS

Equations of a circle:

$(x - h)^2 + (y - k)^2 = r^2$
center (h, k), radius r

$x^2 + y^2 = r^2$
center $(0, 0)$, radius r

REVIEW EXERCISES

Graph each equation.

1. $x^2 + y^2 = 16$

2. $(x - 1)^2 + (y + 2)^2 = 4$

3. Write the equation in standard form and graph it.

$$x^2 + y^2 + 4x - 2y = 4$$

4. ART HISTORY Leonardo da Vinci's *Vitruvian Man* (1492) is one of the most famous pen-and-ink drawings of all time. Use the coordinate system that is superimposed on the drawing to write the equation of the circle in standard form.

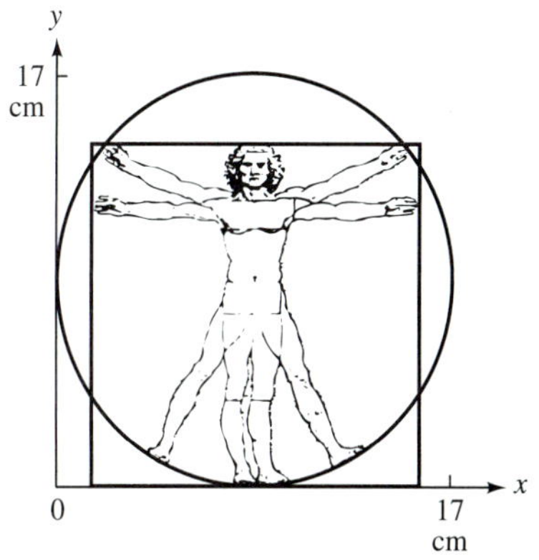

5. Find the center and the radius of the circle whose equation is $(x + 6)^2 + y^2 = 24$.

6. Fill in the blanks: A circle is the set of all points in a plane that are a fixed distance from a point called its ______. The fixed distance is called the ______ of the circle.

Equations of parabolas

General forms:

$y = ax^2 + bx + c$

$a > 0$: up; $a < 0$: down

$x = ay^2 + by + c$

$a > 0$: right; $a < 0$: left

Standard forms:

$y = a(x - h)^2 + k$

$a > 0$: up; $a < 0$: down

Vertex: (h, k)

Axis of symmetry: $x = h$

$x = a(y - k)^2 + h$

$a > 0$: right; $a < 0$: left

Vertex: (h, k)

Axis of symmetry: $y = k$

Graph each parabola and give the coordinates of the vertex.

7. $x = y^2$ **8.** $x = 2(y + 1)^2 - 2$ **9.** $x = -3y^2 + 12y - 7$

10. Find the axis of symmetry of the graph of each equation.

a. $x = (y - 4)^2 + 1$ **b.** $y = 2x^2 - 4x + 5$

11. The axis of symmetry, vertex, and two points on the graph of a parabola are shown. Find the coordinates of two other points on the parabola.

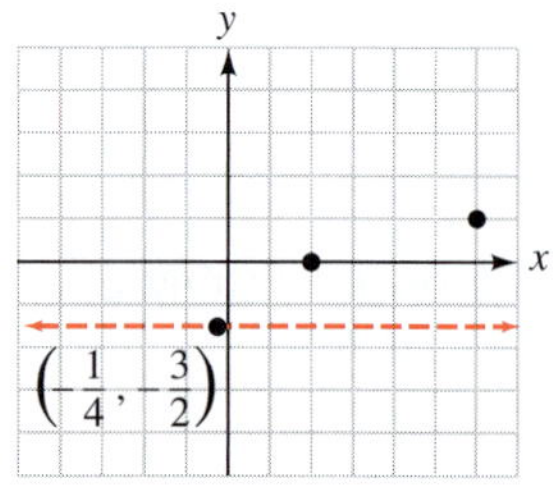

12. LONG JUMP The equation describing the flight path of the long jumper is $y = -\frac{5}{121}(x - 11)^2 + 5$. Show that she will land at a point 22 feet away from the take-off board.

SECTION 10.2 The Ellipse

Equations of an ellipse:

Center at $(0, 0)$

$$\frac{x^2}{a^2} + \frac{y^2}{b^2} = 1$$

Center at (h, k)

$$\frac{(x - h)^2}{a^2} + \frac{(y - k)^2}{b^2} = 1$$

Graph each ellipse.

13. $9x^2 + 16y^2 = 144$ **14.** $\frac{(x - 2)^2}{4} + \frac{(y - 1)^2}{25} = 1$

15. $4(x + 1)^2 + 9(y - 1)^2 = 36$

16. Consider the equation $\frac{x^2}{144} + y^2 = 1$. Write each term on the left-hand side with a denominator that is the square of a number.

17. Consider the equation $\frac{x^2}{9} + \frac{y^2}{4} = 1$. Find two points on the graph of this equation by letting $x = 2$ and finding the corresponding y-coordinates. Express the results as ordered pairs. Give the exact answers and then the approximate answers.

18. SALAMI When a delicatessen slices a cylindrical salami at an angle, the results are elliptical pieces that are larger than circular pieces. Write the equation of the shape of the slice of salami shown in the illustration if it was centered at the origin of a coordinate system.

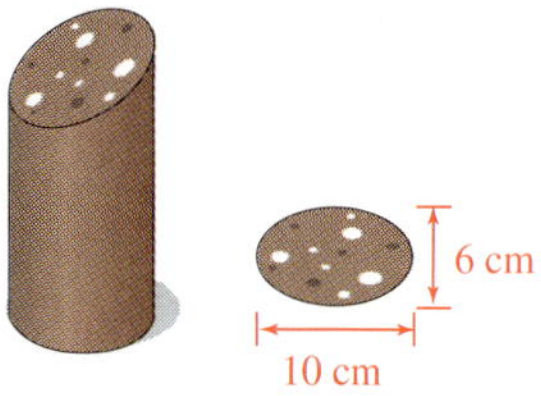

19. Fill in the blanks: An ________ is the set of all points in a plane for which the sum of the distances from two fixed points is a constant. Each of the fixed points is called a ________.

Ellipses have a *reflective* property such that any light or sound originating at one focus is reflected by the interior of the figure to the other focus.

20. CONSTRUCTION Sketch the path of the sound when a person, standing at one focus, whispers something in the whispering gallery dome shown below.

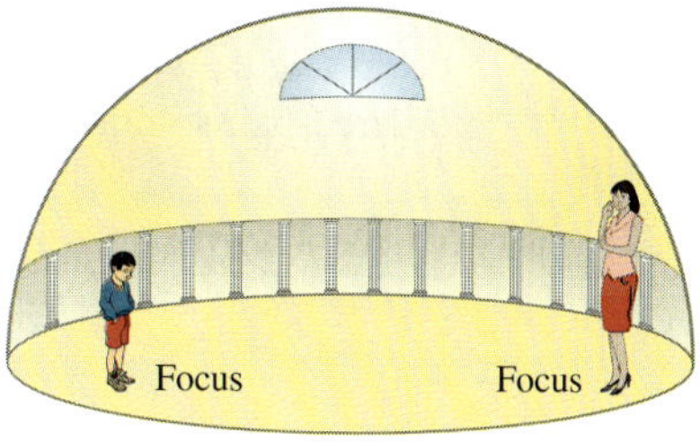

SECTION 10.3 The Hyperbola

Equations of a hyperbola:

Center at (0, 0)

$$\frac{x^2}{a^2} - \frac{y^2}{b^2} = 1$$

$$\frac{y^2}{a^2} - \frac{x^2}{b^2} = 1$$

Center at (h, k)

$$\frac{(x-h)^2}{a^2} - \frac{(y-k)^2}{b^2} = 1$$

$$\frac{(y-k)^2}{a^2} - \frac{(x-h)^2}{b^2} = 1$$

Graph each hyperbola.

21. $\frac{y^2}{9} - \frac{x^2}{1} = 1$

22. $9(x-1)^2 - 4(y+1)^2 = 36$

23. $xy = 9$

24. $y^2 - 4y - x^2 - 2x - 22 = 0$

25. ELECTROSTATIC REPULSION Two similarly charged particles are shot together for an almost head-on collision, as in the illustration. They repel each other and travel the two branches of the hyperbola given by $x^2 - 4y^2 = 4$ on the given coordinate system. How close do they get?

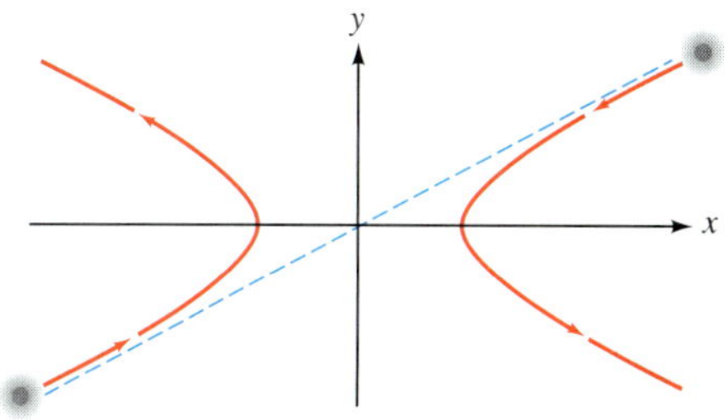

26. Determine whether the equation, when graphed, will be a circle, parabola, ellipse, or hyperbola.

a. $\frac{(x-4)^2}{16} + \frac{y^2}{49} = 1$

b. $x^2 + 6x - y^2 + 2y - 16 = 0$

c. $x = -4y^2 - y + 1$

d. $x^2 + 2x + y^2 - 4y = 40$

SECTION 10.4 Solving Nonlinear Systems of Equations

Systems of nonlinear equations are solved by graphing, by substitution, or by elimination.

27. Check to determine whether $\left(-\sqrt{11}, -3\right)$ is a solution of the system $\begin{cases} x^2 + y^2 = 20 \\ x^2 - y^2 = 2 \end{cases}$.

28. The graphs of $y^2 - x^2 = 9$ and $x^2 + y^2 = 9$ are shown. Estimate the solutions of the system

$$\begin{cases} y^2 - x^2 = 9 \\ x^2 + y^2 = 9 \end{cases}$$

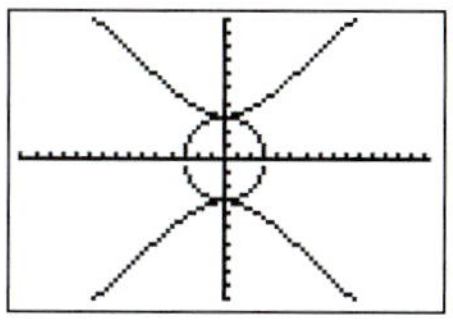

29. Determine the maximum number of solutions there could be for a system of equations consisting of the given curves.

a. a line and an ellipse **b.** two hyperbolas

c. an ellipse and a circle **d.** a parabola and a circle

30. Suppose the x-coordinate of both points of intersection of the circle, defined by $x^2 + y^2 = 1$, and the hyperbola, defined by $4y^2 - x^2 = 4$, is 0. Without graphing, determine the y-coordinates of both points of intersection. Express the answers as ordered-pair solutions.

Solve each system.

31. $\begin{cases} y^2 - x^2 = 16 \\ y + 4 = x^2 \end{cases}$

32. $\begin{cases} y = -x^2 + 2 \\ x^2 - y - 2 = 0 \end{cases}$

33. $\begin{cases} x^2 + 2y^2 = 12 \\ 2x - y = 2 \end{cases}$

34. $\begin{cases} 3x^2 + y^2 = 52 \\ x^2 - y^2 = 12 \end{cases}$

35. $\begin{cases} \dfrac{x^2}{16} + \dfrac{y^2}{12} = 1 \\ x^2 - \dfrac{y^2}{3} = 1 \end{cases}$

36. $\begin{cases} xy = 4 \\ x^2 + \dfrac{y^2}{2} = 9 \end{cases}$

CHAPTER 10 TEST

1. Fill in the blanks: A circle is the set of all points in a plane that are a fixed distance from a point called its _______. The fixed distance is called the _______ of the circle.

2. Find the center and the radius of the circle $x^2 + y^2 = 100$.

3. Find the center and the radius of the circle $x^2 + y^2 + 4x - 6y = 5$

4. TV HISTORY In the early days of television, stations broadcast a black-and-white test pattern like that shown below during the early morning hours. Use the given coordinate system to write an equation of the large, bold circle in the center of the pattern.

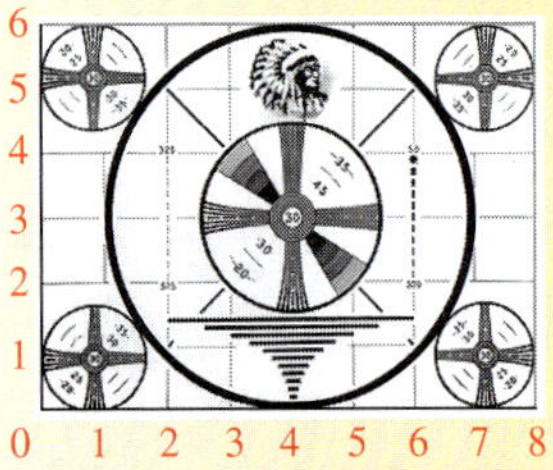

Graph each equation.

5. $(x + 2)^2 + (y - 1)^2 = 9$ **6.** $x = y^2 - 2y + 3$

7. Find the vertex and the axis of symmetry of the graph of $y = -2x^2 - 4x + 5$.

8. SOUND The equation $x = -\frac{1}{10}y^2$ defines a cross-section view of a parabolic dish. Construct a table of values for the equation, plot the ordered pairs, and connect the points with a smooth curve. Then plot the point $(-2.5, 0)$, which locates the microphone that picks up reflected sound waves. Draw a line parallel to the axis of symmetry coming into the dish and striking the dish at $(-0.9, 3)$ and reflecting into the microphone.

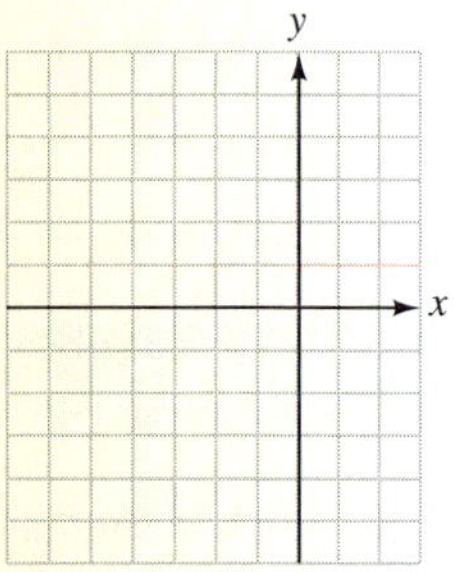

Graph each equation.

9. $9x^2 + 4y^2 = 36$

10. $\dfrac{(x-2)^2}{9} - \dfrac{y^2}{1} = 1$

11. Write the equation in standard form of the ellipse graphed below.

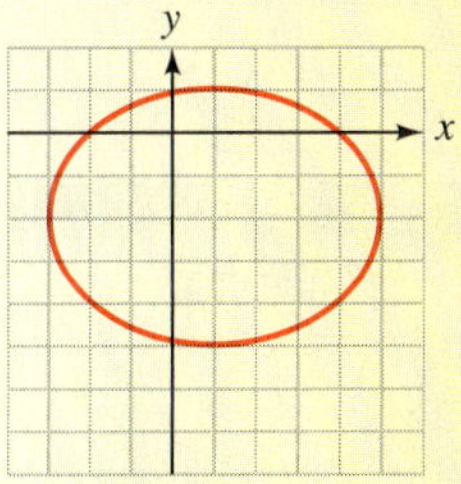

12. Find the center and the vertices of the graph of $25(x+8)^2 + 36(y-10)^2 = 900$.

13. Complete the table of solutions for the equation $\dfrac{x^2}{36} + \dfrac{y^2}{9} = 1$.

x	y
-2	

14. Give an example of the reflective properties of an ellipse. Include a drawing and label it completely.

15. Find the center and the dimensions of the central rectangle of the graph of $x^2 + 2x - y^2 + 2y - 4 = 0$.

16. Graph $xy = -4$.

17. What is the equation in standard form of the hyperbola graphed below?

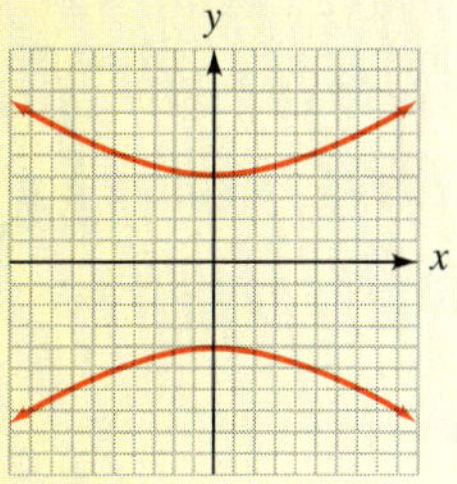

18. Determine whether the equation, when graphed, will be a circle, a parabola, an ellipse, or a hyperbola.

a. $25x^2 + 100y^2 = 400$

b. $x^2 - y^2 = 1$

c. $x^2 + 8x + y^2 - 16y - 1 = 0$

d. $x = 8y^2 - 9y + 4$

Solve the system graphically.

19. $\begin{cases} x^2 + y^2 = 25 \\ y - x = 1 \end{cases}$

Solve each system.

20. $\begin{cases} 2x - y = -2 \\ x^2 + y^2 = 16 + 4y \end{cases}$

21. $\begin{cases} 5x^2 - y^2 - 3 = 0 \\ x^2 + 2y^2 = 5 \end{cases}$

22. $\begin{cases} xy = -\dfrac{9}{2} \\ 3x + 2y = 6 \end{cases}$

Chapter

11 Miscellaneous Topics

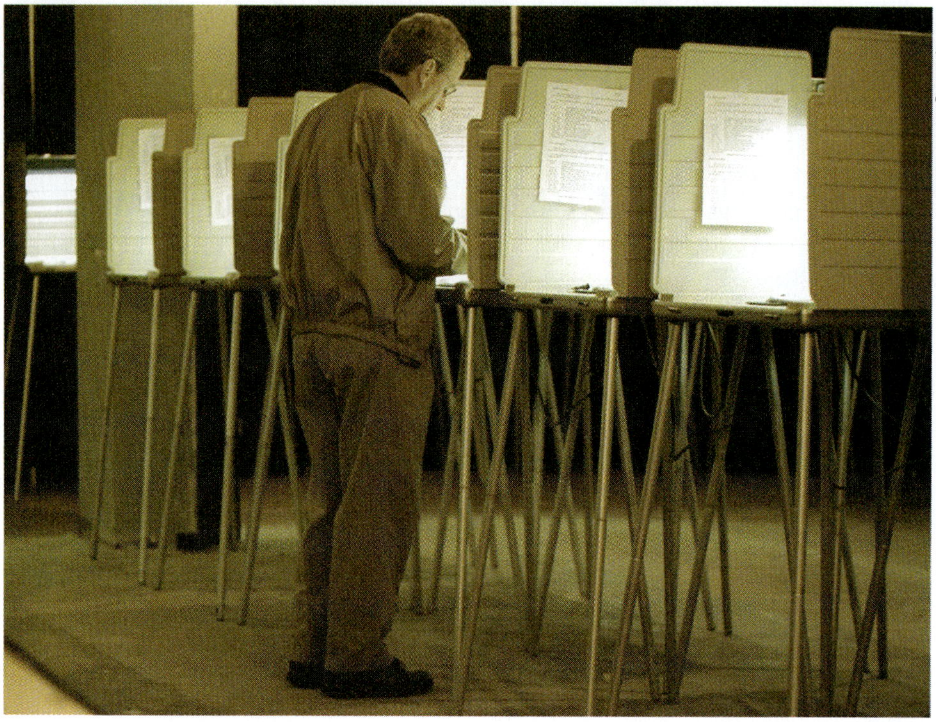
Getty Images News

On a ballot, the names of the candidates running for a given office can be listed in many different orders. For example, they can be presented alphabetically, by party affiliation, or simply in random order. In this chapter, we will learn how to determine the number of ways in which a set of names can be arranged. This counting concept, known as a *permutation,* has many other important applications in areas such as designing license plates, assigning telephone numbers, and listing combinations for locks.

To learn more about permutations, visit *The Learning Equation* on the Internet at http://tle.brookscole.com. (The log-in instructions are in the Preface.) For Chapter 11, the online lesson is:

- *TLE* Lesson 16: Permutations and Combinations

In this chapter, we introduce several topics with applications in advanced mathematics and in certain occupations. The binomial theorem, permutations, and combinations are used in statistics. Arithmetic and geometric sequences are used in finance.

11.1 The Binomial Theorem

- Raising binomials to powers
- Pascal's triangle
- Factorial notation
- The binomial theorem
- Finding a specific term of an expansion

We have discussed how to raise binomials to positive-integer powers. For example, we have learned that

$$(a+b)^2 = a^2 + 2ab + b^2$$

and that

$$\begin{aligned}(a+b)^3 &= (a+b)(a+b)^2\\ &= (a+b)(a^2+2ab+b^2)\\ &= a^3 + 2a^2b + ab^2 + a^2b + 2ab^2 + b^3\\ &= a^3 + 3a^2b + 3ab^2 + b^3\end{aligned}$$

The Language of Algebra

Recall that two-term polynomial expressions such as $a + b$ and $3u - 2v$ are called *binomials.*

In this section, we will learn how to raise binomials to positive-integer powers without performing the multiplications.

RAISING BINOMIALS TO POWERS

To see how to raise binomials to positive-integer powers, we consider the following binomial expansions of $a + b$.

$$\begin{aligned}(a+b)^0 &= 1 && \text{1 term}\\ (a+b)^1 &= a+b && \text{2 terms}\\ (a+b)^2 &= a^2+2ab+b^2 && \text{3 terms}\\ (a+b)^3 &= a^3+3a^2b+3ab^2+b^3 && \text{4 terms}\\ (a+b)^4 &= a^4+4a^3b+6a^2b^2+4ab^3+b^4 && \text{5 terms}\\ (a+b)^5 &= a^5+5a^4b+10a^3b^2+10a^2b^3+5ab^4+b^5 && \text{6 terms}\\ (a+b)^6 &= a^6+6a^5b+15a^4b^2+20a^3b^3+15a^2b^4+6ab^5+b^6 && \text{7 terms}\end{aligned}$$

The Language of Algebra

To *expand* means to increase in size. When we expand a power of a binomial, the result, called a **binomial expansion.** In general, an expansion has more terms than the original binomial.

Several patterns appear in these expansions:

1. Each expansion has one more term than the power of the binomial.
2. For each term of an expansion, the sum of the exponents on a and b is equal to the exponent of the binomial being expanded. For example, in the expansion of $(a + b)^5$, the sum of the exponents in each term is 5:

$$(a+b)^5 = a^5 + \underbrace{5a^4b}_{4+1=5} + \underbrace{10a^3b^2}_{3+2=5} + \underbrace{10a^2b^3}_{2+3=5} + \underbrace{5ab^4}_{1+4=5} + b^5$$

The Language of Algebra

We can state observation 2 in another way: The *degree* of each term of an expansion is equal to the exponent of the binomial that is being expanded.

3. The first term in each expansion is a, raised to the power of the binomial, and the last term in each expansion is b, raised to the power of the binomial.
4. The exponents on a decrease by one in each successive term, ending with $a^0 = 1$ in the last term. The exponents on b, beginning with $b^0 = 1$ in the first term, increase by one in each successive term. For example, the expansion of $(a + b)^4$ could be written as

$$a^4b^0 + 4a^3b^1 + 6a^2b^2 + 4a^1b^3 + a^0b^4$$

Thus, the variables have the pattern

$$a^n, \quad a^{n-1}b, \quad a^{n-2}b^2, \quad \ldots, \quad ab^{n-1}, \quad b^n$$

5. The coefficients of each expansion begin with 1, increase through some values, and then decrease through those same values, back to 1.

PASCAL'S TRIANGLE

To see another pattern, we write the coefficients of each expansion of $a + b$ in a triangular array:

The Language of Algebra

This array of numbers is named *Pascal's triangle* in honor of the French mathematician Blaise Pascal (1623–1662).

1 — Row 0
1 1 — Row 1
1 2 1 — Row 2
1 3 3 1 — Row 3
1 4 6 4 1 — Row 4
1 5 10 10 5 1 — Row 5
1 6 15 20 15 6 1 — Row 6

In this array, called **Pascal's triangle,** each entry between the 1's is the sum of the closest pair of numbers in the line immediately above it. For example, the first 15 in the bottom row is the sum of the 5 and 10 immediately above it. Pascal's triangle continues with the same pattern forever. The next two lines are

1 7 21 35 35 21 7 1 — Row 7
1 8 28 56 70 56 28 8 1 — Row 8

EXAMPLE 1

Expand: $(x + y)^5$.

Solution The first term in the expansion is x^5, and the exponents on x decrease by one in each successive term. A y first appears in the second term, and the exponents on y increase by one in each successive term, concluding when the term y^5 is reached. Thus, the variables in the expansion are

$$x^5, \quad x^4y, \quad x^3y^2, \quad x^2y^3, \quad xy^4, \quad y^5$$

Since the exponent of the binomial that is being expanded is 5, the coefficients of these variables are found in row 5 of Pascal's triangle.

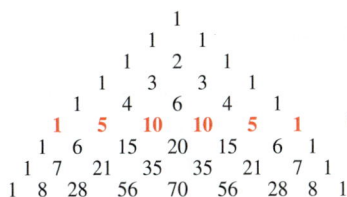

```
                1
              1   1
            1   2   1
          1   3   3   1
        1   4   6   4   1
      1   5  10  10   5   1
    1   6  15  20  15   6   1
  1   7  21  35  35  21   7   1
1   8  28  56  70  56  28   8   1
```

1 5 10 10 5 1

Combining this information gives the following expansion:

$$(x + y)^5 = x^5 + 5x^4y + 10x^3y^2 + 10x^2y^3 + 5xy^4 + y^5$$

Self Check 1 Expand: $(x + y)^4$.

EXAMPLE 2

Expand: $(u - v)^4$.

Solution We note that $(u - v)^4$ can be written in the form $[u + (-v)]^4$. The variables in this expansion are

$$u^4, \quad u^3(-v), \quad u^2(-v)^2, \quad u(-v)^3, \quad (-v)^4$$

and the coefficients are given in row 4 of Pascal's triangle:

1 4 6 4 1

Thus, the required expansion is

$$(u - v)^4 = u^4 + 4u^3(-v) + 6u^2(-v)^2 + 4u(-v)^3 + (-v)^4$$

The Language of Algebra

To *alternate* means to change back and forth. For example, day alternates with night. In this expansion, the signs + and − alternate.

Now we simplify each term. When $-v$ is raised to an even power, the sign is positive, and when $-v$ is raised to an odd power, the sign is negative. This causes the signs of the expansion to alternate between + and −.

$$(u - v)^4 = u^4 - 4u^3v + 6u^2v^2 - 4uv^3 + v^4$$

Self Check 2 Expand: $(x - y)^5$.

FACTORIAL NOTATION

Although Pascal's triangle gives the coefficients of the terms in a binomial expansion, it is not the best way to expand a binomial. To develop another way, we introduce **factorial notation.** The symbol $n!$ (read as "n **factorial**") is defined as follows.

Factorial Notation

$n!$ is the product of consecutively decreasing natural numbers from n to 1.

For any natural number n,

$$n! = n(n - 1)(n - 2)(n - 3) \cdot \cdots \cdot 3 \cdot 2 \cdot 1$$

Zero factorial is defined as

$$0! = 1$$

EXAMPLE 3

Evaluate each expression: **a.** $4!$, **b.** $6!$, **c.** $3! \cdot 2!$, and **d.** $5! \cdot 0!$.

Solution

a. $4! = 4 \cdot 3 \cdot 2 \cdot 1 = 24$ — Read as "4 factorial."

b. $6! = 6 \cdot 5 \cdot 4 \cdot 3 \cdot 2 \cdot 1 = 720$

c. $3! \cdot 2! = (3 \cdot 2 \cdot 1) \cdot (2 \cdot 1) = 6 \cdot 2 = 12$ — Find each factorial first, then multiply the results.

d. $5! \cdot \mathbf{0!} = (5 \cdot 4 \cdot 3 \cdot 2 \cdot 1) \cdot \mathbf{1} = 120$ — $0! = 1$

Self Check 3 Evaluate each expression: **a.** $7!$, **b.** $4! \cdot 3!$, and **c.** $1! \cdot 0!$.

ACCENT ON TECHNOLOGY: FACTORIALS

We can find factorials using a calculator. For example, to find 12! with a scientific calculator, we enter

12 [x!] (You may have to use a [2nd] or [SHIFT] key first.) — display: 479001600

To find 12! on a TI-83 Plus graphing calculator, we enter

12 [MATH] arrow [→] to PRB [4] [ENTER] — display: 12! 479001600

The following property follows from the definition of factorial.

Factorial Property For any natural number n,

$$n(n-1)! = n!$$

We can use this property to simplify certain expressions involving factorials.

EXAMPLE 4

Simplify each expression: **a.** $\dfrac{6!}{5!}$ and **b.** $\dfrac{10!}{8!(10-8)!}$.

Solution

a. If we write 6! as $6 \cdot 5!$, we can simplify the fraction by removing the common factor 5! in the numerator and denominator.

$$\frac{6!}{5!} = \frac{6 \cdot 5!}{5!} = \frac{6 \cdot \overset{1}{\cancel{5!}}}{\underset{1}{\cancel{5!}}} = 6$$

Simplify: $\frac{5!}{5!} = 1$.

b. First, we subtract within the parentheses. Then we write 10! as $10 \cdot 9 \cdot 8!$ and simplify.

$$\frac{10!}{8!\,(\mathbf{10-8})!} = \frac{10!}{8! \cdot \mathbf{2}!} = \frac{10 \cdot 9 \cdot \overset{1}{\cancel{8!}}}{\underset{1}{\cancel{8!}} \cdot 2!} = \frac{5 \cdot \overset{1}{\cancel{2}} \cdot 9}{\underset{1}{\cancel{2}} \cdot 1} = 45$$

Simplify: $\frac{8!}{8!} = 1$. Factor 10 as $5 \cdot 2$ and simplify: $\frac{2}{2} = 1$.

Self Check 4 Simplify: **a.** $\dfrac{4!}{3!}$ and **b.** $\dfrac{7!}{5!(7-5)!}$.

THE BINOMIAL THEOREM

The following theorem brings together our observations about binomial expansions and our work with factorials. Known as the *binomial theorem,* it is the most efficient way to expand a binomial.

The Binomial Theorem For any positive integer n,

$$(a+b)^n = a^n + \frac{n!}{1!(n-1)!}a^{n-1}b + \frac{n!}{2!(n-2)!}a^{n-2}b^2 + \frac{n!}{3!(n-3)!}a^{n-3}b^3 + \cdots + \frac{n!}{r!(n-r)!}a^{n-r}b^r + \cdots + b^n$$

In the binomial theorem, the exponents on the variables follow the familiar pattern:

- The sum of the exponents on a and b in each term is n,
- the exponents on a decrease by 1 in each successive term, and
- the exponents on b increase by 1 in each successive term.

The method of finding the coefficients involves factorials. Except for the first and last terms, the numerator of each coefficient is $n!$. If the exponent on b in a particular term is r, the denominator of the coefficient of that term is $r!(n-r)!$.

EXAMPLE 5 Use the binomial theorem to expand $(a+b)^3$.

Solution We can substitute directly into the binomial theorem and simplify:

$$(a+b)^3 = a^3 + \frac{3!}{1!(3-1)!}a^2b + \frac{3!}{2!(3-2)!}ab^2 + b^3$$

$$= a^3 + \frac{3!}{1!\cdot 2!}a^2b + \frac{3!}{2!\cdot 1!}ab^2 + b^3$$

$$= a^3 + \frac{3\cdot \overset{1}{\cancel{2}}\cdot \overset{1}{\cancel{1}}}{\underset{1}{\cancel{1}}\cdot \underset{1}{\cancel{2}}\cdot 1}a^2b + \frac{3\cdot \overset{1}{\cancel{2}}\cdot \overset{1}{\cancel{1}}}{\underset{1}{\cancel{2}}\cdot \underset{1}{\cancel{1}}\cdot 1}ab^2 + b^3$$

$$= a^3 + 3a^2b + 3ab^2 + b^3$$

Self Check 5 Use the binomial theorem to expand $(a+b)^4$.

EXAMPLE 6

Use the binomial theorem to expand $(x - y)^4$.

Solution We can write $(x - y)^4$ in the form $[x + (-y)]^4$, substitute directly into the binomial theorem, and simplify:

$$
\begin{aligned}
(x - y)^4 &= [x + (-y)]^4 \\
&= x^4 + \frac{4!}{1!(4-1)!}x^3(-y) + \frac{4!}{2!(4-2)!}x^2(-y)^2 + \frac{4!}{3!(4-3)!}x(-y)^3 + (-y)^4 \\
&= x^4 - \frac{4 \cdot 3!}{1! \cdot 3!}x^3y + \frac{4 \cdot 3 \cdot 2!}{2! \cdot 2!}x^2y^2 - \frac{4 \cdot 3!}{3! \cdot 1!}xy^3 + y^4 \\
&= x^4 - 4x^3y + 6x^2y^2 - 4xy^3 + y^4
\end{aligned}
$$

Note the alternating signs.

Self Check 6 Use the binomial theorem to expand $(x - y)^3$.

EXAMPLE 7

Use the binomial theorem to expand $(3u - 2v)^4$.

Solution We write $(3u - 2v)^4$ in the form $[3u + (-2v)]^4$ and let $a = 3u$ and $b = -2v$. Then we can use the binomial theorem to expand $(a + b)^4$.

$$
\begin{aligned}
(a + b)^4 &= a^4 + \frac{4!}{1!(4-1)!}a^3b + \frac{4!}{2!(4-2)!}a^2b^2 + \frac{4!}{3!(4-3)!}ab^3 + b^4 \\
&= a^4 + 4a^3b + 6a^2b^2 + 4ab^3 + b^4
\end{aligned}
$$

Now we can substitute $3u$ for a and $-2v$ for b and simplify:

$$
\begin{aligned}
(3u - 2v)^4 &= (3u)^4 + 4(3u)^3(-2v) + 6(3u)^2(-2v)^2 + 4(3u)(-2v)^3 + (-2v)^4 \\
&= 81u^4 - 216u^3v + 216u^2v^2 - 96uv^3 + 16v^4
\end{aligned}
$$

Self Check 7 Use the binomial theorem to expand $(4a - 5b)^3$.

FINDING A SPECIFIC TERM OF AN EXPANSION

To find a specific term of an expansion, we don't need to write out the entire expansion. A close examination of the binomial theorem and the pattern of the terms suggests the following method for finding a single term of an expansion.

Finding a Specific Term of an Expansion

The $(r + 1)$st term of the expansion of $(a + b)^n$ is

$$\frac{n!}{r!(n-r)!}a^{n-r}b^r$$

EXAMPLE 8

Find the 4th term of the expansion of $(a + b)^9$.

Solution To use the formula for finding a specific term of $(a + b)^9$, we must determine n and r. If we are to find the fourth term, $r + 1 = 4$, and it follows that $r = 3$.

We also see that $n = 9$. We substitute those values into the formula to find the 4th term of the expansion.

Success Tip

r is always 1 less than the number of the term that you are finding.

$$\frac{n!}{r!(n-r)!}a^{n-r}b^r = \frac{9!}{3!(9-3)!}a^{9-3}b^3$$

$$= \frac{9!}{3!6!}a^6b^3 \qquad \frac{9!}{3!6!} = \frac{9 \cdot 8 \cdot 7 \cdot \overset{1}{\cancel{6!}}}{3 \cdot 2 \cdot 1 \cdot \underset{1}{\cancel{6!}}} = 84.$$

$$= 84a^6b^3$$

Self Check 8 Find the 3rd term of the expansion of $(a + b)^9$.

EXAMPLE 9

Find the 6th term of the expansion of $\left(x^2 - \frac{y}{2}\right)^7$.

Solution To use the formula for finding a specific term of $\left(x^2 - \frac{y}{2}\right)^7$, we must determine n, r, a, and b. If we are to find the sixth term, $r + 1 = 6$, and it follows that $r = 5$.

We also see that $a = x^2$, $b = -\frac{y}{2}$, and $n = 7$. We substitute those values into the formula to find the 6th term.

$$\frac{n!}{r!(n-r)!}a^{n-r}b^r = \frac{7!}{5!(7-5)!}(x^2)^{7-5}\left(-\frac{y}{2}\right)^5$$

$$= \frac{7!}{5!2!}(x^2)^2\left(-\frac{y^5}{32}\right) \qquad \frac{7!}{5!2!} = \frac{7 \cdot 6 \cdot \overset{1}{\cancel{5!}}}{\underset{1}{\cancel{5!}} \cdot 2 \cdot 1} = 21.$$

$$= -\frac{21}{32}x^4y^5$$

Self Check 9 Find the 5th term of the expansion of $\left(c^2 - \frac{d}{3}\right)^7$.

Answers to Self Checks

1. $x^4 + 4x^3y + 6x^2y^2 + 4xy^3 + y^4$ **2.** $x^5 - 5x^4y + 10x^3y^2 - 10x^2y^3 + 5xy^4 - y^5$ **3. a.** 5,040, **b.** 144, **c.** 1 **4. a.** 4, **b.** 21 **5.** $a^4 + 4a^3b + 6a^2b^2 + 4ab^3 + b^4$ **6.** $x^3 - 3x^2y + 3xy^2 - y^3$ **7.** $64a^3 - 240a^2b + 300ab^2 - 125b^3$ **8.** $36a^7b^2$ **9.** $\frac{35}{81}c^6d^4$

11.1 STUDY SET

VOCABULARY Fill in the blanks.

1. The two-term polynomial expression $a + b$ is called a ________.
2. $a^4 + 4a^3b + 6a^2b^2 + 4ab^3 + b^4$ is the binomial ________ of $(a + b)^4$.
3. We can use the ________ theorem to raise binomials to positive-integer powers without doing the actual multiplication.
4. The array of numbers that gives the coefficients of the terms of a binomial expansion is called ________ triangle.
5. $n!$ (read as "n ________") is the product of consecutively ________ natural numbers from n to 1.
6. In the expansion $a^3 - 3a^2b + 3ab^2 - b^3$, the signs ________ between + and −.

CONCEPTS Fill in the blanks.

7. Every binomial expansion has ______ more term than the power of the binomial.
8. For each term of the expansion of $(a + b)^8$, the sum of the exponents of a and b is ___.
9. The first term of the expansion of $(r + s)^{20}$ is ___ and the last term is ___.
10. In the expansion of $(m - n)^{15}$, the exponents on m ________ and the exponents on n ________.
11. The coefficients of the terms of the expansion of $(c + d)^{20}$ begin with ___, increase through some values, and then decrease through those same values, back to ___.
12. Complete Pascal's Triangle:

1
1 1
1 2 ___
1 ___ 3 1
1 ___ 6 4 1
1 5 10 10 5 1
1 ___ 15 ___ 15 6 1
___ 7 21 35 ___ 21 7 1
1 8 28 56 70 56 ___ 8 ___

13. $n \cdot$ ______ $= n!$
14. $8! = 8 \cdot$ ___
15. $0! =$ ___
16. According to the binomial theorem, the third term of the expansion of $(a + b)^n$ is ______.
17. The coefficient of the fourth term of the expansion of $(a + b)^9$ is 9! divided by ______.
18. The exponent on a in the fourth term of the expansion of $(a + b)^6$ is ___ and the exponent on b is ___.
19. The exponent on a in the fifth term of the expansion of $(a + b)^6$ is ___ and the exponent on b is ___.
20. The expansion of $(a - b)^4$ is

$$a^4 \;\_\_\; 4a^3b \;\_\_\; 6a^2b^2 \;\_\_\; 4ab^3 \;\_\_\; b^4$$

21. $(x + y)^3$

$$= x^{\_\_} + \frac{\_\_}{1!(3-1)!}x^2y^{\_\_} + \frac{\_\_}{\_\_!(3-2)!}xy^{\_\_} + y^{\_\_}$$

22. Fill in the blanks.
 a. The $(r + 1)$st term of the expansion of $(a + b)^n$ is $\frac{n!}{r!(n - \_\_)!}a^{\_\_ - r}b^{\_\_}$.
 b. To use this formula to find the 6th term of the expansion of $\left(m + \frac{n}{2}\right)^8$, we note that $r =$ ___, $n =$ ___, $a =$ ___, and $b =$ ___.

NOTATION Fill in the blanks.

23. $n! = n \cdot (\_\_\_)(n - 2) \cdots 3 \cdot 2 \cdot 1$
24. The symbol 5! is read as "______ ________" and it means $5 \cdot$ ______.

PRACTICE Evaluate each expression.

25. 3!
26. 7!
27. 5!
28. 6!
29. 3! + 4!
30. 2!(3!)
31. 3!(4!)
32. 4! + 4!
33. 8(7!)
34. 4!(5)
35. $\frac{9!}{11!}$
36. $\frac{13!}{10!}$
37. $\frac{49!}{47!}$
38. $\frac{101!}{100!}$
39. $\frac{9!}{7!0!}$
40. $\frac{7!}{5!0!}$

41. $\dfrac{5!}{1!(5-1)!}$

42. $\dfrac{15!}{14!(15-14)!}$

43. $\dfrac{5!}{3!(5-3)!}$

44. $\dfrac{6!}{4!(6-4)!}$

45. $\dfrac{7!}{5!(7-5)!}$

46. $\dfrac{8!}{6!(8-6)!}$

47. $\dfrac{5!(8-5)!}{4!\cdot 7!}$

48. $\dfrac{6!\cdot 7!}{(8-3)!(7-4)!}$

Use a calculator to evaluate each expression.

49. 11!

50. 13!

51. 20!

52. 55!

Expand each expression.

53. $(x+y)^4$

54. $(a-b)^4$

55. $(c-d)^5$

56. $(c+d)^5$

57. $(s+t)^6$

58. $(s-t)^6$

59. $(a-b)^9$

60. $(a+b)^7$

61. $(2x+y)^3$

62. $(x+2y)^3$

63. $(2t-3)^5$

64. $(2b+1)^4$

65. $(5m-2n)^4$

66. $(2m+3n)^5$

67. $\left(\dfrac{x}{3}+\dfrac{y}{2}\right)^3$

68. $\left(\dfrac{x}{2}-\dfrac{y}{3}\right)^3$

69. $\left(\dfrac{x}{3}-\dfrac{y}{2}\right)^4$

70. $\left(\dfrac{x}{2}+\dfrac{y}{3}\right)^4$

71. $(c^2-d^2)^5$

72. $(u^2-v^3)^5$

Find the indicated term of each binomial expansion.

73. $(x-y)^4$; 4th

74. $(x-y)^5$; 2nd

75. $(r+s)^6$; 5th

76. $(r+s)^7$; 5th

77. $(x-y)^8$; 3rd

78. $(x-y)^9$; 7th

79. $(x-3y)^4$; 2nd

80. $(3x-y)^5$; 3rd

81. $(2t-5)^7$; 4th

82. $(2t+3)^6$; 6th

83. $(2x-3y)^5$; 5th

84. $(3x-2y)^4$; 2nd

85. $\left(\dfrac{c}{2}-\dfrac{d}{3}\right)^4$; 2nd

86. $\left(\dfrac{c}{3}+\dfrac{d}{2}\right)^5$; 4th

87. $(a^2-b^2)^6$; 2nd

88. $(a^2+b^2)^7$; 6th

WRITING

89. Describe how to construct Pascal's triangle.

90. Explain why the signs alternate in the expansion of $(x-y)^9$.

91. Explain why the third term of the expansion of $(m+3n)^9$ could not be $324m^7n^3$.

92. Using your own words, write a definition of $n!$.

REVIEW

Assume that x, y, z, and b represent positive numbers. Use the properties of logarithms to write each expression as the logarithm of a single quantity.

93. $2\log x + \dfrac{1}{2}\log y$

94. $-2\log x - 3\log y + \log z$

95. $\ln(xy+y^2) - \ln(xz+yz) + \ln z$

96. $\log_2(x+1) - \log_2 x$

CHALLENGE PROBLEMS

97. Find the constant term in the expansion of $\left(x+\dfrac{1}{x}\right)^{10}$.

98. Find the coefficient of a^5 in the expansion of $\left(a-\dfrac{1}{a}\right)^9$.

99. a. If we applied the pattern of the coefficients to the coefficient of the first term in a binomial expansion, the coefficient would be $\frac{n!}{0!(n-0)!}$. Show that this expression is 1.

b. If we applied the pattern of the coefficients to the coefficient of the last term in a binomial expansion, the coefficient would be $\frac{n!}{n!(n-n)!}$. Show that this expression is 1.

100. Expand $(i-1)^7$, where $i = \sqrt{-1}$.

11.2 Arithmetic Sequences and Series

- Sequences
- Arithmetic sequences
- Arithmetic means
- The sum of the first n terms
- Summation notation

The word *sequence* is used in everyday conversation when referring to an ordered list. For example, a history instructor might discuss the sequence of events that led up to the sinking of the *Titanic*. In mathematics, a **sequence** is a list of numbers written in a specific order. When we put a + symbol between the numbers in a sequence, the sum is called a *series.*

SEQUENCES

Each number in a sequence is called a **term** of the sequence. **Finite sequences** contain a finite number of terms and **infinite sequences** contain an infinite number of terms. Two examples of sequences are:

Finite sequence: 1, 5, 9, 13, 17, 21, 25

Infinite sequence: 3, 6, 9, 12, 15, . . . The . . . indicates that the sequence goes on forever.

Sequences are defined formally using the terminology of functions.

Finite and Infinite Sequences

A **finite sequence** is a function whose domain is the set of natural numbers $\{1, 2, 3, 4, \ldots, n\}$, for some natural number n.

An **infinite sequence** is a function whose domain is the set of natural numbers: $\{1, 2, 3, 4, \ldots\}$.

Instead of using $f(x)$ notation, we use a_n (read as "a sub n") notation to write the value of a sequence at the number n. For the infinite sequence introduced earlier, we have:

1st term	2nd term	3rd term	4th term	5th term
3,	6,	9,	12,	15, . . .
↑	↑	↑	↑	↑
a_1	a_2	a_3	a_4	a_5

To specifically describe *all* the terms of a sequence we can write a formula for a_n, called the **general term** of the sequence. For the sequence 3, 6, 9, 12, 15, . . . , we note that $a_1 = 3 \cdot 1$, $a_2 = 3 \cdot 2$, $a_3 = 3 \cdot 3$, and so on. In general, the nth term of the sequence is found by multiplying n by 3.

$a_n = 3n$ Read a_n as "a sub n."

We can use this formula to find any term of the sequence. For example, to find the 12th term, we substitute 12 for n.

$$a_{12} = 3(12) = 36$$

EXAMPLE 1

Given an infinite sequence with $a_n = 2n - 3$, find each of the following: **a.** the first four terms and **b.** a_{50}.

Solution

a. To find the first four terms of the sequence, we substitute 1, 2, 3, and 4 for n in $a_n = 2n - 3$ and simplify.

$$a_1 = 2(1) - 3 = -1$$
$$a_2 = 2(2) - 3 = 1$$
$$a_3 = 2(3) - 3 = 3$$
$$a_4 = 2(4) - 3 = 5$$

The first four terms of the sequence are -1, 1, 3, and 5.

b. To find a_{50}, the 50th term of the sequence, we let $n = 50$:

$$a_{50} = 2(50) - 3 = 97$$

Self Check 1 Given an infinite sequence with $a_n = 3n + 5$, find each of the following: **a.** the first three terms and **b.** a_{100}.

ARITHMETIC SEQUENCES

A sequence where each term is found by adding the same number to the previous term is called an *arithmetic sequence.* Two examples are

The Language of Algebra

We pronounce the adjective *arithmetic* in the term *arithmetic* sequence as: air-rith-met′-ic.

5, 12, 19, 26, 33, 40
Add 7

This is a finite arithmetic sequence where each term is found by adding 7 to the previous term.

$3, 1, -1, -3, -5, -7, \ldots$
Add -2

This is an infinite arithmetic sequence where each term is found by adding -2 to the previous term.

Arithmetic Sequence

An **arithmetic sequence** is a sequence of the form

$$a_1, \quad a_1 + d, \quad a_1 + 2d, \quad a_1 + 3d, \quad \ldots, \quad a_1 + (n-1)d, \ldots$$

where a_1 is the **first term** and d is the **common difference.** The nth term is given by

$$a_n = a_1 + (n-1)d$$

We note that the second term of an arithmetic sequence has an addend of $1d$, the third term has an addend of $2d$, the fourth term has an addend of $3d$, and the nth term has an addend of $(n-1)d$. We also note that the *difference between any two consecutive terms in an arithmetic sequence is d.*

EXAMPLE 2

An arithmetic sequence has a first term 5 and a common difference 4. Find the 25th term of the sequence.

Solution Since the first term is $a_1 = 5$ and the common difference is $d = 4$, the arithmetic sequence is defined by the formula

$$a_n = 5 + (n - 1)4 \qquad \text{In } a_n = a_1 + (n-1)d\text{, substitute 5 for } a_1 \text{ and 4 for } d.$$

To find the 25th term, we substitute 25 for n and simplify.

$$\begin{aligned} a_{25} &= 5 + (25 - 1)4 \\ &= 5 + (24)4 \\ &= 101 \end{aligned}$$

The 25th term is 101.

Self Check 2 An arithmetic sequence has a first term 10 and a common difference 8. Find the 30th term of the sequence.

EXAMPLE 3

The first three terms of an arithmetic sequence are 3, 8, and 13. Find the 100th term.

Solution The common difference d is the difference between any two successive terms. Since $a_1 = 3$ and $a_2 = 8$, we can find d using subtraction.

$$d = a_2 - a_1 = 8 - 3 = 5 \qquad \text{Also note that } a_3 - a_2 = 13 - 8 = 5.$$

Success Tip

The common difference d of an arithmetic sequence is defined to be

$$d = a_{n+1} - a_n$$

To find 100th term, we substitute 3 for a_1, 5 for d and 100 for n in the formula for the nth term.

$$\begin{aligned} a_n &= a_1 + (n - 1)d \\ a_{100} &= 3 + (100 - 1)5 \\ &= 3 + (99)5 \\ &= 498 \end{aligned}$$

Self Check 3 The first three terms of an arithmetic sequence are -3, 6, and 15. Find the 99th term.

EXAMPLE 4

The first term of an arithmetic sequence is 12 and the 50th term is 3,099. Write the first six terms of the sequence.

Solution The key is to find the common difference. Because the 50th term of the sequence is 3,099, we substitute 3,099 for a_{50} and $a_1 = 12$ in the formula for the 50th term and solve for d.

$$\begin{aligned} a_{50} &= a_1 + (50 - 1)d && \text{This gives the 50th term of any arithmetic sequence.} \\ 3{,}099 &= 12 + (50 - 1)d && \text{Substitute 3,099 for } a_{50} \text{ and 12 for } a_1. \\ 3{,}099 &= 12 + 49d && \text{Simplify.} \\ 3{,}087 &= 49d && \text{Subtract 12 from both sides.} \\ 63 &= d && \text{Divide both sides by 49.} \end{aligned}$$

Since the first term is 12 and the common difference is 63, the first six terms are

12, 75, 138, 201, 264, 327   Add 63 to a term to get the next term.

Self Check 4 The first term of an arithmetic sequence is 15 and the 12th term is 92. Write the first four terms of the sequence.

ARITHMETIC MEANS

If numbers are inserted between two numbers a and b to form an arithmetic sequence, the inserted numbers are called **arithmetic means** between a and b. If a single number is inserted, it is called **the arithmetic mean** between a and b.

EXAMPLE 5

Insert two arithmetic means between 6 and 27.

Solution The first term is $a_1 = 6$ and the fourth term is $a_4 = 27$. We must find the common difference so that the terms

$$\underset{a_1}{\underset{\uparrow}{6}},\quad \underset{a_2}{\underset{\uparrow}{6+d}},\quad \underset{a_3}{\underset{\uparrow}{6+2d}},\quad \underset{a_4}{\underset{\uparrow}{27}}$$

form an arithmetic sequence. To find d, we substitute 6 for a_1 and 27 for a_4 in the formula for the 4th term:

$a_4 = a_1 + (4 - 1)d$ This gives the 4th term of any arithmetic sequence.

$27 = 6 + (4 - 1)d$ Substitute.

$27 = 6 + 3d$ Simplify.

$21 = 3d$ Subtract 6 from both sides.

$7 = d$ Divide both sides by 3.

The two arithmetic means between 6 and 27 are

$6 + d = 6 + 7$
$= 13$ This is a_2.

or

$6 + 2d = 6 + 2(7)$
$= 6 + 14$
$= 20$ This is a_3.

Two arithmetic means between 6 and 27 are 13 and 20.

Self Check 5 Insert two arithmetic means between 8 and 44.

THE SUM OF THE FIRST n TERMS

When the commas between the terms of a sequence are replaced with + signs, we call the sum a **series.** The sum of the terms of an arithmetic sequence is called an **arithmetic series.** Some examples are

$4 + 8 + 12 + 16 + 20 + 24$ Since this series has a limited number of terms, it is a finite arithmetic series.

$5 + 8 + 11 + 14 + 17 + \cdots$ Since this series has an unlimited number of terms, it is an infinite arithmetic series.

To develop a formula for evaluating the sum of the first n terms of an arithmetic sequence, we let S_n represent the sum of the first n terms of an arithmetic sequence:

$$S_n = a_1 + [a_1 + d] + [a_1 + 2d] + \cdots + [a_1 + (n-1)d]$$

We write the same sum again, but in reverse order:

$$S_n = [a_1 + (n-1)d] + [a_1 + (n-2)d] + [a_1 + (n-3)d] + \cdots + a_1$$

Adding these equations together, term by term, we get

$$2S_n = [2a_1 + (n-1)d] + [2a_1 + (n-1)d] + [2a_1 + (n-1)d] + \cdots + [2a_1 + (n-1)d]$$

Success Tip

An alternate form of the summation formula for arithmetic sequences can be obtained from Equation (1) by combining like terms and dividing both sides by 2.

$$S_n = \frac{n[2a_1 + (n-1)d]}{2}$$

Because there are n equal terms on the right-hand side of the preceding equation, we can write

$$2S_n = n[2a_1 + (n-1)d]$$

(1) $2S_n = n[a_1 + \mathbf{a_1 + (n-1)d}]$ Write $2a_1$ as $a_1 + a_1$.

$2S_n = n(a_1 + \mathbf{a_n})$ Substitute a_n for $a_1 + (n-1)d$.

$S_n = \dfrac{n(a_1 + a_n)}{2}$ Divide both sides by 2.

This reasoning establishes the following formula.

Sum of the First *n* Terms of an Arithmetic Sequence

The sum of the first n terms of an arithmetic sequence is given by the formula

$$S_n = \frac{n(a_1 + a_n)}{2}$$

where a_1 is the first term, a_n is the nth (or last) term, and n is the number of terms in the sequence.

EXAMPLE 6

Find the sum of the first 40 terms of the arithmetic sequence 4, 10, 16, ...

Solution In this example, we let $a_1 = 4$, $n = 40$, $d = 10 - 4 = 6$, and $a_{40} = 4 + (40 - 1)6 = 238$ and substitute these values into the formula for S_n:

$$S_n = \frac{n(a_1 + a_n)}{2}$$

$$S_{40} = \frac{40(4 + 238)}{2} \quad \text{Substitute } a_1 = 4 \text{ and } a_{40} = 238.$$

$$= 20(242)$$

$$= 4{,}840$$

The sum of the first 40 terms is 4,840.

Self Check 6 Find the sum of the first 50 terms of the arithmetic sequence 3, 8, 13, ...

SUMMATION NOTATION

When the general term of a sequence is known, we can use a special notation to write a series. This notation, called **summation notation,** involves the Greek letter Σ (sigma). The expression

$$\sum_{k=1}^{4} 3k \quad \text{Read as "the summation of } 3k \text{ as } k \text{ runs from 1 to 4."}$$

designates the sum of all terms obtained if we successively substitute the numbers 1, 2, 3, and 4 for k, called the **index of the summation.** Thus, we have

$$\begin{aligned}\sum_{k=1}^{4} 3k &= 3(1) + 3(2) + 3(3) + 3(4) \quad (k=1,\ k=2,\ k=3,\ k=4)\\ &= 3 + 6 + 9 + 12\\ &= 30\end{aligned}$$

EXAMPLE 7

Find each sum: **a.** $\sum_{k=1}^{3} (2k + 1)$ and **b.** $\sum_{k=2}^{8} k^2$.

Solution

a.
$$\begin{aligned}\sum_{k=1}^{3} (2k + 1) &= [2(1) + 1] + [2(2) + 1] + [2(3) + 1]\\ &= 3 + 5 + 7\\ &= 15\end{aligned}$$

b. Here, we substitute the integers from 2 to 8 for k and find the sum.
$$\begin{aligned}\sum_{k=2}^{8} k^2 &= 2^2 + 3^2 + 4^2 + 5^2 + 6^2 + 7^2 + 8^2\\ &= 4 + 9 + 16 + 25 + 36 + 49 + 64\\ &= 203\end{aligned}$$

Self Check 7 Find the sum: $\sum_{k=1}^{4} (2k^2 - 2)$.

Answers to Self Checks **1. a.** 8, 11, 14, **b.** 305 **2.** 242 **3.** 879 **4.** 15, 22, 29, 36 **5.** 20, 32 **6.** 6,275 **7.** 52

11.2 STUDY SET

VOCABULARY **Fill in the blanks.**

1. A ________ is a function whose domain is the set of natural numbers.
2. A sequence with an unlimited number of terms is called a(n) ________ sequence.

   A sequence with a specific number of terms is called a(n) ______ sequence.
3. Each term of an ________ sequence is found by adding the same number to the previous term.
4. 5, 15, 25, 35, 45, 55, . . . is an example of an ________ sequence. The first ______ is 5 and the common ________ is 10.
5. If a single number is inserted between a and b to form an arithmetic sequence, the number is called the arithmetic ______ between a and b.

6. The sum of the terms of an arithmetic sequence is called an arithmetic ______.

CONCEPTS

7. Write the first three terms of an arithmetic sequence if $a_1 = 1$ and $d = 6$.
8. Given the arithmetic sequence 4, 7, 10, 13, 16, 19, . . . , find a_5 and d.
9. a. Write the formula for a_n, the general term of an arithmetic sequence.
   b. Write the formula for S_n, the sum of the first n terms of an arithmetic sequence.
10. An infinite arithmetic sequence is of the form

$$a_1, a_1 + d, \underline{\quad\quad}, a_1 + 3d, \underline{\quad\quad}, \ldots$$

NOTATION Fill in the blanks.

11. The notation a_n represents the ______ term of a sequence.
12. To find the common difference of an arithmetic sequence, we use the formula $d = a_{\square} - a_{\square}$.
13. The symbol Σ is the Greek letter ______.
14. In the symbol $\sum_{k=1}^{5} (2k - 5)$, k is called the ______ of summation.
15. We read $\sum_{k=1}^{10} 3k$ as "the ______ of $3k$ as k ______ from 1 to 10."
16. $\sum_{k=1}^{5} k = \square + \square + \square + \square + \square$

PRACTICE Write the first five terms of each sequence.

17. $a_n = 4n - 1$
18. $a_n = 5n - 3$
19. $a_n = -3n + 1$
20. $a_n = -6n + 2$

Write the first five terms of each arithmetic sequence with the given properties.

21. $a_1 = 3, d = 2$
22. $a_1 = -2, d = 3$
23. $a_1 = -5, d = -3$
24. $a_1 = 8, d = -5$
25. $a_1 = 5$, fifth term is 29
26. $a_1 = 4$, sixth term is 39
27. $a_1 = -4$, sixth term is -39
28. $a_1 = -5$, fifth term is -37
29. $d = 7$, sixth term is -83
30. $d = 3$, seventh term is 12
31. $d = -3$, seventh term is 16
32. $d = -5$, seventh term is -12
33. The 19th term is 131 and the 20th term is 138.
34. The 16th term is 70 and the 18th term is 78.
35. Find the 30th term of the arithmetic sequence with $a_1 = 7$ and $d = 12$.
36. Find the 55th term of the arithmetic sequence with $a_1 = -5$ and $d = 4$.
37. Find the 37th term of the arithmetic sequence with a second term of -4 and a third term of -9.
38. Find the 40th term of the arithmetic sequence with a second term of 6 and a fourth term of 16.
39. Find the first term of the arithmetic sequence with a common difference of 11 if its 27th term is 263.
40. Find the common difference of the arithmetic sequence with a first term of -164 if its 36th term is -24.
41. Find the common difference of the arithmetic sequence with a first term of 40 if its 44th term is 556.
42. Find the first term of the arithmetic sequence with a common difference of -5 if its 23rd term is -625.
43. Insert three arithmetic means between 2 and 11.
44. Insert four arithmetic means between 5 and 25.
45. Insert four arithmetic means between 10 and 20.
46. Insert three arithmetic means between 20 and 30.
47. Find the arithmetic mean between 10 and 19.
48. Find the arithmetic mean between -4.5 and 7.

Write the series associated with each summation.

49. $\sum_{k=1}^{4} (3k)$
50. $\sum_{k=1}^{4} (k - 9)$
51. $\sum_{k=2}^{4} k^2$
52. $\sum_{k=3}^{5} (-2k)$

Write the summation notation for each sum.

53. $1 + 4 + 9 + 16 + 25$

54. $2 + 4 + 6 + 8$

55. $3 + 4 + 5 + 6$

56. $-1 - 4 - 9 - 16 - 25 - 36$

Find the sum of the first *n* terms of each arithmetic sequence.

57. $1, 4, 7, \ldots; n = 30$

58. $2, 6, 10, \ldots; n = 28$

59. $-5, -1, 3, \ldots; n = 17$

60. $-7, -1, 5, \ldots; n = 15$

61. Second term is 7, third term is 12; $n = 12$

62. Second term is 5, fourth term is 9; $n = 16$

63. $a_n = 2n + 1$, nth term is 31; n is a natural number

64. $a_n = 4n + 3$, nth term is 23; n is a natural number

65. Find the sum of the first 50 natural numbers.

66. Find the sum of the first 100 natural numbers.

67. Find the sum of the first 50 odd natural numbers.

68. Find the sum of the first 50 even natural numbers.

Find each sum.

69. $\sum_{k=1}^{4} (6k)$

70. $\sum_{k=2}^{5} (3k)$

71. $\sum_{k=3}^{4} k^3$

72. $\sum_{k=2}^{4} (-k^2)$

73. $\sum_{k=3}^{4} (k^2 + 3)$

74. $\sum_{k=2}^{6} (k^2 + 1)$

75. $\sum_{k=4}^{4} (2k + 4)$

76. $\sum_{k=3}^{5} (3k^2 - 7)$

APPLICATIONS

77. SAVING MONEY Yasmeen puts \$60 into a safety deposit box. After each succeeding month, she puts \$50 more in the box. Write the first six terms of an arithmetic sequence that gives the monthly amounts in her savings, and find her savings after 10 years.

78. INSTALLMENT LOANS Maria borrowed \$10,000, interest-free, from her mother. She agreed to pay back the loan in monthly installments of \$275. Write the first six terms of an arithmetic sequence that shows the balance due after each month, and find the balance due after 17 months.

79. DESIGNING PATIOS Each row of bricks in the following triangular patio is to have one more brick than the previous row, ending with the longest row of 150 bricks. How many bricks will be needed?

80. FALLING OBJECTS The equation $s = 16t^2$ represents the distance s in feet that an object will fall in t seconds. After 1 second, the object has fallen 16 feet. After 2 seconds, it has fallen 64 feet, and so on. Find the distance that the object will fall during the second and third seconds.

81. FALLING OBJECTS Refer to Exercise 80. How far will the object fall during the 12th second?

82. INTERIOR ANGLES The sums of the angles of several polygons are given in the table. Assuming that the pattern continues, complete the table.

Figure	Number of sides	Sum of angles
Triangle	3	180°
Quadrilateral	4	360°
Pentagon	5	540°
Hexagon	6	720°
Octagon	8	
Dodecagon	12	

WRITING

83. Explain why 1, 4, 8, 13, 19, 26, . . . is not an arithmetic sequence.

84. What is the difference between a sequence and a series?

85. What is the difference between a_n and S_n?

86. How is the symbol Σ used in this section?

REVIEW

Assume that x, y, z, and b represent positive numbers. Use the properties of logarithms to write each expression in terms of the logarithms of x, y, and z.

87. $\log_2 \dfrac{2x}{y}$

88. $\ln x\sqrt{z}$

89. $\log x^3y^2$

90. $\log x^3y^{1/2}$

CHALLENGE PROBLEMS

91. Show that $\sum_{k=1}^{5} 5k = 5\sum_{k=1}^{5} k$.

92. Show that $\sum_{k=3}^{6} (k^2 + 3k) = \sum_{k=3}^{6} k^2 + \sum_{k=3}^{6} 3k$.

93. Show that $\sum_{k=1}^{n} 3 = 3n$. (*Hint:* Consider 3 to be $3k^0$.)

94. Show that $\sum_{k=1}^{3} \dfrac{k^2}{k} \neq \dfrac{\sum_{k=1}^{3} k^2}{\sum_{k=1}^{3} k}$.

11.3 Geometric Sequences and Series

- Geometric sequences
- Geometric means
- The sum of the first n terms of a geometric sequence
- Infinite geometric series

We have seen that the same number is added to each term of an arithmetic sequence to get the next term. In this section, we will consider another type of sequence where we multiply each term by the same number to get the next term. This type of sequence is called a *geometric sequence.* Two examples are

$2, 8, 32, 128, \ldots$ (Multiply by 4) — This is an infinite geometric sequence where each term is found by multiplying the previous term by 4.

$27, 9, 3, 1, \frac{1}{3}, \frac{1}{9}$ (Multiply by $\frac{1}{3}$) — This is a finite geometric sequence where each term is found by multiplying the previous term by $\frac{1}{3}$.

GEOMETRIC SEQUENCES

Each term of a geometric sequence is found by multiplying the previous term by the same number.

Geometric Sequence

A **geometric sequence** is a sequence of the form

$$a_1, \quad a_1r, \quad a_1r^2, \quad a_1r^3, \quad \ldots, \quad a_1r^{n-1}, \quad \ldots$$

where a_1 is the **first term** and r is the **common ratio.** The nth term is given by

$$a_n = a_1r^{n-1}$$

We note that the second term of a geometric sequence has a factor r^1, the third term has a factor r^2, the fourth term has a factor r^3, and the nth term has a factor r^{n-1}. We also note that *r is the quotient obtained when any term is divided by the previous term.*

EXAMPLE 1

A geometric sequence has a first term 5 and a common ratio 3. **a.** Write the first five terms of the sequence and **b.** find the ninth term.

Solution

a. Because the first term is $a_1 = 5$ and the common ratio is $r = 3$, the first five terms are

$$\underset{a_1}{5},\quad \underset{a_2}{5(3)},\quad \underset{a_3}{5(3^2)},\quad \underset{a_4}{5(3^3)},\quad \underset{a_5}{5(3^4)}$$

Each term is found by multiplying the previous term by 3.

or

5, 15, 45, 135, 405

b. The nth term is a_1r^{n-1} with $a_1 = 5$ and $r = 3$. Because we want the ninth term, we let $n = 9$:

$$\begin{aligned} a_n &= a_1r^{n-1} \\ a_9 &= 5(3)^{9-1} \\ &= 5(3)^8 \\ &= 5(6{,}561) \\ &= 32{,}805 \end{aligned}$$

Success Tip

Note the difference: Each term of an *arithmetic* sequence is found by adding the same number to the previous term. Each term of a *geometric* sequence is found by multiplying the previous term by the same number.

Self Check 1 A geometric sequence has a first term 3 and a common ratio 4. **a.** Write the first four terms and **b.** find the eighth term.

EXAMPLE 2

The first three terms of a geometric sequence are 16, 4, and 1. Find the seventh term.

Solution

The common ratio r is the ratio between any two successive terms. Since $a_1 = 16$ and $a_2 = 4$, we can find r as follows:

$$r = \frac{a_2}{a_1} = \frac{4}{16} = \frac{1}{4}$$

Also note that $\frac{a_3}{a_2} = \frac{1}{4}$.

To find the seventh term, we substitute 16 for a_1, $\frac{1}{4}$ for r, and 7 for n in the formula for the nth term and simplify:

$$\begin{aligned} a_n &= a_1r^{n-1} \\ a_7 &= 16\left(\frac{1}{4}\right)^{7-1} \\ &= 16\left(\frac{1}{4}\right)^6 \\ &= 16\left(\frac{1}{4{,}096}\right) \\ &= \frac{1}{256} \end{aligned}$$

Success Tip

The common ratio r of a geometric sequence is defined to be

$$r = \frac{a_{n+1}}{a_n}$$

Self Check 2 The first three terms of a geometric sequence are 25, 5, and 1. Find the seventh term.

GEOMETRIC MEANS

If numbers are inserted between two numbers a and b to form a geometric sequence, the inserted numbers are called **geometric means** between a and b. If a single number is inserted, that number is called the **geometric mean** between a and b.

EXAMPLE 3

Insert two geometric means between 7 and 1,512.

Solution In this example, the first term is $a_1 = 7$, and the fourth term is $a_4 = 1{,}512$. To find the common ratio r so that the terms

$$\underset{a_1}{7,} \quad \underset{a_2}{7r,} \quad \underset{a_3}{7r^2,} \quad \underset{a_4}{1{,}512}$$

form a geometric sequence, we substitute 4 for n and 7 for a_1 in the formula for the nth term of a geometric sequence and solve for r.

$$\begin{aligned} a_n &= a_1 r^{n-1} \\ a_4 &= 7r^{4-1} \\ 1{,}512 &= 7r^3 \\ 216 &= r^3 && \text{Divide both sides by 7.} \\ 6 &= r && \text{Take the cube root of both sides.} \end{aligned}$$

The two geometric means between 7 and 1,512 are

$$7r = 7(6) = 42 \quad \text{and} \quad 7r^2 = 7(6)^2 = 7(36) = 252$$

The numbers 7, 42, 252, and 1,512 are the first four terms of a geometric sequence.

Self Check 3 Insert three positive geometric means between 1 and 16.

EXAMPLE 4

Find a geometric mean between 2 and 20.

Solution We want to find the middle term of the three-termed geometric sequence

$$\underset{a_1}{2,} \quad \underset{a_2}{2r,} \quad \underset{a_3}{20}$$

with $a_1 = 2$, $a_3 = 20$, and $n = 3$. To find r, we substitute these values into the formula for the nth term of a geometric sequence:

$$\begin{aligned} a_n &= a_1 r^{n-1} \\ a_3 &= 2r^{3-1} \\ 20 &= 2r^2 \\ 10 &= r^2 && \text{Divide both sides by 2.} \\ \pm\sqrt{10} &= r && \text{Use the square root property.} \end{aligned}$$

Because r can be either $\sqrt{10}$ or $-\sqrt{10}$, there are two values for a geometric mean. They are

$$2r = 2\sqrt{10} \quad \text{and} \quad 2r = -2\sqrt{10}$$

The sets of numbers 2, $2\sqrt{10}$, 20 and 2, $-2\sqrt{10}$, 20 both form geometric sequences. The common ratio of the first sequence is $\sqrt{10}$, and the common ratio of the second sequence is $-\sqrt{10}$.

Self Check 4 Find the positive geometric mean between 2 and 200.

THE SUM OF THE FIRST *n* TERMS OF A GEOMETRIC SEQUENCE

When we add the terms of a geometric sequence, we form a **geometric series.** There is a formula that gives the sum of the first n terms of a geometric sequence. To develop this formula, we let S_n represent the sum of the first n terms of a geometric sequence.

(1) $$S_n = a_1 + a_1r + a_1r^2 + a_1r^3 + \dots + a_1r^{n-1}$$

We multiply both sides of Equation 1 by r to get

(2) $$S_nr = \quad a_1r + a_1r^2 + a_1r^3 + \dots + a_1r^{n-1} + a_1r^n$$

We now subtract Equation 2 from Equation 1 and solve for S_n:

$$S_n - S_nr = a_1 - a_1r^n$$

$$S_n(1 - r) = a_1 - a_1r^n \quad \text{Factor out } S_n \text{ from the left-hand side.}$$

$$S_n = \frac{a_1 - a_1r^n}{1 - r} \quad \text{Divide both sides by } 1 - r.$$

This reasoning establishes the following formula.

Success Tip

If the common factor of a_1 in the numerator of $\frac{a_1 - a_1r^n}{1 - r}$ is factored out, the formula can be written in the alternate form: $S_n = \frac{a_1(1 - r^n)}{1 - r}$.

Sum of the First *n* Terms of a Geometric Sequence

The sum of the first n terms of a geometric sequence is given by the formula

$$S_n = \frac{a_1 - a_1r^n}{1 - r} \quad \text{or} \quad S_n = \frac{a_1(1 - r^n)}{1 - r} \quad \text{where } r \neq 1$$

where S_n is the sum, a_1 is the first term, r is the common ratio, and n is the number of terms.

EXAMPLE 5

Find the sum of the first six terms of the geometric sequence 250, 50, 10, ...

Solution In this sequence, $a_1 = 250$, $r = \frac{1}{5}$, and $n = 6$. We substitute these values into the formula for the sum of the first n terms of a geometric sequence and simplify:

$$S_n = \frac{a_1 - a_1r^n}{1 - r}$$

$$S_6 = \frac{250 - 250\left(\frac{1}{5}\right)^6}{1 - \frac{1}{5}}$$

$$= \frac{250 - 250\left(\frac{1}{15{,}625}\right)}{\frac{4}{5}}$$

$$= \frac{5}{4}\left(250 - \frac{250}{15{,}625}\right)$$

$$= 312.48 \quad \text{Use a calculator.}$$

The sum of the first six terms is 312.48.

Self Check 5 Find the sum of the first five terms of the geometric sequence 100, 20, 4, ...

EXAMPLE 6

Inheritances. A father decides to give his son an early inheritance. Each year, on the son's birthday, the father pays the son 15% of what is left of the $100,000 inheritance fund. How much money will be left in the fund after 20 years of payments?

Solution If 15% of the money in the inheritance fund is given to the son each year, 85% of that amount remains after a payment. To find the amount of money that remains in the fund after a payment is made, we multiply the amount that was in the fund by 0.85. Over the years, the amounts of money that are left in the fund after a payment form a geometric sequence.

	Amount of money remaining in fund	
The fund begins with:	100,000	$\leftarrow a_1$
After first payment:	$100{,}000(\mathbf{0.85}) = 100{,}000(0.85)^1$	$\leftarrow a_2$
After second payment:	$\left[100{,}000(0.85)^1\right](\mathbf{0.85}) = 100{,}000(0.85)^2$	$\leftarrow a_3$
After third payment:	$\left[100{,}000(0.85)^2\right](\mathbf{0.85}) = 100{,}000(0.85)^3$	$\leftarrow a_4$
After fourth payment:	$\left[100{,}000(0.85)^3\right](\mathbf{0.85}) = 100{,}000(0.85)^4$	$\leftarrow a_5$
⋮		⋮
After 20th payment:	???	$\leftarrow a_{21}$

The amount of money remaining in the inheritance fund after 20 years is represented by the 21st term of a geometric sequence, where $a_1 = 100{,}000$, $r = 0.85$, and $n = 21$.

$$a_n = a_1 r^{n-1}$$ The formula for the nth term of a geometric sequence.

$$a_{21} = a_1 r^{21-1}$$ Substitute 21 for n.

$$= \mathbf{100{,}000}(\mathbf{0.85})^{21-1}$$ Substitute 100,000 for a_1 and 0.85 for r.

$$= 100{,}000(0.85)^{20}$$

$$\approx 3{,}876$$ Use a calculator. Round to the nearest dollar.

In 20 years, approximately $3,876 of the inheritance fund will be left.

Self Check 6 How much money will be left in the inheritance fund after 30 years of payments?

INFINITE GEOMETRIC SERIES

If we form the sum of the terms of an infinite geometric sequence, we get a series called an **infinite geometric series.** For example, if the common ratio r is 3, we have

Infinite geometric sequence	***Infinite geometric series***
2, 6, 18, 54, 162, . . .	$2 + 6 + 18 + 54 + 162 + \cdots$

As the number of terms of this series gets larger, the value of the series gets larger. We can see that this is true by forming some **partial sums.**

The first partial sum, S_1, of the series is $S_1 = 2$.
The second partial sum, S_2, of the series is $S_2 = 2 + 6 = 8$.
The third partial sum, S_3, of the series is $S_3 = 2 + 6 + 18 = 26$.
The fourth partial sum, S_4, of the series is $S_4 = 2 + 6 + 18 + 54 = 80$.

We can now see that as the number of terms gets infinitely large, the value of the series gets infinitely large. The values of some infinite geometric series get closer and closer to a specific number as the number of terms approaches infinity. One such series is

$$\frac{3}{2} + \frac{3}{4} + \frac{3}{8} + \frac{3}{16} + \frac{3}{32} \cdots \qquad \text{Here, } r = \frac{1}{2}.$$

To see that this is true, we form some partial sums.

The first partial sum is $S_1 = \frac{3}{2} = 1.5$.

The second partial sum is $S_2 = \frac{3}{2} + \frac{3}{4} = \frac{9}{4} = 2.25$.

The third partial sum is $S_3 = \frac{3}{2} + \frac{3}{4} + \frac{3}{8} = \frac{21}{8} = 2.625$.

The fourth partial sum is $S_4 = \frac{3}{2} + \frac{3}{4} + \frac{3}{8} + \frac{3}{16} = \frac{45}{16} = 2.8125$.

The fifth partial sum is $S_5 = \frac{3}{2} + \frac{3}{4} + \frac{3}{8} + \frac{3}{16} + \frac{3}{32} = \frac{93}{32} = 2.90625$.

As the number of terms in this series gets larger, the values of the partial sums approach the number 3. We say that 3 is the **limit** of S_n as n approaches infinity, and we say that 3 is the **sum of the infinite geometric series.**

To develop a formula for finding the sum of an infinite geometric series, we consider the formula that gives the sum of the first n terms.

$$S_n = \frac{a_1 - a_1r^n}{1 - r} \qquad \text{where } r \neq 1$$

If $|r| < 1$ and a_1 is constant, the term a_1r^n in the above formula approaches 0 as n becomes very large. For example,

$$a_1\left(\frac{1}{2}\right)^1 = \frac{1}{2}a_1, \qquad a_1\left(\frac{1}{2}\right)^2 = \frac{1}{4}a_1, \qquad a_1\left(\frac{1}{2}\right)^3 = \frac{1}{8}a_1$$

and so on. Thus, when n is very large, the value of a_1r^n is negligible, and the term a_1r^n in the above formula can be ignored. This reasoning justifies the following formula.

Sum of the Terms of an Infinite Geometric Sequence

If a_1 is the first term and r is the common ratio of an infinite geometric sequence, and if $|r| < 1$, the sum of the terms of the sequence is given by

$$S = \frac{a_1}{1 - r}$$

EXAMPLE 7

Find the sum of the terms of the infinite geometric sequence 125, 25, 5, ...

Solution In this geometric sequence, $a_1 = 125$ and $r = \frac{25}{125} = \frac{1}{5}$. Since $|r| = \left|\frac{1}{5}\right| = \frac{1}{5} < 1$, we can find the sum of the terms of the sequence. We do this by substituting 125 for a_1 and $\frac{1}{5}$ for r in the formula $S = \frac{a_1}{1-r}$ and simplifying:

$$S = \frac{a_1}{1-r} = \frac{125}{1-\frac{1}{5}} = \frac{125}{\frac{4}{5}} = \frac{5}{4}(125) = \frac{625}{4}$$

The sum of the terms of the sequence 125, 25, 5, . . . is 156.25.

Notation

The sum of the terms of an infinite geometric sequence is also denoted S_∞.

Self Check 7 Find the sum of the terms of the infinite geometric sequence 100, 20, 4, ...

EXAMPLE 8

Find the sum of the infinite geometric sequence 64, -4, $\frac{1}{4}$, ...

Solution In this geometric sequence, $a_1 = 64$ and $r = \frac{-4}{64} = -\frac{1}{16}$. Since $|r| = \left|-\frac{1}{16}\right| = \frac{1}{16} < 1$, we can find the sum of all the terms of the sequence. We substitute 64 for a_1 and $-\frac{1}{16}$ for r in the formula $S = \frac{a_1}{1-r}$ and simplify:

$$S = \frac{a_1}{1-r} = \frac{64}{1-\left(-\frac{1}{16}\right)} = \frac{64}{\frac{17}{16}} = 64 \cdot \frac{16}{17} = \frac{1{,}024}{17}$$

The sum of the terms of the geometric sequence 64, -4, $\frac{1}{4}$, . . . is $\frac{1{,}024}{17}$.

Caution

If $|r| \geq 1$ for an infinite geometric sequence, the sum of the terms of the sequence, does not exist.

Self Check 8 Find the sum of the infinite geometric sequence 81, -27, 9, ...

EXAMPLE 9

Change $0.\overline{8}$ to a common fraction.

Solution The decimal $0.\overline{8}$ can be written as the infinite series

$$0.\overline{8} = 0.888\ldots = \frac{8}{10} + \frac{8}{100} + \frac{8}{1{,}000} + \cdots$$

where $a_1 = \frac{8}{10}$ and $r = \frac{1}{10}$. Because $|r| = \left|\frac{1}{10}\right| = \frac{1}{10} < 1$, we can find the sum as follows:

$$S = \frac{a_1}{1-r} = \frac{\frac{8}{10}}{1-\frac{1}{10}} = \frac{\frac{8}{10}}{\frac{9}{10}} = \frac{8}{9}$$

Thus, $0.\overline{8} = \frac{8}{9}$. Long division will verify that $\frac{8}{9} = 0.888\ldots$.

Self Check 9 Change $0.\overline{6}$ to a common fraction.

EXAMPLE 10

Testing steel. One way to measure the hardness of a steel anvil is to drop a ball bearing onto the face of the anvil. The bearing should rebound at least $\frac{4}{5}$ of the distance from which it was dropped. If a bearing is dropped from a height of 10 inches onto a hard forged steel anvil, and if it could bounce forever, what total distance would the bearing travel?

Solution The total distance the ball bearing travels is the sum of two motions, falling and rebounding. The bearing falls 10 inches, then rebounds $\frac{4}{5} \cdot 10 = 8$ inches, and falls 8 inches, and rebounds $\frac{4}{5} \cdot 8 = \frac{32}{5}$ inches, and falls $\frac{32}{5}$ inches, and rebounds $\frac{4}{5} \cdot \frac{32}{5} = \frac{128}{25}$ inches, and so on.

The distance the ball falls is given by the sum

$$10 + 8 + \frac{32}{5} + \frac{128}{25} + \cdots$$ This is an infinite geometric series with $a_1 = 10$ and $r = \frac{4}{5}$.

The distance the ball rebounds is given by the sum

$$8 + \frac{32}{5} + \frac{128}{25} + \cdots$$ This is an infinite geometric series with $a_1 = 8$ and $r = \frac{4}{5}$.

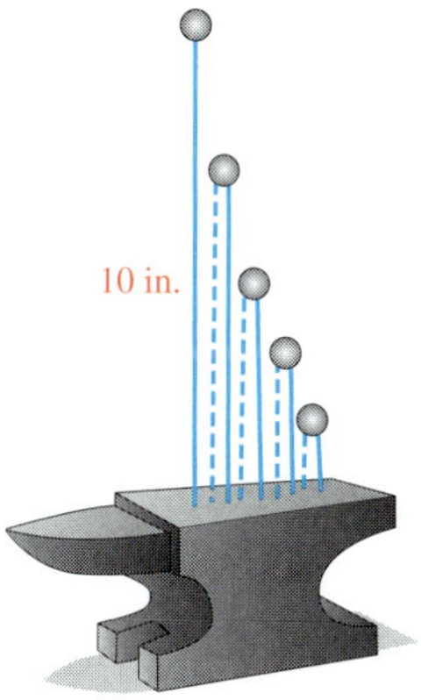

Since each of these is an infinite geometric series with $|r| < 1$, we can use the formula $S = \frac{a_1}{1 - r}$ to find each sum.

$$\text{Falling: } \frac{10}{1 - \frac{4}{5}} = \frac{10}{\frac{1}{5}} = 50 \text{ inches} \qquad \text{Rebounding: } \frac{8}{1 - \frac{4}{5}} = \frac{8}{\frac{1}{5}} = 40 \text{ inches}$$

The total distance the bearing travels is 50 inches + 40 inches = 90 inches.

Self Check 10 If a bearing was dropped from a height of 15 inches onto a hard forged steel anvil, and if it could bounce forever, what total distance would it travel?

Answers to Self Checks **1. a.** 3, 12, 48, 192, **b.** 49,152 **2.** $\frac{1}{625}$ **3.** 2, 4, 8 **4.** 20 **5.** 124.96 **6.** about \$763 **7.** 125 **8.** $\frac{243}{4}$ **9.** $\frac{2}{3}$ **10.** 135 in.

11.3 STUDY SET

VOCABULARY **Fill in the blanks.**

1. Each term of a ________ sequence is found by multiplying the previous term by the same number.

2. 8, 16, 32, 64, 128, . . . is an example of a ________ sequence. The first ______ is 8 and the common ______ is 2.

3. If a single number is inserted between a and b to form a geometric sequence, the number is called the geometric ______ between a and b.

4. The sum of the terms of a geometric sequence is called a geometric ______. The sum of the terms of an infinite geometric sequence is called an ______ geometric series.

CONCEPTS

5. Write the first three terms of a geometric sequence if $a_1 = 16$ and $r = \frac{1}{4}$.
6. Given the geometric sequence 1, 2, 4, 8, 16, 32, ... Find a_5 and r.
7. Write the formula for a_n, the general term of a geometric sequence.
8. **a.** Write the formula for S_n, the sum of the first n terms of a geometric sequence.

   b. Write the formula for S, the sum of the terms of an infinite geometric sequence, where $|r| < 1$.
9. Which of the following values of r satisfy $|r| < 1$?

   a. $r = \frac{2}{3}$ **b.** $r = -3$

   c. $r = 6$ **d.** $r = -\frac{1}{5}$
10. Write $0.\overline{7}$ as an infinite geometric series:

$$0.\overline{7} = \frac{7}{\square} + \frac{7}{\square} + \frac{7}{\square}$$

NOTATION **Fill in the blanks.**

11. An infinite geometric sequence is of the form

$$a_1,\ a_1r,\ \square,\ a_1r^3,\ \square,\ \ldots$$

12. The first four terms of the sequence defined by $a_n = 4(3)^{n-1}$ are $\square$, $\square$, $\square$, $\square$.
13. To find the common ratio of a geometric sequence, we use the formula $r = \frac{a_{\square}}{a_{\square}}$.
14. S_8 represents the sum of the first $\square$ terms of a geometric sequence.

PRACTICE **Write the first five terms of each geometric sequence with the given properties.**

15. $a_1 = 3, r = 2$
16. $a_1 = -2, r = 2$
17. $a_1 = -5, r = \frac{1}{5}$
18. $a_1 = 8, r = \frac{1}{2}$
19. $a_1 = 2, r > 0$, third term is 32
20. $a_1 = 3$, fourth term is 24
21. $a_1 = -3$, fourth term is -192
22. $a_1 = 2, r < 0$, third term is 50
23. $a_1 = -64, r < 0$, fifth term is -4
24. $a_1 = -64, r > 0$, fifth term is -4
25. $a_1 = -64$, sixth term is -2
26. $a_1 = -81$, sixth term is $\frac{1}{3}$
27. The second term is 10, and the third term is 50.
28. The third term is -27, and the fourth term is 81.
29. Find the tenth term of the geometric sequence with $a_1 = 7$ and $r = 2$.
30. Find the 12th term of the geometric sequence with $a_1 = 64$ and $r = \frac{1}{2}$.
31. Find the first term of the geometric sequence with a common ratio -3 and an eighth term -81.
32. Find the first term of the geometric sequence with a common ratio 2 and a tenth term 384.
33. Find the common ratio of the geometric sequence with a first term -8 and a sixth term $-1{,}944$.
34. Find the common ratio of the geometric sequence with a first term 12 and a sixth term $\frac{3}{8}$.
35. Insert three positive geometric means between 2 and 162.
36. Insert four geometric means between 3 and 96.
37. Insert four geometric means between -4 and $-12{,}500$.
38. Insert three geometric means (two positive and one negative) between -64 and $-1{,}024$.
39. Find the negative geometric mean between 2 and 128.
40. Find the positive geometric mean between 3 and 243.
41. Find the positive geometric mean between 10 and 20.
42. Find the negative geometric mean between 5 and 15.
43. Find a geometric mean, if possible, between -50 and 10.
44. Find a negative geometric mean, if possible, between -25 and -5.

Find the sum of the first *n* terms of each geometric sequence.

45. $2, 6, 18, \ldots; n = 6$

46. $2, -6, 18, \ldots; n = 6$

47. $2, -6, 18, \ldots; n = 5$

48. $2, 6, 18, \ldots; n = 5$

49. $3, -6, 12, \ldots; n = 8$

50. $3, 6, 12, \ldots; n = 8$

51. $3, 6, 12, \ldots; n = 7$

52. $3, -6, 12, \ldots; n = 7$

53. The second term is 1 and the third term is $\frac{1}{5}$; $n = 4$.

54. The second term is 1 and the third term is 4; $n = 5$.

55. The third term is -2 and the fourth term is 1; $n = 6$.

56. The third term is -3 and the fourth term is 1; $n = 5$.

Find the sum of each infinite geometric series, if possible.

57. $8 + 4 + 2 + \cdots$

58. $12 + 6 + 3 + \cdots$

59. $54 + 18 + 6 + \cdots$

60. $45 + 15 + 5 + \cdots$

61. $12 - 6 + 3 - \cdots$

62. $8 - 4 + 2 - \cdots$

63. $-45 + 15 - 5 + \cdots$

64. $-54 + 18 - 6 + \cdots$

65. $\frac{9}{2} + 6 + 8 + \cdots$

66. $-112 - 28 - 7 - \cdots$

67. $-\frac{27}{2} - 9 - 6 - \cdots$

68. $\frac{18}{25} + \frac{6}{5} + 2 + \cdots$

Write each decimal in fraction form. Then check the answer by performing a long division.

69. $0.\overline{1}$

70. $0.\overline{2}$

71. $0.\overline{3}$

72. $0.\overline{4}$

73. $0.\overline{12}$

74. $0.\overline{21}$

75. $0.\overline{75}$

76. $0.\overline{57}$

APPLICATIONS **Use a calculator to help solve each problem.**

77. DECLINING SAVINGS John has \$10,000 in a safety deposit box. Each year, he spends 12% of what is left in the box. How much will be in the box after 15 years?

78. SAVINGS GROWTH Sally has \$5,000 in a savings account earning 12% annual interest. How much will be in her account 10 years from now? (Assume that Sally makes no deposits or withdrawals.)

79. HOUSE APPRECIATION A house appreciates by 6% each year. If the house is worth \$70,000 today, how much will it be worth 12 years from now?

80. BOAT DEPRECIATION A boat that cost \$5,000 when new depreciates at a rate of 9% per year. How much will the boat be worth in 5 years?

81. INSCRIBED SQUARES Each inscribed square in the illustration joins the midpoints of the next larger square. The area of the first square, the largest, is 1. Find the area of the 12th square.

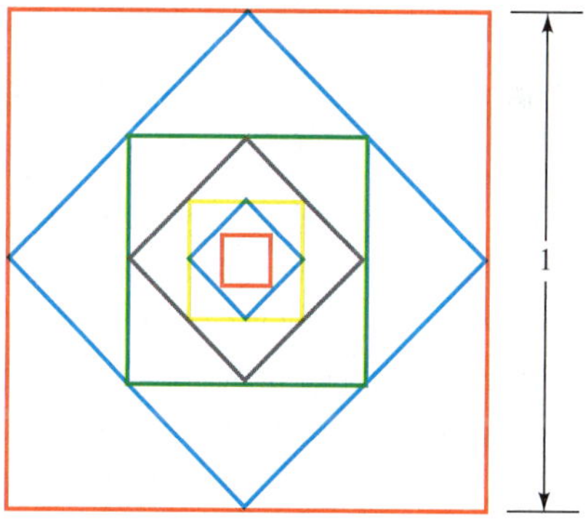

82. GENEALOGY The following family tree spans 3 generations and lists 7 people. How many names would be listed in a family tree that spans 10 generations?

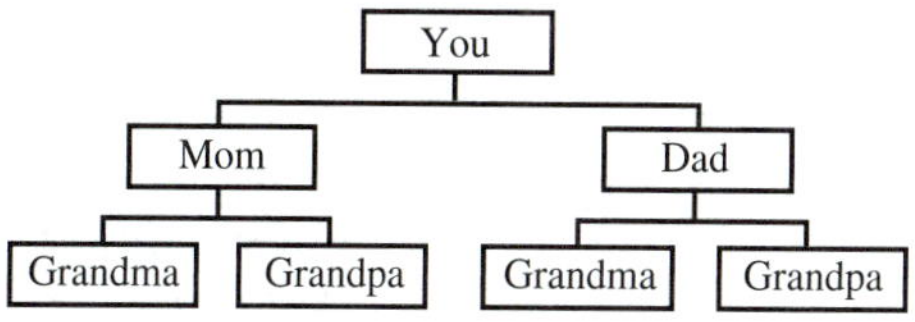

83. BOUNCING BALLS On each bounce, the rubber ball in the illustration rebounds to a height one-half of that from which it fell. Find the total vertical distance the ball travels.

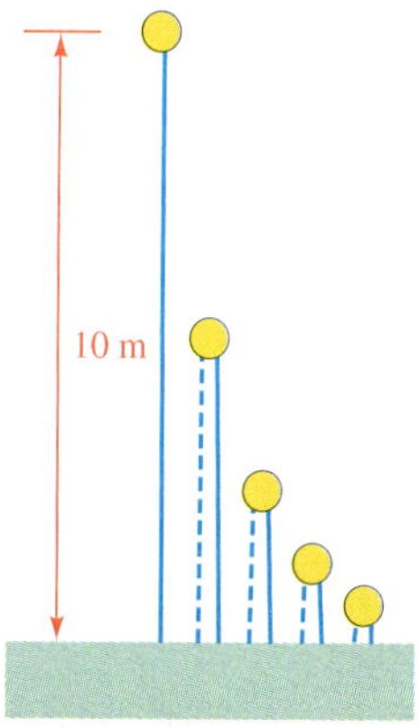

84. BOUNCING BALLS A golf ball is dropped from a height of 12 feet. On each bounce, it returns to a height that is two-thirds of the distance it fell. Find the total vertical distance the ball travels.

85. PEST CONTROL To reduce the population of a destructive moth, biologists release 1,000 sterilized male moths each day into the environment. If 80% of these moths alive one day survive until the next, then after a long time the population of sterile males is the sum of the infinite geometric series

$$1{,}000 + 1{,}000(0.8) + 1{,}000(0.8)^2 + 1{,}000(0.8)^3 + \cdots$$

Find the long-term population.

86. PEST CONTROL If mild weather increases the day-to-day survival rate of the sterile male moths in Exercise 85 to 90%, find the long-term population.

WRITING

87. Describe the real numbers that satisfy $|r| < 1$.

88. Why must the common ratio be less than 1 before an infinite geometric sequence can have a sum?

89. Explain the difference between an arithmetic sequence and a geometric sequence.

90. Why is $1 - \frac{1}{2} + \frac{1}{4} - \frac{1}{8} + \frac{1}{16} - \frac{1}{32} + \ldots$ called an alternating infinite geometric series?

REVIEW **Solve each inequality. Write the solution set using interval notation.**

91. $x^2 - 5x - 6 \le 0$

92. $a^2 - 7a + 12 \ge 0$

93. $\dfrac{x-4}{x+3} > 0$

94. $\dfrac{t^2 + t - 20}{t+2} < 0$

CHALLENGE PROBLEMS

95. If $f(x) = 1 + x + x^2 + x^3 + x^4 + \cdots$, find $f\left(\frac{1}{2}\right)$ and $f\left(-\frac{1}{2}\right)$.

96. Find the sum:

$$\frac{1}{\sqrt{3}} + \frac{1}{3} + \frac{1}{3\sqrt{3}} + \frac{1}{9} + \cdots$$

97. If $a > b > 0$, which is larger: the arithmetic mean between a and b or the geometric mean between a and b?

98. Is there a geometric mean between -5 and 5?

11.4 Permutations and Combinations

- The fundamental counting principle
- Permutations
- Combinations
- Alternative form of the binomial theorem

In this section, we will discuss methods of counting the different ways we can do something like lining up in a row or arranging books on a shelf. These kinds of problems are important in the fields of statistics, insurance, and telecommunications. Although one might think that counting problems are easy to solve, they can be deceptively difficult.

THE FUNDAMENTAL COUNTING PRINCIPLE

When a student goes to the cafeteria for lunch, he has a choice of three different sandwiches (hamburger, hot dog, or ham and cheese) and four different beverages (cola, root beer, water, or milk). His options are shown in the *tree diagram* on the right.

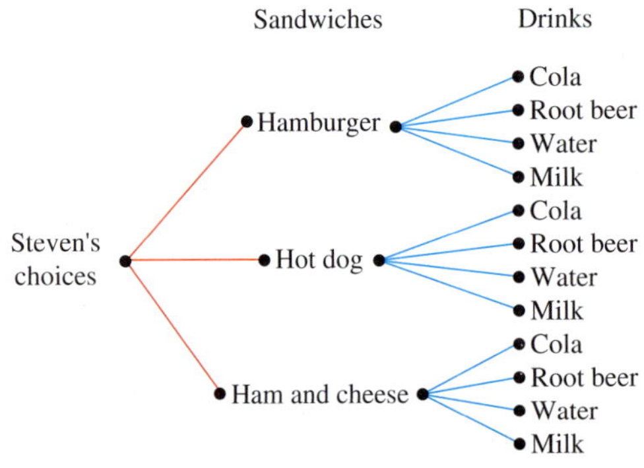

The tree diagram shows that there are a total of 12 different lunches to choose from. One possibility is a hamburger with a cola, and another is a hot dog with milk.

A situation that can have several different outcomes—such as choosing a sandwich—is called an **event.** Choosing a sandwich and choosing a beverage can be thought of as two events. The preceding example illustrates the fundamental counting principle.

Fundamental Counting Principle If event E_1 can be done in m ways, and if (after E_1 has occurred) event E_2 can be done in n ways, then the event "E_1 followed by E_2" can be done in $m \cdot n$ ways.

EXAMPLE 1

Watching television. Before studying for an exam, Taylor plans to watch the evening news and then a situation comedy on television. If she has a choice of 4 news broadcasts and 2 comedies, in how many ways can she choose to watch television?

Solution Let E_1 be the event "watching the news" and E_2 be the event "watching a comedy." Because there are 4 ways to accomplish E_1 and 2 ways to accomplish E_2, the number of choices that Taylor has is $4 \cdot 2 = 8$.

Self Check 1 If Alex has 7 shirts and 5 pairs of pants, how many outfits could he wear?

The fundamental counting principle can be extended to any number of events. In Example 2, we use it to compute the number of ways in which we can arrange objects in a row.

EXAMPLE 2

Arranging books. In how many ways can we arrange 5 books on a shelf?

Solution We can fill the first space with any of the 5 books, the second space with any of the remaining 4 books, the third space with any of the remaining 3 books, the fourth space with any of the remaining 2 books, and the fifth space with the remaining 1 (or last) book. By the fundamental counting principle for events, the number of ways that the books can be arranged is

$$5 \cdot 4 \cdot 3 \cdot 2 \cdot 1 = 120$$

Success Tip

A general guideline is to use the fundamental counting principle to solve problems where consecutive choices are being made.

Self Check 2 In how many ways can 4 people line up in a row?

EXAMPLE 3

Signal flags. If a sailor has 6 flags (each of a different color) to hang on a flagpole, how many different signals can the sailor send by using 4 flags?

Solution The sailor must find the number of arrangements of 4 flags when there are 6 flags to choose from. The sailor can hang any one of the 6 flags in the top position, any one of the remaining 5 flags in the second position, any one of the remaining 4 flags in the third position, and any one of the remaining 3 flags in the lowest position. By the fundamental counting principle for events, the total number of signals that can be sent is

$$6 \cdot 5 \cdot 4 \cdot 3 = 360$$

Self Check 3 How many different signals can the sailor send if each signal uses 3 flags?

PERMUTATIONS

When counting a number of possible arrangements such as books on a shelf or flags on a pole, we are finding the number of **permutations** of those objects. In Example 2, we found that the number of permutations of 5 books, using all 5 of them is 120. In Example 3, we found that the number of permutations of 6 flags, using 4 of them, is 360.

The symbol $P(n, r)$, read as "the number of permutations of n objects taken r at a time," is often used to express permutation problems. In Example 2, we found that $P(5, 5) = 120$. In Example 3, we found that $P(6, 4) = 360$.

EXAMPLE 4

Signal flags. If Sarah has 7 flags (each of a different color) to hang on a flagpole, how many different signals can she send by using 3 flags?

Solution We must find $P(7, 3)$ (the number of permutations of 7 things 3 at a time). In the top position Sarah can hang any of the 7 flags, in the middle position any one of the remaining 6 flags, and in the bottom position any one of the remaining 5 flags. By the fundamental counting principle for events,

$$P(7, 3) = 7 \cdot 6 \cdot 5 = 210$$

Sarah can send 210 signals using only 3 of the 7 flags.

Self Check 4 How many different signals can Sarah send using 4 flags?

Although it is correct to write $P(7, 3) = 7 \cdot 6 \cdot 5$, there is an advantage in changing the form of this answer to obtain a formula for computing $P(7, 3)$:

$$P(7, 3) = 7 \cdot 6 \cdot 5$$

$$= \frac{7 \cdot 6 \cdot 5}{1} \cdot \frac{\mathbf{4 \cdot 3 \cdot 2 \cdot 1}}{\mathbf{4 \cdot 3 \cdot 2 \cdot 1}} \quad \text{Multiply by a form of 1: } \frac{4 \cdot 3 \cdot 2 \cdot 1}{4 \cdot 3 \cdot 2 \cdot 1} = 1.$$

$$= \frac{7!}{4!} \quad \text{Multiply the numerators and the denominators. Use factorial notation.}$$

$$= \frac{7!}{(7-3)!} \quad \text{Write } 4! \text{ as } (7-3)!.$$

The generalization of this idea gives the following formula.

Finding $P(n, r)$ The number of permutations of n objects taken r at a time is given by the formula

$$P(n, r) = \frac{n!}{(n-r)!}$$

EXAMPLE 5

Compute: **a.** $P(8, 2)$, and **b.** $P(n, n)$.

Solution We use the permutation formula $P(n, r) = \dfrac{n!}{(n-r)!}$.

a. $P(8, 2) = \dfrac{8!}{(8-2)!}$ $n = 8$ and $r = 2$.

$= \dfrac{8 \cdot 7 \cdot 6!}{6!}$

$= 8 \cdot 7$

$= 56$

Success Tip

A *permutation* is an ordered arrangement of a given set of objects. To solve counting problems where order is important, use the permutation formula.

b. $P(n, n) = \dfrac{n!}{(n-n)!}$

$= \dfrac{n!}{0!}$

$= \dfrac{n!}{1}$

$= n!$

Self Check 5 Compute: **a.** $P(10, 6)$ and **b.** $P(10, 10)$.

Part b of Example 5 establishes the following formula.

Finding $P(n, n)$ The number of permutations of n objects taken n at a time is $n!$.

$$P(n, n) = n!$$

EXAMPLE 6

Television schedules. **a.** In how many ways can a television executive arrange the Saturday night lineup of six programs if there are 15 programs to choose from? **b.** If there are only six programs to choose from?

Solution **a.** To find the number of permutations of 15 programs 6 at a time, we use the formula $P(n, r) = \frac{n!}{(n-r)!}$ with $n = 15$ and $r = 6$.

$$P(15, 6) = \frac{15!}{(15-6)!}$$

$$= \frac{15 \cdot 14 \cdot 13 \cdot 12 \cdot 11 \cdot 10 \cdot 9!}{9!}$$

$$= 15 \cdot 14 \cdot 13 \cdot 12 \cdot 11 \cdot 10$$

$$= 3{,}603{,}600$$

Success Tip

Problems solved with the permutation formula can also be solved by using the fundamental counting principle.

b. To find the number of permutations of 6 programs 6 at a time, we use the formula $P(n, n) = n!$ with $n = 6$.

$$P(6, 6) = 6! = 720$$

Self Check 6 How many ways are there to arrange the lineup if the executive has 20 programs to choose from?

COMBINATIONS

Suppose that Raul must read 4 books from a reading list of 10 books. The order in which he reads them is not important. For the moment, however, let's assume that order is important and find the number of permutations of 10 things 4 at a time:

$$\begin{aligned} P(10, 4) &= \frac{10!}{(10-4)!} \\ &= \frac{10 \cdot 9 \cdot 8 \cdot 7 \cdot 6!}{6!} \\ &= 10 \cdot 9 \cdot 8 \cdot 7 \\ &= 5{,}040 \end{aligned}$$

If order is important, there are 5,040 ways of choosing 4 books when there are 10 books to choose from. However, because the order in which Raul reads the books does not matter, the previous result of 5,040 is too big. Since there are 24 (or 4!) ways of ordering the 4 books that are chosen, the result of 5,040 is exactly 24 (or 4!) times too big. Therefore, the number of choices that Raul has is the number of permutations of 10 things 4 at a time, divided by 24:

$$\frac{P(10, 4)}{24} = \frac{5{,}040}{24} = 210$$

Raul has 210 ways of choosing 4 books to read from the list of 10 books.

In situations where order is *not* important, we are interested in **combinations,** not permutations. The symbols $C(n, r)$ and $\binom{n}{r}$ both mean the number of combinations of n objects taken r at a time.

Success Tip

A *combination* is a distinct group of objects without regard to their arrangement. To solve counting problems where order is not important, use the combination formula.

If a selection of r books is chosen from a total of n books, the number of possible selections is $C(n, r)$, and there are $r!$ arrangements of the r books in each selection. If we consider the selected books as an ordered grouping, the number of orderings is $P(n, r)$. Therefore, we have

(1) $r! \cdot C(n, r) = P(n, r)$

We can divide both sides of Equation 1 by $r!$ to get the formula for finding $C(n, r)$:

$$C(n, r) = \binom{n}{r} = \frac{P(n, r)}{r!} = \frac{n!}{r!(n-r)!}$$

Finding $C(n, r)$ The number of combinations of n objects taken r at a time is given by

$$C(n, r) = \frac{n!}{r!(n-r)!}$$

EXAMPLE 7

Compute: **a.** $C(8, 5)$ and **b.** $\binom{7}{2}$.

Solution We use the combination formula $C(n, r) = \dfrac{n!}{r!(n-r)!}$.

Notation

The notation $\binom{7}{2}$ means $C(7, 2)$ and is read as "the number of combinations of 7 objects taken 2 at a time."

a. $$C(8, 5) = \frac{8!}{5!(8-5)!} \quad \begin{matrix} n = 8 \\ r = 5 \end{matrix}$$
$$= \frac{8 \cdot 7 \cdot 6 \cdot 5!}{5! \cdot 3!}$$
$$= 8 \cdot 7$$
$$= 56$$

b. $$\binom{7}{2} = \frac{7!}{2!(7-2)!} \quad \begin{matrix} n = 7 \\ r = 2 \end{matrix}$$
$$= \frac{7 \cdot 6 \cdot 5!}{2 \cdot 1 \cdot 5!}$$
$$= 21$$

Self Check 7 Compute: **a.** $C(9, 6)$ and **b.** $C(10, 10)$.

EXAMPLE 8

Choosing committees. If 15 students want to pick a committee of 4 students to plan a party, how many different committees are possible?

Solution Since the ordering of people on each possible committee is not important, we find the number of combinations of 15 people 4 at a time:

$$C(15, 4) = \frac{15!}{4!(15-4)!}$$
$$= \frac{15 \cdot 14 \cdot 13 \cdot 12 \cdot 11!}{4 \cdot 3 \cdot 2 \cdot 1 \cdot 11!}$$
$$= \frac{15 \cdot 14 \cdot 13 \cdot 12}{4 \cdot 3 \cdot 2 \cdot 1}$$
$$= 1{,}365$$

There are 1,365 possible committees.

Success Tip

If, instead, four officers (president, vice president, secretary, and treasurer) were to be selected, order would be important and we would use the permutation formula.

Self Check 8 In how many ways can 20 students pick a committee of 5 students to plan a party?

EXAMPLE 9

Choosing subcommittees. A committee in Congress consists of 10 Democrats and 8 Republicans. In how many ways can a subcommittee be chosen if it is to contain 5 Democrats and 4 Republicans?

Solution There are $C(10, 5)$ ways of choosing the 5 Democrats and $C(8, 4)$ ways of choosing the 4 Republicans. By the fundamental counting principle for events, there are $C(10, 5) \cdot C(8, 4)$ ways of choosing the subcommittee:

$$C(10, 5) \cdot C(8, 4) = \frac{10!}{5!(10-5)!} \cdot \frac{8!}{4!(8-4)!}$$
$$= \frac{10 \cdot 9 \cdot 8 \cdot 7 \cdot 6 \cdot 5!}{120 \cdot 5!} \cdot \frac{8 \cdot 7 \cdot 6 \cdot 5 \cdot 4!}{24 \cdot 4!}$$
$$= \frac{10 \cdot 9 \cdot 8 \cdot 7 \cdot 6}{120} \cdot \frac{8 \cdot 7 \cdot 6 \cdot 5}{24}$$
$$= 17{,}640$$

There are 17,640 possible subcommittees.

Success Tip

To help determine a method of solution, always ask yourself: Does the order in which the objects are chosen matter?

Self Check 9 In how many ways can a subcommittee be chosen if it is to contain 4 members from each party?

ALTERNATIVE FORM OF THE BINOMIAL THEOREM

We have seen that the expansion of $(x + y)^3$ is

$$(x + y)^3 = 1x^3 + 3x^2y + 3xy^2 + 1y^3$$

and that the coefficients can be written as

$$\binom{3}{0} = 1, \quad \binom{3}{1} = 3, \quad \binom{3}{2} = 3, \quad \text{and} \quad \binom{3}{3} = 1$$

The Language of Algebra

Here, the coefficients of the terms of an expansion are expressed in the form $\binom{n}{r}$. In this context, we call $\binom{3}{0}$, $\binom{3}{1}$, $\binom{3}{2}$, and $\binom{3}{3}$ **binomial coefficients.**

Combining these facts gives the following way of writing the expansion of $(x + y)^3$:

$$(x + y)^3 = \binom{3}{0}x^3 + \binom{3}{1}x^2y + \binom{3}{2}xy^2 + \binom{3}{3}y^3$$

Likewise, we have

$$(x + y)^4 = \binom{4}{0}x^4 + \binom{4}{1}x^3y + \binom{4}{2}x^2y^2 + \binom{4}{3}xy^3 + \binom{4}{4}y^4$$

The generalization of this idea enables us to write the binomial theorem in an alternative form using combinatorial notation.

The Binomial Theorem For any positive integer n,

$$(a + b)^n = \binom{n}{0}a^n + \binom{n}{1}a^{n-1}b + \binom{n}{2}a^{n-2}b^2 + \dots + \binom{n}{r}a^{n-r}b^r + \dots + \binom{n}{n}b^n$$

EXAMPLE 10

Use the alternative form of the binomial theorem to expand $(x + y)^6$.

Solution

$$(x + y)^6 = \binom{6}{0}x^6 + \binom{6}{1}x^5y + \binom{6}{2}x^4y^2 + \binom{6}{3}x^3y^3 + \binom{6}{4}x^2y^4 + \binom{6}{5}xy^5 + \binom{6}{6}y^6$$
$$= x^6 + 6x^5y + 15x^4y^2 + 20x^3y^3 + 15x^2y^4 + 6xy^5 + y^6$$

Self Check 10 Use the alternative form of the binomial theorem to expand $(a + b)^5$.

The alternative form for finding a specific term of an expansion is as follows.

Finding a Specific Term The $(r + 1)$st term of the expansion of $(a + b)^n$ is $\binom{n}{r}a^{n-r}b^r$.

EXAMPLE 11

Find the fifth term of the expansion of $(2x - y)^7$.

Solution Since $r + 1 = 5$, it follows that $r = 4$. Also, we see that $n = 7$, $a = 2x$, and $b = -y$. The fifth term of the expansion is

$$\binom{n}{r}a^{n-r}b^r = \binom{7}{4}(2x)^{7-4}(-y)^4$$

$$= \frac{7!}{4!(7-4)!}(2x)^3y^4$$

$$= 280x^3y^4 \qquad \frac{7!}{4!(7-4)!} = 35,\ 2^3 = 8, \text{ and } 35 \cdot 8 = 280.$$

Success Tip

Remember that r is always 1 less than the number of the term that you are finding.

Self Check 11 Find the third term of the expansion of $(2a - 3y)^5$.

Answers to Self Checks **1.** 35 **2.** 24 **3.** 120 **4.** 840 **5. a.** 151,200, **b.** 3,628,800 **6.** 27,907,200 **7. a.** 84, **b.** 1 **8.** 15,504 **9.** 14,700 **10.** $a^5 + 5a^4b + 10a^3b^2 + 10a^2b^3 + 5ab^4 + b^5$ **11.** $720a^3y^2$

11.4 STUDY SET

VOCABULARY Fill in the blanks.

1. A ______ diagram like that shown below can be used to count the number of possible outcomes.

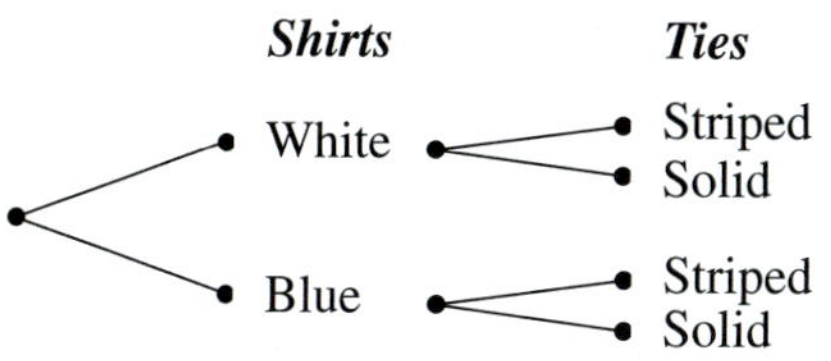

2. Using the fundamental ________ principle, we can determine the number of ways two events can occur.

3. A __________ is an arrangement of objects.

4. When selecting objects when order is not important, we count __________.

CONCEPTS Fill in the blanks.

5. If an event E_1 can be done in p ways and (after it occurs) a second event E_2 can be done in q ways, the event E_1 followed by E_2 can be done in ______ ways.

6. The symbol ______ means the number of permutations of n things taken r at a time.

7. The formula for the number of permutations of n things taken r at a time is ______.

8. $P(n, n) =$ ______

9. The symbol $C(n, r)$ or $\binom{\ }{\ }$ means the number of __________ of n things taken r at a time.

10. The formula for the number of combinations of n things taken r at a time is ______.

11. $0! =$ ______

12. $1! =$ ______

NOTATION Complete each solution.

13. $P(6, 2) = \frac{\square}{(6-2)!}$

$= \frac{6 \cdot 5 \cdot 4!}{\square}$

$= 6 \cdot \square$

$= 30$

14. $C(6, 2) = \frac{\square}{\square(6-2)!}$

$= \frac{6 \cdot 5 \cdot 4!}{2 \cdot 1 \cdot \square}$

$= \square \cdot 5$

$= 15$

PRACTICE Evaluate each permutation or combination.

15. $P(3, 3)$

16. $P(4, 4)$

17. $P(5, 3)$

18. $P(3, 2)$

19. $P(2, 2) \cdot P(3, 3)$

20. $P(3, 2) \cdot P(3, 3)$

21. $\frac{P(5, 3)}{P(4, 2)}$

22. $\frac{P(6, 2)}{P(5, 4)}$

23. $\frac{P(6, 2) \cdot P(7, 3)}{P(5, 1)}$

24. $\frac{P(8, 3)}{P(5, 3) \cdot P(4, 3)}$

25. $C(5, 3)$

26. $C(5, 4)$

27. $\binom{6}{3}$

28. $\binom{6}{4}$

29. $\binom{5}{4}\binom{5}{3}$

30. $\binom{6}{5}\binom{6}{4}$

31. $\dfrac{C(38, 37)}{C(19, 18)}$

32. $\dfrac{C(25, 23)}{C(40, 39)}$

33. $C(12, 0) \cdot C(12, 12)$

34. $\dfrac{C(8, 0)}{C(8, 1)}$

35. $C(n, 2)$

36. $C(n, 3)$

Use the alternative form of the binomial theorem to expand each expression.

37. $(x + y)^4$

38. $(c - d)^5$

39. $(2x + y)^3$

40. $(2x + y)^4$

41. $(3x - 2)^4$

42. $(3 - x^2)^3$

Find the indicated term of each binomial expansion.

43. $(x - 5y)^5$; fourth term

44. $(2x - y)^5$; third term

45. $(x^2 - y^3)^4$; second term

46. $(x^3 - y^2)^4$; fourth term

APPLICATIONS

47. PLANNING AN EVENING Kristy plans to go to dinner and see a movie. In how many ways can she arrange her evening if she has a choice of 5 movies and 7 restaurants?

48. TRAVEL CHOICES Paula has 5 ways to travel from New York to Chicago, 3 ways to travel from Chicago to Denver, and 4 ways to travel from Denver to Los Angeles. How many choices are available if she travels from New York to Los Angeles?

49. LICENSE PLATES How many 6-digit license plates can be manufactured? Note that there are 10 choices—0, 1, 2, 3, 4, 5, 6, 7, 8, 9—for each digit.

50. LICENSE PLATES How many 6-digit license plates can be manufactured if no digit can be repeated?

51. LICENSE PLATES How many 6-digit license plates can be manufactured if no license can begin with 0 and if no digit can be repeated?

52. LICENSE PLATES How many license plates can be manufactured with 2 letters followed by 4 digits?

53. PHONE NUMBERS How many 7-digit phone numbers are available in area code 815 if no phone number can begin with 0 or 1?

54. PHONE NUMBERS How many 10-digit phone numbers are available if area codes of 000 and 911 cannot be used and if no local number can begin with 0 or 1?

55. LINING UP In how many ways can 6 people be placed in a line?

56. ARRANGING BOOKS In how many ways can 7 books be placed on a shelf?

57. ARRANGING BOOKS In how many ways can 4 novels and 5 biographies be arranged on a shelf if the novels are placed first?

58. BALLOTS In how many ways can 6 candidates for mayor and 4 candidates for the county board be arranged on a ballot if all of the candidates for mayor must be placed first?

59. LOCKS How many permutations does a combination lock have if each combination has 3 numbers, no 2 numbers of any combination are equal, and the lock has 25 numbers?

60. LOCKS How many permutations does a combination lock have if each combination has 3 numbers, no 2 numbers of any combination are equal, and the lock has 50 numbers?

61. ARRANGING APPOINTMENTS The receptionist at a dental office has only 3 appointment times available before next Tuesday, and 10 patients have toothaches. In how many ways can the receptionist fill those appointments?

62. COMPUTERS In many computers, a *word* consists of 32 *bits*—a string of thirty-two 1's and 0's. How many different words are possible?

63. PALINDROMES A palindrome is any word, such as *madam* or *radar,* that reads the same backward and forward. How many 5-digit numerical palindromes (like 13531) are there? (*Hint:* A leading 0 would be dropped.)

64. CALL LETTERS The call letters of a U.S. commercial radio station have 3 or 4 letters, and the first is either a W or a K. How many radio stations could this system support?

65. PICNICS A class of 14 students wants to pick a committee of 3 students to plan a picnic. How many committees are possible?

66. CHOOSING BOOKS Jeffrey must read 3 books from a reading list of 15 books. How many choices does he have?

67. COMMITTEES The number of 3-person committees that can be formed from a group of persons is 10. How many persons are in the group?

68. COMMITTEES The number of 3-person committees that can be formed from a group of persons is 20. How many persons are in the group?

69. LOTTERIES In one state lottery, anyone who picks the correct 6 numbers (in any order) wins. With the numbers 0 through 99 available, how many choices are possible?

70. TAKING TESTS The instructions on a test read, *Answer any 10 of the following 15 questions. Then choose one of the remaining questions for homework, and turn in its solution tomorrow.* In how many ways can the questions be chosen?

71. COMMITTEES In how many ways can we select a committee of 2 men and 2 women from a group containing 3 men and 4 women?

72. COMMITTEES In how many ways can we select a committee of 3 men and 2 women from a group containing 5 men and 3 women?

73. CHOOSING CLOTHES In how many ways can we select 2 shirts and 3 neckties from a group of 12 shirts and 10 neckties?

74. CHOOSING CLOTHES In how many ways can we select 5 dresses and 2 coats from a wardrobe containing 9 dresses and 3 coats?

WRITING

75. State the fundamental counting principle.

76. Explain why *permutation lock* would be a better name for a combination lock.

REVIEW

Solve each equation. Give the solution to four decimal places.

77. $2^{x+1} = 3^x$

78. $5^{x-3} = 3^{2x}$

79. $e^{3x} = 9$

80. $8^{x^2} = 9^x$

CHALLENGE PROBLEMS

81. How many ways could 5 people stand in line if 2 people insist on standing together?

82. How many ways could 5 people stand in line if 2 people refuse to stand next to each other?

11.5 Probability

- Probability

The probability that an event will occur is a measure of the likelihood of that event. A tossed coin, for example, can land in two ways, either heads or tails. Because one of these two equally likely outcomes is heads, we expect that out of several tosses, about half will be heads. We say that the probability of obtaining heads in a single toss of the coin is $\frac{1}{2}$.

If records show that out of 100 days with weather conditions like today's, 30 have received rain, the weather service will report, "There is a $\frac{30}{100}$ or 30% probability of rain today."

PROBABILITY

Activities such as tossing a coin, rolling a die, drawing a card, and predicting rain are called **experiments.** For any experiment, a list of all possible outcomes is called a **sample space.** For example, the sample space S for the experiment of tossing two coins is the set

$$S = \{(H, H), (H, T), (T, H), (T, T)\}$$ There are four possible outcomes.

where the ordered pair (H, T) represents the outcome "heads on the first coin and tails on the second coin."

An **event** is a subset of the sample space of an experiment. For example, if E is the event "getting at least one heads" in the experiment of tossing two coins, then

$E = \{(H, H), (H, T), (T, H)\}$ There are 3 ways of getting at least one heads.

Because the outcome of getting at least one heads can occur in 3 out of 4 possible ways, we say that the **probability** of E is $\frac{3}{4}$, and we write

$$P(E) = P(\text{at least one heads}) = \frac{3}{4}$$

Probability of an Event If a sample space of an experiment has n distinct and equally likely outcomes and E is an event that occurs in s of those ways, the **probability of *E*** is

$$P(E) = \frac{s}{n}$$

Since $0 \le s \le n$, it follows that $0 \le \frac{s}{n} \le 1$. This implies that all probabilities have value from 0 to 1. If an event cannot happen, its probability is 0. If an event is certain to happen, its probability is 1.

EXAMPLE 1

List the sample space of the experiment "rolling two dice a single time."

Solution We can list ordered pairs and let the first number be the result on the first die and the second number the result on the second die. The sample space S is the following set of ordered pairs:

(1, 1) (1, 2) (1, 3) (1, 4) (1, 5) (1, 6)
(2, 1) (2, 2) (2, 3) (2, 4) (2, 5) (2, 6)
(3, 1) (3, 2) (3, 3) (3, 4) (3, 5) (3, 6)
(4, 1) (4, 2) (4, 3) (4, 4) (4, 5) (4, 6)
(5, 1) (5, 2) (5, 3) (5, 4) (5, 5) (5, 6)
(6, 1) (6, 2) (6, 3) (6, 4) (6, 5) (6, 6)

By counting, we see that the experiment has 36 equally likely possible outcomes.

Self Check 1 How many pairs in the sample space have a sum of 4?

EXAMPLE 2

Find the probability of the event "rolling a sum of 7 on one roll of two dice."

Solution In the sample space listed in Example 1, the following ordered pairs give a sum of 7:

$$\{(1, 6), (2, 5), (3, 4), (4, 3), (5, 2), (6, 1)\}$$

Since there are 6 ordered pairs whose numbers give a sum of 7 out of a total of 36 equally likely outcomes, we have

$$P(E) = P(\text{rolling a 7}) = \frac{s}{n} = \frac{6}{36} = \frac{1}{6}$$

Self Check 2 Find the probability of rolling a sum of 4.

A standard playing deck of 52 cards has two red suits, hearts and diamonds, and two black suits, clubs and spades. Each suit has 13 cards, including the ace, king, queen, jack, and cards numbered from 2 to 10. We will refer to a standard deck of cards in many examples and exercises.

EXAMPLE 3

Find the probability of drawing an ace on one draw from a standard card deck.

Solution Since there are 4 aces in the deck, the number of favorable outcomes is $s = 4$. Since there are 52 cards in the deck, the total number of possible outcomes is $n = 52$. The probability of drawing an ace is the ratio of the number of favorable outcomes to the number of possible outcomes.

$$P(\text{an ace}) = \frac{s}{n} = \frac{4}{52} = \frac{1}{13}$$

The probability of drawing an ace is $\frac{1}{13}$.

Self Check 3 Find the probability of drawing a red ace on one draw from a standard card deck.

EXAMPLE 4

Find the probability of drawing 5 cards, all hearts, from a standard card deck.

Solution The number of ways we can draw 5 hearts from the 13 hearts is $C(13, 5)$, the number of combinations of 13 things taken 5 at a time. The number of ways to draw 5 cards from the deck is $C(52, 5)$, the number of combinations of 52 things taken 5 at a time. The probability of drawing 5 hearts is the ratio of the number of favorable outcomes to the number of possible outcomes.

$$P(\text{5 hearts}) = \frac{s}{n} = \frac{C(13, 5)}{C(52, 5)}$$

$$P(\text{5 hearts}) = \frac{\frac{13!}{5!8!}}{\frac{52!}{5!47!}}$$

$$= \frac{13!}{\overset{1}{\cancel{5!}}\,8!} \cdot \frac{\overset{1}{\cancel{5!}}\,47!}{52!}$$

$$= \frac{13 \cdot 12 \cdot 11 \cdot 10 \cdot 9 \cdot 8!}{8!} \cdot \frac{47!}{52 \cdot 51 \cdot 50 \cdot 49 \cdot 48 \cdot 47!}$$

$$= \frac{13 \cdot 12 \cdot 11 \cdot 10 \cdot 9}{52 \cdot 51 \cdot 50 \cdot 49 \cdot 48}$$

$$= \frac{33}{66{,}640}$$

The probability of drawing 5 hearts is $\frac{33}{66{,}640}$.

Self Check 4 Find the probability of drawing 6 cards, all diamonds, from a standard card deck.

Answers to Self Checks 1. 3 2. $\frac{1}{12}$ 3. $\frac{1}{26}$ 4. $\frac{33}{391,510}$

11.5 STUDY SET

VOCABULARY **Fill in the blanks.**

1. An __________ is any activity for which the outcome is uncertain.
2. A list of all possible outcomes for an experiment is called a ________ ________.

CONCEPTS **Fill in the blanks.**

3. The probability of an event E is defined as $P(E) = \quad$.
4. If an event is certain to happen, its probability is .
5. If an event cannot happen, its probability is .
6. All probability values are between and , inclusive.

NOTATION **Complete each solution.**

7. Find the probability of drawing a black face card from a standard deck.
   a. The number of black face cards is .
   b. The number of cards in the deck is .
   c. The probability is or .
8. Find the probability of drawing 4 aces from a standard card deck.
   a. The number of ways to draw 4 aces from 4 aces is $C(4, 4) = \quad$.
   b. The number of ways to draw 4 cards from 52 cards is $C(52, 4) = \quad$.
   c. The probability is .

PRACTICE **List the sample space of each experiment.**

9. Rolling one die and tossing one coin
10. Tossing three coins
11. Selecting a letter of the alphabet
12. Picking a one-digit number

An ordinary die is rolled once. Find the probability of each event.

13. Rolling a 2
14. Rolling a number greater than 4
15. Rolling a number larger than 1 but less than 6
16. Rolling an odd number

Balls numbered from 1 to 42 are placed in a container and stirred. If one is drawn at random, find the probability of each result.

17. The number is less than 20.
18. The number is less than 50.
19. The number is a prime number.
20. The number is less than 10 or greater than 40.

Refer to the following spinner. If the spinner is spun, find the probability of each event. Assume that the spinner never stops on a line.

21. The spinner stops on red.
22. The spinner stops on green.
23. The spinner stops on brown.
24. The spinner stops on yellow.

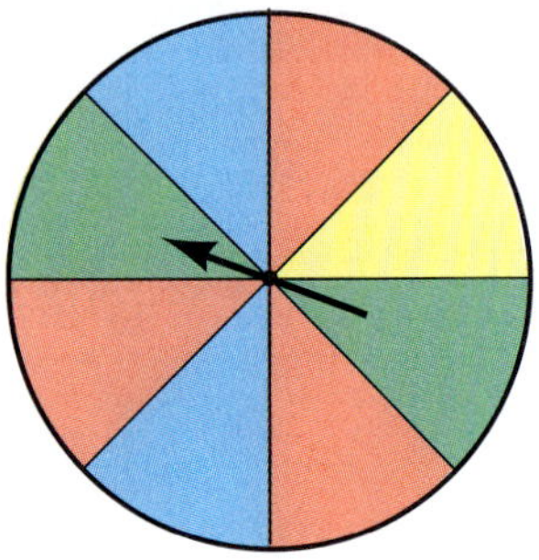

Find the probability of each event.

25. Rolling a sum of 4 on one roll of two dice
26. Drawing a diamond on one draw from a card deck
27. Drawing a red egg from a basket containing 5 red eggs and 7 blue eggs
28. Drawing a yellow egg from a basket containing 5 red eggs and 7 yellow eggs
29. Drawing 6 diamonds from a standard card deck without replacing the cards after each draw
30. Drawing 5 aces from a standard card deck without replacing the cards after each draw

31. Drawing 5 clubs from the black cards in a standard card deck

32. Drawing a face card from a standard card deck

Assume that the probability that an airplane engine will fail a test is $\frac{1}{2}$ and that the aircraft in question has 4 engines. In Exercises 34–38, find each probability.

33. Construct a sample space for the test.

34. All engines will survive the test.

35. Exactly 1 engine will survive.

36. Exactly 2 engines will survive.

37. Exactly 3 engines will survive.

38. No engines will survive.

39. Find the sum of the probabilities in Exercises 34 through 38.

A survey of 282 people is taken to determine the opinions of doctors, teachers, and lawyers on a proposed piece of legislation, with the results shown in the table. A person is chosen at random from those surveyed. Refer to the table to find each probability.

40. The person favors the legislation.

41. A doctor opposes the legislation.

42. A person who opposes the legislation is a lawyer.

	Number that favor	Number that oppose	Number with no opinion	Total
Doctors	70	32	17	119
Teachers	83	24	10	117
Lawyers	23	15	8	46
Total	176	71	35	282

APPLICATIONS

43. QUALITY CONTROL In a batch of 10 tires, 2 are known to be defective. If 4 tires are chosen at random, find the probability that all 4 tires are good.

44. MEDICINE Out of a group of 9 patients treated with a new drug, 4 suffered a relapse. If 3 patients are selected at random from the group of 9, find the probability that none of the 3 patients suffered a relapse.

WRITING

45. Explain why all probability values range from 0 to 1.

46. Explain the concept of probability.

REVIEW **Solve each equation.**

47. $5^{4x} = \frac{1}{125}$

48. $8^{-x+1} = \frac{1}{64}$

49. $2^{x^2-2x} = 8$

50. $3^{x^2-3x} = 81$

51. $3^{x^2+4x} = \frac{1}{81}$

52. $7^{x^2+3x} = \frac{1}{49}$

CHALLENGE PROBLEMS **If $P(A)$ represents the probability of event A, and $P(B \mid A)$ represents the probability that event B will occur after event A, then**

$$P(A \text{ and } B) = P(A) \cdot P(B \mid A)$$

53. In a school, 30% of the students are gifted in math and 10% are gifted in art and math. If a student is gifted in math, find the probability that the student is also gifted in art.

54. The probability that a person owns a luxury car is 0.2, and the probability that the owner of such a car also owns a second car is 0.7. Find the probability that a person chosen at random owns both a luxury car and a second car.

ACCENT ON TEAMWORK

SEATING ARRANGEMENTS

Overview: In this activity, you are to use permutations to determine the number of possible seating arrangements for 4 people.

Instructions: Form groups of 4 students.

a. If possible, arrange your desks in a straight row to help you visualize the following problem: In how many ways can 4 people be seated on a bench?

b. If possible, now arrange your desks in a circle to help you visualize this next situation.

Suppose the same 4 people are seated around a circular table. Some of the possible seating arrangements are really the same, because the relative position of each person to the others is not different. For an example of this, see the illustration below. If we agree not to count any of the duplicate arrangements, in how many different ways can 4 people be seated around a circular table?

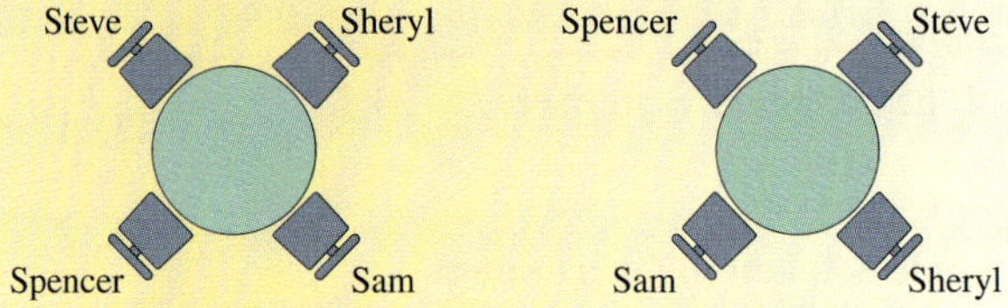

THROWING DICE

Overview: In this activity, you will learn more about probability.

Instructions: Form groups of 2 or 3 students. Your group will need a pair of standard dice.

a. Suppose two dice are rolled, and the sum of the number of dots on the top face is recorded. Complete the table below, which gives the number of ways each sum can occur. For example, there are 3 ways to obtain a sum of 4: a ⚀ on die 1 and a ⚂ on die 2, a ⚂ on die 1 and a ⚀ on die 2, and a ⚁ on die 1 and a ⚁ on die 2.

Sum	2	3	4	5	6	7	8	9	10	11	12
Number of ways			3								

b. Use the table to determine the probability of obtaining a sum of 7 or 11 on one roll of the dice.

c. One-at-a-time, each member of the group should roll the dice until they get a 7 or 11. Keep track of the number of rolls it takes to get one of these outcomes. When finished, compare your results with those of the other members of the class. Which student got 7 or 11 in the least number of rolls? Which student needed the greatest number of rolls to get a 7 or 11?

KEY CONCEPT: THE LANGUAGE OF ALGEBRA

One of the keys to becoming a good algebra student is to know the vocabulary of algebra. Match each instruction in column I with the most appropriate item in column II. Each letter in column II is used only once.

	Column I	*Column II*
___	**1.** Use the FOIL method.	**a.** 2,300,000,000
___	**2.** Apply a rule for exponents to simplify.	**b.** e^3
___	**3.** Add the rational expressions.	**c.** $f(x) = x^2 + 1$ and $g(x) = 5 - 3x$
___	**4.** Rationalize the denominator.	**d.** $-2x(3x^2 - 4x + 8)$
___	**5.** Factor completely.	**e.** $4x - 7 > -3x - 7$
___	**6.** Evaluate the expression for $a = -1$ and $b = -6$.	**f.** $\begin{cases} 2x = y - 5 \\ x + y = -1 \end{cases}$
___	**7.** Express in lowest terms.	**g.** $(x^2 - 5)(x^2 + 3)$
___	**8.** Solve for t.	**h.** $(x + 2)(x - 10) = 0$
___	**9.** Combine like terms.	**i.** $\dfrac{x - 1}{2x^2} + \dfrac{x + 1}{8x}$
___	**10.** Remove parentheses.	**j.** $\sqrt{4x^2}$
___	**11.** Solve the system by graphing.	**k.** $\ln 6 + \ln x$
___	**12.** Find $f(g(x))$.	**l.** $\dfrac{10}{\sqrt{6} - \sqrt{2}}$
___	**13.** Solve using the quadratic formula.	**m.** $2x - 8 + 6y - 14$
___	**14.** Identify the base and the exponent.	**n.** $(3, -2)$ and $(0, -5)$
___	**15.** Write without a radical symbol.	**o.** $x^n \cdot x^{3n}$
___	**16.** Write the equation of the line having the given slope and y-intercept.	**p.** $\dfrac{4x^2y}{16xy}$
___	**17.** Solve the inequality.	**q.** $h(x) = 10^x$
___	**18.** Complete the square to make a perfect square trinomial.	**r.** $\log_2 8 = 3$
___	**19.** Find the slope of the line passing through the given points.	**s.** $m = \dfrac{2}{3}$ and passes through $(0, 2)$
___	**20.** Use a property of logarithms to simplify.	**t.** $2, 6, 18, \ldots$
___	**21.** Set each factor equal to zero and solve for x.	**u.** $3y^3 - 243b^6$
___	**22.** State the solution of the compound inequality using interval notation.	**v.** $x + 7 \geq 0$ and $-x < -1$
___	**23.** Find the inverse function, $h^{-1}(x)$.	**w.** $x^2 - 3x - 4 = 0$
___	**24.** Write using scientific notation.	**x.** $Rt = cd + 2t$
___	**25.** Write the logarithmic statement in exponential form.	**y.** $-2\pi a^2 b - 3b^3$
___	**26.** Find the sum of the first 6 terms of the sequence.	**z.** $x^2 + 4x$

CHAPTER REVIEW

SECTION 11.1 The Binomial Theorem

CONCEPTS

Pascal's triangle gives the coefficients of the terms of the expansion of $(a + b)^n$.

The symbol $n!$ (*n factorial*) is defined as

$n! = n(n - 1)(n - 2) \cdots 2 \cdot 1$

$1! = 1$ and $0! = 1$

$n(n - 1)! = n!$

The binomial theorem:

$$(a + b)^n = a^n + \frac{n!}{1!(n-1)!}a^{n-1}b + \frac{n!}{2!(n-2)!}a^{n-2}b^2 + \cdots + b^n$$

The $(r + 1)$st term of the expansion of $(a + b)^n$ is

$$\frac{n!}{r!(n-r)!}a^{n-r}b^r$$

Remember that r is always 1 less than the number of the term that you are finding.

REVIEW EXERCISES

1. Complete Pascal's triangle. List the row that gives the coefficients for the expansion of $(a + b)^5$.

1
___ 1
1 2 1
1 ___ 3 ___
___ 4 6 4 1
1 5 ___ 10 5 1
1 6 15 20 ___ 6 1

2. Consider the expansion of $(a + b)^{12}$.

a. How many terms does the expansion have?

b. For each term, what is the sum of the exponents on a and b?

c. What is the first term? What is the last term?

d. How do the exponents on a and b change from term to term?

Evaluate each expression.

3. $(4!)(3!)$

4. $\frac{5!}{3!}$

5. $\frac{6!}{2!(6-2)!}$

6. $\frac{12!}{3!(12-3)!}$

7. $(n - n)!$

8. $\frac{8!}{7!}$

Use the binomial theorem to find each expansion.

9. $(x + y)^5$

10. $(x - y)^9$

11. $(4x - y)^3$

12. $\left(\frac{c}{2} + \frac{d}{3}\right)^4$

Find the specified term in each expansion.

13. $(x + y)^4$; third term

14. $(x - y)^6$; fourth term

15. $(3x - 4y)^3$; second term

16. $(u^2 - v^3)^5$; fifth term

SECTION 11.2 Arithmetic Sequences and Series

An *arithmetic sequence:*

$a_1, a_1 + d, a_1 + 2d, \ldots, a_1 + (n-1)d, \ldots$

where a_1 is the first term, d is the common difference, and $a_n = a_1 + (n-1)d$

17. Find the first four terms of the sequence defined by $a_n = 2n - 4$.

18. Find the eighth term of an arithmetic sequence whose first term is 7 and whose common difference is 5.

19. Write the first five terms of the arithmetic sequence whose ninth term is 242 and whose seventh term is 212.

20. The first three terms of an arithmetic sequence are 6, −6, and −18. Find the 101th term.

21. Find the common difference of an arithmetic sequence if its 1st term is −515 and the 23rd term is −625.

If numbers are inserted between two given numbers a and b to form an arithmetic sequence, the inserted numbers are *arithmetic means* between a and b.

22. Find two arithmetic means between 8 and 25.

The sum of the first n terms of an arithmetic sequence is given by

$$S_n = \frac{n(a_1 + a_n)}{2}$$

23. Find the sum of the first ten terms of the sequence $9, 6\frac{1}{2}, 4, \ldots$.

24. Find the sum of the first 28 terms of an arithmetic sequence if the second term is 6 and the sixth term is 22.

Find each sum.

25. $\sum_{k=4}^{6} \frac{1}{2}k$

26. $\sum_{k=2}^{5} 7k^2$

27. $\sum_{k=1}^{4} (3k - 4)$

28. $\sum_{k=10}^{10} 36k$

Summation notation involves the Greek letter sigma Σ. It designates the sum of terms.

29. What is the sum of the first 100 natural numbers?

30. SEATING The illustration shows the first 2 of a total of 30 rows of seats in an amphitheater. The number of seats in each row forms an arithmetic sequence. Find the total number of seats.

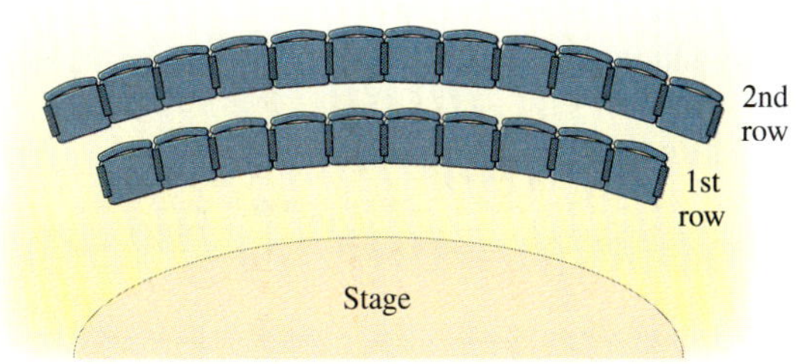

SECTION 11.3 Geometric Sequences and Series

A *geometric sequence:*

$a_1, a_1r, a_1r^2, a_1r^3, \ldots, a_1r^{n-1}, \ldots$

where a_1 is the first term, r is the common ratio, and $a_n = a_1r^{n-1}$.

31. Find the sixth term of a geometric sequence with a first term of $\frac{1}{8}$ and a common ratio of 2.

32. Write the first five terms of the geometric sequence whose fourth term is 3 and whose fifth term is $\frac{3}{2}$.

33. Find the first term of a geometric sequence if it has a common ratio of −3 and the ninth term is 243.

If numbers are inserted between a and b to form a geometric sequence, the inserted numbers are *geometric means* between a and b.

The sum of the first n terms of a geometric sequence:

$$S_n = \frac{a_1 - a_1 r^n}{1 - r} \quad r \neq 1$$

The sum of the terms of an infinite geometric sequence is given by:

$$S = \frac{a_1}{1 - r} \quad |r| < 1$$

34. Find two geometric means between -6 and 384.

35. Find the sum of the first seven terms of the sequence 162, 54, 18, ...

36. Find the sum of the first eight terms of the sequence $\frac{1}{8}, -\frac{1}{4}, \frac{1}{2}, \ldots$.

37. FEEDING BIRDS Tom has 50 pounds of birdseed stored in his garage. Each month, he uses 25% of what is left in the bag to feed the birds in his yard. How much birdseed will be left in 12 months?

38. Find the sum of the infinite geometric sequence 25, 20, 16, ...

39. Change the decimal $0.\overline{05}$ to a common fraction.

40. WHAM-O TOYS Tests have found that 1998 Superballs© rebound $\frac{9}{10}$ of the distance from which they are dropped. If a Superball© is dropped from a height of 10 feet, and if it could bounce forever, what total distance would it travel?

SECTION 11.4 Permutations and Combinations

The fundamental counting principle for events: If E_1 and E_2 are two events, and if E_1 can be done in m ways and E_2 can be done in n ways, then the event "E_1 followed by E_2" can be done in $m \cdot n$ ways.

Formula for permutations:

$$P(n, r) = \frac{n!}{(n - r)!}$$

$$P(n, n) = n!$$

Formula for combinations:

$$C(n, r) = \binom{n}{r} = \frac{n!}{r!(n - r)!}$$

A *permutation* is an ordered arrangement of a given set of objects. To solve counting problems where order is important, use the permutation formula.

41. TRAVEL If there are 17 flights from New York to Chicago and 8 flights from Chicago to San Francisco, in how many different ways could a passenger plan her trip?

42. LICENSE PLATES Refer to the illustration. How many different license plates are possible if they are to have 3 letters followed by 3 digits?

Evaluate each expression.

43. $P(7, 7)$ **44.** $P(7, 0)$ **45.** $P(8, 6)$

46. $\dfrac{P(9, 6)}{P(10, 7)}$ **47.** $C(7, 7)$ **48.** $C(7, 0)$

49. $\dbinom{8}{6}$ **50.** $C(6, 3) \cdot C(7, 3)$ **51.** $\dfrac{C(7, 3)}{C(6, 3)}$

52. Use the alternative form of the binomial theorem to expand $(3y - 2z)^4$.

53. LINING UP In how many ways can 5 persons be arranged in a line?

54. LINING UP In how many ways can 3 men and 5 women be arranged in a line if the women are placed ahead of the men?

A *combination* is a distinct group of objects without regard to their arrangement. To solve counting problems where order is not important, use the combination formula.

55. CHOOSING PEOPLE In how many ways can we pick 3 persons from a group of 10 persons?

56. FORMING COMMITTEES In how many ways can we pick a committee of 2 Democrats and 2 Republicans from a group containing 5 Democrats and 6 Republicans?

SECTION 11.5 Probability

An event that cannot happen has a *probability* of 0. An event that is certain to happen has a probability of 1. All other events have probabilities between 0 and 1.

If S is the *sample space* of an experiment with n distinct and equally likely outcomes, and E is an event that occurs in s of those ways, then the probability of E is

$$P(E) = \frac{s}{n}$$

In Exercises 57–59, assume that a dart is randomly thrown at the colored chart.

57. What is the probability that the dart lands in a blue area?

58. What is the probability that the dart lands in an even-numbered area?

59. What is the probability that the dart lands in an area whose number is greater than 2?

60. Find the probability of rolling an 11 on one roll of two dice.

61. Find the probability of living forever.

62. Find the probability of drawing a 10 from a standard deck of cards.

63. Find the probability of drawing a 5-card poker hand that has exactly 3 aces.

64. Find the probability of drawing 5 cards, all spades, from a standard card deck.

CHAPTER 11 TEST

1. Find the first 4 terms of the sequence defined by $a_n = -6n + 8$.

2. Expand: $(a - b)^6$.

3. Find the third term in the expansion of $(x^2 + 2y)^4$.

4. Find the tenth term of an arithmetic sequence whose first 3 terms are 3, 10, and 17.

5. Find the sum of the first 12 terms of the sequence $-2, 3, 8, \ldots$.

6. Find two arithmetic means between 2 and 98.

7. Find the common difference of an arithmetic sequence if the second term is $\frac{5}{4}$ and the 17th term is 5.

8. Find the sum of the first 27 terms of an arithmetic sequence if the 4th term is -11 and the 20th term is -75.

9. PLUMBING Plastic pipe is stacked so that the bottom row has 25 pipes, the next row has 24 pipes, the next row has 23 pipes, and so on until there is 1 pipe at the top of the stack. If a worker removes the top 15 rows of pipe, how many pieces of pipe will be left in the stack?

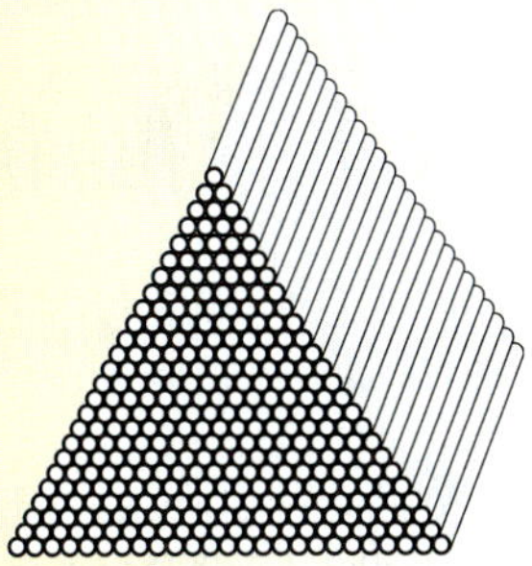

10. Evaluate: $\sum_{k=1}^{3} (2k - 3)$.

11. Find the seventh term of the geometric sequence whose first 3 terms are $-\frac{1}{9}$, $-\frac{1}{3}$, and -1.

12. Find the sum of the first 6 terms of the sequence $\frac{1}{27}$, $\frac{1}{9}$, $\frac{1}{3}$, ...

13. Find the first term of a geometric sequence if the common ratio is $-\frac{2}{3}$ and the fourth term is $-\frac{16}{9}$.

14. Find two geometric means between 3 and 648.

15. Find the sum of all of the terms of the infinite geometric sequence 9, 3, 1, ...

16. FALLING OBJECTS If an object is in free fall, the sequence 16, 48, 80, . . . represents the distance in feet that object falls during the first second, during the second second, during the third second, and so on. How far will the object fall during the first 10 seconds?

Find the value of each expression.

17. $\frac{7!}{4!}$

18. $0!$

19. $P(5, 4)$

20. $C(6, 4)$

21. $\binom{8}{3}$

22. $C(6, 0) \cdot P(3, 3)$

23. PHONE NUMBERS How many 7-digit phone numbers are available in area code 626 if no phone number can begin with 0, 1, or 2?

24. THE SUPREME COURT The last names of the members of the U.S. Supreme Court, as of 2004, are shown in one possible seating arrangement below. How many possible seating arrangements in a line are there?

Bader · Breyer · Kennedy · O'Conner · Rehnquist · Scalia · Souter · Stevens · Thomas

25. CHOOSING PEOPLE In how many ways can we pick 3 persons from a group of 7 persons?

26. CHOOSING COMMITTEES From a group of 5 men and 4 women, how many 3-person committees can be chosen that will include 2 women?

Find each probability.

27. Rolling a 5 on one roll of a die

28. Drawing a jack or a queen from a standard card deck

29. Receiving 5 hearts for a 5-card poker hand

30. Tossing 2 heads in 5 tosses of a fair coin

31. Shade an appropriate number of pie-shaped sections of a circle so that the probability of a spinner stopping on an *unshaded* section is $\frac{9}{16}$.

32. Fill in the blanks:

a. The probability of an event that cannot happen is ____.

b. The probability of an event that is guaranteed to happen is ____.

CHAPTERS 1–11 CUMULATIVE REVIEW EXERCISES

Consider the set $\left\{-\frac{4}{3}, \pi, 5.6, \sqrt{2}, 0, -23, e, 7i\right\}$. List the elements in the set that are

1. whole numbers

2. rational numbers

3. irrational numbers

4. real numbers

5. FINANCIAL PLANNING Anna has some money to invest. Her financial planner tells her that if she can come up with \$3,000 more, she will qualify for an 11% annual interest rate. Otherwise, she will have to invest the money at 7.5% annual interest. The financial planner urges her to invest the larger amount, because the 11% investment would yield twice as much annual income as the 7.5% investment. How much does she originally have on hand to invest?

6. BOATING Use the following graph to determine the average rate of change in the sound level of the engine of a boat in relation to rpm of the engine.

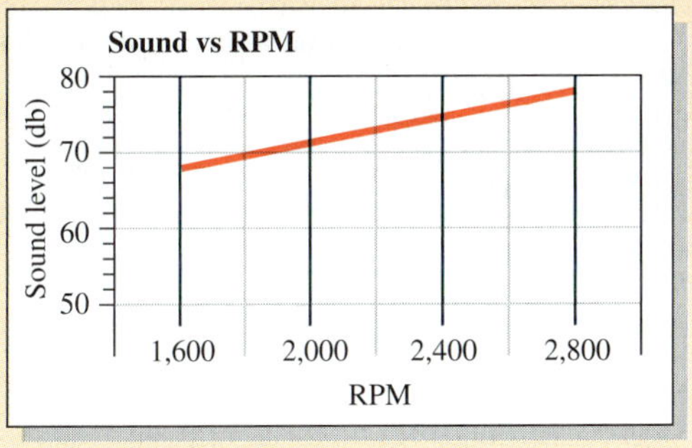

Decide whether the graphs of the equations are parallel or perpendicular.

7. $3x - 4y = 12,\ y = \frac{3}{4}x - 5$

8. $y = 3x + 4,\ x = -3y + 4$

Write the equation of the line with the given properties.

9. $m = -2$, passing through $(0, 5)$

10. Passing through $(8, -5)$ and $(-5, 4)$

11. Use substitution to solve $\begin{cases} 2x - y = -21 \\ 4x + 5y = 7 \end{cases}$.

12. Use addition to solve $\begin{cases} 4y + 5x - 7 = 0 \\ \frac{10}{7}x - \frac{4}{9}y = \frac{17}{21} \end{cases}$.

13. Use Cramer's rule to solve $\begin{cases} 2(x + y) + 1 = 0 \\ 3x + 4y = 0 \end{cases}$.

14. Solve: $\begin{cases} b + 2c = 7 - a \\ a + c = 8 - 2b \\ 2a + b + c = 9 \end{cases}$.

15. The graphs of $y = 4(x - 5) - x - 2$ and $y = -(2x + 6) - 1$ are shown in the illustration. Use the information in the display to solve $4(x - 5) - x - 2 = -(2x + 6) - 1$ graphically.

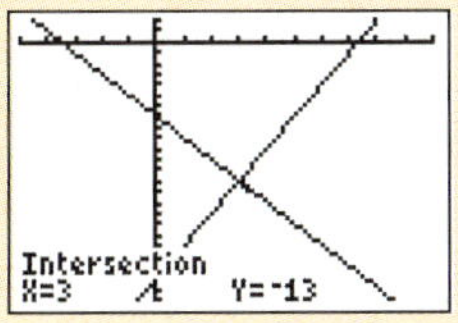

16. MARTIAL ARTS Find the measure of each angle of the triangle shown in the illustration.

17. Explain why the graph does not represent a function.

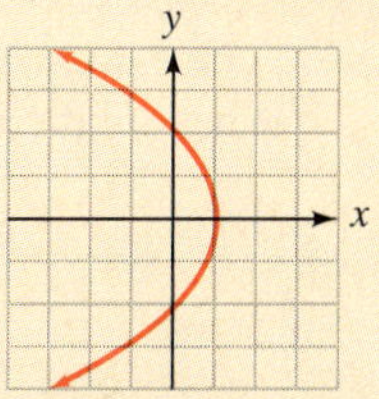

18. If $f(x) = 3x^5 - 2x^2 + 1$, find $f(-1)$ and $f(a)$.

19. Use the graph of function h to find each of the following:

a. $h(-3)$ **b.** $h(4)$

c. The value(s) of x for which $h(x) = 1$

d. The value(s) of x for which $h(x) = 0$

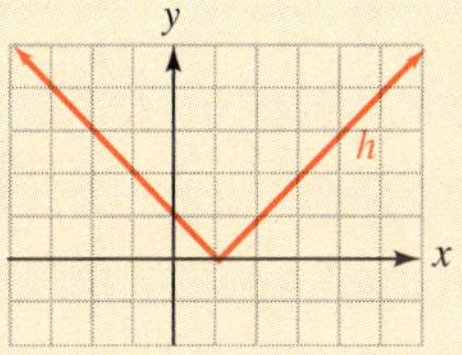

20. Write 173,000,000,000,000 and 0.000000046 in scientific notation.

21. Solve: $\begin{cases} 3x - 2y \le 6 \\ y < -x + 2 \end{cases}$.

Give the solution in interval notation and graph the solution set.

22. Solve: $|5 - 3x| - 14 \le 0$.

23. Solve: $4.5x - 1 < -10$ or $6 - 2x \ge 12$.

Perform the operations.

24. $(x - 3y)(x^2 + 3xy + 9y^2)$

25. $(-2x^2y^3 + 6xy + 5y^2) - (-4x^2y^3 - 7xy + 2y^2)$

26. $(9ab^2 - 4)^2$

27. $ab^{-2}c^{-3}(a^{-4}bc^3 + a^{-3}b^4c^3)$

Factor the expression completely.

28. $3x^3y - 4x^2y^2 - 6x^2y + 8xy^2$

29. $256x^4y^4 - z^8$

30. $12y^6 + 23y^3 + 10$

31. Solve for λ: $\frac{A\lambda}{2} + 1 = 2d + 3\lambda$.

32. Solve: $(x + 7)^2 = -2(x + 7) - 1$.

33. Solve: $x^3 + x^2 = 0$.

34. Complete the table of function values $f(x) = -x^3 - x^2 + 6x$ and then graph the function. What are the x- and y-intercepts of the graph?

x	$f(x)$
−4	
−3	
−2	
−1	
0	
1	
2	
3	

Simplify.

35. $\left(\frac{3x^5y^2}{6x^5y^{-2}}\right)^{-4}$

36. $\frac{6x^2 + 13x + 6}{6 - 5x - 6x^2}$

37. $\frac{p^3 - q^3}{q^2 - p^2} \cdot \frac{q^2 + pq}{p^3 + p^2q + pq^2}$

38. $\frac{2}{a - 2} + \frac{3}{a + 2} - \frac{a - 1}{a^2 - 4}$

39. Solve: $\frac{x - 4}{x - 3} + \frac{x - 2}{x - 3} = x - 3$.

40. Solve: $\frac{1}{R} = \frac{1}{R_1} + \frac{1}{R_2} + \frac{1}{R_3}$ for R.

41. TIRE WEAR See the illustration in the next column.

a. What type of function does it appear would model the relationship between the inflation of a tire and the percent of service it gives?

b. At what percent(s) of inflation will a tire offer only 90% of its possible service?

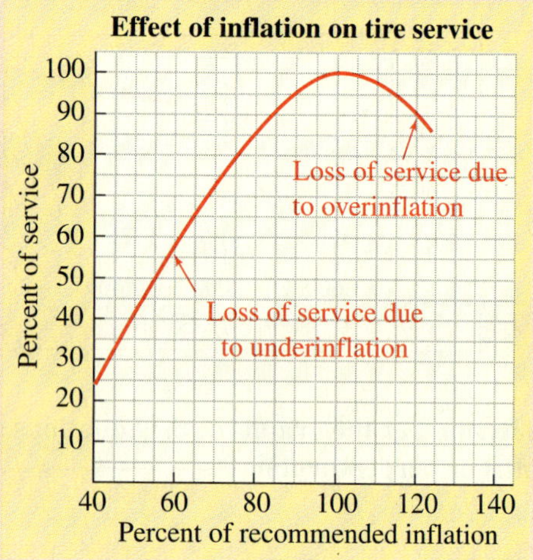

42. CHANGING DIAPERS The following illustration shows how to put a diaper on a baby. If the diaper is a square with sides 16 inches long, what is the largest waist size that this diaper can wrap around, assuming an overlap of 1 inch to pin the diaper?

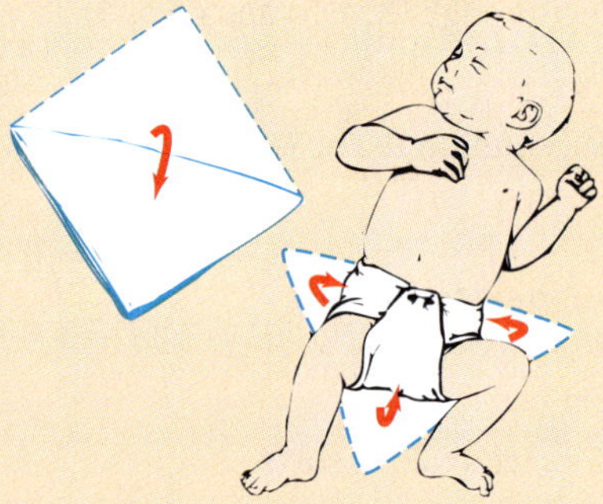

43. Use the long division method to find $(2x^2 + 4x - x^3 + 3) \div (x - 1)$.

44. LIGHT The intensity of a light source is inversely proportional to the square of the distance from the source. If the intensity is 18 lumens at a distance of 4 feet, what is the intensity when the distance is 12 feet?

45. Graph: $f(x) = \sqrt{x} + 2$. Give the domain and range of the function.

Simplify each expression.

46. $\sqrt{98} + \sqrt{8} - \sqrt{32}$

47. $3(\sqrt{5x} - \sqrt{3})^2$

48. $12\sqrt[3]{648x^4} + 3\sqrt[3]{81x^4}$

49. Evaluate: $\left(\frac{25}{49}\right)^{-3/2}$.

Rationalize each denominator.

50. $\frac{\sqrt[3]{4a^2}}{\sqrt[3]{2ab}}$

51. $\frac{3t - 1}{\sqrt{3t} + 1}$

Write each expression in *a* + *bi* form.

52. $(-7 + \sqrt{-81}) - (-2 - \sqrt{-64})$

53. $\frac{2 - 5i}{2 + 5i}$

54. Simplify: i^{42}.

Solve each equation.

55. $\sqrt{3a + 1} = a - 1$

56. $\sqrt{x + 3} - \sqrt{3} = \sqrt{x}$

57. $x^4 + 19x^2 + 18 = 0$

58. $4w^2 + 6w + 1 = 0$

59. $2(2x + 1)^2 - 7(2x + 1) + 6 = 0$

60. $3x^2 - 4x = -2$

61. First determine the coordinates of the vertex and the axis of symmetry of the graph of $f(x) = -6x^2 - 12x - 8$ using the vertex formula. Then determine the *x*- and *y*-intercepts of the graph. Finally, plot several points and complete the graph.

62. If $f(x) = x^2 - 2$ and $g(x) = 2x + 1$, find $(f \circ g)(x)$.

63. Find the inverse function of $f(x) = 2x^3 - 1$.

64. Graph: $f(x) = \left(\frac{1}{2}\right)^x$. Give the domain and range of the function.

65. Graph $y = e^x$ and its inverse on the same coordinate system.

66. Use the properties of logarithms to simplify $\log_6 \frac{36}{x^3}$.

67. Write the expression as a single logarithm:

$$\frac{1}{2}\ln x + \ln y - \ln z$$

68. POPULATION GROWTH As of 2003, the population of Mexico was about 119 million and the annual growth rate was 1.43%. If the growth rate remains the same, estimate the population of Mexico in 25 years.

Find *x*.

69. $\log_x 25 = 2$

70. $\log 1{,}000 = x$

71. $\log_3 x = -3$

72. $\ln e = x$

73. Let $\log 7 = 0.8451$ and $\log 14 = 1.1461$. Evaluate $\log 98$ without using a calculator.

74. Find $\log 0$, if possible.

Solve each equation. Round to four decimal places when necessary.

75. $2^{x+2} = 3^x$

76. $\log x + \log(x + 9) = 1$

77. $5^{4x} = \frac{1}{125}$

78. $\log_3 x = \log_3\left(\frac{1}{x}\right) + 4$

79. Write the equation of the circle that has its center at (1, 3) and passes through (−2, −1).

80. Write the equation in standard form and graph it: $(x - 2)^2 - 9y^2 = 9$.

81. Graph: $\frac{(x-1)^2}{4} + \frac{(y-3)^2}{16} = 1$.

82. Complete the square to write the equation $y^2 + 4x - 6y = -1$ in $x = a(y-k)^2 + h$ form. Determine the vertex and the axis of symmetry of the graph. Then plot several points and complete the graph.

83. Use the binomial theorem to expand $(3a - b)^4$.

84. Find the seventh term of the expansion of $(2x - y)^8$.

85. Find the 20th term of an arithmetic sequence with a first term -11 and a common difference 6.

86. Find the sum of the first 20 terms of an arithmetic sequence with a first term 6 and a common difference 3.

87. Evaluate: $\sum_{k=3}^{5} (2k + 1)$.

88. Find the seventh term of a geometric sequence with a first term $\frac{1}{27}$ and a common ratio 3.

89. BOAT DEPRECIATION How much will a \$9,000 boat be worth after 9 years if it depreciates 12% per year?

90. Find the sum of the first ten terms of the sequence $\frac{1}{64}, \frac{1}{32}, \frac{1}{16}, \ldots$.

91. Find the sum of all the terms of the sequence 9, 3, 1, ...

92. LINING UP In how many ways can 7 people stand in a line?

93. FORMING COMMITTEES In how many ways can a committee of 3 people be chosen from a group containing 9 people?

94. CARDS What is the probability of drawing a face card from a standard deck of cards?

Roots and Powers

n	n^2	$\sqrt{n}$	n^3	$\sqrt[3]{n}$	n	n^2	$\sqrt{n}$	n^3	$\sqrt[3]{n}$
1	1	1.000	1	1.000	51	2,601	7.141	132,651	3.708
2	4	1.414	8	1.260	52	2,704	7.211	140,608	3.733
3	9	1.732	27	1.442	53	2,809	7.280	148,877	3.756
4	16	2.000	64	1.587	54	2,916	7.348	157,464	3.780
5	25	2.236	125	1.710	55	3,025	7.416	166,375	3.803
6	36	2.449	216	1.817	56	3,136	7.483	175,616	3.826
7	49	2.646	343	1.913	57	3,249	7.550	185,193	3.849
8	64	2.828	512	2.000	58	3,364	7.616	195,112	3.871
9	81	3.000	729	2.080	59	3,481	7.681	205,379	3.893
10	100	3.162	1,000	2.154	60	3,600	7.746	216,000	3.915
11	121	3.317	1,331	2.224	61	3,721	7.810	226,981	3.936
12	144	3.464	1,728	2.289	62	3,844	7.874	238,328	3.958
13	169	3.606	2,197	2.351	63	3,969	7.937	250,047	3.979
14	196	3.742	2,744	2.410	64	4,096	8.000	262,144	4.000
15	225	3.873	3,375	2.466	65	4,225	8.062	274,625	4.021
16	256	4.000	4,096	2.520	66	4,356	8.124	287,496	4.041
17	289	4.123	4,913	2.571	67	4,489	8.185	300,763	4.062
18	324	4.243	5,832	2.621	68	4,624	8.246	314,432	4.082
19	361	4.359	6,859	2.668	69	4,761	8.307	328,509	4.102
20	400	4.472	8,000	2.714	70	4,900	8.367	343,000	4.121
21	441	4.583	9,261	2.759	71	5,041	8.426	357,911	4.141
22	484	4.690	10,648	2.802	72	5,184	8.485	373,248	4.160
23	529	4.796	12,167	2.844	73	5,329	8.544	389,017	4.179
24	576	4.899	13,824	2.884	74	5,476	8.602	405,224	4.198
25	625	5.000	15,625	2.924	75	5,625	8.660	421,875	4.217
26	676	5.099	17,576	2.962	76	5,776	8.718	438,976	4.236
27	729	5.196	19,683	3.000	77	5,929	8.775	456,533	4.254
28	784	5.292	21,952	3.037	78	6,084	8.832	474,552	4.273
29	841	5.385	24,389	3.072	79	6,241	8.888	493,039	4.291
30	900	5.477	27,000	3.107	80	6,400	8.944	512,000	4.309
31	961	5.568	29,791	3.141	81	6,561	9.000	531,441	4.327
32	1,024	5.657	32,768	3.175	82	6,724	9.055	551,368	4.344
33	1,089	5.745	35,937	3.208	83	6,889	9.110	571,787	4.362
34	1,156	5.831	39,304	3.240	84	7,056	9.165	592,704	4.380
35	1,225	5.916	42,875	3.271	85	7,225	9.220	614,125	4.397
36	1,296	6.000	46,656	3.302	86	7,396	9.274	636,056	4.414
37	1,369	6.083	50,653	3.332	87	7,569	9.327	658,503	4.431
38	1,444	6.164	54,872	3.362	88	7,744	9.381	681,472	4.448
39	1,521	6.245	59,319	3.391	89	7,921	9.434	704,969	4.465
40	1,600	6.325	64,000	3.420	90	8,100	9.487	729,000	4.481
41	1,681	6.403	68,921	3.448	91	8,281	9.539	753,571	4.498
42	1,764	6.481	74,088	3.476	92	8,464	9.592	778,688	4.514
43	1,849	6.557	79,507	3.503	93	8,649	9.644	804,357	4.531
44	1,936	6.633	85,184	3.530	94	8,836	9.695	830,584	4.547
45	2,025	6.708	91,125	3.557	95	9,025	9.747	857,375	4.563
46	2,116	6.782	97,336	3.583	96	9,216	9.798	884,736	4.579
47	2,209	6.856	103,823	3.609	97	9,409	9.849	912,673	4.595
48	2,304	6.928	110,592	3.634	98	9,604	9.899	941,192	4.610
49	2,401	7.000	117,649	3.659	99	9,801	9.950	970,299	4.626
50	2,500	7.071	125,000	3.684	100	10,000	10.000	1,000,000	4.642

Appendix

II Answers to Selected Exercises

Study Set 1.1 (page 7)

1. equation **3.** expressions **5.** subtraction **7.** formula **9. a.** a line graph **b.** 1 hour; 2 inches **c.** 7 in.; 0 in. **11.** $c = 13u + 24$ **13.** $w = \frac{c}{75}$ **15.** $A = t + 15$ **17.** $c = 12b$ **19.** $b = t - 10$; the height of the base is 10 ft less than the height of the tower **21. a.** expression **b.** equation **c.** equation **d.** expression **e.** expression **f.** expression **23.** 2, 6, 15 **25.** 22.44, 21.43, 0 **27.** 37 in. **29.** 8 yd **31. a.** The rental cost is the product of 10 and the number of hours it is rented, increased by 20. **b.** $C = 10h + 20$ **c.** 30, 40, 50, 60, 100

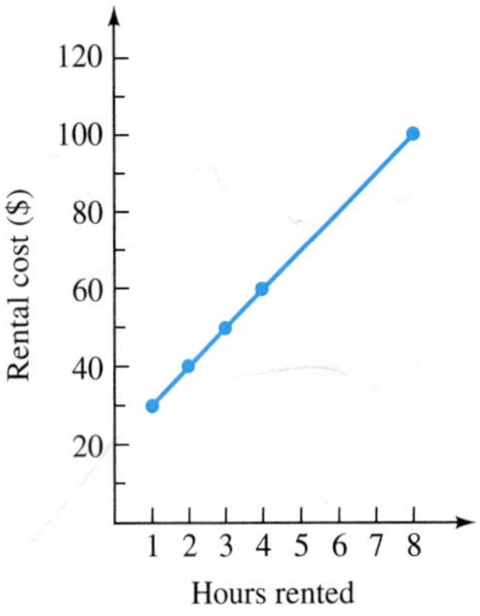

33. 150, 135, 90, 45, 30

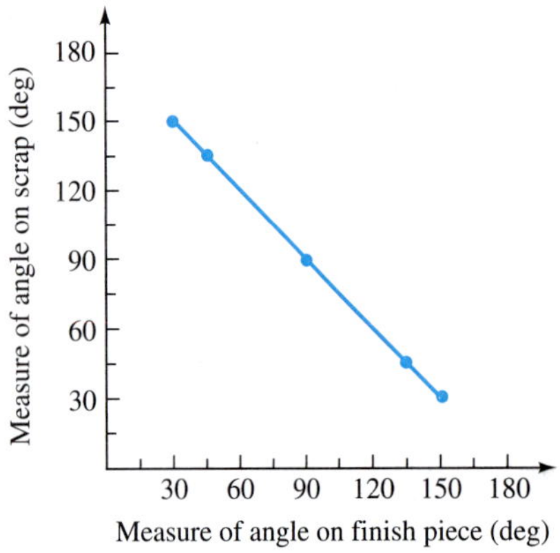

Study Set 1.2 (page 17)

1. rational **3.** absolute value **5.** Irrational **7.** subset **9.** 1, 2, 9 **11.** −3, 0, 1, 2, 9 **13.** $\sqrt{3}, \pi$ **15.** 2 **17.** 2 **19.** 9 **21.** nonrepeating, irrational **23.** repeating, rational **25.** $7 = \frac{7}{1}, -7\frac{3}{5} = \frac{-38}{5}, 0.007 = \frac{7}{1{,}000}, 700.1 = \frac{7{,}001}{10}$ **27.** a, $-a$ **29.** Irrational numbers, Rational numbers, Integers, Whole numbers **31.** is less than **33.** braces **35.** $\left\{\frac{a}{b} \middle| a \text{ and } b \text{ are integers, with } b \neq 0.\right\}$ **37.** 0.875, terminating **39.** $-0.7\overline{3}$, repeating

41.

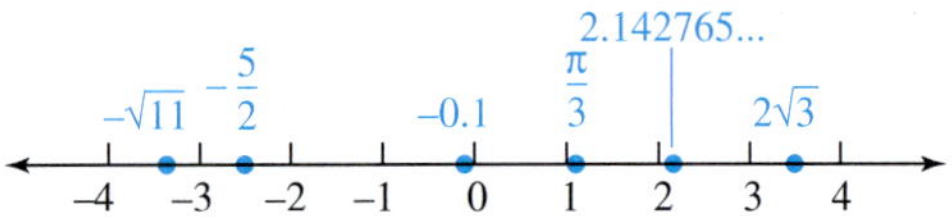

43.

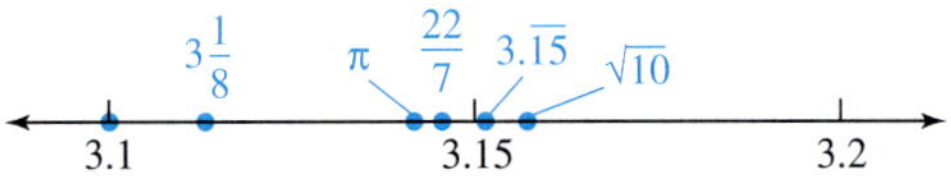

45.

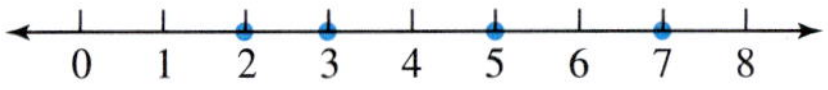

47.

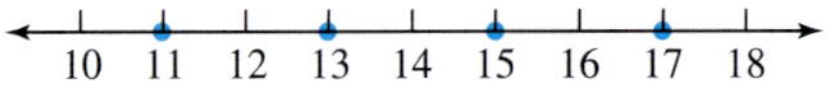

49.

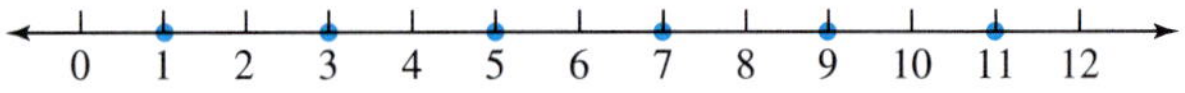

51. < **53.** > **55.** < **57.** > **59.** 12 < 19 **61.** $-5 \geq -6$ **63.** 20 **65.** −6 **67.** 5.9 **69.** $\frac{5}{4}$ **71.** $3\frac{2}{25} = 3.0800, \frac{77}{50} = 1.5400, \frac{15}{16} = 0.9375, 2\frac{5}{8} = 2.6250, \frac{\pi}{4} = 0.7854, \sqrt{8} = 2.8284$ **77.** expression **79.** 2.2, 8.5, 29.1

Study Set 1.3 (page 33)

1. sum, difference **3.** evaluate **5.** squared, cubed **7.** base, exponent **9.** opposite **11. a.** addition first or multiplication first **b.** 12; multiplication is performed before addition **13. a.** area **b.** volume **15. a.** −6 **b.** 6 **17.** radical sign **19.** −8 **21.** −4.3 **23.** −13 **25.** −7 **27.** 0 **29.** −2 **31.** −12 **33.** −1.5 **35.** 60 **37.** −2 **39.** $\frac{1}{6}$ **41.** $\frac{11}{10}$ **43.** $-\frac{6}{7}$ **45.** $\frac{24}{25}$ **47.** 144 **49.** −25 **51.** 64 **53.** 32 **55.** 1.69 **57.** 8 **59.** $-\frac{3}{4}$ **61.** −17 **63.** 64 **65.** 13 **67.** −2 **69.** −8 **71.** 10,000 **73.** −39 **75.** −114 **77.** −32 **79.** 5 **81.** $-\frac{1}{2}$ **83.** 2 **85.** 91,985 **87.** −24 **89.** −2 **91.** 61 **93.** 1 **95.** 10 **97.** 11.3 cm^2

99. 94.35 cm^3 **101.** 775.73 m^3 **103.** 25 ft^2
105. 1st term: area of bottom flap; 2nd term: area of left and right flaps; 3rd term: area of top flap; 4th term: area of face; 42.5 $in.^2$ **107.** \$(967) **109.** yes **111.** 9 **115.** −7 and 3 **117.** {. . . , −2, −1, 0, 1, 2, . . . } **119.** true

Study Set 1.4 (page 44)

1. Like **3.** term **5.** simplify **7.** undefined
9. $(x + y) + z = x + (y + z)$ **11.** $r(s + t) = rs + rt$
13. a. 5 **b.** −5 **15. a.** $\frac{16}{15}$ **b.** $-\frac{1}{20}$ **c.** 2 **d.** $\frac{1}{x}$
17. a. $2x^2, -x, 6$ **b.** 2, −1, 6 **19.** yes, $8x$ **21.** no
23. yes, $-2x^2$ **25.** no **27.** multiplication by −1
29. 7 + 3 **31.** $3 \cdot 2 + 3d$ **33.** c **35.** 1
37. $(8 + 7) + a$ **39.** $2(x + y)$ **41.** assoc. prop. of add.
43. distrib. prop. **45.** $-4t + 12$ **47.** $-t + 3$
49. $-3y + 6$ **51.** $2s - 6$ **53.** $0.7s + 1.4$
55. $4x - 5y + 1$ **57.** $72m$ **59.** $-45q$ **61.** $30bp$
63. $80ry$ **65.** $18x$ **67.** $13x^2$ **69.** 0 **71.** 0
73. $6x$ **75.** 0 **77.** $3.1h$ **79.** $\frac{9}{10}ab$ **81.** $\frac{14}{15}t$
83. $-4y + 36$ **85.** $7z - 20$ **87.** $14c + 62$
89. $-2x^2 + 15x$ **91.** $-2a - A$ **93.** $-6p + 17$
95. $8x - 9$ **97.** $-9x + 87$ **99. a.** $20(x + 6)$ m^2
b. $(20x + 120)$ m^2 **c.** $20(x + 6) = 20x + 120$; distrib. prop.
105. 0 **107.** 1,000 **109.** −3

Study Set 1.5 (page 57)

1. equation **3.** satisfies **5.** identity **7.** c, c, Adding, both **9. a.** $2y + 2$ **b.** 3 **c.** 18 **11.** It clears the equation of decimals. **13.** $-2x$, 14, 14, −2, −2, −17, −10, $\stackrel{?}{=}$, 20, 13 **15.** is possibly equal to **17.** yes
19. no **21.** 6 **23.** $-\frac{6}{5}$ **25.** 2 **27.** −9 **29.** $\frac{2}{3}$
31. 28 **33.** −20 **35.** 2.52 **37.** −30 **39.** −8
41. 0 **43.** 1.395 **45.** 3 **47.** 13 **49.** −11 **51.** −8
53. 2 **55.** −2 **57.** 24 **59.** 0 **61.** 6 **63.** $\frac{21}{19}$
65. −5 **67.** 3 **69.** 24 **71.** $\mathbb{R}$; identity **73.** $\varnothing$; contradiction **75.** $\mathbb{R}$; identity **77.** $B = \frac{3V}{h}$ **79.** $t = \frac{I}{Pr}$
81. $w = \frac{P - 2l}{2}$ **83.** $B = \frac{2A}{h} - b$ or $B = \frac{2A - bh}{h}$
85. $x = \frac{y - b}{m}$ **87.** $v_0 = 2\bar{v} - v$ **89.** $\ell = \frac{a - S + Sr}{r}$
91. $\ell = \frac{2S - na}{n}$ or $\ell = \frac{2S}{n} - a$ **93.** $C = \frac{5(F - 32)}{9}$, 432, −179, 58, −89, 17, −66 **95.** $d = \frac{360A}{\pi(r_1^2 - r_2^2)}$, 140, 160
97. $n = \frac{PV}{R(T + 273)}$; 0.008, 0.090 **99.** $n = \frac{C - 6.50}{0.07}$; 621, 1,000, about 1,692.9 kwh **101.** $h = \frac{A - 2\pi r^2}{2\pi r}$ **107.** $t - 4$
109. $-4b + 32$ **111.** $2.9b$ **113.** t

Study Set 1.6 (page 67)

1. acute **3.** complementary **5.** right **7.** angles
9. $d + 15$, $2d - 10$, $2d + 20$, $\frac{d}{2} - 10$, $2d$ **11. a.** $\frac{2}{3}x$
b. $2x$ **c.** $x + 2x + \frac{2}{3}x$ **13.** $26.5 - x$ **15.** Cheerios: \$663.5 million; Frosted Flakes: \$339.5 million **17.** 20
19. 310 mi **21.** 300 shares of BB, 200 shares of SS
23. 35 \$18 calculators, 50 \$87 calculators **25.** 7 ft, 15 ft
27. 30°, 150° **29.** 10° **31.** 50° **33.** 10 **35.** 60°
37. ii **39.** 156 ft by 312 ft **41.** 10 ft **43.** 6 in.
47. repeating **49.** {. . . , −4, −3, −2, −1, 0, 1, 2, 3, 4, . . . }
51. 0

Study Set 1.7 (page 80)

1. principal **3.** median **5.** amount, base **7.** ______ is ______ % of ______ ? **9. a.** 0.045 **b.** 6%
11. 90, $0.0565x$, $0.07(850 - x)$ **13.** 1, $0.15x$, x, 1, $0.18x$
15. 223.50, $8.25p$, $7.75(p + 30)$ **17.** $x = 0.05 \cdot 10.56$
19. $32.5 = 0.74x$ **21.** $I = Prt$ **23.** $v = np$ **25.** about 415 quadrillion Btu **27.** 20% **29.** \$50 **31.** 9.3%
33. −7.9%, 4.3% **35.** city: mean 43, median 42, mode 42, hwy: mean 48.8, median 49, mode 49 **37.** 84 **39.** CD: \$10,000; Money market: \$2,000 **41.** \$45,000 **43.** \$100,000
45. $\frac{1}{4}$hr = 15 min **47.** $\frac{2}{3}$ hr **49.** 3:30 P.M. **51.** $1\frac{1}{2}$ hr
53. 20 lb of 95¢ candy; 10 lb of \$1.10 candy **55.** 4,000 ft^3 of the premium mix, 2,000 ft^3 of sawdust **57.** 10 oz **59.** 2 gal
65. 0 **67.** 8

Key Concept (page 87)

1. given: 48 states, 4 more lie east of the Miss. River than west; find: how many states lie west **2.** given: one angle is 5° more than twice another, one angle is 90°, the sum of the angles' measures is 180°; find: the measure of the smallest angle
3. Let x = the length of the shortest piece. **4.** Let x = the number of miles he can drive. **5. a.** subtraction
b. multiplication **c.** addition **d.** division
6. a. $d = rt$ **b.** $I = Prt$ **c.** $v = np$ **d.** $P = 2l + 2w$
7. $0.15(15) + 0.50x = 0.40(15 + x)$, 0.15(15), $0.50x$, $0.40(15 + x)$ **8.** $450x + 500x = 2,850$

Chapter Review (page 88)

1. $C = 2t + 15$ **2.** $l = \frac{25}{w}$ **3.** $P = u - 3$
4. 180, 195, 210, 225, 240
5. a.

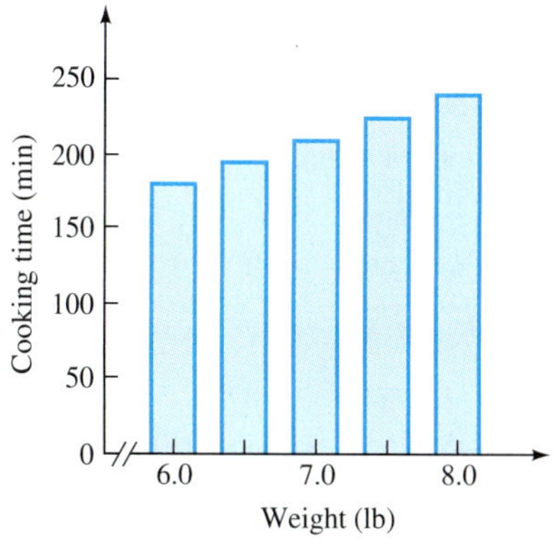

b.

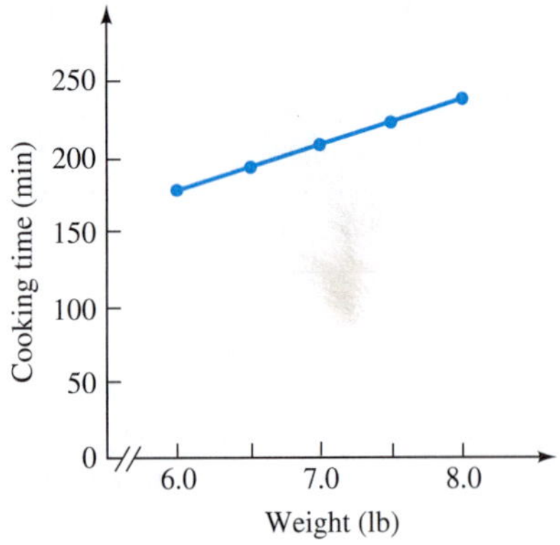

6. 17.5 in. **7.** 7 **8.** 0, 7 **9.** $-5, 0, 7$ **10.** $-5, 0, 2.4, 7, -\frac{2}{3}, -3.\overline{6}, \frac{15}{4}$ **11.** $-\sqrt{3}, \pi, 0.13242368\ldots$ **12.** all **13.** $-5, -\sqrt{3}, -\frac{2}{3}, -3.\overline{6}$ **14.** $2.4, 7, \pi, \frac{15}{4}, 0.13242368\ldots$ **15.** 7 **16.** none **17.** 0 **18.** $-5, 7$ **19. a.** $>$ **b.** $<$ **20. a.** false **b.** true

21.

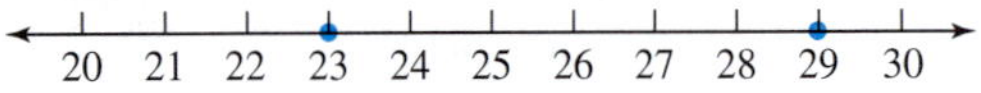

22.

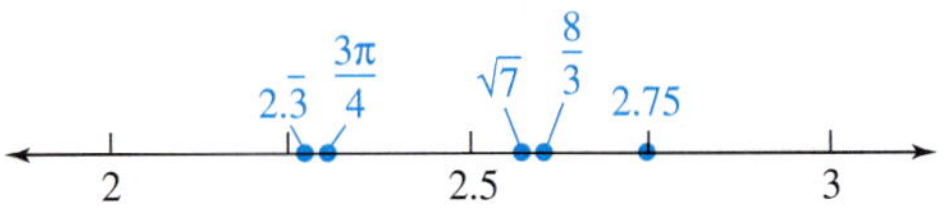

23. 18 **24.** -6.26 **25.** -7 **26.** 10.1 **27.** $-\frac{3}{4}$ **28.** 2 **29.** 12.6 **30.** $-\frac{1}{32}$ **31.** 0.2 **32.** $-\frac{3}{56}$ **33.** -33 **34.** -5.7 **35.** -120 **36.** 1 **37.** -243 **38.** $\frac{4}{9}$ **39.** 0.027 **40.** -25 **41.** 2 **42.** -10 **43.** $\frac{3}{5}$ **44.** 0.8 **45.** 44 **46.** 1 **47.** -12 **48.** 58 **49.** 8 **50.** 3 **51.** 3,000 **52.** -64 **53.** 56 **54.** $-\frac{1}{2}$ **55.** 100 in.^{2} **56.** 251.3 in.^{3} **57.** $3x + 21$ **58.** $5t$ **59.** 0 **60.** $27 + (1 + 99)$ **61.** 1 **62.** m **63.** 1 **64.** 0 **65.** $-3(5 \cdot 2)$ **66.** $(z + t) \cdot t$ **67.** 1 **68.** 0 **69.** -25 **70.** undefined **71.** $8x + 48$ **72.** $-6x + 12$ **73.** $4 - 3y$ **74.** $3.6x - 2.4y$ **75.** $6c^2 - 3c + \frac{3}{4}$ **76.** $2t + 6$ **77.** $48k$ **78.** $70xy$ **79.** $-189p$ **80.** $45a + 7$ **81.** 0 **82.** $3m - 48$ **83.** x **84.** $-24.54l$ **85.** $40a^3 - 16$ **86.** $\frac{1}{4}h + 8$ **87.** yes **88.** no **89.** $\{-225\}$ **90.** $\{7.9\}$ **91.** $\{0.014\}$ **92.** $\{-4\}$ **93.** $-\frac{12}{5}$ **94.** -9 **95.** 8 **96.** $\frac{11}{7}$ **97.** $\frac{88}{17}$ **98.** 12 **99.** 0.06 **100.** -8 **101.** 0 **102.** 3 **103.** $\varnothing$; contradiction **104.** $\mathbb{R}$; identity **105.** $h = \frac{V}{\pi r^2}$ **106.** $g = \frac{m - Y}{2}$ **107.** $x = \frac{T}{ab} - y$ or $x = \frac{T - aby}{ab}$ **108.** $r^3 = \frac{3V}{4\pi}$ **109.** O'Hare: 66.5 million; Atlanta: 76.5 million **110.** $245 - 5c$ **111.** \$5,000; \$1,000x **112.** 42 ft, 45 ft, 48 ft, 51 ft **113.** 50°, 130° **114.** \$18,000 at 10%, \$7,000 at 9% **115. a.** 11.2% **b.** 3.8% **116.** 504; 505; 505 **117.** 3 min **118.** 10 gal **119.** $3.95(x + 3)$

Chapter 1 Test (page 94)

1. $s = T + 10$ **2.** $A = \frac{1}{2}bh$ **3.** $-2, 0, 5$ **4.** $-2, 0, -3\frac{3}{4}, 9.2, \frac{14}{5}, 5$ **5.** $\pi, -\sqrt{7}$ **6.** all

7.

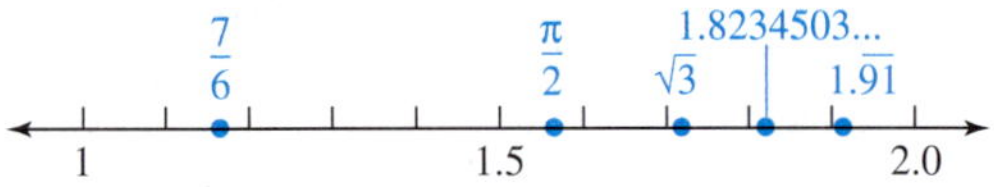

8.

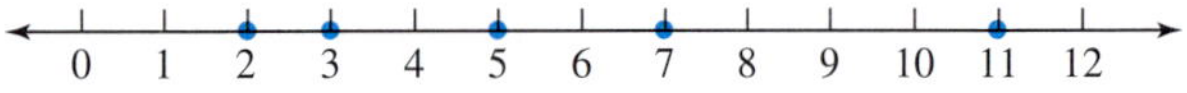

9. -8 **10.** 5.5 **11.** 12.3 **12.** $\frac{4}{15}$ **13.** $\frac{11}{10}$ **14.** -64 **15.** $-\frac{8}{9}$ **16.** -35 **17.** -3 **18.** 100 mg **19.** commut. prop. of add. **20.** assoc. prop. of mult. **21.** $11y$ **22.** -4 **23.** $11h - 17H - 11$ **24.** $x^2 - 1$ **25.** -12 **26.** 6 **27.** $\varnothing$; contradiction **28.** 5.5 **29.** 5.6 **30.** 5.6 **31.** $i = \frac{f(P - L)}{s}$ **32.** 4 cm by 9 cm **33.** 84% **34.** \$4,000 **35.** 10 oz

Study Set 2.1 (page 104)

1. ordered **3.** origin **5.** rectangular **7.** midpoint **9.** origin, right, down **11.** II **13.** yes **15.** capital letters **17.** $x^2 = x \cdot x$; x_2 represents the x-coordinate of a point.

19–25.

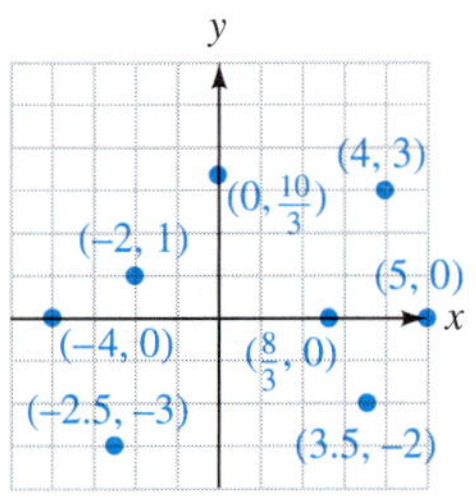

27. (2, 4) **29.** $(-2.5, -1.5)$ **31.** (3, 0) **33.** (0, 0) **35. a.** on the surface **b.** diving **c.** 1,000 ft **d.** 500 ft **37. a.** 1993 **b.** Imports exceeded production by about 3.5 million barrels per day. **39.** (3, 4) **41.** (9, 12) **43.** $\left(\frac{1}{2}, -2\right)$ **45.** $(-4, 0)$ **47.** (4, 1) **49.** $(-20, -3)$ **51.** Jonesville (5, B), Easley (1, B), Hodges (2, E), Union (6, C) **53. a.** $(2, -1)$ **b.** no **c.** yes

55.

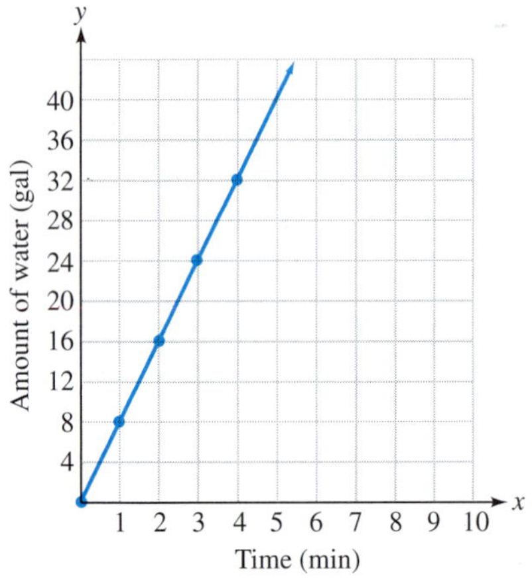

57. a. 6 **b.** 7 strokes **c.** 16th **d.** 18th **59. a.** 83¢ **b.** 23¢ **c.** 3 oz **61. a.** T **b.** R **65.** 20 **67.** $\frac{1}{2}$ **69.** 0.7

Study Set 2.2 (page 118)

1. ordered, pair **3.** linear **5.** vertical **7. a.** yes **b.** no **9. a.** 1, 1 **b.** 2, infinitely many **11.** 1, 1, 1 **13. a.** (−3, 0); (0, 4) **b.** false **15.** $y = -4x - 1$ **17. a.** the y-axis **b.** the x-axis **19.** 5, 4, 2 **21.** 0, −1, −2

23.
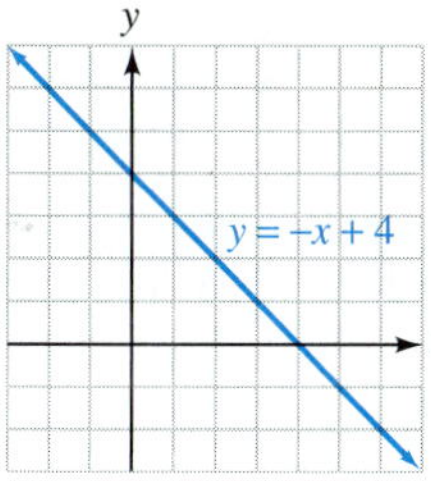

25.
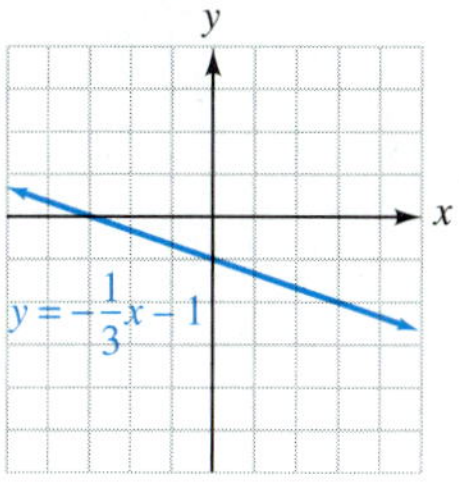

27.
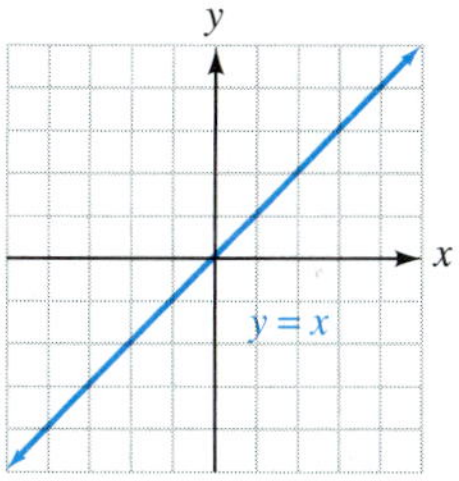

29.
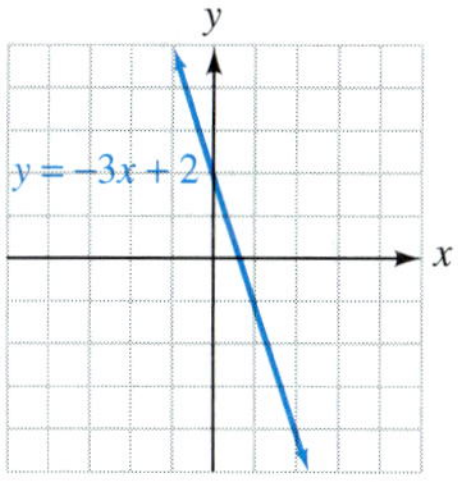

31.
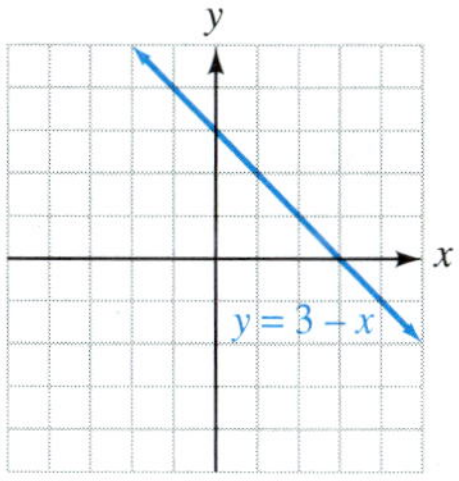

33.
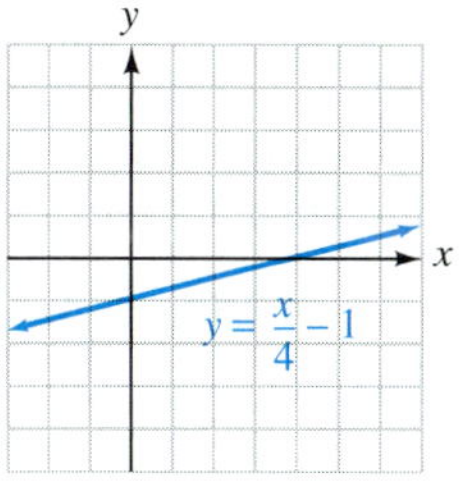

35.
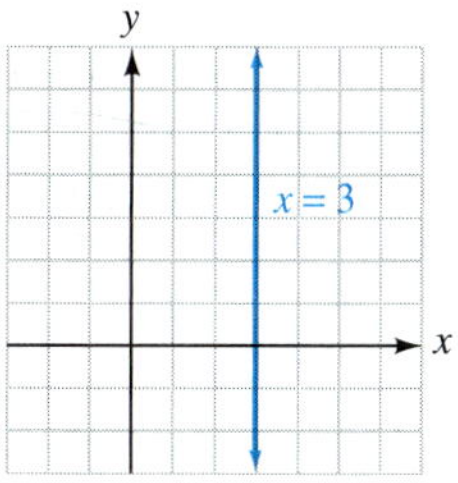

37.
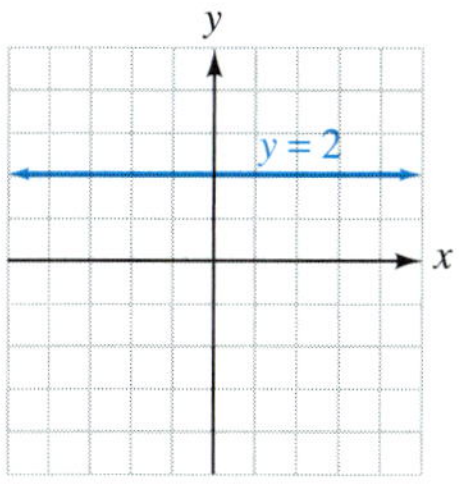

39.
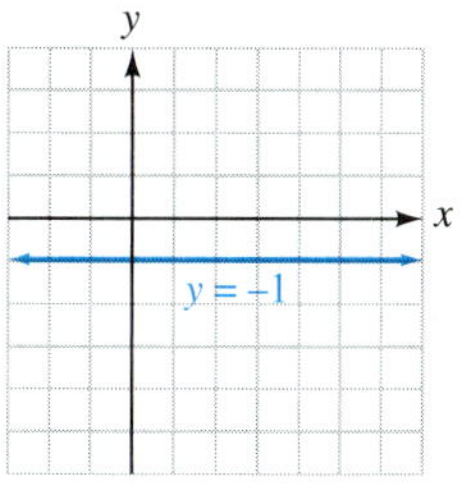

41.
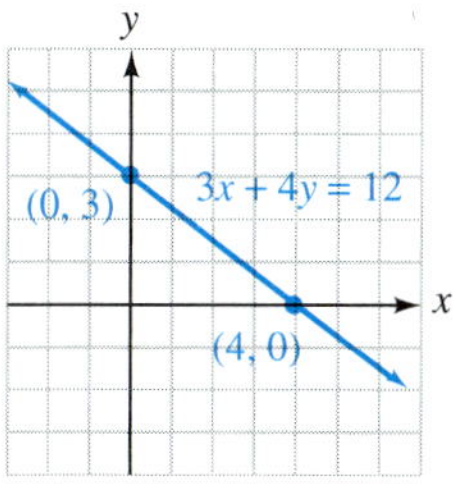

43.
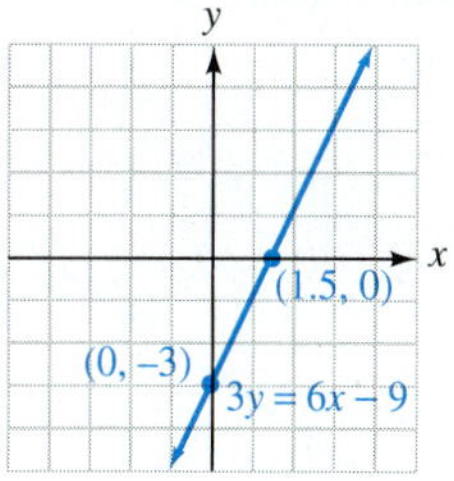

45.
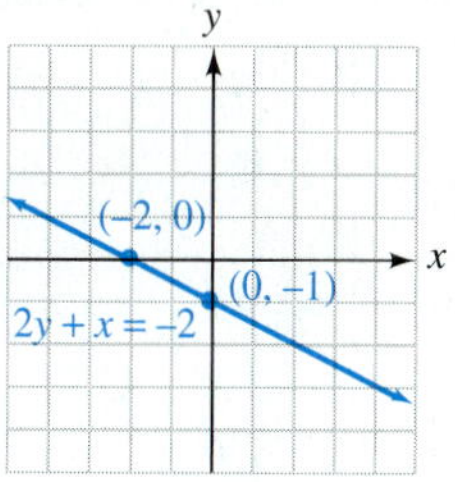

47.
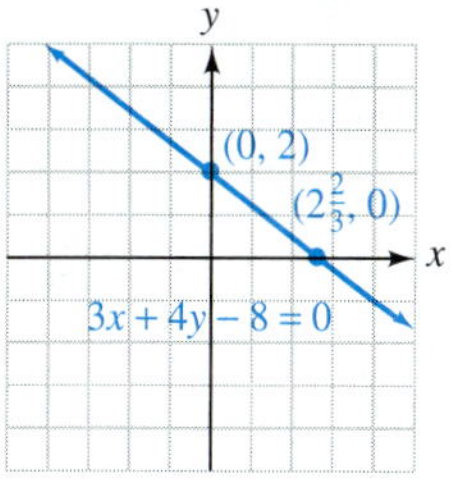

49.
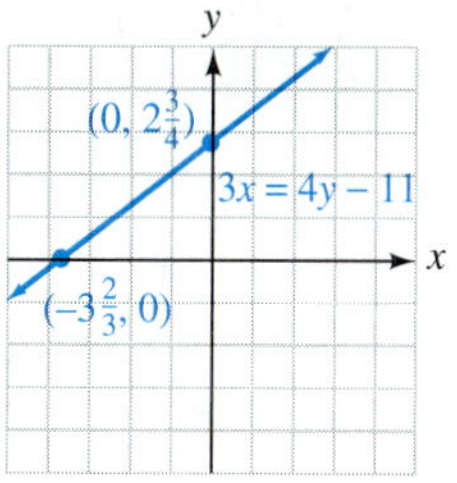

51. 1.22 **53.** 4.67
55. a. $c = 10t + 2$
b. 12, 22, 32, 42
c. $62

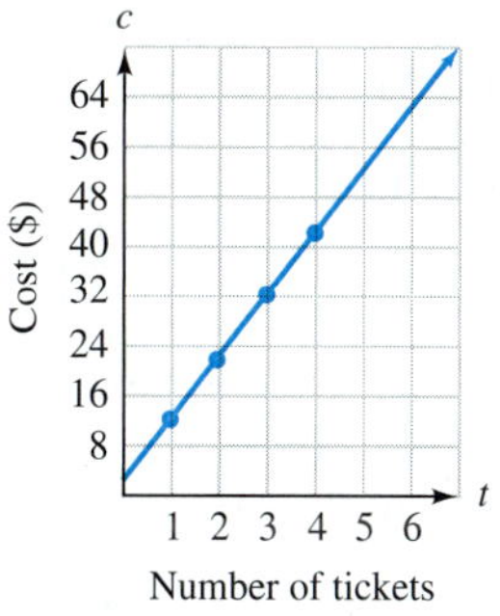

57. a. In 1990, there were 65.5 million swimmers.
b. 54.7 million

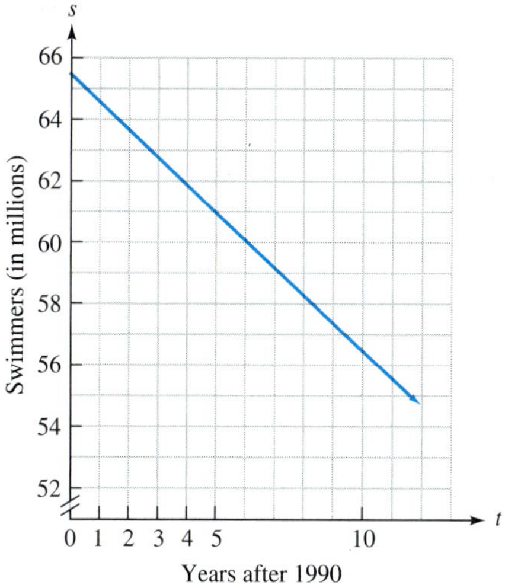

59. a. The life expectancy for someone born in 1980 is 74 years. **b.** 76.3 yr

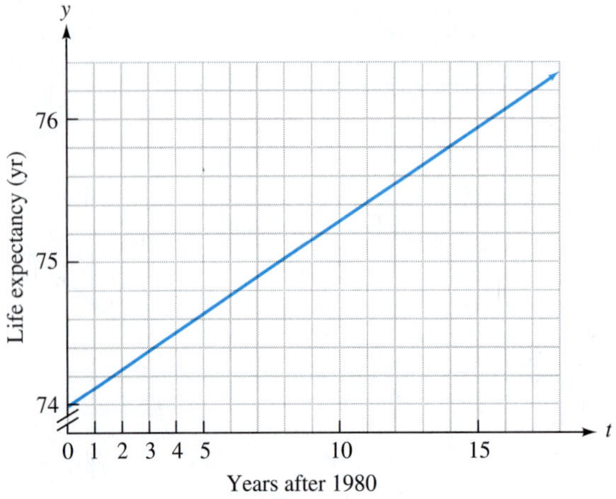

61. In 8 years, the computer will have no value. When new, the computer was worth $3,000. **63.** 200 **67.** 11, 13, 17, 19, 23, 29 **69.** III **71.** 80*s* **73.** $3x + 8$

Study Set 2.3 (page 129)

1. Slope **3.** change **5.** reciprocals **7. a.** l_3; 0 **b.** l_2; undefined **c.** l_1; 2 **d.** l_4; -3 **9. a.** an increase of 73 million units/yr **b.** a decrease of 35 million units/yr **11. a.** $-\frac{4}{3}$ **b.** $-\frac{2}{3}$ **13.** $m = \frac{y_2 - y_1}{x_2 - x_1}$ **15. a.** 6 **b.** 8 **c.** $\frac{3}{4}$ **17.** $-\frac{8}{3}$ **19.** $\frac{6}{7}$ **21.** 3 **23.** -1 **25.** $-\frac{1}{3}$ **27.** 0 **29.** undefined **31.** $\frac{1}{2}$ **33.** -0.5 **35.** -1 **37.** perpendicular **39.** neither **41.** neither **43.** parallel **45.** perpendicular **47.** neither **49.** $\frac{3}{140}, \frac{1}{15}, \frac{1}{20}$; part 2 **51.** $\frac{1}{10}; \frac{1}{4}$ **53.** $\frac{1}{25}$; 4% **55.** brace: $\frac{1}{2}$; support 1: -2; support 2: -1; yes, to support 1 **61.** 40 lb licorice; 20 lb gumdrops

Study Set 2.4 (page 141)

1. $y - y_1 = m(x - x_1)$ **3.** perpendicular **5.** no **7.** $m = \frac{2}{3}$; $y + 3 = \frac{2}{3}(x + 2)$ **9.** $m = -\frac{2}{3}$; $(0, 1)$ **11.** yes **13. a.** $(0, 0)$ **b.** none **15.** No; the slopes are not negative reciprocals. Their product is not -1: $1(-0.9) = -0.9$. **17.** $\frac{1}{3}x$, 2, 2, 1, $\frac{1}{3}$, -1 **19.** $y = 5x + 7$ **21.** $y = -3x + 6$ **23.** $y = x$ **25.** $y = \frac{7}{3}x - 3$ **27.** $y = \frac{2}{3}x + \frac{11}{3}$ **29.** $y = 3x + 17$ **31.** $y = -7x + 54$ **33.** $y = -4$ **35.** $y = -\frac{1}{2}x + 11$ **37.** $\frac{3}{2}$, $(0, -4)$ **39.** $-\frac{1}{3}$, $\left(0, -\frac{5}{6}\right)$

41. 1, $(0, -1)$

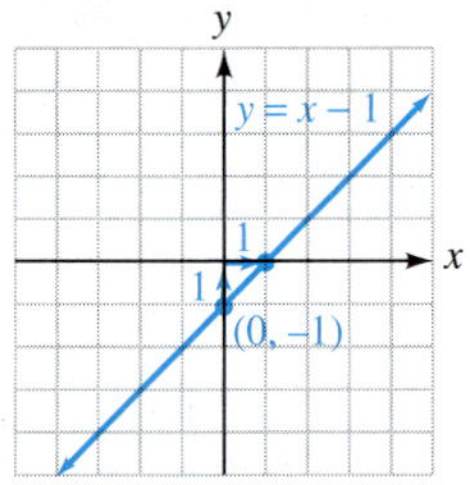

43. $\frac{2}{3}$, $(0, 2)$

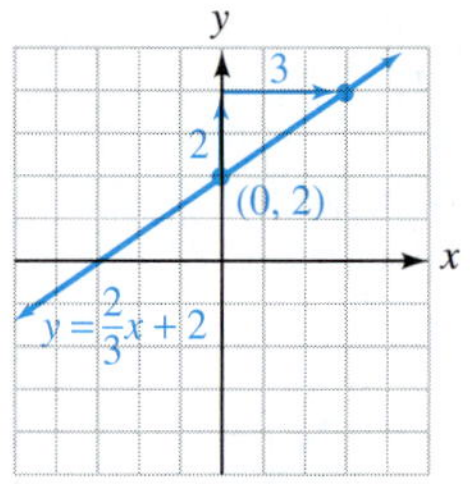

45. $-\frac{3}{4}$, $(0, -2)$

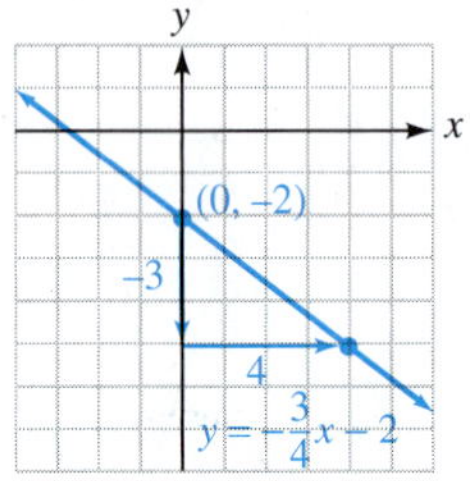

47. parallel **49.** perpendicular **51.** neither **53.** perpendicular **55.** $y = 4x$ **57.** $y = 4x - 3$ **59.** $y = \frac{4}{5}x - \frac{26}{5}$ **61.** $y = -\frac{1}{4}x$ **63.** $y = -\frac{1}{4}x + \frac{11}{2}$ **65.** $y = -\frac{5}{4}x + 3$

65. $y = -\frac{5}{4}x + 3$ **67.** $y = -\frac{950}{3}x + 1{,}750$ **69.** $y = 1{,}811{,}250x + 36{,}225{,}000$ **71. a.** $B = \frac{1}{100}p - 195$ **b.** 905 **73. a.** $E = -\frac{1}{2}t + \frac{21}{2}$ **b.** The number of errors is reduced by 1 for every 2 trials. **c.** On the 21st trial, the rat should make no errors. **75. a.** $y = 1.35x - 31.25$ **b.** When the actual temperature is 0°F, the wind-chill temperature is about -31°F. **81.** $29,100

Study Set 2.5 (page 154)

1. function, one **3.** input, output **5.** value **7.**

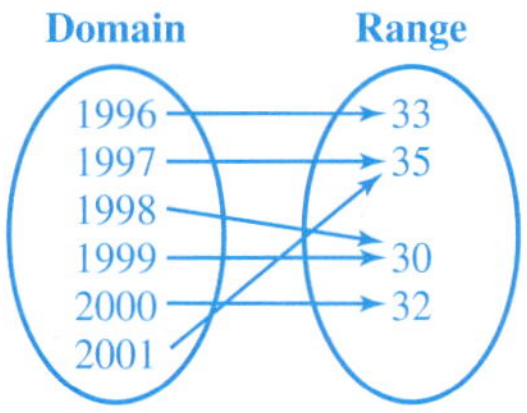

9. $f(8)$ **11.** 17, -1, 7; $(-3, 17)$, $(0, -1)$, $(2, 7)$ **13. a.** $(-2, 4)$, $(-2, -4)$ **b.** No; the *x*-value -2 is assigned more than one *y*-value (4 and -4). **15.** of **17.** $f(x)$ **19.** $f(x)$, y **21.** yes **23.** no; (4, 2), (4, 4), (4, 6) **25.** yes **27.** no; $(-1, 0)$, $(-1, 2)$ **29.** no; $(3, 4)$, $(3, -4)$ or $(4, 3)$, $(4, -3)$ **31.** yes **33.** yes **35.** no **37.** no **39.** 9, -3 **41.** 3, -5 **43.** 22, 2 **45.** 3, 11 **47.** 4, 9 **49.** 7, 26 **51.** 9, 16 **53.** 6, 15 **55.** 4, 4 **57.** 2, 2 **59.** $\frac{1}{5}$, 1 **61.** -2, $\frac{2}{5}$ **63.** 3.7, 1.1, 3.4 **65.** $-\frac{27}{64}$, $\frac{1}{216}$, $\frac{125}{8}$ **67.** $2w$, $2w + 2$ **69.** $3w - 5$, $3w - 2$ **71.** 0 **73.** 1 **75.** D: $\{-2, 4, 6\}$, R: $\{3, 5, 7\}$ **77.** D: the set of all real numbers, R: the set of all real numbers **79.** D: the set of all real numbers, R: the set of all nonnegative real numbers **81.** D: the set of all real numbers, R: the set of all real numbers greater than or equal to 0 **83.** D: the set of all real numbers except 4, R: the set of all real numbers except 0 **85.** not a function **87.** a function **89.** a function **91.** a function

93.
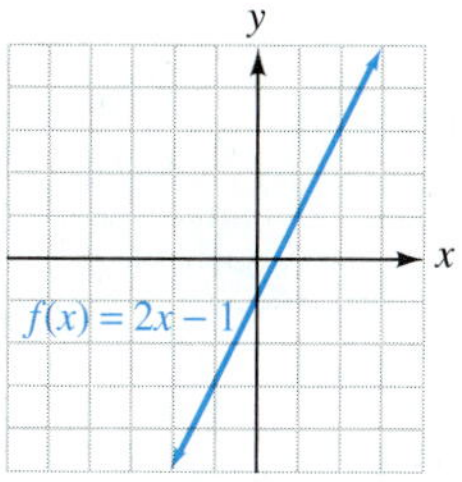

95.
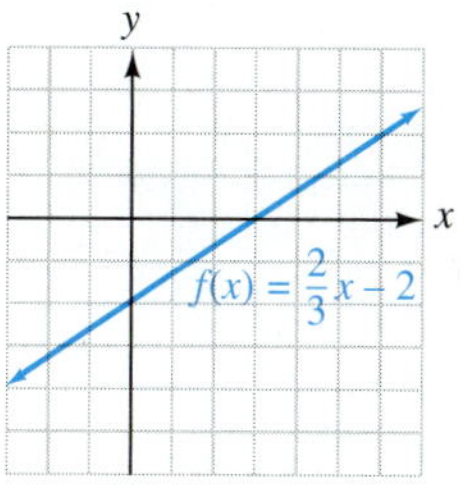

97. between 20°C and 25°C **99. a.** $I(b) = 1.75b - 50$ **b.** \$142.50 **101. a.** (200, 25), (200, 90), (200, 105) **b.** It doesn't pass the vertical line test. **103. a.** \$3,400; the tax on an income of \$25,000 is \$3,400. **b.** $T(a) = 3{,}910 + 0.25(a - 28{,}400)$ **107.** $\frac{-15}{4}$ **109.** $\frac{1}{3}$

Study Set 2.6 (page 166)

1. nonlinear **3.** parabola **5.** cubing **7. a.** −4 **b.** 0 **c.** 2 **d.** −1 **9. a.** 4 **b.** 3 **c.** 0, 2 **d.** 1

11. a.
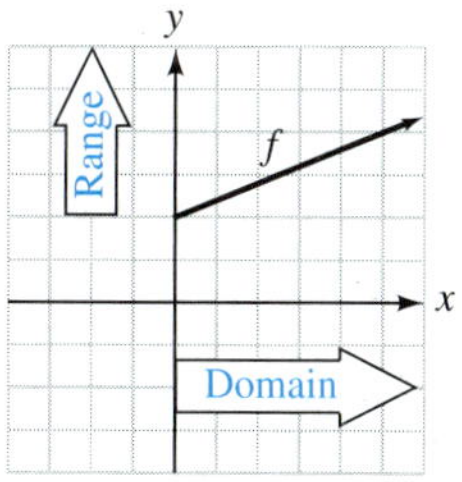

b. D: all nonnegative real numbers, R: all real numbers greater than or equal to 2.

13. 4, 0, −2 **15.**
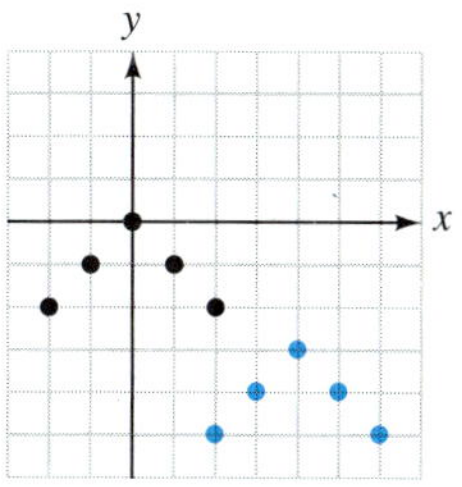

17. 4, left **19.** 5, up

21. D: the set of real numbers, R: the set of all real numbers greater than or equal to −3.
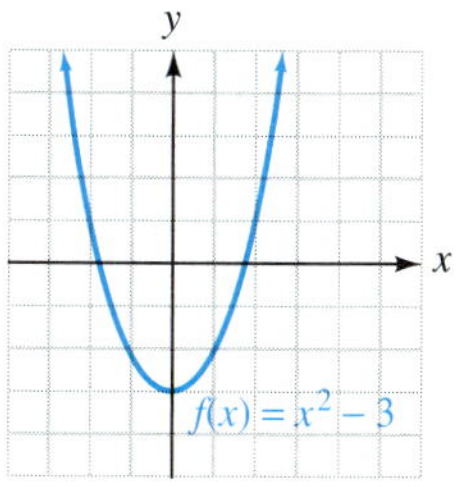

23. D: the set of real numbers, R: the set of real numbers.
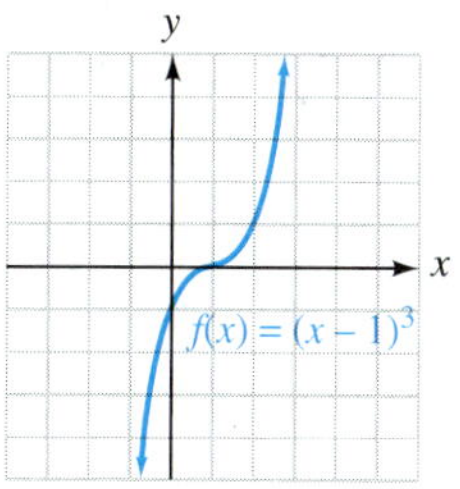

25. D: the set of real numbers, R: the set of all real numbers greater than or equal to −2.
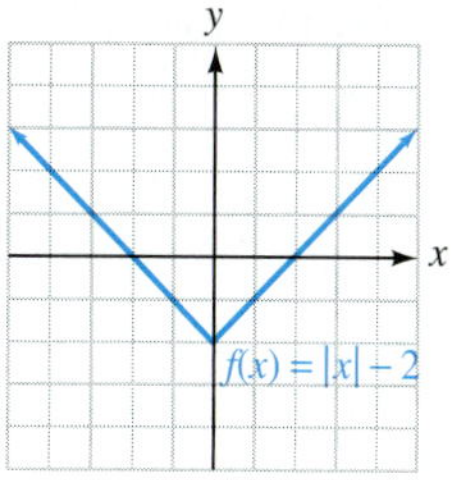

27. D: the set of real numbers, R: the set of real numbers greater than or equal to 0.
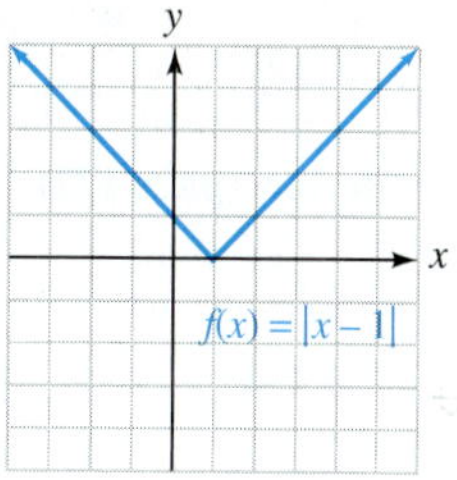

29. D: the set of real numbers, R: the set of real numbers.
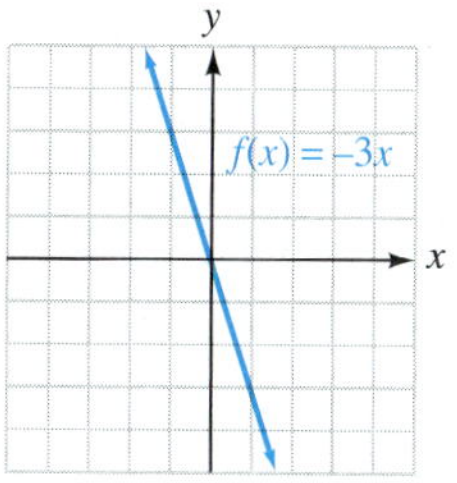

31.
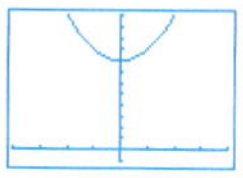

33.
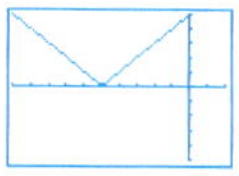

35.
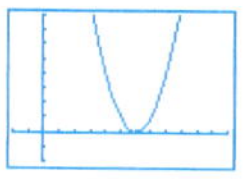

37.
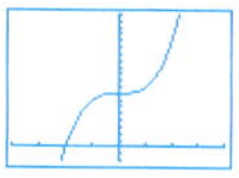

39.
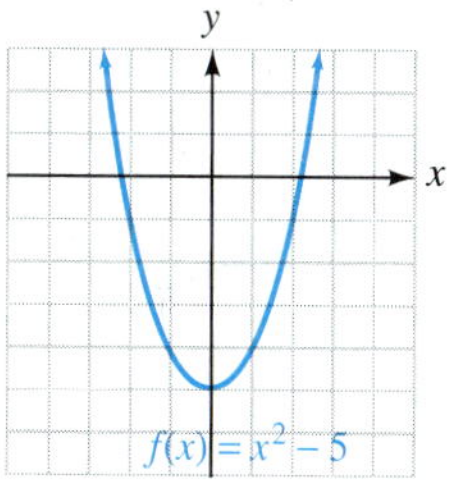

41.
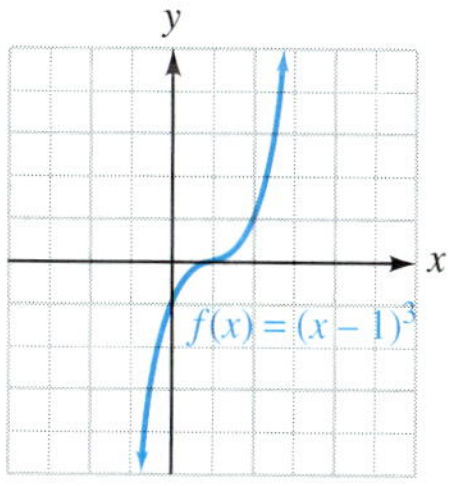

43.
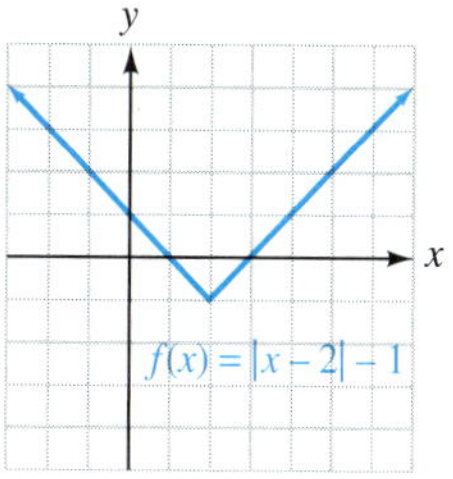

45.
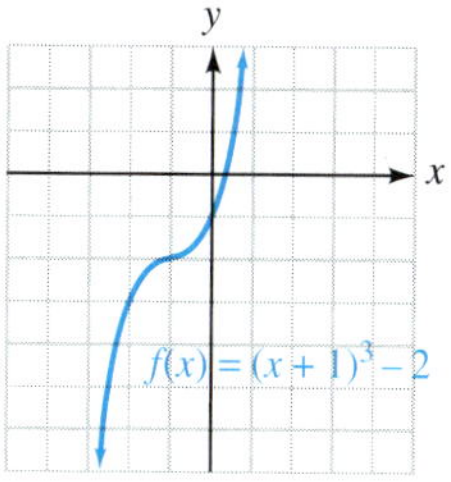

47.
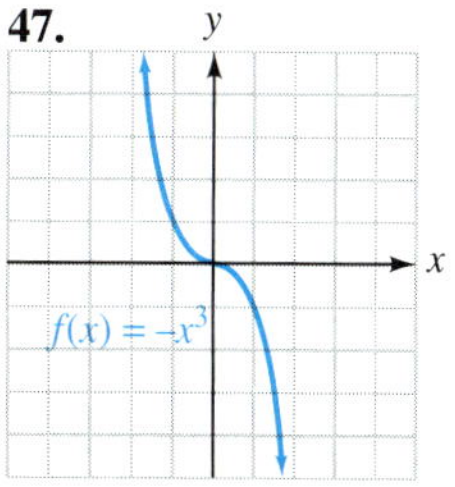

49.
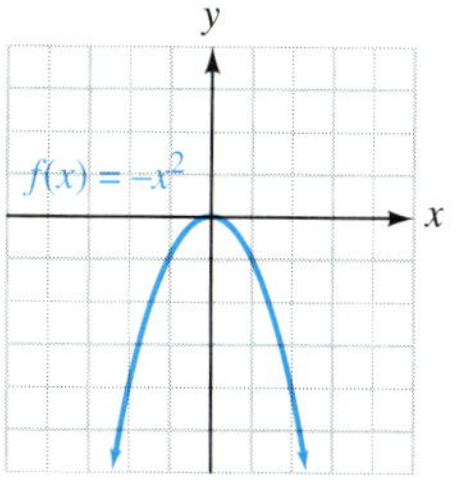

51. $f(x) = |x|$ **53.** a parabola **59.** $W = T - ma$ **61.** $g = \frac{2(s - vt)}{t^2}$ **63.** \$5.4 million

Key Concept (page 171)

1. a. function, independent, dependent **b.** domain, range **3.** -17 **4.** $A = \pi r^2$ **5.** 60 ft **6.** 2 **7.** $f(x) = 2x + 3$, 3 **8.** $A(r) = \pi r^2$ **9.** 0; the projectile will strike the ground 4 seconds after being shot into the air. **10. a.** -2 **b.** 2

Chapter Review (page 172)

1.–5.

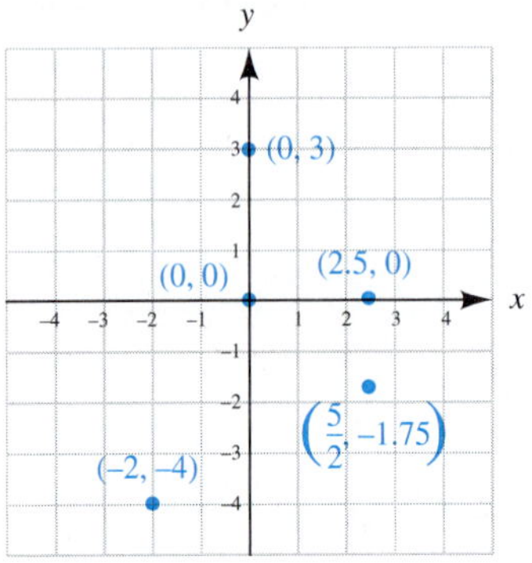

6. a. 1 ft below its normal level **b.** decreased by 3 ft **c.** from day 3 to the beginning of day 4 **7. a.** \$10 increments **b.** \$800 **8.** $(7, -3)$ **9.** yes **10. a.** true **b.** false Complete the table of solutions for each equation.
11. 9, 0, -9 **12.** -4, $-\frac{5}{2}$, -1

13.

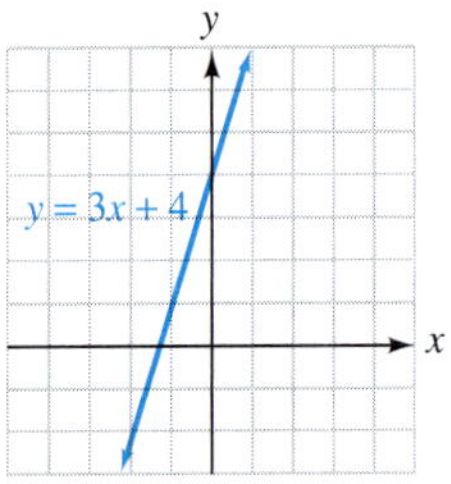

14.

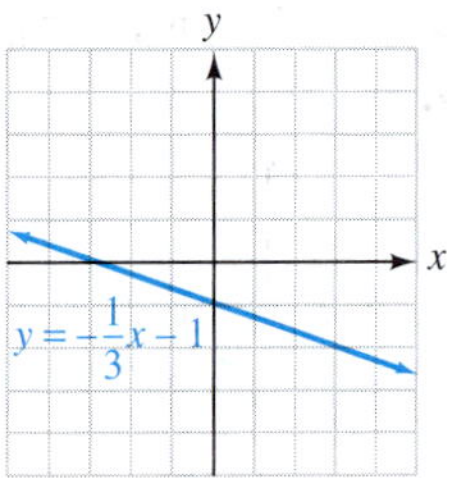

15.

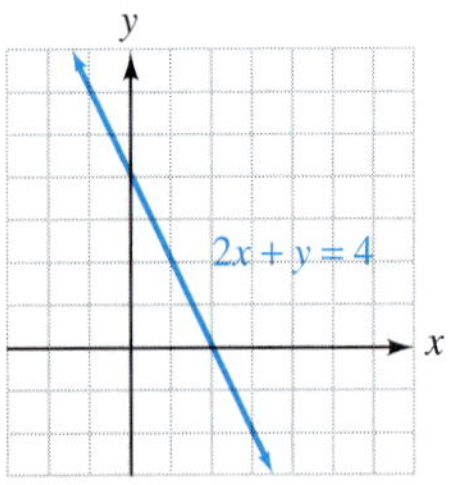

16.

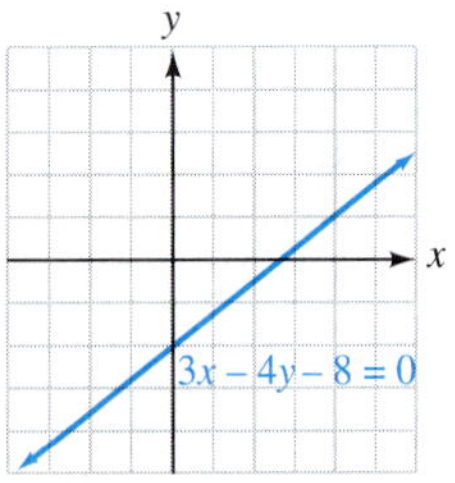

17.

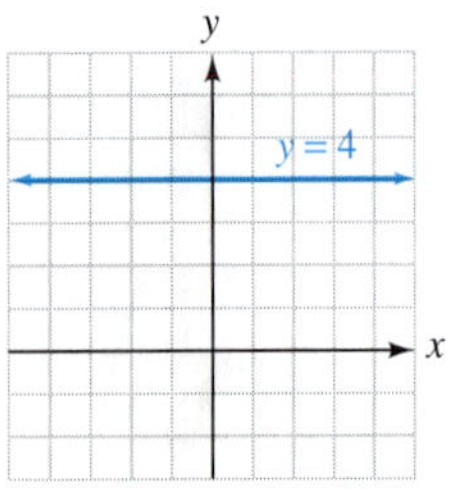

18.

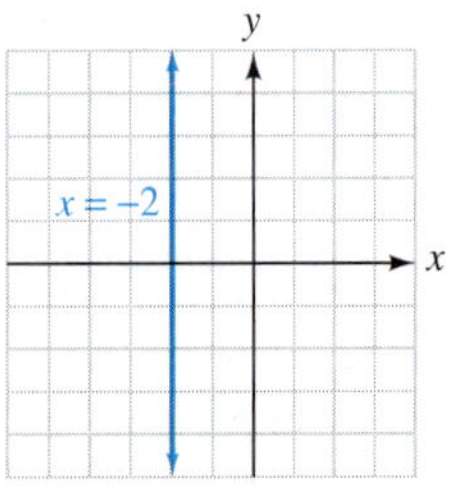

19. 1, 1
20.

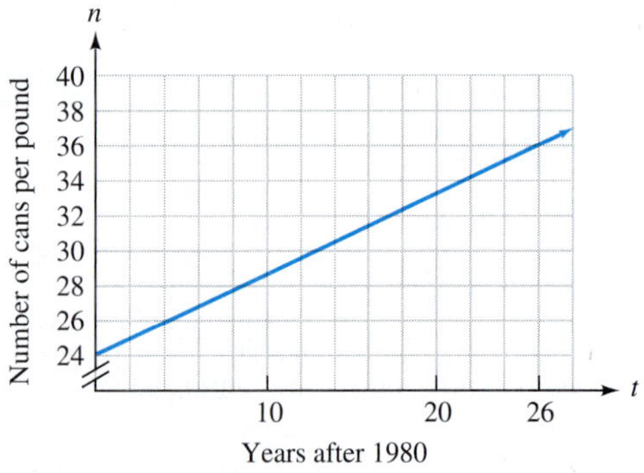

a. In 1980, it took about 24 cans to weigh one pound. **b.** about 35 **21.** slope of $l_1 = \frac{4}{5}$; slope of $l_2 = -\frac{8}{5}$ **22.** 1.37% per yr **23.** 1 **24.** $-\frac{14}{9}$ **25.** 0 **26.** undefined **27.** perpendicular **28.** parallel **29.** perpendicular **30.** 31.5% **31.** $y = 3x + 29$ **32.** $y = -\frac{13}{8}x + \frac{3}{4}$ **33.** $y = \frac{3}{2}x - \frac{1}{2}$ **34.** $y = -\frac{2}{3}x - 7$ **35.** $y = -\frac{3}{4}x - 3$; $m = -\frac{3}{4}$, $(0, -3)$

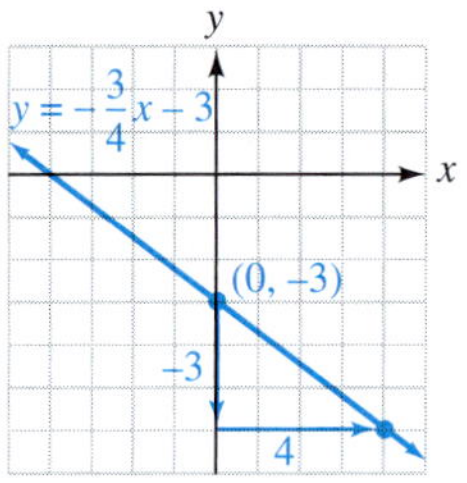

36. $y = 0$ **37.** $x = 0$ **38.** $y = -\frac{4}{5}x + 4$ **39.** $L = \frac{1}{150}p + 110$ **40. a.** $y = -1{,}720x + 8{,}700$ **b.** (0, 8,700); It gives the value of the saw blade when new: \$8,700. **41.** yes **42.** no **43.** yes **44.** yes **45.** no **46.** no **47.** -7 **48.** 18 **49.** 8

50. $3t + 2$ **51.** 3 **52.** $\frac{4}{3}$ **53.** D: the set of real numbers. R: the set of real numbers. **54.** D: the set of real numbers, R: the set of all real numbers greater than or equal to 1. **55.** D: the set of all real numbers except 2, R: the set of all real numbers except 0. **56.** D: the set of real numbers, R: the set of real numbers that are less than or equal to 0. **57.** function **58.** not a function **59.** $f(t) = 1.37t + 21.2$; about 54%
60.

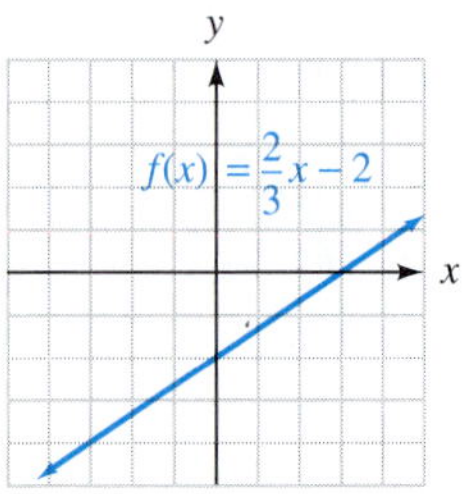

61. a. -4 **b.** 3 **c.** 0

62.

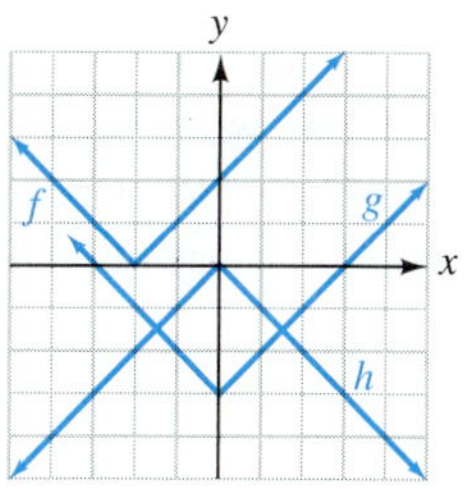

63.

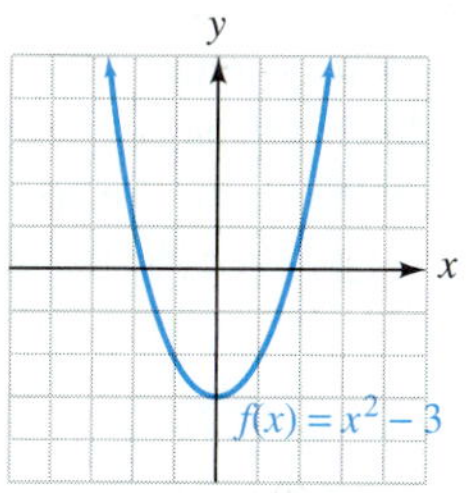

64.

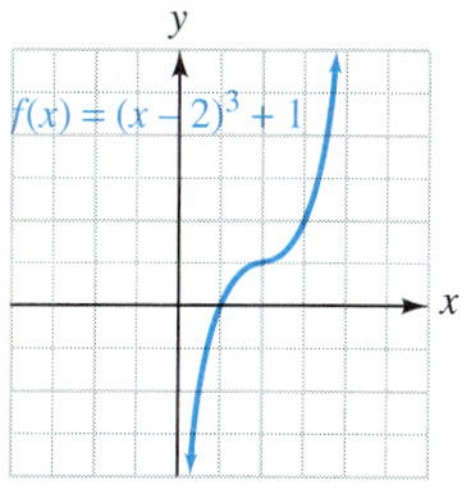

65. D: the set of real numbers, R: the set of real numbers **66.** D: the set of real numbers, R: the set of all real numbers greater than or equal to 1.

Chapter 2 Test (page 177)

1. 240 ft **2.** 1 sec and 7 sec **3.** about 255 ft **4.** 8 sec
5. $\left(\frac{5}{2}, \frac{3}{2}\right)$ **6.**

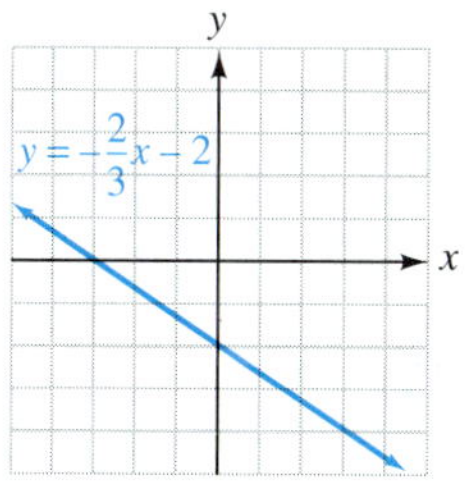

7. (5, 0), (0, −2)

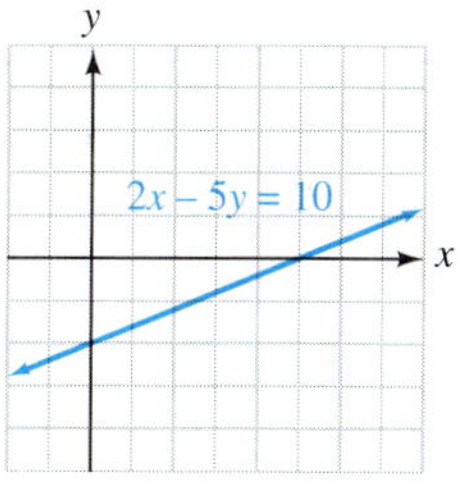

8.

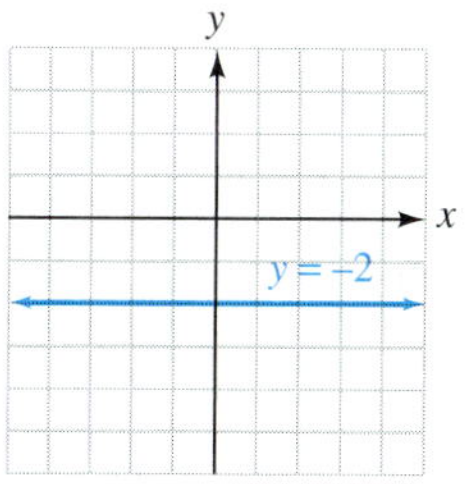

9. 3 **10.** −1.5 degree/hr **11.** $\frac{1}{2}$ **12.** $\frac{2}{3}$ **13.** undefined
14. 0 **15.** $y = 3x + 1$ **16.** $y = 8x + 22$
17. $m = -\frac{1}{3}, \left(0, -\frac{3}{2}\right)$ **18.** neither **19.** $y = -\frac{3}{2}x$
20. a. $v = -600x + 4{,}000$ **b.** (0, 4,000); It gives the value of the copier when new: $4,000 **21.** no **22.** yes
23. D: the set of real numbers, R: the set of nonnegative real numbers. **24.** −20 **25.** 10 **26.** −1 **27.** 3
28. $r^2 - 2r - 1$ **29.** −2 **30.** 2 **31.** function
32. not a function

33.

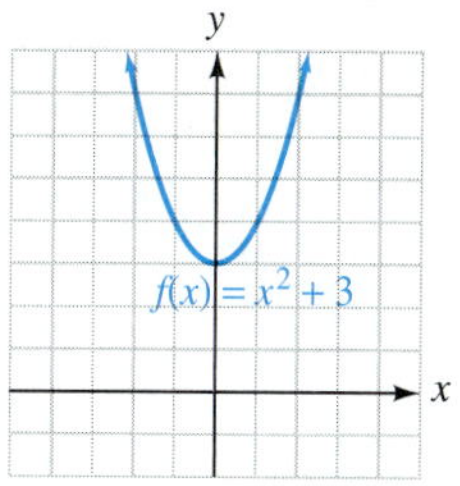

34.

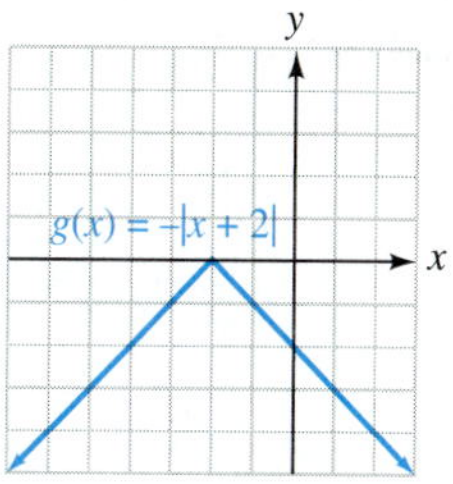

35. D: the set of real numbers, R: the set of all real numbers greater than or equal to −5.

Cumulative Review Exercises, Chapters 1–2 (page 179)

1. 1, 2, 6, 7 **2.** 0, 1, 2, 6, 7 **3.** $-2, 0, 1, 2, \frac{13}{12}, 6, 7$
4. $\sqrt{5}, \pi$ **5.** −2 **6.** $-2, 0, 1, 2, \frac{13}{12}, 6, 7, \sqrt{5}, \pi$
7. 2, 7 **8.** 6 **9.** −2, 0, 2, 6 **10.** 1, 7 **11.** −2
12. −2 **13.** 22 **14.** −2 **15.** $\frac{24}{25}$ **16.** 2 **17.** 4
18. −5 **19.** assoc. prop. of add. **20.** distrib. prop.
21. commut. prop. of add. **22.** assoc. prop. of mult.
23. $-5y$ **24.** $-28st$ **25.** 0 **26.** $z - 4$ **27.** 8
28. −27 **29.** −1 **30.** 6 **31.** $\frac{8}{3}$ **32.** 24
33. $B = \frac{c + Tx}{3y}$ **34.** $h = \frac{2A}{b_1 + b_2}$ **35.** $14,000
36. 39 mph going, 65 mph returning
37.

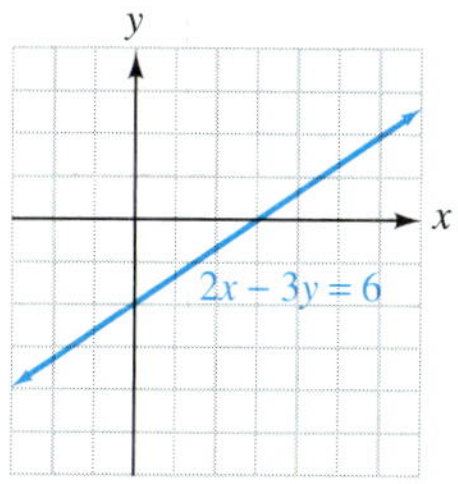

38. $-\frac{5}{6}$ **39.** $y = -\frac{7}{5}x + \frac{11}{5}$
40. $y = -3x - 3$ **41.** 0
42. 2 **43.** yes **44.** no
45. no **46.** −60 **47.** 5
48. −1 **49.** 3 **50.** $3r^2 + 2$

51. It is a function; D: the set of real numbers, R: the set of all real numbers less than or equal to 1.

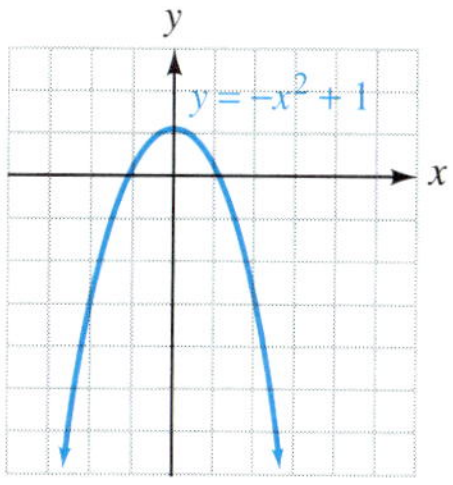

52. It is a function; D: the set of real numbers, R: the set of nonnegative real numbers.

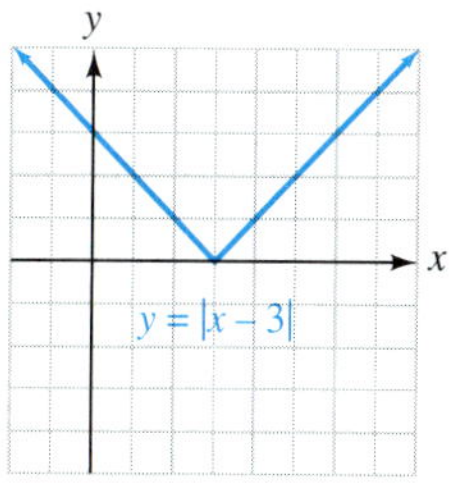

53. The points (2, 20), (3, 40) and (6, 60) do not lie on a straight line. **54. a.** $R = 0.02t + 5.05$ **b.** 7.05 milliohms

Study Set 3.1 (page 190)

1. system **3.** inconsistent **5.** dependent
7. a. true **b.** false **c.** true **d.** true

9. a. −4, (−4, 0), 2, (0, 2), 3, (2, 3) **b.** (−4, 0); (0, 2)
11. no solution; independent **13.** (−2, −1)
15. brace **17.** yes **19.** no

21.
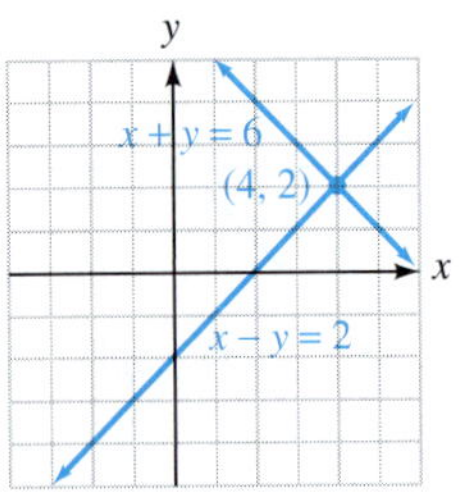

23.
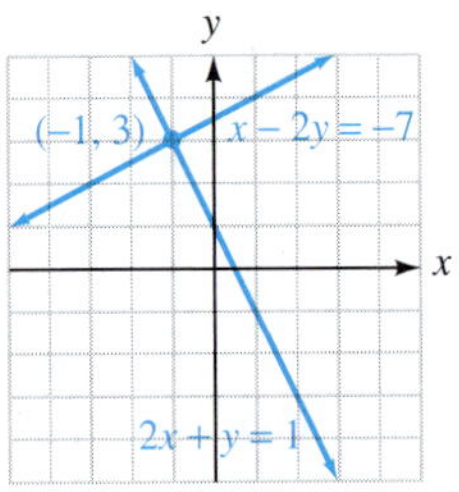

25.
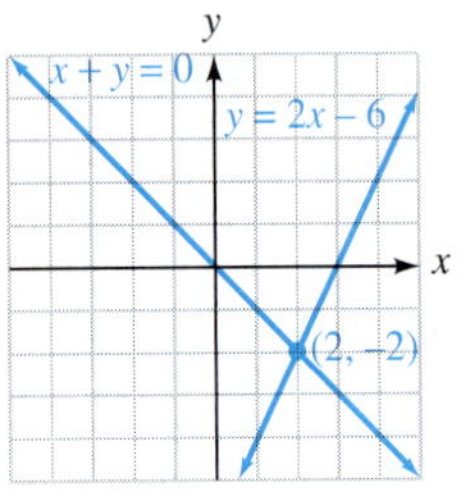

27.
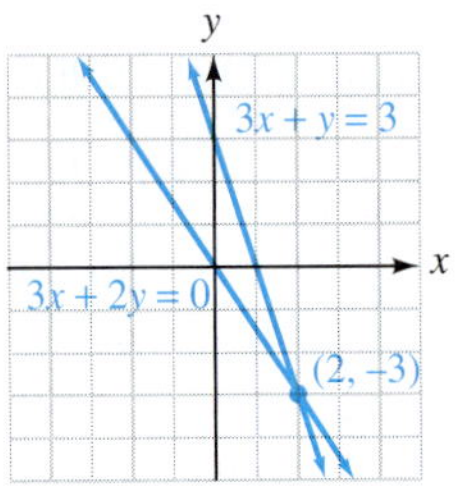

29.
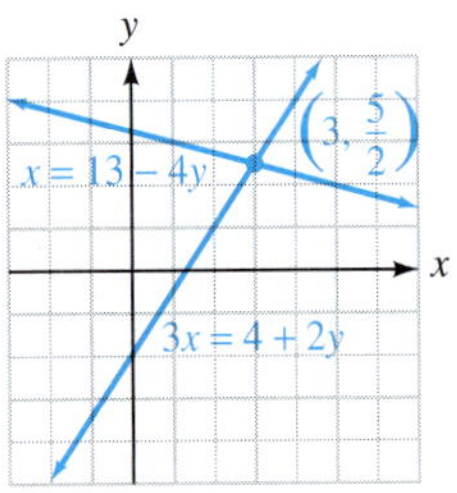

31.
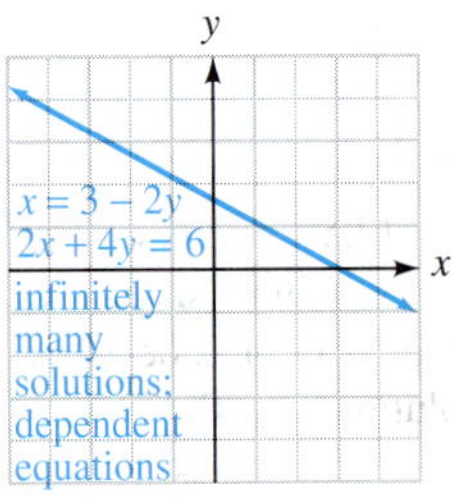

33.
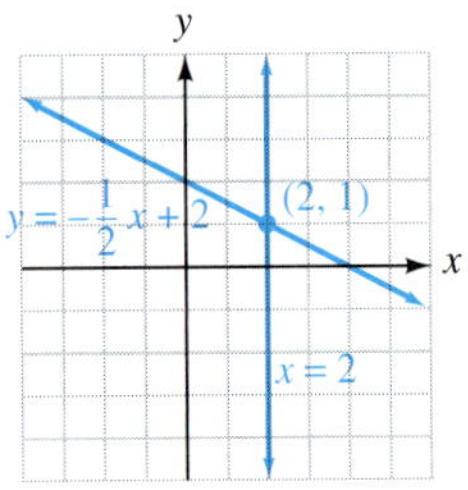

35.
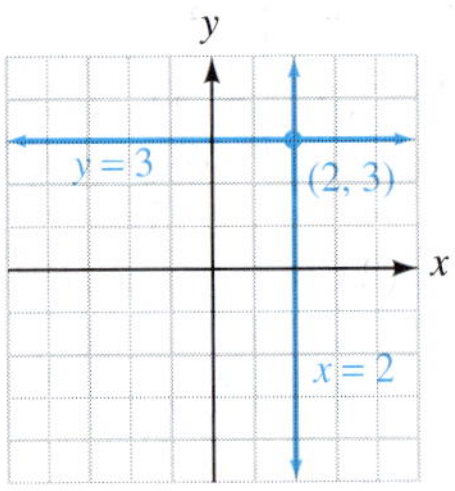

37.
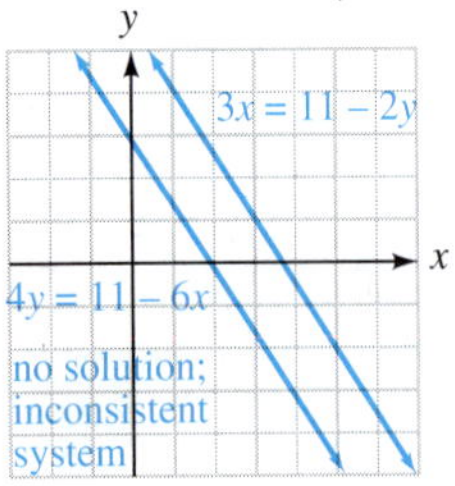

39.
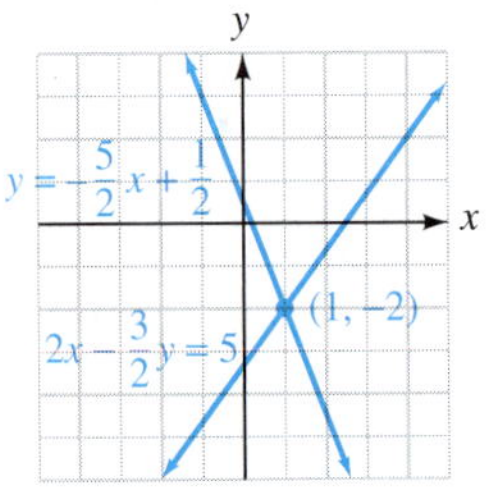

41.
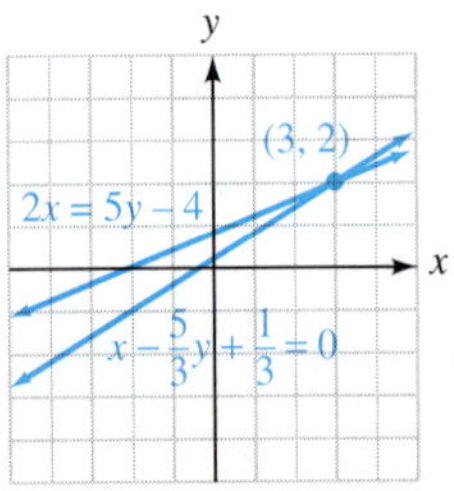

43.
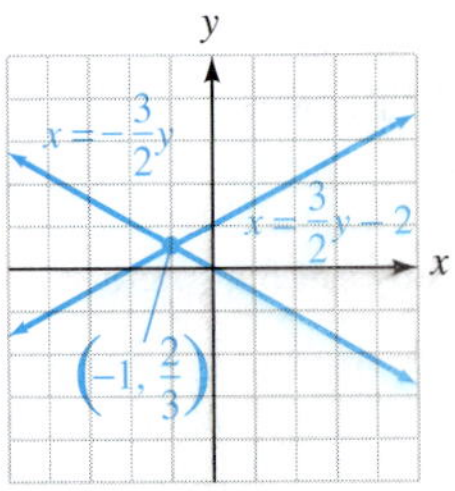

45. (−0.37, −2.69) **47.** (−7.64, 7.04) **49.** 4 **51.** −3
53. Albuquerque **55.** (2,000, 50) **57. a.** 2001; 135 hr
b. 2003; 175 hr **c.** 2007; 110 hr **59. a.** yes
b. (3.75, −0.5) **c.** no **63.** −3 **65.** 0 **67.** D: the set of real numbers, R: the set of all real numbers greater than or equal to −2 **69.** 40.5 cm^2

Study Set 3.2 (page 203)

1. general **3.** eliminated **5. a.** 3; −4 (answers may vary)
b. 2; −3 (answers may vary) **7.** elimination method
9. a. ii **b.** iii **c.** i **11.** $\begin{cases} 7x + 4y = 8 \\ 2x - y = -3 \end{cases}$ **13.** (2, 2)
15. (5, 3) **17.** (−2, 4) **19.** no solution, inconsistent system **21.** (5, 2) **23.** (−4, −2) **25.** (1, 2)
27. $\left(\frac{1}{2}, \frac{2}{3}\right)$ **29.** $\left(5, \frac{3}{2}\right)$ **31.** $\left(-2, \frac{3}{2}\right)$
33. infinitely many solutions, dependent equations **35.** no solution, inconsistent system **37.** (4, 8) **39.** (20, −12)
41. $\left(\frac{2}{3}, \frac{3}{2}\right)$ **43.** $\left(\frac{1}{2}, -3\right)$ **45.** (2, −3)
47. (9, −1) **49.** (2, 3) **51.** $\left(-\frac{1}{3}, 1\right)$ **53.** $475, $800 **55.** dogs: 60 million; cats: 75 million **57.** 16 m by 20 m **59.** 75°, 25° **61.** $3,000 at 10%, $5,000 at 12%
63. 45 mi **65.** 85 racing bikes, 120 mountain bikes
67. 148 g of the 0.2%, 37 g of the 0.7% **69.** Gummy Bears: 45 lb; Jelly Beans: 15 lb **71.** 200 plates **73.** 103
75. a. 590 units per month **b.** 620 units per month
c. A (smaller loss) **83.** $-\frac{5}{2}$ **85.** $-\frac{8}{5}$ **87.** $\frac{4}{3}$

Study Set 3.3 (page 216)

1. system **3.** three **5.** dependent **7. a.** no solution
b. no solution **9.** $x + 2y - 3z = -6$ **11.** yes
13. (1, 1, 2) **15.** (0, 2, 2) **17.** (3, 2, 1) **19.** no solution, inconsistent system **21.** (60, 30, 90) **23.** (2, 4, 8)
25. infinitely many solutions, dependent equations
27. (2, 6, 9) **29.** 30 expensive, 50 middle-priced, 100 inexpensive **31.** 2, 3, 1 **33.** 3 poles, 2 bears, 4 deer
35. 78%, 21%, 1% **37. a.** infinitely many solutions, all lying on the line running down the binding **b.** 3 parallel planes (shelves); no solution **c.** each pair of planes (cards) intersect; no solution **d.** 3 planes (faces of die) intersect at a corner; 1 solution **39.** $y = \frac{1}{2}x^2 - 2x - 1$
41. $x^2 + y^2 - 2x - 2y - 2 = 0$ **43.** $A = 40°$, $B = 60°$, $C = 80°$ **45.** 12, 15, 21

49.
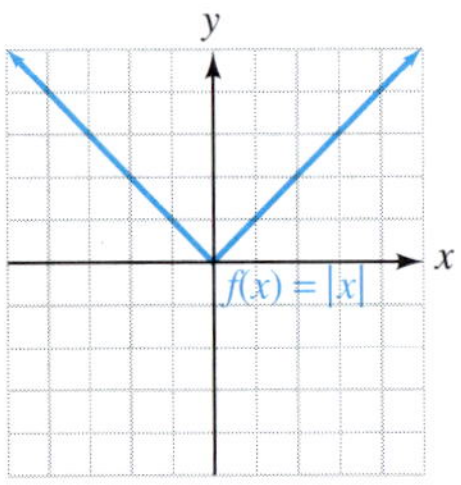

51.
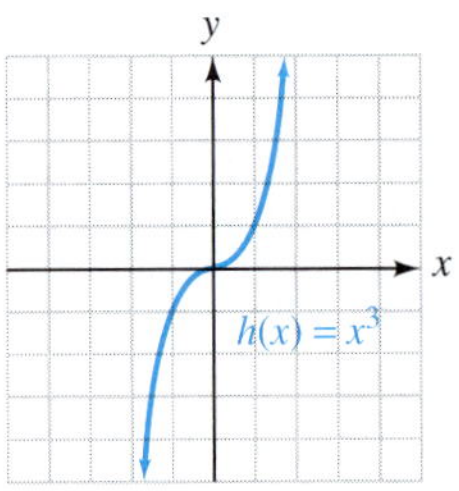

Study Set 3.4 (page 228)

1. matrix **3.** rows, columns **5.** augmented **7. a.** 2×3 **b.** 3×4 **9.** $\begin{cases} x - y = -10 \\ y = 6 \end{cases}$; $(-4, 6)$ **11.** It has no solution. The system is inconsistent. **13. a.** multiply row 1 by $\frac{1}{3}$; $\left[\begin{array}{ccc|c} 1 & 2 & -3 & 0 \\ 1 & 5 & -2 & 1 \\ -2 & 2 & -2 & 5 \end{array}\right]$ **b.** to row 2, add -1 times row 1; $\left[\begin{array}{ccc|c} 1 & 2 & -3 & 0 \\ 0 & 3 & 1 & 1 \\ -2 & 2 & -2 & 5 \end{array}\right]$ **15.** $-1, 1, -5, 2, y, 4$ **17.** $(1, 1)$ **19.** $(2, -3)$ **21.** $(-1, -1)$ **23.** $(0, -3)$ **25.** $(1, 2, 3)$ **27.** $(4, 5, 4)$ **29.** $(2, 1, 0)$ **31.** $(-1, -1, 2)$ **33.** $(-3, 2, 1)$ **35.** $\left(\frac{1}{2}, 1, -2\right)$ **37.** no solution, inconsistent system **39.** infinitely many solutions, dependent equations **41.** $(0, 1, 3)$ **43.** no solution, inconsistent system **45.** $(-4, 8, 5)$ **47.** infinitely many solutions, dependent equations **49.** 22°, 68° **51.** 40°, 65°, 75° **53.** 262,144 **55.** 76°, 104° **57.** founder's circle: 100; box seats: 300; promenade: 400 **61.** $m = \frac{y_2 - y_1}{x_2 - x_1}$ $(x_2 \neq x_1)$ **63.** $y - y_1 = m(x - x_1)$

Study Set 3.5 (page 239)

1. number **3.** minor **5.** rows, columns **7.** dependent, inconsistent **9.** *ad*, *bc* **11.** $\begin{vmatrix} 3 & 4 \\ 2 & -3 \end{vmatrix}$ **13.** $\left(\frac{7}{11}, -\frac{5}{11}\right)$ **15.** 6, 30 **17.** 8 **19.** -2 **21.** 200 **23.** 6 **25.** 1 **27.** 26 **29.** 0 **31.** -79 **33.** $(4, 2)$ **35.** $\left(-\frac{1}{2}, \frac{1}{3}\right)$ **37.** no solution, inconsistent system **39.** $(2, -1)$ **41.** $(1, 1, 2)$ **43.** $(3, 2, 1)$ **45.** $(3, -2, 1)$ **47.** $\left(-\frac{1}{2}, -1, -\frac{1}{2}\right)$ **49.** infinitely many solutions, dependent equations **51.** no solution, inconsistent system **53.** $(-2, 3, 1)$ **55.** 200 of the \$67 phones, 160 of the \$100 phones **57.** \$5,000 in HiTech, \$8,000 in SaveTel, \$7,000 in OilCo **59.** -23 **61.** 26 **67.** no **69.** The graph of *g* is 2 units below the graph of *f*. **71.** *y*-intercept **73.** *x*; *y*

Key Concept (page 243)

1. yes **2.** no **3.** $(-1, 2)$ **4.** $\left(\frac{1}{3}, \frac{1}{4}\right)$ **5.** $(3, 2)$ **6.** $(-1, 0, 2)$ **7.** $(15, 2)$ **8.** $(3, 3, 2)$ **9.** The equations of the system are dependent. There are infinitely many solutions. **10.** The system is inconsistent. There are no solutions.

Chapter Review (page 244)

1. a. $(1, 3), (2, 1), (4, -3)$ (answers may vary) **b.** $(0, -4), (2, -2), (4, 0)$ (answers may vary) **c.** $(3, -1)$ **2.** President Clinton's job approval and disapproval ratings were the same: approximately 47% in 5/94 and approximately 48% in 5/95.

3.

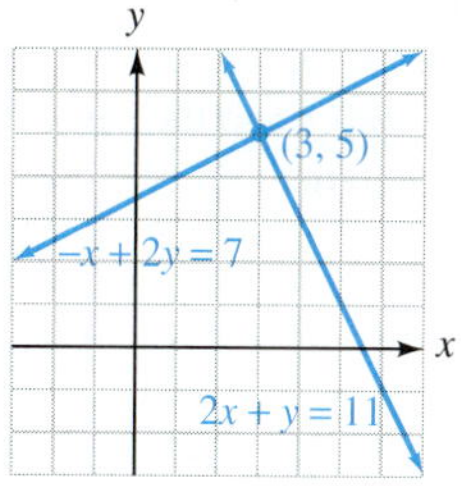

4.

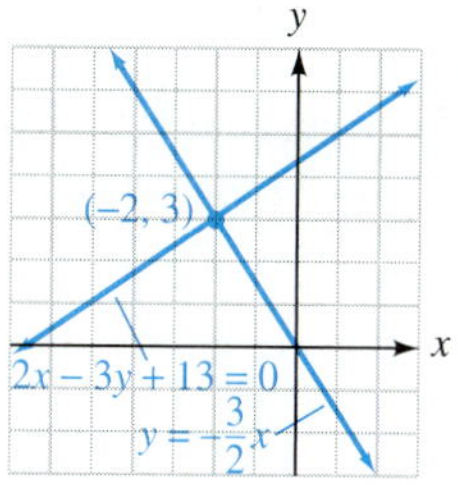

5.

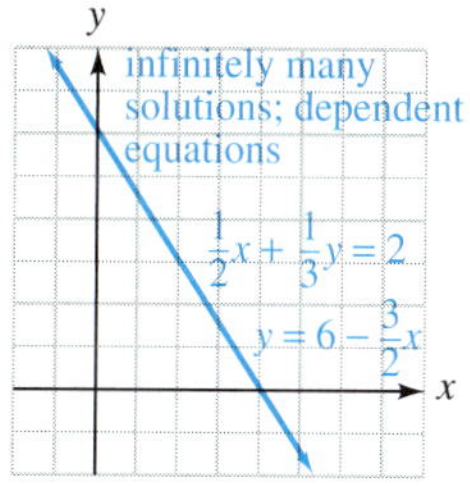

6.

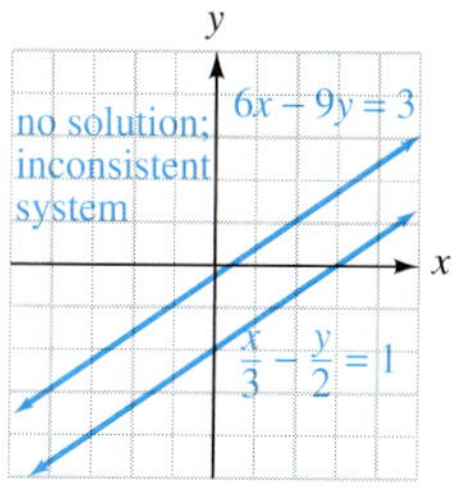

7. 2 **8.** -1 **9.** $(-1, 3)$ **10.** $(-3, -1)$ **11.** $(3, 4)$ **12.** infinitely many solutions, dependent equations **13.** $(-3, 1)$ **14.** no solution, inconsistent system **15.** $(9, -4)$ **16.** $\left(4, \frac{1}{2}\right)$ **17.** Using the elimination method, the computations are easier. **18.** $(-1, 0.7)$ (answers may vary); $\left(-1, \frac{2}{3}\right)$ **19.** A-H: 162 mi, A-SA: 83 mi **20.** 8 mph, 2 mph **21.** 17,500 bottles **22.** no **23.** $(1, 2, 3)$ **24.** no solution, inconsistent system **25.** $(-1, 1, 3)$ **26.** infinitely many solutions, dependent equations **27.** yes; infinitely many solutions **28.** 25 lb peanuts, 10 lb cashews, 15 lb Brazil nuts **29.** $\left[\begin{array}{cc|c} 5 & 4 & 3 \\ 1 & -1 & -3 \end{array}\right]$ **30.** $\left[\begin{array}{ccc|c} 1 & 2 & 3 & 6 \\ 1 & -3 & -1 & 4 \\ 6 & 1 & -2 & -1 \end{array}\right]$ **31.** $(1, -3)$ **32.** $(5, -3, -2)$ **33.** infinitely many solutions, dependent equations **34.** no solution, inconsistent system **35.** \$4,000 at 6%, \$6,000 at 12% **36.** 18 **37.** 38 **38.** -3 **39.** 28 **40.** $(2, 1)$ **41.** no solution, inconsistent system **42.** $(1, -2, 3)$ **43.** $(-3, 2, 2)$ **44.** 2 cups mix A, 1 cup mix B, 1 cup mix C

Chapter Test (page 248)

1.

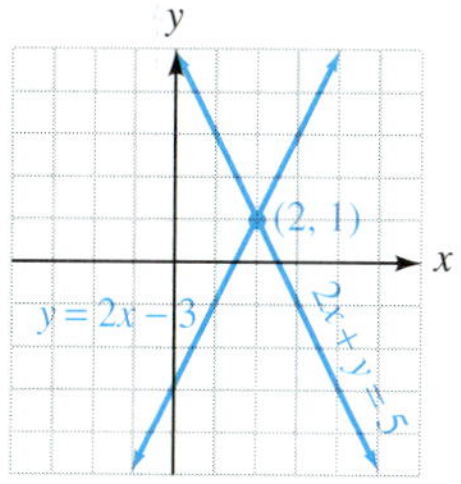

2. $(3, 1)$ **3.** $(7, 0)$ **4.** $(2, -3)$ **5.** dependent **6.** no **7.** $(3, 2, -1)$ **8.** 55, 70 **9.** 15 gal 40%, 5 gal 80% **10.** 1,375 impressions **11.** $(2, 2)$ **12.** no solution, inconsistent system **13.** 22 **14.** 4 **15. a.** $\begin{vmatrix} -6 & -1 \\ -6 & 1 \end{vmatrix}$ **b.** $\begin{vmatrix} 1 & -1 \\ 3 & 1 \end{vmatrix}$ **16.** -3 **17.** 3

18. −1 **19.** C: 60, GA: 30, S: 10 **21.** The system has no solution. **22.** (2035, 25); in the year 2035, the percent of the U.S. population that is children under age 18 and the percent of the U.S. population that is adults 65 and older will be the same, about 25%.

Cumulative Review Exercises, Chapters 1–3 (page 249)

1.

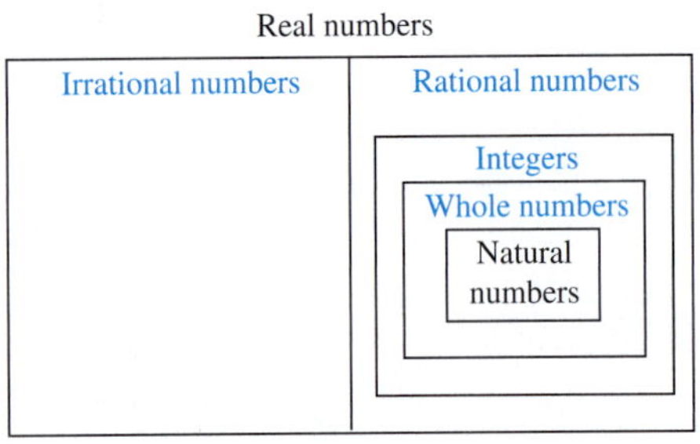

2. $504,000,000,000 **3.** 70 **4.** 2 **5.** $-12.1x^2 + 12.7x$ **6.** −4 **7.** 20 mph **8.** 5 lb apple slices, 5 lb banana chips **9.** −28 **10.** $-\frac{1}{3}$ **11.** −2 **12.** R, identity **13.** $B = \frac{\lambda - Ax}{A}$ **14.** $d_2 = d_1 - vt$

15.

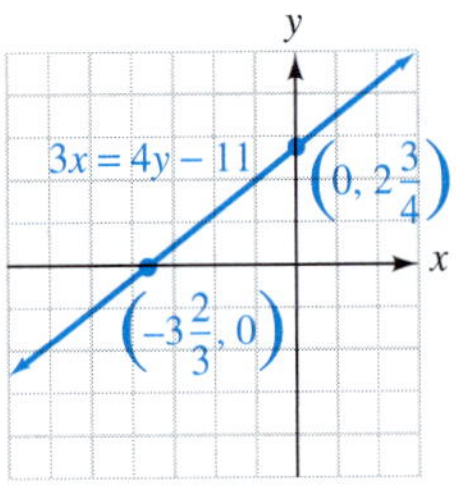

16.

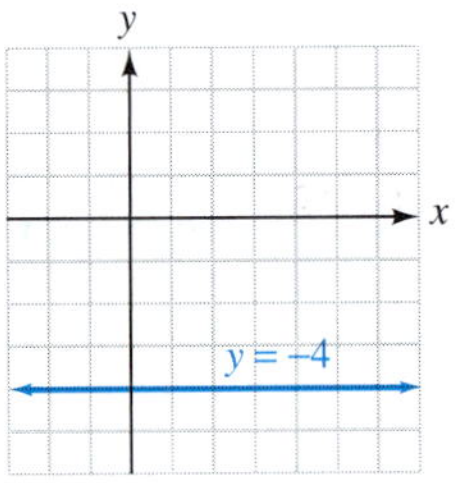

17. $y = -3x + 17$ **18.** $-\frac{4}{5}$ **19.** −105 **20.** −95 **21.** $f(x) = x^3$ (answers may vary) **22.** yes **23.** D: the set of real numbers, R: the set of real numbers greater than or equal to 0.

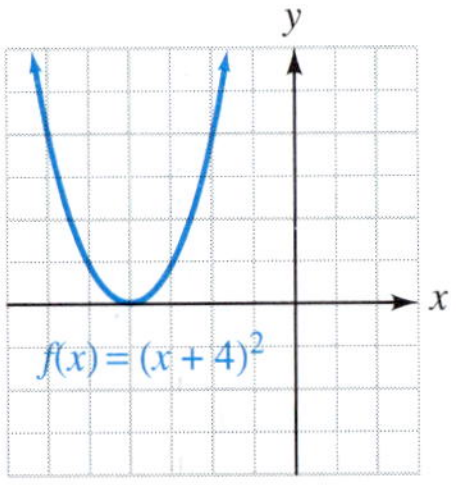

24. a. −1 **b.** 2 **25.** No. It does not pass the vertical line test. **26.** $v = 17.5x + 300$ **27.** (2, −1) **28.** (−1, 0, 2) **29.** 26 **30.** 26

Study Set 4.1 (page 259)

1. inequality **3.** parenthesis **5.** linear **7.** is less than, is greater than or equal to **9. a.** equation **b.** expression **c.** inequality **d.** expression **11.** i and ii **13. a.** $6 > 0$ **b.** $0 > -6$ **c.** $16 > -8$ **d.** $-2 < 1$ **15. a.** yes **b.** no **c.** yes **d.** no **17.** $(-\infty, \infty)$; −1 0 1 **19.** −10, ≤, −5, −∞, $x \leq 2$ **21. a.** iii **b.** i **c.** ii **23.** $(-3, \infty)$ −3 **25.** $[20, \infty)$ 20 **27.** $[60, \infty)$ 60 **29.** $\left(-\frac{10}{3}, \infty\right)$ −10/3 **31.** $(-\infty, 1)$ 1 **33.** $\left[-\frac{2}{5}, \infty\right)$ −2/5 **35.** $[-2, \infty)$ −2 **37.** $(6, \infty)$ 6 **39.** $(-\infty, 3)$ 3 **41.** $(-\infty, \infty)$ −1 0 1 **43.** $(-\infty, -6]$ −6 **45.** $(-\infty, 1.5]$ 1.5 **47.** $(-\infty, 20]$ 20 **49.** $(-\infty, 10)$ 10 **51.** $\left(-\infty, \frac{4}{3}\right]$ 4/3 **53.** $(-\infty, \infty)$ −1 0 1 **55.** ∅ −1 0 1 **57.** $[6, \infty)$ 6 **59.** $[-36, \infty)$ −36 **61.** $\left(-\infty, \frac{45}{7}\right]$ 45/7 **63.** $(-\infty, 3]$ 3 **65.** Midwest, South **67.** $6 + 45 \not> 52$ **68.** *d* is negative, *d* is zero, *d* is positive **69.** 8 hr **71.** 15 **73.** 13 hr **75.** anything over $900 **77.** $x < 1$ **79.** $x \geq -4$ **85.** 4, 5, 3 **87.** 6, −6

Study Set 4.2 (page 271)

1. compound **3.** double **5.** intersection **7.** both **9.** and **11.** union **13. a.** no **b.** yes **15. a.** $[-2, 1)$ **b.** $[2, 2]$ **c.** ∅ **17.** union, intersection **19. a.** $(-3, 3)$ −3 3 **b.** $(-\infty, -3) \cup (3, \infty)$ −3 3 **c.** $(-3, 3]$ −3 3 **21. a.** 2 3 **b.** −2 3 **23.** $(-2, 5]$ −2 5 **25.** $(-10, -9)$ −10 −9 **27.** $(2, 3]$ 2 3 **29.** ∅ −1 0 1 **31.** $[2, \infty)$ 2 **33.** $(-\infty, -15)$ −15 **35.** $(-3, 1)$ 0.8 1.1 **37.** $(3, 9)$ 3 9

39. $(-11, -4)$ 

41. $(-12, -6]$ 

43. $[-1, -1]$ 

45. $\varnothing$ 

47. $(-6, -3)$ 

49. $(-2, 4]$ 

51. $[4, 4]$ 

53. $(-\infty, -2] \cup (6, \infty)$ 

55. $(-\infty, -1) \cup (2, \infty)$ 

57. $(-\infty, 2) \cup (7, \infty)$ 

59. $(-\infty, \infty)$ 

61. $(-\infty, 1)$ 

63. $(-\infty, \infty)$ 

65. a. 128, 192 **b.** $32 \le s \le 48$ **67.** See doctor today. **69. a.** 1999 **b.** 1998, 1999, 2000, 2001 **c.** 1999, 2000 **d.** 1998, 1999, 2000, 2001 **71. a.**  **b.** 

77. 85.7, 86, 86 **79.** 13.3 pts/game

Study Set 4.3 (page 284)

1. absolute value **3.** inequality **5.** opposite **7.** compound **9.** negative **11.** more than **13.** 5 **15. a.** $-2, 2$ **b.** $-1.99, -1, 0, 1, 1.99$ **c.** $-4, -3, -2.01, 2.01, 3, 4$ **17. a.** $x - 7 = 8, x - 7 = -8$, **b.** $x + 10, x - 3, x + 10, -(x - 3)$ **19. a.** $x = 8$ or $x = -8$ **b.** $x \le -8$ or $x \ge 8$ **c.** $-8 \le x \le 8$ **d.** $5x - 1 = x + 3$ or $5x - 1 = -(x + 3)$ **21. a.** ii **b.** iii **c.** i **23.** $(-\infty, -1) \cup (3, \infty)$ **25.** $23, -23$ **27.** $4, -4$ **29.** $9.1, -2.9$ **31.** $\frac{14}{3}, -6$ **33.** $12, -12$ **35.** no solution **37.** $2, -\frac{1}{2}$ **39.** -8 **41.** $-4, -28$ **43.** $40, -20$ **45.** $\frac{12}{11}, -\frac{12}{11}$ **47.** $-2, -\frac{4}{5}$ **49.** $0, -2$ **51.** 0 **53.** $\frac{4}{3}$ **55.** $0, -6$ **57.** $\frac{5}{6}, -\frac{5}{6}$

59. $(-4, 4)$ 

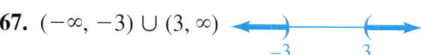

61. $[-21, 3]$ 

63. $\left(-\frac{8}{3}, 4\right)$ 

65. no solution

67. $(-\infty, -3) \cup (3, \infty)$ 

69. $(-\infty, -12) \cup (36, \infty)$ 

71. $\left(-\infty, -\frac{16}{3}\right) \cup (4, \infty)$ 

73. $(-\infty, \infty)$ 

75. $(-\infty, -2] \cup \left[\frac{10}{3}, \infty\right)$ 

77. $(-\infty, -2) \cup (5, \infty)$ 

79. $[-10, 14]$ 

81. $\left(-\frac{5}{3}, 1\right)$ 

83. $(-\infty, -24) \cup (-18, \infty)$ 

85. no solution **87.** $(-3, 1)$ **89.** $\left(-\infty, -\frac{5}{3}\right] \cup [-1, \infty)$ **91.** $70° \le t \le 86°$ **93. a.** $|c - 0.6°| \le 0.5°$ **b.** $[0.1°, 1.1°]$ **95. a.** 26.45%, 24.76% **b.** It is less than or equal to 1%. **101.** 50°, 130°

Study Set 4.4 (page 292)

1. linear, two **3.** edge **5. a.** (3, 1) yes **b.** no **c.** yes **d.** no **7.** $m = 3$; $(0, -1)$ **9.** no **11. a.** $x \ge 2$ 

b. 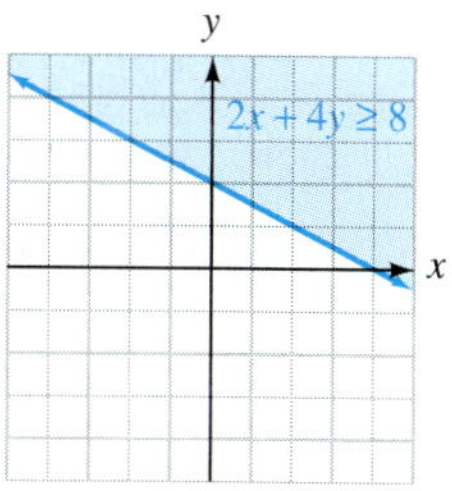

13. 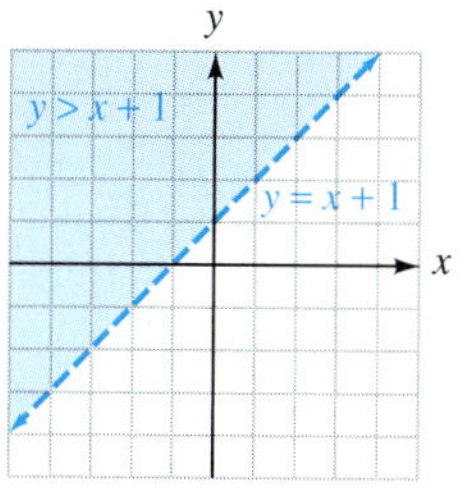

15. 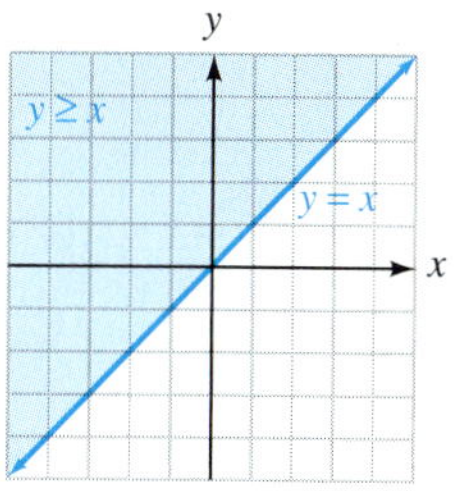

17. 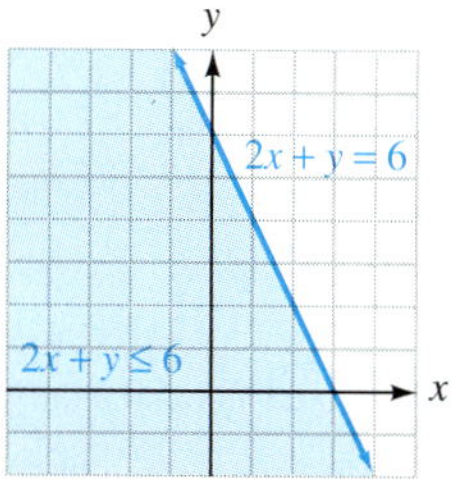

19. 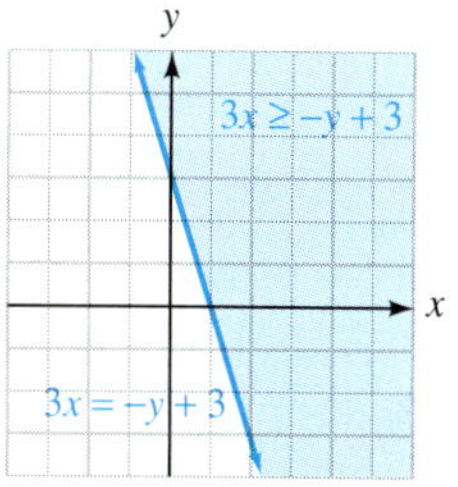

21.

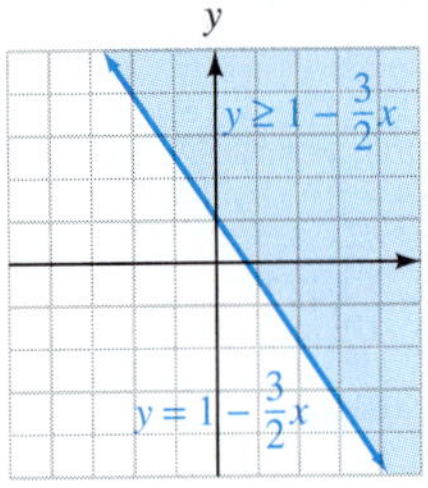

23.

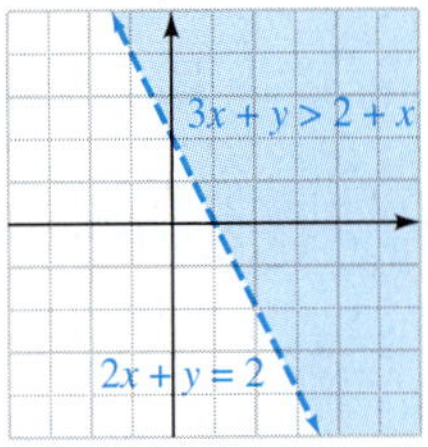

25.

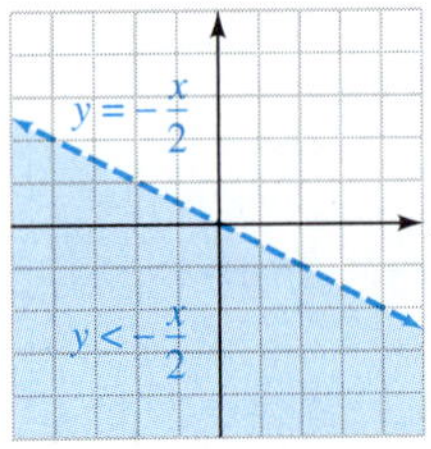

27.

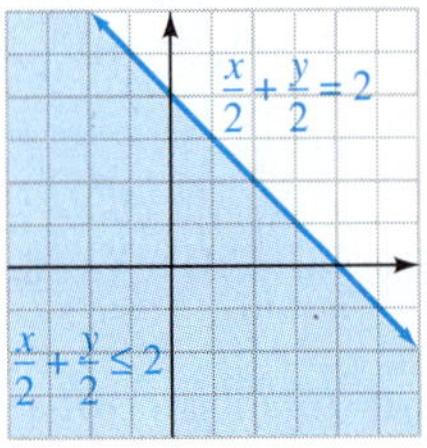

29.

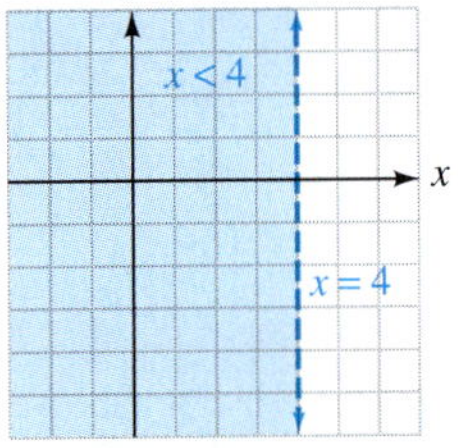

31.

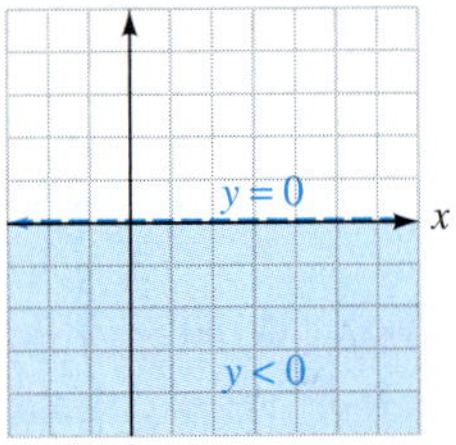

33. $3x + 2y > 6$ **35.** $x \le 3$

37.

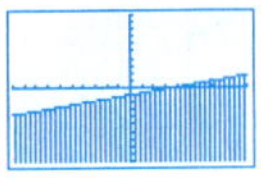

39.

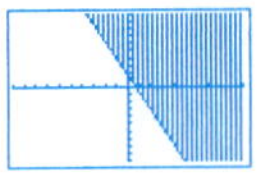

41. a. the Mississippi River **b.** the area of the U.S. west of the Mississippi River

43. $4x + 6y \le 120$; (5, 15), (15, 10), (20, 5)

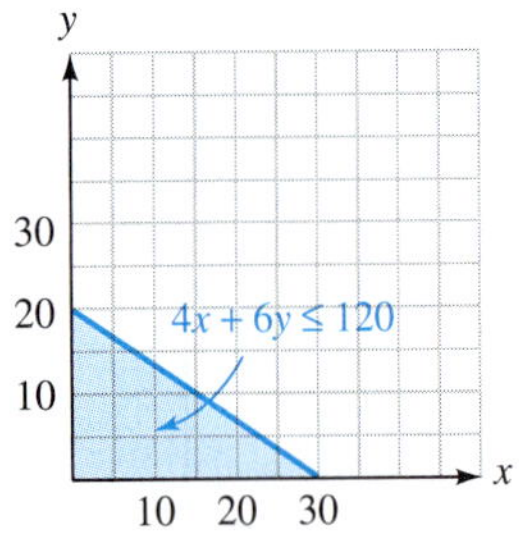

45. $10x + 15y \ge 1{,}200$; (40, 80), (80, 80), (120, 40) **49.** yes **51.** $(-2, -3)$

Study Set 4.5 (page 299)

1. inequalities **3.** intersect **5. a.** yes **b.** no **c.** no **d.** yes **7. a.** false **b.** true **c.** true **d.** false **e.** true **f.** true

9.

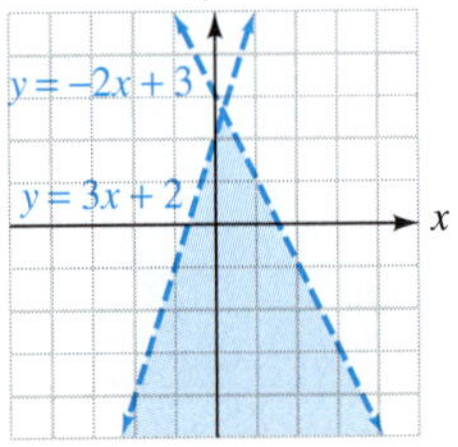

11.

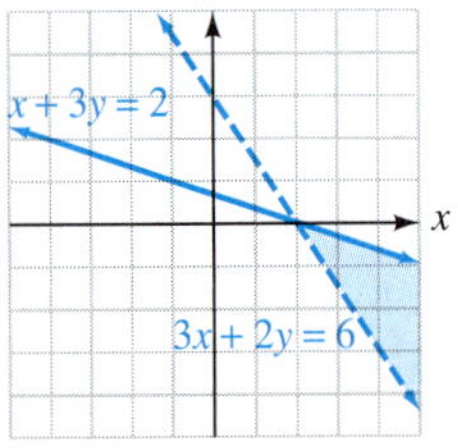

13.

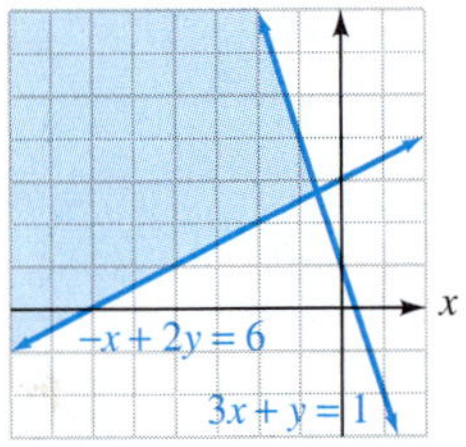

15.

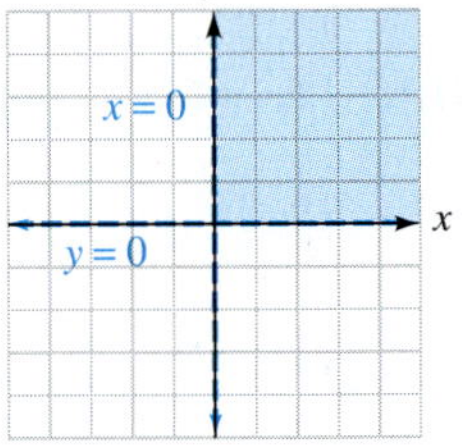

17.

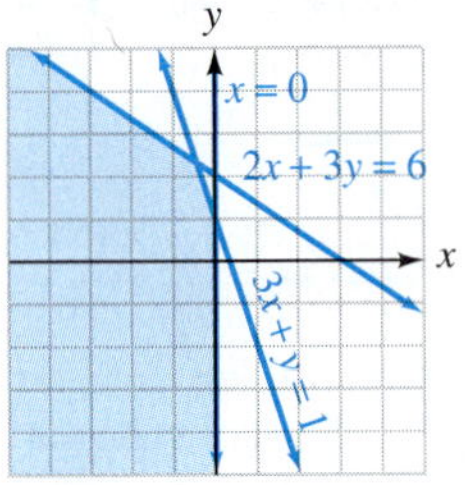

19.

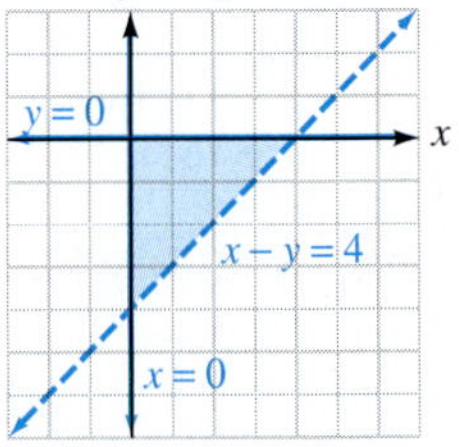

21.

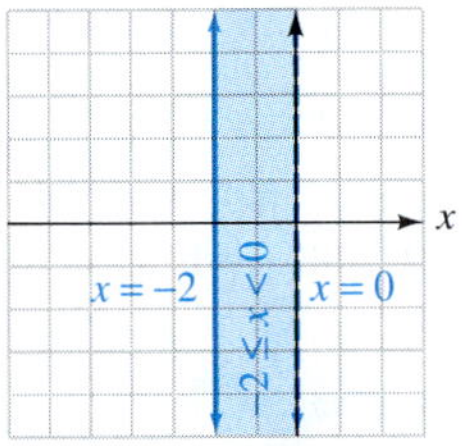

23.

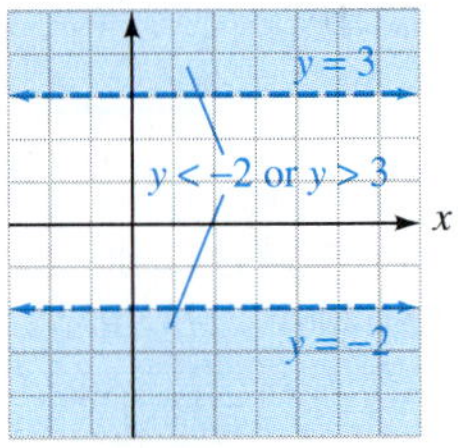

25.

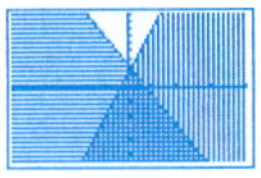

27.

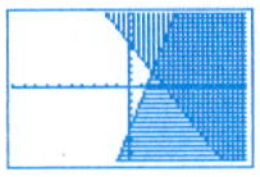

29.

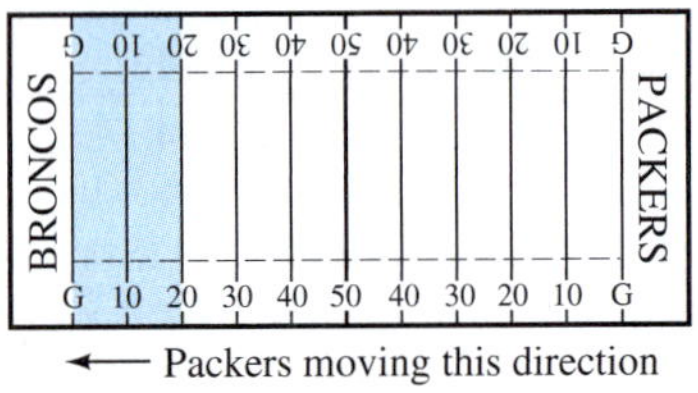

31.

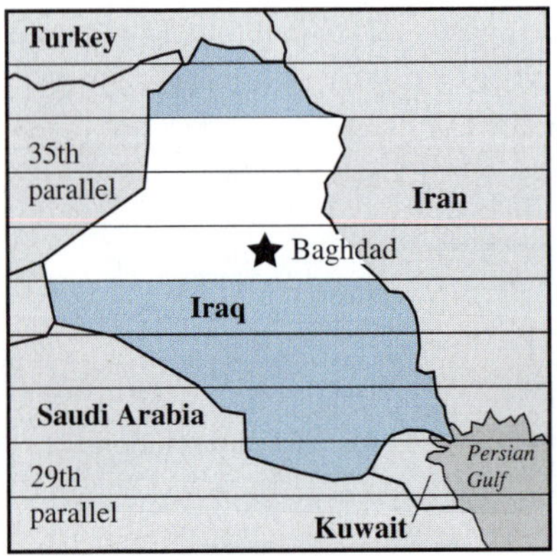

33. 1 \$10 CD and 2 \$15 CDs, 4 \$10 CDs and 1 \$15 CD

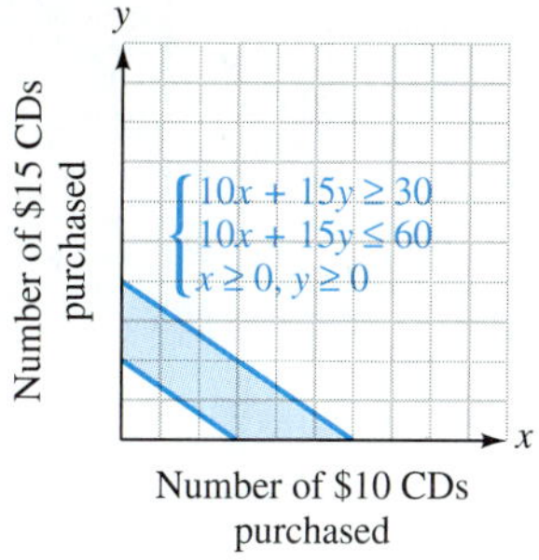

35. 2 desk chairs and 4 side chairs, 1 desk chair and 5 side chairs

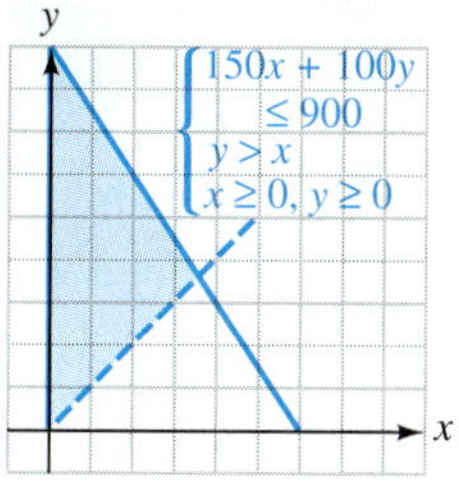

41. IV **43.** II

Key Concept (page 303)

1. compound inequality **2.** system of linear inequalities **3.** absolute value inequality **4.** linear inequality in two variables **5.** linear inequality in one variable **6.** double linear inequality **7.** compound inequality **8.** absolute value inequality **9.** linear inequality in two variables **10.** yes **11.** no **12.** yes **13.** no **14.** yes **15.** yes **16.** no **17.** no **18.** yes **19.**

20.

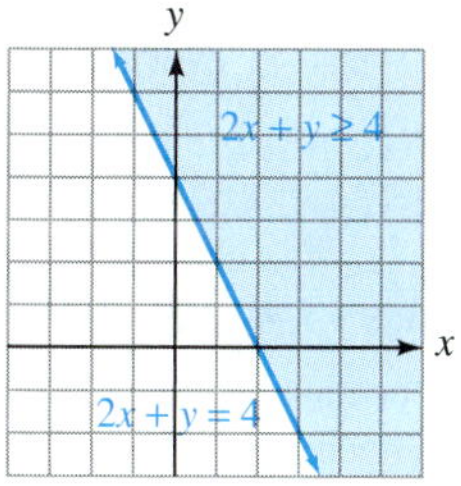

Chapter Review (page 304)

1. $(-\infty, 3]$

2. $[4, \infty)$

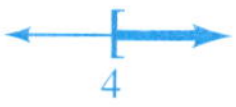

3. $(-\infty, 20)$

4. $\left(-\infty, -\frac{51}{11}\right)$

5. $(-\infty, \infty)$

6. $\varnothing$

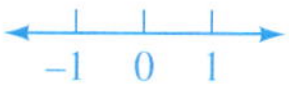

8. \$20,000 or more **9.** yes **10.** no

11.

12.

13. $[-10, -4)$

14. $(-\infty, -11)$

15. $\varnothing$

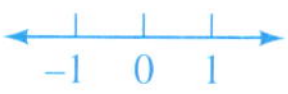

16. $[0, 0]$

17. $\left(-\frac{1}{3}, 2\right)$

18. $[1, 9]$

19. yes **20.** no

21. $(-\infty, -5) \cup (4, \infty)$

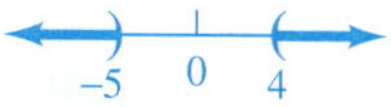

22. $(-\infty, \infty)$

23. $17 \le 4l \le 25$, $4.25 \text{ ft} \le l \le 6.25 \text{ ft}$ **24. a.** ii, iv **b.** i, iii **25.** 2, -2 **26.** 3, $-\frac{11}{3}$ **27.** $\frac{26}{3}$, $-\frac{10}{3}$ **28.** no solution **29.** $\frac{1}{5}$, -5 **30.** $\frac{13}{12}$

31. $[-3, 3]$

32. $(-5, -2)$

33. $\left[-3, \frac{19}{3}\right]$

34. no solution

35. $(-\infty, -1) \cup (1, \infty)$

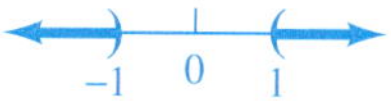

36. $(-\infty, -4] \cup \left[\frac{22}{5}, \infty\right)$

37. $\left(-\infty, \frac{4}{3}\right) \cup (4, \infty)$

38. $(-\infty, \infty)$

39. Since $|0.04x - 8.8|$ is always greater than or equal to 0 for any real number x, this absolute value inequality has no solution. **40.** Since $\left|\frac{3x}{50} + \frac{1}{45}\right|$ is always greater than or equal to 0 for any real number x, this absolute value inequality is true for all real numbers. **41. a.** $|w - 8| \le 2$ **b.** $[6, 10]$ **42.** 3, -3

43.

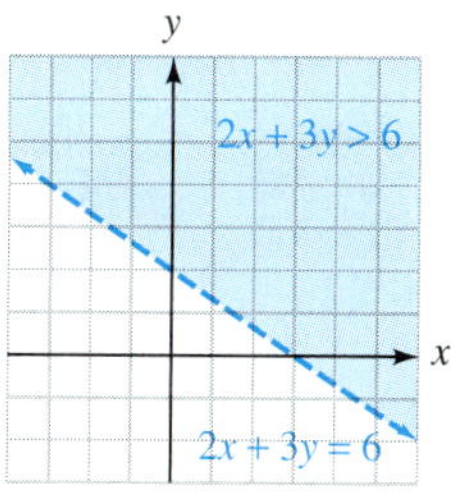

44.

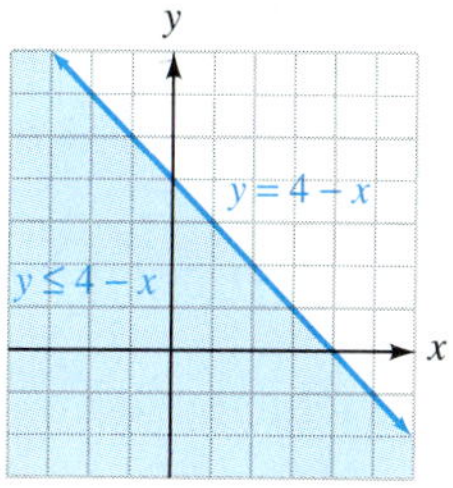

45.

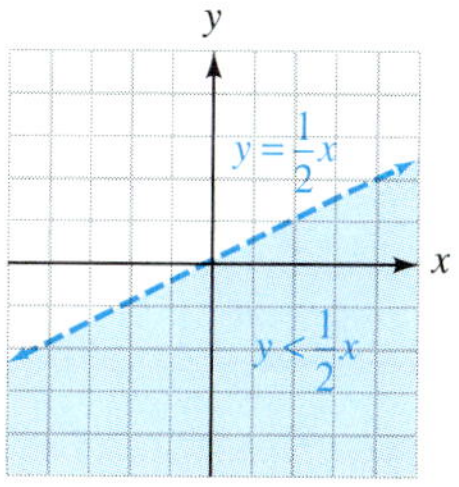

46.

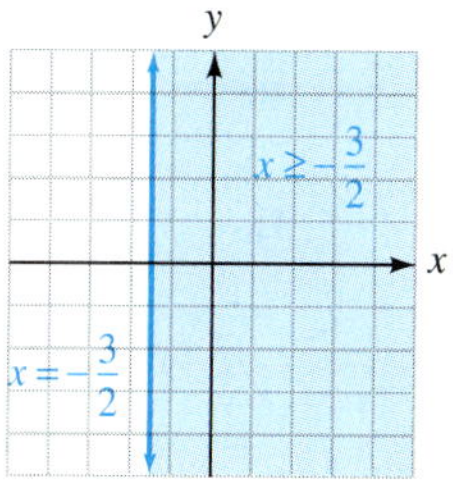

47. $6x + 4y \ge 10{,}200$; (1,800, 0), (1,000, 1,500), (2,000, 2,000)

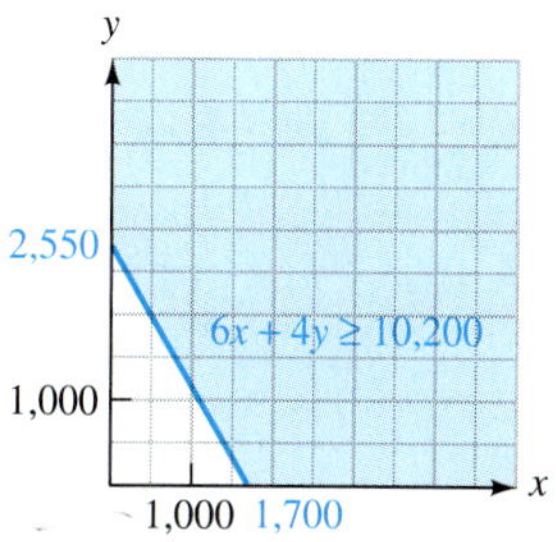

48. $3x - 4y > 12$

49.

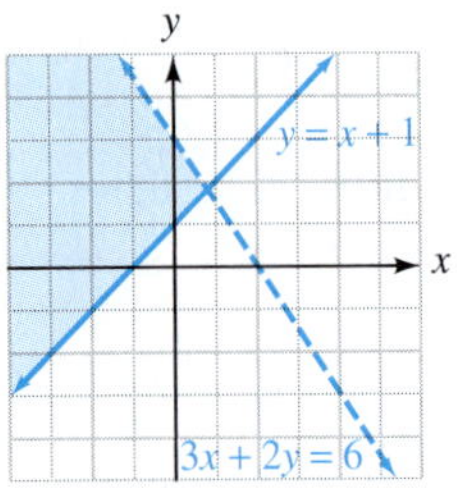

50.

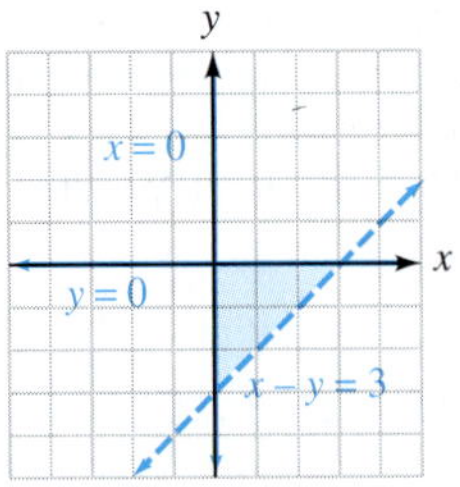

51.

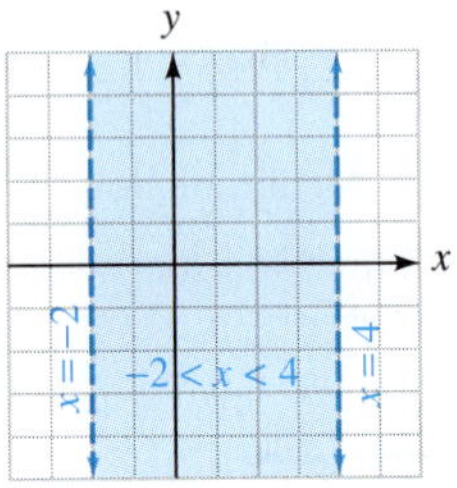

52.

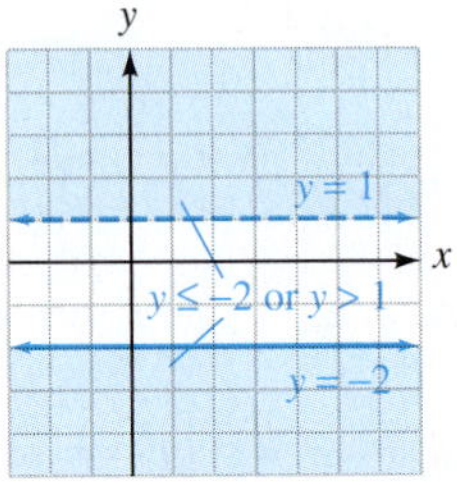

53. ≥, ≤, ≥, ≤ **54. a.** true **b.** false **c.** true **d.** false **e.** true **f.** true

Chapter Test (page 307)

1. false **2.** yes **3.** $(12, \infty)$ 

4. $(-\infty, -5]$ 

5. $(-\infty, -14)$ 

6. $(-\infty, \infty)$ 

7. more than 78

8. $\left[1, \frac{9}{4}\right]$ 

9. $(-\infty, -3) \cup (8, \infty)$ 

10. $(-2, 16)$ 

11. no solution 

12. $-5, \frac{23}{3}$

13. 4, −4 **14.** $\frac{8}{9}, -\frac{8}{9}$ **15.** no solution

16. 

17. 

18. $[-7, 1]$ 

19. $(-\infty, -9) \cup (13, \infty)$ 

20. $(-\infty, 1) \cup (3, \infty)$ 

21. $\left[\frac{4}{3}, \frac{8}{3}\right]$ 

22. $(-\text{in}, \infty)$ 

23. $(-6, -3)$

24.

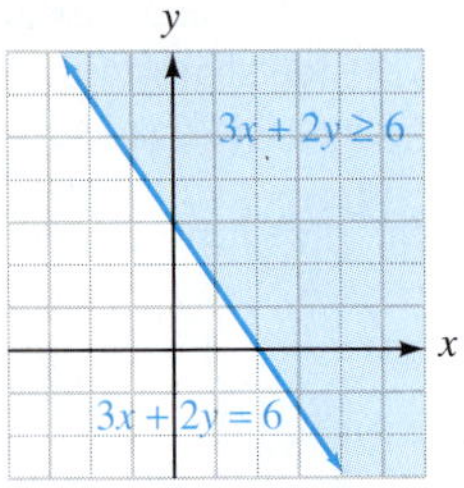

25.

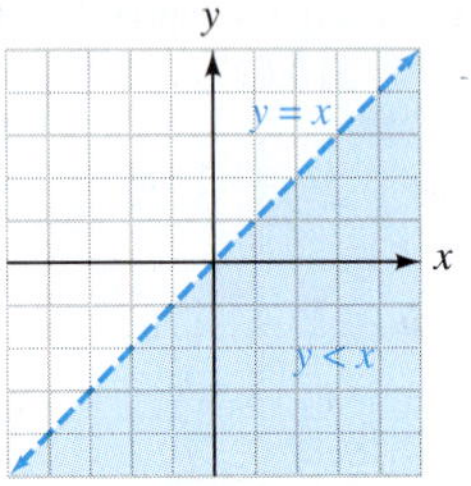

26.

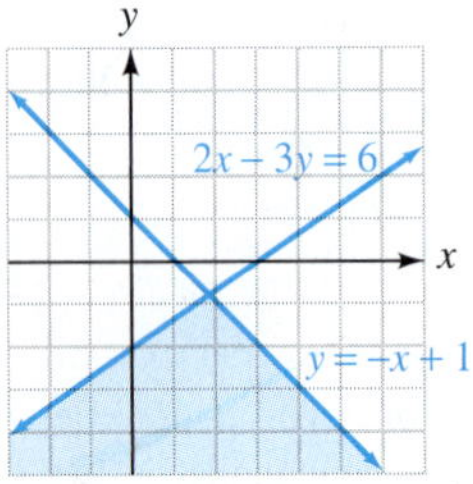

27.

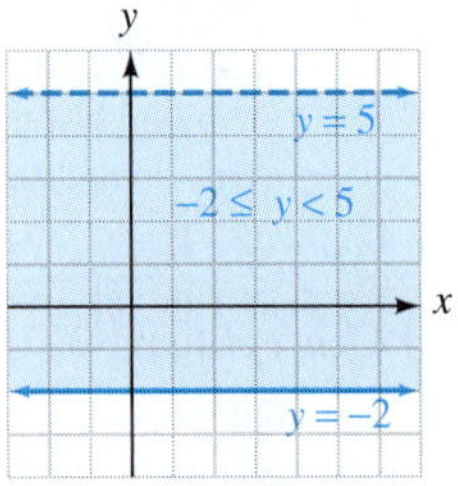

28. (1, 1), (2, 1), (2, 2)

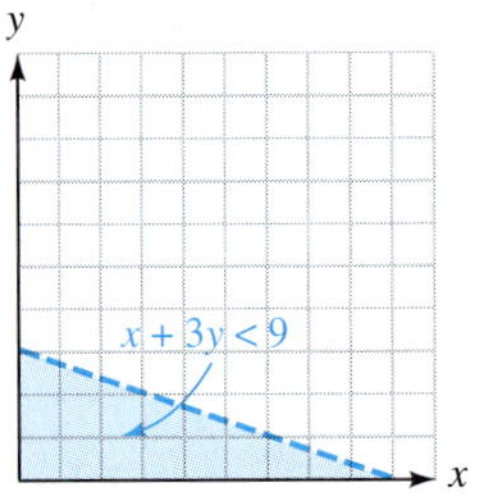

29. a. (3, −4) is a solution of inequality 2. **b.** No; it does not lie in the doubly shaded region. **30.** ≤, ≥, ≥, ≤

Cumulative Review Exercises, Chapters 1–4 (page 309)

1. rational numbers: terminating and repeating decimals; irrational numbers: nonterminating, nonrepeating decimals **2.** 0.125, 0.0625, 0.03125, 0.054125 **3.** 10 **4.** −6 **5.** $-2a + b - 2$ **6.** $8t - 9$ **7.** 201 ft^2 **8.** \$20,000 **9.** 3 **10.** 6 **11.** no solution **12.** −2 **13.** $d = \frac{l - a}{n - 1}$ **14.** parallel **15.** $y = \frac{1}{3}x + \frac{11}{3}$ **16.** $-\frac{8}{5}$ **17.** 9,000 prisoners/yr **18.** 1990-1995, 72,200 prisoners/yr **19.** 10 **20.** 14

21.

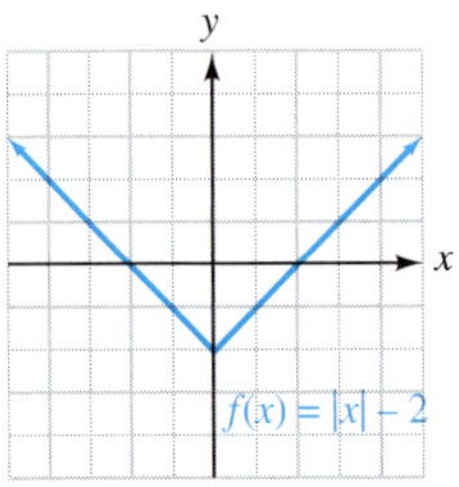

D: the set of real numbers, R: the set of all real numbers greater than or equal to −2

22. yes **23.** (2, 1) **24.** (3, 1) **25.** (1, 1) **26.** (−1, −1, 3) **27.** (−1, −1) **28.** (1, 2, −1) **29.** (7, 23), in 1907 the percent of U.S. workers in white-collar and farming jobs was the same (23%); (45, 42), in 1945 the percent of U.S. workers in white-collar and blue-collar jobs was the same (42%).

30. a. $y = -0.06x + 8.2$ **b.** 2.8 L/min **31.** 750 **32.** 250 \$5 tickets, 375 \$3 tickets, 125 \$2 tickets **33.** $3, -\frac{3}{2}$ **34.** $-5, -\frac{3}{5}$ **35.** $(-\infty, 11]$ 

36. $(-3, 3)$ 

37. $\left[-\frac{2}{3}, 2\right]$ 

38. $(-\infty, -4) \cup (1, \infty)$ 

39.
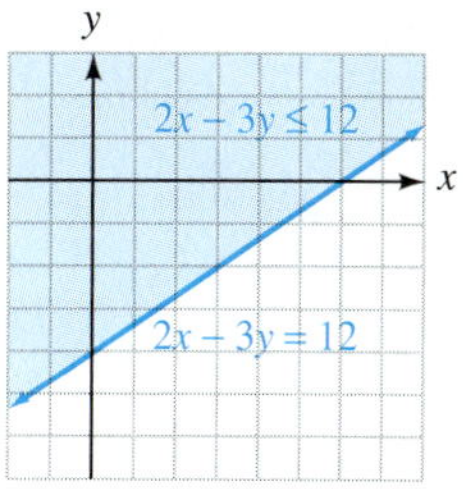

40.
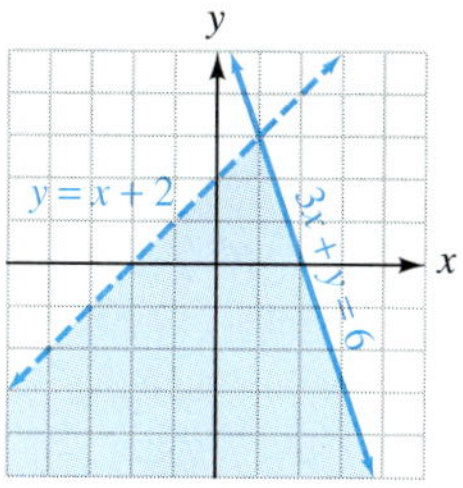

Study Set 5.1 (page 322)

1. exponential **3.** factor **5.** power **7. a.** x^{m+n} **b.** x^{mn} **c.** $x^n y^n$ **d.** $\frac{x^n}{y^n}$ **e.** 1 **f.** $\frac{1}{x^n}$ **g.** x^{m-n} **h.** n **i.** $\frac{y^n}{x^m}$ **9.** multiply **11. a.** 1 **b.** reciprocal **13.** 9, (−2), 11 **15.** $6x$; 3 **17.** x; 5 **19.** b; 6 **21.** $\frac{n}{4}$; 3 **23.** $m - 8$; 6 **25.** −9 **27.** 9 **29.** $\frac{1}{25}$ **31.** $-\frac{1}{81}$ **33.** $\frac{1}{36}$ **35.** −1 **37.** 1 **39.** $\frac{27}{64}$ **41.** 49 **43.** $\frac{1}{16}$ **45.** −48 **47.** $\frac{9}{4}$ **49.** x^5 **51.** y^{12} **53.** x^{20} **55.** h^5 **57.** $\frac{1}{m^{10}}$ **59.** $2a^4b^5$ **61.** $3p^{10}$ **63.** x^5y^4 **65.** $\frac{1}{b^{72}}$ **67.** x^{28} **69.** $-32x^5$ **71.** r^3 **73.** $\frac{1}{m^{10}}$ **75.** $-5r^{13}$ **77.** $\frac{s^3}{r^9}$ **79.** $16a^{20}$ **81.** $-\frac{27}{d^3}$ **83.** $27x^9y^{12}$ **85.** $-\frac{1}{s^6}$ **87.** $\frac{1}{729}m^6n^{12}$ **89.** $\frac{a^{15}}{b^{10}}$ **91.** a^4 **93.** a^{18} **95.** $\frac{a^6}{b^4}$ **97.** a^5 **99.** c^7 **101.** 8 **103.** $\frac{1}{9x^3}$ **105.** $-\frac{3}{8d^3}$ **107.** $\frac{1}{9x^3}$ **109.** $\frac{64b^{12}}{27a^9}$ **111.** 0 **113.** $-\frac{8a^{21}}{b^3}$ **115.** $-\frac{27}{8a^{18}}$ **117.** $\frac{27z^{21}}{64a^{12}b^{12}}$ **123.** $10^{-2}, 10^{-3}, 10^{-4}, 10^{-5}, 10^{-6}, 10^{-7}, 10^{-8}, 10^{-9}$ **125.** $10^3 \cdot 26^3$; 17,576,000 **127. a.** x^6 ft² **b.** x^9 ft³ **131.** $\left(-\infty, -\frac{10}{9}\right]$ 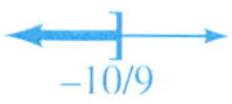

Study Set 5.2 (page 330)

1. scientific, standard **3.** 10^n, integer **5.** left **7. a.** 60.22 is not between 1 and 10. **b.** 0.6022 is not between 1 and 10. **9.** 3.9×10^3 **11.** 7.8×10^{-3} **13.** 1.73×10^{14} **15.** 9.6×10^{-6} **17.** 3.23×10^7 **19.** 6.0×10^{-4} **21.** 5.27×10^3 **23.** 3.17×10^{-4} **25.** 270 **27.** 0.00323 **29.** 796,0000 **31.** 0.00037 **33.** 5.23 **35.** 23,650,000 **37.** 1.817×10^{12} **39.** 5.005×10^8 **41.** 7.2×10^{-6} **43.** 5.0×10^{-8} **45.** 1.9×10^1 **47.** 4.005×10^{20}; 400,500,000,000,000,000,000 **49.** 4.3×10^{-3}; 0.0043 **51.** 6.0×10^3; 6,000 **53.** 3.6×10^{-5}; 0.000036 **55.** 2,600,000 to 1 **57.** $\$1.7 \times 10^{12}$, $\$3.9 \times 10^9$, $\$2.75 \times 10^8$, $\$3.12 \times 10^8$ **59.** 8.5×10^{-28} g **61.** about 9.5×10^{15} m **63. a.** 2.5×10^9 sec = 2,500,000,000 sec **b.** about 79 years **65.** 1.209×10^8 mi **67.** 1.0×10^{21} **73.** $\left[1, \frac{9}{4}\right]$ 

75. $(-\infty, 2) \cup (7, \infty)$ 

Study Set 5.3 (page 341)

1. polynomial **3.** degree **5.** coefficient, degree **7.** like **9.** monomial; 2 **11.** trinomial; 3 **13.** binomial; 2 **15.** monomial; 0 **17.** none of these; 10 **19.** binomial; 9 **21.** like terms, $10x$ **23.** unlike terms **25.** like terms, $-5r^2t^3$ **27.** unlike terms **29.** $9x^2 + 3x - 2$ **31.** 3, 3, 3, −27, −29 **33. a.** $-2x^4 - 5x^2 + 3x + 7$ **b.** $7a^3x^5 - ax^3 - 5a^3x^2 + a^2x$ **35.** 8, 2, 0, 2, 8 **37.** −42, 0, 12, 6, −6, −12, 0, 42

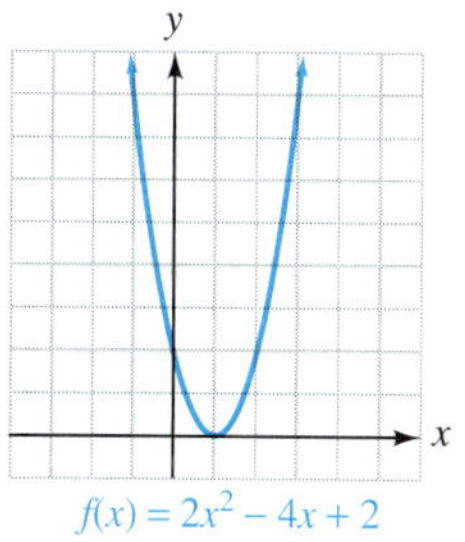

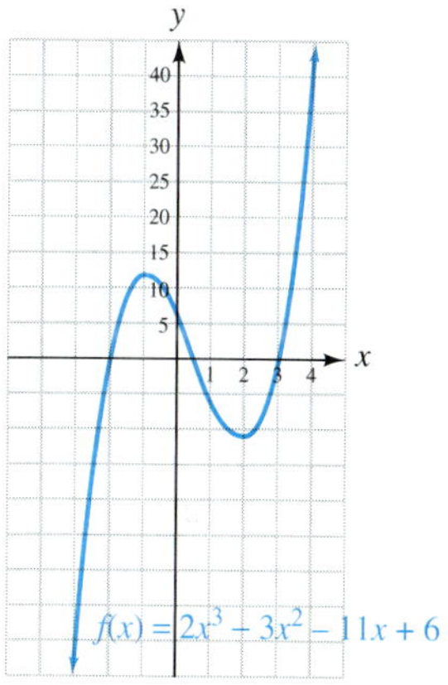

39. 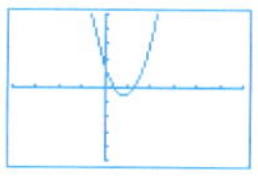

41. $20x^2 + 4x - 5$ **43.** $-10y^3 - 7y - 4$ **45.** $ab^2 + ab - a$ **47.** $8rst^2 - 1$ **49.** $x^2 - 5x + 6$ **51.** $-5a^2 + 4a + 4$ **53.** $5a^2 + 3ab - 9$ **55.** $-y^3 + 4y^2 + 6$ **57.** $2p^2q^2 - 2q$ **59.** $2x^2y^3 + 13xy + 3y^2$ **61.** $4x^2 - 11$ **63.** $\frac{1}{6}y^6 - \frac{2}{3}y^4 - \frac{1}{2}y^2$ **65.** $6x^3 - 6x^2 + 14x - 17$ **67.** $x^2 - 8x + 22$ **69.** $-3y^3 + 18y^2 - 28y + 35$ **71.** $2x^2 + 7a$ **73.** $-2xy^2 + 9x^2$ **75.** $-8x^2 - 2x + 2$ **77.** 20 ft **79.** 872 ft³ **81.** 2,160 in.³ **83. a.** about 6.4 **b.** about 9,000 **91.** $[-5, 5]$ **93.** $(-1, 9)$

Study Set 5.4 (page 352)

1. monomials, binomials **3.** sum **5.** factors **7.** term **9.** $x^2 + 2xy + y^2$ **11.** $x^2 - y^2$ **13.** $x^2 + 2x - 8$ **15.** $16a^2 + 24a + 9$ **17. a.** $2x$, $4x$ **b.** $2x$, −3 **c.** 4, $4x$ **d.** 4, −3 **19.** $-6a^3b$ **21.** $-15a^2b^2c^3$ **23.** $-120a^9b^3$ **25.** $-405x^7y^4$ **27.** $3x + 6$ **29.** $3x^3 + 9x^2$ **31.** $-6x^3 + 6x^2 - 4x$ **33.** $7r^3st + 7rs^3t - 7rst^3$ **35.** $-12m^4n^2 - 12m^3n^3$ **37.** $x^2 + 5x + 6$

39. $6t^2 + 5t - 6$ **41.** $6y^2 - 5yz + z^2$ **43.** $2b^2 + 35b + 48$ **45.** $0.2t^2 - 2.7t + 9$ **47.** $b^4 + b^3 - b - 1$ **49.** $-6t^2u^2 + 11tu - 3$ **51.** $27b^5 - 9b^3c - 3b^2c + c^2$ **53.** $55m^3 + 22m^2n^2 + 15mn^3 + 6n^5$ **55.** $18p^4 + 30p^3 - 72p^2$ **57.** $24m^3y - 20m^2y^2 + 4my^3$ **59.** $x^2 + 4x + 4$ **61.** $9a^2 - 24a + 16$ **63.** $4a^2 + 4ab + b^2$ **65.** $25r^4 + 60r^2 + 36$ **67.** $81a^2b^4 - 72ab^2 + 16$ **69.** $\frac{1}{16}b^2 + b + 4$ **71.** $16k^2 - 10.4k + 1.69$ **73.** $x^2 - 4$ **75.** $y^6 - 4$ **77.** $x^2y^2 - 36$ **79.** $\frac{1}{4}x^2 - 256$ **81.** $5.76 - y^2$ **83.** $x^3 - y^3$ **85.** $6y^3 + 11y^2 + 9y + 2$ **87.** $8a^3 - b^3$ **89.** $a^3 - 3a^2b - ab^2 + 3b^3$ **91.** $2a^2 + ab - b^2 - 3bc - 2c^2$ **93.** $r^4 - 2r^2s^2 + s^4$ **95.** $3x^2 + 12x$ **97.** $-p^2 + 4pq$ **99.** $3x^2 + 3x - 11$ **101.** $5x^2 - 36x + 7$ **103.** $9.2127x^2 - 7.7956x - 36.0315$ **105.** $299.29y^2 - 150.51y + 18.9225$ **107. a.** $(x + y)(x - y)$ **b.** $x(x - y); x^2 - xy$ **c.** $y(x - y); xy - y^2$ **d.** They represent the same area. $(x + y)(x - y) = x^2 - y^2$ **109.** $x(12 - 2x)(12 - 2x)$ in.^{3} $= (144x - 48x^2 + 4x^3)$ in.^{3}

115.

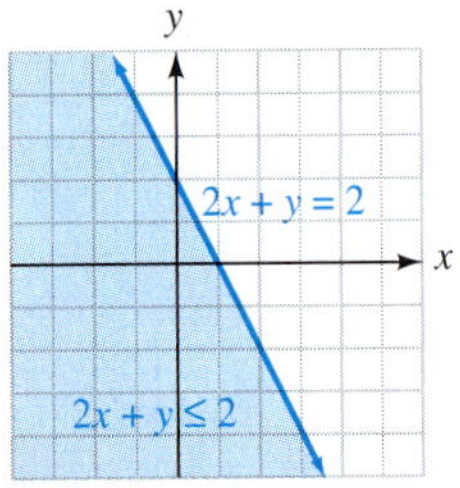

117.

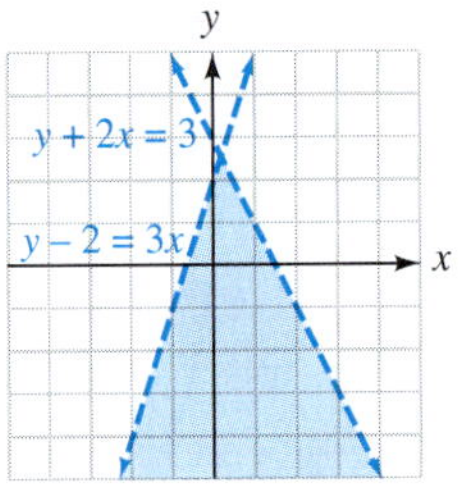

Study Set 5.5 (page 361)

1. factored **3.** greatest common factor **5.** completely, prime **7.** $6xy^2$ **9. a.** The terms within the parentheses have a common factor of 2. **b.** The terms within the parentheses have a common factor of t. **11.** 4 **13.** $x^2, 2, x - 1$ **15.** $2 \cdot 3$ **17.** $3^3 \cdot 5$ **19.** 2^7 **21.** $5^2 \cdot 13$ **23.** 12 **25.** 2 **27.** $4a^2$ **29.** $6xy^2z^2$ **31.** $2(x + 4)$ **33.** $2x(x - 3)$ **35.** prime **37.** $5x^2y(3 - 2y)$ **39.** prime **41.** $3z(9z^2 + 4z + 1)$ **43.** $9x^7y^3(5x^3 - 7y^4 + 9x^3y^7)$ **45.** $\frac{1}{5}x^2(3ax^2 + b - 4ax)$ **47.** $-(a + b)$ **49.** $-(5xy - y + 4)$ **51.** $-(60P^2 + 17)$ **53.** $-3(a + 2)$ **55.** $-x(3x + 1)$ **57.** $-3x(2x + y)$ **59.** $-6ab(3a + 2b)$ **61.** $-7u^2v^3z^2(9uv^3z^7 - 4v^4 + 3uz^2)$ **63.** $(x + y)(4 + t)$ **65.** $(a - b)(r - s)$ **67.** $(m + n + p)(3 + x)$ **69.** $(u + v)(u + v - 1)$ **71.** $-(x + y)(a + b)$ **73.** $2(x^2 + 1)^2(x^2 + 3)$ **75.** $(x + y)(a + b)$ **77.** $(x + y)(x + 1)$ **79.** $(3 - c)(c + d)$ **81.** $(1 - m)(1 - n)$ **83.** $(2x^2 + 1)(a - 4)$ **85.** $(a + b)(a - 4)$ **87.** $(a^2 + b)(x - 1)$ **89.** $(x + y)(x + y + z)$ **91.** $x(m + n)(p + q)$ **93.** $y(x + y)(x + y + 2z)$ **95.** $n(2n - p + 2m)(n^2p - 1)$ **97.** $r_1 = \frac{rr_2}{r_2 - r}$ **99.** $f = \frac{d_1d_2}{d_2 + d_1}$ **101.** $a^2 = \frac{b^2x^2}{b^2 - y^2}$ **103.** $r = \frac{S - a}{S - \ell}$ **105. a.** $\frac{1}{2}b_1h$ **b.** $\frac{1}{2}b_2h$ **c.** $\frac{1}{2}h(b_1 + b_2)$; the formula for the area of a trapezoid **107.** $r^2(4 - \pi)$ **113.** \$4,900

Study Set 5.6 (page 374)

1. trinomial **3.** lead, coefficient, 2 **5.** product, sum **7.** prime **9. a.** positive **b.** negative **c.** positive **11.** descending, GCF, positive **13.** 9, 6, −9, −6 **15.** $-1 + (-12) = -13, -2(-6) = 12, -3 + (-4) = -7$ **17.** 4, −4, 1 **19.** $(x + 2)$ **21.** $(x - 3)$ **23.** $(2a + 1)$ **25.** $(x + 1)^2$ **27.** $(a - 9)^2$ **29.** $(2y + 1)^2$ **31.** $(3b - 2c^2)^2$ **33.** $(y^2 + 5)^2$ **35.** $(5m^4 - 6n)^2$ **37.** $(x - 3)(x - 2)$ **39.** $(x - 2)(x - 5)$ **41.** prime **43.** $(x + 5)(x - 6)$ **45.** $(a + 10)(a - 5)$ **47.** $(x + 3y)(x - 7y)$ **49.** $(s - 2t)(s - 8t)$ **51.** $(y^2 - 10)(y^2 - 3)$ **53.** $(g^3 - 9)(g^3 + 7)$ **55.** $3(x + 7)(x - 3)$ **57.** $x^2(b^2 - 7)(b^2 - 5)$ **59.** $-(a - 8)(a + 4)$ **61.** $-3a^2(x - 3)(x - 2)$ **63.** $-2(p + 2q)(p - q)$ **65.** $(3y + 2)(2y + 1)$ **67.** $(4a - 3)(2a + 3)$ **69.** $(3x - 4y)(2x + y)$ **71.** prime **73.** $(4x - 3)(2x - 1)$ **75.** $4h^4(8h - 1)(2h + 1)$ **77.** $x(3x^2 - 11x + 8)$ **79.** $-(3a + 2b)(a - b)$ **81.** $5(2a - 3b)^2$ **83.** $x^2(7x - 8)(3x + 2)$ **85.** $(3y^3 + 2)(4y^3 + 5)$ **87.** $(m + n)(6a - 5)(a + 3)$ **89.** $(x + a + 1)^2$ **91.** $(a + b + 4)(a + b - 6)$ **93.** $(7q - 7r + 2)(2q - 2r - 3)$ **95.** $x + 3$ **101.** 5 **103.** $\frac{8}{3}$ **105.** $-26p^2 - 6p$

Study Set 5.7 (page 383)

1. squares **3.** 1, 4, 9, 16, 25, 36, 49, 64, 81, 100 **5. a.** $(x + 5)(x + 5) = x^2 + 10x + 25$ **b.** $(x + 5)(x - 5) = x^2 - 25$ **9.** $(p - q)$ **11.** pq **13.** $(p^2 + pq + q^2)$ **15. a.** $x^2 - 4$ **b.** $(x - 4)^2$ **c.** $x^2 + 4$ **d.** $x^3 + 8$ **e.** $(x + 8)^3$ **17.** $(x + 2)(x - 2)$ **19.** $(3y + 8)(3y - 8)$ **21.** prime **23.** $(20 + c)(20 - c)$ **25.** $(25a + 13b^2)(25a - 13b^2)$ **27.** $(9a^2 + 7b)(9a^2 - 7b)$ **29.** $(6x^2y + 7z^2)(6x^2y - 7z^2)$ **31.** $(x + y + z)(x + y - z)$ **33.** $(a - b + c)(a - b - c)$ **35.** $(x^2 + y^2)(x + y)(x - y)$ **37.** $(16x^2y^2 + z^4)(4xy + z^2)(4xy - z^2)$ **39.** $\left(\frac{1}{6} + y^2\right)\left(\frac{1}{6} - y^2\right)$ **41.** $2(x + 12)(x - 12)$ **43.** $2x(x + 4)(x - 4)$ **45.** $5x(x + 5)(x - 5)$ **47.** $t^2(rs + x^2y)(rs - x^2y)$ **49.** $(a + b)(a - b + 1)$ **51.** $(a - b)(a + b + 2)$ **53.** $(2x + y)(1 + 2x - y)$ **55.** $(x - 4)(x + y)(x - y)$ **57.** $(x + 2 + y)(x + 2 - y)$ **59.** $(x + 1 + 3z)(x + 1 - 3z)$ **61.** $(c + 2a - b)(c - 2a + b)$ **63.** $(r + s)(r^2 - rs + s^2)$ **65.** $(x - 2y)(x^2 + 2xy + 4y^2)$ **67.** $(4a - 5b^2)(16a^2 + 20ab^2 + 25b^4)$ **69.** $(5xy^2 + 6z^3)(25x^2y^4 - 30xy^2z^3 + 36z^6)$ **71.** $(x^2 + y^2)(x^4 - x^2y^2 + y^4)$ **73.** $5(x + 5)(x^2 - 5x + 25)$ **75.** $4x^2(x - 4)(x^2 + 4x + 16)$ **77.** $2u^2(4v - t)(16v^2 + 4tv + t^2)$ **79.** $(a + b)(x + 3)(x^2 - 3x + 9)$ **81.** $(x^3 - y^4z^5)(x^6 + x^3y^4z^5 + y^8z^{10})$ **83.** $(a + b + 3)(a^2 + 2ab + b^2 - 3a - 3b + 9)$ **85.** $(y + 1)(y - 1)(y - 3)(y^2 + 3y + 9)$ **87.** $(x + 1)(x^2 - x + 1)(x - 1)(x^2 + x + 1)$ **89.** $(x^2 + y)(x^4 - x^2y + y^2)(x^2 - y)(x^4 + x^2y + y^2)$

91. $(a + 3)(a - 3)(a + 2)(a - 2)$

93. $\frac{4}{3}\pi(r_1 - r_2)(r_1^2 + r_1r_2 + r_2^2)$

97.

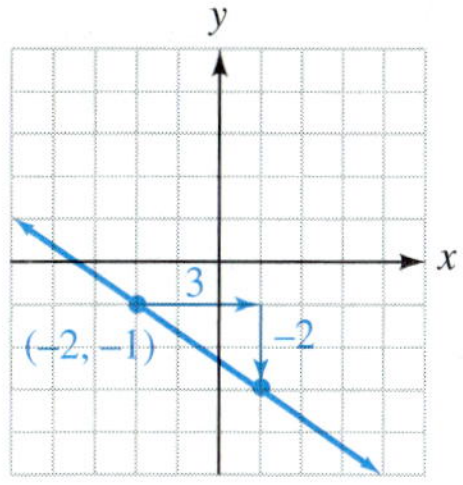

99. $y = -4$

Study Set 5.8 (page 389)

1. completely, prime **3.** cubes, cubes **5.** common **7.** trinomial **9.** Multiply the factors of $y^2z^3(x + 6)(x + 1)$ to see if the product is $x^2y^2z^3 + 7xy^2z^3 + 6y^2z^3$. **11.** $3ab, 2b, 2a$ **13.** $4bc(a - 5d)(a + 6d)$ **15.** $3xy(x + 2y - 4)$ **17.** $(3x + 1)(3x^3 + x^2 + 1)$ **19.** $(5x + 4y)(5x - 4y)$ **21.** prime **23.** $2(3x - 4)(x - 1)$ **25.** $y^2(2x + 1)(2x + 1)$ **27.** $2x^2y^2z^2(2 - 13z)$ **29.** $4(xy + 4)(x^2y^2 - 4xy + 16)$ **31.** $(x^2 + y^2)(a^2 + b^2)$ **33.** $(x - y + 5)[(x - y)^2 - 5(x - y) + 25]$ **35.** $(2a - 2b + 3)(a - b + 1)$ **37.** $(2x - 9)(3x + 7)$ **39.** $(x + 1)(x - 1)(x + 4)(x - 4)$ **41.** $(x + 5 + y^4)(x + 5 - y^4)$ **43.** $(3x - 1 - 5y)(3x - 1 + 5y)$ **45.** $(a + b - 2)(a - b + 2)$ **47.** $2x^8(4x + 3)^2$ **49.** $\left(\frac{9}{4}x^2 + y^{20}\right)\left(\frac{3}{2}x + y^{10}\right)\left(\frac{3}{2}x - y^{10}\right)$ **51.** $16(m^8 + 1)(m^4 + 1)(m^2 + 1)(m + 1)(m - 1)$ **53.** $(3y + 1)(3y^4 + y^3 + 1)$ **57.** 6 **59.** −13

Study Set 5.9 (page 398)

1. quadratic **3.** standard **5.** At least one is 0. **7. a.** yes **b.** no **c.** yes **d.** no **9. a.** add $11x$ to both sides **b.** multiply both sides by 4 **c.** multiply both sides by −1 **d.** distribute the multiplication by m **11. a.** $3, -\frac{5}{2}$ **b.** $0, 1, -\frac{15}{2}$ **13.** 3, −1 **15.** $y + 6, y - 9, -6$ **17.** 0, −2 **19.** 4, −4 **21.** 0, −1 **23.** 0, 5 **25.** −3, −5 **27.** −2, −4 **29.** $-\frac{1}{3}, -3$ **31.** $\frac{1}{2}, 2$ **33.** $1, -\frac{1}{2}$ **35.** 3, 3 **37.** $\frac{1}{4}, -\frac{3}{2}$ **39.** $2, -\frac{5}{6}$ **41.** $\frac{1}{3}, -1$ **43.** $2, \frac{1}{2}$ **45.** $\frac{1}{5}, -\frac{5}{3}$ **47.** 0, 0, −1 **49.** 0, 7, −7 **51.** 0, 7, −3 **53.** 3, −3, 2, −2 **55.** 0, −2, −3 **57.** $0, \frac{5}{6}, -7$ **59.** $\frac{3}{2}, -\frac{7}{3}$ **61.** $2, -\frac{5}{2}$ **63.** −8, −8 **65.** $5, -\frac{7}{2}$ **67.** −2, −1, 2 **69.** 1, 2 **71.** 0, 2, 4 **73.** 2.78, 0.72 **75.** 1 **77.** 16, 18 or −18, −16 **79.** 10 in., 16 in. **81.** 3 ft **83.** 20 ft by 40 ft **85.** 11 sec and 19 sec **87.** 9 sec **89.** 6 m/sec **91.** 4 **99.** 25 ft^2

Key Concept (page 404)

1. $-4x^3 - 4x^2 + 5x - 2$ **2.** $4s^3t - 3s^2 + 3s - 7t$ **3.** $3m^2 + 5m - 12$ **4.** $3r^4st - 6r^2s^2t + 9r^2st^3$ **5.** $a^2 - 4ad + 4d^2$ **6.** $-3x^3 + 12x^2 - 9x$ **7.** $6b^3 + 11b^2 + 9b + 2$ **8.** $-y^2 - 4y - 1$ **9.** 0; after falling for 6 seconds, the squeegee will strike the ground. **10.** $V(x) = 16x^3 + 12x^2 - 4x$; 528 **11.** 861 **12. a.** 6 **b.** 10 **c.** −2, 1, 2 **13.** 9, −9 **14.** 0, 5 **15.** −3, −5 **16.** 0, −1 **17.** $\frac{1}{3}, -1$ **18.** 1, 0, −9

Chapter Review (page 405)

1. 729 **2.** −32 **3.** −64 **4.** $\frac{9}{4}$ **5.** x^6 **6.** $\frac{m^3}{n^5}$ **7.** m^{18} **8.** t^{13} **9.** $9x^4y^6$ **10.** $\frac{x^{16}}{b^4}$ **11.** −3 **12.** $\frac{1}{x^{10}}$ **13.** $\frac{b}{2a}$ **14.** $70x^4$ **15.** $\frac{x^6}{9}$ **16.** $\frac{2}{9x}$ **17.** $-c^{10}$ **18.** $\frac{25}{16}$ **19.** $\frac{1}{y^8}$ **20.** $\frac{-b^3}{8a^{21}}$ **21.** 1.93×10^{10} **22.** 2.735×10^{-8} **23.** 72,770,000 **24.** 0.0000000083 **25.** 7.6×10^2 sec **26.** 1.67248×10^{-18} g **27.** 8.4×10^6 **28.** no **29.** yes **30.** yes **31.** no **32.** binomial, 2 **33.** monomial, 4 **34.** none of these, 4 **35.** trinomial, 8 **36.** 134 in.^3

37.

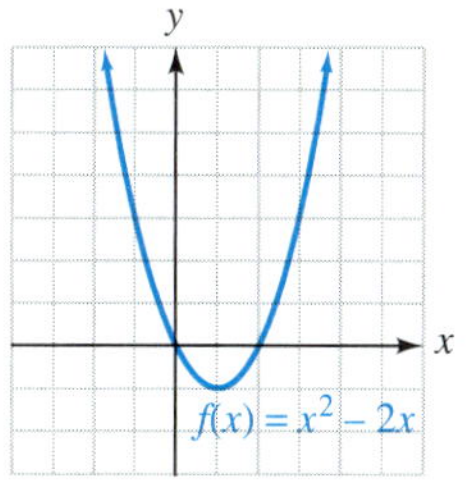

38.

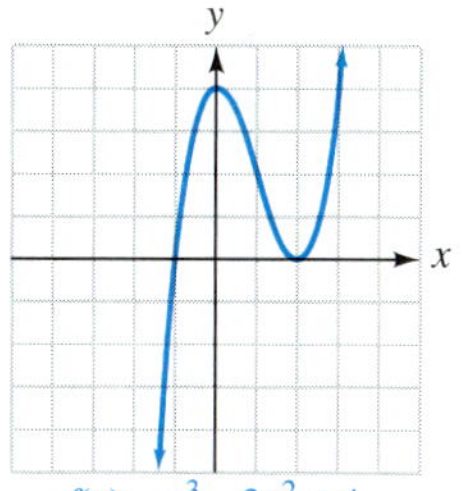

$f(x) = x^3 - 3x^2 + 4$

39. $3x^2y^3 - 8x^2y + 8y$ **40.** $6k^4 - 6k^3 + 9k^2 - 2$ **41.** $4c^2d^2 + 4cd^2$ **42. a.** −1 **b.** −2, −1, 1 **c.** D: $(-\infty, \infty)$; R: $[-1, \infty)$ **43.** $-4a^3$ **44.** $6x^3y^2z^5$ **45.** $2x^4y^3 - 8x^2y^7$ **46.** $a^4b + 2a^3b^2 - a^2b^3$ **47.** $6x^3 - 12x^2 + 4x - 8$ **48.** $25a^2t^2 - 60at + 36$ **49.** $49c^4d^4 - d^2$ **50.** $15x^4 - 22x^3 + 58x^2 - 40x$ **51.** $r^3 - 3r^2s - rs^2 + 3s^3$ **52.** $9c^2 - \frac{9}{2}c + \frac{9}{16}$ **53.** $154x^2 + 27x + 1$ **54. a.** $f(x) = x^3 + 3x^2 + 2x$ **b.** 210 in.^3 **55.** $2 \cdot 5^2 \cdot 7$ **56. a.** 6 **b.** $3xy^3$ **57.** $4(x^4 + 2)$ **58.** $\frac{x}{5}(3x^2 - 6x + 1)$ **59.** prime **60.** $7a^3b(ab + 7)$ **61.** $5x^2(x + y + 1)(1 - 3x)$ **62.** $9x^2y^3z^2(3xz + 9x^2y^2 - 10z^5)$ **63.** $-7(b^3 - 2c)$ **64.** $-7a^2b^2(a - b)^3(7a^2 - 7ab - 9b^2)$ **65.** $(x + 2)(y + 4)$ **66.** $(ry - a + 1)(r - 1)$ **67.** $m_1 = \frac{mm_2}{m_2 - m}$ **68.** $A = 2\pi r(r + h)$ **69.** not factorable **70.** factorable **71.** $(x + 5)^2$ **72.** $(7a^3 + 6b^2)^2$ **73.** $(y + 20)(y + 1)$ **74.** $(z - 5)(z - 6)$ **75.** $-(x + 7)(x - 4)$ **76.** $(a - 8b)(a + 3b)$ **77.** $(4a - 1)(a - 1)$ **78.** prime **79.** $y(y + 2)(y - 1)$ **80.** $9st(3r - 2)(r + 4)$ **81.** $(r + s)(6t - 5)(t + 3)$ **82.** $(v^2 - 7)(v^2 - 6)$ **83.** $(w^4 - 10)(w^4 + 9)$ **84.** $(s + t - 1)^2$ **85.** $(z + 4)(z - 4)$ **86.** $(xy^2 + 8z^3)(xy^2 - 8z^3)$ **87.** prime **88.** $(c + a + b)(c - a - b)$ **89.** $(m^2 + 4)(m + 2)(m - 2)$ **90.** $(m + n)(m - n - 1)$ **91.** $2c(4a^2 + 9b^2)(2a + 3b)(2a - 3b)$

92. $(k + 1 + 3m)(k + 1 - 3m)$ **93.** $(t + 4)(t^2 - 4t + 16)$
94. $(2a + 5b^3)(4a^2 - 10ab^3 + 25b^6)$ **95.** $4rs(q - 5t)(q + 6t)$
96. $(2m + 2n + 3)(m + n - 1)$ **97.** $(z - 2)(z + x + 2)$
98. prime **99.** $(x + 2 + 2p^2)(x + 2 - 2p^2)$
100. $(y + 2)(y + 1 + x)$ **101.** $4c^2(ab + 4)(a^2b^2 - 4ab + 16)$
102. $(a + 3)(a - 3)(a + 2)(a - 2)$
103. $(2x + 3)(2x^3 + 3x^2 + 1)$ **104.** $\frac{\pi}{2}h(r_1 + r_2)(r_1 - r_2)$
105. $0, \frac{3}{4}$ **106.** $6, -6$ **107.** $\frac{1}{2}, -\frac{5}{6}$ **108.** $3, -3, 1, -1$
109. $0, -\frac{2}{3}, \frac{4}{5}$ **110.** $-\frac{2}{3}, 7, 0$ **111.** $-8, -8$
112. 17 m by 20 m **113.** $-\frac{1}{2}, 1$ **114.** 1, 3

Chapter Test (page 410)

1. x^9 **2.** $\frac{-8x^6y^9}{125}$ **3.** $\frac{12}{m^5}$ **4.** $\frac{m^4}{9n^{10}}$ **5.** $\frac{9}{8}$ **6.** $\frac{st}{44}$
7. 4.706×10^{12} **8.** 0.000245 **9.** 1.45×10^{19}
10. 1.116×10^7 mi/min **11.** 5 **12.** 13 **13.** 110 ft
14. $f(x) = 3x^2 + 4x - 4$
15.

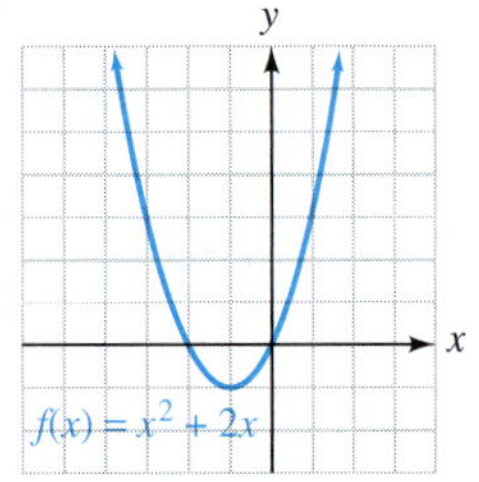

16. a.

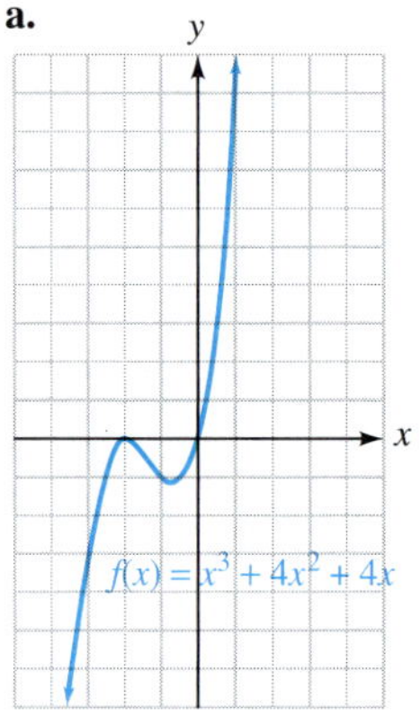

b. $-2, 0$ **17. a.** 0 **b.** 2, 6 **c.** D: $(-\infty, \infty)$; R: $(-\infty, 2]$
18. $-y^3 + 4y^2 + 6$ **19.** $2x^2y^3 + 13xy + 3y^2$
20. $-15a^3b^4 + 10a^3b^5$ **21.** $6y^3 + 11y^2 + 9y + 2$
22. $0.06d^2 + 1.6d - 6$ **23.** $16t^8 - 72t^4 + 81$
24. $2s^3 - 2st^2$ **25.** $3abc(4a^2b - abc + 2c^2)$
26. $(x + y)(x + y + z)$ **27.** $(5m^4 - 6n)^2$
28. $x^2(7x - 8)(3x + 2)$ **29.** $(s + 3)(s - 3)(s + 2)(s - 2)$
30. prime **31.** $5(x + 5)(x^2 - 5x + 25)$
32. $(4a - 5b^2)(16a^2 + 20ab^2 + 25b^4)$
33. $(x - y + 5)(x - y - 2)$ **34.** $(3b + 2c)(2b - c)$
35. $(x + 3 + y)(x + 3 - y)$ **36.** 0, 5 **37.** $\frac{3}{4}, -\frac{3}{2}$
38. $1, 0, -9$ **39.** $-3, -3$ **40.** $v = \frac{v_1v_3}{v_3 + v_1}$
41. 11 ft by 22 ft

Study Set 6.1 (page 421)

1. rational **3.** simplify **5.** opposites **7. a.** 1
b. 0.5 **c.** 0.25 **d.** D: $(0, \infty)$; R: $(0, \infty)$ **9. a.** 1
b. 1 **c.** 1 **d.** -1 **11.** 1: decreases then steadily increases; 2: increases, decreases, then steadily increases; 3: steadily increases; 4: steadily decreases, approaching a cost of \$2.00 per unit **13.** yes, yes, yes, yes, yes

15. 6, 3, 1.5, 1, 0.75, 0.6, 0.5

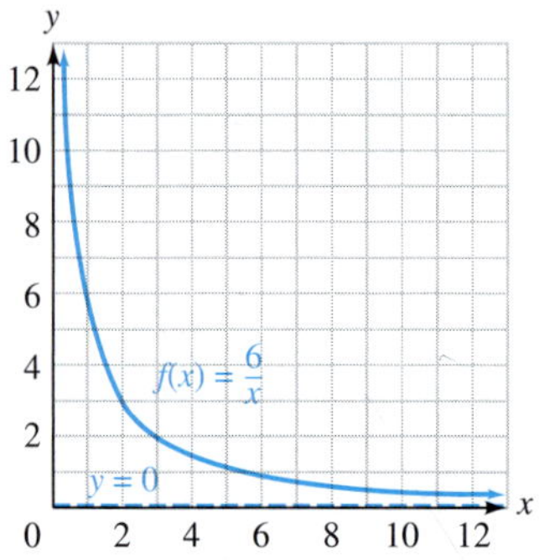

17. 3, 2, 1.5, 1.33, 1.25, 1.2, 1.17

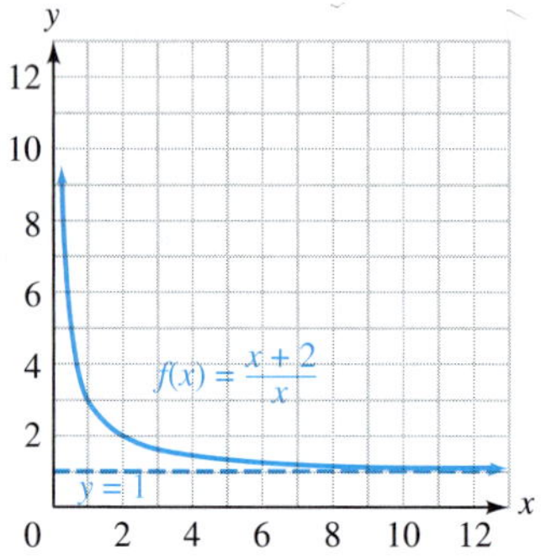

19. $(-\infty, 0) \cup (0, \infty)$ **21.** $(-\infty, -2) \cup (-2, \infty)$
23. $(-\infty, 0) \cup (0, 1) \cup (1, \infty)$ **25.** $(-\infty, -7) \cup (-7, 8) \cup (8, \infty)$
27. $\frac{2}{3}$ **29.** $-\frac{28}{9}$ **31.** $4x^2$ **33.** $-\frac{4y}{3x}$ **35.** $-\frac{x}{2(x - y)}$
37. -1 **39.** $\frac{1}{x - y}$ **41.** $\frac{5}{x - 2}$ **43.** $\frac{-3(x + 2)}{x + 1}$
45. in lowest terms **47.** $x + 2$ **49.** $\frac{x + 1}{x + 3}$ **51.** $\frac{s - 3}{s + 6}$
53. $\frac{x + 4}{2(2x - 3)}$ **55.** $\frac{3(x - y)}{x + 2}$ **57.** $\frac{2x + 1}{2 - x}$ **59.** $\frac{a^2 - 3a + 9}{4(a - 3)}$
61. in lowest terms **63.** $m + n$ **65.** -1
67. $-\frac{m + n}{2m + n}$ or $\frac{-m - n}{2m + n}$ **69.** $\frac{x - y}{x + y}$ **71.** $\frac{10}{3}$ **73.** $\frac{2x - 3}{2y - 3}$
75. $\frac{3a + b}{y + b}$ **77.** 1 **79.** D: $(-\infty, 2) \cup (2, \infty)$; R: $(-\infty, 1) \cup (1, \infty)$
81. D: $(-\infty, -2) \cup (-2, 2); \cup (2, \infty)$; R: $(-\infty, \infty)$
83. a. \$50,000 **b.** \$200,000 **85. a.** $c(n) = 0.09n + 7.50$
b. $c(n) = \frac{0.09n + 7.50}{n}$ **c.** about 10¢ **87. a.** about 2.5 hr
b. about 4.6 hr **91.** $a^3 - 6a^2 + 5a + 6$
93. $-3m^4n^2 + 21m^2n^3 + 6m^3n^2$

Study Set 6.2 (page 431)

1. rational **3.** invert **5.** numerators, denominators, $\frac{AC}{BD}$
7. 1 **9.** $(x - 5)$, $(5x - 25)$, $x(x + 3)$, $5(x - 5)$, x
11. yes, no, yes **13.** $\frac{5}{4}$ **15.** $-\frac{5}{6}$ **17.** $\frac{xy^2d}{2c^2}$
19. $-\frac{x^{10}}{y^2}$ **21.** $x + 1$ **23.** $2y + 16$ or $2(y + 8)$
25. $10h - 30$ or $10(h - 3)$ **27.** $\frac{x + 1}{9}$ **29.** 1
31. $\frac{2(x - 4)}{x + 5}$ **33.** $-\frac{(a + 7)^2(a - 5)}{12x^2}$ **35.** $\frac{t - 1}{t + 1}$ **37.** $\frac{n + 2}{n + 1}$

39. $-\frac{x+y}{x-y}$ **41.** $\frac{1}{x+1}$ **43.** $\frac{(x+1)^2(x+2)}{x+2c}$
45. $\frac{(a+7)^2(a-5)}{12x^2}$ **47.** $\frac{x+y}{x-y}$ **49.** $\frac{a+b}{(x-3)(c+d)}$
51. $-\frac{x+1}{x+3}$ or $\frac{-x-1}{x+3}$ **53.** $x^2(x+3)$ **55.** $\frac{x+2}{x-2}$
57. $\frac{3x}{2}$ **59.** 1 **61.** $\frac{x^2-6x+9}{x^6+8x^3+16}$
63. $\frac{4m^4-4m^3-11m^2+6m+9}{x^4-2x^2+1}$ **65.** $k_1(k_1+2), k_2+6$
71. x^{m+n} **73.** x^ny^n **75.** 1 **77.** x^{m-n} **79.** $\frac{y^n}{x^m}$

Study Set 6.3 (page 442)

1. common **3.** build **5.** numerators, denominator, $A+B$, $A-B$ **7.** factor, greatest **9. a.** ii **b.** adding or subtracting rational expressions **c.** simplifying a rational expression **11. a.** twice **b.** once **13. a.** $2 \cdot 2 \cdot 2 \cdot 5 \cdot x \cdot x$ **b.** $2x(x-3)$ **c.** $(n+8)(n-8)$ **15.** $3x-2, 3, 3x-1, 3$
17. $\frac{11}{4y}$ **19.** 2 **21.** 3 **23.** 3 **25.** $\frac{6x}{(x-3)(x-2)}$
27. $\frac{1}{(x^2+9)(x-3)}$ **29.** $3b-x$ **31.** $36x^2$
33. $x(x+3)(x-3)$ **35.** $(x+3)^2(x^2-3x+9)$
37. $(2x+3)^2(x+1)^2$ **39.** $\frac{17}{12x}$ **41.** $\frac{9a^2-4b^2}{6ab}$
43. $\frac{3a-5b}{a^2b^2}$ **45.** $\frac{3r+2bs}{12b^2}$ **47.** $\frac{2(5a+2b)}{21}$
49. $\frac{2(4x-1)}{(x+2)(x-4)}$ **51.** $\frac{7x+29}{(x+5)(x+7)}$ **53.** $\frac{5(x+2)}{12(x+3)}$
55. $\frac{4x+1}{x}$ **57.** 2 **59.** $\frac{3(a-1)}{3a-2}$ **61.** $\frac{x(2x+1)}{(x+3)(x+2)(x-2)}$
63. $\frac{1}{a-b}$ **65.** $\frac{-x^2+11x+8}{(3x+2)(x+1)(x-3)}$ **67.** $\frac{2x^2+5x+4}{x+1}$
69. $\frac{x^2-5x-5}{x-5}$ **71.** $\frac{-2(2x^2-7x-27)}{x(x+3)(x-3)}$ **73.** $\frac{14s+58}{(s+3)(s+7)}$
75. $\frac{11x^2+7x-3}{(2x-1)(3x+2)}$ **77.** $\frac{2}{x+1}$ **79.** $\frac{2b}{a+b}$
81. $\frac{7mn^2-7n^3-6m^2+3mn+n}{(m-n)^2}$ **83.** $\frac{2}{m-1}$ **85.** $\frac{10r+20}{r}$; $\frac{3t+9}{t}$ **91.** 3, 3 **93.** 0, 0, -1

Study Set 6.4 (page 453)

1. complex, fractions **3.** $\frac{t^2}{t^2}$ **5.** $\div$, $\frac{3}{25m}$, m, 3, 3, m, 10
7. $\div$, **9.** $\frac{2}{3}$ **11.** $-\frac{1}{7}$ **13.** $\frac{2y}{3z}$ **15.** $125b$ **17.** $-\frac{1}{y}$
19. $\frac{y-x}{x^2y^2}$ **21.** $\frac{b+a}{b}$ **23.** $\frac{y+x}{y-x}$ **25.** $\frac{1}{c-d}$ **27.** $y-x$
29. $-\frac{1}{a+b}$ **31.** x^2+x-6 **33.** $\frac{5x^2y^2}{xy+1}$ **35.** -1
37. $\frac{x+2}{x-3}$ **39.** $\frac{a-1}{a+1}$ **41.** $\frac{2x^2+5x}{x^3+x^2+x+1}$ **43.** $\frac{xy^2}{y-x}$
45. $\frac{x^2(xy^2-1)}{y^2(x^2y-1)}$ **47.** $\frac{1}{x-y}$ **49.** $\frac{x-9}{3}$ **51.** $-\frac{h}{4}$
53. $\frac{3a^2+2a}{2a+1}$ **55.** $\frac{(-x^2+2x-2)(3x+2)}{(2-x)(-3x^2-2x+9)}$ **57.** $\frac{k_1k_2}{k_2+k_1}$
59. $\frac{4d-1}{3d-1}$ **65.** 8 **67.** 2, -2, 3, -3

Study Set 6.5 (page 462)

1. monomial, polynomial, binomial **3.** Divisor, Quotient, Dividend, Remainder **5. a.** term **b.** 9, 9 **c.** 6, 6, 6
7. $(2x-1)(x^2+3x-4) = 2x^3+5x^2-11x+4$
9. $2x^2, x, 4$ **11.** $3a^2+5+\frac{6}{3a-2}$ **13.** $\frac{x^2-x-12}{x-4}$, $x-4\overline{)x^2-x-12}$, $(x^2-x-12) \div (x-4)$ **15.** $\frac{y}{2x^3}$
17. $-\frac{3}{4a^2}$ **19.** $2x+3$ **21.** $-\frac{2x}{3}+\frac{x^2}{6}$ **23.** $2xy^2+\frac{x^2y}{6}$
25. $\frac{x^4y^4}{2}-\frac{x^3y^9}{4}+\frac{3}{4xy^2}$ **27.** $x+2$ **29.** $x+7$
31. $3x-5+\frac{3}{2x+3}$ **33.** $3x^2+x+2-\frac{4}{x-1}$
35. $2x^2+5x+3+\frac{4}{3x-2}$ **37.** $3x^2+4x+3$ **39.** $a+1$
41. $2y+2$ **43.** $6x-12$ **45.** $4x^3-3x^2+3x+1$
47. $a^2+a+1+\frac{2}{a-1}$ **49.** $5a^2-3a-4$ **51.** $6y-12$
53. x^4+x^2+4 **55.** x^2+x+1 **57.** x^2+x+2
59. $-3a^2-a-9$ **61.** a^4-a^2+1
63. $2x+3+\frac{20x-13}{3x^2-7x+4}$ **65.** $9.8x+16.4-\frac{36.5}{x-2}$
67. x^2-5x+6 **69.** $3x-2, x+5$ **73.** $8x^2+2x+4$
75. $-2y^3-y^2+10y-14$

Study Set 6.6 (page 470)

1. synthetic **3.** divisor **5.** theorem
7. a. $(5x^3+x-3) \div (x+2)$ **b.** $5x^2-10x+21-\frac{45}{x+2}$
9. $6x^3-x^2-17x+9, x-8$ **11.** 2, 1, 12, 26, 6, 8
13. $x+2$ **15.** $x-3$ **17.** $x+2$ **19.** $x-7+\frac{28}{x+2}$
21. $3x^2-x+2$ **23.** $2x^2+4x+3$
25. $6x^2-x+1+\frac{3}{x+1}$ **27.** $t^2+1+\frac{1}{t+1}$
29. $a^4+a^3+a^2+a+1$ **31.** $-5x^4-11x^3-3x^2-7x-1$
33. $8t^2+2$ **35.** x^3+7x^2-2 **37.** $7.2x-0.66+\frac{0.368}{x-0.2}$
39. $9x^2-513x+29{,}241-\frac{1{,}666{,}762}{x+57}$ **41.** -1 **43.** -37
45. 23 **47.** -1 **49.** 2 **51.** -1 **53.** 18 **55.** 174
57. -8 **59.** 59 **61.** 44 **63.** $\frac{29}{32}$ **65.** yes **67.** no
69. 64 **75.** 0 **77.** 2

Study Set 6.7 (page 482)

1. rational **3.** clear **5. a.** yes **b.** no **7.** $\frac{12}{x}, \frac{12}{x+15}$
9. a. 3, 0 **b.** 3, 0 **c.** 3, 0 **11.** $30y$, $\frac{9}{2y}$, $\frac{10}{3y}$, $7y$, 35
13. 12 **15.** 40 **17.** $\frac{1}{2}$ **19.** no solution **21.** $\frac{17}{25}$
23. 0 **25.** 1 **27.** 2 **29.** -1 **31.** no solution; -3 is extraneous **33.** 2 **35.** $\frac{1}{3}$ **37.** 0 **39.** 2, -5
41. -4, 3 **43.** 6, $\frac{17}{3}$ **45.** 1, -11 **47.** no solution; 0 is extraneous **49.** 1 **51.** $r = \frac{E - IR_L}{I}$ **53.** $r = \frac{S-a}{S-\ell}$
55. $n_2 = \frac{2\mu_R - n_1^2 - n_1}{n_1}$ **57.** $R = \frac{R_1R_2R_3}{R_2R_3+R_1R_3+R_1R_2}$
59. -2, 2, 3 **61.** -7, $-\frac{1}{2}$, 7 **63.** $4\frac{8}{13}$ in.
65. $L = \frac{SN-CN}{V-C}$, 8 years **67. a.** $1\frac{7}{8}$ days **b.** Santos: \$412.50, Mays: \$375 **69.** about 110 sec
71. $2\frac{4}{13}$ weeks **73.** $1\frac{1}{2}$ hours **75.** 6 mph **77.** 35 mph and 45 mph **79.** 5 mph **81.** 9 **89.** 9.0×10^9
91. 4.4×10^{-22}

Study Set 6.8 (page 496)

1. ratio **3.** extremes, means **5.** direct, inverse **7.** Inverse

9.

11.

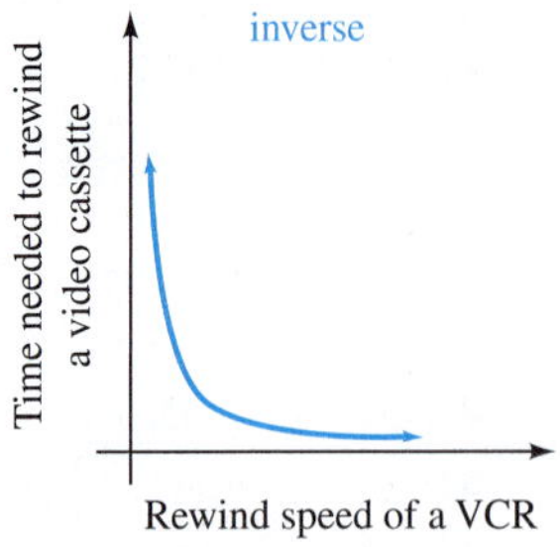

13. −7, 6, 18, −102, −17 **15.** 3 **17.** 5 **19.** 5 **21.** 39 **23.** 2, −2 **25.** 4, −1 **27.** $-\frac{5}{2}$, −1 **29.** no solution **31.** $-\frac{1}{2}$, 0, 5 **33.** $-\frac{5}{2}$, −1, 1 **35.** $A = kp^2$ **37.** $v = \frac{k}{r^2}$ **39.** $P = \frac{ka^2}{j^3}$ **41.** L varies jointly with m and n. **43.** R varies directly with L and inversely with d^2. **45.** 202, 172, 136 **47.** about 2 gal **49.** eye: 49.9 in.; seat: 17.6 in.; elbow; 27.8 in. **51.** 12.5 in. **53.** 555 ft **55.** $46\frac{7}{8}$ ft **57.** 880 ft **59.** 1,600 ft **61.** 25 days **63.** 12 in.^{3} **65.** $9,000 **67.** 3 ohms **69.** 0.275 in. **71.** 12.8 lb **75.** $\frac{1}{c^2}$ **77.** −1

Key Concept (page 502)

1. a. $\frac{3x+2}{4x+3}$ **b.** $2x - 1$ **2. a.** $\frac{d-1}{d+1}$ **b.** $3d + 2$, $2d - 3$, $2d + 3$ **3. a.** $\frac{2(4x-1)}{(x+2)(x-4)}$ **b.** $\frac{x-4}{x-4}, \frac{x+2}{x+2}$ **4. a.** $\frac{3(n^2-n-2)}{n^2}$ **b.** $3n$ **5. a.** 4 **b.** $t(t-2)$ **6. a.** $\frac{1}{6}$ **b.** $(x-1)(2x+3)$ **7. a.** 1 **b.** $(x+2)(x-4)$ **8. a.** $a^2 = \frac{b^2x^2}{b^2+y^2}$ **b.** a^2b^2

Chapter Review (page 503)

1. 8, 4, 2, 1.33, 1, 0.8, 0.67, 0.57, 0.5

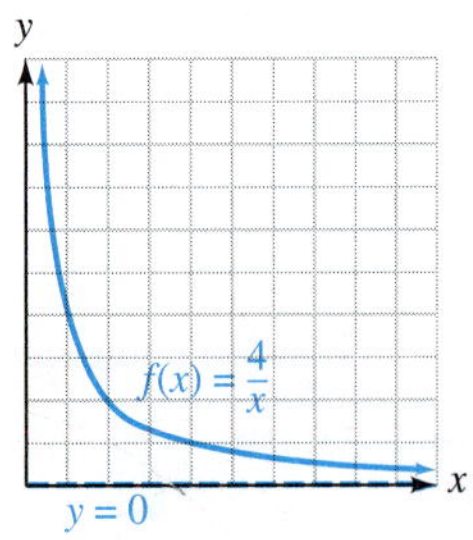

2. a. 1 **b.** 2 **c.** D: $(0, \infty)$; R: $(0, \infty)$ **3.** $(-\infty, -6) \cup (-6, 4) \cup (4, \infty)$ **4.** $y = 3$, $x = 0$; D: $(-\infty, 0) \cup (0, \infty)$, R: $(-\infty, 3) \cup (3, \infty)$ **5.** $\frac{12x}{19y^7}$ **6.** $\frac{x-7}{x+7}$ **7.** $\frac{1}{2(x+2)}$ **8.** $\frac{1}{x-6}$ **9.** $\frac{-a-b}{c+d}$ **10.** $\frac{m+2n}{2m+n}$

11. $\frac{2x+1}{(3x-4)^2}$ **12.** −2 **13.** $\frac{cd}{7x^2}$ **14.** −1 **15.** $\frac{2a-1}{a+2}$ **16.** $\frac{t-2}{t}$ **17.** $\frac{h^2-4h+4}{h^6+8h^3+16}$ **18.** $\frac{a+b}{(m+p)(m-3)}$ **19.** $\frac{2m-n}{m+n}$ **20.** $\frac{3x(x-1)}{(x-3)(x+1)}$ **21.** $\frac{5y-3}{x-y}$ **22.** $\frac{1}{c-d}$ **23.** $-\frac{2}{t-3}$ **24.** $-\frac{1}{p+12}$ **25.** $60a^2h^3$ **26.** $ab^2(b-1)$ **27.** $(x-5)(x+5)(x+1)$ **28.** $(m^2+2m+4)(m-2)^2$ **29.** $\frac{9a+8}{a+1}$ **30.** $\frac{40x+7y^2z}{112z^2}$ **31.** $\frac{4x^2+9x+12}{(x-4)(x+3)}$ **32.** $\frac{12y+20}{15y(x-2)}$ **33.** $\frac{14y+58}{(y+3)(y+7)}$ **34.** $\frac{x^2+26x+3}{(x+3)(x-3)^2}$ **35.** $\frac{2bc^3}{7}$ **36.** $\frac{p-3}{2(p+2)}$ **37.** $\frac{b+2a}{2b-a}$ **38.** $\frac{x-2}{x+3}$ **39.** $\frac{x^2y^2}{(x-y)^2(y^2-x^2)}$ **40.** $\frac{3x^2-18x+2}{x^2-6x+3}$ **41.** $\frac{5h^3}{11k^2}$ **42.** $-\frac{1}{2y^3z^{10}}$ **43.** $6a + \frac{16}{3}$ **44.** $-3x^2y + \frac{3x}{2} + y$ **45.** $b + 4$ **46.** $v^2 - 3v - 10$ **47.** $x^2 - 2x + 4$ **48.** $m^4 - m^2 + 1 + \frac{2}{m^4+2m^2-3}$ **49.** $x + 7$ **50.** $4x^2 - 3x + 6 - \frac{13}{x+2}$ **51.** $-5n^4 - 11n^3 - 3n^2 - 7n - 1$ **52.** 54 **53.** yes **54.** no **55.** 5 **56.** 2 **57.** $-\frac{1}{2}$ **58.** −1, −2 **59.** 2 and −2 are extraneous, no solution **60.** 0 **61.** $y^2 = \frac{x^2b^2 - a^2b^2}{a^2}$ **62.** $b = \frac{Ha}{2a-H}$ **63.** 5 mph **64.** 50 mph **65. a.** 1/10 of the job per hour **b.** $\frac{x}{10}$ of the job is completed **66.** $14\frac{2}{5}$ hr **67.** $18\frac{2}{3}$ days **68.** about 3,500,000 in. lb/rad **69.** 5 **70.** −4, −12 **71.** 70.4 ft **72.** 20 **73.** 66 in. **74.** $5,460 **75.** 1.25 amps **76.** 126.72 lb **77.** inverse variation **78.** 0.2

Chapter Test (page 508)

1. $\frac{2}{3xy}$ **2.** $\frac{2}{x-2}$ **3.** −3 **4.** $\frac{2x+y}{4y}$

5.

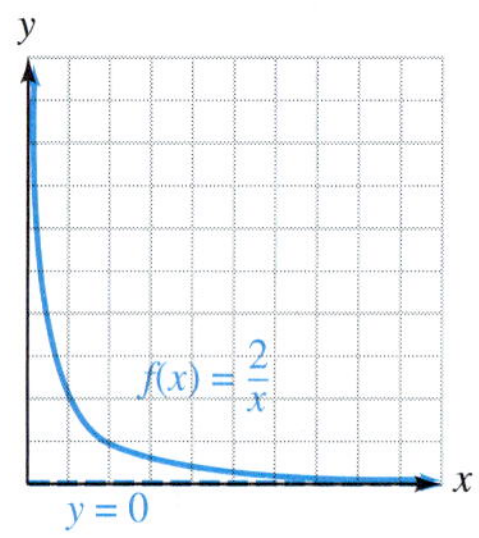

6. $(-\infty, 0) \cup (0, 1) \cup (1, \infty)$ **7.** $\frac{xz}{y^4}$ **8.** 1 **9.** $\frac{13}{x+1}$ **10.** $\frac{(x+y)^2}{2x}$ **11.** −1 **12.** $\frac{(2x-3)^2}{x^6}$ **13.** $\frac{2}{t-4}$ **14.** 2 **15.** $\frac{24b^2-2b-1}{3b+1}$ **16.** $\frac{6a-17}{(a+1)(a-2)(a-3)}$

17. $\frac{u^2}{2vw}$ **18.** $\frac{3k^2+4k+4}{3k^2-9k-9}$ **19.** $-\frac{6x}{y} + \frac{4x^2}{y^2} - \frac{3}{y^3}$ **20.** $y^2 - 2y + 4 - \frac{56}{y+2}$ **21.** 41 **22.** $x + 3$ is a factor of $P(x)$. **23.** 40 **24.** 5; 3 is extraneous. **25.** 6, −1 **26.** 26 **27.** $a^2 = \frac{x^2b^2}{b^2-y^2}$ **28.** $r_2 = \frac{rr_1}{r_1-r}$ **29.** no, $\frac{5}{11}$ of an hour **30.** 3 mph **31.** 80 ft **32.** $77.32 **33.** 25 decibels **34.**

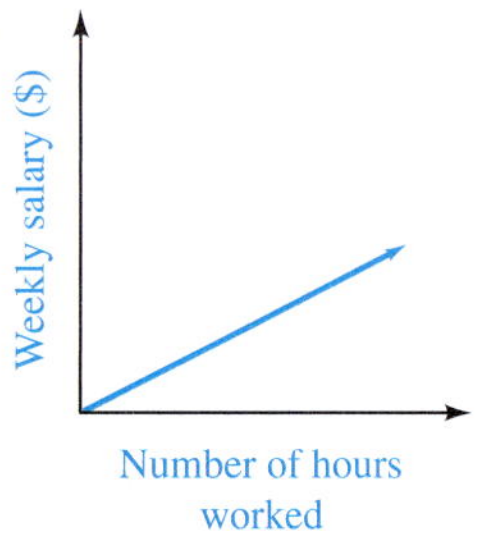

35. Step 1: Factor each denominator. Step 2: The LCD is the product that uses each different factor in step 1 the greatest number of times it appears in any one factorization.

Cumulative Review Exercises, Chapters 1–6 (page 510)

1. $\frac{8}{3}$ **2.** -2 **3.** 14 **4.** Life expectancy will increase 0.1 year each year during this period. **5.** $y = -7x + 54$ **6.** $y = -\frac{11}{6}x - \frac{7}{3}$ **7.** D: $(-\infty, \infty)$; R: $(-\infty, \infty)$

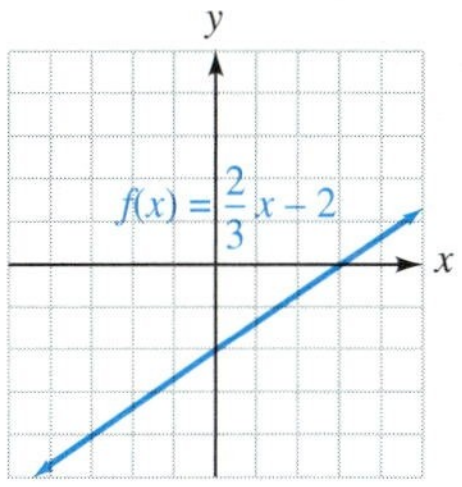

8. $T_1 = 80, T_2 = 60$ **9.** $\left(-\infty, \frac{4}{3}\right]$

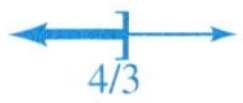

10. $(-\infty, 3] \cup \left[\frac{11}{3}, \infty\right)$

11. $(-\infty, -11)$

12. $(-\infty, 1)$

13. a^8b^4 **14.** $\frac{b^4}{a^4}$ **15.** 81 **16.** $\frac{x^{21}y^3}{8}$

17. 42,500 **18.** 0.000712 **19.** $y = \frac{kxz}{r}$ **20.** 8

21.

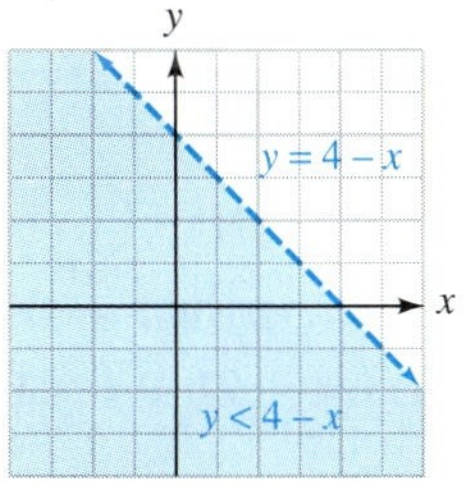

22. D: $(-\infty, \infty)$; R: $[-3, \infty)$

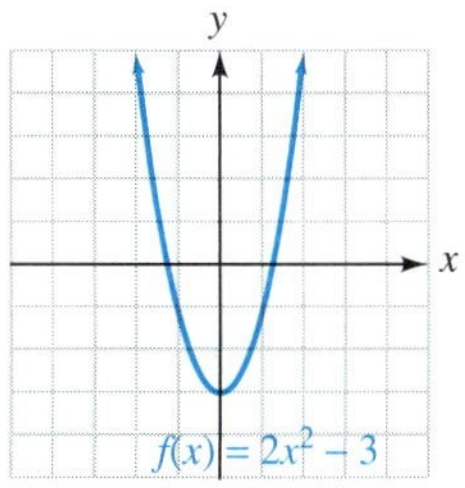

23. 18 **24.** 7 **25. a.** $(-3, 0), (0, -7)$ **b.** $-\frac{7}{3}$ **c.** 1 **d.** $y = -\frac{7}{3}x - 7$ **e.** yes **26.** penny: $\$10^{-2}$, dime: $\$10^{-1}$, one dollar bill: $\$10^0$, one hundred thousand dollar bill: $\$10^5$ **27.** $2x^3 + x^2 + 12$ **28.** $-3x^2 - 3$ **29.** $2a^2 + ab - b^2 - 3bc - 2c^2$ **30.** $4x^6 - 4x^3 + 1$ **31. a.** yes **b.** about 17 **c.** about 14 **32.** 4 **33.** $3rs^3(r - 2s)$ **34.** $(x - y)(5 - a)$ **35.** $(x + y)(u + v)$ **36.** $(9x^2 + 4y^2)(3x + 2y)(3x - 2y)$ **37.** $(2x - 3y^2)(4x^2 + 6xy^2 + 9y^4)$ **38.** $(4x - 3)(2x - 1)$ **39.** $(x + 5 + 4z)(x + 5 - 4z)$ **40.** $(x - y + 5)(x - y - 2)$ **41.** $-\frac{1}{3}, -\frac{7}{2}$ **42.** $0, 2, -2$ **43.** $b^2 = \frac{a^2y^2}{a^2 - x^2}$ **44.** 9 in. by 12 in. **45.** $\frac{2x - 3}{3x - 1}$ **46.** $-\frac{q}{p}$ **47.** $\frac{4}{x - y}$ **48.** $\frac{a^2 + ab^2}{a^2b - b^2}$ **49.** 0 **50.** -17 **51.** $x + 4$ **52.** $-x^2 + x + 5 + \frac{8}{2x - 1}$ **53.** It will rise sharply. **54.** It has dropped. **55.** 13 cups **56.** 1 day

Study Set 7.1 (page

1. square, cube **3.** index, radicand **5.** odd, even **7.** a **9.** two, 5 **11.** x **13.** 3, up **15. a.** 3 **b.** 0 **c.** 6 **d.** D: $[2, \infty)$, R: $[0, \infty)$ **17.** 0, 1, 2, 3, 4; D: $[0, \infty)$; R: $[0, \infty)$

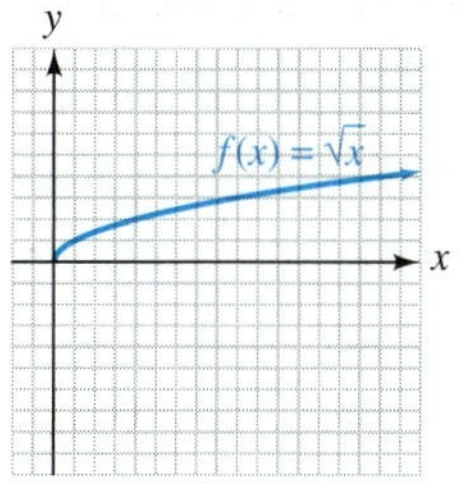

19. $\sqrt{x^2} = |x|$ **21.** $f(x) = \sqrt{x - 5}$ **23.** 11 **25.** -8 **27.** $\frac{1}{3}$ **29.** 0.5 **31.** not real **33.** 4 **35.** 3.4641 **37.** 26.0624 **39.** $2|x|$ **41.** $|t + 5|$ **43.** $5|b|$ **45.** $|a + 3|$ **47.** 0 **49.** 4 **51.** 2 **53.** -10 **55.** 4.1231 **57.** 3.3322 **59.** $0, -1, -2, -3, -4$; D: $[0, \infty)$, R: $(-\infty, 0]$ **61.** D: $[-4, \infty)$, R: $[0, \infty)$

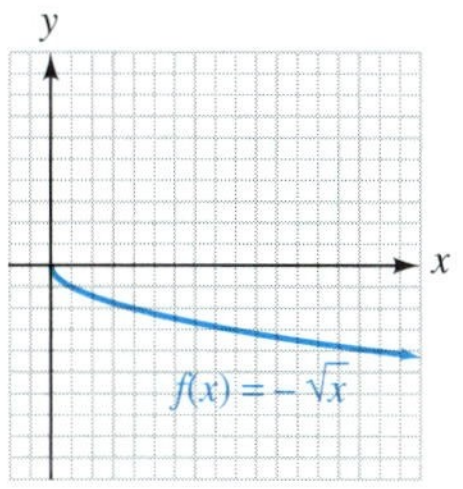

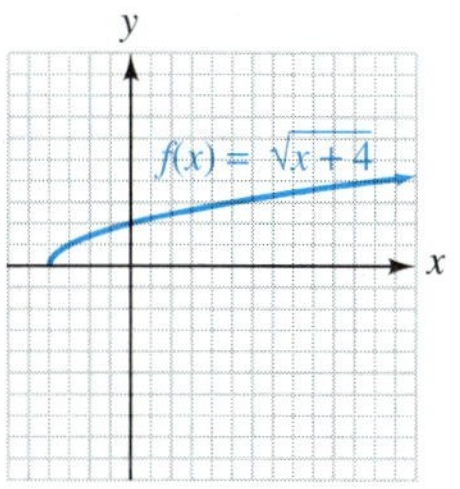

63. D: $(-\infty, \infty)$, R: $(-\infty, \infty)$

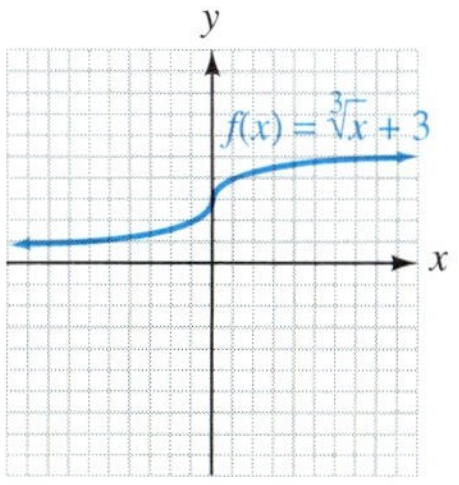

65. 1 **67.** -5 **69.** $-\frac{2}{3}$ **71.** 0.4 **73.** $2a$ **75.** $-10pq$ **77.** $-\frac{1}{2}m^2n$ **79.** $-0.4s^3t^2$ **81.** 3 **83.** -3 **85.** not real **87.** $\frac{2}{5}$ **89.** $\frac{1}{2}$ **91.** $2a$ **93.** $2a$ **95.** k^3 **97.** $\frac{1}{2}m$ **99.** $x + 2$ **101.** 7.0 in. **103.** about 61.3 beats/min **105.** 13.4 ft **107.** 3.5% **113.** 1 **115.** $\frac{3(m^2 + 2m - 1)}{(m + 1)(m - 1)}$

Study Set 7.2 (page 535)

1. rational (or fractional) **3.** index, radicand **5.** $25^{1/5}$, $(\sqrt[3]{-27})^2$, $16^{-3/4}$, $(\sqrt{81})^3$, $-\left(\frac{9}{64}\right)^{1/2}$ **7.**

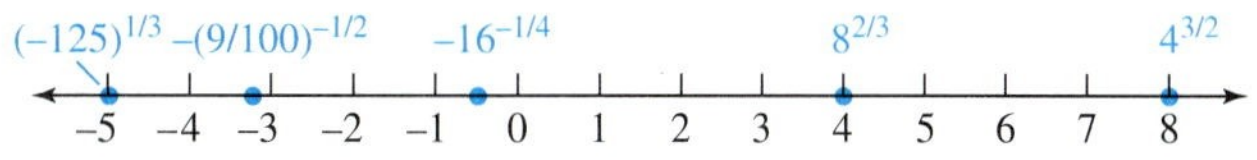

9. $\sqrt[n]{x}$ **11.** $\frac{1}{x^{m/n}}$ **13.** $100a^4, 10a^2$ **15.** $\sqrt[3]{x}$ **17.** $\sqrt[4]{3x}$ **19.** $\sqrt[4]{6x^3y}$ **21.** $\sqrt{x^2 + y^2}$ **23.** $m^{1/2}$ **25.** $(3a)^{1/4}$ **27.** $(8abc)^{1/6}$ **29.** $(a^2 - b^2)^{1/3}$ **31.** 2 **33.** 5 **35.** 3

−6

47. $|x|$ **49.** m^2 **51.** n **53.** $5|y|$ **55.** $2|x|$ **57.** not real **59.** $|x+1|$ **61.** 216 **63.** 27 **65.** 1,728 **67.** $\frac{1}{4}$ **69.** $125x^6$ **71.** $\frac{4x^2}{9}$ **73.** $\frac{1}{2}$ **75.** $\frac{1}{3,125}$ **77.** $\frac{1}{64x^3}$ **79.** $\frac{1}{9y^2}$ **81.** $\frac{16}{81}$ **83.** $-\frac{3}{2x}$ **85.** $9^{5/7}$ **87.** $\frac{1}{36}$ **89.** 3 **91.** a **93.** $a^{2/9}$ **95.** $a^{3/4}b^{1/2}$ **97.** $\frac{1}{3}x$ **99.** $y + y^2$ **101.** $x^2 - x + x^{3/5}$ **103.** $\sqrt{p}$ **105.** $\sqrt{5b}$ **107.** $\sqrt[5]{xy}$ **109.** $\sqrt[18]{c}$ **111.** $\sqrt[15]{7m}$ **113.** 2.47 **115.** 1.01 **117.** 736 ft/sec **119.** 1.96 units **121.** 4,608 in.^{2}, 32 ft^2 **125.** $2\frac{1}{2}$ hr

Study Set 7.3 (page 544)

1. like **3.** factor **5.** $\sqrt[n]{a}\sqrt[n]{b}$, product, roots **7. a.** $\sqrt{4 \cdot 5}$ **b.** $\sqrt{4}\sqrt{5}$ **c.** $\sqrt{4 \cdot 5} = \sqrt{4}\sqrt{5}$ **9. a.** $\sqrt{5}, \sqrt[3]{5}$ (answers may vary); no **b.** $\sqrt{5}, \sqrt{6}$ (answers may vary); no **11.** $8k^3, 8k^3, 4k$ **13.** $2\sqrt{5}$ **15.** $10\sqrt{2}$ **17.** $2\sqrt[3]{10}$ **19.** $-3\sqrt[3]{3}$ **21.** $2\sqrt[4]{2}$ **23.** $-2\sqrt[5]{3}$ **25.** $2\sqrt[6]{5}$ **27.** $\frac{\sqrt{7}}{3}$ **29.** $\frac{\sqrt[3]{7}}{4}$ **31.** $\frac{\sqrt[4]{3}}{10}$ **33.** $\frac{\sqrt[5]{3}}{2}$ **35.** 10 **37.** 2 **39.** $5x\sqrt{2}$ **41.** $4\sqrt{2b}$ **43.** $-4a\sqrt{7a}$ **45.** $5ab\sqrt{7b}$ **47.** $-10\sqrt{3xy}$ **49.** $-3x^2\sqrt[3]{2}$ **51.** $2x^4y\sqrt[3]{2}$ **53.** $2x^3y\sqrt[4]{2}$ **55.** $a\sqrt[5]{a^2}$ **57.** $m\sqrt[6]{m^5}$ **59.** $2t^2\sqrt[5]{t}$ **61.** $3a$ **63.** $7x$ **65.** $\frac{z}{4x}$ **67.** $\frac{\sqrt[4]{5x}}{2z}$ **69.** $10\sqrt{2x}$ **71.** $\sqrt[5]{7a^2}$ **73.** $-\sqrt{2}$ **75.** $2\sqrt{2}$ **77.** $9\sqrt{6}$ **79.** $3\sqrt[3]{3x}$ **81.** $-\sqrt[3]{4}$ **83.** -10 **85.** $-17\sqrt[4]{2}$ **87.** $16\sqrt[4]{2}$ **89.** $-4\sqrt{2}$ **91.** $3\sqrt{2t} + \sqrt{3t}$ **93.** $-11\sqrt[3]{2}$ **95.** $y\sqrt{z}$ **97.** $13y\sqrt{x}$ **99.** $12\sqrt[3]{a}$ **101.** $-7y^2\sqrt{y}$ **103.** $4x\sqrt[5]{xy^2}$ **105.** $8\pi\sqrt{5}$ ft^2; 56.2 ft^2 **107.** $5\sqrt{3}$ amps; 8.7 amps **109.** $(26\sqrt{5} + 10\sqrt{3})$ in.; 75.5 in. **115.** $\frac{-15x^5}{y}$ **117.** $3p + 4 - \frac{5}{2p-5}$

Study Set 7.4 (page 555)

1. FOIL **3.** irrational **5.** perfect **7. a.** $6\sqrt{6}$ **b.** 48 **c.** can't be simplified **d.** $-6\sqrt{6}$ **9.** Any number multiplied by 1 is the same number. $\frac{\sqrt{7}}{\sqrt{7}} = 1.$ **11.** A radical appears in the denominator. **13.** $\sqrt{6}$, 48, 16, 4 **15.** 11 **17.** 7 **19.** 4 **21.** $5\sqrt{2}$ **23.** $6\sqrt{2}$ **25.** 5 **27.** 18 **29.** 8 **31.** 18 **33.** $2\sqrt[3]{3}$ **35.** ab^2 **37.** $5a\sqrt{b}$ **39.** $-20r\sqrt[3]{10s}$ **41.** $x^2(x+3)$ **43.** $9b$ **45.** $a\sqrt[4]{6ab}$ **47.** $2\sqrt[5]{t^2}$ **49.** $12\sqrt{5} - 15$ **51.** $24\sqrt{3} + 6\sqrt{14}$ **53.** $-8x\sqrt{10} + 6\sqrt{15x}$ **55.** $-1 - 2\sqrt{2}$ **57.** $\sqrt[3]{25z^2} + 3\sqrt[3]{15z} + 2\sqrt[3]{9}$ **59.** $3x - 2y$ **61.** $6a + 5\sqrt{3ab} - 3b$ **63.** $18r - 12\sqrt{2r} + 4$ **65.** $-6x - 12\sqrt{x} - 6$ **67.** $4a\sqrt[3]{2a} - 5\sqrt[3]{4a^2} - 3$ **69.** $\frac{\sqrt{7}}{7}$ **71.** $\frac{\sqrt{30}}{5}$ **73.** $\frac{\sqrt{10}}{4}$ **75.** $\frac{\sqrt[3]{4}}{2}$ **77.** $\sqrt[3]{3}$ **79.** $\frac{\sqrt[3]{6}}{3}$ **81.** $\frac{2\sqrt{2xy}}{xy}$ **83.** $\frac{\sqrt{5y}}{y}$ **85.** $\frac{\sqrt[3]{2ab^2}}{b}$ **87.** $\frac{\sqrt[4]{4}}{2}$ **89.** $\frac{\sqrt[5]{2}}{2}$ **91.** $\frac{\sqrt[4]{27st^2}}{3t}$ **93.** $\frac{t\sqrt[5]{9a^4}}{3a}$ **95.** $\frac{3\sqrt{2} - \sqrt{10}}{4}$ **97.** $\frac{9 - 2\sqrt{14}}{5}$ **99.** $\frac{3\sqrt{6} + 4}{2}$ **101.** $\frac{2(\sqrt{x} - 1)}{x - 1}$ or $\frac{2\sqrt{x} - 2}{x - 1}$ **103.** $\sqrt{2z} + 1$ **105.** $\frac{x - 2\sqrt{xy} + y}{x - y}$ **107.** $\frac{6\sqrt{ab} + a + 9b}{a - 9b}$ **109.** $\frac{x - 9}{x(\sqrt{x} - 3)}$ **111.** $\frac{x - y}{\sqrt{x}(\sqrt{x} - \sqrt{y})}$ **113.** $\frac{\sqrt{2\pi}}{2\pi\sigma}$ **115.** $\frac{\sqrt{2}}{2}$ **119.** $\frac{1}{3}$

Study Set 7.5 (page 565)

1. radical **3.** extraneous **5.** n, n **7. a.** Square both sides. **b.** Cube both sides. **c.** Subtract 3 from both sides. **9. a.** x **b.** $x - 5$ **c.** $32x$ **d.** $x + 3$ **11. a.** $x - 6\sqrt{x} + 9$ **b.** $2y + 10\sqrt{2y+1} + 26$ **13.** The principal square root of a number, in this case $\sqrt{8x - 7}$, is never negative. **15.** 6, 2, 2, $3x$, yes **17.** 2 **19.** 4 **21.** 0 **23.** 4 **25.** 8 **27.** 1 **29.** -16 **31.** $\cancel{4}$, no solution **33.** $\frac{5}{2}, \frac{1}{2}$ **35.** 1 **37.** 16 **39.** 14, $\cancel{6}$ **41.** $\cancel{-5}$, 5 **43.** 4, 3 **45.** 2, $\cancel{7}$ **47.** 2, -1 **49.** -1, $\cancel{1}$ **51.** $\cancel{1}$, no solution **53.** 0, $\cancel{4}$ **55.** $-\frac{1}{2}$ **57.** $\frac{1}{3}, \frac{4}{5}$ **59.** $\cancel{-3}$, no solution **61.** 1, 9 **63.** 4, $\cancel{0}$ **65.** 2, $\cancel{142}$ **67.** $\cancel{6}$, no solution **69.** 5 **71.** 1 **73.** $h = \frac{v^2}{2g}$ **75.** $\ell = \frac{8T^2}{\pi^2}$ **77.** $A = P(r + 1)^3$ **79.** $v^2 = c^2\left(1 - \frac{L_A^2}{L_B^2}\right)$ **81.** 178 ft **83.** about 488 watts **85.** 10 lb **87.** \$5 **95.** 2.5 foot-candles **97.** 0.41511 in.

Study Set 7.6 (page 575)

1. hypotenuse **3.** Pythagorean **5.** $a^2 + b^2 = c^2$ **7.** $\sqrt{2}$ **9.** $\sqrt{3}$ **11.** 30°, 60° **13. a.** 8 **b.** $\sqrt{15}$ **c.** $2\sqrt{6}$ **15.** 6, 52, 4, 2 **17.** 10 ft **19.** 80 m **21.** $h = 2\sqrt{2} \approx 2.83, x = 2$ **23.** $x = 5\sqrt{3} \approx 8.66, h = 10$ **25.** $x = 4.69, y = 8.11$ **27.** $x = 12.11, y = 12.11$ **29.** $7\sqrt{2}$ cm **31.** 5 **33.** 13 **35.** 10 **37.** $2\sqrt{26}$ **41.** $(5\sqrt{2}, 0), (0, 5\sqrt{2}), (-5\sqrt{2}, 0), (0, -5\sqrt{2})$; (7.07, 0), (0, 7.07), (−7.07, 0), (0, −7.07) **43.** $10\sqrt{3}$ mm, 17.32 mm **45.** $10\sqrt{181}$ ft, 134.54 ft **47.** about 0.13 ft **49. a.** 21.21 units **b.** 8.25 units **c.** 13.00 units **51.** yes **55.** 7 **57.** 9

Study Set 7.7 (page 586)

1. imaginary **3.** real, imaginary **5.** conjugates **7. a.** $\sqrt{-1}$ **b.** -1 **c.** $-i$ **d.** 1 **9.** real, imaginary **11.** denominator **13. a.** $7i$ **b.** $-\sqrt{6}$ **15.** Complex, Real, Rational, Irrational, Imaginary **17. a.** yes **b.** yes

19. 9, $6i$, 2, 11 **21. a.** true **b.** false **c.** false **d.** false
23. $3i$ **25.** $\sqrt{7}i$ or $i\sqrt{7}$ **27.** $2\sqrt{6}i$ or $2i\sqrt{6}$
29. $-2\sqrt{6}i$ or $-2i\sqrt{6}$ **31.** $45i$ **33.** $\frac{5}{3}i$ **35.** -6
37. $-2\sqrt{3}$ **39.** $\frac{5}{8}$ **41.** -20 **43.** $8 - 2i$ **45.** $3 - 5i$
47. $15 + 2i$ **49.** $15 + 7i$ **51.** $6 - 3i$ **53.** $-25 - 25i$
55. $7 + i$ **57.** $12 + 5i$ **59.** $13 - i$ **61.** $14 - 8i$
63. $8 + \sqrt{2}i$ **65.** $3 + 4i$ **67.** $-9 + 0i$ **69.** $-6 + 0i$
71. $0 - i$ **73.** $0 + \frac{4}{5}i$ **75.** $\frac{1}{8} + 0i$ **77.** $0 + \frac{3}{5}i$
79. $2 + i$ **81.** $-\frac{42}{25} - \frac{6}{25}i$ **83.** $\frac{1}{4} + \frac{3}{4}i$ **85.** $-\frac{4}{13} - \frac{6}{13}i$
87. $\frac{5}{13} - \frac{12}{13}i$ **89.** $\frac{11}{10} + \frac{3}{10}i$ **91.** $\frac{1}{4} - \frac{\sqrt{15}}{4}i$ **93.** i
95. $-i$ **97.** 1 **99.** i **101.** $-1 + i$ **105.** 20 mph

Key Concept (page 589)

1. $-3h^2\sqrt[3]{2}$ **2.** $14\sqrt[3]{e}$ **3.** $-4\sqrt{2}$ **4.** $-20r\sqrt[3]{10s}$
5. $3s - 2t$ **6.** $-\sqrt{21} + \sqrt{15}$ **7.** $18n - 12\sqrt{2n} + 4$
8. $\frac{\sqrt[3]{3k^2}}{k}$ **9.** -3 **10.** 5 **11.** 2; 7 is extraneous
12. no solutions **13.** $3^{1/3}$ **14.** $5\sqrt[5]{a^2}$

Chapter Review (page 590)

1. 7 **2.** -11 **3.** $\frac{15}{7}$ **4.** not real **5.** 0.1 **6.** $5|x|$
7. x^4 **8.** $|x + 2|$ **9.** -3 **10.** -6 **11.** $4x^2y$ **12.** $\frac{x^3}{5}$
13. 5 **14.** -2 **15.** $4x^2|y|$ **16.** $x + 1$ **17.** $-\frac{1}{2}$
18. not real **19.** 0 **20.** 0 **21.** 48 ft **22.** 24 cm^2
23. D: $[-2, \infty)$, R: $[0, \infty)$ **24.** D: $(-\infty, \infty)$, R: $(-\infty, \infty)$

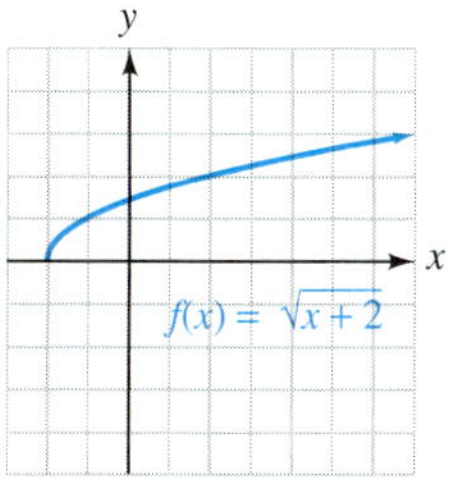

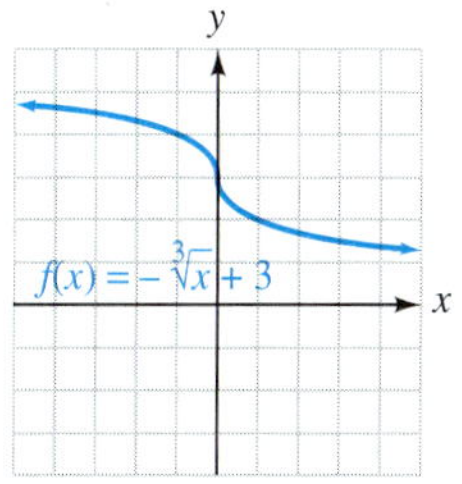

25. $\sqrt{t}$ **26.** $\sqrt[4]{5xy^3}$ **27.** 5 **28.** -6 **29.** not real
30. 1 **31.** $\frac{3}{x}$ **32.** -2 **33.** 5 **34.** $3cd$ **35.** 27
36. $\frac{1}{4}$ **37.** $-16{,}807$ **38.** 10 **39.** $\frac{27}{8}$ **40.** $\frac{1}{3{,}125}$
41. $125x^3y^6$ **42.** $\frac{1}{4u^4v^2}$ **43.** $5^{3/4}$ **44.** $a^{1/7}$ **45.** k^8
46. $3^{2/3}$ **47.** $u - 1$ **48.** $v + v^2$ **49.** $\sqrt{a}$ **50.** $\sqrt[6]{c}$
51. 183 mi **52.** Two true statements result: 32 = 32.
53. $4\sqrt{15}$ **54.** $3\sqrt[3]{2}$ **55.** $2\sqrt[4]{2}$ **56.** $4\sqrt[5]{3}$
57. $2x^2\sqrt{2x}$ **58.** $r^5\sqrt[3]{r^2}$ **59.** $2xy\sqrt[3]{2x^2y}$ **60.** $9j^2\sqrt[3]{jk}$
61. $4x$ **62.** $\frac{\sqrt{17xy}}{8a^2}$ **63.** $3\sqrt{2}$ **64.** $11\sqrt{5}$ **65.** 0
66. $-8a\sqrt[4]{2a}$ **67.** $29x\sqrt{2}$ **68.** $13x\sqrt[3]{2}$
70. $(6\sqrt{2} + 2\sqrt{10})$ in., 14.8 in. **71.** 7 **72.** $6\sqrt{10}$
73. 32 **74.** 72 **75.** $3x$ **76.** $x + 1$ **77.** $-2x^3\sqrt[3]{x}$
78. 3 **79.** $42t + 9t\sqrt{21t}$ **80.** $-20x^3y^3\sqrt[4]{x^2y^2}$
81. $3b + 6\sqrt{b} + 3$ **82.** $\sqrt[3]{9p^2} - \sqrt[3]{6p} - 2\sqrt[3]{4}$
83. $\frac{10\sqrt{3}}{3}$ **84.** $\frac{\sqrt{15xy}}{5xy}$ **85.** $\frac{\sqrt[3]{u^2}}{u^2v^2}$ **86.** $\frac{\sqrt[4]{27ab^2}}{3b}$
87. $2(\sqrt{2} + 1)$ **88.** $\frac{12\sqrt{xz} - 16x - 2z}{z - 16x}$ **89.** $\frac{a - b}{a + \sqrt{ab}}$
90. $r = \frac{\sqrt[3]{6\pi^2V}}{2\pi}$ **91.** 22 **92.** 16, 9 **93.** $\frac{13}{2}$ **94.** $\frac{9}{16}$
95. 2, -4 **96.** ~~x~~, no solution **97.** 1 **98.** $-\frac{3}{2}$, 1
99. $-\frac{1}{2}$, 4 **100.** 2 **101.** $P = \frac{A}{(r + 1)^2}$ **102.** $I = \frac{h^3b}{12}$
103. 17 ft **104.** 88 yd **105.** $7\sqrt{2}$ m **106.** $6\sqrt{3}$ cm, 18 cm **107.** 7.07 in. **108.** 8.66 cm **109.** 13
110. $2\sqrt{2}$ **111.** $5i$ **112.** $12i\sqrt{2}$ **113.** $-i\sqrt{6}$
114. $\frac{3}{8}i$ **115.** Real, Imaginary **116. a.** true **b.** true **c.** false **d.** false **117.** $3 - 6i$ **118.** $-1 + 7i$
119. $0 - 19i$ **120.** $0 + i$ **121.** $8 - 2i$ **122.** $3 - 5i$
123. $3 + 6i$ **124.** $22 + 29i$ **125.** $-3\sqrt{3} + 0i$
126. $-81 + 0i$ **127.** $0 - \frac{3}{4}i$ **128.** $-\frac{5}{13} + \frac{12}{13}i$
129. -1 **130.** i

Chapter Test (page 596)

1. D: $[1, \infty)$, R: $[0, \infty)$

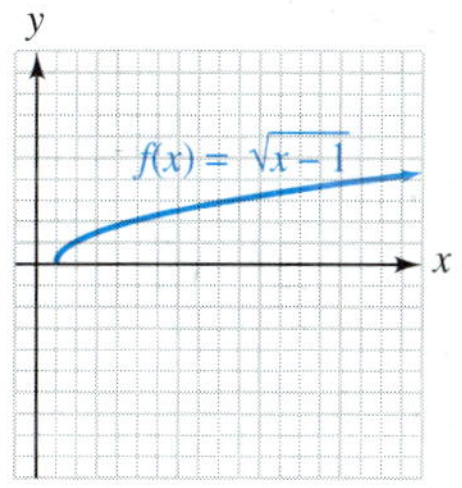

2. 46 feet per second
3. a -1 **b.** 2 **c.** 1
d. D: $(-\infty, \infty)$, R: $(-\infty, \infty)$
4. No real number raised to the fourth power is -16.
5. $7x^2$ **6.** -9 **7.** $\frac{1}{216}$
8. $\frac{25n^4}{4}$ **9.** $2^{4/3}$ **10.** $a^{1/9}$
11. $|x|$ **12.** $|y + 5|$
13. $-4xy^2$ **14.** $\frac{2}{3}a$ **15.** $t + 8$ **16.** $5xy^2\sqrt{10xy}$
17. $2x^5y\sqrt[3]{3}$ **18.** $2\sqrt[7]{2}$ **19.** $2y^2\sqrt{3y}$ **20.** $14\sqrt[3]{5}$
21. $5z^3\sqrt[4]{3z}$ **22.** $-6x\sqrt{y} - 2xy^2$ **23.** $3 - 7\sqrt{6}$
24. $\sqrt[3]{4a^2} + 18\sqrt[3]{2a} + 81$ **25.** $\frac{4\sqrt{10}}{5}$ **26.** $\sqrt{3t} + 1$
27. $\frac{\sqrt[3]{18a^2}}{2a}$ **28.** $\frac{1}{\sqrt{2}(\sqrt{5} - 3)}$ **29.** $\frac{1}{3}$ **30.** 10
31. 4 is extraneous, no solution **32.** 3, -3 is extraneous
33. $\sqrt{x} - 8$ is the principal square root. It cannot be negative.
34. $G = \frac{4\pi^2r^3}{Mt^2}$ **35.** 9.24 cm **36.** 8.67 cm **37.** 25
38. 28 in. **39.** $i\sqrt{5}$ **40.** -1 **41.** $-1 + 11i$
42. $4 - 7i$ **43.** $8 + 6i$ **44.** $-10 - 11i$ **45.** $0 - \frac{\sqrt{2}}{2}i$
46. $\frac{1}{2} + \frac{1}{2}i$

Study Set 8.1 (page 608)

1. quadratic **3.** perfect **5.** $\sqrt{c}, -\sqrt{c}$ **7.** 5
9. It is. **11. a.** 36 **b.** $\frac{25}{4}$ **c.** $\frac{1}{16}$ **d.** $\frac{9}{64}$
13. a. Subtract 35 from both sides. **b.** Add 36 to both sides.

15. $x = -7 \pm \sqrt{6}$ **17. a.** 4 is not a factor of the numerator. Only common factors of the numerator and denominator can be removed. **b.** 5 is not a factor of the numerator. Only common factors of the numerator and denominator can be removed.
19. a. 2; $2\sqrt{5}, -2\sqrt{5}$ **b.** ± 4.47 **21.** $0, -2$ **23.** $5, -5$
25. $-2, -4$ **27.** $2, \frac{1}{2}$ **29.** ± 6 **31.** $\pm\sqrt{5}$
33. $\pm\frac{4\sqrt{3}}{3}$ **35.** $0, -2$ **37.** $4, 10$ **39.** $-5 \pm \sqrt{3}$
41. $-2 \pm 2\sqrt{2}$ **43.** $2, -\frac{4}{3}$ **45.** $\pm 4i$ **47.** $\pm\frac{9}{2}i$
49. $3 \pm i\sqrt{5}$ **51.** $d = \frac{\sqrt{6h}}{2}$ **53.** $c = \frac{\sqrt{Em}}{m}$ **55.** $2, -4$
57. $2, 6$ **59.** $1, -6$ **61.** $-1, 4$ **63.** $-4 \pm \sqrt{10}$
65. $1 \pm 3\sqrt{2}$ **67.** $\frac{7 \pm \sqrt{37}}{2}$ **69.** $\frac{1 \pm \sqrt{13}}{2}$ **71.** $-\frac{1}{2}, 1$
73. $\frac{3 \pm 2\sqrt{3}}{3}$ **75.** $\frac{1 \pm 2\sqrt{2}}{2}$ **77.** $\frac{-5 \pm \sqrt{41}}{4}$
79. $\frac{-7 \pm \sqrt{29}}{10}$ **81.** $-1 \pm i$ **83.** $-4 \pm i\sqrt{2}$
85. $\frac{1}{3} \pm \frac{2\sqrt{2}}{3}i$ **87.** width: $7\frac{1}{4}$ ft; length: $13\frac{3}{4}$ ft
89. 1.6 sec **91.** 1.70 in. **93.** 0.92 in. **95.** 2.9 ft, 6.9 ft
99. $2ab^2\sqrt[3]{5}$ **101.** x^3 **103.** $5ab\sqrt{7b}$

Study Set 8.2 (page 618)

1. quadratic **3. a.** $x^2 + 2x + 5 = 0$ **b.** $3x^2 + 2x - 1 = 0$
5. a. true **b.** true **7. a.** $2, -4$ **b.** $\frac{1 \pm \sqrt{33}}{4}$
9. a. $1 \pm 2\sqrt{2}$ **b.** $\frac{-3 \pm \sqrt{7}}{2}$ **11. a.** The fraction bar wasn't drawn under both parts of the numerator. **b.** A $\pm$ sign wasn't written between b and the radical. **13.** $-1, -2$
15. $-6, -6$ **17.** $\frac{1}{2}, -3$ **19.** $\frac{-5 \pm \sqrt{5}}{10}$ **21.** $-\frac{3}{2}, -\frac{1}{2}$
23. $\frac{1}{4}, -\frac{3}{4}$ **25.** $\frac{4}{3}, -\frac{2}{5}$ **27.** $\frac{-5 \pm \sqrt{17}}{2}$ **29.** $\frac{3 \pm \sqrt{17}}{4}$
31. $5 \pm \sqrt{7}$ **33.** $23, -17$ **35.** $\frac{2}{3}, -\frac{5}{2}$ **37.** $-1 \pm i$
39. $-\frac{1}{4} \pm \frac{\sqrt{7}}{4}i$ **41.** $\frac{2}{3} \pm \frac{\sqrt{2}}{3}i$ **43.** $\frac{1}{3} \pm \frac{2\sqrt{2}}{3}i$
45. $2 \pm 2i$ **47.** $3 \pm i\sqrt{5}$ **49.** $\frac{-3 \pm \sqrt{29}}{10}$ **51.** $\frac{9 \pm \sqrt{89}}{2}$
53. $\frac{10 \pm \sqrt{55}}{30}$ **55.** $-0.68, -7.32$ **57.** $1.22, -0.55$
59. $8.98, -3.98$ **61.** 97 ft by 117 ft **63.** 0.5 mi by 2.5 mi
65. 34 in. **67.** \$4.80 or \$5.20 **69.** 4,000 **71.** 9%
73. early 1976 **77.** $n^{1/2}$ **79.** $(3b)^{1/4}$ **81.** $\sqrt[3]{t}$ **83.** $\sqrt[4]{3t}$

Study Set 8.3 (page 628)

1. discriminant **3.** conjugates **5.** rational, unequal
7. a. x^2 **b.** $\sqrt{x}$ **c.** $x^{1/3}$ **d.** $\frac{1}{x}$ **e.** $x + 1$ **9.** $4ac$, 5, 6, 24, rational **11.** rational, equal **13.** complex conjugates
15. irrational, unequal **17.** rational, unequal **19.** yes
21. $-1, 1, -4, 4$ **23.** $1, -1, \sqrt{5}, -\sqrt{5}$ **25.** $2, -2, i\sqrt{7}, -i\sqrt{7}$ **27.** $-i, i, -3i\sqrt{2}, 3i\sqrt{2}$ **29.** 1 **31.** no solution
33. $16, 4$ **35.** $\frac{9}{4}$ **37.** $-8, -27$ **39.** $-1, 27$
41. $\frac{243}{32}, 1$ **43.** $\frac{1}{4}, \frac{1}{2}$ **45.** $1 \pm 2i$ **47.** $-2\sqrt{3}, 2\sqrt{3}, 0$
49. $-\frac{3}{2}, -\frac{3}{2}$ **51.** $49, 225$ **53.** $-4, \frac{2}{3}$ **55.** $-\frac{5}{7}, 3$
57. $1, 1, -1, -1$ **59.** $1 \pm i$ **61.** $-1, -4$ **63.** $-1, -\frac{27}{13}$
65. $\frac{3 \pm \sqrt{57}}{6}$ **67.** 12.1 in. **69.** 30 mph **71.** 14.3 min
75. $x = 3$ **77.** $y = \frac{2}{3}x$

Study Set 8.4 (page 640)

1. quadratic, parabola **3.** axis of symmetry **5. a.** a parabola **b.** (1, 0), (3, 0) **c.** (0, −3) **d.** (2, 1) **e.** $x = 2$

7.

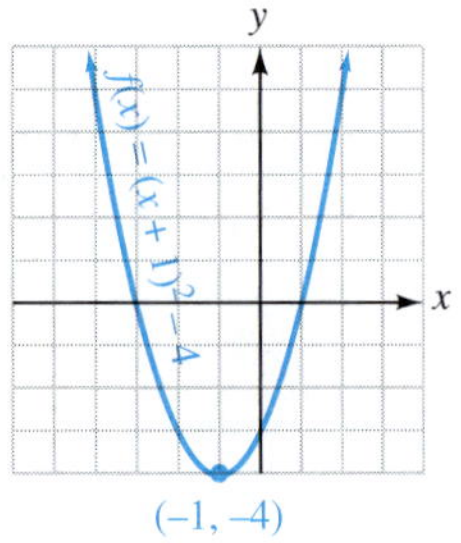

9. a. 2 **b.** 9, 18
11. $-3, 5$ **13.** $h = -1$; $f(x) = 2[x - (-1)]^2 + 6$

15.

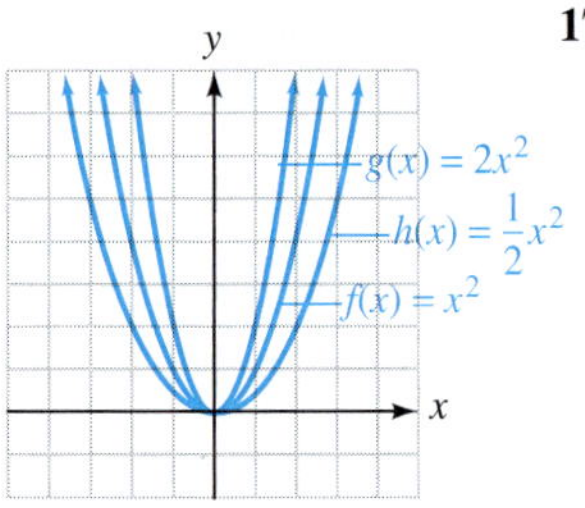

17.

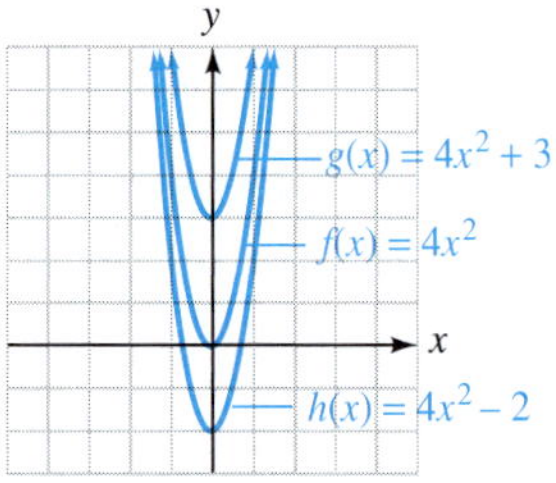

19.

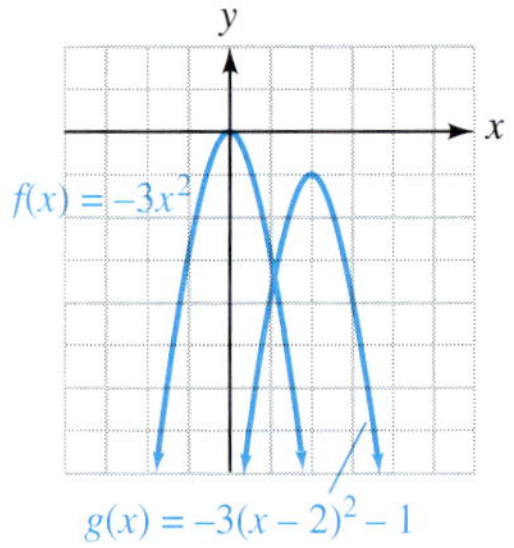

21. (1, 2), $x = 1$, upward
23. (−3, −4), $x = -3$, upward **25.** (7.5, 8.5), $x = 7.5$, upward
27. (1, −2), $x = 1$, upward
29. (2, 21), $x = 2$, downward
31. $\left(-\frac{2}{3}, \frac{2}{3}\right)$, $x = -\frac{2}{3}$, upward

33.

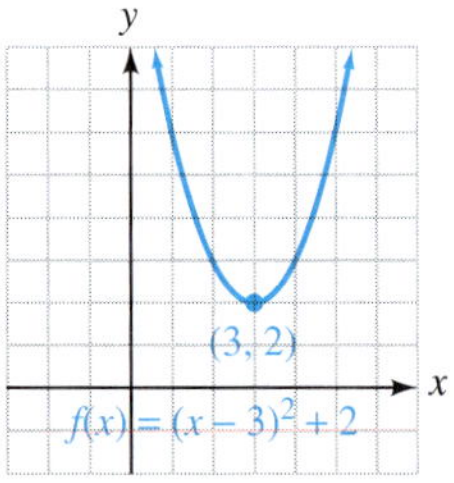

35.

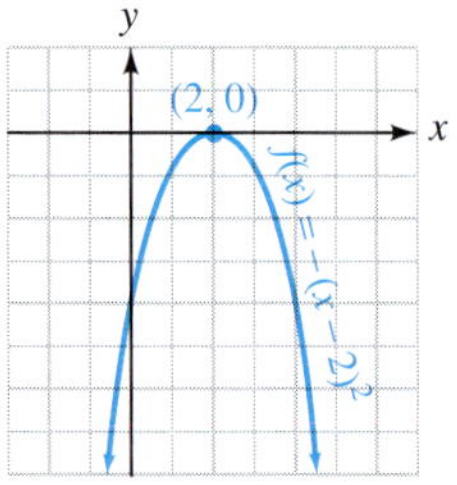

37.

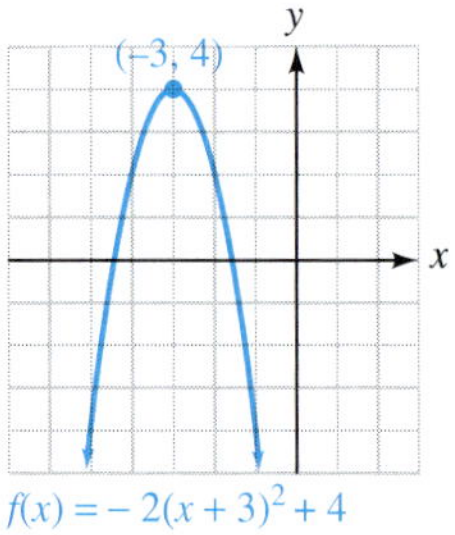

39.

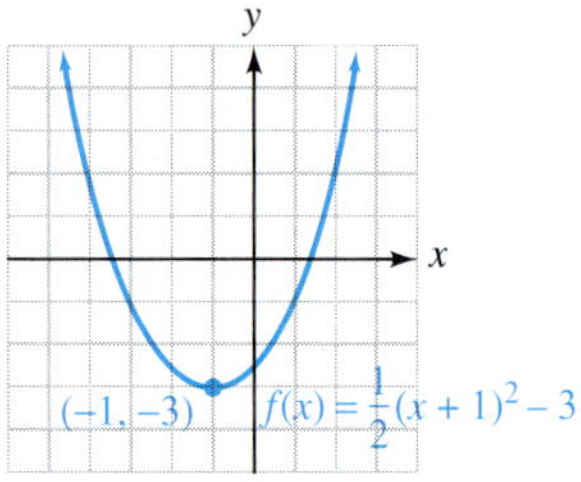

41.

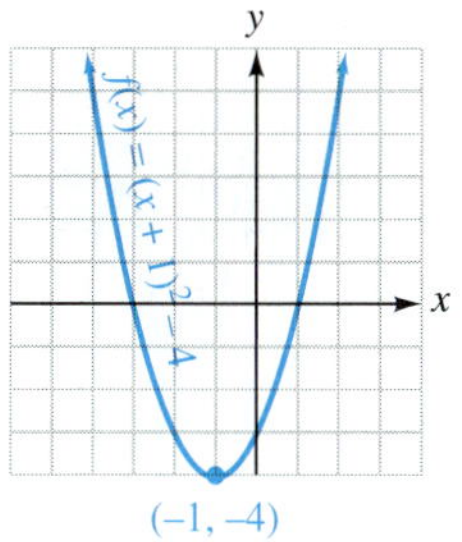

43.

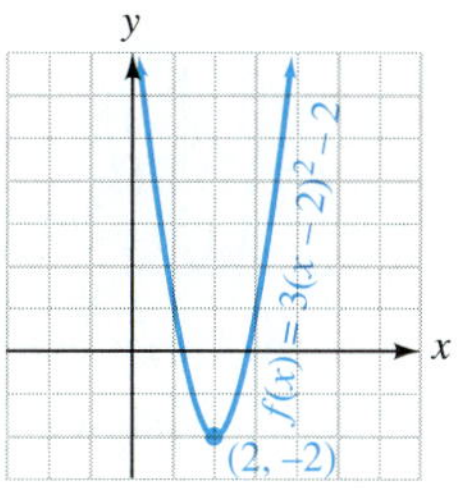

45.

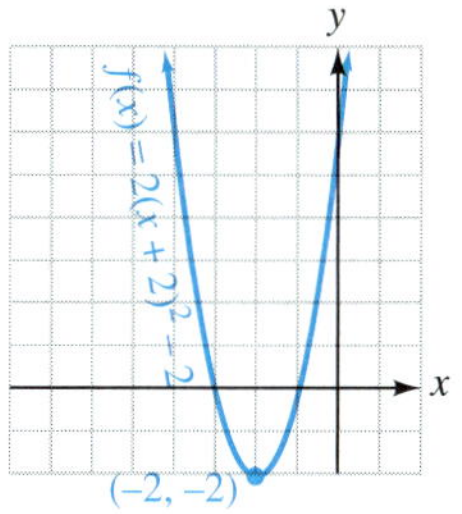

47.

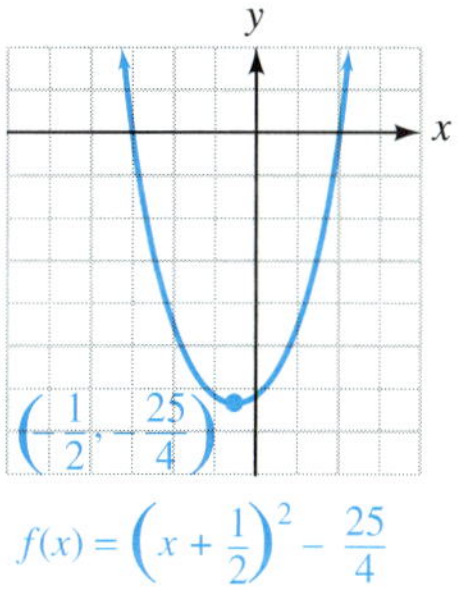

49.

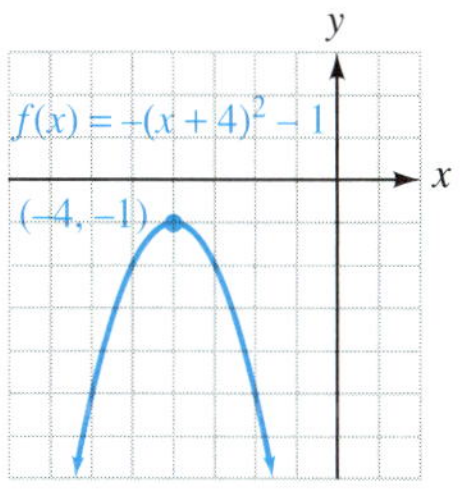

51.

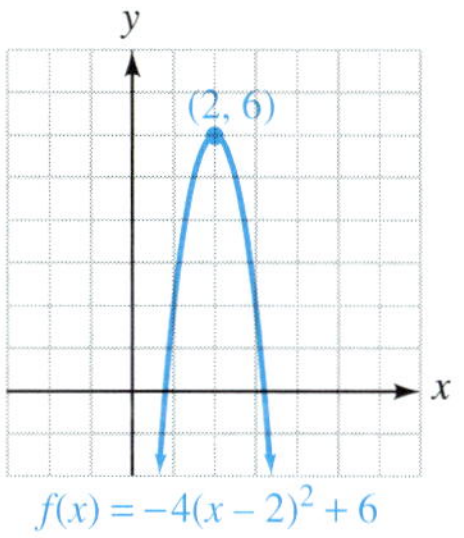

53. (1, 0); (0, 1)

55. (−3, 0), (−7, 0); (0, −21)

57. (−2, 0); (−2, 0); (0, 4)

59. (1, 0); (1, 0); (0, −1)

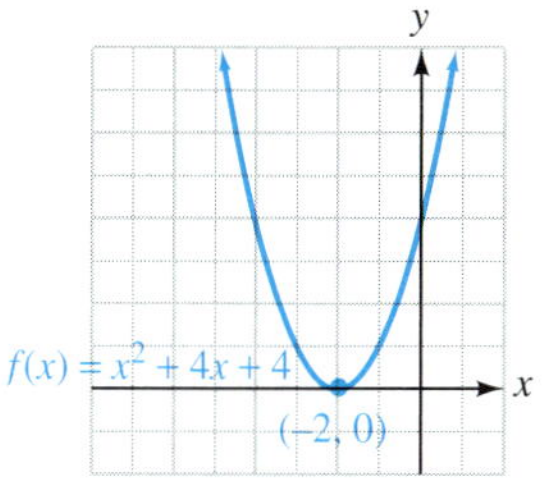

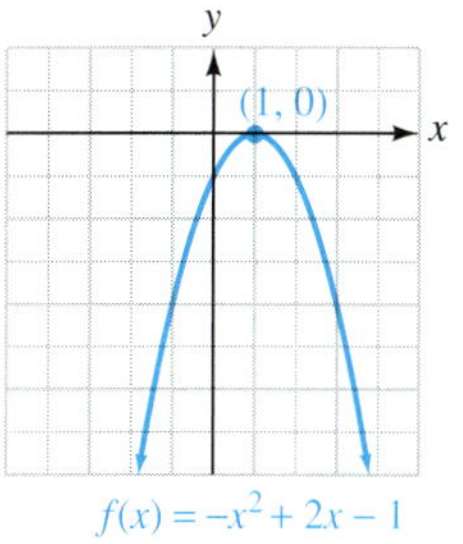

61. (1, −1); (0, 0), (2, 0); (0, 0)

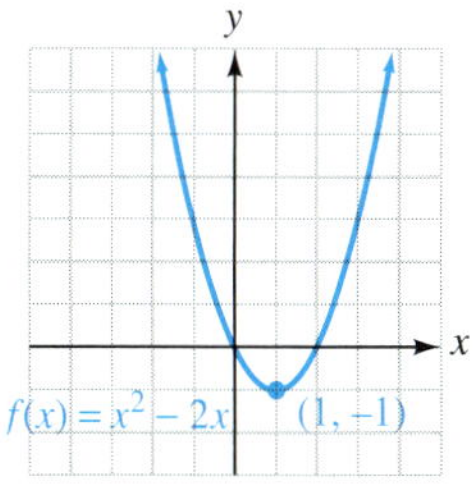

63. $\left(\frac{3}{2}, 0\right)$; $\left(\frac{3}{2}, 0\right)$; (0, 9)

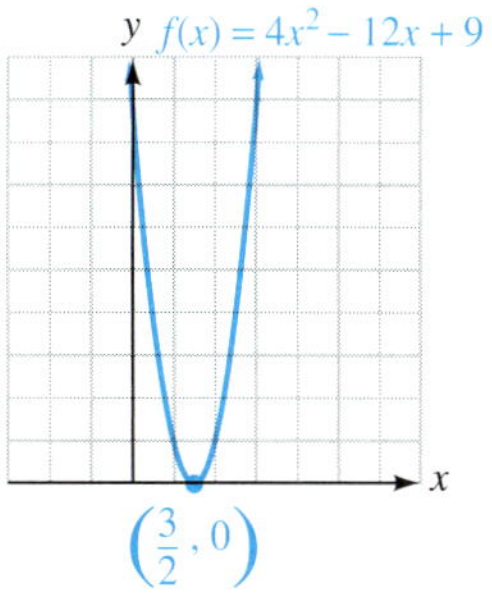

65. (2, −2); (1, 0), (3, 0); (0, 6)

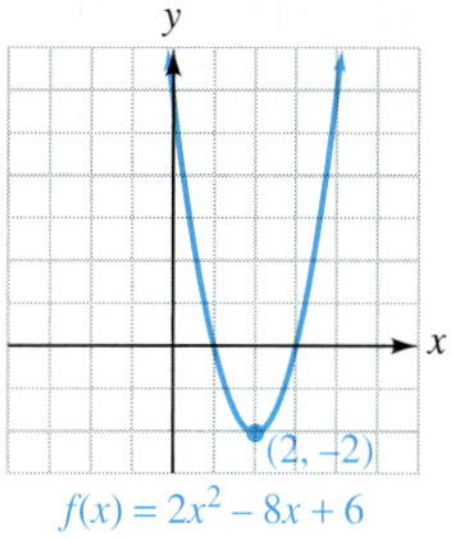

67. (−1, −2); no x-intercept; (0, −8)

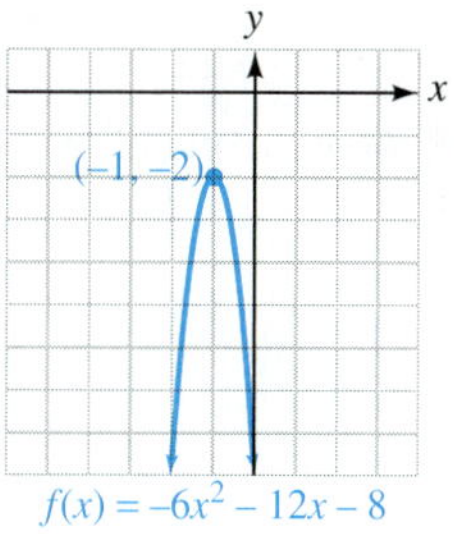

69. (0.25, 0.88) **71.** (0.50, 7.25) **73.** 2, −3

75. −1.85, 3.25

77.

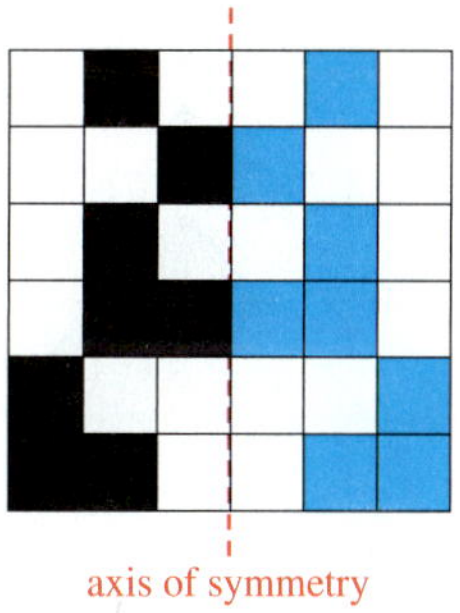

79. 3.75 sec, 225 ft

81. 250 ft by 500 ft

83. 15 min, $160

85. 1968, 1.5 million; the U.S. involvement in the war in Vietnam was at its peak

87. 200, $7,000 **95.** $4a^2\sqrt{b}$

97. $\frac{\sqrt{6}}{10}$

99. $15b - 6\sqrt{15b} + 9$

Study Set 8.5 (page 653)

1. quadratic **3.** two **5.** $(-\infty, -1)$, $(-1, 4)$, $(4, \infty)$

7. a.

b.

9. a. $(-3, 2)$ **b.** $(-\infty, -1] \cup [1, \infty)$ **11. a.** solid **b.** yes **13.** $x^2 - 6x - 7 \geq 0$

15. $(1, 4)$

17. $(-\infty, 3) \cup (5, \infty)$

19. $[-4, 3]$

21. no solutions

23. $(-\infty, -3] \cup [3, \infty)$

25. $(-5, 5)$

27. $(-\infty, 0) \cup \left(\frac{1}{2}, \infty\right)$

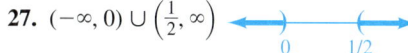

29. $\left(-\infty, -\frac{5}{3}\right) \cup (0, \infty)$

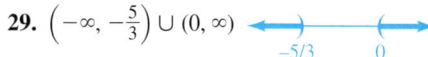

31. $(-\infty, -3) \cup (1, 4)$

33. $\left(-\frac{1}{2}, \frac{1}{3}\right) \cup \left(\frac{1}{2}, \infty\right)$

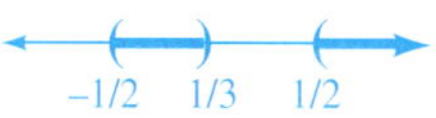

35. $(0, 2) \cup (8, \infty)$

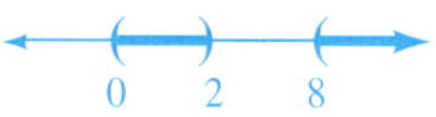

37. $\left[-\frac{34}{5}, -4\right) \cup (3, \infty)$

39. $(-4, -2] \cup (-1, 2]$ 

41. $(-\infty, -2) \cup (-2, \infty)$ 

43. $(-1, 3)$

45. $(-\infty, -3) \cup (2, \infty)$

47.

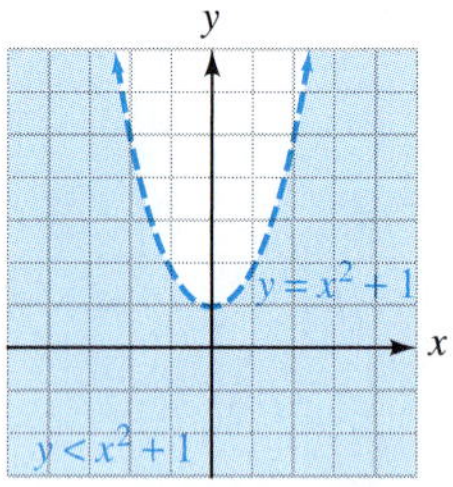

49.

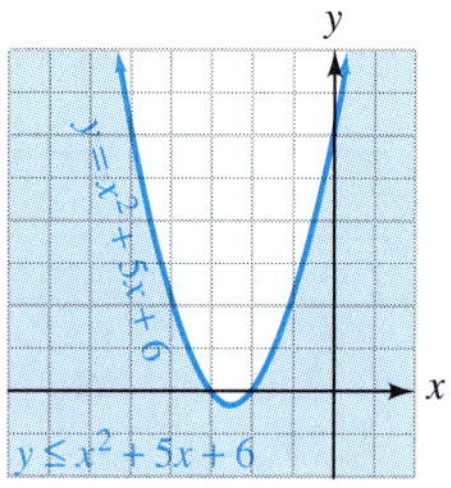

51.

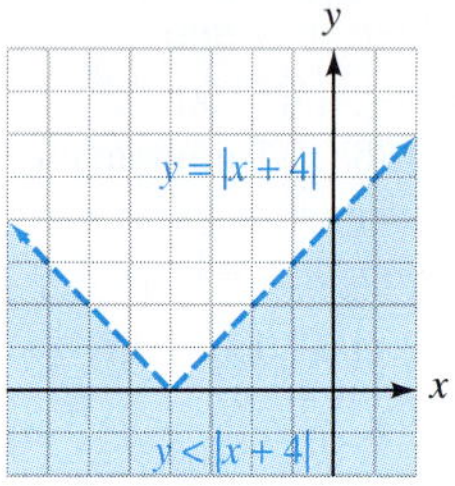

53.

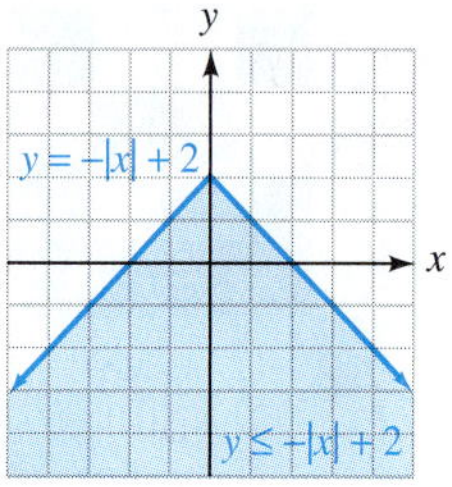

55. $(-2{,}100, -900) \cup (900, 2{,}100)$ **61.** $x = ky$ **63.** $t = kxy$

Key Concept (page 656)

1. 0, −2 **2.** −3, −5 **3.** $-\frac{3}{2}, -1$ **4.** $\pm 2\sqrt{6}$ **5.** 4, 10 **6.** $0 \pm \frac{4\sqrt{3}}{3}i$ **7.** $-5 \pm 4\sqrt{2}$ **8.** $\frac{1 \pm \sqrt{2}}{2}$ **9.** $-1 \pm i$ **10.** $\frac{3 \pm \sqrt{17}}{4}$ **11.** 23, −17 **12.** $-\frac{1}{3} \pm \frac{\sqrt{2}}{3}i$ **13.** 1, −2

Chapter Review (page 657)

1. −5, −4 **2.** $-\frac{1}{3}, -\frac{5}{2}$ **3.** $\pm 2\sqrt{7}$ **4.** 4, −8 **5.** $0, -\frac{11}{5}$ **6.** $\pm\frac{7\sqrt{5}}{5}$ **7.** $\pm 5i$ **8.** $\frac{1}{4}$ **9.** −4, −2 **10.** $\frac{3 \pm \sqrt{3}}{2}$ **11.** $1 \pm 2i\sqrt{3}$ **12.** $r = \frac{\sqrt{\pi A}}{\pi}$ **13.** 2 is not a factor of the numerator. Only common factors of the numerator and denominator can be removed. **14.** 6 seconds before midnight **15.** 0, 10 **16.** $5 \pm \sqrt{7}$ **17.** $\frac{1}{2}, -7$ **18.** $\frac{13 \pm \sqrt{163}}{3}$ **19.** $-\frac{3}{4} \pm \frac{\sqrt{15}}{4}i$ **20.** $\frac{2}{3} \pm \frac{\sqrt{2}}{3}i$ **21.** $\frac{-3 \pm \sqrt{29}}{10}$ **22.** \$24 or \$26 **23.** sides: 1.25 in. wide; top/bottom: 2.5 in. wide **24.** 0.7 sec, 1.8 sec **25.** irrational, unequal **26.** complex conjugates **27.** equal rational numbers **28.** rational, unequal **29.** 1, 144 **30.** 8, −27 **31.** $i, -i, \frac{\sqrt{6}}{3}, -\frac{\sqrt{6}}{3}$ **32.** $1, -\frac{8}{5}$ **33.** $4 \pm i$ **34.** 1, 1, −1, −1 **35.** $-\frac{2}{5}, -\frac{2}{5}$ **36.** about 81 min **37.** \$44.3 billion

38.

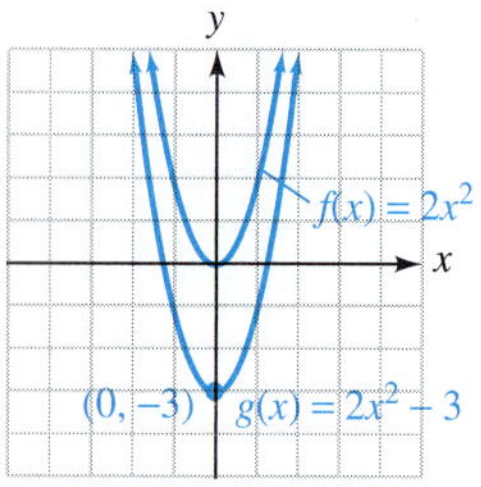

39.

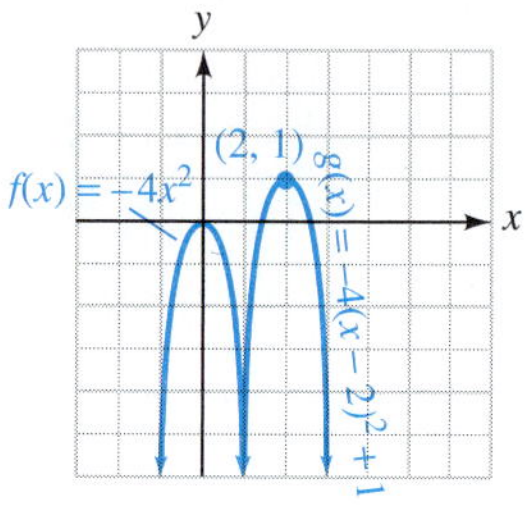

40.

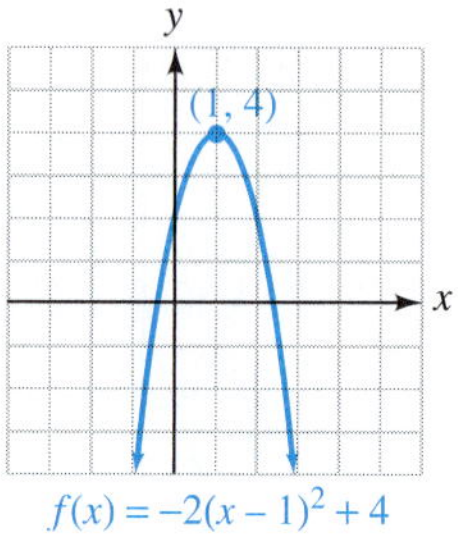

41.

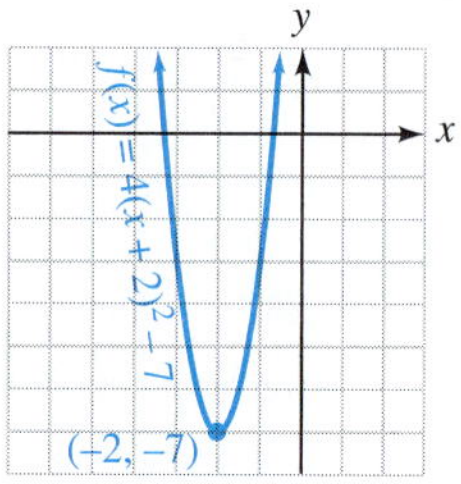

42.

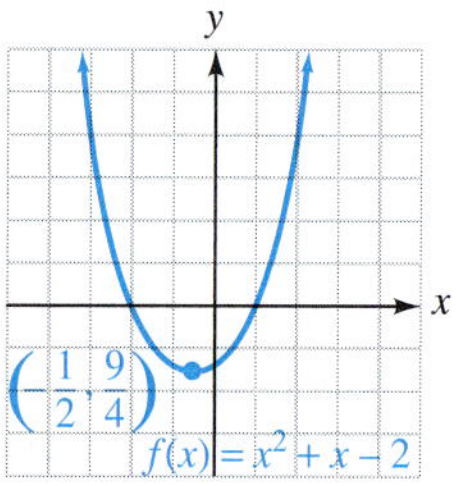

43. 1921; 6,469,326

44. $-2, \frac{1}{3}$

45. $(-\infty, -7) \cup (5, \infty)$

46. $[-9, 9]$

47. $(-\infty, 0) \cup [3/5, \infty)$

48. $(-7/2, 1) \cup (4, \infty)$

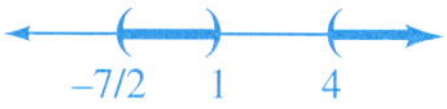

49. $\left[-4, \frac{2}{3}\right]$ **50.** $(-\infty, 0) \cup (1, \infty)$

51.

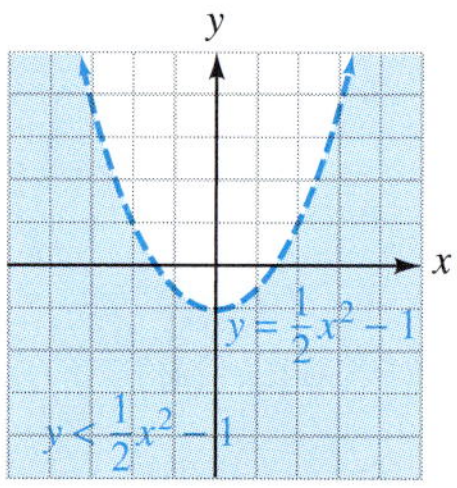

52.

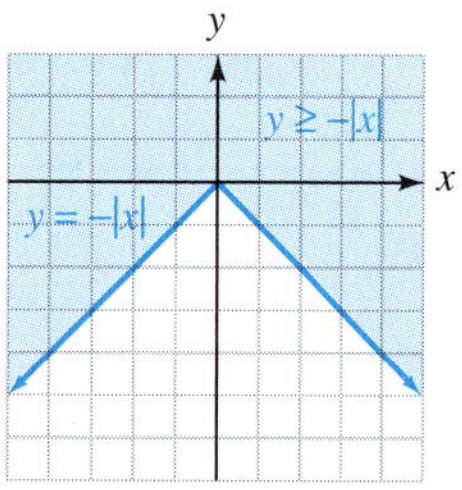

Chapter Test (page 661)

1. 0, −6 **2.** $\pm 2i$ **3.** $-7 \pm 5\sqrt{2}$ **4.** $-\frac{3}{2}, -\frac{5}{3}$ **5.** 144 **6.** $-3, \frac{8}{3}$ **7.** $\frac{4 \pm \sqrt{6}}{2}$ **8.** $1 \pm \sqrt{5}$ **9.** $2 \pm 3i$ **10.** −5, −3 **11.** $1, \frac{1}{4}$ **12.** $-\frac{1}{2} \pm \frac{\sqrt{3}}{2}i$ **13.** $2, -2, i\sqrt{3}, -i\sqrt{3}$ **14.** $-\frac{4}{5}, \frac{4}{7}$ **15.** $c = \frac{\sqrt{Em}}{m}$

16. a. complex conjugates **b.** rational, equal **17.** 4.5 ft by 1,502 ft **18.** about 54 min **19.** 20 in. **20.** iii

21.

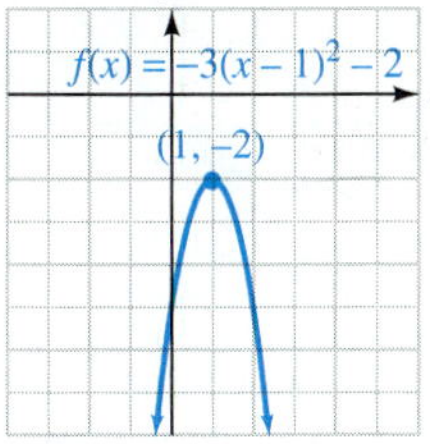

22.

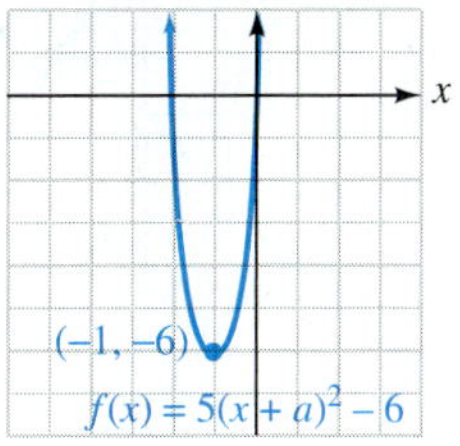

23. $\left(-\frac{1}{4}, -\frac{9}{8}\right)$, $(-1, 0)$, $\left(\frac{1}{2}, 0\right)$; $(0, -1)$ **24.** 211 ft

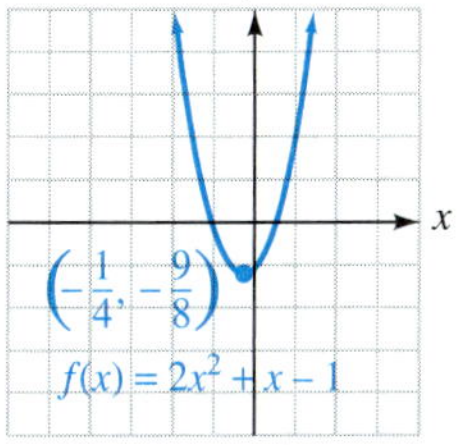

25. $(-\infty, -2) \cup (4, \infty)$

26. $(-3, 2]$

27.

We don't know whether x is positive or negative. When we multiply both sides by x, we don't know whether or not to reverse the inequality symbol.

28.

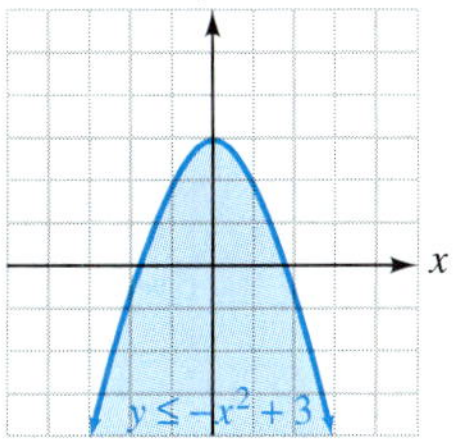

29. $-3, 2$ **30.** $[-3, 2]$

Cumulative Review Exercises, Chapters 1–8 (page 663)

1. $y = 3x + 2$ **2.** $y = -\frac{2}{3}x - 2$ **3.** an increase of about 13,333 a year **4.**

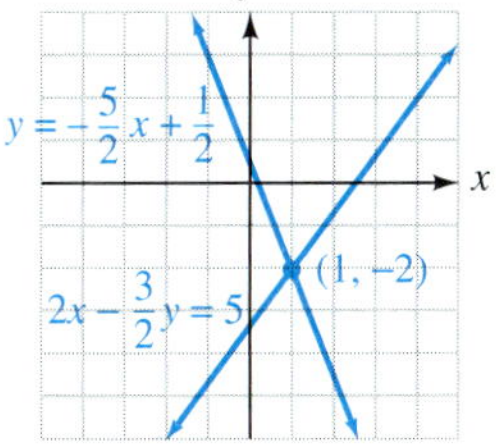

5. $(2, -1, 1)$

6.

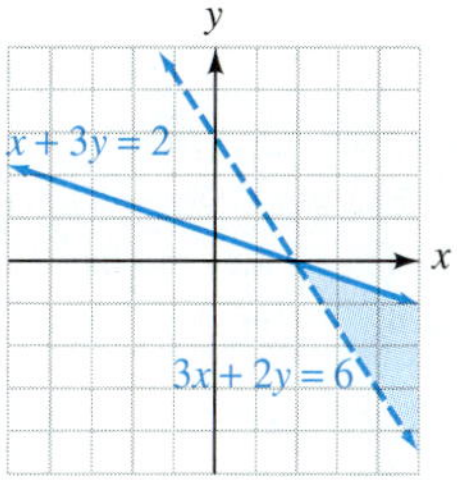

7. $\left(-\infty, -\frac{10}{9}\right)$

8. $(-\infty, -10] \cup [15, \infty)$

9. $\frac{y^8}{9x^2}$ **10. a.** all real numbers from 0 through 24 **b.** 0.5 **c.** 1.5 **d.** about -1.4 **e.** The low tide mark was -2.5 m. **f.** 0, 2, 9, 17 **11.** $3a^3 + 10a^2 + 6a - 4$ **12.** $\frac{x + y}{x - y}$ **13.** 0 **14.** $\frac{1}{2r(r + 2)}$ **15.** $(x^2 + 4y^2)(x + 2y)(x - 2y)$ **16.** $2a^2(3a + 2)(5a - 4)$ **17.** $(x - y)(x - 4)$ **18.** $(2x^2 + 5y)(4x^4 - 10x^2y + 25y^2)$ **19.** $(x + 5 + y^4)(x + 5 - y^4)$ **20.** $(7s^3 - 6n^2)^2$ **21.** $2, -\frac{5}{2}$ **22.** $0, \frac{2}{3}, -\frac{1}{2}$ **23.** 5; 3 is extraneous **24.** $b = \frac{-RTV^2 + aV + PV^3}{PV^2 + a}$ **25.** D: $(-\infty, \infty)$, R: $(-\infty, \infty)$ **26.** D: $(0, \infty)$, R: $(0, \infty)$

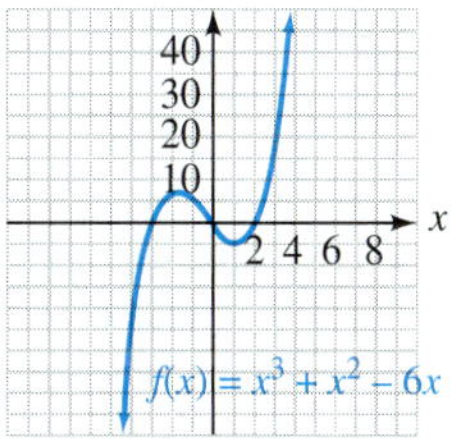

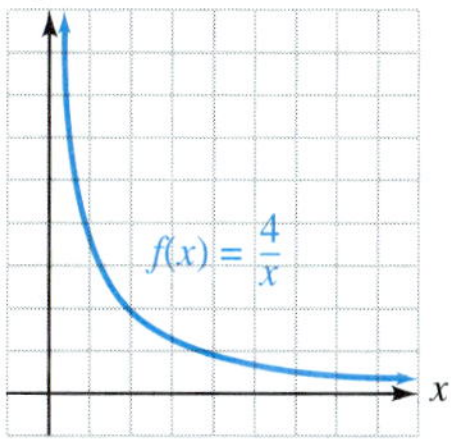

27. 9 **28.** D: $[2, \infty)$, R: $[0, \infty)$

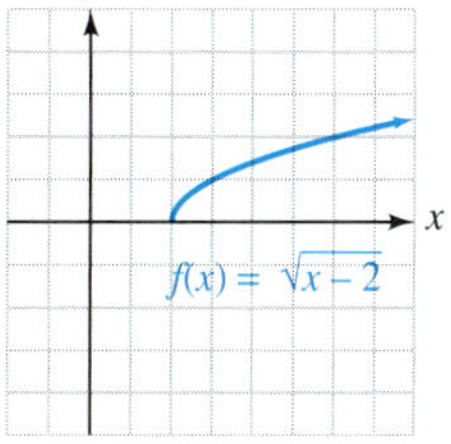

29. $-3x$ **30.** $4t\sqrt{3t}$ **31.** $\frac{1}{16}$ **32.** $x^{17/12}$ **33.** $-12\sqrt[4]{2} + 10\sqrt[4]{3}$ **34.** $-18\sqrt{6}$ **35.** $\frac{x + 3\sqrt{x} + 2}{x - 1}$ **36.** $\frac{5\sqrt[3]{x^2}}{x}$ **37.** 2, 7 **38.** $\frac{1}{4}$ **39.** $3\sqrt{2}$ in. **40.** $2\sqrt{3}$ in. **41.** 10 **42.** $-i$ **43.** $-5 + 17i$ **44.** $\frac{3}{2} + \frac{1}{2}i$ **45.** $3 + 4i$ **46.** $0 - \frac{2}{3}i$ **47.** 9 **48.** $1, -\frac{3}{2}$ **49.** $2 \pm 2i$ **50.** 10 ft by 18 ft **51.** 50 m and 120 m **52.** $(-2, 4)$; $(-4, 0)$, $(0, 0)$; $(0, 0)$

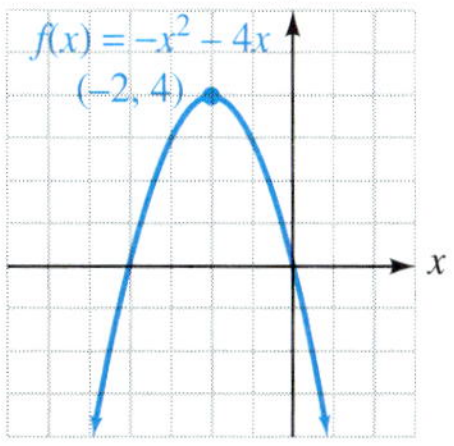

53. 9, 16 **54.** $1, 1, -1, -1$ **55.** $-\frac{3}{4}$ **56.** no solution

Study Set 9.1 (page 672)

1. sum, $f(x) + g(x)$, difference, $f(x) - g(x)$ **3.** domain **5.** identity **7. a.** $g(3)$ **b.** $g(3)$ **9.** $-2, -10, 0$ **11. a.** 7 **b.** 8 **c.** 12 **d.** 0 **13.** $g(x)$, $(3x - 1)$, $9x$, $2x$ **15.** $(f + g)(x) = 7x, (-\infty, \infty)$ **17.** $(f \cdot g)(x) = 12x^2$, $(-\infty, \infty)$ **19.** $(g - f)(x) = x, (-\infty, \infty)$ **21.** $(g/f)(x) = \frac{4}{3}$, $(-\infty, 0) \cup (0, \infty)$ **23.** $(f + g)(x) = 3x - 2, (-\infty, \infty)$ **25.** $(f \cdot g)(x) = 2x^2 - 5x - 3, (-\infty, \infty)$ **27.** $(g - f)(x) = -x - 4, (-\infty, \infty)$ **29.** $(g/f)(x) = \frac{x-3}{2x+1}$, $\left(-\infty, -\frac{1}{2}\right) \cup \left(-\frac{1}{2}, \infty\right)$ **31.** $(f - g)(x) = -2x^2 + 3x - 3$, $(-\infty, \infty)$ **33.** $(f/g)(x) = \frac{3x-2}{2x^2+1}, (-\infty, \infty)$ **35.** $(f - g)(x) = 3, (-\infty, \infty)$ **37.** $(g/f)(x) = \frac{x^2-4}{x^2-1}$, $(-\infty, -1) \cup (-1, 1) \cup (1, \infty)$ **39.** 7 **41.** 24 **43.** -1 **45.** $-\frac{1}{2}$ **47.** $(f \circ g)(x) = 2x^2 - 1$ **49.** $(g \circ f)(2x) = 16x^2 + 8x$ **51.** 58 **53.** 110 **55.** 2 **57.** $(g \circ f)(x) = 9x^2 - 9x + 2$ **59.** 16 **61.** $\frac{1}{9}$ **63.** $(g \circ f)(8x) = 64x^2$ **67.** 500, 503, 1,003 **69.** In 2000, the average combined score on the SAT was 1,019. **71.** $C(t) = \frac{5}{9}(2{,}668 - 200t)$ **73. a.** about \$37.50 **b.** $C(m) = \frac{1.50m}{20} = 0.075m$ **79.** $\frac{1}{c-d}$

Study Set 9.2 (page 683)

1. one-to-one **3.** inverses **5.** images, symmetric **7. a.** once **b.** one-to-one **9.** x **11.** y, interchange **13.** yes, no **15.** yes **17.** 2 **19.** The graphs are not symmetric about the line $y = x$. **21.**

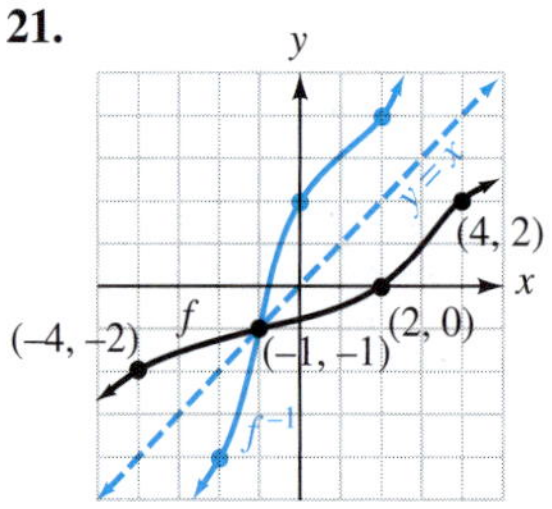

23. y, $2y$, 3, y, $f^{-1}(x)$ **25.** inverse of, inverse **27.** yes **29.** no **31.** no **33.** no **35.** one-to-one **37.** not one-to-one **39.** not one-to-one **41.** $f^{-1}(x) = \frac{x-4}{2}$ **43.** $f^{-1}(x) = 5x - 4$ **45.** $f^{-1}(x) = 5x + 4$ **47.** $f^{-1}(x) = \frac{2}{x} + 3$ **49.** $f^{-1}(x) = \frac{4}{x}$ **51.** $f^{-1}(x) = \sqrt[3]{x - 8}$ **53.** $f^{-1}(x) = x^3$ **55.** $f^{-1}(x) = \sqrt[3]{x} - 10$ **57.** $f^{-1}(x) = \sqrt[3]{\frac{x+3}{2}}$

63.

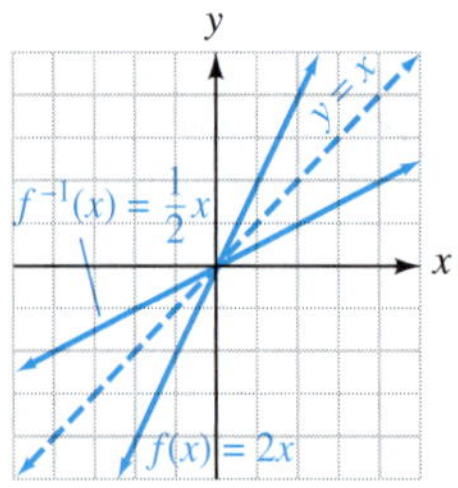

65.

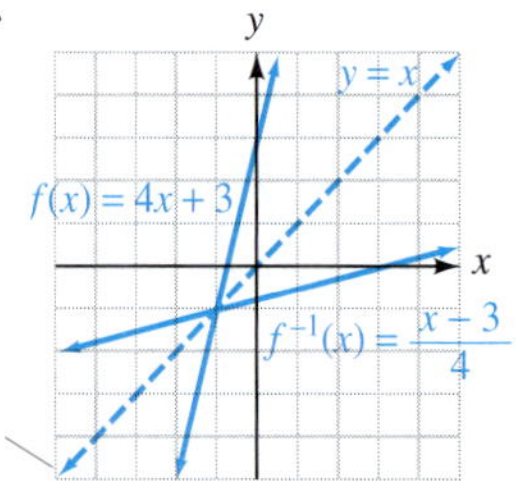

67.

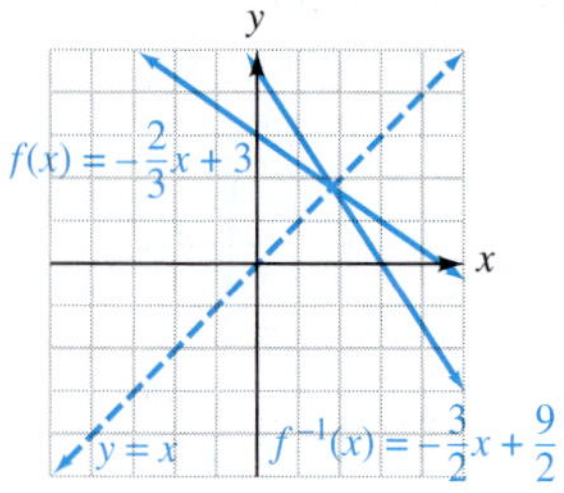

69.

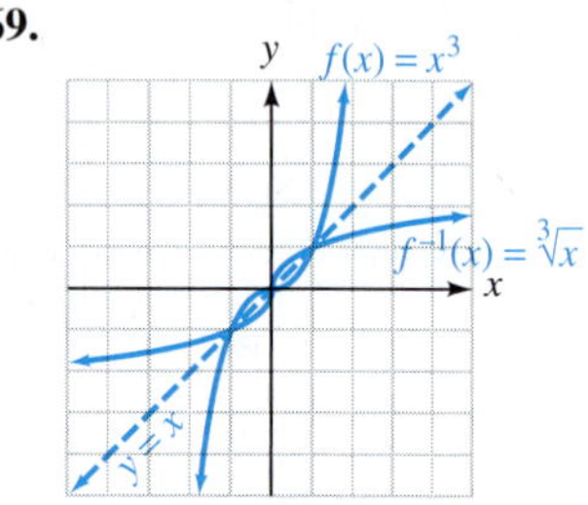

71.

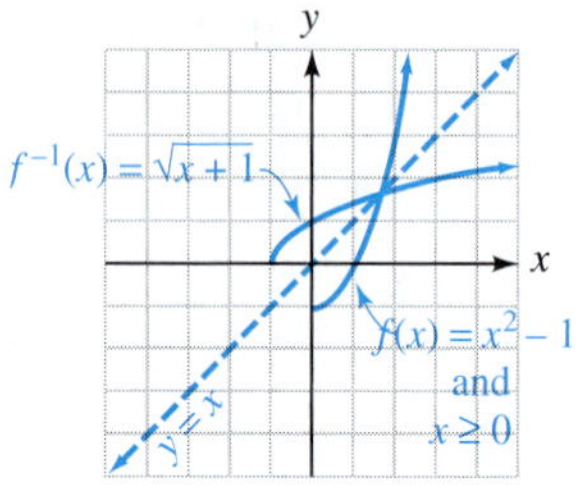

73. a. yes, no **b.** No. Twice during this period, the person's anxiety level was at the maximum threshold value. **79.** $3 - 8i$ **81.** $18 - i$ **83.** $-28 - 96i$

Study Set 9.3 (page 696)

1. exponential **3.** $(0, \infty)$ **5.** yes **7.** increasing **9.**

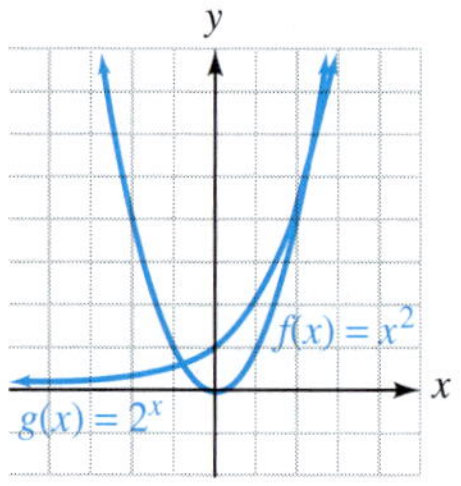

11. $A = P\left(1 + \frac{r}{k}\right)^{kt}$, $FV = PV(1 + i)^n$ **13.** $g(x) = 2^x + 3$; $h(x) = 2^x - 2$ **15.** $\left(1 + \frac{r}{k}\right)$, kt **17.** 2.6651 **19.** 36.5548 **21.** 8 **23.** $7^{3\sqrt{3}}$

25.

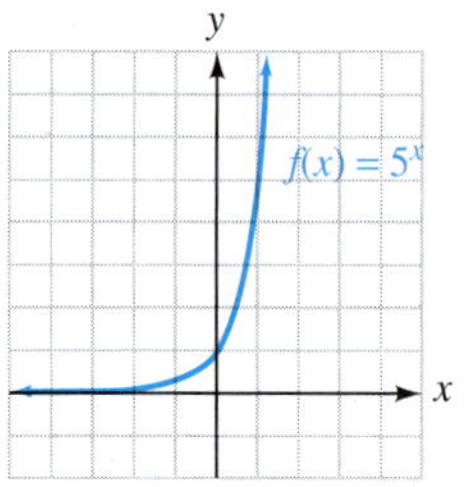

27.

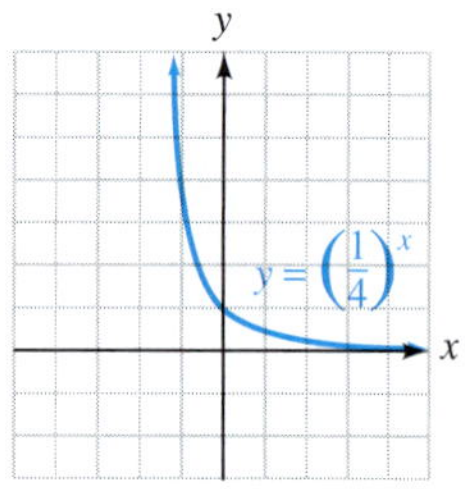

29.

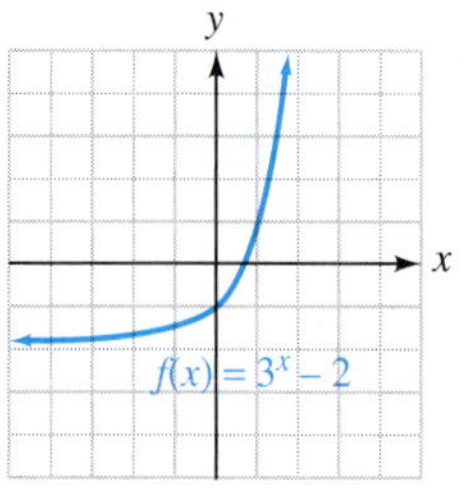

31.

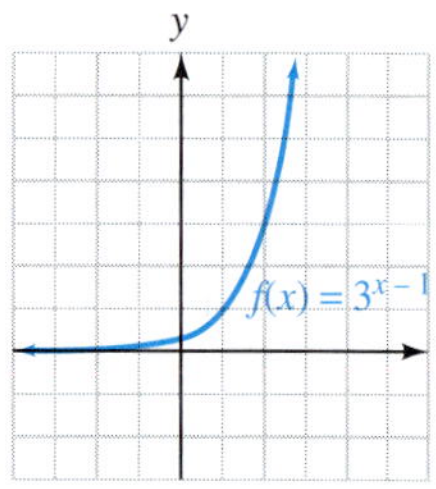

33. increasing

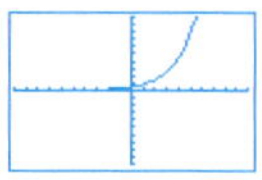

35. decreasing

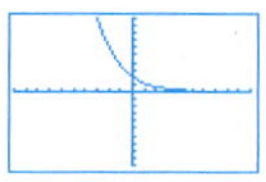

37. \$22,080.40 **39.** \$32.03 **41.** \$2,273,996.13 **43. a.** about 1500, about 1825 **b.** 6.5 billion **c.** exponential **45. a.** at the end of the 2nd year **b.** at the end of the 4th year **c.** during the 7th year

47. 1.679046×10^8 **49.** 5.0421×10^{-5} coulombs
51. $1,115.33 **57.** 40 **59.** 120°

Study Set 9.4 (page 705)

1. the natural exponential function **3.** $(0, \infty)$ **5.** yes
7. increasing **9.** continuous **11.** 2.72
13.

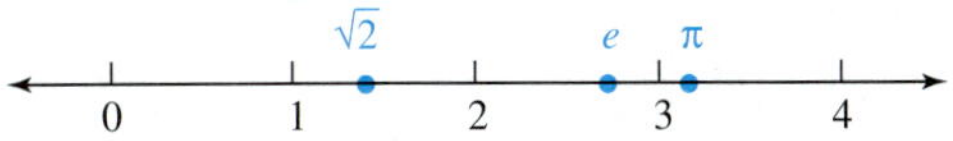

15. an exponential function

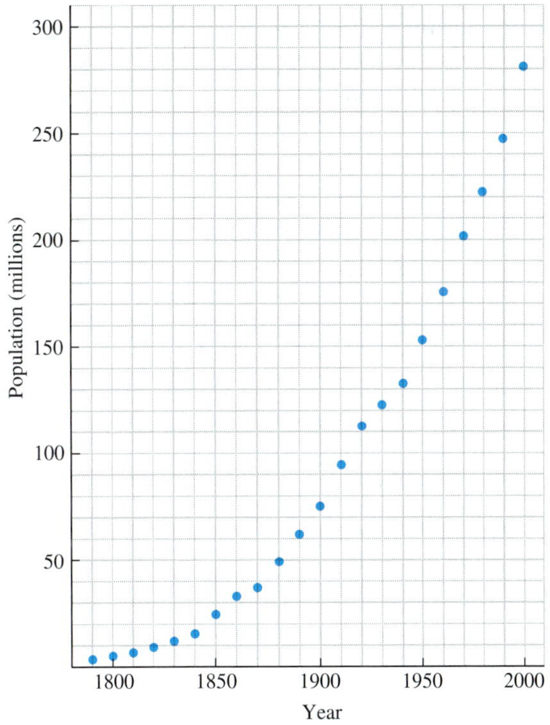

17. 2.7182818 . . .; e **19.** 1,000, 10, 0.9, 1,000
21.

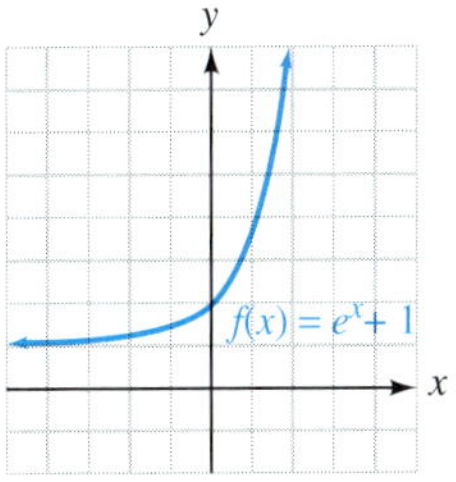

23.

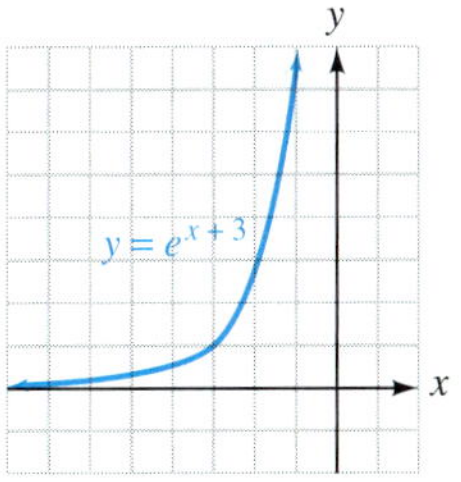

25.

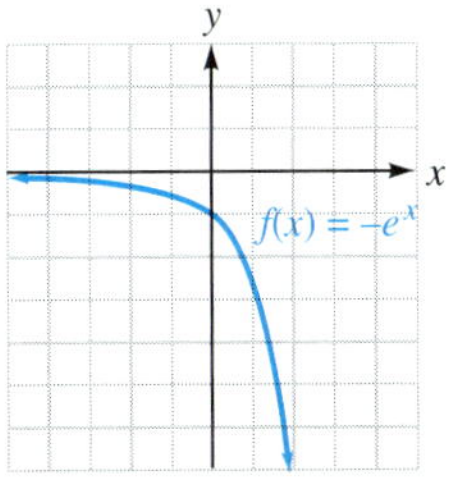

27.

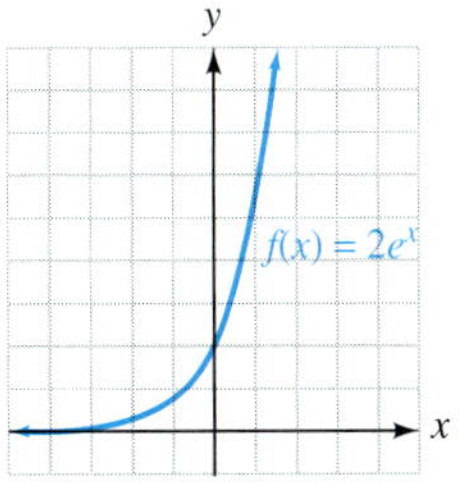

29. $13,375.68 **31.** $7,518.28 from annual compounding, $7,647.95 from continuous compounding **33.** $6,849.16
35. about 9.3 billion **37.** 315 **39.** 51 **41.** 12 hr
43. 49 mps **45.** about 72 yr **51.** $4x^2\sqrt{15x}$ **53.** $10y\sqrt{3y}$

Study Set 9.5 (page 717)

1. logarithmic **3.** $(-\infty, \infty)$ **5.** yes **7.** increasing
9. $x = b^y$ **11.** inverse **13.** $-2, -1, 0, 1, 2$ **15.** none, none, $-3, \frac{1}{2}, 8$ **17.** $-0.30, 0, 0.30, 0.60, 0.78, 0.90, 1$
19. They decrease. **21. a.** 10 **b.** x **23.** $3^4 = 81$
25. $10^1 = 10$ **27.** $4^{-3} = \frac{1}{64}$ **29.** $5^{1/2} = \sqrt{5}$
31. $\log_8 64 = 2$ **33.** $\log_4 \frac{1}{16} = -2$ **35.** $\log_{1/2} 32 = -5$
37. $\log_x z = y$ **39.** 3 **41.** 2 **43.** 5 **45.** $\frac{1}{2}$ **47.** -1
49. 6 **51.** 64 **53.** 5 **55.** $\frac{1}{25}$ **57.** $\frac{1}{6}$ **59.** 10
61. $\frac{2}{3}$ **63.** 5 **65.** $\frac{1}{3}$ **67.** 1,000 **69.** 4 **71.** 1
73. 0.5119 **75.** -2.3307 **77.** 25.25 **79.** 0.02
81. increasing

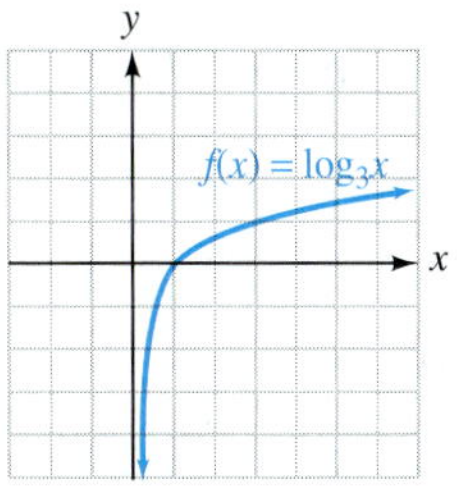

83. decreasing

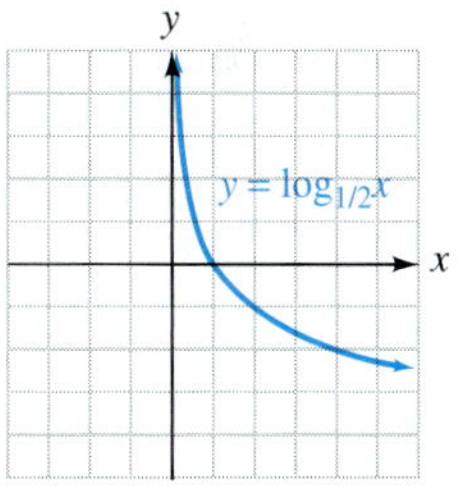

85.

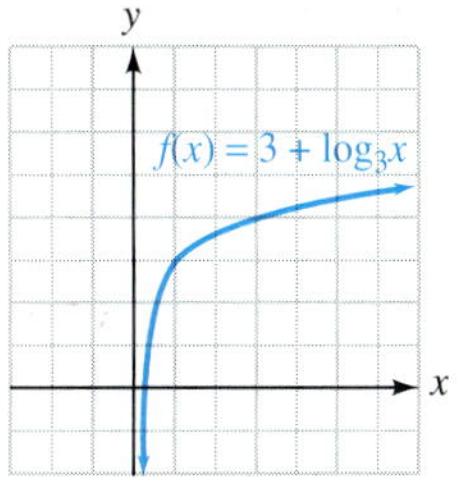

87.

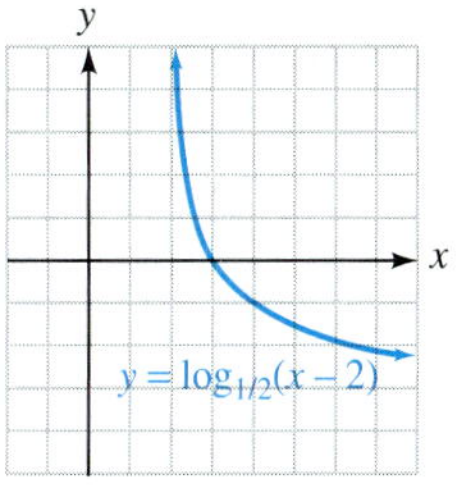

89.

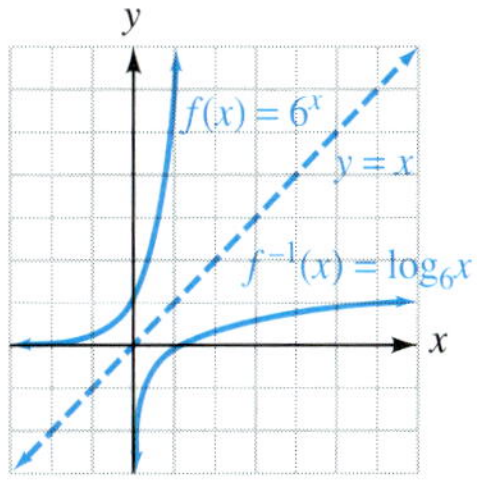

91.

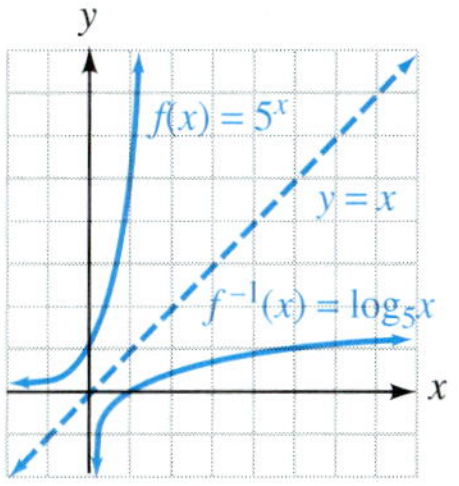

93. 29 db **95.** 49.5 db **97.** 4.4 **99.** 2,500 micrometers
101. 3 yr old **103.** 10.8 yr **107.** 10 **109.** 0; $-\frac{5}{9}$ does not check

Study Set 9.6 (page 724)

1. natural, e **3.** $-0.69, 0, 0.69, 1.39, 1.79, 2.08, 2.30$
5. y-axis **7.** $(-\infty, \infty)$ **9.** 1, asymptote **11.** The logarithm of a negative number or zero is not defined.

13. 2.7182818 . . . , e **15.** ℓ, n **17.** $\frac{\ln 2}{r}$ **19.** 1 **21.** 6
23. -1 **25.** $\frac{1}{4}$ **27.** 3.5596 **29.** -5.3709 **31.** 0.5423
33. undefined **35.** 4.0645 **37.** 69.4079 **39.** 0.0245
41. 2.7210 **43.**

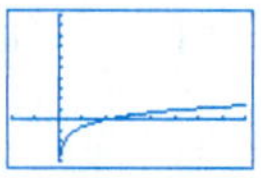

45.

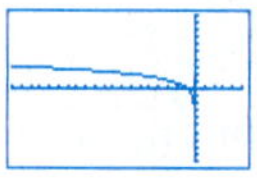

47. 5.8 yr **49.** 9.2 yr **51.** about 3.5 hr **57.** $y = 5x$
59. $y = -\frac{3}{2}x + \frac{13}{2}$ **61.** $x = 2$

Study Set 9.7 (page 735)

1. product **3.** power **5.** 0 **7.** M, N **9.** $-$ **11.** x
13. $\neq$ **15.** $\log_a b$ **17.** $10^0 = 1, 10^1 = 10, 10^2 = 10^2$
19. 0 **21.** 7 **23.** 10 **25.** 2 **27.** 1 **29.** 7
31. rs, t, r, s **39.** $2 + \log_2 5$ **41.** $\log_6 x - 2$
43. $4 \ln y$ **45.** $\frac{1}{2} \log 5$ **47.** $\log x + \log y + \log z$
49. $1 + \log_2 x - \log_2 y$ **51.** $3 \log x + 2 \log y$
53. $\frac{1}{2}(\log_b x + \log_b y)$ **55.** $\frac{1}{3} \log_a x - \frac{1}{4} \log_a y - \frac{1}{4} \log_a z$
57. $\ln x + \frac{1}{2} \ln z$ **59.** $\log_2 \frac{x+1}{x}$ **61.** $\log x^2 y^{1/2}$
63. $\log_b \frac{z^{1/2}}{x^3 y^2}$ **65.** $\ln \frac{\frac{x}{z} + x}{\frac{y}{z} + y} = \ln \frac{x}{y}$ **67.** false **69.** false
71. true **73.** 1.4472 **75.** -1.1972 **77.** 1.1972
79. 1.8063 **81.** 1.7712 **83.** -1.0000 **85.** 1.8928
87. 2.3219 **89.** 4.8 **91.** from 2.5×10^{-8} to 1.6×10^{-7}
95. $-\frac{7}{6}$ **97.** $\left(1, -\frac{1}{2}\right)$

Study Set 9.8 (page 746)

1. exponential **3.** It is a solution. **5.** yes **7.** 5
9. a. $10^2 = x + 1$ **b.** $e^2 = x + 1$ **11.** $\log 9$
13. $x \log 7 = 12$ **15. a.** 1.2920 **b.** 2.2619
17. a. $A_0 2^{-t/h}$ **b.** $P_0 e^{kt}$ **19.** log, log 2 **21.** 8
23. $-\frac{3}{4}$ **25.** $3, -1$ **27.** $-2, -2$ **29.** 1.1610
31. 1.2702 **33.** 1.7095 **35.** 0 **37.** ± 1.0878
39. 0, 1.0566 **41.** 0.7324 **43.** -13.2662 **45.** 1.8
47. 2.1 **49.** 9,998 **51.** -93 **53.** $e \approx 2.7183$
55. 19.0855 **57.** 2 **59.** 3 **61.** -7 **63.** 4
65. $10, -10$ **67.** 50 **69.** 20 **71.** 10 **73.** 10
75. no solution **77.** 6 **79.** 9 **81.** 4 **83.** 1, 7
85. 20 **87.** 8 **89.** 5.1 yr **91.** 42.7 days
93. about 4,200 yr **95.** 5.6 yr **97.** 5.4 yr
99. because $\ln 2 \approx 0.7$ **101.** 25.3 yr **103.** 2.828 times larger **105.** 13.3 **107.** $\frac{1}{3} \ln 0.75$ **113.** $\sqrt{137}$ in.

Key Concept (page 750)

1. no **2.** no **3.** yes **4.** $f^{-1}(x) = -\frac{x+1}{2}$
5. $-2, 1, 3$ **6.** $\log 1{,}000 = 3$ **7.** $2^{-3} = \frac{1}{8}$
8. 2 **9.** $\frac{3}{2}$ **10.** 2.71828 **11.** e **12.** 0.7558
13.

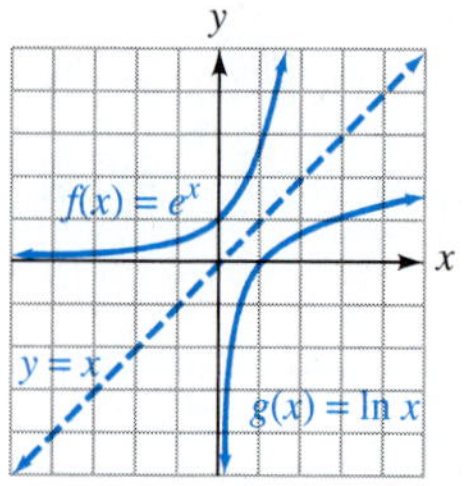

Chapter Review (page 751)

1. $(f + g)(x) = 3x + 1, (-\infty, \infty)$ **2.** $(f - g)(x) = x - 1, (-\infty, \infty)$ **3.** $(f \cdot g)(x) = 2x^2 + 2x, (-\infty, \infty)$
4. $(f/g)(x) = \frac{2x}{x+1}, (-\infty, -1) \cup (-1, \infty)$ **5.** 3 **6.** 5
7. $(f \circ g)(x) = 4x^2 + 4x + 3$ **8.** $(g \circ f)(x) = 2x^2 + 5$
9. a. 12 **b.** 27 **c.** 7 **d.** 0 **10.** $C(m) = \frac{1.85m}{8}$
11. no **12.** yes **13.** yes **14.** no **15.** yes **16.** no
17. $-6, -1, 7, 20$ **18.**

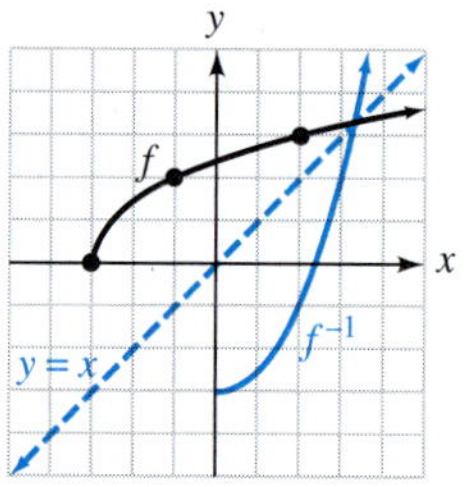

19. $f^{-1}(x) = \frac{x+3}{6}$ **20.** $f^{-1}(x) = \frac{4}{x} + 1$
21. $f^{-1}(x) = \sqrt[3]{x} - 2$ **22.** $f^{-1}(x) = 6x + 1$
23.

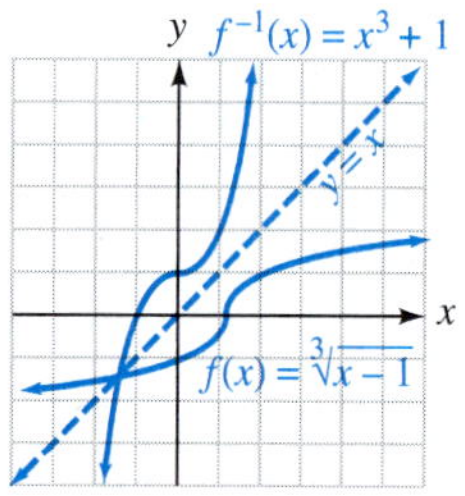

25. $5^{4\sqrt{6}}$ **26.** $2^{2\sqrt{7}}$
27. D: $(-\infty, \infty)$, R: $(0, \infty)$

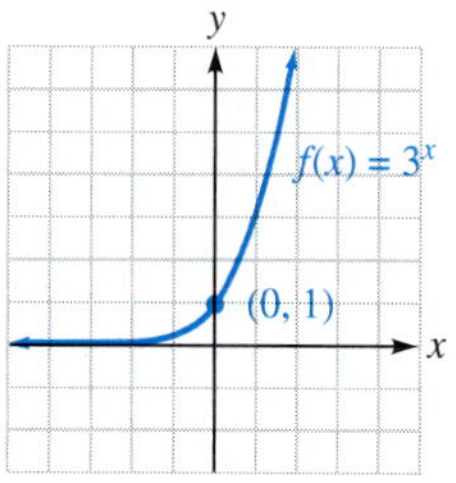

28. D: $(-\infty, \infty)$, R: $(0, \infty)$

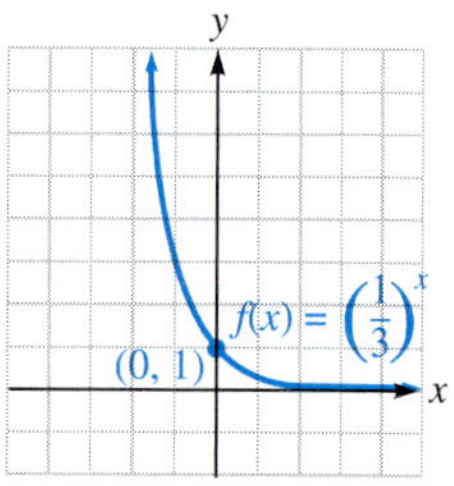

29. D: $(-\infty, \infty)$, R: $(-2, \infty)$

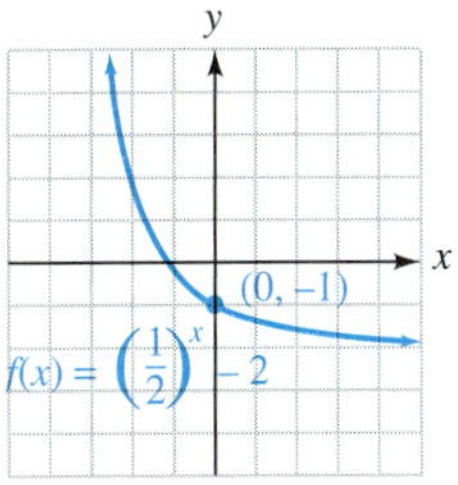

30. D: $(-\infty, \infty)$, R: $(0, \infty)$

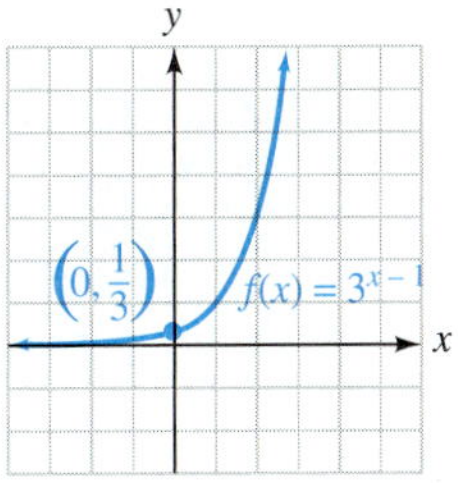

31. the x-axis ($y = 0$)
32. an exponential function

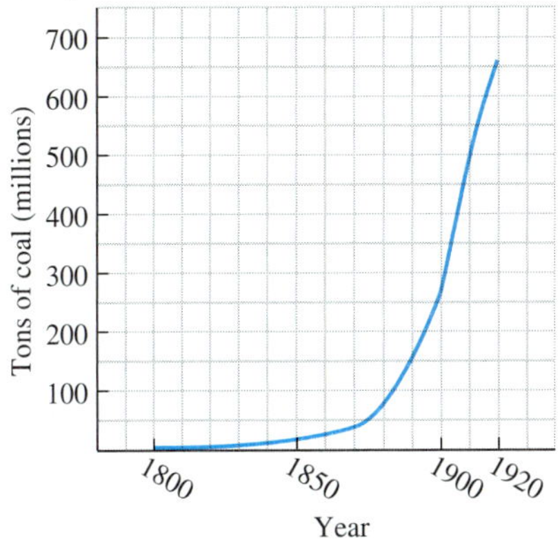

33. \$2,189,703.45 **34.** about \$2,015
35. 8.22% **36.** \$2,324,767.37
37. D: $(-\infty, \infty)$, R: $(1, \infty)$

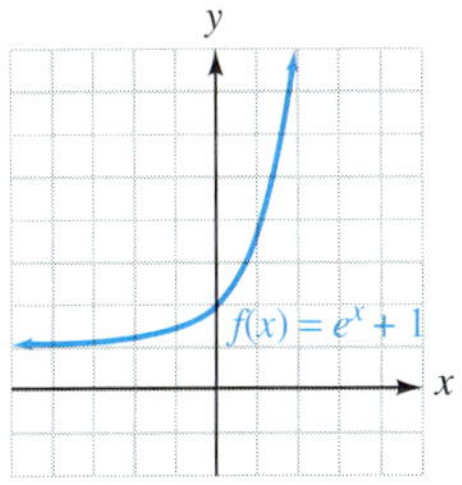

38. D: $(-\infty, \infty)$, R: $(0, \infty)$

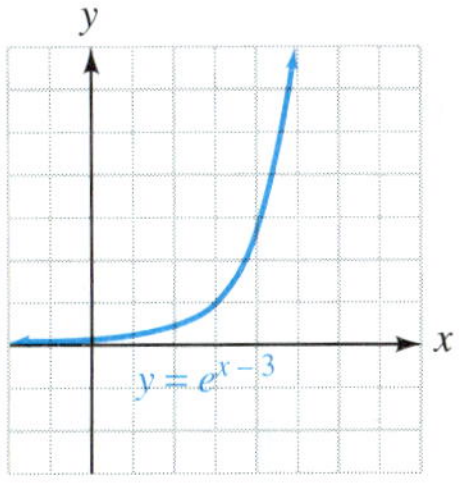

39. about 459,920,041 **40.** The exponent on the base e is negative. **41.** D: $(0, \infty)$, R: $(-\infty, \infty)$ **42.** Since there is no real number such that $10^? = 0$, log 0 is undefined. **43.** $4^3 = 64$
44. $\log_7 \frac{1}{7} = -1$ **45.** 2 **46.** -2 **47.** 0 **48.** not possible **49.** $\frac{1}{2}$ **50.** 3 **51.** 32 **52.** $\frac{1}{81}$ **53.** 4
54. 10 **55.** $\frac{1}{2}$ **56.** 9 **57.** 0.6542 **58.** 26.9153

59.

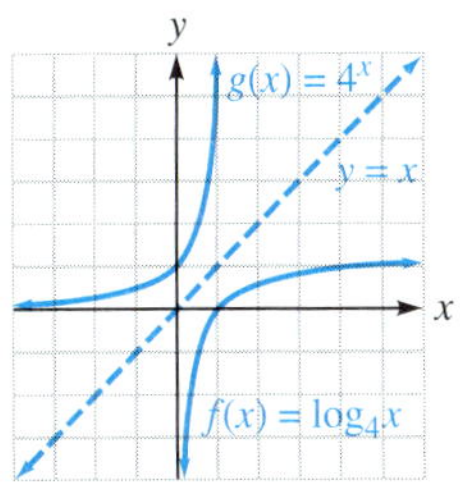

60.

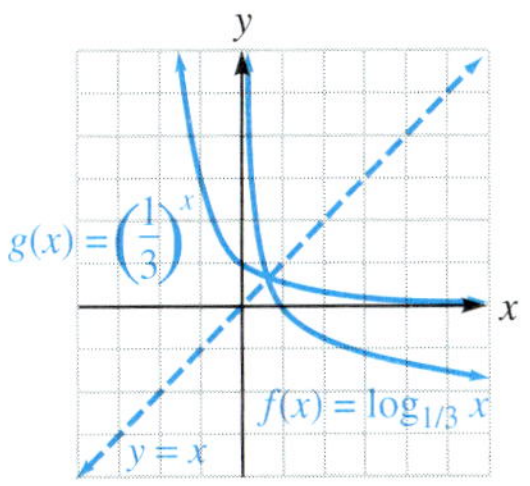

61.

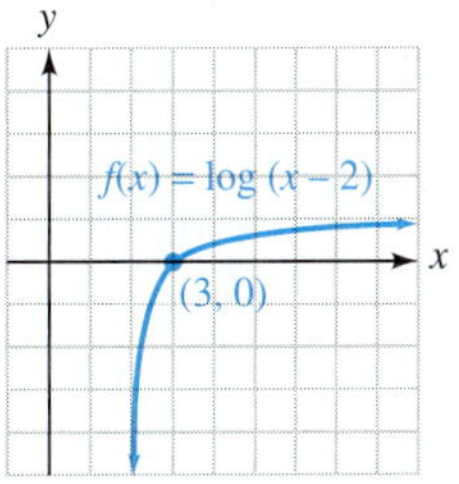

62.

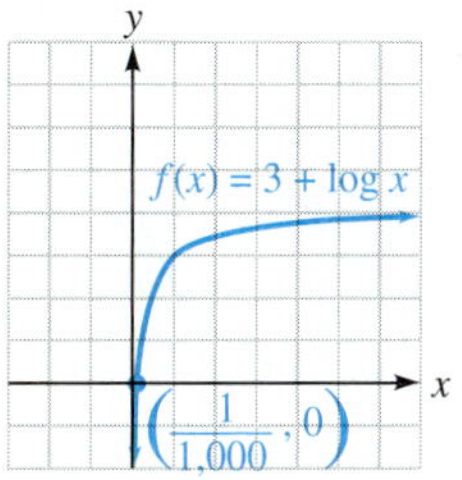

63. about 53 **64.** about 4.4 **65.** 1 **66.** 2 **67.** -5
68. $\frac{1}{2}$ **69.** undefined **70.** undefined **71.** 0 **72.** -7
73. 6.1137 **74.** -0.1625 **75.** 10.3398 **76.** 0.0002
77. $\log x = \log_{10} x$ and $\ln x = \log_e x$ **78.** $f^{-1}(x) = e^x$
79.

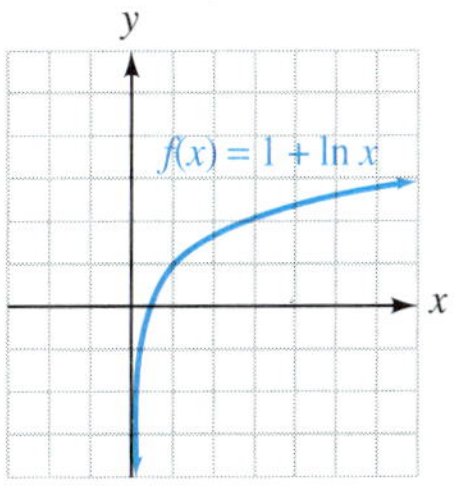

80.

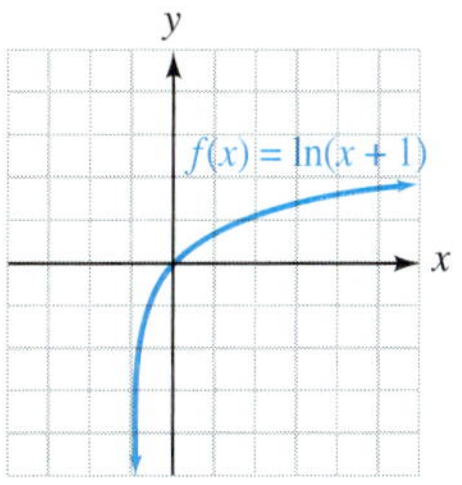

81. about $48\frac{1}{2}$ yr **82.** about 72 in. (6 ft) **83.** 0
84. 1 **85.** 3 **86.** 4 **87.** $3 + \log_3 x$ **88.** $2 - \log x$
89. $\frac{1}{2}\log_5 27$ **90.** $\log_b 10 + \log_b a + 1$
91. $2\log_b x + 3\log_b y - \log_b z$ **92.** $\frac{1}{2}(\ln x - \ln y - 2\ln z)$
93. $\log_2 \frac{x^3 z^7}{y^5}$ **94.** $\log_b \frac{\sqrt{x+2}}{y^3 z^7}$ **95.** 2.6609 **96.** 3.0000
97. 1.7604 **98.** about 7.9×10^{-4} gram-ions/liter
99. 2 **100.** $-3, -1$ **101.** 1.7712 **102.** 2.7095
103. 1.9459 **104.** -8.0472 **105.** 104 **106.** 9
107. 25, 4 **108.** 4 **109.** 4, 3 **110.** 2 **111.** 6
112. 31 **113.** $0.76787 \neq -0.27300$ **114.** about 3,300 yr
115. about 91 days **116.** 2, 5

Chapter Test (page 757)

1. $(f + g)(x) = 4x^2 - 2x + 11, (-\infty, \infty)$
2. $(g/f)(x) = \frac{4x^2 - 3x + 2}{x + 9}, (-\infty, -9) \cup (-9, \infty)$ **3.** 76
4. $(f \circ g)(x) = 32x^2 - 128x + 131$ **5.** -16 **6.** 17 **7.** yes
8. no **9.**

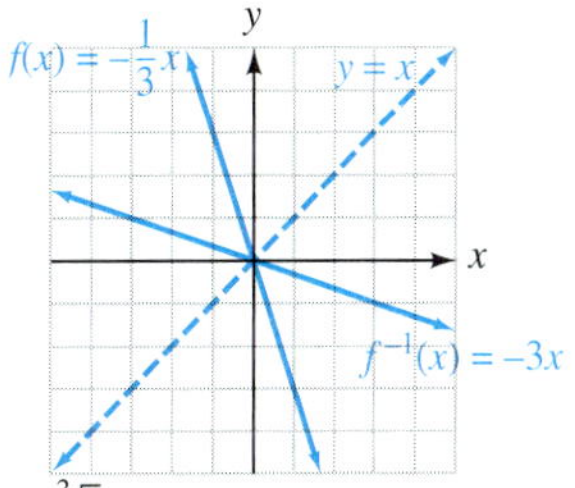

10. $f^{-1}(x) = \sqrt[3]{x} + 15$ **12. a.** yes **b.** yes
c. 80; when the temperature of the tire tread is 260°,

the vehicle is traveling 80 mph

13. D: $(-\infty, \infty)$, R: $(1, \infty)$

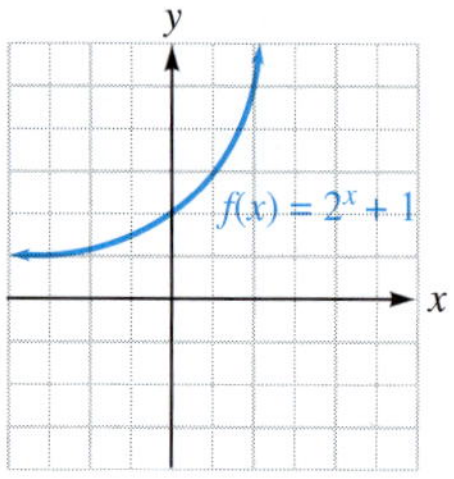

14. D: $(-\infty, \infty)$, R: $(0, \infty)$

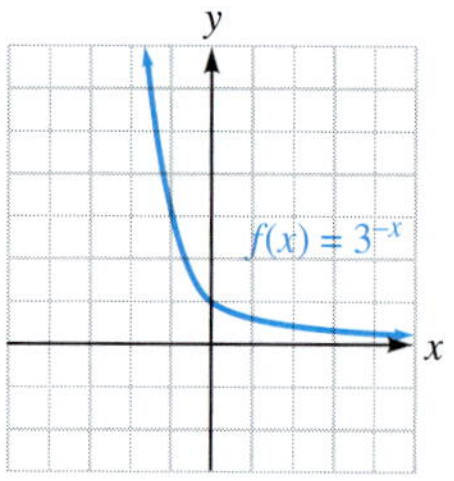

15. $\frac{3}{64}$ g = 0.046875 g **16.** \$1,060.90

17.

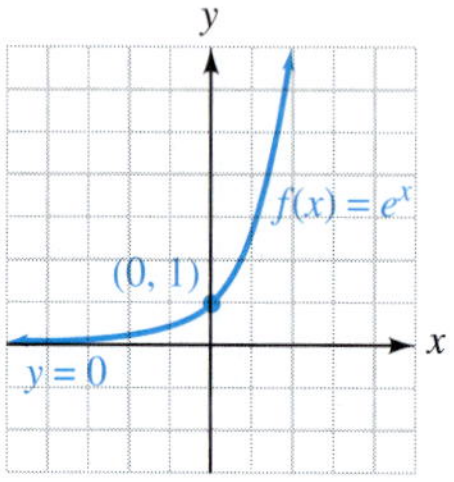

18. about 1,631,973,737 **19.** $6^{-2} = \frac{1}{36}$ **20.** **a.** D: $(0, \infty)$, R: $(-\infty, \infty)$ **b.** $f^{-1}(x) = 10^x$ **21.** 2 **22.** -2 **23.** undefined **24.** -6 **25.** $\frac{1}{2}$ **26.** 0 **27.** 2 **28.** 16 **29.** $\frac{1}{27}$ **30.** e

31.

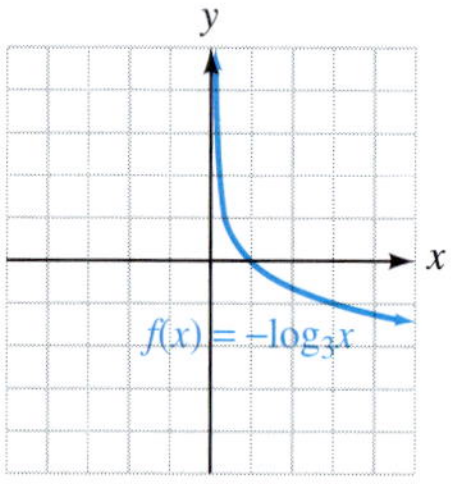

32.

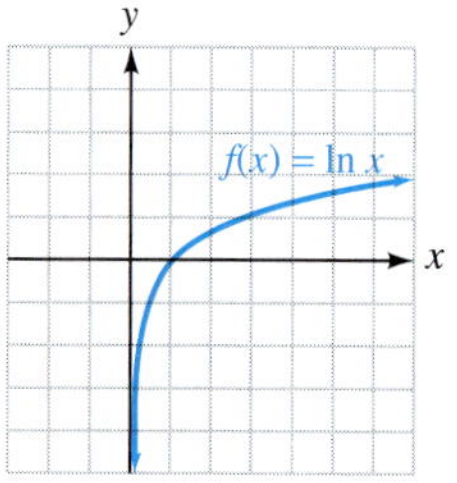

33. 6.4 **34.** about 46 **35.** .0871 **36.** 0.5646 **37.** $2 \log_b a + 1 + 3 \log_b c$ **38.** $\ln \frac{b\sqrt{a+2}}{c^3}$ **39.** 0.6826 **40.** 4 **41.** 1 **42.** 10 **43.** 5 **44.** about 20 min

Study Set 10.1 (page 770)

1. conic sections **3.** circle, radius **5.** **a.** $(x - h)^2 + (y - k)^2 = r^2$ **b.** $x^2 + y^2 = r^2$ **7.** **a.** $(2, -1)$; $r = 4$ **b.** $(x - 2)^2 + (y + 1)^2 = 16$ **9.** **a.** $y = a(x - h)^2 + k$ **b.** $x = a(y - k)^2 + h$ **11.** Determine whether the graph of each equation is a circle or a parabola. **a.** circle **b.** parabola **c.** parabola **d.** circle **13.** 6, -2, 3

15.

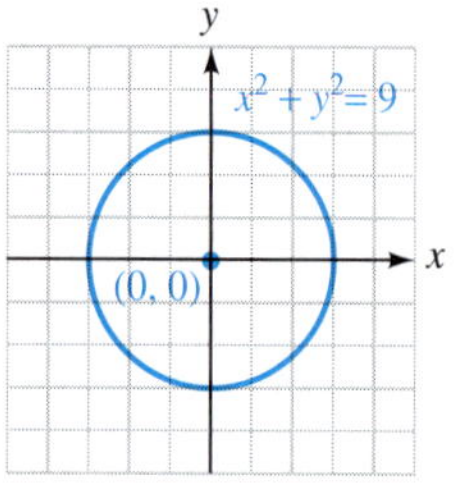

17.

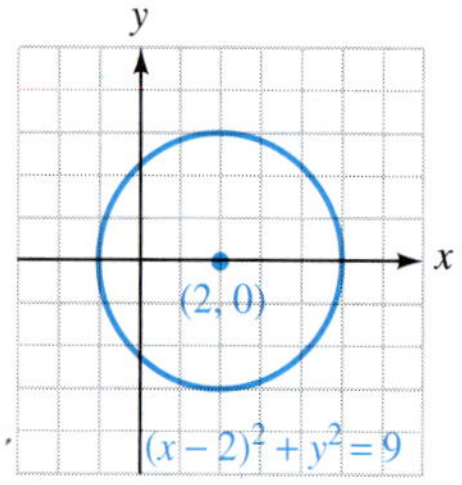

19.

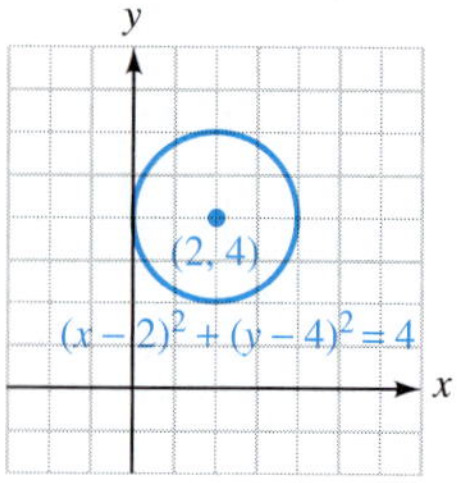

21.

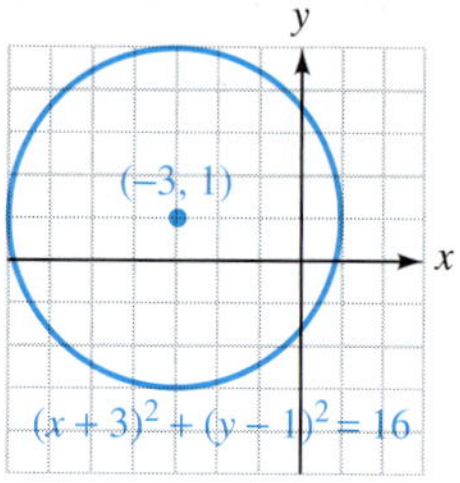

23.

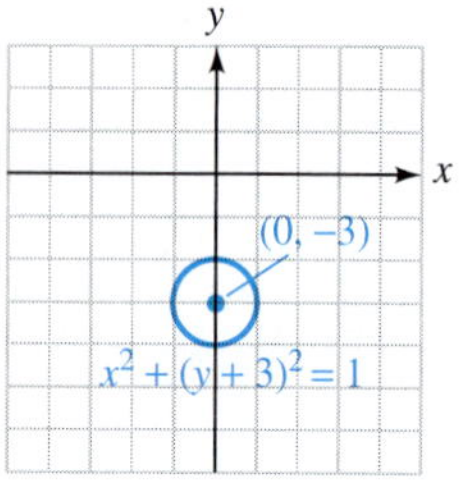

25.

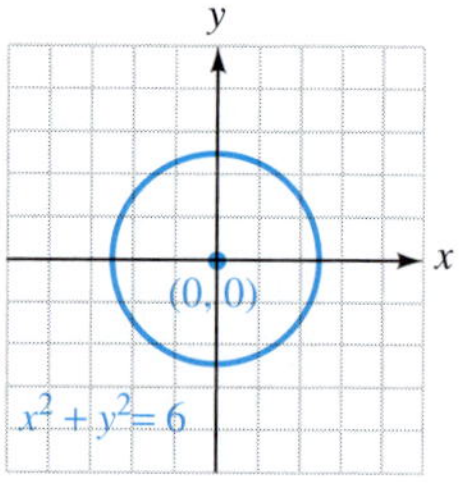

27.

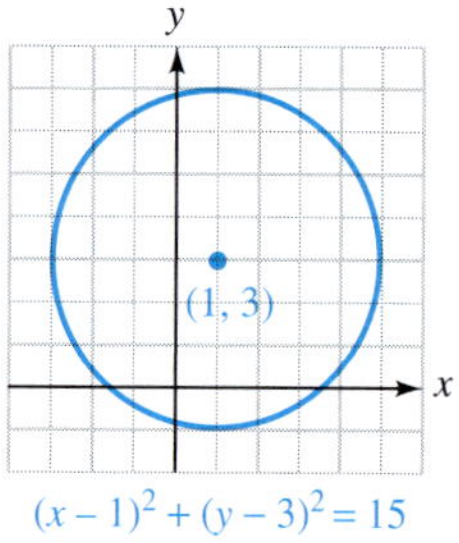

29.

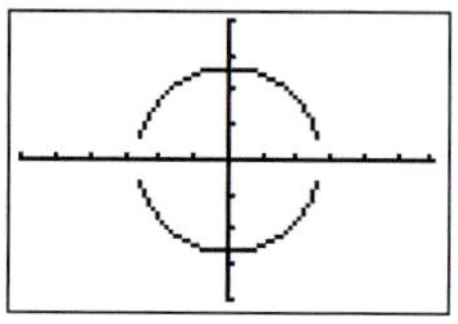

31.

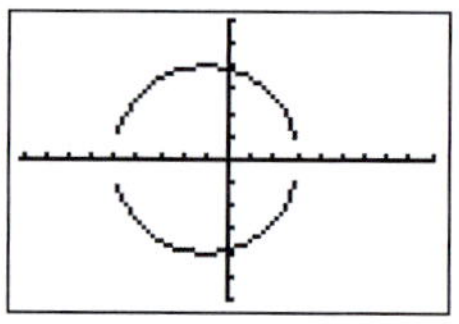

33. $x^2 + y^2 = 1$ **35.** $(x - 6)^2 + (y - 8)^2 = 25$ **37.** $(x + 2)^2 + (y - 6)^2 = 144$ **39.** $x^2 + y^2 = \frac{1}{16}$ **41.** $\left(x - \frac{2}{3}\right)^2 + \left(y + \frac{7}{8}\right)^2 = 2$ **43.** $x^2 + y^2 = 8$

45.

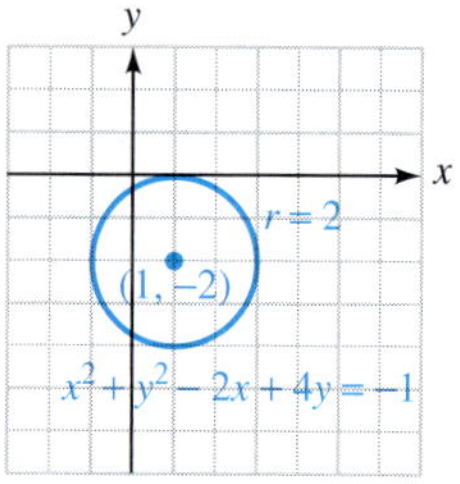

47.

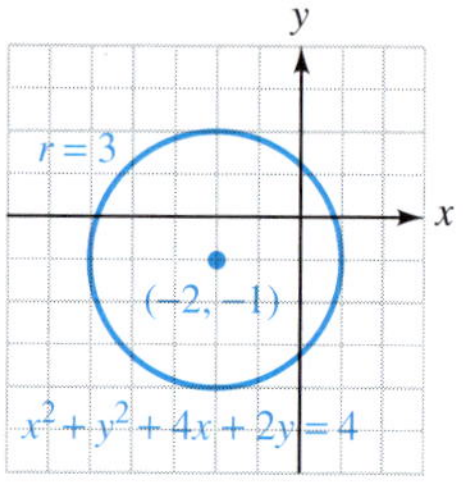

49.

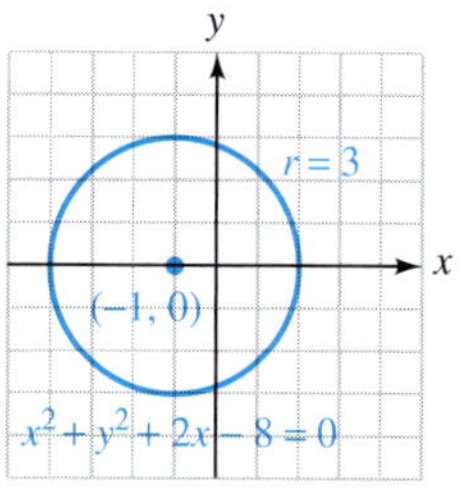

51.

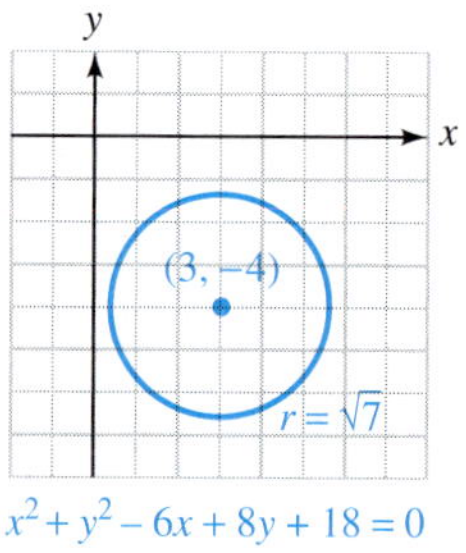

53.

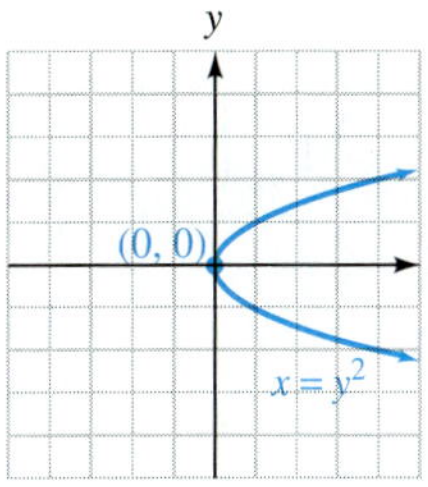

55.

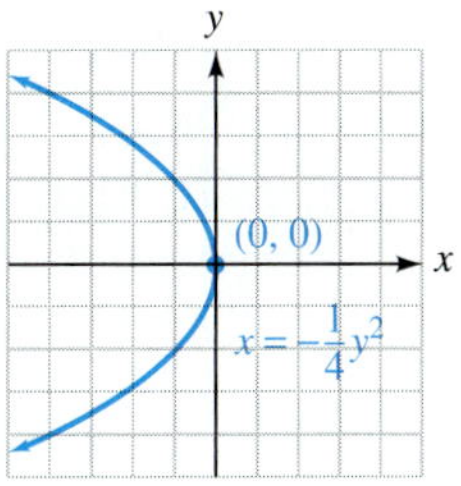

57.

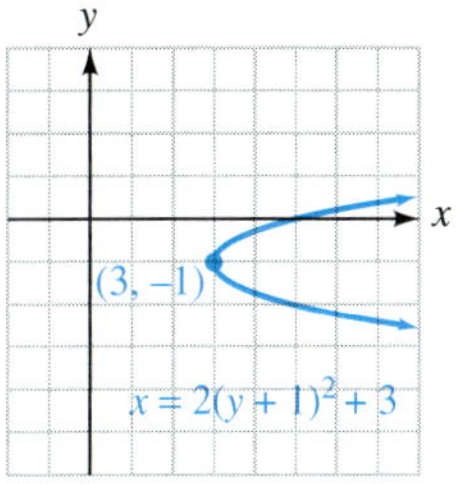

59.

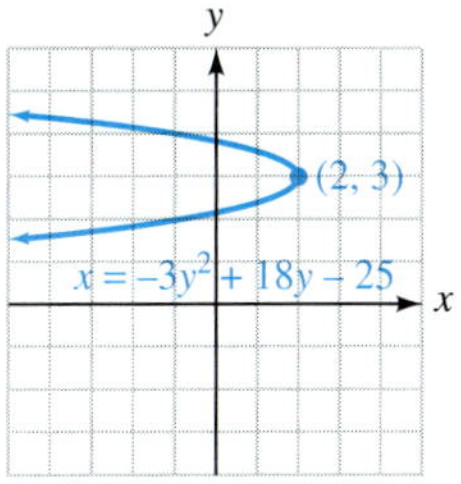

61.

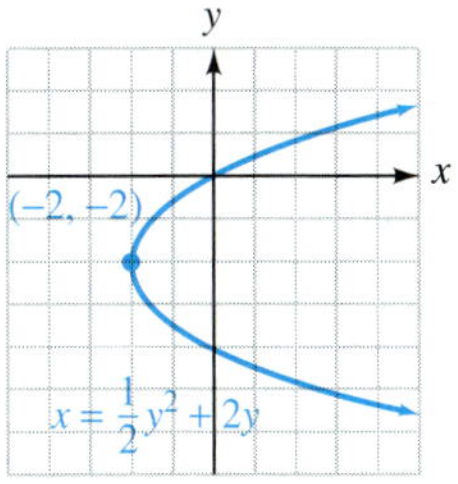

63.

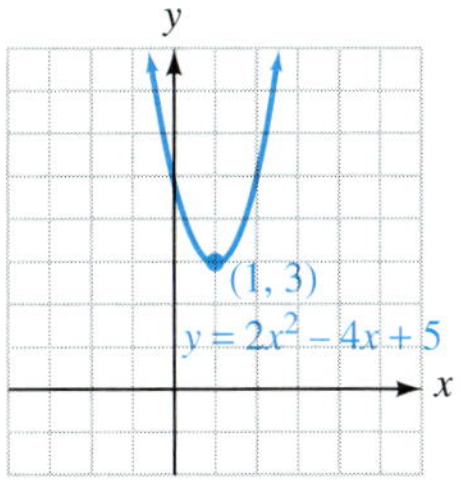

65.

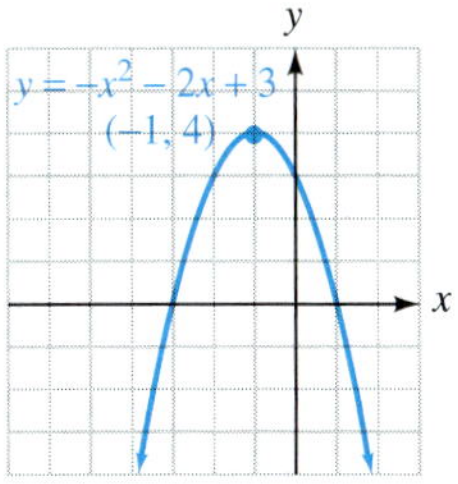

67.

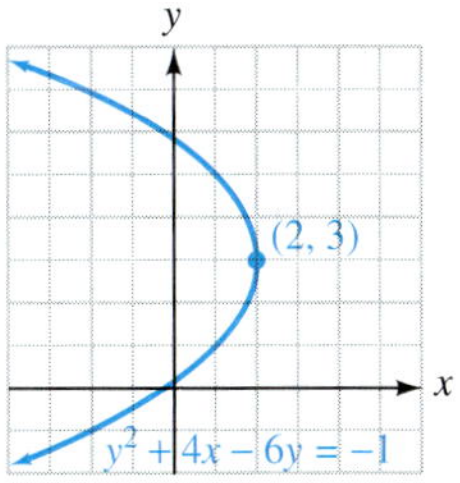

69.

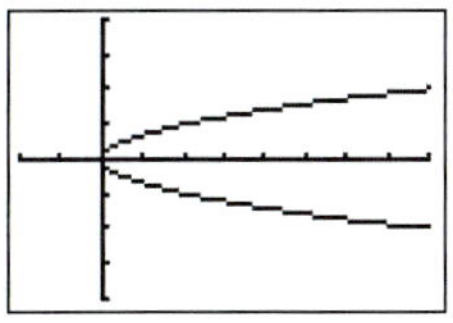

71.

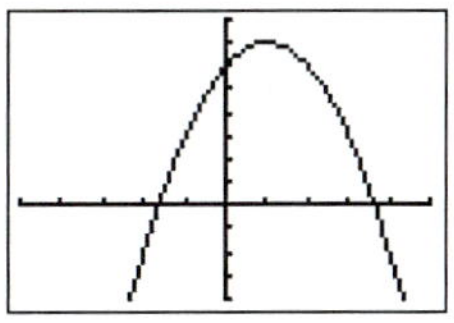

73. $(x - 7)^2 + y^2 = 9$ **75.** no **77.** 5 ft **79.** 2 AU
85. $5, -\frac{7}{3}$ **87.** $3, -\frac{1}{4}$

Study Set 10.2 (page 780)

1. ellipse **3.** foci, focus **5.** major axis **7.** $\frac{x^2}{a^2} + \frac{y^2}{b^2} = 1$
9. x-intercepts: $(a, 0), (-a, 0)$; y-intercepts: $(0, b), (0, -b)$.
11. a. $(-2, 1)$; $a = 2, b = 5$ **b.** vertical
c. $\frac{(x + 2)^2}{4} + \frac{(y - 1)^2}{25} = 1$ **13.** $\frac{(x - 1)^2}{16} + \frac{(y + 5)^2}{1} = 1$
15. $h = -8, k = 6, a = 10, b = 12$

17.

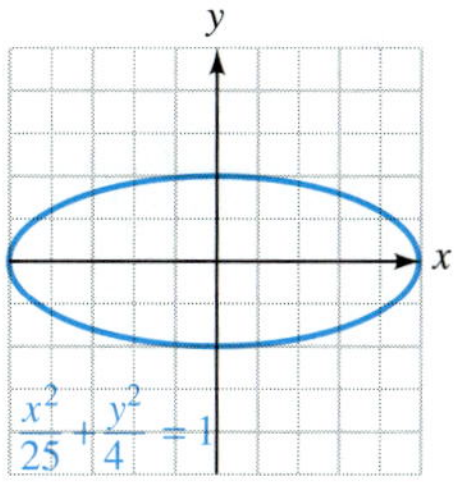

19.

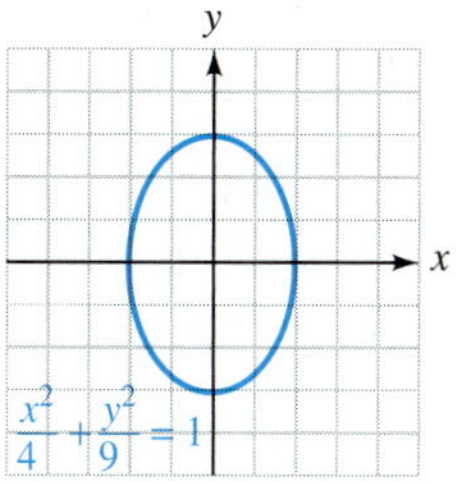

21.

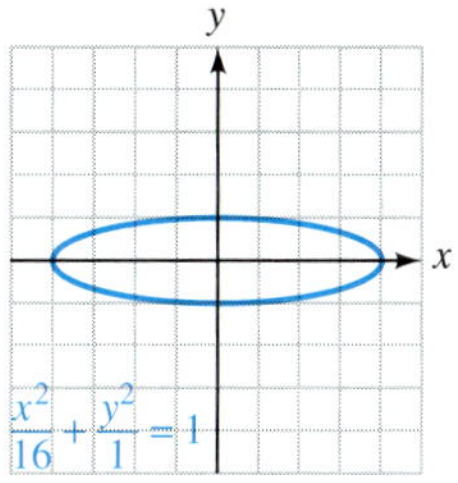

23.

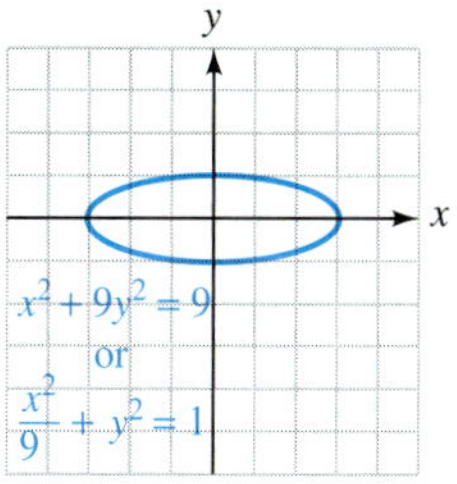

25.

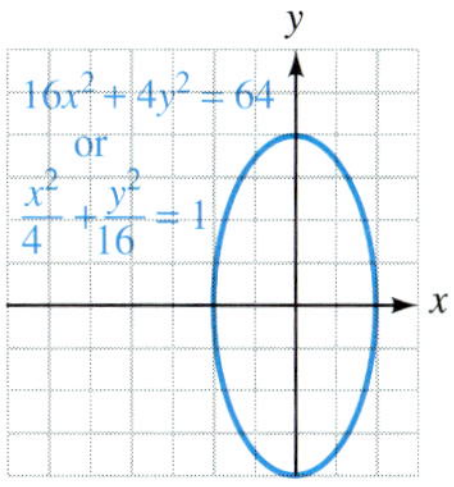

27.

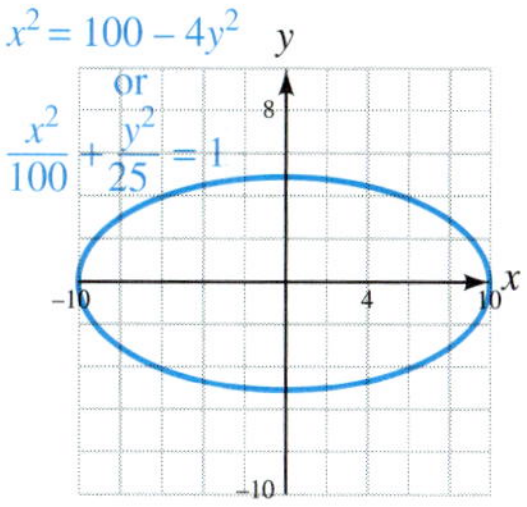

29.

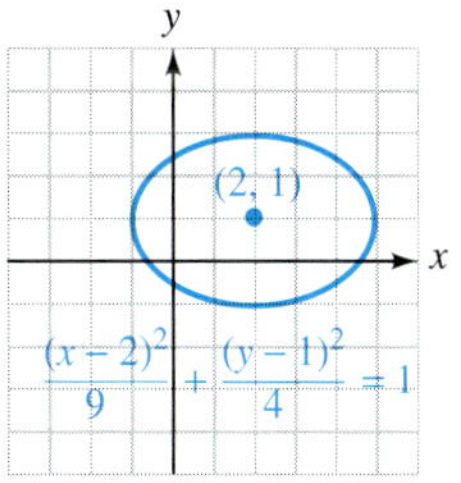

31.

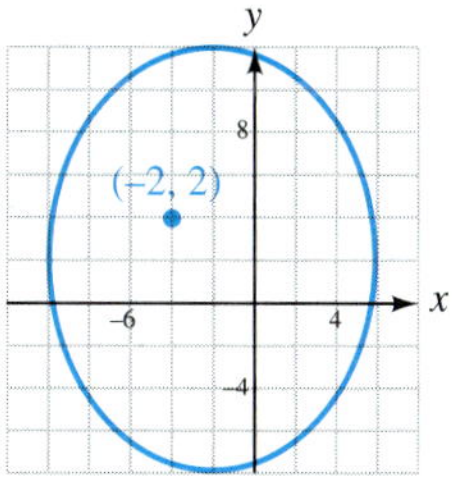

$\frac{(x + 2)^2}{64} + \frac{(y - 2)^2}{100} = 1$

33.

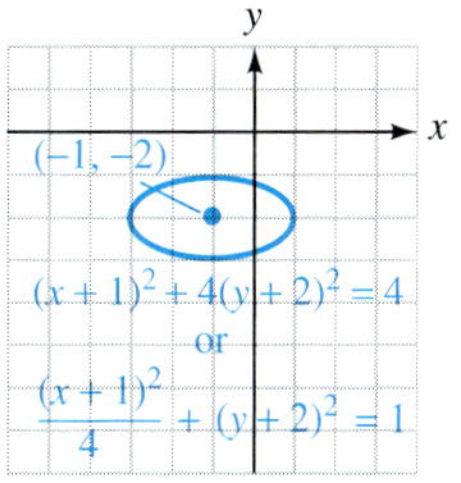

35.

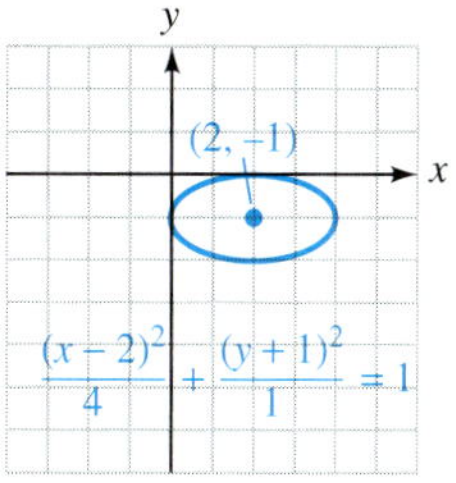

37.

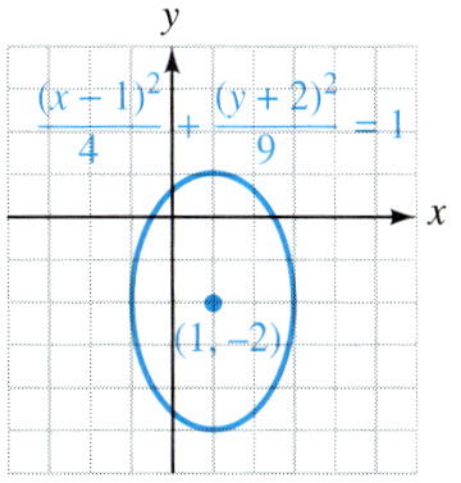

39. 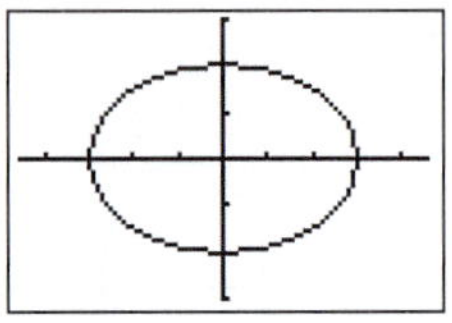

41. 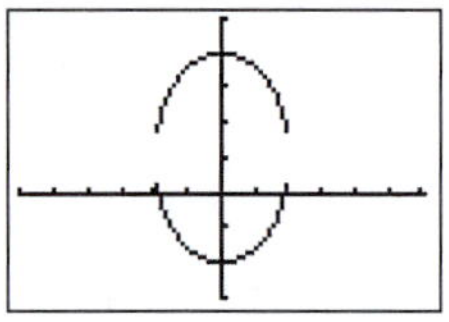

43. $\frac{x^2}{144} + \frac{y^2}{25} = 1$ 45. $5\sqrt{3}$ ft

47. 12π sq. units ≈ 37.7 sq. units 53. $12y^2 + \frac{9}{x^2}$

55. $\frac{y^2 + x^2}{y^2 - x^2}$

Study Set 10.3 (page 790)

1. hyperbola 3. vertices 5. diagonals 7. $\frac{x^2}{a^2} - \frac{y^2}{b^2} = 1$

9. $\frac{(x-h)^2}{a^2} - \frac{(y-k)^2}{b^2} = 1$ 11. a. $(-1, -2)$; $a = 3$, $b = 1$

b. $\frac{(y+2)^2}{9} - \frac{(x+1)^2}{1} = 1$ 13. $\frac{(x+1)^2}{1} - \frac{(y-5)^2}{4} = 1$

15. $h = 5$, $k = -11$, $a = 5$, $b = 6$

17. 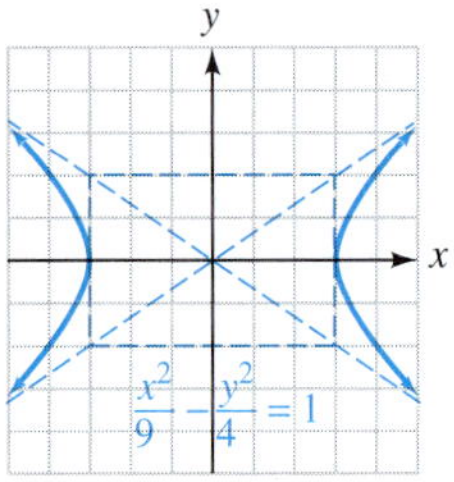

19. 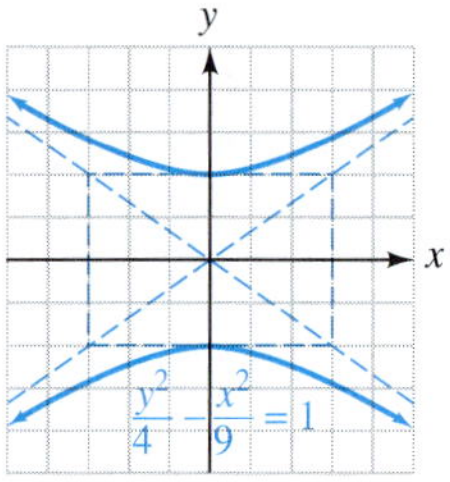

21. 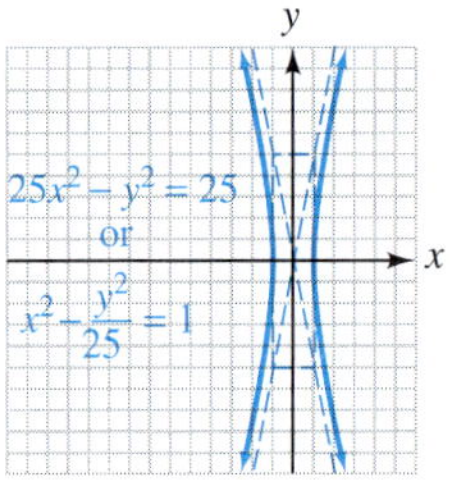

23. 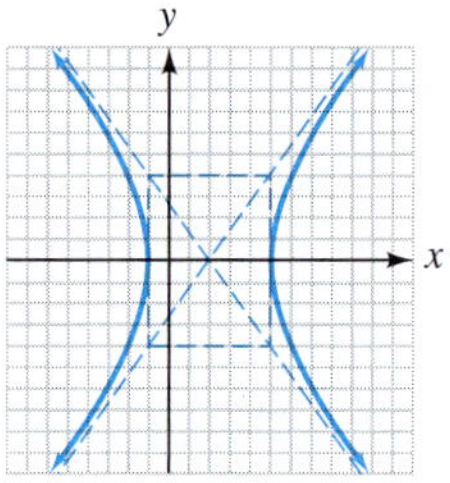

$\frac{(x-2)^2}{9} - \frac{y^2}{16} = 1$

25. 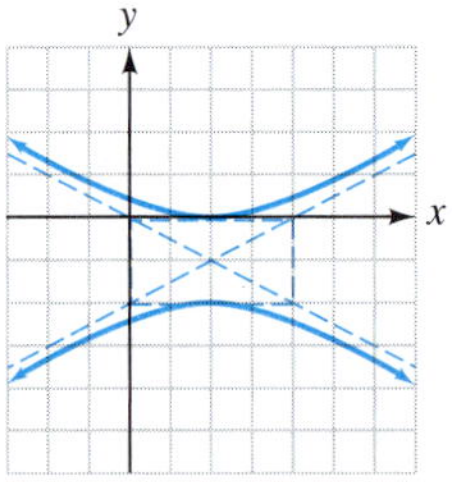

$\frac{(y+1)^2}{1} - \frac{(x-2)^2}{4} = 1$

27. 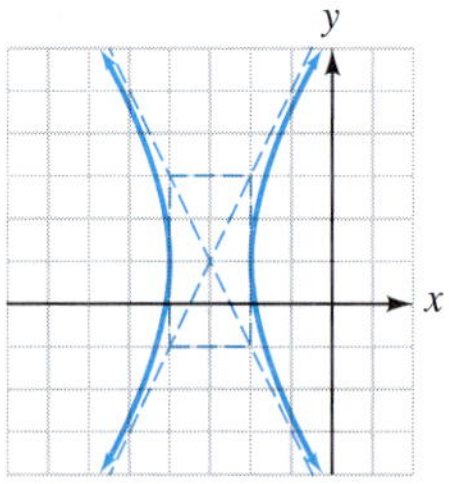

$4(x+3)^2 - (y-1)^2 = 4$
or $(x+3)^2 - \frac{(y-1)^2}{4} = 1$

29. 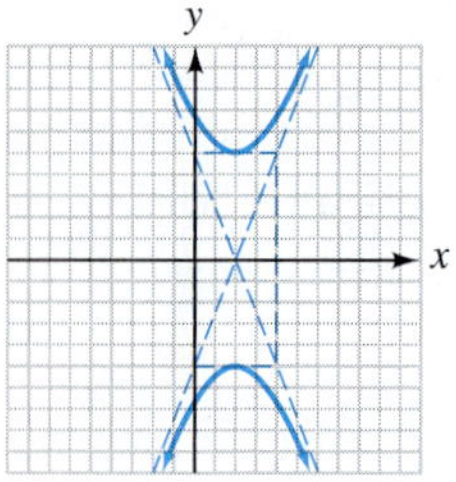

$\frac{y^2}{25} - \frac{(x-2)^2}{4} = 1$

31. 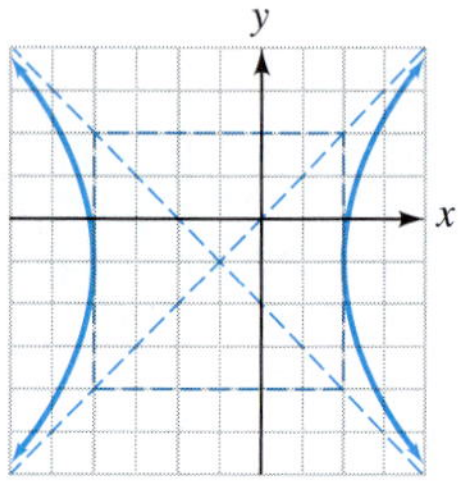

$\frac{(x+1)^2}{9} - \frac{(y+1)^2}{9} = 1$

33. 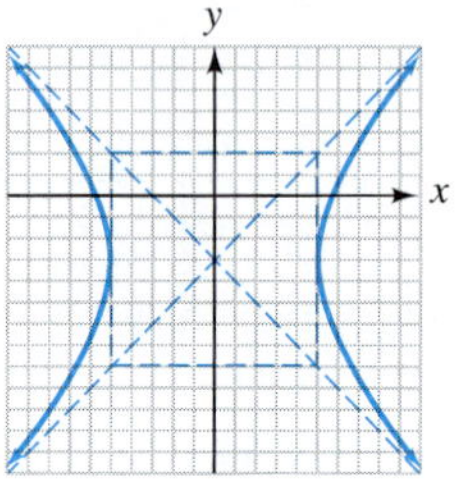

$\frac{x^2}{25} - \frac{(y+3)^2}{25} = 1$

35. 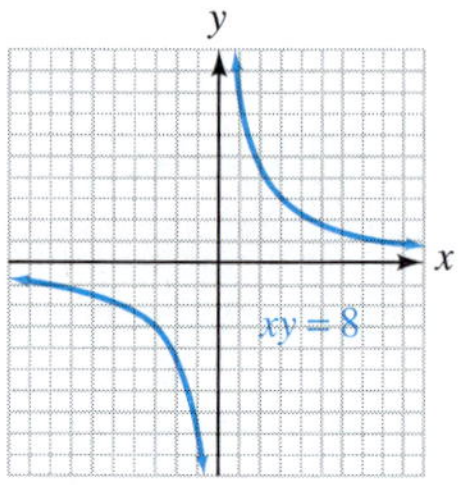

37. 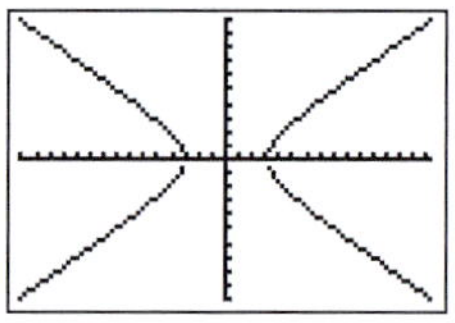

39. 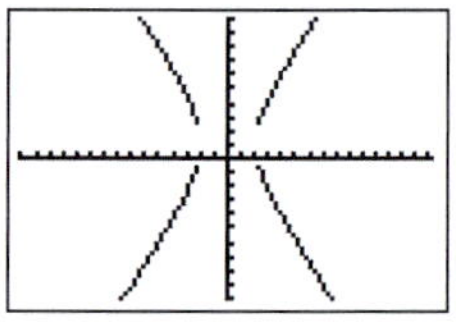

41. 3 units 43. $10\sqrt{3}$ miles 49. 64 51. 3 53. $\frac{3}{2}$

Study Set 10.4 (page 796)

1. system 3. graphing 5. secant 7. a. two b. four c. four d. four 9. $(-3, 2)$, $(3, 2)$, $(-3, -2)$, $(3, -2)$

11. a. -4 b. -2 13. $2x$, 5, 5, 1, 1, -1, -2

15. 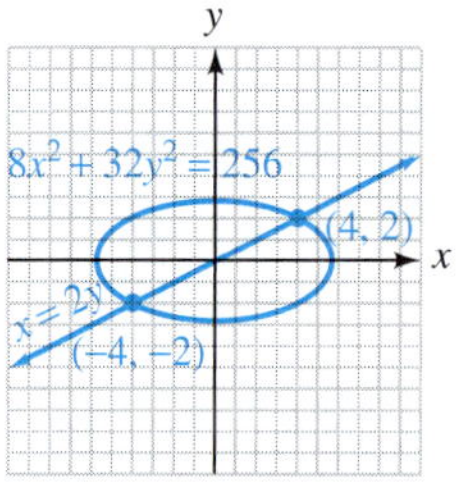

17. 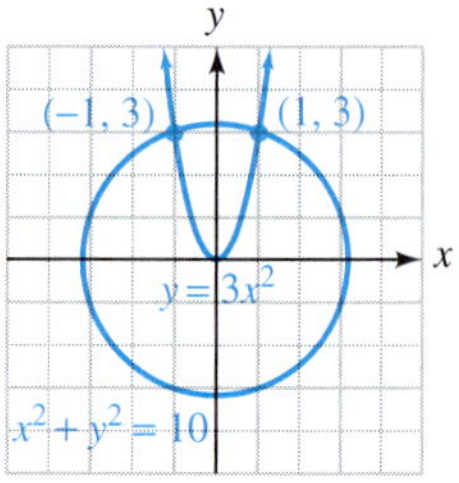

19. 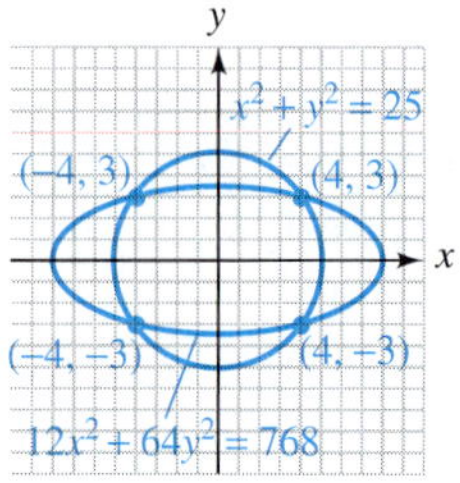

21. 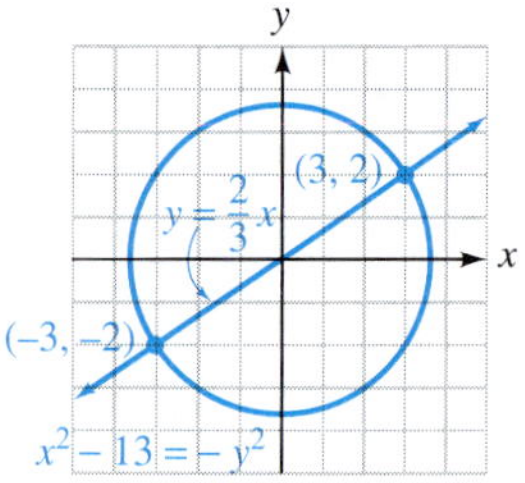

23. (1, 0), (5, 0)

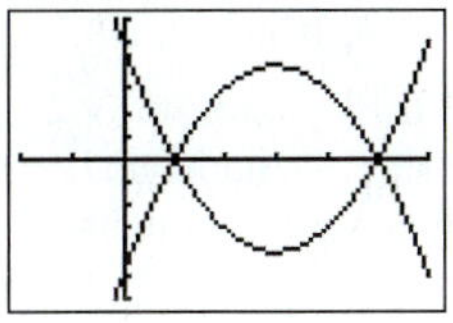

25. (3, 0), (0, 5) **27.** (1, 1) **29.** (1, 2), (2, 1) **31.** (−2, 3), (2, 3) **33.** $(\sqrt{5}, 5), (-\sqrt{5}, 5)$ **35.** (3, 2), (3, −2), (−3, 2), (−3, −2)

37. (2, 4), (2, −4), (−2, 4), (−2, −4) **39.** $(-\sqrt{15}, 5)$, $(\sqrt{15}, 5)$, (−2, −6), (2, −6) **41.** (0, −4), (−3, 5), (3, 5) **43.** (−2, 3), (2, 3), (−2, −3), (2, −3) **45.** (3, 3) **47.** (6, 2), (−6, −2), $(\sqrt{42}, 0)$, $(-\sqrt{42}, 0)$ **49.** $\left(\frac{1}{2}, \frac{1}{3}\right)$, $\left(\frac{1}{3}, \frac{1}{2}\right)$ **51.** 4, 8 **53.** 7 cm by 9 cm **55.** either \$750 at 9% or \$900 at 7.5% **57.** 68 mph, 4.5 hr **61.** 2,000 **63.** 7

Key Concept (page 800)

1. ellipse **2.** circle **3.** parabola **4.** hyperbola **5.** hyperbola **6.** ellipse **7.** ellipse **8.** parabola **9.** circle **10. a.** (−1, 2) **b.** 4 **11. a.** (−2, 1) **b.** right **12. a.** (0, 0) **b.** vertical **c.** (0, 4), (0, −4) **13. a.** (−2, 1) **b.** left and right **c.** 6 units horizontally, 4 units vertically

14.

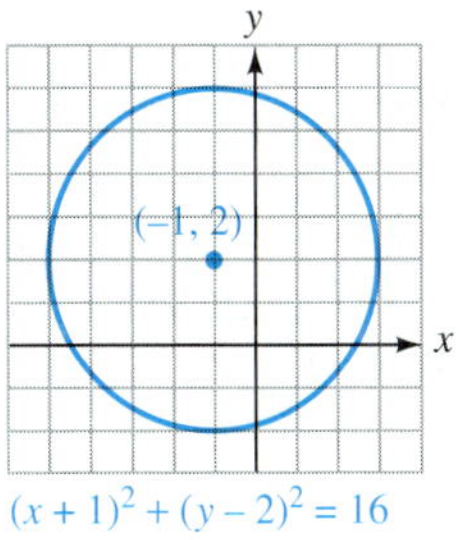

15.

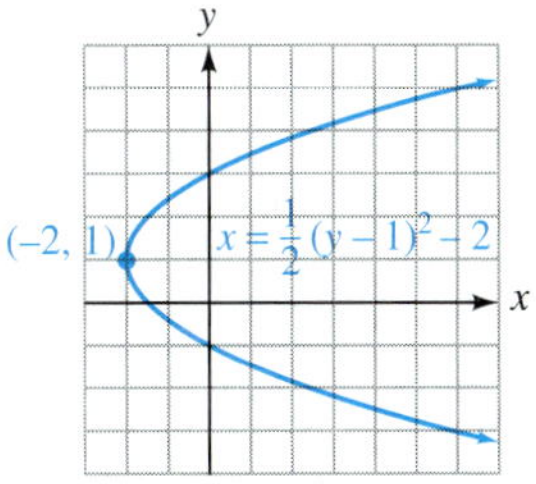

16.

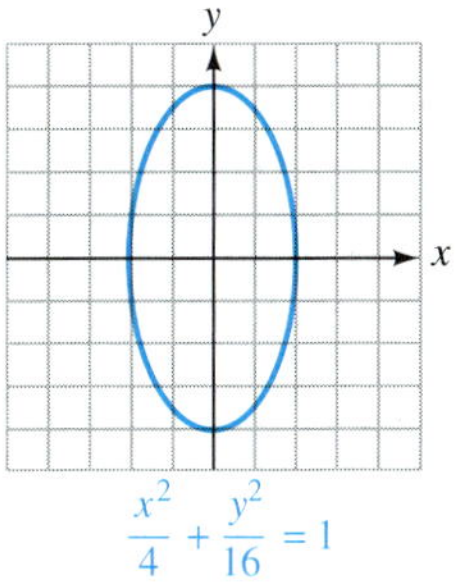

17.

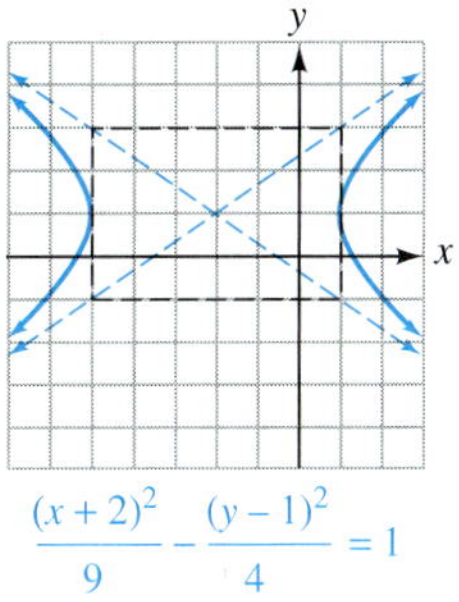

Chapter Review (page 801)

1.

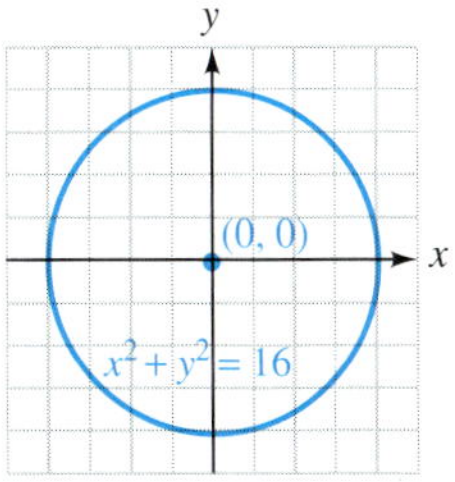

2.

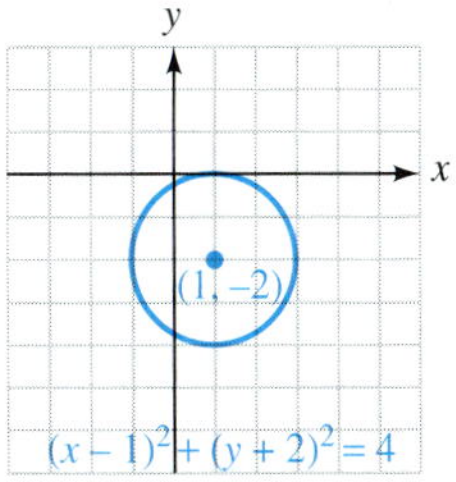

3.

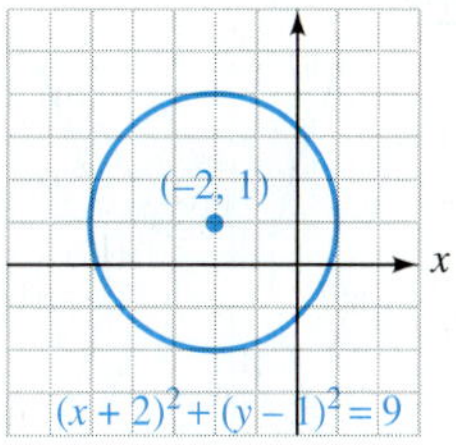

4. $(x - 8.5)^2 + (y - 8.5)^2 = (8.5)^2$
5. (−6, 0); $r = 2\sqrt{6}$
6. center, radius

7.

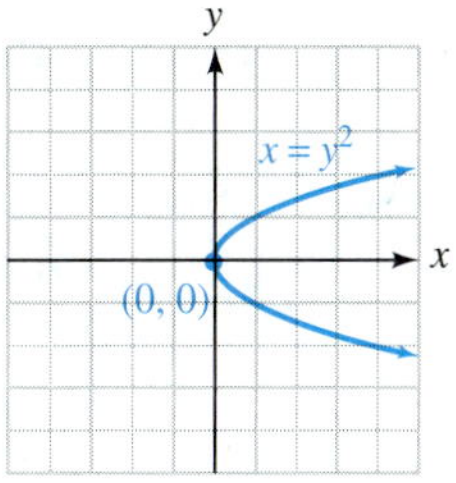

8.

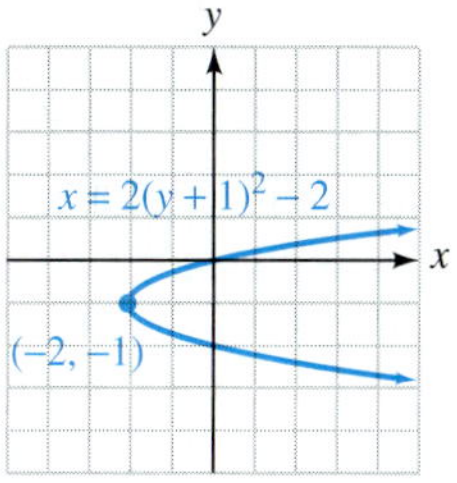

9.

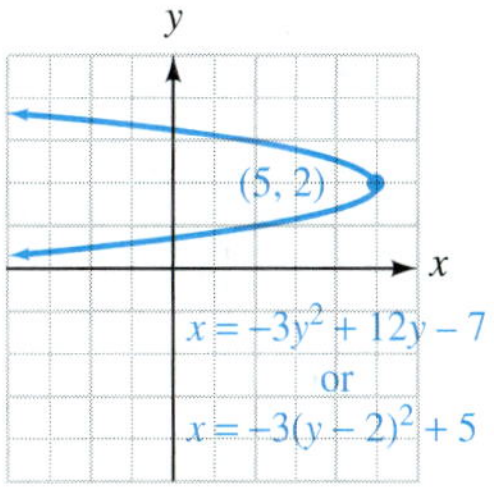

10. a. $y = 4$ **b.** $x = 1$
11. (2, −3), (6, −4)
12. When $x = 22$, $y = 0$: $-\frac{5}{121}(22 - 11)^2 + 5 = 0$

13.

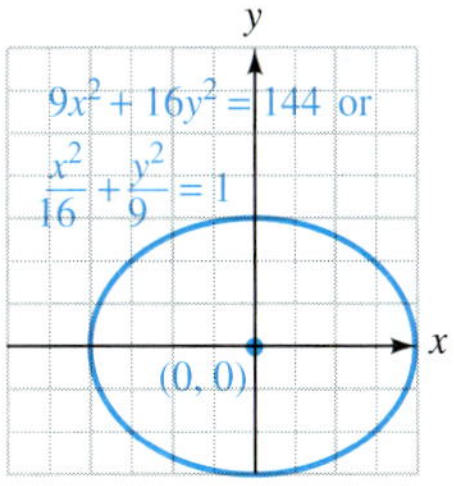

14.

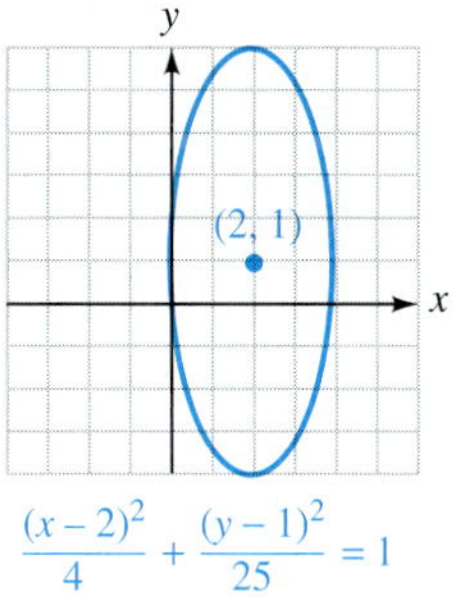

15.

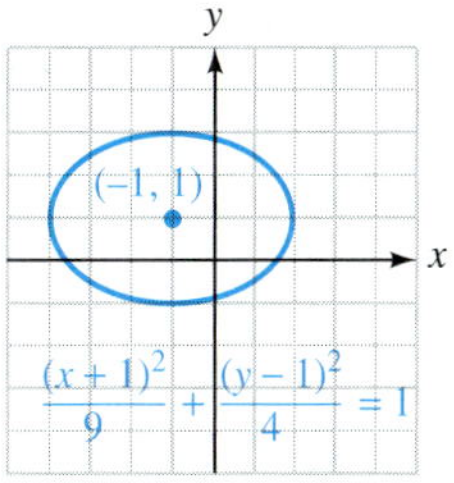

16. $\frac{x^2}{12^2} + \frac{y^2}{1^2} = 1$
17. $\left(2, \frac{2\sqrt{5}}{3}\right), \left(2, -\frac{2\sqrt{5}}{3}\right)$; (2, 1.5), (2, −1.5)
18. $\frac{x^2}{25} + \frac{y^2}{9} = 1$
19. ellipse, focus

20.

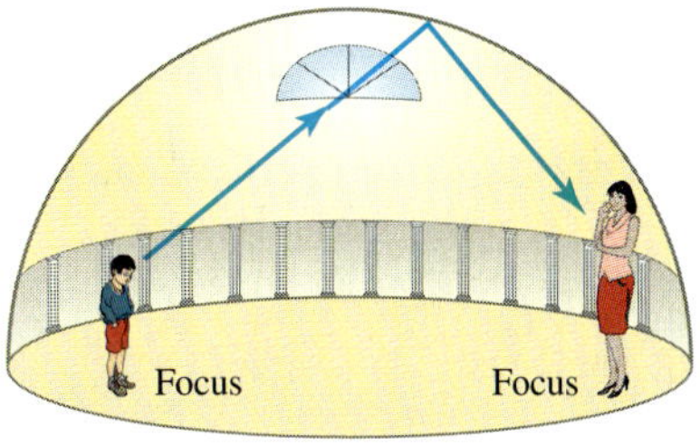

21.

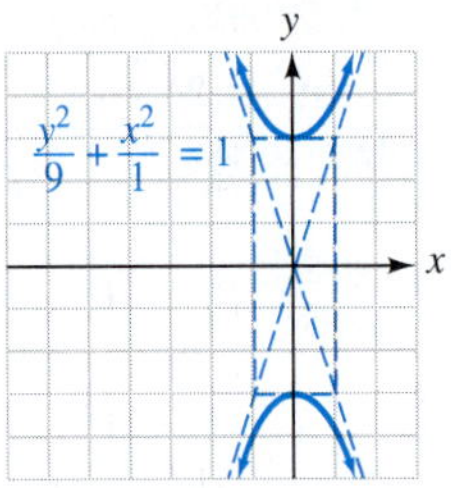

22.

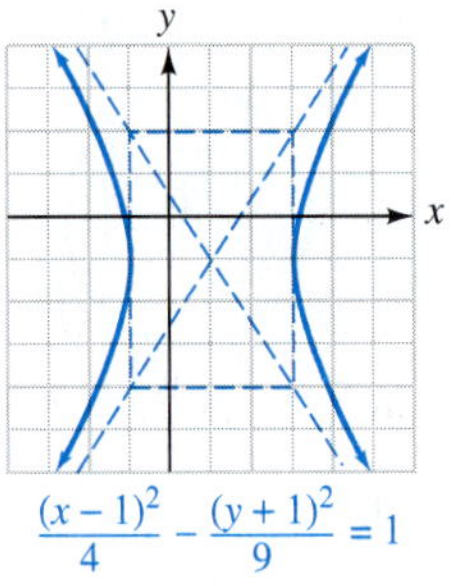

23.

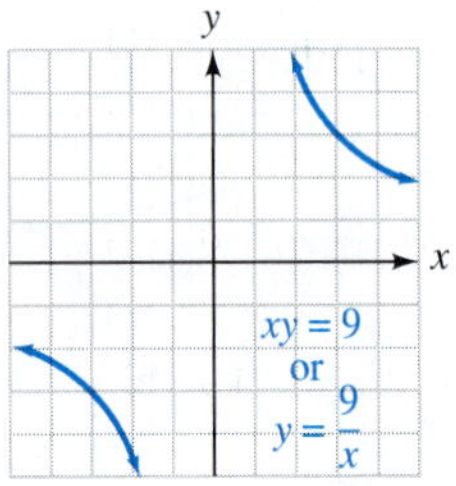

24.

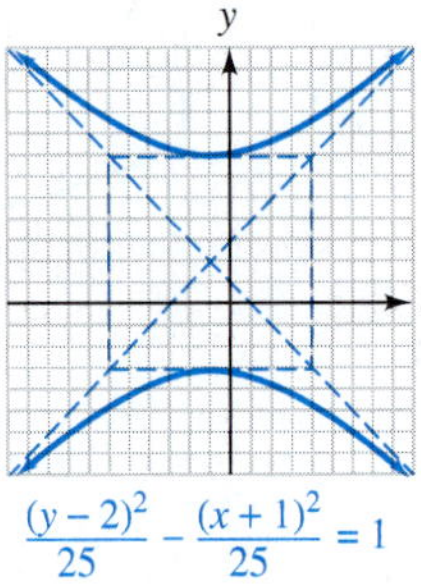

25. 4 units **26. a.** ellipse **b.** hyperbola **c.** parabola **d.** circle **27.** yes **28.** $(0, 3), (0, -3)$ **29. a.** 2 **b.** 4 **c.** 4 **d.** 4 **30.** $(0, 1), (0, -1)$ **31.** $(0, -4), (-3, 5), (3, 5)$ **32.** $(\sqrt{2}, 0), (-\sqrt{2}, 0)$ **33.** $(2, 2), \left(-\frac{2}{9}, -\frac{22}{9}\right)$ **34.** $(4, 2), (4, -2), (-4, 2), (-4, -2)$ **35.** $(2, 3), (2, -3), (-2, 3), (-2, -3)$ **36.** $(2\sqrt{2}, \sqrt{2}), (-2\sqrt{2}, -\sqrt{2}), (1, 4), (-1, -4)$

Chapter Test (page 804)

1. center, radius **2.** $(0, 0); r = 10$ **3.** $(-2, 3), r = 3\sqrt{2}$ **4.** $(x - 4)^2 + (y - 3)^2 = 9$

5.

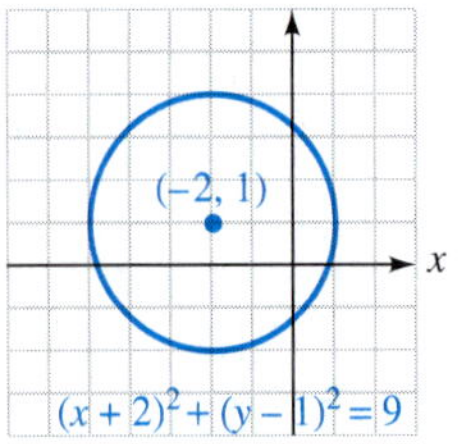

6.

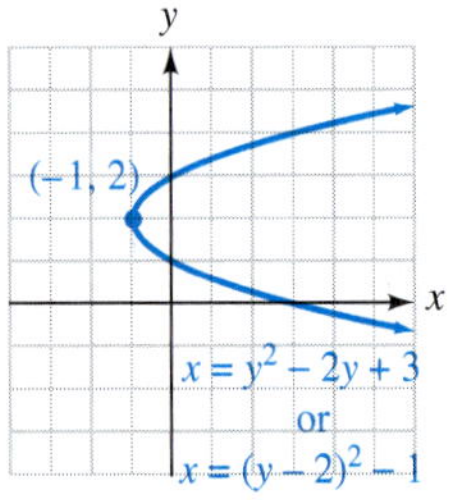

7. $(-1, 7); x = -1$

8.

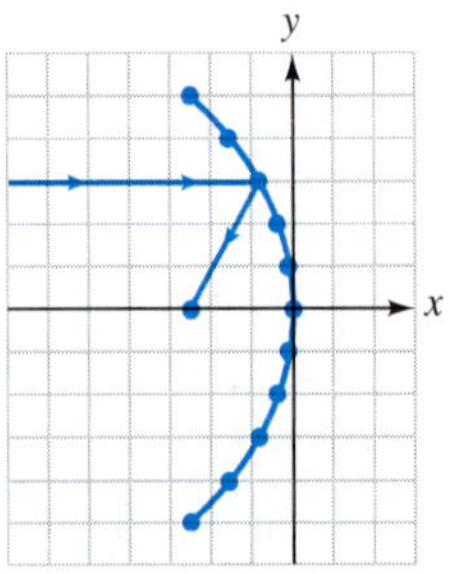

9.

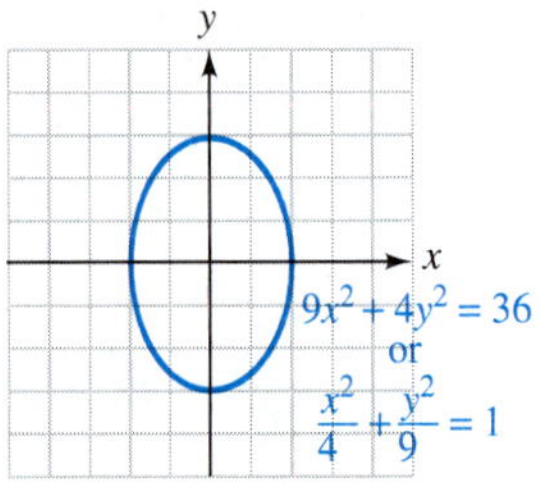

10.

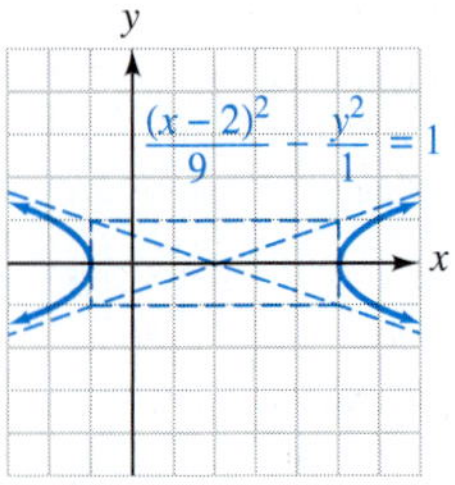

11. $\frac{(x-1)^2}{16} + \frac{(y+2)^2}{9} = 1$ **12.** $(-8, 10); (-2, 10), (-14, 10)$ **13.** $\pm 2\sqrt{2}$ **15.** $(-1, 1)$; horizontal: 4 units, vertical: 4 units

16.

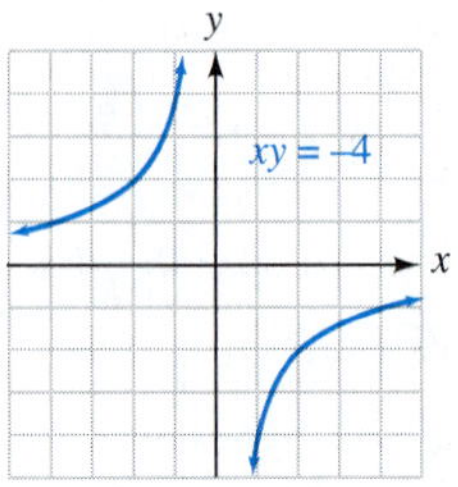

17. $\frac{y^2}{16} - \frac{x^2}{36} = 1$ **18. a.** ellipse **b.** hyperbola **c.** circle **d.** parabola **19.** $(-4, -3), (3, 4)$

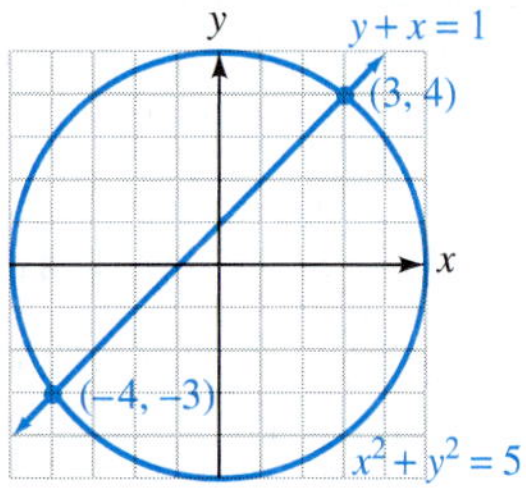

20. $(2, 6), (-2, -2)$ **21.** $(1, \sqrt{2}), (1, -\sqrt{2}), (-1, \sqrt{2}), (-1, -\sqrt{2})$ **22.** $\left(-1, \frac{9}{2}\right), \left(3, -\frac{3}{2}\right)$

Study Set 11.1 (page 814)

1. binomial **3.** binomial **5.** factorial, decreasing **7.** one **9.** r^{20}, s^{20} **11.** 1, 1 **13.** $(n - 1)!$ **15.** 1 **17.** $3!(9 - 3)!$ **19.** 2, 4 **21.** 3, 3!, y, 3!, 2, 2, 3 **23.** $n - 1$ **25.** 6 **27.** 120 **29.** 30 **31.** 144 **33.** 40,320 **35.** $\frac{1}{110}$ **37.** 2,352 **39.** 72 **41.** 5 **43.** 10 **45.** 21 **47.** $\frac{1}{168}$ **49.** 39,916,800 **51.** $2.432902008 \times 10^{18}$ **53.** $x^4 + 4x^3y + 6x^2y^2 + 4xy^3 + y^4$ **55.** $c^5 - 5c^4d + 10c^3d^2 - 10c^2d^3 + 5cd^4 - d^5$ **57.** $s^6 + 6s^5t + 15s^4t^2 + 20s^3t^3 + 15s^2t^4 + 6st^5 + t^6$ **59.** $a^9 - 9a^8b + 36a^7b^2 - 84a^6b^3 + 126a^5b^4 - 126a^4b^5 + 84a^3b^6 - 36a^2b^7 + 9ab^8 - b^9$ **61.** $8x^3 + 12x^2y + 6xy^2 + y^3$ **63.** $32t^5 - 240t^4 + 720t^3 - 1{,}080t^2 + 810t - 243$ **65.** $625m^4 - 1{,}000m^3n + 600m^2n^2 - 160mn^3 + 16n^4$ **67.** $\frac{x^3}{27} + \frac{x^2y}{6} + \frac{xy^2}{4} + \frac{y^3}{8}$ **69.** $\frac{x^4}{81} - \frac{2x^3y}{27} + \frac{x^2y^2}{6} - \frac{xy^3}{6} + \frac{y^4}{16}$ **71.** $c^{10} - 5c^8d^2 + 10c^6d^4 - 10c^4d^6 + 5c^2d^8 - d^{10}$ **73.** $-4xy^3$ **75.** $15r^2s^4$ **77.** $28x^6y^2$ **79.** $-12x^3y$ **81.** $-70{,}000t^4$ **83.** $810xy^4$ **85.** $-\frac{1}{6}c^3d$ **87.** $-6a^{10}b^2$ **93.** $\log x^2y^{1/2}$ **95.** $\ln y$

Study Set 11.2 (page 821)

1. sequence **3.** arithmetic **5.** mean **7.** 1, 7, 13 **9. a.** $a_n = a_1 + (n-1)d$ **b.** $S_n = \frac{n(a_1 + a_n)}{2}$ **11.** nth **13.** sigma **15.** summation, runs **17.** 3, 7, 11, 15, 19 **19.** −2, −5, −8, −11, −14 **21.** 3, 5, 7, 9, 11 **23.** −5, −8, −11, −14, −17 **25.** 5, 11, 17, 23, 29 **27.** −4, −11, −18, −25, −32 **29.** −118, −111, −104, −97, −90 **31.** 34, 31, 28, 25, 22 **33.** 5, 12, 19, 26, 33 **35.** 355 **37.** −179 **39.** −23 **41.** 12 **43.** $\frac{17}{4}, \frac{13}{2}, \frac{35}{4}$ **45.** 12, 14, 16, 18 **47.** $\frac{29}{2}$ **49.** $3 + 6 + 9 + 12$ **51.** $4 + 9 + 16$ **53.** $\sum_{k=1}^{5} k^2$ **55.** $\sum_{k=3}^{6} k$ **57.** 1,335 **59.** 459 **61.** 354 **63.** 255 **65.** 1,275 **67.** 2,500 **69.** 60 **71.** 91 **73.** 31 **75.** 12 **77.** 60, 110, 160, 210, 260, 310; $6,060 **79.** 11,325 **81.** 368 ft **87.** $1 + \log_2 x - \log_2 y$ **89.** $3 \log x + 2 \log y$

Study Set 11.3 (page 831)

1. geometric **3.** mean **5.** 16, 4, 1 **7.** $a_n = a_1 r^{n-1}$ **9. a.** yes **b.** no **c.** no **d.** yes **11.** $a_1 r^2, a_1 r^4$ **13.** $n+1, n$ **15.** 3, 6, 12, 24, 48 **17.** $-5, -1, -\frac{1}{5}, -\frac{1}{25}, -\frac{1}{125}$ **19.** 2, 8, 32, 128, 512 **21.** −3, −12, −48, −192, −768 **23.** −64, 32, −16, 8, −4 **25.** −64, −32, −16, −8, −4 **27.** 2, 10, 50, 250, 1,250 **29.** 3,584 **31.** $\frac{1}{27}$ **33.** 3 **35.** 6, 18, 54 **37.** −20, −100, −500, −2,500 **39.** −16 **41.** $10\sqrt{2}$ **43.** No geometric mean exists. **45.** 728 **47.** 122 **49.** −255 **51.** 381 **53.** $\frac{156}{25}$ **55.** $-\frac{21}{4}$ **57.** 16 **59.** 81 **61.** 8 **63.** $-\frac{135}{4}$ **65.** no sum **67.** $-\frac{81}{2}$ **69.** $\frac{1}{9}$ **71.** $\frac{1}{3}$ **73.** $\frac{4}{33}$ **75.** $\frac{25}{33}$ **77.** $1,469.74 **79.** $140,853.75 **81.** $\left(\frac{1}{2}\right)^{11} \approx 0.0005$ **83.** 30 m **85.** 5,000 **91.** $[-1, 6]$ **93.** $(-\infty, -3) \cup (4, \infty)$

Study Set 11.4 (page 841)

1. tree **3.** permutation **5.** $p \cdot q$ **7.** $P(n, r) = \frac{n!}{(n-r)!}$ **9.** n, r, combinations **11.** 1 **13.** 6!, 4!, 5 **15.** 6 **17.** 60 **19.** 12 **21.** 5 **23.** 1,260 **25.** 10 **27.** 20 **29.** 50 **31.** 2 **33.** 1 **35.** $\frac{n!}{2!(n-2)!}$ **37.** $x^4 + 4x^3y + 6x^2y^2 + 4xy^3 + y^4$ **39.** $8x^3 + 12x^2y + 6xy^2 + y^3$ **41.** $81x^4 - 216x^3 + 216x^2 - 96x + 16$ **43.** $-1{,}250x^2y^3$ **45.** $-4x^6y^3$ **47.** 35 **49.** 1,000,000 **51.** 136,080 **53.** 8,000,000 **55.** 720 **57.** 2,880 **59.** 13,800 **61.** 720 **63.** 900 **65.** 364 **67.** 5 **69.** 1,192,052,400 **71.** 18 **73.** 7,920 **77.** 1.7095 **79.** 0.7324

Study Set 11.5 (page 846)

1. experiment **3.** $\frac{s}{n}$ **5.** 0 **7. a.** 6 **b.** 52 **c.** $\frac{6}{52}, \frac{3}{26}$ **9.** {(1, H), (2, H), (3, H), (4, H), (5, H), (6, H), (1, T), (2, T), (3, T), (4, T), (5, T), (6, T)} **11.** {a, b, c, d, e, f, g, h, i, j, k, l, m, n, o, p, q, r, s, t, u, v, w, x, y, z} **13.** $\frac{1}{6}$ **15.** $\frac{2}{3}$ **17.** $\frac{19}{42}$ **19.** $\frac{13}{42}$ **21.** $\frac{3}{8}$ **23.** 0 **25.** $\frac{1}{12}$ **27.** $\frac{5}{12}$ **29.** $\frac{33}{391{,}510}$ **31.** $\frac{9}{460}$ **35.** $\frac{1}{4}$ **37.** $\frac{1}{4}$ **39.** 1 **41.** $\frac{32}{119}$ **43.** $\frac{1}{3}$ **47.** $-\frac{3}{4}$ **49.** 3, −1 **51.** −2, −2

Key Concept (page 849)

1. g **2.** o **3.** i **4.** l **5.** u **6.** y **7.** p **8.** x **9.** m **10.** d **11.** f **12.** c **13.** w **14.** b **15.** j **16.** s **17.** e **18.** z **19.** n **20.** k **21.** h **22.** v **23.** q **24.** a **25.** r **26.** t

Chapter Review (page 850)

1. 1, 5, 10, 10, 5, 1 **2. a.** 13 **b.** 12 **c.** a^{12}, b^{12} **d.** a: decrease; b: increase **3.** 144 **4.** 20 **5.** 15 **6.** 220 **7.** 1 **8.** 8 **9.** $x^5 + 5x^4y + 10x^3y^2 + 10x^2y^3 + 5xy^4 + y^5$ **10.** $x^9 - 9x^8y + 36x^7y^2 - 84x^6y^3 + 126x^5y^4 - 126x^4y^5 + 84x^3y^6 - 36x^2y^7 + 9xy^8 - y^9$ **11.** $64x^3 - 48x^2y + 12xy^2 - y^3$ **12.** $\frac{c^4}{16} + \frac{c^3d}{6} + \frac{c^2d^2}{6} + \frac{2cd^3}{27} + \frac{d^4}{81}$ **13.** $6x^2y^2$ **14.** $-20x^3y^3$ **15.** $-108x^2y$ **16.** $5u^2v^{12}$ **17.** −2, 0, 2, 4 **18.** 42 **19.** 122, 137, 152, 167, 182 **20.** −1,194 **21.** −5 **22.** $\frac{41}{3}, \frac{58}{3}$ **23.** $-\frac{45}{2}$ **24.** 1,568 **25.** $\frac{15}{2}$ **26.** 378 **27.** 14 **28.** 360 **29.** 5,050 **30.** 1,170 **31.** 4 **32.** 24, 12, 6, 3, $\frac{3}{2}$ **33.** $\frac{1}{27}$ **34.** 24, −96 **35.** $\frac{2{,}186}{9}$ **36.** $-\frac{85}{8}$ **37.** about 1.6 lb **38.** 125 **39.** $\frac{5}{99}$ **40.** 190 ft **41.** 136 **42.** 17,576,000 **43.** 5,040 **44.** 1 **45.** 20,160 **46.** $\frac{1}{10}$ **47.** 1 **48.** 1 **49.** 28 **50.** 700 **51.** $\frac{7}{4}$ **52.** $81y^4 - 216y^3z + 216y^2z^2 - 96yz^3 + 16z^4$ **53.** 120 **54.** 720 **55.** 120 **56.** 150 **57.** $\frac{3}{8}$ **58.** $\frac{1}{2}$ **59.** $\frac{7}{8}$ **60.** $\frac{1}{18}$ **61.** 0 **62.** $\frac{1}{13}$ **63.** $\frac{94}{54{,}145}$ **64.** $\frac{33}{66{,}640}$

Chapter Test (page 853)

1. 2, −4, −10, −16 **2.** $a^6 - 6a^5b + 15a^4b^2 - 20a^3b^3 + 15a^2b^4 - 6ab^5 + b^6$ **3.** $24x^4y^2$ **4.** 66 **5.** 306 **6.** 34, 66 **7.** $\frac{1}{4}$ **8.** −1,377 **9.** 205 **10.** 3 **11.** −81 **12.** $\frac{364}{27}$ **13.** 6 **14.** 18, 108 **15.** $\frac{27}{2}$ **16.** 1,600 ft **17.** 210 **18.** 1 **19.** 120 **20.** 15 **21.** 56 **22.** 6 **23.** 7,000,000 **24.** 362,880 **25.** 35 **26.** 30 **27.** $\frac{1}{6}$ **28.** $\frac{2}{13}$ **29.** $\frac{33}{66{,}640}$ **30.** $\frac{5}{16}$ **31.** shade 7 sections **32. a.** 0 **b.** 1

Cumulative Review Exercises, Chapters 1–11 (page 854)

1. 0 **2.** $-\frac{4}{3}$, 5.6, 0, −23 **3.** $\pi, \sqrt{2}, e$ **4.** $-\frac{4}{3}, \pi$, 5.6, $\sqrt{2}$, 0, −23, e **5.** $8,250 **6.** $\frac{1}{120}$ db/rpm **7.** parallel **8.** perpendicular **9.** $y = -2x + 5$ **10.** $y = -\frac{9}{13}x + \frac{7}{13}$ **11.** $(-7, 7)$ **12.** $\left(\frac{4}{5}, \frac{3}{4}\right)$ **13.** $\left(-2, \frac{3}{2}\right)$ **14.** (3, 2, 1) **15.** 3 **16.** 85°, 80°, 15° **17.** It doesn't pass the vertical line test. **18.** -4; $3a^5 - 2a^2 + 1$

19. a. 4 **b.** 3 **c.** 0, 2 **d.** 1 **20.** 1.73×10^{14}; 4.6×10^{-8}

21.

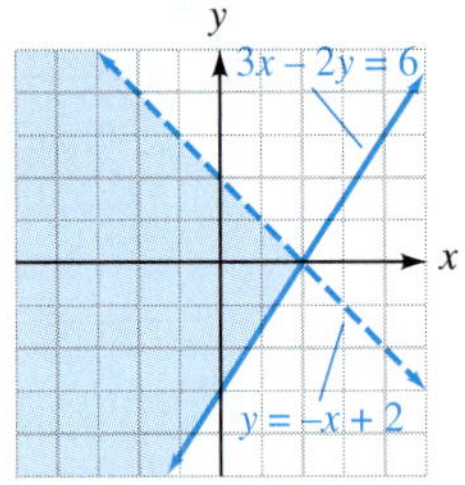

22. $\left[-3, \frac{19}{3}\right]$

23. $(-\infty, -2)$

24. $x^3 - 27y^3$ **25.** $2x^2y^3 + 13xy + 3y^2$

26. $81a^2b^4 - 72ab^2 + 16$ **27.** $\frac{1}{a^3b} + \frac{b^2}{a^2}$

28. $xy(3x - 4y)(x - 2)$ **29.** $(16x^2y^2 + z^4)(4xy + z^2)(4xy - z^2)$

30. $(3y^3 + 2)(4y^3 + 5)$ **31.** $\lambda = \frac{4d - 2}{A - 6}$ **32.** $-8, -8$

33. $0, 0, -1$ **34.** $(-3, 0), (0, 0), (2, 0)$; $(0, 0)$; $24, 0, -8, -6, 0, 4, 0, -18$

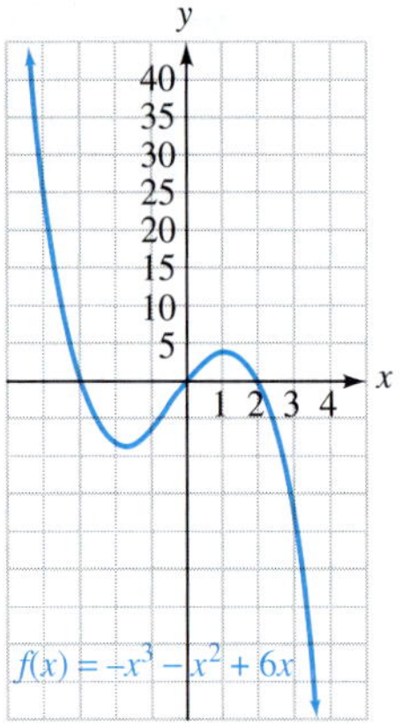

35. $\frac{16}{y^{16}}$ **36.** $-\frac{3x + 2}{3x - 2}$ **37.** $-\frac{q}{p}$ **38.** $\frac{4a - 1}{(a + 2)(a - 2)}$

39. 5; 3 is extraneous **40.** $R = \frac{R_1R_2R_3}{R_2R_3 + R_1R_3 + R_1R_2}$

41. a. a quadratic function **b.** at about 85% and 120% of the suggested inflation **42.** about $21\frac{1}{2}$ in.

43. $-x^2 + x + 5 + \frac{8}{x - 1}$ **44.** 2 lumens

45. D: $[0, \infty)$; R: $[2, \infty)$

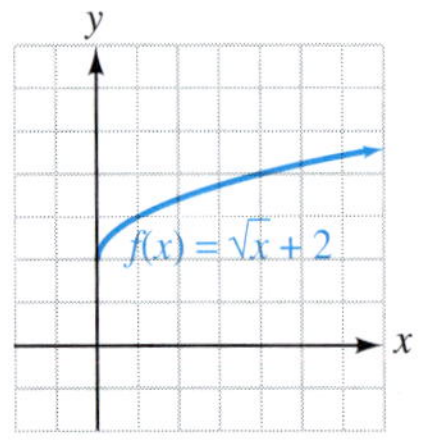

46. $5\sqrt{2}$

47. $15x - 6\sqrt{15x} + 9$

48. $81x\sqrt[3]{3x}$ **49.** $\frac{343}{125}$

50. $\frac{\sqrt[3]{2ab^2}}{b}$ **51.** $\sqrt{3t} - 1$

52. $-5 + 17i$ **53.** $-\frac{21}{29} - \frac{20}{29}i$

54. -1 **55.** 5, 0 does not check **56.** 0

57. $-i, i, -3i\sqrt{2}, 3i\sqrt{2}$ **58.** $\frac{-3 \pm \sqrt{5}}{4}$ **59.** $\frac{1}{4}, \frac{1}{2}$

60. $\frac{2}{3} \pm \frac{\sqrt{2}}{3}i$

61.

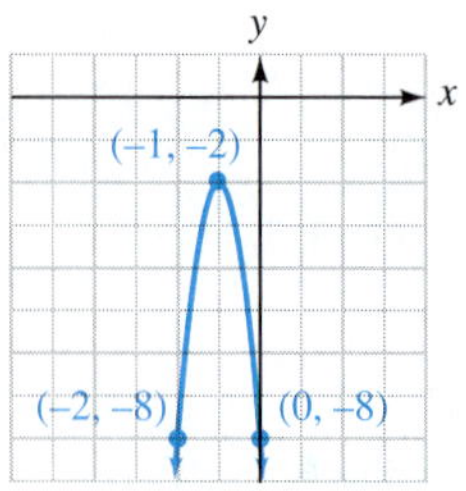

$f(x) = -6x^2 - 12x - 8$

62. $4x^2 + 4x - 1$

63. $f^{-1}(x) = \sqrt[3]{\frac{x + 1}{2}}$

64. D: $(-\infty, \infty)$; R: $(0, \infty)$

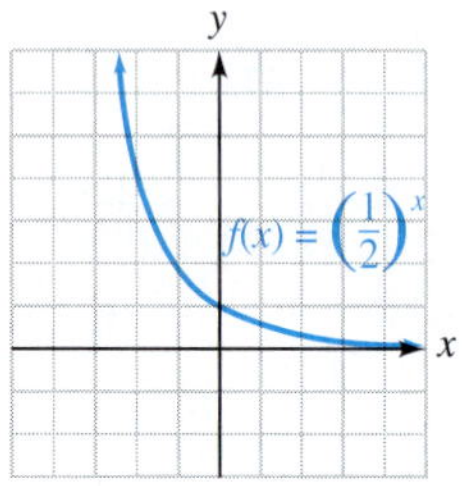

65.

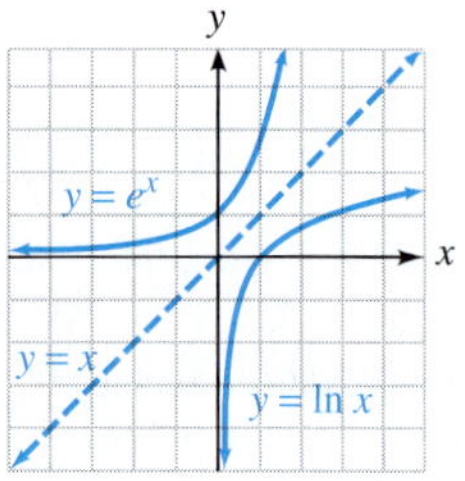

66. $2 - 3\log_6 x$ **67.** $\ln \frac{y\sqrt{x}}{z}$ **68.** about 170 million

69. 5 **70.** 3 **71.** $\frac{1}{27}$ **72.** 1 **73.** 1.9912

74. undefined **75.** 3.4190 **76.** 1, -10 does not check

77. $-\frac{3}{4}$ **78.** 9 **79.** $x^2 + y^2 - 2x - 6y - 15 = 0$

80. $\frac{(x - 2)^2}{9} - \frac{y^2}{1} = 1$

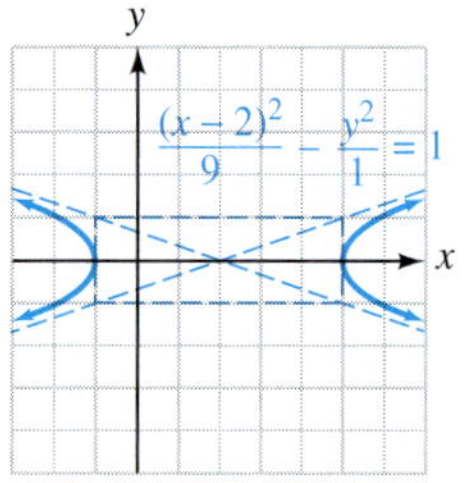

81.

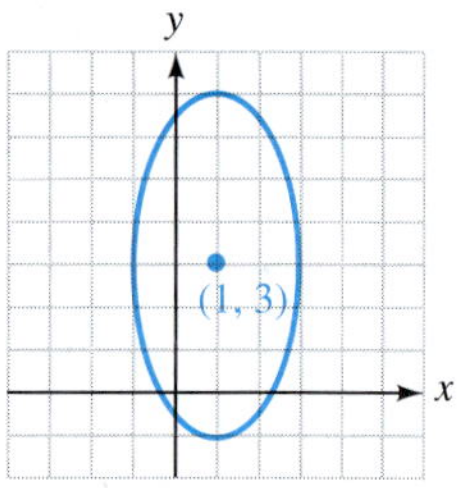

$\frac{(x - 1)^2}{4} + \frac{(y - 3)^2}{16} = 1$

82.

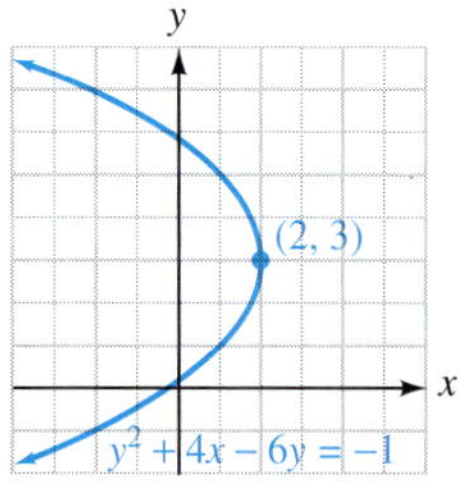

83. $81a^4 - 108a^3b + 54a^2b^2 - 12ab^3 + b^4$ **84.** $112x^2y^6$

85. 103 **86.** 690 **87.** 27 **88.** 27 **89.** \$2,848.31

90. $\frac{1{,}023}{64}$ **91.** $\frac{27}{2}$ **92.** 5,040 **93.** 84 **94.** $\frac{3}{13}$

Index